Some Physical Properties

Air (dry, at 20°C and 1 atm)	
Density	1.21 kg/m³
Specific heat at constant pressure	1010 J/kg·K
Ratio of specific heats	1.40
Speed of sound	343 m/s
Electrical breakdown strength	3×10^6 V/m
Effective molar mass	0.0289 kg/mol
Water	
Density	1000 kg/m³
Speed of sound	1460 m/s
Specific heat at constant pressure	4190 J/kg·K
Heat of fusion (0°C)	333 kJ/kg
Heat of vaporization (100°C)	2260 kJ/kg
Index of refraction (λ = 589 nm)	1.33
Molar mass	0.0180 kg/mol
Earth	
Mass	5.98×10^{24} kg
Mean radius	6.37×10^6 m
Free-fall acceleration at Earth's surface	9.8 m/s²
Standard atmosphere	1.01×10^5 Pa
Period of satellite at 100 km altitude	86.3 min
Radius of the geosynchronous orbit	42 200 km
Escape speed	11.2 km/s
Magnetic dipole moment	8.0×10^{22} A·m²
Mean electric field at surface	150 V/m, down
Distance to	
Moon	3.82×10^8 m
Sun	1.50×10^{11} m
Nearest star	4.04×10^{16} m
Galactic center	2.2×10^{20} m
Andromeda galaxy	2.1×10^{22} m
Edge of the observable universe	$\sim 10^{26}$ m

SI Prefixes*

Factor	Prefix	Symbol	Factor	Prefix	Symbol
10^{24}	yotta	Y	10^{-1}	deci	d
10^{21}	zetta	Z	10^{-2}	centi	c
10^{18}	exa	E	10^{-3}	milli	m
10^{15}	peta	P	10^{-6}	micro	μ
10^{12}	tera	T	10^{-9}	nano	n
10^{9}	giga	G	10^{-12}	pico	p
10^{6}	mega	M	10^{-15}	femto	f
10^{3}	kilo	k	10^{-18}	atto	a
10^{2}	hecto	h	10^{-21}	zepto	z
10^{1}	deka	da	10^{-24}	yocto	y

* In all cases, the first syllable is accented, as in ná-no-mé-ter.

Mathematical Formulas*

Quadratic Formula

If $ax^2 + bx + c = 0$, then $x = \dfrac{-b \pm \sqrt{b^2 - 4ac}}{2a}$

Binomial Theorem

$$(1 + x)^n = 1 + \frac{nx}{1!} + \frac{n(n-1)x^2}{2!} + \ldots \qquad (x^2 < 1)$$

Products of Vectors

Let θ be the smaller of the two angles between $\vec{a}$ and $\vec{b}$. Then

$$\vec{a} \cdot \vec{b} = \vec{b} \cdot \vec{a} = a_x b_x + a_y b_y + a_z b_z = ab \cos \theta$$

$$\vec{a} \times \vec{b} = -\vec{b} \times \vec{a} = \begin{vmatrix} \hat{\mathrm{i}} & \hat{\mathrm{j}} & \hat{\mathrm{k}} \\ a_x & a_y & a_z \\ b_x & b_y & b_z \end{vmatrix}$$

$$= \hat{\mathrm{i}} \begin{vmatrix} a_y & a_z \\ b_y & b_z \end{vmatrix} - \hat{\mathrm{j}} \begin{vmatrix} a_x & a_z \\ b_x & b_z \end{vmatrix} + \hat{\mathrm{k}} \begin{vmatrix} a_x & a_y \\ b_x & b_y \end{vmatrix}$$

$$= (a_y b_z - b_y a_z)\hat{\mathrm{i}} + (a_z b_x - b_z a_x)\hat{\mathrm{j}} + (a_x b_y - b_x a_y)\hat{\mathrm{k}}$$

$$|\vec{a} \times \vec{b}| = ab \sin \theta$$

Trigonometric Identities

$$\sin \alpha \pm \sin \beta = 2 \sin \tfrac{1}{2}(\alpha \pm \beta) \cos \tfrac{1}{2}(\alpha \mp \beta)$$
$$\cos \alpha + \cos \beta = 2 \cos \tfrac{1}{2}(\alpha + \beta) \cos \tfrac{1}{2}(\alpha - \beta)$$

* See Appendix E for a more complete list.

Derivatives and Integrals

$$\frac{d}{dx} \sin x = \cos x \qquad \int \sin x \, dx = -\cos x$$

$$\frac{d}{dx} \cos x = -\sin x \qquad \int \cos x \, dx = \sin x$$

$$\frac{d}{dx} e^x = e^x \qquad \int e^x \, dx = e^x$$

$$\int \frac{dx}{\sqrt{x^2 + a^2}} = \ln(x + \sqrt{x^2 + a^2})$$

$$\int \frac{x\,dx}{(x^2 + a^2)^{3/2}} = -\frac{1}{(x^2 + a^2)^{1/2}}$$

$$\int \frac{dx}{(x^2 + a^2)^{3/2}} = \frac{x}{a^2(x^2 + a^2)^{1/2}}$$

Cramer's Rule

Two simultaneous equations in unknowns x and y,

$$a_1 x + b_1 y = c_1 \quad \text{and} \quad a_2 x + b_2 y = c_2,$$

have the solutions

$$x = \frac{\begin{vmatrix} c_1 & b_1 \\ c_2 & b_2 \end{vmatrix}}{\begin{vmatrix} a_1 & b_1 \\ a_2 & b_2 \end{vmatrix}} = \frac{c_1 b_2 - c_2 b_1}{a_1 b_2 - a_2 b_1}$$

and

$$y = \frac{\begin{vmatrix} a_1 & c_1 \\ a_2 & c_2 \end{vmatrix}}{\begin{vmatrix} a_1 & b_1 \\ a_2 & b_2 \end{vmatrix}} = \frac{a_1 c_2 - a_2 c_1}{a_1 b_2 - a_2 b_1}.$$

The Greek Alphabet

Alpha	A	α	Iota	I	ι	Rho	P	ρ
Beta	B	β	Kappa	K	κ	Sigma	Σ	σ
Gamma	Γ	γ	Lambda	Λ	λ	Tau	T	τ
Delta	Δ	δ	Mu	M	μ	Upsilon	Υ	υ
Epsilon	E	ϵ	Nu	N	ν	Phi	Φ	ϕ, φ
Zeta	Z	ζ	Xi	Ξ	ξ	Chi	X	χ
Eta	H	η	Omicron	O	o	Psi	Ψ	ψ
Theta	Θ	θ	Pi	Π	π	Omega	Ω	ω

SIXTH EDITION

Fundamentals of Physics

EXTENDED

SIXTH EDITION

Fundamentals of Physics

EXTENDED

David Halliday

University of Pittsburgh

Robert Resnick

Rensselaer Polytechnic Institute

Jearl Walker

Cleveland State University

WILEY-
INDIA

Fundamentals of Physics
Extended, Sixth Edition

Authorised reprint by Wiley India (P) Ltd., 4435/7, Ansari Road, Daryaganj, New Delhi-110002.

Edition: 2006

Printed at: *Sanat Printers, Kundli, Haryana*

ISBN: 81-265-0823-X

BRIEF CONTENTS

TABLES

CONTENTS

PART 1

CHAPTER 1

CHAPTER 2

CHAPTER 3

CHAPTER 4

CHAPTER 5

CHAPTER 6

CHAPTER 7

CHAPTER 8

CHAPTER 9

CHAPTER 10

CHAPTER 19

Temperature, Heat, and the First Law of Thermodynamics 425

What thermal protection do bees have against a hornet invasion?

CHAPTER 20

The Kinetic Theory of Gases 454

Why can a cold container of a carbonated drink form a fog when opened?

CHAPTER 21

Entropy and the Second Law of Thermodynamics 482

What in the world gives direction to time?

PART 3

CHAPTER 22

Electric Charge 505

What causes the sparking of a wintergreen Lifesaver?

CHAPTER 23

Electric Fields 520

What causes the huge electrical discharges above some volcanoes?

CHAPTER 24

Gauss' Law 543

How wide is a lightning strike?

CHAPTER 25

Electric Potential 564

What is the danger if your hair suddenly stands up?

CHAPTER 26

Capacitance 588

How do you stop ventricular fibrillation in a remote location?

CHAPTER 27

Current and Resistance 611

What caused the flaming crash of the Zeppelin Hindenburg?

CHAPTER 33

Electromagnetic Oscillations and Alternating Current 768

Why do electrical transmission lines have high potential and not high current?

PART 4

CHAPTER 34

Electromagnetic Waves 801

What shapes the curved dust tail of a comet?

CHAPTER 35

Images 833

What distortion of reality lies in Manet's A Bar at the Folies-Bergère?

CHAPTER 36

Interference 861

What produces the blue-green of a Morpho's wing?

CHAPTER 37

Diffraction 890

Why do the colors in a pointillism painting change with viewing distance?

CHAPTER 38

Relativity 919

Why is special relativity so important in modern navigation?

PART 5

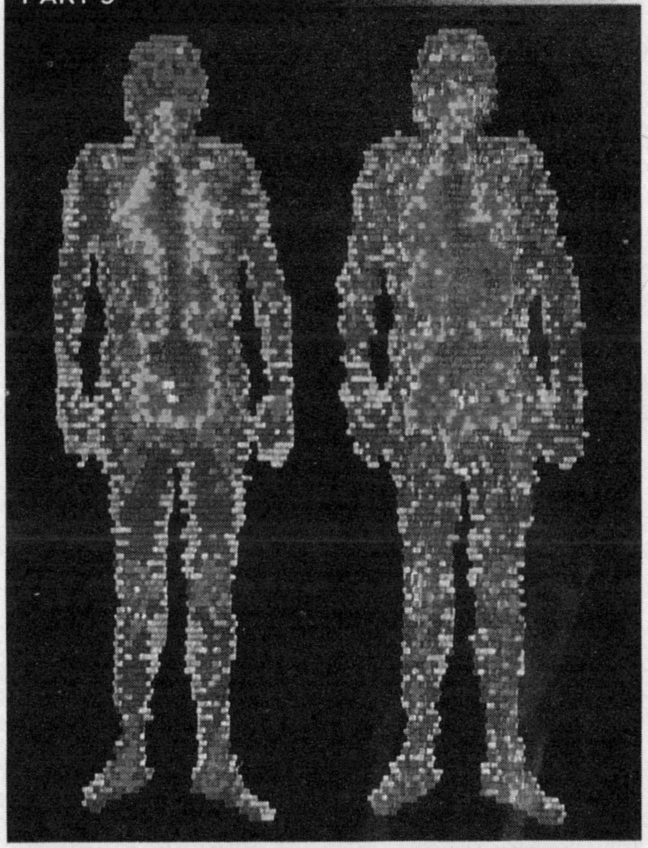

CHAPTER 39

Photons and Matter Waves 953

How can a particle such as an electron be a wave?

CHAPTER 40

More About Matter Waves 979

How can you corral an electron?

CHAPTER 41

All About Atoms 1006

What is so different about light from a laser?

CHAPTER 42

Conduction of Electricity in Solids 1037

Why are "spacesuits" the dress code at the Fab 11 factory in New Mexico?

CHAPTER 43

Nuclear Physics 1062

Why and how do (some) nuclei undergo decay?

CHAPTER 44

Energy from the Nucleus 1092

What physics underlies the image that has horrified the world since World War II?

CHAPTER 45

APPENDICES

PREFACE

This sixth edition of *Fundamentals of Physics* contains a redesign and major rewrites of the widely used fifth edition, while maintaining many elements of the classic text first written by David Halliday and Robert Resnick. Nearly all the changes are based on suggestions from instructors and students using the fifth edition, from reviewers of the manuscripts for the sixth edition, and from research done on the process of learning. You can send suggestions, corrections, and positive or negative comments to John Wiley & Sons (http://www.wiley.com/college/hrw) or Jearl Walker (mail address: Physics Department, Cleveland State University, Cleveland OH 44115 USA; fax number: (USA) (216) 687-2424; or email address: physics@wiley.com). We may not be able to respond to all suggestions, but we keep and study each of them.

Design Changes

➤ ***More open format.*** Previous editions have been printed in a double-column format, which many students and instructors have found cluttered and distracting. In this edition, the narrative is presented in a single-column format with a wide margin for note-taking.

➤ ***Streamlined presentation.*** It is a common complaint of all texts that they cover too much material. As a response to this criticism, the sixth edition has been shortened in two ways.

1. Material regarding special relativity and quantum physics has been moved from the early chapters to the later chapters devoted to those subjects.
2. The essential sample problems have been retained in this book, but the more specialized sample problems have been shifted to the Problem Supplement that is automatically provided with this book. (The Problem Supplement is described on the next page.)

➤ ***Vector notation.*** Vectors are now presented with an overhead arrow (such as $\vec{F}$) instead of as a bold symbol (such as **F**).

➤ ***Emphasis on metric units.*** Except in Chapter 1 (in which various systems of units are employed) and certain problems involving baseball (in which English units are traditional), metric units are used almost exclusively.

➤ ***Structured versus unstructured order of problems.*** The homework problems in this book are still ordered approximately according to their difficulty and grouped under section titles corresponding to the narrative of the chapter. However, many of the homework problems of the fifth edition have been shifted, without order or grouping, to the Problem Supplement. (The total number of problems in this book and in the Problem Supplement exceeds what was available in the fifth edition.)

➤ ***Icons for additional help.*** When worked-out solutions are provided either in print or electronically for certain of the odd-numbered homework problems, the statements for those problems include a trailing icon to alert both student and instructor as to where the solutions are located. An icon guide is provided here and at the beginning of each set of homework problems:

ssm	Solution is in the Student Solutions Manual.
www	Solution is available on the World Wide Web at: http://www.wiley.com/college/hrw
ilw	Solution is available on the Interactive LearningWare.

These resources are described later in this preface.

Pedagogy Changes

➤ ***Reasoning versus plug-and-chug.*** The primary goal of this book is to teach students to reason through challenging situations, from basic principles to a solution. Although some plug-and-chug homework problems remain in this book, most homework problems emphasize reasoning.

➤ ***Key Ideas in the sample problems.*** The solutions to all 360 sample problems in this book and in the Problem Supplement have been rewritten to begin with one or more Key Ideas based on basic principles.

➤ ***Lengthened solutions to sample problems.*** Most of the solutions to the sample problems are now longer because they build step by step from the beginning Key Ideas to an answer, often repeating some of the important reasoning of the narrative preceding the sample problems. For example, see Sample Problem 8-3 on page 148 and Sample Problem 10-2 on pages 200–201.

➤ ***Use of vector-capable calculators.*** When vector calculations in a sample problem can be performed directly on-screen with a vector-capable calculator, the solution of the sample problem indicates that fact but still carries through the traditional component analysis. When vector calculations cannot be performed directly on-screen, the solution explains why.

➤ ***Problems with applied physics,*** based on published research, have been added in many places, either as sample problems or homework problems. For example, see Sample Problem 11-6 on page 229, homework problem 64 on page 71, and homework problem 56 on page 214. For an example of homework problems that build with a continuing story, see problems 4, 32, and 48 (on pages 112, 114, and 115, respectively) in Chapter 6.

Content Changes

➤ ***Chapter 5 on force and motion*** contains clearer explanations of the gravitational force, weight, and normal force (pages 80–82).

➤ ***Chapter 7 on kinetic energy and work*** begins with a rough definition of energy. It then defines kinetic energy, work, and the work–kinetic energy theorem in ways that are more closely tied to Newton's second law than in the fifth edition, while keeping those definitions consistent with thermodynamics (pages 117–120).

➤ ***Chapter 8 on the conservation of energy*** avoids the much criticized definition of work done by a nonconservative force by explaining, instead, the energy transfers that occur due to a nonconservative force (page 153). (The wording still allows an instructor to superimpose a definition of work done by a nonconservative force.)

➤ ***Chapter 10 on collisions*** now presents the general situation of inelastic one-dimensional collisions (pages 198–200) before the special situation of elastic one-dimensional collisions (pages 202–204).

➤ ***Chapters 16, 17, and 18 on SHM and waves*** have been rewritten to better ease a student into these difficult subjects.

➤ ***Chapter 21 on entropy*** now presents a Carnot engine as the ideal heat engine with the greatest efficiency.

Chapter Features

➤ ***Opening puzzlers.*** A curious puzzling situation opens each chapter and is explained somewhere within the chapter, to entice a student to read the chapter.

➤ ***Checkpoints*** are stopping points that effectively ask the student, "Can you answer this question with some reasoning based on the narrative or sample problem that you just read?" If not, then the student should go back over that previous material before traveling deeper into the chapter. For example, see Checkpoint 3 on page 78 and Checkpoint 1 on page 101. **Answers to all checkpoints are in the back of the book.**

➤ ***Sample problems*** have been chosen to help the student organize the basic concepts of the narrative and to develop problem-solving skills. Each sample problem builds step by step from one or more Key Ideas to a solution.

➤ ***Problem-solving tactics*** contain helpful instructions to guide the beginning physics student as to how to solve problems and to avoid common errors.

➤ ***Review & Summary*** is a brief outline of the chapter contents that contains the essential concepts but which is not a substitute for reading the chapter.

➤ ***Questions*** are like the checkpoints and require reasoning and understanding rather than calculations. **Answers to the odd-number questions are in the back of the book.**

➤ ***Exercises & Problems*** are ordered approximately according to difficulty and grouped under section titles. **The odd-numbered ones are answered in the back of the book.** Worked-out solutions to the odd-numbered problems with trailing icons are available either in print or electronically. (See the icon guide at the beginning of the Exercises & Problems.) A problem number with a star indicates an especially challenging problem.

➤ ***Additional Problems*** appear at the end of the Exercises & Problems in certain chapters. They are not sorted according to section titles and many involve applied physics.

Problem Supplement

A problem supplement is automatically provided with this book. The *Problem Supplement #1* (green book) will be provided until May 15, 2002. Thereafter, the *Problem Supplement #2* (blue book) will be provided. The blue book will have a different set of questions and homework problems and will contain more sample problems. The features of both versions of the problem supplement are the following:

➤ ***Additional sample problems*** that were shifted from the main book, plus many new ones. All begin with the basic *Key Ideas* and then build step by step to a solution.

➤ ***Questions*** include:

1. *Checkpoint-style questions,* as in the main book.
2. *Organizing questions,* which request that equations be set up for common situations, as a warm-up for the homework problems.

3. *Discussion questions* from the fourth and earlier editions of the book (back by request).

➤ ***Exercises & Problems.*** More homework problems, including many shifted from the main book. These are *not* ordered according to difficulty, section titles, or appearance of the associated physics in the chapter. Some of the new problems involve applied physics. In some chapters the homework problems end with *Clustered Problems,* in which similar problems are grouped together. In the other chapters, the homework problems end with *Tutorial Problems,* in which solutions are worked out.

Versions of the Text

The sixth edition of *Fundamentals of Physics* is available in a number of different versions, to accommodate the individual needs of instructors and students. The Regular Edition consists of Chapters 1 through 38 (ISBN 0-471-32000-5). The Extended Edition contains seven additional chapters on quantum physics and cosmology (Chapters 1–45) (ISBN 0-471-33236-4). Both editions are available as single, hardcover books, or in the following alternative versions:

➤ ***Volume 1—Chapters 1–21 (Mechanics/Thermodynamics), hardcover, 0-471-33235-6***

➤ ***Volume 2—Chapters 22–45 (E&M and Modern Physics), hardcover, 0-471-36037-6***

➤ ***Part 1—Chapters 1–12, paperback, 0-471-33234-8***

➤ ***Part 2—Chapters 13–21, paperback, 0-471-36041-4***

➤ ***Part 3—Chapters 22–33, paperback, 0-471-36040-6***

➤ ***Part 4—Chapters 34–38, paperback, 0-471-36039-2***

➤ ***Part 5—Chapters 39–45, paperback, 0-471-36038-4***

Supplements

The sixth edition of *Fundamentals of Physics* is supplemented by a comprehensive ancillary package carefully developed to help teachers teach and students learn.

Instructor's Supplements

➤ ***Instructor's Manual*** by J. RICHARD CHRISTMAN, U.S. Coast Guard Academy. This manual contains lecture notes outlining the most important topics of each chapter, demonstration experiments, laboratory and computer projects, film and video sources, answers to all Questions, Exercises & Problems, and Checkpoints, and a correlation guide to the Questions and Exercises & Problems in the previous edition.

➤ ***Instructor's Solutions Manual*** by JAMES WHITENTON, Southern Polytechnic University. This manual provides worked-out solutions for all the exercises and problems found at the end of each chapter within the text and in the Problem Supplement #1. *This supplement is available only to instructors.*

➤ ***Test Bank*** by J. RICHARD CHRISTMAN, U.S. Coast Guard Academy. More than 2200 multiple-choice questions are included in this manual. These items are also available in the Computerized Test Bank (see below).

➤ ***Instructor's Resource CD.*** This CD contains:

- All of the Instructor's Solutions Manual in both LaTex and PDF files.
- Computerized Test Bank in both IBM and Macintosh versions, with full editing features to help instructors customize tests.
- All text illustrations suitable for both classroom presentation and printing.

➤ ***Transparencies.*** More than 200 four-color illustrations from the text are provided in a form suitable for projection in the classroom.

➤ ***On-line Course Management.***

- WebAssign, CAPA, and Wiley eGrade are on-line homework and quizzing programs that give instructors the ability to deliver and grade homework and quizzes over the Internet.
- Instructors will also have access to WebCT course materials. WebCT is a powerful Web site program that allows instructors to set up complete on-line courses with chat rooms, bulletin boards, quizzing, student tracking, etc. Please contact your local Wiley representative for more information.

Student's Supplements

➤ ***A Student Companion*** by J. RICHARD CHRISTMAN, U.S. Coast Guard Academy. This student study guide consists of a traditional print component and an accompanying Web site, which together provide a rich, interactive environment for review and study. The Student Companion Web site includes self-quizzes, simulation exercises, hints for solving end-of-chapter problems, the *Interactive LearningWare* program (see the next page), and links to other Web sites that offer physics tutorial help.

➤ ***Student Solutions Manual*** by J. RICHARD CHRISTMAN, U.S. Coast Guard Academy and EDWARD DERRINGH, Wentworth Institute. This manual provides students with complete worked-out solutions to 30 percent of the exercises and problems found at the end of each chapter within the text. These problems are indicated with an ssm icon in the text.

➤ ***Interactive LearningWare.*** This software guides students through solutions to 200 of the end-of-chapter problems. The solutions process is developed interactively, with appropriate feedback and access to error-specific help for the most common mistakes. These problems are indicated with an ilw icon in the text.

➤ ***CD-Physics, 3.0.*** This CD-ROM based version of *Fundamentals of Physics,* Sixth Edition, contains the complete, extended version of the text, *A Student's Companion,* the *Student's Solutions Manual,* the *Interactive LearningWare,* and numerous simulations all connected with extensive hyperlinking.

➤ ***Take Note!*** This bound notebook lets students take notes directly onto large, black-and-white versions of textbook illustrations. All of the illustrations from the transparency set are included. In-class time spent copying illustrations is substantially reduced by this supplement.

➤ ***Physics Web Site.*** This Web site, **http://www.wiley.com/college/hrw**, was developed specifically for *Fundamentals of Physics,* Sixth Edition, and is designed to further assist students in the study of physics and offers additional physics resources. The site also includes solutions to selected end-of-chapter problems. These problems are identified with a www icon in the text.

ACKNOWLEDGMENTS

A textbook contains far more contributions to the elucidation of a subject than those made by the authors alone. J. Richard Christman, of the U.S. Coast Guard Academy, has once again created many fine supplements for us; his knowledge of our book and his recommendations to students and faculty are invaluable. James Tanner, of Georgia Institute of Technology, and Gary Lewis, of Kennesaw State College, have provided us with innovative software, closely tied to the text exercises and problems. James Whitenton, of Southern Polytechnic State University, and Jerry Shi, of Pasadena City College, performed the Herculean task of working out solutions for every one of the Exercises & Problems in the text. We thank John Merrill, of Brigham Young University, and Edward Derringh, of the Wentworth Institute of Technology, for their many contributions in the past. We also thank George W. Hukle of Oxnard, California, and Frank G. Jacobs of Evanston, Illinois, for their check of the answers for the problems in the book.

At John Wiley publishers, we have been fortunate to receive strong coordination and support from our former editor, Cliff Mills. Cliff guided our efforts and encouraged us along the way. When Cliff moved on to other responsibilities at Wiley, we were ably guided to completion by his successor, Stuart Johnson. Ellen Ford has coordinated the developmental editing and multilayered preproduction process. Sue Lyons, our marketing manager, has been tireless in her efforts on behalf of this edition. Joan Kalkut has built a fine supporting package of ancillary materials. Thomas Hempstead managed the reviews of manuscript and the multiple administrative duties admirably.

We thank Lucille Buonocore, our production director, and Monique Calello, our production editor, for pulling all the pieces together and guiding us through the complex production process. We also thank Maddy Lesure, for her design; Helen Walden for her copyediting; Edward Starr and Anna Melhorn, for managing the illustration program; Georgia Kamvosoulis Mederer, Katrina Avery, and Lilian Brady, for their proofreading; and all other members of the production team.

Hilary Newman and her team of photo researchers were inspired in their search for unusual and interesting photographs that communicate physics principles beautifully. We also owe a debt of gratitude for the line art to the late John Balbalis, whose careful hand and understanding of physics can still be seen in every diagram.

We especially thank Edward Millman for his developmental work on the manuscript. With us, he has read every word, asking many questions from the point of view of a student. Many of his questions and suggested changes have added to the clarity of this volume.

We owe a particular debt of gratitude to the numerous students who used the previous editions of *Fundamentals of Physics* and took the time to fill out the response cards and return them to us. As the ultimate consumers of this text, students are extremely important to us. By sharing their opinions with us, your students help us ensure that we are providing the best possible product and the most value for their textbook dollars. We encourage the users of this book to contact us with their thoughts and concerns so that we can continue to improve this text in the years to come.

Finally, our external reviewers have been outstanding and we acknowledge here our debt to each member of that team:

Edward Adelson
Ohio State University

Mark Arnett
Kirkwood Community College

Arun Bansil
Northeastern University

J. Richard Christman
U.S. Coast Guard Academy

Robert N. Davie, Jr.
St. Petersburg Junior College

Cheryl K. Dellai
Glendale Community College

Eric R. Dietz
California State University at Chico

N. John DiNardo
Drexel University

Harold B. Hart
Western Illinois University

Rebecca Hartzler
Edmonds Community College

Joey Huston
Michigan State University

Hector Jimenez
University of Puerto Rico

Sudhakar B. Joshi
York University

Leonard M. Kahn
University of Rhode Island

Yuichi Kubota
Cornell University

Priscilla Laws
Dickinson College

Edbertho Leal
Polytechnic University of Puerto Rico

Dale Long
Virginia Tech

Andreas Mandelis
University of Toronto

Paul Marquard
Caspar College

James Napolitano
Rensselaer Polytechnic Institute

Des Penny
Southern Utah University

Joe Redish
University of Maryland

Timothy M. Ritter
University of North Carolina at Pembroke

Gerardo A. Rodriguez
Skidmore College

John Rosendahl
University of California at Irvine

Michael Schatz
Georgia Institute of Technology

Michael G. Strauss
University of Oklahoma

Dan Styer
Oberlin College

Marshall Thomsen
Eastern Michigan University

Fred F. Tomblin
New Jersey Institute of Technology

B. R. Weinberger
Trinity College

William M. Whelan
Ryerson Polytechnic University

William Zimmerman, Jr.
University of Minnesota

Reviewers of the Fifth and Previous Editions

Maris A. Abolins
Michigan State University

Barbara Andereck
Ohio Wesleyan University

Albert Bartlett
University of Colorado

Michael E. Browne
University of Idaho

Timothy J. Burns
Leeward Community College

Joseph Buschi
Manhattan College

Philip A. Casabella
Rensselaer Polytechnic Institute

Randall Caton
Christopher Newport College

J. Richard Christman
U.S. Coast Guard Academy

Roger Clapp
University of South Florida

W. R. Conkie
Queen's University

Peter Crooker
University of Hawaii at Manoa

William P. Crummett
Montana College of Mineral Science and Technology

Eugene Dunnam
University of Florida

Robert Endorf
University of Cincinnati

F. Paul Esposito
University of Cincinnati

Jerry Finkelstein
San Jose State University

Alexander Firestone
Iowa State University

Alexander Gardner
Howard University

Andrew L. Gardner
Brigham Young University

John Gieniec
Central Missouri State University

John B. Gruber
San Jose State University

Ann Hanks
American River College

Samuel Harris
Purdue University

Emily Haught
Georgia Institute of Technology

Laurent Hodges
Iowa State University

John Hubisz
North Carolina State University

Joey Huston
Michigan State University

Darrell Huwe
Ohio University

Claude Kacser
University of Maryland

Leonard Kleinman
University of Texas at Austin

Earl Koller
Stevens Institute of Technology

Arthur Z. Kovacs
Rochester Institute of Technology

Kenneth Krane
Oregon State University

Sol Krasner
University of Illinois at Chicago

Peter Loly
University of Manitoba

Robert R. Marchini
Memphis State University

David Markowitz
University of Connecticut

Howard C. McAllister
University of Hawaii at Manoa

W. Scott McCullough
Oklahoma State University

James H. McGuire
Tulane University

David M. McKinstry
Eastern Washington University

Joe P. Meyer
Georgia Institute of Technology

Roy Middleton
University of Pennsylvania

Irvin A. Miller
Drexel University

Eugene Mosca
United States Naval Academy

Michael O'Shea
Kansas State University

Patrick Papin
San Diego State University

George Parker
North Carolina State University

Robert Pelcovits
Brown University

Oren P. Quist
South Dakota State University

Jonathan Reichart
SUNY—Buffalo

Manuel Schwartz
University of Louisville

Darrell Seeley
Milwaukee School of Engineering

Bruce Arne Sherwood
Carnegie Mellon University

John Spangler
St. Norbert College

Ross L. Spencer
Brigham Young University

Harold Stokes
Brigham Young University

Jay D. Strieb
Villanova University

David Toot
Alfred University

J. S. Turner
University of Texas at Austin

T. S. Venkataraman
Drexel University

Gianfranco Vidali
Syracuse University

Fred Wang
Prairie View A & M

Robert C. Webb
Texas A & M University

George Williams
University of Utah

David Wolfe
University of New Mexico

1 Measurement

You can watch the Sun set and disappear over a calm ocean once while you lie on a beach, and then once again if you stand up. Surprisingly, by measuring the time between the two sunsets, you can approximate Earth's radius.

How can such a simple observation be used to measure Earth?

The answer is in this chapter.

1-1 Measuring Things

Physics is based on measurement. We discover physics by learning how to measure the quantities that are involved in physics. Among these quantities are length, time, mass, temperature, pressure, and electric current.

We measure each physical quantity in its own units, by comparison with a **standard.** The **unit** is a unique name we assign to measures of that quantity—for example, meter (or m) for the quantity length. The standard corresponds to exactly 1.0 unit of the quantity. As you will see, the standard for length, which corresponds to exactly 1.0 m, is the distance traveled by light in a vacuum during a certain fraction of a second. We can define a unit and its standard in any way we care to. However, the important thing is to do so in such a way that scientists around the world will agree that our definitions are both sensible and practical.

Once we have set up a standard, say, for length, we must work out procedures by which any length whatever, be it the radius of a hydrogen atom, the wheelbase of a skateboard, or the distance to a star, can be expressed in terms of the standard. Rulers, which approximate our length standard, give us one such procedure for measuring length. However, many of our comparisons must be indirect. You cannot use a ruler, for example, to measure the radius of an atom or the distance to a star.

There are so many physical quantities that it is a problem to organize them. Fortunately, they are not all independent; for example, speed is the ratio of a length to a time. Thus, what we do is pick out—by international agreement—a small number of physical quantities, such as length and time, and assign standards to them alone. We then define all other physical quantities in terms of these *base quantities* and their standards (called *base standards*). Speed, for example, is defined in terms of the base quantities length and time and the associated base standards.

Base standards must be both accessible and invariable. If we define the length standard as the distance between one's nose and the index finger on an outstretched arm, we certainly have an accessible standard—but it will, of course, vary from person to person. The demand for precision in science and engineering pushes us to aim first for invariability. We then exert great effort to make duplicates of the base standards that are accessible to those who need them.

1-2 The International System of Units

In 1971, the 14th General Conference on Weights and Measures picked seven quantities as base quantities, thereby forming the basis of the International System of Units, abbreviated SI from its French name and popularly known as the *metric system.* Table 1-1 shows the units for the three base quantities—length, mass, and time—that we use in the early chapters of this book. These units were defined to be on a "human scale."

TABLE 1-1 Some SI Base Units

Quantity	Unit Name	Unit Symbol
Length	meter	m
Time	second	s
Mass	kilogram	kg

Many SI *derived units* are defined in terms of these base units. For example, the SI unit for power, called the **watt** (symbol: W), is defined in terms of the base units for mass, length, and time. Thus, as you will see in Chapter 7,

$$1 \text{ watt} = 1 \text{ W} = 1 \text{ kg} \cdot \text{m}^2/\text{s}^3, \tag{1-1}$$

where the last collection of unit symbols is read as kilogram–square meter per second–cubed.

To express the very large and very small quantities that we often run into in physics, we use *scientific notation,* which employs powers of 10. In this notation,

$$3\,560\,000\,000 \text{ m} = 3.56 \times 10^9 \text{ m} \tag{1-2}$$

and

$$0.000\,000\,492 \text{ s} = 4.92 \times 10^{-7} \text{ s}. \tag{1-3}$$

TABLE 1-2 Prefixes for SI Units

Factor	Prefix[a]	Symbol
10^{24}	yotta-	Y
10^{21}	zetta-	Z
10^{18}	exa-	E
10^{15}	peta-	P
10^{12}	tera-	T
10^{9}	**giga-**	**G**
10^{6}	**mega-**	**M**
10^{3}	**kilo-**	**k**
10^{2}	hecto-	h
10^{1}	deka-	da
10^{-1}	deci-	d
10^{-2}	**centi-**	**c**
10^{-3}	**milli-**	**m**
10^{-6}	**micro-**	**μ**
10^{-9}	**nano-**	**n**
10^{-12}	**pico-**	**p**
10^{-15}	femto-	f
10^{-18}	atto-	a
10^{-21}	zepto-	z
10^{-24}	yocto-	y

[a]The most commonly used prefixes are shown in bold type.

Scientific notation on computers sometimes takes on an even briefer look, as in 3.56 E9 and 4.92 E–7, where E stands for "exponent of ten." It is briefer still on some calculators, where E is replaced with an empty space.

As a further convenience when dealing with very large or very small measurements, we use the prefixes listed in Table 1-2. As you can see, each prefix represents a certain power of 10, as a factor. Attaching a prefix to an SI unit has the effect of multiplying by the associated factor. Thus, we can express a particular electric power as

$$1.27 \times 10^{9} \text{ watts} = 1.27 \text{ gigawatts} = 1.27 \text{ GW} \tag{1-4}$$

or a particular time interval as

$$2.35 \times 10^{-9} \text{ s} = 2.35 \text{ nanoseconds} = 2.35 \text{ ns}. \tag{1-5}$$

Some prefixes, as used in milliliter, centimeter, kilogram, and megabyte, are probably familiar to you.

1-3 Changing Units

We often need to change the units in which a physical quantity is expressed. We do so by a method called *chain-link conversion*. In this method, we multiply the original measurement by a **conversion factor** (a ratio of units that is equal to unity). For example, because 1 min and 60 s are identical time intervals, we have

$$\frac{1 \text{ min}}{60 \text{ s}} = 1 \quad \text{and} \quad \frac{60 \text{ s}}{1 \text{ min}} = 1.$$

Thus, the ratios (1 min)/(60 s) and (60 s)/(1 min) can be used as conversion factors. This is *not* the same as writing $\frac{1}{60} = 1$ or $60 = 1$; each *number* and its *unit* must be treated together.

Because multiplying any quantity by unity leaves it unchanged, we can introduce such conversion factors wherever we find them useful. In chain-link conversion, we use the factors to cancel unwanted units. For example, to convert 2 min to seconds, we have

$$2 \text{ min} = (2 \text{ min})(1) = (2 \text{ min})\left(\frac{60 \text{ s}}{1 \text{ min}}\right) = 120 \text{ s}. \tag{1-6}$$

If you introduce a conversion factor in such a way that unwanted units do *not* cancel, invert the factor and try again. In conversions, the units obey the same algebraic rules as variables and numbers.

Appendix D and the inside back cover give conversion factors between SI and other systems of units, including non-SI units still used in the United States. However, the conversion factors are written in the style of "1 min = 60 s" rather than as a ratio as used previously. The following sample problem gives an example of how to set up such ratios.

Sample Problem 1-1

When Pheidippides ran from Marathon to Athens in 490 B.C. to bring word of the Greek victory over the Persians, he probably ran at a speed of about 23 rides per hour (rides/h). The ride is an ancient Greek unit for length, as are the stadium and the plethron: 1 ride was defined to be 4 stadia, 1 stadium was defined to be 6 plethra, and, in terms of a modern unit, 1 plethron is 30.8 m. How fast did Pheidippides run in kilometers per second (km/s)?

SOLUTION: The **Key Idea** in chain-link conversions is to write the conversion factors as ratios that will eliminate unwanted units. Here we write

$$23 \text{ rides/h} = \left(23 \frac{\text{rides}}{\text{h}}\right)\left(\frac{4 \text{ stadia}}{1 \text{ ride}}\right)\left(\frac{6 \text{ plethra}}{1 \text{ stadium}}\right)$$
$$\times \left(\frac{30.8 \text{ m}}{1 \text{ plethron}}\right)\left(\frac{1 \text{ km}}{1000 \text{ m}}\right)\left(\frac{1 \text{ h}}{3600 \text{ s}}\right)$$
$$= 4.7227 \times 10^{-3} \text{ km/s} \approx 4.7 \times 10^{-3} \text{ km/s}.$$

(Answer)

Sample Problem 1-2

The cran is a British volume unit for freshly caught herrings: 1 cran = 170.474 liters (L) of fish, about 750 herrings. Suppose that, to be cleared through customs in Saudia Arabia, a shipment of 1255 crans must be declared in terms of cubic covidos, where the covido is an Arabic unit of length: 1 covido = 48.26 cm. What is the required declaration?

SOLUTION: From Appendix D we see that 1 L is equivalent to 1000 cm^3. A **Key Idea** then helps: To convert from *cubic* centimeters to *cubic* covidos, we must *cube* the conversion ratio between centimeters and covidos. Thus, we write the following chain-link conversion:

$$\begin{aligned}1255 \text{ crans} &= (1255 \text{ crans})\left(\frac{170.474 \text{ L}}{1 \text{ cran}}\right)\left(\frac{1000 \text{ cm}^3}{1 \text{ L}}\right)\left(\frac{1 \text{ covido}}{48.26 \text{ cm}}\right)^3 \\ &= 1.903 \times 10^3 \text{ covidos}^3. \qquad \text{(Answer)}\end{aligned}$$

PROBLEM-SOLVING TACTICS

Tactic 1: *Significant Figures and Decimal Places*

If you calculated the answer to Sample Problem 1-1 without your calculator automatically rounding it off, the number $4.722\,666\,666\,67 \times 10^{-3}$ might have appeared in the display. The precision implied by this number is meaningless. We rounded the answer to 4.7×10^{-3} km/s so as not to imply that it is more precise than the given data. The given speed of 23 rides/h consists of two digits, called **significant figures.** Thus, we rounded the answer to two significant figures. In this book, final results of calculations are often rounded to match the least number of significant figures in the given data. (However, sometimes an extra significant figure is kept.) When the leftmost of the digits to be discarded is 5 or more, the last remaining digit is rounded up; otherwise it is retained as is. For example, 11.3516 is rounded to three significant figures as 11.4 and 11.3279 is rounded to three significant figures as 11.3. (The answers to sample problems in this book are usually presented with the symbol = instead of ≈ even if rounding off is involved.)

When a number such as 3.15 or 3.15×10^3 is provided in a problem, the number of significant figures is apparent, but how about the number 3000? Is it known to only one significant figure (could it be written as 3×10^3)? Or is it known to as many as four significant figures (could it be written as 3.000×10^3)? In this book, we assume that all the zeros in such given numbers as 3000 are significant, but you had better not make that assumption elsewhere.

Don't confuse *significant figures* with *decimal places*. Consider the lengths 35.6 mm, 3.56 m, and 0.00356 m. They all have three significant figures but they have one, two, and five decimal places, respectively.

1-4 Length

In 1792, the newborn Republic of France established a new system of weights and measures. Its cornerstone was the meter, defined to be one ten-millionth of the distance from the north pole to the equator. Later, for practical reasons, this Earth standard was abandoned and the meter came to be defined as the distance between two fine lines engraved near the ends of a platinum–iridium bar, the **standard meter bar,** which was kept at the International Bureau of Weights and Measures near Paris. Accurate copies of the bar were sent to standardizing laboratories throughout the world. These **secondary standards** were used to produce other, still more accessible standards so that ultimately every measuring device derived its authority from the standard meter bar through a complicated chain of comparisons.

Eventually, modern science and technology required a more precise standard than the distance between two fine scratches on a metal bar. In 1960, a new standard for the meter, based on the wavelength of light, was adopted. Specifically, the standard for the meter was redefined to be 1 650 763.73 wavelengths of a particular orange-red light emitted by atoms of krypton-86 (a particular isotope, or type, of krypton) in a gas discharge tube. This awkward number of wavelengths was chosen so that the new standard would be close to the old meter-bar standard.

By 1983, however, the demand for higher precision had reached such a point that even the krypton-86 standard could not meet it, and in that year a bold step was taken. The meter was redefined as the distance traveled by light in a specified time interval. In the words of the 17th General Conference on Weights and Measures:

TABLE 1-3 Some Approximate Lengths

Measurement	Length in Meters
Distance to the first galaxies formed	2×10^{26}
Distance to the Andromeda galaxy	2×10^{22}
Distance to the nearest star (Proxima Centauri)	4×10^{16}
Distance to Pluto	6×10^{12}
Radius of Earth	6×10^{6} m
Height of Mt. Everest	9×10^{3}
Thickness of this page	1×10^{-4}
Length of a typical virus	1×10^{-8}
Radius of a hydrogen atom	5×10^{-11} 0.5 A°
Radius of a proton	1×10^{-15} 1 fm

The meter is the length of the path traveled by light in a vacuum during a time interval of 1/299 792 458 of a second.

This time interval was chosen so that the speed of light c is exactly

$$c = 299\,792\,458 \text{ m/s}.$$

Measurements of the speed of light had become extremely precise, so it made sense to adopt the speed of light as a defined quantity and to use it to redefine the meter.

Table 1-3 shows a wide range of lengths, from that of the universe to those of some very small objects.

PROBLEM-SOLVING TACTICS

Tactic 2: *Order of Magnitude*

The *order of magnitude* of a number is the power of ten when the number is expressed in scientific notation. For example, if $A = 2.3 \times 10^4$ and $B = 7.8 \times 10^4$, then the orders of magnitude of both A and B are 4.

Often, engineering and science professionals will estimate the result of a calculation to the *nearest* order of magnitude. For our example, the nearest order of magnitude is 4 for A and 5 for B. Such estimation is common when detailed or precise data required in the calculation are not known or easily found. Sample Problem 1-3 gives an example.

Sample Problem 1-3

The world's largest ball of string is about 2 m in radius. To the nearest order of magnitude, what is the total length L of the string in the ball?

SOLUTION: We could, of course, take the ball apart and measure the total length L, but that would take great effort and make the ball's builder most unhappy. A **Key Idea** here is that, because we want only the nearest order of magnitude, we can estimate any quantities required in the calculation.

Let us assume the ball is spherical with radius $R = 2$ m. The string in the ball is not closely packed (there are uncountable gaps between nearby sections of string). To allow for these gaps, let us somewhat overestimate the cross-sectional area of the string by assuming the cross section is square, with an edge length $d = 4$ mm. Then, with a cross-sectional area of d^2 and a length L, the string occupies a total volume of

$$V = (\text{cross-sectional area})(\text{length}) = d^2 L.$$

This is approximately equal to the volume of the ball, given by $\frac{4}{3}\pi R^3$, which is about $4R^3$ because π is about 3. Thus, we have

$$d^2 L = 4R^3,$$

or

$$L = \frac{4R^3}{d^2} = \frac{4(2\text{ m})^3}{(4 \times 10^{-3}\text{ m})^2}$$

$$= 2 \times 10^6 \text{ m} \approx 10^6 \text{ m} = 10^3 \text{ km}. \qquad \text{(Answer)}$$

(Note that you do not need a calculator for such a simplified calculation.) Thus, to the nearest order of magnitude, the ball contains about 1000 km of string!

TABLE 1-4 Some Approximate Time Intervals

Measurement	Time Interval in Seconds
Lifetime of the proton (predicted)	1×10^{39}
Age of the universe	5×10^{17}
Age of the pyramid of Cheops	1×10^{11}
Human life expectancy	2×10^{9}
Length of a day	9×10^{4}
Time between human heartbeats	8×10^{-1}
Lifetime of the muon	2×10^{-6}
Shortest lab light pulse	6×10^{-15}
Lifetime of the most unstable particle	1×10^{-23}
The Planck time[a]	1×10^{-43}

[a]This is the earliest time after the big bang at which the laws of physics as we know them can be applied.

Fig. 1-1 When the metric system was proposed in 1792, the hour was redefined to provide a 10-hour day. The idea did not catch on. The maker of this 10-hour watch wisely provided a small dial that kept conventional 12-hour time. Do the two dials indicate the same time?

1-5 Time

Time has two aspects. For civil and some scientific purposes, we want to know the time of day so that we can order events in sequence. In much scientific work, we want to know how long an event lasts. Thus, any time standard must be able to answer two questions: "*When* did it happen?" and "What is its *duration*?" Table 1-4 shows some time intervals.

Any phenomenon that repeats itself is a possible time standard. Earth's rotation, which determines the length of the day, has been used in this way for centuries; Fig. 1-1 shows one novel example of a watch based on that rotation. A quartz clock, in which a quartz ring is made to vibrate continuously, can be calibrated against Earth's rotation via astronomical observations and used to measure time intervals in the laboratory. However, the calibration cannot be carried out with the accuracy called for by modern scientific and engineering technology.

To meet the need for a better time standard, atomic clocks have been developed. An atomic clock at the National Institute of Standards and Technology (NIST) in Boulder, Colorado, is the standard for Coordinated Universal Time (UTC) in the United States. Its time signals are available by shortwave radio (stations WWV and WWVH) and by telephone (303-499-7111). Time signals (and related information) are also available from the United States Naval Observatory at Web site http://tycho.usno.navy.mil/time.html. (To set a clock extremely accurately at your particular location, you would have to account for the travel time that is required for these signals to reach you.)

Figure 1-2 shows variations in the length of one day on Earth over a 4-year period, as determined by comparison with a cesium (atomic) clock. Because the variation displayed by Fig. 1-2 is seasonal and repetitious, we suspect the rotating Earth when there is a difference between Earth and atom as timekeepers. The variation is probably due to tidal effects caused by the Moon and to large-scale winds.

The 13th General Conference on Weights and Measures in 1967 adopted a standard second based on the cesium clock:

> One second is the time taken by 9 192 631 770 oscillations of the light (of a specified wavelength) emitted by a cesium-133 atom.

Atomic clocks are so consistent that, in principle, two cesium clocks would have to run for 6000 years before their readings would differ by more than 1 s. Even such accuracy pales in comparison to that of clocks currently being developed; their precision may be 1 part in 10^{18}—that is, 1 s in 1×10^{18} s (about 3×10^{10} y).

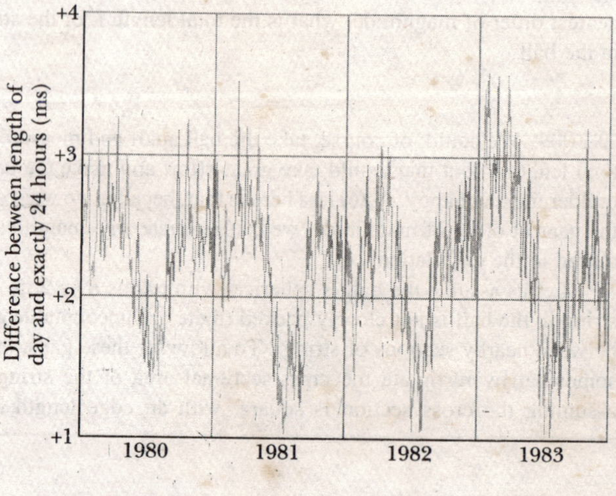

Fig. 1-2 Variations in the length of the day over a 4-year period. Note that the entire vertical scale amounts to only 3 ms (3 milliseconds = 0.003 s).

Sample Problem 1-4*

Suppose that while lying on a beach watching the Sun set over a calm ocean, you start a stopwatch just as the top of the Sun disappears. You then stand, elevating your eyes by a height $h = 1.70$ m, and stop the watch when the top of the Sun again disappears. If the elapsed time on the watch is $t = 11.1$ s, what is the radius r of Earth?

SOLUTION: A **Key Idea** here is that just as the Sun disappears, your line of sight to the top of the Sun is tangent to Earth's surface. Two such lines of sight are shown in Fig. 1-3. There your eyes are located at point A while you are lying, and at height h above point A while you are standing. For the latter situation, the line of sight is tangent to Earth's surface at point B. Let d represent the distance between point B and the location of your eyes when you are standing, and draw radii r as shown in Fig. 1-3. From the Pythagorean theorem, we then have

$$d^2 + r^2 = (r + h)^2 = r^2 + 2rh + h^2,$$

or

$$d^2 = 2rh + h^2. \tag{1-7}$$

Because the height h is so much smaller than Earth's radius r, the term h^2 is negligible compared to the term $2rh$, and we can rewrite Eq. 1-7 as

$$d^2 = 2rh. \tag{1-8}$$

In Fig. 1-3, the angle between the radii to the two tangent points A and B is θ, which is also the angle through which the Sun moves about Earth during the measured time $t = 11.1$ s. During a full day, which is approximately 24 h, the Sun moves through an angle of 360° about Earth. This allows us to write

$$\frac{\theta}{360°} = \frac{t}{24\ \text{h}},$$

*Adapted from "Doubling Your Sunsets, or How Anyone Can Measure the Earth's Size with a Wristwatch and Meter Stick," by Dennis Rawlins, *American Journal of Physics*, Feb. 1979, Vol. 47, pp. 126–128. This technique works best at the equator.

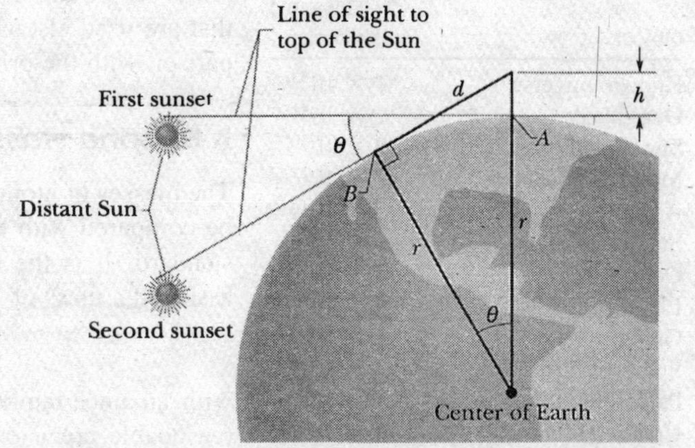

Fig. 1-3 Sample Problem 1-4. Your line of sight to the top of the setting Sun rotates through the angle θ when you stand up at point A and elevate your eyes by a distance h. (Angle θ and distance h are exaggerated here for clarity.)

which, with $t = 11.1$ s, gives us

$$\theta = \frac{(360°)(11.1\ \text{s})}{(24\ \text{h})(60\ \text{min/h})(60\ \text{s/min})} = 0.04625°.$$

Again in Fig. 1-3, we see that $d = r \tan \theta$. Substituting this for d in Eq. 1-8 gives us

$$r^2 \tan^2\theta = 2rh,$$

or

$$r = \frac{2h}{\tan^2\theta}.$$

Substituting $\theta = 0.04625°$ and $h = 1.70$ m, we find

$$r = \frac{(2)(1.70\ \text{m})}{\tan^2 0.04625°} = 5.22 \times 10^6\ \text{m}, \qquad \text{(Answer)}$$

which is within 20% of the accepted value (6.37×10^6 m) for the (mean) radius of Earth.

1-6 Mass

The Standard Kilogram

The SI standard of mass is a platinum–iridium cylinder (Fig. 1-4) kept at the International Bureau of Weights and Measures near Paris and assigned, by international agreement, a mass of 1 kilogram. Accurate copies have been sent to standardizing laboratories in other countries, and the masses of other bodies can be determined by balancing them against a copy. Table 1-5 shows some masses expressed in kilograms, ranging over about 83 orders of magnitude.

Fig. 1-4 The international 1 kg standard of mass, a platinum–iridium cylinder 3.9 cm in height and in diameter.

TABLE 1-5 Some Approximate Masses

Object	Mass in Kilograms
Known universe	1×10^{53}
Our galaxy	2×10^{41}
Sun	2×10^{30}
Moon	7×10^{22}
Asteroid Eros	5×10^{15}
Small mountain	1×10^{12}
Ocean liner	7×10^{7}
Elephant	5×10^{3}
Grape	3×10^{-3}
Speck of dust	7×10^{-10}
Penicillin molecule	5×10^{-17}
Uranium atom	4×10^{-25}
Proton	2×10^{-27}
Electron	9×10^{-31}

The U.S. copy of the standard kilogram is housed in a vault at NIST. It is removed, no more than once a year, for the purpose of checking duplicate copies that are used elsewhere. Since 1889, it has been taken to France twice for recomparison with the primary standard.

A Second Mass Standard

The masses of atoms can be compared with each other more precisely than they can be compared with the standard kilogram. For this reason, we have a second mass standard. It is the carbon-12 atom, which, by international agreement, has been assigned a mass of 12 **atomic mass units** (u). The relation between the two units is

$$1 \text{ u} = 1.6605402 \times 10^{-27} \text{ kg}, \tag{1-9}$$

with an uncertainty of ± 10 in the last two decimal places. Scientists can, with reasonable precision, experimentally determine the masses of other atoms relative to the mass of carbon-12. What we presently lack is a reliable means of extending that precision to more common units of mass, such as a kilogram.

REVIEW & SUMMARY

Measurement in Physics Physics is based on measurement of physical quantities. Certain physical quantities have been chosen as **base quantities** (such as length, time, and mass); each has been defined in terms of a **standard** and given a **unit** of measure (such as meter, second, and kilogram). Other physical quantities are defined in terms of the base quantities and their standards and units.

SI Units The unit system emphasized in this book is the International System of Units (SI). The three physical quantities displayed in Table 1-1 are used in the early chapters. Standards, which must be both accessible and invariable, have been established for these base quantities by international agreement. These standards are used in all physical measurement, for both the base quantities and the quantities derived from them. Scientific notation and the prefixes of Table 1-2 can be used to simplify measurement notation in many cases.

Changing Units Conversion of units from one system to another (for example, from miles per hour to kilometers per second) may be performed by using *chain-link conversions* in which the original data are multiplied successively by conversion factors written as unity and the units are manipulated like algebraic quantities until only the desired units remain.

Length The unit of length—the meter—is defined as the distance traveled by light during a precisely specified time interval.

Time The unit of time—the second—was formerly defined in terms of the rotation of Earth. It is now defined in terms of the oscillations of light emitted by an atomic (cesium-133) source. Accurate time signals are sent worldwide by radio signals keyed to atomic clocks in standardizing laboratories.

Mass The unit of mass—the kilogram—is defined in terms of a particular platinum–iridium prototype kept near Paris, France. For measurements on an atomic scale, the atomic mass unit, defined in terms of the atom carbon-12, is usually used.

EXERCISES & PROBLEMS

ssm Solution is in the Student Solutions Manual.
www Solution is available on the World Wide Web at: http://www.wiley.com/college/hrw
ilw Solution is available on the Interactive LearningWare.

SEC. 1-4 Length

1E. The micrometer (1 μm) is often called the *micron.* (a) How many microns make up 1.0 km? (b) What fraction of a centimeter equals 1.0 μm? (c) How many microns are in 1.0 yd? **ssm**

2E. Two types of *barrel* units were in use in the 1920s in the United States. The apple barrel had a legally set volume of 7056 cubic inches; the cranberry barrel, 5826 cubic inches. If a merchant sells 20 cranberry barrels of goods to a customer who thinks he is receiving apple barrels, what is the discrepancy in the shipment volume in liters?

3E. Horses are to race over a certain English meadow for a distance of 4.0 furlongs. What is the race distance in units of (a) rods and (b) chains? (1 furlong = 201.168 m, 1 rod = 5.0292 m, and 1 chain = 20.117 m.) **ssm**

4E. Spacing in this book was generally done in units of points and picas: 12 points = 1 pica, and 6 picas = 1 inch. If a figure was misplaced in the page proofs by 0.80 cm, what was the misplacement in (a) points and (b) picas?

5E. Earth is approximately a sphere of radius 6.37×10^6 m. What are (a) its circumference in kilometers, (b) its surface area in square kilometers, and (c) its volume in cubic kilometers? ssm

6E. An old manuscript reveals that a landowner in the time of King Arthur held 3.00 acres of plowed land plus a livestock area of 25.0 perches by 4.00 perches. What was the total area in (a) the old unit of roods and (b) the more modern unit of square meters? Here, 1 acre is an area of 40 perches by 4 perches, 1 rood is 40 perches by 1 perch, and 1 perch is 16.5 ft.

7P. Antarctica is roughly semicircular, with a radius of 2000 km (Fig. 1-5). The average thickness of its ice cover is 3000 m. How many cubic centimeters of ice does Antarctica contain? (Ignore the curvature of Earth.) ssm

Fig. 1-5 Problem 7.

8P. In the United States, a doll house has the scale of 1:12 of a real house (that is, each length of the doll house is $\frac{1}{12}$ that of the real house) and a miniature house (a doll house to fit within a doll house) has the scale of 1:144 of a real house. Suppose a real house (Fig. 1-6) has a front length of 20 m, a depth of 12 m, a height of 6.0 m, and a standard sloped roof (vertical triangular faces on the ends) of height 3.0 m. In cubic meters, what are the volumes of the corresponding (a) doll house and (b) miniature house?

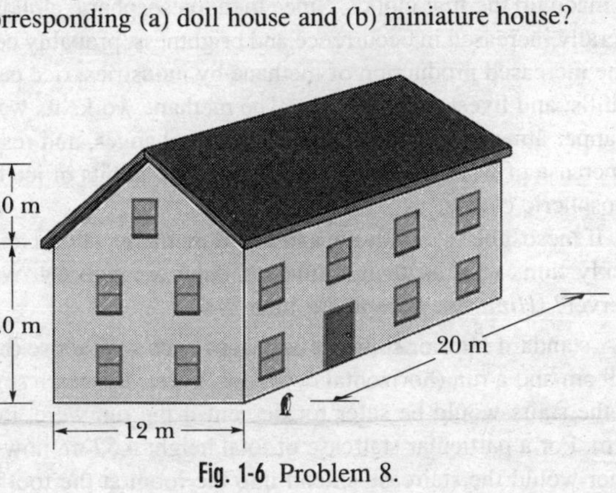

Fig. 1-6 Problem 8.

9P. Hydraulic engineers in the United States often use, as a unit of volume of water, the *acre-foot*, defined as the volume of water that will cover 1 acre of land to a depth of 1 ft. A severe thunderstorm dumped 2.0 in. of rain in 30 min on a town of area 26 km². What volume of water, in acre-feet, fell on the town? ssm ilw www

SEC. 1-5 Time

10E. Physicist Enrico Fermi once pointed out that a standard lecture period (50 min) is close to 1 microcentury. (a) How long is a microcentury in minutes? (b) Using

$$\text{percentage difference} = \left(\frac{\text{actual} - \text{approximation}}{\text{actual}}\right)100,$$

find the percentage difference from Fermi's approximation.

11E. Express the speed of light, 3.0×10^8 m/s, in (a) feet per nanosecond and (b) millimeters per picosecond. ssm

12E. A unit of time sometimes used in microscopic physics is the *shake*. One shake equals 10^{-8} s. (a) Are there more shakes in a second than there are seconds in a year? (b) Humans have existed for about 10^6 years, whereas the universe is about 10^{10} years old. If the age of the universe now is taken to be 1 "universe day," for how many "universe seconds" have humans existed?

13P. Five clocks are being tested in a laboratory. Exactly at noon, as determined by the WWV time signal, on successive days of a week the clocks read as in the following table. Rank the five clocks according to their relative value as good timekeepers, best to worst. Justify your choice. ssm

Clock	Sun.	Mon.	Tues.	Wed.	Thurs.	Fri.	Sat.
A	12:36:40	12:36:56	12:37:12	12:37:27	12:37:44	12:37:59	12:38:14
B	11:59:59	12:00:02	11:59:57	12:00:07	12:00:02	11:59:56	12:00:03
C	15:50:45	15:51:43	15:52:41	15:53:39	15:54:37	15:55:35	15:56:33
D	12:03:59	12:02:52	12:01:45	12:00:38	11:59:31	11:58:24	11:57:17
E	12:03:59	12:02:49	12:01:54	12:01:52	12:01:32	12:01:22	12:01:12

14P. Three digital clocks *A*, *B*, and *C* run at different rates and do not have simultaneous readings of zero. Figure 1-7 shows simultaneous readings on pairs of the clocks for four occasions. (At the earliest occasion, for example, *B* reads 25.0 s and *C* reads 92.0 s.) If two events are 600 s apart on clock *A*, how far apart are they on (a) clock *B* and (b) clock *C*? (c) When clock *A* reads 400 s, what does clock *B* read? (d) When clock *C* reads 15.0 s, what does clock *B* read? (Assume negative readings for prezero times.)

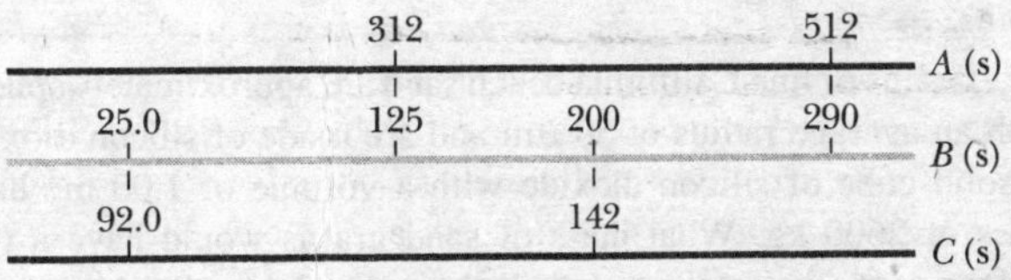

Fig. 1-7 Problem 14.

15P. An astronomical unit (AU) is the average distance of Earth from the Sun, approximately 1.50×10^8 km. The speed of light is about 3.0×10^8 m/s. Express the speed of light in terms of astronomical units per minute. ssm

16P. Until 1883, every city and town in the United States kept its own local time. Today, travelers reset their watches only when the time change equals 1.0 h. How far, on the average, must you travel in degrees of longitude until your watch must be reset by 1.0 h? (*Hint:* Earth rotates 360° in about 24 h.)

17P. Assuming the length of the day uniformly increases by 0.0010 s per century, calculate the cumulative effect on the measure of time over 20 centuries. (Such slowing of Earth's rotation is indicated by observations of the occurrences of solar eclipses during this period.) ssm www

18P. Time standards are now based on atomic clocks. A promising second standard is based on *pulsars,* which are rotating neutron stars (highly compact stars consisting only of neutrons). Some rotate at a rate that is highly stable, sending out a radio beacon that sweeps briefly across Earth once with each rotation, like a lighthouse beacon. Pulsar PSR 1937+21 is an example; it rotates once every 1.557 806 448 872 75 ± 3 ms, where the trailing ±3 indicates the uncertainty in the last decimal place (it does *not* mean ±3 ms). (a) How many times does PSR 1937+21 rotate in 7.00 days? (b) How much time does the pulsar take to rotate 1.0×10^6 times and (c) what is the associated uncertainty?

SEC. 1-6 Mass

19E. Earth has a mass of 5.98×10^{24} kg. The average mass of the atoms that make up Earth is 40 u. How many atoms are there in Earth? ssm

20P. Gold, which has a mass of 19.32 g for each cubic centimeter of volume, is the most ductile metal and can be pressed into a thin leaf or drawn out into a long fiber. (a) If 1.000 oz of gold, with a mass of 27.63 g, is pressed into a leaf of 1.000 μm thickness, what is the area of the leaf? (b) If, instead, the gold is drawn out into a cylindrical fiber of radius 2.500 μm, what is the length of the fiber?

21P (a) Assuming that each cubic centimeter of water has a mass of exactly 1 g, find the mass of one cubic meter of water in kilograms. (b) Suppose that it takes 10.0 h to drain a container of 5700 m^3 of water. What is the "mass flow rate," in kilograms per second, of water from the container? ssm

22P. What mass of water fell on the town in Problem 9 during the thunderstorm? One cubic meter of water has a mass of 10^3 kg.

23P. Iron has a mass of 7.87 g per cubic centimeter of volume, and the mass of an iron atom is 9.27×10^{-26} kg. If the atoms are spherical and tightly packed, (a) what is the volume of an iron atom and (b) what is the distance between the centers of adjacent atoms? ssm

24P. Grains of fine California beach sand are approximately spheres with an average radius of 50 μm and are made of silicon dioxide. A solid cube of silicon dioxide with a volume of 1.00 m^3 has a mass of 2600 kg. What mass of sand grains would have a total surface area (the total area of all the individual spheres) equal to the surface area of a cube 1 m on an edge?

Additional Problems

25. Harvard Bridge, which connects MIT with its fraternities across the Charles River, has a length of 364.4 Smoots plus one ear. The unit of one Smoot is based on the length of Oliver Reed Smoot, Jr., class of 1962, who was carried or dragged length by length across the bridge so that other pledge members of the Lambda Chi Alpha fraternity could mark off (with paint) 1-Smoot lengths along the bridge. The marks have been repainted biannually by fraternity pledges since the initial measurement, usually during times of traffic congestion so that the police could not easily interfere. (Presumably, the police were originally upset because a Smoot is not an SI base unit, but these days they seem to have accepted the unit.) Figure 1-8 shows three parallel paths, measured in Smoots (S), Willies (W), and Zeldas (Z). What is the length of 50.0 Smoots in (a) Willies and (b) Zeldas?

Fig. 1-8 Problem 25.

26. An old English children's rhyme states, "Little Miss Muffet sat on her tuffet, eating her curds and whey, when along came a spider who sat down beside her. . . ." The spider sat down not because of the curds and whey but because Miss Muffet had a stash of 11 tuffets of dried flies. The volume measure of a tuffet is given by 1 tuffet = 2 pecks = 0.50 bushel, where 1 Imperial (British) bushel = 36.3687 liters (L). What was Miss Muffet's stash in (a) pecks, (b) bushels, and (c) liters?

27. During the summers at high latitudes, ghostly, silver-blue clouds occasionally appear after sunset when common clouds are in Earth's shadow and are no longer visible. The ghostly clouds have been called *noctilucent clouds* (NLC), which means "luminous night clouds," but now are often called *mesospheric clouds,* after the *mesosphere,* the name of the atmosphere at the altitude of the clouds.

These clouds were first seen in June 1885, after dust and water from the massive 1883 volcanic explosion of Krakatoa Island (near Java in the Southeast Pacific) reached the high altitudes in the Northern Hemisphere. In the low temperatures of the mesosphere, the water collected and froze on the volcanic dust (and perhaps on comet and meteor dust already present there) to form the particles that made up the first clouds. Since then, mesospheric clouds have generally increased in occurrence and brightness, probably because of the increased production of methane by industries, rice paddies, landfills, and livestock flatulence. The methane works its way into the upper atmosphere, undergoes chemical changes, and results in an increase of water molecules there, and also in bits of ice for the mesospheric clouds.

If mesospheric clouds are spotted 38 min after sunset and then quickly dim, what is their altitude if they are directly over the observer? (*Hint:* See Sample Problem 1-4.)

28. A standard interior staircase has steps each with a rise (height) of 19 cm and a run (horizontal depth) of 23 cm. Research suggests that the stairs would be safer for descent if the run were, instead, 28 cm. For a particular staircase of total height 4.57 m, how much farther would the staircase extend into the room at the foot of the stairs if this change in run were made?

29. As a contrast between the old and the modern and between the large and the small, consider the following: In old rural England 1 hide (between 100 and 120 acres) was the area of land needed to sustain one family with a single plough for one year. (An area of 1 acre is equal to 4047 m^2.) Also, 1 wapentake was the area of land needed by 100 such families. In quantum physics, the cross-sectional area of a nucleus (defined in terms of the chance of a particle hitting and being absorbed by it) is measured in units of barns, where 1 barn is 1×10^{-28} m^2. (In nuclear physics jargon, if a nucleus is "large," then shooting a particle at it is like shooting a bullet at a barn door, which can hardly be missed.) What is the ratio of 25 wapentakes to 11 barns?

2 Motion Along a Straight Line

On September 26, 1993, Dave Munday, a diesel mechanic by trade, went over the Canadian edge of Niagara Falls for the second time, freely falling 48 m to the water (and rocks) below. On this attempt, he rode in a steel ball with a hole for air. Munday, keen on surviving this plunge that had killed four other stuntmen, had done considerable research on the physics and engineering aspects of the plunge.

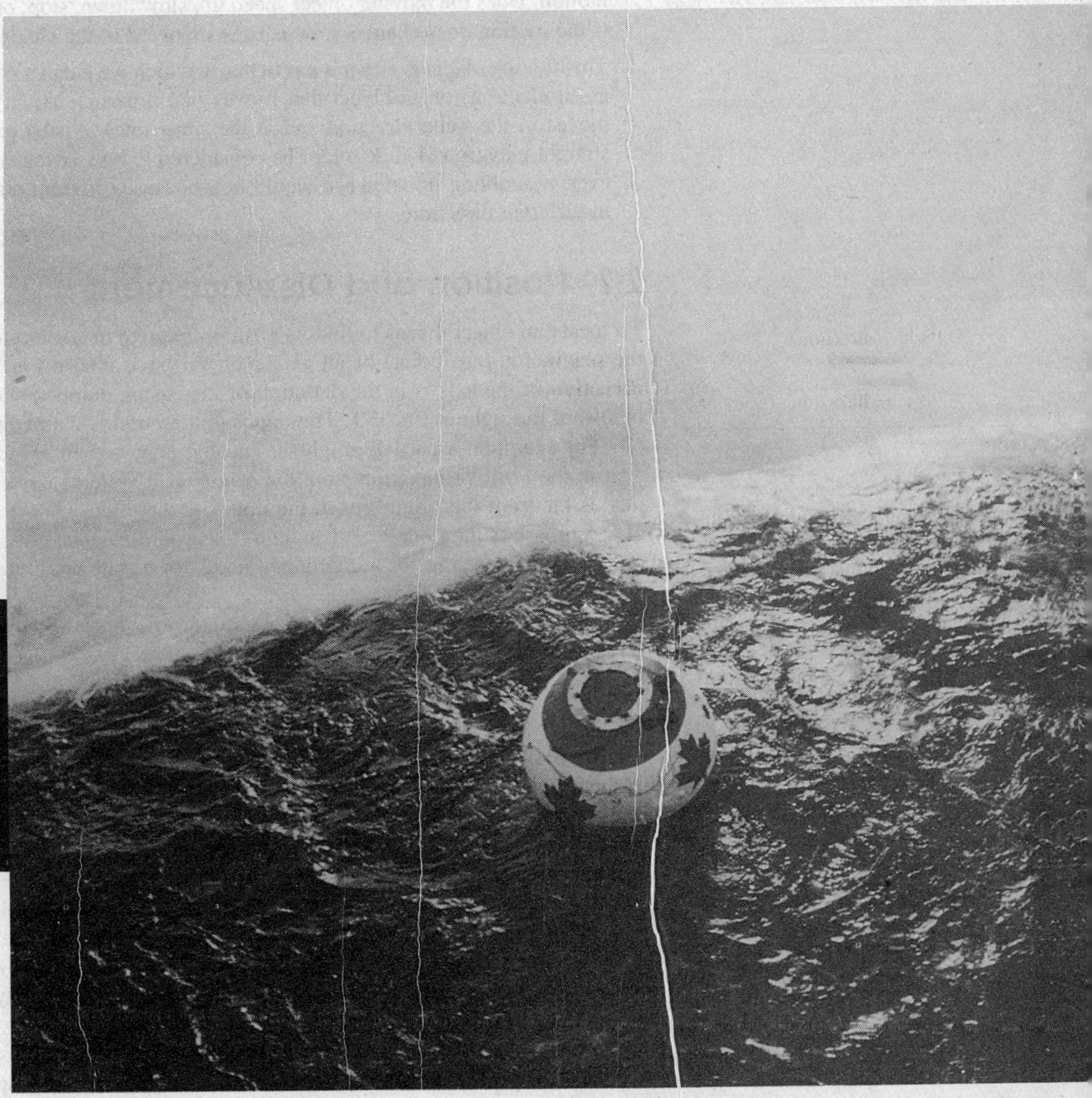

How long would he fall, and at what speed would he hit the turbulent water at the bottom?

The answer is in this chapter.

2-1 Motion

The world, and everything in it, moves. Even seemingly stationary things, such as a roadway, move with Earth's rotation, Earth's orbit around the Sun, the Sun's orbit around the center of the Milky Way galaxy, and that galaxy's migration relative to other galaxies. The classification and comparison of motions (called **kinematics**) is often challenging. What exactly do you measure, and how do you compare?

Before we attempt an answer, we shall examine some general properties of motion that is restricted in three ways.

1. The motion is along a straight line only. The line may be vertical (that of a falling stone), horizontal (that of a car on a level highway), or slanted, but it must be straight.
2. Forces (pushes and pulls) cause motion, but forces will not be discussed until Chapter 5. In this chapter you study only the motion itself, and changes in the motion. Does the moving object speed up, slow down, stop, or reverse direction? If the motion does change, how is time involved in the change?
3. The moving object is either a **particle** (by which we mean a pointlike object such as an electron) or an object that moves like a particle (such that every portion moves in the same direction and at the same rate). A stiff pig slipping down a straight playground slide might be considered to be moving like a particle; however, a tumbling tumbleweed would not, because different points inside it move in different directions.

2-2 Position and Displacement

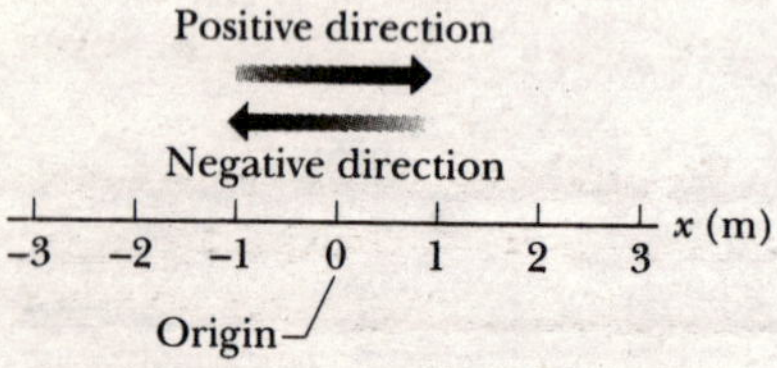

Fig. 2-1 Position is determined on an axis that is marked in units of length (here meters) and that extends indefinitely in opposite directions. The axis label, here x, is always on the positive side of the origin.

To locate an object means to find its position relative to some reference point, often the **origin** (or zero point) of an axis such as the x axis in Fig. 2-1. The **positive direction** of the axis is in the direction of increasing numbers (coordinates), which is toward the right in Fig. 2-1. The opposite direction is the **negative direction.**

For example, a particle might be located at $x = 5$ m, which means that it is 5 m in the positive direction from the origin. If it were at $x = -5$ m, it would be just as far from the origin but in the opposite direction. On the axis, a coordinate of -5 m is less than one of -1 m, and both coordinates are less than a coordinate of $+5$ m. A plus sign for a coordinate need not be shown, but a minus sign must always be shown.

A change from one position x_1 to another position x_2 is called a **displacement** Δx, where

$$\Delta x = x_2 - x_1. \qquad (2\text{-}1)$$

(The symbol Δ, the Greek uppercase delta, represents a change in a quantity, and it means the final value of that quantity minus the initial value.) When numbers are inserted for the position values x_1 and x_2, a displacement in the positive direction (toward the right in Fig. 2-1) always comes out positive, and one in the opposite direction (left in the figure), negative. For example, if the particle moves from $x_1 = 5$ m to $x_2 = 12$ m, then $\Delta x = (12 \text{ m}) - (5 \text{ m}) = +7$ m. The positive result indicates that the motion is in the positive direction. If the particle then returns to $x = 5$ m, the displacement for the full trip is zero. The actual number of meters covered for the full trip is irrelevant; displacement involves only the original and final positions.

A plus sign for a displacement need not be shown, but a minus sign must always be shown. If we ignore the sign (and thus the direction) of a displacement, we are

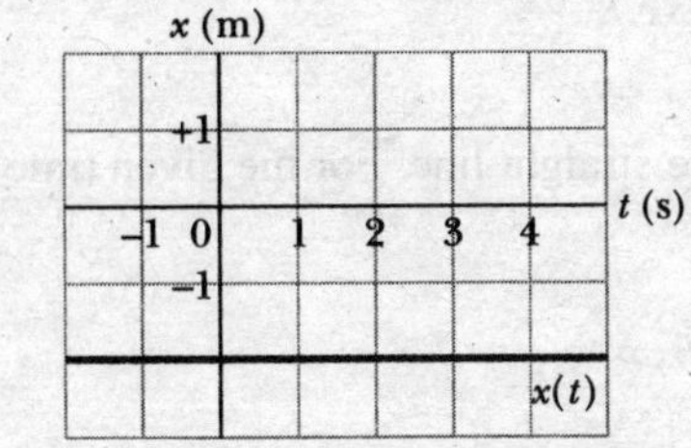

Fig. 2-2 The graph of $x(t)$ for an armadillo that is stationary at $x = -2$ m. The value of x is -2 m for all times t.

left with the **magnitude** (or absolute value) of the displacement. In the previous example, the magnitude of Δx is 7 m.

Displacement is an example of a **vector quantity,** which is a quantity that has both a direction and a magnitude. We explore vectors more fully in Chapter 3 (in fact, some of you may have already read that chapter), but here all we need is the idea that displacement has two features: (1) Its *magnitude* is the distance (such as the number of meters) between the original and final positions. (2) Its *direction*, from an original position to a final position, can be represented by a plus sign or a minus sign if the motion is along a single axis.

What follows is the first of many checkpoints that you will see in this book. Each consists of one or more questions whose answers require some reasoning or a mental calculation, and each gives you a quick check of your understanding. The answers are listed in the back of the book.

CHECKPOINT 1: Here are three pairs of initial and final positions, respectively, along an x axis. Which pairs give a negative displacement: (a) -3 m, $+5$ m; (b) -3 m, -7 m; (c) 7 m, -3 m?

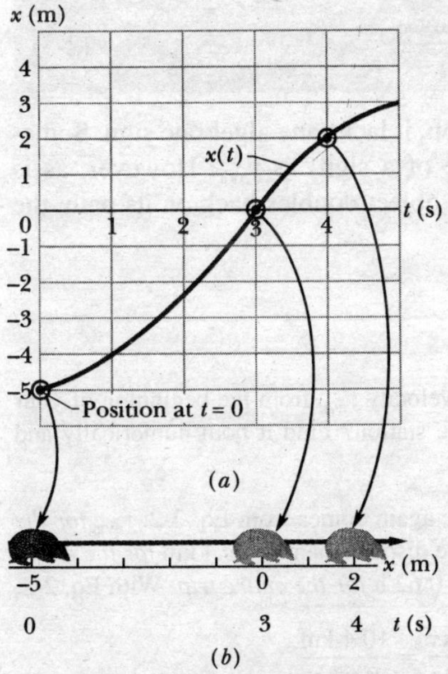

Fig. 2-3 (*a*) The graph of $x(t)$ for a moving armadillo. (*b*) The path associated with the graph. The scale below the x axis shows the times at which the armadillo reaches various x values.

2-3 Average Velocity and Average Speed

A compact way to describe position is with a graph of position x plotted as a function of time t—a graph of $x(t)$. (The notation $x(t)$ represents a function x of t, not the product x times t.) As a simple example, Fig. 2-2 shows the position function $x(t)$ for a stationary armadillo (which we treat as a particle) at $x = -2$ m.

Figure 2-3*a*, also for an armadillo, is more interesting, because it involves motion. The armadillo is apparently first noticed at $t = 0$ when it is at the position $x = -5$ m. It moves toward $x = 0$, passes through that point at $t = 3$ s, and then moves on to increasingly larger positive values of x.

Figure 2-3*b* depicts the actual straight-line motion of the armadillo and is something like what you would see. The graph in Fig. 2-3*a* is more abstract and quite unlike what you would see, but it is richer in information. It also reveals how fast the armadillo moves.

Actually, several quantities are associated with the phrase "how fast." One of them is the **average velocity** v_{avg}, which is the ratio of the displacement Δx that occurs during a particular time interval Δt to that interval:

$$v_{avg} = \frac{\Delta x}{\Delta t} = \frac{x_2 - x_1}{t_2 - t_1}. \tag{2-2}$$

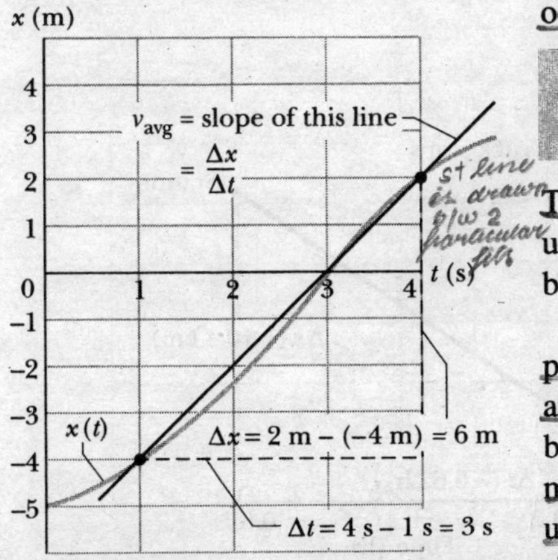

Fig. 2-4 Calculation of the average velocity between $t = 1$ s and $t = 4$ s as the slope of the line that connects the points (on the $x(t)$ curve) representing those times.

The notation means that the position is x_1 at time t_1 and then x_2 at time t_2. A common unit for v_{avg} is the meter per second (m/s). You may see other units in the problems, but they are always in the form of length/time.

On a graph of x versus t, v_{avg} is the **slope** of the straight line that connects two particular points on the $x(t)$ curve: one is the point that corresponds to x_2 and t_2, and the other is the point that corresponds to x_1 and t_1. Like displacement, v_{avg} has both magnitude and direction (it is another vector quantity). Its magnitude is the magnitude of the line's slope. A positive v_{avg} (and slope) tells us that the line slants upward to the right; a negative v_{avg} (and slope), that the line slants downward to the right. The average velocity v_{avg} always has the same sign as the displacement Δx because Δt in Eq. 2-2 is always positive.

Figure 2-4 shows how to find v_{avg} for the armadillo of Fig. 2-3, for the time interval $t = 1$ s to $t = 4$ s. We draw the straight line that connects the point on the position curve at the beginning of the interval and the point on the curve at the end of the interval. Then we find the slope $\Delta x/\Delta t$ of the straight line. For the given time interval, the average velocity is

$$v_{avg} = \frac{6\text{ m}}{3\text{ s}} = 2\text{ m/s}.$$

Average speed s_{avg} is a different way of describing "how fast" a particle moves. Whereas the average velocity involves the particle's displacement Δx, the average speed involves the total distance covered (for example, the number of meters moved), independent of direction; that is,

$$s_{avg} = \frac{\text{total distance}}{\Delta t}. \tag{2-3}$$

Because average speed does *not* include direction, it lacks any algebraic sign. Sometimes s_{avg} is the same (except for the absence of a sign) as v_{avg}. However, as is demonstrated in Sample Problem 2-1, when an object doubles back on its path, the two can be quite different.

Sample Problem 2-1

You drive a beat-up pickup truck along a straight road for 8.4 km at 70 km/h, at which point the truck runs out of gasoline and stops. Over the next 30 min, you walk another 2.0 km farther along the road to a gasoline station.

(a) What is your overall displacement from the beginning of your drive to your arrival at the station?

SOLUTION: Assume, for convenience, that you move in the positive direction of an x axis, from a first position of $x_1 = 0$ to a second position of x_2 at the station. That second position must be at $x_2 = 8.4\text{ km} + 2.0\text{ km} = 10.4\text{ km}$. Then the **Key Idea** here is that your displacement Δx along the x axis is the second position minus the first position. From Eq. 2-1, we have

$$\Delta x = x_2 - x_1 = 10.4\text{ km} - 0 = 10.4\text{ km}. \quad \text{(Answer)}$$

Thus, your overall displacement is 10.4 km in the positive direction of the x axis.

(b) What is the time interval Δt from the beginning of your drive to your arrival at the station?

SOLUTION: We already know the time interval Δt_{wlk} (= 0.50 h) for the walk, but we lack the time interval Δt_{dr} for the drive. However, we know that for the drive the displacement Δx_{dr} is 8.4 km and the average velocity $v_{avg,dr}$ is 70 km/h. A **Key Idea** to use here comes from Eq. 2-2: This average velocity is the ratio of the displacement for the drive to the time interval for the drive:

$$v_{avg,dr} = \frac{\Delta x_{dr}}{\Delta t_{dr}}.$$

Rearranging and substituting data then give us

$$\Delta t_{dr} = \frac{\Delta x_{dr}}{v_{avg,dr}} = \frac{8.4\text{ km}}{70\text{ km/h}} = 0.12\text{ h}.$$

So,

$$\Delta t = \Delta t_{dr} + \Delta t_{wlk}$$
$$= 0.12\text{ h} + 0.50\text{ h} = 0.62\text{ h}. \quad \text{(Answer)}$$

(c) What is your average velocity v_{avg} from the beginning of your drive to your arrival at the station? Find it both numerically and graphically.

SOLUTION: The **Key Idea** here again comes from Eq. 2-2: v_{avg} *for the entire trip* is the ratio of the displacement of 10.4 km *for the entire trip* to the time interval of 0.62 h *for the entire trip*. With Eq. 2-2, we find it is

$$v_{avg} = \frac{\Delta x}{\Delta t} = \frac{10.4\text{ km}}{0.62\text{ h}}$$
$$= 16.8\text{ km/h} \approx 17\text{ km/h}. \quad \text{(Answer)}$$

To find v_{avg} graphically, first we graph $x(t)$ as shown in Fig. 2-5, where the beginning and arrival points on the graph are the origin and the point labeled as "Station." The **Key Idea** here is that your average velocity is the slope of the straight line connecting those

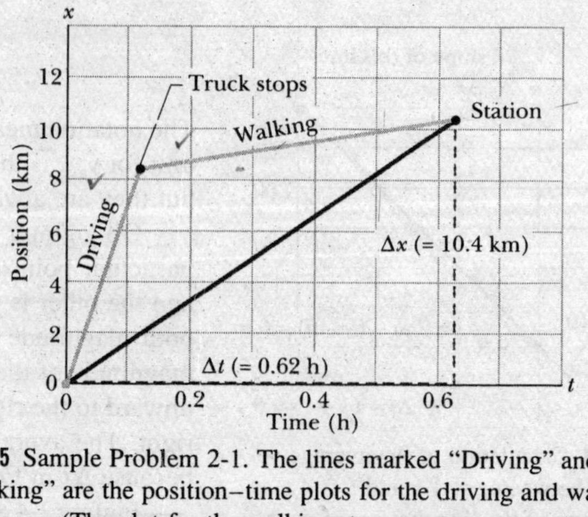

Fig. 2-5 Sample Problem 2-1. The lines marked "Driving" and "Walking" are the position–time plots for the driving and walking stages. (The plot for the walking stage assumes a constant rate of walking.) The slope of the straight line joining the origin and the point labeled "Station" is the average velocity for the trip, from the beginning to the station.

points; that is, it is the ratio of the *rise* ($\Delta x = 10.4$ km) to the *run* ($\Delta t = 0.62$ h), which gives us $v_{avg} = 16.8$ km/h.

(d) Suppose that to pump the gasoline, pay for it, and walk back to the truck takes you another 45 min. What is your average speed from the beginning of your drive to your return to the truck with the gasoline?

SOLUTION: The **Key Idea** here is that your average speed is the ratio of the total distance you move to the total time interval you take to make that move. The total distance is 8.4 km + 2.0 km + 2.0 km = 12.4 km. The total time interval is 0.12 h + 0.50 h + 0.75 h = 1.37 h. Thus, Eq. 2-3 gives us

$$s_{avg} = \frac{12.4 \text{ km}}{1.37 \text{ h}} = 9.1 \text{ km/h}. \qquad \text{(Answer)}$$

CHECKPOINT 2: In this sample problem, suppose that right after refueling the truck, you drive back to x_1 at 35 km/h. What is your average velocity for your entire trip?

PROBLEM-SOLVING TACTICS

Tactic 1: *Do You Understand the Problem?*
For beginning problem solvers, no difficulty is more common than simply not understanding the problem. The best test of understanding is this: Can you explain the problem in your own words?

Write down the given data, with units, using the symbols of the chapter. (In Sample Problem 2-1, the given data allow you to find your net displacement Δx in part (a) and the corresponding time interval Δt in part (b).) Identify the unknown and its symbol. (In the sample problem, the unknown in part (c) is your average velocity v_{avg}.) Then find the connection between the unknown and the data. (The connection is Eq. 2-2, the definition of average velocity.)

Tactic 2: *Are the Units OK?*
Be sure to use a consistent set of units when putting numbers into the equations. In Sample Problem 2-1, the logical units in terms of the given data are kilometers for distances, hours for time intervals, and kilometers per hour for velocities. You may sometimes need to make conversions.

Tactic 3: *Is Your Answer Reasonable?*
Does your answer make sense? Is it far too large or far too small? Is the sign correct? Are the units appropriate? In part (c) of Sample Problem 2-1, for example, the correct answer is 17 km/h. If you find 0.00017 km/h, −17 km/h, 17 km/s, or 17,000 km/h, you should realize at once that you have done something wrong. The error may lie in your method, in your algebra, or in your keystroking of numbers on a calculator.

Tactic 4: *Reading a Graph*
Figures 2-2, 2-3*a*, 2-4, and 2-5 are graphs that you should be able to read easily. In each graph, the variable on the horizontal axis is the time t, with the direction of increasing time to the right. In each, the variable on the vertical axis is the position x of the moving particle with respect to the origin, with the positive direction of x upward. Always note the units (seconds or minutes; meters or kilometers) in which the variables are expressed.

2-4 Instantaneous Velocity and Speed

You have now seen two ways to describe how fast something moves: average velocity and average speed, both of which are measured over a time interval Δt. However, the phrase "how fast" more commonly refers to how fast a particle is moving at a given instant—and that is its **instantaneous velocity** (or simply **velocity**) v.

The velocity at any instant is obtained from the average velocity by shrinking the time interval Δt closer and closer to 0. As Δt dwindles, the average velocity approaches a limiting value, which is the velocity at that instant:

$$v = \lim_{\Delta t \to 0} \frac{\Delta x}{\Delta t} = \frac{dx}{dt}. \qquad (2\text{-}4)$$

This equation displays two features of the instantaneous velocity v. First, v is the rate at which the particle's position x is changing with time at a given instant; that is, v is the derivative of x with respect to t. Second, v at any instant is the slope of the particle's position–time curve at the point representing that instant. Velocity is another vector quantity and thus has an associated direction.

Speed is the magnitude of velocity; that is, speed is velocity that has been stripped of any indication of direction, either in words or via an algebraic sign. (*Caution:* Speed and average speed can be quite different.) A velocity of +5 m/s and one of −5 m/s both have an associated speed of 5 m/s. The speedometer in a car measures the speed, not the velocity, because it cannot determine the direction.

Sample Problem 2-2

Figure 2-6*a* is an $x(t)$ plot for an elevator that is initially stationary, then moves upward (which we take to be the positive direction of x), and then stops. Plot v as a function of time.

SOLUTION: The **Key Idea** here is that we can find the velocity at any time from the slope of the curve of $x(t)$ at that time. The slope of $x(t)$, and so also the velocity, is zero in the intervals from 0 to 1 s and from 9 s on, so then the cab is stationary. During the interval *bc* the slope is constant and nonzero; so then the cab moves with constant velocity. We calculate the slope of $x(t)$ then as

$$\frac{\Delta x}{\Delta t} = v = \frac{24\text{ m} - 4.0\text{ m}}{8.0\text{ s} - 3.0\text{ s}} = +4.0\text{ m/s}.$$

The plus sign indicates that the cab is moving in the positive x direction. These intervals (where $v = 0$ and $v = 4$ m/s) are plotted in Fig. 2-6*b*. In addition, as the cab initially begins to move and then later slows to a stop, v varies as indicated in the intervals 1 s to 3 s and 8 s to 9 s. Thus, Fig. 2-6*b* is the required plot. (Figure 2-6*c* is considered in Section 2-5.)

Given a $v(t)$ graph such as Fig. 2-6*b*, we could "work backward" to produce the shape of the associated $x(t)$ graph (Fig. 2-6*a*). However, we would not know the actual values for x at various times, because the $v(t)$ graph indicates only *changes* in x. To find such a change in x during any interval, we must, in the language of calculus, calculate the area "under the curve" on the $v(t)$ graph for that interval. For example, during the interval 3 s to 8 s in which the cab has a velocity of 4.0 m/s, the change in x is

$$\Delta x = (4.0\text{ m/s})(8.0\text{ s} - 3.0\text{ s}) = +20\text{ m}.$$

(This area is positive because the $v(t)$ curve is above the t axis.) Figure 2-6*a* shows that x does indeed increase by 20 m in that interval. However, Fig. 2-6*b* does not tell us the *values* of x at the beginning and end of the interval. For that we need additional information such as the value of x for some given instant.

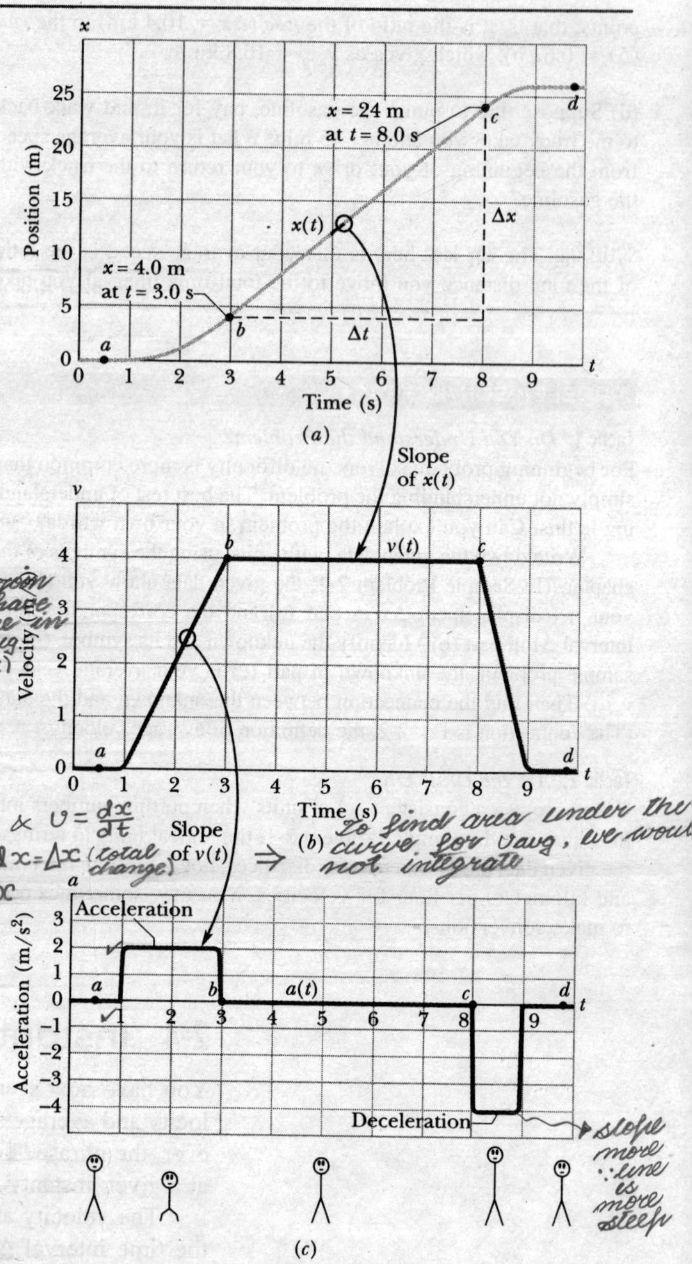

Fig. 2-6 Sample Problem 2-2. (*a*) The $x(t)$ curve for an elevator cab that moves upward along an x axis. (*b*) The $v(t)$ curve for the cab. Note that it is the derivative of the $x(t)$ curve ($v = dx/dt$). (*c*) The $a(t)$ curve for the cab. It is the derivative of the $v(t)$ curve ($a = dv/dt$). The stick figures along the bottom suggest how a passenger's body might feel during the accelerations.

Sample Problem 2-3

The position of a particle moving on an x axis is given by

$$x = 7.8 + 9.2t - 2.1t^3, \quad (2\text{-}5)$$

with x in meters and t in seconds. What is its velocity at t = 3.5 s? Is the velocity constant, or is it continuously changing?

SOLUTION: For simplicity, the units have been omitted from Eq. 2-5, but you can insert them if you like by changing the coefficients to 7.8 m, 9.2 m/s, and -2.1 m/s³. The **Key Idea** here is that velocity is the first derivative (with respect to time) of the position function $x(t)$. Thus, we write

$$v = \frac{dx}{dt} = \frac{d}{dt}(7.8 + 9.2t - 2.1t^3),$$

which becomes

$$v = 0 + 9.2 - (3)(2.1)t^2 = 9.2 - 6.3t^2. \quad (2\text{-}6)$$

At t = 3.5 s,

$$v = 9.2 - (6.3)(3.5)^2 = -68\text{ m/s}. \quad \text{(Answer)}$$

At t = 3.5 s, the particle is moving in the negative direction of x (note the minus sign) with a speed of 68 m/s. Since the quantity t appears in Eq. 2-6, the velocity v depends on t and so is continuously changing.

CHECKPOINT 3: The following equations give the position $x(t)$ of a particle in four situations (in each equation, x is in meters, t is in seconds, and $t > 0$): (1) $x = 3t - 2$; (2) $x = -4t^2 - 2$; (3) $x = 2/t^2$; and (4) $x = -2$. (a) In which situation is the velocity v of the particle constant? (b) In which is v in the negative x direction?

2-5 Acceleration

When a particle's velocity changes, the particle is said to undergo **acceleration** (or to accelerate). For motion along an axis, the **average acceleration** a_{avg} over a time interval Δt is

$$a_{avg} = \frac{v_2 - v_1}{t_2 - t_1} = \frac{\Delta v}{\Delta t}, \tag{2-7}$$

where the particle has velocity v_1 at time t_1 and then velocity v_2 at time t_2. The **instantaneous acceleration** (or simply **acceleration**) is the derivative of the velocity with respect to time:

$$a = \frac{dv}{dt}. \tag{2-8}$$

In words, the acceleration of a particle at any instant is the rate at which its velocity is changing at that instant. Graphically, the acceleration at any point is the slope of the curve of $v(t)$ at that point.

We can combine Eq. 2-8 with Eq. 2-4 to write

$$a = \frac{dv}{dt} = \frac{d}{dt}\left(\frac{dx}{dt}\right) = \frac{d^2x}{dt^2}. \tag{2-9}$$

In words, the acceleration of a particle at any instant is the second derivative of its position $x(t)$ with respect to time.

A common unit of acceleration is the meter per second per second: m/(s · s) or m/s^2. You will see other units in the problems, but they will each be in the form of length/(time · time) or length/time^2. Acceleration has both magnitude and direction (it is yet another vector quantity). Its algebraic sign represents its direction on an axis just as for displacement and velocity; that is, acceleration with a positive value is in the positive direction of an axis, and acceleration with a negative value is in the negative direction.

Figure 2-6c is a plot of the acceleration of the elevator cab discussed in Sample Problem 2-2. Compare this $a(t)$ curve with the $v(t)$ curve—each point on the $a(t)$ curve shows the derivative (slope) of the $v(t)$ curve at the corresponding time. When v is constant (at either 0 or 4 m/s), the derivative is zero and so also is the acceleration. When the cab first begins to move, the $v(t)$ curve has a positive derivative (the slope is positive), which means that $a(t)$ is positive. When the cab slows to a stop, the derivative and slope of the $v(t)$ curve are negative; that is, $a(t)$ is negative.

Next compare the slopes of the $v(t)$ curve during the two acceleration periods. The slope associated with the cab's slowing down (commonly called "deceleration") is steeper, because the cab stops in half the time it took to get up to speed. The steeper slope means that the magnitude of the deceleration is larger than that of the acceleration, as indicated in Fig. 2-6c.

The sensations you would feel while riding in the cab of Fig. 2-6 are indicated by the sketched figures. When the cab first accelerates, you feel as though you are pressed downward; when later the cab is braked to a stop, you seem to be stretched upward. In between, you feel nothing special. Your body reacts to accelerations (it is an accelerometer) but not to velocities (it is not a speedometer). When you are in

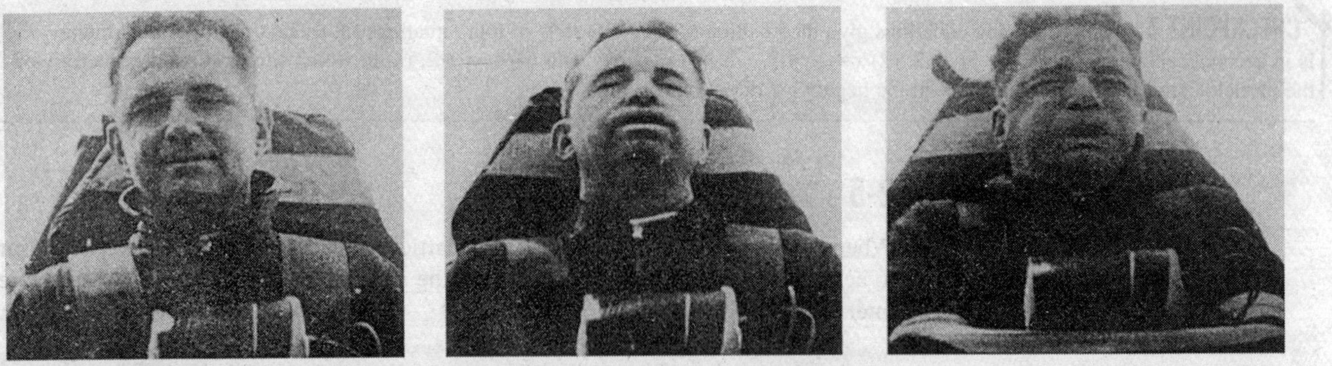

Fig. 2-7 Colonel J. P. Stapp in a rocket sled as it is brought up to high speed (acceleration out of the page) and then very rapidly braked (acceleration into the page).

a car traveling at 90 km/h or an airplane traveling at 900 km/h, you have no bodily awareness of the motion. However, if the car or plane quickly changes velocity, you may become keenly aware of the change, perhaps even frightened by it. Part of the thrill of an amusement park ride is due to the quick changes of velocity that you undergo (you pay for the accelerations, not for the speed). A more extreme example is shown in the photographs of Fig. 2-7, which were taken while a rocket sled was rapidly accelerated along a track and then rapidly braked to a stop.

Large accelerations are sometimes expressed in terms of g units, with

$$1g = 9.8 \text{ m/s}^2 \qquad (g \text{ unit}). \tag{2-10}$$

(As we shall discuss in Section 2-8, g is the magnitude of the acceleration of a falling object near Earth's surface.) On a roller coaster, you may experience brief accelerations up to $3g$, which is $(3)(9.8 \text{ m/s}^2)$ or about 29 m/s^2, more than enough to justify the cost of the ride.

PROBLEM-SOLVING TACTICS

Tactic 5: *An Acceleration's Sign*

In common language, the sign of an acceleration has a nonscientific meaning: positive acceleration means that the speed of an object is increasing, and negative acceleration means that the speed is decreasing (the object is decelerating). In this book, however, the sign of an acceleration indicates a direction, not whether an object's speed is increasing or decreasing.

For example, if a car with an initial velocity $v = -25$ m/s is braked to a stop in 5.0 s, then $a_{avg} = +5.0 \text{ m/s}^2$. The acceleration is *positive*, but the car's speed has decreased. The reason is the difference in signs: the direction of the acceleration is opposite that of the velocity.

Here then is the proper way to interpret the signs:

> If the signs of the velocity and acceleration of a particle are the same, the speed of the particle increases. If the signs are opposite, the speed decreases.

CHECKPOINT 4: A wombat moves along an x axis. What is the sign of its acceleration if it is moving (a) in the positive direction with increasing speed, (b) in the positive direction with decreasing speed, (c) in the negative direction with increasing speed, and (d) in the negative direction with decreasing speed?

Sample Problem 2-4

A particle's position on the x axis of Fig. 2-1 is given by

$$x = 4 - 27t + t^3,$$

with x in meters and t in seconds.

(a) Find the particle's velocity function $v(t)$ and acceleration function $a(t)$.

SOLUTION: One **Key Idea** is that to get the velocity function $v(t)$, we differentiate the position function $x(t)$ with respect to time. Here we find

$$v = -27 + 3t^2, \qquad \text{(Answer)}$$

with v in meters per second.

Another **Key Idea** is that to get the acceleration function $a(t)$,

Fig. 2-7 *Continued*

we differentiate the velocity function $v(t)$ with respect to time. This gives us

$$a = +6t, \qquad \text{(Answer)}$$

with a in meters per second squared.

(b) Is there ever a time when $v = 0$?

SOLUTION: Setting $v(t) = 0$ yields

$$0 = -27 + 3t^2,$$

which has the solution

$$t = \pm 3 \text{ s.} \qquad \text{(Answer)}$$

Thus, the velocity is zero both 3 s before and 3 s after the clock reads 0.

(c) Describe the particle's motion for $t \geq 0$.

SOLUTION: The **Key Idea** is to examine the expressions for $x(t)$, $v(t)$, and $a(t)$.

At $t = 0$, the particle is at $x(0) = +4$ m and is moving with a velocity of $v(0) = -27$ m/s—that is, in the negative direction of the x axis. Its acceleration is $a(0) = 0$, because just then the particle's velocity is not changing.

For $0 < t < 3$ s, the particle still has a negative velocity, so it continues to move in the negative direction. However, its acceleration is no longer 0 but is increasing and positive. Because the signs of the velocity and the acceleration are opposite, the particle must be slowing.

Indeed, we already know that it stops momentarily at $t = 3$ s. Just then the particle is as far to the left of the origin in Fig. 2-1 as it will ever get. Substituting $t = 3$ s into the expression for $x(t)$, we find that the particle's position just then is $x = -50$ m. Its acceleration is still positive.

For $t > 3$ s, the particle moves to the right on the axis. Its acceleration remains positive and grows progressively larger in magnitude. The velocity is now positive, and it too grows progressively larger in magnitude.

2-6 Constant Acceleration: A Special Case

In many types of motion, the acceleration is either constant or approximately so. For example, you might accelerate a car at an approximately constant rate when a traffic light turns from red to green. Then graphs of your position, velocity, and acceleration would resemble those in Fig. 2-8. (Note that $a(t)$ in Fig. 2-8c is constant, which requires that $v(t)$ in Fig. 2-8b have a constant slope.) Later when you brake the car to a stop, the deceleration might also be approximately constant.

Such cases are so common that a special set of equations has been derived for dealing with them. One approach to the derivation of these equations is given in this section. A second approach is given in the next section. Throughout both sections and later when you work on the homework problems, keep in mind that *these equations are valid only for constant acceleration (or situations in which you can approximate the acceleration as being constant)*.

When the acceleration is constant, the average acceleration and instantaneous acceleration are equal and we can write Eq. 2-7, with some changes in notation, as

$$a = a_{avg} = \frac{v - v_0}{t - 0}.$$

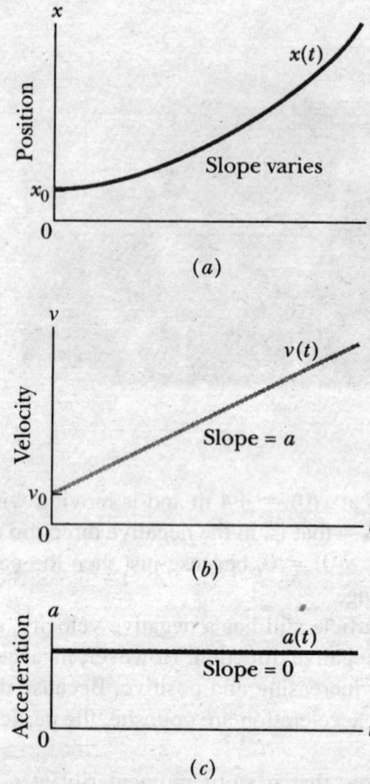

Fig. 2-8 (*a*) The position $x(t)$ of a particle moving with constant acceleration. (*b*) Its velocity $v(t)$, given at each point by the slope of the curve in (*a*). (*c*) Its (constant) acceleration, equal to the (constant) slope of the curve of $v(t)$.

Here v_0 is the velocity at time $t = 0$, and v is the velocity at any later time t. We can recast this equation as

$$v = v_0 + at. \tag{2-11}$$

As a check, note that this equation reduces to $v = v_0$ for $t = 0$, as it must. As a further check, take the derivative of Eq. 2-11. Doing so yields $dv/dt = a$, which is the definition of a. Figure 2-8*b* shows a plot of Eq. 2-11, the $v(t)$ function; the function is linear and thus the plot is a straight line.

In a similar manner we can rewrite Eq. 2-2 (with a few changes in notation) as

$$v_{avg} = \frac{x - x_0}{t - 0}$$

and then as

$$x = x_0 + v_{avg}t, \tag{2-12}$$

in which x_0 is the position of the particle at $t = 0$, and v_{avg} is the average velocity between $t = 0$ and a later time t.

For the linear velocity function in Eq. 2-11, the *average* velocity over any time interval (say, from $t = 0$ to a later time t) is the average of the velocity at the beginning of the interval ($= v_0$) and the velocity at the end of the interval ($= v$). For the interval from $t = 0$ to the later time t then, the average velocity is

$$v_{avg} = \tfrac{1}{2}(v_0 + v). \tag{2-13}$$

Substituting the right side of Eq. 2-11 for v yields, after a little rearrangement,

$$v_{avg} = v_0 + \tfrac{1}{2}at. \tag{2-14}$$

Finally, substituting Eq. 2-14 into Eq. 2-12 yields

$$x - x_0 = v_0t + \tfrac{1}{2}at^2. \tag{2-15}$$

As a check, note that putting $t = 0$ yields $x = x_0$, as it must. As a further check, taking the derivative of Eq. 2-15 yields Eq. 2-11, again as it must. Figure 2-8*a* shows a plot of Eq. 2-15; the function is quadratic and thus the plot is curved.

Equations 2-11 and 2-15 are the *basic equations for constant acceleration*; they can be used to solve any constant acceleration problem in this book. However, we can derive other equations that might prove useful in certain specific situations. First, note that five quantities can possibly be involved in any problem regarding constant acceleration, namely, $x - x_0$, v, t, a, and v_0. Usually, one of these quantities is *not* involved in the problem, *either as a given or as an unknown*. We are then presented with three of the remaining quantities and asked to find the fourth.

Equations 2-11 and 2-15 each contain four of these quantities, but not the same four. In Eq. 2-11, the "missing ingredient" is the displacement, $x - x_0$. In Eq. 2-15, it is the velocity v. These two equations can also be combined in three ways to yield three additional equations, each of which involves a different "missing variable." First, we can eliminate t to obtain

$$v^2 = v_0^2 + 2a(x - x_0). \tag{2-16}$$

This equation is useful if we do not know t and are not required to find it. Second, we can eliminate the acceleration a between Eqs. 2-11 and 2-15 to produce an equation in which a does not appear:

$$x - x_0 = \tfrac{1}{2}(v_0 + v)t. \tag{2-17}$$

TABLE 2-1 Equations for Motion with Constant Acceleration[a]

Equation Number	Equation	Missing Quantity
2-11	$v = v_0 + at$	$x - x_0$
2-15	$x - x_0 = v_0 t + \frac{1}{2}at^2$	v
2-16	$v^2 = v_0^2 + 2a(x - x_0)$	t
2-17	$x - x_0 = \frac{1}{2}(v_0 + v)t$	a
2-18	$x - x_0 = vt - \frac{1}{2}at^2$	v_0

[a] Make sure that the acceleration is indeed constant before using the equations in this table.

Finally, we can eliminate v_0, obtaining

$$x - x_0 = vt - \tfrac{1}{2}at^2. \tag{2-18}$$

Note the subtle difference between this equation and Eq. 2-15. One involves the initial velocity v_0; the other involves the velocity v at time t.

Table 2-1 lists the basic constant acceleration equations (Eqs. 2-11 and 2-15) as well as the specialized equations that we have derived. To solve a simple constant acceleration problem, you can usually use an equation from this list (*if* you have the list). Choose an equation for which the only unknown variable is the variable requested in the problem. A simpler plan is to remember only Eqs. 2-11 and 2-15, and then solve them as simultaneous equations whenever needed. An example is given in Sample Problem 2-5.

CHECKPOINT 5: The following equations give the position $x(t)$ of a particle in four situations: (1) $x = 3t - 4$; (2) $x = -5t^3 + 4t^2 + 6$; (3) $x = 2/t^2 - 4/t$; (4) $x = 5t^2 - 3$. To which of these situations do the equations of Table 2-1 apply?

Sample Problem 2-5

Spotting a police car, you brake a Porsche from a speed of 100 km/h to a speed of 80.0 km/h during a displacement of 88.0 m, at a constant acceleration.

(a) What is that acceleration?

SOLUTION: Assume that the motion is along the positive direction of an x axis. For simplicity, let us take the beginning of the braking to be at time $t = 0$, at position x_0. The **Key Idea** here is that, with the acceleration constant, we can relate the car's acceleration to its velocity and displacement via the basic constant acceleration equations (Eqs. 2-11 and 2-15). The initial velocity is $v_0 = 100\text{ km/h} = 27.78\text{ m/s}$, the displacement is $x - x_0 = 88.0$ m, and the velocity at the end of that displacement is $v = 80.0\text{ km/h} = 22.22\text{ m/s}$. However, we do not know the acceleration a and time t, which appear in both basic equations. So, we must solve those equations simultaneously.

To eliminate the unknown t, we use Eq. 2-11 to write

$$t = \frac{v - v_0}{a}, \tag{2-19}$$

and then we substitute this expression into Eq. 2-15 to write

$$x - x_0 = v_0\left(\frac{v - v_0}{a}\right) + \tfrac{1}{2}a\left(\frac{v - v_0}{a}\right)^2.$$

Solving for a and substituting known data then yield

$$a = \frac{v^2 - v_0^2}{2(x - x_0)} = \frac{(22.22\text{ m/s})^2 - (27.78\text{ m/s})^2}{2(88.0\text{ m})}$$

$$= -1.58\text{ m/s}^2. \qquad \text{(Answer)}$$

Note that we could have used Eq. 2-16 instead to solve for a because the unknown t is the missing variable in that equation.

(b) How much time is required for the given decrease in speed?

SOLUTION: Now that we know a, we can use Eq. 2-19 to solve for t:

$$t = \frac{v - v_0}{a} = \frac{22.22\text{ m/s} - 27.78\text{ m/s}}{-1.58\text{ m/s}^2}$$

$$= 3.519\text{ s} \approx 3.52\text{ s}. \qquad \text{(Answer)}$$

If you are initially speeding and trying to slow to the speed limit, this is plenty of time for the police officer to measure your speed.

PROBLEM-SOLVING TACTICS

Tactic 6: *Check the Dimensions*
The dimension of a velocity is L/T—that is, length L divided by time T—and the dimension of an acceleration is L/T^2. In any equation, the dimensions of all terms must be the same. If you are in doubt about an equation, check its dimensions.

To check the dimensions of Eq. 2-15 ($x - x_0 = v_0t + \frac{1}{2}at^2$), we note that every term must be a length, because that is the dimension of x and of x_0. The dimension of the term v_0t is $(L/T)(T)$, which is L. The dimension of $\frac{1}{2}at^2$ is $(L/T^2)(T^2)$, which is also L. Thus, this equation checks out.

2-7 Another Look at Constant Acceleration*

The first two equations in Table 2-1 are the basic equations from which the others are derived. Those two can be obtained by integration of the acceleration with the condition that a is constant. To find Eq. 2-11, we rewrite the definition of acceleration (Eq. 2-8) as

$$dv = a\,dt.$$

We next write the *indefinite integral* (or *antiderivative*) of both sides:

$$\int dv = \int a\,dt.$$

Since acceleration a is a constant, it can be taken outside the integration. Then we obtain

$$\int dv = a\int dt$$

or

$$v = at + C. \tag{2-20}$$

To evaluate the constant of integration C, we let $t = 0$, at which time $v = v_0$. Substituting these values into Eq. 2-20 (which must hold for all values of t, including $t = 0$) yields

$$v_0 = (a)(0) + C = C.$$

Substituting this into Eq. 2-20 gives us Eq. 2-11.

To derive Eq. 2-15, we rewrite the definition of velocity (Eq. 2-4) as

$$dx = v\,dt$$

and then take the indefinite integral of both sides to obtain

$$\int dx = \int v\,dt.$$

Generally v is not constant, so we cannot move it outside the integration. However, we can substitute for v with Eq. 2-11:

$$\int dx = \int (v_0 + at)\,dt.$$

Since v_0 is a constant, as is the acceleration a, this can be rewritten as

$$\int dx = v_0\int dt + a\int t\,dt.$$

Integration now yields

$$x = v_0t + \tfrac{1}{2}at^2 + C', \tag{2-21}$$

where C' is another constant of integration. At time $t = 0$, we have $x = x_0$. Substituting these values in Eq. 2-21 yields $x_0 = C'$. Replacing C' with x_0 in Eq. 2-21 gives us Eq. 2-15.

* This section is intended for students who have had integral calculus.

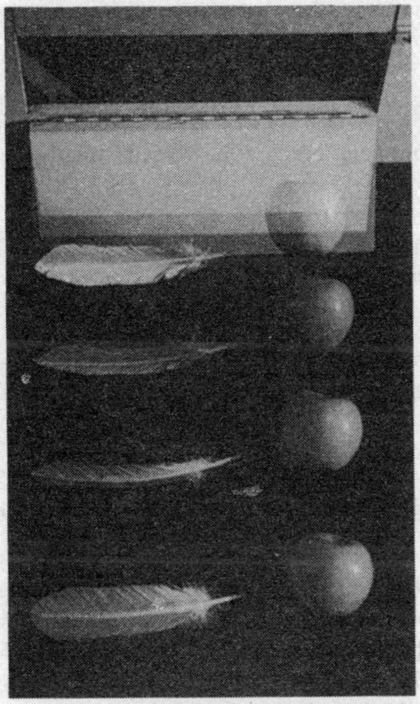

Fig. 2-9 A feather and an apple, undergoing free fall in a vacuum, move downward with the same magnitude of acceleration g. The acceleration causes the increase in distance between successive images during the fall, but note that, in the absence of air, the feather and apple fall the same distance each time.

2-8 Free-Fall Acceleration

If you tossed an object either up or down and could somehow eliminate the effects of air on its flight, you would find that the object accelerates downward at a certain constant rate. That rate is called the **free-fall acceleration,** and its magnitude is represented by g. The acceleration is independent of the object's characteristics, such as mass, density, or shape; it is the same for all objects.

Two examples of free-fall acceleration are shown in Fig. 2-9, which is a series of stroboscopic photos of a feather and an apple. As these objects fall, they accelerate downward—both at the same rate g. Their speeds increase together.

The value of g varies slightly with latitude and with elevation. At sea level in Earth's midlatitudes the value is $9.8\ \text{m/s}^2$ (or $32\ \text{ft/s}^2$), which is what you should use for the problems in this chapter.

The equations of motion in Table 2-1 for constant acceleration also apply to free fall near Earth's surface; that is, they apply to an object in vertical flight, either up or down, when the effects of the air can be neglected. However, note that for free fall: (1) The directions of motion are now along a vertical y axis instead of the x axis, with the positive direction of y upward. (This is important for later chapters when combined horizontal and vertical motions are examined.) (2) The free-fall acceleration is now negative—that is, downward on the y axis, toward Earth's center—and so it has the value $-g$ in the equations.

> The free-fall acceleration near Earth's surface is $a = -g = -9.8\ \text{m/s}^2$, and the *magnitude* of the acceleration is $g = 9.8\ \text{m/s}^2$. Do not substitute $-9.8\ \text{m/s}^2$ for g.

Suppose that you toss a tomato directly upward with an initial (positive) velocity v_0 and then catch it when it returns to the release level. During its *free-fall flight* (just after its release and just before it is caught), the equations of Table 2-1 apply to its motion. The acceleration is always $a = -g = -9.8\ \text{m/s}^2$, negative and thus downward. The velocity, however, changes, as indicated by Eqs. 2-11 and 2-16: during the ascent, the magnitude of the positive velocity decreases, until it momentarily becomes zero. Because the tomato has then stopped, it is at its maximum height. During the descent, the magnitude of the (now negative) velocity increases.

Sample Problem 2-6

Let us return to the opening story about Dave Munday's Niagara free fall in a steel ball. He fell 48 m. Let us assume that his initial velocity was zero and neglect the effect of the air on the ball during the fall.

(a) How long did Munday fall to reach the water surface below the falls?

SOLUTION: The **Key Idea** here is that, because Munday's fall was a free fall, the equations of Table 2-1 apply. Let us place a y axis along the path of Munday's fall, with $y = 0$ at his starting point and the positive direction upward along the axis (Fig. 2-10). Then the acceleration is $a = -g$ along that axis and the water level is at $y = -48$ m (negative because it is below $y = 0$). Let the fall begin at time $t = 0$, with initial velocity $v_0 = 0$.

Then from Table 2-1 we choose Eq. 2-15 (but in y notation) because it contains the requested time t and all the other variables

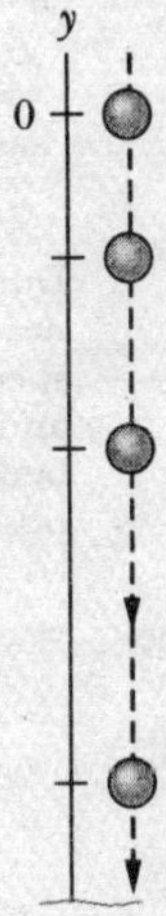

t (s)	y (m)	v (m/s)	a (m/s^2)
0	0	0	−9.8
1	−4.9	−9.8	−9.8
2	−19.6	−19.6	−9.8
3	−44.1	−29.4	−9.8
	−48.0		−9.8

Fig. 2-10 Sample Problem 2-6. The position, velocity, and acceleration of a freely falling object, here the steel ball ridden by Dave Munday over Niagara Falls.

have known values. We find

$$y - y_0 = v_0 t - \tfrac{1}{2}gt^2,$$

$$-48 \text{ m} - 0 = 0t - \tfrac{1}{2}(9.8 \text{ m/s}^2)t^2,$$

$$t^2 = 48/4.9,$$

and

$$t = 3.1 \text{ s}. \qquad \text{(Answer)}$$

Note that Munday's displacement $y - y_0$ is a negative quantity—Munday fell down, in the *negative direction* of the y axis (he did not fall up!). Also note that 48/4.9 has two square roots: 3.1 and −3.1. Here we choose the positive root because Munday obviously reaches the water surface *after* he begins to fall at $t = 0$.

(b) Munday could count off the three seconds of free fall but could not see how far he had fallen with each count. Determine his position at each full second.

SOLUTION: We again use Eq. 2-15 but now we substitute, in turn, the values t = 1.0 s, 2.0 s, and 3.0 s, and solve for Munday's position y. The results are shown in Fig. 2-10.

(c) What was Munday's velocity as he reached the water surface?

SOLUTION: To find the velocity from the original data without using the time of fall from (a), we rewrite Eq. 2-16 in y notation and then substitute known data:

$$v^2 = v_0^2 - 2g(y - y_0) = 0 - (2)(9.8 \text{ m/s}^2)(-48 \text{ m}),$$

so

$$v = -30.67 \text{ m/s} \approx -31 \text{ m/s} = -110 \text{ km/h}. \qquad \text{(Answer)}$$

We chose the negative root here, because the velocity was in the negative direction.

(d) What was Munday's velocity at each count of one full second? Was Munday aware of this increasing speed?

SOLUTION: To find the velocities from the original data without using the positions from (b), we let $a = -g$ in Eq. 2-11 and then substitute, in turn, the values t = 1.0 s, 2.0 s, and 3.0 s. Here is an example:

$$v = v_0 - gt$$
$$= 0 - (9.8 \text{ m/s}^2)(1.0 \text{ s}) = -9.8 \text{ m/s}. \qquad \text{(Answer)}$$

The other results are shown in Fig. 2-10.

Once he was in free fall, Munday was unaware of the increasing speed because the acceleration during the fall was always -9.8 m/s^2, as noted in the last column of Fig. 2-10. He was, of course, sharply aware of hitting the water because then the acceleration abruptly changed. (Munday survived the fall but then faced stiff legal fines for his daredevil action.)

Sample Problem 2-7

In Fig. 2-11, a pitcher tosses a baseball up along a y axis, with an initial speed of 12 m/s.

(a) How long does the ball take to reach its maximum height?

SOLUTION: One **Key Idea** here is that once the ball leaves the pitcher and before it returns, its acceleration is the free-fall acceleration $a = -g$. Because this is constant, Table 2-1 applies to the motion. A second **Key Idea** is that the velocity v at the maxium height must be 0. So, knowing v, a, and the initial velocity v_0 = 12 m/s, and seeking t, we solve Eq. 2-11, which contains those four variables. Rearranging yields

$$t = \frac{v - v_0}{a} = \frac{0 - 12 \text{ m/s}}{-9.8 \text{ m/s}^2} = 1.2 \text{ s}. \qquad \text{(Answer)}$$

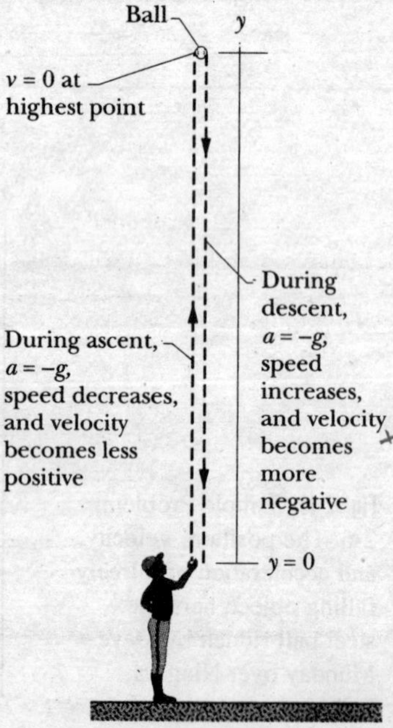

Fig. 2-11 Sample Problem 2-7. A pitcher tosses a baseball straight up into the air. The equations of free fall apply for rising as well as for falling objects, provided any effects from the air can be neglected.

(b) What is the ball's maximum height above its release point?

SOLUTION: We can take the ball's release point to be $y_0 = 0$. We can then write Eq. 2-16 in y notation, set $y - y_0 = y$ and $v = 0$ (at the maximum height), and solve for y. We get

$$y = \frac{v^2 - v_0^2}{2a} = \frac{0 - (12 \text{ m/s})^2}{2(-9.8 \text{ m/s}^2)} = 7.3 \text{ m}. \qquad \text{(Answer)}$$

(c) How long does the ball take to reach a point 5.0 m above its release point?

SOLUTION: We know v_0, $a = -g$, and the displacement $y - y_0$ = 5.0 m, and we want t, so we choose Eq. 2-15. Rewriting it for y and setting $y_0 = 0$ give us

$$y = v_0 t - \tfrac{1}{2}gt^2,$$

or

$$5.0 \text{ m} = (12 \text{ m/s})t - (\tfrac{1}{2})(9.8 \text{ m/s}^2)t^2.$$

If we temporarily omit the units (having noted that they are consistent), we can rewrite this as

$$4.9t^2 - 12t + 5.0 = 0.$$

Solving this quadratic equation for t yields

$$t = 0.53 \text{ s} \quad \text{and} \quad t = 1.9 \text{ s}. \qquad \text{(Answer)}$$

There are two such times! This is not really surprising because the ball passes twice through $y = 5.0$ m, once on the way up and once on the way down.

CHECKPOINT 6: (a) In this sample problem, what is the sign of the ball's displacement for the ascent, from the release point to the highest point? (b) What is it for the descent, from the highest point back to the release point? (c) What is the ball's acceleration at its highest point?

PROBLEM-SOLVING TACTICS

Tactic 7: *Meanings of Minus Signs*

In Sample Problems 2-6 and 2-7, many answers emerged automatically with minus signs. It is important to know what these signs mean. For these two falling-body problems, we established a vertical axis (the y axis) and we chose—quite arbitrarily—its upward direction to be positive.

We then chose the origin of the y axis (that is, the $y = 0$ position) to suit the problem. In Sample Problem 2-6, the origin was at the top of the falls, and in Sample Problem 2-7 it was at the pitcher's hand. A negative value of y then means that the body is below the chosen origin. A negative velocity means that the body is moving in the negative direction of the y axis—that is, downward. This is true no matter where the body is located.

We take the acceleration to be negative (-9.8 m/s^2) in all problems dealing with falling bodies. A negative acceleration means that, as time goes on, the velocity of the body becomes either less positive or more negative. This is true no matter where the body is located and no matter how fast or in what direction it is moving. In Sample Problem 2-7, the acceleration of the ball is negative (downward) throughout its flight, whether the ball is rising or falling.

Tactic 8: *Unexpected Answers*

Mathematics often generates answers that you might not have thought of as possibilities, as in Sample Problem 2-7c. If you get more answers than you expect, do not automatically discard the ones that do not seem to fit. Examine them carefully for physical meaning. If time is your variable, even a negative value can mean something; negative time simply refers to time before $t = 0$, the (arbitrary) time at which you decided to start your stopwatch.

REVIEW & SUMMARY

Position The *position* x of a particle on an x axis locates the particle with respect to the **origin,** or zero point, of the axis. The position is either positive or negative, according to which side of the origin the particle is on, or zero if the particle is at the origin. The **positive direction** on an axis is the direction of increasing positive numbers; the opposite direction is the **negative direction**.

Displacement The *displacement* Δx of a particle is the change in its position:

$$\Delta x = x_2 - x_1. \tag{2-1}$$

Displacement is a vector quantity. It is positive if the particle has moved in the positive direction of the x axis, and negative if the particle has moved in the negative direction.

Average Velocity When a particle has moved from position x_1 to x_2 during a time interval $\Delta t = t_2 - t_1$, its *average velocity* during that interval is

$$v_{avg} = \frac{\Delta x}{\Delta t} = \frac{x_2 - x_1}{t_2 - t_1}. \tag{2-2}$$

The algebraic sign of v_{avg} indicates the direction of motion (v_{avg} is a vector quantity). Average velocity does not depend on the actual distance a particle moves, but instead depends on its original and final positions.

On a graph of x versus t, the average velocity for a time interval Δt is the slope of the straight line connecting the points on the curve that represent the ends of the interval.

Average Speed The *average speed* s_{avg} of a particle during a time interval Δt depends on the total distance the particle moves in that time interval:

$$s_{avg} = \frac{\text{total distance}}{\Delta t}. \tag{2-3}$$

Instantaneous Velocity The *instantaneous velocity* (or simply **velocity**) v of a moving particle is

$$v = \lim_{\Delta t \to 0} \frac{\Delta x}{\Delta t} = \frac{dx}{dt}, \tag{2-4}$$

where Δx and Δt are defined by Eq. 2-2. The instantaneous velocity (at a particular time) may be found as the slope (at that particular time) of the graph of x versus t. **Speed** is the magnitude of instantaneous velocity.

Average Acceleration *Average acceleration* is the ratio of a change in velocity Δv to the time interval Δt in which the change occurs:

$$a_{avg} = \frac{\Delta v}{\Delta t}. \tag{2-7}$$

The algebraic sign indicates the direction of a_{avg}.

Instantaneous Acceleration *Instantaneous acceleration* (or simply **acceleration**) a is the rate of change of velocity with time and the second derivative of position $x(t)$ with respect to time:

$$a = \frac{dv}{dt} = \frac{d^2x}{dt^2}. \qquad (2\text{-}8, 2\text{-}9)$$

On a graph of v versus t, the acceleration a at any time t is the slope of the curve at the point that represents t.

Constant Acceleration The five equations in Table 2-1 describe the motion of a particle with constant acceleration:

$$v = v_0 + at, \qquad (2\text{-}11)$$

$$x - x_0 = v_0 t + \tfrac{1}{2}at^2, \qquad (2\text{-}15)$$

$$v^2 = v_0^2 + 2a(x - x_0), \qquad (2\text{-}16)$$

$$x - x_0 = \tfrac{1}{2}(v_0 + v)t, \qquad (2\text{-}17)$$

$$x - x_0 = vt - \tfrac{1}{2}at^2. \qquad (2\text{-}18)$$

These are *not* valid when the acceleration is not constant.

Free-Fall Acceleration An important example of straight-line motion with constant acceleration is that of an object rising or falling freely near Earth's surface. The constant acceleration equations describe this motion, but we make two changes in notation: (1) we refer the motion to the vertical y axis with $+y$ vertically *up*; (2) we replace a with $-g$, where g is the magnitude of the free-fall acceleration. Near Earth's surface, $g = 9.8\ \text{m/s}^2$ $(= 32\ \text{ft/s}^2)$.

QUESTIONS

1. Figure 2-12 shows four paths along which objects move from a starting point to a final point, all in the same time. The paths pass over a grid of equally spaced straight lines. Rank the paths according to (a) the average velocity of the objects and (b) the average speed of the objects, greatest first.

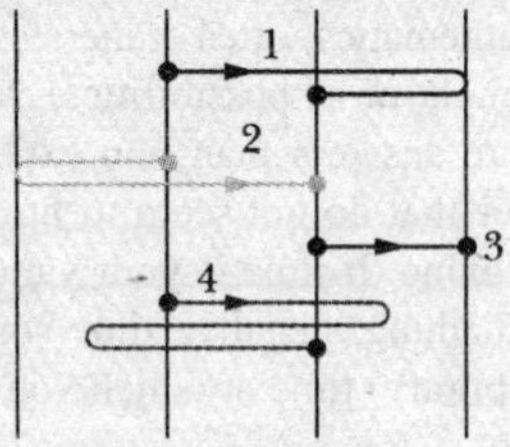

Fig. 2-12 Question 1.

2. Figure 2-13 gives the velocity of a particle moving on an x axis. What are (a) the initial and (b) the final directions of travel? (c) Does the particle stop momentarily? (d) Is the acceleration positive or negative? (e) Is it constant or varying?

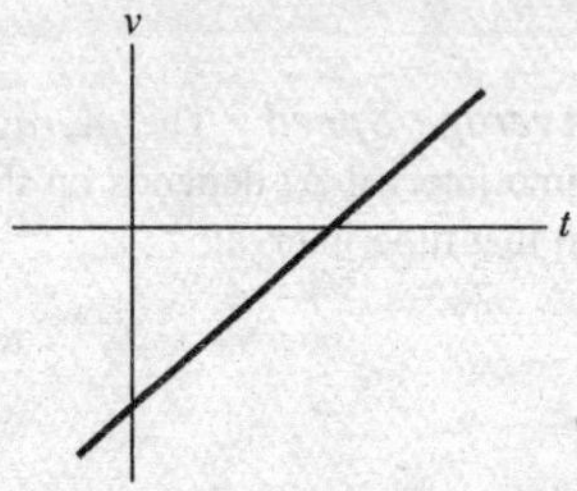

Fig. 2-13 Question 2.

3. Figure 2-14 gives the acceleration $a(t)$ of a Chihuahua as it chases a German shepherd along an axis. In which of the time periods indicated does the Chihuahua move at constant speed?

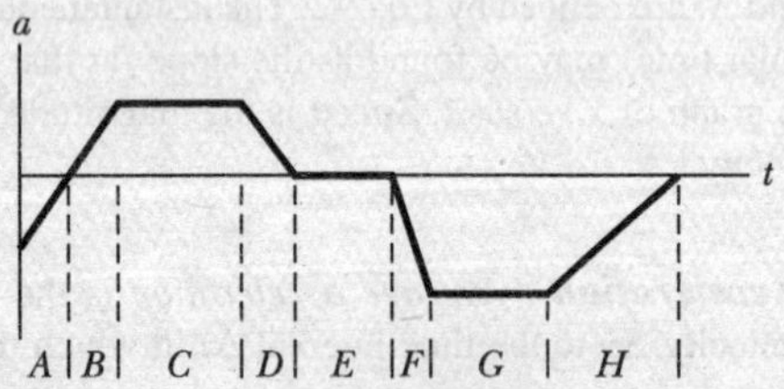

Fig. 2-14 Question 3.

4. At $t = 0$, a particle moving along an x axis is at position $x_0 = -20$ m. The signs of the particle's initial velocity v_0 (at time t_0) and constant acceleration a are, respectively, for four situations: (1) +, +; (2) +, −; (3) −, +; (4) −, −. In which situation will the particle (a) undergo a momentary stop, (b) definitely pass through the origin (given enough time), and (c) definitely not pass through the origin?

5. The following equations give the velocity $v(t)$ of a particle in four situations: (a) $v = 3$; (b) $v = 4t^2 + 2t - 6$; (c) $v = 3t - 4$; (d) $v = 5t^2 - 3$. To which of these situations do the equations of Table 2-1 apply?

6. The driver of a blue car, moving at a speed of 80 km/h, suddenly realizes that she is about to rear-end a red car, moving at a speed of 60 km/h. To avoid a collision, what is the maximum speed the blue car can have just as it reaches the red car? (Warm-up for Problem 38)

7. At $t = 0$ and $x = 0$, an initially stationary blue car begins to accelerate at the constant rate of $2.0\ \text{m/s}^2$ in the positive direction of the x axis. At $t = 2$ s, a red car traveling in an adjacent lane and in the same direction, passes $x = 0$ with a speed of 8.0 m/s and a constant acceleration of $3.0\ \text{m/s}^2$. What pair of simultaneous equations should be solved to find when the red car passes the blue car? (Warm-up for Problem 36)

8. In Fig. 2-15, a cream tangerine is thrown directly upward past three evenly spaced windows of equal heights. Rank the windows according to (a) the average speed of the cream tangerine while passing them, (b) the time the cream tangerine takes to pass them, (c) the magnitude of the acceleration of the cream tangerine while passing them, and (d) the change Δv in the speed of the cream tangerine during the passage, greatest first.

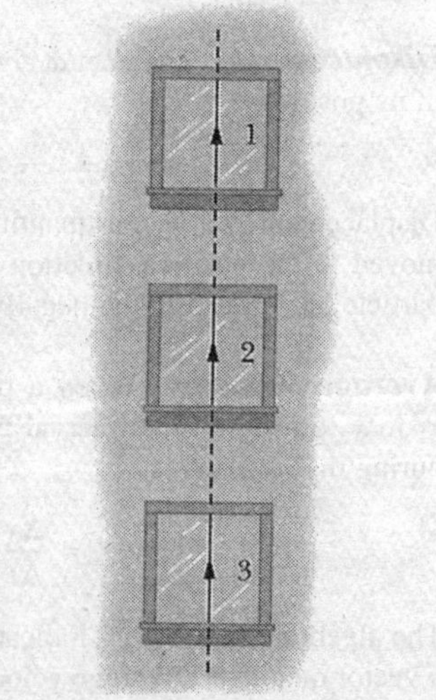

Fig. 2-15 Question 8.

9. You throw a ball straight up from the edge of a cliff, and it lands on the ground below the cliff. If you had, instead, thrown the ball down from the cliff edge with the same speed, would the ball's speed just before landing be larger than, smaller than, or the same as previously? (*Hint:* Consider Eq. 2-16.)

EXERCISES & PROBLEMS

ssm Solution is in the Student Solutions Manual.
www Solution is available on the World Wide Web at: http://www.wiley.com/college/hrw
ilw Solution is available on the Interactive LearningWare.

In several of the problems that follow you are asked to graph position, velocity, and acceleration versus time. Usually a sketch will suffice, appropriately labeled and with straight and curved portions apparent. If you have a computer or graphing calculator, you might use it to produce the graph.

SEC. 2-3 Average Velocity and Average Speed

1E. If a baseball pitcher throws a fastball at a horizontal speed of 160 km/h, how long does the ball take to reach home plate 18.4 m away? ssm

2E. A world speed record for bicycles was set in 1992 by Chris Huber riding Cheetah, a high-tech bicycle built by three mechanical engineering graduates. The record (average) speed was 110.6 km/h through a measured length of 200.0 m on a desert road. At the end of the run, Huber commented, "Cogito ergo zoom!" (I think, therefore I go fast!) What was Huber's elapsed time through the 200.0 m?

3E. An automobile travels on a straight road for 40 km at 30 km/h. It then continues in the same direction for another 40 km at 60 km/h. (a) What is the average velocity of the car during this 80 km trip? (Assume that it moves in the positive x direction.) (b) What is the average speed? (c) Graph x versus t and indicate how the average velocity is found on the graph. ssm

4P. A top-gun pilot, practicing radar avoidance maneuvers, is manually flying horizontally at 1300 km/h, just 35 m above the level ground. Suddenly, the plane encounters terrain that slopes gently upward at 4.3°, an amount difficult to detect visually (Fig. 2-16). How much times does the pilot have to make a correction to avoid flying into the ground?

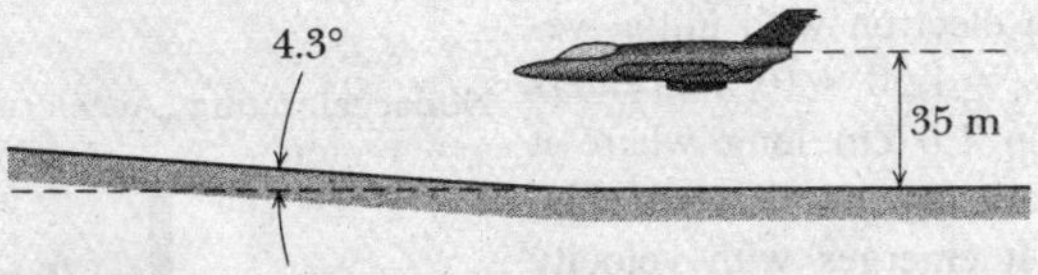

Fig. 2-16 Problem 4.

5P. You drive on Interstate 10 from San Antonio to Houston, half the *time* at 55 km/h and the other half at 90 km/h. On the way back you travel half the *distance* at 55 km/h and the other half at 90 km/h. What is your average speed (a) from San Antonio to Houston, (b) from Houston back to San Antonio, and (c) for the entire trip? (d) What is your average velocity for the entire trip? (e) Sketch x versus t for (a), assuming the motion is all in the positive x direction. Indicate how the average velocity can be found on the sketch. ilw

6P. Compute your average velocity in the following two cases: (a) You walk 73.2 m at a speed of 1.22 m/s and then run 73.2 m at a speed of 3.05 m/s along a straight track. (b) You walk for 1.00 min at a speed of 1.22 m/s and then run for 1.00 min at 3.05 m/s along a straight track. (c) Graph x versus t for both cases and indicate how the average velocity is found on the graph.

7P. The position of an object moving along an x axis is given by $x = 3t - 4t^2 + t^3$, where x is in meters and t in seconds. (a) What is the position of the object at t = 1, 2, 3, and 4 s? (b) What is the object's displacement between $t = 0$ and $t = 4$ s? (c) What is its average velocity for the time interval from $t = 2$ s to $t = 4$ s? (d) Graph x versus t for $0 \leq t \leq 4$ s and indicate how the answer for (c) can be found on the graph. ssm www

8P. Two trains, each having a speed of 30 km/h, are headed at each other on the same straight track. A bird that can fly 60 km/h flies off the front of one train when they are 60 km apart and heads directly for the other train. On reaching the other train it flies directly back to the first train, and so forth. (We have no idea *why* a bird would behave in this way.) What is the total distance the bird travels?

9P. On two *different* tracks, the winners of the one-kilometer race ran their races in 2 min, 27.95 s and 2 min, 28.15 s. In order to conclude that the runner with the shorter time was indeed faster, how much longer can the other track be in *actual* length? ilw

SEC. 2-4 Instantaneous Velocity and Speed

10E. The graph in Fig. 2-17 is for an armadillo that scampers left (negative direction of x) and right along an x axis. (a) When, if ever, is the animal to the left of the origin on the axis? When, if ever, is its velocity (b) negative, (c) positive, or (d) zero?

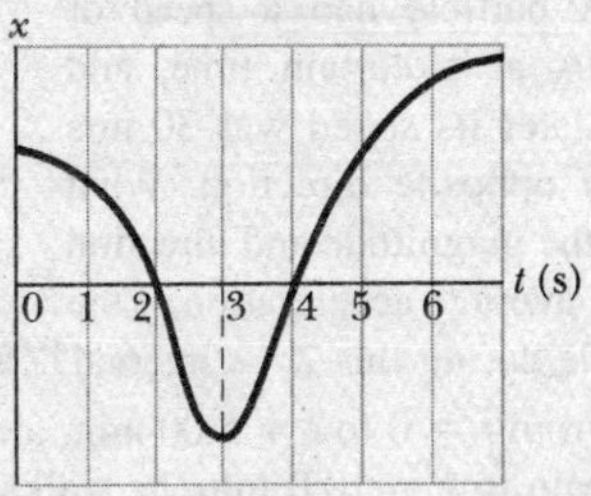

Fig. 2-17 Exercise 10.

11E. (a) If a particle's position is given by $x = 4 - 12t + 3t^2$ (where t is in seconds and x is in meters), what is its velocity at $t = 1$ s? (b) Is it moving in the positive or negative direction of x just then? (c) What is its speed just then? (d) Is the speed larger or smaller at later times? (Try answering the next two questions without further calculation.) (e) Is there ever an instant when the velocity is zero? (f) Is there a time after $t = 3$ s when the particle is moving in the negative direction of x?

12P. The position of a particle moving along the x axis is given in centimeters by $x = 9.75 + 1.50t^3$, where t is in seconds. Calculate (a) the average velocity during the time interval $t = 2.00$ s to $t = 3.00$ s; (b) the instantaneous velocity at $t = 2.00$ s; (c) the instantaneous velocity at $t = 3.00$ s; (d) the instantaneous velocity at $t = 2.50$ s; and (e) the instantaneous velocity when the particle is midway between its positions at $t = 2.00$ s and $t = 3.00$ s. (f) Graph x versus t and indicate your answers graphically.

13P. How far does the runner whose velocity–time graph is shown in Fig. 2-18 travel in 16 s? ilw

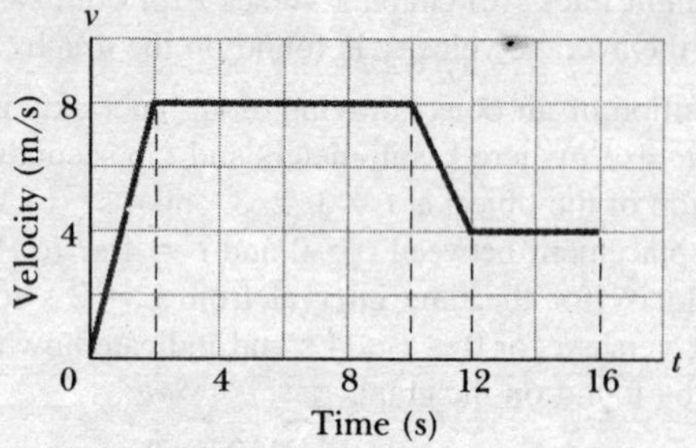

Fig. 2-18 Problem 13.

SEC. 2-5 Acceleration

14E. Sketch a graph that is a possible description of position as a function of time for a particle that moves along the x axis and, at $t = 1$ s, has (a) zero velocity and positive acceleration; (b) zero velocity and negative acceleration; (c) negative velocity and positive acceleration; (d) negative velocity and negative acceleration. (e) For which of these situations is the speed of the particle increasing at $t = 1$ s?

15E. What do the quantities (a) $(dx/dt)^2$ and (b) d^2x/dt^2 represent? (c) What are their SI units?

16E. A frightened ostrich moves in a straight line with velocity described by the velocity–time graph of Fig. 2-19. Sketch acceleration versus time.

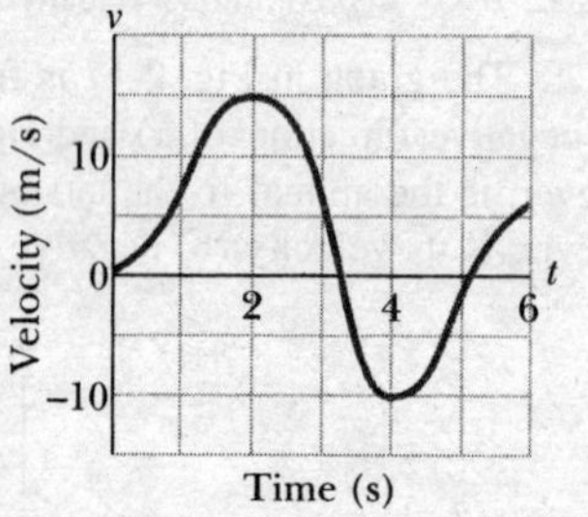

Fig. 2-19 Exercise 16.

17E. A particle had a speed of 18 m/s at a certain time, and 2.4 s later its speed was 30 m/s in the opposite direction. What were the magnitude and direction of the average acceleration of the particle during this 2.4 s interval? ssm

18P. From $t = 0$ to $t = 5.00$ min, a man stands still, and from $t = 5.00$ min to $t = 10.0$ min, he walks briskly in a straight line at a constant speed of 2.20 m/s. What are (a) his average velocity v_{avg} and (b) his average acceleration a_{avg} in the time interval 2.00 min to 8.00 min? What are (c) v_{avg} and (d) a_{avg} in the time interval 3.00 min to 9.00 min? (e) Sketch x versus t and v versus t, and indicate how the answers to (a) through (d) can be obtained from the graphs.

19P. A proton moves along the x axis according to the equation $x = 50t + 10t^2$, where x is in meters and t is in seconds. Calculate (a) the average velocity of the proton during the first 3.0 s of its motion, (b) the instantaneous velocity of the proton at $t = 3.0$ s, and (c) the instantaneous acceleration of the proton at $t = 3.0$ s. (d) Graph x versus t and indicate how the answer to (a) can be obtained from the plot. (e) Indicate the answer to (b) on the graph. (f) Plot v versus t and indicate on it the answer to (c). ssm

20P. A electron moving along the x axis has a position given by $x = 16te^{-t}$ m, where t is in seconds. How far is the electron from the origin when it momentarily stops?

21P. The position of a particle moving along the x axis depends on the time according to the equation $x = ct^2 - bt^3$, where x is in meters and t in seconds. (a) What units must c and b have? Let their numerical values be 3.0 and 2.0, respectively. (b) At what time does the particle reach its maximum positive x position? From $t = 0.0$ s to $t = 4.0$ s, (c) what distance does the particle move and (d) what is its displacement? At $t = 1.0$, 2.0, 3.0, and 4.0 s, what are (e) its velocities and (f) its accelerations? ssm

SEC. 2-6 Constant Acceleration: A Special Case

22E. An automobile driver increases the speed at a constant rate from 25 km/h to 55 km/h in 0.50 min. A bicycle rider speeds up at a constant rate from rest to 30 km/h in 0.50 min. Calculate their accelerations.

23E. A muon (an elementary particle) enters a region with a speed of 5.00×10^6 m/s and then is slowed at the rate of 1.25×10^{14} m/s^2. (a) How far does the muon take to stop? (b) Graph x versus t and v versus t for the muon. ssm

24E. The head of a rattlesnake can accelerate at 50 m/s^2 in striking a victim. If a car could do as well, how long would it take to reach a speed of 100 km/h from rest?

25E. An electron has a constant acceleration of +3.2 m/s^2. At a certain instant its velocity is +9.6 m/s. What is its velocity (a) 2.5 s earlier and (b) 2.5 s later? ssm

26E. The speed of a bullet is measured to be 640 m/s as the bullet emerges from a barrel of length 1.20 m. Assuming constant acceleration, find the time that the bullet spends in the barrel after it is fired.

27E. Suppose a rocket ship in deep space moves with constant acceleration equal to 9.8 m/s^2, which gives the illusion of normal gravity during the flight. (a) If it starts from rest, how long will it take to acquire a speed one-tenth that of light, which travels at 3.0×10^8 m/s? (b) How far will it travel in so doing? ssm

28E. A jumbo jet must reach a speed of 360 km/h on the runway for takeoff. What is the least constant acceleration needed for takeoff from a 1.80 km runway?

29E. An electron with initial velocity $v_0 = 1.50 \times 10^5$ m/s enters a region 1.0 cm long where it is electrically accelerated (Fig. 2-20). It emerges with velocity $v = 5.70 \times 10^6$ m/s. What is its acceleration, assumed constant? (Such a process occurs in conventional television sets.) ssm

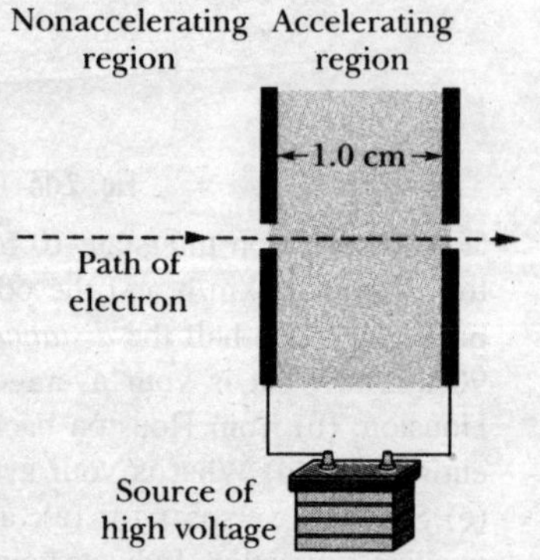

Fig. 2-20 Exercise 29.

30E. A world's land speed record was set by Colonel John P. Stapp when in March 1954 he rode a rocket-propelled sled that moved along a track at 1020 km/h. He and the sled were brought to a stop in 1.4 s. (See Fig. 2-7.) In g units, what acceleration did he experience while stopping?

31E. The brakes on your automobile are capable of creating a deceleration of 5.2 m/s^2. (a) If you are going 137 km/h and suddenly see a state trooper, what is the minimum time in which you can get your car under the 90 km/h speed limit? (The answer reveals the futility of braking to keep your high speed from being detected with a radar or laser gun.) (b) Graph x versus t and v versus t for such a deceleration. ssm www

32E. Figure 2-21 depicts the motion of a particle moving along an x axis with a constant acceleration. What are the magnitude and direction of the particle's acceleration?

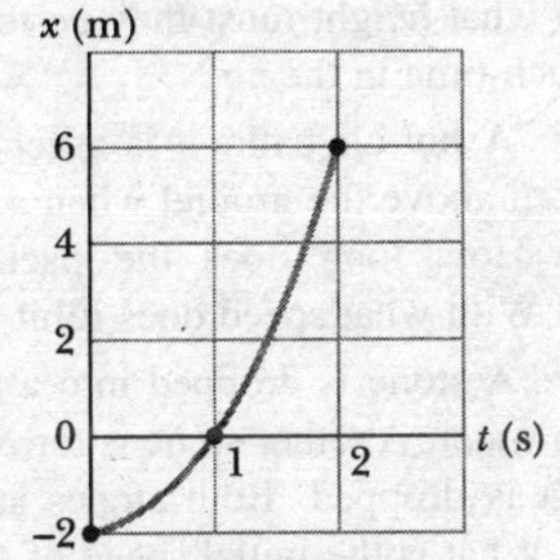

Fig. 2-21 Exercise 32.

33P. A car traveling 56.0 km/h is 24.0 m from a barrier when the driver slams on the brakes. The car hits the barrier 2.00 s later. (a) What is the car's constant deceleration before impact? (b) How fast is the car traveling at impact? ssm ilw

34P. A red train traveling at 72 km/h and a green train traveling at 144 km/h are headed toward one another along a straight, level track. When they are 950 m apart, each engineer sees the other's train and applies the brakes. The brakes decelerate each train at the rate of 1.0 m/s^2. Is there a collision? If so, what is the speed of each train at impact? If not, what is the separation between the trains when they stop?

35P. A car moving with constant acceleration covered the distance between two points 60.0 m apart in 6.00 s. Its speed as it passes the second point was 15.0 m/s. (a) What was the speed at the first point? (b) What was the acceleration? (c) At what prior distance from the first point was the car at rest? (d) Graph x versus t and v versus t for the car, from rest ($t = 0$). ssm www

36P. At the instant the traffic light turns green, an automobile starts with a constant acceleration a of 2.2 m/s^2. At the same instant a truck, traveling with a constant speed of 9.5 m/s, overtakes and passes the automobile. (a) How far beyond the traffic signal will the automobile overtake the truck? (b) How fast will the car be traveling at that instant?

37P. To stop a car, first you require a certain reaction time to begin braking; then the car slows under the constant braking deceleration. Suppose that the total distance moved by your car during these two phases is 56.7 m when its initial speed is 80.5 km/h, and 24.4 m when its initial speed is 48.3 km/h. What are (a) your reaction time and (b) the magnitude of the deceleration? ssm

38P. When a high-speed passenger train traveling at 161 km/h rounds a bend, the engineer is shocked to see that a locomotive has improperly entered onto the track from a siding and is a distance D = 676 m ahead (Fig. 2-22). The locomotive is moving at 29.0 km/h. The engineer of the high-speed train immediately applies the brakes. (a) What must be the magnitude of the resulting constant deceleration if a collision is to be just avoided? (b) Assume that the engineer is at $x = 0$ when, at $t = 0$, he first spots the locomotive. Sketch the $x(t)$ curves representing the locomotive and high-speed train for the situations in which a collision is just avoided and is not quite avoided.

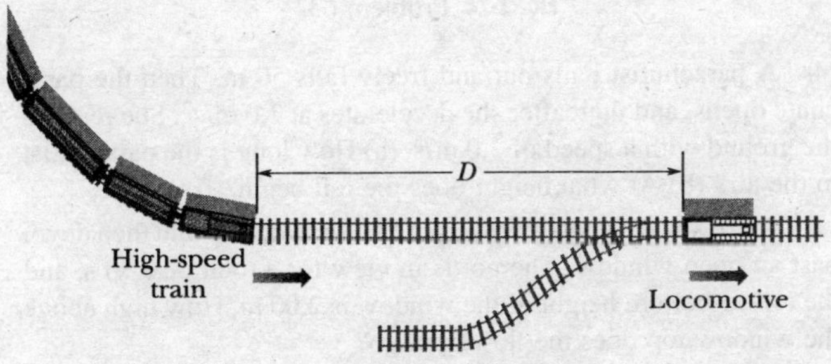

Fig. 2-22 Problem 38.

39P. An elevator cab in the New York Marquis Marriott has a total run of 190 m. Its maximum speed is 305 m/min. Its acceleration and deceleration both have a magnitude of 1.22 m/s^2. (a) How far does the cab move while accelerating to full speed from rest? (b) How long does it take to make the nonstop 190 m run, starting and ending at rest? ilw

SEC. 2-8 Free-Fall Acceleration

40E. Raindrops fall 1700 m from a cloud to the ground. (a) If they were not slowed by air resistance, how fast would the drops be moving when they struck the ground? (b) Would it be safe to walk outside during a rainstorm?

41E. At a construction site a pipe wrench struck the ground with a speed of 24 m/s. (a) From what height was it inadvertently dropped? (b) How long was it falling? (c) Sketch graphs of y, v, and a versus t for the wrench. ssm

42E. A hoodlum throws a stone vertically downward with an initial speed of 12.0 m/s from the roof of a building, 30.0 m above the ground. (a) How long does it take the stone to reach the ground? (b) What is the speed of the stone at impact?

43E. (a) With what speed must a ball be thrown vertically from ground level to rise to a maximum height of 50 m? (b) How long will it be in the air? (c) Sketch graphs of y, v, and a versus t for the ball. On the first two graphs, indicate the time at which 50 m is reached. ssm

44E. The Zero Gravity Research Facility at the NASA Lewis Research Center includes a 145 m drop tower. This is an evacuated vertical tower through which, among other possibilities, a 1 m diameter sphere containing an experimental package can be dropped. (a) How long is the sphere in free fall? (b) What is its speed just as it reaches a catching device at the bottom of the tower? (c) When caught, the sphere experiences an average deceleration of $25g$ as its speed is reduced to zero. Through what distance does it travel during the deceleration?

45E. A rock is dropped from a 100-m-high cliff. How long does it take to fall (a) the first 50 m and (b) the second 50 m? ssm

46P. A ball is thrown *down* vertically with an initial *speed* of v_0 from a height of h. (a) What is its speed just before it strikes the ground? (b) How long does the ball take to reach the ground? What would be the answers to (c) part a and (d) part b if the ball were thrown *upward* from the same height and with the same initial speed? Before solving any equations, decide whether the answers to (c) and (d) should be greater than, less than, or the same as in (a) and (b).

47P. A startled armadillo leaps upward, rising 0.544 m in the first 0.200 s. (a) What is its initial speed as it leaves the ground? (b) What is its speed at the height of 0.544 m? (c) How much higher does it go? ssm www

48P. A rock is dropped (from rest) from the top of a 60-m-tall building. How far above the ground is the rock 1.2 s before it reaches the ground?

49P. A key falls from a bridge that is 45 m above the water. It falls directly into a model boat, moving with constant velocity, that is 12 m from the point of impact when the key is released. What is the speed of the boat? ssm ilw

50P. A ball is thrown vertically downward from the top of a 36.6-m-tall building. The ball passes the top of a window that is 12.2 m above the ground 2.00 s after being thrown. What is the speed of the ball as it passes the top of the window?

51P. A ball of moist clay falls 15.0 m to the ground. It is in contact with the ground for 20.0 ms before stopping. What is the average acceleration of the ball during the time it is in contact with the ground? (Treat the ball as a particle.) ssm

52P. A model rocket fired vertically from the ground ascends with a constant vertical acceleration of 4.00 m/s^2 for 6.00 s. Its fuel is then exhausted, so it continues upward as a free-fall particle and then falls back down. (a) What is the maximum altitude reached? (b) What is the total time elapsed from takeoff until the rocket strikes the ground?

53P. To test the quality of a tennis ball, you drop it onto the floor from a height of 4.00 m. It rebounds to a height of 2.00 m. If the ball is in contact with the floor for 12.0 ms, what is its average acceleration during that contact? ssm

54P. A basketball player, standing near the basket to grab a rebound, jumps 76.0 cm vertically. How much (total) time does the player spend (a) in the top 15.0 cm of this jump and (b) in the bottom 15.0 cm? Does this help explain why such players seem to hang in the air at the tops of their jumps?

55P. Water drips from the nozzle of a shower onto the floor 200 cm below. The drops fall at regular (equal) intervals of time, the first drop striking the floor at the instant the fourth drop begins to fall. Find the locations of the second and third drops when the first strikes the floor.

56P. A ball is shot vertically upward from the surface of a planet in a distant solar system. A plot of y versus t for the ball is shown in Fig. 2-23, where y is the height of the ball above its starting point and $t = 0$ at the instant the ball is shot. What are the magnitudes of (a) the free-fall acceleration on the planet and (b) the initial velocity of the ball?

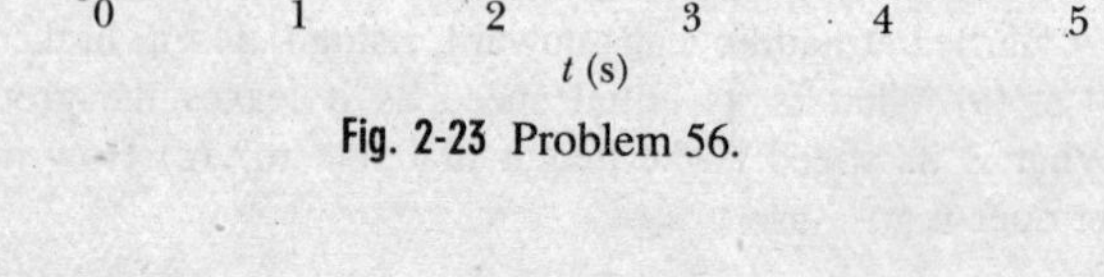

Fig. 2-23 Problem 56.

57P. Two diamonds begin a free fall from rest from the same height, 1.0 s apart. How long after the first diamond begins to fall will the two diamonds be 10 m apart? ssm

58P. A certain juggler usually tosses balls vertically to a height H. To what height must they be tossed if they are to spend twice as much time in the air?

59P. A hot-air balloon is ascending at the rate of 12 m/s and is 80 m above the ground when a package is dropped over the side. (a) How long does the package take to reach the ground? (b) With what speed does it hit the ground? ssm

60P. A stone is dropped into a river from a bridge 43.9 m above the water. Another stone is thrown vertically down 1.00 s after the first is dropped. Both stones strike the water at the same time. (a) What is the initial speed of the second stone? (b) Plot velocity versus time on a graph for each stone, taking zero time as the instant the first stone is released.

61P. An elevator without a ceiling is ascending with a constant speed of 10 m/s. A boy on the elevator shoots a ball directly upward, from a height of 2.0 m above the elevator floor, just as the elevator floor is 28 m above the ground. The initial speed of the ball with respect to the elevator is 20 m/s. (a) What maximum height above the ground does the ball reach? (b) How long does the ball take to return to the elevator floor? ssm

62P. A stone is thrown vertically upward. On its way up it passes point A with speed v, and point B, 3.00 m higher than A, with speed $\frac{1}{2}v$. Calculate (a) the speed v and (b) the maximum height reached by the stone above point B.

63P. Figure 2-24 shows a simple device for measuring your reaction time. It consists of a cardboard strip marked with a scale and two large dots. A friend holds the strip *vertically*, with thumb and forefinger at the dot on the right in Fig. 2-24. You then position your thumb and forefinger at the other dot (on the left in Fig. 2-24), being careful not to touch the strip. Your friend releases the strip, and you try to pinch it as soon as possible after you see it begin to fall. The mark at the place where you pinch the strip gives your reaction time. (a) How far from the lower dot should you place the 50.0 ms mark? (b) How much higher should the marks for 100, 150, 200, and 250 ms be? (For example, should the 100 ms marker be two times as far from the dot as the 50 ms marker? Can you find any pattern in the answers?)

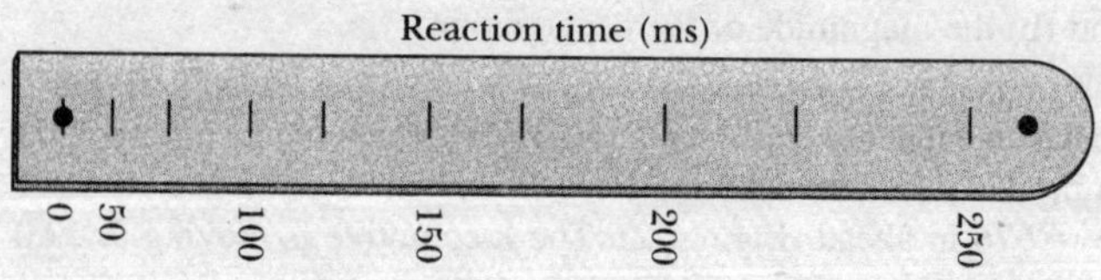

Fig. 2-24 Problem 63.

64P. A parachutist bails out and freely falls 50 m. Then the parachute opens, and thereafter she decelerates at 2.0 m/s^2. She reaches the ground with a speed of 3.0 m/s. (a) How long is the parachutist in the air? (b) At what height does the fall begin?

65P. A drowsy cat spots a flowerpot that sails first up and then down past an open window. The pot is in view for a total of 0.50 s, and the top-to-bottom height of the window is 2.00 m. How high above the window top does the flowerpot go?

3 Vectors

For two decades spelunking teams crawled, climbed, and squirmed through 200 km of Mammoth Cave and the Flint Ridge cave system, seeking a connection. The photograph shows Richard Zopf pushing his pack through the Tight Tube, far inside the Flint Ridge system. After 12 hours of "caving" along a labyrinthine route, Zopf and six others waded through a stretch of chilling water and found themselves in Mammoth Cave. Their breakthrough established the Mammoth–Flint cave system as the longest cave in the world.

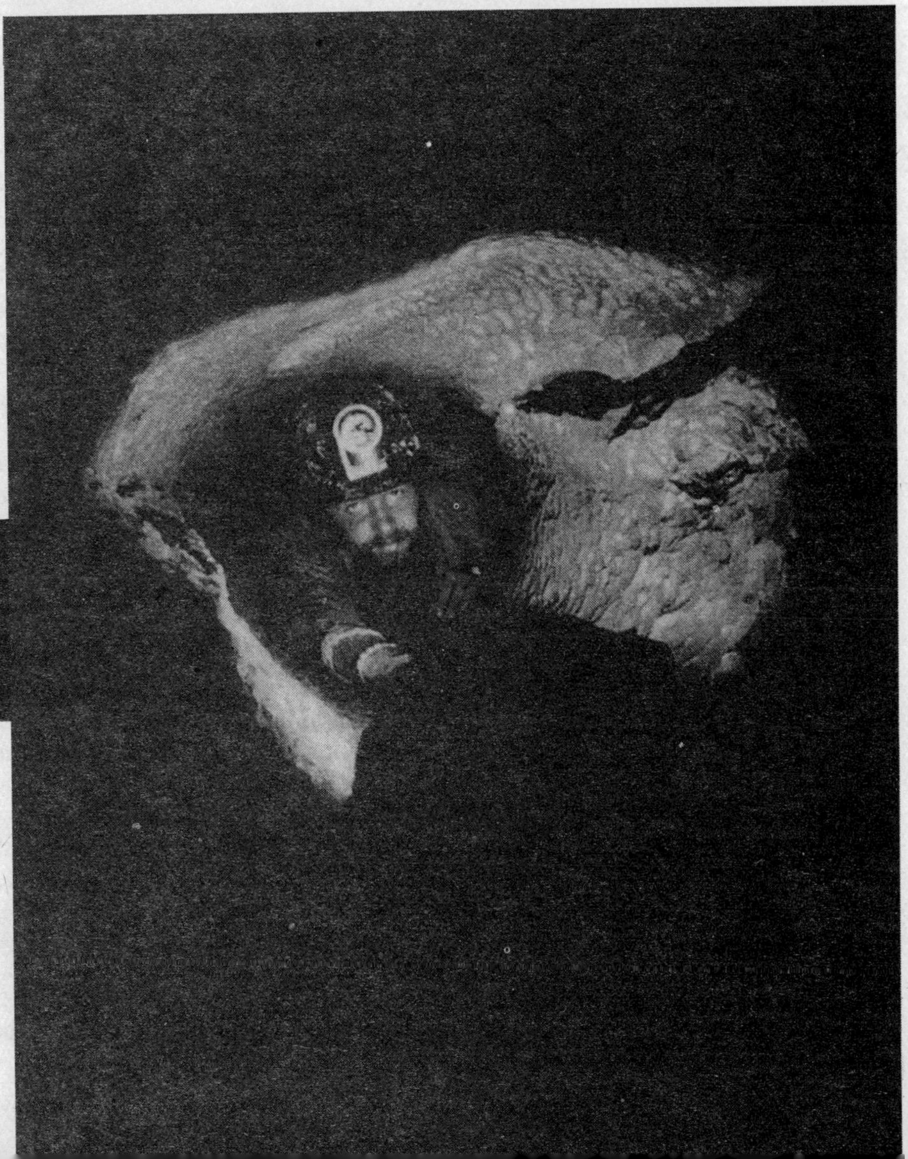

How does their final point relate to their initial point other than in terms of the actual route they covered?

The answer is in this chapter.

3-1 Vectors and Scalars

A particle moving along a straight line can move in only two directions. We can take its motion to be positive in one of these directions and negative in the other. For a particle moving in three dimensions, however, a plus sign or minus sign is no longer enough to indicate the direction of the motion. Instead, we must use a *vector*.

A **vector** has magnitude as well as direction, and vectors follow certain (vector) rules of combination, which we examine in this chapter. A **vector quantity** is a quantity that has both a magnitude and a direction and thus can be represented with a vector. Some physical quantities that are vector quantities are displacement, velocity, and acceleration. You will see many more throughout this book, so learning the rules of vector combination now will help you greatly in later chapters.

Not all physical quantities involve a direction. Temperature, pressure, energy, mass, and time, for example, do not "point" in the spatial sense. We call such quantities **scalars**, and we deal with them by the rules of ordinary algebra. A single value, with a sign (as in a temperature of $-40°F$), specifies a scalar.

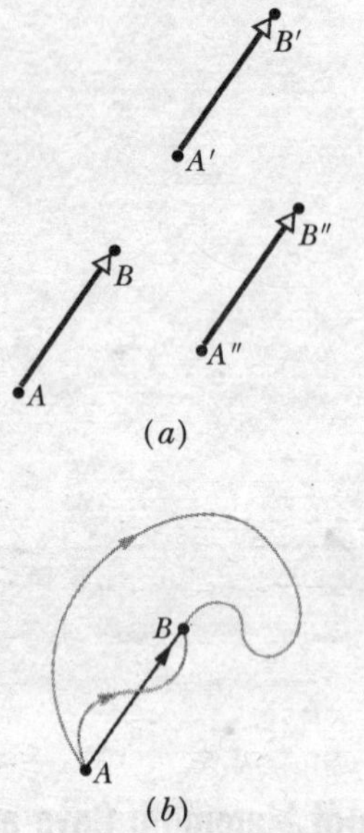

Fig. 3-1 (*a*) All three arrows have the same magnitude and direction and thus represent the same displacement. (*b*) All three paths connecting the two points correspond to the same displacement vector.

The simplest vector quantity is displacement, or change of position. A vector that represents a displacement is called, reasonably, a **displacement vector.** (Similarly, we have velocity vectors and acceleration vectors.) If a particle changes its position by moving from A to B in Fig. 3-1*a*, we say that it undergoes a displacement from A to B, which we represent with an arrow pointing from A to B. The arrow specifies the vector graphically. To distinguish vector symbols from other kinds of arrows in this book, we use the outline of a triangle as the arrowhead.

In Fig. 3-1*a*, the arrows from A to B, from A' to B', and from A'' to B'' have the same magnitude and direction. Thus, they specify identical displacement vectors and represent the same *change of position* for the particle. A vector can be shifted without changing its value, *if* its magnitude (length) and direction are not changed.

The displacement vector tells us nothing about the actual path that the particle takes. In Fig. 3-1*b*, for example, all three paths connecting points A and B correspond to the same displacement vector, that of Fig. 3-1*a*. Displacement vectors represent only the overall effect of the motion, not the motion itself.

3-2 Adding Vectors Geometrically

Suppose that, as in the vector diagram of Fig. 3-2*a*, a particle moves from A to B and then later from B to C. We can represent its overall displacement (no matter what its actual path) with two successive displacement vectors, AB and BC. The *net* displacement of these two displacements is a single displacement from A to C. We call AC the **vector sum** (or **resultant**) of the vectors AB and BC. This sum is not the usual algebraic sum.

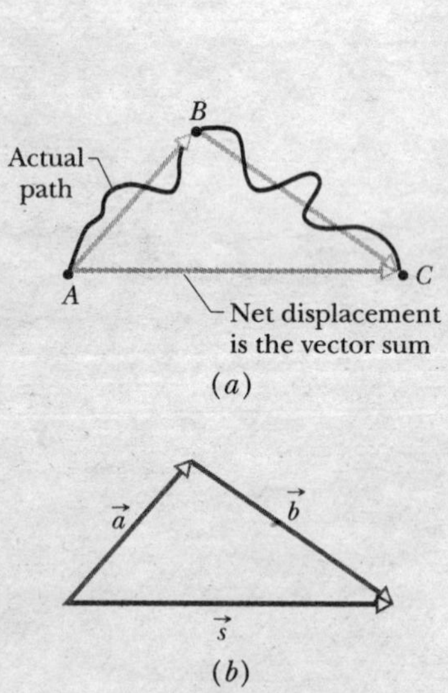

Fig. 3-2 (*a*) AC is the vector sum of the vectors AB and BC. (*b*) The same vectors relabeled.

In Fig. 3-2*b*, we redraw the vectors of Fig. 3-2*a* and relabel them in the way that we shall use from now on, namely, with an arrow over an italic symbol, as in $\vec{a}$. If we want to indicate only the magnitude of the vector (a quantity that lacks a sign or direction), we shall use the italic symbol, as in a, b, and s. (You can use just a handwritten symbol.) A symbol with an overhead arrow always implies both properties of a vector, magnitude and direction.

We can represent the relation among the three vectors in Fig. 3-2*b* with the *vector equation*

$$\vec{s} = \vec{a} + \vec{b}, \tag{3-1}$$

which says that the vector $\vec{s}$ is the vector sum of vectors $\vec{a}$ and $\vec{b}$. The symbol $+$ in Eq. 3-1 and the words "sum" and "add" have different meanings for vectors than they do in the usual algebra because they involve both magnitude *and* direction.

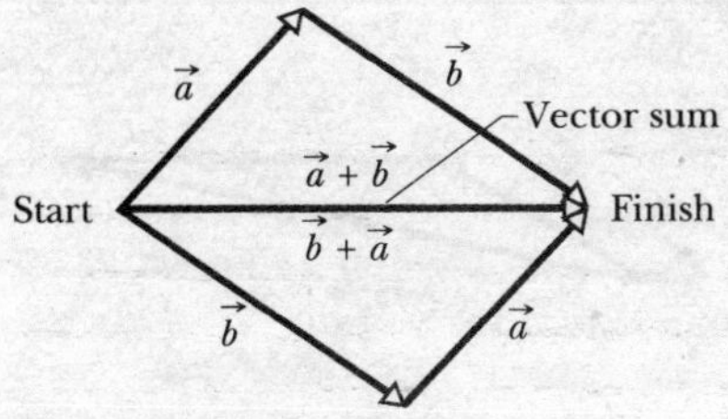

Fig. 3-3 The two vectors $\vec{a}$ and $\vec{b}$ can be added in either order; see Eq. 3-2.

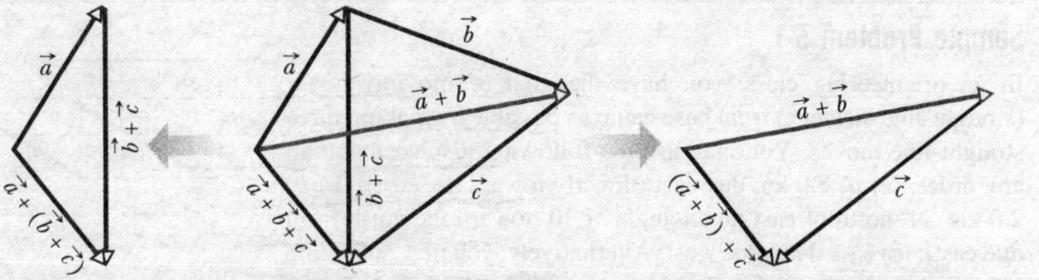

Fig. 3-4 The three vectors $\vec{a}$, $\vec{b}$, and $\vec{c}$ can be grouped in any way as they are added; see Eq. 3-3.

Figure 3-2 suggests a procedure for adding two-dimensional vectors $\vec{a}$ and $\vec{b}$ geometrically. (1) On paper, sketch vector $\vec{a}$ to some convenient scale and at the proper angle. (2) Sketch vector $\vec{b}$ to the same scale, with its tail at the head of vector $\vec{a}$, again at the proper angle. (3) The vector sum $\vec{s}$ is the vector that extends from the tail of $\vec{a}$ to the head of $\vec{b}$.

Vector addition, defined in this way, has two important properties. First, the order of addition does not matter. Adding $\vec{a}$ to $\vec{b}$ gives the same result as adding $\vec{b}$ to $\vec{a}$ (Fig. 3-3); that is,

$$\vec{a}+\vec{b}=\vec{b}+\vec{a} \quad \text{(commutative law).} \tag{3-2}$$

Second, when there are more than two vectors, we can group them in any order as we add them. Thus, if we want to add vectors $\vec{a}$, $\vec{b}$, and $\vec{c}$, we can add $\vec{a}$ and $\vec{b}$ first and then add their vector sum to $\vec{c}$. We can also add $\vec{b}$ and $\vec{c}$ first and then add *that* sum to $\vec{a}$. We get the same result either way, as shown in Fig. 3-4. That is,

$$(\vec{a}+\vec{b})+\vec{c}=\vec{a}+(\vec{b}+\vec{c}) \quad \text{(associative law).} \tag{3-3}$$

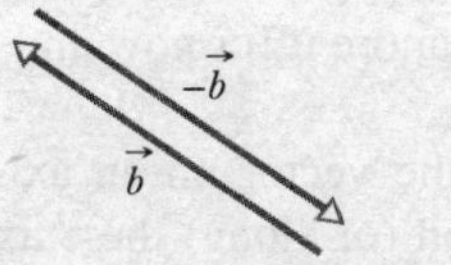

Fig. 3-5 The vectors $\vec{b}$ and $-\vec{b}$ have the same magnitude and opposite directions.

The vector $-\vec{b}$ is a vector with the same magnitude as $\vec{b}$ but the opposite direction (see Fig. 3-5). Adding the two vectors in Fig. 3-5 would yield

$$\vec{b}+(-\vec{b})=0.$$

Thus, adding $-\vec{b}$ has the effect of subtracting $\vec{b}$. We use this property to define the difference between two vectors: let $\vec{d}=\vec{a}-\vec{b}$. Then

$$\vec{d}=\vec{a}-\vec{b}=\vec{a}+(-\vec{b}) \quad \text{(vector subtraction);} \tag{3-4}$$

that is, we find the difference vector $\vec{d}$ by adding the vector $-\vec{b}$ to the vector $\vec{a}$. Figure 3-6 shows how this is done geometrically.

As in the usual algebra, we can move a term that includes a vector symbol from one side of a vector equation to the other, but we must change its sign. For example, if we are given Eq. 3-4 and need to solve for $\vec{a}$, we can rearrange the equation as

$$\vec{d}+\vec{b}=\vec{a} \quad \text{or} \quad \vec{a}=\vec{d}+\vec{b}.$$

Remember, although we have used displacement vectors here, the rules for addition and subtraction hold for vectors of all kinds, whether they represent velocities, accelerations, or any other vector quantity. However, we can add only vectors of the same kind. For example, we can add two displacements, or two velocities, but adding a displacement and a velocity makes no sense. In the arithmetic of scalars, that would be like trying to add 21 s and 12 m.

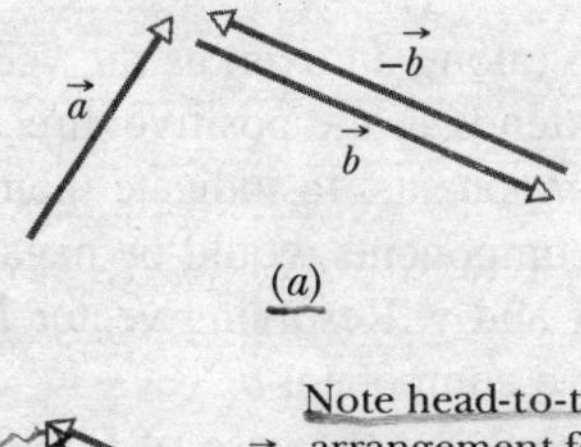

(a)

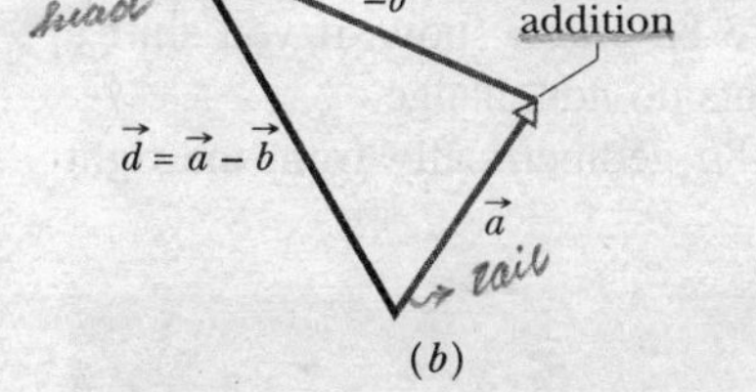

(b)

Fig. 3-6 (a) Vectors $\vec{a}$, $\vec{b}$, and $-\vec{b}$. (b) To subtract vector $\vec{b}$ from vector $\vec{a}$, add vector $-\vec{b}$ to vector $\vec{a}$.

CHECKPOINT 1: The magnitudes of displacements $\vec{a}$ and $\vec{b}$ are 3 m and 4 m, respectively, and $\vec{c}=\vec{a}+\vec{b}$. Considering various orientations of $\vec{a}$ and $\vec{b}$, what is (a) the maximum possible magnitude for $\vec{c}$ and (b) the minimum possible magnitude?

a & b 7m in same direc 1m

Sample Problem 3-1

In an orienteering class, you have the goal of moving as far (straight-line distance) from base camp as possible by making three straight-line moves. You may use the following displacements in any order: (a) $\vec{a}$, 2.0 km due east (directly toward the east); (b) $\vec{b}$, 2.0 km 30° north of east (at an angle of 30° toward the north from due east); (c) $\vec{c}$, 1.0 km due west. Alternatively, you may substitute either $-\vec{b}$ for $\vec{b}$ or $-\vec{c}$ for $\vec{c}$. What is the greatest distance you can be from base camp at the end of the third displacement?

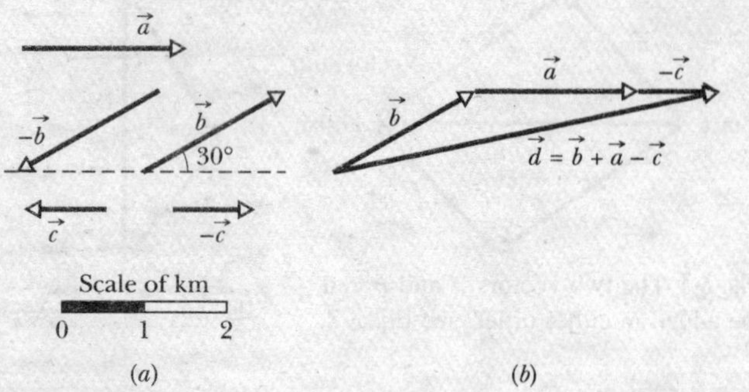

Fig. 3-7 Sample Problem 3-1. (*a*) Displacement vectors; three are to be used. (*b*) Your distance from base camp is greatest if you undergo displacements $\vec{a}$, $\vec{b}$, and $-\vec{c}$, in any order. One choice of order is shown; it gives vector sum $\vec{d} = \vec{b} + \vec{a} - \vec{c}$.

SOLUTION: Using a convenient scale, we draw vectors $\vec{a}$, $\vec{b}$, $\vec{c}$, $-\vec{b}$, and $-\vec{c}$ as in Fig. 3-7*a*. We then mentally slide the vectors over the page, connecting three of them at a time in head-to-tail arrangements to find their vector sum $\vec{d}$. The tail of the first vector represents base camp. The head of the third vector represents the point at which you stop. The vector sum $\vec{d}$ extends from the tail of the first vector to the head of the third vector. Its magnitude d is your distance from base camp.

We find that distance d is greatest for a head-to-tail arrangement of vectors $\vec{a}$, $\vec{b}$, and $-\vec{c}$. They can be in any order, because their vector sum is the same for any order. The order shown in Fig. 3-7*b* is for the vector sum

$$\vec{d} = \vec{b} + \vec{a} + (-\vec{c}).$$

Using the scale given in Fig. 3-7*a*, we measure the length d of this vector sum, finding

$$d = 4.8 \text{ m}. \qquad \text{(Answer)}$$

3-3 Components of Vectors

Adding vectors geometrically can be tedious. A neater and easier technique involves algebra but requires that the vectors be placed on a rectangular coordinate system. The x and y axes are usually drawn in the plane of the page, as in Fig. 3-8*a*. The z axis comes directly out of the page at the origin; we ignore it for now and deal only with two-dimensional vectors.

A **component** of a vector is the projection of the vector on an axis. In Fig. 3-8*a*, for example, a_x is the component of vector $\vec{a}$ on (or along) the x axis and a_y is the component along the y axis. To find the projection of a vector along an axis, we draw perpendicular lines from the two ends of the vector to the axis, as shown. The projection of a vector on an x axis is its x *component*, and similarly the projection on the y axis is the y *component*. The process of finding the components of a vector is called **resolving the vector.**

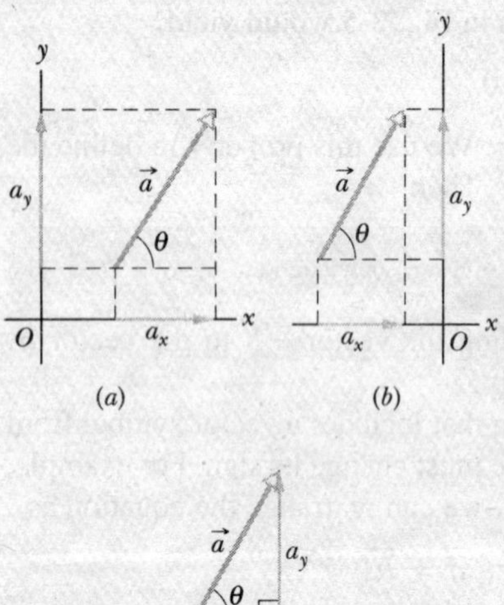

Fig. 3-8 (*a*) The components a_x and a_y of vector $\vec{a}$. (*b*) The components are unchanged if the vector is shifted, as long as the magnitude and orientation are maintained. (*c*) The components form the legs of a right triangle whose hypotenuse is the magnitude of the vector.

A component of a vector has the same direction (along an axis) as the vector. In Fig. 3-8, a_x and a_y are both positive because $\vec{a}$ extends in the positive direction of both axes. (Note the small arrowheads on the components, to indicate their direction.) If we were to reverse vector $\vec{a}$, then both components would be negative and their arrowheads would point toward negative x and y. Resolving vector $\vec{b}$ in Fig. 3-9 yields a positive component b_x and a negative component b_y.

In general, a vector has three components, although for the case of Fig. 3-8*a* the component along the z axis is zero. As Figs. 3-8*a* and *b* show, if you shift a vector without changing its direction, its components do not change.

We can find the components of $\vec{a}$ in Fig. 3-8*a* geometrically from the right triangle there:

$$a_x = a \cos \theta \quad \text{and} \quad a_y = a \sin \theta, \qquad (3\text{-}5)$$

where θ is the angle that the vector $\vec{a}$ makes with the positive direction of the x axis, and a is the magnitude of $\vec{a}$. Figure 3-8*c* shows that $\vec{a}$ and its x and y components

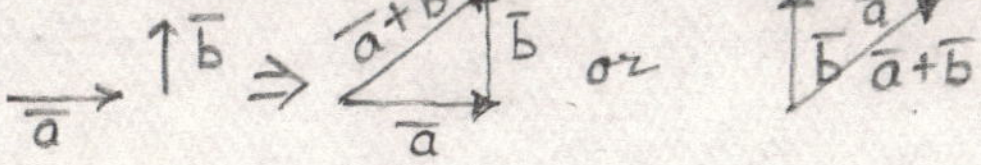

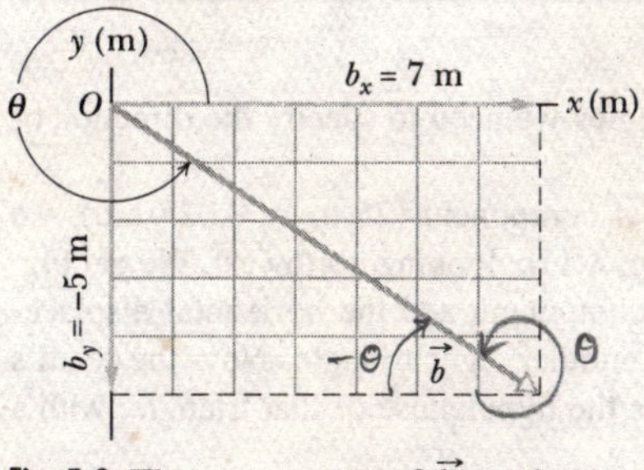

Fig. 3-9 The component of $\vec{b}$ on the x axis is positive, and that on the y axis is negative.

form a right triangle. It also shows how we can reconstruct a vector from its components: we arrange those components *head to tail*. Then we complete a right triangle with the vector forming the hypotenuse, from the tail of one component to the head of the other component.

Once a vector has been resolved into its components along a set of axes, the components themselves can be used in place of the vector. For example, $\vec{a}$ in Fig. 3-8*a* is given (completely determined) by a and θ. It can also be given by its components a_x and a_y. Both pairs of values contain the same information. If we know a vector in *component notation* (a_x and a_y) and want it in *magnitude-angle notation* (a and θ), we can use the equations

$$a = \sqrt{a_x^2 + a_y^2} \quad \text{and} \quad \tan\theta = \frac{a_y}{a_x} \tag{3-6}$$

to transform it.

In the more general three-dimensional case, we need a magnitude and two angles (say, a, θ, and ϕ) or three components (a_x, a_y, and a_z) to specify a vector.

CHECKPOINT 2: In the figure, which of the indicated methods for combining the x and y components of vector $\vec{a}$ are proper to determine that vector?

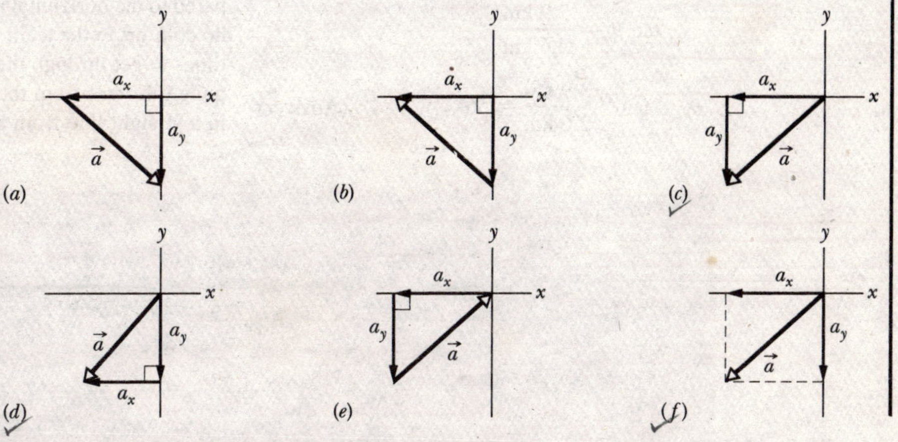

Sample Problem 3-2

A small airplane leaves an airport on an overcast day and is later sighted 215 km away, in a direction making an angle of 22° east of north. How far east and north is the airplane from the airport when sighted?

SOLUTION: The **Key Idea** here is that we are given the magnitude (215 km) and the angle (22° east of north) of a vector and need to find the components of the vector. We draw an xy coordinate system with the positive direction of x due east and that of y due north (Fig. 3-10). For convenience, the origin is placed at the airport. The airplane's displacement vector $\vec{d}$ points from the origin to where the airplane is sighted.

To find the components of $\vec{d}$, we use Eq. 3-5 with $\theta = 68°$ $(= 90° - 22°)$:

$$d_x = d\cos\theta = (215 \text{ km})(\cos 68°)$$
$$= 81 \text{ km} \qquad \text{(Answer)}$$
$$d_y = d\sin\theta = (215 \text{ km})(\sin 68°)$$
$$= 199 \text{ km}. \qquad \text{(Answer)}$$

Thus, the airplane is 81 km east and 199 km north of the airport.

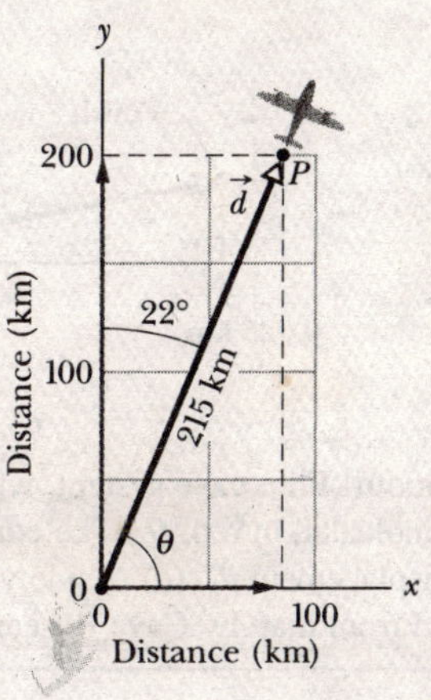

Fig. 3-10 Sample Problem 3-2. A plane takes off from an airport at the origin and is later sighted at P.

Sample Problem 3-3

The 1972 team that connected the Mammoth–Flint cave system went from Austin Entrance in the Flint Ridge system to Echo River in Mammoth Cave (Fig. 3-11*a*), traveling a net 2.6 km westward, 3.9 km southward, and 25 m upward. What was their displacement vector from start to finish?

SOLUTION: The **Key Idea** here is that we have the components of a three-dimensional vector, and we need to find the vector's magnitude and two angles to specify the vector's direction. We first draw the components as in Fig. 3-11*b*. The horizontal components (2.6 km west and 3.9 km south) form the legs of a horizontal right triangle. The team's horizontal displacement forms the hypotenuse of the triangle, and its magnitude d_h is given by the Pythagorean theorem:

$$d_h = \sqrt{(2.6\text{ km})^2 + (3.9\text{ km})^2} = 4.69\text{ km}.$$

Also from the horizontal triangle in Fig. 3-11*b*, we see that this horizontal displacement is directed south of due west by an angle θ_h given by

$$\tan\theta_h = \frac{3.9\text{ km}}{2.6\text{ km}},$$

so

$$\theta_h = \tan^{-1}\frac{3.9\text{ km}}{2.6\text{ km}} = 56°, \qquad \text{(Answer)}$$

which is one of the two angles we need to specify the direction of the overall displacement.

To include the vertical component (25 m = 0.025 km), we now take a side view of Fig. 3-11*b*, looking northwest. We get Fig. 3-11*c*, where the vertical component and the horizontal displacement d_h form the legs of another right triangle. Now the team's overall displacement forms the hypotenuse of that triangle, with a magnitude d given by

$$d = \sqrt{(4.69\text{ km})^2 + (0.025\text{ km})^2} = 4.69\text{ km}$$
$$\approx 4.7\text{ km}. \qquad \text{(Answer)}$$

This displacement is directed upward from the horizontal displacement by the angle

$$\theta_v = \tan^{-1}\frac{0.025\text{ km}}{4.69\text{ km}} = 0.3°. \qquad \text{(Answer)}$$

Thus, the team's displacement vector had a magnitude of 4.7 km and was at an angle of 56° south of west and at an angle of 0.3° upward. The net vertical motion was, of course, insignificant compared to the horizontal motion. However, that fact would have been no comfort to the team, which had to climb up and down countless times to get through the cave. The route they actually covered was quite different from the displacement vector, which merely points in a straight line from start to finish.

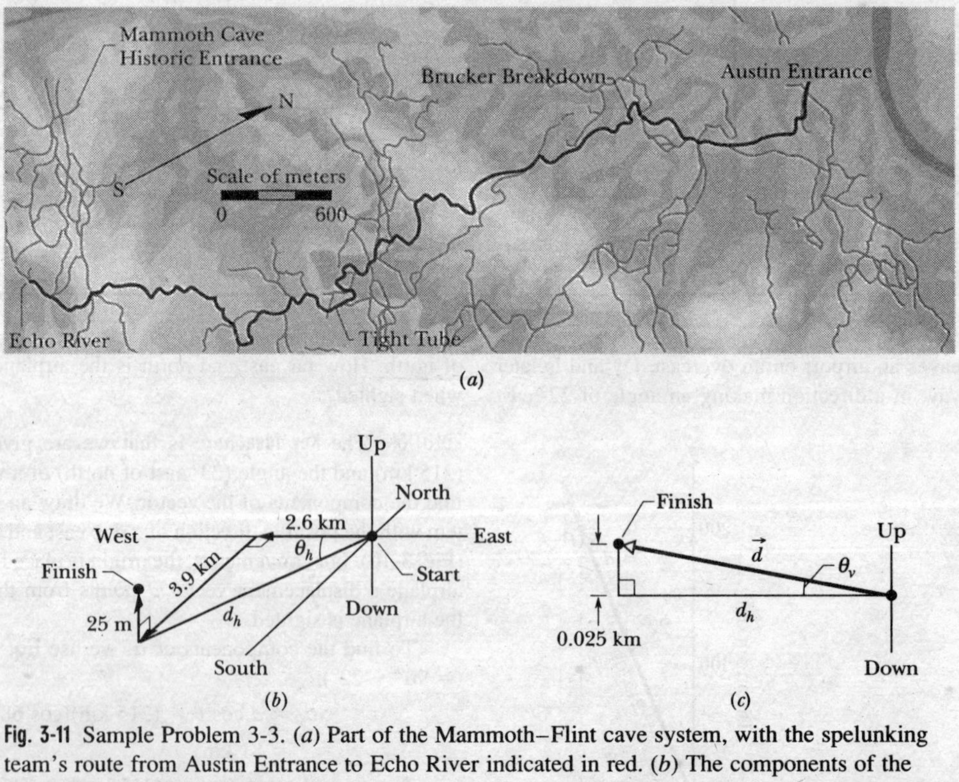

Fig. 3-11 Sample Problem 3-3. (*a*) Part of the Mammoth–Flint cave system, with the spelunking team's route from Austin Entrance to Echo River indicated in red. (*b*) The components of the team's overall displacement and their horizontal displacement d_h. (*c*) A side view showing d_h and the team's overall displacement vector $\vec{d}$. (Adapted from map by Cave Research Foundation.)

PROBLEM-SOLVING TACTICS

Tactic 1: *Angles—Degrees and Radians*
Angles that are measured relative to the positive direction of the x axis are positive if they are measured in the counterclockwise direction, and negative if measured clockwise. For example, 210° and −150° are the same angle.

Angles may be measured in degrees or radians (rad). You can relate the two measures by remembering that one full circle is equivalent to 360° and to 2π rad. If you needed to convert, say, 40° to radians, you would write

$$40^\circ \frac{2\pi \text{ rad}}{360^\circ} = 0.70 \text{ rad}.$$

Tactic 2: *Trig Functions*
You need to know the definitions of the common trigonometric functions—sine, cosine, and tangent—because they are part of the language of science and engineering. They are given in Fig. 3-12 in a form that does not depend on how the triangle is labeled.

You should also be able to sketch how the trig functions vary with angle, as in Fig. 3-13, in order to be able to judge whether a calculator result is reasonable. Even knowing the signs of the functions in the various quadrants can be of help.

Tactic 3: *Inverse Trig Functions*
When the inverse trig functions $\sin^{-1}$, $\cos^{-1}$, and $\tan^{-1}$ are taken on a calculator, you must consider the reasonableness of the answer you get, because there is usually another possible answer that the calculator does not give. The range of operation for a calculator in taking each inverse trig function is indicated in Fig. 3-13. As an example, $\sin^{-1} 0.5$ has associated angles of 30° (which is displayed by the calculator, since 30° falls within its range of operation) and 150°. To see both values, draw a horizontal line through 0.5 in Fig. 3-13*a* and note where it cuts the sine curve.

How do you distinguish a correct answer? It is the one that seems more reasonable for the given situation. As an example, reconsider the calculation of θ_h in Sample Problem 3-3, where $\tan \theta_h = 3.9/2.6 = 1.5$. Taking $\tan^{-1} 1.5$ on your calculator tells you that $\theta_h = 56°$, but $\theta_h = 236°$ $(= 180° + 56°)$ also has a tangent of 1.5. Which is correct? From the physical situation (Fig. 3-11*b*), 56° is reasonable and 236° is clearly not.

$$\sin\theta = \frac{\text{leg opposite } \theta}{\text{hypotenuse}}$$

$$\cos\theta = \frac{\text{leg adjacent to } \theta}{\text{hypotenuse}}$$

$$\tan\theta = \frac{\text{leg opposite } \theta}{\text{leg adjacent to } \theta}$$

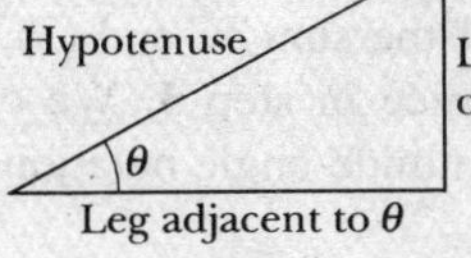

Fig. 3-12 A triangle used to define the trigonometric functions. See also Appendix E.

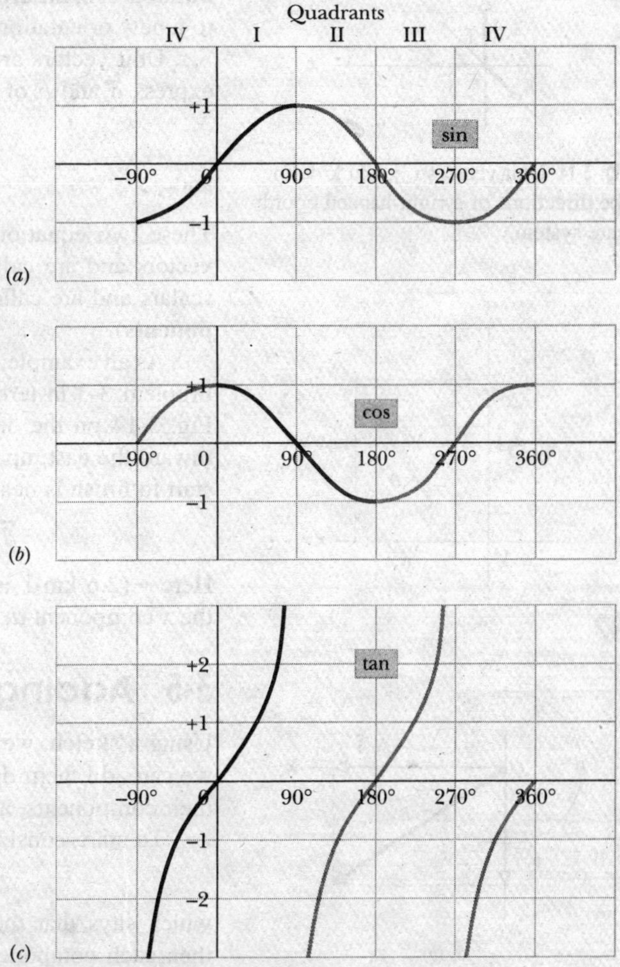

Fig. 3-13 Three useful curves to remember. A calculator's range of operation for taking *inverse* trig functions is indicated by the darker portions of the colored curves.

Tactic 4: *Measuring Vector Angles*
The equations for $\cos\theta$ and $\sin\theta$ in Eq. 3-5 and the equation for $\tan\theta$ in Eq. 3-6 are valid only if the angle is measured relative to the positive direction of the x axis. If it is measured relative to some other direction, then the trig functions in Eq. 3-5 may have to be interchanged, and the ratio in Eq. 3-6 may have to be inverted. A safer method is to convert the given angle into one that is measured from the positive direction of the x axis.

3-4 Unit Vectors

A **unit vector** is a vector that has a magnitude of exactly 1 and points in a particular direction. It lacks both dimension and unit. Its sole purpose is to point—that is, to specify a direction. The unit vectors in the positive directions of the x, y, and z axes are labeled $\hat{\mathrm{i}}$, $\hat{\mathrm{j}}$, and $\hat{\mathrm{k}}$, where the hat ^ is used instead of an overhead arrow as for

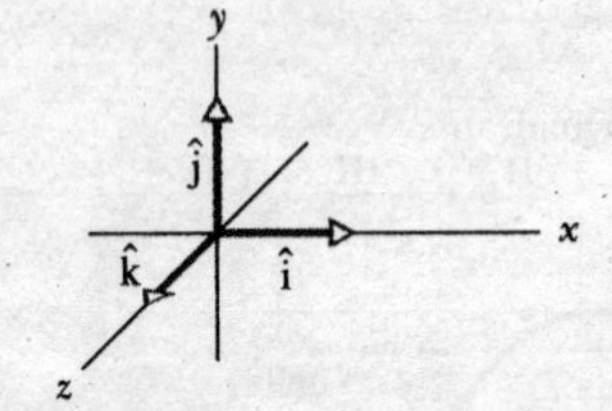

Fig. 3-14 Unit vectors î, ĵ, and k̂ define the directions of a right-handed coordinate system.

other vectors (Fig. 3-14). The arrangement of axes in Fig. 3-14 is said to be a **right-handed coordinate system.** The system remains right-handed if it is rotated rigidly to a new orientation. We use such coordinate systems exclusively in this book.

Unit vectors are very useful for expressing other vectors; for example, we can express $\vec{a}$ and $\vec{b}$ of Figs. 3-8 and 3-9 as

$$\vec{a} = a_x\hat{\mathrm{i}} + a_y\hat{\mathrm{j}} \tag{3-7}$$

and

$$\vec{b} = b_x\hat{\mathrm{i}} + b_y\hat{\mathrm{j}}. \tag{3-8}$$

These two equations are illustrated in Fig. 3-15. The quantities $a_x\hat{\mathrm{i}}$ and $a_y\hat{\mathrm{j}}$ are vectors and are called the **vector components** of $\vec{a}$. The quantities a_x and a_y are scalars and are called the **scalar components** of $\vec{a}$ (or, as before, simply its **components**).

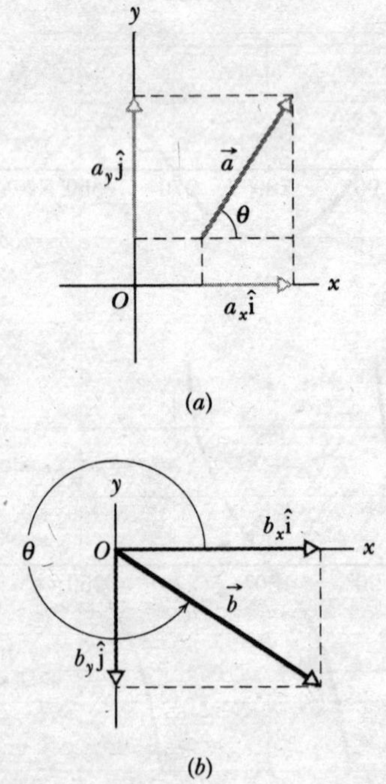

Fig. 3-15 (*a*) The vector components of vector $\vec{a}$. (*b*) The vector components of vector $\vec{b}$.

As an example, let us write the displacement $\vec{d}$ of the spelunking team of Sample Problem 3-3 in terms of unit vectors. First, superimpose the coordinate system of Fig. 3-14 on the one shown in Fig. 3-11*b*. Then the direction of î, ĵ, and k̂ are toward the east, up, and toward the south, respectively. Thus, displacement $\vec{d}$ from start to finish is neatly expressed in unit-vector notation as

$$\vec{d} = -(2.6\text{ km})\hat{\mathrm{i}} + (0.025\text{ km})\hat{\mathrm{j}} + (3.9\text{ km})\hat{\mathrm{k}}. \tag{3-9}$$

Here $-(2.6\text{ km})\hat{\mathrm{i}}$ is the vector component $d_x\hat{\mathrm{i}}$ along the x axis, and $-(2.6\text{ km})$ is the x component d_x.

3-5 Adding Vectors by Components

Using a sketch, we can add vectors geometrically. On a vector-capable calculator, we can add them directly on the screen. A third way to add vectors is to combine their components, axis by axis.

To start, consider the statement

$$\vec{r} = \vec{a} + \vec{b}, \tag{3-10}$$

which says that the vector $\vec{r}$ is the same as the vector $(\vec{a} + \vec{b})$. If that is so, then each component of $\vec{r}$ must be the same as the corresponding component of $(\vec{a} + \vec{b})$:

$$r_x = a_x + b_x \tag{3-11}$$

$$r_y = a_y + b_y \tag{3-12}$$

$$r_z = a_z + b_z. \tag{3-13}$$

In other words, two vectors must be equal if their corresponding components are equal. Equations 3-10 to 3-13 tell us that to add vectors $\vec{a}$ and $\vec{b}$, we must (1) resolve the vectors into their scalar components; (2) combine these scalar components, axis by axis, to get the components of the sum $\vec{r}$; and (3) combine the components of $\vec{r}$ to get $\vec{r}$ itself. We have a choice in step 3. We can express $\vec{r}$ in unit-vector notation (as in Eq. 3-9) or in magnitude-angle notation (as in the answer to Sample Problem 3-3).

This procedure for adding vectors by components also applies to vector subtractions. Recall that a subtraction such as $\vec{d} = \vec{a} - \vec{b}$ can be rewritten as an addition $\vec{d} = \vec{a} + (-\vec{b})$. To subtract we simply add $\vec{a}$ and $-\vec{b}$ by components, to get

$$d_x = a_x - b_x, \quad d_y = a_y - b_y, \quad \text{and} \quad d_z = a_z - b_z,$$

where

$$\vec{d} = d_x\hat{\mathrm{i}} + d_y\hat{\mathrm{j}} + d_z\hat{\mathrm{k}}.$$

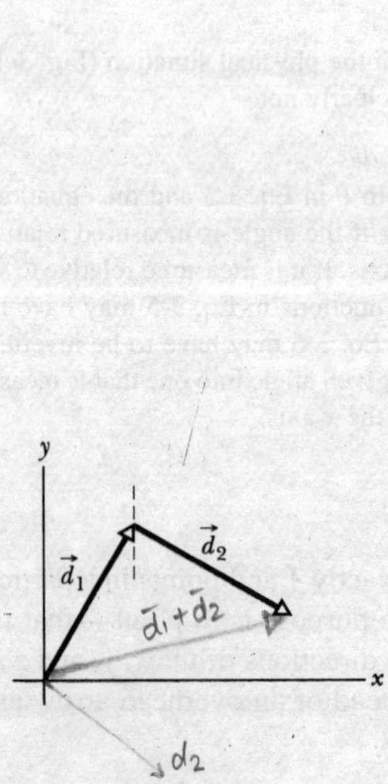

CHECKPOINT 3: (a) In the figure here, what are the signs of the x components of $\vec{d}_1$ and $\vec{d}_2$? (b) What are the signs of the y components of $\vec{d}_1$ and $\vec{d}_2$? (c) What are the signs of the x and y components of $\vec{d}_1 + \vec{d}_2$?

Sample Problem 3-4

Figure 3-16*a* shows the following three vectors:

$$\vec{a} = (4.2\text{ m})\hat{i} - (1.5\text{ m})\hat{j},$$

$$\vec{b} = (-1.6\text{ m})\hat{i} + (2.9\text{ m})\hat{j},$$

and $$\vec{c} = (-3.7\text{ m})\hat{j}.$$

What is their vector sum $\vec{r}$, which is also shown?

SOLUTION: The **Key Idea** here is that we can add the three vectors by components, axis by axis. For the x axis, we add the x components of $\vec{a}$, $\vec{b}$, and $\vec{c}$ to get the x component of $\vec{r}$:

$$r_x = a_x + b_x + c_x$$
$$= 4.2\text{ m} - 1.6\text{ m} + 0 = 2.6\text{ m}.$$

Similarly, for the y axis,

$$r_y = a_y + b_y + c_y$$
$$= -1.5\text{ m} + 2.9\text{ m} - 3.7\text{ m} = -2.3\text{ m}.$$

Another **Key Idea** is that we can combine these components of $\vec{r}$ to write the vector in unit-vector notation:

$$\vec{r} = (2.6\text{ m})\hat{i} - (2.3\text{ m})\hat{j}, \quad \text{(Answer)}$$

where $(2.6\text{ m})\hat{i}$ is the vector component of $\vec{r}$ along the x axis and $-(2.3\text{ m})\hat{j}$ is that along the y axis. Figure 3-16*b* shows one way to arrange these vector components to form $\vec{r}$. (Can you sketch the other way?)

A third **Key Idea** is that we can also answer the question by giving the magnitude and an angle for $\vec{r}$. From Eq. 3-6, the magnitude is

$$r = \sqrt{(2.6\text{ m})^2 + (-2.3\text{ m})^2} \approx 3.5\text{ m} \quad \text{(Answer)}$$

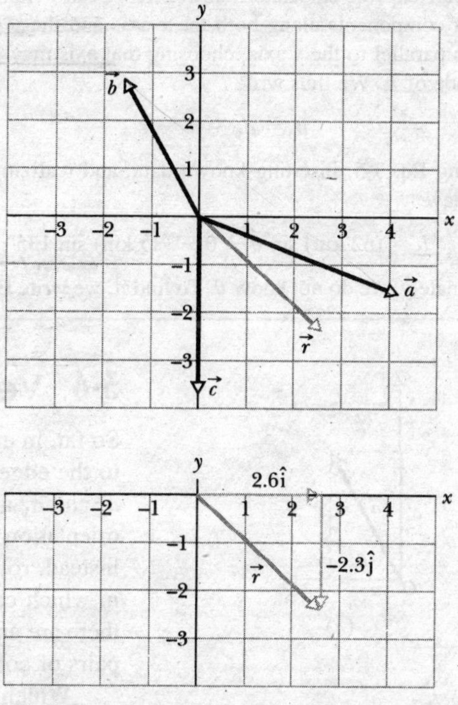

Fig. 3-16 Sample Problem 3-4. Vector $\vec{r}$ is the vector sum of the other three vectors.

and the angle (measured from the positive direction of x) is

$$\theta = \tan^{-1}\left(\frac{-2.3\text{ m}}{2.6\text{ m}}\right) = -41°, \quad \text{(Answer)}$$

where the minus sign means that the angle is measured clockwise.

Sample Problem 3-5

Figure 3-17 gives an incomplete map of a road rally. From the starting point (at the origin), you must use available roads to go through the following displacements:

(1) $\vec{a}$ to checkpoint Able, magnitude 36 km, due east,
(2) $\vec{b}$ to checkpoint Baker, due north,
(3) $\vec{c}$ to checkpoint Charlie, magnitude 25 km, at the angle shown.

Your net displacement $\vec{d}$ from the starting point is 62.0 km. What is the magnitude b of $\vec{b}$?

SOLUTION: The **Key Idea** here is that the net displacement $\vec{d}$ is the vector sum of the three individual displacements, so we can write

$$\vec{d} = \vec{a} + \vec{b} + \vec{c},$$

which gives us

$$\vec{b} = \vec{d} - \vec{a} - \vec{c}. \quad (3\text{-}14)$$

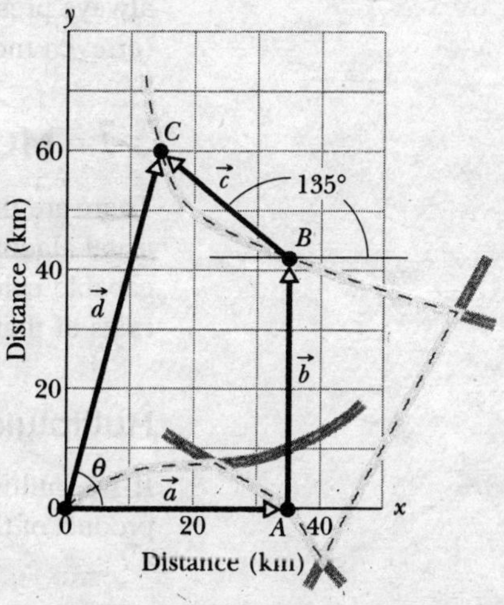

Fig. 3-17 Sample Problem 3-5. A rally route, showing the origin, checkpoints Able (A), Baker (B), and Charlie (C), and the roads.

Although we know both magnitude and direction for $\vec{a}$ and $\vec{c}$, we do not know both for $\vec{d}$, so we cannot directly solve for $\vec{b}$

on a vector-capable calculator. However, we can write Eq. 3-14 in terms of components along both the x axis and the y axis. Since $\vec{b}$ points parallel to the y axis, choosing that axis may give us the magnitude of $\vec{b}$. We thus write

$$b_y = d_y - a_y - c_y. \tag{3-15}$$

Following Eq. 3-5, inserting known data, and realizing that $b = b_y$, we have

$$b = (62\text{ km}) \sin \theta - 0 - (25\text{ km}) \sin 135°. \tag{3-16}$$

Unfortunately, we do not know θ. To find it, we write Eq. 3-14 for components along the x axis:

$$b_x = d_x - a_x - c_x, \tag{3-17}$$

which gives us

$$0 = (62\text{ km}) \cos \theta - 36\text{ km} - (25\text{ km}) \cos 135°$$

and $$\theta = \cos^{-1} \frac{36 + (25)(\cos 135°)}{62} = 72.81°.$$

Inserting this into Eq. 3-16, we find

$$b \approx 42\text{ km}. \qquad \text{(Answer)}$$

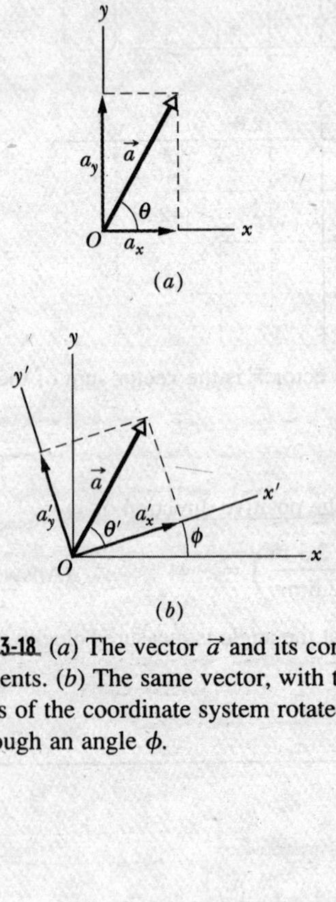

Fig. 3-18 (*a*) The vector $\vec{a}$ and its components. (*b*) The same vector, with the axes of the coordinate system rotated through an angle ϕ.

3-6 Vectors and the Laws of Physics

So far, in every figure that includes a coordinate system, the x and y axes are parallel to the edges of the book page. Thus, when a vector $\vec{a}$ is included, its components a_x and a_y are also parallel to the edges (as in Fig. 3-18*a*). The only reason for that orientation of the axes is that it looks "proper": there is no deeper reason. We could, instead, rotate the axes (but not the vector $\vec{a}$) through an angle ϕ as in Fig. 3-18*b*, in which case the components would have new values, call them a'_x and a'_y. Since there are an infinite number of choices of ϕ, there are an infinite number of different pairs of components for $\vec{a}$.

Which then is the "right" pair of components? The answer is that they are all equally valid because each pair (with its axes) just gives us a different way of describing the same vector $\vec{a}$; all produce the same magnitude and direction for the vector. In Fig. 3-18 we have

$$a = \sqrt{a_x^2 + a_y^2} = \sqrt{a_x'^2 + a_y'^2} \tag{3-18}$$

and $$\theta = \theta' + \phi. \tag{3-19}$$

The point is that we have great freedom in choosing a coordinate system, because the relations among vectors (including, for example, the vector addition of Eq. 3-1) do not depend on the location of the origin of the coordinate system or on the orientation of the axes. This is also true of the relations of physics; they are all independent of the choice of coordinate system. Add to that the simplicity and richness of the language of vectors and you can see why the laws of physics are almost always presented in that language: one equation, like Eq. 3-10, can represent three (or even more) relations, like Eqs. 3-11, 3-12, and 3-13.

3-7 Multiplying Vectors*

There are three ways in which vectors can be multiplied, but none is exactly like the usual algebraic multiplication. As you read this section, keep in mind that a vector-capable calculator will help you multiply vectors only if you understand the basic rules of that multiplication.

Multiplying a Vector by a Scalar

If we multiply a vector $\vec{a}$ by a scalar s, we get a new vector. Its magnitude is the product of the magnitude of $\vec{a}$ and the absolute value of s. Its direction is the direction

* This material will not be employed until later (Chapter 7 for scalar products and Chapter 12 for vector products), and so your instructor may wish to postpone assignment of the section.

of $\vec{a}$ if s is positive, but the opposite direction if s is negative. To divide $\vec{a}$ by s, we multiply $\vec{a}$ by $1/s$.

Multiplying a Vector by a Vector

There are two ways to multiply a vector by a vector: one way produces a scalar (called the *scalar product*), and the other produces a new vector (called the *vector product*). Students commonly confuse the two ways, and so starting now, you should carefully distinguish between them.

The Scalar Product

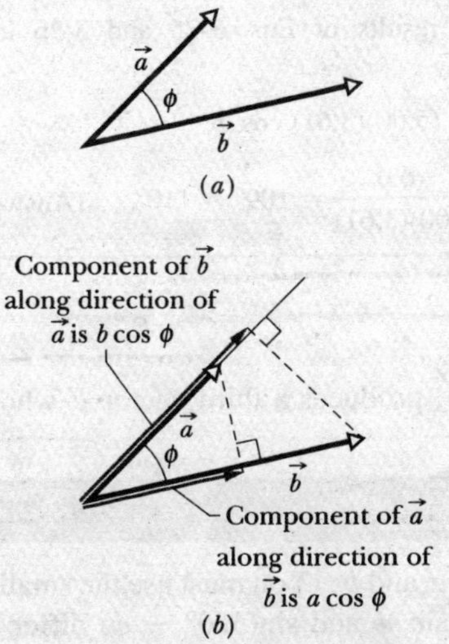

Fig. 3-19 (a) Two vectors $\vec{a}$ and $\vec{b}$, with an angle ϕ between them. (b) Each vector has a component along the direction of the other vector.

The **scalar product** of the vectors $\vec{a}$ and $\vec{b}$ in Fig. 3-19*a* is written as $\vec{a} \cdot \vec{b}$ and defined to be

$$\vec{a} \cdot \vec{b} = ab \cos \phi, \tag{3-20}$$

where a is the magnitude of $\vec{a}$, b is the magnitude of $\vec{b}$, and ϕ is the angle between $\vec{a}$ and $\vec{b}$ (or, more properly, between the directions of $\vec{a}$ and $\vec{b}$). There are actually two such angles: ϕ and $360° - \phi$. Either can be used in Eq. 3-20, because their cosines are the same.

Note that there are only scalars on the right side of Eq. 3-20 (including the value of $\cos \phi$). Thus $\vec{a} \cdot \vec{b}$ on the left side represents a *scalar* quantity. Because of the notation, $\vec{a} \cdot \vec{b}$ is also known as the **dot product** and is spoken as "a dot b."

A dot product can be regarded as the product of two quantities: (1) the magnitude of one of the vectors and (2) the scalar component of the second vector along the direction of the first vector. For example, in Fig. 3-19*b*, $\vec{a}$ has a scalar component $a \cos \phi$ along the direction of $\vec{b}$; note that a perpendicular dropped from the head of $\vec{a}$ to $\vec{b}$ determines that component. Similarly, $\vec{b}$ has a scalar component $b \cos \phi$ along the direction of $\vec{a}$.

> **If the angle ϕ between two vectors is 0°, the component of one vector along the other is maximum, and so also is the dot product of the vectors. If, instead, ϕ is 90°, the component of one vector along the other is zero, and so is the dot product.**

Equation 3-20 can be rewritten as follows to emphasize the components:

$$\vec{a} \cdot \vec{b} = (a \cos \phi)(b) = (a)(b \cos \phi). \tag{3-21}$$

The commutative law applies to a scalar product, so we can write

$$\vec{a} \cdot \vec{b} = \vec{b} \cdot \vec{a}.$$

When two vectors are in unit-vector notation, we write their dot product as

$$\vec{a} \cdot \vec{b} = (a_x\hat{\text{i}} + a_y\hat{\text{j}} + a_z\hat{\text{k}}) \cdot (b_x\hat{\text{i}} + b_y\hat{\text{j}} + b_z\hat{\text{k}}), \tag{3-22}$$

which we can expand according to the distributive law: Each vector component of the first vector is to be dotted with each vector component of the second vector. By doing so, we can show that

$$\vec{a} \cdot \vec{b} = a_x b_x + a_y b_y + a_z b_z. \tag{3-23}$$

CHECKPOINT 4: Vectors $\vec{C}$ and $\vec{D}$ have magnitudes of 3 units and 4 units, respectively. What is the angle between the directions of $\vec{C}$ and $\vec{D}$ if $\vec{C} \cdot \vec{D}$ equals (a) zero, (b) 12 units, and (c) -12 units?

Sample Problem 3-6

What is the angle ϕ between $\vec{a} = 3.0\hat{\mathrm{i}} - 4.0\hat{\mathrm{j}}$ and $\vec{b} = -2.0\hat{\mathrm{i}} + 3.0\hat{\mathrm{k}}$?

SOLUTION: First, a caution: Although many of the following steps can be bypassed with a vector-capable calculator, you will learn more about scalar products if, at least here, you use these steps.

One **Key Idea** here is that the angle between the directions of two vectors is included in the definition of their scalar product (Eq. 3-20):

$$\vec{a} \cdot \vec{b} = ab \cos \phi. \tag{3-24}$$

In this equation, a is the magnitude of $\vec{a}$, or

$$a = \sqrt{3.0^2 + (-4.0)^2} = 5.00, \tag{3-25}$$

and b is the magnitude of $\vec{b}$, or

$$b = \sqrt{(-2.0)^2 + 3.0^2} = 3.61. \tag{3-26}$$

A second **Key Idea** is that we can separately evaluate the left side of Eq. 3-24 by writing the vectors in unit-vector notation and using the distributive law:

$$\begin{aligned}\vec{a} \cdot \vec{b} &= (3.0\hat{\mathrm{i}} - 4.0\hat{\mathrm{j}}) \cdot (-2.0\hat{\mathrm{i}} + 3.0\hat{\mathrm{k}}) \\ &= (3.0\hat{\mathrm{i}}) \cdot (-2.0\hat{\mathrm{i}}) + (3.0\hat{\mathrm{i}}) \cdot (3.0\hat{\mathrm{k}}) \\ &\quad + (-4.0\hat{\mathrm{j}}) \cdot (-2.0\hat{\mathrm{i}}) + (-4.0\hat{\mathrm{j}}) \cdot (3.0\hat{\mathrm{k}}).\end{aligned}$$

We next apply Eq. 3-20 to each term in this last expression. The angle between the vectors in the first term ($3.0\hat{\mathrm{i}}$ and $-2.0\hat{\mathrm{i}}$) is 0°, and in the other terms it is 90°. We then have

$$\vec{a} \cdot \vec{b} = -(6.0)(1) + (9.0)(0) + (8.0)(0) - (12)(0) = -6.0.$$

Substituting this and the results of Eqs. 3-25 and 3-26 into Eq. 3-24 yields

$$-6.0 = (5.00)(3.61) \cos \phi,$$

so

$$\phi = \cos^{-1} \frac{-6.0}{(5.00)(3.61)} = 109° \approx 110°. \quad \text{(Answer)}$$

The Vector Product

The **vector product** of $\vec{a}$ and $\vec{b}$, written $\vec{a} \times \vec{b}$, produces a third vector $\vec{c}$ whose magnitude is

$$c = ab \sin \phi, \tag{3-27}$$

where ϕ is the *smaller* of the two angles between $\vec{a}$ and $\vec{b}$. (You must use the smaller of the two angles between the vectors because $\sin \phi$ and $\sin(360° - \phi)$ differ in algebraic sign.) Because of the notation, $\vec{a} \times \vec{b}$ is also known as the **cross product**, and in speech it is "a cross b."

If $\vec{a}$ and $\vec{b}$ are parallel or antiparallel, $\vec{a} \times \vec{b} = 0$. The magnitude of $\vec{a} \times \vec{b}$, which can be written as $|\vec{a} \times \vec{b}|$, is maximum when $\vec{a}$ and $\vec{b}$ are perpendicular to each other.

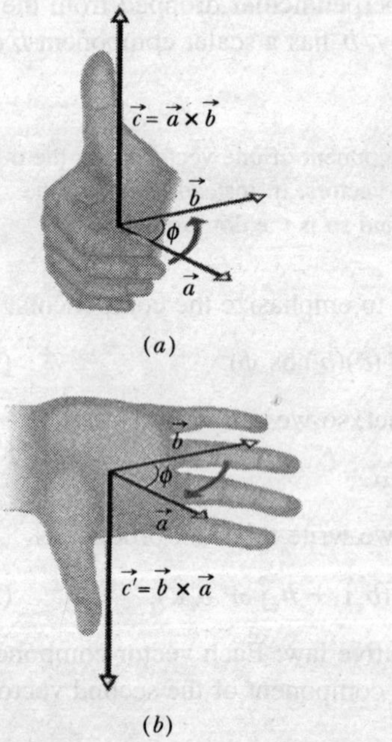

Fig. 3-20 Illustration of the right-hand rule for vector products. (*a*) Sweep vector $\vec{a}$ into vector $\vec{b}$ with the fingers of your right hand. Your outstretched thumb shows the direction of vector $\vec{c} = \vec{a} \times \vec{b}$. (*b*) Showing that $\vec{a} \times \vec{b}$ is the reverse of $\vec{b} \times \vec{a}$.

The direction of $\vec{c}$ is perpendicular to the plane that contains $\vec{a}$ and $\vec{b}$. Figure 3-20*a* shows how to determine the direction of $\vec{c} = \vec{a} \times \vec{b}$ with what is known as the **right-hand rule**. Place the vectors $\vec{a}$ and $\vec{b}$ tail to tail without altering their orientations, and imagine a line that is perpendicular to their plane where they meet. Pretend to place your *right* hand around that line in such a way that your fingers would sweep $\vec{a}$ into $\vec{b}$ through the smaller angle between them. Your outstretched thumb points in the direction of $\vec{c}$.

The order of the vector multiplication is important. In Fig. 3-20*b*, we are determining the direction of $\vec{c}' = \vec{b} \times \vec{a}$, so the fingers are placed to sweep $\vec{b}$ into $\vec{a}$ through the smaller angle. The thumb ends up in the opposite direction from previously, and so it must be that $\vec{c}' = -\vec{c}$; that is,

$$\vec{b} \times \vec{a} = -(\vec{a} \times \vec{b}). \tag{3-28}$$

In other words, the commutative law does not apply to a vector product.

In unit-vector notation, we write

$$\vec{a} \times \vec{b} = (a_x\hat{\mathrm{i}} + a_y\hat{\mathrm{j}} + a_z\hat{\mathrm{k}}) \times (b_x\hat{\mathrm{i}} + b_y\hat{\mathrm{j}} + b_z\hat{\mathrm{k}}), \tag{3-29}$$

which can be expanded according to the distributive law; that is, each component of the first vector is to be crossed with each component of the second vector. The cross products of unit vectors are given in Appendix E (see "Products of Vectors").

For example, in the expansion of Eq. 3-29, we have

$$a_x\hat{\mathrm{i}} \times b_x\hat{\mathrm{i}} = a_x b_x(\hat{\mathrm{i}} \times \hat{\mathrm{i}}) = 0,$$

because the two unit vectors $\hat{\mathrm{i}}$ and $\hat{\mathrm{i}}$ are parallel and thus have a zero cross product. Similarly, we have

$$a_x\hat{\mathrm{i}} \times b_y\hat{\mathrm{j}} = a_x b_y(\hat{\mathrm{i}} \times \hat{\mathrm{j}}) = a_x b_y\hat{\mathrm{k}}.$$

In the last step we used Eq. 3-27 to evaluate the magnitude of $\hat{\mathrm{i}} \times \hat{\mathrm{j}}$ as unity. (The vectors $\hat{\mathrm{i}}$ and $\hat{\mathrm{j}}$ each have a magnitude of unity, and the angle between them is 90°.) Also, we used the right-hand rule to get the direction of $\hat{\mathrm{i}} \times \hat{\mathrm{j}}$ as being in the positive direction of the z axis (thus in the direction of $\hat{\mathrm{k}}$).

Continuing to expand Eq. 3-29, you can show that

$$\vec{a} \times \vec{b} = (a_y b_z - b_y a_z)\hat{\mathrm{i}} + (a_z b_x - b_z a_x)\hat{\mathrm{j}} + (a_x b_y - b_x a_y)\hat{\mathrm{k}}. \quad (3\text{-}30)$$

You can also evaluate a cross product by setting up and evaluating a determinant (as shown in Appendix E) or by using a vector-capable calculator.

To check whether any xyz coordinate system is a right-handed coordinate system, use the right-hand rule for the cross product $\hat{\mathrm{i}} \times \hat{\mathrm{j}} = \hat{\mathrm{k}}$ with that system. If your fingers sweep $\hat{\mathrm{i}}$ (positive direction of x) into $\hat{\mathrm{j}}$ (positive direction of y) with the outstretched thumb pointing in the positive direction of z, then the system is right-handed.

CHECKPOINT 5: Vectors $\vec{C}$ and $\vec{D}$ have magnitudes of 3 units and 4 units, respectively. What is the angle between the directions of $\vec{C}$ and $\vec{D}$ if the magnitude of the vector product $\vec{C} \times \vec{D}$ is (a) zero and (b) 12 units?

Sample Problem 3-7

In Fig. 3-21, vector $\vec{a}$ lies in the xy plane, has a magnitude of 18 units, and points in a direction 250° from the positive direction of x. Also, vector $\vec{b}$ has a magnitude of 12 units and points along the positive direction of z. What is the vector product $\vec{c} = \vec{a} \times \vec{b}$?

SOLUTION: One **Key Idea** is that when we have two vectors in magnitude-angle notation, we find the magnitude of their cross product (that is, the vector that results from taking their cross product) with Eq. 3-27. Here that means the magnitude of $\vec{c}$ is

$$c = ab \sin \phi = (18)(12)(\sin 90°) = 216. \quad \text{(Answer)}$$

A second **Key Idea** is that with two vectors in magnitude-angle notation, we find the direction of their cross product with the right-hand rule of Fig. 3-20. In Fig. 3-21, imagine placing the fingers of your right hand around a line perpendicular to the plane of $\vec{a}$ and $\vec{b}$ (the line on which $\vec{c}$ is shown) such that your fingers sweep $\vec{a}$ into $\vec{b}$. Your outstretched thumb then gives the direction of $\vec{c}$. Thus, as shown in Fig. 3-21, $\vec{c}$ lies in the xy plane. Because its direction is perpendicular to the direction of $\vec{a}$, it is at an angle of

$$250° - 90° = 160° \quad \text{(Answer)}$$

from the positive direction of x.

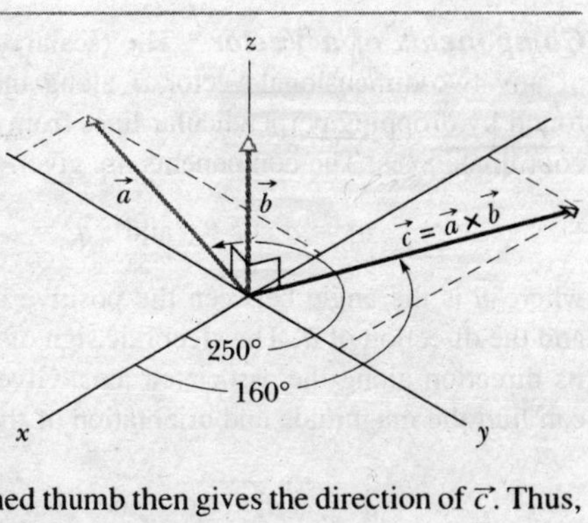

Fig. 3-21 Sample Problem 3-7. Vector $\vec{c}$ (in the xy plane) is the vector (or cross) product of vectors $\vec{a}$ and $\vec{b}$.

Sample Problem 3-8

If $\vec{a} = 3\hat{\mathrm{i}} - 4\hat{\mathrm{j}}$ and $\vec{b} = -2\hat{\mathrm{i}} + 3\hat{\mathrm{k}}$, what is $\vec{c} = \vec{a} \times \vec{b}$?

SOLUTION: The **Key Idea** is that when two vectors are in unit-vector notation, we can find their cross product by using the distributive law. Here that means we can write

$$\vec{c} = (3\hat{\mathrm{i}} - 4\hat{\mathrm{j}}) \times (-2\hat{\mathrm{i}} + 3\hat{\mathrm{k}})$$
$$= 3\hat{\mathrm{i}} \times (-2\hat{\mathrm{i}}) + 3\hat{\mathrm{i}} \times 3\hat{\mathrm{k}} + (-4\hat{\mathrm{j}}) \times (-2\hat{\mathrm{i}}) + (-4\hat{\mathrm{j}}) \times 3\hat{\mathrm{k}}.$$

We next evaluate each term with Eq. 3-27, determining the direction with the right-hand rule. For the first term here, the angle ϕ between the two vectors being crossed is 0. For the other terms, ϕ is 90°. We find

$$\vec{c} = 6(0) + 9(-\hat{\mathrm{j}}) + 8(-\hat{\mathrm{k}}) - 12\hat{\mathrm{i}}$$
$$= -12\hat{\mathrm{i}} - 9\hat{\mathrm{j}} - 8\hat{\mathrm{k}}. \quad \text{(Answer)}$$

This vector $\vec{c}$ is perpendicular to both $\vec{a}$ and $\vec{b}$, a fact you can check by showing that $\vec{c} \cdot \vec{a} = 0$ and $\vec{c} \cdot \vec{b} = 0$; that is, there is no component of $\vec{c}$ along the direction of either $\vec{a}$ or $\vec{b}$.

PROBLEM-SOLVING TACTICS

Tactic 5: *Common Errors with Cross Products*
Several errors are common in finding a cross product. (1) Failure to arrange vectors tail to tail is tempting when an illustration presents them head to tail: you must mentally shift (or better, redraw) one vector to the proper arrangement without changing its orientation. (2) Failing to use the right hand in applying the right-hand rule is easy when the right hand is occupied with a calculator or pencil. (3) Failure to sweep the first vector of the product into the second vector can occur when the orientations of the vectors require an awkward twisting of your hand to apply the right-hand rule. Sometimes that happens when you try to make the sweep mentally rather than actually using your hand. (4) Failure to work with a right-handed coordinate system results when you forget how to draw such a system (see Fig. 3-14).

REVIEW & SUMMARY

Scalars and Vectors *Scalars,* such as temperature, have magnitude only. They are specified by a number with a unit (10°C) and obey the rules of arithmetic and ordinary algebra. *Vectors,* such as displacement, have both magnitude and direction (5 m, north) and obey the special rules of vector algebra.

Adding Vectors Geometrically Two vectors $\vec{a}$ and $\vec{b}$ may be added geometrically by drawing them to a common scale and placing them head to tail. The vector connecting the tail of the first to the head of the second is the vector sum $\vec{s}$. To subtract $\vec{b}$ from $\vec{a}$, reverse the direction of $\vec{b}$ to get $-\vec{b}$; then add $-\vec{b}$ to $\vec{a}$. Vector addition is commutative and obeys the associative law.

Components of a Vector The (scalar) *components* a_x and a_y of any two-dimensional vector $\vec{a}$ along the coordinate axes are found by dropping perpendicular lines from the ends of $\vec{a}$ onto the coordinate axes. The components are given by

$$a_x = a\cos\theta \quad \text{and} \quad a_y = a\sin\theta, \tag{3-5}$$

where θ is the angle between the positive direction of the x axis and the direction of $\vec{a}$. The algebraic sign of a component indicates its direction along the associated axis. Given its components, we can find the magnitude and orientation of the vector $\vec{a}$ with

$$a = \sqrt{a_x^2 + a_y^2} \quad \text{and} \quad \tan\theta = \frac{a_y}{a_x}. \tag{3-6}$$

Unit-Vector Notation Unit vectors $\hat{\text{i}}$, $\hat{\text{j}}$, and $\hat{\text{k}}$ have magnitudes of unity and are directed in the positive directions of the x, y, and z axes, respectively, in a right-handed coordinate system. We can write a vector $\vec{a}$ in terms of unit vectors as

$$\vec{a} = a_x\hat{\text{i}} + a_y\hat{\text{j}} + a_z\hat{\text{k}}, \tag{3-7}$$

in which $a_x\hat{\text{i}}$, $a_y\hat{\text{j}}$, and $a_z\hat{\text{k}}$ are the **vector components** of $\vec{a}$ and a_x, a_y, and a_z are its **scalar components.**

Adding Vectors in Component Form To add vectors in component form, we use the rules

$$r_x = a_x + b_x \quad r_y = a_y + b_y \quad r_z = a_z + b_z. \tag{3-11 to 3-13}$$

Here $\vec{a}$ and $\vec{b}$ are the vectors to be added, and $\vec{r}$ is the vector sum.

Vectors and Physical Laws Any physical situation involving vectors can be described using many possible coordinate systems. We usually choose the one that most simplifies the situation. However, the relationship between the vector quantities does not depend on our choice of coordinates. The laws of physics are also independent of that choice.

Product of a Scalar and a Vector The product of a scalar s and a vector $\vec{v}$ is a new vector whose magnitude is sv and whose direction is the same as that of $\vec{v}$ if s is positive, and opposite that of $\vec{v}$ if s is negative. To divide $\vec{v}$ by s, multiply $\vec{v}$ by $(1/s)$.

The Scalar Product The **scalar** (or **dot**) **product** of two vectors $\vec{a}$ and $\vec{b}$ is written $\vec{a}\cdot\vec{b}$ and is the *scalar* quantity given by

$$\vec{a}\cdot\vec{b} = ab\cos\phi, \tag{3-20}$$

in which ϕ is the angle between the directions of $\vec{a}$ and $\vec{b}$. The scalar product may be positive, zero, or negative, depending on the value of ϕ. A scalar product is the product of the magnitude of one vector and the component of the second vector along the direction of the first vector.

In unit-vector notation,

$$\vec{a}\cdot\vec{b} = (a_x\hat{\text{i}} + a_y\hat{\text{j}} + a_z\hat{\text{k}})\cdot(b_x\hat{\text{i}} + b_y\hat{\text{j}} + b_z\hat{\text{k}}), \tag{3-22}$$

which may be expanded according to the distributive law. Note that $\vec{a}\cdot\vec{b} = \vec{b}\cdot\vec{a}$.

The Vector Product The **vector** (or **cross**) **product** of two vectors $\vec{a}$ and $\vec{b}$ is written $\vec{a}\times\vec{b}$ and is a *vector* $\vec{c}$ whose magnitude c is given by

$$c = ab\sin\phi, \tag{3-27}$$

in which ϕ is the smaller of the angles between the directions of $\vec{a}$ and $\vec{b}$. The direction of $\vec{c}$ is perpendicular to the plane defined by $\vec{a}$ and $\vec{b}$ and is given by a right-hand rule, as shown in Fig. 3-20. Note that $\vec{a}\times\vec{b} = -(\vec{b}\times\vec{a})$. In unit-vector notation,

$$\vec{a}\times\vec{b} = (a_x\hat{\text{i}} + a_y\hat{\text{j}} + a_z\hat{\text{k}})\times(b_x\hat{\text{i}} + b_y\hat{\text{j}} + b_z\hat{\text{k}}), \tag{3-29}$$

which we may expand with the distributive law.

QUESTIONS

1. Displacement $\vec{D}$ points from coordinates (5 m, 3 m) to coordinates (7 m, 6 m) in the xy plane. Which of the following displacement vectors are equivalent to $\vec{D}$: vector $\vec{A}$, which points from (−6 m, −5 m) to (−4 m, −2 m); vector $\vec{B}$, which points from (−6 m, 1 m) to (−4 m, 4 m); and vector $\vec{C}$, which points from (−8 m, −6 m) to (−10 m, −9 m)?

2. Can the magnitude of the difference between two vectors ever be greater than (a) the magnitude of one of the vectors, (b) the magnitudes of both vectors, and (c) the magnitude of their sum?

3. Equation 3-2 shows that the addition of two vectors $\vec{a}$ and $\vec{b}$ is commutative. Does that mean subtraction is commutative, so that $\vec{a} - \vec{b} = \vec{b} - \vec{a}$?

4. If $\vec{d} = \vec{a} + \vec{b} + (-\vec{c})$, does (a) $\vec{a} + (-\vec{d}) = \vec{c} + (-\vec{b})$, (b) $\vec{a} = (-\vec{b}) + \vec{d} + \vec{c}$, and (c) $\vec{c} + (-\vec{d}) = \vec{a} + \vec{b}$?

5. Describe two vectors $\vec{a}$ and $\vec{b}$ such that
(a) $\vec{a} + \vec{b} = \vec{c}$ and $a + b = c$;
(b) $\vec{a} + \vec{b} = \vec{a} - \vec{b}$;
(c) $\vec{a} + \vec{b} = \vec{c}$ and $a^2 + b^2 = c^2$.

6. In Fig. 3-22, are (a) the x component and (b) the y component of vector $\vec{A}$ positive or negative? Are (c) the x component and (d) the y component of the vector combination $\vec{A} - \vec{B}$ positive or negative?

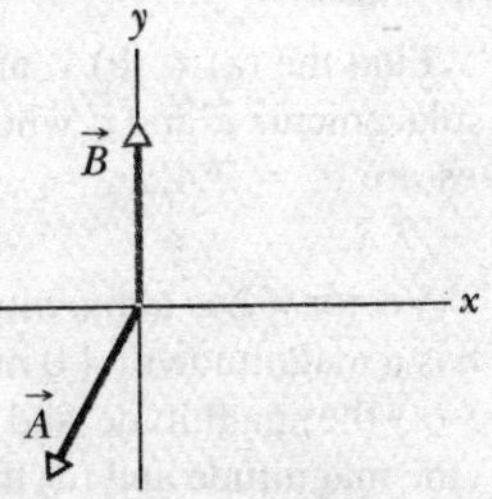

Fig. 3-22 Question 6.

7. Which of the arrangements of axes in Fig. 3-23 can be labeled "right-handed coordinate system"? As usual, each axis label indicates the positive side of the axis.

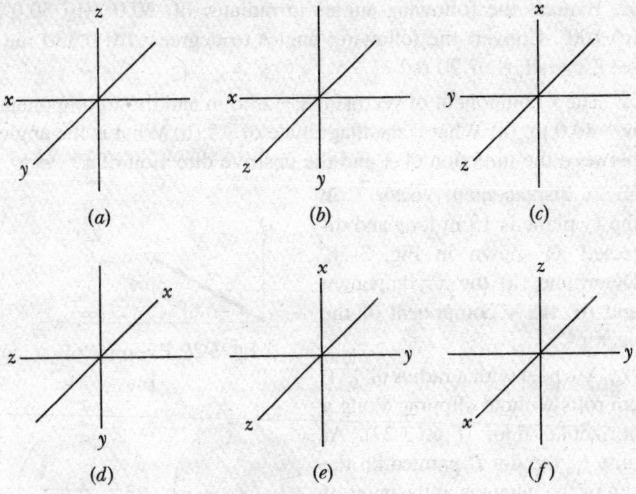

Fig. 3-23 Question 7.

8. If $\vec{a} \cdot \vec{b} = \vec{a} \cdot \vec{c}$, must $\vec{b}$ equal $\vec{c}$?

9. If $\vec{A} = 2\hat{i} + 4\hat{j}$, what is $\vec{A} \times \vec{B}$ when (a) $\vec{B} = 8\hat{i} + 16\hat{j}$ and (b) $\vec{B} = -8\hat{i} - 16\hat{j}$? (This question can be answered without computation.)

10. Figure 3-24 shows vector $\vec{A}$ and four other vectors that have the same magnitude but differ in orientation. (a) Which of those other four vectors have the same dot product with $\vec{A}$? (b) Which have a negative dot product with $\vec{A}$?

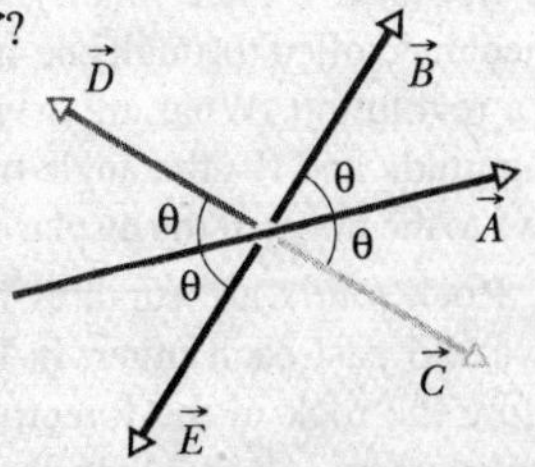

Fig. 3-24 Question 10.

EXERCISES & PROBLEMS

ssm Solution is in the Student Solutions Manual.
www Solution is available on the World Wide Web at: http://www.wiley.com/college/hrw
ilw Solution is available on the Interactive LearningWare.

SEC. 3-2 Adding Vectors Geometrically

1E. Consider two displacements, one of magnitude 3 m and another of magnitude 4 m. Show how the displacement vectors may be combined to get a resultant displacement of magnitude (a) 7 m, (b) 1 m, and (c) 5 m.

2P. A bank in downtown Boston is robbed (see the map in Fig. 3-25). To elude police, the robbers escape by helicopter, making three successive flights described by the following displacements: 32 km, 45° south of east; 53 km, 26° north of west; 26 km, 18° east of south. At the end of the third flight they are captured. In what town are they apprehended? (Use the geometrical method to add these displacements on the map.)

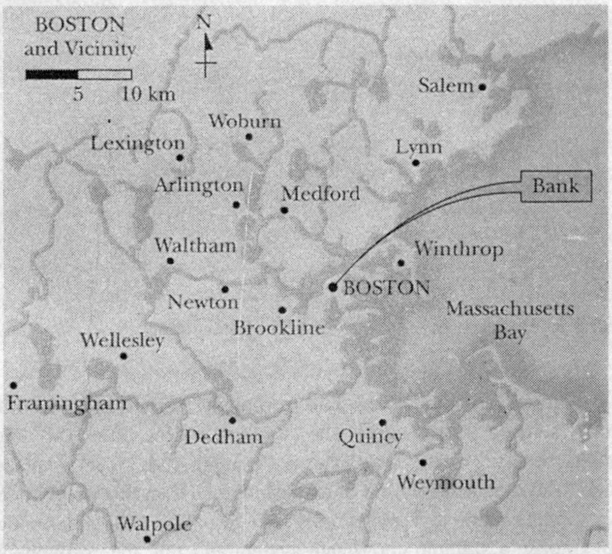

Fig. 3-25 Problem 2.

SEC. 3-3 Components of Vectors

3E. What are (a) the x component and (b) the y component of a vector $\vec{a}$ in the xy plane if its direction is 250° counterclockwise from the positive direction of the x axis and its magnitude is 7.3 m? ssm

4E. Express the following angles in radians: (a) 20.0°, (b) 50.0°, (c) 100°. Convert the following angles to degrees: (d) 0.330 rad, (e) 2.10 rad, (f) 7.70 rad.

5E. The x component of vector $\vec{A}$ is -25.0 m and the y component is $+40.0$ m. (a) What is the magnitude of $\vec{A}$? (b) What is the angle between the direction of $\vec{A}$ and the positive direction of x? ssm

6E. A displacement vector $\vec{r}$ in the xy plane is 15 m long and directed as shown in Fig. 3-26. Determine (a) the x component and (b) the y component of the vector.

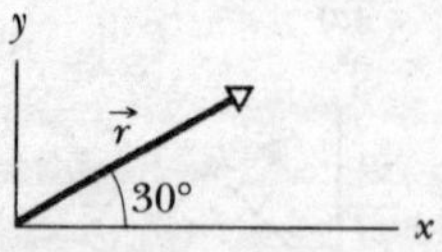

Fig. 3-26 Exercise 6.

7P. A wheel with a radius of 45.0 cm rolls without slipping along a horizontal floor (Fig. 3-27). At time t_1, the dot P painted on the rim of the wheel is at the point of contact between the wheel and the floor. At a later time t_2, the wheel has rolled through one-half of a revolution. What are (a) the magnitude and (b) the angle (relative to the floor) of the displacement of P during this interval? ssm

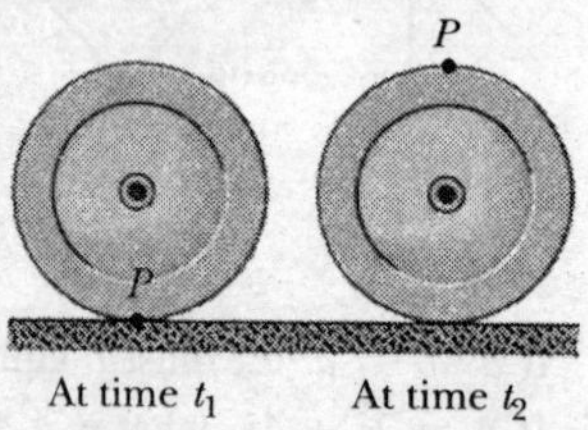

Fig. 3-27 Problem 7.

8P. Rock *faults* are ruptures along which opposite faces of rock have slid past each other. In Fig. 3-28, points A and B coincided before the rock in the foreground slid down to the right. The net displacement $\overrightarrow{AB}$ is along the plane of the fault. The horizontal component of $\overrightarrow{AB}$ is the *strike-slip AC*. The component of $\overrightarrow{AB}$ that is directly down the plane of the fault is the *dip-slip AD*. (a) What is the magnitude of the net displacement $\overrightarrow{AB}$ if the strike-slip is 22.0 m and the dip-slip is 17.0 m? (b) If the plane of the fault is inclined 52.0° to the horizontal, what is the vertical component of $\overrightarrow{AB}$?

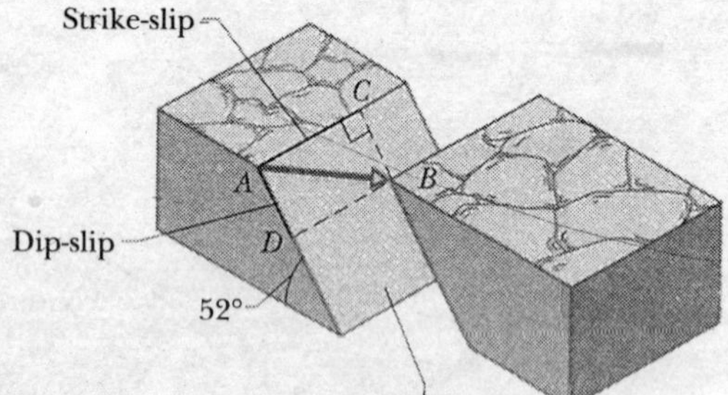

Fig. 3-28 Problem 8.

9P A room has dimensions 3.00 m (height) × 3.70 m × 4.30 m. A fly starting at one corner flies around, ending up at the diagonally opposite corner. (a) What is the magnitude of its displacement? (b) Could the length of its path be less than this magnitude? (c) Greater than this magnitude? (d) Equal to this magnitude? (e) Choose a suitable coordinate system and find the components of the displacement vector in that system. (f) If the fly walks rather than flies, what is the length of the shortest path it can take? (*Hint:* This can be answered without calculus. The room is like a box. Unfold its walls to flatten them into a plane.) ssm www

SEC. 3-5 Adding Vectors by Components

10E. A car is driven east for a distance of 50 km, then north for 30 km, and then in a direction 30° east of north for 25 km. Sketch the vector diagram and determine (a) the magnitude and (b) the angle of the car's total displacement from its starting point.

11E. A woman walks 250 m in the direction 30° east of north, then 175 m directly east. Find (a) the magnitude and (b) the angle of her final displacement from the starting point. (c) Find the distance she walks. (d) Which is greater, that distance or the magnitude of her displacement? ssm

12E. A person walks in the following pattern: 3.1 km north, then 2.4 km west, and finally 5.2 km south. (a) Sketch the vector diagram that represents this motion. (b) How far and (c) in what direction would a bird fly in a straight line from the same starting point to the same final point?

13E. (a) In unit-vector notation, what is the sum of

$$\vec{a} = (4.0\text{ m})\hat{\text{i}} + (3.0\text{ m})\hat{\text{j}} \quad\text{and}\quad \vec{b} = (-13.0\text{ m})\hat{\text{i}} + (7.0\text{ m})\hat{\text{j}}?$$

What are (b) the magnitude and (c) the direction of $\vec{a} + \vec{b}$ (relative to $\hat{\text{i}}$)? ssm

14E. Find the (a) x, (b) y, and (c) z components of the sum $\vec{r}$ of the displacements $\vec{c}$ and $\vec{d}$ whose components in meters along the three axes are $c_x = 7.4$, $c_y = -3.8$, $c_z = -6.1$; $d_x = 4.4$, $d_y = -2.0$, $d_z = 3.3$.

15E. Vector $\vec{a}$ has a magnitude of 5.0 m and is directed east. Vector $\vec{b}$ has a magnitude of 4.0 m and is directed 35° west of north. What are (a) the magnitude and (b) the direction of $\vec{a} + \vec{b}$? What are (c) the magnitude and (d) the direction of $\vec{b} - \vec{a}$? (e) Draw a vector diagram for each combination. ssm

16E. For the vectors

$$\vec{a} = (3.0\text{ m})\hat{\text{i}} + (4.0\text{ m})\hat{\text{j}} \quad\text{and}\quad \vec{b} = (5.0\text{ m})\hat{\text{i}} + (-2.0\text{ m})\hat{\text{j}},$$

give $\vec{a} + \vec{b}$ in (a) unit-vector notation, and as (b) a magnitude and (c) an angle (relative to $\hat{\text{i}}$). Now give $\vec{b} - \vec{a}$ in (d) unit-vector notation, and as (e) a magnitude and (f) an angle.

17E. Two vectors are given by

$$\vec{a} = (4.0\text{ m})\hat{\text{i}} - (3.0\text{ m})\hat{\text{j}} + (1.0\text{ m})\hat{\text{k}}$$

and

$$\vec{b} = (-1.0\text{ m})\hat{\text{i}} + (1.0\text{ m})\hat{\text{j}} + (4.0\text{ m})\hat{\text{k}}.$$

In unit-vector notation, find (a) $\vec{a} + \vec{b}$, (b) $\vec{a} - \vec{b}$, and (c) a third vector $\vec{c}$ such that $\vec{a} - \vec{b} + \vec{c} = 0$. ssm

18P. Here are two vectors:

$$\vec{a} = (4.0\text{ m})\hat{\text{i}} - (3.0\text{ m})\hat{\text{j}} \quad\text{and}\quad \vec{b} = (6.0\text{ m})\hat{\text{i}} + (8.0\text{ m})\hat{\text{j}}.$$

What are (a) the magnitude and (b) the angle (relative to $\hat{\text{i}}$) of $\vec{a}$? What are (c) the magnitude and (d) the angle of $\vec{b}$? What are (e) the magnitude and (f) the angle of $\vec{a} + \vec{b}$; (g) the magnitude and (h) the angle of $\vec{b} - \vec{a}$; and (i) the magnitude and (j) the angle of $\vec{a} - \vec{b}$? (k) What is the angle between the directions of $\vec{b} - \vec{a}$ and $\vec{a} - \vec{b}$?

19P. Three vectors $\vec{a}$, $\vec{b}$, and $\vec{c}$ each have a magnitude of 50 m and lie in an xy plane. Their directions relative to the positive direction of the x axis are 30°, 195°, and 315°, respectively. What are (a) the magnitude and (b) the angle of the vector $\vec{a} + \vec{b} + \vec{c}$, and (c) the magnitude and (d) the angle of $\vec{a} - \vec{b} + \vec{c}$? What are (e) the magnitude and (f) the angle of a fourth vector $\vec{d}$ such that $(\vec{a} + \vec{b}) - (\vec{c} + \vec{d}) = 0$? ilw

20P. What is the sum of the following four vectors in (a) unit-vector notation and (b) magnitude-angle notation? For the latter, give the angle in both degrees and radians. Positive angles are counterclockwise from the positive direction of the x axis; negative angles are clockwise.

$\vec{E}$: 6.00 m at +0.900 rad $\vec{F}$: 5.00 m at −75.0°

$\vec{G}$: 4.00 m at +1.20 rad $\vec{H}$: 6.00 m at −210°

21P. The two vectors $\vec{a}$ and $\vec{b}$ in Fig. 3-29 have equal magnitudes of 10.0 m. Find (a) the x component and (b) the y component of their vector sum $\vec{r}$, (c) the magnitude of $\vec{r}$, and (d) the angle $\vec{r}$ makes with the positive direction of the x axis. ssm ilw www

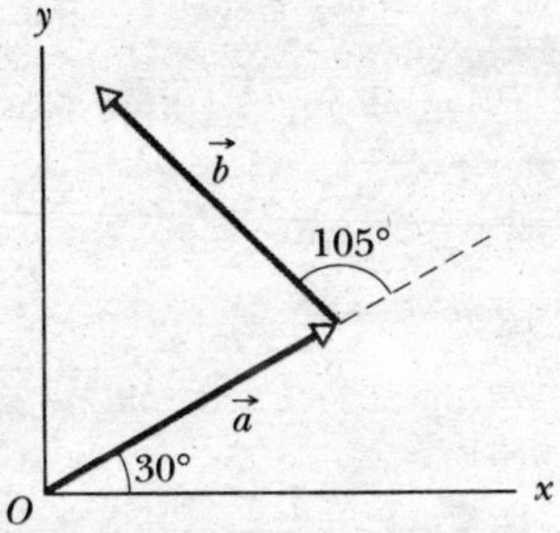

Fig. 3-29 Problem 21.

22P. In the sum $\vec{A} + \vec{B} = \vec{C}$, vector $\vec{A}$ has a magnitude of 12.0 m and is angled 40.0° counterclockwise from the $+x$ direction, and vector $\vec{C}$ has a magnitude of 15.0 m and is angled 20.0° counterclockwise from the $-x$ direction. What are (a) the magnitude and (b) the angle (relative to $+x$) of $\vec{B}$?

23P. Prove that two vectors must have equal magnitudes if their sum is perpendicular to their difference. ssm

24P. Find the sum of the following four vectors in (a) unit-vector notation, and as (b) a magnitude and (c) an angle relative to $+x$.

$\vec{P}$: 10.0 m, at 25.0° counterclockwise from $+x$

$\vec{Q}$: 12.0 m, at 10.0° counterclockwise from $+y$

$\vec{R}$: 8.00 m, at 20.0° clockwise from $-y$

$\vec{S}$: 9.00 m, at 40.0° counterclockwise from $-y$

25P. Two vectors of magnitudes a and b make an angle θ with each other when placed tail to tail. Prove, by taking components along two perpendicular axes, that

$$r = \sqrt{a^2 + b^2 + 2ab\cos\theta}$$

gives the magnitude of the sum $\vec{r}$ of the two vectors. ssm

26P. What is the sum of the following four vectors in (a) unit-vector notation, and as (b) a magnitude and (c) an angle? Positive angles are counterclockwise from the positive direction of the x axis; negative angles are clockwise.

$\vec{A} = (2.00\text{ m})\hat{\imath} + (3.00\text{ m})\hat{\jmath}$ $\vec{B}$: 4.00 m, at +65.0°

$\vec{C} = (-4.00\text{ m})\hat{\imath} - (6.00\text{ m})\hat{\jmath}$ $\vec{D}$: 5.00 m, at −235°

27P. (a) Using unit vectors, write expressions for the four body diagonals (the straight lines from one corner to another through the center) of a cube in terms of its edges, which have length a. (b) Determine the angles that the body diagonals make with the adjacent edges. (c) Determine the length of the body diagonals in terms of a. ssm

SEC. 3-6 Vectors and the Laws of Physics

28E. $\vec{A}$ has the magnitude 12.0 m and is angled 60.0° counterclockwise from the positive direction of the x axis of an xy coordinate system. Also, $\vec{B} = (12.0\text{ m})\hat{\imath} + (8.00\text{ m})\hat{\jmath}$ on that same coordinate system. We now rotate the system counterclockwise about the origin by 20.0° to form an $x'y'$ system. On this new system, what are (a) $\vec{A}$ and (b) $\vec{B}$, both in unit-vector notation?

SEC. 3-7 Multiplying Vectors

29E. A vector $\vec{a}$ of magnitude 10 units and another vector $\vec{b}$ of magnitude 6.0 units differ in directions by 60°. Find (a) the scalar product of the two vectors and (b) the magnitude of the vector product $\vec{a} \times \vec{b}$. ssm

30E. Derive Eq. 3-23 for a scalar product in unit-vector notation.

31P. Use the definition of scalar product, $\vec{a} \cdot \vec{b} = ab\cos\theta$, and the fact that $\vec{a} \cdot \vec{b} = a_xb_x + a_yb_y + a_zb_z$ (see Exercise 30) to calculate the angle between the two vectors given by $\vec{a} = 3.0\hat{\imath} + 3.0\hat{\jmath} + 3.0\hat{k}$ and $\vec{b} = 2.0\hat{\imath} + 1.0\hat{\jmath} + 3.0\hat{k}$. ssm ilw www

32P. Derive Eq. 3-30 for a vector product in unit-vector notation.

33P. Show that the area of the triangle contained between $\vec{a}$ and $\vec{b}$ and the red line in Fig. 3-30 is $\frac{1}{2}|\vec{a} \times \vec{b}|$. ssm

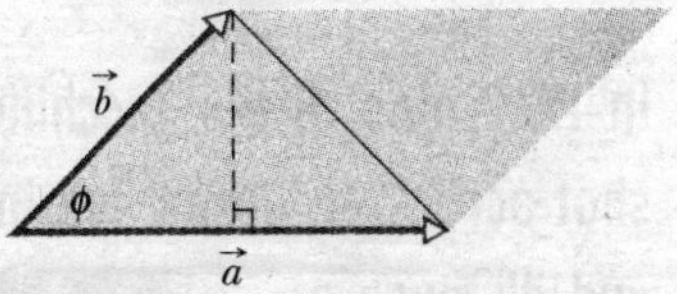

Fig. 3-30 Problem 33.

34P. In the product $\vec{F} = q\vec{v} \times \vec{B}$, take $q = 2$,

$$\vec{v} = 2.0\hat{\imath} + 4.0\hat{\jmath} + 6.0\hat{k} \quad \text{and} \quad \vec{F} = 4.0\hat{\imath} - 20\hat{\jmath} + 12\hat{k}.$$

What then is $\vec{B}$ in unit-vector notation if $B_x = B_y$?

35P. (a) Show that $\vec{a} \cdot (\vec{b} \times \vec{a})$ is zero for all vectors $\vec{a}$ and $\vec{b}$. (b) What is the magnitude of $\vec{a} \times (\vec{b} \times \vec{a})$ if there is an angle ϕ between the directions of $\vec{a}$ and $\vec{b}$? ssm

36P. For the following three vectors, what is $3\vec{C} \cdot (2\vec{A} \times \vec{B})$?

$$\vec{A} = 2.00\hat{\imath} + 3.00\hat{\jmath} - 4.00\hat{k}$$

$$\vec{B} = -3.00\hat{\imath} + 4.00\hat{\jmath} + 2.00\hat{k} \quad \vec{C} = 7.00\hat{\imath} - 8.00\hat{\jmath}$$

37P. The three vectors in Fig. 3-31 have magnitudes $a = 3.00$ m, $b = 4.00$ m, and $c = 10.0$ m. What are (a) the x component and (b) the y component of $\vec{a}$; (c) the x component and (d) the y component of $\vec{b}$; and (e) the x component and (f) the y component of $\vec{c}$? If $\vec{c} = p\vec{a} + q\vec{b}$, what are the values of (g) p and (h) q? ilw

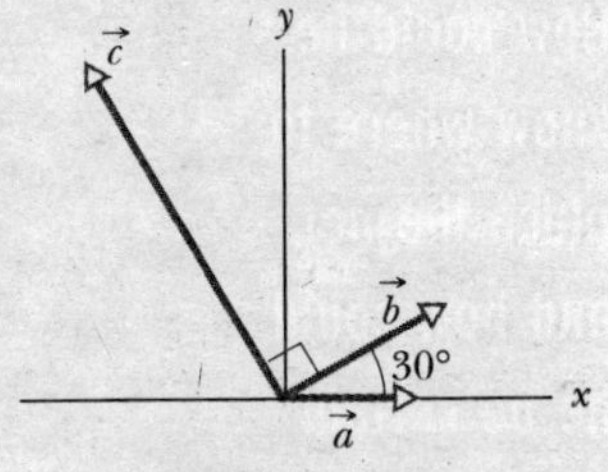

Fig. 3-31 Problem 37.

38P. Two vectors $\vec{a}$ and $\vec{b}$ have the components, in meters, $a_x = 3.2$, $a_y = 1.6$, $b_x = 0.50$, $b_y = 4.5$. (a) Find the angle between the directions of $\vec{a}$ and $\vec{b}$. There are two vectors in the xy plane that are perpendicular to $\vec{a}$ and have a magnitude of 5.0 m. One, vector $\vec{c}$, has a positive x component and the other, vector $\vec{d}$, a negative x component. What are (b) the x component and (c) the y component of $\vec{c}$, and (d) the x component and (e) the y component of vector $\vec{d}$?

4 Motion in Two and Three Dimensions

In 1922, one of the Zacchinis, a famous family of circus performers, was the first human cannonball to be shot across an arena and into a net. To increase the excitement, the family gradually increased the height and distance of the flight until, in 1939 or 1940, Emanuel Zacchini soared over three Ferris wheels and through a horizontal distance of 69 m.

How could he know where to place the net, and how could he be certain he would clear the Ferris wheels?

The answer is in this chapter.

4-1 Moving in Two or Three Dimensions

This chapter extends the material of the preceding two chapters to two and three dimensions. Many of the ideas of Chapter 2, such as position, velocity, and acceleration, are used here, but they are now a little more complex because of the extra dimensions. To keep the notation manageable, we use the vector algebra of Chapter 3. As you read this chapter, you might want to thumb back to those previous chapters to refresh your memory.

4-2 Position and Displacement

One general way of locating a particle (or particle-like object) is with a **position vector** $\vec{r}$, which is a vector that extends from a reference point (usually the origin of a coordinate system) to the particle. In the unit-vector notation of Section 3-4, $\vec{r}$ can be written

$$\vec{r} = x\hat{i} + y\hat{j} + z\hat{k}, \tag{4-1}$$

where $x\hat{i}$, $y\hat{j}$, and $z\hat{k}$ are the vector components of $\vec{r}$, and the coefficients x, y, and z are its scalar components.

The coefficients x, y, and z give the particle's location along the coordinate axes and relative to the origin; that is, the particle has the rectangular coordinates (x, y, z). For instance, Fig. 4-1 shows a particle with position vector

$$\vec{r} = (-3 \text{ m})\hat{i} + (2 \text{ m})\hat{j} + (5 \text{ m})\hat{k}$$

and rectangular coordinates $(-3 \text{ m}, 2 \text{ m}, 5 \text{ m})$. Along the x axis the particle is 3 m from the origin, in the $-\hat{i}$ direction. Along the y axis it is 2 m from the origin, in the $+\hat{j}$ direction. Along the z axis it is 5 m from the origin, in the $+\hat{k}$ direction.

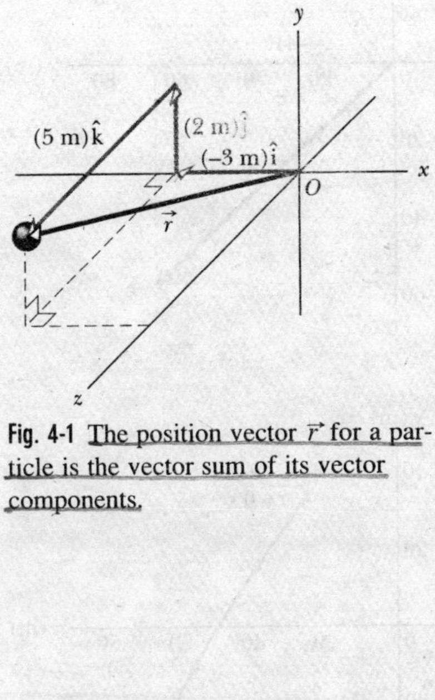

Fig. 4-1 The position vector $\vec{r}$ for a particle is the vector sum of its vector components.

As a particle moves, its position vector changes in such a way that the vector always extends to the particle from the reference point (the origin). If the position vector changes, say, from $\vec{r}_1$ to $\vec{r}_2$ during a certain time interval, then the particle's **displacement** $\Delta\vec{r}$ during that time interval is

$$\Delta\vec{r} = \vec{r}_2 - \vec{r}_1. \tag{4-2}$$

Using the unit-vector notation of Eq. 4-1, we can rewrite this displacement as

$$\Delta\vec{r} = (x_2\hat{i} + y_2\hat{j} + z_2\hat{k}) - (x_1\hat{i} + y_1\hat{j} + z_1\hat{k})$$

or as

$$\Delta\vec{r} = (x_2 - x_1)\hat{i} + (y_2 - y_1)\hat{j} + (z_2 - z_1)\hat{k}, \tag{4-3}$$

where coordinates (x_1, y_1, z_1) correspond to position vector $\vec{r}_1$ and coordinates (x_2, y_2, z_2) correspond to position vector $\vec{r}_2$. We can also rewrite the displacement by substituting Δx for $(x_2 - x_1)$, Δy for $(y_2 - y_1)$, and Δz for $(z_2 - z_1)$:

$$\Delta\vec{r} = \Delta x\hat{i} + \Delta y\hat{j} + \Delta z\hat{k}. \tag{4-4}$$

Sample Problem 4-1

In Fig. 4-2, the position vector for a particle is initially

$$\vec{r}_1 = (-3.0 \text{ m})\hat{i} + (2.0 \text{ m})\hat{j} + (5.0 \text{ m})\hat{k}$$

and then later is

$$\vec{r}_2 = (9.0 \text{ m})\hat{i} + (2.0 \text{ m})\hat{j} + (8.0 \text{ m})\hat{k}.$$

What is the particle's displacement $\Delta\vec{r}$ from $\vec{r}_1$ to $\vec{r}_2$?

SOLUTION: The **Key Idea** is that the displacement $\Delta\vec{r}$ is obtained by subtracting the initial position vector $\vec{r}_1$ from the later position vector $\vec{r}_2$. That is most easily done by components:

$$\begin{aligned}\Delta\vec{r} &= \vec{r}_2 - \vec{r}_1\\ &= [9.0 - (-3.0)]\hat{i} + [2.0 - 2.0]\hat{j} + [8.0 - 5.0]\hat{k}\\ &= (12 \text{ m})\hat{i} + (3.0 \text{ m})\hat{k}. \qquad \text{(Answer)}\end{aligned}$$

This displacement vector is parallel to the xz plane, because it lacks any y component, a fact that is easier to see in the numerical result than in Fig. 4-2.

CHECKPOINT 1: (a) If a wily bat flies from xyz coordinates $(-2\text{ m}, 4\text{ m}, -3\text{ m})$ to coordinates $(6\text{ m}, -2\text{ m}, -3\text{ m})$, what is its displacement $\Delta\vec{r}$ in unit-vector notation? (b) Is $\Delta\vec{r}$ parallel to one of the three coordinate planes? If so, which plane?

8î - 6ĵ   xy

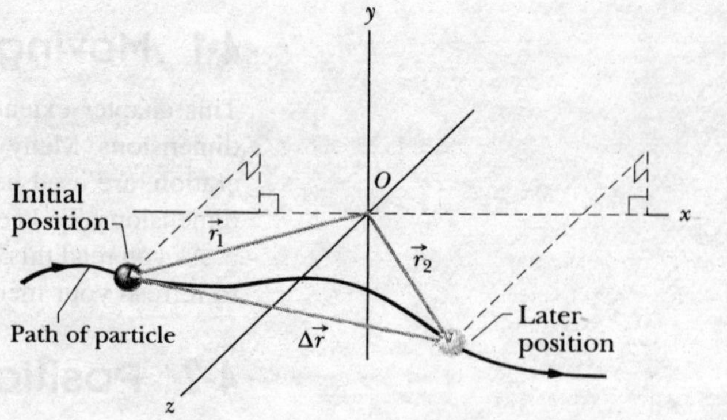

Fig. 4-2 Sample Problem 4-1. The displacement $\Delta\vec{r} = \vec{r}_2 - \vec{r}_1$ extends from the head of the initial position vector $\vec{r}_1$ to the head of the later position vector $\vec{r}_2$.

Sample Problem 4-2

A rabbit runs across a parking lot on which a set of coordinate axes has, strangely enough, been drawn. The coordinates of the rabbit's position as functions of time t are given by

$$x = -0.31t^2 + 7.2t + 28 \qquad (4\text{-}5)$$

and

$$y = 0.22t^2 - 9.1t + 30, \qquad (4\text{-}6)$$

with t in seconds and x and y in meters.

(a) At $t = 15$ s, what is the rabbit's position vector $\vec{r}$ in unit-vector notation and as a magnitude and an angle?

SOLUTION: The **Key Idea** here is that the x and y coordinates of the rabbit's position, as given by Eqs. 4-5 and 4-6, are the scalar components of the rabbit's position vector $\vec{r}$. Thus, we can write

$$\vec{r}(t) = x(t)\hat{\mathrm{i}} + y(t)\hat{\mathrm{j}}. \qquad (4\text{-}7)$$

(We write $\vec{r}(t)$ rather than $\vec{r}$ because the components are functions of t, and thus $\vec{r}$ is also.)

At $t = 15$ s, the scalar components are

$$x = (-0.31)(15)^2 + (7.2)(15) + 28 = 66\text{ m}$$

and

$$y = (0.22)(15)^2 - (9.1)(15) + 30 = -57\text{ m}.$$

Thus, at $t = 15$ s,

$$\vec{r} = (66\text{ m})\hat{\mathrm{i}} - (57\text{ m})\hat{\mathrm{j}}, \qquad \text{(Answer)}$$

which is drawn in Fig. 4-3a.

To get the magnitude and angle of $\vec{r}$, we can use a vector-capable calculator, or we can be guided by Eq. 3-6 to write

$$r = \sqrt{x^2 + y^2} = \sqrt{(66\text{ m})^2 + (-57\text{ m})^2}$$
$$= 87\text{ m}, \qquad \text{(Answer)}$$

and

$$\theta = \tan^{-1}\frac{y}{x} = \tan^{-1}\left(\frac{-57\text{ m}}{66\text{ m}}\right) = -41°. \qquad \text{(Answer)}$$

(Although $\theta = 139°$ has the same tangent as $-41°$, study of the signs of the components of $\vec{r}$ rules out 139°.)

(b) Graph the rabbit's path for $t = 0$ to $t = 25$ s.

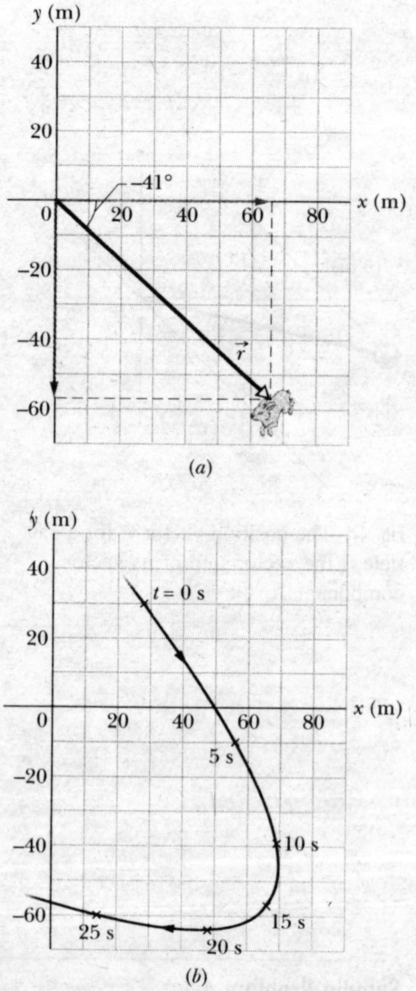

Fig. 4-3 Sample Problem 4-2. (a) A rabbit's position vector $\vec{r}$ at time $t = 15$ s. The scalar components of $\vec{r}$ are shown along the axes. (b) The rabbit's path and its position at five values of t.

SOLUTION: We can repeat part (a) for several values of t and then plot the results. Figure 4-3b shows the plots for five values of t and the path connecting them. We can also use a graphing calculator to make a *parametric graph;* that is, we would have the calculator plot y versus x, where these coordinates are given by Eqs. 4-5 and 4-6 as functions of time t.

$\tan^{-1}\left(\frac{57}{66}\right) = 41°$

IV quad $\Rightarrow -41°$

4-3 Average Velocity and Instantaneous Velocity

If a particle moves through a displacement $\Delta\vec{r}$ in a time interval Δt, then its **average velocity** $\vec{v}_{avg}$ is

$$\text{average velocity} = \frac{\text{displacement}}{\text{time interval}},$$

or

$$\vec{v}_{avg} = \frac{\Delta\vec{r}}{\Delta t}. \tag{4-8}$$

This tells us the direction of $\vec{v}_{avg}$ (the vector on the left side of Eq. 4-8) must be the same as that of the displacement $\Delta\vec{r}$ (the vector on the right side). Using Eq. 4-4, we can write Eq. 4-8 in vector components as

$$\vec{v}_{avg} = \frac{\Delta x\hat{\mathrm{i}} + \Delta y\hat{\mathrm{j}} + \Delta z\hat{\mathrm{k}}}{\Delta t} = \frac{\Delta x}{\Delta t}\hat{\mathrm{i}} + \frac{\Delta y}{\Delta t}\hat{\mathrm{j}} + \frac{\Delta z}{\Delta t}\hat{\mathrm{k}}. \tag{4-9}$$

For example, if the particle in Sample Problem 4-1 moves from its initial position to its later position in 2.0 s, then its average velocity during that move is

$$\vec{v}_{avg} = \frac{\Delta\vec{r}}{\Delta t} = \frac{(12\text{ m})\hat{\mathrm{i}} + (3.0\text{ m})\hat{\mathrm{k}}}{2.0\text{ s}} = (6.0\text{ m/s})\hat{\mathrm{i}} + (1.5\text{ m/s})\hat{\mathrm{k}}.$$

When we speak of the **velocity** of a particle, we usually mean the particle's **instantaneous velocity** $\vec{v}$ at some instant. This $\vec{v}$ is the value that $\vec{v}_{avg}$ approaches in the limit as we shrink the time interval Δt to 0 about that instant. Using the language of calculus, we may write $\vec{v}$ as the derivative

$$\vec{v} = \frac{d\vec{r}}{dt}. \tag{4-10}$$

Figure 4-4 shows the path of a particle that is restricted to the xy plane. As the particle travels to the right along the curve, its position vector sweeps to the right. During time interval Δt, the position vector changes from $\vec{r}_1$ to $\vec{r}_2$, and the particle's displacement is $\Delta\vec{r}$.

To find the instantaneous velocity of the particle at, say, instant t_1 (when the particle is at position 1), we shrink interval Δt to 0 about t_1. Three things happen as we do so: (1) Position vector $\vec{r}_2$ in Fig. 4-4 moves toward $\vec{r}_1$ so that $\Delta\vec{r}$ shrinks toward zero. (2) The direction of $\Delta\vec{r}/\Delta t$ (thus of $\vec{v}_{avg}$) approaches the direction of the tangent line to the particle's path at position 1. (3) The average velocity $\vec{v}_{avg}$ approaches the instantaneous velocity $\vec{v}$ at t_1.

In the limit as $\Delta t \to 0$, we have $\vec{v}_{avg} \to \vec{v}$ and, most important here, $\vec{v}_{avg}$ takes on the direction of the tangent line. Thus, $\vec{v}$ has that direction as well:

> The direction of the instantaneous velocity $\vec{v}$ of a particle is always tangent to the particle's path at the particle's position.

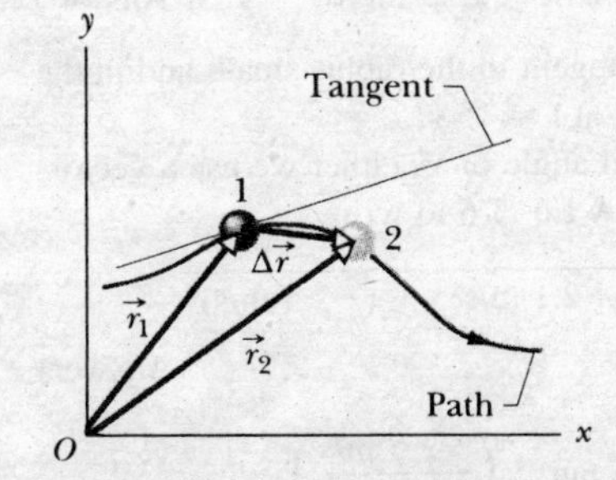

Fig. 4-4 The displacement $\Delta\vec{r}$ of a particle during a time interval Δt, from position 1 with position vector $\vec{r}_1$ at time t_1 to position 2 with position vector $\vec{r}_2$ at time t_2. The tangent to the particle's path at position 1 is shown.

The result is the same in three dimensions: $\vec{v}$ is always tangent to the particle's path.

To write Eq. 4-10 in unit-vector form, we substitute for $\vec{r}$ from Eq. 4-1:

$$\vec{v} = \frac{d}{dt}(x\hat{\mathrm{i}} + y\hat{\mathrm{j}} + z\hat{\mathrm{k}}) = \frac{dx}{dt}\hat{\mathrm{i}} + \frac{dy}{dt}\hat{\mathrm{j}} + \frac{dz}{dt}\hat{\mathrm{k}}.$$

This equation can be simplified somewhat by writing it as

$$\vec{v} = v_x\hat{\mathrm{i}} + v_y\hat{\mathrm{j}} + v_z\hat{\mathrm{k}}, \tag{4-11}$$

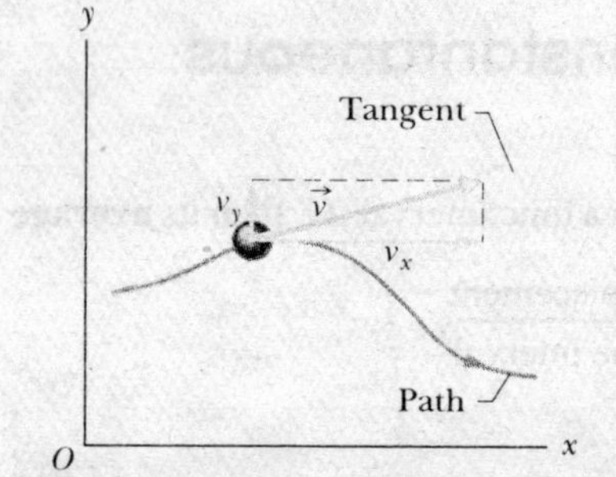

Fig. 4-5 The velocity $\vec{v}$ of a particle, along with the scalar components of $\vec{v}$.

where the scalar components of $\vec{v}$ are

$$v_x = \frac{dx}{dt}, \quad v_y = \frac{dy}{dt}, \quad \text{and} \quad v_z = \frac{dz}{dt}. \tag{4-12}$$

For example, dx/dt is the scalar component of $\vec{v}$ along the x axis. Thus, we can find the scalar components of $\vec{v}$ by differentiating the scalar components of $\vec{r}$.

Figure 4-5 shows a velocity vector $\vec{v}$ and its scalar x and y components. Note that $\vec{v}$ is tangent to the particle's path at the particle's position. *Caution:* When a position vector is drawn as in Figs. 4-1 through 4-4, it is an arrow that extends from one point (a "here") to another point (a "there"). However, when a velocity vector is drawn as in Fig. 4-5, it does *not* extend from one point to another. Rather, it shows the instantaneous direction of travel of a particle located at the tail, and its length (representing the velocity magnitude) can be drawn to any scale.

CHECKPOINT 2: The figure shows a circular path taken by a particle. If the instantaneous velocity of the particle is $\vec{v} = (2 \text{ m/s})\hat{i} - (2 \text{ m/s})\hat{j}$, through which quadrant is the particle moving when it is traveling (a) clockwise and (b) counterclockwise around the circle? For both cases, draw $\vec{v}$ on the figure.

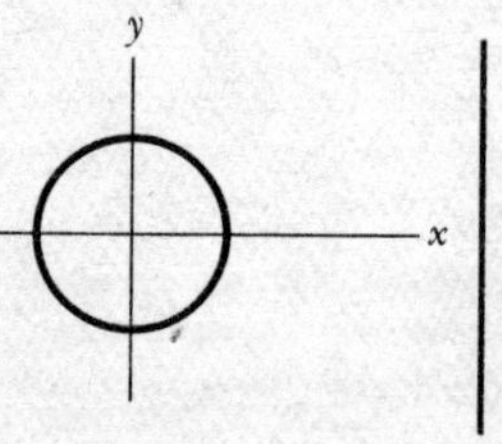

Sample Problem 4-3

For the rabbit in Sample Problem 4-2, find the velocity $\vec{v}$ at time $t = 15$ s, in unit-vector notation and as a magnitude and an angle.

SOLUTION: There are two **Key Ideas** here: (1) We can find the rabbit's velocity $\vec{v}$ by first finding the velocity components. (2) We can find those components by taking derivatives of the components of the rabbit's position vector. Applying the first of Eqs. 4-12 to Eq. 4-5, we find the x component of $\vec{v}$ to be

$$v_x = \frac{dx}{dt} = \frac{d}{dt}(-0.31t^2 + 7.2t + 28)$$
$$= -0.62t + 7.2. \tag{4-13}$$

At $t = 15$ s, this gives $v_x = -2.1$ m/s. Similarly, applying the second of Eqs. 4-12 to Eq. 4-6, we find that the y component is

$$v_y = \frac{dy}{dt} = \frac{d}{dt}(0.22t^2 - 9.1t + 30)$$
$$= 0.44t - 9.1. \tag{4-14}$$

At $t = 15$ s, this gives $v_y = -2.5$ m/s. Equation 4-11 then yields

$$\vec{v} = -(2.1 \text{ m/s})\hat{i} - (2.5 \text{ m/s})\hat{j}, \quad \text{(Answer)}$$

which is shown in Fig. 4-6, tangent to the rabbit's path and in the direction the rabbit is running at $t = 15$ s.

To get the magnitude and angle of $\vec{v}$, either we use a vector-capable calculator or we follow Eq. 3-6 to write

$$v = \sqrt{v_x^2 + v_y^2} = \sqrt{(-2.1 \text{ m/s})^2 + (-2.5 \text{ m/s})^2}$$
$$= 3.3 \text{ m/s} \quad \text{(Answer)}$$

and

$$\theta = \tan^{-1}\frac{v_y}{v_x} = \tan^{-1}\left(\frac{-2.5 \text{ m/s}}{-2.1 \text{ m/s}}\right)$$
$$= \tan^{-1} 1.19 = -130°. \quad \text{(Answer)}$$

(Although 50° has the same tangent as −130°, inspection of the signs of the velocity components indicates that the desired angle is in the third quadrant, given by 50° − 180° = −130°.)

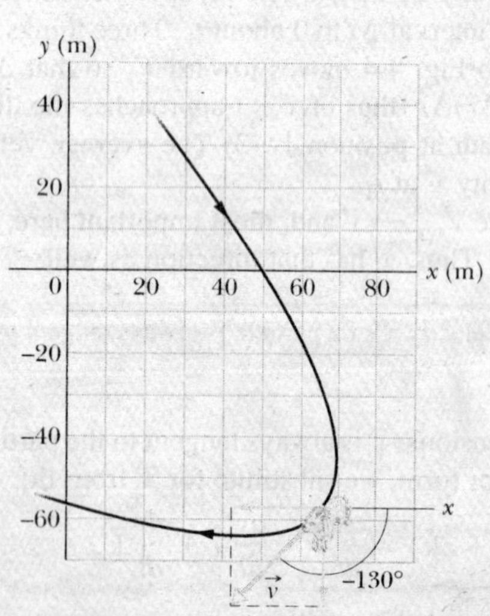

Fig. 4-6 Sample Problem 4-3. The rabbit's velocity $\vec{v}$ at $t = 15$ s. The velocity vector is tangent to the path at the rabbit's position at that instant. The scalar components of $\vec{v}$ are shown.

4-4 Average Acceleration and Instantaneous Acceleration

When a particle's velocity changes from $\vec{v}_1$ to $\vec{v}_2$ in a time interval Δt, its **average acceleration** $\vec{a}_{avg}$ during Δt is

$$\text{average acceleration} = \frac{\text{change in velocity}}{\text{time interval}},$$

or

$$\vec{a}_{avg} = \frac{\vec{v}_2 - \vec{v}_1}{\Delta t} = \frac{\Delta \vec{v}}{\Delta t}. \tag{4-15}$$

If we shrink Δt to zero about some instant, then in the limit $\vec{a}_{avg}$ approaches the **instantaneous acceleration** (or **acceleration**) $\vec{a}$ at that instant; that is,

$$\vec{a} = \frac{d\vec{v}}{dt}. \tag{4-16}$$

If the velocity changes in *either* magnitude *or* direction (or both), the particle must have an acceleration.

We can write Eq. 4-16 in unit-vector form by substituting for $\vec{v}$ from Eq. 4-11 to obtain

$$\vec{a} = \frac{d}{dt}(v_x\hat{\mathrm{i}} + v_y\hat{\mathrm{j}} + v_z\hat{\mathrm{k}})$$

$$= \frac{dv_x}{dt}\hat{\mathrm{i}} + \frac{dv_y}{dt}\hat{\mathrm{j}} + \frac{dv_z}{dt}\hat{\mathrm{k}}.$$

We can rewrite this as

$$\vec{a} = a_x\hat{\mathrm{i}} + a_y\hat{\mathrm{j}} + a_z\hat{\mathrm{k}}, \tag{4-17}$$

where the scalar components of $\vec{a}$ are

$$a_x = \frac{dv_x}{dt}, \quad a_y = \frac{dv_y}{dt}, \quad \text{and} \quad a_z = \frac{dv_z}{dt}. \tag{4-18}$$

Thus, we can find the scalar components of $\vec{a}$ by differentiating the scalar components of $\vec{v}$.

Figure 4-7 shows an acceleration vector $\vec{a}$ and its scalar components for a particle moving in two dimensions. *Caution:* When an acceleration vector is drawn as in Fig. 4-7, it does *not* extend from one position to another. Rather, it shows the direction of acceleration for a particle located at its tail, and its length (representing the acceleration magnitude) can be drawn to any scale.

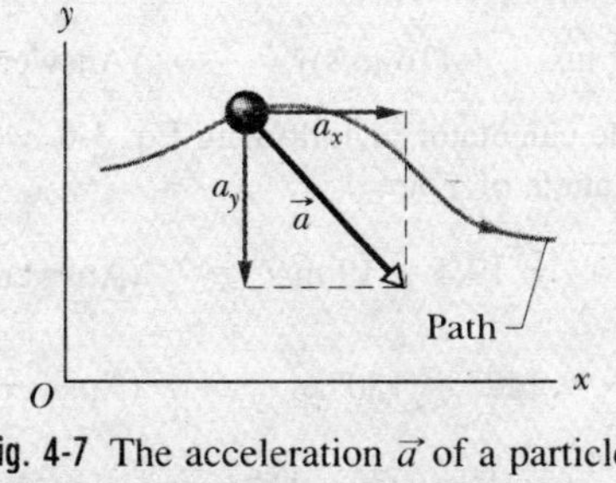

Fig. 4-7 The acceleration $\vec{a}$ of a particle and the scalar components of $\vec{a}$.

Sample Problem 4-4

For the rabbit in Sample Problems 4-2 and 4-3, find the acceleration $\vec{a}$ at time $t = 15$ s, in unit-vector notation and as a magnitude and an angle.

SOLUTION: There are two **Key Ideas** here: (1) We can find the rabbit's acceleration $\vec{a}$ by first finding the acceleration components. (2) We can find those components by taking derivatives of the rabbit's velocity components. Applying the first of Eqs. 4-18 to Eq. 4-13, we find the x component of $\vec{a}$ to be

$$a_x = \frac{dv_x}{dt} = \frac{d}{dt}(-0.62t + 7.2) = -0.62 \text{ m/s}^2.$$

Similarly, applying the second of Eqs. 4-18 to Eq. 4-14 yields the y component as

$$a_y = \frac{dv_y}{dt} = \frac{d}{dt}(0.44t - 9.1) = 0.44 \text{ m/s}^2.$$

We see that the acceleration does not vary with time (it is a constant) because the time variable t does not appear in the expression for either acceleration component. Equation 4-17 then yields

$$\vec{a} = (-0.62 \text{ m/s}^2)\hat{\mathrm{i}} + (0.44 \text{ m/s}^2)\hat{\mathrm{j}}, \quad \text{(Answer)}$$

which is shown superimposed on the rabbit's path in Fig. 4-8.

To get the magnitude and angle of $\vec{a}$, either we use a vector-capable calculator or we follow Eq. 3-6. For the magnitude we have

$$a = \sqrt{a_x^2 + a_y^2} = \sqrt{(-0.62\ \text{m/s}^2)^2 + (0.44\ \text{m/s}^2)^2}$$
$$= 0.76\ \text{m/s}^2. \qquad \text{(Answer)}$$

For the angle we have

$$\theta = \tan^{-1}\frac{a_y}{a_x} = \tan^{-1}\left(\frac{0.44\ \text{m/s}^2}{-0.62\ \text{m/s}^2}\right) = -35°.$$

However, this last result, which is displayed on a calculator, indicates that $\vec{a}$ is directed to the right and downward in Fig. 4-8. Yet, we know from the components above that $\vec{a}$ must be directed to the left and upward. To find the other angle that has the same tangent as $-35°$, but which is not displayed on a calculator, we add 180°:

$$-35° + 180° = 145°. \qquad \text{(Answer)}$$

This *is* consistent with the components of $\vec{a}$. Note that $\vec{a}$ has the same magnitude and direction throughout the rabbit's run because, as we noted previously, the acceleration is constant.

CHECKPOINT 3: Here are four descriptions of the position (in meters) of a hockey puck as it moves in the xy plane:

(1) $x = -3t^2 + 4t - 2$ and $y = 6t^2 - 4t$
(2) $x = -3t^3 - 4t$ and $y = -5t^2 + 6$
(3) $\vec{r} = 2t^2\hat{i} - (4t + 3)\hat{j}$
(4) $\vec{r} = (4t^3 - 2t)\hat{i} + 3\hat{j}$

For each description, determine whether the x and y components of the puck's acceleration are constant, and whether the acceleration $\vec{a}$ is constant.

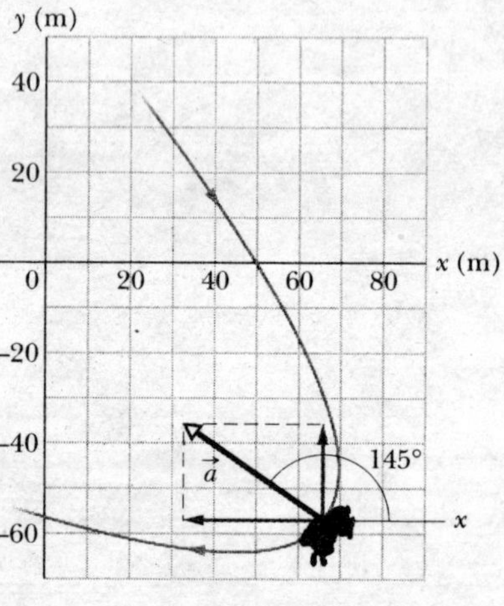

Fig. 4-8 Sample Problem 4-4. The acceleration $\vec{a}$ of the rabbit at $t = 15$ s. The rabbit happens to have this same acceleration at all points of its path.

Sample Problem 4-5

A particle with velocity $\vec{v}_0 = -2.0\hat{i} + 4.0\hat{j}$ (in meters per second) at $t = 0$ undergoes a constant acceleration $\vec{a}$ of magnitude $a = 3.0\ \text{m/s}^2$ at an angle $\theta = 130°$ from the positive direction of the x axis. What is the particle's velocity $\vec{v}$ at $t = 5.0$ s, in unit-vector notation and as a magnitude and an angle?

SOLUTION: We first note that this is two-dimensional motion, in the xy plane. Then there are two **Key Ideas** here. One is that, because the acceleration is constant, Eq. 2-11 ($v = v_0 + at$) applies. The second is that, because Eq. 2-11 applies only to straight-line motion, we must apply it separately for motion parallel to the x axis and motion parallel to the y axis; that is, we must find the velocity components v_x and v_y from the equations

$$v_x = v_{0x} + a_x t \quad \text{and} \quad v_y = v_{0y} + a_y t.$$

In these equations, v_{0x} ($= -2.0$ m/s) and v_{0y} ($= 4.0$ m/s) are the x and y components of $\vec{v}_0$, and a_x and a_y are the x and y components of $\vec{a}$. To find a_x and a_y, we resolve $\vec{a}$ either with a vector-capable calculator or with Eqs. 3-5:

$$a_x = a\cos\theta = (3.0\ \text{m/s}^2)(\cos 130°) = -1.93\ \text{m/s}^2,$$
$$a_y = a\sin\theta = (3.0\ \text{m/s}^2)(\sin 130°) = +2.30\ \text{m/s}^2.$$

When these values are inserted into the equations for v_x and v_y, we find that, at time $t = 5.0$ s,

$$v_x = -2.0\ \text{m/s} + (-1.93\ \text{m/s}^2)(5.0\ \text{s}) = -11.65\ \text{m/s},$$
$$v_y = 4.0\ \text{m/s} + (2.30\ \text{m/s}^2)(5.0\ \text{s}) = 15.50\ \text{m/s}.$$

Thus, at $t = 5.0$ s, we have, after rounding,

$$\vec{v} = (-12\ \text{m/s})\hat{i} + (16\ \text{m/s})\hat{j}. \qquad \text{(Answer)}$$

Either using a vector-capable calculator or following Eq. 3-6, we find that the magnitude and angle of $\vec{v}$ are

$$v = \sqrt{v_x^2 + v_y^2} = 19.4 \approx 19\ \text{m/s} \qquad \text{(Answer)}$$

and

$$\theta = \tan^{-1}\frac{v_y}{v_x} = 127° \approx 130°. \qquad \text{(Answer)}$$

Check the last line with your calculator. Does 127° appear on the display, or does $-53°$ appear? Now sketch the vector $\vec{v}$ with its components to see which angle is reasonable.

CHECKPOINT 4: If the position of a hobo's marble is given by $\vec{r} = (4t^3 - 2t)\hat{i} + 3\hat{j}$, with $\vec{r}$ in meters and t in seconds, what must be the units of the coefficients 4, -2, and 3?

4-5 Projectile Motion

We next consider a special case of two-dimensional motion: A particle moves in a vertical plane with some initial velocity $\vec{v}_0$ but its acceleration is always the free-fall acceleration $\vec{g}$, which is downward. Such a particle is called a **projectile** (meaning that it is projected or launched) and its motion is called **projectile motion.** A projectile might be a golf ball (Fig. 4-9) or a baseball in flight, but it is not an airplane or a duck in flight. Our goal here is to analyze projectile motion using the tools for two-dimensional motion in Sections 4-2 through 4-4 and making the assumption that air has no effect on the projectile.

Figure 4-10, which is analyzed in the next section, shows the path followed by a projectile when the air has no effect. The projectile is launched with an initial velocity $\vec{v}_0$ that can be written as

$$\vec{v}_0 = v_{0x}\hat{\text{i}} + v_{0y}\hat{\text{j}}. \qquad (4\text{-}19)$$

The components v_{0x} and v_{0y} can then be found if we know the angle θ_0 between $\vec{v}_0$ and the positive x direction:

$$v_{0x} = v_0 \cos\theta_0 \quad \text{and} \quad v_{0y} = v_0 \sin\theta_0. \qquad (4\text{-}20)$$

During its two-dimensional motion, the projectile's position vector $\vec{r}$ and velocity vector $\vec{v}$ change continuously, but its acceleration vector $\vec{a}$ is constant and *always* directed vertically downward. The projectile has *no* horizontal acceleration.

Projectile motion, like that in Figs. 4-9 and 4-10, looks complicated, but we have the following simplifying feature (known from experiment):

> In projectile motion, the horizontal motion and the vertical motion are independent of each other; that is, neither motion affects the other.

This feature allows us to break up a problem involving two-dimensional motion into two separate and easier one-dimensional problems, one for the horizontal motion (with *zero acceleration*) and one for the vertical motion (with *constant downward acceleration*). Here are two experiments that show that the horizontal motion and the vertical motion are independent.

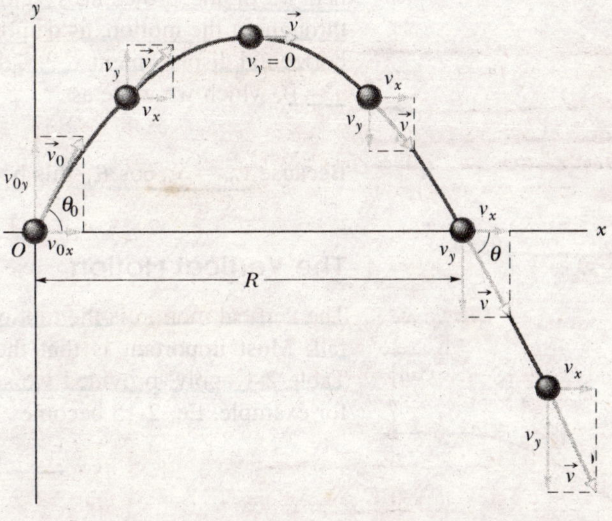

Fig. 4-10 The path of a projectile that is launched at $x_0 = 0$ and $y_0 = 0$, with an initial velocity $\vec{v}_0$. The initial velocity and the velocities at various points along its path are shown, along with their components. Note that the horizontal velocity component remains constant but the vertical velocity component changes continuously. The *range R* is the horizontal distance the projectile has traveled *when it returns to its launch height*.

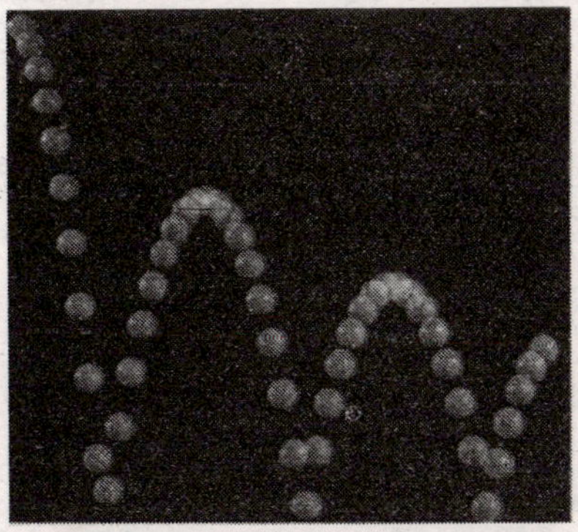

Fig. 4-9 A stroboscopic photograph of an orange golf ball bouncing off a hard surface. Between impacts, the ball has projectile motion.

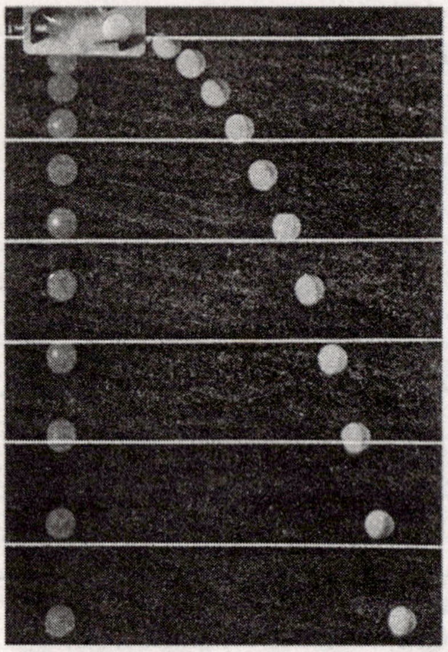

Fig. 4-11 One ball is released from rest at the same instant that another ball is shot horizontally to the right. Their vertical motions are identical.

⇒ acceleration due to gravity acts similarly on G and can, w.r.t inertial ref frame eg an observer on the ground & both fall same distance down ∴ G would always hit the can

Two Golf Balls

Figure 4-11 is a stroboscopic photograph of two golf balls, one simply released and the other shot horizontally by a spring. The golf balls have the same vertical motion, both falling through the same vertical distance in the same interval of time. *The fact that one ball is moving horizontally while it is falling has no effect on its vertical motion;* that is, the horizontal and vertical motions are independent.

A Great Student Rouser

Figure 4-12 shows a demonstration that has enlivened many a physics lecture. It involves a blow gun G, using a ball as a projectile. The target is a can suspended from a magnet M, and the tube of the blow gun is aimed directly at the can. The experiment is arranged so that the magnet releases the can just as the ball leaves the blow gun.

If g (the magnitude of the free-fall acceleration) were zero, the ball would follow the straight line shown in Fig. 4-12 and the can would float in place after the magnet released it. The ball would certainly hit the can.

However, g is *not* zero. The ball *still* hits the can! As Fig. 4-12 shows, during the time of flight of the ball, both ball and can fall the same distance h from their zero-g locations. The harder the demonstrator blows, the greater the ball's initial speed, the shorter the time of flight, and the smaller the value of h.

4-6 Projectile Motion Analyzed

Now we are ready to analyze projectile motion, horizontally and vertically.

The Horizontal Motion

Because there is *no acceleration* in the horizontal direction, the horizontal component v_x of the projectile's velocity remains unchanged from its initial value v_{0x} throughout the motion, as demonstrated in Fig. 4-13. At any time t, the projectile's horizontal displacement $x - x_0$ from an initial position x_0 is given by Eq. 2-15 with $a = 0$, which we write as

$$x - x_0 = v_{0x}t.$$

Because $v_{0x} = v_0 \cos \theta_0$, this becomes

$$x - x_0 = (v_0 \cos \theta_0)t. \tag{4-21}$$

The Vertical Motion

The vertical motion is the motion we discussed in Section 2-8 for a particle in free fall. Most important is that the acceleration is constant. Thus, the equations of Table 2-1 apply, provided we substitute $-g$ for a and switch to y notation. Then, for example, Eq. 2-15 becomes

$$\begin{aligned} y - y_0 &= v_{0y}t - \tfrac{1}{2}gt^2 \\ &= (v_0 \sin \theta_0)t - \tfrac{1}{2}gt^2, \end{aligned} \tag{4-22}$$

where the initial vertical velocity component v_{0y} is replaced with the equivalent $v_0 \sin \theta_0$. Similarly, Eqs. 2-11 and 2-16 become

$$v_y = v_0 \sin \theta_0 - gt \tag{4-23}$$

and

$$v_y^2 = (v_0 \sin \theta_0)^2 - 2g(y - y_0). \tag{4-24}$$

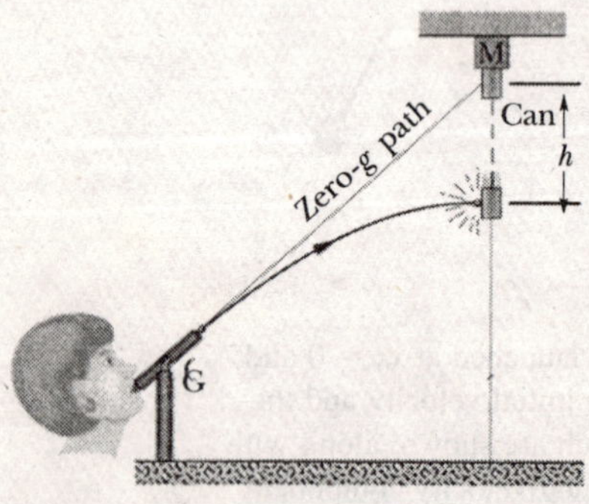

Fig. 4-12 The projectile ball always hits the falling can. Each falls a distance h from where it would be were there no free-fall acceleration.

As is illustrated in Fig. 4-10 and Eq. 4-23, the vertical velocity component behaves just as for a ball thrown vertically upward. It is directed upward initially,

Fig. 4-13 The vertical component of this skateboarder's velocity is changing, but not the horizontal component, which matches the skateboard's velocity. As a result, the skateboard stays underneath him, allowing him to land on it.

and its magnitude steadily decreases to zero, *which marks the maximum height of the path.* The vertical velocity component then reverses direction, and its magnitude becomes larger with time.

The Equation of the Path

We can find the equation of the projectile's path (its **trajectory**) by eliminating t between Eqs. 4-21 and 4-22. Solving Eq. 4-21 for t and substituting into Eq. 4-22, we obtain, after a little rearrangement,

$$y = (\tan \theta_0)x - \frac{gx^2}{2(v_0 \cos \theta_0)^2} \quad \text{(trajectory).} \tag{4-25}$$

This is the equation of the path shown in Fig. 4-10. In deriving it, for simplicity we let $x_0 = 0$ and $y_0 = 0$ in Eqs. 4-21 and 4-22, respectively. Because g, θ_0, and v_0 are constants, Eq. 4-25 is of the form $y = ax + bx^2$, in which a and b are constants. This is the equation of a parabola, so the path is *parabolic.*

The Horizontal Range

The *horizontal range R* of the projectile, as Fig. 4-10 shows, is the *horizontal* distance the projectile has traveled when it returns to its initial (launch) height. To find range R, let us put $x - x_0 = R$ in Eq. 4-21 and $y - y_0 = 0$ in Eq. 4-22, obtaining

$$R = (v_0 \cos \theta_0)t$$

and

$$0 = (v_0 \sin \theta_0)t - \tfrac{1}{2}gt^2.$$

Eliminating t between these two equations yields

$$R = \frac{2v_0^2}{g} \sin \theta_0 \cos \theta_0.$$

Using the identity $\sin 2\theta_0 = 2 \sin \theta_0 \cos \theta_0$ (see Appendix E), we obtain

$$R = \frac{v_0^2}{g} \sin 2\theta_0. \tag{4-26}$$

Caution: This equation does *not* give the horizontal distance traveled by a projectile when the final height is not the launch height.

Note that R in Eq. 4-26 has its maximum value when $\sin 2\theta_0 = 1$, which corresponds to $2\theta_0 = 90°$ or $\theta_0 = 45°$.

➤ The horizontal range R is maximum for a launch angle of 45°.

The Effects of the Air

We have assumed that the air through which the projectile moves has no effect on its motion. However, in many situations, the disagreement between our calculations and the actual motion of the projectile can be large because the air resists (or opposes) the motion. Figure 4-14, for example, shows two paths for a fly ball that leaves the

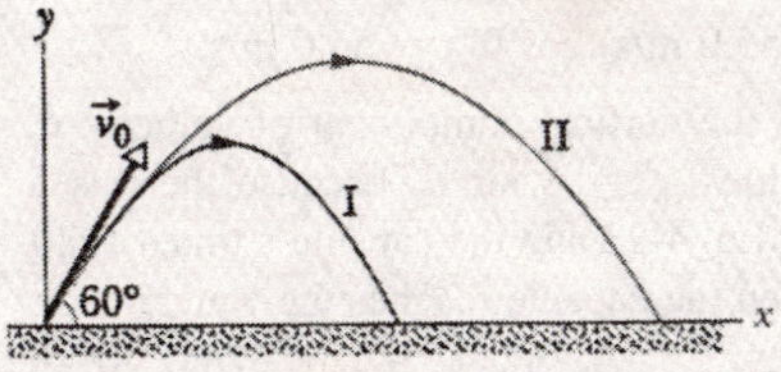

Fig. 4-14 (I) The path of a fly ball, calculated by taking air resistance into account. (II) The path the ball would follow in a vacuum, calculated by the methods of this chapter. See Table 4-1 for corresponding data. (Adapted from "The Trajectory of a Fly Ball," by Peter J. Brancazio, *The Physics Teacher*, January 1985.)

TABLE 4-1 Two Fly Balls[a]

	Path I (Air)	Path II (Vacuum)
Range	98.5 m	177 m
Maximum height	53.0 m	76.8 m
Time of flight	6.6 s	7.9 s

[a]See Fig. 4-14. The launch angle is 60° and the launch speed is 44.7 m/s.

bat at an angle of 60° with the horizontal and an initial speed of 44.7 m/s. Path I (the baseball player's fly ball) is a calculated path that approximates normal conditions of play, in air. Path II (the physics professor's fly ball) is the path that the ball would follow in a vacuum.

CHECKPOINT 5: A fly ball is hit to the outfield. During its flight (ignore the effects of the air), what happens to its (a) horizontal and (b) vertical components of velocity? What are the (c) horizontal and (d) vertical components of its acceleration during its ascent and its descent, and at the topmost point of its flight?

Sample Problem 4-6

In Fig. 4-15, a rescue plane flies at 198 km/h (= 55.0 m/s) and a constant elevation of 500 m toward a point directly over a boating accident victim struggling in the water. The pilot wants to release a rescue capsule so that it hits the water very close to the victim.

(a) What should be the angle ϕ of the pilot's line of sight to the victim when the release is made?

SOLUTION: The **Key Idea** here is that, once released, the capsule is a projectile, so its horizontal and vertical motions are independent and can be considered separately (we need not consider the actual curved path of the capsule). Figure 4-15 includes a coordinate system with its origin at the point of release, and we see there that ϕ is given by

$$\phi = \tan^{-1}\frac{x}{h}, \tag{4-27}$$

where x is the horizontal coordinate of the victim at release (and of the capsule when it hits the water), and h is the height of the plane. That height is 500 m, so we need only x in order to find ϕ. We should be able to find x with Eq. 4-21:

$$x - x_0 = (v_0 \cos\theta_0)t. \tag{4-28}$$

Here we know $x_0 = 0$ because the origin is placed at the point of release. Because the capsule is *released* and not shot from the plane, its initial velocity $\vec{v}_0$ is equal to the plane's velocity. Thus, we know also that the initial velocity has magnitude $v_0 = 55.0$ m/s and angle $\theta_0 = 0°$ (measured relative to the positive direction of the x axis). However, we do not know the time t the capsule takes to move from the plane to the victim.

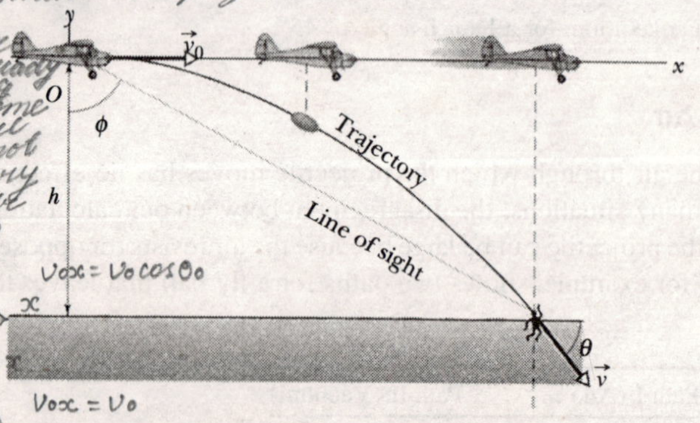

Fig. 4-15 Sample Problem 4-6. A plane drops a rescue capsule while moving at constant velocity in level flight. While the capsule is falling, its horizontal velocity component remains equal to the velocity of the plane.

To find t, we next consider the vertical motion and specifically Eq. 4-22:

$$y - y_0 = (v_0 \sin\theta_0)t - \tfrac{1}{2}gt^2. \tag{4-29}$$

Here the vertical displacement $y - y_0$ of the capsule is -500 m (the negative value indicates that the capsule moves *downward*). Putting this and other known values into Eq. 4-29 gives us

$$-500 \text{ m} = (55.0 \text{ m/s})(\sin 0°)t - \tfrac{1}{2}(9.8 \text{ m/s}^2)t^2.$$

Solving for t, we find $t = 10.1$ s. Using that value in Eq. 4-28 yields

$$x - 0 = (55.0 \text{ m/s})(\cos 0°)(10.1 \text{ s}),$$

or

$$x = 555.5 \text{ m}.$$

Then Eq. 4-27 gives us

$$\phi = \tan^{-1}\frac{555.5 \text{ m}}{500 \text{ m}} = 48°. \quad \text{(Answer)}$$

(b) As the capsule reaches the water, what is its velocity $\vec{v}$ in unit-vector notation and as a magnitude and an angle?

SOLUTION: Again, we need the **Key Idea** that the horizontal and vertical motions of the capsule are independent during the capsule's flight. In particular, the horizontal and vertical components of the capsule's velocity are independent of each other.

A second **Key Idea** is that the horizontal component of velocity v_x does not change from its initial value $v_{0x} = v_0 \cos\theta_0$ because there is no horizontal acceleration. Thus, when the capsule reaches the water,

$$v_x = v_0 \cos\theta_0 = (55.0 \text{ m/s})(\cos 0°) = 55.0 \text{ m/s}.$$

A third **Key Idea** is that the vertical component of velocity v_y changes from its initial value $v_{0y} = v_0 \sin\theta_0$ because there is a vertical acceleration. Using Eq. 4-23 and the capsule's time of fall $t = 10.1$ s, we find that when the capsule reaches the water,

$$\begin{aligned} v_y &= v_0 \sin\theta_0 - gt \\ &= (55.0 \text{ m/s})(\sin 0°) - (9.8 \text{ m/s}^2)(10.1 \text{ s}) \\ &= -99.0 \text{ m/s}. \end{aligned}$$

Thus, when the capsule reaches the water it has the velocity

$$\vec{v} = (55.0 \text{ m/s})\hat{\text{i}} - (99.0 \text{ m/s})\hat{\text{j}}. \quad \text{(Answer)}$$

Using either Eq. 3-6 as a guide or a vector-capable calculator, we find that the magnitude and the angle of $\vec{v}$ are

$$v = 113 \text{ m/s} \quad \text{and} \quad \theta = -61°. \quad \text{(Answer)}$$

Sample Problem 4-7

Figure 4-16 shows a pirate ship 560 m from a fort defending the harbor entrance of an island. A defense cannon, located at sea level, fires balls at initial speed $v_0 = 82$ m/s.

(a) At what angle θ_0 from the horizontal must a ball be fired to hit the ship?

SOLUTION: The **Key Idea** here is obvious: A fired cannonball is a projectile and the projectile equations apply. We want an equation that relates the launch angle θ_0 to the horizontal displacement of the ball from the cannon to the ship.

A second **Key Idea** is that, because the cannon and the ship are at the same height, the horizontal displacement is the range. We can then relate the launch angle θ_0 to the range R with Eq. 4-26,

$$R = \frac{v_0^2}{g}\sin 2\theta_0, \tag{4-30}$$

which gives us

$$2\theta_0 = \sin^{-1}\frac{gR}{v_0^2} = \sin^{-1}\frac{(9.8\ \text{m/s}^2)(560\ \text{m})}{(82\ \text{m/s})^2} = \sin^{-1} 0.816. \tag{4-31}$$

The inverse function $\sin^{-1}$ always has two possible solutions. One solution (here, 54.7°) is displayed by a calculator; we subtract it from 180° to get the other solution (here, 125.3°). Thus, Eq. 4-31 gives us

$$\theta_0 = \tfrac{1}{2}(54.7°) \approx 27° \quad \text{(Answer)}$$

and

$$\theta_0 = \tfrac{1}{2}(125.3°) \approx 63°. \quad \text{(Answer)}$$

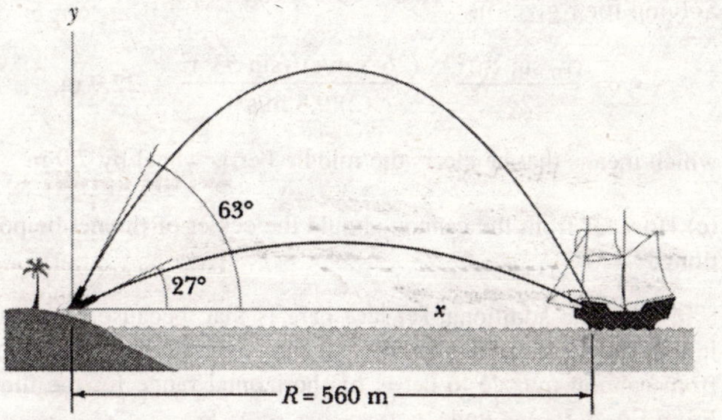

Fig. 4-16 Sample Problem 4-7. At this range, the cannon can hit the pirate ship at two elevation angles of the cannon.

The commandant of the fort can elevate the cannon to either of these two angles and (if only there were no intervening air!) hit the pirate ship.

(b) How far should the pirate ship be from the cannon if it is to be beyond the maximum range of the cannonballs?

SOLUTION: We have seen that maximum range corresponds to an elevation angle θ_0 of 45°. Thus, from Eq. 4-30 with $\theta_0 = 45°$,

$$R = \frac{v_0^2}{g}\sin 2\theta_0 = \frac{(82\ \text{m/s})^2}{9.8\ \text{m/s}^2}\sin(2\times 45°) = 686\ \text{m} \approx 690\ \text{m}. \quad \text{(Answer)}$$

As the pirate ship sails away, the two elevation angles at which the ship can be hit draw together, eventually merging at $\theta_0 = 45°$ when the ship is 690 m away. Beyond that distance the ship is safe.

Sample Problem 4-8

Figure 4-17 illustrates the flight of Emanuel Zacchini over three Ferris wheels, located as shown and each 18 m high. Zacchini is launched with speed $v_0 = 26.5$ m/s, at an angle $\theta_0 = 53°$ up from the horizontal and with an initial height of 3.0 m above the ground. The net in which he is to land is at the same height.

(a) Does he clear the first Ferris wheel?

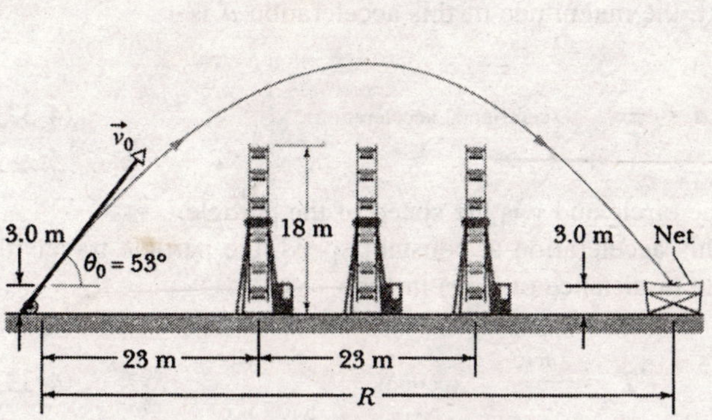

Fig. 4-17 Sample Problem 4-8. The flight of a human cannonball over three Ferris wheels and into a net.

SOLUTION: A **Key Idea** here is that Zacchini is a human projectile, so we can use the projectile equations. To do so, we place the origin of an xy coordinate system at the cannon muzzle. Then $x_0 = 0$ and $y_0 = 0$ and we want his height y when $x = 23$ m, but we do not know the time t when he reaches that height. To relate y to x without t, we use Eq. 4-25:

$$y = (\tan\theta_0)x - \frac{gx^2}{2(v_0\cos\theta_0)^2}$$
$$= (\tan 53°)(23\ \text{m}) - \frac{(9.8\ \text{m/s}^2)(23\ \text{m})^2}{2(26.5\ \text{m/s})^2(\cos 53°)^2}$$
$$= 20.3\ \text{m}.$$

Since he begins 3.0 m off the ground, he clears the Ferris wheel by about 5.3 m.

(b) If he reaches his maximum height when he is over the middle Ferris wheel, what is his clearance above it?

SOLUTION: A **Key Idea** here is that the vertical component v_y of his velocity is zero when he reaches his maximum height. Since Eq. 4-24 relates v_y and his height y, we write

$$v_y^2 = (v_0\sin\theta_0)^2 - 2gy = 0.$$

Solving for y gives us

$$y = \frac{(v_0 \sin \theta_0)^2}{2g} = \frac{(26.5 \text{ m/s})^2(\sin 53°)^2}{(2)(9.8 \text{ m/s}^2)} = 22.9 \text{ m},$$

which means that he clears the middle Ferris wheel by 7.9 m.

(c) How far from the cannon should the center of the net be positioned?

SOLUTION: The additional **Key Idea** here is that, because Zacchini's initial and landing heights are the same, the horizontal distance from cannon muzzle to net is his horizontal range for the flight. From Eq. 4-26, we find

$$R = \frac{v_0^2}{g} \sin 2\theta_0 = \frac{(26.5 \text{ m/s})^2}{9.8 \text{ m/s}^2} \sin 2(53°)$$

$$= 69 \text{ m.} \qquad \text{(Answer)}$$

We can now answer the questions that opened this chapter: How could Zacchini know where to place the net, and how could he be certain he would clear the Ferris wheels? He (or someone) did the calculations as we have here. Although he could not take into account the complicated effects of the air on his flight, Zacchini knew that the air would slow him and thus decrease his range from the calculated value, so he used a wide net and biased it toward the cannon. He was then relatively safe whether the effects of the air in a particular flight happened to slow him considerably or very little. Still, the variability of this factor of air effects must have played on his imagination before each flight.

Zacchini still faced a subtle danger: Even for shorter flights, his propulsion through the cannon was so severe that he underwent a momentary blackout. If he landed during the blackout, he could break his neck. To avoid this, he had trained himself to awake quickly. Indeed, not waking up in time presents the only real danger to a human cannonball in the short flights today.

PROBLEM-SOLVING TACTICS

Tactic 1: *Numbers Versus Algebra*
One way to avoid rounding errors and other numerical errors is to solve problems algebraically, substituting numbers only in the final step. That is easy to do in Sample Problems 4-6 to 4-8, and that is the way experienced problem solvers operate. In these early chapters, however, we prefer to solve most problems in parts, to give you a firmer numerical grasp of what is going on. Later we shall stick to the algebra longer.

4-7 Uniform Circular Motion

A particle is in **uniform circular motion** if it travels around a circle or a circular arc at constant (*uniform*) speed. Although the speed does not vary, *the particle is accelerating*. That fact may be surprising because we often think of acceleration (a change in velocity) as an increase or decrease in speed. However, actually velocity is a vector, not a scalar. Thus, even if a velocity changes only in direction, there is still an acceleration, and that is what happens in uniform circular motion.

Figure 4-18 shows the relation between the velocity and acceleration vectors at various stages during uniform circular motion. Both vectors have constant magnitude as the motion progresses, but their directions change continuously. The velocity is always directed tangent to the circle in the direction of motion. The acceleration is always directed *radially inward*. Because of this, the acceleration associated with uniform circular motion is called a **centripetal** (meaning "center seeking") **acceleration**. As we prove next, the magnitude of this acceleration $\vec{a}$ is

$$a = \frac{v^2}{r} \quad \text{(centripetal acceleration),} \tag{4-32}$$

where r is the radius of the circle and v is the speed of the particle.

In addition, during this acceleration at constant speed, the particle travels the circumference of the circle (a distance of $2\pi r$) in time

$$T = \frac{2\pi r}{v} \quad \text{(period).} \tag{4-33}$$

T is called the *period of revolution*, or simply the *period*, of the motion. It is, in general, the time for a particle to go around a closed path exactly once.

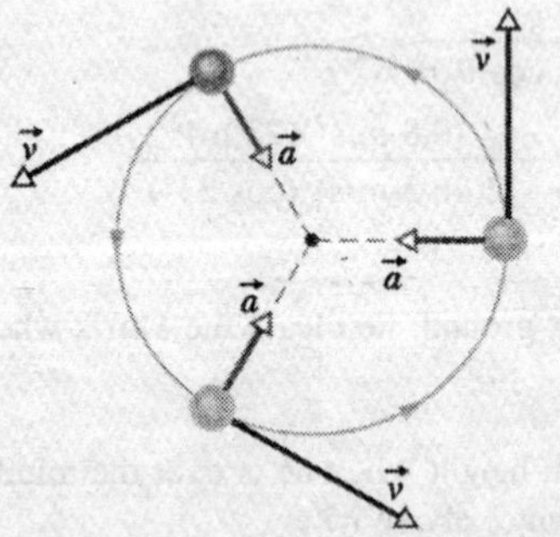

Fig. 4-18 Velocity and acceleration vectors for a particle in counterclockwise uniform circular motion. Both have constant magnitude but vary continuously in direction.

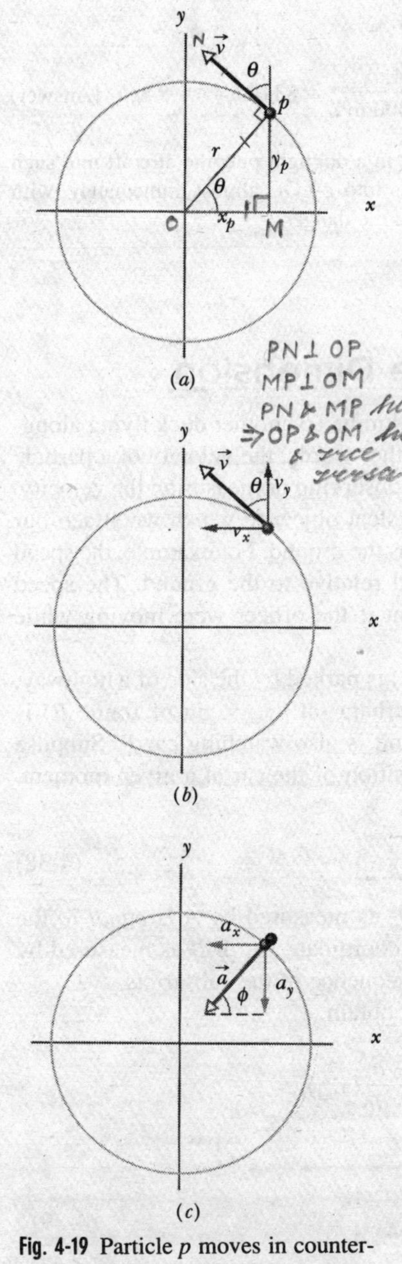

Fig. 4-19 Particle p moves in counterclockwise uniform circular motion. (a) Its position and velocity $\vec{v}$ at a certain instant. (b) Velocity $\vec{v}$ and its components. (c) The particle's acceleration $\vec{a}$ and its components.

Proof of Eq. 4-32

To find the magnitude and direction of the acceleration for uniform circular motion, we consider Fig. 4-19. In Fig. 4-19a, particle p moves at constant speed v around a circle of radius r. At the instant shown, p has coordinates x_p and y_p.

Recall from Section 4-3 that the velocity $\vec{v}$ of a moving particle is always tangent to the particle's path at the particle's position. In Fig. 4-19a, that means $\vec{v}$ is perpendicular to a radius r drawn to the particle's position. Then the angle θ that $\vec{v}$ makes with a vertical at p equals the angle θ that radius r makes with the x axis.

The scalar components of $\vec{v}$ are shown in Fig. 4-19b. With them, we can write the velocity $\vec{v}$ as

$$\vec{v} = v_x\hat{i} + v_y\hat{j} = (-v \sin\theta)\hat{i} + (v\cos\theta)\hat{j}. \tag{4-34}$$

Now, using the right triangle in Fig. 4-19a, we can replace $\sin\theta$ with y_p/r and $\cos\theta$ with x_p/r to write

$$\vec{v} = \left(-\frac{vy_p}{r}\right)\hat{i} + \left(\frac{vx_p}{r}\right)\hat{j}. \tag{4-35}$$

To find the acceleration $\vec{a}$ of particle p, we must take the time derivative of this equation. Noting that speed v and radius r do not change with time, we obtain

$$\vec{a} = \frac{d\vec{v}}{dt} = \left(-\frac{v}{r}\frac{dy_p}{dt}\right)\hat{i} + \left(\frac{v}{r}\frac{dx_p}{dt}\right)\hat{j}. \tag{4-36}$$

Now note that the rate dy_p/dt at which y_p changes is equal to the velocity component v_y. Similarly, $dx_p/dt = v_x$, and, again from Fig. 4-19b, we see that $v_x = -v\sin\theta$ and $v_y = v\cos\theta$. Making these substitutions in Eq. 4-36, we find

$$\vec{a} = \left(-\frac{v^2}{r}\cos\theta\right)\hat{i} + \left(-\frac{v^2}{r}\sin\theta\right)\hat{j}. \tag{4-37}$$

This vector and its components are shown in Fig. 4-19c. Following Eq. 3-6, we find that the magnitude of $\vec{a}$ is

$$a = \sqrt{a_x^2 + a_y^2} = \frac{v^2}{r}\sqrt{(\cos\theta)^2 + (\sin\theta)^2} = \frac{v^2}{r},$$

as we wanted to prove. To orient $\vec{a}$, we can find the angle ϕ shown in Fig. 4-19c:

$$\tan\phi = \frac{a_y}{a_x} = \frac{-(v^2/r)\sin\theta}{-(v^2/r)\cos\theta} = \tan\theta.$$

Thus, $\phi = \theta$, which means that $\vec{a}$ is directed along the radius r of Fig. 4-19a, toward the circle's center, as we wanted to prove.

CHECKPOINT 6: An object moves at constant speed along a circular path in a horizontal xy plane, with the center at the origin. When the object is at $x = -2$ m, its velocity is $-(4\text{ m/s})\hat{j}$. Give the object's (a) velocity and (b) acceleration when it is at $y = 2$ m.

Sample Problem 4-9

"Top gun" pilots have long worried about taking a turn too tightly. As a pilot's body undergoes centripetal acceleration, with the head toward the center of curvature, the blood pressure in the brain decreases, leading to loss of brain function.

There are several warning signs to signal a pilot to ease up: when the centripetal acceleration is $2g$ or $3g$, the pilot feels heavy. At about $4g$, the pilot's vision switches to black and white and narrows to "tunnel vision." If that acceleration is sustained or increased, vision ceases and, soon after, the pilot is unconscious—a condition known as g-LOC for "g-induced loss of consciousness."

What is the centripetal acceleration, in g units, of a pilot flying an F-22 at speed $v = 2500$ km/h (694 m/s) through a circular arc with radius of curvature $r = 5.80$ km?

SOLUTION: The **Key Idea** here is that although the pilot's speed is constant, the circular path requires a (centripetal) acceleration, with magnitude given by Eq. 4-32:

$$a = \frac{v^2}{r} = \frac{(694\ \text{m/s})^2}{5800\ \text{m}} = 83.0\ \text{m/s}^2 = 8.5g. \quad \text{(Answer)}$$

If an unwary pilot caught in a dogfight puts the aircraft into such a tight turn, the pilot goes into g-LOC almost immediately, with no warning signs to signal the danger.

4-8 Relative Motion in One Dimension

Suppose you see a duck flying north at, say, 30 km/h. To another duck flying alongside, the first duck seems to be stationary. In other words, the velocity of a particle depends on the **reference frame** of whoever is observing or measuring the velocity. For our purposes, a reference frame is the physical object to which we attach our coordinate system. In everyday life, that object is the ground. For example, the speed listed on a speeding ticket is always measured relative to the ground. The speed relative to the police officer would be different if the officer were moving while making the speed measurement.

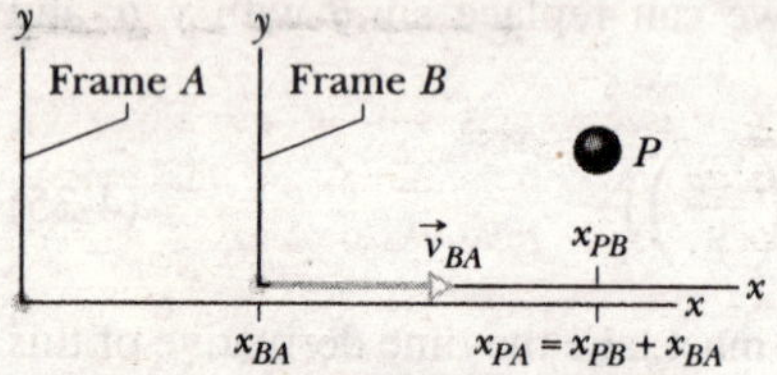

Fig. 4-20 Alex (frame A) and Barbara (frame B) watch car P, as both B and P move at different velocities along the common x axes of the two frames. At the instant shown, x_{BA} is the coordinate of B in the A frame. Also, P is at coordinate x_{PB} in the B frame and coordinate $x_{PA} = x_{PB} + x_{BA}$ in the A frame.

Suppose that Alex (at the origin of frame A) is parked by the side of a highway, watching car P (the "particle") speed past. Barbara (at the origin of frame B) is driving along the highway at constant speed and is also watching car P. Suppose that, as in Fig. 4-20, they both measure the position of the car at a given moment. From the figure we see that

$$x_{PA} = x_{PB} + x_{BA}. \tag{4-38}$$

The equation is read: "The coordinate x_{PA} of P as measured by A *is equal to* the coordinate x_{PB} of P as measured by B *plus* the coordinate x_{BA} of B as measured by A." Note how this reading is supported by the sequence of the subscripts.

Taking the time derivative of Eq. 4-38, we obtain

$$\frac{d}{dt}(x_{PA}) = \frac{d}{dt}(x_{PB}) + \frac{d}{dt}(x_{BA}),$$

or (because $v = dx/dt$)

$$v_{PA} = v_{PB} + v_{BA}. \tag{4-39}$$

This equation is read: "The velocity v_{PA} of P as measured by A *is equal to* the velocity v_{PB} of P as measured by B *plus* the velocity v_{BA} of B as measured by A." The term v_{BA} is the velocity of frame B relative to frame A. (Because the motions are along a single axis, we can use components along that axis in Eq. 4-39 and omit overhead vector arrows.)

Here we consider only frames that move at constant velocity relative to each other. In our example, this means that Barbara (frame B) will drive always at constant velocity v_{BA} relative to Alex (frame A). Car P (the moving particle), however, may speed up, slow down, come to rest, or reverse direction (that is, it can accelerate).

To relate an acceleration of P as measured by Barbara and by Alex, we take the time derivative of Eq. 4-39:

$$\frac{d}{dt}(v_{PA}) = \frac{d}{dt}(v_{PB}) + \frac{d}{dt}(v_{BA}).$$

Because v_{BA} is constant, the last term is zero and we have

$$a_{PA} = a_{PB}. \tag{4-40}$$

In other words,

> Observers on different frames of reference (that move at constant velocity relative to each other) will measure the same acceleration for a moving particle.

Situation	v_{BA}	v_{PA}	v_{PB}
(a)	+50	+50	
(b)	+30		+40
(c)		+60	−20

CHECKPOINT 7: The table here gives velocities (km/h) for Barbara and car P of Fig. 4-20 for three situations. For each, what is the missing value and how is the distance between Barbara and car P changing?

Sample Problem 4-10

For the situation of Fig. 4-20 and this section, Barbara's velocity relative to Alex is a constant $v_{BA} = 52$ km/h and car P is moving in the negative direction of the x axis.

(a) If Alex measures a constant velocity $v_{PA} = -78$ km/h for car P, what velocity v_{PB} will Barbara measure?

SOLUTION: **The Key Idea** here is that we can attach a frame of reference A to Alex and another frame of reference B to Barbara. Further, because these two frames move at constant velocity relative to each other, along a single axis, we can use Eq. 4-39 to relate v_{PB} and v_{PA}. We find

$$v_{PA} = v_{PB} + v_{BA}$$

or

$$-78 \text{ km/h} = v_{PB} + 52 \text{ km/h}.$$

Thus,

$$v_{PB} = -130 \text{ km/h}. \qquad \text{(Answer)}$$

If car P were connected to Barbara's car by a cord wound on a spool, the cord would be unwinding at a speed of 130 km/h as the two cars separated.

(b) If car P brakes to a stop relative to Alex (and thus the ground) in time $t = 10$ s at constant acceleration, what is its acceleration a_{PA} relative to Alex?

SOLUTION: **The Key Idea** here is that, to calculate the acceleration of car P *relative to Alex*, we must use the car's velocities *relative to Alex*. Because the acceleration is constant, we can use Eq. 2-11 ($v = v_0 + at$) to relate the acceleration to the initial and final velocities of P. The initial velocity of P relative to Alex is $v_{PA} = -78$ km/h and the final velocity is 0. Thus, Eq. 2-11 gives us

$$a_{PA} = \frac{v - v_0}{t} = \frac{0 - (-78 \text{ km/h})}{10 \text{ s}} \frac{1 \text{ m/s}}{3.6 \text{ km/h}}$$
$$= 2.2 \text{ m/s}^2. \qquad \text{(Answer)}$$

(c) What is the acceleration a_{PB} of car P relative to Barbara during the braking?

SOLUTION: **The Key Idea** is that now, to calculate the acceleration of car P *relative to Barbara*, we must use the car's velocities *relative to Barbara*. We know the initial velocity of P relative to Barbara from part (a) ($v_{PB} = -130$ km/h). The final velocity of P relative to Barbara is -52 km/h (this is the velocity of the stopped car relative to the moving Barbara). Thus, again from Eq. 2-11,

$$a_{PB} = \frac{v - v_0}{t} = \frac{-52 \text{ km/h} - (-130 \text{ km/h})}{10 \text{ s}} \frac{1 \text{ m/s}}{3.6 \text{ km/h}}$$
$$= 2.2 \text{ m/s}^2. \qquad \text{(Answer)}$$

We should have foreseen this result: Because Alex and Barbara have a constant relative velocity, they must measure the same acceleration for the car.

4-9 Relative Motion in Two Dimensions

Now we turn from relative motion in one dimension to relative motion in two (and, by extension, in three) dimensions. In Fig. 4-21, our two observers are again watching a moving particle P from the origins of reference frames A and B, while B moves at a constant velocity $\vec{v}_{BA}$ relative to A. (The corresponding axes of these two frames remain parallel.)

Figure 4-21 shows a certain instant during the motion. At that instant, the position vector of B relative to A is $\vec{r}_{BA}$. Also, the position vectors of particle P are

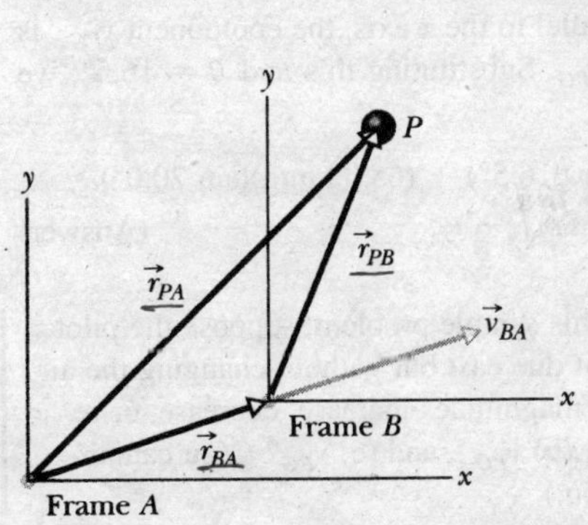

Fig. 4-21 Frame B has the constant two-dimensional velocity $\vec{v}_{BA}$ relative to frame A. The position vector of B relative to A is $\vec{r}_{BA}$. The position vectors of particle P are $\vec{r}_{PA}$ relative to A and $\vec{r}_{PB}$ relative to B.

$\vec{r}_{PA}$ relative to A and $\vec{r}_{PB}$ relative to B. From the arrangement of heads and tails of those three position vectors, we can relate the vectors with

$$\vec{r}_{PA} = \vec{r}_{PB} + \vec{r}_{BA}. \tag{4-41}$$

By taking the time derivative of this equation, we can relate the velocities $\vec{v}_{PA}$ and $\vec{v}_{PB}$ of particle P relative to our observers. We get

$$\vec{v}_{PA} = \vec{v}_{PB} + \vec{v}_{BA}. \tag{4-42}$$

By taking the time derivative of this relation, we can relate the accelerations $\vec{a}_{PA}$ and $\vec{a}_{PB}$ of the particle P relative to our observers. However, note that because $\vec{v}_{BA}$ is constant, its time derivative is zero. Thus, we get

$$\vec{a}_{PA} = \vec{a}_{PB}. \tag{4-43}$$

As for one-dimensional motion, we have the following rule: Observers on different frames of reference that move at constant velocity relative to each other will measure the *same* acceleration for a moving particle.

Sample Problem 4-11

In Fig. 4-22*a*, a plane moves due east (directly toward the east) while the pilot points the plane somewhat south of east, toward a steady wind that blows to the northeast. The plane has velocity $\vec{v}_{PW}$ relative to the wind, with an airspeed (speed relative to the wind) of 215 km/h, directed at angle θ south of east. The wind has velocity $\vec{v}_{WG}$ relative to the ground, with a speed of 65.0 km/h, directed 20.0° east of north. What is the magnitude of the velocity $\vec{v}_{PG}$ of the plane relative to the ground, and what is θ?

SOLUTION: The **Key Idea** is that the situation is like the one in Fig. 4-21. Here the moving particle P is the plane, frame A is attached to the ground (call it G), and frame B is "attached" to the wind (call it W). We need to construct a vector diagram like that in Fig. 4-21 but this time using the three velocity vectors.

First construct a sentence that relates the three vectors:

velocity of plane relative to ground (PG)	=	velocity of plane relative to wind (PW)	+	velocity of wind relative to ground. (WG)

This relation can be drawn as in Fig. 4-22*b* and written in vector notation as

$$\vec{v}_{PG} = \vec{v}_{PW} + \vec{v}_{WG}. \tag{4-44}$$

We want the magnitude of the first vector and the direction of the second vector. With unknowns in two vectors, we cannot solve Eq. 4-44 directly on a vector-capable calculator. Instead, we need to resolve the vectors into components on the coordinate system of Fig. 4-22*b*, and then solve Eq. 4-44 axis by axis (see Section 3-5). For the y components, we find

$$v_{PG,y} = v_{PW,y} + v_{WG,y}$$

or $\quad 0 = -(215 \text{ km/h}) \sin\theta + (65.0 \text{ km/h})(\cos 20.0°).$

Solving for θ gives us

$$\theta = \sin^{-1}\frac{(65.0 \text{ km/h})(\cos 20.0°)}{215 \text{ km/h}} = 16.5°. \qquad \text{(Answer)}$$

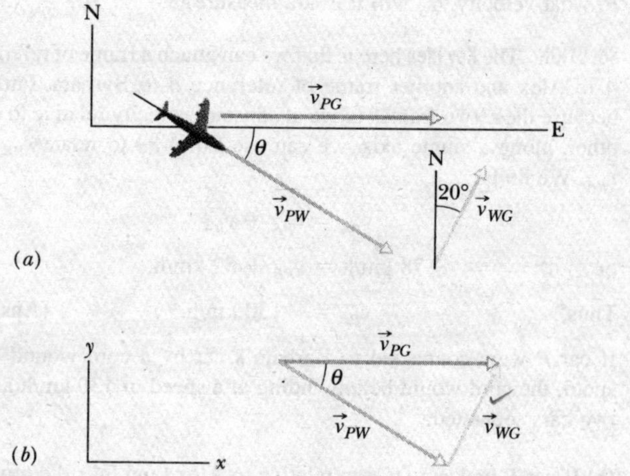

Fig. 4-22 Sample Problem 4-11. To travel due east, the plane must head somewhat into the wind.

Similarly, for the x components we find

$$v_{PG,x} = v_{PW,x} + v_{WG,x}.$$

Here, because $\vec{v}_{PG}$ is parallel to the x axis, the component $v_{PG,x}$ is equal to the magnitude v_{PG}. Substituting this and $\theta = 16.5°$, we find

$$v_{PG} = (215 \text{ km/h})(\cos 16.5°) + (65.0 \text{ km/h})(\sin 20.0°)$$
$$= 228 \text{ km/h}. \qquad \text{(Answer)}$$

CHECKPOINT 8: In this sample problem, suppose the pilot turns the plane to point it due east but without changing the air speed. Do the following magnitudes increase, decrease, or remain the same: (a) $v_{PG,y}$, (b) $v_{PG,x}$, and (c) v_{PG}? (You can answer without computation.)

REVIEW & SUMMARY

Position Vector The location of a particle relative to the origin of a coordinate system is given by a *position vector* $\vec{r}$, which in unit-vector notation is

$$\vec{r} = x\hat{i} + y\hat{j} + z\hat{k}. \tag{4-1}$$

Here $x\hat{i}$, $y\hat{j}$, and $z\hat{k}$ are the *vector components* of position vector $\vec{r}$, and x, y, and z are its *scalar components* (as well as the coordinates of the particle). A position vector is described by a magnitude and one or two angles for orientation, or by its vector or scalar components.

Displacement If a particle moves so that its position vector changes from $\vec{r}_1$ to $\vec{r}_2$, then the particle's *displacement* $\Delta\vec{r}$ is

$$\Delta\vec{r} = \vec{r}_2 - \vec{r}_1. \tag{4-2}$$

The displacement can also be written as

$$\Delta\vec{r} = (x_2 - x_1)\hat{i} + (y_2 - y_1)\hat{j} + (z_2 - z_1)\hat{k} \tag{4-3}$$

$$= \Delta x\hat{i} + \Delta y\hat{j} + \Delta z\hat{k}, \tag{4-4}$$

where coordinates (x_1, y_1, z_1) correspond to position vector $\vec{r}_1$ and coordinates (x_2, y_2, z_2) correspond to position vector $\vec{r}_2$.

Average Velocity and (Instantaneous) Velocity If a particle undergoes a displacement $\Delta\vec{r}$ in time Δt, its *average velocity* $\vec{v}_{avg}$ for that time interval is

$$\vec{v}_{avg} = \frac{\Delta\vec{r}}{\Delta t}. \tag{4-8}$$

As Δt in Eq. 4-8 is shrunk to 0, $\vec{v}_{avg}$ reaches a limit called the *velocity* or *instantaneous velocity* $\vec{v}$:

$$\vec{v} = \frac{d\vec{r}}{dt}, \tag{4-10}$$

which can be rewritten in unit-vector notation as

$$\vec{v} = v_x\hat{i} + v_y\hat{j} + v_z\hat{k}, \tag{4-11}$$

where $v_x = dx/dt$, $v_y = dy/dt$, and $v_z = dz/dt$. The instantaneous velocity $\vec{v}$ of a particle is always directed along the tangent to the particle's path at the particle's position.

Average Acceleration and (Instantaneous) Acceleration If a particle's velocity changes from $\vec{v}_1$ to $\vec{v}_2$ in time interval Δt, its *average acceleration* during Δt is

$$\vec{a}_{avg} = \frac{\vec{v}_2 - \vec{v}_1}{\Delta t} = \frac{\Delta\vec{v}}{\Delta t}. \tag{4-15}$$

As Δt in Eq. 4-15 is shrunk to 0, $\vec{a}_{avg}$ reaches a limiting value called the *acceleration* or *instantaneous acceleration* $\vec{a}$:

$$\vec{a} = \frac{d\vec{v}}{dt}. \tag{4-16}$$

In unit-vector notation,

$$\vec{a} = a_x\hat{i} + a_y\hat{j} + a_z\hat{k}, \tag{4-17}$$

where $a_x = dv_x/dt$, $a_y = dv_y/dt$, and $a_z = dv_z/dt$.

Projectile Motion *Projectile motion* is the motion of a particle that is launched with an initial velocity $\vec{v}_0$. During its flight, the particle's horizontal acceleration is zero and its vertical acceleration is the free-fall acceleration $-g$. (Upward is taken to be a positive direction.) If $\vec{v}_0$ is expressed as a magnitude (the speed v_0) and an angle θ_0, the particle's equations of motion along the horizontal x axis and vertical y axis are

$$x - x_0 = (v_0 \cos\theta_0)t, \tag{4-21}$$

$$y - y_0 = (v_0 \sin\theta_0)t - \tfrac{1}{2}gt^2, \tag{4-22}$$

$$v_y = v_0 \sin\theta_0 - gt, \tag{4-23}$$

$$v_y^2 = (v_0 \sin\theta_0)^2 - 2g(y - y_0). \tag{4-24}$$

The **trajectory** (path) of a particle in projectile motion is parabolic and is given by

$$y = (\tan\theta_0)x - \frac{gx^2}{2(v_0 \cos\theta_0)^2}, \tag{4-25}$$

where the origin has been chosen so that x_0 and y_0 of Eqs. 4-21 to 4-24 are zero. The particle's **horizontal range** R, which is the horizontal distance from the launch point to the point at which the particle returns to the launch height, is

$$R = \frac{v_0^2}{g}\sin 2\theta_0. \tag{4-26}$$

Uniform Circular Motion If a particle travels along a circle or circular arc with radius r at constant speed v, it is in *uniform circular motion* and has an acceleration $\vec{a}$ of magnitude

$$a = \frac{v^2}{r}. \tag{4-32}$$

The direction of $\vec{a}$ is toward the center of the circle or circular arc, and $\vec{a}$ is said to be *centripetal*. The time for the particle to complete a circle is

$$T = \frac{2\pi r}{v}. \tag{4-33}$$

T is called the *period of revolution*, or simply the *period*, of the motion.

Relative Motion When two frames of reference A and B are moving relative to each other at constant velocity, the velocity of a particle P as measured by an observer in frame A usually differs from that measured from frame B. The two measured velocities are related by

$$\vec{v}_{PA} = \vec{v}_{PB} + \vec{v}_{BA}, \tag{4-42}$$

in which $\vec{v}_{BA}$ is the velocity of B with respect to A. Both observers measure the same acceleration for the particle; that is,

$$\vec{a}_{PA} = \vec{a}_{PB}. \tag{4-43}$$

QUESTIONS

1. Figure 4-23 shows the initial position i and the final position f of a particle. What are the particle's (a) initial position vector $\vec{r}_i$ and (b) final position vector $\vec{r}_f$, both in unit-vector notation? (c) What is the x component of the particle's displacement $\Delta\vec{r}$?

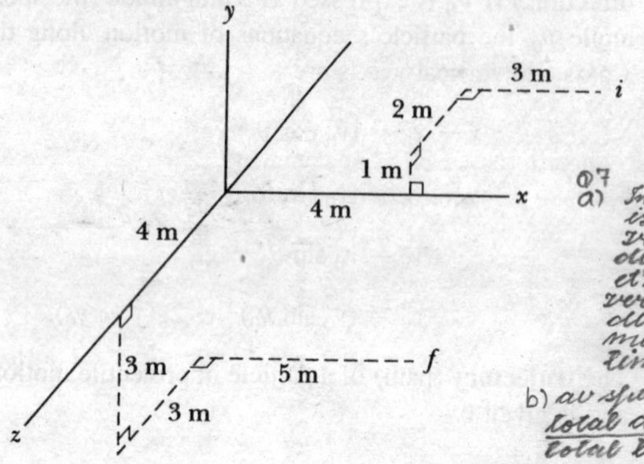

Fig. 4-23 Question 1.

2. Here are four descriptions for the velocity of a hockey puck in the xy plane, all in meters per second:
(1) $v_x = -3t^2 + 4t - 2$ and $v_y = 6t - 4$
(2) $v_x = -3$ and $v_y = -5t^2 + 6$
(3) $\vec{v} = 2t^2\hat{i} - (4t + 3)\hat{j}$
(4) $\vec{v} = -2t\hat{i} + 3\hat{j}$
(a) For each description, are the x and y components of the acceleration constant, and is the acceleration vector $\vec{a}$ constant? (b) In description (4), if $\vec{v}$ is in meters per second and t is in seconds, what must be the units of the coefficients -2 and 3?

3. Figure 4-24 shows three situations in which identical projectiles are launched from the ground (at the same level) at identical initial speeds and angles. The projectiles do not land on the same terrain, however. Rank the situations according to the final speeds of the projectiles just before they land, greatest first.

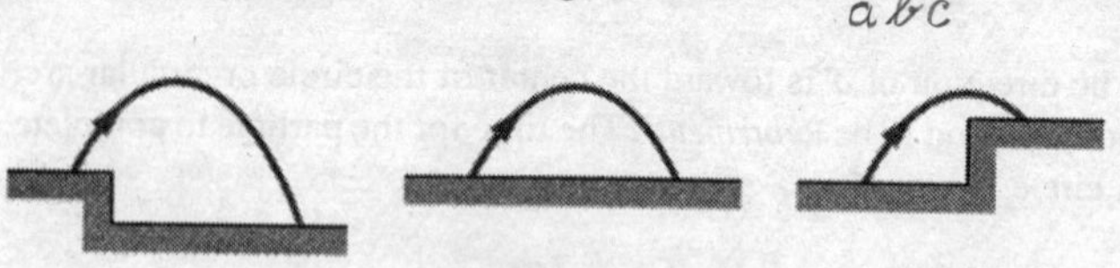

Fig. 4-24 Question 3.

4. At a certain instant, a fly ball has velocity $\vec{v} = 25\hat{i} - 4.9\hat{j}$ (the x axis is horizontal, the y axis is upward, and $\vec{v}$ is in meters per second). Has the ball passed the highest point of its trajectory?

5. You are to launch a rocket, from just above the ground, with one of the following initial velocity vectors: (1) $\vec{v}_0 = 20\hat{i} + 70\hat{j}$, (2) $\vec{v}_0 = -20\hat{i} + 70\hat{j}$, (3) $\vec{v}_0 = 20\hat{i} - 70\hat{j}$, (4) $\vec{v}_0 = -20\hat{i} - 70\hat{j}$. In your coordinate system, x runs along level ground and y increases upward. (a) Rank the vectors according to the launch speed of the projectile, greatest first. (b) Rank the vectors according to the time of flight of the projectile, greatest first.

6. A mud ball is launched 2 m above the ground with initial velocity $\vec{v}_0 = (2\hat{i} + 4\hat{j})$ m/s. What is its velocity just before it lands on a surface that is 2 m above the ground?

7. In Fig. 4-25, a cream tangerine is thrown up past windows 1, 2, and 3, which are identical in size and regularly spaced vertically. Rank those three windows according to (a) the time the cream tangerine takes to pass them and (b) the average speed of the cream tangerine during the passage, greatest first.

The cream tangerine then moves down past windows 4, 5, and 6, which are identical in size and irregularly spaced horizontally. Rank those three windows according to (c) the time the cream tangerine takes to pass them and (d) the average speed of the cream tangerine during the passage, greatest first.

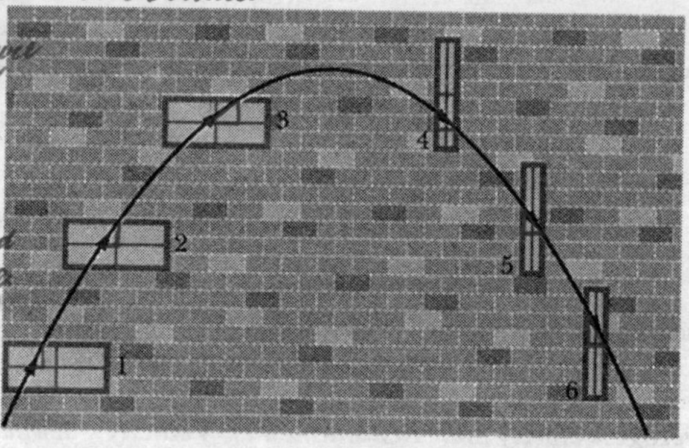

Fig. 4-25 Question 7.

8. An airplane flying horizontally at a constant speed of 350 km/h over level ground releases a bundle of food supplies. Ignore the effect of the air on the bundle. What are the bundle's initial (a) vertical and (b) horizontal components of velocity? (c) What is its horizontal component of velocity just before hitting the ground? (d) If the airplane's speed were, instead, 450 km/h, would the time of fall be larger, smaller, or the same?

9. You throw a ball with a launch velocity of $\vec{v}_i = (3 \text{ m/s})\hat{i} + (4 \text{ m/s})\hat{j}$ toward a wall, where it hits at height h_1 in time t_1 after the launch (Fig. 4-26). Suppose that the launch velocity were, instead, $\vec{v}_i = (5 \text{ m/s})\hat{i} + (4 \text{ m/s})\hat{j}$. (a) Would the time taken by the ball to reach the wall be greater than, less than, or equal to t_1, or is the question unanswerable without more information? (b) Would the height at which the ball hits be greater than, less than, or equal to h_1, or is the question unanswerable?

Suppose, instead, that the launch velocity were $\vec{v}_i = (3 \text{ m/s})\hat{i} + (5 \text{ m/s})\hat{j}$. (c) Would the time taken by the ball to reach the wall be greater than, less than, or equal to t_1, or is the question unanswerable? (d) Would the height at which the ball hits be greater than, less than, or equal to h_1, or is the question unanswerable?

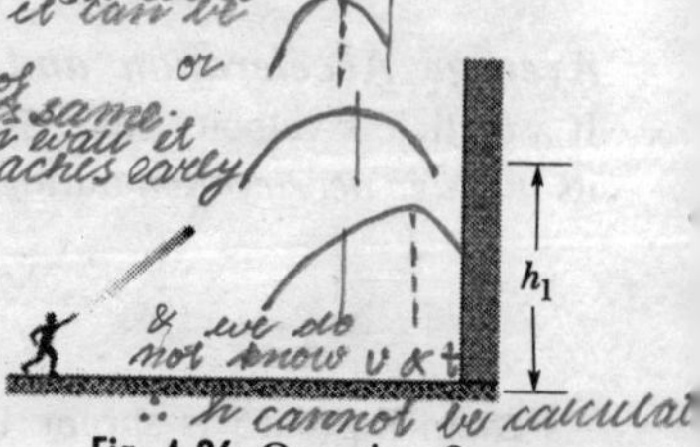

Fig. 4-26 Question 9.

10. Figure 4-27 shows three paths for a football kicked from ground level. Ignoring the effects of air

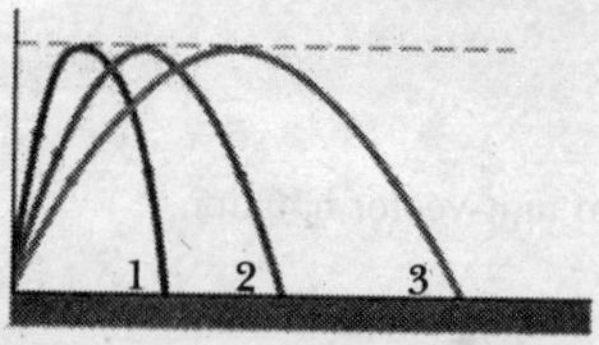

Fig. 4-27 Question 10.

on the flight, rank the paths according to (a) time of flight, (b) initial vertical velocity component, (c) initial horizontal velocity component, and (d) initial speed, greatest first.

11. Figure 4-28 shows the velocity and acceleration of a particle at a particular instant in three situations. In which situation is (a) the speed of the particle increasing, (b) the speed decreasing, (c) the speed not changing, (d) $\vec{v} \cdot \vec{a}$ positive, (e) $\vec{v} \cdot \vec{a}$ negative, and (f) $\vec{v} \cdot \vec{a} = 0$?

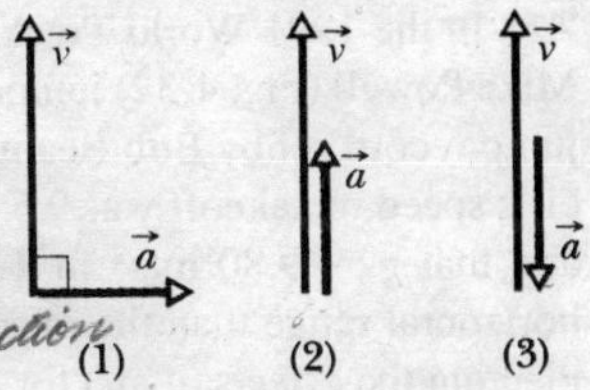

Fig. 4-28 Question 11.

12. Figure 4-29 shows four tracks (either half- or quarter-circles) that can be taken by a train, which moves at a constant speed. Rank the tracks according to the magnitude of a train's acceleration on the curved portion, greatest first.

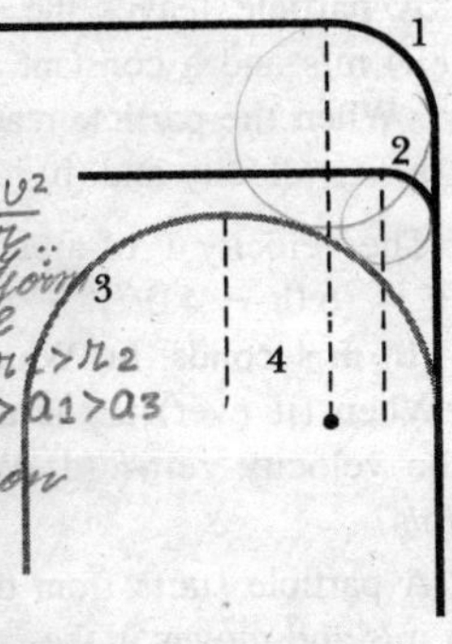

Fig. 4-29 Question 12.

13. (a) Is it possible to be accelerating while traveling at constant speed? Is it possible to round a curve with (b) zero acceleration and (c) a constant magnitude of acceleration?

EXERCISES & PROBLEMS

ssm Solution is in the Student Solutions Manual.
www Solution is available on the World Wide Web at: http://www.wiley.com/college/hrw
ilw Solution is available on the Interactive LearningWare.

SEC. 4-2 Position and Displacement

1E. A watermelon seed has the following coordinates: $x = -5.0$ m, $y = 8.0$ m, and $z = 0$ m. Find its position vector (a) in unit-vector notation and as (b) a magnitude and (c) an angle relative to the positive direction of the x axis. (d) Sketch the vector on a right-handed coordinate system. If the seed is moved to the xyz coordinates (3.00 m, 0 m, 0 m), what is its displacement (e) in unit-vector notation and as (f) a magnitude and (g) an angle relative to the positive direction of the x axis?

2E. The position vector for an electron is $\vec{r} = (5.0\text{ m})\hat{i} - (3.0\text{ m})\hat{j} + (2.0\text{ m})\hat{k}$. (a) Find the magnitude of $\vec{r}$. (b) Sketch the vector on a right-handed coordinate system.

3E. The position vector for a proton is initially $\vec{r} = 5.0\hat{i} - 6.0\hat{j} + 2.0\hat{k}$ and then later is $\vec{r} = -2.0\hat{i} + 6.0\hat{j} + 2.0\hat{k}$, all in meters. (a) What is the proton's displacement vector, and (b) to what plane is that vector parallel?

4P. A radar station detects an airplane approaching directly from the east. At first observation, the range to the plane is 360 m at 40° above the horizon. The airplane is tracked for another 123° in the vertical east–west plane, the range at final contact being 790 m. See Fig. 4-30. Find the displacement of the airplane during the period of observation.

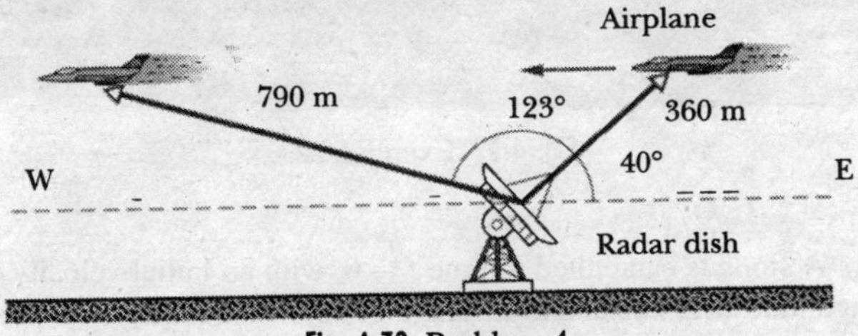

Fig. 4-30 Problem 4.

SEC. 4-3 Average Velocity and Instantaneous Velocity

5E. A train moving at a constant speed of 60.0 km/h moves east for 40.0 min, then in a direction 50.0° east of north for 20.0 min, and finally west for 50.0 min. What is the average velocity of the train during this trip? **ssm**

6E. An ion's position vector is initially $\vec{r} = 5.0\hat{i} - 6.0\hat{j} + 2.0\hat{k}$, and 10 s later it is $\vec{r} = -2.0\hat{i} + 8.0\hat{j} - 2.0\hat{k}$, all in meters. What is its average velocity during the 10 s?

7P. The position of an electron is given by $\vec{r} = 3.00t\hat{i} - 4.00t^2\hat{j} + 2.00\hat{k}$, with t in seconds and $\vec{r}$ in meters. (a) What is the electron's velocity $\vec{v}(t)$? At $t = 2.00$ s, what is $\vec{v}$ (b) in unit-vector notation and as (c) a magnitude and (d) an angle relative to the positive direction of the x axis?

8P. Oasis A is 90 km west of oasis B. A camel leaves oasis A and during a 50 h period walks 75 km in a direction 37° north of east. The camel then walks toward the south a distance of 65 km in a 35 h period, after which it rests for 5.0 h. (a) What is the camel's displacement with respect to oasis A after resting? (b) What is the camel's average velocity from the time it leaves oasis A until it finishes resting? (c) What is the camel's average speed from the time it leaves oasis A until it finishes resting? (d) If the camel is able to go without water for five days (120 h), what must its average velocity be after resting if it is to reach oasis B just in time?

SEC. 4-4 Average Acceleration and Instantaneous Acceleration

9E. A particle moves so that its position (in meters) as a function of time (in seconds) is $\vec{r} = \hat{i} + 4t^2\hat{j} + t\hat{k}$. Write expressions for (a) its velocity and (b) its acceleration as functions of time. **ssm**

10E. A proton initially has $\vec{v} = 4.0\hat{i} - 2.0\hat{j} + 3.0\hat{k}$ and then 4.0 s later has $\vec{v} = -2.0\hat{i} - 2.0\hat{j} + 5.0\hat{k}$ (in meters per second). For that 4.0 s, what is the proton's average acceleration $\vec{a}_{avg}$ (a) in unit-vector notation and (b) as a magnitude and a direction?

11E. The position $\vec{r}$ of a particle moving in an xy plane is given by $\vec{r} = (2.00t^3 - 5.00t)\hat{i} + (6.00 - 7.00t^4)\hat{j}$, with $\vec{r}$ in meters and t in seconds. Calculate (a) $\vec{r}$, (b) $\vec{v}$, and (c) $\vec{a}$ for $t = 2.00$ s. (d) What is the orientation of a line that is tangent to the particle's path at $t = 2.00$ s?

12E. An iceboat sails across the surface of a frozen lake with constant acceleration produced by the wind. At a certain instant the boat's velocity is $(6.30\hat{i} - 8.42\hat{j})$ m/s. Three seconds later, because of a wind shift, the boat is instantaneously at rest. What is its average acceleration for this 3 s interval?

13P. A particle leaves the origin with an initial velocity $\vec{v} = (3.00\hat{i})$ m/s and a constant acceleration $\vec{a} = (-1.00\hat{i} - 0.500\hat{j})$ m/s^2. When the particle reaches its maximum x coordinate, what are (a) its velocity and (b) its position vector? ssm ilw www

14P. The velocity $\vec{v}$ of a particle moving in the xy plane is given by $\vec{v} = (6.0t - 4.0t^2)\hat{i} + 8.0\hat{j}$, with $\vec{v}$ in meters per second and t (> 0) in seconds. (a) What is the acceleration when $t = 3.0$ s? (b) When (if ever) is the acceleration zero? (c) When (if ever) is the velocity zero? (d) When (if ever) does the speed equal 10 m/s?

15P. A particle starts from the origin at $t = 0$ with a velocity of $8.0\hat{j}$ m/s and moves in the xy plane with a constant acceleration of $(4.0\hat{i} + 2.0\hat{j})$ m/s^2. At the instant the particle's x coordinate is 29 m, what are (a) its y coordinate and (b) its speed?

16P. Particle A moves along the line $y = 30$ m with a constant velocity $\vec{v}$ of magnitude 3.0 m/s and directed parallel to the positive x axis (Fig. 4-31). Particle B starts at the origin with zero speed and constant acceleration $\vec{a}$ (of magnitude 0.40 m/s^2) at the same instant that particle A passes the y axis. What angle θ between $\vec{a}$ and the positive y axis would result in a collision between these two particles? (If your computation involves an equation with a term such as t^4, substitute $u = t^2$ and then consider solving the resulting quadratic equation to get u.)

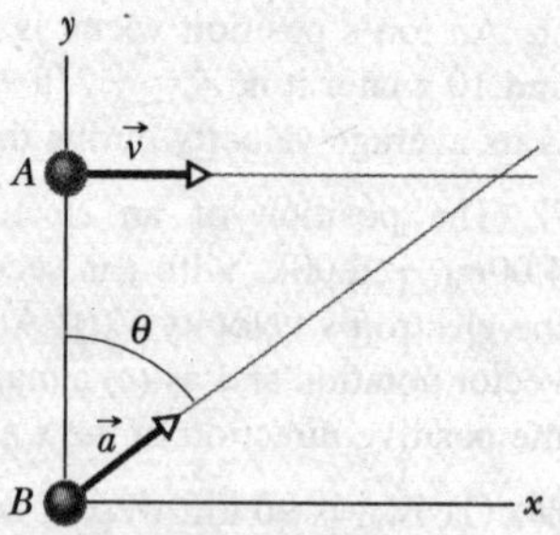

Fig. 4-31 Problem 16.

SEC. 4-6 Projectile Motion Analyzed

In some of these problems, exclusion of the effects of the air is unwarranted but helps simplify the calculations.

17E. A rifle is aimed horizontally at a target 30 m away. The bullet hits the target 1.9 cm below the aiming point. What are (a) the bullet's time of flight and (b) its speed as it emerges from the rifle? ssm

18E. A small ball rolls horizontally off the edge of a tabletop that is 1.20 m high. It strikes the floor at a point 1.52 m horizontally away from the edge of the table. (a) How long is the ball in the air? (b) What is its speed at the instant it leaves the table?

19E. A baseball leaves a pitcher's hand horizontally at a speed of 161 km/h. The distance to the batter is 18.3 m. (Ignore the effect of air resistance.) (a) How long does the ball take to travel the first half of that distance? (b) The second half? (c) How far does the ball fall freely during the first half? (d) During the second half? (e) Why aren't the quantities in (c) and (d) equal?

20E. A dart is thrown horizontally with an initial speed of 10 m/s toward point P, the bull's-eye on a dart board. It hits at point Q on the rim, vertically below P, 0.19 s later. (a) What is the distance PQ? (b) How far away from the dart board is the dart released?

21E. An electron, with an initial horizontal velocity of magnitude 1.00×10^9 cm/s, travels into the region between two horizontal metal plates that are electrically charged. In that region, it travels a horizontal distance of 2.00 cm and has a constant downward acceleration of magnitude 1.00×10^{17} cm/s^2 due to the charged plates. Find (a) the time required by the electron to travel the 2.00 cm and (b) the vertical distance it travels during that time. Also find the magnitudes of the (c) horizontal and (d) vertical velocity components of the electron as it emerges. ssm

22E. In the 1991 World Track and Field Championships in Tokyo, Mike Powell (Fig. 4-32) jumped 8.95 m, breaking the 23-year long-jump record set by Bob Beamon by a full 5 cm. Assume that Powell's speed on takeoff was 9.5 m/s (about equal to that of a sprinter) and that $g = 9.80$ m/s^2 in Tokyo. How much less was Powell's horizontal range than the maximum possible horizontal range (neglecting the effects of air) for a particle launched at the same speed of 9.5 m/s?

Fig. 4-32 Exercise 22.

23E. A stone is catapulted at time $t = 0$, with an initial velocity of magnitude 20.0 m/s and at an angle of 40.0° above the horizontal. What are the magnitudes of the (a) horizontal and (b) vertical components of its displacement from the catapult site at $t = 1.10$ s? Repeat for the (c) horizontal and (d) vertical components at $t = 1.80$ s, and for the (e) horizontal and (f) vertical components at $t = 5.00$ s. ssm

24P. A golf ball is struck at ground level. The speed of the golf ball as a function of the time is shown in Fig. 4-33, where $t = 0$ at the instant the ball is struck. (a) How far does the golf ball travel horizontally before returning to ground level? (b) What is the maximum height above ground level attained by the ball?

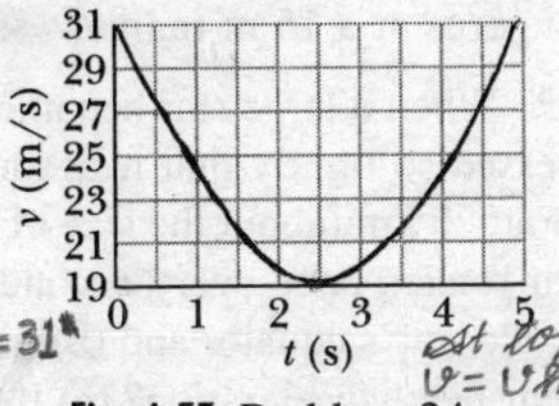

Fig. 4-33 Problem 24.

25P. A rifle that shoots bullets at 460 m/s is to be aimed at a target 45.7 m away and level with the rifle. How high above the target must the rifle barrel be pointed so that the bullet hits the target? ssm

26P. The pitcher in a slow-pitch softball game releases the ball at a point 3.0 ft above ground level. A stroboscopic plot of the position of the ball is shown in Fig. 4-34, where the readings are 0.25 s apart and the ball is released at $t = 0$. (a) What is the initial speed of the ball? (b) What is the speed of the ball at the instant it reaches its maximum height above ground level? (c) What is that maximum height?

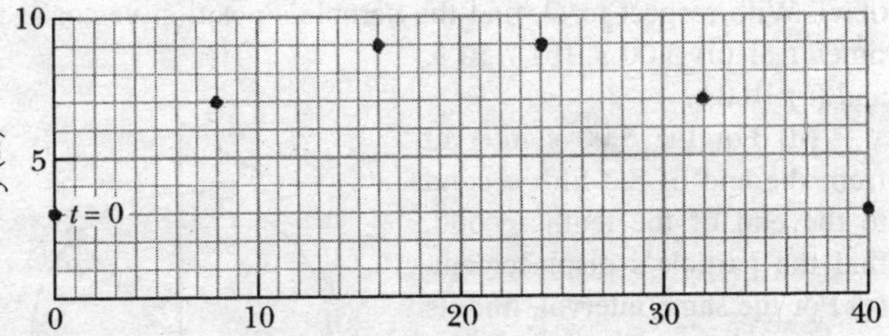

Fig. 4-34 Problem 26.

27P. Show that the maximum height reached by a projectile is $y_{max} = (v_0 \sin \theta_0)^2/2g$. ssm www

28P. You throw a ball toward a wall with a speed of 25.0 m/s and at an angle of 40.0° above the horizontal (Fig. 4-35). The wall is 22.0 m from the release point of the ball. (a) How far above the release point does the ball hit the wall? (b) What are the horizontal and vertical components of its velocity as it hits the wall? (c) When it hits, has it passed the highest point on its trajectory?

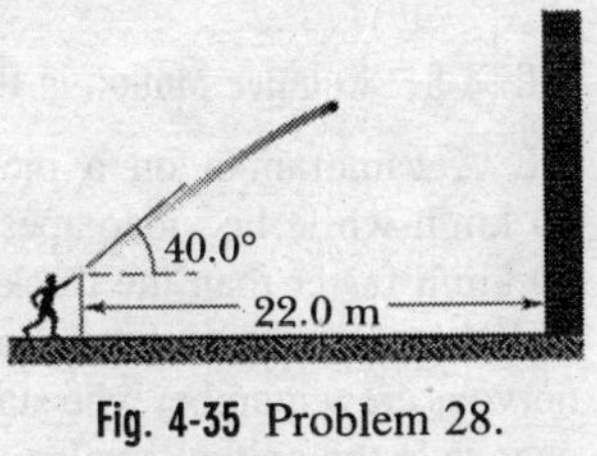

Fig. 4-35 Problem 28.

29P. A ball is shot from the ground into the air. At a height of 9.1 m, its velocity is observed to be $\vec{v} = 7.6\hat{i} + 6.1\hat{j}$ in meters per second ($\hat{i}$ horizontal, $\hat{j}$ upward). (a) To what maximum height does the ball rise? (b) What total horizontal distance does the ball travel? What are (c) the magnitude and (d) the direction of the ball's velocity just before it hits the ground? ilw

30P. Two seconds after being projected from ground level, a projectile is displaced 40 m horizontally and 53 m vertically above its point of projection. What are the (a) horizontal and (b) vertical components of the initial velocity of the projectile? (c) At the instant the projectile achieves its maximum height above ground level, how far is it displaced horizontally from its point of projection?

31P. A football player punts the football so that it will have a "hang time" (time of flight) of 4.5 s and land 46 m away. If the ball leaves the player's foot 150 cm above the ground, what must be (a) the magnitude and (b) the direction of the ball's initial velocity? ssm

32P. A lowly high diver pushes off horizontally with a speed of 2.00 m/s from the edge of a platform that is 10.0 m above the surface of the water. (a) At what horizontal distance from the edge of the platform is the diver 0.800 s after pushing off? (b) At what vertical distance above the surface of the water is the diver just then? (c) At what horizontal distance from the edge of the platform does the diver strike the water?

33P. A certain airplane has a speed of 290.0 km/h and is diving at an angle of 30.0° below the horizontal when the pilot releases a radar decoy (Fig. 4-36). The horizontal distance between the release point and the point where the decoy strikes the ground is 700 m. (a) How long is the decoy in the air? (b) How high was the released point? ilw

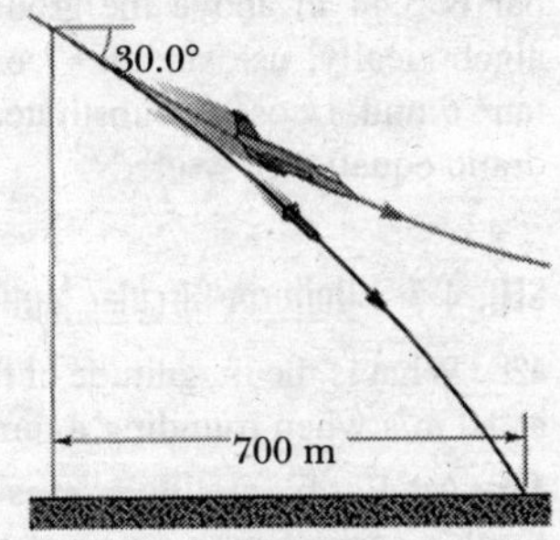

Fig. 4-36 Problem 33.

34P. The launching speed of a certain projectile is five times the speed it has at its maximum height. Calculate the elevation angle θ_0 at launching.

35P. A ball rolls horizontally off the top of a stairway with a speed of 1.52 m/s. The steps are 20.3 cm high and 20.3 cm wide. Which step does the ball hit first? ssm

36P. A soccer ball is kicked from the ground with an initial speed of 19.5 m/s at an upward angle of 45°. A player 55 m away in the direction of the kick starts running to meet the ball at that instant. What must be his average speed if he is to meet the ball just before it hits the ground? Neglect air resistance.

37P. An airplane, diving at an angle of 53.0° with the vertical, releases a projectile at an altitude of 730 m. The projectile hits the ground 5.00 s after being released. (a) What is the speed of the aircraft? (b) How far did the projectile travel horizontally during its flight? What were the (c) horizontal and (d) vertical components of its velocity just before striking the ground? ssm

38P. For women's volleyball the top of the net is 2.24 m above the floor and the court measures 9.0 m by 9.0 m on each side of the net. Using a jump serve, a player strikes the ball at a point that is 3.0 m above the floor and a horizontal distance of 8.0 m from the net. If the initial velocity of the ball is horizontal, (a) what minimum magnitude must it have if the ball is to clear the net and (b) what maximum magnitude can it have if the ball is to strike the floor inside the back line on the other side of the net?

39P. A batter hits a pitched ball when the center of the ball is 1.22 m above the ground. The ball leaves the bat at an angle of 45° with the ground. With that launch, the ball should have a horizontal range (returning to the *launch* level) of 107 m. (a) Does the ball clear a 7.32-m-high fence that is 97.5 m horizontally from the launch point? (b) Either way, find the distance between the top of the fence and the center of the ball when the ball reaches the fence. ssm www

40P. During a tennis match, a player serves the ball at 23.6 m/s, with the center of the ball leaving the racquet horizontally 2.37 m above the court surface. The net is 12 m away and 0.90 m high. When the ball reaches the net, (a) does the ball clear it and (b) what is the distance between the center of the ball and the top of the net? Suppose that, instead, the ball is served as before but now it leaves the racquet at 5.00° below the horizontal. When the ball reaches the net, (c) does the ball clear it and (d) what now is the distance between the center of the ball and the top of the net?

41P. A football kicker can give the ball an initial speed of 25 m/s. Within what two elevation angles must he kick the ball to score a field goal from a point 50 m in front of goalposts whose horizontal bar is 3.44 m above the ground? (If you want to work this out algebraically, use $\sin^2\theta + \cos^2\theta = 1$ to get a relation between $\tan^2\theta$ and $1/\cos^2\theta$, substitute, and then solve the resulting quadratic equation.) ssm

SEC. 4-7 Uniform Circular Motion

42E. What is the magnitude of the acceleration of a sprinter running at 10 m/s when rounding a turn with a radius of 25 m?

43E. An Earth satellite moves in a circular orbit 640 km above Earth's surface with a period of 98.0 min. What are (a) the speed and (b) the magnitude of the centripetal acceleration of the satellite? ssm

44E. A rotating fan completes 1200 revolutions every minute. Consider the tip of a blade, at a radius of 0.15 m. (a) Through what distance does the tip move in one revolution? What are (b) the tip's speed and (c) the magnitude of its acceleration? (d) What is the period of the motion?

45E. An astronaut is rotated in a horizontal centrifuge at a radius of 5.0 m. (a) What is the astronaut's speed if the centripetal acceleration has a magnitude of $7.0g$? (b) How many revolutions per minute are required to produce this acceleration? (c) What is the period of the motion? ssm

46P. A carnival merry-go-round rotates about a vertical axis at a constant rate. A passenger standing on the edge of the merry-go-round has a constant speed of 3.66 m/s. For each of the following instantaneous situations, state how far the passenger is from the center of the merry-go-round, and in which direction. (a) The passenger has an acceleration of 1.83 m/s^2, east. (b) The passenger has an acceleration of 1.83 m/s^2, south.

47P. (a) What is the magnitude of the centripetal acceleration of an object on Earth's equator owing to the rotation of Earth? (b) What would the period of rotation of Earth have to be for objects on the equator to have a centripetal acceleration with a magnitude of 9.8 m/s^2? ssm www

48P. The fast French train known as the TGV (Train à Grande Vitesse) has a scheduled average speed of 216 km/h. (a) If the train goes around a curve at that speed and the magnitude of the acceleration experienced by the passengers is to be limited to $0.050g$, what is the smallest radius of curvature for the track that can be tolerated? (b) At what speed must the train go around a curve with a 1.00 km radius to be at the acceleration limit?

49P. A carnival Ferris wheel has a 15 m radius and completes five turns about its horizontal axis every minute. (a) What is the period of the motion? What is the centripetal acceleration of a passenger at (b) the highest point and (c) the lowest point, assuming the passenger is at a 15 m radius? ssm ilw

50P. When a large star becomes a *supernova,* its core may be compressed so tightly that it becomes a *neutron star,* with a radius of about 20 km (about the size of the San Francisco area). If a neutron star rotates once every second, (a) what is the speed of a particle on the star's equator and (b) what is the magnitude of the particle's centripetal acceleration? (c) If the neutron star rotates faster, do the answers to (a) and (b) increase, decrease, or remain the same?

51P. A boy whirls a stone in a horizontal circle of radius 1.5 m and at height 2.0 m above level ground. The string breaks, and the stone flies off horizontally and strikes the ground after traveling a horizontal distance of 10 m. What is the magnitude of the centripetal acceleration of the stone while in circular motion? ssm

52P. A particle P travels with constant speed on a circle of radius $r = 3.00$ m (Fig. 4-37) and completes one revolution in 20.0 s. The particle passes through O at time $t = 0$. State the following vectors in magnitude-angle notation (angle relative to the positive direction of x). With respect to O, find the particle's position vector at the times t of (a) 5.00 s, (b) 7.50 s, and (c) 10.0 s.

(d) For the 5.00 s interval from the end of the fifth second to the end of the tenth second, find the particle's displacement. (e) For the same interval, find its average velocity. Find its velocity at (f) the beginning and (g) the end of that 5.00 s interval. Next, find the acceleration at (h) the beginning and (i) the end of that interval.

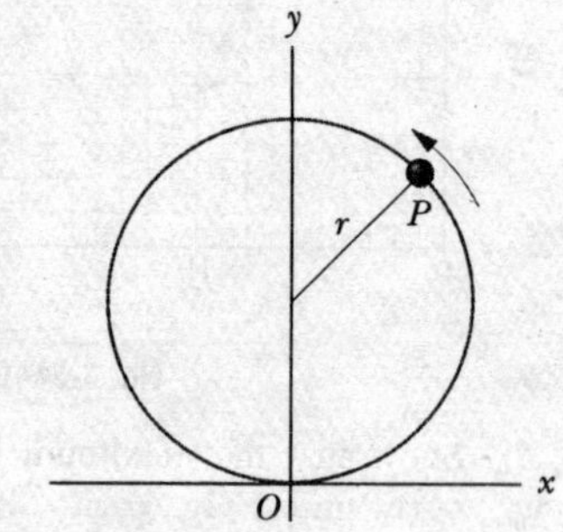

Fig. 4-37 Problem 52.

SEC. 4-8 Relative Motion in One Dimension

53E. A cameraman on a pickup truck is traveling westward at 20 km/h while he videotapes a cheetah that is moving westward 30 km/h faster than the truck. Suddenly, the cheetah stops, turns, and then runs at 45 km/h eastward, as measured by a suddenly nervous crew member who stands alongside the cheetah's path. The change in the animal's velocity takes 2.0 s. What is its acceleration from the perspective of (a) the cameraman and (b) the nervous crew member?

54E. A boat is traveling upstream at 14 km/h with respect to the water of a river. The water is flowing at 9 km/h with respect to the ground. (a) What is the velocity of the boat with respect to the ground? (b) A child on the boat walks from front to rear at 6 km/h with respect to the boat. What is the child's velocity with respect to the ground?

55P. A person walks up a stalled 15-m-long escalator in 90 s. When standing on the same escalator, now moving, the person is carried up in 60 s. How much time would it take that person to walk up the moving escalator? Does the answer depend on the length of the escalator? ssm

SEC. 4-9 Relative Motion in Two Dimensions

56E. In rugby a player can legally pass the ball to a teammate as long as the pass is not "forward" (it must not have a velocity component parallel to the length of the field and directed toward the other team's goal). Suppose a player runs parallel to the field's length and toward the other team's goal with a speed of 4.0 m/s while he passes the ball with a speed of 6.0 m/s relative to himself. What is the smallest angle from the forward direction that keeps the pass legal?

57E. Snow is falling vertically at a constant speed of 8.0 m/s. At what angle from the vertical do the snowflakes appear to be falling as viewed by the driver of a car traveling on a straight, level road with a speed of 50 km/h? ssm

58E. Two highways intersect as shown in Fig. 4-38. At the instant shown, a police car P is 800 m from the intersection and moving at 80 km/h. Motorist M is 600 m from the intersection and moving at 60 km/h. (a) In unit-vector notation, what is the velocity of the motorist with respect to the police car? (b) For the instant shown in Fig. 4-38, how does the direction of the velocity found in (a) compare to the line of sight between the two cars? (c) If the cars maintain their velocities, do the answers to (a) and (b) change as the cars move nearer the intersection?

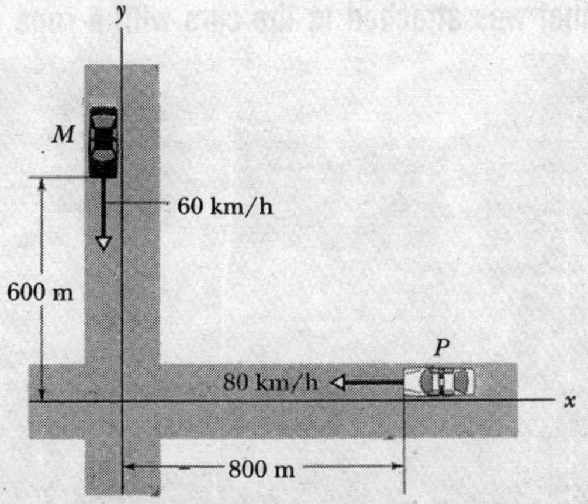

Fig. 4-38 Exercise 58.

59P. A train travels due south at 30 m/s (relative to the ground) in a rain that is blown toward the south by the wind. The path of each raindrop makes an angle of 70° with the vertical, as measured by an observer stationary on the ground. An observer on the train, however, sees the drops fall perfectly vertically. Determine the speed of the raindrops relative to the ground. ssm

60P. Ship A is located 4.0 km north and 2.5 km east of ship B. Ship A has a velocity of 22 km/h toward the south and ship B has a velocity of 40 km/h in a direction 37° north of east. (a) What is the velocity of A relative to B? (Express your answer in terms of the unit vectors $\hat{\mathrm{i}}$ and $\hat{\mathrm{j}}$, where $\hat{\mathrm{i}}$ is toward the east.) (b) Write an expression (in terms of $\hat{\mathrm{i}}$ and $\hat{\mathrm{j}}$) for the position of A relative to B as a function of t, where $t = 0$ when the ships are in the positions described above. (c) At what time is the separation between the ships least? (d) What is that least separation?

61P. Two ships, A and B, leave port at the same time. Ship A travels northwest at 24 knots and ship B travels at 28 knots in a direction 40° west of south. (1 knot = 1 nautical mile per hour; see Appendix D.) (a) What are the magnitude and direction of the velocity of ship A relative to B? (b) After what time will the ships be 160 nautical miles apart? (c) What will be the bearing of B (the direction of the position of B) relative to A at that time? ssm ilw

62P. A wooden boxcar is moving along a straight railroad track at speed v_1. A sniper fires a bullet (initial speed v_2) at it from a high-powered rifle. The bullet passes through both lengthwise walls of the car, its entrance and exit holes being exactly opposite each other as viewed from within the car. From what direction, relative to the track, is the bullet fired? Assume that the bullet is not deflected upon entering the car, but that its speed decreases by 20%. Take $v_1 = 85$ km/h and $v_2 = 650$ m/s. (Why don't you need to know the width of the boxcar?)

63P. A 200-m-wide river has a uniform flow speed of 1.1 m/s through a jungle and toward the east. An explorer wishes to leave a small clearing on the south bank and cross the river in a powerboat that moves at a constant speed of 4.0 m/s with respect to the water. There is a clearing on the north bank 82 m upstream from a point directly opposite the clearing on the south bank. (a) In what direction must the boat be pointed in order to travel in a straight line and land in the clearing on the north bank? (b) How long will the boat take to cross the river and land in the clearing?

Additional Problem

64. *Curtain of death.* A large metallic asteroid strikes Earth and quickly digs a crater into the rocky material below ground level by launching rocks upward and outward. The following table gives five pairs of launch speeds and angles (from the horizontal) for such rocks, based on a model of crater formation. (Other rocks, with intermediate speeds and angles, are also launched.) Suppose that you are at $x = 20$ km when the asteroid strikes the ground at time $t = 0$ and position $x = 0$ (Fig. 4-39). (a) At $t = 20$ s, what are the x and y coordinates of the rocks headed in your direction from launches A through E? (b) Plot these coordinates and then sketch a curve through the points to include rocks with intermediate launch speeds and angles. The curve should give you an idea of what you would see as you look up into the approaching rocks and what dinosaurs must have seen during asteroid strikes long ago.

Launch	Speed (m/s)	Angle (degrees)
A	520	14.0
B	630	16.0
C	750	18.0
D	870	20.0
E	1000	22.0

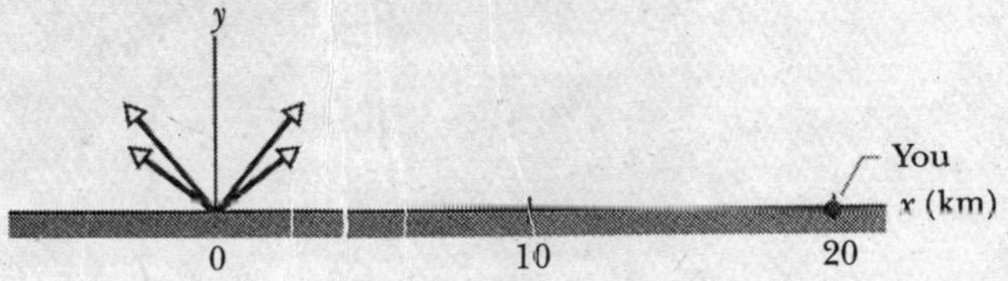

Fig. 4-39 Problem 64.

5 Force and Motion—I

On April 4, 1974, John Massis of Belgium managed to move two passenger cars belonging to New York's Long Island Railroad. He did so by clamping his teeth down on a bit that was attached to the cars with a rope and then leaning backward while pressing his feet against the railway ties. The cars together weighed about 80 tons.

Did Massis have to pull with superhuman force to accelerate them?

The answer is in this chapter.

5-1 What Causes an Acceleration?

If you see the velocity of a particle-like body change in either magnitude or direction, you know that something must have *caused* that change (that acceleration). Indeed, out of common experience, you know that the change in velocity must be due to an interaction between the body and something in its surroundings. For example, if you see a hockey puck that is sliding across an ice rink suddenly stop or suddenly change direction, you will suspect that the puck encountered a slight ridge on the ice surface.

An interaction that can cause an acceleration of a body is called a **force**, which is, loosely speaking, a push or pull on the body—the force is said to *act* on the body. For example, the bump on the hockey puck by the ridge is a push on the puck, causing an acceleration. The relationship between a force and the acceleration it causes was first understood by Isaac Newton (1642–1727) and is the subject of this chapter. The study of that relationship, as Newton presented it, is called *Newtonian mechanics*. We shall focus on its three primary laws of motion.

Newtonian mechanics does not apply to all situations. If the speeds of the interacting bodies are very large—an appreciable fraction of the speed of light—we must replace Newtonian mechanics with Einstein's special theory of relativity, which holds at any speed, including those near the speed of light. If the interacting bodies are on the scale of atomic structure (for example, they might be electrons within an atom), we must replace Newtonian mechanics with quantum mechanics. Physicists now view Newtonian mechanics as a special case of these two more comprehensive theories. Still, it is a very important special case because it applies to the motion of objects ranging in size from the very small (almost on the scale of atomic structure) to astronomical (objects such as galaxies and clusters of galaxies).

5-2 Newton's First Law

Before Newton formulated his mechanics, it was thought that some influence, a "force," was needed to keep a body moving at constant velocity. Similarly, a body was thought to be in its "natural state" when it was at rest. For it to move with constant velocity, it seemingly had to be propelled in some way, by a push or a pull. Otherwise, it would "naturally" stop moving.

These ideas were reasonable. If you send a puck sliding across a wooden floor, it does indeed slow and then stop. If you want to make it move across the floor with constant velocity, you have to continuously pull or push it.

Send a puck sliding over the ice of a skating rink, however, and it goes a lot farther. You can imagine longer and more slippery surfaces, over which the puck would slide farther and farther. In the limit you can think of a long, extremely slippery surface (said to be a **frictionless surface**), over which the puck would hardly slow. (We can in fact come close to this situation in the laboratory, by sending a puck sliding over a horizontal air table, across which it moves on a film of air.)

From these observations, we can conclude that a body will keep moving with constant velocity if no force acts on it. That leads us to the first of Newton's three laws of motion:

> **Newton's First Law:** If no force acts on a body, then the body's velocity cannot change; that is, the body cannot accelerate.

In other words, if the body is at rest, it stays at rest. If it is moving, it will continue to move with the same velocity (same magnitude *and* same direction).

5-3 Force

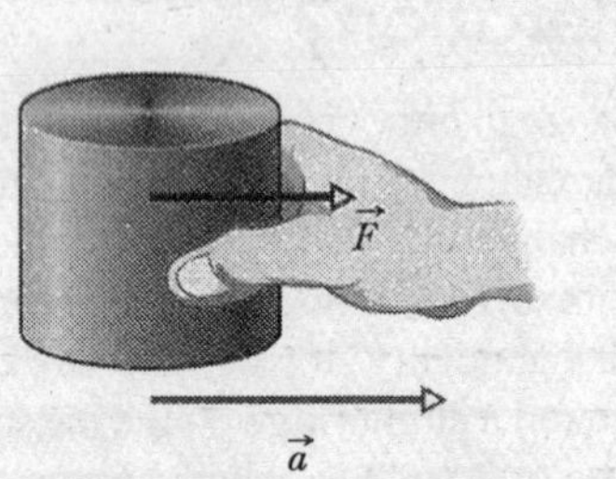

Fig. 5-1 A force $\vec{F}$ on the standard kilogram gives that body an acceleration $\vec{a}$.

We now wish to define the unit of force. We know that a force can cause the acceleration of a body. Thus, we shall define the unit of force in terms of the acceleration that a force gives to a standard reference body. As the standard body, we shall use (or rather imagine that we use) the standard kilogram of Fig. 1-4. This body has been assigned, exactly and by definition, a mass of 1 kg.

We put the standard body on a horizontal frictionless table and pull the body to the right (Fig. 5-1) so that by trial and error, it eventually experiences a measured acceleration of 1 m/s^2. We then declare, as a matter of definition, that the force we are exerting on the standard body has a magnitude of 1 newton (abbreviated N).

We can exert a 2 N force on our standard body by pulling it so that its measured acceleration is 2 m/s^2, and so on. Thus in general, if our standard body of 1 kg mass has an acceleration of magnitude a, we know that force F must be acting on it and that the magnitude of the force (in newtons) is equal to the magnitude of the acceleration (in meters per second per second).

Thus, a force is measured by the acceleration it produces. However, acceleration is a vector quantity, with both magnitude and direction. Is force also a vector quantity? We can easily assign a direction to a force (just assign the direction of the acceleration), but that is not sufficient. We must prove by experiment that forces are vector quantities. Actually, that has been done: forces are indeed vector quantities; they have magnitudes and directions and they combine according to the vector rules of Chapter 3.

This means that when two or more forces act on a body, we can find their **net force** or **resultant force** by adding the individual forces vectorially. A single force with the magnitude and direction of the net force has the same effect on the body as all the individual forces together. This fact is called the **principle of superposition for forces.** The world would be quite strange if, for example, you and a friend were to pull on the standard body in the same direction, each with a force of 1 N, and yet somehow the net pull was 14 N.

In this book, forces are most often represented with a vector symbol such as $\vec{F}$, and a net force is represented with the vector symbol $\vec{F}_{net}$. As with other vectors, a force or a net force can have components along coordinate axes. When forces act only along a single axis, they are single-component forces. Then we can drop the overhead arrows on the force symbols and just use signs to indicate the directions of the forces along that axis.

Instead of what was previously given, the proper statement of Newton's First Law is in terms of a *net* force:

> **Newton's First Law:** If no *net* force acts on a body ($\vec{F}_{net} = 0$), then the body's velocity cannot change; that is, the body cannot accelerate.

There may be multiple forces acting on a body, but if their net (or resultant) force is zero, then the body cannot accelerate.

Inertial Reference Frames

Newton's first law is not true in all reference frames, but we can always find reference frames in which it (and the rest of Newtonian mechanics) is true. Such frames are called **inertial reference frames,** or simply **inertial frames.**

> An inertial reference frame is one in which Newton's laws hold.

For example, we can assume that the ground is an inertial frame provided that we can neglect Earth's actual astronomical motions (such as its rotation).

That assumption works well if we, say, send a puck sliding along a short strip of frictionless ice—an observer on the ground would find that the puck's motion obeys the laws of Newtonian mechanics. However, suppose we made the strip very long, extending, say, from north to south. Then an observer on the ground would find that the puck accelerates slightly toward the west as it moves south (Fig. 5-2*a*); yet, that observer would not be able to find a force that causes the westward acceleration. In this case, the ground is a **noninertial frame** because, for the puck's long path, Earth's rotation cannot be neglected. The surprising westward acceleration of the sliding puck relative to the ground is actually due to the eastward rotation of the ground beneath the puck (Fig. 5-2*b*).

In this book we usually assume that the ground is an inertial frame and that measurements of forces and accelerations are made from it. If measurements are made in, say, an elevator that is accelerating relative to the ground, then the measurements are being made in a noninertial frame and the results can be surprising. We see an example of this in Sample Problem 5-8.

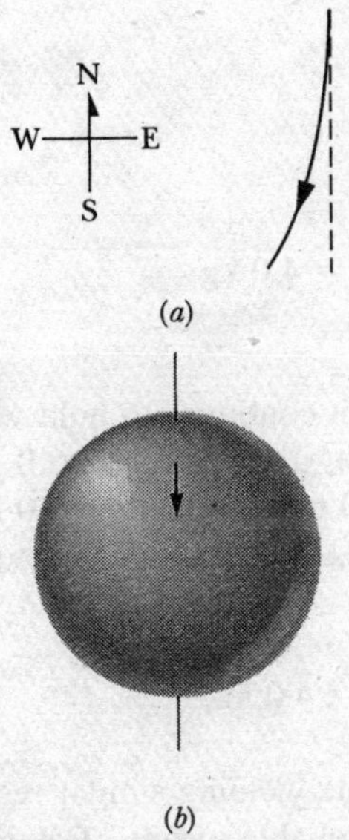

Fig. 5-2 (*a*) The path of a puck that is sent sliding due south along a long strip of frictionless ice, as seen by an observer on the ground. (*b*) The ground beneath the southward sliding puck rotates to the east as Earth rotates.

CHECKPOINT 1: Which of the figure's six arrangements correctly show the vector addition of forces $\vec{F}_1$ and $\vec{F}_2$ to yield the third vector, which is meant to represent their net force $\vec{F}_{net}$?

F1 & F2 must be head to tail
Fnet must be from tail of one of them to head of another

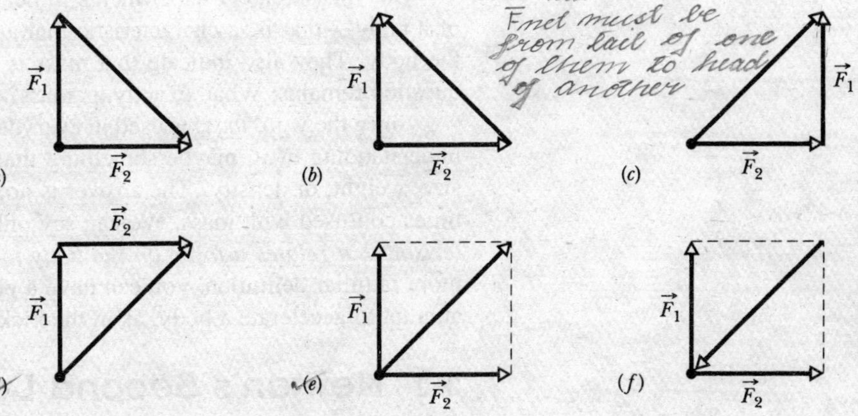

5-4 Mass

Everyday experience tells us that a given force produces different magnitudes of acceleration for different bodies. Put a baseball and a bowling ball on the floor and give both the same sharp kick. Even if you don't actually do this, you know the result: The baseball receives a noticeably larger acceleration than the bowling ball. The two accelerations differ because the mass of the baseball differs from the mass of the bowling ball—but what, exactly, is mass?

We can explain how to measure mass by imagining a series of experiments in an inertial frame. In the first experiment we exert a force on our standard body, whose mass m_0 is defined to be 1.0 kg. Suppose that the standard body accelerates at 1.0 m/s^2. We can then say the force on that body is 1.0 N.

We next apply that same force (we would need some way of being certain it is the same force) to a second body, body X, whose mass is not known. Suppose we find that this body X accelerates at 0.25 m/s^2. We know that a *less massive* baseball receives a *greater acceleration* than a more massive bowling ball when the same force (kick) is applied to both. Let us then make the following conjecture: The ratio of the masses of two bodies is equal to the inverse of the ratio of their accelerations when the same force is applied to both. For body X and the standard body, this tells us that

$$\frac{m_X}{m_0} = \frac{a_0}{a_X}.$$

Solving for m_X yields

$$m_X = m_0 \frac{a_0}{a_X} = (1.0\text{ kg})\frac{1.0\text{ m/s}^2}{0.25\text{ m/s}^2} = 4.0\text{ kg}.$$

Our conjecture will be useful, of course, only if it continues to hold when we change the applied force to other values. For example, if we apply an 8.0 N force to the standard body, we obtain an acceleration of 8.0 m/s^2. When the 8.0 N force is applied to body X, we obtain an acceleration of 2.0 m/s^2. Our conjecture then gives us

$$m_X = m_0 \frac{a_0}{a_X} = (1.0\text{ kg})\frac{8.0\text{ m/s}^2}{2.0\text{ m/s}^2} = 4.0\text{ kg},$$

consistent with our first experiment. Many experiments yielding similar results indicate that our conjecture provides a consistent and reliable means of assigning a mass to any given body.

Our measurement experiments indicate that mass is an *intrinsic* characteristic of a body—that is, a characteristic that automatically comes with the existence of the body. They also indicate that mass is a scalar quantity. However, the nagging question remains: What, exactly, is mass?

Since the word *mass* is used in everyday English, we should have some intuitive understanding of it, maybe something that we can physically sense. Is it a body's size, weight, or density? The answer is no, although those characteristics are sometimes confused with mass. We can say only that *the mass of a body is the characteristic that relates a force on the body to the resulting acceleration.* Mass has no more familiar definition; you can have a physical sensation of mass only when you attempt to accelerate a body, as in the kicking of a baseball or a bowling ball.

5-5 Newton's Second Law

All the definitions, experiments, and observations that we have discussed so far can be summarized in one neat statement:

Newton's Second Law: The net force on a body is equal to the product of the body's mass and the acceleration of the body.

In equation form,

$$\vec{F}_{\text{net}} = m\vec{a} \quad \text{(Newton's second law).} \tag{5-1}$$

This equation is simple, but we must use it cautiously. First, we must be certain about which body we are applying it to. Then $\vec{F}_{\text{net}}$ must be the vector sum of *all* the forces that act on *that* body. Only forces that act on *that* body are to be included in the vector sum, not forces acting on other bodies that might be involved in the given situation. For example, if you are in a rugby scrum, the net force on *you* is the vector sum of all the pushes and pulls on *your* body. It does not include any push or pull on another player from you.

Like other vector equations, Eq. 5-1 is equivalent to three component equations, one written for each axis of an *xyz* coordinate system:

$$F_{\text{net},x} = ma_x, \quad F_{\text{net},y} = ma_y, \quad \text{and} \quad F_{\text{net},z} = ma_z. \tag{5-2}$$

TABLE 5-1 Units in Newton's Second Law (Eqs. 5-1 and 5-2)

System	Force	Mass	Acceleration
SI	newton (N)	kilogram (kg)	m/s^2
CGS[a]	dyne	gram (g)	cm/s^2
British[b]	pound (lb)	slug	ft/s^2

[a] 1 dyne = 1 $g \cdot cm/s^2$.
[b] 1 lb = 1 $slug \cdot ft/s^2$.

Each of these equations relates the net force component along an axis to the acceleration along that same axis. For example, the first equation tells us that the sum of all the force components along the x axis causes the x component a_x of the body's acceleration, but causes no acceleration in the y and z directions. Turned around, the acceleration component a_x is caused only by the sum of the force components along the x axis. In general,

> The acceleration component along a given axis is caused only by the sum of the force components along that *same* axis, and not by force components along any other axis.

Equation 5-1 tells us that if the net force on a body is zero, the body's acceleration $\vec{a} = 0$. If the body is at rest, it stays at rest; if it is moving, it continues to move at constant velocity. In such cases, any forces on the body *balance* one another, and both the forces and the body can be said to be in *equilibrium*. Commonly, the forces are also said to *cancel* one another, but the term "cancel" is tricky. It does *not* mean that the forces cease to exist (canceling forces is not like canceling dinner reservations). The forces still act on the body.

For SI units, Eq. 5-1 tells us that

$$1 \text{ N} = (1 \text{ kg})(1 \text{ m/s}^2) = 1 \text{ kg} \cdot \text{m/s}^2. \tag{5-3}$$

Some force units in other systems of units are given in Table 5-1 and Appendix D.

To solve problems with Newton's second law, we often draw a **free-body diagram** in which the only body shown is the one for which we are summing forces. A sketch of the body itself is preferred by some teachers but, to save space in these chapters, we shall usually represent the body with a dot. Also, each force on the body is drawn as a vector arrow with its tail on the body. A coordinate system is usually included, and the acceleration of the body is sometimes shown with a vector arrow (labeled as an acceleration).

A collection of two or more bodies is called a **system,** and any force on the bodies inside the system from bodies outside the system is called an **external force.** If the bodies are rigidly connected, then we can treat the system as one composite body, and the net force $\vec{F}_{net}$ on it is the vector sum of all external forces. (We do not include **internal forces**—that is, forces between two bodies inside the system.) For example, a connected railroad engine and car form a system. If, say, a tow line pulls on the front of the engine, then the force due to the tow line acts on the whole engine–car system. Just as for a single body, we can relate the net external force on a system to its acceleration with Newton's second law, $\vec{F}_{net} = m\vec{a}$, where m is the total mass of the system.

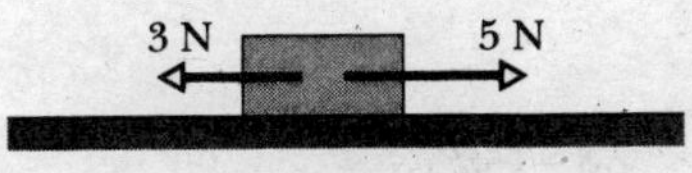

CHECKPOINT 2: The figure here shows two horizontal forces acting on a block that is on a frictionless floor. Assume that a third horizontal force $\vec{F}_3$ also acts on the block. What are the magnitude and direction of $\vec{F}_3$ when the block is (a) stationary and (b) moving to the left with a constant speed of 5 m/s?

Sample Problem 5-1

In Figs. 5-3*a* to *c*, one or two forces act on a puck that moves over frictionless ice and along an x axis, in one-dimensional motion. The puck's mass is $m = 0.20$ kg. Forces $\vec{F}_1$ and $\vec{F}_2$ are directed along the axis and have magnitudes $F_1 = 4.0$ N and $F_2 = 2.0$ N. Force $\vec{F}_3$ is directed at angle $\theta = 30°$ and has magnitude $F_3 = 1.0$ N. In each situation, what is the acceleration of the puck?

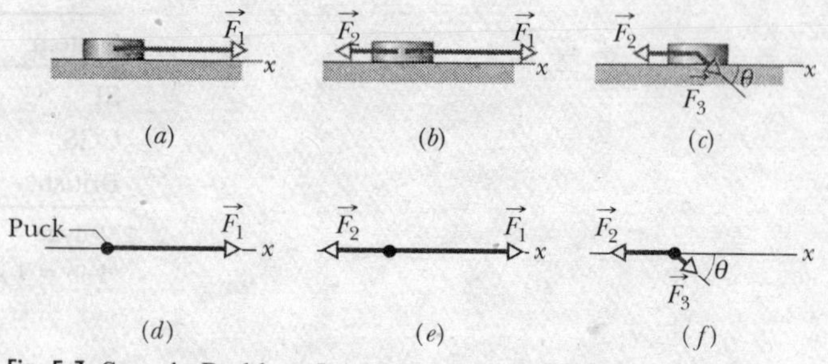

Fig. 5-3 Sample Problem 5-1. (*a*)–(*c*) In three situations, forces act on a puck that moves on frictionless ice along an x axis. (*d*)–(*f*) Free-body diagrams for the three situations.

SOLUTION: The **Key Idea** in each situation is that we can relate the acceleration $\vec{a}$ to the net force $\vec{F}_{net}$ acting on the puck with Newton's second law, $\vec{F}_{net} = m\vec{a}$. However, because the motion is along only the x axis, we can simplify each situation by writing the second law for x components only:

$$F_{net,x} = ma_x. \quad (5\text{-}4)$$

The free-body diagrams for the three situations are given in Figs. 5-3*d* to *f*, where the puck is represented by a dot.

For Fig. 5-3*d*, where only one horizontal force acts, Eq. 5-4 gives us

$$F_1 = ma_x,$$

which, with given data, yields

$$a_x = \frac{F_1}{m} = \frac{4.0\text{ N}}{0.20\text{ kg}} = 20\text{ m/s}^2. \quad \text{(Answer)}$$

The positive answer indicates that the acceleration is in the positive direction of the x axis.

In Fig. 5-3*e*, two horizontal forces act on the puck, $\vec{F}_1$ in the positive direction of x and $\vec{F}_2$ in the negative direction. Now Eq. 5-4 gives us

$$F_1 - F_2 = ma_x,$$

which, with given data, yields

$$a_x = \frac{F_1 - F_2}{m} = \frac{4.0\text{ N} - 2.0\text{ N}}{0.20\text{ kg}} = 10\text{ m/s}^2. \quad \text{(Answer)}$$

Thus, the net force accelerates the puck in the positive direction of the x axis.

In Fig. 5-3*f*, force $\vec{F}_3$ is not directed along the direction of the puck's acceleration; only the x component $F_{3,x}$ is. (Force $\vec{F}_3$ is two-dimensional but the motion is only one-dimensional.) Thus, we write Eq. 5-4 as

$$F_{3,x} - F_2 = ma_x.$$

From the figure, we see that $F_{3,x} = F_3 \cos\theta$. Solving for the acceleration and substituting for $F_{3,x}$ yield

$$a_x = \frac{F_{3,x} - F_2}{m} = \frac{F_3\cos\theta - F_2}{m}$$

$$= \frac{(1.0\text{ N})(\cos 30°) - 2.0\text{ N}}{0.20\text{ kg}} = -5.7\text{ m/s}^2. \quad \text{(Answer)}$$

Thus, the net force accelerates the puck in the negative direction of the x axis.

CHECKPOINT 3: The figure shows *overhead* views of four situations in which two forces accelerate the same block across a frictionless floor. Rank the situations according to the magnitudes of (a) the net force on the block and (b) the acceleration of the block, greatest first.

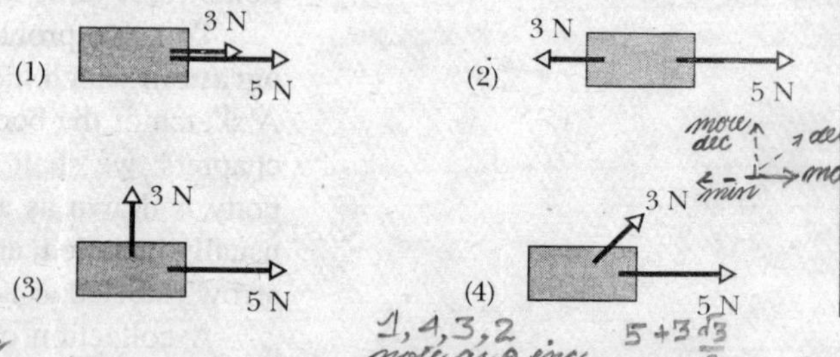

Sample Problem 5-2

In the overhead view of Fig. 5-4*a*, a 2.0 kg cookie tin is accelerated at 3.0 m/s² in the direction shown by $\vec{a}$, over a frictionless horizontal surface. The acceleration is caused by three horizontal forces, only two of which are shown: $\vec{F}_1$ of magnitude 10 N and $\vec{F}_2$ of magnitude 20 N. What is the third force $\vec{F}_3$ in unit-vector notation and as a magnitude and an angle?

SOLUTION: One **Key Idea** here is that the net force $\vec{F}_{net}$ on the tin is the sum of the three forces and is related to the acceleration $\vec{a}$ of the tin via Newton's second law ($\vec{F}_{net} = m\vec{a}$). Thus,

$$\vec{F}_1 + \vec{F}_2 + \vec{F}_3 = m\vec{a},$$

which gives us

$$\vec{F}_3 = m\vec{a} - \vec{F}_1 - \vec{F}_2. \quad (5\text{-}5)$$

A second **Key Idea** is that this is a two-dimensional problem; so we *cannot* find $\vec{F}_3$ merely by substituting the magnitudes for the vector quantities on the right side of Eq. 5-5. Instead, we must vectorially add $m\vec{a}$, $-\vec{F}_1$ (the reverse of $\vec{F}_1$), and $-\vec{F}_2$ (the reverse of $\vec{F}_2$), as is shown in Fig. 5-4*b*. This addition can be done directly on a vector-capable calculator because we know both magnitude and angle for each of these three vectors. However, here we shall evaluate the right side of Eq. 5-5 in terms of components, first along the x axis and then along the y axis.

Along the x axis we have

$$F_{3,x} = ma_x - F_{1,x} - F_{2,x}$$
$$= m(a\cos 50°) - F_1\cos(-150°) - F_2\cos 90°.$$

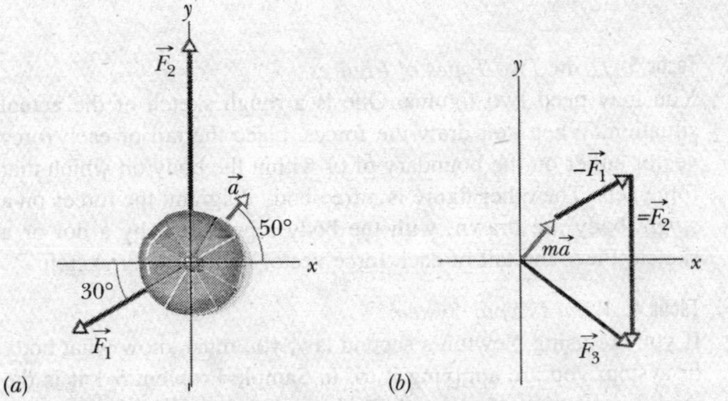

Fig. 5-4 Sample Problem 5-2. (*a*) An overhead view of two of three horizontal forces that act on a cookie tin, resulting in acceleration $\vec{a}$. $\vec{F}_3$ is not shown. (*b*) An arrangement of vectors $m\vec{a}$, $-\vec{F}_1$, and $-\vec{F}_2$ to find force $\vec{F}_3$.

Then, substituting known data, we find

$$\begin{aligned} F_{3,x} &= (2.0\ \text{kg})(3.0\ \text{m/s}^2)\cos 50^\circ - (10\ \text{N})\cos(-150^\circ) \\ &\quad - (20\ \text{N})\cos 90^\circ \\ &= 12.5\ \text{N}. \end{aligned}$$

Similarly, along the y axis we find

$$\begin{aligned} F_{3,y} &= ma_y - F_{1,y} - F_{2,y} \\ &= m(a\sin 50^\circ) - F_1\sin(-150^\circ) - F_2\sin 90^\circ \\ &= (2.0\ \text{kg})(3.0\ \text{m/s}^2)\sin 50^\circ - (10\ \text{N})\sin(-150^\circ) \\ &\quad - (20\ \text{N})\sin 90^\circ \\ &= -10.4\ \text{N}. \end{aligned}$$

Thus, in unit-vector notation, we have

$$\begin{aligned} \vec{F}_3 &= F_{3,x}\hat{\mathrm{i}} + F_{3,y}\hat{\mathrm{j}} = (12.5\ \text{N})\hat{\mathrm{i}} - (10.4\ \text{N})\hat{\mathrm{j}} \\ &\approx (13\ \text{N})\hat{\mathrm{i}} - (10\ \text{N})\hat{\mathrm{j}}. \end{aligned} \qquad \text{(Answer)}$$

We can now use a vector-capable calculator to get the magnitude and the angle of $\vec{F}_3$. We can also use Eq. 3-6 to obtain the magnitude and the angle (from the positive direction of the x axis) as

$$F_3 = \sqrt{F_{3,x}^2 + F_{3,y}^2} = 16\ \text{N}$$

and

$$\theta = \tan^{-1}\frac{F_{3,y}}{F_{3,x}} = -40^\circ. \qquad \text{(Answer)}$$

Sample Problem 5-3

In a two-dimensional tug-of-war, Alex, Betty, and Charles pull horizontally on an automobile tire at the angles shown in the overhead view of Fig. 5-5*a*. The tire remains stationary in spite of the three pulls. Alex pulls with force $\vec{F}_A$ of magnitude 220 N, and Charles pulls with force $\vec{F}_C$ of magnitude 170 N. The direction of $\vec{F}_C$ is not given. What is the magnitude of Betty's force $\vec{F}_B$?

SOLUTION: Because the three forces pulling on the tire do not accelerate the tire, the tire's acceleration is $\vec{a} = 0$ (that is, the forces are in equilibrium). The **Key Idea** here is that we can relate that acceleration to the net force $\vec{F}_{\text{net}}$ on the tire with Newton's second law ($\vec{F}_{\text{net}} = m\vec{a}$), which we can write as

$$\vec{F}_A + \vec{F}_B + \vec{F}_C = m(0) = 0,$$

or

$$\vec{F}_B = -\vec{F}_A - \vec{F}_C. \qquad (5\text{-}6)$$

The free-body diagram for the tire is shown in Fig. 5-5*b*, where we have conveniently centered a coordinate system on the tire and assigned ϕ to the angle of $\vec{F}_C$.

We want to solve for the magnitude of $\vec{F}_B$. Although we know both magnitude and direction for $\vec{F}_A$, we know only the magnitude of $\vec{F}_C$ and not its direction. Thus, with unknowns on both sides of Eq. 5-6, we cannot directly solve it on a vector-capable calculator. Instead we must rewrite Eq. 5-6 in terms of components for either the x or the y axis. Since $\vec{F}_B$ is directed along the y axis, we choose that axis and write

$$F_{By} = -F_{Ay} - F_{Cy}.$$

Evaluating these components with their angles and using the angle 133° (= 180° − 47.0°) for $\vec{F}_A$, we obtain

$$F_B\sin(-90^\circ) = -F_A\sin 133^\circ - F_C\sin\phi,$$

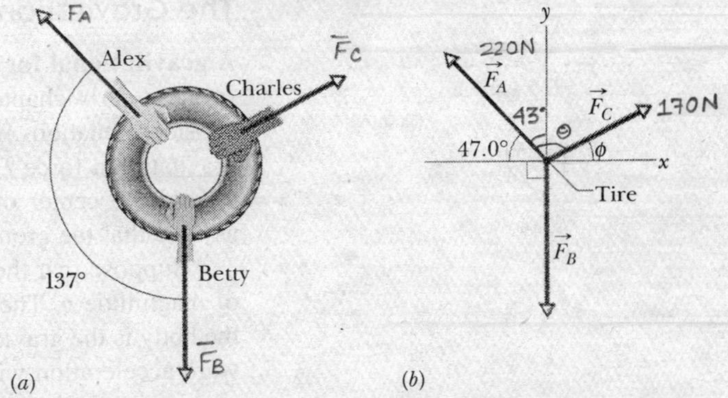

Fig. 5-5 Sample Problem 5-3. (*a*) An overhead view of three people pulling on a tire. (*b*) A free-body diagram for the tire.

which, with the given data for the magnitudes, yields

$$-F_B = -(220\ \text{N})(\sin 133^\circ) - (170\ \text{N})\sin\phi. \qquad (5\text{-}7)$$

However, we do not know ϕ.

We can find it by rewriting Eq. 5-6 for the x axis as

$$F_{Bx} = -F_{Ax} - F_{Cx}$$

and then as

$$F_B\cos(-90^\circ) = -F_A\cos 133^\circ - F_C\cos\phi,$$

which gives us

$$0 = -(220\ \text{N})(\cos 133^\circ) - (170\ \text{N})\cos\phi$$

and

$$\phi = \cos^{-1} -\frac{(220\ \text{N})(\cos 133^\circ)}{170\ \text{N}} = 28.04^\circ.$$

Inserting this into Eq. 5-7, we find

$$F_B = 241\ \text{N}. \qquad \text{(Answer)}$$

PROBLEM-SOLVING TACTICS

Tactic 1: *Dimensions and Vectors*

Many students fail to grasp the second **Key Idea** in Sample Problem 5-2, and that failure haunts them through the rest of this book. When you are dealing with forces, you cannot just add or subtract their magnitudes to find their net force unless they happen to be directed *along the same axis*. If they are not, you must use vector addition, either by means of a vector-capable calculator or by finding components along axes, as is done in Sample Problem 5-2.

Tactic 2: *Reading Force Problems*

Read the problem statement several times until you have a good mental picture of what the situation is, what data are given, and what is requested. If you know what the problem is about but don't know what to do next, put the problem aside and reread the text. If you are hazy about Newton's second law, reread that section. Study the sample problems. And remember, solving physics problems (like repairing cars and designing computer chips) takes training—you were not born with the ability.

Tactic 3: *Draw Two Types of Figures*

You may need two figures. One is a rough sketch of the actual situation. When you draw the forces, place the tail of each force vector either on the boundary of or within the body on which that force acts. The other figure is a free-body diagram: the forces on a *single* body are drawn, with the body represented by a dot or a sketch. Place the tail of each force vector on the dot or sketch.

Tactic 4: *What Is Your System?*

If you are using Newton's second law, you must know what body or system you are applying it to. In Sample Problem 5-1 it is the puck (not the ice). In Sample Problem 5-2, it is the cookie tin. In Sample Problem 5-3, it is the tire (not the people).

Tactic 5: *Choose Your Axes Wisely*

In Sample Problem 5-3, we saved a lot of work by choosing one of our coordinate axes to coincide with one of the forces (the y axis with $\vec{F}_B$).

5-6 Some Particular Forces

The Gravitational Force

A **gravitational force** $\vec{F}_g$ on a body is a pull that is directed toward a second body. In these early chapters, we do not discuss the nature of this force and we usually consider situations in which the second body is Earth. Thus, when we speak of *the* gravitational force $\vec{F}_g$ on a body, we usually mean a force that pulls on it directly toward the center of Earth—that is, directly down toward the ground. We shall assume that the ground is an inertial frame.

Suppose that the body, of mass m, is in free fall with the free-fall acceleration of magnitude g. Then, if we neglect the effects of the air, the only force acting on the body is the gravitational force $\vec{F}_g$. We can relate this downward force and downward acceleration with Newton's second law ($\vec{F} = m\vec{a}$). We place a vertical y axis along the body's path, with the positive direction upward. For this axis, Newton's second law can be written in the form $F_{\text{net},y} = ma_y$, which, in our situation, becomes

$$-F_g = m(-g)$$

or

$$F_g = mg. \tag{5-8}$$

In other words, the magnitude of the gravitational force is equal to the product mg.

This same gravitational force, with the same magnitude, still acts on the body even when the body is not in free fall but, say, at rest on a pool table or moving across the table. (For the gravitational force to disappear, Earth would have to disappear.)

We can write Newton's second law for the gravitational force in these vector forms:

$$\vec{F}_g = -F_g\hat{\mathrm{j}} = -mg\hat{\mathrm{j}} = m\vec{g}, \tag{5-9}$$

where $\hat{\mathrm{j}}$ is the unit vector that points upward along a y axis, directly away from the ground, and $\vec{g}$ is the free-fall acceleration (written as a vector), directed downward.

Weight

The **weight** W of a body is the magnitude of the net force required to prevent the body from falling freely, as measured by someone on the ground. For example, to

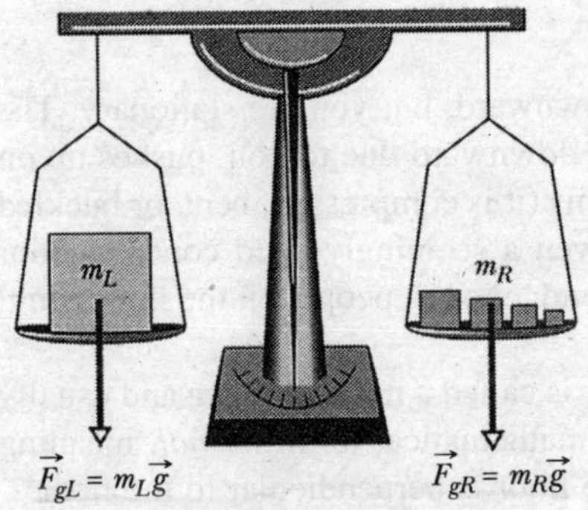

Fig. 5-6 An equal-arm balance. When the device is in balance, the gravitational force $\vec{F}_{gL}$ on the body being weighed (on the left pan) and the total gravitational force $\vec{F}_{gR}$ on the reference bodies (on the right pan) are equal. Thus, the mass of the body being weighed is equal to the total mass of the reference bodies.

keep a ball at rest in your hand while you stand on the ground, you must provide an upward force to balance the gravitational force on the ball from Earth. Suppose the magnitude of the gravitational force is 2.0 N. Then the magnitude of your upward force must be 2.0 N, and thus, the weight W of the ball is 2.0 N. We also say that the ball *weighs* 2.0 N, and speak about the ball *weighing* 2.0 N.

A ball with a weight of 3.0 N would require a greater force from you—namely, a 3.0 N force—to keep it at rest. The reason is that the gravitational force you must balance has a greater magnitude—namely, 3.0 N. We say that this second ball is *heavier* than the first ball.

Now let us generalize the situation. Consider a body that has an acceleration $\vec{a}$ of zero relative to the ground, which we again assume to be an inertial frame. Two forces act on the body: a downward gravitational force $\vec{F}_g$ and a balancing upward force of magnitude W. We can write Newton's second law for a vertical y axis, with the positive direction upward, as

$$F_{net,y} = ma_y.$$

In our situation, this becomes

$$W - F_g = m(0) \tag{5-10}$$

or

$$W = F_g \quad \text{(weight, with ground as inertial frame).} \tag{5-11}$$

This equation tells us (assuming the ground is an inertial frame) that

> The weight W of a body is equal to the magnitude F_g of the gravitational force on the body.

Substituting mg for F_g from Eq. 5-8, we find

$$\boxed{W = mg} \quad \text{(weight),} \tag{5-12}$$

which relates a body's weight to its mass.

To *weigh* a body means to measure its weight. One way to do this is to place the body on one of the pans of an equal-arm balance (Fig. 5-6) and then add reference bodies (whose masses are known) on the other pan until we strike a balance (so that the gravitational forces on the two sides match). The masses on the pans then match, and we know the mass m of the body. If we know the value of g for the location of the balance, we can find the weight of the body with Eq. 5-12.

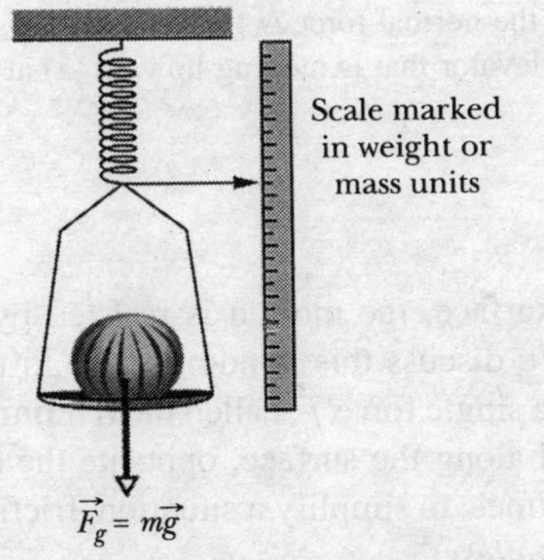

Fig. 5-7 A spring scale. The reading is proportional to the *weight* of the object placed on the pan, and the scale gives that weight if marked in weight units. If, instead, it is marked in mass units, the reading is accurate only where the free-fall acceleration g is the same as where the scale was calibrated.

We can also weigh a body with a spring scale (Fig. 5-7). The body stretches a spring, moving a pointer along a scale that has been calibrated and marked in either mass or weight units. (Most bathroom scales in the United States work this way and are marked in the force unit pounds.) If the scale is marked in mass units, it is accurate only where the value of g is the same as where the scale was calibrated.

The weight of a body must be measured when the body is not accelerating vertically relative to the ground. For example, you can measure your weight on a scale in your bathroom or on a fast train. However, if you repeat the measurement with the scale in an accelerating elevator, the reading on the scale differs from your weight because of the acceleration. Such a measurement is called an *apparent weight.*

Caution: The weight of a body is not the mass of the body. Weight is the magnitude of a force and is related to the mass by Eq. 5-12. If you move a body to a point where the value of g is different, the body's mass (an intrinsic property of the body) is not different but the weight is. For example, the weight of a bowling ball with a mass of 7.2 kg is 71 N on Earth but would be only 12 N on the Moon. The mass is the same on Earth and Moon, but the free-fall acceleration on the Moon is only 1.7 m/s^2.

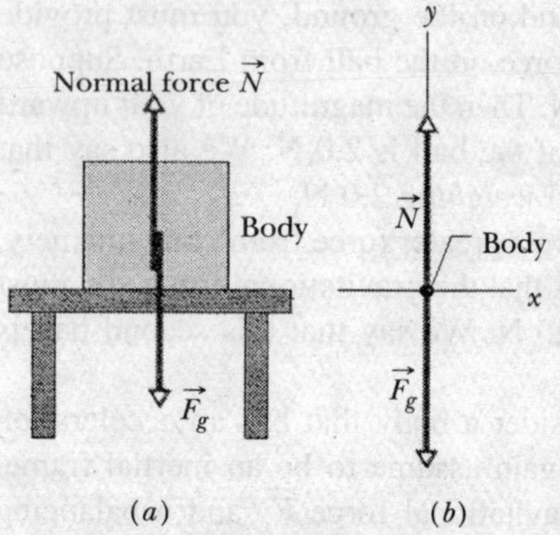

Fig. 5-8 (*a*) A body resting on a tabletop experiences a normal force $\vec{N}$ perpendicular to the tabletop. (*b*) The corresponding free-body diagram for the body.

The Normal Force

If you stand on a mattress, Earth pulls you downward, but you are stationary. The reason is that the mattress, because it deforms downward due to you, pushes up on you. Similarly, if you stand on a floor, it deforms (it is compressed, bent, or buckled ever so slightly), and it pushes up on you. Even a seemingly rigid concrete floor does this (if it is not sitting directly on the ground, enough people on the floor could actually break it).

The push on you from the mattress or floor is called a **normal force** and usually symbolized as $\vec{N}$. The name comes from the mathematical term *normal,* meaning perpendicular: The force on you from, say, the floor is perpendicular to the floor.

> When a body presses against a surface, the surface (even a seemingly rigid surface) deforms and pushes on the body with a normal force $\vec{N}$ that is perpendicular to the surface.

Figure 5-8*a* shows an example. A block of mass m lies on a table's horizontal surface and presses down on the table, deforming the table somewhat because of the gravitational force $\vec{F}_g$ on the block. The table pushes up on the block with normal force $\vec{N}$. The free-body diagram for the block is given in Fig. 5-8*b*. Forces $\vec{F}_g$ and $\vec{N}$ are the only two forces on the block and they are vertical. Thus, for the block we can write Newton's second law for a positive-upward y axis ($F_{\text{net},y} = ma_y$) as

$$N - F_g = ma_y.$$

From Eq. 5-8, we substitute mg for F_g, finding

$$N - mg = ma_y.$$

Then the magnitude of the normal force is

$$N = mg + ma_y = m(g + a_y) \tag{5-13}$$

for any vertical acceleration a_y of the table and block (they might be in an accelerating elevator). If the table and block are not accelerating relative to the ground, then $a_y = 0$ and Eq. 5-13 yields

$$N = mg. \tag{5-14}$$

CHECKPOINT 4: In Fig. 5-8, is the magnitude of the normal force $\vec{N}$ greater than, less than, or equal to mg if the body and table are in an elevator that is moving upward (a) at constant speed and (b) at increasing speed?

Friction

If we slide or attempt to slide a body over a surface, the motion is resisted by a bonding between the body and the surface. (We discuss this bonding more in the next chapter.) The resistance is considered to be a single force $\vec{f}$, called the **frictional force,** or simply **friction.** This force is directed along the surface, opposite the direction of the intended motion (Fig. 5-9). Sometimes, to simplify a situation, friction is assumed to be negligible (the surface is *frictionless*).

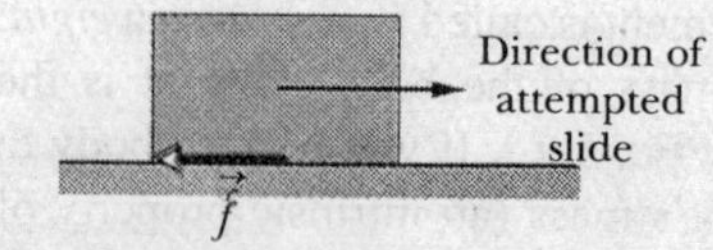

Fig. 5-9 A frictional force $\vec{f}$ opposes the attempted slide of a body over a surface.

Tension

When a cord (or a rope, cable, or other such object) is attached to a body and pulled taut, the cord pulls on the body with a force $\vec{T}$ directed away from the body and along the cord (Fig. 5-10*a*). The force is often called a *tension force* because the

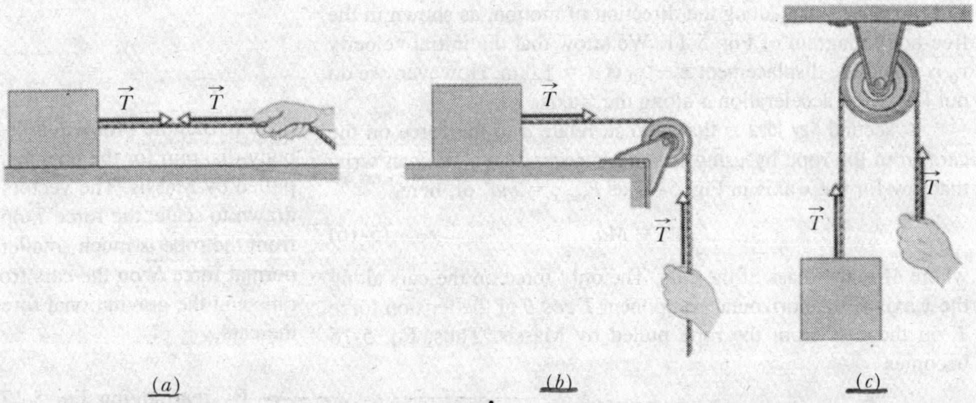

Fig. 5-10 (*a*) The cord, pulled taut, is under tension. If its mass is negligible, it pulls on the body and the hand with force $\vec{T}$, even if it runs around a massless, frictionless pulley as in (*b*) and (*c*).

cord is said to be in a state of *tension* (or to be *under tension*), which means that it is being pulled taut. The *tension in the cord* is the magnitude T of the force on the body. For example, if the force on the body has magnitude $T = 50$ N, then the tension in the cord is 50 N.

A cord is often said to be *massless* (meaning its mass is negligible compared to the body's mass) and unstretchable. The cord then exists only as a connection between two bodies. It pulls on both bodies with the same magnitude T, even if the bodies and the cord are accelerating and even if the cord runs around a *massless, frictionless pulley* (Figs. 5-10*b* and *c*). Such a pulley has negligible mass compared to the bodies and has negligible friction on its axle opposing its rotation. If the cord wraps halfway around a pulley, as in Fig. 5-10*c*, then the net force on the pulley from the cord has the magnitude $2T$.

CHECKPOINT 5: The body that is suspended by a rope in Fig. 5-10*c* has a weight of 75 N. Is T equal to, greater than, or less than 75 N when the body is moving upward (a) at constant speed, (b) at increasing speed, and (c) at decreasing speed?

PROBLEM-SOLVING TACTICS

Tactic 6: *Normal Force*

Equation 5-14 for the normal force on a body holds only when $\vec{N}$ is directed upward and the body's vertical acceleration is zero, so we do *not* apply it for other orientations of $\vec{N}$ or when the vertical acceleration is not zero. Instead, we must derive a new expression for $\vec{N}$ from Newton's second law.

We are free to move $\vec{N}$ around in a figure as long as we maintain its orientation. For example, in Fig. 5-8*a* we can slide it downward so that its head is at the boundary of the body and the tabletop. However, $\vec{N}$ is least likely to be misinterpreted if its tail is at that boundary or somewhere within the body (as shown). An even better technique is to draw a free-body diagram as in Fig. 5-8*b*, with the tail of $\vec{N}$ directly on the dot or sketch representing the body.

Sample Problem 5-4

Let us return to John Massis and the railroad cars, and assume that Massis pulled (with his teeth) on his end of the rope with a constant force that was 2.5 times his body weight, at an angle θ of 30° from the horizontal. His mass m was 80 kg. The weight W of the cars was 700 kN, and he moved them 1.0 m along the rails. Assume that the rolling wheels encountered no retarding force from the rails. What was the speed of the cars at the end of the pull?

SOLUTION: A **Key Idea** here is that, from Newton's second law, the constant horizontal force on the cars from Massis causes a constant horizontal acceleration of the cars. Because the acceleration is constant and the motion is one-dimensional, we can use the equations of Table 2-1 to find the velocity v at the end of the pulling distance $d = 1.0$ m. We need an equation that contains v. Let us try Eq. 2-16,

$$v^2 = v_0^2 + 2a(x - x_0), \tag{5-15}$$

and place an x axis along the direction of motion, as shown in the free-body diagram of Fig. 5-11. We know that the initial velocity v_0 is 0 and the displacement $x - x_0$ is $d = 1.0$ m. However, we do not know the acceleration a along the x axis.

A second **Key Idea** is that we can relate a to the force on the cars from the rope by using Newton's second law. We can write that law for the x axis in Fig. 5-11 as $F_{net,x} = ma_x$ or, here,

$$F_{net,x} = Ma, \tag{5-16}$$

where M is the mass of the cars. The only force on the cars along the x axis is the horizontal component $T \cos \theta$ of the tension force $\vec{T}$ on the cars from the rope pulled by Massis. Thus, Eq. 5-16 becomes

$$T \cos \theta = Ma. \tag{5-17}$$

We know that T is 2.5 times the body weight of Massis. From Eq. 5-12, his weight is equal to mg, so we have

$$T = 2.5mg = (2.5)(80 \text{ kg})(9.8 \text{ m/s}^2) = 1960 \text{ N},$$

which is the force a good middle-weight weight lifter can produce—and far from a superhuman force.

We still need M in order to evaluate Eq. 5-17 for a. To find M, we again use Eq. 5-12, but now with the weight W of the cars:

$$M = \frac{W}{g} = \frac{7.0 \times 10^5 \text{ N}}{9.8 \text{ m/s}^2} = 7.143 \times 10^4 \text{ kg}.$$

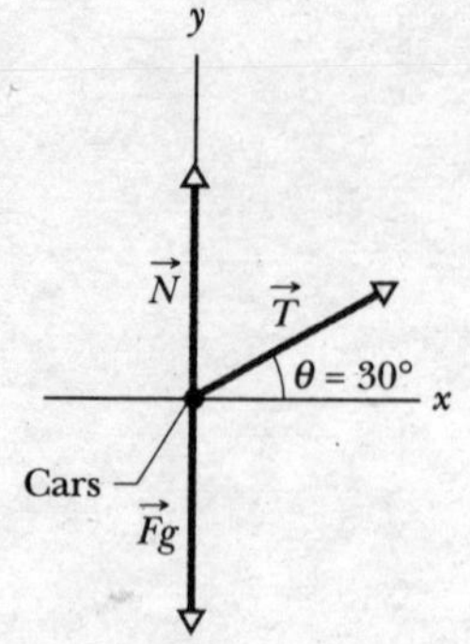

Fig. 5-11 Sample Problem 5-4. Free-body diagram for the passenger cars pulled by Massis. The vectors are not drawn to scale; the force $\vec{T}$ on the cars from the rope is much smaller than the normal force $\vec{N}$ on the cars from the rails and the gravitational force $\vec{F}_g$ on the cars.

By rearranging Eq. 5-17 and substituting for T, M, and θ, we get

$$a = \frac{T \cos \theta}{M} = \frac{(1960 \text{ N})(\cos 30°)}{7.143 \times 10^4 \text{ kg}} = 0.02376 \text{ m/s}^2.$$

Substituting this value and the other known values into Eq. 5-15 now gives us

$$v^2 = 0 + 2(0.02376 \text{ m/s}^2)(1.0 \text{ m})$$

and

$$v = 0.22 \text{ m/s}. \qquad \text{(Answer)}$$

Massis would have done better if the rope had been attached higher on the car, so that it was horizontal. Can you see why?

5-7 Newton's Third Law

Two bodies are said to *interact* when they push or pull on each other—that is, when a force acts on each body due to the other body. For example, suppose that you position a book B so it leans against a crate C (Fig. 5-12a). Then the book and crate interact: There is a horizontal force $\vec{F}_{BC}$ on the book from the crate (or due to the crate) and a horizontal force $\vec{F}_{CB}$ on the crate from the book (or due to the book). This pair of forces is shown in Fig. 5-12b. Newton's third law states that

> **Newton's Third Law:** When two bodies interact, the forces on the bodies from each other are always equal in magnitude and opposite in direction.

For the book and crate, we can write this law as the scalar relation

$$F_{BC} = F_{CB} \quad \text{(equal magnitudes)}$$

or as the vector relation

$$\vec{F}_{BC} = -\vec{F}_{CB} \quad \text{(equal magnitudes and opposite directions)},$$

where the minus sign means that these two forces are in opposite directions. We can call the forces between two interacting bodies a **third-law force pair.** When any two bodies interact in any situation, a third-law force pair is present. The book and crate in Fig. 5-12a are stationary, but the third law would still hold if they were moving and even if they were accelerating.

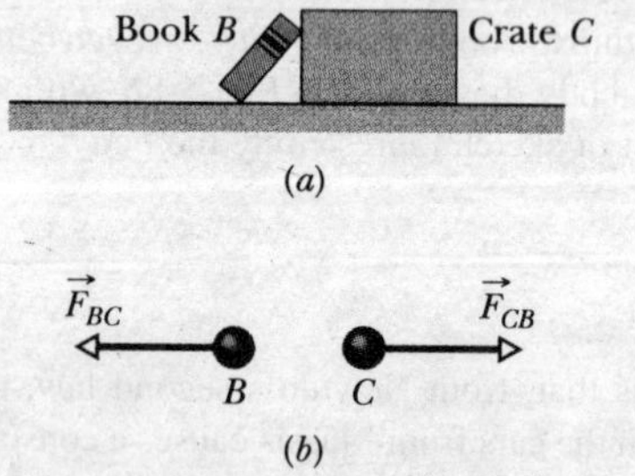

Fig. 5-12 (a) Book B leans against crate C. (b) According to Newton's third law, the force $\vec{F}_{BC}$ on the book from the crate has the same magnitude but the opposite direction of the force $\vec{F}_{CB}$ on the crate from the book.

As another example, let us find the third-law force pairs involving the cantaloupe in Fig. 5-13a, which lies on a table that stands on Earth. The cantaloupe interacts with the table and with Earth (this time, there are three bodies whose interactions we must sort out).

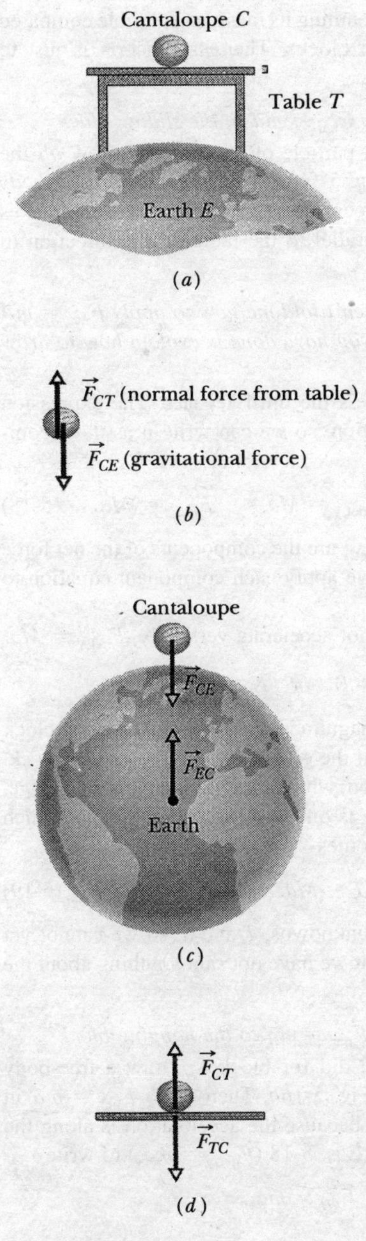

Fig. 5-13 (a) A cantaloupe lies on a table that stands on Earth. (b) The forces *on the cantaloupe* are $\vec{F}_{CT}$ and $\vec{F}_{CE}$. (c) The third-law force pair for the cantaloupe–Earth interaction. (d) The third-law force pair for the cantaloupe–table interaction.

Let's first focus only on the cantaloupe (Fig. 5-13b). Force $\vec{F}_{CT}$ is the normal force on the cantaloupe from the table, and force $\vec{F}_{CE}$ is the gravitational force on the cantaloupe due to Earth. Are they a third-law force pair? No, they are forces on a single body, the cantaloupe, and not on two interacting bodies.

To find a third-law pair, we should focus not on the cantaloupe but on the interaction between the cantaloupe and one of the other two bodies. First, in the cantaloupe–Earth interaction (Fig. 5-13c), Earth pulls on the cantaloupe with a gravitational force $\vec{F}_{CE}$ and the cantaloupe pulls on Earth with a gravitational force $\vec{F}_{EC}$. Are these forces a third-law force pair? Yes, they are forces on two interacting bodies, the force on each due to the other. Thus, by Newton's third law,

$$\vec{F}_{CE} = -\vec{F}_{EC} \quad \text{(cantaloupe–Earth interaction)}.$$

Next, in the cantaloupe–table interaction, the force on the cantaloupe from the table is $\vec{F}_{CT}$ and the force on the table from the cantaloupe is $\vec{F}_{TC}$ (Fig. 5-13d). These forces are also a third-law force pair, and so

$$\vec{F}_{CT} = -\vec{F}_{TC} \quad \text{(cantaloupe–table interaction)}.$$

✔CHECKPOINT 6: Suppose that the cantaloupe and table of Fig. 5-13 are in an elevator cab that begins to accelerate upward. (a) Do the magnitudes of forces $\vec{F}_{TC}$ and $\vec{F}_{CT}$ increase, decrease, or stay the same? (b) Are those two forces still equal in magnitude and opposite in direction? (c) Do the magnitudes of forces $\vec{F}_{CE}$ and $\vec{F}_{EC}$ increase, decrease, or stay the same? (d) Are those two forces still equal in magnitude and opposite in direction? Yes

Yes

Table like weight machine
weight-force ↑

5-8 Applying Newton's Laws

The rest of this chapter consists of sample problems. You should pore over them, learning not just their particular answers but, instead, the procedures they show for attacking a problem. Especially important is knowing how to translate a sketch of a situation into a free-body diagram with appropriate axes, so that Newton's laws can be applied. We begin with a sample problem that is worked out in exhaustive detail, using a question-and-answer format.

Sample Problem 5-5

Figure 5-14 shows a block S (the *sliding block*) with mass M = 3.3 kg. The block is free to move along a horizontal frictionless surface such as an air table. This first block is connected by a cord that wraps over a frictionless pulley to a second block H (the *hanging block*), with mass m = 2.1 kg. The cord and pulley have negligible masses compared to the blocks (they are "massless"). The hanging block H falls as the sliding block S accelerates to the right. Find (a) the acceleration of the sliding block, (b) the acceleration of the hanging block, and (c) the tension in the cord.

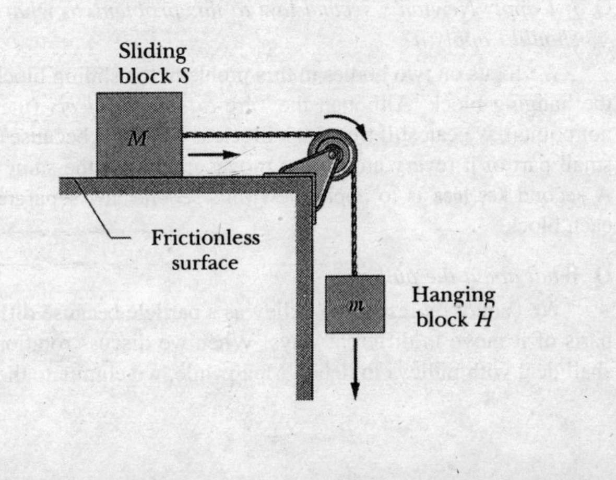

Fig. 5-14 Sample Problem 5-5. A block S of mass M is connected to a block H of mass m by a cord that wraps over a pulley.

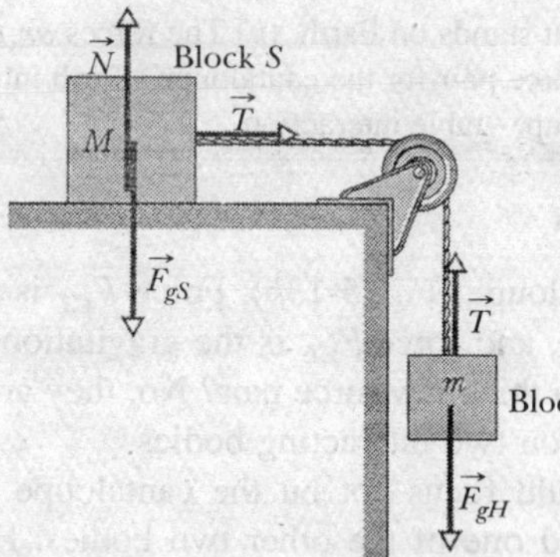

Fig. 5-15 The forces acting on the two blocks of Fig. 5-14.

Q *What is this problem all about?*

You are given two bodies, the sliding block and the hanging block, *and also Earth,* which pulls on both bodies. (Without Earth, nothing would happen here.) A total of five forces act on the blocks, as shown in Fig. 5-15:

1. The cord pulls to the right on sliding block S with a force of magnitude T.
2. The cord pulls upward on hanging block H with a force of the same magnitude T. This upward force keeps the hanging block from falling freely.
3. Earth pulls down on sliding block S with the gravitational force $\vec{F}_{gS}$, which has a magnitude equal to Mg.
4. Earth pulls down on hanging block H with the gravitational force $\vec{F}_{gH}$, which has a magnitude equal to mg.
5. The table pushes up on sliding block S with a normal force $\vec{N}$.

There is another thing that you should note. We assume that the cord does not stretch, so that if block H falls 1 mm in a certain time, block S moves 1 mm to the right in that same time. This means that the blocks move together and their accelerations have the same magnitude a.

Q *How do I classify this problem? Should it suggest a particular law of physics to me?*

Yes. Forces, masses, and accelerations are involved, and they should suggest Newton's second law of motion, $\vec{F}_{\text{net}} = m\vec{a}$. That is our starting **Key Idea.**

Q *If I apply Newton's second law to this problem, to what body should I apply it?*

We focus on two bodies in this problem, the sliding block and the hanging block. Although they are *extended objects* (they are not points), we can still treat each block as a particle because every small part of it (every atom, say) moves in exactly the same way. A second **Key Idea** is to apply Newton's second law separately to each block.

Q *What about the pulley?*

We cannot represent the pulley as a particle because different parts of it move in different ways. When we discuss rotation, we shall deal with pulleys in detail. Meanwhile, we eliminate the pulley from consideration by assuming its mass is negligible compared with the masses of the two blocks. Then its function is just to change the cord's orientation.

Q *OK. Now how do I apply $\vec{F}_{\text{net}} = m\vec{a}$ to the sliding block?*

Represent block S as a particle of mass M and draw *all* the forces that act *on* it, as in Fig. 5-16*a*. This is the block's *free-body diagram*. There are three forces. Next, draw a set of axes. It makes sense to draw the x axis parallel to the table, in the direction in which the block moves.

Q *Thanks, but you still haven't told me how to apply $\vec{F}_{\text{net}} = m\vec{a}$ to the sliding block. All you have done is explain how to draw a free-body diagram.*

You are right, and here's the third **Key Idea:** The expression $\vec{F}_{\text{net}} = M\vec{a}$ is a vector equation, so we can write it as three component equations:

$$F_{\text{net},x} = Ma_x \qquad F_{\text{net},y} = Ma_y \qquad F_{\text{net},z} = Ma_z \qquad (5\text{-}18)$$

in which $F_{\text{net},x}$, $F_{\text{net},y}$, and $F_{\text{net},z}$ are the components of the net force along the three axes. Now we apply each component equation to its corresponding direction.

Because block S does not accelerate vertically, $F_{\text{net},y} = Ma_y$ becomes

$$N - F_{gS} = 0 \quad \text{or} \quad N = F_{gS}.$$

Thus in the y direction, the magnitude of the normal force on block S is equal to the magnitude of the gravitational force on that block. No force acts in the z direction, which is perpendicular to the page.

In the x direction, there is only one force component, which is T. Thus, $F_{\text{net},x} = Ma_x$ becomes

$$T = Ma. \qquad (5\text{-}19)$$

This equation contains two unknowns, T and a, so we cannot yet solve it. Recall, however, that we have not said anything about the hanging block.

Q *I agree. How do I apply $\vec{F}_{\text{net}} = m\vec{a}$ to the hanging block?*

We apply it just as we did for block S: Draw a free-body diagram for block H, as in Fig. 5-16*b*. Then apply $\vec{F}_{\text{net}} = m\vec{a}$ in component form. This time, because the acceleration is along the y axis, we use the second of Eqs. 5-18 ($F_{\text{net},y} = ma_y$) to write

$$T - F_{gH} = ma_y.$$

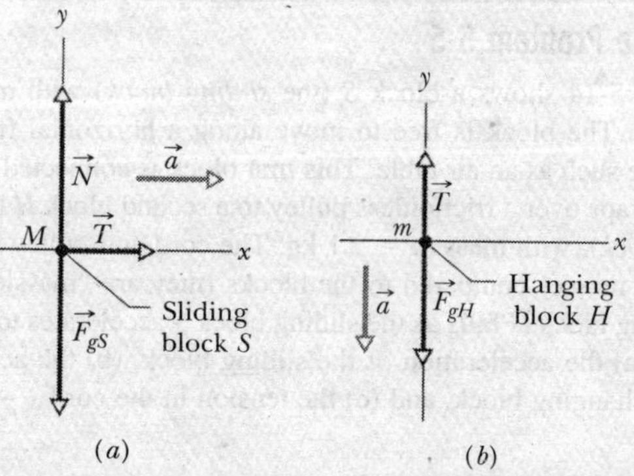

Fig. 5-16 (*a*) A free-body diagram for block S of Fig. 5-14. (*b*) A free-body diagram for the hanging block H of Fig. 5-14.

We can now substitute mg for F_{gH} and $-a$ for a_y (negative because block H accelerates downward, in the negative direction of the y axis). We find

$$T - mg = -ma. \tag{5-20}$$

Now note that Eqs. 5-19 and 5-20 are simultaneous equations with the same two unknowns, T and a. Subtracting these equations eliminates T. Then solving for a yields

$$a = \frac{m}{M + m} g. \tag{5-21}$$

Substituting this result into Eq. 5-19 yields

$$T = \frac{Mm}{M + m} g. \tag{5-22}$$

Putting in the numbers gives, for these two quantities,

$$a = \frac{m}{M + m} g = \frac{2.1 \text{ kg}}{3.3 \text{ kg} + 2.1 \text{ kg}} (9.8 \text{ m/s}^2)$$

$$= 3.8 \text{ m/s}^2 \qquad \text{(Answer)}$$

and

$$T = \frac{Mm}{M + m} g = \frac{(3.3 \text{ kg})(2.1 \text{ kg})}{3.3 \text{ kg} + 2.1 \text{ kg}} (9.8 \text{ m/s}^2)$$

$$= 13 \text{ N}. \qquad \text{(Answer)}$$

Q *The problem is now solved, right?*

That's a fair question, but the problem is not really finished until we have examined the results to see whether they make sense. (If you made these calculations on the job, wouldn't you want to see whether they made sense before you turned them in?)

Look first at Eq. 5-21. Note that it is dimensionally correct and that the acceleration a will always be less than g. This is as it must be, because the hanging block is not in free fall. The cord pulls upward on it.

Look now at Eq. 5-22, which we can rewrite in the form

$$T = \frac{M}{M + m} mg. \tag{5-23}$$

In this form, it is easier to see that this equation is also dimensionally correct, because both T and mg have dimensions of forces. Equation 5-23 also lets us see that the tension in the cord is always less than mg, and thus always less than the gravitational force on the hanging block. That is a comforting thought because, if T were *greater* than mg, the hanging block would accelerate upward.

We can also check the results by studying special cases, in which we can guess what the answers must be. A simple example is to put $g = 0$, as if the experiment were carried out in interstellar space. We know that in that case, the blocks would not move from rest, there would be no forces on the ends of the cord, and so there would be no tension in the cord. Do the formulas predict this? Yes, they do. If you put $g = 0$ in Eqs. 5-21 and 5-22, you find $a = 0$ and $T = 0$. Two more special cases that you might try are $M = 0$ and $m \to \infty$.

Sample Problem 5-6

In Fig. 5-17a, a block B of mass $M = 15.0$ kg hangs by a cord from a knot K of mass m_K, which hangs from a ceiling by means of two other cords. The cords have negligible mass, and the magnitude of the gravitational force on the knot is negligible compared to the gravitational force on the block. What are the tensions in the three cords?

SOLUTION: Let's start with the block because it has only one attached cord. The free-body diagram in Fig. 5-17b shows the forces on the block: gravitational force $\vec{F}_g$ (with a magnitude of Mg) and force T_3 from the attached cord. A **Key Idea** is that we can relate these forces to the acceleration of the block via Newton's second law ($\vec{F}_{net} = m\vec{a}$). Because the forces are both vertical, we choose the vertical component version of the law ($F_{net,y} = ma_y$) and write

$$T_3 - F_g = Ma_y.$$

Substituting Mg for F_g and 0 for the block's acceleration a_y, we find

$$T_3 - Mg = M(0) = 0.$$

This means that the two forces on the block are in equilibrium. Substituting for M (= 15.0 kg) and g and solving for T_3 yield

$$T_3 = 147 \text{ N}. \qquad \text{(Answer)}$$

We next consider the knot in the free-body diagram of Fig. 5-17c, where the negligible gravitational force on the knot is not included. The **Key Idea** here is that we can relate the three other forces acting on the knot to the acceleration of the knot via New-

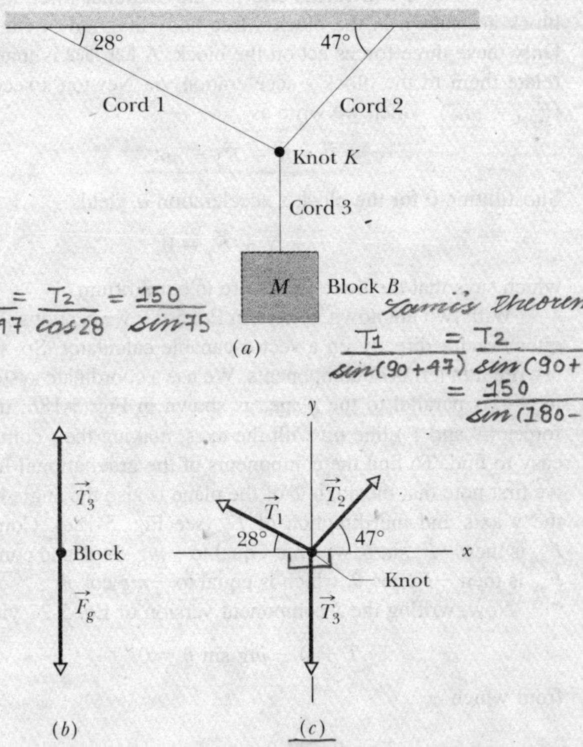

Fig. 5-17 Sample Problem 5-6. (a) A block of mass M hangs from three cords by means of a knot. (b) A free-body diagram for the block. (c) A free-body diagram for the knot.

ton's second law ($\vec{F}_{net} = m\vec{a}$) by writing

$$\vec{T}_1 + \vec{T}_2 + \vec{T}_3 = m_K\vec{a}_K.$$

Substituting 0 for the knot's acceleration $\vec{a}_K$ yields

$$\vec{T}_1 + \vec{T}_2 + \vec{T}_3 = 0, \tag{5-24}$$

which means that the three forces on the knot are in equilibrium. Although we know both magnitude and angle for $\vec{T}_3$, we know only the angles and not the magnitudes for $\vec{T}_1$ and $\vec{T}_2$. With unknowns in two vectors, we cannot solve Eq. 5-24 for $\vec{T}_1$ or $\vec{T}_2$ directly on a vector-capable calculator.

Instead we rewrite Eq. 5-24 in terms of components along the x and y axes. For the x axis, we have

$$T_{1x} + T_{2x} + T_{3x} = 0,$$

which, using the given data, yields

$$-T_1 \cos 28° + T_2 \cos 47° + 0 = 0. \tag{5-25}$$

(For the first term, we have two choices, either the one shown or the equivalent $T_1 \cos 152°$, where 152° is the angle from the positive direction of the x axis.)

Similarly, for the y axis we rewrite Eq. 5-24 as

$$T_{1y} + T_{2y} + T_{3y} = 0$$

or

$$T_1 \sin 28° + T_2 \sin 47° - T_3 = 0.$$

Substituting our previous result for T_3 then gives us

$$T_1 \sin 28° + T_2 \sin 47° - 147 \text{ N} = 0. \tag{5-26}$$

We cannot solve Eq. 5-25 or Eq. 5-26 separately because each contains two unknowns, but we can solve them simultaneously because they contain the same two unknowns. Doing so (either by substitution, by adding or subtracting the equations appropriately, or by using the equation-solving capability of a calculator), we discover

$$T_1 = 104 \text{ N} \quad\text{and}\quad T_2 = 134 \text{ N}. \qquad \text{(Answer)}$$

Thus, the tensions in the cords are 104 N in cord 1, 134 N in cord 2, and 147 N in cord 3.

Sample Problem 5-7

In Fig. 5-18a, a cord holds stationary a block of mass $m = 15$ kg, on a frictionless plane that is inclined at angle $\theta = 27°$.

(a) What are the magnitudes of the force $\vec{T}$ on the block from the cord and the normal force $\vec{N}$ on the block from the plane?

SOLUTION: Those two forces and the gravitational force $\vec{F}_g$ on the block are shown in the block's free-body diagram of Fig. 5-18b. Only these three forces act on the block. A **Key Idea** is that we can relate them to the block's acceleration via Newton's second law ($\vec{F}_{net} = m\vec{a}$), which we write as

$$\vec{T} + \vec{N} + \vec{F}_g = m\vec{a}.$$

Substituting 0 for the block's acceleration $\vec{a}$ yields

$$\vec{T} + \vec{N} + \vec{F}_g = 0, \tag{5-27}$$

which says that the three forces are in equilibrium.

With two unknown vectors in Eq. 5-27, we cannot solve it for either vector directly on a vector-capable calculator. So, we must rewrite it in terms of components. We use a coordinate system with its x axis parallel to the plane, as shown in Fig. 5-18b; then two forces ($\vec{N}$ and $\vec{T}$) line up with the axes, making their components easy to find. To find the components of the gravitational force $\vec{F}_g$, we first note that the angle θ of the plane is also the angle between the y axis and the direction of $\vec{F}_g$ (see Fig. 5-18c). Component F_{gx} is then $-F_g \sin\theta$, which is equal to $-mg\sin\theta$; and component F_{gy} is then $-F_g\cos\theta$, which is equal to $-mg\cos\theta$.

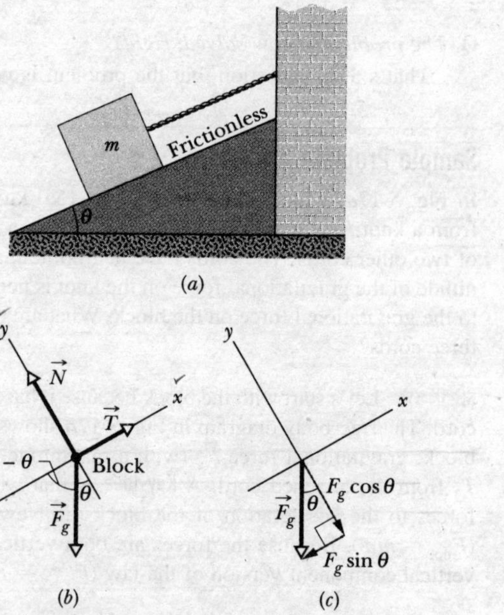

Fig. 5-18 Sample Problem 5-7 (a) A block of mass m held stationary by a cord. (b) A free-body diagram for the block. (c) The x and y components of $\vec{F}_g$.

Now, writing the x component version of Eq. 5-27 yields

$$T + 0 - mg\sin\theta = 0,$$

from which

$$\begin{aligned} T &= mg\sin\theta \\ &= (15 \text{ kg})(9.8 \text{ m/s}^2)(\sin 27°) \\ &= 67 \text{ N}. \end{aligned} \qquad \text{(Answer)}$$

Similarly, for the y axis, Eq. 5-27 yields

$$0 + N - mg\cos\theta = 0$$

or

$$\begin{aligned} N &= mg\cos\theta \\ &= (15 \text{ kg})(9.8 \text{ m/s}^2)(\cos 27°) \\ &= 131 \text{ N} \approx 130 \text{ N}. \end{aligned} \qquad \text{(Answer)}$$

(b) We now cut the cord. As the block then slides down the inclined plane, does it accelerate? If so, what is its acceleration?

SOLUTION: Cutting the cord removes force $\vec{T}$ from the block. Along the y axis, the normal force and component F_{gy} are still in equilibrium. However, along the x axis, only force component F_{gx} acts on the block; because it is directed down the plane (along the x axis), that component must cause the block to accelerate down the plane. Our **Key Idea** here is that we can relate F_{gx} to the acceleration a that it produces with Newton's second law written for x components ($F_{\text{net},x} = ma_x$). We get

$$F_{gx} = ma$$

or

$$-mg \sin \theta = ma,$$

which gives us

$$a = -g \sin \theta. \tag{5-28}$$

Substituting known data then yields

$$a = -(9.8 \text{ m/s}^2)(\sin 27°) = -4.4 \text{ m/s}^2. \quad \text{(Answer)}$$

The magnitude of this acceleration a is less than the magnitude 9.8 m/s^2 of the free-fall acceleration because only a component of $\vec{F}_g$ (the component that is directed down the plane) is producing acceleration a.

CHECKPOINT 7: In the figure, horizontal force $\vec{F}$ is applied to a block on a ramp. (a) Is the component of $\vec{F}$ that is perpendicular to the ramp $F \cos \theta$ or $F \sin \theta$? (b) Does the presence of $\vec{F}$ increase or decrease the magnitude of the normal force on the block from the ramp?

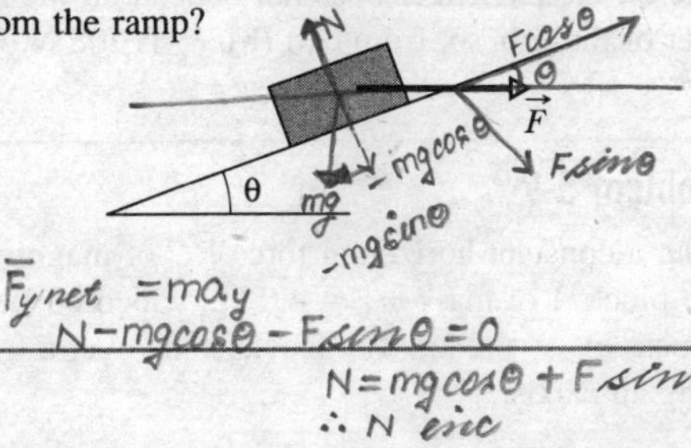

Sample Problem 5-8

In Fig. 5-19*a*, a passenger of mass $m = 72.2$ kg stands on a platform scale in an elevator cab. We are concerned with the scale readings when the cab is stationary, and when it is moving up or down.

(a) Find a general solution for the scale reading, whatever the vertical motion of the cab.

SOLUTION: One **Key Idea** here is that the scale reading is equal to the magnitude of the normal force $\vec{N}$ on the passenger from the scale. The only other force acting on the passenger is the gravitational force $\vec{F}_g$, as shown in the free-body diagram of the passenger in Fig. 5-19*b*.

A second **Key Idea** is that we can relate the forces on the passenger to the acceleration $\vec{a}$ of the passenger with Newton's second law ($\vec{F}_{\text{net}} = m\vec{a}$). However, recall that we can use this law only in an inertial frame. If the cab accelerates, then it is *not* an inertial frame. So, we choose the ground to be our inertial frame and make any measure of the passenger's acceleration relative to it.

Because the two forces on the passenger and the passenger's acceleration are all directed vertically, along the y axis shown in Fig. 5-19*b*, we can use Newton's second law written for y components ($F_{\text{net},y} = ma_y$) to get

$$N - F_g = ma$$

or

$$N = F_g + ma. \tag{5-29}$$

This tells us that the scale reading, which is equal to N, depends on the vertical acceleration a of the cab. Substituting mg for F_g gives us

$$N = m(g + a) \quad \text{(Answer)} \tag{5-30}$$

for any choice of acceleration a.

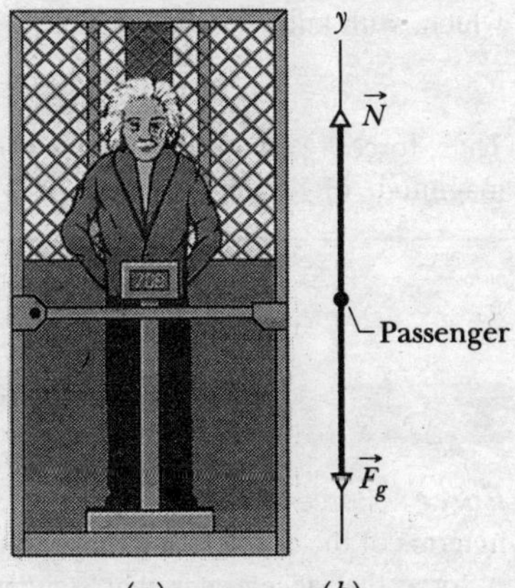

(*a*) (*b*)

Fig. 5-19 Sample Problem 5-8. (*a*) A passenger stands on a platform scale that indicates his weight or apparent weight. (*b*) The free-body diagram for the passenger, showing the normal force $\vec{N}$ on him from the scale and the gravitational force $\vec{F}_g$.

(b) What does the scale read if the cab is stationary or moving upward at a constant 0.50 m/s?

SOLUTION: The **Key Idea** here is that for any constant velocity (zero or otherwise), the acceleration a of the passenger is zero. Substituting this and other known values into Eq. 5-30, we find

$$N = (72.2 \text{ kg})(9.8 \text{ m/s}^2 + 0) = 708 \text{ N}. \quad \text{(Answer)}$$

This is the weight of the passenger and is equal to the magnitude F_g of the gravitational force on him.

(c) What does the scale read if the cab accelerates upward at 3.20 m/s^2 and downward at 3.20 m/s^2?

SOLUTION: For $a = 3.20 \text{ m/s}^2$, Eq. 5-30 gives

$$N = (72.2 \text{ kg})(9.8 \text{ m/s}^2 + 3.20 \text{ m/s}^2) = 939 \text{ N}, \quad \text{(Answer)}$$

and for $a = -3.20 \text{ m/s}^2$, it gives

$$N = (72.2 \text{ kg})(9.8 \text{ m/s}^2 - 3.20 \text{ m/s}^2) = 477 \text{ N} \quad \text{(Answer)}$$

So, for an upward acceleration (either the cab's upward speed is increasing or its downward speed is decreasing), the scale reading is greater than the passenger's weight. That reading is a measurement of an apparent weight, because it is made in a noninertial frame. Similarly, for a downward acceleration (either the cab's upward speed is decreasing or its downward speed is increasing), the scale reading is less than the passenger's weight.

(d) During the upward acceleration in part (c), what is the magnitude F_{net} of the net force on the passenger, and what is the magnitude $a_{p,cab}$ of the passenger's acceleration as measured in the frame of the cab? Does $\vec{F}_{net} = m\vec{a}_{p,cab}$?

SOLUTION: One **Key Idea** here is that the magnitude F_g of the gravitational force on the passenger does not depend on the motion of the passenger or the cab, so, from part (b), F_g is 708 N. From part (c), the magnitude N of the normal force on the passenger during the upward acceleration is the 939 N reading on the scale. Thus, the net force on the passenger is

$$F_{net} = N - F_g = 939 \text{ N} - 708 \text{ N} = 231 \text{ N}, \quad \text{(Answer)}$$

during the upward acceleration. However, the acceleration $a_{p,cab}$ of the passenger relative to the frame of the cab is zero. Thus, in the noninertial frame of the accelerating cab, F_{net} is not equal to $ma_{p,cab}$, and Newton's second law does not hold.

CHECKPOINT 8: In this sample problem what does the scale read if the elevator cable breaks, so that the cab falls freely; that is, what is the apparent weight of the passenger in free fall?

Sample Problem 5-9

In Fig. 5-20*a*, a constant horizontal force $\vec{F}_{ap}$ of magnitude 20 N is applied to block *A* of mass $m_A = 4.0$ kg, which pushes against block *B* of mass $m_B = 6.0$ kg. The blocks slide over a frictionless surface, along an *x* axis.

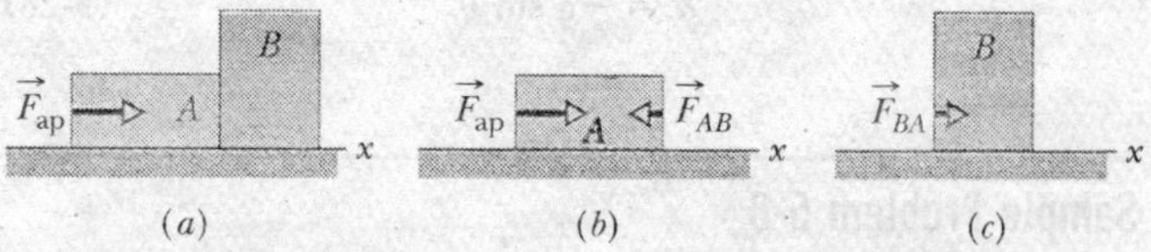

Fig. 5-20 Sample Problem 5-9. (*a*) A constant horizontal force $\vec{F}_{ap}$ is applied to block *A*, which pushes against block *B*. (*b*) Two horizontal forces act on block *A*: applied force $\vec{F}_{ap}$ and force $\vec{F}_{AB}$ from block *B*. (*c*) Only one horizontal force acts on block *B*: force $\vec{F}_{BA}$ from block *A*.

(a) What is the acceleration of the blocks?

SOLUTION: We shall first examine a solution with a serious error, then a dead-end solution, and then a successful solution.

Serious Error: Because force $\vec{F}_{ap}$ is applied directly to block *A*, we use Newton's second law to relate that force to the acceleration $\vec{a}$ of block *A*. Because the motion is along the *x* axis, we use that law for *x* components ($F_{net,x} = ma_x$), writing it as

$$F_{ap} = m_A a.$$

However, this is seriously wrong because $\vec{F}_{ap}$ is not the only horizontal force acting on block *A*. There is also the force $\vec{F}_{AB}$ from block *B* (as shown in Fig. 5-20*b*).

Dead-End Solution: Let us now include force $\vec{F}_{AB}$ by writing, again for the *x* axis,

$$F_{ap} - F_{AB} = m_A a.$$

(We use the minus sign to include the direction of $\vec{F}_{AB}$.) However, F_{AB} is a second unknown, so we cannot solve this equation for the desired acceleration *a*.

Successful Solution: The **Key Idea** here is that, because of the direction in which force $\vec{F}_{ap}$ is applied, the two blocks form a rigidly connected system. We can relate the net force *on the system* to the acceleration *of the system* with Newton's second law. Here, once again for the *x* axis, we can write that law as

$$F_{ap} = (m_A + m_B)a,$$

where now we properly apply $\vec{F}_{ap}$ to the system with total mass $m_A + m_B$. Solving for *a* and substituting known values, we find

$$a = \frac{F_{ap}}{m_A + m_B} = \frac{20 \text{ N}}{4.0 \text{ kg} + 6.0 \text{ kg}} = 2.0 \text{ m/s}^2. \quad \text{(Answer)}$$

Thus, the acceleration of the system and of each block is in the positive direction of the *x* axis and has the magnitude 2.0 m/s^2.

(b) What is the force $\vec{F}_{BA}$ on block *B* from block *A* (Fig. 5-20*c*)?

SOLUTION: The **Key Idea** here is that we can relate the net force on block *B* to the block's acceleration with Newton's second law. Here we can write that law, still for components along the *x* axis, as

$$F_{BA} = m_B a,$$

which, with known values, gives

$$F_{BA} = (6.0 \text{ kg})(2.0 \text{ m/s}^2) = 12 \text{ N}. \quad \text{(Answer)}$$

Thus, force $\vec{F}_{BA}$ is in the positive direction of the *x* axis and has a magnitude of 12 N.

REVIEW & SUMMARY

Newtonian Mechanics The velocity of a particle or a particle-like body can change (the particle can accelerate) when the particle is acted on by one or more **forces** (pushes or pulls) from other objects. *Newtonian mechanics* relates accelerations and forces.

Force Forces are vector quantities. Their magnitudes are defined in terms of the acceleration they would give the standard kilogram. A force that accelerates that standard body by exactly 1 m/s^2 is defined to have a magnitude of 1 N. The direction of a force is the direction of the acceleration it causes. Forces are combined according to the rules of vector algebra. The **net force** on a body is the vector sum of all the forces acting on it.

Mass The **mass** of a body is the characteristic of that body that relates the body's acceleration to the force (or net force) causing the acceleration. Masses are scalar quantities.

Newton's First Law If there is no net force on a body, the body must remain at rest if it is initially at rest, or move in a straight line at constant speed if it is in motion.

Inertial Reference Frames Reference frames in which Newtonian mechanics holds are called *inertial reference frames* or simply *inertial frames*. We can approximate the ground as an inertial frame if Earth's motions can be neglected. Reference frames in which Newtonian mechanics does not hold are called *noninertial reference frames* or simply *noninertial frames*. An elevator accelerating relative to the ground is a noninertial frame.

Newton's Second Law The net force $\vec{F}_{net}$ on a body with mass m is related to the body's acceleration $\vec{a}$ by

$$\vec{F}_{net} = m\vec{a}, \tag{5-1}$$

which may be written in the component versions

$$F_{net,x} = ma_x \quad F_{net,y} = ma_y \quad \text{and} \quad F_{net,z} = ma_z. \tag{5-2}$$

The second law indicates that in SI units

$$1\ \text{N} = 1\ \text{kg}\cdot\text{m/s}^2. \tag{5-3}$$

A **free-body diagram** is helpful in solving problems with the second law: It is a stripped-down diagram in which only *one* body is considered. That body is represented by a sketch or simply a dot. The external forces on the body are drawn, and a coordinate system is superimposed, oriented so as to simplify the solution.

Some Particular Forces A **gravitational force** $\vec{F}_g$ on a body is a pull by another body. In most situations in this book, the other body is Earth or some other astronomical body. For Earth, the force is directed down toward the ground, which is assumed to be an inertial frame. With that assumption, the magnitude of the force is

$$F_g = mg, \tag{5-8}$$

where m is the body's mass and g is the magnitude of the free-fall acceleration.

The **weight** W of a body is the magnitude of the upward force needed to balance the gravitational force on the body due to Earth (or another astronomical body). It is related to the body's mass by

$$W = mg. \tag{5-12}$$

A **normal force** $\vec{N}$ is the force on a body from a surface against which the body presses. The normal force is always perpendicular to the surface.

A **frictional force** $\vec{f}$ is the force on a body when the body slides or attempts to slide along a surface. The force is always parallel to the surface and directed so as to oppose the motion of the body. On a *frictionless surface*, the frictional force is negligible.

When a cord is under **tension**, it pulls on a body at each of its ends. The pull is directed along the cord, away from the point of attachment to each body. For a *massless* cord (a cord with negligible mass), the pulls at both ends of the cord have the same magnitude T, even if the cord runs around a *massless, frictionless pulley* (a pulley with negligible mass and negligible friction on its axle to oppose its rotation).

Newton's Third Law If a force $\vec{F}_{BC}$ acts on body B due to body C, then there is a force $\vec{F}_{CB}$ on body C due to body B. The forces are equal in magnitude and opposite in direction:

$$\vec{F}_{BC} = -\vec{F}_{CB}.$$

QUESTIONS

1. Two horizontal forces,

$$\vec{F}_1 = (3\ \text{N})\hat{\text{i}} - (4\ \text{N})\hat{\text{j}} \quad \text{and} \quad \vec{F}_2 = -(1\ \text{N})\hat{\text{i}} - (2\ \text{N})\hat{\text{j}},$$

pull a banana split across a frictionless lunch counter. Without using a calculator, determine which of the vectors in the free-body diagram of Fig. 5-21 best represent (a) $\vec{F}_1$ and (b) $\vec{F}_2$. What is the net-force component along (c) the x axis and (d) the y axis? Into which quadrants do (e) the net-force vector and (f) the split's acceleration vector point?

2. At time $t = 0$, a single force $\vec{F}$ of constant magnitude begins to act on a rock that is moving along an x axis through deep space. The rock continues to move along that axis. (a) For time $t > 0$, which of the following is a possible function $x(t)$ for the rock's position: (1) $x = 4t - 3$, (2) $x = -4t^2 + 6t - 3$, (3) $x = 4t^2 + 6t - 3$? (b) For which function is $\vec{F}$ directed opposite the rock's initial direction of motion?

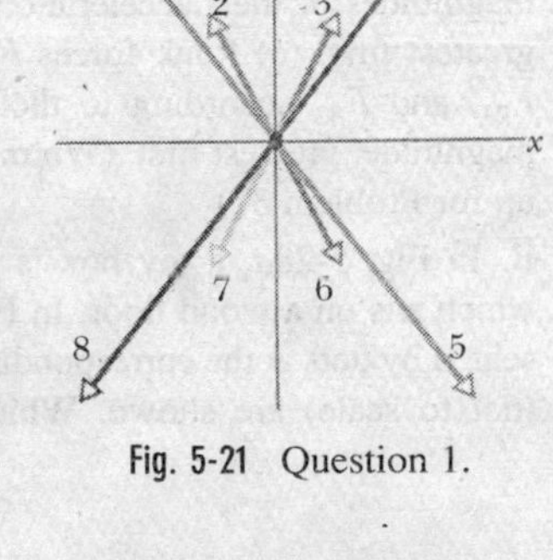

Fig. 5-21 Question 1.

3. Figure 5-22 shows overhead views of four situations in which forces act on a block that lies on a frictionless floor. If the force

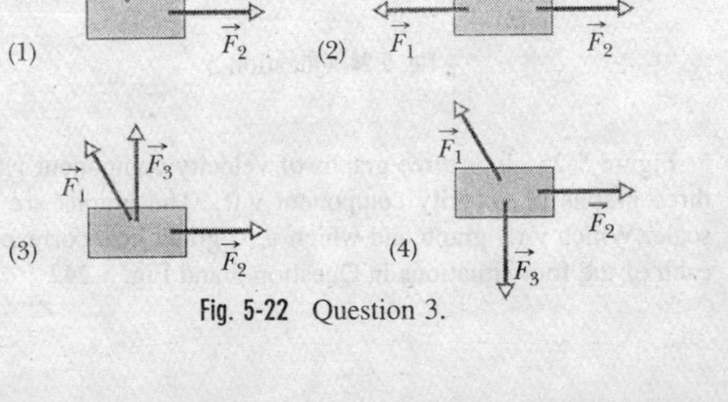

Fig. 5-22 Question 3.

magnitudes are chosen properly, in which situations is it possible that the block is (a) stationary and (b) moving with a constant velocity?

4. In Fig. 5-23, two forces $\vec{F}_1$ and $\vec{F}_2$ act on a "Rocky and Bullwinkle" lunch box as the lunch box slides at constant velocity over a frictionless lunchroom floor. We are to decrease the angle θ of $\vec{F}_1$ without changing the magnitude of $\vec{F}_1$. To keep the lunch box sliding at constant velocity, should we increase, decrease, or maintain the magnitude of $\vec{F}_2$?

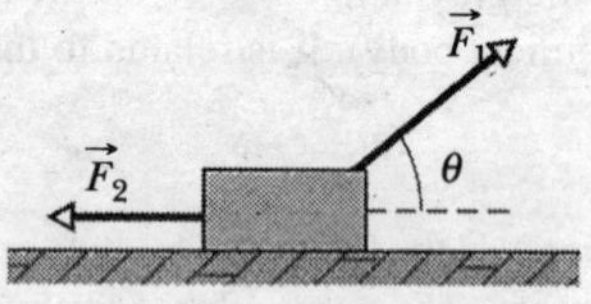

Fig. 5-23 Question 4.

5. Figure 5-24 gives the free-body diagram for four situations in which an object is pulled by several forces across a frictionless floor, as seen from overhead. In which situations does the object's acceleration $\vec{a}$ have (a) an x component and (b) a y component? (c) In each situation, give the direction of $\vec{a}$ by naming either a quadrant or a direction along an axis. (This can be done with a few mental calculations.)

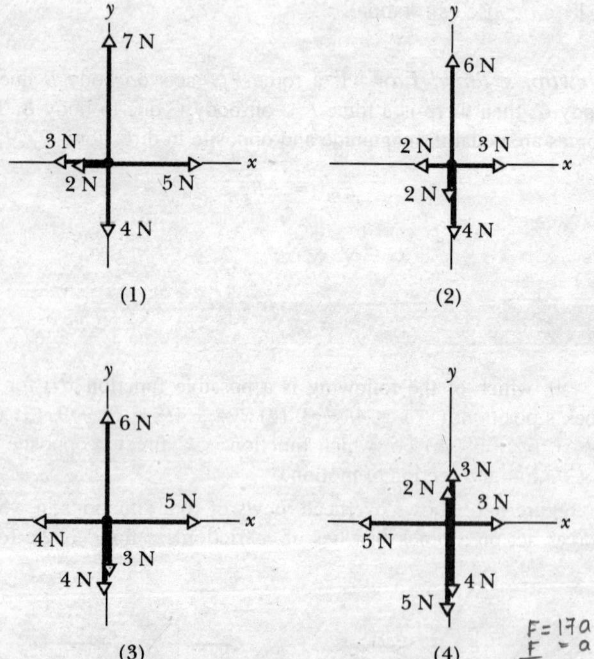

Fig. 5-24 Question 5.

6. Figure 5-25 gives three graphs of velocity component $v_x(t)$ and three graphs of velocity component $v_y(t)$. The graphs are not to scale. Which $v_x(t)$ graph and which $v_y(t)$ graph best correspond to each of the four situations in Question 5 and Fig. 5-24?

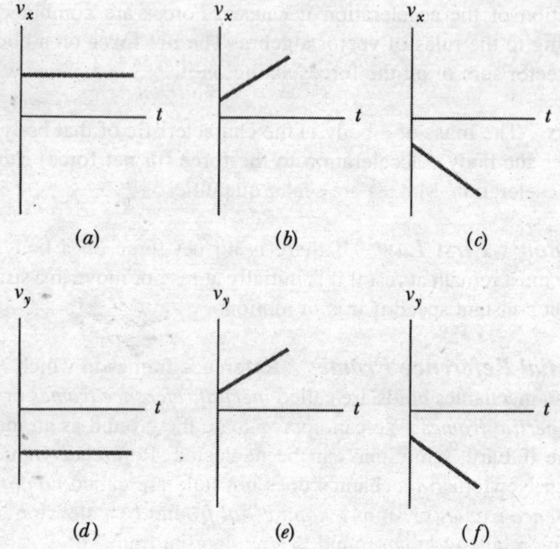

Fig. 5-25 Question 6.

7. The body that is suspended by a rope in Fig. 5-10c has a weight of 75 N. Is T equal to, greater than, or less than 75 N when the body is moving downward at (a) increasing speed and (b) decreasing speed?

8. A vertical force $\vec{F}$ is applied to a block of mass m that lies on a floor. What happens to the magnitude of the normal force $\vec{N}$ on the block from the floor as magnitude F is increased from zero if force $\vec{F}$ is (a) downward and (b) upward?

9. Figure 5-26 shows a train of four blocks being pulled across a frictionless floor by force $\vec{F}$. What total mass is accelerated to the right by (a) force $\vec{F}$, (b) cord 3, and (c) cord 1? (d) Rank the blocks according to their accelerations, greatest first. (e) Rank the cords according to their tension, greatest first. (Warm-up for Problems 34 and 36)

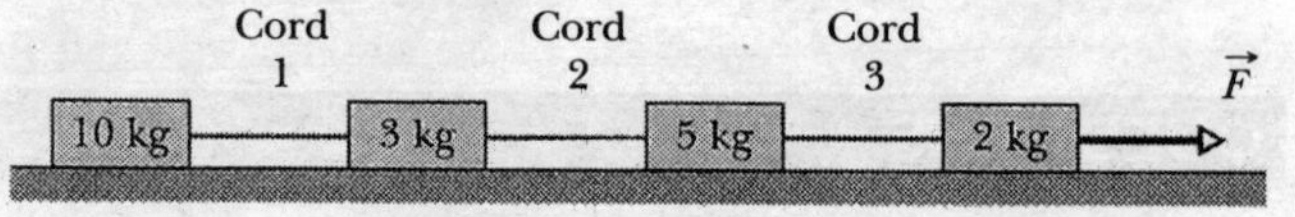

Fig. 5-26 Question 9.

10. Figure 5-27 shows a group of three blocks being pushed across a frictionless floor by horizontal force $\vec{F}$. What total mass is accelerated to the right by (a) force $\vec{F}$, (b) force $\vec{F}_{21}$ on block 2 from block 1, and (c) force $\vec{F}_{32}$ on block 3 from block 2? (d) Rank the blocks according to the magnitudes of their acceleration, greatest first. (e) Rank forces $\vec{F}$, $\vec{F}_{21}$, and $\vec{F}_{32}$ according to their magnitude, greatest first. (Warm-up for Problem 31)

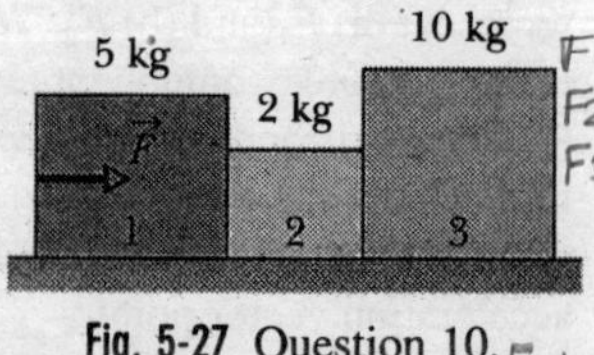

Fig. 5-27 Question 10.

11. In Fig. 5-28a, a toy box is on top of a (heavier) dog house, which sits on a wood floor. In Fig. 5-28b, these objects are represented by dots at the corresponding heights, and six vertical vectors (not to scale) are shown. Which of the vectors best represents

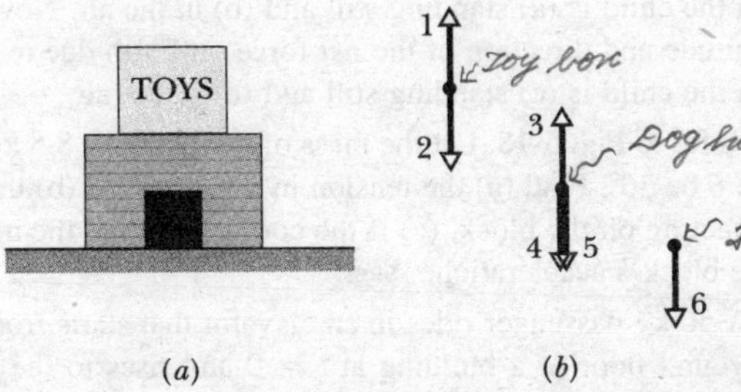

Fig. 5-28 Question 11.

(a) the gravitational force on the dog house, (b) the gravitational force on the toy box, (c) the force on the toy box from the dog house, (d) the force on the dog house from the toy box, (e) the force on the dog house from the floor, and (f) the force on the floor from the dog house? (g) Which of the forces are equal in magnitude? Which are (h) greatest and (i) least in magnitude?

12. In Fig. 5-29*a*, a block is attached by a rope to a bar that is itself rigidly attached to a ramp. Determine whether the magnitudes of the following increase, decrease, or remain the same as the angle θ of the ramp is increased from zero: (a) the component of the gravitational force $\vec{F}_g$ on the block that is along the ramp, (b) the tension in the cord, (c) the component of $\vec{F}_g$ that is perpendicular to the ramp, and (d) the normal force on the block from the ramp. (e) Which of the curves in Fig. 5-29*b* corresponds to each of the quantities in parts (a) through (d)?

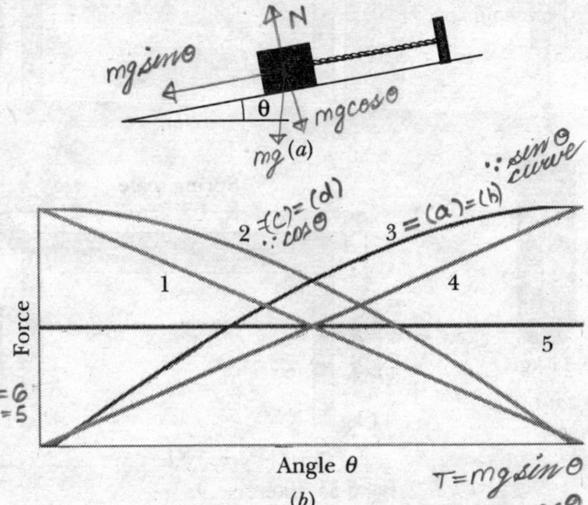

Fig. 5-29 Question 12.

EXERCISES & PROBLEMS

ssm Solution is in the Student Solutions Manual.
www Solution is available on the World Wide Web at: http://www.wiley.com/college/hrw
ilw Solution is available on the Interactive LearningWare.

SEC. 5-5 Newton's Second Law

1E. If the 1 kg standard body has an acceleration of 2.00 m/s^2 at 20° to the positive direction of the x axis, then what are (a) the x component and (b) the y component of the net force on it, and (c) what is the net force in unit-vector notation?

2E. Two horizontal forces act on a 2.0 kg chopping block that can slide over a frictionless kitchen counter, which lies in an xy plane. One force is $\vec{F}_1 = (3.0\text{ N})\hat{i} + (4.0\text{ N})\hat{j}$. Find the acceleration of the chopping block in unit-vector notation when the other force is (a) $\vec{F}_2 = (-3.0\text{ N})\hat{i} + (-4.0\text{ N})\hat{j}$, (b) $\vec{F}_2 = (-3.0\text{ N})\hat{i} + (4.0\text{ N})\hat{j}$, and (c) $\vec{F}_2 = (3.0\text{ N})\hat{i} + (-4.0\text{ N})\hat{j}$.

3E. Only two horizontal forces act on a 3.0 kg body. One force is 9.0 N, acting due east, and the other is 8.0 N, acting 62° north of west. What is the magnitude of the body's acceleration?

4E. While two forces act on it, a particle is to move at the constant velocity $\vec{v} = (3\text{ m/s})\hat{i} - (4\text{ m/s})\hat{j}$. One of the forces is $\vec{F}_1 = (2\text{ N})\hat{i} + (-6\text{ N})\hat{j}$. What is the other force?

5E. Three forces act on a particle that moves with unchanging velocity $\vec{v} = (2\text{ m/s})\hat{i} - (7\text{ m/s})\hat{j}$. Two of the forces are $\vec{F}_1 = (2\text{ N})\hat{i} + (3\text{ N})\hat{j} + (-2\text{ N})\hat{k}$ and $\vec{F}_2 = (-5\text{ N})\hat{i} + (8\text{ N})\hat{j} + (-2\text{ N})\hat{k}$. What is the third force?

6P. Three astronauts, propelled by jet backpacks, push and guide a 120 kg asteroid toward a processing dock, exerting the forces shown in Fig. 5-30. What is the asteroid's acceleration (a) in unit-vector notation and as (b) a magnitude and (c) a direction?

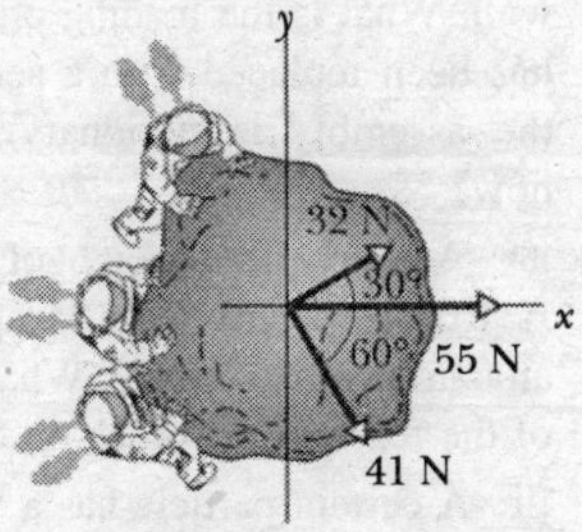

Fig. 5-30 Problem 6.

7P. There are two forces on the 2.0 kg box in the overhead view of Fig. 5-31 but only one is shown. The figure also shows the acceleration of the box. Find the second force (a) in unit-vector notation and as (b) a magnitude and (c) a direction. **ssm**

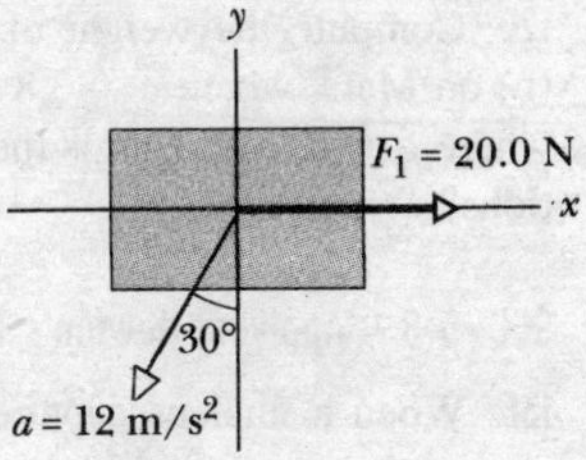

Fig. 5-31 Problem 7.

8P. Figure 5-32 is an overhead view of a 12 kg tire that is to be pulled by three ropes. One force ($\vec{F}_1$, with magnitude 50 N) is indicated. Orient the other two forces $\vec{F}_2$ and $\vec{F}_3$ so that the magnitude of the resulting acceleration of the tire is least, and find that magnitude if (a) $F_2 = 30$ N, $F_3 = 20$ N; (b) $F_2 = 30$ N, $F_3 = 10$ N; and (c) $F_2 = F_3 = 30$ N.

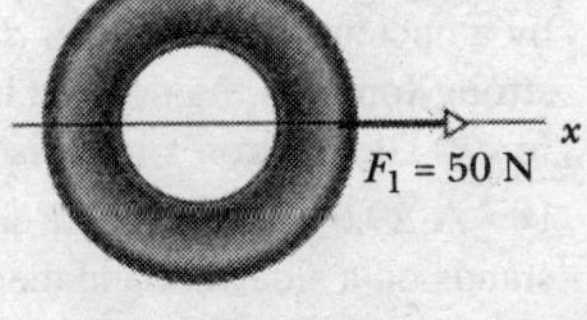

Fig. 5-32 Problem 8.

SEC. 5-6 Some Particular Forces

9E. (a) An 11.0 kg salami is supported by a cord that runs to a spring scale, which is supported

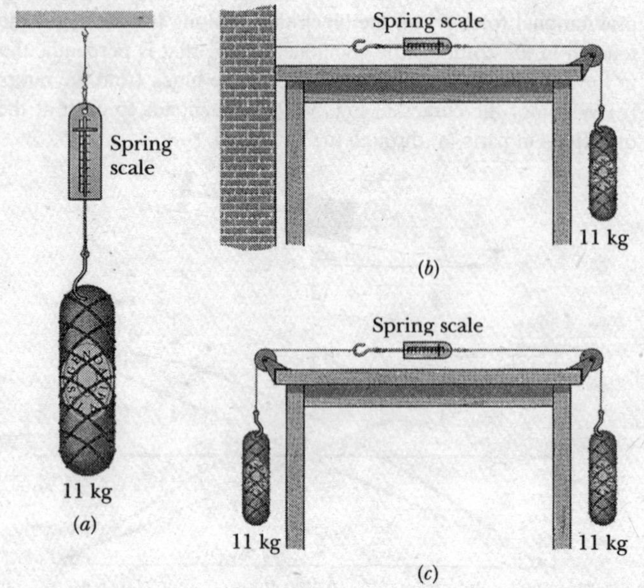

Fig. 5-33 Exercise 9.

by another cord from the ceiling (Fig. 5-33*a*). What is the reading on the scale, which is marked in weight units? (b) In Fig. 5-33*b* the salami is supported by a cord that runs around a pulley and to a scale. The opposite end of the scale is attached by a cord to a wall. What is the reading on the scale? (c) In Fig. 5-33*c* the wall has been replaced with a second 11.0 kg salami on the left, and the assembly is stationary. What is the reading on the scale now? ssm www

10E. A block with a weight of 3.0 N is at rest on a horizontal surface. A 1.0 N upward force is applied to the block by means of an attached vertical string. What are the magnitude and the direction of the force of the block on the horizontal surface?

11E. A certain particle has a weight of 22 N at a point where $g = 9.8\ \text{m/s}^2$. What are its (a) weight and (b) mass at a point where $g = 4.9\ \text{m/s}^2$? What are its (c) weight and (d) mass if it is moved to a point in space where $g = 0$? ssm

12E. Compute the weight of a 75 kg space ranger (a) on Earth, (b) on Mars, where $g = 3.8\ \text{m/s}^2$, and (c) in interplanetary space, where $g = 0$. (d) What is the ranger's mass at each of these locations?

SEC. 5-8 Applying Newton's Laws

13E. When a nucleus captures a stray neutron, it must bring the neutron to a stop within the diameter of the nucleus by means of the *strong force*. That force, which "glues" the nucleus together, is approximately zero outside the nucleus. Suppose that a stray neutron with an initial speed of 1.4×10^7 m/s is just barely captured by a nucleus with diameter $d = 1.0 \times 10^{-14}$ m. Assuming that the strong force on the neutron is constant, find the magnitude of that force. The neutron's mass is 1.67×10^{-27} kg. ssm

14E. A 29.0 kg child, with a 4.50 kg backpack on his back, first stands on a sidewalk and then jumps up into the air. Find the magnitude and direction of the force on the sidewalk from the child when the child is (a) standing still and (b) in the air. Now find the magnitude and direction of the *net* force on Earth due to the child when the child is (c) standing still and (d) in the air.

15E. Refer to Fig. 5-18. Let the mass of the block be 8.5 kg and the angle θ be 30°. Find (a) the tension in the cord and (b) the normal force acting on the block. (c) If the cord is cut, find the magnitude of the block's acceleration. ssm

16E. A 50 kg passenger rides in an elevator that starts from rest on the ground floor of a building at $t = 0$ and rises to the top floor during a 10 s interval. The acceleration of the elevator as a function of the time is shown in Fig. 5-34, where positive values of the acceleration mean that it is directed upward. Give the magnitude and direction of the following forces: (a) the maximum force on the passenger from the floor, (b) the minimum force on the passenger from the floor, and (c) the maximum force on the floor from the passenger.

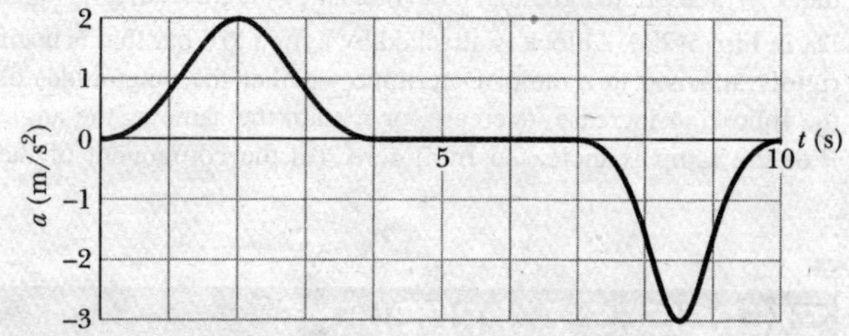

Fig. 5-34 Exercise 16.

17E. *Sunjamming*. A "sun yacht" is a spacecraft with a large sail that is pushed by sunlight. Although such a push is tiny in everyday circumstances, it can be large enough to send the spacecraft outward from the Sun on a cost-free but slow trip. Suppose that the spacecraft has a mass of 900 kg and receives a push of 20 N. (a) What is the magnitude of the resulting acceleration? If the craft starts from rest, (b) how far will it travel in 1 day and (c) how fast will it then be moving?

18E. The tension at which a fishing line snaps is commonly called the line's "strength." What minimum strength is needed for a line that is to stop a salmon of weight 85 N in 11 cm if the fish is initially drifting at 2.8 m/s? Assume a constant deceleration.

19E. An experimental rocket sled can be accelerated at a constant rate from rest to 1600 km/h in 1.8 s. What is the magnitude of the required net force if the sled has a mass of 500 kg? ssm

20E. A car that weighs 1.30×10^4 N is initially moving at a speed of 40 km/h when the brakes are applied and the car is brought to a stop in 15 m. Assuming that the force that stops the car is constant, find (a) the magnitude of that force and (b) the time required for the change in speed. If the initial speed is doubled, and the car experiences the same force during the braking, by what factors are (c) the stopping distance and (d) the stopping time multiplied? (There could be a lesson here about the danger of driving at high speeds.)

21E. An electron with a speed of 1.2×10^7 m/s moves horizontally into a region where a constant vertical force of 4.5×10^{-16} N acts on it. The mass of the electron is 9.11×10^{-31} kg. Determine the vertical distance the electron is deflected during the time it has moved 30 mm horizontally. ssm

22E. A car traveling at 53 km/h hits a bridge abutment. A passenger in the car moves forward a distance of 65 cm (with respect to the road) while being brought to rest by an inflated air bag. What magnitude of force (assumed constant) acts on the passenger's upper torso, which has a mass of 41 kg?

23E. Tarzan, who weighs 820 N, swings from a cliff at the end of a 20 m vine that hangs from a high tree limb and initially makes an angle of 22° with the vertical. Immediately after Tarzan steps off the cliff, the tension in the vine is 760 N. Choose a coordinate system for which the x axis points horizontally away from the edge of the cliff and the y axis points upward. (a) What is the force of the vine on Tarzan in unit-vector notation? (b) What is the net force acting on Tarzan in unit-vector notation? What are (c) the magnitude and (d) the direction of the net force acting on Tarzan? What are (e) the magnitude and (f) the direction of Tarzan's acceleration?

24P. A 50 kg skier is pulled up a frictionless ski slope that makes an angle of 8.0° with the horizontal by holding onto a tow rope that moves parallel to the slope. Determine the magnitude of the force of the rope on the skier at an instant when (a) the rope is moving with a constant speed of 2.0 m/s and (b) the rope is moving with a speed of 2.0 m/s but that speed is increasing at a rate of 0.10 m/s^2.

25P. A 40 kg girl and an 8.4 kg sled are on the frictionless ice of a frozen lake, 15 m apart but connected by a rope of negligible mass. The girl exerts a horizontal 5.2 N force on the rope. (a) What is the acceleration of the sled? (b) What is the acceleration of the girl? (c) How far from the girl's initial position do they meet? ssm

26P. You pull a short refrigerator with a constant force $\vec{F}$ across a greased (frictionless) floor, either with $\vec{F}$ horizontal (case 1) or with $\vec{F}$ tilted upward at an angle θ (case 2). (a) What is the ratio of the refrigerator's speed in case 2 to its speed in case 1 if you pull for a certain time t? (b) What is this ratio if you pull for a certain distance d?

27P. A firefighter with a weight of 712 N slides down a vertical pole with an acceleration of 3.00 m/s^2, directed downward. What are the magnitudes and directions of the vertical forces (a) on the firefighter from the pole and (b) on the pole from the firefighter? ilw

28P. For sport, a 12 kg armadillo runs onto a large pond of level, frictionless ice with an initial velocity of 5.0 m/s along the positive direction of an x axis. Take its initial position on the ice as being the origin. It slips over the ice while being pushed by a wind with a force of 17 N in the positive direction of the y axis. In unit-vector notation, what are the animal's (a) velocity and (b) position vector when it has slid for 3.0 s?

29P. A sphere of mass 3.0×10^{-4} kg is suspended from a cord. A steady horizontal breeze pushes the sphere so that the cord makes a constant angle of 37° with the vertical. Find (a) the magnitude of that push and (b) the tension in the cord. ssm ilw www

30P. A 40 kg skier comes directly down a frictionless ski slope that is inclined at an angle of 10° with the horizontal while a strong wind blows parallel to the slope. Determine the magnitude and direction of the force of the wind on the skier if (a) the magnitude of the skier's velocity is constant, (b) the magnitude of the skier's velocity is increasing at a rate of 1.0 m/s^2, and (c) the magnitude of the skier's velocity is increasing at a rate of 2.0 m/s^2.

31P. Two blocks are in contact on a frictionless table. A horizontal force is applied to the larger block, as shown in Fig. 5-35. (a) If $m_1 = 2.3$ kg, $m_2 = 1.2$ kg, and $F = 3.2$ N, find the magnitude of the force between the two blocks. (b) Show that if a force of the same magnitude F is applied to the smaller block but in the opposite direction, the magnitude of the force between the blocks is 2.1 N, which is not the same value calculated in (a). (c) Explain the difference. ssm ilw

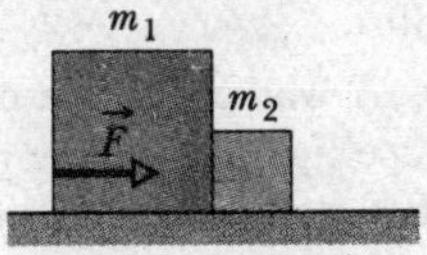

Fig. 5-35 Problem 31.

32P. A 1400 kg jet engine is fastened to the fuselage of a passenger jet by just three bolts (this is the usual practice). Assume that each bolt supports one-third of the load. (a) Calculate the force on each bolt as the plane waits in line for clearance to take off. (b) During flight, the plane encounters turbulence, which suddenly imparts an upward vertical acceleration of 2.6 m/s^2 to the plane. Calculate the force on each bolt now.

33P. An elevator and its load have a combined mass of 1600 kg. Find the tension in the supporting cable when the elevator, originally moving downward at 12 m/s, is brought to rest with constant acceleration in a distance of 42 m. ssm

34P. Figure 5-36 shows four penguins that are being playfully pulled along very slippery (frictionless) ice by a curator. The masses of three penguins and the tension in two of the cords are given. Find the penguin mass that is not given.

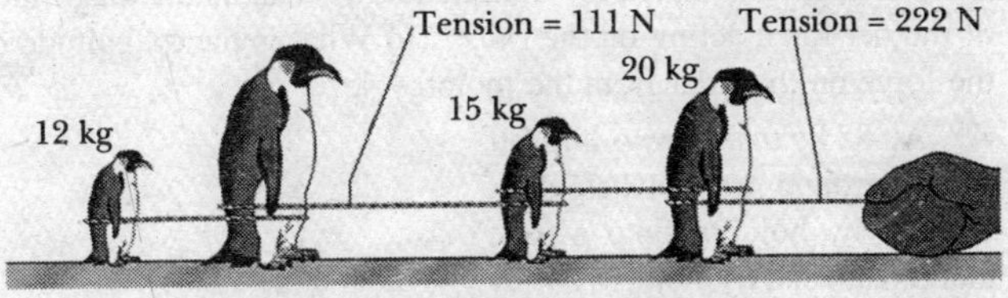

Fig. 5-36 Problem 34.

35P. An 80 kg person is parachuting and experiencing a downward acceleration of 2.5 m/s^2. The mass of the parachute is 5.0 kg. (a) What is the upward force on the open parachute from the air? (b) What is the downward force on the parachute from the person? ssm

36P. In Fig. 5-37, three blocks are connected and pulled to the right on a horizontal frictionless table by a force with a magnitude of $T_3 = 65.0$ N. If $m_1 = 12.0$ kg, $m_2 = 24.0$ kg, and $m_3 = 31.0$ kg, calculate (a) the acceleration of the system and the tensions (b) T_1 and (c) T_2 in the interconnecting cords.

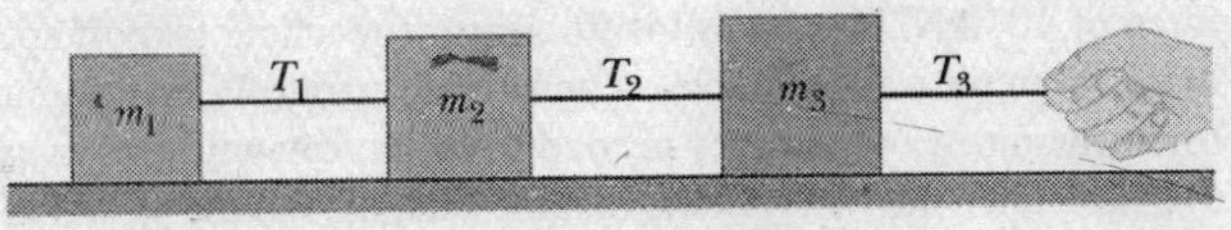

Fig. 5-37 Problem 36.

37P. Imagine a landing craft approaching the surface of Callisto, one of Jupiter's moons. If the engine provides an upward force (thrust) of 3260 N, the craft descends at constant speed; if the engine provides only 2200 N, the craft accelerates downward at 0.39 m/s^2. (a) What is the weight of the landing craft in the vicinity of Callisto's surface? (b) What is the mass of the craft? (c) What is the magnitude of the free-fall acceleration near the surface of Callisto? ssm

38P. A worker drags a crate across a factory floor by pulling on a rope tied to the crate (Fig. 5-38). The worker exerts a force of 450 N on the rope, which is inclined at 38° to the horizontal, and the floor exerts a horizontal force of 125 N that opposes the motion. Calculate the magnitude of the acceleration of the crate if (a) its mass is 310 kg and (b) its weight is 310 N.

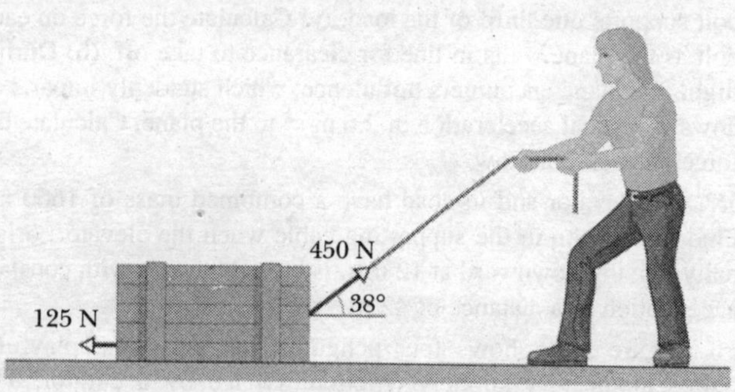

Fig. 5-38 Problem 38.

39P. A motorcycle and 60.0 kg rider accelerate at 3.0 m/s^2 up a ramp inclined 10° above the horizontal. (a) What is the magnitude of the net force acting on the rider? (b) What is the magnitude of the force on the rider from the motorcycle?

40P. An 85 kg man lowers himself to the ground from a height of 10.0 m by holding onto a rope that runs over a frictionless pulley to a 65 kg sandbag. With what speed does the man hit the ground if he started from rest?

41P. In Fig. 5-39, a chain consisting of five links, each of mass 0.100 kg, is lifted vertically with a constant acceleration of 2.50 m/s^2. Find the magnitudes of (a) the force on link 1 from link 2, (b) the force on link 2 from link 3, (c) the force on link 3 from link 4, and (d) the force on link 4 from link 5. Then find the magnitudes of (e) the force $\vec{F}$ on the top link from the person lifting the chain and (f) the *net* force accelerating each link. ssm

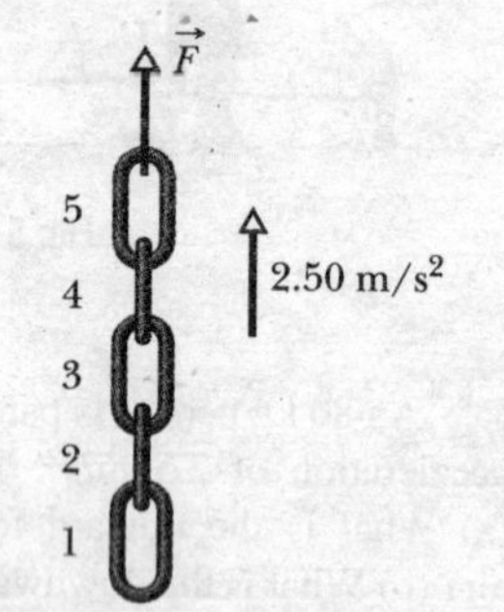

Fig. 5-39 Problem 41.

42P. A Navy jet (Fig. 5-40) with a weight of 231 kN requires an airspeed of 85 m/s for liftoff. The engine develops a maximum force of 107 kN, but that is insufficient for reaching takeoff speed in the 90 m runway available on an aircraft carrier. What minimum force (assumed constant) is needed from the catapult that is used to help launch the jet? Assume that the catapult and the jet's engine each exert a constant force over the 90 m distance used for takeoff.

Fig. 5-40 Problem 42.

43P. A block of mass $m_1 = 3.70$ kg on a frictionless inclined plane of angle 30.0° is connected by a cord over a massless, frictionless pulley to a second block of mass $m_2 = 2.30$ kg hanging vertically (Fig. 5-41). What are (a) the magnitude of the acceleration of each block and (b) the direction of the acceleration of the hanging block? (c) What is the tension in the cord? ssm ilw www

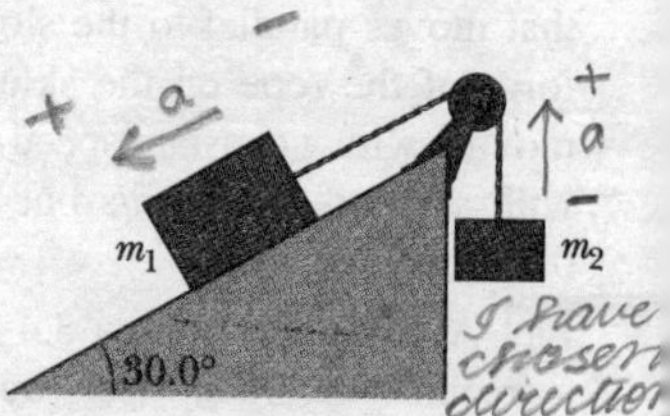

Fig. 5-41 Problem 43.

44P. In Fig. 5-42, a 1.0 kg pencil box on a 30° frictionless incline is connected to a 3.0 kg pen box on a horizontal frictionless surface. The pulley is frictionless and massless. (a) If the magnitude of $\vec{F}$ is 2.3 N, what is the tension in the connecting cord? (b) What is the largest value that the magnitude of $\vec{F}$ may have without the connecting cord becoming slack?

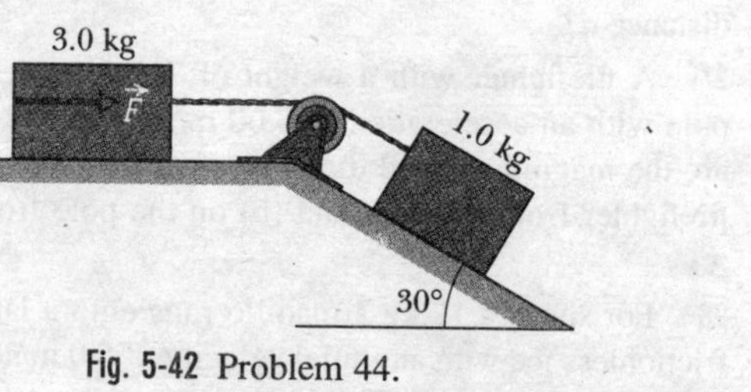

Fig. 5-42 Problem 44.

45P. A block is projected up a frictionless inclined plane with initial speed $v_0 = 3.50$ m/s. The angle of incline is $\theta = 32.0°$. (a) How far up the plane does it go? (b) How long does it take to get there? (c) What is its speed when it gets back to the bottom? ssm

46P. An interstellar ship has a mass of 1.20×10^6 kg and is initially at rest relative to a star system. (a) What constant acceleration is needed to bring the ship up to a speed of $0.10c$ (where c is the speed of light, 3.0×10^8 m/s) relative to the star system in 3.0 days? (b) What is that acceleration in g units? (c) What force is required for the acceleration? (d) If the engines are shut down when $0.10c$ is reached (the speed then remains constant), how long does the ship take (start to finish) to journey 5.0 light-months, the distance that light travels in 5.0 months?

47P. A 10 kg monkey climbs up a massless rope that runs over a frictionless tree limb and back down to a 15 kg package on the ground (Fig. 5-43). (a) What is the magnitude of the least acceleration the monkey must have if it is to lift the package off the ground? If, after the package has been lifted, the monkey stops its climb and holds onto the rope, what are (b) the magnitude and (c) the direction of the monkey's acceleration, and (d) what is the tension in the rope? ssm

Fig. 5-43 Problem 47.

48P. In earlier days, horses pulled barges down canals in the manner shown in Fig. 5-44. Suppose that the horse pulls on the rope with a force of 7900 N at an angle of 18° to the direction of motion of the barge, which is headed straight along the canal. The mass of the barge is 9500 kg, and its acceleration is 0.12 m/s^2. What are (a) the magnitude and (b) the direction of the force on the barge from the water?

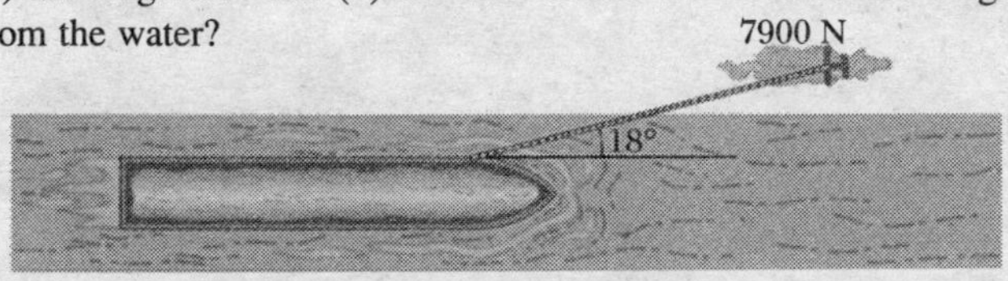

Fig. 5-44 Problem 48.

49P. In Fig. 5-45, a 5.00 kg block is pulled along a horizontal frictionless floor by a cord that exerts a force of magnitude $F = 12.0$ N at an angle $\theta = 25.0°$ above the horizontal. (a) What is the magnitude of the block's acceleration? (b) The force magnitude F is slowly increased. What is its value just before the block is lifted (completely) off the floor? (c) What is the magnitude of the block's acceleration just before it is lifted (completely) off the floor? ssm

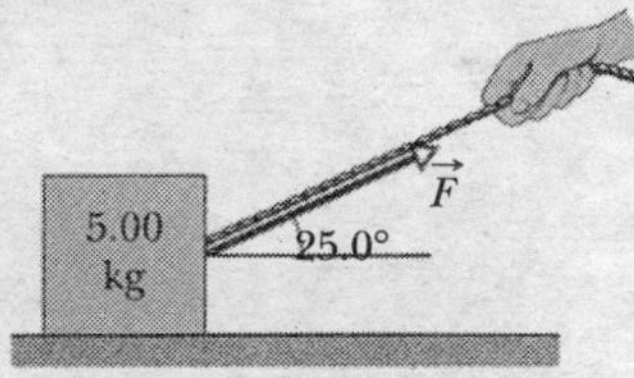

Fig. 5-45 Problem 49.

50P. Figure 5-46 shows a man sitting in a bosun's chair that dangles from a massless rope, which runs over a massless, frictionless pulley and back down to the man's hand. The combined mass of man and chair is 95.0 kg. With what force magnitude must the man pull on the rope if he is to rise (a) with a constant velocity and (b) with an upward acceleration of 1.30 m/s^2? (*Hint:* A free-body diagram can really help.) Problem continues, next column.

Fig. 5-46 Problem 50.

Suppose, instead, that the rope on the right extends to the ground, where it is pulled by a co-worker. With what force magnitude must the co-worker pull for the man to rise (c) with a constant velocity and (d) with an upward acceleration of 1.30 m/s^2? What is the magnitude of the force on the ceiling from the pulley system in (e) part a (f) part b, (g) part c, and (h) part d?

51P. A block of mass M is pulled along a horizontal frictionless surface by a rope of mass m, as shown in Fig. 5-47. A horizontal force $\vec{F}$ is applied to one end of the rope. (a) Show that the rope *must* sag, even if only by an imperceptible amount. Then, assuming that the sag is negligible, find (b) the acceleration of rope and block, (c) the force on the block from the rope, and (d) the tension in the rope at its midpoint. ssm

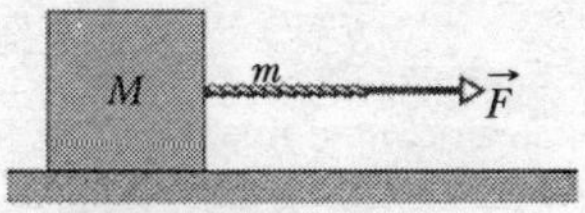

Fig. 5-47 Problem 51.

52P. In Fig. 5-48, a 100 kg crate is pushed at constant speed up the frictionless 30.0° ramp by a horizontal force $\vec{F}$. What are the magnitudes of (a) $\vec{F}$ and (b) the force on the crate from the ramp?

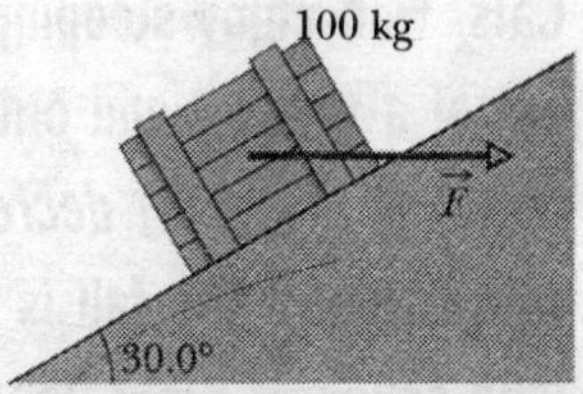

Fig. 5-48 Problem 52.

53P. A hot-air balloon of mass M is descending vertically with downward acceleration of magnitude a. How much mass (ballast) must be thrown out to give the balloon an upward acceleration of magnitude a (same magnitude but opposite direction)? Assume that the upward force from the air (the lift) does not change because of the decrease in mass. ssm ilw www

54P. Figure 5-49 shows a section of an alpine cable-car system. The maximum permissible mass of each car with occupants is 2800 kg. The cars, riding on a support cable, are pulled by a second cable attached to each pylon (support tower); assume the cables are straight. What is the difference in tension between adjacent sections of pull cable if the cars are at the maximum permissible mass and are being accelerated up the 35° incline at 0.81 m/s^2?

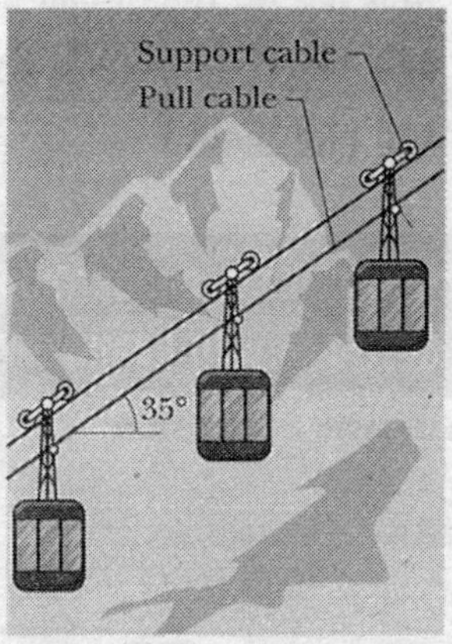

Fig. 5-49 Problem 54.

55P. An elevator with a weight of 27.8 kN is given an upward acceleration of 1.22 m/s^2 by a cable. (a) Calculate the tension in the cable. (b) What is the tension when the elevator is decelerating at the rate of 1.22 m/s^2 but is still moving upward?

56P A lamp hangs vertically from a cord in a descending elevator that decelerates at 2.4 m/s^2. (a) If the tension in the cord is 89 N, what is the lamp's mass? (b) What is the cord's tension when the elevator ascends with an upward acceleration of 2.4 m/s^2?

6 Force and Motion — II

Cats, who enjoy sleeping on window sills, are often kept in apartment buildings. When a cat accidentally falls out of a window and onto a sidewalk, the extent of injury (such as the number of fractured bones or the certainty of death) *decreases* with height if the fall is more than seven or eight floors. (There is even a record of a cat who fell 32 floors and suffered only slight damage to its thorax and one tooth.)

How can the danger possibly decrease with height?

The answer is in this chapter.

6-1 Friction

Frictional forces are unavoidable in our daily lives. If we were not able to counteract them, they would stop every moving object and bring to a halt every rotating shaft. About 20% of the gasoline used in an automobile is needed to counteract friction in the engine and in the drive train. On the other hand, if friction were totally absent, we could not get an automobile to go anywhere, and we could not walk or ride a bicycle. We could not hold a pencil and, if we could, it would not write. Nails and screws would be useless, woven cloth would fall apart, and knots would untie.

Here we deal with the frictional forces that exist between dry solid surfaces, either stationary relative to each other or moving across each other at slow speeds. Consider three simple thought experiments:

1. Send a book sliding across a long horizontal counter. As expected, the book slows and then stops. This means the book must have an acceleration parallel to the counter surface, in the direction opposite the book's velocity. From Newton's law, then, a force must act on the book parallel to the counter surface, in the direction opposite its velocity. That force is a frictional force.
2. Push horizontally on the book to make it travel at constant velocity along the counter. Can the force from you be the only horizontal force on the book? No, because then the book would accelerate. From Newton's second law, there must be a second force, directed opposite your force but with the same magnitude, so that the two forces balance. That second force is a frictional force, directed parallel to the counter.
3. Push horizontally on a heavy crate. The crate does not move. From Newton's second law, a second force must also act on the crate to counteract your force. Moreover, it must be directed opposite your force and have the same magnitude as your force, so that the two forces balance. That second force is a frictional force. Push even harder. The crate still does not move. Apparently the frictional force can change in magnitude so that the two forces still balance. Now push with all your strength. The crate begins to slide. Evidently, there is a maximum magnitude of the frictional force. When you exceed that maximum magnitude, the crate slides.

Figure 6-1 shows a similar situation in detail. In Fig. 6-1*a*, a block rests on a tabletop, with the gravitational force $\vec{F}_g$ balanced by a normal force $\vec{N}$. In Fig. 6-1*b*, you exert a force $\vec{F}$ on the block, attempting to pull it to the left. In response, a frictional force $\vec{f}_s$ is directed to the right, exactly balancing your force. The force $\vec{f}_s$ is called the **static frictional force.** The block does not move.

Figures 6-1*c* and 6-1*d* show that as you increase the magnitude of your applied force, the magnitude of the static frictional force $\vec{f}_s$ also increases and the block remains at rest. When the applied force reaches a certain magnitude, however, the block "breaks away" from its intimate contact with the tabletop and accelerates leftward (Fig. 6-1*e*). The frictional force that then opposes the motion is called the **kinetic frictional force** $\vec{f}_k$.

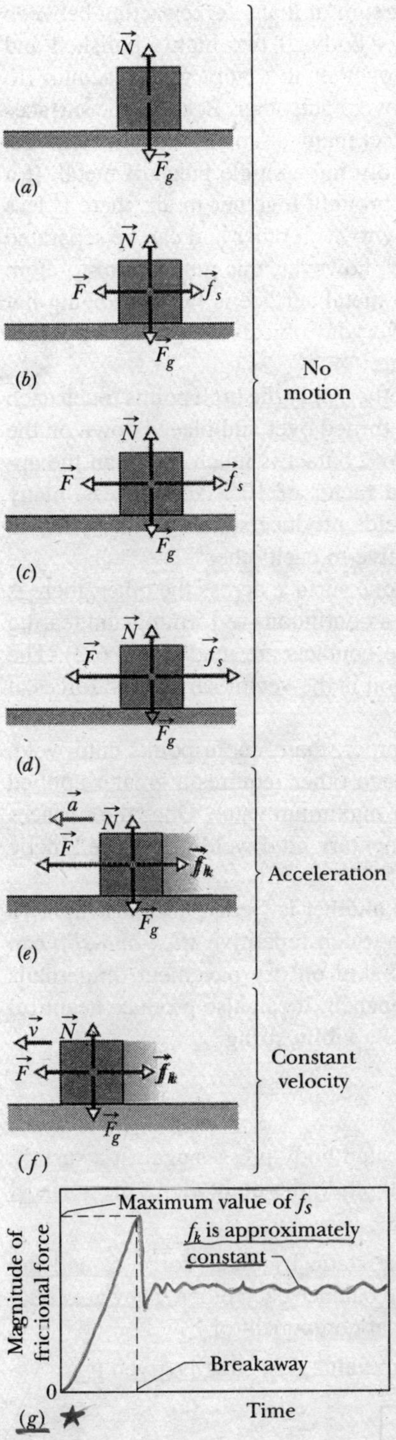

Fig. 6-1 (*a*) The forces on a stationary block. (*b–d*) An external force $\vec{F}$, applied to the block, is balanced by a static frictional force $\vec{f}_s$. As F is increased, f_s also increases, until f_s reaches a certain maximum value. (*e*) The block then "breaks away," accelerating suddenly in the direction of $\vec{F}$. (*f*) If the block is now to move with constant velocity, F must be reduced from the maximum value it had just before the block broke away. (*g*) Some experimental results for the sequence (*a*) through (*f*).

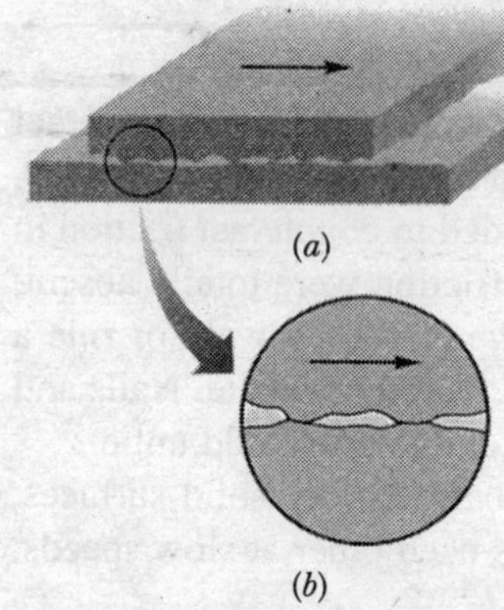

Fig. 6-2 The mechanism of sliding friction. (*a*) The upper surface is sliding to the right over the lower surface in this enlarged view. (*b*) A detail, showing two spots where cold-welding has occurred. Force is required to break the welds and maintain the motion.

★ Usually, the magnitude of the kinetic frictional force, which acts when there is motion, is less than the maximum magnitude of the static frictional force, which acts when there is no motion. Thus, if you wish the block to move across the surface with a constant speed, you must usually decrease the magnitude of the applied force once the block begins to move, as in Fig. 6-1*f*. As an example, Fig. 6-1*g* shows the results of an experiment in which the force on a block was slowly increased until breakaway occurred. Note the reduced force needed to keep the block moving at constant speed after breakaway.

A frictional force is, in essence, the vector sum of many forces acting between surface atoms of one body and those of another body. If two highly polished and carefully cleaned metal surfaces are brought together in a very good vacuum (to keep them clean), they cannot be made to slide over each other. Because the surfaces are so smooth, many atoms of one surface contact many atoms of the other surface, and the surfaces *cold-weld* together instantly, forming a single piece of metal. If a machinist's specially polished gage blocks are brought together in air, there is less atom-to-atom contact but the blocks stick firmly to each other and can be separated only by means of a wrenching motion. Usually, however, this much atom-to-atom contact is not possible. Even a highly polished metal surface is far from being flat on the atomic scale. Moreover, the surfaces of everyday objects have layers of oxides and other contaminants that reduce cold-welding.

When two ordinary surfaces are placed together, only the high points touch each other. (It is like having the Alps of Switzerland turned over and placed down on the Alps of Austria.) The actual *micro*scopic area of contact is much less than the apparent *macro*scopic contact area, perhaps by a factor of 10^4. Nonetheless, many contact points do cold-weld together. These welds produce static friction when an applied force attempts to slide the surfaces relative to each other.

If the applied force is great enough to pull one surface across the other, there is first a tearing of welds (at breakaway) and then a continuous re-forming and tearing apart of welds as movement occurs and chance contacts are made (Fig. 6-2). The kinetic frictional force $\vec{f}_k$ that opposes the motion is the vector sum of the forces at those many chance contacts.

★ If the two surfaces are pressed together harder, many more points cold-weld. Then, getting the surfaces to slide relative to each other requires a greater applied force: The static frictional force $\vec{f}_s$ has a greater maximum value. Once the surfaces are sliding, there are many more points of momentary cold-welding, so the kinetic frictional force $\vec{f}_k$ also has a greater magnitude.

Often, the sliding motion of one surface on another is "jerky," because the two surfaces alternately stick together and then slip. Such repetitive *stick-and-slip* can produce squeaking or squealing, as when tires skid on dry pavement, fingernails scratch along a chalkboard, or a rusty hinge is opened. It can also produce beautiful sounds, as when a bow is drawn properly across a violin string.

6-2 Properties of Friction

Experiment shows that when a dry and unlubricated body presses against a surface in the same condition, and a force $\vec{F}$ attempts to slide the body along the surface, the resulting frictional force has three properties:

★ **Property 1.** If the body does not move, then the static frictional force $\vec{f}_s$ and the component of $\vec{F}$ that is parallel to the surface balance each other. They are equal in magnitude, and $\vec{f}_s$ is directed opposite that component of $\vec{F}$.

★ **Property 2.** The magnitude of $\vec{f}_s$ has a maximum value $f_{s,\max}$ that is given by

$$f_{s,\max} = \mu_s N, \tag{6-1}$$

where μ_s is the **coefficient of static friction** and N is the magnitude of the normal force on the body from the surface. If the magnitude of the component of $\vec{F}$ that is parallel to the surface exceeds $f_{s,\max}$, then the body begins to slide along the surface.

Property 3. If the body begins to slide along the surface, the magnitude of the frictional force rapidly decreases to a value f_k given by

$$f_k = \mu_k N, \tag{6-2}$$

where μ_k is the **coefficient of kinetic friction.** Thereafter, during the sliding, a kinetic frictional force $\vec{f}_k$ with magnitude given by Eq. 6-2 opposes the motion.

The magnitude N of the normal force appears in properties 2 and 3 as a measure of how firmly the body presses against the surface. If the body presses harder, then, by Newton's third law, N is greater. Properties 1 and 2 are worded in terms of a single applied force $\vec{F}$, but they also hold for the net force of several applied forces acting on the body. Equations 6-1 and 6-2 are *not* vector equations; the direction of $\vec{f}_s$ or $\vec{f}_k$ is always parallel to the surface and opposed to the attempted sliding, and the normal force $\vec{N}$ is perpendicular to the surface.

The coefficients μ_s and μ_k are dimensionless and must be determined experimentally. Their values depend on certain properties of both the body and the surface; hence, they are usually referred to with the preposition "between," as in "the value of μ_s *between* an egg and a Teflon-coated skillet is 0.04, but that *between* rock-climbing shoes and rock is as much as 1.2." We assume that the value of μ_k does not depend on the speed at which the body slides along the surface.

CHECKPOINT 1: A block lies on a floor. (a) What is the magnitude of the frictional force on it from the floor? (b) If a horizontal force of 5 N is now applied to the block, but the block does not move, what is the magnitude of the frictional force on it? (c) If the maximum value $f_{s,\max}$ of the static frictional force on the block is 10 N, will the block move if the magnitude of the horizontally applied force is 8 N? (d) If the magnitude is 12 N? (e) What is the magnitude of the frictional force in part (c)?

Sample Problem 6-1

If a car's wheels are "locked" (kept from rolling) during emergency braking, the car slides along the road. Ripped-off bits of tire and small melted sections of road form the "skid marks" that reveal that cold-welding occurred during the slide. The record for the longest skid marks on a public road was reportedly set in 1960 by a Jaguar on the M1 highway in England (Fig. 6-3*a*)—the marks were 290 m long! Assuming that $\mu_k = 0.60$ and the car's acceleration was constant during the braking, how fast was the car going when the wheels became locked?

SOLUTION: One **Key Idea** here is that, because the acceleration is assumed constant, we can use the equations of Table 2-1 to find the car's initial speed v_0. Let us try Eq. 2-16,

$$v^2 = v_0^2 + 2a(x - x_0), \tag{6-3}$$

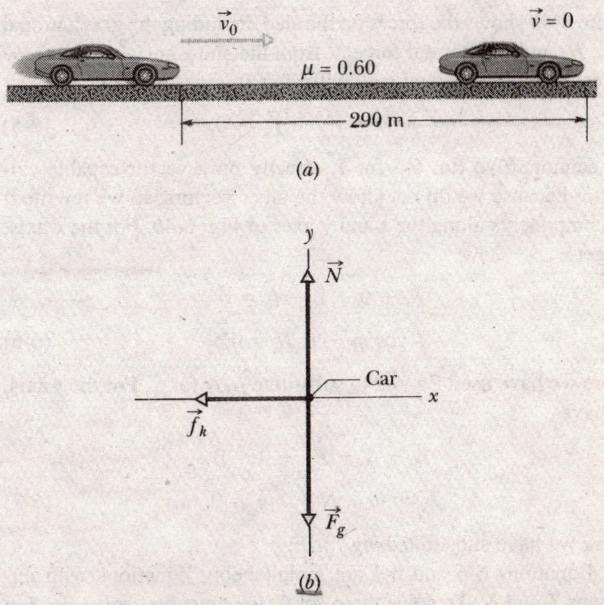

Fig. 6-3 Sample Problem 6-1. (*a*) A car, sliding to the right and finally stopping after a displacement of 290 m. (*b*) A free-body diagram for the car.

assuming that the car moved in the positive direction of an x axis. We know that the displacement $x - x_0$ was 290 m, we assume that the final speed v was 0, and we want v_0. However, we do not know the car's acceleration a.

To find a, we use another **Key Idea**: If we neglect the effects of the air on the car, the acceleration a was due only to a kinetic frictional force $\vec{f}_k$ on the car from the road, directed opposite the direction of the car's motion (Fig. 6-3b). We can relate this force to the acceleration by writing Newton's second law for x components ($F_{\text{net},x} = ma_x$) as

$$-f_k = ma,$$

where m is the car's mass. The minus sig[illegible] of the kinetic frictional force.

From Eq. 6-2, that frictional force [illegible] $\mu_k N$, where N is the magnitude of the norm[illegible] the road. Because the car is not accelerat[illegible]

from Fig. 6-3b and Newton's second law that the magnitude of $\vec{N}$ is equal to the magnitude of the gravitational force $\vec{F}_g$ on the car, which is mg. Thus, we have $N = mg$.

Now solving Eq. 6-4 for a and substituting $f_k = \mu_k N = \mu_k mg$ for f_k yield

$$a = -\frac{f_k}{m} = -\frac{\mu_k mg}{m} = -\mu_k g,$$

where the minus sign indicates that the acceleration is in the negative direction of the x axis, opposite the direction of the velocity. [illegible] and solving for [illegible]

[illegible] (290 m)

(Answer)

[illegible] arks. Actually, [illegible] ad after 290 m. [illegible] ore.

When a car is moving, every atom of car or of a person inside experience same acceleration but force on each is diff. If we approximate that every atom of car is same, then F on each is same every atom of
Frictional force b/w the surface of (any) tyre is μN where N is the TOTAL NORMAL REAC AS IF CAR WERE A POINT OBJECT

Another approach
Net force along $-\hat{x}$ direction is balanced by net force along $+\hat{x}$
∵ Two tyres support the car, by symmetry, it is as if each tyre supports half the car
Net force of friction $\mu g\left(\frac{M}{2}+\frac{M}{2}\right)$

Sample Problem 6-2

In Fig. 6-4a, a woman pulls a loaded sled o[illegible] kg along a horizontal surface at constant velocity. The coefficient of kinetic friction μ_k between the runners and the snow is 0.10, and the angle ϕ is 42°.

(a) What is the magnitude of the force $\vec{T}$ on the sled from the rope?

SOLUTION: We need three **Key Ideas** here:

1. Because the sled's velocity is constant, its acceleration is zero in spite of the woman's pull.
2. Acceleration is prevented by a kinetic frictional force $\vec{f}_k$ on the sled from the snow.
3. We can relate the (zero) acceleration of the sled to the forces on the sled, including the desired $\vec{T}$, with Newton's second law ($\vec{F}_{\text{net}} = m\vec{a}$).

Figure 6-4b shows the forces on the sled, including the gravitational force $\vec{F}_g$ and the normal force $\vec{N}$ from the snow surface. For these forces, Newton's second law, with $\vec{a} = 0$, gives us

$$\vec{T} + \vec{N} + \vec{F}_g + \vec{f}_k = 0. \tag{6-5}$$

We cannot solve Eq. 6-5 for $\vec{T}$ directly on a vector-capable calculator because we do not know the other vectors, so we rewrite it for components along the x and y axes of Fig. 6-4b. For the x axis, we get

$$T_x + 0 + 0 - f_k = 0$$

or

$$T\cos\phi - \mu_k N = 0, \tag{6-6}$$

where we have used Eq. 6-2 to substitute $\mu_k N$ for f_k. For the y axis, we have

$$T_y + N - F_g + 0 = 0$$

or

$$T\sin\phi + N - mg = 0, \tag{6-7}$$

where we have substituted mg for F_g.

Equations 6-6 and 6-7 are simultaneous equations with unknowns T and N. To solve them for T, we must first solve Eq. 6-6 for N and then substitute that expression into Eq. 6-7 to get

$$T = \frac{\mu_k mg}{\cos\phi + \mu_k \sin\phi}$$

$$= \frac{(0.10)(75\text{ kg})(9.8\text{ m/s}^2)}{\cos 42° + (0.10)(\sin 42°)}$$

$$= 91\text{ N}. \qquad \text{(Answer)}$$

(Instead, we could substitute known data into Eqs. 6-6 and 6-7 and then solve them by using a simultaneous-equations solver on a calculator.)

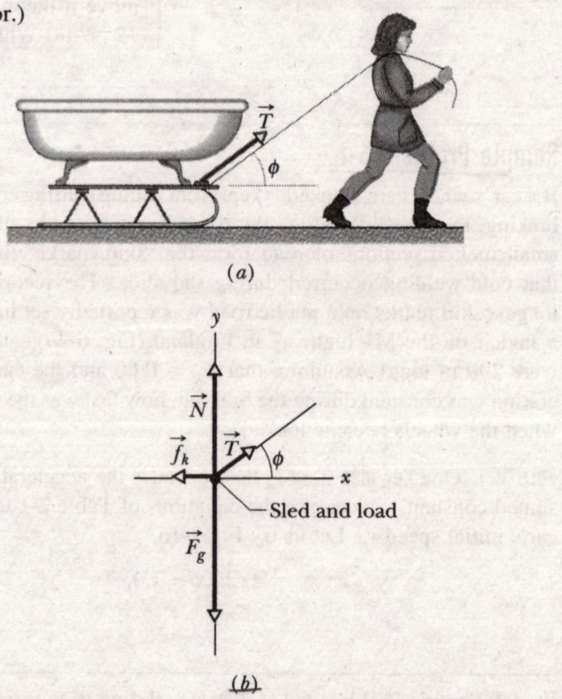

Fig. 6-4 Sample Problem 6-2. (a) A woman pulls a loaded sled at constant velocity via a force $\vec{T}$ on the sled from a rope. (b) A free-body diagram for the loaded sled.

(b) If the woman increases her pull on the rope, so that T is greater than 91 N, is the magnitude f_k of the frictional force greater than, less than, or the same as in (a)?

SOLUTION: The **Key Idea** here is that, by Eq. 6-2, the magnitude of f_k depends directly on the magnitude N of the normal force. Thus, we can answer if we find a relation between N and T. Equation 6-7 is such a relation. Rewriting it as

$$N = mg - T \sin \phi, \tag{6-8}$$

we note that if T is increased, then N will decrease. (The physical reason is that the upward component of the rope's pull is greater, and thus the force on the sled from the snow is less.) Because $f_k = \mu_k N$, we see that f_k will be less than it was.

CHECKPOINT 2: In the figure, horizontal force $\vec{F}_1$ of magnitude 10 N is applied to a box on a floor, but the box does not slide. Then, as the magnitude of vertically applied force $\vec{F}_2$ is increased from zero but before the box begins to slide, do the following quantities increase, decrease, or stay the same: (a) the magnitude of the frictional force on the box; (b) the magnitude of the normal force on the box from the floor; (c) the maximum value $f_{s,\max}$ of the static frictional force on the box?

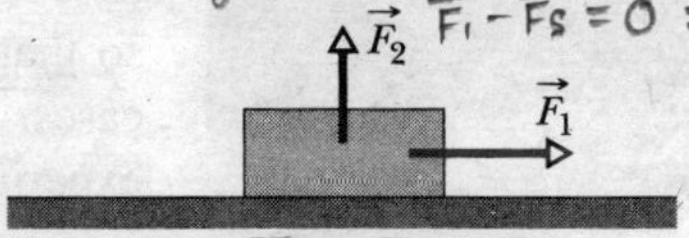

Sample Problem 6-3

Figure 6-5*a* shows a coin of mass m at rest on a book that has been tilted at an angle θ with the horizontal. By experimenting, you find that when θ is increased to 13°, the coin is on the *verge* of sliding down the book, which means that even a slight increase beyond 13° produces sliding. What is the coefficient of static friction μ_s between the coin and the book?

SOLUTION: If the book were frictionless, the coin would surely slide down it for any tilt of the book because of the gravitational force on the coin. Thus, one **Key Idea** here is that a frictional force $\vec{f}_s$ must be holding the coin in place. A second **Key Idea** is that, because the coin is *on the verge* of sliding *down* the book, that force is at its *maximum* magnitude $f_{s,\max}$ and is directed *up* the book. Also, from Eq. 6-1, we know that $f_{s,\max} = \mu_s N$, where N is the magnitude of the normal force $\vec{N}$ on the coin from the book. Thus,

$$f_s = f_{s,\max} = \mu_s N,$$

from which

$$\mu_s = \frac{f_s}{N}. \tag{6-9}$$

To evaluate this equation, we need to find the force magnitudes f_s and N. To do that, we use another **Key Idea**: When the coin is on the verge of sliding, it is stationary and thus its acceleration $\vec{a}$ is zero. We can relate this acceleration to the forces on the coin with Newton's second law ($\vec{F}_{\text{net}} = m\vec{a}$). As shown in the free-body diagram of the coin in Fig. 6-5*b*, those forces are (1) the frictional force $\vec{f}_s$, (2) the normal force $\vec{N}$, and (3) the gravitational force $\vec{F}_g$ on the coin, with magnitude equal to mg. Then, from Newton's second law with $\vec{a} = 0$, we have

$$\vec{f}_s + \vec{N} + \vec{F}_g = 0. \tag{6-10}$$

To find f_s and N, we rewrite Eq. 6-10 for components along the x and y axes of the tilted coordinate system in Fig. 6-5*b*. For the x axis and with mg substituted for F_g, we have

$$f_s + 0 - mg \sin \theta = 0,$$

so

$$f_s = mg \sin \theta. \tag{6-11}$$

Similarly, for the y axis we have

$$0 + N - mg \cos \theta = 0,$$

so

$$N = mg \cos \theta. \tag{6-12}$$

Substituting Eqs. 6-11 and 6-12 into Eq. 6-9 produces

$$\mu_s = \frac{mg \sin \theta}{mg \cos \theta} = \tan \theta, \tag{6-13}$$

which here means

$$\mu_s = \tan 13° = 0.23. \quad \text{(Answer)}$$

Actually you do not need to measure θ to get μ_s. Instead, measure the two lengths shown in Fig. 5-6*a* and then substitute h/d for $\tan \theta$ in Eq. 6-13.

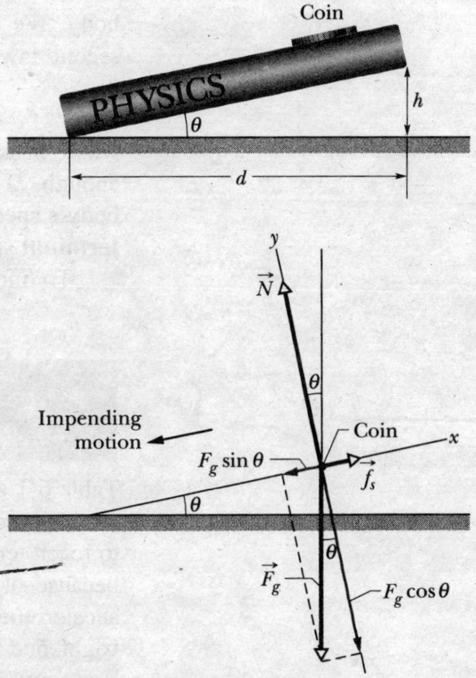

Fig. 6-5 Sample Problem 6-3. (*a*) A coin on the verge of sliding down a book. (*b*) A free-body diagram for the coin, showing the three forces (drawn to scale) that act on it. The gravitational force $\vec{F}_g$ is shown resolved into its components along the x and the y axes, whose orientations are chosen to simplify the problem. Component $F_g \sin \theta$ tends to slide the coin down the book. Component $F_g \cos \theta$ presses the coin onto the book.

Fig. 6-6 This skier crouches in an "egg position" so as to minimize her effective cross-sectional area and thus the air drag acting on her.

6-3 The Drag Force and Terminal Speed

A **fluid** is anything that can flow—generally either a gas or a liquid. When there is a relative velocity between a fluid and a body (either because the body moves through the fluid or because the fluid moves past the body), the body experiences a **drag force** $\vec{D}$ that opposes the relative motion and points in the direction in which the fluid flows relative to the body.

Here we examine only cases in which air is the fluid, the body is blunt (like a baseball) rather than slender (like a javelin), and the relative motion is fast enough so that the air becomes turbulent (breaks up into swirls) behind the body. In such cases, the magnitude of the drag force $\vec{D}$ is related to the relative speed v by an experimentally determined **drag coefficient** C according to

$$D = \tfrac{1}{2}C\rho Av^2, \tag{6-14}$$

where ρ is the air density (mass per volume) and A is the **effective cross-sectional area** of the body (the area of a cross section taken perpendicular to the velocity $\vec{v}$). The drag coefficient C (typical values range from 0.4 to 1.0) is not truly a constant for a given body, because if v varies significantly, the value of C can vary as well. Here, we ignore such complications.

Downhill speed skiers know well that drag depends on A and v^2. To reach high speeds a skier must reduce D as much as possible by, for example, riding the skis in the "egg position" (Fig. 6-6) to minimize A.

When a blunt body falls from rest through air, the drag force $\vec{D}$ is directed upward; its magnitude gradually increases from zero as the speed of the body increases. This upward force $\vec{D}$ opposes the downward gravitational force $\vec{F}_g$ on the body. We can relate these forces to the body's acceleration by writing Newton's second law for a vertical y axis ($F_{net,y} = ma_y$) as

$$D - F_g = ma, \tag{6-15}$$

where m is the mass of the body. As suggested in Fig. 6-7, if the body falls long enough, D eventually equals F_g. From Eq. 6-15, this means that $a = 0$ and so the body's speed no longer increases. The body then falls at a constant speed, called the **terminal speed** v_t.

To find v_t, we set $a = 0$ in Eq. 6-15 and substitute for D from Eq. 6-14, obtaining

$$\tfrac{1}{2}C\rho Av_t^2 - F_g = 0,$$

which gives

$$v_t = \sqrt{\frac{2F_g}{C\rho A}}. \tag{6-16}$$

Table 6-1 gives values of v_t for some common objects.

According to calculations* based on Eq. 6-14, a cat must fall about six floors to reach terminal speed. Until it does so, $F_g > D$ and the cat accelerates downward because of the net downward force. Recall from Chapter 2 that your body is an accelerometer, not a speedometer. Because the cat also senses the acceleration, it is frightened and keeps its feet underneath its body, its head tucked in, and its spine bent upward, making A small, v_t large, and injury on landing likely.

However, if the cat does reach v_t, the acceleration vanishes and the cat relaxes somewhat, stretching its legs and neck horizontally outward and straightening its spine (it then resembles a flying squirrel). These actions increase area A and thus

*W. O. Whitney and C. J. Mehlhaff, "High-Rise Syndrome in Cats." *The Journal of the American Veterinary Medical Association*, 1987, Vol. 191, pp. 1399–1403.

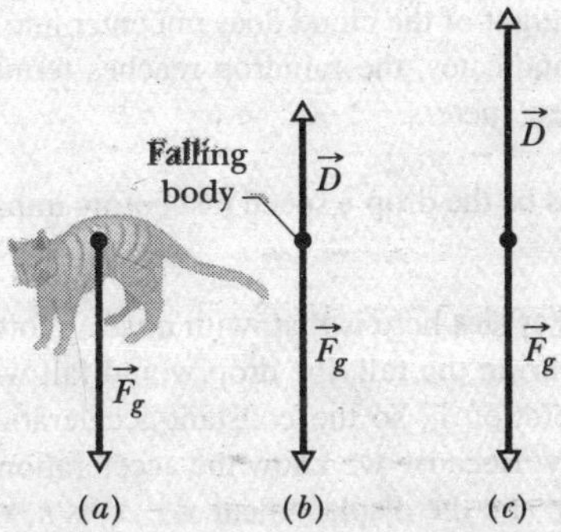

Fig. 6-7 The forces that act on a body falling through air: (a) the body when it has just begun to fall and (b) the free-body diagram a little later, after a drag force has developed. (c) The drag force has increased until it balances the gravitational force on the body. The body now falls at its constant terminal speed.

TABLE 6-1 Some Terminal Speeds in Air

Object	Terminal Speed (m/s)	95% Distance[a] (m)
Shot (from shot put)	145	2500
Sky diver (typical)	60	430
Baseball	42	210
Tennis ball	31	115
Basketball	20	47
Ping-Pong ball	9	10
Raindrop (radius = 1.5 mm)	7	6
Parachutist (typical)	5	3

[a]This is the distance through which the body must fall from rest to reach 95% of its terminal speed.

Source: Adapted from Peter J. Brancazio, *Sport Science*, 1984, Simon & Schuster, New York.

Fig. 6-8 A sky diver in a horizontal "spread eagle" maximizes the air drag.

also, by Eq. 6-14, the drag D. The cat begins to slow because now $D > F_g$ (the net force is upward), until a new, smaller v_t is reached. The decrease in v_t reduces the possibility of serious injury on landing. Just before the end of the fall, when it sees it is nearing the ground, the cat pulls its legs back beneath its body to prepare for the landing.

Humans often fall from great heights for the fun of skydiving. However, in April 1987, during a jump, sky diver Gregory Robertson noticed that fellow sky diver Debbie Williams had been knocked unconscious in a collision with a third sky diver and was unable to open her parachute. Robertson, who was well above Williams at the time and who had not yet opened his parachute for the 4 km plunge, reoriented his body head-down so as to minimize A and maximize his downward speed. Reaching an estimated v_t of 320 km/h, he caught up with Williams and then went into a horizontal "spread eagle" (as in Fig. 6-8) to increase D so that he could grab her. He opened her parachute and then, after releasing her, his own, a scant 10 s before impact. Williams received extensive internal injuries due to her lack of control on landing but survived.

Sample Problem 6-4

If a falling cat reaches a first terminal speed of 97 km/h while it is tucked in and then stretches out, doubling A, how fast is it falling when it reaches a new terminal speed?

SOLUTION: The **Key Idea** here is that the terminal speeds of the cat depend on (among other things) the effective cross-sectional areas A of the cat, according to Eq. 6-16. Thus, we can use that equation to set up a ratio of speeds. We let v_{to} and v_{tn} represent the original and new terminal speeds, and A_o and A_n the original and new areas. Then by Eq. 6-16,

$$\frac{v_{tn}}{v_{to}} = \frac{\sqrt{2F_g/C\rho A_n}}{\sqrt{2F_g/C\rho A_o}} = \sqrt{\frac{A_o}{A_n}} = \sqrt{\frac{A_o}{2A_o}} = \sqrt{0.5} \approx 0.7,$$

which means that $v_{tn} \approx 0.7 v_{to}$, or about 68 km/h.

Sample Problem 6-5

A raindrop with radius $R = 1.5$ mm falls from a cloud that is at height $h = 1200$ m above the ground. The drag coefficient C for the drop is 0.60. Assume that the drop is spherical throughout its fall. The density of water ρ_w is 1000 kg/m^3, and the density of air ρ_a is 1.2 kg/m^3.

(a) What is the terminal speed of the drop?

SOLUTION: The **Key Idea** here is that the drop reaches a terminal speed v_t when the gravitational force on it is balanced by the air drag force on it, so its acceleration is zero. We could then apply Newton's second law and the drag force equation to find v_t, but Eq. 6-16 does all that for us.

To use Eq. 6-16, we need the drop's effective cross-sectional area A and the magnitude F_g of the gravitational force. Because the drop is spherical, A is the area of a circle (πR^2) with the same radius as the sphere. To find F_g, we use three facts: (1) $F_g = mg$, where m is the drop's mass; (2) the (spherical) drop's volume is $V = \frac{4}{3}\pi R^3$; and (3) the density of the water in the drop is the mass per volume, or $\rho_w = m/V$. Thus, we find

$$F_g = V\rho_w g = \tfrac{4}{3}\pi R^3 \rho_w g.$$

We next substitute this, the expression for A, and the given data into Eq. 6-16. Being careful to distinguish between the air density ρ_a and the water density ρ_w, we obtain

$$v_t = \sqrt{\frac{2F_g}{C\rho_a A}} = \sqrt{\frac{8\pi R^3 \rho_w g}{3C\rho_a \pi R^2}} = \sqrt{\frac{8R\rho_w g}{3C\rho_a}}$$

$$= \sqrt{\frac{(8)(1.5\times10^{-3}\text{ m})(1000\text{ kg/m}^3)(9.8\text{ m/s}^2)}{(3)(0.60)(1.2\text{ kg/m}^3)}}$$

$$= 7.4\text{ m/s} \approx 27\text{ km/h}. \qquad \text{(Answer)}$$

Note that the height of the cloud does not enter into the calculation. As Table 6-1 indicates, the raindrop reaches terminal speed after falling just a few meters.

(b) What would be the drop's speed just before impact if there were no drag force?

SOLUTION: The **Key Idea** here is that, with no drag force to reduce the drop's speed during the fall, the drop would fall with the constant free-fall acceleration g, so the constant-acceleration equations of Table 2-1 apply. Because we know the acceleration is g, the initial velocity v_0 is 0, and the displacement $x - x_0$ is h, we use Eq. 2-16 to find v:

$$v = \sqrt{2gh} = \sqrt{(2)(9.8\text{ m/s}^2)(1200\text{ m})}$$

$$= 153\text{ m/s} \approx 550\text{ km/h}. \qquad \text{(Answer)}$$

For that speed, Shakespeare would scarcely have written, "it droppeth as the gentle rain from heaven, upon the place beneath."

CHECKPOINT 3: Near the ground, is the speed of large raindrops greater than, less than, or the same as the speed of small raindrops, assuming that all raindrops are spherical and have the same drag coefficient?

6-4 Uniform Circular Motion

From Section 4-7, recall that when a body moves in a circle (or a circular arc) at constant speed v, it is said to be in uniform circular motion. Also recall that the body has a centripetal acceleration (directed toward the center of the circle), of constant magnitude given by

$$a = \frac{v^2}{R} \quad \text{(centripetal acceleration),} \qquad (6\text{-}17)$$

where R is the radius of the circle.

Let us examine two examples of uniform circular motion:

1. *Rounding a curve in a car.* You are sitting in the center of the rear seat of a car moving at a constant high speed along a flat road. When the driver suddenly turns left, rounding a corner in a circular arc, you slide across the seat toward the right and then jam against the car wall for the rest of the turn. What is going on?

   While the car moves in the circular arc, it is in uniform circular motion; that is, it has an acceleration that is directed toward the center of the circle. By Newton's second law, a force must cause this acceleration. Moreover, the force must also be directed toward the center of the circle. Thus, it is a **centripetal force**, where the adjective indicates the direction. In this example, the centripetal force is a frictional force on the tires from the road; it makes the turn possible.

   If you are to move in uniform circular motion along with the car, there must also be a centripetal force on you. However, apparently the frictional force on you from the seat was not great enough to make you go in a circle with the car. Thus, the seat slid beneath you, until the right wall of the car jammed into you. Then, its push on you provided the needed centripetal force on you, and you joined the car's uniform circular motion.

2. *Orbiting Earth.* This time you are a passenger in the space shuttle *Atlantis*. As it and you orbit Earth, you float through your cabin. What is going on?

Both you and the shuttle are in uniform circular motion and have accelerations directed toward the center of the circle. Again by Newton's second law, centripetal forces must cause these accelerations. This time the centripetal forces are gravitational pulls (the pull on you and the pull on the shuttle) by Earth, radially inward, toward the center of Earth.

In both car and shuttle you are in uniform circular motion, acted on by a centripetal force—yet your sensations in the two situations are quite different. In the car, jammed up against the wall, you are aware of being compressed by the wall. In the orbiting shuttle, however, you are floating around with no sensation of any force acting on you. Why this difference?

The difference is due to the nature of the two centripetal forces. In the car, the centripetal force is the push on the part of your body touching the car wall. You can sense the compression on that part of your body. In the shuttle, the centripetal force is Earth's gravitational pull on every atom of your body. Thus, there is no compression (or pull) on any one part of your body and no sensation of a force acting on you. (The sensation is said to be one of "weightlessness," but that description is tricky. The pull on you by Earth has certainly not disappeared and, in fact, is only a little less than it would be with you on the ground.)

Another example of a centripetal force is shown in Fig. 6-9. There a hockey puck moves around in a circle at constant speed v while tied to a string looped around a central peg. This time the centripetal force is the radially inward pull on the puck from the string. Without that force, the puck would slide off in a straight line instead of moving in a circle.

Note again that a centripetal force is not a new kind of force. The name merely indicates the direction of the force. It can, in fact, be a frictional force, a gravitational force, the force from a car wall or a string, or any other force. For any situation;

> A centripetal force accelerates a body by changing the direction of the body's velocity without changing the body's speed.

From Newton's second law and Eq. 6-17 ($a = v^2/R$), we can write the magnitude F of a centripetal force (or a net centripetal force) as

$$F = m\frac{v^2}{R} \quad \text{(magnitude of centripetal force).} \tag{6-18}$$

Because the speed v here is constant, so are also the magnitudes of the acceleration and the force. → in uniform circular motion, not in nonuniform circular motion

However, the directions of the centripetal acceleration and force are not constant; they vary continuously so as to always point toward the center of the circle. For this reason, the force and acceleration vectors are sometimes drawn along a radial axis r that moves with the body and always extends from the center of the circle to the body, as in Fig. 6-9. The positive direction of the axis is radially outward, but the acceleration and force vectors point radially inward.

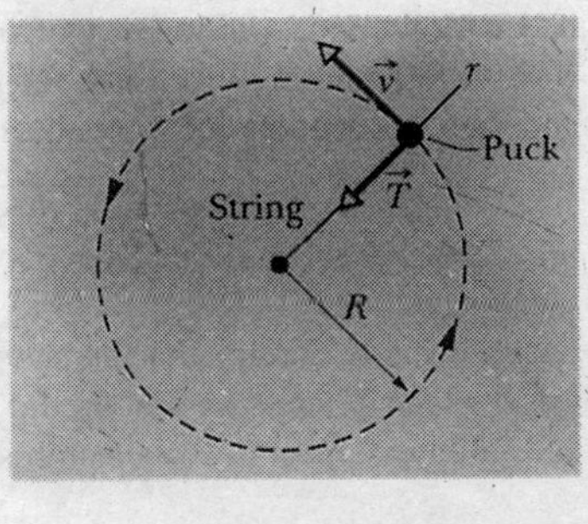

Fig. 6-9 An overhead view of a hockey puck of mass m moving with constant speed v in a circular path of radius R on a horizontal frictionless surface. The centripetal force on the puck is $\vec{T}$, the pull from the string, directed inward along the radial axis r extending through the puck.

Sample Problem 6-6

Igor is a cosmonaut-engineer on the *International Space Station*, in a circular orbit around Earth, at an altitude h of 520 km and with a constant speed v of 7.6 km/s. Igor's mass m is 79 kg.

(a) What is his acceleration?

SOLUTION: The **Key Idea** here is that Igor is in uniform circular motion and thus has a centripetal acceleration of magnitude given by Eq. 6-17 ($a = v^2/R$). The radius R of Igor's motion is $R_E + h$, where R_E is Earth's radius (6.37×10^6 m, from Appendix C). Thus,

$$a = \frac{v^2}{R} = \frac{v^2}{R_E + h}$$

$$= \frac{(7.6 \times 10^3\ \text{m/s})^2}{6.37 \times 10^6\ \text{m} + 0.52 \times 10^6\ \text{m}}$$

$$= 8.38\ \text{m/s}^2 \approx 8.4\ \text{m/s}^2. \qquad \text{(Answer)}$$

This is the value of the free-fall acceleration at Igor's altitude. If he were lifted to that altitude and released, instead of being put into orbit there, he would fall toward Earth's center, starting out with that value for his acceleration. The difference in the two situations is that when he orbits Earth, he always has a "sideways" motion as well: As he falls, he also moves to the side, so that he ends up moving along a curved path around Earth.

(b) What force does Earth exert on Igor?

SOLUTION: There are two **Key Ideas** here. First, there must be a centripetal force on Igor if he is to be in uniform circular motion. Second, that force is the gravitational force $\vec{F}_g$ on him from Earth, directed toward his center of rotation (at the center of Earth). From Newton's second law, written along the radial axis r, this force has the magnitude

$$F_g = ma = (79\ \text{kg})(8.38\ \text{m/s}^2)$$

$$= 662\ \text{N} \approx 660\ \text{N}. \qquad \text{(Answer)}$$

If Igor were to stand on a scale placed on the top of a tower with height h = 520 km, the scale would read 660 N. In orbit, the scale (if Igor could "stand" on it) would read zero because he and the scale are in free fall together, and therefore his feet do not actually press against it.

Sample Problem 6-7

In a 1901 circus performance, Allo "Dare Devil" Diavolo introduced the stunt of riding a bicycle in a loop-the-loop (Fig. 6-10*a*). Assuming that the loop is a circle with radius R = 2.7 m, what is the least speed v Diavolo could have at the top of the loop to remain in contact with it there?

SOLUTION: A **Key Idea** in analyzing Diavolo's stunt is to assume that he and his bicycle travel through the top of the loop as a single particle in uniform circular motion. Thus, at the top, the acceleration $\vec{a}$ of this particle must have the magnitude $a = v^2/R$ given by Eq. 6-17 and be directed downward, toward the center of the circular loop.

The forces on the particle when it is at the top of the loop are shown in the free-body diagram of Fig 6-10*b*. The gravitational force $\vec{F}_g$ is directed downward along a y axis. The normal force $\vec{N}$ on the particle from the loop is also directed downward. Thus, Newton's second law for y components ($F_{\text{net},y} = ma_y$) gives us

$$-N - F_g = m(-a).$$

This becomes

$$-N - mg = m\left(-\frac{v^2}{R}\right). \qquad (6\text{-}19)$$

Another **Key Idea** is that if the particle has the *least speed v* needed to remain in contact, then it is on the *verge of losing contact* with the loop (falling away from the loop), which means that $N = 0$. Substituting this value for N into Eq. 6-19, solving for v, and then substituting known values give us

$$v = \sqrt{gR} = \sqrt{(9.8\ \text{m/s}^2)(2.7\ \text{m})}$$

$$= 5.1\ \text{m/s}. \qquad \text{(Answer)}$$

Diavolo made certain that his speed at the top of the loop was greater than 5.1 m/s so that he did not lose contact with the loop and fall away from it. Note that this speed requirement is independent of the mass of Diavolo and his bicycle. Had he feasted on, say, pierogies before his performance, he still would have had to exceed only 5.1 m/s.

(*a*)

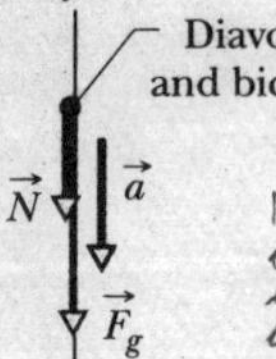

(*b*)

Fig. 6-10 Sample Problem 6-7. (*a*) Contemporary advertisement for Diavolo and (*b*) free-body diagram for the performer at the top of the loop.

CHECKPOINT 4: When you ride in a Ferris wheel at constant speed, what are the directions of your acceleration $\vec{a}$ and the normal force $\vec{N}$ on you (from the always upright seat) as you pass through (a) the highest point and (b) the lowest point of the ride?

Sample Problem 6-8

Even some seasoned roller-coaster riders blanch at the thought of riding the Rotor, which is essentially a large, hollow cylinder that is rotated rapidly around its central axis (Fig. 6-11). Before the ride begins, a rider enters the cylinder through a door on the side and stands on a floor, up against a canvas-covered wall. The door is closed, and as the cylinder begins to turn, the rider, wall, and floor move in unison. When the rider's speed reaches some predetermined value, the floor abruptly and alarmingly falls away. The rider does not fall with it but instead is pinned to the wall while the cylinder rotates, as if an unseen (and somewhat unfriendly) agent is pressing the body to the wall. Later, the floor is eased back to the rider's feet, the cylinder slows, and the rider sinks a few centimeters to regain footing on the floor. (Some riders consider all this to be fun.)

Suppose that the coefficient of static friction μ_s between the rider's clothing and the canvas is 0.40 and that the cylinder's radius R is 2.1 m.

(a) What minimum speed v must the cylinder and rider have if the rider is not to fall when the floor drops?

SOLUTION: We start with a question: What force can keep the rider from falling and how is it related to the speed v of her and the cylinder? To answer, we use three **Key Ideas**:

1. The gravitational force $\vec{F}_g$ on the rider tends to slide her down the wall, but she does not move because a frictional force from the wall acts upward on her (Fig 6-11).
2. If she is to be on the verge of sliding down, that upward force must be a *static* frictional force $\vec{f}_s$ at its maximum value $\mu_s N$, where N is the magnitude of the normal force $\vec{N}$ on her from the cylinder (Fig. 6-11).
3. This normal force is directed horizontally toward the central axis of the cylinder and is the centripetal force that causes the rider to move in a circular path, with centripetal acceleration of magnitude $a = v^2/R$.

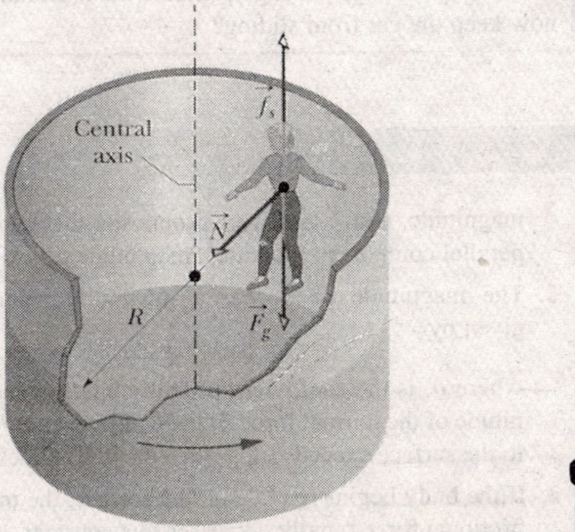

Fig. 6-11 Sample Problem 6-8. A Rotor in an amusement park, showing the forces on a rider. The centripetal force is the normal force with which the wall pushes inward on the rider.

We want speed v in that last expression, for the condition that the rider is on the verge of sliding.

We first place a vertical y axis through the rider, with the positive direction upward. For **Key Idea** 1, we can then apply Newton's second law to the rider, writing it for y components ($F_{net,y} = ma_y$) as

$$f_s - mg = m(0),$$

where m is the rider's mass and mg is the magnitude of $\vec{F}_g$. For **Key Idea** 2, we substitute the maximum value $\mu_s N$ for f_s in this equation, getting

$$\mu_s N - mg = 0,$$

or

$$N = \frac{mg}{\mu_s}. \tag{6-20}$$

Next we place a radial r axis through the rider, with the positive direction outward. For **Key Idea** 3, we can then write Newton's second law for components along that axis as

$$-N = m\left(-\frac{v^2}{R}\right). \tag{6-21}$$

Substituting Eq. 6-20 for N and then solving for v, we find

$$v = \sqrt{\frac{gR}{\mu_s}} = \sqrt{\frac{(9.8\ \text{m/s}^2)(2.1\ \text{m})}{0.40}}$$

$$= 7.17\ \text{m/s} \approx 7.2\ \text{m/s}. \qquad \text{(Answer)}$$

Note that the result is independent of the rider's mass; it holds for anyone riding the Rotor, from a child to a sumo wrestler, which is why no one has to "weigh in" to ride the Rotor.

(b) If the rider's mass is 49 kg, what is the magnitude of the centripetal force on her?

SOLUTION: According to Eq. 6-21,

$$N = m\frac{v^2}{R} = (49\ \text{kg})\frac{(7.17\ \text{m/s})^2}{2.1\ \text{m}}$$

$$\approx 1200\ \text{N}. \qquad \text{(Answer)}$$

Although this force is directed toward the central axis, the rider has an overwhelming sensation that the force pinning her against the wall is directed radially outward. Her sensation stems from the fact that she is in a noninertial frame (she and it are accelerating). As measured from such frames, forces can be illusionary. The illusion is part of the Rotor's attraction.

CHECKPOINT 5: If the Rotor initially moves at the minimum required speed for the rider not to fall and then its speed is increased in steps, do the following increase, decrease, or remain the same: (a) the magnitude of $\vec{f}_s$; (b) the magnitude of $\vec{N}$; (c) the value of $f_{s,max}$?

Sample Problem 6-9

Figure 6-12*a* represents a stock car of mass $m = 1600$ kg traveling at a constant speed $v = 20$ m/s around a flat, circular track of radius $R = 190$ m. For what value of μ_s between the track and the tires of the car will the car be on the verge of sliding off the track?

SOLUTION: We need to relate μ_s to the circular motion of the car. We start with four **Key Ideas**, all related to one force on the car:

1. A centripetal force must act on the car if the car is moving around a circular path; the force must be directed horizontally toward the center of the circle.
2. The only horizontal force acting on the car is a frictional force on the tires from the road. So, the required centripetal force is a frictional force.
3. Because the car is not sliding, the frictional force must be a *static* frictional force, the $\vec{f}_s$ shown in Fig 6-12*a*.
4. If the car is just on the verge of sliding, the magnitude f_s of the frictional force is just equal to the maximum value $f_{s,\max} = \mu_s N$, where N is the magnitude of the normal force on the car from the track.

Thus, to find μ_s, we start with the centripetal force on the car. Figure 6-12*b* is a free-body diagram for the car, shown on a radial axis r that always extends from the center of the circle through the car as the car moves. The centripetal force $\vec{f}_s$ is directed inward along that axis, in the negative direction of the axis, and so is the car's centripetal acceleration $\vec{a}$ (of magnitude v^2/R). We can relate this force and acceleration by writing Newton's second law for components along the r axis ($F_{\text{net},r} = ma_r$) as

$$-f_s = m\left(-\frac{v^2}{R}\right). \tag{6-22}$$

Substituting $f_{s,\max} = \mu_s N$ for f_s and solving for μ_s give us

$$\mu_s = \frac{mv^2}{NR}. \tag{6-23}$$

Because the car does not accelerate vertically, the two vertical forces acting on it (see Fig. 6-12*b*) must balance; that is, the magnitude N of the normal force must equal the magnitude mg of the gravitational force. Substituting $N = mg$ into Eq. 6-23, we find

$$\mu_s = \frac{mv^2}{mgR} = \frac{v^2}{gR} \tag{6-24}$$

$$= \frac{(20\text{ m/s})^2}{(9.8\text{ m/s}^2)(190\text{ m})} = 0.21. \quad \text{(Answer)}$$

This means that if $\mu_s = 0.21$, the car is on the verge of sliding off the track; if $\mu_s > 0.21$, the car is in no danger of sliding off; and if $\mu_s < 0.21$, the car will certainly slide off.

Equation 6-24 contains two important lessons for road engineers. First, the value of μ_s (required to prevent sliding) depends on the *square* of v. Much more friction is required when the turning speed is increased. You may have noted this effect if you have ever taken a flat turn too fast and suddenly felt the tires slip. Second, the mass m dropped out in our derivation of Eq. 6-24. Thus, Eq. 6-24 holds for a vehicle of any mass, from a kiddy car to a bicycle to a heavy truck.

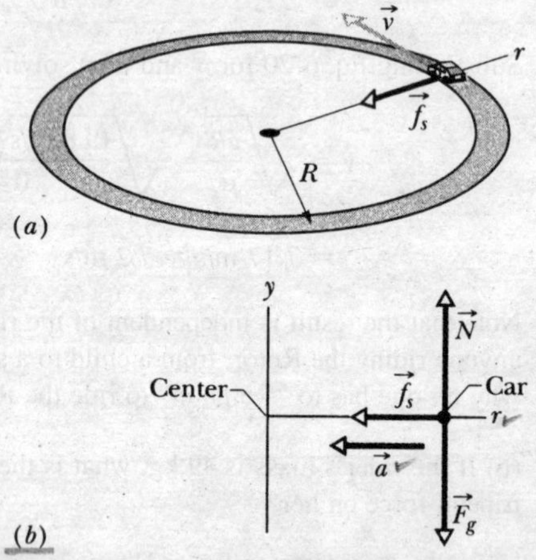

Fig. 6-12 Sample Problem 6-9. (*a*) A car moves around a flat curved road at constant speed v. The frictional force $\vec{f}_s$ provides the necessary centripetal force along a radial axis r. (*b*) A free-body diagram (not to scale) for the car, in the vertical plane containing r.

CHECKPOINT 6: In Fig. 6-12, suppose the car is on the verge of sliding when the radius of the circle is R_1. (a) If we double the car's speed, what is the least radius that would now keep the car from sliding? (b) If we also double the mass of the car (say by adding sandbags), what is the least radius that would now keep the car from sliding?

REVIEW & SUMMARY

Friction When a force $\vec{F}$ tends to slide a body along a surface, a **frictional force** from the surface acts on the body. The frictional force is parallel to the surface and directed so as to oppose the sliding. It is due to bonding between the body and the surface.

If the body does not slide, the frictional force is a **static frictional force** $\vec{f}_s$. If there is sliding, the frictional force is a **kinetic frictional force** $\vec{f}_k$.

Three Properties of Friction

1. If the body does not move, then the static frictional force $\vec{f}_s$ and the component of $\vec{F}$ that is parallel to the surface are equal in magnitude, and $\vec{f}_s$ is directed opposite that component. If that parallel component increases, magnitude f_s also increases.
2. The magnitude of $\vec{f}_s$ has a maximum value $f_{s,\max}$ that is given by
$$f_{s,\max} = \mu_s N, \tag{6-1}$$
where μ_s is the **coefficient of static friction** and N is the magnitude of the normal force. If the component of $\vec{F}$ that is parallel to the surface exceeds $f_{s,\max}$, then the body slides on the surface.
3. If the body begins to slide on the surface, the magnitude of the frictional force rapidly decreases to a constant value f_k given by
$$f_k = \mu_k N, \tag{6-2}$$
where μ_k is the **coefficient of kinetic friction.**

Drag Force When there is a relative motion between air (or some other fluid) and a body, the body experiences a **drag force** $\vec{D}$ that opposes the relative motion and points in the direction in which the fluid flows relative to the body. The magnitude of $\vec{D}$ is related to the relative speed v by an experimentally determined **drag coefficient** C according to

$$D = \tfrac{1}{2}C\rho A v^2, \tag{6-14}$$

where ρ is the fluid density (mass per volume) and A is the **effective cross-sectional area of the body** (the area of a cross section taken perpendicular to the relative velocity $\vec{v}$).

Terminal Speed When a blunt object has fallen far enough through air, the magnitudes of the drag force $\vec{D}$ and the gravitational force $\vec{F}_g$ on the body become equal. The body then falls at a constant **terminal speed** v_t given by

$$v_t = \sqrt{\frac{2F_g}{C\rho A}}. \tag{6-16}$$

Uniform Circular Motion If a particle moves in a circle or a circular arc with radius R at constant speed v, it is said to be in **uniform circular motion.** It then has a **centripetal acceleration** $\vec{a}$ with magnitude given by

$$a = \frac{v^2}{R}. \tag{6-17}$$

This acceleration is due to a net **centripetal force** on the particle, with magnitude given by

$$F = \frac{mv^2}{R}, \tag{6-18}$$

where m is the particle's mass. The vector quantities $\vec{a}$ and $\vec{F}$ are directed toward the center of curvature of the particle's path.

QUESTIONS

1 In three experiments, three different horizontal forces are applied to the same block lying on the same countertop. The force magnitudes are $F_1 = 12$ N, $F_2 = 8$ N, and $F_3 = 4$ N. In each experiment, the block remains stationary in spite of the applied force. Rank the forces according to (a) the magnitude f_s of the static frictional force on the block from the countertop and (b) the maximum value $f_{s,\max}$ of that force, greatest first.

2 In Fig. 6-13*a*, a "Batman" thermos is sent sliding leftward across a long plastic tray. What are the directions of the kinetic frictional forces on (a) the thermos and (b) the tray from each other? (c) Does the former increase or decrease the speed of the thermos relative to the floor? In Fig. 6-13*b*, the tray is now sent sliding leftward beneath the thermos. What now are the directions of the kinetic frictional forces on (d) the thermos and (e) the tray from each other? (f) Does the former increase or decrease the speed of the thermos relative to the floor? (g) *Do kinetic frictional forces always slow objects?*

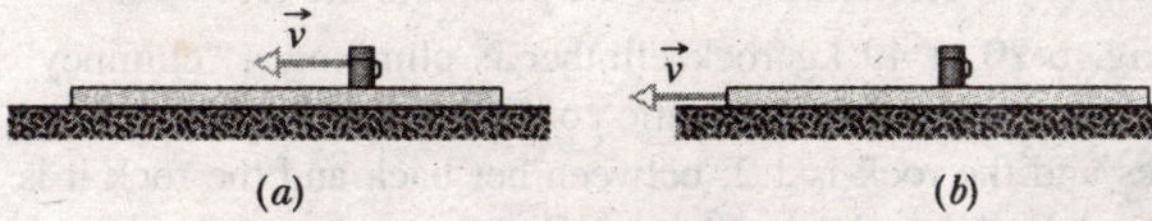

Fig. 6-13 Question 2.

3 In Fig. 6-14, horizontal force $\vec{F}_1$ of magnitude 10 N is applied to a box on a floor, but the box does not slide. Then, as the magnitude of vertical force $\vec{F}_2$ is increased from zero, do the following quantities increase, decrease, or stay the same: (a) the magnitude of the frictional force $\vec{f}_s$ on the box; (b) the magnitude of the normal force $\vec{N}$ on the box from the floor; (c) the maximum value $f_{s,\max}$ of the magnitude of the static frictional force on the box? (d) Does the box eventually slide?

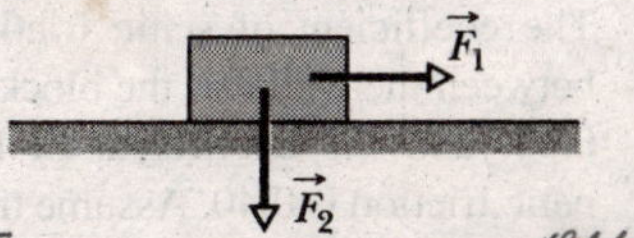

Fig. 6-14 Question 3.

4 If you press an apple crate against a wall so hard that the crate cannot slide down the wall, what is the direction of (a) the static frictional force $\vec{f}_s$ on the crate from the wall and (b) the normal force $\vec{N}$ on the crate from the wall? If you increase your push, what happens to (c) f_s, (d) N, and (e) $f_{s,\max}$?

5 In Fig. 6-15, if the box is stationary and the angle θ of force $\vec{F}$ is increased, do the following quantities increase, decrease, or remain the same: (a) F_x; (b) f_s; (c) N; (d) $f_{s,\max}$? (e) If, instead, the box is sliding and θ is increased, does the magnitude of the frictional force on the box increase, decrease, or remain the same?

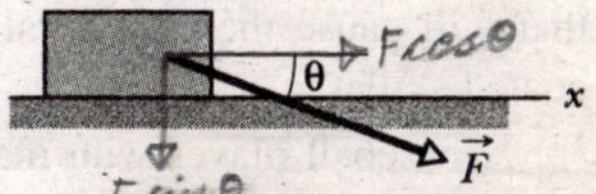

Fig. 6-15 Question 5.

6 Repeat Question 5 for force $\vec{F}$ angled upward instead of downward as drawn.

7 Figure 6-16 shows a block of mass m on a slab of mass M and a horizontal force $\vec{F}$ applied to the block, causing it to slide over the slab. There is friction between the block and the slab (but not between the slab and the floor). (a) What mass determines the magnitude of the frictional force between the block and the slab? (b) At the block–slab interface, is the magnitude of the frictional force acting on the block greater than, less than, or equal to that of the frictional force acting on the slab? (c) What are the directions of those two frictional forces? (d) If we write Newton's second law for the slab, what mass should be multiplied by the acceleration of the slab? (Warm-up for Problem 27)

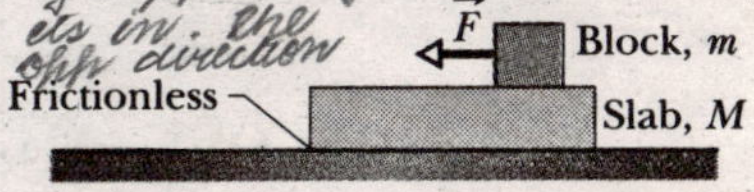

Fig. 6-16 Question 7.

8 *Follow-up to Problem 25.* Suppose the larger of the two blocks in Fig. 6-33 is, instead, fixed to the surface below it. Again, the smaller block is not to slip down the face of the larger block. (a) Is the magnitude of the frictional force between the blocks then greater than, less than, or the same as in the original problem? (b) Is the required minimum magnitude of the horizontal force $\vec{F}$ then greater than, less than, or the same as in the original problem?

9. Figure 6-17 shows the path of a park ride that travels at constant speed through five circular arcs of radii R_0, $2R_0$, and $3R_0$. Rank the arcs according to the magnitude of the centripetal force on a rider traveling in the arcs, greatest first.

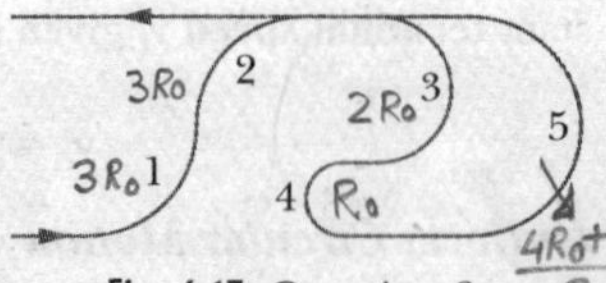

Fig. 6-17 Question 9.

10. A person riding a Ferris wheel moves through positions at (1) the top, (2) the bottom, and (3) midheight. If the wheel rotates at a constant rate, rank these three positions according to (a) the magnitude of the person's centripetal acceleration, (b) the magnitude of the net centripetal force on the person, and (c) the magnitude of the normal force on the person, greatest first.

EXERCISES & PROBLEMS

ssm Solution is in the Student Solutions Manual.
www Solution is available on the World Wide Web at: http://www.wiley.com/college/hrw
ilw Solution is available on the Interactive LearningWare.

SEC. 6-2 Properties of Friction

1E. A bedroom bureau with a mass of 45 kg, including drawers and clothing, rests on the floor. (a) If the coefficient of static friction between the bureau and the floor is 0.45, what is the magnitude of the minimum horizontal force that a person must apply to start the bureau moving? (b) If the drawers and clothing, with 17 kg mass, are removed before the bureau is pushed, what is the new minimum magnitude? ssm

2E. The coefficient of static friction between Teflon and scrambled eggs is about 0.04. What is the smallest angle from the horizontal that will cause the eggs to slide across the bottom of a Teflon-coated skillet?

3E. A baseball player with mass $m = 79$ kg, sliding into second base, is retarded by a frictional force of magnitude 470 N. What is the coefficient of kinetic friction μ_k between the player and the ground? ssm

4E. *The mysterious sliding stones.* Along the remote Racetrack Playa in Death Valley, California, stones sometimes gouge out prominent trails in the desert floor, as if they had been migrating (Fig. 6-18). For years curiosity mounted about why the stones moved. One explanation was that strong winds during the occasional rainstorms would drag the rough stones over ground softened by rain. When the desert dried out, the trails behind the stones were hard-baked in place. According to measurements, the coefficient of kinetic friction between the stones and the wet playa ground is about 0.80. What horizontal force is needed on a stone of typical mass 20 kg to maintain the stone's motion once a gust has started it moving? (Story continues with Exercise 32.)

Fig. 6-18 Exercise 4.

5E. A person pushes horizontally with a force of 220 N on a 55 kg crate to move it across a level floor. The coefficient of kinetic friction is 0.35. (a) What is the magnitude of the frictional force? (b) What is the magnitude of the crate's acceleration? ssm ilw

6E. A house is built on the top of a hill with a nearby 45° slope (Fig. 6-19). An engineering study indicates that the slope angle should be reduced because the top layers of soil along the slope might slip past the lower layers. If the static coefficient of friction between two such layers is 0.5, what is the least angle ϕ through which the present slope should be reduced to prevent slippage?

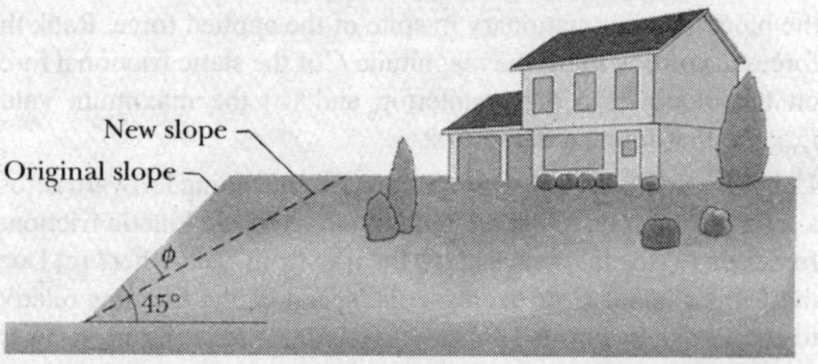

Fig. 6-19 Exercise 6.

7E. A 110 g hockey puck sent sliding over ice is stopped in 15 m by the frictional force on it from the ice. (a) If its initial speed is 6.0 m/s, what is the magnitude of the frictional force? (b) What is the coefficient of friction between the puck and the ice? ssm

8E. In Fig. 6-20, a 49 kg rock climber is climbing a "chimney" between two rock slabs. The static coefficient of friction between her shoes and the rock is 1.2; between her back and the rock it is 0.80. She has reduced her push against the rock until her back and her shoes are on the verge of slipping. (a) Draw a free-body diagram of the climber. (b) What is her push against the rock? (c) What fraction of her weight is supported by the frictional force on her shoes?

Fig. 6-20 Exercise 8.

9P. A 12 N horizontal force $\vec{F}$ pushes a block weighing 5.0 N against a vertical wall (Fig. 6-21). The coefficient of static friction between the wall and the block is 0.60, and the coefficient of kinetic friction is 0.40. Assume that the block is not moving initially.

(a) Will the block move? (b) In unit-vector notation, what is the force on the block from the wall? ssm www

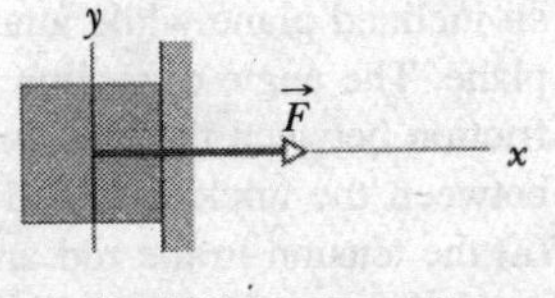

Fig. 6-21 Problem 9.

10P. A 2.5 kg block is initially at rest on a horizontal surface. A 6.0 N horizontal force and a vertical force $\vec{P}$ are applied to the block as shown in Fig. 6-22. The coefficients of friction for the block and surface are $\mu_s = 0.40$ and $\mu_k = 0.25$. Determine the magnitude and direction of the frictional force acting on the block if the magnitude of $\vec{P}$ is (a) 8.0 N, (b) 10 N, and (c) 12 N.

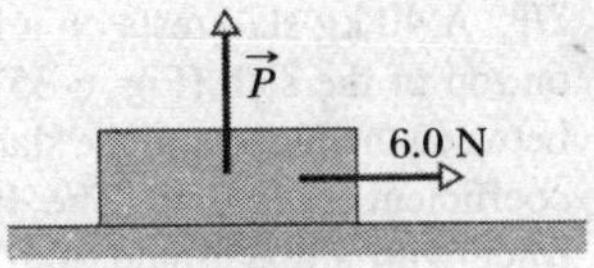

Fig. 6-22 Problem 10.

11P. A worker wishes to pile a cone of sand onto a circular area in his yard. The radius of the circle is R, and no sand is to spill onto the surrounding area (Fig. 6-23). If μ_s is the static coefficient of friction between each layer of sand along the slope and the sand beneath it (along which it might slip), show that the greatest volume of sand that can be stored in this manner is $\pi\mu_s R^3/3$. (The volume of a cone is $Ah/3$, where A is the base area and h is the cone's height.) ssm

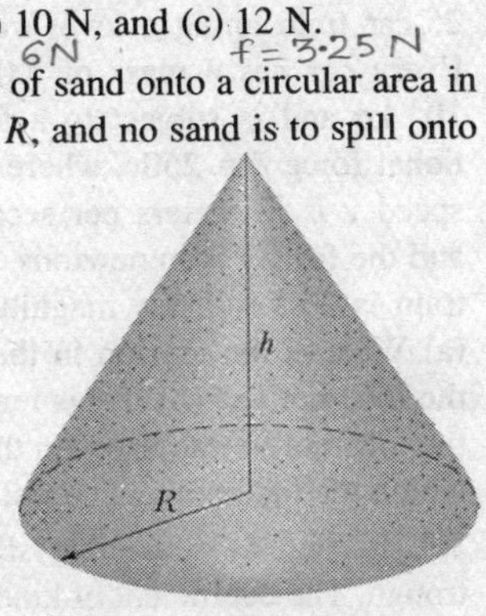

Fig. 6-23 Problem 11.

12P. A worker pushes horizontally on a 35 kg crate with a force of magnitude 110 N. The coefficient of static friction between the crate and the floor is 0.37. (a) What is the frictional force on the crate from the floor? (b) What is the maximum magnitude $f_{s,\max}$ of the static frictional force under the circumstances? (c) Does the crate move? (d) Suppose, next, that a second worker pulls directly upward on the crate to help out. What is the least vertical pull that will allow the first worker's 110 N push to move the crate? (e) If, instead, the second worker pulls horizontally to help out, what is the least pull that will get the crate moving?

13P. A 68 kg crate is dragged across a floor by pulling on a rope attached to the crate and inclined 15° above the horizontal. (a) If the coefficient of static friction is 0.50, what minimum force magnitude is required from the rope to start the crate moving? (b) If $\mu_k = 0.35$, what is the magnitude of the initial acceleration of the crate? ssm

Fig. 6-24 Problem 14.

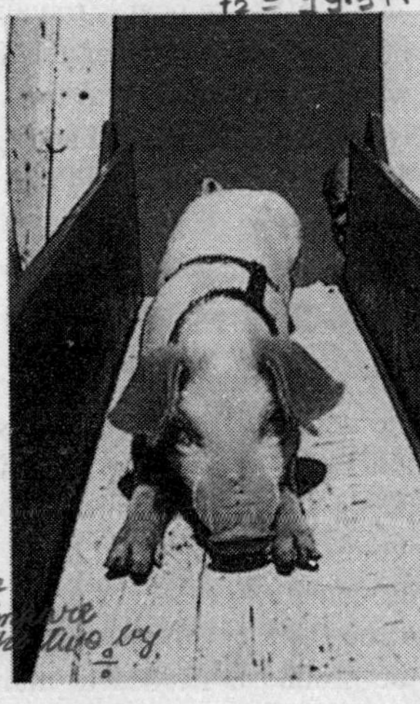

14P. A slide-loving pig slides down a certain 35° slide (Fig. 6-24) in twice the time it would take to slide down a frictionless 35° slide. What is the coefficient of kinetic friction between the pig and the slide?

15P. In Fig. 6-25, blocks A and B have weights of 44 N and 22 N, respectively. (a) Determine the minimum weight of block C to keep A from sliding if μ_s between A and the table is 0.20. (b) Block C suddenly is lifted off A. What is the acceleration of block A if μ_k between A and the table is 0.15? ssm

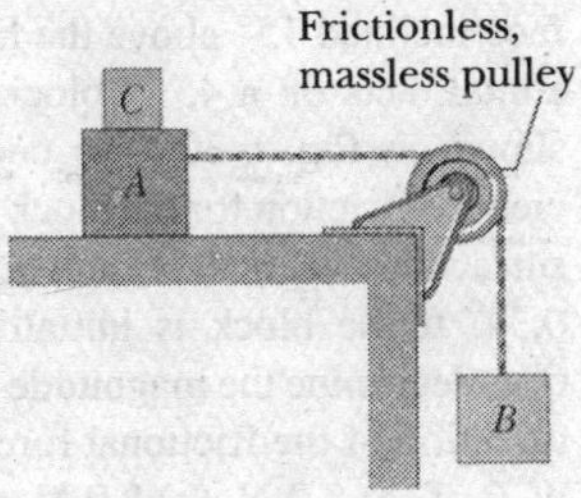

Fig. 6-25 Problem 15.

16P. A 3.5 kg block is pushed along a horizontal floor by a force $\vec{F}$ of magnitude 15 N at an angle $\theta = 40°$ with the horizontal (Fig. 6-26). The coefficient of kinetic friction between the block and the floor is 0.25. Calculate the magnitudes of (a) the frictional force on the block from the floor and (b) the acceleration of the block.

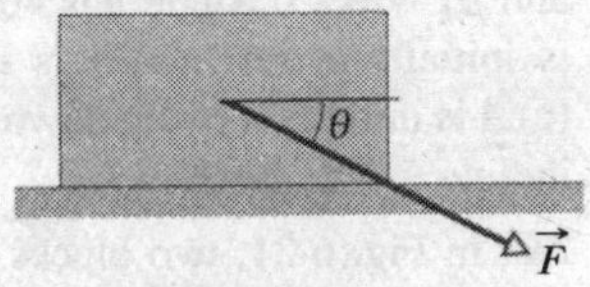

Fig. 6-26 Problem 16.

17P. Figure 6-27 shows the cross section of a road cut into the side of a mountain. The solid line AA' represents a weak bedding plane along which sliding is possible. Block B directly above the highway is separated from uphill rock by a large crack (called a *joint*), so that only friction between the block and the bedding plane prevents sliding. The mass of the block is 1.8×10^7 kg, the *dip angle* θ of the bedding plane is 24°, and the coefficient of static friction between block and plane is 0.63. (a) Show that the block will not slide. (b) Water seeps into the joint and expands upon freezing, exerting on the block a force $\vec{F}$ parallel to AA'. What minimum value of F will trigger a slide?

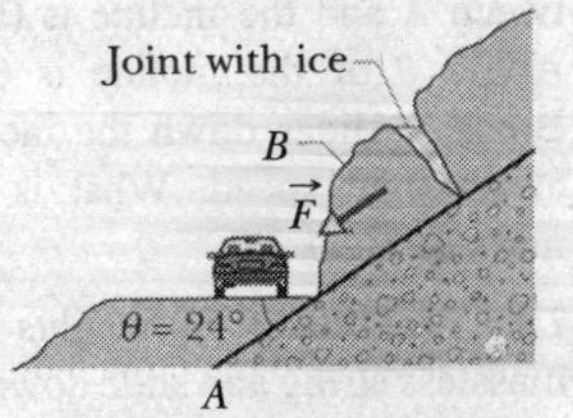

Fig. 6-27 Problem 17.

18P. A loaded penguin sled weighing 80 N rests on a plane inclined at 20° to the horizontal (Fig. 6-28). Between the sled and the plane, the coefficient of static friction is 0.25, and the coefficient of kinetic friction is 0.15. (a) What is the minimum magnitude of the force $\vec{F}$, parallel to the plane, that will prevent the sled from slipping down the plane? (b) What is the minimum magnitude F that will start the sled moving up the plane? (c) What value of F is required to move the sled up the plane at constant velocity?

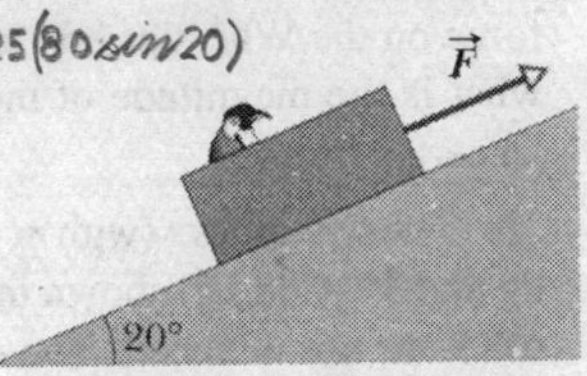

Fig. 6-28 Problem 18.

19P. Block B in Fig. 6-29 weighs 711 N. The coefficient of static friction between block and table is 0.25; assume that the cord between B and the knot is horizontal. Find the maximum weight of block A for which the system will be stationary. ssm

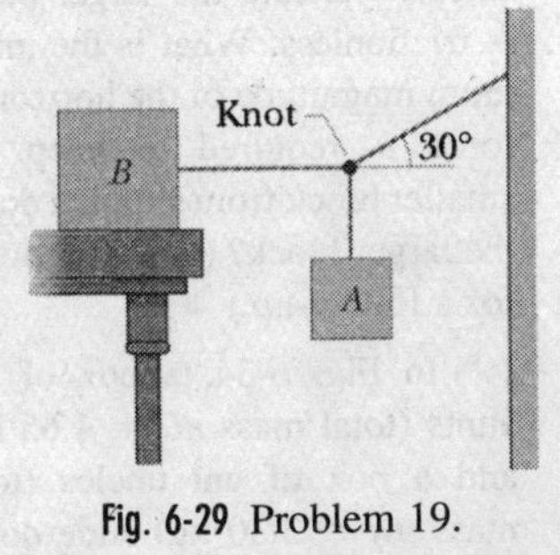

Fig. 6-29 Problem 19.

20P. A force $\vec{P}$, parallel to a surface inclined 15° above the horizontal, acts on a 45 N block, as shown in Fig. 6-30. The coefficients of friction for the block and surface are $\mu_s = 0.50$ and $\mu_k = 0.34$. If the block is initially at rest, determine the magnitude and direction of the frictional force acting on the block for magnitudes of $\vec{P}$ of (a) 5.0 N, (b) 8.0 N, and (c) 15 N.

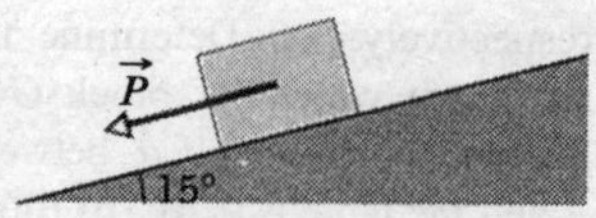

Fig. 6-30 Problem 20.

21P. Body A in Fig. 6-31 weighs 102 N, and body B weighs 32 N. The coefficients of friction between A and the incline are $\mu_s = 0.56$ and $\mu_k = 0.25$. Angle θ is 40°. Find the acceleration of A if (a) A is initially at rest, (b) A is initially moving up the incline, and (c) A is initially moving down the incline. ssm www

22P. In Fig. 6-31, two blocks are connected over a pulley. The mass of block A is 10 kg and the coefficient of kinetic friction between A and the incline is 0.20. Angle θ of the incline is 30°. Block A slides down the incline at constant speed. What is the mass of block B?

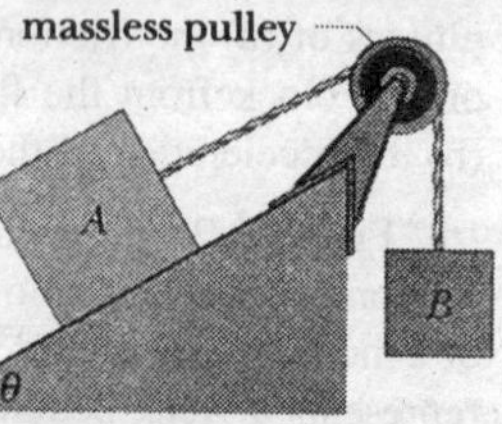

Fig. 6-31 Problems 21 and 22.

23P. Two blocks, of weights 3.6 N and 7.2 N, are connected by a massless string and slide down a 30° inclined plane. The coefficient of kinetic friction between the lighter block and the plane is 0.10; that between the heavier block and the plane is 0.20. Assuming that the lighter block leads, find (a) the magnitude of the acceleration of the blocks and (b) the tension in the string. (c) Describe the motion if, instead, the heavier block leads. ssm

24P. In Fig. 6-32, a box of Cheerios and a box of Wheaties are accelerated across a horizontal surface by a horizontal force $\vec{F}$ applied to the Cheerios box. The magnitude of the frictional force on the Cheerios box is 2.0 N, and the magnitude of the frictional force on the Wheaties box is 4.0 N. If the magnitude of $\vec{F}$ is 12 N, what is the magnitude of the force on the Wheaties box from the Cheerios box?

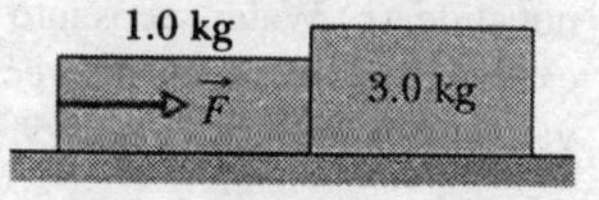

Fig. 6-32 Problem 24.

25P. The two blocks (with $m = 16$ kg and $M = 88$ kg) shown in Fig. 6-33 are not attached. The coefficient of static friction between the blocks is $\mu_s = 0.38$, but the surface beneath the larger block is frictionless. What is the minimum magnitude of the horizontal force $\vec{F}$ required to keep the smaller block from slipping down the larger block? (See Question 8 for a follow-up.) ilw

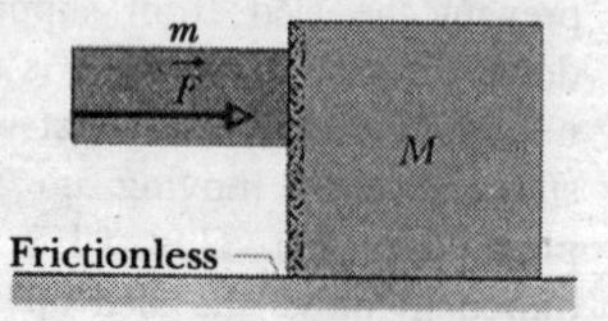

Fig. 6-33 Problem 25.

26P. In Fig. 6-34, a box of ant aunts (total mass $m_1 = 1.65$ kg) and a box of ant uncles (total mass $m_2 = 3.30$ kg) slide down an inclined plane while attached by a massless rod parallel to the plane. The angle of incline is $\theta = 30°$. The coefficient of kinetic friction between the aunt box and the incline is $\mu_1 = 0.226$; that between the uncle box and the incline is $\mu_2 = 0.113$. Compute (a) the tension in the rod and (b) the common acceleration of the two boxes. (c) How would the answers to (a) and (b) change if the uncles trailed the aunts?

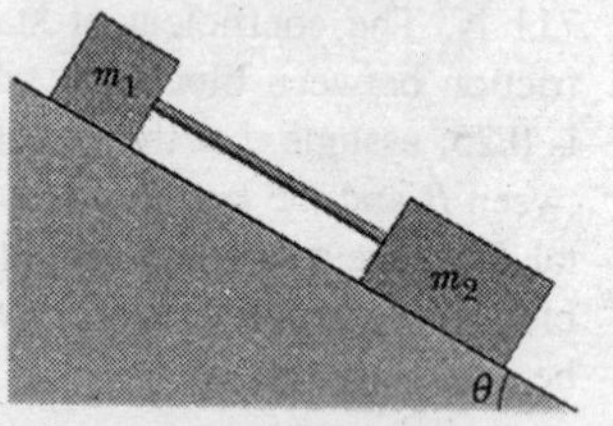

Fig. 6-34 Problem 26.

27P. A 40 kg slab rests on a frictionless floor. A 10 kg block rests on top of the slab (Fig. 6-35). The coefficient of static friction μ_s between the block and the slab is 0.60, whereas their kinetic friction coefficient μ_k is 0.40. The 10 kg block is pulled by a horizontal force with a magnitude of 100 N. What are the resulting accelerations of (a) the block and (b) the slab? ssm www

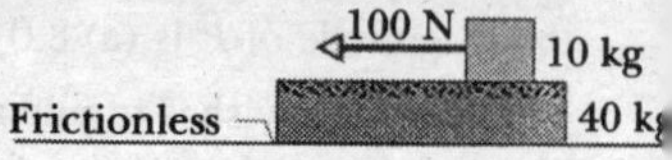

Fig. 6-35 Problem 27.

28P. A locomotive accelerates a 25-car train along a level track. Every car has a mass of 5.0×10^4 kg and is subject to a frictional force $f = 250v$, where the speed v is in meters per second and the force f is in newtons. At the instant when the speed of the train is 30 km/h, the magnitude of its acceleration is 0.20 m/s^2. (a) What is the tension in the coupling between the first car and the locomotive? (b) If this tension is equal to the maximum force the locomotive can exert on the train, what is the steepest grade up which the locomotive can pull the train at 30 km/h?

29P. In Fig. 6-36, a crate slides down an inclined right-angled trough. The coefficient of kinetic friction between the crate and the trough is μ_k. What is the acceleration of the crate in terms of μ_k, θ, and g? ssm

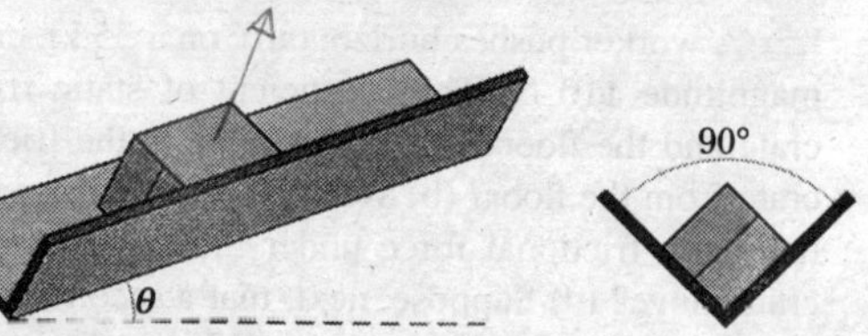

Fig. 6-36 Problem 29.

30P. An initially stationary box of sand is to be pulled across a floor by means of a cable in which the tension should not exceed 1100 N. The coefficient of static friction between the box and the floor is 0.35. (a) What should be the angle between the cable and the horizontal in order to pull the greatest possible amount of sand, and (b) what is the weight of the sand and box in that situation?

31P. A 1000 kg boat is traveling at 90 km/h when its engine is shut off. The magnitude of the frictional force $\vec{f}_k$ between boat and water is proportional to the speed v of the boat: $f_k = 70v$, where v is in meters per second and f_k is in newtons. Find the time required for the boat to slow to 45 km/h. ssm

SEC. 6-3 The Drag Force and Terminal Speed

32E. *Continuation of Exercise 4.* First reread the explanation of how the wind might drag desert stones across the playa. Now assume that Eq. 6-14 gives the magnitude of the air drag force on the typical 20 kg stone, which presents a vertical cross-sectional area to the wind of 0.040 m^2 and has a drag coefficient C of 0.80. Take the air density to be 1.21 kg/m^3, and the coefficient of kinetic friction to be 0.80. (a) In kilometers per hour, what wind speed V along the ground is needed to maintain the stone's motion once it has started moving? Because winds along the ground are retarded by the ground, the wind speeds reported for storms are often measured at a height of 10 m. Assume wind speeds are 2.00 times those along the ground. (b) For your answer to (a), what wind speed would be reported for the storm and is that value reasonable for a high-speed wind in a storm? (Story continues with Problem 51.)

33E. Calculate the drag force on a missile 53 cm in diameter cruising with a speed of 250 m/s at low altitude, where the density of air is 1.2 kg/m³. Assume $C = 0.75$. ssm

34E. The terminal speed of a sky diver is 160 km/h in the spread-eagle position and 310 km/h in the nosedive position. Assuming that the diver's drag coefficient C does not change from one position to the other, find the ratio of the effective cross-sectional area A in the slower position to that in the faster position.

35P. Calculate the ratio of the drag force on a passenger jet flying with a speed of 1000 km/h at an altitude of 10 km to the drag force on a prop-driven transport flying at half the speed and half the altitude of the jet. At 10 km the density of air is 0.38 kg/m³, and at 5.0 km it is 0.67 kg/m³. Assume that the airplanes have the same effective cross-sectional area and the same drag coefficient C.

SEC. 6-4 Uniform Circular Motion

36E. During an Olympic bobsled run, the Jamaican team makes a turn of radius 7.6 m at a speed of 96.6 km/h. What is their acceleration in g-units?

37E. Suppose the coefficient of static friction between the road and the tires on a Formula One car is 0.6 during a Grand Prix auto race. What speed will put the car on the verge of sliding as it rounds a level curve of 30.5 m radius? ssm

38E. A roller-coaster car has a mass of 1200 kg when fully loaded with passengers. As the car passes over the top of a circular hill of radius 18 m, its speed is not changing. What are the magnitude and direction of the force of the track on the car at the top of the hill if the car's speed is (a) 11 m/s and (b) 14 m/s?

39E. What is the smallest radius of an unbanked (flat) track around which a bicyclist can travel if her speed is 29 km/h and the coefficient of static friction between tires and track is 0.32? ilw

40P. An amusement park ride consists of a car moving in a vertical circle on the end of a rigid boom of negligible mass. The combined weight of the car and riders is 5.0 kN, and the radius of the circle is 10 m. What are the magnitude and direction of the force of the boom on the car at the top of the circle if the car's speed there is (a) 5.0 m/s and (b) 12 m/s?

41P. A puck of mass m slides on a frictionless table while attached to a hanging cylinder of mass M by a cord through a hole in the table (Fig. 6-37). What speed keeps the cylinder at rest? ssm

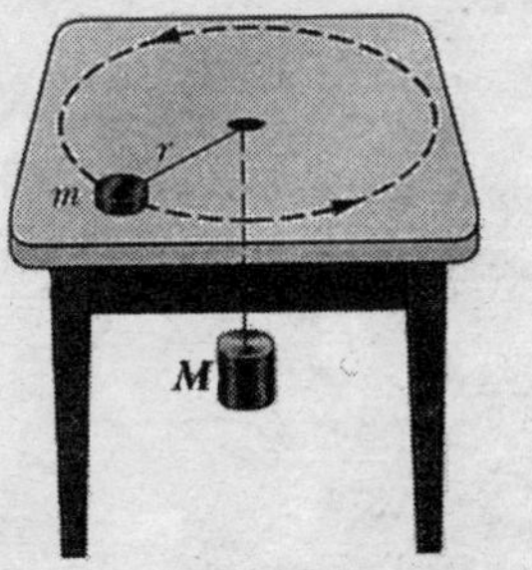

Fig. 6-37 Problem 41.

42P. A bicyclist travels in a circle of radius 25.0 m at a constant speed of 9.00 m/s. The bicycle–rider mass is 85.0 kg. Calculate the magnitudes of (a) the force of friction on the bicycle from the road and (b) the *net* force on the bicycle from the road.

43P. A student of weight 667 N rides a steadily rotating Ferris wheel (the student sits upright). At the highest point, the magnitude of the normal force $\vec{N}$ on the student from the seat is 556 N. (a) Does the student feel "light" or "heavy" there? (b) What is the magnitude of $\vec{N}$ at the lowest point? (c) What is the magnitude N if the wheel's speed is doubled? ssm ilw www

44P. An old streetcar rounds a flat corner of radius 9.1 m, at 16 km/h. What angle with the vertical will be made by the loosely hanging hand straps?

45P. An airplane is flying in a horizontal circle at a speed of 480 km/h. If its wings are tilted 40° to the horizontal, what is the radius of the circle in which the plane is flying? (See Fig. 6-38.) Assume that the required force is provided entirely by an "aerodynamic lift" that is perpendicular to the wing surface. ssm

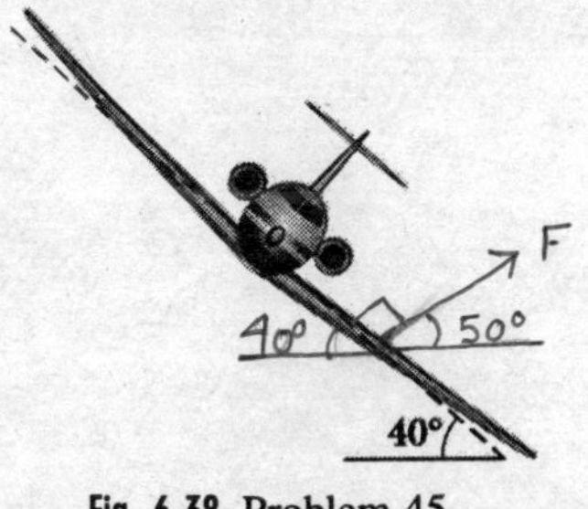

Fig. 6-38 Problem 45.

46P. A high-speed railway car goes around a flat, horizontal circle of radius 470 m at a constant speed. The magnitudes of the horizontal and vertical components of the force of the car on a 51.0 kg passenger are 210 N and 500 N, respectively. (a) What is the magnitude of the net force (of *all* the forces) on the passenger? (b) What is the speed of the car?

47P. As shown in Fig. 6-39, a 1.34 kg ball is connected by means of two massless strings to a vertical, rotating rod. The strings are tied to the rod and are taut. The tension in the upper string is 35 N. (a) Draw the free-body diagram for the ball. What are (b) the tension in the lower string, (c) the net force on the ball, and (d) the speed of the ball? ssm ilw

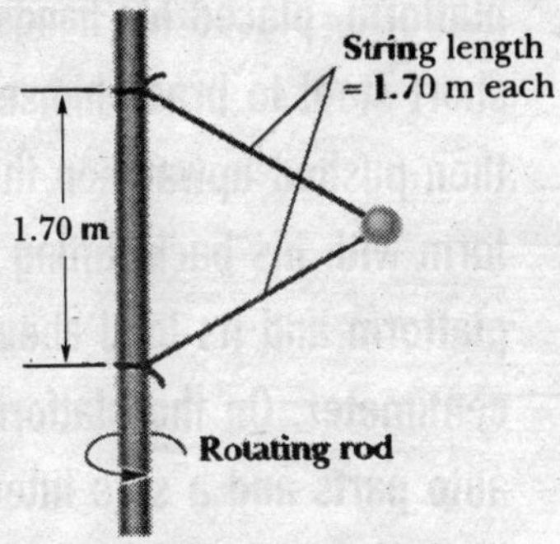

Fig. 6-39 Problem 47.

Additional Problem

48. *Continuation of Exercises 4 and 32*. Another explanation is that the stones move only when the water dumped on the playa during a storm freezes into a large, thin sheet. The stones are trapped in place in the sheet. Then, as air flows across the sheet during a wind, the air drag forces on the sheet and on the stones move them, with the stones gouging out the trails. The magnitude of the air drag force on this horizontal "ice sail" is given by $D_{ice} = 4C_{ice}\rho A_{ice}v^2$, where C_{ice} is the drag coefficient (about 2.0×10^{-3}), ρ is the air density (1.21 kg/m³), A_{ice} is the horizontal area of the ice, and v is the wind speed along the ice.

Assume the following: The sheet measures 400 m by 500 m by 4.0 mm and has a coefficient of kinetic friction of 0.10 with the ground and a density of 917 kg/m³. Also assume that 100 stones identical to the one in Exercise 4 are trapped in the ice. To maintain the motion of the sheet, what are the required wind speeds (a) near the sheet and (b) at a height of 10 m? (c) Are these reasonable values for high-speed winds in a storm?

7 Kinetic Energy and Work

In the weight-lifting competition of the 1996 Olympics, Andrey Chemerkin lifted a record-breaking 260.0 kg from the floor to over his head (about 2 m). In 1957 Paul Anderson stooped beneath a reinforced wood platform, placed his hands on a short stool to brace himself, and then pushed upward on the platform with his back, lifting the platform and its load about a centimeter. On the platform were auto parts and a safe filled with lead; the composite weight of the load was 27 900 N (6270 lb)!

Who did more work on the objects he lifted — Chemerkin or Anderson?

The answer is in this chapter.

7-1 Energy

Newton's laws of motion allow us to analyze many kinds of motion. However, the analysis is often complicated, requiring details about the motion that we simply do not know. Here is an example: A puck is sent sliding along an inclined frictionless track that includes several ups and downs (hills and valleys) of various shapes. The puck's initial speed is 4.0 m/s, and its initial height is 0.46 m. Using Newton's second law, can you calculate the puck's speed when it reaches the end of the track, at zero height? No, not without the details of how the incline varies all along the track, and then the calculations can be very complicated.

Long ago, scientists and engineers gradually began to realize that there is another, sometimes more powerful technique for analyzing motion. Moreover, this technique could be, and eventually was, extended to other situations, such as chemical reactions, geological processes, and biological functions, that do not involve motion. This other technique involves **energy**, which comes in a great many *forms* (or types). In fact, the term *energy* is so broad that a clear definition for it is difficult to write. Technically, energy is a scalar quantity that is associated with a state (or condition) of one or more objects. However, this definition is too vague to be of help to us now.

A looser definition might at least get us started. Energy is a number that we associate with a system of one or more objects. If a force changes one of the objects by, say, making it move, then the number changes. After countless experiments, scientists and engineers realized that if the scheme by which we assign these energy numbers is planned carefully, then the numbers can be used to predict the outcomes of experiments. (For example, they would allow us to easily find the puck's speed in the previous example.) However, learning how to use the numbering scheme is not easy and the scheme is not obvious. Thus, in this chapter we focus on only one form of energy—kinetic energy. Other forms of energy will show up throughout this book and your work in science or engineering.

Kinetic energy K is energy associated with the *state of motion* of an object. The faster the object moves, the greater is its kinetic energy. When the object is stationary, its kinetic energy is zero.

For an object of mass m whose speed v is well below the speed of light, we define kinetic energy as

$$K = \tfrac{1}{2}mv^2 \quad \text{(kinetic energy).} \tag{7-1}$$

For example, a 3.0 kg duck flying past us at 2.0 m/s has a kinetic energy of $6.0\ \text{kg}\cdot\text{m}^2/\text{s}^2$; that is, we associate that number with the duck's motion.

The SI unit of kinetic energy (and every other type of energy) is the **joule** (J), named for James Prescott Joule, an English scientist of the 1800s. It is defined directly from Eq. 7-1 in terms of the units for mass and velocity:

$$1\ \text{joule} = 1\ \text{J} = 1\ \text{kg}\cdot\text{m}^2/\text{s}^2. \tag{7-2}$$

Thus, the flying duck has a kinetic energy of 6.0 J.

Sample Problem 7-1

In 1896 in Waco, Texas, William Crush of the "Katy" railroad parked two locomotives at opposite ends of a 6.4-km-long track, fired them up, tied their throttles open, and then allowed them to crash head-on at full speed (Fig. 7-1) in front of 30,000 spectators. Hundreds of people were hurt by flying debris; several were killed. Assuming each locomotive weighed 1.2×10^6 N and its acceleration along the track was a constant $0.26\ \text{m/s}^2$, what was the total kinetic energy of the two locomotives just before the collision?

Fig. 7-1 Sample Problem 7-1. The aftermath of an 1896 crash of two locomotives.

SOLUTION: One **Key Idea** here is to find the kinetic energy of each locomotive with Eq. 7-1, but that means we need each locomotive's speed just before the collision and its mass. A second **Key Idea** is that, because we can assume each locomotive had constant acceleration, we can use the equations in Table 2-1 to find its speed v just before the collision. We choose Eq. 2-16 because we know values for all the variables except v:

$$v^2 = v_0^2 + 2a(x - x_0).$$

With $v_0 = 0$ and $x - x_0 = 3.2 \times 10^3$ m (half the initial separation), this yields

$$v^2 = 0 + 2(0.26 \text{ m/s}^2)(3.2 \times 10^3 \text{ m}),$$

or

$$v = 40.8 \text{ m/s}$$

(about 150 km/h).

A third **Key Idea** is that we can find the mass of each locomotive by dividing its given weight by g:

$$m = \frac{1.2 \times 10^6 \text{ N}}{9.8 \text{ m/s}^2} = 1.22 \times 10^5 \text{ kg}.$$

Now, using Eq. 7-1, we find the total kinetic energy of the two locomotives just before the collision as

$$K = 2(\tfrac{1}{2}mv^2) = (1.22 \times 10^5 \text{ kg})(40.8 \text{ m/s})^2$$
$$= 2.0 \times 10^8 \text{ J}. \qquad \text{(Answer)}$$

Sitting near this collision was like sitting near an exploding bomb.

7-2 Work

If you accelerate an object to a greater speed by applying a force to the object, you increase the kinetic energy $K\ (= \frac{1}{2}mv^2)$ of the object. Similarly, if you decelerate the object to a lesser speed by applying a force, you decrease the kinetic energy of the object. We account for these changes in kinetic energy by saying that your force has transferred energy *to* the object from yourself or *from* the object to yourself.

In such a transfer of energy via a force, **work** W is said to be *done on the object by the force*. More formally, we define work as follows:

> Work W is energy transferred to or from an object by means of a force acting on the object. Energy transferred to the object is positive work, and energy transferred from the object is negative work.

"Work," then, is transferred energy; "doing work" is the act of transferring the energy. Work has the same units as energy and is a scalar quantity.

The term *transfer* can be misleading. It does not mean that anything material flows into or out of the object; that is, the transfer is not like a flow of water. Rather it is like the electronic transfer of money between two bank accounts: The number in one account goes up while the number in the other account goes down, with nothing material passing between the two accounts.

Note that we are not concerned here with the common meaning of the word "work," which implies that *any* physical or mental labor is work. For example, if you push hard against a wall, you tire because of the continuously repeated muscle contractions that are required, and you are, in the common sense, working. However, such effort does not cause an energy transfer to or from the wall and thus is not work done on the wall as defined here.

To avoid confusion in this chapter, we shall use the symbol W only for work and shall represent a weight with its equivalent mg.

7-3 Work and Kinetic Energy

Finding an Expression for Work

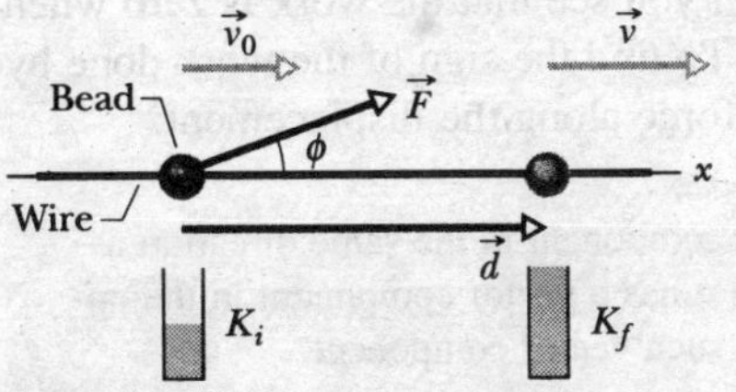

Fig. 7-2 A constant force $\vec{F}$ directed at angle ϕ to the displacement $\vec{d}$ of a bead on a wire accelerates the bead along the wire, changing the velocity of the bead from $\vec{v}_0$ to $\vec{v}$. A "kinetic energy gauge" indicates the resulting change in the kinetic energy of the bead, from the value K_i to K_f.

Let us find an expression for work by considering a bead that can slide along a frictionless wire, which is stretched along a horizontal x axis (Fig. 7-2). A constant force $\vec{F}$, directed at an angle ϕ to the wire, accelerates the bead along the wire. We can relate the force and the acceleration with Newton's second law, written for components along the x axis:

$$F_x = ma_x, \tag{7-3}$$

where m is the bead's mass. As the bead moves through a displacement $\vec{d}$, the force changes the bead's velocity from an initial value $\vec{v}_0$ to another value $\vec{v}$. Because the force is constant, we know that the acceleration is also constant. Thus, we can use Eq. 2-16 (one of the basic constant-acceleration equations of Chapter 2) to write, for components along the x axis,

$$v^2 = v_0^2 + 2a_x d. \tag{7-4}$$

Solving this equation for a_x, substituting into Eq. 7-3, and rearranging then give us

$$\tfrac{1}{2}mv^2 - \tfrac{1}{2}mv_0^2 = F_x d. \tag{7-5}$$

The first term on the left side of the equation is the kinetic energy K_f of the bead at the end of the displacement d, and the second term is the kinetic energy K_i of the bead at the start of the displacement. Thus, the left side of Eq. 7-5 tells us the kinetic energy has been changed by the force, and the right side tells us the change is equal to $F_x d$. Therefore, the work W done on the bead by the force (the energy transfer due to the force) is

$$W = F_x d. \tag{7-6}$$

If we know values for F_x and d, we can use this equation to calculate the work W done on the bead by the force.

> To calculate the work done on an object by a force during a displacement, we use only the force component along the object's displacement. The force component perpendicular to the displacement does zero work.

From Fig. 7-2, we see that we can write F_x as $F\cos\phi$, where ϕ is the angle between the directions of the displacement $\vec{d}$ and the force $\vec{F}$. We can rewrite Eq. 7-6 in a more general form as

$$W = Fd\cos\phi \quad \text{(work done by a constant force).} \tag{7-7}$$

This equation is useful for calculating the work if we know values for F, d, and ϕ. Because the right side of this equation is equivalent to the scalar (or dot) product $\vec{F}\cdot\vec{d}$, we can also write

$$W = \vec{F}\cdot\vec{d} \quad \text{(work done by a constant force).} \tag{7-8}$$

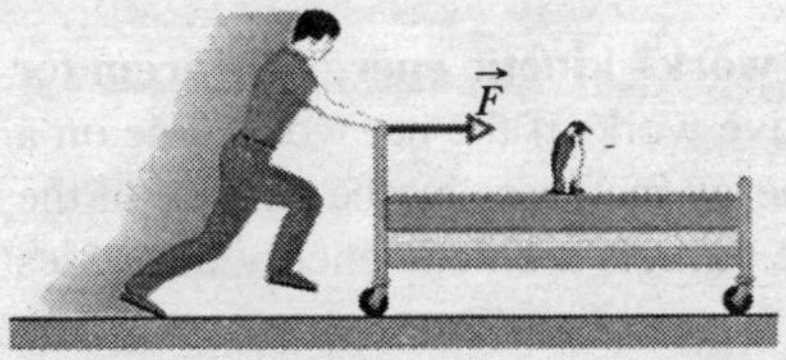

Fig. 7-3 A contestant in a bed race. We can approximate the bed and its rider as being a particle for the purpose of calculating the work done on them by the force applied by the student.

(You may wish to review the discussion of scalar products in Section 3-7.) Equation 7-8 is especially useful for calculating the work when $\vec{F}$ and $\vec{d}$ are given in unit-vector notation.

Cautions: There are two restrictions to using Eqs. 7-6 through 7-8 to calculate work done on an object by a force. First, the force must be a *constant force*; that is, it must not change in magnitude or direction as the object moves. (Later, we shall discuss what to do with a *variable force* that changes in magnitude.) Second, the object must be *particle-like*. This means that the object must be *rigid*; all parts of the object must move together, in the same direction. In this chapter we consider only particle-like objects, such as the bed and its rider being pushed in Fig. 7-3.

Signs for work. The work done on an object by a force can be either positive work or negative work. For example, if the angle ϕ in Eq. 7-7 is less than 90°, then $\cos\phi$ is positive and thus so is the work. If ϕ is greater than 90° (up to 180°), then $\cos\phi$ is negative and thus so is the work. (Can you see that the work is zero when $\phi = 90°$?) These results lead to a simple rule. To find the sign of the work done by a force, consider the vector component of the force along the displacement:

> A force does positive work when it has a vector component in the same direction as the displacement, and it does negative work when it has a vector component in the opposite direction. It does zero work when it has no such vector component.

Units for work. Work has the SI unit of the joule, the same as kinetic energy. However, from Eqs. 7-6 and 7-7 we can see that an equivalent unit is the newton-meter (N · m). The corresponding unit in the British system is the foot-pound (ft · lb). Extending Eq. 7-2, we have

$$1\ \text{J} = \text{kg}\cdot\text{m}^2/\text{s}^2 = 1\ \text{N}\cdot\text{m} = 0.738\ \text{ft}\cdot\text{lb}. \tag{7-9}$$

Net work done by several forces. When two or more forces act on an object, the **net work** done on the object is the sum of the works done by the individual forces. We can calculate the net work in two ways. (1) We can find the work done by each force and then sum those works. (2) Alternatively, we can first find the net force $\vec{F}_{\text{net}}$ of those forces. Then we can use Eq. 7-7, substituting the magnitude F_{net} for F, and the angle between the directions of $\vec{F}_{\text{net}}$ and the displacement for ϕ; or we can use Eq. 7-8 with $\vec{F}_{\text{net}}$ substituted for $\vec{F}$.

Work–Kinetic Energy Theorem

Equation 7-5 relates the change in kinetic energy of the bead (from an initial $K_i = \frac{1}{2}mv_0^2$ to a later $K_f = \frac{1}{2}mv^2$) to the work W $(= F_x d)$ done on the bead. For such particle-like objects, we can generalize that equation. Let ΔK be the change in the kinetic energy of the object, and W the net work done on it. Then we can write

$$\Delta K = K_f - K_i = W, \tag{7-10}$$

which says that

$$\left(\begin{array}{c}\text{change in the kinetic}\\ \text{energy of a particle}\end{array}\right) = \left(\begin{array}{c}\text{net work done on}\\ \text{the particle}\end{array}\right).$$

We can also write

$$K_f = K_i + W, \tag{7-11}$$

which says that

$$\left(\begin{array}{c}\text{kinetic energy after}\\ \text{the net work is done}\end{array}\right) = \left(\begin{array}{c}\text{kinetic energy}\\ \text{before the net work}\end{array}\right) + \left(\begin{array}{c}\text{the net}\\ \text{work done}\end{array}\right).$$

These statements are known traditionally as the **work–kinetic energy theorem** for particles. They hold for both positive and negative work: If the net work done on a particle is positive, then the particle's kinetic energy increases by the amount of the work. If the net work done is negative, then the particle's kinetic energy decreases by the amount of the work.

For example, if the kinetic energy is initially 5 J and there is a net transfer of 2 J to the particle (positive net work), then the final kinetic energy is 7 J. If, instead, there is a net transfer of 2 J from the particle (negative net work), then the final kinetic energy is 3 J.

CHECKPOINT 1: A particle moves along an x axis. Does the kinetic energy of the particle increase, decrease, or remain the same if the particle's velocity changes (a) from -3 m/s to -2 m/s and (b) from -2 m/s to 2 m/s? (c) In each situation, is the work done on the particle positive, negative, or zero?

Sample Problem 7-2

Figure 7-4*a* shows two industrial spies sliding an initially stationary 225 kg floor safe a displacement $\vec{d}$ of magnitude 8.50 m, straight toward their truck. The push $\vec{F}_1$ of Spy 001 is 12.0 N, directed at an angle of 30° downward from the horizontal; the pull $\vec{F}_2$ of Spy 002 is 10.0 N, directed at 40° above the horizontal. The magnitudes and directions of these forces do not change as the safe moves, and the floor and safe make frictionless contact.

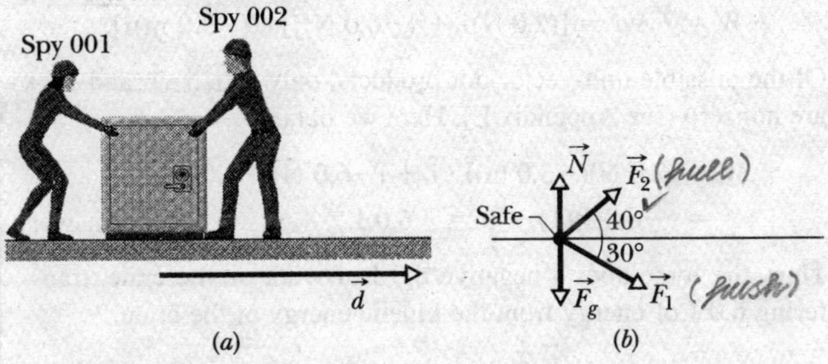

Fig. 7-4 Sample Problem 7-2. (*a*) Two spies move a floor safe through displacement $\vec{d}$. (*b*) A free-body diagram for the safe.

(a) What is the net work done on the safe by forces $\vec{F}_1$ and $\vec{F}_2$ during the displacement $\vec{d}$?

SOLUTION: We use two **Key Ideas** here. First, the net work W done on the safe by the two forces is the sum of the works they do individually. Second, because we can treat the safe as a particle and the forces are constant in both magnitude and direction, we can use either Eq. 7-7 ($W = Fd \cos \phi$) or Eq. 7-8 ($W = \vec{F} \cdot \vec{d}$) to calculate those works. Since we know the magnitudes and directions of the forces, we choose Eq. 7-7. From it and the free-body diagram for the safe in Fig. 7-4*b*, the work done by $\vec{F}_1$ is

$$W_1 = F_1 d \cos \phi_1 = (12.0\ \text{N})(8.50\ \text{m})(\cos 30°)$$
$$= 88.33\ \text{J},$$

and the work done by $\vec{F}_2$ is

$$W_2 = F_2 d \cos \phi_2 = (10.0\ \text{N})(8.50\ \text{m})(\cos 40°)$$
$$= 65.11\ \text{J}.$$

Thus, the net work W is

$$W = W_1 + W_2 = 88.33\ \text{J} + 65.11\ \text{J}$$
$$= 153.4\ \text{J} \approx 153\ \text{J}. \quad \text{(Answer)}$$

During the 8.50 m displacement, therefore, the spies transfer 153 J of energy to the kinetic energy of the safe.

(b) During the displacement, what is the work W_g done on the safe by the gravitational force $\vec{F}_g$ and what is the work W_N done on the safe by the normal force $\vec{N}$ from the floor?

SOLUTION: The **Key Idea** is that, because these forces are constant in both magnitude and direction, we can find the work they do with Eq. 7-7. Thus, with mg as the magnitude of the gravitational force, we write

$$W_g = mgd \cos 90° = mgd(0) = 0 \quad \text{(Answer)}$$

and

$$W_N = Nd \cos 90° = Nd(0) = 0. \quad \text{(Answer)}$$

We should have known this result. Because these forces are perpendicular to the displacement of the safe, they do zero work on the safe and do not transfer any energy to or from it.

(c) The safe is initially stationary. What is its speed v_f at the end of the 8.50 m displacement?

SOLUTION: Here the **Key Idea** is that the speed of the safe changes because its kinetic energy is changed when energy is transferred to it by $\vec{F}_1$ and $\vec{F}_2$. We relate the speed to the work done by combining Eqs. 7-10 and 7-1:

$$W = K_f - K_i = \tfrac{1}{2}mv_f^2 - \tfrac{1}{2}mv_i^2.$$

The initial speed v_i is zero, and we now know that the work done is 153.4 J. Solving for v_f and then substituting known data, we find that

$$v_f = \sqrt{\frac{2W}{m}} = \sqrt{\frac{2(153.4\ \text{J})}{225\ \text{kg}}}$$
$$= 1.17\ \text{m/s}. \quad \text{(Answer)}$$

Sample Problem 7-3

During a storm, a crate of crepe is sliding across a slick, oily parking lot through a displacement $\vec{d} = (-3.0\ \text{m})\hat{\mathrm{i}}$ while a steady wind pushes against the crate with a force $\vec{F} = (2.0\ \text{N})\hat{\mathrm{i}} + (-6.0\ \text{N})\hat{\mathrm{j}}$. The situation and coordinate axes are shown in Fig. 7-5.

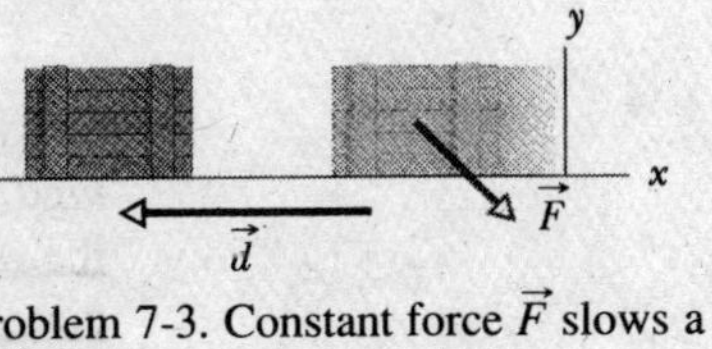

Fig. 7-5 Sample Problem 7-3. Constant force $\vec{F}$ slows a crate during displacement $\vec{d}$.

(a) How much work does this force from the wind do on the crate during the displacement?

SOLUTION: The **Key Idea** here is that, because we can treat the crate as a particle and because the wind force is constant ("steady") in both magnitude and direction during the displacement, we can use either Eq. 7-7 ($W = Fd\cos\phi$) or Eq. 7-8 ($W = \vec{F}\cdot\vec{d}$) to calculate the work. Since we know $\vec{F}$ and $\vec{d}$ in unit-vector notation, we choose Eq. 7-8 and write

$$W = \vec{F}\cdot\vec{d} = [(2.0\text{ N})\hat{i} + (-6.0\text{ N})\hat{j}]\cdot[(-3.0\text{ m})\hat{i}].$$

Of the possible unit-vector dot products, only $\hat{i}\cdot\hat{i}$, $\hat{j}\cdot\hat{j}$, and $\hat{k}\cdot\hat{k}$ are nonzero (see Appendix E). Here we obtain

$$W = (2.0\text{ N})(-3.0\text{ m})\hat{i}\cdot\hat{i} + (-6.0\text{ N})(-3.0\text{ m})\hat{j}\cdot\hat{i}$$
$$= (-6.0\text{ J})(1) + 0 = -6.0\text{ J}. \qquad \text{(Answer)}$$

Thus, the force does a negative 6.0 J of work on the crate, transfering 6.0 J of energy from the kinetic energy of the crate.

(b) If the crate has a kinetic energy of 10 J at the beginning of displacement $\vec{d}$, what is its kinetic energy at the end of $\vec{d}$?

SOLUTION: The **Key Idea** here is that, because the force does negative work on the crate, it reduces the crate's kinetic energy. Using the work–kinetic energy theorem in the form of Eq. 7-11, we have

$$K_f = K_i + W = 10\text{ J} + (-6.0\text{ J}) = 4.0\text{ J}. \qquad \text{(Answer)}$$

Because the kinetic energy is decreased to 4.0 J, the crate has been slowed.

CHECKPOINT 2: The figure shows four situations in which a force acts on a box while the box slides rightward a distance d across a frictionless floor. The magnitudes of the forces are identical; their orientations are as shown. Rank the situations according to the work done on the box during the displacement, from most positive to most negative.

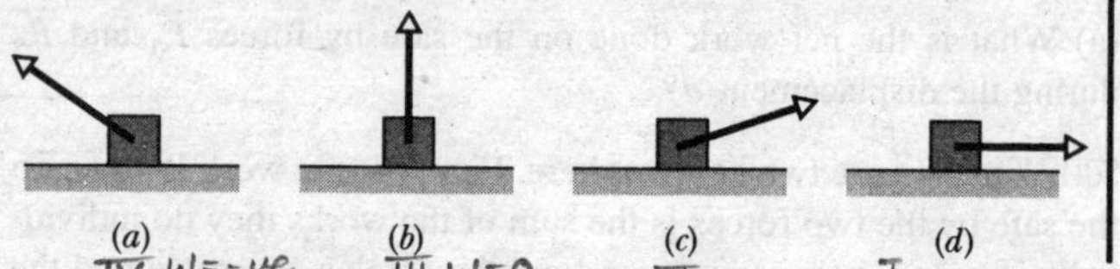

7-4 Work Done by a Gravitational Force

We next examine the work done on an object by a particular type of force—namely, the gravitational force acting on it. Figure 7-6 shows a particle-like tomato of mass m that is thrown upward with initial speed v_0 and thus with initial kinetic energy $K_i = \frac{1}{2}mv_0^2$. As the tomato rises, it is slowed by a gravitational force $\vec{F}_g$; that is, the tomato's kinetic energy decreases because $\vec{F}_g$ does work on the tomato as it rises.

Because we can treat the tomato as a particle, we can use Eq. 7-7 ($W = Fd\cos\phi$) to express the work done during a displacement $\vec{d}$. For the force magnitude F, we use mg as the magnitude of $\vec{F}_g$. Thus, the work W_g done by the gravitational force $\vec{F}_g$ is

$$W_g = mgd\cos\phi \quad \text{(work done by gravitational force).} \qquad (7\text{-}12)$$

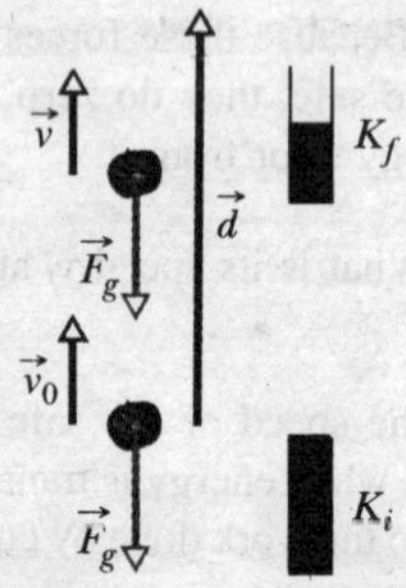

Fig. 7-6 A particle-like tomato of mass m thrown upward slows from velocity $\vec{v}_0$ to velocity $\vec{v}$ during displacement $\vec{d}$ because the gravitational force $\vec{F}_g$ acts on it. A kinetic energy gauge indicates the resulting change in the kinetic energy of the object, from K_i ($= \frac{1}{2}mv_0^2$) to K_f ($= \frac{1}{2}mv^2$).

For a rising object, force $\vec{F}_g$ is directed opposite the displacement $\vec{d}$, as indicated in Fig. 7-6. Thus, $\phi = 180°$ and

$$W_g = mgd\cos 180° = mgd(-1) = -mgd. \qquad (7\text{-}13)$$

The minus sign tells us that during the object's rise, the gravitational force on the object transfers energy in the amount mgd from the kinetic energy of the object. This is consistent with the slowing of the object as it rises.

After the object has reached its maximum height and is falling back down, the angle ϕ between force $\vec{F}_g$ and displacement $\vec{d}$ is zero. Thus,

$$W_g = mgd\cos 0° = mgd(+1) = +mgd. \qquad (7\text{-}14)$$

The plus sign tells us that the gravitational force now transfers energy in the amount mgd to the kinetic energy of the object. This is consistent with the speeding up of the object as it falls. (Actually, as we shall see in Chapter 8, energy transfers associated with lifting and lowering an object involve not just the object, but the full object–Earth system. Without Earth, of course, "lifting" would be meaningless.)

Work Done in Lifting and Lowering an Object

Now suppose we lift a particle-like object by applying a vertical force $\vec{F}$ to it. During the upward displacement, our applied force does positive work W_a on the object while the gravitational force does negative work W_g on it. Our force tends to transfer energy to the object while the gravitational force tends to transfer energy from it. By Eq. 7-10, the change ΔK in the kinetic energy of the object due to these two energy transfers is

$$\Delta K = K_f - K_i = W_a + W_g, \qquad (7\text{-}15)$$

in which K_f is the kinetic energy at the end of the displacement and K_i is that at the start of the displacement. This equation also applies if we lower the object, but then the gravitational force tends to transfer energy *to* the object while our force tends to transfer energy *from* it.

In one common situation the object is stationary before and after the lift—for example, when you lift a book from the floor to a shelf. Then K_f and K_i are both zero, and Eq. 7-15 reduces to

$$W_a + W_g = 0$$

or

$$W_a = -W_g. \qquad (7\text{-}16)$$

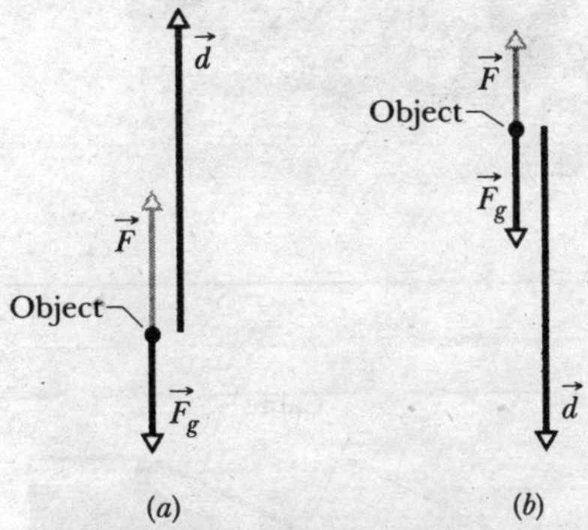

Fig. 7-7 (*a*) An applied force $\vec{F}$ lifts an object. The object's displacement $\vec{d}$ makes an angle $\phi = 180°$ with the gravitational force $\vec{F}_g$ on the object. The applied force does positive work on the object. (*b*) An applied force $\vec{F}$ lowers an object. The displacement $\vec{d}$ of the object makes an angle $\phi = 0°$ with the gravitational force $\vec{F}_g$. The applied force does negative work on the object.

Note that we get the same result if K_f and K_i are not zero but are still equal. Either way, the result means that the work done by the applied force is the negative of the work done by the gravitational force; that is, the applied force transfers the same amount of energy to the object as the gravitational force transfers from the object. Using Eq. 7-12, we can rewrite Eq. 7-16 as

$$W_a = -mgd\cos\phi \quad \text{(work in lifting and lowering; } K_f = K_i\text{)}, \qquad (7\text{-}17)$$

with ϕ being the angle between $\vec{F}_g$ and $\vec{d}$. If the displacement is vertically upward (Fig. 7-7*a*), then $\phi = 180°$ and the work done by our force equals mgd. If the displacement is vertically downward (Fig. 7-7*b*), then $\phi = 0°$ and the work done by the applied force equals $-mgd$.

Equations 7-16 and 7-17 apply to any situation in which an object is lifted or lowered, with the object stationary before and after the lift. They are independent of the magnitude of the force used. For example, when Chemerkin made his record-breaking lift, his force on the object he lifted varied considerably during the lift. Still, because the object was stationary before and after the lift, the work he did is given by Eqs. 7-16 and 7-17, where, in Eq. 7-17, mg is the weight of the object he lifted and d is the distance he lifted it.

Sample Problem 7-4

Let us return to the lifting feats of Andrey Chemerkin and Paul Anderson.

(a) Chemerkin made his record-breaking lift with rigidly connected objects (a barbell and disk weights) having a total mass $m = 260.0$ kg; he lifted them a distance of 2.0 m. During the lift, how much work was done on the objects by the gravitational force $\vec{F}_g$ acting on them?

SOLUTION: The **Key Idea** here is that we can treat the rigidly connected objects as a single particle and thus use Eq. 7-12 ($W_g = mgd\cos\phi$) to find the work W_g done on them by $\vec{F}_g$. The total weight mg was 2548 N, the magnitude d of the displacement was 2.0 m, and the angle ϕ between the directions of the downward gravitational force and the upward displacement was 180°. Thus,

$$W_g = mgd\cos\phi = (2548\text{ N})(2.0\text{ m})(\cos 180°)$$
$$= -5100\text{ J}. \qquad \text{(Answer)}$$

(b) How much work was done on the objects by Chemerkin's force during the lift?

SOLUTION: We do not have an expression for Chemerkin's force on the object, and even if we did, his force was certainly not constant. Thus, one **Key Idea** here is that we *cannot* just substitute his force into Eq. 7-7 to find his work. However, we know that the objects were stationary at the start and end of the lift. Therefore, as a second **Key Idea**, we know that the work W_{AC} done by Chemerkin's applied force was the negative of the work W_g done by the gravitational force $\vec{F}_g$. Equation 7-16 expresses this fact and gives us

$$W_{AC} = -W_g = +5100 \text{ J.} \qquad \text{(Answer)}$$

(c) While Chemerkin held the objects stationary above his head, how much work was done on them by his force?

SOLUTION: The **Key Idea** is that when he supported the objects, they were stationary. Thus, their displacement $d = 0$ and, by Eq. 7-7, the work done on them was zero (even though supporting them was a very tiring task).

(d) How much work was done by the force Paul Anderson applied to lift objects with a total weight of 27 900 N, a distance of 1.0 cm?

SOLUTION: Following the argument of parts (a) and (b), but now with $mg = 27\,900$ N and $d = 1.0$ cm, we find

$$W_{PA} = -W_g = -mgd\cos\phi = -mgd\cos 180°$$
$$= -(27\,900 \text{ N})(0.010 \text{ m})(-1) = 280 \text{ J.} \qquad \text{(Answer)}$$

Anderson's lift required a tremendous upward force but only a small energy transfer of 280 J, because of the short displacement involved. This photo shows another of his lifts.

Sample Problem 7-5

An initially stationary 15.0 kg crate of cheese wheels is pulled, via a cable, a distance $L = 5.70$ m up a frictionless ramp, to a height h of 2.50 m, where it stops (Fig. 7-8*a*).

(a) How much work W_g is done on the crate by the gravitational force $\vec{F}_g$ during the lift?

SOLUTION: A **Key Idea** is that we can treat the crate as a particle and thus use Eq. 7-12 ($W_g = mgd\cos\phi$) to find the work W_g done by $\vec{F}_g$. However, we do not know the angle ϕ between the directions of $\vec{F}_g$ and displacement $\vec{d}$. From the crate's free-body diagram in Fig. 7-8*b*, we find that ϕ is $\theta + 90°$, where θ is the (unknown) angle of the ramp. Equation 7-12 then gives us

$$W_g = mgd\cos(\theta + 90°) = -mgd\sin\theta, \qquad (7\text{-}18)$$

where we have used a trigonometic identity to simplify the expression. The result seems to be useless because θ is unknown. But (continuing with physics courage) we see from Fig. 7-8*a* that $d\sin\theta = h$, where h is a known quantity (2.50 m). With this substitution, Eq. 7-18 then gives us

$$W_g = -mgh \qquad (7\text{-}19)$$
$$= -(15.0 \text{ kg})(9.8 \text{ m/s}^2)(2.50 \text{ m})$$
$$= -368 \text{ J.} \qquad \text{(Answer)}$$

Note that Eq. 7-19 tells us that the work W_g done by the gravitational force depends on the vertical displacement but (surprisingly) not on the horizontal displacement. (We return to this point in Chapter 8.)

(b) How much work W_T is done on the crate by the force $\vec{T}$ from the cable during the lift?

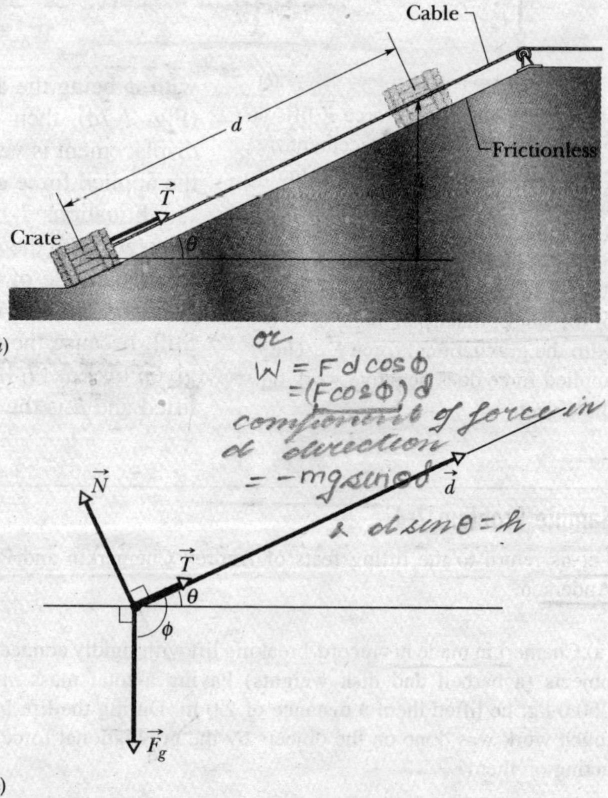

Fig. 7-8 Sample Problem 7-5. (*a*) A crate is pulled up a frictionless ramp by a force $\vec{T}$ parallel to the ramp. (*b*) A free-body diagram for the crate, showing also the displacement $\vec{d}$.

SOLUTION: We cannot just substitute the force magnitude T for F in Eq. 7-7 ($W = Fd\cos\phi$) because we do not know the value of T. However, a **Key Idea** to get us going is that we can treat the crate as a particle and then apply the work–kinetic energy theorem ($\Delta K = W$) to it. Because the crate is stationary before and after the lift, the change ΔK in its kinetic energy is zero. For the net work W done on the crate, we must sum the works done by all three forces acting on the crate. From (a), the work W_g done by the gravitational force $\vec{F}_g$ is -368 J. The work W_N done by the normal force $\vec{N}$ on the crate from the ramp is zero because $\vec{N}$ is perpendicular to the displacement. We want the work W_T done by $\vec{T}$. Thus, the work–kinetic energy theorem gives us

$$\Delta K = W_T + W_g + W_N$$

or

$$0 = W_T - 368\text{ J} + 0,$$

and so

$$W_T = 368\text{ J}. \qquad \text{(Answer)}$$

CHECKPOINT 3: Suppose we raise the crate by the same height h but with a longer ramp. (a) Is the work done by force $\vec{T}$ now greater, smaller, or the same as before? (b) Is the magnitude of $\vec{T}$ needed to move the crate now greater, smaller, or the same as before?

Sample Problem 7-6

An elevator cab of mass $m = 500$ kg is descending with speed $v_i = 4.0$ m/s when its supporting cable begins to slip, allowing it to fall with constant acceleration $\vec{a} = \vec{g}/5$ (Fig. 7-9*a*).

(a) During the fall through a distance $d = 12$ m, what is the work W_g done on the cab by the gravitational force $\vec{F}_g$?

SOLUTION: The **Key Idea** here is that we can treat the cab as a particle and thus use Eq. 7-12 ($W_g = mgd\cos\phi$) to find the work W_g done on it by $\vec{F}_g$. From Fig. 7-9*b* (a free-body diagram of the cab with the cab's displacement $\vec{d}$ included), we see that the angle between the directions of $\vec{F}_g$ and the cab's displacement $\vec{d}$ is 0°. Then, from Eq. 7-12, we find

$$W_g = mgd\cos 0° = (500\text{ kg})(9.8\text{ m/s}^2)(12\text{ m})(1)$$
$$= 5.88 \times 10^4\text{ J} \approx 59\text{ kJ}. \qquad \text{(Answer)}$$

(b) During the 12 m fall, what is the work W_T done on the cab by the upward pull $\vec{T}$ of the elevator cable?

SOLUTION: A **Key Idea** here is that we can calculate the work W_T with Eq. 7-7 ($W = Fd\cos\phi$) if we first find an expression for the magnitude T of the cable's pull. A second **Key Idea** is that we can find that expression by writing Newton's second law for components along the y axis in Fig. 7-9*b* ($F_{\text{net},y} = ma_y$). We get

$$T - F_g = ma.$$

Solving for T, substituting mg for F_g, and then substituting the result in Eq. 7-7, we obtain

$$W_T = Td\cos\phi = m(a + g)d\cos\phi.$$

Next, substituting $-g/5$ for the (downward) acceleration a and then 180° for the angle ϕ between the directions of forces $\vec{T}$ and $m\vec{g}$, we find

$$W_T = m\left(-\frac{g}{5} + g\right)d\cos\phi = \frac{4}{5}mgd\cos\phi$$
$$= \frac{4}{5}(500\text{ kg})(9.8\text{ m/s}^2)(12\text{ m})\cos 180°$$
$$= -4.70 \times 10^4\text{ J} \approx -47\text{ kJ}. \qquad \text{(Answer)}$$

Now note that W_T is not simply the negative of W_g, which we found in (a). The reason is that, because the cab accelerates during the fall, its speed changes during the fall, and thus its kinetic energy also changes. Therefore, Eq. 7-16 (which assumes that the initial and final kinetic energies are equal) does *not* apply here.

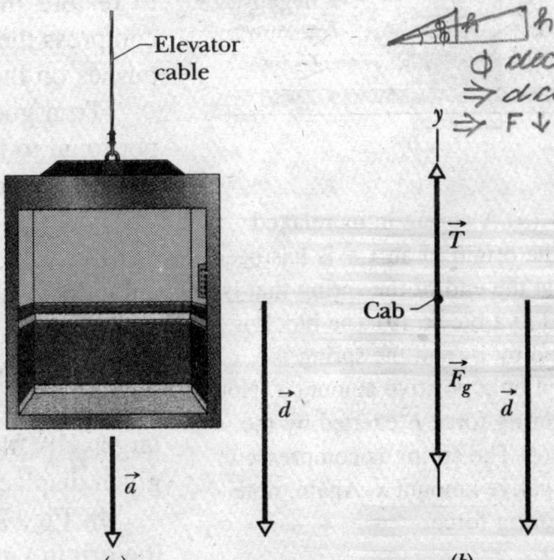

Fig. 7-9 Sample Problem 7-6. An elevator cab, descending with speed v_i, suddenly begins to accelerate downward. (*a*) It moves through a displacement $\vec{d}$ with constant acceleration $\vec{a} = \vec{g}/5$. (*b*) A free-body diagram for the cab, displacement included.

(c) What is the net work W done on the cab during the fall?

SOLUTION: The **Key Idea** here is that the net work is the sum of the works done by the forces acting on the cab:

$$W = W_g + W_T = 5.88 \times 10^4\text{ J} - 4.70 \times 10^4\text{ J}$$
$$= 1.18 \times 10^4\text{ J} \approx 12\text{ kJ}. \qquad \text{(Answer)}$$

(d) What is the cab's kinetic energy at the end of the 12 m fall?

SOLUTION: The **Key Idea** here is that the kinetic energy changes *because* of the net work done on the cab, according to Eq. 7-11 ($K_f = K_i + W$). From Eq. 7-1, we can write the kinetic energy at the start of the fall as $K_i = \frac{1}{2}mv_i^2$. We can then write Eq. 7-11 as

$$K_f = K_i + W = \tfrac{1}{2}mv_i^2 + W$$
$$= \tfrac{1}{2}(500\text{ kg})(4.0\text{ m/s})^2 + 1.18 \times 10^4\text{ J}$$
$$= 1.58 \times 10^4\text{ J} \approx 16\text{ kJ}. \qquad \text{(Answer)}$$

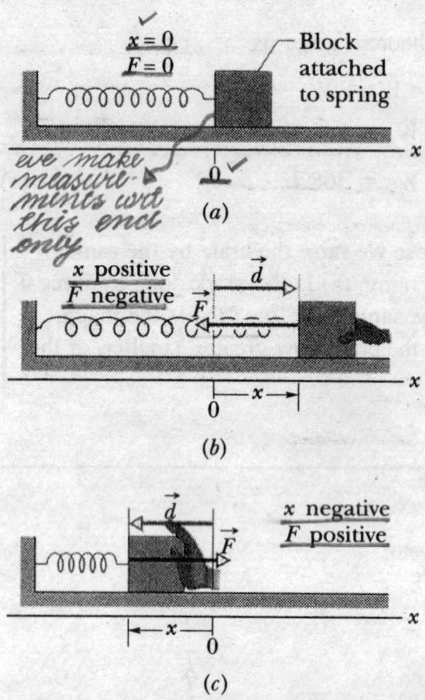

Fig. 7-10 (*a*) A spring in its relaxed state. The origin of an x axis has been placed at the end of the spring that is attached to a block. (*b*) The block is displaced by $\vec{d}$, and the spring is stretched by a positive amount x. Note the restoring force $\vec{F}$ exerted by the spring. (*c*) The spring is compressed by a negative amount x. Again, note the restoring force.

7-5 Work Done by a Spring Force

We next want to examine the work done on a particle-like object by a particular type of *variable force*—namely, a **spring force**, the force from a spring. Many forces in nature have the same mathematical form as the spring force. Thus, by examining this one force, you can gain an understanding of many others.

The Spring Force

Figure 7-10*a* shows a spring in its **relaxed state**—that is, neither compressed nor extended. One end is fixed, and a particle-like object, say, a block, is attached to the other, free end. If we stretch the spring by pulling the block to the right as in Fig. 7-10*b*, the spring pulls on the block toward the left. (Because a spring's force acts to restore the relaxed state, it is sometimes said to be a *restoring force*.) If we compress the spring by pushing the block to the left as in Fig. 7-10*c*, the spring now pushes on the block toward the right.

To a good approximation for many springs, the force $\vec{F}$ from a spring is proportional to the displacement $\vec{d}$ of the free end from its position when the spring is in the relaxed state. The *spring force* is given by

$$\vec{F} = -k\vec{d} \quad \text{(Hooke's law)}, \tag{7-20}$$

which is known as **Hooke's law** after Robert Hooke, an English scientist of the late 1600s. The minus sign in Eq. 7-20 indicates that the spring force is always opposite in direction from the displacement of the free end. The constant k is called the **spring constant** (or **force constant**) and is a measure of the stiffness of the spring. The larger k is, the stiffer the spring; that is, the stronger will be its pull or push for a given displacement. The SI unit for k is the newton per meter.

In Fig. 7-10 an x axis has been placed parallel to the length of a spring, with the origin ($x = 0$) at the position of the free end when the spring is in its relaxed state. For this common arrangement, we can write Eq. 7-20 as

$$F = -kx \quad \text{(Hooke's law)}. \tag{7-21}$$

If x is positive (the spring is stretched toward the right on the x axis), then F is negative (it is a pull toward the left). If x is negative (the spring is compressed toward the left), then F is positive (it is a push toward the right).

Note that a spring force is a *variable force* because its magnitude and direction depend on the position x of the free end; F can be symbolized as $F(x)$. Also note that Hooke's law is a *linear* relationship between F and x.

The Work Done by a Spring Force

To find an expression for the work done by the spring force as the block in Fig. 7-10*a* moves, let us make two simplifying assumptions about the spring. (1) It is *massless*; that is, its mass is negligible compared to the block's mass. (2) It is an *ideal spring*; that is, it obeys Hooke's law exactly. Let us also assume that the contact between the block and the floor is frictionless and that the block is particle-like.

We give the block a rightward jerk to get it moving, and then leave it alone. As the block moves rightward, the spring force $\vec{F}$ does work on the block, decreasing the kinetic energy and slowing the block. However, we *cannot* find this work by using Eq. 7-7 ($W = Fd\cos\phi$) because that equation assumes a constant force. The spring force is a variable force.

To find the work done by the spring, we use calculus. Let the block's initial position be x_i and its later position x_f. Then divide the distance between those two positions into many segments, each of tiny length Δx. Label these segments, starting from x_i, as segments 1, 2, and so on. As the block moves through a segment, the spring force hardly varies because the segment is so short that x hardly varies. Thus, we can approximate the force magnitude as being constant within the segment. Label these magnitudes as F_1 in segment 1, F_2 in segment 2, and so on.

With the force now constant in each segment, we *can* find the work done within each segment by using Eq. 7-7 ($W = Fd \cos \phi$). Here $\phi = 0$, so $\cos \phi = 1$. Then the work done is $F_1 \Delta x$ in segment 1, $F_2 \Delta x$ in segment 2, and so on. The net work W_s done by the spring, from x_i to x_f, is the sum of all these works:

$$W_s = \sum F_j \Delta x, \tag{7-22}$$

where j labels the segments. In the limit as Δx goes to zero, Eq. 7-22 becomes

W = F · d = constant force
W = ∫F · dl = variable force

$$W_s = \int_{x_i}^{x_f} F\, dx. \tag{7-23}$$

Substituting for F from Eq. 7-21, we find

$$\begin{aligned} W_s &= \int_{x_i}^{x_f} (-kx)\, dx = -k \int_{x_i}^{x_f} x\, dx \\ &= (-\tfrac{1}{2}k)\, [x^2]_{x_i}^{x_f} = (-\tfrac{1}{2}k)(x_f^2 - x_i^2). \end{aligned} \tag{7-24}$$

Ws = -1/2 k (xf² - xi²)

Multiplied out, this yields

$$W_s = \tfrac{1}{2}kx_i^2 - \tfrac{1}{2}kx_f^2 \quad \text{(work by a spring force).} \tag{7-25}$$

This work W_s done by the spring force can have a positive or negative value, depending on whether the *net* transfer of energy is to or from the block as the block moves from x_i to x_f. *Caution:* The final position x_f appears in the *second* term on the right side of Eq. 7-25. Therefore, Eq. 7-25 tells us:

➤ Work W_s is positive if the block ends up closer to the relaxed position ($x = 0$) than it was initially. It is negative if the block ends up farther away from $x = 0$. It is zero if the block ends up at the same distance from $x = 0$.

If $x_i = 0$ and if we call the final position x, then Eq. 7-25 becomes

$$W_s = -\tfrac{1}{2}kx^2 \quad \text{(work by a spring force).} \tag{7-26}$$

The Work Done by an Applied Force

Now suppose that we displace the block along the x axis while continuing to apply a force $\vec{F}_a$ to it. During the displacement, our applied force does work W_a on the block while the spring force does work W_s. By Eq. 7-10, the change ΔK in the kinetic energy of the block due to these two energy transfers is

Work-KE theorem holds for constant as well as variable forces

$$\Delta K = K_f - K_i = W_a + W_s, \tag{7-27}$$

in which K_f is the kinetic energy at the end of the displacement and K_i is that at the start of the displacement. If the block is stationary before and after the displacement, then K_f and K_i are both zero and Eq. 7-27 reduces to

$$W_a = -W_s. \tag{7-28}$$

➤ If a block that is attached to a spring is stationary before and after a displacement, then the work done on it by the applied force displacing it is the negative of the work done on it by the spring force.

Caution: If the block is not stationary before and after the displacement, then this statement is *not* true.

CHECKPOINT 4: For three situations, the initial and final positions, respectively, along the x axis for the block in Fig. 7-10 are (a) -3 cm, 2 cm; (b) 2 cm, 3 cm; and (c) -2 cm, 2 cm. In each situation is the work done by the spring force on the block positive, negative, or zero?

Sample Problem 7-7

A package of spicy Cajun pralines lies on a frictionless floor, attached to the free end of a spring in the arrangement of Fig. 7-10*a*. An applied force of magnitude $F_a = 4.9$ N would be needed to hold the package stationary at $x_1 = 12$ mm.

(a) How much work does the spring force do on the package if the package is pulled rightward from $x_0 = 0$ to $x_2 = 17$ mm?

SOLUTION: A **Key Idea** here is that as the package moves from one position to another, the spring force does work on it as given by Eq. 7-25 or Eq. 7-26. We know that the initial position x_i is 0 and the final position x_f is 17 mm, but we do not know the spring constant k.

We can probably find k with Eq. 7-21 (Hooke's law), but we need a second **Key Idea** to use it: Were the package held stationary at $x_1 = 12$ mm, the spring force would have to balance the applied force (according to Newton's second law). Thus, the spring force F would have to be -4.9 N (toward the left in Fig. 7-10*b*), so Eq. 7-21 ($F = -kx$) gives us

$$k = -\frac{F}{x_1} = -\frac{-4.9\text{ N}}{12 \times 10^{-3}\text{ m}} = 408\text{ N/m}.$$

Now, with the package at $x_2 = 17$ mm, Eq. 7-26 yields

$$W_s = -\tfrac{1}{2}kx_2^2 = -\tfrac{1}{2}(408\text{ N/m})(17 \times 10^{-3}\text{ m})^2$$
$$= -0.059\text{ J}. \qquad \text{(Answer)}$$

(b) Next, the package is moved leftward to $x_3 = -12$ mm. How much work does the spring force do on the package during this displacement? Explain the sign of this work.

SOLUTION: The **Key Idea** here is the first one we noted in part (a). Now $x_i = +17$ mm and $x_f = -12$ mm, and Eq. 7-25 yields

$$W_s = \tfrac{1}{2}kx_i^2 - \tfrac{1}{2}kx_f^2 = \tfrac{1}{2}k(x_i^2 - x_f^2)$$
$$= \tfrac{1}{2}(408\text{ N/m})[(17 \times 10^{-3}\text{ m})^2 - (-12 \times 10^{-3}\text{ m})^2]$$
$$= 0.030\text{ J} = 30\text{ mJ}. \qquad \text{(Answer)}$$

This work done on the block by the spring force is positive because the spring force does more positive work as the block moves from $x_i = +17$ mm to the spring's relaxed position than it does negative work as the block moves from the spring's relaxed position to $x_f = -12$ mm.

Sample Problem 7-8

In Fig. 7-11, a cumin canister of mass $m = 0.40$ kg slides across a horizontal frictionless counter with speed $v = 0.50$ m/s. It then runs into and compresses a spring of spring constant $k = 750$ N/m. When the canister is momentarily stopped by the spring, by what distance d is the spring compressed?

SOLUTION: There are three **Key Ideas** here:

1. The work W_s done on the canister by the spring force is related to the requested distance d by Eq. 7-26 ($W_s = -\tfrac{1}{2}kx^2$), with d replacing x.

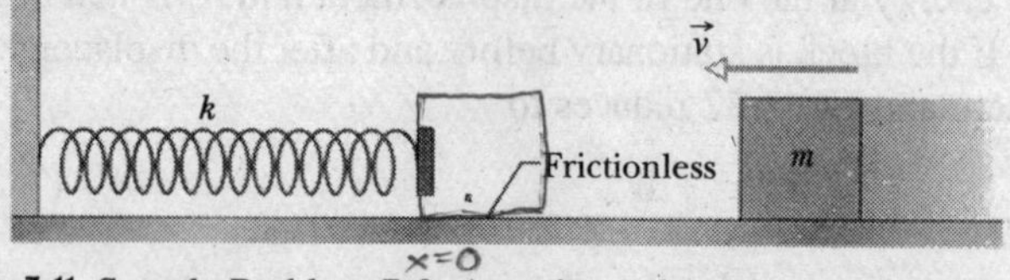

Fig. 7-11 Sample Problem 7-8. A canister of mass m moves at velocity $\vec{v}$ toward a spring with spring constant k.

2. The work W_s is also related to the kinetic energy of the canister by Eq. 7-10 ($K_f - K_i = W$).
3. The canister's kinetic energy has an initial value of $K = \tfrac{1}{2}mv^2$ and a value of zero when the canister is momentarily at rest.

Putting the first two of these ideas together, we write the work–kinetic energy theorem for the canister as

$$K_f - K_i = -\tfrac{1}{2}kd^2.$$

Substituting according to the third idea makes this

$$0 - \tfrac{1}{2}mv^2 = -\tfrac{1}{2}kd^2.$$

Simplifying, solving for d, and substituting known data then give us

$$d = v\sqrt{\frac{m}{k}} = (0.50\text{ m/s})\sqrt{\frac{0.40\text{ kg}}{750\text{ N/m}}}$$
$$= 1.2 \times 10^{-2}\text{ m} = 1.2\text{ cm}. \qquad \text{(Answer)}$$

7-6 Work Done by a General Variable Force

One-Dimensional Analysis

Let us return to the situation of Fig. 7-2 but now consider the force to be directed along the x axis and the force magnitude to vary with position x. Thus, as the bead (particle) moves, the magnitude of the force doing work on it changes. Only the magnitude of this variable force changes, not its direction, and the magnitude at any position does not change with time.

Figure 7-12a shows a plot of such a *one-dimensional variable force*. We want an expression for the work done on the particle by this force as the particle moves from an initial point x_i to a final point x_f. However, we *cannot* use Eq. 7-7 because it applies only for a constant force $\vec{F}$. Here, again, we shall use calculus. We divide the area under the curve of Fig. 7-12a into a number of narrow strips of width Δx (Fig. 7-12b). We choose Δx small enough to permit us to take the force $F(x)$ as being reasonably constant over that interval. We let $F_{j,\text{avg}}$ be the average value of $F(x)$ within the jth interval. Then in Fig. 7-12b, $F_{j,\text{avg}}$ is the height of the jth strip.

With $F_{j,\text{avg}}$ considered constant, the increment (small amount) of work ΔW_j done by the force in the jth interval is now approximately given by Eq. 7-7 and is

$$\Delta W_j = F_{j,\text{avg}}\,\Delta x. \tag{7-29}$$

In Fig. 7-12b, ΔW_j is then equal to the area of the jth rectangular, shaded strip.

To approximate the total work W done by the force as the particle moves from x_i to x_f, we add the areas of all the strips between x_i and x_f in Fig. 7-12b:

$$W = \sum \Delta W_j = \sum F_{j,\text{avg}}\,\Delta x. \tag{7-30}$$

Equation 7-30 is an approximation because the broken "skyline" formed by the tops of the rectangular strips in Fig. 7-12b only approximates the actual curve of $F(x)$.

We can make the approximation better by reducing the strip width Δx and using more strips, as in Fig. 7-12c. In the limit, we let the strip width approach zero; the number of strips then becomes infinitely large and we have, as an exact result,

$$W = \lim_{\Delta x \to 0} \sum F_{j,\text{avg}}\,\Delta x. \tag{7-31}$$

This limit is exactly what we mean by the integral of the function $F(x)$ between the limits x_i and x_f. Thus, Eq. 7-31 becomes

$$W = \int_{x_i}^{x_f} F(x)\,dx \qquad \text{(work: variable force).} \tag{7-32}$$

If we know the function $F(x)$, we can substitute it into Eq. 7-32, introduce the proper limits of integration, carry out the integration, and thus find the work. (Appendix E contains a list of common integrals.) Geometrically, the work is equal to the area between the $F(x)$ curve and the x axis, between the limits x_i and x_f (shaded in Fig. 7-12d).

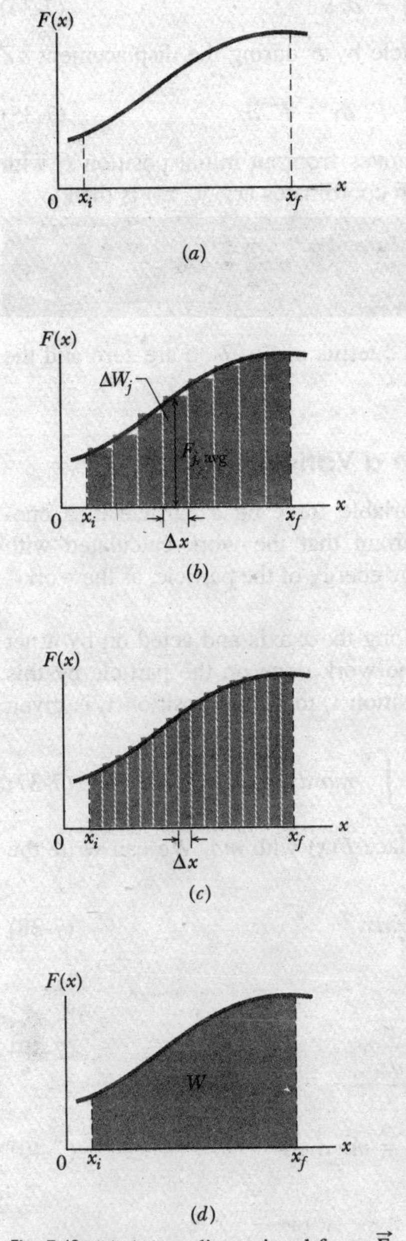

Fig. 7-12 (a) A one-dimensional force $\vec{F}$ plotted against the displacement x of a particle on which it acts. The particle moves from x_i to x_f. (b) Same as (a) but with the area under the curve divided into narrow strips. (c) Same as (b) but with the area divided into narrower strips. (d) The limiting case. The work done by the force is given by Eq. 7-32 and is represented by the shaded area between the curve and the x axis and between x_i and x_f.

Three-Dimensional Analysis

Consider now a particle that is acted on by a three-dimensional force

$$\vec{F} = F_x\hat{\mathrm{i}} + F_y\hat{\mathrm{j}} + F_z\hat{\mathrm{k}}, \tag{7-33}$$

in which the components F_x, F_y, and F_z can depend on the position of the particle; that is, they can be functions of that position. However, we make three simplifications: F_x may depend on x but not on y or z, F_y may depend on y but not on x or z, and F_z may depend on z but not on x or y. Now let the particle move through an incremental displacement

$$d\vec{r} = dx\hat{\mathrm{i}} + dy\hat{\mathrm{j}} + dz\hat{\mathrm{k}}. \tag{7-34}$$

The increment of work dW done on the particle by $\vec{F}$ during the displacement $d\vec{r}$ is, by Eq. 7-8,

$$dW = \vec{F} \cdot d\vec{r} = F_x\, dx + F_y\, dy + F_z\, dz. \tag{7-35}$$

The work W done by $\vec{F}$ while the particle moves from an initial position r_i with coordinates (x_i, y_i, z_i) to a final position r_f with coordinates (x_f, y_f, z_f) is then

$$W = \int_{r_i}^{r_f} dW = \int_{x_i}^{x_f} F_x\, dx + \int_{y_i}^{y_f} F_y\, dy + \int_{z_i}^{z_f} F_z\, dz. \tag{7-36}$$

If $\vec{F}$ has only an x component, then the y and z terms in Eq. 7-36 are zero and the equation reduces to Eq. 7-32.

Work–Kinetic Energy Theorem with a Variable Force

Equation 7-32 gives the work done by a variable force on a particle in a one-dimensional situation. Let us now make certain that the work calculated with Eq. 7-32 is indeed equal to the change in kinetic energy of the particle, as the work–kinetic energy theorem states.

Consider a particle of mass m, moving along the x axis and acted on by a net force $F(x)$ that is directed along that axis. The work done on the particle by this force as the particle moves from an initial position x_i to a final position x_f is given by Eq. 7-32 as

$$W = \int_{x_i}^{x_f} F(x)\, dx = \int_{x_i}^{x_f} ma\, dx, \tag{7-37}$$

in which we use Newton's second law to replace $F(x)$ with ma. We can write the quantity $ma\, dx$ in Eq. 7-37 as

$$ma\, dx = m\frac{dv}{dt}\, dx. \tag{7-38}$$

From the "chain rule" of calculus, we have

$$\frac{dv}{dt} = \frac{dv}{dx}\frac{dx}{dt} = \frac{dv}{dx}v, \tag{7-39}$$

and Eq. 7-38 becomes

$$ma\, dx = m\frac{dv}{dx}v\, dx = mv\, dv. \tag{7-40}$$

Substituting Eq. 7-40 into Eq. 7-37 yields

$$\begin{aligned} W &= \int_{v_i}^{v_f} mv\, dv = m\int_{v_i}^{v_f} v\, dv \\ &= \tfrac{1}{2}mv_f^2 - \tfrac{1}{2}mv_i^2. \end{aligned} \tag{7-41}$$

Note that when we change the variable from x to v we are required to express the limits on the integral in terms of the new variable. Note also that because the mass m is a constant, we are able to move it outside the integral.

Recognizing the terms on the right side of Eq. 7-41 as kinetic energies allows us to write this equation as

$$W = K_f - K_i = \Delta K,$$

which is the work–kinetic energy theorem.

$W = \int \vec{F} \cdot d\vec{l} = \int (3x^2 \hat{i} + 4\hat{j}) \cdot (dx\hat{i} + dy\hat{j}) = \int 3x^2 dx + \int 4 dy$

Sample Problem 7-9

Force $\vec{F} = (3x^2 \text{ N})\hat{i} + (4 \text{ N})\hat{j}$, with x in meters, acts on a particle, changing only the kinetic energy of the particle. How much work is done on the particle as it moves from coordinates (2 m, 3 m) to (3 m, 0 m)? Does the speed of the particle increase, decrease, or remain the same?

SOLUTION: The **Key Idea** here is that the force is a variable force because its x component depends on the value of x. Thus, we cannot use Eqs. 7-7 and 7-8 to find the work done. Instead, we must use Eq. 7-36 to integrate the force:

$$W = \int_2^3 3x^2\,dx + \int_3^0 4\,dy = 3\int_2^3 x^2\,dx + 4\int_3^0 dy$$
$$= 3[\tfrac{1}{3}x^3]_2^3 + 4[y]_3^0 = [3^3 - 2^3] + 4[0 - 3]$$
$$= 7.0 \text{ J}. \qquad \text{(Answer)}$$

The positive result means that energy is transferred to the particle by force $\vec{F}$. Thus, the kinetic energy of the particle increases, and so must its speed.

7-7 Power

A contractor wishes to lift a load of bricks from the sidewalk to the top of a building by means of a winch. We can now calculate how much work the force applied by the winch must do on the load to make the lift. The contractor, however, is much more interested in the *rate* at which that work is done. Will the job take 5 minutes (acceptable) or a week (unacceptable)?

The time rate at which work is done by a force is said to be the **power** due to the force. If an amount of work W is done in an amount of time Δt by a force, the **average power** due to the force during that time interval is

$$P_{avg} = \frac{W}{\Delta t} \quad \text{(average power)}. \tag{7-42}$$

The **instantaneous power** P is the instantaneous time rate of doing work, which we can write as

$$P = \frac{dW}{dt} \quad \text{(instantaneous power)}. \tag{7-43}$$

Instantaneous Power
$P = \vec{F} \cdot \vec{v}$ → const force
$\vec{v}$ is particle's (particle-like object) velocity

$P_{net} = P_1 + P_2$

Suppose we know the work $W(t)$ done by a force as a function of time. Then to get the instantaneous power P at, say, time $t = 3.0$ s during the work, we would first take the time derivative of $W(t)$, and then evaluate the result for $t = 3.0$ s.

The SI unit of power is the joule per second. This unit is used so often that it has a special name, the **watt** (W), after James Watt, who greatly improved the rate at which steam engines could do work. In the British system, the unit of power is the foot-pound per second. Often the horsepower is used. Some relations among these units are

$$1 \text{ watt} = 1 \text{ W} = 1 \text{ J/s} = 0.738 \text{ ft} \cdot \text{lb/s} \tag{7-44}$$

and

$$1 \text{ horsepower} = 1 \text{ hp} = 550 \text{ ft} \cdot \text{lb/s} = 746 \text{ W}. \tag{7-45}$$

Inspection of Eq. 7-42 shows that work can be expressed as power multiplied by time, as in the common unit, the kilowatt-hour. Thus,

$$1 \text{ kilowatt-hour} = 1 \text{ kW} \cdot \text{h} = (10^3 \text{ W})(3600 \text{ s}) = 3.60 \times 10^6 \text{ J} = 3.60 \text{ MJ}. \tag{7-46}$$

Perhaps because they appear on our utility bills, the watt and the kilowatt-hour have become identified as electrical units. They can be used equally well as units for other examples of power and work or energy. Thus, if you pick up this book from the floor and put it on a tabletop, you are free to report the work that you have done as 4×10^{-6} kW · h (or more conveniently as 4 mW · h).

We can also express the rate at which a force does work on a particle (or particle-like object) in terms of that force and the particle's velocity. For a particle that is moving along a straight line (say, the x axis) and is acted on by a constant force $\vec{F}$ directed at some angle ϕ to that line, Eq. 7-43 becomes

$$P = \frac{dW}{dt} = \frac{F\cos\phi\,dx}{dt} = F\cos\phi\left(\frac{dx}{dt}\right),$$

or

$$P = Fv\cos\phi. \tag{7-47}$$

Reorganizing the right side of Eq. 7-47 as the dot product $\vec{F}\cdot\vec{v}$, we may also write Eq. 7-47 as

$$P = \vec{F}\cdot\vec{v} \quad \text{(instantaneous power).} \tag{7-48}$$

For example, the truck in Fig. 7-13 exerts a force $\vec{F}$ on the trailing load, which has velocity $\vec{v}$ at some instant. The instantaneous power due to $\vec{F}$ is the rate at which $\vec{F}$ does work on the load at that instant and is given by Eqs. 7-47 and 7-48. Saying that this power is "the power of the truck" is often acceptable, but we should keep in mind what is meant: Power is the rate at which the applied *force* does work.

CHECKPOINT 5: A block moves with uniform circular motion because a cord tied to the block is anchored at the center of a circle. Is the power due to the force on the block from the cord positive, negative, or zero?

Fig. 7-13 The power due to the truck's applied force on the trailing load is the rate at which that force does work on the load.

Sample Problem 7-10

Figure 7-14 shows constant forces $\vec{F}_1$ and $\vec{F}_2$ acting on a box as the box slides rightward across a frictionless floor. Force $\vec{F}_1$ is horizontal, with magnitude 2.0 N; force $\vec{F}_2$ is angled upward by 60° to the floor and has magnitude 4.0 N. The speed v of the box at a certain instant is 3.0 m/s.

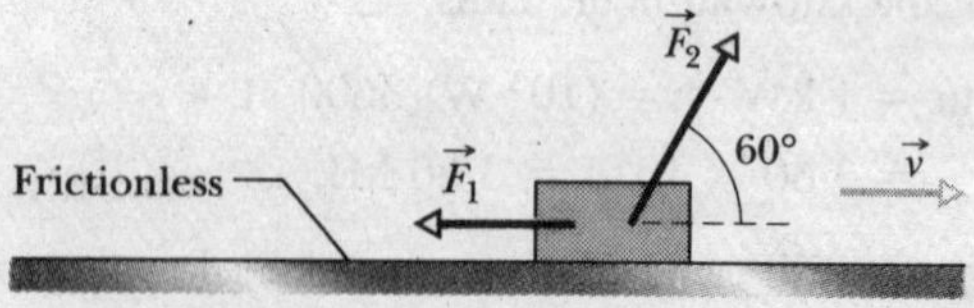

Fig. 7-14 Sample Problem 7-10. Two forces $\vec{F}_1$ and $\vec{F}_2$ act on a box that slides rightward across a frictionless floor. The velocity of the box is $\vec{v}$.

(a) What is the power due to each force acting on the box at that instant, and what is the net power? Is the net power changing at that instant?

SOLUTION: A **Key Idea** here is that we want an instantaneous power, not an average power over a time period. Also, we know the particle's velocity (rather than the work done on it). Therefore, we use Eq. 7-47 for each force. For force $\vec{F}_1$, at angle $\phi_1 = 180°$ to velocity

$\vec{v}$, we have

$$P_1 = F_1 v \cos \phi_1 = (2.0 \text{ N})(3.0 \text{ m/s}) \cos 180°$$
$$= -6.0 \text{ W}. \quad \text{(Answer)}$$

This result tells us that force $\vec{F}_1$ is transferring energy *from* the box at the rate of 6.0 J/s.

For force $\vec{F}_2$, at angle $\phi_2 = 60°$ to velocity $\vec{v}$, we have

$$P_2 = F_2 v \cos \phi_2 = (4.0 \text{ N})(3.0 \text{ m/s}) \cos 60°$$
$$= 6.0 \text{ W}. \quad \text{(Answer)}$$

This result tells us that force $\vec{F}_2$ is transferring energy *to* the box at the rate of 6.0 J/s.

A second **Key Idea** is that the net power is the sum of the individual powers:

$$P_{net} = P_1 + P_2$$
$$= -6.0 \text{ W} + 6.0 \text{ W} = 0, \quad \text{(Answer)}$$

which tells us that the net rate of transfer of energy to or from the box is zero. Thus, the kinetic energy ($K = \frac{1}{2}mv^2$) of the box is not changing, and so the speed of the box will remain at 3.0 m/s. With neither the forces $\vec{F}_1$ and $\vec{F}_2$ nor the velocity $\vec{v}$ changing, we see from Eq. 7-48 that P_1 and P_2 are constant and thus so is P_{net}.

(b) If the magnitude of $\vec{F}_2$ is, instead, 6.0 N, what now is the net power, and is it changing?

SOLUTION: The same **Key Ideas** as above give us, for the power now due to $\vec{F}_2$,

$$P_2 = F_2 v \cos \phi_2 = (6.0 \text{ N})(3.0 \text{ m/s}) \cos 60°$$
$$= 9.0 \text{ W}.$$

The power of force $\vec{F}_1$ is still $P_1 = -6.0$ W, so the net power is now

$$P_{net} = P_1 + P_2 = -6.0 \text{ W} + 9.0 \text{ W}$$
$$= 3.0 \text{ W}, \quad \text{(Answer)}$$

which tells us that the net rate of transfer of energy to the box has a positive value. Thus, the kinetic energy of the box is increasing, and so also is the speed of the box. With the speed increasing, we see from Eq. 7-48 that the values of P_1 and P_2, and thus also of P_{net}, will be changing. Hence, this net power of 3.0 W is the net power only at the instant the speed is the given 3.0 m/s.

REVIEW & SUMMARY

Kinetic Energy The **kinetic energy** K associated with the motion of a particle of mass m and speed v, where v is well below the speed of light, is

$$K = \tfrac{1}{2}mv^2 \quad \text{(kinetic energy)}. \qquad (7\text{-}1)$$

Work Work W is energy transferred to or from an object via a force acting on the object. Energy transferred to the object is positive work, and energy transferred from the object is negative work.

Work Done by a Constant Force The work done on a particle by a constant force $\vec{F}$ during displacement $\vec{d}$ of the particle is

$$W = Fd \cos \phi = \vec{F} \cdot \vec{d} \quad \text{(work, constant force)}, \qquad (7\text{-}7, 7\text{-}8)$$

in which ϕ is the constant angle between the directions of $\vec{F}$ and $\vec{d}$. Only the component of $\vec{F}$ that is along the displacement $\vec{d}$ can do work on the object. When two or more forces act on an object, their **net work** is the sum of the individual works by the forces, which is also equal to the work that would be done on the object by the net force $\vec{F}_{net}$ of those forces.

Work and Kinetic Energy We can relate a change ΔK in kinetic energy of a particle to the net work W done on the particle with

$$\Delta K = K_f - K_i = W \quad \text{(work–kinetic energy theorem)}, \qquad (7\text{-}10)$$

in which K_i is the initial kinetic energy of the object and K_f is the kinetic energy after the work is done. Equation 7-10 rearranged gives us

$$K_f = K_i + W. \qquad (7\text{-}11)$$

Work Done by the Gravitational Force The work W_g done by the gravitational force $\vec{F}_g$ on a particle-like object of mass m during a displacement $\vec{d}$ of the object is given by

$$W_g = mgd \cos \phi, \qquad (7\text{-}12)$$

in which ϕ is the angle between $\vec{F}_g$ and $\vec{d}$.

Work Done in Lifting and Lowering an Object The work W_a done by an applied force during a lifting or lowering of a particle-like object is related to the work W_g done by the gravitational force and the change ΔK in the object's kinetic energy by

$$\Delta K = K_f - K_i = W_a + W_g. \qquad (7\text{-}15)$$

If the kinetic energy at the beginning of a lift equals that at the end of the lift, then Eq. 7-15 reduces to

$$W_a = -W_g, \qquad (7\text{-}16)$$

which tells us that the applied force transfers as much energy to the object as the gravitational force transfers from the object.

Spring Force The force $\vec{F}$ from a spring is

$$\vec{F} = -k\vec{d} \quad \text{(Hooke's law)}, \qquad (7\text{-}20)$$

where $\vec{d}$ is the displacement of its free end from the position when the spring is in its **relaxed state** (neither compressed nor extended), and k is the **spring constant** (a measure of the spring's stiffness). If an x axis lies along the spring, with the origin at the location of the spring's free end when the spring is in its relaxed state, Eq. 7-20 can be written as

$$F = -kx \quad \text{(Hooke's law)}. \qquad (7\text{-}21)$$

A spring force is thus a variable force: It varies with the displacement of the spring's free end.

Work Done by a Spring Force If an object is attached to the spring's free end, the work W_s done on the object by the spring force when the object is moved from an initial position x_i to a final position x_f is

$$W_s = \tfrac{1}{2}kx_i^2 - \tfrac{1}{2}kx_f^2. \qquad (7\text{-}25)$$

If $x_i = 0$ and $x_f = x$, then Eq. 7-25 becomes

$$W_s = -\tfrac{1}{2}kx^2. \qquad (7\text{-}26)$$

Work Done by a Variable Force When the force $\vec{F}$ on a particle-like object depends on the position of the object, the work done by $\vec{F}$ on the object while the object moves from an initial position r_i with coordinates (x_i, y_i, z_i) to a final position r_f with coordinates (x_f, y_f, z_f) must be found by integrating the force. If we assume that component F_x may depend on x but not on y or z, component F_y may depend on y but not on x or z, and component F_z may depend on z but not on x or y, then the work is

$$W = \int_{x_i}^{x_f} F_x\,dx + \int_{y_i}^{y_f} F_y\,dy + \int_{z_i}^{z_f} F_z\,dz. \qquad (7\text{-}36)$$

If $\vec{F}$ has only an x component, then Eq. 7-36 reduces to

$$W = \int_{x_i}^{x_f} F(x)\,dx. \qquad (7\text{-}32)$$

Power The **power** due to a force is the *rate* at which that force does work on an object. If the force does work W during a time interval Δt, the *average power* due to the force over that time interval is

$$P_{avg} = \frac{W}{\Delta t}. \qquad (7\text{-}42)$$

Instantaneous power is the instantaneous rate of doing work:

$$P = \frac{dW}{dt}. \qquad (7\text{-}43)$$

If the direction of a force $\vec{F}$ is at an angle ϕ to the direction of travel of the object, the instantaneous power is

$$P = Fv\cos\phi = \vec{F}\cdot\vec{v}, \qquad (7\text{-}47, 7\text{-}48)$$

in which $\vec{v}$ is the instantaneous velocity of the object.

QUESTIONS

1. Rank the following velocities according to the kinetic energy a particle will have with each velocity, greatest first: (a) $\vec{v} = 4\hat{i} + 3\hat{j}$, (b) $\vec{v} = -4\hat{i} + 3\hat{j}$, (c) $\vec{v} = -3\hat{i} + 4\hat{j}$, (d) $\vec{v} = 3\hat{i} - 4\hat{j}$, (e) $\vec{v} = 5\hat{i}$, and (f) $v = 5$ m/s at 30° to the horizontal.

2. Is positive or negative work done by a constant force $\vec{F}$ on a particle during a straight-line displacement $\vec{d}$ if (a) the angle between $\vec{F}$ and $\vec{d}$ is 30°; (b) the angle is 100°; (c) $\vec{F} = 2\hat{i} - 3\hat{j}$ and $\vec{d} = -4\hat{i}$?

3. Figure 7-15 shows six situations in which two forces act simultaneously on a box after the box has been sent sliding over a frictionless surface, either to the left or to the right. The forces are either 1 N or 2 N in magnitude, as indicated by the vector lengths. For each situation, is the work done on the box by the net force during the indicated displacement $\vec{d}$ positive, negative, or zero?

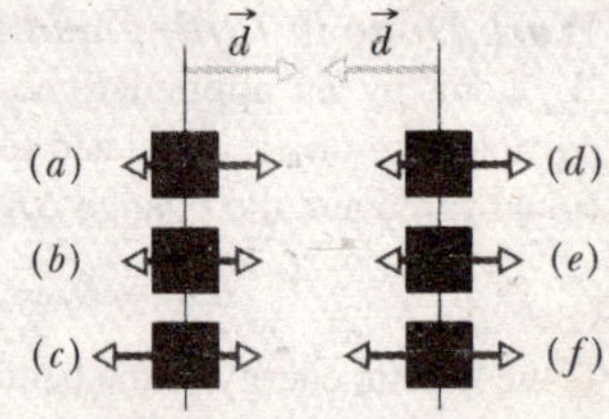

Fig. 7-15 Question 3.

4. Figure 7-16 shows the values of a force $\vec{F}$, directed along an x axis, that will act on a particle at the corresponding values of x. If the particle begins at rest at $x = 0$, what is the particle's coordinate when it has (a) its greatest kinetic energy, (b) its greatest speed, and (c) zero speed? (d) What is the particle's direction of travel after it reaches $x = 6$ m?

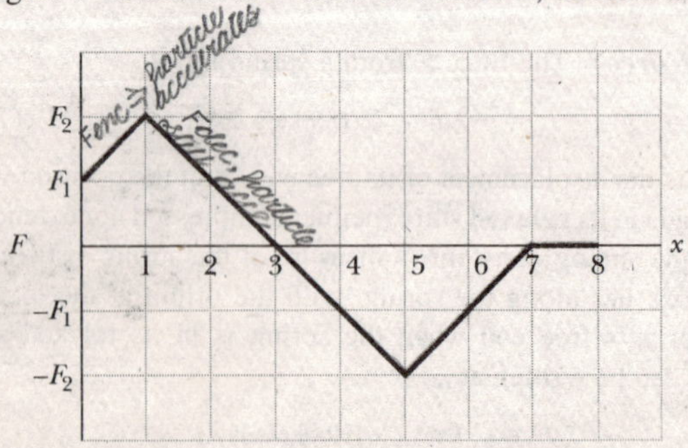

Fig. 7-16 Question 4.

5. Figure 7-17 gives, for three situations, the position versus time of a box of contraband pulled by applied forces directed along an x axis on a frictionless surface. Line B is straight; the others are curved. Rank the situations according to the kinetic energy of the box at (a) time t_1 and (b) time t_2, greatest first. (c) Now rank them according to the net work done on the box by the applied forces during the time period t_1 to t_2, greatest first.

(d) For each situation, which of the following best describes the net work done by the applied forces during the period t_1 to t_2?

(1) Energy is transferred to the box.

(2) Energy is transferred from the box.

(3) The net work is zero.

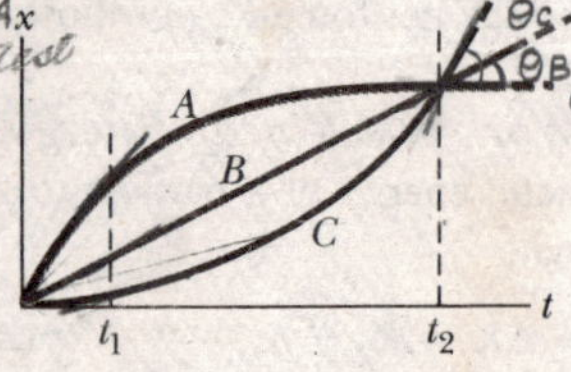

Fig. 7-17 Question 5.

6. Figure 7-18 shows four graphs (drawn to the same scale) of the x component of a variable force $\vec{F}$ (directed along an x axis) versus the position x of a particle on which the force acts. Rank the graphs according to the work done by $\vec{F}$ on the particle from $x = 0$ to x_1, from most positive work to most negative work.

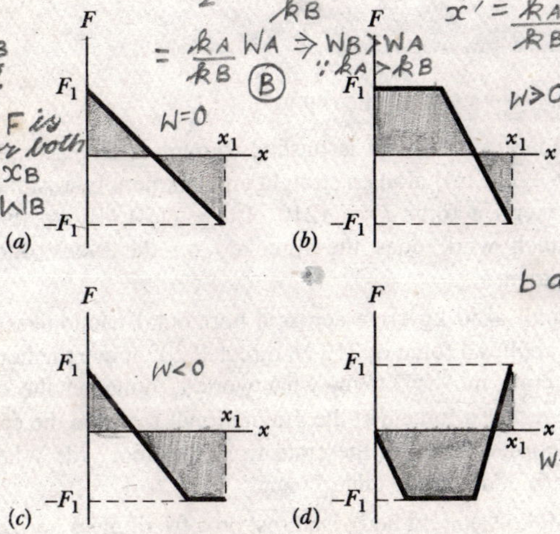

Fig. 7-18 Question 6.

7. In Fig. 7-19, a greased pig has a choice of three frictionless slides along which to slide to the ground. Rank the slides according to how much work the gravitational force does on the pig during the descent, greatest first.

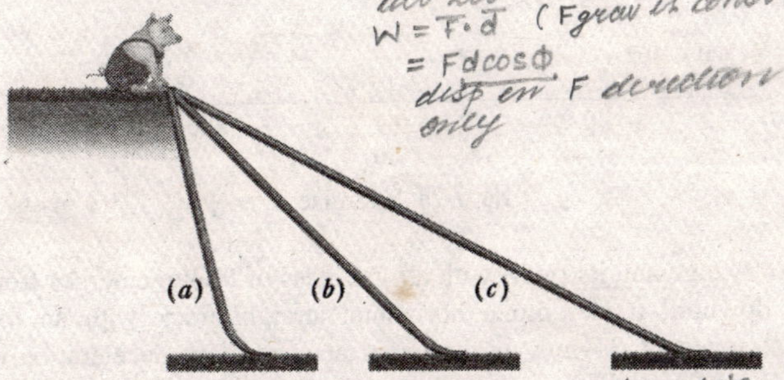

Fig. 7-19 Question 7.

8. You lift an armadillo to a shelf. Does the work done by your force on the armadillo depend on (a) the mass of the armadillo, (b) the weight of the armadillo, (c) the height of the shelf, (d) the time you take, or (e) whether you move the armadillo sideways or directly upward?

9. Figure 7-20 shows a bundle of magazines that is lifted by a cord through distance d. The table shows six pairs of values of the initial speed v_0 and final speed v (in meters per second) of the bundle at the beginning and end of distance d. Rank the pairs according to the work done by the cord's force on the bundle over distance d, most positive first, most negative last.

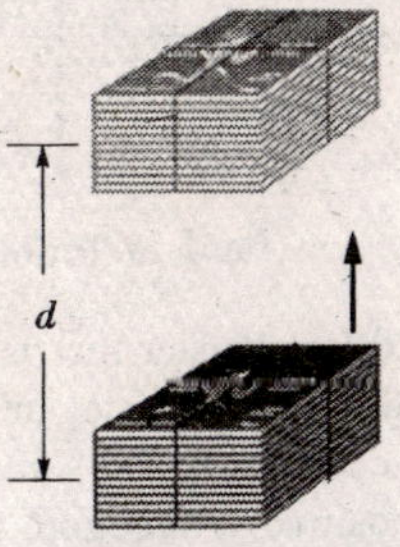

	a	b	c	d	e	f
$v=$	0	2	2	2	0	1
$v_0=$	0	2	0	1	2	2

Fig. 7-20 Question 9.

10. Spring A is stiffer than spring B; that is, $k_A > k_B$. The spring force of which spring does more work if the springs are compressed (a) the same distance and (b) by the same applied force?

11. A block is attached to a relaxed spring as in Fig. 7-21a. The spring constant k of the spring is such that for a certain rightward displacement $\vec{d}$ of the block, the spring force acting on the block has magnitude F_1 and has done work W_1 on the block. A second, identical spring is then attached on the opposite side of the block, as shown in Fig. 7-21b; both springs are in their relaxed state in the figure. If the block is again displaced by $\vec{d}$, (a) what is the magnitude of the net force on it from both springs and (b) how much work has been done on it by the spring forces?

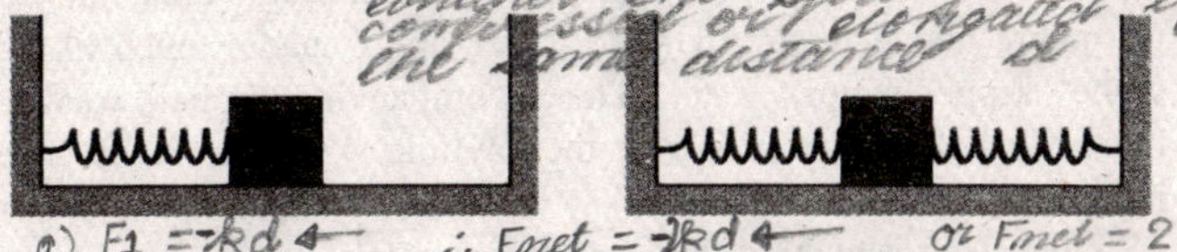

Fig. 7-21 Question 11.

12. Figure 7-22 gives the velocity versus time of a scooter car being moved along an axis by a varying applied force. The time axis shows six periods: $\Delta t_1 = \Delta t_2 = \Delta t_3 = \Delta t_6 = 2\Delta t_4 = \frac{2}{3}\Delta t_5$. (a) During which of the time periods is energy transferred *from* the scooter car by the applied force? (b) Rank the time periods according to the work done on the scooter car by the applied force during the period, most positive work first, most negative work last. (c) Rank the periods according to the rate at which the applied force transfers energy, greatest rate of transfer *to* the scooter car first, greatest rate of transfer *from* it last.

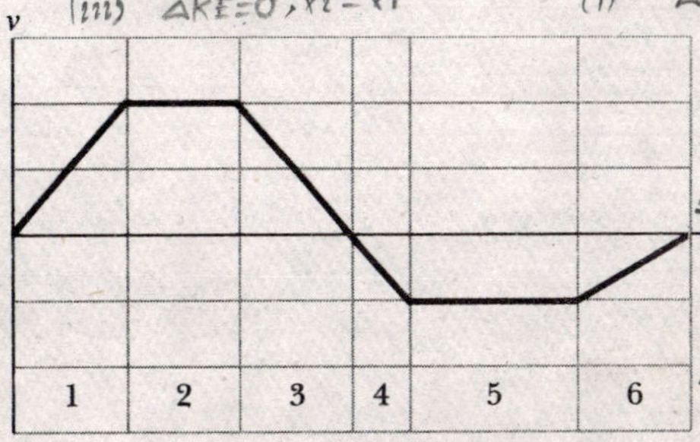

Fig. 7-22 Question 12.

13. In three situations, an initially stationary can of crayons is sent sliding over a frictionless floor by a different applied force. Plots of the resulting acceleration versus time for the three situations are given in Fig. 7-23. Rank the plots according to the work done by the applied force during the acceleration period, greatest first.

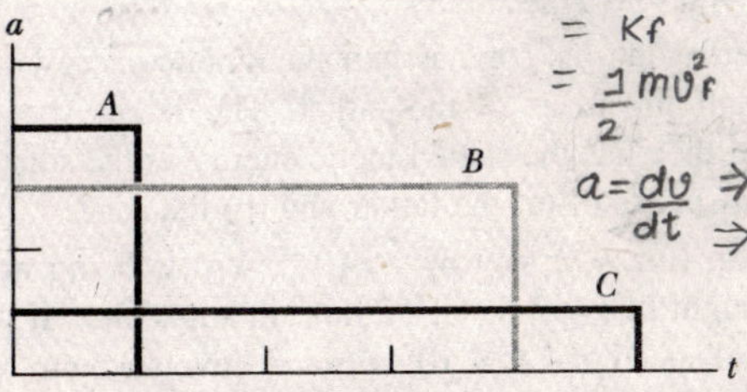

Fig. 7-23 Question 13.

EXERCISES & PROBLEMS

ssm Solution is in the Student Solutions Manual.
www Solution is available on the World Wide Web at: http://www.wiley.com/college/hrw
ilw Solution is available on the Interactive LearningWare.

SEC. 7-1 Energy

1E. If an electron (mass $m = 9.11 \times 10^{-31}$ kg) in copper near the lowest possible temperature has a kinetic energy of 6.7×10^{-19} J, what is the speed of the electron? ssm

2E. On August 10, 1972, a large meteorite skipped across the atmosphere above western United States and Canada, much like a stone skipped across water. The accompanying fireball was so bright that it could be seen in the daytime sky (Fig. 7-24). The meteorite's mass was about 4×10^6 kg; its speed was about 15 km/s. Had it entered the atmosphere vertically, it would have hit Earth's surface with about the same speed. (a) Calculate the meteorite's loss of kinetic energy (in joules) that would have been associated with the vertical impact. (b) Express the energy as a multiple of the explosive energy of 1 megaton of TNT, which is 4.2×10^{15} J. (c) The energy associated with the atomic bomb explosion over Hiroshima was equivalent to 13 kilotons of TNT. To how many Hiroshima bombs would the meteorite impact have been equivalent?

Fig. 7-24 Exercise 2. A large meteorite skips across the atmosphere in the sky above the mountains (upper right).

3E. Calculate the kinetic energies of the following objects moving at the given speeds: (a) a 110 kg football linebacker running at 8.1 m/s; (b) a 4.2 g bullet at 950 m/s; (c) the aircraft carrier *Nimitz*, 91,400 tons at 32 knots.

4P. A father racing his son has half the kinetic energy of the son, who has half the mass of the father. The father speeds up by 1.0 m/s and then has the same kinetic energy as the son. What are the original speeds of (a) the father and (b) the son?

5P. A proton (mass $m = 1.67 \times 10^{-27}$ kg) is being accelerated along a straight line at 3.6×10^{15} m/s^2 in a machine. If the proton has an initial speed of 2.4×10^7 m/s and travels 3.5 cm, what then is (a) its speed and (b) the increase in its kinetic energy? ssm

SEC. 7-3 Work and Kinetic Energy

6E. A floating ice block is pushed through a displacement $\vec{d} = (15\text{ m})\hat{i} - (12\text{ m})\hat{j}$ along a straight embankment by rushing water, which exerts a force $\vec{F} = (210\text{ N})\hat{i} - (150\text{ N})\hat{j}$ on the block. How much work does the force do on the block during the displacement?

7E. To pull a 50 kg crate across a horizontal frictionless floor, a worker applies a force of 210 N, directed 20° above the horizontal. As the crate moves 3.0 m, what work is done on the crate by (a) the worker's force, (b) the gravitational force on the crate, and (c) the normal force on the crate from the floor? (d) What is the total work done on the crate? ssm

8E. A 1.0 kg standard body is at rest on a frictionless horizontal air track when a constant horizontal force $\vec{F}$ acting in the positive direction of an x axis along the track is applied to the body. A stroboscopic graph of the position of the body as it slides to the right is shown in Fig. 7-25. The force $\vec{F}$ is applied to the body at $t = 0$, and the graph records the position of the body at 0.50 s intervals. How much work is done on the body by the applied force $\vec{F}$ between $t = 0$ and $t = 2.0$ s?

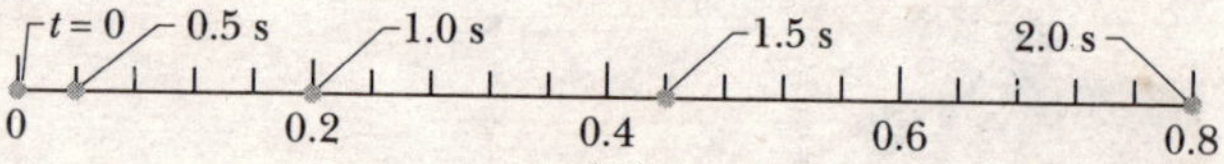

Fig. 7-25 Exercise 8.

9E. A luge and its rider, with a total mass of 85 kg, emerges from a downhill track onto a horizontal straight track with an initial speed of 37 m/s. If they stop at a constant deceleration of 2.0 m/s^2, (a) what magnitude F is required for the decelerating force, (b) what distance d do they travel while decelerating, and (c) what work W is done on them by the decelerating force? What are (d) F, (e) d, and (f) W for a deceleration of 4.0 m/s^2? ilw

10P. A force acts on a 3.0 kg particle-like object in such a way that the position of the object as a function of time is given by $x = 3.0t - 4.0t^2 + 1.0t^3$, with x in meters and t in seconds. Find the work done on the object by the force from $t = 0$ to $t = 4.0$ s. (*Hint:* What are the speeds at those times?)

11P. Figure 7-26 shows three forces applied to a trunk that moves leftward by 3.00 m over a frictionless floor. The force magnitudes are $F_1 = 5.00$ N, $F_2 = 9.00$ N, and $F_3 = 3.00$ N. During the displacement, (a) what is the net work done on the trunk by the three forces and (b) does the kinetic energy of the trunk increase or decrease? ssm

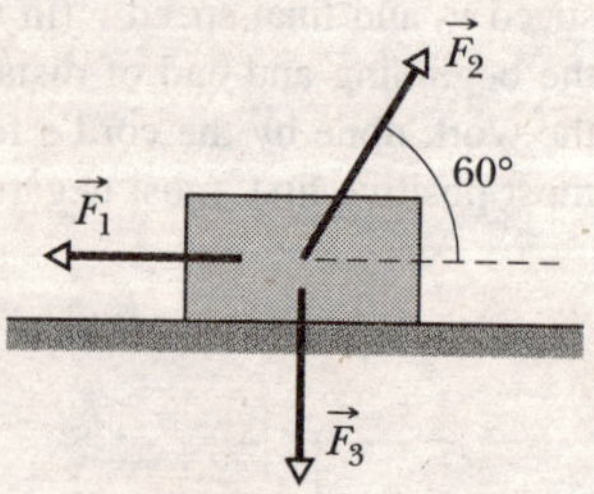

Fig. 7-26 Problem 11.

12P. The only force acting on a 2.0 kg canister that is moving in an xy plane has a magnitude of 5.0 N. The canister initially has a velocity of 4.0 m/s in the positive x direction, and some time later has a velocity of 6.0 m/s in the positive y direction. How much work is done on the canister by the 5.0 N force during this time?

13P. Figure 7-27 shows an overhead view of three horizontal forces acting on a cargo canister that was initially stationary but that now moves across a frictionless floor. The force magnitudes are $F_1 = 3.00$ N, $F_2 = 4.00$ N, and $F_3 = 10.0$ N. What is the net work done on the canister by the three forces during the first 4.00 m of displacement? ssm

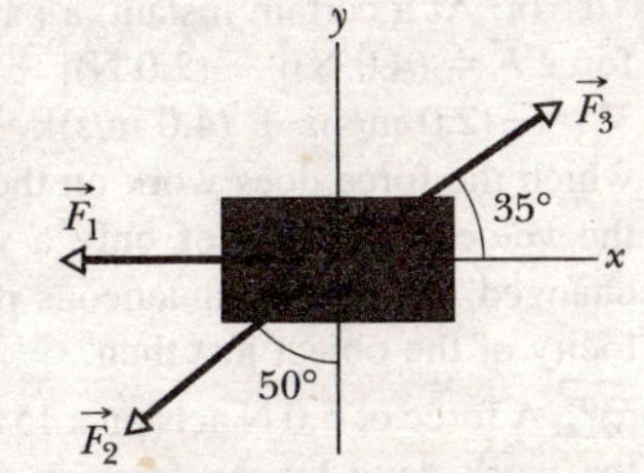

Fig. 7-27 Problem 13.

SEC. 7-4 Work Done by a Gravitational Force

14E. (a) In 1975 the roof of Montreal's Velodrome, with a weight of 360 kN, was lifted by 10 cm so that it could be centered. How much work was done on the roof by the forces making the lift? (b) In 1960 Mrs. Maxwell Rogers of Tampa, Florida, reportedly raised one end of a car that had fallen onto her son when a jack failed. If her panic lift effectively raised 4000 N (about $\frac{1}{4}$ of the car's weight) by 5.0 cm, how much work did her force do on the car?

15E. In Fig. 7-28, a cord runs around two massless, frictionless pulleys; a canister with mass m = 20 kg hangs from one pulley; and you exert a force $\vec{F}$ on the free end of the cord. (a) What must be the magnitude of $\vec{F}$ if you are to lift the canister at a constant speed? (b) To lift the canister by 2.0 cm, how far must you pull the free end of the cord? During that lift, what is the work done on the canister by (c) your force (via the cord) and (d) the gravitational force on the canister? (*Hint:* When a cord loops around a pulley as shown, it pulls on the pulley with a net force that is twice the tension in the cord.)

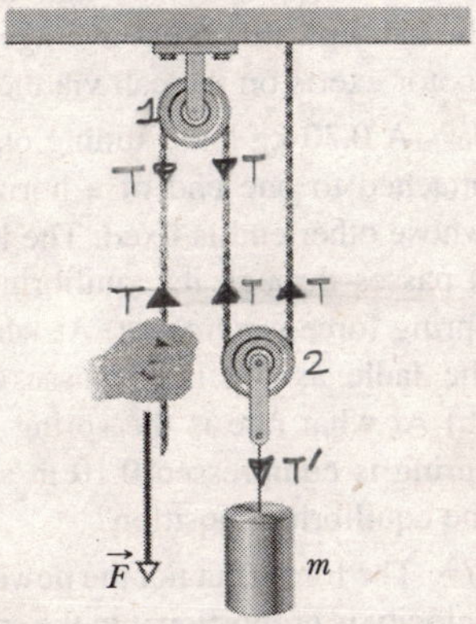

Fig. 7-28 Exercise 15.

16E. A 45 kg block of ice slides down a frictionless incline 1.5 m long and 0.91 m high. A worker pushes up against the ice, parallel to the incline, so that the block slides down at constant speed. (a) Find the magnitude of the worker's force. How much work is done on the block by (b) the worker's force, (c) the gravitational force on the block, (d) the normal force on the block from the surface of the incline, and (e) the net force on the block?

17P. A helicopter lifts a 72 kg astronaut 15 m vertically from the ocean by means of a cable. The acceleration of the astronaut is $g/10$. How much work is done on the astronaut by (a) the force from the helicopter and (b) the gravitational force on her? What are (c) the kinetic energy and (d) the speed of the astronaut just before she reaches the helicopter? ssm www

18P. A cave rescue team lifts an injured spelunker directly upward and out of a sinkhole by means of a motor-driven cable. The lift is performed in three stages, each requiring a vertical distance of 10.0 m: (a) the initially stationary spelunker is accelerated to a speed of 5.00 m/s; (b) he is then lifted at the constant speed of 5.00 m/s; (c) finally he is decelerated to zero speed. How much work is done on the 80.0 kg rescuee by the force lifting him during each stage?

19P. A cord is used to vertically lower an initially stationary block of mass M at a constant downward acceleration of $g/4$. When the block has fallen a distance d, find (a) the work done by the cord's force on the block, (b) the work done by the gravitational force on the block, (c) the kinetic energy of the block, and (d) the speed of the block. ssm

SEC. 7-5 Work Done by a Spring Force

20E. During spring semester at MIT, residents of the parallel buildings of the East Campus dorms battle one another with large catapults that are made with surgical hose mounted on a window frame. A balloon filled with dyed water is placed in a pouch attached to the hose, which is then stretched through the width of the room. Assume that the stretching of the hose obeys Hooke's law with a spring constant of 100 N/m. If the hose is stretched by 5.00 m and then released, how much work does the force from the hose do on the balloon in the pouch by the time the hose reaches its relaxed length?

21E. A spring with a spring constant of 15 N/cm has a cage attached to one end (Fig. 7-29). (a) How much work does the spring force do on the cage when the spring is stretched from its relaxed length by 7.6 mm? (b) How much additional work is done by the spring force when the spring is stretched by an additional 7.6 mm? ssm ilw www

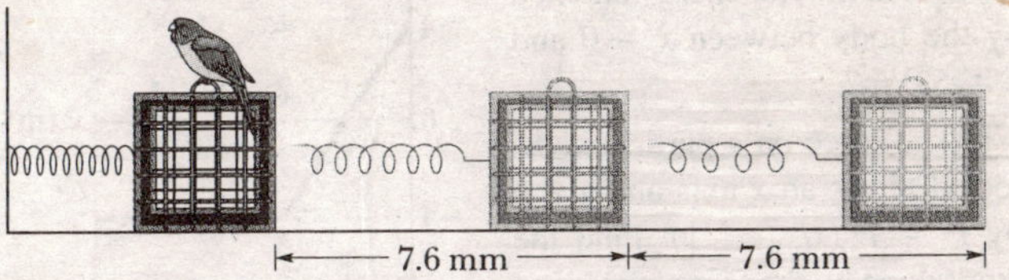

Fig. 7-29 Exercise 21.

22P. A 250 g block is dropped onto a relaxed vertical spring that has a spring constant of k = 2.5 N/cm (Fig. 7-30). The block becomes attached to the spring and compresses the spring 12 cm before momentarily stopping. While the spring is being compressed, what work is done on the block by (a) the gravitational force on it and (b) the spring force? (c) What is the speed of the block just before it hits the spring? (Assume that friction is negligible.) (d) If the speed at impact is doubled, what is the maximum compression of the spring?

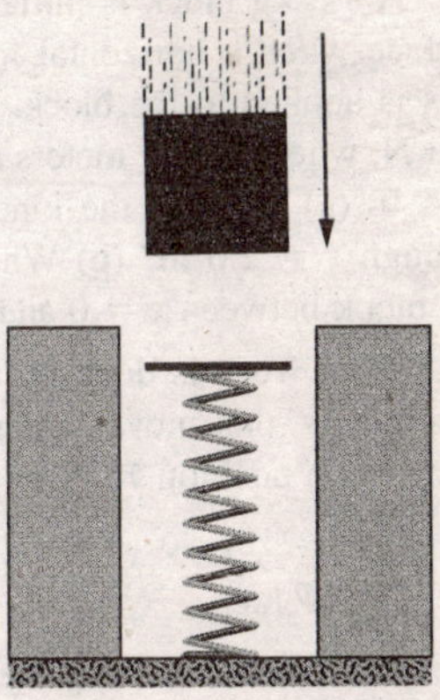
Fig. 7-30 Problem 22.

23P. The only force acting on a 2.0 kg body as it moves along the positive x axis has an x component $F_x = -6x$ N, where x is in meters. The velocity of the body at x = 3.0 m is 8.0 m/s. (a) What is the velocity of the body at x = 4.0 m? (b) At what positive value of x will the body have a velocity of 5.0 m/s? ssm

SEC. 7-6 Work Done by a General Variable Force

24E. A 5.0 kg block moves in a straight line on a horizontal frictionless surface under the influence of a force that varies with position as shown in Fig. 7-31. How much work is done by the force as the block moves from the origin to $x = 8.0$ m?

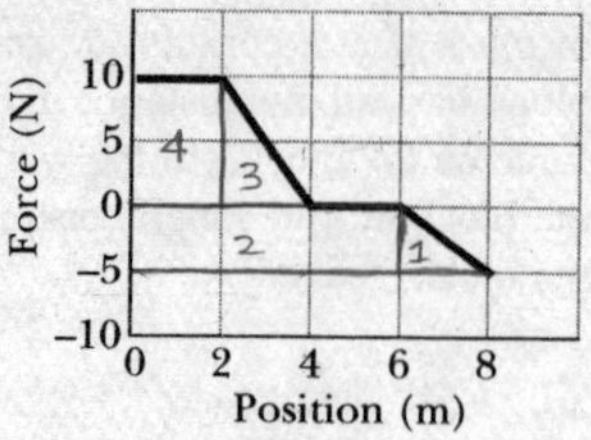

Fig. 7-31 Exercise 24.

25E. A 10 kg brick moves along an x axis. Its acceleration as a function of its position is shown in Fig. 7-32. What is the net work performed on the brick by the force causing the acceleration as the brick moves from $x = 0$ to $x = 8.0$ m? ssm ilw

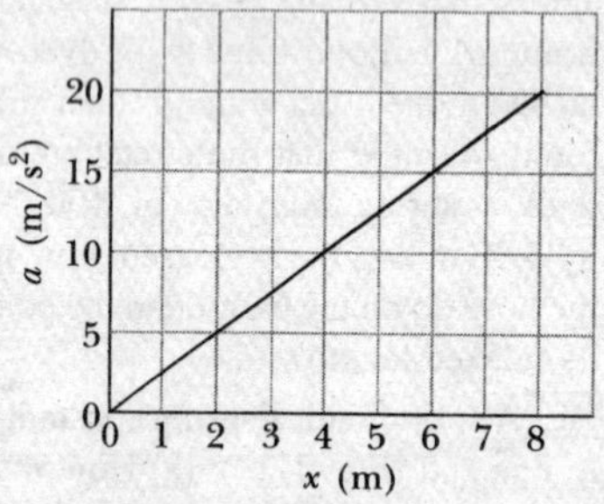

Fig. 7-32 Exercise 25.

26P. The only force acting on a 2.0 kg body as the body moves along the x axis varies as shown in Fig. 7-33. The velocity of the body at $x = 0$ is 4.0 m/s. (a) What is the kinetic energy of the body at $x = 3.0$ m? (b) At what value of x will the body have a kinetic energy of 8.0 J? (c) What is the maximum kinetic energy attained by the body between $x = 0$ and $x = 5.0$ m?

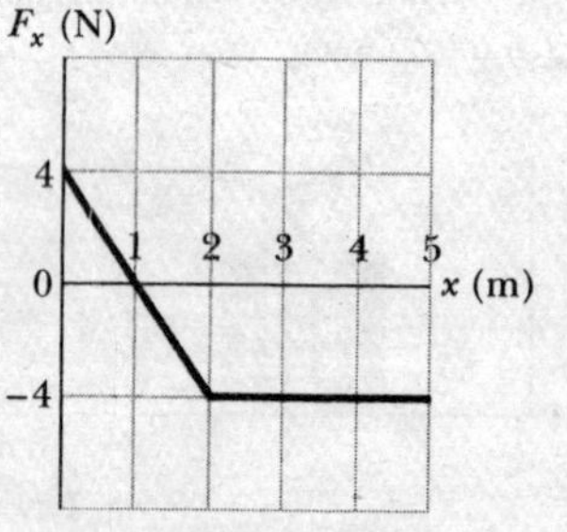

Fig. 7-33 Problem 26.

27P. The force on a particle is directed along an x axis and given by $F = F_0(x/x_0 - 1)$. Find the work done by the force in moving the particle from $x = 0$ to $x = 2x_0$ by (a) plotting $F(x)$ and measuring the work from the graph and (b) integrating $F(x)$. ssm

28P. A 1.5 kg block is initially at rest on a horizontal frictionless surface when a horizontal force in the positive direction of an x axis is applied to the block. The force is given by $\vec{F}(x) = (2.5 - x^2)\hat{\imath}$ N, where x is in meters and the initial position of the block is $x = 0$. (a) What is the kinetic energy of the block as it passes through $x = 2.0$ m? (b) What is the maximum kinetic energy of the block between $x = 0$ and $x = 2.0$ m?

29P. What work is done by a force $\vec{F} = (2x \text{ N})\hat{\imath} + (3 \text{ N})\hat{\jmath}$, with x in meters, that moves a particle from a position $\vec{r}_i = (2 \text{ m})\hat{\imath} + (3 \text{ m})\hat{\jmath}$ to a position $\vec{r}_f = -(4 \text{ m})\hat{\imath} - (3 \text{ m})\hat{\jmath}$? ssm

SEC. 7-7 Power

30E. The loaded cab of an elevator has a mass of 3.0×10^3 kg and moves 210 m up the shaft in 23 s at constant speed. At what average rate does the force from the cable do work on the cab?

31E. A 100 kg block is pulled at a constant speed of 5.0 m/s across a horizontal floor by an applied force of 122 N directed 37° above the horizontal. What is the rate at which the force does work on the block? ssm ilw

32E. (a) At a certain instant, a particle-like object is acted on by a force $\vec{F} = (4.0 \text{ N})\hat{\imath} - (2.0 \text{ N})\hat{\jmath} + (9.0 \text{ N})\hat{k}$ while having a velocity $\vec{v} = -(2.0 \text{ m/s})\hat{\imath} + (4.0 \text{ m/s})\hat{k}$. What is the instantaneous rate at which the force does work on the object? (b) At some other time, the velocity consists of only a y component. If the force is unchanged, and the instantaneous power is -12 W, what is the velocity of the object just then?

33P. A force of 5.0 N acts on a 15 kg body initially at rest. Compute the work done by the force in (a) the first, (b) the second, and (c) the third seconds and (d) the instantaneous power due to the force at the end of the third second. ssm

34P. A skier is pulled by a tow rope up a frictionless ski slope that makes an angle of 12° with the horizontal. The rope moves parallel to the slope with a constant speed of 1.0 m/s. The force of the rope does 900 J of work on the skier as the skier moves a distance of 8.0 m up the incline. (a) If the rope moved with a constant speed of 2.0 m/s, how much work would the force of the rope do on the skier as the skier moved a distance of 8.0 m up the incline? At what rate is the force of the rope doing work on the skier when the rope moves with a speed of (b) 1.0 m/s and (c) 2.0 m/s?

35P. A fully loaded, slow-moving freight elevator has a cab with a total mass of 1200 kg, which is required to travel upward 54 m in 3.0 min, starting and ending at rest. The elevator's counterweight has a mass of only 950 kg, so the elevator motor must help pull the cab upward. What average power is required of the force the motor exerts on the cab via the cable? ssm www

36P. A 0.30 kg ladle sliding on a horizontal frictionless surface is attached to one end of a horizontal spring (with $k = 500$ N/m) whose other end is fixed. The ladle has a kinetic energy of 10 J as it passes through its equilibrium position (the point at which the spring force is zero). (a) At what rate is the spring doing work on the ladle as the ladle passes through its equilibrium position? (b) At what rate is the spring doing work on the ladle when the spring is compressed 0.10 m and the ladle is moving away from the equilibrium position?

37P. The force (but not the power) required to tow a boat at constant velocity is proportional to the speed. If a speed of 4.0 km/h requires 7.5 kW, how much power does a speed of 12 km/h require? ssm

38P. Boxes are transported from one location to another in a warehouse by means of a conveyor belt that moves with a constant speed of 0.50 m/s. At a certain location the conveyor belt moves for 2.0 m up an incline that makes an angle of 10° with the horizontal, then for 2.0 m horizontally, and finally for 2.0 m down an incline that makes an angle of 10° with the horizontal. Assume that a 2.0 kg box rides on the belt without slipping. At what rate is the force of the conveyor belt doing work on the box (a) as the box moves up the 10° incline, (b) as the box moves horizontally, and (c) as the box moves down the 10° incline?

39P. A horse pulls a cart with a force of 40 lb at an angle of 30° above the horizontal and moves along at a speed of 6.0 mi/h. (a) How much work does the force do in 10 min? (b) What is the average power (in horsepower) of the force?

40P. An initially stationary 2.0 kg object accelerates horizontally and uniformly to a speed of 10 m/s in 3.0 s. (a) In that 3.0 s interval, how much work is done on the object by the force accelerating it? What is the instantaneous power due to that force (b) at the end of the interval and (c) at the end of the first half of the interval?

8 Potential Energy and Conservation of Energy

The prehistoric people of Easter Island carved hundreds of giant stone statues in their quarry and then moved them to sites all over the island. How they managed to move them by as much as 10 km without the use of sophisticated machines has been a hotly debated subject, with many fanciful theories about the source of the required energy.

How much energy *was* required to move one of the statues using only primitive means?

The answer is in this chapter.

8-1 Potential Energy

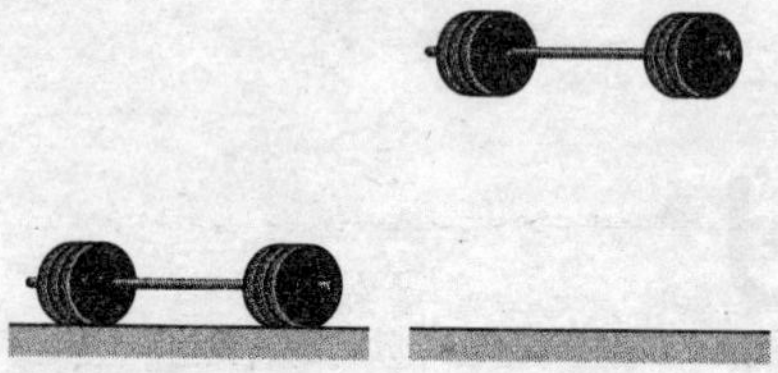

Fig. 8-1 When Chemerkin lifted the weights above his head, he increased the separation between the weights and Earth and thus changed the configuration of the weights–Earth system from that in (*a*) to that in (*b*).

In this chapter we continue the discussion of energy that we began in Chapter 7. To do so, we define a second form of energy: **potential energy** U is energy that can be associated with the configuration (or arrangement) of a system of objects that exert forces on one another. If the configuration of the system changes, then the potential energy of the system can also change.

One type of potential energy is the **gravitational potential energy** that is associated with the state of separation between objects, which attract one another via the gravitational force. For example, when Andrey Chemerkin lifted the record-breaking weights above his head in the 1996 Olympics, he increased the separation between the weights and Earth. The work his force did changed the gravitational potential energy of the weights–Earth system because it changed the configuration of the system—that is, the force shifted the relative locations of the weights and Earth (Fig. 8-1).

Another type of potential energy is **elastic potential energy**, which is associated with the state of compression or extension of an elastic (springlike) object. If you compress or extend a spring, you do work to change the relative locations of the coils within the spring. The result of the work done by your force is an increase in the elastic potential energy of the spring.

The idea of potential energy can be an enormously powerful tool in understanding situations involving the motion of an object. With it we can, in fact, easily solve problems that would require careful computer programming if we used only the ideas of earlier chapters.

Work and Potential Energy

In Chapter 7 we discussed the relation between work and a change in kinetic energy. Here we discuss the relation between work and a change in potential energy.

Let us throw a tomato upward (Fig. 8-2). We already know that as the tomato rises, the work W_g done on the tomato by the gravitational force is negative because the force transfers energy *from* the kinetic energy of the tomato. We can now finish the story by saying that this energy is transferred by the gravitational force *to* the gravitational potential energy of the tomato–Earth system.

The tomato slows, stops, and then begins to fall back down because of the gravitational force. During the fall, the transfer is reversed: The work W_g done on the tomato by the gravitational force is now positive—that force transfers energy *from* the gravitational potential energy of the tomato–Earth system *to* the kinetic energy of the tomato.

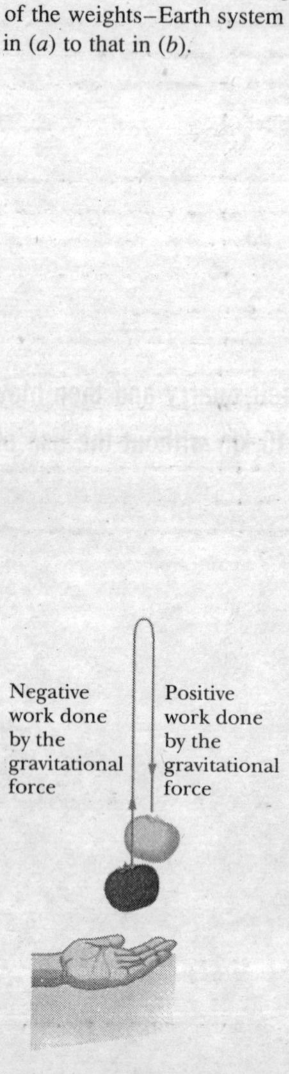

Fig. 8-2 A tomato is thrown upward. As it rises, the gravitational force does negative work on it, decreasing its kinetic energy. As the tomato descends, the gravitational force does positive work on it, increasing its kinetic energy.

For either rise or fall, the change ΔU in gravitational potential energy is defined to equal the negative of the work done on the tomato by the gravitational force. Using the general symbol W for work, we write this as

$$\Delta U = -W. \tag{8-1}$$

This equation also applies to a block–spring system, as in Fig. 8-3. If we abruptly shove the block to send it moving rightward, the spring force acts leftward and thus does negative work on the block, transferring energy from the kinetic energy of the block to the elastic potential energy of the spring. The block slows and eventually stops, and then begins to move leftward because the spring force is still leftward. The transfer of energy is then reversed—it is from potential energy of the spring to kinetic energy of the block.

Conservative and Nonconservative Forces

Let us list the key elements of the two situations we just discussed:

1. The *system* consists of two or more objects.
2. A *force* acts between a particle-like object (tomato or block) in the system and the rest of the system.
3. When the system configuration changes, the force does *work* (call it W_1) on the particle-like object, transferring energy between the kinetic energy K of the object and some other form of energy of the system.
4. When the configuration change is reversed, the force reverses the energy transfer, doing work W_2 in the process.

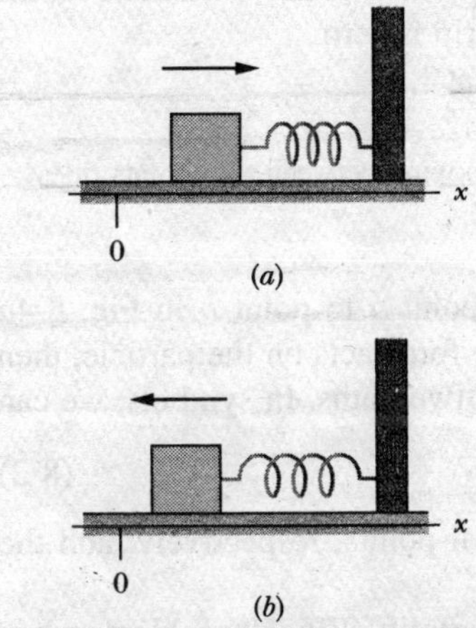

Fig. 8-3 A block, attached to a spring and initially at rest at $x = 0$, is set in motion toward the right. (*a*) As the block moves rightward (as indicated by the arrow), the spring force does negative work on it. (*b*) Then, as the block moves back toward $x = 0$, the spring force does positive work on it.

In a situation in which $W_1 = -W_2$ is always true, the other form of energy is a potential energy, and the force is said to be a **conservative force.** As you might suspect, the gravitational force and the spring force are both conservative (since otherwise we could not have spoken of gravitational potential energy and elastic potential energy, as we did previously).

A force that is not conservative is called a **nonconservative force.** The kinetic frictional force and drag force are nonconservative. For an example, let us send a block sliding across a floor that is not frictionless. During the sliding, a kinetic frictional force from the floor does negative work on the block, slowing the block by transferring energy from its kinetic energy to a form of energy called *thermal energy* (which has to do with the random motions of atoms and molecules). We know from experiment that this energy transfer cannot be reversed (thermal energy cannot be transferred back to kinetic energy of the block by the kinetic frictional force). Thus, although we have a system (made up of the block and the floor), a force that acts between parts of the system, and a transfer of energy by the force, the force is not conservative. Therefore, thermal energy is not a potential energy.

When only conservative forces act on a particle-like object, we can greatly simplify otherwise difficult problems involving motion of the object. The next section, in which we develop a test for identifying conservative forces, provides one means for simplifying such problems.

8-2 Path Independence of Conservative Forces

The primary test for determining whether a force is conservative or nonconservative is this: Let the force act on a particle that moves along any *closed path,* beginning at some initial position and eventually returning to that position (so that the particle makes a *round trip* beginning and ending at the initial position). The force is conservative only if the total energy it transfers to and from the particle during the round trip along this and any other closed path is zero. In other words:

> The net work done by a conservative force on a particle moving around every closed path is zero.

We know from experiment that the gravitational force passes this *closed-path test.* An example is the tossed tomato of Fig. 8-2. The tomato leaves the launch point with speed v_0 and kinetic energy $\frac{1}{2}mv_0^2$. The gravitational force acting on the tomato slows it, stops it, and then causes it to fall back down. When the tomato returns to the launch point, it again has speed v_0 and kinetic energy $\frac{1}{2}mv_0^2$. Thus, the gravita-

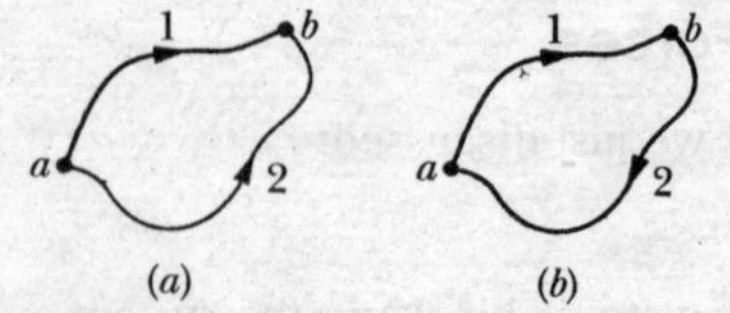

(a) (b)

Fig. 8-4 (*a*) As a conservative force acts on it, a particle can move from point *a* to point *b* along either path 1 or path 2. (*b*) The particle moves in a round trip, from point *a* to point *b* along path 1 and then back to point *a* along path 2.

tional force transfers as much energy *from* the tomato during the ascent as it transfers *to* the tomato during the descent back to the launch point. The net work done on the tomato by the gravitational force during the round trip is zero.

An important result of the closed-path test is that

> The work done by a conservative force on a particle moving between two points does not depend on the path taken by the particle.

For example, suppose that a particle moves from point *a* to point *b* in Fig. 8-4*a* along either path 1 or path 2. If only a conservative force acts on the particle, then the work done on the particle is the same along the two paths. In symbols, we can write this result as

$$W_{ab,1} = W_{ab,2}, \tag{8-2}$$

where the subscript *ab* indicates the initial and final points, respectively, and the subscripts 1 and 2 indicate the path.

This result is powerful, because it allows us to simplify difficult problems when only a conservative force is involved. Suppose you need to calculate the work done by a conservative force along a given path between two points, and the calculation is difficult or even impossible without additional information. You can find the work by substituting some other path between those two points for which the calculation is easier and possible. Sample Problem 8-1 gives an example, but first we need to prove Eq. 8-2.

Proof of Equation 8-2

Figure 8-4*b* shows an arbitrary round trip for a particle that is acted upon by a single force. The particle moves from an initial point *a* to point *b* along path 1, and then back to point *a* along path 2. The force does work on the particle as the particle moves along each path. Without worrying about where positive work is done and where negative work is done, let us just represent the work done from *a* to *b* along path 1 as $W_{ab,1}$ and the work done from *b* back to *a* along path 2 as $W_{ba,2}$. If the force is conservative, then the net work done during the round trip must be zero:

$$W_{ab,1} + W_{ba,2} = 0,$$

and thus

$$W_{ab,1} = -W_{ba,2}. \tag{8-3}$$

In words, the work done along the outward path must be the negative of the work done along the path back.

Let us now consider the work $W_{ab,2}$ done on the particle by the force when the particle moves from *a* to *b* along path 2, as indicated in Fig. 8-4*a*. If the force is conservative, that work is the negative of $W_{ba,2}$:

$$W_{ab,2} = -W_{ba,2}. \tag{8-4}$$

Substituting $W_{ab,2}$ for $-W_{ba,2}$ in Eq. 8-3, we obtain

$$W_{ab,1} = W_{ab,2},$$

which is what we set out to prove.

CHECKPOINT 1: The figure shows three paths connecting points *a* and *b*. A single force $\vec{F}$ does the indicated work on a particle moving along each path in the indicated direction. On the basis of this information, is force $\vec{F}$ conservative?

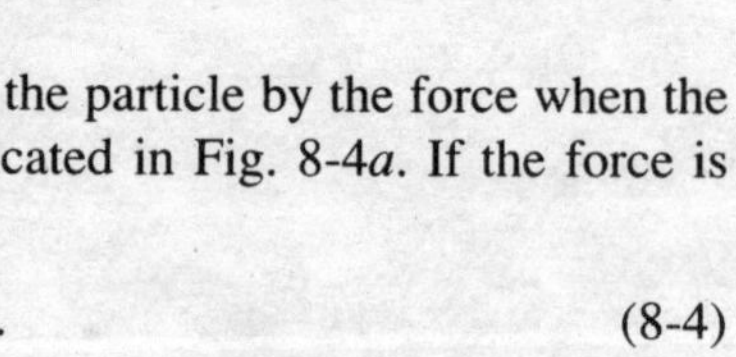

Sample Problem 8-1

Figure 8-5*a* shows a 2.0 kg block of slippery cheese that slides along a frictionless track from point *a* to point *b*. The cheese travels through a total distance of 2.0 m along the track, and a net vertical distance of 0.80 m. How much work is done on the cheese by the gravitational force during the slide?

SOLUTION: A **Key Idea** here is that we *cannot* use Eq. 7-12 ($W_g = mgd \cos \phi$) to calculate the work done by the gravitational force $\vec{F}_g$ as the cheese moves along the track. The reason is that the angle ϕ between the directions of $\vec{F}_g$ and the displacement $\vec{d}$ varies along the track in an unknown way. (Even if we did know the shape of the track and could calculate ϕ along it, the calculation could be very difficult.)

A second **Key Idea** is that because $\vec{F}_g$ is a conservative force, we can find the work by choosing some other path between *a* and *b*—one that makes the calculation easy. Let us choose the dashed path in Fig. 8-5*b*; it consists of two straight segments. Along the horizontal segment, the angle ϕ is a constant 90°. Even though we do not know the displacement along that horizontal segment, Eq. 7-12 tells us that the work W_h done there is

$$W_h = mgd \cos 90° = 0.$$

Along the vertical segment, the displacement *d* is 0.80 m and, with $\vec{F}_g$ and $\vec{d}$ both downward, the angle ϕ is a constant 0°. Thus, Eq. 7-12 gives us, for the work W_v done along the vertical part of the dashed path,

$$W_v = mgd \cos 0°$$
$$= (2.0\ \text{kg})(9.8\ \text{m/s}^2)(0.80\ \text{m})(1) = 15.7\ \text{J}.$$

The total work done on the cheese by $\vec{F}_g$ as the cheese moves from point *a* to point *b* along the dashed path is then

$$W = W_h + W_v = 0 + 15.7\ \text{J} \approx 16\ \text{J}. \qquad \text{(Answer)}$$

This is also the work done as the cheese moves along the track from *a* to *b*.

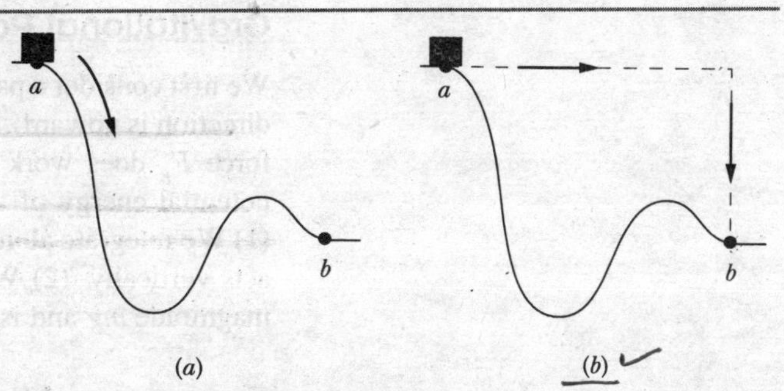

Fig. 8-5 Sample Problem 8-1. (*a*) A block of cheese slides along a frictionless track from point *a* to point *b*. (*b*) Finding the work done on the cheese by the gravitational force is easier along the dashed path than along the actual path taken by the cheese; the result is the same for both paths.

8-3 Determining Potential Energy Values

Here we find equations that give the value of the two types of potential energy discussed in this chapter: gravitational potential energy and elastic potential energy. However, first we must find a general relation between a conservative force and the associated potential energy.

Consider a particle-like object that is part of a system in which a conservative force $\vec{F}$ acts. When that force does work W on the object, the change ΔU in the potential energy associated with the system is the negative of the work done. We wrote this fact as Eq. 8-1 ($\Delta U = -W$). For the most general case, in which the force may vary with position, we may write the work W as in Eq. 7-32:

$$W = \int_{x_i}^{x_f} F(x)\, dx. \qquad (8\text{-}5)$$

This equation gives the work done by the force when the object moves from point x_i to point x_f, changing the configuration of the system. (Because the force is conservative, the work is the same for all paths between those two points.)

Substituting Eq. 8-5 into Eq. 8-1, we find that the change in potential energy due to the change in configuration is

$$\Delta U = -\int_{x_i}^{x_f} F(x)\, dx. \qquad (8\text{-}6)$$

This is the general relation we sought. Let's put it to use.

Gravitational Potential Energy

We first consider a particle with mass m moving vertically along a y axis (the positive direction is upward). As the particle moves from point y_i to point y_f, the gravitational force $\vec{F}_g$ does work on it. To find the corresponding change in the gravitational potential energy of the particle–Earth system, we use Eq. 8-6 with two changes: (1) We integrate along the y axis instead of the x axis, because the gravitational force acts vertically (2) We substitute $-mg$ for the force symbol F, because $\vec{F}_g$ has the magnitude mg and is directed down the y axis. We then have

$$\Delta U = -\int_{y_i}^{y_f} (-mg)\, dy = mg \int_{y_i}^{y_f} dy = mg\Big[\, y \,\Big]_{y_i}^{y_f},$$

which yields

$$\Delta U = mg(y_f - y_i) = mg\, \Delta y. \tag{8-7}$$

Only *changes* ΔU in gravitational potential energy (or any other type of potential energy) are physically meaningful. However, to simplify a calculation or a discussion, we sometimes would like to say that a certain gravitational potential value U is associated with a certain particle–Earth system when the particle is at a certain height y. To do so, we rewrite Eq. 8-7 as

$$U - U_i = mg(y - y_i). \tag{8-8}$$

Then we take U_i to be the gravitational potential energy of the system when it is in a **reference configuration** in which the particle is at a **reference point** y_i. Usually we take $U_i = 0$ and $y_i = 0$. Doing this changes Eq. 8-8 to

$$U(y) = mgy \qquad \text{(gravitational potential energy).} \tag{8-9}$$

This equation tells us:

➤ **The gravitational potential energy associated with a particle–Earth system depends only on the vertical position y (or height) of the particle relative to the reference position y = 0, not on the horizontal position.**

Elastic Potential Energy

We next consider the block–spring system shown in Fig. 8-3, with the block moving on the end of a spring of spring constant k. As the block moves from point x_i to point x_f, the spring force $F = -kx$ does work on the block. To find the corresponding change in the elastic potential energy of the block–spring system, we substitute $-kx$ for $F(x)$ in Eq. 8-6. We then have

$$\Delta U = -\int_{x_i}^{x_f} (-kx)\, dx = k\int_{x_i}^{x_f} x\, dx = \tfrac{1}{2}k\Big[\, x^2 \,\Big]_{x_i}^{x_f},$$

or $\quad W = -\frac{1}{2}k(x_2^2 - x_1^2)$

$$\Delta U = \tfrac{1}{2}kx_f^2 - \tfrac{1}{2}kx_i^2. \tag{8-10}$$

To associate a potential energy value U with the block at position x, we choose the reference configuration to be when the spring is at its relaxed length and the block is at $x_i = 0$. Then the elastic potential energy U_i is 0, and Eq. 8-10 becomes

$$U - 0 = \tfrac{1}{2}kx^2 - 0,$$

which gives us

$$U(x) = \tfrac{1}{2}kx^2 \qquad \text{(elastic potential energy).} \tag{8-11}$$

CHECKPOINT 2: A particle is to move along the x axis from $x = 0$ to x_1 while a conservative force, directed along the x axis, acts on the particle. The figure shows three situations in which the x component of that force varies with x. The force has the same maximum magnitude F_1 in all three situations. Rank the situations according to the change in the associated potential energy during the particle's motion, most positive first.

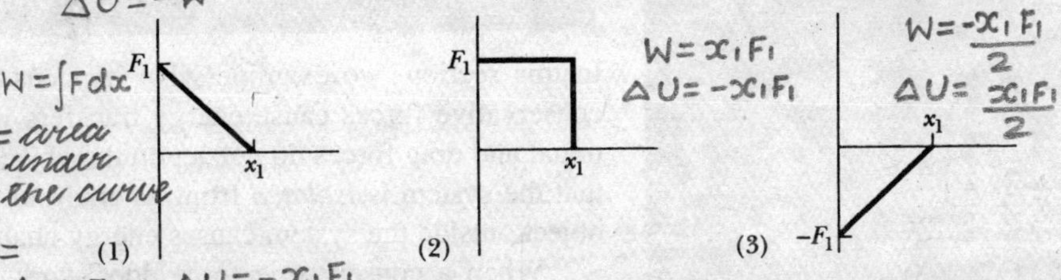

PROBLEM-SOLVING TACTICS

Tactic 1: *Using the Term "Potential Energy"*

A potential energy is associated with a system as a whole. However, you might see statements that associate it with only part of the system. For example, you might read, "An apple hanging in a tree has a gravitational potential energy of 30 J." Such statements are often acceptable, but you should always keep in mind that the potential energy is actually associated with a system—here the apple–Earth system. Also keep in mind that assigning a particular potential energy value, such as 30 J here, to an object or even a system makes sense *only* if the reference potential energy value is known, as explored in Sample Problem 8-2.

Sample Problem 8-2

A 2.0 kg sloth hangs 5.0 m above the ground (Fig. 8-6).

(a) What is the gravitational potential energy U of the sloth–Earth system if we take the reference point $y = 0$ to be (1) at the ground, (2) at a balcony floor that is 3.0 m above the ground, (3) at the limb, and (4) 1.0 m above the limb? Take the gravitational potential energy to be zero at $y = 0$.

SOLUTION: The **Key Idea** here is that once we have chosen the reference point for $y = 0$, we can calculate the gravitational potential energy U of the system *relative to that reference point* with Eq. 8-9. For example, for choice (1) the sloth is at $y = 5.0$ m, and

$$U = mgy = (2.0\text{ kg})(9.8\text{ m/s}^2)(5.0\text{ m}) = 98\text{ J.} \quad \text{(Answer)}$$

For the other choices, the values of U are

(2) $U = mgy = mg(2.0\text{ m}) = 39$ J,
(3) $U = mgy = mg(0) = 0$ J,
(4) $U = mgy = mg(-1.0\text{ m}) = -19.6\text{ J} \approx -20$ J. (Answer)

(b) The sloth drops to the ground. For each choice of reference point, what is the change ΔU in the potential energy of the sloth–Earth system due to the fall?

SOLUTION: The **Key Idea** here is that the *change* in potential energy does not depend on the choice of the reference point for $y = 0$; instead, it depends on the change in height Δy. For all four situations, we have the same $\Delta y = -5.0$ m. Thus, for (1) to (4), Eq. 8-7 tells us that

$$\Delta U = mg\,\Delta y = (2.0\text{ kg})(9.8\text{ m/s}^2)(-5.0\text{ m}) = -98\text{ J.} \quad \text{(Answer)}$$

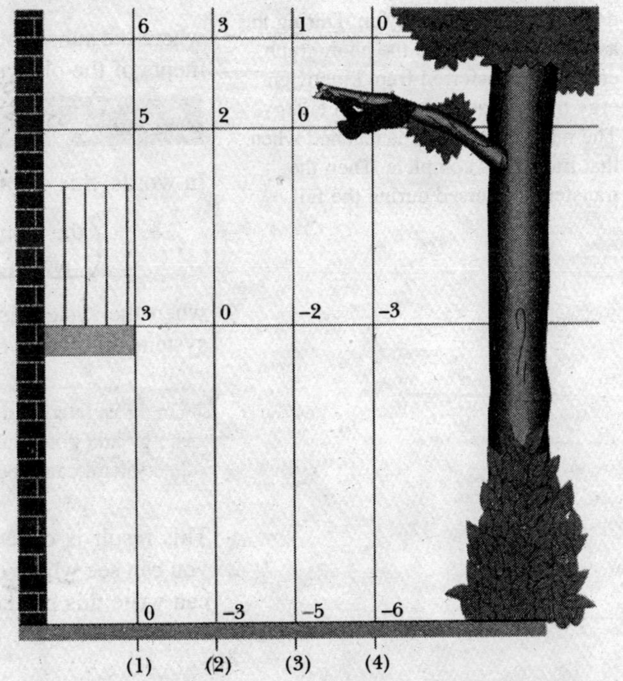

Fig. 8-6 Sample Problem 8-2. Four choices of reference point $y = 0$. Each y axis is marked in units of meters. The choice affects the value of the potential energy U of the sloth–Earth system. However, it does not affect the change ΔU in potential energy of the system if the sloth moves by, say, falling.

In olden days, a native Alaskan would be tossed via a blanket to be able to see farther over the flat terrain. Nowadays, it is done just for fun. During the ascent of the child in the photograph, energy is transferred from kinetic energy to gravitational potential energy. The maximum height is reached when that transfer is complete. Then the transfer is reversed during the fall.

8-4 Conservation of Mechanical Energy

The **mechanical energy** E_{mec} of a system is the sum of its potential energy U and the kinetic energy K of the objects within it:

$$E_{mec} = K + U \quad \text{(mechanical energy).} \tag{8-12}$$

In this section, we examine what happens to this mechanical energy when only conservative forces cause energy transfers within the system—that is, when frictional and drag forces do not act on the objects in the system. Also, we shall assume that the system is *isolated* from its environment; that is, no *external force* from an object outside the system causes energy changes inside the system.

When a conservative force does work W on an object within the system, it transfers energy between kinetic energy K of the object and potential energy U of the system. From Eq. 7-10, the change ΔK in kinetic energy is

$$\Delta K = W \tag{8-13}$$

and from Eq. 8-1, the change ΔU in potential energy is

$$\Delta U = -W. \tag{8-14}$$

Combining Eqs. 8-13 and 8-14, we find that

$$\Delta K = -\Delta U. \tag{8-15}$$

In words, one of these energies increases exactly as much as the other decreases.

We can rewrite Eq. 8-15 as

$$K_2 - K_1 = -(U_2 - U_1), \tag{8-16}$$

where the subscripts refer to two different instants and thus to two different arrangements of the objects in the system. Rearranging Eq. 8-16 yields

$$K_2 + U_2 = K_1 + U_1 \quad \text{(conservation of mechanical energy).} \tag{8-17}$$

In words, this equation says:

$$\left(\begin{array}{c}\text{the sum of } K \text{ and } U \text{ for}\\ \text{any state of a system}\end{array}\right) = \left(\begin{array}{c}\text{the sum of } K \text{ and } U \text{ for}\\ \text{any other state of the system}\end{array}\right),$$

when the system is isolated and only conservative forces act on the objects in the system. In other words:

> In an isolated system where only conservative forces cause energy changes, the kinetic energy and potential energy can change, but their sum, the mechanical energy E_{mec} of the system, cannot change.

This result is called the **principle of conservation of mechanical energy.** (Now you can see where *conservative* forces got their name.) With the aid of Eq. 8-15, we can write this principle in one more form, as

$$\Delta E_{mec} = \Delta K + \Delta U = 0. \tag{8-18}$$

The principle of conservation of mechanical energy allows us to solve problems that would be quite difficult to solve using only Newton's laws:

> When the mechanical energy of a system is conserved, we can relate the sum of kinetic energy and potential energy at one instant to that at another instant *without considering the intermediate motion* and *without finding the work done by the forces involved.*

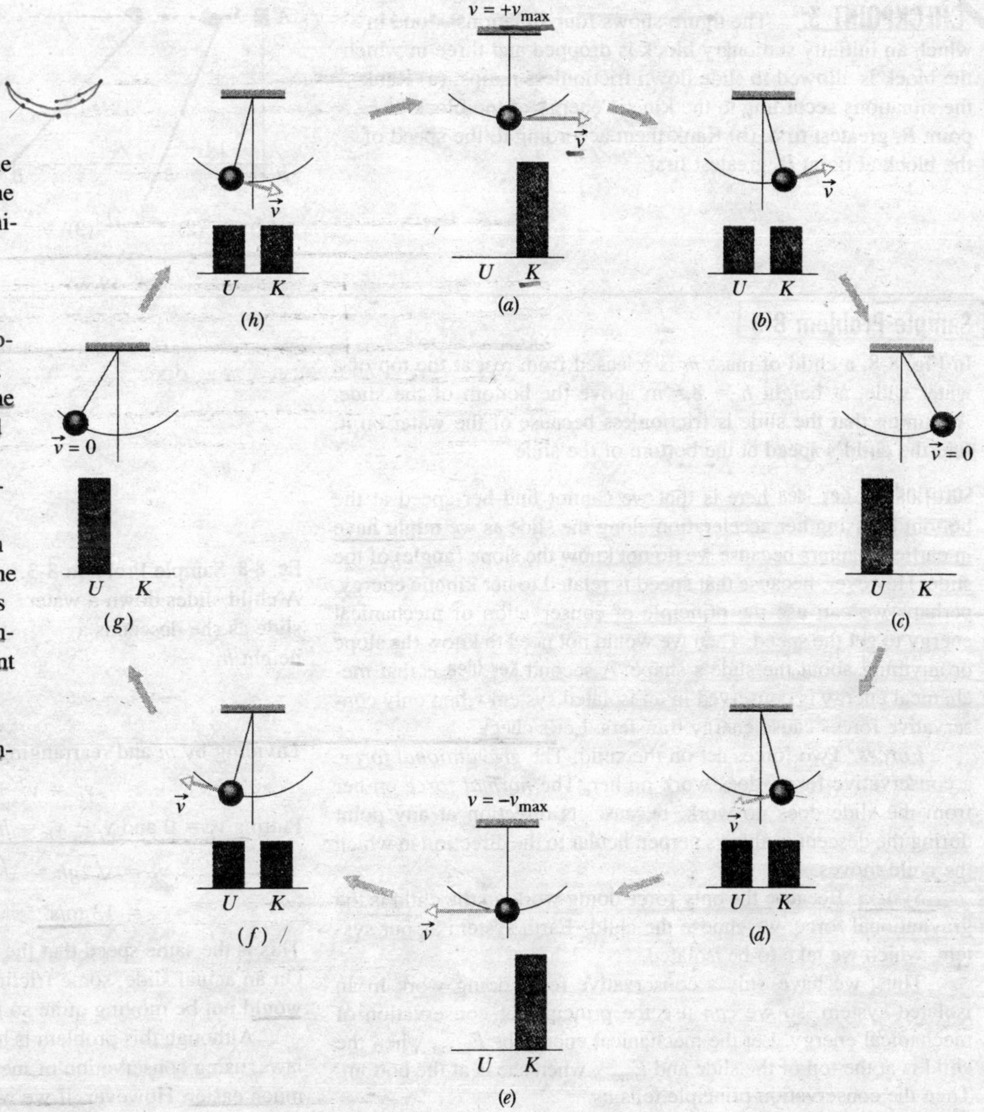

Fig. 8-7 A pendulum, with its mass concentrated in a bob at the lower end, swings back and forth. One full cycle of the motion is shown. During the cycle the values of the potential and kinetic energies of the pendulum–Earth system vary as the bob rises and falls, but the mechanical energy E_{mec} of the system remains constant. The energy E_{mec} can be described as continuously shifting between the kinetic and potential forms. In stages (*a*) and (*e*), all the energy is kinetic energy. The bob then has its greatest speed and is at its lowest point. In stages (*c*) and (*g*), all the energy is potential energy. The bob then has zero speed and is at its highest point. In stages (*b*), (*d*), (*f*), and (*h*), half the energy is kinetic energy and half is potential energy. If the swinging involved a frictional force at the point where the pendulum is attached to the ceiling, or a drag force due to the air, then E_{mec} would not be conserved, and eventually the pendulum would stop.

Figure 8-7 shows an example in which the principle of conservation of mechanical energy can be applied: As a pendulum swings, the energy of the pendulum–Earth system is transferred back and forth between kinetic energy K and gravitational potential energy U, with the sum $K + U$ being constant. If we know the gravitational potential energy when the pendulum bob is at its highest point (Fig. 8-7*c*), Eq. 8-17 gives us the kinetic energy of the bob at the lowest point (Fig. 8-7*e*).

For example, let us choose the lowest point as the reference point, with the gravitational potential energy $U_2 = 0$. Suppose then that the potential energy at the highest point is $U_1 = 20$ J relative to the reference point. Because the bob momentarily stops at its highest point, the kinetic energy there is $K_1 = 0$. Substituting these values into Eq. 8-17 gives us the kinetic energy K_2 at the lowest point:

$$K_2 + 0 = 0 + 20 \text{ J} \quad \text{or} \quad K_2 = 20 \text{ J}.$$

Note that we get this result without considering the motion between the highest and lowest points (such as in Fig. 8-7*d*) and without finding the work done by any forces involved in the motion.

CHECKPOINT 3: The figure shows four situations—one in which an initially stationary block is dropped and three in which the block is allowed to slide down frictionless ramps. (a) Rank the situations according to the kinetic energy of the block at point B, greatest first. (b) Rank them according to the speed of the block at point B, greatest first.

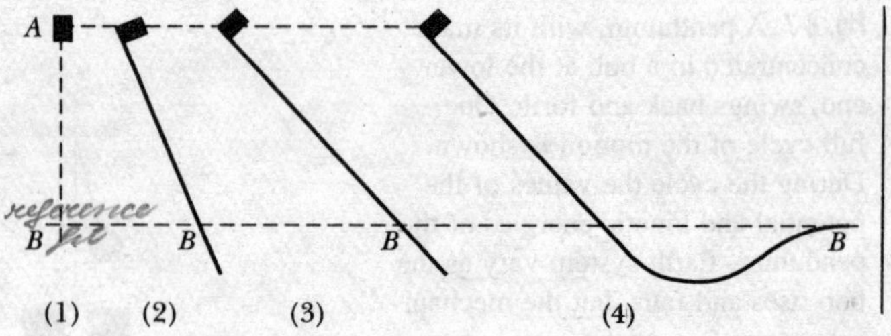

Sample Problem 8-3

In Fig. 8-8, a child of mass m is released from rest at the top of a water slide, at height h = 8.5 m above the bottom of the slide. Assuming that the slide is frictionless because of the water on it, find the child's speed at the bottom of the slide.

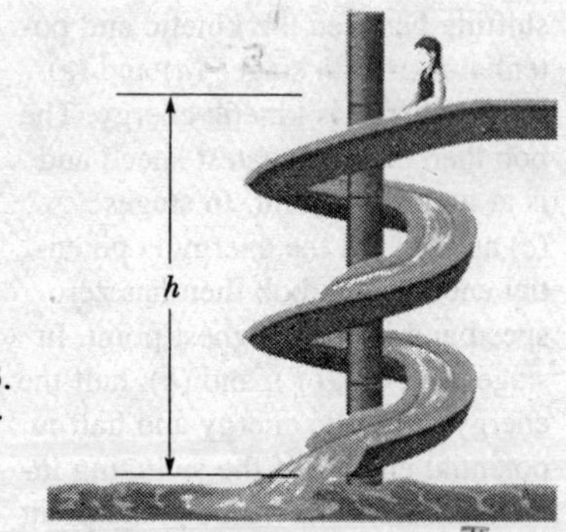

Fig. 8-8 Sample Problem 8-3. A child slides down a water slide as she descends a height h.

SOLUTION: A **Key Idea** here is that we cannot find her speed at the bottom by using her acceleration along the slide as we might have in earlier chapters because we do not know the slope (angle) of the slide. However, because that speed is related to her kinetic energy, perhaps we can use the principle of conservation of mechanical energy to get the speed. Then we would not need to know the slope or anything about the slide's shape. A second **Key Idea** is that mechanical energy is conserved in an isolated system when only conservative forces cause energy transfers. Let's check.

Forces: Two forces act on the child. The *gravitational force*, a conservative force, does work on her. The *normal force* on her from the slide does no work, because its direction at any point during the descent is always perpendicular to the direction in which the child moves.

System: Because the only force doing work on the child is the gravitational force, we choose the child–Earth system as our system, which we take to be isolated.

Thus, we have only a conservative force doing work in an isolated system, so we *can* use the principle of conservation of mechanical energy. Let the mechanical energy be $E_{\text{mec},t}$ when the child is at the top of the slide and $E_{\text{mec},b}$ when she is at the bottom. Then the conservation principle tells us

$$E_{\text{mec},b} = E_{\text{mec},t}.$$

Expanding this to show both kinds of mechanical energy, we have

$$K_b + U_b = K_t + U_t,$$

or

$$\tfrac{1}{2}mv_b^2 + mgy_b = \tfrac{1}{2}mv_t^2 + mgy_t.$$

Dividing by m and rearranging yield

$$v_b^2 = v_t^2 + 2g(y_t - y_b).$$

Putting $v_t = 0$ and $y_t - y_b = h$ leads to

$$v_b = \sqrt{2gh} = \sqrt{(2)(9.8\ \text{m/s}^2)(8.5\ \text{m})}$$
$$= 13\ \text{m/s}. \qquad \text{(Answer)}$$

This is the same speed that the child would reach if she fell 8.5 m. On an actual slide, some frictional forces would act and the child would not be moving quite so fast.

Although this problem is hard to solve directly with Newton's laws, using conservation of mechanical energy makes the solution much easier. However, if we were asked to find the time taken for the child to reach the bottom of the slide, energy methods would be of no use; we would need to know the shape of the slide, and we would have a difficult problem.

Now that we have worked this sample problem, return to the puck example in the first paragraph of Chapter 7. See if you can show that the puck's speed at the end of the track is 5.0 m/s.

Sample Problem 8-4

A 61.0 kg bungee-cord jumper is on a bridge 45.0 m above a river. The elastic bungee cord has a relaxed length of L = 25.0 m. Assume that the cord obeys Hooke's law, with a spring constant of 160 N/m. If the jumper stops before reaching the water, what is the height h of her feet above the water at her lowest point?

SOLUTION: Figure 8-9 shows the jumper at the lowest point, with her feet at height h and with the cord stretched by distance d from its relaxed length. If we knew d, we could find h. One **Key Idea** is that perhaps we can solve for d by applying the principle of conservation of mechanical energy, between her initial point (on the bridge) and her lowest point. In that case, a second **Key Idea** is that mechanical energy is conserved in an isolated system when only conservative forces cause energy transfers. Let's check.

Forces: The gravitational force does work on the jumper throughout her fall. Once the bungee cord becomes taut, the spring-like force from it does work on her, transferring energy to elastic potential energy of the cord. The force from the cord also pulls on the bridge, which is attached to Earth. The gravitational force and the spring-like force are conservative.

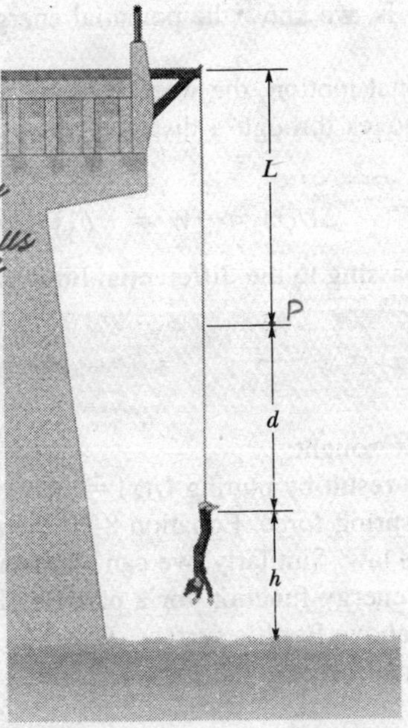

Fig. 8-9 Sample Problem 8-4. A bungee-cord jumper at the lowest point of the jump.

System: The jumper–Earth–cord system includes all these forces and energy transfers, and we can take it to be isolated. Thus, we *can* apply the principle of conservation of mechanical energy to the system. From Eq. 8-18, we can write the principle as

$$\Delta K + \Delta U_e + \Delta U_g = 0, \qquad (8\text{-}19)$$

where ΔK is the change in the jumper's kinetic energy, ΔU_e is the change in the elastic potential energy of the bungee cord, and ΔU_g is the change in the jumper's gravitational potential energy. All these changes must be computed between her initial point and her lowest point. Because she is stationary (at least momentarily) both initially and at her lowest point, $\Delta K = 0$. From Fig. 8-9, we see that the change Δy in her height is $-(L + d)$, so we have

$$\Delta U_g = mg\,\Delta y = -mg(L + d),$$

where m is her mass. Also from Fig. 8-9, we see that the bungee cord is stretched by distance d. Thus, we also have

$$\Delta U_e = \tfrac{1}{2}kd^2.$$

Inserting these expressions and the given data into Eq. 8-19, we obtain

$$0 + \tfrac{1}{2}kd^2 - mg(L + d) = 0$$

or

$$\tfrac{1}{2}kd^2 - mgL - mgd = 0$$

and then

$$\tfrac{1}{2}(160\ \text{N/m})d^2 - (61.0\ \text{kg})(9.8\ \text{m/s}^2)(25.0\ \text{m}) - (61.0\ \text{kg})(9.8\ \text{m/s}^2)d = 0.$$

Solving this quadratic equation yields

$$d = 17.9\ \text{m}.$$

The jumper's feet are then a distance of $(L + d) = 42.9$ m below their initial height. Thus,

$$h = 45.0\ \text{m} - 42.9\ \text{m} = 2.1\ \text{m}. \qquad \text{(Answer)}$$

PROBLEM-SOLVING TACTICS

Tactic 2: *Conservation of Mechanical Energy*
Asking the following questions will help you to solve problems involving the conservation of mechanical energy.

For what system is mechanical energy conserved? You should be able to separate your system from its environment. Imagine drawing a closed surface such that whatever is inside is your system and whatever is outside is the environment of that system. In Sample Problem 8-3 the system is the *child + Earth* and in Sample Problem 8-4, it is the *jumper + Earth + cord*.

Is friction or drag present? If friction or drag is present, mechanical energy is not conserved.

Is your system isolated? Conservation of mechanical energy applies only to isolated systems. That means that no *external forces* (forces exerted by objects outside the system) should do work on the objects in the system.

What are the initial and final states of your system? The system changes from some initial state (or configuration) to some final state. You apply the principle of conservation of mechanical energy by saying that E_{mec} has the same value in both these states. Be very clear about what these two states are.

8-5 Reading a Potential Energy Curve

Once again we consider a particle that is part of a system in which a conservative force acts. This time suppose that the particle is constrained to move along an x axis while the conservative force does work on it. We can learn a lot about the motion of the particle from a plot of the system's potential energy $U(x)$. However, before we discuss such plots, we need one more relationship.

Finding the Force Analytically

Equation 8-6 tells us how to find the change ΔU in potential energy between two points in a one-dimensional situation if we know the force $F(x)$. Now we want to go the other way; that is, we know the potential energy function $U(x)$ and want to find the force.

For one-dimensional motion, the work W done by a force that acts on a particle as the particle moves through a distance Δx is $F(x)\,\Delta x$. We can then write Eq. 8-1 as

$$\Delta U(x) = -W = -F(x)\,\Delta x.$$

Solving for $F(x)$ and passing to the differential limit yield

$$F(x) = -\frac{dU(x)}{dx} \quad \text{(one-dimensional motion),} \qquad (8\text{-}20)$$

which is the relation we sought.

We can check this result by putting $U(x) = \frac{1}{2}kx^2$, which is the elastic potential energy function for a spring force. Equation 8-20 then yields, as expected, $F(x) = -kx$, which is Hooke's law. Similarly, we can substitute $U(x) = mgx$, which is the gravitational potential energy function for a particle–Earth system, with a particle of mass m at height x above Earth's surface. Equation 8-20 then yields $F = -mg$, which is the gravitational force on the particle.

The Potential Energy Curve

Figure 8-10*a* is a plot of a potential energy function $U(x)$ for a system in which a particle is in one-dimensional motion while a conservative force $F(x)$ does work on it. We can easily find $F(x)$ by (graphically) taking the slope of the $U(x)$ curve at various points. (Equation 8-20 tells us that $F(x)$ is negative the slope of the $U(x)$ curve.) Figure 8-10*b* is a plot of $F(x)$ found in this way.

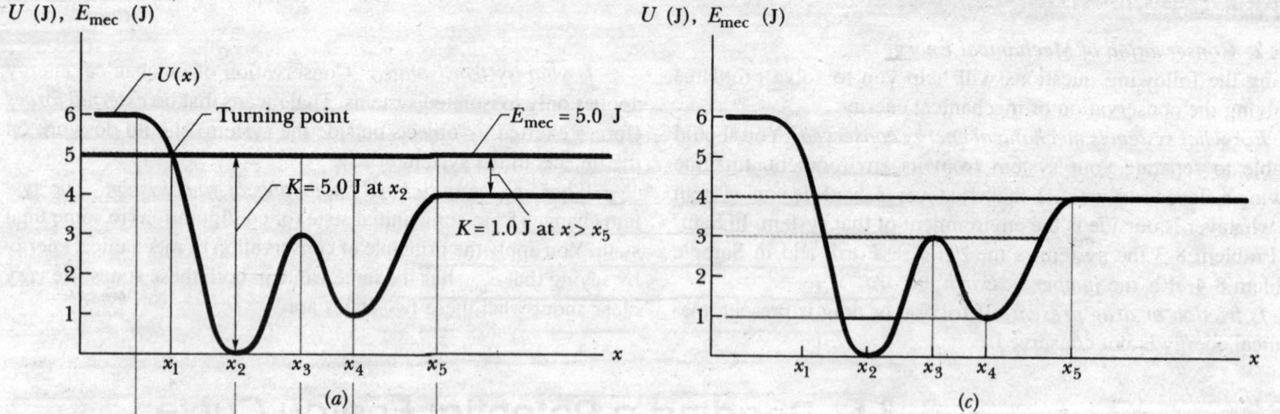

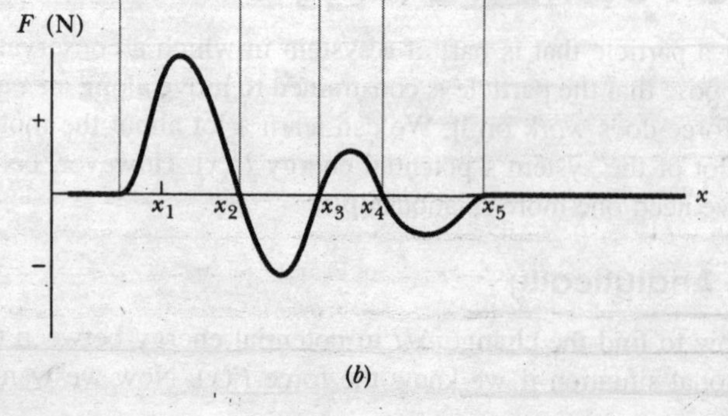

Fig. 8-10 (*a*) A plot of $U(x)$, the potential energy function of a system containing a particle confined to move along the x axis. There is no friction, so mechanical energy is conserved. (*b*) A plot of the force $F(x)$ acting on the particle, derived from the potential energy plot by taking its slope at various points. (*c*) The $U(x)$ plot of (*a*) with three different possible values of E_{mec} shown.

Turning Points

In the absence of a nonconservative force, the mechanical energy E of the system has a constant value given by

$$U(x) + K(x) = E_{\text{mec}}. \tag{8-21}$$

Here $K(x)$ is the *kinetic energy function* of the particle (this $K(x)$ gives the kinetic energy as a function of the particle's location x). We may rewrite Eq. 8-21 as

$$K(x) = E_{\text{mec}} - U(x). \tag{8-22}$$

Suppose that E_{mec} (which has a constant value, remember) happens to be 5.0 J. It would be represented in Fig. 8-10a by a horizontal line that runs through the value 5.0 J on the energy axis. (It is, in fact, shown there.)

Equation 8-22 tells us how to determine the kinetic energy K for any location x of the particle: On the $U(x)$ curve, find U for that location x and then subtract U from E_{mec}. For example, if the particle is at any point to the right of x_5, then $K =$ 1.0 J. The value of K is greatest (5.0 J) when the particle is at x_2, and least (0 J) when the particle is at x_1.

Since K can never be negative (because v^2 is always positive), the particle can never move to the left of x_1, where $E_{\text{mec}} - U$ is negative. Instead, as the particle moves toward x_1 from x_2, K decreases (the particle slows) until $K = 0$ at x_1 (the particle stops there).

Note that when the particle reaches x_1, the force on the particle, given by Eq. 8-20, is positive (because the slope dU/dx is negative). This means that the particle does not remain at x_1 but instead begins to move to the right, opposite its earlier motion. Hence x_1 is a **turning point,** a place where $K = 0$ (because $U = E$) and the particle changes direction. There is no turning point (where $K = 0$) on the right side of the graph. When the particle heads to the right, it will continue indefinitely.

Check at points where mech energy line cuts PE curve ie $E_{mech} = U \Rightarrow K=0$

+ve slope

−ve slope

Equilibrium Points

Figure 8-10c shows three different values for E_{mec} superposed on the plot of the same potential energy function $U(x)$. Let us see how they would change the situation. If $E_{\text{mec}} = 4.0$ J (purple line), the turning point shifts from x_1 to a point between x_1 and x_2. Also, at any point to the right of x_5, the system's mechanical energy is equal to its potential energy; thus, the particle has no kinetic energy and (by Eq. 8-20) no force acts on it, and so it must be stationary. A particle at such a position is said to be in **neutral equilibrium.** (A marble placed on a horizontal tabletop is in that state.)

For equil slope = 0, we talk abt a point

Particle slightly displaced, remains stationary

If $E_{\text{mec}} = 3.0$ J (pink line), there are two turning points: One is between x_1 and x_2, and the other is between x_4 and x_5. In addition, x_3 is a point at which $K = 0$. If the particle is located exactly there, the force on it is also zero, and the particle remains stationary. However, if it is displaced even slightly in either direction, a nonzero force pushes it farther in the same direction, and the particle continues to move. A particle at such a position is said to be in **unstable equilibrium.** (A marble balanced on top of a bowling ball is an example.)

Next consider the particle's behavior if $E_{\text{mec}} = 1.0$ J (green line). If we place it at x_4, it is stuck there. It cannot move left or right on its own because to do so would require a negative kinetic energy. If we push it slightly left or right, a restoring force appears that moves it back to x_4. A particle at such a position is said to be in **stable equilibrium.** (A marble placed at the bottom of a hemispherical bowl is an example.) If we place the particle in the cuplike *potential well* centered at x_2, it is between two turning points. It can still move somewhat, but only partway to x_1 or x_3. on its own

CHECKPOINT 4: The figure gives the potential energy function $U(x)$ for a system in which a particle is in one-dimensional motion. (a) Rank regions AB, BC, and CD according to the magnitude of the force on the particle, greatest first. (b) What is the direction of the force when the particle is in region AB?

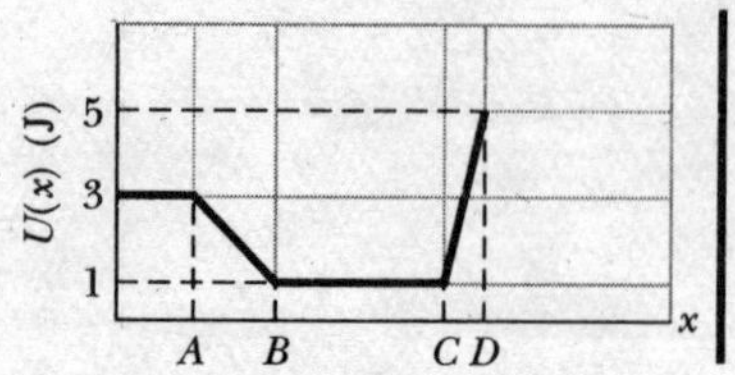

8-6 Work Done on a System by an External Force

In Chapter 7, we defined work as being energy transferred to or from an object by means of a force acting on the object. We can now extend that definition to an external force acting on a system of objects.

> Work is energy transferred to or from a system by means of an external force acting on that system.

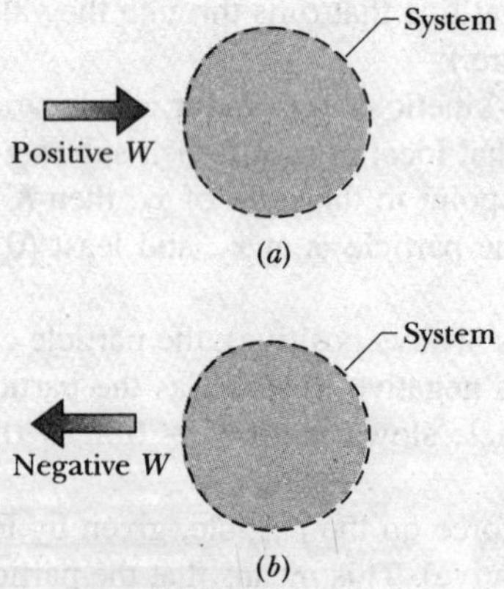

Fig. 8-11 (*a*) Positive work W on an arbitrary system means a transfer of energy to the system. (*b*) Negative work W means a transfer of energy from the system.

Figure 8-11*a* represents positive work (a transfer of energy *to* a system), and Fig. 8-11*b* represents negative work (a transfer of energy *from* a system). When more than one force acts on a system, their *net work* is the energy transferred to or from the system.

These transfers are like transfers of money to and from a bank account. If a system consists of a single particle or particle-like object, as in Chapter 7, the work done on the system by a force can change only the kinetic energy of the system. The energy statement for such transfers is the work–kinetic energy theorem of Eq. 7-10 ($\Delta K = W$); that is, a single particle has only one energy account, called kinetic energy. External forces can transfer energy into or out of that account. If a system is more complicated, however, an external force can change other forms of energy (such as potential energy); that is, a more complicated system can have multiple energy accounts.

Let us find energy statements for such systems by examining two basic situations, one that does not involve friction and one that does.

No Friction Involved

To compete in a bowling-ball hurling contest, you first squat and cup your hands under the ball on the floor. Then you rapidly straighten up while also pulling your hands up sharply, launching the ball upward at about face level. During your upward motion, your applied force on the ball obviously does work; that is, it is an external force that transfers energy, but to what system?

To answer, we check to see which energies change. There is a change ΔK in the ball's kinetic energy and, because the ball and Earth become more separated, there is a change ΔU in the gravitational potential energy of the ball–Earth system. To include both changes, we need to consider the ball–Earth system. Then your force is an external force doing work on that system, and the work is

$$W = \Delta K + \Delta U, \tag{8-23}$$

or

$$W = \Delta E_{mec} \quad \text{(work done on system, no friction involved),} \tag{8-24}$$

where ΔE_{mec} is the change in the mechanical energy of the system. These two equations, which are represented in Fig. 8-12, are equivalent energy statements for work done on a system by an external force when friction is not involved.

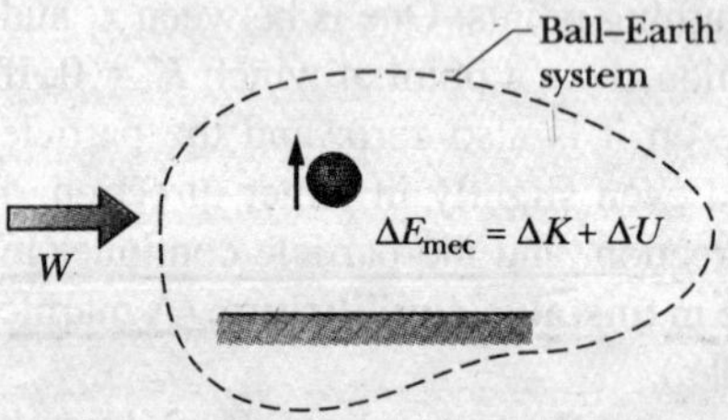

Fig. 8-12 Positive work W is done on a system of a bowling ball and Earth, causing a change ΔE_{mec} in the mechanical energy of the system, a change ΔK in the ball's kinetic energy, and a change ΔU in the system's gravitational potential energy.

Friction Involved

We next consider the example in Fig. 8-13*a*. A constant horizontal force $\vec{F}$ pulls a block along an x axis and through a displacement of magnitude d, increasing the block's velocity from $\vec{v}_0$ to $\vec{v}$. During the motion, a constant kinetic frictional force $\vec{f}_k$ from the floor acts on the block. Let us first choose the block as our system and apply Newton's second law to it. We can write that law for components along the x axis ($F_{\text{net},x} = ma_x$) as

$$F - f_k = ma. \qquad (8\text{-}25)$$

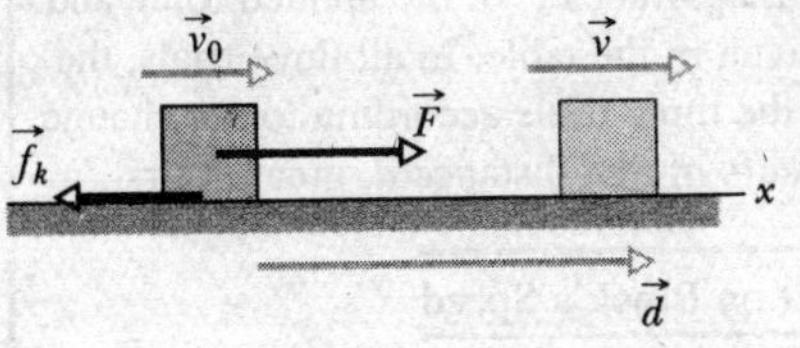

(*a*)

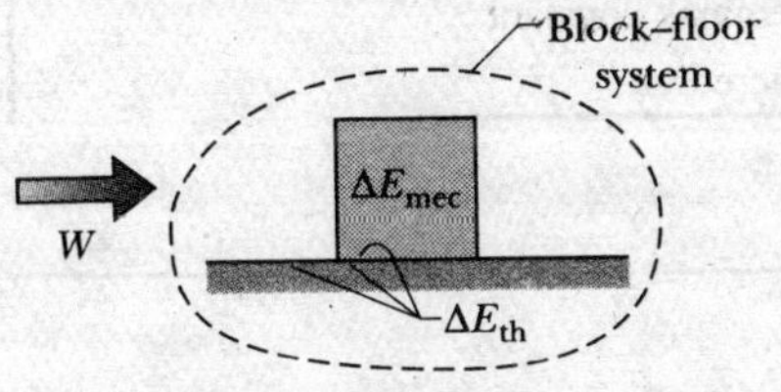

(*b*)

Fig. 8-13 (*a*) A block is pulled across a floor by force $\vec{F}$ while a kinetic frictional force opposes the motion. The block has velocity $\vec{v}_0$ at the start of a displacement $\vec{d}$ and a velocity $\vec{v}$ at the end of the displacement. (*b*) Positive work W is done on the block–floor system by force $\vec{F}$, resulting in a change ΔE_{mec} in the block's mechanical energy and a change ΔE_{th} in the thermal energy of the block and floor.

Because the forces are constant, the acceleration $\vec{a}$ is also constant. Thus, we can use Eq. 2-16 to write

$$v^2 = v_0^2 + 2ad.$$

Solving this equation for a, substituting the result into Eq. 8-25, and rearranging then give us

$$Fd = \tfrac{1}{2}mv^2 - \tfrac{1}{2}mv_0^2 + f_k d \qquad (8\text{-}26)$$

or, because $\frac{1}{2}mv^2 - \frac{1}{2}mv_0^2 = \Delta K$ for the block,

$$Fd = \Delta K + f_k d. \qquad (8\text{-}27)$$

In a more general situation (say, one in which the block is moving up a ramp), there can be a change in potential energy. To include such a possible change, we generalize Eq. 8-27 by writing

$$Fd = \Delta E_{\text{mec}} + f_k d. \qquad (8\text{-}28)$$

By experiment we find that the block and the portion of the floor along which it slides become warmer as the block slides. As we shall discuss in Chapter 19, the temperature of an object is related to the object's thermal energy E_{th} (the energy associated with the random motion of the atoms and molecules in the object). Here, the thermal energy of the block and floor increase because (1) there is friction between them and (2) there is sliding. Recall that friction is due to the cold-welding between two surfaces. As the block slides over the floor, the sliding causes repeated tearing and reforming of the welds between the block and the floor, which makes the block and floor warmer. Thus, the sliding increases their thermal energy E_{th}.

Through experiment, we find that the increase ΔE_{th} in thermal energy is equal to the product of the magnitudes f_k and d:

$$\Delta E_{\text{th}} = f_k d \quad \text{(increase in thermal energy by sliding).} \qquad (8\text{-}29)$$

Thus, we can rewrite Eq. 8-28 as

$$Fd = \Delta E_{\text{mec}} + \Delta E_{\text{th}}. \qquad (8\text{-}30)$$

Fd is the work W done by the external force $\vec{F}$ (the energy transferred by the force), but on which system is the work done (where are the energy transfers made)? To answer, we check to see which energies change. The block's mechanical energy changes, and the thermal energies of the block and floor also change. Therefore, the work done by force $\vec{F}$ is done on the block–floor system. That work is

$$W = \Delta E_{\text{mec}} + \Delta E_{\text{th}} \quad \text{(work done on system, friction involved).} \qquad (8\text{-}31)$$

This equation, which is represented in Fig. 8-13*b*, is the energy statement for the work done on a system by an external force when friction is involved.

CHECKPOINT 5: In three trials, a block is pushed by a horizontal applied force across a floor that is not frictionless, as in Fig. 8-13*a*. The magnitudes F of the applied force and the result of the pushing on the block's speed are given in the table. In all three trials, the block is pushed through the same distance d. Rank the three trials according to the change in the thermal energy of the block and floor that occurs in that distance d, greatest first.

$\Delta E_{th} = f_k d$ constant for a given system of block & floor ∴ all tie

Trial	F	Result on Block's Speed
a	5.0 N	decreases
b	7.0 N	remains constant
c	8.0 N	increases

Sample Problem 8-5

The giant stone statues of Easter Island were most likely moved by the prehistoric islanders by cradling each statue in a wooden sled and then pulling the sled over a "runway" consisting of almost identical logs acting as rollers. In a modern reenactment of this technique, 25 men were able to move a 9000 kg Easter Island-type statue 45 m over level ground in 2 min.

(a) Estimate the work the net force $\vec{F}$ from the men did during the 45 m displacement of the statue, and determine the system on which that force did the work.

SOLUTION: One **Key Idea** is that we can calculate the work done with Eq. 7-7 ($W = Fd\cos\phi$). Here d is the distance 45 m, F is the magnitude of the net force on the statue from the 25 men, and $\phi = 0°$. Let us estimate that each man pulled with a force magnitude equal to twice his weight, which we take to be the same value mg for all the men. Thus, the magnitude of the net force was $F = (25)(2mg) = 50mg$. Estimating a man's mass as 80 kg, we can then write Eq. 7-7 as

$$\begin{aligned} W &= Fd\cos\phi = 50mgd\cos\phi \\ &= (50)(80\text{ kg})(9.8\text{ m/s}^2)(45\text{ m})\cos 0° \\ &= 1.8 \times 10^6\text{ J} \approx 2\text{ MJ}. \qquad \text{(Answer)} \end{aligned}$$

The **Key Idea** in determining the system on which the work is done is to see which energies change. Because the statue moved, there was certainly a change ΔK in its kinetic energy during the motion. We can easily guess that there must have been considerable kinetic friction between the sled, logs, and ground, resulting in a change ΔE_{th} in their thermal energies. Thus, the system on which the work was done consisted of the statue, sled, logs, and ground.

(b) What was the increase ΔE_{th} in the thermal energy of the system during the 45 m displacement?

SOLUTION: The **Key Idea** here is that we can relate ΔE_{th} to the work W done by $\vec{F}$ with the energy statement of Eq. 8-31 for a system that involves friction:

$$W = \Delta E_{mec} + \Delta E_{th}.$$

We know the value of W from (a). The change ΔE_{mec} in the crate's mechanical energy was zero because the statue was stationary at the beginning and the end of the move and did not change in elevation. Thus, we find

$$\Delta E_{th} = W = 1.8 \times 10^6\text{ J} \approx 2\text{ MJ}. \qquad \text{(Answer)}$$

(c) Estimate the work that would have been done by the 25 men if they had moved the statue 10 km across level ground on Easter Island. Also estimate the total change ΔE_{th} that would have occurred in the statue–sled–logs–ground system.

SOLUTION: The **Key Ideas** here are the same as in (a) and (b). Thus we calculate W as in (a), but with 1×10^4 m now substituted for d. Also, we again equate ΔE_{th} to W. We get

$$W = \Delta E_{th} = 3.9 \times 10^8\text{ J} \approx 400\text{ MJ}. \qquad \text{(Answer)}$$

This would have been a staggering amount of energy for the men to have transferred during the movement of a statue. Still, the 25 men *could* have moved the statue 10 km, and the required energy does not suggest some mysterious source.

Sample Problem 8-6

A food shipper pushes a wood crate of cabbage heads (total mass $m = 14$ kg) across a concrete floor with a constant horizontal force $\vec{F}$ of magnitude 40 N. In a straight-line displacement of magnitude $d = 0.50$ m, the speed of the crate decreases from $v_0 = 0.60$ m/s to $v = 0.20$ m/s.

(a) How much work is done by force $\vec{F}$, and on what system does it do the work?

SOLUTION: One **Key Idea** is that Eq. 7-7 holds here: The work W done by $\vec{F}$ can be calculated as

includes work done on crate & floor both

$$\begin{aligned} W &= Fd\cos\phi = (40\text{ N})(0.50\text{ m})\cos 0° \\ &= 20\text{ J}. \qquad \text{(Answer)} \end{aligned}$$

The **Key Idea** in determining the system on which the work is done is to see which energies change. Because the crate's speed changes, there is certainly a change ΔK in the crate's kinetic energy. Is there friction between the floor and the crate, and thus a change in thermal energy? Note that $\vec{F}$ and the crate's velocity have the same direction. Thus, a **Key Idea** here is that if there is no friction, then $\vec{F}$ should be accelerating the crate to a *greater* speed. However, the crate is *slowing*, so there must be friction and a change ΔE_{th} in thermal energy of the crate and the floor. Therefore, the system on which the work is done is the crate–floor system, because both energy changes occur in that system.

(b) What is the increase ΔE_{th} in the thermal energy of the crate and floor?

SOLUTION: The **Key Idea** here is that we can relate ΔE_{th} to the work W done by $\vec{F}$ with the energy statement of Eq. 8-31 for a system that involves friction:

$$W = \Delta E_{mec} + \Delta E_{th}. \tag{8-32}$$

We know the value of W from (a). The change ΔE_{mec} in the crate's mechanical energy is just the change in its kinetic energy because no potential energy changes occur, so we have

$$\Delta E_{mec} = \Delta K = \tfrac{1}{2}mv^2 - \tfrac{1}{2}mv_0^2.$$

Substituting this into Eq. 8-32 and solving for ΔE_{th}, we find

$$\begin{aligned}\Delta E_{th} &= W - (\tfrac{1}{2}mv^2 - \tfrac{1}{2}mv_0^2) = W - \tfrac{1}{2}m(v^2 - v_0^2)\\ &= 20\text{ J} - \tfrac{1}{2}(14\text{ kg})[(0.20\text{ m/s})^2 - (0.60\text{ m/s})^2]\\ &= 22.2\text{ J} \approx 22\text{ J}. \qquad \text{(Answer)}\end{aligned}$$

8-7 Conservation of Energy

We now have discussed several situations in which energy is transferred to or from objects and systems, much like money is transferred between accounts. In each situation we assume that the energy that was involved could always be accounted for; that is, energy could not magically appear or disappear. In more formal language, we assumed (correctly) that energy obeys a law called the **law of conservation of energy,** which is concerned with the **total energy** E of a system. That total is the sum of the system's mechanical energy, thermal energy, and any form of *internal energy* in addition to thermal energy. (We have not yet discussed other forms of internal energy.) The law states that

> The total energy E of a system can change only by amounts of energy that are transferred to or from the system.

The only type of energy transfer that we have considered is work W done on a system. Thus, for us at this point, this law states that

$$W = \Delta E = \Delta E_{mec} + \Delta E_{th} + \Delta E_{int}, \tag{8-33}$$

where ΔE_{mec} is any change in the mechanical energy of the system, ΔE_{th} is any change in the thermal energy of the system, and ΔE_{int} is any change in any other form of internal energy of the system. Included in ΔE_{mec} are changes ΔK in kinetic energy and changes ΔU in potential energy (elastic, gravitational, or any other form we might find).

This law of conservation of energy is *not* something we have derived from basic physics principles. Rather, it is a law based on countless experiments. Scientists and engineers have never found an exception to it.

Isolated System

If a system is isolated from its environment, then there can be no energy transfers to or from it. For that case, the law of conservation of energy states:

> The total energy E of an isolated system cannot change.

Fig. 8-14 To descend, the rock climber must transfer energy from the gravitational potential energy of a system consisting of her, her gear, and Earth. She has wrapped the rope around metal rings so that the rope rubs against the rings. This allows most of the transferred energy to go to the thermal energy of the rope and rings rather than to her kinetic energy.

Many energy transfers may be going on *within* an isolated system, between, say, kinetic energy and a potential energy or kinetic energy and thermal energy. However, the total of all the forms of energy in the system cannot change.

We can use the rock climber in Fig. 8-14 as an example, approximating her, her gear, and Earth as an isolated system. As she rappels down the rock face, changing the configuration of the system, she needs to control the transfer of energy from the gravitational potential energy of the system. (That energy cannot just disappear.) Some of it is transferred to her kinetic energy. However, she obviously does not want very much transferred to that form or she will be moving too quickly, so she has wrapped the rope around metal rings to produce friction between the rope and the rings as she moves down. The sliding of the rings on the rope then transfers the gravitational potential energy of the system to thermal energy of the rings and rope in a way that she can control. The total energy of the climber–gear–Earth system (the total of its gravitational potential energy, kinetic energy, and thermal energy) does not change during her descent.

For an isolated system, the law of conservation of energy can be written in two ways. First, by setting $W = 0$ in Eq. 8-33, we get

$$\Delta E_{mec} + \Delta E_{th} + \Delta E_{int} = 0 \quad \text{(isolated system).} \tag{8-34}$$

We can also let $\Delta E_{mec} = E_{mec,2} - E_{mec,1}$, where the subscripts 1 and 2 refer to two different instants, say before and after a certain process has occurred. Then Eq. 8-34 becomes

$$E_{mec,2} = E_{mec,1} - \Delta E_{th} - \Delta E_{int}. \tag{8-35}$$

Equation 8-35 tells us:

> In an isolated system, we can relate the total energy at one instant to the total energy at another instant *without considering the energies at intermediate times.*

This fact can be a very powerful tool in solving problems about isolated systems when you need to relate energies of a system before and after a certain process occurs in the system.

In Section 8-4, we discussed a special situation for isolated systems—namely, the situation in which nonconservative forces (such as a kinetic frictional force) do not act within them. In that special situation, ΔE_{th} and ΔE_{int} are both zero, and so Eq. 8-35 reduces to Eq. 8-18. In other words, the mechanical energy of an isolated system is conserved when nonconservative forces do not act within it.

Power

Now that you have seen how energy can be transferred from one form to another, we can expand the definition of power given in Section 7-7. There it is the rate at which work is done by a force. In a more general sense, power P is the rate at which energy is transferred by a force from one form to another. If an amount of energy ΔE is transferred in an amount of time Δt, the **average power** due to the force is

$$P_{avg} = \frac{\Delta E}{\Delta t}. \tag{8-36}$$

Similarly, the **instantaneous power** due to the force is

$$P = \frac{dE}{dt}. \tag{8-37}$$

Sample Problem 8-7

In Fig. 8-15, a 2.0 kg package of tamale slides along a floor with speed $v_1 = 4.0$ m/s. It then runs into and compresses a spring, until the package momentarily stops. Its path to the initially relaxed spring is frictionless, but as it compresses the spring, a kinetic frictional force from the floor, of magnitude 15 N, acts on it. The spring constant is 10,000 N/m. By what distance d is the spring compressed when the package stops?

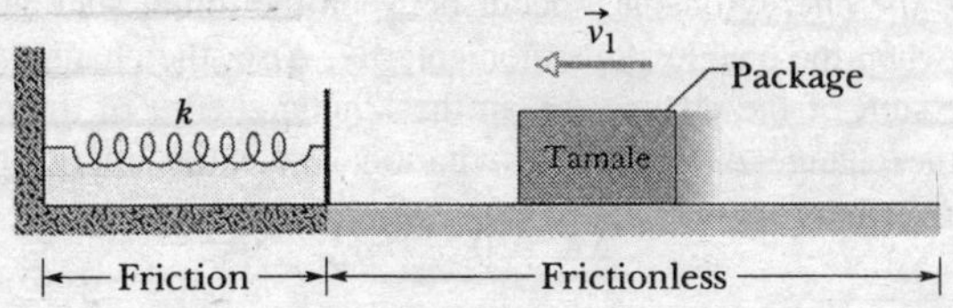

Fig. 8-15 Sample Problem 8-7. A package slides across a frictionless floor with velocity $\vec{v}_1$ toward a spring of spring constant k. When the package reaches the spring, a frictional force from the floor acts on it.

SOLUTION: A starting **Key Idea** is to examine all the forces acting on the package, and then to determine whether we have an isolated system or a system on which an external force is doing work.

Forces: The normal force on the package from the floor does no work on the package, because its direction is always perpendicular to that of the package's displacement. For the same reason, the gravitational force on the package does no work. As the spring is compressed, however, a spring force does work on the package, transferring energy to elastic potential energy of the spring. The spring force also pushes against a rigid wall. Because there is friction between the package and the floor, the sliding of the package across the floor increases their thermal energies.

System: The package–spring–floor–wall system includes all these forces and energy transfers in one isolated system. Therefore, a second **Key Idea** is that, because the system is isolated, its total energy cannot change. We can then apply the law of conservation of energy in the form of Eq. 8-35 to the system:

$$E_{mec,2} = E_{mec,1} - \Delta E_{th}. \qquad (8\text{-}38)$$

Let subscript 1 correspond to the initial state of the sliding package, and subscript 2 correspond to the state in which the package is momentarily stopped and the spring is compressed by distance d. For both states the mechanical energy of the system is the sum of the package's kinetic energy ($K = \frac{1}{2}mv^2$) and the spring's potential energy ($U = \frac{1}{2}kx^2$). For state 1, $U = 0$ (because the spring is not compressed), and the package's speed is v_1. Thus, we have

$$E_{mec,1} = K_1 + U_1 = \tfrac{1}{2}mv_1^2 + 0.$$

For state 2, $K = 0$ (because the package is stopped), and the compression distance is d. Therefore, we have

$$E_{mec,2} = K_2 + U_2 = 0 + \tfrac{1}{2}kd^2.$$

Finally, by Eq. 8-29, we can substitute $f_k d$ for the change ΔE_{th} in the thermal energy of the package and the floor. We can now rewrite Eq. 8-38 as

$$\tfrac{1}{2}kd^2 = \tfrac{1}{2}mv_1^2 - f_k d.$$

Rearranging and substituting known data give us

$$5000d^2 + 15d - 16 = 0.$$

Solving this quadratic equation yields

$$d = 0.055 \text{ m} = 5.5 \text{ cm}. \qquad \text{(Answer)}$$

$\Delta E_{mech} + \Delta U + \Delta E_{th} = 0$
$\Rightarrow \Delta K + \Delta U + \Delta E_{th} = 0$
$\Rightarrow (0 - \frac{1}{2}mv_i^2) + (\frac{1}{2}kx^2 - 0) + (f_k d - 0) = 0 \Rightarrow -\frac{1}{2}mv_i^2 + \frac{1}{2}kx^2 + f_k d = 0$, $d = x$

Sample Problem 8-8

In Fig. 8-16, a circus beagle of mass $m = 6.0$ kg runs onto the left end of a curved ramp with speed $v_0 = 7.8$ m/s at height $y_0 = 8.5$ m above the floor. It then slides to the right and comes to a momentary stop when it reaches a height $y = 11.1$ m from the floor. The ramp is not frictionless. What is the increase ΔE_{th} in the thermal energy of the beagle and ramp because of the sliding?

Fig. 8-16 Sample Problem 8-8. A beagle slides along a curved ramp, starting with speed v_0 at height y_0, and reaching a height y at which it momentarily stops.

SOLUTION: A **Key Idea** to get us started is to examine all the forces on the beagle, and then see if we have an isolated system or a system on which an external force is doing work.

① *Forces:* The normal force on the beagle from the ramp does no work on the beagle, because its direction is always perpendicular to that of the beagle's displacement. The gravitational force on the beagle does do work as the beagle's elevation changes. Because there is friction between the beagle and the ramp, the sliding increases their thermal energy.

② * *System:* The beagle–ramp–Earth system includes all these forces and energy transfers in one isolated system. Then a second **Key Idea** is that, because the system is isolated, its total energy cannot change. We can apply the law of conservation of energy in the form of Eq. 8-34 to this system:

$$\Delta E_{mec} + \Delta E_{th} = 0, \qquad (8\text{-}39)$$

$\Delta K + \Delta U + \Delta E_{th} = 0$
$\Delta E_{th} = \frac{1}{2} m v_0^2 - mg(y - y_0)$

where the energy changes occur between the initial state and the state when the beagle stops momentarily. Also, the change ΔE_{mec} is the sum of the change ΔK in the kinetic energy of the beagle and the change ΔU in the gravitational potential energy of the system, where

$$\Delta K = 0 - \tfrac{1}{2} m v_0^2$$

and

$$\Delta U = mgy - mgy_0.$$

Substituting these expressions into Eq. 8-39 and solving for ΔE_{th} yield

$$\begin{aligned} \Delta E_{th} &= \tfrac{1}{2} m v_0^2 - mg(y - y_0) \\ &= \tfrac{1}{2}(6.0\ \text{kg})(7.8\ \text{m/s})^2 \\ &\quad - (6.0\ \text{kg})(9.8\ \text{m/s}^2)(11.1\ \text{m} - 8.5\ \text{m}) \\ &\approx 30\ \text{J}. \end{aligned} \qquad \text{(Answer)}$$

REVIEW & SUMMARY

Conservative Forces A force is a **conservative force** if the net work it does on a particle moving around every closed path, from an initial point and then back to that point, is zero. Equivalently, it is conservative if the net work it does on a particle moving between two points does not depend on the path taken by the particle. The gravitational force and the spring force are conservative forces; the kinetic frictional force is a **nonconservative force.**

Potential Energy A **potential energy** is energy that is associated with the configuration of a system in which a conservative force acts. When the conservative force does work W on a particle within the system, the change ΔU in the potential energy of the system is

$$\Delta U = -W. \qquad (8\text{-}1)$$

If the particle moves from point x_i to point x_f, the change in the potential energy of the system is

$$\Delta U = -\int_{x_i}^{x_f} F(x)\, dx. \qquad (8\text{-}6)$$

Gravitational Potential Energy The potential energy associated with a system consisting of Earth and a nearby particle is **gravitational potential energy.** If the particle moves from height y_i to height y_f, the change in the gravitational potential energy of the particle–Earth system is

$$\Delta U = mg(y_f - y_i) = mg\,\Delta y. \qquad (8\text{-}7)$$

If the **reference position** of the particle is set as $y_i = 0$ and the corresponding gravitational potential energy of the system is set as $U_i = 0$, then the gravitational potential energy U when the particle is at any height y is

$$U(y) = mgy. \qquad (8\text{-}9)$$

Elastic Potential Energy **Elastic potential energy** is the energy associated with the state of compression or extension of an elastic object. For a spring that exerts a spring force $F = -kx$ when its free end has displacement x, the elastic potential energy is

$$U(x) = \tfrac{1}{2}kx^2. \qquad (8\text{-}11)$$

The reference configuration has the spring at its relaxed length, at which $x = 0$ and $U = 0$.

Mechanical Energy The **mechanical energy** E_{mec} of a system is the sum of its kinetic energy K and its potential energy U:

$$E_{mec} = K + U. \qquad (8\text{-}12)$$

An *isolated system* is one in which no *external force* causes energy changes. If only conservative forces do work within an isolated system, then the mechanical energy E_{mec} of the system cannot change. This **principle of conservation of mechanical energy** is written as

$$K_2 + U_2 = K_1 + U_1, \qquad (8\text{-}17)$$

in which the subscripts refer to different instants during an energy transfer process. This conservation principle can also be written as

$$\Delta E_{mec} = \Delta K + \Delta U = 0. \qquad (8\text{-}18)$$

Potential Energy Curves If we know the **potential energy function** $U(x)$ for a system in which a one-dimensional force F acts on a particle, we can find the force as

$$F(x) = -\frac{dU(x)}{dx}. \qquad (8\text{-}20)$$

If $U(x)$ is given on a graph, then at any value of x, the force F is the negative of the slope of the curve there and the kinetic energy of the particle is given by

$$K(x) = E_{mec} - U(x), \qquad (8\text{-}22)$$

where E_{mec} is the mechanical energy of the system. A **turning point** is a point x at which the particle reverses its motion (there, $K = 0$). The particle is in **equilibrium** at points where the slope of the $U(x)$ curve is zero (there, $F(x) = 0$).

Work Done on a System by an External Force Work W is energy transferred to or from a system by means of an external force acting on the system. When more than one force acts on a system, their *net work* is the transferred energy. When friction is not involved, the work done on the system and the change ΔE_{mec} in the mechanical energy of the system are equal:

$$W = \Delta E_{mec} = \Delta K + \Delta U. \qquad (8\text{-}24, 8\text{-}23)$$

When a kinetic frictional force acts within the system, then the thermal energy E_{th} of the system changes. (This energy is associated with the random motion of atoms and molecules in the system.) The work done on the system is then

$$W = \Delta E_{mec} + \Delta E_{th}. \qquad (8\text{-}31)$$

The change ΔE_{th} is related to the magnitude f_k of the frictional force and the magnitude d of the displacement caused by the external force by

$$\Delta E_{th} = f_k d. \qquad (8\text{-}29)$$

Conservation of Energy The total energy E of a system (the sum of its mechanical energy and its internal energies, including thermal energy) can change only by amounts of energy that are transferred to or from the system. This experimental fact is known as the **law of conservation of energy.** If work W is done on the system, then

$$W = \Delta E = \Delta E_{mec} + \Delta E_{th} + \Delta E_{int}. \tag{8-33}$$

If the system is isolated ($W = 0$), this gives

$$\Delta E_{mec} + \Delta E_{th} + \Delta E_{int} = 0 \tag{8-34}$$

and

$$E_{mec,2} = E_{mec,1} - \Delta E_{th} - \Delta E_{int}, \tag{8-35}$$

where the subscripts 1 and 2 refer to two different instants.

Power The **power** due to a force is the *rate* at which that force transfers energy. If an amount of energy ΔE is transferred by a force in an amount of time Δt, the **average power** of the force is

$$P_{avg} = \frac{\Delta E}{\Delta t}. \tag{8-36}$$

The **instantaneous power** due to a force is

$$P = \frac{dE}{dt}. \tag{8-37}$$

QUESTIONS

1. Figure 8-17 shows one direct path and four indirect paths from point i to point f. Along the direct path and three of the indirect paths, only a conservative force F_c acts on a certain object. Along the fourth indirect path, both F_c and a nonconservative force F_{nc} act on the object. The change ΔE_{mec} in the object's mechanical energy (in joules) in going from i to f is indicated along each straight line segment of the indirect paths. (a) What is ΔE_{mec} in moving from i to f along the direct path? (b) What is ΔE_{mec} due to F_{nc} along the one path where it acts?

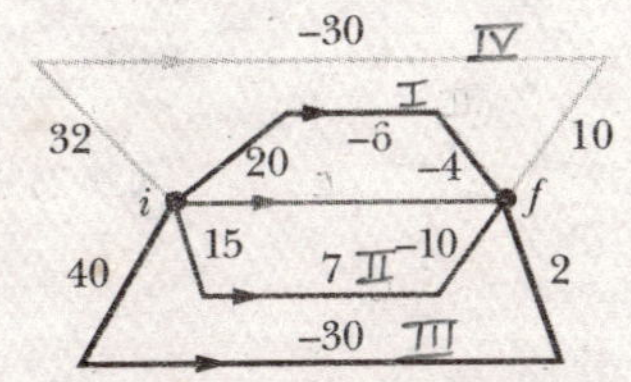

Fig. 8-17 Question 1.

2. A spring is initially stretched by 3.0 cm from its relaxed length. Here are four choices for changing the initial stretch: (a) to a stretch of 2.0 cm, (b) to a compression of 2.0 cm, (c) to a compression of 4.0 cm, and (d) to a stretch of 4.0 cm. Rank the choices according to the changes they make in the elastic potential energy of the spring, most positive first, most negative last.

3. A coconut is thrown from a cliff edge toward a wide, flat valley, with initial speed $v_0 = 8$ m/s. Rank the following choices for the launch direction according to (a) the initial kinetic energy of the coconut and (b) its kinetic energy just before hitting the valley bottom, greatest first: (1) $\vec{v}_0$ almost vertically upward, (2) $\vec{v}_0$ angled upward by 45°, (3) $\vec{v}_0$ horizontal, (4) $\vec{v}_0$ angled downward by 45°, and (5) $\vec{v}_0$ almost vertically downward.

4. In Fig. 8-18, a brave skater slides down three slopes of frictionless ice whose vertical heights d are identical. Rank the slopes according to (a) the work done on the skater by the gravitational force during the descent on each slope and (b) the change in her kinetic energy produced along the slope, greatest first.

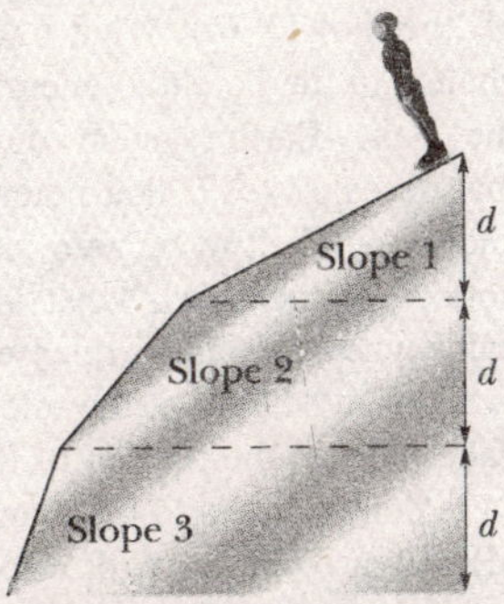

Fig. 8-18 Question 4.

5. In Fig. 8-19, a small, initially stationary block is released on a frictionless ramp at a height of 3.0 m. Hill heights along the ramp are as shown. The hills have identical circular tops (assume that the block does not fly off any hill). (a) Which hill is the first the block cannot cross? (b) What does it do after failing to cross that hill? On which hilltop is (c) the centripetal acceleration of the block greatest and (d) the normal force on the block least?

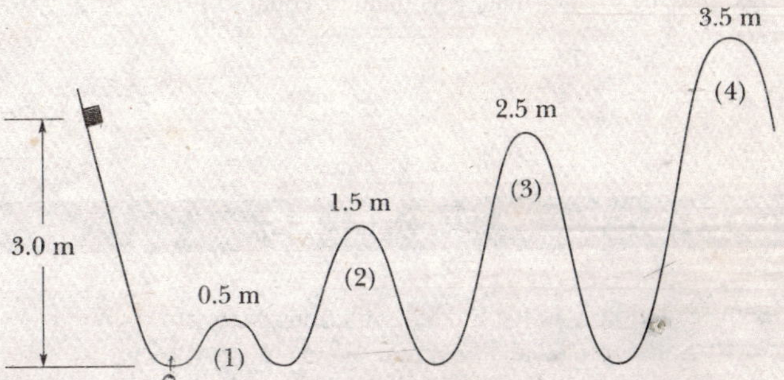

Fig. 8-19 Question 5.

6. Figure 8-20 gives the potential energy function of a particle. (a) Rank regions AB, BC, CD, and DE according to the magnitude of the force on the particle, greatest first. What value must the mechanical energy E_{mec} of the particle not exceed if the particle is to be (b) trapped in the potential well at the left, (c) trapped in the

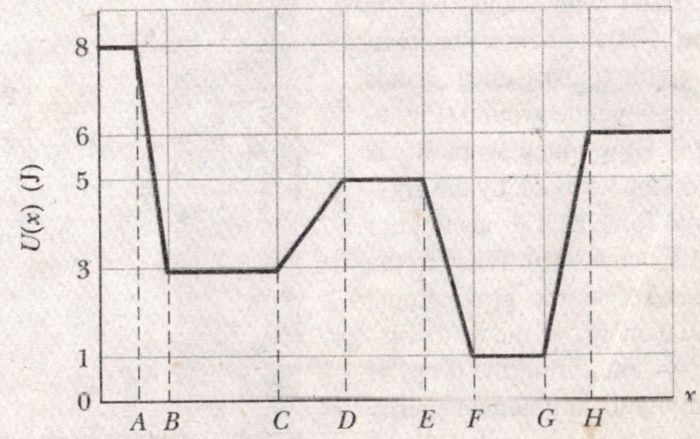

Fig. 8-20 Question 6.

potential well at the right, and (d) able to move between the two potential wells but not to the right of point H? For the situation of (d), in which of regions BC, DE, and FG will the particle have (e) the greatest kinetic energy and (f) the least speed?

7 In Fig. 8-21, a block is released from rest on a track with an initial gravitational potential energy U_i. The curved portions on the track are frictionless, but the horizontal portion, of length L, produces a frictional force f on the block. (a) How much energy is transferred to thermal energy if the block passes once through length L? How many times does the block pass through that length if the initial potential energy U_i is equal to (b) $0.50fL$, (c) $1.25fL$, and (d) $2.25fL$? (e) For each of those three situations, does the block come to a stop at the center of the horizontal portion, to the left of center, or to the right of center? (Warm-up for Problem 63)

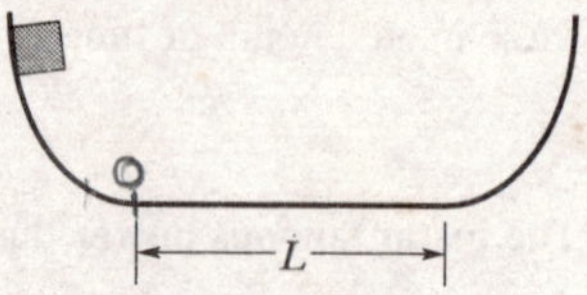

Fig. 8-21 Question 7.

8 In Fig. 8-22, a block slides along a track that descends through distance h. The track is frictionless except for the lower section. There the block slides to a stop in a certain distance D because of friction. (a) If we decrease h, will the block now slide to a stop in a distance that is greater than, less than, or equal to D? (b) If, instead, we increase the mass of the block, will the stopping distance now be greater than, less than, or equal to D?

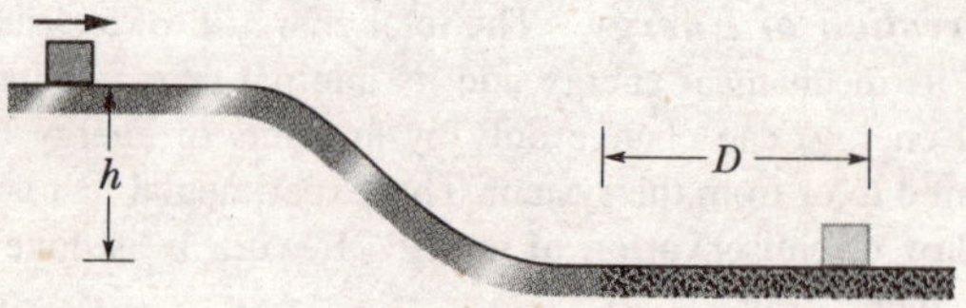

Fig. 8-22 Question 8.

9 In Fig. 8-23, a block slides from A to C along a frictionless ramp, and then it passes through horizontal region CD, where a frictional force acts on it. Is the block's kinetic energy increasing, decreasing, or constant in (a) region AB, (b) region BC, and (c) region CD? (d) Is the block's mechanical energy increasing, decreasing, or constant in those regions?

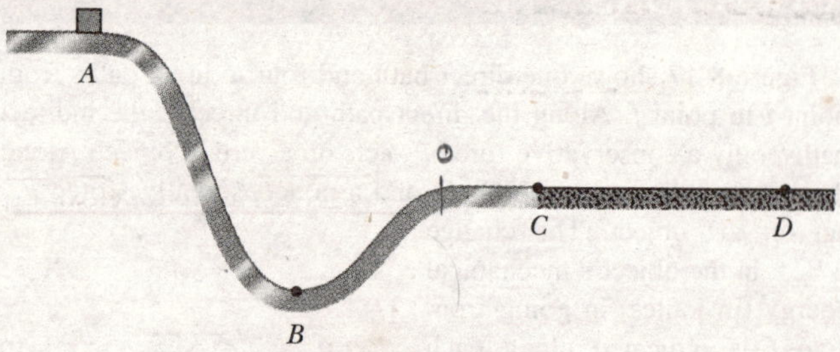

Fig. 8-23 Question 9.

EXERCISES & PROBLEMS

ssm Solution is in the Student Solutions Manual.
www Solution is available on the World Wide Web at: http://www.wiley.com/college/hrw
ilw Solution is available on the Interactive LearningWare.

SEC. 8-3 Determining Potential Energy Values

1E. What is the spring constant of a spring that stores 25 J of elastic potential energy when compressed by 7.5 cm from its relaxed length? **ssm**

2E. You drop a 2.00 kg textbook to a friend who stands on the ground 10.0 m below the textbook with outstretched hands 1.50 m above the ground (Fig. 8-24). (a) How much work W_g is done on the textbook by the gravitational force as it drops to your friend's hands? (b) What is the change ΔU in the gravitational potential energy of the textbook–Earth system during the drop? If the gravitational potential energy U of that system is taken to be zero at ground level, what is U when the textbook (c) is released and (d) reaches the hands? Now take U to be 100 J at ground level and again find (e) W_g, (f) ΔU, (g) U at the release point, and (h) U at the hands.

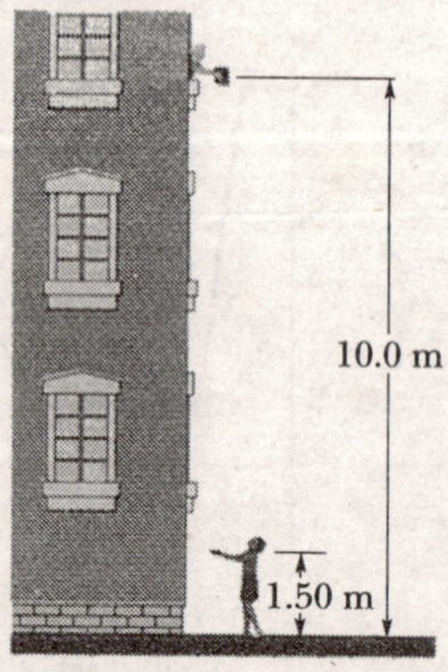

Fig. 8-24 Exercises 2 and 10.

3E. In Fig. 8-25, a 2.00 g ice flake is released from the edge of a hemispherical bowl whose radius r is 22.0 cm. The flake–bowl contact is frictionless. (a) How much work is done on the flake by the gravitational force during the flake's descent to the bottom of the bowl? (b) What is the change in the potential energy of the flake–Earth system during that descent? (c) If that potential energy is taken to be zero at the bottom of the bowl, what is its value when the flake is released? (d) If, instead, the potential energy is taken to be zero at the release point, what is its value when the flake reaches the bottom of the bowl? (e) If the mass of the flake were doubled, would the magnitudes of the answers to (a) through (d) increase, decrease, or remain the same? **ssm**

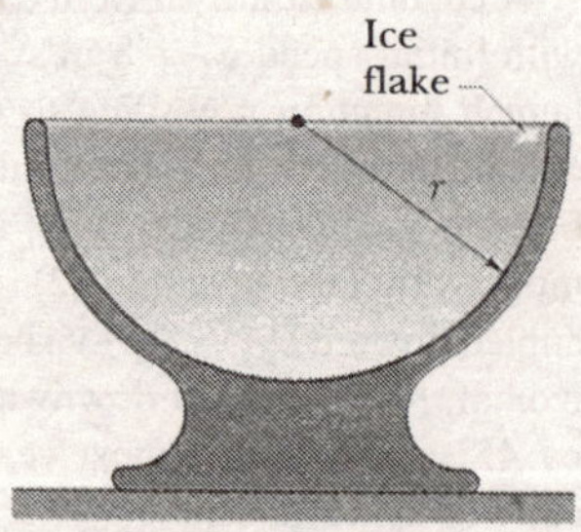

Fig. 8-25 Exercises 3 and 9.

4E. In Fig. 8-26, a frictionless roller coaster of mass m tops the first hill with speed v_0. How much work does the gravitational force do on it from that point to (a) point A, (b) point B, and (c) point C? If the gravitational potential energy of the coaster–Earth system is taken to be zero at point C, what is its value when the coaster is at (d) point B and (e) point A? (f) If mass m were doubled, would the change in the gravitational potential energy of the system between points A and B increase, decrease, or remain the same?

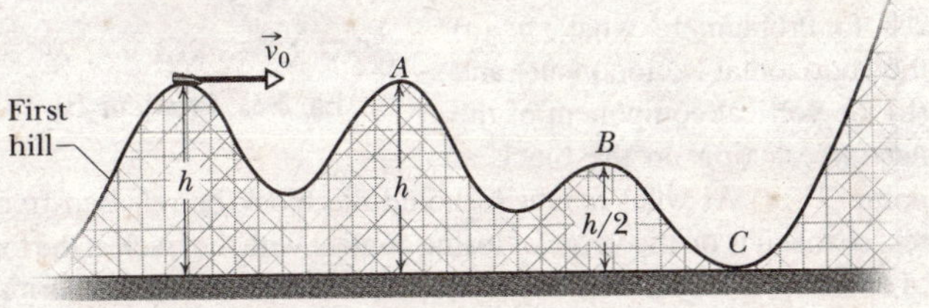

Fig. 8-26 Exercises 4 and 12.

5E. Figure 8-27 shows a ball with mass m attached to the end of a thin rod with length L and negligible mass. The other end of the rod is pivoted so that the ball can move in a vertical circle. The rod is held in the horizontal position as shown and then given enough of a downward push to cause the ball to swing down and around and just reach the vertically upward position, with zero speed there. How much work is done on the ball by the gravitational force from the initial point to (a) the lowest point, (b) the highest point, and (c) the point on the right at which the ball is level with the initial point? If the gravitational potential energy of the ball–Earth system is taken to be zero at the initial point, what is its value when the ball reaches (d) the lowest point, (e) the highest point, and (f) the point on the right that is level with the initial point? (g) Suppose the rod were pushed harder so that the ball passed through the highest point with a nonzero speed. Would the change in the gravitational potential energy from the lowest point to the highest point then be greater, less, or the same? ssm

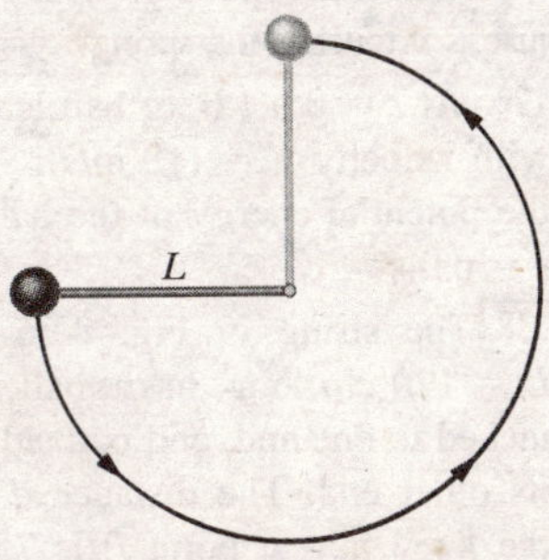

Fig. 8-27 Exercises 5 and 11.

6P. In Fig. 8-28, a small block of mass m can slide along the frictionless loop-the-loop. The block is released from rest at point P, at height $h = 5R$ above the bottom of the loop. How much work does the gravitational force do on the block as the block travels from point P to (a) point Q and (b) the top of the loop? If the gravitational potential energy of the block–Earth system is taken to be zero at the bottom of the loop, what is that potential energy when the block is (c) at point P, (d) at point Q, and (e) at the top of the loop? (f) If, instead of being released, the block is given some initial speed downward along the track, do the answers to (a) through (e) increase, decrease, or remain the same?

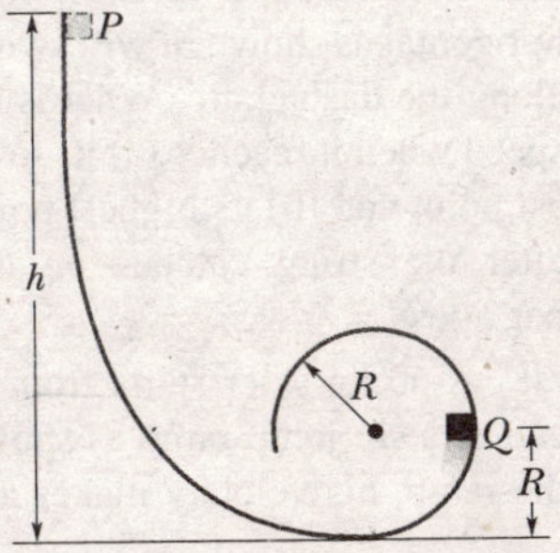

Fig. 8-28 Problems 6 and 20.

7P. A 1.50 kg snowball is fired from a cliff 12.5 m high with an initial velocity of 14.0 m/s, directed 41.0° above the horizontal. (a) How much work is done on the snowball by the gravitational force during its flight to the flat ground below the cliff? (b) What is the change in the gravitational potential energy of the snowball–Earth system during the flight? (c) If that gravitational potential energy is taken to be zero at the height of the cliff, what is its value when the snowball reaches the ground? ssm

8P. Figure 8-29 shows a thin rod, of length L and negligible mass, that can pivot about one end to rotate in a vertical circle. A heavy ball of mass m is attached to the other end. The rod is pulled aside through an angle θ and released. As the ball descends to its lowest point, (a) how much work does the gravitational force do on it and (b) what is the change in the gravitational potential energy of the ball–Earth system? (c) If the gravitational potential energy is taken to be zero at the lowest point, what is its value just as the ball is released? (d) Do the magnitudes of the answers to (a) through (c) increase, decrease, or remain the same if angle θ is increased?

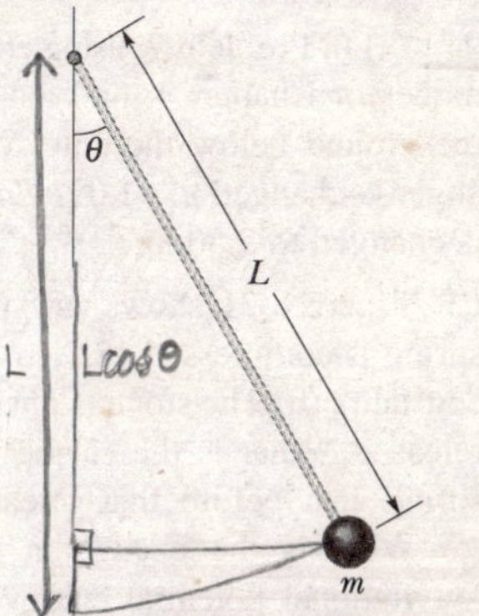

Fig. 8-29 Problems 8 and 14.

SEC. 8-4 Conservation of Mechanical Energy

9E. (a) In Exercise 3, what is the speed of the flake when it reaches the bottom of the bowl? (b) If we substituted a second flake with twice the mass, what would its speed be? (c) If, instead, we gave the flake an initial downward speed along the bowl, would the answer to (a) increase, decrease, or remain the same? ssm www

10E. (a) In Exercise 2, what is the speed of the textbook when it reaches the hands? (b) If we substituted a second textbook with twice the mass, what would its speed be? (c) If, instead, the textbook were thrown down, would the answer to (a) increase, decrease, or remain the same?

11E. (a) In Exercise 5, what initial speed must be given the ball so that it reaches the vertically upward position with zero speed? What then is its speed at (b) the lowest point and (c) the point on the right at which the ball is level with the initial point? (d) If the ball's mass were doubled, would the answers to (a) through (c) increase, decrease, or remain the same? ssm

12E. In Exercise 4, what is the speed of the coaster at (a) point A, (b) point B, and (c) point C? (d) How high will it go on the last hill, which is too high for it to cross? (e) If we substitute a second coaster with twice the mass, what then are the answers to (a) through (d)?

13E. In Fig. 8-30, a runaway truck with failed brakes is moving downgrade at 130 km/h just before the driver steers the truck up a

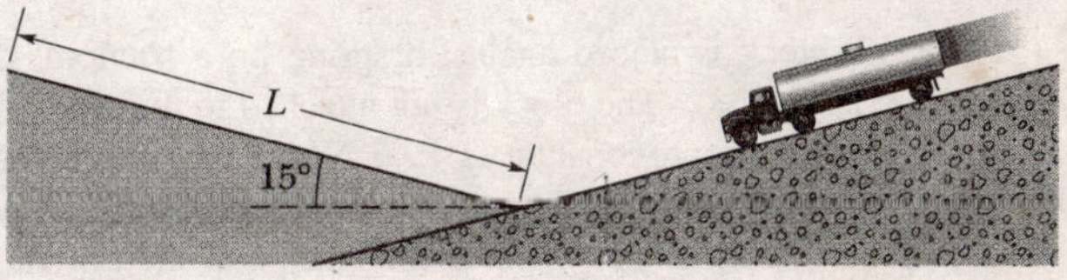

Fig. 8-30 Exercise 13.

frictionless emergency escape ramp with an inclination of 15°. The truck's mass is 5000 kg. (a) What minimum length L must the ramp have if the truck is to stop (momentarily) along it? (Assume the truck is a particle, and justify that assumption.) Does the minimum length L increase, decrease, or remain the same if (b) the truck's mass is decreased and (c) its speed is decreased? ssm

14P. (a) In Problem 8, what is the speed of the ball at the lowest point if $L = 2.00$ m, $\theta = 30.0°$, and $m = 5.00$ kg? (b) Does the speed increase, decrease, or remain the same if the mass is increased?

15P (a) In Problem 7, using energy techniques rather than the techniques of Chapter 4, find the speed of the snowball as it reaches the ground below the cliff. What is that speed (b) if the launch angle is changed to 41.0° *below* the horizontal and (c) if the mass is changed to 2.50 kg? ssm

16P. Figure 8-31 shows an 8.00 kg stone at rest on a spring. The spring is compressed 10.0 cm by the stone. (a) What is the spring constant? (b) The stone is pushed down an additional 30.0 cm and released. What is the elastic potential energy of the compressed spring just before that release? (c) What is the change in the gravitational potential energy of the stone–Earth system when the stone moves from the release point to its maximum height? (d) What is that maximum height, measured from the release point?

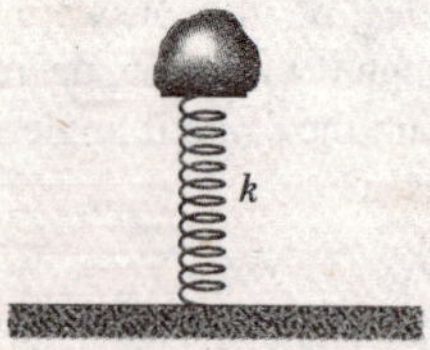

Fig. 8-31 Problem 16.

17P. A 5.0 g marble is fired vertically upward using a spring gun. The spring must be compressed 8.0 cm if the marble is to just reach a target 20 m above the marble's position on the compressed spring. (a) What is the change ΔU_g in the gravitational potential energy of the marble–Earth system during the 20 m ascent? (b) What is the change ΔU_s in the elastic potential energy of the spring during its launch of the marble? (c) What is the spring constant of the spring? ssm www

18P. Figure 8-32 shows a pendulum of length L. Its bob (which effectively has all the mass) has speed v_0 when the cord makes an angle θ_0 with the vertical. (a) Derive an expression for the speed of the bob when it is in its lowest position. What is the least value that v_0 can have if the pendulum is to swing down and then up (b) to a horizontal position, and (c) to a vertical position with the cord remaining straight? (d) Do the answers to (b) and (c) increase, decrease, or remain the same if θ_0 is increased by a few degrees?

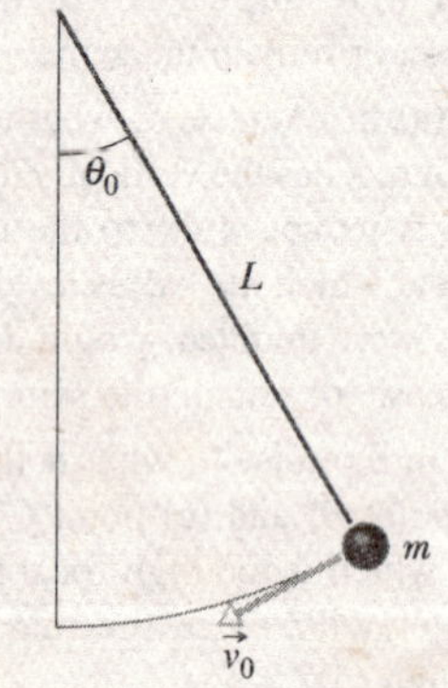

Fig. 8-32 Problem 18.

19P. A 2.00 kg block is placed against a spring on a frictionless 30.0° incline (Fig. 8-33). (The block is not attached to the spring.) The spring, whose spring constant is 19.6 N/cm, is compressed 20.0 cm and then released. (a) What is the elastic potential energy of the compressed spring? (b) What is the change in the gravitational potential energy of the block–Earth system as the block moves from the release point to its highest point on the incline? (c) How far along the incline is the highest point from the release point? ilw

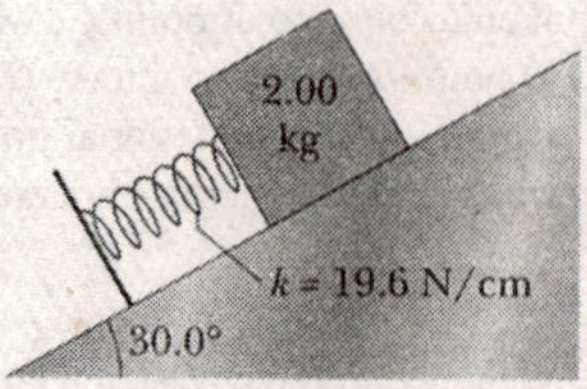

Fig. 8-33 Problem 19.

20P. In Problem 6, what are (a) the horizontal component and (b) the vertical component of the *net* force acting on the block at point Q? (c) At what height h should the block be released from rest so that it is on the verge of losing contact with the track at the top of the loop? (*On the verge of losing contact* means that the normal force on the block from the track has just then become zero.) (d) Graph the magnitude of the normal force on the block at the top of the loop versus initial height h, for the range $h = 0$ to $h = 6R$.

21P. In Fig. 8-34, a 12 kg block is released from rest on a 30° frictionless incline. Below the block is a spring that can be compressed 2.0 cm by a force of 270 N. The block momentarily stops when it compresses the spring by 5.5 cm. (a) How far does the block move down the incline from its rest position to this stopping point? (b) What is the speed of the block just as it touches the spring? ssm

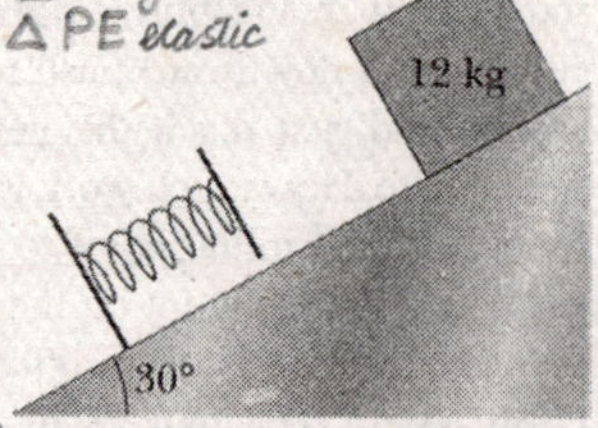

Fig. 8-34 Problem 21.

22P. At $t = 0$ a 1.0 kg ball is thrown from the top of a tall tower with velocity $\vec{v} = (18 \text{ m/s})\hat{i} + (24 \text{ m/s})\hat{j}$. What is the change in the potential energy of the ball–Earth system between $t = 0$ and $t = 6.0$ s?

23P. The string in Fig. 8-35 is $L = 120$ cm long, has a ball attached to one end, and is fixed at its other end. The distance d to the fixed peg at point P is 75.0 cm. When the initially stationary ball is released with the string horizontal as shown, it will swing along the dashed arc. What is its speed when it reaches (a) its lowest point and (b) its highest point after the string catches on the peg? ilw

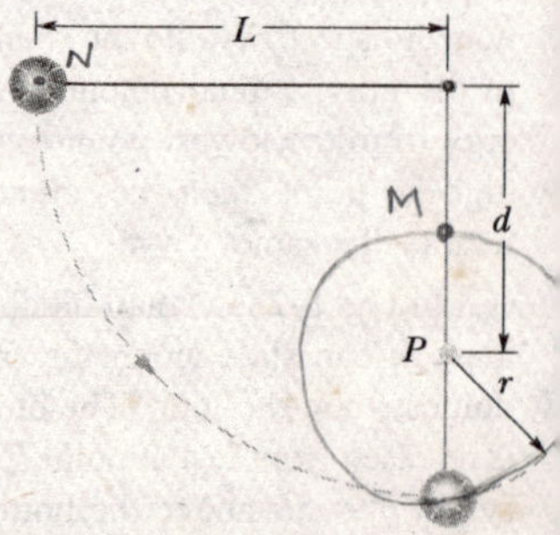

Fig. 8-35 Problems 23 and 29.

24P. A 60 kg skier starts from rest at a height of 20 m above the end of a ski-jump ramp as shown in Fig. 8-36. As the skier leaves the ramp, his velocity makes an angle of 28° with the horizontal.

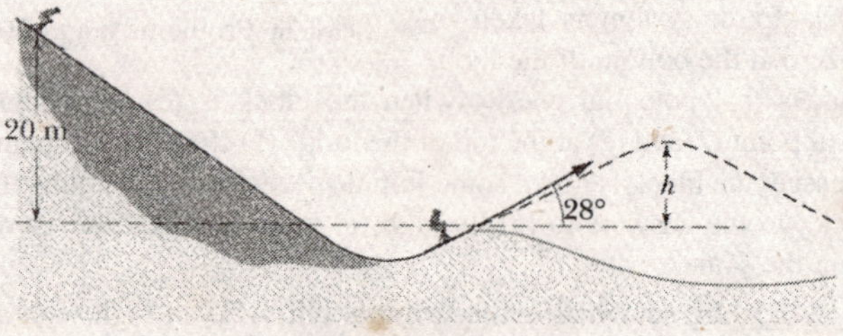

Fig. 8-36 Problem 24.

Neglect the effects of air resistance and assume the ramp is frictionless. (a) What is the maximum height h of his jump above the end of the ramp? (b) If he increased his weight by putting on a backpack, would h then be greater, less, or the same?

25P. A 2.0 kg block is dropped from a height of 40 cm onto a spring of spring constant k = 1960 N/m (Fig. 8-37). Find the maximum distance the spring is compressed. ssm

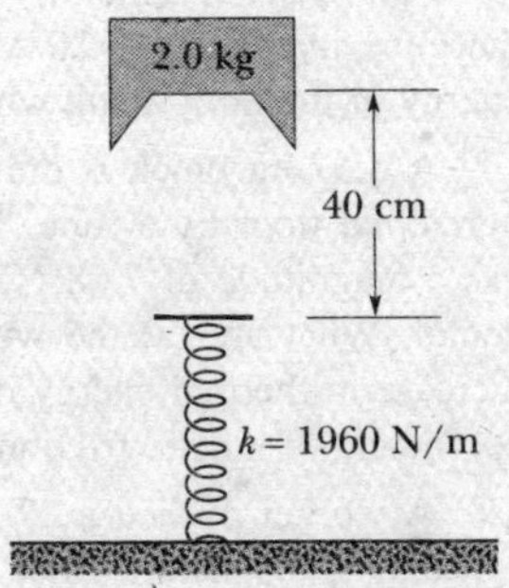

Fig. 8-37 Problem 25.

26P. Tarzan, who weighs 688 N, swings from a cliff at the end of a convenient vine that is 18 m long (Fig. 8-38). From the top of the cliff to the bottom of the swing, he descends by 3.2 m. The vine will break if the force on it exceeds 950 N. (a) Does the vine break? (b) If no, what is the greatest force on it during the swing? If yes, at what angle with the vertical does it break?

Fig. 8-38 Problem 26.

27P. Two children are playing a game in which they try to hit a small box on the floor with a marble fired from a spring-loaded gun that is mounted on a table. The target box is 2.20 m horizontally from the edge of the table; see Fig. 8-39. Bobby compresses the spring 1.10 cm, but the center of the marble falls 27.0 cm short of the center of the box. How far should Rhoda compress the spring to score a direct hit? Assume that neither the spring nor the ball encounters friction in the gun. ssm

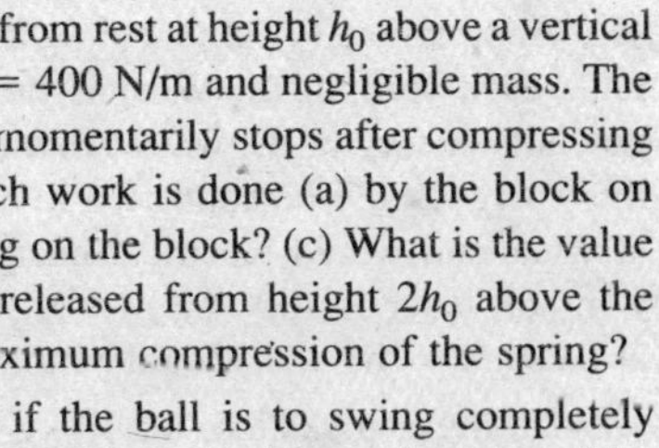

Fig. 8-39 Problem 27

28P. A 700 g block is released from rest at height h_0 above a vertical spring with spring constant k = 400 N/m and negligible mass. The block sticks to the spring and momentarily stops after compressing the spring 19.0 cm. How much work is done (a) by the block on the spring and (b) by the spring on the block? (c) What is the value of h_0? (d) If the block were released from height $2h_0$ above the spring, what would be the maximum compression of the spring?

29P. In Fig. 8-35 show that, if the ball is to swing completely around the fixed peg, then $d > 3L/5$. (*Hint:* The ball must still be moving at the top of its swing. Do you see why?) ssm www

30P. To make a pendulum, a 300 g ball is attached to one end of a string that has a length of 1.4 m and negligible mass. (The other end of the string is fixed.) The ball is pulled to one side until the string makes an angle of 30.0° with the vertical; then (with the string taut) the ball is released from rest. Find (a) the speed of the ball when the string makes an angle of 20.0° with the vertical and (b) the maximum speed of the ball. (c) What is the angle between the string and the vertical when the speed of the ball is one-third its maximum value?

31P. A rigid rod of length L and negligible mass has a ball with mass m attached to one end and its other end fixed, to form a pendulum. The pendulum is inverted, with the rod straight up, and then released. At the lowest point, what are (a) the ball's speed and (b) the tension in the rod? (c) The pendulum is next released at rest from a horizontal position. At what angle from the vertical does the tension in the rod equal the weight of the ball? ssm

32P. In Fig. 8-40, a spring with spring constant k = 170 N/m is at the top of a 37.0° frictionless incline. The lower end of the incline is 1.00 m from the end of the spring, which is at its relaxed length. A 2.00 kg canister is pushed against the spring until the spring is compressed 0.200 m and released from rest. (a) What is the speed of the canister at the instant the spring returns to its relaxed length (which is when the canister loses contact with the spring)? (b) What is the speed of the canister when it reaches the lower end of the incline?

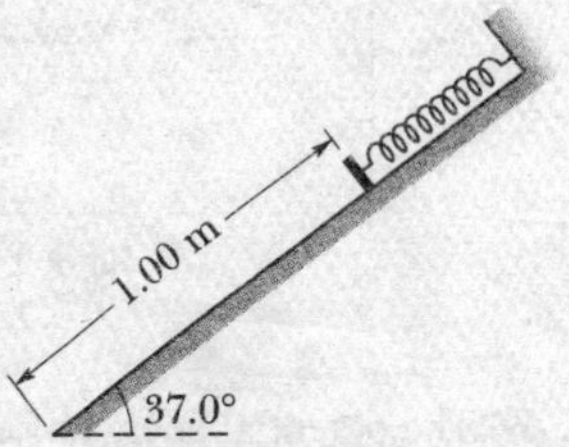

Fig. 8-40 Problem 32.

33P. In Fig. 8-41, a chain is held on a frictionless table with one-fourth of its length hanging over the edge. If the chain has length L and mass m, how much work is required to pull the hanging part back onto the table? ssm

Fig. 8-41 Problem 33.

34P. A spring with spring constant k = 400 N/m is placed in a vertical orientation with its lower end supported by a horizontal surface. The upper end is depressed 25.0 cm, and a block with a weight of 40.0 N is placed (unattached) on the depressed spring. The system is then released from rest. Assume the gravitational potential energy U_g of the block is zero at the release point ($y = 0$) and calculate the gravitational potential energy, the elastic potential energy U_e, and the kinetic energy K of the block for y equal to (a) 0, (b) 5.00 cm, (c) 10.0 cm, (d) 15.0 cm, (e) 20.0 cm, (f) 25.0 cm, and (g) 30.0 cm. Also, (h) how far above its point of release does the block rise?

35P* A boy is seated on the top of a hemispherical mound of ice (Fig. 8-42). He is given a very small push and starts sliding down the ice. Show that he leaves the ice at a point whose height is $2R/3$ if the ice is frictionless. (*Hint:* The normal force vanishes as he leaves the ice.) ssm

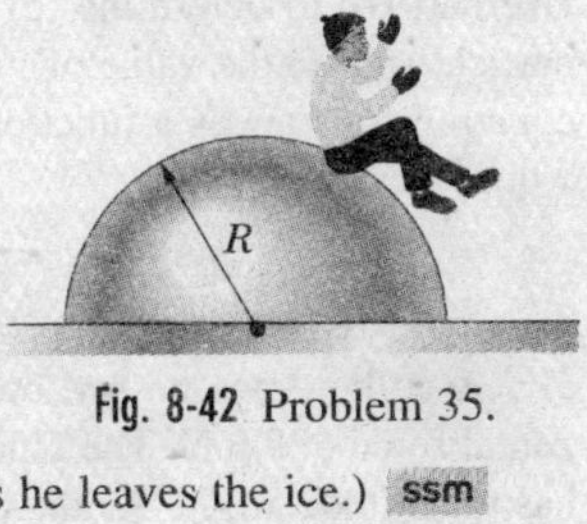

Fig. 8-42 Problem 35.

SEC. 8-5 Reading a Potential Energy Curve

36E. A conservative force $F(x)$ acts on a 2.0 kg particle that moves along the x axis. The potential energy $U(x)$ associated with $F(x)$ is graphed in Fig. 8-43. When the particle is at $x = 2.0$ m, its velocity is -1.5 m/s. (a) What are the magnitude and direction of $F(x)$ at this position? (b) Between what limits of x does the particle move? (c) What is its speed at $x = 7.0$ m?

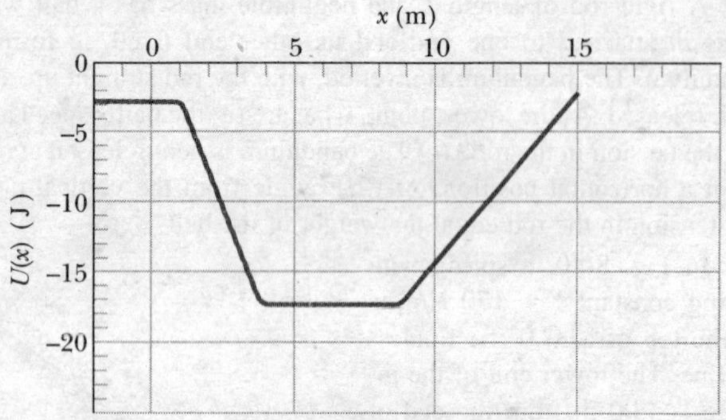

Fig. 8-43 Exercise 36.

37P. The potential energy of a diatomic molecule (a two-atom system like H_2 or O_2) is given by

$$U = \frac{A}{r^{12}} - \frac{B}{r^6},$$

where r is the separation of the two atoms of the molecule and A and B are positive constants. This potential energy is associated with the force that binds the two atoms together. (a) Find the *equilibrium separation,* that is, the distance between the atoms at which the force on each atom is zero. Is the force repulsive (the atoms are pushed apart) or attractive (they are pulled together) if their separation is (b) smaller and (c) larger than the equilibrium separation? ssm

38P. A single conservative force $F(x)$ acts on a 1.0 kg particle that moves along an x axis. The potential energy $U(x)$ associated with $F(x)$ is given by

$$U(x) = -4x\, e^{-x/4}\ \text{J},$$

where x is in meters. At $x = 5.0$ m the particle has a kinetic energy of 2.0 J. (a) What is the mechanical energy of the system? (b) Make a plot of $U(x)$ as a function of x for $0 \leq x \leq 10$ m, and on the same graph draw the line that represents the mechanical energy of the system. Use part (b) to determine (c) the least value of x and (d) the greatest value of x between which the particle can move. Use part (b) to determine (e) the maximum kinetic energy of the particle and (f) the value of x at which it occurs. (g) Determine the equation for $F(x)$ as a function of x. (h) For what (finite) value of x does $F(x) = 0$?

SEC. 8-6 Work Done on a System by an External Force

39E. A collie drags its bed box across a floor by applying a horizontal force of 8.0 N. The kinetic frictional force acting on the box has magnitude 5.0 N. As the box is dragged through 0.70 m along the way, what are (a) the work done by the collie's applied force and (b) the increase in thermal energy of the bed and floor?

40E. The temperature of a plastic cube is monitored while the cube is pushed 3.0 m across a floor at constant speed by a horizontal force of 15 N. The monitoring reveals that the thermal energy of the cube increases by 20 J. What is the increase in the thermal energy of the floor along which the cube slides?

41P. A 3.57 kg block is drawn at constant speed 4.06 m along a horizontal floor by a rope. The force on the block from the rope has a magnitude of 7.68 N and is directed 15.0° above the horizontal. What are (a) the work done by the rope's force, (b) the increase in thermal energy of the block–floor system, and (c) the coefficient of kinetic friction between the block and floor? ssm

42P. A worker pushed a 27 kg block 9.2 m along a level floor at constant speed with a force directed 32° below the horizontal. If the coefficient of kinetic friction between block and floor was 0.20, what were (a) the work done by the worker's force and (b) the increase in thermal energy of the block–floor system?

SEC. 8-7 Conservation of Energy

43E. A 25 kg bear slides, from rest, 12 m down a lodgepole pine tree, moving with a speed of 5.6 m/s just before hitting the ground. (a) What change occurs in the gravitational potential energy of the bear–Earth system during the slide? (b) What is the kinetic energy of the bear just before hitting the ground? (c) What is the average frictional force that acts on the sliding bear? ssm ilw

44E. A 30 g bullet, with a horizontal velocity of 500 m/s, comes to a stop 12 cm within a solid wall. (a) What is the change in its mechanical energy? (b) What is the magnitude of the average force from the wall stopping it?

45E. A 60 kg skier leaves the end of a ski-jump ramp with a velocity of 24 m/s directed 25° above the horizontal. Suppose that as a result of air drag the skier returns to the ground with a speed of 22 m/s, landing 14 m vertically below the end of the ramp. From the launch to the return to the ground, by how much is the mechanical energy of the skier–Earth system reduced because of air drag?

46E. A 75 g Frisbee is thrown from a point 1.1 m above the ground with a speed of 12 m/s. When it has reached a height of 2.1 m, its speed is 10.5 m/s. What was the reduction in the mechanical energy of the Frisbee–Earth system because of air drag?

47E. An outfielder throws a baseball with an initial speed of 81.8 mi/h. Just before an infielder catches the ball at the same level, the ball's speed is 110 ft/s. In foot-pounds, by how much is the mechanical energy of the ball–Earth system reduced because of air drag? (The weight of a baseball is 9.0 oz.)

48E. Approximately 5.5×10^6 kg of water fall 50 m over Niagara Falls each second. (a) What is the decrease in the gravitational potential energy of the water–Earth system each second? (b) If all this energy could be converted to electrical energy (it cannot be), at what rate would electrical energy be supplied? (The mass of 1 m^3 of water is 1000 kg.) (c) If the electrical energy were sold at 1 cent/kW · h, what would be the yearly cost?

49E. During a rockslide, a 520 kg rock slides from rest down a hillside that is 500 m long and 300 m high. The coefficient of kinetic friction between the rock and the hill surface is 0.25. (a) If the gravitational potential energy U of the rock–Earth system is zero at the bottom of the hill, what is the value of U just before the slide? (b) How much energy is transferred to thermal energy during the slide? (c) What is the kinetic energy of the rock as it reaches the bottom of the hill? (d) What is its speed then?

50P. You push a 2.0 kg block against a horizontal spring, compressing the spring by 15 cm. Then you release the block, and the spring sends it sliding across a tabletop. It stops 75 cm from where you released it. The spring constant is 200 N/m. What is the coefficient of kinetic friction between the block and the table?

51P. As Fig. 8-44 shows, a 3.5 kg block is accelerated by a compressed spring whose spring constant is 640 N/m. After leaving the spring at the spring's relaxed length, the block travels over a horizontal surface, with a coefficient of kinetic friction of 0.25, for a distance of 7.8 m before stopping. (a) What is the increase in the thermal energy of the block–floor system? (b) What is the maximum kinetic energy of the block? (c) Through what distance is the spring compressed before the block begins to move? ssm www

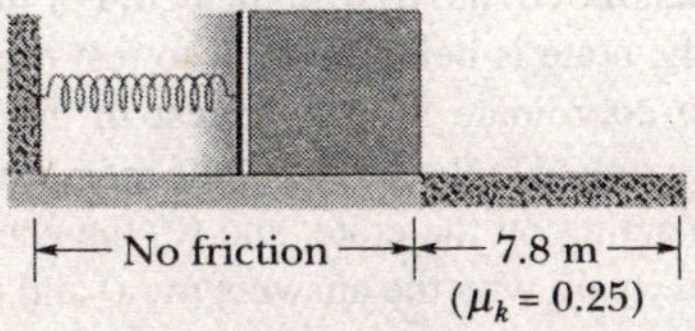

Fig. 8-44 Problem 51.

52P. In Fig. 8-45, a block is moved down an incline a distance of 5.0 m from point A to point B by a force $\vec{F}$ that is parallel to the incline and has magnitude 2.0 N. The magnitude of the frictional force acting on the block is 10 N. If the kinetic energy of the block increases by 35 J between A and B, how much work is done on the block by the gravitational force as the block moves from A to B?

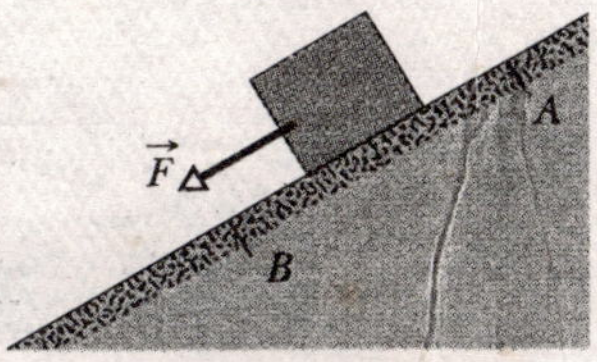

Fig. 8-45 Problem 52.

53P. A certain spring is found *not* to conform to Hooke's law. The force (in newtons) it exerts when stretched a distance x (in meters) is found to have magnitude $52.8x + 38.4x^2$ in the direction opposing the stretch. (a) Compute the work required to stretch the spring from $x = 0.500$ m to $x = 1.00$ m. (b) With one end of the spring fixed, a particle of mass 2.17 kg is attached to the other end of the spring when it is extended by an amount $x = 1.00$ m. If the particle is then released from rest, what is its speed at the instant the spring has returned to the configuration in which the extension is $x = 0.500$ m? (c) Is the force exerted by the spring conservative or nonconservative? Explain. ssm

54P. A 4.0 kg bundle starts up a 30° incline with 128 J of kinetic energy. How far will it slide up the incline if the coefficient of kinetic friction between bundle and incline is 0.30?

55P. Two snowy peaks are 850 m and 750 m above the valley between them. A ski run extends down from the top of the higher peak and then back up to the top of the lower one, with a total length of 3.2 km and an average slope of 30° (Fig. 8-46). (a) A skier starts from rest at the top of the higher peak. At what speed will he arrive at the top of the lower peak if he coasts without using ski poles? Ignore friction. (b) Approximately what coefficient of kinetic friction between snow and skis would make him stop just at the top of the lower peak? ssm

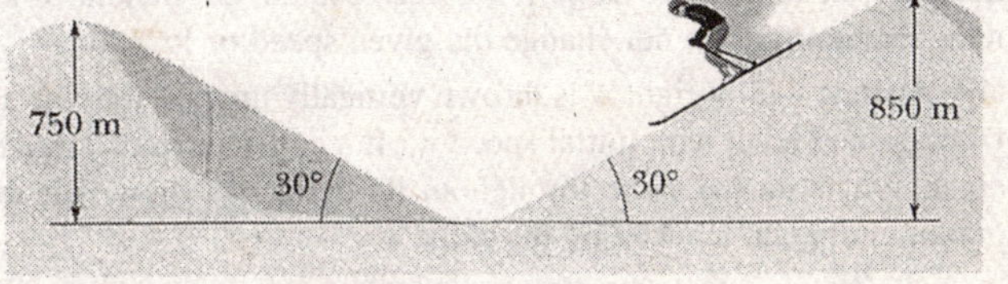

Fig. 8-46 Problem 55.

56P. A girl whose weight is 267 N slides down a 6.1 m playground slide that makes an angle of 20° with the horizontal. The coefficient of kinetic friction between slide and child is 0.10. (a) How much energy is transferred to thermal energy? (b) If the girl starts at the top with a speed of 0.457 m/s, what is her speed at the bottom?

57P. In Fig. 8-47, a 2.5 kg block slides head on into a spring with a spring constant of 320 N/m. When the block stops, it has compressed the spring by 7.5 cm. The coefficient of kinetic friction between the block and the horizontal surface is 0.25. While the block is in contact with the spring and being brought to rest, what are (a) the work done by the spring force and (b) the increase in thermal energy of the block–floor system? (c) What is the block's speed just as the block reaches the spring? ilw

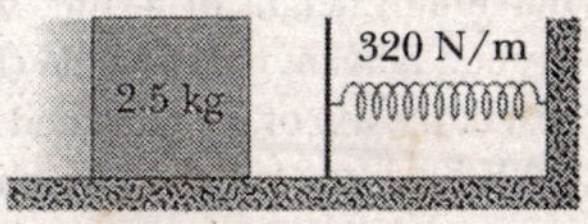

Fig. 8-47 Problem 57.

58P. A factory worker accidentally releases a 180 kg crate that was being held at rest at the top of a 3.7 m-long-ramp inclined at 39° to the horizontal. The coefficient of kinetic friction between the crate and the ramp, and between the crate and the horizontal factory floor, is 0.28. (a) How fast is the crate moving as it reaches the bottom of the ramp? (b) How far will it subsequently slide across the factory floor? (Assume that the crate's kinetic energy does not change as it moves from the ramp onto the floor.) (c) Do the answers to (a) and (b) increase, decrease, or remain the same if we halve the mass of the crate?

59P. In Fig. 8-48, a block slides along a track from one level to a higher level, by moving through an intermediate valley. The track is frictionless until the block reaches the higher level. There a frictional force stops the block in a distance d. The block's initial speed v_0 is 6.0 m/s; the height difference h is 1.1 m; and the coefficient of kinetic friction μ is 0.60. Find d.

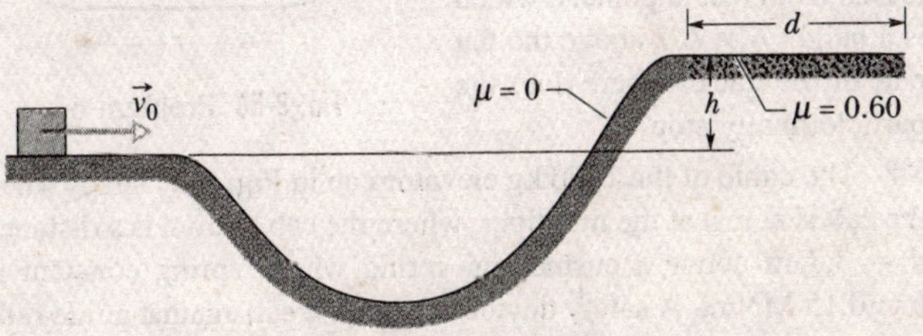

Fig. 8-48 Problem 59.

60P. A cookie jar is moving up a 40° incline. At a point 55 cm from the bottom of the incline (measured along the incline), it has a speed of 1.4 m/s. The coefficient of kinetic friction between jar and incline is 0.15. (a) How much farther up the incline will the jar move? (b) How fast will it be going when it has slid back to the bottom of the incline? (c) Do the answers to (a) and (b) increase, decrease, or remain the same if we decrease the coefficient of kinetic friction (but do not change the given speed or location)?

61P. A stone with weight w is thrown vertically upward into the air from ground level with initial speed v_0. If a constant force f due to air drag acts on the stone throughout its flight, (a) show that the maximum height reached by the stone is

$$h = \frac{v_0^2}{2g(1 + f/w)}.$$

(b) Show that the stone's speed is

$$v = v_0\left(\frac{w - f}{w + f}\right)^{1/2}$$

just before impact with the ground. ssm

62P. A playground slide is in the form of an arc of a circle with a maximum height of 4.0 m, with a radius of 12 m, and with the ground tangent to the circle (Fig. 8-49). A 25 kg child starts from rest at the top of the slide and has a speed of 6.2 m/s at the bottom. (a) What is the length of the slide? (b) What average frictional force acts on the child over this distance? If, instead of the ground, a vertical line through the *top of the slide* is tangent to the circle, what are (c) the length of the slide and (d) the average frictional force on the child?

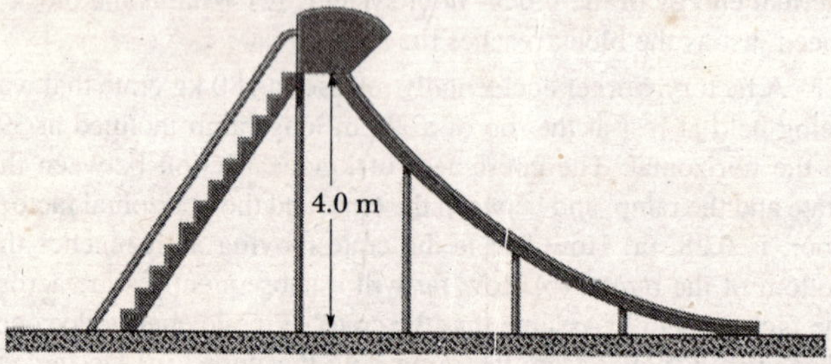

Fig. 8-49 Problem 62.

63P. A particle can slide along a track with elevated ends and a flat central part, as shown in Fig. 8-50. The flat part has length L. The curved portions of the track are frictionless, but for the flat part the coefficient of kinetic friction is $\mu_k = 0.20$. The particle is released from rest at point A, which is a height $h = L/2$ above the flat part of the track. Where does the particle finally stop?

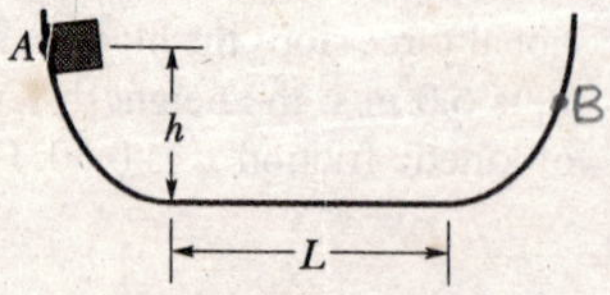

Fig. 8-50 Problem 63.

64P. The cable of the 1800 kg elevator cab in Fig. 8-51 snaps when the cab is at rest at the first floor, where the cab bottom is a distance $d = 3.7$ m above a cushioning spring whose spring constant is $k = 0.15$ MN/m. A safety device clamps the cab against guide rails so that a constant frictional force of 4.4 kN opposes the cab's motion. (a) Find the speed of the cab just before it hits the spring. (b) Find the maximum distance x that the spring is compressed (the frictional force still acts during this compression). (c) Find the distance that the cab will bounce back up the shaft. (d) Using conservation of energy, find the approximate total distance that the cab will move before coming to rest. (Assume that the frictional force on the cab is negligible when the cab is stationary.)

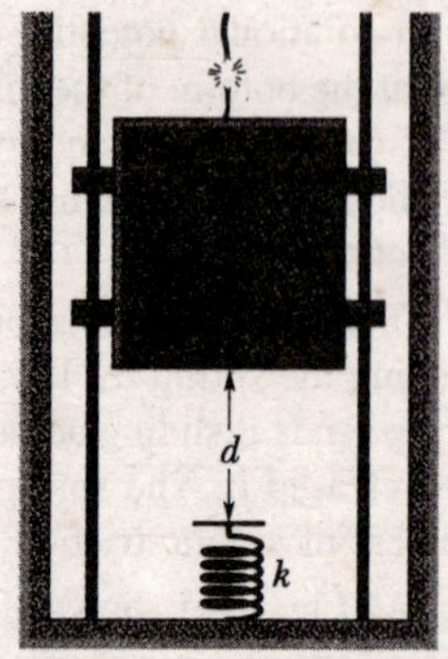

Fig. 8-51 Problem 64.

65P. At a certain factory, 300 kg crates are dropped vertically from a packing machine onto a conveyor belt moving at 1.20 m/s (Fig. 8-52). (A motor maintains the belt's constant speed.) The coefficient of kinetic friction between the belt and each crate is 0.400. After a short time, slipping between the belt and the crate ceases, and the crate then moves along with the belt. For the period of time during which the crate is being brought to rest relative to the belt, calculate, for a coordinate system at rest in the factory, (a) the kinetic energy supplied to the crate, (b) the magnitude of the kinetic frictional force acting on the crate, and (c) the energy supplied by the motor. (d) Explain why the answers to (a) and (c) are different.

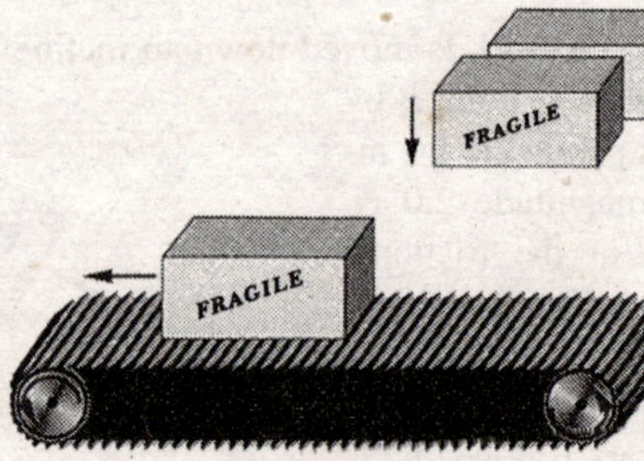

Fig. 8-52 Problem 65.

Additional Problems

66. The maximum force that you can exert on an object with one of your back teeth is about 750 N. Suppose that as you gradually bite on a clump of elastic licorice, the licorice resists its compression by one of the teeth with a spring-like force of spring constant 2.5×10^5 N/m. Find (a) the distance the licorice is compressed by your tooth and (b) the work your tooth does on the licorice during the compression. (c) Plot the magnitude of your force versus the compression distance. (d) If there is a potential energy associated with this compression, plot it versus compression distance.

In the 1990s the pelvis of a particular *Triceratops* dinosaur was found to have deep bite marks. The shape of the marks suggested that they were made by a *Tyrannosaurus rex* dinosaur. To test the idea, researchers made a replica of a *T. rex* tooth from bronze and aluminum and then used a hydraulic press to gradually drive the replica into cow bone to the depth seen in the *Triceratops* bone. A graph of the force required for the penetration versus the depth of penetration is given in Fig. 8-53 for one trial; the required force increased with depth because, as the nearly conical tooth penetrated the bone, more of the tooth came in contact with the bone. (e) How much work was done by the hydraulic press—and thus, presumably, by the *T. rex*—in such a penetration? (f) Is there a potential energy associated with this penetration? The large biting force and energy expenditure attributed to the *T. rex* by this research (and unsurpassed by modern animals) suggest that the animal was a predator and not a scavenger (as had been argued by some researchers).

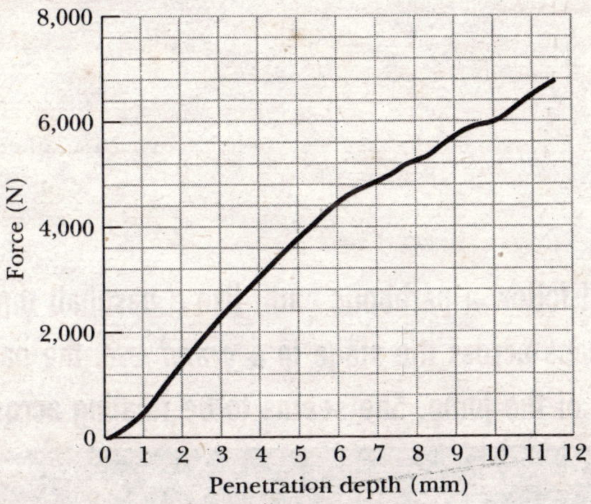

Fig. 8-53 Problem 66.

67. *Fly-Fishing and Speed Amplification.* If you throw a loose fishing fly, it will travel horizontally only about 1 m. However, if you throw that fly attached to fishing line by casting the line with a rod, the fly will easily travel horizontally to the full length of the line, say, 20 m.

The cast is depicted in Fig. 8-54: Initially (Fig. 8-54*a*) the line of length L is extended horizontally leftward and moving rightward with a speed v_0. As the fly at the end of the line moves forward, the line doubles over, with the upper section still moving and the lower section stationary (Fig. 8-54*b*). The upper section decreases in length as the lower section increases in length (Fig. 8-54*c*), until the line is extended horizontally rightward and there is only a lower section (Fig. 8-54*d*). If air drag is neglected, the initial kinetic energy of the line in Fig. 8-54*a* becomes progressively concentrated in the fly and the decreasing portion of the line that is still moving, resulting in an amplification (increase) in the speed of the fly and that portion.

(a) Using the x axis indicated, show that when the fly position is x, the length of the still-moving (upper) section of line is $(L - x)/2$. (b) Assuming that the line is uniform with a linear density ρ (mass per unit length), what is the mass of the still-moving section? Next, let m_f represent the mass of the fly, and assume that the kinetic energy of the moving section does not change from its initial value (when the moving section had length L and speed v_0) even though the length of the moving section is decreasing during the cast. (c) Find an expression for the speed of the still-moving section and the fly.

Assume that initial speed $v_0 = 6.0$ m/s, line length $L = 20$ m, fly mass $m_f = 0.80$ g, and linear density $\rho = 1.3$ g/m. (d) Plot the fly's speed v versus its position x. (e) What is the fly's speed just as the line approaches its final horizontal orientation and the fly is about to flip over and stop? (In more realistic calculations, air drag reduces this final speed.)

Speed amplification can also be produced with a bullwhip and even a rolled-up, wet towel that is popped against a victim in a common locker-room prank. (Adapted from "The Mechanics of Flycasting: The Flyline," by Graig A. Spolek, *American Journal of Physics*, Sept. 1986, Vol. 54, No. 9, pp. 832–836.)

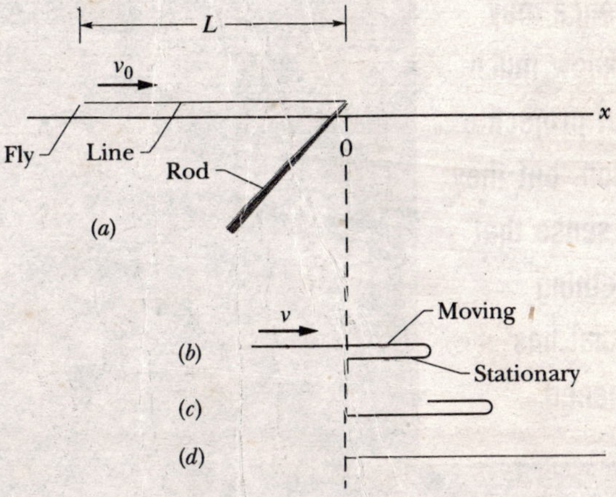

Fig. 8-54 Problem 67.

9 Systems of Particles

If you leap forward, chances are that your head and torso will follow a parabolic path, like a baseball thrown in from the outfield. However, when a skilled ballet dancer leaps across the stage in a *grand jeté*, the path taken by her head and torso is nearly horizontal during much of the jump. She seems to be floating across the stage. The audience may not know much about projectile motion, but they still sense that something unusual has happened.

How does the ballerina seemingly "turn off" the gravitational force?

The answer is in this chapter.

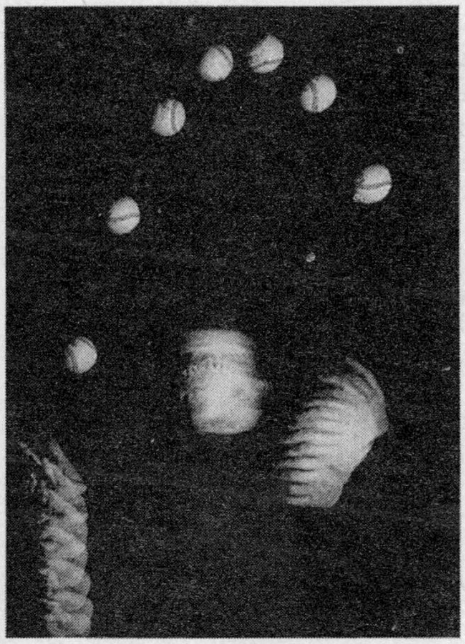

(a)

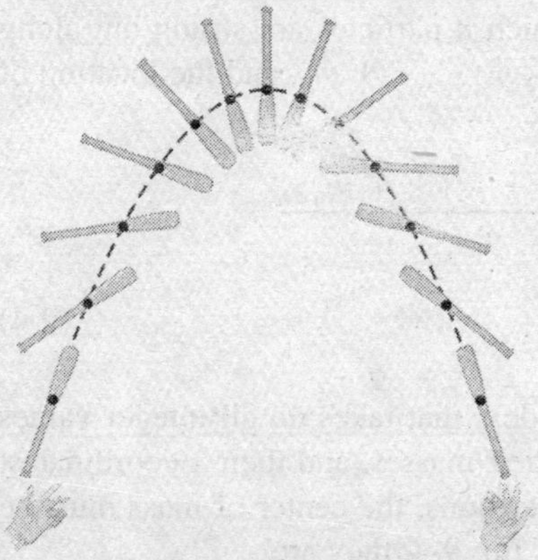

(b)

Fig. 9-1 (a) A ball tossed into the air follows a parabolic path. (b) The center of mass (the black dot) of a baseball bat that is flipped into the air does also, but all other points of the bat follow more complicated curved paths.

9-1 A Special Point

Physicists love to look at something complicated and find in it something simple and familiar. Here is an example. If you flip a baseball bat into the air, its motion as it turns is clearly more complicated than that of, say, a nonspinning tossed ball (Fig. 9-1*a*), which moves like a particle. Every part of the bat moves in a different way from every other part, so you cannot represent the bat as a tossed particle; instead, it is a system of particles.

However, if you look closely, you will find that one special point of the bat moves in a simple parabolic path, just as a particle would if tossed into the air (Fig. 9-1*b*). In fact, that special point moves as though (1) the bat's total mass were concentrated there and (2) the gravitational force on the bat acted only there. That special point is said to be the **center of mass** of the bat. In general:

> The center of mass of a body or a system of bodies is the point that moves as though all of the mass were concentrated there and all external forces were applied there.

The center of mass of a baseball bat lies along the bat's central axis. You can locate it by balancing the bat horizontally on an outstretched finger: The center of mass is on the bat's axis just above your finger.

9-2 The Center of Mass

We shall now spend some time determining how to find the center of mass in various systems. We start with a system of a few particles, and then we consider a system of a great many particles (as in a baseball bat).

Systems of Particles

Figure 9-2*a* shows two particles of masses m_1 and m_2 separated by a distance d. We have arbitrarily chosen the origin of the x axis to coincide with the particle of mass m_1. We *define* the position of the center of mass (com) of this two-particle system to be

$$x_{com} = \frac{m_2}{m_1 + m_2} d. \tag{9-1}$$

Suppose, as an example, that $m_2 = 0$. Then there is only one particle, of mass m_1, and the center of mass must lie at the position of that particle; Eq. 9-1 dutifully

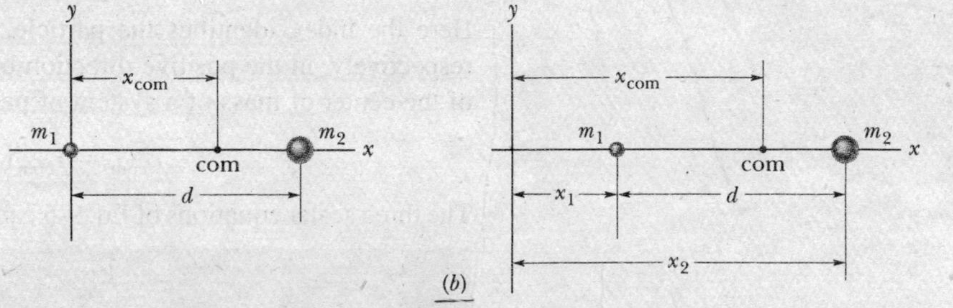

Fig. 9-2 (*a*) Two particles of masses m_1 and m_2 are separated by a distance d. The dot labeled com shows the position of the center of mass, calculated from Eq. 9-1. (*b*) The same as (*a*) except that the origin is located farther from the particles. The position of the center of mass is calculated from Eq. 9-2. The location of the center of mass (with respect to the particles) is the same in both cases.

reduces to $x_{com} = 0$. If $m_1 = 0$, there is again only one particle (of mass m_2), and we have, as we expect, $x_{com} = d$. If $m_1 = m_2$, the masses of the particles are equal and the center of mass should be halfway between them; Eq. 9-1 reduces to $x_{com} = \frac{1}{2}d$, again as we expect. Finally, Eq. 9-1 tells us that if neither m_1 nor m_2 is zero, x_{com} can have only values that lie between zero and d; that is, the center of mass must lie somewhere between the two particles.

Figure 9-2b shows a more generalized situation, in which the coordinate system has been shifted leftward. The position of the center of mass is now defined as

$$x_{com} = \frac{m_1 x_1 + m_2 x_2}{m_1 + m_2}. \tag{9-2}$$

Note that if we put $x_1 = 0$, then x_2 becomes d and Eq. 9-2 reduces to Eq. 9-1, as it must. Note also that in spite of the shift of the coordinate system, the center of mass is still the same distance from each particle.

We can rewrite Eq. 9-2 as

$$x_{com} = \frac{m_1 x_1 + m_2 x_2}{M}, \tag{9-3}$$

in which M is the total mass of the system. (Here, $M = m_1 + m_2$.) We can extend this equation to a more general situation in which n particles are strung out along the x axis. Then the total mass is $M = m_1 + m_2 + \cdots + m_n$, and the location of the center of mass is

$$x_{com} = \frac{m_1 x_1 + m_2 x_2 + m_3 x_3 + \cdots + m_n x_n}{M} = \frac{1}{M}\sum_{i=1}^{n} m_i x_i. \tag{9-4}$$

Here the subscript i is a running number, or index, that takes on all integer values from 1 to n. It identifies the various particles, their masses, and their x coordinates.

If the particles are distributed in three dimensions, the center of mass must be identified by three coordinates. By extension of Eq. 9-4, they are

$$x_{com} = \frac{1}{M}\sum_{i=1}^{n} m_i x_i, \qquad y_{com} = \frac{1}{M}\sum_{i=1}^{n} m_i y_i, \qquad z_{com} = \frac{1}{M}\sum_{i=1}^{n} m_i z_i. \tag{9-5}$$

We can also define the center of mass with the language of vectors. First recall that the position of a particle at coordinates x_i, y_i, and z_i is given by a position vector:

$$\vec{r}_i = x_i\hat{\mathrm{i}} + y_i\hat{\mathrm{j}} + z_i\hat{\mathrm{k}}. \tag{9-6}$$

Here the index identifies the particle, and $\hat{\mathrm{i}}$, $\hat{\mathrm{j}}$, and $\hat{\mathrm{k}}$ are unit vectors pointing, respectively, in the positive direction of the x, y, and z axes. Similarly, the position of the center of mass of a system of particles is given by a position vector:

$$\vec{r}_{com} = x_{com}\hat{\mathrm{i}} + y_{com}\hat{\mathrm{j}} + z_{com}\hat{\mathrm{k}}. \tag{9-7}$$

The three scalar equations of Eq. 9-5 can now be replaced by a single vector equation,

$$\vec{r}_{com} = \frac{1}{M}\sum_{i=1}^{n} m_i \vec{r}_i, \tag{9-8}$$

where again M is the total mass of the system. You can check that this equation is correct by substituting Eqs. 9-6 and 9-7 into it, and then separating out the x, y, and z components. The scalar relations of Eq. 9-5 result.

Solid Bodies

An ordinary object, such as a baseball bat, contains so many particles (atoms) that we can best treat it as a continuous distribution of matter. The "particles" then become differential mass elements *dm*, the sums of Eq. 9-5 become integrals, and the coordinates of the center of mass are defined as

$$x_{com} = \frac{1}{M}\int x\,dm, \qquad y_{com} = \frac{1}{M}\int y\,dm, \qquad z_{com} = \frac{1}{M}\int z\,dm, \tag{9-9}$$

where *M* is now the mass of the object.

Evaluating these integrals for most common objects (like a television set or a moose) would be difficult, so here we shall consider only *uniform* objects. Such an object has *uniform density*, or mass per unit volume; that is, the density ρ (Greek letter rho) is the same for any given element of the object as for the whole object:

$$\rho = \frac{dm}{dV} = \frac{M}{V}, \tag{9-10}$$

where dV is the volume occupied by a mass element dm, and V is the total volume of the object. If we substitute $dm = M/V\,dV$ from Eq. 9-10 into Eq. 9-9, we find that

$$x_{com} = \frac{1}{V}\int x\,dV, \qquad y_{com} = \frac{1}{V}\int y\,dV, \qquad z_{com} = \frac{1}{V}\int z\,dV. \tag{9-11}$$

You can bypass one or more of these integrals if an object has a point, a line, or a plane of symmetry. The center of mass of such an object then lies at that point, on that line, or in that plane. For example, the center of mass of a uniform sphere (which has a point of symmetry) is at the center of the sphere (which is the point of symmetry). The center of mass of a uniform cone (whose axis is a line of symmetry) lies on the axis of the cone. The center of mass of a banana (which has a plane of symmetry that splits it into two equal parts) lies somewhere in that plane.

The center of mass of an object need not lie within the object. There is no dough at the center of mass of a doughnut, and no iron at the center of mass of a horseshoe.

CHECKPOINT 1: The figure shows a uniform square plate from which four identical squares at the corners will be removed. (a) Where is the center of mass of the plate originally? Where is it after the removal of (b) square 1; (c) squares 1 and 2; (d) squares 1 and 3; (e) squares 1, 2, and 3; (f) all four squares? Answer in terms of quadrants, axes, or points (without calculation, of course).

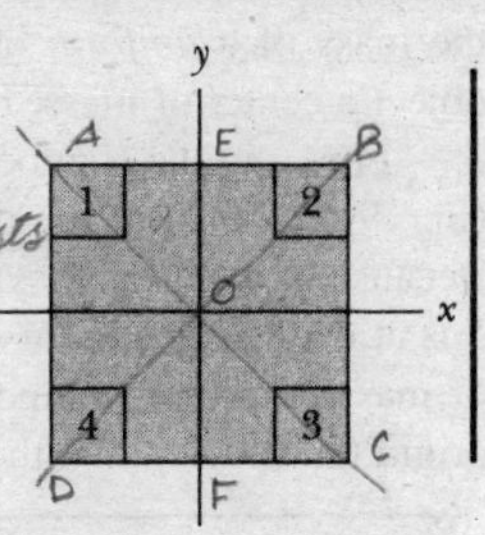

Sample Problem 9-1

Three particles of masses $m_1 = 1.2$ kg, $m_2 = 2.5$ kg, and $m_3 = 3.4$ kg form an equilateral triangle of edge length $a = 140$ cm. Where is the center of mass of this three-particle system?

SOLUTION: A **Key Idea** to get us started is that we are dealing with particles instead of an extended solid body, so we can use Eq. 9-5 to locate their center of mass. The particles are in the plane of the equilateral triangle, so we need only the first two equations. A second **Key Idea** is that we can simplify the calculations by choosing the *x* and *y* axes so that one of the particles is located at the origin and the *x* axis coincides with one of the triangle's sides (Fig. 9-3). The three particles then have the following coordinates:

Particle	Mass (kg)	*x* (cm)	*y* (cm)
1	1.2	0	0
2	2.5	140	0
3	3.4	70	121

The total mass *M* of the system is 7.1 kg.

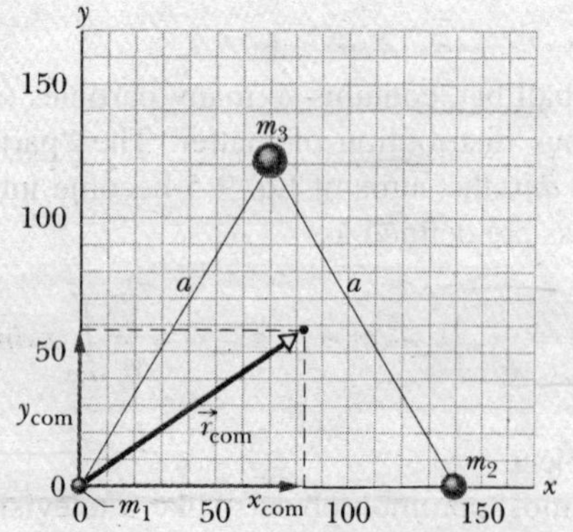

Fig. 9-3 Sample Problem 9-1. Three particles form an equilateral triangle of edge length a. The center of mass is located by the position vector $\vec{r}_{com}$.

From Eq. 9-5, the coordinates of the center of mass are

$$x_{com} = \frac{1}{M}\sum_{i=1}^{3} m_i x_i = \frac{m_1x_1 + m_2x_2 + m_3x_3}{M}$$

$$= \frac{(1.2\text{ kg})(0) + (2.5\text{ kg})(140\text{ cm}) + (3.4\text{ kg})(70\text{ cm})}{7.1\text{ kg}}$$

$$= 83\text{ cm} \qquad \text{(Answer)}$$

and

$$y_{com} = \frac{1}{M}\sum_{i=1}^{3} m_i y_i = \frac{m_1y_1 + m_2y_2 + m_3y_3}{M}$$

$$= \frac{(1.2\text{ kg})(0) + (2.5\text{ kg})(0) + (3.4\text{ kg})(121\text{ cm})}{7.1\text{ kg}}$$

$$= 58\text{ cm}. \qquad \text{(Answer)}$$

In Fig. 9-3, the center of mass is located by the position vector $\vec{r}_{com}$, which has components x_{com} and y_{com}.

Sample Problem 9-2

Figure 9-4a shows a uniform metal plate P of radius $2R$ from which a disk of radius R has been stamped out (removed) in an assembly line. Using the xy coordinate system shown, locate the center of mass com_P of the plate.

SOLUTION: First, let us roughly locate the center of plate P by using the **Key Idea** of symmetry. We note that the plate is symmetric about the x axis (we get the portion below that axis by rotating the upper portion about the axis). Thus, com_P must be on the x axis. The plate (with the disk removed) is not symmetric about the y axis. However, because there is somewhat more mass on the right of the y axis, com_P must be somewhat to the right of that axis. Thus, the location of com_P should be roughly as indicated in Fig. 9-4a.

Another **Key Idea** here is that plate P is an extended solid body, so we can use Eqs. 9-11 to find the actual coordinates of com_P. However, that procedure is difficult. A much easier way is to use this **Key Idea**: In working with centers of mass, we can assume that the mass of a *uniform* object is concentrated in a particle at the object's center of mass. Here is how we do so:

First, put the stamped-out disk (call it disk S) back into place (Fig. 9-4b) to form the original composite plate (call it plate C). Because of its circular symmetry, the center of mass com_S for disk S is at the center of S, at $x = -R$ (as shown). Similarly, the center of mass com_C for composite plate C is at the center of C, at the origin (as shown). We then have the following:

Plate	Center of Mass	Location of com	Mass
P	com_P	$x_P = ?$	m_P
S	com_S	$x_S = -R$	m_S
C	com_C	$x_C = 0$	$m_C = m_S + m_P$

Now we use the **Key Idea** of concentrated mass: Assume that mass m_S of disk S is concentrated in a particle at $x_S = -R$, and mass m_P is concentrated in a particle at x_P (Fig. 9-4c). Next treat these two particles as a two-particle system, using Eq. 9-2 to find their

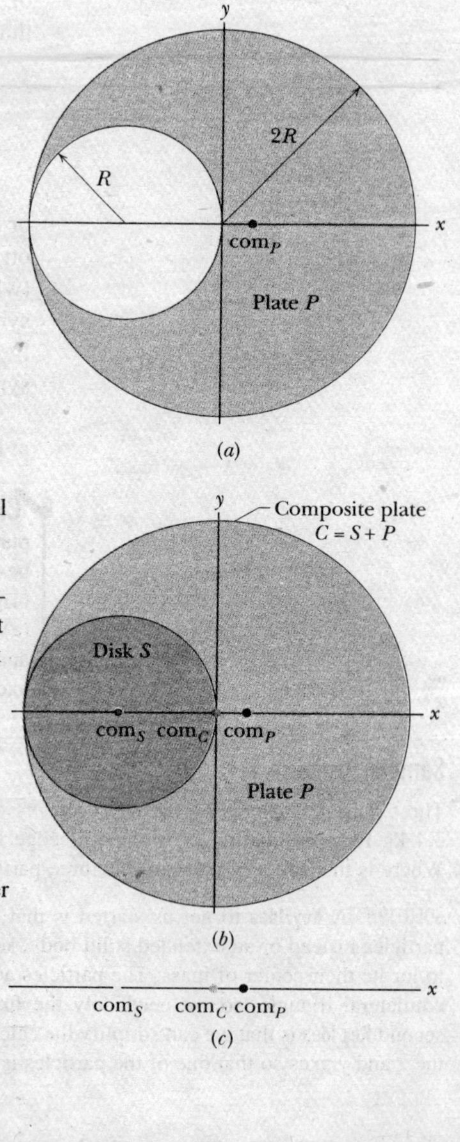

Fig. 9-4 Sample Problem 9-2. (a) Plate P is a metal plate of radius $2R$, with a circular hole of radius R. The center of mass of P is at point com_P. (b) Disk S has been put back into place to form a composite plate C. The center of mass com_S of disk S and the center of mass com_C of plate C are shown. (c) The center of mass com_{S+P} of the combination of S and P coincides with com_C, which is at $x = 0$.

center of mass x_{S+P}. We get

$$x_{S+P} = \frac{m_S x_S + m_P x_P}{m_S + m_P}. \tag{9-12}$$

Next note that the combination of disk S and plate P is composite plate C. Thus, the position x_{S+P} of com$_{S+P}$ must coincide with the position x_C of com$_C$, which is at the origin; so $x_{S+P} = x_C = 0$. Substituting this into Eq. 9-12 and solving for x_P, we get

$$x_P = -x_S \frac{m_S}{m_P}. \tag{9-13}$$

Now we seem to have a problem, because we do not know the masses in Eq. 9-13. However, we can relate the masses to the face areas of S and P by noting that

$$\begin{aligned} \text{mass} &= \text{density} \times \text{volume} \\ &= \text{density} \times \text{thickness} \times \text{area}. \end{aligned}$$

Then $$\frac{m_S}{m_P} = \frac{\text{density}_S}{\text{density}_P} \times \frac{\text{thickness}_S}{\text{thickness}_P} \times \frac{\text{area}_S}{\text{area}_P}.$$

Because the plate is uniform, the densities and thicknesses are equal; we are left with

$$\frac{m_S}{m_P} = \frac{\text{area}_S}{\text{area}_P} = \frac{\text{area}_S}{\text{area}_C - \text{area}_S} = \frac{\pi R^2}{\pi(2R)^2 - \pi R^2} = \frac{1}{3}.$$

Substituting this and $x_S = -R$ into Eq. 9-13, we have

$$x_P = \tfrac{1}{3}R. \qquad \text{(Answer)}$$

PROBLEM-SOLVING TACTICS

Tactic 1: *Center-of-Mass Problems* ★

Sample Problems 9-1 and 9-2 provide three strategies for simplifying center-of-mass problems. (1) Make full use of the symmetry of the object, be it about a point, a line, or a plane. (2) If the object can be divided into several parts, treat each of these parts as a particle, located at its own center of mass. (3) Choose your axes wisely: If your system is a group of particles, choose one of the particles as your origin. If your system is a body with a line of symmetry, let that be your x or y axis. The choice of origin is completely arbitrary; the location of the center of mass is the same regardless of the origin from which it is measured.

9-3 Newton's Second Law for a System of Particles

If you roll a cue ball at a second billiard ball that is at rest, you expect that the two-ball system will continue to have some forward motion after impact. You would be surprised, for example, if both balls came back toward you or if both moved to the right or to the left.

What continues to move forward, its steady motion completely unaffected by the collision, is the center of mass of the two-ball system. If you focus on this point—which is always halfway between these bodies because they have identical masses—you can easily convince yourself by trial at a billiard table that this is so. No matter whether the collision is glancing, head on, or somewhere in between, the center of mass continues to move forward, as if the collision had never occurred. Let us look into this center-of-mass motion in more detail.

To do so, we replace the pair of billiard balls with an assemblage of n particles of (possibly) different masses. We are interested not in the individual motions of these particles but *only* in the motion of their center of mass. Although the center of mass is just a point, it moves like a particle whose mass is equal to the total mass of the system; we can assign a position, a velocity, and an acceleration to it. We state (and shall prove next) that the (vector) equation that governs the motion of the center of mass of such a system of particles is

$$\vec{F}_{\text{net}} = M\vec{a}_{\text{com}} \quad \text{(system of particles).} \tag{9-14}$$

This equation is Newton's second law for the motion of the center of mass of a system of particles. Note that it has the same form ($\vec{F}_{\text{net}} = m\vec{a}$) that holds for the motion of a single particle. However, the three quantities that appear in Eq. 9-14

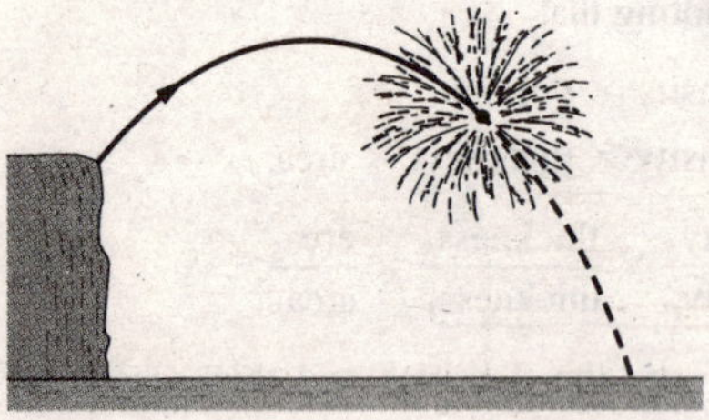

Fig. 9-5 A fireworks rocket explodes in flight. In the absence of air drag, the center of mass of the fragments would continue to follow the original parabolic path, until fragments began to hit the ground.

must be evaluated with some care:

1. $\vec{F}_{net}$ is the net force of *all external forces* that act on the system. Forces on one part of the system from another (*internal forces*) are not included in Eq. 9-14.
2. M is the *total mass* of the system. We assume that no mass enters or leaves the system as it moves, so that M remains constant. The system is said to be **closed.**
3. $\vec{a}_{com}$ is the acceleration of the *center of mass* of the system. Equation 9-14 gives no information about the acceleration of any other point of the system.

Equation 9-14 is equivalent to three equations involving the components of $\vec{F}_{net}$ and $\vec{a}_{com}$ along the three coordinate axes. These equations are

$$F_{net,x} = Ma_{com,x} \qquad F_{net,y} = Ma_{com,y} \qquad F_{net,z} = Ma_{com,z} \tag{9-15}$$

Now we can go back and examine the behavior of the billiard balls. Once the cue ball has begun to roll, no net external force acts on the (two-ball) system. Thus, because $\vec{F}_{net} = 0$, Eq. 9-14 tells us that $\vec{a}_{com} = 0$ also. Because acceleration is the rate of change of velocity, we conclude that the velocity of the center of mass of the system of two balls does not change. When the two balls collide, the forces that come into play are *internal* forces, on one ball from the other. Such forces do not contribute to the net force $\vec{F}_{net}$, which remains zero. Thus, the center of mass of the system, which was moving forward before the collision, must continue to move forward after the collision, with the same speed and in the same direction.

Equation 9-14 applies not only to a system of particles but also to a solid body, such as the bat of Fig. 9-1*b*. In that case, M in Eq. 9-14 is the mass of the bat and $\vec{F}_{net}$ is the gravitational force on the bat. Equation 9-14 then tells us that $\vec{a}_{com} = \vec{g}$. In other words, the center of mass of the bat moves as if the bat were a single particle of mass M, with force $\vec{F}_g$ acting on it.

Figure 9-5 shows another interesting case. Suppose that at a fireworks display, a rocket is launched on a parabolic path. At a certain point, it explodes into fragments. If the explosion had not occurred, the rocket would have continued along the trajectory shown in the figure. The forces of the explosion are *internal* to the system (first the rocket and then its fragments); that is, they are forces on parts of the system from other parts. If we ignore air drag, the net *external* force $\vec{F}_{net}$ acting on the system is the gravitational force on the system, regardless of whether the rocket explodes. Thus, from Eq. 9-14, the acceleration $\vec{a}_{com}$ of the center of mass of the fragments (while they are in flight) remains equal to $\vec{g}$. This means that the center of mass of the fragments follows the same parabolic trajectory that the rocket would have followed had it not exploded.

When a ballet dancer leaps across the stage in a grand jeté, she raises her arms and stretches her legs out horizontally as soon as her feet leave the stage (Fig. 9-6). These actions shift her center of mass upward through her body. Although the shifting center of mass faithfully follows a parabolic path across the stage, its movement relative to the body decreases the height that is attained by her head and torso, relative to that of a normal jump. The result is that the head and torso follow a nearly horizontal path, giving an illusion that the dancer is floating.

Proof of Equation 9-14

Now let us prove this important equation. From Eq. 9-8 we have, for a system of n particles,

$$M\vec{r}_{com} = m_1\vec{r}_1 + m_2\vec{r}_2 + m_3\vec{r}_3 + \cdots + m_n\vec{r}_n, \tag{9-16}$$

Fig. 9-6 A grand jeté. (Adapted from *The Physics of Dance*, by Kenneth Laws, Schirmer Books, 1984.)

in which M is the system's total mass and $\vec{r}_{com}$ is the vector locating the position of the system's center of mass.

Differentiating Eq. 9-16 with respect to time gives

$$M\vec{v}_{com} = m_1\vec{v}_1 + m_2\vec{v}_2 + m_3\vec{v}_3 + \cdots + m_n\vec{v}_n. \tag{9-17}$$

Here $\vec{v}_i$ $(= d\vec{r}_i/dt)$ is the velocity of the ith particle, and $\vec{v}_{com}$ $(= d\vec{r}_{com}/dt)$ is the velocity of the center of mass.

Differentiating Eq. 9-17 with respect to time leads to

$$M\vec{a}_{com} = m_1\vec{a}_1 + m_2\vec{a}_2 + m_3\vec{a}_3 + \cdots + m_n\vec{a}_n. \tag{9-18}$$

Here $\vec{a}_i$ $(= d\vec{v}_i/dt)$ is the acceleration of the ith particle, and $\vec{a}_{com}$ $(= d\vec{v}_{com}/dt)$ is the acceleration of the center of mass. Although the center of mass is just a geometrical point, it has a position, a velocity, and an acceleration, as if it were a particle.

From Newton's second law, $m_i\vec{a}_i$ is equal to the resultant force $\vec{F}_i$ that acts on the ith particle. Thus, we can rewrite Eq. 9-18 as

$$M\vec{a}_{com} = \vec{F}_1 + \vec{F}_2 + \vec{F}_3 + \cdots + \vec{F}_n. \tag{9-19}$$

Among the forces that contribute to the right side of Eq. 9-19 will be forces that the particles of the system exert on each other (internal forces) and forces exerted on the particles from outside the system (external forces). By Newton's third law, the internal forces form third-law force pairs and cancel out in the sum that appears on the right side of Eq. 9-19. What remains is the vector sum of all the *external* forces that act on the system. Equation 9-19 then reduces to Eq. 9-14, the relation that we set out to prove.

✔CHECKPOINT 2: Two skaters on frictionless ice hold opposite ends of a pole of negligible mass. An axis runs along the pole, and the origin of the axis is at the center of mass of the two-skater system. One skater, Fred, weighs twice as much as the other skater, Ethel. Where do the skaters meet if (a) Fred pulls hand over hand along the pole so as to draw himself to Ethel, (b) Ethel pulls hand over hand to draw herself to Fred, and (c) both skaters pull hand over hand?

Sample Problem 9-3

The three particles in Fig. 9-7*a* are initially at rest. Each experiences an *external* force due to bodies outside the three-particle system. The directions are indicated, and the magnitudes are $F_1 = 6.0$ N, $F_2 = 12$ N, and $F_3 = 14$ N. What is the acceleration of the center of mass of the system, and in what direction does it move?

SOLUTION: The position of the center of mass, calculated by the method of Sample Problem 9-1, is marked by a dot in the figure. One **Key Idea** here is that we can treat the center of mass as if it were a real particle, with a mass equal to the system's total mass $M = 16$ kg. We can also treat the three external forces as if they act at the center of mass (Fig. 9-7*b*).

A second **Key Idea** is that we can now apply Newton's second law ($\vec{F}_{net} = m\vec{a}$) to the center of mass, writing

$$\vec{F}_{net} = M\vec{a}_{com} \tag{9-20}$$

or

$$\vec{F}_1 + \vec{F}_2 + \vec{F}_3 = M\vec{a}_{com}$$

so

$$\vec{a}_{com} = \frac{\vec{F}_1 + \vec{F}_2 + \vec{F}_3}{M}. \tag{9-21}$$

Equation 9-20 tells us that the acceleration $\vec{a}_{com}$ of the center of mass is in the same direction as the net external force $\vec{F}_{net}$ on the system (Fig. 9-7*b*). Because the particles are initially at rest, the center of mass must also be at rest. As the center of mass then begins to accelerate, it must move off in the common direction of $\vec{a}_{com}$ and $\vec{F}_{net}$.

We can evaluate the right side of Eq. 9-21 directly on a vector-capable calculator, or we can rewrite Eq. 9-21 in component form, find the components of $\vec{a}_{com}$, and then find $\vec{a}_{com}$. Along the x axis, we have

$$a_{com,x} = \frac{F_{1x} + F_{2x} + F_{3x}}{M}$$

$$= \frac{-6.0 \text{ N} + (12 \text{ N}) \cos 45° + 14 \text{ N}}{16 \text{ kg}} = 1.03 \text{ m/s}^2.$$

Along the y axis, we have

$$a_{com,y} = \frac{F_{1y} + F_{2y} + F_{3y}}{M}$$

$$= \frac{0 + (12 \text{ N}) \sin 45° + 0}{16 \text{ kg}} = 0.530 \text{ m/s}^2.$$

From these components, we find that $\vec{a}_{com}$ has the magnitude

$$a_{com} = \sqrt{(a_{com,x})^2 + (a_{com,y})^2}$$

$$= 1.16 \text{ m/s}^2 \approx 1.2 \text{ m/s}^2 \qquad \text{(Answer)}$$

and the angle (from the positive direction of the x axis)

$$\theta = \tan^{-1} \frac{a_{com,y}}{a_{com,x}} = 27°. \qquad \text{(Answer)}$$

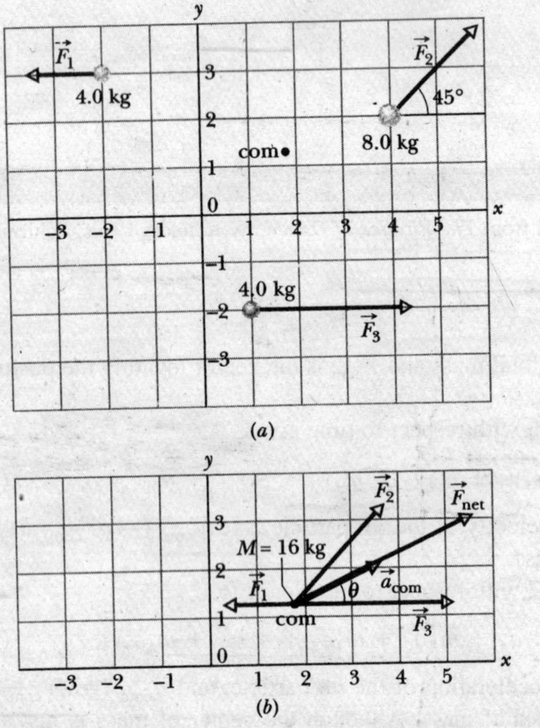

Fig. 9-7 Sample Problem 9-3. (*a*) Three particles, initially at rest in the positions shown, are acted on by the external forces shown. The center of mass (com) of the system is marked. (*b*) The forces are now transferred to the center of mass of the system, which behaves like a particle with a mass M equal to the total mass of the system. The net external force $\vec{F}_{net}$ and the acceleration $\vec{a}_{com}$ of the center of mass are shown.

9-4 Linear Momentum

Momentum is a word that has several meanings in everyday language but only a single precise meaning in physics. The **linear momentum** of a particle is a vector $\vec{p}$, defined as

$$\vec{p} = m\vec{v} \quad \text{(linear momentum of a particle),} \tag{9-22}$$

in which m is the mass of the particle and $\vec{v}$ is its velocity. (The adjective *linear* is often dropped, but it serves to distinguish $\vec{p}$ from *angular* momentum, which is introduced in Chapter 12 and which is associated with rotation.) Since m is always a positive scalar quantity, Eq. 9-22 tells us that $\vec{p}$ and $\vec{v}$ have the same direction. From Eq. 9-22, the SI unit for momentum is the kilogram-meter per second.

Newton actually expressed his second law of motion in terms of momentum:

> The time rate of change of the momentum of a particle is equal to the net force acting on the particle and is in the direction of that force.

In equation form this becomes

$$\vec{F}_{\text{net}} = \frac{d\vec{p}}{dt}. \tag{9-23}$$

Substituting for $\vec{p}$ from Eq. 9-22 gives

$$\vec{F}_{\text{net}} = \frac{d\vec{p}}{dt} = \frac{d}{dt}(m\vec{v}) = m\frac{d\vec{v}}{dt} = m\vec{a}.$$

Thus, the relations $\vec{F}_{\text{net}} = d\vec{p}/dt$ and $\vec{F}_{\text{net}} = m\vec{a}$ are equivalent expressions of Newton's second law of motion for a particle.

CHECKPOINT 3: The figure gives the linear momentum versus time for a particle moving along an axis. A force directed along the axis acts on the particle. (a) Rank the four regions indicated according to the magnitude of the force, greatest first. (b) In which region is the particle slowing?

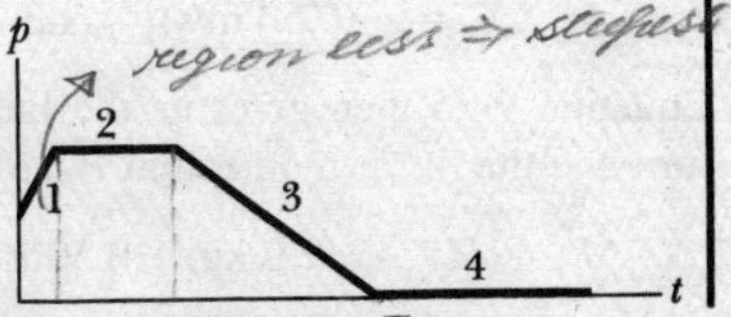

9-5 The Linear Momentum of a System of Particles

Now consider a system of n particles, each with its own mass, velocity, and linear momentum. The particles may interact with each other, and external forces may act on them as well. The system as a whole has a total linear momentum $\vec{P}$, which is defined to be the vector sum of the individual particles' linear momenta. Thus,

$$\begin{aligned}\vec{P} &= \vec{p}_1 + \vec{p}_2 + \vec{p}_3 + \cdots + \vec{p}_n \\ &= m_1\vec{v}_1 + m_2\vec{v}_2 + m_3\vec{v}_3 + \cdots + m_n\vec{v}_n.\end{aligned} \tag{9-24}$$

If we compare this equation with Eq. 9-17, we see that

$$\vec{P} = M\vec{v}_{\text{com}} \quad \text{(linear momentum, system of particles)}, \tag{9-25}$$

which gives us another way to define the linear momentum of a system of particles:

> The linear momentum of a system of particles is equal to the product of the total mass M of the system and the velocity of the center of mass.

If we take the time derivative of Eq. 9-25, we find

$$\frac{d\vec{P}}{dt} = M\frac{d\vec{v}_{\text{com}}}{dt} = M\vec{a}_{\text{com}}. \tag{9-26}$$

Comparing Eqs. 9-14 and 9-26 allows us to write Newton's second law for a system of particles in the equivalent form

$$\vec{F}_{net} = \frac{d\vec{P}}{dt} \quad \text{(system of particles),} \tag{9-27}$$

where $\vec{F}_{net}$ is the net external force acting on the system. This equation is the generalization of the single-particle equation $\vec{F}_{net} = d\vec{p}/dt$ to a system of many particles.

Sample Problem 9-4

Figure 9-8a shows a 2.0 kg toy race car before and after taking a turn on a track. Its speed is 0.50 m/s before the turn and 0.40 m/s after the turn. What is the change $\Delta\vec{P}$ in the linear momentum of the car due to the turn?

SOLUTION: We treat the car as a system of particles. Then a **Key Idea** is that to get $\Delta\vec{P}$, we need the car's linear momenta before and after the turn. However, that means we first need its velocity $\vec{v}_i$ before the turn and its velocity $\vec{v}_f$ after the turn. Using the coordinate system of Fig. 9-8*a*, we write $\vec{v}_i$ and $\vec{v}_f$ as

$$\vec{v}_i = -(0.50 \text{ m/s})\hat{j} \quad \text{and} \quad \vec{v}_f = (0.40 \text{ m/s})\hat{i}.$$

Equation 9-25 then gives us the linear momentum $\vec{P}_i$ before the turn and the linear momentum $\vec{P}_f$ after the turn:

$$\vec{P}_i = M\vec{v}_i = (2.0 \text{ kg})(-0.50 \text{ m/s})\hat{j} = (-1.0 \text{ kg}\cdot\text{m/s})\hat{j}$$

and $$\vec{P}_f = M\vec{v}_f = (2.0 \text{ kg})(0.40 \text{ m/s})\hat{i} = (0.80 \text{ kg}\cdot\text{m/s})\hat{i}.$$

A second **Key Idea** is that, because these linear momenta are not directed along the same axis, we *cannot* find the change in linear momentum $\Delta\vec{P}$ by merely subtracting the magnitude of $\vec{P}_i$ from the magnitude of $\vec{P}_f$. Instead, we must write the change in linear momentum as the vector equation

$$\Delta\vec{P} = \vec{P}_f - \vec{P}_i \tag{9-28}$$

and then as

$$\Delta\vec{P} = (0.80 \text{ kg}\cdot\text{m/s})\hat{i} - (-1.0 \text{ kg}\cdot\text{m/s})\hat{j}$$
$$= (0.8\hat{i} + 1.0\hat{j}) \text{ kg}\cdot\text{m/s}. \quad \text{(Answer)}$$

Figure 9-8*b* shows $\Delta\vec{P}$, $\vec{P}_f$, and $-\vec{P}_i$. Note that we subtract $\vec{P}_i$ from $\vec{P}_f$ in Eq. 9-28 by adding $-\vec{P}_i$ to $\vec{P}_f$.

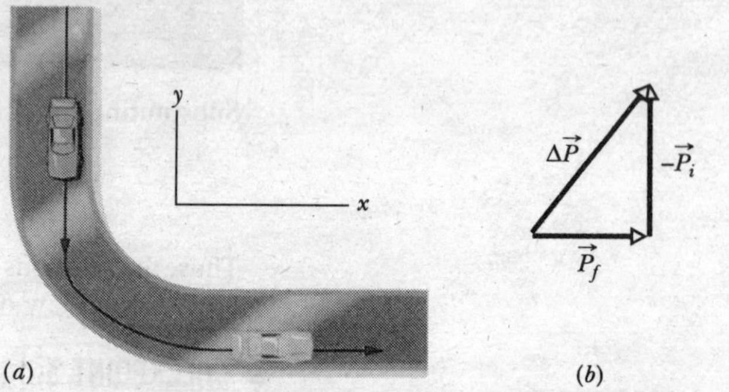

Fig. 9-8 Sample Problem 9-4. (*a*) A toy car takes a turn. (*b*) The change $\Delta\vec{P}$ in the car's linear momentum is the vector sum of its final linear momentum $\vec{P}_f$ and the negative of its initial linear momentum $\vec{P}_i$.

9-6 Conservation of Linear Momentum

Suppose that the net external force acting on a system of particles is zero (the system is isolated) and that no particles leave or enter the system (the system is closed). Putting $\vec{F}_{net} = 0$ in Eq. 9-27 then yields $d\vec{P}/dt = 0$, or

$$\vec{P} = \text{constant} \quad \text{(closed, isolated system).} \tag{9-29}$$

In words,

> If no net external force acts on a system of particles, the total linear momentum $\vec{P}$ of the system cannot change.

This result is called the **law of conservation of linear momentum.** It can also be written as

$$\vec{P}_i = \vec{P}_f \quad \text{(closed, isolated system).} \tag{9-30}$$

In words, this equation says that, for a closed, isolated system,

$$\begin{pmatrix}\text{total linear momentum}\\ \text{at some initial time } t_i\end{pmatrix} = \begin{pmatrix}\text{total linear momentum}\\ \text{at some later time } t_f\end{pmatrix}.$$

Equations 9-29 and 9-30 are vector equations and, as such, each is equivalent to three equations corresponding to the conservation of linear momentum in three mutually perpendicular directions as in, say, the xyz coordinate system. Depending on the forces acting on a system, linear momentum might be conserved in one or two directions but not in all directions. However,

> If the component of the net *external* force on a closed system is zero along an axis, then the component of the linear momentum of the system along that axis cannot change.

As an example, suppose that you toss a grapefruit across a room. During its flight, the only external force acting on the grapefruit (which we take as the system) is the gravitational force $\vec{F}_g$, which is directed vertically downward. Thus, the vertical component of the linear momentum of the grapefruit changes, but since no horizontal external force acts on the grapefruit, the horizontal component of the linear momentum cannot change.

Note that we focus on the external forces acting on a closed system. Although internal forces can change the linear momentum of portions of the system, they cannot change the total linear momentum of the entire system.

CHECKPOINT 4: An initially stationary device lying on a frictionless floor explodes into two pieces, which then slide across the floor. One piece slides in the positive direction of an x axis. (a) What is the sum of the momenta of the two pieces after the explosion? (b) Can the second piece move at an angle to the x axis? (c) What is the direction of the momentum of the second piece?

Sample Problem 9-5

A ballot box with mass $m = 6.0$ kg slides with speed $v = 4.0$ m/s across a frictionless floor in the positive direction of an x axis. It suddenly explodes into two pieces. One piece, with mass $m_1 = 2.0$ kg, moves in the positive direction of the x axis with speed $v_1 = 8.0$ m/s. What is the velocity of the second piece, with mass m_2?

SOLUTION: There are two **Key Ideas** here. First, we could get the velocity of the second piece if we knew its momentum, because we already know its mass is $m_2 = m - m_1 = 4.0$ kg. Second, we can relate the momenta of the two pieces to the original momentum of the box if momentum is conserved. Let's check.

Our reference frame will be that of the floor. Our system, which consists initially of the box and then of the two pieces, is closed but is not isolated, because the box and pieces each experience a normal force from the floor and a gravitational force. However, those forces are both vertical and thus cannot change the horizontal component of the momentum of the system. Neither can the forces produced by the explosion, because those forces are internal to the system. Thus, the horizontal component of the momentum of the system is conserved, and we can apply Eq. 9-30 along the x axis.

The initial momentum of the system is that of the box:

$$\vec{P}_i = m\vec{v}.$$

Similarly, we can write the final momenta of the two pieces as

$$\vec{P}_{f1} = m_1\vec{v}_1 \quad \text{and} \quad \vec{P}_{f2} = m_2\vec{v}_2.$$

The final total momentum $\vec{P}_f$ of the system is the vector sum of the momenta of the two pieces:

$$\vec{P}_f = \vec{P}_{f1} + \vec{P}_{f2} = m_1\vec{v}_1 + m_2\vec{v}_2.$$

Since all the velocities and momenta in this problem are vectors along the x axis, we can write them in terms of their x components. Doing so while applying Eq. 9-30, we now obtain

$$P_i = P_f$$

or

$$mv = m_1v_1 + m_2v_2.$$

Inserting known data, we find

$$(6.0 \text{ kg})(4.0 \text{ m/s}) = (2.0 \text{ kg})(8.0 \text{ m/s}) + (4.0 \text{ kg})v_2$$

and thus

$$v_2 = 2.0 \text{ m/s}. \qquad \text{(Answer)}$$

Since the result is positive, the second piece moves in the positive direction of the x axis.

Sample Problem 9-6

Figure 9-9*a* shows a space hauler and cargo module, of total mass M traveling along an x axis in deep space. They have an initial velocity $\vec{v}_i$ of magnitude 2100 km/h relative to the Sun. With a small explosion, the hauler ejects the cargo module, of mass $0.20M$ (Fig. 9-9*b*). The hauler then travels 500 km/h faster than the module along the x axis; that is, the relative speed v_{rel} between the hauler and the module is 500 km/h. What then is the velocity $\vec{v}_{HS}$ of the hauler relative to the Sun?

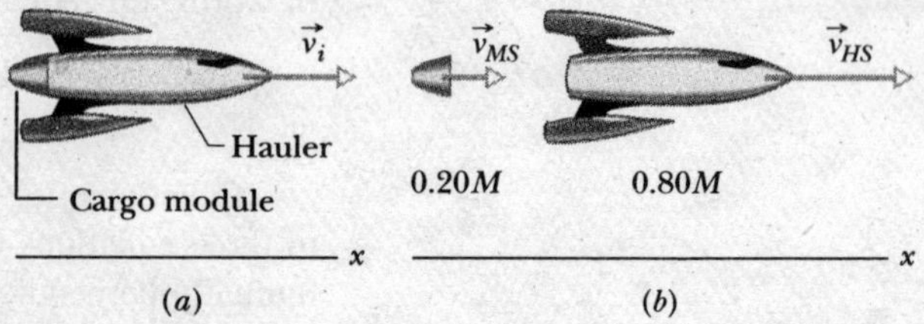

Fig. 9-9 Sample Problem 9-6. (*a*) A space hauler, with a cargo module, moving at initial velocity $\vec{v}_i$. (*b*) The hauler has ejected the cargo module. Now the module has velocity $\vec{v}_{MS}$ and the hauler has velocity $\vec{v}_{HS}$.

SOLUTION: The **Key Idea** here is that, because the hauler–module system is closed and isolated, its total linear momentum is conserved; that is,

$$\vec{P}_i = \vec{P}_f, \tag{9-31}$$

where the subscripts i and f refer to values before and after the ejection, respectively. Because the motion is along a single axis, we can write momenta and velocities in terms of their x components. Before the ejection, we have

$$P_i = Mv_i. \tag{9-32}$$

Let v_{MS} be the velocity of the ejected module relative to the Sun. The total linear momentum of the system after the ejection is then

$$P_f = (0.20M)v_{MS} + (0.80M)v_{HS}, \tag{9-33}$$

where the first term on the right is the linear momentum of the module and the second term is that of the hauler.

We do not know the velocity v_{MS} of the module relative to the Sun, but we can relate it to the known velocities with

$$\begin{pmatrix}\text{velocity of}\\ \text{hauler relative}\\ \text{to Sun}\end{pmatrix} = \begin{pmatrix}\text{velocity of}\\ \text{hauler relative}\\ \text{to module}\end{pmatrix} + \begin{pmatrix}\text{velocity of}\\ \text{module relative}\\ \text{to Sun}\end{pmatrix}.$$

In symbols, this gives us

$$v_{HS} = v_{rel} + v_{MS} \tag{9-34}$$

or

$$v_{MS} = v_{HS} - v_{rel}.$$

Substituting this expression for v_{MS} into Eq. 9-33, and then substituting Eqs. 9-32 and 9-33 into Eq. 9-31, we find

$$Mv_i = 0.20M(v_{HS} - v_{rel}) + 0.80Mv_{HS},$$

which gives us

$$v_{HS} = v_i + 0.20v_{rel},$$

or

$$v_{HS} = 2100 \text{ km/h} + (0.20)(500 \text{ km/h})$$
$$= 2200 \text{ km/h}. \quad \text{(Answer)}$$

CHECKPOINT 5: The table gives velocities of the hauler and module (after ejection and relative to the Sun), and the relative speed between the hauler and the module for three situations. What are the missing values?

	Velocities (km/h)		Relative Speed
	Module	Hauler	(km/h)
(a)	1500	2000	
(b)		3000	400
(c)	1000		600

Sample Problem 9-7

A firecracker placed inside a coconut of mass M, initially at rest on a frictionless floor, blows the coconut into three pieces that slide across the floor. An overhead view is shown in Fig. 9-10*a*. Piece C, with mass $0.30M$, has final speed $v_{fC} = 5.0$ m/s.

(a) What is the speed of piece B, with mass $0.20M$?

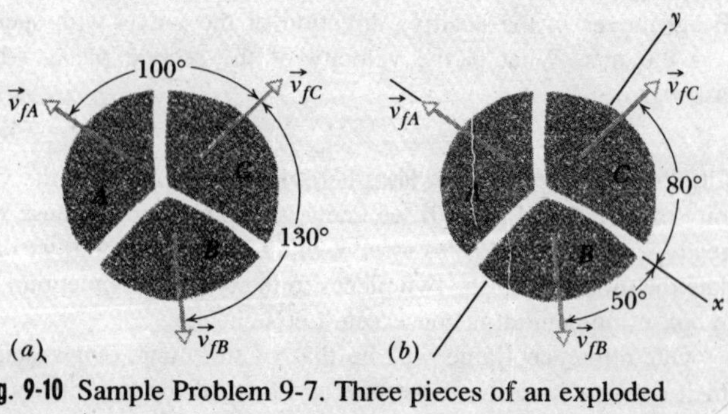

Fig. 9-10 Sample Problem 9-7. Three pieces of an exploded coconut move off in three directions along a frictionless floor. (*a*) An overhead view of the event. (*b*) The same with a two-dimensional axis system imposed.

SOLUTION: A **Key Idea** here is to see whether linear momentum is conserved. We note that (1) the coconut and its pieces form a closed system, (2) the explosion forces are internal to that system, and (3) no net external force acts on the system. Therefore, the linear momentum of the system is conserved.

To get started, we superimpose an xy coordinate system as shown in Fig. 9-10*b*, with the negative direction of the x axis coinciding with the direction of $\vec{v}_{fA}$. The x axis is at 80° with the direction of $\vec{v}_{fC}$ and 50° with the direction of $\vec{v}_{fB}$.

A second **Key Idea** is that linear momentum is conserved separately along both x and y axes. Let's use the y axis and write

$$P_{iy} = P_{fy}, \tag{9-35}$$

where subscript i refers to the initial value (before the explosion), and subscript y refers to the y component of $\vec{P}_i$ or $\vec{P}_f$.

The component P_{iy} of the initial linear momentum is zero, because the coconut is initially at rest. To get an expression for P_{fy}, we find the y component of the final linear momentum of each piece, using the y-component version of Eq. 9-22 ($p_y = mv_y$):

$$p_{fA,y} = 0,$$

$$p_{fB,y} = -0.20Mv_{fB,y} = -0.20Mv_{fB} \sin 50°,$$

$$p_{fC,y} = 0.30Mv_{fC,y} = 0.30Mv_{fC} \sin 80°.$$

(Note that $p_{fA,y} = 0$ because of our choice of axes.) Equation 9-35 can now be written as

$$P_{iy} = P_{fy} = p_{fA,y} + p_{fB,y} + p_{fC,y}.$$

Then, with $v_{fC} = 5.0$ m/s, we have

$$0 = 0 - 0.20Mv_{fB} \sin 50° + (0.30M)(5.0 \text{ m/s}) \sin 80°,$$

from which we find

$$v_{fB} = 9.64 \text{ m/s} \approx 9.6 \text{ m/s}. \qquad \text{(Answer)}$$

(b) What is the speed of piece A?

SOLUTION: Because linear momentum is also conserved along the x axis, we have

$$P_{ix} = P_{fx}, \qquad (9\text{-}36)$$

where $P_{ix} = 0$ because the coconut is initially at rest. To get P_{fx}, we find the x components of the final momenta, using the fact that piece A must have a mass of $0.50M$ ($= M - 0.20M - 0.30M$):

$$p_{fA,x} = -0.50Mv_{fA},$$

$$p_{fB,x} = 0.20Mv_{fB,x} = 0.20Mv_{fB} \cos 50°,$$

$$p_{fC,x} = 0.30Mv_{fC,x} = 0.30Mv_{fC} \cos 80°.$$

Equation 9-36 can now be written as

$$P_{ix} = P_{fx} = p_{fA,x} + p_{fB,x} + p_{fC,x}.$$

Then, with $v_{fC} = 5.0$ m/s and $v_{fB} = 9.64$ m/s, we have

$$0 = -0.50Mv_{fA} + 0.20M(9.64 \text{ m/s}) \cos 50° + 0.30M(5.0 \text{ m/s}) \cos 80°,$$

from which we find

$$v_{fA} = 3.0 \text{ m/s}. \qquad \text{(Answer)}$$

CHECKPOINT 6: Suppose that the exploding coconut is accelerating in the negative direction of y in Fig. 9-10 (say it is on a ramp that slants downward in that direction). Is linear momentum conserved along (a) the x axis (as stated in Eq. 9-36) and (b) the y axis (as stated in Eq. 9-35)?

PROBLEM-SOLVING TACTICS

Tactic 2: *Conservation of Linear Momentum*

For problems involving the conservation of linear momentum, first make sure that you have chosen a closed, isolated system. *Closed* means that no matter (no particles) passes through the system boundary in any direction. *Isolated* means that the net external force acting on the system is zero. If it is not isolated, then remember that each component of linear momentum is conserved separately if the corresponding component of the net external force is zero. So, you might conserve one component and not another.

Next, select two appropriate states of the system (which you may choose to call the initial state and the final state) and write expressions for the linear momentum of the system in each of these two states. In writing these expressions, make sure that you know what inertial reference frame you are using, and make sure also that you include the entire system, not missing any part of it and not including objects that do not belong to your system.

Finally, set your expressions for $\vec{P}_i$ and $\vec{P}_f$ equal to each other and solve for what is requested.

9-7 Systems with Varying Mass: A Rocket

In the systems we have dealt with so far, we have assumed that the total mass of the system remains constant. Sometimes, as in a rocket (Fig. 9-11), it does not. Most of the mass of a rocket on its launching pad is fuel, all of which will eventually be burned and ejected from the nozzle of the rocket engine.

We handle the variation of the mass of the rocket as the rocket accelerates by applying Newton's second law, not to the rocket alone but to the rocket and its ejected combustion products taken together. The mass of *this* system does *not* change as the rocket accelerates.

Finding the Acceleration

Assume that we are at rest relative to an inertial reference frame, watching a rocket accelerate through deep space with no gravitational or atmospheric drag forces acting

Fig. 9-11 Liftoff of Project Mercury spacecraft.

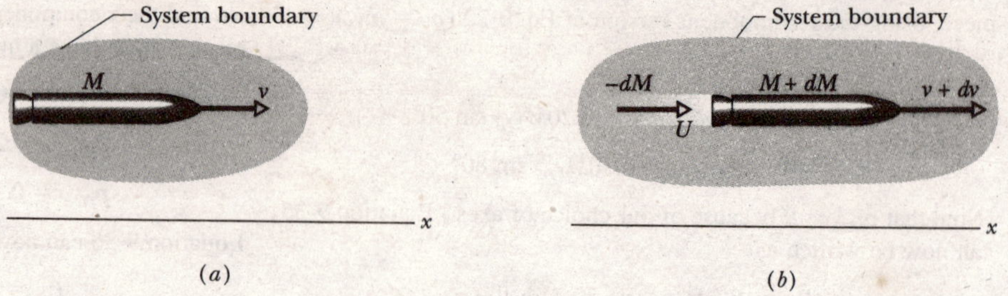

Fig. 9-12 (*a*) An accelerating rocket of mass M at time t, as seen from an inertial reference frame. (*b*) The same but at time $t + dt$. The exhaust products released during interval dt are shown.

on it. For this one-dimensional motion, let M be the mass of the rocket and v its velocity at an arbitrary time t (see Fig. 9-12*a*).

Figure 9-12*b* shows how things stand a time interval dt later. The rocket now has velocity $v + dv$ and mass $M + dM$, where the change in mass dM is a *negative quantity*. The exhaust products released by the rocket during interval dt have mass $-dM$ and velocity U relative to our inertial reference frame.

Our system consists of the rocket and the exhaust products released during interval dt. The system is closed and isolated, so the linear momentum of the system must be conserved during dt; that is,

$$P_i = P_f, \tag{9-37}$$

where the subscripts i and f indicate the values at the beginning and end of time interval dt. We can rewrite Eq. 9-37 as

$$Mv = -dM\,U + (M + dM)(v + dv), \tag{9-38}$$

where the first term on the right is the linear momentum of the exhaust products released during interval dt and the second term is the linear momentum of the rocket at the end of interval dt.

We can simplify Eq. 9-38 by using the relative speed v_{rel} between the rocket and the exhaust products, which is related to the velocities relative to the frame with

$$\begin{pmatrix}\text{velocity of rocket}\\ \text{relative to frame}\end{pmatrix} = \begin{pmatrix}\text{velocity of rocket}\\ \text{relative to products}\end{pmatrix} + \begin{pmatrix}\text{velocity of products}\\ \text{relative to frame}\end{pmatrix}.$$

In symbols, this means

$$(v + dv) = v_{rel} + U,$$

or

$$U = v + dv - v_{rel}. \tag{9-39}$$

Substituting this result for U into Eq. 9-38 yields, with a little algebra,

$$-dM\, v_{rel} = M\, dv. \tag{9-40}$$

Dividing each side by dt gives us

$$-\frac{dM}{dt} v_{rel} = M \frac{dv}{dt}. \tag{9-41}$$

We replace dM/dt (the rate at which the rocket loses mass) by $-R$, where R is the (positive) mass rate of fuel consumption, and we recognize that dv/dt is the acceleration of the rocket. With these changes, Eq. 9-41 becomes

$$Rv_{rel} = Ma \quad \text{(first rocket equation).} \tag{9-42}$$

Equation 9-42 holds at any instant, with the mass M, the fuel consumption rate R, and the acceleration a evaluated at that instant.

The left side of Eq. 9-42 has the dimensions of a force ($\text{kg} \cdot \text{m/s}^2 = \text{N}$) and depends only on design characteristics of the rocket engine, namely, the rate R at which it consumes fuel mass and the speed v_{rel} with which that mass is ejected relative to the rocket. We call this term Rv_{rel} the **thrust** of the rocket engine and represent it with T. Newton's second law emerges clearly if we write Eq. 9-42 as $T = Ma$, in which a is the acceleration of the rocket at the time that its mass is M.

Finding the Velocity

How will the velocity of a rocket change as it consumes its fuel? From Eq. 9-40 we have

$$dv = -v_{rel}\frac{dM}{M}.$$

Integrating leads to

$$\int_{v_i}^{v_f} dv = -v_{rel}\int_{M_i}^{M_f}\frac{dM}{M},$$

in which M_i is the initial mass of the rocket and M_f its final mass. Evaluating the integrals then gives

$$v_f - v_i = v_{rel}\ln\frac{M_i}{M_f} \quad \text{(second rocket equation)} \qquad (9\text{-}43)$$

for the increase in the speed of the rocket during the change in mass from M_i to M_f. (The symbol "ln" in Eq. 9-43 means the *natural logarithm.*) We see here the advantage of multistage rockets, in which M_f is reduced by discarding successive stages when their fuel is depleted. An ideal rocket would reach its destination with only its payload remaining.

Sample Problem 9-8

A rocket whose initial mass M_i is 850 kg consumes fuel at the rate $R = 2.3$ kg/s. The speed v_{rel} of the exhaust gases relative to the rocket engine is 2800 m/s.

(a) What thrust does the rocket engine provide?

SOLUTION: The **Key Idea** here is that the thrust T is equal to the product of the fuel consumption rate R and the relative speed v_{rel} at which exhaust gases are expelled:

$$T = Rv_{rel} = (2.3 \text{ kg/s})(2800 \text{ m/s})$$
$$= 6440 \text{ N} \approx 6400 \text{ N}. \qquad \text{(Answer)}$$

(b) What is the initial acceleration of the rocket?

SOLUTION: We can relate the thrust T of a rocket to the magnitude a of the resulting acceleration with $T = Ma$, where M is the rocket's mass. The **Key Idea**, however, is that M decreases and a increases as fuel is consumed. Because we want the initial value of a here, we must use the initial value M_i of the mass, finding that

$$a = \frac{T}{M_i} = \frac{6440 \text{ N}}{850 \text{ kg}} = 7.6 \text{ m/s}^2. \qquad \text{(Answer)}$$

To be launched from Earth's surface, a rocket must have an initial acceleration greater than $g = 9.8$ m/s^2. Put another way, the thrust T of the rocket engine must exceed the initial gravitational force on the rocket, which here has the magnitude $M_i g$, which gives us (850 kg)(9.8 m/s^2), or 8330 N. Because the acceleration or thrust requirement is not met (here $T = 6400$ N), our rocket could not be launched from Earth's surface by itself; it would require another, more powerful, rocket.

(c) Suppose, instead, that the rocket is launched from a spacecraft already in deep space, where we can neglect any gravitational force acting on it. The mass M_f of the rocket when its fuel is exhausted is 180 kg. What is its speed relative to the spacecraft at that time? Assume that the spacecraft is so massive that the launch does not alter its speed.

SOLUTION: The **Key Idea** here is that the rocket's final speed v_f (when the fuel is exhausted) depends on the ratio M_i/M_f of its initial mass to its final mass, as given by Eq. 9-43. With the initial speed $v_i = 0$, we have

$$v_f = v_{rel}\ln\frac{M_i}{M_f}$$
$$= (2800 \text{ m/s})\ln\frac{850 \text{ kg}}{180 \text{ kg}}$$
$$= (2800 \text{ m/s})\ln 4.72 \approx 4300 \text{ m/s}. \qquad \text{(Answer)}$$

Note that the ultimate speed of the rocket can exceed the exhaust speed v_{rel}.

9-8 External Forces and Internal Energy Changes

As the ice skater in Fig. 9-13*a* pushes herself away from a railing, there is a force $\vec{F}$ on her from the railing at angle ϕ to the horizontal. This force accelerates her, increasing her speed until she leaves the railing (Fig. 9-13*b*). Thus, her kinetic energy is increased via the force. This example differs in two ways from earlier examples in which an object's kinetic energy is changed via a force:

1. Previously, each part of an object moved rigidly in the same direction. Here the skater's arm does not move like the rest of her body.
2. Previously, energy was transferred between the object (or system) and its environment via an external force; that is, the force did work. Here the energy is transferred internally (from one part of the system to another) via the external force $\vec{F}$. In particular, the energy is transferred from internal biochemical energy of the skater's muscles to kinetic energy of her body as a whole.

We want to relate the external force $\vec{F}$ to that internal energy transfer.

In spite of the two differences listed here, we can relate $\vec{F}$ to the change in kinetic energy much as we did for a particle in Section 7-3. To do so, we first mentally concentrate the mass M of the skater at her center of mass so that we can treat her as a particle at the center of mass. Then we treat the external force $\vec{F}$ as acting on the particle (Fig. 9-13*c*). The horizontal force component $F \cos \phi$ accelerates that particle, resulting in a change ΔK in its kinetic energy during a displacement of magnitude d. As we prove later, we can relate these quantities with

$$\Delta K = Fd \cos \phi. \tag{9-44}$$

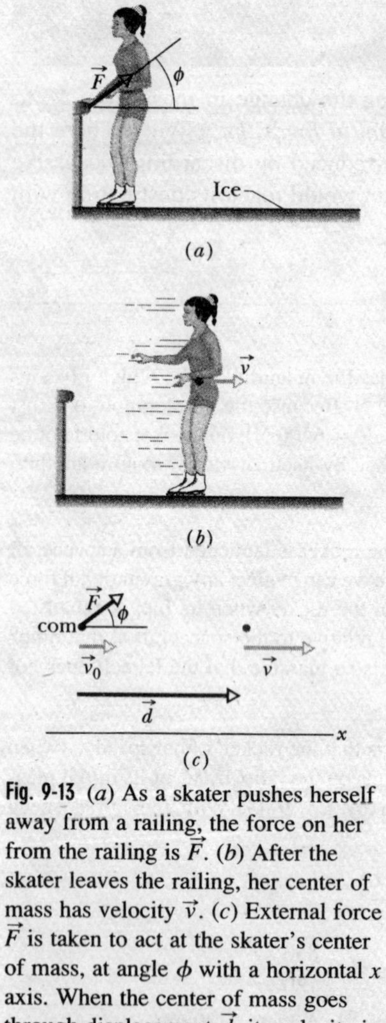

Fig. 9-13 (*a*) As a skater pushes herself away from a railing, the force on her from the railing is $\vec{F}$. (*b*) After the skater leaves the railing, her center of mass has velocity $\vec{v}$. (*c*) External force $\vec{F}$ is taken to act at the skater's center of mass, at angle ϕ with a horizontal x axis. When the center of mass goes through displacement $\vec{d}$, its velocity is changed from $\vec{v}_0$ to $\vec{v}$ by the horizontal component of $\vec{F}$.

We can imagine cases in which the external force would also change the height of the skater's center of mass and thus cause a change ΔU in the gravitational potential energy of the skater–Earth system. To include such a change ΔU, we can rewrite Eq. 9-44 as

$$\Delta K + \Delta U = Fd \cos \phi. \tag{9-45}$$

The left side here is ΔE_{mec}, the change in the mechanical energy of the system. Thus, in the general case, we have

$$\Delta E_{\text{mec}} = Fd \cos \phi \quad \text{(external force, change in } E_{\text{mec}}\text{).} \tag{9-46}$$

Next let us consider the energy transfers in the ice skater, which we now take as the system. Although an external force acts on the system, the force does not transfer energy to or from the system. Thus, the total energy E of the system cannot change: $\Delta E = 0$. We know that when the ice skater pushes off from the rail, there is a change not only in the mechanical energy of her center of mass, but also in the energy of her muscles. Without going into the details of the change in muscular energy, let us just represent it with ΔE_{int} (for a change in internal energy). Then we can write $\Delta E = 0$ as

$$\Delta E_{\text{int}} + \Delta E_{\text{mec}} = 0, \tag{9-47}$$

or

$$\Delta E_{\text{int}} = -\Delta E_{\text{mec}}. \tag{9-48}$$

This equation means that as E_{mec} increases for the ice skater, E_{int} decreases by just as much. Substituting for ΔE_{mec} in Eq. 9-48 from Eq. 9-46, we have

$$\Delta E_{\text{int}} = -Fd \cos \phi \quad \text{(external force, internal energy change).} \tag{9-49}$$

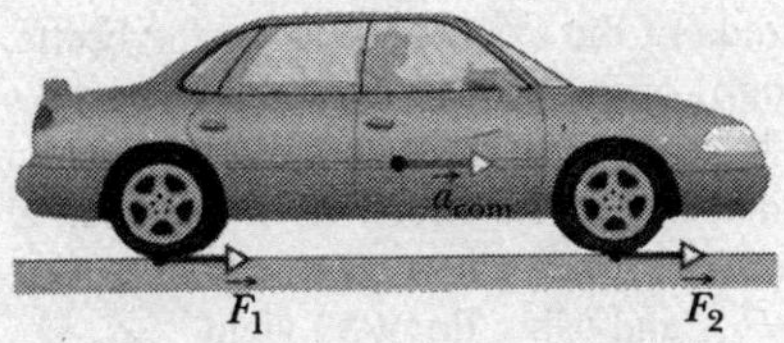

Fig. 9-14 A vehicle accelerates to the right using four-wheel drive. The road exerts four frictional forces (two of them shown) on the bottom surfaces of the tires. Taken together, these four forces make up the net external force $\vec{F}$ acting on the car.

This equation relates the change ΔE_{int} in internal energy that is made via the external force $\vec{F}$. If $\vec{F}$ is not constant in magnitude, we can replace the symbol F in Eqs. 9-46 and 9-49 with the symbol F_{avg} for the average magnitude of $\vec{F}$.

Although we have used an ice skater to derive Eqs. 9-46 and 9-49, the equations hold for other objects in which a change ΔE_{int} in the internal energy of a system is made via an external force. For example, consider a vehicle with four-wheel drive (all four wheels are made to turn by the engine) as its speed is increased. During the acceleration, the engine causes the tires to push backward on the road's surface. This push produces frictional forces that act on each tire in the forward direction (Fig. 9-14). The net external force $\vec{F}$, which is the sum of these frictional forces, gives the car's center of mass an acceleration $\vec{a}$. Thus, there is a transfer of energy from the internal energy stored in the fuel to the kinetic energy of the car. If $\vec{F}$ is constant, then for a given displacement $\vec{d}$ of the car's center of mass along level road, we can relate the change ΔK in the car's kinetic energy to the external force $\vec{F}$ with Eq. 9-45, with $\Delta U = 0$ and $\phi = 0$.

If the driver applies the brakes, Eq. 9-45 still holds. Now $\vec{F}$ due to the frictional forces is toward the rear, and $\phi = 180°$. Energy is now transferred from kinetic energy of the car's center of mass to thermal energy of the brakes.

Proof of Equation 9-44

Let us return to the ice skater and Fig. 9-13. Suppose that during displacement $\vec{d}$ of her center of mass, the velocity of her center of mass changes from $\vec{v}_0$ to $\vec{v}$. Then the magnitude of $\vec{v}$ is, by Eq. 2-16,

$$v^2 = v_0^2 + 2a_x d, \tag{9-50}$$

where a_x is her acceleration. Multiplying both sides of Eq. 9-50 by the skater's mass M and rearranging yield

$$\tfrac{1}{2}Mv^2 - \tfrac{1}{2}Mv_0^2 = Ma_x d. \tag{9-51}$$

The left side of Eq. 9-51 is the difference between the final kinetic energy K_f of the center of mass and the initial kinetic energy K_i. This difference is the change ΔK in the kinetic energy of the center of mass due to the force $\vec{F}$. Making that substitution, and substituting the product $F \cos\phi$ for the product Ma_x according to Newton's second law, we get

$$\Delta K = Fd \cos\phi,$$

as we intended to prove.

Sample Problem 9-9

When a click beetle is upside down on its back, it jumps upward by suddenly arching its back, transferring energy stored in a muscle to mechanical energy. This launching mechanism produces an audible click, giving the beetle its name. Videotape of a certain click-beetle jump shows that the center of mass of a click beetle of mass $m = 4.0 \times 10^{-6}$ kg moved directly upward by 0.77 mm during the launch and then to a maximum height of $h = 0.30$ m. What was the average magnitude of the external force $\vec{F}$ on the beetle's back from the floor during the launch?

SOLUTION: The **Key Idea** here is that during the launch, energy is transferred from an internal energy of the beetle to the mechanical energy of the beetle–Earth system, in the amount ΔE_{mec}. The transfer is made via the external force $\vec{F}$. To find the magnitude of the force with Eq. 9-46, we first need an expression for ΔE_{mec}.

Let the system's mechanical energy be $E_{mec,0}$ just before the launch and $E_{mec,1}$ at the end of the launch. Then the change ΔE_{mec} is

$$\Delta E_{mec} = E_{mec,1} - E_{mec,0}. \tag{9-52}$$

We must now find expressions for $E_{mec,0}$ and $E_{mec,1}$. Let the gravitational potential energy of the beetle–Earth system be $U_0 = 0$ when the beetle is on the floor. Also, just before the launch, the kinetic energy of the beetle's center of mass is $K_0 = 0$, so the beginning mechanical energy $E_{mec,0}$ is 0. Unfortunately, when we try to find $E_{mec,1}$, we get stuck because we do not know the beetle's kinetic energy K_1 or speed v_1 at the end of the launch.

We get unstuck with a second **Key Idea**: The mechanical energy of the system does not change from the time the beetle is launched until it reaches its maximum height. We can find this mechanical energy $E_{mec,1}$ at the maximum height because we know the beetle's speed ($v = 0$) and height ($y = h$) there. Thus, we have

$$E_{mec,1} = K + U = \tfrac{1}{2}mv^2 + mgy = 0 + mgh = mgh.$$

Substituting this and $E_{mec,0} = 0$ into Eq. 9-52 now gives us

$$\Delta E_{mec} = mgh - 0 = mgh. \tag{9-53}$$

We can now use Eq. 9-46 to relate this change in mechanical energy to the external force. We write

$$\Delta E_{mec} = F_{avg} d \cos\phi. \tag{9-54}$$

F_{avg} is the average magnitude of the external force on the beetle, d is the magnitude (0.77 mm) of the displacement of the beetle's center of mass during the launch (when the external force acts on it), and ϕ (= 0°) is the angle between the directions of the external force and the displacement.

Solving Eq. 9-54 for F_{avg} and using Eq. 9-53 yield

$$F_{avg} = \frac{\Delta E_{mec}}{d\cos\phi} = \frac{mgh}{d\cos\phi} = \frac{(4.0\times10^{-6}\text{ kg})(9.8\text{ m/s}^2)(0.30\text{ m})}{(7.7\times10^{-4}\text{ m})(\cos 0°)} = 1.5\times10^{-2}\text{ N}. \quad \text{(Answer)}$$

This force magnitude may seem small, but to the click beetle it is enormous because, as you can show, it gives the beetle an acceleration of over 380g during the launch.

REVIEW & SUMMARY

Center of Mass The **center of mass** of a system of n particles is defined to be the point whose coordinates are given by

$$x_{com} = \frac{1}{M}\sum_{i=1}^{n} m_i x_i, \quad y_{com} = \frac{1}{M}\sum_{i=1}^{n} m_i y_i, \quad z_{com} = \frac{1}{M}\sum_{i=1}^{n} m_i z_i, \tag{9-5}$$

or

$$\vec{r}_{com} = \frac{1}{M}\sum_{i=1}^{n} m_i \vec{r}_i, \tag{9-8}$$

where M is the total mass of the system. If the mass is continuously distributed, the center of mass is given by

$$x_{com} = \frac{1}{M}\int x\,dm, \quad y_{com} = \frac{1}{M}\int y\,dm, \quad z_{com} = \frac{1}{M}\int z\,dm. \tag{9-9}$$

If the density (mass per unit volume) is uniform, then Eq. 9-9 can be rewritten as

$$x_{com} = \frac{1}{V}\int x\,dV, \quad y_{com} = \frac{1}{V}\int y\,dV, \quad z_{com} = \frac{1}{V}\int z\,dV, \tag{9-11}$$

where V is the volume occupied by M.

Newton's Second Law for a System of Particles The motion of the center of mass of any system of particles is governed by **Newton's second law for a system of particles**, which is

$$\vec{F}_{net} = M\vec{a}_{com}. \tag{9-14}$$

Here $\vec{F}_{net}$ is the net force of all the *external* forces acting on the system, M is the total mass of the system, and $\vec{a}_{com}$ is the acceleration of the system's center of mass.

Linear Momentum and Newton's Second Law For a single particle, we define a quantity $\vec{p}$ called its **linear momentum** as

$$\vec{p} = m\vec{v}, \tag{9-22}$$

and can write Newton's second law in terms of this momentum:

$$\vec{F}_{net} = \frac{d\vec{p}}{dt}. \tag{9-23}$$

For a system of particles these relations become

$$\vec{P} = M\vec{v}_{com} \quad \text{and} \quad \vec{F}_{net} = \frac{d\vec{P}}{dt}. \tag{9-25, 9-27}$$

Conservation of Linear Momentum If a system is isolated so that no net *external* force acts on the system, the linear momentum $\vec{P}$ of the system remains constant:

$$\vec{P} = \text{constant} \quad \text{(closed, isolated system)}. \tag{9-29}$$

This can also be written as

$$\vec{P}_i = \vec{P}_f \quad \text{(closed, isolated system)}, \tag{9-30}$$

where the subscripts refer to the values of $\vec{P}$ at some initial time and at a later time. Equations 9-29 and 9-30 are equivalent statements of the **law of conservation of linear momentum.**

Variable-Mass Systems If a system has varying mass, we redefine the system, enlarging its boundaries until they encompass a larger system whose mass *does* remain constant; then we apply the law of conservation of linear momentum. For a rocket, this means that the system includes both the rocket and its exhaust gases. Analysis of such a system shows that in the absence of external forces a rocket accelerates at an instantaneous rate given by

$$Rv_{rel} = Ma \quad \text{(first rocket equation)}, \tag{9-42}$$

in which M is the rocket's instantaneous mass (including unexpended fuel), R is the fuel consumption rate, and v_{rel} is the fuel's exhaust speed relative to the rocket. The term Rv_{rel} is the **thrust** of the rocket engine. For a rocket with constant R and v_{rel}, whose speed changes from v_i to v_f when its mass changes from M_i to M_f,

$$v_f - v_i = v_{rel} \ln \frac{M_i}{M_f} \quad \text{(second rocket equation).} \qquad (9\text{-}43)$$

External Forces and Internal Energy Changes Energy can be transferred inside a system between an internal energy and mechanical energy via an external force $\vec{F}$. The change ΔE_{int} in the internal energy is related by

$$\Delta E_{int} = -Fd \cos \phi \qquad (9\text{-}49)$$

to the external force, the displacement $\vec{d}$ of the system's center of mass, and the angle ϕ between the directions of $\vec{F}$ and $\vec{d}$. The change in the mechanical energy is

$$\Delta E_{mec} = \Delta K + \Delta U = Fd \cos \phi \qquad (9\text{-}46, 9\text{-}45)$$

QUESTIONS

1 Figure 9-15 shows four uniform square metal plates that have had a section removed. The origin of the x and y axes is at the center of the original plate, and in each case the center of mass of the removed section was at the origin. In each case, where is the center of mass of the remaining section of the plate? Answer in terms of quadrants, lines, or points.

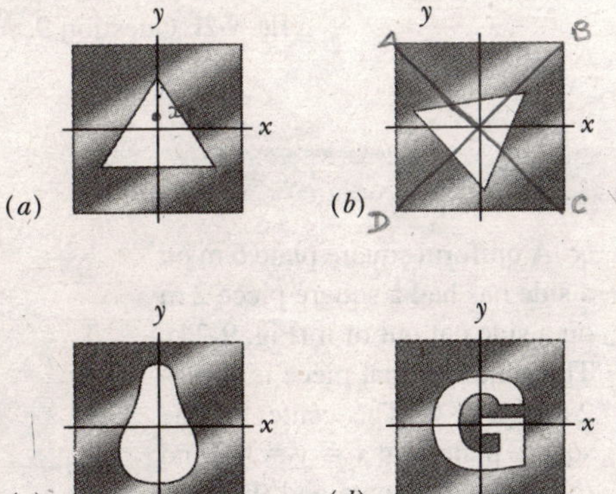

Fig. 9-15 Question 1.

2 Some skilled basketball players seem to hang in midair during a jump at the basket, allowing them more time to shift the ball from hand to hand and then into the basket. If a player raises arms or legs during a jump, does the player's time in the air increase, decrease, or stay the same?

3 In Fig. 9-16, a penguin stands at the left edge of a uniform sled of length L, which lies on frictionless ice. The sled and penguin have equal masses. (a) Where is the center of mass of the sled? (b) How far and in what direction is the center of the sled from the center of mass of the sled–penguin system?

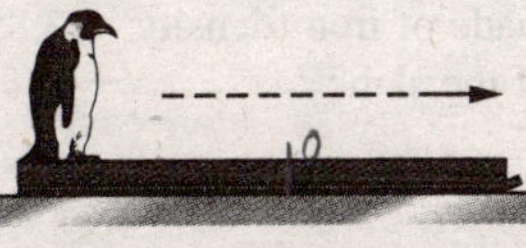

Fig. 9-16 Question 3.

The penguin then waddles to the right edge of the sled, and the sled slides on the ice. (c) Does the center of mass of the sled–penguin system move leftward, rightward, or not at all? (d) Now how far and in what direction is the center of the sled from the center of mass of the sled–penguin system? (e) How far does the penguin move relative to the sled? Relative to the center of mass of the sled–penguin system, how far does (f) the center of the sled move and (g) the penguin move? (Warm-up for Problem 19)

4 In Question 3 and Fig. 9-16, suppose that the sled and penguin are initially moving rightward at speed v_0. (a) As the penguin waddles to the right edge of the sled, is the speed v of the sled less than, greater than, or equal to v_0? (b) If the penguin then waddles back to the left edge, during that motion is the speed v of the sled less than, greater than, or equal to v_0?

5 Figure 9-17 shows an overhead view of four particles of equal mass sliding over a frictionless surface at constant velocity. The directions of the velocities are indicated; their magnitudes are equal. Consider pairing the particles. Which pairs form a system with a center of mass that (a) is stationary, (b) is stationary and at the origin, and (c) passes through the origin?

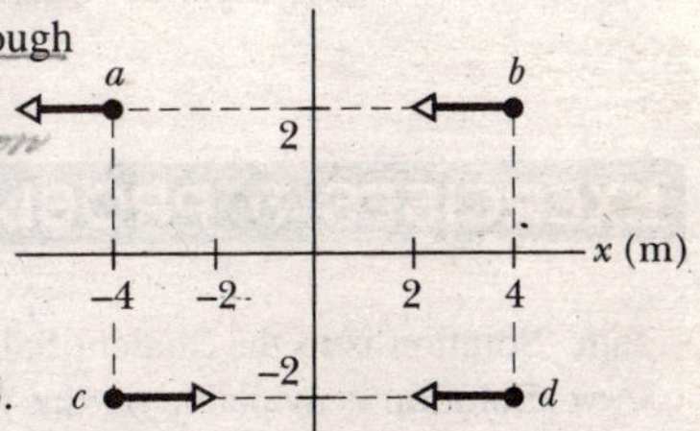

Fig. 9-17 Question 5.

6 Figure 9-18 shows an overhead view of three particles on which external forces act. The magnitudes and directions of the forces on two of the particles are indicated. What are the magnitude and direction of the force acting on the third particle if the center of mass of the three-particle system is (a) stationary, (b) moving at a constant velocity rightward, and (c) accelerating rightward?

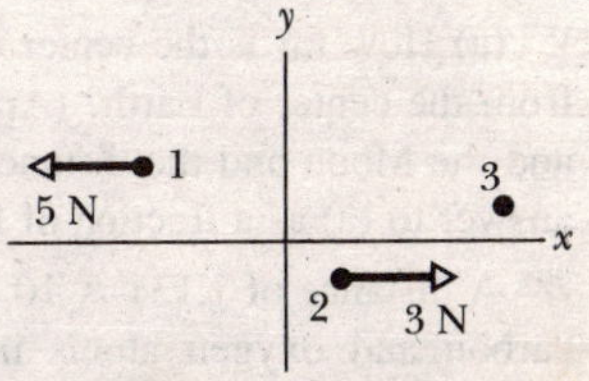

Fig. 9-18 Question 6.

7 A container sliding along an x axis on a frictionless surface explodes into three pieces. The pieces then move along the x axis in the directions indicated in Fig. 9-19. The following table gives four sets of magnitudes (in kg · m/s) for the momenta $\vec{p}_1$, $\vec{p}_2$, and $\vec{p}_3$ of the pieces. Rank the sets according to the initial speed of the container, greatest first.

	p_1	p_2	p_3		p_1	p_2	p_3
(a)	10	2	6	(b)	10	6	2
(c)	2	10	6	(d)	6	2	10

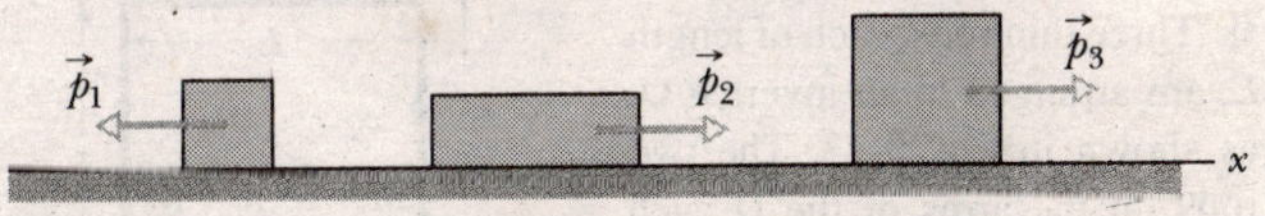

Fig. 9-19 Question 7.

8 A spaceship that is moving along an x axis separates into two parts, like the hauler in Fig. 9-9. (a) Which of the graphs in

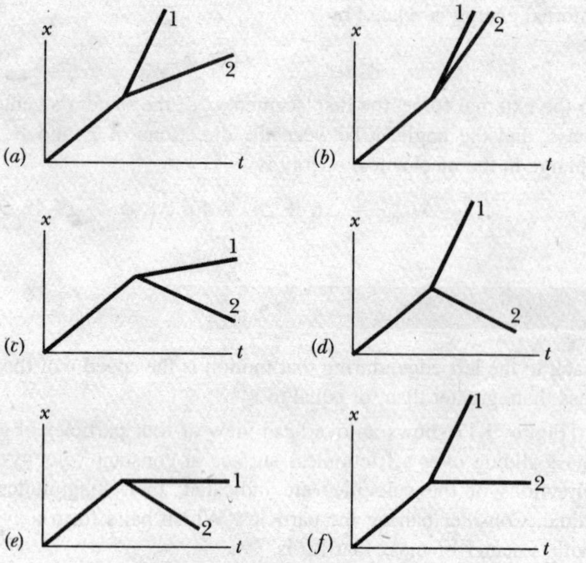

Fig. 9-20 Question 8.

Fig. 9-20 could possibly give the position versus time for the ship and the two parts? (b) Which of the numbered lines pertains to the trailing part? (c) Rank the possible graphs according to the relative speed between the parts, greatest first.

9. Consider a box, like that in Sample Problem 9-5, which explodes into two pieces while moving with a constant positive velocity along an x axis. If one piece, with mass m_1, ends up with positive velocity $\vec{v}_1$, then the second piece, with mass m_2, could end up with (a) a positive velocity $\vec{v}_2$ (Fig. 9-21*a*), (b) a negative velocity $\vec{v}_2$ (Fig. 9-21*b*), or (c) zero velocity (Fig. 9-21*c*). Rank those three possible results for the second piece according to the corresponding magnitude of $\vec{v}_1$, greatest first.

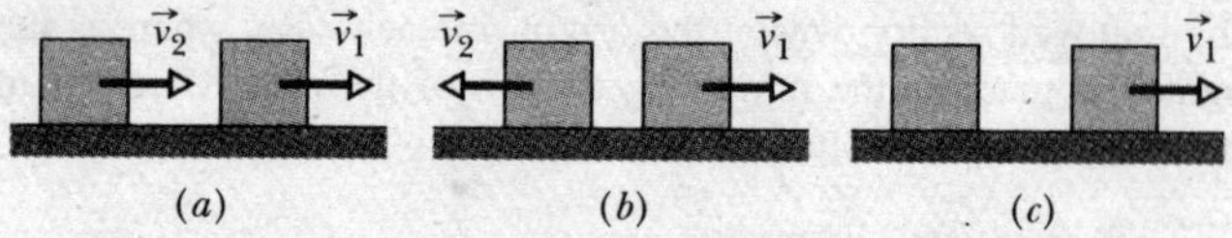

Fig. 9-21 Question 9.

EXERCISES & PROBLEMS

ssm Solution is in the Student Solutions Manual.
www Solution is available on the World Wide Web at: http://www.wiley.com/college/hrw
ilw Solution is available on the Interactive LearningWare.

SEC. 9-2 The Center of Mass

1E. (a) How far is the center of mass of the Earth–Moon system from the center of Earth? (Appendix C gives the masses of Earth and the Moon and the distance between the two.) (b) Express the answer to (a) as a fraction of Earth's radius R_e. **ssm**

2E. A distance of 1.131×10^{-10} m lies between the centers of the carbon and oxygen atoms in a carbon monoxide (CO) gas molecule. Locate the center of mass of a CO molecule relative to the carbon atom. (Find the masses of C and O in Appendix F.)

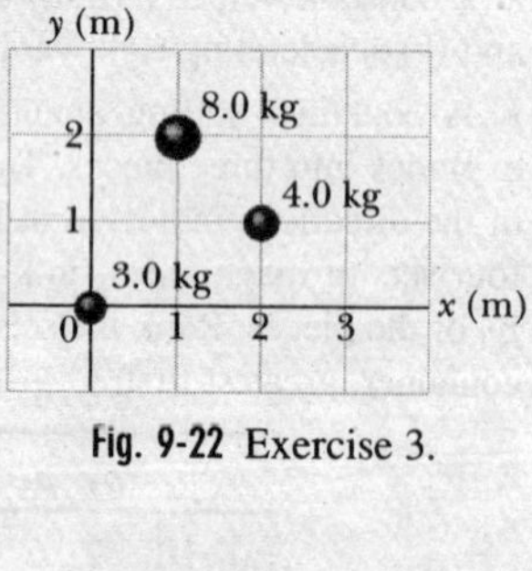

Fig. 9-22 Exercise 3.

3E. What are (a) the x coordinate and (b) the y coordinate of the center of mass of the three-particle system shown in Fig. 9-22? (c) What happens to the center of mass as the mass of the topmost particle is gradually increased? **ssm**

4E. Three thin rods, each of length L, are arranged in an inverted U, as shown in Fig. 9-23. The two rods on the arms of the U each have mass M; the third rod has mass $3M$. Where is the center of mass of the assembly?

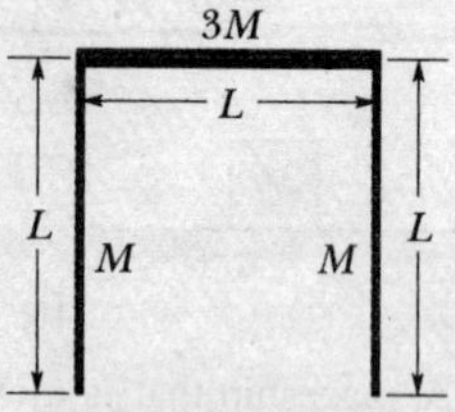

Fig. 9-23 Exercise 4.

5E. A uniform square plate 6 m on a side has had a square piece 2 m on a side cut out of it (Fig. 9-24). The center of that piece is at $x = 2$ m, $y = 0$. The center of the square plate is at $x = y = 0$. Find (a) the x coordinate and (b) the y coordinate of the center of mass of the remaining piece.

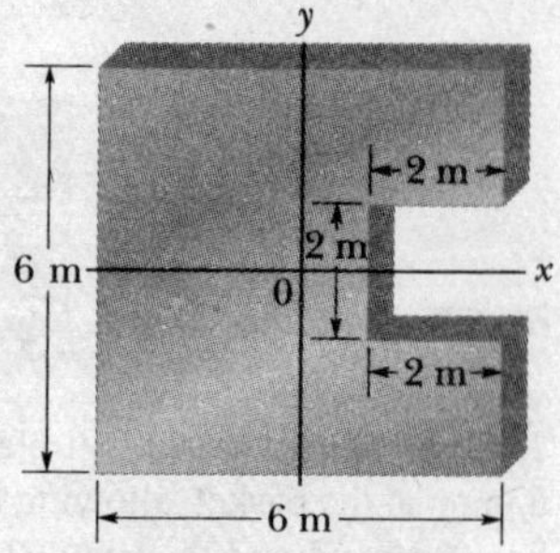

Fig. 9-24 Exercise 5.

6P. Figure 9-25 shows the dimensions of a composite slab; half the slab is made of aluminum (density = 2.70 g/cm³) and half is made of iron (density = 7.85 g/cm³). Where is the center of mass of the slab?

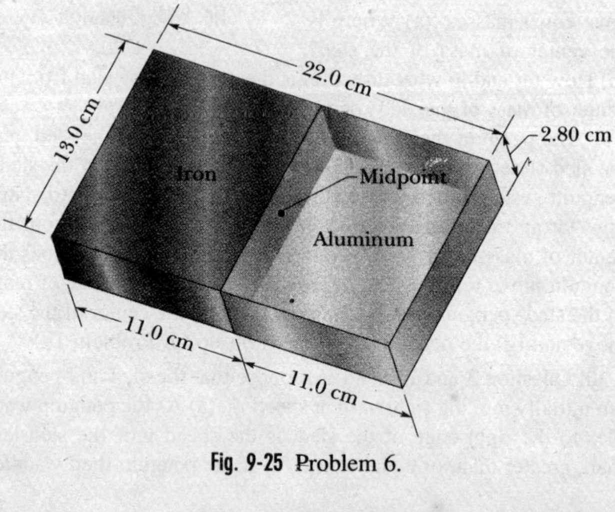

Fig. 9-25 Problem 6.

7P. In the ammonia (NH_3) molecule (see Fig. 9-26), the three hydrogen (H) atoms form an equilateral triangle; the center of the triangle is 9.40×10^{-11} m from each hydrogen atom. The nitrogen (N) atom is at the apex of a pyramid, with the three hydrogen atoms forming the base. The nitrogen-to-hydrogen atomic mass ratio is 13.9, and the nitrogen-to-hydrogen distance is 10.14×10^{-11} m. Locate the center of mass of the molecule relative to the nitrogen atom. ilw

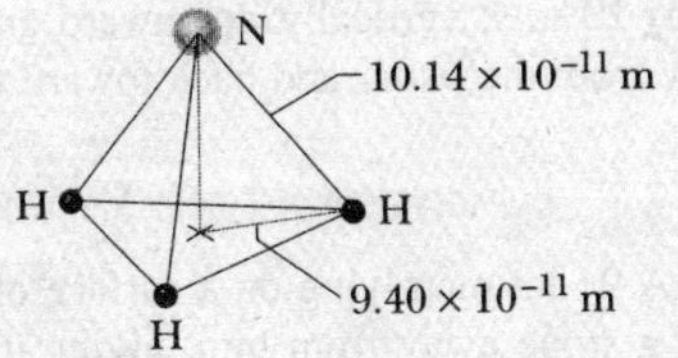

Fig. 9-26 Problem 7.

8P. Figure 9-27 shows a cubical box that has been constructed from metal plate of uniform density and negligible thickness. The box is open at the top and has edge length 40 cm. Find (a) the x coordinate, (b) the y coordinate, and (c) the z coordinate of the center of mass of the box.

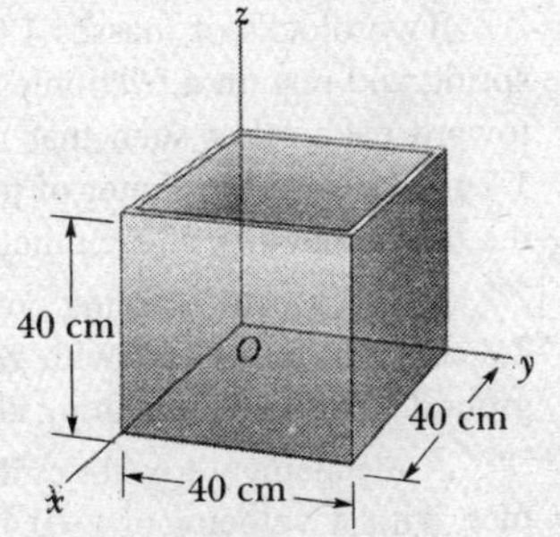

Fig. 9-27 Problem 8.

9P. A right cylindrical can with mass M, height H, and uniform density is initially filled with soda of mass m (Fig. 9-28). We punch small holes in the top and bottom to drain the soda; we then consider the height h of the center of mass of the can and any soda within it. What is h (a) initially and (b) when all the soda has drained? (c) What happens to h during the draining of the soda? (d) If x is the height of the remaining soda at any given instant, find x (in terms of M, H, and m) when the center of mass reaches its lowest point. ssm

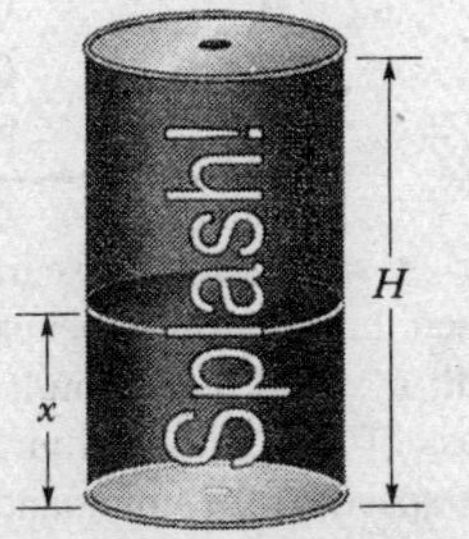

Fig. 9-28 Problem 9.

SEC. 9-3 Newton's Second Law for a System of Particles

10E. Two skaters, one with mass 65 kg and the other with mass 40 kg, stand on an ice rink holding a pole of length 10 m and negligible mass. Starting from the ends of the pole, the skaters pull themselves along the pole until they meet. How far does the 40 kg skater move?

11E. An old Chrysler with mass 2400 kg is moving along a straight stretch of road at 80 km/h. It is followed by a Ford with mass 1600 kg moving at 60 km/h. How fast is the center of mass of the two cars moving? ssm

12E. A man of mass m clings to a rope ladder suspended below a balloon of mass M; see Fig. 9-29. The balloon is stationary with respect to the ground. (a) If the man begins to climb the ladder at speed v (with respect to the ladder), in what direction and with what speed (with respect to the ground) will the balloon move? (b) What is the state of the motion after the man stops climbing?

Fig. 9-29 Exercise 12.

13P. A stone is dropped at $t = 0$. A second stone, with twice the mass of the first, is dropped from the same point at $t = 100$ ms. (a) How far below the release point is the center of mass of the two stones at $t = 300$ ms? (Neither stone has yet reached the ground.) (b) How fast is the center of mass of the two-stone system moving at that time? ilw

14P. A 1000 kg automobile is at rest at a traffic signal. At the instant the light turns green, the automobile starts to move with a constant acceleration of 4.0 m/s^2. At the same instant a 2000 kg truck, traveling at a constant speed of 8.0 m/s, overtakes and passes the automobile. (a) How far is the center of mass of the automobile–truck system from the traffic light at $t = 3.0$ s? (b) What is the speed of the center of mass of the automobile–truck system then?

15P. A shell is shot with an initial velocity $\vec{v}_0$ of 20 m/s, at an angle of 60° with the horizontal. At the top of the trajectory, the shell explodes into two fragments of equal mass (Fig. 9-30). One fragment, whose speed immediately after the explosion is zero, falls vertically. How far from the gun does the other fragment land, assuming that the terrain is level and that air drag is negligible? ssm

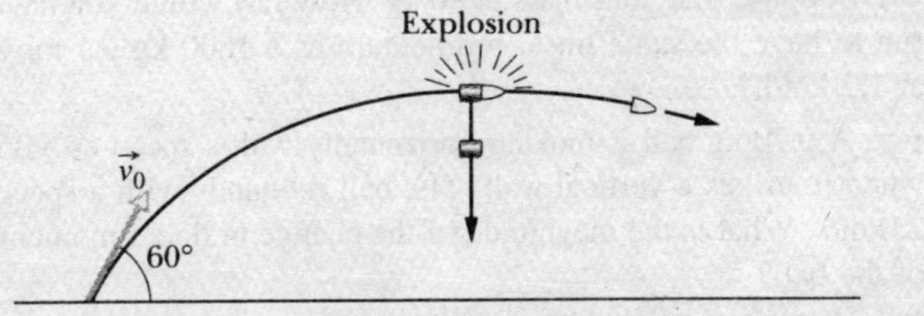

Fig. 9-30 Problem 15.

16P. A big olive ($m = 0.50$ kg) lies at the origin and a big Brazil nut ($M = 1.5$ kg) lies at the point (1.0, 2.0) m in an xy plane. At $t = 0$, a force $\vec{F}_o = (2\hat{i} + 3\hat{j})$ N begins to act on the olive, and a force $\vec{F}_n = (-3\hat{i} - 2\hat{j})$ N begins to act on the nut. In unit-vector notation, what is the displacement of the center of mass of the olive–nut system at $t = 4.0$ s, with respect to its position at $t = 0$?

17P. Two identical containers of sugar are connected by a massless cord that passes over a massless, frictionless pulley with a diameter of 50 mm (Fig. 9-31). The two containers are at the same level. Each originally has a mass of 500 g. (a) What is the horizontal position of their center of mass? (b) Now 20 g of sugar is transferred from one container to the other, but the containers are prevented from moving. What is the

Fig. 9-31 Problem 17.

new horizontal position of their center of mass, relative to the central axis through the lighter container? (c) The two containers are now released. In what direction does the center of mass move? (d) What is its acceleration? ssm

18P. Ricardo, of mass 80 kg, and Carmelita, who is lighter, are enjoying Lake Merced at dusk in a 30 kg canoe. When the canoe is at rest in the placid water, they exchange seats, which are 3.0 m apart and symmetrically located with respect to the canoe's center. Ricardo notices that the canoe moves 40 cm relative to a submerged log during the exchange and calculates Carmelita's mass, which she has not told him. What is it?

19P. In Fig. 9-32*a*, a 4.5 kg dog stands on an 18 kg flatboat and is 6.1 m from the shore. He walks 2.4 m along the boat toward shore and then stops. Assuming there is no friction between the boat and the water, find how far the dog is then from the shore. (*Hint:* See Fig. 9-32*b*. The dog moves leftward and the boat moves rightward, but does the center of mass of the *boat* + *dog* system move?) ssm www

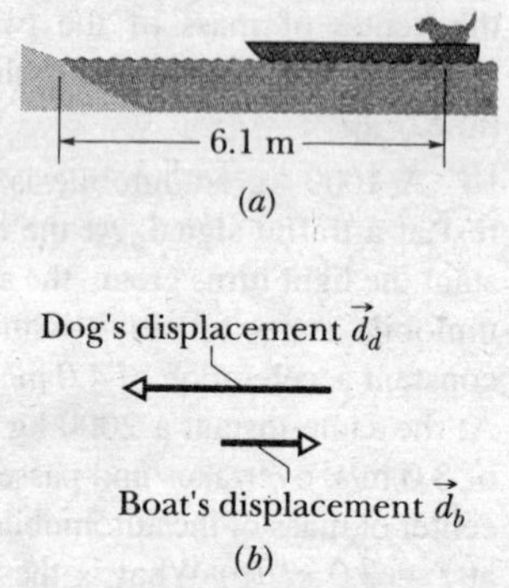

Fig. 9-32 Problem 19.

SEC. 9-5 The Linear Momentum of a System of Particles

20E. How fast must an 816 kg VW Beetle travel (a) to have the same linear momentum as a 2650 kg Cadillac going 16 km/h and (b) to have the same kinetic energy?

21E. Suppose that your mass is 80 kg. How fast would you have to run to have the same linear momentum as a 1600 kg car moving at 1.2 km/h?

22E. A 0.70 kg ball is moving horizontally with a speed of 5.0 m/s when it strikes a vertical wall. The ball rebounds with a speed of 2.0 m/s. What is the magnitude of the change in linear momentum of the ball?

23P. A 2100 kg truck traveling north at 41 km/h turns east and accelerates to 51 km/h. (a) What is the change in the kinetic energy of the truck? What are (b) the magnitude and (c) the direction of the change in the linear momentum of the truck? ilw

24P. A 0.165 kg cue ball with an initial speed of 2.00 m/s bounces off the rail in a game of pool, as shown from an overhead view in Fig. 9-33. For x and y axes located as shown, the bounce reverses the y component of the ball's velocity but does not alter the x component. (a) What is θ in Fig. 9-33? (b) What is the change in the ball's linear momentum in unit-vector notation? (The fact that the ball rolls is not relevant to either question.)

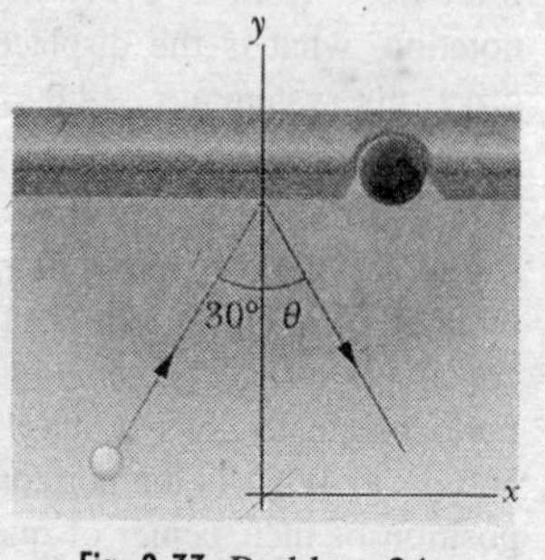

Fig. 9-33 Problem 24.

25P. An object is tracked by a radar station and found to have a position vector given by $\vec{r} = (3500 - 160t)\hat{\text{i}} + 2700\hat{\text{j}} + 300\hat{\text{k}}$, with $\vec{r}$ in meters and t in seconds. The radar station's x axis points east, its y axis north, and its z axis vertically up. If the object is a 250 kg meteorological missile, what are (a) its linear momentum, (b) its direction of motion, and (c) the net force on it?

26P. A 0.30 kg softball has a velocity of 15 m/s at an angle of 35° below the horizontal just before making contact with the bat. What is the magnitude of the change in momentum of the ball while it is in contact with the bat if the ball leaves the bat with a velocity of (a) 20 m/s, vertically downward and (b) 20 m/s, horizontally away from the batter and back toward the pitcher?

SEC. 9-6 Conservation of Linear Momentum

27E. A 91 kg man lying on a surface of negligible friction shoves a 68 g stone away from him, giving it a speed of 4.0 m/s. What velocity does the man acquire as a result? ssm

28E. Two blocks of masses 1.0 kg and 3.0 kg are connected by a spring and rest on a frictionless surface. They are given velocities toward each other such that the 1.0 kg block travels initially at 1.7 m/s toward the center of mass, which remains at rest. What is the initial velocity of the other block?

29E. A 75 kg man is riding on a 39 kg cart traveling at a speed of 2.3 m/s. He jumps off with zero horizontal speed relative to the ground. What is the resulting change in the speed of the cart? ssm

30E. A mechanical toy slides along an x axis on a frictionless surface with a velocity of $(-0.40 \text{ m/s})\hat{\text{i}}$ when two internal springs separate the toy into three parts, as given in the table. What is the velocity of part A?

Part	Mass (kg)	Velocity (m/s)
A	0.50	?
B	0.60	$0.20\hat{\text{i}}$
C	0.20	$0.30\hat{\text{i}}$

31E. A space vehicle is traveling at 4300 km/h relative to Earth when the exhausted rocket motor is disengaged and sent backward with a speed of 82 km/h relative to the command module. The mass of the motor is four times the mass of the module. What is the speed of the command module relative to Earth just after the separation?

32E. A railroad flatcar of weight W can roll without friction along a straight horizontal track. Initially, a man of weight w is standing on the car, which is moving to the right with speed v_0 (see Fig. 9-34). What is the change in velocity of the car if the man runs to the left (in the figure) so that his speed relative to the car is v_{rel}?

Fig. 9-34 Exercise 32.

33P. The last stage of a rocket, which is traveling at a speed of 7600 m/s, consists of two parts that are clamped together: a rocket case with a mass of 290.0 kg and a payload capsule with a mass of 150.0 kg. When the clamp is released, a compressed spring causes the two parts to separate with a relative speed of 910.0 m/s. What are the speeds of (a) the rocket case and (b) the payload after they have separated? Assume that all velocities are along the same line. Find the total kinetic energy of the two parts (c) before and (d) after they separate; account for any difference. ssm

34P. A 4.0 kg mess kit sliding on a frictionless surface explodes into two 2.0 kg parts, one moving at 3.0 m/s, due north, and the other at 5.0 m/s, 30° north of east. What is the original speed of the mess kit?

35P A certain radioactive nucleus can transform to another nucleus by emitting an electron and a neutrino. (The *neutrino* is one of the fundamental particles of physics.) Suppose that in such a transformation, the initial nucleus is stationary, the electron and neutrino are emitted along perpendicular paths, and the magnitudes of the linear momenta are 1.2×10^{-22} kg · m/s for the electron and 6.4×10^{-23} kg · m/s for the neutrino. As a result of the emissions, the new nucleus moves (recoils). (a) What is the magnitude of its linear momentum? What is the angle between its path and the path of (b) the electron and (c) the neutrino? (d) What is its kinetic energy if its mass is 5.8×10^{-26} kg? ilw

36P. Particle *A* and particle *B* are held together with a compressed spring between them. When they are released, the spring pushes them apart and they then fly off in opposite directions, free of the spring. The mass of *A* is 2.00 times the mass of *B*, and the energy stored in the spring was 60 J. Assume that the spring has negligible mass and that all its stored energy is transferred to the particles. Once that transfer is complete, what are the kinetic energies of (a) particle *A* and (b) particle *B*?

37P. A 20.0 kg body is moving in the positive x direction with a speed of 200 m/s when, owing to an internal explosion, it breaks into three parts. One part, with a mass of 10.0 kg, moves away from the point of explosion with a speed of 100 m/s in the positive y direction. A second fragment, with a mass of 4.00 kg, moves in the negative x direction with a speed of 500 m/s. (a) What is the velocity of the third (6.00 kg) fragment? (b) How much energy is released in the explosion? Ignore effects due to the gravitational force. ssm ilw www

38P. An object, with mass m and speed v relative to an observer, explodes into two pieces, one three times as massive as the other; the explosion takes place in deep space. The less massive piece stops relative to the observer. How much kinetic energy is added to the system in the explosion, as measured in the observer's reference frame?

39P. A vessel at rest explodes, breaking into three pieces. Two pieces, having equal mass, fly off perpendicular to one another with the same speed of 30 m/s. The third piece has three times the mass of each other piece. What are the magnitude and direction of its velocity immediately after the explosion? ssm

40P. An 8.0 kg body is traveling at 2.0 m/s with no external force acting on it. At a certain instant an internal explosion occurs, splitting the body into two chunks of 4.0 kg mass each. The explosion gives the chunks an additional 16 J of kinetic energy. Neither chunk leaves the line of original motion. Determine the speed and direction of motion of each of the chunks after the explosion.

SEC. 9-7 Systems with Varying Mass: A Rocket

41E. A 6090 kg space probe, moving nose-first toward Jupiter at 105 m/s relative to the Sun, fires its rocket engine, ejecting 80.0 kg of exhaust at a speed of 253 m/s relative to the space probe. What is the final velocity of the probe? ssm

42E. A rocket is moving away from the solar system at a speed of 6.0×10^3 m/s. It fires its engine, which ejects exhaust with a speed of 3.0×10^3 m/s relative to the rocket. The mass of the rocket at this time is 4.0×10^4 kg, and its acceleration is 2.0 m/s^2. (a) What is the thrust of the engine? (b) At what rate, in kilograms per second, is exhaust ejected during the firing?

43E. A rocket, which is in deep space and initially at rest relative to an inertial reference frame, has a mass of 2.55×10^5 kg, of which 1.81×10^5 kg is fuel. The rocket engine is then fired for 250 s, during which fuel is consumed at the rate of 480 kg/s. The speed of the exhaust products relative to the rocket is 3.27 km/s. (a) What is the rocket's thrust? After the 250 s firing, what are (b) the mass and (c) the speed of the rocket? ssm ilw

44E. Consider a rocket that is in deep space and at rest relative to an inertial reference frame. The rocket's engine is to be fired for a certain interval. What must be the rocket's *mass ratio* (ratio of initial to final mass) over that interval if the rocket's original speed relative to the inertial frame is to be equal to (a) the exhaust speed (speed of the exhaust products relative to the rocket) and (b) 2.0 times the exhaust speed?

45E. During a lunar mission, it is necessary to increase the speed of a spacecraft by 2.2 m/s when it is moving at 400 m/s relative to the Moon. The speed of the exhaust products from the rocket engine is 1000 m/s relative to the spacecraft. What fraction of the initial mass of the spacecraft must be burned and ejected to accomplish the speed increase?

46E. A railroad car moves at a constant speed of 3.20 m/s under a grain elevator. Grain drops into it at the rate of 540 kg/min. What is the magnitude of the force needed to keep the car moving at constant speed if friction is negligible?

47P. In Fig. 9-35, two long barges are moving in the same direction in still water, one with a speed of 10 km/h and the other with a speed of 20 km/h. While they are passing each other, coal is shov-

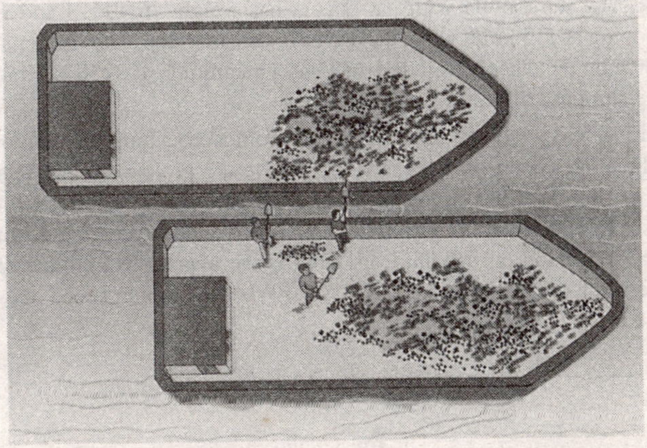

Fig. 9-35 Problem 47.

eled from the slower to the faster one at a rate of 1000 kg/min. How much additional force must be provided by the driving engines of (a) the fast barge and (b) the slow barge if neither is to change speed? Assume that the shoveling is always perfectly sideways and that the frictional forces between the barges and the water do not depend on the mass of the barges. ssm

48P. A 6100 kg rocket is set for vertical firing from the ground. If the exhaust speed is 1200 m/s, how much gas must be ejected each second if the thrust (a) is to equal the magnitude of the gravitational force on the rocket and (b) is to give the rocket an initial upward acceleration of 21 m/s^2?

SEC. 9-8 External Forces and Internal Energy Changes

49E. In 1981, Daniel Goodwin climbed 443 m up the *exterior* of the Sears Building in Chicago using suction cups and metal clips. (a) Approximate his mass and then compute how much energy he had to transfer from biomechanical (internal) energy to the gravitational potential energy of the Earth–Goodwin system to lift his center of mass to that height. (b) How much energy would he have had to transfer if he had, instead, taken the stairs inside the building (to the same height)?

50E. The summit of Mount Everest is 8850 m above sea level. (a) How much energy would a 90 kg climber expend against the gravitational force on him in climbing to the summit from sea level? (b) How many candy bars, at 1.25 MJ per bar, would supply an energy equivalent to this? Your answer should suggest that work done against the gravitational force is a very small part of the energy expended in climbing a mountain.

51E. A sprinter who weighs 670 N runs the first 7.0 m of a race in 1.6 s, starting from rest and accelerating uniformly. What are the sprinter's (a) speed and (b) kinetic energy at the end of the 1.6 s? (c) What average power does the sprinter generate during the 1.6 s interval?

52E. The luxury liner *Queen Elizabeth 2* has a diesel-electric powerplant with a maximum power of 92 MW at a cruising speed of 32.5 knots. What forward force is exerted on the ship at this speed? (1 knot = 1.852 km/h.)

53E. A swimmer moves through the water at an average speed of 0.22 m/s. The average drag force opposing this motion is 110 N. What average power is required of the swimmer?

54E. An automobile with passengers has weight 16,400 N and is moving at 113 km/h when the driver brakes to a stop. The frictional force on the wheels from the road has a magnitude of 8230 N. Find the stopping distance.

55E. A 55 kg woman leaps vertically from a crouching position in which her center of mass is 40 cm above the ground. As her feet leave the floor her center of mass is 90 cm above the ground; it rises to 120 cm at the top of her leap. (a) As she is pressing down on the ground during the leap, what is the average magnitude of the force on her from the ground? (b) What maximum speed does she attain? ssm www

56P. A 1500 kg automobile starts from rest on a horizontal road and gains a speed of 72 km/h in 30 s. (a) What is the kinetic energy of the auto at the end of the 30 s? (b) What is the average power required of the car during the 30 s interval? (c) What is the instantaneous power at the end of the 30 s interval, assuming that the acceleration is constant?

57P. A locomotive with a power capability of 1.5 MW can accelerate a train from a speed of 10 m/s to 25 m/s in 6.0 min. (a) Calculate the mass of the train. Find (b) the speed of the train and (c) the force accelerating the train as functions of time (in seconds) during the 6.0 min interval. (d) Find the distance moved by the train during the interval.

58P. Resistance to the motion of an automobile consists of road friction, which is almost independent of speed, and air drag, which is proportional to speed-squared. For a certain car with a weight of 12,000 N, the total resistant force F is given by $F = 300 + 1.8v^2$, where F is in newtons and v is in meters per second. Calculate the power (in horsepower) required to accelerate the car at 0.92 m/s^2 when the speed is 80 km/h.

Additional Problem

59. Some solidified lava contains a pattern of horizontal bubble layers separated vertically with few intermediate bubbles. (Researchers must slice open solidified lava to see these bubbles.) Apparently, as the lava was cooling, bubbles rising from the bottom of the lava separated into these layers and then were locked into place when the lava solidified. Similar layering of bubbles has been studied in certain creamy stouts poured fresh from a tap into a clear glass. The rising bubbles quickly become sorted into layers (Fig. 9-36). The bubbles trapped within a layer rise at speed v_t; the free bubbles between the layers rise at a greater speed v_f. Bubbles breaking free from the top of one layer rise to join the bottom of the next layer. Assume that the rate at which a layer loses height at its top is $dy/dt = v_f$ and the rate at which it gains height at its bottom is $dy/dt = v_f$. Also assume that $v_f = 2.0v_t = 1.0$ cm/s. What are the speed and direction of motion of the layer's center of mass?

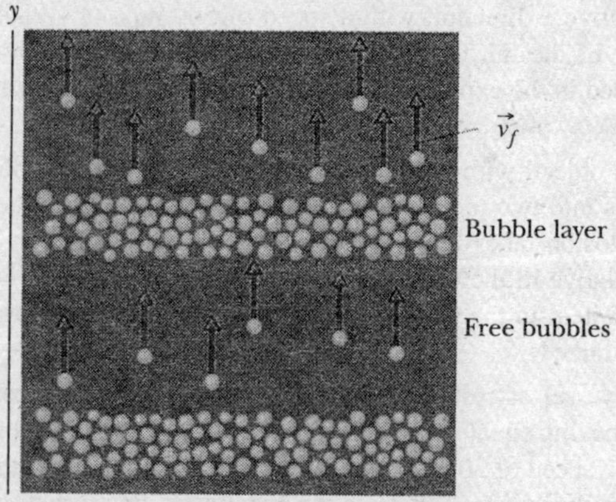

Fig. 9-36 Problem 59.

10 Collisions

Ronald McNair, a physicist and one of the astronauts killed in the explosion of the *Challenger* space shuttle, held a black belt in karate. Here he breaks several concrete slabs with one blow. In such karate demonstrations, a pine board or a concrete "patio block" is typically used. When struck, the board or block bends, storing energy like a stretched spring does, until a critical energy is reached. Then the object breaks. The energy necessary to break the board is about three times that for the block.

Why then is the board considerably easier to break?

The answer is in this chapter.

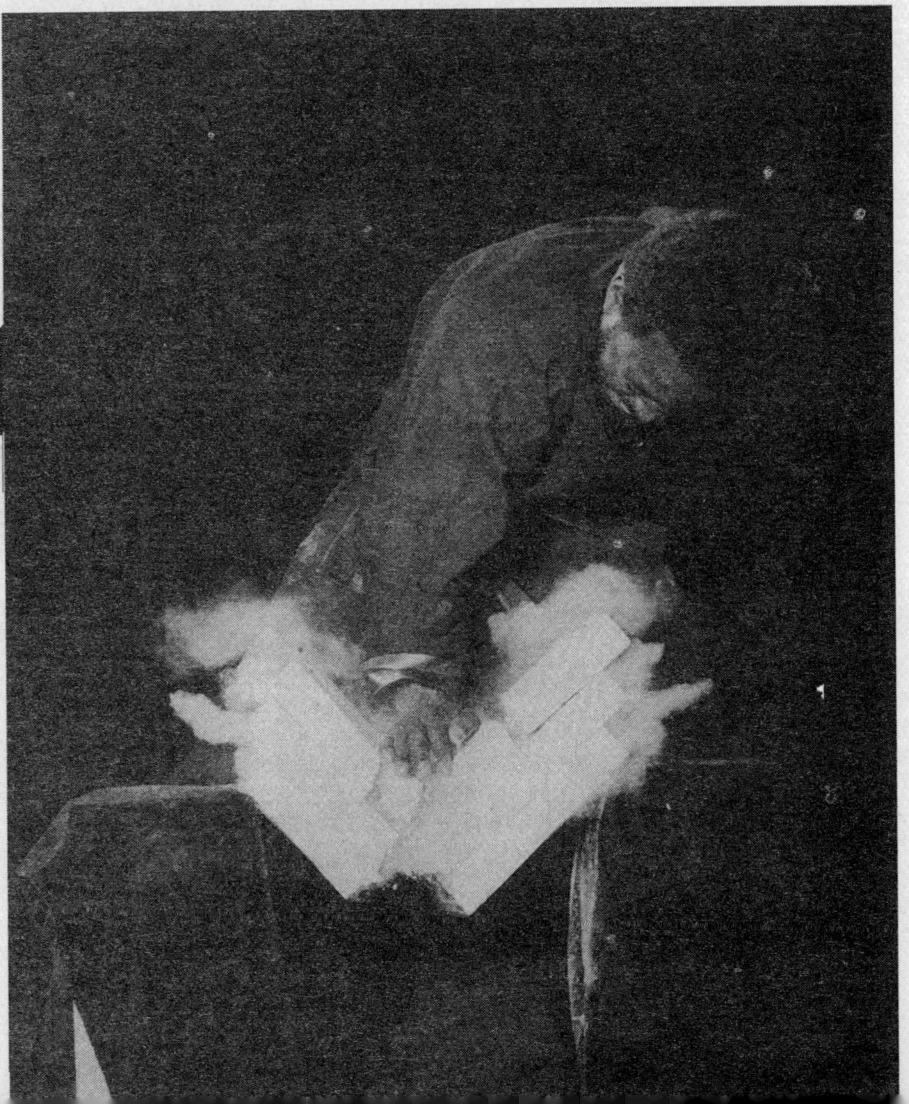

10-1 What Is a Collision?

In everyday language, a *collision* occurs when objects crash into each other. Although we will refine that definition, it conveys the meaning well enough and covers common collisions, such as those between billiard balls, a hammer and a nail, and—too commonly—automobiles. Figure 10-1*a* shows the lasting result of a collision (an impressive crash) that occurred about 20,000 years ago. Collisions range from the microscopic scale of subatomic particles (Fig. 10-1*b*) to the astronomic scale of colliding stars and colliding galaxies. Even when they occur on a human scale, they are often too brief to be visible, although they involve significant distortion of the colliding bodies (Fig. 10-1*c*).

We shall use the following more formal definition of collision:

> A **collision** is an isolated event in which two or more bodies (the colliding bodies) exert relatively strong forces on each other for a relatively short time.

We must be able to distinguish times that are *before, during,* and *after* a collision, as suggested in Fig. 10-2. That figure shows a system of two colliding bodies and indicates that the forces the bodies exert on each other are internal to the system.

Note that our formal definition of collision does not require the "crash" of our informal definition. When a space probe swings around a large planet to pick up speed (a *slingshot* encounter), that too is a collision. The probe and planet do not actually "touch," but a collision does not require contact, and a collision force does not have to be a force involving contact; it can just as easily be a gravitational force, as in this case.

(*a*)

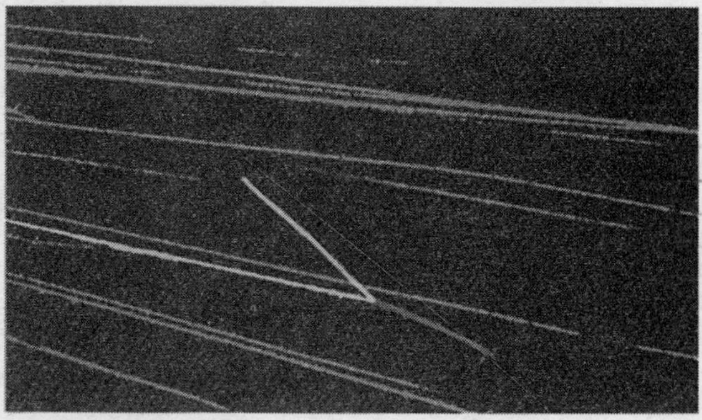

(*b*)

(*c*)

Fig. 10-1 Collisions range widely in scale. (*a*) Meteor Crater in Arizona is about 1200 m wide and 200 m deep. (*b*) An alpha particle coming in from the left (whose trail is colored yellow in this false-color photograph) bounces off a nitrogen nucleus that had been stationary and that now moves toward the bottom right (red trail). (*c*) In a tennis match, the ball is in contact with the racquet for about 4 ms in each collision (for a cumulative time of only 1 s per set).

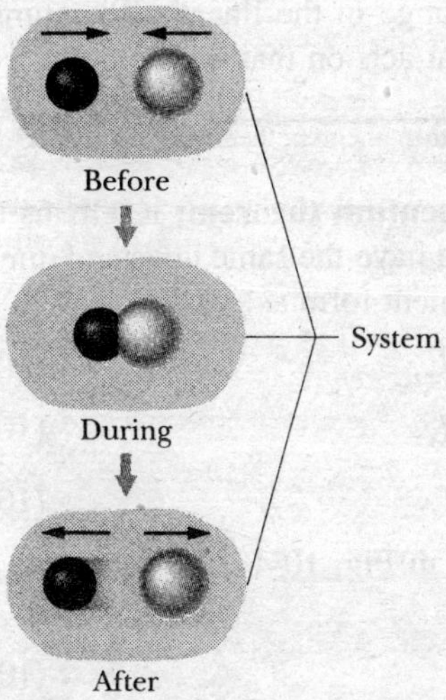

Fig. 10-2 A flowchart showing the system in which a collision occurs.

Many physicists today spend their time playing what we can call "the collision game." A principal goal of this game is to find out as much as possible about the forces that act during a collision, from knowledge of the state of the particles before and after the collision. Virtually all our understanding of the subatomic world—electrons, protons, neutrons, muons, quarks, and the like—comes from experiments involving collisions. The rules of the game are the laws of conservation of momentum and of energy.

10-2 Impulse and Linear Momentum

Single Collision

Figure 10-3 shows the third-law force pair, $\vec{F}(t)$ and $-\vec{F}(t)$, that acts during a simple head-on collision between two particle-like bodies of different masses. These forces will change the linear momentum of both bodies; the amount of the change will depend not only on the average values of the forces, but also on the time Δt during which they act. To see this quantitatively, let us apply Newton's second law in the form $\vec{F} = d\vec{p}/dt$ to, say, body R on the right in Fig. 10-3. We have

$$d\vec{p} = \vec{F}(t)\,dt, \tag{10-1}$$

in which $\vec{F}(t)$ is a time-varying force with magnitude given by the curve in Fig. 10-4a. Let us integrate Eq. 10-1 over the interval Δt—that is, from an initial time t_i (just before the collision) to a final time t_f (just after the collision). We obtain

$$\int_{\vec{p}_i}^{\vec{p}_f} d\vec{p} = \int_{t_i}^{t_f} \vec{F}(t)\,dt. \tag{10-2}$$

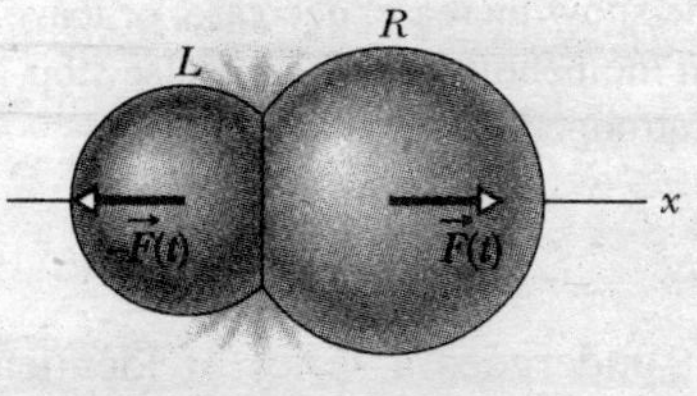

Fig. 10-3 Two particle-like bodies L and R collide with each other. During the collision, body L exerts force $\vec{F}(t)$ on body R, and body R exerts force $-\vec{F}(t)$ on body L. Forces $\vec{F}(t)$ and $-\vec{F}(t)$ are a third-law force pair. Their magnitudes vary with time during the collision, but at any given instant those magnitudes are equal.

The left side of this equation is $\vec{p}_f - \vec{p}_i$, the change in linear momentum of body R. The right side, which is a measure of both the strength and the duration of the collision force, is called the **impulse** $\vec{J}$ of the collision. Thus,

$$\vec{J} = \int_{t_i}^{t_f} \vec{F}(t)\,dt \quad \text{(impulse defined).} \tag{10-3}$$

Equation 10-3 tells us that the magnitude of the impulse is equal to the area under the $F(t)$ curve of Fig. 10-4a. Because $F(t)$ on body R and $-F(t)$ on body L are third-law force pairs, their impulses have the same magnitudes but opposite directions.

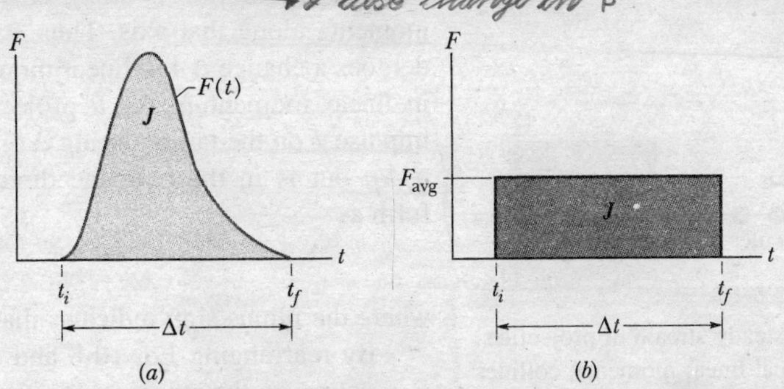

Fig. 10-4 (a) The curve shows the magnitude of the time-varying force $F(t)$ that acts on body R during the collision of Fig. 10-3. The area under the curve is equal to the magnitude of the impulse $\vec{J}$ on body R in the collision. (b) The height of the rectangle represents the average force F_{avg} acting on body R over the time interval Δt. The area within the rectangle is equal to the area under the curve in (a) and thus is also equal to the magnitude of the impulse $\vec{J}$ in the collision.

From Eqs. 10-2 and 10-3 we see that the change in the linear momentum of each body in a collision is equal to the impulse that acts on that body:

$$\vec{p}_f - \vec{p}_i = \Delta\vec{p} = \vec{J} \quad \text{(impulse–linear momentum theorem).} \tag{10-4}$$

Equation 10-4 is called the **impulse–linear momentum theorem;** it tells us that impulse and linear momentum are both vectors and have the same units and dimensions. Equation 10-4 can also be written in component form as

$$p_{fx} - p_{ix} = \Delta p_x = J_x, \tag{10-5}$$

$$p_{fy} - p_{iy} = \Delta p_y = J_y, \tag{10-6}$$

and

$$p_{fz} - p_{iz} = \Delta p_z = J_z. \tag{10-7}$$

If F_{avg} is the average magnitude of the force in Fig. 10-4*a*, we can write the magnitude of the impulse as

$$J = F_{avg}\,\Delta t, \tag{10-8}$$

where Δt is the duration of the collision. The value of F_{avg} must be such that the area within the rectangle of Fig. 10-4*b* is equal to the area under the actual $F(t)$ curve of Fig. 10-4*a*.

CHECKPOINT 1: A paratrooper whose chute fails to open lands in snow; he is hurt slightly. Had he landed on bare ground, the stopping time would have been 10 times shorter and the collision lethal. Does the presence of the snow increase, decrease, or leave unchanged the values of (a) the paratrooper's change in momentum, (b) the impulse stopping the paratrooper, and (c) the force stopping the paratrooper?

Series of Collisions

Now let's consider the force on a body when it undergoes a series of identical, repeated collisions. For example, as a prank, we might adjust one of those machines that fires tennis balls to fire them at a rapid rate directly at a wall. Each collision would produce a force on the wall, but that is not the force we are seeking. We want the average force F_{avg} on the wall during the bombardment—that is, the average force during a large number of collisions.

In Fig. 10-5, a steady stream of projectile bodies, with identical mass m and linear momenta $m\vec{v}$, moves along an x axis and collides with a target body that is fixed in place. Let n be the number of projectiles that collide in a time interval Δt. Because the motion is along only the x axis, we can use the components of the momenta along that axis. Thus, each projectile has initial momentum mv and undergoes a change Δp in linear momentum because of the collision. The total change in linear momentum for n projectiles during interval Δt is $n\,\Delta p$. The resulting impulse J on the target during Δt is along the x axis and has the same magnitude of $n\,\Delta p$ but is in the opposite direction. We can write this relation in component form as

$$J = -n\,\Delta p, \tag{10-9}$$

where the minus sign indicates that J and Δp have opposite directions.

By rearranging Eq. 10-8 and substituting Eq. 10-9, we find the average force F_{avg} acting on the target during the collisions:

$$F_{avg} = \frac{J}{\Delta t} = -\frac{n}{\Delta t}\Delta p = -\frac{n}{\Delta t} m\,\Delta v. \tag{10-10}$$

This equation gives us F_{avg} in terms of $n/\Delta t$, the rate at which the projectiles collide with the target, and Δv, the change in the velocity of those projectiles.

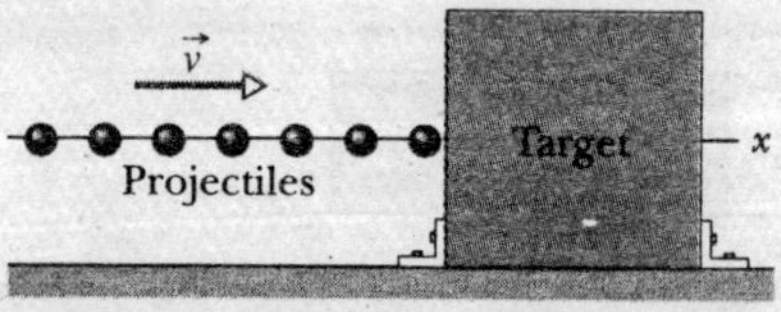

Fig. 10-5 A steady stream of projectiles, with identical linear momenta, collides with a target, which is fixed in place. The average force F_{avg} on the target is to the right and has a magnitude that depends on the rate at which the projectiles collide or, equivalently, the rate at which mass collides.

If the projectiles stop upon impact, then in Eq. 10-10 we can substitute, for Δv,

$$\Delta v = v_f - v_i = 0 - v = -v, \tag{10-11}$$

where v_i (= v) and v_f (= 0) are the velocities before and after the collision, respectively. If, instead, the projectiles bounce (rebound) directly backward from the target with no change in speed, then $v_f = -v$ and we can substitute

$$\Delta v = v_f - v_i = -v - v = -2v. \tag{10-12}$$

In time interval Δt, an amount of mass $\Delta m = nm$ collides with the target. With this result, we can rewrite Eq. 10-10 as

$$F_{\text{avg}} = -\frac{\Delta m}{\Delta t}\Delta v. \tag{10-13}$$

This equation gives the average force F_{avg} in terms of $\Delta m/\Delta t$, the rate at which mass collides with the target. Here again we can substitute for Δv from Eq. 10-11 or 10-12 depending on what the projectiles do.

CHECKPOINT 2: The figure shows an overhead view of a ball bouncing from a vertical wall without any change in its speed. Consider the change $\Delta\vec{p}$ in the ball's linear momentum. (a) Is Δp_x positive, negative, or zero? (b) Is Δp_y positive, negative, or zero? (c) What is the direction of $\Delta\vec{p}$?

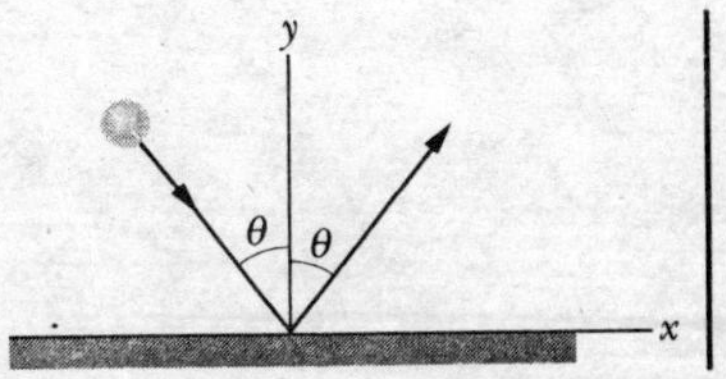

Sample Problem 10-1

A pitched 140 g baseball, in horizontal flight with a speed v_i of 39.0 m/s, is struck by a bat. After leaving the bat, the ball travels in the opposite direction with speed v_f, also 39.0 m/s.

(a) What impulse J acts on the ball while it is in contact with the bat during the collision?

SOLUTION: The **Key Idea** here is that we can calculate the impulse from the change it produces in the ball's linear momentum, using Eq. 10-4 for one-dimensional motion. Let us choose the direction in which the ball is initially moving to be the negative direction. From Eq. 10-4 we have

$$\begin{aligned} J &= p_f - p_i = mv_f - mv_i \\ &= (0.140\ \text{kg})(39.0\ \text{m/s}) - (0.140\ \text{kg})(-39.0\ \text{m/s}) \\ &= 10.9\ \text{kg}\cdot\text{m/s}. \end{aligned} \quad \text{(Answer)}$$

With our sign convention, the initial velocity of the ball is negative and the final velocity is positive. The impulse turns out to be positive, which tells us that the direction of the impulse vector acting on the ball is the direction in which the bat is swinging.

(b) The impact time Δt for the baseball–bat collision is 1.20 ms. What average force acts on the baseball?

SOLUTION: The **Key Idea** here is that the average force of the collision is the ratio of the impulse J to the duration Δt of the collision (see Eq. 10-8). Thus,

$$\begin{aligned} F_{\text{avg}} &= \frac{J}{\Delta t} = \frac{10.9\ \text{kg}\cdot\text{m/s}}{0.00120\ \text{s}} \\ &= 9080\ \text{N}. \end{aligned} \quad \text{(Answer)}$$

Note that this is the *average* force; the *maximum* force is larger. The sign of the average force on the ball from the bat is positive, which means that the direction of the force vector is the same as that of the impulse vector.

In defining a collision, we assumed that no net external force acts on the colliding bodies. That is not true in this case, because the gravitational force always acts on the ball, whether the ball is in flight or in contact with the bat. However, this force, with a magnitude of $mg = 1.37$ N, is negligible compared to the average force exerted by the bat, which has a magnitude of 9080 N. We are quite safe in treating the collision as "isolated."

(c) Now suppose the collision is not head-on, and the ball leaves the bat with a speed v_f of 45.0 m/s at an upward angle of 30.0° (Fig. 10-6). What now is the impulse on the ball?

SOLUTION: The **Key Idea** here is that now the collision is two-dimensional because the ball's outward path is not along the same axis as its incoming path. Thus, we must use vectors to find the impulse

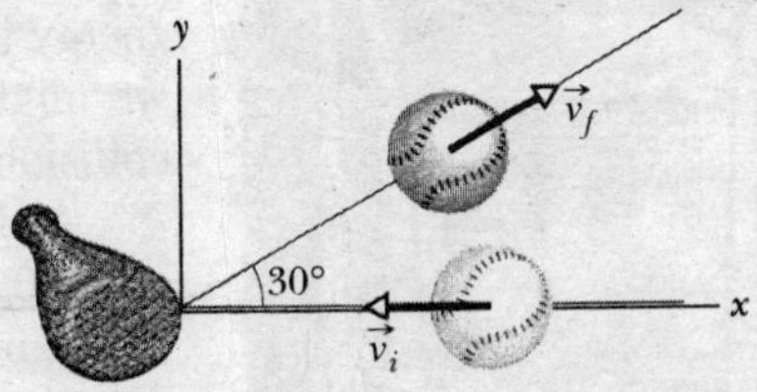

Fig. 10-6 Sample Problem 10-1. A bat collides with a pitched baseball, sending the ball off at an angle of 30° up from the horizontal.

$\vec{J}$. From Eq. 10-4, we can write

$$\vec{J} = \Delta\vec{p} = \vec{p}_f - \vec{p}_i = m\vec{v}_f - m\vec{v}_i.$$

Thus,

$$\vec{J} = m(\vec{v}_f - \vec{v}_i). \tag{10-14}$$

We can evaluate the right side of this equation directly on a vector-capable calculator, since we know that the mass m is 0.140 kg, the final velocity $\vec{v}_f$ is 45.0 m/s at 30.0°, and the initial velocity $\vec{v}_i$ is 39.0 m/s at 180°.

Instead, we can evaluate Eq. 10-14 in component form. To do so, we first place an xy coordinate system as shown in Fig. 10-6. Then along the x axis we have

$$J_x = p_{fx} - p_{ix} = m(v_{fx} - v_{ix})$$
$$= (0.140 \text{ kg})[(45.0 \text{ m/s})(\cos 30.0°) - (-39.0 \text{ m/s})]$$
$$= 10.92 \text{ kg} \cdot \text{m/s}.$$

Along the y axis,

$$J_y = p_{fy} - p_{iy} = m(v_{fy} - v_{iy})$$
$$= (0.140 \text{ kg})[(45.0 \text{ m/s})(\sin 30.0°) - 0]$$
$$= 3.150 \text{ kg} \cdot \text{m/s}.$$

The impulse is then

$$\vec{J} = (10.9\hat{\text{i}} + 3.15\hat{\text{j}}) \text{ kg} \cdot \text{m/s}, \quad \text{(Answer)}$$

and the magnitude and direction of $\vec{J}$ are

$$J = \sqrt{J_x^2 + J_y^2} = 11.4 \text{ kg} \cdot \text{m/s}$$

and

$$\theta = \tan^{-1}\frac{J_y}{J_x} = 16°. \quad \text{(Answer)}$$

When two bodies collide head on, the ratio of ratio of their relative velocities after collision to their relative velocities before collision is called coefficient of restitution

$$e = -\left(\frac{v_1 - v_2}{u_1 - u_2}\right)$$

10-3 Momentum and Kinetic Energy in Collisions

Consider a system of two colliding bodies. If there is to be a collision, then at least one of the bodies must be moving, so the system has a certain kinetic energy and a certain linear momentum before the collision. During the collision, the kinetic energy and linear momentum of each body are changed by the impulse from the other body. For the rest of this chapter we shall discuss these changes—and also the changes in the kinetic energy and linear momentum of the system as a whole—without knowing the details of the impulses that determine the changes. The discussion will be limited to collisions in systems that are **closed** (no mass enters or leaves them) and **isolated** (no net external forces act on the bodies within the system).

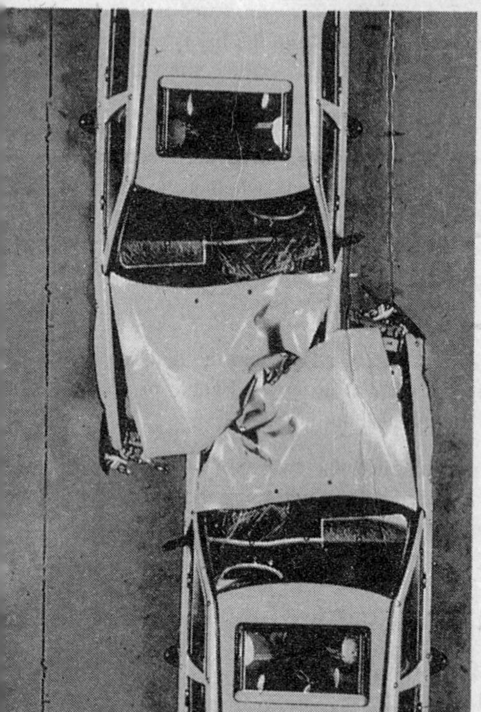

Fig. 10-7 Two cars after an almost head-on, almost completely inelastic collision.

Kinetic Energy

If the total kinetic energy of the system of two colliding bodies is unchanged by the collision, then the kinetic energy of the system is *conserved* (it is the same before and after the collision). Such a collision is called an **elastic collision.** In everyday collisions of common bodies, such as two cars or a ball and a bat, some energy is always transferred from kinetic energy to other forms of energy, such as thermal energy or energy of sound. Thus, the kinetic energy of the system is *not* conserved. Such a collision is called an **inelastic collision.**

However, in some situations, we can *approximate* a collision of common bodies as elastic. Suppose that you drop a Superball onto a hard floor. If the collision between the ball and floor (or Earth) were elastic, the ball would lose no kinetic energy because of the collision and would rebound to its original height. However, the actual rebound height is somewhat short, showing that at least some kinetic energy is lost in the collision and thus that the collision is somewhat inelastic. Still, we might choose to neglect that small loss of kinetic energy to approximate the collision as elastic. eg billiard balls when collide produce sound, still we consider collision to be elastic

A dropped golf ball will lose more of its kinetic energy and will rebound to only 60% of its original height. This collision is noticeably inelastic and cannot be approximated as elastic. If you drop a ball of wet putty onto the floor, it sticks to the floor and does not rebound at all. Because the putty sticks, this collision is called a **completely inelastic collision.** Figure 10-7 shows a more dramatic example of a completely inelastic collision. In such collisions, the bodies always stick together and lose kinetic energy.

Linear Momentum

Regardless of the details of the impulses in a collision, and regardless of what happens to the total kinetic energy of the system, the total linear momentum $\vec{P}$ of a closed, isolated system *cannot* change. The reason is that $\vec{P}$ can be changed only by external forces (from outside the system), but the forces in the collision are internal forces (inside the system). Thus, we have this important rule:

> In a closed, isolated system containing a collision, the linear momentum of each colliding body may change but the total linear momentum $\vec{P}$ of the system cannot change, whether the collision is elastic or inelastic.

This is actually another statement of the **law of conservation of linear momentum** that we first discussed in Section 9-6. In the next two sections we apply this law to some specific collisions, first inelastic and then elastic.

10-4 Inelastic Collisions in One Dimension

One-Dimensional Collision

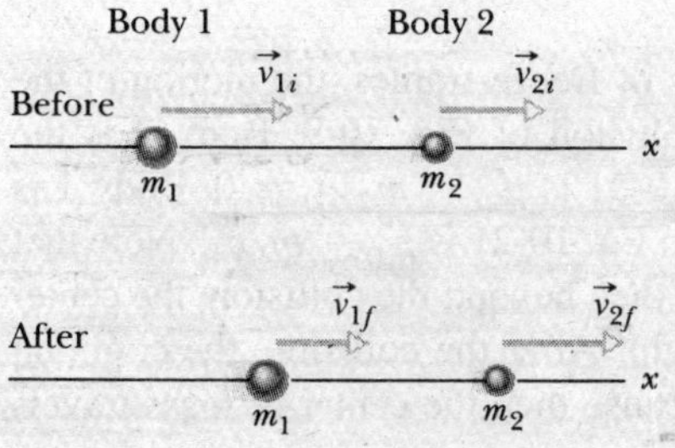

Fig. 10-8 Bodies 1 and 2 move along an x axis, before and after they have an inelastic collision.

Figure 10-8 shows two bodies just before and just after they have a *one-dimensional collision* (meaning that the motions before and after the collision are along a single axis). The velocities before the collision (subscript i) and after the collision (subscript f) are indicated. The two bodies form our system, which is closed and isolated. We can write the law of conservation of linear momentum for this two-body system as

$$\begin{pmatrix}\text{total momentum } \vec{P}_i\\ \text{before the collision}\end{pmatrix} = \begin{pmatrix}\text{total momentum } \vec{P}_f\\ \text{after the collision}\end{pmatrix},$$

which we can symbolize as

$$\vec{p}_{1i} + \vec{p}_{2i} = \vec{p}_{1f} + \vec{p}_{2f} \quad \text{(conservation of linear momentum).} \tag{10-15}$$

Because the motion is one-dimensional, we can drop the overhead arrows for vectors and use only components along the axis. Thus, from $p = mv$, we can rewrite Eq. 10-15 as

$$m_1 v_{1i} + m_2 v_{2i} = m_1 v_{1f} + m_2 v_{2f}. \tag{10-16}$$

If we know values for, say, the masses, the initial velocities, and one of the final velocities, we can find the other final velocity with Eq. 10-16.

Completely Inelastic Collision

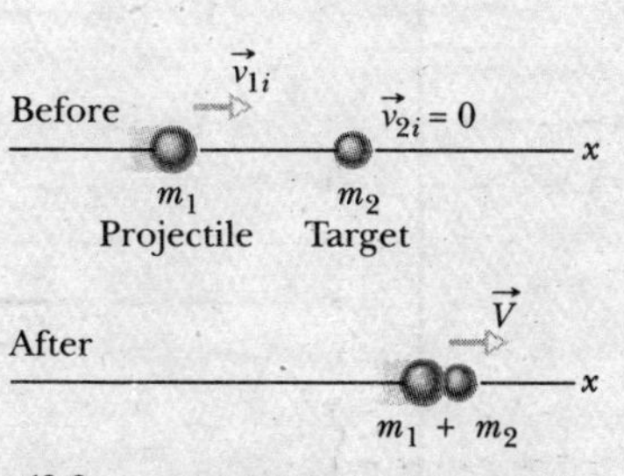

Fig. 10-9 A completely inelastic collision between two bodies. Before the collision, the body with mass m_2 is at rest and the body with mass m_1 moves directly toward it. After the collision, the stuck-together bodies move with the same velocity $\vec{V}$.

Figure 10-9 shows two bodies before and after they have a completely inelastic collision (meaning they stick together). The body with mass m_2 happens to be initially at rest ($v_{2i} = 0$). We can refer to that body as the *target*, and the incoming body as the *projectile*. After the collision, the stuck-together bodies move with velocity V. For this situation, we can rewrite Eq. 10-16 as

$$m_1 v_{1i} = (m_1 + m_2)V \tag{10-17}$$

or

$$V = \frac{m_1}{m_1 + m_2} v_{1i}. \tag{10-18}$$

If we know values for, say, the masses and the initial velocity v_{1i} of the projectile, we can find the final velocity V with Eq. 10-18. Note that V must be less than v_{1i}, because the mass ratio $m_1/(m_1 + m_2)$ must be less than unity.

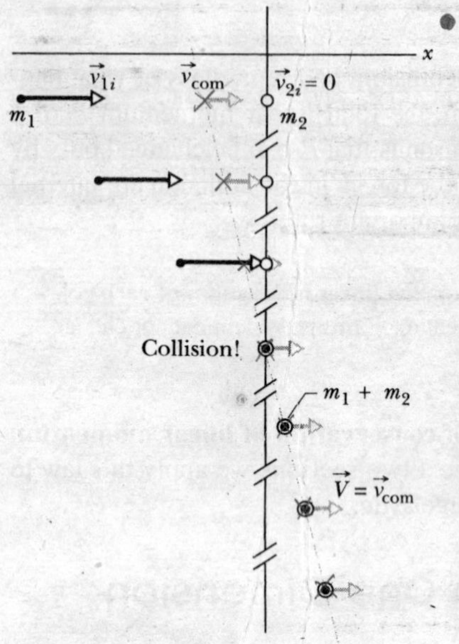

Fig. 10-10 Some freeze-frames of the two-body system in Fig. 10-9, which undergoes a completely inelastic collision. The system's center of mass is shown in each freeze-frame. The velocity $\vec{v}_{com}$ of the center of mass is unaffected by the collision. Because the bodies stick together after the collision, their common velocity $\vec{V}$ must be equal to $\vec{v}_{com}$.

Velocity of Center of Mass

In a closed, isolated system, the velocity $\vec{v}_{com}$ of the center of mass of the system cannot be changed by a collision because, with the system isolated, there is no net external force to change it. To get an expression for $\vec{v}_{com}$, let us return to the two-body system and one-dimensional collision of Fig. 10-8. From Eq. 9-25 ($\vec{P} = M\vec{v}_{com}$), we can relate $\vec{v}_{com}$ to the total linear momentum $\vec{P}$ of that two-body system by writing

$$\vec{P} = M\vec{v}_{com} = (m_1 + m_2)\vec{v}_{com}. \tag{10-19}$$

The total linear momentum $\vec{P}$ is conserved during the collision; so it is given by either side of Eq. 10-15. Let us use the left side to write

$$\vec{P} = \vec{p}_{1i} + \vec{p}_{2i}. \tag{10-20}$$

Substituting this for $\vec{P}$ in Eq. 10-19 and solving for $\vec{v}_{com}$ give us

$$\vec{v}_{com} = \frac{\vec{P}}{m_1 + m_2} = \frac{\vec{p}_{1i} + \vec{p}_{2i}}{m_1 + m_2}. \tag{10-21}$$

The right side of this equation is a constant, and $\vec{v}_{com}$ has that same constant value before and after the collision.

For example, Fig. 10-10 shows, in a series of freeze-frames, the motion of the center of mass for the completely inelastic collision of Fig. 10-9. Body 2 is the target, and its initial linear momentum in Eq. 10-21 is $\vec{p}_{2i} = m_2\vec{v}_{2i} = 0$. Body 1 is the projectile, and its initial linear momentum in Eq. 10-21 is $\vec{p}_{1i} = m_1\vec{v}_{1i}$. Note that as the series of freeze-frames progresses to and then beyond the collision, the center of mass moves at a constant velocity to the right. After the collision, the common final speed V of the bodies is equal to v_{com} because then the center of mass travels with the stuck-together bodies.

CHECKPOINT 3: Body 1 and body 2 are in a completely inelastic one-dimensional collision. What is their final momentum if their initial momenta are, respectively, (a) 10 kg · m/s and 0; (b) 10 kg · m/s and 4 kg · m/s; (c) 10 kg · m/s and −4 kg · m/s?

Sample Problem 10-2

The *ballistic pendulum* was used to measure the speeds of bullets before electronic timing devices were developed. The version shown in Fig. 10-11 consists of a large block of wood of mass $M = 5.4$ kg, hanging from two long cords. A bullet of mass $m = 9.5$ g is fired into the block, coming quickly to rest. The *block + bullet* then swing upward, their center of mass rising a vertical distance $h = 6.3$ cm before the pendulum comes momentarily to rest at the end of its arc. What is the speed of the bullet just prior to the collision?

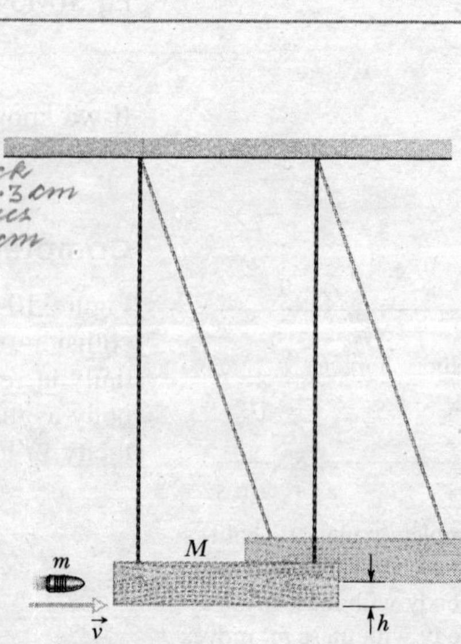

Fig. 10-11 Sample Problem 10-2. A ballistic pendulum, used to measure the speeds of bullets.

SOLUTION: We can see that the bullet's speed v must determine the rise height h. However, a **Key Idea** is that we cannot use the conservation of mechanical energy to relate these two quantities because surely energy is transferred from mechanical energy to other forms (such as thermal energy and energy to break apart the wood) as the bullet penetrates the block. Another **Key Idea** helps—we can split this complicated motion into two steps that we can separately analyze: (1) the bullet–block collision and (2) the bullet–block rise, during which mechanical energy *is* conserved.

Step 1. Because the collision within the bullet–block system is so brief, we can make two important assumptions: (1) During the collision, the gravitational force on the block and the force on the block from the cords are still balanced. Thus, during the collision, the net external impulse on the bullet–block system is zero. Therefore, the system is isolated and its total linear momentum is conserved. (2) The collision is one-dimensional in the sense that the direction of the bullet and block *just after the collision* is in the bullet's original direction of motion.

Because the collision is one-dimensional, the block is initially at rest, and the bullet sticks in the block, we use Eq. 10-18 to express the conservation of linear momentum. Replacing the symbols there with the corresponding symbols here, we have

$$V = \frac{m}{m + M} v. \tag{10-22}$$

Step 2. As the bullet and block now swing up together, the mechanical energy of the bullet–block–Earth system is conserved. (This mechanical energy is not changed by the force of the cords on the block, because that force is always directed perpendicular to the block's direction of travel.) Let's take the block's initial level as our reference level of zero gravitational potential energy. Then conservation of mechanical energy means that the system's kinetic energy at the start of the swing must equal its gravitational potential energy at the highest point of the swing. Because the speed of the bullet and block at the start of the swing is the speed V immediately after the collision, we may write this conservation as

$$\tfrac{1}{2}(m + M)V^2 = (m + M)gh.$$

Substituting for V from Eq. 10-22 leads to

$$v = \frac{m + M}{m}\sqrt{2gh}$$
$$= \left(\frac{0.0095 \text{ kg} + 5.4 \text{ kg}}{0.0095 \text{ kg}}\right)\sqrt{(2)(9.8 \text{ m/s}^2)(0.063 \text{ m})}$$
$$= 630 \text{ m/s}. \quad \text{(Answer)}$$

The ballistic pendulum is a kind of "transformer," exchanging the high speed of a light object (the bullet) for the low—and thus more easily measurable—speed of a massive object (the block).

Sample Problem 10-3

A karate expert strikes downward with his fist (of mass $m_1 = 0.70$ kg), breaking a 0.14 kg board (Fig. 10-12*a*). He then does the same to a 3.2 kg concrete block. The spring constants k for bending are 4.1×10^4 N/m for the board and 2.6×10^6 N/m for the block. Breaking occurs at a deflection d of 16 mm for the board and 1.1 mm for the block (Fig. 10-12*c*). (The data are taken from "The Physics of Karate," by S. R. Wilk, R. E. McNair, and M. S. Feld, *American Journal of Physics*, September 1983.)

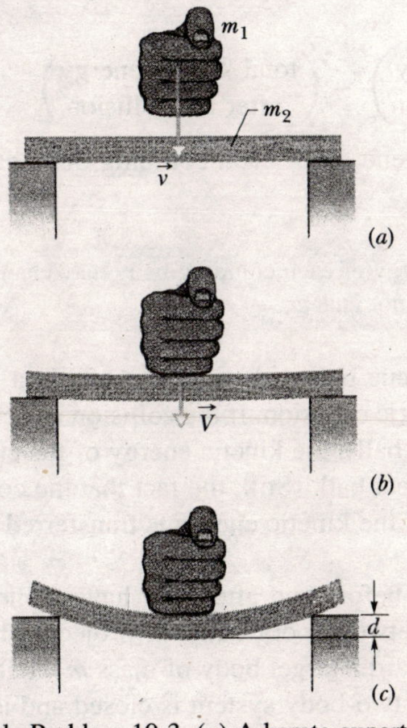

Fig. 10-12 Sample Problem 10-3. (*a*) A karate expert strikes at a flat object with speed v. (*b*) Fist and object undergo a completely inelastic collision, and bending begins. The *fist + object* then have speed V. (*c*) The object breaks when its center has been deflected by an amount d.

(a) Just before the object (board or block) breaks, what is the energy stored in it?

SOLUTION: The **Key Idea** here is that we can treat the bending as the compression of a spring for which Hooke's law applies. The stored potential energy is then, from Eq. 8-11, $U = \frac{1}{2}kd^2$. For the board,

$$U = \tfrac{1}{2}(4.1 \times 10^4 \text{ N/m})(0.016 \text{ m})^2$$
$$= 5.248 \text{ J} \approx 5.2 \text{ J}. \quad \text{(Answer)}$$

For the block,

$$U = \tfrac{1}{2}(2.6 \times 10^6 \text{ N/m})(0.0011 \text{ m})^2$$
$$= 1.573 \text{ J} \approx 1.6 \text{ J}. \quad \text{(Answer)}$$

(b) What is the lowest fist speed v_{fist} required to break the object (board or block)? Assume the following: The collisions are completely inelastic collisions of only the fist and the object. Bending begins just after the collision. Mechanical energy is conserved from the beginning of the bending until just before the object breaks. The speed of the fist and object is negligible at that point.

SOLUTION: The **Key Idea** here is that we can split up this complicated motion into three steps that we can separately analyze:

1. The completely inelastic one-dimensional collision of the fist and the object transfers energy to the kinetic energy of the fist–object system.
2. That energy is then transferred to the potential energy U stored by the bending.

3. The object breaks when U reaches the value calculated in part (a).

In step 1, we can use Eq. 10-18 to relate the fist speed v_{fist} just before the collision to the fist–object speed V_{fo} just after the collision and as the bending begins. With the notation we are using here, Eq. 10-18 becomes

$$V_{\text{fo}} = \frac{m_1}{m_1 + m_2} v_{\text{fist}}. \tag{10-23}$$

In step 2, the mechanical energy of the fist–object system is conserved during the bending (up to the break). (Because the downward deflection of the object is small, the change in the gravitational potential energies of fist and object during the deflection are small enough to be neglected.) We can write the conservation of mechanical energy during the bending as

$$\begin{pmatrix}\text{kinetic energy at}\\ \text{start of bending}\end{pmatrix} = \begin{pmatrix}\text{potential energy of bending}\\ \text{just before the break}\end{pmatrix},$$

or

$$\tfrac{1}{2}(m_1 + m_2)V_{\text{fo}}^2 = U. \tag{10-24}$$

Substituting for V_{fo} from Eq. 10-23 and solving for v_{fist}, we find

$$v_{\text{fist}} = \frac{1}{m_1}\sqrt{2U(m_1 + m_2)}. \tag{10-25}$$

For step 3, we substitute the proper mass value and the breaking value U as obtained in part (a). We find that for the board,

$$v_{\text{fist}} = 4.2 \text{ m/s} \qquad \text{(Answer)}$$

and for the block,

$$v_{\text{fist}} = 5.0 \text{ m/s}. \qquad \text{(Answer)}$$

Thus, from the answer to (a) we see that breaking a board requires the greater energy. However, from the answers to (b) we see why a board is easier to break: The required fist speed is less. The reason lies in Eq. 10-23. If we decrease the target mass in the collision, we increase the speed V_{fo} given to the object. Thus, we also increase the fraction of the fist's energy that is transferred to the object. (One reason why breaking a pencil in the arrangement of Fig. 10-12 would be easy is because of the pencil's small mass.)

In such cases we see proportion
$U_{block} < U_{board}$
but extent of energy transfer to block >> s.t. it overcomes this diff

10-5 Elastic Collisions in One Dimension

Stationary Target

→ headon collision ∵ no b

As we discussed in Section 10-3, everyday collisions are inelastic but we can approximate some of them as being elastic; that is, we can approximate that the total kinetic energy of the colliding bodies is conserved and is not transferred to other forms of energy:

$$\begin{pmatrix}\text{total kinetic energy}\\ \text{before the collision}\end{pmatrix} = \begin{pmatrix}\text{total kinetic energy}\\ \text{after the collision}\end{pmatrix}.$$

This does not mean that the kinetic energy of each colliding body cannot change. Rather, it means this:

> In an elastic collision, the kinetic energy of each colliding body may change, but the total kinetic energy of the system does not change.

For example, the collision of a cue ball with an object ball in a game of pool can be approximated as being an elastic collision. If the collision is head-on (the cue ball heads directly toward the object ball), the kinetic energy of the cue ball can be transferred almost entirely to the object ball. (Still, the fact that the collision makes a sound means that at least a little of the kinetic energy is transferred to the energy of the sound.)

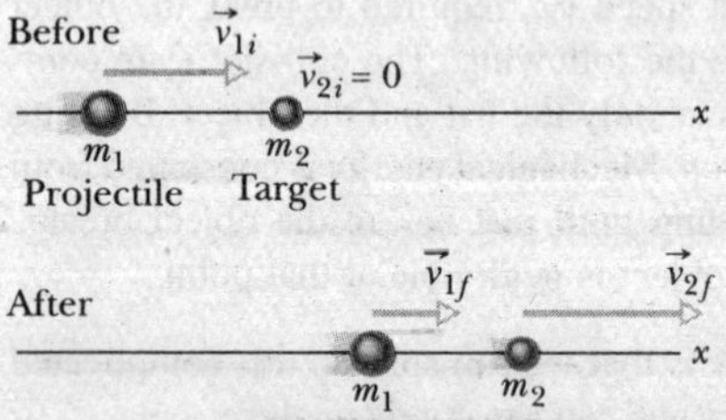

Fig. 10-13 Body 1 moves along an x axis before having an elastic collision with body 2, which is initially at rest. Both bodies move along that axis after the collision.

Figure 10-13 shows two bodies before and after they have a one-dimensional collision, like a head-on collision between pool balls. A projectile body of mass m_1 and initial velocity v_{1i} moves toward a target body of mass m_2 that is initially at rest ($v_{2i} = 0$). Let's assume that this two-body system is closed and isolated. Then the net linear momentum of the system is conserved, and from Eq. 10-15 we can write that conservation as

$$m_1 v_{1i} = m_1 v_{1f} + m_2 v_{2f} \qquad \text{(linear momentum).} \tag{10-26}$$

If the collision is also elastic, then the total kinetic energy is conserved and we can write that conservation as

$$\tfrac{1}{2}m_1v_{1i}^2 = \tfrac{1}{2}m_1v_{1f}^2 + \tfrac{1}{2}m_2v_{2f}^2 \qquad \text{(kinetic energy).} \tag{10-27}$$

In each of these equations, the subscript i identifies the initial velocities and the subscript f the final velocities of the bodies. If we know the masses of the bodies and if we also know v_{1i}, the initial velocity of body 1, the only unknown quantities are v_{1f} and v_{2f}, the final velocities of the two bodies. With two equations at our disposal, we should be able to find these two unknowns.

To do so, we rewrite Eq. 10-26 as

$$m_1(v_{1i} - v_{1f}) = m_2v_{2f} \tag{10-28}$$

and Eq. 10-27 as*

$$m_1(v_{1i} - v_{1f})(v_{1i} + v_{1f}) = m_2v_{2f}^2. \tag{10-29}$$

After dividing Eq. 10-29 by Eq. 10-28 and doing some more algebra, we obtain

$$v_{1f} = \frac{m_1 - m_2}{m_1 + m_2}v_{1i} \tag{10-30}$$

and

$$v_{2f} = \frac{2m_1}{m_1 + m_2}v_{1i}. \tag{10-31}$$

We note from Eq. 10-31 that v_{2f} is always positive (the target body with mass m_2 always moves forward). From Eq. 10-30 we see that v_{1f} may be of either sign (the projectile body with mass m_1 moves forward if $m_1 > m_2$ but rebounds if $m_1 < m_2$). Let us look at a few special situations.

1. ***Equal masses*** If $m_1 = m_2$, Eqs. 10-30 and 10-31 reduce to

$$v_{1f} = 0 \quad \text{and} \quad v_{2f} = v_{1i},$$

which we might call a pool player's result. It predicts that after a head-on collision of bodies with equal masses, body 1 (initially moving) stops dead in its tracks and body 2 (initially at rest) takes off with the initial speed of body 1. In head-on collisions, bodies of equal mass simply exchange velocities. This is true even if the target particle (body 2) is not initially at rest.

2. ***A massive target*** In Fig. 10-13, a massive target means that $m_2 \gg m_1$. For example, we might fire a golf ball at a cannonball. Equations 10-30 and 10-31 then reduce to

$$v_{1f} \approx -v_{1i} \quad \text{and} \quad v_{2f} \approx \left(\frac{2m_1}{m_2}\right)v_{1i}. \tag{10-32}$$

This tells us that body 1 (the golf ball) simply bounces back along its incoming path, its speed essentially unchanged. Body 2 (the cannonball) moves forward at a low speed, because the quantity in parentheses in Eq. 10-32 is much less than unity. All this is what we should expect.

3. ***A massive projectile*** This is the opposite case; that is, $m_1 \gg m_2$. This time, we fire a cannonball at a golf ball. Equations 10-30 and 10-31 reduce to

$$v_{1f} \approx v_{1i} \quad \text{and} \quad v_{2f} \approx 2v_{1i}. \tag{10-33}$$

*In this step, we use the identity $a^2 - b^2 = (a - b)(a + b)$. It reduces the amount of algebra needed to solve the simultaneous equations, Eqs. 10-28 and 10-29.

Equation 10-33 tells us that body 1 (the cannonball) simply keeps on going, scarcely slowed by the collision. Body 2 (the golf ball) charges ahead at twice the speed of the cannonball.

You may wonder: "Why twice the speed?" As a starting point in thinking about the matter, recall the collision described by Eq. 10-32, in which the velocity of the incident light body (the golf ball) changed from $+v$ to $-v$, a velocity *change* of $2v$. The same *change* in velocity (but now from zero to $2v$) occurs in this example also.

Moving Target

Now that we have examined the elastic collision of a projectile and a stationary target, let us examine the situation in which both bodies are moving before they undergo an elastic collision.

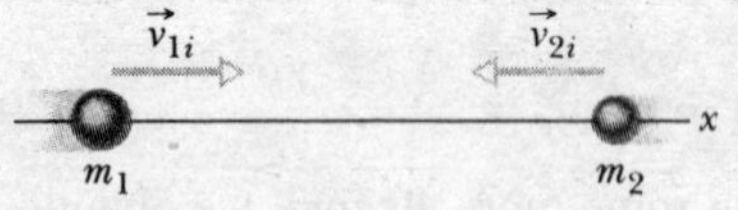

Fig. 10-14 Two bodies headed for a one-dimensional elastic collision.

For the situation of Fig. 10-14, the conservation of linear momentum is written as

$$m_1 v_{1i} + m_2 v_{2i} = m_1 v_{1f} + m_2 v_{2f}, \tag{10-34}$$

and the conservation of kinetic energy is written as

$$\tfrac{1}{2} m_1 v_{1i}^2 + \tfrac{1}{2} m_2 v_{2i}^2 = \tfrac{1}{2} m_1 v_{1f}^2 + \tfrac{1}{2} m_2 v_{2f}^2. \tag{10-35}$$

To solve these simultaneous equations for v_{1f} and v_{2f}, we first rewrite Eq. 10-34 as

$$m_1(v_{1i} - v_{1f}) = -m_2(v_{2i} - v_{2f}), \tag{10-36}$$

and Eq. 10-35 as

$$m_1(v_{1i} - v_{1f})(v_{1i} + v_{1f}) = -m_2(v_{2i} - v_{2f})(v_{2i} + v_{2f}). \tag{10-37}$$

After dividing Eq. 10-37 by Eq. 10-36 and doing some more algebra, we obtain

$$v_{1f} = \frac{m_1 - m_2}{m_1 + m_2} v_{1i} + \frac{2m_2}{m_1 + m_2} v_{2i} \tag{10-38}$$

and

$$v_{2f} = \frac{2m_1}{m_1 + m_2} v_{1i} + \frac{m_2 - m_1}{m_1 + m_2} v_{2i}. \tag{10-39}$$

Note that the assignment of subscripts 1 and 2 to the bodies is arbitrary. If we exchange those subscripts in Fig. 10-14 and in Eqs. 10-38 and 10-39, we end up with the same set of equations. Note also that if we set $v_{2i} = 0$, body 2 becomes a stationary target as in Fig. 10-13, and Eqs. 10-38 and 10-39 reduce to Eqs. 10-30 and 10-31, respectively.

CHECKPOINT 4: What is the final linear momentum of the target in Fig. 10-13 if the initial linear momentum of the projectile is 6 kg · m/s and the final linear momentum of the projectile is (a) 2 kg · m/s and (b) −2 kg · m/s? (c) What is the final kinetic energy of the target if the initial and final kinetic energies of the projectile are, respectively, 5 J and 2 J?

Sample Problem 10-4

Two metal spheres, suspended by vertical cords, initially just touch, as shown in Fig. 10-15. Sphere 1, with mass $m_1 = 30$ g, is pulled to the left to height $h_1 = 8.0$ cm, and then released from rest. After swinging down, it undergoes an elastic collision with sphere 2, whose mass $m_2 = 75$ g. What is the velocity v_{1f} of sphere 1 just after the collision?

SOLUTION: A first **Key Idea** is that we can split this complicated motion into two steps that we can separately analyze: (1) the descent of sphere 1 and (2) the two-sphere collision.

Step 1. The **Key Idea** here is that as sphere 1 swings down, the mechanical energy of the sphere–Earth system is conserved.

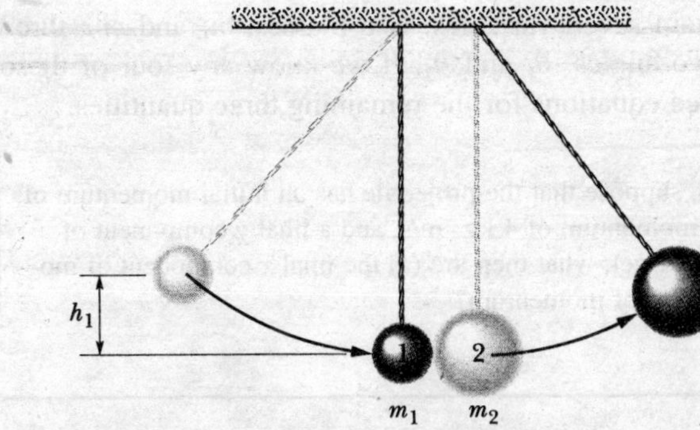

Fig. 10-15 Sample Problem 10-4. Two metal spheres suspended by cords just touch when they are at rest. Sphere 1, with mass m_1, is pulled to the left to height h_1 and then released.

(The mechanical energy is not changed by the force of the cord on sphere 1 because that force is always directed perpendicular to the sphere's direction of travel.) Let's take the lowest level as our reference level of zero gravitational potential energy. Then the kinetic energy of sphere 1 at the lowest level must equal the gravitational potential energy of the system when sphere 1 is at the initial height. Thus,

$$\tfrac{1}{2}m_1v_{1i}^2 = m_1gh_1,$$

which we solve for the speed v_{1i} of sphere 1 just before the collision:

$$v_{1i} = \sqrt{2gh_1} = \sqrt{(2)(9.8\ \text{m/s}^2)(0.080\ \text{m})} = 1.252\ \text{m/s}.$$

Step 2. Here we can make two assumptions in addition to the assumption that the collision is elastic. First, we can assume that the collision is one-dimensional because the motions of the spheres are approximately horizontal from just before the collision to just after it. Second, because the collision is so brief, we can assume that the two-sphere system is closed and isolated. This gives the **Key Idea** that the total linear momentum of the system is conserved. Thus, we can use Eq. 10-30 to find the velocity of sphere 1 just after the collision:

$$v_{1f} = \frac{m_1 - m_2}{m_1 + m_2}v_{1i} = \frac{0.030\ \text{kg} - 0.075\ \text{kg}}{0.030\ \text{kg} + 0.075\ \text{kg}}(1.252\ \text{m/s})$$
$$= -0.537\ \text{m/s} \approx -0.54\ \text{m/s}. \quad \text{(Answer)}$$

The minus sign tells us that sphere 1 moves to the left just after the collision.

10-6 Collisions in Two Dimensions

When two bodies collide, the impulses of one on the other determine the directions in which they then travel. In particular, when the collision is not head-on, the bodies do not end up traveling along their initial axis. For such two-dimensional collisions in a closed, isolated system, the total linear momentum must still be conserved:

$$\vec{P}_{1i} + \vec{P}_{2i} = \vec{P}_{1f} + \vec{P}_{2f} \tag{10-40}$$

If the collision is also elastic (a special case), then the total kinetic energy is also conserved:

$$K_{1i} + K_{2i} = K_{1f} + K_{2f}. \tag{10-41}$$

Equation 10-40 is often more useful for analyzing a two-dimensional collision if we write it in terms of components on an xy coordinate system. For example, Fig. 10-16 shows a *glancing collision* (it is not head-on) between a projectile body and a target body initially at rest. The impulses between the bodies have sent the bodies off at angles θ_1 and θ_2 to the x axis, along which the projectile initially traveled. In this situation we would rewrite Eq. 10-40 for components along the x axis as

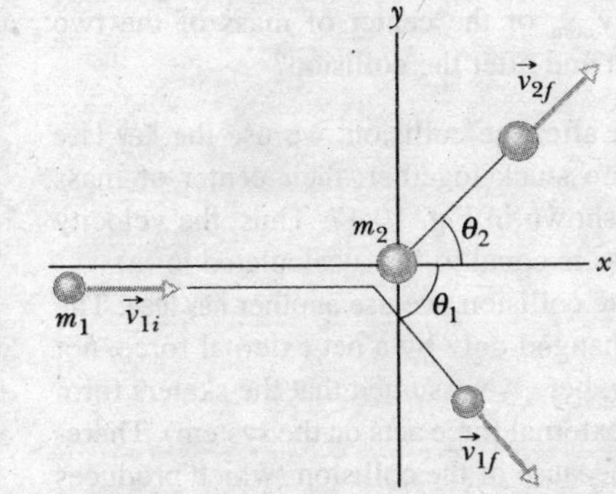

Fig. 10-16 An elastic collision between two bodies in which the collision is not head-on. The body with mass m_2 (the target) is initially at rest.

$$m_1v_{1i} = m_1v_{1f}\cos\theta_1 + m_2v_{2f}\cos\theta_2, \tag{10-42}$$

and along the y axis as

$$0 = -m_1v_{1f}\sin\theta_1 + m_2v_{2f}\sin\theta_2. \tag{10-43}$$

We can also write Eq. 10-41 (for the special case of an elastic collision) in terms of speeds:

$$\tfrac{1}{2}m_1v_{1i}^2 = \tfrac{1}{2}m_1v_{1f}^2 + \tfrac{1}{2}m_2v_{2f}^2 \quad \text{(kinetic energy).} \tag{10-44}$$

Equations 10-42 to 10-44 contain seven variables: two masses, m_1 and m_2; three speeds, v_{1i}, v_{1f}, and v_{2f}; and two angles, θ_1 and θ_2. If we know any four of these quantities, we can solve the three equations for the remaining three quantities.

CHECKPOINT 5: In Fig. 10-16, suppose that the projectile has an initial momentum of 6 kg · m/s, a final x component of momentum of 4 kg · m/s, and a final y component of momentum of −3 kg · m/s. For the target, what then are (a) the final x component of momentum and (b) the final y component of momentum?

→ $0 + \bar{P}_{1i} = \bar{P}_{1f} + \bar{P}_{2f}$
→ $6 = 4 + P_{2fx}$
$2 = P_{2fx}$
→ $0 = -3 + P_{2fy}$
$3 = P_{2fy}$

Sample Problem 10-5

Two skaters collide and embrace, in a completely inelastic collision. Thus, they stick together after impact, as suggested by Fig. 10-17, where the origin is placed at the point of collision. Alfred, whose mass m_A is 83 kg, is originally moving east with speed $v_A = 6.2$ km/h. Barbara, whose mass m_B is 55 kg, is originally moving north with speed $v_B = 7.8$ km/h.

(a) What is the velocity $\vec{V}$ of the couple after they collide?

SOLUTION: One **Key Idea** here is the assumption that the two skaters form a closed, isolated system; that is, we assume no *net* external force acts on them. In particular, we neglect any frictional force on their skates from the ice. With that assumption, we can apply the conservation of the total linear momentum $\vec{P}$ to the system by writing $\vec{P}_i = \vec{P}_f$ as

1) we cannot write $V = \frac{m_1 V_{1i}}{m_1+m_2}$ ∵ motion is not 1D

$$m_A\vec{v}_A + m_B\vec{v}_B = (m_A + m_B)\vec{V}. \quad (10\text{-}45)$$

Solving for $\vec{V}$ gives us

2) vel of CM also has same expression $\vec{V}_{com} = \frac{m_A\vec{V}_A + m_B\vec{V}_B}{m_A+m_B}$

$$\vec{V} = \frac{m_A\vec{v}_A + m_B\vec{v}_B}{m_A + m_B}.$$

We can solve this directly on a vector-capable calculator by substituting given data for the symbols on the right side. We can also solve it by applying a second **Key Idea** (one we have used before) and then some algebra: The idea is that the total linear momentum of the system is conserved separately for components along the x axis and y axis shown in Fig. 10-17. Writing Eq. 10-45 in component form for the x axis yields

$$m_Av_A + m_B(0) = (m_A + m_B)V\cos\theta, \quad (10\text{-}46)$$

and for the y axis

$$m_A(0) + m_Bv_B = (m_A + m_B)V\sin\theta. \quad (10\text{-}47)$$

We cannot solve either of these equations separately because they both contain two unknowns (V and θ), but we can solve them simultaneously by dividing Eq. 10-47 by Eq. 10-46. We get

$$\tan\theta = \frac{m_Bv_B}{m_Av_A} = \frac{(55\text{ kg})(7.8\text{ km/h})}{(83\text{ kg})(6.2\text{ km/h})} = 0.834.$$

Thus,

$$\theta = \tan^{-1} 0.834 = 39.8° \approx 40°. \quad \text{(Answer)}$$

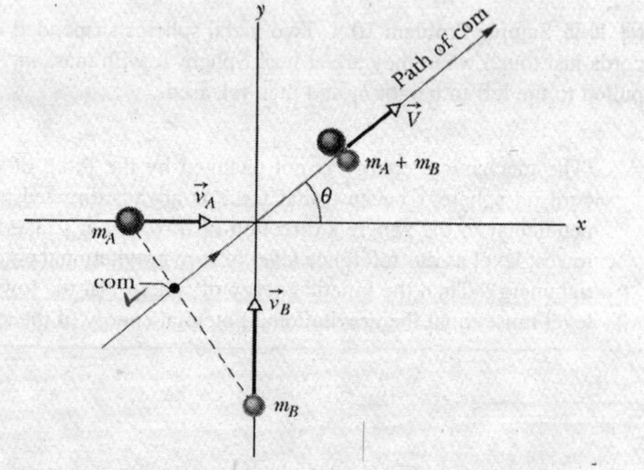

Fig. 10-17 Sample Problem 10-5. Two skaters, Alfred (A) and Barbara (B), represented by spheres in this simplified overhead view, have a completely inelastic collision. Afterward, they move off together at angle θ, with speed V. The path of their center of mass is shown. The position of the center of mass for the indicated positions of the skaters before the collision is also shown.

From Eq. 10-47, with $m_A + m_B = 138$ kg, we then have

$$V = \frac{m_Bv_B}{(m_A + m_B)\sin\theta} = \frac{(55\text{ kg})(7.8\text{ km/h})}{(138\text{ kg})(\sin 39.8°)}$$
$$= 4.86\text{ km/h} \approx 4.9\text{ km/h}. \quad \text{(Answer)}$$

(b) What is the velocity $\vec{v}_{com}$ of the center of mass of the two skaters before the collision and after the collision?

SOLUTION: For the situation after the collision, we use the **Key Idea** that because the skaters are stuck together, their center of mass must travel with them, as shown in Fig. 10-17. Thus, the velocity $\vec{v}_{com}$ of their center of mass is equal to $\vec{V}$, as calculated in (a).

To find $\vec{v}_{com}$ before the collision, we use another **Key Idea**: The $\vec{v}_{com}$ of a system can be changed only by a net external force, not an internal force. However, here we assumed that the skaters form an isolated system (no net external force acts on the system). Therefore, $\vec{v}_{com}$ cannot change because of the collision (which produces only internal forces). Thus, before and after the collision, we have

$$\vec{v}_{com} = \vec{V}. \quad \text{(Answer)}$$

OR

$V_{com,x} = \frac{m_AV_{Ax} + 0}{m_A+m_B} = \frac{83 \times 6.2}{138} = 3.73$

$V_{com,y} = 0 + \frac{m_BV_{By}}{m_A+m_B} = \frac{55 \times 7.8}{138} = 3.10$

$V_{com} = \sqrt{3.73^2 + 3.10^2} = 4.85$ km/h

$\tan\theta = +\frac{3.10}{3.73} \Rightarrow \theta = 39.72$ wrt $+x$ axis

y is + ∵ ↑

REVIEW & SUMMARY

Collisions In a **collision,** two bodies exert strong forces on each other for a relatively short time. These forces are internal to the two-body system and are significantly larger than any external force during the collision.

Impulse and Linear Momentum Applying Newton's second law in momentum form to a particle-like body involved in a collision leads to the **impulse–linear momentum theorem:**

$$\vec{p}_f - \vec{p}_i = \Delta\vec{p} = \vec{J}, \tag{10-4}$$

where $\vec{p}_f - \vec{p}_i = \Delta\vec{p}$ is the change in the body's linear momentum, and $\vec{J}$ is the **impulse** due to the force $\vec{F}(t)$ exerted on the body by the other body in the collision:

$$\vec{J} = \int_{t_i}^{t_f} \vec{F}(t)\, dt. \tag{10-3}$$

If F_{avg} is the average magnitude of $\vec{F}(t)$ during the collision and Δt is the duration of the collision, then for one-dimensional motion

$$J = F_{avg}\, \Delta t. \tag{10-8}$$

When a steady stream of bodies, each with mass m and speed v, collides with a body whose position is fixed, the average force on the fixed body is

$$F_{avg} = -\frac{n}{\Delta t}\Delta p = -\frac{n}{\Delta t} m\, \Delta v, \tag{10-10}$$

where $n/\Delta t$ is the rate at which the bodies collide with the fixed body, and Δv is the change in velocity of each colliding body. This average force can also be written as

$$F_{avg} = -\frac{\Delta m}{\Delta t}\Delta v, \tag{10-13}$$

where $\Delta m/\Delta t$ is the rate at which mass collides with the fixed body. In Eqs. 10-10 and 10-13, $\Delta v = -v$ if the bodies stop upon impact, or $\Delta v = -2v$ if they bounce directly backward with no change in their speed.

Inelastic Collision—One Dimension In an *inelastic collision* of two bodies, the kinetic energy of the two-body system is not conserved. If the system is closed and isolated, then the total linear momentum of the system *must* be conserved, which we can write in vector form as

$$\vec{p}_{1i} + \vec{p}_{2i} = \vec{p}_{1f} + \vec{p}_{2f}, \tag{10-15}$$

where subscripts i and f refer to values just before and just after the collision, respectively.

If the motion of the bodies is along a single axis, the collision is one-dimensional and we can write Eq. 10-15 in terms of velocity components along that axis:

$$m_1v_{1i} + m_2v_{2i} = m_1v_{1f} + m_2v_{2f}. \tag{10-16}$$

If the bodies stick together, the collision is a *completely inelastic collision* and the bodies have the same final velocity V (because they *are* stuck together).

Motion of the Center of Mass The center of mass of a closed, isolated system of two colliding bodies is not affected by the collision. In particular, the velocity $\vec{v}_{com}$ of the center of mass cannot be changed by the collision and is related to the constant total momentum $\vec{P}$ of the system by

$$\vec{v}_{com} = \frac{\vec{P}}{m_1 + m_2} = \frac{\vec{p}_{1i} + \vec{p}_{2i}}{m_1 + m_2}. \tag{10-21}$$

Elastic Collisions—One Dimension An *elastic collision* is a special type of collision in which the kinetic energy of the system of colliding bodies is conserved. Some collisions in the everyday world can be approximated as being elastic collisions. If the system is closed and isolated, its linear momentum is also conserved. For a one-dimensional collision in which body 2 is a target and body 1 is an incoming projectile, conservation of kinetic energy and linear momentum yield the following expressions for the velocities immediately after the collision:

$$v_{1f} = \frac{m_1 - m_2}{m_1 + m_2} v_{1i} \tag{10-30}$$

and

$$v_{2f} = \frac{2m_1}{m_1 + m_2} v_{1i}. \tag{10-31}$$

If both bodies are moving prior to the collision, their velocities immediately after the collision are given by

$$v_{1f} = \frac{m_1 - m_2}{m_1 + m_2} v_{1i} + \frac{2m_2}{m_1 + m_2} v_{2i} \tag{10-38}$$

and

$$v_{2f} = \frac{2m_1}{m_1 + m_2} v_{1i} + \frac{m_2 - m_1}{m_1 + m_2} v_{2i}. \tag{10-39}$$

Note the symmetry of subscripts 1 and 2 in Eqs. 10-38 and 10-39.

Collisions in Two Dimensions If two bodies collide and their motion is not along a single axis (the collision is not head-on), then the collision is two-dimensional. If the two-body system is closed and isolated, then the law of conservation of momentum applies to the collision and can be written as

$$\vec{P}_{1i} + \vec{P}_{2i} = \vec{P}_{1f} + \vec{P}_{2f}, \tag{10-40}$$

In component form, the law gives two equations that describe the collision (one equation for each of the two dimensions). If the collision is also elastic (a special case), then the conservation of kinetic energy during the collision gives a third equation:

$$K_{1i} + K_{2i} = K_{1f} + K_{2f}. \tag{10-41}$$

QUESTIONS

1 Figure 10-18 shows three graphs of force magnitude versus time for a body involved in a collision. Rank the graphs according to the magnitude of the impulse on the body, greatest first.

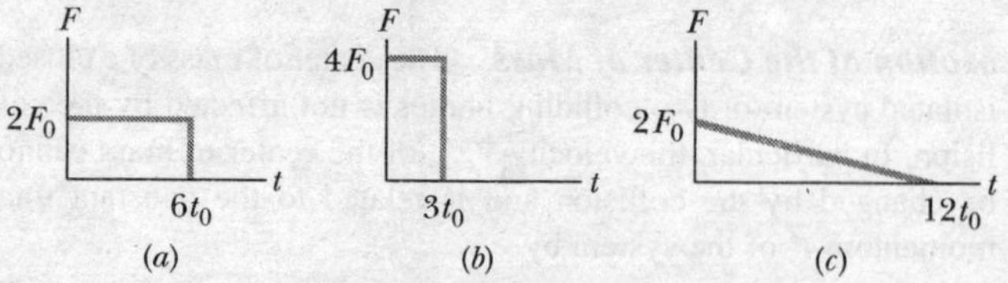

Fig. 10-18 Question 1.

2. Two objects that are moving along an xy plane on a frictionless floor collide. Assume that they form a closed, isolated system. The following table gives some of the momentum components (in kilogram-meters per second) before and after the collision. What are the missing values?

		Before		After	
Situation	Object	p_x	p_y	p_x	p_y
1	A	3	4	7	2
	B	2	2		
2	C	−4	5	3	
	D		−2	4	2
3	E	−6			3
	F	6	2	−4	−3

3 The following table gives, for three situations, the masses (in kilograms) and velocities (in meters per second) of the two bodies in Fig. 10-14. For which situations is the center of mass of the two-body system stationary?

Situation	m_1	v_1	m_2	v_2
a	2	3	4	−3
b	6	2	3	−4
c	4	3	4	−3

4 Figure 10-19 shows four graphs of position versus time for two bodies and their center of mass. The two bodies form a closed, isolated system and undergo a completely inelastic, one-dimensional collision along an x axis. In graph 1, are (a) the two bodies and (b) the center of mass moving in the positive or negative direction of the x axis? (c) Which graphs correspond to a physically impossible situation? Explain.

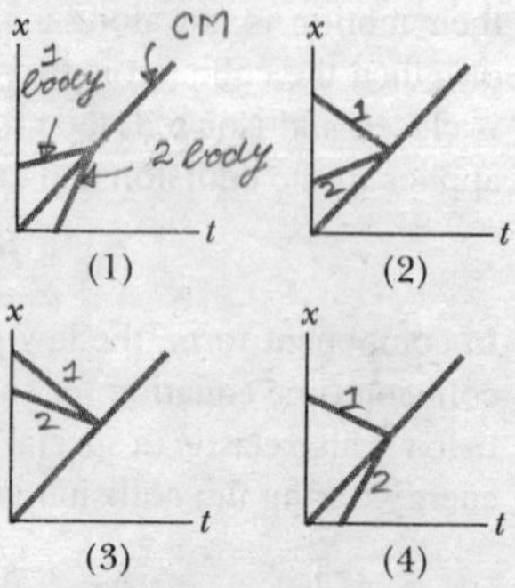

Fig. 10-19 Question 4.

5 In Fig. 10-20, blocks A and B have linear momenta with directions as shown and with magnitudes of 9 kg · m/s and 4 kg · m/s, respectively. (a) What is the direction of motion of the center of mass of the two-block system over the frictionless floor? (b) If the blocks stick together during their collision, in what direction do they move? (c) If, instead, block A ends up moving to the left, is the magnitude of its momentum then smaller than, more than, or the same as that of block B?

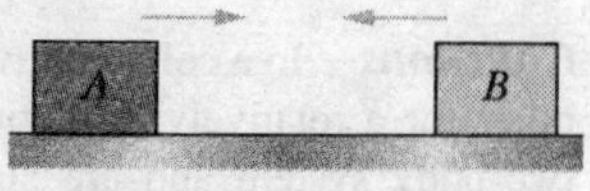

Fig. 10-20 Question 5.

6 A projectile body moving in the positive direction of an x axis on a frictionless floor runs into an initially stationary target body (as in Fig. 10-13) in a one-dimensional collision. Assume the particles form a closed, isolated system. Nine choices for a graph of the momenta of the bodies versus time (before and after the collision) are given in Fig. 10-21. Determine which choices represent physically impossible situations and explain why.

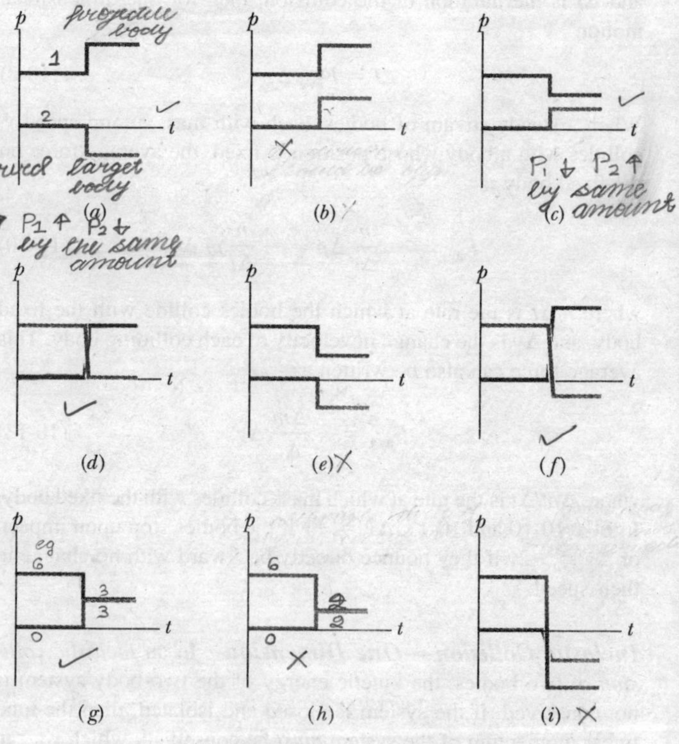

Fig. 10-21 Question 6.

7 Two bodies have undergone an elastic one-dimensional collision along an x axis. Figure 10-22 is a graph of position versus time for those bodies and for their center of mass. (a) Were both bodies initially moving, or was one initially stationary? Which line segment corresponds to the motion of the center of mass (b) before the collision and (c) after the col-

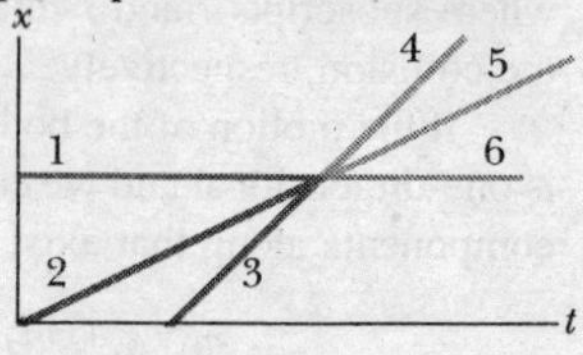

Fig. 10-22 Question 7.

lision? (d) Is the mass of the body that was moving faster before the collision greater than, less than, or equal to that of the other body?

8. Drop, in succession, a baseball and a basketball from about shoulder height above a hard floor, and note how high each rebounds. Then align the baseball above the basketball (with a small separation as in Fig. 10-23a) and drop them simultaneously. (Be prepared to duck, and guard your face.) (a) Is the rebound height of the basketball now higher or lower than before (Fig. 10-23b)? (b) Is the rebound height of the baseball less than or greater than the sum of the individual baseball and basketball rebound heights? (See also Problem 45.)

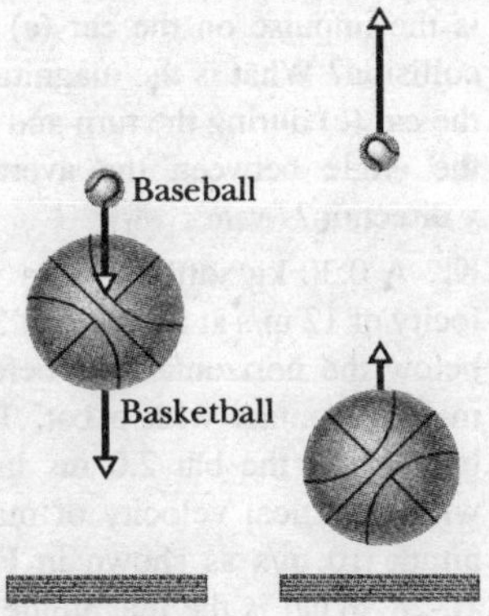

Fig. 10-23 Question 8 and Problem 45.

9. A projectile hockey puck A, with initial momentum 5 kg · m/s along an x axis, collides with an initially stationary hockey puck B. The pucks slide over frictionless ice and are shown in an overhead view in Fig. 10-24. Also shown are three general choices for the path taken by puck A after the collision. Which choice is appropriate if the momentum of puck B after the collision has an x component of (a) 5 kg · m/s, (b) more than 5 kg · m/s, and (c) less than 5 kg · m/s? Can that x component be (d) 1 kg · m/s or (e) −1 kg · m/s?

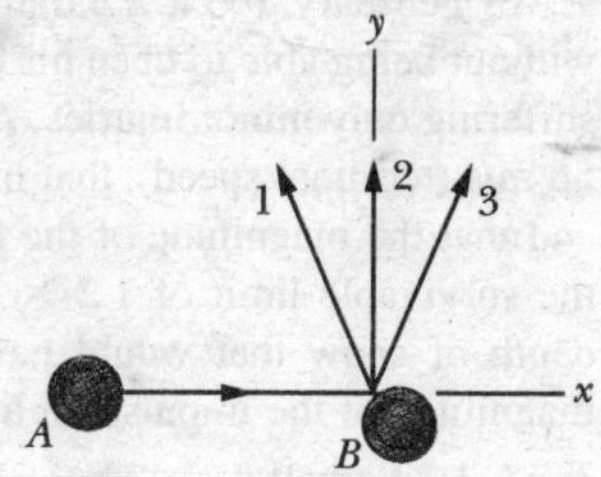

Fig. 10-24 Question 9.

10. Two bodies that form a closed, isolated system undergo a collision along a frictionless floor. Which of the three choices in Fig. 10-25 best represents the paths of those bodies and the path of their center of mass as seen from overhead?

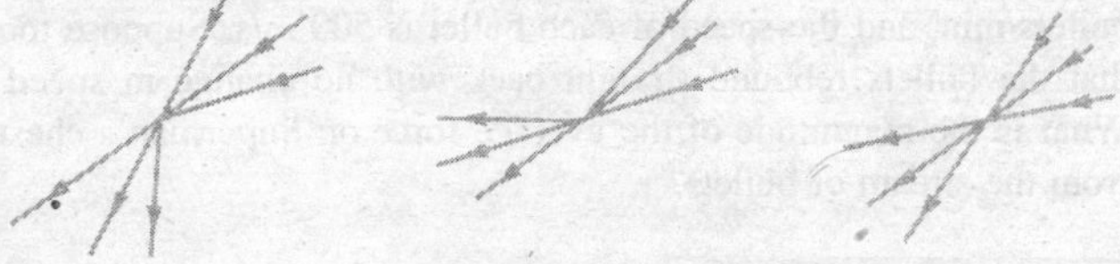

Fig. 10-25 Question 10.

EXERCISES & PROBLEMS

ssm Solution is in the Student Solutions Manual.
www Solution is available on the World Wide Web at: http://www.wiley.com/college/hrw
ilw Solution is available on the Interactive LearningWare.

SEC. 10-2 Impulse and Linear Momentum

1E. A cue stick strikes a stationary pool ball, with an average force of 50 N over a time of 10 ms. If the ball has mass 0.20 kg, what speed does it have just after impact? **ssm**

2E. The National Transportation Safety Board is testing the crashworthiness of a new car. The 2300 kg vehicle, moving at 15 m/s, is allowed to collide with a bridge abutment, which stops it in 0.56 s. What is the magnitude of the average force that acts on the car during the impact?

3E. A 150 g baseball pitched at a speed of 40 m/s is hit straight back to the pitcher at a speed of 60 m/s. What is the magnitude of the average force on the ball from the bat if the bat is in contact with the ball for 5.0 ms?

4E. Until he was in his seventies, Henri LaMothe excited audiences by belly-flopping from a height of 12 m into 30 cm of water (Fig. 10-26). Assuming that he stops just as he reaches the bottom of the water and estimating his mass, find the magnitudes of (a) the average force and (b) the average impulse on him from the water.

5E. A force that averages 1200 N is applied to a 0.40 kg steel ball moving at 14 m/s in a collision lasting 27 ms. If the force is in a direction opposite the initial velocity of the ball, find the final speed and direction of the ball. **ssm**

Fig. 10-26 Exercise 4.

6E. In February 1955, a paratrooper fell 370 m from an airplane without being able to open his chute but happened to land in snow, suffering only minor injuries. Assume that his speed at impact was 56 m/s (terminal speed), that his mass (including gear) was 85 kg, and that the magnitude of the force on him from the snow was at the survivable limit of 1.2×10^5 N. What are (a) the minimum depth of snow that would have stopped him safely and (b) the magnitude of the impulse on him from the snow?

7E. A 1.2 kg ball drops vertically onto a floor, hitting with a speed of 25 m/s. It rebounds with an initial speed of 10 m/s. (a) What impulse acts on the ball during the contact? (b) If the ball is in contact with the floor for 0.020 s, what is the magnitude of the average force on the floor from the ball?

8P. It is well known that bullets and other missiles fired at Superman simply bounce off his chest (Fig. 10-27). Suppose that a gangster sprays Superman's chest with 3 g bullets at the rate of 100 bullets/min, and the speed of each bullet is 500 m/s. Suppose too that the bullets rebound straight back with no change in speed. What is the magnitude of the average force on Superman's chest from the stream of bullets?

Fig. 10-27 Problem 8.

9P. A 1400 kg car moving at 5.3 m/s is initially traveling north in the positive y direction. After completing a 90° right-hand turn to the positive x direction in 4.6 s, the inattentive operator drives into a tree, which stops the car in 350 ms. In unit-vector notation, what is the impulse on the car (a) due to the turn and (b) due to the collision? What is the magnitude of the average force that acts on the car (c) during the turn and (d) during the collision? (e) What is the angle between the average force in (c) and the positive x direction? ssm www

10P. A 0.30 kg softball has a velocity of 12 m/s at an angle of 35° below the horizontal just before making contact with a bat. The ball leaves the bat 2.0 ms later with a vertical velocity of magnitude 10 m/s as shown in Fig. 10-28. What is the magnitude of the average force of the bat on the ball during the ball–bat contact?

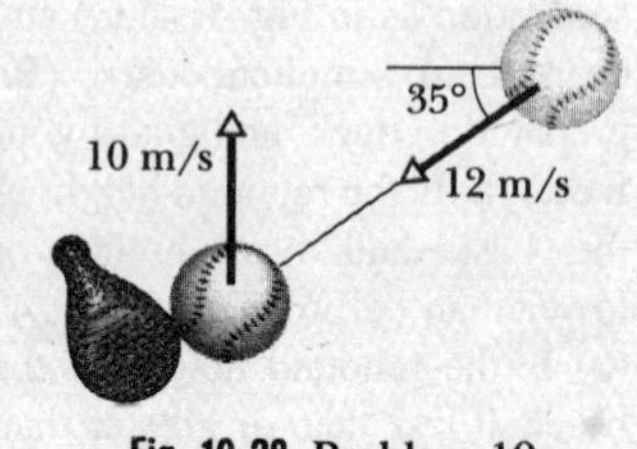

Fig. 10-28 Problem 10.

11P. The magnitude of an unbalanced force on a 10 kg object increases at a constant rate from zero to 50 N in 4.0 s, causing the initially stationary object to move. What is the object's speed at the end of the 4.0 s? ssm

12P. During a violent thunderstorm, hail of diameter 1.0 cm falls directly downward at a speed of 25 m/s. There are estimated to be 120 hailstones per cubic meter of air. (a) What is the mass of each hailstone (density = 0.92 g/cm^3)? (b) Assuming that the hail does not bounce, find the magnitude of the average force on a flat roof measuring 10 m × 20 m due to the impact of the hail. (*Hint:* During impact, the force on a hailstone from the roof is approximately equal to the net force on the hailstone, because the gravitational force on it is small.)

13P. A pellet gun fires ten 2.0 g pellets per second with a speed of 500 m/s. The pellets are stopped by a rigid wall. What are (a) the momentum of each pellet, (b) the kinetic energy of each pellet, and (c) the magnitude of the average force on the wall from the stream of pellets? (d) If each pellet is in contact with the wall for 0.6 ms, what is the magnitude of the average force on the wall from each pellet during contact? (e) Why is this average force so different from the average force calculated in (c)? ssm

14P. Figure 10-29 shows an approximate plot of force magnitude versus time during the collision of a 58 g Superball with a wall. The initial velocity of the ball is 34 m/s perpendicular to the wall; it rebounds directly back with approximately the same speed, also perpendicular to the wall. What is F_{max}, the maximum magnitude of the force on the ball from the wall during the collision?

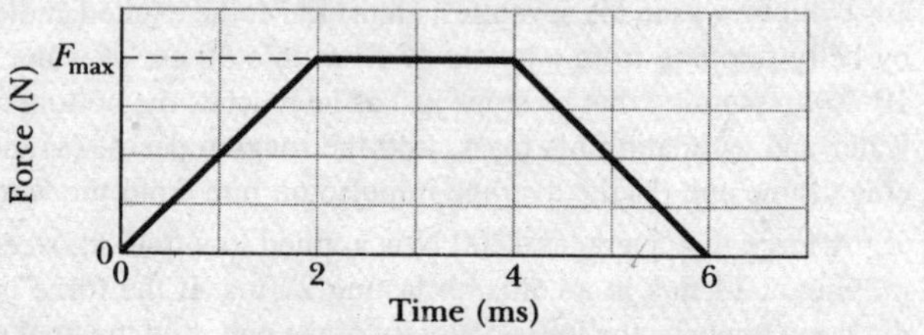

Fig. 10-29 Problem 14.

15P. A spacecraft is separated into two parts by detonating the explosive bolts that hold them together. The masses of the parts are 1200 kg and 1800 kg; the magnitude of the impulse on each part from the bolts is 300 N · s. With what relative speed do the two parts separate because of the detonation? ssm

16P. A ball having a mass of 150 g strikes a wall with a speed of 5.2 m/s and rebounds with only 50% of its initial kinetic energy. (a) What is the speed of the ball immediately after rebounding? (b) What is the magnitude of the impulse on the wall from the ball? (c) If the ball was in contact with the wall for 7.6 ms, what was the magnitude of the average force on the ball from the wall during this time interval?

17P. In the overhead view of Fig. 10-30, a 300 g ball with a speed v of 6.0 m/s strikes a wall at an angle θ of 30° and then rebounds with the same speed and angle. It is in contact with the wall for 10 ms. (a) What is the impulse on the ball from the wall? (b) What is the average force on the wall from the ball?

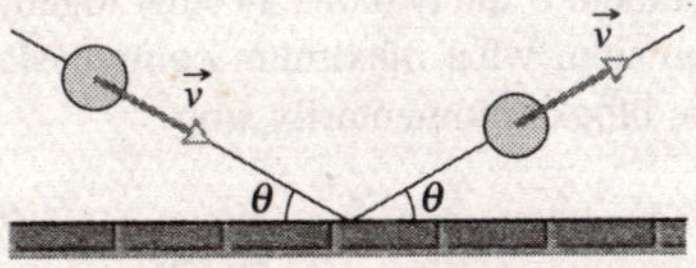

Fig. 10-30 Problem 17.

18P. A 2500 kg unmanned space probe is moving in a straight line at a constant speed of 300 m/s. Control rockets on the space probe execute a burn in which a thrust of 3000 N acts for 65.0 s. (a) What is the change in the magnitude of the probe's linear momentum if the thrust is backward, forward, or directly sideways? (b) What is the change in kinetic energy under the same three conditions? Assume that the mass of the ejected burn products is negligible compared to the mass of the space probe.

19P. A soccer player kicks a soccer ball of mass 0.45 kg that is initially at rest. The player's foot is in contact with the ball for 3.0×10^{-3} s, and the force of the kick is given by

$$F(t) = [(6.0 \times 10^6)t - (2.0 \times 10^9)t^2]\ \text{N},$$

for $0 \le t \le 3.0 \times 10^{-3}$ s, where t is in seconds. Find the magnitudes of the following: (a) the impulse on the ball due to the kick, (b) the average force on the ball from the player's foot during the period of contact, (c) the maximum force on the ball from the player's foot during the period of contact, and (d) the ball's speed immediately after it loses contact with the player's foot. ssm

SEC. 10-4 Inelastic Collisions in One Dimension

20E. A 5.20 g bullet moving at 672 m/s strikes a 700 g wooden block at rest on a frictionless surface. The bullet emerges, traveling in the same direction with its speed reduced to 428 m/s. (a) What is the resulting speed of the block? (b) What is the speed of the bullet–block center of mass?

21E. A 6.0 kg box sled is coasting across frictionless ice at a speed of 9.0 m/s when a 12 kg package is dropped into it from above. What is the new speed of the sled? ssm

22E. A bullet of mass 10 g strikes a ballistic pendulum of mass 2.0 kg. The center of mass of the pendulum rises a vertical distance of 12 cm. Assuming that the bullet remains embedded in the pendulum, calculate the bullet's initial speed.

23E. Meteor Crater in Arizona (Fig. 10-1a) is thought to have been formed by the impact of a meteor with Earth some 20,000 years ago. The mass of the meteor is estimated at 5×10^{10} kg, and its speed at 7200 m/s. What speed would such a meteor give Earth in a head-on collision? ssm

24E. A bullet of mass 4.5 g is fired horizontally into a 2.4 kg wooden block at rest on a horizontal surface. The coefficient of kinetic friction between block and surface is 0.20. The bullet stops in the block, which slides straight ahead for 1.8 m (without rotation). (a) What is the speed of the block immediately after the bullet stops relative to it? (b) At what speed is the bullet fired?

25P. Two cars A and B slide on an icy road as they attempt to stop at a traffic light. The mass of A is 1100 kg, and the mass of B is 1400 kg. The coefficient of kinetic friction between the locked wheels of either car and the road is 0.13. Car A succeeds in stopping at the light, but car B cannot stop and rear-ends car A. After the collision, A stops 8.2 m ahead of its position at impact, and B 6.1 m ahead; see Fig. 10-31. Both drivers had their brakes locked throughout the incident. From the distance each car moved after the collision, find the speed of (a) car A and (b) car B immediately after impact. (c) Use conservation of linear momentum to find the speed at which car B struck car A. On what grounds can the use of linear momentum conservation be criticized here?

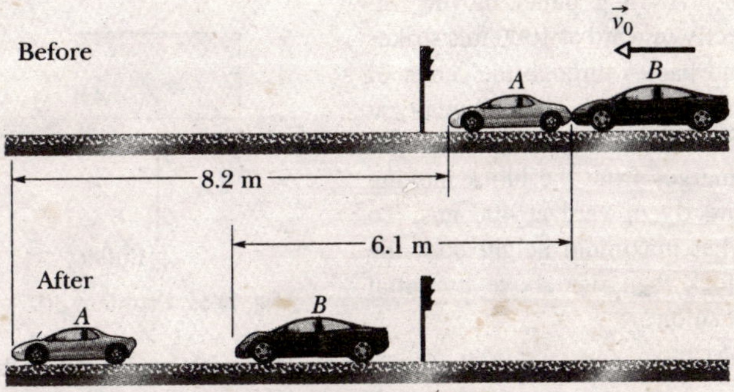

Fig. 10-31 Problem 25.

26P. In Fig. 10-32a, a 3.50 g bullet is fired horizontally at two blocks at rest on a frictionless tabletop. The bullet passes through the first block, with mass 1.20 kg, and embeds itself in the second, with mass 1.80 kg. Speeds of 0.630 m/s and 1.40 m/s, respectively, are thereby given to the blocks (Fig. 10-32b). Neglecting the mass

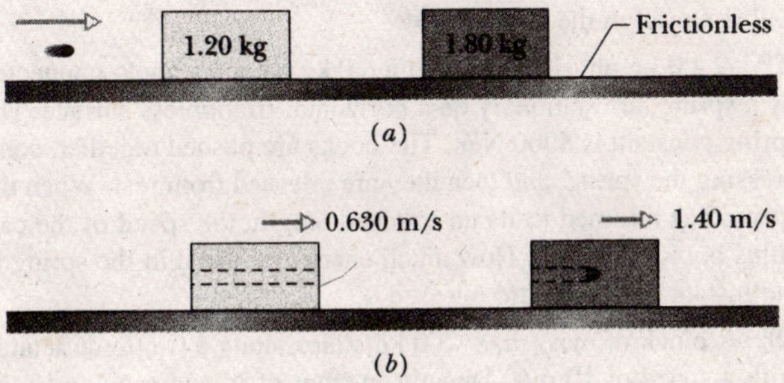

Fig. 10-32 Problem 26.

removed from the first block by the bullet, find (a) the speed of the bullet immediately after it emerges from the first block and (b) the bullet's original speed.

27P. A box is put on a scale that is marked in units of mass and adjusted to read zero when the box is empty. A stream of marbles is then poured into the box from a height h above its bottom at a rate of R (marbles per second). Each marble has mass m. (a) If the collisions between the marbles and the box are completely inelastic, find the scale reading at time t after the marbles begin to fill the box. (b) Determine a numerical answer when $R = 100\ s^{-1}$, $h = 7.60$ m, $m = 4.50$ g, and $t = 10.0$ s. ssm

28P. A 5.0 kg block with a speed of 3.0 m/s collides with a 10 kg block that has a speed of 2.0 m/s in the same direction. After the collision, the 10 kg block is observed to be traveling in the original direction with a speed of 2.5 m/s. (a) What is the velocity of the 5.0 kg block immediately after the collision? (b) By how much does the total kinetic energy of the system of two blocks change because of the collision? (c) Suppose, instead, that the 10 kg block ends up with a speed of 4.0 m/s. What then is the change in the total kinetic energy? (d) Account for the result you obtained in (c). Ilw

29P. A railroad freight car of mass 3.18×10^4 kg collides with a stationary caboose car. They couple together, and 27.0% of the initial kinetic energy is transferred to thermal energy, sound, vibrations, and so on. Find the mass of the caboose. ssm www

30P. A 10 g bullet moving directly upward at 1000 m/s strikes and passes through the center of mass of a 5.0 kg block initially at rest (Fig. 10-33). The bullet emerges from the block moving directly upward at 400 m/s. To what maximum height does the block then rise above its initial position?

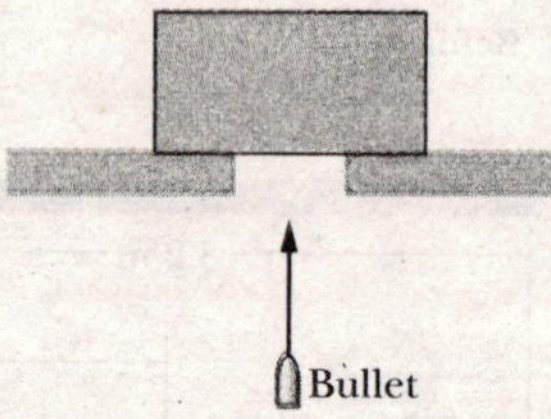

Fig. 10-33 Problem 30.

31P. In Fig. 10-34, a ball of mass m is shot with speed v_i into the barrel of a spring gun of mass M initially at rest on a frictionless surface. The ball sticks in the barrel at the point of maximum compression of the spring. Assume that the increase in thermal energy due to friction between the ball and the barrel is negligible. (a) What is the speed of the spring gun after the ball stops in the barrel? (b) What fraction of the initial kinetic energy of the ball is stored in the spring? ssm

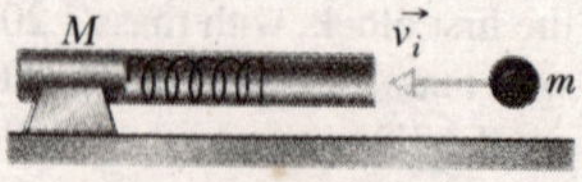

Fig. 10-34 Problem 31.

32P. A 4.0 kg physics book and a 6.0 kg calculus book, connected by a spring, are stationary on a horizontal frictionless surface. The spring constant is 8000 N/m. The books are pushed together, compressing the spring, and then they are released from rest. When the spring has returned to its unstretched length, the speed of the calculus book is 4.0 m/s. How much energy is stored in the spring at the instant the books are released?

33P. A block of mass $m_1 = 2.0$ kg slides along a frictionless table with a speed of 10 m/s. Directly in front of it, and moving in the same direction, is a block of mass $m_2 = 5.0$ kg moving at 3.0 m/s. A massless spring with spring constant $k = 1120$ N/m is attached to the near side of m_2, as shown in Fig. 10-35. When the blocks collide, what is the maximum compression of the spring? (*Hint:* At the moment of maximum compression of the spring, the two blocks move as one. Find the velocity by noting that the collision is completely inelastic at this point.) Ilw

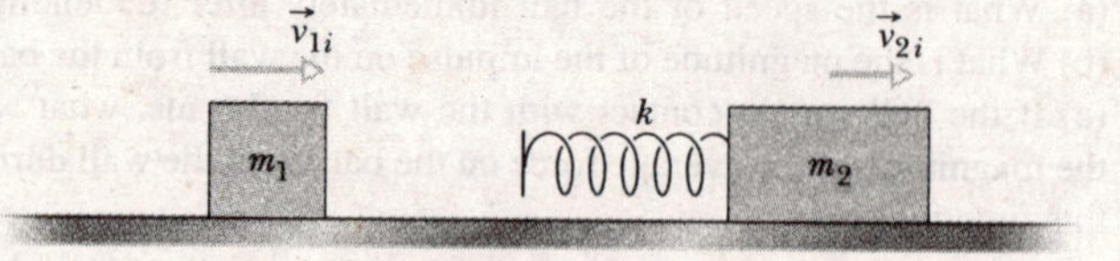

Fig. 10-35 Problem 33.

34P. A 1.0 kg block at rest on a horizontal frictionless surface is connected to an unstretched spring ($k = 200$ N/m) whose other end is fixed (Fig. 10-36). A 2.0 kg block moving at 4.0 m/s collides with the 1.0 kg block. If the two blocks stick together after the one-dimensional collision, what maximum compression of the spring occurs when the blocks momentarily stop?

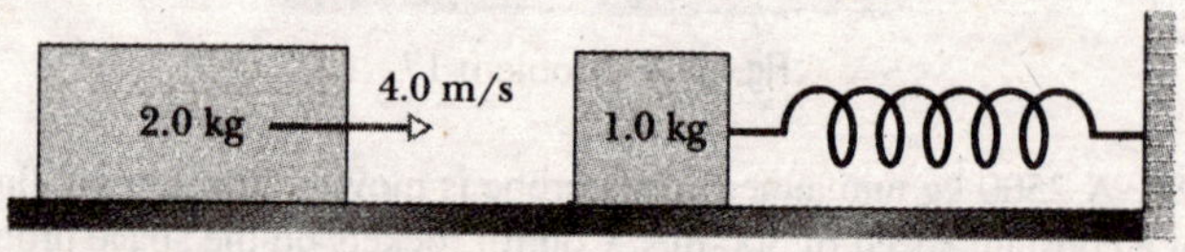

Fig. 10-36 Problem 34.

SEC. 10-5 Elastic Collisions in One Dimension

35E. The blocks in Fig. 10-37 slide without friction. (a) What is the velocity $\vec{v}$ of the 1.6 kg block after the collision? (b) Is the collision elastic? (c) Suppose the initial velocity of the 2.4 kg block is the reverse of what is shown. Can the velocity $\vec{v}$ of the 1.6 kg block after the collision be in the direction shown? ssm

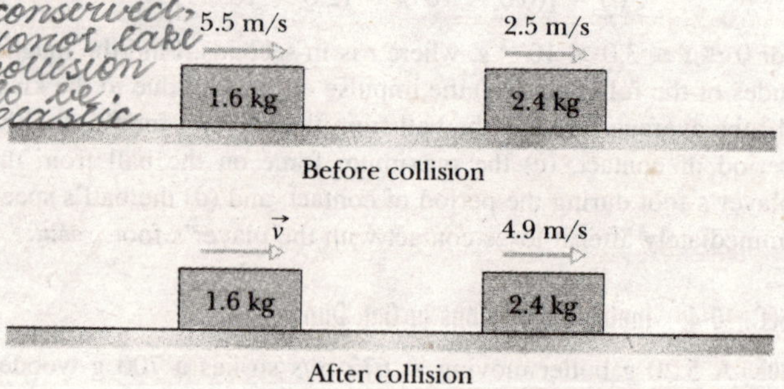

Fig. 10-37 Exercise 35.

36E. An electron undergoes a one-dimensional elastic collision with an initially stationary hydrogen atom. What percentage of the electron's initial kinetic energy is transferred to kinetic energy of the hydrogen atom? (The mass of the hydrogen atom is 1840 times the mass of the electron.)

37E. A cart with mass 340 g moving on a frictionless linear air track at an initial speed of 1.2 m/s undergoes an elastic collision

$m_1 v_{1i} + m_2 v_{2i} = (m_1 + m_2) v$

$\frac{1}{2} m_1 v_{1i}^2 = \frac{1}{2}(m_1 + m_2) v^2 + \frac{1}{2} k x_f^2$

with an initially stationary cart of unknown mass. After the collision, the first cart continues in its original direction at 0.66 m/s. (a) What is the mass of the second cart? (b) What is its speed after impact? (c) What is the speed of the two-cart center of mass? ssm

38E. Spacecraft *Voyager 2* (of mass m and speed v relative to the Sun) approaches the planet Jupiter (of mass M and speed V_J relative to the Sun) as shown in Fig. 10-38. The spacecraft rounds the planet and departs in the opposite direction. What is its speed, relative to the Sun, after this slingshot encounter, which can be analyzed as a collision? Assume $v = 12$ km/s and $V_J = 13$ km/s (the orbital speed of Jupiter). The mass of Jupiter is very much greater than the mass of the spacecraft ($M \gg m$).

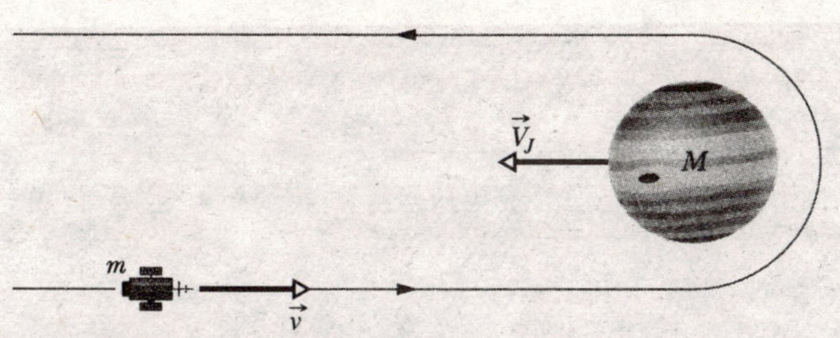

Fig. 10-38 Exercise 38.

39E. An alpha particle (mass 4 u) experiences an elastic head-on collision with a gold nucleus (mass 197 u) that is originally at rest. (The symbol u represents the atomic mass unit.) What percentage of its original kinetic energy does the alpha particle lose? ilw

40P. A steel ball of mass 0.500 kg is fastened to a cord that is 70.0 cm long and fixed at the far end. The ball is then released when the cord is horizontal (Fig. 10-39). At the bottom of its path, the ball strikes a 2.50 kg steel block initially at rest on a frictionless surface. The collision is elastic. Find (a) the speed of the ball and (b) the speed of the block, both just after the collision.

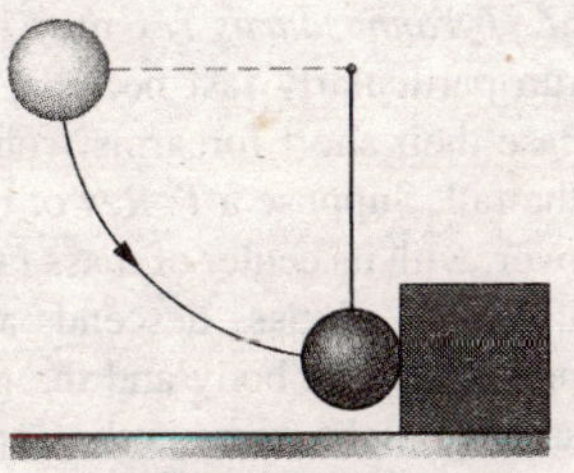

Fig. 10-39 Problem 40.

41P. A body of mass 2.0 kg makes an elastic collision with another body at rest and continues to move in the original direction but with one-fourth of its original speed. (a) What is the mass of the other body? (b) What is the speed of the two-body center of mass if the initial speed of the 2.0 kg body was 4.0 m/s? ssm www

42P. In the two-sphere arrangement of Sample Problem 10-4, assume that sphere 1 has a mass of 50 g and an initial height of 9.0 cm and that sphere 2 has a mass of 85 g. After the collision, what height is reached by (a) sphere 1 and (b) sphere 2? After the next (elastic) collision, what height is reached by (c) sphere 1 and (d) sphere 2? (*Hint:* Do not use rounded-off values.)

43P. Two titanium spheres approach each other head-on with the same speed and collide elastically. After the collision, one of the spheres, whose mass is 300 g, remains at rest. (a) What is the mass of the other sphere? (b) What is the speed of the two-sphere center of mass if the initial speed of each sphere is 2.0 m/s? ssm

44P. In Fig. 10-40, block 1 of mass m_1 is at rest on a long frictionless table that is up against a wall. Block 2 of mass m_2 is placed between block 1 and the wall and sent sliding to the left, toward block 1, with constant speed v_{2i}. Assuming that all collisions are elastic, find the value of m_2 (in terms of m_1) for which both blocks move with the same velocity after block 2 has collided once with block 1 and once with the wall. Assume the wall to have infinite mass.

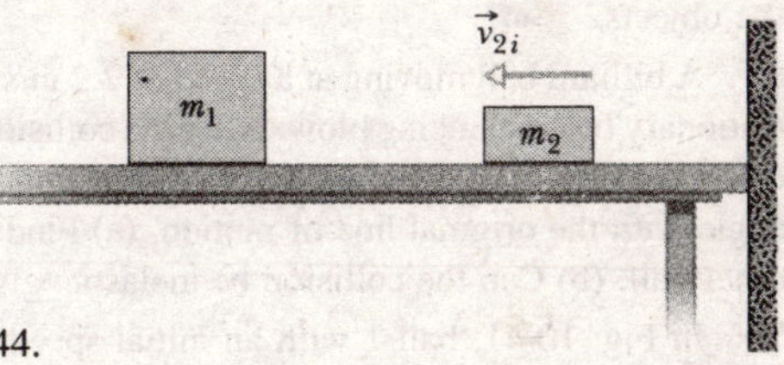

Fig. 10-40 Problem 44.

45P. A small ball of mass m is aligned above a larger ball of mass M (with a slight separation, as in Fig. 10-23a), and the two are dropped simultaneously from height h. (Assume the radius of each ball is negligible compared to h.) (a) If the larger ball rebounds elastically from the floor and then the small ball rebounds elastically from the larger ball, what ratio m/M results in the larger ball stopping upon its collision with the small ball? (The answer is approximately the mass ratio of a baseball to a basketball, as in Question 8.) (b) What height does the small ball then reach? ssm

SEC. 10-6 Collisions in Two Dimensions

46E. Two 2.0 kg bodies, A and B, collide. The velocities before the collision are $\vec{v}_A = 15\hat{i} + 30\hat{j}$ and $\vec{v}_B = -10\hat{i} + 5.0\hat{j}$. After the collision, $\vec{v}'_A = -5.0\hat{i} + 20\hat{j}$. All speeds are given in meters per second. (a) What is the final velocity of B? (b) How much kinetic energy is gained or lost in the collision?

47E. An alpha particle collides with an oxygen nucleus that is initially at rest. The alpha particle is scattered at an angle of 64.0° from its initial direction of motion, and the oxygen nucleus recoils at an angle of 51.0° on the opposite side of that initial direction. The final speed of the nucleus is 1.20×10^5 m/s. Find (a) the final speed and (b) the initial speed of the alpha particle. (In atomic mass units, the mass of an alpha particle is 4.0 u, and the mass of an oxygen nucleus is 16 u.) ilw

48E. A proton with a speed of 500 m/s collides elastically with another proton initially at rest. The projectile and target protons then move along perpendicular paths, with the projectile path at 60° from the original direction. After the collision, what are the speeds of (a) the target proton and (b) the projectile proton?

49E. In a game of pool, the cue ball strikes another ball of the same mass and initially at rest. After the collision, the cue ball moves at 3.50 m/s along a line making an angle of 22.0° with its original direction of motion, and the second ball has a speed of 2.00 m/s. Find (a) the angle between the direction of motion of the second ball and the original direction of motion of the cue ball and (b) the original speed of the cue ball. (c) Is kinetic energy (of the centers of mass, don't consider the rotation) conserved? ssm

50P. Two balls A and B, having different but unknown masses, collide. Initially, A is at rest and B has speed v. After the collision, B has speed $v/2$ and moves perpendicularly to its original motion. (a) Find the direction in which ball A moves after the collision. (b) Show that you cannot determine the speed of A from the information given.

51P. After a completely inelastic collision, two objects of the same mass and same initial speed are found to move away together at $\frac{1}{2}$ their initial speed. Find the angle between the initial velocities of the objects. ssm

52P. A billiard ball moving at a speed of 2.2 m/s strikes an identical stationary ball a glancing blow. After the collision, one ball is found to be moving at a speed of 1.1 m/s in a direction making a 60° angle with the original line of motion. (a) Find the velocity of the other ball. (b) Can the collision be inelastic, given these data?

53P. In Fig. 10-41, ball 1 with an initial speed of 10 m/s collides elastically with stationary balls 2 and 3, whose centers are on a line perpendicular to the initial velocity of ball 1 and that are initially in contact with each other. The three balls are identical. Ball 1 is aimed directly at the contact point, and all motion is frictionless. After the collision, what are the velocities of (a) ball 2, (b) ball 3, and (c) ball 1? (*Hint:* With friction absent, each impulse is directed along the line connecting the centers of the colliding balls, normal to the colliding surfaces.) ssm

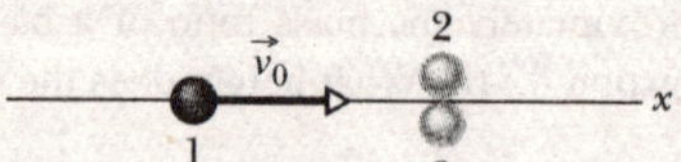

Fig. 10-41 Problem 53.

54P. Two 30 kg children, each with a speed of 4.0 m/s, are sliding on a frictionless frozen pond when they collide and stick together because they have Velcro straps on their jackets. The two children then collide and stick to a 75 kg man who was sliding at 2.0 m/s. After this collision, the three-person composite is stationary. What is the angle between the initial velocity vectors of the two children?

55P. Show that if a neutron is scattered through 90° in an elastic collision with an initially stationary deuteron, the neutron loses $\frac{2}{3}$ of its initial kinetic energy to the deuteron. (In atomic mass units, the mass of a neutron is 1.0 u and the mass of a deuteron is 2.0 u.) ssm

Additional Problems

56. Basilisk lizards can run across the top of a water surface (Fig. 10-42). With each step, a lizard first slaps its foot against the water and then pushes it down into the water rapidly enough to form an air cavity around the top of the foot. To avoid having to pull the foot back up against water drag in order to complete the step, the lizard withdraws the foot before water can flow into the air cavity. During this full action of slap, downward push, and withdrawal, the average upward impulse on the lizard must match the downward impulse due to the gravitational force if the lizard is not to sink. Suppose that the mass of a basilisk lizard is 90.0 g, the mass of each foot is 3.00 g, the speed of a foot as it slaps the water is 1.50 m/s, and the time for a single step is 0.600 s. (a) What is the magnitude of the impulse on the lizard during the slap? (Assume this impulse is directly upward.) (b) During the 0.600 s duration of a step, what is the downward impulse on the lizard due to the gravitational force? (c) Which action, the slap or the push, provides the primary support for the lizard, or are they approximately equal in their support?

Fig. 10-42 Problem 56. Basilisk lizard running across water.

57. *Tyrannosaurus rex* may have known from experience not to run particularly fast because of the danger of tripping, in which case their short forearms would have been no help in cushioning the fall. Suppose a *T. Rex* of mass m trips while walking, toppling over, with its center of mass falling freely a distance of 1.5 m. Then its center of mass descends an additional 0.30 m owing to compression of its body and the ground. (a) In multiples of the dinosaur's weight, what is the approximate magnitude of the average vertical force on the dinosaur during its collision with the ground (during the descent of 0.30 m)? Now assume that the dinosaur is running at a speed of 19 m/s (fast) when it trips, falls to the ground, and then slides to a stop with a coefficient of kinetic friction of 0.6. Assume also that the average vertical force during the collision and sliding is that in (a). What, approximately, are (b) the magnitude of the average total force on it from the ground (again in multiples of its weight) and (c) the sliding distance? The force magnitudes of (a) and (b) strongly suggest that the collision would injure the torso of the dinosaur. The head, which would fall farther, would suffer even greater injury.

11 Rotation

In judo, a weaker and smaller fighter who understands physics can defeat a stronger and larger fighter who does not. This fact is demonstrated by the basic "hip throw," in which a fighter rotates the fighter's opponent around his hip and—if the throw is successful—onto the mat. Without the proper use of physics, the throw requires considerable strength and can easily fail.

What is the advantage offered by physics?

The answer is in this chapter.

(a)

(b)

Fig. 11-1 Figure skater Michelle Kwan in motion of (a) pure translation in a fixed direction and (b) pure rotation about a vertical axis.

11-1 Translation and Rotation

The graceful movement of figure skaters can be used to illustrate two kinds of pure, or unmixed, motion. Figure 11-1*a* shows a skater gliding across the ice in a straight line with constant speed. Her motion is one of pure **translation.** Figure 11-1*b* shows her spinning at a constant rate about a vertical axis, in a motion of pure **rotation.**

Translation is motion along a straight line, which has been our focus up to now. Rotation is the motion of wheels, gears, motors, planets, the hands of clocks, the rotors of jet engines, and the blades of helicopters. It is our focus in this chapter.

11-2 The Rotational Variables

We wish to examine the rotation of a rigid body about a fixed axis. A **rigid body** is a body that can rotate with all its parts locked together and without any change in its shape. A **fixed axis** means that the rotation occurs about an axis that does not move. Thus, we shall not examine an object like the Sun, because the parts of the Sun (a ball of gas) are not locked together. We also shall not examine an object like a bowling ball rolling along a bowling alley, because the ball rotates about an axis that moves (the ball's motion is a mixture of rotation and translation).

Figure 11-2 shows a rigid body of arbitrary shape in rotation about a fixed axis, called the **axis of rotation** or the **rotation axis.** Every point of the body moves in a circle whose center lies on the axis of rotation, and every point moves through the same angle during a particular time interval. In pure translation, every point of the body moves in a straight line, and every point moves through the same *linear distance* during a particular time interval. (Comparisons between angular and linear motion will appear throughout this chapter.)

We deal now—one at a time—with the angular equivalents of the linear quantities position, displacement, velocity, and acceleration.

Angular Position

Figure 11-2 shows a *reference line*, fixed in the body, perpendicular to the rotation axis, and rotating with the body. The **angular position** of this line is the angle of the line relative to a fixed direction, which we take as the **zero angular position.** In Fig. 11-3, the angular position θ is measured relative to the positive direction of the x axis. From geometry, we know that θ is given by

$$\theta = \frac{s}{r} \quad \text{(radian measure).} \tag{11-1}$$

Here s is the length of arc (or the arc distance) along a circle and between the x axis (the zero angular position) and the reference line; r is the radius of that circle.

An angle defined in this way is measured in **radians** (rad) rather than in revolutions (rev) or degrees. The radian, being the ratio of two lengths, is a pure number and thus has no dimension. Because the circumference of a circle of radius r is $2\pi r$, there are 2π radians in a complete circle:

$$1 \text{ rev} = 360° = \frac{2\pi r}{r} = 2\pi \text{ rad,} \tag{11-2}$$

and thus

$$1 \text{ rad} = 57.3° = 0.159 \text{ rev.} \tag{11-3}$$

We do *not* reset θ to zero with each complete rotation of the reference line about the rotation axis. If the reference line completes two revolutions from the zero angular position, then the angular position θ of the line is $\theta = 4\pi$ rad.

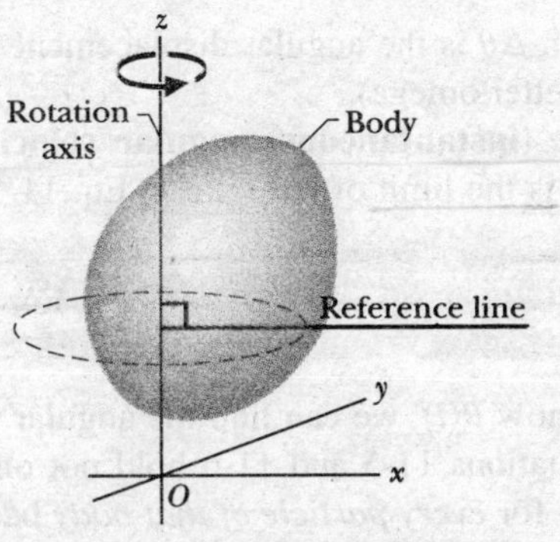

Fig. 11-2 A rigid body of arbitrary shape in pure rotation about the z axis of a coordinate system. The position of the *reference line* with respect to the rigid body is arbitrary, but it is perpendicular to the rotation axis. It is fixed in the body and rotates with the body.

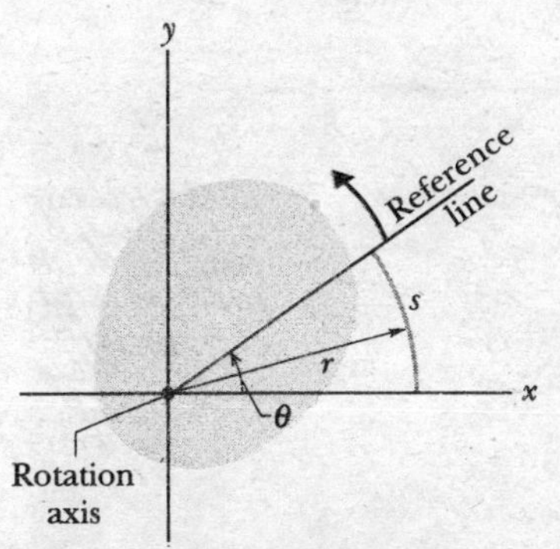

Fig. 11-3 The rotating rigid body of Fig. 11-2 in cross section, viewed from above. The plane of the cross section is perpendicular to the rotation axis, which now extends out of the page, toward you. In this position of the body, the reference line makes an angle θ with the x axis.

For pure translational motion along the x direction, we can know all there is to know about a moving body if we are given $x(t)$, its position as a function of time. Similarly, for pure rotation, we can know all there is to know about a rotating body if we are given $\theta\,(t)$, the angular position of the body's reference line as a function of time.

Angular Displacement

If the body of Fig. 11-3 rotates about the rotation axis as in Fig. 11-4, changing the angular position of the reference line from θ_1 to θ_2, the body undergoes an **angular displacement** $\Delta\theta$ given by

$$\Delta\theta = \theta_2 - \theta_1. \tag{11-4}$$

This definition of angular displacement holds not only for the rigid body as a whole but also for *every particle within that body* because the particles are all locked together.

If a body is in translational motion along an x axis, its displacement Δx is either positive or negative, depending on whether the body is moving in the positive or negative direction of the axis. Similarly, the angular displacement $\Delta\theta$ of a rotating body is either positive or negative, according to the following rule:

> An angular displacement in the counterclockwise direction is positive, and one in the clockwise direction is negative.

The phrase "*clocks are negative*" can help you remember this rule.

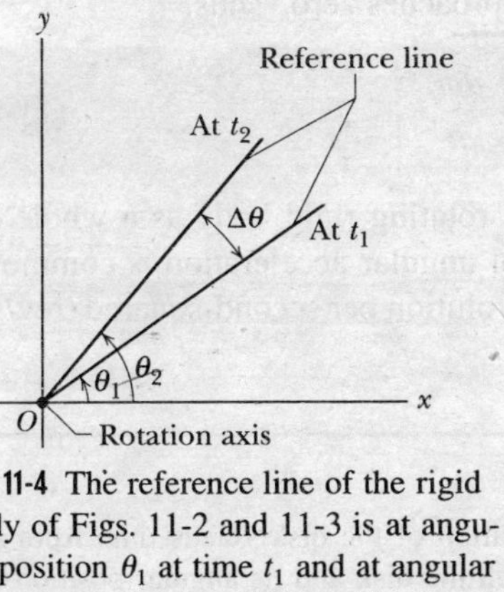

Fig. 11-4 The reference line of the rigid body of Figs. 11-2 and 11-3 is at angular position θ_1 at time t_1 and at angular position θ_2 at a later time t_2. The quantity $\Delta\theta$ (= $\theta_2 - \theta_1$) is the angular displacement that occurs during the interval Δt (= $t_2 - t_1$). The body itself is not shown.

Angular Velocity

Suppose (see Fig. 11-4) that our rotating body is at angular position θ_1 at time t_1 and at angular position θ_2 at time t_2. We define the **average angular velocity** of the body in the time interval Δt from t_1 to t_2 to be

$$\omega_{\text{avg}} = \frac{\theta_2 - \theta_1}{t_2 - t_1} = \frac{\Delta\theta}{\Delta t}, \tag{11-5}$$

in which $\Delta\theta$ is the angular displacement that occurs during Δt (ω is the lowercase Greek letter omega).

The **(instantaneous) angular velocity** ω, with which we shall be most concerned, is the limit of the ratio in Eq. 11-5 as Δt approaches zero. Thus,

Whenever we say angular vel we mean $\frac{d\theta}{dt}$ only

$$\omega = \lim_{\Delta t \to 0} \frac{\Delta\theta}{\Delta t} = \frac{d\theta}{dt}. \tag{11-6}$$

If we know $\theta(t)$, we can find the angular velocity ω by differentiation.

Equations 11-5 and 11-6 hold not only for the rotating rigid body as a whole but also for *every particle of that body* because the particles are all locked together. The unit of angular velocity is commonly the radian per second (rad/s) or the revolution per second (rev/s). Another measure of angular velocity was used during at least the first three decades of rock: Music was produced by vinyl (phonograph) records that were played on turntables at "$33\frac{1}{3}$ rpm" or "45 rpm," meaning at $33\frac{1}{3}$ rev/min or 45 rev/min.

If a particle moves in translation along an x axis, its linear velocity v is either positive or negative, depending on whether the particle is moving in the positive or negative direction of the axis. Similarly, the angular velocity ω of a rotating rigid body is either positive or negative, depending on whether the body is rotating counterclockwise (positive) or clockwise (negative). ("Clocks are negative" still works.) The magnitude of an angular velocity is called the **angular speed,** which is also represented with ω.

Angular Acceleration

If the angular velocity of a rotating body is not constant, then the body has an angular acceleration. Let ω_2 and ω_1 be its angular velocities at times t_2 and t_1, respectively. The **average angular acceleration** of the rotating body in the interval from t_1 to t_2 is defined as

$$\alpha_{avg} = \frac{\omega_2 - \omega_1}{t_2 - t_1} = \frac{\Delta\omega}{\Delta t}, \tag{11-7}$$

in which $\Delta\omega$ is the change in the angular velocity that occurs during the time interval Δt. The **(instantaneous) angular acceleration** α, with which we shall be most concerned, is the limit of this quantity as Δt approaches zero. Thus,

$$\alpha = \lim_{\Delta t \to 0} \frac{\Delta\omega}{\Delta t} = \frac{d\omega}{dt}. \tag{11-8}$$

Equations 11-7 and 11-8 hold not only for the rotating rigid body as a whole but also for *every particle of that body*. The unit of angular acceleration is commonly the radian per second-squared (rad/s^2) or the revolution per second-squared (rev/s^2).

Sample Problem 11-1

The disk in Fig. 11-5*a* is rotating about its central axis like a merry-go-round. The angular position $\theta(t)$ of a reference line on the disk is given by

$$\theta = -1.00 - 0.600t + 0.250t^2, \tag{11-9}$$

with t in seconds, θ in radians, and the zero angular position as indicated in the figure.

(a) Graph the angular position of the disk versus time from $t = -3.0$ s to $t = 6.0$ s. Sketch the disk and its angular position reference line at $t = -2.0$ s, 0 s, and 4.0 s, and when the curve crosses the t axis.

SOLUTION: The **Key Idea** here is that the angular position of the disk is the angular position $\theta(t)$ of its reference line, which is given by

Eq. 11-9 as a function of time. So we graph Eq. 11-9; the result is shown in Fig. 11-5*b*.

To sketch the disk and its reference line at a particular time, we need to determine θ for that time. To do so, we substitute the time into Eq. 11-9. For $t = -2.0$ s, we get

$$\theta = -1.00 - (0.600)(-2.0) + (0.250)(-2.0)^2$$
$$= 1.2 \text{ rad} = 1.2 \text{ rad}\,\frac{360°}{2\pi \text{ rad}} = 69°.$$

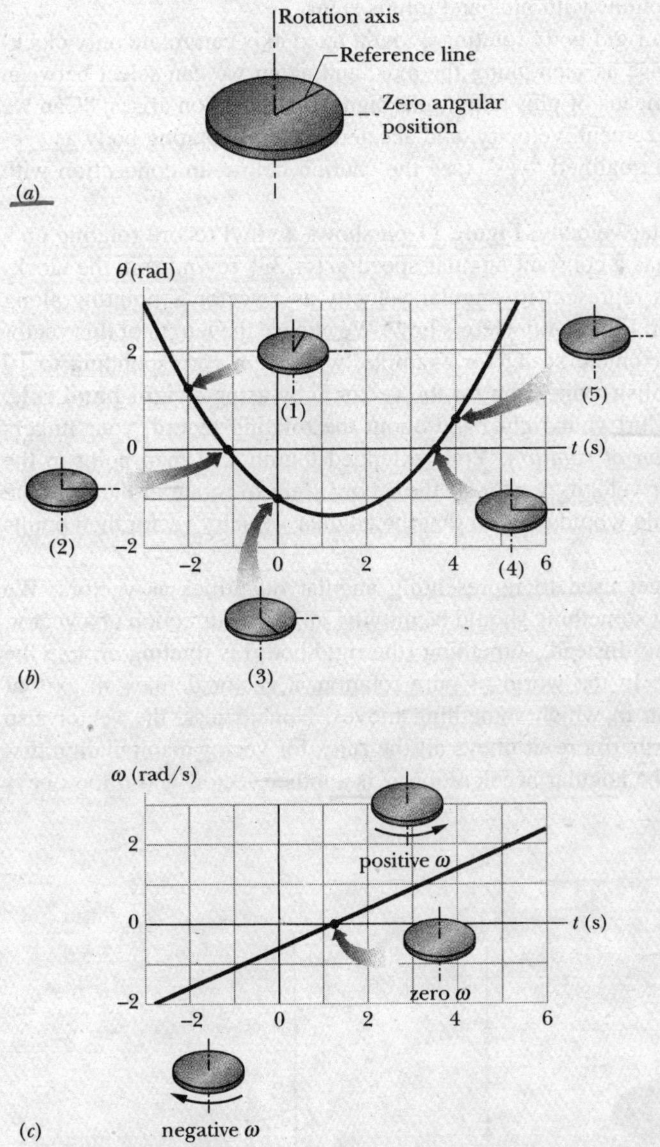

Fig. 11-5 Sample Problem 11-1. (*a*) A rotating disk. (*b*) A plot of the disk's angular position $\theta(t)$. Five sketches indicate the angular position of the reference line on the disk for five points on the curve. (*c*) A plot of the disk's angular velocity $\omega(t)$. Positive values of ω correspond to counterclockwise rotation, and negative values to clockwise rotation.

This means that at $t = -2.0$ s the reference line on the disk is rotated counterclockwise from the zero angular position by 1.2 rad or 69° (counterclockwise because θ is positive). Sketch 1 in Fig. 11-5*b* shows this angular position of the reference line.

Similarly, for $t = 0$, we find $\theta = -1.00$ rad $= -57°$, which means that the reference line is rotated clockwise from the zero angular position by 1.0 rad or 57°, as shown in sketch 3. For $t = 4.0$ s, we find $\theta = 0.60$ rad $= 34°$ (sketch 5). Drawing sketches for when the curve crosses the t axis is easy, because then $\theta = 0$ and the reference line is momentarily aligned with the zero angular position (sketches 2 and 4).

(b) At what time t_{min} does $\theta(t)$ reach the minimum value shown in Fig. 11-5*b*? What is that minimum value?

SOLUTION: The **Key Idea** here is that to find the extreme value (here the minimum) of a function, we take the first derivative of the function and set the result to zero. The first derivative of $\theta(t)$ is

$$\frac{d\theta}{dt} = -0.600 + 0.500t. \qquad (11\text{-}10)$$

Setting this to zero and solving for t give us the time at which $\theta(t)$ is minimum:

$$t_{min} = 1.20 \text{ s.} \qquad \text{(Answer)}$$

To get the minimum value of θ, we next substitute t_{min} into Eq. 11-9, finding

$$\theta = -1.36 \text{ rad} \approx -77.9°. \qquad \text{(Answer)}$$

This *minimum* of $\theta(t)$ (the bottom of the curve in Fig. 11-5*b*) corresponds to the *maximum clockwise* rotation of the disk from the zero angular position, somewhat more than is shown in sketch 3.

(c) Graph the angular velocity ω of the disk versus time from $t = -3.0$ s to $t = 6.0$ s. Sketch the disk and indicate the direction of turning and the sign of ω at $t = -2.0$ s and 4.0 s, and also at t_{min}.

SOLUTION: The **Key Idea** here is that, from Eq. 11-6, the angular velocity ω is equal to $d\theta/dt$ as given in Eq. 11-10. So, we have

$$\omega = -0.600 + 0.500t. \qquad (11\text{-}11)$$

The graph of this function $\omega(t)$ is shown in Fig. 11-5*c*.

To sketch the disk at $t = -2.0$ s, we substitute that value into Eq. 11-11, obtaining

$$\omega = -1.6 \text{ rad/s.} \qquad \text{(Answer)}$$

The minus sign tells us that at $t = -2.0$ s, the disk is turning clockwise, as suggested by the lowest sketch in Fig. 11-5*c*.

Substituting $t = 4.0$ s into Eq. 11-11 gives us

$$\omega = 1.4 \text{ rad/s.} \qquad \text{(Answer)}$$

The implied plus sign tells us that at $t = 4.0$ s, the disk is turning counterclockwise (the highest sketch in Fig. 11-5*c*).

For t_{min}, we already know that $d\theta/dt = 0$. So, we must also have $\omega = 0$. That is, the disk momentarily stops when the reference line reaches the minimum value of θ in Fig. 11-5*b*, as suggested by the center sketch in Fig. 11-5*c*.

(d) Use the results in parts (a) through (c) to describe the motion of the disk from $t = -3.0$ s to $t = 6.0$ s.

SOLUTION: When we first observe the disk at $t = -3.0$ s, it has a positive angular position and is turning clockwise but slowing. It stops at angular position $\theta = -1.36$ rad and then begins to turn counterclockwise, with its angular position eventually becoming positive again.

CHECKPOINT 1 A disk can rotate about its central axis like the one in Fig. 11-5*a*. Which of the following pairs of values for its initial and final angular positions, respectively, give a negative angular displacement: (a) -3 rad, $+5$ rad, (b) -3 rad, -7 rad, (c) 7 rad, -3 rad?

11-3 Are Angular Quantities Vectors?

We can describe the position, velocity, and acceleration of a single particle by means of vectors. If the particle is confined to a straight line, however, we do not really need vector notation. Such a particle has only two directions available to it, and we can indicate these directions with plus and minus signs.

In the same way, a rigid body rotating about a fixed axis can rotate only clockwise or counterclockwise as seen along the axis, and again we can select between the two directions by means of plus and minus signs. The question arises: "Can we treat the angular displacement, velocity, and acceleration of a rotating body as vectors?" The answer is a qualified "yes" (see the caution below, in connection with angular displacements).

Consider the angular velocity. Figure 11-6*a* shows a vinyl record rotating on a turntable. The record has a constant angular speed ω ($= 33\frac{1}{3}$ rev/min) in the clockwise direction. We can represent its angular velocity as a vector $\vec{\omega}$ pointing along the axis of rotation, as in Fig. 11-6*b*. Here's how: We choose the length of this vector according to some convenient scale, for example, with 1 cm corresponding to 10 rev/min. Then we establish a direction for the vector $\vec{\omega}$ by using a **right-hand rule,** as Fig. 11-6*c* shows: Curl your right hand about the rotating record, your fingers pointing *in the direction of rotation*. Your extended thumb will then point in the direction of the angular velocity vector. If the record were to rotate in the opposite sense, the right-hand rule would tell you that the angular velocity vector then points in the opposite direction.

It is not easy to get used to representing angular quantities as vectors. We instinctively expect that something should be moving *along* the direction of a vector. That is not the case here. Instead, something (the rigid body) is rotating *around* the direction of the vector. In the world of pure rotation, a vector defines an axis of rotation, not a direction in which something moves. Nonetheless, the vector also defines the motion. Furthermore, it obeys all the rules for vector manipulation discussed in Chapter 3. The angular acceleration $\vec{\alpha}$ is another vector, and it too obeys those rules.

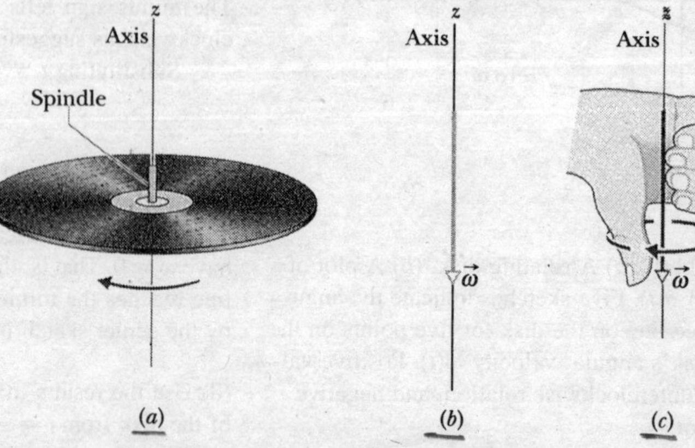

Fig. 11-6 (*a*) A record rotating about a vertical axis that coincides with the axis of the spindle. (*b*) The angular velocity of the rotating record can be represented by the vector $\vec{\omega}$, lying along the axis and pointing down, as shown. (*c*) We establish the direction of the angular velocity vector as downward by using the right-hand rule. When the fingers of the right hand curl around the record and point the way it is moving, the extended thumb points in the direction of $\vec{\omega}$.

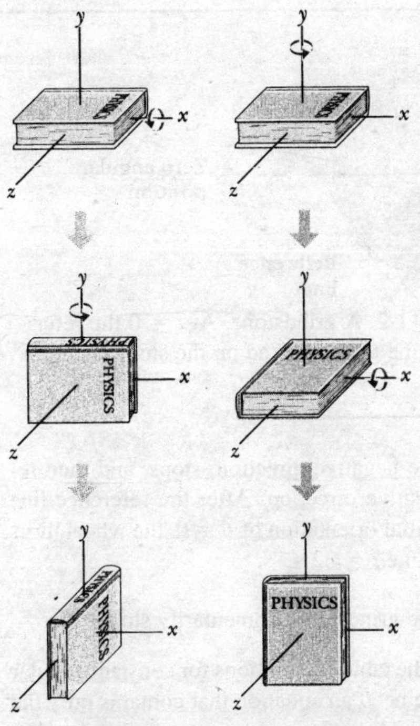

Fig. 11-7 (*a*) From its initial position, at the top, the book is given two successive 90° rotations, first about the (horizontal) *x* axis and then about the (vertical) *y* axis. (*b*) The book is given the same rotations, but in the reverse order.

In this chapter we consider only rotations that are about a fixed axis. For such situations, we need not consider vectors—we can represent angular velocity with ω and angular acceleration with α, and we can indicate direction with an implied plus sign for counterclockwise or an explicit minus sign for clockwise.

Now for the caution: Angular *displacements* (unless they are very small) *cannot* be treated as vectors. Why not? We can certainly give them both magnitude and direction, as we did for the angular velocity vector in Fig. 11-6. However, to be represented as a vector, a quantity must *also* obey the rules of vector addition, one of which says that if you add two vectors, the order in which you add them does not matter. Angular displacements fail this test.

Figure 11-7 gives an example. An initially horizontal book is given two 90° angular displacements, first in the order of Fig. 11-7*a* and then in the order of Fig. 11-7*b*. Although the two angular displacements are identical, their order is not, and the book ends up with different orientations. Thus, the addition of the two angular displacements depends on their order and they cannot be vectors.

11-4 Rotation with Constant Angular Acceleration

In pure translation, motion with a *constant linear acceleration* (for example, that of a falling body) is an important special case. In Table 2-1, we displayed a series of equations that hold for such motion.

In pure rotation, the case of *constant angular acceleration* is also important, and a parallel set of equations holds for this case also. We shall not derive them here, but simply write them from the corresponding linear equations, substituting equivalent angular quantities for the linear ones. This is done in Table 11-1, which lists both sets of equations (Eqs. 2-11 and 2-15 to 2-18; 11-12 to 11-16).

Recall that Eqs. 2-11 and 2-15 are basic equations for constant linear acceleration—the other equations in the Linear list can be derived from them. Similarly, Eqs. 11-12 and 11-13 are the basic equations for constant angular acceleration, and the other equations in the Angular list can be derived from them. To solve a simple problem involving constant angular acceleration, you can usually use an equation from the Angular list (*if* you have the list). Choose an equation for which the only unknown variable will be the variable requested in the problem. A better plan is to remember only Eqs. 11-12 and 11-13, and then solve them as simultaneous equations whenever needed. An example is given in Sample Problem 11-3.

CHECKPOINT 2: In four situations, a rotating body has angular position $\theta(t)$ given by (a) $\theta = 3t - 4$, (b) $\theta = -5t^3 + 4t^2 + 6$, (c) $\theta = 2/t^2 - 4/t$, and (d) $\theta = 5t^2 - 3$. To which situations do the angular equations of Table 11-1 apply?

TABLE 11-1 Equations of Motion for Constant Linear Acceleration and for Constant Angular Acceleration

Equation Number	Linear Equation	Missing Variable		Angular Equation	Equation Number
(2-11)	$v = v_0 + at$	$x - x_0$	$\theta - \theta_0$	$\omega = \omega_0 + \alpha t$	(11-12)
(2-15)	$x - x_0 = v_0 t + \frac{1}{2}at^2$	v	ω	$\theta - \theta_0 = \omega_0 t + \frac{1}{2}\alpha t^2$	(11-13)
(2-16)	$v^2 = v_0^2 + 2a(x - x_0)$	t	t	$\omega^2 = \omega_0^2 + 2\alpha(\theta - \theta_0)$	(11-14)
(2-17)	$x - x_0 = \frac{1}{2}(v_0 + v)t$	a	α	$\theta - \theta_0 = \frac{1}{2}(\omega_0 + \omega)t$	(11-15)
(2-18)	$x - x_0 = vt - \frac{1}{2}at^2$	v_0	ω_0	$\theta - \theta_0 = \omega t - \frac{1}{2}\alpha t^2$	(11-16)

Sample Problem 11-2

A grindstone (Fig. 11-8) rotates at constant angular acceleration $\alpha = 0.35\ \text{rad/s}^2$. At time $t = 0$, it has an angular velocity of $\omega_0 = -4.6$ rad/s and a reference line on it is horizontal, at the angular position $\theta_0 = 0$.

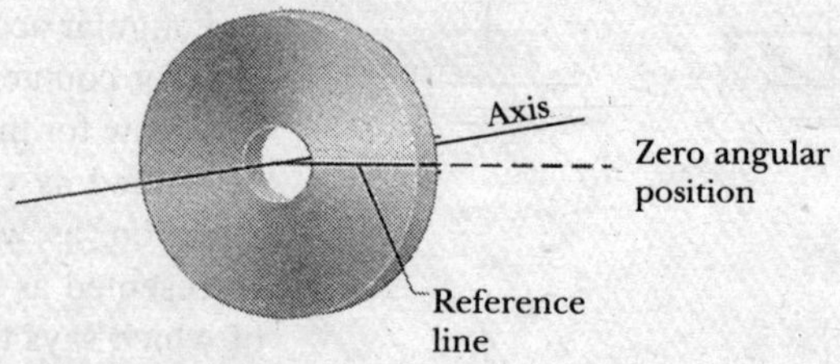

Fig. 11-8 Sample Problem 11-2. A grindstone. At $t = 0$ the reference line (which we imagine to be marked on the stone) is horizontal.

(a) At what time after $t = 0$ is the reference line at the angular position $\theta = 5.0$ rev?

SOLUTION: The **Key Idea** here is that the angular acceleration is constant, so we can use the rotation equations of Table 11-1. We choose Eq. 11-13,

∵ ω is missing in this eq

$$\theta - \theta_0 = \omega_0 t + \tfrac{1}{2}\alpha t^2,$$

because the only unknown variable it contains is the desired time t. Substituting known values and setting $\theta_0 = 0$ and $\theta = 5.0$ rev $= 10\pi$ rad give us

$$10\pi\ \text{rad} = (-4.6\ \text{rad/s})t + \tfrac{1}{2}(0.35\ \text{rad/s}^2)t^2.$$

(We converted 5.0 rev to 10π rad to keep the units consistent.) Solving this quadratic equation for t, we find

$$t = 32\ \text{s}. \qquad \text{(Answer)}$$

(b) Describe the grindstone's rotation between $t = 0$ and $t = 32$ s.

SOLUTION: The wheel is initially rotating in the negative (clockwise) direction with angular velocity $\omega_0 = -4.6$ rad/s, but its angular acceleration α is positive. This initial opposition of the signs of angular velocity and angular acceleration means that the wheel slows in its rotation in the negative direction, stops, and then reverses to rotate in the positive direction. After the reference line comes back through its initial orientation of $\theta = 0$, the wheel turns an additional 5.0 rev by time $t = 32$ s.

(c) At what time t does the grindstone momentarily stop?

SOLUTION: We again go to the table of equations for constant angular acceleration, and again we need an equation that contains only the desired unknown variable t. However, now we use another **Key Idea**. The equation must also contain the variable ω, so that we can set it to 0 and then solve for the corresponding time t. We choose Eq. 11-12, which yields

$$t = \frac{\omega - \omega_0}{\alpha} = \frac{0 - (-4.6\ \text{rad/s})}{0.35\ \text{rad/s}^2} = 13\ \text{s}. \qquad \text{(Answer)}$$

NOTE: ω=0 at a point ⇏ α=0 or $\frac{d\omega}{dt}=0$ ∵ change from any previous point to that point may not be equal to zero.

Sample Problem 11-3

While you are operating a Rotor (the rotating cylindrical ride discussed in Sample Problem 6-8), you spot a passenger in acute distress and decrease the angular speed of the cylinder from 3.40 rad/s to 2.00 rad/s in 20.0 rev, at constant angular acceleration. (The passenger is obviously more of a "translation person" than a "rotation person.")

(a) What is the constant angular acceleration during this decrease in angular speed?

SOLUTION: Assume that the rotation is in the counterclockwise direction, and let the acceleration begin at time $t = 0$, at angular position θ_0. The **Key Idea** here is that, because the angular acceleration is constant, we can relate the cylinder's angular acceleration to its angular velocity and angular displacement via the basic equations for constant angular acceleration (Eqs. 11-12 and 11-13). The initial angular velocity is $\omega_0 = 3.40$ rad/s, the angular displacement is $\theta - \theta_0 = 20.0$ rev, and the angular velocity at the end of that displacement is $\omega = 2.00$ rad/s. But we do not know the angular acceleration α and time t, which are in both basic equations.

To eliminate the unknown t, we use Eq. 11-12 to write

$$t = \frac{\omega - \omega_0}{\alpha},$$

which we then substitute into Eq. 11-13 to write

$$\theta - \theta_0 = \omega_0\left(\frac{\omega - \omega_0}{\alpha}\right) + \tfrac{1}{2}\alpha\left(\frac{\omega - \omega_0}{\alpha}\right)^2.$$

Solving for α, substituting known data, and converting 20 rev to 125.7 rad, we find

$$\alpha = \frac{\omega^2 - \omega_0^2}{2(\theta - \theta_0)} = \frac{(2.00\ \text{rad/s})^2 - (3.40\ \text{rad/s})^2}{2(125.7\ \text{rad})}$$
$$= -0.0301\ \text{rad/s}^2. \qquad \text{(Answer)}$$

(b) How much time did the speed decrease take?

SOLUTION: Now that we know α, we can use Eq. 11-12 to solve for t:

∵ $\theta - \theta_0 = \omega_0 t + \frac{1}{2}\alpha t^2$ is hard

$$t = \frac{\omega - \omega_0}{\alpha} = \frac{2.00\ \text{rad/s} - 3.40\ \text{rad/s}}{-0.0301\ \text{rad/s}^2}$$
$$= 46.5\ \text{s}. \qquad \text{(Answer)}$$

This sample problem is actually an angular version of Sample Problem 2-5. The data and symbols differ, but the techniques of solution are identical.

11-5 Relating the Linear and Angular Variables

In Section 4-7, we discussed uniform circular motion, in which a particle travels at constant linear speed v along a circle and around an axis of rotation. When a rigid body, such as a merry-go-round, rotates around an axis, each particle in the body moves in its own circle around that axis. Since the body is rigid, all the particles make one revolution in the same amount of time; that is, they all have the same angular speed ω.

However, the farther a particle is from the axis, the greater the circumference of its circle is, and so the faster its linear speed v must be. You can notice this on a merry-go-round. You turn with the same angular speed ω regardless of your distance from the center, but your linear speed v increases noticeably if you move to the outside edge of the merry-go-round.

We often need to relate the linear variables s, v, and a for a particular point in a rotating body to the angular variables θ, ω, and α for that body. The two sets of variables are related by r, the *perpendicular distance* of the point from the rotation axis. This perpendicular distance is the distance between the point and the rotation axis, measured along a perpendicular to the axis. It is also the radius r of the circle traveled by the point around the axis of rotation.

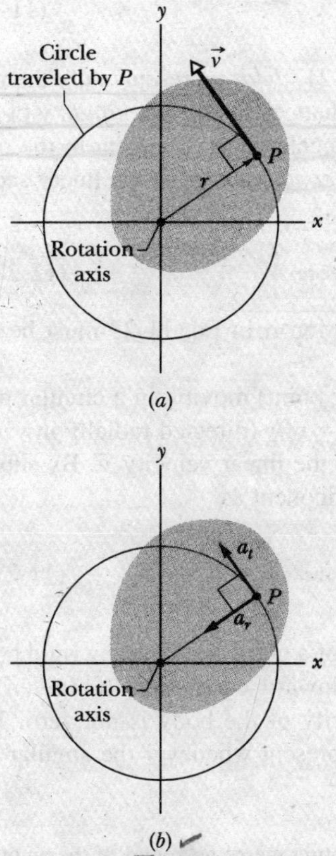

Fig. 11-9 The rotating rigid body of Fig. 11-2, shown in cross section viewed from above. Every point of the body (such as P) moves in a circle around the rotation axis. (a) The linear velocity $\vec{v}$ of every point is tangent to the circle in which the point moves. (b) The linear acceleration $\vec{a}$ of the point has (in general) two components: a tangential component a_t and a radial component a_r.

The Position

If a reference line on a rigid body rotates through an angle θ, a point within the body at a position r from the rotation axis moves a distance s along a circular arc, where s is given by Eq. 11-1:

$$s = \theta r \quad \text{(radian measure).} \tag{11-17}$$

This is the first of our linear–angular relations. *Caution:* The angle θ here must be measured in radians because Eq. 11-17 is itself the definition of angular measure in radians.

The Speed

Differentiating Eq. 11-17 with respect to time—with r held constant—leads to

$$\frac{ds}{dt} = \frac{d\theta}{dt} r.$$

However, ds/dt is the linear speed (the magnitude of the linear velocity) of the point in question, and $d\theta/dt$ is the angular speed ω of the rotating body. So

$$v = \omega r \quad \text{(radian measure).} \tag{11-18}$$

Caution: The angular speed ω must be expressed in radian measure.

Equation 11-18 tells us that since all points within the rigid body have the same angular speed ω, points with greater radius r have greater linear speed v. Figure 11-9a reminds us that the linear velocity is always tangent to the circular path of the point in question.

If the angular speed ω of the rigid body is constant, then Eq. 11-18 tells us that the linear speed v of any point within it is also constant. Thus, each point within the body undergoes uniform circular motion. The period of revolution T for the motion of each point and for the rigid body itself is given by Eq. 4-33:

$$T = \frac{2\pi r}{v}. \tag{11-19}$$

This equation tells us that the time for one revolution is the distance $2\pi r$ traveled in one revolution divided by the speed at which that distance is traveled. Substituting for v from Eq. 11-18 and canceling r, we find also that

$$T = \frac{2\pi}{\omega} \quad \text{(radian measure).} \tag{11-20}$$

This equivalent equation says that the time for one revolution is the angular distance 2π rad traveled in one revolution divided by the angular speed (or rate) at which that angle is traveled.

The Acceleration

Differentiating Eq. 11-18 with respect to time—again with r held constant—leads to

$$\frac{dv}{dt} = \frac{d\omega}{dt} r. \tag{11-21}$$

$a = \sqrt{a_t^2 + a_n^2}$

Here we run up against a complication. In Eq. 11-21, dv/dt represents only the part of the linear acceleration that is responsible for changes in the *magnitude* v of the linear velocity $\vec{v}$. Like $\vec{v}$, that part of the linear acceleration is tangent to the path of the point in question. We call it the *tangential component* a_t of the linear acceleration of the point, and we write

$$a_t = \alpha r \quad \text{(radian measure),} \tag{11-22}$$

where $\alpha = d\omega/dt$. *Caution:* The angular acceleration α in Eq. 11-22 must be expressed in radian measure.

In addition, as Eq. 4-32 tells us, a particle (or point) moving in a circular path has a *radial component* of linear acceleration, $a_r = v^2/r$ (directed radially inward), that is responsible for changes in the *direction* of the linear velocity $\vec{v}$. By substituting for v from Eq. 11-18, we can write this component as

$$a_r = \frac{v^2}{r} = \omega^2 r \quad \text{(radian measure).} \tag{11-23}$$

Thus, as Fig. 11-9*b* shows, the linear acceleration of a point on a rotating rigid body has, in general, two components. The radially inward component a_r (given by Eq. 11-23) is present whenever the angular velocity of the body is not zero. The tangential component a_t (given by Eq. 11-22) is present whenever the angular acceleration is not zero.

CHECKPOINT 3: A cockroach rides the rim of a rotating merry-go-round. If the angular speed of this system (*merry-go-round + cockroach*) is constant, does the cockroach have (a) radial acceleration and (b) tangential acceleration? If the angular speed is decreasing, does the cockroach have (c) radial acceleration and (d) tangential acceleration?

Sample Problem 11-4

Figure 11-10 shows a centrifuge used to accustom astronaut trainees to high accelerations. The radius r of the circle traveled by an astronaut is 15 m.

(a) At what constant angular speed must the centrifuge rotate if the astronaut is to have a linear acceleration of magnitude $11g$?

Fig. 11-10 Sample Problem 11-4. A centrifuge is used to accustom astronauts to the large acceleration experienced during a liftoff.

SOLUTION: The **Key Idea** is this: Because the angular speed is constant, the angular acceleration α (= $d\omega/dt$) is zero and so is the tangential component of the linear acceleration ($a_t = \alpha r$). This leaves only the radial component. From Eq. 11-23 ($a_r = \omega^2 r$), with $a_r = 11g$, we have

$$\omega = \sqrt{\frac{a_r}{r}} = \sqrt{\frac{(11)(9.8\ \text{m/s}^2)}{15\ \text{m}}}$$
$$= 2.68\ \text{rad/s} \approx 26\ \text{rev/min}. \qquad \text{(Answer)}$$

(b) What is the tangential acceleration of the astronaut if the centrifuge accelerates at a constant rate from rest to the angular speed of (a) in 120 s?

SOLUTION: The **Key Idea** here is that the tangential acceleration a_t, which is the linear acceleration along the circular path, is related to the angular acceleration α by Eq. 11-22 ($a_t = \alpha r$). Also, because the angular acceleration is constant, we can use Eq. 11-12 ($\omega = \omega_0 + \alpha t$) from Table 11-1 to find α from the given angular speeds. Putting these two equations together, we find

$$a_t = \alpha r = \frac{\omega - \omega_0}{t} r$$
$$= \frac{2.68\ \text{rad/s} - 0}{120\ \text{s}}(15\ \text{m}) = 0.34\ \text{m/s}^2$$
$$= 0.034g. \qquad \text{(Answer)}$$

Although the final radial acceleration $a_r = 11g$ is large (and alarming), the astronaut's tangential acceleration a_t during the speed-up is not.

PROBLEM-SOLVING TACTICS

Tactic 1: *Units for Angular Variables*

In Eq. 11-1 ($\theta = s/r$), we began the use of radian measure for all angular variables whenever we are using equations that contain both angular and linear variables. Thus, we must express angular displacements in radians, angular velocities in rad/s and rad/min, and angular accelerations in rad/s^2 and rad/min^2. Equations 11-17, 11-18, 11-20, 11-22, and 11-23 are marked to emphasize this. The only exceptions to this rule are equations that involve *only* angular variables, such as the angular equations listed in Table 11-1. Here you are free to use any unit you wish for the angular variables; that is, you may use radians, degrees, or revolutions, as long as you use them consistently.

In equations where radian measure must be used, you need not keep track of the unit "radian" (rad) algebraically, as you must do for other units. You can add or delete it at will, to suit the context. In Sample Problem 11-4a the unit was added to the answer; in Sample Problem 11-4b it was omitted from the answer.

11-6 Kinetic Energy of Rotation

The rapidly rotating blade of a table saw certainly has kinetic energy due to that rotation. How can we express the energy? We cannot apply the familiar formula $K = \frac{1}{2}mv^2$ to the saw as a whole because that would only give us the kinetic energy of the saw's center of mass, which is zero.

Instead, we shall treat the table saw (and any other rotating rigid body) as a collection of particles with different speeds. We can then add up the kinetic energies of all the particles to find the kinetic energy of the body as a whole. In this way we obtain, for the kinetic energy of a rotating body,

$$K = \tfrac{1}{2}m_1v_1^2 + \tfrac{1}{2}m_2v_2^2 + \tfrac{1}{2}m_3v_3^2 + \cdots$$
$$= \sum \tfrac{1}{2}m_iv_i^2, \qquad (11\text{-}24)$$

in which m_i is the mass of the ith particle and v_i is its speed. The sum is taken over all the particles in the body.

The problem with Eq. 11-24 is that v_i is not the same for all particles. We solve this problem by substituting for v from Eq. 11-18 ($v = \omega r$), so that we have

$$K = \sum \tfrac{1}{2} m_i (\omega r_i)^2 = \tfrac{1}{2}\left(\sum m_i r_i^2\right)\omega^2, \qquad (11\text{-}25)$$

in which ω *is* the same for all particles.

The quantity in parentheses on the right side of Eq. 11-25 tells us how the mass of the rotating body is distributed about its axis of rotation. We call that quantity the **rotational inertia** (or **moment of inertia**) I of the body with respect to the axis of rotation. It is a constant for a particular rigid body and a particular rotation axis. (That axis must always be specified if the value of I is to be meaningful.)

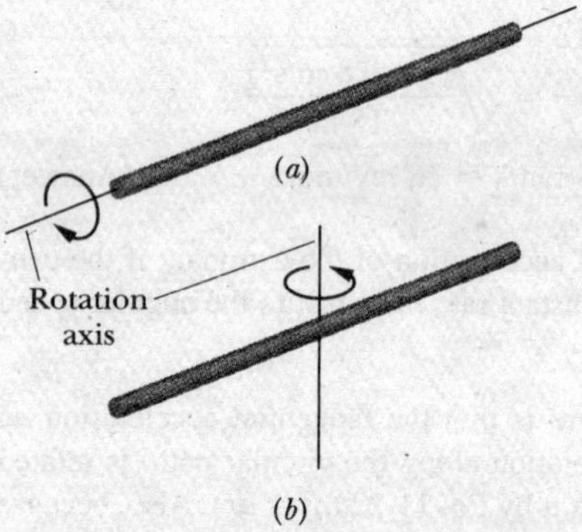

Fig. 11-11 A long rod is much easier to rotate about (*a*) its central (longitudinal) axis than about (*b*) an axis through its center and perpendicular to its length because the mass is distributed closer to the rotation axis in (*a*) than in (*b*).

We may now write

$$I = \sum m_i r_i^2 \quad \text{(rotational inertia)} \qquad (11\text{-}26)$$

and substitute into Eq. 11-25, obtaining

$$K = \tfrac{1}{2} I \omega^2 \quad \text{(radian measure)} \qquad (11\text{-}27)$$

as the expression we seek. Because we have used the relation $v = \omega r$ in deriving Eq. 11-27, ω must be expressed in radian measure. The SI unit for I is the kilogram–square meter ($kg \cdot m^2$).

Equation 11-27, which gives the kinetic energy of a rigid body in pure rotation, is the angular equivalent of the formula $K = \tfrac{1}{2} M v_{com}^2$, which gives the kinetic energy of a rigid body in pure translation. In both formulas there is a factor of $\tfrac{1}{2}$. Where mass M appears in one equation, I (which involves both mass and its distribution) appears in the other. Finally, each equation contains as a factor the square of a speed—translational or rotational as appropriate. The kinetic energies of translation and of rotation are not different kinds of energy. They are both kinetic energy, expressed in ways that are appropriate to the motion at hand.

We noted previously that the rotational inertia of a rotating body involves not only its mass but also how that mass is distributed. Here is an example that you can literally feel. Rotate a long, fairly heavy rod (a pole, a length of lumber, or something similar), first around its central (longitudinal) axis (Fig. 11-11*a*) and then around an axis perpendicular to the rod and through the center (Fig. 11-11*b*). Both rotations involve the very same mass, but the first rotation is much easier than the second. The reason is that the mass is distributed much closer to the rotation axis in the first rotation. As a result, the rotational inertia of the rod is much smaller in Fig. 11-11*a* than in Fig. 11-11*b*. In general, smaller rotational inertia means easier rotation.

point mass

CHECKPOINT 4: The figure shows three small spheres that rotate about a vertical axis. The perpendicular distance between the axis and the center of each sphere is given. Rank the three spheres according to their rotational inertia about that axis, greatest first.

I $m_1 r_1^2 = 36$
II $m_2 r_2^2 = 9 \times 2^2 = 36$
III $m_3 r_3^2 = 4 \times 3^2 = 36$ all tie

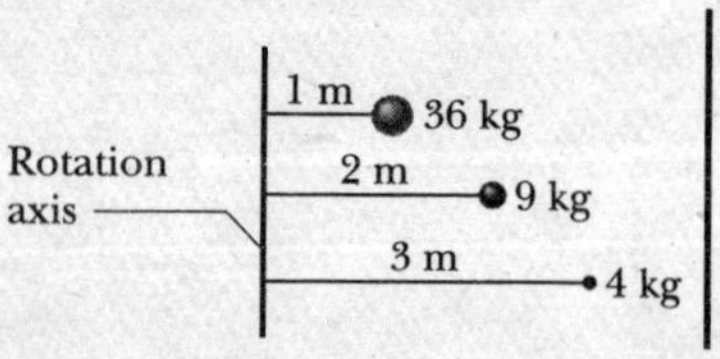

11-7 Calculating the Rotational Inertia

If a rigid body consists of a few particles, we can calculate its rotational inertia about a given rotation axis with Eq. 11-26 ($I = \sum m_i r_i^2$), that is, we can find the product mr^2 for each particle and then sum the products. (Recall that r is the perpendicular distance a particle is from the given rotation axis.)

TABLE 11-2 Some Rotational Inertias

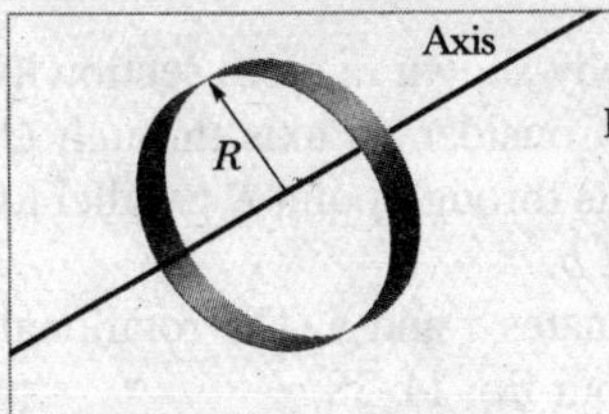 Hoop about central axis $I = MR^2$ (*a*)	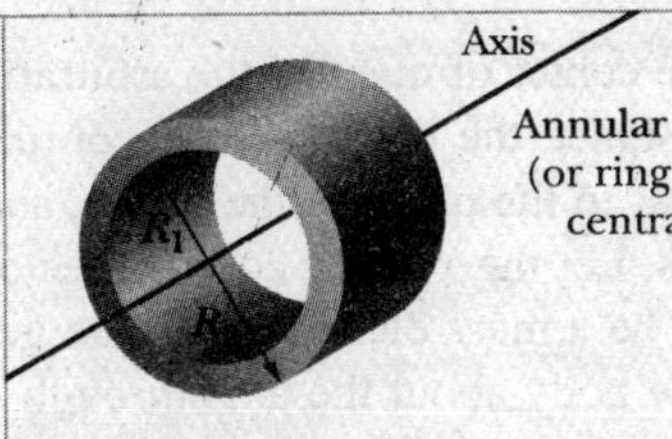 Annular cylinder (or ring) about central axis $I = \frac{1}{2}M(R_1^2 + R_2^2)$ (*b*)	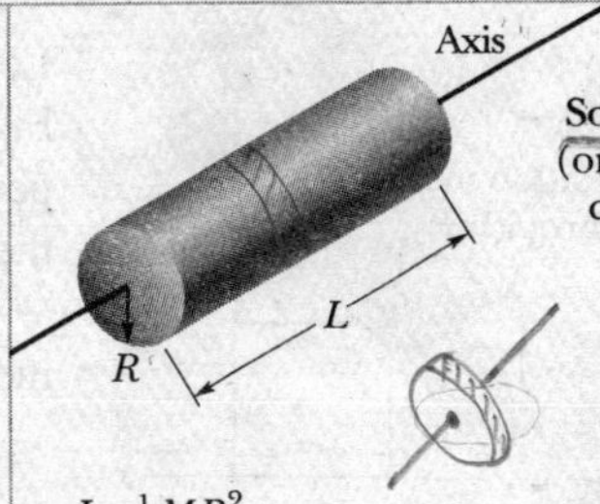 Solid cylinder (or disk) about central axis $I = \frac{1}{2}MR^2$ (*c*)
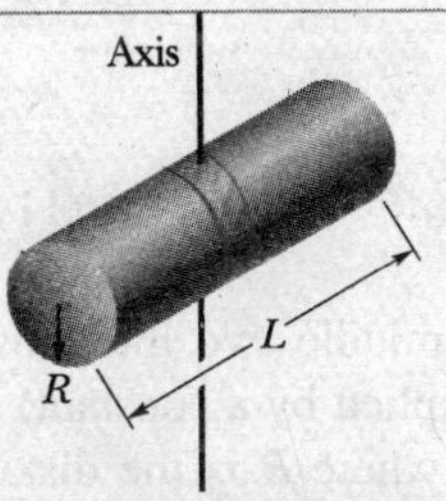 Solid cylinder (or disk) about central diameter $I = \frac{1}{4}MR^2 + \frac{1}{12}ML^2$ (*d*)	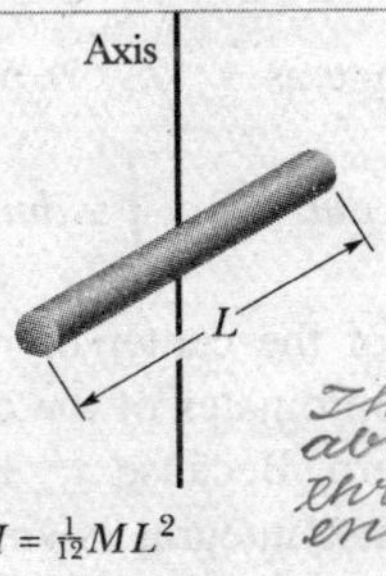 Thin rod about axis through center perpendicular to length Thin rod abt an axis (⊥) thro' one of the ends $\frac{ML^2}{3}$ $I = \frac{1}{12}ML^2$ (*e*)	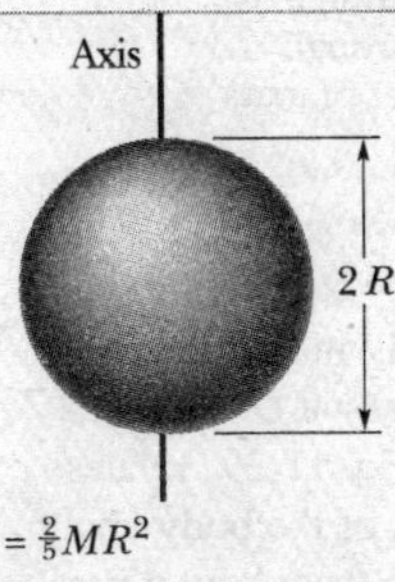 Solid sphere about any diameter $I = \frac{2}{5}MR^2$ (*f*)
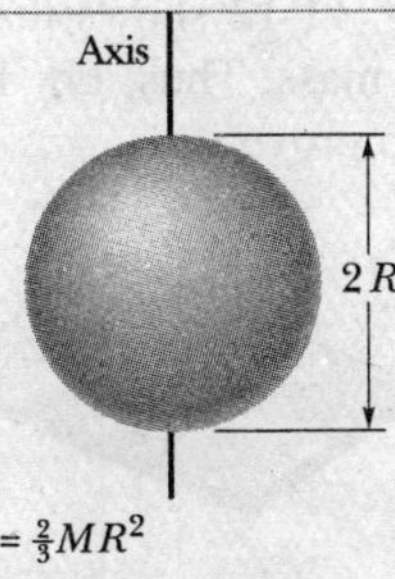 Thin spherical shell about any diameter $I = \frac{2}{3}MR^2$ (*g*)	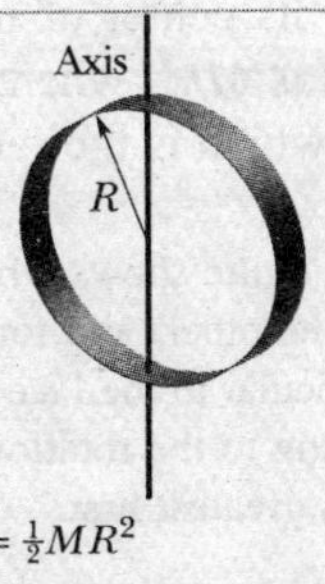 Hoop about any diameter $I = \frac{1}{2}MR^2$ (*h*)	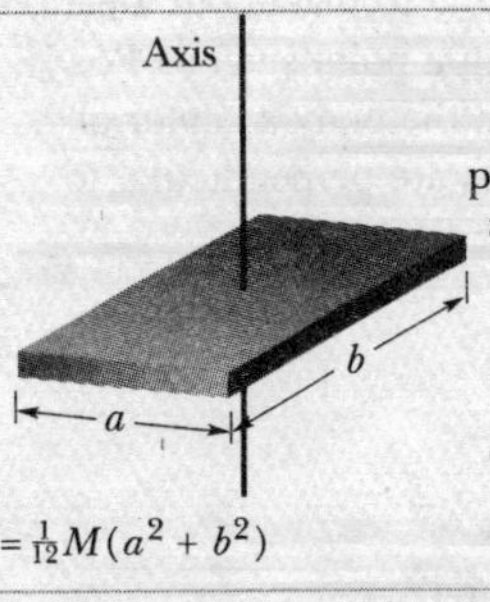 Uniform Slab about perpendicular axis through center (⊥ to large faces) $I = \frac{1}{12}M(a^2 + b^2)$ (*i*)

Perpendicular Axis Theorem

MI of a plane lamina about an axis perpendicular to the plane of lamina is equal to the sum of the MI of the lamina about any two mutually perpendicular axis in its own plane and intersecting each other at the point where the perpendicular axis passes through it

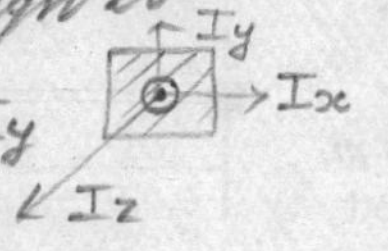

$I_z = I_x + I_y$

If a rigid body consists of a great many adjacent particles (it is *continuous*, like a Frisbee), using Eq. 11-26 would require a computer. Thus, instead, we replace the sum in Eq. 11-26 with an integral and define the rotational inertia of the body as

$$I = \int r^2\,dm \quad \text{(rotational inertia, continuous body).} \tag{11-28}$$

Table 11-2 gives the results of such integration for nine common body shapes and the indicated axes of rotation.

Parallel-Axis Theorem

Suppose we want to find the rotational inertia I of a body of mass M about a given axis. In principle, we can always find I with the integration of Eq. 11-28. However, there is a shortcut if we happen to already know the rotational inertia I_{com} of the body about a *parallel* axis that extends through the body's center of mass. Let h be the perpendicular distance between the given axis and the axis through the center of mass (remember these two axes must be parallel). Then the rotational inertia I about the given axis is

$$I = I_{com} + Mh^2 \quad \text{(parallel-axis theorem).} \tag{11-29}$$

This equation is known as the **parallel-axis theorem.** We shall now prove it and then put it to use in Checkpoint 5 and Sample Problem 11-5.

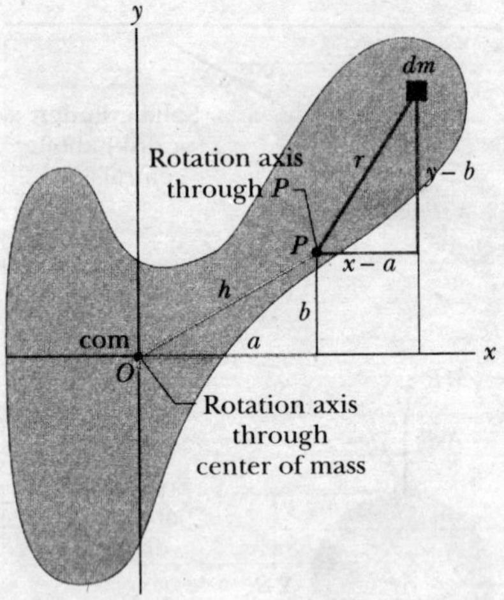

Fig. 11-12 A rigid body in cross section, with its center of mass at O. The parallel-axis theorem (Eq. 11-29) relates the rotational inertia of the body about an axis through O to that about a parallel axis through a point such as P, a distance h from the body's center of mass. Both axes are perpendicular to the plane of the figure.

Proof of the Parallel-Axis Theorem

Let O be the center of mass of the arbitrarily shaped body shown in cross section in Fig. 11-12. Place the origin of the coordinates at O. Consider an axis through O perpendicular to the plane of the figure, and another axis through point P parallel to the first axis. Let the x and y coordinates of P be a and b.

Let dm be a mass element with the general coordinates x and y. The rotational inertia of the body about the axis through P is then, from Eq. 11-28,

$$I = \int r^2\, dm = \int [(x-a)^2 + (y-b)^2]\, dm,$$

which we can rearrange as

$$I = \int (x^2 + y^2)\, dm - 2a \int x\, dm - 2b \int y\, dm + \int (a^2 + b^2)\, dm. \qquad (11\text{-}30)$$

From the definition of the center of mass (Eq. 9-9), the middle two integrals of Eq. 11-30 give the coordinates of the center of mass (multiplied by a constant) and thus must each be zero. Because $x^2 + y^2$ is equal to R^2, where R is the distance from O to dm, the first integral is simply I_{com}, the rotational inertia of the body about an axis through its center of mass. Inspection of Fig. 11-12 shows that the last term in Eq. 11-30 is Mh^2, where M is the body's total mass. Thus, Eq. 11-30 reduces to Eq. 11-29, which is the relation that we set out to prove.

CHECKPOINT 5: The figure shows a booklike object (one side is longer than the other) and four choices of rotation axes, all perpendicular to the face of the object. Rank the choices according to the rotational inertia of the object about the axis, greatest first.

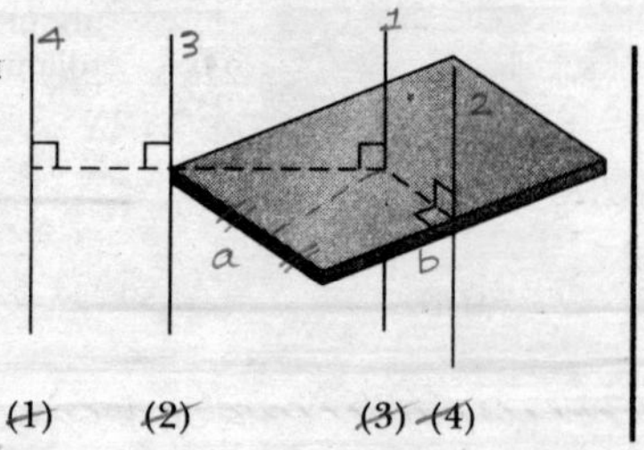

$I_1 = I_O = \frac{1}{12} M(a^2+b^2)$

$I_4 > I_3 > I_2 > I_O$

$I_2 = I_O + M\left(\frac{a}{2}\right)^2 = I_O + \frac{Ma^2}{4}$

$I_3 = I_O + M\left(\left[\frac{a}{2}\right]^2 + \left[\frac{b}{2}\right]^2\right) = I_O + \frac{M}{4}(a^2+b^2)$

$I_4 = I_O + Mh^2$ where $h^2 > \frac{a^2}{4} + \frac{b^2}{4}$

Sample Problem 11-5

Figure 11-13*a* shows a rigid body consisting of two particles of mass m connected by a rod of length L and negligible mass.

(a) What is the rotational inertia I_{com} of this body about an axis through its center of mass, perpendicular to the rod as shown?

SOLUTION: The **Key Idea** is that because we have only two particles with mass, we can find the body's rotational inertia I_{com} by using Eq. 11-26 rather than by integration. For the two particles, each at perpendicular distance $\frac{1}{2}L$ from the rotation axis, we have

$$I = \sum m_i r_i^2 = (m)(\tfrac{1}{2}L)^2 + (m)(\tfrac{1}{2}L)^2$$
$$= \tfrac{1}{2}mL^2. \qquad \text{(Answer)}$$

(b) What is the rotational inertia I of the body about an axis through the left end of the rod and parallel to the first axis (Fig. 11-13*b*)?

SOLUTION: This situation is simple enough that we can find I using either of two **Key Ideas**. The first is identical to the one we used in (a). The only difference here is that the perpendicular distance r_i is zero for the particle on the left, and L for the particle on the right. Now Eq. 11-26 gives us

$$I = m(0)^2 + mL^2 = mL^2. \qquad \text{(Answer)}$$

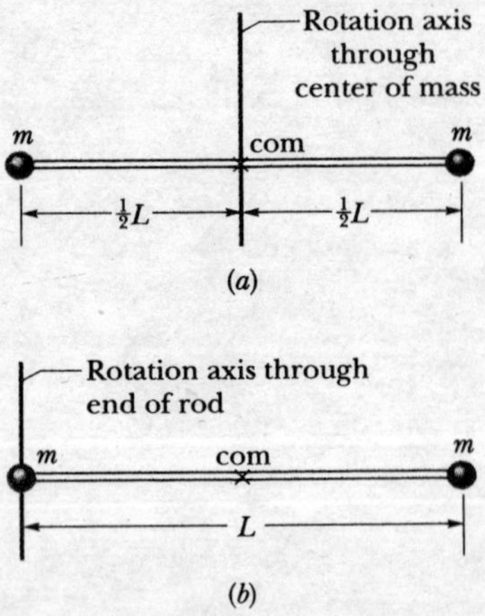

Fig. 11-13 Sample Problem 11-5. A rigid body of two particles of mass m, which are joined by a rod of negligible mass.

The second **Key Idea** provides a more powerful technique: Because we already know I_{com} about an axis through the center of mass and because the axis here is parallel to that "com axis," we can apply the parallel-axis theorem (Eq. 11-29). We find

$$I = I_{com} + Mh^2 = \tfrac{1}{2}mL^2 + (2m)(\tfrac{1}{2}L)^2$$
$$= mL^2. \qquad \text{(Answer)}$$

Sample Problem 11-6

Large machine components that undergo prolonged, high-speed rotation are first examined for the possibility of failure in a *spin test system*. In this system, a component is *spun up* (brought up to high speed) while inside a cylindrical arrangement of lead bricks and containment liner, all within a steel shell that is closed by a lid clamped into place. If the rotation causes the component to shatter, the soft lead bricks are supposed to catch the pieces so that the failure can then be analyzed.

In early 1985, Test Devices, Inc. (www.testdevices.com) was spin testing a sample of a solid steel rotor (a disk) of mass M = 272 kg and radius R = 38.0 cm. When the sample reached an angular speed ω of 14 000 rev/min, the test engineers heard a dull thump from the test system, which was located one floor down and one room over from them. Investigating, they found that lead bricks had been thrown out in the hallway leading to the test room, a door to the room had been hurled into the adjacent parking lot, one lead brick had shot from the test site through the wall of a neighbor's kitchen, the structural beams of the test building had been damaged, the concrete floor beneath the spin chamber had been shoved downward by about 0.5 cm, and the 900 kg lid had been blown upward through the ceiling and had then crashed back onto the test equipment (Fig. 11-14). The exploding pieces had not penetrated the room of the test engineers only by luck.

How much energy was released in the explosion of the rotor?

SOLUTION: The **Key Idea** here is that this released energy was equal to the rotational kinetic energy K of the rotor just as it reached the angular speed of 14 000 rev/min. We can find K with Eq. 11-27 ($K = \frac{1}{2}I\omega^2$), but first we need an expression for the rotational inertia I. Because the rotor was a disk that rotated like a merry-go-round, I is given by the expression in Table 11-2*c* ($I = \frac{1}{2}MR^2$).

Fig. 11-14 Sample Problem 11-6. Some of the destruction caused by the explosion of a rapidly rotating steel disk.

Thus, we have

$$I = \tfrac{1}{2}MR^2 = \tfrac{1}{2}(272 \text{ kg})(0.38 \text{ m})^2 = 19.64 \text{ kg}\cdot\text{m}^2.$$

The angular speed of the rotor was

$$\omega = (14\,000 \text{ rev/min})(2\pi \text{ rad/rev})\left(\frac{1 \text{ min}}{60 \text{ s}}\right)$$
$$= 1.466 \times 10^3 \text{ rad/s}.$$

Now we can use Eq. 11-27 to write

$$K = \tfrac{1}{2}I\omega^2 = \tfrac{1}{2}(19.64 \text{ kg}\cdot\text{m}^2)(1.466 \times 10^3 \text{ rad/s})^2$$
$$= 2.1 \times 10^7 \text{ J}. \quad \text{(Answer)}$$

Being near this explosion was like being near an exploding bomb.

11-8 Torque

A doorknob is located as far as possible from the door's hinge line for a good reason. If you want to open a heavy door, you must certainly apply a force; that alone, however, is not enough. Where you apply that force and in what direction you push are also important. If you apply your force nearer to the hinge line than the knob, or at any angle other than 90° to the plane of the door, you must use a greater force to move the door than if you apply the force at the knob and perpendicular to the door's plane.

Figure 11-15*a* shows a cross section of a body that is free to rotate about an axis passing through O and perpendicular to the cross section. A force $\vec{F}$ is applied at point P, whose position relative to O is defined by a position vector $\vec{r}$. The directions of vectors $\vec{F}$ and $\vec{r}$ make an angle ϕ with each other. (For simplicity, we consider only forces that have no component parallel to the rotation axis; thus, $\vec{F}$ is in the plane of the page.)

To determine how $\vec{F}$ results in a rotation of the body around the rotation axis, we resolve $\vec{F}$ into two components (Fig. 11-15*b*). One component, called the *radial component* F_r, points along $\vec{r}$. This component does not cause rotation, because it acts along a line that extends through O. (If you pull on a door parallel to the plane of the door, you do not rotate the door.) The other component of $\vec{F}$, called the

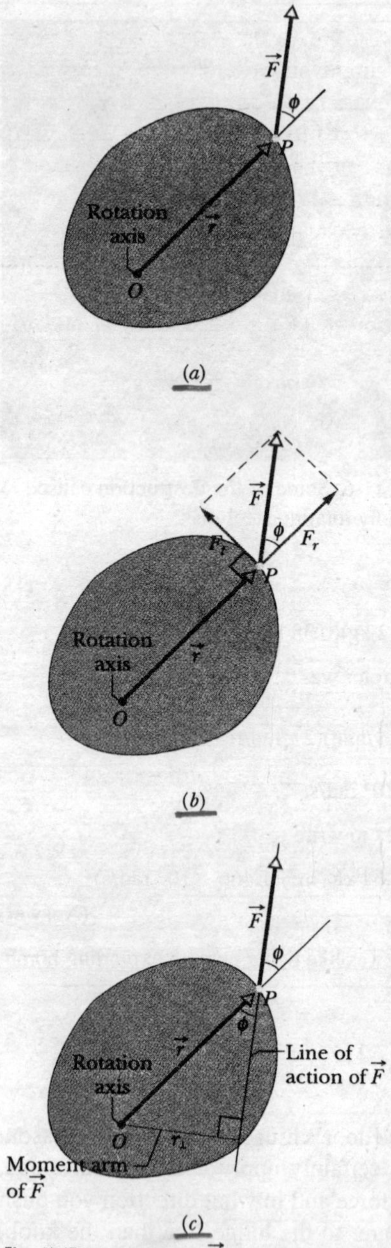

Fig. 11-15 (*a*) A force $\vec{F}$ acts at point *P* on a rigid body that is free to rotate about an axis through *O*; the axis is perpendicular to the plane of the cross section shown here. (*b*) The torque due to this force is $(r)(F \sin \phi)$. We can also write it as rF_t, where F_t is the tangential component of $\vec{F}$. (*c*) The torque can also be written as $r_\perp F$, where $r_\perp$ is the moment arm of $\vec{F}$.

tangential component F_t, is perpendicular to $\vec{r}$ and has magnitude $F_t = F \sin \phi$. This component *does* cause rotation. (If you pull on a door perpendicular to its plane, you can rotate the door.)

The ability of $\vec{F}$ to rotate the body depends not only on the magnitude of its tangential component F_t, but also on just how far from *O* the force is applied. To include both these factors, we define a quantity called **torque** τ as the product of the two factors and write it as

$$\tau = (r)(F \sin \phi). \tag{11-31}$$

Two equivalent ways of computing the torque are

$$\tau = (r)(F \sin \phi) = rF_t \tag{11-32}$$

and

$$\tau = (r \sin \phi)(F) = r_\perp F, \tag{11-33}$$

where $r_\perp$ is the perpendicular distance between the rotation axis at *O* and an extended line running through the vector $\vec{F}$ (Fig. 11-15*c*). This extended line is called the **line of action** of $\vec{F}$, and $r_\perp$ is called the **moment arm** of $\vec{F}$. Figure 11-15*b* shows that we can describe *r*, the magnitude of $\vec{r}$, as being the moment arm of the force component F_t.

Torque, which comes from the Latin word meaning "to twist," may be loosely identified as the turning or twisting action of the force $\vec{F}$. When you apply a force to an object—such as a screwdriver or torque wrench—with the purpose of turning that object, you are applying a torque. The SI unit of torque is the newton-meter (N · m). *Caution*: The newton-meter is also the unit of work. Torque and work, however, are quite different quantities and must not be confused. Work is often expressed in joules (1 J = 1 N · m), but torque never is.

In the next chapter we shall discuss torque in a general way as being a vector quantity. Here, however, because we consider only rotation around a single axis, we do not need vector notation. Instead, a torque has either a positive or negative value depending on the direction of rotation it would give a body initially at rest: If the body would rotate counterclockwise, the torque is positive. If the object would rotate clockwise, the torque is negative. (The phrase "clocks are negative" from Section 11-2 still works.)

Torques obey the superposition principle that we discussed in Chapter 5 for forces: When several torques act on a body, the **net torque** (or **resultant torque**) is the sum of the individual torques. The symbol for net torque is τ_{net}.

In stick's plane

CHECKPOINT 6: The figure shows an overhead view of a meter stick that can pivot about the dot at the position marked 20 (for 20 cm). All five horizontal forces on the stick have the same magnitude. Rank those forces according to the magnitude of the torque that they produce, greatest first.

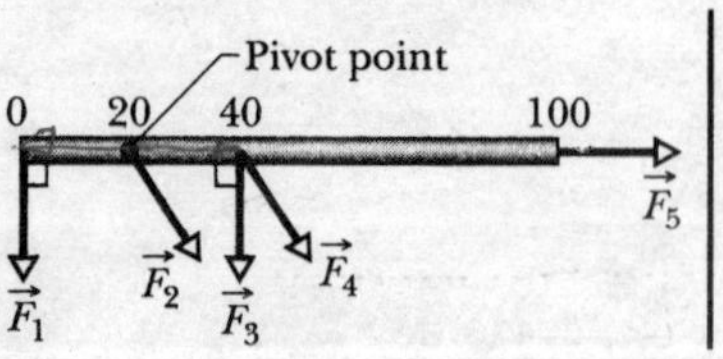

11-9 Newton's Second Law for Rotation

A torque can cause rotation of a rigid body, as when you use a torque to rotate a door. Here we want to relate the net torque τ_{net} on a rigid body to the angular acceleration α it causes about a rotation axis. We do so by analogy with Newton's second law ($F_{net} = ma$) for the acceleration *a* of a body of mass *m* due to a net force F_{net} along a coordinate axis. We replace F_{net} with τ_{net}, *m* with *I*, and *a* with α, writing

$$\tau_{net} = I\alpha \quad \text{(Newton's second law for rotation)}, \tag{11-34}$$

where α must be in radian measure.

we consider F in CGS units too

$\tau = r F \sin\phi$

$\tau_1 = 20 F \sin 90° = 20F$ ∵ r is dis from rotation axis

$\tau_2 = 0$

$\tau_3 = 20F \sin 90° = 20F$

~~$\tau_4 = 20F \sin(90+\theta) = 20F\cos\theta$~~ — angle is measured b/w direction of $\bar{F}$ & $\bar{r}$

$\tau_5 = 80F \times 0 = 0$

3 & 1, 4, 5 & 2

$\theta < 90°$, $\sin\theta < \sin 90$

Proof of Equation 11-34

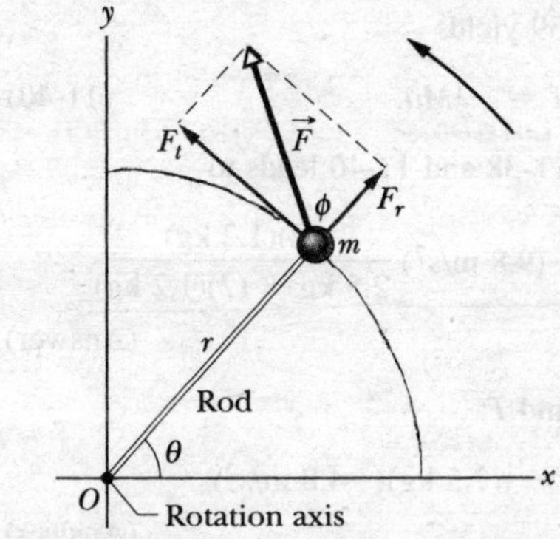

Fig. 11-16 A simple rigid body, free to rotate about an axis through O, consists of a particle of mass m fastened to the end of a rod of length r and negligible mass. An applied force $\vec{F}$ causes the body to rotate.

We prove Eq. 11-34 by first considering the simple situation shown in Fig. 11-16. The rigid body there consists of a particle of mass m on one end of a massless rod of length r. The rod can move only by rotating about its other end, around a rotation axis (an axle) that is perpendicular to the plane of the page. Thus, the particle can move only in a circular path that has the rotation axis at its center.

A force $\vec{F}$ acts on the particle. However, because the particle can move only along the circular path, only the tangential component F_t of the force (the component that is tangent to the circular path) can accelerate the particle along the path. We can relate F_t to the particle's tangential acceleration a_t along the path with Newton's second law, writing

$$F_t = ma_t.$$

The torque acting on the particle is, from Eq. 11-32,

$$\tau = F_t r = ma_t r.$$

From Eq. 11-22 ($a_t = \alpha r$) we can write this as

$$\tau = m(\alpha r)r = (mr^2)\alpha. \tag{11-35}$$

The quantity in parentheses on the right side of Eq. 11-35 is the rotational inertia of the particle about the rotation axis (see Eq. 11-26). Thus, Eq. 11-35 reduces to

$$\tau = I\alpha \quad \text{(radian measure).} \tag{11-36}$$

For the situation in which more than one force is applied to the particle, we can generalize Eq. 11-36 as

$$\tau_{net} = I\alpha \quad \text{(radian measure),} \tag{11-37}$$

which we set out to prove. We can extend this equation to any rigid body rotating about a fixed axis, because any such body can always be analyzed as an assembly of single particles.

CHECKPOINT 7: The figure shows an overhead view of a meter stick that can pivot about the point indicated, which is to the left of the stick's midpoint. Two horizontal forces, $\vec{F}_1$ and $\vec{F}_2$, are applied to the stick. Only $\vec{F}_1$ is shown. Force $\vec{F}_2$ is perpendicular to the stick and is applied at the right end. If the stick is not to turn, (a) what should be the direction of $\vec{F}_2$, and (b) should F_2 be greater than, less than, or equal to F_1?

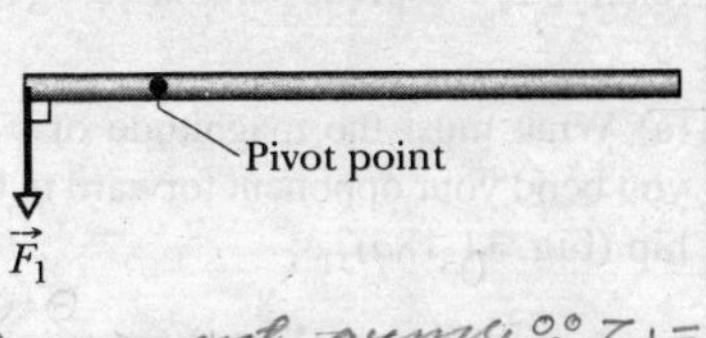

(a) downward in the fig
(b) $F_2 < F_1$ ∵ $\tau = r_\perp F$ considering moment arms ∵ $\tau_1 = \tau_2$
$F_1 x = F_2(L - x)$
$L - x > x \Rightarrow F_2 < F_1$

Sample Problem 11-7

Figure 11-17*a* shows a uniform disk, with mass $M = 2.5$ kg and radius $R = 20$ cm, mounted on a fixed horizontal axle. A block with mass $m = 1.2$ kg hangs from a massless cord that is wrapped around the rim of the disk. Find the acceleration of the falling block, the angular acceleration of the disk, and the tension in the cord. The cord does not slip, and there is no friction at the axle.

SOLUTION: One **Key Idea** here is that, taking the block as a system, we can relate its acceleration a to the forces acting on it with Newton's second law ($\vec{F}_{net} = m\vec{a}$). Those forces are shown in the block's free-body diagram in Fig. 11-17*b*: The force from the cord is $\vec{T}$ and the gravitational force is $\vec{F}_g$, of magnitude mg. We can

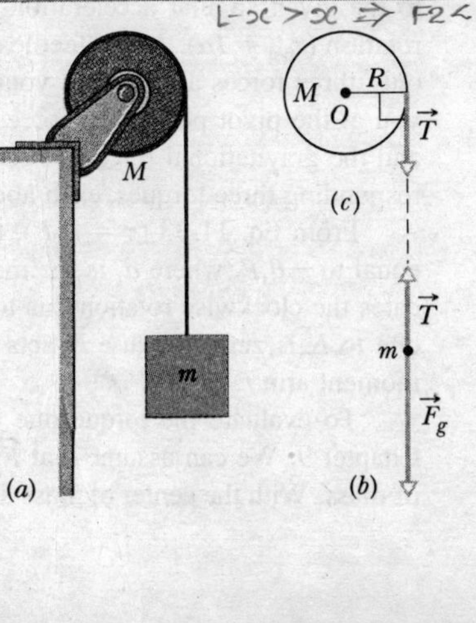

Fig. 11-17 Sample Problems 11-7 and 11-9. (*a*) The falling block causes the disk to rotate. (*b*) A free-body diagram for the block. (*c*) An incomplete free-body diagram for the disk.

now write Newton's second law for components along a vertical y axis ($F_{\text{net},y} = ma_y$) as

$$T - mg = ma. \tag{11-38}$$

However, we cannot solve this equation for a because it also contains the unknown T.

Previously, when we got stuck on the y axis, we would switch to the x axis. Here, we switch to the rotation of the disk and use this **Key Idea**: Taking the disk as a system, we can relate its angular acceleration α to the torque acting on it with Newton's second law for rotation ($\tau_{\text{net}} = I\alpha$). To calculate the torques and the rotational inertia I, we take the rotation axis to be perpendicular to the disk and through its center, at point O in Fig. 11-17c.

The torques are then given by Eq. 11-32 ($\tau = rF_t$). The gravitational force on the disk and the force on the disk from the axle both act at the center of the disk and thus at distance $r = 0$, so their torques are zero. The force $\vec{T}$ on the disk due to the cord acts at distance $r = R$ and is tangent to the rim of the disk. Therefore, its torque is $-RT$, negative because the torque rotates the disk clockwise from rest. From Table 11-2c, the rotational inertia I of the disk is $\frac{1}{2}MR^2$. Thus we can write $\tau_{\text{net}} = I\alpha$ as

$$-RT = \tfrac{1}{2}MR^2\alpha. \tag{11-39}$$

This equation seems useless because it has two unknowns, α and T, neither of which is the desired a. However, mustering physics courage, we can make it useful with a third **Key Idea**: Because the cord does not slip, the linear acceleration a of the block and the (tangential) linear acceleration a_t of the rim of the disk are equal. Then, by Eq. 11-22 ($a_t = \alpha r$) we see that here $\alpha = a/R$. Substituting this in Eq. 11-39 yields

$$T = -\tfrac{1}{2}Ma. \tag{11-40}$$

Now combining Eqs. 11-38 and 11-40 leads to

$$a = -g\frac{2m}{M + 2m} = -(9.8\text{ m/s}^2)\frac{(2)(1.2\text{ kg})}{2.5\text{ kg} + (2)(1.2\text{ kg})}$$
$$= -4.8\text{ m/s}^2. \qquad \text{(Answer)}$$

We then use Eq. 11-40 to find T:

$$T = -\tfrac{1}{2}Ma = -\tfrac{1}{2}(2.5\text{ kg})(-4.8\text{ m/s}^2)$$
$$= 6.0\text{ N}. \qquad \text{(Answer)}$$

As we should expect, the acceleration of the falling block is less than g, and the tension in the cord (= 6.0 N) is less than the gravitational force on the hanging block (= mg = 11.8 N). We see also that the acceleration of the block and the tension depend on the mass of the disk but not on its radius. As a check, we note that the formulas derived above predict $a = -g$ and $T = 0$ for the case of a massless disk ($M = 0$). This is what we would expect; the block simply falls as a free body, trailing the string behind it.

From Eq. 11-22, the angular acceleration of the disk is

$$\alpha = \frac{a}{R} = \frac{-4.8\text{ m/s}^2}{0.20\text{ m}} = -24\text{ rad/s}^2. \qquad \text{(Answer)}$$

Sample Problem 11-8

To throw an 80 kg opponent with a basic judo hip throw, you intend to pull his uniform with a force $\vec{F}$ and a moment arm $d_1 = 0.30$ m from a pivot point (rotation axis) on your right hip (Fig. 11-18). You wish to rotate him about the pivot point with an angular acceleration α of -6.0 rad/s²—that is, with an angular acceleration that is *clockwise* in the figure. Assume that his rotational inertia I relative to the pivot point is 15 kg · m².

(a) What must the magnitude of $\vec{F}$ be if, before you throw him, you bend your opponent forward to bring his center of mass to your hip (Fig. 11-18a)?

SOLUTION: One **Key Idea** here is that we can relate your pull $\vec{F}$ on him to the given angular acceleration α via Newton's second law for rotation ($\tau_{\text{net}} = I\alpha$). As his feet leave the floor, we can assume that only three forces act on him: your pull $\vec{F}$, a force $\vec{N}$ on him from you at the pivot point (this force is not indicated in Fig. 11-18), and the gravitational force $\vec{F}_g$. To use $\tau_{\text{net}} = I\alpha$, we need the corresponding three torques, each about the pivot point.

From Eq. 11-33 ($\tau = r_\perp F$), the torque due to your pull $\vec{F}$ is equal to $-d_1F$, where d_1 is the moment arm $r_\perp$ and the sign indicates the clockwise rotation this torque tends to cause. The torque due to $\vec{N}$ is zero, because $\vec{N}$ acts at the pivot point and thus has moment arm $r_\perp = 0$.

To evaluate the torque due to $\vec{F}_g$, we need a **Key Idea** from Chapter 9: We can assume that $\vec{F}_g$ acts at your opponent's center of mass. With the center of mass at the pivot point, $\vec{F}_g$ has moment arm $r_\perp = 0$ and thus the torque due to $\vec{F}_g$ is zero. Thus, the only torque on your opponent is due to your pull $\vec{F}$, and we can write $\tau_{\text{net}} = I\alpha$ as

$$-d_1F = I\alpha.$$

We then find

$$F = \frac{-I\alpha}{d_1} = \frac{-(15\text{ kg}\cdot\text{m}^2)(-6.0\text{ rad/s}^2)}{0.30\text{ m}}$$
$$= 300\text{ N}. \qquad \text{(Answer)}$$

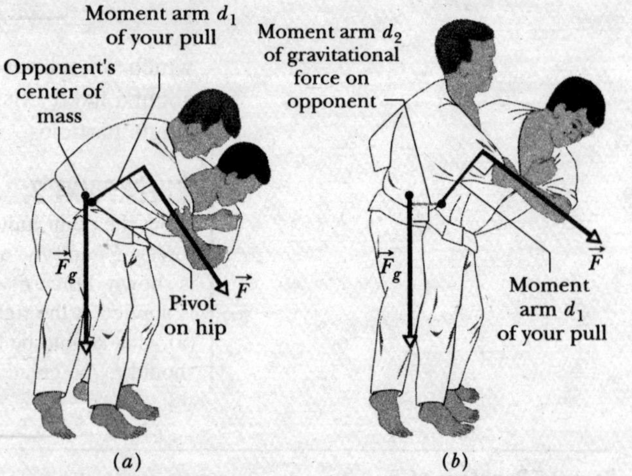

Fig. 11-18 Sample Problem 11-8. A judo hip throw (a) correctly executed and (b) incorrectly executed.

(b) What must the magnitude of $\vec{F}$ be if your opponent remains upright before you throw him, so that $\vec{F}_g$ has a moment arm $d_2 = 0.12$ m from the pivot point (Fig. 11-18*b*)?

SOLUTION: The **Key Ideas** we need here are similar to those in (a) with one exception: Because the moment arm for $\vec{F}_g$ is no longer zero, the torque due to $\vec{F}_g$ is now equal to d_2mg, and is positive because the torque attempts counterclockwise rotation. Now we write $\tau_{net} = I\alpha$ as

$$-d_1F + d_2mg = I\alpha,$$

which gives

$$F = -\frac{I\alpha}{d_1} + \frac{d_2mg}{d_1}.$$

From (a), we know that the first term on the right is equal to 300 N. Substituting this and the given data, we have

$$F = 300\ \text{N} + \frac{(0.12\ \text{m})(80\ \text{kg})(9.8\ \text{m/s}^2)}{0.30\ \text{m}}$$

$$= 613.6\ \text{N} \approx 610\ \text{N}. \qquad \text{(Answer)}$$

The results indicate that you will have to pull much harder if you do not initially bend your opponent to bring his center of mass to your hip. A good judo fighter knows this lesson from physics. (An analysis of the physics of judo and aikido is given in "The Amateur Scientist" by J. Walker, *Scientific American,* July 1980, Vol. 243, pp. 150–161.)

11-10 Work and Rotational Kinetic Energy

As we discussed in Chapter 7, when a force F causes a rigid body of mass m to accelerate along a coordinate axis, it does work W on the body. Thus, the body's kinetic energy ($K = \frac{1}{2}mv^2$) can change. Suppose it is the only energy of the body that changes. Then we relate the change ΔK in kinetic energy to the work W with the work–kinetic energy theorem (Eq. 7-10), writing

$$\Delta K = K_f - K_i = \tfrac{1}{2}mv_f^2 - \tfrac{1}{2}mv_i^2 = W \qquad \text{(work–kinetic energy theorem).} \qquad (11\text{-}41)$$

For motion confined to an x axis, we can calculate the work with Eq. 7-32,

$$W = \int_{x_i}^{x_f} F\,dx \qquad \text{(work, one-dimensional motion).} \qquad (11\text{-}42)$$

This reduces to $W = Fd$ when F is constant and the body's displacement is d. The rate at which the work is done is the power, which we can find with Eqs. 7-43 and 7-48,

$$P = \frac{dW}{dt} = Fv \qquad \text{(power, one-dimensional motion).} \qquad (11\text{-}43)$$

Now let us consider a rotational situation that is similar. When a torque accelerates a rigid body in rotation about a fixed axis, it does work W on the body. Therefore, the body's rotational kinetic energy ($K = \frac{1}{2}I\omega^2$) can change. Suppose that it is the only energy of the body that changes. Then we can still relate the change ΔK in kinetic energy to the work W with the work–kinetic energy theorem, except now the kinetic energy is a rotational kinetic energy:

$$\Delta K = K_f - K_i = \tfrac{1}{2}I\omega_f^2 - \tfrac{1}{2}I\omega_i^2 = W \qquad \text{(work–kinetic energy theorem).} \qquad (11\text{-}44)$$

Here, I is the rotational inertia of the body about the fixed axis and ω_i and ω_f are the angular speeds of the body before and after the work is done, respectively.

Also, we can calculate the work with a rotational equivalent of Eq. 11-42,

$$W = \int_{\theta_i}^{\theta_f} \tau\,d\theta \qquad \text{(work, rotation about fixed axis),} \qquad (11\text{-}45)$$

where τ is the torque doing the work W, and θ_i and θ_f are the body's angular positions before and after the work is done, respectively. When τ is constant, Eq. 11-45 reduces to

$$W = \tau(\theta_f - \theta_i) \qquad \text{(work, constant torque).} \qquad (11\text{-}46)$$

The rate at which the work is done is the power, which we can find with the rotational equivalent of Eq. 11-43,

$$P = \frac{dW}{dt} = \tau\omega \quad \text{(power, rotation about fixed axis).} \qquad (11\text{-}47)$$

Table 11-3 summarizes the equations that apply to the rotation of a rigid body about a fixed axis and the corresponding equations for translational motion.

Proof of Eqs. 11-44 through 11-47

Let us again consider the situation of Fig. 11-16, in which force $\vec{F}$ rotates a rigid body consisting of a single particle of mass m fastened to the end of a massless rod. During the rotation, force $\vec{F}$ does work on the body. Let us assume that the only energy of the body that is changed by $\vec{F}$ is the kinetic energy. Then we can apply the work–kinetic energy theorem of Eq. 11-41:

$$\Delta K = K_f - K_i = W. \qquad (11\text{-}48)$$

Using $K = \frac{1}{2}mv^2$ and Eq. 11-18 ($v = \omega r$), we can rewrite Eq. 11-48 as

$$\Delta K = \tfrac{1}{2}mr^2\omega_f^2 - \tfrac{1}{2}mr^2\omega_i^2 = W. \qquad (11\text{-}49)$$

From Eq. 11-26, the rotational inertia for this one-particle body is $I = mr^2$. Substituting this into Eq. 11-49 yields

$$\Delta K = \tfrac{1}{2}I\omega_f^2 - \tfrac{1}{2}I\omega_i^2 = W,$$

which is Eq. 11-44. We derived it for a rigid body with one particle, but it holds for any rigid body rotated about a fixed axis.

We next relate the work W done on the body in Fig. 11-16 to the torque τ on the body due to force $\vec{F}$. When the particle moves a distance ds along its circular path, only the tangential component F_t of the force accelerates the particle along the path. Therefore, only F_t does work on the particle. We write that work dW as $F_t\,ds$. However, we can replace ds with $r\,d\theta$, where $d\theta$ is the angle through which the particle moves. Thus we have

$$dW = F_t r\,d\theta. \qquad (11\text{-}50)$$

From Eq. 11-32, we see that the product $F_t r$ is equal to the torque τ, so we can rewrite Eq. 11-50 as

$$dW = \tau\,d\theta. \qquad (11\text{-}51)$$

TABLE 11-3 Some Corresponding Relations for Translational and Rotational Motion

Pure Translation (Fixed Direction)		Pure Rotation (Fixed Axis)	
Position	x	Angular position	θ
Velocity	$v = dx/dt$	Angular velocity	$\omega = d\theta/dt$
Acceleration	$a = dv/dt$	Angular acceleration	$\alpha = d\omega/dt$
Mass	m	Rotational inertia	I
Newton's second law	$F_{net} = ma$	Newton's second law	$\tau_{net} = I\alpha$
Work	$W = \int F\,dx$	Work	$W = \int \tau\,d\theta$
Kinetic energy	$K = \frac{1}{2}mv^2$	Kinetic energy	$K = \frac{1}{2}I\omega^2$
Power (constant force)	$P = Fv$	Power (constant torque)	$P = \tau\omega$
Work–kinetic energy theorem	$W = \Delta K$	Work–kinetic energy theorem	$W = \Delta K$

The work done during a finite angular displacement from θ_i to θ_f is then

$$W = \int_{\theta_i}^{\theta_f} \tau\, d\theta,$$

which is Eq. 11-45. It holds for any rigid body rotating about a fixed axis. Equation 11-46 comes directly from Eq. 11-45.

We can find the power P for rotational motion from Eq. 11-51:

$$P = \frac{dW}{dt} = \tau\frac{d\theta}{dt} = \tau\omega,$$

which is Eq. 11-47.

Sample Problem 11-9

Let the disk in Sample Problem 11-7 and Fig. 11-17 start from rest at time $t = 0$. What is its rotational kinetic energy K at $t = 2.5$ s?

SOLUTION: We can find K with Eq. 11-27 ($K = \frac{1}{2}I\omega^2$). We already know that I is equal to $\frac{1}{2}MR^2$, but we do not yet know ω at $t =$ 2.5 s. A **Key Idea**, though, is that the angular acceleration α has the constant value of -24 rad/s^2, so we can apply the equations for constant angular acceleration in Table 11-1. Because we want ω and know α and ω_0 ($= 0$), we use Eq. 11-12:

$$\omega = \omega_0 + \alpha t = 0 + \alpha t = \alpha t.$$

Substituting $\omega = \alpha t$ and $I = \frac{1}{2}MR^2$ into Eq. 11-27, we find

$$\begin{aligned} K &= \tfrac{1}{2}I\omega^2 = \tfrac{1}{2}(\tfrac{1}{2}MR^2)(\alpha t)^2 = \tfrac{1}{4}M(R\alpha t)^2 \\ &= \tfrac{1}{4}(2.5\ \text{kg})[(0.20\ \text{m})(-24\ \text{rad/s}^2)(2.5\ \text{s})]^2 \\ &= 90\ \text{J}. \end{aligned}$$ (Answer)

We can also get this answer with a different **Key Idea**: We can find the disk's kinetic energy from the work done on the disk. First, we relate the *change* in the kinetic energy of the disk to the net work W done on the disk, using the work–kinetic energy theorem of Eq. 11-44 ($K_f - K_i = W$). With K substituted for K_f and 0 for K_i, we get

$$K = K_i + W = 0 + W = W. \tag{11-52}$$

Next we want to find the work W. We can relate W to the torques acting on the disk with Eq. 11-45 or 11-46. The only torque causing angular acceleration and doing work is the torque due to force $\vec{T}$ on the disk from the cord. From Sample Problem 11-7, this torque is equal to $-TR$. Another **Key Idea** is that because α is constant, this torque also must be constant. Thus, we can use Eq. 11-46 to write

$$W = \tau(\theta_f - \theta_i) = -TR(\theta_f - \theta_i). \tag{11-53}$$

We need one more **Key Idea**: Because α is constant, we can use Eq. 11-13 to find $\theta_f - \theta_i$. With $\omega_i = 0$, we have

$$\theta_f - \theta_i = \omega_i t + \tfrac{1}{2}\alpha t^2 = 0 + \tfrac{1}{2}\alpha t^2 = \tfrac{1}{2}\alpha t^2.$$

Now we substitute this into Eq. 11-53 and then substitute the result into Eq. 11-52. With $T = 6.0$ N and $\alpha = -24$ rad/s^2 (from Sample Problem 11-7), we have

$$\begin{aligned} K = W &= -TR(\theta_f - \theta_i) = -TR(\tfrac{1}{2}\alpha t^2) = -\tfrac{1}{2}TR\alpha t^2 \\ &= -\tfrac{1}{2}(6.0\ \text{N})(0.20\ \text{m})(-24\ \text{rad/s}^2)(2.5\ \text{s})^2 \\ &= 90\ \text{J}. \end{aligned}$$ (Answer)

Sample Problem 11-10

A rigid sculpture consists of a thin hoop (of mass m and radius $R = 0.15$ m) and a thin radial rod (of mass m and length $L = 2.0R$), arranged as shown in Fig. 11-19. The sculpture can pivot around a horizontal axis in the plane of the hoop, passing through its center.

(a) In terms of m and R, what is the sculpture's rotational inertia I about the rotation axis?

SOLUTION: A **Key Idea** here is that we can separately find the rotational inertias of the hoop and the rod and then add the results to get the sculpture's total rotational inertia I. From Table 11-2*h*, the hoop has rotational inertia $I_{hoop} = \frac{1}{2}mR^2$ about its diameter. From Table 11-2*e*, the rod has rotational inertia $I_{com} = mL^2/12$ about an axis through its center of mass and parallel to the sculpture's rotation axis. To find its rotational inertia I_{rod} about that rotation axis, we

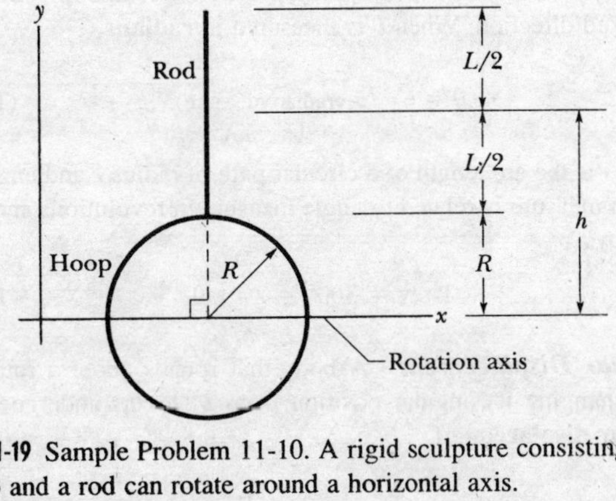

Fig. 11-19 Sample Problem 11-10. A rigid sculpture consisting of a hoop and a rod can rotate around a horizontal axis.

use Eq. 11-29, the parallel-axis theorem:

$$I_{rod} = I_{com} + mh_{com}^2 = \frac{mL^2}{12} + m\left(R + \frac{L}{2}\right)^2$$
$$= 4.33mR^2,$$

where we have used the fact that $L = 2.0R$ and where the perpendicular distance between the rod's center of mass and the rotation axis is $h = R + L/2$. Thus, the rotational inertia I of the sculpture about the rotation axis is

$$I = I_{hoop} + I_{rod} = \tfrac{1}{2}mR^2 + 4.33mR^2$$
$$= 4.83mR^2 \approx 4.8mR^2. \quad \text{(Answer)}$$

(b) Starting from rest, the sculpture rotates around the rotation axis from the initial upright orientation of Fig. 11-19. What is its angular speed ω about the axis when it is inverted?

SOLUTION: Three **Key Ideas** are required here:

1. We can relate the sculpture's speed ω to its rotational kinetic energy K with Eq. 11-27 ($K = \frac{1}{2}I\omega^2$).
2. We can relate K to the gravitational potential energy U of the sculpture via the conservation of the sculpture's mechanical energy E during the rotation. Thus, during the rotation, E does not change ($\Delta E = 0$) as energy is transferred from U to K.
3. For the gravitational potential energy we can treat the rigid sculpture as a particle located at the center of mass, with the total mass $2m$ concentrated there.

We can write the conservation of mechanical energy ($\Delta E = 0$) as

$$\Delta K + \Delta U = 0. \tag{11-54}$$

As the sculpture rotates from its initial position at rest to its inverted position, when the angular speed is ω, the change ΔK in its kinetic energy is

$$\Delta K = K_f - K_i = \tfrac{1}{2}I\omega^2 - 0 = \tfrac{1}{2}I\omega^2. \tag{11-55}$$

From Eq. 8-7 ($\Delta U = mg\,\Delta y$), the corresponding change ΔU in the gravitational potential energy is

$$\Delta U = (2m)g\,\Delta y_{com}, \tag{11-56}$$

where $2m$ is the sculpture's total mass, and Δy_{com} is the vertical displacement of its center of mass during the rotation.

To find Δy_{com}, we first find the initial location y_{com} of the center of mass in Fig. 11-19. The hoop (with mass m) is centered at $y = 0$. The rod (with mass m) is centered at $y = R + L/2$. Thus, from Eq. 9-5, the sculpture's center of mass is at

$$y_{com} = \frac{m(0) + m(R + L/2)}{2m} = \frac{0 + m(R + 2R/2)}{2m} = R.$$

When the sculpture is inverted, the center of mass is this same distance R from the rotation axis but *below* it. Therefore, the vertical displacement of the center of mass from the initial position to the inverted position is $\Delta y_{com} = -2R$.

Now let's pull these results together. Substituting Eqs. 11-55 and 11-56 into 11-54 gives us

$$\tfrac{1}{2}I\omega^2 + (2m)g\,\Delta y_{com} = 0.$$

Substituting $I = 4.83mR^2$ from (a) and $\Delta y_{com} = -2R$ from above and solving for ω, we find

$$\omega = \sqrt{\frac{8g}{4.83R}} = \sqrt{\frac{(8)(9.8\text{ m/s}^2)}{(4.83)(0.15\text{ m})}}$$
$$= 10\text{ rad/s}. \quad \text{(Answer)}$$

REVIEW & SUMMARY

Angular Position To describe the rotation of a rigid body about a fixed axis, called the **rotation axis,** we assume a **reference line** is fixed in the body, perpendicular to that axis and rotating with the body. We measure the **angular position** θ of this line relative to a fixed direction. When θ is measured in **radians,**

$$\theta = \frac{s}{r} \quad \text{(radian measure)}, \tag{11-1}$$

where s is the arc length of a circular path of radius r and angle θ. Radian measure is related to angle measure in revolutions and degrees by

$$1\text{ rev} = 360° = 2\pi\text{ rad}. \tag{11-2}$$

Angular Displacement A body that rotates about a rotation axis, changing its angular position from θ_1 to θ_2, undergoes an **angular displacement**

$$\Delta\theta = \theta_2 - \theta_1, \tag{11-4}$$

where $\Delta\theta$ is positive for counterclockwise rotation and negative for clockwise rotation.

Angular Velocity and Speed If a body rotates through an angular displacement $\Delta\theta$ in a time interval Δt, its **average angular velocity** ω_{avg} is

$$\omega_{avg} = \frac{\Delta\theta}{\Delta t}. \tag{11-5}$$

The **(instantaneous) angular velocity** ω of the body is

$$\omega = \frac{d\theta}{dt}. \tag{11-6}$$

Both ω_{avg} and ω are vectors, with directions given by the **right-hand rule** of Fig. 11-6. They are positive for counterclockwise rotation and negative for clockwise rotation. The magnitude of the body's angular velocity is the **angular speed.**

Angular Acceleration If the angular velocity of a body changes from ω_1 to ω_2 in a time interval $\Delta t = t_2 - t_1$, the **average angular acceleration** α_{avg} of the body is

$$\alpha_{avg} = \frac{\omega_2 - \omega_1}{t_2 - t_1} = \frac{\Delta\omega}{\Delta t}. \qquad (11\text{-}7)$$

The **(instantaneous) angular acceleration** α of a body is

$$\alpha = \frac{d\omega}{dt}. \qquad (11\text{-}8)$$

Both α_{avg} and α are vectors.

The Kinematic Equations for Constant Angular Acceleration *Constant angular acceleration* (α = constant) is an important special case of rotational motion. The appropriate kinematic equations, given in Table 11-1, are

$$\omega = \omega_0 + \alpha t, \qquad (11\text{-}12)$$

$$\theta - \theta_0 = \omega_0 t + \tfrac{1}{2}\alpha t^2, \qquad (11\text{-}13)$$

$$\omega^2 = \omega_0^2 + 2\alpha(\theta - \theta_0), \qquad (11\text{-}14)$$

$$\theta - \theta_0 = \tfrac{1}{2}(\omega_0 + \omega)t, \qquad (11\text{-}15)$$

$$\theta - \theta_0 = \omega t - \tfrac{1}{2}\alpha t^2. \qquad (11\text{-}16)$$

Linear and Angular Variables Related A point in a rigid rotating body, at a *perpendicular distance r* from the rotation axis, moves in a circle with radius r. If the body rotates through an angle θ, the point moves along an arc with length s given by

$$s = \theta r \qquad \text{(radian measure)}, \qquad (11\text{-}17)$$

where θ is in radians.

The linear velocity $\vec{v}$ of the point is tangent to the circle; the point's linear speed v is given by

$$v = \omega r \qquad \text{(radian measure)}, \qquad (11\text{-}18)$$

where ω is the angular speed (in radians per second) of the body.

The linear acceleration $\vec{a}$ of the point has both *tangential* and *radial* components. The tangential component is

$$a_t = \alpha r \qquad \text{(radian measure)}, \qquad (11\text{-}22)$$

where α is the magnitude of the angular acceleration (in radians per second-squared) of the body. The radial component of $\vec{a}$ is

$$a_r = \frac{v^2}{r} = \omega^2 r \qquad \text{(radian measure)}. \qquad (11\text{-}23)$$

If the point moves in uniform circular motion, the period T of the motion for the point and the body is

$$T = \frac{2\pi r}{v} = \frac{2\pi}{\omega} \qquad \text{(radian measure)}. \qquad (11\text{-}19, 11\text{-}20)$$

Rotational Kinetic Energy and Rotational Inertia The kinetic energy K of a rigid body rotating about a fixed axis is given by

$$K = \tfrac{1}{2}I\omega^2 \qquad \text{(radian measure)}, \qquad (11\text{-}27)$$

in which I is the **rotational inertia** of the body, defined as

$$I = \sum m_i r_i^2 \qquad (11\text{-}26)$$

for a system of discrete particles and as

$$I = \int r^2\, dm \qquad (11\text{-}28)$$

for a body with continuously distributed mass. The r and r_i in these expressions represent the perpendicular distance from the axis of rotation to each mass element in the body.

The Parallel-Axis Theorem The *parallel-axis theorem* relates the rotational inertia I of a body about any axis to that of the same body about a parallel axis through the center of mass:

$$I = I_{com} + Mh^2. \qquad (11\text{-}29)$$

Here h is the perpendicular distance between the two axes.

Torque *Torque* is a turning or twisting action on a body about a rotation axis due to a force $\vec{F}$. If $\vec{F}$ is exerted at a point given by the position vector $\vec{r}$ relative to the axis, then the magnitude of the torque is

$$\tau = rF_t = r_\perp F = rF\sin\phi, \qquad (11\text{-}32, 11\text{-}33, 11\text{-}31)$$

where F_t is the component of $\vec{F}$ perpendicular to $\vec{r}$, and ϕ is the angle between $\vec{r}$ and $\vec{F}$. The quantity $r_\perp$ is the perpendicular distance between the rotation axis and an extended line running through the $\vec{F}$ vector. This line is called the **line of action** of $\vec{F}$, and $r_\perp$ is called the **moment arm** of $\vec{F}$. Similarly, r is the moment arm of F_t.

The SI unit of torque is the newton-meter (N · m). A torque τ is positive if it tends to rotate a body at rest counterclockwise and negative if it tends to rotate the body in the clockwise direction.

Newton's Second Law in Angular Form The rotational analog of Newton's second law is

$$\tau_{net} = I\alpha, \qquad (11\text{-}37)$$

where τ_{net} is the net torque acting on a particle or rigid body, I is the rotational inertia of the particle or body about the rotation axis, and α is the resulting angular acceleration about that axis.

Work and Rotational Kinetic Energy The equations used for calculating work and power in rotational motion correspond to equations used for translational motion and are

$$W = \int_{\theta_i}^{\theta_f} \tau\, d\theta \qquad (11\text{-}45)$$

and

$$P = \frac{dW}{dt} = \tau\omega. \qquad (11\text{-}47)$$

When τ is constant, Eq. 11-45 reduces to

$$W = \tau(\theta_f - \theta_i). \qquad (11\text{-}46)$$

The form of the work–kinetic energy theorem used for rotating bodies is

$$\Delta K = K_f - K_i = \tfrac{1}{2}I\omega_f^2 - \tfrac{1}{2}I\omega_i^2 = W. \qquad (11\text{-}44)$$

QUESTIONS

1. Figure 11-20*b* is a graph of the angular position of the rotating disk of Fig. 11-20*a*. Is the angular velocity of the disk positive, negative, or zero at (a) $t = 1$ s, (b) $t = 2$ s, and (c) $t = 3$ s? (d) Is the angular acceleration positive or negative?

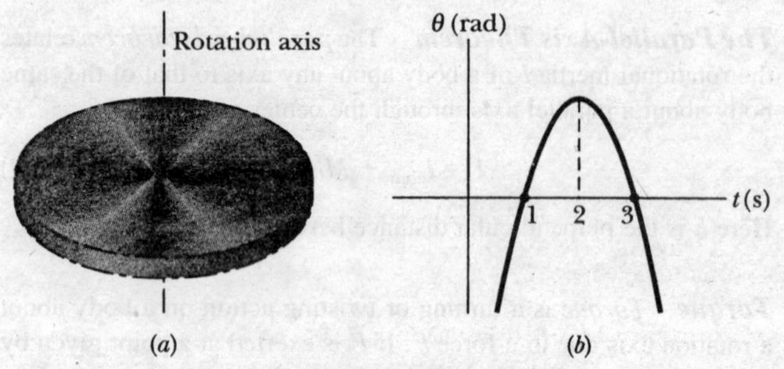

Fig. 11-20 Question 1.

2. Figure 11-21 is a graph of the angular velocity of the rotating disk of Fig. 11-20*a*. What are the (a) initial and (b) final directions of rotation? (c) Does the disk momentarily stop? (d) Is the angular acceleration positive or negative? (e) Is the angular acceleration constant or varying?

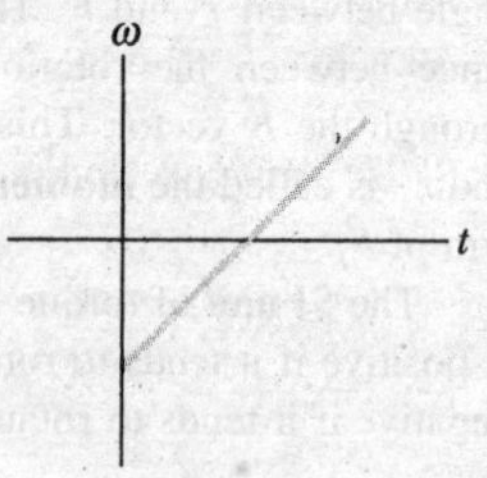

Fig. 11-21 Question 2.

3. Hold your right arm downward, palm toward your thigh. Keeping your wrist rigid, (1) lift the arm until it is horizontal and forward, (2) move it horizontally until it is pointed toward the right, and (3) then bring it down to your side. Your palm faces forward. If you start over, but reverse the steps, why does your palm *not* face forward?

4. For which of the following expressions for $\omega(t)$ of a rotating object do the angular equations of Table 11-1 apply? (a) $\omega = 3$; (b) $\omega = 4t^2 + 2t - 6$; (c) $\omega = 3t - 4$; (d) $\omega = 5t^2 - 3$, all with ω in radians per second and t in seconds.

5. Figure 11-22 is a graph of the angular velocity versus time for the rotating disk of Fig. 11-20*a*. For a point on the rim of the disk, rank the four instants *a*, *b*, *c*, and *d* according to the magnitude of (a) the tangential acceleration and (b) the radial acceleration, greatest first.

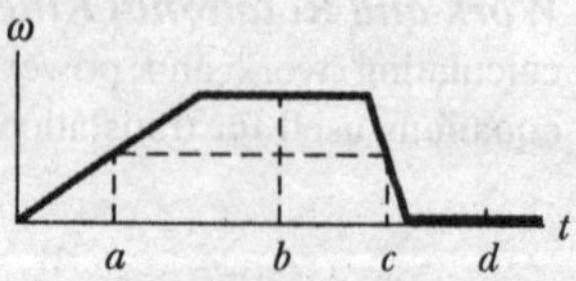

Fig. 11-22 Question 5.

6. The overhead view of Fig. 11-23 is a snapshot of a disk turning counterclockwise like a merry-go-round. The angular speed ω of the disk is decreasing (the disk is turning counterclockwise slower and slower). The figure shows the position of a cockroach that rides the rim of the disk. At the instant of the snapshot, what are the directions of (a) the cockroach's radial acceleration and (b) its tangential acceleration?

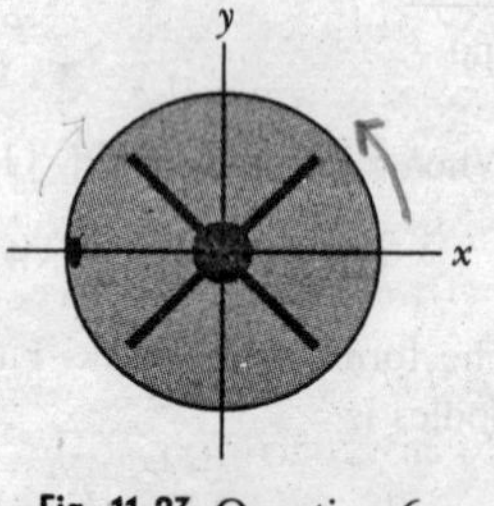

Fig. 11-23 Question 6.

7. Figure 11-24 shows an assembly of three small spheres of the same mass that are attached to a massless rod with the indicated spacings. Consider the rotational inertia I of the assembly about each sphere, in turn. Then rank the spheres according to the rotational inertia about them, greatest first.

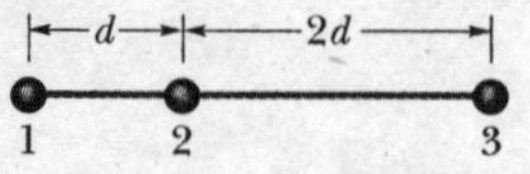

Fig. 11-24 Question 7.

8. Figure 11-25*a* shows an overhead view of a horizontal bar that can pivot about the point indicated. Two horizontal forces act on the bar, but the bar is stationary. If the angle between the bar and force $\vec{F}_2$ is now decreased from the initial 90° and the bar is still not to turn, should the magnitude of $\vec{F}_2$ be made larger, made smaller, or left the same?

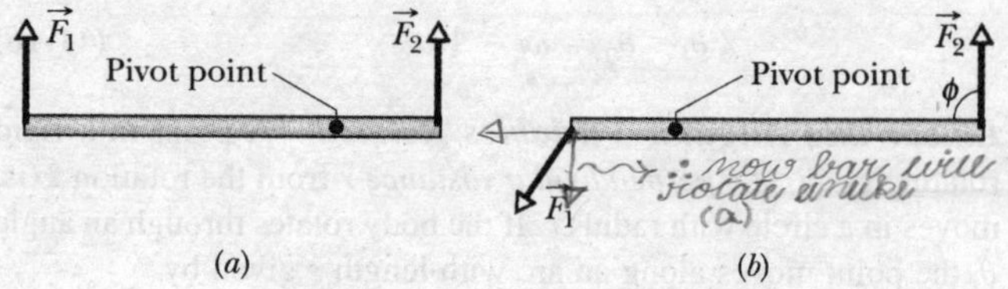

Fig. 11-25 Questions 8 and 9.

9. Figure 11-25*b* shows an overhead view of a horizontal bar that is rotated about the pivot point by two horizontal forces, $\vec{F}_1$ and $\vec{F}_2$, at opposite ends of the bar. The direction of $\vec{F}_2$ is at angle ϕ to the bar. Rank the following values of ϕ according to the magnitude of the angular acceleration of the bar, greatest first: 90°, 70°, and 110°.

10. In the overhead view of Fig. 11-26, five forces of the same magnitude act on a merry-go-round for the strange; it is a square that can rotate about point P at midlength along one of the edges. Rank the forces acting on it according to the magnitude of the torque they create about point P, greatest first.

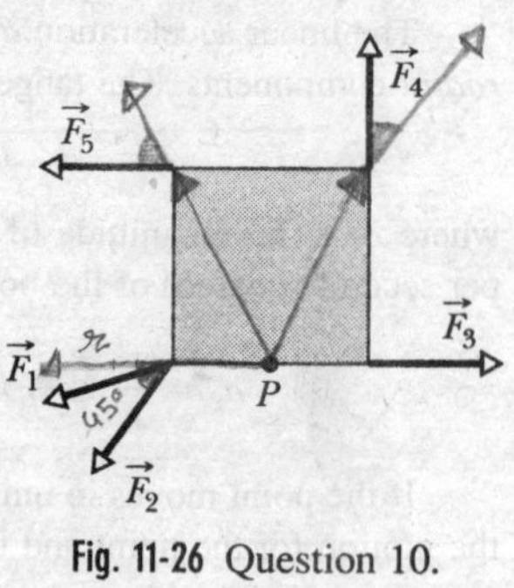

Fig. 11-26 Question 10.

11. In Fig. 11-27, two forces $\vec{F}_1$ and $\vec{F}_2$ act on a disk that turns about its center like a merry-go-round. The forces maintain the indicated angles during the rotation, which is counterclockwise and at a constant rate. However, we are to decrease the angle θ of $\vec{F}_1$ without changing the magnitude of $\vec{F}_1$. (a) To keep the angular speed constant, should we increase, decrease, or maintain

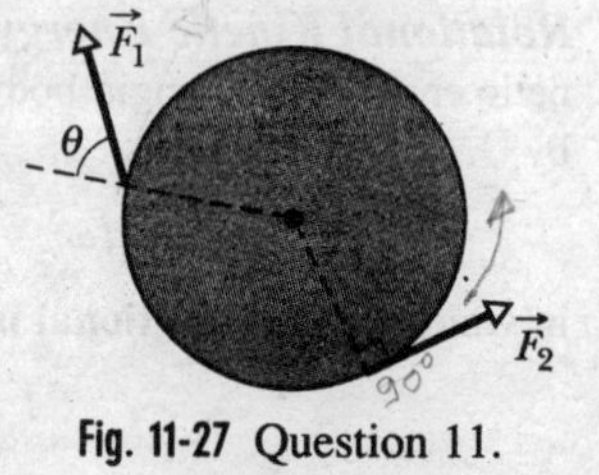

Fig. 11-27 Question 11.

the magnitude of $\vec{F}_2$? Do forces (b) $\vec{F}_1$ and (c) $\vec{F}_2$ tend to rotate the disk clockwise or counterclockwise?

12. Figure 11-28 gives the angular velocity ω versus time t for a merry-go-round being turned by a varying applied force. The magnitudes of the curve's slopes in time intervals 1, 3, 4, and 6 are equal. (a) During which of the time intervals is energy transferred *from* the merry-go-round by the applied force? (b) Rank the time intervals according to the work done on the merry-go-round by the applied force during the interval, most positive work first, most negative work last. (c) Rank the intervals according to the rate at which the applied force transfers energy, greatest rate of transfer *to* the merry-go-round first, greatest rate of transfer *from* it last.

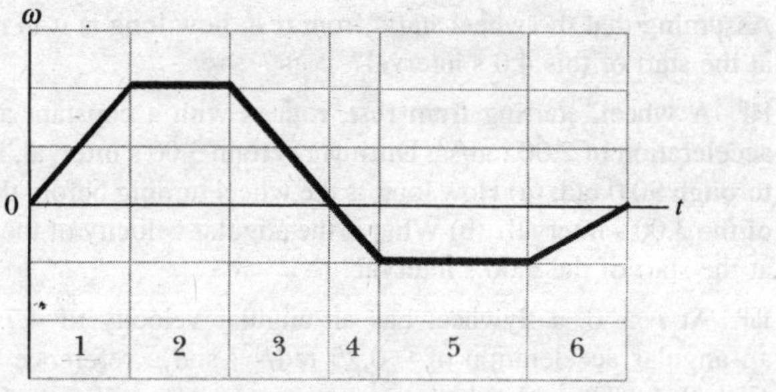

Fig. 11-28 Question 12.

EXERCISES & PROBLEMS

ssm Solution is in the Student Solutions Manual.
www Solution is available on the World Wide Web at: http://www.wiley.com/college/hrw
ilw Solution is available on the Interactive LearningWare.

SEC. 11-2 The Rotational Variables

1E. During a time interval t the flywheel of a generator turns through the angle $\theta = at + bt^3 - ct^4$, where a, b, and c are constants. Write expressions for the wheel's (a) angular velocity and (b) angular acceleration.

2E. What is the angular speed of (a) the second hand, (b) the minute hand, and (c) the hour hand of a smoothly running analog watch? Answer in radians per second.

3E. Our Sun is 2.3×10^4 ly (light-years) from the center of our Milky Way galaxy and is moving in a circle around that center at a speed of 250 km/s. (a) How long does it take the Sun to make one revolution about the galactic center? (b) How many revolutions has the Sun completed since it was formed about 4.5×10^9 years ago? **ssm**

4E. The angular position of a point on the rim of a rotating wheel is given by $\theta = 4.0t - 3.0t^2 + t^3$, where θ is in radians and t is in seconds. What are the angular velocities at (a) $t = 2.0$ s and (b) $t = 4.0$ s? (c) What is the average angular acceleration for the time interval that begins at $t = 2.0$ s and ends at $t = 4.0$ s? What are the instantaneous angular accelerations at (d) the beginning and (e) the end of this time interval?

5E. The angular position of a point on a rotating wheel is given by $\theta = 2 + 4t^2 + 2t^3$, where θ is in radians and t is in seconds. At $t = 0$, what are (a) the point's angular position and (b) its angular velocity? (c) What is its angular velocity at $t = 4.0$ s? (d) Calculate its angular acceleration at $t = 2.0$ s. (e) Is its angular acceleration constant? **ssm**

6P. The wheel in Fig. 11-29 has eight equally spaced spokes and a radius of 30 cm. It is mounted on a fixed axle and is spinning at 2.5 rev/s. You want to shoot a 20-cm-long arrow parallel to this axle and through the wheel without hitting any of the spokes. Assume that the arrow and the spokes are very thin. (a) What minimum speed must the arrow have? (b) Does it matter where between the axle and rim of the wheel you aim? If so, what is the best location?

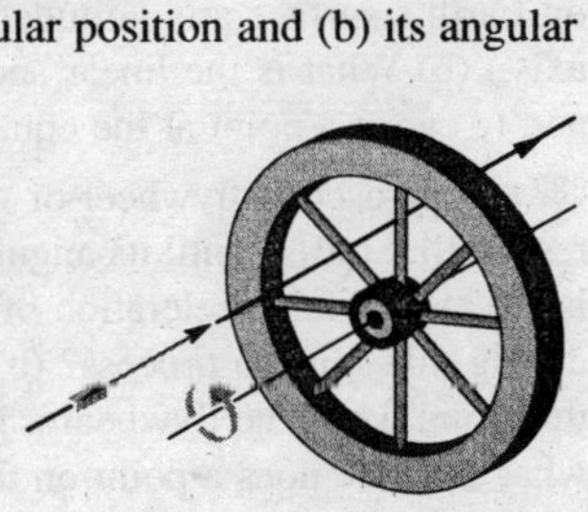
Fig. 11-29 Problem 6.

7P. A diver makes 2.5 revolutions on the way from a 10-m-high platform to the water. Assuming zero initial vertical velocity, find the diver's average angular velocity during a dive. **ilw**

SEC. 11-4 Rotation with Constant Angular Acceleration

8E. The angular speed of an automobile engine is increased at a constant rate from 1200 rev/min to 3000 rev/min in 12 s. (a) What is its angular acceleration in revolutions per minute-squared? (b) How many revolutions does the engine make during this 12 s interval?

9E. A record turntable rotating at $33\frac{1}{3}$ rev/min slows down and stops in 30 s after the motor is turned off. (a) Find its (constant) angular acceleration in revolutions per minute-squared. (b) How many revolutions does it make in this time? **ssm**

10E. A disk, initially rotating at 120 rad/s, is slowed down with a constant angular acceleration of magnitude 4.0 rad/s^2. (a) How much time does the disk take to stop? (b) Through what angle does the disk rotate during that time?

11E. A heavy flywheel rotating on its central axis is slowing down because of friction in its bearings. At the end of the first minute of slowing, its angular speed is 0.90 of its initial angular speed of 250 rev/min. Assuming a constant angular acceleration, find its angular speed at the end of the second minute.

12E. Starting from rest, a disk rotates about its central axis with constant angular acceleration. In 5.0 s, it rotates 25 rad. During that time, what are the magnitudes of (a) the angular acceleration and (b) the average angular velocity? (c) What is the instantaneous angular velocity of the disk at the end of the 5.0 s? (d) With the angular acceleration unchanged, through what additional angle will the disk turn during the next 5.0 s?

13P. A wheel has a constant angular acceleration of 3.0 rad/s^2. During a certain 4.0 s interval, it turns through an angle of 120 rad.

Assuming that the wheel starts from rest, how long is it in motion at the start of this 4.0 s interval? ssm www

14P. A wheel, starting from rest, rotates with a constant angular acceleration of 2.00 rad/s^2. During a certain 3.00 s interval, it turns through 90.0 rad. (a) How long is the wheel turning before the start of the 3.00 s interval? (b) What is the angular velocity of the wheel at the start of the 3.00 s interval?

15P. At $t = 0$, a flywheel has an angular velocity of 4.7 rad/s, an angular acceleration of -0.25 rad/s^2, and a reference line at $\theta_0 = 0$. (a) Through what maximum angle θ_{max} will the reference line turn in the positive direction? At what times t will the reference line be at (b) $\theta = \frac{1}{2}\theta_{max}$ and (c) $\theta = -10.5$ rad (consider both positive and negative values of t)? (d) Graph θ versus t, and indicate the answers to (a), (b), and (c) on the graph.

16P. A disk rotates about its central axis starting from rest and accelerates with constant angular acceleration. At one time it is rotating at 10 rev/s; 60 revolutions later, its angular speed is 15 rev/s. Calculate (a) the angular acceleration, (b) the time required to complete the 60 revolutions, (c) the time required to reach the 10 rev/s angular speed, and (d) the number of revolutions from rest until the time the disk reaches the 10 rev/s angular speed.

17P. A flywheel turns through 40 rev as it slows from an angular speed of 1.5 rad/s to a stop. (a) Assuming a constant angular acceleration, find the time for it to come to rest. (b) What is its angular acceleration? (c) How much time is required for it to complete the first 20 of the 40 revolutions? ilw

★ $w_2^2 - w_1^2 = 2\alpha S$; keep unsolved α & put in. Now $w_2 = w_1 + \alpha t$, simplify for final ans (many quantities will get cut)

18P. A wheel rotating about a fixed axis through its center has a constant angular acceleration of 4.0 rad/s^2. In a certain 4.0 s interval the wheel turns through an angle of 80 rad. (a) What is the angular velocity of the wheel at the start of the 4.0 s interval? (b) Assuming that the wheel starts from rest, how long is it in motion at the start of the 4.0 s interval?

★ 1 rev = 2π rad

SEC. 11-5 Relating the Linear and Angular Variables

19E. What is the linear acceleration of a point on the rim of a 30-cm-diameter record rotating at a constant angular speed of $33\frac{1}{3}$ rev/min? ssm

20E. A vinyl record on a turntable rotates at $33\frac{1}{3}$ rev/min. (a) What is its angular speed in radians per second? What is the linear speed of a point on the record at the needle when the needle is (b) 15 cm and (c) 7.4 cm from the turntable axis?

21E. What is the angular speed of a car traveling at 50 km/h and rounding a circular turn of radius 110 m? SSM

22E. A flywheel with a diameter of 1.20 m is rotating at an angular speed of 200 rev/min. (a) What is the angular speed of the flywheel in radians per second? (b) What is the linear speed of a point on the rim of the flywheel? (c) What constant angular acceleration (in revolutions per minute-squared) will increase the wheel's angular speed to 1000 rev/min in 60 s? (d) How many revolutions does the wheel make during that 60 s?

23E. An astronaut is being tested in a centrifuge. The centrifuge has a radius of 10 m and, in starting, rotates according to $\theta = 0.30t^2$, where t is in seconds and θ is in radians. When $t = 5.0$ s, what are the magnitudes of the astronaut's (a) angular velocity, (b) linear velocity, (c) tangential acceleration, and (d) radial acceleration?

24E. What are the magnitudes of (a) the angular velocity, (b) the radial acceleration, and (c) the tangential acceleration of a spaceship taking a circular turn of radius 3220 km at a speed of 29 000 km/h?

25P. An early method of measuring the speed of light makes use of a rotating slotted wheel. A beam of light passes through a slot at the outside edge of the wheel, as in Fig. 11-30, travels to a distant mirror, and returns to the wheel just in time to pass through the next slot in the wheel. One such slotted wheel has a radius of 5.0 cm and 500 slots at its edge. Measurements taken when the mirror is $L = 500$ m from the wheel indicate a speed of light of 3.0×10^5 km/s. (a) What is the (constant) angular speed of the wheel? (b) What is the linear speed of a point on the edge of the wheel? ssm

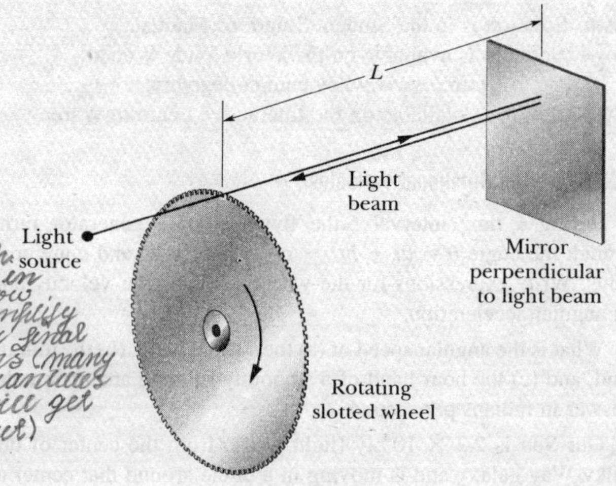

Fig. 11-30 Problem 25.

26P. The flywheel of a steam engine runs with a constant angular velocity of 150 rev/min. When steam is shut off, the friction of the bearings and of the air stops the wheel in 2.2 h. (a) What is the constant angular acceleration, in revolutions per minute-squared, of the wheel during the slowdown? (b) How many rotations does the wheel make before stopping? (c) At the instant the flywheel is turning at 75 rev/min, what is the tangential component of the linear acceleration of a flywheel particle that is 50 cm from the axis of rotation? (d) What is the magnitude of the net linear acceleration of the particle in (c)?

27P. (a) What is the angular speed ω about the polar axis of a point on Earth's surface at a latitude of 40° N? (Earth rotates about that axis.) (b) What is the linear speed v of the point? What are (c) ω and (d) v for a point at the equator? ssm

28P. A gyroscope flywheel of radius 2.83 cm is accelerated from rest at 14.2 rad/s^2 until its angular speed is 2760 rev/min. (a) What is the tangential acceleration of a point on the rim of the flywheel during this spin-up process? (b) What is the radial acceleration of this point when the flywheel is spinning at full speed? (c) Through what distance does a point on the rim move during the spin-up?

29P. In Fig. 11-31, wheel A of radius $r_A = 10$ cm is coupled by belt B to wheel C of radius $r_C = 25$ cm. The angular speed of wheel A is increased from rest at a constant rate of 1.6 rad/s^2. Find the time for wheel C to reach a rotational speed of 100 rev/min, assuming the belt does not slip. (*Hint:* If the belt does not slip, the linear speeds at the rims of the two wheels must be equal.) ssm www

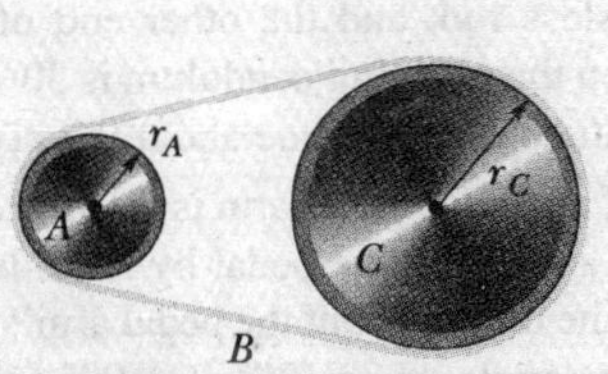

Fig. 11-31 Problem 29.

30P. An object rotates about a fixed axis, and the angular position of a reference line on the object is given by $\theta = 0.40e^{2t}$, where θ is in radians and t is in seconds. Consider a point on the object that is 4.0 cm from the axis of rotation. At $t = 0$, what are the magnitudes of the point's (a) tangential component of acceleration and (b) radial component of acceleration?

31P. A pulsar is a rapidly rotating neutron star that emits a radio beam like a lighthouse emits a light beam. We receive a radio pulse for each rotation of the star. The period T of rotation is found by measuring the time between pulses. The pulsar in the Crab nebula (Fig. 11-32) has a period of rotation of $T = 0.033$ s that is increasing at the rate of 1.26×10^{-5} s/y. (a) What is the pulsar's angular acceleration? (b) If its angular acceleration is constant, how many years from now will the pulsar stop rotating? (c) The pulsar originated in a supernova explosion seen in the year 1054. What was the initial T for the pulsar? (Assume constant angular acceleration since the pulsar originated.) ssm

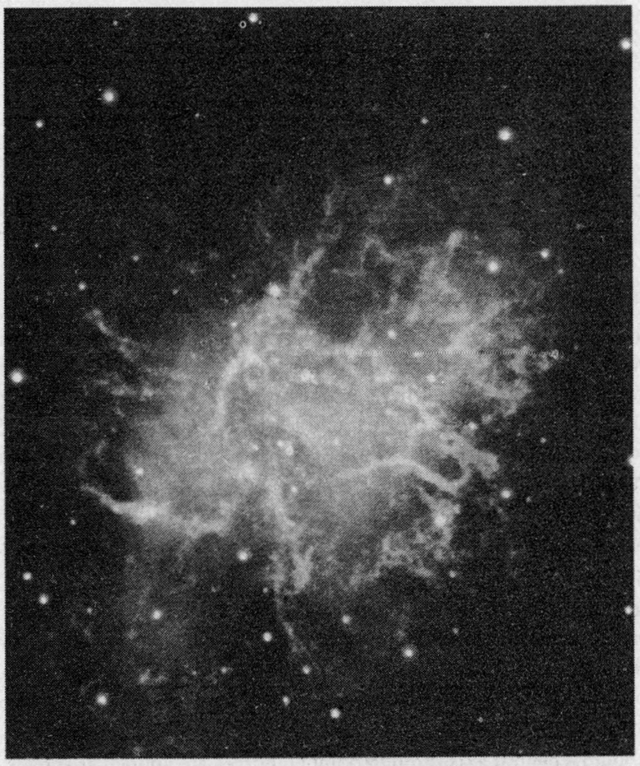

Fig. 11-32 Problem 31. The Crab nebula resulted from a star whose explosion was seen in 1054. In addition to the gaseous debris seen here, the explosion left a spinning neutron star at its center. The star has a diameter of only 30 km.

32P. A record turntable is rotating at $33\frac{1}{3}$ rev/min. A watermelon seed is on the turntable 6.0 cm from the axis of rotation. (a) Calculate the acceleration of the seed, assuming that it does not slip. (b) What is the minimum value of the coefficient of static friction between the seed and the turntable if the seed is not to slip? (c) Suppose that the turntable achieves its angular speed by starting from rest and undergoing a constant angular acceleration for 0.25 s. Calculate the minimum coefficient of static friction required for the seed not to slip during the acceleration period.

SEC. 11-6 . Kinetic Energy of Rotation

33E. Calculate the rotational inertia of a wheel that has a kinetic energy of 24 400 J when rotating at 602 rev/min. ssm

34P. The oxygen molecule O_2 has a mass of 5.30×10^{-26} kg and a rotational inertia of 1.94×10^{-46} kg $\cdot$ m^2 about an axis through the center of the line joining the atoms and perpendicular to that line. Suppose the center of mass of an O_2 molecule in a gas has a translational speed of 500 m/s and the molecule has a rotational kinetic energy that is $\frac{2}{3}$ of the translational kinetic energy of its center of mass. What then is the molecule's angular speed about the center of mass?

SEC. 11-7 Calculating the Rotational Inertia

35E. Two uniform solid cylinders, each rotating about its central (longitudinal) axis, have the same mass of 1.25 kg and rotate with the same angular speed of 235 rad/s, but they differ in radius. What is the rotational kinetic energy of (a) the smaller cylinder, of radius 0.25 m, and (b) the larger cylinder, of radius 0.75 m? ssm

36E. A communications satellite is a solid cylinder with mass 1210 kg, diameter 1.21 m, and length 1.75 m. Prior to launching from the shuttle cargo bay, it is set spinning at 1.52 rev/s about the cylinder axis (Fig. 11-33). Calculate the satellite's (a) rotational inertia about the rotation axis and (b) rotational kinetic energy.

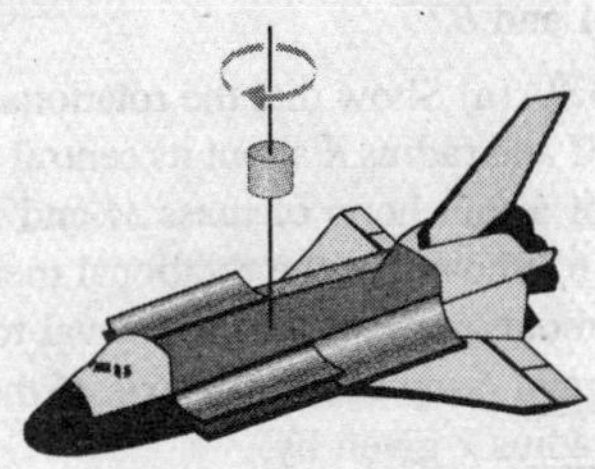

Fig. 11-33 Exercise 36.

37E. In Fig. 11-34, two particles, each with mass m, are fastened to each other, and to a rotation axis at O, by two thin rods, each with length d and mass M. The combination rotates around the rotation axis with angular velocity ω. In terms of these symbols, and measured about O, what are the combination's (a) rotational inertia and (b) kinetic energy?

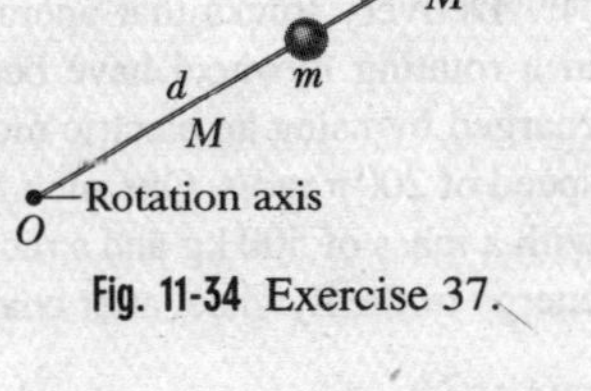

Fig. 11-34 Exercise 37.

38E. Each of the three helicopter rotor blades shown in Fig. 11-35 is 5.20 m long and has a mass of 240 kg. The rotor is rotating at 350 rev/min. (a) What is the rotational inertia of the rotor assembly about the axis of rotation? (Each blade can be considered to be a thin rod rotated about one end.) (b) What is the total kinetic energy of rotation?

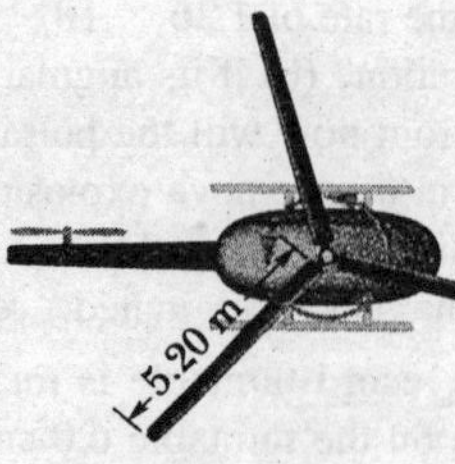

Fig. 11-35 Exercise 38.

39E. Calculate the rotational inertia of a meter stick, with mass 0.56 kg, about an axis perpendicular to the stick and located at the 20 cm mark. (Treat the stick as a thin rod.) ssm

40P. Four identical particles of mass 0.50 kg each are placed at the vertices of a 2.0 m × 2.0 m square and held there by four massless rods, which form the sides of the square. What is the rotational inertia of this rigid body about an axis that (a) passes through the midpoints of opposite sides and lies in the plane of the square, (b) passes through the midpoint of one of the sides and is perpendicular to the plane of the square, and (c) lies in the plane of the square and passes through two diagonally opposite particles?

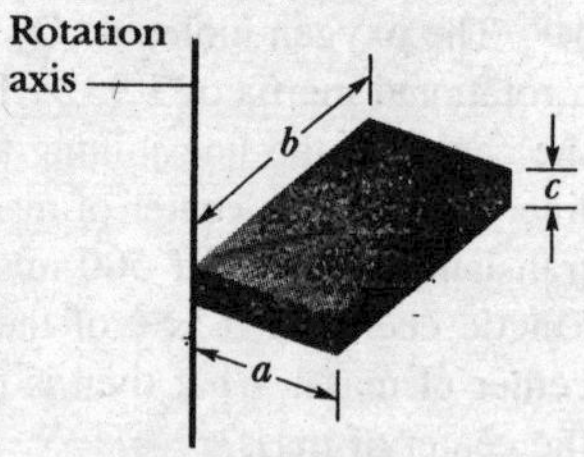

Fig. 11-36 Problem 41.

41P. The uniform solid block in Fig. 11-36 has mass M and edge lengths a, b, and c. Calculate its rotational inertia about an axis through one corner and perpendicular to the large faces. ssm

42P. The masses and coordinates of four particles are as follows: 50 g, $x = 2.0$ cm, $y = 2.0$ cm; 25 g, $x = 0$, $y = 4.0$ cm; 25 g, $x = -3.0$ cm, $y = -3.0$ cm; 30 g, $x = -2.0$ cm, $y = 4.0$ cm. What are the rotational inertias of this collection about the (a) x, (b) y, and (c) z axes? (d) Suppose the answers to (a) and (b) are A and B, respectively. Then what is the answer to (c) in terms of A and B?

43P. (a) Show that the rotational inertia of a solid cylinder of mass M and radius R about its central axis is equal to the rotational inertia of a thin hoop of mass M and radius $R/\sqrt{2}$ about its central axis. (b) Show that the rotational inertia I of any given body of mass M about any given axis is equal to the rotational inertia of an *equivalent hoop* about that axis, if the hoop has the same mass M and a radius k given by

$$k = \sqrt{\frac{I}{M}}.$$

The radius k of the equivalent hoop is called the *radius of gyration* of the given body. ssm

44P. Delivery trucks that operate by making use of energy stored in a rotating flywheel have been used in Europe. The trucks are charged by using an electric motor to get the flywheel up to its top speed of 200π rad/s. One such flywheel is a solid, uniform cylinder with a mass of 500 kg and a radius of 1.0 m. (a) What is the kinetic energy of the flywheel after charging? (b) If the truck operates with an average power requirement of 8.0 kW, for how many minutes can it operate between chargings?

SEC. 11-8 Torque

45E. A small ball of mass 0.75 kg is attached to one end of a 1.25-m-long massless rod, and the other end of the rod is hung from a pivot. When the resulting pendulum is 30° from the vertical, what is the magnitude of the torque about the pivot? ssm

46E. The length of a bicycle pedal arm is 0.152 m, and a downward force of 111 N is applied to the pedal by the rider's foot. What is the magnitude of the torque about the pedal arm's pivot point when the arm makes an angle of (a) 30°, (b) 90°, and (c) 180° with the vertical?

47P. The body in Fig. 11-37 is pivoted at O, and two forces act on it as shown. (a) Find an expression for the net torque on the body about the pivot. (b) If $r_1 = 1.30$ m, $r_2 = 2.15$ m, $F_1 = 4.20$ N, $F_2 = 4.90$ N, $\theta_1 = 75.0°$, and $\theta_2 = 60.0°$, what is the net torque about the pivot? ssm ilw

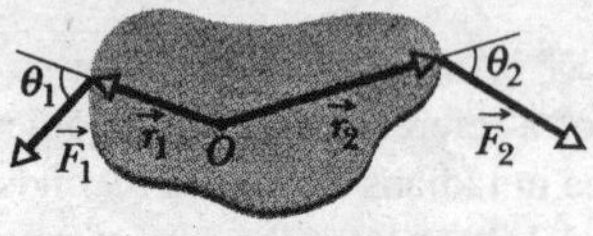

Fig. 11-37 Problem 47.

48P. The body in Fig. 11-38 is pivoted at O. Three forces act on it in the directions shown: $F_A = 10$ N at point A, 8.0 m from O; $F_B = 16$ N at point B, 4.0 m from O; and $F_C = 19$ N at point C, 3.0 m from O. What is the net torque about O?

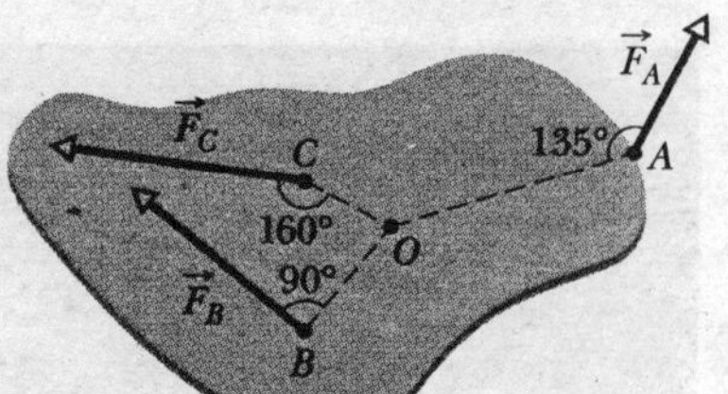

Fig. 11-38 Problem 48.

SEC. 11-9 Newton's Second Law for Rotation

49E. During the launch from a board, a diver's angular speed about her center of mass changes from zero to 6.20 rad/s in 220 ms. Her rotational inertia about her center of mass is 12.0 kg · m^2. During the launch, what are the magnitudes of (a) her average angular acceleration and (b) the average external torque on her from the board? ssm ilw

50E. A torque of 32.0 N · m on a certain wheel causes an angular acceleration of 25.0 rad/s^2. What is the wheel's rotational inertia?

51E. A thin spherical shell has a radius of 1.90 m. An applied torque of 960 N · m gives the shell an angular acceleration of 6.20 rad/s^2 about an axis through the center of the shell. What are (a) the rotational inertia of the shell about that axis and (b) the mass of the shell? ssm

52E. In Fig. 11-39, a cylinder having a mass of 2.0 kg can rotate about its central axis through point O. Forces are applied as shown: $F_1 = 6.0$ N, $F_2 = 4.0$ N, $F_3 = 2.0$ N, and $F_4 = 5.0$ N. Also, $R_1 = 5.0$ cm and $R_2 = 12$ cm. Find the magnitude and direction of the angular acceleration of the cylinder. (During the rotation, the forces maintain their same angles relative to the cylinder.)

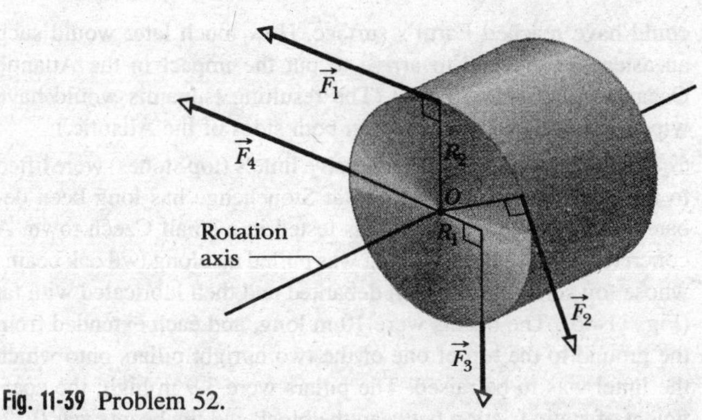

Fig. 11-39 Problem 52.

53P. Figure 11-40 shows the massive shield door at a neutron test facility at Lawrence Livermore Laboratory; this is the world's heaviest hinged door. The door has a mass of 44,000 kg, a rotational inertia about a vertical axis through its huge hinges of 8.7×10^4 kg · m², and a (front) face width of 2.4 m. Neglecting friction, what steady force, applied at its outer edge and perpendicular to the plane of the door, can move it from rest through an angle of 90° in 30 s?

Fig. 11-40 Problem 53.

54P. A wheel of radius 0.20 m is mounted on a frictionless horizontal axis. The rotational inertia of the wheel about the axis is 0.050 kg · m². A massless cord wrapped around the wheel is attached to a 2.0 kg block that slides on a horizontal frictionless surface. If a horizontal force of magnitude $P = 3.0$ N is applied to the block as shown in Fig. 11-41, what is the magnitude of the angular acceleration of the wheel? Assume the string does not slip on the wheel.

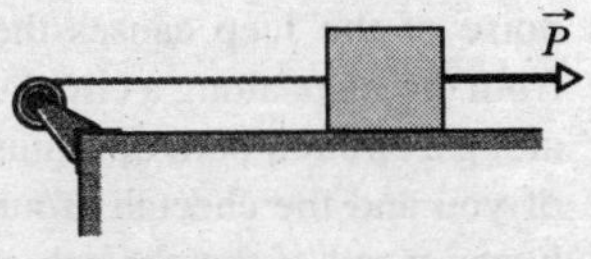

Fig. 11-41 Problem 54.

55P. In Fig. 11-42, one block has mass $M = 500$ g, the other has mass $m = 460$ g, and the pulley, which is mounted in horizontal frictionless bearings, has a radius of 5.00 cm. When released from rest, the heavier block falls 75.0 cm in 5.00 s (without the cord slipping on the pulley). (a) What is the magnitude of the blocks' acceleration? What is the tension in the part of the cord that supports (b) the heavier block and (c) the lighter block? (d) What is the magnitude of the pulley's angular acceleration? (e) What is its rotational inertia? ssm www

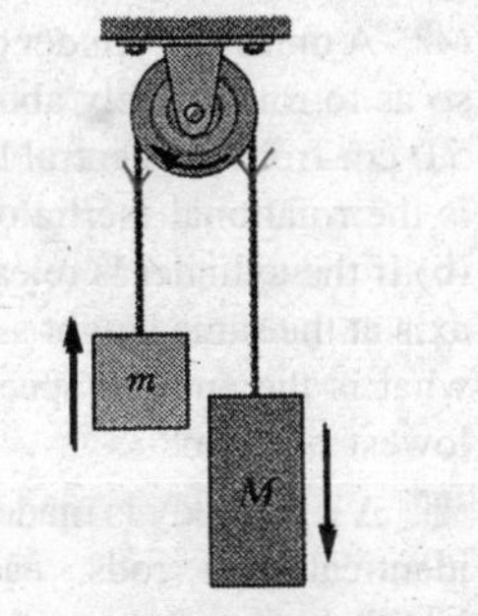

Fig. 11-42 Problem 55.

56P. A pulley, with a rotational inertia of 1.0×10^{-3} kg · m² about its axle and a radius of 10 cm, is acted on by a force applied tangentially at its rim. The force magnitude varies in time as $F = 0.50t + 0.30t^2$, with F in newtons and t in seconds. The pulley is initially at rest. At $t = 3.0$ s what are (a) its angular acceleration and (b) its angular speed?

57P. Figure 11-43 shows two blocks, each of mass m, suspended from the ends of a rigid massless rod of length $L_1 + L_2$, with $L_1 = 20$ cm and $L_2 = 80$ cm. The rod is held horizontally on the fulcrum and then released. What are the magnitudes of the initial accelerations of (a) the block closer to the fulcrum and (b) the other block?

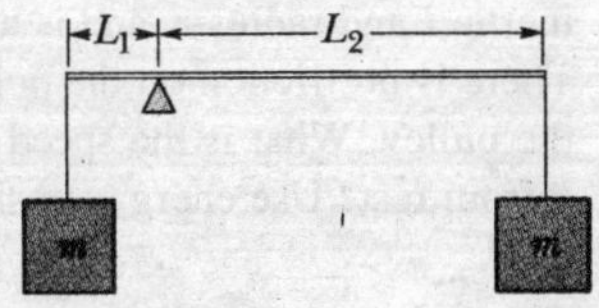

Fig. 11-43 Problem 57.

SEC. 11-10 Work and Rotational Kinetic Energy

58E. (a) If $R = 12$ cm, $M = 400$ g, and $m = 50$ g in Fig. 11-17, find the speed of the block after it has descended 50 cm starting from rest. Solve the problem using energy conservation principles. (b) Repeat (a) with $R = 5.0$ cm.

59E. An automobile crankshaft transfers energy from the engine to the axle at the rate of 100 hp (= 74.6 kW) when rotating at a speed of 1800 rev/min. What torque (in newton-meters) does the crankshaft deliver?

60E. A 32.0 kg wheel, essentially a thin hoop with radius 1.20 m, is rotating at 280 rev/min. It must be brought to a stop in 15.0 s. (a) How much work must be done to stop it? (b) What is the required average power?

61P. A thin rod of length L and mass m is suspended freely from one end. It is pulled to one side and then allowed to swing like a pendulum, passing through its lowest position with angular speed ω. In terms of these symbols and g, and neglecting friction and air resistance, find (a) the rod's kinetic energy at its lowest position and (b) how far above that position the center of mass rises.

62P. Calculate (a) the torque, (b) the energy, and (c) the average power required to accelerate Earth in 1 day from rest to its present angular speed about its axis.

63P. A meter stick is held vertically with one end on the floor and is then allowed to fall. Find the speed of the other end when it hits the floor, assuming that the end on the floor does not slip. (*Hint:* Consider the stick to be a thin rod and use the conservation of energy principle.) ssm ilw

64P. A uniform cylinder of radius 10 cm and mass 20 kg is mounted so as to rotate freely about a horizontal axis that is parallel to and 5.0 cm from the central longitudinal axis of the cylinder. (a) What is the rotational inertia of the cylinder about the axis of rotation? (b) If the cylinder is released from rest with its central longitudinal axis at the same height as the axis about which the cylinder rotates, what is the angular speed of the cylinder as it passes through its lowest position?

65P. A rigid body is made of three identical thin rods, each with length L, fastened together in the form of a letter **H** (Fig. 11-44). The body is free to rotate about a horizontal axis that runs along the length of one of the legs of the **H**. The body is allowed to fall from rest from a position in which the plane of the **H** is horizontal. What is the angular speed of the body when the plane of the **H** is vertical? ssm www

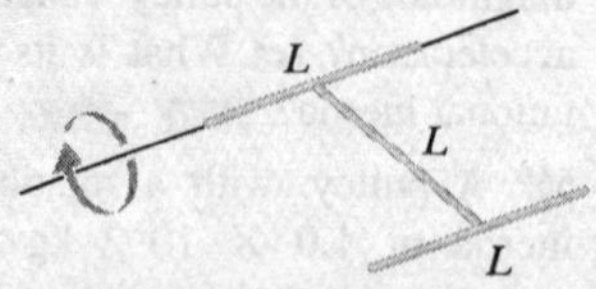

Fig. 11-44 Problem 65.

66P. A uniform spherical shell of mass M and radius R rotates about a vertical axis on frictionless bearings (Fig. 11-45). A massless cord passes around the equator of the shell, over a pulley of rotational inertia I and radius r, and is attached to a small object of mass m. There is no friction on the pulley's axle; the cord does not slip on the pulley. What is the speed of the object after it falls a distance h from rest? Use energy considerations.

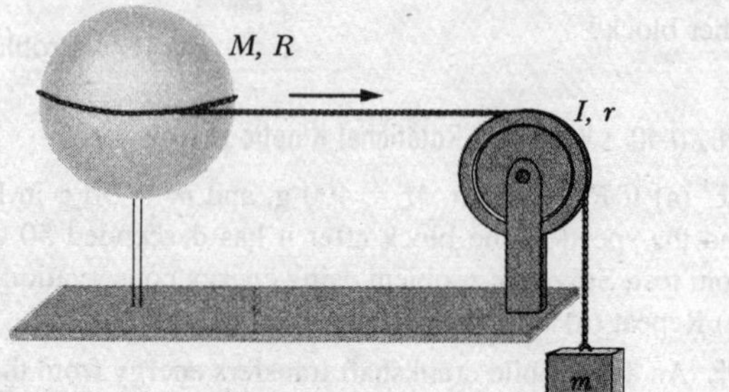

Fig. 11-45 Problem 66.

67P. A tall, cylinder-shaped chimney falls over when its base is ruptured. Treat the chimney as a thin rod of length H, and let θ be the angle the chimney makes with the vertical. In terms of these symbols and g, express the following: (a) the angular speed of the chimney, (b) the radial acceleration of the chimney's top, and (c) the tangential acceleration of the top. (*Hint:* Use energy considerations, not a torque. In part (c), recall that $\alpha = d\omega/dt$.) (d) At what angle θ does the tangential acceleration equal g? ssm

Additional Problems

68. At 7:14 A.M. on June 30, 1908, a huge explosion occurred above remote central Siberia, at latitude 61° N and longitude 102° E; the fireball thus created was the brightest flash seen by anyone before nuclear weapons. The *Tunguska Event,* which according to one chance witness "covered an enormous part of the sky," was probably the explosion of a *stony asteroid* about 140 m wide. (a) Considering only Earth's rotation, determine how much later the asteroid would have had to arrive to put the explosion above Helsinki at longitude 25° E. This would have obliterated the city. (b) If the asteroid had, instead, been a *metallic asteroid,* it could have reached Earth's surface. How much later would such an asteroid have had to arrive to put the impact in the Atlantic Ocean at longitude 20° W? (The resulting tsunamis would have wiped out coastal civilization on both sides of the Atlantic.)

69. The method in which the massive lintels (top stones) were lifted to the top of the upright stones at Stonehenge has long been debated. One possible method was tested in a small Czech town. A concrete block of mass 5124 kg was pulled up along two oak beams whose top surfaces had been debarked and then lubricated with fat (Fig. 11-46). The beams were 10 m long, and each extended from the ground to the top of one of the two upright pillars onto which the lintel was to be raised. The pillars were 3.9 m high; the coefficient of static friction between the block and the beams was 0.22. The pull on the block was via ropes that wrapped around the block and also around the top ends of two spruce logs of length 4.5 m. A riding platform was strung at the opposite end of each log. When enough workers sat or stood on a riding platform, the corresponding spruce log would pivot about the top of its upright pillar and pull one side of the block a short distance up a beam. For each log, the ropes were approximately perpendicular to the log; the distance between the pivot point and the point where the ropes wrapped around the log was 0.70 m. Assuming that each worker had a mass of 85 kg, find the least number of workers needed on the two platforms so that the block begins to move up along the beams. (About half this number could actually move the block by moving first one side of the block and then the other side.)

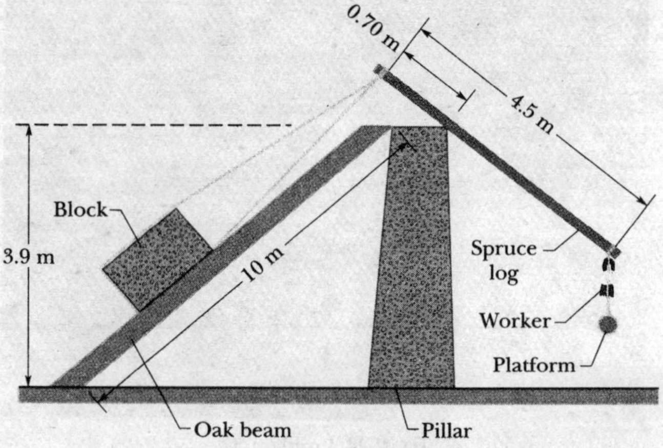

Fig. 11-46 Problem 69.

70. Cheetahs in full run have been reported to have an astounding speed of 114 km/h (about 71 mi/h) by observers moving along with the cheetahs in Jeep-like vehicles. Imagine trying to measure a cheetah's speed by keeping your Jeep abreast of the animal while also glancing at your speedometer, which is registering 114 km/h. You keep the Jeep at a constant 8.0 m from the cheetah, but the noise of the Jeep causes the cheetah to continuously veer away from the Jeep, along a circular path of radius 92 m. Thus, you travel along a circular path of radius 100 m. (a) What is the angular speed of you and the cheetah around the circular paths? (b) What is the linear speed of the cheetah along its path? (If you did not account for the circular motion, you would conclude erroneously that the cheetah's speed is 114 km/h, and that type of error was apparently made in the published reports.)

12 Rolling, Torque, and Angular Momentum

In 1897, a European "aerialist" made the first triple somersault during the flight from a swinging trapeze to the hands of a partner. For the next 85 years aerialists attempted to complete a *quadruple* somersault, but not until 1982 was it done before an audience: Miguel Vazquez of the Ringling Bros. and Barnum & Bailey Circus rotated his body in four complete circles in midair before his brother Juan caught him. Both were stunned by their success.

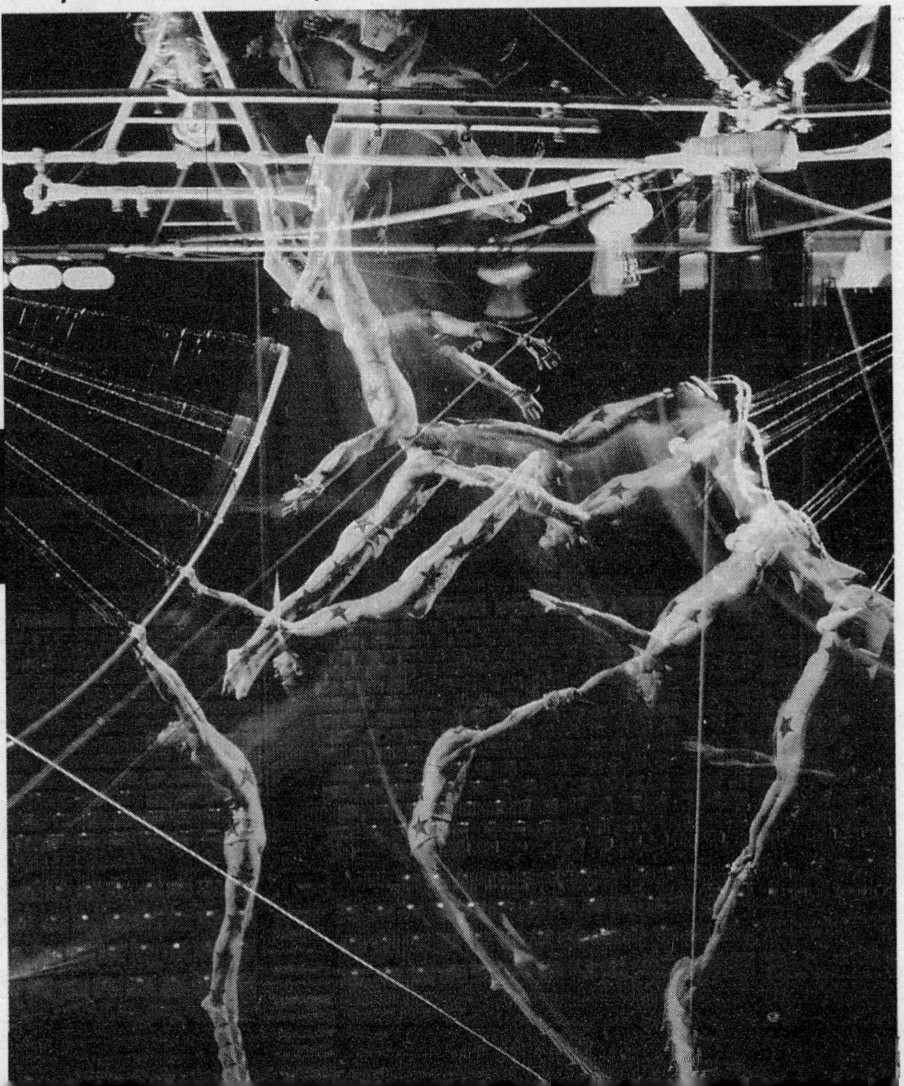

Why was the feat so difficult, and what feature of physics made it (finally) possible?

The answer is in this chapter.

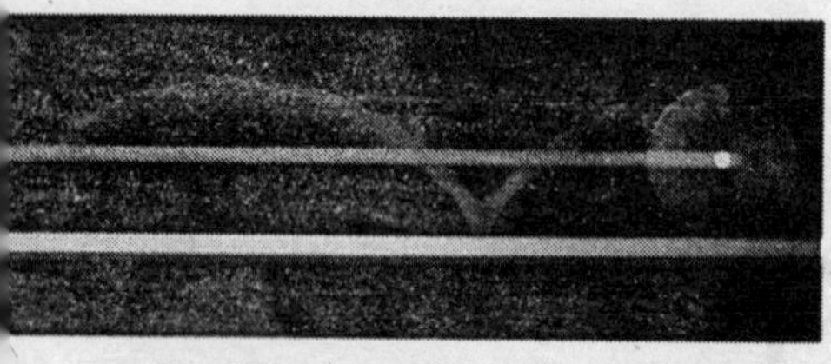

Fig. 12-1 A time exposure photograph of a rolling disk. Small lights have been attached to the disk, one at its center and one at its edge. The latter traces out a curve called a *cycloid*.

12-1 Rolling

When a bicycle moves along a straight track, the center of each wheel moves forward in pure translation. A point on the rim of the wheel, however, traces out a more complex path, as Fig. 12-1 shows. In what follows, we analyze the motion of a rolling wheel first by viewing it as a combination of pure translation and pure rotation, and then by viewing it as rotation alone.

Rolling as Rotation and Translation Combined

Imagine that you are watching the wheel of a bicycle, which passes you at constant speed while *rolling smoothly* (that is, without sliding) along a street. As shown in Fig. 12-2, the center of mass O of the wheel moves forward at constant speed v_{com}. The point P on the street where the wheel makes contact also moves forward at speed v_{com}, so that it always remains directly below O.

During a time interval t, you see both O and P move forward by a distance s. The bicycle rider sees the wheel rotate through an angle θ about the center of the wheel, with the point of the wheel that was touching the street at the beginning of t moving through arc length s. Equation 11-17 relates the arc length s to the rotation angle θ:

$$s = \theta R, \tag{12-1}$$

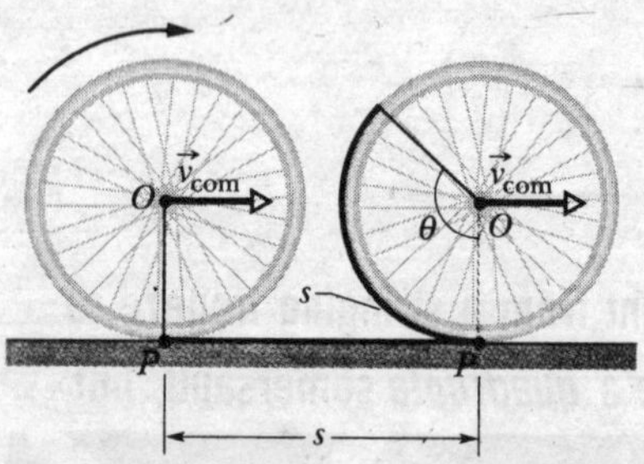

Fig. 12-2 The center of mass O of a rolling wheel moves a distance s at velocity $\vec{v}_{com}$, while the wheel rotates through angle θ. The point P at which the wheel makes contact with the surface over which the wheel rolls also moves a distance s.

where R is the radius of the wheel. The linear speed v_{com} of the center of the wheel (the center of mass of this uniform wheel) is ds/dt. The angular speed ω of the wheel about its center is $d\theta/dt$. Thus, differentiating Eq. 12-1 with respect to time (with R held constant) gives us

$$v_{com} = \omega R \quad \text{(smooth rolling motion)}. \tag{12-2}$$

Figure 12-3 shows that the rolling motion of a wheel is a combination of purely translational and purely rotational motions. Figure 12-3*a* shows the purely rotational motion (as if the rotation axis through the center were stationary): Every point on the wheel rotates about the center with angular speed ω. (This is the type of motion we considered in Chapter 11.) Every point on the outside edge of the wheel has linear speed v_{com} given by Eq. 12-2. Figure 12-3*b* shows the purely translational

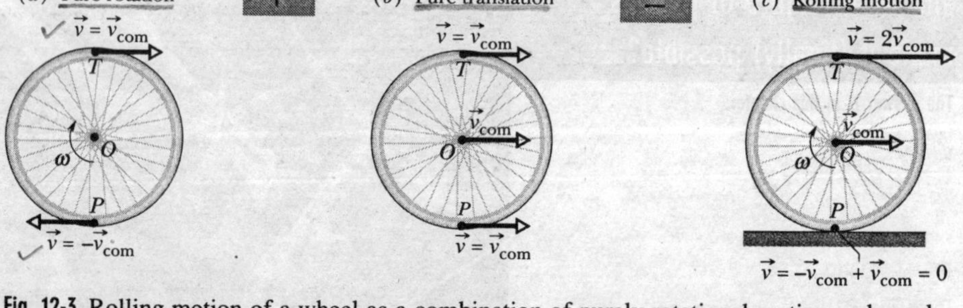

Fig. 12-3 Rolling motion of a wheel as a combination of purely rotational motion and purely translational motion. (*a*) The purely rotational motion: All points on the wheel move with the same angular speed ω. Points on the outside edge of the wheel all move with the same linear speed $v = v_{com}$. The linear velocities $\vec{v}$ of two such points, at top (T) and bottom (P) of the wheel, are shown. (*b*) The purely translational motion: All points on the wheel move to the right with the same linear velocity $\vec{v}_{com}$ as the center of the wheel. (*c*) The rolling motion of the wheel is the combination of (*a*) and (*b*).

motion (as if the wheel did not rotate at all): Every point on the wheel moves to the right with speed v_{com}.

The combination of Figs. 12-3*a* and 12-3*b* yields the actual rolling motion of the wheel, Fig. 12-3*c*. Note that in this combination of motions, the portion of the wheel at the bottom (at point *P*) is stationary and the portion at the top (at point *T*) is moving at speed $2v_{com}$, faster than any other portion of the wheel. These results are demonstrated in Fig. 12-4, which is a time exposure of a rolling bicycle wheel. You can tell that the wheel is moving faster near its top than near its bottom because the spokes are more blurred at the top than at the bottom.

The motion of any round body rolling smoothly over a surface can be separated into purely rotational and purely translational motions, as in Figs. 12-3*a* and 12-3*b*.

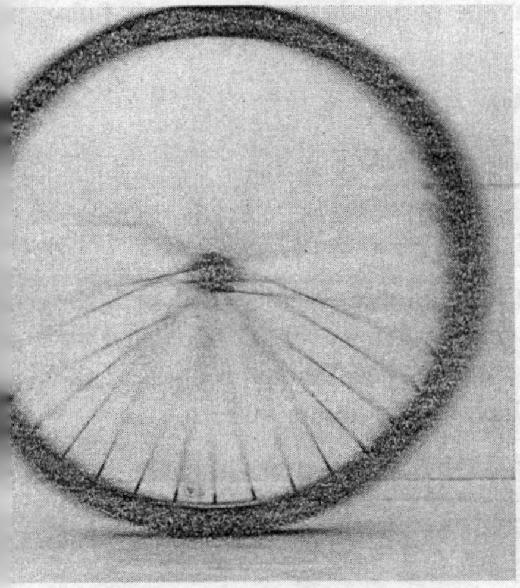

Fig. 12-4 A photograph of a rolling bicycle wheel. The spokes near the top of the wheel are more blurred than those near the bottom of the wheel because they are moving faster, as Fig. 12-3*c* shows.

Rolling as Pure Rotation

Figure 12-5 suggests another way to look at the rolling motion of a wheel—namely, as pure rotation about an axis that always extends through the point where the wheel contacts the street as the wheel moves. We consider the rolling motion to be pure rotation about an axis passing through point *P* in Fig. 12-3*c* and perpendicular to the plane of the figure. The vectors in Fig. 12-5 then represent the instantaneous velocities of points on the rolling wheel.

Question: What angular speed about this new axis will a stationary observer assign to a rolling bicycle wheel?

Answer: The same angular speed ω that the rider assigns to the wheel as she or he observes it in pure rotation about an axis through its center of mass.

To verify this answer, let us use it to calculate the linear speed of the top of the rolling wheel from the point of view of a stationary observer. If we call the wheel's radius *R*, the top is a distance 2*R* from the axis through *P* in Fig. 12-5, so the linear speed at the top should be (using Eq. 12-2)

$$v_{top} = (\omega)(2R) = 2(\omega R) = 2v_{com},$$

in exact agreement with Fig. 12-3*c*. You can similarly verify the linear speeds shown for the portion of the wheel at points *O* and *P* in Fig. 12-3*c*.

CHECKPOINT 1 The rear wheel on a clown's bicycle has twice the radius of the front wheel. (a) When the bicycle is moving, is the linear speed at the very top of the rear wheel greater than, less than, or the same as that of the front wheel? (b) Is the angular speed of the rear wheel greater than, less than, or the same as that of the front wheel?

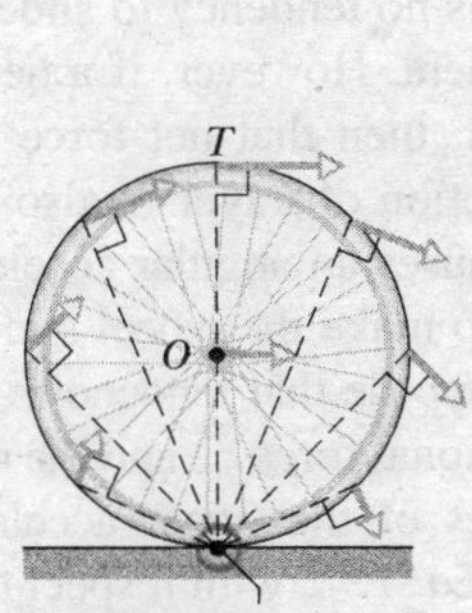

Rotation axis at *P*

Fig. 12-5 Rolling can be viewed as pure rotation, with angular speed ω, about an axis that always extends through *P*. The vectors show the instantaneous linear velocities of selected points on the rolling wheel. You can obtain the vectors by combining the translational and rotational motions as in Fig. 12-3.

12-2 The Kinetic Energy of Rolling

Let us now calculate the kinetic energy of the rolling wheel as measured by the stationary observer. If we view the rolling as pure rotation about an axis through *P* in Fig. 12-5, then from Eq. 11-27 we have

$$K = \tfrac{1}{2} I_P \omega^2, \tag{12-3}$$

in which ω is the angular speed of the wheel and I_P is the rotational inertia of the wheel about the axis through *P*. From the parallel-axis theorem of Eq. 11-29 ($I = I_{com} + Mh^2$), we have

$$I_P = I_{com} + MR^2, \tag{12-4}$$

in which *M* is the mass of the wheel, I_{com} is its rotational inertia about an axis through its center of mass, and R (the wheel's radius) is the perpendicular distance h. Substituting Eq. 12-4 into Eq. 12-3, we obtain

$$K = \tfrac{1}{2}I_{com}\omega^2 + \tfrac{1}{2}MR^2\omega^2,$$

and using the relation $v_{com} = \omega R$ (Eq. 12-2) yields

$$K = \tfrac{1}{2}I_{com}\omega^2 + \tfrac{1}{2}Mv_{com}^2. \qquad (12\text{-}5)$$

We can interpret the term $\frac{1}{2}I_{com}\omega^2$ as the kinetic energy associated with the rotation of the wheel about an axis through its center of mass (Fig. 12-3*a*), and the term $\frac{1}{2}Mv_{com}^2$ as the kinetic energy associated with the translational motion of the wheel's center of mass (Fig. 12-3*b*). Thus, we have the following rule:

> A rolling object has two types of kinetic energy: a rotational kinetic energy ($\frac{1}{2}I_{com}\omega^2$) due to its rotation about its center of mass and a translational kinetic energy ($\frac{1}{2}Mv_{com}^2$) due to translation of its center of mass.

Sample Problem 12-1

A uniform solid cylindrical disk, of mass $M = 1.4$ kg and radius $R = 8.5$ cm, rolls smoothly across a horizontal table at a speed of 15 cm/s. What is its kinetic energy K?

SOLUTION: Equation 12-5 gives the kinetic energy of a rolling object, but we need three **Key Ideas** to use it.

1. When we speak of the speed of a rolling object, we always mean the speed of the center of mass, so here $v_{com} = 15$ cm/s.
2. Equation 12-5 requires the angular speed ω of the rolling object, which we can relate to v_{com} with Eq. 12-2, writing $\omega = v_{com}/R$.
3. Equation 12-5 also requires the rotational inertia I_{com} of the object about its center of mass. From Table 11-2*c*, we find that for a solid disk $I_{com} = \frac{1}{2}MR^2$.

Now Eq. 12-5 gives us

$$K = \tfrac{1}{2}I_{com}\omega^2 + \tfrac{1}{2}Mv_{com}^2$$
$$= (\tfrac{1}{2})(\tfrac{1}{2}MR^2)(v_{com}/R)^2 + \tfrac{1}{2}Mv_{com}^2 = \tfrac{3}{4}Mv_{com}^2$$
$$= \tfrac{3}{4}(1.4\text{ kg})(0.15\text{ m/s})^2$$
$$= 0.024\text{ J} = 24\text{ mJ}. \qquad \text{(Answer)}$$

12-3 The Forces of Rolling

Friction and Rolling

If a wheel rolls at constant speed, as in Fig. 12-2, it has no tendency to slide at the point of contact P, and thus no frictional force acts there. However, if a net force acts on the rolling wheel to speed it up or to slow it, then that net force causes acceleration $\vec{a}_{com}$ of the center of mass along the direction of travel. It also causes the wheel to rotate faster or slower, which means it causes an angular acceleration α about the center of mass. These accelerations tend to make the wheel slide at P. Thus, a frictional force must act on the wheel at P to oppose that tendency.

If the wheel *does not* slide, the force is a *static* frictional force $\vec{f}_s$ and the motion is smooth rolling. We can then relate the magnitudes of the linear acceleration $\vec{a}_{com}$ and the angular acceleration α by differentiating Eq. 12-2 with respect to time (with R held constant). On the left side, dv_{com}/dt is a_{com}, and on the right side $d\omega/dt$ is α. So, for smooth rolling we have

$$a_{com} = \alpha R \quad \text{(smooth rolling motion)}. \qquad (12\text{-}6)$$

If the wheel *does* slide when the net force acts on it, the frictional force that acts at P in Fig. 12-2 is a *kinetic* frictional force $\vec{f}_k$. The motion then is not smooth rolling, and Eq. 12-6 does not apply to the motion. In this chapter we discuss only smooth rolling motion.

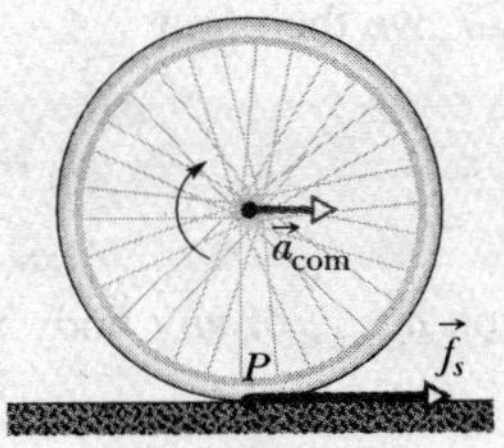

Fig. 12-6 A wheel rolls horizontally without sliding while accelerating with linear acceleration $\vec{a}_{com}$. A static frictional force $\vec{f}_s$ acts on the wheel at P, opposing its tendency to slide.

Figure 12-6 shows an example in which a wheel is being made to rotate faster while rolling along a flat surface, as on a bicycle wheel at the start of a race. The faster rotation tends to make the bottom of the wheel slide to the left at point P. A frictional force at P, directed to the right, opposes this tendency to slide. If the wheel does not slide, that frictional force is a static frictional force $\vec{f}_s$ (as shown), the motion is smooth rolling, and Eq. 12-6 applies to the motion. (Without friction, bicycle races would be stationary and very boring.)

If the wheel in Fig. 12-6 were made to rotate slower, as on a slowing bicycle, we would change the figure in two ways: The directions of the center-of-mass acceleration $\vec{a}_{com}$ and the frictional force $\vec{f}_s$ at point P would now be to the left.

Rolling down a Ramp

Figure 12-7 shows a round uniform body of mass M and radius R rolling smoothly down a ramp at angle θ, along an x axis. We want to find expressions for the body's acceleration $a_{com,x}$ down the ramp. We do this by using Newton's second law in both its linear version ($F_{net} = Ma$) and its angular version ($\tau_{net} = I\alpha$).

We start by drawing the forces on the body as shown in Fig. 12-7:

1. The gravitational force $\vec{F}_g$ on the body is directed downward. The tail of the vector is placed at the center of mass of the body. The component along the ramp is $F_g \sin\theta$, which is equal to $Mg \sin\theta$.
2. A normal force $\vec{N}$ is perpendicular to the ramp. It acts at the point of contact P, but in Fig. 12-7 the vector has been shifted along its direction until its tail is at the body's center of mass.
3. A static frictional force $\vec{f}_s$ acts at the point of contact P and is directed up the ramp. (Do you see why? If the body were to slide at P, it would slide *down* the ramp. Thus, the frictional force opposing the sliding must be *up* the ramp.)

We can write Newton's second law for components along the x axis in Fig. 12-7 ($F_{net,x} = ma_x$) as

$$f_s - Mg\sin\theta = Ma_{com,x}. \tag{12-7}$$

This equation contains two unknowns, f_s and $a_{com,x}$. (We should *not* assume that f_s is at its maximum value $f_{s,max}$. All we know is that the value of f_s is just right for the body to roll smoothly down the ramp, without sliding.)

We now wish to apply Newton's second law in angular form to the body's rotation about its center of mass. First, we shall use Eq. 11-33 ($\tau = r_\perp F$) to write the torques on the body about that point. The frictional force $\vec{f}_s$ has moment arm R and thus produces a torque Rf_s, which is positive because it tends to rotate the body counterclockwise in Fig. 12-7. Forces $\vec{F}_g$ and $\vec{N}$ have zero moment arms about the center of mass and thus produce zero torques. So we can write the angular form of Newton's second law ($\tau_{net} = I\alpha$) about an axis through the body's center of mass as

$$Rf_s = I_{com}\alpha. \tag{12-8}$$

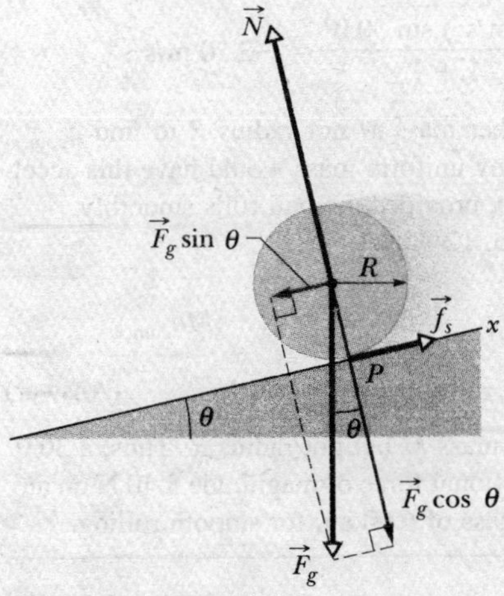

Fig. 12-7 A round uniform body of radius R rolls down a ramp. The forces that act on the body are the gravitational force $\vec{F}_g$, a normal force $\vec{N}$, and a frictional force $\vec{f}_s$ pointing up the ramp. (For clarity, vector $\vec{N}$ has been shifted in the direction it points until its tail is at the center of the body.)

This equation contains two unknowns, f_s and α.

Because the body is rolling smoothly, we can use Eq. 12-6 ($a_{com} = \alpha R$) to relate the unknowns $a_{com,x}$ and α. But we must be cautious because here $a_{com,x}$ is negative (in the negative direction of the x axis) and α is positive (counterclockwise). Thus we substitute $-a_{com,x}/R$ for α in Eq. 12-8. Then, solving for f_s, we obtain

$$f_s = -I_{com}\frac{a_{com,x}}{R^2}. \tag{12-9}$$

Substituting the right side of Eq. 12-9 for f_s in Eq. 12-7, we then find

$$a_{com,x} = -\frac{g \sin\theta}{1 + I_{com}/MR^2}. \tag{12-10}$$

We can use this equation to find the linear acceleration $a_{com,x}$ of any body rolling along an incline of angle θ with the horizontal.

CHECKPOINT 2: Disks A and B are identical and roll across a floor with equal speeds. Then disk A rolls up an incline, reaching a maximum height h, and disk B moves up an incline that is identical except that it is frictionless. Is the maximum height reached by disk B greater than, less than, or equal to h?

Sample Problem 12-2

A uniform ball, of mass $M = 6.00$ kg and radius R, rolls smoothly from rest down a ramp at angle $\theta = 30.0°$ (Fig. 12-7).

(a) The ball descends a vertical height $h = 1.20$ m to reach the bottom of the ramp. What is its speed at the bottom?

SOLUTION: One **Key Idea** here is that we can relate the ball's speed at the bottom to its kinetic energy K_f there. A second **Key Idea** is that the mechanical energy E of the ball–Earth system is conserved as the ball rolls down the ramp. The reason is that the only force doing work on the ball is the gravitational force, a conservative force. The normal force on the ball from the ramp does zero work because it is perpendicular to the ball's path. The frictional force on the ball from the ramp does not transfer any energy to thermal energy because the ball does not slide (it *rolls smoothly*).

Therefore, we can write the conservation of mechanical energy ($E^f = E^i$) as

$$K_f + U_f = K_i + U_i, \tag{12-11}$$

where subscripts f and i refer to the final values (at the bottom) and initial values (at rest), respectively. The gravitational potential energy is initially $U_i = Mgh$ (where M is the ball's mass) and finally $U_f = 0$. The kinetic energy is initially $K_i = 0$. For the final kinetic energy K_f, we need a third **Key Idea**: Because the ball rolls, the kinetic energy involves both translation *and* rotation, so we include them both with the right side of Eq. 12-5. Substituting all these expressions into Eq. 12-11 gives us

$$(\tfrac{1}{2}I_{com}\omega^2 + \tfrac{1}{2}Mv_{com}^2) + 0 = 0 + Mgh, \tag{12-12}$$

where I_{com} is the ball's rotational inertia about an axis through its center of mass, v_{com} is the requested speed at the bottom, and ω is the angular speed there.

Because the ball rolls smoothly, we can use Eq. 12-2 to substitute v_{com}/R for ω to reduce the unknowns in Eq. 12-12. Doing so, substituting $\frac{2}{5}MR^2$ for I_{com} (from Table 11-2f), and then solving for v_{com} give us

$$v_{com} = \sqrt{(\tfrac{10}{7})gh} = \sqrt{(\tfrac{10}{7})(9.8\ \text{m/s}^2)(1.2\ \text{m})}$$
$$= 4.1\ \text{m/s}. \qquad \text{(Answer)}$$

Note that the answer does not depend on the mass M or radius R of the ball.

(b) What are the magnitude and direction of the friction force on the ball as it rolls down the ramp?

SOLUTION: The **Key Idea** here is that, because the ball rolls smoothly, Eq. 12-9 gives the frictional force on the ball. However, first we need the ball's acceleration $a_{com,x}$ from Eq. 12-10:

$$a_{com,x} = -\frac{g\sin\theta}{1 + I_{com}/MR^2} = -\frac{g\sin\theta}{1 + \frac{2}{5}MR^2/MR^2}$$
$$= -\frac{(9.8\ \text{m/s}^2)\sin 30.0°}{1 + \frac{2}{5}} = -3.50\ \text{m/s}^2.$$

Note that we needed neither mass M nor radius R to find $a_{com,x}$. Thus, any size ball with any uniform mass would have this acceleration down a 30.0° ramp, provided the ball rolls smoothly.

We can now solve Eq. 12-9 as

$$f_s = -I_{com}\frac{a_{com,x}}{R^2} = -\tfrac{2}{5}MR^2\frac{a_{com,x}}{R^2} = -\tfrac{2}{5}Ma_{com,x}$$
$$= -\tfrac{2}{5}(6.00\ \text{kg})(-3.50\ \text{m/s}^2) = 8.40\ \text{N}. \qquad \text{(Answer)}$$

Note that here we needed mass M but not radius R. Thus, a 30.0° ramp would produce a frictional force of magnitude 8.40 N on any size ball with a uniform mass of 6.00 kg, for smooth rolling.

12-4 The Yo-Yo

A yo-yo is a physics lab that you can fit in your pocket. If a yo-yo rolls down its string for a distance h, it loses potential energy in amount mgh but gains kinetic energy in both translational ($\frac{1}{2}mv_{com}^2$) and rotational ($\frac{1}{2}I_{com}\omega^2$) forms. As it climbs back up, it loses kinetic energy and regains potential energy.

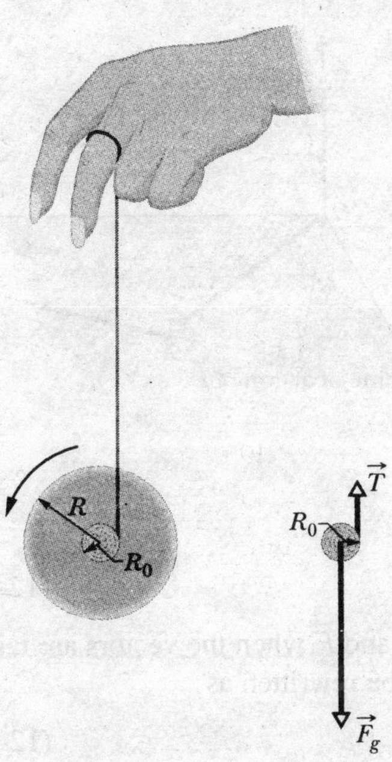

(a) (b)

Fig. 12-8 (a) A yo-yo, shown in cross section. The string, of assumed negligible thickness, is wound around an axle of radius R_0. (b) A free-body diagram for the falling yo-yo. Only the axle is shown.

In a modern yo-yo, the string is not tied to the axle but is looped around it. When the yo-yo "hits" the bottom of its string, an upward force on the axle from the string stops the descent. The yo-yo then spins, axle inside loop, with only rotational kinetic energy. The yo-yo keeps spinning ("sleeping") until you "wake it" by jerking on the string, causing the string to catch on the axle and the yo-yo to climb back up. The rotational kinetic energy of the yo-yo at the bottom of its string (and thus the sleeping time) can be considerably increased by throwing the yo-yo downward so that it starts down the string with initial speeds v_{com} and ω instead of rolling down from rest.

To find an expression for the linear acceleration a_{com} of a yo-yo rolling down a string, we could use Newton's second law just as we did for the body rolling down a ramp in Fig. 12-7. The analysis is the same except for the following:

1. Instead of rolling down a ramp at angle θ with the horizontal, the yo-yo rolls down a string at angle $\theta = 90°$ with the horizontal.
2. Instead of rolling on its outer surface at radius R, the yo-yo rolls on an axle of radius R_0 (Fig. 12-8a).
3. Instead of being slowed by frictional force $\vec{f}_s$, the yo-yo is slowed by the force $\vec{T}$ on it from the string (Fig. 12-8b).

The analysis would again lead us to Eq. 12-10. Therefore, let us just change the notation in Eq. 12-10 and set $\theta = 90°$ to write the linear acceleration as

$$a_{com} = -\frac{g}{1 + I_{com}/MR_0^2}, \tag{12-13}$$

where I_{com} is the yo-yo's rotational inertia about its center, and M is its mass. A yo-yo has the same downward acceleration when it is climbing back up the string, because the forces on it are still those shown in Fig. 12-8b.

12-5 Torque Revisited

In Chapter 11 we defined torque τ for a rigid body that can rotate around a fixed axis, with each particle in the body forced to move in a path that is a circle about that axis. We now expand the definition of torque to apply it to an individual particle that moves along any path relative to a fixed *point* (rather than a fixed axis). The path need no longer be a circle, and we must write the torque as a vector $\vec{\tau}$ that may have any direction.

Figure 12-9a shows such a particle at point A in the xy plane. A single force $\vec{F}$ in that plane acts on the particle, and the particle's position relative to the origin O is given by position vector $\vec{r}$. The torque $\vec{\tau}$ acting on the particle relative to the fixed point O is a vector quantity defined as

$$\vec{\tau} = \vec{r} \times \vec{F} \quad \text{(torque defined).} \tag{12-14}$$

We can evaluate the vector (or cross) product in this definition of $\vec{\tau}$ by using the rules for such products given in Section 3-7. To find the direction of $\vec{\tau}$, we slide the vector $\vec{F}$ (without changing its direction) until its tail is at the origin O, so that the two vectors in the vector product are tail to tail as in Fig. 12-9b. We then use the right-hand rule for vector products in Fig. 3-20a, sweeping the fingers of the right hand from $\vec{r}$ (the first vector in the product) into $\vec{F}$ (the second vector). The outstretched right thumb then gives the direction of $\vec{\tau}$. In Fig. 12-9b, the direction of $\vec{\tau}$ is in the positive direction of the z axis.

To determine the magnitude of $\vec{\tau}$, we apply the general result of Eq. 3-27

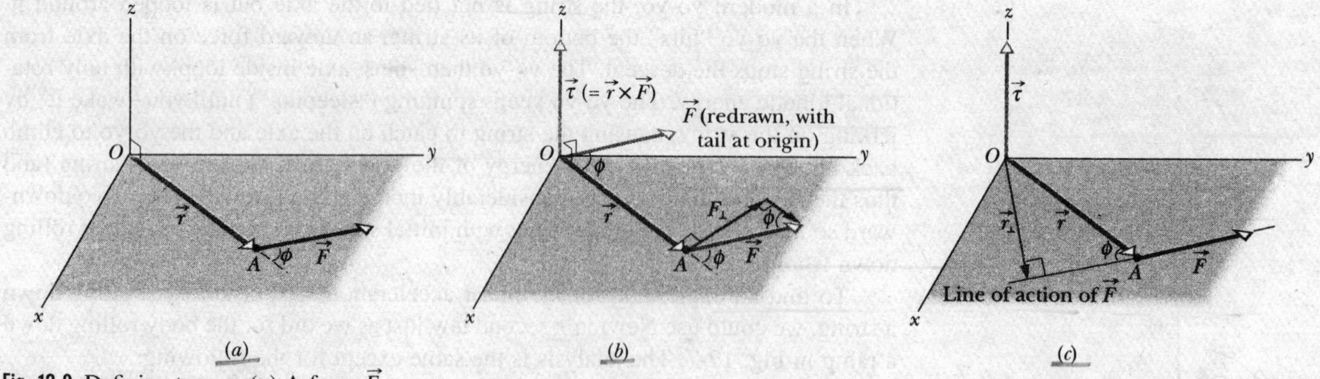

Fig. 12-9 Defining torque. (*a*) A force $\vec{F}$, lying in the *xy* plane, acts on a particle at point *A*. (*b*) This force produces a torque $\vec{\tau}\ (= \vec{r} \times \vec{F})$ on the particle with respect to the origin *O*. By the right-hand rule for vector (cross) products, the torque vector points in the positive direction of *z*. Its magnitude is given by $rF_\perp$ in (*b*) and by $r_\perp F$ in (*c*).

($c = ab \sin \phi$), finding

$$\tau = rF \sin \phi, \tag{12-15}$$

where ϕ is the angle between the directions of $\vec{r}$ and $\vec{F}$ when the vectors are tail to tail. From Fig. 12-9*b*, we see that Eq. 12-15 can be rewritten as

$$\tau = rF_\perp, \tag{12-16}$$

where $F_\perp\ (= F \sin \phi)$ is the component of $\vec{F}$ perpendicular to $\vec{r}$. From Fig. 12-9*c*, we see that Eq. 12-15 can also be rewritten as

$$\tau = r_\perp F, \tag{12-17}$$

where $r_\perp\ (= r \sin \phi)$ is the moment arm of $\vec{F}$ (the perpendicular distance between *O* and the line of action of $\vec{F}$).

Sample Problem 12-3

In Fig. 12-10*a*, three forces, each of magnitude 2.0 N, act on a particle. The particle is in the *xz* plane at point *A* given by position vector $\vec{r}$, where $r = 3.0$ m and $\theta = 30°$. Force $\vec{F}_1$ is parallel to the *x* axis, force $\vec{F}_2$ is parallel to the *z* axis, and force $\vec{F}_3$ is parallel to the *y* axis. What is the torque, about the origin *O*, due to each force?

SOLUTION: The Key Idea here is that, because the three force vectors do not lie in a plane, we cannot evaluate their torques as in Chapter 11. Instead, we must use vector (or cross) products, with magnitudes given by Eq. 12-15 ($\tau = rF \sin \phi$) and directions given by the right-hand rule for vector products.

Because we want the torques with respect to the origin *O*, the vector $\vec{r}$ required for each cross product is the given position vector. To determine the angle ϕ between the direction of $\vec{r}$ and the direction of each force, we shift the force vectors of Fig. 12-10*a*, each in turn, so that their tails are at the origin. Figures 12-10*b*, *c*,

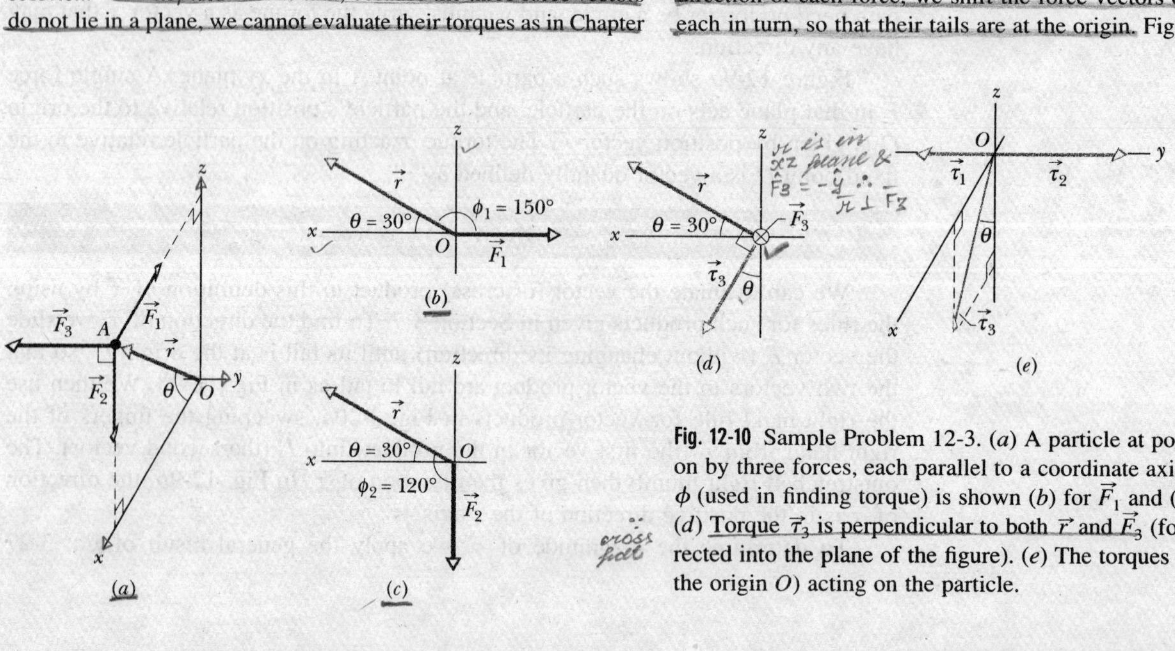

Fig. 12-10 Sample Problem 12-3. (*a*) A particle at point *A* is acted on by three forces, each parallel to a coordinate axis. The angle ϕ (used in finding torque) is shown (*b*) for $\vec{F}_1$ and (*c*) for $\vec{F}_2$. (*d*) Torque $\vec{\tau}_3$ is perpendicular to both $\vec{r}$ and $\vec{F}_3$ (force $\vec{F}_3$ is directed into the plane of the figure). (*e*) The torques (relative to the origin *O*) acting on the particle.

and d, which are direct views of the xz plane, show the shifted force vectors $\vec{F}_1$, $\vec{F}_2$, and $\vec{F}_3$, respectively. (Note how much easier the angles are to see.) In Fig. 12-10d, the angle between the directions of $\vec{r}$ and $\vec{F}_3$ is 90° and the symbol ⊗ means $\vec{F}_3$ is directed into the page. If it were directed out of the page, it would be represented with the symbol ⊙.

Now, applying Eq. 12-15 for each force, we find the magnitudes of the torques to be

$$\tau_1 = rF_1 \sin \phi_1 = (3.0 \text{ m})(2.0 \text{ N})(\sin 150°) = 3.0 \text{ N}\cdot\text{m},$$

$$\tau_2 = rF_2 \sin \phi_2 = (3.0 \text{ m})(2.0 \text{ N})(\sin 120°) = 5.2 \text{ N}\cdot\text{m},$$

and

$$\tau_3 = rF_3 \sin \phi_3 = (3.0 \text{ m})(2.0 \text{ N})(\sin 90°) = 6.0 \text{ N}\cdot\text{m}. \quad \text{(Answer)}$$

To find the directions of these torques, we use the right-hand rule, placing the fingers of the right hand so as to rotate $\vec{r}$ into $\vec{F}$ through the *smaller* of the two angles between their directions. The thumb points in the direction of the torque. Thus $\vec{\tau}_1$ is directed into the page in Fig. 12-10b; $\vec{\tau}_2$ is directed out of the page in Fig. 12-10c; and $\vec{\tau}_3$ is directed as shown in Fig. 12-10d. All three torque vectors are shown in Fig. 12-10e.

CHECKPOINT 3: The position vector $\vec{r}$ of a particle points along the positive direction of the z axis. If the torque on the particle is (a) zero, (b) in the negative direction of x, and (c) in the negative direction of y, in what direction is the force producing the torque?

PROBLEM-SOLVING TACTICS

Tactic 1: *Vector Products and Torques*
Equation 12-15 for torques is our first application of the vector (or cross) product. You might want to review Section 3-7, where the rules for the vector product are given. In that section, Problem-Solving Tactic 5 lists many common errors in finding the direction of a vector product.

Keep in mind that a torque is calculated *with respect to* (or *about*) a point, which must be known if the value of the torque is to be meaningful. Changing the point can change the torque in both magnitude and direction. For example, in Sample Problem 12-3, the torques due to the three forces are calculated about the origin O. You can show that the torques due to the same three forces are all zero if they are calculated about point A (at the position of the particle), because then $r = 0$ for each force.

12-6 Angular Momentum

NOTE: The unit of P is kgm/s & L = r×p has unit kg m²/s

Recall that the concept of linear momentum $\vec{p}$ and the principle of conservation of linear momentum are extremely powerful tools. They allow us to predict the outcome of, say, a collision of two cars without knowing the details of the collision. Here we begin a discussion of the angular counterpart of $\vec{p}$. We shall end the chapter by discussing the angular counterpart of the conservation principle.

Figure 12-11 shows a particle of mass m with linear momentum $\vec{p}\ (= m\vec{v})$ as it passes through point A in the xy plane. The **angular momentum** $\vec{\ell}$ of this particle with respect to the origin O is a vector quantity defined as

$$\vec{\ell} = \vec{r} \times \vec{p} = m(\vec{r} \times \vec{v}) \quad \text{(angular momentum defined)}, \tag{12-18}$$

where $\vec{r}$ is the position vector of the particle with respect to O. As the particle moves relative to O in the direction of its momentum $\vec{p}\ (= m\vec{v})$, position vector $\vec{r}$ rotates around O. Note carefully that to have angular momentum about O, the particle does *not* itself have to rotate around O. Comparison of Eqs. 12-14 and 12-18 shows that angular momentum bears the same relation to linear momentum that torque does to force. The SI unit of angular momentum is the kilogram-meter-squared per second ($\text{kg}\cdot\text{m}^2/\text{s}$), equivalent to the joule-second ($\text{J}\cdot\text{s}$).

To find the direction of the angular momentum vector $\vec{\ell}$ in Fig. 12-11, we slide the vector $\vec{p}$ until its tail is at the origin O. Then we use the right-hand rule for

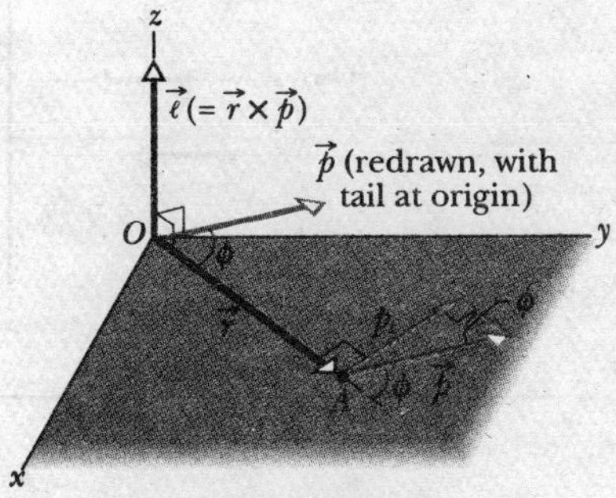

(a)

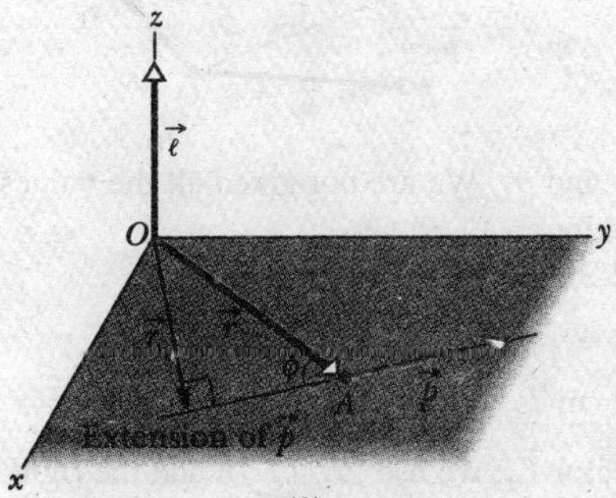

(b)

Fig. 12-11 Defining angular momentum. A particle passing through point A has linear momentum $\vec{p}\ (= m\vec{v})$, with the vector $\vec{p}$ lying in the xy plane. The particle has angular momentum $\vec{\ell}\ (= \vec{r} \times \vec{p})$ with respect to the origin O. By the right-hand rule, the angular momentum vector points in the positive direction of z. (a) The magnitude of $\vec{\ell}$ is given by $\ell = rp_\perp = rmv_\perp$. ($b$) The magnitude of $\vec{\ell}$ is also given by $\ell = r_\perp p = r_\perp mv$.

vector products, sweeping the fingers from $\vec{r}$ into $\vec{p}$. The outstretched thumb then shows that the direction of $\vec{\ell}$ is in the positive direction of the z axis in Fig. 12-11. This positive direction is consistent with the counterclockwise rotation of the particle's position vector $\vec{r}$ about the z axis, as the particle continues to move. (A negative direction of $\vec{\ell}$ would be consistent with a clockwise rotation of $\vec{r}$ about the z axis.)

To find the magnitude of $\vec{\ell}$, we use the general result of Eq. 3-27, to write

$$\ell = rmv \sin \phi, \tag{12-19}$$

where ϕ is the angle between $\vec{r}$ and $\vec{p}$ when these two vectors are tail to tail. From Fig. 12-11a, we see that Eq. 12-19 can be rewritten as

$$\ell = rp_\perp = rmv_\perp, \tag{12-20}$$

where $p_\perp$ is the component of $\vec{p}$ perpendicular to $\vec{r}$ and $v_\perp$ is the component of $\vec{v}$ perpendicular to $\vec{r}$. From Fig. 12-11b, we see that Eq. 12-19 can also be rewritten as

$$\ell = r_\perp p = r_\perp mv, \tag{12-21}$$

where $r_\perp$ is the perpendicular distance between O and the extension of $\vec{p}$.

Just as is true for torque, angular momentum has meaning only with respect to a specified origin. Moreover, if the particle in Fig. 12-11 did not lie in the xy plane, or if the linear momentum $\vec{p}$ of the particle did not also lie in that plane, the angular momentum $\vec{\ell}$ would not be parallel to the z axis. The direction of the angular momentum vector is always perpendicular to the plane formed by the position and linear momentum vectors $\vec{r}$ and $\vec{p}$.

CHECKPOINT 4: In part a of the figure, particles 1 and 2 move around point O in opposite directions, in circles with radii 2 m and 4 m. In part b, particles 3 and 4 travel in the same direction, along straight lines at perpendicular distances of 4 m and 2 m from point O. Particle 5 moves directly away from O. All five particles have the same mass and the same constant speed.
(a) Rank the particles according to the magnitudes of their angular momentum about point O, greatest first.
(b) Which particles have negative angular momentum about point O?

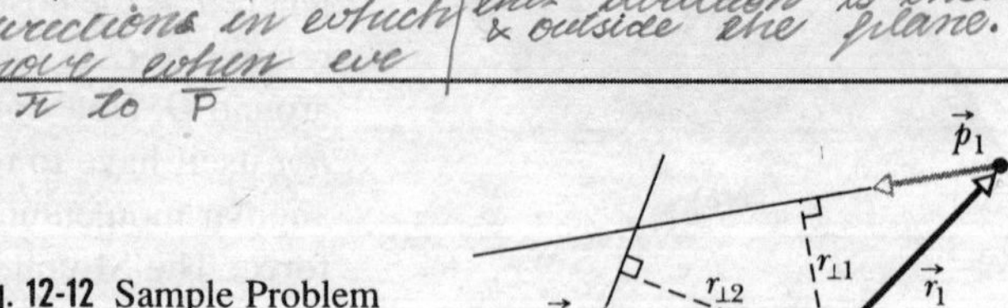

Sample Problem 12-4

Figure 12-12 shows an overhead view of two particles moving at constant momentum along horizontal paths. Particle 1, with momentum magnitude $p_1 = 5.0$ kg · m/s, has position vector $\vec{r}_1$ and will pass 2.0 m from point O. Particle 2, with momentum magnitude $p_2 = 2.0$ kg · m/s, has position vector $\vec{r}_2$ and will pass 4.0 m from point O. What is the net angular momentum $\vec{L}$ about point O of the two-particle system?

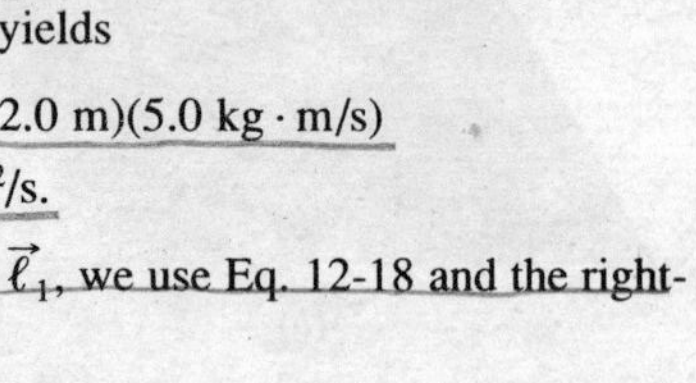

Fig. 12-12 Sample Problem 12-4. Two particles pass near point O.

SOLUTION: The **Key Idea** here is that to find $\vec{L}$, we can first find the individual angular momenta $\vec{\ell}_1$ and $\vec{\ell}_2$ and then add them. To evaluate their magnitudes, we can use any one of Eqs. 12-18 through 12-21. However, Eq. 12-21 is easiest, because we are given the perpendicular distances $r_{\perp 1}$ (= 2.0 m) and $r_{\perp 2}$ (= 4.0 m) and the momentum magnitudes p_1 and p_2. We are not given all the values required for the other equations.

For particle 1, Eq. 12-21 yields

$$\ell_1 = r_{\perp 1} p_1 = (2.0 \text{ m})(5.0 \text{ kg} \cdot \text{m/s}) = 10 \text{ kg} \cdot \text{m}^2/\text{s}.$$

To find the direction of vector $\vec{\ell}_1$, we use Eq. 12-18 and the right-hand rule for vector products. For $\vec{r}_1 \times \vec{p}_1$, the vector product is out of the page, perpendicular to the plane of Fig. 12-12. This is the positive direction, consistent with the counterclockwise rotation of the particle's position vector $\vec{r}_1$ around O as particle 1 moves. Thus, the angular momentum vector for particle 1 is

$$\ell_1 = +10 \text{ kg} \cdot \text{m}^2/\text{s}.$$

Similarly, the magnitude of $\vec{\ell}_2$ is

$$\ell_2 = r_{\perp 2} p_2 = (4.0 \text{ m})(2.0 \text{ kg} \cdot \text{m/s}) = 8.0 \text{ kg} \cdot \text{m}^2/\text{s},$$

and the vector product $\vec{r}_2 \times \vec{p}_2$ is into the page, which is the negative direction, consistent with the clockwise rotation of $\vec{r}_2$ around O as particle 2 moves. Thus, the angular momentum vector for particle 2 is

$$\ell_2 = -8.0 \text{ kg} \cdot \text{m}^2/\text{s}.$$

The net angular momentum for the two-particle system is then

$$L = \ell_1 + \ell_2 = +10 \text{ kg} \cdot \text{m}^2/\text{s} + (-8.0 \text{ kg} \cdot \text{m}^2/\text{s}) = +2.0 \text{ kg} \cdot \text{m}^2/\text{s}. \quad \text{(Answer)}$$

The plus sign means that the system's net angular momentum about point O is out of the page.

12-7 Newton's Second Law in Angular Form

Newton's second law written in the form

$$\vec{F}_{\text{net}} = \frac{d\vec{p}}{dt} \quad \text{(single particle)} \tag{12-22}$$

expresses the close relation between force and linear momentum for a single particle. We have seen enough of the parallelism between linear and angular quantities to be pretty sure that there is also a close relation between torque and angular momentum. Guided by Eq. 12-22, we can even guess that it must be

$$\vec{\tau}_{\text{net}} = \frac{d\vec{\ell}}{dt} \quad \text{(single particle)}. \tag{12-23}$$

Equation 12-23 is indeed an angular form of Newton's second law for a single particle:

The (vector) sum of all the torques acting on a particle is equal to the time rate of change of the angular momentum of that particle.

Equation 12-23 has no meaning unless the torques $\vec{\tau}$ and the angular momentum $\vec{\ell}$ are defined with respect to the same origin.

Proof of Equation 12-23

We start with Eq. 12-18, the definition of the angular momentum of a particle:

$$\vec{\ell} = m(\vec{r} \times \vec{v}),$$

where $\vec{r}$ is the position vector of the particle and $\vec{v}$ is the velocity of the particle. Differentiating* each side with respect to time t yields

$$\frac{d\vec{\ell}}{dt} = m\left(\vec{r} \times \frac{d\vec{v}}{dt} + \frac{d\vec{r}}{dt} \times \vec{v}\right). \tag{12-24}$$

However, $d\vec{v}/dt$ is the acceleration $\vec{a}$ of the particle, and $d\vec{r}/dt$ is its velocity $\vec{v}$. Thus, we can rewrite Eq. 12-24 as

$$\frac{d\vec{\ell}}{dt} = m(\vec{r} \times \vec{a} + \vec{v} \times \vec{v}).$$

*In differentiating a vector product, be sure not to change the order of the two quantities (here $\vec{r}$ and $\vec{v}$) that form that product. (See Eq. 3-28.)

Now $\vec{v} \times \vec{v} = 0$ (the vector product of any vector with itself is zero because the angle between the two vectors is necessarily zero). This leads to

$$\frac{d\vec{\ell}}{dt} = m(\vec{r} \times \vec{a}) = \vec{r} \times m\vec{a}.$$

We now use Newton's second law ($\vec{F}_{net} = m\vec{a}$) to replace $m\vec{a}$ with its equal, the vector sum of the forces that act on the particle, obtaining

$$\frac{d\vec{\ell}}{dt} = \vec{r} \times \vec{F}_{net} = \sum(\vec{r} \times \vec{F}). \tag{12-25}$$

Here the symbol Σ indicates that we must sum the vector products $\vec{r} \times \vec{F}$ for all the forces. However, from Eq. 12-14, we know that each one of those vector products is the torque associated with one of the forces. Therefore, Eq. 12-25 tells us that

$$\vec{\tau}_{net} = \frac{d\vec{\ell}}{dt}.$$

This is Eq. 12-23, the relation that we set out to prove.

CHECKPOINT 5: The figure shows the position vector $\vec{r}$ of a particle at a certain instant, and four choices for the direction of a force that is to accelerate the particle. All four choices lie in the xy plane. (a) Rank the choices according to the magnitude of the time rate of change ($d\vec{\ell}/dt$) they produce in the angular momentum of the particle about point O, greatest first. (b) Which choice results in a negative rate of change about O?

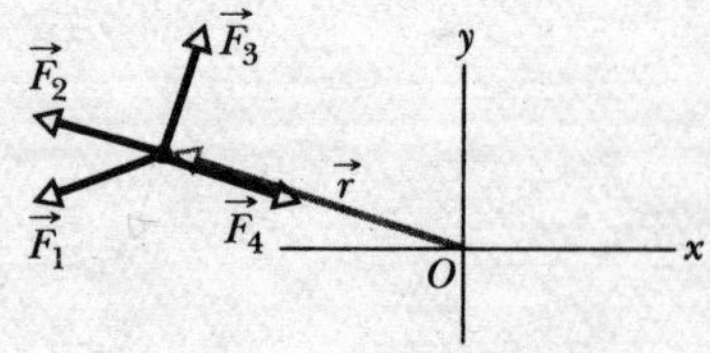

Sample Problem 12-5

In Fig. 12-13, a penguin of mass m falls from rest at point A, a horizontal distance D from the origin O of an xyz coordinate system. (The positive direction of the z axis is directly outward from the plane of the figure.)

(a) What is the angular momentum $\vec{\ell}$ of the falling penguin about O?

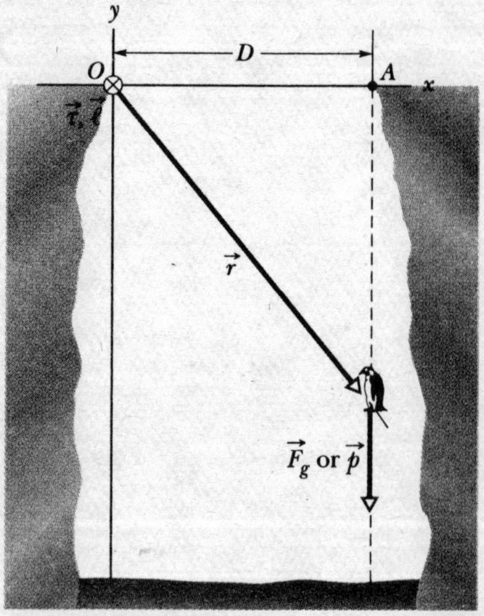

Fig. 12-13 Sample Problem 12-5. A penguin falls vertically from point A. The torque $\vec{\tau}$ and the angular momentum $\vec{\ell}$ of the falling penguin with respect to the origin O are directed into the plane of the figure at O.

SOLUTION: One **Key Idea** here is that we can treat the penguin as a particle, and thus its angular momentum $\vec{\ell}$ is given by Eq. 12-18 ($\vec{\ell} = \vec{r} \times \vec{p}$), where $\vec{r}$ is the penguin's position vector (extending from O to the penguin) and $\vec{p}$ is the penguin's linear momentum. The second **Key Idea** is that the penguin has *angular* momentum about O even though it moves in a straight line, because $\vec{r}$ rotates about O as the penguin falls.

To find the magnitude of $\vec{\ell}$, we can use any one of the scalar equations derived from Eq. 12-18—namely, Eqs. 12-19 through 12-21. However, Eq. 12-21 ($\ell = r_\perp mv$) is easiest because the perpendicular distance $r_\perp$ between O and an extension of vector $\vec{p}$ is the given distance D. A third **Key Idea** is an old one: The speed of an object that has fallen from rest for a time t is $v = gt$. We can now write Eq. 12-21 in terms of given quantities as

$$\ell = r_\perp mv = Dmgt. \quad \text{(Answer)}$$

To find the direction of $\vec{\ell}$, we use the right-hand rule for the vector product $\vec{r} \times \vec{p}$ in Eq. 12-18. Mentally shift $\vec{p}$ until its tail is at the origin, and then use the fingers of your right hand to rotate $\vec{r}$ into $\vec{p}$ through the smaller angle between the two vectors. Your outstretched thumb then points into the plane of the figure, indicating that the product $\vec{r} \times \vec{p}$ and thus also $\vec{\ell}$ are directed into that plane, in the negative direction of the z axis. We represent $\vec{\ell}$ with an encircled cross ⊗ at O. The vector $\vec{\ell}$ changes with time in magnitude only; its direction remains unchanged.

(b) About the origin O, what is the torque $\vec{\tau}$ on the penguin due to the gravitational force $\vec{F}_g$?

SOLUTION: One **Key Idea** here is that the torque is given by Eq. 12-14 ($\vec{\tau} = \vec{r} \times \vec{F}$), where now the force is $\vec{F}_g$. An associated **Key Idea** is that $\vec{F}_g$ causes a torque on the penguin, even though the penguin moves in a straight line, because $\vec{r}$ rotates about O as the penguin moves.

To find the magnitude of $\vec{\tau}$, we can use any one of the scalar equations derived from Eq. 12-14—namely, Eqs. 12-15 through 12-17. However, Eq. 12-17 ($\tau = r_{\perp}F$) is easiest because the perpendicular distance $r_{\perp}$ between O and the line of action of $\vec{F}_g$ is the given distance D. So, substituting D and using mg for the magnitude of $\vec{F}_g$, we can write Eq. 12-17 as

$$\tau = DF_g = Dmg. \qquad \text{(Answer)}$$

Using the right-hand rule for the vector product $\vec{r} \times \vec{F}$ in Eq. 12-14, we find that the direction of $\vec{\tau}$ is the negative direction of the z axis, the same as $\vec{\ell}$.

The results we obtained in parts (a) and (b) must be consistent with Newton's second law in the angular form of Eq. 12-23 ($\vec{\tau}_{net} = d\vec{\ell}/dt$). To check the magnitudes we got, we write Eq. 12-23 in component form for the z axis and then substitute our result $\ell = Dmgt$. We find

$$\tau = \frac{d\ell}{dt} = \frac{d(Dmgt)}{dt} = Dmg,$$

which is the magnitude we found for $\vec{\tau}$. To check the directions, we note that Eq. 12-23 tells us that $\vec{\tau}$ and $d\vec{\ell}/dt$ must have the same direction. So $\vec{\tau}$ and $\vec{\ell}$ must also have the same direction, which is what we found.

12-8 The Angular Momentum of a System of Particles

Now we turn our attention to the angular momentum of a system of particles with respect to an origin. The total angular momentum $\vec{L}$ of the system is the (vector) sum of the angular momenta $\vec{\ell}$ of the individual particles:

$$\vec{L} = \vec{\ell}_1 + \vec{\ell}_2 + \vec{\ell}_3 + \cdots + \vec{\ell}_n = \sum_{i=1}^{n} \vec{\ell}_i, \qquad (12\text{-}26)$$

in which i (= 1, 2, 3, . . .) labels the particles.

With time, the angular momenta of individual particles may change, either because of interactions within the system (between the individual particles) or because of influences that may act on the system from the outside. We can find the change in $\vec{L}$ as these changes take place by taking the time derivative of Eq. 12-26. Thus,

$$\frac{d\vec{L}}{dt} = \sum_{i=1}^{n} \frac{d\vec{\ell}_i}{dt}. \qquad (12\text{-}27)$$

From Eq. 12-23, we see that $d\vec{\ell}_i/dt$ is equal to the net torque $\vec{\tau}_{net,i}$ on the ith particle. We can rewrite Eq. 12-27 as

$$\frac{d\vec{L}}{dt} = \sum_{i=1}^{n} \vec{\tau}_{net,i}. \qquad (12\text{-}28)$$

That is, the rate of change of the system's angular momentum $\vec{L}$ is equal to the vector sum of the torques on its individual particles. Those torques include *internal torques* (due to forces between the particles) and *external torques* (due to forces on the particles from bodies external to the system). However, the forces between the particles always come in third-law force pairs so their torques sum to zero. Thus, the only torques that can change the total angular momentum $\vec{L}$ of the system are the external torques acting on the system.

Let $\vec{\tau}_{net}$ represent the net external torque, the vector sum of all external torques on all particles in the system. Then we can write Eq. 12-28 as

$$\vec{\tau}_{net} = \frac{d\vec{L}}{dt} \quad \text{(system of particles)}. \qquad (12\text{-}29)$$

This equation is Newton's second law for rotation in angular form, for a system of particles. It says:

> The net external torque $\vec{\tau}_{net}$ acting on a system of particles is equal to the time rate of change of the system's total angular momentum $\vec{L}$.

Equation 12-29 is analogous to $\vec{F}_{net} = d\vec{P}/dt$ (Eq. 9-23) but requires extra caution: Torques and the system's angular momentum must be measured relative to the same origin. If the center of mass of the system is not accelerating relative to an inertial frame, that origin can be any point. However, if the center of mass of the system *is* accelerating, the origin can be only at that center of mass. As an example, consider a wheel as the system of particles. If the wheel is rotating about an axis that is fixed relative to the ground, then the origin for applying Eq. 12-29 can be any point that is stationary relative to the ground. However, if the wheel is rotating about an axis that is accelerating (such as when the wheel rolls down a ramp), then the origin can be only at the wheel's center of mass.

12-9 The Angular Momentum of a Rigid Body Rotating About a Fixed Axis

We next evaluate the angular momentum of a system of particles that form a rigid body that rotates about a fixed axis. Figure 12-14*a* shows such a body. The fixed axis of rotation is the z axis, and the body rotates about it with constant angular speed ω. We wish to find the angular momentum of the body about that axis.

We can find the angular momentum by summing the z components of the angular momenta of the mass elements in the body. In Fig. 12-14*a*, a typical mass element, of mass Δm_i, moves around the z axis in a circular path. The position of the mass element is located relative to the origin O by position vector $\vec{r}_i$. The radius of the mass element's circular path is $r_{\perp i}$, the perpendicular distance between the element and the z axis.

The magnitude of the angular momentum $\vec{\ell}_i$ of this mass element, with respect to O, is given by Eq. 12-19:

$$\ell_i = (r_i)(p_i)(\sin 90°) = (r_i)(\Delta m_i\, v_i),$$

where p_i and v_i are the linear momentum and linear speed of the mass element, and 90° is the angle between $\vec{r}_i$ and $\vec{p}_i$. The angular momentum vector $\vec{\ell}_i$ for the mass element in Fig. 12-14*a* is shown in Fig. 12-14*b*; its direction must be perpendicular to those of $\vec{r}_i$ and $\vec{p}_i$.

We are interested in the component of $\vec{\ell}_i$ that is parallel to the rotation axis, here the z axis. That z component is

$$\ell_{iz} = \ell_i \sin\theta = (r_i \sin\theta)(\Delta m_i\, v_i) = r_{\perp i}\, \Delta m_i\, v_i.$$

The z component of the angular momentum for the rotating rigid body as a whole is found by adding up the contributions of all the mass elements that make up the body. Thus, because $v = \omega r_\perp$, we may write

$$L_z = \sum_{i=1}^{n} \ell_{iz} = \sum_{i=1}^{n} \Delta m_i\, v_i r_{\perp i} = \sum_{i=1}^{n} \Delta m_i(\omega r_{\perp i}) r_{\perp i}$$

$$= \omega\left(\sum_{i=1}^{n} \Delta m_i\, r_{\perp i}^2\right). \tag{12-30}$$

We can remove ω from the summation here because it has the same value for all points of the rotating rigid body.

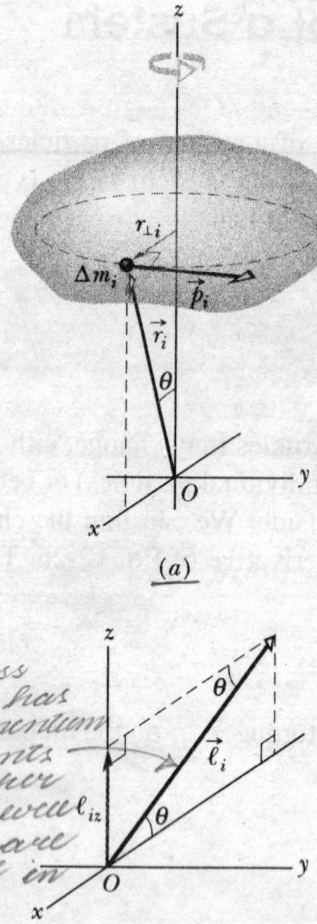

Fig. 12-14 (*a*) A rigid body rotates about the z axis with angular speed ω. A mass element of mass Δm_i within the body moves about the z axis in a circle with radius $r_{\perp i}$. The mass element has linear momentum $\vec{p}_i$, and it is located relative to the origin O by position vector $\vec{r}_i$. Here the mass element is shown when $r_{\perp i}$ is parallel to the x axis. (*b*) The angular momentum $\vec{\ell}_i$, with respect to O, of the mass element in (*a*). The z component ℓ_{iz} is also shown.

TABLE 12-1 More Corresponding Variables and Relations for Translational and Rotational Motion[a]

Translational		Rotational	
Force	$\vec{F}$	Torque	$\vec{\tau}\ (= \vec{r} \times \vec{F})$
Linear momentum	$\vec{p}$	Angular momentum	$\vec{\ell}\ (= \vec{r} \times \vec{p})$
Linear momentum[b]	$\vec{P}\ (= \Sigma \vec{p}_i)$	Angular momentum[b]	$\vec{L}\ (= \Sigma \vec{\ell}_i)$
Linear momentum[b]	$\vec{P} = M\vec{v}_{com}$	Angular momentum[c]	$L = I\omega$
Newton's second law[b]	$\vec{F}_{net} = \dfrac{d\vec{P}}{dt}$	Newton's second law[b]	$\vec{\tau}_{net} = \dfrac{d\vec{L}}{dt}$
Conservation law[d]	$\vec{P}$ = a constant	Conservation law[d]	$\vec{L}$ = a constant

[a] See also Table 11-3. [b] For systems of particles, including rigid bodies.
[c] For a rigid body about a fixed axis, with L being the component along that axis.
[d] For a closed, isolated system.

The quantity $\Sigma\,\Delta m_i\, r_{\perp i}^2$ in Eq. 12-30 is the rotational inertia I of the body about the fixed axis (see Eq. 11-26). Thus Eq. 12-30 reduces to

$$L = I\omega \qquad \text{(rigid body, fixed axis)}. \tag{12-31}$$

We have dropped the subscript z, but you must remember that the angular momentum defined by Eq. 12-31 is the angular momentum about the rotation axis. Also, I in that equation is the rotational inertia about that same axis.

Table 12-1, which supplements Table 11-3, extends our list of corresponding linear and angular relations.

CHECKPOINT 6: In the figure, a disk, a hoop, and a solid sphere are made to spin about fixed central axes (like a top) by means of strings wrapped around them, with the strings producing the same constant tangential force $\vec{F}$ on all three objects. The three objects have the same mass and radius, and they are initially stationary. Rank the objects according to (a) their angular momentum about their central axes and (b) their angular speed, greatest first, when the strings have been pulled for a certain time t.

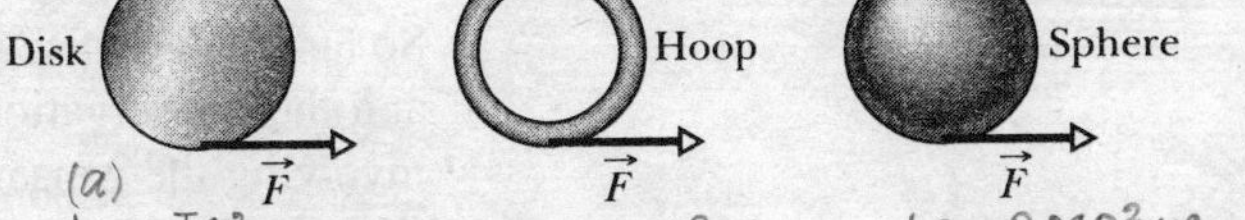

Sample Problem 12-6

George Washington Gale Ferris, Jr., a civil engineering graduate from Rensselaer Polytechnic Institute, built the original Ferris wheel (Fig. 12-15) for the 1893 World's Columbian Exposition in Chicago. The wheel, an astounding engineering construction at the time, carried 36 wooden cars, each holding as many as 60 passengers, around a circle of radius $R = 38$ m. The mass of each car was about 1.1×10^4 kg. The mass of the wheel's structure was about 6.0×10^5 kg, which was mostly in the circular grid from which the cars were suspended. The cars were loaded 6 at a time, and once all 36 cars were full, the wheel made a complete rotation at an angular speed ω_F in about 2 min.

Fig. 12-15 Sample Problem 12-6. The original Ferris wheel, built in 1893 near the University of Chicago, towered over the surrounding buildings.

(a) Estimate the magnitude L of the angular momentum of the wheel and its passengers while the wheel rotated at ω_F.

SOLUTION: The **Key Idea** here is that we can treat the wheel, cars, and passengers as a rigid object rotating about a fixed axis, at the wheel's axle. Then Eq. 12-31 ($L = I\omega$) gives the magnitude of the angular momentum of that object. We need to find the rotational inertia I of this object and the angular speed ω_F.

To find I, let us start with the loaded cars. Because we can treat them as particles, at distance R from the axis of rotation, we know from Eq. 11-26 that their rotational inertia is $I_{pc} = M_{pc}R^2$, where M_{pc} is their total mass. Let us assume that the 36 cars are each filled with 60 passengers, each of mass 70 kg. Then their total mass is

$$M_{pc} = 36[1.1 \times 10^4 \text{ kg} + 60(70 \text{ kg})] = 5.47 \times 10^5 \text{ kg}$$

and their rotational inertia is

$$I_{pc} = M_{pc}R^2 = (5.47 \times 10^5 \text{ kg})(38 \text{ m})^2 = 7.90 \times 10^8 \text{ kg}\cdot\text{m}^2.$$

Next we consider the structure of the wheel. Let us assume that the rotational inertia of the structure is due mainly to the circular grid suspending the cars. Further, let us assume that the grid forms a hoop of radius R, with a mass M_{hoop} of 3.0×10^5 kg (half the wheel's mass). From Table 11-2*a*, the rotational inertia of the hoop is

$$I_{hoop} = M_{hoop}R^2 = (3.0 \times 10^5 \text{ kg})(38 \text{ m})^2 = 4.33 \times 10^8 \text{ kg}\cdot\text{m}^2.$$

The combined rotational inertia I of the cars, passengers, and hoop is then

$$I = I_{pc} + I_{hoop} = 7.90 \times 10^8 \text{ kg}\cdot\text{m}^2 + 4.33 \times 10^8 \text{ kg}\cdot\text{m}^2 = 1.22 \times 10^9 \text{ kg}\cdot\text{m}^2.$$

To find the rotational speed ω_F, we use Eq. 11-5 ($\omega_{avg} = \Delta\theta/\Delta t$). Here the wheel goes through an angular displacement of $\Delta\theta = 2\pi$ rad in a time period $\Delta t = 2$ min. Thus, we have

$$\omega_F = \frac{2\pi \text{ rad}}{(2 \text{ min})(60 \text{ s/min})} = 0.0524 \text{ rad/s}.$$

Now we can find the magnitude L of the angular momentum with Eq. 12-31:

$$L = I\omega_F = (1.22 \times 10^9 \text{ kg}\cdot\text{m}^2)(0.0524 \text{ rad/s}) = 6.39 \times 10^7 \text{ kg}\cdot\text{m}^2/\text{s} \approx 6.4 \times 10^7 \text{ kg}\cdot\text{m}^2/\text{s}. \quad \text{(Answer)}$$

(b) Assume that the fully loaded wheel is rotated from rest to ω_F in a time period $\Delta t_1 = 5.0$ s. What is the magnitude τ_{avg} of the average net external torque acting on it during Δt_1?

SOLUTION: The **Key Idea** here is that the average net external torque is related to the change ΔL in the angular momentum of the loaded wheel by Eq. 12-29 ($\vec{\tau}_{net} = d\vec{L}/dt$). Because the wheel rotates about a fixed axis to reach angular speed ω_F in time period Δt_1, we can rewrite Eq. 12-29 as $\tau_{avg} = \Delta L/\Delta t_1$. The change ΔL is from zero to the answer for part (a). Thus, we have

$$\tau_{avg} = \frac{\Delta L}{\Delta t_1} = \frac{6.39 \times 10^7 \text{ kg}\cdot\text{m}^2/\text{s} - 0}{5.0 \text{ s}} \approx 1.3 \times 10^7 \text{ N}\cdot\text{m}. \quad \text{(Answer)}$$

12-10 Conservation of Angular Momentum

So far we have discussed two powerful conservation laws, the conservation of energy and the conservation of linear momentum. Now we meet a third law of this type, involving the conservation of angular momentum. We start from Eq. 12-29 ($\vec{\tau}_{net} = d\vec{L}/dt$), which is Newton's second law in angular form. If no net external torque acts on the system, this equation becomes $d\vec{L}/dt = 0$, or

$$\vec{L} = \text{a constant} \quad \text{(isolated system)}. \tag{12-32}$$

This result, called the **law of conservation of angular momentum,** can also be written as

$$\begin{pmatrix}\text{net angular momentum} \\ \text{at some initial time } t_i\end{pmatrix} = \begin{pmatrix}\text{net angular momentum} \\ \text{at some later time } t_f\end{pmatrix},$$

or

$$\vec{L}_i = \vec{L}_f \quad \text{(isolated system)}. \tag{12-33}$$

Equations 12-32 and 12-33 tell us:

> If the net external torque acting on a system is zero, the angular momentum $\vec{L}$ of the system remains constant, no matter what changes take place within the system.

Equations 12-32 and 12-33 are vector equations; as such, they are equivalent to three component equations corresponding to the conservation of angular momentum in three mutually perpendicular directions. Depending on the torques acting on a system, the angular momentum of the system might be conserved in only one or two directions but not in all directions:

> If the component of the net *external* torque on a system along a certain axis is zero, then the component of the angular momentum of the system along that axis cannot change, no matter what changes take place within the system.

We can apply this law to the isolated body in Fig. 12-14, which rotates around the z axis. Suppose that the initially rigid body somehow redistributes its mass relative to that rotation axis, changing its rotational inertia about that axis. Equations 12-32 and 12-33 state that the angular momentum of the body cannot change. Substituting Eq. 12-31 (for the angular momentum along the rotational axis) into Eq. 12-33, we write this conservation law as

$$I_i\omega_i = I_f\omega_f. \tag{12-34}$$

Here the subscripts refer to the values of the rotational inertia I and angular speed ω before and after the redistribution of mass.

Like the other two conservation laws that we have discussed, Eqs. 12-32 and 12-33 hold beyond the limitations of Newtonian mechanics. They hold for particles whose speeds approach that of light (where the theory of special relativity reigns), and they remain true in the world of subatomic particles (where quantum physics reigns). No exceptions to the law of conservation of angular momentum have ever been found.

We now discuss four examples involving this law.

1. ***The spinning volunteer*** Figure 12-16 shows a student seated on a stool that can rotate freely about a vertical axis. The student, who has been set into rotation at a modest initial angular speed ω_i, holds two dumbbells in his outstretched hands. His angular momentum vector $\vec{L}$ lies along the vertical rotation axis, pointing upward.

The instructor now asks the student to pull in his arms; this action reduces his rotational inertia from its initial value I_i to a smaller value I_f because he moves mass closer to the rotation axis. His rate of rotation increases markedly, from ω_i to ω_f. The student can then slow down by extending his arms once more.

No net external torque acts on the system consisting of the student, stool, and dumbbells. Thus, the angular momentum of that system about the rotation

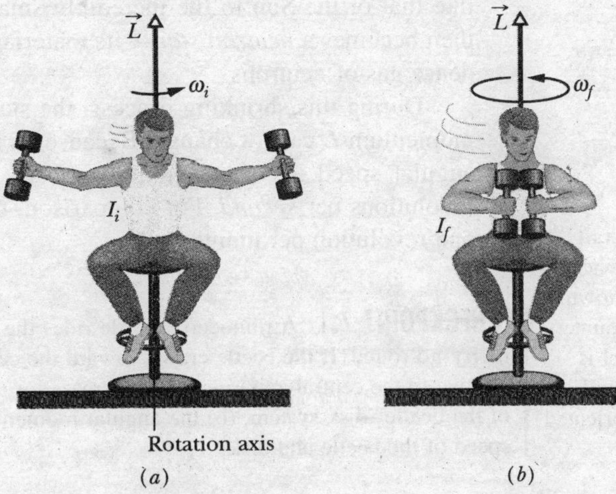

Fig. 12-16 (*a*) The student has a relatively large rotational inertia about the rotation axis and a relatively small angular speed. (*b*) By decreasing his rotational inertia, the student automatically increases his angular speed. The angular momentum $\vec{L}$ of the rotating system remains unchanged.

Fig. 12-17 The diver's angular momentum $\vec{L}$ is constant throughout the dive, being represented by the tail ⊗ of an arrow that is perpendicular to the plane of the figure. Note also that her center of mass (see the dots) follows a parabolic path.

axis must remain constant, no matter how the student maneuvers the dumbbells. In Fig. 12-16*a*, the student's angular speed ω_i is relatively low and his rotational inertia I_i is relatively high. According to Eq. 12-34, his angular speed in Fig. 12-16*b* must be greater to compensate for the decreased rotational inertia.

2. ***The springboard diver*** Figure 12-17 shows a diver doing a forward one-and-a-half-somersault dive. As you should expect, her center of mass follows a parabolic path. She leaves the springboard with a definite angular momentum $\vec{L}$ about an axis through her center of mass, represented by a vector pointing into the plane of Fig. 12-17, perpendicular to the page. When she is in the air, no net external torque acts on her about her center of mass, so her angular momentum about her center of mass cannot change. By pulling her arms and legs into the closed *tuck position,* she can considerably reduce her rotational inertia about the same axis and thus, according to Eq. 12-34, considerably increase her angular speed. Pulling out of the tuck position (into the *open layout position*) at the end of the dive increases her rotational inertia and thus slows her rotation rate so she can enter the water with little splash. Even in a more complicated dive involving both twisting and somersaulting, the angular momentum of the diver must be conserved, in both magnitude *and* direction, throughout the dive.

3. ***Spacecraft orientation*** Figure 12-18, which represents a spacecraft with a rigidly mounted flywheel, suggests a scheme (albeit crude) for orientation control. The *spacecraft + flywheel* form an isolated system. Therefore, if the system's total angular momentum $\vec{L}$ is zero because neither spacecraft nor flywheel is turning, it must remain zero (as long as the system remains isolated).

   To change the orientation of the spacecraft, the flywheel is made to rotate (Fig. 12-18*a*). The spacecraft will start to rotate in the opposite sense to maintain the system's angular momentum at zero. When the flywheel is then brought to rest, the spacecraft will also stop rotating but will have changed its orientation (Fig. 12-18*b*). Throughout, the angular momentum of the system *spacecraft + flywheel* never differs from zero.

   Interestingly, the spacecraft *Voyager 2*, on its 1986 flyby of the planet Uranus, was set into unwanted rotation by this flywheel effect every time its tape recorder was turned on at high speed. The ground staff at the Jet Propulsion Laboratory had to program the on-board computer to turn on counteracting thruster jets every time the tape recorder was turned on or off.

4. ***The incredible shrinking star*** When the nuclear fire in the core of a star burns low, the star may eventually begin to collapse, building up pressure in its interior. The collapse may go so far as to reduce the radius of the star from something like that of the Sun to the incredibly small value of a few kilometers. The star then becomes a *neutron star*—its material has been compressed to an incredibly dense gas of neutrons.

   During this shrinking process, the star is an isolated system and its angular momentum $\vec{L}$ cannot change. Because its rotational inertia is greatly reduced, its angular speed is correspondingly greatly increased, to as much as 600 to 800 revolutions per *second.* For comparison, the Sun, a typical star, rotates at about one revolution per month.

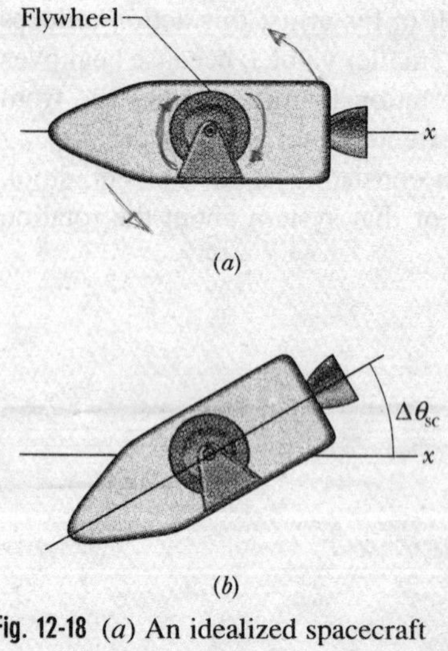

Fig. 12-18 (*a*) An idealized spacecraft containing a flywheel. If the flywheel is made to rotate clockwise as shown, the spacecraft itself will rotate counterclockwise. (*b*) When the flywheel is braked to a stop, the spacecraft will also stop rotating but will be reoriented by the angle $\Delta\theta_{sc}$.

CHECKPOINT 7: A rhinoceros beetle rides the rim of a small disk that rotates like a merry-go-round. If the beetle crawls toward the center of the disk, do the following (each relative to the central axis) increase, decrease, or remain the same: (a) the rotational inertia of the beetle–disk system, (b) the angular momentum of the system, and (c) the angular speed of the beetle and disk?

Sample Problem 12-7

Figure 12-19*a* shows a student, again sitting on a stool that can rotate freely about a vertical axis. The student, initially at rest, is holding a bicycle wheel whose rim is loaded with lead and whose rotational inertia I_{wh} about its central axis is 1.2 kg · m². The wheel is rotating at an angular speed ω_{wh} of 3.9 rev/s; as seen from overhead, the rotation is counterclockwise. The axis of the wheel is vertical, and the angular momentum $\vec{L}_{wh}$ of the wheel points vertically upward. The student now inverts the wheel (Fig. 12-19*b*) so that, as seen from overhead, it is rotating clockwise. Its angular momentum is then $-\vec{L}_{wh}$. The inversion results in the student, the stool, and the wheel's center rotating together as a composite rigid body about the stool's rotation axis, with rotational inertia I_b = 6.8 kg · m². (The fact that the wheel is also rotating about its center does not affect the mass distribution of this composite body; thus, I_b has the same value whether or not the wheel rotates.) With what angular speed ω_b and in what direction does the composite body rotate after the inversion of the wheel?

SOLUTION: The **Key Ideas** here are these:

1. The angular speed ω_b we seek is related to the final angular momentum $\vec{L}_b$ of the composite body about the stool's rotation axis by Eq. 12-31 ($L = I\omega$).
2. The initial angular speed ω_{wh} of the wheel is related to the angular momentum $\vec{L}_{wh}$ of the wheel's rotation about its center by the same equation.
3. The vector addition of $\vec{L}_b$ and $\vec{L}_{wh}$ gives the total angular momentum $\vec{L}_{tot}$ of the system of student, stool, and wheel.
4. As the wheel is inverted, no net *external* torque acts on that system to change $\vec{L}_{tot}$ about any vertical axis. (Torques due to forces between the student and the wheel as the student inverts the wheel are *internal* to the system.) So, the system's total angular momentum is conserved about any vertical axis.

The conservation of $\vec{L}_{tot}$ is represented with vectors in Fig. 12-19*c*. We can also write it in terms of components along a vertical axis as

$$L_{b,f} + L_{wh,f} = L_{b,i} + L_{wh,i}, \qquad (12\text{-}35)$$

where *i* and *f* refer to the initial state (before inversion of the wheel) and the final state (after inversion). Because inversion of the wheel inverted the angular momentum vector of the wheel's rotation, we substitute $-L_{wh,i}$ for $L_{wh,f}$. Then, if we set $L_{b,i} = 0$ (because the student, the stool, and the wheel's center were initially at rest),

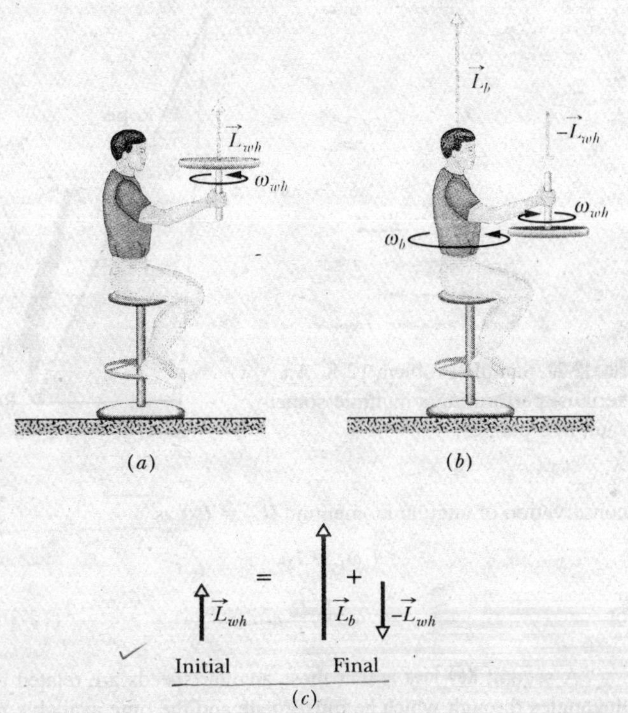

Fig. 12-19 Sample Problem 12-7. (*a*) A student holds a bicycle wheel rotating around the vertical. (*b*) The student inverts the wheel, setting himself into rotation. (*c*) The net angular momentum of the system must remain the same in spite of the inversion.

Eq. 12-35 yields

$$L_{b,f} = 2L_{wh,i}.$$

Using Eq. 12-31, we next substitute $I_b\omega_b$ for $L_{b,f}$ and $I_{wh}\omega_{wh}$ for $L_{wh,i}$ and solve for ω_b, finding

$$\omega_b = \frac{2I_{wh}}{I_b}\,\omega_{wh} = \frac{(2)(1.2\ \text{kg}\cdot\text{m}^2)(3.9\ \text{rev/s})}{6.8\ \text{kg}\cdot\text{m}^2} = 1.4\ \text{rev/s}. \quad \text{(Answer)}$$

This positive result tells us that the student rotates counterclockwise about the stool axis as seen from overhead. If the student wishes to stop rotating, he has only to invert the wheel once more.

Sample Problem 12-8

During a jump to his partner, an aerialist is to make a quadruple somersault lasting a time t = 1.87 s. For the first and last quarter-revolution, he is in the extended orientation shown in Fig. 12-20, with rotational inertia I_1 = 19.9 kg · m² around his center of mass (the dot). During the rest of the flight he is in a tight tuck, with rotational inertia I_2 = 3.93 kg · m². What must be his angular speed ω_2 around his center of mass during the tuck?

SOLUTION: Obviously he must turn fast enough to complete the 4.0 rev required for a quadruple somersault in the given 1.87 s. To do so, he increases his angular speed to ω_2 by tucking. We can relate ω_2 to his initial angular speed ω_1 with this **Key Idea:** His angular momentum about his center of mass is conserved throughout the free flight because there is no net external torque about his center of mass to change it. From Eq. 12-34, we can write the

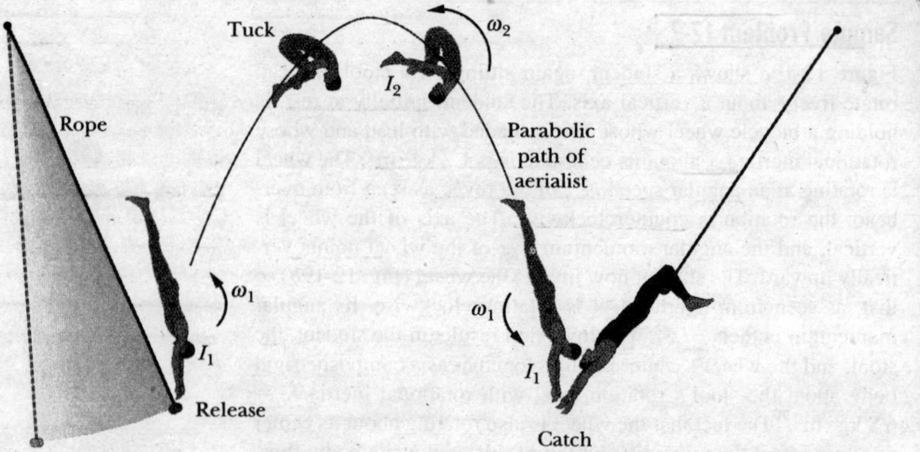

Fig. 12-20 Sample Problem 12-8. An aerialist performing a multiple somersault to a partner.

conservation of angular momentum ($L_1 = L_2$) as

$$I_1\omega_1 = I_2\omega_2$$

or

$$\omega_1 = \frac{I_2}{I_1}\omega_2. \tag{12-36}$$

A second **Key Idea** is that these angular speeds are related to the angles through which he must rotate and the time available to do so. At the start and at the end, he must rotate in the extended orientation for a total angle of $\theta_1 = 0.500$ rev (two quarter-turns) in a time we shall call t_1. In the tuck, he must rotate through an angle of $\theta_2 = 3.50$ rev in a time t_2. From Eq. 11-5 ($\omega_{avg} = \Delta\theta/\Delta t$), we can write

$$t_1 = \frac{\theta_1}{\omega_1} \quad \text{and} \quad t_2 = \frac{\theta_2}{\omega_2}.$$

Thus, his total flight time is

$$t = t_1 + t_2 = \frac{\theta_1}{\omega_1} + \frac{\theta_2}{\omega_2}, \tag{12-37}$$

which we know to be 1.87 s. Now substituting from Eq. 12-36 for ω_1 yields

$$t = \frac{\theta_1 I_1}{\omega_2 I_2} + \frac{\theta_2}{\omega_2} = \frac{1}{\omega_2}\left(\theta_1\frac{I_1}{I_2} + \theta_2\right).$$

Inserting the known data, we obtain

$$1.87\text{ s} = \frac{1}{\omega_2}\left((0.500\text{ rev})\frac{19.9\text{ kg}\cdot\text{m}^2}{3.93\text{ kg}\cdot\text{m}^2} + 3.50\text{ rev}\right),$$

which gives us

$$\omega_2 = 3.23\text{ rev/s}. \qquad \text{(Answer)}$$

This angular speed is so fast that the aerialist cannot clearly see his surroundings or fine-tune his rotation by adjusting his tuck. The possibility of an aerialist making a four-and-a-half somersault flight, which would require a greater value of ω_2 and thus a smaller I_2 via a tighter tuck, seems very small.

Sample Problem 12-9

(This final sample problem of the chapter is long and challenging, but it is helpful because it pulls together many ideas of Chapters 11 and 12.) In the overhead view of Fig. 12-21, four thin, uniform rods, each of mass M and length $d = 0.50$ m, are rigidly connected to a vertical axle to form a turnstile. The turnstile rotates clockwise about the axle, which is attached to a floor, with initial angular velocity $\omega_i = -2.0$ rad/s. A mud ball of mass $m = \frac{1}{3}M$ and initial speed $v_i = 12$ m/s is thrown along the path shown and sticks to the end of one rod. What is the final angular velocity ω_f of the ball–turnstile system?

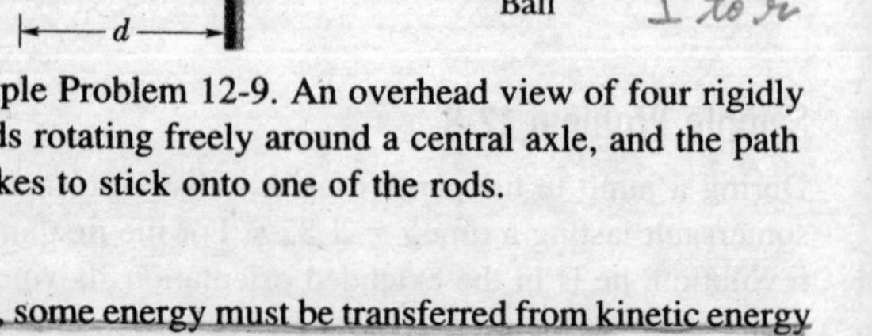

Fig. 12-21 Sample Problem 12-9. An overhead view of four rigidly connected rods rotating freely around a central axle, and the path a mud ball takes to stick onto one of the rods.

SOLUTION: A **Key Idea** here can be stated in a question-and-answer format. The question is this: Does the system have a quantity that is conserved during the collision and that involves angular velocity, so that we can solve for ω_f? To answer, let us check the conservation possibilities:

1. The total kinetic energy K is *not* conserved, because the collision between ball and rod is completely inelastic (the ball sticks). So, some energy must be transferred from kinetic energy to other types of energy (such as thermal energy). For the same reason, total mechanical energy is not conserved.
2. The total linear momentum $\vec{P}$ is also *not* conserved, because during the collision an external force acts on the turnstile at the

attachment of the axle to the floor. (This is the force that keeps the turnstile from moving across the floor when it is hit by the mud ball.)

3. The total angular momentum $\vec{L}$ of the system about the axle *is* conserved because there is no net external torque to change $\vec{L}$. (The forces in the collision produce only internal torques; the external force on the turnstile acts at the axle, has zero moment arm, and thus does not produce an external torque.)

We can write the conservation of the system's total angular momentum ($L_f = L_i$) about the axle as

$$L_{ts,f} + L_{ball,f} = L_{ts,i} + L_{ball,i}, \qquad (12\text{-}38)$$

where ts stands for "turnstile." The final angular velocity ω_f is contained in the terms $L_{ts,f}$ and $L_{ball,f}$ because those final angular momenta depend on how fast the turnstile and ball are rotating. To find ω_f, we consider first the turnstile and then the ball, and then we return to Eq. 12-38.

Turnstile: The **Key Idea** here is that, because the turnstile is a rotating rigid object, Eq. 12-31 ($L = I\omega$) gives its angular momentum. Thus, we can write its final and initial angular momenta about the axle as

$$L_{ts,f} = I_{ts}\omega_f \quad \text{and} \quad L_{ts,i} = I_{ts}\omega_i. \qquad (12\text{-}39)$$

Because the turnstile consists of four rods, each rotating around an end, the rotational inertia I_{ts} of the turnstile is four times the rotational inertia I_{rod} of each rod about its end. From Table 11-2*e*, we know that the rotational inertia I_{com} of a rod about its center is $\frac{1}{12}Md^2$, where M is its mass and d is its length. To get I_{rod} we use the parallel-axis theorem of Eq. 11-29 ($I = I_{com} + Mh^2$). Here perpendicular distance h is $d/2$. Thus, we find

$$I_{rod} = \tfrac{1}{12}Md^2 + M\left(\frac{d}{2}\right)^2 = \tfrac{1}{3}Md^2.$$

With four rods in the turnstile, we then have

$$I_{ts} = \tfrac{4}{3}Md^2. \qquad (12\text{-}40)$$

Ball: Before the collision, the ball is like a particle moving along a straight line, as in Fig. 12-11. So, to find the ball's initial angular momentum $L_{ball,i}$ about the axle, we can use any of Eqs. 12-18 through 12-21, but Eq. 12-20 ($\ell = rmv_\perp$) is easiest. Here ℓ is $L_{ball,i}$; just before the ball hits, its radial distance r from the axle is d and the component $v_\perp$ of the ball's velocity perpendicular to r is $v_i \cos 60°$.

To give a sign to this angular momentum, we mentally draw a position vector from the turnstile's axle to the ball. As the ball approaches the turnstile, this position vector rotates counterclockwise about the axle, so the ball's angular momentum is a positive quantity. We can now rewrite $\ell = rmv_\perp$ as

$$L_{ball,i} = mdv_i \cos 60° \qquad (12\text{-}41)$$

After the collision, the ball is like a particle rotating in a circle of radius d. So, from Eq. 11-26 ($I = \Sigma m_i r_i^2$), we have $I_{ball} = md^2$ about the axle. Then from Eq. 12-31 ($L = I\omega$), we can write the final angular momentum of the ball about the axle as

$$L_{ball,f} = I_{ball}\omega_f = md^2\omega_f. \qquad (12\text{-}42)$$

Return to Eq. 12-38: Substituting from Eqs. 12-39 through 12-42 into Eq.12-38, we have

$$\tfrac{4}{3}Md^2\omega_f + md^2\omega_f = \tfrac{4}{3}Md^2\omega_i + mdv_i \cos 60°.$$

Substituting $M = 3m$ and solving for ω_f, we find

$$\omega_f = \frac{1}{5d}(4d\omega_i + v_i \cos 60°)$$

$$= \frac{1}{5(0.50\text{ m})}[4(0.50\text{ m})(-2.0\text{ rad/s}) + (12\text{ m/s})(\cos 60°)]$$

$$= 0.80\text{ rad/s}. \qquad \text{(Answer)}$$

Thus, the turnstile is now turning counterclockwise.

REVIEW & SUMMARY

Rolling Bodies For a wheel of radius R that is rolling smoothly (no sliding),

$$v_{com} = \omega R, \qquad (12\text{-}2)$$

where v_{com} is the linear speed of the wheel's center and ω is the angular speed of the wheel about its center. The wheel may also be viewed as rotating instantaneously about the point P of the "road" that is in contact with the wheel. The angular speed of the wheel about this point is the same as the angular speed of the wheel about its center. The rolling wheel has kinetic energy

$$K = \tfrac{1}{2}I_{com}\omega^2 + \tfrac{1}{2}Mv_{com}^2, \qquad (12\text{-}5)$$

where I_{com} is the rotational moment of the wheel about its center and M is the mass of the wheel. If the wheel is being accelerated but is still rolling smoothly, the acceleration of the center of mass $\vec{a}_{com}$ is related to the angular acceleration α about the center with

$$a_{com} = \alpha R. \qquad (12\text{-}6)$$

If the wheel rolls smoothly down a ramp of angle θ, its acceleration along an x axis extending up the ramp is

$$a_{com,x} = -\frac{g \sin\theta}{1 + I_{com}/MR^2}. \qquad (12\text{-}10)$$

Torque as a Vector In three dimensions, *torque* $\vec{\tau}$ is a vector quantity defined relative to a fixed point (usually an origin); it is

$$\vec{\tau} = \vec{r} \times \vec{F}, \qquad (12\text{-}14)$$

where $\vec{F}$ is a force applied to a particle and $\vec{r}$ is a position vector locating the particle relative to the fixed point (or origin). The magnitude of $\vec{\tau}$ is given by

$$\tau = rF \sin \phi = rF_{\perp} = r_{\perp}F, \qquad (12\text{-}15, 12\text{-}16, 12\text{-}17)$$

where ϕ is the angle between $\vec{F}$ and $\vec{r}$, $F_{\perp}$ is the component of $\vec{F}$ perpendicular to $\vec{r}$, and $r_{\perp}$ is the moment arm of $\vec{F}$. The direction of $\vec{\tau}$ is given by the right-hand rule for cross products.

Angular Momentum of a Particle The *angular momentum* $\vec{\ell}$ of a particle with linear momentum $\vec{p}$, mass m, and linear velocity $\vec{v}$ is a vector quantity defined relative to a fixed point (usually an origin); it is

$$\vec{\ell} = \vec{r} \times \vec{p} = m(\vec{r} \times \vec{v}). \qquad (12\text{-}18)$$

The magnitude of $\vec{\ell}$ is given by

$$\ell = rmv \sin \phi \qquad (12\text{-}19)$$
$$= rp_{\perp} = rmv_{\perp} \qquad (12\text{-}20)$$
$$= r_{\perp}p = r_{\perp}mv, \qquad (12\text{-}21)$$

where ϕ is the angle between $\vec{r}$ and $\vec{p}$, $p_{\perp}$ and $v_{\perp}$ are the components of $\vec{p}$ and $\vec{v}$ perpendicular to $\vec{r}$, and $r_{\perp}$ is the perpendicular distance between the fixed point and the extension of $\vec{p}$. The direction of $\vec{\ell}$ is given by the right-hand rule for cross products.

Newton's Second Law in Angular Form Newton's second law for a particle can be written in angular form as

$$\vec{\tau}_{\text{net}} = \frac{d\vec{\ell}}{dt}, \qquad (12\text{-}23)$$

where $\vec{\tau}_{\text{net}}$ is the net torque acting on the particle, and $\vec{\ell}$ is the angular momentum of the particle.

Angular Momentum of a System of Particles The angular momentum $\vec{L}$ of a system of particles is the vector sum of the angular momenta of the individual particles:

$$\vec{L} = \vec{\ell}_1 + \vec{\ell}_2 + \cdots + \vec{\ell}_n = \sum_{i=1}^{n} \vec{\ell}_i. \qquad (12\text{-}26)$$

The time rate of change of this angular momentum is equal to the net external torque on the system (the vector sum of the torques due to interactions of the particles of the system with particles external to the system):

$$\vec{\tau}_{\text{net}} = \frac{d\vec{L}}{dt} \quad \text{(system of particles).} \qquad (12\text{-}29)$$

Angular Momentum of a Rigid Body For a rigid body rotating about a fixed axis, the component of its angular momentum parallel to the rotation axis is

$$L = I\omega \quad \text{(rigid body, fixed axis).} \qquad (12\text{-}31)$$

Conservation of Angular Momentum The angular momentum $\vec{L}$ of a system remains constant if the net external torque acting on the system is zero:

$$\vec{L} = \text{a constant} \quad \text{(isolated system)} \qquad (12\text{-}32)$$

or

$$\vec{L}_i = \vec{L}_f \quad \text{(isolated system).} \qquad (12\text{-}33)$$

This is the **law of conservation of angular momentum.** It is one of the fundamental conservation laws of nature, having been verified even in situations (involving high-speed particles or subatomic dimensions) in which Newton's laws are not applicable.

QUESTIONS

1. In Fig. 12-22, a block slides down a frictionless ramp and a sphere rolls without sliding down a ramp of the same angle θ. The block and sphere have the same mass, start from rest at point A, and descend through point B. (a) In that descent, is the work done by the gravitational force on the block greater than, less than, or the same as the work done by the gravitational force on the sphere? At B, which object has more (b) translational kinetic energy and (c) speed down the ramp?

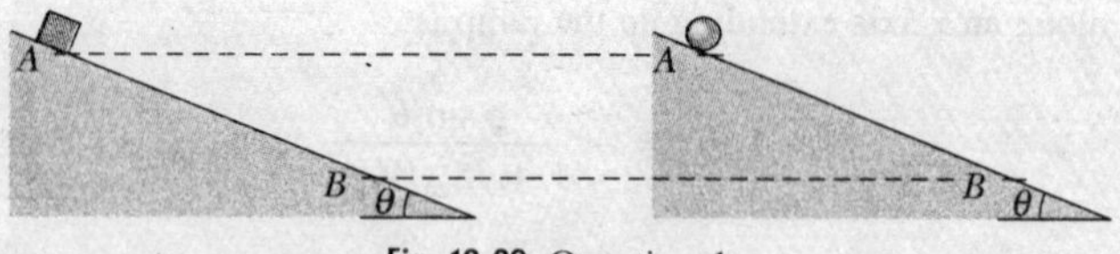

Fig. 12-22 Question 1.

2. A cannonball rolls down an incline without sliding. If the roll is now repeated with an incline that is less steep but of the same height as the first incline, are (a) the ball's time to reach the bottom and (b) its translational kinetic energy at the bottom greater than, less than, or the same as previously?

3. In Fig. 12-23, a woman rolls a cylindrical drum by pushing a board across its top. The drum moves through the distance $L/2$, which is half the board's length. The drum rolls smoothly without sliding or bouncing, and the board does not slide over the drum. (a) What length of board rolls over the top of the drum? (b) How far does the woman walk?

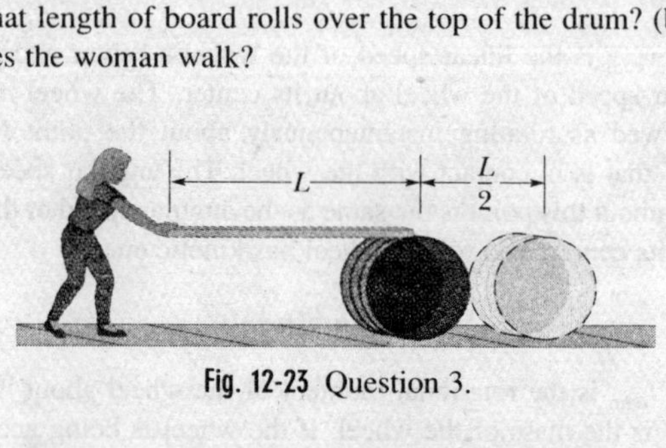

Fig. 12-23 Question 3.

4 The position vector $\vec{r}$ of a particle relative to a certain point has a magnitude of 3 m, and the force $\vec{F}$ on the particle has a magnitude of 4 N. What is the angle between the directions of $\vec{r}$ and $\vec{F}$ if the magnitude of the associated torque equals (a) zero and (b) 12 N · m?

5 Figure 12-24 shows a particle moving at constant velocity $\vec{v}$ and five points with their xy coordinates. Rank the points according to the magnitude of the angular momentum of the particle measured about them, greatest first.

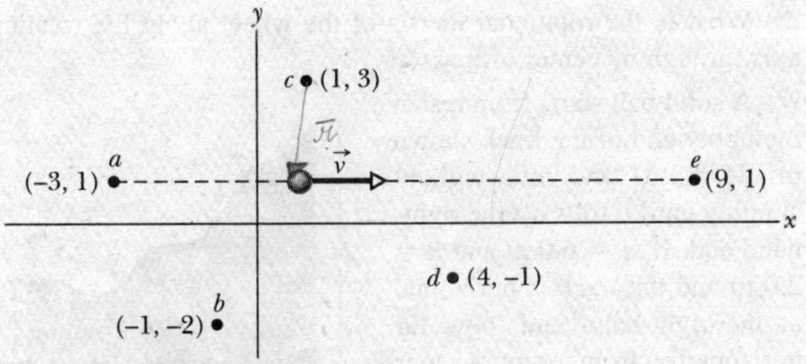

Fig. 12-24 Question 5.

6 (a) In Checkpoint 4, what is the torque on particles 1 and 2 about point O due to the centripetal forces that cause those particles to circle at constant speed? (b) As particles 3, 4, and 5 move from the left to the right of point O, do their individual angular momenta increase, decrease, or stay the same?

7 Figure 12-25 shows three particles of the same mass and the same constant speed moving as indicated by the velocity vectors. Points a, b, c, and d form a square, with point e at the center. Rank the points according to the magnitude of the net angular momentum of the three-particle system when measured about the points, greatest first.

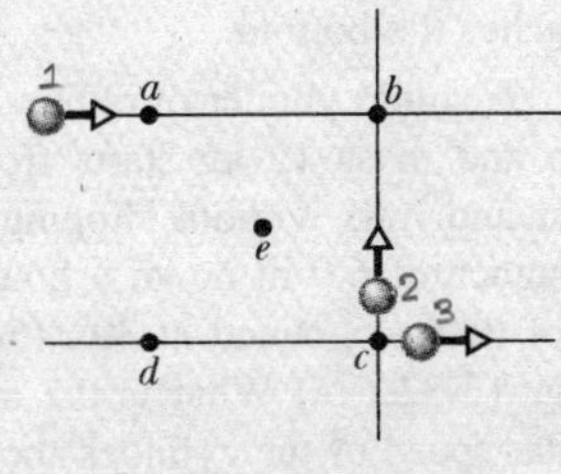

Fig. 12-25 Question 7.

8 A bola, which consists of three heavy balls connected to a common point by identical lengths of sturdy string, is readied for launch by holding one of the balls overhead and rotating the wrist, causing the other two balls to rotate in a horizontal circle about the hand. The bola is then released, and its configuration rapidly changes from that in the overhead view of Fig. 12-26a to that of Fig. 12-26b. Thus, the rotation is initially around axis 1 through the ball that was held. Then it is around axis 2 through the center of mass. Are (a) the angular momentum and (b) the angular speed around axis 2 greater than, less than, or the same as that around axis 1?

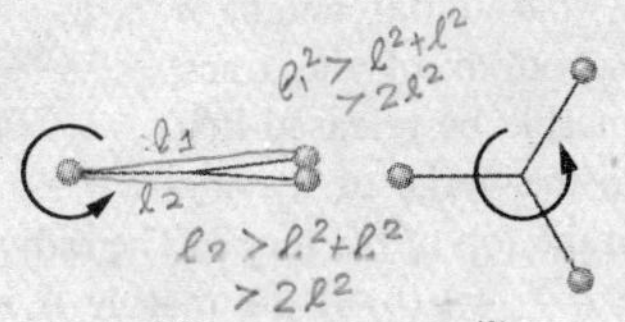

Fig. 12-26 Question 8. (*a*) (*b*)

9 A rhinoceros beetle rides the rim of a horizontal disk rotating counterclockwise like a merry-go-round. If it then walks along the rim in the direction of the rotation, will the magnitudes of the following increase, decrease, or remain the same: (a) the angular momentum of the beetle–disk system, (b) the angular momentum and angular velocity of the beetle, and (c) the angular momentum and angular velocity of the disk? (d) What are your answers if the beetle walks in the direction opposite the rotation?

10 Figure 12-27 shows an overhead view of a rectangular slab that can spin like a merry-go-round about its center at O. Also shown are seven paths along which wads of bubble gum can be thrown (all with the same speed and mass) to stick onto the stationary slab. (a) Rank the paths according to the angular speed that the slab (and gum) will have after the gum sticks, greatest first. (b) For which paths will the angular momentum of the slab (and gum) about O be negative from the view of Fig. 12-27?

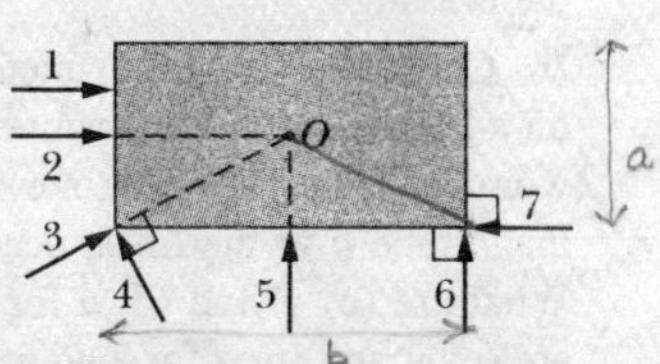

Fig. 12-27 Question 10.

11 In Fig. 12-28, three forces of the same magnitude are applied to a particle at the origin ($\vec{F}_1$ acts directly into the plane of the figure). Rank the forces according to the magnitudes of the torques they create about (a) point P_1, (b) point P_2, and (c) point P_3, greatest first.

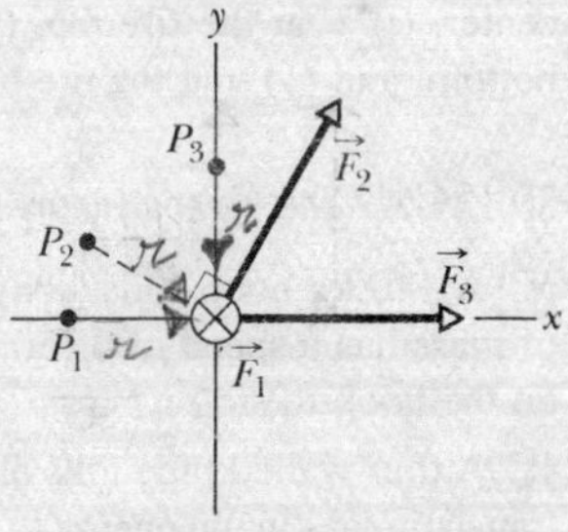

Fig. 12-28 Question 11.

12. Figure 12-29 shows two particles A and B at xyz coordinates (1 m, 1 m, 0) and (1 m, 0, 1 m). Acting on each particle are three numbered forces, all of the same magnitude and each directed parallel to an axis. (a) Which of the forces produce a torque about the origin that is directed parallel to y? (b) Rank the forces according to the magnitudes of the torques they produce on the particles about the origin, greatest first.

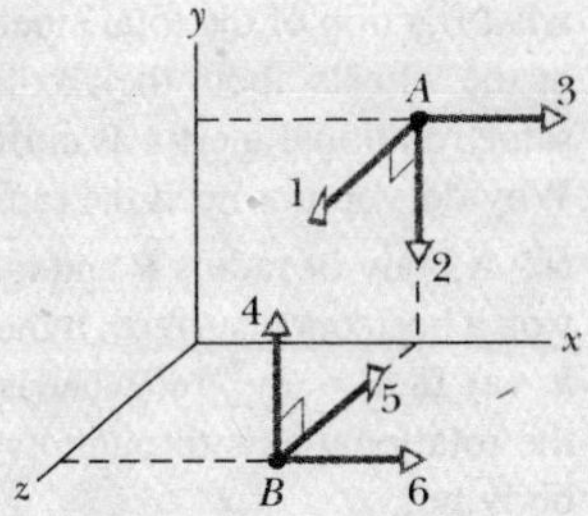

Fig. 12-29 Question 12.

13 Figure 11-24 in Chapter 11 shows an assembly of three small spheres of the same mass that are attached to a massless rod with the indicated spacings. The assembly is to be rotated at 3.0 rad/s about an axis through one of the spheres and perpendicular to the plane of the page. There are, of course, three such choices of the axis. Rank those choices according to (a) the magnitude of the angular momentum the assembly will have about the chosen rotation axis and (b) the rotational kinetic energy the assembly will have, greatest first.

EXERCISES & PROBLEMS

ssm Solution is in the Student Solutions Manual.
www Solution is available on the World Wide Web at:
http://www.wiley.com/college/hrw
ilw Solution is available on the Interactive LearningWare.

SEC. 12-1 Rolling

1E. An automobile traveling 80.0 km/h has tires of 75.0 cm diameter. (a) What is the angular speed of the tires about their axles? (b) If the car is brought to a stop uniformly in 30.0 complete turns of the tires (without skidding), what is the magnitude of the angular acceleration of the wheels? (c) How far does the car move during the braking?

2P. Consider a 66-cm-diameter tire on a car traveling at 80 km/h on a level road in the positive direction of an x axis. Relative to a woman in the car, what are (a) the linear velocity $\vec{v}$ and (b) the magnitude a of the linear acceleration of the center of the wheel? What are (c) $\vec{v}$ and (d) a for a point at the top of the tire? What are (e) $\vec{v}$ and (f) a for a point at the bottom of the tire?

Now repeat the questions relative to a hitchhiker sitting near the road: What are (g) $\vec{v}$ at the wheel's center, (h) a at the wheel's center, (i) $\vec{v}$ at the tire top, (j) a at the tire top, (k) $\vec{v}$ at the tire bottom, and (l) a at the tire bottom?

SEC. 12-2 The Kinetic Energy of Rolling

3E. A 140 kg hoop rolls along a horizontal floor so that its center of mass has a speed of 0.150 m/s. How much work must be done on the hoop to stop it? ssm

4E. A thin-walled pipe rolls along the floor. What is the ratio of its translational kinetic energy to its rotational kinetic energy about an axis parallel to its length and through its center of mass?

5E. A 1000 kg car has four 10 kg wheels. When the car is moving, what fraction of the total kinetic energy of the car is due to rotation of the wheels about their axles? Assume that the wheels have the same rotational inertia as uniform disks of the same mass and size. Why do you not need the radius of the wheels? ssm ilw www

6P. A body of radius R and mass m is rolling smoothly with speed v on a horizontal surface. It then rolls up a hill to a maximum height h. (a) If $h = 3v^2/4g$, what is the body's rotational inertia about the rotational axis through its center of mass? (b) What might the body be?

SEC. 12-3 The Forces of Rolling

7E. A uniform solid sphere rolls down an incline. (a) What must be the incline angle if the linear acceleration of the center of the sphere is to have a magnitude of $0.10g$? (b) If a frictionless block were to slide down the incline at that angle, would its acceleration magnitude be more than, less than, or equal to $0.10g$? Why?

8P. A constant horizontal force of magnitude 10 N is applied to a wheel of mass 10 kg and radius 0.30 m as shown in Fig. 12-30. The wheel rolls smoothly on the horizontal surface, and the acceleration of its center of mass has magnitude 0.60 m/s^2. (a) What are the magnitude and direction of the frictional force on the wheel?

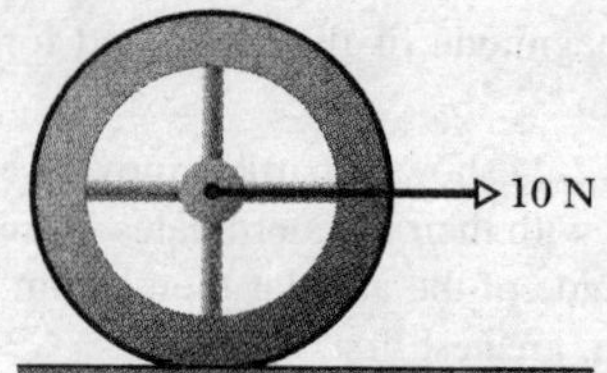

Fig. 12-30 Problem 8.

(b) What is the rotational inertia of the wheel about the rotation axis through its center of mass?

9P. A solid ball starts from rest at the upper end of the track shown in Fig. 12-31 and rolls without slipping until it rolls off the right-hand end. If $H = 6.0$ m and $h = 2.0$ m and the track is horizontal at the right-hand end, how far horizontally from point A does the ball land on the floor? ssm

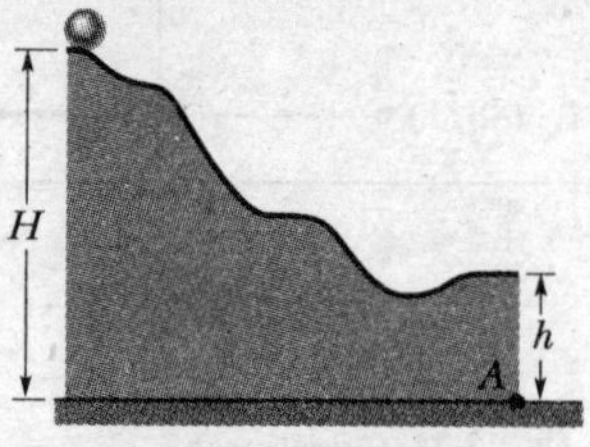

Fig. 12-31 Problem 9.

10P. A small sphere, with radius r and mass m, rolls without slipping on the inside of a large fixed hemisphere with radius R and a vertical axis of symmetry. It starts at the top from rest. (a) What is its kinetic energy at the bottom? (b) What fraction of its kinetic energy at the bottom is associated with rotation about an axis through its center of mass? (c) Assuming $r \ll R$, find the magnitude of the normal force on the hemisphere from the ball when the ball reaches the bottom. $\omega = \frac{2}{R}\sqrt{\frac{8(H-h)}{3}}$, $S = \sqrt{\frac{8}{3}h(H-h)}$

11P. A solid cylinder of radius 10 cm and mass 12 kg starts from rest and rolls without slipping a distance of 6.0 m down a house roof that is inclined at 30°. (See Fig. 12-32.) (a) What is the angular speed of the cylinder about its center as it leaves the house roof? (b) The roof's edge is 5.0 m high. How far horizontally from the roof's edge does the cylinder hit the level ground? ilw

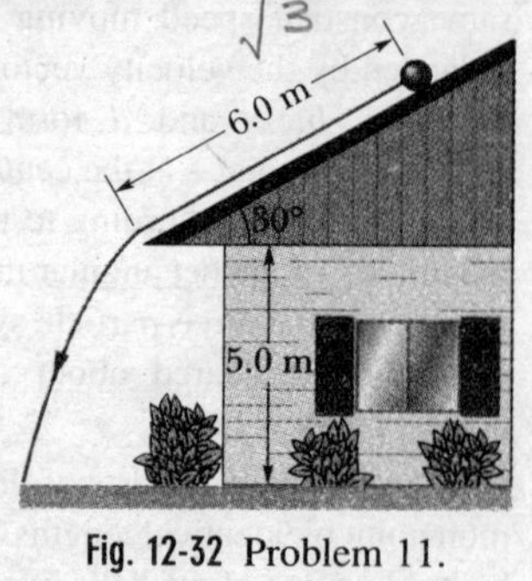

Fig. 12-32 Problem 11.

•12P. A small solid marble of mass m and radius r will roll without slipping along the loop-the-loop track shown in Fig. 12-33 if it is released from rest somewhere on the straight section of track. (a) From what initial height h above the bottom of the track must the marble be released if it is to be on the verge of leaving the track at the top of the loop? (The radius of the loop-the-loop is R; assume $R \gg r$.) (b) If the marble is released from height $6R$ above the bottom of the track, what is the horizontal component of the force acting on it at point Q?

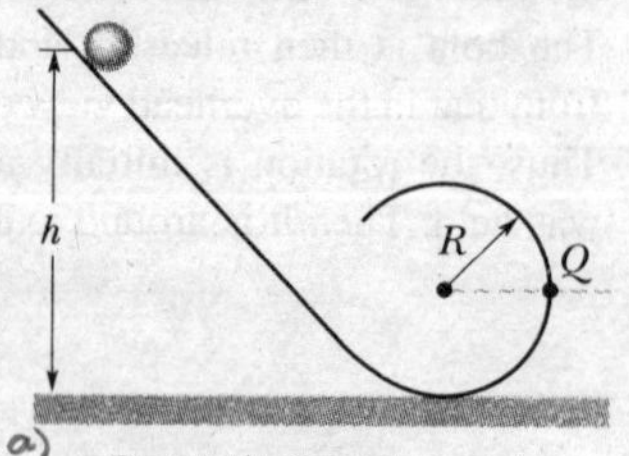

Fig. 12-33 Problem 12.

a) $h = \frac{R}{2}$

b) $N = \frac{mv^2}{R} \Rightarrow N = \frac{mgR}{R} = mg$

13P. A hollow sphere of radius 0.15 m, with rotational inertia $I = 0.040\ \text{kg}\cdot\text{m}^2$ about a line through its center of mass, rolls without slipping up a surface inclined at 30° to the horizontal. At a certain initial position, the sphere's total kinetic energy is 20 J. (a) How much of this initial kinetic energy is rotational? (b) What is the speed of the center of mass of the sphere at the initial position? What are (c) the total kinetic energy of the sphere and (d) the speed of its center of mass after it has moved 1.0 m up along the incline from its initial position?

14P. A bowler throws a bowling ball of radius $R = 11$ cm along a lane. The ball slides on the lane, with initial speed $v_{com,0} = 8.5$ m/s and initial angular speed $\omega_0 = 0$. The coefficient of kinetic friction between the ball and the lane is 0.21. The kinetic frictional force $\vec{f}_k$ acting on the ball (Fig. 12-34) causes a linear acceleration of the ball while producing a torque that causes an angular acceleration of the ball. When speed v_{com} has decreased enough and angular speed ω has increased enough, the ball stops sliding and then rolls smoothly. (a) What then is v_{com} in terms of ω? During the sliding, what are the ball's (b) linear acceleration and (c) angular acceleration? (d) How long does the ball slide? (e) How far does the ball slide? (f) What is the speed of the ball when smooth rolling begins?

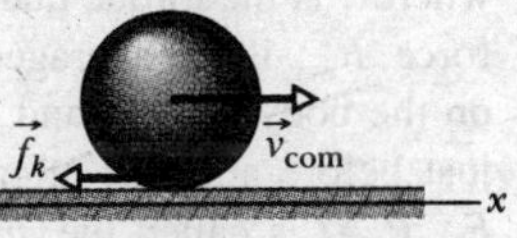

Fig. 12-34 Problem 14.

SEC. 12-4 The Yo-Yo

15E. A yo-yo has a rotational inertia of $950\ \text{g}\cdot\text{cm}^2$ and a mass of 120 g. Its axle radius is 3.2 mm, and its string is 120 cm long. The yo-yo rolls from rest down to the end of the string. (a) What is the magnitude of its linear acceleration? (b) How long does it take to reach the end of the string? As it reaches the end of the string, what are its (c) linear speed, (d) translational kinetic energy, (e) rotational kinetic energy, and (f) angular speed? ssm

16P. Suppose that the yo-yo in Exercise 15, instead of rolling from rest, is thrown so that its initial speed down the string is 1.3 m/s. (a) How long does the yo-yo take to reach the end of the string? As it reaches the end of the string, what are its (b) total kinetic energy, (c) linear speed, (d) translational kinetic energy, (e) angular speed, and (f) rotational kinetic energy?

SEC. 12-5 Torque Revisited

17E. Show that, if $\vec{r}$ and $\vec{F}$ lie in a given plane, the torque $\vec{\tau} = \vec{r} \times \vec{F}$ has no component in that plane.

18E. What are the magnitude and direction of the torque about the origin on a plum located at coordinates (−2.0 m, 0, 4.0 m) due to force $\vec{F}$ whose only component is (a) $F_x = 6.0$ N, (b) $F_x = -6.0$ N, (c) $F_z = 6.0$ N, and (d) $F_z = -6.0$ N?

19E. What are the magnitude and direction of the torque about the origin on a particle located at coordinates (0, −4.0 m, 3.0 m) due to (a) force $\vec{F}_1$ with components $F_{1x} = 2.0$ N and $F_{1y} = F_{1z} = 0$, and (b) force $\vec{F}_2$ with components $F_{2x} = 0$, $F_{2y} = 2.0$ N, and $F_{2z} = 4.0$ N?

20P. Force $\vec{F} = (2.0\ \text{N})\hat{i} - (3.0\ \text{N})\hat{k}$ acts on a pebble with position vector $\vec{r} = (0.50\ \text{m})\hat{j} - (2.0\ \text{m})\hat{k}$, relative to the origin. What is the resulting torque acting on the pebble about (a) the origin and (b) a point with coordinates (2.0 m, 0, −3.0 m)?

21P. Force $\vec{F} = (-8.0\ \text{N})\hat{i} + (6.0\ \text{N})\hat{j}$ acts on a particle with position vector $\vec{r} = (3.0\ \text{m})\hat{i} + (4.0\ \text{m})\hat{j}$. What are (a) the torque on the particle about the origin and (b) the angle between the directions of $\vec{r}$ and $\vec{F}$? ssm

22P. What is the torque about the origin on a jar of jalapeño peppers located at coordinates (3.0 m, −2.0 m, 4.0 m) due to (a) force $\vec{F}_1 = (3.0\ \text{N})\hat{i} - (4.0\ \text{N})\hat{j} + (5.0\ \text{N})\hat{k}$, (b) force $\vec{F}_2 = (-3.0\ \text{N})\hat{i} - (4.0\ \text{N})\hat{j} - (5.0\ \text{N})\hat{k}$, and (c) the vector sum of $\vec{F}_1$ and $\vec{F}_2$? (d) Repeat part (c) about a point with coordinates (3.0 m, 2.0 m, 4.0 m) instead of about the origin.

SEC. 12-6 Angular Momentum

23E. Two objects are moving as shown in Fig. 12-35. What is their total angular momentum about point O? ilw

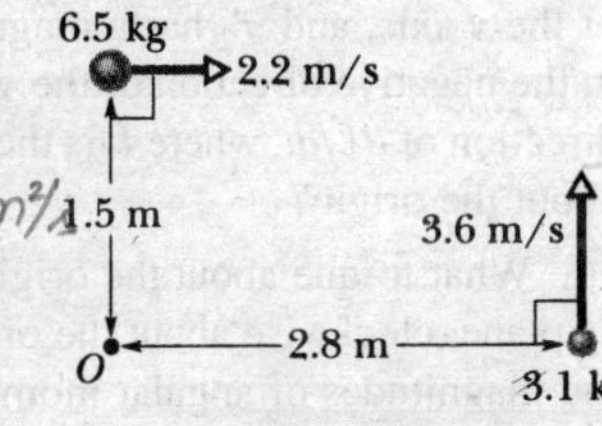

Fig. 12-35 Exercise 23.

24E. In Fig. 12-36, a particle P with mass 2.0 kg has position vector $\vec{r}$ of magnitude 3.0 m and velocity $\vec{v}$ of magnitude 4.0 m/s. A force $\vec{F}$ of magnitude 2.0 N acts on the particle. All three vectors lie in the xy plane oriented as shown. About the origin, what are (a) the angular momentum of the particle and (b) the torque acting on the particle?

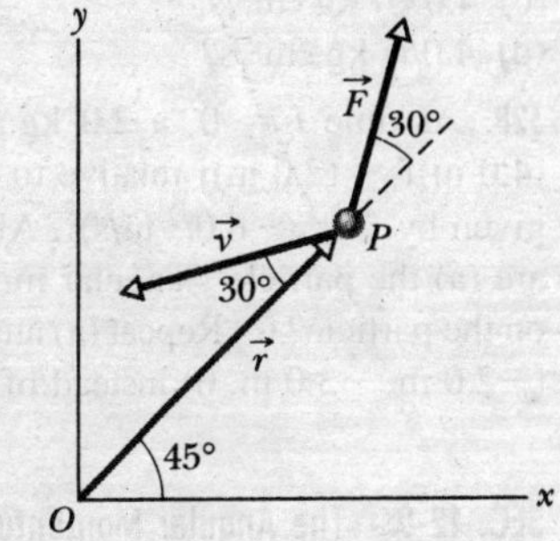

Fig. 12-36 Exercise 24.

25E. At a certain time, a 0.25 kg object has a position vector $\vec{r} = 2.0\hat{i} - 2.0\hat{k}$, in meters. At that instant, its velocity in meters per second is $\vec{v} = -5.0\hat{i} + 5.0\hat{k}$, and the force in newtons acting on it is $\vec{F} = 4.0\hat{j}$. (a) What is the angular momentum of the object about the origin? (b) What torque acts on it? ssm

26P. A 2.0 kg particle-like object moves in a plane with velocity components $v_x = 30$ m/s and $v_y = 60$ m/s as it passes through the point with (x, y) coordinates of (3.0, −4.0) m. Just then, what is its angular momentum relative to (a) the origin and (b) the point (−2.0, −2.0) m?

27P. Two particles, each of mass m and speed v, travel in opposite directions along parallel lines separated by a distance d. (a) In terms of m, v, and d, find an expression for the magnitude L of the angular momentum of the two-particle system around a point midway between the two lines. (b) Does the expression change if the point about which L is calculated is not midway between the lines? (c) Now reverse the direction of travel for one of the particles and repeat (a) and (b). ssm

28P. A 4.0 kg particle moves in an xy plane. At the instant when the particle's position and velocity are $\vec{r} = (2.0\hat{i} + 4.0\hat{j})$ m and $\vec{v} = -4.0\hat{j}$ m/s, the force on the particle is $\vec{F} = -3.0\hat{i}$ N. At this instant, determine (a) the particle's angular momentum about the origin, (b) the particle's angular momentum about the point $x = 0$,

$y = 4.0$ m, (c) the torque acting on the particle about the origin, and (d) the torque acting on the particle about the point $x = 0$, $y = 4.0$ m.

SEC. 12-7 Newton's Second Law in Angular Form

29E. A 3.0 kg particle with velocity $\vec{v} = (5.0 \text{ m/s})\hat{i} - (6.0 \text{ m/s})\hat{j}$ is at $x = 3.0$ m, $y = 8.0$ m. It is pulled by a 7.0 N force in the negative x direction. (a) What is the angular momentum of the particle about the origin? (b) What torque about the origin acts on the particle? (c) At what rate is the angular momentum of the particle changing with time? ssm ilw

30E. A particle is acted on by two torques about the origin: $\vec{\tau}_1$ has a magnitude of 2.0 N · m and is directed in the positive direction of the x axis, and $\vec{\tau}_2$ has a magnitude of 4.0 N · m and is directed in the negative direction of the y axis. What are the magnitude and direction of $d\vec{\ell}/dt$, where $\vec{\ell}$ is the angular momentum of the particle about the origin?

31E. What torque about the origin acts on a particle moving in the xy plane, clockwise about the origin, if the particle has the following magnitudes of angular momentum about the origin:
(a) $4.0 \text{ kg} \cdot \text{m}^2/\text{s}$,
(b) $4.0t^2 \text{ kg} \cdot \text{m}^2/\text{s}$,
(c) $4.0\sqrt{t} \text{ kg} \cdot \text{m}^2/\text{s}$,
(d) $4.0/t^2 \text{ kg} \cdot \text{m}^2/\text{s}$?

32P. At time $t = 0$, a 2.0 kg particle has position vector $\vec{r} = (4.0 \text{ m})\hat{i} - (2.0 \text{ m})\hat{j}$ relative to the origin. Its velocity just then is given by $\vec{v} = (-6.0t^2 \text{ m/s})\hat{i}$. About the origin and for $t > 0$, what are (a) the particle's angular momentum and (b) the torque acting on the particle? (c) Repeat (a) and (b) about a point with coordinates $(-2.0 \text{ m}, -3.0 \text{ m}, 0)$ instead of about the origin.

SEC. 12-9 The Angular Momentum of a Rigid Body Rotating About a Fixed Axis

33E. The angular momentum of a flywheel having a rotational inertia of $0.140 \text{ kg} \cdot \text{m}^2$ about its central axis decreases from 3.00 to $0.800 \text{ kg} \cdot \text{m}^2/\text{s}$ in 1.50 s. (a) What is the magnitude of the average torque acting on the flywheel about its central axis during this period? (b) Assuming a constant angular acceleration, through what angle does the flywheel turn? (c) How much work is done on the wheel? (d) What is the average power of the flywheel? ssm

34E. A sanding disk with rotational inertia $1.2 \times 10^{-3} \text{ kg} \cdot \text{m}^2$ is attached to an electric drill whose motor delivers a torque of 16 N · m. Find (a) the angular momentum of the disk about its central axis and (b) the angular speed of the disk 33 ms after the motor is turned on.

35E. Three particles, each of mass m, are fastened to each other and to a rotation axis at O by three massless strings, each with length d as shown in Fig. 12-37. The combination rotates around the rotational axis with angular velocity ω in such a way that the particles remain in a straight line. In terms of m, d, and ω, and relative to point O, what are (a) the rotational inertia of the combination, (b) the angular momentum of the middle particle, and (c) the total angular momentum of the three particles? ssm

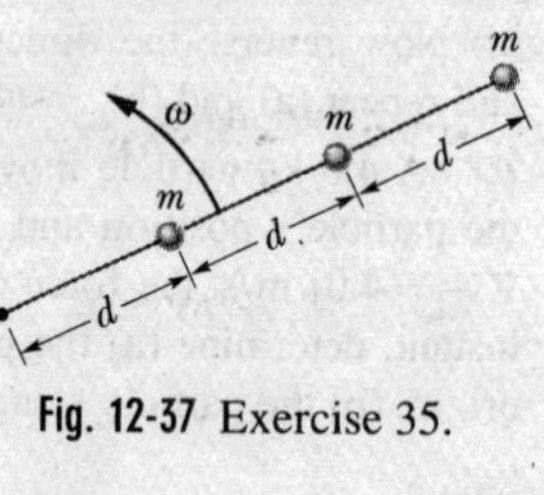

Fig. 12-37 Exercise 35.

36P. An impulsive force $F(t)$ acts for a short time Δt on a rotating rigid body with rotational inertia I. Show that

$$\int \tau\, dt = F_{avg} R\, \Delta t = I(\omega_f - \omega_i),$$

where τ is the torque due to the force, R is the moment arm of the force, F_{avg} is the average value of the force during the time it acts on the body, and ω_i and ω_f are the angular velocities of the body just before and just after the force acts. (The quantity $\int \tau\, dt = F_{avg} R\, \Delta t$ is called the *angular impulse,* in analogy with $F_{avg}\, \Delta t$, the linear impulse.)

37P*. Two cylinders having radii R_1 and R_2 and rotational inertias I_1 and I_2 about their central axes are supported by axles perpendicular to the plane of Fig. 12-38. The large cylinder is initially rotating clockwise with angular velocity ω_0. The small cylinder is moved to the right until it touches the large cylinder and is caused to rotate by the frictional force between the two. Eventually, slipping ceases, and the two cylinders rotate at constant rates in opposite directions. Find the final angular velocity ω_2 of the small cylinder in terms of I_1, I_2, R_1, R_2, and ω_0. (*Hint:* Neither angular momentum nor kinetic energy is conserved. Apply the angular impulse equation of Problem 36.) ssm www

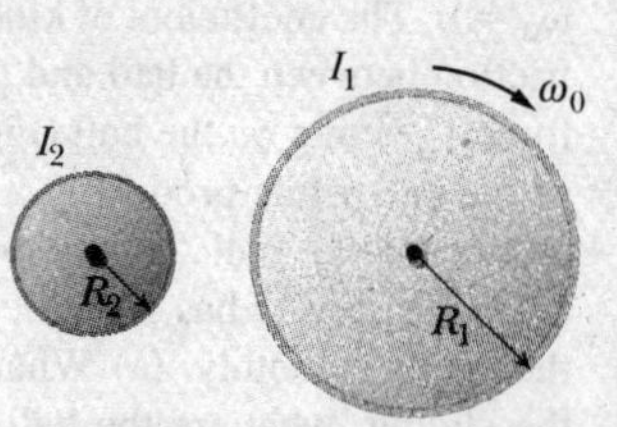

Fig. 12-38 Problem 37.

38P. Figure 12-39 shows a rigid structure consisting of a circular hoop of radius R and mass m, and a square made of four thin bars, each of length R and mass m. The rigid structure rotates at a constant speed about a vertical axis, with a period of rotation of 2.5 s. Assuming $R = 0.50$ m and $m = 2.0$ kg, calculate (a) the structure's rotational inertia about the axis of rotation and (b) its angular momentum about that axis.

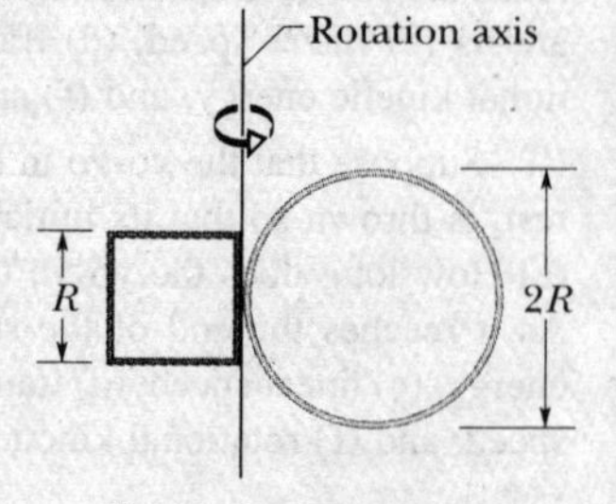

Fig. 12-39 Problem 38.

SEC. 12-10 Conservation of Angular Momentum

39E. A man stands on a platform that is rotating (without friction) with an angular speed of 1.2 rev/s; his arms are outstretched and he holds a brick in each hand. The rotational inertia of the system consisting of the man, bricks, and platform about the central axis is $6.0 \text{ kg} \cdot \text{m}^2$. If by moving the bricks the man decreases the rotational inertia of the system to $2.0 \text{ kg} \cdot \text{m}^2$, (a) what is the resulting angular speed of the platform and (b) what is the ratio of the new kinetic energy of the system to the original kinetic energy? (c) What provided the added kinetic energy? ssm

40E. The rotor of an electric motor has rotational inertia $I_m = 2.0 \times 10^{-3} \text{ kg} \cdot \text{m}^2$ about its central axis. The motor is used to change the orientation of the space probe in which it is mounted.

The motor axis is mounted parallel to the axis of the probe, which has rotational inertia $I_p = 12\ \text{kg}\cdot\text{m}^2$ about its axis. Calculate the number of revolutions of the rotor required to turn the probe through 30° about its axis.

41E. A wheel is rotating freely at angular speed 800 rev/min on a shaft whose rotational inertia is negligible. A second wheel, initially at rest and with twice the rotational inertia of the first, is suddenly coupled to the same shaft. (a) What is the angular speed of the resultant combination of the shaft and two wheels? (b) What fraction of the original rotational kinetic energy is lost? ssm ilw

42E. Two disks are mounted on low-friction bearings on the same axle and can be brought together so that they couple and rotate as one unit. (a) The first disk, with rotational inertia $3.3\ \text{kg}\cdot\text{m}^2$ about its central axis, is set spinning at 450 rev/min. The second disk, with rotational inertia $6.6\ \text{kg}\cdot\text{m}^2$ about its central axis, is set spinning at 900 rev/min in the same direction as the first. They then couple together. What is their angular speed after coupling? (b) If instead the second disk is set spinning at 900 rev/min in the direction opposite the first disk's rotation, what is their angular speed and direction of rotation after coupling?

43E. In a playground, there is a small merry-go-round of radius 1.20 m and mass 180 kg. Its radius of gyration (see Problem 43 of Chapter 11) is 91.0 cm. A child of mass 44.0 kg runs at a speed of 3.00 m/s along a path that is tangent to the rim of the initially stationary merry-go-round and then jumps on. Neglect friction between the bearings and the shaft of the merry-go-round. Calculate (a) the rotational inertia of the merry-go-round about its axis of rotation, (b) the magnitude of the angular momentum of the running child about the axis of rotation of the merry-go-round, and (c) the angular speed of the merry-go-round and child after the child has jumped on. ssm

44E. The rotational inertia of a collapsing spinning star changes to $\frac{1}{3}$ its initial value. What is the ratio of the new rotational kinetic energy to the initial rotational kinetic energy?

45P. A track is mounted on a large wheel that is free to turn with negligible friction about a vertical axis (Fig. 12-40). A toy train of mass m is placed on the track and, with the system initially at rest, the electrical power is turned on. The train reaches a steady speed v with respect to the track. What is the angular speed of the wheel if its mass is M and its radius is R? (Treat the wheel as a hoop, and neglect the mass of the spokes and hub.) ssm www

Fig. 12-40 Problem 45.

46P. In Fig. 12-41, two skaters, each of mass 50 kg, approach each other along parallel paths separated by 3.0 m. They have opposite velocities of 1.4 m/s each. One skater carries one end of a long pole with negligible mass, and the other skater grabs the other end of it as she passes. Assume frictionless ice. (a) Describe quantitatively the motion of the skaters after they have become connected by the pole. (b) What is the kinetic energy of the two-skater system?

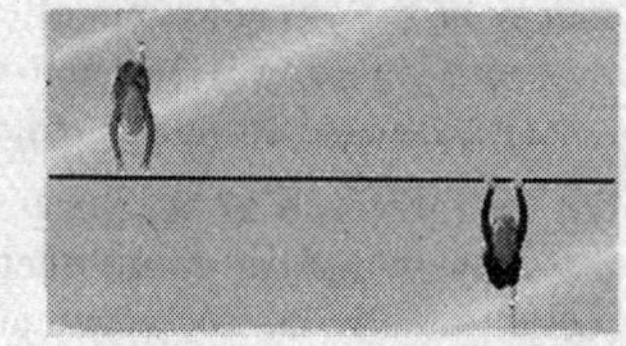

Fig. 12-41 Problem 46.

Next, the skaters each pull along the pole so as to reduce their separation to 1.0 m. What then are (c) their angular speed and (d) the kinetic energy of the system? (e) Explain the source of the increased kinetic energy.

47P. A cockroach of mass m runs counterclockwise around the rim of a lazy Susan (a circular dish mounted on a vertical axle) of radius R and rotational inertia I and having frictionless bearings. The cockroach's speed (relative to the ground) is v, whereas the lazy Susan turns clockwise with angular speed ω_0. The cockroach finds a bread crumb on the rim and, of course, stops. (a) What is the angular speed of the lazy Susan after the cockroach stops? (b) Is mechanical energy conserved as they stop?

48P. A girl of mass M stands on the rim of a frictionless merry-go-round of radius R and rotational inertia I that is not moving. She throws a rock of mass m horizontally in a direction that is tangent to the outer edge of the merry-go-round. The speed of the rock, relative to the ground, is v. Afterward, what are (a) the angular speed of the merry-go-round and (b) the linear speed of the girl?

49P. A horizontal vinyl record of mass 0.10 kg and radius 0.10 m rotates freely about a vertical axis through its center with an angular speed of 4.7 rad/s. The rotational inertia of the record about its axis of rotation is $5.0 \times 10^{-4}\ \text{kg}\cdot\text{m}^2$. A wad of wet putty of mass 0.020 kg drops vertically onto the record from above and sticks to the edge of the record. What is the angular speed of the record immediately after the putty sticks to it?

50P. A uniform thin rod of length 0.50 m and mass 4.0 kg can rotate in a horizontal plane about a vertical axis through its center. The rod is at rest when a 3.0 g bullet traveling in the horizontal plane of the rod is fired into one end of the rod. As viewed from above, the direction of the bullet's velocity makes an angle of 60° with the rod (Fig. 12-42). If the bullet lodges in the rod and the angular velocity of the rod is 10 rad/s immediately after the collision, what is the bullet's speed just before impact?

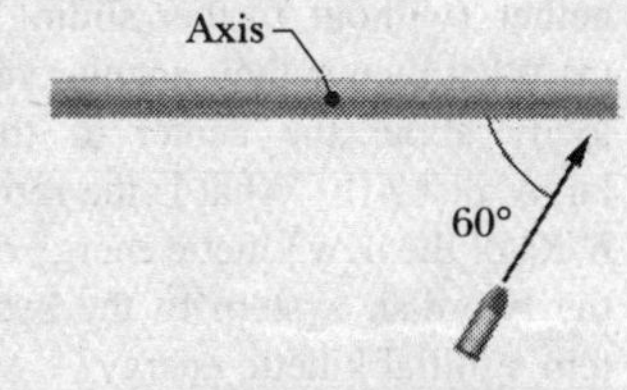

Fig. 12-42 Problem 50.

51P*. Two 2.00 kg balls are attached to the ends of a thin rod of negligible mass, 50.0 cm long. The rod is free to rotate in a vertical plane without friction about a horizontal axis through its center. With the rod initially horizontal (Fig. 12-43), a 50.0 g wad of wet putty drops onto one of the balls, hitting it with a speed of 3.00 m/s and then sticking to it. (a) What is the angular speed of the system just after the putty wad hits? (b) What is the ratio of the kinetic

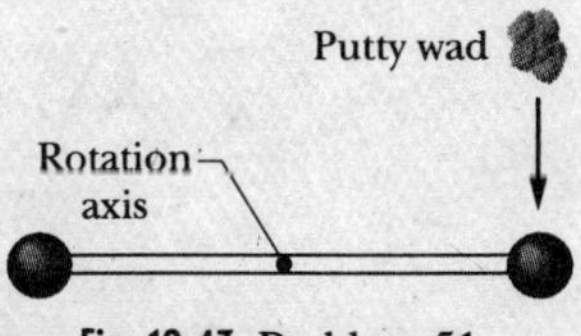

Fig. 12-43 Problem 51.

energy of the entire system after the collision to that of the putty wad just before? (c) Through what angle will the system rotate until it momentarily stops? **ssm**

52P. A cockroach of mass m lies on the rim of a uniform disk of mass $10.0m$ that can rotate freely about its center like a merry-go-round. Initially the cockroach and disk rotate together with an angular velocity of ω_0. Then the cockroach walks halfway to the center of the disk. (a) What is the change $\Delta\omega$ in the angular velocity of the cockroach–disk system? (b) What is the ratio K/K_0 of the new kinetic energy of the system to its initial kinetic energy? (c) What accounts for the change in the kinetic energy?

53P. If Earth's polar ice caps fully melted and the water returned to the oceans, the oceans would be deeper by about 30 m. What effect would this have on Earth's rotation? Make an estimate of the resulting change in the length of the day. (Concern has been expressed that warming of the atmosphere resulting from industrial pollution could cause the ice caps to melt.)

54P. A horizontal platform in the shape of a circular disk rotates on a frictionless bearing about a vertical axle through the center of the disk. The platform has a mass of 150 kg, a radius of 2.0 m, and a rotational inertia of 300 kg · m^2 about the axis of rotation. A 60 kg student walks slowly from the rim of the platform toward the center. If the angular speed of the system is 1.5 rad/s when the student starts at the rim, what is the angular speed when she is 0.50 m from the center?

55P. A uniform disk of mass $10m$ and radius $3.0r$ can rotate freely about its fixed center like a merry-go-round. A smaller uniform disk of mass m and radius r lies on top of the larger disk, concentric with it. Initially the two disks rotate together with an angular velocity of 20 rad/s. Then a slight disturbance causes the smaller disk to slide outward across the larger disk, until the outer edge of the smaller disk catches on the outer edge of the larger disk. Afterward, the two disks again rotate together (without further sliding). (a) What then is their angular velocity about the center of the larger disk? (b) What is the ratio K/K_0 of the new kinetic energy of the two-disk system to the system's initial kinetic energy?

56P. A 30 kg child stands on the edge of a stationary merry-go-round of mass 100 kg and radius 2.0 m. The rotational inertia of the merry-go-round about its axis of rotation is 150 kg · m^2. The child catches a ball of mass 1.0 kg thrown by a friend. Just before the ball is caught, it has a horizontal velocity of 12 m/s that makes an angle of 37° with a line tangent to the outer edge of the merry-go-round, as shown in the overhead view of Fig. 12-44. What is the angular speed of the merry-go-round just after the ball is caught?

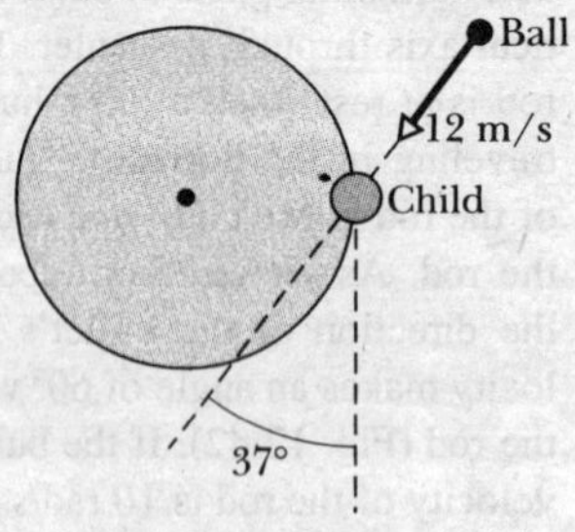

Fig. 12-44 Problem 56.

57P. In Fig. 12-45, a 1.0 g bullet is fired into a 0.50 kg block that is mounted on the end of a 0.60 m nonuniform rod of mass 0.50 kg. The block–rod–bullet system then rotates about a fixed axis at point A. The rotational inertia of the rod alone about A is 0.060 kg · m^2. Assume the block is small enough to treat as a particle on the end of the rod. (a) What is the rotational inertia of the block–rod–bullet system about point A? (b) If the angular speed of the system about A just after the bullet's impact is 4.5 rad/s, what is the speed of the bullet just before the impact?

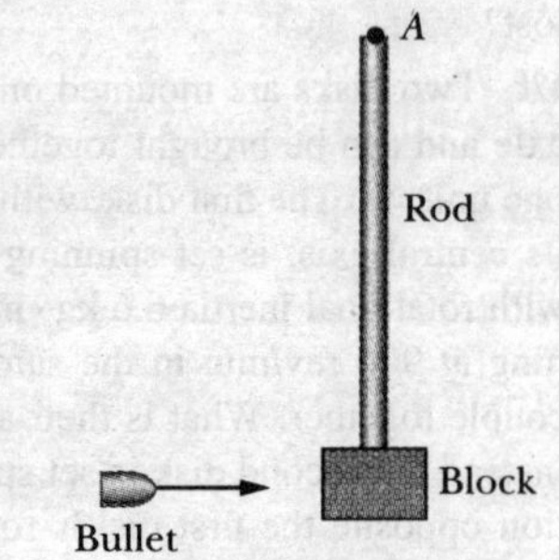

Fig. 12-45 Problem 57.

58P. In Fig. 12-46, a uniform rod (length = 0.60 m, mass = 1.0 kg) rotates about an axis through one end, with a rotational inertia of 0.12 kg · m^2. As the rod swings through its lowest position, the end of the rod collides with a small 0.20 kg putty wad that sticks to the end of the rod. If the angular speed of the rod just before the collision is 2.4 rad/s, what is the angular speed of the rod–putty system immediately after the collision?

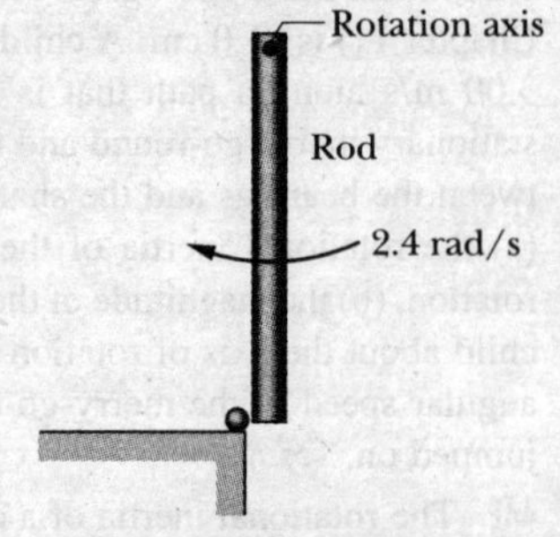

Fig. 12-46 Problem 58.

59P*. The particle of mass m in Fig. 12-47 slides down the frictionless surface through height h and collides with the uniform vertical rod (of mass M and length d), sticking to it. The rod pivots about point O through the angle θ before momentarily stopping. Find θ.

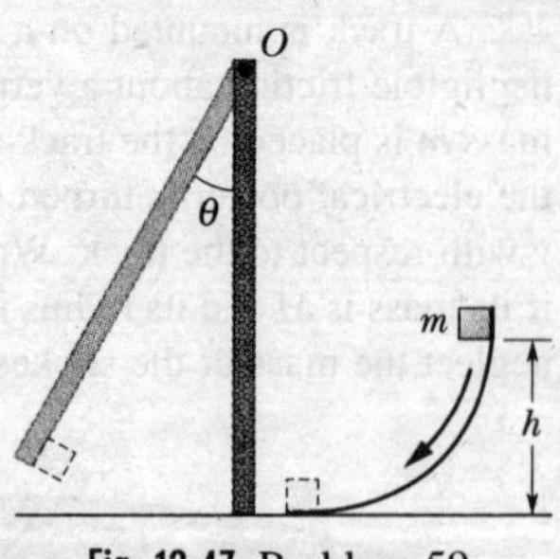

Fig. 12-47 Problem 59.

13 Equilibrium and Elasticity

Rock climbing may be the ultimate physics exam. Failure can mean death, and even "partial credit" can mean severe injury. For example, in a long chimney climb, in which your torso is pressed against one wall of a wide vertical fissure and your feet are pressed against the opposite wall, you need to rest occasionally or you will fall due to exhaustion. Here the exam consists of a single question: What can you do to relax your push on the walls in order to rest? If you relax without considering the physics, the walls will not hold you up.

What is the answer to this life-and-death, one-question exam?

The answer is in this chapter.

Fig. 13-1 A balancing rock near Petrified Forest National Park in Arizona. Although its perch seems precarious, the rock is in static equilibrium.

13-1 Equilibrium

Consider these objects: (1) a book resting on a table, (2) a hockey puck sliding across a frictionless surface with constant velocity, (3) the rotating blades of a ceiling fan, and (4) the wheel of a bicycle that is traveling along a straight path at constant speed. For each of these four objects:

1. The linear momentum $\vec{P}$ of its center of mass is constant.
2. Its angular momentum $\vec{L}$ about its center of mass, or about any other point, is also constant.

We say that such objects are in **equilibrium.** The two requirements for equilibrium are then

$$\vec{P} = \text{a constant} \quad \text{and} \quad \vec{L} = \text{a constant}. \tag{13-1}$$

Our concern in this chapter is with situations in which the constants in Eq. 13-1 are in fact zero; that is, we are concerned largely with objects that are not moving in any way—either in translation or in rotation—in the reference frame from which we observe them. Such objects are in **static equilibrium.** Of the four objects mentioned at the beginning of this section, only one—the book resting on the table—is in static equilibrium.

The balancing rock of Fig. 13-1 is another example of an object that, for the present at least, is in static equilibrium. It shares this property with countless other structures, such as cathedrals, houses, filing cabinets, and taco stands, that remain stationary over time.

As we discussed in Section 8-5, if a body returns to a state of static equilibrium after having been displaced from it by a force, the body is said to be in *stable* static equilibrium. A marble placed at the bottom of a hemispherical bowl is an example. However, if a small force can displace the body and end the equilibrium, the body is in *unstable* static equilibrium.

For example, suppose we balance a domino with the domino's center of mass vertically above the supporting edge, as in Fig. 13-2*a*. The torque about the supporting edge due to the gravitational force $\vec{F}_g$ on the domino is zero, because the line of action of $\vec{F}_g$ is through that edge. Thus, the domino is in equilibrium. Of course, even a slight force on it due to some chance disturbance ends the equilibrium. As the line of action of $\vec{F}_g$ moves to one side of the supporting edge (as in Fig. 13-2*b*), the torque due to $\vec{F}_g$ increases the rotation of the domino. Therefore, the domino in Fig. 13-2*a* is in unstable static equilibrium.

The domino in Fig. 13-2*c* is not quite as unstable. To topple this domino, a force would have to rotate it through and then beyond the balance position of Fig. 13-2*a*, in which the center of mass is above a supporting edge. A slight force will not topple this domino, but a vigorous flick of the finger against the domino certainly will. (If we arrange a chain of such upright dominos, a finger flick against the first can cause the whole chain to fall.)

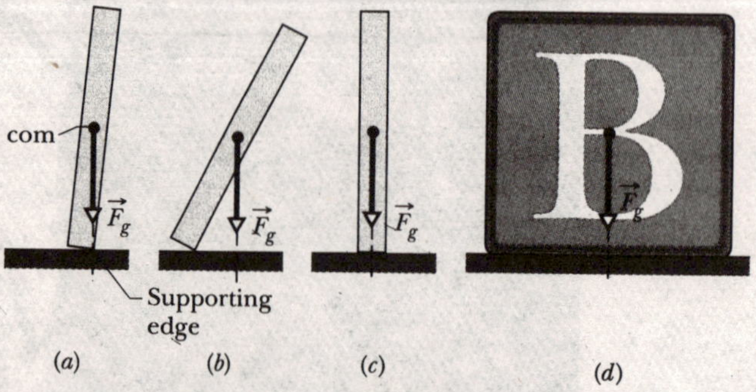

Fig. 13-2 (*a*) A domino balanced on one edge, with its center of mass vertically above that edge. The gravitational force $\vec{F}_g$ on the domino is directed through the supporting edge. (*b*) If the domino is rotated even slightly from the balanced orientation, then $\vec{F}_g$ causes a torque that increases the rotation. (*c*) A domino upright on a narrow side is somewhat more stable than the domino in (*a*). (*d*) A square block is even more stable.

Fig. 13-3 A construction worker balanced above New York City is in static equilibrium but is more stable parallel to the beam than perpendicular to it.

The child's square block in Fig. 13-2*d* is even more stable because its center of mass would have to be moved even farther to get it to pass above a supporting edge. A flick of the finger may not topple the block. (This is why you never see a chain of toppling square blocks.) The worker in Fig. 13-3 is like both the domino and the square block: Parallel to the beam, his stance is wide and he is stable; perpendicular to the beam, his stance is narrow and he is unstable (and at the mercy of a chance gust of wind).

The analysis of static equilibrium is very important in engineering practice. The design engineer must isolate and identify all the external forces and torques that may act on a structure and, by good design and wise choice of materials, ensure that the structure will remain stable under these loads. Such analysis is necessary to ensure, for example, that bridges do not collapse under their traffic and wind loads, and that the landing gear of aircraft will survive the shock of rough landings.

13-2 The Requirements of Equilibrium

The translational motion of a body is governed by Newton's second law in its linear momentum form, given by Eq. 9-27 as

$$\vec{F}_{\text{net}} = \frac{d\vec{P}}{dt}. \tag{13-2}$$

If the body is in translational equilibrium—that is, if $\vec{P}$ is a constant—then $d\vec{P}/dt = 0$ and we must have

$$\vec{F}_{\text{net}} = 0 \quad \text{(balance of forces).} \tag{13-3}$$

The rotational motion of a body is governed by Newton's second law in its angular momentum form, given by Eq. 12-29 as

$$\vec{\tau}_{\text{net}} = \frac{d\vec{L}}{dt}. \tag{13-4}$$

If the body is in rotational equilibrium—that is, if $\vec{L}$ is a constant—then $d\vec{L}/dt = 0$ and we must have

$$\vec{\tau}_{\text{net}} = 0 \quad \text{(balance of torques).} \tag{13-5}$$

Thus, the two requirements for a body to be in equilibrium are as follows:

1. **The vector sum of all the external forces that act on the body must be zero.**
2. **The vector sum of all the external torques that act on the body, measured about *any* possible point, must also be zero.**

These requirements obviously hold for *static* equilibrium. They also hold for the more general equilibrium in which $\vec{P}$ and $\vec{L}$ are constant but not zero.

Equations 13-3 and 13-5, as vector equations, are each equivalent to three independent component equations, one for each direction of the coordinate axes:

Balance of forces	Balance of torques
$F_{\text{net},x} = 0$	$\tau_{\text{net},x} = 0$
$F_{\text{net},y} = 0$	$\tau_{\text{net},y} = 0$
$F_{\text{net},z} = 0$	$\tau_{\text{net},z} = 0$

(13-6)

We shall simplify matters by considering only situations in which the forces that act on the body lie in the xy plane. This means that the only torques that can act on the body must tend to cause rotation around an axis parallel to the z axis. With this assumption, we eliminate one force equation and two torque equations from Eqs. 13-6, leaving

$$F_{\text{net},x} = 0 \quad \text{(balance of forces)}, \tag{13-7}$$

$$F_{\text{net},y} = 0 \quad \text{(balance of forces)}, \tag{13-8}$$

$$\tau_{\text{net},z} = 0 \quad \text{(balance of torques)}. \tag{13-9}$$

Here, $\tau_{\text{net},z}$ is the net torque that the external forces produce either about the z axis or about *any* axis parallel to it.

A hockey puck sliding at constant velocity over ice satisfies Eqs. 13-7, 13-8, and 13-9 and is thus in equilibrium *but not in static equilibrium*. For static equilibrium, the linear momentum $\vec{P}$ of the puck must be not only constant but also zero; the puck must be at rest on the ice. Thus, there is another requirement for static equilibrium:

> **3. The linear momentum $\vec{P}$ of the body must be zero.**

CHECKPOINT 1: The figure gives six overhead views of a uniform rod on which two or more forces act perpendicularly to the rod. If the magnitudes of the forces are adjusted properly (but kept nonzero), in which situations can the rod be in static equilibrium?

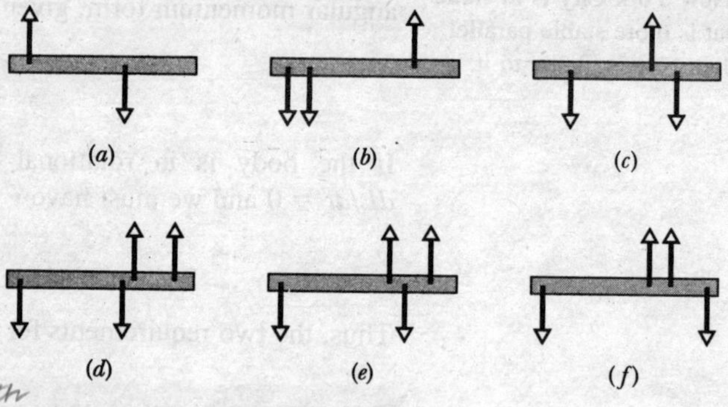

13-3 The Center of Gravity

The gravitational force on an extended body is the vector sum of the gravitational forces acting on the individual elements (the atoms) of the body. Instead of considering all those individual elements, we can say:

> **The gravitational force $\vec{F}_g$ on a body effectively acts at a single point, called the center of gravity (cog) of the body.**

Here the word "effectively" means that if the forces on the individual elements were somehow turned off and force $\vec{F}_g$ at the center of gravity were turned on, the net force and the net torque (about any point) acting on the body would not change.

Until now, we have assumed that the gravitational force $\vec{F}_g$ acts at the center of mass (com) of the body. This is equivalent to assuming that the center of gravity is

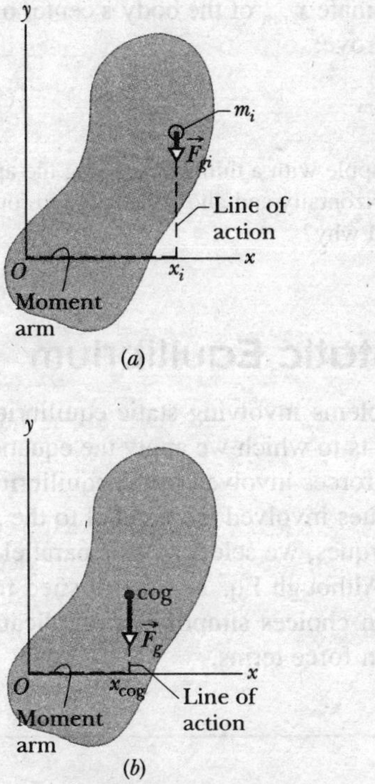

Fig. 13-4 (*a*) An element of mass m_i in an extended body. The gravitational force $\vec{F}_{gi}$ on it has moment arm x_i about the origin O of the coordinate system. (*b*) The gravitational force $\vec{F}_g$ on a body is said to act at the center of gravity (cog) of the body. Here it has moment arm x_{cog} about origin O.

at the center of mass. Recall that, for a body of mass M, the force $\vec{F}_g$ is equal to $M\vec{g}$, where $\vec{g}$ is the acceleration that the force would produce if the body were to fall freely. In the proof that follows, we show that

> If $\vec{g}$ is the same for all elements of a body, then the body's center of gravity (cog) is coincident with the body's center of mass (com).

This is approximately true for everyday objects because $\vec{g}$ varies only a little along Earth's surface and decreases in magnitude only slightly with altitude. Thus, for objects like a mouse or a moose, we have been justified in assuming that the gravitational force acts at the center of mass. After the following proof, we shall resume that assumption.

Proof

First, we consider the individual elements of the body. Figure 13-4*a* shows an extended body, of mass M, and one of its elements, of mass m_i. A gravitational force $\vec{F}_{gi}$ acts on each such element and is equal to $m_i\vec{g}_i$. The subscript on $\vec{g}_i$ means $\vec{g}_i$ is the gravitational acceleration *at the location of the element* (it can be different for other elements).

In Fig. 13-4*a*, each force $\vec{F}_{gi}$ produces a torque τ_i on the element about the origin O, with moment arm x_i. Using Eq. 11-33 ($\tau = r_\perp F$), we can write torque τ_i as

$$\tau_i = x_i F_{gi}. \tag{13-10}$$

The net torque on all the elements of the body is then

$$\tau_{net} = \sum \tau_i = \sum x_i F_{gi}. \tag{13-11}$$

Next, we consider the body as a whole. Figure 13-4*b* shows the gravitational force $\vec{F}_g$ acting at the body's center of gravity. This force produces a torque τ on the body about O, with moment arm x_{cog}. Again using Eq. 11-33, we can write this torque as

$$\tau = x_{cog} F_g. \tag{13-12}$$

The gravitational force $\vec{F}_g$ on the body is equal to the sum of the gravitational forces $\vec{F}_{gi}$ on all its elements, so we can substitute ΣF_{gi} for F_g in Eq. 13-12 to write

$$\tau = x_{cog} \sum F_{gi}. \tag{13-13}$$

Now recall that the torque due to force $\vec{F}_g$ acting at the center of gravity is equal to the net torque due to all the forces $\vec{F}_{gi}$ acting on all the elements of the body. (That is how we defined the cog.) Therefore, τ in Eq. 13-13 is equal to τ_{net} in Eq. 13-11. Putting those two equations together, we can write

$$x_{cog} \sum F_{gi} = \sum x_i F_{gi}.$$

Substituting $m_i g_i$ for F_{gi} gives us

$$x_{cog} \sum m_i g_i = \sum x_i m_i g_i.$$

Now here is a key idea: If the accelerations g_i at all the locations of the elements are the same, we can cancel g_i from this equation to write

$$x_{cog} \sum m_i = \sum x_i m_i. \tag{13-14}$$

The sum Σm_i of the masses of all the elements is the mass M of the body. Therefore, we can rewrite Eq. 13-14 as

$$x_{cog} = \frac{1}{M} \sum x_i m_i. \tag{13-15}$$

The right side of this equation gives the coordinate x_{com} of the body's center of mass (Eq. 9-4). We now have what we sought to prove:

$$x_{cog} = x_{com}. \tag{13-16}$$

✔CHECKPOINT 2: Suppose that you skewer an apple with a thin rod, missing the apple's center of gravity. When you hold the rod horizontally and allow the apple to rotate freely, where does the center of gravity end up and why?

13-4 Some Examples of Static Equilibrium

In this section we examine four sample problems involving static equilibrium. In each, we select a system of one or more objects to which we apply the equations of equilibrium (Eqs. 13-7, 13-8, and 13-9). The forces involved in the equilibrium are all in the *xy* plane, which means that the torques involved are parallel to the *z* axis. Thus, in applying Eq. 13-9, the balance of torques, we select an axis parallel to the *z* axis about which to calculate the torques. Although Eq. 13-9 is satisfied for *any* such choice of axis, you will see that certain choices simplify the application of Eq. 13-9 by eliminating one or more unknown force terms.

Sample Problem 13-1

In Fig. 13-5*a*, a uniform beam, of length L and mass $m = 1.8$ kg, is at rest with its ends on two scales. A uniform block, with mass $M = 2.7$ kg, is at rest on the beam, with its center a distance $L/4$ from the beam's left end. What do the scales read?

SOLUTION: The first steps in the solution of *any* problem about static equilibrium are these: Clearly define the system to be analyzed and then draw a free-body diagram of it, indicating all the forces on the system. Here, let us choose the system as the beam and block taken together. Then the forces on the system are shown in the free-body diagram of Fig. 13-5*b*. (Choosing the system takes experience and often there can be more than one good choice; see Problem Solving Tactic 1 below.)

The normal forces on the beam from the scales are $\vec{F}_l$ on the left and $\vec{F}_r$ on the right. The scale readings that we want are equal to the magnitudes of those forces. The gravitational force $\vec{F}_{g,beam}$ on the beam acts at the beam's center of mass and is equal to $m\vec{g}$. Similarly, the gravitational force $\vec{F}_{g,block}$ on the block acts at the block's center of mass and is equal to $M\vec{g}$. However, to simplify Fig. 13-5*b*, the block is represented by a dot within the boundary of the beam and the vector $\vec{F}_{g,block}$ is drawn with its tail on that dot. (This downward shift of vector $\vec{F}_{g,block}$ along its line of action does not alter the torque due to $\vec{F}_{g,block}$ about any axis perpendicular to the figure.)

The **Key Idea** here is that, because the system is in static equilibrium, we can apply the balance of forces equations (Eqs. 13-7 and 13-8) and the balance of torques equation (13-9) to it. The forces have no x components, so Eq. 13-7 ($F_{net,x} = 0$) provides no information. For the y components, Eq. 13-8 ($F_{net,y} = 0$) gives us

$$F_l + F_r - Mg - mg = 0. \tag{13-17}$$

This equation contains two unknowns, the forces F_l and F_r, so we also need to use Eq. 13-9, the balance of torques equation.

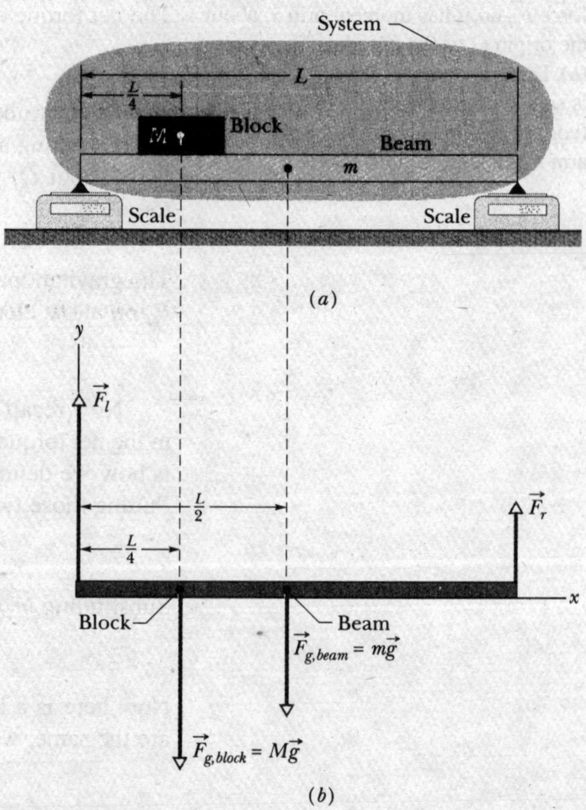

Fig. 13-5 Sample Problem 13-1. (*a*) A beam of mass m supports a block of mass M. (*b*) A free-body diagram, showing the forces that act on the system *beam + block*.

We can apply it to *any* rotation axis perpendicular to the plane of Fig. 13-5. Let us choose a rotation axis through the left end of the beam. We shall also use our general rule for assigning signs to torques: If a torque would cause an initially stationary body to rotate clockwise about the rotation axis, the torque is negative. If the rotation would be counterclockwise, the torque is positive. Finally, we shall write the torques in the form $r_\perp F$, where the moment arm $r_\perp$ is 0 for $\vec{F}_l$, $L/4$ for $M\vec{g}$, $L/2$ for $m\vec{g}$, and L for $\vec{F}_r$.

We now can write the balancing equation ($\tau_{\text{net},z} = 0$) as

$$(0)(F_l) - (L/4)(Mg) - (L/2)(mg) + (L)(F_r) = 0,$$

which gives us

$$\begin{aligned} F_r &= \tfrac{1}{4}Mg + \tfrac{1}{2}mg \\ &= \tfrac{1}{4}(2.7\text{ kg})(9.8\text{ m/s}^2) + \tfrac{1}{2}(1.8\text{ kg})(9.8\text{ m/s}^2) \\ &= 15.44\text{ N} \approx 15\text{ N}. \qquad \text{(Answer)} \end{aligned}$$

Now, solving Eq. 13-17 for F_l and substituting this result, we find

$$\begin{aligned} F_l &= (M + m)g - F_r \\ &= (2.7\text{ kg} + 1.8\text{ kg})(9.8\text{ m/s}^2) - 15.44\text{ N} \\ &= 28.66\text{ N} \approx 29\text{ N}. \qquad \text{(Answer)} \end{aligned}$$

Notice the strategy in the solution: When we wrote an equation for the balance of force components, we got stuck with two unknowns. If we had written an equation for the balance of torques around some *arbitrary* axis, we would have again gotten stuck with those two unknowns. However, because we chose the axis to pass through the point of application of one of the unknown forces, here $\vec{F}_l$, we did not get stuck. Our choice neatly eliminated that force from the torque equation, allowing us to solve for the other unknown force magnitude F_r. Then we returned to the equation for the balance of force components to find the remaining unknown force magnitude.

CHECKPOINT 3: The figure gives an overhead view of a uniform rod in static equilibrium. (a) Can you find the magnitudes of unknown forces $\vec{F}_1$ and $\vec{F}_2$ by balancing the forces? (b) If you wish to find the magnitude of force $\vec{F}_2$ by using a single equation, where should you place a rotational axis? (c) The magnitude of $\vec{F}_2$ turns out to be 65 N. What then is the magnitude of $\vec{F}_1$?

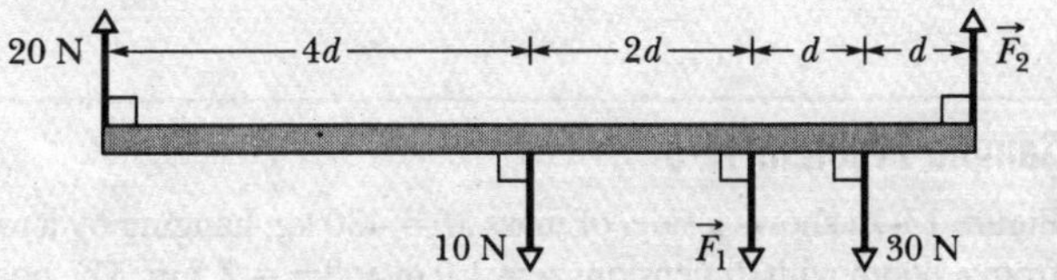

Sample Problem 13-2

In Fig. 13-6*a*, a ladder of length $L = 12$ m and mass $m = 45$ kg leans against a slick (frictionless) wall. Its upper end is at height $h = 9.3$ m above the pavement on which the lower end rests (the pavement is not frictionless). The ladder's center of mass is $L/3$ from the lower end. A firefighter of mass $M = 72$ kg climbs the ladder until her center of mass is $L/2$ from the lower end. What then are the magnitudes of the forces on the ladder from the wall and the pavement?

SOLUTION: First, we choose our system as being the firefighter and ladder, together, and then we draw the free-body diagram of Fig. 13-6*b*. The firefighter is represented with a dot within the boundary of the ladder. The gravitational force on her is represented with its equivalent $M\vec{g}$, and that vector has been shifted along its line of action, so that its tail is on the dot. (The shift does not alter a torque due to $M\vec{g}$ about any axis perpendicular to the figure.)

The only force on the ladder from the wall is the horizontal force $\vec{F}_w$ (there cannot be a frictional force along a frictionless wall). The force $\vec{F}_p$ on the ladder from the pavement has a horizontal component $\vec{F}_{px}$ that is a static frictional force and a vertical component $\vec{F}_{py}$ that is a normal force.

A **Key Idea** here is that the system is in static equilibrium, so the balancing equations (Eqs. 13-7 through 13-9) apply to it. Let us start with Eq. 13-9 ($\tau_{\text{net},z} = 0$). To choose an axis about which to calculate the torques, note that we have unknown forces ($\vec{F}_w$ and $\vec{F}_p$) at the two ends of the ladder. To eliminate, say, $\vec{F}_p$ from the calculation, we place the axis at point O, perpendicular to the figure. We also place the origin of an xy coordinate system at O. We can find torques about O with any of Eqs. 11-31 through 11-33, but Eq. 11-33 ($\tau = r_\perp F$) is easiest to use here.

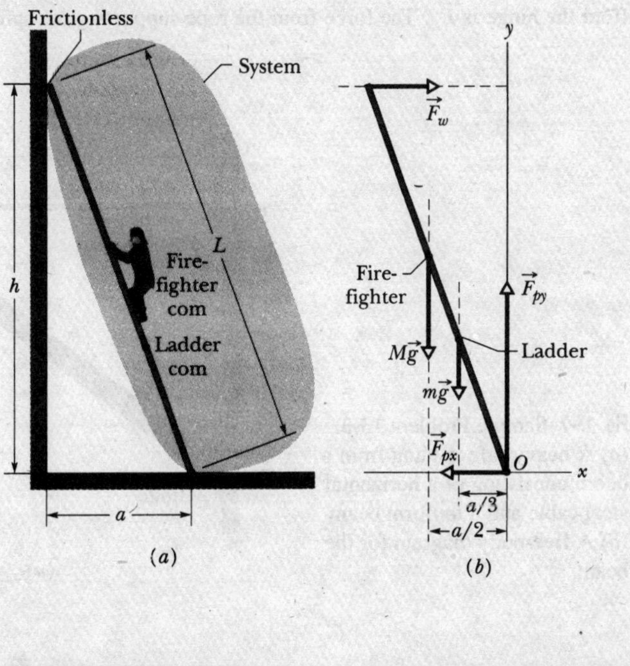

Fig. 13-6 Sample Problem 13-2 (*a*) A firefighter climbs halfway up a ladder that is leaning against a frictionless wall. The pavement beneath the ladder is not frictionless. (*b*) A free-body diagram, showing the forces that act on the firefighter–ladder system. The origin O of a coordinate system is placed at the point of application of the unknown force $\vec{F}_p$ (whose vector components $\vec{F}_{px}$ and $\vec{F}_{py}$ are shown).

To find the moment arm $r_\perp$ of $\vec{F}_w$, we draw a line of action through that vector (Fig. 13-6*b*). Then $r_\perp$ is the perpendicular distance between O and the line of action. In Fig. 13-6*b*, it extends along the y axis and is equal to the height h. We similarly draw lines of action for $M\vec{g}$ and $m\vec{g}$ and see that their moment arms extend along the x axis. For the distance a shown in Fig. 13-6*a*, the moment arms are $a/2$ (the firefighter is halfway up the ladder) and $a/3$ (the ladder's center of mass is one-third of the way up the ladder), respectively. The moment arms for $\vec{F}_{px}$ and $\vec{F}_{py}$ are zero.

Now, with torques written in the form $r_\perp F$, the balancing equation $\tau_{\text{net},z} = 0$ becomes

$$-(h)(F_w) + (a/2)(Mg) + (a/3)(mg) + (0)(F_{px}) + (0)(F_{py}) = 0. \quad (13\text{-}18)$$

(Recall our rule: A positive torque corresponds to counterclockwise rotation and a negative torque corresponds to clockwise rotation.)

Using the Pythagorean theorem, we find that

$$a = \sqrt{L^2 - h^2} = 7.58 \text{ m}.$$

Then Eq. 13-18 gives us

$$F_w = \frac{ga(M/2 + m/3)}{h}$$
$$= \frac{(9.8 \text{ m/s}^2)(7.58 \text{ m})(72/2 \text{ kg} + 45/3 \text{ kg})}{9.3 \text{ m}}$$
$$= 407 \text{ N} \approx 410 \text{ N}. \quad \text{(Answer)}$$

Now we need to use the force balancing equations. The equation $F_{\text{net},x} = 0$ gives us

$$F_w - F_{px} = 0, \quad (13\text{-}19)$$

so

$$F_{px} = F_w = 410 \text{ N}. \quad \text{(Answer)}$$

The equation $F_{\text{net},y} = 0$ gives us

$$F_{py} - Mg - mg = 0, \quad (13\text{-}20)$$

so

$$F_{py} = (M + m)g = (72 \text{ kg} + 45 \text{ kg})(9.8 \text{ m/s}^2)$$
$$= 1146.6 \text{ N} \approx 1100 \text{ N}. \quad \text{(Answer)}$$

Sample Problem 13-3

Figure 13-7*a* shows a safe, of mass $M = 430$ kg, hanging by a rope from a boom with dimensions $a = 1.9$ m and $b = 2.5$ m. The boom consists of a hinged beam and a horizontal cable that connects the beam to a wall. The uniform beam has a mass m of 85 kg; the mass of the cable and rope are negligible.

(a) What is the tension T_c in the cable? In other words, what is the magnitude of the force $\vec{T}_c$ on the beam from the cable?

SOLUTION: The system here is the beam alone, and the forces on it are shown in the free-body diagram of Fig. 13-7*b*. The force from the cable is $\vec{T}_c$. The gravitational force on the beam acts at the beam's center of mass (at the beam's center) and is represented by its equivalent $m\vec{g}$. The vertical component of the force on the beam from the hinge is $\vec{F}_v$, and the horizontal component of the force from the hinge is $\vec{F}_h$. The force from the rope supporting the safe is $\vec{T}_r$. Because beam, rope, and safe are stationary, the magnitude of $\vec{T}_r$ is equal to the weight of the safe: $T_r = Mg$. We place the origin O of an xy coordinate system at the hinge.

One **Key Idea** here is that our system is in static equilibrium, so the balancing equations apply to it. Let us start with Eq. 13-9 ($\tau_{\text{net},z} = 0$). Note that we are asked for the magnitude of force $\vec{T}_c$ and not of forces $\vec{F}_h$ and $\vec{F}_v$ acting at the hinge, at point O. Thus, a second **Key Idea** is that, to eliminate $\vec{F}_h$ and $\vec{F}_v$ from the torque calculation, we should calculate torques about an axis that is perpendicular to the figure at point O. Then $\vec{F}_h$ and $\vec{F}_v$ will have moment arms of zero. The lines of action for $\vec{T}_c$, $\vec{T}_r$, and $m\vec{g}$ are dashed in Fig. 13-7*b*. The corresponding moment arms are a, b, and $b/2$.

Writing torques in the form of $r_\perp F$ and using our rule about signs for torques, the balancing equation $\tau_{\text{net},z} = 0$ becomes

$$(a)(T_c) - (b)(T_r) - (\tfrac{1}{2}b)(mg) = 0.$$

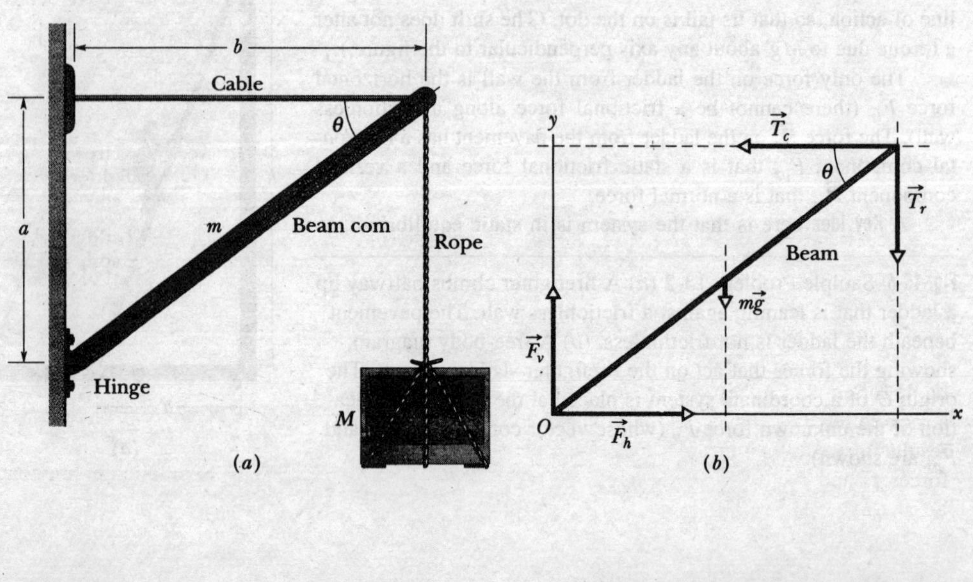

Fig. 13-7 Sample Problem 13-3. (*a*) A heavy safe is hung from a boom consisting of a horizontal steel cable and a uniform beam. (*b*) A free-body diagram for the beam.

Substituting Mg for T_r and solving for T_c, we find that

$$T_c = \frac{gb(M + \frac{1}{2}m)}{a}$$
$$= \frac{(9.8\ \text{m/s}^2)(2.5\ \text{m})(430\ \text{kg} + 85/2\ \text{kg})}{1.9\ \text{m}}$$
$$= 6093\ \text{N} \approx 6100\ \text{N}. \qquad \text{(Answer)}$$

(b) Find the magnitude F of the net force on the beam from the hinge.

SOLUTION: Now we want F_h and F_v so we can combine them to get F. Because we know T_c, our **Key Idea** here is to apply the force balancing equations to the beam. For the horizontal balance, we write $F_{\text{net},x} = 0$ as

$$F_h - T_c = 0,$$

and so

$$F_h = T_c = 6093\ \text{N}.$$

For the vertical balance, we write $F_{\text{net},y} = 0$ as

$$F_v - mg - T_r = 0.$$

Substituting Mg for T_r and solving for F_v, we find that

$$F_v = (m + M)g = (85\ \text{kg} + 430\ \text{kg})(9.8\ \text{m/s}^2)$$
$$= 5047\ \text{N}.$$

From the Pythagorean theorem, we now have

$$F = \sqrt{F_h^2 + F_v^2}$$
$$= \sqrt{(6093\ \text{N})^2 + (5047\ \text{N})^2} \approx 7900\ \text{N}. \qquad \text{(Answer)}$$

Note that F is substantially greater than either the combined weights of the safe and the beam, 5000 N, or the tension in the horizontal wire, 6100 N.

CHECKPOINT 4: In the figure, a stationary 5 kg rod AC is held against a wall by a rope and friction between rod and wall. The uniform rod is 1 m long, and angle $\theta = 30°$. (a) If you are to find the magnitude of the force $\vec{T}$ on the rod from the rope with a single equation, at what labeled point should a rotational axis be placed? With that choice of axis and counterclockwise torques positive, what is the sign of (b) the torque τ_w due to the rod's weight and (c) the torque τ_r due to the pull on the rod by the rope? (d) Is the magnitude of τ_r greater than, less than, or equal to the magnitude of τ_w?

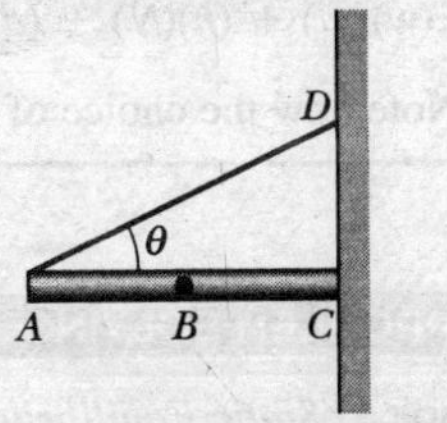

Sample Problem 13-4

In Fig. 13-8, a rock climber with mass $m = 55$ kg rests during a "chimney climb," pressing only with her shoulders and feet against the walls of a fissure of width $w = 1.0$ m. Her center of mass is a horizontal distance $d = 0.20$ m from the wall against which her shoulders are pressed. The coefficient of static friction between her shoes and the wall is $\mu_1 = 1.1$, and between her shoulders and the wall it is $\mu_2 = 0.70$. To rest, the climber wants to minimize her horizontal push on the walls. The minimum occurs when her feet and her shoulders are both on the verge of sliding.

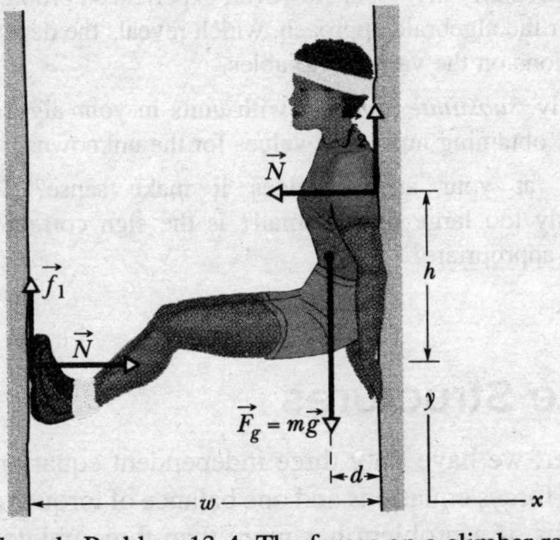

Fig. 13-8 Sample Problem 13-4. The forces on a climber resting in a rock chimney. The push of the climber on the chimney walls results in rise to the normal forces $\vec{N}$ and the static frictional forces $\vec{f}_1$ and $\vec{f}_2$.

(a) What is that minimum horizontal push on the walls?

SOLUTION: Our system is the climber, and Fig. 13-8 shows the forces that act on her. The only horizontal forces are the normal forces $\vec{N}$ on her from the walls, at her feet and shoulders. The static frictional forces on her are $\vec{f}_1$ and $\vec{f}_2$, directed upward. The gravitational force $\vec{F}_g = m\vec{g}$ acts at her center of mass.

A **Key Idea** is that, because the system is in static equilibrium, we can apply the force balancing equations (Eqs. 13-7 and 13-8) to it. The equation $F_{\text{net},x} = 0$ tells us that the two normal forces on her must be equal in magnitude and opposite in direction. We seek the magnitude N of these two forces, which is also the magnitude of her push against either wall.

The balancing equation $F_{\text{net},y} = 0$ gives us

$$f_1 + f_2 - mg = 0. \qquad (13\text{-}21)$$

We want the climber to be on the verge of sliding at both her feet and her shoulders. That means we want the static frictional forces there to be at their maximum values. Those maximum values are, from Eq. 6-1 ($f_{s,\max} = \mu_s N$),

$$f_1 = \mu_1 N \quad \text{and} \quad f_2 = \mu_2 N. \qquad (13\text{-}22)$$

Substituting these expressions into Eq. 13-21 and solving for N give us

$$N = \frac{mg}{\mu_1 + \mu_2} = \frac{(55\ \text{kg})(9.8\ \text{m/s}^2)}{1.1 + 0.70} = 299\ \text{N} \approx 300\ \text{N}.$$

Thus, her minimum horizontal push must be about 300 N.

(b) For that push, what must be the vertical distance h between her feet and her shoulders if she is to be stable?

SOLUTION: A **Key Idea** here is that the climber will be stable if the torque balancing equation ($\tau_{\text{net},z} = 0$) applies to her. This means that the forces on her must not produce a net torque about *any* rotation axis. Another **Key Idea** is that we are free to choose a rotation axis that helps simplify the calculation. We shall write the torques in the form $r_\perp F$, where $r_\perp$ is the moment arm of force F. In Fig. 13-8, we choose a rotation axis at her shoulders, perpendicular to the figure's plane. Then the moment arms of the forces acting there ($\vec{N}$ and $\vec{f}_2$) are zero. Frictional force $\vec{f}_1$, the normal force $\vec{N}$ at her feet, and the gravitational force $\vec{F}_g = m\vec{g}$ have the corresponding moment arms w, h, and d.

Recalling our rule about the signs of torques and the corresponding directions, we can now write $\tau_{\text{net},z} = 0$ as

$$-(w)(f_1) + (h)(N) + (d)(mg) + (0)(f_2) + (0)(N) = 0. \quad (13\text{-}23)$$

(Note how the choice of rotation axis neatly eliminates f_2 from the calculation.) Next, solving Eq. 13-23 for h, setting $f_1 = \mu_1 N$, and substituting $N = 299$ N and other known values, we find that

$$h = \frac{f_1 w - mgd}{N} = \frac{\mu_1 N w - mgd}{N} = \mu_1 w - \frac{mgd}{N}$$

$$= (1.1)(1.0\text{ m}) - \frac{(55\text{ kg})(9.8\text{ m/s}^2)(0.20\text{ m})}{299\text{ N}}$$

$$= 0.739\text{ m} \approx 0.74\text{ m}. \quad \text{(Answer)}$$

We would find the same required value of h if we wrote the torques about any other rotation axis perpendicular to the page, such as one at her feet.

If h is more than *or* less than 0.74 m, she must exert a force greater than 299 N on the walls to be stable. Here, then, is the advantage of knowing the physics before you climb a chimney. When you need to rest, you will avoid the (dire) error of novice climbers who place their feet too high or too low. Instead, you will know that there is a "best" distance between shoulders and feet, requiring the least push, and giving you a good chance to rest.

PROBLEM-SOLVING TACTICS

Tactic 1: *Static Equilibrium Problems*
Here is a list of steps for solving static equilibrium problems:

1. Draw a *sketch* of the problem.
2. Select the *system* to which you will apply the laws of equilibrium, drawing a closed curve around it on your sketch to fix it clearly in your mind. In some situations you can select a single object as the system; it is the object you wish to be in equilibrium (such as the rock climber in Sample Problem 13-4). In other situations, you might include additional objects in the system *if* their inclusion simplifies the calculations for equilibrium. For example, suppose in Sample Problem 13-2 you select only the ladder as the system. Then in Fig. 13-6*b* you will have to account for additional unknown forces exerted on the ladder by the hands and feet of the firefighter. These additional unknowns complicate the equilibrium calculations. The system of Fig. 13-6 was chosen to include the firefighter so that those unknown forces are *internal* to the system and thus need not be found in order to solve Sample Problem 13-2.
3. Draw a *free-body diagram* of the system. Show all the forces that act on the system, labeling them and making sure that their points of application and lines of action are correctly shown.
4. Draw in the *x and y axes* of a coordinate system. Choose them so that at least one axis is parallel to one or more unknown force. Resolve into components the forces that do not lie along one of the axes. In all our sample problems it made sense to choose the x axis horizontal and the y axis vertical.
5. Write the two *balance of forces equations,* using symbols throughout.
6. Choose one or more rotation axes perpendicular to the plane of the figure and write the *balance of torques equation* for each axis. If you choose an axis that passes through the line of action of an unknown force, the equation will be simplified because that force will not appear in it.
7. *Solve* your equations *algebraically* for the unknowns. Some students feel more confident in substituting numbers with units in the independent equations at this stage, especially if the algebra is particularly involved. However, experienced problem solvers prefer the algebraic approach, which reveals the dependence of solutions on the various variables.
8. Finally, *substitute numbers* with units in your algebraic solutions, obtaining numerical values for the unknowns.
9. Look at your answer—does it make sense? Is it obviously too large or too small? Is the sign correct? Are the units appropriate?

13-5 Indeterminate Structures

For the problems of this chapter, we have only three independent equations at our disposal, usually two balance of forces equations and one balance of torques equation about a given rotation axis. Thus, if a problem has more than three unknowns, we cannot solve it.

It is easy to find such problems. In Sample Problem 13-2, for example, we could have assumed that there is friction between the wall and the top of the ladder. Then there would have been a vertical frictional force acting where the ladder touches the wall, making a total of four unknown forces. With only three equations, we could not have solved this problem.

Consider also an unsymmetrically loaded car. What are the forces—all different—on the four tires? Again, we cannot find them because we have only three independent equations with which to work. Similarly, we can solve an equilibrium problem for a table with three legs but not for one with four legs. Problems like these, in which there are more unknowns than equations, are called **indeterminate.**

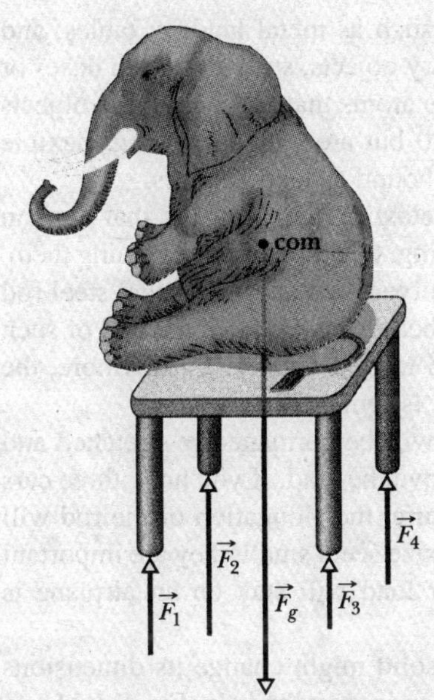

Fig. 13-9 The table is an indeterminate structure. The four forces on the table legs are different in magnitude and cannot be found from the laws of static equilibrium alone.

Yet solutions to indeterminate problems exist in the real world. If you rest the tires of the car on four platform scales, each scale will register a definite reading, the sum of the readings being the weight of the car. What is eluding us in our efforts to find the individual forces by solving equations?

The problem is that we have assumed—without making a great point of it—that the bodies to which we apply the equations of static equilibrium are perfectly rigid. By this we mean that they do not deform when forces are applied to them. Strictly, there are no such bodies. The tires of the car, for example, deform easily under load until the car settles into a position of static equilibrium.

We have all had experience with a wobbly restaurant table, which we usually level by putting folded paper under one of the legs. If a big enough elephant sat on such a table, however, you may be sure that if the table did not collapse, it would deform just like the tires of a car. Its legs would all touch the floor, the forces acting upward on the table legs would all assume definite (and different) values as in Fig. 13-9, and the table would no longer wobble. How do we find the values of those forces acting on the legs?

To solve such indeterminate equilibrium problems, we must supplement equilibrium equations with some knowledge of *elasticity,* the branch of physics and engineering that describes how real bodies deform when forces are applied to them. The next section provides an introduction to this subject.

CHECKPOINT 5: A horizontal uniform bar of weight 10 N is to hang from a ceiling by two wires that exert upward forces $\vec{F}_1$ and $\vec{F}_2$ on the bar. The figure shows four arrangements for the wires. Which arrangements, if any, are indeterminate (so that we cannot solve for numerical values of $\vec{F}_1$ and $\vec{F}_2$)?

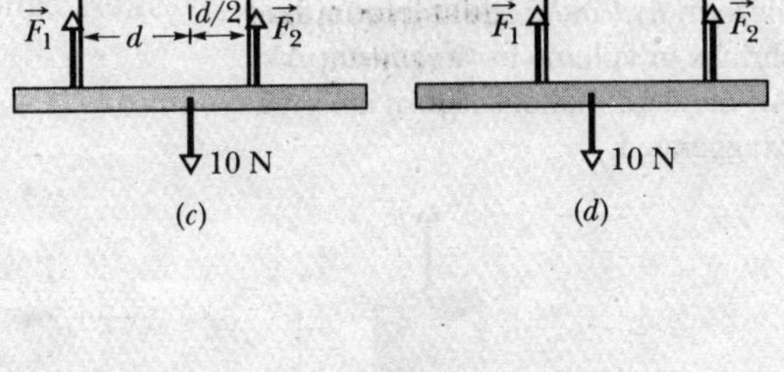

13-6 Elasticity

When a large number of atoms come together to form a metallic solid, such as an iron nail, they settle into equilibrium positions in a three-dimensional *lattice,* a repetitive arrangement in which each atom has a well-defined equilibrium distance from its nearest neighbors. The atoms are held together by interatomic forces that are modeled as tiny springs in Fig. 13-10. The lattice is remarkably rigid, which is another way of saying that the "interatomic springs" are extremely stiff. It is for this

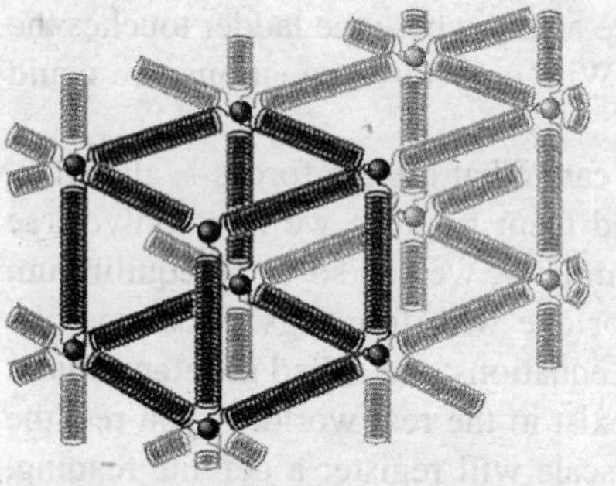

Fig. 13-10 The atoms of a metallic solid are distributed on a repetitive three-dimensional lattice. The springs represent interatomic forces.

reason that we perceive many ordinary objects such as metal ladders, tables, and spoons as perfectly rigid. Of course, some ordinary objects, such as garden hoses or rubber gloves, do not strike us as rigid at all. The atoms that make up these objects *do not* form a rigid lattice like that of Fig. 13-10 but are aligned in long, flexible molecular chains, each chain being only loosely bound to its neighbors.

All real "rigid" bodies are to some extent **elastic,** which means that we can change their dimensions slightly by pulling, pushing, twisting, or compressing them. To get a feeling for the orders of magnitude involved, consider a vertical steel rod 1 m long and 1 cm in diameter. If you hang a subcompact car from the end of such a rod, the rod will stretch, but only by about 0.5 mm, or 0.05%. Furthermore, the rod will return to its original length when the car is removed.

If you hang two cars from the rod, the rod will be permanently stretched and will not recover its original length when you remove the load. If you hang three cars from the rod, the rod will break. Just before rupture, the elongation of the rod will be less than 0.2%. Although deformations of this size seem small, they are important in engineering practice. (Whether a wing under load will stay on an airplane is obviously important.)

Figure 13-11 shows three ways in which a solid might change its dimensions when forces act on it. In Fig. 13-11*a*, a cylinder is stretched. In Fig. 13-11*b*, a cylinder is deformed by a force perpendicular to its axis, much as we might deform a pack of cards or a book. In Fig. 13-11*c*, a solid object, placed in a fluid under high pressure, is compressed uniformly on all sides. What the three deformation types have in common is that a **stress,** or deforming force per unit area, produces a **strain,** or unit deformation. In Fig. 13-11, *tensile stress* (associated with stretching) is illustrated in (*a*), *shearing stress* in (*b*), and *hydraulic stress* in (*c*).

The stresses and the strains take different forms in the three situations of Fig. 13-11, but—over the range of engineering usefulness—stress and strain are proportional to each other. The constant of proportionality is called a **modulus of elasticity,** so that

$$\text{stress} = \text{modulus} \times \text{strain}. \tag{13-24}$$

In a standard test of tensile properties, the tensile stress on a test cylinder (like that in Fig. 13-12) is slowly increased from zero to the point at which the cylinder fractures, and the strain is carefully measured and plotted. The result is a graph of stress versus strain like that in Fig. 13-13. For a substantial range of applied stresses, the stress–strain relation is linear, and the specimen recovers its original dimensions when the stress is removed; it is here that Eq. 13-24 applies. If the stress is increased beyond the **yield strength** S_y of the specimen, the specimen becomes permanently deformed. If the stress continues to increase, the specimen eventually ruptures, at a stress called the **ultimate strength** S_u.

Fig. 13-11 (*a*) A cylinder subject to *tensile stress* stretches by an amount ΔL. (*b*) A cylinder subject to *shearing stress* deforms by an amount Δx, somewhat like a pack of playing cards would. (*c*) A solid sphere subject to uniform *hydraulic stress* from a fluid shrinks in volume by an amount ΔV. All the deformations shown are greatly exaggerated.

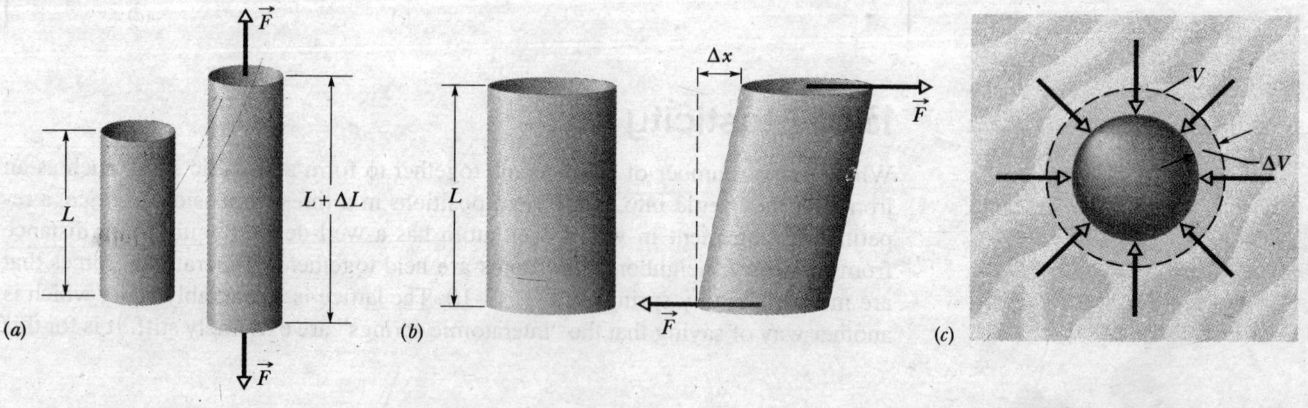

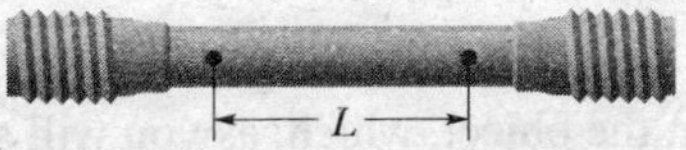

Fig. 13-12 A test specimen, used to determine a stress–strain curve such as that of Fig. 13-13. The change ΔL that occurs in a certain length L is measured in a tensile stress–strain test.

Tension and Compression

For simple tension or compression, the stress on an object is defined as F/A, where F is the magnitude of the force applied perpendicularly to the area A on the object. The strain, or unit deformation, is then the dimensionless quantity $\Delta L/L$, the fractional (or sometimes percentage) change in the length of the specimen. If the specimen is a long rod and the stress does not exceed the yield strength, then not only the entire rod but also every section of it experiences the same strain when a given stress is applied. Because the strain is dimensionless, the modulus in Eq. 13-24 has the same dimensions as the stress, namely, force per unit area.

The modulus for tensile and compressive stresses is called the **Young's modulus** and is represented in engineering practice by the symbol E. Equation 13-24 becomes

$$\frac{F}{A} = E\frac{\Delta L}{L}. \tag{13-25}$$

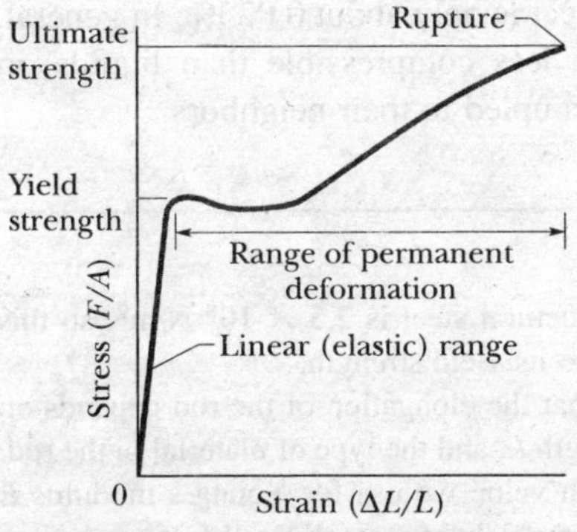

Fig. 13-13 A stress–strain curve for a steel test specimen such as that of Fig. 13-12. The specimen deforms permanently when the stress is equal to the *yield strength* of the material. It ruptures when the stress is equal to the *ultimate strength* of the material.

The strain $\Delta L/L$ in a specimen can often be measured conveniently with a *strain gauge* (Fig. 13-14). This simple and useful device, which can be attached directly to operating machinery with an adhesive, is based on the principle that its electrical properties are dependent on the strain it undergoes.

Although the Young's modulus for an object may be almost the same for tension and compression, the object's ultimate strength may well be different for the two types of stress. Concrete, for example, is very strong in compression but is so weak in tension that it is almost never used in that manner. Table 13-1 shows the Young's modulus and other elastic properties for some materials of engineering interest.

Shearing

In the case of shearing, the stress is also a force per unit area, but the force vector lies in the plane of the area rather than perpendicular to it. The strain is the dimensionless ratio $\Delta x/L$, with the quantities defined as shown in Fig. 13-11*b*. The corresponding modulus, which is given the symbol G in engineering practice, is called the **shear modulus.** For shearing, Eq. 13-24 is written as

$$\frac{F}{A} = G\frac{\Delta x}{L}. \tag{13-26}$$

Shearing stresses play a critical role in the buckling of shafts that rotate under load and in bone fractures caused by bending.

Fig. 13-14 A strain gauge of overall dimensions 9.8 mm by 4.6 mm. The gauge is fastened with adhesive to the object whose strain is to be measured; it experiences the same strain as the object. The electrical resistance of the gauge varies with the strain, permitting strains up to 3% to be measured.

TABLE 13-1 Some Elastic Properties of Selected Materials of Engineering Interest

Material	Density ρ (kg/m^3)	Young's Modulus E ($10^9\ N/m^2$)	Ultimate Strength S_u ($10^6\ N/m^2$)	Yield Strength S_y ($10^6\ N/m^2$)
Steel[a]	7860	200	400	250
Aluminum	2710	70	110	95
Glass	2190	65	50[b]	—
Concrete[c]	2320	30	40[b]	—
Wood[d]	525	13	50[b]	—
Bone	1900	9[b]	170[b]	—
Polystyrene	1050	3	48	—

[a]Structural steel (ASTM-A36). [b]In compression. [c]High strength. [d]Douglas fir.

Hydraulic Stress

In Fig. 13-11*c*, the stress is the fluid pressure p on the object, which, as you will see in Chapter 15, is a force per unit area. The strain is $\Delta V/V$, where V is the original volume of the specimen and ΔV is the absolute value of the change in volume. The corresponding modulus, with symbol B, is called the **bulk modulus** of the material. The object is said to be under *hydraulic compression,* and the pressure can be called the *hydraulic stress*. For this situation, we write Eq. 13-24 as

$$p = B\frac{\Delta V}{V}. \tag{13-27}$$

The bulk modulus is 2.2×10^9 N/m² for water and 16×10^{10} N/m² for steel. The pressure at the bottom of the Pacific Ocean, at its average depth of about 4000 m, is 4.0×10^7 N/m². The fractional compression $\Delta V/V$ of a volume of water due to this pressure is 1.8%; that for a steel object is only about 0.025%. In general, solids—with their rigid atomic lattices—are less compressible than liquids, in which the atoms or molecules are less tightly coupled to their neighbors.

Sample Problem 13-5

A structural steel rod has a radius R of 9.5 mm and a length L of 81 cm. A 62 kN force $\vec{F}$ stretches it along its length. What are the stress on the rod and the elongation and strain of the rod?

SOLUTION: The first **Key Idea** here has to do with what is meant by the second sentence in the problem statement. We assume the rod is held stationary by, say, a clamp or vise at one end. Then force $\vec{F}$ is applied at the other end, parallel to the length of the rod and thus perpendicular to the end face there. Therefore, the situation is like that in Fig. 13-11*a*.

The next **Key Idea** is that we assume the force is applied uniformly across the end face and thus over an area $A = \pi R^2$. Then the stress on the rod is given by the left side of Eq. 13-25:

$$\text{stress} = \frac{F}{A} = \frac{F}{\pi R^2} = \frac{6.2 \times 10^4 \text{ N}}{(\pi)(9.5 \times 10^{-3} \text{ m})^2}$$
$$= 2.2 \times 10^8 \text{ N/m}^2. \qquad \text{(Answer)}$$

The yield strength for structural steel is 2.5×10^8 N/m², so this rod is dangerously close to its yield strength.

Another **Key Idea** is that the elongation of the rod depends on the stress, the original length L, and the type of material in the rod. The last determines which value we use for Young's modulus E (from Table 13-1). Using the value for steel, Eq. 13-25 gives us

$$\Delta L = \frac{(F/A)L}{E} = \frac{(2.2 \times 10^8 \text{ N/m}^2)(0.81 \text{ m})}{2.0 \times 10^{11} \text{ N/m}^2}$$
$$= 8.9 \times 10^{-4} \text{ m} = 0.89 \text{ mm}. \qquad \text{(Answer)}$$

The last **Key Idea** we need here is that strain is the ratio of the change in length to the original length, so we have

$$\frac{\Delta L}{L} = \frac{8.9 \times 10^{-4} \text{ m}}{0.81 \text{ m}}$$
$$= 1.1 \times 10^{-3} = 0.11\%. \qquad \text{(Answer)}$$

Sample Problem 13-6

A table has three legs that are 1.00 m in length and a fourth leg that is longer by $d = 0.50$ mm, so that the table wobbles slightly. A heavy steel cylinder with mass $M = 290$ kg is placed upright on the table (with a mass much less than M) so that all four legs are compressed and the table no longer wobbles. The legs are wooden cylinders with cross-sectional area $A = 1.0$ cm². The Young's modulus E for the wood is 1.3×10^{10} N/m². Assume that the tabletop remains level and that the legs do not buckle. What are the magnitudes of the forces on the legs from the floor?

SOLUTION: We take the table plus steel cylinder as our system. The situation is like that in Fig. 13-9, except now we have a steel cylinder on the table. One **Key Idea** is that if the tabletop remains level, the legs must be compressed in the following ways: Each of the short legs must be compressed by the same amount (call it ΔL_3) and thus by the same force of magnitude F_3. The single long leg must be compressed by a larger amount ΔL_4 and thus by a force with a larger magnitude F_4. In other words, for a level tabletop, we must have

$$\Delta L_4 = \Delta L_3 + d. \tag{13-28}$$

A second **Key Idea** is that, from Eq. 13-25, we can relate a change in length to the force causing the change with $\Delta L = FL/AE$, where L is the original length of a leg. We can use this relation to replace ΔL_4 and ΔL_3 in Eq. 13-28. However, note that we can approximate the original length L as being the same for all four legs. The replacements then give us

$$\frac{F_4 L}{AE} = \frac{F_3 L}{AE} + d. \tag{13-29}$$

We cannot solve this equation because it has two unknowns, F_4 and F_3.

To get a second equation containing F_4 and F_3, we can use a vertical y axis and then write the balance of vertical forces ($F_{net,y} = 0$) as

$$3F_3 + F_4 - Mg = 0, \qquad (13\text{-}30)$$

where Mg is equal to the magnitude of the gravitational force on the system. (*Three* legs have force $\vec{F}_3$ on them.) To solve the simultaneous equations 13-29 and 13-30 for, say, F_3, we first use Eq. 13-30 to find that $F_4 = Mg - 3F_3$. Substituting that into Eq. 13-29 then yields, after some algebra,

$$F_3 = \frac{Mg}{4} - \frac{dAE}{4L}$$

$$= \frac{(290\text{ kg})(9.8\text{ m/s}^2)}{4}$$

$$- \frac{(5.0 \times 10^{-4}\text{ m})(10^{-4}\text{ m}^2)(1.3 \times 10^{10}\text{ N/m}^2)}{(4)(1.00\text{ m})}$$

$$= 548\text{ N} \approx 550\text{ N}. \qquad \text{(Answer)}$$

From Eq. 13-30 we then find

$$F_4 = Mg - 3F_3 = (290\text{ kg})(9.8\text{ m/s}^2) - 3(548\text{ N})$$

$$\approx 1200\text{ N}. \qquad \text{(Answer)}$$

You can show that to reach their equilibrium configuration, the three short legs are each compressed by 0.42 mm and the single long leg by 0.92 mm.

✔CHECKPOINT 6: The figure shows a horizontal block that is suspended by two wires, A and B, which are identical except for their original lengths. The center of mass of the block is closer to wire B than to wire A. (a) Measuring torques about the block's center of mass, state whether the magnitude of the torque due to wire A is greater than, less than, or equal to the magnitude of the torque due to wire B. (b) Which wire exerts more force on the block? (c) If the wires are now equal in length, which one was originally shorter?

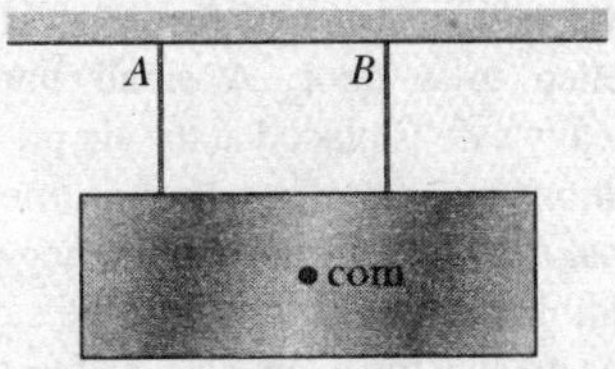

REVIEW & SUMMARY

Static Equilibrium A rigid body at rest is said to be in **static equilibrium.** For such a body, the vector sum of the external forces acting on it is zero:

$$\vec{F}_{net} = 0 \quad \text{(balance of forces)}. \qquad (13\text{-}3)$$

If all the forces lie in the xy plane, this vector equation is equivalent to two component equations:

$$F_{net,x} = 0 \quad \text{and} \quad F_{net,y} = 0 \quad \text{(balance of forces)}. \qquad (13\text{-}7, 13\text{-}8)$$

Static equilibrium also implies that the vector sum of the external torques acting on the body about *any* point is zero, or

$$\vec{\tau}_{net} = 0 \quad \text{(balance of torques)}. \qquad (13\text{-}5)$$

If the forces lie in the xy plane, all torque vectors are parallel to the z axis, and Eq. 13-5 is equivalent to the single component equation

$$\tau_{net,z} = 0 \quad \text{(balance of torques)}. \qquad (13\text{-}9)$$

Center of Gravity The gravitational force acts individually on each element of a body. The net effect of all individual actions may be found by imagining an equivalent total gravitational force $\vec{F}_g$ acting at a specific point called the **center of gravity.** If the gravitational acceleration $\vec{g}$ is the same for all the elements of the body, the center of gravity is at the center of mass.

Elastic Moduli Three **elastic moduli** are used to describe the elastic behavior (deformations) of objects as they respond to forces that act on them. The **strain** (fractional change in length) is linearly related to the applied **stress** (force per unit area) by the proper modulus, according to the general relation

$$\text{stress} = \text{modulus} \times \text{strain}. \qquad (13\text{-}24)$$

Tension and Compression When an object is under tension or compression, Eq. 13-24 is written as

$$\frac{F}{A} = E\frac{\Delta L}{L}, \qquad (13\text{-}25)$$

where $\Delta L/L$ is the tensile or compressive strain of the object, F is the magnitude of the applied force $\vec{F}$ causing the strain, A is the cross-sectional area over which $\vec{F}$ is applied (perpendicular to A, as in Fig. 13-11*a*), and E is the **Young's modulus** for the object. The stress is F/A.

Shearing When an object is under a shearing stress, Eq. 13-24 is written as

$$\frac{F}{A} = G\frac{\Delta x}{L}, \qquad (13\text{-}26)$$

where $\Delta x/L$ is the shearing strain of the object, Δx is the displacement of one end of the object in the direction of the applied force $\vec{F}$ (as in Fig. 13-11*b*), and G is the **shear modulus** of the object. The stress is F/A.

Hydraulic Stress When an object undergoes *hydraulic compression* due to a stress exerted by a surrounding fluid, Eq. 13-24 is written as

$$p = B\frac{\Delta V}{V}, \qquad (13\text{-}27)$$

where p is the pressure (*hydraulic stress*) on the object due to the fluid, $\Delta V/V$ (the strain) is the absolute value of the fractional change in the object's volume due to that pressure, and B is the **bulk modulus** of the object.

QUESTIONS

1. Figure 13-15 shows an overhead view of a uniform stick on which four forces act. Suppose we choose a rotational axis through point O, calculate the torques about that axis due to the forces, and find that these torques balance. Will the torques balance if, instead, the rotational axis is chosen to be at (a) point A, (b) point B, or (c) point C? (d) Suppose, instead, that we find that the torques about point O do not balance. Is there another point about which the torques will balance?

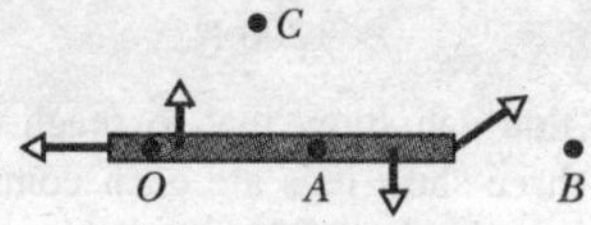

Fig. 13-15 Question 1.

2. In Fig. 13-16, a rigid beam is attached to two posts that are fastened to a floor. A small but heavy safe is placed at the six positions indicated, in turn. Assume that the mass of the beam is negligible compared to that of the safe. (a) Rank the positions according to the force on post A due to the safe, greatest compression first, greatest tension last, and indicate where, if anywhere, the force is zero. (b) Now rank them according to the force on post B.

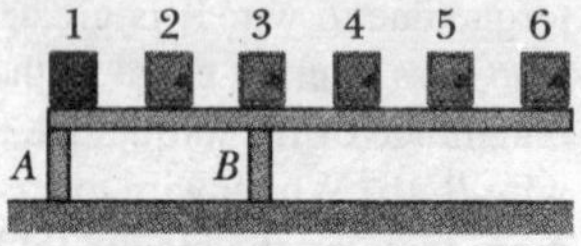

Fig. 13-16 Question 2.

3. Figure 13-17 shows four overhead views of rotating uniform disks that are sliding across a frictionless floor. Three forces, of magnitude F, $2F$, or $3F$, act on each disk, either at the rim, at the center, or halfway between rim and center. The force vectors rotate along with the disks, and, in the "snapshots" of Fig. 13-17, point left or right. Which disks are in equilibrium?

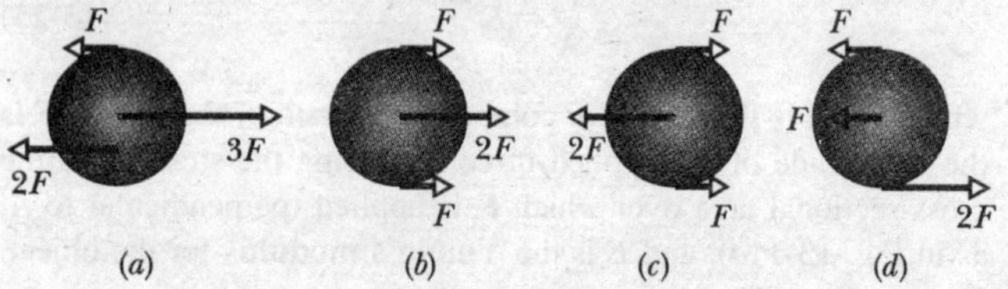

Fig. 13-17 Question 3.

4. Figure 13-18 shows overhead views of two structures on which three forces act. The directions of the forces are as indicated. If the magnitudes of the forces are adjusted properly (but kept nonzero), which structure can be in static equilibrium?

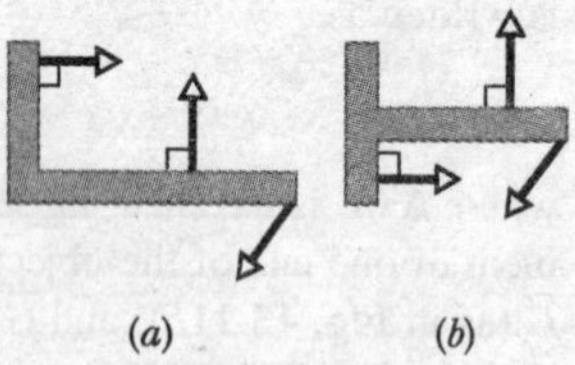

Fig. 13-18 Question 4.

5. Figure 13-19 shows a mobile of toy penguins hanging from a ceiling. Each crossbar is horizontal, has negligible mass, and extends three times as far to the right of the wire supporting it as to the left. Penguin 1 has mass $m_1 = 48$ kg. What are the masses of the other penguins?

Fig. 13-19 Question 5.

6. A ladder leans against a frictionless wall but is prevented from falling because of friction between it and the ground. Suppose you shift the base of the ladder toward the wall. Determine whether the following become larger, smaller, or stay the same (in magnitude): (a) the normal force on the ladder from the ground, (b) the force on the ladder from the wall, (c) the static frictional force on the ladder from the ground, and (d) the maximum value $f_{s,\max}$ of the static frictional force.

7. Three piñatas hang from the (stationary) assembly of massless pulleys and cords seen in Fig. 13-20. One long cord runs from the ceiling at the right to the lower pulley at the left. Several shorter cords suspend pulleys from the ceiling or piñatas from the pulleys. The weights (in newtons) of two piñatas are given. (a) What is the weight of the third piñata? (*Hint:* When a cord loops halfway around a pulley, it pulls on the pulley with a net force that is twice the tension in the cord.) (b) What is the tension in the short cord labeled with T?

Fig. 13-20 Question 7.

8. (a) In Checkpoint 4, to express τ_r in terms of T, should you use $\sin\theta$ or $\cos\theta$? (b) If angle θ is decreased (by shortening the rope but still keeping the rod horizontal), does the torque τ_r required for equilibrium become larger, smaller, or stay the same? (c) Does the corresponding force magnitude T become larger, smaller, or stay the same?

9. The table gives the areas of three surfaces and the magnitude of a force that is applied perpendicular to the surface and uniformly across it. Rank the surfaces according to the stress on them, greatest first.

	Area	Force
Surface A	$0.5A_0$	$2F_0$
Surface B	$2A_0$	$4F_0$
Surface C	$3A_0$	$6F_0$

10. Four cylindrical rods are stretched as in Fig. 13-11*a*. The force magnitudes, the areas of the end faces, the changes in length, and the initial lengths are given here. Rank the rods according to their Young's moduli, greatest first.

Rod	Force	Area	Length Change	Initial Length
1	F	A	ΔL	L
2	$2F$	$2A$	$2\Delta L$	L
3	F	$2A$	$2\Delta L$	$2L$
4	$2F$	A	ΔL	$2L$

EXERCISES & PROBLEMS

ssm Solution is in the Student Solutions Manual.
www Solution is available on the World Wide Web at: http://www.wiley.com/college/hrw
ilw Solution is available on the Interactive LearningWare.

SEC. 13-4 Some Examples of Static Equilibrium

1E. A physics Brady Bunch, whose weights in newtons are indicated in Fig. 13-21, is balanced on a seesaw. What is the number of the person who causes the largest torque, about the rotation axis at *fulcrum f*, directed (a) out of the page and (b) into the page?

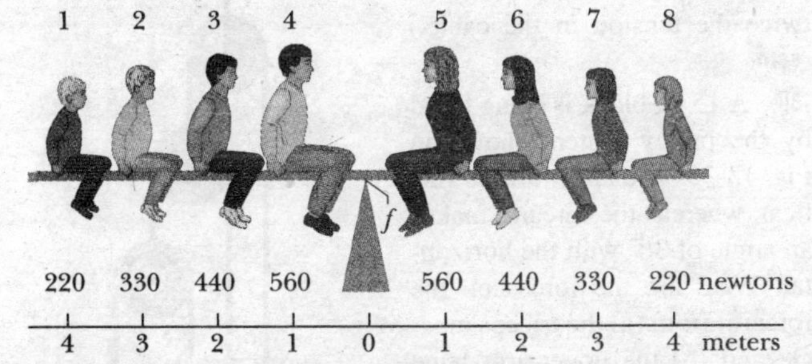

Fig. 13-21 Exercise 1.

2E. The leaning Tower of Pisa (Fig. 13-22) is 55 m high and 7.0 m in diameter. The top of the tower is displaced 4.5 m from the vertical. Treat the tower as a uniform, circular cylinder. (a) What additional displacement, measured at the top, would bring the tower to the verge of toppling? (b) What angle would the tower then make with the vertical?

Fig. 13-22 Exercise 2.

3E. A particle is acted on by forces given, in newtons, by $\vec{F}_1 = 10\hat{i} - 4\hat{j}$ and $\vec{F}_2 = 17\hat{i} + 2\hat{j}$. (a) What force $\vec{F}_3$ balances these forces? (b) What direction does $\vec{F}_3$ have relative to the x axis? ssm

4E. A bow is drawn at its midpoint until the tension in the string is equal to the force exerted by the archer. What is the angle between the two halves of the string?

5E. A rope of negligible mass is stretched horizontally between two supports that are 3.44 m apart. When an object of weight 3160 N is hung at the center of the rope, the rope is observed to sag by 35.0 cm. What is the tension in the rope? ilw

6E. A scaffold of mass 60 kg and length 5.0 m is supported in a horizontal position by a vertical cable at each end. A window washer of mass 80 kg stands at a point 1.5 m from one end. What is the tension in (a) the nearer cable and (b) the farther cable?

7E. In Fig. 13-23, a uniform sphere of mass m and radius r is held in place by a massless rope attached to a frictionless wall a distance L above the center of the sphere. Find (a) the tension in the rope and (b) the force on the sphere from the wall. ssm

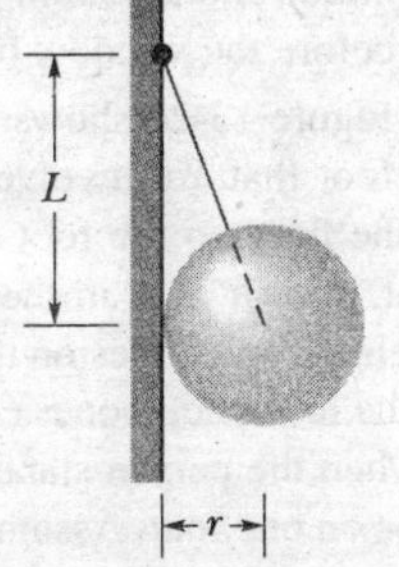

Fig. 13-23 Exercise 7.

8E. An automobile with a mass of 1360 kg has 3.05 m between the front and rear axles. Its center of gravity is located 1.78 m behind the front axle. With the automobile on level ground, determine the magnitude of the force from the ground on (a) each front wheel (assuming equal forces on the front wheels) and (b) each rear wheel (assuming equal forces on the rear wheels).

9E. A diver of weight 580 N stands at the end of a 4.5 m diving board of negligible mass (Fig. 13-24). The board is attached to two pedestals 1.5 m apart. What are the magnitude and direction of the force on the board from (a) the left pedestal and (b) the right pedestal? (c) Which pedestal is being stretched, and (d) which compressed? ssm

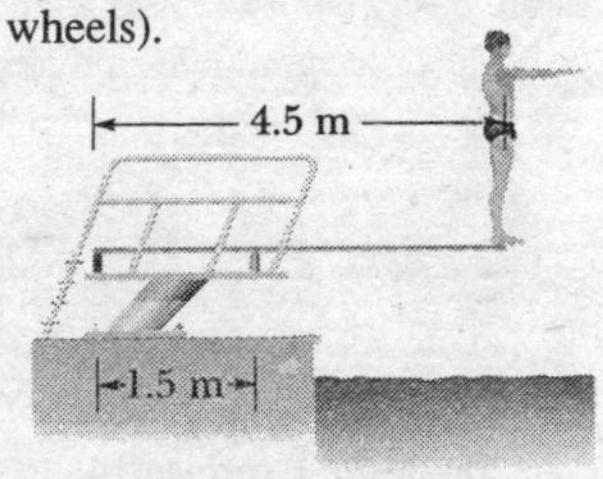

Fig. 13-24 Exercise 9.

10E. In Fig. 13-25, a man is trying to get his car out of mud on the shoulder of a road. He ties one end of a rope tightly around the front bumper and the other end tightly around a utility pole 18 m away. He then pushes sideways on the rope at its midpoint with a force of 550 N, displacing the center of the rope 0.30 m from its previous position, and the car barely moves. What is the magnitude of the force on the car from the rope? (The rope stretches somewhat.)

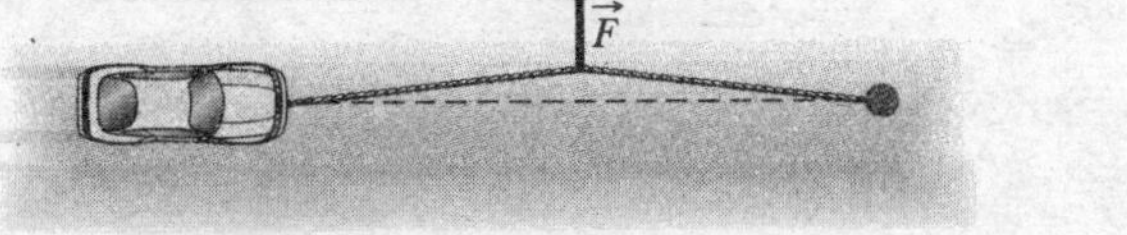

Fig. 13-25 Exercise 10.

11E. A meter stick balances horizontally on a knife-edge at the 50.0 cm mark. With two 5.0 g coins stacked over the 12.0 cm mark, the stick is found to balance at the 45.5 cm mark. What is the mass of the meter stick? ssm

12E. A uniform cubical crate is 0.750 m on each side and weighs 500 N. It rests on a floor with one edge against a very small, fixed obstruction. At what least height above the floor must a horizontal force of magnitude 350 N be applied to the crate to tip it?

13E. A 75 kg window cleaner uses a 10 kg ladder that is 5.0 m long. He places one end on the ground 2.5 m from a wall, rests the upper end against a cracked window, and climbs the ladder. He is 3.0 m up along the ladder when the window breaks. Neglecting friction between the ladder and window and assuming that the base of the ladder does not slip, find (a) the magnitude of the force on the window from the ladder just before the window breaks and (b) the magnitude and direction of the force on the ladder from the ground just before the window breaks. ssm

14E. Figure 13-26 shows the anatomical structures in the lower leg and foot that are involved in standing tiptoe with the heel raised off the floor so the foot effectively contacts the floor at only one point, shown as P in the figure. Calculate, in terms of a person's weight W, the forces on the foot from (a) the calf muscle (at A) and (b) the lower-leg bones (at B) when the person stands tiptoe on one foot. Assume that $a = 5.0$ cm and $b = 15$ cm.

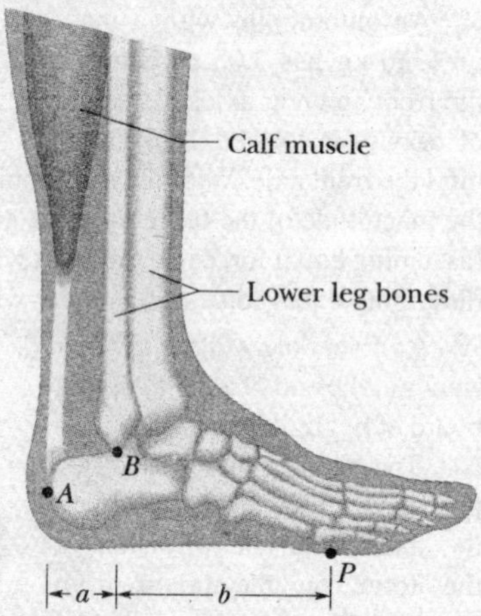

Fig. 13-26 Exercise 14.

15P. In Fig. 13-27, an 817 kg construction bucket is suspended by a cable A that is attached at O to two other cables B and C, making angles of 51.0° and 66.0° with the horizontal. Find the tensions in (a) cable A, (b) cable B, and (c) cable C. (*Hint:* To avoid solving two equations in two unknowns, position the axes as shown in the figure.) ssm

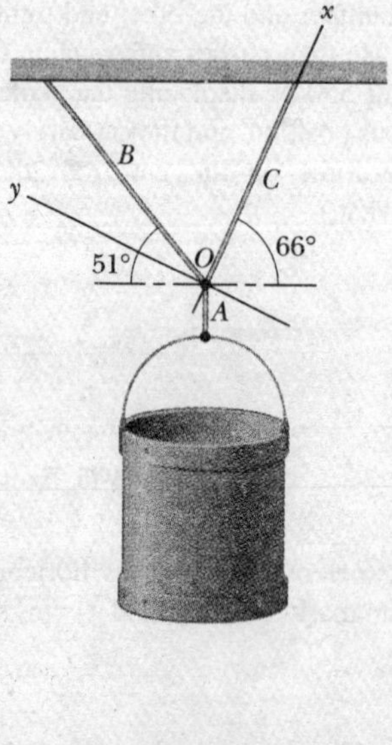

Fig. 13-27 Problem 15.

16P. The system in Fig. 13-28 is in equilibrium, with the string in the center exactly horizontal. Find (a) tension T_1, (b) tension T_2, (c) tension T_3, and (d) angle θ.

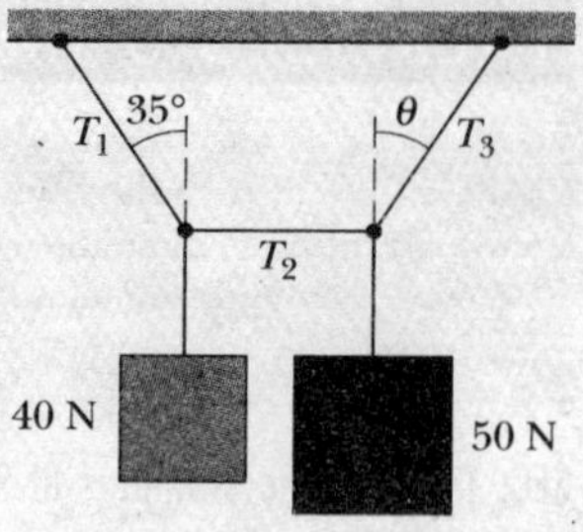

Fig. 13-28 Problem 16.

17P. The force $\vec{F}$ in Fig. 13-29 keeps the 6.40 kg block and the pulleys in equilibrium. The pulleys have negligible mass and friction. Calculate the tension T in the upper cable. (*Hint:* When a cable wraps halfway around a pulley as here, the magnitude of its net force on the pulley is twice the tension in the cable.) ssm ilw

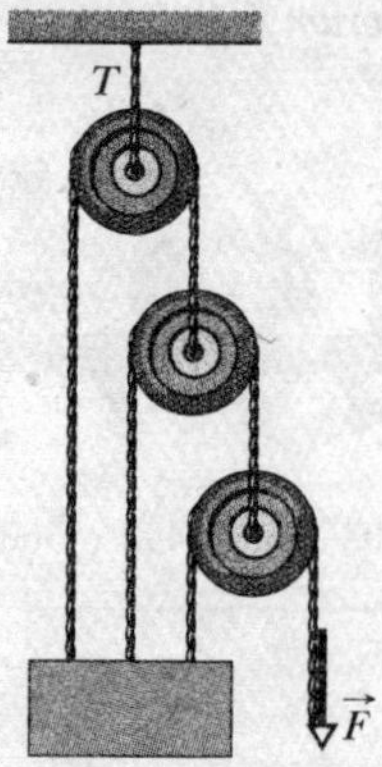

Fig. 13-29 Problem 17.

18P. A 15 kg block is being lifted by the pulley system shown in Fig. 13-30. The upper arm is vertical, whereas the forearm makes an angle of 30° with the horizontal. What are the forces on the forearm from (a) the triceps muscle and (b) the upper-arm bone (the humerus)? The forearm and hand together have a mass of 2.0 kg with a center of mass 15 cm (measured along the arm) from the point where the forearm and upper-arm bones are in contact. The triceps muscle pulls vertically upward at a point 2.5 cm behind that contact point.

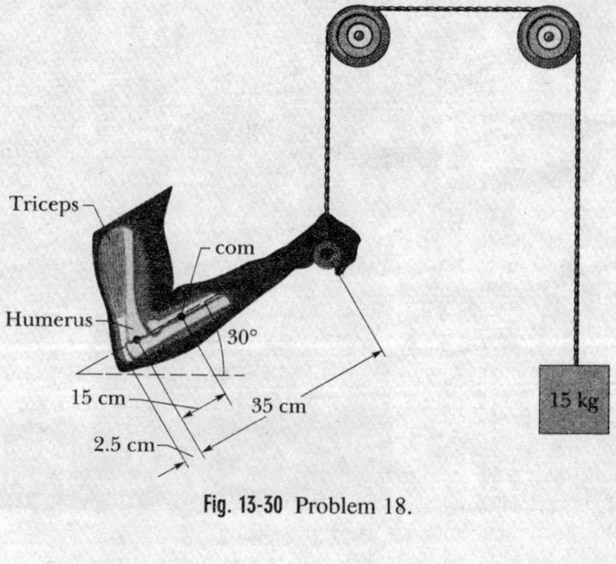

Fig. 13-30 Problem 18.

19P. Forces $\vec{F}_1$, $\vec{F}_2$, and $\vec{F}_3$ act on the structure of Fig. 13-31, shown in an overhead view. We wish to put the structure in equilibrium by applying a fourth force, at a point such as P. The fourth force has vector components $\vec{F}_h$ and $\vec{F}_v$. We are given that $a = 2.0$ m,

$b = 3.0$ m, $c = 1.0$ m, $F_1 = 20$ N, $F_2 = 10$ N, and $F_3 = 5.0$ N. Find (a) F_h, (b) F_v, and (c) d. ilw

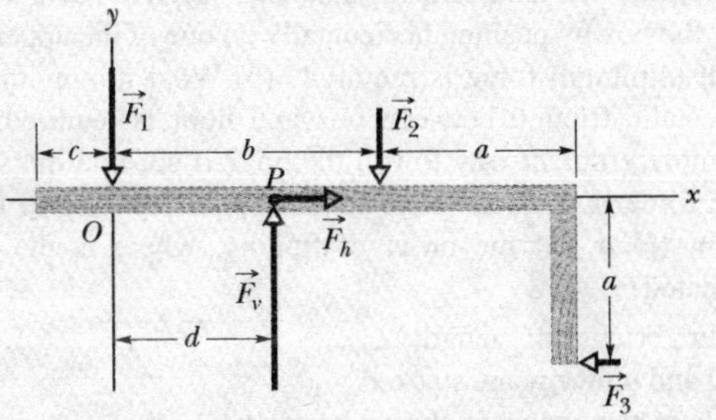

Fig. 13-31 Problem 19.

20P. In Fig. 13-32, a 50.0 kg uniform square sign, 2.00 m on a side, is hung from a 3.00 m horizontal rod of negligible mass. A cable is attached to the end of the rod and to a point on the wall 4.00 m above the point where the rod is hinged to the wall. (a) What is the tension in the cable? What are the magnitudes and directions of the (b) horizontal and (c) vertical components of the force on the rod from the wall?

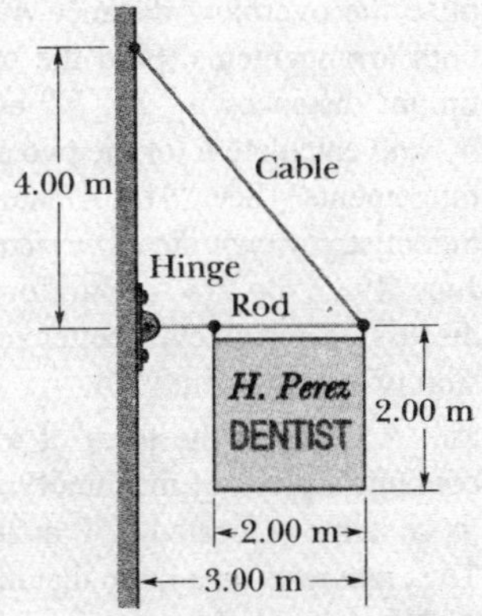

Fig. 13-32 Problem 20.

21P. In Fig. 13-33, what magnitude of force $\vec{F}$ applied horizontally at the axle of the wheel is necessary to raise the wheel over an obstacle of height h? The wheel's radius is r and its mass is m. ssm www

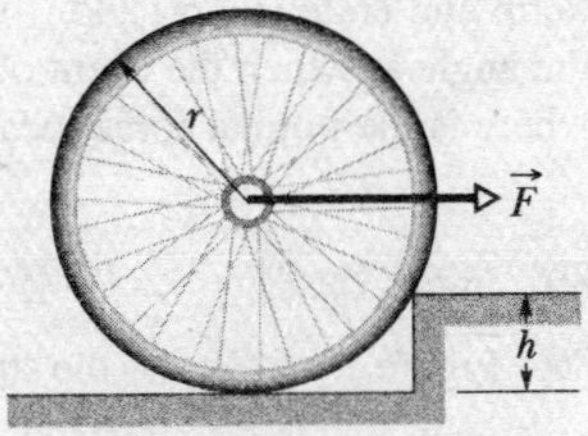

Fig. 13-33 Problem 21.

22P. In Fig. 13-34, a 55 kg rock climber is in a lie-back climb along a fissure, with hands pulling on one side of the fissure and feet pressed against the opposite side. The fissure has width $w = 0.20$ m, and the center of mass of the climber is a horizontal distance $d = 0.40$ m from the fissure. The coefficient of static friction between hands and rock is $\mu_1 = 0.40$, and between boots and rock it is $\mu_2 = 1.2$. (a) What is the least horizontal pull by the hands and push by the feet that will keep the climber stable? (b) For the horizontal pull of (a), what must be the vertical distance h between hands and feet? (c) If the climber encounters wet rock, so that μ_1 and μ_2 are reduced, what happens to the answers to (a) and (b), respectively?

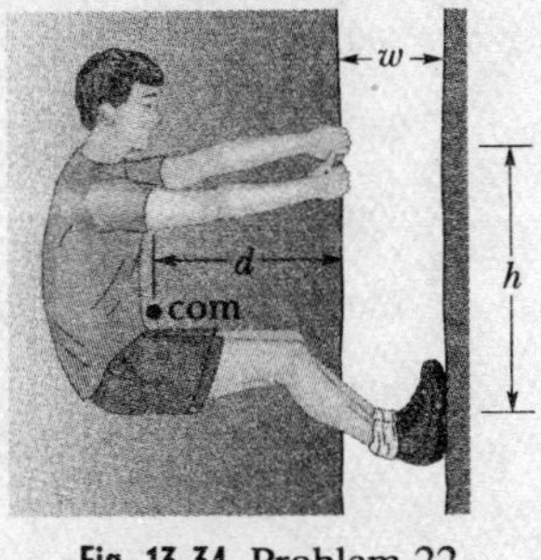

Fig. 13-34 Problem 22.

23P. In Fig. 13-35, one end of a uniform beam that weighs 222 N is attached to a wall with a hinge. The other end is supported by a wire. (a) Find the tension in the wire. What are the (b) horizontal and (c) vertical components of the force of the hinge on the beam? ssm

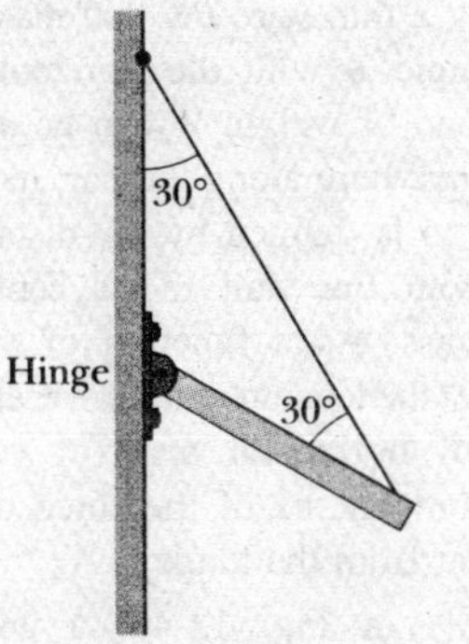

Fig. 13-35 Problem 23.

24P. Four bricks of length L, identical and uniform, are stacked on top of one another (Fig. 13-36) in such a way that part of each extends beyond the one beneath. Find, in terms of L, the maximum values of (a) a_1, (b) a_2, (c) a_3, (d) a_4, and (e) h, such that the stack is in equilibrium.

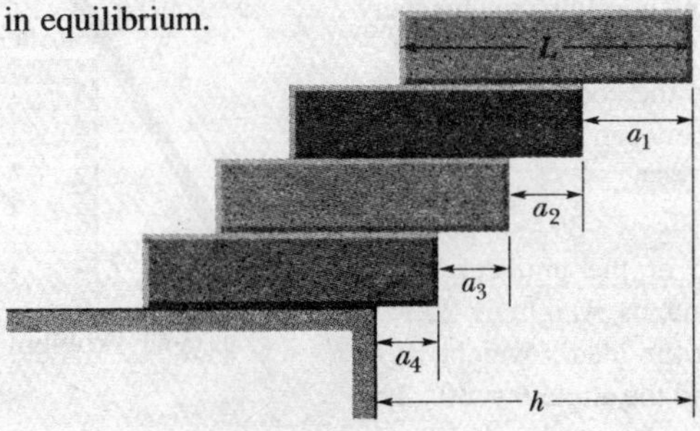

Fig. 13-36 Problem 24.

25P. The system in Fig. 13-37 is in equilibrium. A concrete block of mass 225 kg hangs from the end of the uniform strut whose mass is 45.0 kg. Find (a) the tension T in the cable and the (b) horizontal and (c) vertical force components on the strut from the hinge. ilw

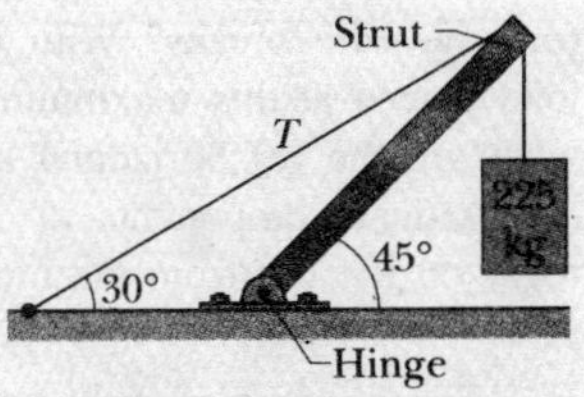

Fig. 13-37 Problem 25.

26P. A door 2.1 m high and 0.91 m wide has a mass of 27 kg. A hinge 0.30 m from the top and another 0.30 m from the bottom each support half the door's mass. Assume that the center of gravity is at the geometrical center of the door, and determine the (a) vertical and (b) horizontal components of the force from each hinge on the door.

27P. A nonuniform bar is suspended at rest in a horizontal position by two massless cords as shown in Fig. 13-38. One cord makes the angle $\theta = 36.9°$ with the vertical; the other makes the angle $\phi = 53.1°$ with the vertical. If the length L of the bar is 6.10 m, compute the distance x from the left-hand end of the bar to its center of mass. ssm

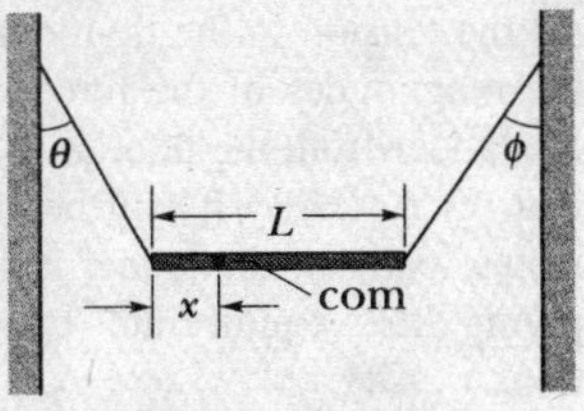

Fig. 13-38 Problem 27.

28P. In Fig. 13-39, a thin horizontal bar AB of negligible weight and length L is hinged to a vertical wall at A and supported at B

by a thin wire BC that makes an angle θ with the horizontal. A load of weight W can be moved anywhere along the bar; its position is defined by the distance x from the wall to its center of mass. As a function of x, find (a) the tension in the wire, and the (b) horizontal and (c) vertical components of the force on the bar from the hinge at A.

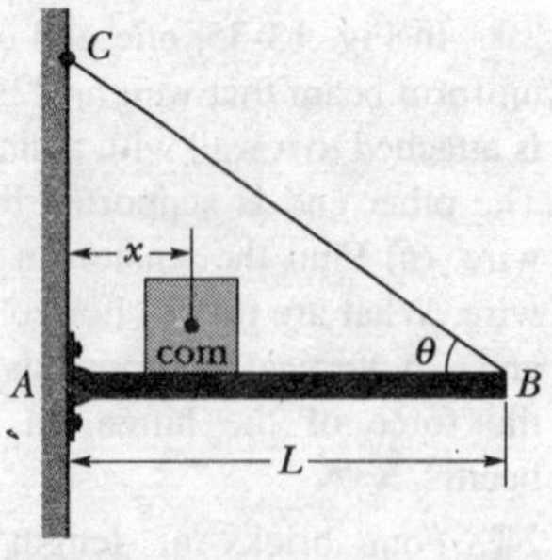

Fig. 13-39 Problems 28 and 30.

29P. In Fig. 13-40, a uniform plank, with a length L of 6.10 m and a weight of 445 N, rests on the ground and against a frictionless roller at the top of a wall of height $h = 3.05$ m. The plank remains in equilibrium for any value of $\theta \geq 70°$ but slips if $\theta < 70°$. Find the coefficient of static friction between the plank and the ground. ssm

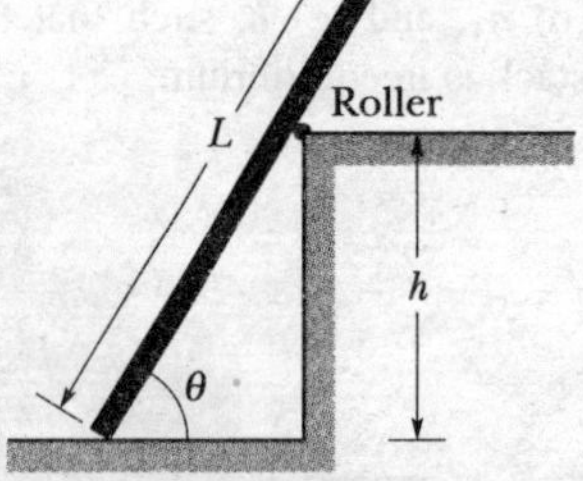

Fig. 13-40 Problem 29.

30P. In Fig. 13-39, suppose the length L of the uniform bar is 3.0 m and its weight is 200 N. Also, let the load's weight $W =$ 300 N and the angle $\theta = 30°$. The wire can withstand a maximum tension of 500 N. (a) What is the maximum possible distance x before the wire breaks? With the load placed at this maximum x, what are the (b) horizontal and (c) vertical components of the force on the bar from the hinge at A?

31P. For the stepladder shown in Fig. 13-41, sides AC and CE are each 2.44 m long and hinged at C. Bar BD is a tie-rod 0.762 m long, halfway up. A man weighing 854 N climbs 1.80 m along the ladder. Assuming that the floor is frictionless and neglecting the mass of the ladder, find (a) the tension in the tie-rod and the magnitudes of the forces on the ladder from the floor at (b) A and (c) E. (*Hint:* It will help to isolate parts of the ladder in applying the equilibrium conditions.) ssm

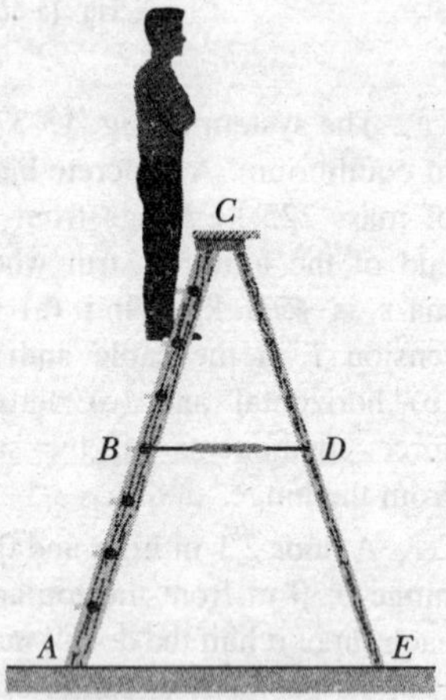

Fig. 13-41 Problem 31.

32P. Two uniform beams, A and B, are attached to a wall with hinges and then loosely bolted together as in Fig. 13-42. Find the x and y components of the force on (a) beam A due to its hinge, (b) beam A due to the bolt, (c) beam B due to its hinge, and (d) beam B due to the bolt.

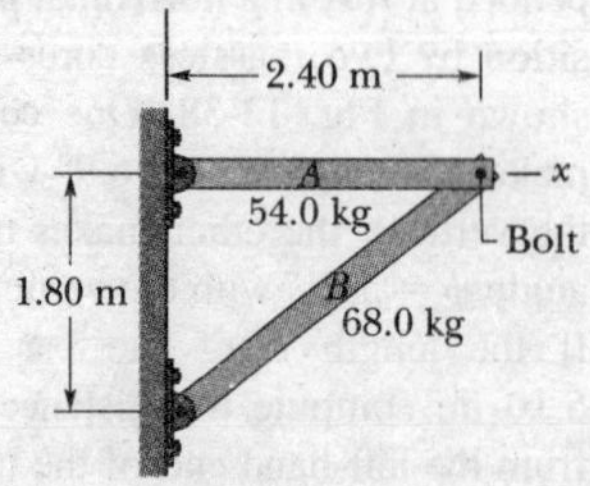

Fig. 13-42 Problem 32.

33P. A cubical box is filled with sand and weighs 890 N. We wish to "roll" the box by pushing horizontally on one of the upper edges. (a) What minimum force is required? (b) What minimum coefficient of static friction between box and floor is required? (c) Is there a more efficient way to roll the box? If so, find the smallest possible force that would have to be applied directly to the box to roll it. (*Hint:* At the onset of tipping, where is the normal force located?) ssm

34P. Four bricks of length L, identical and uniform, are stacked on a table in two ways, as shown in Fig. 13-43 (compare with Problem 24). We seek to maximize the overhang distance h in both arrangements. Find the optimum distances a_1, a_2, b_1, and b_2, and calculate h for the two arrangements. (See "The Amateur Scientist," *Scientific American*, June 1985, pp. 133–134, for a discussion and an even better version of arrangement (*b*).)

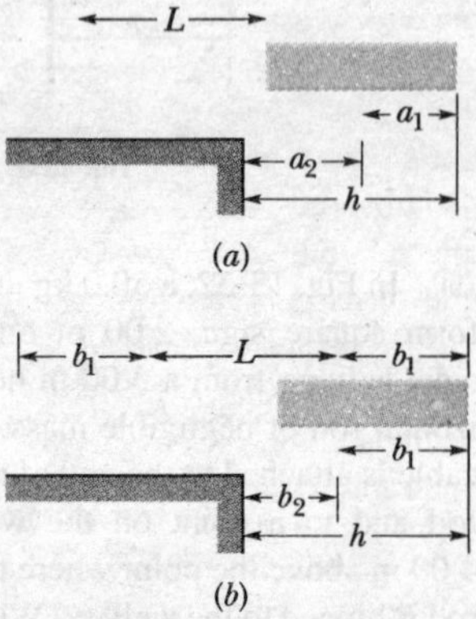

Fig. 13-43 Problem 34.

35P. A crate, in the form of a cube with edge lengths of 1.2 m, contains a piece of machinery; the center of mass of the crate and its contents is located 0.30 m above the crate's geometrical center. The crate rests on a ramp that makes an angle θ with the horizontal. As θ is increased from zero, an angle will be reached at which the crate will either start to slide down the ramp or tip over. Which event will occur (a) when the coefficient of static friction between ramp and crate is 0.60 and (b) when it is 0.70? In each case, give the angle at which the event occurs. (*Hint:* At the onset of tipping, where is the normal force located?) ssm www

SEC. 13-6 Elasticity

36E. Figure 13-44 shows the stress–strain curve for quartzite. What

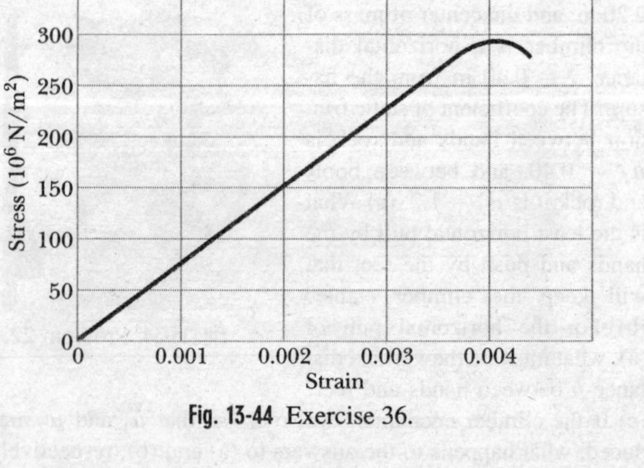

Fig. 13-44 Exercise 36.

are (a) the Young's modulus and (b) the approximate yield strength for this material?

37E. A horizontal aluminum rod 4.8 cm in diameter projects 5.3 cm from a wall. A 1200 kg object is suspended from the end of the rod. The shear modulus of aluminum is 3.0×10^{10} N/m^2. Neglecting the rod's mass, find (a) the shear stress on the rod and (b) the vertical deflection of the end of the rod. ssm ilw

38P. In Fig. 13-45, a lead brick rests horizontally on cylinders A and B. The areas of the top faces of the cylinders are related by $A_A = 2A_B$; the Young's moduli of the cylinders are related by $E_A = 2E_B$. The cylinders had identical lengths before the brick was placed on them. What fraction of the brick's mass is supported (a) by cylinder A and (b) by cylinder B? The horizontal distances between the center of mass of the brick and the centerlines of the cylinders are d_A for cylinder A and d_B for cylinder B. (c) What is the ratio d_A/d_B?

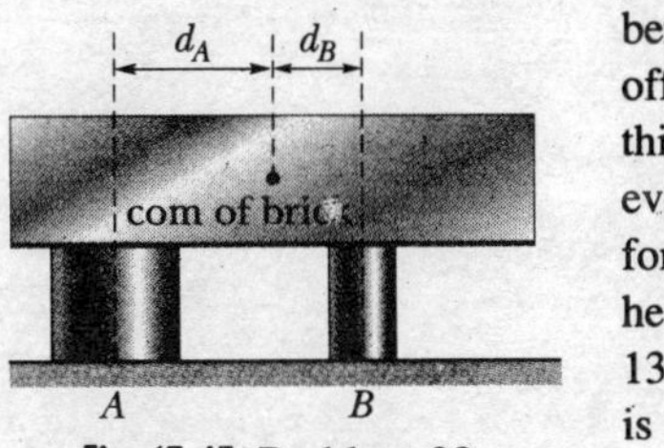

Fig. 13-45 Problem 38.

39P. In Fig. 13-46, a 103 kg uniform log hangs by two steel wires, A and B, both of radius 1.20 mm. Initially, wire A was 2.50 m long and 2.00 mm shorter than wire B. The log is now horizontal. What are the magnitudes of the forces on it from (a) wire A and (b) wire B? (c) What is the ratio d_A/d_B? ssm www

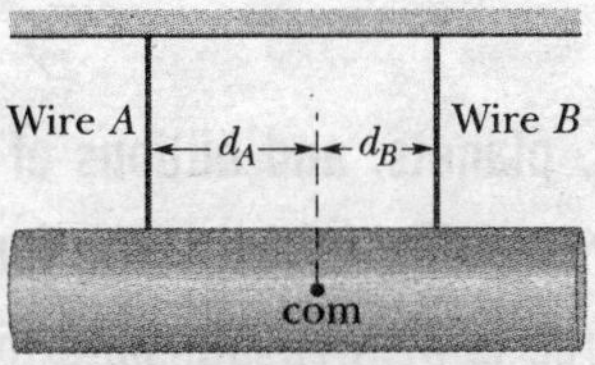

Fig. 13-46 Problem 39.

40P. A tunnel 150 m long, 7.2 m high, and 5.8 m wide (with a flat roof) is to be constructed 60 m beneath the ground. (See Fig. 13-47.) The tunnel roof is to be supported entirely by square steel columns, each with a cross-sectional area of 960 cm^2. The density of the ground material is 2.8 g/cm^3. (a) What is the total mass of the material that the columns must support? (b) How many columns are needed to keep the compressive stress on each column at one-half its ultimate strength?

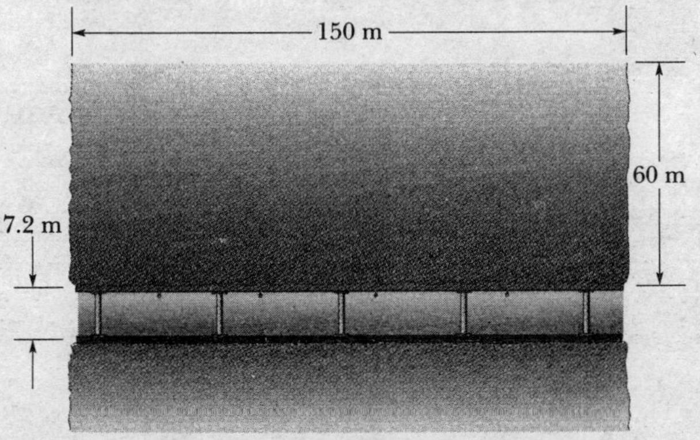

Fig. 13-47 Problem 40.

Additional Problems

41. Here is a way to move a heavy log through a tropical forest. Find a young tree in the general direction of travel; find a vine that hangs from the top of the tree down to ground level; pull the vine over to the log; wrap the vine around a limb on the log; pull hard enough on the vine to bend the tree over; and then tie off the vine on the limb. Repeat this procedure with several trees; eventually the net force of the vines on the log moves the log forward. Although tedious, this technique allowed workers to move heavy logs long before modern machinery was available. Figure 13-48 shows the essentials of the technique. There, a single vine is shown attached to a branch at one end of a uniform log of mass M. The coefficient of static friction between the log and the ground is 0.80. If the log is on the verge of sliding, with the left end raised slightly by the vine, what are (a) the angle θ and (b) the magnitude T of the force on the log from the vine?

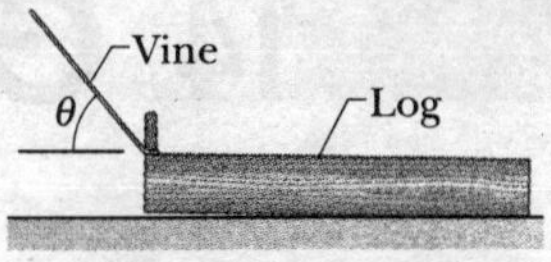

Fig. 13-48 Problem 41.

42. You have been hired to build a large sand mound in an indoor playground and must be careful about the stress that the sand will put on the floor. Consulting research literature, your are surprised to find that the greatest stress occurs, not directly beneath the apex (top) of the mound, but at points that are a distance r_m from that central point (Fig. 13-49a). This outward displacement of the maximum stress is presumably due to the sand grains forming arches within the mound. For a mound of height $H = 3.00$ m and angle $\theta = 33°$, and with sand of density $\rho = 1800$ kg/m^3, Fig. 13-49b gives the stress σ as a function of radius r from the central point of the mound's base. In that figure, $\sigma_0 = 40\,000$ N/m^2, $\sigma_m = 40\,024$ N/m^2, and $r_m = 1.82$ m.

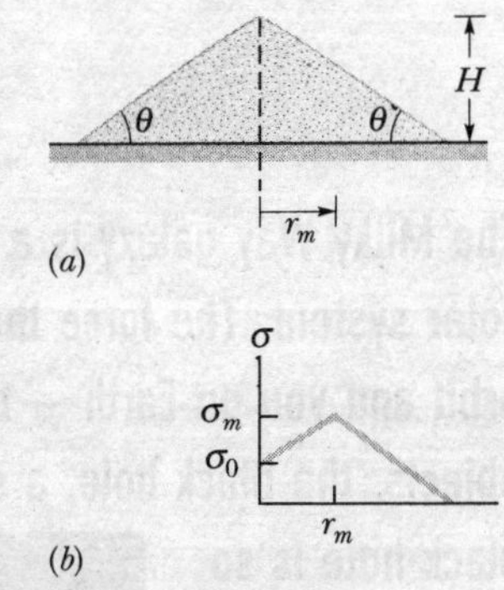

Fig. 13-49 Problem 42.

(a) What is the volume of sand contained in the mound for $r \leq r_m/2$? (*Hint:* The volume is that of a vertical cylinder plus a cone on top of the cylinder. The volume of the cone is $\pi R^2 h/3$, where R is the cone's radius and h is the cone's height.) (b) What is the weight W of that volume of sand? (c) Use Fig. 13-49b to write an expression for the stress σ on the floor as a function of radius r, for $r \leq r_m$. (d) On the floor, what is the area dA of a thin ring of radius r centered on the mound's central axis and with radial width dr? (e) What then is the magnitude dF of the downward force on the ring due to the sand? (f) What is the magnitude F of the net downward force on the floor due to all the sand contained in the mound for $r \leq r_m/2$? (*Hint:* Integrate the expression of (e) from $r = 0$ to $r = r_m/2$.) Now note the surprise: This force magnitude F on the floor is less than the weight W of the sand above the floor, as found in (b). (g) By what fraction is F reduced from W; that is, what is $(F - W)/W$?

14 Gravitation

The Milky Way galaxy is a disk-shaped collection of dust, planets, and billions of stars, including our Sun and solar system. The force that binds it or any other galaxy together is the same force that holds the Moon in orbit and you on Earth—the gravitational force. That force is also responsible for one of nature's strangest objects, the black hole, a star that has completely collapsed onto itself. The gravitational force near a black hole is so strong that not even light can escape it.

If that is the case, how can a black hole be detected?

The answer is in this chapter.

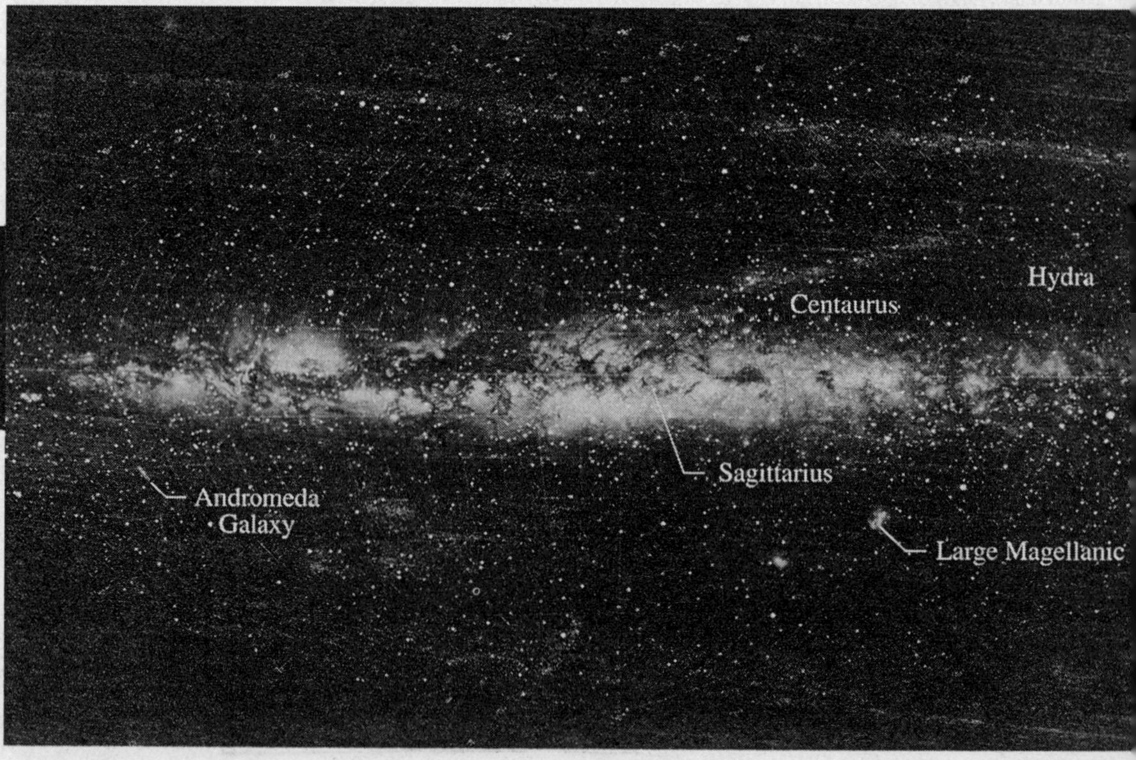

Fig. 14-1 The Andromeda galaxy. Located 2.3×10^6 light-years from us, and faintly visible to the naked eye, it is very similar to our home galaxy, the Milky Way.

14-1 The World and the Gravitational Force

The drawing that opens this chapter shows our view of the Milky Way galaxy. We are near the edge of the disk of the galaxy, about 26 000 light-years (2.5×10^{20} m) from its center, which in the drawing lies in the star collection known as Sagittarius. Our galaxy is a member of the Local Group of galaxies, which includes the Andromeda galaxy (Fig. 14-1) at a distance of 2.3×10^6 light-years, and several closer dwarf galaxies, such as the Large Magellanic Cloud shown in the opening drawing.

The Local Group is part of the Local Supercluster of galaxies. Measurements taken during and since the 1980s suggest that the Local Supercluster and the supercluster consisting of the clusters Hydra and Centaurus are all moving toward an exceptionally massive region called the Great Attractor. This region appears to be about 300 million light-years away, on the opposite side of the Milky Way from us, past the clusters Hydra and Centaurus.

The force that binds together these progressively larger structures, from star to galaxy to supercluster, and may be drawing them all toward the Great Attractor, is the gravitational force. That force not only holds you on Earth but also reaches out across intergalactic space.

14-2 Newton's Law of Gravitation

Physicists like to study seemingly unrelated phenomena to show that a relationship can be found if they are examined closely enough. This search for unification has been going on for centuries. In 1665, the 23-year-old Isaac Newton made a basic contribution to physics when he showed that the force that holds the Moon in its orbit is the same force that makes an apple fall. We take this so much for granted now that it is not easy for us to comprehend the ancient belief that the motions of earthbound bodies and heavenly bodies were different in kind and were governed by different laws.

Newton concluded that not only does Earth attract an apple and the Moon but every body in the universe attracts every other body; this tendency of bodies to move toward each other is called **gravitation.** Newton's conclusion takes a little getting used to, because the familiar attraction of Earth for earthbound bodies is so great that it overwhelms the attraction that earthbound bodies have for each other. For example, Earth attracts an apple with a force magnitude of about 0.8 N. You also attract a nearby apple (and it attracts you), but the force of attraction has less magnitude than the weight of a speck of dust.

Quantitatively, Newton proposed a *force law* that we call **Newton's law of gravitation:** Every particle attracts any other particle with a **gravitational force** whose magnitude is given by

$$F = G\frac{m_1 m_2}{r^2} \quad \text{(Newton's law of gravitation).} \qquad (14\text{-}1)$$

Here m_1 and m_2 are the masses of the particles, r is the distance between them, and G is the **gravitational constant,** with a value that is now known to be

$$G = 6.67 \times 10^{-11}\ \mathrm{N \cdot m^2/kg^2}$$
$$= 6.67 \times 10^{-11}\ \mathrm{m^3/kg \cdot s^2}. \qquad (14\text{-}2)$$

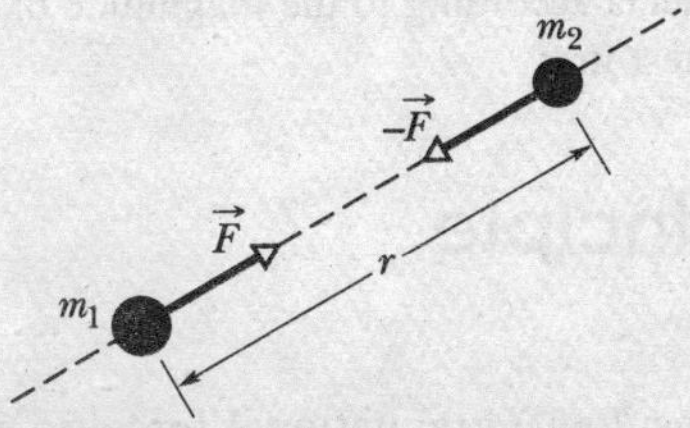

Fig. 14-2 Two particles, of masses m_1 and m_2 and with separation r, attract each other according to Newton's law of gravitation, Eq. 14-1. The forces of attraction, $\vec{F}$ and $-\vec{F}$, are equal in magnitude and in opposite directions.

As Fig. 14-2 shows, a particle m_2 attracts a particle m_1 with a gravitational force $\vec{F}$ that is directed toward particle m_2, and particle m_1 attracts particle m_2 with a gravitational force $-\vec{F}$ that is directed toward m_1. The forces $\vec{F}$ and $-\vec{F}$ form a third-law force pair; they are opposite in direction but equal in magnitude. They depend on the separation of the two particles, but not on their location: the particles could be in a deep cave or in deep space. Also, forces $\vec{F}$ and $-\vec{F}$ are not altered by the presence of other bodies, even if those bodies lie between the two particles we are considering.

The strength of the gravitational force—that is, how strongly two particles with given masses at a given separation attract each other—depends on the value of the gravitational constant G. If G—by some miracle—were suddenly multiplied by a factor of 10, you would be crushed to the floor by Earth's attraction. If G were divided by this factor, Earth's attraction would be weak enough that you could jump over a building.

Although Newton's law of gravitation applies strictly to particles, we can also apply it to real objects as long as the sizes of the objects are small compared to the distance between them. The Moon and Earth are far enough apart so that, to a good approximation, we can treat them both as particles—but what about an apple and Earth? From the point of view of the apple, the broad and level Earth, stretching out to the horizon beneath the apple, certainly does not look like a particle.

Newton solved the apple–Earth problem by proving an important theorem called the *shell theorem:*

> A uniform spherical shell of matter attracts a particle that is outside the shell as if all the shell's mass were concentrated at its center.

Earth can be thought of as a nest of such shells, one within another, and each attracting a particle outside Earth's surface as if the mass of that shell were at the center of the shell. Thus, from the apple's point of view, Earth *does* behave like a particle, one that is located at the center of Earth and has a mass equal to that of Earth.

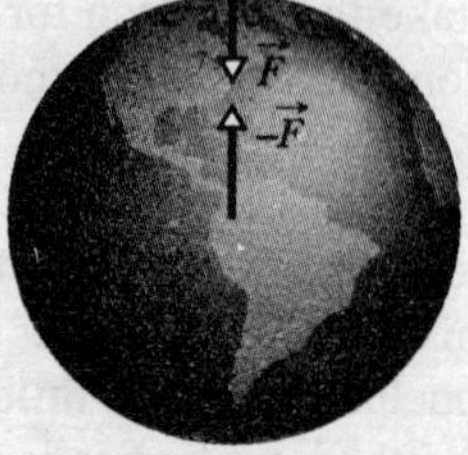

Fig. 14-3 The apple pulls up on Earth just as much as Earth pulls down on the apple.

Suppose, as in Fig. 14-3, that Earth pulls down on an apple with a force of magnitude 0.80 N. The apple must then pull up on Earth with a force of magnitude 0.80 N, which we take to act at the center of Earth. Although the forces are matched in magnitude, they produce different accelerations when the apple is released. For the apple, the acceleration is about 9.8 m/s^2, the familiar acceleration of a falling body near Earth's surface. For Earth, the acceleration measured in a reference frame attached to the center of mass of the apple–Earth system is only about 1×10^{-25} m/s^2.

CHECKPOINT 1: A particle is to be placed, in turn, outside four objects, each of mass m: (1) a large uniform solid sphere, (2) a large uniform spherical shell, (3) a small uniform solid sphere, and (4) a small uniform shell. In each situation, the distance between the particle and the center of the object is d. Rank the objects according to the magnitude of the gravitational force they exert on the particle, greatest first.

14-3 Gravitation and the Principle of Superposition

Given a group of particles, we find the net (or resultant) gravitational force on any one of them from the others by using the **principle of superposition.** This is a general principle that says a net effect is the sum of the individual effects. Here, the principle means that we first compute the gravitational force that acts on our selected particle due to each of the other particles, in turn. We then find the net force by adding these forces vectorially, as usual.

For n interacting particles, we can write the principle of superposition for gravitational forces as

$$\vec{F}_{1,\text{net}} = \vec{F}_{12} + \vec{F}_{13} + \vec{F}_{14} + \vec{F}_{15} + \cdots + \vec{F}_{1n}. \quad (14\text{-}3)$$

Here $\vec{F}_{1,\text{net}}$ is the net force on particle 1 and, for example, $\vec{F}_{13}$ is the force on particle 1 from particle 3. We can express this equation more compactly as a vector sum:

$$\vec{F}_{1,\text{net}} = \sum_{i=2}^{n} \vec{F}_{1i}. \quad (14\text{-}4)$$

What about the gravitational force on a particle from a real extended object? The force is found by dividing the object into parts small enough to treat as particles and then using Eq. 14-4 to find the vector sum of the forces on the particle from all the parts. In the limiting case, we can divide the extended object into differential parts of mass dm, each of which produces only a differential force $d\vec{F}$ on the particle. In this limit, the sum of Eq. 14-4 becomes an integral and we have

$$\vec{F}_1 = \int d\vec{F}, \quad (14\text{-}5)$$

in which the integral is taken over the entire extended object and we drop the subscript "net." If the object is a uniform sphere or a spherical shell, we can avoid the integration of Eq. 14-5 by assuming that the object's mass is concentrated at the object's center and using Eq. 14-1.

If mass of shells making the sphere vary with r only, then also it is allowed

Sample Problem 14-1

Figure 14-4*a* shows an arrangement of three particles, particle 1 having mass $m_1 = 6.0$ kg and particles 2 and 3 having mass $m_2 = m_3 = 4.0$ kg, and with distance $a = 2.0$ cm. What is the net gravitational force $\vec{F}_1$ that acts on particle 1 due to the other particles?

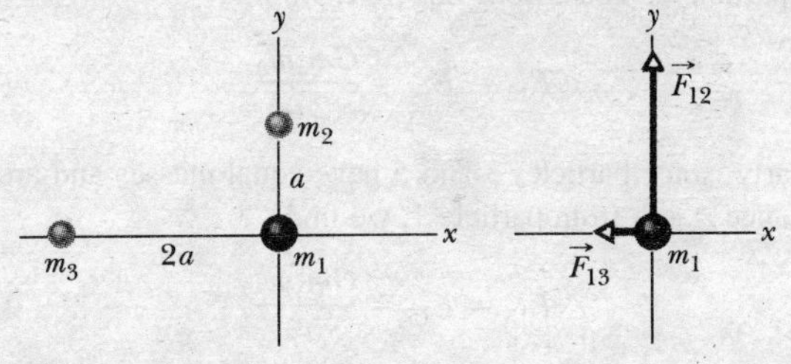

Fig. 14-4 Sample Problem 14-1. (*a*) An arrangement of three particles. (*b*) The forces acting on the particle of mass m_1 due to the other particles.

SOLUTION: One **Key Idea** here is that, because we have particles, the magnitude of the gravitational force on particle 1 due to either of the other particles is given by Eq. 14-1 ($F = Gm_1m_2/r^2$). Thus, the magnitude of the force $\vec{F}_{12}$ on particle 1 from particle 2 is

$$F_{12} = \frac{Gm_1m_2}{a^2} = \frac{(6.67 \times 10^{-11}\ \text{m}^3/\text{kg}\cdot\text{s}^2)(6.0\ \text{kg})(4.0\ \text{kg})}{(0.020\ \text{m})^2} = 4.00 \times 10^{-6}\ \text{N}.$$

Similarly, the magnitude of force $\vec{F}_{13}$ on particle 1 from particle 3 is

$$F_{13} = \frac{Gm_1m_3}{(2a)^2} = \frac{(6.67 \times 10^{-11}\ \text{m}^3/\text{kg}\cdot\text{s}^2)(6.0\ \text{kg})(4.0\ \text{kg})}{(0.040\ \text{m})^2} = 1.00 \times 10^{-6}\ \text{N}.$$

To determine the directions of $\vec{F}_{12}$ and $\vec{F}_{13}$, we use this **Key Idea**: Each force on particle 1 is directed toward the particle responsible for that force. Thus, $\vec{F}_{12}$ is directed in the positive direction of y (Fig. 14-4*b*) and has only the y component F_{12}. Similarly, $\vec{F}_{13}$ is directed in the negative direction of x and has only the x component $-F_{13}$.

To find the net force $\vec{F}_{1,\text{net}}$ on particle 1, we first use this very important **Key Idea**: Because the forces are not directed along the same line, we *cannot* simply add or subtract their magnitudes or their components to get their net force. Instead, we must add them as vectors.

We can do so on a vector-capable calculator. However, here we note that $-F_{13}$ and F_{12} are actually the x and y components of $\vec{F}_{1,\text{net}}$. Therefore, we shall follow the guide of Eq. 3-6 to find first the magnitude and then the direction of $\vec{F}_{1,\text{net}}$. The magnitude is

$$F_{1,\text{net}} = \sqrt{(F_{12})^2 + (-F_{13})^2} = \sqrt{(4.00 \times 10^{-6}\ \text{N})^2 + (-1.00 \times 10^{-6}\ \text{N})^2} = 4.1 \times 10^{-6}\ \text{N}. \quad \text{(Answer)}$$

Relative to the positive direction of the x axis, Eq. 3-6 gives the direction of $\vec{F}_{1,\text{net}}$ as

$$\theta = \tan^{-1}\frac{F_{12}}{-F_{13}} = \tan^{-1}\frac{4.00 \times 10^{-6}\ \text{N}}{-1.00 \times 10^{-6}\ \text{N}} = -76^\circ.$$

Is this a reasonable direction? No, the direction of $\vec{F}_{1,\text{net}}$ must be between the directions of $\vec{F}_{12}$ and $\vec{F}_{13}$. Recall from Chapter 3 (Tactic 3) that a calculator displays only one of the two possible answers to a $\tan^{-1}$ function. We find the other answer by adding 180°. That gives us

$$-76^\circ + 180^\circ = 104^\circ, \qquad \text{(Answer)}$$

which *is* a reasonable direction for $\vec{F}_{1,\text{net}}$.

CHECKPOINT 2: The figure shows four arrangements of three particles of equal masses. (a) Rank the arrangements according to the magnitude of the net gravitational force on the particle labeled m, greatest first. (b) In arrangement 2, is the direction of the net force closer to the line of length d or to the line of length D?

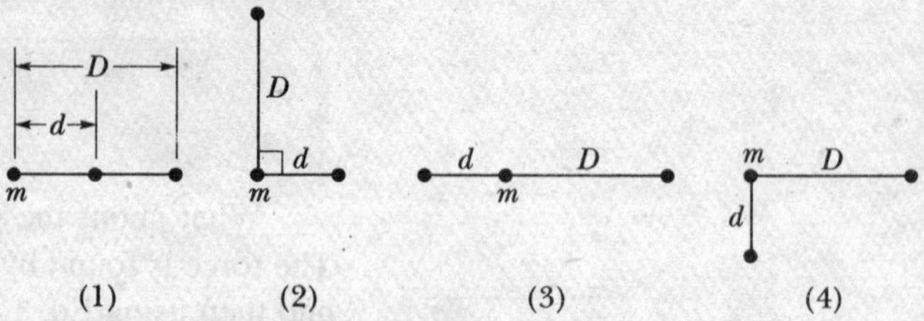

Sample Problem 14-2

Figure 14-5*a* shows an arrangement of five particles, with masses $m_1 = 8.0$ kg, $m_2 = m_3 = m_4 = m_5 = 2.0$ kg, and with $a = 2.0$ cm and $\theta = 30^\circ$. What is the net gravitational force $\vec{F}_{1,\text{net}}$ on particle 1 due to the other particles?

SOLUTION: Our **Key Ideas** are the same as in Sample Problem 14-1. However, this problem has a lot of symmetry that can help simplify the solution.

For the magnitudes of the forces on particle 1, first note that particles 2 and 4 have equal masses and equal distances of $r = 2a$ from particle 1. Thus, from Eq. 14-1, we find

$$F_{12} = F_{14} = \frac{Gm_1m_2}{(2a)^2}. \qquad (14\text{-}6)$$

Similarly, since particles 3 and 5 have equal masses and are both a distance $r = a$ from particle 1, we find

$$F_{13} = F_{15} = \frac{Gm_1m_3}{a^2}. \qquad (14\text{-}7)$$

We can now substitute known data into these two equations to evaluate the magnitudes of the forces. Then we can indicate the directions of the forces on the free-body diagram of Fig. 14-5*b*, and find the net force in either of two basic ways: We could resolve the vectors into x and y components, find the net x component and the net y component, and then vectorially combine those net components. We could also add the vectors directly on a vector-capable calculator.

Instead, however, we shall make further use of the symmetry of the problem. First, we note that $\vec{F}_{12}$ and $\vec{F}_{14}$ are equal in magnitude but opposite in direction; thus, those forces *cancel*. Inspection of Fig. 14-5*b* and Eq. 14-7 reveals that the x components of $\vec{F}_{13}$ and $\vec{F}_{15}$ also *cancel*, and that their y components are identical in magnitude and both act in the positive direction of y. Thus, $\vec{F}_{1,\text{net}}$ acts in that same direction, and its magnitude is twice the y component of $\vec{F}_{13}$:

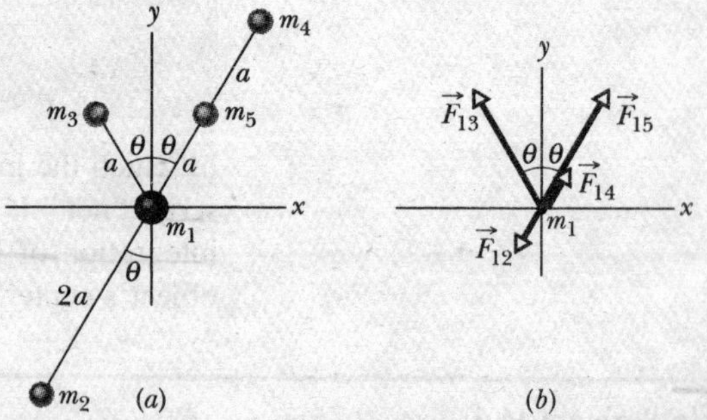

Fig. 14-5 Sample Problem 14-2. (*a*) An arrangement of five particles. (*b*) The forces acting on the particle of mass m_1 due to the other four particles.

$$F_{1,\text{net}} = 2F_{13}\cos\theta = 2\frac{Gm_1m_3}{a^2}\cos\theta$$

$$= 2\frac{(6.67 \times 10^{-11}\ \text{m}^3/\text{kg}\cdot\text{s}^2)(8.0\ \text{kg})(2.0\ \text{kg})}{(0.020\ \text{m})^2}\cos 30^\circ$$

$$= 4.6 \times 10^{-6}\ \text{N}. \qquad \text{(Answer)}$$

Note that the presence of particle 5 along the line between particles 1 and 4 does not alter the gravitational force on particle 1 from particle 4.

CHECKPOINT 3: In the figure here, what is the direction of the net gravitational force on the particle of mass m_1 due to the other particles, each of mass m, that are arranged symmetrically relative to the y axis?

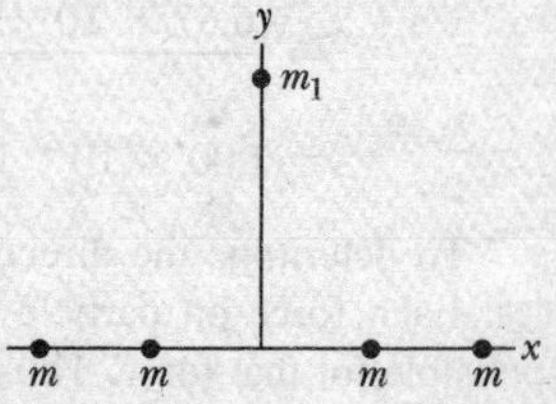

PROBLEM-SOLVING TACTICS

Tactic 1: *Drawing Gravitational Force Vectors*

When you are given a diagram of particles, such as Fig. 14-4*a*, and asked to find the net gravitational force on one of them, you should usually draw a free-body diagram showing only the particle of concern and the forces on *it alone*, as in Fig. 14-4*b*. If, instead, you choose to superimpose the force vectors on the given diagram, be sure to draw the vectors with either their tails (preferably) or their heads on the particle experiencing those forces. If you draw the vectors elsewhere, you invite confusion—and confusion is guaranteed if you draw the vectors on the particles *causing* the forces on the particle of concern.

Tactic 2: *Simplifying a Sum of Forces with Symmetry*
In Sample Problem 14-2 we used the symmetry of the situation: By realizing that particles 2 and 4 are positioned symmetrically about particle 1, and thus that $\vec{F}_{12}$ and $\vec{F}_{14}$ cancel, we avoided calculating either force. By realizing that the x components of $\vec{F}_{13}$ and $\vec{F}_{15}$ cancel and that their y components are identical and add, we saved even more effort.

14-4 Gravitation Near Earth's Surface

Let us assume that Earth is a uniform sphere of mass M. The magnitude of the gravitational force from Earth on a particle of mass m, located outside Earth a distance r from Earth's center, is then given by Eq. 14-1 as

$$F = G\frac{Mm}{r^2}. \tag{14-8}$$

If the particle is released, it will fall toward the center of Earth, as a result of the gravitational force $\vec{F}$, with an acceleration we shall call the **gravitational acceleration** $\vec{a}_g$. Newton's second law tells us that magnitudes F and a_g are related by

$$F = ma_g. \tag{14-9}$$

Now, substituting F from Eq. 14-8 into Eq. 14-9 and solving for a_g, we find

$$a_g = \frac{GM}{r^2}. \tag{14-10}$$

Table 14-1 shows values of a_g computed for various altitudes above Earth's surface.

Since Section 5-6, we have assumed that Earth is an inertial frame by neglecting its actual rotation. This simplification has allowed us to assume that the actual free-fall acceleration g of a particle is the same as the gravitational acceleration (which we now call a_g). Furthermore, we assumed that g has the constant value of 9.8 m/s^2 over Earth's surface. However, the g we would measure differs from the a_g we would calculate with Eq. 14-10 for three reasons: (1) Earth is not uniform, (2) it is not a perfect sphere, and (3) it rotates. Moreover, because g differs from a_g, the measured weight mg of the particle differs from the magnitude of the gravitational force on the particle as given by Eq. 14-8 for the same three reasons. Let us now examine those reasons.

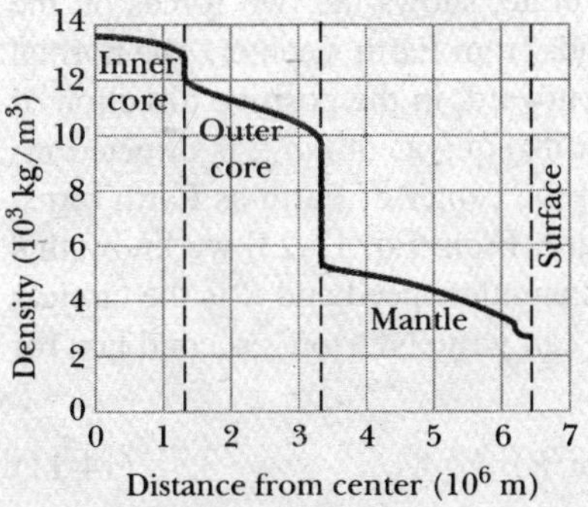

Fig. 14-6 The density of Earth as a function of distance from the center. The limits of the solid inner core, the largely liquid outer core, and the solid mantle are shown, but the crust of Earth is too thin to show clearly on this plot.

1. ***Earth is not uniform.*** The density (mass per unit volume) of Earth varies radially as shown in Fig. 14-6, and the density of the crust (or outer section) of Earth varies from region to region over Earth's surface. Thus, g varies from region to region over the surface.

TABLE 14-1 Variation of a_g with Altitude

Altitude (km)	a_g (m/s^2)	Altitude Example
0	9.83	Mean Earth surface
8.8	9.80	Mt. Everest
36.6	9.71	Highest manned balloon
400	8.70	Space shuttle orbit
35 700	0.225	Communications satellite

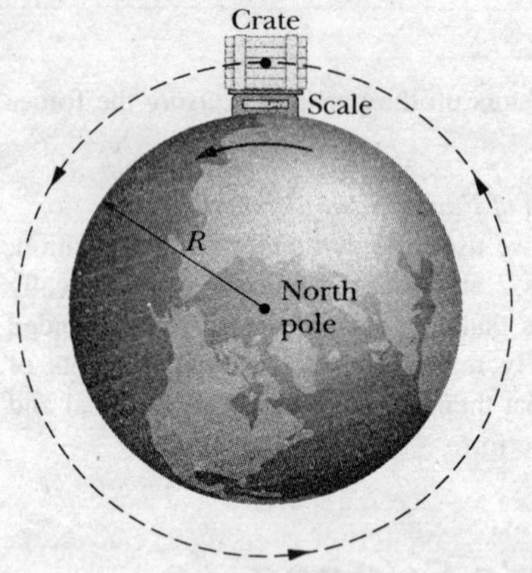

(a)

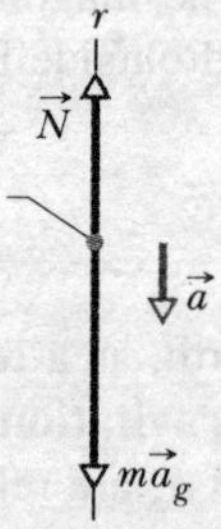

(b)

Fig. 14-7 (a) A crate lies on a scale at Earth's equator, as seen along Earth's rotation axis from above the north pole. (b) A free-body diagram for the crate, with a radially outward r axis. The gravitational force on the crate is represented with its equivalent $m\vec{a}_g$. The normal force on the crate from the scale is $\vec{N}$. Because of Earth's rotation, the crate has a centripetal acceleration $\vec{a}$ that is directed toward Earth's center.

2. ***Earth is not a sphere.*** Earth is approximately an ellipsoid, flattened at the poles and bulging at the equator. Its equatorial radius is greater than its polar radius by 21 km. Thus, a point at the poles is closer to the dense core of Earth than is a point on the equator. This is one reason the free-fall acceleration g increases as one proceeds, at sea level, from the equator toward either pole.
3. ***Earth is rotating.*** The rotation axis runs through the north and south poles of Earth. An object located on Earth's surface anywhere except at those poles must rotate in a circle about the rotation axis and thus must have a centripetal acceleration directed toward the center of the circle. This centripetal acceleration requires a centripetal net force that is also directed toward that center.

To see how Earth's rotation causes g to differ from a_g, let us analyze a simple situation in which a crate of mass m is on a scale at the equator. Figure 14-7a shows this situation as viewed from a point in space above the north pole.

Figure 14-7b, a free-body diagram for the crate, shows the two forces on the crate, both acting along a radial axis r that extends from Earth's center. The normal force $\vec{N}$ on the crate from the scale is directed outward, in the positive direction of axis r. The gravitational force, represented with its equivalent $m\vec{a}_g$, is directed inward. Because the crate travels in a circle about the center of Earth as Earth turns, it has a centripetal acceleration $\vec{a}$ directed inward. From Eq. 11-23, we know this acceleration is equal to $\omega^2 R$, where ω is Earth's angular speed and R is the circle's radius (approximately Earth's radius). Thus, we can write Newton's second law for the r axis ($F_{\text{net},r} = ma_r$) as

$$N - ma_g = m(-\omega^2 R). \quad (14\text{-}11)$$

The magnitude N of the normal force is equal to the weight mg read on the scale. With mg substituted for N, Eq. 14-11 gives us

$$mg = ma_g - m(\omega^2 R), \quad (14\text{-}12)$$

which says

$$\begin{pmatrix}\text{measured}\\ \text{weight}\end{pmatrix} = \begin{pmatrix}\text{magnitude of}\\ \text{gravitational force}\end{pmatrix} - \begin{pmatrix}\text{mass times}\\ \text{centripetal acceleration}\end{pmatrix}.$$

Thus, the measured weight is actually less than the magnitude of the gravitational force on the crate, because of Earth's rotation.

To find a corresponding expression for g and a_g, we cancel m from Eq. 14-12 to write

$$g = a_g - \omega^2 R, \quad (14\text{-}13)$$

which says

$$\begin{pmatrix}\text{free-fall}\\ \text{acceleration}\end{pmatrix} = \begin{pmatrix}\text{gravitational}\\ \text{acceleration}\end{pmatrix} - \begin{pmatrix}\text{centripetal}\\ \text{acceleration}\end{pmatrix}.$$

Thus, the measured free-fall acceleration is actually less than the gravitational acceleration, because of Earth's rotation.

The difference between accelerations g and a_g is equal to $\omega^2 R$ and is greatest on the equator (for one reason, the radius of the circle traveled by the crate is greatest there). To find the difference, we can use Eq. 11-5 ($\omega = \Delta\theta/\Delta t$) and Earth's radius $R = 6.37 \times 10^6$ m. For one rotation of Earth, θ is 2π rad and the time period Δt is about 24 h. Using these values (and converting hours to seconds), we find that g is less than a_g by only about 0.034 m/s^2 (compared to 9.8 m/s^2). Therefore, neglecting the difference in accelerations g and a_g is often justified. Similarly, neglecting the difference between weight and the magnitude of the gravitational acceleration is also often justified.

Sample Problem 14-3

(a) An astronaut whose height h is 1.70 m floats "feet down" in an orbiting space shuttle at a distance $r = 6.77 \times 10^6$ m from the center of Earth. What is the difference between the gravitational acceleration at her feet and that at her head?

SOLUTION: One **Key Idea** here is that we can approximate Earth as a uniform sphere of mass M_E. Then, from Eq. 14-10, the gravitational acceleration at any distance r from the center of Earth is

$$a_g = \frac{GM_E}{r^2}. \tag{14-14}$$

We might simply apply Eq. 14-14 twice, first with, say, $r = 6.77 \times 10^6$ m for the feet and then with $r = 6.77 \times 10^6$ m + 1.70 m for the head. However, a calculator may give us the same value for a_g twice, and thus a difference of zero, because h is so small compared to r. A second **Key Idea** helps here: Because we have a differential change dr in r between the astronaut's feet and head, let us differentiate Eq. 14-14 with respect to r. That gives us

$$da_g = -2\frac{GM_E}{r^3}\,dr, \tag{14-15}$$

where da_g is the differential change in the gravitational acceleration due to the differential change dr in r. For the astronaut, $dr = h$ and $r = 6.77 \times 10^6$ m. Substituting data into Eq. 14-15, we find

$$da_g = -2\frac{(6.67 \times 10^{-11}\ \text{m}^3/\text{kg}\cdot\text{s}^2)(5.98 \times 10^{24}\ \text{kg})}{(6.77 \times 10^6\ \text{m})^3}(1.70\ \text{m})$$

$$= -4.37 \times 10^{-6}\ \text{m/s}^2. \qquad \text{(Answer)}$$

This result means that the gravitational acceleration of the astronaut's feet toward Earth is slightly greater than the gravitational acceleration of her head toward Earth. This difference in acceleration tends to stretch her body, but the difference is so small that the stretching is unnoticeable.

(b) If the astronaut is now "feet down" at the same orbital radius r of 6.77×10^6 m about a black hole of mass $M_h = 1.99 \times 10^{31}$ kg (which is 10 times our Sun's mass), what is now the difference between the gravitational acceleration at her feet and that at her head? The black hole has a surface (called its *event horizon*) of radius $R_h = 2.95 \times 10^4$ m. Nothing, not even light, can escape from that surface or anywhere inside it. Note that the astronaut is (wisely) well outside the surface (at $r = 229R_h$).

SOLUTION: The **Key Idea** here is that we again have a differential change dr in r between the astronaut's feet and head, so we can again use Eq. 14-15. However, now we substitute $M_h = 1.99 \times 10^{31}$ kg for M_E. We find

$$da_g = -2\frac{(6.67 \times 10^{-11}\ \text{m}^3/\text{kg}\cdot\text{s}^2)(1.99 \times 10^{31}\ \text{kg})}{(6.77 \times 10^6\ \text{m})^3}(1.70\ \text{m})$$

$$= -14.5\ \text{m/s}^2. \qquad \text{(Answer)}$$

This means that the gravitational acceleration of the astronaut's feet toward the black hole is noticeably larger than that of her head. The resulting tendency to stretch her body would be bearable but quite painful. If she drifted closer to the black hole, the stretching tendency would increase drastically.

14-5 Gravitation Inside Earth

Newton's shell theorem can also be applied to a situation in which a particle is located *inside* a uniform shell, to show the following:

> A uniform shell of matter exerts no *net* gravitational force on a particle located inside it. (spherical condition not here unlike first shell theorem)

Caution: This statement does *not* mean that the gravitational forces on the particle from the various elements of the shell magically disappear. Rather, it means that the *sum* of the force vectors on the particle from all the elements is zero.

If the density of Earth were uniform, the gravitational force acting on a particle would be a maximum at Earth's surface and would decrease as the particle moved outward. If the particle were to move inward, perhaps down a deep mine shaft, the gravitational force would change for two reasons. (1) It would tend to increase because the particle would be moving closer to the center of Earth. (2) It would tend to decrease because the thickening shell of material lying outside the particle's radial position would not exert any net force on the particle.

For a uniform Earth, the second influence would prevail and the force on the particle would steadily decrease to zero as the particle approached the center of Earth. However, for the real (nonuniform) Earth, the force on the particle actually increases as the particle begins to descend. The force reaches a maximum at a certain depth; only then does it begin to decrease as the particle descends farther.

Sample Problem 14-4

In *Pole to Pole*, an early science fiction story by George Griffith, three explorers attempt to travel by capsule through a naturally formed (and, of course, fictional) tunnel directly from the south pole to the north pole (Fig. 14-8). According to the story, as the capsule approaches Earth's center, the gravitational force on the explorers becomes alarmingly large and then, exactly at the center, it suddenly but only momentarily disappears. Then the capsule travels through the second half of the tunnel, to the north pole.

Check Griffith's description by finding the gravitational force on the capsule of mass m when it reaches a distance r from Earth's center. Assume that Earth is a sphere of uniform density ρ (mass per unit volume).

SOLUTION: Newton's shell theorem gives us three **Key Ideas** here:

1. When the capsule is at a radius r from Earth's center, the portion of Earth that lies outside a sphere of radius r does *not* produce a net gravitational force on the capsule.
2. The portion that lies inside that sphere *does* produce a net gravitational force on the capsule.
3. We can treat the mass M_{ins} of that inside portion of Earth as being the mass of a particle located at Earth's center.

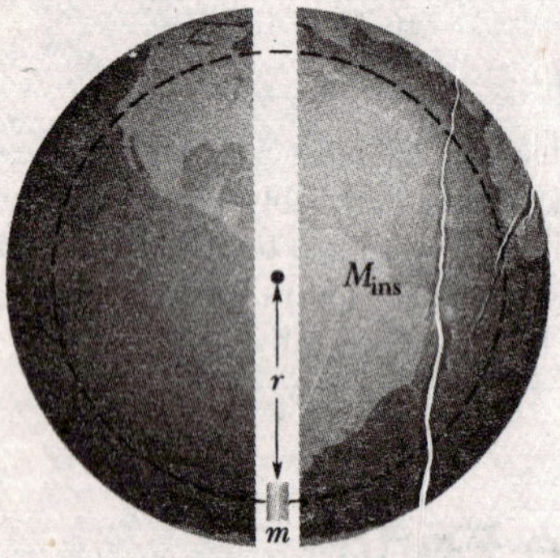

Fig. 14-8 Sample Problem 14-4. A capsule of mass m falls from rest through a tunnel that connects Earth's south and north poles. When the capsule is at distance r from Earth's center, the portion of Earth's mass that is contained in a sphere of that radius is M_{ins}.

All three ideas tell us that we can write Eq. 14-1, for the magnitude of the gravitational force on the capsule, as

$$F = \frac{GmM_{ins}}{r^2}. \qquad (14\text{-}16)$$

To write the mass M_{ins} in terms of the radius r, we note that the volume V_{ins} containing this mass is $\frac{4}{3}\pi r^3$. Also, its density is Earth's density ρ. Thus, we have

$$M_{ins} = \rho V_{ins} = \rho\frac{4\pi r^3}{3}. \qquad (14\text{-}17)$$

Then, after substituting this expression into Eq. 14-16 and canceling, we have

$$F = \frac{4\pi G m\rho}{3}r. \qquad \text{(Answer)} \qquad (14\text{-}18)$$

This equation tells us that the force magnitude F depends linearly on the capsule's distance r from Earth's center. Thus, as r decreases, F also decreases (opposite of Griffith's description), until it is zero at Earth's center. At least Griffith got zero-at-the-center correct.

Equation 14-18 can also be written in terms of the force vector $\vec{F}$ and the capsule's position vector $\vec{r}$ along a radial axis extending from Earth's center. Let K represent the collection of constants $4\pi Gm\rho/3$. Then, Eq. 14-18 becomes

$$\vec{F} = -K\vec{r}, \qquad (14\text{-}19)$$

in which we have inserted a minus sign to indicate that $\vec{F}$ and $\vec{r}$ have opposite directions. Equation 14-19 has the form of Hooke's law (Eq. 7-20). Thus, under the idealized conditions of the story, the capsule would oscillate like a block on a spring, with the center of the oscillation at Earth's center. After the capsule had fallen from the south pole to Earth's center, it would travel from the center to the north pole (as Griffith said) and then back again.

14-6 Gravitational Potential Energy

In Section 8-3, we discussed the gravitational potential energy of a particle–Earth system. We were careful to keep the particle near Earth's surface, so that we could regard the gravitational force as constant. We then chose some reference configuration of the system as having a gravitational potential energy of zero. Often, in this configuration the particle was on Earth's surface. For particles not on Earth's surface, the gravitational potential energy decreased when the separation between the particle and Earth decreased.

Here, we broaden our view and consider the gravitational potential energy U of two particles, of masses m and M, separated by a distance r. We again choose a reference configuration with U equal to zero. However, to simplify the equations, the separation distance r in the reference configuration is now large enough to be approximated as *infinite*. As before, the gravitational potential energy decreases when

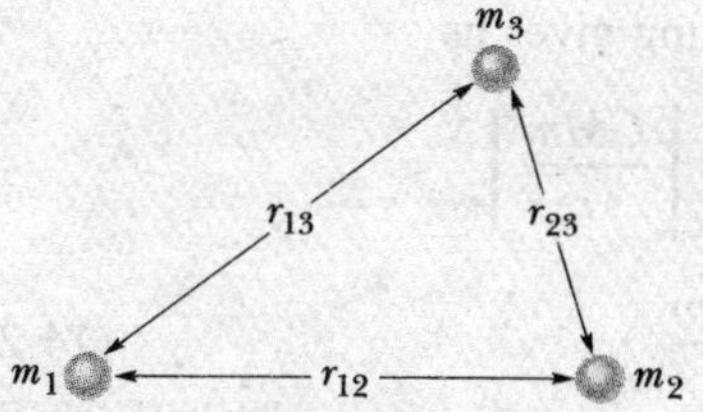

Fig. 14-9 Three particles form a system. (The separation for each pair of particles is labeled with a double subscript to indicate the particles.) The gravitational potential energy *of the system* is the sum of the gravitational potential energies of all three pairs of particles.

the separation decreases. Since $U = 0$ for $r = \infty$, the potential energy is negative for any finite separation and becomes progressively more negative as the particles move closer together.

With these facts in mind and as we shall justify next, we take the gravitational potential energy of the two-particle system to be

$$U = -\frac{GMm}{r} \quad \text{(gravitational potential energy).} \tag{14-20}$$

Note that $U(r)$ approaches zero as r approaches infinity and that for any finite value of r, the value of $U(r)$ is negative.

The potential energy given by Eq. 14-20 is a property of the system of two particles rather than of either particle alone. There is no way to divide this energy and say that so much belongs to one particle and so much to the other. However, if $M \gg m$, as is true for Earth (mass M) and a baseball (mass m), we often speak of "the potential energy of the baseball." We can get away with this because, when a baseball moves in the vicinity of Earth, changes in the potential energy of the baseball–Earth system appear almost entirely as changes in the kinetic energy of the baseball, since changes in the kinetic energy of Earth are too small to be measured. Similarly, in Section 14-8 we shall speak of "the potential energy of an artificial satellite" orbiting Earth, because the satellite's mass is so much smaller than Earth's mass. When we speak of the potential energy of bodies of comparable mass, however, we have to be careful to treat them as a system.

If our system contains more than two particles, we consider each pair of particles in turn, calculate the gravitational potential energy of that pair with Eq. 14-20 as if the other particles were not there, and then algebraically sum the results. Applying Eq. 14-20 to each of the three pairs of Fig. 14-9, for example, gives the potential energy of the system as

$$U = -\left(\frac{Gm_1m_2}{r_{12}} + \frac{Gm_1m_3}{r_{13}} + \frac{Gm_2m_3}{r_{23}}\right). \tag{14-21}$$

Proof of Equation 14-20

Let us shoot a baseball directly away from Earth along the path in Fig. 14-10. We want to find an expression for the gravitational potential energy U of the ball at point P along its path, at radial distance R from Earth's center. To do so, we first find the work W done on the ball by the gravitational force as the ball travels from point P to a great (infinite) distance from Earth. Because the gravitational force $\vec{F}(r)$ is a variable force (its magnitude depends on r), we must use the techniques of Section 7-6 to find the work. In vector notation, we can write

$$W = \int_R^\infty \vec{F}(r) \cdot d\vec{r}. \tag{14-22}$$

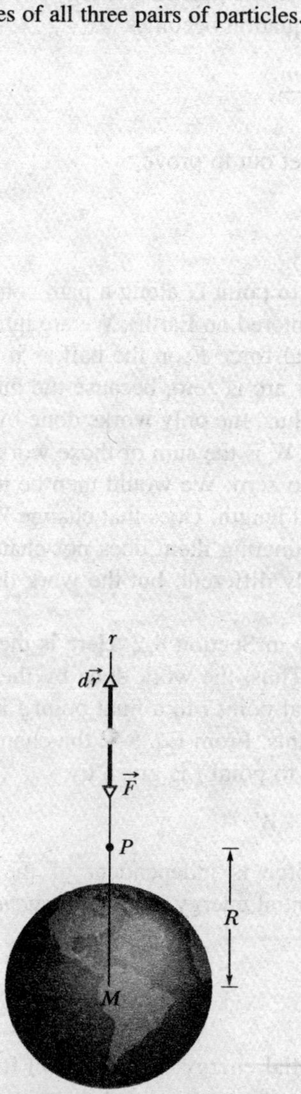

Fig. 14-10 A baseball is shot directly away from Earth, through point P at radial distance R from Earth's center. The gravitational force $\vec{F}$ on the ball and a differential displacement vector $d\vec{r}$ are shown, both directed along a radial r axis.

The integral contains the scalar (or dot) product of the force $\vec{F}(r)$ and the differential displacement vector $d\vec{r}$ along the ball's path. We can expand that product as

$$\vec{F}(r) \cdot d\vec{r} = F(r)\, dr \cos\phi, \tag{14-23}$$

where ϕ is the angle between the directions of $\vec{F}(r)$ and $d\vec{r}$. When we substitute 180° for ϕ and Eq. 14-1 for $F(r)$, Eq. 14-23 becomes

$$\vec{F}(r) \cdot d\vec{r} = -\frac{GMm}{r^2}\, dr,$$

where M is Earth's mass and m is the mass of the ball.

Substituting this into Eq. 14-22 and integrating gives us

$$W = -GMm \int_R^\infty \frac{1}{r^2}\,dr = \left[\frac{GMm}{r}\right]_R^\infty$$

$$= 0 - \frac{GMm}{R} = -\frac{GMm}{R}. \qquad (14\text{-}24)$$

W in Eq. 14-24 is the work required to move the ball from point P (at distance R) to infinity. Equation 8-1 ($\Delta U = -W$) tells us that we can also write that work in terms of potential energies as

$$U_\infty - U = -W.$$

The potential energy U_∞ at infinity is zero, and U is the potential energy at P. Thus, with Eq. 14-24 substituted for W, the previous equation becomes

$$U = W = -\frac{GMm}{R}.$$

Switching R to r gives us Eq. 14-20, which we set out to prove.

Path Independence

In Fig. 14-11, we move a baseball from point A to point G along a path consisting of three radial lengths and three circular arcs (centered on Earth). We are interested in the total work W done by Earth's gravitational force $\vec{F}$ on the ball as it moves from A to G. The work done along each circular arc is zero, because the direction of $\vec{F}$ is perpendicular to the arc at every point. Thus, the only works done by $\vec{F}$ are along the three radial lengths, and the total work W is the sum of those works.

Now, suppose we mentally shrink the arcs to zero. We would then be moving the ball directly from A to G along a single radial length. Does that change W? No. Because no work was done along the arcs, eliminating them does not change the work. The path taken from A to G now is clearly different, but the work done by $\vec{F}$ is the same.

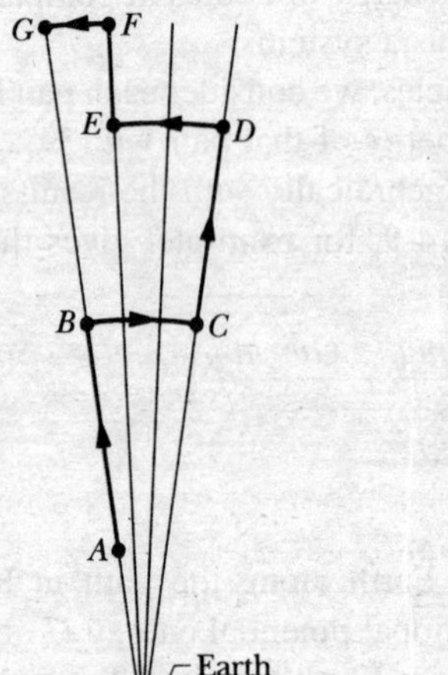

Fig. 14-11 Near Earth, a baseball is moved from point A to point G along a path consisting of radial lengths and circular arcs.

We discussed such a result in a general way in Section 8-2. Here is the point: The gravitational force is a conservative force. Thus, the work done by the gravitational force on a particle moving from an initial point i to a final point f is independent of the actual path taken between the points. From Eq. 8-1, the change ΔU in the gravitational potential energy from point i to point f is given by

$$\boxed{\Delta U = U_f - U_i = -W.} \qquad (14\text{-}25)$$

Since the work W done by a conservative force is independent of the actual path taken, the change ΔU in gravitational potential energy is *also independent* of the actual path taken.

Potential Energy and Force

In the proof of Eq. 14-20, we derived the potential energy function $U(r)$ from the force function $\vec{F}(r)$. We should be able to go the other way—that is, to start from the potential energy function and derive the force function. Guided by Eq. 8-20, we can write

$$F = -\frac{dU}{dr} = -\frac{d}{dr}\left(-\frac{GMm}{r}\right)$$

$$= -\frac{GMm}{r^2}. \qquad (14\text{-}26)$$

This is Newton's law of gravitation (Eq. 14-1). The minus sign indicates that the force on mass m points radially inward, toward mass M.

Escape Speed

If you fire a projectile upward, usually it will slow, stop momentarily, and return to Earth. There is, however, a certain minimum initial speed that will cause it to move upward forever, theoretically coming to rest only at infinity. This initial speed is called the (Earth) **escape speed.**

Consider a projectile of mass m, leaving the surface of a planet (or some other astronomical body or system) with escape speed v. It has a kinetic energy K given by $\frac{1}{2}mv^2$ and a potential energy U given by Eq. 14-20:

$$U = -\frac{GMm}{R},$$

in which M is the mass of the planet, and R is its radius.

When the projectile reaches infinity, it stops and thus has no kinetic energy. It also has no potential energy because this is our zero-potential-energy configuration. Its total energy at infinity is therefore zero. From the principle of conservation of energy, its total energy at the planet's surface must also have been zero, so

$$K + U = \tfrac{1}{2}mv^2 + \left(-\frac{GMm}{R}\right) = 0.$$

This yields

$$v = \sqrt{\frac{2GM}{R}}. \qquad (14\text{-}27)$$

The escape speed v does not depend on the direction in which a projectile is fired from a planet. However, attaining that speed is easier if the projectile is fired in the direction the launch site is moving as the planet rotates about its axis. For example, rockets are launched eastward at Cape Canaveral to take advantage of the Cape's eastward speed of 1500 km/h due to Earth's rotation.

Equation 14-27 can be applied to find the escape speed of a projectile from any astronomical body, provided we substitute the mass of the body for M and the radius of the body for R. Table 14-2 shows escape speeds from some astronomical bodies.

TABLE 14-2 Some Escape Speeds

Body	Mass (kg)	Radius (m)	Escape Speed (km/s)
Ceres[a]	1.17×10^{21}	3.8×10^{5}	0.64
Earth's moon[a]	7.36×10^{22}	1.74×10^{6}	2.38
Earth	5.98×10^{24}	6.37×10^{6}	11.2
Jupiter	1.90×10^{27}	7.15×10^{7}	59.5
Sun	1.99×10^{30}	6.96×10^{8}	618
Sirius B[b]	2×10^{30}	1×10^{7}	5200
Neutron star[c]	2×10^{30}	1×10^{4}	2×10^{5}

[a] The most massive of the asteroids.

[b] A *white dwarf* (a star in a final stage of evolution) that is a companion of the bright star Sirius.

[c] The collapsed core of a star that remains after that star has exploded in a *supernova* event.

CHECKPOINT 4: You move a ball of mass m away from a sphere of mass M. (a) Does the gravitational potential energy of the ball–sphere system increase or decrease? (b) Is positive or negative work done by the gravitational force between the ball and the sphere?

Sample Problem 14-5

An asteroid, headed directly toward Earth, has a speed of 12 km/s relative to the planet when it is at a distance of 10 Earth radii from Earth's center. Neglecting the effects of Earth's atmosphere on the asteroid, find the asteroid's speed v_f when it reaches Earth's surface.

SOLUTION: One **Key Idea** is that, because we are to neglect the effects of the atmosphere on the asteroid, the mechanical energy of the asteroid–Earth system is conserved during the fall. Thus, the final mechanical energy (when the asteroid reaches Earth's surface) is equal to the initial mechanical energy. We can write this as

$$K_f + U_f = K_i + U_i, \tag{14-28}$$

where K is kinetic energy and U is gravitational potential energy.

A second **Key Idea** is that, if we assume the system is isolated, the system's linear momentum must be conserved during the fall. Therefore, the momentum change of the asteroid and that of Earth must be equal in magnitude and opposite in sign. However, because Earth's mass is so great relative to the asteroid's mass, the change in Earth's speed is negligible relative to the change in the asteroid's speed. So, the change in Earth's kinetic energy is also negligible. Thus, we can assume that the kinetic energies in Eq. 14-28 are those of the asteroid alone.

Let m represent the asteroid's mass and M represent Earth's mass (5.98×10^{24} kg). The asteroid is initially at the distance $10R_E$ and finally at the distance R_E, where R_E is Earth's radius (6.37×10^6 m). Substituting Eq. 14-20 for U and $\frac{1}{2}mv^2$ for K, we rewrite Eq. 14-28 as

$$\tfrac{1}{2}mv_f^2 - \frac{GMm}{R_E} = \tfrac{1}{2}mv_i^2 - \frac{GMm}{10R_E}.$$

Rearranging and substituting known values, we find

$$\begin{aligned} v_f^2 &= v_i^2 + \frac{2GM}{R_E}\left(1 - \frac{1}{10}\right) \\ &= (12 \times 10^3\ \text{m/s})^2 \\ &\quad + \frac{2(6.67 \times 10^{-11}\ \text{m}^3/\text{kg}\cdot\text{s}^2)(5.98 \times 10^{24}\ \text{kg})}{6.37 \times 10^6\ \text{m}}\ 0.9 \\ &= 2.567 \times 10^8\ \text{m}^2/\text{s}^2, \end{aligned}$$

and thus

$$v_f = 1.60 \times 10^4\ \text{m/s} = 16\ \text{km/s}. \qquad \text{(Answer)}$$

At this speed, the asteroid would not have to be particularly large to do considerable damage at impact. As an example, if it were only 5 m across, the impact could release about as much energy as the nuclear explosion at Hiroshima. Alarmingly, about 500 million asteroids of this size are near Earth's orbit, and in 1994 one of them apparently penetrated Earth's atmosphere and exploded at an altitude of 20 km near a remote South Pacific island (setting off nuclear-explosion warnings on six military satellites). The impact of an asteroid 500 m across (there may be a million of them near Earth's orbit) could end modern civilization and almost eliminate humans worldwide.

14-7 Planets and Satellites: Kepler's Laws

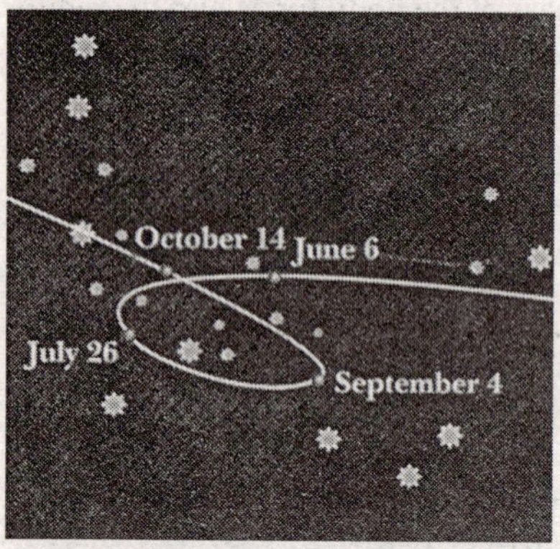

Fig. 14-12 The path of the planet Mars as it moved against a background of the constellation Capricorn during 1971. Its position on four selected days is marked. Both Mars and Earth are moving in orbits around the Sun so that we see the position of Mars relative to us; this sometimes results in an apparent loop in the path of Mars.

The motions of the planets, as they seemingly wander against the background of the stars, have been a puzzle since the dawn of history. The "loop-the-loop" motion of Mars, shown in Fig. 14-12, was particularly baffling. Johannes Kepler (1571–1630), after a lifetime of study, worked out the empirical laws that govern these motions. Tycho Brahe (1546–1601), the last of the great astronomers to make observations without the help of a telescope, compiled the extensive data from which Kepler was able to derive the three laws of planetary motion that now bear his name. Later, Newton (1642–1727) showed that his law of gravitation leads to Kepler's laws.

In this section we discuss each of Kepler's laws in turn. Although here we apply the laws to planets orbiting the Sun, they hold equally well for satellites, either natural or artificial, orbiting Earth or any other massive central body.

1. THE LAW OF ORBITS: All planets move in elliptical orbits, with the Sun at one focus.

Figure 14-13 shows a planet of mass m moving in such an orbit around the Sun, whose mass is M. We assume that $M \gg m$, so that the center of mass of the planet–Sun system is approximately at the center of the Sun.

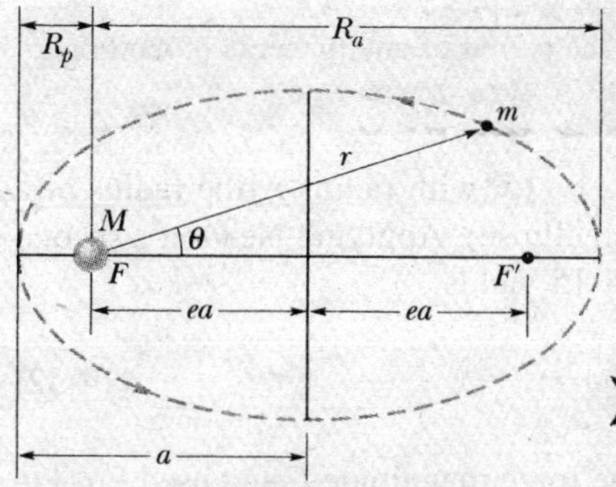

Fig. 14-13 A planet of mass m moving in an elliptical orbit around the Sun. The Sun, of mass M, is at one focus F of the ellipse. The other focus is F', which is located in empty space. Each focus is a distance ea from the ellipse's center, with e being the eccentricity of the ellipse. The semimajor axis a of the ellipse, the perihelion (nearest the Sun) distance R_p, and the aphelion (farthest from the Sun) distance R_a are also shown.

The orbit in Fig. 14-13 is described by giving its **semimajor axis** a and its **eccentricity** e, the latter defined so that ea is the distance from the center of the ellipse to either focus F or F'. *An eccentricity of zero corresponds to a circle*, in which the two foci merge to a single central point. The eccentricities of the planetary orbits are not large, so—sketched on paper—the orbits look circular. The eccentricity of the ellipse of Fig. 14-13, which has been exaggerated for clarity, is 0.74. The eccentricity of Earth's orbit is only 0.0167.

2. THE LAW OF AREAS: A line that connects a planet to the Sun sweeps out equal areas in the plane of the planet's orbit in equal times; that is, the rate dA/dt at which it sweeps out area A is constant.

Qualitatively, this second law tells us that the planet will move most slowly when it is farthest from the Sun and most rapidly when it is nearest to the Sun. As it turns out, Kepler's second law is totally equivalent to the law of conservation of angular momentum. Let us prove it. (for planet-sun system)

The area of the shaded wedge in Fig. 14-14a closely approximates the area swept out in time Δt by a line connecting the Sun and the planet, which are separated by a distance r. The area ΔA of the wedge is approximately the area of a triangle with base $r\,\Delta\theta$ and height r. Since the area of a triangle is one-half of the base times the height, $\Delta A \approx \frac{1}{2}r^2\,\Delta\theta$. This expression for ΔA becomes more exact as Δt (hence $\Delta\theta$) approaches zero. The instantaneous rate at which area is being swept out is then

$$\frac{dA}{dt} = \frac{1}{2}r^2\frac{d\theta}{dt} = \frac{1}{2}r^2\omega, \tag{14-29}$$

dA/dt for any 2 different planets is also same provided both orbit same body & lie in the same plane

in which ω is the angular speed of the rotating line connecting Sun and planet.

Figure 14-14b shows the linear momentum $\vec{p}$ of the planet, along with its radial and perpendicular components. From Eq. 12-20 ($L = rp_\perp$), the magnitude of the angular momentum $\vec{L}$ of the planet about the Sun is given by the product of r and $p_\perp$, the component of $\vec{p}$ perpendicular to r. Here, for a planet of mass m,

$$L = rp_\perp = (r)(mv_\perp) = (r)(m\omega r) = mr^2\omega, \tag{14-30}$$

where we have replaced $v_\perp$ with its equivalent ωr (Eq. 11-18). Eliminating $r^2\omega$ between Eqs. 14-29 and 14-30 leads to

$$\frac{dA}{dt} = \frac{L}{2m}. \tag{14-31}$$

If dA/dt is constant, as Kepler said it is, then Eq. 14-31 means that L must also be constant—angular momentum is conserved. Kepler's second law is indeed equivalent to the law of conservation of angular momentum.

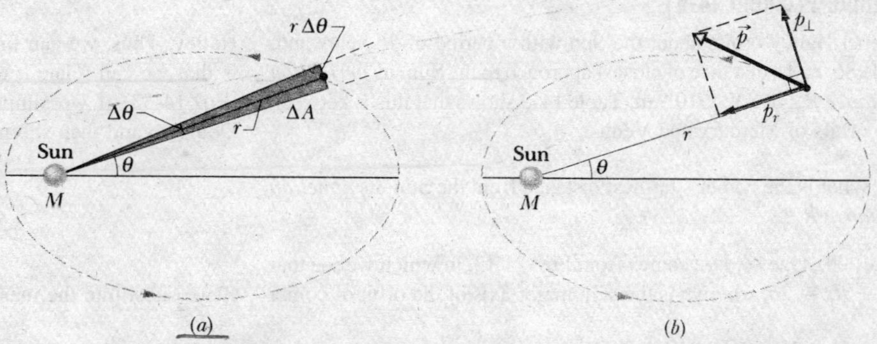

Fig. 14-14 (a) In time Δt, the line r connecting the planet to the Sun (of mass M) sweeps through an angle $\Delta\theta$, sweeping out an area ΔA (shaded). (b) The linear momentum $\vec{p}$ of the planet and its components.

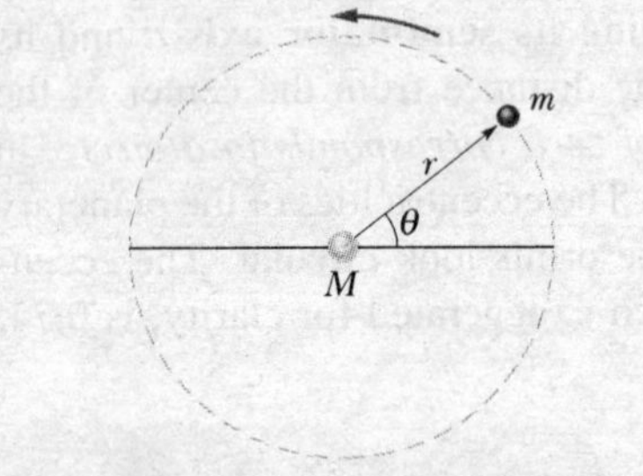

Fig. 14-15 A planet of mass m moving around the Sun in a circular orbit of radius r.

3. THE LAW OF PERIODS: The square of the period of any planet is proportional to the cube of the semimajor axis of its orbit.

To see this, consider the circular orbit of Fig. 14-15, with radius r (the radius of a circle is equivalent to the semimajor axis of an ellipse). Applying Newton's second law ($F = ma$) to the orbiting planet in Fig. 14-15 yields

$$\frac{GMm}{r^2} = (m)(\omega^2 r). \tag{14-32}$$

Here we have substituted from Eq. 14-1 for the force magnitude F and used Eq. 11-23 to substitute $\omega^2 r$ for the centripetal acceleration. If we use Eq. 11-20 to replace ω with $2\pi/T$, where T is the period of the motion, we obtain Kepler's third law:

$$T^2 = \left(\frac{4\pi^2}{GM}\right) r^3 \quad \text{(law of periods).} \tag{14-33}$$

The quantity in parentheses is a constant that depends only on the mass M of the central body about which the planet orbits.

Equation 14-33 holds also for elliptical orbits, provided we replace r with a, the semimajor axis of the ellipse. This law predicts that the ratio T^2/a^3 has essentially the same value for every planetary orbit around a given massive body. Table 14-3 shows how well it holds for the orbits of the planets of the solar system.

TABLE 14-3 Kepler's Law of Periods for the Solar System

Planet	Semimajor Axis a (10^{10} m)	Period T (y)	T^2/a^3 (10^{-34} y^2/m^3)
Mercury	5.79	0.241	2.99
Venus	10.8	0.615	3.00
Earth	15.0	1.00	2.96
Mars	22.8	1.88	2.98
Jupiter	77.8	11.9	3.01
Saturn	143	29.5	2.98
Uranus	287	84.0	2.98
Neptune	450	165	2.99
Pluto	590	248	2.99

CHECKPOINT 5: Satellite 1 is in a certain circular orbit about a planet, while satellite 2 is in a larger circular orbit. Which satellite has (a) the longer period and (b) the greater speed?

Sample Problem 14-6

Comet Halley orbits about the Sun with a period of 76 years and, in 1986, had a distance of closest approach to the Sun, its *perihelion distance* R_p, of 8.9×10^{10} m. Table 14-3 shows that this is between the orbits of Mercury and Venus.

(a) What is the comet's farthest distance from the Sun, its *aphelion distance* R_a?

SOLUTION: One Key Idea comes from Fig. 14-13, in which we see that $R_a + R_p = 2a$, where a is the semimajor axis of the orbit of comet Halley. Thus, we can find R_a if we first find a. A second Key Idea is that we can relate a to the given period via the law of periods (Eq. 14-33) if we simply substitute the semimajor axis a for r. Doing so and then solving for a, we have

$$a = \left(\frac{GMT^2}{4\pi^2}\right)^{1/3}. \tag{14-34}$$

If we substitute the mass M of the Sun, 1.99×10^{30} kg, and the period T of the comet, 76 years or 2.4×10^9 s, into Eq. 14-34, we find that $a = 2.7 \times 10^{12}$ m. Now we have

$$R_a = 2a - R_p = (2)(2.7 \times 10^{12}\text{ m}) - 8.9 \times 10^{10}\text{ m} = 5.3 \times 10^{12}\text{ m.} \quad \text{(Answer)}$$

Table 14-3 shows that this is a little less than the semimajor axis of the orbit of Pluto. Thus, the comet does not get farther from the Sun than Pluto.

(b) What is the eccentricity e of the orbit of comet Halley?

SOLUTION: The **Key Idea** here is that we can relate e, a, and R_p via Fig. 14-13. We see there that $ea = a - R_p$, or

$$e = \frac{a - R_p}{a} = 1 - \frac{R_p}{a} = 1 - \frac{8.9 \times 10^{10}\text{ m}}{2.7 \times 10^{12}\text{ m}} = 0.97. \quad \text{(Answer)}$$

This cometary orbit, with an eccentricity approaching unity, is a long thin ellipse.

Sample Problem 14-7

Hunting a black hole. Observations of the light from a certain star indicate that it is part of a binary (two-star) system. This visible star has orbital speed $v = 270$ km/s, orbital period $T = 1.70$ days, and approximate mass $m_1 = 6M_s$, where M_s is the Sun's mass, 1.99×10^{30} kg. Assuming that the visible star and its companion star, which is dark and unseen, are both in circular orbits (see Fig. 14-16), determine the approximate mass m_2 of the dark star.

SOLUTION: Some of the **Key Ideas** in this challenging problem are as follows:

1. The two stars are in circular orbits, not about each other, but about the center of mass of this two-star system.
2. As with the two-particle systems of Section 9-2, the center of mass of the two-star system must lie along a line connecting the centers of the stars—that is, at point O in Fig. 14-16. The visible star orbits at radius r_1, the dark star at radius r_2.
3. The center of mass of the system is not even approximately at the center of a central, massive object (like the Sun). Therefore, Kepler's law of periods, Eq. 14-33, does *not* apply here and we cannot easily find mass m_2 with it.
4. The centripetal force causing each star to move in a circle is the gravitational force due to the other star. The magnitude of the force is Gm_1m_2/r^2, where r is the distance between the centers of the stars.
5. From Eq. 4-32, the centripetal acceleration a of the visible star is v^2/r_1.

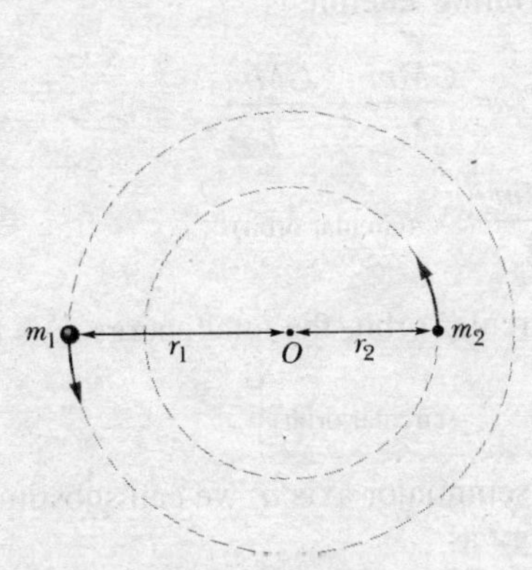

Fig. 14-16 Sample Problem 14-7. A visible star with mass m_1 and a dark, unseen star with mass m_2 orbit around the center of mass of the two-star system at O.

These ideas lead us to write Newton's second law ($F = ma$) for the visible star as

$$\frac{Gm_1m_2}{r^2} = m_1\frac{v^2}{r_1}. \tag{14-35}$$

This equation contains the required mass m_2, but to find it we need expressions for r and r_1. (Note that m_1 cancels out.)

We start by locating the center of mass relative to the visible star, using Eq. 9-1. That star is at distance zero relative to itself, the center of mass is at distance r_1, and the dark star is at distance r. Equation 9-2 then becomes

$$r_1 = \frac{m_1(0) + m_2r}{m_1 + m_2}, \tag{14-36}$$

which yields

$$r = r_1\frac{m_1 + m_2}{m_2}. \tag{14-37}$$

To find an expression for r_1, we note that the visible star is moving in a circle of radius r_1, at speed v, and with period T. Thus, from Eq. 4-33, we have $v = 2\pi r_1/T$ or

$$r_1 = \frac{vT}{2\pi}. \tag{14-38}$$

Substituting this for r_1 in Eq. 14-37 results in

$$r = \frac{vT}{2\pi}\frac{m_1 + m_2}{m_2}. \tag{14-39}$$

Now we return to Eq. 14-35 and substitute for r with Eq. 14-39, for r_1 with Eq. 14-38, and for m_1 with the given $6M_s$. Rearranging the result and substituting known data then give us

$$\frac{m_2^3}{(6M_s + m_2)^2} = \frac{v^3T}{2\pi G} = \frac{(2.7 \times 10^5\text{ m/s})^3(1.70\text{ days})(86\,400\text{ s/day})}{(2\pi)(6.67 \times 10^{-11}\text{ N}\cdot\text{m}^2/\text{kg}^2)} = 6.90 \times 10^{30}\text{ kg},$$

or

$$\frac{m_2^3}{(6M_s + m_2)^2} = 3.47M_s. \tag{14-40}$$

We can solve this cubic equation for m_2 with a polynomial solver on a calculator. Instead, since we are working with approximate masses anyway, we can substitute integer multiples of M_s for m_2 until we find one that makes Eq. 14-40 nearly true. This occurs for

$$m_2 \approx 9M_s. \qquad \text{(Answer)}$$

The data here approximate those for the binary system LMC X-3 in the Large Magellanic Cloud (shown in the figure that begins this chapter). From other data, the dark object is known to be especially compact: It may be a star that collapsed under its own gravitational pull to become a neutron star or a black hole. Since a neutron star cannot have a mass larger than about $2M_s$, the result $m_2 \approx 9M_s$ strongly suggests that the dark object is a black hole.

Thus, we can detect the presence of a black hole provided it is part of a binary system with a visible star whose mass, orbital speed, and orbital period can be measured.

14-8 Satellites: Orbits and Energy

On February 7, 1984, at a height of 102 km above Hawaii and with a speed of about 29 000 km/h, Bruce McCandless stepped (untethered) into space from a space shuttle and became the first human satellite.

As a satellite orbits Earth on its elliptical path, both its speed, which fixes its kinetic energy K, and its distance from the center of Earth, which fixes its gravitational potential energy U, fluctuate with fixed periods. However, the mechanical energy E of the satellite remains constant. (Since the satellite's mass is so much smaller than Earth's mass, we assign U and E for the Earth–satellite system to the satellite alone.)

The potential energy of the system is given by Eq. 14-20 and is

$$U = -\frac{GMm}{r}$$

(with $U = 0$ for infinite separation). Here r is the radius of the orbit, assumed for the time being to be circular, and M and m are the masses of Earth and the satellite, respectively.

To find the kinetic energy of a satellite in a circular orbit, we write Newton's second law ($F = ma$) as

$$\frac{GMm}{r^2} = m\frac{v^2}{r}, \qquad (14\text{-}41)$$

where v^2/r is the centripetal acceleration of the satellite. Then, from Eq. 14-41, the kinetic energy is

$$K = \tfrac{1}{2}mv^2 = \frac{GMm}{2r}, \qquad (14\text{-}42)$$

which shows us that for a satellite in a circular orbit,

$$K = -\frac{U}{2} \quad \text{(circular orbit)}. \qquad (14\text{-}43)$$

The total mechanical energy of the orbiting satellite is

$$E = K + U = \frac{GMm}{2r} - \frac{GMm}{r}$$

or

$$E = -\frac{GMm}{2r} \quad \text{(circular orbit)}. \qquad (14\text{-}44)$$

This tells us that for a satellite in a circular orbit, the total energy E is the negative of the kinetic energy K:

$$E = -K \quad \text{(circular orbit)}. \qquad (14\text{-}45)$$

For a satellite in an elliptical orbit of semimajor axis a, we can substitute a for r in Eq. 14-44 to find the mechanical energy as

$$E = -\frac{GMm}{2a} \quad \text{(elliptical orbit)}. \qquad (14\text{-}46)$$

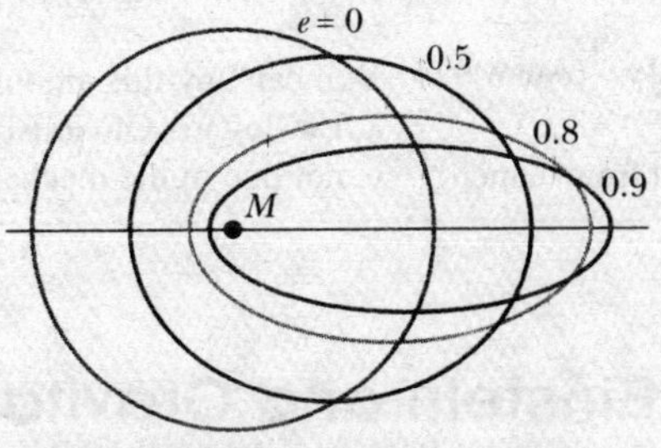

Fig. 14-17 Four orbits about an object of mass M. All four orbits have the same semimajor axis a and thus correspond to the same total mechanical energy E. Their eccentricities e are marked.

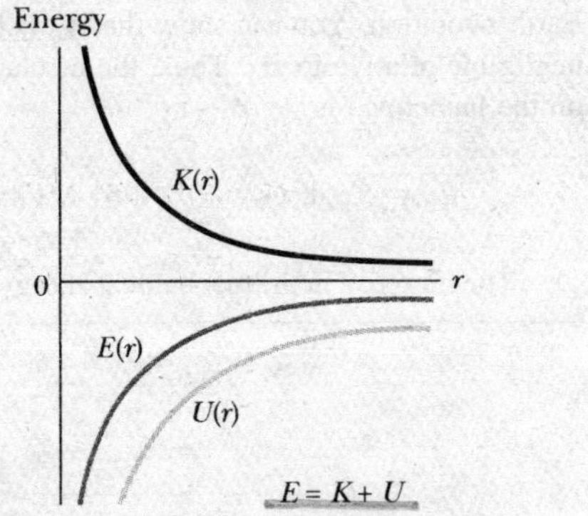

Fig. 14-18 The variation of kinetic energy K, potential energy U, and total energy E with radius r for a satellite in a circular orbit. For any value of r, the values of U and E are negative, the value of K is positive, and $E = -K$. As $r \to \infty$, all three energy curves approach a value of zero.

Equation 14-46 tells us that the total energy of an orbiting satellite depends only on the semimajor axis of its orbit and not on its eccentricity e. For example, four orbits with the same semimajor axis are shown in Fig. 14-17; the same satellite would have the same total mechanical energy E in all four orbits. Figure 14-18 shows the variation of K, U, and E with r for a satellite moving in a circular orbit about a massive central body.

(a) Path 1 Decreased 'E' ⇒ more negative gives decreased 'a'

(b) less Decreased 'a' gives decreased 'T'

CHECKPOINT 6: In the figure here, a space shuttle is initially in a circular orbit of radius r about Earth. At point P, the pilot briefly fires a forward-pointing thruster to decrease the shuttle's kinetic energy K and mechanical energy E. (a) Which of the dashed elliptical orbits shown in the figure will the shuttle then take? (b) Is the orbital period T of the shuttle (the time to return to P) then greater than, less than, or the same as in the circular orbit?

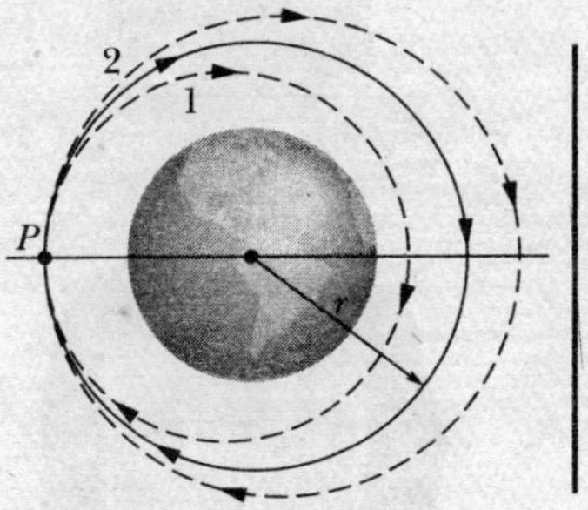

Sample Problem 14-8

A playful astronaut releases a bowling ball, of mass $m = 7.20$ kg, into circular orbit about Earth at an altitude h of 350 km. (is measured from surface of earth)

(a) What is the mechanical energy E of the ball in its orbit?

SOLUTION: The **Key Idea** here is that we can get E from the orbital energy, given by Eq. 14-44 ($E = -GMm/2r$), if we first find the orbital radius r. That radius must be

$$r = R + h = 6370 \text{ km} + 350 \text{ km} = 6.72 \times 10^6 \text{ m},$$

in which R is the radius of Earth. Then, from Eq. 14-44, the mechanical energy is

$$E = -\frac{GMm}{2r}$$
$$= -\frac{(6.67 \times 10^{-11} \text{ N}\cdot\text{m}^2/\text{kg}^2)(5.98 \times 10^{24} \text{ kg})(7.20 \text{ kg})}{(2)(6.72 \times 10^6 \text{ m})}$$
$$= -2.14 \times 10^8 \text{ J} = -214 \text{ MJ}. \quad \text{(Answer)}$$

(b) What is the mechanical energy E_0 of the ball on the launchpad at Cape Canaveral? From there to the orbit, what is the change ΔE in the ball's mechanical energy?

SOLUTION: The **Key Idea** here is that, on the launchpad, the ball is *not* in orbit and thus Eq. 14-44 does *not* apply. Instead, we must find $E_0 = K_0 + U_0$, where K_0 is the ball's kinetic energy and U_0 is the gravitational potential energy of the ball–Earth system. To find U_0, we use Eq. 14-20 to write

$$U_0 = -\frac{GMm}{R}$$
$$= -\frac{(6.67 \times 10^{-11} \text{ N}\cdot\text{m}^2/\text{kg}^2)(5.98 \times 10^{24} \text{ kg})(7.20 \text{ kg})}{6.37 \times 10^6 \text{ m}}$$
$$= -4.51 \times 10^8 \text{ J} = -451 \text{ MJ}.$$

The kinetic energy K_0 of the ball is due to the ball's motion with

Earth's rotation. You can show that K_0 is less than 1 MJ, which is negligible relative to U_0. Thus, the mechanical energy of the ball on the launchpad is

$$E_0 = K_0 + U_0 \approx 0 - 451 \text{ MJ} = -451 \text{ MJ}. \quad \text{(Answer)}$$

The *increase* in the mechanical energy of the ball from launchpad to orbit is

$$\Delta E = E - E_0 = (-214 \text{ MJ}) - (-451 \text{ MJ}) = 237 \text{ MJ}. \quad \text{(Answer)}$$

You can buy this amount of energy from your utility company for a few dollars. Obviously the high cost of placing objects into orbit is not due to the mechanical energy those objects require.

14-9 Einstein and Gravitation

Principle of Equivalence

Albert Einstein once said: "I was . . . in the patent office at Bern when all of a sudden a thought occurred to me: 'If a person falls freely, he will not feel his own weight.' I was startled. This simple thought made a deep impression on me. It impelled me toward a theory of gravitation."

Thus Einstein tells us how he began to form his **general theory of relativity.** The fundamental postulate of this theory about gravitation (the gravitating of objects toward each other) is called the **principle of equivalence,** which says that gravitation and acceleration are equivalent. If a physicist were locked up in a small box as in Fig. 14-19, he would not be able to tell whether the box was at rest on Earth (and subject only to Earth's gravitational force), as in Fig. 14-19*a*, or accelerating through interstellar space at 9.8 m/s^2 (and subject only to the force producing that acceleration), as in Fig. 14-19*b*. In both situations he would feel the same and would read the same value for his weight on a scale. Moreover, if he watched an object fall past him, the object would have the same acceleration relative to him in both situations.

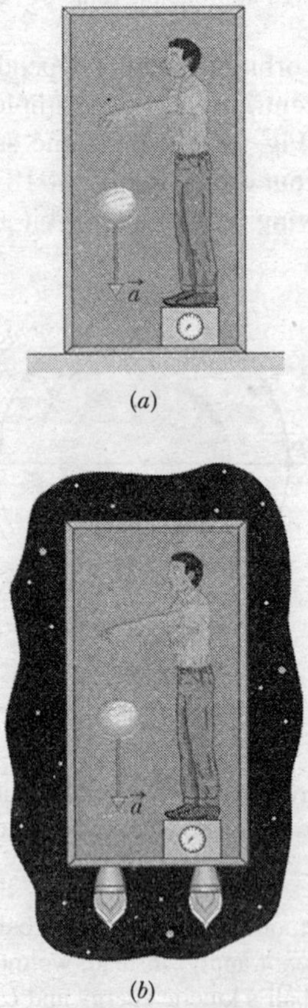

Fig. 14-19 (*a*) A physicist in a box resting on Earth sees a cantaloupe falling with acceleration $a = 9.8$ m/s^2. (*b*) If he and the box accelerate in deep space at 9.8 m/s^2, the cantaloupe has the same acceleration relative to him. It is not possible, by doing experiments within the box, for the physicist to tell which situation he is in. For example, the platform scale on which he stands reads the same weight in both situations.

Curvature of Space

We have thus far explained gravitation as due to a force between masses. Einstein showed that, instead, gravitation is due to a curvature (or shape) of space that is caused by the masses. (As is discussed later in this book, space and time are entangled, so the curvature of which Einstein spoke is really a curvature of *spacetime,* the combined four dimensions of our universe.)

Picturing how space (such as vacuum) can have curvature is difficult. An analogy might help: Suppose that from orbit we watch a race in which two boats begin on the equator with a separation of 20 km and head due south (Fig. 14-20*a*). To the sailors, the boats travel along flat, parallel paths. However, with time the boats draw together until, nearer the south pole, they touch. The sailors in the boats can interpret this drawing together in terms of a force acting on the boats. However, we can see that the boats draw together simply because of the curvature of Earth's surface. We can see this because we are viewing the race from "outside" that surface.

Figure 14-20*b* shows a similar race: Two horizontally separated apples are dropped from the same height above Earth. Although the apples may appear to travel along parallel paths, they actually move toward each other because they both fall toward Earth's center. We can interpret the motion of the apples in terms of the gravitational force on the apples from Earth. We can also interpret the motion in terms of a curvature of the space near Earth, due to the presence of Earth's mass. This time we cannot see the curvature because we cannot get "outside" the curved space, as we got "outside" the curved Earth in the boat example. However, we can depict the curvature with a drawing like Fig. 14-20*c*; there the apples would move along a surface that curves toward Earth because of Earth's mass.

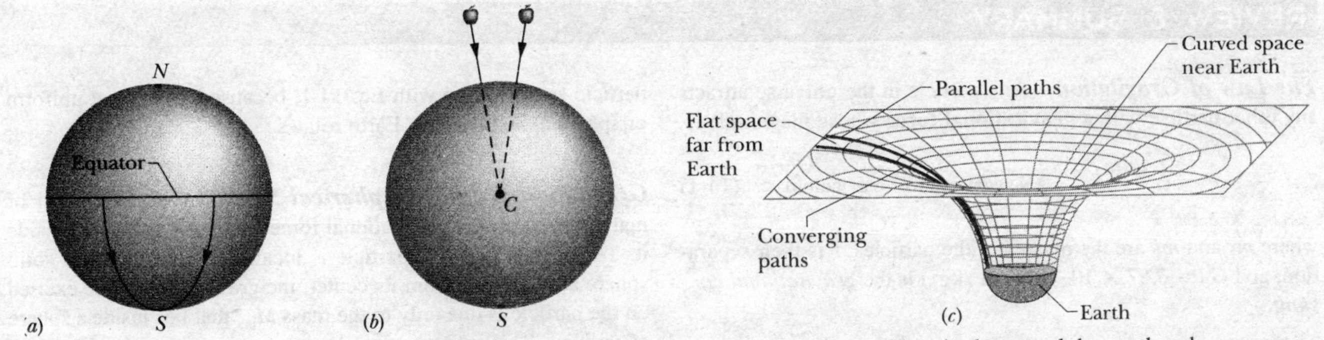

Fig. 14-20 (*a*) Two objects moving along lines of longitude toward the south pole converge because of the curvature of Earth's surface. (*b*) Two objects falling freely near Earth move along lines that converge toward the center of Earth because of the curvature of space near Earth. (*c*) Far from Earth (and other masses), space is flat and parallel paths remain parallel. Close to Earth, the parallel paths begin to converge because space is curved by Earth's mass.

When light passes near Earth, its path bends slightly because of the curvature of space there, an effect called *gravitational lensing*. When it passes a more massive structure, like a galaxy or a black hole having large mass, its path can be bent more. If such a massive structure is between us and a quasar (an extremely bright, extremely distant source of light), the light from the quasar can bend around the massive structure and toward us (Fig. 14-21*a*). Then, because the light seems to be coming to us from a number of slightly different directions in the sky, we see the same quasar in all those different directions. In some situations, the quasars we see blend together to form a giant luminous arc, which is called an *Einstein ring* (Fig. 14-21*b*).

Should we attribute gravitation to the curvature of spacetime due to the presence of masses or to a force between masses? Or should we attribute it to the actions of a type of fundamental particle called a *graviton*, as conjectured in some modern physics theories? We do not know.

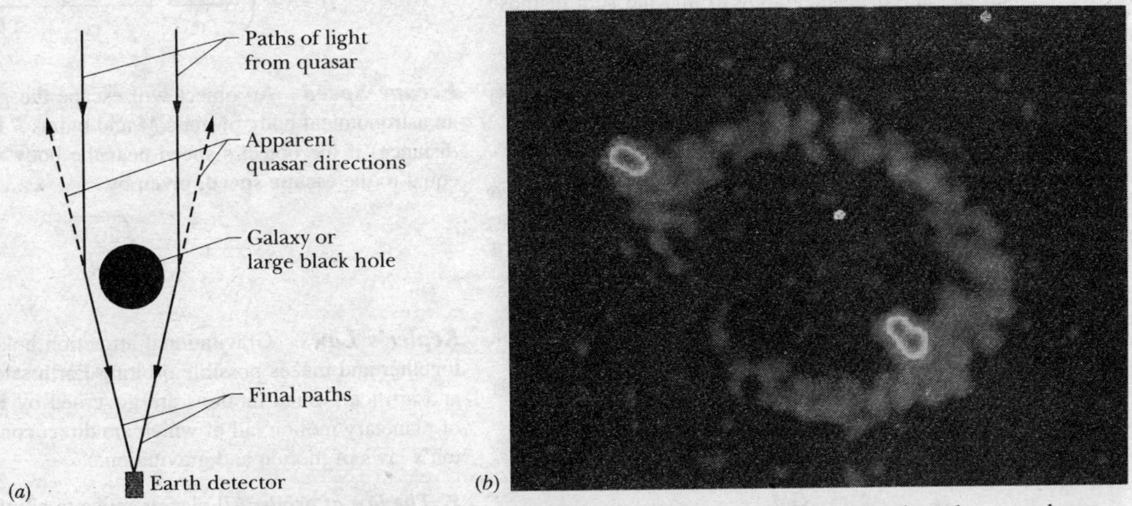

Fig. 14-21 (*a*) Light from a distant quasar follows curved paths around a galaxy or a large black hole because the mass of the galaxy or black hole has curved the adjacent space. If the light is detected, it appears to have originated along the backward extensions of the final paths (dashed lines). (*b*) The Einstein ring known as MG1131+0456 on the computer screen of a telescope. The source of the light (actually, radio waves, which are a form of invisible light) is far behind the large, unseen galaxy that produces the ring; a portion of the source appears as the two bright spots seen along the ring.

REVIEW & SUMMARY

The Law of Gravitation Any particle in the universe attracts any other particle with a **gravitational force** whose magnitude is

$$F = G\frac{m_1 m_2}{r^2} \quad \text{(Newton's law of gravitation)}, \tag{14-1}$$

where m_1 and m_2 are the masses of the particles, r is their separation, and G (= $6.67 \times 10^{-11}\ \text{N} \cdot \text{m}^2/\text{kg}^2$) is the *gravitational constant.*

Gravitational Behavior of Uniform Spherical Shells Equation 14-1 holds only for particles. The gravitational force between extended bodies must generally be found by adding (integrating) the individual forces on individual particles within the bodies. However, if either of the bodies is a uniform spherical shell or a spherically symmetric solid, the net gravitational force it exerts on an *external* object may be computed as if all the mass of the shell or body were located at its center.

Superposition Gravitational forces obey the **principle of superposition;** that is, if n particles interact, the net force $\vec{F}_{1,\text{net}}$ on a particle labeled as particle 1 is the sum of the forces on it from all the other particles taken one at a time:

$$\vec{F}_{1,\text{net}} = \sum_{i=2}^{n} \vec{F}_{1i}, \tag{14-4}$$

in which the sum is a vector sum of the forces $\vec{F}_{1i}$ on particle 1 from particles 2, 3, . . . , n. The gravitational force $\vec{F}_1$ on a particle from an extended body is found by dividing the body into units of differential mass dm, each of which produces a differential force $d\vec{F}$ on the particle, and then integrating to find the sum of those forces:

$$\vec{F}_1 = \int d\vec{F}. \tag{14-5}$$

Gravitational Acceleration The *gravitational acceleration* a_g of a particle (of mass m) is due solely to the gravitational force acting on it. When the particle is at distance r from the center of a uniform, spherical body of mass M, the magnitude F of the gravitational force on the particle is given by Eq. 14-1. Thus, by Newton's second law,

$$F = ma_g, \tag{14-9}$$

which gives

$$a_g = \frac{GM}{r^2}. \tag{14-10}$$

Free-Fall Acceleration and Weight The actual free-fall acceleration $\vec{g}$ of a particle near Earth differs slightly from the gravitational acceleration $\vec{a}_g$, and the particle's weight (equal to mg) differs from the magnitude of the gravitational force acting on the particle as computed with Eq. 14-1, because Earth is not uniform or spherical and because Earth rotates.

Gravitation Within a Spherical Shell A uniform shell of matter exerts no net gravitational force on a particle located inside it. This means that if a particle is located inside a uniform solid sphere at distance r from its center, the gravitational force exerted on the particle is due only to the mass M_{ins} that lies inside a sphere of radius r. This mass is given by

$$M_{\text{ins}} = \rho\frac{4\pi r^3}{3}, \tag{14-17}$$

where ρ is the density of the sphere.

Gravitational Potential Energy The gravitational potential energy $U(r)$ of a system of two particles, with masses M and m and separated by a distance r, is the negative of the work that would be done by the gravitational force of either particle acting on the other if the separation between the particles were changed from infinite (very large) to r. This energy is

$$U = -\frac{GMm}{r} \quad \text{(gravitational potential energy)}. \tag{14-20}$$

Potential Energy of a System If a system contains more than two particles, its total gravitational potential energy U is the sum of terms representing the potential energies of all the pairs. As an example, for three particles, of masses m_1, m_2, and m_3,

$$U = -\left(\frac{Gm_1m_2}{r_{12}} + \frac{Gm_1m_3}{r_{13}} + \frac{Gm_2m_3}{r_{23}}\right). \tag{14-21}$$

Escape Speed An object will escape the gravitational pull of an astronomical body of mass M and radius R (to reach an infinite distance) if the object's speed near the body's surface is at least equal to the **escape speed,** given by

$$v = \sqrt{\frac{2GM}{R}}. \tag{14-27}$$

Kepler's Laws Gravitational attraction holds the solar system together and makes possible orbiting Earth satellites, both natural and artificial. Such motions are governed by Kepler's three laws of planetary motion, all of which are direct consequences of Newton's laws of motion and gravitation:

1. ***The law of orbits.*** All planets move in elliptical orbits with the Sun at one focus.
2. ***The law of areas.*** A line joining any planet to the Sun sweeps out equal areas in equal times as the planet orbits the Sun. (This statement is equivalent to conservation of angular momentum.)
3. ***The law of periods.*** The square of the period T of any planet about the Sun is proportional to the cube of the semimajor axis

a of the orbit. For circular orbits with radius *r*, the semimajor axis *a* is replaced by *r* and the law is written as

$$T^2 = \left(\frac{4\pi^2}{GM}\right) r^3 \quad \text{(law of periods)}, \qquad (14\text{-}33)$$

where *M* is the mass of the attracting body—the Sun in the case of the solar system. This equation is generally valid for elliptical planetary orbits, when the semimajor axis *a* is inserted in place of the circular radius *r*.

Energy in Planetary Motion When a planet or satellite with mass *m* moves in a circular orbit with radius *r*, its potential energy *U* and kinetic energy *K* are given by

$$U = -\frac{GMm}{r} \quad \text{and} \quad K = \frac{GMm}{2r}. \qquad (14\text{-}20, 14\text{-}42)$$

The mechanical energy $E = K + U$ is then

$$E = -\frac{GMm}{2r}. \qquad (14\text{-}44)$$

For an elliptical orbit of semimajor axis *a*,

$$E = -\frac{GMm}{2a}. \qquad (14\text{-}46)$$

Einstein's View of Gravitation Einstein pointed out that gravitation and acceleration are equivalent. This **principle of equivalence** led him to a theory of gravitation (the **general theory of relativity**) that explains gravitational effects in terms of a curvature of space.

QUESTIONS

1. In Fig. 14-22, two particles, of masses *m* and 2*m*, are fixed in place on an axis. (a) Where on the axis can a third particle of mass 3*m* be placed (other than at infinity) so that the net gravitational force on it from the first two particles is zero: to the left of the first two particles, to their right, between them but closer to the more massive particle, or between them but closer to the less massive particle? (b) Does the answer change if the third particle has, instead, a mass of 16*m*? (c) Is there a point off the axis at which the net force on the third particle would be zero?

Fig. 14-22 Question 1.

2. In Fig. 14-23, a central particle is surrounded by two circular rings of particles, at radii *r* and *R*, with $R > r$. All the particles have mass *m*. What are the magnitude and direction of the net gravitational force on the central particle due to the particles in the rings?

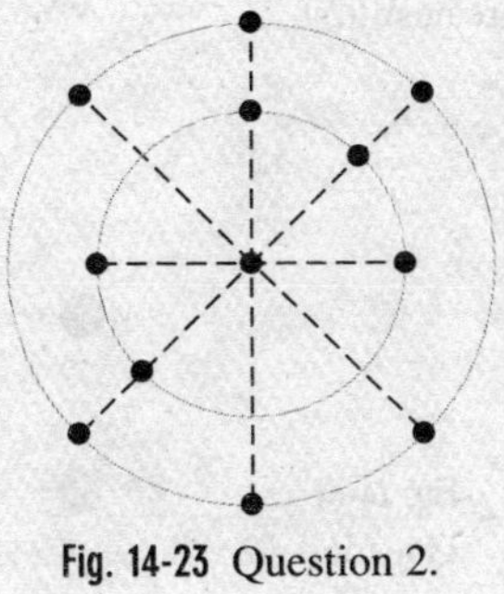

Fig. 14-23 Question 2.

3. In Fig. 14-24, a central particle of mass *M* is surrounded by a square array of other particles, separated by either distance *d* or distance *d*/2 along the perimeter of the square. What are the magnitude and direction of the net gravitational force on the central particle due to the other particles?

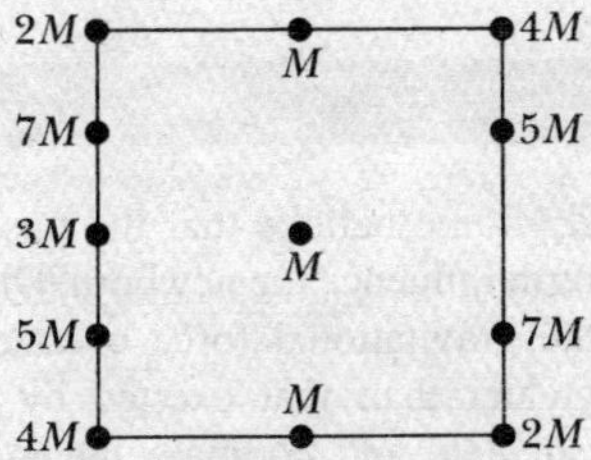

Fig. 14-24 Question 3.

4. Figure 14-25 shows four arrangements of a particle of mass *m* and one or more uniform rods of mass *M* and length *L*, each a distance *d* from the particle. Rank the arrangements according to the magnitude of the net gravitational force on the particle from the rods, greatest first.

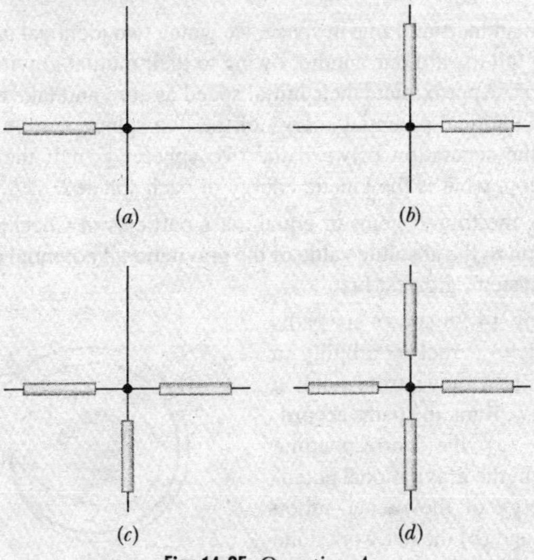

Fig. 14-25 Question 4.

5. Figure 14-26 gives the gravitational acceleration a_g for four planets as a function of the radial distance *r* from the center of the

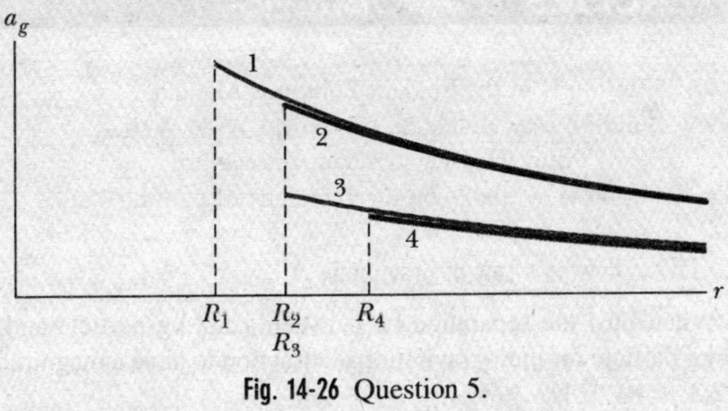

Fig. 14-26 Question 5.

planet, starting at the surface of the planet (at radius R_1, R_2, R_3, or R_4). Plots 1 and 2 coincide for $r \geq R_2$; plots 3 and 4 coincide for $r \geq R_4$. Rank the four planets according to (a) their mass and (b) their density, greatest first.

6. Figure 14-27 shows three uniform spherical planets that are identical in size and mass. The periods of rotation T for the planets are given, and six lettered points are indicated—three points are on the equators of the planets and three points are on the north poles. Rank the points according to the value of the free-fall acceleration g at them, greatest first.

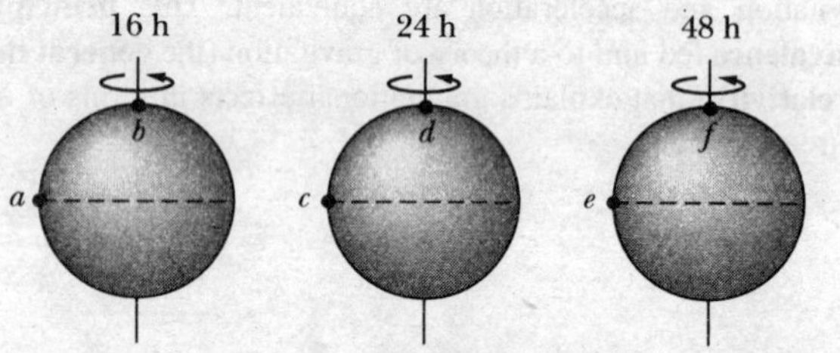

Fig. 14-27 Question 6.

7. From an inertial frame in space, we watch two identical uniform spheres fall toward one another owing to their mutual gravitational attraction. Approximate their initial speed as zero and take the initial gravitational potential energy of the two-sphere system as U_i. When the separation between the two spheres is half the initial separation, what is the kinetic energy of each sphere?

8. Rank the four systems of equal-mass particles in Checkpoint 2 according to the absolute value of the gravitational potential energy of the system, greatest first.

9. Figure 14-28 shows six paths by which a rocket orbiting a moon might move from point a to point b. Rank the paths according to (a) the corresponding change in the gravitational potential energy of the rocket–moon system and (b) the net work done on the rocket by the gravitational force from the moon, greatest first.

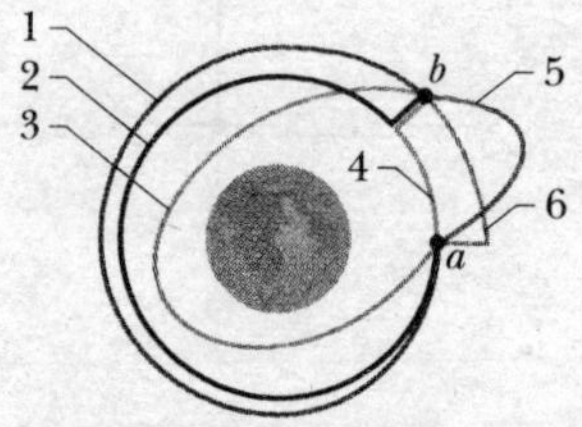

Fig. 14-28 Question 9.

10. In Fig. 14-29, a particle of mass m (not shown) is to be moved from an infinite distance to one of the three possible locations a, b, and c. Two other particles, of masses m and $2m$, are fixed in place. Rank the three possible locations according to the work done by the net gravitational force on the moving particle due to the fixed particles, greatest first.

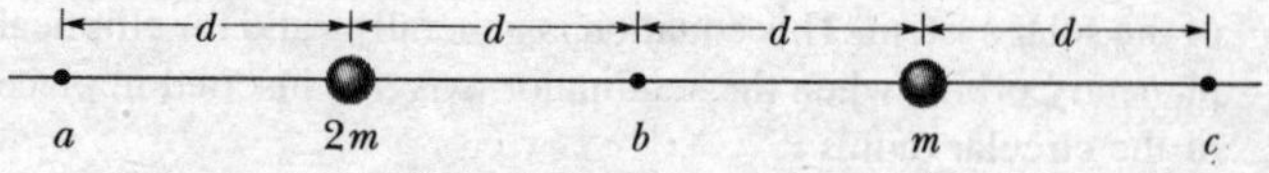

Fig. 14-29 Question 10.

11. In Fig. 14-30, a particle of mass m is initially at point A, at distance d from the center of one uniform sphere and distance $4d$ from the center of another uniform sphere, both of mass $M \gg m$. State whether, if you moved the particle to point D, the following would be positive, negative, or zero: (a) the change in the gravitational potential energy of the particle, (b) the work done by the net gravitational force on the particle, (c) the work done by your force. (d) What are the answers if, instead, the move were from point B to point C?

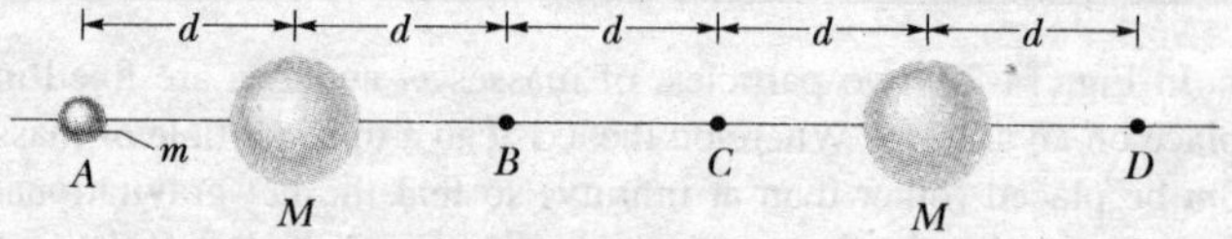

Fig. 14-30 Question 11.

12. Figure 14-31 gives the masses and separations for three pairs of stars that each form a binary star system. (a) Locate the point about which the stars of each pair orbit. (b) Rank the pairs according to the magnitude of the centripetal acceleration of the stars, greatest first.

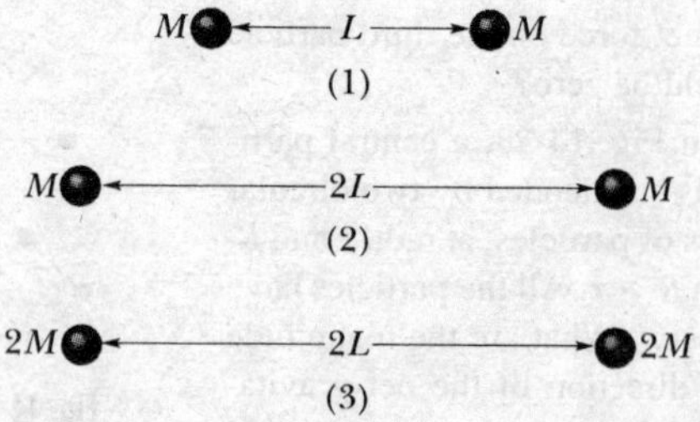

Fig. 14-31 Question 12.

EXERCISES & PROBLEMS

ssm Solution is in the Student Solutions Manual.
www Solution is available on the World Wide Web at: http://www.wiley.com/college/hrw
ilw Solution is available on the Interactive LearningWare.

SEC. 14-2 Newton's Law of Gravitation

1E. What must the separation be between a 5.2 kg particle and a 2.4 kg particle for their gravitational attraction to have a magnitude of 2.3×10^{-12} N? ssm

2E. Some believe that the positions of the planets at the time of birth influence the newborn. Others deride this belief and claim that the gravitational force exerted on a baby by the obstetrician is greater than that exerted by the planets. To check this claim, calculate and compare the magnitude of the gravitational force exerted on a 3 kg baby (a) by a 70 kg obstetrician who is 1 m away and roughly approximated as a point mass, (b) by the massive planet Jupiter ($m = 2 \times 10^{27}$ kg) at its closest approach to Earth ($= 6 \times 10^{11}$ m), and (c) by Jupiter at its greatest distance from Earth ($= 9 \times 10^{11}$ m). (d) Is the claim correct?

3E. One of the *Echo* satellites consisted of an inflated spherical aluminum balloon 30 m in diameter and of mass 20 kg. Suppose a meteor having a mass of 7.0 kg passes within 3.0 m of the surface of the satellite. What is the magnitude of the gravitational force on the meteor from the satellite at the closest approach? ssm

4E. The Sun and Earth each exert a gravitational force on the Moon. What is the ratio F_{Sun}/F_{Earth} of these two forces? (The average Sun–Moon distance is equal to the Sun–Earth distance.)

5E. A mass M is split into two parts, m and $M - m$, which are then separated by a certain distance. What ratio m/M maximizes the magnitude of the gravitational force between the parts? ilw

SEC. 14-3 Gravitation and the Principle of Superposition

6E. A spaceship is on a straight-line path between Earth and its moon. At what distance from Earth is the net gravitational force on the spaceship zero?

7E. How far from Earth must a space probe be along a line toward the Sun so that the Sun's gravitational pull on the probe balances Earth's pull? ssm

8P. Three 5.0 kg spheres are located in the xy plane as shown in Fig. 14-32. What is the magnitude of the net gravitational force on the sphere at the origin due to the other two spheres?

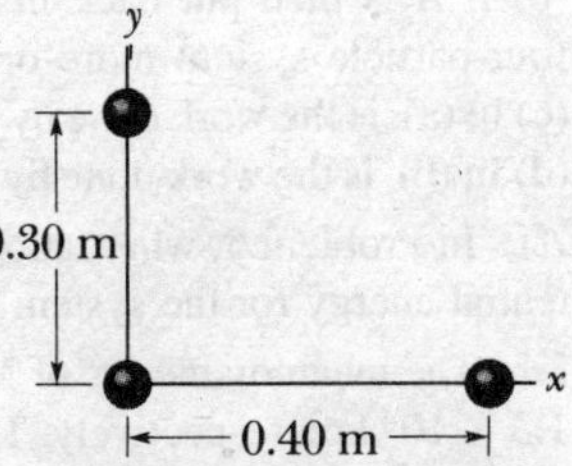

Fig. 14-32 Problem 8.

9P. In Fig. 14-33*a*, four spheres form the corners of a square whose side is 2.0 cm long. What are the magnitude and direction of the net gravitational force from them on a central sphere with mass $m_5 = 250$ kg?

10P. In Fig. 14-33*b*, two spheres of mass m and a third sphere of mass M form an equilateral triangle, and a fourth sphere of mass m_4 is at the center of the triangle. The net gravitational force on that central sphere from the three other spheres is zero. (a) What is M in terms of m? (b) If we double the value of m_4, what then is the magnitude of the net gravitational force on the central sphere?

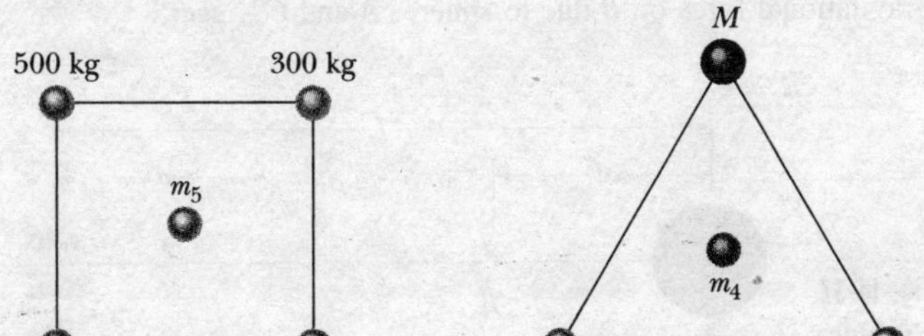

Fig. 14-33 Problems 9 and 10.

11P. The masses and coordinates of three spheres are as follows: 20 kg, $x = 0.50$ m, $y = 1.0$ m; 40 kg, $x = -1.0$ m, $y = -1.0$ m; 60 kg, $x = 0$ m, $y = -0.50$ m. What is the magnitude of the gravitational force on a 20 kg sphere located at the origin due to the other spheres? ilw

12P. Four uniform spheres, with masses $m_A = 400$ kg, $m_B = 350$ kg, $m_C = 2000$ kg, and $m_D = 500$ kg, have (x, y) coordinates of (0, 50 cm), (0, 0), (−80 cm, 0), and (40 cm, 0), respectively. What is the net gravitational force on sphere B due to the other spheres?

13P. Figure 14-34 shows a spherical hollow inside a lead sphere of radius R; the surface of the hollow passes through the center of the sphere and "touches" the right side of the sphere. The mass of the sphere before hollowing was M. With what gravitational force does the hollowed-out lead sphere attract a small sphere of mass m that lies at a distance d from the center of the lead sphere, on the straight line connecting the centers of the spheres and of the hollow? ssm

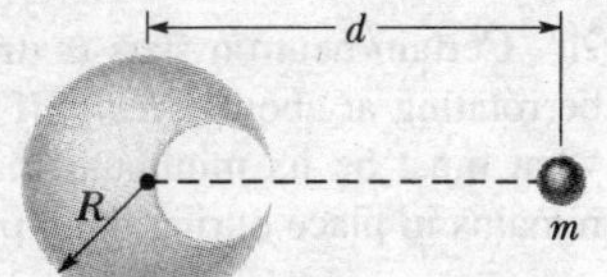

Fig. 14-34 Problem 13.

SEC. 14-4 Gravitation Near Earth's Surface

14E. You weigh 530 N at sidewalk level outside the World Trade Center in New York City. Suppose that you ride from this level to the top of one of its 410 m towers. Ignoring Earth's rotation, how much less would you weigh there (because you are slightly farther from the center of Earth)?

15E. At what altitude above Earth's surface would the gravitational acceleration be 4.9 m/s^2? ssm

16E. (a) What will an object weigh on the Moon's surface if it weighs 100 N on Earth's surface? (b) How many Earth radii must this same object be from the center of Earth if it is to weigh the same as it does on the Moon?

17P. The fastest possible rate of rotation of a planet is that for which the gravitational force on material at the equator just barely provides the centripetal force needed for the rotation. (Why?) (a) Show that the corresponding shortest period of rotation is

$$T = \sqrt{\frac{3\pi}{G\rho}},$$

where ρ is the uniform density of the spherical planet. (b) Calculate the rotation period assuming a density of 3.0 g/cm^3, typical of many planets, satellites, and asteroids. No astronomical object has ever been found to be spinning with a period shorter than that determined by this analysis. ssm

18P. One model for a certain planet has a core of radius R and mass M surrounded by an outer shell of inner radius R, outer radius $2R$, and mass $4M$. If $M = 4.1 \times 10^{24}$ kg and $R = 6.0 \times 10^6$ m, what is the gravitational acceleration of a particle at points (a) R and (b) $3R$ from the center of the planet?

19P. A body is suspended from a spring scale in a ship sailing along the equator with speed v. (a) Show that the scale reading will be very close to $W_0(1 \pm 2\omega v/g)$, where ω is the angular speed of Earth and W_0 is the scale reading when the ship is at rest. (b) Explain the $\pm$ sign. ssm www

20P. The radius R_h and mass M_h of a black hole are related by $R_h = 2GM_h/c^2$, where c is the speed of light. Assume that the gravitational acceleration a_g of an object at a distance r_o =

$1.001R_h$ from the center of a black hole is given by Eq. 14-10 (it is, for large black holes). (a) Find an expression for a_g at r_o in terms of M_h. (b) Does a_g at r_o increase or decrease with an increase of M_h? (c) What is a_g at r_o for a very large black hole whose mass is 1.55×10^{12} times the solar mass of 1.99×10^{30} kg? (d) If the astronaut of Sample Problem 14-3 is at r_o with her feet toward this black hole, what is the difference in gravitational acceleration between her head and her feet? (e) Is the tendency to stretch the astronaut severe?

21P. Certain neutron stars (extremely dense stars) are believed to be rotating at about 1 rev/s. If such a star has a radius of 20 km, what must be its minimum mass so that material on its surface remains in place during the rapid rotation? **ilw**

SEC. 14-5 Gravitation Inside Earth

22E. Two concentric shells of uniform density having masses M_1 and M_2 are situated as shown in Fig. 14-35. Find the magnitude of the net gravitational force on a particle of mass m, due to the shells, when the particle is located at (a) point A, at distance $r = a$ from the center, (b) point B at $r = b$, and (c) point C at $r = c$. The distance r is measured from the center of the shells.

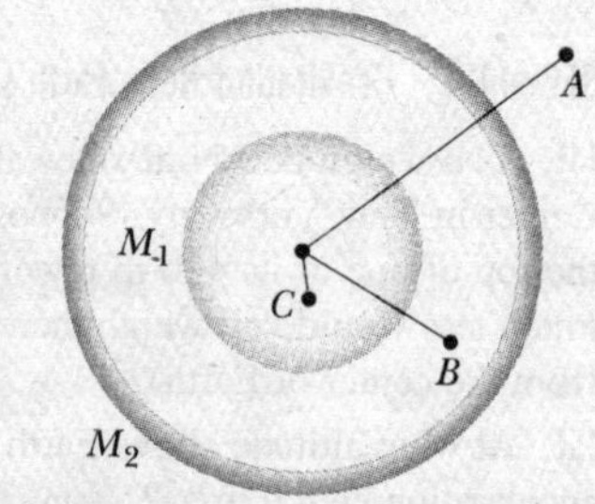

Fig. 14-35 Exercise 22.

23P. A solid sphere of uniform density has a mass of 1.0×10^4 kg and a radius of 1.0 m. What is the magnitude of the gravitational force due to the sphere on a particle of mass m located at a distance of (a) 1.5 m and (b) 0.50 m from the center of the sphere? (c) Write a general expression for the magnitude of the gravitational force on the particle at a distance $r \leq 1.0$ m from the center of the sphere.

24P. A uniform solid sphere of radius R produces a gravitational acceleration of a_g on its surface. At what two distances from the center of the sphere is the gravitational acceleration $a_g/3$? (*Hint:* Consider distances both inside and outside the sphere.)

25P. Figure 14-36 shows, not to scale, a cross section through the interior of Earth. Rather than being uniform throughout, Earth is divided into three zones: an outer *crust*, a *mantle*, and an inner *core*. The dimensions of these zones and the masses contained within them are shown on the figure. Earth has a total mass of 5.98×10^{24} kg and a radius of 6370 km. Ignore rotation and assume that Earth is spherical. (a) Calculate a_g at the surface. (b) Suppose that a bore hole (the *Mohole*) is driven to the crust–mantle interface at a depth of 25 km; what would be the value of a_g at the bottom of the hole? (c) Suppose that Earth were a uniform sphere with the same total mass and size. What would be the value of a_g at a depth of 25 km? (Precise measurements of a_g are sensitive probes of the interior structure of Earth, although results can be clouded by local density variations.) **ssm**

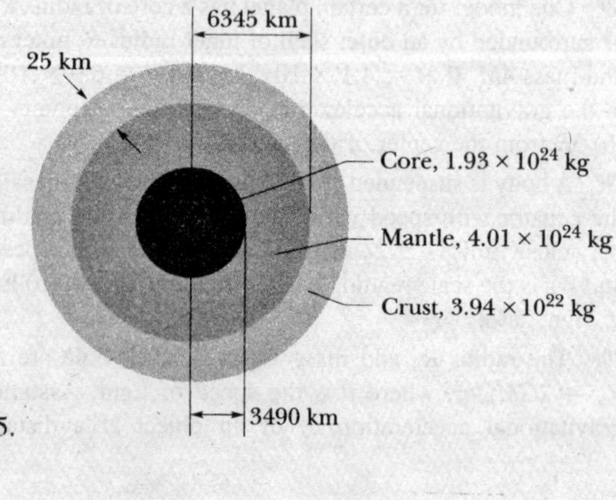

Fig. 14-36 Problem 25.

SEC. 14-6 Gravitational Potential Energy

26E. (a) What is the gravitational potential energy of the two-particle system in Exercise 1? If you triple the separation between the particles, how much work is done (b) by the gravitational force between the particles and (c) by you?

27E. (a) In Problem 12, remove sphere A and calculate the gravitational potential energy of the remaining three-particle system. (b) If A is then put back in place, is the potential energy of the four-particle system more or less than that of the system in (a)? (c) In (a), is the work done by you to remove A positive or negative? (d) In (b), is the work done by you to replace A positive or negative?

28E. In Problem 5, what ratio m/M gives the least gravitational potential energy for the system?

29E. The mean diameters of Mars and Earth are 6.9×10^3 km and 1.3×10^4 km, respectively. The mass of Mars is 0.11 times Earth's mass. (a) What is the ratio of the mean density of Mars to that of Earth? (b) What is the value of the gravitational acceleration on Mars? (c) What is the escape speed on Mars? **ssm**

30E. Calculate the amount of energy required to escape from (a) Earth's moon and (b) Jupiter relative to that required to escape from Earth.

31P. The three spheres in Fig. 14-37, with masses $m_A = 800$ g, $m_B = 100$ g, and $m_C = 200$ g, have their centers on a common line, with $L = 12$ cm and $d = 4.0$ cm. You move sphere B along the line until its center-to-center separation from C is $d = 4.0$ cm. How much work is done on sphere B (a) by you and (b) by the net gravitational force on B due to spheres A and C? **ssm**

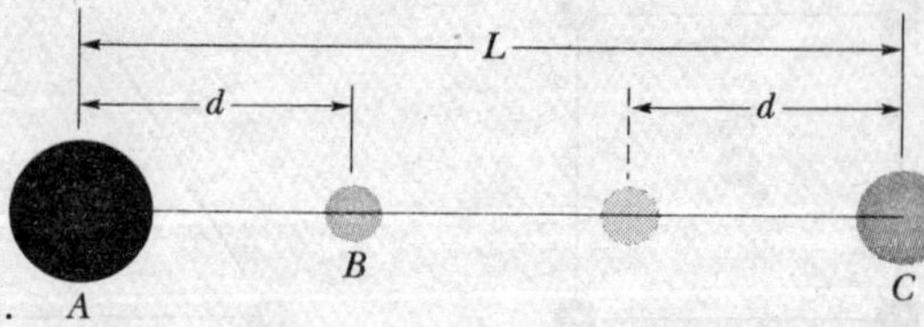

Fig. 14-37 Problem 31.

32P. Zero, a hypothetical planet, has a mass of 5.0×10^{23} kg, a radius of 3.0×10^6 m, and no atmosphere. A 10 kg space probe is to be launched vertically from its surface. (a) If the probe is launched with an initial energy of 5.0×10^7 J, what will be its kinetic energy when it is 4.0×10^6 m from the center of Zero? (b) If the probe is to achieve a maximum distance of 8.0×10^6 m from the center of Zero, with what initial kinetic energy must it be launched from the surface of Zero?

33P. A rocket is accelerated to speed $v = 2\sqrt{gR_e}$ near Earth's surface (where Earth's radius is R_e), and it then coasts upward. (a) Show that it will escape from Earth. (b) Show that very far from Earth its speed will be $v = \sqrt{2gR_e}$. ssm

34P. Planet Roton, with a mass of 7.0×10^{24} kg and a radius of 1600 km, gravitationally attracts a meteorite that is initially at rest relative to the planet, at a great enough distance to take as infinite. The meteorite falls toward the planet. Assuming the planet is airless, find the speed of the meteorite when it reaches the planet's surface.

35P. (a) What is the escape speed on a spherical asteroid whose radius is 500 km and whose gravitational acceleration at the surface is 3.0 m/s^2? (b) How far from the surface will a particle go if it leaves the asteroid's surface with a radial speed of 1000 m/s? (c) With what speed will an object hit the asteroid if it is dropped from 1000 km above the surface? ssm www

36P. A 150.0 kg rocket moving radially outward from Earth has a speed of 3.70 km/s when its engine shuts off 200 km above Earth's surface. (a) Assuming negligible air drag, find the rocket's kinetic energy when the rocket is 1000 km above Earth's surface. (b) What maximum height above the surface is reached by the rocket?

37P. Two neutron stars are separated by a distance of 10^{10} m. They each have a mass of 10^{30} kg and a radius of 10^5 m. They are initially at rest with respect to each other. As measured from that rest frame, how fast are they moving when (a) their separation has decreased to one-half its initial value and (b) they are about to collide? ssm www

38P. In deep space, sphere A of mass 20 kg is located at the origin of an x axis and sphere B of mass 10 kg is located on the axis at $x = 0.80$ m. Sphere B is released from rest while sphere A is held at the origin. (a) What is the gravitational potential energy of the two-sphere system as B is released? (b) What is the kinetic energy of B when it has moved 0.20 m toward A?

39P. A projectile is fired vertically from Earth's surface with an initial speed of 10 km/s. Neglecting air drag, how far above the surface of Earth will it go? ilw

SEC. 14-7 Planets and Satellites: Kepler's Laws

40E. The mean distance of Mars from the Sun is 1.52 times that of Earth from the Sun. From Kepler's law of periods, calculate the number of years required for Mars to make one revolution about the Sun; compare your answer with the value given in Appendix C.

41E. The Martian satellite Phobos travels in an approximately circular orbit of radius 9.4×10^6 m with a period of 7 h 39 min. Calculate the mass of Mars from this information. ssm

42E. Determine the mass of Earth from the period T (27.3 days) and the radius r (3.82×10^5 km) of the Moon's orbit about Earth. Assume the Moon orbits the center of Earth rather than the center of mass of the Earth–Moon system.

43E. Our Sun, with mass 2.0×10^{30} kg, revolves about the center of the Milky Way galaxy, which is 2.2×10^{20} m away, once every 2.5×10^8 years. Assuming that each of the stars in the galaxy has a mass equal to that of our Sun, that the stars are distributed uniformly in a sphere about the galactic center, and that our Sun is essentially at the edge of that sphere, estimate roughly the number of stars in the galaxy ssm

44E. A satellite is placed in a circular orbit about Earth with a radius equal to one-half the radius of the Moon's orbit. What is its period of revolution in lunar months? (A lunar month is the period of revolution of the Moon.)

45E. (a) What linear speed must an Earth satellite have to be in a circular orbit at an altitude of 160 km? (b) What is the period of revolution? ssm

46E. The Sun's center is at one focus of Earth's orbit. How far from this focus is the other focus, (a) in meters and (b) in terms of the solar radius, 6.96×10^8 m? The eccentricity of Earth's orbit is 0.0167, and the semimajor axis is 1.50×10^{11} m.

47E. A satellite, moving in an elliptical orbit, is 360 km above Earth's surface at its farthest point and 180 km above at its closest point. Calculate (a) the semimajor axis and (b) the eccentricity of the orbit. (*Hint:* See Sample Problem 14-6.) ssm

48E. A satellite hovers over a certain spot on the equator of (rotating) Earth. What is the altitude of its orbit (called a *geosynchronous orbit*)?

49E. A comet that was seen in April 574 by Chinese astronomers on a day known by them as the Woo Woo day was spotted again in May 1994. Assume the time between observations is the period of the Woo Woo day comet and take its eccentricity as 0.11. What are (a) the semimajor axis of the comet's orbit and (b) its greatest distance from the Sun in terms of the mean orbital radius R_P of Pluto?

50E. In 1993 the spacecraft *Galileo* sent home an image (Fig. 14-38) of asteroid 243 Ida and a tiny orbiting moon (now known as Dactyl), the first confirmed example of an asteroid–moon system. In the image, the moon, which is 1.5 km wide, is 100 km from the center of the asteroid, which is 55 km long. The shape of the moon's orbit is not well known; assume it is circular with a period of 27 h. (a) What is the mass of the asteroid? (b) The volume of the asteroid, measured from the *Galileo* images, is 14 100 km^3. What is the density of the asteroid?

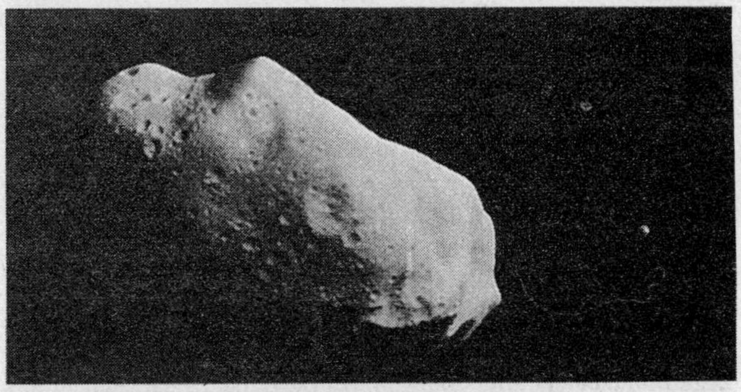

Fig. 14-38 Exercise 50. An image from the spacecraft *Galileo* shows a tiny moon orbiting asteroid 243 Ida.

51P. In 1610, Galileo used his telescope to discover four prominent moons around Jupiter. Their mean orbital radii a and periods T are as follows:

Name	a (10^8 m)	T (days)
Io	4.22	1.77
Europa	6.71	3.55
Ganymede	10.7	7.16
Callisto	18.8	16.7

(a) Plot log a (y axis) against log T (x axis) and show that you get a straight line. (b) Measure the slope of the line and compare it with the value that you expect from Kepler's third law. (c) Find the mass of Jupiter from the intercept of this line with the y axis.

52P. A 20 kg satellite has a circular orbit with a period of 2.4 h and a radius of 8.0×10^6 m around a planet of unknown mass. If the magnitude of the gravitational acceleration on the surface of the planet is 8.0 m/s^2, what is the radius of the planet?

53P. In a certain binary-star system, each star has the same mass as our Sun, and they revolve about their center of mass. The distance between them is the same as the distance between Earth and the Sun. What is their period of revolution in years? itw

54P. A certain triple-star system consists of two stars, each of mass m, revolving about a central star of mass M in the same circular orbit of radius r (Fig. 14-39). The two stars are always at opposite ends of a diameter of the circular orbit. Derive an expression for the period of revolution of the stars.

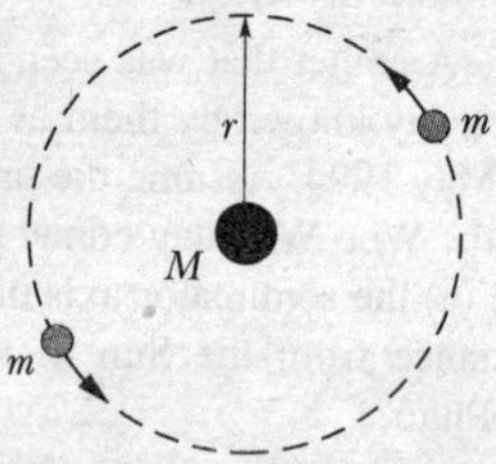

Fig. 14-39 Problem 54.

55P*. Three identical stars of mass M are located at the vertices of an equilateral triangle with side L. At what speed must they move if they all revolve under the influence of one another's gravitational force in a circular orbit circumscribing the triangle while still preserving the equilateral triangle? ssm www

SEC. 14-8 Satellites: Orbits and Energy

56E. Consider two satellites, A and B, both of mass m, moving in the same circular orbit of radius r around Earth, of mass M_E, but in opposite senses of rotation and therefore on a collision course (see Fig. 14-40). (a) In terms of G, M_E, m, and r, find the total mechanical energy $E_A + E_B$ of the two-satellite-plus-Earth system before collision. (b) If the collision is completely inelastic so that the wreckage remains as one piece of tangled material (mass = $2m$), find the total mechanical energy immediately after collision. (c) Describe the subsequent motion of the wreckage.

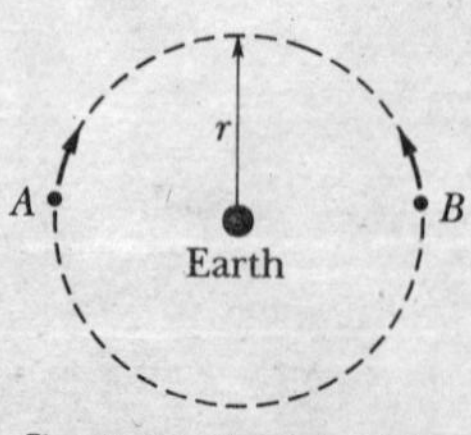

Fig. 14-40 Exercise 56.

57E. An asteroid, whose mass is 2.0×10^{-4} times the mass of Earth, revolves in a circular orbit around the Sun at a distance that is twice Earth's distance from the Sun. (a) Calculate the period of revolution of the asteroid in years. (b) What is the ratio of the kinetic energy of the asteroid to that of Earth? ssm

58P. Two Earth satellites, A and B, each of mass m, are to be launched into circular orbits about Earth's center. Satellite A is to orbit at an altitude of 6370 km. Satellite B is to orbit at an altitude of 19 110 km. The radius of Earth R_E is 6370 km. (a) What is the ratio of the potential energy of satellite B to that of satellite A, in orbit? (b) What is the ratio of the kinetic energy of satellite B to that of satellite A, in orbit? (c) Which satellite has the greater total energy if each has a mass of 14.6 kg? By how much?

59P. Show that if an object is in an elliptical orbit with semimajor axis a about a planet of mass M, then its distance r from the planet and speed v are related by

$$v^2 = GM\left(\frac{2}{r} - \frac{1}{a}\right).$$

(*Hint:* Use the law of conservation of mechanical energy and Eq. 14-46.) ssm

60P. Use the result of Problem 59 and data contained in Sample Problem 14-6 to calculate (a) the speed v_p of comet Halley at perihelion and (b) its speed v_a at aphelion. (c) Using the law of conservation of angular momentum relative to the Sun, find the ratio of the comet's perihelion distance R_p to its aphelion distance R_a in terms of v_p and v_a.

61P. (a) Does it take more energy to get a satellite up to 1500 km above Earth than to put it in circular orbit once it is there? (Take Earth's radius to be 6370 km.) (b) What about 3185 km? (c) What about 4500 km?

62P. One way to attack a satellite in Earth orbit is to launch a swarm of pellets in the same orbit as the satellite but in the opposite direction. Suppose a satellite in a circular orbit 500 km above Earth's surface collides with a pellet having mass 4.0 g. (a) What is the kinetic energy of the pellet in the reference frame of the satellite just before the collision? (b) What is the ratio of this kinetic energy to the kinetic energy of a 4.0 g bullet from a modern army rifle with a muzzle speed of 950 m/s?

63P. What are (a) the speed and (b) the period of a 220 kg satellite in an approximately circular orbit 640 km above the surface of Earth? Suppose the satellite loses mechanical energy at the average rate of 1.4×10^5 J per orbital revolution. Adopting the reasonable approximation that the satellite's orbit becomes a "circle of slowly diminishing radius," determine the satellite's (c) altitude, (d) speed, and (e) period at the end of its 1500th revolution. (f) What is the magnitude of the average retarding force on the satellite? Is angular momentum around Earth's center conserved for (g) the satellite and (h) the satellite–Earth system? ssm

SEC. 14-9 Einstein and Gravitation

64E. In Fig. 14-19b, the scale on which the 60 kg physicist stands reads 220 N. How long will the cantaloupe take to reach the floor if the physicist drops it from rest (relative to himself), 2.1 m from the floor?

15 Fluids

The force exerted by water on the body of a descending diver increases noticeably, even for a relatively shallow descent to the bottom of a swimming pool. However, in 1975, using scuba gear with a special gas mixture for breathing, William Rhodes emerged from a chamber that had been lowered 300 m into the Gulf of Mexico, and he then swam to a record depth of 350 m. Strangely, a novice scuba diver practicing in a swimming pool might be in more danger from the force exerted by the water than was Rhodes. Occasionally, novice scuba divers die because they have neglected that danger.

What is this potentially lethal risk?

The answer is in this chapter.

15-1 Fluids and the World Around Us

Fluids—which include both liquids and gases—play a central role in our daily lives. We breathe and drink them, and a rather vital fluid circulates in the human cardiovascular system. There are the fluid ocean and the fluid atmosphere.

In a car, there are fluids in the tires, the gas tank, the radiator, the combustion chambers of the engine, the exhaust manifold, the battery, the air-conditioning system, the windshield wiper reservoir, the lubrication system, and the hydraulic system. (*Hydraulic* means operated via a liquid.) The next time you see a large piece of earthmoving machinery, count the hydraulic cylinders that permit the machine to do its work. Large jet planes have scores of them.

We use the kinetic energy of a moving fluid in windmills, and the potential energy of another fluid in hydroelectric power plants. Given time, fluids carve the landscape. We often travel great distances just to watch fluids move. Perhaps it is time to see what physics can tell us about fluids.

15-2 What Is a Fluid?

A **fluid**, in contrast to a solid, is a substance that can flow. Fluids conform to the boundaries of any container in which we put them. They do so because a fluid cannot sustain a force that is tangential to its surface. (In the more formal language of Section 13-6, a fluid is a substance that flows because it cannot withstand a shearing stress. It can, however, exert a force in the direction perpendicular to its surface.) Some materials, such as pitch, take a long time to conform to the boundaries of a container, but they do so eventually; thus, we classify them as fluids.

You may wonder why we lump liquids and gases together and call them fluids. After all (you may say), liquid water is as different from steam as it is from ice. Actually, it is not. Ice, like other crystalline solids, has its constituent atoms organized in a fairly rigid three-dimensional array called a crystalline lattice. In neither steam nor liquid water, however, is there any such orderly long-range arrangement.

15-3 Density and Pressure

When we discuss rigid bodies, we are concerned with particular lumps of matter, such as wooden blocks, baseballs, or metal rods. Physical quantities that we find useful, and in whose terms we express Newton's laws, are *mass* and *force*. We might speak, for example, of a 3.6 kg block acted on by a 25 N force.

With fluids, we are more interested in the extended substance, and in properties that can vary from point to point in that substance. It is more useful to speak of **density** and **pressure** than of mass and force.

Density

To find the density ρ of a fluid at any point, we isolate a small volume element ΔV around that point and measure the mass Δm of the fluid contained within that element. The **density** is then

$$\rho = \frac{\Delta m}{\Delta V}. \tag{15-1}$$

In theory, the density at any point in a fluid is the limit of this ratio as the volume element ΔV at that point is made smaller and smaller. In practice, we assume that a fluid sample is large compared to atomic dimensions and thus is "smooth" (with uniform density), rather than "lumpy" with atoms. This assumption allows us to

TABLE 15-1 Some Densities

Material or Object		Density (kg/m³)
Interstellar space		10^{-20}
Best laboratory vacuum		10^{-17}
Air:	20°C and 1 atm pressure	1.21
	20°C and 50 atm	60.5
Styrofoam		1×10^2
Ice		0.917×10^3
Water:	20°C and 1 atm	0.998×10^3
	20°C and 50 atm	1.000×10^3
Seawater:	20°C and 1 atm	1.024×10^3
Whole blood		1.060×10^3
Iron		7.9×10^3
Mercury (the metal)		13.6×10^3
Earth:	average	5.5×10^3
	core	9.5×10^3
	crust	2.8×10^3
Sun:	average	1.4×10^3
	core	1.6×10^5
White dwarf star (core)		10^{10}
Uranium nucleus		3×10^{17}
Neutron star (core)		10^{18}
Black hole (1 solar mass)		10^{19}

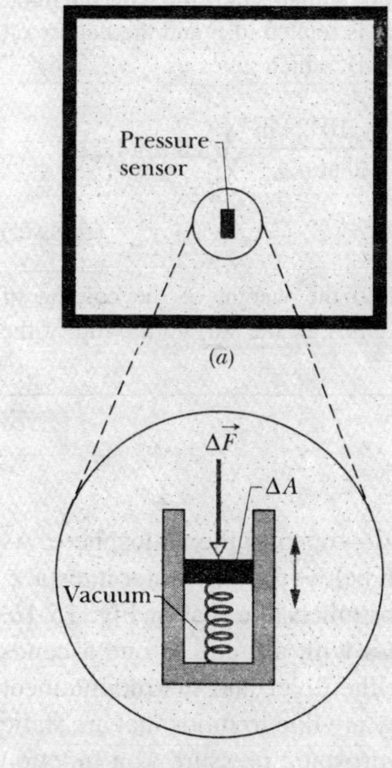

Fig. 15-1 (*a*) A fluid-filled vessel containing a small pressure sensor, shown in (*b*). The pressure is measured by the relative position of the movable piston in the sensor.

write Eq. 15-1 as

$$\rho = \frac{m}{V} \quad \text{(uniform density)}, \tag{15-2}$$

where m and V are the mass and volume of the sample.

Density is a scalar property; its SI unit is the kilogram per cubic meter. Table 15-1 shows the densities of some substances and the average densities of some objects. Note that the density of a gas (see Air in the table) varies considerably with pressure, but the density of a liquid (see Water) does not; that is, gases are readily *compressible* but liquids are not.

Pressure

Let a small pressure-sensing device be suspended inside a fluid-filled vessel, as in Fig. 15-1*a*. The sensor (Fig. 15-1*b*) consists of a piston of area ΔA riding in a close-fitting cylinder and resting against a spring. A readout arrangement allows us to record the amount by which the (calibrated) spring is compressed by the surrounding fluid, thus indicating the magnitude ΔF of the force that acts normal to the piston. We define the **pressure** on the piston from the fluid as

$$p = \frac{\Delta F}{\Delta A}. \tag{15-3}$$

In theory, the pressure at any point in the fluid is the limit of this ratio as the area ΔA of the piston, centered on that point, is made smaller and smaller. However, if the force is uniform over a flat area A, we can write Eq. 15-3 as

$$p = \frac{F}{A} \quad \text{(pressure of uniform (normal) force on flat area)}, \tag{15-4}$$

TABLE 15-2 Some Pressures

	Pressure (Pa)
Center of the Sun	2×10^{16}
Center of Earth	4×10^{11}
Highest sustained laboratory pressure	1.5×10^{10}
Deepest ocean trench (bottom)	1.1×10^{8}
Spike heels on a dance floor	1×10^{6}
Automobile tire[a]	2×10^{5}
Atmosphere at sea level	1.0×10^{5}
Normal blood pressure[a,b]	1.6×10^{4}
Best laboratory vacuum	10^{-12}

[a]Pressure in excess of atmospheric pressure.

[b]The systolic pressure, corresponding to 120 torr on the physician's pressure gauge.

where F is the magnitude of the normal force on area A. (When we say a force is uniform over an area, we mean that it is evenly distributed over every point of the area.)

We find by experiment that at a given point in a fluid at rest, the pressure p defined by Eq. 15-3 has the same value no matter how the pressure sensor is oriented. Pressure is a scalar, having no directional properties. It is true that the force acting on the piston of our pressure sensor is a vector, but Eq. 15-3 involves only the *magnitude* of that force, a scalar quantity.

The SI unit of pressure is the newton per square meter, which is given a special name, the **pascal** (Pa). In metric countries, tire pressure gauges are calibrated in kilopascals. The pascal is related to some other common (non-SI) pressure units as follows:

$$1 \text{ atm} = 1.01 \times 10^5 \text{ Pa} = 760 \text{ torr} = 14.7 \text{ lb/in.}^2.$$

The *atmosphere* (atm) is, as the name suggests, the approximate average pressure of the atmosphere at sea level. The *torr* (named for Evangelista Torricelli, who invented the mercury barometer in 1674) was formerly called the *millimeter of mercury* (mm Hg). The pound per square inch is often abbreviated psi. Table 15-2 shows some pressures.

Sample Problem 15-1

Density of air at 1 atm & 20°C is 1.21

A living room has floor dimensions of 3.5 m and 4.2 m and a height of 2.4 m.

(a) What does the air in the room weigh when the air pressure is 1.0 atm?

SOLUTION: The **Key Ideas** here are these: (1) The air's weight is equal to mg, where m is its mass. (2) Mass m is related to the air density ρ and the air's volume V by Eq. 15-2 ($\rho = m/V$). Putting these two ideas together and taking the density of air at 1.0 atm from Table 15-1, we find

$$\begin{aligned} mg &= (\rho V)g \\ &= (1.21 \text{ kg/m}^3)(3.5 \text{ m} \times 4.2 \text{ m} \times 2.4 \text{ m})(9.8 \text{ m/s}^2) \\ &= 418 \text{ N} \approx 420 \text{ N}. \qquad \text{(Answer)} \end{aligned}$$

This is the weight of about 110 cans of Pepsi.

(b) What is the magnitude of the atmosphere's force on the floor of the room?

SOLUTION: The **Key Idea** here is that the atmosphere pushes down on the floor with a force of magnitude F that is uniform over the floor. Thus, it produces a pressure that is related to F and the flat area A of the floor by Eq. 15-4 ($p = F/A$), which gives us

$$\begin{aligned} F = pA &= (1.0 \text{ atm})\left(\frac{1.01 \times 10^5 \text{ N/m}^2}{1.0 \text{ atm}}\right)(3.5 \text{ m})(4.2 \text{ m}) \\ &= 1.5 \times 10^6 \text{ N}. \qquad \text{(Answer)} \end{aligned}$$

This enormous force is equal to the weight of the column of air that covers the floor and extends all the way to the top of the atmosphere. ? Why (a) & (b) are not equal

15-4 Fluids at Rest

Figure 15-2*a* shows a tank of water—or other liquid—open to the atmosphere. As every diver knows, the pressure *increases* with depth below the air–water interface. The diver's depth gauge, in fact, is a pressure sensor much like that of Fig. 15-1*b*. As every mountaineer knows, the pressure *decreases* with altitude as one ascends into the atmosphere. The pressures encountered by the diver and the mountaineer are usually called *hydrostatic pressures*, because they are due to fluids that are static (at rest). Here we want to find an expression for hydrostatic pressure as a function of depth or altitude.

Let us look first at the increase in pressure with depth below the water's surface. We set up a vertical y axis in the tank, with its origin at the air–water interface and the positive direction upward. We next consider a water sample contained in an imaginary right circular cylinder of horizontal base (or face) area A, such that y_1 and

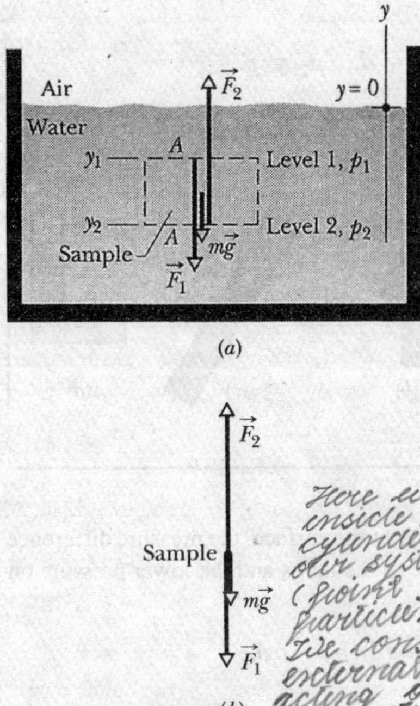

Here water inside the cylinder is our system (point particle). We consider only external forces acting on such a system

Fig. 15-2 (*a*) A tank of water in which a sample of water is contained in an imaginary cylinder of horizontal base area *A*. Force $\vec{F}_1$ acts at the top surface of the cylinder; force $\vec{F}_2$ acts at the bottom surface of the cylinder; the gravitational force on the water in the cylinder is represented by $m\vec{g}$. (*b*) A free-body diagram of the water sample.

y_2 (both of which are *negative* numbers) are the depths below the surface of the upper and lower cylinder faces, respectively.

Figure 15-2*b* shows a free-body diagram for the water in the cylinder. The water is in *static equilibrium;* that is, it is stationary and the forces on it balance. Three forces act on it vertically: Force $\vec{F}_1$ acts at the top surface of the cylinder and is due to the water above the cylinder. Similarly, force $\vec{F}_2$ acts at the bottom surface of the cylinder and is due to the water below the cylinder. The gravitational force on the water in the cylinder is represented by $m\vec{g}$, where m is the mass of the water in the cylinder. The balance of these forces is written as

$$F_2 = F_1 + mg. \tag{15-5}$$

We want to transform Eq. 15-5 into an equation involving pressures. From Eq. 15-4, we know that

$$F_1 = p_1 A \quad \text{and} \quad F_2 = p_2 A. \tag{15-6}$$

The mass m of the water in the cylinder is, from Eq. 15-2, $m = \rho V$, where the cylinder's volume V is the product of its face area A and its height $y_1 - y_2$. Thus, m is equal to $\rho A(y_1 - y_2)$. Substituting this and Eq. 15-6 into Eq. 15-5, we find

$$p_2 A = p_1 A + \rho A g(y_1 - y_2)$$

or

$$p_2 = p_1 + \rho g(y_1 - y_2). \tag{15-7}$$

This equation can be used to find pressure both in a liquid (as a function of depth) and in the atmosphere (as a function of altitude or height). For the former, suppose we seek the pressure p at a depth h below the liquid surface. Then we choose level 1 to be the surface, level 2 to be a distance h below it (as in Fig. 15-3), and p_0 to represent the atmospheric pressure on the surface. We then substitute

$$y_1 = 0, \quad p_1 = p_0 \quad \text{and} \quad y_2 = -h, \quad p_2 = p$$

into Eq. 15-7, which becomes

$$p = p_0 + \rho g h \quad \text{(pressure at depth } h\text{)}. \tag{15-8}$$

Note that the pressure at a given depth in the liquid depends on that depth but not on any horizontal dimension.

> The pressure at a point in a fluid in static equilibrium depends on the depth of that point but not on any horizontal dimension of the fluid or its container.

Thus, Eq. 15-8 holds no matter what the shape of the container. If the bottom surface of the container is at depth h, then Eq. 15-8 gives the pressure p there.

In Eq. 15-8, p is said to be the total pressure or **absolute pressure** at level 2. To see why, note in Fig. 15-3 that the pressure p at level 2 consists of two contributions: (1) p_0, the pressure due to the atmosphere, which bears down on the liquid, and (2) $\rho g h$, the pressure due to the liquid above level 2, which bears down on level 2. In general, the difference between an absolute pressure and an atmospheric pressure is called the **gauge pressure.** (The name comes from the use of a gauge to measure this difference in pressures.) For the situation of Fig. 15-3, the gauge pressure is $\rho g h$.

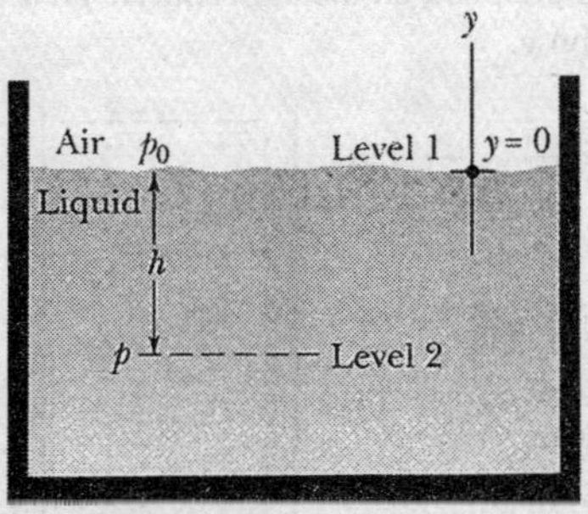

Fig. 15-3 The pressure p increases with depth h below the liquid surface according to Eq. 15-8.

Equation 15-7 also holds above the liquid surface: It gives the atmospheric pressure at a given distance above level 1 in terms of the atmospheric pressure p_1 at level 1 (*assuming* that the atmospheric density is uniform over that distance). For example, to find the atmospheric pressure at a distance d above level 1 in

Fig. 15-3, we substitute

$$y_1 = 0, \quad p_1 = p_0 \quad \text{and} \quad y_2 = d, \quad p_2 = p.$$

Then with $\rho = \rho_{air}$, we obtain

$$p = p_0 - \rho_{air} g d.$$

These are all different containers of fluid in static equilibrium. For pressure to be same at a given depth, p_0, ρ, g, L must be equal

CHECKPOINT 1: The figure shows four containers of olive oil. Rank them according to the pressure at depth h, greatest first.

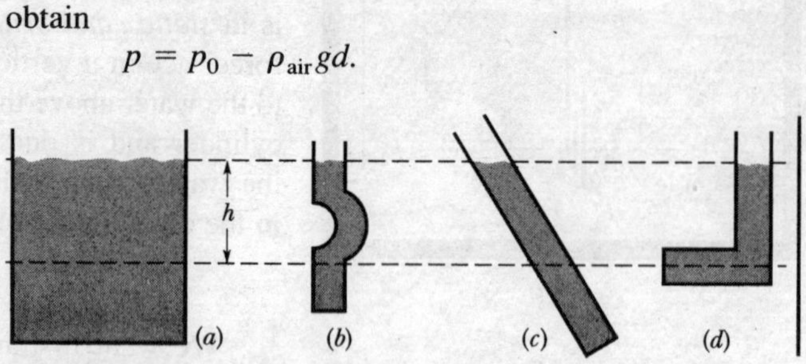

Sample Problem 15-2

A novice scuba diver practicing in a swimming pool takes enough air from his tank to fully expand his lungs before abandoning the tank at depth L and swimming to the surface. He ignores instructions and fails to exhale during his ascent. When he reaches the surface, the difference between the external pressure on him and the air pressure in his lungs is 9.3 kPa. From what depth does he start? What potentially lethal danger does he face?

SOLUTION: The **Key Idea** here is that when the diver fills his lungs at depth L, the external pressure on him (and thus the air pressure within his lungs) is greater than normal and given by Eq. 15-8 as

$$p = p_0 + \rho g L,$$

where p_0 is atmospheric pressure and ρ is the water's density (998 kg/m^3, from Table 15-1). As he ascends, the external pressure on him decreases, until it is atmospheric pressure p_0 at the surface. His blood pressure also decreases, until it is normal. However, because he does not exhale, the air pressure in his lungs remains at the value it had at depth L. At the surface, the pressure difference between the higher pressure in his lungs and the lower pressure on his chest is

$$\Delta p = p - p_0 = \rho g L,$$

from which we find

$$L = \frac{\Delta p}{\rho g} = \frac{9300 \text{ Pa}}{(998 \text{ kg/m}^3)(9.8 \text{ m/s}^2)} = 0.95 \text{ m.} \qquad \text{(Answer)}$$

This is not deep! Yet, the pressure difference of 9.3 kPa (about 9% of atmospheric pressure) is sufficient to rupture the diver's lungs and force air from them into the depressurized blood, which then carries the air to the heart, killing the diver. If the diver follows instructions and gradually exhales as he ascends, he allows the pressure in his lungs to equalize with the external pressure, and then there is no danger.

Sample Problem 15-3

The U-tube in Fig. 15-4 contains two liquids in static equilibrium: Water of density ρ_w (= 998 kg/m^3) is in the right arm, and oil of unknown density ρ_x is in the left. Measurement gives l = 135 mm and d = 12.3 mm. What is the density of the oil?

SOLUTION: One **Key Idea** here is that the pressure p_{int} at the oil–water interface in the left arm depends on the density ρ_x and height of the oil above the interface. A second **Key Idea** is that the water in the right arm *at the same level* must be at the same pressure p_{int}. The reason is that, because the water is in static equilibrium, pressures at points in the water at the same level must be the same even if the points are separated horizontally.

In the right arm, the interface is a distance l below the free surface of the *water* and we have, from Eq. 15-8,

$$p_{int} = p_0 + \rho_w g l \qquad \text{(right arm).}$$

In the left arm, the interface is a distance $l + d$ below the free surface of the *oil* and we have, again from Eq. 15-8,

$$p_{int} = p_0 + \rho_x g (l + d) \qquad \text{(left arm).}$$

Equating these two expressions and solving for the unknown density yield

$$\rho_x = \rho_w \frac{l}{l + d} = (998 \text{ kg/m}^3) \frac{135 \text{ mm}}{135 \text{ mm} + 12.3 \text{ mm}} = 915 \text{ kg/m}^3. \qquad \text{(Answer)}$$

Note that the answer does not depend on the atmospheric pressure p_0 or the free-fall acceleration g.

This is one container of fluid/fluids in static equilibrium. Pressure at a given level must be the same, no matter what

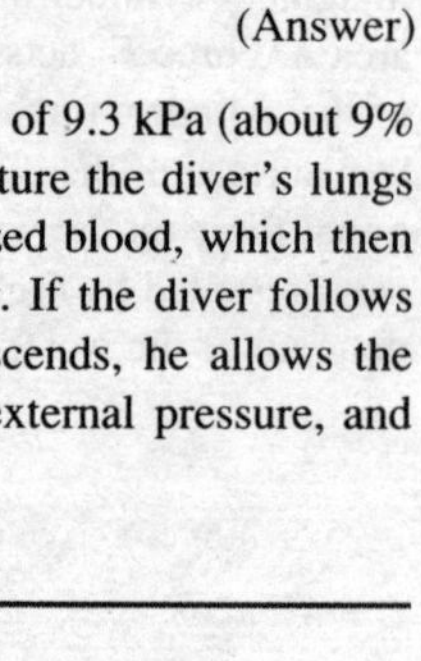

Fig. 15-4 Sample Problem 15-3. The oil in the left arm stands higher than the water in the right arm because the oil is less dense than the water. Both fluid columns produce the same pressure p_{int} at the level of the interface.

15-5 Measuring Pressure

The Mercury Barometer

Figure 15-5*a* shows a very basic *mercury barometer,* a device used to measure the pressure of the atmosphere. The long glass tube is filled with mercury and inverted with its open end in a dish of mercury, as the figure shows. The space above the mercury column contains only mercury vapor, whose pressure is so small at ordinary temperatures that it can be neglected. (nearly vacuum)

We can use Eq. 15-7 to find the atmospheric pressure p_0 in terms of the height h of the mercury column. We choose level 1 of Fig. 15-2 to be that of the air–mercury interface and level 2 to be that of the top of the mercury column, as labeled in Fig. 15-5*a*. We then substitute

$$y_1 = 0, \quad p_1 = p_0 \quad \text{and} \quad y_2 = h, \quad p_2 = 0$$

into Eq. 15-7, finding that

$$p_0 = \rho g h, \tag{15-9}$$

where ρ is the density of the mercury.

For a given pressure, the height h of the mercury column does not depend on the cross-sectional area of the vertical tube. The fanciful mercury barometer of Fig. 15-5*b* gives the same reading as that of Fig. 15-5*a*; all that counts is the vertical distance h between the mercury levels.

Equation 15-9 shows that, for a given pressure, the height of the column of mercury depends on the value of g at the location of the barometer and on the density of mercury, which varies with temperature. The column height (in millimeters) is numerically equal to the pressure (in torr) *only* if the barometer is at a place where g has its accepted standard value of 9.80665 m/s^2 *and* the temperature of the mercury is 0°C. If these conditions do not prevail (and they rarely do), small corrections must be made before the height of the mercury column can be transformed into a pressure.

The Open-Tube Manometer

An *open-tube manometer* (Fig. 15-6) measures the gauge pressure p_g of a gas. It consists of a U-tube containing a liquid, with one end of the tube connected to the

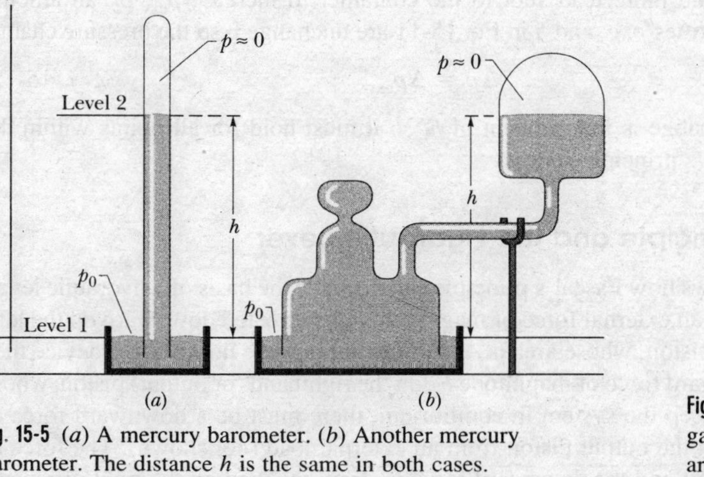

Fig. 15-5 (*a*) A mercury barometer. (*b*) Another mercury barometer. The distance h is the same in both cases.

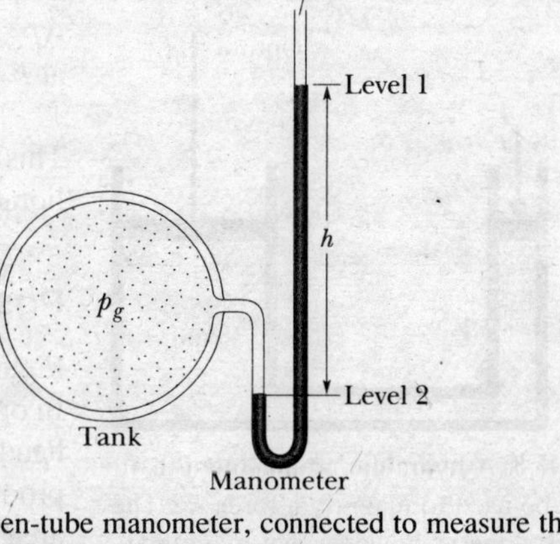

Fig. 15-6 An open-tube manometer, connected to measure the gauge pressure of the gas in the tank on the left. The right arm of the U-tube is open to the atmosphere.

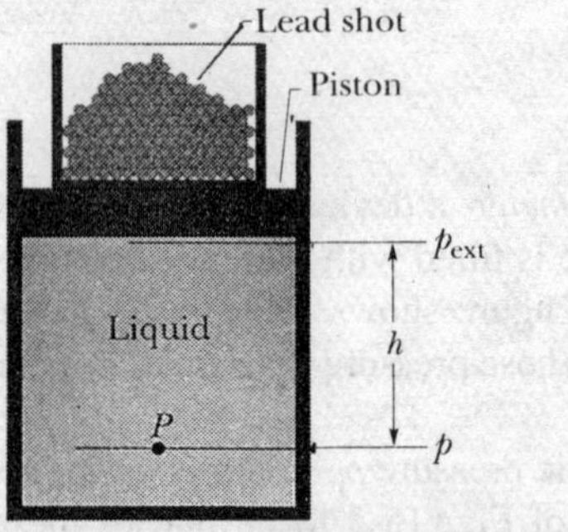

Fig. 15-7 Lead shot (small balls of lead) loaded onto the piston create a pressure p_{ext} at the top of the enclosed (incompressible) liquid. If p_{ext} is increased, by adding more lead shot, the pressure increases by the same amount at all points within the liquid.

vessel whose gauge pressure we wish to measure and the other end open to the atmosphere. We can use Eq. 15-7 to find the gauge pressure in terms of the height h shown in Fig. 15-6. Let us choose levels 1 and 2 as shown in Fig. 15-6. We then substitute

$$y_1 = 0, \quad p_1 = p_0 \quad \text{and} \quad y_2 = -h, \quad p_2 = p$$

into Eq. 15-7, finding that

$$p_g = p - p_0 = \rho g h, \tag{15-10}$$

where ρ is the density of the liquid in the tube. The gauge pressure p_g is directly proportional to h.

The gauge pressure can be positive or negative, depending on whether $p > p_0$ or $p < p_0$. In inflated tires or the human circulatory system, the (absolute) pressure is greater than atmospheric pressure, so the gauge pressure is a positive quantity, sometimes called the *overpressure*. If you suck on a straw to pull fluid up the straw, the (absolute) pressure in your lungs is actually less than atmospheric pressure. The gauge pressure in your lungs is then a negative quantity.

15-6 Pascal's Principle

When you squeeze one end of a tube to get toothpaste out the other end, you are watching **Pascal's principle** in action. This principle is also the basis for the Heimlich maneuver, in which a sharp pressure increase properly applied to the abdomen is transmitted to the throat, forcefully ejecting food lodged there. The principle was first stated clearly in 1652 by Blaise Pascal (for whom the unit of pressure is named):

> A change in the pressure applied to an enclosed incompressible fluid is transmitted undiminished to every portion of the fluid and to the walls of its container.

Demonstrating Pascal's Principle

Consider the case in which the incompressible fluid is a liquid contained in a tall cylinder, as in Fig. 15-7. The cylinder is fitted with a piston on which a container of lead shot rests. The atmosphere, container, and shot put pressure p_{ext} on the piston and thus on the liquid. The pressure p at any point P in the liquid is then

$$p = p_{ext} + \rho g h. \tag{15-11}$$

Let us add a little more lead shot to the container to increase p_{ext} by an amount Δp_{ext}. The quantities ρ, g, and h in Eq. 15-11 are unchanged, so the pressure change at P is

$$\Delta p = \Delta p_{ext}. \tag{15-12}$$

This pressure change is independent of h, so it must hold for all points within the liquid, as Pascal's principle states.

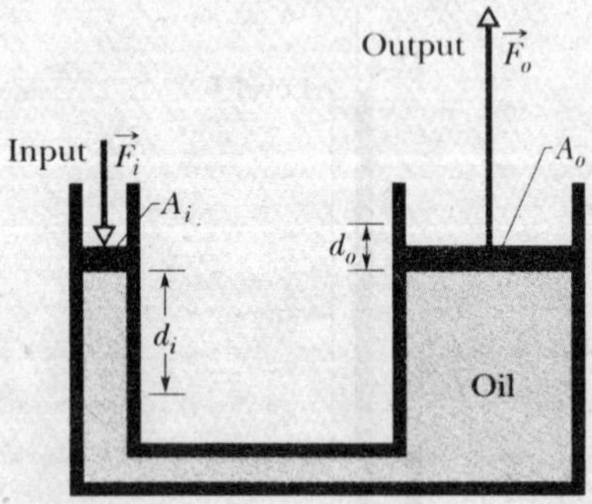

Fig. 15-8 A hydraulic arrangement that can be used to magnify a force $\vec{F}_i$. The work done is, however, not magnified and is the same for both the input and output forces.

Pascal's Principle and the Hydraulic Lever

Figure 15-8 shows how Pascal's principle can be made the basis of a hydraulic lever. In operation, let an external force of magnitude F_i be directed downward on the left-hand (or input) piston, whose area is A_i. An incompressible liquid in the device then produces an upward force of magnitude F_o on the righthand (or output) piston, whose area is A_o. To keep the system in equilibrium, there must be a downward force of magnitude F_o on the output piston from an external load (not shown). The force $\vec{F}_i$ applied on the left and the downward force $\vec{F}_o$ from the load on the right produce a change Δp in the pressure of the liquid that is given by

$$\Delta p = \frac{F_i}{A_i} = \frac{F_o}{A_o},$$

so

$$F_o = F_i \frac{A_o}{A_i}. \tag{15-13}$$

Equation 15-13 shows that the output force F_o on the load must be greater than the input force F_i if $A_o > A_i$, as is the case in Fig. 15-8.

If we move the input piston downward a distance d_i, the output piston moves upward a distance d_o, such that the same volume V of the incompressible liquid is displaced at both pistons. Then

$$V = A_i d_i = A_o d_o,$$

which we can write as

$$d_o = d_i \frac{A_i}{A_o}. \tag{15-14}$$

This shows that, if $A_o > A_i$ (as in Fig. 15-8), the output piston moves a smaller distance than the input piston moves.

From Eqs. 15-13 and 15-14 we can write the output work as

$$W = F_o d_o = \left(F_i \frac{A_o}{A_i}\right)\left(d_i \frac{A_i}{A_o}\right) = F_i d_i, \tag{15-15}$$

which shows that the work W done *on* the input piston by the applied force is equal to the work W done *by* the output piston in lifting the load placed on it.

The advantage of a hydraulic lever is this:

> With a hydraulic lever, a given force applied over a given distance can be transformed to a greater force applied over a smaller distance.

The product of force and distance remains unchanged so that the same work is done. However, there is often tremendous advantage in being able to exert the larger force. Most of us, for example, cannot lift an automobile directly but can with a hydraulic jack, even though we have to pump the handle farther than the automobile rises. In this device, the displacement d_i is accomplished not in a single stroke but over a series of small strokes.

15-7 Archimedes' Principle

Fig. 15-9 A thin-walled plastic sack of water is in static equilibrium in the pool. The gravitational force on it must be balanced by a net upward force on the sack from the surrounding water.

Figure 15-9 shows a student in a swimming pool, manipulating a very thin plastic sack (of negligible mass) that is filled with water. She finds that the sack and its contained water are in static equilibrium, tending neither to rise nor to sink. The downward gravitational force $\vec{F}_g$ on the contained water must be balanced by a net upward force from the water surrounding the sack.

This net upward force is a **buoyant force** $\vec{F}_b$. It exists because the pressure in the surrounding water increases with depth below the surface. Thus, the pressure near the bottom of the sack is greater than the pressure near the top. Then the forces on the sack due to this pressure are greater in magnitude near the bottom of the sack than near the top. Some of the forces are represented in Fig. 15-10*a*, where the space occupied by the sack has been left empty. Note that the force vectors drawn near the bottom of that space (with upward components) have longer lengths than those drawn near the top of the sack (with downward components). If we vectorially add all the forces on the sack from the water, the horizontal components cancel and the vertical components add to yield the upward buoyant force $\vec{F}_b$ on the sack. (Force $\vec{F}_b$ is shown to the right of the pool in Fig. 15-10*a*.)

Because the sack of water is in static equilibrium, the magnitude of $\vec{F}_b$ is equal to the magnitude $m_f g$ of the gravitational force $\vec{F}_g$ on the sack of water: $F_b = m_f g$. (Subscript *f* refers to *fluid*, here the water.) In words, the magnitude of the buoyant force is equal to the weight of the water in the sack.

In Fig. 15-10*b*, we have replaced the sack of water with a stone that exactly fills the hole in Fig. 15-10*a*. The stone is said to *displace* the water, meaning that it occupies space that would otherwise be occupied by water. We have changed nothing about the shape of the hole, so the forces at the hole's surface must be the same as when the water-filled sack was in place. Thus, the same upward buoyant force that acted on the water-filled sack now acts on the stone; that is, the magnitude F_b of the buoyant force is equal to $m_f g$, the weight of the water displaced by the stone.

Unlike the water-filled sack, the stone is not in static equilibrium. The downward gravitational force $\vec{F}_g$ on the stone is greater in magnitude than the upward buoyant force, as is shown in the free-body diagram to the right of the pool in Fig. 15-10*b*. The stone thus accelerates downward, sinking to the bottom of the pool.

Let us next exactly fill the hole in Fig. 15-10*a* with a block of light-weight wood, as in Fig. 15-10*c*. Again, nothing has changed about the forces at the hole's surface, so the magnitude F_b of the buoyant force is still equal to $m_f g$, the weight of the displaced water. Like the stone, the block is not in static equilibrium. However, this time the gravitational force $\vec{F}_g$ is lesser in magnitude than the buoyant force (as shown to the right of the pool), and so the block accelerates upward, rising to the top surface of the water.

Our results with the sack, stone, and block apply to all fluids and are summarized in **Archimedes' principle:** volume of the fluid displaced under water = volume of body

> When a body is fully or partially submerged in a fluid, a buoyant force $\vec{F}_b$ from the surrounding fluid acts on the body. The force is directed upward and has a magnitude equal to the weight $m_f g$ of the fluid that has been displaced by the body.

The buoyant force on a body in a fluid has the magnitude

$$F_b = m_f g \quad \text{(buoyant force)}, \tag{15-16}$$

where m_f is the mass of the fluid that is displaced by the body.

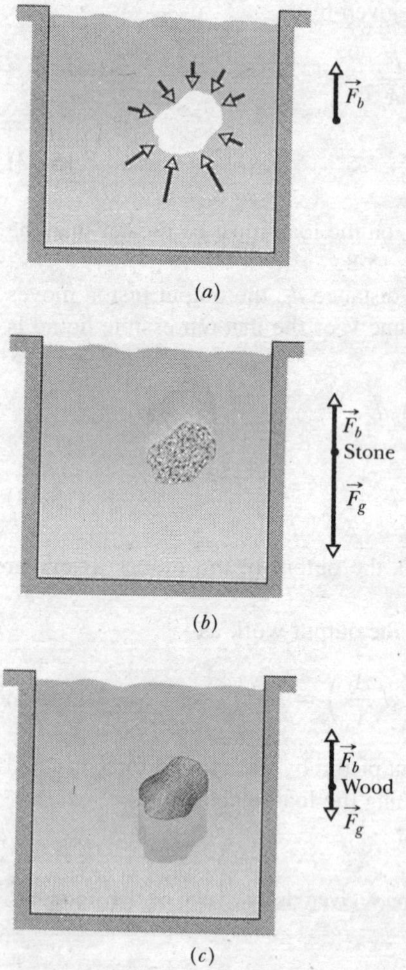

Fig. 15-10 (*a*) The water surrounding the hole in the water produces a net upward buoyant force on whatever fills the hole. (*b*) For a stone of the same volume as the hole, the gravitational force exceeds the buoyant force in magnitude. (*c*) For a lump of wood of the same volume, the gravitational force is less than the buoyant force in magnitude.

In the late evening of August 21, 1986, something (possibly a volcanic tremor) disturbed Cameroon's Lake Nyos, which has a high concentration of dissolved carbon dioxide. The disturbance caused that gas to form bubbles. Being lighter than the surrounding fluid (the water), those bubbles were buoyed to the surface, where they released the carbon dioxide. The gas, being heavier than the surrounding fluid (now the air), rushed down the mountainside like a river, asphyxiating 1700 persons and the scores of animals seen here.

Floating

When we release a block of light-weight wood just above the water in a pool, it moves into the water because the gravitational force on it pulls it downward. As the block displaces more and more water, the magnitude F_b of the upward buoyant force acting on it increases. Eventually, F_b is large enough to equal the magnitude F_g of the downward gravitational force on the block, and the block comes to rest. The block is then in static equilibrium and is said to be *floating* in the water. In general,

> **When a body floats in a fluid, the magnitude F_b of the buoyant force on the body is equal to the magnitude F_g of the gravitational force on the body.**

We can write this statement as

$$F_b = F_g \quad \text{(floating).} \tag{15-17}$$

From Eq. 15-16, we know that $F_b = m_f g$. Thus,

> **When a body floats in a fluid, the magnitude F_g of the gravitational force on the body is equal to the weight $m_f g$ of the fluid that has been displaced by the body.**

We can write this statement as

$$F_g = m_f g \quad \text{(floating).} \tag{15-18}$$

In other words, a floating body displaces its own weight of fluid.

Apparent Weight in a Fluid

If we place a stone on a scale that is calibrated to measure weight, then the reading on the scale is the stone's weight. However, if we do this underwater, the upward buoyant force on the stone from the water decreases the reading. That reading is then an apparent weight. In general, an apparent weight is related to the actual weight of a body and the buoyant force on the body by

$$\begin{pmatrix}\text{apparent}\\ \text{weight}\end{pmatrix} = \begin{pmatrix}\text{actual}\\ \text{weight}\end{pmatrix} - \begin{pmatrix}\text{magnitude of}\\ \text{buoyant force}\end{pmatrix},$$

which we can write as

$$\text{weight}_{\text{app}} = \text{weight} - F_b \quad \text{(apparent weight).} \tag{15-19}$$

If, in some strange test of strength, you had to lift a heavy stone, you could do it more easily with the stone underwater. Then your applied force would need to exceed only the stone's apparent weight, not its larger actual weight, because the upward buoyant force would help you lift the stone.

The magnitude of the buoyant force on a floating body is equal to the body's weight. Equation 15-19 thus tells us that a floating body has an apparent weight of zero—the body would produce a reading of zero on a scale. (When astronauts prepare to perform a complex task in space, they practice the task floating underwater, where their apparent weight is zero as it is in space.)

CHECKPOINT 2: A penguin floats first in a fluid of density ρ_0, then in a fluid of density $0.95\rho_0$, and then in a fluid of density $1.1\rho_0$. (a) Rank the densities according to the magnitude of the buoyant force on the penguin, greatest first. (b) Rank the densities according to the amount of fluid displaced by the penguin, greatest first.

(a) For floating
$F_b = F_g$
all tie
(b) $F_b = m_f g$
F_b, g constants
$m_f = \rho_f V$
m_f const
amount of fluid $\propto$ volume V
$\therefore 0.95\rho_0 > \rho_0 > 1.1\rho_0$

Sample Problem 15-4

What fraction of the volume of an iceberg floating in seawater is visible?

SOLUTION: Let V_i be the total volume of the iceberg. The nonvisible portion is below water and thus is equal to the volume V_f of the fluid (the seawater) displaced by the iceberg. We seek the fraction (call it frac)

$$\text{frac} = \frac{V_i - V_f}{V_i} = 1 - \frac{V_f}{V_i}, \qquad (15\text{-}20)$$

but we know neither volume. A **Key Idea** here is that, because the iceberg is floating, Eq. 15-18 ($F_g = m_f g$) applies. We can write that equation as

$$m_i g = m_f g,$$

from which we see that $m_i = m_f$. Thus, the mass of the iceberg is equal to the mass of the displaced fluid (seawater). Although we know neither mass, we can relate them to the densities of ice and seawater given in Table 15-1 by using Eq. 15-2 ($\rho = m/V$). Because $m_i = m_f$, we can write

$$\rho_i V_i = \rho_f V_f$$

or

$$\frac{V_f}{V_i} = \frac{\rho_i}{\rho_f}.$$

Substituting this into Eq. 15-20 and then using the known densities, we find

$$\text{frac} = 1 - \frac{\rho_i}{\rho_f} = 1 - \frac{917 \text{ kg/m}^3}{1024 \text{ kg/m}^3}$$

$$= 0.10 \text{ or } 10\%. \qquad \text{(Answer)}$$

Sample Problem 15-5

A spherical, helium-filled balloon has a radius R of 12.0 m. The balloon, support cables, and basket have a mass m of 196 kg. What maximum load M can the balloon support while it floats at an altitude at which the helium density ρ_{He} is 0.160 kg/m^3 and the air density ρ_{air} is 1.25 kg/m^3? Assume that the volume of air displaced by the load, support cables, and basket is negligible.

SOLUTION: The **Key Idea** here is that the balloon, cables, basket, load, *and* the contained helium form a floating body, with a total mass of $m + M + m_{He}$, where m_{He} is the mass of the contained helium. Then the magnitude of the total gravitational force on this body must be equal to the weight of the air displaced by the body (the air is the fluid in which this body floats). Let m_{air} be the mass of the air displaced by the body. From Eq. 15-18 ($F_g = m_f g$), we have

$$(m + M + m_{He})g = m_{air} g$$

or

$$M = m_{air} - m_{He} - m. \qquad (15\text{-}21)$$

We do not know m_{He} and m_{air} but we do know the corresponding densities, so we can use Eq. 15-2 ($\rho = m/V$) to rewrite Eq. 15-21 in terms of those densities. First we note that, because the load, support cables, and basket displace a negligible amount of air, the volume of the displaced air is equal to the volume V ($= \frac{4}{3}\pi R^3$) of the spherical balloon. Then Eq. 15-21 becomes

$$M = \rho_{air} V - \rho_{He} V - m$$

$$= (\tfrac{4}{3}\pi R^3)(\rho_{air} - \rho_{He}) - m$$

$$= (\tfrac{4}{3}\pi)(12.0 \text{ m})^3(1.25 \text{ kg/m}^3 - 0.160 \text{ kg/m}^3) - 196 \text{ kg}$$

$$= 7694 \text{ kg} \approx 7690 \text{ kg}. \qquad \text{(Answer)}$$

Fig. 15-11 At a certain point, the rising flow of smoke and heated gas changes from steady to turbulent.

15-8 Ideal Fluids in Motion

The motion of *real fluids* is very complicated and not yet fully understood. Instead, we shall discuss the motion of an **ideal fluid,** which is simpler to handle mathematically and yet provides useful results. Here are four assumptions that we make about our ideal fluid; they all are concerned with *flow:*

1. ***Steady flow*** In *steady* (or *laminar*) *flow*, the velocity of the moving fluid at any fixed point does not change with time, either in magnitude or in direction. The gentle flow of water near the center of a quiet stream is steady; that in a chain of rapids is not. Figure 15-11 shows a transition from steady flow to *nonsteady* (or *turbulent*) flow for a rising stream of smoke. The speed of the smoke particles increases as they rise and, at a certain critical speed, the flow changes from steady to nonsteady (that is, from laminar to *nonlaminar* flow).
2. ***Incompressible flow*** We assume, as we have already done for fluids at rest, that our ideal fluid is incompressible; that is, its density has a constant, uniform value.
3. ***Nonviscous flow*** Roughly speaking, the viscosity of a fluid is a measure of how resistive the fluid is to flow. For example, thick honey is more resistive to flow than water, and so honey is said to be more viscous than water. Viscosity is the

Fig. 15-12 The steady flow of a fluid around a cylinder, as revealed by a dye tracer that was injected into the fluid upstream of the cylinder.

Fig. 15-13 Smoke reveals streamlines in airflow past a car in a wind-tunnel test.

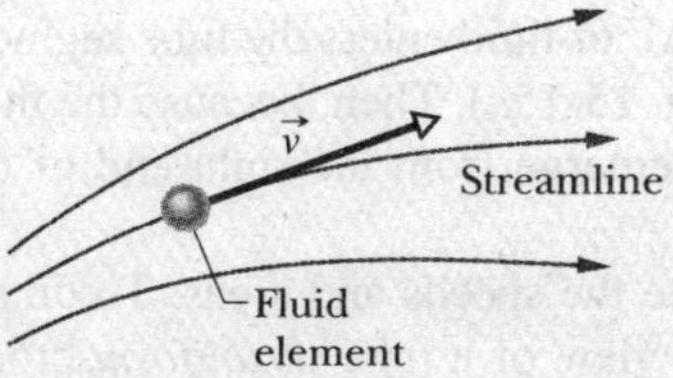

Fig. 15-14 A fluid element P traces out a streamline as it moves. The velocity vector of the element is tangent to the streamline at every point.

fluid analog of friction between solids; both are mechanisms by which the kinetic energy of moving objects can be transferred to thermal energy. In the absence of friction, a block could glide at constant speed along a horizontal surface. In the same way, an object moving through a nonviscous fluid would experience no *viscous drag force*—that is, no resistive force due to viscosity; it could move at constant speed through the fluid. The British scientist Lord Rayleigh noted that in an ideal fluid a ship's propeller would not work but, on the other hand, a ship (once set into motion) would not need a propeller!

4. ***Irrotational flow*** Although it need not concern us further, we also assume that the flow is *irrotational*. To test for this property, let a tiny grain of dust move with the fluid. Although this test body may (or may not) move in a circular path, in irrotational flow the test body will not rotate about an axis through its own center of mass. For a loose analogy, the motion of a Ferris wheel is rotational; that of its passengers is irrotational.

We can make the flow of a fluid visible by adding a *tracer*. This might be a dye injected into many points across a liquid stream (Fig. 15-12) or smoke particles added to a gas flow (Figs. 15-11 and 15-13). Each bit of a tracer follows a *streamline*, which is the path that a tiny element of the fluid would take as the fluid flows. Recall from Chapter 2 that the velocity of a particle is always tangent to the path taken by the particle. Here the particle is the fluid element, and its velocity $\vec{v}$ is always tangent to a streamline (Fig. 15-14). For this reason, two streamlines can never intersect; if they did, then an element arriving at their intersection would have two different velocities simultaneously, an impossibility.

15-9 The Equation of Continuity

You may have noticed that you can increase the speed of the water emerging from a garden hose by partially closing the hose opening with your thumb. Apparently the speed v of the water depends on the cross-sectional area A through which the water flows.

Here we wish to derive an expression that relates v and A for the steady flow of an ideal fluid through a tube with varying cross section, like that in Fig. 15-15. The flow there is toward the right, and the tube segment shown (part of a longer

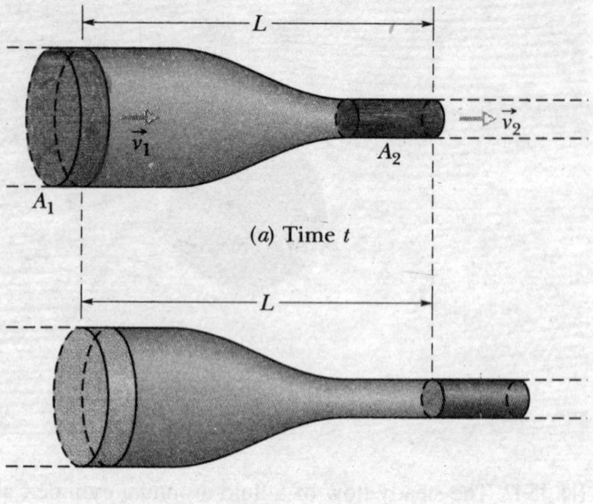

Fig. 15-15 Fluid flows from left to right at a steady rate through a tube segment of length L. The fluid's speed is v_1 at the left side and v_2 at the right side. The tube's cross-sectional area is A_1 at the left side and A_2 at the right side. From time t in (*a*) to time $t + \Delta t$ in (*b*), the amount of fluid shown in purple enters at the left side and the equal amount of fluid shown in green emerges at the right side.

tube) has length L. The fluid has speeds v_1 at the left end of the segment and v_2 at the right end. The tube has cross-sectional areas A_1 at the left end and A_2 at the right end. Suppose that in a time interval Δt a volume ΔV of fluid enters the tube segment at its left end (that volume is colored purple in Fig. 15-15*a*). Then, because the fluid is incompressible, an identical volume ΔV must emerge from the right end of the segment (it is colored green in Fig. 15-15*b*).

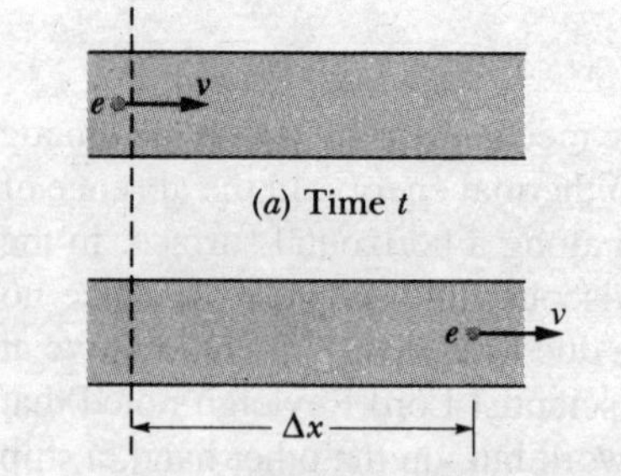

Fig. 15-16 Fluid flows at a constant speed v through a tube. (*a*) At time t, fluid element e is about to pass the dashed line. (*b*) At time $t + \Delta t$, element e is a distance $\Delta x = v\,\Delta t$ from the dashed line.

We can use this common volume ΔV to relate the speeds and areas. To do so, we first consider Fig. 15-16, which shows a side view of a tube of *uniform* cross-sectional area A. In Fig. 15-16*a*, a fluid element e is about to pass through the dashed line drawn across the tube width. The element's speed is v, so during a time interval Δt, the element moves along the tube a distance $\Delta x = v\,\Delta t$. The volume ΔV of fluid that has passed through the dashed line in that time interval Δt is

$$\Delta V = A\,\Delta x = Av\,\Delta t. \tag{15-22}$$

Applying Eq. 15-22 to both the left and right ends of the tube segment in Fig. 15-15, we have

$$\Delta V = A_1 v_1\,\Delta t = A_2 v_2\,\Delta t$$

or

$$A_1 v_1 = A_2 v_2 \quad \text{(equation of continuity).} \tag{15-23}$$

This relation between speed and cross-sectional area is called the **equation of continuity** for the flow of an ideal fluid. It tells us that the flow speed increases when we decrease the cross-sectional area through which the fluid flows (as when we partially close off a garden hose with a thumb).

Equation 15-23 applies not only to an actual tube but also to any so-called *tube of flow*, or imaginary tube whose boundary consists of streamlines. Such a tube acts like a real tube because no fluid element can cross a streamline; thus, all the fluid within a tube of flow must remain within its boundary. Figure 15-17 shows a tube of flow in which the cross-sectional area increases from area A_1 to area A_2 along the flow direction. From Eq. 15-23 we know that, with the increase in area, the speed must decrease, as is indicated by the greater spacing between streamlines at the right in Fig. 15-17. Similarly, you can see that in Fig. 15-12 the speed of the flow is greatest just above and just below the cylinder.

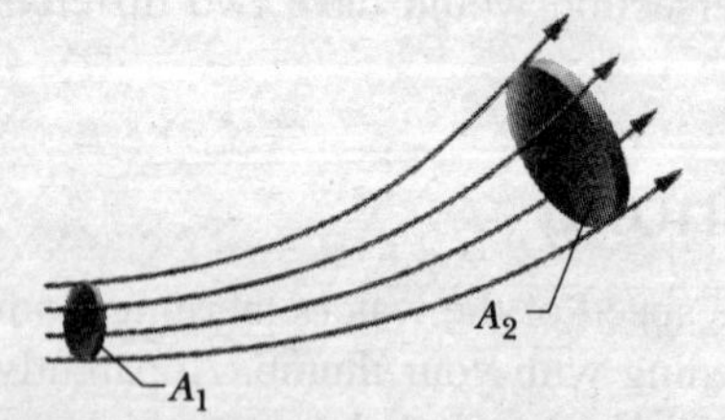

Fig. 15-17 A tube of flow is defined by the streamlines that form the boundary. The volume flow rate must be the same for all cross sections of the tube of flow.

We can rewrite Eq. 15-23 as

$$R_V = Av = \text{a constant} \quad \text{(volume flow rate, equation of continuity),} \tag{15-24}$$

in which R_V is the **volume flow rate** of the fluid (volume per unit time). Its SI unit is the cubic meter per second (m^3/s). If the density ρ of the fluid is uniform, we can multiply Eq. 15-24 by that density to get the **mass flow rate** R_m (mass per unit time):

$$R_m = \rho R_V = \rho A v = \text{a constant} \quad \text{(mass flow rate)}. \tag{15-25}$$

The SI unit of mass flow rate is the kilogram per second (kg/s). Equation 15-25 says that the mass that flows into the tube segment of Fig. 15-15 each second must be equal to the mass that flows out of that segment each second.

CHECKPOINT 3: The figure shows a pipe and gives the volume flow rate (in cm^3/s) and the direction of flow for all but one section. What are the volume flow rate and the direction of flow for that section?

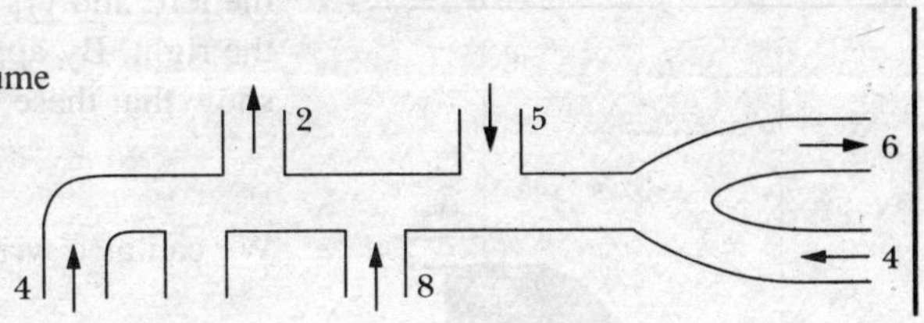

Sample Problem 15-6

The cross-sectional area A_0 of the aorta (the major blood vessel emerging from the heart) of a normal resting person is 3 cm^2, and the speed v_0 of the blood through it is 30 cm/s. A typical capillary (diameter $\approx 6\ \mu m$) has a cross-sectional area A of $3 \times 10^{-7}\ cm^2$ and a flow speed v of 0.05 cm/s. How many capillaries does such a person have?

SOLUTION: The **Key Idea** here is that all the blood that passes through the capillaries must have passed through the aorta. Therefore, the volume flow rate through the aorta must equal the total volume flow rate through the capillaries. Let us assume that the capillaries are identical, with the given cross-sectional area A and flow speed v. Then, from Eq. 15-24 we have

$$A_0 v_0 = nAv,$$

where n is the number of capillaries. Solving for n yields

$$n = \frac{A_0 v_0}{Av} = \frac{(3\ cm^2)(30\ cm/s)}{(3 \times 10^{-7}\ cm^2)(0.05\ cm/s)}$$
$$= 6 \times 10^9 \text{ or 6 billion.} \quad \text{(Answer)}$$

You can easily show that the combined cross-sectional area of the capillaries is about 600 times the cross-sectional area of the aorta.

Sample Problem 15-7

Figure 15-18 shows how the stream of water emerging from a faucet "necks down" as it falls. The indicated cross-sectional areas are $A_0 = 1.2\ cm^2$ and $A = 0.35\ cm^2$. The two levels are separated by a vertical distance $h = 45$ mm. What is the volume flow rate from the tap?

SOLUTION: The **Key Idea** here is simply that the volume flow rate through the higher cross section must be the same as that through the lower cross section. Thus, from Eq. 15-24, we have

$$A_0 v_0 = Av, \tag{15-26}$$

where v_0 and v are the water speeds at the levels corresponding to A_0 and A. From Eq. 2-16 we can also write, because the water is falling freely with acceleration g,

$$v^2 = v_0^2 + 2gh. \tag{15-27}$$

Eliminating v between Eqs. 15-26 and 15-27 and solving for v_0, we obtain

$$v_0 = \sqrt{\frac{2ghA^2}{A_0^2 - A^2}}$$
$$= \sqrt{\frac{(2)(9.8\ m/s^2)(0.045\ m)(0.35\ cm^2)^2}{(1.2\ cm^2)^2 - (0.35\ cm^2)^2}}$$
$$= 0.286\ m/s = 28.6\ cm/s.$$

From Eq. 15-24, the volume flow rate R_V is then

$$R_V = A_0 v_0 = (1.2\ cm^2)(28.6\ cm/s)$$
$$= 34\ cm^3/s. \quad \text{(Answer)}$$

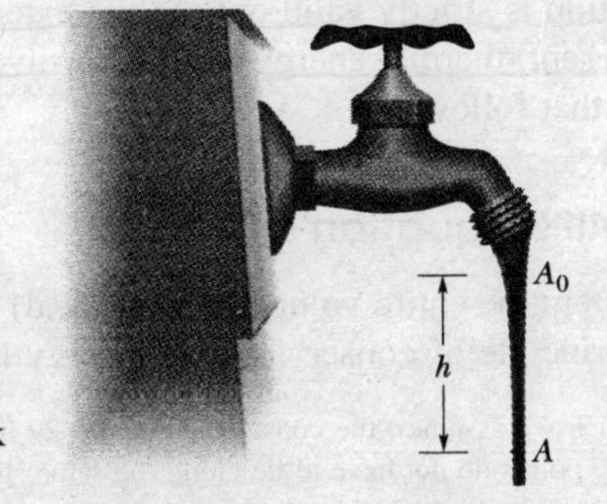

Fig. 15-18 Sample Problem 15-7. As water falls from a tap, its speed increases. Because the flow rate must be the same at all cross sections, the stream must "neck down."

15-10 Bernoulli's Equation

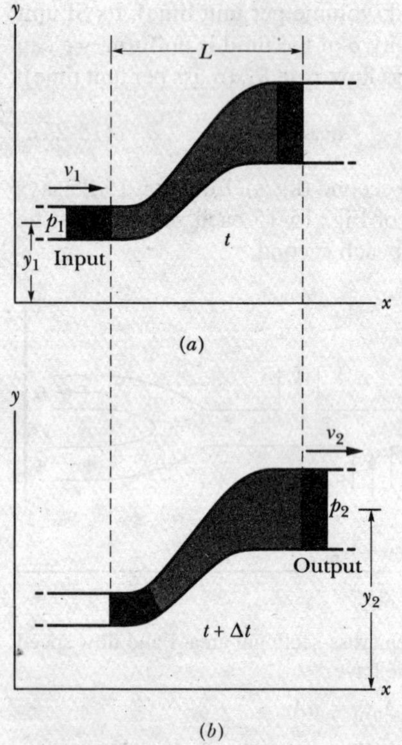

Fig. 15-19 Fluid flows at a steady rate through a length L of a tube, from the input end at the left to the output end at the right. From time t in (a) to time $t + \Delta t$ in (b), the amount of fluid shown in purple enters the input end and the equal amount shown in green emerges from the output end.

Figure 15-19 represents a tube through which an ideal fluid is flowing at a steady rate. In a time interval Δt, suppose that a volume of fluid ΔV, colored purple in Fig. 15-19a, enters the tube at the left (or input) end and an identical volume, colored green in Fig. 15-19b, emerges at the right (or output) end. The emerging volume must be the same as the entering volume because the fluid is incompressible, with an assumed constant density ρ.

Let y_1, v_1, and p_1 be the elevation, speed, and pressure of the fluid entering at the left, and y_2, v_2, and p_2 be the corresponding quantities for the fluid emerging at the right. By applying the principle of conservation of energy to the fluid, we shall show that these quantities are related by

$$p_1 + \tfrac{1}{2}\rho v_1^2 + \rho g y_1 = p_2 + \tfrac{1}{2}\rho v_2^2 + \rho g y_2. \qquad (15\text{-}28)$$

We can also write this equation as

$$p + \tfrac{1}{2}\rho v^2 + \rho g y = \text{a constant} \quad \text{(Bernoulli's equation).} \qquad (15\text{-}29)$$

Equations 15-28 and 15-29 are equivalent forms of **Bernoulli's equation,** after Daniel Bernoulli, who studied fluid flow in the 1700s.* Like the equation of continuity (Eq. 15-24), Bernoulli's equation is not a new principle but simply the reformulation of a familiar principle in a form more suitable to fluid mechanics. As a check, let us apply Bernoulli's equation to fluids at rest, by putting $v_1 = v_2 = 0$ in Eq. 15-28. The result is

$$p_2 = p_1 + \rho g(y_1 - y_2),$$

which is Eq. 15-7 with a slight change in notation.

A major prediction of Bernoulli's equation emerges if we take y to be a constant ($y = 0$, say) so that the fluid does not change elevation as it flows. Equation 15-28 then becomes

$$p_1 + \tfrac{1}{2}\rho v_1^2 = p_2 + \tfrac{1}{2}\rho v_2^2, \qquad (15\text{-}30)$$

which tells us that:

> If the speed of a fluid element increases as it travels along a horizontal streamline, the pressure of the fluid must decrease, and conversely.

Put another way, where the streamlines are relatively close together (that is, where the velocity is relatively great), the pressure is relatively low, and conversely.

The link between a change in speed and a change in pressure makes sense if you consider a fluid element. When the element nears a narrow region, the higher pressure behind it accelerates it so that it then has a greater speed in the narrow region. When it nears a wide region, the higher pressure ahead of it decelerates it so that it then has a lesser speed in the wide region.

Bernoulli's equation is strictly valid only to the extent that the fluid is ideal. If viscous forces are present, thermal energy will be involved. We take no account of this in the derivation that follows.

Proof of Bernoulli's Equation

Let us take as our system the entire volume of the (ideal) fluid shown in Fig. 15-19. We shall apply the principle of conservation of energy to this system as it moves

*For irrotational flow (which we assume), the constant in Eq. 15-29 has the same value for all points within the tube of flow; the points do not have to lie along the same streamline. Similarly, the points 1 and 2 in Eq. 15-28 can lie anywhere within the tube of flow.

from its initial state (Fig. 15-19*a*) to its final state (Fig. 15-19*b*). The fluid lying between the two vertical planes separated by a distance L in Fig. 15-19 does not change its properties during this process; we need be concerned only with changes that take place at the input and output ends.

We apply energy conservation in the form of the work–kinetic energy theorem,

$$W = \Delta K, \tag{15-31}$$

which tells us that the change in the kinetic energy of our system must equal the net work done on the system. The change in kinetic energy results from the change in speed between the ends of the tube and is

$$\begin{aligned}\Delta K &= \tfrac{1}{2}\Delta m\, v_2^2 - \tfrac{1}{2}\Delta m\, v_1^2 \\ &= \tfrac{1}{2}\rho\, \Delta V\, (v_2^2 - v_1^2),\end{aligned} \tag{15-32}$$

in which $\Delta m\ (= \rho\, \Delta V)$ is the mass of the fluid that enters at the input end and leaves at the output end during a small time interval Δt.

The work done on the system arises from two sources. The work W_g done by the gravitational force ($\Delta m\, \vec{g}$) on the fluid of mass Δm during the vertical lift of the mass from the input level to the output level is

$$\begin{aligned}W_g &= -\Delta m\, g(y_2 - y_1) \\ &= -\rho g\, \Delta V\, (y_2 - y_1).\end{aligned} \tag{15-33}$$

This work is negative because the upward displacement and the downward gravitational force have opposite directions.

Work must also be done *on* the system (at the input end) to push the entering fluid into the tube and *by* the system (at the output end) to push forward the fluid that is located ahead of the emerging fluid. In general, the work done by a force of magnitude F, acting on a fluid sample contained in a tube of area A to move the fluid through a distance Δx, is

$$F\, \Delta x = (pA)(\Delta x) = p(A\, \Delta x) = p\, \Delta V.$$

The work done on the system is then $p_1\, \Delta V$, and the work done by the system is $-p_2\, \Delta V$. Their sum W_p is

$$\begin{aligned}W_p &= -p_2\, \Delta V + p_1\, \Delta V \\ &= -(p_2 - p_1)\, \Delta V.\end{aligned} \tag{15-34}$$

The work–kinetic energy theorem of Eq. 15-31 now becomes

$$W = W_g + W_p = \Delta K.$$

Substituting from Eqs. 15-32, 15-33, and 15-34 yields

$$-\rho g\, \Delta V\, (y_2 - y_1) - \Delta V\, (p_2 - p_1) = \tfrac{1}{2}\rho\, \Delta V\, (v_2^2 - v_1^2).$$

This, after a slight rearrangement, matches Eq. 15-28, which we set out to prove.

CHECKPOINT 4: Water flows smoothly through the pipe shown in the figure, descending in the process. Rank the four numbered sections of pipe according to (a) the volume flow rate R_V through them, (b) the flow speed v through them, and (c) the water pressure p within them, greatest first.

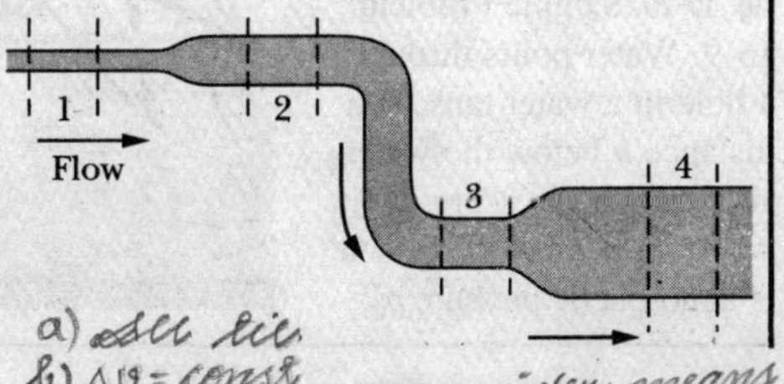

$P_4 > P_3 > P_2 > P_1$ wider & slower means more pressure

a) all tie
b) AV = const
$v_1 > v_2 = v_3 > v_4$
c) 4>3>2>1

wider means slower

Sample Problem 15-8

Ethanol of density $\rho = 791\ kg/m^3$ flows smoothly through a horizontal pipe that tapers (as in Fig. 15-15) in cross-sectional area from $A_1 = 1.20 \times 10^{-3}\ m^2$ to $A_2 = A_1/2$. The pressure difference between the wide and narrow sections of pipe is 4120 Pa. What is the volume flow rate R_V of the ethanol?

SOLUTION: One **Key Idea** here is that, because the fluid flowing through the wide section of pipe must entirely pass through the narrow section, the volume flow rate R_V must be the same in the two sections. Thus, from Eq. 15-24,

$$R_V = v_1 A_1 = v_2 A_2. \tag{15-35}$$

However, with two unknown speeds, we cannot evaluate this equation for R_V.

A second **Key Idea** is that, because the flow is smooth, we can apply Bernoulli's equation. From Eq. 15-28, we can write

$$p_1 + \tfrac{1}{2}\rho v_1^2 + \rho g y = p_2 + \tfrac{1}{2}\rho v_2^2 + \rho g y, \tag{15-36}$$

where subscripts 1 and 2 refer to the wide and narrow sections of pipe, respectively, and y is their common elevation. This equation hardly seems to help because it does not contain the desired volume flow R_V and it contains the unknown speeds v_1 and v_2.

However, there is a neat way to make it work for us: First, we can use Eq. 15-35 and the fact that $A_2 = A_1/2$ to write

$$v_1 = \frac{R_V}{A_1} \quad \text{and} \quad v_2 = \frac{R_V}{A_2} = \frac{2R_V}{A_1}. \tag{15-37}$$

Then we can substitute these expressions into Eq. 15-36 to eliminate the unknown speeds and introduce the desired volume flow rate. Doing this and solving for R_V yield

$$R_V = A_1\sqrt{\frac{2(p_1 - p_2)}{3\rho}}. \tag{15-38}$$

We still have a decision to make: We know that the pressure difference between the two sections is 4120 Pa, but does that mean that $p_1 - p_2$ is 4120 Pa or −4120 Pa? We could guess the former is true, or otherwise the square root in Eq. 15-38 would give us an imaginary number. Instead of guessing, however, let's try some reasoning. From Eq. 15-35 we see that speed v_2 in the narrow section (small A_2) must be greater than speed v_1 in the wider section (larger A_1). Recall that if the speed of a fluid increases as it travels along a horizontal path (as here), the pressure of the fluid must decrease. Thus, p_1 is greater than p_2, and $p_1 - p_2 = 4120$ Pa. Inserting this and known data into Eq. 15-38 gives

$$R_V = 1.20 \times 10^{-3}\ m^2 \sqrt{\frac{(2)(4120\ Pa)}{(3)(791\ kg/m^3)}}$$
$$= 2.24 \times 10^{-3}\ m^3/s. \quad \text{(Answer)}$$

Sample Problem 15-9

In the old West, a desperado fires a bullet into an open water tank (Fig. 15-20), creating a hole a distance h below the water surface. What is the speed v of the water emerging from the hole?

SOLUTION: A **Key Idea** is that this situation is essentially that of water moving (downward) with speed v_0 through a wide pipe (the tank) of cross-sectional area A and then moving (horizontally) with speed v through a narrow pipe (the hole) of cross-sectional area a. Another **Key Idea** is that, because the water flowing through the wide pipe must entirely pass through the narrow pipe, the volume flow rate R_V must be the same in the two "pipes." Then from Eq. 15-24,

$$R_V = av = Av_0$$

and thus

$$v_0 = \frac{a}{A}v.$$

Because $a \ll A$, we see that $v_0 \ll v$.

A third **Key Idea** is that we can also relate v to v_0 (and to h) through Bernoulli's equation (Eq. 15-28). We take the level of the hole as our reference level for measuring elevations (and thus gravitational potential energy). Noting that the pressure at the top of the tank and at the bullet hole is the atmospheric pressure p_0 (because both places are exposed to the atmosphere), we write Eq. 15-28 as

$$p_0 + \tfrac{1}{2}\rho v_0^2 + \rho g h = p_0 + \tfrac{1}{2}\rho v^2 + \rho g(0). \tag{15-39}$$

(Here the top of the tank is represented by the left side of the equation, and the hole by the right side. The zero on the right indicates that the hole is at our reference level.) Before we solve Eq. 15-39 for v, we can use our result that $v_0 \ll v$ to simplify it: We assume that v_0^2, and thus the term $\tfrac{1}{2}\rho v_0^2$ in Eq. 15-39, is negligible compared to the other terms, and we drop it. Solving the remaining equation for v then yields

$$v = \sqrt{2gh}. \quad \text{(Answer)}$$

This is the same speed that an object would have when falling a height h from rest.

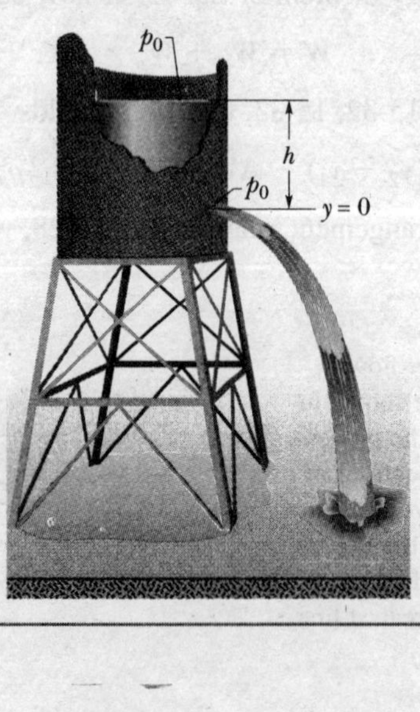

Fig. 15-20 Sample Problem 15-9. Water pours through a hole in a water tank, at a distance h below the water surface. The pressure at the water surface and at the hole is atmospheric pressure p_0.

REVIEW & SUMMARY

Density The **density** ρ of any material is defined as its mass per unit volume:

$$\rho = \frac{\Delta m}{\Delta V}. \quad (15\text{-}1)$$

Usually, where a material sample is large compared with atomic dimensions, we can write Eq. 15-1 as

$$\rho = \frac{m}{V}. \quad (15\text{-}2)$$

Fluid Pressure A **fluid** is a substance that can flow; it conforms to the boundaries of its container because it cannot withstand shearing stress. It can, however, exert a force perpendicular to its surface. That force is described in terms of **pressure** p:

$$p = \frac{\Delta F}{\Delta A}, \quad (15\text{-}3)$$

in which ΔF is the force acting on a surface element of area ΔA. If the force is uniform over a flat area, Eq. 15-3 can be written as

$$p = \frac{F}{A}. \quad (15\text{-}4)$$

The force resulting from fluid pressure at a particular point in a fluid has the same magnitude in all directions. *Gauge pressure* is the difference between the actual pressure (or *absolute pressure*) at a point and the atmospheric pressure.

Pressure Variation with Height and Depth Pressure in a fluid at rest varies with vertical position y. For y measured positive upward,

$$p_2 = p_1 + \rho g(y_1 - y_2). \quad (15\text{-}7)$$

The pressure in a fluid is the same for all points at the same level. If h is the *depth* of a fluid sample below some reference level at which the pressure is p_0, Eq. 15-7 becomes

$$p = p_0 + \rho gh, \quad (15\text{-}8)$$

where p is the pressure in the sample.

Pascal's Principle *Pascal's principle,* which can be derived from Eq. 15-7, states that a change in the pressure applied to an enclosed fluid is transmitted undiminished to every portion of the fluid and to the walls of the containing vessel.

Archimedes' Principle When a body is fully or partially submerged in a fluid, a buoyant force $\vec{F}_b$ from the surrounding fluid acts on the body. The force is directed upward and has a magnitude given by

$$F_b = m_f g, \quad (15\text{-}16)$$

where m_f is the mass of the fluid that has been displaced by the body.

When a body floats in a fluid, the magnitude F_b of the (upward) buoyant force on the body is equal to the magnitude F_g of the (downward) gravitational force on the body. The apparent weight of a body on which a buoyant force acts is related to its actual weight by

$$\text{weight}_{\text{app}} = \text{weight} - F_b. \quad (15\text{-}19)$$

Flow of Ideal Fluids An *ideal fluid* is incompressible and lacks viscosity, and its flow is steady and irrotational. A **streamline** is the path followed by an individual fluid particle. A *tube of flow* is a bundle of streamlines. The flow within any tube of flow obeys the **equation of continuity:**

$$R_V = Av = \text{a constant}, \quad (15\text{-}24)$$

in which R_V is the **volume flow rate**, A is the cross-sectional area of the tube of flow at any point, and v is the speed of the fluid at that point, assumed to be constant across A. The **mass flow rate** R_m is

$$R_m = \rho R_V = \rho Av = \text{a constant}. \quad (15\text{-}25)$$

Bernoulli's Equation Applying the principle of conservation of mechanical energy to the flow of an ideal fluid leads to **Bernoulli's equation:**

$$p + \tfrac{1}{2}\rho v^2 + \rho gy = \text{a constant} \quad (15\text{-}29)$$

along any tube of flow.

QUESTIONS

1. Figure 15-21 shows a tank filled with water. Five horizontal floors and ceilings are indicated; all have the same area and are located at distances L, $2L$, or $3L$ below the top of the tank. Rank the floors and ceilings according to the force on them due to the water, greatest first.

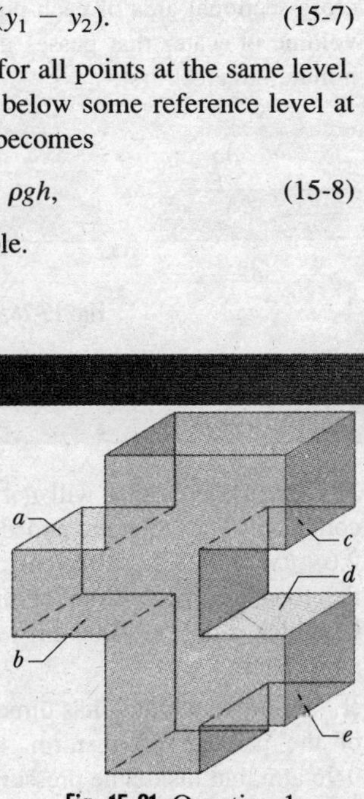

Fig. 15-21 Question 1.

2. *The Teapot Effect:* When water is poured slowly from a teapot spout, it can double back under the spout for a considerable distance before detaching and falling. (The water layer is held against the underside of the spout by atmospheric pressure.) In Fig. 15-22, within the water layer in the spout, point a is at the top of the layer

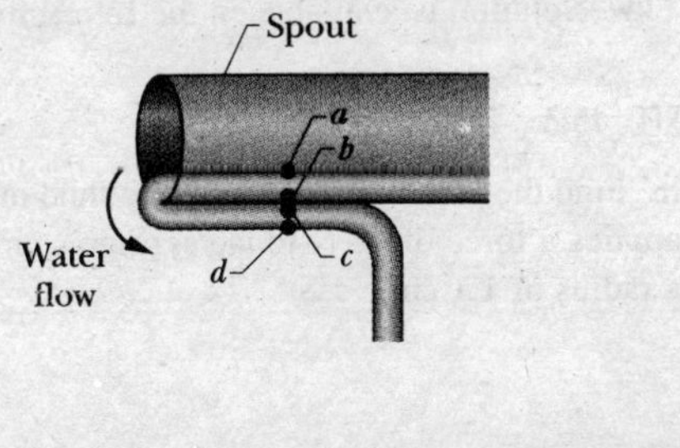

Fig. 15-22 Question 2.

and point b is at the bottom of the layer; within the water layer below the spout, point c is at the top of the layer and point d is at the bottom of the layer. Rank those four points according to the gauge pressure in the water there, most positive first, most negative last.

3. Figure 15-23 shows four situations in which a red liquid and a gray liquid are in a U-tube. In one situation the liquids cannot be in static equilibrium. (a) Which situation is that? (b) For the other three situations, assume static equilibrium. For each, is the density of the red liquid greater than, less than, or equal to the density of the gray liquid?

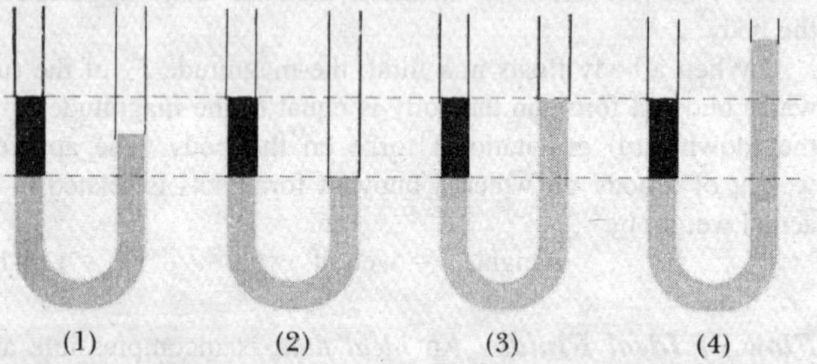

Fig. 15-23 Question 3.

4. Three hydraulic levers such as that in Fig. 15-8 are used to lift identical loads (at the output side) by identical distances. The levels are identical on the input side but differ in the area of the output piston: Lever 1 has a piston of area A, lever 2 has a piston of area $2A$, and lever 3 has a piston of area $3A$. For the lift, rank the levers according to (a) the required work at the input side, (b) the required magnitude of the force (assumed to be constant) at the input side, and (c) the displacement of the piston at the input side, greatest first.

5. We fully submerge an irregular 3 kg lump of material in a certain fluid. The fluid that would have been in the space now occupied by the lump has a mass of 2 kg. (a) When we release the lump, does it move upward, move downward, or remain in place? (b) If we next fully submerge the lump in a less dense fluid and again release it, what does it do?

6. Figure 15-24 shows four solid objects floating in corn syrup. Rank the objects according to their density, greatest first.

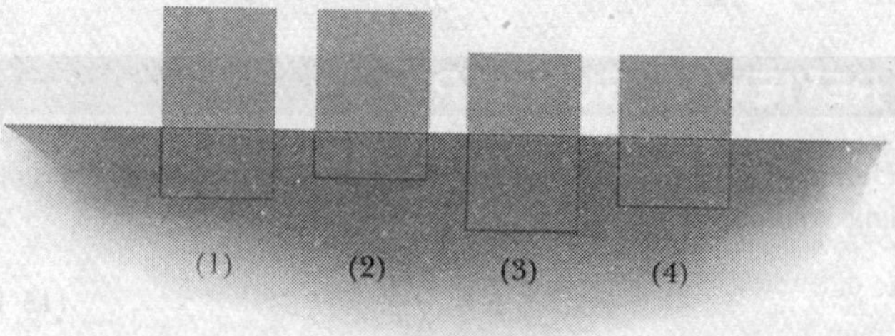

Fig. 15-24 Question 6.

7. Figure 15-25 shows three identical open-top containers filled to the brim with water; toy ducks float in two of them. Rank the containers and contents according to their weight, greatest first.

Fig. 15-25 Question 7.

8. A block floats in a pail of water in a stationary elevator. Does the block float higher, lower, or at the same level when the elevator cab (a) moves upward at constant speed, (b) moves downward at constant speed, (c) accelerates upward, and (d) accelerates downward with an acceleration magnitude less than g?

9. A boat with an anchor on board floats in a swimming pool that is somewhat wider than the boat. Does the water level in the pool move upward, move downward, or remain the same if the anchor is (a) dropped into the water or (b) thrown onto the surrounding ground? (c) Does the water level in the pool move upward, move downward, or remain the same if, instead, a cork is dropped from the boat into the water, where it floats?

10. Figure 15-26 shows three straight pipes through which water flows. The figure gives the speed of the water in each pipe and the cross-sectional area of each pipe. Rank the pipes according to the volume of water that passes through the cross-sectional area per minute, greatest first.

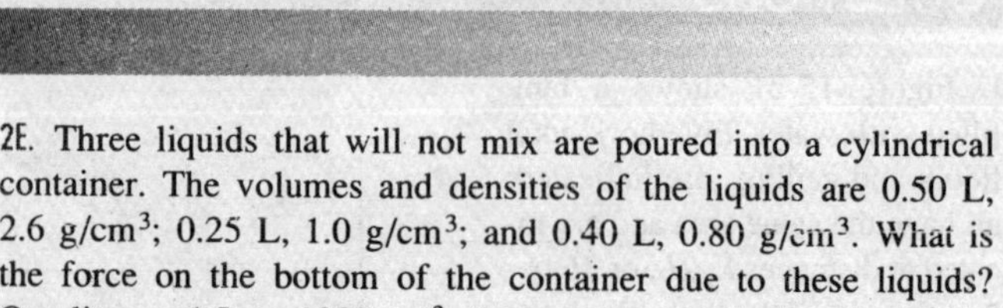

Fig. 15-26 Question 10.

EXERCISES & PROBLEMS

ssm Solution is in the Student Solutions Manual.
www Solution is available on the World Wide Web at: http://www.wiley.com/college/hrw
ilw Solution is available on the Interactive LearningWare.

SEC. 15-3 Density and Pressure

1E. Find the pressure increase in the fluid in a syringe when a nurse applies a force of 42 N to the syringe's circular piston, which has a radius of 1.1 cm. **ssm**

2E. Three liquids that will not mix are poured into a cylindrical container. The volumes and densities of the liquids are 0.50 L, 2.6 g/cm^3; 0.25 L, 1.0 g/cm^3; and 0.40 L, 0.80 g/cm^3. What is the force on the bottom of the container due to these liquids? One liter = 1 L = 1000 cm^3. (Ignore the contribution due to the atmosphere.)

3E. An office window has dimensions 3.4 m by 2.1 m. As a result of the passage of a storm, the outside air pressure drops to 0.96 atm, but inside the pressure is held at 1.0 atm. What net force pushes out on the window? **ssm**

4E. You inflate the front tires on your car to 28 psi. Later, you measure your blood pressure, obtaining a reading of 120/80, the readings being in mm Hg. In metric countries (which is to say, most of the world), these pressures are customarily reported in kilopascals (kPa). In kilopascals, what are (a) your tire pressure and (b) your blood pressure?

5E. A fish maintains its depth in fresh water by adjusting the air content of porous bone or air sacs to make its average density the same as that of the water. Suppose that with its air sacs collapsed, a fish has a density of 1.08 g/cm^3. To what fraction of its expanded body volume must the fish inflate the air sacs to reduce its density to that of water? **itw**

6P. An airtight container having a lid with negligible mass and an area of 77 cm^2 is partially evacuated. If a 480 N force is required to pull the lid off the container and the atmospheric pressure is 1.0×10^5 Pa, what is the air pressure in the container before it is opened? **ssm**

7P. In 1654 Otto von Guericke, inventor of the air pump, gave a demonstration before the noblemen of the Holy Roman Empire in which two teams of eight horses could not pull apart two evacuated brass hemispheres. (a) Assuming that the hemispheres have thin walls, so that R in Fig. 15-27 may be considered both the inside and outside radius, show that the force $\vec{F}$ required to pull apart the hemispheres has magnitude $F = \pi R^2 \Delta p$, where Δp is the difference between the pressures outside and inside the sphere. (b) Taking R as 30 cm, the inside pressure as 0.10 atm, and the outside pressure as 1.00 atm, find the force magnitude the teams of horses would have had to exert to pull apart the hemispheres. (c) Explain why one team of horses could have proved the point just as well if the hemispheres were attached to a sturdy wall. **ssm** **www**

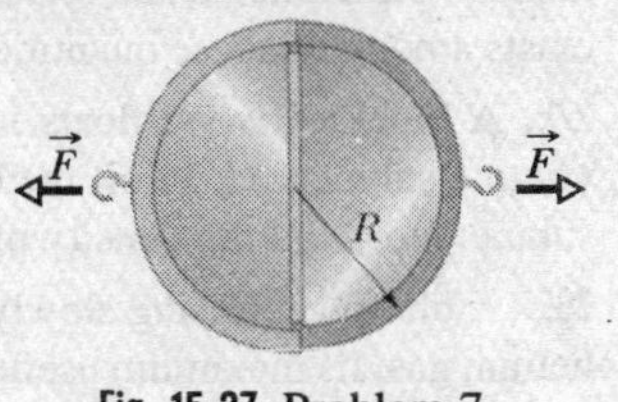

Fig. 15-27 Problem 7.

SEC. 15-4 Fluids at Rest

8E. Calculate the hydrostatic difference in blood pressure between the brain and the foot in a person of height 1.83 m. The density of blood is 1.06×10^3 kg/m^3.

9E. The sewage outlet of a house constructed on a slope is 8.2 m below street level. If the sewer is 2.1 m below street level, find the minimum pressure difference that must be created by the sewage pump to transfer waste of average density 900 kg/m^3 from outlet to sewer. **ssm**

10E. Figure 15-28 displays the *phase diagram* of carbon, showing the ranges of temperature and pressure in which carbon will crystallize either as diamond or graphite. What is the minimum depth at which diamonds can form if the temperature at that depth is 1000°C and the rocks there have density 3.1 g/cm^3? Assume that, as in a fluid, the pressure at any level is due to the gravitational force on the material lying above that level.

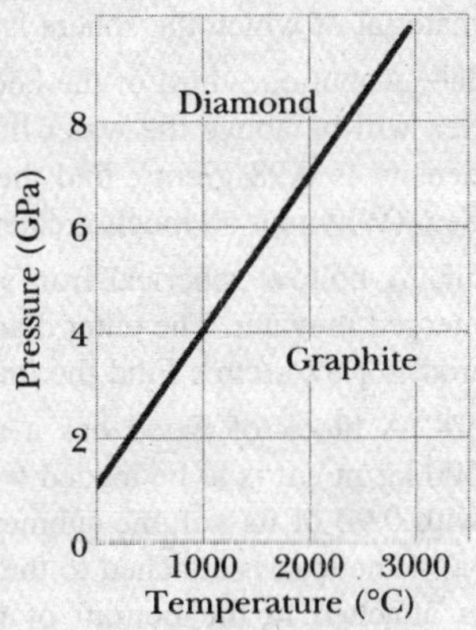

Fig. 15-28 Exercise 10.

11E. A swimming pool has the dimensions 24 m × 9.0 m × 2.5 m. When it is filled with water, what is the force (resulting from the water alone) on (a) the bottom, (b) each short side, and (c) each long side? (d) If you are concerned with the possibility that the concrete walls and floor will collapse, is it appropriate to take the atmospheric pressure into account? Why? **itw**

12E. (a) Assuming the density of seawater is 1.03 g/cm^3, find the total weight of water on top of a nuclear submarine at a depth of 200 m if its (horizontal cross-sectional) hull area is 3000 m^2. (b) In atmospheres, what water pressure would a diver experience at this depth? Do you think that occupants of a damaged submarine at this depth could escape without special equipment?

13E. Crew members attempt to escape from a damaged submarine 100 m below the surface. What force must be applied to a pop-out hatch, which is 1.2 m by 0.60 m, to push it out at that depth? Assume that the density of the ocean water is 1025 kg/m^3. **ssm**

14E. A cylindrical barrel has a narrow tube fixed to the top, as shown (with dimensions) in Fig. 15-29. The vessel is filled with water to the top of the tube. Calculate the ratio of the hydrostatic force on the bottom of the barrel to the gravitational force on the water contained inside the barrel. Why is that ratio not equal to one? (You need not consider the atmospheric pressure.)

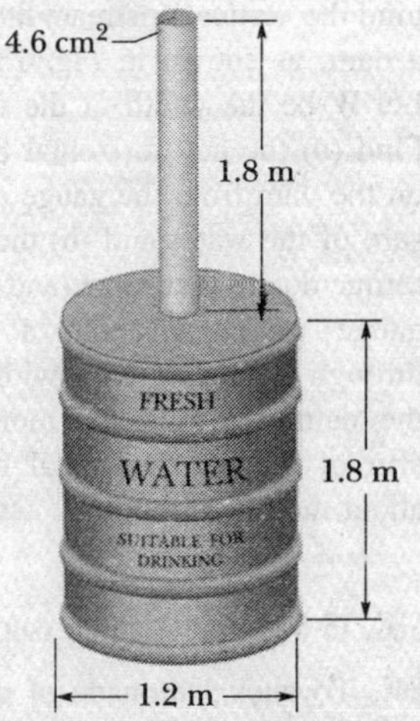

Fig. 15-29 Exercise 14.

15P. Two identical cylindrical vessels with their bases at the same level each contain a liquid of density ρ. The area of each base is A, but in one vessel the liquid height is h_1, and in the other it is h_2. Find the work done by the gravitational force in equalizing the levels when the two vessels are connected. **ssm**

16P. In analyzing certain geological features, it is often appropriate to assume that the pressure at some horizontal *level of compensation,* deep inside Earth, is the same over a large region and is equal to the pressure due to the gravitational force on the overlying material. Thus, the pressure on the level of compensation is given by the fluid pressure formula. This model requires, for one thing, that mountains have *roots* of continental rock extending into the denser mantle (Fig. 15-30). Consider a mountain

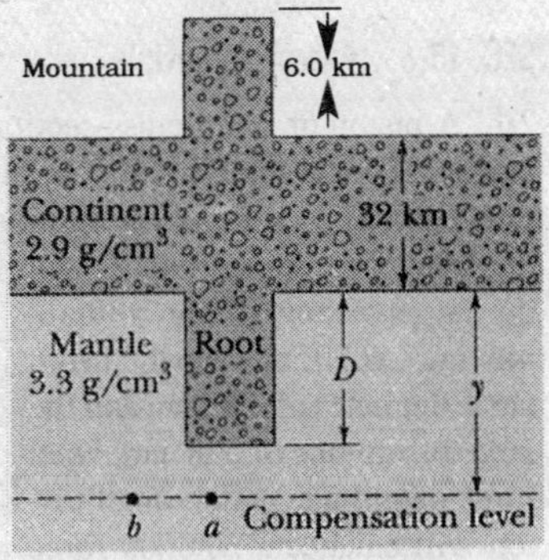

Fig. 15-30 Problem 16.

6.0 km high. The continental rocks have a density of 2.9 g/cm^3, and beneath the continent the mantle has a density of 3.3 g/cm^3. Calculate the depth D of the root. (*Hint:* Set the pressure at points a and b equal; the depth y of the level of compensation will cancel out.)

17P. Figure 15-31 shows the juncture of ocean and continent. Find the depth h of the ocean using the level-of-compensation technique presented in Problem 16. ssm www

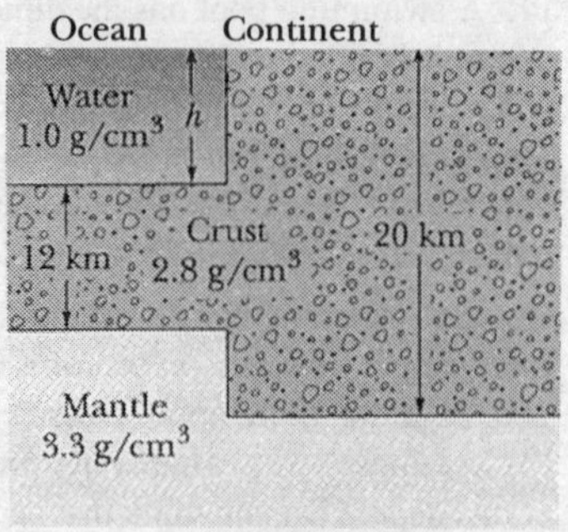

Fig. 15-31 Problem 17.

18P. The L-shaped tank shown in Fig. 15-32 is filled with water and is open at the top. If $d = 5.0$ m, what are (a) the force on face A and (b) the force on face B due to the water?

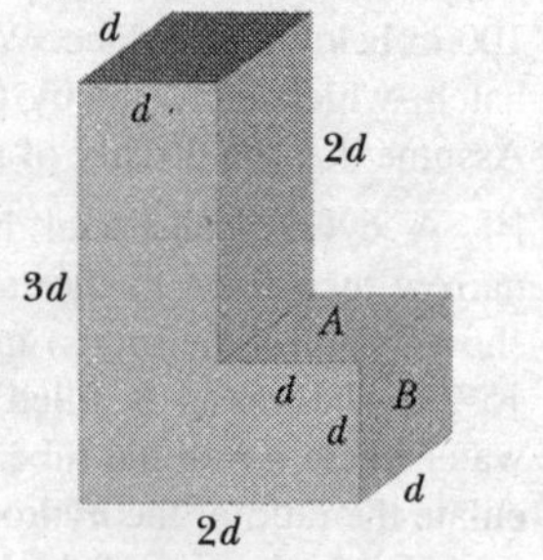

Fig. 15-32 Problem 18.

19P. Water stands at a depth D behind the vertical upstream face of a dam, as shown in Fig. 15-33. Let W be the width of the dam. Find (a) the net horizontal force on the dam from the gauge pressure of the water and (b) the net torque due to that force (and thus gauge pressure) about a line through O parallel to the width of the dam. (c) Find the moment arm of the net horizontal force about the line through O. ssm

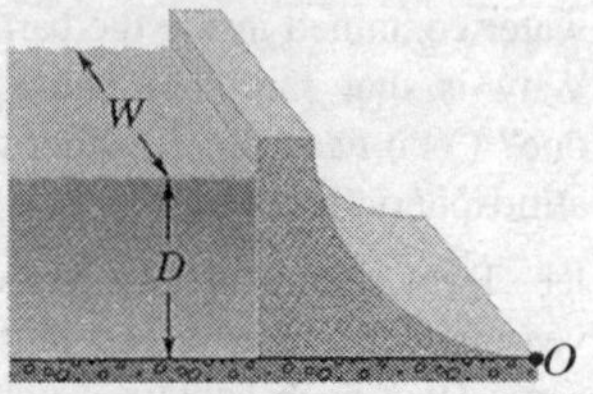

Fig. 15-33 Problem 19.

SEC. 15-5 Measuring Pressure

20E. To suck lemonade of density 1000 kg/m^3 up a straw to a maximum height of 4.0 cm, what minimum gauge pressure (in atmospheres) must you produce in your lungs?

21E. What would be the height of the atmosphere if the air density (a) were uniform and (b) decreased linearly to zero with height? Assume that at sea level the air pressure is 1.0 atm and the air density is 1.3 kg/m^3. ssm

SEC. 15-6 Pascal's Principle

22E. A piston of small cross-sectional area a is used in a hydraulic press to exert a small force $\vec{f}$ on the enclosed liquid. A connecting pipe leads to a larger piston of cross-sectional area A (Fig. 15-34). (a) What force magnitude F will the larger piston sustain without moving? (b) If the small piston has a diameter of 3.80 cm and the large piston one of 53.0 cm, what force magnitude on the small piston will balance a 20.0 kN force on the large piston? ssm

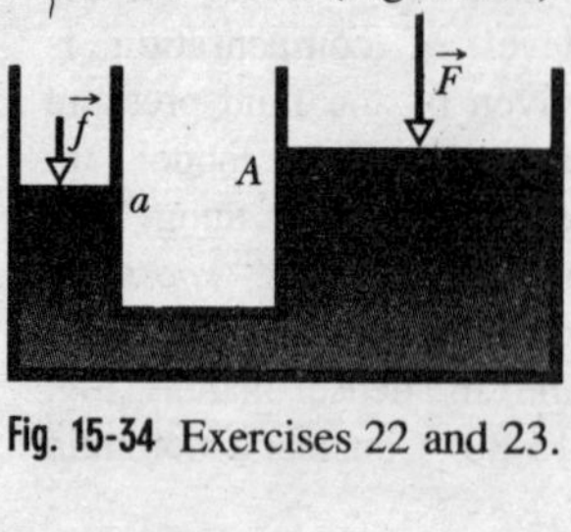

Fig. 15-34 Exercises 22 and 23.

23E. In the hydraulic press of Exercise 22, through what distance must the large piston be moved to raise the small piston a distance of 0.85 m? ssm

SEC. 15-7 Archimedes' Principle

24E. A boat floating in fresh water displaces water weighing 35.6 kN. (a) What is the weight of the water that this boat would displace if it were floating in salt water with a density of 1.10×10^3 kg/m^3? (b) Would the volume of the displaced water change? If so, by how much?

25E. An iron anchor of density 7870 kg/m^3 appears 200 N lighter in water than in air. (a) What is the volume of the anchor? (b) How much does it weigh in air? ssm

26E. In Fig. 15-35, a cubical object of dimensions $L = 0.600$ m on a side and with a mass of 450 kg is suspended by a rope in an open tank of liquid of density 1030 kg/m^3. (a) Find the magnitude of the total downward force on the top of the object from the liquid and the atmosphere, assuming that atmospheric pressure is 1.00 atm. (b) Find the magnitude of the total upward force on the bottom of the object. (c) Find the tension in the rope. (d) Calculate the magnitude of the buoyant force on the object using Archimedes' principle. What relation exists among all these quantities?

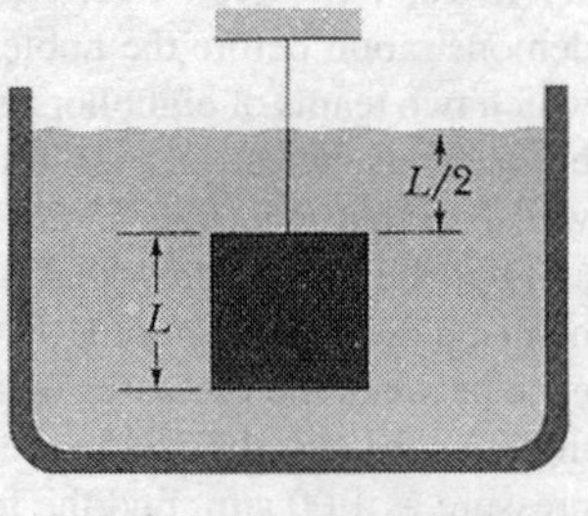

Fig. 15-35 Exercise 26.

27E. A block of wood floats in fresh water with two-thirds of its volume submerged. In oil the block floats with 0.90 of its volume submerged. Find the density of (a) the wood and (b) the oil. ssm

28E. A blimp is cruising slowly at low altitude, filled as usual with helium gas. Its maximum useful payload, including crew and cargo, is 1280 kg. The volume of the helium-filled interior space is 5000 m^3. The density of helium gas is 0.16 kg/m^3, and the density of hydrogen is 0.081 kg/m^3. How much more payload could the blimp carry if you replaced the helium with hydrogen? (Why not do it?)

29P. A hollow sphere of inner radius 8.0 cm and outer radius 9.0 cm floats half-submerged in a liquid of density 800 kg/m^3. (a) What is the mass of the sphere? (b) Calculate the density of the material of which the sphere is made. ssm www

30P. About one-third of the body of a person floating in the Dead Sea will be above the water line. Assuming that the human body density is 0.98 g/cm^3, find the density of the water in the Dead Sea. (Why is it so much greater than 1.0 g/cm^3?)

31P. A hollow spherical iron shell floats almost completely submerged in water. The outer diameter is 60.0 cm, and the density of iron is 7.87 g/cm^3. Find the inner diameter. ilw

32P. A block of wood has a mass of 3.67 kg and a density of 600 kg/m^3. It is to be loaded with lead so that it will float in water with 0.90 of its volume submerged. What mass of lead is needed (a) if the lead is attached to the top of the wood and (b) if the lead is attached to the bottom of the wood? The density of lead is 1.13×10^4 kg/m^3.

33P. An iron casting containing a number of cavities weighs 6000 N in air and 4000 N in water. What is the total volume of all the cavities in the casting? The density of iron (that is, a sample with no cavities) is 7.87 g/cm^3. ssm

34P. Assume the density of brass weights to be 8.0 g/cm^3 and that of air to be 0.0012 g/cm^3. What percent error arises from neglecting the buoyancy of air in weighing an object of mass m and density ρ on a beam balance, as in Fig. 5-6?

35P. (a) What is the minimum area of the top surface of a slab of ice 0.30 m thick floating on fresh water that will hold up an automobile of mass 1100 kg? (b) Does it matter where the car is placed on the block of ice? ssm

36P. Three children, each of weight 356 N, make a log raft by lashing together logs of diameter 0.30 m and length 1.80 m. How many logs will be needed to keep them afloat in fresh water? Take the density of the logs to be 800 kg/m^3.

37P. A metal rod of length 80 cm and mass 1.6 kg has a uniform cross-sectional area of 6.0 cm^2. Due to a nonuniform density, the center of mass of the rod is 20 cm from one end of the rod. The rod is suspended in a horizontal position in water by ropes attached to both ends (Fig. 15-36). (a) What is the tension in the rope closer to the center of mass? (b) What is the tension in the rope farther from the center of mass? (*Hint:* The buoyancy force on the rod effectively acts at the rod's center.) ssm

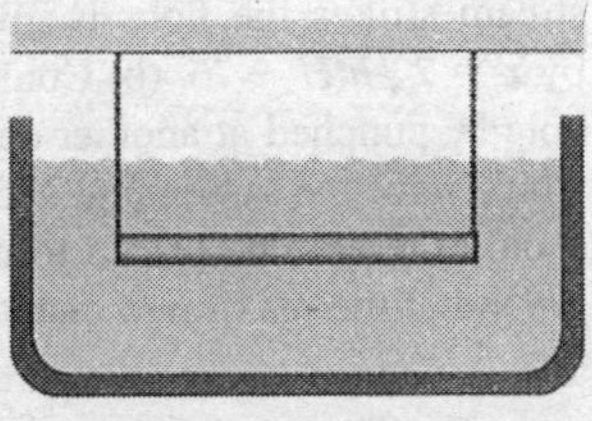

Fig. 15-36 Problem 37.

38P. A car has a total mass of 1800 kg. The volume of air space in the passenger compartment is 5.00 m^3. The volume of the motor and front wheels is 0.750 m^3, and the volume of the rear wheels, gas tank, and trunk is 0.800 m^3; water cannot enter these areas. The car is parked on a hill; the handbrake cable snaps and the car rolls down the hill into a lake (Fig. 15-37). (a) At first, no water enters the passenger compartment. How much of the car, in cubic meters, is below the water surface with the car floating as shown? (b) As water slowly enters, the car sinks. How many cubic meters of water are in the car as it disappears below the water surface? (The car, with a heavy load in the trunk, remains horizontal.)

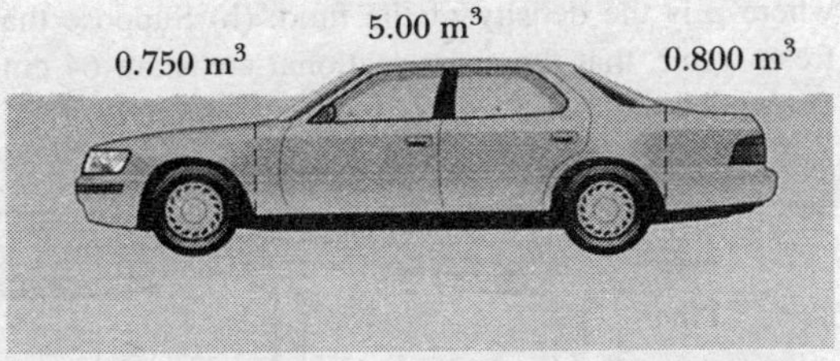

Fig. 15-37 Problem 38.

SEC. 15-9 The Equation of Continuity

39E. A garden hose with an internal diameter of 1.9 cm is connected to a (stationary) lawn sprinkler that consists merely of an enclosure with 24 holes, each 0.13 cm in diameter. If the water in the hose has a speed of 0.91 m/s, at what speed does it leave the sprinkler holes? ssm

40E. Figure 15-38 shows the merging of two streams to form a river. One stream has a width of 8.2 m, depth of 3.4 m, and current speed of 2.3 m/s. The other stream is 6.8 m wide and 3.2 m deep, and flows at 2.6 m/s. The width of the river is 10.5 m, and the current speed is 2.9 m/s. What is its depth?

Fig. 15-38 Exercise 40.

41P. Water is pumped steadily out of a flooded basement at a speed of 5.0 m/s through a uniform hose of radius 1.0 cm. The hose passes out through a window 3.0 m above the waterline. What is the power of the pump? ssm

42E. The water flowing through a 1.9 cm (inside diameter) pipe flows out through three 1.3 cm pipes. (a) If the flow rates in the three smaller pipes are 26, 19, and 11 L/min, what is the flow rate in the 1.9 cm pipe? (b) What is the ratio of the speed of water in the 1.9 cm pipe to that in the pipe carrying 26 L/min?

SEC. 15-10 Bernoulli's Equation

43E. Water is moving with a speed of 5.0 m/s through a pipe with a cross-sectional area of 4.0 cm^2. The water gradually descends 10 m as the pipe increases in area to 8.0 cm^2. (a) What is the speed at the lower level? (b) If the pressure at the upper level is 1.5×10^5 Pa, what is the pressure at the lower level? ssm

44E. Models of torpedoes are sometimes tested in a horizontal pipe of flowing water, much as a wind tunnel is used to test model airplanes. Consider a circular pipe of internal diameter 25.0 cm and a torpedo model, aligned along the axis of the pipe, with a diameter of 5.00 cm. The model is to be tested with water flowing past it at 2.50 m/s. (a) With what speed must the water flow in the part of the pipe that is unconstricted by the model? (b) What will the pressure difference be between the constricted and unconstricted parts of the pipe?

45E. A water pipe having a 2.5 cm inside diameter carries water into the basement of a house at a speed of 0.90 m/s and a pressure of 170 kPa. If the pipe tapers to 1.2 cm and rises to the second floor 7.6 m above the input point, what are (a) the speed and (b) the water pressure at the second floor? ilw

46E. A water intake at a pump storage reservoir (Fig. 15-39) has a cross-sectional area of $0.74\ m^2$. The water flows in at a speed of 0.40 m/s. At the generator building 180 m below the intake point, the cross-sectional area is smaller than at the intake and the water flows out at 9.5 m/s. What is the difference in pressure, in megapascals, between inlet and outlet?

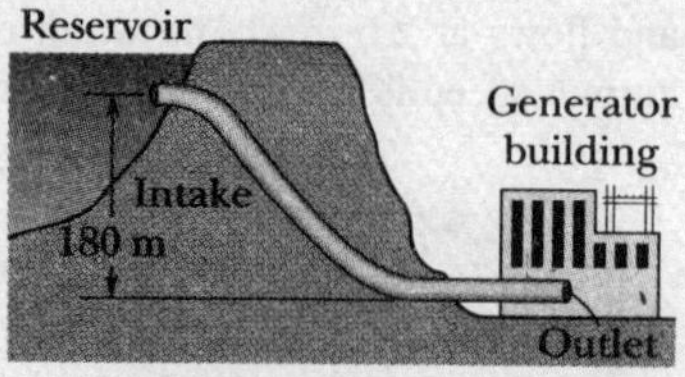

Fig. 15-39 Exercise 46.

47E. A tank of large area is filled with water to a depth $D = 0.30$ m. A hole of cross-sectional area $A = 6.5\ cm^2$ in the bottom of the tank allows water to drain out. (a) What is the rate at which water flows out, in cubic meters per second? (b) At what distance below the bottom of the tank is the cross-sectional area of the stream equal to one-half the area of the hole? ssm

48E. Air flows over the top of an airplane wing of area A with speed v_t and past the underside of the wing (also of area A) with speed v_u. Show that in this simplified situation Bernoulli's equation predicts that the magnitude L of the upward lift force on the wing will be

$$L = \tfrac{1}{2}\rho A(v_t^2 - v_u^2),$$

where ρ is the density of the air.

49E. If the speed of flow past the lower surface of an airplane wing is 110 m/s, what speed of flow over the upper surface will give a pressure difference of 900 Pa between upper and lower surfaces? Take the density of air to be $1.30 \times 10^{-3}\ g/cm^3$, and see Exercise 48. ssm

50E. Suppose that two tanks, 1 and 2, each with a large opening at the top, contain different liquids. A small hole is made in the side of each tank at the same depth h below the liquid surface, but the hole in tank 1 has half the cross-sectional area of the hole in tank 2. (a) What is the ratio ρ_1/ρ_2 of the densities of the liquids if the mass flow rate is the same for the two holes? (b) What is the ratio of the volume flow rates from the two tanks? (c) To what height above the hole in the second tank should liquid be added or drained to equalize the volume flow rates?

51P. In Fig. 15-40, water flows through a horizontal pipe, and then out into the atmosphere at a speed of 15 m/s. The diameters of the left and right sections of the pipe are 5.0 cm and 3.0 cm, respectively. (a) What volume of water flows into the atmosphere during a 10 min period? In the left section of the pipe, what are (b) the speed v_2, and (c) the gauge pressure?

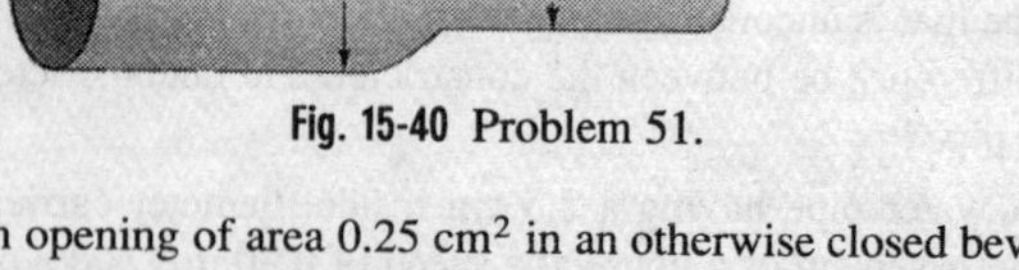

Fig. 15-40 Problem 51.

52P. An opening of area $0.25\ cm^2$ in an otherwise closed beverage keg is 50 cm below the level of the liquid (of density $1.0\ g/cm^3$) in the keg. What is the speed of the liquid flowing through the opening if the gauge pressure in the air space above the liquid is (a) zero and (b) 0.40 atm?

53P. The fresh water behind a reservoir dam is 15 m deep. A horizontal pipe 4.0 cm in diameter passes through the dam 6.0 m below the water surface, as shown in Fig. 15-41. A plug secures the pipe opening. (a) Find the magnitude of the frictional force between plug and pipe wall. (b) The plug is removed. What volume of water flows out of the pipe in 3.0 h? ilw

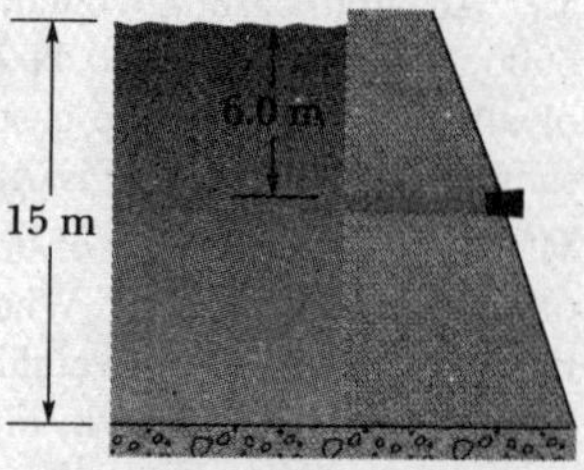

Fig. 15-41 Problem 53.

54P. A tank is filled with water to a height H. A hole is punched in one of the walls at a depth h below the water surface (Fig. 15-42). (a) Show that the distance x from the base of the tank to the point at which the resulting stream strikes the floor is given by $x = 2\sqrt{h(H-h)}$. (b) Could a hole be punched at another depth to produce a second stream that would have the same range? If so, at what depth? (c) At what depth should the hole be placed to make the emerging stream strike the ground at the maximum distance from the base of the tank?

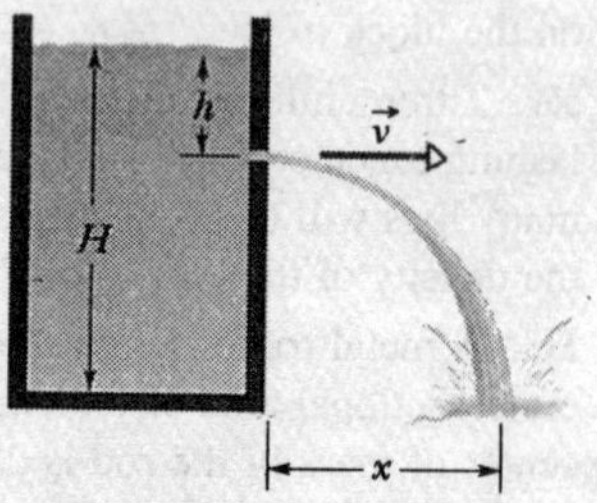

Fig. 15-42 Problem 54.

55P. A *venturi meter* is used to measure the flow speed of a fluid in a pipe. The meter is connected between two sections of the pipe (Fig. 15-43); the cross-sectional area A of the entrance and exit of the meter matches the pipe's cross-sectional area. Between the entrance and exit, the fluid flows from the pipe with speed V and then through a narrow "throat" of cross-sectional area a with speed v. A manometer connects the wider portion of the meter to the narrower portion. The change in the fluid's speed is accompanied by a change Δp in the fluid's pressure, which causes a height difference h of the liquid in the two arms of the manometer. (Here Δp means pressure in the throat minus pressure in the pipe.) (a) By applying Bernoulli's equation and the equation of continuity to points 1 and 2 in Fig. 15-43, show that

$$V = \sqrt{\frac{2a^2\,\Delta p}{\rho(a^2 - A^2)}},$$

where ρ is the density of the fluid. (b) Suppose that the fluid is fresh water, that the cross-sectional areas are $64\ cm^2$ in the pipe

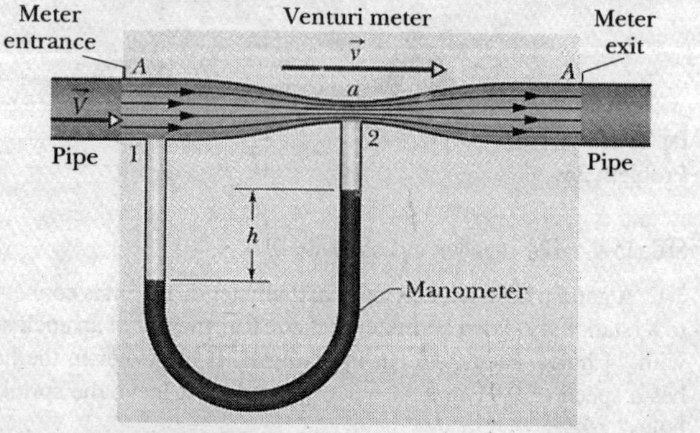

Fig. 15-43 Problems 55 and 56.

and 32 cm^2 in the throat, and that the pressure is 55 kPa in the pipe and 41 kPa in the throat. What is the rate of water flow in cubic meters per second? ssm www

56P. Consider the venturi tube of Problem 55 and Fig. 15-43 without the manometer. Let A equal $5a$. Suppose that the pressure p_1 at A is 2.0 atm. Compute the values of (a) V at A and (b) v at a that would make the pressure p_2 at a equal to zero. (c) Compute the corresponding volume flow rate if the diameter at A is 5.0 cm. The phenomenon that occurs at a when p_2 falls to nearly zero is known as cavitation. The water vaporizes into small bubbles.

57P. A pitot tube (Fig. 15-44) is used to determine the airspeed of an airplane. It consists of an outer tube with a number of small holes B (four are shown) that allow air into the tube; that tube is connected to one arm of a U-tube. The other arm of the U-tube is connected to hole A at the front end of the device, which points in the direction the plane is headed. At A the air becomes stagnant so that $v_A = 0$. At B, however, the speed of the air presumably equals the airspeed v of the aircraft. (a) Use Bernoulli's equation to show that

$$v = \sqrt{\frac{2\rho g h}{\rho_{air}}},$$

where ρ is the density of the liquid in the U-tube and h is the difference in the fluid levels in that tube. (b) Suppose that the tube contains alcohol and indicates a level difference h of 26.0 cm. What is the plane's speed relative to the air? The density of the air is 1.03 kg/m^3 and that of alcohol is 810 kg/m^3.

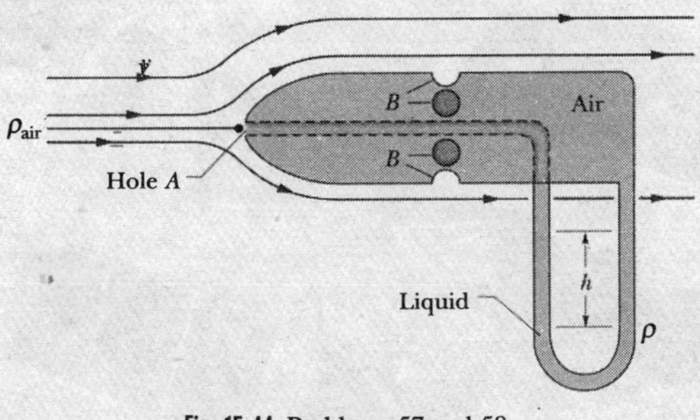

Fig. 15-44 Problems 57 and 58.

58P. A pitot tube (see Problem 57) on a high-altitude aircraft measures a differential pressure of 180 Pa. What is the airspeed if the density of the air is 0.031 kg/m^3?

Additional Problems

59. The dinosaur *Diplodocus* was enormous, with long neck and tail and a mass that was great enough to test its leg strength. According to conjecture, *Diplodocus* waded in water, perhaps up to its head, so that buoyancy could offset its weight and lighten the load on its legs. To check the conjecture, take the density of *Diplodocus* to be 0.90 that of water, and assume that its mass was the published estimate of 1.85×10^4 kg. (a) What then would be its actual weight? Find its apparent weight when it had the following fractions of its volume submerged: (b) 0.50, (c) 0.80, and (d) 0.90. When almost fully submerged, with only its head above water, its lungs would have been about 8.0 m below the water surface. (e) At that depth, what would be the difference between the (external) water pressure and the pressure of the air in its lungs? For the dinosaur to breathe in, its lung muscles would have had to expand its lungs against this pressure difference. It probably could not do so against a pressure difference of more than 8 kPa. (f) Did *Diplodocus* wade as conjectured?

60. When you cough, you expel air at high speed through the trachea and upper bronchi so that the air will remove excess mucus lining the pathway. You produce the high speed by this procedure: You breathe in a large amount of air, trap it by closing the glottis (the narrow opening in the larynx), increase the air pressure by contracting the lungs, partially collapse the trachea and upper bronchi to narrow the pathway, and then expel the air through the pathway by suddenly reopening the glottis. Assume that during the expulsion the volume flow rate is 7.0×10^{-3} m^3/s. In terms of the speed of sound $v_s = 343$ m/s, what is the air speed through the trachea if the trachea diameter (a) remains its normal diameter of 14 mm and (b) contracts to a diameter of 5.2 mm?

61. Suppose that your body has a uniform density of 0.95 times that of water. (a) If you float in a swimming pool, what fraction of your body's volume is above the water surface?

Quicksand is a fluid that is produced when water is forced up into sand, moving the sand grains away from one another so they are no longer locked together by friction. Pools of quicksand can occur when water drains underground from hills into valleys with sand pockets. (b) If you step into a deep pool of quicksand with a density 1.6 times that of water, what fraction of your body's volume is then above the quicksand surface? (c) In particular, are you submerged enough to be unable to breathe? The viscosity of quicksand dramatically increases during any quick movements (the fluid is said to be *thixotropic*). Thus, if you struggle to escape from quicksand, the quicksand holds you tighter. (d) How might you escape without someone helping?

62. In a sink with a flat bottom, turn on a sink faucet so that a smoothly flowing (laminar) stream strikes the bottom. The water spreads from the impact point in a shallow layer but then, at a certain radius r_J from the impact point, it suddenly increases in depth. This depth change, called a *hydraulic jump*, forms a prominent circle around the impact point. Inside the circle, the speed v_1 of the spreading water is constant and is equal to its speed in the falling stream just before impact.

In a certain experiment, the radius of the falling stream is 1.3 mm just before impact, the volume flow rate R_V is 7.9 cm^3/s, the jump radius r_J is 2.0 cm, and the depth just after the jump is 2.0 mm. (a) What is speed v_1? (b) For $r < r_J$, express the water depth d as a function of the radial distance r from the impact point. (c) Does the depth of the water increase or decrease with r? (d) What is the depth just before the water reaches the hydraulic jump? (e) What is the speed v_2 of the water just after the jump? What are the kinetic energy densities (f) just before and (g) just after the jump? (h) What is the change in pressure on the sink due to the jump? (i) Does Bernoulli's equation apply to a streamline through the jump?

16 Oscillations

On September 19, 1985, seismic waves from an earthquake that originated along the west coast of Mexico caused terrible and widespread damage in Mexico City, about 400 km from the origin.

Why did the seismic waves cause such extensive damage in Mexico City but relatively little on the way there?

The answer is in this chapter.

16-1 Oscillations

We are surrounded by oscillations—motions that repeat themselves. There are swinging chandeliers, boats bobbing at anchor, and the surging pistons in the engines of cars. There are oscillating guitar strings, drums, bells, diaphragms in telephones and speaker systems, and quartz crystals in wristwatches. Less evident are the oscillations of the air molecules that transmit the sensation of sound, the oscillations of the atoms in a solid that convey the sensation of temperature, and the oscillations of the electrons in the antennas of radio and TV transmitters that convey information.

Oscillations in the real world are usually *damped;* that is, the motion dies out gradually, transferring mechanical energy to thermal energy by the action of frictional forces. Although we cannot totally eliminate such loss of mechanical energy, we can replenish the energy from some source. As an example, you know that by swinging your legs or torso you can "pump" a swing to maintain or increase the oscillations. In doing this, you transfer biochemical energy to mechanical energy of the oscillating system.

16-2 Simple Harmonic Motion

Figure 16-1*a* shows a sequence of "snapshots" of a simple oscillating system, a particle moving repeatedly back and forth about the origin of an x axis. In this section we simply describe the motion. Later, we shall discuss how to attain such motion.

One important property of oscillatory motion is its **frequency,** or number of oscillations that are completed each second. The symbol for frequency is f, and its SI unit is the **hertz** (abbreviated Hz), where

$$1 \text{ hertz} = 1 \text{ Hz} = 1 \text{ oscillation per second} = 1 \text{ s}^{-1} \tag{16-1}$$

Related to the frequency is the **period** T of the motion, which is the time for one complete oscillation (or **cycle**); that is,

$$T = \frac{1}{f}. \tag{16-2}$$

Any motion that repeats itself at regular intervals is called **periodic motion** or **harmonic motion.** We are interested here in motion that repeats itself in a particular way—namely, like that in Fig. 16-1*a*. For such motion the displacement x of the

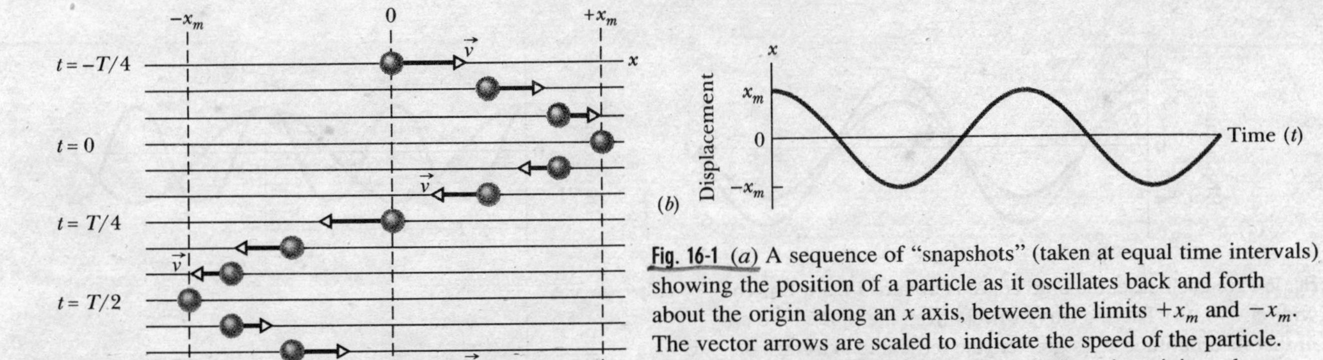

Fig. 16-1 (*a*) A sequence of "snapshots" (taken at equal time intervals) showing the position of a particle as it oscillates back and forth about the origin along an x axis, between the limits $+x_m$ and $-x_m$. The vector arrows are scaled to indicate the speed of the particle. The speed is maximum when the particle is at the origin and zero when it is at $\pm x_m$. If the time t is chosen to be zero when the particle is at $+x_m$, then the particle returns to $+x_m$ at $t = T$, where T is the period of the motion. The motion is then repeated. (*b*) A graph of x as a function of time for the motion of (*a*).

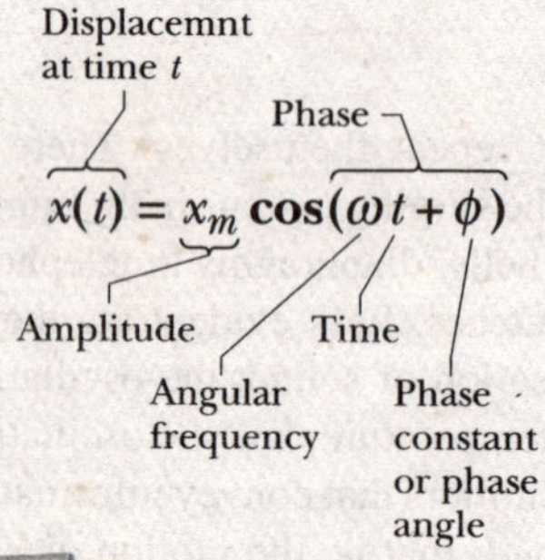

Fig. 16-2 A handy reference to the quantities in Eq. 16-3 for simple harmonic motion.

particle from the origin is given as a function of time by

$$x(t) = x_m \cos(\omega t + \phi) \quad \text{(displacement)}, \tag{16-3}$$

in which x_m, ω, and ϕ are constants. This motion is called **simple harmonic motion** (SHM), a term that means the periodic motion is a sinusoidal function of time. Equation 16-3, in which the sinusoidal function is a cosine function, is graphed in Fig. 16-1*b*. (You can get that graph by rotating Fig. 16-1*a* counterclockwise by 90° and then connecting the successive locations of the particle with a curve.) The quantities that determine the shape of the graph are displayed in Fig. 16-2 with their names. We now shall define those quantities.

The quantity x_m, called the **amplitude** of the motion, is a positive constant whose value depends on how the motion was started. The subscript m stands for *maximum* because the amplitude is the magnitude of the maximum displacement of the particle in either direction. The cosine function in Eq. 16-3 varies between the limits ± 1, so the displacement $x(t)$ varies between the limits $\pm x_m$.

The time-varying quantity $(\omega t + \phi)$ in Eq. 16-3 is called the **phase** of the motion, and the constant ϕ is called the **phase constant** (or **phase angle**). The value of ϕ depends on the displacement and velocity of the particle at time $t = 0$. For the $x(t)$ plots of Fig. 16-3*a*, the phase constant ϕ is zero.

To interpret the constant ω, called the **angular frequency** of the motion, we first note that the displacement $x(t)$ must return to its initial value after one period T of the motion; that is, $x(t)$ must equal $x(t + T)$ for all t. To simplify this analysis, let us put $\phi = 0$ in Eq. 16-3. From that equation we then can write

$$x_m \cos \omega t = x_m \cos \omega(t + T). \tag{16-4}$$

The cosine function first repeats itself when its argument (the phase) has increased by 2π rad, so Eq. 16-4 gives us

$$\omega(t + T) = \omega t + 2\pi$$

or

$$\omega T = 2\pi.$$

Thus, from Eq. 16-2 the angular frequency is

$$\omega = \frac{2\pi}{T} = 2\pi f. \tag{16-5}$$

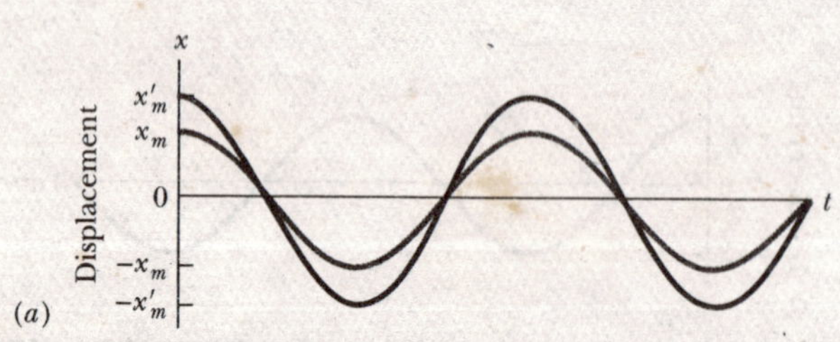

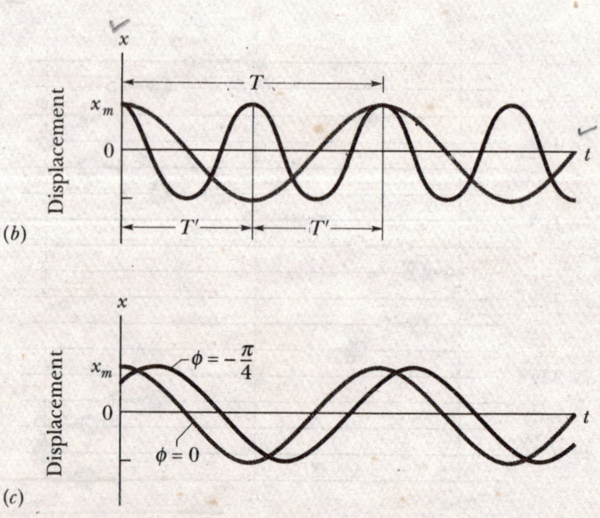

Fig. 16-3 In each case, the blue curve is obtained from Eq. 16-3 with $\phi = 0$. (*a*) The red curve differs from the blue curve *only* in that its amplitude x'_m is greater (the red-curve extremes of displacement are higher and lower). (*b*) The red curve differs from the blue curve *only* in that its period is $T' = T/2$ (the red curve is compressed horizontally). (*c*) The red curve differs from the blue curve *only* in that $\phi = -\pi/4$ rad rather than zero (the negative value of ϕ shifts the red curve to the right).

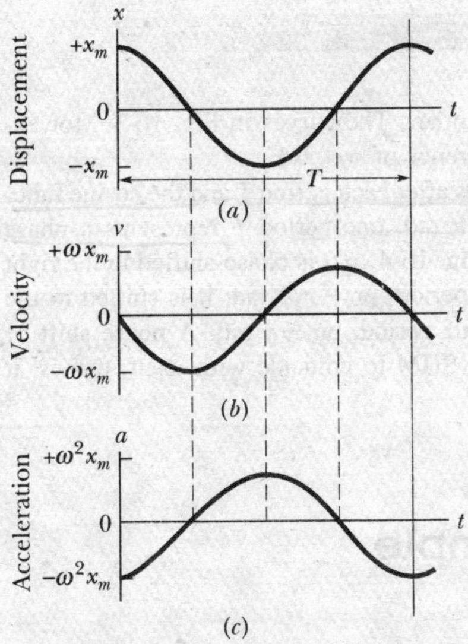

Fig. 16-4 (a) The displacement $x(t)$ of a particle oscillating in SHM with phase angle ϕ equal to zero. The period T marks one complete oscillation. (b) The velocity $v(t)$ of the particle. (c) The acceleration $a(t)$ of the particle.

The SI unit of angular frequency is the radian per second. (To be consistent, then, ϕ must be in radians.) Figure 16-3 compares $x(t)$ for two simple harmonic motions that differ either in amplitude, in period (and thus in frequency and angular frequency), or in phase constant.

CHECKPOINT 1: A particle undergoing simple harmonic oscillation of period T (like that in Fig. 16-1) is at $-x_m$ at time $t = 0$. Is it at $-x_m$, at $+x_m$, at 0, between $-x_m$ and 0, or between 0 and $+x_m$ when (a) $t = 2.00T$, (b) $t = 3.50T$, and (c) $t = 5.25T$?

The Velocity of SHM

By differentiating Eq. 16-3, we can find an expression for the velocity of a particle moving with simple harmonic motion; that is,

$$v(t) = \frac{dx(t)}{dt} = \frac{d}{dt}[x_m \cos(\omega t + \phi)]$$

or

$$v(t) = -\omega x_m \sin(\omega t + \phi) \quad \text{(velocity).} \tag{16-6}$$

Figure 16-4a is a plot of Eq. 16-3 with $\phi = 0$. Figure 16-4b shows Eq. 16-6, also with $\phi = 0$. Analogous to the amplitude x_m in Eq. 16-3, the positive quantity ωx_m in Eq. 16-6 is called the **velocity amplitude** v_m. As you can see in Fig. 16-4b, the velocity of the oscillating particle varies between the limits $\pm v_m = \pm \omega x_m$. Note also in that figure that the curve of $v(t)$ is *shifted* (to the left) from the curve of $x(t)$ by one-quarter period; when the magnitude of the displacement is greatest (that is, $x(t) = x_m$), the magnitude of the velocity is least (that is, $v(t) = 0$). When the magnitude of the displacement is least (that is, zero), the magnitude of the velocity is greatest (that is, $v_m = \omega x_m$).

The Acceleration of SHM

Knowing the velocity $v(t)$ for simple harmonic motion, we can find an expression for the acceleration of the oscillating particle by differentiating once more. Thus, we have, from Eq. 16-6,

$$a(t) = \frac{dv(t)}{dt} = \frac{d}{dt}[-\omega x_m \sin(\omega t + \phi)]$$

or

$$a(t) = -\omega^2 x_m \cos(\omega t + \phi) \quad \text{(acceleration).} \tag{16-7}$$

Figure 16-4c is a plot of Eq. 16-7 for the case $\phi = 0$. The positive quantity $\omega^2 x_m$ in Eq. 16-7 is called the **acceleration amplitude** a_m; that is, the acceleration of the particle varies between the limits $\pm a_m = \pm \omega^2 x_m$, as Fig. 16-4c shows. Note also that the curve of $a(t)$ is shifted (to the left) by $\frac{1}{4}T$ relative to the curve of $v(t)$.

We can combine Eqs. 16-3 and 16-7 to yield

$$a(t) = -\omega^2 x(t), \tag{16-8}$$

which is the hallmark of simple harmonic motion:

> In SHM, the acceleration is proportional to the displacement but opposite in sign, and the two quantities are related by the square of the angular frequency.

Thus, as Fig. 16-4 shows, when the displacement has its greatest positive value, the acceleration has its greatest negative value, and conversely. When the displacement is zero, the acceleration is also zero.

PROBLEM-SOLVING TACTICS

Tactic 1: *Phase Angles*

Note the effect of the phase angle ϕ on a plot of $x(t)$. When $\phi = 0$, $x(t)$ has a graph like that in Fig. 16-4*a*, a typical cosine curve. Increasing ϕ shifts the curve leftward along the t axis. (You might remember this with the symbol $\leftarrow\!\!\uparrow\phi$, where the up arrow indicates an increase in ϕ and the left arrow indicates the resulting shift in the curve.) Decreasing ϕ shifts the curve rightward, as in Fig. 16-3*c* for $\phi = -\pi/4$.

Two plots of SHM with different phase angles are said to have a *phase difference*; each is said to be *phase-shifted* from the other, or *out of phase* with the other. The curves in Fig. 16-3*c*, for example, have a phase difference of $\pi/4$ rad.

Because SHM repeats after each period T and the cosine function repeats after each 2π rad, one period T represents a phase difference of 2π rad. In Fig. 16-4, $x(t)$ is phase-shifted to the right from $v(t)$ by one-quarter period, or $-\pi/2$ rad; it is shifted to the right from $a(t)$ by one-half period, or $-\pi$ rad. A phase shift of 2π rad causes a curve of SHM to coincide with itself; that is, it looks unchanged.

16-3 The Force Law for Simple Harmonic Motion

Once we know how the acceleration of a particle varies with time, we can use Newton's second law to learn what force must act on the particle to give it that acceleration. If we combine Newton's second law and Eq. 16-8, we find, for simple harmonic motion,

$$F = ma = -(m\omega^2)x. \tag{16-9}$$

This result—a restoring force that is proportional to the displacement but opposite in sign—is familiar. It is Hooke's law,

$$F = -kx, \tag{16-10}$$

for a spring, the spring constant here being

$$\boxed{k = m\omega^2.} \tag{16-11}$$

We can in fact take Eq. 16-10 as an alternative definition of simple harmonic motion. It says:

> **Simple harmonic motion is the motion executed by a particle of mass m subject to a force that is proportional to the displacement of the particle but opposite in sign.**

The block–spring system of Fig. 16-5 forms a **linear simple harmonic oscillator** (linear oscillator, for short), where "linear" indicates that F is proportional to x rather than to some other power of x. The angular frequency ω of the simple harmonic motion of the block is related to the spring constant k and the mass m of the block by Eq. 16-11, which yields

$$\omega = \sqrt{\frac{k}{m}} \quad \text{(angular frequency)}. \tag{16-12}$$

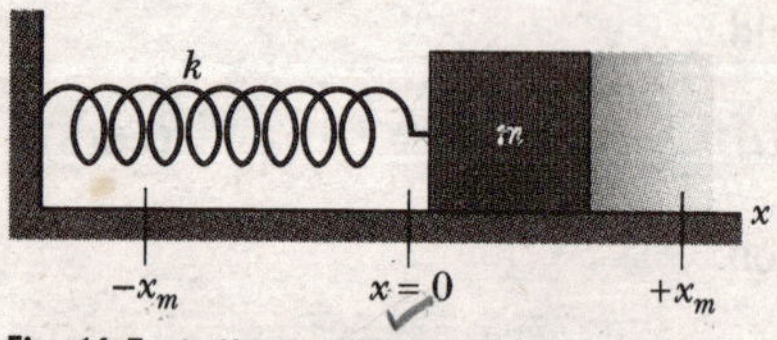

Fig. 16-5 A linear simple harmonic oscillator. The surface is frictionless. Like the particle of Fig. 16-2, the block moves in simple harmonic motion once it has been pulled to the side and released. Its displacement is then given by Eq. 16-3.

By combining Eqs. 16-5 and 16-12, we can write, for the **period** of the linear oscillator of Fig. 16-5,

$$T = 2\pi\sqrt{\frac{m}{k}} \quad \text{(period)}. \tag{16-13}$$

Equations 16-12 and 16-13 tell us that a large angular frequency (and thus a small period) goes with a stiff spring (large k) and a light block (small m).

Every oscillating system, be it the linear oscillator of Fig. 16-5, a diving board, or a violin string, has some element of "springiness" and some element of "inertia" or mass, and thus resembles a linear oscillator. In the linear oscillator of Fig. 16-5, these elements are located in separate parts of the system: The springiness is entirely in the spring, which we assume to be massless, and the inertia is entirely in the block, which we assume to be rigid. In a violin string, however, the two elements are both within the string itself, as you will see in Chapter 17.

✔CHECKPOINT 2: Which of the following relationships between the force F on a particle and the particle's position x implies simple harmonic oscillation: (a) $F = -5x$, (b) $F = -400x^2$, (c) $F = 10x$, (d) $F = 3x^2$?

Sample Problem 16-1

A block whose mass m is 680 g is fastened to a spring whose spring constant k is 65 N/m. The block is pulled a distance $x = 11$ cm from its equilibrium position at $x = 0$ on a frictionless surface and released from rest at $t = 0$.

(a) What are the angular frequency, the frequency, and the period of the resulting motion?

SOLUTION: The **Key Idea** here is that the block–spring system forms a linear simple harmonic oscillator, with the block undergoing SHM. Then the angular frequency is given by Eq. 16-12:

$$\omega = \sqrt{\frac{k}{m}} = \sqrt{\frac{65\text{ N/m}}{0.68\text{ kg}}} = 9.78\text{ rad/s}$$
$$\approx 9.8\text{ rad/s.} \qquad \text{(Answer)}$$

The frequency follows from Eq. 16-5, which yields

$$f = \frac{\omega}{2\pi} = \frac{9.78\text{ rad/s}}{2\pi\text{ rad}} = 1.56\text{ Hz} \approx 1.6\text{ Hz.} \qquad \text{(Answer)}$$

The period follows from Eq. 16-2, which yields

$$T = \frac{1}{f} = \frac{1}{1.56\text{ Hz}} = 0.64\text{ s} = 640\text{ ms.} \qquad \text{(Answer)}$$

(b) What is the amplitude of the oscillation?

SOLUTION: The **Key Idea** here is that, with no friction involved, the mechanical energy of the spring–block system is conserved. The block is released from rest 11 cm from its equilibrium position, with zero kinetic energy and the elastic potential energy of the system at a maximum. Thus, the block will have zero kinetic energy whenever it is again 11 cm from its equilibrium position, which means it will never be farther than 11 cm from that position. Its maximum displacement is 11 cm:

$$x_m = 11\text{ cm.} \qquad \text{(Answer)}$$

(c) What is the maximum speed v_m of the oscillating block, and where is the block when it occurs?

SOLUTION: The **Key Idea** here is that the maximum speed v_m is the velocity amplitude ωx_m in Eq. 16-6; that is,

$$v_m = \omega x_m = (9.78\text{ rad/s})(0.11\text{ m})$$
$$= 1.1\text{ m/s.} \qquad \text{(Answer)}$$

This maximum speed occurs when the oscillating block is rushing through the origin; compare Figs. 16-4*a* and 16-4*b*, where you can see that the speed is a maximum whenever $x = 0$.

(d) What is the magnitude a_m of the maximum acceleration of the block?

SOLUTION: The **Key Idea** this time is that the magnitude a_m of the maximum acceleration is the acceleration amplitude $\omega^2 x_m$ in Eq. 16-7; that is,

$$a_m = \omega^2 x_m = (9.78\text{ rad/s})^2(0.11\text{ m})$$
$$= 11\text{ m/s}^2. \qquad \text{(Answer)}$$

This maximum acceleration occurs when the block is at the ends of its path. At those points, the force acting on the block has its maximum magnitude; compare Figs. 16-4*a* and 16-4*c*, where you can see that the magnitudes of the displacement and acceleration are maximum at the same times.

(e) What is the phase constant ϕ for the motion?

SOLUTION: Here the **Key Idea** is that Eq. 16-3 gives the displacement of the block as a function of time. We know that at time $t = 0$, the block is located at $x = x_m$. Substituting these *initial conditions*, as they are called, into Eq. 16-3 and canceling x_m give us

$$1 = \cos\phi. \qquad (16\text{-}14)$$

Taking the inverse cosine then yields

$$\phi = 0\text{ rad.} \qquad \text{(Answer)}$$

(Any angle that is an integer multiple of 2π rad also satisfies Eq. 16-14; we chose the smallest angle.)

(f) What is the displacement function $x(t)$ for the spring–block system?

SOLUTION: The **Key Idea** here is that $x(t)$ is given in general form by Eq. 16-3. Substituting known quantities into that equation gives us

$$x(t) = x_m\cos(\omega t + \phi)$$
$$= (0.11\text{ m})\cos[(9.8\text{ rad/s})t + 0]$$
$$= 0.11\cos(9.8t), \qquad \text{(Answer)}$$

where x is in meters and t is in seconds.

Sample Problem 16-2

At $t = 0$, the displacement $x(0)$ of the block in a linear oscillator like that of Fig. 16-5 is -8.50 cm. (Read $x(0)$ as "x at time zero.") The block's velocity $v(0)$ then is -0.920 m/s, and its acceleration $a(0)$ is $+47.0\ \text{m/s}^2$.

(a) What is the angular frequency ω of this system?

SOLUTION: A **Key Idea** here is that, with the block in SHM, Eqs. 16-3, 16-6, and 16-7 give its displacement, velocity, and acceleration, respectively, and each contains ω. Let's substitute $t = 0$ into each to see whether we can solve any one of them for ω. We find

$$x(0) = x_m \cos \phi, \tag{16-15}$$

$$v(0) = -\omega x_m \sin \phi, \tag{16-16}$$

and

$$a(0) = -\omega^2 x_m \cos \phi. \tag{16-17}$$

In Eq. 16-15, ω has disappeared. In Eqs. 16-16 and 16-17, we know values for the left sides, but we do not know x_m and ϕ. However, if we divide Eq. 16-17 by Eq. 16-15, we neatly eliminate both x_m and ϕ and can then solve for ω as

$$\omega = \sqrt{-\frac{a(0)}{x(0)}} = \sqrt{-\frac{47.0\ \text{m/s}^2}{-0.0850\ \text{m}}}$$

$$= 23.5\ \text{rad/s}. \qquad \text{(Answer)}$$

(b) What are the phase constant ϕ and amplitude x_m?

SOLUTION: The same **Key Idea** as in part (a) also applies here, as do Eqs. 16-15 through 16-17. Now, however, we know ω and want ϕ and x_m. If we divide Eq. 16-16 by Eq. 16-15, we find

$$\frac{v(0)}{x(0)} = \frac{-\omega x_m \sin \phi}{x_m \cos \phi} = -\omega \tan \phi.$$

Solving for $\tan \phi$, we find

$$\tan \phi = -\frac{v(0)}{\omega x(0)} = -\frac{-0.920\ \text{m/s}}{(23.5\ \text{rad/s})(-0.0850\ \text{m})}$$

$$= -0.461.$$

This equation has two solutions:

$$\phi = -25^\circ \quad \text{and} \quad \phi = 180^\circ + (-25^\circ) = 155^\circ.$$

(Normally only the first solution here is displayed by a calculator.) A **Key Idea** in choosing the proper solution is to test them both by using them to compute values for the amplitude x_m. From Eq. 16-15, we find that if $\phi = -25^\circ$, then

$$x_m = \frac{x(0)}{\cos \phi} = \frac{-0.0850\ \text{m}}{\cos(-25^\circ)} = -0.094\ \text{m}.$$

We find similarly that if $\phi = 155^\circ$, then $x_m = 0.094$ m. Because the amplitude of SHM must be a positive constant, the correct phase constant and amplitude here are

$$\phi = 155^\circ \quad \text{and} \quad x_m = 0.094\ \text{m} = 9.4\ \text{cm}. \qquad \text{(Answer)}$$

PROBLEM-SOLVING TACTICS

Tactic 2: *Identifying SHM*

In linear SHM the acceleration a and displacement x of the system are related by an equation of the form

$$a = -(\text{a positive constant})x,$$

which says that the acceleration is proportional to the displacement from the equilibrium position but is in the opposite direction. Once you find such an expression for an oscillating system, you can immediately compare it to Eq. 16-8, identify the positive constant as being equal to ω^2, and so quickly get an expression for the angular frequency of the motion. With Eq. 16-5 you then can find the period T and the frequency f.

In some problems you might derive an expression for the force F as a function of displacement x. If the motion is linear SHM, the force and displacement are related by

$$F = -(\text{a positive constant})x,$$

which says that the force is proportional to the displacement but is in the opposite direction. Once you have found such an expression, you can immediately compare it to Eq. 16-10 and identify the positive constant as being k. If you know the mass that is involved, you can then use Eqs. 16-12, 16-13, and 16-5 to find the angular frequency ω, period T, and frequency f.

16-4 Energy in Simple Harmonic Motion

In Chapter 8 we saw that the energy of a linear oscillator transfers back and forth between kinetic energy and potential energy, while the sum of the two—the mechanical energy E of the oscillator—remains constant. We now consider this situation quantitatively.

The potential energy of a linear oscillator like that of Fig. 16-5 is associated entirely with the spring. Its value depends on how much the spring is stretched or compressed—that is, on $x(t)$. We can use Eqs. 8-11 and 16-3 to find

$$U(t) = \tfrac{1}{2}kx^2 = \tfrac{1}{2}kx_m^2 \cos^2(\omega t + \phi). \tag{16-18}$$

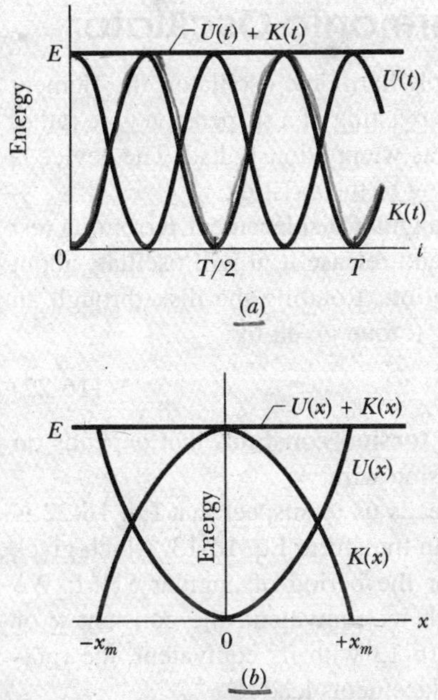

Fig. 16-6 (*a*) Potential energy $U(t)$, kinetic energy $K(t)$, and mechanical energy E as functions of time t for a linear harmonic oscillator. Note that all energies are positive and that the potential energy and the kinetic energy peak twice during every period. (*b*) Potential energy $U(x)$, kinetic energy $K(x)$, and mechanical energy E as functions of position x for a linear harmonic oscillator with amplitude x_m. For $x = 0$ the energy is all kinetic, and for $x = \pm x_m$ it is all potential.

Note carefully that a function written in the form $\cos^2 A$ (as here) means $(\cos A)^2$ and is *not* the same as one written $\cos A^2$, which means $\cos (A^2)$.

The kinetic energy of the system of Fig. 16-5 is associated entirely with the block. Its value depends on how fast the block is moving—that is, on $v(t)$. We can use Eq. 16-6 to find

$$K(t) = \tfrac{1}{2}mv^2 = \tfrac{1}{2}m\omega^2 x_m^2 \sin^2(\omega t + \phi). \tag{16-19}$$

If we use Eq. 16-12 to substitute k/m for ω^2, we can write Eq. 16-19 as

$$K(t) = \tfrac{1}{2}mv^2 = \tfrac{1}{2}kx_m^2 \sin^2(\omega t + \phi). \tag{16-20}$$

The mechanical energy follows from Eqs. 16-18 and 16-20 and is

$$\begin{aligned} E &= U + K \\ &= \tfrac{1}{2}kx_m^2 \cos^2(\omega t + \phi) + \tfrac{1}{2}kx_m^2 \sin^2(\omega t + \phi) \\ &= \tfrac{1}{2}kx_m^2[\cos^2(\omega t + \phi) + \sin^2(\omega t + \phi)]. \end{aligned}$$

For any angle α,

$$\cos^2 \alpha + \sin^2 \alpha = 1.$$

Thus, the quantity in the square brackets above is unity and we have

$$E = U + K = \tfrac{1}{2}kx_m^2. \tag{16-21}$$

The mechanical energy of a linear oscillator is indeed constant and independent of time. The potential energy and kinetic energy of a linear oscillator are shown as functions of time t in Fig. 16-6*a*, and as functions of displacement x in Fig. 16-6*b*.

You might now understand why an oscillating system normally contains an element of springiness and an element of inertia: The former stores its potential energy and the latter stores its kinetic energy.

CHECKPOINT 3: In Fig. 16-5, the block has a kinetic energy of 3 J and the spring has an elastic potential energy of 2 J when the block is at $x = +2.0$ cm. (a) What is the kinetic energy when the block is at $x = 0$? What are the elastic potential energies when the block is at (b) $x = -2.0$ cm and (c) $x = -x_m$?

a) 5 J
b) 2 J
c) 5 J

Sample Problem 16-3

(a) What is the mechanical energy E of the linear oscillator of Sample Problem 16-1? (Initially, the block's position is $x = 11$ cm and its speed is $v = 0$. Spring constant k is 65 N/m.)

SOLUTION: The Key Idea here is that the mechanical energy E (the sum of the kinetic energy $K = \tfrac{1}{2}mv^2$ of the block and the potential energy $U = \tfrac{1}{2}kx^2$ of the spring) is constant throughout the motion of the oscillator. Thus, we can evaluate E at any point during the motion. Because we are given the initial conditions of the oscillator as $x = 11$ cm and $v = 0$, let us evaluate E for those conditions. We find

$$\begin{aligned} E &= K + U = \tfrac{1}{2}mv^2 + \tfrac{1}{2}kx^2 = 0 + \tfrac{1}{2}(65\ \text{N/m})(0.11\ \text{m})^2 \\ &= 0.393\ \text{J} \approx 0.39\ \text{J}. \qquad \text{(Answer)} \end{aligned}$$

(b) What are the potential energy U and kinetic energy K of the oscillator when the block is at $x = \tfrac{1}{2}x_m$? What are they when the block is at $x = -\tfrac{1}{2}x_m$?

SOLUTION: The Key Idea here is that, because we are given the location of the block, we can easily find the spring's potential energy with $U = \tfrac{1}{2}kx^2$. For $x = \tfrac{1}{2}x_m$, we have

$$U = \tfrac{1}{2}kx^2 = \tfrac{1}{2}k(\tfrac{1}{2}x_m)^2 = (\tfrac{1}{2})(\tfrac{1}{4})kx_m^2.$$

We can substitute for k and x_m, or we can use the Key Idea that the total mechanical energy, which we know from part (a), is $\tfrac{1}{2}kx_m^2$. That idea allows us to write, from the above equation,

$$U = \tfrac{1}{4}(\tfrac{1}{2}kx_m^2) = \tfrac{1}{4}E = \tfrac{1}{4}(0.393\ \text{J}) = 0.098\ \text{J}. \qquad \text{(Answer)}$$

Now, using the Key Idea of (a) (namely, $E = K + U$), we can write

$$K = E - U = 0.393\ \text{J} - 0.098\ \text{J} \approx 0.30\ \text{J}. \qquad \text{(Answer)}$$

By repeating these calculations for $x = -\tfrac{1}{2}x_m$, we would find the same answers for that displacement, consistent with the left–right symmetry of Fig. 16-6*b*.

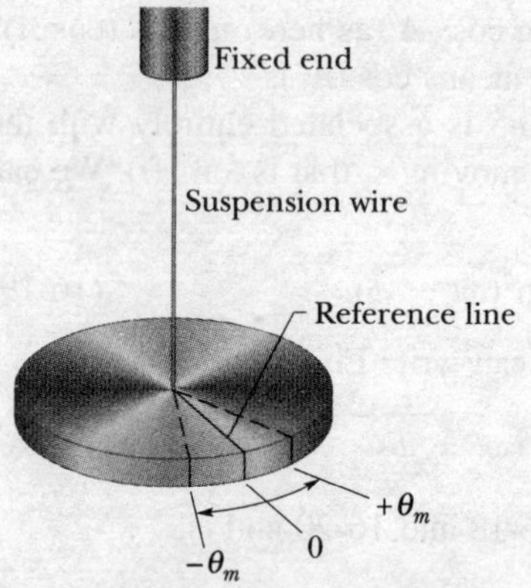

Fig. 16-7 An angular simple harmonic oscillator, or torsion pendulum, is an angular version of the linear simple harmonic oscillator of Fig. 16-5. The disk oscillates in a horizontal plane; the reference line oscillates with angular amplitude θ_m. The twist in the suspension wire stores potential energy as a spring does and provides the restoring torque.

16-5 An Angular Simple Harmonic Oscillator

Figure 16-7 shows an angular version of a simple harmonic oscillator; the element of springiness or elasticity is associated with the twisting of a suspension wire rather than the extension and compression of a spring as we previously had. The device is called a **torsion pendulum,** with *torsion* referring to the twisting.

If we rotate the disk in Fig. 16-7 by some angular displacement θ from its rest position (where the reference line is at $\theta = 0$) and release it, it will oscillate about that position in **angular simple harmonic motion.** Rotating the disk through an angle θ in either direction introduces a restoring torque given by

$$\tau = -\kappa\theta. \tag{16-22}$$

Here κ (Greek *kappa*) is a constant, called the **torsion constant,** that depends on the length, diameter, and material of the suspension wire.

Comparison of Eq. 16-22 with Eq. 16-10 leads us to suspect that Eq. 16-22 is the angular form of Hooke's law, and that we can transform Eq. 16-13, which gives the period of linear SHM, into an equation for the period of angular SHM: We replace the spring constant k in Eq. 16-13 with its equivalent, the constant κ of Eq. 16-22, and we replace the mass m in Eq. 16-13 with *its* equivalent, the rotational inertia I of the oscillating disk. These replacements lead to

$$T = 2\pi\sqrt{\frac{I}{\kappa}} \quad \text{(torsion pendulum)}, \tag{16-23}$$

which is the correct equation for the period of an angular simple harmonic oscillator, or torsion pendulum.

PROBLEM-SOLVING TACTICS

Tactic 3: *Identifying Angular SHM*

When a system undergoes angular simple harmonic motion, its angular acceleration α and angular displacement θ are related by an equation of the form

$$\alpha = -(\text{a positive constant})\theta.$$

This equation is the angular equivalent of Eq. 16-8 ($a = -\omega^2 x$). It says that the angular acceleration α is proportional to the angular displacement θ from the equilibrium position but tends to rotate the system in the direction opposite the displacement. If you have an expression of this form, you can identify the positive constant as being ω^2, and then you can determine ω, f, and T.

You can also identify angular SHM if you have an expression for the torque τ in terms of the angular displacement, because that expression must be in the form of Eq. 16-22 ($\tau = -\kappa\theta$) or

$$\tau = -(\text{a positive constant})\theta.$$

This equation is the angular equivalent of Eq. 16-10 ($F = -kx$). It says that the torque τ is proportional to the angular displacement θ from the equilibrium position but tends to rotate the system in the opposite direction. If you have an expression of this form, then you can identify the positive constant as being the system's torsion constant κ. If you know the rotational inertia I of the system, you can then determine T with Eq. 16-23.

Sample Problem 16-4

Figure 16-8*a* shows a thin rod whose length L is 12.4 cm and whose mass m is 135 g, suspended at its midpoint from a long wire. Its period T_a of angular SHM is measured to be 2.53 s. An irregularly shaped object, which we call object X, is then hung from the same wire, as in Fig. 16-8*b*, and its period T_b is found to be 4.76 s. What is the rotational inertia of object X about its suspension axis?

SOLUTION: The **Key Idea** here is that the rotational inertia of either the rod or object X is related to the measured period by Eq. 16-23. In Table 11-2*e*, the rotational inertia of a thin rod about a perpendic-

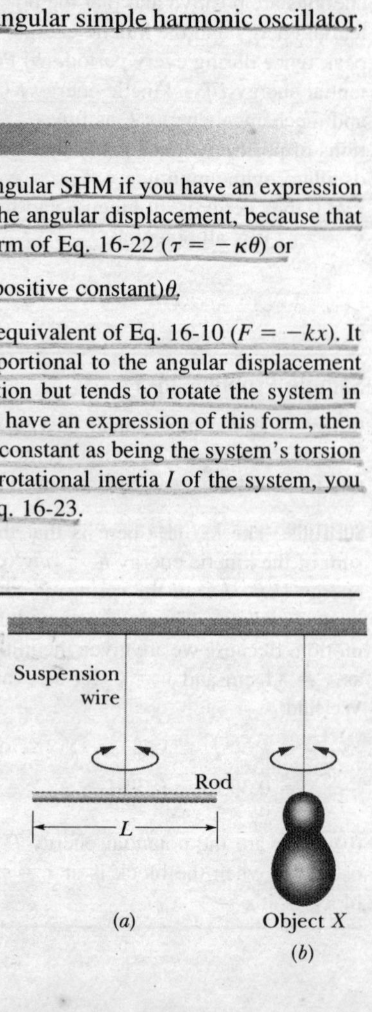

Fig. 16-8 Sample Problem 16-4. Two torsion pendulums, consisting of (*a*) a wire and a rod and (*b*) the same wire and an irregularly shaped object.

ular axis through its midpoint is given as $\frac{1}{12}mL^2$. Thus, we have, for the rod in Fig. 16-8*a*,

$$I_a = \tfrac{1}{12}mL^2 = (\tfrac{1}{12})(0.135\ \text{kg})(0.124\ \text{m})^2 = 1.73 \times 10^{-4}\ \text{kg}\cdot\text{m}^2.$$

Now let us write Eq. 16-23 twice, once for the rod and once for object *X*:

$$T_a = 2\pi\sqrt{\frac{I_a}{\kappa}} \quad \text{and} \quad T_b = 2\pi\sqrt{\frac{I_b}{\kappa}}.$$

The constant κ, which is a property of the wire, is the same for both figures; only the periods and the rotational inertias differ.

Let us square each of these equations, divide the second by the first, and solve the resulting equation for I_b. The result is

$$I_b = I_a\frac{T_b^2}{T_a^2} = (1.73 \times 10^{-4}\ \text{kg}\cdot\text{m}^2)\frac{(4.76\ \text{s})^2}{(2.53\ \text{s})^2} = 6.12 \times 10^{-4}\ \text{kg}\cdot\text{m}^2. \quad \text{(Answer)}$$

16-6 Pendulums

We turn now to a class of simple harmonic oscillators in which the springiness is associated with the gravitational force rather than with the elastic properties of a twisted wire or a compressed or stretched spring.

The Simple Pendulum

If you hang an apple at the end of a long thread fixed at its upper end, and then set the apple swinging back and forth a small distance, you easily see that the apple's motion is periodic. Is it, in fact, simple harmonic motion? If so, what is the period T? To answer, we consider a **simple pendulum,** which consists of a particle of mass m (called the *bob* of the pendulum) suspended from one end of an unstretchable, massless string of length L that is fixed at the other end, as in Fig. 16-9*a*. The bob is free to swing back and forth in the plane of the page, to the left and right of a vertical line through the pendulum's pivot point.

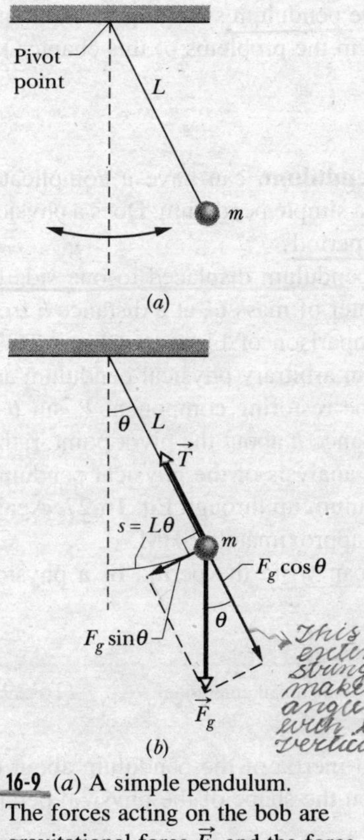

Fig. 16-9 (*a*) A simple pendulum. (*b*) The forces acting on the bob are the gravitational force $\vec{F}_g$ and the force $\vec{T}$ from the string. The tangential component $F_g \sin\theta$ of the gravitational force is a restoring force that tends to bring the pendulum back to its central position.

The forces acting on the bob are the force $\vec{T}$ from the string and the gravitational force $\vec{F}_g$, as shown in Fig. 16-9*b* where the string makes an angle θ with the vertical. We resolve $\vec{F}_g$ into a radial component $F_g \cos\theta$ and a component $F_g \sin\theta$ that is tangent to the path taken by the bob. This tangential component produces a restoring torque about the pendulum's pivot point, because it always acts opposite the displacement of the bob so as to bring the bob back toward its central location. That location is called the *equilibrium position* ($\theta = 0$), because the pendulum would be at rest there were it not swinging.

From Eq. 11-33 ($\tau = r_\perp F$), we can write this restoring torque as

$$\tau = -L(F_g \sin\theta), \quad (16\text{-}24)$$

where the minus sign indicates that the torque acts to reduce θ, and L is the moment arm of the force component $F_g \sin\theta$ about the pivot point. Substituting Eq. 16-24 into Eq. 11-36 ($\tau = I\alpha$) and then substituting mg as the magnitude of F_g, we obtain

$$-L(mg \sin\theta) = I\alpha, \quad (16\text{-}25)$$

where I is the pendulum's rotational inertia about the pivot point and α is its angular acceleration about that point.

We can simplify Eq. 16-25 if we assume the angle θ is small, for then we can approximate $\sin\theta$ with θ (expressed in radian measure). (As an example, if $\theta = 5.00° = 0.0873$ rad, then $\sin\theta = 0.0872$, a difference of only about 0.1%.) With that approximation and some rearranging, we then have

$$\alpha = -\frac{mgL}{I}\theta. \quad (16\text{-}26)$$

This equation is the angular equivalent of Eq. 16-8, the hallmark of SHM. It tells us that the angular acceleration α of the pendulum is proportional to the angular displacement θ but opposite in sign. Thus, as the pendulum bob moves to, say, the right as in Fig. 16-9*a*, its acceleration *to the left* increases until it stops and begins moving to the left. Then, when it is on the left, its acceleration to the right tends to return it to the right, and so on, as it swings back and forth in SHM. More precisely, the motion of a *simple pendulum swinging through only small angles* is approximately SHM. We can state this restriction to small angles another way: The **angular amplitude** θ_m of the motion (the maximum angle of swing) must be small.

Comparing Eqs. 16-26 and Eq. 16-8, we see that the angular frequency of the pendulum is $\omega = \sqrt{mgL/I}$. Next, if we substitute this expression for ω into Eq. 16-5 ($\omega = 2\pi/T$), we see that the period of the pendulum may be written as

$$T = 2\pi\sqrt{\frac{I}{mgL}}. \tag{16-27}$$

All the mass of a simple pendulum is concentrated in the mass m of the particle-like bob, which is at radius L from the pivot point. Thus, we can use Eq. 11-26 ($I = mr^2$) to write $I = mL^2$ for the rotational inertia of the pendulum. Substituting this into Eq. 16-27 and simplifying then yield

$$T = 2\pi\sqrt{\frac{L}{g}} \quad \text{(simple pendulum, small amplitude)} \tag{16-28}$$

as a simpler expression for the period of a simple pendulum swinging through only small angles. (We assume small-angle swinging in the problems of this chapter.)

The Physical Pendulum

A real pendulum, usually called a **physical pendulum,** can have a complicated distribution of mass, much different from that of a simple pendulum. Does a physical pendulum also undergo SHM? If so, what is its period?

Figure 16-10 shows an arbitrary physical pendulum displaced to one side by angle θ. The gravitational force $\vec{F}_g$ acts at its center of mass C, at a distance h from the pivot point O. In spite of their shapes, comparison of Figs. 16-10 and 16-9*b* reveals only one important difference between an arbitrary physical pendulum and a simple pendulum. For a physical pendulum the restoring component $F_g \sin\theta$ of the gravitational force has a moment arm of distance h about the pivot point, rather than of string length L. In all other respects, an analysis of the physical pendulum would duplicate our analysis of the simple pendulum up through Eq. 16-27. Again, (for small θ_m) we would find that the motion is approximately SHM.

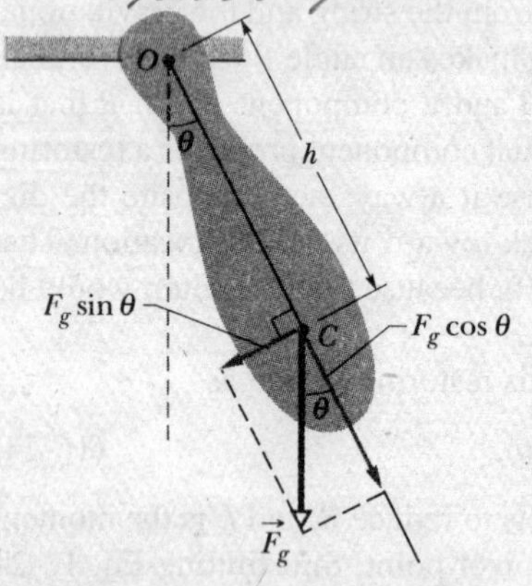

Fig. 16-10 A physical pendulum. The restoring torque is $hF_g \sin\theta$. When $\theta = 0$, center of mass C hangs directly below pivot point O.

If we replace L with h in Eq. 16-27, we can write the period of a physical pendulum as

$$T = 2\pi\sqrt{\frac{I}{mgh}} \quad \text{(physical pendulum, small amplitude).} \tag{16-29}$$

As with the simple pendulum, I is the rotational inertia of the pendulum about O. However, now I is not simply mL^2 (it depends on the shape of the physical pendulum), but it is still proportional to m.

A physical pendulum will not swing if it pivots at its center of mass. Formally, this corresponds to putting $h = 0$ in Eq. 16-29. That equation then predicts $T \rightarrow \infty$, which implies that such a pendulum will never complete one swing.

Corresponding to any physical pendulum that oscillates about a given pivot point O with period T is a simple pendulum of length L_0 with the same period T. We can find L_0 with Eq. 16-28. The point along the physical pendulum at distance L_0 from point O is called the *center of oscillation* of the physical pendulum for the given suspension point.

Measuring g

We can use a physical pendulum to measure the free-fall acceleration g at a particular location on Earth's surface. (Countless thousands of such measurements have been made during geophysical prospecting.)

To analyze a simple case, take the pendulum to be a uniform rod of length L, suspended from one end. For such a pendulum, h in Eq. 16-29, the distance between the pivot point and the center of mass, is $\frac{1}{2}L$. Table 11-2*e* tells us that the rotational inertia of this pendulum about a perpendicular axis through its center of mass is $\frac{1}{12}mL^2$. From the parallel-axis theorem of Eq. 11-29 ($I = I_{com} + Mh^2$), we then find that the rotational inertia about a perpendicular axis through one end of the rod is

$$I = I_{com} + mh^2 = \tfrac{1}{12}mL^2 + m(\tfrac{1}{2}L)^2 = \tfrac{1}{3}mL^2. \tag{16-30}$$

If we put $h = \frac{1}{2}L$ and $I = \frac{1}{3}mL^2$ in Eq. 16-29 and solve for g, we find

$$g = \frac{8\pi^2 L}{3T^2}. \tag{16-31}$$

Thus, by measuring L and the period T, we can find the value of g at the pendulum's location. (If precise measurements are to be made, a number of refinements are needed, such as swinging the pendulum in an evacuated chamber.)

Sample Problem 16-5

In Fig. 16-11*a*, a meter stick swings about a pivot point at one end, at distance h from its center of mass.

(a) What is its period of oscillation T?

SOLUTION: One **Key Idea** here is that the stick is not a simple pendulum because its mass is not concentrated in a bob at the end opposite the pivot point—so the stick is a physical pendulum. Then its period is given by Eq. 16-29, for which we need the rotational inertia I of the stick about the pivot point. We can treat the stick as a uniform rod of length L and mass m. Then Eq. 16-30 tells us that $I = \frac{1}{3}mL^2$, and the distance h in Eq. 16-29 is $\frac{1}{2}L$. Substituting these quantities into Eq. 16-29, we find

$$T = 2\pi\sqrt{\frac{I}{mgh}} = 2\pi\sqrt{\frac{\frac{1}{3}mL^2}{mg(\frac{1}{2}L)}} = 2\pi\sqrt{\frac{2L}{3g}} \tag{16-32}$$

$$= 2\pi\sqrt{\frac{(2)(1.00\text{ m})}{(3)(9.8\text{ m/s}^2)}} = 1.64\text{ s.} \qquad \text{(Answer)}$$

Note that the result is independent of the pendulum's mass m.

(b) What is the distance L_0 between the pivot point O of the stick and the center of oscillation of the stick?

SOLUTION: The **Key Idea** here is that we want the length L_0 of the simple pendulum (drawn in Fig. 16-11*b*) that has the same period as the physical pendulum (the stick) of Fig. 16-11*a*. Setting

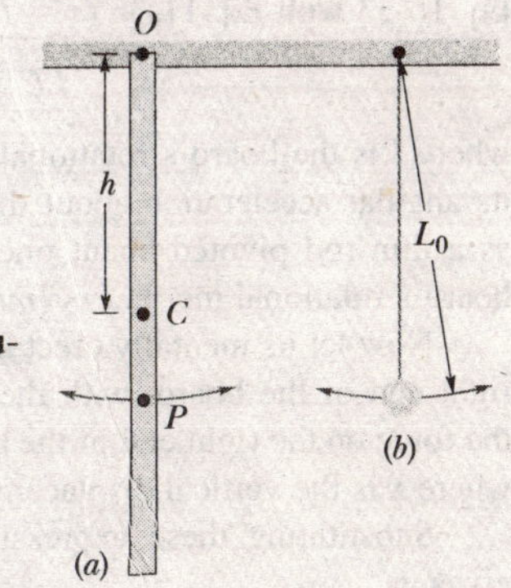

Fig. 16-11 Sample Problem 16-5. (*a*) A meter stick suspended from one end as a physical pendulum. (*b*) A simple pendulum whose length L_0 is chosen so that the periods of the two pendulums are equal. Point P on the pendulum of (*a*) marks the center of oscillation.

Eqs. 16-28 and 16-32 equal yields

$$T = 2\pi\sqrt{\frac{L_0}{g}} = 2\pi\sqrt{\frac{2L}{3g}}.$$

You can see by inspection that

$$L_0 = \tfrac{2}{3}L = (\tfrac{2}{3})(100\text{ cm}) = 66.7\text{ cm.} \qquad \text{(Answer)}$$

In Fig. 16-11*a*, point P marks this distance from suspension point O. Thus, point P is the stick's center of oscillation for the given suspension point.

CHECKPOINT 4: Three physical pendulums, of masses m_0, $2m_0$, and $3m_0$, have the same shape and size and are suspended at the same point. Rank the masses according to the periods of the pendulums, greatest period first.

Sample Problem 16-6

In Fig. 16-12, a penguin (obviously skilled in aquatic sports) dives from a uniform board that is hinged at the left and attached to a spring at the right. The board has length $L = 2.0$ m and mass $m = 12$ kg; the spring constant k is 1300 N/m. When the penguin dives, it leaves the board and spring oscillating with a small amplitude. Assume that the board is stiff enough not to bend, and find the period T of the oscillations.

SOLUTION: Since a spring is involved, we might guess that the oscillations are SHM, but we shall not assume that. Instead, we use the following **Key Idea**: If the board is in SHM, then the acceleration and displacement of the oscillating end of the board must be related by an expression in the form of Eq. 16-8 ($a = -\omega^2 x$). If so, we shall be able to find ω and then the desired T from the expression. Let us check by finding the relation between the acceleration and displacement of the board's right end.

Because the board rotates about the hinge as one end oscillates, we are concerned with a torque $\vec{\tau}$ on the board about the hinge. That torque is due to the force $\vec{F}$ on the board from the spring. Because $\vec{F}$ varies with time, $\vec{\tau}$ must also. However, at any given instant we can relate the magnitudes of $\vec{\tau}$ and $\vec{F}$ with Eq. 11-31 ($\tau = rF \sin \phi$). Here we have

$$\tau = LF \sin 90°, \tag{16-33}$$

where L is the moment arm of force $\vec{F}$ and 90° is the angle between the moment arm and the force's line of action. Combining Eq. 16-33 with Eq. 11-36 ($\tau = I\alpha$) gives us

$$LF = I\alpha, \tag{16-34}$$

where I is the board's rotational inertia about the hinge, and α is its angular acceleration about that point. We may treat the board as a thin rod pivoted about one end. Then, from Eq. 16-30, the board's rotational inertia I is $\frac{1}{3}mL^2$.

Now let us mentally erect a vertical x through the oscillating right end of the board, with the positive direction upward. Then the force on the right end of the board from the spring is $F = -kx$, where x is the vertical displacement of the right end.

Substituting these expressions for I and F into Eq. 16-34 gives us

$$-Lkx = \frac{mL^2\alpha}{3}. \tag{16-35}$$

We now have a mixture of linear displacement x (vertically) and rotational acceleration α (about the hinge). We can replace α in Eq. 16-35 with the (linear) acceleration a along the x axis by substituting according to Eq. 11-22 ($a_t = \alpha r$) for tangential acceleration. Here the tangential acceleration is a and the radius of rotation r is L, so $\alpha = a/L$. With that substitution, Eq. 16-35 becomes

$$-Lkx = \frac{mL^2 a}{3L},$$

which yields

$$a = -\frac{3k}{m}x. \tag{16-36}$$

Equation 16-36 is, in fact, of the same form as Eq. 16-8 ($a = -\omega^2 x$). Therefore, the board does indeed undergo SHM, and comparison of Eqs. 16-36 and 16-8 shows that

$$\omega^2 = \frac{3k}{m},$$

which gives $\omega = \sqrt{3k/m}$. Using Eq. 16-5 ($\omega = 2\pi/T$) to find T then gives us

$$T = 2\pi\sqrt{\frac{m}{3k}} = 2\pi\sqrt{\frac{12 \text{ kg}}{3(1300 \text{ N/m})}}$$
$$= 0.35 \text{ s.} \qquad \text{(Answer)}$$

Perhaps surprisingly, the period is independent of the board's length L.

Fig. 16-12 Sample Problem 16-6. The dive by the penguin causes the board and spring to oscillate; the board pivots about the hinge at the left.

16-7 Simple Harmonic Motion and Uniform Circular Motion

In 1610, Galileo, using his newly constructed telescope, discovered the four principal moons of Jupiter. Over weeks of observation, each moon seemed to him to be moving back and forth relative to the planet in what today we would call simple harmonic motion; the disk of the planet was the midpoint of the motion. The record of Galileo's observations, written in his own hand, is still available. A. P. French of MIT used Galileo's data to work out the position of the moon Callisto relative to

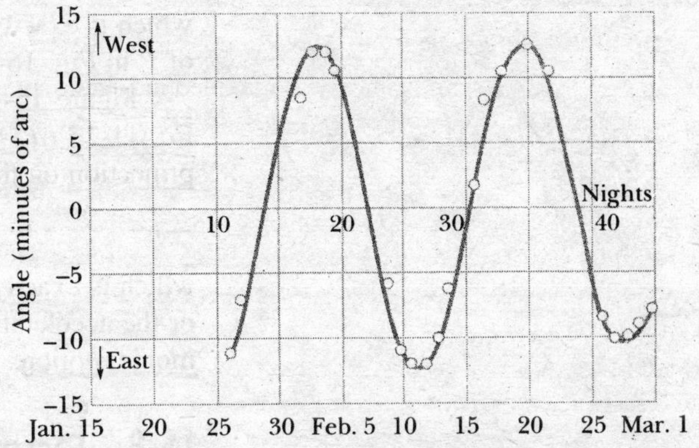

Fig. 16-13 The angle between Jupiter and its moon Callisto as seen from Earth. The circles are based on Galileo's 1610 measurements. The curve is a best fit, strongly suggesting simple harmonic motion. At Jupiter's mean distance, 10 minutes of arc corresponds to about 2×10^6 km. (Adapted from A. P. French, *Newtonian Mechanics,* W. W. Norton & Company, New York, 1971, p. 288.)

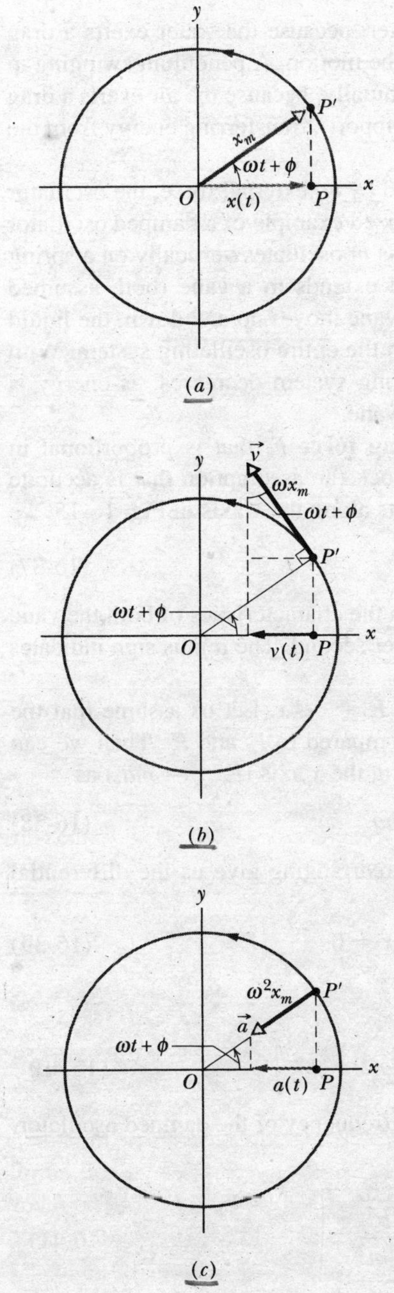

Jupiter. In the results shown in Fig. 16-13, the circles are based on Galileo's observations and the curve is a best fit to the data. The curve strongly suggests Eq. 16-3, the displacement function for SHM. A period of about 16.8 days can be measured from the plot.

Actually, Callisto moves with essentially constant speed in an essentially circular orbit around Jupiter. Its true motion—far from being simple harmonic—is uniform circular motion. What Galileo saw—and what you can see with a good pair of binoculars and a little patience—is the projection of this uniform circular motion on a line in the plane of the motion. We are led by Galileo's remarkable observations to the conclusion that simple harmonic motion is uniform circular motion viewed edge-on. In more formal language:

> Simple harmonic motion is the projection of uniform circular motion on a diameter of the circle in which the latter motion occurs.

Figure 16-14*a* gives an example. It shows a *reference particle P'* moving in uniform circular motion with (constant) angular speed ω in a *reference circle.* The radius x_m of the circle is the magnitude of the particle's position vector. At any time t, the angular position of the particle is $\omega t + \phi$, where ϕ is its angular position at $t = 0$.

The projection of particle P' onto the x axis is a point P, which we take to be a second particle. The projection of the position vector of particle P' onto the x axis gives the location $x(t)$ of P. Thus, we find

$$x(t) = x_m \cos(\omega t + \phi),$$

which is precisely Eq. 16-3. Our conclusion is correct. If reference particle P' moves in uniform circular motion, its projection particle P moves in simple harmonic motion along a diameter of the circle.

Figure 16-14*b* shows the velocity $\vec{v}$ of the reference particle. From Eq. 11-18 ($v = \omega r$), the magnitude of the velocity vector is ωx_m; its projection on the x axis is

$$v(t) = -\omega x_m \sin(\omega t + \phi),$$

Fig. 16-14 (*a*) A reference particle P' moving with uniform circular motion in a reference circle of radius x_m. Its projection P on the x axis executes simple harmonic motion. (*b*) The projection of the velocity $\vec{v}$ of the reference particle is the velocity of SHM. (*c*) The projection of the radial acceleration $\vec{a}$ of the reference particle is the acceleration of SHM.

which is exactly Eq. 16-6. The minus sign appears because the velocity component of P in Fig. 16-14b is directed to the left, in the negative direction of x.

Figure 16-14c shows the radial acceleration $\vec{a}$ of the reference particle. From Eq. 11-23 ($a_r = \omega^2 r$), the magnitude of the radial acceleration vector is $\omega^2 x_m$; its projection on the x axis is

$$a(t) = -\omega^2 x_m \cos(\omega t + \phi),$$

which is exactly Eq. 16-7. Thus, whether we look at the displacement, the velocity, or the acceleration, the projection of uniform circular motion is indeed simple harmonic motion.

16-8 Damped Simple Harmonic Motion

A pendulum will swing only briefly under water, because the water exerts a drag force on the pendulum that quickly eliminates the motion. A pendulum swinging in air does better, but still the motion dies out eventually, because the air exerts a drag force on the pendulum (and friction acts at its support), transferring energy from the pendulum's motion.

When the motion of an oscillator is reduced by an external force, the oscillator and its motion are said to be **damped.** An idealized example of a damped oscillator is shown in Fig. 16-15, where a block with mass m oscillates vertically on a spring with spring constant k. From the block, a rod extends to a vane (both assumed massless) that is submerged in a liquid. As the vane moves up and down, the liquid exerts an inhibiting drag force on it and thus on the entire oscillating system. With time, the mechanical energy of the block–spring system decreases, as energy is transferred to thermal energy of the liquid and vane.

Let us assume the liquid exerts a **damping force** $\vec{F}_d$ that is proportional in magnitude to the velocity $\vec{v}$ of the vane and block (an assumption that is accurate if the vane moves slowly). Then, for components along the x axis in Fig. 16-15, we have

$$F_d = -bv, \tag{16-37}$$

where b is a **damping constant** that depends on the characteristics of both the vane and the liquid and has the SI unit of kilogram per second. The minus sign indicates that $\vec{F}_d$ opposes the motion.

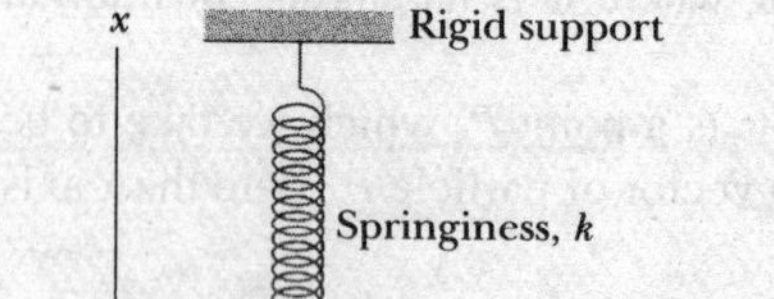

Fig. 16-15 An idealized damped simple harmonic oscillator. A vane immersed in a liquid exerts a damping force on the block as the block oscillates parallel to the x axis.

The force on the block from the spring is $F_s = -kx$. Let us assume that the gravitational force on the block is negligible compared to F_d and F_s. Then we can write Newton's second law for components along the x axis ($F_{\text{net},x} = ma_x$) as

$$-bv - kx = ma. \tag{16-38}$$

Substituting dx/dt for v and d^2x/dt^2 for a and rearranging give us the differential equation

$$m\frac{d^2x}{dt^2} + b\frac{dx}{dt} + kx = 0. \tag{16-39}$$

The solution of this equation is

$$x(t) = x_m e^{-bt/2m} \cos(\omega' t + \phi), \tag{16-40}$$

where x_m is the amplitude and ω' is the angular frequency of the damped oscillator. This angular frequency is given by

$$\omega' = \sqrt{\frac{k}{m} - \frac{b^2}{4m^2}}. \tag{16-41}$$

If $b = 0$ (there is no damping), then Eq. 16-41 reduces to Eq. 16-12 ($\omega = \sqrt{k/m}$) for the angular frequency of an undamped oscillator, and Eq. 16-40 reduces to Eq. 16-3 for the displacement of an undamped oscillator. If the damping constant is small but not zero (so that $b \ll \sqrt{km}$), then $\omega' \approx \omega$.

We can regard Eq. 16-40 as a cosine function whose amplitude, which is $x_m e^{-bt/2m}$, gradually decreases with time, as Fig. 16-16 suggests. For an undamped oscillator, the mechanical energy is constant and is given by Eq. 16-21 ($E = \frac{1}{2}kx_m^2$). If the oscillator is damped, the mechanical energy is not constant but decreases with time. If the damping is small, we can find $E(t)$ by replacing x_m in Eq. 16-21 with $x_m e^{-bt/2m}$, the amplitude of the damped oscillations. By doing so, we find that

$$E(t) \approx \tfrac{1}{2}kx_m^2 e^{-bt/m}, \tag{16-42}$$

which tells us that, like the amplitude, the mechanical energy decreases exponentially with time.

CHECKPOINT 5: Here are three sets of values for the spring constant, damping constant, and mass for the damped oscillator of Fig. 16-15. Rank the sets according to the time required for the mechanical energy to decrease to one-fourth of its initial value, greatest first.

Set 1	$2k_0$	b_0	m_0
Set 2	k_0	$6b_0$	$4m_0$
Set 3	$3k_0$	$3b_0$	m_0

Sample Problem 16-7

For the damped oscillator of Fig. 16-15, $m = 250$ g, $k = 85$ N/m, and $b = 70$ g/s.

(a) What is the period of the motion?

SOLUTION: The **Key Idea** here is that because $b \ll \sqrt{km} = 4.6$ kg/s, the period is approximately that of the undamped oscillator. From Eq. 16-13, we then have

$$T = 2\pi\sqrt{\frac{m}{k}} = 2\pi\sqrt{\frac{0.25\text{ kg}}{85\text{ N/m}}} = 0.34\text{ s.} \quad \text{(Answer)}$$

(b) How long does it take for the amplitude of the damped oscillations to drop to half its initial value?

SOLUTION: Now the **Key Idea** is that the amplitude at time t is displayed in Eq. 16-40 as $x_m e^{-bt/2m}$. It has the value x_m at $t = 0$. Thus, we must find the value of t for which

$$x_m e^{-bt/2m} = \tfrac{1}{2}x_m.$$

Canceling x_m and taking the natural logarithm of the equation that remains, we have $\ln \frac{1}{2}$ on the right side and

$$\ln(e^{-bt/2m}) = -bt/2m$$

on the left side. Thus,

$$t = \frac{-2m \ln \frac{1}{2}}{b} = \frac{-(2)(0.25\text{ kg})(\ln \frac{1}{2})}{0.070\text{ kg/s}}$$

$$= 5.0\text{ s.} \quad \text{(Answer)}$$

Because $T = 0.34$ s, this is about 15 periods of oscillation.

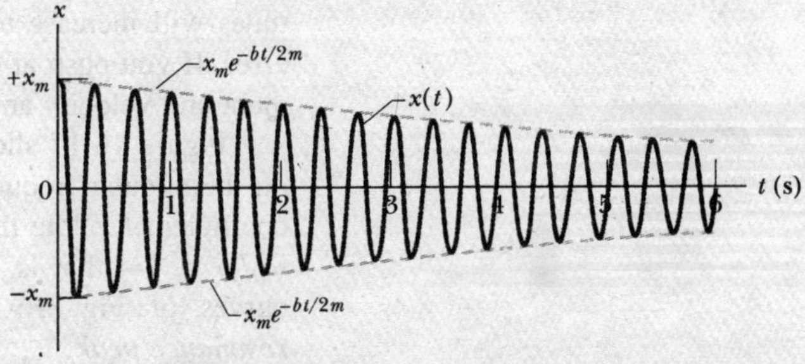

Fig. 16-16 The displacement function $x(t)$ for the damped oscillator of Fig. 16-15, with values given in Sample Problem 16-7. The amplitude, which is $x_m e^{-bt/2m}$, decreases exponentially with time.

(c) How long does it take for the mechanical energy to drop to one-half its initial value?

SOLUTION: Here the **Key Idea** is that, from Eq. 16-42, the mechanical energy at time t is $\frac{1}{2}kx_m^2 e^{-bt/m}$. It has the value $\frac{1}{2}kx_m^2$ at $t = 0$. Thus, we must find the value of t for which

$$\tfrac{1}{2}kx_m^2 e^{-bt/m} = \tfrac{1}{2}(\tfrac{1}{2}kx_m^2).$$

If we divide both sides of this equation by $\frac{1}{2}kx_m^2$ and solve for t as we did above, we find

$$t = \frac{-m \ln \frac{1}{2}}{b} = \frac{-(0.25\text{ kg})(\ln \frac{1}{2})}{0.070\text{ kg/s}} = 2.5\text{ s.} \quad \text{(Answer)}$$

This is exactly half the time we calculated in (b), or about 7.5 periods of oscillation. Figure 16-16 was drawn to illustrate this sample problem.

16-9 Forced Oscillations and Resonance

A person swinging in a swing without anyone pushing it is an example of *free oscillation*. However, if someone pushes the swing periodically, the swing has *forced*, or *driven, oscillations*. Two angular frequencies are associated with a system undergoing driven oscillations: (1) the *natural* angular frequency ω of the system, which is the angular frequency at which it would oscillate if it were suddenly disturbed and then left to oscillate freely, and (2) the angular frequency ω_d of the external driving force causing the driven oscillations.

We can use Fig. 16-15 to represent an idealized forced simple harmonic oscillator if we allow the structure marked "rigid support" to move up and down at a variable angular frequency ω_d. Such a forced oscillator oscillates at the angular frequency ω_d of the driving force, and its displacement $x(t)$ is given by

$$x(t) = x_m \cos(\omega_d t + \phi), \tag{16-43}$$

where x_m is the amplitude of the oscillations.

How large the displacement amplitude x_m is depends on a complicated function of ω_d and ω. The velocity amplitude v_m of the oscillations is easier to describe: it is greatest when

$$\omega_d = \omega \quad \text{(resonance)}, \tag{16-44}$$

a condition called **resonance.** Equation 16-44 is also *approximately* the condition at which the displacement amplitude x_m of the oscillations is greatest. Thus, if you push a swing at its natural angular frequency, the displacement and velocity amplitudes will increase to large values, a fact that children learn quickly by trial and error. If you push at other angular frequencies, either higher or lower, the displacement and velocity amplitudes will be smaller.

Figure 16-17 shows how the displacement amplitude of an oscillator depends on the angular frequency ω_d of the driving force, for three values of the damping coefficient b. Note that for all three the amplitude is approximately greatest when $\omega_d/\omega = 1$—that is, when the resonance condition of Eq. 16-44 is satisfied. The curves of Fig. 16-17 show that less damping gives a taller and narrower *resonance peak*.

All mechanical structures have one or more natural angular frequencies, and if a structure is subjected to a strong external driving force that matches one of these

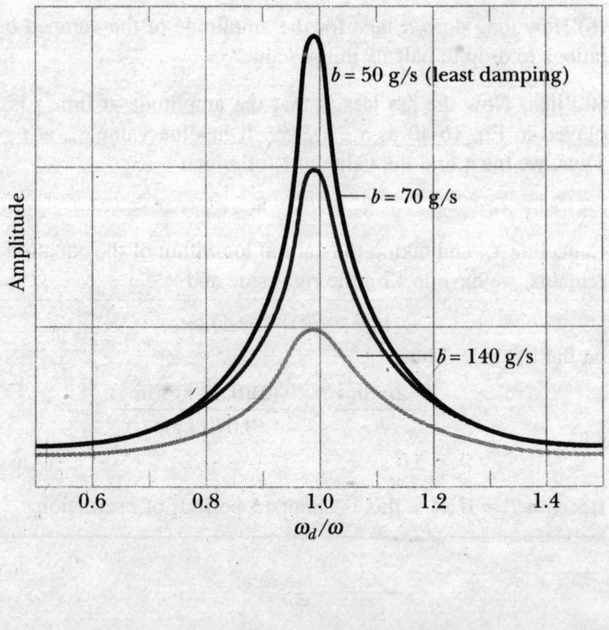

Fig. 16-17 The displacement amplitude x_m of a forced oscillator varies as the angular frequency ω_d of the driving force is varied. The amplitude is greatest *approximately* at $\omega_d/\omega = 1$, the resonance condition. The curves here correspond to three values of the damping constant b.

angular frequencies, the resulting oscillations of the structure may rupture it. Thus, for example, aircraft designers must make sure that none of the natural angular frequencies at which a wing can oscillate matches the angular frequency of the engines in flight. A wing that flaps violently at certain engine speeds would obviously be dangerous.

Mexico's earthquake in September 1985 was a major earthquake (8.1 on the Richter scale), but the seismic waves from it should have been too weak to cause extensive damage when they reached Mexico City about 400 km away. However, Mexico City is largely built on an ancient lake bed, where the soil is still soft with water. Although the amplitude of the seismic waves was weak in the firmer ground en route to Mexico City, their amplitude substantially increased in the loose soil of the city. Acceleration amplitudes of the waves were as much as $0.20g$, and the angular frequency was (surprisingly) concentrated around 3 rad/s. Not only was the ground severely oscillated, but many of the buildings with intermediate height had resonant angular frequencies of about 3 rad/s. Most of those intermediate-height buildings collapsed during the shaking, while shorter buildings (with higher resonant angular frequencies) and taller buildings (with lower resonant angular frequencies) remained standing.

REVIEW & SUMMARY

Frequency The *frequency* f of periodic or oscillatory motion is the number of oscillations per second. In the SI system, it is measured in hertz:

$$1 \text{ hertz} = 1 \text{ Hz} = 1 \text{ oscillation per second} = 1 \text{ s}^{-1}. \quad (16\text{-}1)$$

Period The *period* T is the time required for one complete oscillation, or **cycle.** It is related to the frequency by

$$T = \frac{1}{f}. \quad (16\text{-}2)$$

Simple Harmonic Motion In *simple harmonic motion* (SHM), the displacement $x(t)$ of a particle from its equilibrium position is described by the equation

$$x = x_m \cos(\omega t + \phi) \quad \text{(displacement)}, \quad (16\text{-}3)$$

in which x_m is the **amplitude** of the displacement, the quantity $(\omega t + \phi)$ is the **phase** of the motion, and ϕ is the **phase constant.** The **angular frequency** ω is related to the period and frequency of the motion by

$$\omega = \frac{2\pi}{T} = 2\pi f \quad \text{(angular frequency)}. \quad (16\text{-}5)$$

Differentiating Eq. 16-3 leads to equations for the particle's velocity and acceleration during SHM as functions of time:

$$v = -\omega x_m \sin(\omega t + \phi) \quad \text{(velocity)} \quad (16\text{-}6)$$

and

$$a = -\omega^2 x_m \cos(\omega t + \phi) \quad \text{(acceleration)}. \quad (16\text{-}7)$$

In Eq. 16-6, the positive quantity ωx_m is the **velocity amplitude** v_m of the motion. In Eq. 16-7, the positive quantity $\omega^2 x_m$ is the **acceleration amplitude** a_m of the motion.

The Linear Oscillator A particle with mass m that moves under the influence of a Hooke's law restoring force given by $F = -kx$ exhibits simple harmonic motion with

$$\omega = \sqrt{\frac{k}{m}} \quad \text{(angular frequency)} \quad (16\text{-}12)$$

and

$$T = 2\pi\sqrt{\frac{m}{k}} \quad \text{(period)}. \quad (16\text{-}13)$$

Such a system is called a **linear simple harmonic oscillator.**

Energy A particle in simple harmonic motion has, at any time, kinetic energy $K = \frac{1}{2}mv^2$ and potential energy $U = \frac{1}{2}kx^2$. If no friction is present, the mechanical energy $E = K + U$ remains constant even though K and U change.

Pendulums Examples of devices that undergo simple harmonic motion are the **torsion pendulum** of Fig. 16-7, the **simple pendulum** of Fig. 16-9, and the **physical pendulum** of Fig. 16-10. Their periods of oscillation for small oscillations are, respectively,

$$T = 2\pi\sqrt{I/\kappa}, \quad (16\text{-}23)$$

$$T = 2\pi\sqrt{L/g}, \quad (16\text{-}28)$$

and

$$T = 2\pi\sqrt{I/mgh}. \quad (16\text{-}29)$$

Simple Harmonic Motion and Uniform Circular Motion Simple harmonic motion is the projection of uniform circular motion onto the diameter of the circle in which the latter motion occurs. Figure 16-14 shows that all parameters of circular motion (position, velocity, and acceleration) project to the corresponding values for simple harmonic motion.

Damped Harmonic Motion The mechanical energy E in a real oscillating system decreases during the oscillations because external forces, such as a drag force, inhibit the oscillations and transfer mechanical energy to thermal energy. The real oscillator and its motion are then said to be **damped.** If the **damping force** is given by $\vec{F}_d = -b\vec{v}$, where v is the velocity of the oscillator and b is a **damping constant,** then the displacement of the oscillator is given by

$$x(t) = x_m e^{-bt/2m} \cos(\omega' t + \phi), \tag{16-40}$$

where ω', the angular frequency of the damped oscillator, is given by

$$\omega' = \sqrt{\frac{k}{m} - \frac{b^2}{4m^2}}. \tag{16-41}$$

If the damping constant is small ($b \ll \sqrt{km}$), then $\omega' \approx \omega$, where ω is the angular frequency of the undamped oscillator. For small b, the mechanical energy E of the oscillator is given by

$$E(t) \approx \tfrac{1}{2}kx_m^2 e^{-bt/m}. \tag{16-42}$$

Forced Oscillations and Resonance If an external driving force with angular frequency ω_d acts on an oscillating system with *natural* angular frequency ω, the system oscillates with angular frequency ω_d. The velocity amplitude v_m of the system is greatest when

$$\omega_d = \omega, \tag{16-44}$$

a condition called **resonance.** The amplitude x_m of the system is (approximately) greatest under the same condition.

QUESTIONS

1. Which of the following relationships between the acceleration a and the displacement x of a particle involve SHM: (a) $a = 0.5x$, (b) $a = 400x^2$, (c) $a = -20x$, (d) $a = -3x^2$?

2. Given $x = (2.0 \text{ m}) \cos(5t)$ for SHM and needing to find the velocity at $t = 2$ s, should you substitute for t and then differentiate with respect to t or vice versa?

3. The acceleration $a(t)$ of a particle undergoing SHM is graphed in Fig. 16-18. (a) Which of the labeled points corresponds to the particle at $-x_m$? (b) At point 4, is the velocity of the particle positive, negative, or zero? (c) At point 5, is the particle at $-x_m$, at $+x_m$, at 0, between $-x_m$ and 0, or between 0 and $+x_m$?

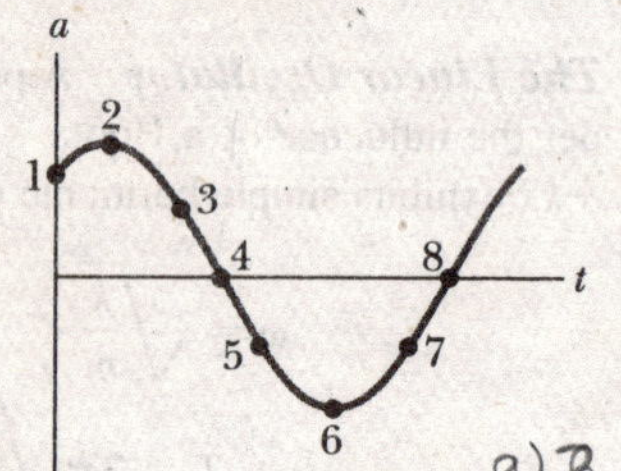

Fig. 16-18 Question 3.

4. Which of the following describe ϕ for the SHM of Fig. 16-19a:

(a) $-\pi < \phi < -\pi/2$, (b) $\pi < \phi < 3\pi/2$, (c) $-3\pi/2 < \phi < -\pi$?

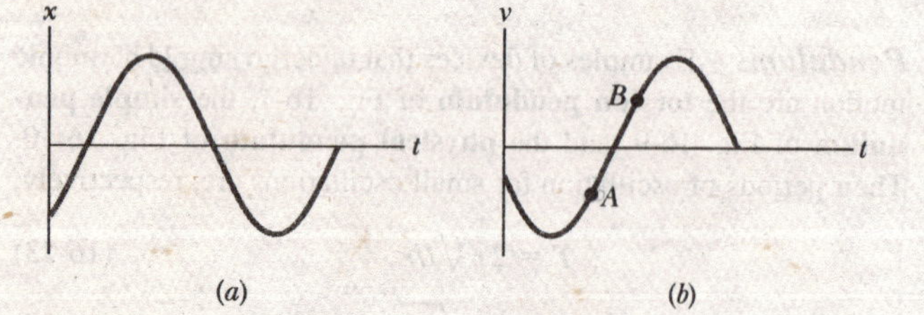

Fig. 16-19 Questions 4 and 5.

5. The velocity $v(t)$ of a particle undergoing SHM is graphed in Fig. 16-19b. Is the particle momentarily stationary, headed toward $-x_m$, or headed toward $+x_m$ at (a) point A on the graph and (b) point B? Is the particle at $-x_m$, at $+x_m$, at 0, between $-x_m$ and 0, or between 0 and $+x_m$ when its velocity is represented by (c) point A and (d) point B? Is the speed of the particle increasing or decreasing at (e) point A and (f) point B?

6. Figure 16-20 gives, for three situations, the displacements $x(t)$ of a pair of simple harmonic oscillators (A and B) that are identical except for phase. For each pair, what phase shift (in radians and in degrees) is needed to shift the curve for A to coincide with the curve for B? Of the many possible answers, choose the shift with the smallest absolute magnitude.

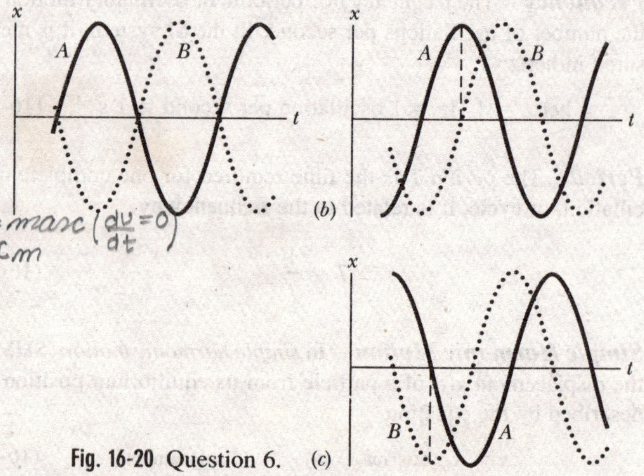

Fig. 16-20 Question 6.

7. Figures 16-21a and b show the positions of four linear oscillators with identical masses and spring constants, in snapshots at the same instant. What is the phase difference of the two linear oscil-

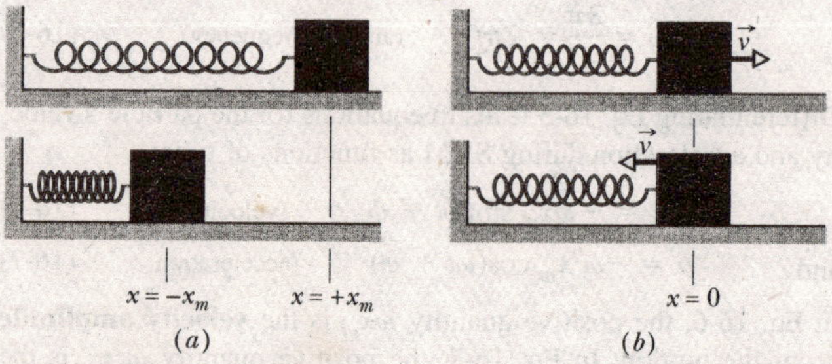

Fig. 16-21 Question 7.

lators in (a) Fig. 16-21*a* and (b) Fig. 16-21*b*? (c) What is the phase difference between the red oscillator in Fig. 16-21*a* and the green oscillator in Fig. 16-21*b*?

8. (a) Which curve in Fig. 16-22*a* gives the acceleration $a(t)$ versus displacement $x(t)$ of a simple harmonic oscillator? (b) Which curve in Fig. 16-22*b* gives the velocity $v(t)$ versus $x(t)$?

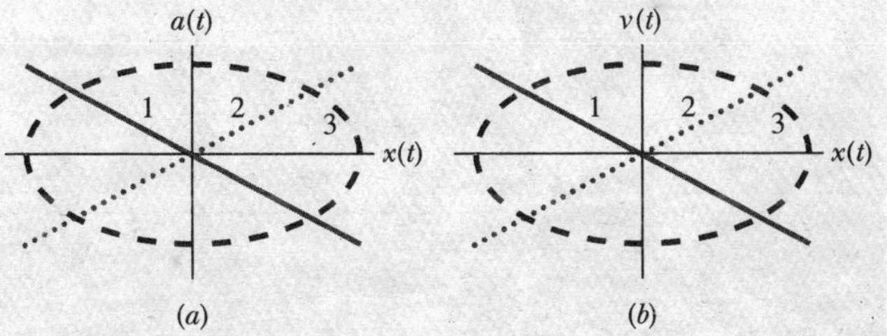

Fig. 16-22 Question 8.

9. In Fig. 16-23, a small block *A* sits on a large block *B* with a certain nonzero coefficient of static friction between the two blocks. Block *B*, which lies on a frictionless surface, is initially at $x = 0$, with the spring at its relaxed length; then we pull the block a distance *d* to the right and release it. As the spring–blocks system undergoes SHM, with amplitude x_m, block *A* is on the verge of slipping over *B*.

(a) Is the acceleration of block *A* constant or does it vary? (b) Is the magnitude of the frictional force accelerating *A* constant or does it vary? (c) Is *A* more likely to slip at $x = 0$ or at $x = \pm x_m$? (d) If the SHM began with an initial displacement that was greater than *d*, would slippage then be more likely or less likely? (Warm-up for Problem 16)

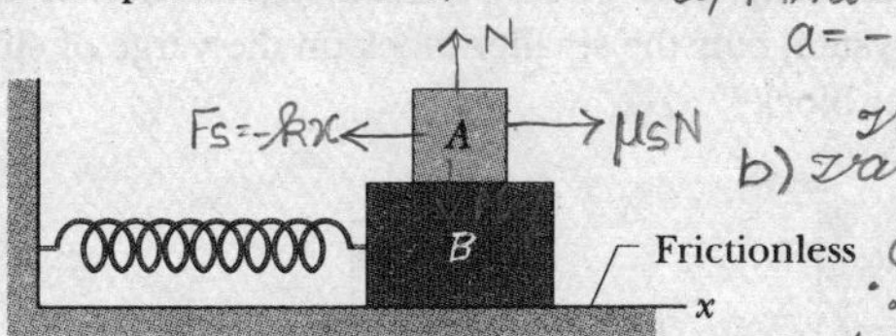

Fig. 16-23 Question 9.

10. In Fig. 16-24, a spring–block system is put into SHM in two experiments. In the first, the block is pulled from the equilibrium position through a displacement d_1 and then released. In the second, it is pulled from the equilibrium position through a greater displacement d_2 and then released. Are the (a) amplitude, (b) period, (c) frequency, (d) maximum kinetic energy, and (e) maximum potential energy in the second experiment greater than, less than, or the same as those in the first experiment?

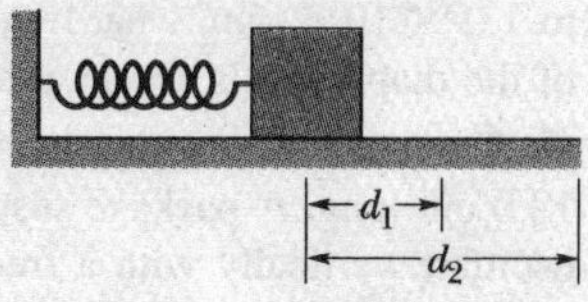

Fig. 16-24 Question 10.

11. Figure 16-25 shows three physical pendulums consisting of identical uniform spheres of the same mass that are rigidly connected by identical rods of negligible mass. Each pendulum is vertical and can pivot about suspension point *O*. Rank the pendulums according to period of oscillation, greatest first.

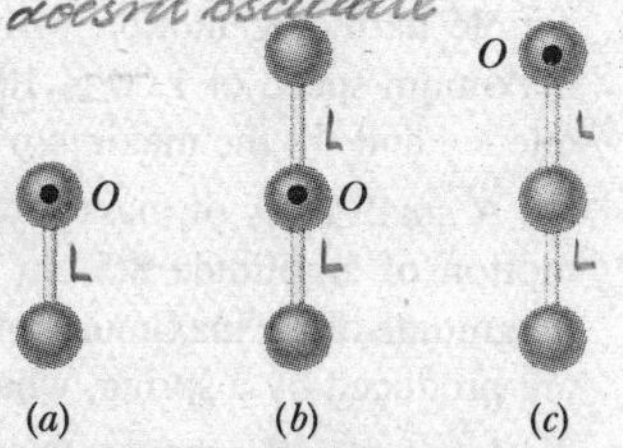

Fig. 16-25 Question 11.

12. Follow-up to Exercise 36: If the speed of the bullet were greater, would the following quantities of the resulting SHM then be greater, less, or the same: (a) amplitude, (b) period, (c) maximum potential energy?

13. You are to build the oscillation transfer device shown in Fig. 16-26. It consists of two spring–block systems hanging from a flexible rod. When the spring of system 1 is stretched and then released, the resulting SHM of system 1 at frequency f_1 oscillates the rod. The rod then provides a driving force on system 2, at the same frequency f_1. You can choose from four springs with spring constants *k* of 1600, 1500, 1400, and 1200 N/m, and four blocks with masses *m* of 800, 500, 400, and 200 kg. Mentally determine which spring should go with which block in each of the two systems to maximize the amplitude of oscillations in system 2.

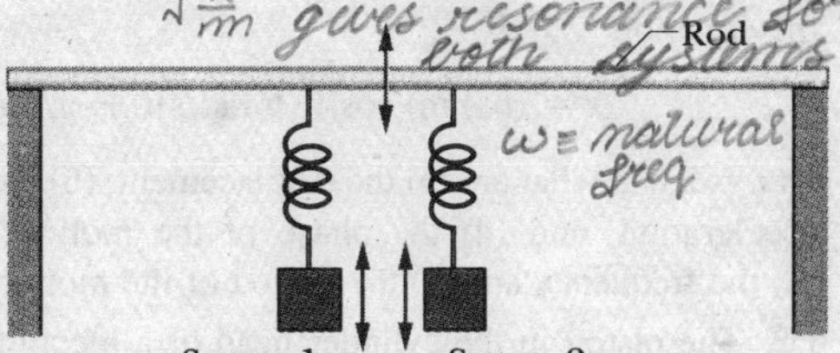

Fig. 16-26 Question 13.

EXERCISES & PROBLEMS

ssm Solution is in the Student Solutions Manual.
www Solution is available on the World Wide Web at: http://www.wiley.com/college/hrw
ilw Solution is available on the Interactive LearningWare.

SEC. 16-3 The Force Law for Simple Harmonic Motion

1E. An object undergoing simple harmonic motion takes 0.25 s to travel from one point of zero velocity to the next such point. The distance between those points is 36 cm. Calculate (a) the period, (b) the frequency, and (c) the amplitude of the motion.

2E. An oscillating block–spring system takes 0.75 s to begin repeating its motion. Find (a) the period, (b) the frequency in hertz, and (c) the angular frequency in radians per second.

3E. An oscillator consists of a block of mass 0.500 kg connected to a spring. When set into oscillation with amplitude 35.0 cm, the oscillator repeats its motion every 0.500 s. Find (a) the period, (b) the frequency, (c) the angular frequency, (d) the spring constant, (e) the maximum speed, and (f) the magnitude of the maximum force on the block from the spring. ssm

4E. What is the maximum acceleration of a platform that oscillates with an amplitude of 2.20 cm at a frequency of 6.60 Hz?

5E. A loudspeaker produces a musical sound by means of the oscillation of a diaphragm. If the amplitude of oscillation is limited to 1.0×10^{-3} mm, what frequencies will result in the magnitude of the diaphragm's acceleration exceeding g? ssm

6E. The scale of a spring balance that reads from 0 to 15.0 kg is 12.0 cm long. A package suspended from the balance is found to oscillate vertically with a frequency of 2.00 Hz. (a) What is the spring constant? (b) How much does the package weigh?

7E. A particle with a mass of 1.00×10^{-20} kg is oscillating with simple harmonic motion with a period of 1.00×10^{-5} s and a maximum speed of 1.00×10^{3} m/s. Calculate (a) the angular frequency and (b) the maximum displacement of the particle. ssm

8E. A small body of mass 0.12 kg is undergoing simple harmonic motion of amplitude 8.5 cm and period 0.20 s. (a) What is the magnitude of the maximum force acting on it? (b) If the oscillations are produced by a spring, what is the spring constant?

9E. In an electric shaver, the blade moves back and forth over a distance of 2.0 mm in simple harmonic motion, with frequency 120 Hz. Find (a) the amplitude, (b) the maximum blade speed, and (c) the magnitude of the maximum blade acceleration. ssm

10E. A loudspeaker diaphragm is oscillating in simple harmonic motion with a frequency of 440 Hz and a maximum displacement of 0.75 mm. What are (a) the angular frequency, (b) the maximum speed, and (c) the magnitude of the maximum acceleration?

11E. An automobile can be considered to be mounted on four identical springs as far as vertical oscillations are concerned. The springs of a certain car are adjusted so that the oscillations have a frequency of 3.00 Hz. (a) What is the spring constant of each spring if the mass of the car is 1450 kg and the mass is evenly distributed over the springs? (b) What will be the oscillation frequency if five passengers, averaging 73.0 kg each, ride in the car? (Again, consider an even distribution of mass.)

12E. A body oscillates with simple harmonic motion according to the equation

$$x = (6.0 \text{ m}) \cos[(3\pi \text{ rad/s})t + \pi/3 \text{ rad}].$$

At $t = 2.0$ s, what are (a) the displacement, (b) the velocity, (c) the acceleration, and (d) the phase of the motion? Also, what are (e) the frequency and (f) the period of the motion?

13E. The piston in the cylinder head of a locomotive has a stroke (twice the amplitude) of 0.76 m. If the piston moves with simple harmonic motion with an angular frequency of 180 rev/min, what is its maximum speed? ssm

14P. Figure 16-27 shows an astronaut on a body-mass measuring device (BMMD). Designed for use on orbiting space vehicles, its purpose is to allow astronauts to measure their mass in the "weightless" conditions in Earth orbit. The BMMD is a spring-mounted chair; an astronaut measures his or her period of oscillation in the chair; the mass follows from the formula for the period of an oscillating block–spring system. (a) If M is the mass of the astronaut and m the effective mass of that part of the BMMD that also oscillates, show that

$$M = (k/4\pi^2)T^2 - m,$$

where T is the period of oscillation and k is the spring constant. (b) The spring constant was $k = 605.6$ N/m for the BMMD on

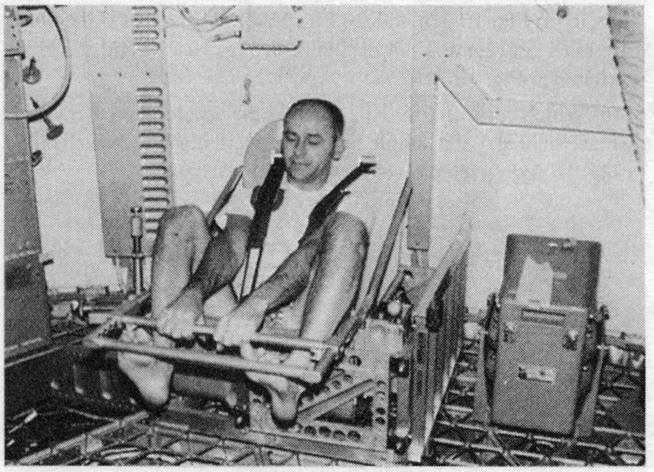

Fig. 16-27 Problem 14.

Skylab Mission Two; the period of oscillation of the empty chair was 0.90149 s. Calculate the effective mass of the chair. (c) With an astronaut in the chair, the period of oscillation became 2.08832 s. Calculate the mass of the astronaut.

15P. At a certain harbor, the tides cause the ocean surface to rise and fall a distance d (from highest level to lowest level) in simple harmonic motion, with a period of 12.5 h. How long does it take for the water to fall a distance $d/4$ from its highest level?

16P. In Fig. 16-28, two blocks ($m = 1.0$ kg and $M = 10$ kg) and a spring ($k = 200$ N/m) are arranged on a horizontal, frictionless surface. The coefficient of static friction between the two blocks is 0.40. What amplitude of simple harmonic motion of the spring–blocks system puts the smaller block on the verge of slipping over the larger block?

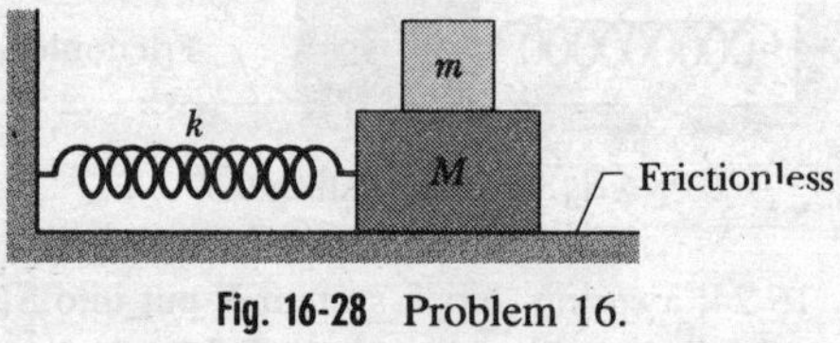

Fig. 16-28 Problem 16.

17P. A block is on a horizontal surface (a shake table) that is moving back and forth horizontally with simple harmonic motion of frequency 2.0 Hz. The coefficient of static friction between block and surface is 0.50. How great can the amplitude of the SHM be if the block is not to slip along the surface? ssm www

18P. A block rides on a piston that is moving vertically with simple harmonic motion. (a) If the SHM has period 1.0 s, at what amplitude of motion will the block and piston separate? (b) If the piston has an amplitude of 5.0 cm, what is the maximum frequency for which the block and piston will be in contact continuously?

19P. An oscillator consists of a block attached to a spring ($k = 400$ N/m). At some time t, the position (measured from the system's equilibrium location), velocity, and acceleration of the block are $x = 0.100$ m, $v = -13.6$ m/s, and $a = -123$ m/s^2. Calculate (a) the frequency of oscillation, (b) the mass of the block, and ? (c) the amplitude of the motion. ilw

$|a| = \omega^2 x$ find ω
$k = m\omega^2$ find m

20P. A simple harmonic oscillator consists of a block of mass 2.00 kg attached to a spring of spring constant 100 N/m. When $t = 1.00$ s, the position and velocity of the block are $x = 0.129$ m and $v = 3.415$ m/s. (a) What is the amplitude of the oscillations? What were the (b) position and (c) velocity of the block at $t = 0$ s?

21P. A massless spring hangs from the ceiling with a small object attached to its lower end. The object is initially held at rest in a position y_i such that the spring is at its rest length. The object is then released from y_i and oscillates up and down, with its lowest position being 10 cm below y_i. (a) What is the frequency of the oscillation? (b) What is the speed of the object when it is 8.0 cm below the initial position? (c) An object of mass 300 g is attached to the first object, after which the system oscillates with half the original frequency. What is the mass of the first object? (d) Relative to y_i, where is the new equilibrium (rest) position with both objects attached to the spring? ssm

22P. Two particles execute simple harmonic motion of the same amplitude and frequency along close parallel lines. They pass each other moving in opposite directions each time their displacement is half their amplitude. What is their phase difference?

23P. Two particles oscillate in simple harmonic motion along a common straight-line segment of length A. Each particle has a period of 1.5 s, but they differ in phase by $\pi/6$ rad. (a) How far apart are they (in terms of A) 0.50 s after the lagging particle leaves one end of the path? (b) Are they then moving in the same direction, toward each other, or away from each other? ssm

24P. In Fig. 16-29, two identical springs of spring constant k are attached to a block of mass m and to fixed supports. Show that the block's frequency of oscillation on the frictionless surface is

$$f = \frac{1}{2\pi}\sqrt{\frac{2k}{m}}.$$

25P. Suppose that the two springs in Fig. 16-29 have different spring constants k_1 and k_2. Show that the frequency f of oscillation of the block is then given by

$$f = \sqrt{f_1^2 + f_2^2},$$

where f_1 and f_2 are the frequencies at which the block would oscillate if connected only to spring 1 or only to spring 2. ilw

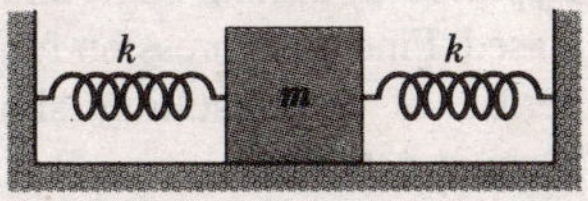

Fig. 16-29 Problems 24 and 25.

26P. The end of one of the prongs of a tuning fork that executes simple harmonic motion of frequency 1000 Hz has an amplitude of 0.40 mm. Find (a) the magnitude of the maximum acceleration and (b) the maximum speed of the end of the prong. Find (c) the magnitude of the acceleration and (d) the speed of the end of the prong when the end has a displacement of 0.20 mm.

27P. In Fig. 16-30, two springs are joined and connected to a block of mass m. The surface is frictionless. If the springs both have spring constant k, show that

$$f = \frac{1}{2\pi}\sqrt{\frac{k}{2m}}$$

gives the block's frequency of oscillation. ssm

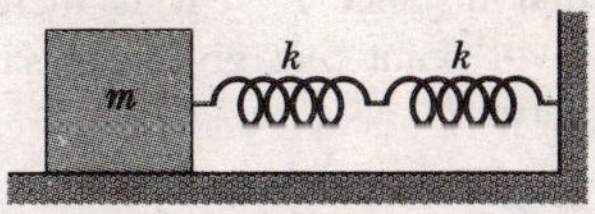

Fig. 16-30 Problem 27.

28P. In Fig. 16-31, a block weighing 14.0 N, which slides without friction on a 40.0° incline, is connected to the top of the incline by a massless spring of unstretched length 0.450 m and spring constant 120 N/m. (a) How far from the top of the incline does the block stop? (b) If the block is pulled slightly down the incline and released, what is the period of the resulting oscillations?

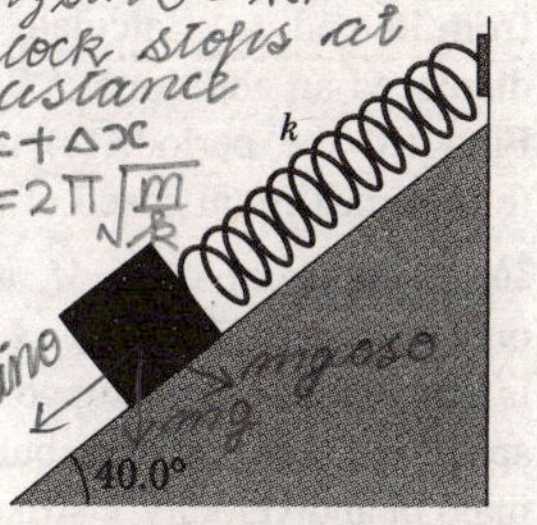

Fig. 16-31 Problem 28.

29P. A uniform spring with unstretched length L and spring constant k is cut into two pieces of unstretched lengths L_1 and L_2, with $L_1 = nL_2$. What are the corresponding spring constants (a) k_1 and (b) k_2 in terms of n and k? If a block is attached to the original spring, as in Fig. 16-5, it oscillates with frequency f. If the spring is replaced with the piece L_1 or L_2, the corresponding frequency is f_1 or f_2. Find (c) f_1 and (d) f_2 in terms of f. ssm www

30P. In Fig. 16-32, three 10 000 kg ore cars are held at rest on a 30° incline on a mine railway using a cable that is parallel to the incline. The cable stretches 15 cm just before the coupling between the two lower cars breaks, detaching the lowest car. Assuming that the cable obeys Hooke's law, find (a) the frequency and (b) the amplitude of the resulting oscillations of the remaining two cars.

Fig. 16-32 Problem 30.

SEC. 16-4 Energy in Simple Harmonic Motion

31E. Find the mechanical energy of a block–spring system having a spring constant of 1.3 N/cm and an oscillation amplitude of 2.4 cm. ssm

32E. An oscillating block–spring system has a mechanical energy of 1.00 J, an amplitude of 10.0 cm, and a maximum speed of 1.20 m/s. Find (a) the spring constant, (b) the mass of the block and (c) the frequency of oscillation.

33E. A 5.00 kg object on a horizontal frictionless surface is attached to a spring with spring constant 1000 N/m. The object is displaced from equilibrium 50.0 cm horizontally and given an initial velocity of 10.0 m/s back toward the equilibrium position. (a) What is the frequency of the motion? What are (b) the initial potential energy of the block–spring system, (c) the initial kinetic energy, and (d) the amplitude of the oscillation? ilw

34E. A (hypothetical) large slingshot is stretched 1.50 m to launch a 130 g projectile with speed sufficient to escape from Earth (11.2 km/s). Assume the elastic bands of the slingshot obey Hooke's law. (a) What is the spring constant of the device, if all the elastic potential energy is converted to kinetic energy? (b) Assume that an average person can exert a force of 220 N. How many people are required to stretch the elastic bands?

35E. A vertical spring stretches 9.6 cm when a 1.3 kg block is hung from its end. (a) Calculate the spring constant. This block is then displaced an additional 5.0 cm downward and released from rest. Find (b) the period, (c) the frequency, (d) the amplitude, and (e) the maximum speed of the resulting SHM. ssm

•36E. A block of mass M, at rest on a horizontal frictionless table, is attached to a rigid support by a spring of constant k. A bullet of mass m and velocity $\vec{v}$ strikes the block as shown in Fig. 16-33. The bullet is embedded in the block. Determine (a) the speed of the block immediately after the collision and (b) the amplitude of the resulting simple harmonic motion.

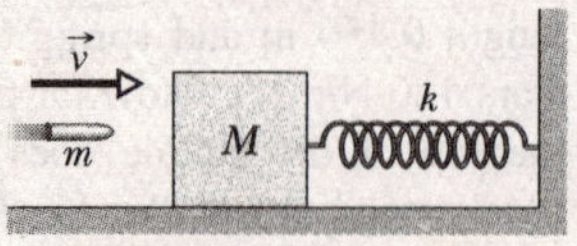

Fig. 16-33 Exercise 36.

37E. When the displacement in SHM is one-half the amplitude x_m, what fraction of the total energy is (a) kinetic energy and (b) potential energy? (c) At what displacement, in terms of the amplitude, is the energy of the system half kinetic energy and half potential energy? ssm

38P. A 10 g particle is undergoing simple harmonic motion with an amplitude of 2.0×10^{-3} m and a maximum acceleration of magnitude 8.0×10^3 m/s^2. The phase constant is $-\pi/3$ rad. (a) Write an equation for the force on the particle as a function of time. (b) What is the period of the motion? (c) What is the maximum speed of the particle? (d) What is the total mechanical energy of this simple harmonic oscillator?

39P*. A 4.0 kg block is suspended from a spring with a spring constant of 500 N/m. A 50 g bullet is fired into the block from directly below with a speed of 150 m/s and becomes embedded in the block. (a) Find the amplitude of the resulting simple harmonic motion. (b) What fraction of the original kinetic energy of the bullet is transferred to mechanical energy of the harmonic oscillator? ssm www

SEC. 16-5 An Angular Simple Harmonic Oscillator

40E. A flat uniform circular disk has a mass of 3.00 kg and a radius of 70.0 cm. It is suspended in a horizontal plane by a vertical wire attached to its center. If the disk is rotated 2.50 rad about the wire, a torque of 0.0600 N·m is required to maintain that orientation. Calculate (a) the rotational inertia of the disk about the wire, (b) the torsion constant, and (c) the angular frequency of this torsion pendulum when it is set oscillating.

41P. The balance wheel of a watch oscillates with an angular amplitude of π rad and a period of 0.500 s. Find (a) the maximum angular speed of the wheel, (b) the angular speed of the wheel when its displacement is $\pi/2$ rad, and (c) the magnitude of the angular acceleration of the wheel when its displacement is $\pi/4$ rad. ssm

SEC. 16-6 Pendulums

42E. In Fig. 16-34, a 2500 kg demolition ball swings from the end of a crane. The length of the swinging segment of cable is 17 m. (a) Find the period of the swinging, assuming that the system can be treated as a simple pendulum. (b) Does the period depend on the ball's mass?

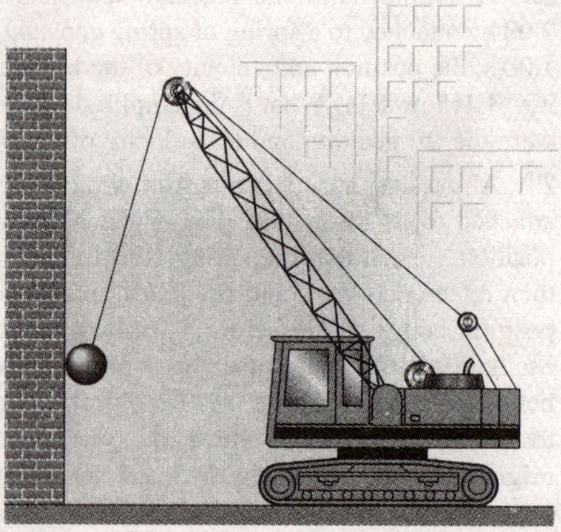
Fig. 16-34 Exercise 42.

43E. What is the length of a simple pendulum that marks seconds by completing a full swing from left to right and then back again every 2.0 s? ssm

44E. A performer seated on a trapeze is swinging back and forth with a period of 8.85 s. If she stands up, thus raising the center of mass of the *trapeze + performer* system by 35.0 cm, what will be the new period of the system? Treat *trapeze + performer* as a simple pendulum.

45E. A physical pendulum consists of a meter stick that is pivoted at a small hole drilled through the stick a distance d from the 50 cm mark. The period of oscillation is 2.5 s. Find d. ilw

46E. In Fig. 16-35, a physical pendulum consists of a uniform solid disk (of mass M and radius R) supported in a vertical plane by a pivot located a distance d from the center of the disk. The disk is displaced by a small angle and released. Find an expression for the period of the resulting simple harmonic motion.

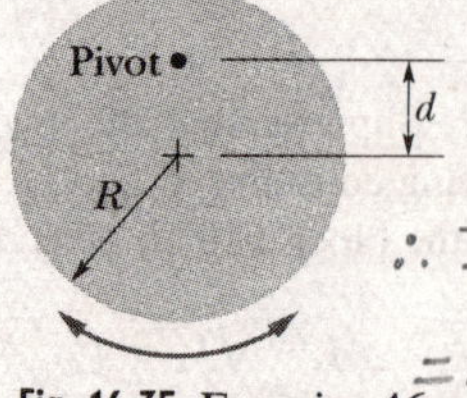

Fig. 16-35 Exercise 46.

47E. A pendulum is formed by pivoting a long thin rod of length L and mass m about a point on the rod that is a distance d above the center of the rod. (a) Find the period of this pendulum in terms of d, L, m, and g, assuming small-amplitude swinging. What happens to the period if (b) d is decreased, (c) L is increased, or (d) m is increased? ssm

48E. A uniform circular disk whose radius R is 12.5 cm is suspended as a physical pendulum from a point on its rim. (a) What is its period? (b) At what radial distance $r < R$ is there a pivot point that gives the same period?

•49E. The pendulum in Fig. 16-36 consists of a uniform disk with

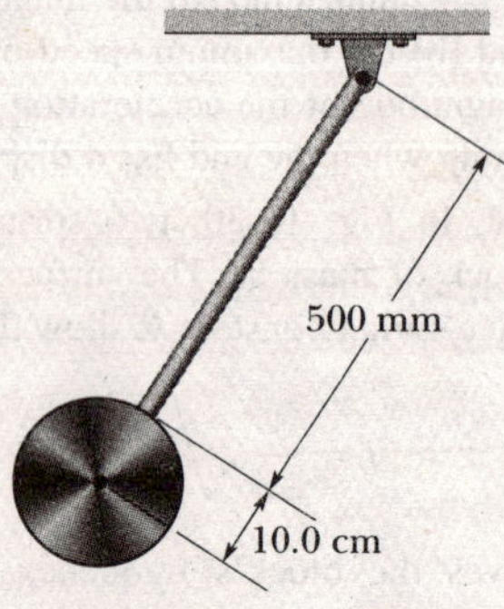

Fig. 16-36 Exercise 49.

radius 10.0 cm and mass 500 g attached to a uniform rod with length 500 mm and mass 270 g. (a) Calculate the rotational inertia of the pendulum about the pivot point. (b) What is the distance between the pivot point and the center of mass of the pendulum? (c) Calculate the period of oscillation. ssm

50E. (a) If the physical pendulum of Sample Problem 16-5 is inverted and suspended at point P, what is its period of oscillation? (b) Is the period now greater than, less than, or equal to its previous value?

51E. In Sample Problem 16-5, we saw that a physical pendulum has a center of oscillation at distance $2L/3$ from its point of suspension. Show that the distance between the point of suspension and the center of oscillation for a physical pendulum of any form is I/mh, where I and h have the meanings assigned to them in Eq. 16-29, and m is the mass of the pendulum.

52P. A stick with length L oscillates as a physical pendulum, pivoted about point O in Fig. 16-37. (a) Derive an expression for the period of the pendulum in terms of L and x, the distance from the pivot point to the center of mass of the pendulum. (b) For what value of x/L is the period a minimum? (c) Show that if L = 1.00 m and g = 9.80 m/s^2, this minimum is 1.53 s.

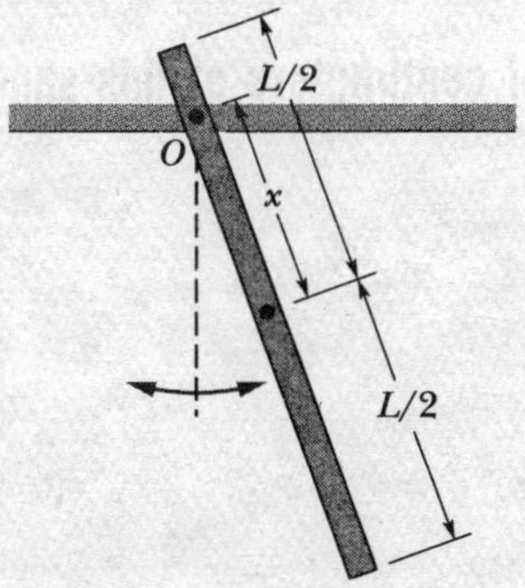

Fig. 16-37 Problem 52.

53P. In the overhead view of Fig. 16-38, a long uniform rod of length L and mass m is free to rotate in a horizontal plane about a vertical axis through its center. A spring with force constant k is connected horizontally between one end of the rod and a fixed wall. When the rod is in equilibrium, it is parallel to the wall. What is the period of the small oscillations that result when the rod is rotated slightly and released? ssm ilw www

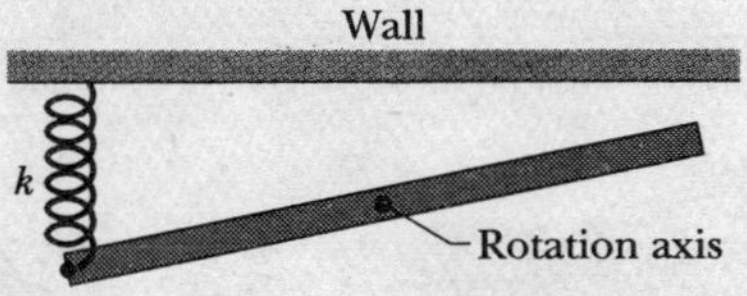

Fig. 16-38 Problem 53.

54P. A simple pendulum of length L and mass m is suspended in a car that is traveling with constant speed v around a circle of radius R. If the pendulum undergoes small oscillations in a radial direction about its equilibrium position, what will be its frequency of oscillation?

55P. What is the frequency of a simple pendulum 2.0 m long (a) in a room, (b) in an elevator accelerating upward at a rate of 2.0 m/s^2, and (c) in free fall? ssm

56P. For a simple pendulum, find the angular amplitude θ_m at which the restoring torque required for simple harmonic motion deviates from the actual restoring torque by 1.0%. (See "Trigonometric Expansions" in Appendix E.)

57P. The bob on a simple pendulum of length R moves in an arc of a circle. (a) By considering that the radial acceleration of the bob as it moves through its equilibrium position is that for uniform circular motion (v^2/R), show that the tension in the string at that position is $mg(1 + \theta_m^2)$ if the angular amplitude θ_m is small. (See "Trigonometric Expansions" in Appendix E.) (b) Is the tension at other positions of the bob greater, smaller, or the same?

58P. A wheel is free to rotate about its fixed axle. A spring is attached to one of its spokes a distance r from the axle, as shown in Fig. 16-39. (a) Assuming that the wheel is a hoop of mass m and radius R, obtain the angular frequency of small oscillations of this system in terms of m, R, r, and the spring constant k. How does the result change if (b) $r = R$ and (c) $r = 0$?

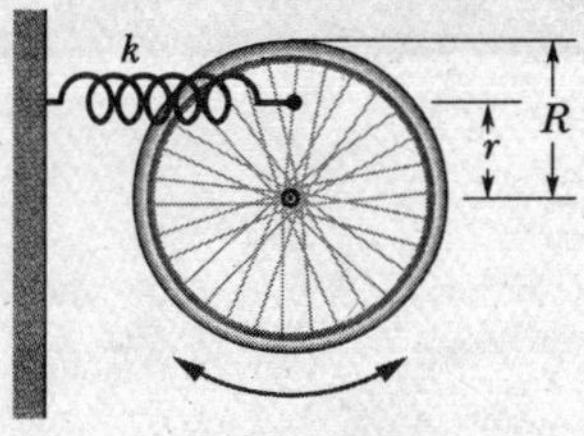

Fig. 16-39 Problem 58.

SEC. 16-8 Damped Simple Harmonic Motion

59E. In Sample Problem 16-7, what is the ratio of the amplitude of the damped oscillations to the initial amplitude when 20 full oscillations have elapsed? ssm

60E. The amplitude of a lightly damped oscillator decreases by 3.0% during each cycle. What fraction of the mechanical energy of the oscillator is lost in each full oscillation?

61E. For the system shown in Fig. 16-15, the block has a mass of 1.50 kg and the spring constant is 8.00 N/m. The damping force is given by $-b(dx/dt)$, where b = 230 g/s. Suppose that the block is initially pulled down a distance 12.0 cm and released. (a) Calculate the time required for the amplitude of the resulting oscillations to fall to one-third of its initial value. (b) How many oscillations are made by the block in this time? ssm

62P. Assume that you are examining the oscillation characteristics of the suspension system of a 2000 kg automobile. The suspension "sags" 10 cm when the entire automobile is placed on it. Also, the amplitude of oscillation decreases by 50% during one complete oscillation. Estimate the values of (a) the spring constant k and (b) the damping constant b for the spring and shock absorber system of one wheel, assuming each wheel supports 500 kg.

SEC. 16-9 Forced Oscillations and Resonance

63E. For Eq. 16-43, suppose the amplitude x_m is given by

$$x_m = \frac{F_m}{[m^2(\omega_d^2 - \omega^2)^2 + b^2\omega_d^2]^{1/2}},$$

where F_m is the (constant) amplitude of the external oscillating force exerted on the spring by the rigid support in Fig. 16-15. At resonance, what are (a) the amplitude and (b) the velocity amplitude of the oscillating object?

64P. A 1000 kg car carrying four 82 kg people travels over a rough "washboard" dirt road with corrugations 4.0 m apart, which cause the car to bounce on its spring suspension. The car bounces with maximum amplitude when its speed is 16 km/h. The car now stops, and the four people get out. By how much does the car body rise on its suspension owing to this decrease in mass?

17 Waves—I

When a beetle moves along the sand within a few tens of centimeters of this sand scorpion, the scorpion immediately turns toward the beetle and dashes to it (for lunch). The scorpion can do this without seeing (it is nocturnal) or hearing the beetle.

How can the scorpion so precisely locate its prey?

The answer is in this chapter.

17-1 Waves and Particles

Two ways to get in touch with a friend in a distant city are to write a letter and to use the telephone.

The first choice (the letter) involves the concept of "particle": A material object moves from one point to another, carrying with it information and energy. Most of the preceding chapters deal with particles or with systems of particles.

The second choice (the telephone) involves the concept of "wave," the subject of this chapter and the next. In a wave, information and energy move from one point to another but no material object makes that journey. In your telephone call, a sound wave carries your message from your vocal cords to the telephone. There, an electromagnetic wave takes over, passing along a copper wire or an optical fiber or through the atmosphere, possibly by way of a communications satellite. At the receiving end there is another sound wave, from a telephone to your friend's ear. Although the message is passed, nothing that you have touched reaches your friend. Leonardo da Vinci understood about waves when he wrote of water waves: "It often happens that the wave flees the place of its creation, while the water does not; like the waves made in a field of grain by the wind, where we see the waves running across the field while the grain remains in place."

Particle and *wave* are the two great concepts in classical physics, in the sense that we seem able to associate almost every branch of the subject with one or the other. The two concepts are quite different. The word *particle* suggests a tiny concentration of matter capable of transmitting energy. The word *wave* suggests just the opposite—namely, a broad distribution of energy, filling the space through which it passes. The job at hand is to put aside particles for a while and to learn something about waves.

17-2 Types of Waves

Waves are of three main types:

1. ***Mechanical waves.*** These waves are most familiar because we encounter them almost constantly; common examples include water waves, sound waves, and seismic waves. All these waves have certain central features: They are governed by Newton's laws, and they can exist only within a material medium, such as water, air, and rock.

2. ***Electromagnetic waves.*** These waves are less familiar, but you use them constantly; common examples include visible and ultraviolet light, radio and television waves, microwaves, x rays, and radar waves. These waves require no material medium to exist. Light waves from stars, for example, travel through the vacuum of space to reach us. All electromagnetic waves travel through a vacuum at the same speed c, given by

$$c = 299\,792\,458 \text{ m/s} \qquad \text{(speed of light).} \qquad (17\text{-}1)$$

3. ***Matter waves.*** Although these waves are commonly used in modern technology, their type is probably very unfamiliar to you. These waves are associated with electrons, protons, and other fundamental particles, and even atoms and molecules. Because we commonly think of these things as constituting matter, such waves are called matter waves.

Much of what we discuss in this chapter applies to waves of all kinds. However, for specific examples we shall refer to mechanical waves.

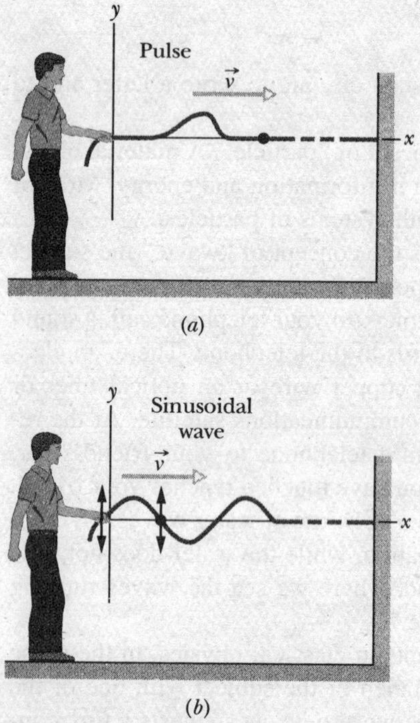

Fig. 17-1 (*a*) A single pulse is sent along a stretched string. A typical string element (marked with a dot) moves up once and then down as the pulse passes. The element's motion is perpendicular to the wave's direction of travel, so the pulse is a *transverse wave*. (*b*) A sinusoidal wave is sent along the string. A typical string element moves up and down continuously as the wave passes. This too is a transverse wave.

17-3 Transverse and Longitudinal Waves

A wave sent along a stretched, taut string is the simplest mechanical wave. If you give one end of a stretched string a single up-and-down jerk, a wave in the form of a single *pulse* travels along the string, as in Fig. 17-1*a*. This pulse and its motion can occur because the string is under tension. When you pull your end of the string upward, it begins to pull upward on the adjacent section of the string via tension between the two sections. As the adjacent section moves upward, it begins to pull the next section upward, and so on. Meanwhile, you have pulled down on your end of the string. As each section moves upward in turn, it begins to be pulled back downward by neighboring sections that are already on the way down. The net result is that a distortion in the string's shape (the pulse) moves along the string at some velocity $\vec{v}$.

If you move your hand up and down in continuous simple harmonic motion, a continuous wave travels along the string at velocity $\vec{v}$. Because the motion of your hand is a sinusoidal function of time, the wave has a sinusoidal shape at any given instant, as in Fig. 17-1*b*; that is, the wave has the shape of a sine curve or a cosine curve.

We consider here only an "ideal" string, in which no frictionlike forces within the string cause the wave to die out as it travels along the string. In addition, we assume that the string is so long that we need not consider a wave rebounding from the far end.

One way to study the waves of Fig. 17-1 is to monitor the **wave forms** (shapes of the waves) as they move to the right. Alternatively, we could monitor the motion of an element of the string as the element oscillates up and down while a wave passes through it. We would find that the displacement of every such oscillating string element is *perpendicular* to the direction of travel of the wave, as indicated in Fig. 17-1. This motion is said to be **transverse,** and the wave is said to be a **transverse wave.**

Figure 17-2 shows how a sound wave can be produced by a piston in a long, air-filled pipe. If you suddenly move the piston rightward and then leftward, you can send a pulse of sound along the pipe. The rightward motion of the piston moves the elements of air next to it rightward, changing the air pressure there. The increased air pressure then pushes rightward on the elements of air somewhat farther along the pipe. Moving the piston leftward reduces the air pressure next to it. Once they have moved rightward, the nearest elements, and then farther elements, move back leftward. Thus, the motion of the air and the change in air pressure travel rightward along the pipe as a pulse.

If you push and pull on the piston in simple harmonic motion, as is being done in Fig. 17-2, a sinusoidal wave travels along the pipe. Because the motion of the elements of air is parallel to the direction of the wave's travel, the motion is said to be **longitudinal,** and the wave is said to be a **longitudinal wave.** In this chapter we concentrate on transverse waves, and string waves in particular; in Chapter 18 we shall concentrate on longitudinal waves, and sound waves in particular.

Both a transverse wave and a longitudinal wave are said to be **traveling waves** because they both travel from one point to another, as from one end of the string to the other end in Fig. 17-1 or from one end of the pipe to the other end in Fig. 17-2. Note that it is the wave that moves from end to end, not the material (string or air) through which the wave moves.

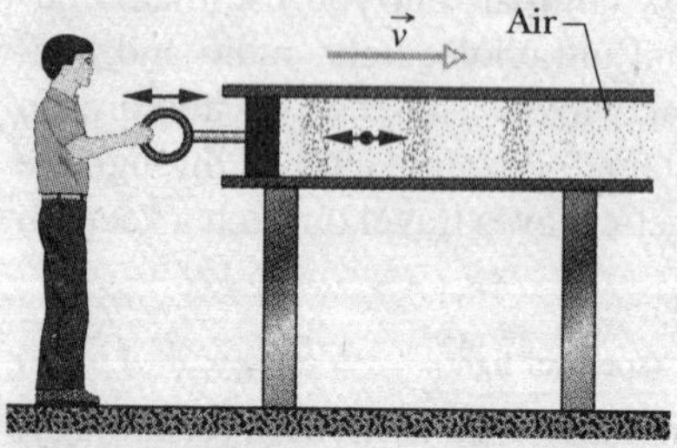

Fig. 17-2 A sound wave is set up in an air-filled pipe by moving a piston back and forth. Because the oscillations of an element of the air (represented by the black dot) are parallel to the direction in which the wave travels, the wave is a *longitudinal wave*.

The sand scorpion shown in the photograph opening this chapter uses waves of both transverse and longitudinal motion to locate its prey. When a beetle even slightly disturbs the sand, it sends pulses along the sand's surface (Fig. 17-3). One

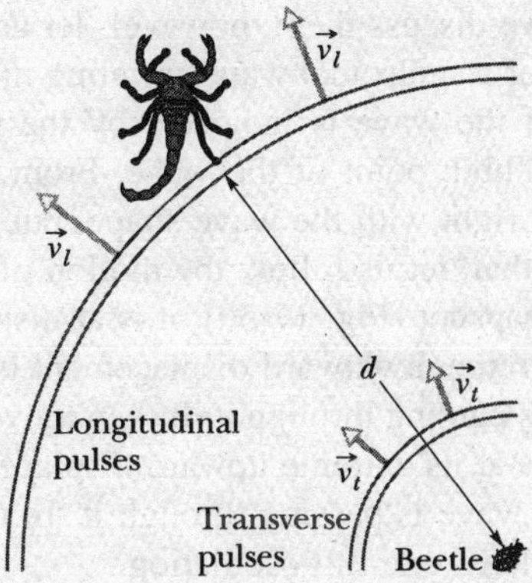

Fig. 17-3 A beetle's motion sends fast longitudinal pulses and slower transverse pulses along the sand's surface. The sand scorpion first intercepts the longitudinal pulses; here, it is the rearmost right leg that senses the pulses earliest.

set of pulses is longitudinal, traveling with speed $v_l = 150$ m/s. A second set is transverse, traveling with speed $v_t = 50$ m/s.

The scorpion, with its eight legs spread roughly in a circle about 5 cm in diameter, intercepts the faster longitudinal pulses first and learns the direction of the beetle; it is in the direction of whichever leg is disturbed earliest by the pulses. The scorpion then senses the time interval Δt between that first interception and the interception of the slower transverse waves and uses it to determine the distance d to the beetle. This distance is given by

$$\Delta t = \frac{d}{v_t} - \frac{d}{v_l},$$

and it turns out to be

$$d = (75 \text{ m/s})\, \Delta t.$$

For example, if $\Delta t = 4.0$ ms, then $d = 30$ cm, which gives the scorpion a perfect fix on the beetle.

17-4 Wavelength and Frequency

To completely describe a wave on a string (and the motion of any element along its length), we need a function that gives the shape of the wave. This means that we need a relation in the form $y = h(x, t)$, in which y is the transverse displacement of any string element as a function h of the time t and the position x of the element along the string. In general, a sinusoidal shape like the wave in Fig. 17-1b can be described with h being either a sine function or a cosine function; both give the same general shape for the wave. In this chapter we use the sine function.

Imagine a sinusoidal wave like that of Fig. 17-1b traveling in the positive direction of an x axis. As the wave sweeps through succeeding elements (that is, very short sections) of the string, the elements oscillate parallel to the y axis. At time t, the displacement y of the element located at position x is given by

$$y(x, t) = y_m \sin(kx - \omega t). \tag{17-2}$$

Because this equation is written in terms of position x, it can be used to find the displacements of all the elements of the string as a function of time. Thus, it can tell us the shape of the wave at any given time and how that shape changes as the wave moves along the string. The names of the quantities in Eq. 17-2 are displayed in Fig. 17-4 and defined next.

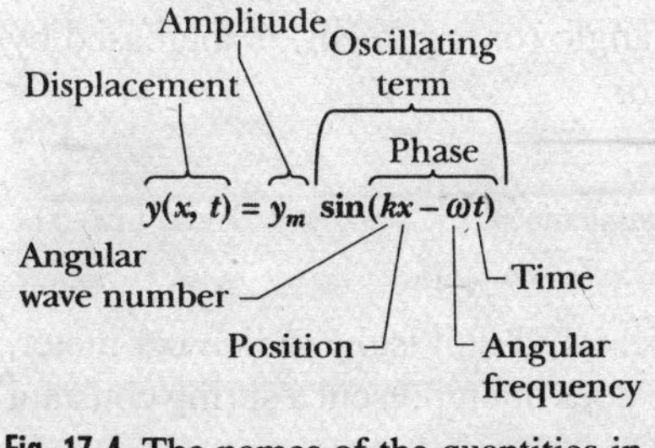

Fig. 17-4 The names of the quantities in Eq. 17-2, for a transverse sinusoidal wave.

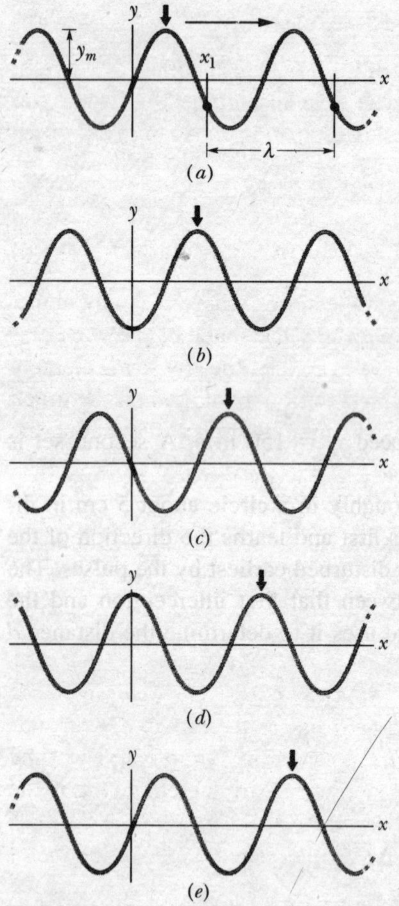

Fig. 17-5 Five "snapshots" of a string wave traveling in the positive direction of an x axis. The amplitude y_m is indicated. A typical wavelength λ, measured from an arbitrary position x_1, is also indicated.

Before we discuss them, however, let us examine Fig. 17-5, which shows five "snapshots" of a sinusoidal wave traveling in the positive direction of an x axis. The movement of the wave is indicated by the rightward progress of the short arrow pointing to a high point of the wave. From snapshot to snapshot, the short arrow moves to the right with the wave shape, but the string moves *only* parallel to the y axis. To see that, let us follow the motion of the red-dyed string element at $x = 0$. In the first snapshot (Fig. 17-5*a*), it is at displacement $y = 0$. In the next snapshot, it is at its extreme downward displacement because a *valley* (or extreme low point) of the wave is passing through it. It then moves back up through $y = 0$. In the fourth snapshot, it is at its extreme upward displacement because a *peak* (or extreme high point) of the wave is passing through it. In the fifth snapshot, it is again at $y = 0$, having completed one full oscillation.

Amplitude and Phase

The **amplitude** y_m of a wave, such as that in Fig. 17-5, is the magnitude of the maximum displacement of the elements from their equilibrium positions as the wave passes through them. (The subscript m stands for maximum.) Because y_m is a magnitude, it is always a positive quantity, even if it is measured downward in Fig. 17-5*a*, instead of upward as drawn.

The **phase** of the wave is the *argument* $kx - \omega t$ of the sine in Eq. 17-2. As the wave sweeps through a string element at a particular position x, the phase changes linearly with time t. This means that the sine also changes, oscillating between $+1$ and -1. Its extreme positive value $(+1)$ corresponds to a peak of the wave moving through the element; then, the value of y at position x is y_m. Its extreme negative value (-1) corresponds to a valley of the wave moving through the element; then, the value of y at position x is $-y_m$. Thus, the sine function and the time-dependent phase of a wave correspond to the oscillation of a string element, and the amplitude of the wave determines the extremes of the element's displacement.

Wavelength and Angular Wave Number

The **wavelength** λ of a wave is the distance (parallel to the direction of the wave's travel) between repetitions of the shape of the wave (or *wave shape*). A typical wavelength is marked in Fig. 17-5*a*, which is a snapshot of the wave at time $t = 0$. At that time, Eq. 17-2 gives, for the description of the wave shape,

$$y(x, 0) = y_m \sin kx. \tag{17-3}$$

By definition, the displacement y is the same at both ends of this wavelength—that is, at $x = x_1$ and $x = x_1 + \lambda$. Thus, by Eq. 17-3,

$$\begin{aligned} y_m \sin kx_1 &= y_m \sin k(x_1 + \lambda) \\ &= y_m \sin(kx_1 + k\lambda). \end{aligned} \tag{17-4}$$

A sine function begins to repeat itself when its angle (or argument) is increased by 2π rad, so in Eq. 17-4 we must have $k\lambda = 2\pi$, or

$$k = \frac{2\pi}{\lambda} \quad \text{(angular wave number).} \tag{17-5}$$

We call k the **angular wave number** of the wave; its SI unit is the radian per meter, or the inverse meter. (Note that the symbol k here does *not* represent a spring constant as previously.)

Notice that the wave in Fig. 17-5 moves to the right by $\frac{1}{4}\lambda$ from one snapshot to the next. Thus, by the fifth snapshot, it has moved to the right by 1λ.

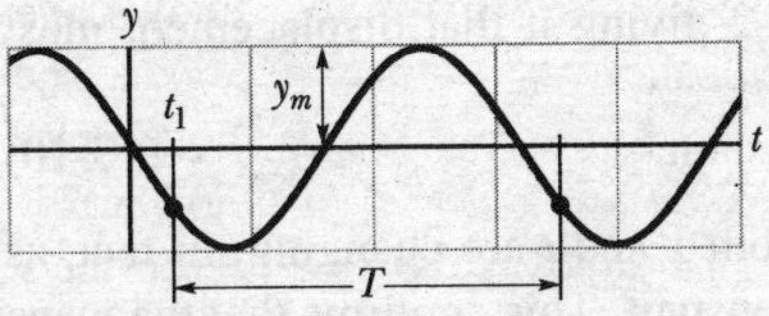

Fig. 17-6 A graph of the displacement of the string element at $x = 0$ as a function of time, as the sinusoidal wave of Fig. 17-5 passes through it. The amplitude y_m is indicated. A typical period T, measured from an arbitrary time t_1, is also indicated.

Period, Angular Frequency, and Frequency

Figure 17-6 shows a graph of the displacement y of Eq. 17-2 versus time t at a certain position along the string, taken to be $x = 0$. If you were to monitor the string, you would see that the single element of the string at that position moves up and down in simple harmonic motion given by Eq. 17-2 with $x = 0$:

$$\begin{aligned} y(0, t) &= y_m \sin(-\omega t) \\ &= -y_m \sin \omega t \qquad (x = 0). \end{aligned} \tag{17-6}$$

Here we have made use of the fact that $\sin(-\alpha) = -\sin \alpha$, where α is any angle. Figure 17-6 is a graph of this equation; it *does not* show the shape of the wave.

We define the **period** of oscillation T of a wave to be the time any string element takes to move through one full oscillation. A typical period is marked on the graph of Fig. 17-6. Applying Eq. 17-6 to both ends of this time interval and equating the results yield

$$\begin{aligned} -y_m \sin \omega t_1 &= -y_m \sin \omega(t_1 + T) \\ &= -y_m \sin(\omega t_1 + \omega T). \end{aligned} \tag{17-7}$$

This can be true only if $\omega T = 2\pi$, or if

$$\omega = \frac{2\pi}{T} \qquad \text{(angular frequency).} \tag{17-8}$$

We call ω the **angular frequency** of the wave; its SI unit is the radian per second.

Look back at the five snapshots of a traveling wave in Fig. 17-5. The time between snapshots is $\frac{1}{4}T$. Thus, by the fifth snapshot, every string element has made one full oscillation.

The **frequency** f of a wave is defined as $1/T$ and is related to the angular frequency ω by

$$f = \frac{1}{T} = \frac{\omega}{2\pi} \qquad \text{(frequency).} \tag{17-9}$$

Like the frequency of simple harmonic motion in Chapter 16, this frequency f is a number of oscillations per unit time—here, the number made by a string element as the wave moves through it. As in Chapter 16, f is usually measured in hertz or its multiples, such as kilohertz.

CHECKPOINT 1: The figure is a composite of three snapshots, each of a wave traveling along a particular string. The phases for the waves are given by (a) $2x - 4t$, (b) $4x - 8t$, and (c) $8x - 16t$. Which phase corresponds to which wave in the figure?

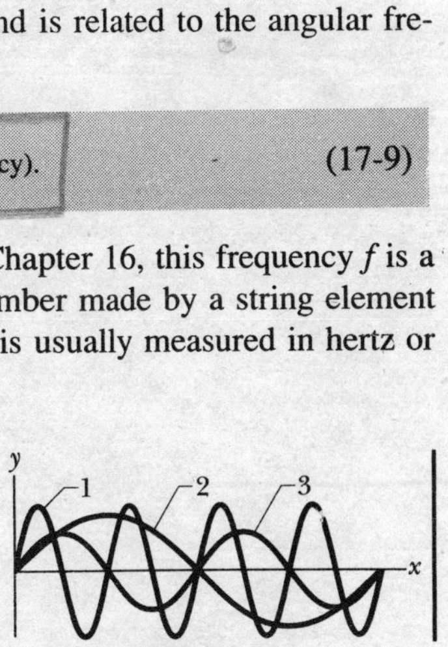

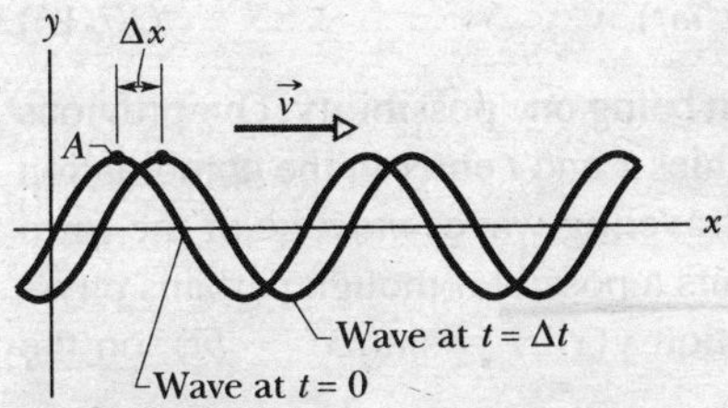

Fig. 17-7 Two snapshots of the wave of Fig. 17-5, at time $t = 0$ and then at time $t = \Delta t$. As the wave moves to the right at velocity $\vec{v}$, the entire curve shifts a distance Δx during Δt. Point A "rides" with the wave form but the string elements move only up and down.

17-5 The Speed of a Traveling Wave

Figure 17-7 shows two snapshots of the wave of Eq. 17-2, taken a small time interval Δt apart. The wave is traveling in the positive direction of x (to the right in Fig. 17-7), the entire wave pattern moving a distance Δx in that direction during the interval Δt. The ratio $\Delta x/\Delta t$ (or, in the differential limit, dx/dt) is the **wave speed** v. How can we find its value?

As the wave in Fig. 17-7 moves, each point of the moving wave form, such as point A marked on a peak, retains its displacement y. (Points on the string do not retain their displacement, but points on the wave *form* do.) If point A retains its displacement as it moves, the phase in Eq. 17-2 giving it that displacement must remain a constant:

$$kx - \omega t = \text{a constant.} \tag{17-10}$$

Note that although this argument is constant, both x and t are changing. In fact, as t increases, x must also, to keep the argument constant. This confirms that the wave pattern is moving in the positive direction of x.

To find the wave speed v, we take the derivative of Eq. 17-10, getting

$$k\frac{dx}{dt} - \omega = 0$$

or

$$\frac{dx}{dt} = v = \frac{\omega}{k}. \tag{17-11}$$

Using Eq. 17-5 ($k = 2\pi/\lambda$) and Eq. 17-8 ($\omega = 2\pi/T$), we can rewrite the wave speed as

$$v = \frac{\omega}{k} = \frac{\lambda}{T} = \lambda f \quad \text{(wave speed).} \tag{17-12}$$

The equation $v = \lambda/T$ tells us that the wave speed is one wavelength per period; the wave moves a distance of one wavelength in one period of oscillation.

Equation 17-2 describes a wave moving in the positive direction of x. We can find the equation of a wave traveling in the opposite direction by replacing t in Eq. 17-2 with $-t$. This corresponds to the condition

$$kx + \omega t = \text{a constant,} \tag{17-13}$$

which (compare Eq. 17-10) requires that x *decrease* with time. Thus, a wave traveling in the negative direction of x is described by the equation

$$y(x, t) = y_m \sin(kx + \omega t). \tag{17-14}$$

If you analyze the wave of Eq. 17-14 as we have just done for the wave of Eq. 17-2, you will find for its velocity

$$\frac{dx}{dt} = -\frac{\omega}{k}. \tag{17-15}$$

The minus sign (compare Eq. 17-11) verifies that the wave is indeed moving in the negative direction of x and justifies our switching the sign of the time variable.

Consider now a wave of arbitrary shape, given by

$$y(x, t) = h(kx \pm \omega t), \tag{17-16}$$

where h represents *any* function, the sine function being one possibility. Our previous analysis shows that all waves in which the variables x and t enter in the combination $kx \pm \omega t$ are traveling waves. Furthermore, all traveling waves *must* be of the form of Eq. 17-16. Thus, $y(x, t) = \sqrt{ax + bt}$ represents a possible (though perhaps physically a little bizarre) traveling wave. The function $y(x, t) = \sin(ax^2 - bt)$, on the other hand, does *not* represent a traveling wave.

Sample Problem 17-1

A wave traveling along a string is described by

$$y(x, t) = 0.00327 \sin(72.1x - 2.72t), \tag{17-17}$$

in which the numerical constants are in SI units (0.00327 m, 72.1 rad/m, and 2.72 rad/s).

(a) What is the amplitude of this wave?

SOLUTION: The **Key Idea** is that Eq. 17-17 is of the same form as Eq. 17-2,

$$y = y_m \sin(kx - \omega t), \qquad (17\text{-}18)$$

so we have a sinusoidal wave. By comparing the two equations, we see that the amplitude is

$$y_m = 0.00327 \text{ m} = 3.27 \text{ mm}. \qquad \text{(Answer)}$$

(b) What are the wavelength, period, and frequency of this wave?

SOLUTION: By comparing Eqs. 17-17 and 17-18, we see that the angular wave number and angular frequency are

$$k = 72.1 \text{ rad/m} \quad \text{and} \quad \omega = 2.72 \text{ rad/s}.$$

We then relate wavelength λ to k via Eq. 17-5:

$$\lambda = \frac{2\pi}{k} = \frac{2\pi \text{ rad}}{72.1 \text{ rad/m}}$$
$$= 0.0871 \text{ m} = 8.71 \text{ cm}. \qquad \text{(Answer)}$$

Next, we relate T to ω with Eq. 17-8:

$$T = \frac{2\pi}{\omega} = \frac{2\pi \text{ rad}}{2.72 \text{ rad/s}} = 2.31 \text{ s}, \qquad \text{(Answer)}$$

and from Eq. 17-9 we have

$$f = \frac{1}{T} = \frac{1}{2.31 \text{ s}} = 0.433 \text{ Hz}. \qquad \text{(Answer)}$$

(c) What is the velocity of this wave?

SOLUTION: The speed of the wave is given by Eq. 17-12:

$$v = \frac{\omega}{k} = \frac{2.72 \text{ rad/s}}{72.1 \text{ rad/m}} = 0.0377 \text{ m/s}$$
$$= 3.77 \text{ cm/s}. \qquad \text{(Answer)}$$

Because the phase in Eq. 17-17 contains the position variable x, the wave is moving along the x axis. Also, because the wave equation is written in the form of Eq. 17-2, the *minus* sign in front of the ωt term indicates that the wave is moving in the *positive* direction of the x axis. (Note that the quantities calculated in (b) and (c) are independent of the amplitude of the wave.)

(d) What is the displacement y at $x = 22.5$ cm and $t = 18.9$ s?

SOLUTION: The **Key Idea** here is that Eq. 17-17 gives the displacement as a function of position x and time t. Substituting the given values into the equation yields

$$y = 0.00327 \sin(72.1 \times 0.225 - 2.72 \times 18.9)$$
$$= (0.00327 \text{ m}) \sin(-35.1855 \text{ rad})$$
$$= (0.00327 \text{ m})(0.588)$$
$$= 0.00192 \text{ m} = 1.92 \text{ mm}. \qquad \text{(Answer)}$$

Thus, the displacement is positive. (Be sure to change your calculator mode to radians before evaluating the sine.)

Sample Problem 17-2

In Sample Problem 17-1d, we showed that at $t = 18.9$ s the transverse displacement y of the element of the string at $x = 0.255$ m due to the wave of Eq. 17-17 is 1.92 mm.

(a) What is u, the transverse velocity of the same element of the string, at that time? (This speed, which is associated with the transverse oscillation of an element of the string, is in the y direction. Do not confuse it with v, the constant velocity at which the *wave form* travels along the x axis.)

SOLUTION: The **Key Idea** here is that the transverse velocity u is the rate at which the displacement y of the element is changing. In general, that displacement is given by

$$y(x, t) = y_m \sin(kx - \omega t). \qquad (17\text{-}19)$$

For an element at a certain location x, we find the rate of change of y by taking the derivative of Eq. 17-19 with respect to t while treating x as a constant. A derivative taken while one (or more) of the variables is treated as a constant is called a *partial derivative* and is represented by the symbol $\partial/\partial x$ rather than d/dx. Here we have

$$u = \frac{\partial y}{\partial t} = -\omega y_m \cos(kx - \omega t). \qquad (17\text{-}20)$$

Next, substituting numerical values from Sample Problem 17-1, we obtain

$$u = (-2.72 \text{ rad/s})(3.27 \text{ mm}) \cos(-35.1855 \text{ rad})$$
$$= 7.20 \text{ mm/s}. \qquad \text{(Answer)}$$

Thus, at $t = 18.9$ s, the element of string at $x = 22.5$ cm is moving in the positive direction of y, with a velocity of 7.20 mm/s.

(b) What is the transverse acceleration a_y of the same element at that time?

SOLUTION: The **Key Idea** here is that the transverse acceleration a_y is the rate at which the transverse velocity of the element is changing. From Eq. 17-20, again treating x as a constant but allowing t to vary, we find

$$a_y = \frac{\partial u}{\partial t} = -\omega^2 y_m \sin(kx - \omega t).$$

Comparison with Eq. 17-19 shows that we can write this as

$$a_y = -\omega^2 y.$$

We see that the transverse acceleration of an oscillating string element is proportional to its transverse displacement but opposite in sign. This is completely consistent with the action of the element itself—namely, that it is moving transversely in simple harmonic motion. Substituting numerical values yields

$$a_y = -(2.72 \text{ rad/s})^2(1.92 \text{ mm})$$
$$= -14.2 \text{ mm/s}^2. \qquad \text{(Answer)}$$

Thus, at $t = 18.9$ s, the element of string at $x = 22.5$ cm is displaced from its equilibrium position by 1.92 mm in the positive y direction and has an acceleration of magnitude 14.2 mm/s^2 in the negative y direction.

✔CHECKPOINT 2: Here are the equations of three waves:
(1) $y(x, t) = 2\sin(4x - 2t)$, (2) $y(x, t) = \sin(3x - 4t)$, (3) $y(x, t) = 2\sin(3x - 3t)$.
Rank the waves according to their (a) wave speed and (b) maximum transverse speed, greatest first.

PROBLEM-SOLVING TACTICS

Tactic 1: *Evaluating Large Phases*

Sometimes, as in Sample Problems 17-1d and 17-2, an angle much greater than 2π rad (or 360°) crops up and you are asked to find its sine or cosine. Adding or subtracting an integral multiple of 2π rad to such an angle does not change the value of any of its trigonometric functions. In Sample Problem 17-1d, for example, the angle is -35.1855 rad. Adding $(6)(2\pi \text{ rad})$ to this angle yields

$$-35.1855 \text{ rad} + (6)(2\pi \text{ rad}) = 2.51361 \text{ rad},$$

an angle of less than 2π rad that has the same trigonometric functions as -35.1855 rad (Fig. 17-8). As an example, the sine of both 2.51361 rad and -35.1855 rad is 0.588.

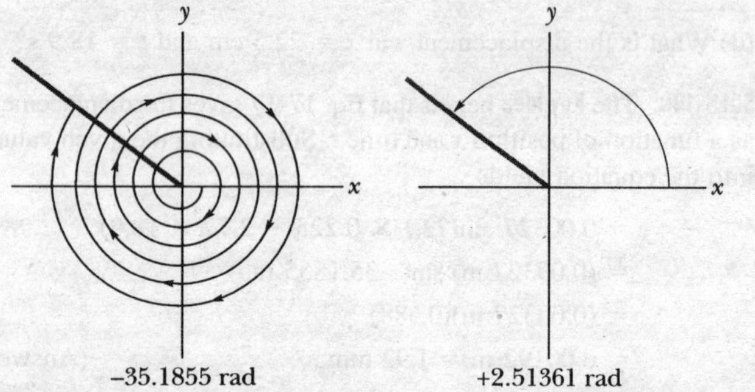

Fig. 17-8 These two angles are different but all their trigonometric functions are identical.

Your calculator will reduce such large angles for you automatically. *Caution:* Do not round off large angles if you intend to take their sines or cosines. In taking the sine of a very large angle, you are throwing away most of the angle and taking the sine of what is left over. If, for example, you were to round -35.1855 rad to -35 rad (a change of 0.5% and normally a reasonable step), you would be changing the sine of the angle by 27%. Also, if you change a large angle from degrees to radians, be sure to use an exact conversion factor (such as $180° = \pi$ rad) rather than an approximate one (such as $57.3° \approx 1$ rad).

17-6 Wave Speed on a Stretched String

The speed of a wave is related to the wave's wavelength and frequency by Eq. 17-12, but *it is set by the properties of the medium.* If a wave is to travel through a medium such as water, air, steel, or a stretched string, it must cause the particles of that medium to oscillate as it passes. For that to happen, the medium must possess both mass (so that there can be kinetic energy) and elasticity (so that there can be potential energy). Thus, the medium's mass and elasticity properties determine how fast the wave can travel in the medium. Conversely, it should be possible to calculate the speed of the wave through the medium in terms of these properties. We do so now for a stretched string, in two ways.

Dimensional Analysis

In dimensional analysis we carefully examine the dimensions of all the physical quantities that enter into a given situation to determine the quantities they produce. In this case, we examine mass and elasticity to find a speed v, which has the dimension of length divided by time, or LT^{-1}.

For the mass, we use the mass of a string element, which is represented by the mass m of the string divided by the length l of the string. We call this ratio the *linear density* μ of the string. Thus, $\mu = m/l$, its dimension being mass divided by length, ML^{-1}.

You cannot send a wave along a string unless the string is under tension, which means that it has been stretched and pulled taut by forces at its two ends. The tension τ in the string is equal to the common magnitude of those two forces. As a wave travels along the string, it displaces elements of the string by causing additional stretching, with adjacent sections of string pulling on each other because of the tension. Thus, we can associate the tension in the string with the stretching (elasticity) of the string. The tension and the stretching forces it produces have the dimension of a force—namely, MLT^{-2} (from $F = ma$).

The goal here is to combine μ (dimension ML^{-1}) and τ (dimension MLT^{-2}) in such a way as to generate v (dimension LT^{-1}). A little juggling of various combinations suggests

$$v = C\sqrt{\frac{\tau}{\mu}}, \tag{17-21}$$

in which C is a dimensionless constant that cannot be determined with dimensional analysis. In our second approach to determining wave speed, you will see that Eq. 17-21 is indeed correct and that $C = 1$.

Derivation from Newton's Second Law

Instead of the sinusoidal wave of Fig. 17-1*b*, let us consider a single symmetrical pulse such as that of Fig. 17-9, moving from left to right along a string with speed v. For convenience, we choose a reference frame in which the pulse remains stationary; that is, we run along with the pulse, keeping it constantly in view. In this frame, the string appears to move past us, from right to left in Fig. 17-9, with speed v.

Consider a small string element within the pulse, of length Δl, forming an arc of a circle of radius R and subtending an angle 2θ at the center of that circle. A force $\vec{\tau}$ with a magnitude equal to the tension in the string pulls tangentially on this element at each end. The horizontal components of these forces cancel, but the vertical components add to form a radial restoring force $\vec{F}$. In magnitude,

$$F = 2(\tau \sin \theta) \approx \tau (2\theta) = \tau \frac{\Delta l}{R} \quad \text{(force)}, \tag{17-22}$$

where we have approximated $\sin \theta$ as θ for the small angles θ in Fig. 17-9. From that figure, we have also used $2\theta = \Delta l / R$.

The mass of the element is given by

$$\Delta m = \mu \, \Delta l \quad \text{(mass)}, \tag{17-23}$$

where μ is the string's linear density.

At the moment shown in Fig. 17-9, the string element Δl is moving in an arc of a circle. Thus, it has a centripetal acceleration toward the center of that circle, given by

$$a = \frac{v^2}{R} \quad \text{(acceleration)}. \tag{17-24}$$

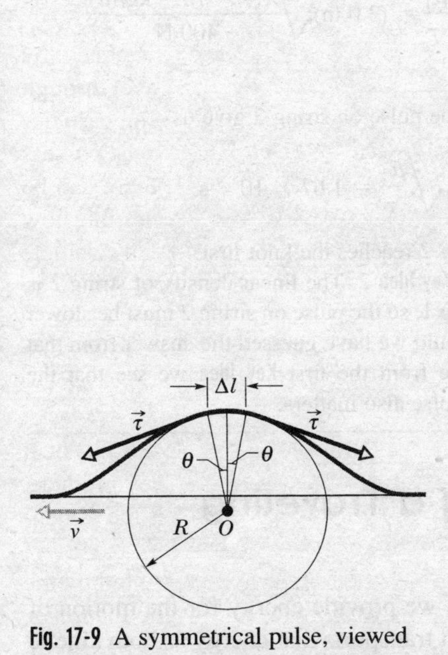

Fig. 17-9 A symmetrical pulse, viewed from a reference frame in which the pulse is stationary and the string appears to move right to left with speed v. We find speed v by applying Newton's second law to a string element of length Δl, located at the top of the pulse.

Equations 17-22, 17-23, and 17-24 contain the elements of Newton's second law. Combining them in the form

$$\text{force} = \text{mass} \times \text{acceleration}$$

gives

$$\frac{\tau \, \Delta l}{R} = (\mu \, \Delta l) \frac{v^2}{R}.$$

Solving this equation for the speed v yields

$$v = \sqrt{\frac{\tau}{\mu}} \quad \text{(speed)}, \tag{17-25}$$

in exact agreement with Eq. 17-21 if the constant C in that equation is given the value unity. Equation 17-25 gives the speed of the pulse in Fig. 17-9 and the speed of *any* other wave on the same string under the same tension.

Equation 17-25 tells us:

> The speed of a wave along a stretched ideal string depends only on the tension and linear density of the string and not on the frequency of the wave.

The *frequency* of the wave is fixed entirely by whatever generates the wave (for example, the person in Fig. 17-1*b*). The *wavelength* of the wave is then fixed by Eq. 17-12 in the form $\lambda = v/f$.

CHECKPOINT 3: You send a traveling wave along a particular string by oscillating one end. If you increase the frequency of the oscillations, do (a) the speed of the wave and (b) the wavelength of the wave increase, decrease, or remain the same? If, instead, you increase the tension in the string, do (c) the speed of the wave and (d) the wavelength of the wave increase, decrease, or remain the same?

Sample Problem 17-3

In Fig. 17-10, two strings have been tied together with a knot and then stretched between two rigid supports. The strings have linear densities $\mu_1 = 1.4 \times 10^{-4}$ kg/m and $\mu_2 = 2.8 \times 10^{-4}$ kg/m. Their lengths are $L_1 = 3.0$ m and $L_2 = 2.0$ m, and string 1 is under a tension of 400 N. Simultaneously, on each string a pulse is sent from the rigid support end, toward the knot. Which pulse reaches the knot first?

SOLUTION: We need several **Key Ideas** here:

1. The time t taken by a pulse to travel a length L is $t = L/v$, where v is the constant speed of the pulse.
2. The speed of a pulse on a stretched string depends on the string's tension τ and linear density μ, and is given by Eq. 17-25 ($v = \sqrt{\tau/\mu}$).
3. Because the two strings are stretched together, they must both be under the same tension τ (= 400 N).

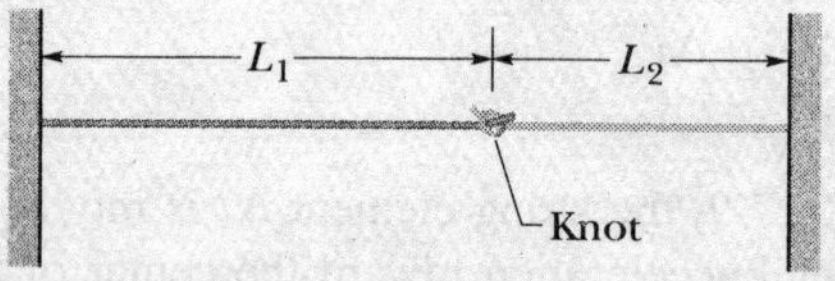

Fig. 17-10 Sample Problem 17-3. Two strings, of lengths L_1 and L_2, tied together with a knot and stretched between two rigid supports.

Putting these three ideas together gives us, as the time for the pulse on string 1 to reach the knot,

$$t_1 = \frac{L_1}{v_1} = L_1\sqrt{\frac{\mu_1}{\tau}} = (3.0 \text{ m})\sqrt{\frac{1.4 \times 10^{-4} \text{ kg/m}}{400 \text{ N}}}$$
$$= 1.77 \times 10^{-3} \text{ s.}$$

Similarly, the data for the pulse on string 2 give us

$$t_2 = L_2\sqrt{\frac{\mu_2}{\tau}} = 1.67 \times 10^{-3} \text{ s.}$$

Thus, the pulse on string 2 reaches the knot first.

Now look back at **Key Idea** 2. The linear density of string 2 is greater than that of string 1, so the pulse on string 2 must be slower than that on string 1. Could we have guessed the answer from that fact alone? No, because from the first **Key Idea** we see that the distance traveled by a pulse also matters.

17-7 Energy and Power of a Traveling String Wave

When we set up a wave on a stretched string, we provide energy for the motion of the string. As the wave moves away from us, it transports that energy as both kinetic energy and elastic potential energy. Let us consider each form in turn.

Kinetic Energy

An element of the string of mass dm, oscillating transversely in simple harmonic motion as the wave passes through it, has kinetic energy associated with its transverse

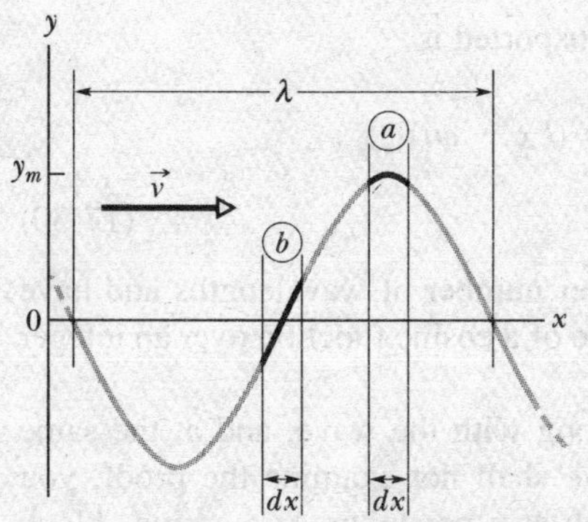

Fig. 17-11 A snapshot of a traveling wave on a string at time $t = 0$. String element a is at displacement $y = y_m$, and string element b is at displacement $y = 0$. The kinetic energy of the string element at each position depends on the transverse velocity of the element. The potential energy depends on the amount by which the string element is stretched as the wave passes through it.

velocity $\vec{u}$. When the element is rushing through its $y = 0$ position (element b in Fig. 17-11), its transverse velocity—and thus its kinetic energy—is a maximum. When the element is at its extreme position $y = y_m$ (as is element a), its transverse velocity—and thus its kinetic energy—is zero.

Elastic Potential Energy

To send a sinusoidal wave along a previously straight string, the wave must necessarily stretch the string. As a string element of length dx oscillates transversely, its length must increase and decrease in a periodic way if the string element is to fit the sinusoidal wave form. Elastic potential energy is associated with these length changes, just as for a spring.

When the string element is at its $y = y_m$ position (element b in Fig. 17-11), its length has its normal undisturbed value dx, so its elastic potential energy is zero. However, when the element is rushing through its $y = 0$ position, it is stretched to its maximum extent, and its elastic potential energy then is a maximum.

Energy Transport

The oscillating string element thus has both its maximum kinetic energy and its maximum elastic potential energy at $y = 0$. In the snapshot of Fig. 17-11, the regions of the string at maximum displacement have no energy, and the regions at zero displacement have maximum energy. As the wave travels along the string, forces due to the tension in the string continuously do work to transfer energy from regions with energy to regions with no energy.

Suppose we set up a wave on a string stretched along a horizontal x axis so that Eq. 17-2 describes the string's displacement. We might send a wave along the string by continuously oscillating one end of the string, as in Fig. 17-1b. In doing so, we continuously provide energy for the motion and stretching of the string—as the string sections oscillate perpendicularly to the x axis, they have kinetic energy and elastic potential energy. As the wave moves into sections that were previously at rest, energy is transferred into those new sections. Thus, we say that the wave *transports* the energy along the string.

The Rate of Energy Transmission

The kinetic energy dK associated with a string element of mass dm is given by

$$dK = \tfrac{1}{2}\, dm\, u^2, \qquad (17\text{-}26)$$

where u is the transverse speed of the oscillating string element. To find u, we differentiate Eq. 17-2 with respect to time while holding x constant:

$$u = \frac{\partial y}{\partial t} = -\omega y_m \cos(kx - \omega t). \qquad (17\text{-}27)$$

Using this relation and putting $dm = \mu\, dx$, we rewrite Eq. 17-26 as

$$dK = \tfrac{1}{2}(\mu\, dx)(-\omega y_m)^2 \cos^2(kx - \omega t). \qquad (17\text{-}28)$$

Dividing Eq. 17-28 by dt gives the rate at which the kinetic energy of a string element changes, and thus the rate at which kinetic energy is carried along by the wave. The ratio dx/dt that then appears on the right of Eq. 17-28 is the wave speed v, so we obtain

$$\frac{dK}{dt} = \tfrac{1}{2}\mu v \omega^2 y_m^2 \cos^2(kx - \omega t). \qquad (17\text{-}29)$$

The *average* rate at which kinetic energy is transported is

$$\left(\frac{dK}{dt}\right)_{avg} = \tfrac{1}{2}\mu v\omega^2 y_m^2[\cos^2(kx - \omega t)]_{avg}$$

$$= \tfrac{1}{4}\mu v\omega^2 y_m^2. \tag{17-30}$$

Here we have taken the average over an integer number of wavelengths and have used the fact that the average value of the square of a cosine function over an integer number of periods is $\frac{1}{2}$.

Elastic potential energy is also carried along with the wave, and at the same average rate given by Eq. 17-30. Although we shall not examine the proof, you should recall that, in an oscillating system such as a pendulum or a spring–block system, the average kinetic energy and the average potential energy are indeed equal.

The **average power,** which is the average rate at which energy of both kinds is transmitted by the wave, is then

$$P_{avg} = 2\left(\frac{dK}{dt}\right)_{avg} \tag{17-31}$$

or, from Eq. 17-30,

$$P_{avg} = \tfrac{1}{2}\mu v\omega^2 y_m^2 \quad \text{(average power).} \tag{17-32}$$

The factors μ and v in this equation depend on the material and tension of the string. The factors ω and y_m depend on the process that generates the wave. The dependence of the average power of a wave on the square of its amplitude and also on the square of its angular frequency is a general result, true for waves of all types.

Sample Problem 17-4

A stretched string has linear density $\mu = 525$ g/m and is under tension $\tau = 45$ N. We send a sinusoidal wave with frequency $f =$ 120 Hz and amplitude $y_m = 8.5$ mm along the string from one end. At what average rate does the wave transport energy?

SOLUTION: The **Key Idea** here is that the average rate of energy transport is the average power P_{avg} as given by Eq. 17-32. To use that equation, however, we first must calculate the angular frequency ω and the wave speed v. From Eq. 17-9,

$$\omega = 2\pi f = (2\pi)(120 \text{ Hz}) = 754 \text{ rad/s}.$$

From Eq. 17-25 we have

$$v = \sqrt{\frac{\tau}{\mu}} = \sqrt{\frac{45 \text{ N}}{0.525 \text{ kg/m}}} = 9.26 \text{ m/s}.$$

Equation 17-32 then yields

$$P_{avg} = \tfrac{1}{2}\mu v\omega^2 y_m^2$$

$$= (\tfrac{1}{2})(0.525 \text{ kg/m})(9.26 \text{ m/s})(754 \text{ rad/s})^2(0.0085 \text{ m})^2$$

$$\approx 100 \text{ W}. \qquad \text{(Answer)}$$

✔CHECKPOINT 4: For the string and wave of this sample problem, we can adjust three parameters: the tension in the string, the frequency of the wave, and the amplitude of the wave. Does the average rate at which the wave transports energy along the string increase, decrease, or remain the same if we increase (a) the tension, (b) the frequency, and (c) the amplitude?

17-8 The Principle of Superposition for Waves

It often happens that two or more waves pass simultaneously through the same region. When we listen to a concert, for example, sound waves from many instruments fall simultaneously on our eardrums. The electrons in the antennas of our radio and television receivers are set in motion by the net effect of many electromagnetic waves from many different broadcasting centers. The water of a lake or harbor may be churned up by waves in the wakes of many boats.

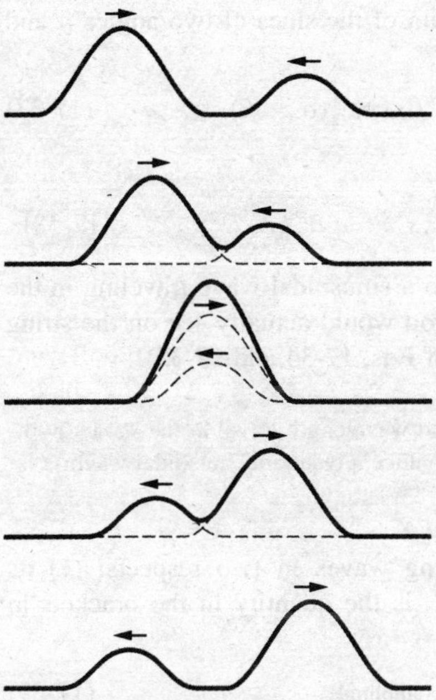

Fig. 17-12 A series of snapshots that show two pulses traveling in opposite directions along a stretched string. The superposition principle applies as the pulses move through each other.

Suppose that two waves travel simultaneously along the same stretched string. Let $y_1(x, t)$ and $y_2(x, t)$ be the displacements that the string would experience if each wave traveled alone. The displacement of the string when the waves overlap is then the algebraic sum

$$y'(x, t) = y_1(x, t) + y_2(x, t). \tag{17-33}$$

This summation of displacements along the string means that

Overlapping waves algebraically add to produce a **resultant wave** (or **net wave**).

This is another example of the **principle of superposition,** which says that when several effects occur simultaneously, their net effect is the sum of the individual effects.

Figure 17-12 shows a sequence of snapshots of two pulses traveling in opposite directions on the same stretched string. When the pulses overlap, the resultant pulse is their sum. Moreover, each pulse moves through the other, as if the other were not present:

Overlapping waves do not in any way alter the travel of each other.

17-9 Interference of Waves

Suppose we send two sinusoidal waves of the same wavelength and amplitude in the same direction along a stretched string. The superposition principle applies. What resultant wave does it predict for the string?

The resultant wave depends on the extent to which the waves are *in phase* (in step) with respect to each other—that is, how much one wave form is shifted from the other wave form. If the waves are exactly in phase (so that the peaks and valleys of one are exactly aligned with those of the other), they combine to double the displacement of either wave acting alone. If they are exactly out of phase (the peaks of one are exactly aligned with the valleys of the other), they combine to cancel everywhere, and the string remains straight. We call this phenomenon of combining waves **interference,** and the waves are said to **interfere.** (These terms refer only to the displacements of the waves; the travel of the waves is unaffected.)

Let one wave traveling along a stretched string be given by

$$y_1(x, t) = y_m \sin(kx - \omega t) \tag{17-34}$$

and another, shifted from the first, by

$$y_2(x, t) = y_m \sin(kx - \omega t + \phi). \tag{17-35}$$

These waves have the same angular frequency ω (and thus the same frequency f), the same angular wave number k (and thus the same wavelength λ), and the same amplitude y_m. They both travel in the positive direction of the x axis, with the same speed, given by Eq. 17-25. They differ only by a constant angle ϕ, which we call the **phase constant.** These waves are said to be *out of phase* by ϕ or to have a *phase difference* of ϕ, or one wave is said to be *phase-shifted* from the other by ϕ.

From the principle of superposition (Eq. 17-33), the resultant wave is the algebraic sum of the two interfering waves and has displacement

$$\begin{aligned} y'(x, t) &= y_1(x, t) + y_2(x, t) \\ &= y_m \sin(kx - \omega t) + y_m \sin(kx - \omega t + \phi). \end{aligned} \tag{17-36}$$

Displacement

$y'(x,t) = [2y_m \cos \tfrac{1}{2}\phi] \sin(kx - \omega t + \tfrac{1}{2}\phi)$

Amplitude — Oscillating term

Fig. 17-13 The resultant wave of Eq. 17-38, due to the interference of two sinusoidal transverse waves, is also a sinusoidal transverse wave, with an amplitude and an oscillating term.

In Appendix E we see that we can write the sum of the sines of two angles α and β as

$$\sin \alpha + \sin \beta = 2 \sin \tfrac{1}{2}(\alpha + \beta) \cos \tfrac{1}{2}(\alpha - \beta). \qquad (17\text{-}37)$$

Applying this relation to Eq. 17-36 leads to

$$y'(x, t) = [2y_m \cos \tfrac{1}{2}\phi] \sin(kx - \omega t + \tfrac{1}{2}\phi). \qquad (17\text{-}38)$$

As Fig. 17-13 shows, the resultant wave is also a sinusoidal wave traveling in the direction of increasing x. It is the only wave you would actually see on the string (you would *not* see the two interfering waves of Eqs. 17-34 and 17-35).

➤ If two sinusoidal waves of the same amplitude and wavelength travel in the *same* direction along a stretched string, they interfere to produce a resultant sinusoidal wave traveling in that direction.

The resultant wave differs from the interfering waves in two respects: (1) its phase constant is $\frac{1}{2}\phi$, and (2) its amplitude y'_m is the quantity in the brackets in Eq. 17-38:

$$y'_m = 2y_m \cos \tfrac{1}{2}\phi \quad \text{(amplitude).} \qquad (17\text{-}39)$$

If $\phi = 0$ rad (or 0°), the two interfering waves are exactly in phase, as in Fig. 17-14*a*. Then Eq. 17-38 reduces to

$$y'(x, t) = 2y_m \sin(kx - \omega t) \qquad (\phi = 0). \qquad (17\text{-}40)$$

This resultant wave is plotted in Fig. 17-14*d*. Note from both that figure and Eq. 17-40 that the amplitude of the resultant wave is twice the amplitude of either interfering wave. That is the greatest amplitude the resultant wave can have, because the cosine term in Eqs. 17-38 and 17-39 has its greatest value (unity) when $\phi = 0$. Interference that produces the greatest possible amplitude is called *fully constructive interference*.

If $\phi = \pi$ rad (or 180°), the interfering waves are exactly out of phase as in Fig. 17-14*b*. Then $\cos \frac{1}{2}\phi$ becomes $\cos \pi/2 = 0$, and the amplitude of the resultant

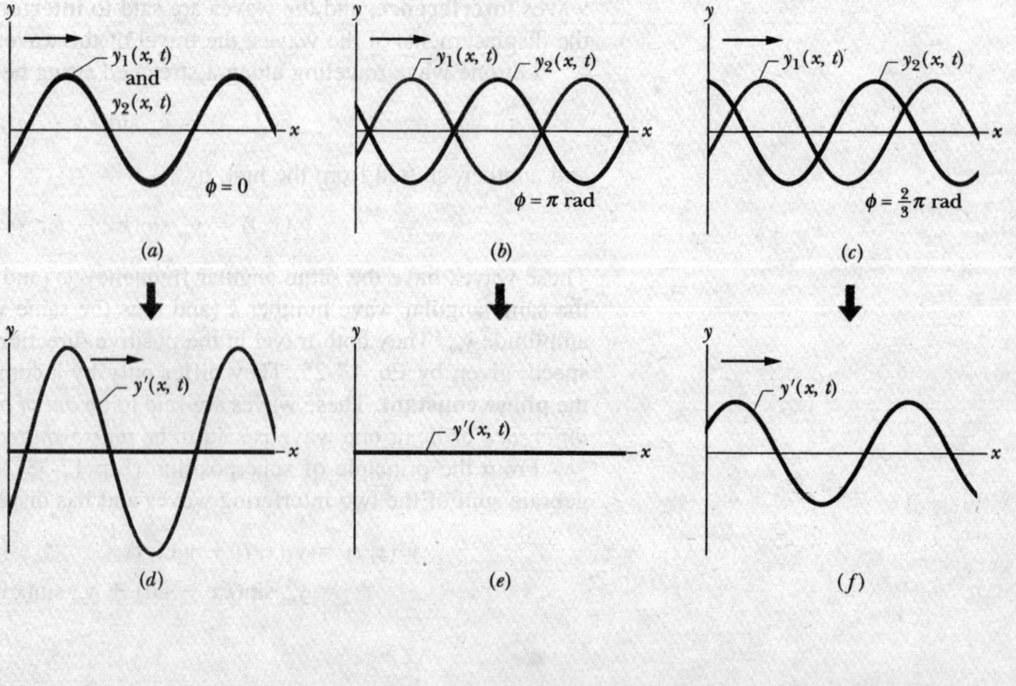

Fig. 17-14 Two identical sinusoidal waves, $y_1(x, t)$ and $y_2(x, t)$, travel along a string in the positive direction of an x axis. They interfere to give a resultant wave $y'(x, t)$. The resultant wave is what is actually seen on the string. The phase difference ϕ between the two interfering waves is (*a*) 0 rad or 0°, (*b*) π rad or 180°, and (*c*) $\frac{2}{3}\pi$ rad or 120°. The corresponding resultant waves are shown in (*d*), (*e*), and (*f*).

TABLE 17-1 Phase Differences and Resulting Interference Types[a]

Phase Difference, in Degrees	Radians	Wavelengths	Amplitude of Resultant Wave	Type of Interference
0	0	0	$2y_y$	Fully constructive
120	$\frac{2}{3}\pi$	0.33	y_m	Intermediate
180	π	0.50	0	Fully destructive
240	$\frac{4}{3}\pi$	0.67	y_m	Intermediate
360	2π	1.00	$2y_m$	Fully constructive
865	15.1	2.40	$0.60y_m$	Intermediate

[a]The phase difference is between two otherwise identical waves, with amplitude y_m, moving in the same direction.

wave as given by Eq. 17-39 is zero. We then have, for all values of x and t,

$$y'(x, t) = 0 \qquad (\phi = \pi \text{ rad}). \tag{17-41}$$

The resultant wave is plotted in Fig. 17-14*e*. Although we sent two waves along the string, we see no motion of the string. This type of interference is called *fully destructive interference*.

Because a sinusoidal wave repeats its shape every 2π rad, a phase difference $\phi = 2\pi$ rad (or 360°) corresponds to a shift of one wave relative to the other wave by a distance equivalent to one wavelength. Thus, phase differences can be described in terms of wavelengths as well as angles. For example, in Fig. 17-14*b* the waves may be said to be 0.50 wavelength out of phase. Table 17-1 shows some other examples of phase differences and the interference they produce. Note that when interference is neither fully constructive nor fully destructive, it is called *intermediate interference*. The amplitude of the resultant wave is then intermediate between 0 and $2y_m$. For example, from Table 17-1, if the interfering waves have a phase difference of 120° ($\phi = \frac{2}{3}\pi$ rad $= 0.33$ wavelength), then the resultant wave has an amplitude of y_m, the same as the interfering waves (see Figs. 17-14*c* and *f*).

Two waves with the same wavelength are in phase if their phase difference is zero or any integer number of wavelengths. Thus, the integer part of any phase difference *expressed in wavelengths* may be discarded. For example, a phase difference of 0.40 wavelength is equivalent in every way to one of 2.40 wavelengths, and so the simpler of the two numbers can be used in computations.

Sample Problem 17-5

Two identical sinusoidal waves, moving in the same direction along a stretched string, interfere with each other. The amplitude y_m of each wave is 9.8 mm, and the phase difference ϕ between them is 100°.

(a) What is the amplitude y'_m of the resultant wave due to the interference of these two waves, and what type of interference occurs?

SOLUTION: The **Key Idea** here is that these are identical sinusoidal waves traveling in the *same direction* along a string, so they interfere to produce a sinusoidal traveling wave. Because they are identical, they have the *same amplitude*. Thus, the amplitude y'_m of the resultant wave is given by Eq. 17-39:

$$y'_m = 2y_m \cos \tfrac{1}{2}\phi = (2)(9.8 \text{ mm}) \cos(100°/2)$$
$$= 13 \text{ mm}. \qquad \text{(Answer)}$$

We can tell that the interference is *intermediate* in two ways. The phase difference is between 0 and 180° and, correspondingly, amplitude y'_m is between 0 and $2y_m$ (= 19.6 mm).

(b) What phase difference, in radians and wavelengths, will give the resultant wave an amplitude of 4.9 mm?

SOLUTION: The same **Key Idea** applies here as in part (a), but now we are given y'_m and seek ϕ. From Eq. 17-39,

$$y'_m = 2y_m \cos \tfrac{1}{2}\phi.$$

We now have

$$4.9\text{ mm} = (2)(9.8\text{ mm})\cos\tfrac{1}{2}\phi,$$

which gives us (with a calculator in the radian mode)

$$\phi = 2\cos^{-1}\frac{4.9\text{ mm}}{(2)(9.8\text{ mm})}$$
$$= \pm 2.636\text{ rad} \approx \pm 2.6\text{ rad.} \quad \text{(Answer)}$$

There are two solutions because we can obtain the same resultant wave by letting the first wave *lead* (travel ahead of) or *lag* (travel behind) the second wave by 2.6 rad. In wavelengths, the phase difference is

$$\frac{\phi}{2\pi\text{ rad/wavelength}} = \frac{\pm 2.636\text{ rad}}{2\pi\text{ rad/wavelength}}$$
$$= \pm 0.42\text{ wavelength.} \quad \text{(Answer)}$$

CHECKPOINT 5: Here are four other possible phase differences between the two waves of this sample problem, expressed in wavelengths: 0.20, 0.45, 0.60, and 0.80. Rank them according to the amplitude of the resultant wave, greatest first.

17-10 Phasors

We can represent a string wave (or any other type of wave) vectorially with a **phasor.** In essence, a phasor is a vector that has a magnitude equal to the amplitude of the wave and that rotates around an origin; the angular speed of the phasor is equal to the angular frequency ω of the wave. For example, the wave

$$y_1(x, t) = y_{m1}\sin(kx - \omega t) \tag{17-42}$$

is represented by the phasor shown in Fig. 17-15*a*. The magnitude of the phasor is the amplitude y_{m1} of the wave. As the phasor rotates around the origin at angular speed ω, its projection y_1 on the vertical axis varies sinusoidally, from a maximum of y_{m1} through zero to a minimum of $-y_{m1}$ and then back to y_{m1}. This variation corresponds to the sinusoidal variation in the displacement y_1 of any point along the string as the wave passes through it.

When two waves travel along the same string in the same direction, we can represent them and their resultant wave in a *phasor diagram.* The phasors in Fig. 17-15*b* represent the wave of Eq. 17-42 and a second wave given by

$$y_2(x, t) = y_{m2}\sin(kx - \omega t + \phi). \tag{17-43}$$

This second wave is phase shifted from the first wave by phase constant ϕ. Because the phasors rotate at the same angular speed ω, the angle between the two phasors is always ϕ. If ϕ is a *positive* quantity, then the phasor for wave 2 *lags* the phasor for wave 1 as they rotate, as drawn in Fig. 17-15*b*. If ϕ is a negative quantity, then the phasor for wave 2 *leads* the phasor for wave 1.

Because waves y_1 and y_2 have the same angular wave number k and angular frequency ω, we know from Eq. 17-38 that their resultant is of the form

$$y'(x, t) = y'_m\sin(kx - \omega t + \beta), \tag{17-44}$$

where y'_m is the amplitude of the resultant wave and β is its phase constant. To find the values of y'_m and β, we would have to sum the two combining waves, as we did to obtain Eq. 17-38.

To do this on a phasor diagram, we vectorially add the two phasors at any instant during their rotation, as in Fig. 17-15*c* where phasor y_{m2} has been shifted to the

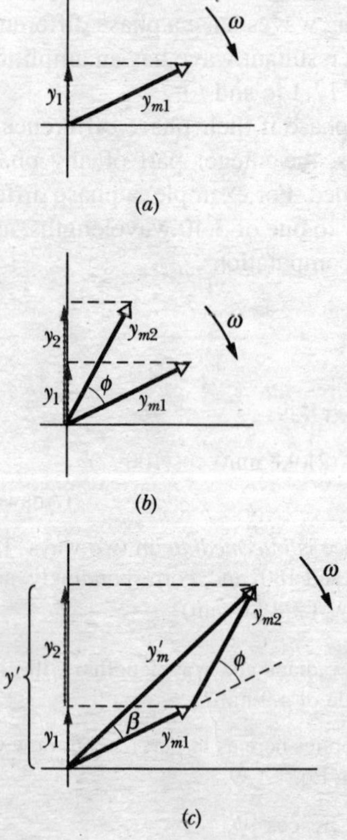

Fig. 17-15 (*a*) A phasor of magnitude y_{m1} rotating about an origin at angular speed ω represents a sinusoidal wave. The phasor's projection y_1 on the vertical axis represents the displacement of a point through which the wave passes. (*b*) A second phasor, also of angular speed ω but of magnitude y_{m2} and rotating at a constant angle ϕ from the first phasor, represents a second wave, with a phase constant ϕ. (*c*) The resultant wave of the two waves is represented by the vector sum y'_m of the two phasors. The projection y' on the vertical axis represents the displacement of a point as that resultant wave passes through it.

head of phasor y_{m1}. The magnitude of the vector sum equals the amplitude y'_m in Eq. 17-44. The angle between the vector sum and the phasor for y_1 equals the phase constant β in Eq. 17-44.

Note that, in contrast to the method of Section 17-9:

> We can use phasors to combine waves *even if their amplitudes are different.*

Sample Problem 17-6

Two sinusoidal waves $y_1(x, t)$ and $y_2(x, t)$ have the same wavelength and travel together in the same direction along a string. Their amplitudes are $y_{m1} = 4.0$ mm and $y_{m2} = 3.0$ mm, and their phase constants are 0 and $\pi/3$ rad, respectively. What are the amplitude y'_m and phase constant β of the resultant wave? Write the resultant wave in the form of Eq. 17-44.

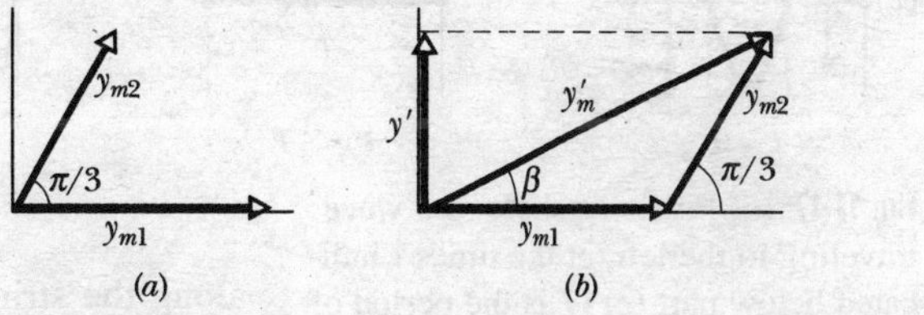

Fig. 17-16 Sample Problem 17-6. (*a*) Two phasors of magnitudes y_{m1} and y_{m2} and with phase difference $\pi/3$. (*b*) Vector addition of these phasors at any instant during their rotation gives the magnitude y'_m of the phasor for the resultant wave.

SOLUTION: One **Key Idea** here is that the two waves have a number of properties in common: Because they travel along the same string, they must have the same speed v, as set by the tension and linear density of the string according to Eq. 17-25. With the same wavelength λ, they must have the same angular wave number $k\ (= 2\pi/\lambda)$. Also, with the same wave number k and speed v, they must have the same angular frequency $\omega\ (= kv)$.

A second **Key Idea** is that the waves (call them waves 1 and 2) can be represented by phasors rotating at the same angular speed ω about an origin. Because the phase constant for wave 2 is *greater* than that for wave 1 by $\pi/3$, phasor 2 must *lag* phasor 1 by $\pi/3$ rad in their clockwise rotation, as shown in Fig. 17-16*a*. The resultant wave due to the interference of waves 1 and 2 can then be represented by a phasor that is the vector sum of phasors 1 and 2.

To simplify the vector summation, we drew phasors 1 and 2 in Fig. 17-16*a* at the instant when phasor 1 lies along the horizontal axis. We then drew lagging phasor 2 at positive angle $\pi/3$ rad. In Fig. 17-16*b* we shifted phasor 2 so its tail is at the head of phasor 1. Then we can draw the phasor y'_m of the resultant wave from the tail of phasor 1 to the head of phasor 2. The phase constant β is the angle it makes with phasor 1.

To find values for y'_m and β, we can sum phasors 1 and 2 directly on a vector-capable calculator, by adding a vector of magnitude 4.0 and angle 0 rad to a vector of magnitude 3.0 and angle $\pi/3$ rad, or we can add the vectors by components. For the horizontal components we have

$$y'_{mh} = y_{m1} \cos 0 + y_{m2} \cos \pi/3$$
$$= 4.0 \text{ mm} + (3.0 \text{ mm}) \cos \pi/3 = 5.50 \text{ mm}.$$

For the vertical components we have

$$y'_{mv} = y_{m1} \sin 0 + y_{m2} \sin \pi/3$$
$$= 0 + (3.0 \text{ mm}) \sin \pi/3 = 2.60 \text{ mm}.$$

Thus, the resultant wave has an amplitude of

$$y'_m = \sqrt{(5.50 \text{ mm})^2 + (2.60 \text{ mm})^2}$$
$$= 6.1 \text{ mm} \qquad \text{(Answer)}$$

and a phase constant of

$$\beta = \tan^{-1} \frac{2.60 \text{ mm}}{5.50 \text{ mm}} = 0.44 \text{ rad}. \qquad \text{(Answer)}$$

From Fig. 17-16*b*, phase constant β is a *positive* angle relative to phasor 1. Thus, the resultant wave *lags* wave 1 in their travel by phase constant $\beta = +0.44$ rad. From Eq. 17-44, we can write the resultant wave as

$$y'(x, t) = (6.1 \text{ mm}) \sin(kx - \omega t + 0.44 \text{ rad}). \qquad \text{(Answer)}$$

17-11 Standing Waves

In the preceding two sections, we discussed two sinusoidal waves of the same wavelength and amplitude traveling *in the same direction* along a stretched string. What if they travel in opposite directions? We can again find the resultant wave by applying the superposition principle.

Figure 17-17 suggests the situation graphically. It shows the two combining waves, one traveling to the left in Fig. 17-17*a*, the other to the right in Fig. 17-17*b*. Figure 17-17*c* shows their sum, obtained by applying the superposition principle graphically. The outstanding feature of the resultant wave is that there are places

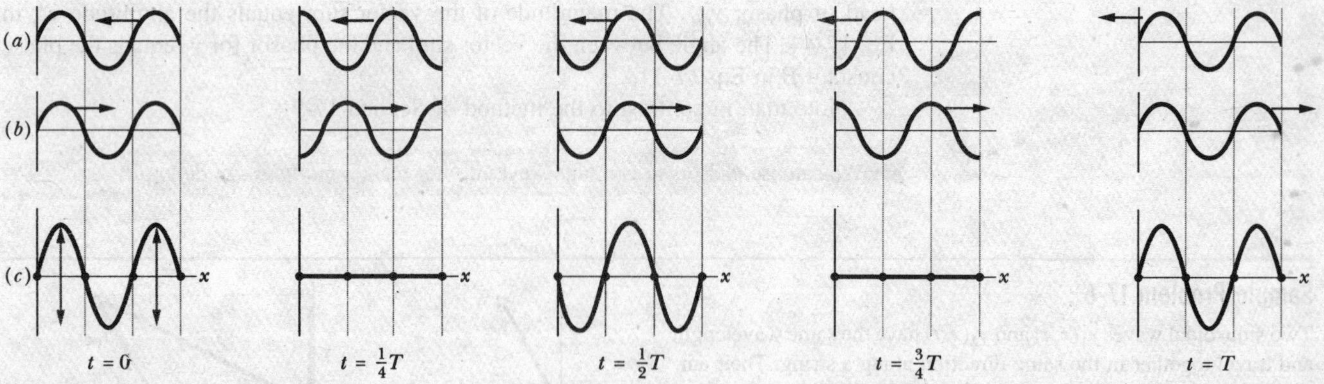

Fig. 17-17 (*a*) Five snapshots of a wave traveling to the left, at the times *t* indicated below part (*c*) (*T* is the period of oscillation). (*b*) Five snapshots of a wave identical to that in (*a*) but traveling to the right, at the same times *t*. (*c*) Corresponding snapshots for the superposition of the two waves on the same string. At $t = 0$, $\frac{1}{2}T$, and T, fully constructive interference occurs because of the alignment of peaks with peaks and valleys with valleys. At $t = \frac{1}{4}T$ and $\frac{3}{4}T$, fully destructive interference occurs because of the alignment of peaks with valleys. Some points (the nodes, marked with dots) never oscillate; some points (the antinodes) oscillate the most.

along the string, called **nodes**, where the string never moves. Four such nodes are marked by dots in Fig. 17-17*c*. Halfway between adjacent nodes are **antinodes**, where the amplitude of the resultant wave is a maximum. Wave patterns such as that of Fig. 17-17*c* are called **standing waves** because the wave patterns do not move left or right; the locations of the maxima and minima do not change.

▶ If two sinusoidal waves of the same amplitude and wavelength travel in *opposite* directions along a stretched string, their interference with each other produces a standing wave.

To analyze a standing wave, we represent the two combining waves with the equations

$$y_1(x, t) = y_m \sin(kx - \omega t) \tag{17-45}$$

and

$$y_2(x, t) = y_m \sin(kx + \omega t). \tag{17-46}$$

The principle of superposition gives, for the combined wave,

$$y'(x, t) = y_1(x, t) + y_2(x, t) = y_m \sin(kx - \omega t) + y_m \sin(kx + \omega t).$$

Applying the trigonometric relation of Eq. 17-37 leads to

$$y'(x, t) = [2y_m \sin kx] \cos \omega t, \tag{17-47}$$

which is displayed in Fig. 17-18. This equation does not describe a traveling wave because it is not of the form of Eq. 17-16. Instead, it describes a standing wave.

The quantity $2y_m \sin kx$ in the brackets of Eq. 17-47 can be viewed as the amplitude of oscillation of the string element that is located at position x. However, since an amplitude is always positive and $\sin kx$ can be negative, we take the absolute value of the quantity $2y_m \sin kx$ to be the amplitude at x.

In a traveling sinusoidal wave, the amplitude of the wave is the same for all string elements. That is not true for a standing wave, in which the amplitude *varies with position.* In the standing wave of Eq. 17-47, for example, the amplitude is zero for values of kx that give $\sin kx = 0$. Those values are

$$kx = n\pi, \quad \text{for } n = 0, 1, 2, \ldots . \tag{17-48}$$

Substituting $k = 2\pi/\lambda$ in this equation and rearranging, we get

$$x = n\frac{\lambda}{2}, \quad \text{for } n = 0, 1, 2, \ldots \quad \text{(nodes)}, \tag{17-49}$$

as the positions of zero amplitude—the nodes—for the standing wave of Eq. 17-47. Note that adjacent nodes are separated by $\lambda/2$, half a wavelength.

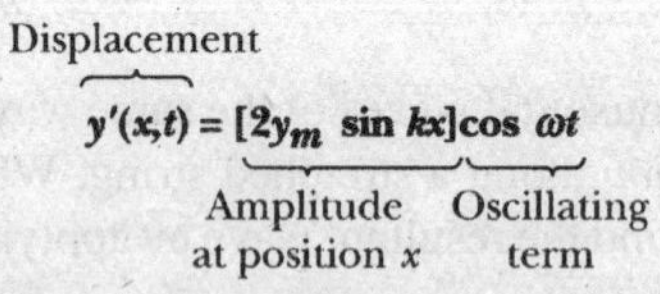

Fig. 17-18 The resultant wave of Eq. 17-47 is a standing wave and is due to the interference of two sinusoidal waves of the same amplitude and wavelength that travel in opposite directions.

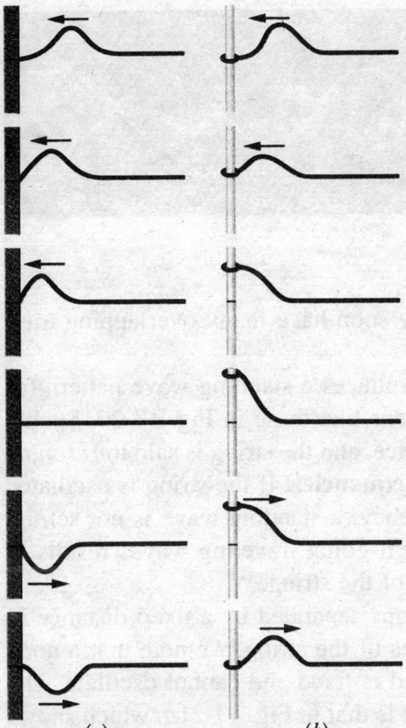

Fig. 17-19 (*a*) A pulse incident from the right is reflected at the left end of the string, which is tied to a wall. Note that the reflected pulse is inverted from the incident pulse. (*b*) Here the left end of the string is tied to a ring that can slide without friction up and down the rod. Now the pulse is not inverted by the reflection.

The amplitude of the standing wave of Eq. 17-47 has a maximum value of $2y$, which occurs for values of kx that give $|\sin kx| = 1$. Those values are

$$\begin{aligned} kx &= \tfrac{1}{2}\pi, \tfrac{3}{2}\pi, \tfrac{5}{2}\pi, \ldots \\ &= (n + \tfrac{1}{2})\pi, \qquad \text{for } n = 0, 1, 2, \ldots. \end{aligned} \tag{17-50}$$

Substituting $k = 2\pi/\lambda$ in Eq. 17-50 and rearranging, we get

$$x = \left(n + \frac{1}{2}\right)\frac{\lambda}{2}, \qquad \text{for } n = 0, 1, 2, \ldots \quad \text{(antinodes)}, \tag{17-51}$$

as the positions of maximum amplitude—the antinodes—of the standing wave of Eq. 17-47. The antinodes are separated by $\lambda/2$ and are located halfway between pairs of nodes.

Reflections at a Boundary

We can set up a standing wave in a stretched string by allowing a traveling wave to be reflected from the far end of the string so that it travels back through itself. The incident (original) wave and the reflected wave can then be described by Eqs. 17-45 and 17-46, respectively, and they can combine to form a pattern of standing waves.

In Fig. 17-19, we use a single pulse to show how such reflections take place. In Fig. 17-19*a*, the string is fixed at its left end. When the pulse arrives at that end, it exerts an upward force on the support (the wall). By Newton's third law, the support exerts an opposite force of equal magnitude on the string. This second force generates a pulse at the support, which travels back along the string in the direction opposite that of the incident pulse. In a "hard" reflection of this kind, there must be a node at the support because the string is fixed there. The reflected and incident pulses must have opposite signs, so as to cancel each other at that point.

In Fig. 17-19*b*, the left end of the string is fastened to a light ring that is free to slide without friction along a rod. When the incident pulse arrives, the ring moves up the rod. As the ring moves, it pulls on the string, stretching the string and producing a reflected pulse with the same sign and amplitude as the incident pulse. Thus, in such a "soft" reflection, the incident and reflected pulses reinforce each other, creating an antinode at the end of the string; the maximum displacement of the ring is twice the amplitude of either of these pulses.

✓CHECKPOINT 6: Two waves with the same amplitude and wavelength interfere in three different situations to produce resultant waves with the following equations:
(1) $y'(x, t) = 4 \sin(5x - 4t)$
(2) $y'(x, t) = 4 \sin(5x) \cos(4t)$
(3) $y'(x, t) = 4 \sin(5x + 4t)$
In which situation are the two combining waves traveling (a) toward positive x, (b) toward negative x, and (c) in opposite directions?

17-12 Standing Waves and Resonance

Consider a string, such as a guitar string, that is stretched between two clamps. Suppose we send a continuous sinusoidal wave of a certain frequency along the string, say, toward the right. When the wave reaches the right end, it reflects and begins to travel back to the left. That left-going wave then overlaps the wave that is still traveling to the right. When the left-going wave reaches the left end, it reflects again and the newly reflected wave begins to travel to the right, overlapping the left-

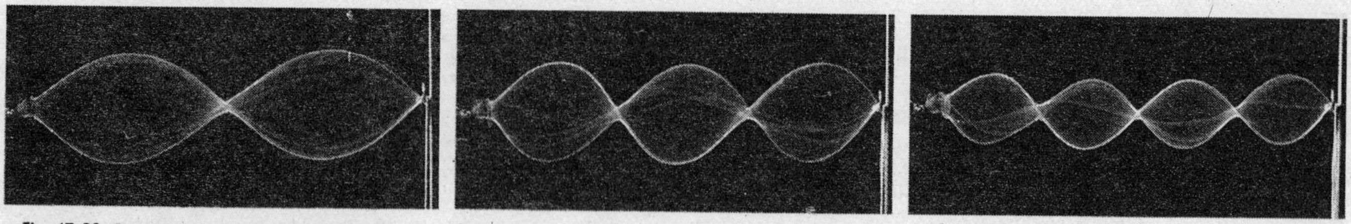

Fig. 17-20 Stroboscopic photographs reveal (imperfect) standing wave patterns on a string being made to oscillate by a vibrator at the left end. The patterns occur at certain frequencies of oscillation.

going and right-going waves. In short, we very soon have many overlapping traveling waves, which interfere with one another.

For certain frequencies, the interference produces a standing wave pattern (or **oscillation mode**) with nodes and large antinodes like those in Fig. 17-20. Such a standing wave is said to be produced at **resonance,** and the string is said to *resonate* at these certain frequencies, called **resonant frequencies.** If the string is oscillated at some frequency other than a resonant frequency, a standing wave is not set up. Then the interference of the right-going and left-going traveling waves results in only small (perhaps imperceptible) oscillations of the string.

Let a string be stretched between two clamps separated by a fixed distance L. To find expressions for the resonant frequencies of the string, we note that a node must exist at each of its ends, because each end is fixed and cannot oscillate. The simplest pattern that meets this key requirement is that in Fig. 17-21*a*, which shows the string at both its extreme displacements (one solid and one dashed, together forming a single "loop"). There is only one antinode, which is at the center of the string. Note that half a wavelength spans the length L, which we take to be the string's length. Thus, for this pattern, $\lambda/2 = L$. This condition tells us that if the left-going and right-going traveling waves are to set up this pattern by their interference, they must have the wavelength $\lambda = 2L$.

A second simple pattern meeting the requirement of nodes at the fixed ends is shown in Fig. 17-21*b*. This pattern has three nodes and two antinodes and is said to be a two-loop pattern. For the left-going and right-going waves to set it up, they must have a wavelength $\lambda = L$. A third pattern is shown in Fig. 17-21*c*. It has four nodes, three antinodes, and three loops, and the wavelength is $\lambda = \frac{2}{3}L$. We could continue this progression by drawing increasingly more complicated patterns. In each step of the progression, the pattern would have one more node and one more antinode than the preceding step, and an additional $\lambda/2$ would be fitted into the distance L.

Thus, a standing wave can be set up on a string of length L by a wave with a wavelength equal to one of the values

$$\lambda = \frac{2L}{n}, \qquad \text{for } n = 1, 2, 3, \ldots. \tag{17-52}$$

The resonant frequencies that correspond to these wavelengths follow from Eq. 17-12:

$$f = \frac{v}{\lambda} = n\frac{v}{2L}, \qquad \text{for } n = 1, 2, 3, \ldots. \tag{17-53}$$

Here v is the speed of traveling waves on the string.

Equation 17-53 tells us that the resonant frequencies are integer multiples of the lowest resonant frequency, $f = v/2L$, which corresponds to $n = 1$. The oscillation mode with that lowest frequency is called the *fundamental mode* or the *first harmonic*. The *second harmonic* is the oscillation mode with $n = 2$, the *third harmonic* is that with $n = 3$, and so on. The frequencies associated with these modes are often

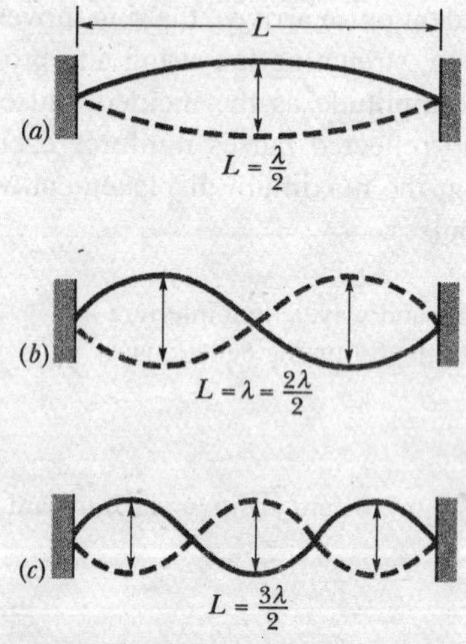

Fig. 17-21 A string, stretched between two clamps, is made to oscillate in standing wave patterns. (*a*) The simplest possible pattern consists of one *loop,* which refers to the composite shape formed by the string in its extreme displacements (the solid and dashed lines). (*b*) The next simplest pattern has two loops. (*c*) The next has three loops.

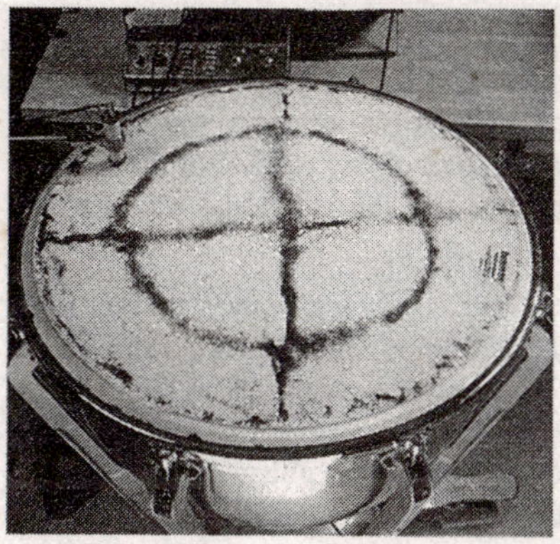

Fig. 17-22 One of many possible standing wave patterns for a kettledrum head, made visible by dark powder sprinkled on the drumhead. As the head is set into oscillation at a single frequency by a mechanical vibrator at the upper left of the photograph, the powder collects at the nodes, which are circles and straight lines in this two-dimensional example.

labeled f_1, f_2, f_3, and so on. The collection of all possible oscillation modes is called the **harmonic series,** and n is called the **harmonic number** of the nth harmonic.

The phenomenon of resonance is common to all oscillating systems and can occur in two and three dimensions. For example, Fig. 17-22 shows a two-dimensional standing wave pattern on the oscillating head of a kettledrum.

✔CHECKPOINT 7: In the following series of resonant frequencies, one frequency (lower than 400 Hz) is missing: 150, 225, 300, 375 Hz. (a) What is the missing frequency? (b) What is the frequency of the seventh harmonic?

Sample Problem 17-7

In Fig. 17-23, a string, tied to a sinusoidal vibrator at P and running over a support at Q, is stretched by a block of mass m. The separation L between P and Q is 1.2 m, the linear density of the string is 1.6 g/m, and the frequency f of the vibrator is fixed at 120 Hz. The amplitude of the motion at P is small enough for that point to be considered a node. A node also exists at Q.

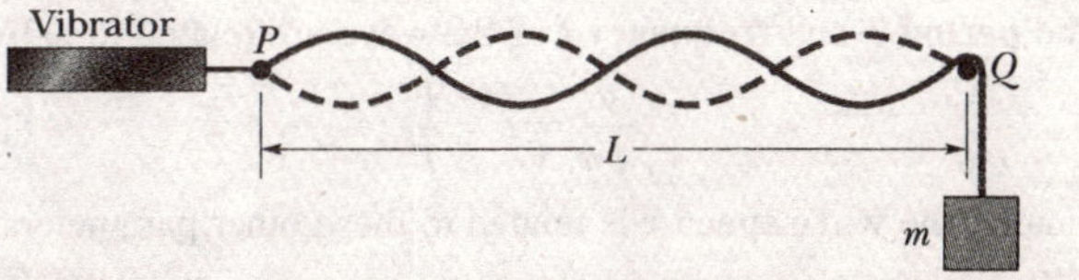

Fig. 17-23 Sample Problem 17-7. A string under tension connected to a vibrator. For a fixed vibrator frequency, standing wave patterns will occur for certain values of the string tension.

(a) What mass m allows the vibrator to set up the fourth harmonic on the string?

SOLUTION: One **Key Idea** here is that the string will resonate at only certain frequencies, determined by the wave speed v on the string and the length L of the string. From Eq. 17-53, these resonant frequencies are

$$f = n\frac{v}{2L}, \qquad \text{for } n = 1, 2, 3, \ldots. \tag{17-54}$$

To set up the fourth harmonic (for which $n = 4$), we need to adjust the right side of this equation, with $n = 4$, so that the left side equals the frequency of the vibrator (120 Hz).

We cannot adjust L in Eq. 17-54; it is set. However, a second **Key Idea** is that we *can* adjust v, because it depends on how much mass m we hang on the string. According to Eq. 17-25, wave speed $v = \sqrt{\tau/\mu}$. Here the tension τ in the string is equal to the weight mg of the block. Thus,

$$v = \sqrt{\frac{\tau}{\mu}} = \sqrt{\frac{mg}{\mu}}. \tag{17-55}$$

Substituting v from Eq. 17-55 into Eq. 17-54, setting $n = 4$ for the fourth harmonic, and solving for m give us

$$m = \frac{4L^2f^2\mu}{n^2g} \tag{17-56}$$

$$= \frac{(4)(1.2\text{ m})^2(120\text{ Hz})^2(0.0016\text{ kg/m})}{(4)^2(9.8\text{ m/s}^2)}$$

$$= 0.846\text{ kg} \approx 0.85\text{ kg}. \qquad \text{(Answer)}$$

(b) What standing wave mode is set up if $m = 1.00$ kg?

SOLUTION: If we insert this value of m into Eq. 17-56 and solve for n, we find that $n = 3.7$. A **Key Idea** here is that n must be an integer, so $n = 3.7$ is impossible. Thus, with $m = 1.00$ kg, the vibrator cannot set up a standing wave on the string, and any oscillation of the string will be small, perhaps even imperceptible.

PROBLEM-SOLVING TACTICS

Tactic 2: *Harmonics on a String*

When you need to obtain information about a certain harmonic on a stretched string of given length L, first draw that harmonic (as in Fig. 17-21). If you are asked about, say, the fifth harmonic, you need to draw five loops between the fixed support points. That would mean that five loops, each of length $\lambda/2$, occupy the length L of the string. Thus, $5(\lambda/2) = L$, and $\lambda = 2L/5$. You can then use Eq. 17-12 ($f = v/\lambda$) to find the frequency of the harmonic.

Keep in mind that the wavelength of a harmonic is set only by the length L of the string, but the frequency depends also on the wave speed v, which is set by the tension and the linear density of the string via Eq. 17-25.

REVIEW & SUMMARY

Transverse and Longitudinal Waves Mechanical waves can exist only in material media and are governed by Newton's laws. **Transverse** mechanical waves, like those on a stretched string, are waves in which the particles of the medium oscillate perpendicular to the wave's direction of travel. Waves in which the particles of the medium oscillate parallel to the wave's direction of travel are **longitudinal** waves.

Sinusoidal Waves A sinusoidal wave moving in the positive x direction has the mathematical form

$$y(x, t) = y_m \sin(kx - \omega t), \tag{17-2}$$

where y_m is the **amplitude** of the wave, k is the **angular wave number,** ω is the **angular frequency,** and $kx - \omega t$ is the **phase.** The **wavelength** λ is related to k by

$$k = \frac{2\pi}{\lambda}. \tag{17-5}$$

The **period** T and **frequency** f of the wave are related to ω by

$$\frac{\omega}{2\pi} = f = \frac{1}{T}. \tag{17-9}$$

Finally, the **wave speed** v is related to these other parameters by

$$v = \frac{\omega}{k} = \frac{\lambda}{T} = \lambda f. \tag{17-12}$$

Equation of a Traveling Wave Any function of the form

$$y(x, t) = h(kx \pm \omega t) \tag{17-16}$$

can represent a **traveling wave** with a wave speed given by Eq. 17-12 and a wave shape given by the mathematical form of h. The plus sign denotes a wave traveling in the negative x direction, and the minus sign a wave traveling in the positive x direction.

Wave Speed on Stretched String The speed of a wave on a stretched string is set by properties of the string. The speed on a string with tension τ and linear density μ is

$$v = \sqrt{\frac{\tau}{\mu}}. \tag{17-25}$$

Power The **average power,** or average rate at which energy is transmitted by a sinusoidal wave on a stretched string, is given by

$$P_{avg} = \tfrac{1}{2}\mu v \omega^2 y_m^2. \tag{17-32}$$

Superposition of Waves When two or more waves traverse the same medium, the displacement of any particle of the medium is the sum of the displacements that the individual waves would give it.

Interferenee of Waves Two sinusoidal waves on the same string exhibit **interference,** adding or canceling according to the principle of superposition. If the two are traveling in the same direction and have the same amplitude y_m and frequency (hence the same wavelength) but differ in phase by a **phase constant** ϕ, the result is a single wave with this same frequency:

$$y'(x, t) = [2y_m \cos \tfrac{1}{2}\phi] \sin(kx - \omega t + \tfrac{1}{2}\phi). \tag{17-38}$$

If $\phi = 0$, the waves are exactly in phase and their interference is fully constructive; if $\phi = \pi$ rad, they are exactly out of phase and their interference is fully destructive.

Phasors A wave $y(x, t)$ can be represented with a *phasor*. This is a vector that has a magnitude equal to the amplitude y_m of the wave and that rotates about an origin with an angular speed equal to the angular frequency ω of the wave. The projection of the rotating phasor on a vertical axis gives the displacement y of a point along the wave's travel.

Standing Waves The interference of two identical sinusoidal waves moving in opposite directions produces **standing waves.** For a string with fixed ends, the standing wave is given by

$$y'(x, t) = [2y_m \sin kx] \cos \omega t. \tag{17-47}$$

Standing waves are characterized by fixed locations of zero displacement called **nodes** and fixed locations of maximum displacement called **antinodes.**

Resonance Standing waves on a string can be set up by reflection of traveling waves from the ends of the string. If an end is fixed, it must be the position of a node. This limits the frequencies at which standing waves will occur on a given string. Each possible frequency is a **resonant frequency,** and the corresponding standing wave pattern is an **oscillation mode.** For a stretched string of length L with fixed ends, the resonant frequencies are

$$f = \frac{v}{\lambda} = n\frac{v}{2L}, \quad \text{for } n = 1, 2, 3, \ldots. \tag{17-53}$$

The oscillation mode corresponding to $n = 1$ is called the *fundamental mode* or the *first harmonic*; the mode corresponding to $n = 2$ is the *second harmonic*; and so on.

QUESTIONS

1. What is the wavelength of the (strange) wave in Fig. 17-24, where each segment of the wave has length d?

Fig. 17-24 Question 1.

2. Figure 17-25*a* gives a snapshot of a wave traveling in the direction of positive x along a string under tension. Four string elements are indicated by the lettered points. For each of those elements, determine whether, at the instant of the snapshot, the element is moving upward or downward or is momentarily at rest. (*Hint:* Imagine the wave as it moves through the four string elements.)

Figure 17-25*b* gives the displacement of a string element

located at, say, $x = 0$ as a function of time. At the lettered times, is the element moving upward or downward or is it momentarily at rest?

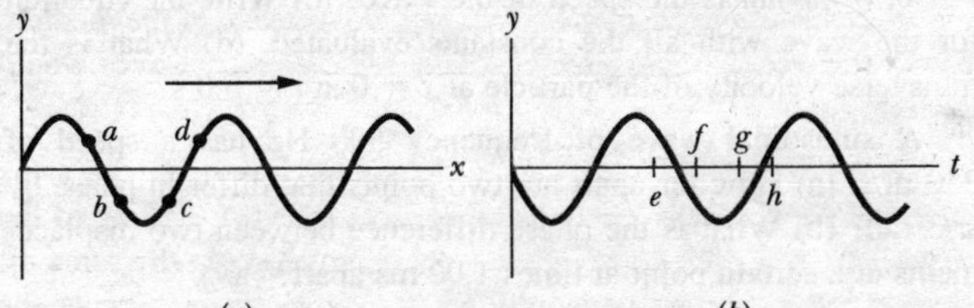

Fig. 17-25 Question 2.

3. In Fig. 17-26, five points are indicated on a snapshot of a sinusoidal wave. What is the phase difference between point 1 and (a) point 2, (b) point 3, (c) point 4, and (d) point 5? Answer in radians and in terms of the wavelength of the wave. The snapshot shows a point of zero displacement at $x = 0$. In terms of the period T of the wave, when will (e) a peak and (f) the next point of zero displacement reach $x = 0$?

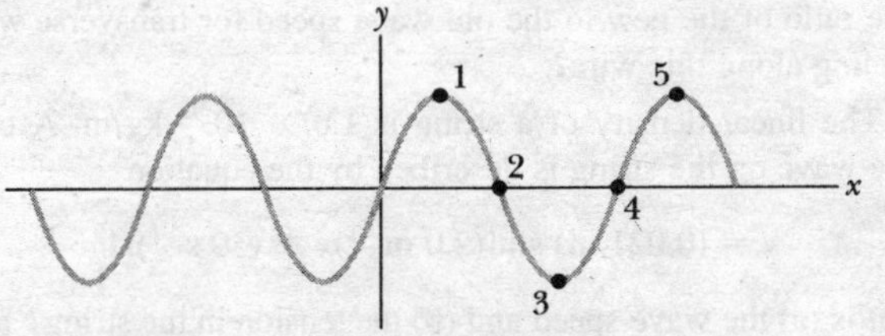

Fig. 17-26 Question 3.

4. The following four waves are sent along strings with the same linear densities (x is in meters and t is in seconds). Rank the waves according to (a) their wave speed and (b) the tension in the strings along which they travel, greatest first:

(1) $y_1 = (3 \text{ mm}) \sin(x - 3t)$, (3) $y_3 = (1 \text{ mm}) \sin(4x - t)$,

(2) $y_2 = (6 \text{ mm}) \sin(2x - t)$, (4) $y_4 = (2 \text{ mm}) \sin(x - 2t)$.

5. In Fig. 17-27, wave 1 consists of a rectangular peak of height 4 units and width d, and a rectangular valley of depth 2 units and width d. The wave travels rightward along an x axis. Choices 2, 3, and 4 are similar waves, with the same heights, depths, and widths, that will travel leftward along that axis and through wave 1. With which choice will the interference give, for an instant, (a) the deepest valley, (b) a flat line, and (c) a level peak $2d$ wide?

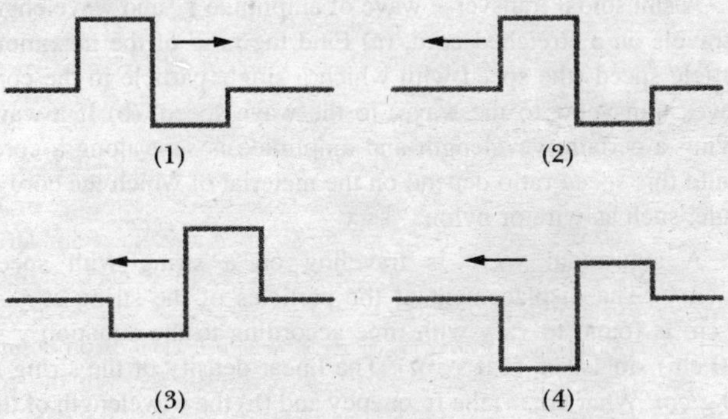

Fig. 17-27 Question 5.

6. If you start with two sinusoidal waves of the same amplitude traveling in phase on a string and then somehow phase-shift one of them by 5.4 wavelengths, what type of interference will occur on the string?

7. The amplitudes and phase differences for four pairs of waves of equal wavelengths are (a) 2 mm, 6 mm, and π rad; (b) 3 mm, 5 mm, and π rad; (c) 7 mm, 9 mm, and π rad; (d) 2 mm, 2 mm, and 0 rad. Each pair travels in the same direction along the same string. Without written calculation, rank the four pairs according to the amplitude of their resultant wave, greatest first. (*Hint:* Construct phasor diagrams.)

8. If you set up the seventh harmonic on a string, (a) how many nodes are present, and (b) is there a node, antinode, or some intermediate state at the midpoint? If you next set up the sixth harmonic, (c) is its resonant wavelength longer or shorter than that for the seventh harmonic, and (d) is the resonant frequency higher or lower?

9. Strings A and B have identical lengths and linear densities, but string B is under greater tension than string A. Figure 17-28 shows four situations, (*a*) through (*d*), in which standing wave patterns exist on the two strings. In which situations is there the possibility that strings A and B are oscillating at the same resonant frequency?

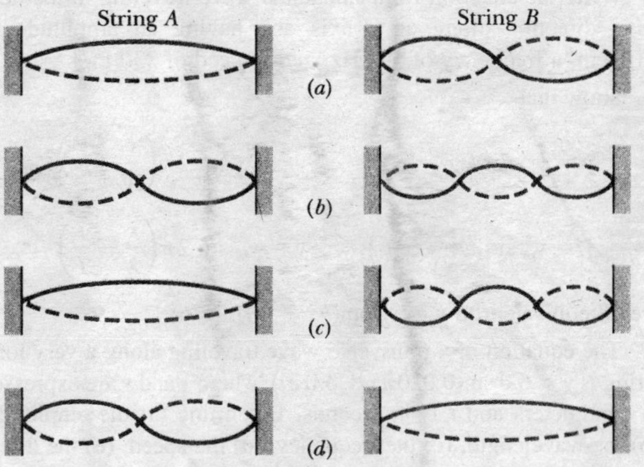

Fig. 17-28 Question 9.

10. (a) If a standing wave on a string is given by

$$y'(t) = (3 \text{ mm}) \sin(5x) \cos(4t),$$

is there a node or an antinode of the oscillations of the string at $x = 0$? (b) If the standing wave is given by

$$y'(t) = (3 \text{ mm}) \sin(5x + \pi/2) \cos(4t),$$

is there a node or an antinode at $x = 0$?

11. (a) In Sample Problem 17-7 and Fig. 17-23, if we gradually increase the mass of the block (the frequency remains fixed), new resonant modes appear. Do the harmonic numbers of the new resonant modes increase or decrease from one to the next? (b) Is the shift from one resonant mode to the next gradual, or does each resonant mode disappear well before the next one appears?

EXERCISES & PROBLEMS

ssm Solution is in the Student Solutions Manual.
www Solution is available on the World Wide Web at: http://www.wiley.com/college/hrw
ilw Solution is available on the Interactive LearningWare.

SEC. 17-5 The Speed of a Traveling Wave

1E. A wave has an angular frequency of 110 rad/s and a wavelength of 1.80 m. Calculate (a) the angular wave number and (b) the speed of the wave.

2E. The speed of electromagnetic waves (which include visible light, radio, and x rays) in vacuum is 3.0×10^8 m/s. (a) Wavelengths of visible light waves range from about 400 nm in the violet to about 700 nm in the red. What is the range of frequencies of these waves? (b) The range of frequencies for shortwave radio (for example, FM radio and VHF television) is 1.5 to 300 MHz. What is the corresponding wavelength range? (c) X ray wavelengths range from about 5.0 nm to about 1.0×10^{-2} nm. What is the frequency range for x rays?

3E. A sinusoidal wave travels along a string. The time for a particular point to move from maximum displacement to zero is 0.170 s. What are the (a) period and (b) frequency? (c) The wavelength is 1.40 m; what is the wave speed? ssm

4E. Write the equation for a sinusoidal wave traveling in the negative direction along an x axis and having an amplitude of 0.010 m, a frequency of 550 Hz, and a speed of 330 m/s.

5E. Show that

$$y = y_m \sin k(x - vt), \qquad y = y_m \sin 2\pi\left(\frac{x}{\lambda} - ft\right),$$

$$y = y_m \sin \omega\left(\frac{x}{v} - t\right), \qquad y = y_m \sin 2\pi\left(\frac{x}{\lambda} - \frac{t}{T}\right)$$

are all equivalent to $y = y_m \sin(kx - \omega t)$. ssm

6P. The equation of a transverse wave traveling along a very long string is $y = 6.0 \sin(0.020\pi x + 4.0\pi t)$, where x and y are expressed in centimeters and t is in seconds. Determine (a) the amplitude, (b) the wavelength, (c) the frequency, (d) the speed, (e) the direction of propagation of the wave, and (f) the maximum transverse speed of a particle in the string. (g) What is the transverse displacement at $x = 3.5$ cm when $t = 0.26$ s?

7P. (a) Write an equation describing a sinusoidal transverse wave traveling on a cord in the $+x$ direction with a wavelength of 10 cm, a frequency of 400 Hz, and an amplitude of 2.0 cm. (b) What is the maximum speed of a point on the cord? (c) What is the speed of the wave? ssm

8P. A transverse sinusoidal wave of wavelength 20 cm is moving along a string in the positive x direction. The transverse displacement of the string particle at $x = 0$ as a function of time is shown in Fig. 17-29. (a) Make a rough sketch of one wavelength of the wave (the portion between $x = 0$ and $x = 20$ cm) at time $t = 0$. (b) What is the speed of the wave? (c) Write the equation for the wave with all the constants evaluated. (d) What is the transverse velocity of the particle at $x = 0$ at $t = 5.0$ s?

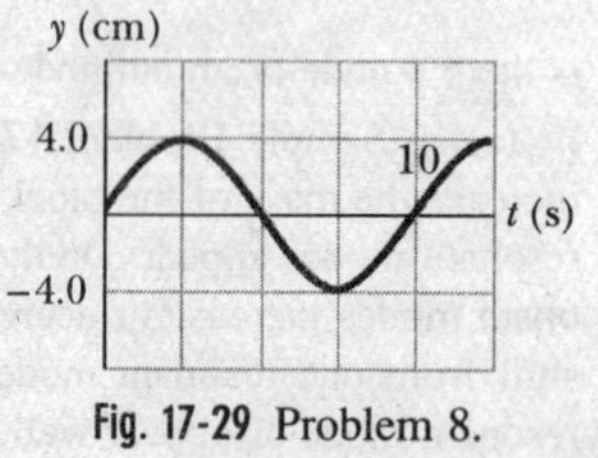

Fig. 17-29 Problem 8.

9P. A sinusoidal wave of frequency 500 Hz has a speed of 350 m/s. (a) How far apart are two points that differ in phase by $\pi/3$ rad? (b) What is the phase difference between two displacements at a certain point at times 1.00 ms apart? ilw

SEC. 17-6 Wave Speed on a Stretched String

10E. The heaviest and lightest strings on a certain violin have linear densities of 3.0 and 0.29 g/m. What is the ratio of the diameter of the heaviest string to that of the lightest string, assuming that the strings are of the same material?

11E. What is the speed of a transverse wave in a rope of length 2.00 m and mass 60.0 g under a tension of 500 N? ssm

12E. The tension in a wire clamped at both ends is doubled without appreciably changing the wire's length between the clamps. What is the ratio of the new to the old wave speed for transverse waves traveling along this wire?

13E. The linear density of a string is 1.6×10^{-4} kg/m. A transverse wave on the string is described by the equation

$$y = (0.021 \text{ m}) \sin[(2.0 \text{ m}^{-1})x + (30 \text{ s}^{-1})t].$$

What is (a) the wave speed and (b) the tension in the string? ssm

14E. The equation of a transverse wave on a string is

$$y = (2.0 \text{ mm}) \sin[(20 \text{ m}^{-1})x - (600 \text{ s}^{-1})t].$$

The tension in the string is 15 N. (a) What is the wave speed? (b) Find the linear density of this string in grams per meter.

15P. A stretched string has a mass per unit length of 5.0 g/cm and a tension of 10 N. A sinusoidal wave on this string has an amplitude of 0.12 mm and a frequency of 100 Hz and is traveling in the negative direction of x. Write an equation for this wave. ssm

16P. What is the fastest transverse wave that can be sent along a steel wire? For safety reasons, the maximum tensile stress to which steel wires should be subjected is 7.0×10^8 N/m^2. The density of steel is 7800 kg/m^3. Show that your answer does not depend on the diameter of the wire.

17P. A sinusoidal transverse wave of amplitude y_m and wavelength λ travels on a stretched cord. (a) Find the ratio of the maximum particle speed (the speed with which a single particle in the cord moves transverse to the wave) to the wave speed. (b) If a wave having a certain wavelength and amplitude is sent along a cord, would this speed ratio depend on the material of which the cord is made, such as wire or nylon? ssm

18P. A sinusoidal wave is traveling on a string with speed 40 cm/s. The displacement of the particles of the string at $x = 10$ cm is found to vary with time according to the equation $y = (5.0 \text{ cm}) \sin[1.0 - (4.0 \text{ s}^{-1})t]$. The linear density of the string is 4.0 g/cm. What are (a) the frequency and (b) the wavelength of the wave? (c) Write the general equation giving the transverse displacement of the particles of the string as a function of position and time. (d) Calculate the tension in the string.

19P. A sinusoidal transverse wave is traveling along a string in the negative direction of an x axis. Figure 17-30 shows a plot of the displacement as a function of position at time $t = 0$; the y intercept is 4.0 cm. The string tension is 3.6 N, and its linear density is 25 g/m. Find (a) the amplitude, (b) the wavelength, (c) the wave speed, and (d) the period of the wave. (e) Find the maximum transverse speed of a particle in the string. (f) Write an equation describing the traveling wave. ssm ilw

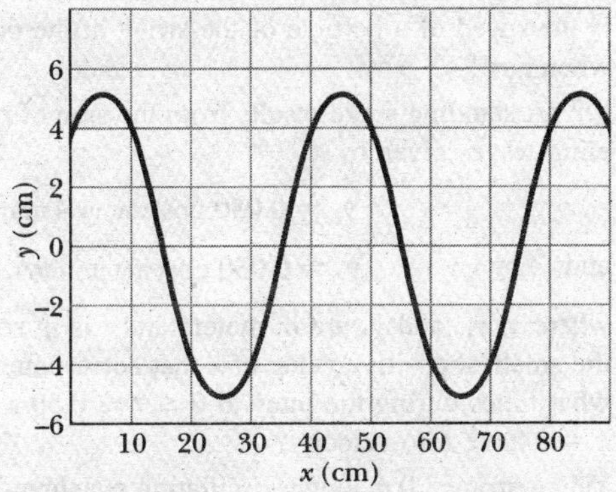

Fig. 17-30 Problem 19.

20P. In Fig. 17-31a, string 1 has a linear density of 3.00 g/m, and string 2 has a linear density of 5.00 g/m. They are under tension owing to the hanging block of mass $M = 500$ g. Calculate the wave speed on (a) string 1 and (b) string 2. (*Hint:* When a string loops halfway around a pulley, it pulls on the pulley with a net force that is twice the tension in the string.) Next the block is divided into two blocks (with $M_1 + M_2 = M$) and the apparatus is rearranged as shown in Fig. 17-31b. Find (c) M_1 and (d) M_2 such that the wave speeds in the two strings are equal.

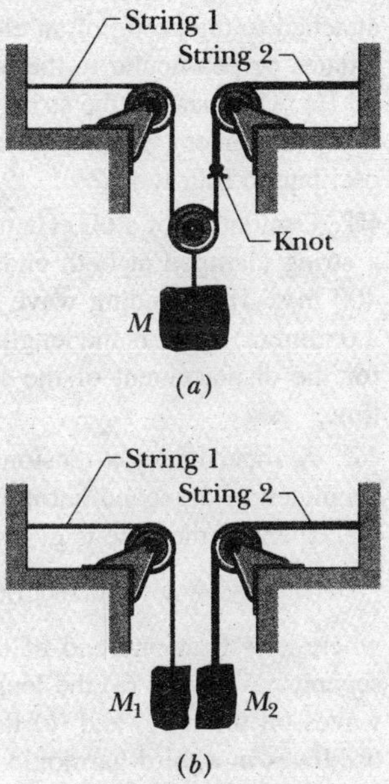

Fig. 17-31 Problem 20.

21P. A wire 10.0 m long and having a mass of 100 g is stretched under a tension of 250 N. If two pulses, separated in time by 30.0 ms, are generated, one at each end of the wire, where will the pulses first meet? ssm ilw

22P. The type of rubber band used inside some baseballs and golf balls obeys Hooke's law over a wide range of elongation of the band. A segment of this material has an unstretched length ℓ and a mass m. When a force F is applied, the band stretches an additional length $\Delta\ell$. (a) What is the speed (in terms of m, $\Delta\ell$, and the spring constant k) of transverse waves on this stretched rubber band? (b) Using your answer to (a), show that the time required for a transverse pulse to travel the length of the rubber band is proportional to $1/\sqrt{\Delta\ell}$ if $\Delta\ell \ll \ell$ and is constant if $\Delta\ell \gg \ell$.

23P*. A uniform rope of mass m and length L hangs from a ceiling. (a) Show that the speed of a transverse wave on the rope is a function of y, the distance from the lower end, and is given by $v = \sqrt{gy}$. (b) Show that the time a transverse wave takes to travel the length of the rope is given by $t = 2\sqrt{L/g}$. ssm

SEC. 17-7 Energy and Power of a Traveling String Wave

24E. A string along which waves can travel is 2.70 m long and has a mass of 260 g. The tension in the string is 36.0 N. What must be the frequency of traveling waves of amplitude 7.70 mm for the average power to be 85.0 W?

25P. A transverse sinusoidal wave is generated at one end of a long, horizontal string by a bar that moves up and down through a distance of 1.00 cm. The motion is continuous and is repeated regularly 120 times per second. The string has linear density 120 g/m and is kept under a tension of 90.0 N. Find the maximum value of (a) the transverse speed u and (b) the transverse component of the tension τ. (*Hint:* That component is $\tau \sin\theta$, where θ is the angle the string makes with the horizontal. You will need to relate angle θ to dy/dx.)

(c) Show that the two maximum values calculated above occur at the same phase values for the wave. What is the transverse displacement y of the string at these phases? (d) What is the maximum rate of energy transfer along the string? (e) What is the transverse displacement y when this maximum transfer occurs? (f) What is the minimum rate of energy transfer along the string? (g) What is the transverse displacement y when this minimum transfer occurs? ssm www

SEC. 17-9 Interference of Waves

26E. What phase difference between two otherwise identical traveling waves, moving in the same direction along a stretched string, will result in the combined wave having an amplitude 1.50 times that of the common amplitude of the two combining waves? Express your answer in (a) degrees, (b) radians, and (c) wavelengths.

27E. Two identical traveling waves, moving in the same direction, are out of phase by $\pi/2$ rad. What is the amplitude of the resultant wave in terms of the common amplitude y_m of the two combining waves? ssm

28P. Two sinusoidal waves, identical except for phase, travel in the same direction along a string and interfere to produce a resultant wave given by $y'(x, t) = (3.0 \text{ mm}) \sin(20x - 4.0t + 0.820 \text{ rad})$, with x in meters and t in seconds. What are (a) the wavelength λ of the two waves, (b) the phase difference between them, and (c) their amplitude y_m?

SEC. 17-10 Phasors

29E. Determine the amplitude of the resultant wave when two sinusoidal string waves having the same frequency and traveling in the same direction on the same string are combined, if their amplitudes are 3.0 cm and 4.0 cm and they have phase constants of 0 and $\pi/2$ rad, respectively. ssm

30P. Two sinusoidal waves of the same period, with amplitudes of 5.0 and 7.0 mm, travel in the same direction along a stretched string; they produce a resultant wave with an amplitude of 9.0 mm. The phase constant of the 5.0 mm wave is 0. What is the phase constant of the 7.0 mm wave?

31P. Three sinusoidal waves of the same frequency travel along a string in the positive direction of an x axis. Their amplitudes are y_1, $y_1/2$, and $y_1/3$, and their phase constants are 0, $\pi/2$, and π, respectively. What are (a) the amplitude and (b) the phase constant of the resultant wave? (c) Plot the wave form of the resultant wave at $t = 0$, and discuss its behavior as t increases. ssm www

SEC. 17-12 Standing Waves and Resonance

32E. A string under tension τ_i oscillates in the third harmonic at frequency f_3, and the waves on the string have wavelength λ_3. If the tension is increased to $\tau_f = 4\tau_i$ and the string is again made to oscillate in the third harmonic, what then are (a) the frequency of oscillation in terms of f_3 and (b) the wavelength of the waves in terms of λ_3?

33E. A nylon guitar string has a linear density of 7.2 g/m and is under a tension of 150 N. The fixed supports are 90 cm apart. The string is oscillating in the standing wave pattern shown in Fig. 17-32. Calculate the (a) speed, (b) wavelength, and (c) frequency of the traveling waves whose superposition gives this standing wave. ilw

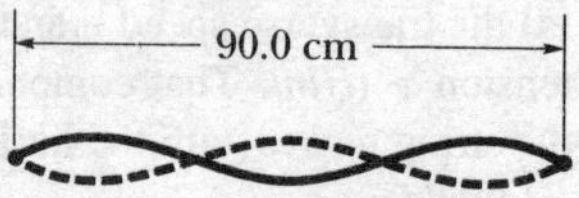

Fig. 17-32 Exercise 33.

34E. Two sinusoidal waves with identical wavelengths and amplitudes travel in opposite directions along a string with a speed of 10 cm/s. If the time interval between instants when the string is flat is 0.50 s, what is the wavelength of the waves?

35E. A string fixed at both ends is 8.40 m long and has a mass of 0.120 kg. It is subjected to a tension of 96.0 N and set oscillating. (a) What is the speed of the waves on the string? (b) What is the longest possible wavelength for a standing wave? (c) Give the frequency of that wave. ssm

36E. A 125 cm length of string has a mass of 2.00 g. It is stretched with a tension of 7.00 N between fixed supports. (a) What is the wave speed for this string? (b) What is the lowest resonant frequency of this string?

37E. What are the three lowest frequencies for standing waves on a wire 10.0 m long having a mass of 100 g, which is stretched under a tension of 250 N? ssm

38E. String A is stretched between two clamps separated by distance L. String B, with the same linear density and under the same tension as string A, is stretched between two clamps separated by distance $4L$. Consider the first eight harmonics of string B. Which, if any, has a resonant frequency that matches a resonant frequency of string A?

39P. A string that is stretched between fixed supports separated by 75.0 cm has resonant frequencies of 420 and 315 Hz, with no intermediate resonant frequencies. What are (a) the lowest resonant frequency and (b) the wave speed? ssm ilw www

40P. In Fig. 17-33, two pulses travel along a string in opposite directions. The wave speed v is 2.0 m/s and the pulses are 6.0 cm apart at $t = 0$. (a) Sketch the wave patterns when t is equal to 5.0, 10, 15, 20, and 25 ms. (b) In what form (or type) is the energy of the pulses at $t = 15$ ms?

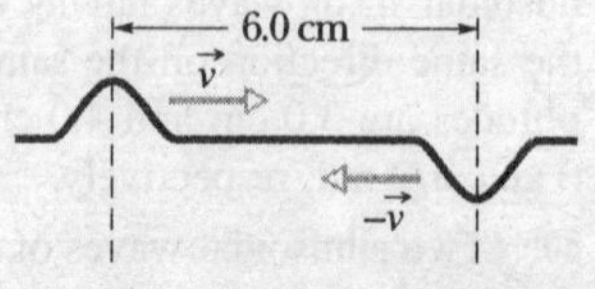

Fig. 17-33 Problem 40.

41P. A string oscillates according to the equation

$$y' = (0.50 \text{ cm}) \sin\left[\left(\frac{\pi}{3} \text{ cm}^{-1}\right)x\right] \cos[(40\pi \text{ s}^{-1})t].$$

What are (a) the amplitude and (b) the speed of the two waves (identical except for direction of travel) whose superposition gives this oscillation? (c) What is the distance between nodes? (d) What is the speed of a particle of the string at the position $x = 1.5$ cm when $t = \frac{9}{8}$ s? ssm

42P. A standing wave results from the sum of two transverse traveling waves given by

$$y_1 = 0.050 \cos(\pi x - 4\pi t)$$

and

$$y_2 = 0.050 \cos(\pi x + 4\pi t),$$

where x, y_1, and y_2 are in meters and t is in seconds. (a) What is the smallest positive value of x that corresponds to a node? (b) At what times during the interval $0 \le t \le 0.50$ s will the particle at $x = 0$ have zero velocity?

43P. A string 3.0 m long is oscillating as a three-loop standing wave with an amplitude of 1.0 cm. The wave speed is 100 m/s. (a) What is the frequency? (b) Write equations for two waves that, when combined, will result in this standing wave. ssm

44P. In an experiment on standing waves, a string 90 cm long is attached to the prong of an electrically driven tuning fork that oscillates perpendicular to the length of the string at a frequency of 60 Hz. The mass of the string is 0.044 kg. What tension must the string be under (weights are attached to the other end) if it is to oscillate in four loops?

45P. Oscillation of a 600 Hz tuning fork sets up standing waves in a string clamped at both ends. The wave speed for the string is 400 m/s. The standing wave has four loops and an amplitude of 2.0 mm. (a) What is the length of the string? (b) Write an equation for the displacement of the string as a function of position and time. ssm

46P. A rope, under a tension of 200 N and fixed at both ends, oscillates in a second-harmonic standing wave pattern. The displacement of the rope is given by

$$y = (0.10 \text{ m})(\sin \pi x/2) \sin 12\pi t,$$

where $x = 0$ at one end of the rope, x is in meters, and t is in seconds. What are (a) the length of the rope, (b) the speed of the waves on the rope, and (c) the mass of the rope? (d) If the rope oscillates in a third-harmonic standing wave pattern, what will be the period of oscillation?

47P. A generator at one end of a very long string creates a wave given by

$$y = (6.0 \text{ cm}) \cos \frac{\pi}{2} [(2.0 \text{ m}^{-1})x + (8.0 \text{ s}^{-1})t],$$

and one at the other end creates the wave

$$y = (6.0 \text{ cm}) \cos \frac{\pi}{2} [(2.0 \text{ m}^{-1})x - (8.0 \text{ s}^{-1})t].$$

Calculate (a) the frequency, (b) the wavelength, and (c) the speed of each wave. At what x values are (d) the nodes and (e) the antinodes? ssm

48P. A standing wave pattern on a string is described by

$$y(x, t) = 0.040 \sin 5\pi x \cos 40\pi t,$$

where x and y are in meters and t is in seconds. (a) Determine the location of all nodes for $0 \leq x \leq 0.40$ m. (b) What is the period of the oscillatory motion of any (nonnode) point on the string? What are (c) the speed and (d) the amplitude of the two traveling waves that interfere to produce this wave? (e) At what times for $0 \leq t \leq 0.050$ s will all the points on the string have zero transverse velocity?

49P. Show that the maximum kinetic energy in each loop of a standing wave produced by two traveling waves of identical amplitudes is $2\pi^2 \mu y_m^2 f v$. ssm www

50P. For a certain transverse standing wave on a long string, an antinode is at $x = 0$ and a node is at $x = 0.10$ m. The displacement $y(t)$ of the string particle at $x = 0$ is shown in Fig. 17-34. When $t = 0.50$ s, what are the displacements of the string particles at (a) $x = 0.20$ m and (b) $x = 0.30$ m? At $x = 0.20$ m, what are the transverse velocities of the string particles at (c) $t = 0.50$ s and (d) $t = 1.0$ s? (e) Sketch the standing wave at $t = 0.50$ s for the range $x = 0$ to $x = 0.40$ m.

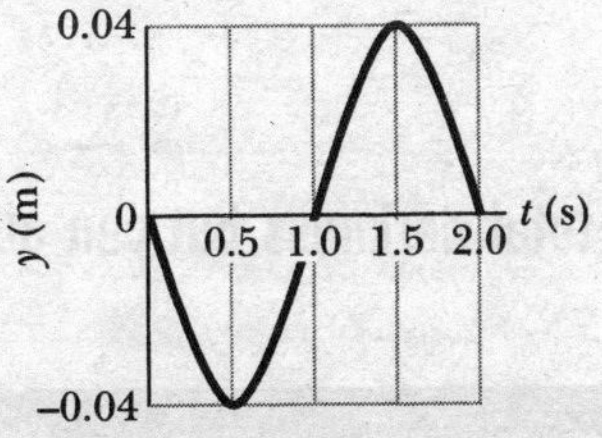

Fig. 17-34 Problem 50.

51P. In Fig. 17-35, an aluminum wire, of length $L_1 = 60.0$ cm, cross-sectional area 1.00×10^{-2} cm^2, and density 2.60 g/cm^3, is joined to a steel wire, of density 7.80 g/cm^3 and the same cross-sectional area. The compound wire, loaded with a block of mass $m = 10.0$ kg, is arranged so that the distance L_2 from the joint to the supporting pulley is 86.6 cm. Transverse waves are set up in the wire by using an external source of variable frequency; a node is located at the pulley. (a) Find the lowest frequency of excitation for which standing waves are observed such that the joint in the wire is one of the nodes. (b) How many nodes are observed at this frequency? ssm

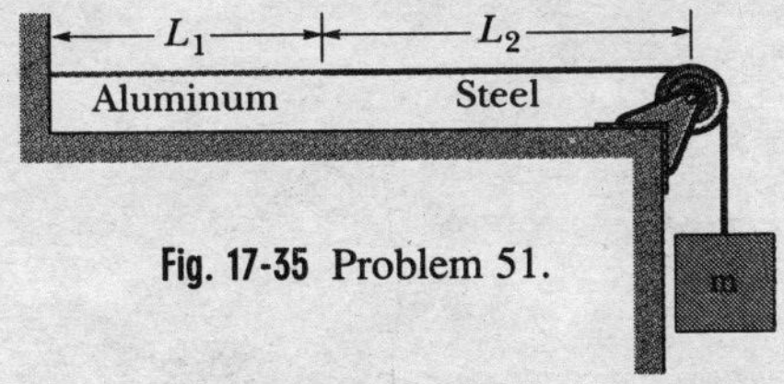

Fig. 17-35 Problem 51.

Additional Problem

52. *Body armor.* When a high-speed projectile such as a bullet or bomb fragment strikes modern body armor, the fabric of the armor stops the projectile and prevents penetration by quickly spreading the projectile's energy over a large area. This spreading is done by longitudinal and transverse pulses that move *radially* from the impact point, where the projectile pushes a cone-shaped dent into the fabric. The longitudinal pulse, racing along the fibers of the fabric at speed v_l ahead of the denting, causes the fibers to thin and stretch, with material flowing radially inward into the dent. One such radial fiber is shown in Fig. 17-36*a*. Part of the projectile's energy goes into this motion and stretching. The transverse pulse, moving at a slower speed v_t, is due to the denting. As the projectile increases the dent's depth, the dent increases in radius, causing the material in the fibers to move in the same direction as the projectile (perpendicular to the transverse pulse's direction of travel). The rest of the projectile's energy goes into this motion. All the energy that does not eventually go into permanently deforming the fibers ends up as thermal energy.

Figure 17-36*b* is a graph of speed v versus time t for a bullet of mass 10.2 g fired from a .38 Special revolver directly into body armor. Take $v_l = 2000$ m/s, and assume that the half-angle θ of the conical dent is 60°. At the end of the collision, what are the radii of (a) the thinned region and (b) the dent (assuming that the person wearing the armor remains stationary)?

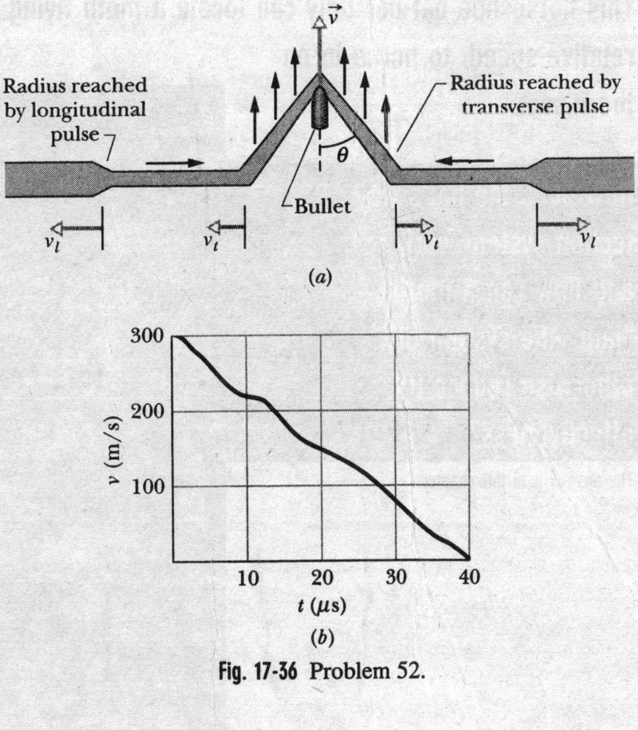

Fig. 17-36 Problem 52.

18 Waves—II

This horseshoe bat not only can locate a moth flying in total darkness but can also determine the moth's relative speed, to home in on the insect.

How does the bat's detection system work, and how can a moth "jam" the system or otherwise reduce its effectiveness?

The answer is in this chapter.

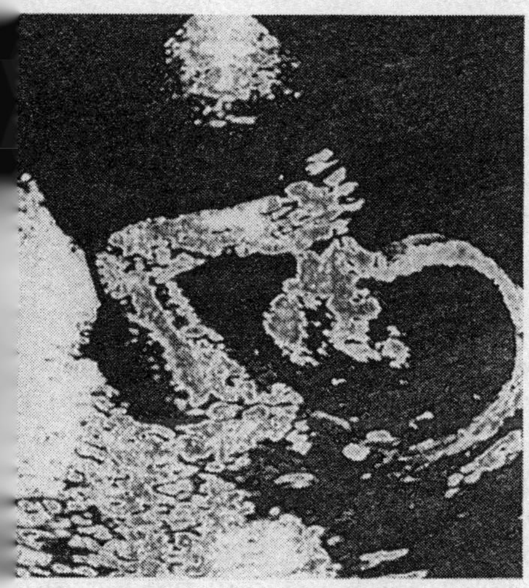

Fig. 18-1 An image of a fetus searching for a thumb to suck; the image is made with ultrasound (which has a frequency above your hearing range).

18-1 Sound Waves

As we saw in Chapter 17, mechanical waves are waves that require a material medium to exist. There are two types of mechanical waves: *Transverse waves* involve oscillations perpendicular to the direction in which the wave travels; *longitudinal waves* involve oscillations parallel to the direction of wave travel.

In this book, a **sound wave** is defined roughly as any longitudinal wave. Seismic prospecting teams use such waves to probe Earth's crust for oil. Ships carry sound-ranging gear (sonar) to detect underwater obstacles. Submarines use sound waves to stalk other submarines, largely by listening for the characteristic noises produced by the propulsion system. Figure 18-1, a computer-processed image of a fetal head and arm, shows how sound waves can be used to explore the soft tissues of the human body. In this chapter we shall focus on sound waves that travel through the air and that are audible to people.

Figure 18-2 illustrates several ideas that we shall use in our discussions. Point S represents a tiny sound source, called a *point source*, that emits sound waves in all directions. The *wavefronts* and *rays* indicate the direction of travel and the spread of the sound waves. **Wavefronts** are surfaces over which the oscillations of the air due to the sound wave have the same value; such surfaces are represented by whole or partial circles in a two-dimensional drawing for a point source. **Rays** are directed lines perpendicular to the wavefronts that indicate the direction of travel of the wavefronts. The short double arrows superimposed on the rays of Fig. 18-2 indicate that the longitudinal oscillations of the air are parallel to the rays.

Near a point source like that of Fig. 18-2, the wavefronts are spherical and spread out in three dimensions, and there the waves are said to be *spherical*. As the wavefronts move outward and their radii become larger, their curvature decreases. Far from the source, we approximate the wavefronts as planes (or lines on two-dimensional drawings), and the waves are said to be *planar*.

18-2 The Speed of Sound

The speed of any mechanical wave, transverse or longitudinal, depends on both an inertial property of the medium (to store kinetic energy) and an elastic property of the medium (to store potential energy). Thus, we can generalize Eq. 17-25, which gives the speed of a transverse wave along a stretched string, by writing

$$v = \sqrt{\frac{\tau}{\mu}} = \sqrt{\frac{\text{elastic property}}{\text{inertial property}}}, \tag{18-1}$$

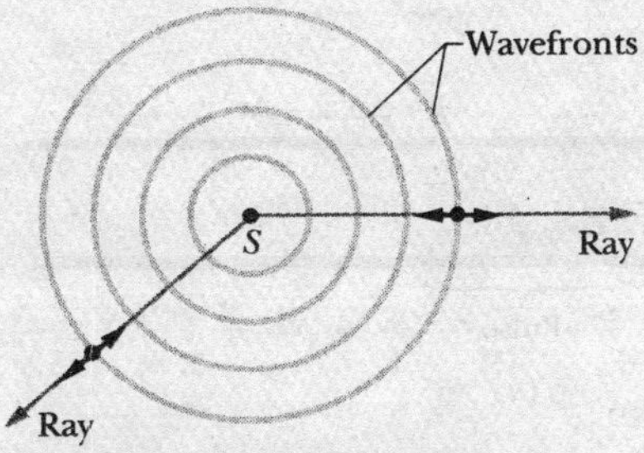

Fig. 18-2 A sound wave travels from a point source S through a three-dimensional medium. The wavefronts form spheres centered on S; the rays are radial to S. The short, double-headed arrows indicate that elements of the medium oscillate parallel to the rays.

where (for transverse waves) τ is the tension in the string and μ is the string's linear density. If the medium is air and the wave is longitudinal, we can guess that the inertial property, corresponding to μ, is the volume density ρ of air. What shall we put for the elastic property?

In a stretched string, potential energy is associated with the periodic stretching of the string elements as the wave passes through them. As a sound wave passes through air, potential energy is associated with periodic compressions and expansions of small volume elements of the air. The property that determines the extent to which an element of a medium changes in volume when the pressure (force per unit area) on it changes is the **bulk modulus** B, defined (from Eq. 13-27) as

$$B = -\frac{\Delta p}{\Delta V/V} \quad \text{(definition of bulk modulus).} \tag{18-2}$$

Here $\Delta V/V$ is the fractional change in volume produced by a change in pressure Δp. As explained in Section 15-3, the SI unit for pressure is the newton per square meter,

TABLE 18-1 The Speed of Sound[a]

Medium	Speed (m/s)
Gases	
Air (0°C)	331
Air (20°C)	343
Helium	965
Hydrogen	1284
Liquids	
Water (0°C)	1402
Water (20°C)	1482
Seawater[b]	1522
Solids	
Aluminum	6420
Steel	5941
Granite	6000

[a]At 0°C and 1 atm pressure, except where noted.

[b]At 20°C and 3.5% salinity.

which is given a special name, the *pascal* (Pa). From Eq. 18-2 we see that the unit for B is also the pascal. The signs of Δp and ΔV are always opposite: When we increase the pressure on an element (Δp is positive), its volume decreases (ΔV is negative). We include a minus sign in Eq. 18-2 so that B is always a positive quantity. Now substituting B for τ and ρ for μ in Eq. 18-1 yields

$$v = \sqrt{\frac{B}{\rho}} \quad \text{(speed of sound)} \tag{18-3}$$

as the speed of sound in a medium with bulk modulus B and density ρ. This is actually the correct equation, as we shall derive shortly. Table 18-1 lists the speed of sound in various media.

The density of water is almost 1000 times greater than the density of air. If this were the only relevant factor, we would expect from Eq. 18-3 that the speed of sound in water would be considerably less than the speed of sound in air. However, Table 18-1 shows us that the reverse is true. We conclude (again from Eq. 18-3) that the bulk modulus of water must be more than 1000 times greater than that of air. This is indeed the case. Water is much more incompressible than air, which (see Eq. 18-2) is another way of saying that its bulk modulus is much greater.

Formal Derivation of Eq. 18-3

We now derive Eq. 18-3 by direct application of Newton's laws. Let a single pulse in which air is compressed travel (from right to left) with speed v through the air in a long tube, like that in Fig. 17-2. Let us run along with the pulse at that speed, so that the pulse appears to stand still in our reference frame. Figure 18-3*a* shows the situation as it is viewed from that frame. The pulse is standing still, and air is moving at speed v through it from left to right.

Let the pressure of the undisturbed air be p and the pressure inside the pulse be $p + \Delta p$, where Δp is positive owing to the compression. Consider a slice of air of thickness Δx and face area A, moving toward the pulse at speed v. As this element of air enters the pulse, its leading face encounters a region of higher pressure, which slows it to speed $v + \Delta v$, in which Δv is negative. This slowing is complete when the rear face reaches the pulse, which requires time interval

$$\Delta t = \frac{\Delta x}{v}. \tag{18-4}$$

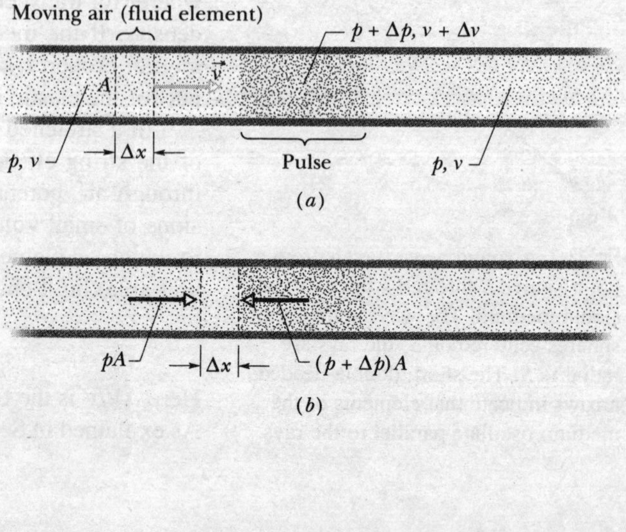

Fig. 18-3 A compression pulse is sent down a long air-filled tube. The reference frame of the figure is chosen so that the pulse is at rest and the air moves from left to right. (*a*) A slice of air of width Δx moves toward the pulse with speed v. (*b*) The leading face of the slice enters the pulse. The forces acting on the leading and trailing faces (due to air pressure) are shown.

Let us apply Newton's second law to the element. During Δt, the average force on the element's trailing face is pA toward the right, and the average force on the leading face is $(p + \Delta p)A$ toward the left (Fig. 18-3*b*). Therefore, the average net force on the element during Δt is

$$F = pA - (p + \Delta p)A$$
$$= -\Delta p\, A \qquad \text{(net force).} \tag{18-5}$$

The minus sign indicates that the net force on the fluid element is directed to the left in Fig. 18-3*b*. The volume of the element is $A\, \Delta x$, so with the aid of Eq. 18-4, we can write its mass as

$$\Delta m = \rho A\, \Delta x = \rho A v\, \Delta t \qquad \text{(mass).} \tag{18-6}$$

Then the average acceleration of the element during Δt is

$$a = \frac{\Delta v}{\Delta t} \qquad \text{(acceleration).} \tag{18-7}$$

From Newton's second law ($F = ma$), we have, from Eqs. 18-5, 18-6, and 18-7,

$$-\Delta p\, A = (\rho A v\, \Delta t)\, \frac{\Delta v}{\Delta t},$$

which we can write as

$$\rho v^2 = -\frac{\Delta p}{\Delta v / v}. \tag{18-8}$$

The air that occupies a volume V $(= Av\, \Delta t)$ outside the pulse is compressed by an amount ΔV $(= A\, \Delta v\, \Delta t)$ as it enters the pulse. Thus,

$$\frac{\Delta V}{V} = \frac{A\, \Delta v\, \Delta t}{Av\, \Delta t} = \frac{\Delta v}{v}. \tag{18-9}$$

Substituting Eq. 18-9 and then Eq. 18-2 into Eq. 18-8 leads to

$$\rho v^2 = -\frac{\Delta p}{\Delta v / v} = -\frac{\Delta p}{\Delta V / V} = B.$$

Solving for v yields Eq. 18-3 for the speed of the air toward the right in Fig. 18-3, and thus for the actual speed of the pulse toward the left.

Sample Problem 18-1

One clue used by your brain to determine the direction of a source of sound is the time delay Δt between the arrival of the sound at the ear closer to the source and the arrival at the farther ear. Assume that the source is distant so that a wavefront from it is approximately planar when it reaches you, and let D represent the separation between your ears.

(a) Find an expression that gives Δt in terms of D and the angle θ between the direction of the source and the forward direction.

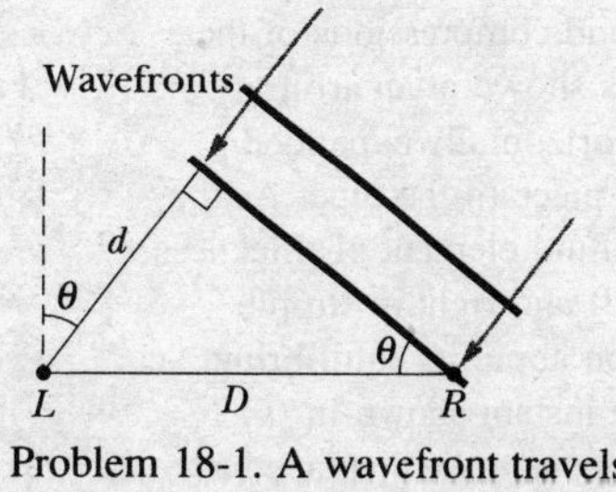

Fig. 18-4 Sample Problem 18-1. A wavefront travels a distance d $(= D \sin \theta)$ farther to reach the left ear (L) than to reach the right ear (R).

SOLUTION: The situation is shown (from an overhead view) in Fig. 18-4, where wavefronts approach you from a source that is located in front of you and to your right. The **Key Idea** here is that the time delay Δt is due to the distance d that each wavefront must travel to reach your left ear (L) after it reaches your right ear (R). From Fig. 18-4, we find

$$\Delta t = \frac{d}{v} = \frac{D \sin \theta}{v}, \quad \text{(Answer)} \quad (18\text{-}10)$$

where v is the speed of sound in air. Based on a lifetime of experience, your brain correlates each detected value of Δt (from zero to the maximum value) with a value of θ (from zero to 90°) for the direction of the sound source.

(b) Suppose that you are submerged in water at 20°C when a wavefront arrives from directly to your right. Based on the time-delay clue, at what angle θ from the forward direction does the source seem to be?

SOLUTION: The **Key Idea** here is that the speed of the sound is now the speed v_w in water, so in Eq. 18-10 we substitute v_w for v and 90° for θ, finding that

$$\Delta t_w = \frac{D \sin 90°}{v_w} = \frac{D}{v_w}. \quad (18\text{-}11)$$

Since v_w is about four times v, delay Δt_w is about one-fourth the maximum time delay in air. Based on experience, your brain will process the water time delay as if it occurred in air. Thus, the sound source appears to be at an angle θ smaller than 90°. To find that apparent angle, we substitute the time delay D/v_w from Eq. 18-11 for Δt in Eq. 18-10, obtaining

$$\frac{D}{v_w} = \frac{D \sin \theta}{v}. \quad (18\text{-}12)$$

Then, to solve for θ we substitute $v = 343$ m/s and $v_w = 1482$ m/s (from Table 18-1) into Eq. 18-12, finding

$$\sin \theta = \frac{v}{v_w} = \frac{343 \text{ m/s}}{1482 \text{ m/s}} = 0.231$$

and thus

$$\theta = 13°. \quad \text{(Answer)}$$

18-3 Traveling Sound Waves

Here we examine the displacements and pressure variations associated with a sinusoidal sound wave traveling through air. Figure 18-5*a* displays such a wave traveling rightward through a long air-filled tube. Recall from Chapter 17 that we can produce such a wave by sinusoidally moving a piston at the left end of the tube (as in Fig. 17-2). The piston's rightward motion moves the element of air next to it and compresses that air; the piston's leftward motion allows the element of air to move back to the left and the pressure to decrease. As each element of air pushes on the next element in turn, the right–left motion of the air and the change in its pressure travel along the tube as a sound wave.

Consider a thin element of air of thickness Δx, located at a position x along the tube. As the wave travels through x, the element of air oscillates left and right in simple harmonic motion about its equilibrium position (Fig. 18-5*b*). Thus, the oscillations of each air element due to the traveling sound wave are like those of a string element due to a transverse wave, except that the air element oscillates *longitudinally* rather than *transversely*. Because string elements oscillate parallel to the y axis, we write their displacements in the form $y(x, t)$. Similarly, because air ele-

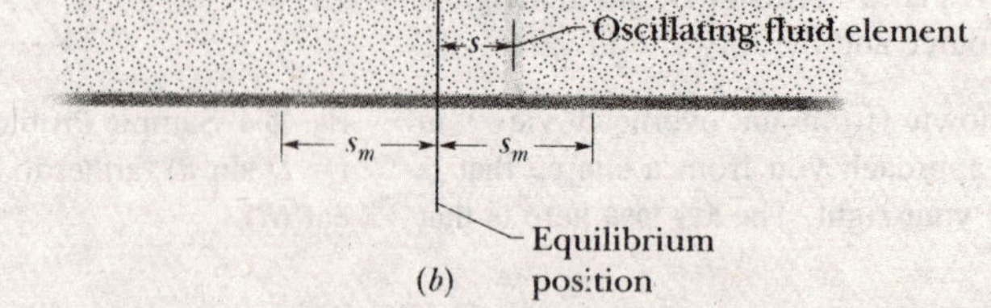

Fig. 18-5 (*a*) A sound wave, traveling through a long air-filled tube with speed v, consists of a moving, periodic pattern of expansions and compressions of the air. The wave is shown at an arbitrary instant. (*b*) A horizontally expanded view of a short piece of the tube. As the wave passes, a fluid element of thickness Δx oscillates left and right in simple harmonic motion about its equilibrium position. At the instant shown in (*b*), the element happens to be displaced a distance s to the right of its equilibrium position. Its maximum displacement, either right or left, is s_m.

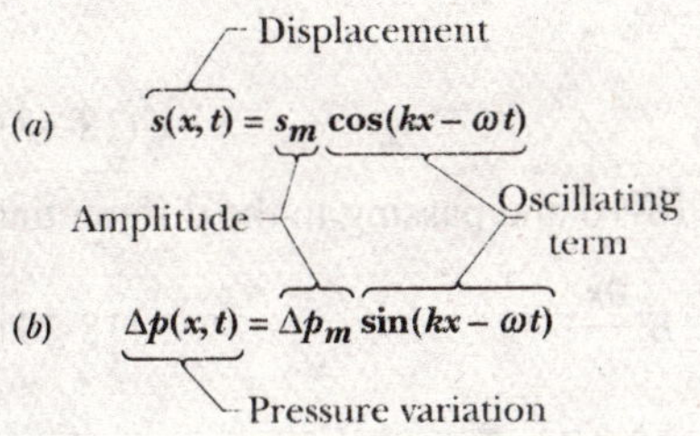

Fig. 18-6 (*a*) The displacement function and (*b*) the pressure-variation function of a traveling sound wave consist of an amplitude and an oscillating term.

ments oscillate parallel to the x axis, we could write their displacements in the form $x(x, t)$. However, we shall avoid that confusing notation and use $s(x, t)$ instead.

To show that the displacements $s(x, t)$ are sinusoidal functions of x and t, we can use either a sine function or a cosine function. In this chapter we use a cosine function, writing

$$s(x, t) = s_m \cos(kx - \omega t). \tag{18-13}$$

Figure 18-6*a* shows the major parts of this equation. In it, s_m is the **displacement amplitude**—that is, the maximum displacement of the air element to either side of its equilibrium position (see Fig. 18-5*b*). The angular wave number k, angular frequency ω, frequency f, wavelength λ, speed v, and period T for a sound (longitudinal) wave are defined and interrelated exactly as for a transverse wave, except that λ is now the distance (again along the direction of travel) in which the pattern of compression and expansion due to the wave begins to repeat itself (see Fig. 18-5*a*). (We assume s_m is much less than λ.)

As the wave moves, the air pressure at any position x in Fig. 18-5*a* varies sinusoidally, as we prove next. To describe this variation we write

$$\Delta p(x, t) = \Delta p_m \sin(kx - \omega t). \tag{18-14}$$

Figure 18-6*b* shows the major parts of this equation. A negative value of Δp in Eq. 18-14 corresponds to an expansion of the air, and a positive value to a compression. Here Δp_m is the **pressure amplitude,** which is the maximum increase or decrease in pressure due to the wave; Δp_m is normally very much less than the pressure p present when there is no wave. As we shall prove, the pressure amplitude Δp_m is related to the displacement amplitude s_m in Eq. 18-13 by

$$\Delta p_m = (v\rho\omega)s_m. \tag{18-15}$$

Figure 18-7 shows plots of Eqs. 18-13 and 18-14 at $t = 0$; with time, the two curves would move rightward along the horizontal axes. Note that the displacement and pressure variation are $\pi/2$ rad (or 90°) out of phase. Thus, for example, the pressure variation Δp at any point along the wave is zero when the displacement there is a maximum.

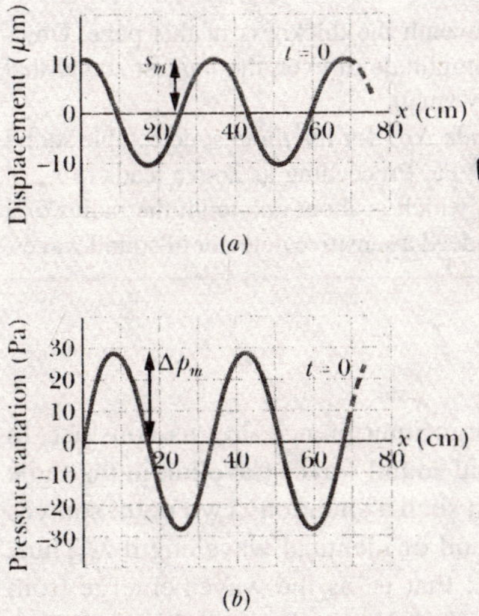

Fig. 18-7 (*a*) A plot of the displacement function (Eq. 18-13) for $t = 0$. (*b*) A similar plot of the pressure-variation function (Eq. 18-14). Both plots are for a 1000 Hz sound wave whose pressure amplitude is at the threshold of pain, see Sample Problem 18-2.

✔CHECKPOINT 1: When the oscillating fluid element in Fig. 18-5*b* is moving rightward through the point of zero displacement, is the pressure in the element at its equilibrium value, just beginning to increase, or just beginning to decrease?

Derivation of Eqs. 18-14 and 18-15

Figure 18-5*b* shows an oscillating element of air of cross-sectional area A and thickness Δx, with its center displaced from its equilibrium position by distance s.

From Eq. 18-2 we can write, for the pressure variation in the displaced element,

$$\Delta p = -B\frac{\Delta V}{V}. \tag{18-16}$$

The quantity V in Eq. 18-16 is the volume of the element, given by

$$V = A\,\Delta x. \tag{18-17}$$

The quantity ΔV in Eq. 18-16 is the change in volume that occurs when the element is displaced. This volume change comes about because the displacements of the two faces of the element are not quite the same, differing by some amount Δs. Thus, we can write the change in volume as

$$\Delta V = A\,\Delta s. \tag{18-18}$$

Substituting Eqs. 18-17 and 18-18 into Eq. 18-16 and passing to the differential limit yield

$$\Delta p = -B\frac{\Delta s}{\Delta x} = -B\frac{\partial s}{\partial x}. \tag{18-19}$$

The symbols ∂ indicate that the derivative in Eq. 18-19 is a *partial derivative*, which tells us how s changes with x when the time t is fixed. From Eq. 18-13 we then have, treating t as a constant,

$$\frac{\partial s}{\partial x} = \frac{\partial}{\partial x}[s_m \cos(kx - \omega t)] = -ks_m \sin(kx - \omega t).$$

Substituting this quantity for the partial derivative in Eq. 18-19 yields

$$\Delta p = Bks_m \sin(kx - \omega t).$$

Setting $\Delta p_m = Bks_m$, this yields Eq. 18-14, which we set out to prove.

Using Eq. 18-3, we can now write

$$\Delta p_m = (Bk)s_m = (v^2\rho k)s_m.$$

Equation 18-15, which we also promised to prove, follows at once if we substitute ω/v for k from Eq. 17-12.

Sample Problem 18-2

The maximum pressure amplitude Δp_m that the human ear can tolerate in loud sounds is about 28 Pa (which is very much less than the normal air pressure of about 10^5 Pa). What is the displacement amplitude s_m for such a sound in air of density $\rho = 1.21\ \text{kg/m}^3$, at a frequency of 1000 Hz and a speed of 343 m/s?

SOLUTION: The **Key Idea** is that the displacement amplitude s_m of a sound wave is related to the pressure amplitude Δp_m of the wave according to Eq. 18-15. Solving that equation for s_m yields

$$s_m = \frac{\Delta p_m}{v\rho\omega} = \frac{\Delta p_m}{v\rho(2\pi f)}.$$

Substituting known data then gives us

$$s_m = \frac{28\ \text{Pa}}{(343\ \text{m/s})(1.21\ \text{kg/m}^3)(2\pi)(1000\ \text{Hz})}$$

$$= 1.1 \times 10^{-5}\ \text{m} = 11\ \mu\text{m}. \quad \text{(Answer)}$$

That is only about one-seventh the thickness of this page. Obviously, the displacement amplitude of even the loudest sound that the ear can tolerate is very small.

The pressure amplitude Δp_m for the *faintest* detectable sound at 1000 Hz is 2.8×10^{-5} Pa. Proceeding as above leads to $s_m = 1.1 \times 10^{-11}$ m or 11 pm, which is about one-tenth the radius of a typical atom. The ear is indeed a sensitive detector of sound waves.

18-4 Interference

Like transverse waves, sound waves can undergo interference. Let us consider, in particular, the interference between two identical sound waves traveling in the same direction. Figure 18-8 shows how we can set up such a situation: Two point sources S_1 and S_2 emit sound waves that are in phase and of identical wavelength λ. Thus, the sources themselves are said to be in phase; that is, as the waves emerge from the sources, their displacements are always identical. We are interested in the waves that then travel through point P in Fig. 18-8. We assume that the distance to P is much greater than the distance between the sources so that we can approximate the waves as traveling in the same direction at P.

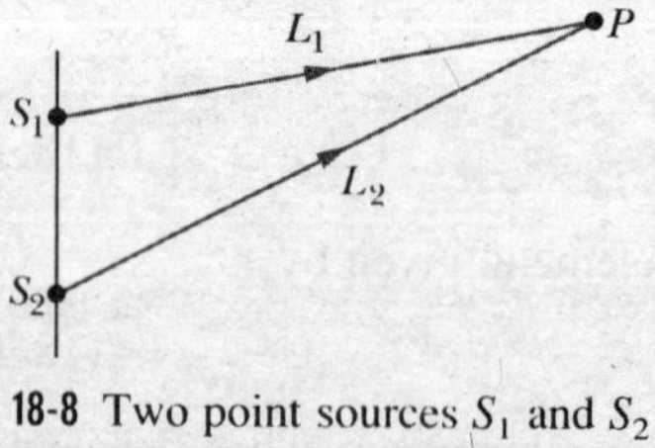

Fig. 18-8 Two point sources S_1 and S_2 emit spherical sound waves in phase. The rays indicate that the waves pass through a common point P.

If the waves traveled along paths with identical lengths to reach point P, they would be in phase there. As with transverse waves, this means that they would undergo fully constructive interference there. However, in Fig. 18-8, path L_2 traveled by the wave from S_2 is longer than path L_1 traveled by the wave from S_1. The difference in path lengths means that the waves may not be in phase at point P. In other words, their phase difference ϕ at P depends on their **path length difference** $\Delta L = |L_2 - L_1|$.

To relate phase difference ϕ to path length difference ΔL, we recall (from Section 17-4) that a phase difference of 2π rad corresponds to one wavelength. Thus, we can write the proportion

$$\frac{\phi}{2\pi} = \frac{\Delta L}{\lambda}, \qquad (18\text{-}20)$$

from which

$$\phi = \frac{\Delta L}{\lambda} 2\pi. \qquad (18\text{-}21)$$

Fully constructive interference occurs when ϕ is zero, 2π, or any integer multiple of 2π. We can write this condition as

$$\phi = m(2\pi), \quad \text{for } m = 0, 1, 2, \ldots \quad \text{(fully constructive interference).} \qquad (18\text{-}22)$$

From Eq. 18-21, this occurs when the ratio $\Delta L/\lambda$ is

$$\frac{\Delta L}{\lambda} = 0, 1, 2, \ldots \quad \text{(fully constructive interference).} \qquad (18\text{-}23)$$

For example, if the path length difference $\Delta L = |L_2 - L_1|$ in Fig. 18-8 is equal to 2λ, then $\Delta L/\lambda = 2$ and the waves undergo fully constructive interference at point P. The interference is fully constructive because the wave from S_2 is phase-shifted relative to the wave from S_1 by 2λ, putting the two waves *exactly in phase* at P.

Fully destructive interference occurs when ϕ is an odd multiple of π, a condition we can write as

$$\phi = (2m + 1)\pi, \quad \text{for } m = 0, 1, 2, \ldots \quad \text{(fully destructive interference).} \qquad (18\text{-}24)$$

From Eq. 18-21, this occurs when the ratio $\Delta L/\lambda$ is

$$\frac{\Delta L}{\lambda} = 0.5, 1.5, 2.5, \ldots \quad \text{(fully destructive interference).} \qquad (18\text{-}25)$$

For example, if the path length difference $\Delta L = |L_2 - L_1|$ in Fig. 18-8 is equal to 2.5λ, then $\Delta L/\lambda = 2.5$ and the waves undergo fully destructive interference at point P. The interference is fully destructive because the wave from S_2 is phase-shifted relative to the wave from S_1 by 2.5 wavelengths, which puts the two waves *exactly out of phase* at P.

Of course, two waves could produce intermediate interference as, say, when $\Delta L/\lambda = 1.2$. This would be closer to fully constructive interference ($\Delta L/\lambda = 1.0$) than to fully destructive interference ($\Delta L/\lambda = 1.5$).

Sample Problem 18-3

In Fig. 18-9*a*, two point sources S_1 and S_2, which are in phase and separated by distance $D = 1.5\lambda$, emit identical sound waves of wavelength λ.

(a) What is the path length difference of the waves from S_1 and S_2 at point P_1, which lies on the perpendicular bisector of distance D, at a distance greater than D from the sources? What type of interference occurs at P_1?

SOLUTION: The **Key Idea** here is that, because the waves travel identical distances to reach P_1, their path length difference is

$$\Delta L = 0. \qquad \text{(Answer)}$$

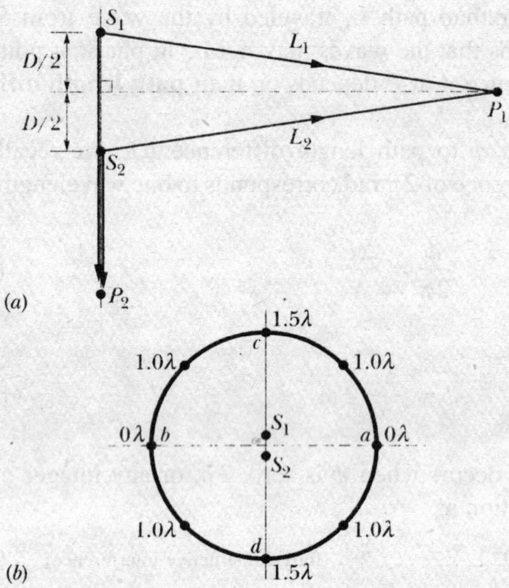

Fig. 18-9 Sample Problem 18-3. (*a*) Two point sources S_1 and S_2, separated by distance D, emit spherical sound waves in phase. The waves travel equal distances to reach point P_1. Point P_2 is on the line extending through S_1 and S_2. (*b*) The path length difference (in terms of wavelength) between the waves from S_1 and S_2, at eight points on a large circle around the sources.

From Eq. 18-23, this means that the waves undergo fully constructive interference at P_1.

(b) What are the path length difference and type of interference at point P_2 in Fig. 18-9*a*?

SOLUTION: Now the **Key Idea** is that the wave from S_1 travels the extra distance D ($= 1.5\lambda$) to reach P_2. Thus, the path length difference is

$$\Delta L = 1.5\lambda. \qquad \text{(Answer)}$$

From Eq. 18-25, this means that the waves are exactly out of phase at P_2 and undergo fully destructive interference there.

(c) Figure 18-9*b* shows a circle with a radius much greater than D, centered on the midpoint between sources S_1 and S_2. What is the number of points N around this circle at which the interference is fully constructive?

SOLUTION: Imagine that, starting at point a, we move clockwise along the circle to point d. One **Key Idea** here is that as we move to point d, the path length difference ΔL increases and so the type of interference changes. From (a), we know that the path length difference is $\Delta L = 0\lambda$ at point a. From (b), we know that $\Delta L = 1.5\lambda$ at point d. Thus, there must be one point along the circle between a and d at which $\Delta L = \lambda$, as indicated in Fig. 18-9*b*. From Eq. 18-23, fully constructive interference occurs at that point. Also, there can be no other point along the way from point a to point d at which fully constructive interference occurs, because there is no other integer than 1 between 0 and 1.5.

Another **Key Idea** here is to use symmetry to locate the other points of fully constructive interference along the rest of the circle. Symmetry about line cd gives us point b, at which $\Delta L = 0\lambda$. Also, there are three more points at which $\Delta L = \lambda$. In all we have

$$N = 6. \qquad \text{(Answer)}$$

CHECKPOINT 2: In this sample problem, if the distance D between sources S_1 and S_2 were, instead, equal to 4λ, what would be the path length difference and what type of interference would occur at (a) point P_1 and (b) point P_2?

18-5 Intensity and Sound Level

If you have ever tried to sleep while someone played loud music nearby, you are well aware that there is more to sound than frequency, wavelength, and speed. There is also intensity. The **intensity** I of a sound wave at a surface is the average rate per unit area at which energy is transferred by the wave through or onto the surface. We can write this as

$$I = \frac{P}{A}, \qquad (18\text{-}26)$$

where P is the time rate of energy transfer (the power) of the sound wave, and A is the area of the surface intercepting the sound. As we shall derive shortly, the intensity I is related to the displacement amplitude s_m of the sound wave by

$$I = \tfrac{1}{2}\rho v \omega^2 s_m^2. \qquad (18\text{-}27)$$

Variation of Intensity with Distance

How intensity varies with distance from a real sound source is often complex. Some real sources (like loudspeakers) may transmit sound only in particular directions,

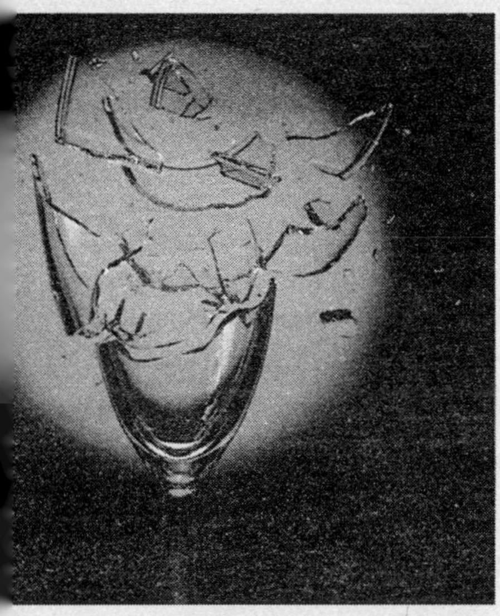

Sound can cause the wall of a drinking glass to oscillate. If the sound produces a standing wave of oscillations and if the intensity of the sound is large enough, the glass will shatter.

and the environment usually produces echoes (reflected sound waves) that overlap the direct sound waves. In some situations, however, we can ignore echoes and assume that the sound source is a point source that emits the sound *isotropically*—that is, with equal intensity in all directions. The wavefronts spreading from such an isotropic point source S at a particular instant are shown in Fig. 18-10.

Let us assume that the mechanical energy of the sound waves is conserved as they spread from this source. Let us also center an imaginary sphere of radius r on the source, as shown in Fig. 18-10. All the energy emitted by the source must pass through the surface of the sphere. Thus, the time rate at which energy is transferred through the surface by the sound waves must equal the time rate at which energy is emitted by the source (that is, the power P_s of the source). From Eq. 18-26, the intensity I at the sphere must then be

$$I = \frac{P_s}{4\pi r^2}, \qquad (18\text{-}28)$$

where $4\pi r^2$ is the area of the sphere. Equation 18-28 tells us that the intensity of sound from an isotropic point source decreases with the square of the distance r from the source.

✔CHECKPOINT 3: The figure indicates three small patches 1, 2, and 3 that lie on the surfaces of two imaginary spheres; the spheres are centered on an isotropic point source S of sound. The rates at which energy is transmitted through the three patches by the sound waves are equal. Rank the patches according to (a) the intensity of the sound on them and (b) their area, greatest first.

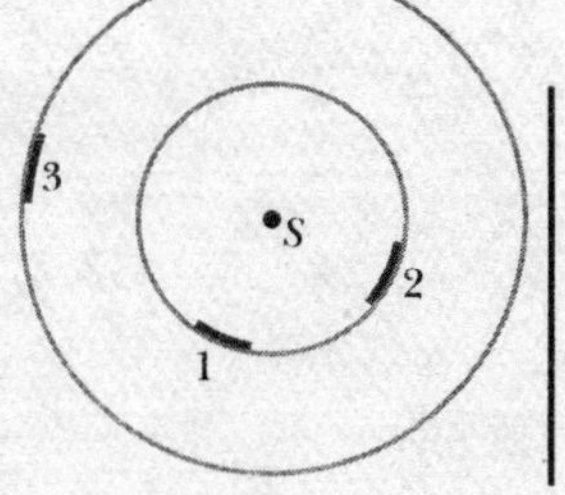

The Decibel Scale

You saw in Sample Problem 18-2 that the displacement amplitude at the human ear ranges from about 10^{-5} m for the loudest tolerable sound to about 10^{-11} m for the faintest detectable sound, a ratio of 10^6. From Eq. 18-27 we see that the intensity of a sound varies as the *square* of its amplitude, so the ratio of intensities at these two limits of the human auditory system is 10^{12}. Humans can hear over an enormous range of intensities.

We deal with such an enormous range of values by using logarithms. Consider the relation

$$y = \log x,$$

in which x and y are variables. It is a property of this equation that if we *multiply* x by 10, then y increases by 1. To see this, we write

$$y' = \log(10x) = \log 10 + \log x = 1 + y.$$

Similarly, if we multiply x by 10^{12}, y increases by only 12.

Thus, instead of speaking of the intensity I of a sound wave, it is much more convenient to speak of its **sound level** β, defined as

$$\beta = (10\text{ dB}) \log \frac{I}{I_0}. \qquad (18\text{-}29)$$

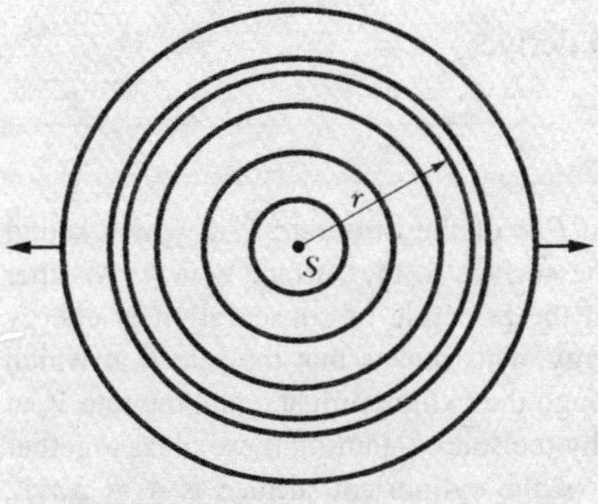

Fig. 18-10 A point source S emits sound waves uniformly in all directions. The waves pass through an imaginary sphere of radius r that is centered on S.

Here dB is the abbreviation for **decibel,** the unit of sound level, a name that was chosen to recognize the work of Alexander Graham Bell. I_0 in Eq. 18-29 is a standard reference intensity ($= 10^{-12}$ W/m^2), chosen because it is near the lower limit of the human range of hearing. For $I = I_0$, Eq. 18-29 gives $\beta = 10 \log 1 = 0$, so our

TABLE 18-2 Some Sound Levels (dB)

Hearing threshold	0
Rustle of leaves	10
Conversation	60
Rock concert	110
Pain threshold	120
Jet engine	130

standard reference level corresponds to zero decibels. Then β increases by 10 dB every time the sound intensity increases by an order of magnitude (a factor of 10). Thus, $\beta = 40$ corresponds to an intensity that is 10^4 times the standard reference level. Table 18-2 lists the sound levels for a variety of environments.

Derivation of Eq. 18-27

Consider, in Fig. 18-5*a*, a thin slice of air of thickness dx, area A, and mass dm, oscillating back and forth as the sound wave of Eq. 18-13 passes through it. The kinetic energy dK of the slice of air is

$$dK = \tfrac{1}{2} dm\, v_s^2. \tag{18-30}$$

Here v_s is not the speed of the wave but the speed of the oscillating element of air, obtained from Eq. 18-13 as

$$v_s = \frac{\partial s}{\partial t} = -\omega s_m \sin(kx - \omega t).$$

Using this relation and putting $dm = \rho A\, dx$ allow us to rewrite Eq. 18-30 as

$$dK = \tfrac{1}{2}(\rho A\, dx)(-\omega s_m)^2 \sin^2(kx - \omega t). \tag{18-31}$$

Dividing Eq. 18-31 by dt gives the rate at which kinetic energy moves along with the wave. As we saw in Chapter 17 for transverse waves, dx/dt is the wave speed v, so we have

$$\frac{dK}{dt} = \tfrac{1}{2}\rho A v \omega^2 s_m^2 \sin^2(kx - \omega t). \tag{18-32}$$

The *average* rate at which kinetic energy is transported is

$$\left(\frac{dK}{dt}\right)_{\text{avg}} = \tfrac{1}{2}\rho A v \omega^2 s_m^2 [\sin^2(kx - \omega t)]_{\text{avg}}$$

$$= \tfrac{1}{4}\rho A v \omega^2 s_m^2. \tag{18-33}$$

To obtain this equation, we have used the fact that the average value of the square of a sine (or a cosine) function over one full oscillation is $\frac{1}{2}$.

We assume that *potential* energy is carried along with the wave at this same average rate. The wave intensity I, which is the average rate per unit area at which energy of both kinds is transmitted by the wave, is then, from Eq. 18-33,

$$I = \frac{2(dK/dt)_{\text{avg}}}{A} = \tfrac{1}{2}\rho v \omega^2 s_m^2,$$

which is Eq. 18-27, the equation we set out to derive.

Sample Problem 18-4

An electric spark jumps along a straight line of length $L = 10$ m, emitting a pulse of sound that travels radially outward from the spark. (The spark is said to be a *line source* of sound.) The power of the emission is $P_s = 1.6 \times 10^4$ W.

(a) What is the intensity I of the sound when it reaches a distance $r = 12$ m from the spark?

SOLUTION: Let us center an imaginary cylinder of radius $r = 12$ m and length $L = 10$ m (open at both ends) on the spark, as shown in Fig. 18-11. One **Key Idea** here is that the intensity I at the cylindrical surface is the ratio P/A of the time rate P at which sound energy passes through the surface to the surface area A. Another **Key Idea** is to assume that the principle of conservation of energy applies to the sound energy. This means that the rate P at which energy is transferred through the cylinder must equal the rate P_s at which energy is emitted by the source. Putting these ideas together and noting that the area of the cylindrical surface is $A = 2\pi rL$, we have

$$I = \frac{P}{A} = \frac{P_s}{2\pi r L}. \tag{18-34}$$

This tells us that the intensity of the sound from a line source

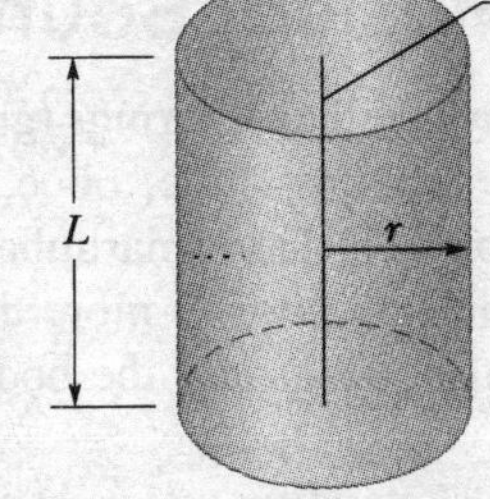

Fig. 18-11 Sample Problem 18-4. A spark along a straight line of length L emits sound waves radially outward. The waves pass through an imaginary cylinder of radius r and length L that is centered on the spark.

decreases with distance r (and not with the square of distance r as for a point source). Substituting the given data, we find

$$I = \frac{1.6 \times 10^4 \text{ W}}{2\pi(12 \text{ m})(10 \text{ m})} = 21.2 \text{ W/m}^2 \approx 21 \text{ W/m}^2. \quad \text{(Answer)}$$

(b) At what time rate P_d is sound energy intercepted by an acoustic detector of area $A_d = 2.0 \text{ cm}^2$, aimed at the spark and located a distance $r = 12$ m from the spark?

SOLUTION: Applying the first **Key Idea** of part (a), we know that the intensity of sound at the detector is the ratio of the energy transfer rate P_d there to the detector's area A_d:

$$I = \frac{P_d}{A_d}. \tag{18-35}$$

We can imagine that the detector lies on the cylindrical surface of (a). Then the sound intensity at the detector is the intensity I (= 21.2 W/m^2) at the cylindrical surface. Solving Eq. 18-35 for P_d gives us

$$P_d = (21.2 \text{ W/m}^2)(2.0 \times 10^{-4} \text{ m}^2) = 4.2 \text{ mW}. \quad \text{(Answer)}$$

Sample Problem 18-5

In 1976, the Who set a record for the loudest concert—the sound level 46 m in front of the speaker systems was $\beta_2 = 120$ dB. What is the ratio of the intensity I_2 of the band at that spot to the intensity I_1 of a jackhammer operating at sound level $\beta_1 = 92$ dB?

SOLUTION: The **Key Idea** here is that for both the Who and the jackhammer, the sound level β is related to the intensity by the definition of sound level in Eq. 18-29. For the Who, we have

$$\beta_2 = (10 \text{ dB}) \log \frac{I_2}{I_0},$$

and for the jackhammer, we have

$$\beta_1 = (10 \text{ dB}) \log \frac{I_1}{I_0}.$$

The difference in the sound levels is

$$\beta_2 - \beta_1 = (10 \text{ dB})\left(\log \frac{I_2}{I_0} - \log \frac{I_1}{I_0}\right). \tag{18-36}$$

Using the identity

$$\log \frac{a}{b} - \log \frac{c}{d} = \log \frac{ad}{bc},$$

we can rewrite Eq. 18-36 as

$$\beta_2 - \beta_1 = (10 \text{ dB}) \log \frac{I_2}{I_1}. \tag{18-37}$$

Rearranging and substituting the known sound levels now yield

$$\log \frac{I_2}{I_1} = \frac{\beta_2 - \beta_1}{10 \text{ dB}} = \frac{120 \text{ dB} - 92 \text{ dB}}{10 \text{ dB}} = 2.8.$$

Taking the antilog of the far left and far right sides of this equation (the antilog key on your calculator is probably marked as 10^x), we find

$$\frac{I_2}{I_1} = \log^{-1} 2.8 = 630. \quad \text{(Answer)}$$

Thus, the Who was *very* loud.

Temporary exposure to sound intensities as great as those of a jackhammer and the 1976 Who concert results in a temporary reduction of hearing. Repeated or prolonged exposure can result in permanent reduction of hearing (Fig. 18-12). Loss of hearing is a clear risk for anyone continually listening to, say, heavy metal at high volume, especially on headphones.

Fig. 18-12 Sample Problem 18-5. Peter Townshend of the Who, playing in front of a speaker system. He suffered a permanent reduction in his hearing ability due to his exposure to high-intensity sound, not so much during on-stage performances as from wearing headphones in recording studios and at home.

18-6 Sources of Musical Sound

Fig. 18-13 The air column within a fujara oscillates when that traditional Slovakian instrument is played.

Musical sounds can be set up by oscillating strings (guitar, piano, violin), membranes (kettledrum, snare drum), air columns (flute, oboe, pipe organ, and the fujara of Fig. 18-13), wooden blocks or steel bars (marimba, xylophone), and many other oscillating bodies. Most instruments involve more than a single oscillating part. In the violin, for example, both the strings and the body of the instrument participate in producing the music.

Recall from Chapter 17 that standing waves can be set up on a stretched string that is fixed at both ends. They arise because waves traveling along the string are reflected back onto the string at each end. If the wavelength of the waves is suitably matched to the length of the string, the superposition of waves traveling in opposite directions produces a standing wave pattern (or oscillation mode). The wavelength required of the waves for such a match is one that corresponds to a *resonant frequency* of the string. The advantage of setting up standing waves is that the string then oscillates with a large, sustained amplitude, pushing back and forth against the surrounding air and thus generating a noticeable sound wave with the same frequency as the oscillations of the string. This production of sound is of obvious importance to, say, a guitarist.

We can set up standing waves of sound in an air-filled pipe in a similar way. As sound waves travel through the air in the pipe, they are reflected at each end and travel back through the pipe. (The reflection occurs even if an end is open, but the reflection is not as complete as when the end is closed.) If the wavelength of the sound waves is suitably matched to the length of the pipe, the superposition of waves traveling in opposite directions through the pipe sets up a standing wave pattern. The wavelength required of the sound waves for such a match is one that corresponds to a resonant frequency of the pipe. The advantage of such a standing wave is that the air in the pipe oscillates with a large, sustained amplitude, emitting at any open end a sound wave that has the same frequency as the oscillations in the pipe. This emission of sound is of obvious importance to, say, an organist.

Many other aspects of standing sound wave patterns are similar to those of string waves: The closed end of a pipe is like the fixed end of a string in that there must be a node (zero displacement) there, and the open end of a pipe is like the end of a string attached to a freely moving ring, as in Fig. 17-19*b*, in that there must be an antinode there. (Actually, the antinode for the open end of a pipe is located slightly beyond the end, but we shall not dwell on that detail here.)

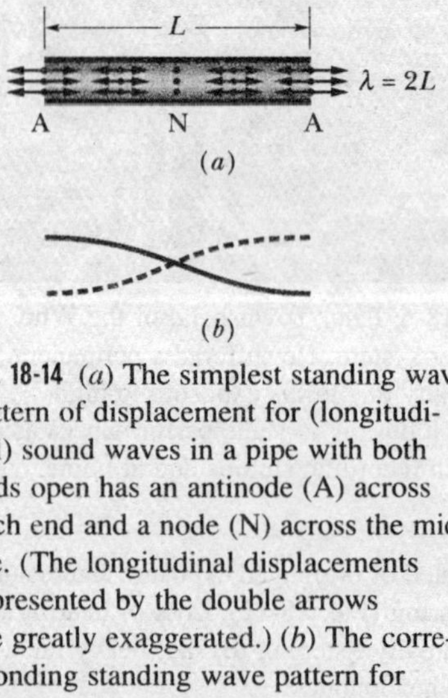

Fig. 18-14 (*a*) The simplest standing wave pattern of displacement for (longitudinal) sound waves in a pipe with both ends open has an antinode (A) across each end and a node (N) across the middle. (The longitudinal displacements represented by the double arrows are greatly exaggerated.) (*b*) The corresponding standing wave pattern for (transverse) string waves.

The simplest standing wave pattern that can be set up in a pipe with two open ends is shown in Fig. 18-14*a*. There is an antinode across each open end, as required. There is also a node across the middle of the pipe. An easier way of representing this standing longitudinal sound wave is shown in Fig. 18-14*b*—by drawing it as a standing transverse string wave.

The standing wave pattern of Fig. 18-14*a* is called the *fundamental mode* or *first harmonic*. For it to be set up, the sound waves in a pipe of length L must have a wavelength given by $L = \lambda/2$, so that $\lambda = 2L$. Several more standing sound wave patterns for a pipe with two open ends are shown in Fig. 18-15*a* using string wave representations. The *second harmonic* requires sound waves of wavelength $\lambda = L$, the *third harmonic* requires wavelength $\lambda = 2L/3$, and so on.

More generally, the resonant frequencies for a pipe of length L with two open ends correspond to the wavelengths

$$\lambda = \frac{2L}{n}, \qquad \text{for } n = 1, 2, 3, \ldots, \tag{18-38}$$

where n is called the *harmonic number*. The resonant frequencies for a pipe with

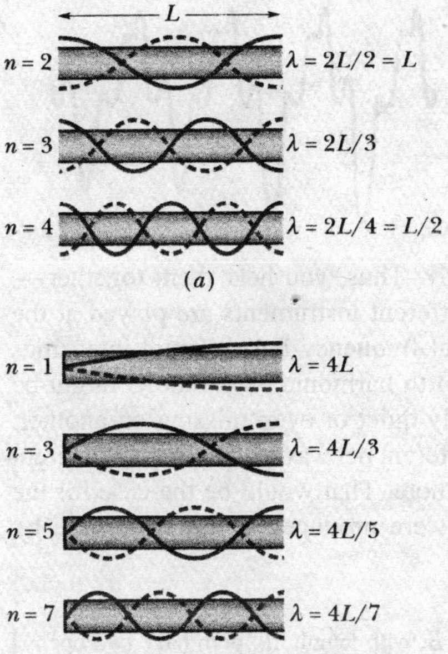

Fig. 18-15 Standing wave patterns for string waves superimposed on pipes to represent standing sound wave patterns in the pipes. (*a*) With *both* ends of the pipe open, any harmonic can be set up in the pipe. (*b*) With only *one* end open, only odd harmonics can be set up.

two open ends are then given by

$$f = \frac{v}{\lambda} = \frac{nv}{2L}, \quad \text{for } n = 1, 2, 3, \ldots \quad \text{(pipe, two open ends)}. \qquad (18\text{-}39)$$

where v is the speed of sound.

Figure 18-15*b* shows (using string wave representations) some of the standing sound wave patterns that can be set up in a pipe with only one open end. As required, across the open end there is an antinode and across the closed end there is a node. The simplest pattern requires sound waves having a wavelength given by $L = \lambda/4$, so that $\lambda = 4L$. The next simplest pattern requires a wavelength given by $L = 3\lambda/4$, so that $\lambda = 4L/3$, and so on.

More generally, the resonant frequencies for a pipe of length L with only one open end correspond to the wavelengths

$$\lambda = \frac{4L}{n}, \quad \text{for } n = 1, 3, 5, \ldots, \qquad (18\text{-}40)$$

in which the harmonic number n *must be an odd number*. The resonant frequencies are then given by

$$f = \frac{v}{\lambda} = \frac{nv}{4L}, \quad \text{for } n = 1, 3, 5, \ldots \quad \text{(pipe, one open end)}. \qquad (18\text{-}41)$$

Note again that only odd harmonics can exist in a pipe with one open end. For example, the second harmonic, with $n = 2$, cannot be set up in such a pipe. Note also that for such a pipe the adjective in a phrase such as "the third harmonic" still refers to the harmonic number n (and not to, say, the third possible harmonic).

The length of a musical instrument reflects the range of frequencies over which the instrument is designed to function, and smaller length implies higher frequencies. Figure 18-16, for example, shows the saxophone and violin families, with their frequency ranges suggested by the piano keyboard. Note that, for every instrument, there is overlap with its higher- and lower-frequency neighbors.

In any oscillating system that gives rise to a musical sound, whether it is a violin string or the air in an organ pipe, the fundamental and one or more of the higher

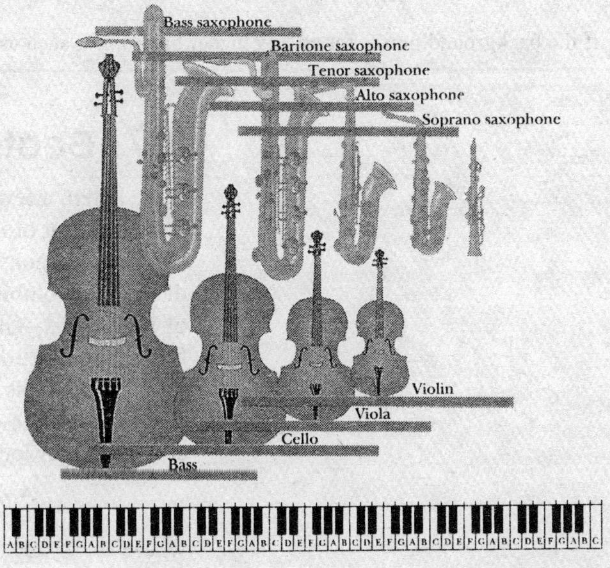

Fig. 18-16 The saxophone and violin families, showing the relations between instrument length and frequency range. The frequency range of each instrument is indicated by a horizontal bar along a frequency scale suggested by the keyboard at the bottom; the frequency increases toward the right.

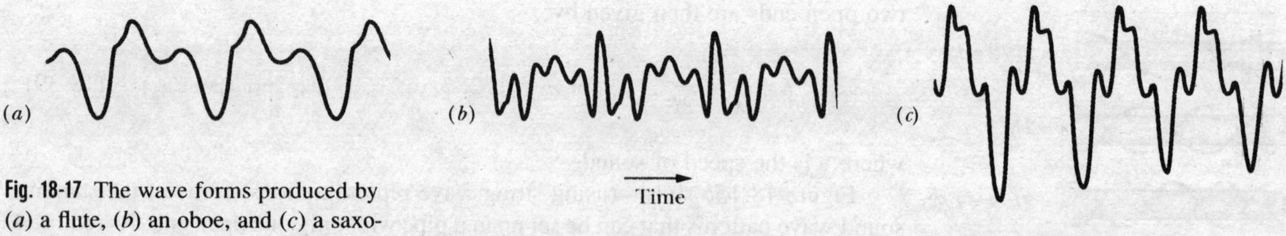

Fig. 18-17 The wave forms produced by (*a*) a flute, (*b*) an oboe, and (*c*) a saxophone when they all play the same note, with the same first harmonic frequency.

harmonics are usually generated simultaneously. Thus, you hear them together—that is, superimposed as a net wave. When different instruments are played at the same note, they produce the same fundamental frequency but different intensities for the higher harmonics. For example, the fourth harmonic of middle C might be relatively loud on one instrument and relatively quiet or even missing on another. Thus, because different instruments produce different net waves, they sound different to you even when they are played at the same note. That would be the case for the three net waves shown in Fig. 18-17, which were produced at the same note by different instruments.

✔**CHECKPOINT 4:** Pipe A, with length L, and pipe B, with length $2L$, both have two open ends. Which harmonic of pipe B has the same frequency as the fundamental of pipe A?

Sample Problem 18-6

Weak background noises from a room set up the fundamental standing wave in a cardboard tube of length $L = 67.0$ cm with two open ends. Assume that the speed of sound in the air within the tube is 343 m/s.

(a) What frequency do you hear from the tube?

SOLUTION: The **Key Idea** here is that, with both pipe ends open, we have a symmetric situation in which the standing wave has an antinode at each end of the tube. The standing wave pattern (in string wave style) is that of Fig. 18-14*b*. The frequency is given by Eq. 18-39 with $n = 1$ for the fundamental mode:

$$f = \frac{nv}{2L} = \frac{(1)(343 \text{ m/s})}{2(0.670 \text{ m})} = 256 \text{ Hz}. \qquad \text{(Answer)}$$

If the background noises set up any higher harmonics, such as the second harmonic, you must also hear frequencies that are *integer* multiples of 256 Hz.

(b) If you jam your ear against one end of the tube, what fundamental frequency do you hear from the tube?

SOLUTION: The **Key Idea** now is that, with your ear effectively closing one end of the tube, we have an asymmetric situation—an antinode still exists at the open end but a node is now at the other (closed) end. The standing wave pattern is the top one in Fig. 18-15*b*. The frequency is given by Eq. 18-41 with $n = 1$ for the fundamental mode:

$$f = \frac{nv}{4L} = \frac{(1)(343 \text{ m/s})}{4(0.670 \text{ m})} = 128 \text{ Hz}. \qquad \text{(Answer)}$$

If the background noises set up any higher harmonics, they will be *odd* multiples of 128 Hz. That means that the frequency of 256 Hz (which is an even multiple) cannot now occur.

18-7 Beats

If we listen, a few minutes apart, to two sounds whose frequencies are, say, 552 and 564 Hz, most of us cannot tell one from the other. However, if the sounds reach our ears simultaneously, what we hear is a sound whose frequency is 558 Hz, the *average* of the two combining frequencies. We also hear a striking variation in the intensity of this sound—it increases and decreases in slow, wavering **beats** that repeat at a frequency of 12 Hz, the *difference* between the two combining frequencies. Figure 18-18 shows this beat phenomenon.

Let the time-dependent variations of the displacements due to two sound waves at a particular location be

$$s_1 = s_m \cos \omega_1 t \quad \text{and} \quad s_2 = s_m \cos \omega_2 t, \qquad (18\text{-}42)$$

where $\omega_1 > \omega_2$. We have assumed, for simplicity, that the waves have the same

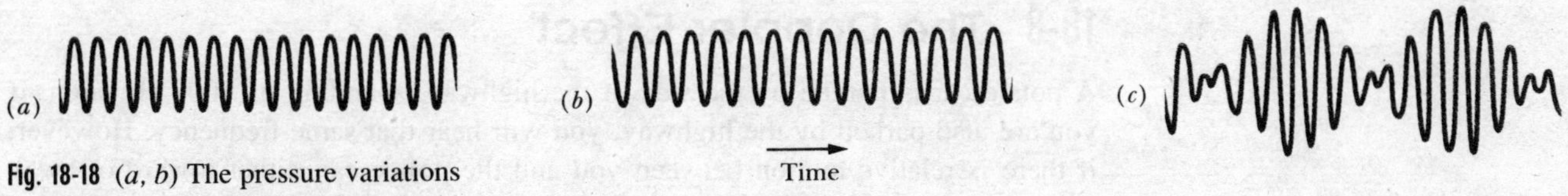

Fig. 18-18 (*a, b*) The pressure variations Δp of two sound waves as they would be detected separately. The frequencies of the waves are nearly equal. (*c*) The resultant pressure variation if the two waves are detected simultaneously.

amplitude. According to the superposition principle, the resultant displacement is

$$s = s_1 + s_2 = s_m(\cos \omega_1 t + \cos \omega_2 t).$$

Using the trigonometric identity (see Appendix E)

$$\cos \alpha + \cos \beta = 2 \cos \tfrac{1}{2}(\alpha - \beta) \cos \tfrac{1}{2}(\alpha + \beta)$$

allows us to write the resultant displacement as

$$s = 2s_m \cos \tfrac{1}{2}(\omega_1 - \omega_2)t \cos \tfrac{1}{2}(\omega_1 + \omega_2)t. \tag{18-43}$$

If we write

$$\omega' = \tfrac{1}{2}(\omega_1 - \omega_2) \quad \text{and} \quad \omega = \tfrac{1}{2}(\omega_1 + \omega_2), \tag{18-44}$$

we can then write Eq. 18-43 as

$$s(t) = [2s_m \cos \omega' t] \cos \omega t. \tag{18-45}$$

We now assume that the angular frequencies ω_1 and ω_2 of the combining waves are almost equal, which means that $\omega \gg \omega'$ in Eq. 18-44. We can then regard Eq. 18-45 as a cosine function whose angular frequency is ω and whose amplitude (which is not constant but varies with angular frequency ω') is the quantity in the brackets.

A maximum amplitude will occur whenever $\cos \omega' t$ in Eq. 18-45 has the value $+1$ or -1, which happens twice in each repetition of the cosine function. Because $\cos \omega' t$ has angular frequency ω', the angular frequency ω_{beat} at which beats occur is $\omega_{beat} = 2\omega'$. Then, with the aid of Eq. 18-44, we can write

$$\omega_{beat} = 2\omega' = (2)(\tfrac{1}{2})(\omega_1 - \omega_2) = \omega_1 - \omega_2.$$

Because $\omega = 2\pi f$, we can recast this as

$$f_{beat} = f_1 - f_2 \quad \text{(beat frequency)}. \tag{18-46}$$

Musicians use the beat phenomenon in tuning their instruments. If an instrument is sounded against a standard frequency (for example, the lead oboe's reference A) and tuned until the beat disappears, then the instrument is in tune with that standard. In musical Vienna, concert A (440 Hz) is available as a telephone service for the benefit of the city's many professional and amateur musicians.

Sample Problem 18-7

You wish to tune the note A_3 on a piano to its proper frequency of 220 Hz. You have available a tuning fork whose frequency is 440 Hz. How should you proceed?

SOLUTION: We need two **Key Ideas** here: (1) The two frequencies are too far apart to produce beats. (2) However, the piano string will oscillate not only in its fundamental mode (at 220 Hz when tuned) but also in its second harmonic mode (at 440 Hz when in tune). Thus, with the string somewhat out of tune, the frequency of its second harmonic will beat against the 440 Hz of the tuning fork. To tune the string, you can listen for those beats and then either tighten or loosen the string to decrease the beat frequency until the beating disappears.

CHECKPOINT 5: In this sample problem, you tighten the string and the beat frequency increases from 6 Hz. Should you continue to tighten the string or should you loosen the string to put the string in tune?

18-8 The Doppler Effect

A police car is parked by the side of the highway, sounding its 1000 Hz siren. If you are also parked by the highway, you will hear that same frequency. However, if there is relative motion between you and the police car, either toward or away from each other, you will hear a different frequency. For example, if you are driving *toward* the police car at 120 km/h (about 75 mi/h), you will hear a *higher* frequency (1096 Hz, an *increase* of 96 Hz). If you are driving *away from* the police car at that same speed, you will hear a *lower* frequency (904 Hz, a *decrease* of 96 Hz).

These motion-related frequency changes are examples of the **Doppler effect.** The effect was proposed (although not fully worked out) in 1842 by Austrian physicist Johann Christian Doppler. It was tested experimentally in 1845 by Buys Ballot in Holland, "using a locomotive drawing an open car with several trumpeters."

The Doppler effect holds not only for sound waves but also for electromagnetic waves, including microwaves, radio waves, and visible light. Here, however, we shall consider only sound waves, and we shall take as a reference frame the body of air through which these waves travel. This means that we shall measure the speeds of a source S of sound waves and a detector D of those waves *relative to that body of air*. (Unless otherwise stated, the body of air is stationary relative to the ground, so the speeds can also be measured relative to the ground.) We shall assume that S and D move either directly toward or directly away from each other, at speeds less than the speed of sound.

If either the detector or the source is moving, or both are moving, the emitted frequency f and the detected frequency f' are related by

$$f' = f\frac{v \pm v_D}{v \pm v_S} \quad \text{(general Doppler effect)}, \qquad (18\text{-}47)$$

where v is the speed of sound through the air, v_D is the detector's speed relative to the air, and v_S is the source's speed relative to the air. The choice of plus or minus signs is set by this rule:

When the motion of detector or source is toward the other, the sign on its speed must give an upward shift in frequency. When the motion of detector or source is away from the other, the sign on its speed must give a downward shift in frequency.

In short, *toward* means *shift up*, and *away* means *shift down*.

Here are some examples of the rule. If the detector moves toward the source, use the plus sign in the numerator of Eq. 18-47 to get a shift up in the frequency. If it moves away, use the minus sign in the numerator to get a shift down. If it is stationary, substitute 0 for v_D. If the source moves toward the detector, use the minus sign in the denominator of Eq. 18-47 to get a shift up in the frequency. If it moves away, use the plus sign in the denominator to get a shift down. If the source is stationary, substitute 0 for v_S.

Next, we derive equations for the Doppler effect for the following two specific situations and then derive Eq. 18-47 for the general situation.

1. When the detector moves relative to the air and the source is stationary relative to the air, the motion changes the frequency at which the detector intercepts wavefronts and thus the detected frequency of the sound wave.
2. When the source moves relative to the air and the detector is stationary relative to the air, the motion changes the wavelength of the sound wave and thus the detected frequency (recall that frequency is related to wavelength).

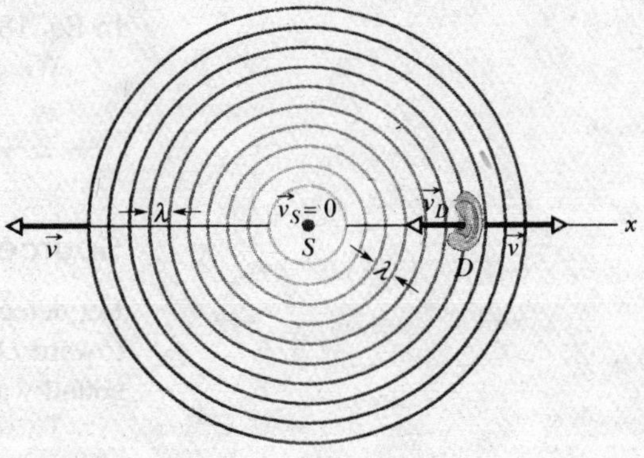

Fig. 18-19 A stationary source of sound S emits spherical wavefronts, shown one wavelength apart, that expand outward at speed v. A sound detector D, represented by an ear, moves with velocity $\vec{v}_D$ toward the source. The detector senses a higher frequency because of its motion.

Detector Moving; Source Stationary

In Fig. 18-19, a detector D (represented by an ear) is moving at speed v_D toward a stationary source S that emits spherical wavefronts, of wavelength λ and frequency f, moving at the speed v of sound in air. The wavefronts are drawn one wavelength apart. The frequency detected by detector D is the rate at which D intercepts wavefronts (or individual wavelengths). If D were stationary, that rate would be f, but since D is moving into the wavefronts, the rate of interception is greater, and thus the detected frequency f' is greater than f.

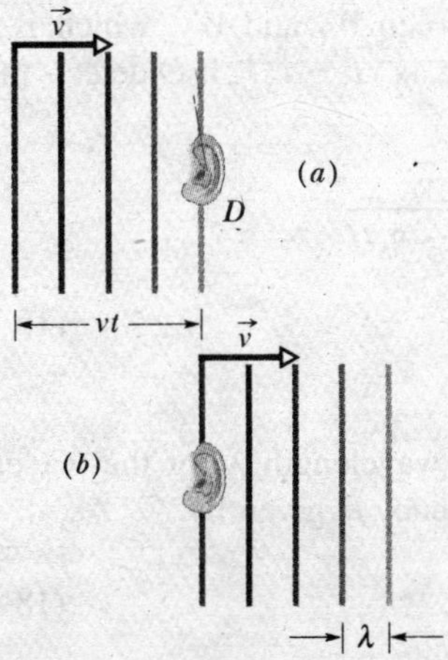

Fig. 18-20 Wavefronts of Fig. 18-19, assumed planar, (*a*) reach and (*b*) pass a stationary detector D; they move a distance vt to the right in time t.

Let us for the moment consider the situation in which D is stationary (Fig. 18-20). In time t, the wavefronts move to the right a distance vt. The number of wavelengths in that distance vt is the number of wavelengths intercepted by D in time t, and that number is vt/λ. The rate at which D intercepts wavelengths, which is the frequency f detected by D, is

$$f = \frac{vt/\lambda}{t} = \frac{v}{\lambda}. \tag{18-48}$$

In this situation, with D stationary, there is no Doppler effect—the frequency detected by D is the frequency emitted by S.

Now let us again consider the situation in which D moves opposite the wavefronts (Fig. 18-21). In time t, the wavefronts move to the right a distance vt as previously, but now D moves to the left a distance $v_D t$. Thus, in this time t, the distance moved by the wavefronts relative to D is $vt + v_D t$. The number of wavelengths in this relative distance $vt + v_D t$ is the number of wavelengths intercepted by D in time t, and is $(vt + v_D t)/\lambda$. The *rate* at which D intercepts wavelengths in this situation is the frequency f', given by

$$f' = \frac{(vt + v_D t)/\lambda}{t} = \frac{v + v_D}{\lambda}. \tag{18-49}$$

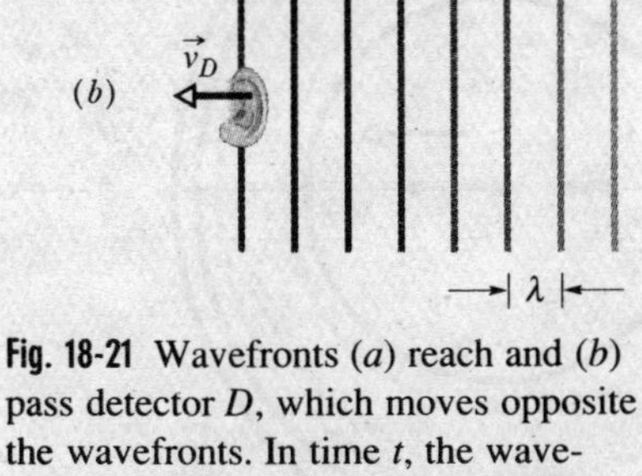

Fig. 18-21 Wavefronts (*a*) reach and (*b*) pass detector D, which moves opposite the wavefronts. In time t, the wavefronts move a distance vt to the right and D moves a distance $v_D t$ to the left.

From Eq. 18-48, we have $\lambda = v/f$. Then Eq. 18-49 becomes

$$f' = \frac{v + v_D}{v/f} = f\frac{v + v_D}{v}. \tag{18-50}$$

Note that in Eq. 18-50, f' must be greater than f unless $v_D = 0$ (the detector is stationary).

Similarly, we can find the frequency detected by D if D moves away from the source. In this situation, the wavefronts move a distance $vt - v_D t$ relative to D in time t, and f' is given by

$$f' = f\frac{v - v_D}{v}. \tag{18-51}$$

In Eq. 18-51, f' must be less than f unless $v_D = 0$.

We can summarize Eqs. 18-50 and 18-51 with

$$f' = f\frac{v \pm v_D}{v} \quad \text{(detector moving; source stationary).} \tag{18-52}$$

Source Moving; Detector Stationary

Let detector D be stationary with respect to the body of air, and let source S move toward D at speed v_S (Fig. 18-22). The motion of S changes the wavelength of the sound waves it emits, and thus the frequency detected by D.

To see this change, let T $(= 1/f)$ be the time between the emission of any pair of successive wavefronts W_1 and W_2. During T, wavefront W_1 moves a distance vT and the source moves a distance $v_S T$. At the end of T, wavefront W_2 is emitted. In the direction in which S moves, the distance between W_1 and W_2, which is the wavelength λ' of the waves moving in that direction, is $vT - v_S T$. If D detects those waves, it detects frequency f' given by

$$\begin{aligned} f' &= \frac{v}{\lambda'} = \frac{v}{vT - v_S T} = \frac{v}{v/f - v_S/f} \\ &= f\frac{v}{v - v_S}. \end{aligned} \tag{18-53}$$

Note that f' must be greater than f unless $v_S = 0$.

In the direction opposite that taken by S, the wavelength λ' of the waves is $vT + v_S T$. If D detects those waves, it detects frequency f' given by

$$f' = f\frac{v}{v + v_S}. \tag{18-54}$$

Now f' must be less than f unless $v_S = 0$.

We can summarize Eqs. 18-53 and 18-54 with

$$f' = f\frac{v}{v \pm v_S} \quad \text{(source moving; detector stationary).} \tag{18-55}$$

General Doppler Effect Equation

We can now derive the general Doppler effect equation by replacing f in Eq. 18-55 (the frequency of the source) with f' of Eq. 18-52 (the frequency associated with motion of the detector). The result is Eq. 18-47 for the general Doppler effect.

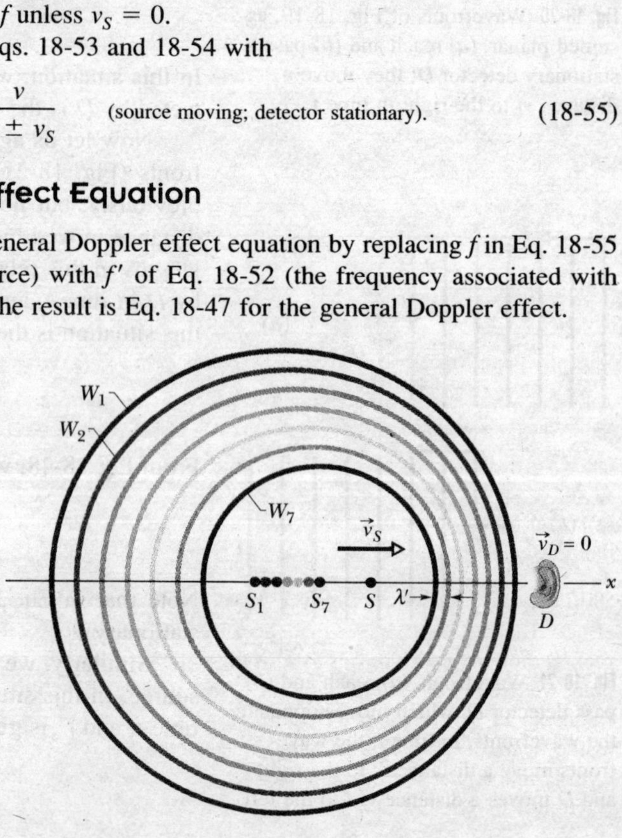

Fig. 18-22 A detector D is stationary, and a source S is moving toward it at speed v_S. Wavefront W_1 was emitted when the source was at S_1, wavefront W_7 when it was at S_7. At the moment depicted, the source is at S. The detector senses a higher frequency because the moving source, chasing its own wavefronts, emits a reduced wavelength λ' in the direction of its motion.

That general equation holds not only when both detector and source are moving but also in the two specific situations we just discussed. For the situation in which the detector is moving and the source is stationary, substitution of $v_S = 0$ into Eq. 18-47 gives us Eq. 18-52, which we previously found. For the situation in which the source is moving and the detector is stationary, substitution of $v_D = 0$ into Eq. 18-47 gives us Eq. 18-55, which we previously found. Thus, Eq. 18-47 is the equation to remember.

Bat Navigation

Bats navigate and search out prey by emitting, and then detecting reflections of, ultrasonic waves. These are sound waves with frequencies greater than can be heard by a human. For example, a horseshoe bat emits ultrasonic waves at 83 kHz, well above the 20 kHz limit of human hearing.

After the sound is emitted through the bat's nostrils, it might reflect (echo) from a moth, and then return to the bat's ears. The motions of the bat and the moth relative to the air cause the frequency heard by the bat to differ by a few kilohertz from the frequency it emitted. The bat automatically translates this difference into a relative speed between itself and the moth, so it can zero in on the moth.

Some moths evade capture by flying away from the direction in which they hear ultrasonic waves. That choice of flight path reduces the frequency difference between what the bat emits and what it hears, and then the bat may not notice the echo. Some moths avoid capture by clicking to produce their own ultrasonic waves, thus "jamming" the detection system and confusing the bat. (Surprisingly, moths and bats do all this without first studying physics.)

✔CHECKPOINT 6: The figure indicates the directions of motion of a sound source and a detector for six situations in stationary air. For each situation, is the detected frequency greater than or less than the emitted frequency, or can't we tell without more information about the actual speeds?

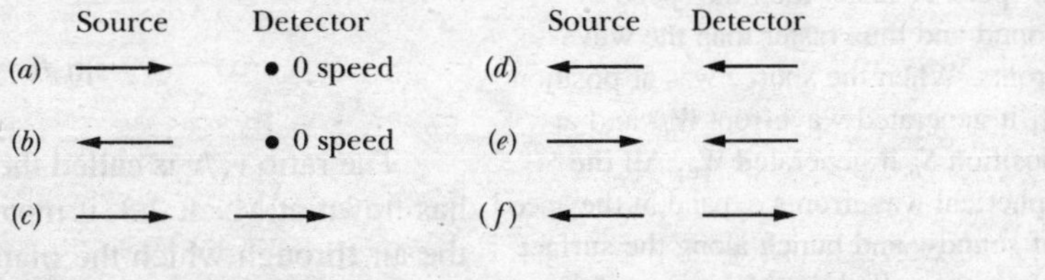

Sample Problem 18-8

A rocket moves at a speed of 242 m/s directly toward a stationary pole (through stationary air) while emitting sound waves at frequency $f = 1250$ Hz.

(a) What frequency f' is measured by a detector that is attached to the pole?

SOLUTION: We can find f' with Eq. 18-47 for the general Doppler effect. The **Key Idea** here is that, because the sound source (the rocket) moves through the air *toward* the stationary detector on the pole, we need to choose the sign on v_S that gives a *shift up* in the frequency of the sound. Thus, in Eq. 18-47 we use the minus sign in the denominator. We then substitute 0 for the detector speed v_D, 242 m/s for the source speed v_S, 343 m/s for the speed of sound v (from Table 18-1), and 1250 Hz for the emitted frequency f. We find

$$f' = f\frac{v \pm v_D}{v \pm v_S} = (1250 \text{ Hz})\frac{343 \text{ m/s} \pm 0}{343 \text{ m/s} - 242 \text{ m/s}}$$

$$= 4245 \text{ Hz} \approx 4250 \text{ Hz}, \qquad \text{(Answer)}$$

which, indeed, is a greater frequency than the emitted frequency.

(b) Some of the sound reaching the pole reflects back to the rocket as an echo. What frequency f'' does a detector on the rocket detect for the echo?

SOLUTION: Two **Key Ideas** here are the following:

1. The pole is now the source of sound (because it is the source of the echo), and the rocket's detector is now the detector (because it detects the echo).

2. The frequency of the sound emitted by the source (the pole) is equal to f', the frequency of the sound the pole intercepts and reflects.

We can rewrite Eq. 18-47 in terms of the source frequency f' and the detected frequency f'' as

$$f'' = f' \frac{v \pm v_D}{v \pm v_S}. \qquad (18\text{-}56)$$

A third **Key Idea** here is that, because the detector (on the rocket) moves through the air *toward* the stationary source, we need to use the sign on v_D that gives a *shift up* in the frequency of the sound. Thus, we use the plus sign in the numerator of Eq. 18-56. Also, we substitute $v_D = 242$ m/s, $v_S = 0$, $v = 343$ m/s, and $f' = 4245$ Hz. We find

$$f'' = (4245 \text{ Hz}) \frac{343 \text{ m/s} + 242 \text{ m/s}}{343 \text{ m/s} \pm 0}$$
$$= 7240 \text{ Hz}, \qquad \text{(Answer)}$$

which, indeed, is greater than the frequency of the sound reflected by the pole.

✔CHECKPOINT 7: If the air in this sample problem is moving toward the pole at speed 20 m/s, (a) what value for the source speed v_S should be used in the solution of part (a), and (b) what value for the detector speed v_D should be used in the solution of part (b)?

18-9 Supersonic Speeds; Shock Waves

If a source is moving toward a stationary detector at a speed equal to the speed of sound—that is, if $v_S = v$—Eqs. 18-47 and 18-55 predict that the detected frequency f' will be infinitely great. This means that the source is moving so fast that it keeps pace with its own spherical wavefronts, as Fig. 18-23*a* suggests. What happens when the speed of the source *exceeds* the speed of sound?

For such *supersonic* speeds, Eqs. 18-47 and 18-55 no longer apply. Figure 18-23*b* depicts the spherical wavefronts that originated at various positions of the source. The radius of any wavefront in this figure is vt, where v is the speed of sound and t is the time that has elapsed since the source emitted that wavefront. Note that all the wavefronts bunch along a V-shaped envelope in the two-dimensional drawing of Fig. 18-23*b*. The wavefronts actually extend in three dimensions, and the bunching actually forms a cone called the *Mach cone*. A *shock wave* is said to exist along the surface of this cone, because the bunching of wavefronts causes an abrupt rise and fall of air pressure as the surface passes through any point. From Fig. 18-23*b*, we see that the half-angle θ of the cone, called the *Mach cone angle*, is given by

$$\sin \theta = \frac{vt}{v_S t} = \frac{v}{v_S} \qquad \text{(Mach cone angle).} \qquad (18\text{-}57)$$

The ratio v_S/v is called the *Mach number*. When you hear that a particular plane has flown at Mach 2.3, it means that its speed was 2.3 times the speed of sound in the air through which the plane was flying. The shock wave generated by a super-

Fig. 18-23 (*a*) A source of sound S moves at speed v_S equal to the speed of sound and thus as fast as the wavefronts it generates. (*b*) A source S moves at speed v_S faster than the speed of sound and thus faster than the wavefronts. When the source was at position S_1 it generated wavefront W_1, and at position S_6 it generated W_6. All the spherical wavefronts expand at the speed of sound v and bunch along the surface of a cone called the Mach cone, forming a shock wave. The surface of the cone has half-angle θ and is tangent to all the wavefronts.

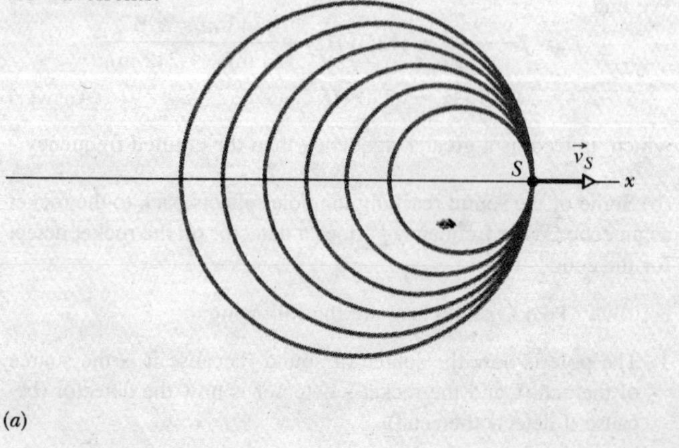

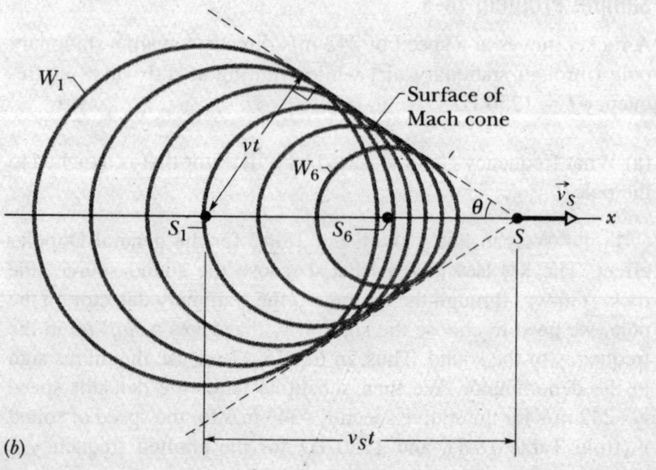

sonic aircraft (Fig. 18-24) or projectile produces a burst of sound, called a *sonic boom,* in which the air pressure first suddenly increases and then suddenly decreases below normal before returning to normal. Part of the sound that is heard when a rifle is fired is the sonic boom produced by the bullet. A sonic boom can also be heard from a long bullwhip when it is snapped quickly: Near the end of the whip's motion, its tip is moving faster than sound and produces a small sonic boom—the *crack* of the whip.

Fig. 18-24 Shock waves produced by the wings of a Navy FA 18 jet. They are visible because the sudden decrease in air pressure in the shock waves caused water molecules in the air to condense, forming a fog.

REVIEW & SUMMARY

Sound Waves Sound waves are longitudinal mechanical waves that can travel through solids, liquids, or gases. The speed v of a sound wave in a medium having **bulk modulus** B and density ρ is

$$v = \sqrt{\frac{B}{\rho}} \quad \text{(speed of sound).} \tag{18-3}$$

In air at 20°C, the speed of sound is 343 m/s.

A sound wave causes a longitudinal displacement s of a mass element in a medium as given by

$$s = s_m \cos(kx - \omega t), \tag{18-13}$$

where s_m is the **displacement amplitude** (maximum displacement) from equilibrium, $k = 2\pi/\lambda$, and $\omega = 2\pi f$, λ and f being the wavelength and frequency, respectively, of the sound wave. The sound wave also causes a pressure change Δp of the medium from the equilibrium pressure:

$$\Delta p = \Delta p_m \sin(kx - \omega t), \tag{18-14}$$

where the **pressure amplitude** is

$$\Delta p_m = (v\rho\omega)s_m. \tag{18-15}$$

Interference The interference of two sound waves with identical wavelengths passing through a common point depends on their phase difference ϕ there. If the sound waves were emitted in phase and are traveling in approximately the same direction, ϕ is given by

$$\phi = \frac{\Delta L}{\lambda} 2\pi, \tag{18-21}$$

where ΔL is their **path length difference** (the difference in the distances traveled by the waves to reach the common point). Fully constructive interference occurs when ϕ is an integer multiple of 2π,

$$\phi = m(2\pi), \quad \text{for } m = 0, 1, 2, \ldots, \tag{18-22}$$

and, equivalently, when ΔL is related to wavelength λ by

$$\frac{\Delta L}{\lambda} = 0, 1, 2, \ldots. \tag{18-23}$$

Fully destructive interference occurs when ϕ is an odd multiple of π,

$$\phi = (2m + 1)\pi, \quad \text{for } m = 0, 1, 2, \ldots, \tag{18-24}$$

and, equivalently, when ΔL is related to λ by

$$\frac{\Delta L}{\lambda} = 0.5, 1.5, 2.5, \ldots. \tag{18-25}$$

Sound Intensity The **intensity** I of a sound wave at a surface is the average rate per unit area at which energy is transferred by the wave through or onto the surface:

$$I = \frac{P}{A}, \tag{18-26}$$

where P is the time rate of energy transfer (power) of the sound wave and A is the area of the surface intercepting the sound. The intensity I is related to the displacement amplitude s_m of the sound wave by

$$I = \tfrac{1}{2}\rho v \omega^2 s_m^2. \tag{18-27}$$

The intensity at a distance r from a point source that emits sound waves of power P_s is

$$I = \frac{P_s}{4\pi r^2}. \tag{18-28}$$

Sound Level in Decibels The *sound level* β in *decibels* (dB) is defined as

$$\beta = (10 \text{ dB}) \log \frac{I}{I_0}, \tag{18-29}$$

where I_0 (= 10^{-12} W/m^2) is a reference intensity level to which all intensities are compared. For every factor-of-10 increase in intensity, 10 dB is added to the sound level.

Standing Wave Patterns in Pipes Standing sound wave patterns can be set up in pipes. A pipe open at both ends will resonate at frequencies

$$f = \frac{v}{\lambda} = \frac{nv}{2L}, \qquad n = 1, 2, 3, \ldots, \tag{18-39}$$

where v is the speed of sound in the air in the pipe. For a pipe closed at one end and open at the other, the resonant frequencies are

$$f = \frac{v}{\lambda} = \frac{nv}{4L}, \qquad n = 1, 3, 5, \ldots. \tag{18-41}$$

Beats *Beats* arise when two waves having slightly different frequencies, f_1 and f_2, are detected together. The beat frequency is

$$f_{beat} = f_1 - f_2. \tag{18-46}$$

The Doppler Effect The *Doppler effect* is a change in the observed frequency of a wave when the source or the detector moves relative to the transmitting medium (such as air). For sound the observed frequency f' is given in terms of the source frequency f by

$$f' = f\frac{v \pm v_D}{v \pm v_S} \quad \text{(general Doppler effect)}, \tag{18-47}$$

where v_D is the speed of the detector relative to the medium, v_S is that of the source, and v is the speed of sound in the medium. The signs are chosen such that f' tends to be *greater* for motion (of detector or source) "toward" and *less* for motion "away."

Shock Wave If the speed of a source relative to the medium exceeds the speed of sound in the medium, the Doppler equation no longer applies. In such a case, shock waves result. The half angle θ of the Mach cone is given by

$$\sin\theta = \frac{v}{v_S} \quad \text{(Mach cone angle)}. \tag{18-57}$$

QUESTIONS

1. Figure 18-25 shows the paths taken by two pulses of sound that begin simultaneously and then race each other through equal distances in air. The only difference between the paths is that a region of hot (low density) air lies along path 2. Which pulse wins the race?

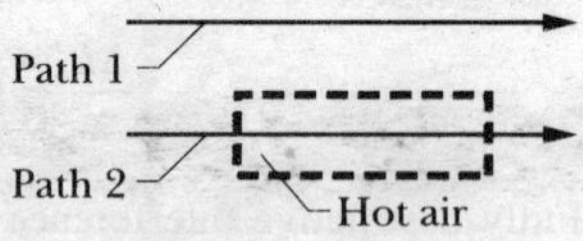

Fig. 18-25 Question 1.

2. A sound wave of wavelength λ and displacement amplitude s_m begins to travel down a passageway (a tube, the opening of the ear, etc.). When a small device in the passageway detects this wave, it issues a second sound wave (said to be *antisound*) that is able to cancel the first wave, so that nothing is heard at the far end of the passageway. For such cancellation, what must be (a) the direction of travel, (b) the wavelength, and (c) the displacement amplitude of the second wave? (d) What must be the phase difference between the two waves? (Such antisound devices are used to eliminate unwanted sound in a noisy environment.)

3. In Fig. 18-26, two point sources S_1 and S_2, which are in phase, emit identical sound waves of wavelength 2.0 m. In terms of wavelengths, what is the phase difference between the waves arriving at point P if (a) L_1 = 38 m and L_2 = 34 m, and (b) L_1 = 39 m and L_2 = 36 m? (c) Assuming that the source separation is much smaller than L_1 and L_2, what type of interference occurs at P in situations (a) and (b), respectively?

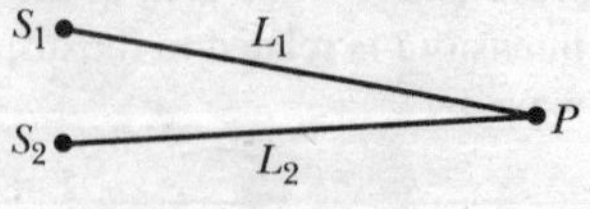

Fig. 18-26 Question 3.

4. In Fig. 18-27, sound waves of wavelength λ are emitted by a point source S and travel to a detector D directly along path 1 and via reflection from a panel along path 2. Initially, the panel is almost along path 1 and the waves arriving at D along the two paths are almost exactly in phase. Then the panel is moved away from path 1 as shown until the waves arriving at D are exactly out of phase. What then is the path length difference ΔL of the waves along the two paths?

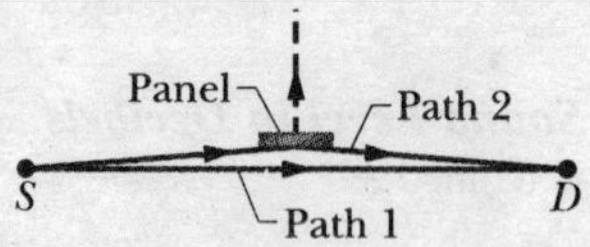

Fig. 18-27 Question 4.

5. In Fig. 18-28, two point sources S_1 and S_2, which are in phase, emit identical sound waves of wavelength λ, and point P is at equal distances from them. Then S_2 is moved directly away from P by a distance equal to $\lambda/4$. Are the waves at P then exactly in phase, exactly out of phase, or do they have some intermediate phase relation if (a) S_1 is moved directly toward P by a distance equal to $\lambda/4$ and (b) S_1 is moved directly away from P by a distance equal to $3\lambda/4$?

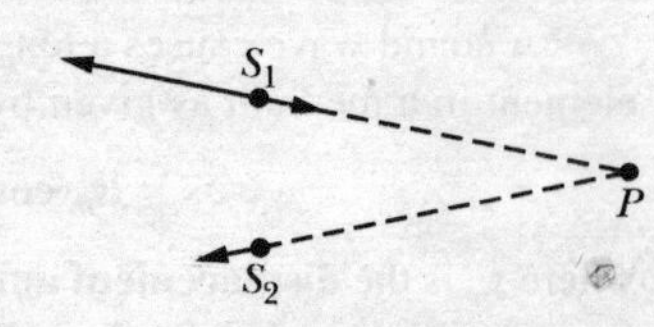

Fig. 18-28 Question 5.

6. In Sample Problem 18-3 and Fig. 18-9*a*, the waves arriving at point P_1 on the perpendicular bisector are exactly in phase; that is, the waves from S_1 and S_2 always tend to move an element of air at P_1 in the same direction. Let the intersection of the perpendicular bisector and the line through S_1 and S_2 be point P_3. (a) Are the waves arriving at P_3 exactly in phase, exactly out of phase, or do they have some intermediate relation? (b) What is the answer if we increase the separation between the sources to 1.7λ?

7. A standing sound wave in a pipe has five nodes and five antinodes. (a) How many open ends does the pipe have? (b) What is the harmonic number n for this standing wave?

8. The sixth harmonic is set up in a pipe. (a) How many open ends does the pipe have (it has at least one)? (b) Is there a node, antinode, or some intermediate state at the midpoint?

9. (a) When an orchestra warms up, the players' warm breath increases the temperature of the air within the wind instruments (and thus decreases the density of that air). Do the resonant frequencies

of those instruments increase or decrease? (b) When the slide of a slide trombone is pushed outward, do the resonant frequencies of the instrument increase or decrease?

10. For a particular tube, here are four of the six harmonic frequencies below 1000 Hz: 300, 600, 750, and 900 Hz. What two frequencies are missing from the list?

11. Pipe A has length L and one open end. Pipe B has length $2L$ and two open ends. Which harmonics of pipe B have a frequency that matches a resonant frequency of pipe A?

12. Figure 18-29 shows a stretched string of length L and pipes a, b, c, and d of lengths L, $2L$, $L/2$, and $L/2$, respectively. The string's tension is adjusted until the speed of waves on the string equals the speed of sound waves in the air. The fundamental mode of oscillation is then set up on the string. In which pipe will the sound produced by the string cause resonance, and what oscillation mode will that sound set up?

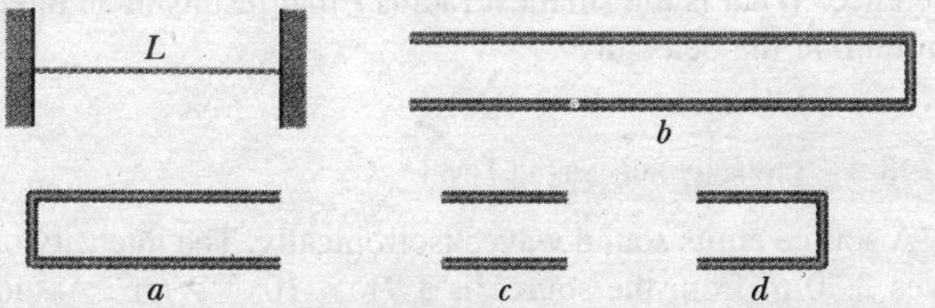

Fig. 18-29 Question 12.

13. Sound waves of frequency f are reflected by a fluid moving through a narrow tube along an x axis (Fig. 18-30a). The tube's inside diameter varies with x. The change in frequency Δf of the sound, due to the Doppler effect, also varies with x, as shown in Fig. 18-30b. Rank the five indicated regions in terms of the tube's inside diameter, greatest first. (*Hint:* See Section 15-10.)

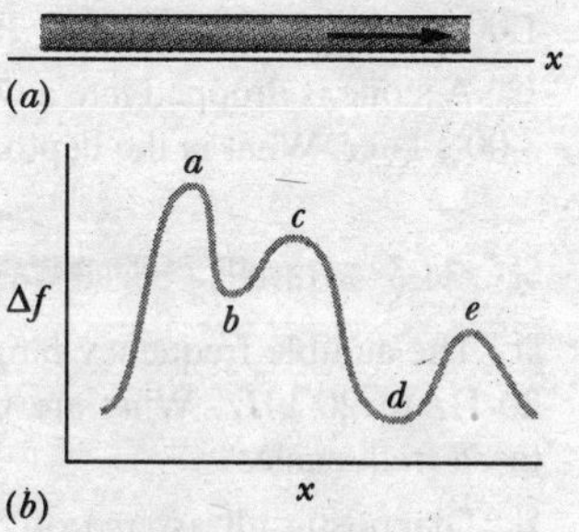

Fig. 18-30 Question 13.

14. A friend rides, in turn, the rims of three fast merry-go-rounds while holding a sound source that emits isotropically at a certain frequency. You stand far from each merry-go-round. The frequency you hear for each of your friend's three rides varies as the merry-go-round rotates. The variations in frequency for the three rides are given by the three curves in Fig. 18-31. Rank the curves according to (a) the linear speed v of the sound source, (b) the angular speeds ω of the merry-go-rounds, and (c) the radii r of the merry-go-rounds, greatest first.

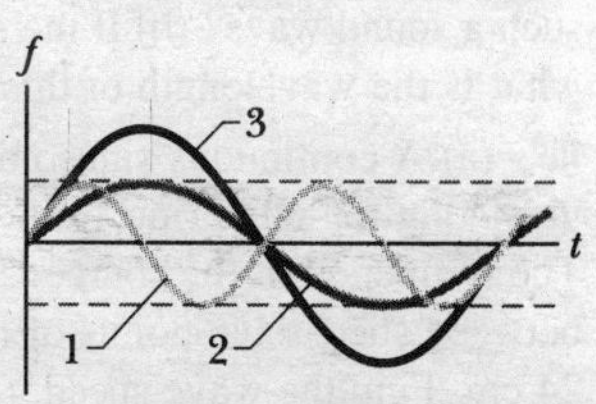

Fig. 18-31 Question 14.

EXERCISES & PROBLEMS

ssm Solution is in the Student Solutions Manual.
www Solution is available on the World Wide Web at: http://www.wiley.com/college/hrw
ltw Solution is available on the Interactive LearningWare.

Where needed in the problems, use

speed of sound in air = 343 m/s

and density of air = 1.21 kg/m^3

unless otherwise specified.

SEC. 18-2 The Speed of Sound

1E. Devise a rule for finding your distance in kilometers from a lightning flash by counting the seconds from the time you see the flash until you hear the thunder. Assume that the sound travels to you along a straight line.

2E. You are at a large outdoor concert, seated 300 m from the speaker system. The concert is also being broadcast live via satellite (at the speed of light, 3.0×10^8 m/s). Consider a listener 5000 km away who receives the broadcast. Who hears the music first, you or the listener, and by what time difference?

3E. Two spectators at a soccer game in Montjuic Stadium see, and a moment later hear, the ball being kicked on the playing field. The time delay for one spectator is 0.23 s and for the other 0.12 s. Sight lines from the two spectators to the player kicking the ball meet at an angle of 90°. (a) How far is each spectator from the player? (b) How far are the spectators from each other? **ssm**

4E. A column of soldiers, marching at 120 paces per minute, keep in step with the beat of a drummer at the head of the column. It is observed that the soldiers in the rear end of the column are striding forward with the left foot when the drummer is advancing with the right. What is the approximate length of the column?

5P. Earthquakes generate sound waves inside Earth. Unlike a gas, Earth can experience both transverse (S) and longitudinal (P) sound waves. Typically, the speed of S waves is about 4.5 km/s, and that of P waves 8.0 km/s. A seismograph records P and S waves from an earthquake. The first P waves arrive 3.0 min before the first S waves (Fig. 18-32). Assuming the waves travel in a straight line, how far away does the earthquake occur? **ssm** **ltw**

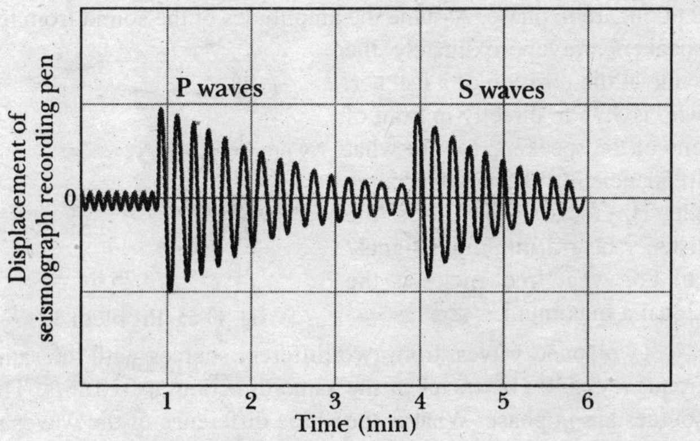

Fig. 18-32 Problem 5.

6P. The speed of sound in a certain metal is V. One end of a long pipe of that metal of length L is struck a hard blow. A listener at the other end hears two sounds, one from the wave that travels along the pipe and the other from the wave that travels through the air. (a) If v is the speed of sound in air, what time interval t elapses between the arrivals of the two sounds? (b) Suppose that $t = 1.00$ s and the metal is steel. Find the length L.

7P. A stone is dropped into a well. The sound of the splash is heard 3.00 s later. What is the depth of the well? ssm

SEC. 18-3 Traveling Sound Waves

8E. The audible frequency range for normal hearing is from about 20 Hz to 20 kHz. What are the wavelengths of sound waves at these frequencies?

9E. Diagnostic ultrasound of frequency 4.50 MHz is used to examine tumors in soft tissue. (a) What is the wavelength in air of such a sound wave? (b) If the speed of sound in tissue is 1500 m/s, what is the wavelength of this wave in tissue? ssm

10P. (a) A continuous sinusoidal longitudinal wave is sent along a very long coiled spring from an oscillating source attached to it. The frequency of the source is 25 Hz, and at any time the distance between successive points of maximum expansion in the spring is 24 cm. Find the wave speed. (b) If the maximum longitudinal displacement of a particle in the spring is 0.30 cm and the wave moves in the negative direction of an x axis, write the equation for the wave. Place $x = 0$ at the source and take the displacement there to be zero when $t = 0$.

11P. The pressure in a traveling sound wave is given by the equation

$$\Delta p = (1.50 \text{ Pa}) \sin \pi[(0.900 \text{ m}^{-1})x - (315 \text{ s}^{-1})t].$$

Find (a) the pressure amplitude, (b) the frequency, (c) the wavelength, and (d) the speed of the wave.

SEC. 18-4 Interference

12P. Two point sources of sound waves of identical wavelength λ and amplitude are separated by distance $D = 2.0\lambda$. The sources are in phase. (a) How many points of maximum signal (that is, maximum constructive interference) lie along a large circle around the sources? (b) How many points of minimum signal (destructive interference) lie around the circle?

13P. In Fig. 18-33, two loudspeakers, separated by a distance of 2.00 m, are in phase. Assume the amplitudes of the sound from the speakers are approximately the same at the position of a listener, who is 3.75 m directly in front of one of the speakers. (a) For what frequencies in the audible range (20 Hz to 20 kHz) does the listener hear a minimum signal? (b) For what frequencies is the signal a maximum? ssm www

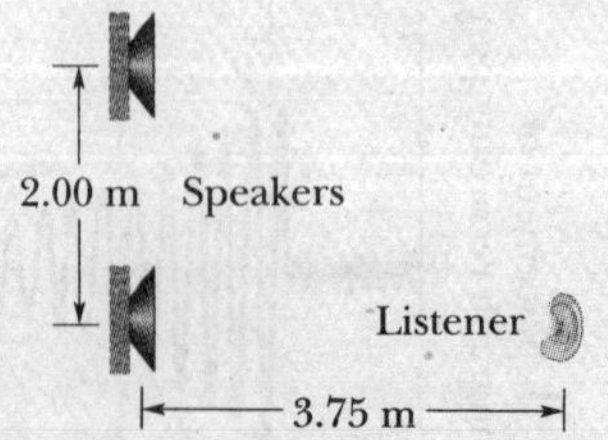

Fig. 18-33 Problem 13.

14P. Two sound waves, from two different sources with the same frequency, 540 Hz, travel in the same direction at 330 m/s. The sources are in phase. What is the phase difference of the waves at a point that is 4.40 m from one source and 4.00 m from the other?

15P. Two loudspeakers are located 3.35 m apart on an outdoor stage. A listener is 18.3 m from one and 19.5 m from the other. During the sound check, a signal generator drives the two speakers in phase with the same amplitude and frequency. The transmitted frequency is swept through the audible range (20 Hz to 20 kHz). (a) What are the three lowest frequencies at which the listener will hear a minimum signal because of destructive interference? (b) What are the three lowest frequencies at which the listener will hear a maximum signal? ilw

16P. In Fig. 18-34, sound with a 40.0 cm wavelength travels rightward from a source and through a tube that consists of a straight portion and a half-circle. Part of the sound wave travels through the half-circle and then rejoins the rest of the wave, which goes directly through the straight portion. This rejoining results in interference. What is the smallest radius r that results in an intensity minimum at the detector?

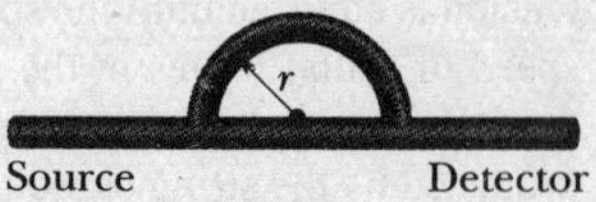

Fig. 18-34 Problem 16.

SEC. 18-5 Intensity and Sound Level

17E. A source emits sound waves isotropically. The intensity of the waves 2.50 m from the source is 1.91×10^{-4} W/m^2. Assuming that the energy of the waves is conserved, find the power of the source. ssm

18E. A 1.0 W point source emits sound waves isotropically. Assuming that the energy of the waves is conserved, find the intensity (a) 1.0 m from the source and (b) 2.5 m from the source.

19E. A sound wave of frequency 300 Hz has an intensity of 1.00 μW/m^2. What is the amplitude of the air oscillations caused by this wave? ssm

20E. Two sounds differ in sound level by 1.00 dB. What is the ratio of the greater intensity to the smaller intensity?

21E. A certain sound source is increased in sound level by 30 dB. By what multiple is (a) its intensity increased and (b) its pressure amplitude increased? ssm

22E. The source of a sound wave has a power of 1.00 μW. If it is a point source, (a) what is the intensity 3.00 m away and (b) what is the sound level in decibels at that distance?

23E. (a) If two sound waves, one in air and one in (fresh) water, are equal in intensity, what is the ratio of the pressure amplitude of the wave in water to that of the wave in air? Assume the water and the air are at 20°C. (See Table 15-1.) (b) If the pressure amplitudes are equal instead, what is the ratio of the intensities of the waves? ssm

24P. Assume that a noisy freight train on a straight track emits a cylindrical, expanding sound wave, and that the air absorbs no energy. How does the amplitude s_m of the wave depend on the perpendicular distance r from the source?

25P. (a) Show that the intensity I of a wave is the product of the wave's energy per unit volume u and its speed v. (b) Radio waves travel at a speed of 3.00×10^8 m/s. Find u for a radio wave 480 km from a 50 000 W source, assuming the wavefronts are spherical. ssm

26P. Find the ratios (greater to smaller) of (a) the intensities,

(b) the pressure amplitudes, and (c) the particle displacement amplitudes for two sounds whose sound levels differ by 37 dB.

27P. A sound wave travels out uniformly in all directions from a point source. (a) Justify the following expression for the displacement s of the transmitting medium at any distance r from the source:

$$s = \frac{b}{r} \sin k(r - vt),$$

where b is a constant. Consider the speed, direction of propagation, periodicity, and intensity of the wave. (b) What is the dimension of the constant b? ssm www

28P. A point source emits 30.0 W of sound isotropically. A small microphone intercepts the sound in an area of 0.750 cm^2, 200 m from the source. Calculate (a) the sound intensity there and (b) the power intercepted by the microphone.

29P*. Figure 18-35 shows an air-filled, acoustic interferometer, used to demonstrate the interference of sound waves. Sound source S is an oscillating diaphragm; D is a sound detector, such as the ear or a microphone. Path SBD can be varied in length, but path SAD is fixed. At D, the sound wave coming along path SBD interferes with that coming along path SAD. In one demonstration, the sound intensity at D has a minimum value of 100 units at one position of the movable arm and continuously climbs to a maximum value of 900 units when that arm is shifted by 1.65 cm. Find (a) the frequency of the sound emitted by the source and (b) the ratio of the amplitude at D of the SAD wave to that of the SBD wave. (c) How can it happen that these waves have different amplitudes, considering that they originate at the same source? ssm

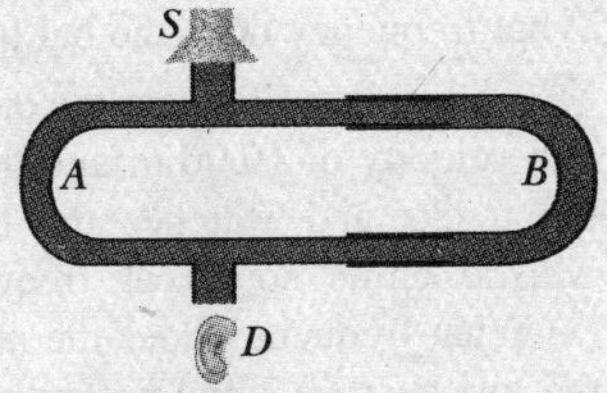

Fig. 18-35 Problem 29.

SEC. 18-6 Sources of Musical Sound

30E. A violin string 15.0 cm long and fixed at both ends oscillates in its $n = 1$ mode. The speed of waves on the string is 250 m/s, and the speed of sound in air is 348 m/s. What are (a) the frequency and (b) the wavelength of the emitted sound wave?

31E. Organ pipe A, with both ends open, has a fundamental frequency of 300 Hz. The third harmonic of organ pipe B, with one end open, has the same frequency as the second harmonic of pipe A. How long are (a) pipe A and (b) pipe B?

32E. The water level in a vertical glass tube 1.00 m long can be adjusted to any position in the tube. A tuning fork vibrating at 686 Hz is held just over the open top end of the tube, to set up a standing wave of sound in the air-filled top portion of the tube. (That air-filled top portion acts as a tube with one end closed and the other end open.) At what positions of the water level is there resonance?

33E. (a) Find the speed of waves on a violin string of mass 800 mg and length 22.0 cm if the fundamental frequency is 920 Hz. (b) What is the tension in the string? For the fundamental, what is the wavelength of (c) the waves on the string and (d) the sound waves emitted by the string? ssm ilw

34P. A certain violin string is 30 cm long between its fixed ends and has a mass of 2.0 g. The "open" string (no applied finger) sounds an A note (440 Hz). (a) To play a C note (523 Hz), how far down the string must one place a finger? (b) What is the ratio of the wavelength of the string waves required for an A note to that required for a C note? (c) What is the ratio of the wavelength of the sound wave for an A note to that for a C note?

35P. In Fig. 18-36, S is a small loudspeaker driven by an audio oscillator and amplifier, adjustable in frequency from 1000 to 2000 Hz only. Tube D is a piece of cylindrical sheet-metal pipe 45.7 cm long and open at both ends. (a) If the speed of sound in air is 344 m/s at the existing temperature, at what frequencies will resonance occur in the pipe when the frequency emitted by the speaker is varied from 1000 Hz to 2000 Hz? (b) Sketch the standing wave (using the style of Fig. 18-14b) for each resonant frequency. ssm www

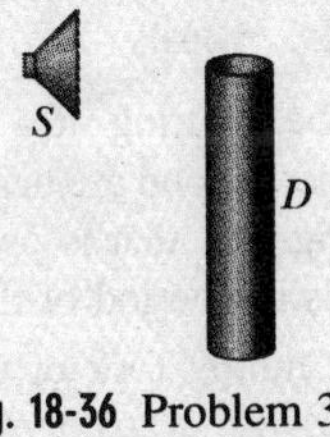

Fig. 18-36 Problem 35.

36P. A string on a cello has length L, for which the fundamental frequency is f. (a) By what length l must the string be shortened by fingering to change the fundamental frequency to rf? (b) What is l if $L = 0.80$ m and $r = 1.2$? (c) For $r = 1.2$, what is the ratio of the wavelength of the new sound wave emitted by the string to that of the wave emitted before fingering?

37P. A well with vertical sides and water at the bottom resonates at 7.00 Hz and at no lower frequency. (The air-filled portion of the well acts as a tube with one closed end and one open end.) The air in the well has a density of 1.10 kg/m^3 and a bulk modulus of 1.33×10^5 Pa. How far down in the well is the water surface? ssm

38P. A tube 1.20 m long is closed at one end. A stretched wire is placed near the open end. The wire is 0.330 m long and has a mass of 9.60 g. It is fixed at both ends and oscillates in its fundamental mode. By resonance, it sets the air column in the tube into oscillation at that column's fundamental frequency. Find (a) that frequency and (b) the tension in the wire.

39P. The period of a pulsating variable star may be estimated by considering the star to be executing *radial* longitudinal pulsations in the fundamental standing wave mode; that is, the star's radius varies periodically with time, with a displacement antinode at the star's surface. (a) Would you expect the center of the star to be a displacement node or antinode? (b) By analogy with a pipe with one open end, show that the period of pulsation T is given by

$$T = \frac{4R}{v},$$

where R is the equilibrium radius of the star and v is the average sound speed in the material of the star. (c) Typical white dwarf stars are composed of material with a bulk modulus of 1.33×10^{22} Pa and a density of 10^{10} kg/m^3. They have radii equal to 9.0×10^{-3} solar radius. What is the approximate pulsation period of a white dwarf? ssm

40P. Pipe A, which is 1.2 m long and open at both ends, oscillates at its third lowest harmonic frequency. It is filled with air for which the speed of sound is 343 m/s. Pipe B, which is closed at one end, oscillates at its second lowest harmonic frequency. The frequencies of pipes A and B happen to match. (a) If an x axis extends along the interior of pipe A, with $x = 0$ at one end, where along the axis are the displacement nodes? (b) How long is pipe B? (c) What is the lowest harmonic frequency of pipe A?

41P. A violin string 30.0 cm long with linear density 0.650 g/m is placed near a loudspeaker that is fed by an audio oscillator of variable frequency. It is found that the string is set into oscillation only at the frequencies 880 and 1320 Hz as the frequency of the oscillator is varied over the range 500–1500 Hz. What is the tension in the string? ssm

SEC. 18-7 Beats

42E. The A string of a violin is a little too tightly stretched. Four beats per second are heard when the string is sounded together with a tuning fork that is oscillating accurately at concert A (440 Hz). What is the period of the violin string oscillation?

43E. A tuning fork of unknown frequency makes three beats per second with a standard fork of frequency 384 Hz. The beat frequency decreases when a small piece of wax is put on a prong of the first fork. What is the frequency of this fork? ssm

44P. You have five tuning forks that oscillate at close but different frequencies. What are the (a) maximum and (b) minimum number of different beat frequencies you can produce by sounding the forks two at a time, depending on how the frequencies differ?

45P. Two identical piano wires have a fundamental frequency of 600 Hz when kept under the same tension. What fractional increase in the tension of one wire will lead to the occurrence of 6 beats/s when both wires oscillate simultaneously? ssm

SEC. 18-8 The Doppler Effect

46E. Trooper B is chasing speeder A along a straight stretch of road. Both are moving at a speed of 160 km/h. Trooper B, failing to catch up, sounds his siren again. Take the speed of sound in air to be 343 m/s and the frequency of the source to be 500 Hz. What is the Doppler shift in the frequency heard by speeder A?

47E. The 16 000 Hz whine of the turbines in the jet engines of an aircraft moving with speed 200 m/s is heard at what frequency by the pilot of a second craft trying to overtake the first at a speed of 250 m/s? ssm

48E. An ambulance with a siren emitting a whine at 1600 Hz overtakes and passes a cyclist pedaling a bike at 2.44 m/s. After being passed, the cyclist hears a frequency of 1590 Hz. How fast is the ambulance moving?

49P. A whistle of frequency 540 Hz moves in a circle of radius 60.0 cm at an angular speed of 15.0 rad/s. What are (a) the lowest and (b) the highest frequencies heard by a listener a long distance away, at rest with respect to the center of the circle? ilw

50P. A stationary motion detector sends sound waves of frequency 0.150 MHz toward a truck approaching at a speed of 45.0 m/s. What is the frequency of the waves reflected back to the detector?

51P. A French submarine and a U.S. submarine move toward each other during maneuvers in motionless water in the North Atlantic (Fig. 18-37). The French sub moves at 50.0 km/h, and the U.S. sub at 70.0 km/h. The French sub sends out a sonar signal (sound wave in water) at 1000 Hz. Sonar waves travel at 5470 km/h. (a) What is the signal's frequency as detected by the U.S. sub? (b) What frequency is detected by the French sub in the signal reflected back to it by the U.S. sub?

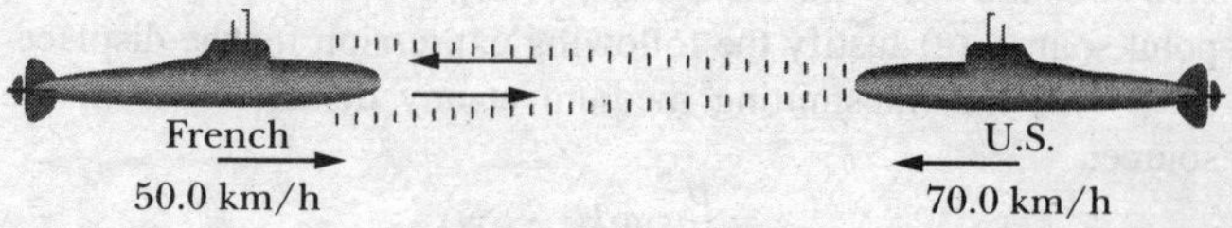

Fig. 18-37 Problem 51.

52P. A sound source A and a reflecting surface B move directly toward each other. Relative to the air, the speed of source A is 29.9 m/s, the speed of surface B is 65.8 m/s, and the speed of sound is 329 m/s. The source emits waves at frequency 1200 Hz as measured in the source frame. In the reflector frame, what are (a) the frequency and (b) the wavelength of the arriving sound waves? In the source frame, what are (c) the frequency and (d) the wavelength of the sound waves reflected back to the source?

53P. An acoustic burglar alarm consists of a source emitting waves of frequency 28.0 kHz. What is the beat frequency between the source waves and the waves reflected from an intruder walking at an average speed of 0.950 m/s directly away from the alarm? ilw

54P. A bat is flitting about in a cave, navigating via ultrasonic bleeps. Assume that the sound emission frequency of the bat is 39 000 Hz. During one fast swoop directly toward a flat wall surface, the bat is moving at 0.025 times the speed of sound in air. What frequency does the bat hear reflected off the wall?

55P. A girl is sitting near the open window of a train that is moving at a velocity of 10.00 m/s to the east. The girl's uncle stands near the tracks and watches the train move away. The locomotive whistle emits sound at frequency 500.0 Hz. The air is still. (a) What frequency does the uncle hear? (b) What frequency does the girl hear? A wind begins to blow from the east at 10.00 m/s. (c) What frequency does the uncle now hear? (d) What frequency does the girl now hear? ssm www

56P. A 2000 Hz siren and a civil defense official are both at rest with respect to the ground. What frequency does the official hear if the wind is blowing at 12 m/s (a) from source to official and (b) from official to source?

57P. Two trains are traveling toward each other at 30.5 m/s relative to the ground. One train is blowing a whistle at 500 Hz. (a) What frequency is heard on the other train in still air? (b) What frequency is heard on the other train if the wind is blowing at 30.5 m/s toward the whistle and away from the listener? (c) What frequency is heard if the wind direction is reversed?

SEC. 18-9 Supersonic Speeds; Shock Waves

58E. A bullet is fired with a speed of 685 m/s. Find the angle made by the shock cone with the line of motion of the bullet.

59P. A jet plane passes over you at a height of 5000 m and a speed of Mach 1.5. (a) Find the Mach cone angle. (b) How long after the jet passes directly overhead does the shock wave reach you? Use 331 m/s for the speed of sound. ssm

60P. A plane flies at 1.25 times the speed of sound. Its sonic boom reaches a man on the ground 1.00 min after the plane passes directly overhead. What is the altitude of the plane? Assume the speed of sound to be 330 m/s.

19 Temperature, Heat, and the First Law of Thermodynamics

The giant hornet *Vespa mandarinia japonica* preys on Japanese bees. However, if one of the hornets attempts to invade a bee hive, several hundred of the bees quickly form a compact ball around the hornet to stop it. After about 20 minutes the hornet is dead, although the bees do not sting, bite, crush, or suffocate it.

Why, then, does the hornet die?

The answer is in this chapter.

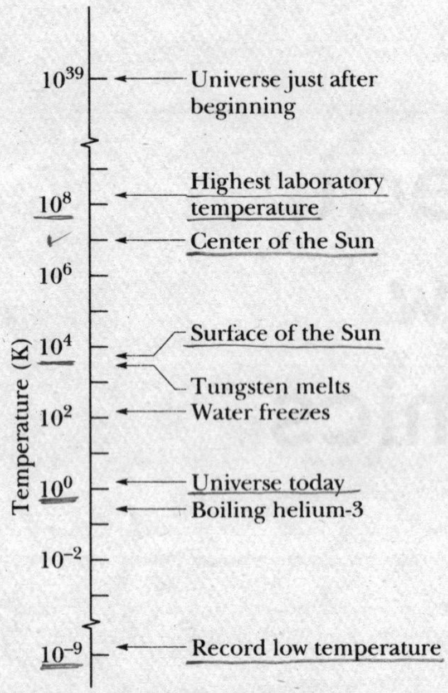

Fig. 19-1 Some temperatures on the Kelvin scale. Temperature $T = 0$ corresponds to $10^{-\infty}$ and cannot be plotted on this logarithmic scale.

19-1 Thermodynamics

In this and the next two chapters we focus on a new subject—**thermodynamics,** the study of the *thermal energy* (often called the *internal energy*) of systems. The central concept of thermodynamics is temperature. This word is so familiar that most of us—because of our built-in sense of hot and cold—tend to be overconfident in our understanding of it. Our "temperature sense" is in fact not always reliable. On a cold winter day, for example, an iron railing seems much colder to the touch than a wooden fence post, yet both are at the same temperature. This error in our perception comes about because iron removes energy from our fingers more quickly than wood does. Here, we shall develop the concept of temperature from its foundations, without relying in any way on our temperature sense.

Temperature is one of the seven SI base quantities. Physicists measure temperature on the **Kelvin scale,** which is marked in units called *kelvins*. Although the temperature of a body apparently has no upper limit, it does have a lower limit; this limiting low temperature is taken as the zero of the Kelvin temperature scale. Room temperature is about 290 kelvins, or 290 K as we write it, above this *absolute zero*. Figure 19-1 shows a wide range of temperatures, either measured or conjectured.

When the universe began, some 10 to 20 billion years ago, its temperature was about 10^{39} K. As the universe expanded it cooled, and it has now reached an average temperature of about 3 K. We on Earth are a little warmer than that because we happen to live near a star. Without our Sun, we too would be at 3 K (or, rather, we could not exist).

19-2 The Zeroth Law of Thermodynamics

The properties of many bodies change as we alter their temperature, perhaps by moving them from a refrigerator to a warm oven. To give a few examples: As their temperatures increase, the volume of a liquid increases, a metal rod grows a little longer, and the electrical resistance of a wire increases, as does the pressure exerted by a confined gas. We can use any one of these properties as the basis of an instrument that will help us to pin down the concept of temperature.

Figure 19-2 shows such an instrument. Any resourceful engineer could design and construct it, using any one of the properties listed above. The instrument is fitted with a digital readout display and has the following properties: If you heat it (say, with a Bunsen burner), the displayed number starts to increase; if you then put it into a refrigerator, the displayed number starts to decrease. The instrument is not calibrated in any way, and the numbers have (as yet) no physical meaning. The device is a *thermoscope* but not (as yet) a *thermometer*.

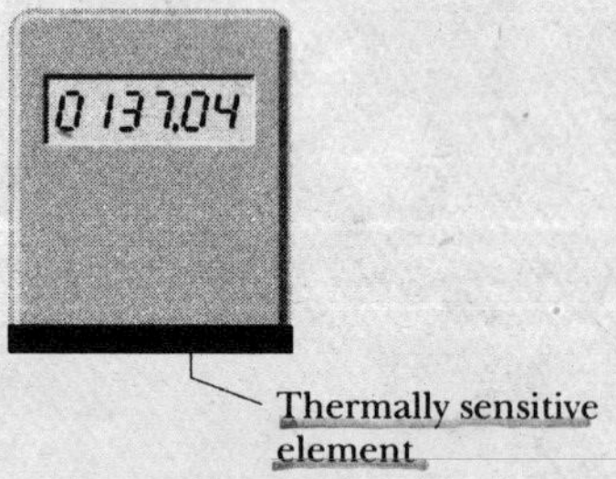

Fig. 19-2 A thermoscope. The numbers increase when the device is heated and decrease when it is cooled. The thermally sensitive element could be—among many possibilities—a coil of wire whose electrical resistance is measured and displayed.

Suppose that, as in Fig. 19-3*a*, we put the thermoscope (which we shall call body T) into intimate contact with another body (body A). The entire system is confined within a thick-walled insulating box. The numbers displayed by the thermoscope roll by until, eventually, they come to rest (let us say the reading is "137.04") and no further change takes place. In fact, we suppose that every measurable property of body T and of body A has assumed a stable, unchanging value. Then we say that the two bodies are in *thermal equilibrium* with each other. Even though the displayed readings for body T have not been calibrated, we conclude that bodies T and A must be at the same (unknown) temperature.

Suppose that we next put body T in intimate contact with body B (Fig. 19-3*b*) and find that the two bodies come to thermal equilibrium *at the same reading of the thermoscope*. Then bodies T and B must be at the same (still unknown) temperature. If we now put bodies A and B into intimate contact (Fig. 19-3*c*), are they immediately in thermal equilibrium with each other? Experimentally, we find that they are.

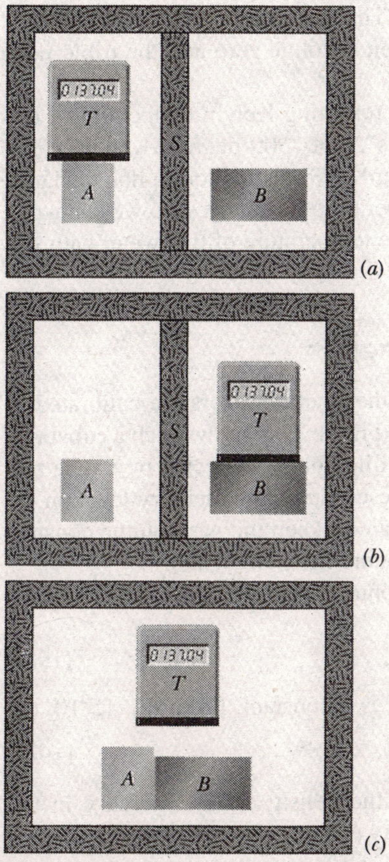

Fig. 19-3 (*a*) Body T (a thermoscope) and body A are in thermal equilibrium. (Body S is a thermally insulating screen.) (*b*) Body T and body B are also in thermal equilibrium, at the same reading of the thermoscope. (*c*) If (*a*) and (*b*) are true, the zeroth law of thermodynamics states that body A and body B are also in thermal equilibrium.

The experimental fact shown in Fig. 19-3 is summed up in the **zeroth law of thermodynamics:**

> If bodies A and B are each in thermal equilibrium with a third body T, then they are in thermal equilibrium with each other.

In less formal language, the message of the zeroth law is: "Every body has a property called **temperature.** When two bodies are in thermal equilibrium, their temperatures are equal. And vice versa." We can now make our thermoscope (the third body T) into a thermometer, confident that its readings will have physical meaning. All we have to do is calibrate it.

We use the zeroth law constantly in the laboratory. If we want to know whether the liquids in two beakers are at the same temperature, we measure the temperature of each with a thermometer. We do not need to bring the two liquids into intimate contact and observe whether they are or are not in thermal equilibrium.

The zeroth law, which has been called a logical afterthought, came to light only in the 1930s, long after the first and second laws of thermodynamics had been discovered and numbered. Because the concept of temperature is fundamental to those two laws, the law that establishes temperature as a valid concept should have the lowest number—hence the zero.

19-3 Measuring Temperature

Here we first define and measure temperatures on the Kelvin scale. Then we calibrate a thermoscope so as to make it a thermometer.

The Triple Point of Water

To set up a temperature scale, we pick some reproducible thermal phenomenon and, quite arbitrarily, assign a certain Kelvin temperature to its environment; that is, we select a *standard fixed point* and give it a standard fixed-point *temperature*. We could, for example, select the freezing point or the boiling point of water but, for various technical reasons, we select instead the **triple point of water.**

Liquid water, solid ice, and water vapor (gaseous water) can coexist, in thermal equilibrium, at only one set of values of pressure and temperature. Figure 19-4 shows a triple-point cell, in which this so-called triple point of water can be achieved in the laboratory. By international agreement, the triple point of water has been assigned a value of 273.16 K as the standard fixed-point temperature for the calibration of thermometers; that is,

$$T_3 = 273.16 \text{ K} \quad \text{(triple-point temperature)}, \tag{19-1}$$

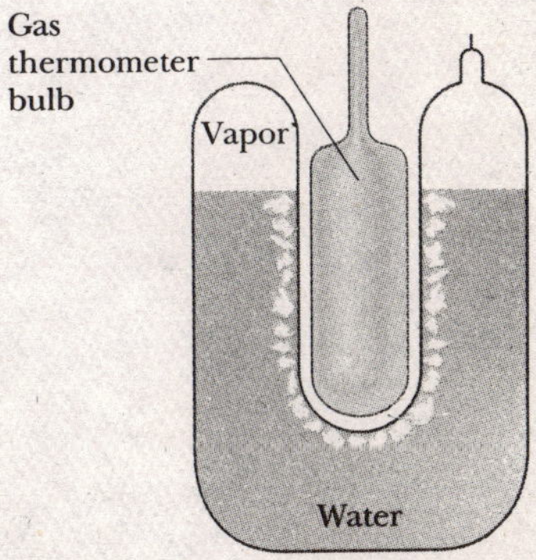

Fig. 19-4 A triple-point cell, in which solid ice, liquid water, and water vapor coexist in thermal equilibrium. By international agreement, the temperature of this mixture has been defined to be 273.16 K. The bulb of a constant-volume gas thermometer is shown inserted into the well of the cell.

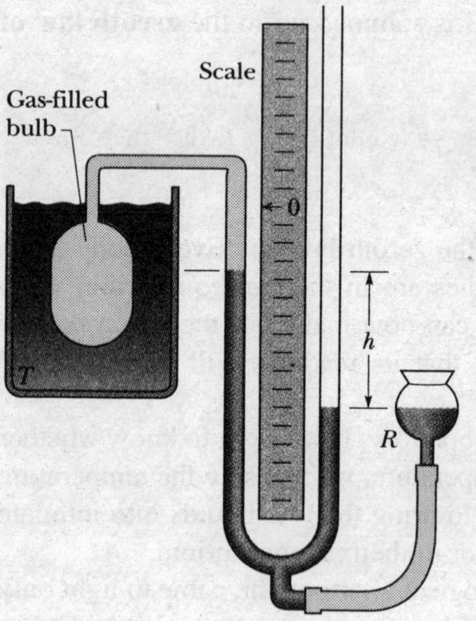

Fig. 19-5 A constant-volume gas thermometer, its bulb immersed in a liquid whose temperature T is to be measured.

in which the subscript 3 means "triple point." This agreement also sets the size of the kelvin as 1/273.16 of the difference between absolute zero and the triple-point temperature of water.

Note that we do not use a degree mark in reporting Kelvin temperatures. It is 300 K (not 300°K), and it is read "300 kelvins" (not "300 degrees Kelvin"). The usual SI prefixes apply. Thus, 0.0035 K is 3.5 mK. No distinction in nomenclature is made between Kelvin temperatures and temperature differences, so we can write, "the boiling point of sulfur is 717.8 K" and "the temperature of this water bath was raised by 8.5 K."

The Constant-Volume Gas Thermometer

The standard thermometer, against which all other thermometers are calibrated, is based on the pressure of a gas in a fixed volume. Figure 19-5 shows such a **constant-volume gas thermometer;** it consists of a gas-filled bulb connected by a tube to a mercury manometer. By raising and lowering reservoir R, the mercury level on the left can always be brought to the zero of the scale to keep the gas volume constant (variations in the gas volume can affect temperature measurements).

The temperature of any body in thermal contact with the bulb (like the liquid in Fig. 19-5) is then defined to be

$$T = Cp, \tag{19-2}$$

in which p is the pressure within the gas and C is a constant. From Eq. 15-10, the pressure p is

$$p = p_0 - \rho g h, \tag{19-3}$$

in which p_0 is the atmospheric pressure, ρ is the density of the mercury in the manometer, and h is the measured difference between the mercury levels in the two arms of the tube.*

If we next put the bulb in a triple-point cell (Fig. 19-4), the temperature now being measured is

$$T_3 = Cp_3, \tag{19-4}$$

in which p_3 is the gas pressure now. Eliminating C between Eqs. 19-2 and 19-4 gives us the temperature as

$$T = T_3\left(\frac{p}{p_3}\right) = (273.16\ \text{K})\left(\frac{p}{p_3}\right) \quad \text{(provisional)}. \tag{19-5}$$

We still have a problem with this thermometer. If we use it to measure, say, the boiling point of water, we find that different gases in the bulb give slightly different results. However, as we use smaller and smaller amounts of gas to fill the bulb, the readings converge nicely to a single temperature, no matter what gas we use. Figure 19-6 shows this convergence for three gases.

Thus the recipe for measuring a temperature with a gas thermometer is

$$T = (273.16\ \text{K})\left(\lim_{\text{gas}\to 0} \frac{p}{p_3}\right). \tag{19-6}$$

The recipe instructs us to measure an unknown temperature T as follows: Fill the thermometer bulb with an arbitrary amount of *any* gas (for example, nitrogen) and measure p_3 (using a triple-point cell) and p, the gas pressure at the temperature being

*For pressure units, we shall use units introduced in Section 15-3. The SI unit for pressure is the newton per square meter, which is called the pascal (Pa). The pascal is related to other common pressure units by

$$1\ \text{atm} = 1.01 \times 10^5\ \text{Pa} = 760\ \text{torr} = 14.7\ \text{lb/in.}^2.$$

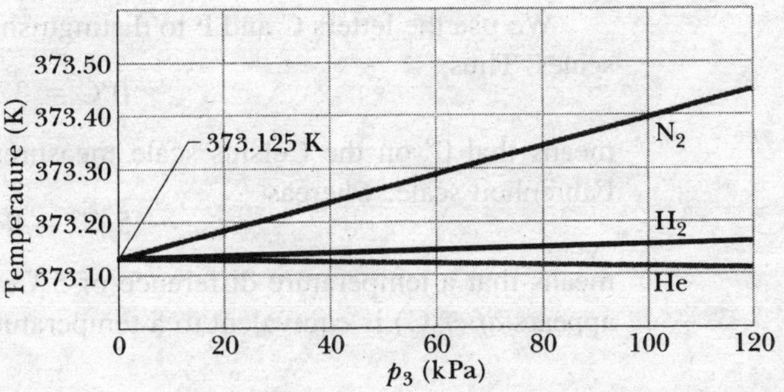

Fig. 19-6 Temperatures measured by a constant-volume gas thermometer, with its bulb immersed in boiling water. For temperature calculations using Eq. 19-5, pressure p_3 was measured at the triple point of water. Three different gases in the thermometer bulbs gave generally different results at different gas pressures, but as the amount of gas was decreased (decreasing p_3), all three curves converged to 373.125 K.

measured. (Keep the gas volume the same.) Calculate the ratio p/p_3. Then repeat both measurements with a smaller amount of gas in the bulb, and again calculate this ratio. Continue this way, using smaller and smaller amounts of gas, until you can extrapolate to the ratio p/p_3 that you would find if there were approximately no gas in the bulb. Calculate the temperature T by substituting that extrapolated ratio into Eq. 19-6. (The temperature is called the *ideal gas temperature*.)

TABLE 19-1 Some Corresponding Temperatures

Temperature	°C	°F
Boiling point of water[a]	100	212
Normal body temperature	37.0	98.6
Accepted comfort level	20	68
Freezing point of water[a]	0	32
Zero of Fahrenheit scale	≈ −18	0
Scales coincide	−40	−40

[a] Strictly, the boiling point of water on the Celsius scale is 99.975°C, and the freezing point is 0.00°C. Thus, there is slightly less than 100 C° between those two points.

19-4 The Celsius and Fahrenheit Scales

So far, we have discussed only the Kelvin scale, used in basic scientific work. In nearly all countries of the world, the Celsius scale (formerly called the centigrade scale) is the scale of choice for popular and commercial use and much scientific use. Celsius temperatures are measured in degrees, and the Celsius degree has the same size as the kelvin. However, the zero of the Celsius scale is shifted to a more convenient value than absolute zero. If T_C represents a Celsius temperature and T a Kelvin temperature, then

$$T_C = T - 273.15°. \qquad (19\text{-}7)$$

In expressing temperatures on the Celsius scale, the degree symbol is commonly used. Thus, we write 20.00°C for a Celsius reading but 293.15 K for a Kelvin reading.

The Fahrenheit scale, used in the United States, employs a smaller degree than the Celsius scale and a different zero of temperature. You can easily verify both these differences by examining an ordinary room thermometer on which both scales are marked. The relation between the Celsius and Fahrenheit scales is

$$T_F = \tfrac{9}{5}T_C + 32°, \qquad (19\text{-}8)$$

where T_F is Fahrenheit temperature. Transferring between these two scales can be done easily by remembering a few corresponding points, such as the freezing and boiling points of water (see Table 19-1). Figure 19-7 compares the Kelvin, Celsius, and Fahrenheit scales.

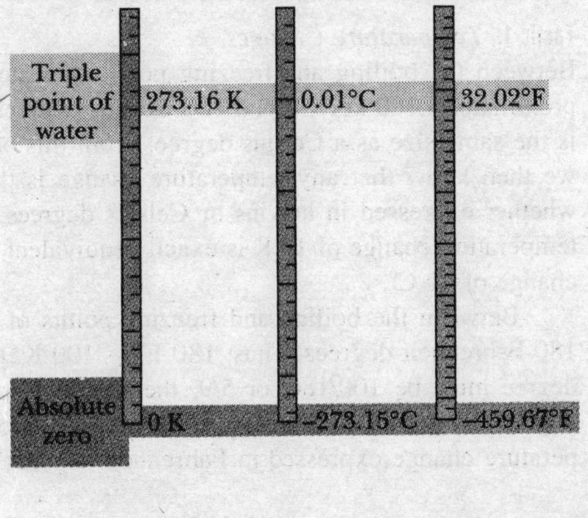

Fig. 19-7 The Kelvin, Celsius, and Fahrenheit temperature scales compared.

We use the letters C and F to distinguish measurements and degrees on the two scales. Thus,

$$0°\text{C} = 32°\text{F}$$

means that 0° on the Celsius scale measures the same temperature as 32° on the Fahrenheit scale, whereas

$$5\ \text{C}° = 9\ \text{F}°$$

means that a temperature difference of 5 Celsius degrees (note the degree symbol appears *after* C) is equivalent to a temperature difference of 9 Fahrenheit degrees.

Sample Problem 19-1

Suppose you come across old scientific notes that describe a temperature scale called Z on which the boiling point of water is 65.0°Z and the freezing point is −14.0°Z. To what temperature on the Fahrenheit scale would a temperature of $T = -98.0$°Z correspond? Assume that the Z scale is linear; that is, the size of a Z degree is the same everywhere on the Z scale.

SOLUTION: One **Key Idea** here is to relate the given temperature T to *either* of the two known temperatures on the Z scale. Since $T = -98.0$°Z is closer to the freezing point of −14.0°Z, we use that point for simplicity. Then we note that T is *below the freezing point* by the difference −14.0°Z − (−98.0°Z) = 84.0 Z° (Fig. 19-8). (Read this difference as "84.0 Z degrees.")

Another **Key Idea** is to set up a conversion factor between the Z and Fahrenheit scales to convert this difference. To do so, we use *both* known temperatures on the Z scale and the corresponding temperatures on the Fahrenheit scale. On the Z scale, the difference between the boiling and freezing points is 65.0°Z − (−14.0°Z) = 79.0 Z°. On the Fahrenheit scale, it is 212°F − 32.0°F = 180 F°. Thus, a temperature difference of 79.0 Z° is equivalent to a temperature difference of 180 F° (Fig. 19-8), and we can use the ratio (180 F°)/(79.0 Z°) as our conversion factor.

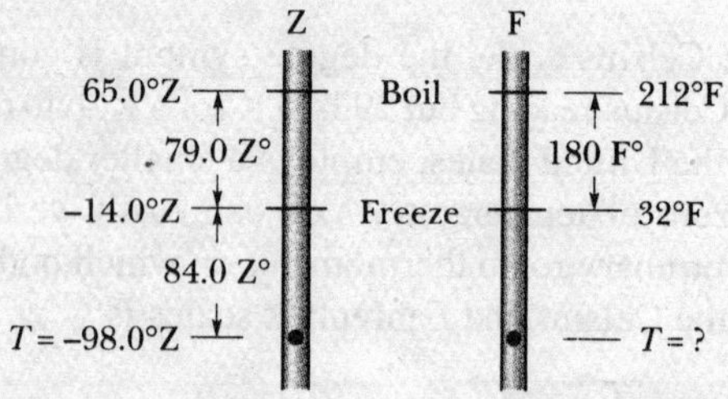

Fig. 19-8 Sample Problem 19-1. An unknown temperature scale compared to the Fahrenheit temperature scale.

Now, since T is below the freezing point by 84.0 Z°, it must also be below the freezing point by

$$(84.0\ \text{Z}°)\frac{180\ \text{F}°}{79.0\ \text{Z}°} = 191\ \text{F}°.$$

Because the freezing point is at 32.0°F, this means that

$$T = 32.0°\text{F} - 191\ \text{F}° = -159°\text{F}. \qquad \text{(Answer)}$$

✔CHECKPOINT 1: The figure here shows three temperature scales with the freezing and boiling points of water indicated. (a) Rank the degrees on these scales by size, greatest first. (b) Rank the following temperatures, highest first: 50°X, 50°W, and 50°Y.

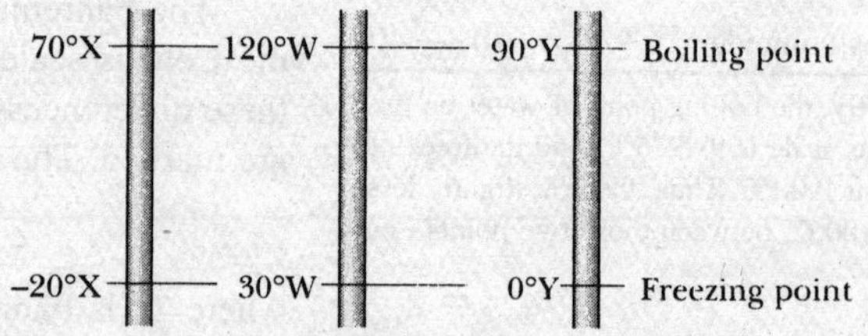

PROBLEM-SOLVING TACTICS

Tactic 1: *Temperature Changes*

Between the boiling and freezing points of water, there are (approximately) 100 kelvins and 100 Celsius degrees. Thus, a kelvin is the same size as a Celsius degree. From this or from Eq. 19-7, we then know that any temperature change is the same number whether expressed in kelvins or Celsius degrees. For example, a temperature change of 10 K is exactly equivalent to a temperature change of 10 C°.

Between the boiling and freezing points of water, there are 180 Fahrenheit degrees. Thus, 180 F° = 100 K, and a Fahrenheit degree must be 100/180, or 5/9, the size of a kelvin or Celsius degree. From this or from Eq. 19-8, we then know that any temperature change expressed in Fahrenheit degrees must be $\frac{9}{5}$ times that same temperature change expressed in either kelvins or Celsius degrees. For example, in Fahrenheit degrees, a temperature change of 10 K is (9/5)(10 K), or 18 F°.

You should take care not to confuse a *temperature* with a temperature *change* or *difference*. A temperature of 10 K is certainly not the same as one of 10°C or 18°F but, as above, a temperature *change* of 10 K is the same as one of 10 C° or 18 F°. This distinction is very important in an equation containing a temperature T instead of a temperature change or difference such as $T_2 - T_1$: A temperature T by itself should generally be in kelvins and not degrees Celsius or Fahrenheit. In short, beware the "bare T."

Fig. 19-9 Railroad tracks in Asbury Park, New Jersey, distorted because of thermal expansion on a very hot July day.

19-5 Thermal Expansion

You can often loosen a tight metal jar lid by holding it under a stream of hot water. Both the metal of the lid and the glass of the jar expand as the hot water adds energy to their atoms. (With the added energy, the atoms can move a bit farther from each other than usual, against the spring-like interatomic forces that hold every solid together.) However, because the atoms in the metal move farther apart than those in the glass, the lid expands more than the jar and thus is loosened.

Such **thermal expansion** is not always desirable, as Fig. 19-9 suggests. To preclude buckling, therefore, expansion slots are placed in bridges to accommodate roadway expansion on hot days. Dental materials used for fillings must be matched in their thermal expansion properties to those of tooth enamel (otherwise consuming hot coffee or cold ice cream would be quite painful). In aircraft manufacture, however, rivets and other fasteners are often cooled in dry ice before insertion and then allowed to expand to a tight fit.

Thermometers and thermostats may be based on the differences in expansion between the components of a *bimetal strip* (Fig. 19-10). Also, the familiar liquid-in-glass thermometers are based on the fact that liquids such as mercury and alcohol expand to a different (greater) extent than their glass containers.

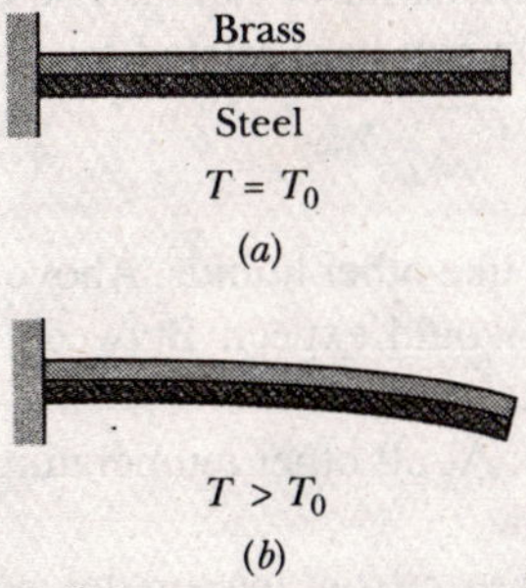

Fig. 19-10 (*a*) A bimetal strip, consisting of a strip of brass and a strip of steel welded together, at temperature T_0. (*b*) The strip bends as shown at temperatures above this reference temperature. Below the reference temperature the strip bends the other way. Many thermostats operate on this principle, making and breaking an electrical contact as the temperature rises and falls.

Linear Expansion

If the temperature of a metal rod of length L is raised by an amount ΔT, its length is found to increase by an amount

$$\Delta L = L\alpha\, \Delta T, \tag{19-9}$$

in which α is a constant called the **coefficient of linear expansion.** The coefficient α has the unit "per degree" or "per kelvin" and depends on the material. Although

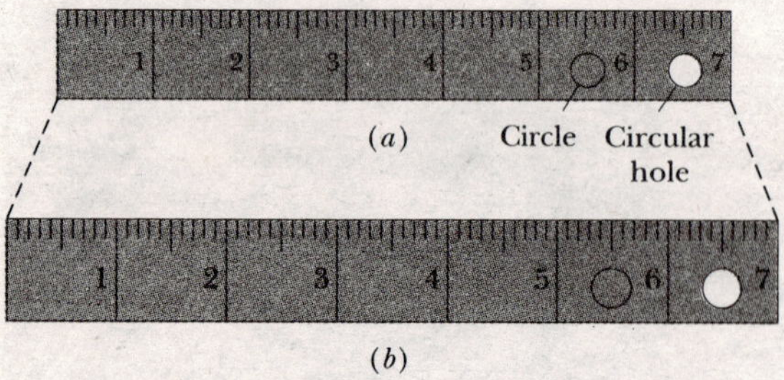

Fig. 19-11 The same steel ruler at two different temperatures. When it expands, the scale, the numbers, the thickness, and the diameters of the circle and circular hole are all increased by the same factor. (The expansion has been exaggerated for clarity.)

TABLE 19-2 Some Coefficients of Linear Expansion[a]

Substance	α (10^{-6}/C°)	Substance	α (10^{-6}/C°)
Ice (at 0°C)	51	Steel	11
Lead	29	Glass (ordinary)	9
Aluminum	23	Glass (Pyrex)	3.2
Brass	19	Diamond	1.2
Copper	17	Invar[b]	0.7
Concrete	12	Fused quartz	0.5

[a]Room temperature values except for the listing for ice.

[b]This alloy was designed to have a low coefficient of expansion. The word is a shortened form of "invariable."

α varies somewhat with temperature, for most practical purposes it can be taken as constant for a particular material. Table 19-2 shows some coefficients of linear expansion. Note that the unit C° there could be replaced with the unit K.

The thermal expansion of a solid is like (three-dimensional) photographic enlargement. Figure 19-11*b* shows the (exaggerated) expansion of a steel ruler after its temperature is increased from that of Fig. 19-11*a*. Equation 19-9 applies to every linear dimension of the ruler, including its edge, thickness, diagonals, and the diameters of the circle etched on it and the circular hole cut in it. If the disk cut from that hole originally fits snugly in the hole, it will continue to fit snugly if it undergoes the same temperature increase as the ruler.

Volume Expansion

If all dimensions of a solid expand with temperature, the volume of that solid must also expand. For liquids, volume expansion is the only meaningful expansion parameter. If the temperature of a solid or liquid whose volume is V is increased by an amount ΔT, the increase in volume is found to be

$$\Delta V = V\beta\,\Delta T, \tag{19-10}$$

where β is the **coefficient of volume expansion** of the solid or liquid. The coefficients of volume expansion and linear expansion for a solid are related by

$$\beta = 3\alpha. \tag{19-11}$$

The most common liquid, water, does not behave like other liquids. Above about 4°C, water expands as the temperature rises, as we would expect. Between 0 and about 4°C, however, water *contracts* with increasing temperature. Thus, at about 4°C, the density of water passes through a maximum. At all other temperatures, the density of water is less than this maximum value.

This behavior of water is the reason why lakes freeze from the top down rather than from the bottom up. As water on the surface is cooled from, say, 10°C toward the freezing point, it becomes denser ("heavier") than lower water and sinks to the bottom. Below 4°C, however, further cooling makes the water then on the surface *less* dense ("lighter") than the lower water, so it stays on the surface until it freezes. Thus the surface freezes while the lower water is still liquid. If lakes froze from the bottom up, the ice so formed would tend not to melt completely during the summer, because it would be insulated by the water above. After a few years, many bodies of open water in the temperate zones of Earth would be frozen solid all year round—and aquatic life as we know it could not exist.

CHECKPOINT 2: The figure here shows four rectangular metal plates, with sides of L, $2L$, or $3L$. They are all made of the same material, and their temperature is to be increased by the same amount. Rank the plates according to the expected increase in (a) their vertical heights and (b) their areas, greatest first.

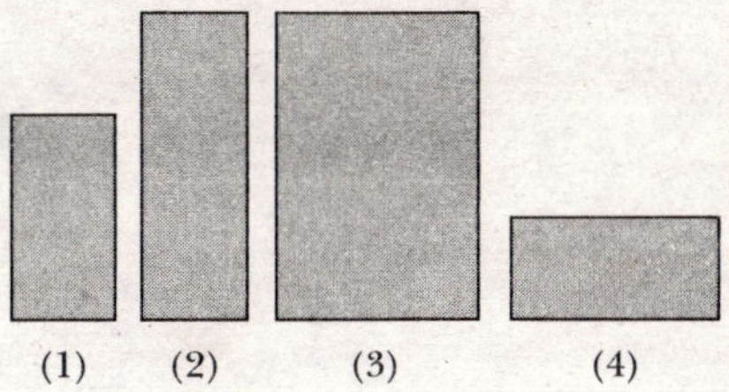

Sample Problem 19-2

On a hot day in Las Vegas, an oil trucker loaded 37,000 L of diesel fuel. He encountered cold weather on the way to Payson, Utah, where the temperature was 23.0 K lower than in Las Vegas, and where he delivered his entire load. How many liters did he deliver? The coefficient of volume expansion for diesel fuel is $9.50 \times 10^{-4}/\text{C}°$, and the coefficient of linear expansion for his steel truck tank is $11 \times 10^{-6}/\text{C}°$.

SOLUTION: The **Key Idea** here is that the volume of the diesel fuel depends directly on the temperature. Thus, because the temperature decreased, the volume of the fuel did also. From Eq. 19-10, the volume change is

$$\Delta V = V\beta\,\Delta T$$
$$= (37{,}000\text{ L})(9.50 \times 10^{-4}/\text{C}°)(-23.0\text{ K}) = -808\text{ L}.$$

Thus, the amount delivered was

$$V_{\text{del}} = V + \Delta V = 37{,}000\text{ L} - 808\text{ L}$$
$$= 36{,}190\text{ L}. \qquad \text{(Answer)}$$

Note that the thermal expansion of the steel tank has nothing to do with the problem. Question: Who paid for the "missing" diesel fuel?

19-6 Temperature and Heat

If you take a can of cola from the refrigerator and leave it on the kitchen table, its temperature will rise—rapidly at first but then more slowly—until the temperature of the cola equals that of the room (the two are then in thermal equilibrium). In the same way, the temperature of a cup of hot coffee, left sitting on the table, will fall until it also reaches room temperature.

In generalizing this situation, we describe the cola or the coffee as a *system* (with temperature T_S) and the relevant parts of the kitchen as the *environment* (with temperature T_E) of that system. Our observation is that if T_S is not equal to T_E, then T_S will change (T_E may also change some) until the two temperatures are equal and thus thermal equilibrium is reached.

Such a change in temperature is due to the transfer of energy between the thermal energy of the system and the system's environment. (*Thermal energy* is an internal energy that consists of the kinetic and potential energies associated with the random motions of the atoms, molecules, and other microscopic bodies within an object.) The transferred energy is called **heat** and is symbolized Q. Heat is *positive* when energy is transferred to a system's thermal energy from its environment (we say that heat is absorbed). Heat is *negative* when energy is transferred from a system's thermal energy to its environment (we say that heat is released or lost).

This transfer of energy is shown in Fig. 19-12. In the situation of Fig. 19-12*a*, in which $T_S > T_E$, energy is transferred from the system to the environment, so Q is negative. In Fig. 19-12*b*, in which $T_S = T_E$, there is no such transfer, Q is zero, and heat is neither released nor absorbed. In Fig. 19-12*c*, in which $T_S < T_E$, the transfer is to the system from the environment, so Q is positive.

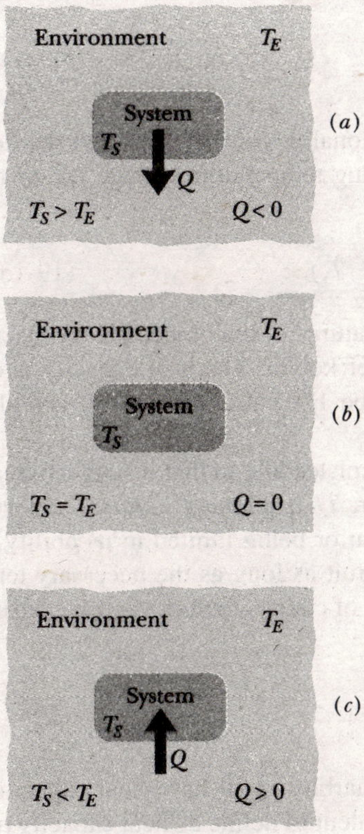

Fig. 19-12 If the temperature of a system exceeds that of its environment as in (*a*), heat Q is lost by the system to the environment until thermal equilibrium (*b*) is established. (*c*) If the temperature of the system is below that of the environment, heat is absorbed by the system until thermal equilibrium is established.

We are led then to this definition of heat:

Heat is the energy that is transferred between a system and its environment because of a temperature difference that exists between them.

Recall that energy can also be transferred between a system and its environment as *work W* via a force acting on a system. Heat and work, unlike temperature, pressure, and volume, are not intrinsic properties of a system. They have meaning only as they describe the transfer of energy into or out of a system. Thus, it is proper to say: "During the last 3 min, 15 J of heat was transferred to the system from its environment" or "During the last minute, 12 J of work was done on the system by its environment." It is meaningless to say: "This system contains 450 J of heat" or "This system contains 385 J of work."

Before scientists realized that heat is transferred energy, heat was measured in terms of its ability to raise the temperature of water. Thus, the **calorie** (cal) was defined as the amount of heat that would raise the temperature of 1 g of water from 14.5°C to 15.5°C. In the British system, the corresponding unit of heat was the **British thermal unit** (Btu), defined as the amount of heat that would raise the temperature of 1 lb of water from 63°F to 64°F.

In 1948, the scientific community decided that since heat (like work) is transferred energy, the SI unit for heat should be the one we use for energy, namely, the **joule.** The calorie is now defined to be 4.1860 J (exactly), with no reference to the heating of water. (The "calorie" used in nutrition, sometimes called the Calorie (Cal), is really a kilocalorie.) The relations among the various heat units are

$$1 \text{ cal} = 3.969 \times 10^{-3} \text{ Btu} = 4.1860 \text{ J}. \qquad (19\text{-}12)$$

19-7 The Absorption of Heat by Solids and Liquids

Heat Capacity

The **heat capacity** C of an object is the proportionality constant between the heat Q that the object absorbs or loses and the resulting temperature change ΔT of the object; that is,

$$Q = C\,\Delta T = C(T_f - T_i), \qquad (19\text{-}13)$$

in which T_i and T_f are the initial and final temperatures of the object. Heat capacity C has the unit of energy per degree or energy per kelvin. The heat capacity C of, say, a marble slab used in a bun warmer might be 179 cal/C°, which we can also write as 179 cal/K or as 749 J/K.

The word "capacity" in this context is really misleading in that it suggests analogy with the capacity of a bucket to hold water. *That analogy is false,* and you should not think of the object as "containing" heat or being limited in its ability to absorb heat. Heat transfer can proceed without limit as long as the necessary temperature difference is maintained. The object may, of course, melt or vaporize during the process.

Specific Heat

Two objects made of the same material—say, marble—will have heat capacities proportional to their masses. It is therefore convenient to define a "heat capacity per

TABLE 19-3 Specific Heats of Some Substances at Room Temperature

	Specific Heat		Molar Specific Heat
Substance	cal/g · K	J/kg · K	J/mol · K
Elemental Solids			
Lead	0.0305	128	26.5
Tungsten	0.0321	134	24.8
Silver	0.0564	236	25.5
Copper	0.0923	386	24.5
Aluminum	0.215	900	24.4
Other Solids			
Brass	0.092	380	
Granite	0.19	790	
Glass	0.20	840	
Ice (−10°C)	0.530	2220	
Liquids			
Mercury	0.033	140	
Ethyl alcohol	0.58	2430	
Seawater	0.93	3900	
Water	1.00	4190	

unit mass" or **specific heat** c that refers not to an object but to a unit mass of the material of which the object is made. Equation 19-13 then becomes

$$Q = cm\,\Delta T = cm(T_f - T_i). \tag{19-14}$$

Through experiment we would find that although the heat capacity of a particular marble slab might be 179 cal/C° (or 749 J/K), the specific heat of marble itself (in that slab or in any other marble object) is 0.21 cal/g · C° (or 880 J/kg · K).

From the way the calorie and the British thermal unit were initially defined, the specific heat of water is

$$c = 1 \text{ cal/g} \cdot \text{C}° = 1 \text{ Btu/lb} \cdot \text{F}° = 4190 \text{ J/kg} \cdot \text{K}. \tag{19-15}$$

Table 19-3 shows the specific heats of some substances at room temperature. Note that the value for water is relatively high. The specific heat of any substance actually depends somewhat on temperature, but the values in Table 19-3 apply reasonably well in a range of temperatures near room temperature.

CHECKPOINT 3: A certain amount of heat Q will warm 1 g of material A by 3 C° and 1 g of material B by 4 C°. Which material has the greater specific heat?

Molar Specific Heat

In many instances the most convenient unit for specifying the amount of a substance is the mole (mol), where

$$1 \text{ mol} = 6.02 \times 10^{23} \text{ elementary units}$$

of *any* substance. Thus 1 mol of aluminum means 6.02×10^{23} atoms (the atom being the elementary unit), and 1 mol of aluminum oxide means 6.02×10^{23} molecules of the oxide (because the molecule is the elementary unit of a compound).

When quantities are expressed in moles, specific heats must also involve moles (rather than a mass unit); they are then called **molar specific heats.** Table 19-3 shows the values for some elemental solids (each consisting of a single element) at room temperature.

An Important Point

In determining and then using the specific heat of any substance, we need to know the conditions under which energy is transferred as heat. For solids and liquids, we usually assume that the sample is under constant pressure (usually atmospheric) during the transfer. It is also conceivable that the sample is held at constant volume while the heat is absorbed. This means that thermal expansion of the sample is prevented by applying external pressure. For solids and liquids, this is very hard to arrange experimentally but the effect can be calculated, and it turns out that the specific heats under constant pressure and constant volume for any solid or liquid differ usually by no more than a few percent. Gases, as you will see, have quite different values for their specific heats under constant-pressure conditions and under constant-volume conditions.

Heats of Transformation

When energy is absorbed as heat by a solid or liquid, the temperature of the sample does not necessarily rise. Instead, the sample may change from one *phase*, or *state*, to another. Matter can exist in three common states: In the *solid state*, the molecules of a sample are locked into a fairly rigid structure by their mutual attraction. In the

TABLE 19-4 Some Heats of Transformation

	Melting		Boiling	
Substance	Melting Point (K)	Heat of Fusion L_F (kJ/kg)	Boiling Point (K)	Heat of Vaporization L_V (kJ/kg)
Hydrogen	14.0	58.0	20.3	455
Oxygen	54.8	13.9	90.2	213
Mercury	234	11.4	630	296
Water	273	333	373	2256
Lead	601	23.2	2017	858
Silver	1235	105	2323	2336
Copper	1356	207	2868	4730

liquid state, the molecules have more energy and move about more. They may form brief clusters, but the sample does not have a rigid structure and can flow or settle into a container. In the *gas* or *vapor state*, the molecules have even more energy, are free of one another, and can fill up the full volume of a container.

To *melt* a solid means to change it from the solid state to the liquid state. The process requires energy because the molecules of the solid must be freed from their rigid structure. Melting an ice cube to form liquid water is a common example. To *freeze* a liquid to form a solid is the reverse of melting and requires that energy be removed from the liquid, so that the molecules can settle into a rigid structure.

To *vaporize* a liquid means to change it from the liquid state to the vapor or gas state. This process, like melting, requires energy because the molecules must be freed from their clusters. Boiling liquid water to transfer it to water vapor (or steam—a gas of individual water molecules) is a common example. *Condensing* a gas to form a liquid is the reverse of vaporizing; it requires that energy be removed from the gas, so that the molecules can cluster instead of flying away from one another.

The amount of energy per unit mass that must be transferred as heat when a sample completely undergoes a phase change is called the **heat of transformation** *L*. Thus, when a sample of mass *m* completely undergoes a phase change, the total energy transferred is

$$Q = Lm. \qquad (19\text{-}16)$$

When the phase change is from liquid to gas (then the sample must absorb heat) or from gas to liquid (then the sample must release heat), the heat of transformation is called the **heat of vaporization** L_V. For water at its normal boiling or condensation temperature,

$$L_V = 539 \text{ cal/g} = 40.7 \text{ kJ/mol} = 2256 \text{ kJ/kg}. \qquad (19\text{-}17)$$

When the phase change is from solid to liquid (then the sample must absorb heat) or from liquid to solid (then the sample must release heat), the heat of transformation is called the **heat of fusion** L_F. For water at its normal freezing or melting temperature,

$$L_F = 79.5 \text{ cal/g} = 6.01 \text{ kJ/mol} = 333 \text{ kJ/kg}. \qquad (19\text{-}18)$$

Table 19-4 shows the heats of transformation for some substances.

Sample Problem 19-3

(a) How much heat must be absorbed by ice of mass $m = 720$ g at $-10°C$ to take it to liquid state at 15°C?

SOLUTION: The first **Key Idea** is that the heating process is accomplished in three steps.

Step 1. The **Key Idea** here is that the ice cannot melt at a temperature below the freezing point—so initially, any energy transferred to the ice as heat can only increase the temperature of the ice. The heat Q_1 needed to increase that temperature from the initial value $T_i = -10°C$ to a final value $T_f = 0°C$ (so that the ice can then melt) is given by Eq. 19-14 ($Q = cm\,\Delta T$). Using the specific heat of ice c_{ice} in Table 19-3 gives us

$$\begin{aligned} Q_1 &= c_{ice}m(T_f - T_i) \\ &= (2220\ \text{J/kg}\cdot\text{K})(0.720\ \text{kg})[0°\text{C} - (-10°\text{C})] \\ &= 15{,}984\ \text{J} \approx 15.98\ \text{kJ}. \end{aligned}$$

Step 2. The next **Key Idea** is that the temperature cannot increase from 0°C until all the ice melts—so any energy transferred to the ice as heat now can only change ice to liquid water. The heat Q_2 needed to melt all the ice is given by Eq. 19-16 ($Q = Lm$). Here L is the heat of fusion L_F, with the value given in Eq. 19-18 and Table 19-4. We find

$$Q_2 = L_F m = (333\ \text{kJ/kg})(0.720\ \text{kg}) \approx 239.8\ \text{kJ}.$$

Step 3. Now we have liquid water at 0°C. The next **Key Idea** is that the energy transferred to the liquid water as heat now can only increase the temperature of the liquid water. The heat Q_3 needed to increase the temperature of the water from the initial value $T_i = 0°C$ to the final value $T_f = 15°C$ is given by Eq. 19-14 (with the specific heat of liquid water c_{liq}):

$$\begin{aligned} Q_3 &= c_{liq}m(T_f - T_i) \\ &= (4190\ \text{J/kg}\cdot\text{K})(0.720\ \text{kg})(15°\text{C} - 0°\text{C}) \\ &= 45{,}252\ \text{J} \approx 45.25\ \text{kJ}. \end{aligned}$$

The total required heat Q_{tot} is the sum of the amounts required in the three steps:

$$\begin{aligned} Q_{tot} &= Q_1 + Q_2 + Q_3 \\ &= 15.98\ \text{kJ} + 239.8\ \text{kJ} + 45.25\ \text{kJ} \\ &\approx 300\ \text{kJ}. \end{aligned} \qquad \text{(Answer)}$$

Note that the heat required to melt the ice is much greater than the heat required to raise the temperature of either the ice or the liquid water.

(b) If we supply the ice with a total energy of only 210 kJ (as heat), what then are the final state and temperature of the water?

SOLUTION: From step 1, we know that 15.98 kJ is needed to raise the temperature of the ice to the melting point. The remaining heat Q_{rem} is then 210 kJ − 15.98 kJ, or about 194 kJ. From step 2, we can see that this amount of heat is insufficient to melt all the ice. Then this **Key Idea** becomes important: Because the melting of the ice is incomplete, we must end up with a mixture of ice and liquid; the temperature of the mixture must be the freezing point, 0°C.

We can find the mass m of ice that is melted by the available energy Q_{rem} by using Eq. 19-16 with L_F:

$$m = \frac{Q_{rem}}{L_F} = \frac{194\ \text{kJ}}{333\ \text{kJ/kg}} = 0.583\ \text{kg} \approx 580\ \text{g}.$$

Thus, the mass of the ice that remains is 720 g − 580 g, or 140 g, and we have

580 g water and 140 g ice, at 0°C. (Answer)

Sample Problem 19-4

A copper slug whose mass m_c is 75 g is heated in a laboratory oven to a temperature T of 312°C. The slug is then dropped into a glass beaker containing a mass $m_w = 220$ g of water. The heat capacity C_b of the beaker is 45 cal/K. The initial temperature T_i of the water and the beaker is 12°C. Assuming that the slug, beaker, and water are an isolated system and the water does not vaporize, find the final temperature T_f of the system at thermal equilibrium.

SOLUTION: One **Key Idea** here is that, with the system isolated, only internal transfers of energy can occur. There are three such transfers, all as heat. The slug loses energy, the water gains energy, and the beaker gains energy. Another **Key Idea** is that, because these transfers do not involve a phase change, the energy transfers can only change the temperatures. To relate the transfers to the temperature changes, we can use Eqs. 19-13 and 19-14 to write

$$\text{for the water: } Q_w = c_w m_w (T_f - T_i); \qquad (19\text{-}19)$$

$$\text{for the beaker: } Q_b = C_b (T_f - T_i); \qquad (19\text{-}20)$$

$$\text{for the copper: } Q_c = c_c m_c (T_f - T). \qquad (19\text{-}21)$$

A third **Key Idea** is that, with the system isolated, the total energy of the system cannot change. This means that the sum of these three energy transfers is zero:

$$Q_w + Q_b + Q_c = 0. \qquad (19\text{-}22)$$

Substituting Eqs. 19-19 through 19-21 into Eq. 19-22 yields

$$c_w m_w (T_f - T_i) + C_b (T_f - T_i) + c_c m_c (T_f - T) = 0. \qquad (19\text{-}23)$$

Temperatures are contained in Eq. 19-23 only as differences. Thus, because the differences on the Celsius and Kelvin scales are identical, we can use either of those scales in this equation. Solving it for T_f, we obtain

$$T_f = \frac{c_c m_c T + C_b T_i + c_w m_w T_i}{c_w m_w + C_b + c_c m_c}.$$

Using Celsius temperatures and taking values for c_c and c_w from Table 19-3, we find the numerator to be

$$(0.0923\ \text{cal/g}\cdot\text{K})(75\ \text{g})(312°\text{C}) + (45\ \text{cal/K})(12°\text{C}) + (1.00\ \text{cal/g}\cdot\text{K})(220\ \text{g})(12°\text{C}) = 5339.8\ \text{cal},$$

and the denominator to be

$$(1.00\ \text{cal/g}\cdot\text{K})(220\ \text{g}) + 45\ \text{cal/K} + (0.0923\ \text{cal/g}\cdot\text{K})(75\ \text{g}) = 271.9\ \text{cal/C°}.$$

We then have

$$T_f = \frac{5339.8\ \text{cal}}{271.9\ \text{cal/C°}} = 19.6°\text{C} \approx 20°\text{C}. \qquad \text{(Answer)}$$

From the given data you can show that

$$Q_w \approx 1670\ \text{cal}, \quad Q_b \approx 342\ \text{cal}, \quad Q_c \approx -2020\ \text{cal}.$$

Apart from rounding errors, the algebraic sum of these three heat transfers is indeed zero, as Eq. 19-22 requires.

19-8 A Closer Look at Heat and Work

Here we look in some detail at how energy can be transferred as heat and work between a system and its environment. Let us take as our system a gas confined to a cylinder with a movable piston, as in Fig. 19-13. The upward force on the piston due to the pressure of the confined gas is equal to the weight of lead shot loaded onto the top of the piston. The walls of the cylinder are made of insulating material that does not allow any transfer of energy as heat. The bottom of the cylinder rests on a reservoir for thermal energy, a *thermal reservoir* (perhaps a hot plate) whose temperature T you can control by turning a knob.

The system (the gas) starts from an *initial state i*, described by a pressure p_i, a volume V_i, and a temperature T_i. You want to change the system to a *final state f*, described by a pressure p_f, a volume V_f, and a temperature T_f. The procedure by which you change the system from its initial state to its final state is called a *thermodynamic process*. During such a process, energy may be transferred into the system from the thermal reservoir (positive heat) or vice versa (negative heat). Also, work can be done by the system to raise the loaded piston (positive work) or lower it (negative work). We assume that all such changes occur slowly, with the result that the system is always in (approximate) thermal equilibrium (that is, every part of the system is always in thermal equilibrium with every other part).

Suppose that you remove a few lead shot from the piston of Fig. 19-13, allowing the gas to push the piston and remaining shot upward through a differential displacement $d\vec{s}$ with an upward force $\vec{F}$. Since the displacement is tiny, we can assume that $\vec{F}$ is constant during the displacement. Then $\vec{F}$ has a magnitude that is equal to pA, where p is the pressure of the gas and A is the face area of the piston. The differential work dW done by the gas during the displacement is

$$dW = \vec{F} \cdot d\vec{s} = (pA)(ds) = p(A\,ds)$$
$$= p\,dV, \qquad (19\text{-}24)$$

in which dV is the differential change in the volume of the gas owing to the movement of the piston. When you have removed enough shot to allow the gas to change its volume from V_i to V_f, the total work done by the gas is

$$W = \int dW = \int_{V_i}^{V_f} p\,dV. \qquad (19\text{-}25)$$

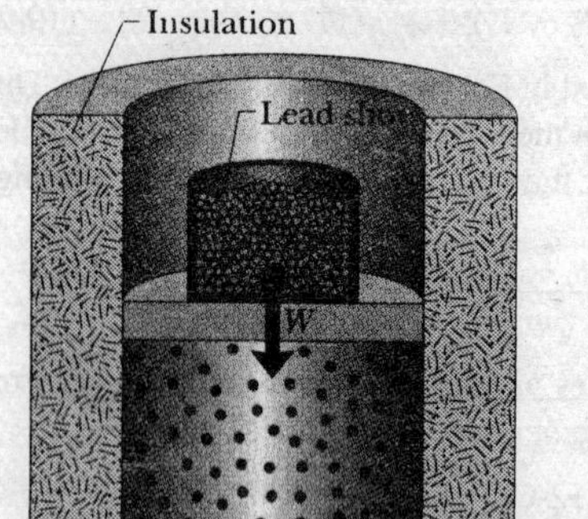

Fig. 19-13 A gas is confined to a cylinder with a movable piston. Heat Q can be added to, or withdrawn from, the gas by regulating the temperature T of the adjustable thermal reservoir. Work W can be done by the gas by raising or lowering the piston.

During the change in volume, the pressure and temperature of the gas may also change. To evaluate the integral in Eq. 19-25 directly, we would need to know how pressure varies with volume for the actual process by which the system changes from state i to state f.

There are actually many ways to take the gas from state i to state f. One way is shown in Fig. 19-14*a*, which is a plot of the pressure of the gas versus its volume and which is called a *p-V diagram*. In Fig. 19-14*a*, the curve indicates that the pressure decreases as the volume increases. The integral Eq. 19-25 (and thus the work W done by the gas) is represented by the shaded area under the curve between points i and f. Regardless of what exactly we do to take the gas along the curve, that work is positive, owing to the fact that the gas increases its volume by forcing the piston upward.

Another way to get from state i to state f is shown in Fig. 19-14*b*. There the change takes place in two steps—the first from state i to state a, and the second from state a to state f.

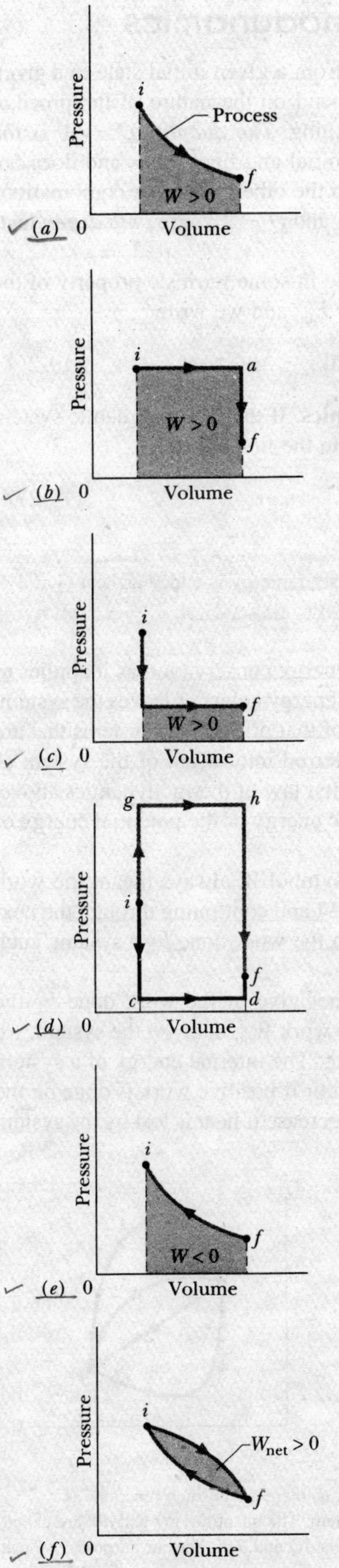

Fig. 19-14 (*a*) The shaded area represents the work W done by a system as it goes from an initial state i to a final state f. Work W is positive because the system's volume increases. (*b*) W is still positive, but now greater. (*c*) W is still positive, but now smaller. (*d*) W can be even smaller (path *icdf*) or larger (path *ighf*). (*e*) Here the system goes from state f to state i, as the gas is compressed to less volume by an external force. The work W done *by* the system is now negative. (*f*) The net work W_{net} done by the system during a complete cycle is represented by the shaded area.

Step *ia* of this process is carried out at constant pressure, which means that you leave undisturbed the lead shot that ride on top of the piston in Fig. 19-13. You cause the volume to increase (from V_i to V_f) by slowly turning up the temperature control knob, raising the temperature of the gas to some higher value T_a. (Increasing the temperature increases the force from the gas on the piston, moving it upward.) During this step, positive work is done by the expanding gas (to lift the loaded piston) and heat is absorbed by the system from the thermal reservoir (in response to the arbitrarily small temperature differences that you create as you turn up the temperature). This heat is positive because it is added to the system.

Step *af* of the process of Fig. 19-14*b* is carried out at constant volume, so you must wedge the piston, preventing it from moving. Then as you use the control knob to decrease the temperature, you find that the pressure drops from p_a to its final value p_f. During this step, heat is lost by the system to the thermal reservoir.

For the overall process *iaf*, the work W, which is positive and is carried out only during step *ia*, is represented by the shaded area under the curve. Energy is transferred as heat during both steps *ia* and *af*, with a net energy transfer Q.

Figure 19-14*c* shows a process in which the previous two steps are carried out in reverse order. The work W in this case is smaller than for Fig. 19-14*b*, as is the net heat absorbed. Figure 19-14*d* suggests that you can make the work done by the gas as small as you want (by following a path like *icdf*) or as large as you want (by following a path like *ighf*).

To sum up: A system can be taken from a given initial state to a given final state by an infinite number of processes. Heat may or may not be involved, and in general, the work W and the heat Q will have different values for different processes. We say that heat and work are *path-dependent* quantities.

Figure 19-14*e* shows an example in which negative work is done by a system as some external force compresses the system, reducing its volume. The absolute value of the work done is still equal to the area beneath the curve, but because the gas is *compressed*, the work done by the gas is negative.

Figure 19-14*f* shows a *thermodynamic cycle* in which the system is taken from some initial state i to some other state f and then back to i. The net work done by the system during the cycle is the sum of the *positive* work done during the expansion and the *negative* work done during the compression. In Fig. 19-14*f*, the net work is positive because the area under the expansion curve (i to f) is greater than the area under the compression curve (f to i).

CHECKPOINT 4: The p-V diagram here shows six curved paths (connected by vertical paths) that can be followed by a gas. Which two of them should be part of a closed cycle if the net work done by the gas is to be at its maximum positive value?

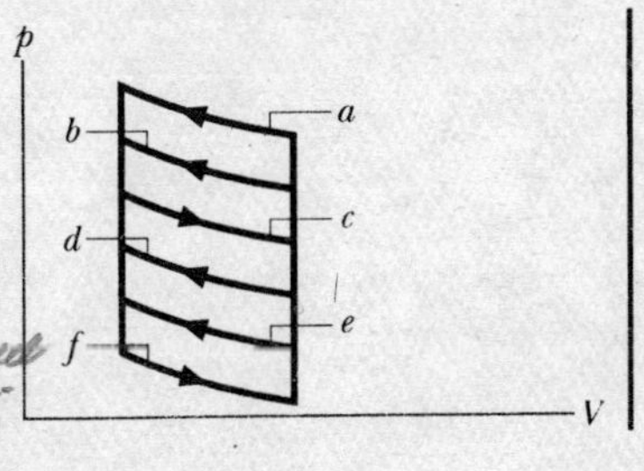

19-9 The First Law of Thermodynamics

You have just seen that when a system changes from a given initial state to a given final state, both the work W and the heat Q depend on the nature of the process. Experimentally, however, we find a surprising thing. *The quantity $Q - W$ is the same for all processes.* It depends only on the initial and final states and does not depend at all on how the system gets from one to the other. All other combinations of Q and W, including Q alone, W alone, $Q + W$, and $Q - 2W$, are *path dependent;* only the quantity $Q - W$ is not.

The quantity $Q - W$ must represent a change in some intrinsic property of the system. We call this property the *internal energy* E_{int} and we write

$$\Delta E_{\text{int}} = E_{\text{int},f} - E_{\text{int},i} = Q - W \quad \text{(first law).} \tag{19-26}$$

Equation 19-26 is the **first law of thermodynamics.** If the thermodynamic system undergoes only a differential change, we can write the first law as*

$$dE_{\text{int}} = dQ - dW \quad \text{(first law).} \tag{19-27}$$

> The internal energy E_{int} of a system tends to increase if energy is added as heat Q and tends to decrease if energy is lost as work W done by the system.

In Chapter 8, we discussed the principle of energy conservation as it applies to isolated systems—that is, to systems in which no energy enters or leaves the system. The first law of thermodynamics is an extension of that principle to systems that are *not* isolated. In such cases, energy may be transferred into or out of the system as either work W or heat Q. In our statement of the first law of thermodynamics above, we assume that there are no changes in the kinetic energy or the potential energy of the system as a whole; that is, $\Delta K = \Delta U = 0$.

Before this chapter, the term *work* and the symbol W always meant the work done *on* a system. However, starting with Eq. 19-24 and continuing through the next two chapters about thermodynamics, we focus on the work done *by* a system, such as the gas in Fig. 19-13.

The work done *on* a system is always the negative of the work done *by* the system, so if we rewrite Eq. 19-26 in terms of the work W_{on} done *on* the system, we have $\Delta E_{\text{int}} = Q + W_{\text{on}}$. This tells us the following: The internal energy of a system tends to increase if heat is absorbed by the system or if positive work is done *on* the system. Conversely, the internal energy tends to decrease if heat is lost by the system or if negative work is done *on* the system.

CHECKPOINT 5: The figure here shows four paths on a p-V diagram along which a gas can be taken from state i to state f. Rank the paths according to (a) the change ΔE_{int}, (b) the work W done by the gas, and (c) the magnitude of the energy transferred as heat Q, greatest first.

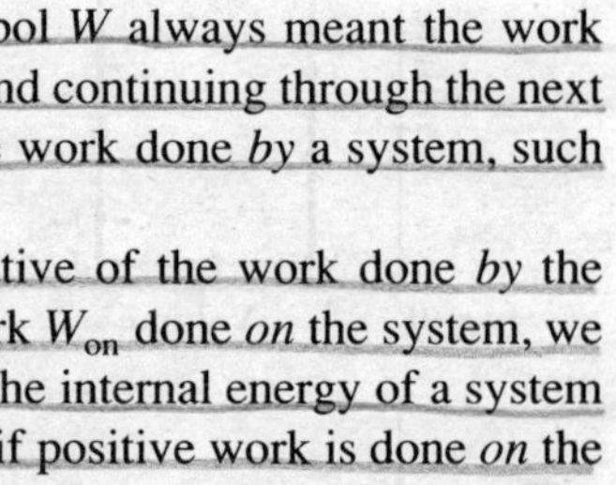

a) all tie ∵ ΔEint depends only on i & f

(b) V ↑ in all cases ⇒ work is done by the gas in all cases
we compare area under the curves (even for 1)
⇒ W4 > W3 > W2 > W1

(c) Q = ΔEint + W
⇒ Q4 > Q3 > Q2 > Q1

*Here dQ and dW, unlike dE_{int}, are not true differentials; that is, there are no such functions as $Q(p, V)$ and $W(p, V)$ that depend only on the state of the system. The quantities dQ and dW are called *inexact differentials* and are usually represented by the symbols $\bar{d}Q$ and $\bar{d}W$. For our purposes, we can treat them simply as infinitesimally small energy transfers.

19-10 Some Special Cases of the First Law of Thermodynamics

Here we look at four different thermodynamic processes, in each of which a certain restriction is imposed on the system. We then see what consequences follow when we apply the first law of thermodynamics to the process. The results are summarized in Table 19-5.

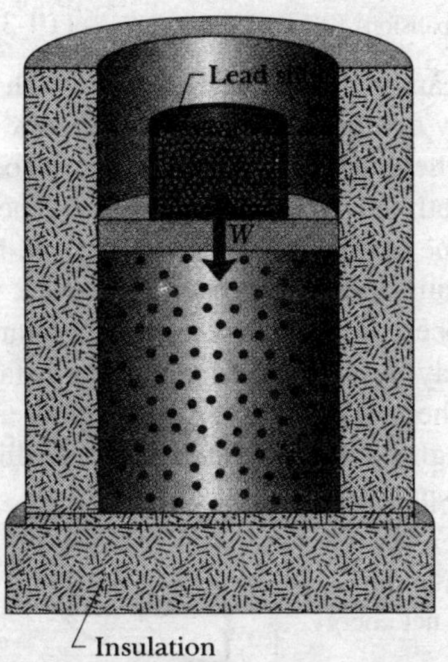

Fig. 19-15 An adiabatic expansion can be carried out by slowly removing lead shot from the top of the piston. Adding lead shot reverses the process at any stage.

1. ***Adiabatic processes.*** An adiabatic process is one that occurs so rapidly or occurs in a system that is so well insulated that *no transfer of energy as heat* occurs between the system and its environment. Putting $Q = 0$ in the first law (Eq. 19-26) yields

$$\Delta E_{\text{int}} = -W \qquad \text{(adiabatic process).} \tag{19-28}$$

This tells us that if work is done *by* the system (that is, if W is positive), the internal energy of the system decreases by the amount of work. Conversely, if work is done *on* the system (that is, if W is negative), the internal energy of the system increases by that amount.

Figure 19-15 shows an idealized adiabatic process. Heat cannot enter or leave the system because of the insulation. Thus, the only way energy can be transferred between the system and its environment is by work. If we remove shot from the piston and allow the gas to expand, the work done by the system (the gas) is positive and the internal energy of the gas decreases. If, instead, we add shot and compress the gas, the work done by the system is negative and the internal energy of the gas increases.

2. ***Constant-volume processes.*** If the volume of a system (such as a gas) is held constant, that system can do no work. Putting $W = 0$ in the first law (Eq. 19-26) yields

$$\Delta E_{\text{int}} = Q \qquad \text{(constant-volume process).} \tag{19-29}$$

Thus, if heat is absorbed by a system (that is, if Q is positive), the internal energy of the system increases. Conversely, if heat is lost during the process (that is, if Q is negative), the internal energy of the system must decrease.

3. ***Cyclical processes.*** There are processes in which, after certain interchanges of heat and work, the system is restored to its initial state. In that case, no intrinsic property of the system—including its internal energy—can possibly change. Putting $\Delta E_{\text{int}} = 0$ in the first law (Eq. 19-26) yields

$$Q = W \qquad \text{(cyclical process).} \tag{19-30}$$

Thus, the net work done during the process must exactly equal the net amount of energy transferred as heat; the store of internal energy of the system remains unchanged. Cyclical processes form a closed loop on a p-V plot, as shown in Fig. 19-14f. We shall discuss such processes in some detail in Chapter 21.

TABLE 19-5 The First Law of Thermodynamics: Four Special Cases

The Law: $\Delta E_{\text{int}} = Q - W$ (Eq. 19-26)

Process	Restriction	Consequence
Adiabatic	$Q = 0$	$\Delta E_{\text{int}} = -W$
Constant volume	$W = 0$	$\Delta E_{\text{int}} = Q$
Closed cycle	$\Delta E_{\text{int}} = 0$	$Q = W$
Free expansion	$Q = W = 0$	$\Delta E_{\text{int}} = 0$

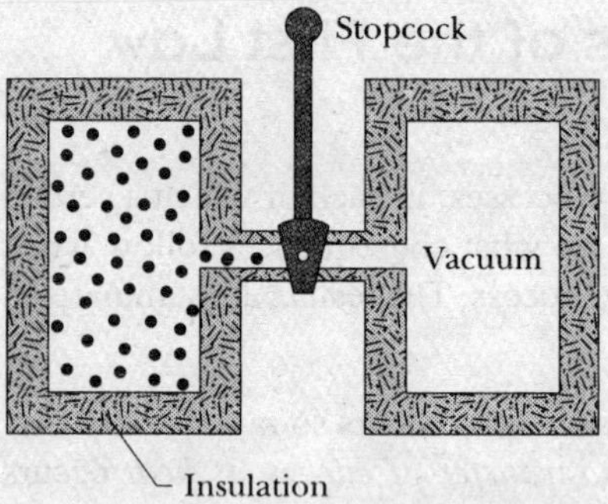

Fig. 19-16 The initial stage of a free-expansion process. After the stopcock is opened, the gas fills both chambers and eventually reaches an equilibrium state.

4. ***Free expansions.*** These are adiabatic processes in which no transfer of heat occurs between the system and its environment and no work is done on or by the system. Thus, $Q = W = 0$ and the first law requires that

$$\Delta E_{\text{int}} = 0 \quad \text{(free expansion)}. \tag{19-31}$$

Figure 19-16 shows how such an expansion can be carried out. A gas, which is in thermal equilibrium within itself, is initially confined by a closed stopcock to one half of an insulated double chamber; the other half is evacuated. The stopcock is opened, and the gas expands freely to fill both halves of the chamber. No heat is transferred to or from the gas because of the insulation. No work is done by the gas because it rushes into a vacuum and thus does not meet any pressure.

A free expansion differs from all other processes we have considered because it cannot be done slowly and in a controlled way. As a result, at any given instant during the sudden expansion, the gas is not in thermal equilibrium and its pressure is not the same everywhere. Therefore, although we can plot the initial and final states on a p-V diagram, we cannot plot the expansion itself.

CHECKPOINT 6: For one complete cycle as shown in the p-V diagram here, are (a) ΔE_{int} for the gas and (b) the net energy transferred as heat Q positive, negative, or zero?

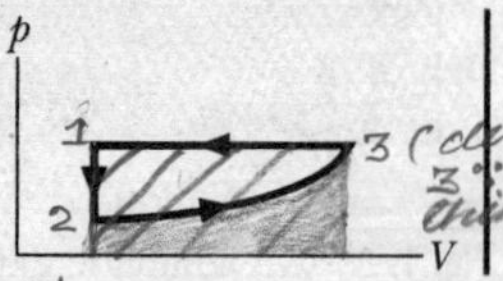

Sample Problem 19-5

Let 1.00 kg of liquid water at 100°C be converted to steam at 100°C by boiling at standard atmospheric pressure (which is 1.00 atm or 1.01×10^5 Pa) in the arrangement of Fig. 19-17. The volume of that water changes from an initial value of 1.00×10^{-3} m^3 as a liquid to 1.671 m^3 as steam.

(a) How much work is done by the system during this process?

SOLUTION: The **Key Idea** here is that the system must do positive work because the volume increases. In the general case we would calculate the work W done by integrating the pressure with respect to the volume (Eq. 19-25). However, here the pressure is constant at 1.01×10^5 Pa, so we can take p outside the integral. We then have

$$W = \int_{V_i}^{V_f} p\, dV = p \int_{V_i}^{V_f} dV = p(V_f - V_i)$$
$$= (1.01 \times 10^5 \text{ Pa})(1.671 \text{ m}^3 - 1.00 \times 10^{-3} \text{ m}^3)$$
$$= 1.69 \times 10^5 \text{ J} = 169 \text{ kJ}. \qquad \text{(Answer)}$$

(b) How much energy is transferred as heat during the process?

SOLUTION: The **Key Idea** here is that the heat causes only a phase change and not a change in temperature, so it is given fully by Eq. 19-16 ($Q = Lm$). Because the change is from liquid to gaseous phase, L is the heat of vaporization L_V, with the value given in Eq. 19-17 and Table 19-4. We find

$$Q = L_V m = (2256 \text{ kJ/kg})(1.00 \text{ kg})$$
$$= 2256 \text{ kJ} \approx 2260 \text{ kJ}. \qquad \text{(Answer)}$$

(c) What is the change in the system's internal energy during the process?

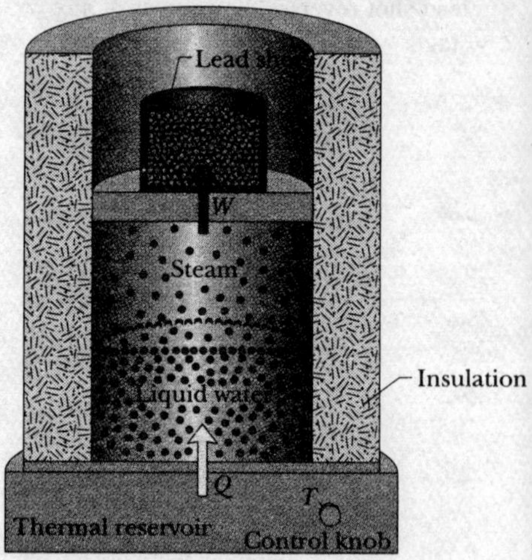

Fig. 19-17 Sample Problem 19-5. Water boiling at constant pressure. Energy is transferred from the thermal reservoir as heat until the liquid water has changed completely into steam. Work is done by the expanding gas as it lifts the loaded piston.

SOLUTION: The **Key Idea** here is that the change in the system's internal energy is related to the heat (here, this is energy transferred into the system) and the work (here, this is energy transferred out of the system) by the first law of thermodynamics (Eq. 19-26). Thus, we can write

$$\Delta E_{\text{int}} = Q - W = 2256 \text{ kJ} - 169 \text{ kJ}$$
$$\approx 2090 \text{ kJ} = 2.09 \text{ MJ}. \qquad \text{(Answer)}$$

This quantity is positive, indicating that the internal energy of the system has increased during the boiling process. This energy goes into separating the H_2O molecules, which strongly attract each other in the liquid state. We see that, when water is boiled, about 7.5% (= 169 kJ/2260 kJ) of the heat goes into the work of pushing back the atmosphere. The rest of the heat goes into the system's internal energy.

19-11 Heat Transfer Mechanisms

We have discussed the transfer of energy as heat between a system and its environment, but we have not yet described how that transfer takes place. There are three transfer mechanisms: conduction, convection, and radiation.

Conduction

If you leave the end of a metal poker in a fire for enough time, its handle will get hot. Energy is transferred from the fire to the handle by (thermal) **conduction** along the length of the poker. The vibration amplitudes of the atoms and electrons of the metal at the fire end of the poker become relatively large because of the high temperature of their environment. These increased vibrational amplitudes, and thus the associated energy, are passed along the poker, from atom to atom, during collisions between adjacent atoms. In this way, a region of rising temperature extends itself along the poker to the handle.

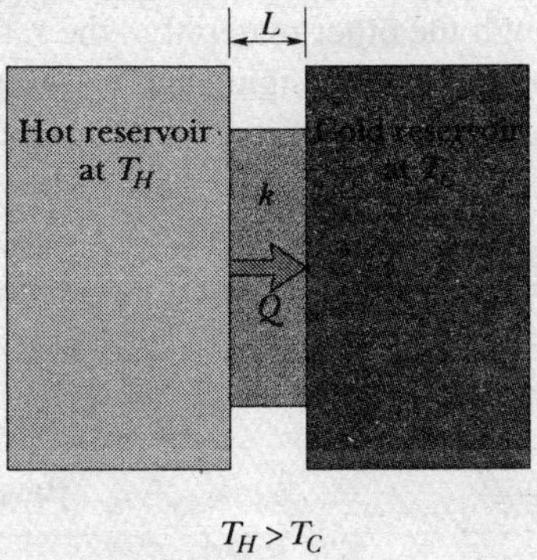

Fig. 19-18 Thermal conduction. Energy is transferred as heat from a reservoir at temperature T_H to a cooler reservoir at temperature T_C through a conducting slab of thickness L and thermal conductivity k.

Consider a slab of face area A and thickness L, whose faces are maintained at temperatures T_H and T_C by a hot reservoir and a cold reservoir, as in Fig. 19-18. Let Q be the energy that is transferred as heat through the slab, from its hot face to its cold face, in time t. Experiment shows that the *conduction rate* P_{cond} (the amount of energy transferred per unit time) is

$$P_{\text{cond}} = \frac{Q}{t} = kA\,\frac{T_H - T_C}{L}, \tag{19-32}$$

in which k, called the *thermal conductivity*, is a constant that depends on the material of which the slab is made. A material that readily transfers energy by conduction is a *good thermal conductor* and has a high value of k. Table 19-6 gives the thermal conductivities of some common metals, gases, and building materials.

TABLE 19-6 Some Thermal Conductivities[a]

Substance	k (W/m · K)
Metals	
Stainless steel	14
Lead	35
Aluminum	235
Copper	401
Silver	428
Gases	
Air (dry)	0.026
Helium	0.15
Hydrogen	0.18
Building Materials	
Polyurethane form	0.024
Rock wool	0.043
Fiberglass	0.048
White pine	0.11
Window glass	1.0

Thermal Resistance to Conduction (R-Value)

If you are interested in insulating your house or in keeping cola cans cold on a picnic, you are more concerned with poor heat conductors than with good ones. For this reason, the concept of *thermal resistance R* has been introduced into engineering practice. The R-value of a slab of thickness L is defined as

$$R = \frac{L}{k}. \tag{19-33}$$

The lower the thermal conductivity of the material of which a slab is made, the higher the R-value of the slab, so something that has a high R-value is a *poor thermal conductor* and thus a *good thermal insulator*.

Note that R is a property attributed to a slab of a specified thickness, not to a material. The commonly used unit for R (which, in the United States at least, is almost never stated) is the square foot–Fahrenheit degree–hour per British thermal unit ($\text{ft}^2 \cdot \text{F}^\circ \cdot \text{h/Btu}$). (Now you know why the unit is rarely stated.)

Conduction Through a Composite Slab

Figure 19-19 shows a composite slab, consisting of two materials having different thicknesses L_1 and L_2 and different thermal conductivities k_1 and k_2. The temperatures of the outer surfaces of the slab are T_H and T_C. Each face of the slab has area A. Let us derive an expression for the conduction rate through the slab under the assumption that the transfer is a *steady-state* process; that is, the temperatures everywhere in the slab and the rate of energy transfer do not change with time.

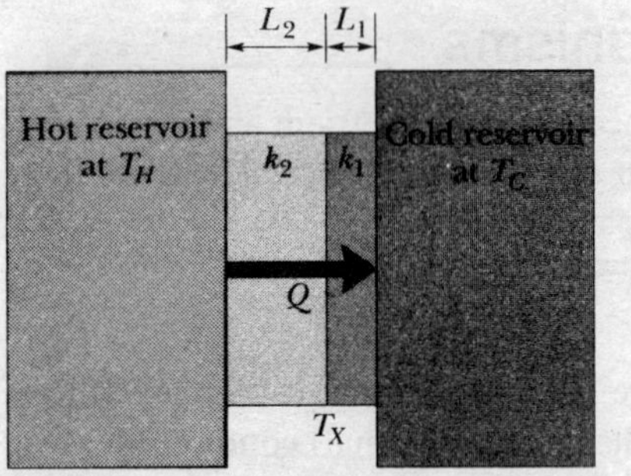

Fig. 19-19 Heat is transferred at a steady rate through a composite slab made up of two different materials with different thicknesses and different thermal conductivities. The steady-state temperature at the interface of the two materials is T_X.

In the steady state, the conduction rates through the two materials must be equal. This is the same as saying that the energy transferred through one material in a certain time must be equal to that transferred through the other material in the same time. If this were not true, temperatures in the slab would be changing and we would not have a steady-state situation. Letting T_X be the temperature of the interface between the two materials, we can now use Eq. 19-32 to write

$$P_{cond} = \frac{k_2 A(T_H - T_X)}{L_2} = \frac{k_1 A(T_X - T_C)}{L_1}. \tag{19-34}$$

Solving Eq. 19-34 for T_X yields, after a little algebra,

$$T_X = \frac{k_1 L_2 T_C + k_2 L_1 T_H}{k_1 L_2 + k_2 L_1}. \tag{19-35}$$

Substituting this expression for T_X into either equality of Eq. 19-34 yields

$$P_{cond} = \frac{A(T_H - T_C)}{L_1/k_1 + L_2/k_2}. \tag{19-36}$$

We can extend Eq. 19-36 to apply to any number n of materials making up a slab:

$$P_{cond} = \frac{A(T_H - T_C)}{\Sigma\,(L/k)}. \tag{19-37}$$

The summation sign in the denominator tells us to add the values of L/k for all the materials.

CHECKPOINT 7: The figure shows the face and interface temperatures of a composite slab consisting of four materials, of identical thicknesses, through which the heat transfer is steady. Rank the materials according to their thermal conductivities, greatest first.

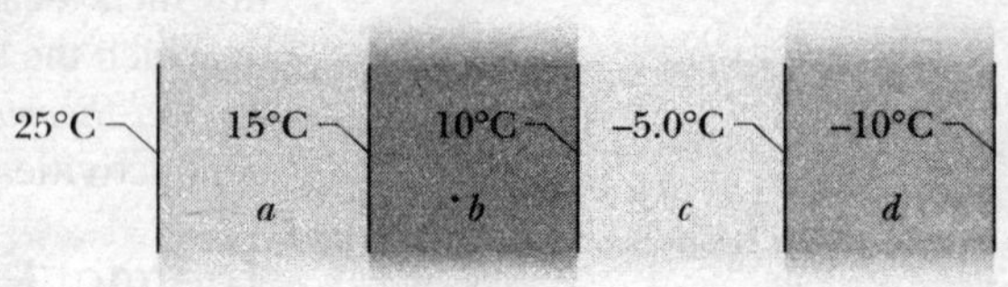

Convection

When you look at the flame of a candle or a match, you are watching thermal energy being transported upward by **convection.** Such energy transfer occurs when a fluid, such as air or water, comes in contact with an object whose temperature is higher than that of the fluid. The temperature of the part of the fluid that is in contact with the hot object increases, and (in most cases) that fluid expands and thus becomes less dense. Because this expanded fluid is now lighter than the surrounding cooler fluid, buoyant forces cause it to rise. Some of the surrounding cooler fluid then flows so as to take the place of the rising warmer fluid, and the process can then continue.

Convection is part of many natural processes. Atmospheric convection plays a fundamental role in determining global climate patterns and daily weather variations. Glider pilots and birds alike seek rising thermals (convection currents of warm air) that keep them aloft. Huge energy transfers take place within the oceans by the same process. Finally, energy is transported to the surface of the Sun from the nuclear furnace at its core by enormous cells of convection, in which hot gas rises to the surface along the cell core and cooler gas around the core descends below the surface.

Radiation

The third method by which an object and its environment can exchange energy as heat is via electromagnetic waves (visible light is one kind of electromagnetic wave).

Fig. 19-20 A false-color thermogram reveals the rate at which energy is radiated by houses along a street. The rates, from largest to smallest, are color coded as white, red, pink, blue, and black. You can tell where there is insulation in the walls, a heavy curtain over a window, and a higher air temperature at the ceiling on the second floor.

Energy transferred in this way is often called **thermal radiation** to distinguish it from electromagnetic *signals* (as in, say, television broadcasts) and from nuclear radiation (energy and particles emitted by nuclei). (To "radiate" generally means to emit.) When you stand in front of a big fire, you are warmed by absorbing thermal radiation from the fire; that is, your thermal energy increases as the fire's thermal energy decreases. No medium is required for heat transfer via radiation—the radiation can travel through vacuum from, say, the Sun to you.

The rate P_{rad} at which an object emits energy via electromagnetic radiation depends on the object's surface area A and the temperature T of that area in kelvins and is given by

$$P_{rad} = \sigma \varepsilon A T^4. \tag{19-38}$$

Here $\sigma = 5.6703 \times 10^{-8}\ \mathrm{W/m^2 \cdot K^4}$ is called the *Stefan–Boltzmann constant* after Josef Stefan (who discovered Eq. 19-38 experimentally in 1879) and Ludwig Boltzmann (who derived it theoretically soon after). The symbol ε represents the *emissivity* of the object's surface, which has a value between 0 and 1, depending on the composition of the surface. A surface with the maximum emissivity of 1.0 is said to be a *blackbody radiator,* but such a surface is an ideal limit and does not occur in nature. Note again that the temperature in Eq. 19-38 must be in kelvins so that a temperature of absolute zero corresponds to no radiation. Note also that every object whose temperature is above 0 K—including you—emits thermal radiation. (See Fig. 19-20.)

The rate P_{abs} at which an object absorbs energy via thermal radiation from its environment, which we take to be at uniform temperature T_{env} (in kelvins), is

$$P_{abs} = \sigma \varepsilon A T_{env}^4. \tag{19-39}$$

The emissivity ε in Eq. 19-39 is the same as that in Eq. 19-38. An idealized blackbody radiator, with $\varepsilon = 1$, will absorb all the radiated energy it intercepts (rather than sending a portion back away from itself through reflection or scattering).

Because an object will radiate energy to the environment while it absorbs energy from the environment, the object's net rate P_{net} of energy exchange due to thermal radiation is

$$P_{net} = P_{abs} - P_{rad} = \sigma \varepsilon A (T_{env}^4 - T^4). \tag{19-40}$$

P_{net} is positive if net energy is being absorbed via radiation, and negative if it is being lost via radiation.

Sample Problem 19-6

Figure 19-21 shows the cross section of a wall made of white pine of thickness L_a and brick of thickness L_d $(= 2.0L_a)$, sandwiching two layers of unknown material with identical thicknesses and thermal conductivities. The thermal conductivity of the pine is k_a and that of the brick is k_d $(= 5.0k_a)$. The face area A of the wall is unknown. Thermal conduction through the wall has reached the steady state; the only known interface temperatures are $T_1 = 25°C$, $T_2 = 20°C$, and $T_5 = -10°C$. What is interface temperature T_4?

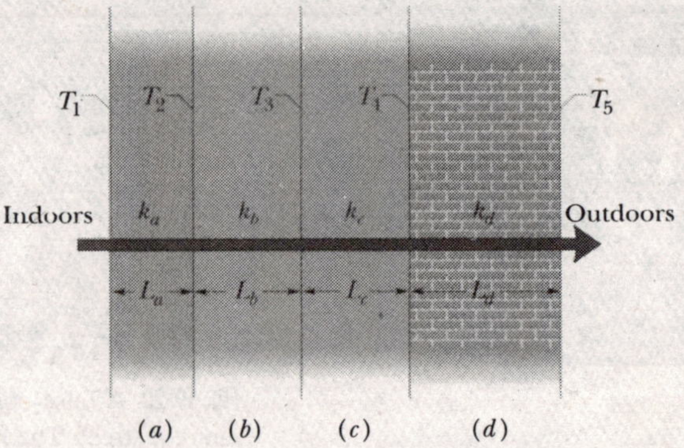

Fig. 19-21 Sample Problem 19-6. A wall of four layers through which there is steady-state heat transfer.

SOLUTION: One **Key Idea** here is that temperature T_4 helps determine the rate P_d at which energy is conducted through the brick, as given by Eq. 19-32. However, we lack enough data to solve Eq. 19-32 for T_4. A second **Key Idea** is that because the conduction is steady, the conduction rate P_d through the brick must equal the conduction rate P_a through the pine. From Eq. 19-32 and Fig. 19-21, we can write

$$P_a = k_aA\frac{T_1 - T_2}{L_a} \quad \text{and} \quad P_d = k_dA\frac{T_4 - T_5}{L_d}.$$

Setting $P_a = P_d$ and solving for T_4 yield

$$T_4 = \frac{k_aL_d}{k_dL_a}(T_1 - T_2) + T_5.$$

Letting $L_d = 2.0L_a$ and $k_d = 5.0k_a$, and inserting the known temperatures, we find

$$T_4 = \frac{k_a(2.0L_a)}{(5.0k_a)L_a}(25°C - 20°C) + (-10°C)$$
$$= -8.0°C. \qquad \text{(Answer)}$$

Sample Problem 19-7

When hundreds of Japanese bees form a compact ball around a giant hornet that attempts to invade their hive, they can quickly raise their body temperature from the normal 35°C to 47°C or 48°C. That higher temperature is lethal to the hornet but not to the bees (Fig. 19-22). Assume the following: 500 bees form a ball of radius $R = 2.0$ cm for a time $t = 20$ min, the primary loss of energy by the ball is by thermal radiation, the ball's surface has emissivity $\varepsilon = 0.80$, and the ball has a uniform temperature. On average, how much additional energy must each bee produce during the 20 min to maintain 47°C?

SOLUTION: The **Key Idea** here is that, because the surface temperature of the bee ball increases after the ball forms, the rate at which energy is radiated by the ball also increases. Thus, the bees lose an additional amount of energy to thermal radiation. We can relate the surface temperature to the rate of radiation (energy per unit time) with Eq. 19-38 ($P_{rad} = \sigma\varepsilon AT^4$), in which A is the ball's surface area and T is the ball's surface temperature in kelvins. This rate is an energy per unit time; that is,

$$P_{rad} = \frac{E}{t}.$$

Thus, the amount of energy E radiated in time t is $E = P_{rad}t$.

At the normal temperature $T_1 = 35°C$, the radiation rate would be P_{r1} and the amount of energy radiated in time t would be $E_1 = P_{r1}t$. At the increased temperature $T_2 = 47°C$, the (greater) radiation rate is P_{r2} and the (greater) amount of energy radiated in time t is $E_2 = P_{r2}t$. Thus, in maintaining the ball at T_2 for time t, the bees must (together) provide an additional energy of $\Delta E = E_2 - E_1$.

We can now write

$$\Delta E = E_2 - E_1 = P_{r2}t - P_{r1}t$$
$$= (\sigma\varepsilon AT_2^4)t - (\sigma\varepsilon AT_1^4)t = \sigma\varepsilon At(T_2^4 - T_1^4). \qquad (19\text{-}41)$$

The temperatures here *must* be in kelvins; thus, we write them as

$$T_2 = 47°C + 273\ C° = 320\ K$$

and

$$T_1 = 35°C + 273\ C° = 308\ K.$$

The surface area A of the ball is

$$A = 4\pi R^2 = (4\pi)(0.020\ m)^2 = 5.027 \times 10^{-3}\ m^2,$$

and the time t is 20 min = 1200 s. Substituting these and other known values into Eq. 19-41, we find

$$\Delta E = (5.6703 \times 10^{-8}\ W/m^2 \cdot K^4)(0.80)(5.027 \times 10^{-3}\ m^2)$$
$$\times (1200\ s)[(320\ K)^4 - (308\ K)^4] = 406.8\ J.$$

Thus, with 500 bees in the ball, each bee must produce an additional energy of

$$\frac{\Delta E}{500} = \frac{406.8\ J}{500} = 0.81\ J. \qquad \text{(Answer)}$$

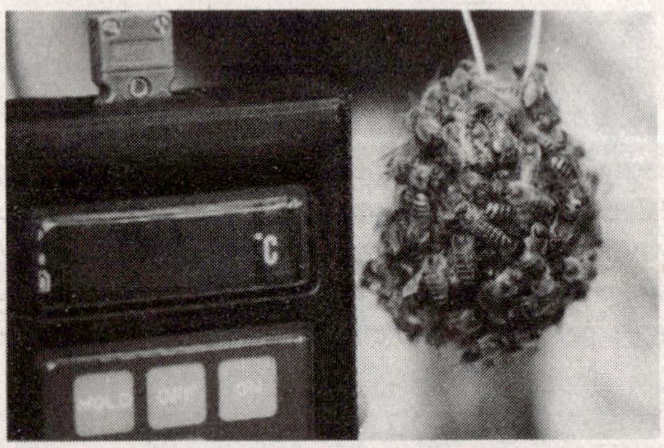

Fig. 19-22 Sample Problem 19-7. The bees were unharmed by their increased body temperature, which the hornet could not withstand.

REVIEW & SUMMARY

Temperature; Thermometers Temperature is an SI base quantity related to our sense of hot and cold. It is measured with a thermometer, which contains a working substance with a measurable property, such as length or pressure, that changes in a regular way as the substance becomes hotter or colder.

Zeroth Law of Thermodynamics When a thermometer and some other object are placed in contact with each other, they eventually reach thermal equilibrium. The reading of the thermometer is then taken to be the temperature of the other object. The process provides consistent and useful temperature measurements because of the **zeroth law of thermodynamics:** If bodies A and B are each in thermal equilibrium with a third body C (the thermometer), then A and B are in thermal equilibrium with each other.

The Kelvin Temperature Scale In the SI system, temperature is measured on the **Kelvin scale,** which is based on the *triple point* of water (273.16 K). Other temperatures are then defined by use of a *constant-volume gas thermometer,* in which a sample of gas is maintained at constant volume so its pressure is proportional to its temperature. We define the *temperature* T as measured with a gas thermometer to be

$$T = (273.16\text{ K})\left(\lim_{\text{gas}\to 0} \frac{p}{p_3}\right). \tag{19-6}$$

Here T is measured in kelvins, and p_3 and p are the pressures of the gas at 273.16 K and the measured temperature, respectively.

Celsius and Fahrenheit Scales The Celsius temperature scale is defined by

$$T_C = T - 273.15°, \tag{19-7}$$

with T in kelvins. The Fahrenheit temperature scale is defined by

$$T_F = \tfrac{9}{5}T_C + 32°. \tag{19-8}$$

Thermal Expansion All objects change size with changes in temperature. For a temperature change ΔT, a change ΔL in any linear dimension L is given by

$$\Delta L = L\alpha\,\Delta T, \tag{19-9}$$

in which α is the **coefficient of linear expansion.** The change ΔV in the volume V of a solid or liquid is

$$\Delta V = V\beta\,\Delta T. \tag{19-10}$$

Here $\beta = 3\alpha$ is the material's **coefficient of volume expansion.**

Heat Heat Q is energy that is transferred between a system and its environment because of a temperature difference between them. It can be measured in **joules** (J), **calories** (cal), **kilocalories** (Cal or kcal), or **British thermal units** (Btu), with

$$1\text{ cal} = 3.969 \times 10^{-3}\text{ Btu} = 4.1860\text{ J}. \tag{19-12}$$

Heat Capacity and Specific Heat If heat Q is absorbed by an object, the object's temperature change $T_f - T_i$ is related to Q by

$$Q = C(T_f - T_i), \tag{19-13}$$

in which C is the **heat capacity** of the object. If the object has mass m, then

$$Q = cm(T_f - T_i), \tag{19-14}$$

where c is the **specific heat** of the material making up the object. The **molar specific heat** of a material is the heat capacity per mole, or per 6.02×10^{23} elementary units of the material.

Heat of Transformation Heat absorbed by a material may change the material's physical state or phase—for example, from solid to liquid or from liquid to gas. The amount of energy required per unit mass to change the phase (but not the temperature) of a particular material is its **heat of transformation** L. Thus,

$$Q = Lm. \tag{19-16}$$

The **heat of vaporization** L_V is the amount of energy per unit mass that must be added to vaporize a liquid or that must be removed to condense a gas. The **heat of fusion** L_F is the amount of energy per unit mass that must be added to melt a solid or that must be removed to freeze a liquid.

Work Associated with Volume Change A gas may exchange energy with its surroundings through work. The amount of work W done *by* a gas as it expands or contracts from an initial volume V_i to a final volume V_f is given by

$$W = \int dW = \int_{V_i}^{V_f} p\,dV. \tag{19-25}$$

The integration is necessary because the pressure p may vary during the volume change.

First Law of Thermodynamics The principle of conservation of energy for a thermodynamic process is expressed in the **first law of thermodynamics**, which may assume either of the forms

$$\Delta E_{\text{int}} = E_{\text{int},f} - E_{\text{int},i} = Q - W \quad \text{(first law)} \tag{19-26}$$

or

$$dE_{\text{int}} = dQ - dW \quad \text{(first law)}. \tag{19-27}$$

E_{int} represents the internal energy of the material, which depends only on its state (temperature, pressure, and volume). Q represents the energy exchanged as heat between the system and its surroundings; Q is positive if the system absorbs heat, and negative if the system loses heat. W is the work done *by* the system; W is positive if the system expands against some external force exerted by the surroundings, and negative if the system contracts because of some external force. *Both Q and W are path dependent; ΔE_{int} is path independent.*

Applications of the First Law The first law of thermodynamics finds application in several special cases:

adiabatic processes: $Q = 0,\quad \Delta E_{\text{int}} = -W$

constant-volume processes: $W = 0,\quad \Delta E_{\text{int}} = Q$

cyclical processes: $\Delta E_{\text{int}} = 0,\quad Q = W$

free expansions: $Q = W = \Delta E_{\text{int}} = 0$

Conduction, Convection, and Radiation The rate P_{cond} at which energy is *conducted* through a slab whose faces are maintained at temperatures T_H and T_C is

$$P_{cond} = \frac{Q}{t} = kA\frac{T_H - T_C}{L}, \tag{19-32}$$

in which A and L are the face area and length of the slab, and k is the thermal conductivity of the material.

Convection occurs when temperature differences cause an energy transfer by motion within a fluid. *Radiation* is an energy transfer via the emission of electromagnetic energy. The rate P_{rad} at which an object emits energy via thermal radiation is

$$P_{rad} = \sigma\varepsilon AT^4, \tag{19-38}$$

where σ (= 5.6703×10^{-8} W/m$^2 \cdot$ K^4) is the Stefan–Boltzmann constant, ε is the emissivity of the object's surface, A is its surface area, and T is its surface temperature (in kelvins). The rate P_{abs} at which an object absorbs energy via thermal radiation from its environment, which is at the uniform temperature T_{env} (in kelvins), is

$$P_{abs} = \sigma\varepsilon AT_{env}^4. \tag{19-39}$$

QUESTIONS

1. Figure 19-23 shows three linear temperature scales with the freezing and boiling points of water indicated. Rank changes of 25 R°, 25 S°, and 25 U° according to the corresponding changes in temperature, greatest first.

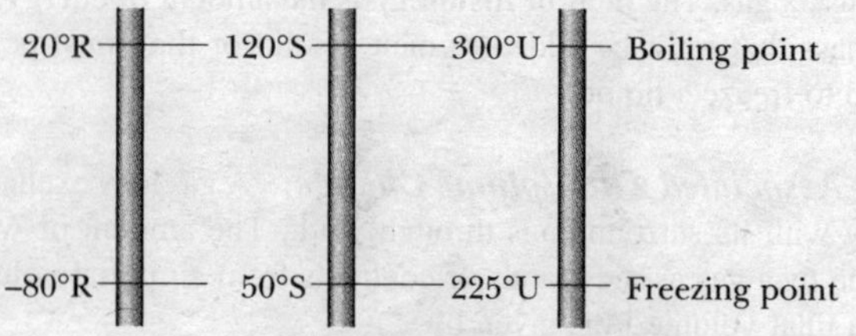

Fig. 19-23 Question 1.

2. The table gives the initial length L, change in temperature ΔT, and change in length ΔL of four rods. Rank the rods according to their coefficients of thermal expansion, greatest first.

Rod	L (m)	ΔT (C°)	ΔL (m)
a	2	10	4×10^{-4}
b	1	20	4×10^{-4}
c	2	10	8×10^{-4}
d	4	5	4×10^{-4}

3. In a thermally isolated container, material A of mass m is placed against material B, also of mass m but at a higher temperature. When thermal equilibrium is reached, the temperature changes ΔT_A and ΔT_B of A and B are recorded. Then the experiment is repeated, using A with other materials, all of the same mass m. The results are given in the table. Rank the four materials according to their specific heats, greatest first.

Experiment	Temperature Changes	
1	$\Delta T_A = +50$ C°	$\Delta T_B = -50$ C°
2	$\Delta T_A = +10$ C°	$\Delta T_C = -20$ C°
3	$\Delta T_A = +2$ C°	$\Delta T_D = -40$ C°

4. Materials A, B, and C are solids that are at their melting temperatures. Material A requires 200 J to melt 4 kg, material B requires 300 J to melt 5 kg, and material C requires 300 J to melt 6 kg. Rank the materials according to their heats of fusion, greatest first.

5. Figure 19-24 shows two closed cycles on p-V diagrams for a gas. The three parts of cycle 1 are of the same length and shape as those of cycle 2. For each cycle, should the cycle be traversed clockwise or counterclockwise if (a) the net work W done by the gas is to be positive and (b) the net energy transferred by the gas as heat Q is to be positive?

6. For which cycle in Fig. 19-24, traversed clockwise, is (a) W greater and (b) Q greater?

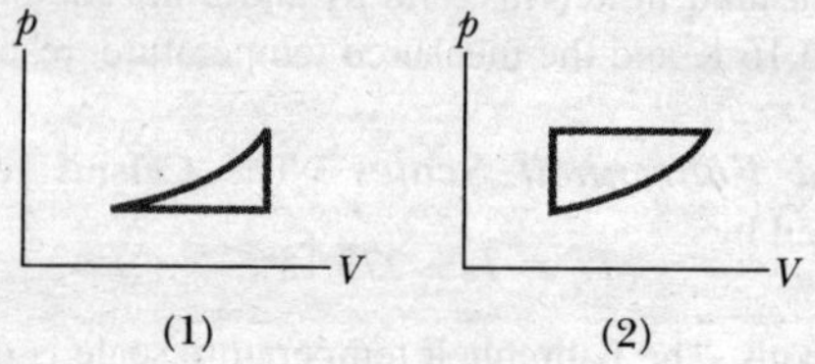

Fig. 19-24 Questions 5 and 6.

7. Figure 19-25 shows a composite slab of three different materials, a, b, and c, with identical thicknesses and with thermal conductivities $k_b > k_a > k_c$. The transfer of energy through them as heat is nonzero and steady. Rank the materials according to the temperature difference ΔT across them, greatest first.

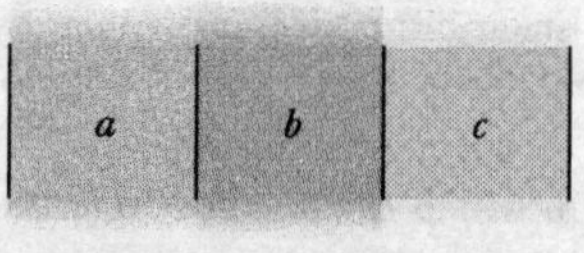

Fig. 19-25 Question 7.

8. During an icicle's growth, its outer surface is covered with a thin sheath of liquid water that slowly seeps downward to form drops, one at a time, that hang at the icicle's tip (Fig. 19-26). Each drop straddles a thin tube of liquid water that extends up into the icicle, toward (but not all the way to) its *root* (its top). As the water at the top of the tube gradually freezes, energy is released. Is that energy conducted radially outward through the ice, downward through the water to the hanging

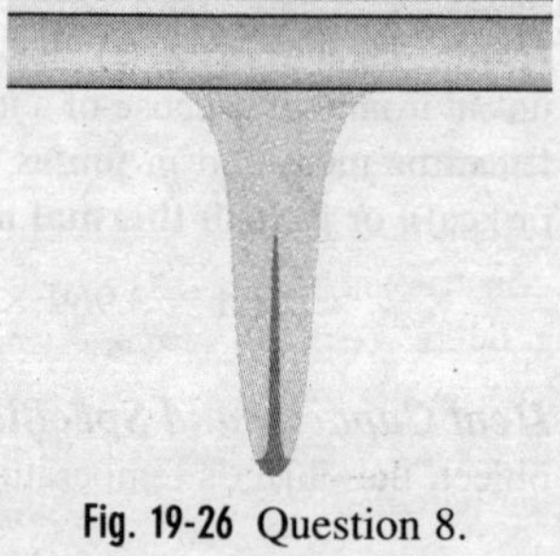

Fig. 19-26 Question 8.

drop, or upward toward the root? (Assume that the air temperature is below 0°C.)

9. The following solid objects, made of the same material, are maintained at a temperature of 300 K in an environment whose temperature is 350 K: a cube of edge length r, a sphere of radius r, and a hemisphere of radius r. Rank the objects according to the net rate at which thermal radiation is exchanged with the environment, greatest first.

10. Three different materials of identical masses are placed, in turn, in a special freezer that can extract energy from a material at a certain constant rate. During the cooling process, each material begins in the liquid state and ends in the solid state; Fig. 19-27 shows graphs of the temperature T versus time t for the three materials. (a) For material 1, is the specific heat for the liquid state greater than or less than that for the solid state? Rank the materials according to (b) their freezing-point temperatures, (c) their specific heats in the liquid state, (d) their specific heats in the solid state, and (e) their heats of fusion, all greatest first.

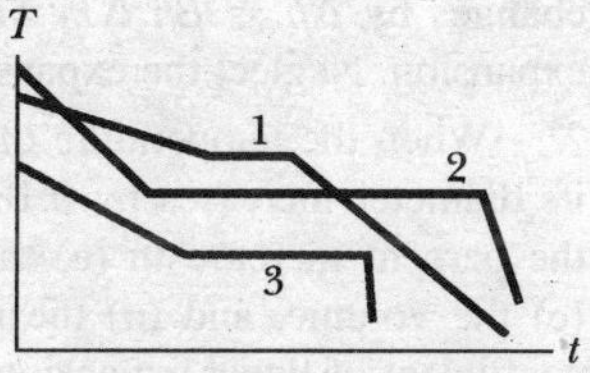

Fig. 19-27 Question 10.

11. A sample A of liquid water and a sample B of ice, of identical masses, are placed in a thermally isolated (insulated) container and allowed to come to thermal equilibrium. Figure 19-28a is a sketch of the temperature T of the samples versus time t. (a) Is the equilibrium temperature above, below, or at the freezing point of water? (b) In reaching equilibrium, does the liquid partly freeze or fully freeze, or does it undergo no freezing? (c) Does the ice partly melt or fully melt, or does it undergo no melting?

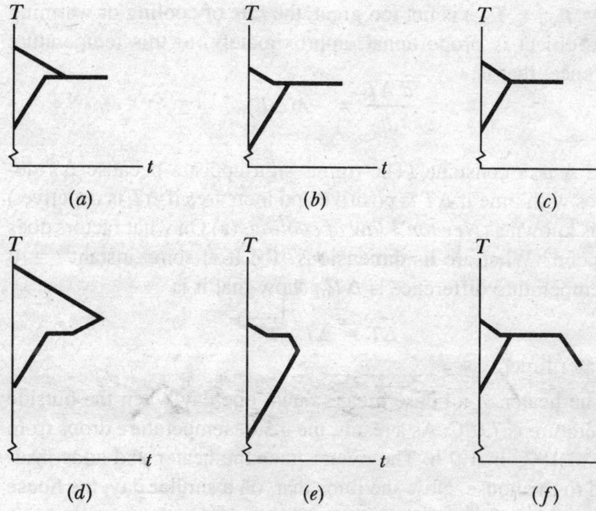

Fig. 19-28 Questions 11 and 12.

12. Question 11 continued: Figure 19-28 gives additional sketches of T versus t, of which one or more are impossible to produce. (a) Which sketch is impossible and why? (b) In the possible ones, is the equilibrium temperature above, below, or at the freezing point of water? (c) As the possible situations reach equilibrium, does the liquid partly freeze or fully freeze, or does it undergo no freezing? Does the ice partly melt or fully melt, or undergo no melting?

EXERCISES & PROBLEMS

ssm Solution is in the Student Solutions Manual.
www Solution is available on the World Wide Web at: http://www.wiley.com/college/hrw
ilw Solution is available on the Interactive LearningWare.

SEC. 19-3 Measuring Temperature

1E. Two constant-volume gas thermometers are assembled, one with nitrogen and the other hydrogen. Both contain enough gas so that $p_3 = 80$ kPa. What is the difference between the pressures in the two thermometers if both bulbs are inserted into boiling water? (*Hint:* See Fig. 19-6.) Which gas is at higher pressure? **ssm**

2E. Suppose the temperature of a gas at the boiling point of water is 373.15 K. What then is the limiting value of the ratio of the pressure of the gas at that boiling point to its pressure at the triple point of water? (Assume the volume of the gas is the same at both temperatures.)

3E. A particular gas thermometer is constructed of two gas-containing bulbs, each of which is put into a water bath, as shown in Fig. 19-29. The pressure difference between the two bulbs is measured by a mercury manometer as shown. Appropriate reservoirs, not shown in the diagram, maintain constant gas volume in the two bulbs. There is no difference in pressure when both baths are at the triple point of water. The pressure difference is 120 torr when one bath is at the triple point and the other is at the boiling point of water. It is 90.0 torr when one bath is at the triple point and the other is at an unknown temperature to be measured. What is the unknown temperature? **ssm**

Fig. 19-29 Exercise 3.

SEC. 19-4 The Celsius and Fahrenheit Scales

4E. At what temperature is the Fahrenheit scale reading equal to (a) twice that of the Celsius and (b) half that of the Celsius?

5E. At what temperature do the following pairs of scales read the same, if ever: (a) Fahrenheit and Celsius (verify the listing in Table 19-1), (b) Fahrenheit and Kelvin, and (c) Celsius and Kelvin? **ssm**

6E. (a) In 1964, the temperature in the Siberian village of Oymyakon reached −71°C. What temperature is this on the Fahrenheit scale? (b) The highest officially recorded temperature in the continental United States was 134°F in Death Valley, California. What is this temperature on the Celsius scale?

7P. It is an everyday observation that hot and cold objects cool down or warm up to the temperature of their surroundings. If the temperature difference ΔT between an object and its surroundings

($\Delta T = T_{obj} - T_{sur}$) is not too great, the rate of cooling or warming of the object is proportional, approximately, to this temperature difference; that is,

$$\frac{d\,\Delta T}{dt} = -A(\Delta T),$$

where A is a constant. (The minus sign appears because ΔT decreases with time if ΔT is positive and increases if ΔT is negative.) This is known as *Newton's law of cooling*. (a) On what factors does A depend? What are its dimensions? (b) If at some instant $t = 0$ the temperature difference is ΔT_0, show that it is

$$\Delta T = \Delta T_0 e^{-At}$$

at a later time t. ssm

8P. The heater of a house breaks down one day when the outside temperature is 7.0°C. As a result, the inside temperature drops from 22°C to 18°C in 1.0 h. The owner fixes the heater and adds insulation to the house. Now she finds that, on a similar day, the house takes twice as long to drop from 22°C to 18°C when the heater is not operating. What is the ratio of the new value of constant A in Newton's law of cooling (see Problem 7) to the previous value?

9P. Suppose that on a linear temperature scale X, water boils at −53.5°X and freezes at −170°X. What is a temperature of 340 K on the X scale? ilw

SEC. 19-5 Thermal Expansion

10E. An aluminum flagpole is 33 m high. By how much does its length increase as the temperature increases by 15 C°?

11E. The Pyrex glass mirror in the telescope at the Mt. Palomar Observatory has a diameter of 200 in. The temperature ranges from −10°C to 50°C on Mt. Palomar. In micrometers, what is the maximum change in the diameter of the mirror, assuming that the glass can freely expand and contract? ssm

12E. An aluminum-alloy rod has a length of 10.000 cm at 20.000°C and a length of 10.015 cm at the boiling point of water. (a) What is the length of the rod at the freezing point of water? (b) What is the temperature if the length of the rod is 10.009 cm?

13E. A circular hole in an aluminum plate is 2.725 cm in diameter at 0.000°C. What is its diameter when the temperature of the plate is raised to 100.0°C? ilw

14E. What is the volume of a lead ball at 30°C if the ball's volume at 60°C is 50 cm^3?

15E. Find the change in volume of an aluminum sphere with an initial radius of 10 cm when the sphere is heated from 0.0°C to 100°C. ssm

16E. The area A of a rectangular plate is ab. Its coefficient of linear expansion is α. After a temperature rise ΔT, side a is longer by Δa and side b is longer by Δb (Fig. 19-30). Show that if the small quantity $(\Delta a\,\Delta b)/ab$ is neglected, then $\Delta A = 2\alpha A\,\Delta T$.

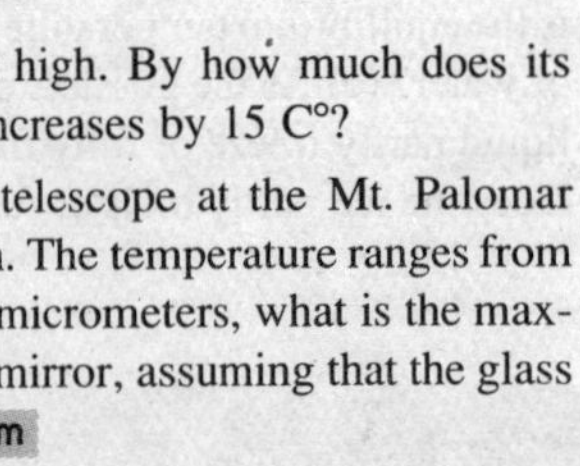

Fig. 19-30 Exercise 16.

17E. An aluminum cup of 100 cm^3 capacity is completely filled with glycerin at 22°C. How much glycerin, if any, will spill out of the cup if the temperature of both the cup and glycerin is increased to 28°C? (The coefficient of volume expansion of glycerin is 5.1×10^{-4}/C°.) ssm

18P. At 20°C, a rod is exactly 20.05 cm long on a steel ruler. Both the rod and the ruler are placed in an oven at 270°C, where the rod now measures 20.11 cm on the same ruler. What is the coefficient of thermal expansion for the material of which the rod is made?

19P. A steel rod is 3.000 cm in diameter at 25°C. A brass ring has an interior diameter of 2.992 cm at 25°C. At what common temperature will the ring just slide onto the rod? ssm ilw www

20P. When the temperature of a metal cylinder is raised from 0.0°C to 100°C, its length increases by 0.23%. (a) Find the percent change in density. (b) What is the metal? Use Table 19-2.

21P. Show that when the temperature of a liquid in a barometer changes by ΔT and the pressure is constant, the liquid's height h changes by $\Delta h = \beta h\,\Delta T$, where β is the coefficient of volume expansion. Neglect the expansion of the glass tube. ssm

22P. When the temperature of a copper coin is raised by 100 C°, its diameter increases by 0.18%. To two significant figures, give the percent increase in (a) the area of a face, (b) the thickness, (c) the volume, and (d) the mass of the coin. (e) Calculate the coefficient of linear expansion of the coin.

23P. A pendulum clock with a pendulum made of brass is designed to keep accurate time at 20°C. If the clock operates at 0.0°C, what is the magnitude of its error, in seconds per hour, and does the clock run fast or slow? ilw

24P. In a certain experiment, a small radioactive source must move at selected, extremely slow speeds. This motion is accomplished by fastening the source to one end of an aluminum rod and heating the central section of the rod in a controlled way. If the effective heated section of the rod in Fig. 19-31 is 2.00 cm, at what constant rate must the temperature of the rod be changed if the source is to move at a constant speed of 100 nm/s?

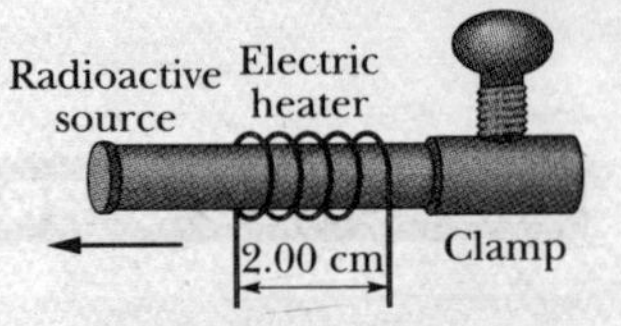

Fig. 19-31 Problem 24.

25P. As a result of a temperature rise of 32 C°, a bar with a crack at its center buckles upward (Fig. 19-32). If the fixed distance L_0 is 3.77 m and the coefficient of linear expansion of the bar is 25×10^{-6}/C°, find the rise x of the center. ssm ilw

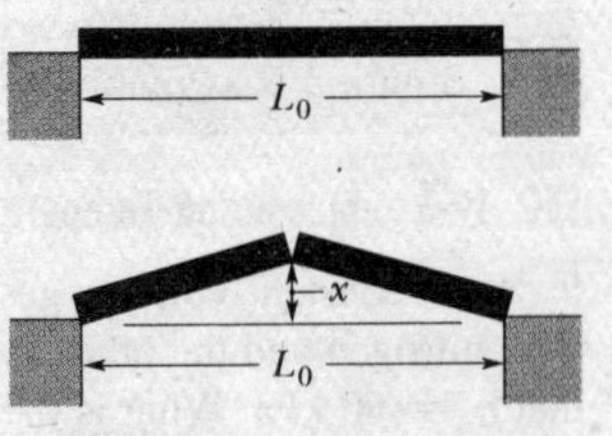

Fig. 19-32 Problem 25.

SEC. 19-7 The Absorption of Heat by Solids and Liquids

26E. A certain diet doctor encourages people to diet by drinking ice water. His theory is that the body must burn off enough fat to raise the temperature of the water from 0.00°C to the body temperature of 37.0°C. How many liters of ice water would have to be consumed to burn off 454 g (about 1 lb) of fat, assuming that this much fat burning requires 3500 Cal be transferred to the ice water? Why is it not advisable to follow this diet? (One liter = 10^3 cm^3. The density of water is 1.00 g/cm^3.)

27E. A certain substance has a mass per mole of 50 g/mol. When 314 J is added as heat to a 30.0 g sample, the sample's temperature rises from 25.0°C to 45.0°C. What are (a) the specific heat and (b) the molar specific heat of this substance? (c) How many moles are present? ssm

28E. How much water remains unfrozen after 50.2 kJ is transferred as heat from 260 g of liquid water initially at its freezing point?

29E. Calculate the minimum amount of energy, in joules, required to completely melt 130 g of silver initially at 15.0°C. ssm

30E. A room is lighted by four 100 W incandescent lightbulbs. (The power of 100 W is the rate at which a bulb converts electrical energy to heat and the energy of visible light.) Assuming that 90% of the energy is converted to heat, how much heat does the room receive in 1.00 h?

31E. An energetic athlete can use up all the energy from a diet of 4000 Cal/day. If he were to use up this energy at a steady rate, how would his rate of energy use compare with the power of a 100 W bulb? (The power of 100 W is the rate at which the bulb converts electrical energy to heat and the energy of visible light.) ilw

32E. How many grams of butter, which has a usable energy content of 6.0 Cal/g (= 6000 cal/g), would be equivalent to the change in gravitational potential energy of a 73.0 kg man who ascends from sea level to the top of Mt. Everest, at elevation 8.84 km? Assume that the average value of g is 9.80 m/s^2.

33E. A power of 0.400 hp is required for 2.00 min to drill a hole in a 1.60-lb copper block. (a) If the full power is the rate at which thermal energy is generated, how much is generated in Btu? (b) What is the rise in temperature of the copper if the copper absorbs 75.0% of this energy? (Use the energy conversion 1 ft · lb = 1.285×10^{-3} Btu.) ssm

34E. One way to keep the contents of a garage from becoming too cold on a night when a severe subfreezing temperature is forecast is to put a tub of water in the garage. If the mass of the water is 125 kg and its initial temperature is 20°C, (a) how much energy must the water transfer to its surroundings in order to freeze completely and (b) what is the lowest possible temperature of the water and its surroundings until that happens?

35E. A small electric immersion heater is used to heat 100 g of water for a cup of instant coffee. The heater is labeled "200 watts," which means that it converts electrical energy to thermal energy at this rate. Calculate the time required to bring all this water from 23°C to 100°C, ignoring any heat losses. ssm

36P. A 150 g copper bowl contains 220 g of water, both at 20.0°C. A very hot 300 g copper cylinder is dropped into the water, causing the water to boil, with 5.00 g being converted to steam. The final temperature of the system is 100°C. Neglect energy transfers with the environment. (a) How much energy (in calories) is transferred to the water as heat? (b) How much to the bowl? (c) What is the original temperature of the cylinder?

37P A chef, on finding his stove out of order, decides to boil the water for his wife's coffee by shaking it in a thermos flask. Suppose that he uses tap water at 15°C and that the water falls 30 cm each shake, the chef making 30 shakes each minute. Neglecting any loss of thermal energy by the flask, how long must he shake the flask until the water reaches 100°C? ssm www

38P. *Nonmetric version:* How long does a 2.0×10^5 Btu/h water heater take to raise the temperature of 40 gal of water from 70°F to 100°F? *Metric version:* How long does a 59 kW water heater take to raise the temperature of 150 L of water from 21°C to 38°C?

39P. Ethyl alcohol has a boiling point of 78°C, a freezing point of −114°C, a heat of vaporization of 879 kJ/kg, a heat of fusion of 109 kJ/kg, and a specific heat of 2.43 kJ/kg · K. How much energy must be removed from 0.510 kg of ethyl alcohol that is initially a gas at 78°C so that it becomes a solid at −114°C?

40P. A 1500 kg Buick moving at 90 km/h brakes to a stop, at uniform deceleration and without skidding, over a distance of 80 m. At what average rate is mechanical energy transferred to thermal energy in the brake system?

41P. The specific heat of a substance varies with temperature according to $c = 0.20 + 0.14T + 0.023T^2$, with T in °C and c in cal/g · K. Find the energy required to raise the temperature of 2.0 g of this substance from 5.0°C to 15°C.

42P. In a solar water heater, energy from the Sun is gathered by water that circulates through tubes in a rooftop collector. The solar radiation enters the collector through a transparent cover and warms the water in the tubes; this water is pumped into a holding tank. Assume that the efficiency of the overall system is 20% (that is, 80% of the incident solar energy is lost from the system). What collector area is necessary to raise the temperature of 200 L of water in the tank from 20°C to 40°C in 1.0 h when the intensity of incident sunlight is 700 W/m^2?

43P. What mass of steam at 100°C must be mixed with 150 g of ice at its melting point, in a thermally insulated container, to produce liquid water at 50°C? ilw

44P. A person makes a quantity of iced tea by mixing 500 g of hot tea (essentially water) with an equal mass of ice at its melting point. If the initial hot tea is at a temperature of (a) 90°C and (b) 70°C, what are the temperature and mass of the remaining ice when the tea and ice reach a common temperature? Neglect energy transfers with the environment.

45P. (a) Two 50 g ice cubes are dropped into 200 g of water in a thermally insulated container. If the water is initially at 25°C, and the ice comes directly from a freezer at −15°C, what is the final temperature of the drink when the drink reaches thermal equilibrium? (b) What is the final temperature if only one ice cube is used? ssm

46P. An insulated Thermos contains 130 cm^3 of hot coffee, at a temperature of 80.0°C. You put in a 12.0 g ice cube at its melting point to cool the coffee. By how many degrees has your coffee cooled once the ice has melted? Treat the coffee as though it were pure water and neglect energy transfers with the environment.

47P. A 20.0 g copper ring has a diameter of 2.54000 cm at its temperature of 0.000°C. An aluminum sphere has a diameter of 2.54508 cm at its temperature of 100.0°C. The sphere is placed on top of the ring (Fig. 19-33), and the two are allowed to come to thermal equilibrium, with no heat lost to the surroundings. The sphere just passes through the ring at the equilibrium temperature. What is the mass of the sphere? ssm

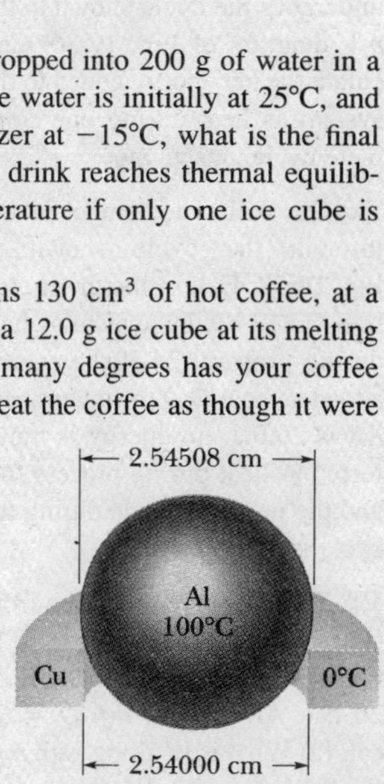

Fig. 19-33 Problem 47.

SEC. 19-10 Some Special Cases of the First Law of Thermodynamics

48E. Consider that 200 J of work is done on a system and 70.0 cal is extracted from the system as heat. In the sense of the first law of thermodynamics, what are the values (including algebraic signs) of (a) W, (b) Q, and (c) ΔE_{int}?

49E. A sample of gas expands from 1.0 m^3 to 4.0 m^3 while its pressure decreases from 40 Pa to 10 Pa. How much work is done by the gas if its pressure changes with volume via each of the three paths shown in the p-V diagram in Fig. 19-34? ssm www

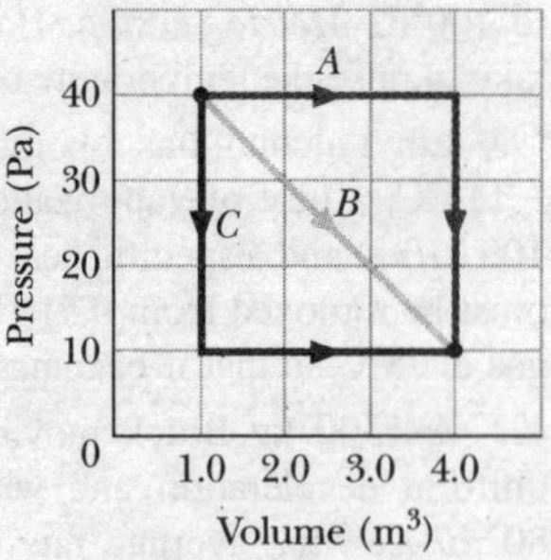

Fig. 19-34 Exercise 49.

50E. A thermodynamic system is taken from an initial state A to another state B and back again to A, via state C, as shown by path $ABCA$ in the p-V diagram of Fig. 19-35a. (a) Complete the table in Fig. 19-35b by filling in either + or − for the sign of each thermodynamic quantity associated with each step of the cycle. (b) Calculate the numerical value of the work done by the system for the complete cycle $ABCA$.

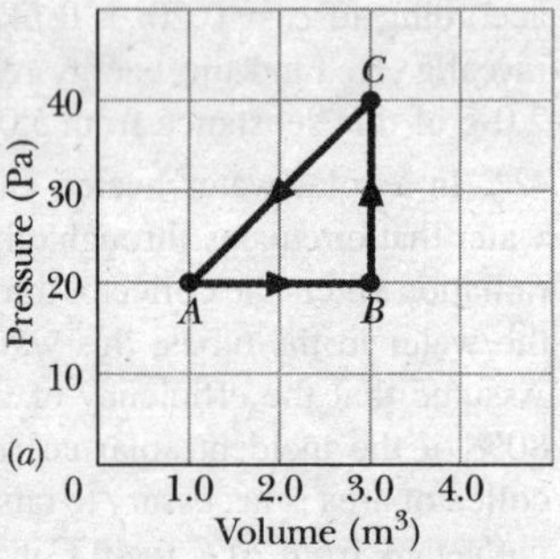

(b)

	Q	W	ΔE_{int}
$A \rightarrow B$			+
$B \rightarrow C$	+		
$C \rightarrow A$			

Fig. 19-35 Exercise 50.

51E. Gas within a closed chamber undergoes the cycle shown in the p-V diagram of Fig. 19-36. Calculate the net energy added to the system as heat during one complete cycle. ssm ilw

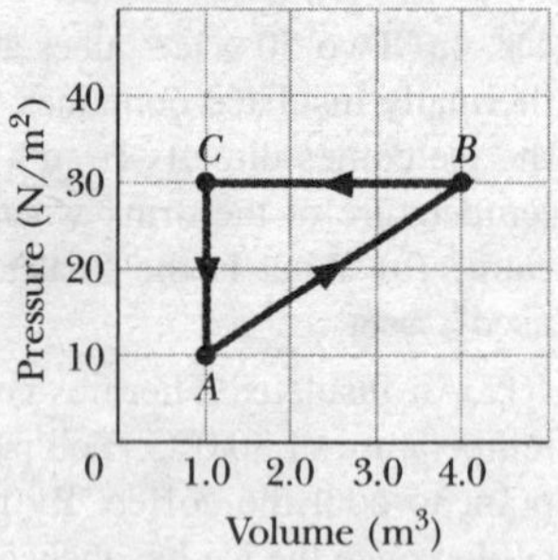

Fig. 19-36 Exercise 51.

52E. Gas within a chamber passes through the cycle shown in Fig. 19-37. Determine the energy transferred by the system as heat during process CA if the energy added as heat Q_{AB} during process AB is 20.0 J, no energy is transferred as heat during process BC, and the net work done during the cycle is 15.0 J.

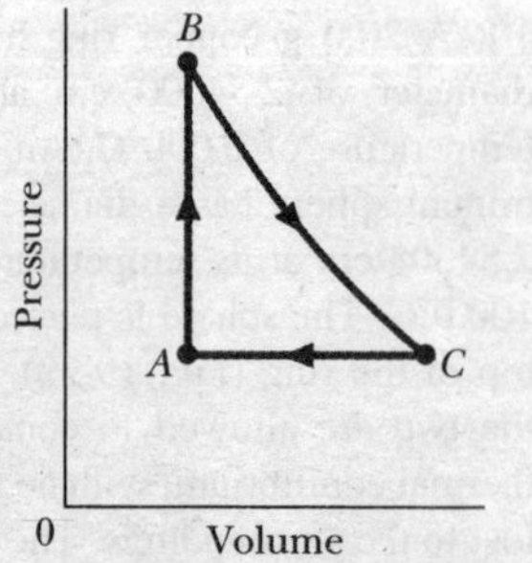

Fig. 19-37 Exercise 52.

53P. When a system is taken from state i to state f along path iaf in Fig. 19-38, Q = 50 cal and W = 20 cal. Along path ibf, Q = 36 cal. (a) What is W along path ibf? (b) If $W = -13$ cal for the return path fi, what is Q for this path? (c) Take $E_{int,i}$ = 10 cal. What is $E_{int,f}$? (d) If $E_{int,b}$ = 22 cal, what are the values of Q for path ib and path bf? ssm

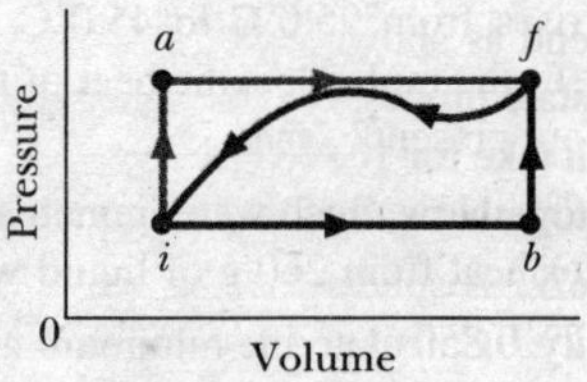

Fig. 19-38 Problem 53.

SEC. 19-11 Heat Transfer Mechanisms

54E. The average rate at which energy is conducted outward through the ground surface in North America is 54.0 mW/m^2, and the average thermal conductivity of the near-surface rocks is 2.50 W/m · K. Assuming a surface temperature of 10.0°C, find the temperature at a depth of 35.0 km (near the base of the crust). Ignore the heat generated by the presence of radioactive elements.

55E. The ceiling of a single-family dwelling in a cold climate should have an R-value of 30. To give such insulation, how thick would a layer of (a) polyurethane foam and (b) silver have to be?

56E. (a) Calculate the rate at which body heat is conducted through the clothing of a skier in a steady-state process, given the following data: the body surface area is 1.8 m^2 and the clothing is 1.0 cm thick; the skin surface temperature is 33°C and the outer surface of the clothing is at 1.0°C; the thermal conductivity of the clothing is 0.040 W/m · K. (b) How would the answer to (a) change if, after a fall, the skier's clothes became soaked with water of thermal conductivity 0.60 W/m · K?

57E. Consider the slab shown in Fig. 19-18. Suppose that L = 25.0 cm, A = 90.0 cm^2, and the material is copper. If T_H = 125°C, T_C = 10.0°C, and a steady state is reached, find the conduction rate through the slab. ssm

58E. If you were to walk briefly in space without a spacesuit while far from the Sun (as an astronaut does in the movie *2001*), you would feel the cold of space—while you radiated energy, you would absorb almost none from your environment. (a) At what rate would you lose energy? (b) How much energy would you lose in 30 s? Assume that your emissivity is 0.90, and estimate other data needed in the calculations.

59E. A cylindrical copper rod of length 1.2 m and cross-sectional area 4.8 cm^2 is insulated to prevent heat loss through its surface. The ends are maintained at a temperature difference of 100 C° by having one end in a water–ice mixture and the other in boiling water and steam. (a) Find the rate at which energy is conducted along the rod. (b) Find the rate at which ice melts at the cold end. ilw

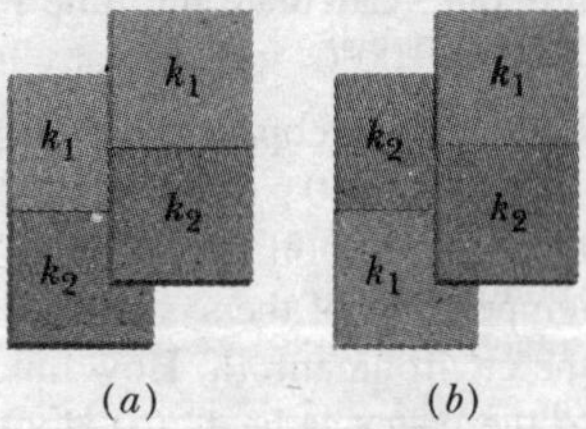

Fig. 19-39 Exercise 60.

60E. Four square pieces of insulation of two different materials, all with the same thickness and area A, are available to cover an opening of area $2A$. This can be done in either of the two ways shown in Fig. 19-39. Which arrangement, (a) or (b), gives the lower energy flow if $k_2 \neq k_1$?

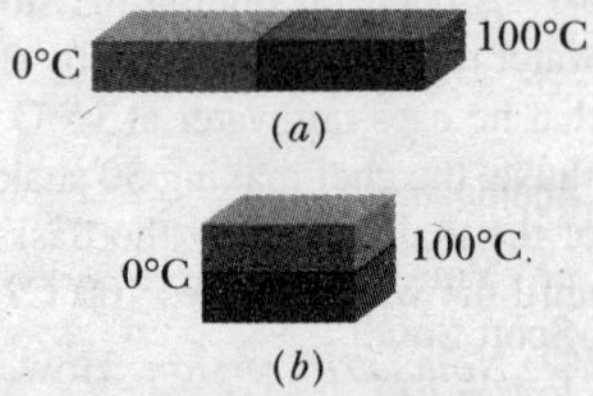

Fig. 19-40 Problem 61.

61P. Two identical rectangular rods of metal are welded end to end as shown in Fig. 19-40*a*, and 10 J is conducted (in a steady-state process) through the rods as heat in 2.0 min. How long would it take for 10 J to be conducted through the rods if they were welded together as shown in Fig. 19-40*b*?

62P. A sphere of radius 0.500 m, temperature 27.0°C, and emissivity 0.850 is located in an environment of temperature 77.0°C. At what rate does the sphere (a) emit and (b) absorb thermal radiation? (c) What is the sphere's net rate of energy exchange?

63P. (a) What is the rate of energy loss in watts per square meter through a glass window 3.0 mm thick if the outside temperature is −20°F and the inside temperature is +72°F? (b) A storm window having the same thickness of glass is installed parallel to the first window, with an air gap of 7.5 cm between the two windows. What now is the rate of energy loss if conduction is the only important energy-loss mechanism? ilw

64P. Figure 19-41 shows (in cross section) a wall that consists of four layers. The thermal conductivities are $k_1 = 0.060$ W/m · K, $k_3 = 0.040$ W/m · K, and $k_4 = 0.12$ W/m · K (k_2 is not known). The layer thicknesses are $L_1 = 1.5$ cm, $L_3 = 2.8$ cm, and $L_4 = 3.5$ cm (L_2 is not known). Energy transfer through the wall is steady. What is the temperature of the interface indicated?

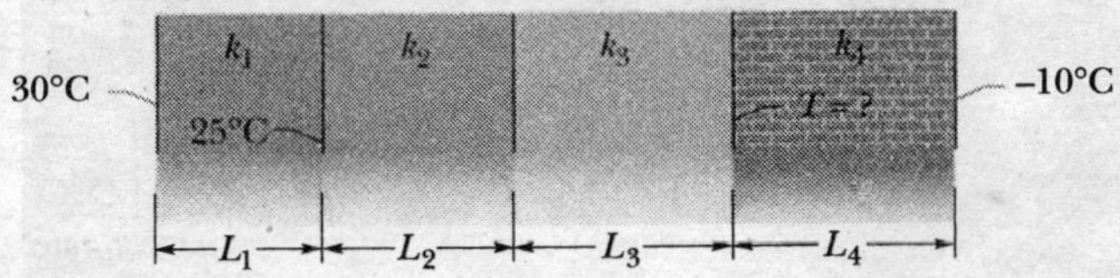

Fig. 19-41 Problem 64.

65P. A tank of water has been outdoors in cold weather, and a slab of ice 5.0 cm thick has formed on its surface (Fig. 19-42). The air above the ice is at −10°C. Calculate the rate of formation of ice (in centimeters per hour) on the ice slab. Take the thermal conductivity and density of ice to be 0.0040 cal/s · cm · C° and 0.92 g/cm³. Assume that energy is not transferred through the walls or bottom of the tank. ssm www

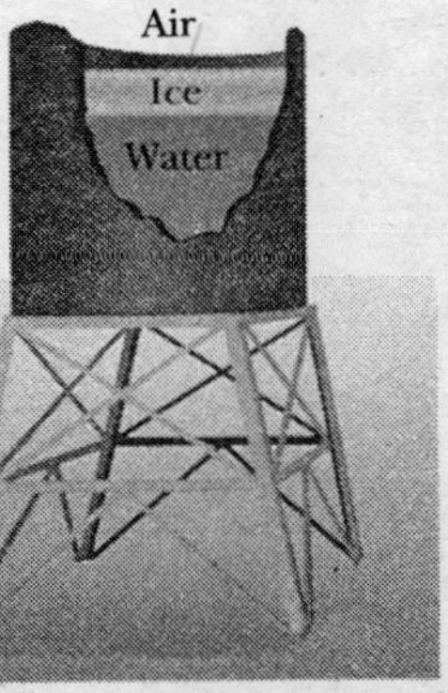

Fig. 19-42 Problem 65.

66P. Ice has formed on a shallow pond and a steady state has been reached, with the air above the ice at −5.0°C and the bottom of the pond at 4.0°C. If the total depth of *ice* + *water* is 1.4 m, how thick is the ice? (Assume that the thermal conductivities of ice and water are 0.40 and 0.12 cal/m · C° · s, respectively.)

Additional Problems

67. You can join the semi-secret "300 F" club at the Amundsen–Scott South Pole Station only when the outside temperature is below −70°C. On such a day, you first bask in a hot sauna and then run outside wearing only your shoes. (This is, of course, extremely dangerous, but the rite is effectively a protest against the constant danger of the winter cold at the south pole.)

Assume that when you step out of the sauna, your skin temperature is 102°F and the walls, ceiling, and floor of the sauna room have a temperature of 30°C. Estimate your surface area, and take your skin emissivity to be 0.80. (a) What is the approximate net rate P_{net} at which you lose energy via thermal radiation exchanges with the room? Next, assume that when you are outside half your surface area exchanges thermal radiation with the sky at a temperature of −25°C and the other half exchanges thermal radiation with the snow and ground at a temperature of −80°C. What is the approximate net rate at which you lose energy via thermal radiation exchanges with (b) the sky and (c) the snow and ground?

68. Emperor penguins (Fig. 19-43), those large penguins that resemble stuffy English butlers, breed and hatch their young even during severe Antarctic winters. Once an egg is laid, the father balances the egg on his feet to prevent the egg from freezing. He must do this for the full incubation period of 105 to 115 days, during which he cannot eat because his food is in the water. He can survive this long without food only if he can reduce his consumption of internal energy significantly. If he is alone, he consumes that energy too quickly to stay warm, and eventually abandons the egg in order to eat. To protect themselves and each other from the cold so as to reduce the consumption of internal energy, penguin fathers huddle closely together, in groups of perhaps several thousand. In addition to providing other benefits, the huddling reduces the rate at which the penguins thermally radiate energy to their surroundings.

Assume that a penguin father is a circular cylinder with top surface area a, height h, surface temperature T, and emissivity ε. (a) Find an expression for the rate P_i at which an individual father would radiate energy to the environment from his top surface and his side surface were he alone with his egg.

If N identical fathers were well apart from one another, the total rate of energy loss via radiation would be NP_i. Suppose, instead, that they huddle closely to form a *huddled cylinder* with top surface area Na and height h. (b) Find an expression for the rate P_h at which energy is radiated by the top surface and the side surface of the huddled cylinder.

(c) Assuming $a = 0.34$ m² and $h = 1.1$ m and using the expressions you obtained for P_i and P_h, graph the ratio P_h/NP_i versus N. Of course, the penguins know nothing about algebra or graphing, but their instinctive huddling reduces this ratio so that more of their eggs survive to the hatching stage. From the graphs (as you will see, you probably need more than one version), approximate how many penguins must huddle so that P_h/NP_i is reduced to (d) 0.5, (e) 0.4, (f) 0.3, (g) 0.2, and (h) 0.15. (i) For the assumed data, what is the lower limiting value for P_h/NP_i?

Fig. 19-43 Problem 68.

20 The Kinetic Theory of Gases

When a container of cold champagne, soda pop, or any other carbonated drink is opened, a slight fog forms around the opening and some of the liquid sprays outward. (In the photograph, the fog is the white cloud that surrounds the stopper, and the spray has formed streaks within the cloud.)

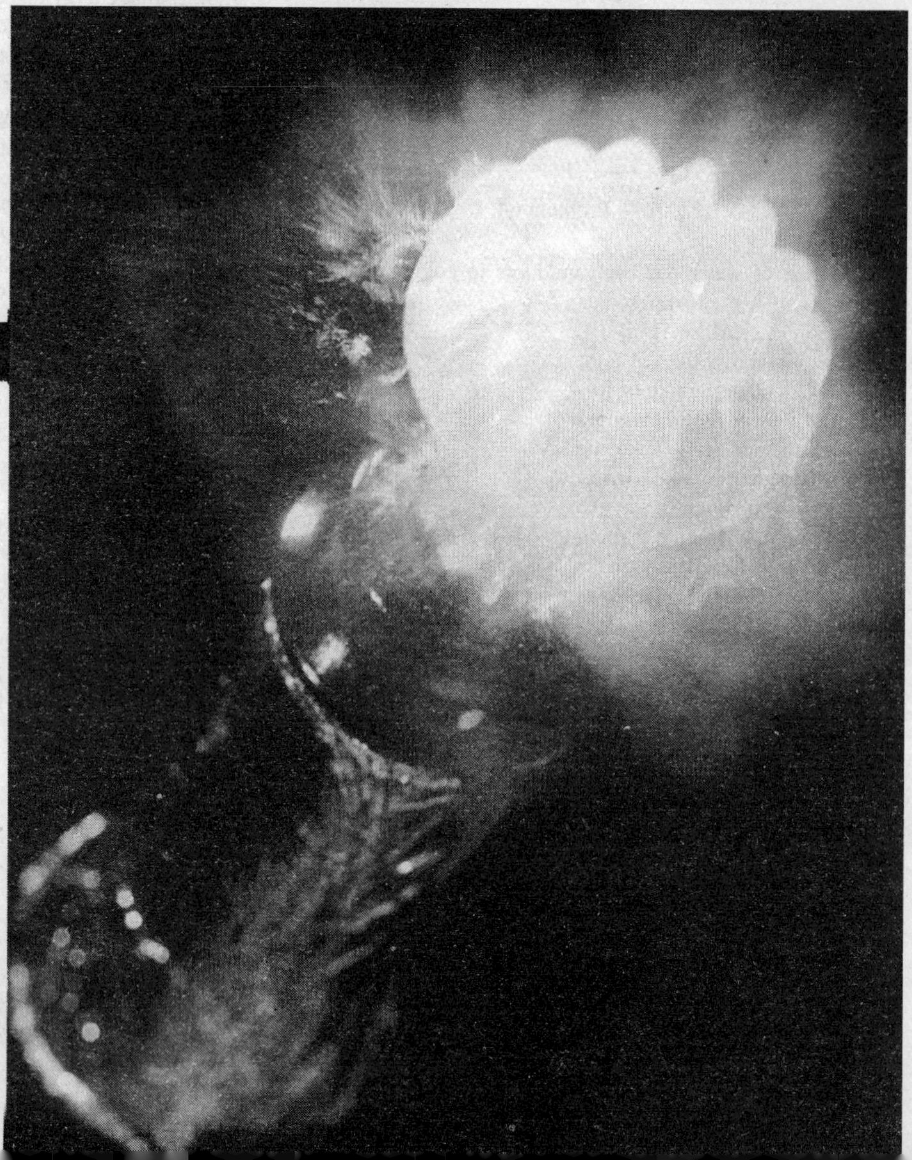

What causes the fog?

The answer is in this chapter.

20-1 A New Way to Look at Gases

Classical thermodynamics—the subject of the last chapter—has nothing to say about atoms. Its laws are concerned only with such macroscopic variables as pressure, volume, and temperature. However, we know that a gas is made up of moving atoms or molecules (groups of atoms bound together). The pressure exerted by a gas must surely be related to the collisions of its molecules with the walls of its container. The ability of a gas to fill the volume of its container must surely be due to the freedom of motion of its molecules, and the temperature and internal energy of a gas must surely be related to the kinetic energy of these molecules. Thus, we can learn something about gases by approaching the subject from this direction. We call this molecular approach the **kinetic theory of gases.** It is the subject of this chapter.

20-2 Avogadro's Number

When our thinking is slanted toward molecules, it makes sense to measure the sizes of our samples in moles. If we do so, we can be certain that we are comparing samples that contain the same number of atoms or molecules. The *mole* is one of the seven SI base units and is defined as follows:

➤ One mole is the number of atoms in a 12 g sample of carbon-12.

The obvious question now is: "How many atoms or molecules are there in a mole?" The answer is determined experimentally and, as you saw in Chapter 19, is

$$N_A = 6.02 \times 10^{23}\ \text{mol}^{-1} \quad \text{(Avogadro's number)}, \tag{20-1}$$

where mol^{-1} represents the inverse mole or "per mole," and mol is the abbreviation for mole. The number N_A is called **Avogadro's number** after Italian scientist Amadeo Avogadro (1776–1856), who suggested that all gases contain the same number of atoms or molecules when they occupy the same volume under the same conditions of temperature and pressure.

The number of moles n contained in a sample of any substance is equal to the ratio of the number of molecules N in the sample to the number of molecules N_A in 1 mole:

$$n = \frac{N}{N_A}. \tag{20-2}$$

(*Caution:* The three symbols in this equation can easily be confused with one another, so you should sort them with their meanings now, before you end in "N-confusion.") We can find the number of moles in a sample from the mass M_{sam} of the sample and either the *molar mass* M (the mass of 1 mol) or the molecular mass m (the mass of one molecule):

$$n = \frac{M_{sam}}{M} = \frac{M_{sam}}{mN_A}. \tag{20-3}$$

In Eq. 20-3, we used the fact that the mass M of 1 mol is the product of the mass m of one molecule and the number of molecules N_A in 1 mol:

$$M = mN_A. \tag{20-4}$$

PROBLEM-SOLVING TACTICS

Tactic 1: *Avogadro's Number of What?*
In Eq. 20-1, Avogadro's number is expressed in terms of mol^{-1}, which is the inverse mole, or 1/mol. We could instead explicitly state the elementary unit involved in a given situation. For example, we might write $N_A = 6.02 \times 10^{23}$ atoms/mole if the elementary unit is an atom. If, instead, the elementary unit is a molecule, then we might write $N_A = 6.02 \times 10^{23}$ molecules/mole.

20-3 Ideal Gases

Our goal in this chapter is to explain the macroscopic properties of a gas—such as its pressure and its temperature—in terms of the behavior of the molecules that make it up. However, there is an immediate problem: which gas? Should it be hydrogen or oxygen, or methane, or perhaps uranium hexafluoride? They are all different. Experimenters have found, though, that if we confine 1 mole samples of various gases in boxes of identical volume and hold the gases at the same temperature, then their measured pressures are nearly—though not exactly—the same. If we repeat the measurements at lower gas densities, then these small differences in the measured pressures tend to disappear. Further experiments show that, at low enough densities, all real gases tend to obey the relation

$$pV = nRT \quad \text{(ideal gas law)}, \tag{20-5}$$

in which p is the absolute (not gauge) pressure, n is the number of moles of gas present, and T is the temperature in kelvins. The symbol R is a constant called the **gas constant** that has the same value for all gases—namely,

$$R = 8.31 \text{ J/mol} \cdot \text{K}. \tag{20-6}$$

Equation 20-5 is called the **ideal gas law.** Provided the gas density is low, this law holds for any single gas or for any mixture of different gases. (For a mixture, n is the total number of moles in the mixture.)

We can rewrite Eq. 20-5 in an alternative form, in terms of a constant called the **Boltzmann constant** k, which is defined as

$$k = \frac{R}{N_A} = \frac{8.31 \text{ J/mol} \cdot \text{K}}{6.02 \times 10^{23} \text{ mol}^{-1}} = 1.38 \times 10^{-23} \text{ J/K}. \tag{20-7}$$

This allows us to write $R = kN_A$. Then, with Eq. 20-2 ($n = N/N_A$), we see that

$$nR = Nk. \tag{20-8}$$

Substituting this into Eq. 20-5 gives a second expression for the ideal gas law:

$$pV = NkT \quad \text{(ideal gas law)}. \tag{20-9}$$

(*Caution:* Note the difference between the two expressions for the ideal gas law—Eq. 20-5 involves the number of moles n and Eq. 20-9 involves the number of molecules N.)

You may well ask, "What is an *ideal gas* and what is so 'ideal' about one?" The answer lies in the simplicity of the law (Eqs. 20-5 and 20-9) that governs its macroscopic properties. Using this law—as you will see—we can deduce many properties of the ideal gas in a simple way. Although there is no such thing in nature as a truly ideal gas, *all real* gases approach the ideal state at low enough densities—that is, under conditions in which their molecules are far enough apart that they do not interact with one another. Thus, the ideal gas concept allows us to gain useful insights into the limiting behavior of real gases.

Work Done by an Ideal Gas at Constant Temperature

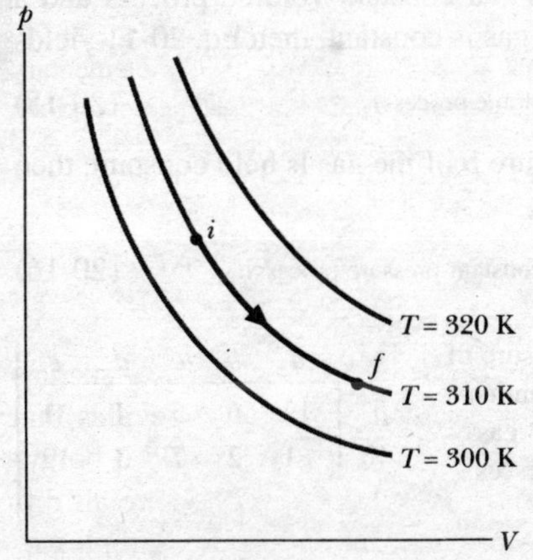

Fig. 20-1 Three isotherms on a p-V diagram. The path shown along the middle isotherm represents an isothermal expansion of a gas from an initial state i to a final state f. The path from f to i along the isotherm would represent the reverse process, that is, an isothermal compression.

∴ V is increasing

Suppose we put an ideal gas in a piston–cylinder arrangement like those in Chapter 19. Suppose also that we allow the gas to expand from an initial volume V_i to a final volume V_f while we keep the temperature T of the gas constant. Such a process, at *constant temperature,* is called an **isothermal expansion** (and the reverse is called an **isothermal compression**).

On a p-V diagram, an *isotherm* is a curve that connects points that have the same temperature. Thus, it is a graph of pressure versus volume for a gas whose temperature T is held constant. For n moles of an ideal gas, it is a graph of the equation

$$p = nRT\frac{1}{V} = (\text{a constant})\,\frac{1}{V}. \qquad (20\text{-}10)$$

$y = mx$ here $y = m\frac{1}{x}$

Figure 20-1 shows three isotherms, each corresponding to a different (constant) value of T. (Note that the values of T for the isotherms increase upward to the right.) Superimposed on the middle isotherm is the path followed by a gas during an isothermal expansion from state i to state f at a constant temperature of 310 K.

To find the work done by an ideal gas during an isothermal expansion, we start with Eq. 19-25,

$$W = \int_{V_i}^{V_f} p\, dV. \qquad (20\text{-}11)$$

This is a general expression for the work done during any change in volume of any gas. For an ideal gas, we can use Eq. 20-5 to substitute for p, obtaining

$$W = \int_{V_i}^{V_f} \frac{nRT}{V}\, dV. \qquad (20\text{-}12)$$

Because we are considering an isothermal expansion, T is constant and we can move it in front of the integral sign to write

$$W = nRT \int_{V_i}^{V_f} \frac{dV}{V} = nRT \Big[\ln V \Big]_{V_i}^{V_f}. \qquad (20\text{-}13)$$

By evaluating the expression in brackets at the limits and then using the relationship $\ln a - \ln b = \ln (a/b)$, we find that

$$W = nRT \ln \frac{V_f}{V_i} \quad \text{(ideal gas, isothermal process).} \qquad (20\text{-}14)$$

Recall that the symbol ln specifies a *natural* logarithm, which has base e.

For an expansion, V_f is greater than V_i, so the ratio V_f/V_i in Eq. 20-14 is greater than unity. The natural logarithm of a quantity greater than unity is positive, and so the work W done by an ideal gas during an isothermal expansion is positive, as we expect. For a compression, V_f is less than V_i, so the ratio of volumes in Eq. 20-14 is less than unity. The natural logarithm in that equation—hence the work W—is negative, again as we expect.

Work Done at Constant Volume and at Constant Pressure

Equation 20-14 does not give the work W done by an ideal gas during *every* thermodynamic process. Instead, it gives the work only for a process in which the temperature is held constant. If the temperature varies, then the symbol T in Eq. 20-12 cannot be moved in front of the integral symbol as in Eq. 20-13, and thus we do not end up with Eq. 20-14.

However, we can go back to Eq. 20-11 to find the work W done by an ideal gas (or any other gas) during two more processes—a constant-volume process and a constant-pressure process. If the volume of the gas is constant, then Eq. 20-11 yields

$$W = 0 \quad \text{(constant-volume process).} \tag{20-15}$$

If, instead, the volume changes while the pressure p of the gas is held constant, then Eq. 20-11 becomes

$$W = p(V_f - V_i) = p\,\Delta V \quad \text{(constant-pressure process).} \tag{20-16}$$

CHECKPOINT 1: An ideal gas has an initial pressure of 3 pressure units and an initial volume of 4 volume units. The table gives the final pressure and volume of the gas (in those same units) in five processes. Which processes start and end on the same isotherm?

	a	b	c	d	e
p	12	6	5	4	1
V	1	2	7	3	12

Sample Problem 20-1

A cylinder contains 12 L of oxygen at 20°C and 15 atm. The temperature is raised to 35°C, and the volume is reduced to 8.5 L. What is the final pressure of the gas in atmospheres? Assume that the gas is ideal.

SOLUTION: The **Key Idea** here is that, because the gas is ideal, its pressure, volume, temperature, and number of moles are related by the ideal gas law, both in the initial state i and in the final state f (after the changes). Thus, from Eq. 20-5 we can write

$$p_iV_i = nRT_i \quad \text{and} \quad p_fV_f = nRT_f.$$

Dividing the second equation by the first equation and solving for p_f yields

$$p_f = \frac{p_iT_fV_i}{T_iV_f}. \tag{20-17}$$

Note here that if we converted the given initial and final volumes from liters to the proper units of cubic meters, the multiplying conversion factors would cancel out of Eq. 20-17. The same would be true for conversion factors that convert the pressures from atmospheres to the proper pascals. However, to convert the given temperatures to kelvins requires the addition of an amount that would not cancel and thus must be included. Hence, we must write

$$T_i = (273 + 20)\text{ K} = 293\text{ K}$$

and

$$T_f = (273 + 35)\text{ K} = 308\text{ K}.$$

Inserting the given data into Eq. 20-17 then yields

$$p_f = \frac{(15\text{ atm})(308\text{ K})(12\text{ L})}{(293\text{ K})(8.5\text{ L})} = 22\text{ atm.} \quad \text{(Answer)}$$

Sample Problem 20-2

One mole of oxygen (assume it to be an ideal gas) expands at a constant temperature T of 310 K from an initial volume V_i of 12 L to a final volume V_f of 19 L. How much work is done by the gas during the expansion?

SOLUTION: The **Key Idea** is this: Generally we find the work by integrating the gas pressure with respect to the gas volume, using Eq. 20-11. However, because the gas here is ideal and the expansion is isothermal, that integration leads to Eq. 20-14. Therefore, we can write

$$W = nRT \ln \frac{V_f}{V_i}$$

$$= (1\text{ mol})(8.31\text{ J/mol}\cdot\text{K})(310\text{ K}) \ln \frac{19\text{ L}}{12\text{ L}}$$

$$= 1180\text{ J.} \quad \text{(Answer)}$$

The expansion is graphed in the p-V diagram of Fig. 20-2. The work done by the gas during the expansion is represented by the area beneath the curve if.

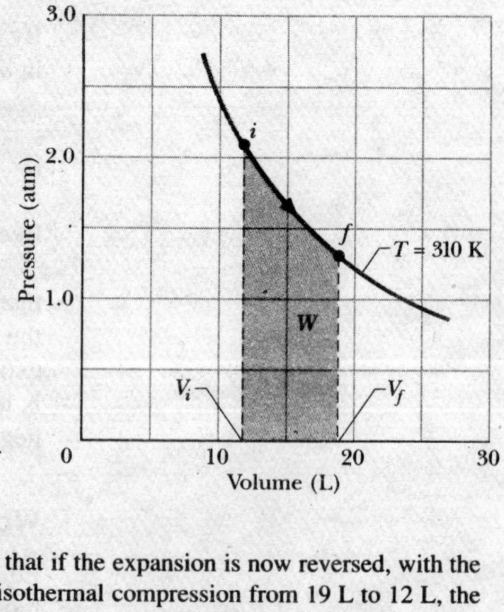

Fig. 20-2 Sample Problem 20-2. The shaded area represents the work done by 1 mol of oxygen in expanding from V_i to V_f at a constant temperature T of 310 K.

You can show that if the expansion is now reversed, with the gas undergoing an isothermal compression from 19 L to 12 L, the work done by the gas will be −1180 J. Thus, an external force would have to do 1180 J of work on the gas to compress it.

20-4 Pressure, Temperature, and RMS Speed

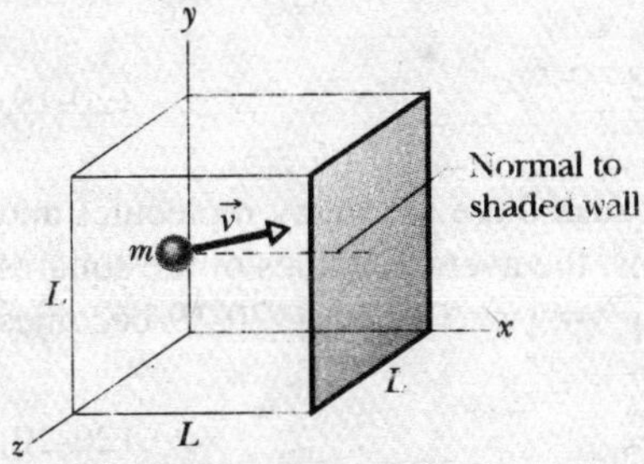

Fig. 20-3 A cubical box of edge L, containing n moles of an ideal gas. A molecule of mass m and velocity $\vec{v}$ is about to collide with the shaded wall of area L^2. A normal to that wall is shown.

Here is our first kinetic theory problem. Let n moles of an ideal gas be confined in a cubical box of volume V, as in Fig. 20-3. The walls of the box are held at temperature T. What is the connection between the pressure p exerted by the gas on the walls and the speeds of the molecules?

The molecules of gas in the box are moving in all directions and with various speeds, bumping into each other and bouncing from the walls of the box like balls in a racquetball court. We ignore (for the time being) collisions of the molecules with one another and consider only elastic collisions with the walls.

Figure 20-3 shows a typical gas molecule, of mass m and velocity $\vec{v}$, that is about to collide with the shaded wall. Because we assume that any collision of a molecule with a wall is elastic, when this molecule collides with the shaded wall, the only component of its velocity that is changed is the x component, and that component is reversed. This means that the only change in the particle's momentum is along the x axis, and that change is

$$\Delta p_x = (-mv_x) - (mv_x) = -2mv_x.$$

Hence, the momentum Δp_x delivered to the wall by the molecule during the collision is $+2mv_x$. (Because in this book the symbol p represents both momentum and pressure, we must be careful to note that here p represents momentum and is a vector quantity.)

The molecule of Fig. 20-3 will hit the shaded wall repeatedly. The time Δt between collisions is the time the molecule takes to travel to the opposite wall and back again (a distance of $2L$) at speed v_x. Thus, Δt is equal to $2L/v_x$. (Note that this result holds even if the molecule bounces off any of the other walls along the way, because those walls are parallel to x and so cannot change v_x.) Therefore, the average rate at which momentum is delivered to the shaded wall by this single molecule is

$$\frac{\Delta p_x}{\Delta t} = \frac{2mv_x}{2L/v_x} = \frac{mv_x^2}{L}.$$

From Newton's second law ($\vec{F} = d\vec{p}/dt$), the rate at which momentum is delivered to the wall is the force acting on that wall. To find the total force, we must add up the contributions of all the molecules that strike the wall, allowing for the possibility that they all have different speeds. Dividing the magnitude of the total force F_x by the area of the wall ($= L^2$) then gives the pressure p on that wall, where now and in the rest of this discussion, p represents pressure. Thus, using the expression for $\Delta p_x/\Delta t$, we can write this pressure as

$$p = \frac{F_x}{L^2} = \frac{mv_{x1}^2/L + mv_{x2}^2/L + \cdots + mv_{xN}^2/L}{L^2}$$

$$= \left(\frac{m}{L^3}\right)(v_{x1}^2 + v_{x2}^2 + \cdots + v_{xN}^2), \qquad (20\text{-}18)$$

where N is the number of molecules in the box.

Since $N = nN_A$, there are nN_A terms in the second parentheses of Eq. 20-18. We can replace that quantity by $nN_A(v_x^2)_{avg}$, where $(v_x^2)_{avg}$ is the average value of the square of the x components of all the molecular speeds. Equation 20-18 then becomes

$$p = \frac{nmN_A}{L^3}(v_x^2)_{avg}.$$

However, mN_A is the molar mass M of the gas (that is, the mass of 1 mol of the gas). Also, L^3 is the volume of the box, so

$$p = \frac{nM(v_x^2)_{avg}}{V}. \tag{20-19}$$

For any molecule, $v^2 = v_x^2 + v_y^2 + v_z^2$. Because there are many molecules and because they are all moving in random directions, the average values of the squares of their velocity components are equal, so that $v_x^2 = \frac{1}{3}v^2$. Thus, Eq. 20-19 becomes

$$p = \frac{nM(v^2)_{avg}}{3V}. \tag{20-20}$$

The square root of $(v^2)_{avg}$ is a kind of average speed, called the **root-mean-square speed** of the molecules and symbolized by v_{rms}. Its name describes it rather well: You *square* each speed, you find the *mean* (that is, the average) of all these squared speeds, and then you take the square *root* of that mean. With $\sqrt{(v^2)_{avg}} = v_{rms}$, we can then write Eq. 20-20 as

$$p = \frac{nMv_{rms}^2}{3V}. \tag{20-21}$$

Equation 20-21 is very much in the spirit of kinetic theory. It tells us how the pressure of the gas (a purely macroscopic quantity) depends on the speed of the molecules (a purely microscopic quantity).

We can turn Eq. 20-21 around and use it to calculate v_{rms}. Combining Eq. 20-21 with the ideal gas law ($pV = nRT$) leads to

$$v_{rms} = \sqrt{\frac{3RT}{M}}. \tag{20-22}$$

Table 20-1 shows some rms speeds calculated from Eq. 20-22. The speeds are surprisingly high. For hydrogen molecules at room temperature (300 K), the rms speed is 1920 m/s or 4300 mi/h—faster than a speeding bullet! On the surface of the Sun, where the temperature is 2×10^6 K, the rms speed of hydrogen molecules would be 82 times greater than at room temperature were it not for the fact that at such high speeds, the molecules cannot survive collisions among themselves. Remember too that the rms speed is only a kind of average speed; many molecules move much faster than this, and some much slower.

TABLE 20-1 Some Molecular Speeds at Room Temperature (T = 300 K)[a]

Gas	Molar Mass (10^{-3} kg/mol)	v_{rms} (m/s)
Hydrogen (H_2)	2.02	1920
Helium (He)	4.0	1370
Water vapor (H_2O)	18.0	645
Nitrogen (N_2)	28.0	517
Oxygen (O_2)	32.0	483
Carbon dioxide (CO_2)	44.0	412
Sulfur dioxide (SO_2)	64.1	342

[a]For convenience, we often set room temperature = 300 K even though (at 27°C or 81°F) that represents a fairly warm room.

The speed of sound in a gas is closely related to the rms speed of the molecules of that gas. In a sound wave, the disturbance is passed on from molecule to molecule by means of collisions. The wave cannot move any faster than the "average" speed of the molecules. In fact, the speed of sound must be somewhat less than this "average" molecular speed because not all molecules are moving in exactly the same direction as the wave. As examples, at room temperature, the rms speeds of hydrogen and nitrogen molecules are 1920 m/s and 517 m/s, respectively. The speeds of sound in these two gases at this temperature are 1350 m/s and 350 m/s, respectively.

A question often arises: If molecules move so fast, why does it take as long as a minute or so before you can smell perfume when someone opens a bottle across a room? The answer is that, as we shall discuss in Section 20-6, each perfume molecule moves away from the bottle only very slowly because its repeated collisions with other molecules prevent it from moving directly across the room to you.

Sample Problem 20-3

Here are five numbers: 5, 11, 32, 67, and 89.

(a) What is the average value n_{avg} of these numbers?

SOLUTION: We find this from

$$n_{avg} = \frac{5 + 11 + 32 + 67 + 89}{5} = 40.8. \quad \text{(Answer)}$$

(b) What is the rms value n_{rms} of these numbers?

SOLUTION: We find this from

$$n_{rms} = \sqrt{\frac{5^2 + 11^2 + 32^2 + 67^2 + 89^2}{5}} = 52.1. \quad \text{(Answer)}$$

The rms value is greater than the average value because the larger numbers—being squared—are relatively more important in forming the rms value. To test this, let us replace 89 in our set of five numbers by 300. The average value of the new set of five numbers (as you should show) is 2.0 times the previous average value. The rms value, however, is 2.7 times the previous rms value.

20-5 Translational Kinetic Energy

We again consider a single molecule of an ideal gas as it moves around in the box of Fig. 20-3, but we now assume that its speed changes when it collides with other molecules. Its translational kinetic energy at any instant is $\frac{1}{2}mv^2$. Its *average* translational kinetic energy over the time that we watch it is

$$K_{avg} = (\tfrac{1}{2}mv^2)_{avg} = \tfrac{1}{2}m(v^2)_{avg} = \tfrac{1}{2}mv_{rms}^2, \qquad (20\text{-}23)$$

in which we make the assumption that the average speed of the molecule during our observation is the same as the average speed of all the molecules at any given time. (Provided the total energy of the gas is not changing and we observe our molecule for long enough, this assumption is appropriate.) Substituting for v_{rms} from Eq. 20-22 leads to

$$K_{avg} = (\tfrac{1}{2}m)\frac{3RT}{M}.$$

However, M/m, the molar mass divided by the mass of a molecule, is simply Avogadro's number. Thus,

$$K_{avg} = \frac{3RT}{2N_A}.$$

Using Eq. 20-7 ($k = R/N_A$), we can then write

$$K_{avg} = \tfrac{3}{2}kT. \qquad (20\text{-}24)$$

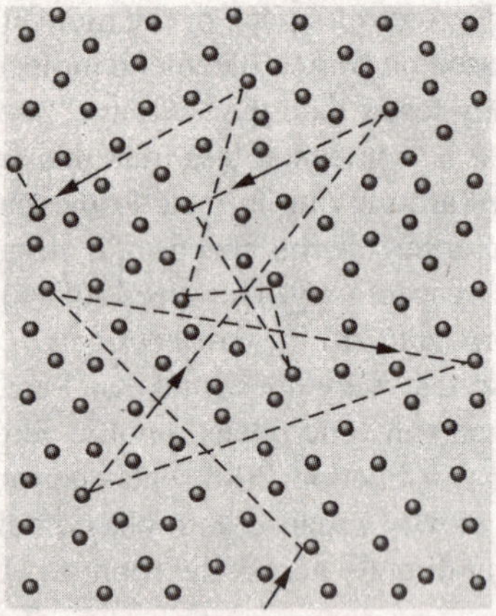

Fig. 20-4 A molecule traveling through a gas, colliding with other gas molecules in its path. Although the other molecules are shown as stationary, they are also moving in a similar fashion.

This equation tells us something unexpected:

At a given temperature T, all ideal gas molecules—no matter what their mass—have the same average translational kinetic energy, namely, $\frac{3}{2}kT$. When we measure the temperature of a gas, we are also measuring the average translational kinetic energy of its molecules.

CHECKPOINT 2: A gas mixture consists of molecules of types 1, 2, and 3, with molecular masses $m_1 > m_2 > m_3$. Rank the three types according to (a) average kinetic energy and (b) rms speed, greatest first.

20-6 Mean Free Path

We continue to examine the motion of molecules in an ideal gas. Figure 20-4 shows the path of a typical molecule as it moves through the gas, changing both speed and direction abruptly as it collides elastically with other molecules. Between collisions, our typical molecule moves in a straight line at constant speed. Although the figure shows all the other molecules as stationary, they are moving similarly.

One useful parameter to describe this random motion is the **mean free path** λ of the molecules. As its name implies, λ is the average distance traversed by a molecule between collisions. We expect λ to vary inversely with N/V, the number of molecules per unit volume (or density of molecules). The larger N/V is, the more collisions there should be and the smaller the mean free path. We also expect λ to vary inversely with the size of the molecules, say, with their diameter d. (If the molecules were points, as we have assumed them to be, they would never collide and the mean free path would be infinite.) Thus, the larger the molecules are, the smaller the mean free path. We can even predict that λ should vary (inversely) as the *square* of the molecular diameter because the cross section of a molecule—not its diameter—determines its effective target area.

The expression for the mean free path does, in fact, turn out to be

$$\lambda = \frac{1}{\sqrt{2}\pi d^2 N/V} \quad \text{(mean free path).} \tag{20-25}$$

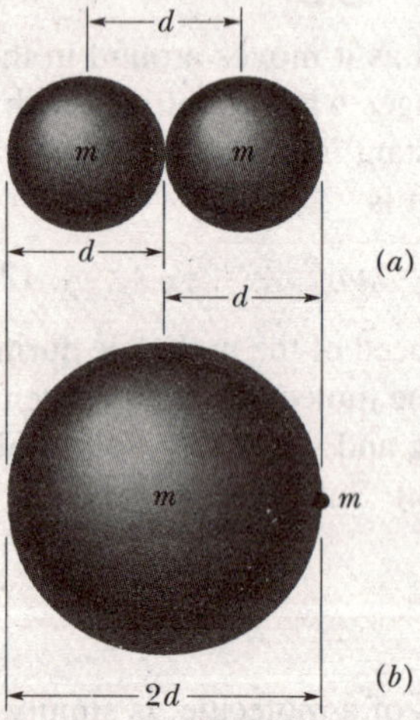

Fig. 20-5 (*a*) A collision occurs when the centers of two molecules come within a distance d of each other, d being the molecular diameter. (*b*) An equivalent but more convenient representation is to think of the moving molecule as having a *radius* d and all other molecules as being points. The condition for a collision is unchanged.

To justify Eq. 20-25, we focus attention on a single molecule and assume—as Fig. 20-4 suggests—that our molecule is traveling with a constant speed v and that all the other molecules are at rest. Later, we shall relax this assumption.

We assume further that the molecules are spheres of diameter d. A collision will then take place if the centers of the molecules come within a distance d of each other, as in Fig. 20-5*a*. Another, more helpful way to look at the situation is to consider our single molecule to have a *radius* of d and all the other molecules to be *points,* as in Fig. 20-5*b*. This does not change our criterion for a collision.

As our single molecule zigzags through the gas, it sweeps out a short cylinder of cross-sectional area πd^2 between successive collisions. If we watch this molecule for a time interval Δt, it moves a distance $v\,\Delta t$, where v is its assumed speed. Thus, if we align all the short cylinders swept out in interval Δt, we form a composite cylinder (Fig. 20-6) of length $v\,\Delta t$ and volume $(\pi d^2)(v\,\Delta t)$. The number of collisions that occur in time Δt is then equal to the number of (point) molecules that lie within this cylinder.

Since N/V is the number of molecules per unit volume, the number of molecules in the cylinder is N/V times the volume of the cylinder, or $(N/V)(\pi d^2 v\,\Delta t)$. This is also the number of collisions in time Δt. The mean free path is the length of the

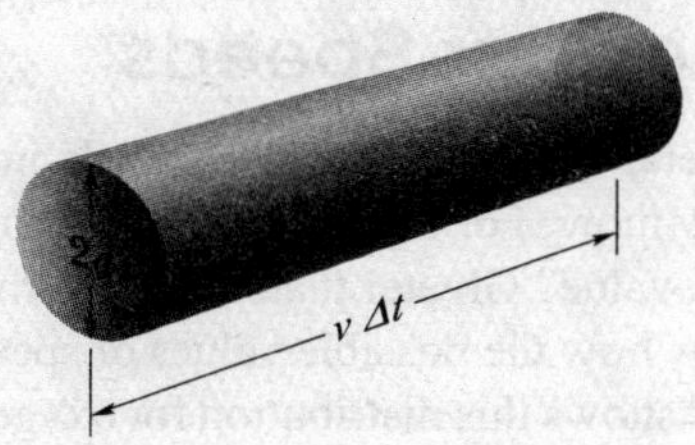

Fig. 20-6 In time Δt the moving molecule effectively sweeps out a cylinder of length $v\,\Delta t$ and radius d.

path (and of the cylinder) divided by this number:

$$\lambda = \frac{\text{length of path during } \Delta t}{\text{number of collisions in } \Delta t} \approx \frac{v\,\Delta t}{\pi d^2 v\,\Delta t\,N/V}$$

$$= \frac{1}{\pi d^2\,N/V}. \tag{20-26}$$

This equation is only approximate because it is based on the assumption that all the molecules except one are at rest. In fact, *all* the molecules are moving; when this is taken properly into account, Eq. 20-25 results. Note that it differs from the (approximate) Eq. 20-26 only by a factor of $1/\sqrt{2}$.

We can even get a glimpse of what is "approximate" about Eq. 20-26. The v in the numerator and that in the denominator are—strictly—not the same. The v in the numerator is v_{avg}, the mean speed of the molecule *relative to the container*. The v in the denominator is v_{rel}, the mean speed of our single molecule *relative to the other molecules,* which are moving. It is this latter average speed that determines the number of collisions. A detailed calculation, taking into account the actual speed distribution of the molecules, gives $v_{rel} = \sqrt{2}\,v_{avg}$ and thus the factor $\sqrt{2}$.

The mean free path of air molecules at sea level is about 0.1 μm. At an altitude of 100 km, the density of air has dropped to such an extent that the mean free path rises to about 16 cm. At 300 km, the mean free path is about 20 km. A problem faced by those who would study the physics and chemistry of the upper atmosphere in the laboratory is the unavailability of containers large enough to hold gas samples that simulate upper atmospheric conditions. Yet studies of the concentrations of Freon, carbon dioxide, and ozone in the upper atmosphere are of vital public concern.

Sample Problem 20-4

(a) What is the mean free path λ for oxygen molecules at temperature $T = 300$ K and pressure $p = 1.0$ atm? Assume that the molecular diameter is $d = 290$ pm and the gas is ideal.

SOLUTION: The **Key Idea** here is that each oxygen molecule moves among other *moving* oxygen molecules in a zigzag path due to the resulting collisions. Thus, we use Eq. 20-25 for the mean free path, for which we need the number of molecules per unit volume, N/V. Because we assume the gas is ideal, we can use the ideal gas law of Eq. 20-9 ($pV = NkT$) to write $N/V = p/kT$. Substituting this into Eq. 20-25, we find

$$\lambda = \frac{1}{\sqrt{2}\pi d^2\,N/V} = \frac{kT}{\sqrt{2}\pi d^2 p}$$

$$= \frac{(1.38 \times 10^{-23}\text{ J/K})(300\text{ K})}{\sqrt{2}\pi(2.9 \times 10^{-10}\text{ m})^2(1.01 \times 10^5\text{ Pa})}$$

$$= 1.1 \times 10^{-7}\text{ m}. \qquad \text{(Answer)}$$

This is about 380 molecular diameters.

(b) Assume the average speed of the oxygen molecules is $v = 450$ m/s. What is the average time t between successive collisions for any given molecule? At what rate does the molecule collide; that is, what is the frequency f of its collisions?

SOLUTION: To find the time t between collisions, we use this **Key Idea**: Between collisions, the molecule travels, on average, the mean free path λ at speed v. Thus, the average time between collisions is

$$t = \frac{\text{distance}}{\text{speed}} = \frac{\lambda}{v} = \frac{1.1 \times 10^{-7}\text{ m}}{450\text{ m/s}}$$

$$= 2.44 \times 10^{-10}\text{ s} \approx 0.24\text{ ns}. \qquad \text{(Answer)}$$

This tells us that, on average, any given oxygen molecule has less than a nanosecond between collisions.

To find the frequency f of the collisions, we use this **Key Idea**: The average rate or frequency at which the collisions occur is the inverse of the time t between collisions. Thus,

$$f = \frac{1}{t} = \frac{1}{2.44 \times 10^{-10}\text{ s}} = 4.1 \times 10^9\text{ s}^{-1}. \qquad \text{(Answer)}$$

This tells us that, on average, any given oxygen molecule makes about 4 billion collisions per second.

CHECKPOINT 3: One mole of gas A, with molecular diameter $2d_0$ and average molecular speed v_0, is placed inside a certain container. One mole of gas B, with molecular diameter d_0 and average molecular speed $2v_0$ (the molecules of B are smaller but faster), is placed in an identical container. Which gas has the greater average collision rate within its container?

20-7 The Distribution of Molecular Speeds

The root-mean-square speed v_{rms} gives us a general idea of molecular speeds in a gas at a given temperature. We often want to know more. For example, what fraction of the molecules have speeds greater than the rms value? Greater than twice the rms value? To answer such questions, we need to know how the possible values of speed are distributed among the molecules. Figure 20-7*a* shows this distribution for oxygen molecules at room temperature ($T = 300$ K); Fig. 20-7*b* compares it with the distribution at $T = 80$ K.

In 1852, Scottish physicist James Clerk Maxwell first solved the problem of finding the speed distribution of gas molecules. His result, known as **Maxwell's speed distribution law,** is

$$P(v) = 4\pi\left(\frac{M}{2\pi RT}\right)^{3/2} v^2 e^{-Mv^2/2RT}. \quad (20\text{-}27)$$

Here v is the molecular speed, T is the gas temperature, M is the molar mass of the gas, and R is the gas constant. It is this equation that is plotted in Fig. 20-7*a*,*b*. The quantity $P(v)$ in Eq. 20-27 and Fig. 20-7 is a *probability distribution function:* For any speed v, the product $P(v)\,dv$ (a dimensionless quantity) is the fraction of molecules whose speeds lie in the interval of width dv centered on speed v.

As Fig. 20-7*a* shows, this fraction is equal to the area of a strip with height $P(v)$ and width dv. The total area under the distribution curve corresponds to the fraction of the molecules whose speeds lie between zero and infinity. All molecules fall into this category, so the value of this total area is unity; that is,

$$\int_0^\infty P(v)\,dv = 1. \quad (20\text{-}28)$$

The fraction (frac) of molecules with speeds in an interval of, say, v_1 to v_2 is then

$$\text{frac} = \int_{v_1}^{v_2} P(v)\,dv. \quad (20\text{-}29)$$

Average, RMS, and Most Probable Speeds

In principle, we can find the **average speed** v_{avg} of the molecules in a gas with the following procedure: We *weight* each value of v in the distribution; that is, we

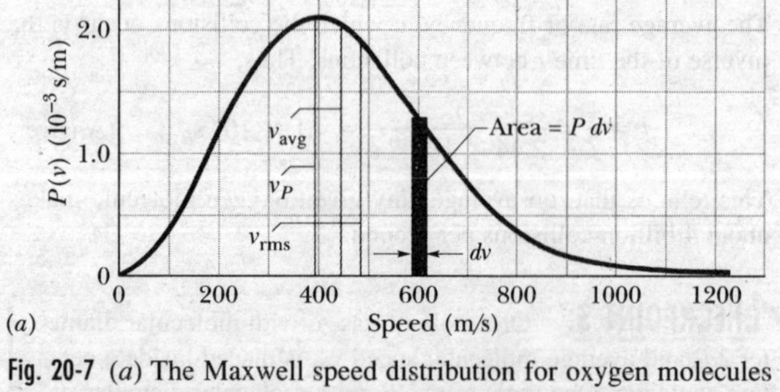

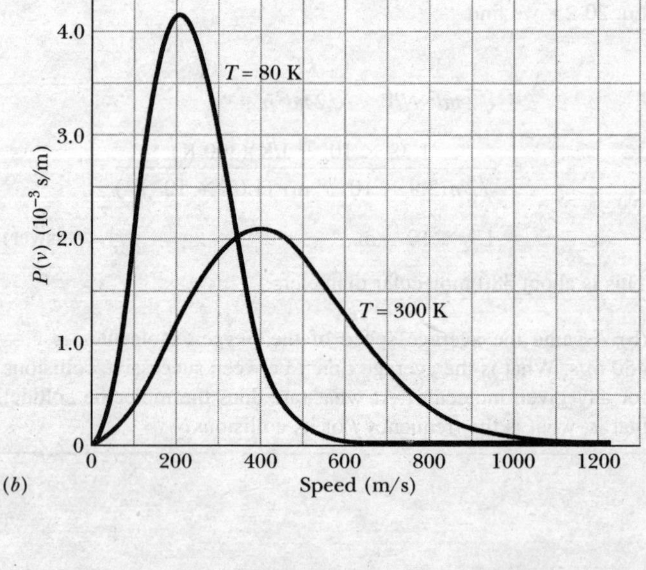

Fig. 20-7 (*a*) The Maxwell speed distribution for oxygen molecules at $T = 300$ K. The three characteristic speeds are marked. (*b*) The curves for 300 K and 80 K. Note that the molecules move more slowly at the lower temperature. Because these are probability distributions, the area under each curve has a numerical value of unity.

multiply it by the fraction $P(v)\,dv$ of molecules with speeds in a differential interval dv centered on v. Then we add up all these values of $v\,P(v)\,dv$. The result is v_{avg}. In practice, we do all this by evaluating

$$v_{avg} = \int_0^\infty v\,P(v)\,dv. \tag{20-30}$$

Substituting for $P(v)$ from Eq. 20-27 and using generic integral 20 from the list of integrals in Appendix E, we find

$$v_{avg} = \sqrt{\frac{8RT}{\pi M}} \quad \text{(average speed).} \tag{20-31}$$

Similarly, we can find the average of the square of the speeds $(v^2)_{avg}$ with

$$(v^2)_{avg} = \int_0^\infty v^2\,P(v)\,dv. \tag{20-32}$$

Substituting for $P(v)$ from Eq. 20-27 and using generic integral 16 from the list of integrals in Appendix E, we find

$$(v^2)_{avg} = \frac{3RT}{M}. \tag{20-33}$$

The square root of $(v^2)_{avg}$ is the **root-mean-square speed** v_{rms}. Thus,

$$v_{rms} = \sqrt{\frac{3RT}{M}} \quad \text{(rms speed),} \tag{20-34}$$

which agrees with Eq. 20-22.

The **most probable speed** v_P is the speed at which $P(v)$ is maximum (see Fig. 20-7*a*). To calculate v_P, we set $dP/dv = 0$ (the slope of the curve in Fig. 20-7*a* is zero at the maximum of the curve) and then solve for v. Doing so, we find

$$v_P = \sqrt{\frac{2RT}{M}} \quad \text{(most probable speed).} \tag{20-35}$$

A molecule is more likely to have speed v_P than any other speed, but some molecules will have speeds that are many times v_P. These molecules lie in the *high-speed tail* of a distribution curve like that in Fig. 20-7*a*. We should be thankful for these few, higher speed molecules because they make possible both rain and sunshine (without which we could not exist). We next see why.

Rain: The speed distribution of water molecules in, say, a pond at summertime temperatures can be represented by a curve similar to that of Fig. 20-7*a*. Most of the molecules do not have nearly enough kinetic energy to escape from the water through its surface. However, small numbers of very fast molecules with speeds far out in the tail of the curve can do so. It is these water molecules that evaporate, making clouds and rain a possibility.

As the fast water molecules leave the surface, carrying energy with them, the temperature of the remaining water is maintained by heat transfer from the surroundings. Other fast molecules—produced in particularly favorable collisions—quickly take the place of those that have left, and the speed distribution is maintained.

Sunshine: Let the distribution curve of Fig. 20-7*a* now refer to protons in the core of the Sun. The Sun's energy is supplied by a nuclear fusion process that starts with the merging of two protons. However, protons repel each other because of their electrical charges, and protons of average speed do not have enough kinetic energy to overcome the repulsion and get close enough to merge. Very fast protons with speeds in the tail of the distribution curve can do so, however, and for that reason the Sun can shine.

Sample Problem 20-5

A container is filled with oxygen gas maintained at room temperature (300 K). What fraction of the molecules have speeds in the interval 599 to 601 m/s? The molar mass M of oxygen is 0.0320 kg/mol.

SOLUTION: The **Key Ideas** here are

1. The speeds of the molecules are distributed over a wide range of values, with the distribution $P(v)$ of Eq. 20-27.
2. The fraction of the molecules with speeds in a differential interval dv is $P(v)\,dv$.
3. For a larger interval, the fraction is found by integrating $P(v)$ over the interval.
4. However, the interval $\Delta v = 2$ m/s here is small compared to the speed $v = 600$ m/s on which it is centered.

Thus, we can avoid the integration by approximating the fraction as

$$\text{frac} = P(v)\,\Delta v = 4\pi \left(\frac{M}{2\pi RT}\right)^{3/2} v^2 e^{-Mv^2/2RT}\,\Delta v.$$

The function $P(v)$ is plotted in Fig. 20-7a. The total area between the curve and the horizontal axis represents the total fraction of molecules (unity). The area of the thin gold strip represents the fraction we seek.

To evaluate frac in parts, we can write

$$\text{frac} = (4\pi)(A)(v^2)(e^B)(\Delta v). \tag{20-36}$$

A and B are

$$A = \left(\frac{M}{2\pi RT}\right)^{3/2} = \left(\frac{0.0320 \text{ kg/mol}}{(2\pi)(8.31 \text{ J/mol}\cdot\text{K})(300 \text{ K})}\right)^{3/2}$$
$$= 2.92 \times 10^{-9} \text{ s}^3/\text{m}^3$$

and $$B = -\frac{Mv^2}{2RT} = -\frac{(0.0320 \text{ kg/mol})(600 \text{ m/s})^2}{(2)(8.31 \text{ J/mol}\cdot\text{K})(300 \text{ K})} = -2.31.$$

Substituting A and B into Eq. 20-36 yields

$$\text{frac} = (4\pi)(A)(v^2)(e^B)(\Delta v)$$
$$= (4\pi)(2.92 \times 10^{-9} \text{ s}^3/\text{m}^3)(600 \text{ m/s})^2(e^{-2.31})(2 \text{ m/s})$$
$$= 2.62 \times 10^{-3}. \quad \text{(Answer)}$$

Thus, at room temperature, 0.262% of the oxygen molecules will have speeds that lie in the narrow range between 599 and 601 m/s. If the gold strip of Fig. 20-7a were drawn to the scale of this problem, it would be a very thin strip indeed.

Sample Problem 20-6

The molar mass M of oxygen is 0.0320 kg/mol.

(a) What is the average speed v_{avg} of oxygen gas molecules at $T =$ 300 K?

SOLUTION: The **Key Idea** here is that to find the average speed, we must weight speed v with the distribution function $P(v)$ of Eq. 20-27 and then integrate the resulting expression over the range of possible speeds (0 to ∞). That leads to Eq. 20-31, which gives us

$$v_{avg} = \sqrt{\frac{8RT}{\pi M}}$$
$$= \sqrt{\frac{8(8.31 \text{ J/mol}\cdot\text{K})(300 \text{ K})}{\pi(0.0320 \text{ kg/mol})}} = 445 \text{ m/s}. \quad \text{(Answer)}$$

This result is plotted in Fig. 20-7a.

(b) What is the root-mean-square speed v_{rms} at 300 K?

SOLUTION: The **Key Idea** here is that to find v_{rms}, we must first find $(v^2)_{avg}$ by weighting v^2 with the distribution function $P(v)$ of Eq. 20-27 and then integrating the expression over the range of possible speeds. Then we must take the square root of the result. That leads to Eq. 20-34, which gives us

$$v_{rms} = \sqrt{\frac{3RT}{M}}$$
$$= \sqrt{\frac{3(8.31 \text{ J/mol}\cdot\text{K})(300 \text{ K})}{0.0320 \text{ kg/mol}}} = 483 \text{ m/s}. \quad \text{(Answer)}$$

This result is plotted in Fig. 20-7a. It is greater than v_{avg} because the greater speed values influence the calculation more when we integrate the v^2 values than when we integrate the v values.

(c) What is the most probable speed v_P at 300 K?

SOLUTION: The **Key Idea** here is that v_P corresponds to the maximum of the distribution function $P(v)$, which we obtain by setting the derivative $dP/dv = 0$ and solving the result for v. That leads to Eq. 20-35, which gives us

$$v_P = \sqrt{\frac{2RT}{M}}$$
$$= \sqrt{\frac{2(8.31 \text{ J/mol}\cdot\text{K})(300 \text{ K})}{0.0320 \text{ kg/mol}}} = 395 \text{ m/s}. \quad \text{(Answer)}$$

This result is also plotted in Fig. 20-7a.

20-8 The Molar Specific Heats of an Ideal Gas

In this section, we want to derive from molecular considerations an expression for the internal energy E_{int} of an ideal gas. In other words, we want an expression for the energy associated with the random motions of the atoms or molecules in the gas. We shall then use that expression to derive the molar specific heats of an ideal gas.

Internal Energy E_{int}

Let us first assume that our ideal gas is a *monatomic gas* (which has individual atoms rather than molecules), such as helium, neon, or argon. Let us also assume that the internal energy E_{int} of our ideal gas is simply the sum of the translational kinetic energies of its atoms. (As explained by quantum theory, individual atoms do not have rotational kinetic energy.)

The average translational kinetic energy of a single atom depends only on the gas temperature and is given by Eq. 20-24 as $K_{avg} = \frac{3}{2}kT$. A sample of n moles of such a gas contains nN_A atoms. The internal energy E_{int} of the sample is then

$$E_{int} = (nN_A)K_{avg} = (nN_A)(\tfrac{3}{2}kT). \tag{20-37}$$

Using Eq. 20-7 ($k = R/N_A$), we can rewrite this as

$$E_{int} = \tfrac{3}{2}nRT \quad \text{(monatomic ideal gas).} \tag{20-38}$$

Thus,

> The internal energy E_{int} of an ideal gas is a function of the gas temperature *only*; it does not depend on any other variable.

With Eq. 20-38 in hand, we are now able to derive an expression for the molar specific heat of an ideal gas. Actually, we shall derive two expressions. One is for the case in which the volume of the gas remains constant as energy is transferred to or from it as heat. The other is for the case in which the pressure of the gas remains constant as energy is transferred to or from it as heat. The symbols for these two molar specific heats are C_V and C_p, respectively. (By convention, the capital letter C is used in both cases, even though C_V and C_p represent types of specific heat and not heat capacities.)

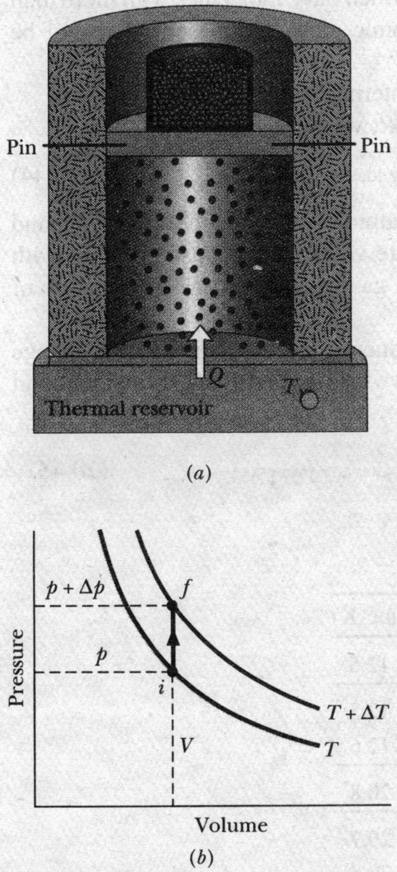

Fig. 20-8 (*a*) The temperature of an ideal gas is raised from T to $T + \Delta T$ in a constant-volume process. Heat is added, but no work is done. (*b*) The process on a *p-V* diagram.

Molar Specific Heat at Constant Volume

Figure 20-8*a* shows n moles of an ideal gas at pressure p and temperature T, confined to a cylinder of fixed volume V. This *initial state i* of the gas is marked on the *p-V* diagram of Fig. 20-8*b*. Suppose now that you add a small amount of energy to the gas as heat Q by slowly turning up the temperature of the thermal reservoir. The gas temperature rises a small amount to $T + \Delta T$, and its pressure rises to $p + \Delta p$, bringing the gas to *final state f*.

In such experiments, we would find that the heat Q is related to the temperature change ΔT by

$$Q = nC_V\,\Delta T \quad \text{(constant volume),} \tag{20-39}$$

where C_V is a constant called the **molar specific heat at constant volume.** Substituting this expression for Q into the first law of thermodynamics as given by

Eq. 19-26 ($\Delta E_{int} = Q - W$) yields

$$\Delta E_{int} = nC_V \, \Delta T - W. \tag{20-40}$$

With the volume held constant, the gas cannot expand and thus cannot do any work. Therefore, $W = 0$, and Eq. 20-40 gives us

$$C_V = \frac{\Delta E_{int}}{n \, \Delta T}. \tag{20-41}$$

From Eq. 20-38 we know that $E_{int} = \frac{3}{2}nRT$, so the change in internal energy must be

$$\Delta E_{int} = \tfrac{3}{2}nR \, \Delta T. \tag{20-42}$$

Substituting this result into Eq. 20-41 yields

$$C_V = \tfrac{3}{2}R = 12.5 \text{ J/mol} \cdot \text{K} \quad \text{(monatomic gas)}. \tag{20-43}$$

As Table 20-2 shows, this prediction of the kinetic theory (for ideal gases) agrees very well with experiment for real monatomic gases, the case that we have assumed. The (predicted and) experimental values of C_V for *diatomic gases* (which have molecules with two atoms) and *polyatomic gases* (which have molecules with more than two atoms) are greater than those for monatomic gases for reasons that will be suggested in Section 20-9.

We can now generalize Eq. 20-38 for the internal energy of any ideal gas by substituting C_V for $\frac{3}{2}R$; we get

$$E_{int} = nC_VT \quad \text{(any ideal gas)}. \tag{20-44}$$

This equation applies not only to an ideal monatomic gas but also to diatomic and polyatomic ideal gases, provided the appropriate value of C_V is used. Just as with Eq. 20-38, we see that the internal energy of a gas depends on the temperature of the gas but not on its pressure or density.

When an ideal gas that is confined to a container undergoes a temperature change ΔT, then from either Eq. 20-41 or Eq. 20-44 we can write the resulting change in its internal energy as

$$\Delta E_{int} = nC_V \, \Delta T \quad \text{(ideal gas, any process)}. \tag{20-45}$$

TABLE 20-2 Molar Specific Heats at Constant Volume

Molecule		Example	C_V (J/mol · K)
Monatomic	Ideal		$\frac{3}{2}R = 12.5$
	Real	He	12.5
		Ar	12.6
Diatomic	Ideal		$\frac{5}{2}R = 20.8$
	Real	N_2	20.7
		O_2	20.8
Polyatomic	Ideal		$3R = 24.9$
	Real	NH_4	29.0
		CO_2	29.7

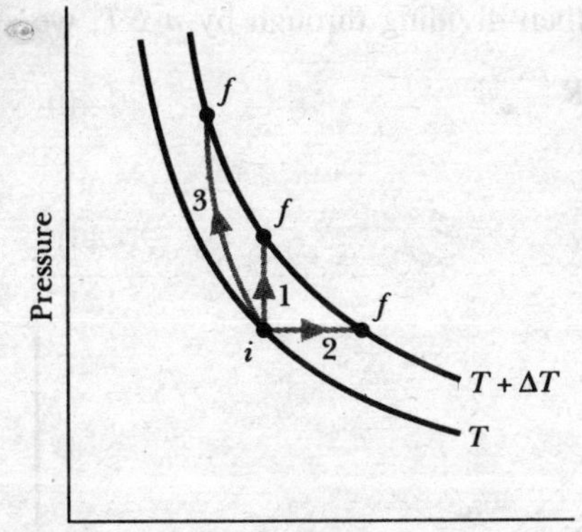

Fig. 20-9 Three paths for three different processes that take an ideal gas from an initial state i at temperature T to some final state f at temperature $T + \Delta T$. The change ΔE_{int} in the internal energy of the gas is the same for these three processes and for any others that result in the same change of temperature.

This equation tells us:

A change in the internal energy E_{int} of a confined ideal gas depends on the change in the gas temperature only; it does *not* depend on what type of process produces the change in the temperature.

As examples, consider the three paths between the two isotherms in the p-V diagram of Fig. 20-9. Path 1 represents a constant-volume process. Path 2 represents a constant-pressure process (that we are about to examine). Path 3 represents a process in which no heat is exchanged with the system's environment (we discuss this in Section 20-11). Although the values of heat Q and work W associated with these three paths differ, as do p_f and V_f, the values of ΔE_{int} associated with the three paths are identical and are all given by Eq. 20-45, because they all involve the same temperature change ΔT. Therefore, no matter what path is actually taken between T and $T + \Delta T$, we can *always* use path 1 and Eq. 20-45 to compute ΔE_{int} easily.

Molar Specific Heat at Constant Pressure

We now assume that the temperature of the ideal gas is increased by the same small amount ΔT as previously, but that the necessary energy (heat Q) is added with the gas under constant pressure. An experiment for doing this is shown in Fig. 20-10a; the p-V diagram for the process is plotted in Fig. 20-10b. From such experiments we find that the heat Q is related to the temperature change ΔT by

$$Q = nC_p \, \Delta T \quad \text{(constant pressure)}, \tag{20-46}$$

where C_p is a constant called the **molar specific heat at constant pressure.** This C_p is *greater* than the molar specific heat at constant volume C_V, because energy must now be supplied not only to raise the temperature of the gas but also for the gas to do work—that is, to lift the weighted piston of Fig. 20-10a.

To relate molar specific heats C_p and C_V, we start with the first law of thermodynamics (Eq. 19-26):

$$\Delta E_{\text{int}} = Q - W. \tag{20-47}$$

We next replace each term in Eq. 20-47. For ΔE_{int}, we substitute from Eq. 20-45. For Q, we substitute from Eq. 20-46. To replace W, we first note that since the pressure remains constant, Eq. 20-16 tells us that $W = p \, \Delta V$. Then we note that, using the ideal gas equation ($pV = nRT$), we can write

$$W = p \, \Delta V = nR \, \Delta T. \tag{20-48}$$

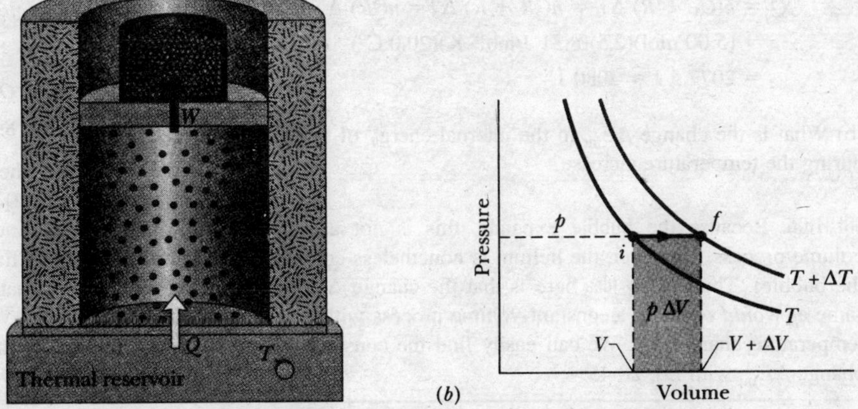

Fig. 20-10 (a) The temperature of an ideal gas is raised from T to $T + \Delta T$ in a constant-pressure process. Heat is added and work is done in lifting the loaded piston. (b) The process on a p-V diagram. The work $p \, \Delta V$ is given by the shaded area.

Making these substitutions in Eq. 20-47, and then dividing through by $n\ \Delta T$, we find

$$C_V = C_p - R$$

and then

$$C_p = C_V + R. \tag{20-49}$$

This prediction of kinetic theory agrees well with experiment, not only for monatomic gases but for gases in general, as long as their density is low enough so that we may treat them as ideal.

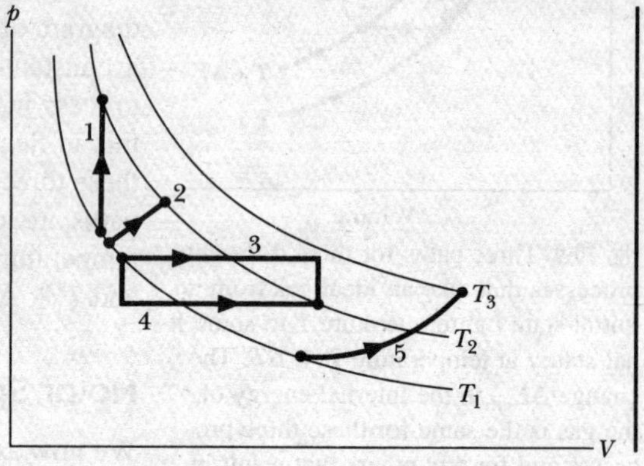

CHECKPOINT 4: The figure here shows five paths traversed by a gas on a p-V diagram. Rank the paths according to the change in internal energy of the gas, greatest first.

Sample Problem 20-7

A bubble of 5.00 mol of helium is submerged at a certain depth in liquid water when the water (and thus the helium) undergoes a temperature increase ΔT of 20.0 C° at constant pressure. As a result, the bubble expands. The helium is monatomic and ideal.

(a) How much energy is added as heat to the helium during the increase and expansion?

SOLUTION: One **Key Idea** here is that the heat Q is related to the temperature change ΔT by a molar specific heat of the gas. Because the pressure p is held constant during the addition of energy, we use the molar specific heat at constant pressure C_p and Eq. 20-46,

$$Q = nC_p\ \Delta T, \tag{20-50}$$

to find Q. To evaluate C_p we go to Eq. 20-49, which tells us that for any ideal gas, $C_p = C_V + R$. Then from Eq. 20-43, we know that for any *monatomic* gas (like the helium here), $C_V = \frac{3}{2}R$. Thus, Eq. 20-50 gives us

$$\begin{aligned} Q &= n(C_V + R)\ \Delta T = n(\tfrac{3}{2}R + R)\ \Delta T = n(\tfrac{5}{2}R)\ \Delta T \\ &= (5.00\ \text{mol})(2.5)(8.31\ \text{J/mol}\cdot\text{K})(20.0\ \text{C}^\circ) \\ &= 2077.5\ \text{J} \approx 2080\ \text{J}. \end{aligned}$$ (Answer)

(b) What is the change ΔE_{int} in the internal energy of the helium during the temperature increase?

SOLUTION: Because the bubble expands, this is not a constant-volume process. However, the helium is nonetheless confined (to the bubble). Thus, a **Key Idea** here is that the change ΔE_{int} is the same as *would occur* in a constant-volume process with the same temperature change ΔT. We can easily find the constant-volume change ΔE_{int} with Eq. 20-45:

$$\begin{aligned} \Delta E_{int} &= nC_V\ \Delta T = n(\tfrac{3}{2}R)\ \Delta T \\ &= (5.00\ \text{mol})(1.5)(8.31\ \text{J/mol}\cdot\text{K})(20.0\ \text{C}^\circ) \\ &= 1246.5\ \text{J} \approx 1250\ \text{J}. \end{aligned}$$ (Answer)

(c) How much work W is done by the helium as it expands against the pressure of the surrounding water during the temperature increase?

SOLUTION: One **Key Idea** here is that the work done by *any* gas expanding against the pressure from its environment is given by Eq. 20-11, which tells us to integrate $p\ dV$. When the pressure is constant (as here), we can simplify that to $W = p\ \Delta V$. When the gas is *ideal* (as here), we can use the ideal gas law (Eq. 20-5) to write $p\ \Delta V = nR\ \Delta T$. We end up with

$$\begin{aligned} W &= nR\ \Delta T \\ &= (5.00\ \text{mol})(8.31\ \text{J/mol}\cdot\text{K})(20.0\ \text{C}^\circ) \\ &= 831\ \text{J}. \end{aligned}$$ (Answer)

Because we happen to know Q and ΔE_{int}, we can work this problem another way: The **Key Idea** now is that we can account for the energy changes of the gas with the first law of thermodynamics, writing

$$\begin{aligned} W &= Q - \Delta E_{int} = 2077.5\ \text{J} - 1246.5\ \text{J} \\ &= 831\ \text{J}. \end{aligned}$$ (Answer)

Note that during the temperature increase, only a portion (1250 J) of the energy (2080 J) that is transferred to the helium as heat goes to increasing the internal energy of the helium and thus the temperature of the helium. The rest (831 J) is transferred out of the helium as work that the helium does during the expansion. If the water were frozen, it would not allow that expansion. Then the same temperature increase of 20.0 C° would require only 1250 J of heat, because no work would be done by the helium.

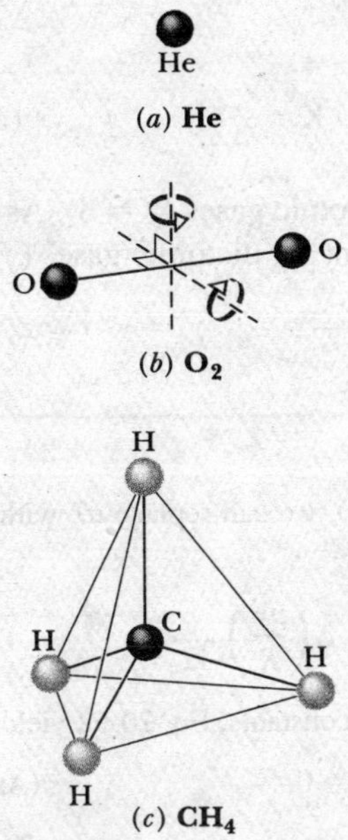

Fig. 20-11 Models of molecules as used in kinetic theory: (*a*) helium, a typical monatomic molecule; (*b*) oxygen, a typical diatomic molecule; and (*c*) methane, a typical polyatomic molecule. The spheres represent atoms, and the lines between them represent bonds. Two rotation axes are shown for the oxygen molecule.

20-9 Degrees of Freedom and Molar Specific Heats

As Table 20-2 shows, the prediction that $C_V = \frac{3}{2}R$ agrees with experiment for monatomic gases but fails for diatomic and polyatomic gases. Let us try to explain the discrepancy by considering the possibility that molecules with more than one atom can store internal energy in forms other than translational motion.

Figure 20-11 shows common models of helium (a *monatomic* molecule, containing a single atom), oxygen (a *diatomic* molecule, containing two atoms), and methane (a *polyatomic* molecule). From such models, we would assume that all three types of molecules can have translational motions (say, moving left–right and up–down) and rotational motions (spinning about an axis like a top). In addition, we would assume that the diatomic and polyatomic molecules can have oscillatory motions, with the atoms oscillating slightly toward and away from one another, as if attached to opposite ends of a spring.

To keep account of the various ways in which energy can be stored in a gas, James Clerk Maxwell introduced the theorem of the **equipartition of energy:**

> Every kind of molecule has a certain number f of *degrees of freedom*, which are independent ways in which the molecule can store energy. Each such degree of freedom has associated with it—on average—an energy of $\frac{1}{2}kT$ per molecule (or $\frac{1}{2}RT$ per mole).

Let us apply the theorem to the translational and rotational motions of the molecules in Fig. 20-11. (We discuss oscillatory motion in the next section.) For the translational motion, superimpose an xyz coordinate system on any gas. The molecules will, in general, have velocity components along all three axes. Thus, gas molecules of all types have three degrees of translational freedom (three ways to move in translation) and, on average, an associated energy of $3(\frac{1}{2}kT)$ per molecule.

For the rotational motion, imagine the origin of our xyz coordinate system at the center of each molecule in Fig. 20-11. In a gas, each molecule should be able to rotate with an angular velocity component along each of the three axes, so each gas should have three degrees of rotational freedom and, on average, an additional energy of $3(\frac{1}{2}kT)$ per molecule. *However*, experiment shows this is true only for the polyatomic molecules. As explained by quantum theory, a monatomic gas molecule does not rotate and so has no rotational energy (a single atom cannot rotate like a top). A diatomic molecule can rotate like a top only about axes perpendicular to the line connecting the atoms (the axes are shown in Fig. 20-11*b*) and not about that line itself. Therefore, a diatomic molecule can have only two degrees of rotational freedom and a rotational energy of only $2(\frac{1}{2}kT)$ per molecule.

To extend our analysis of molar specific heats (C_p and C_V, in Section 20-8) to ideal diatomic and polyatomic gases, it is necessary to retrace the derivations of that analysis in detail. First, we replace Eq. 20-38 ($E_{int} = \frac{3}{2}nRT$) with $E_{int} = (f/2)nRT$, where f is the number of degrees of freedom listed in Table 20-3. Doing so leads to

TABLE 20-3 Degrees of Freedom for Various Molecules

		Degrees of Freedom			Predicted Molar Specific Heats	
Molecule	Example	Translational	Rotational	Total (f)	C_V (Eq. 20-51)	$C_p = C_V + R$
Monatomic	He	3	0	3	$\frac{3}{2}R$	$\frac{5}{2}R$
Diatomic	O_2	3	2	5	$\frac{5}{2}R$	$\frac{7}{2}R$
Polyatomic	CH_4	3	3	6	$3R$	$4R$

the prediction

$$C_V = \left(\frac{f}{2}\right)R = 4.16f \text{ J/mol} \cdot \text{K}, \tag{20-51}$$

which agrees—as it must—with Eq. 20-43 for monatomic gases ($f = 3$). As Table 20-2 shows, this prediction also agrees with experiment for diatomic gases ($f = 5$), but it is too low for polyatomic gases.

Sample Problem 20-8

A cabin of volume V is filled with air (which we consider to be an ideal diatomic gas) at an initial low temperature T_1. After you light a wood stove, the air temperature increases to T_2. What is the resulting change ΔE_{int} in the internal energy of the air in the cabin?

SOLUTION: As the air temperature increases, the air pressure p cannot change but must always be equal to the air pressure outside the room. The reason is that, because the room is not air-tight, the air is not confined. As the temperature increases, air molecules leave through various openings and thus the number of moles n of air in the room decreases. Thus, one **Key Idea** here is that we *cannot* use Eq. 20-45 ($\Delta E_{int} = nC_V\,\Delta T$) to find ΔE_{int}, because it requires constant n.

A second **Key Idea** is that we *can* relate the internal energy E_{int} at any instant to n and the temperature T with Eq. 20-44 ($E_{int} = nC_VT$). From that equation we can then write

$$\Delta E_{int} = \Delta(nC_VT) = C_V\,\Delta(nT).$$

Next, using Eq. 20-5 ($pV = nRT$), we can replace nT with pV/R, obtaining

$$\Delta E_{int} = C_V\,\Delta\left(\frac{pV}{R}\right). \tag{20-52}$$

Now, because p, V, and R are all constants, Eq. 20-52 yields

$$\Delta E_{int} = 0, \qquad \text{(Answer)}$$

even though the temperature changes.

Why does the cabin feel more comfortable at the higher temperature? There are at least two factors involved: (1) You exchange electromagnetic radiation (thermal radiation) with surfaces inside the room, and (2) you exchange energy with air molecules that collide with you. When the room temperature is increased, (1) the amount of thermal radiation emitted by the surfaces and absorbed by you is increased, and (2) the amount of energy you gain through the collisions of air molecules with you is increased.

20-10 A Hint of Quantum Theory

We can improve the agreement of kinetic theory with experiment by including the oscillations of the atoms in a gas of diatomic or polyatomic molecules. For example, the two atoms in the O_2 molecule of Fig. 20-11b can oscillate toward and away from each other, with the interconnecting bond acting like a spring. However, experiment shows that such oscillations occur only at relatively high temperatures of the gas—the motion is "turned on" only when the gas molecules have relatively large energies. Rotational motion is also subject to such "turning on," but at a lower temperature.

Figure 20-12 is of help in seeing this turning on of rotational motion and oscillatory motion. The ratio C_V/R for diatomic hydrogen gas (H_2) is plotted there against temperature, with the temperature scale logarithmic to cover several orders of magnitude. Below about 80 K, we find that $C_V/R = 1.5$. This result implies that only the three translational degrees of freedom of hydrogen are involved in the specific heat.

⇒ C_V is not const as predicted by kinetic theory. It is not independent of temp (implicit dependence)

As the temperature increases, the value of C_V/R gradually increases to 2.5, implying that two additional degrees of freedom have become involved. Quantum theory shows that these two degrees of freedom are associated with the rotational motion of the hydrogen molecules and that this motion requires a certain minimum amount of energy. At very low temperatures (below 80 K), the molecules do not have enough energy to rotate. As the temperature increases from 80 K, first a few molecules and then more and more obtain enough energy to rotate, and C_V/R increases, until all of them are rotating and $C_V/R = 2.5$.

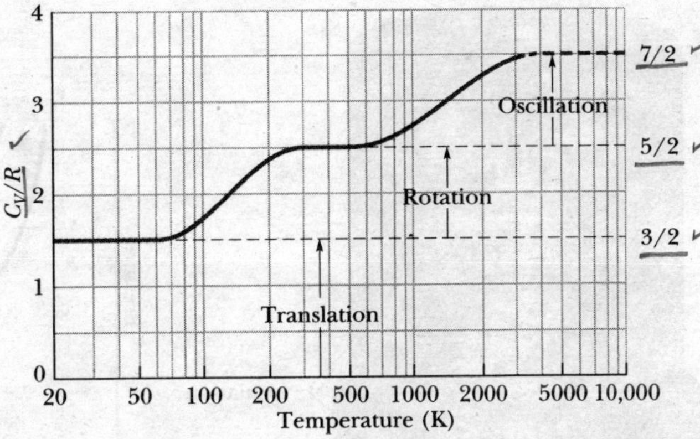

Fig. 20-12 A plot of C_V/R verses temperature for (diatomic) hydrogen gas. Because rotational and oscillatory motions begin at certain energies, only translation is possible at very low temperatures. As the temperature increases, rotational motion can begin. At still higher temperatures, oscillatory motion can begin.

Similarly, quantum theory shows that oscillatory motion of the molecules requires a certain (higher) minimum amount of energy. This minimum amount is not met until the molecules reach a temperature of about 1000 K, as shown in Fig. 20-12. As the temperature increases beyond 1000 K, the number of molecules with enough energy to oscillate increases, and C_V/R increases, until all of them are oscillating and $C_V/R = 3.5$. (In Fig. 20-12, the plotted curve stops at 3200 K because at that temperature, the atoms of a hydrogen molecule oscillate so much that they overwhelm their bond, and the molecule then *dissociates* into two separate atoms.)

20-11 The Adiabatic Expansion of an Ideal Gas

We saw in Section 18-3 that sound waves are propagated through air and other gases as a series of compressions and expansions; these variations in the transmission medium take place so rapidly that there is no time for energy to be transferred from one part of the medium to another as heat. As we saw in Section 19-10, a process for which $Q = 0$ is an *adiabatic process*. We can ensure that $Q = 0$ either by carrying out the process very quickly (as in sound waves) or by doing it (at any rate) in a well-insulated container. Let us see what the kinetic theory has to say about adiabatic processes.

Figure 20-13*a* shows our usual insulated cylinder, now containing an ideal gas and resting on an insulating stand. By removing mass from the piston, we can allow the gas to expand adiabatically. As the volume increases, both the pressure and the temperature drop. We shall prove next that the relation between the pressure and the volume during such an adiabatic process is

$$pV^{\gamma} = \text{a constant} \quad \text{(adiabatic process)}, \tag{20-53}$$

in which $\gamma = C_p/C_V$, the ratio of the molar specific heats for the gas. On a p-V diagram such as that in Fig. 20-13*b*, the process occurs along a line (called an *adiabat*) that has the equation $p = (\text{a constant})/V^{\gamma}$. Since the gas goes from an initial state i to a final state f, we can rewrite Eq. 20-53 as

$$p_i V_i^{\gamma} = p_f V_f^{\gamma} \quad \text{(adiabatic process)}. \tag{20-54}$$

We can also write an equation for an adiabatic process in terms of T and V. To do so, we use the ideal gas equation ($pV = nRT$) to eliminate p from Eq. 20-53, finding

$$\left(\frac{nRT}{V}\right)V^{\gamma} = \text{a constant.}$$

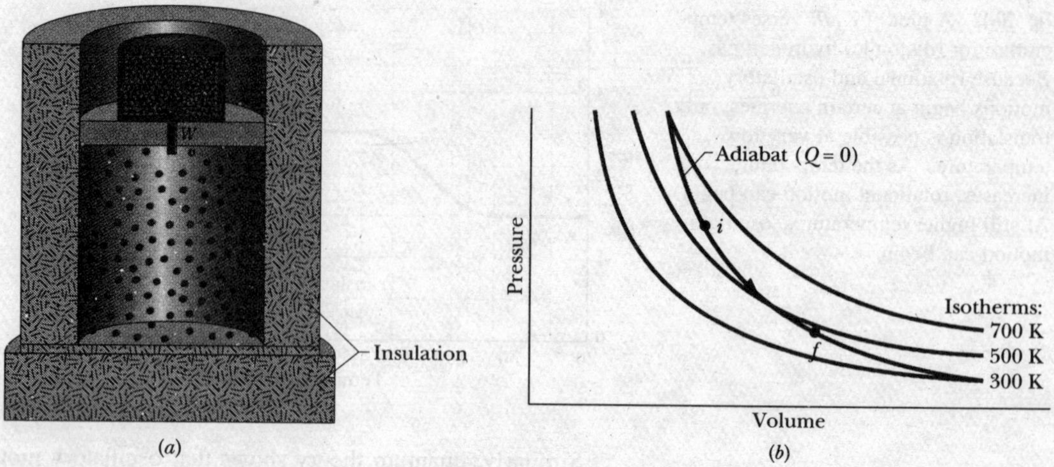

Fig. 20-13 (*a*) The volume of an ideal gas is increased by removing mass from the piston. The process is adiabatic ($Q = 0$). (*b*) The process proceeds from i to f along an adiabat on a p-V diagram.

Because n and R are constants, we can rewrite this in the alternative form

$$TV^{\gamma-1} = \text{a constant} \quad \text{(adiabatic process)}, \tag{20-55}$$

in which the constant is different from that in Eq. 20-53. When the gas goes from an initial state i to a final state f, we can rewrite Eq. 20-55 as

$$T_i V_i^{\gamma-1} = T_f V_f^{\gamma-1} \quad \text{(adiabatic process)}. \tag{20-56}$$

We can now answer the question that opens this chapter. At the top of an unopened carbonated drink, there is a gas of carbon dioxide and water vapor. Because the pressure of the gas is greater than atmospheric pressure, the gas expands out into the atmosphere when the container is opened. Thus, the gas increases its volume, but that means it must do work to push against the atmosphere. Because the expansion is so rapid, it is adiabatic and the only source of energy for the work is the internal energy of the gas. Because the internal energy decreases, the temperature of the gas must also decrease, which can cause the water vapor in the gas to condense into tiny drops, forming the fog. (Note that Eq. 20-56 also tells us that the temperature must decrease during an adiabatic expansion: Since V_f is greater than V_i, then T_f must be less than T_i.)

Proof of Eq. 20-53

Suppose that you remove some shot from the piston of Fig. 20-13*a*, allowing the ideal gas to push the piston and the remaining shot upward and thus to increase the volume by a differential amount dV. Since the volume change is tiny, we may assume that the pressure p of the gas on the piston is constant during the change. This assumption allows us to say that the work dW done by the gas during the volume increase is equal to $p\,dV$. From Eq. 19-27, the first law of thermodynamics can then be written as

$$dE_{\text{int}} = Q - p\,dV. \tag{20-57}$$

Since the gas is thermally insulated (and thus the expansion is adiabatic), we substitute 0 for Q. Then we use Eq. 20-45 to substitute $nC_V\,dT$ for dE_{int}. With these substitutions, and after some rearranging, we have

$$n\,dT = -\left(\frac{p}{C_V}\right) dV. \tag{20-58}$$

Now from the ideal gas law ($pV = nRT$) we have

$$p\,dV + V\,dp = nR\,dT. \tag{20-59}$$

Replacing R with its equal, $C_p - C_V$, in Eq. 20-59 yields

$$n\,dT = \frac{p\,dV + V\,dp}{C_p - C_V}. \tag{20-60}$$

Equating Eqs. 20-58 and 20-60 and rearranging then give

$$\frac{dp}{p} + \left(\frac{C_p}{C_V}\right)\frac{dV}{V} = 0.$$

Replacing the ratio of the molar specific heats with γ and integrating (see integral 5 in Appendix E) yield

$$\ln p + \gamma \ln V = \text{a constant}.$$

Rewriting the left side as $\ln pV^{\gamma}$ and then taking the antilog of both sides, we find

$$pV^{\gamma} = \text{a constant}, \tag{20-61}$$

which is what we set out to prove.

Free Expansions

Recall from Section 19-10 that a free expansion of a gas is an adiabatic process that involves no work done on or by the gas, and no change in the internal energy of the gas. A free expansion is thus quite different from the type of adiabatic process described by Eqs. 20-53 through 20-61, in which work is done and the internal energy changes. Those equations then do *not* apply to a free expansion, even though such an expansion is adiabatic.

Also recall that in a free expansion, a gas is in equilibrium only at its initial and final points; thus, we can plot only those points, but not the expansion itself, on a p-V diagram. In addition, because $\Delta E_{\text{int}} = 0$, the temperature of the final state must be that of the initial state. Thus, the initial and final points on a p-V diagram must be on the same isotherm, and instead of Eq. 20-56 we have

$$T_i = T_f \qquad \text{(free expansion)}. \tag{20-62}$$

If we next assume that the gas is ideal (so that $pV = nRT$), then because there is no change in temperature, there can be no change in the product pV. Thus, instead of Eq. 20-53 a free expansion involves the relation

$$p_iV_i = p_fV_f \qquad \text{(free expansion)}. \tag{20-63}$$

Sample Problem 20-9

In Sample Problem 20-2, 1 mol of oxygen (assumed to be an ideal gas) expands isothermally (at 310 K) from an initial volume of 12 L to a final volume of 19 L.

(a) What would be the final temperature if the gas had expanded adiabatically to this same final volume? Oxygen (O_2) is diatomic and here has rotation but not oscillation.

SOLUTION: The **Key Ideas** here are

1. When a gas expands against the pressure of its environment, it must do work.
2. When the process is adiabatic (no energy is transferred as heat), then the energy required for the work can come only from the internal energy of the gas.
3. Because the internal energy decreases, the temperature T must also decrease.

We can relate the initial and final temperatures and volumes with Eq. 20-56:

$$T_iV_i^{\gamma-1} = T_fV_f^{\gamma-1}. \tag{20-64}$$

Because the molecules are diatomic and have rotation but not oscillation, we can take the molar specific heats from Table 20-3.

Thus,

$$\gamma = \frac{C_p}{C_V} = \frac{\frac{7}{2}R}{\frac{5}{2}R} = 1.40.$$

Solving Eq. 20-64 for T_f and inserting known data then yield

$$T_f = \frac{T_i V_i^{\gamma-1}}{V_f^{\gamma-1}} = \frac{(310\text{ K})(12\text{ L})^{1.40-1}}{(19\text{ L})^{1.40-1}}$$
$$= (310\text{ K})(\tfrac{12}{19})^{0.40} = 258\text{ K}. \qquad \text{(Answer)}$$

(b) What would be the final temperature and pressure if, instead, the gas had expanded freely to the new volume, from an initial pressure of 2.0 Pa?

SOLUTION: Here the **Key Idea** is that the temperature does not change in a free expansion:

$$T_f = T_i = 310\text{ K}. \qquad \text{(Answer)}$$

We find the new pressure using Eq. 20-63, which gives us

$$p_f = p_i \frac{V_i}{V_f} = (2.0\text{ Pa})\frac{12\text{ L}}{19\text{ L}} = 1.3\text{ Pa}. \qquad \text{(Answer)}$$

PROBLEM-SOLVING TACTICS

Tactic 2: *A Graphical Summary of Four Gas Processes*
In this chapter we have discussed four special processes that an ideal gas can undergo. An example of each is shown in Fig. 20-14, and some associated characteristics are given in Table 20-4, including two process names (isobaric and isochoric) that we have not used but that you might see in other courses.

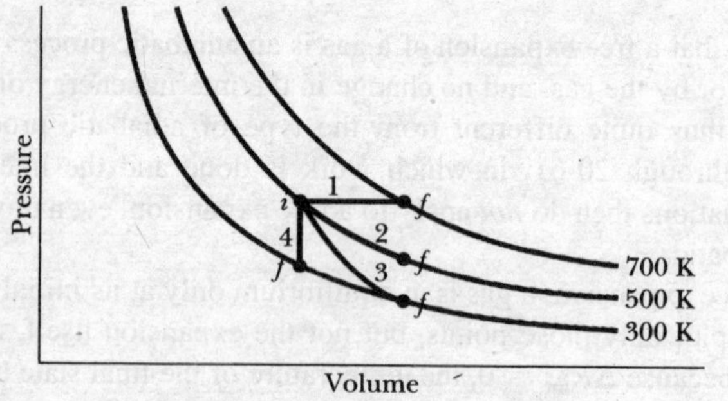

Fig. 20-14 A *p-V* diagram representing four special processes for an ideal gas. Table 20-4 explains the processes.

✔CHECKPOINT 5: Rank paths 1, 2, and 3 in Fig. 20-14 according to the heat transfer to the gas, greatest first.

TABLE 20-4 Four Special Processes

Path in Fig. 20-14	Constant Quantity	Process Type	Some Special Results ($\Delta E_{int} = Q - W$ and $\Delta E_{int} = nC_V\,\Delta T$ for all paths)
1	p	Isobaric	$Q = nC_p\,\Delta T$; $W = p\,\Delta V$
2	T	Isothermal	$Q = W = nRT\ln(V_f/V_i)$; $\Delta E_{int} = 0$
3	pV^γ, $TV^{\gamma-1}$	Adiabatic	$Q = 0$; $W = -\Delta E_{int}$
4	V	Isochoric	$Q = \Delta E_{int} = nC_V\,\Delta T$; $W = 0$

REVIEW & SUMMARY

Kinetic Theory of Gases The *kinetic theory of gases* relates the *macroscopic* properties of gases (for example, pressure and temperature) to the *microscopic* properties of gas molecules (for example, speed and kinetic energy).

Avogadro's Number One mole of a substance contains N_A (*Avogadro's number*) elementary units (usually atoms or molecules), where N_A is found experimentally to be

$$N_A = 6.02 \times 10^{23}\text{ mol}^{-1} \quad \text{(Avogadro's number).} \qquad (20\text{-}1)$$

One molar mass M of any substance is the mass of one mole of the substance. It is related to the mass m of the individual molecules of the substance by

$$M = mN_A. \qquad (20\text{-}4)$$

The number of moles n contained in a sample of mass M_{sam}, consisting of N molecules, is given by

$$n = \frac{N}{N_A} = \frac{M_{sam}}{M} = \frac{M_{sam}}{mN_A}. \qquad (20\text{-}2, 20\text{-}3)$$

Ideal Gas An *ideal gas* is one for which the pressure p, volume V, and temperature T are related by

$$pV = nRT \quad \text{(ideal gas law).} \qquad (20\text{-}5)$$

Here n is the number of moles of the gas present and R is a constant (8.31 J/mol · K) called the **gas constant.** The ideal gas law can also be written as

$$pV = NkT, \qquad (20\text{-}9)$$

where the **Boltzmann constant** k is

$$k = \frac{R}{N_A} = 1.38 \times 10^{-23} \text{ J/K}. \quad (20\text{-}7)$$

Work in an Isothermal Volume Change The work done *by* an ideal gas during an **isothermal** (constant-temperature) change from volume V_i to volume V_f is

$$W = nRT \ln \frac{V_f}{V_i} \quad \text{(ideal gas, isothermal process)}. \quad (20\text{-}14)$$

Pressure, Temperature, and Molecular Speed The pressure exerted by n moles of an ideal gas, in terms of the speed of its molecules, is

$$p = \frac{nMv_{\text{rms}}^2}{3V}, \quad (20\text{-}21)$$

where $v_{\text{rms}} = \sqrt{(v^2)_{\text{avg}}}$ is the **root-mean-square speed** of the molecules of the gas. With Eq. 20-5 this gives

$$v_{\text{rms}} = \sqrt{\frac{3RT}{M}}. \quad (20\text{-}22)$$

Temperature and Kinetic Energy The average translational kinetic energy K_{avg} per molecule of an ideal gas is

$$K_{\text{avg}} = \tfrac{3}{2}kT. \quad (20\text{-}24)$$

Mean Free Path The *mean free path* λ of a gas molecule is its average path length between collisions and is given by

$$\lambda = \frac{1}{\sqrt{2}\pi d^2 N/V}, \quad (20\text{-}25)$$

where N/V is the number of molecules per unit volume and d is the molecular diameter.

Maxwell Speed Distribution The *Maxwell speed distribution* $P(v)$ is a function such that $P(v)\,dv$ gives the *fraction* of molecules with speeds in the interval dv centered on speed v:

$$P(v) = 4\pi\left(\frac{M}{2\pi RT}\right)^{3/2} v^2 e^{-Mv^2/2RT}. \quad (20\text{-}27)$$

Three measures of the distribution of speeds among the molecules of a gas are

$$v_{\text{avg}} = \sqrt{\frac{8RT}{\pi M}} \quad \text{(average speed)}, \quad (20\text{-}31)$$

$$v_P = \sqrt{\frac{2RT}{M}} \quad \text{(most probable speed)}, \quad (20\text{-}35)$$

and the rms speed defined above in Eq. 20-22.

Molar Specific Heats The molar specific heat C_V of a gas at constant volume is defined as

$$C_V = \frac{1}{n}\frac{Q}{\Delta T} = \frac{1}{n}\frac{\Delta E_{\text{int}}}{\Delta T}, \quad (20\text{-}39, 20\text{-}41)$$

in which Q is the energy transferred as heat to or from a sample of n moles of the gas, ΔT is the resulting temperature change of the gas, and ΔE_{int} is the resulting change in the internal energy of the gas. For an ideal monatomic gas,

$$C_V = \tfrac{3}{2}R = 12.5 \text{ J/mol}\cdot\text{K}. \quad (20\text{-}43)$$

The molar specific heat C_p of a gas at constant pressure is defined to be

$$C_p = \frac{1}{n}\frac{Q}{\Delta T}, \quad (20\text{-}46)$$

in which Q, n, and ΔT are defined as above. C_p is also given by

$$C_p = C_V + R. \quad (20\text{-}49)$$

For n moles of an ideal gas,

$$E_{\text{int}} = nC_VT \quad \text{(ideal gas)}. \quad (20\text{-}44)$$

If n moles of a confined ideal gas undergo a temperature change ΔT due to *any* process, the change in the internal energy of the gas is

$$\Delta E_{\text{int}} = nC_V\,\Delta T \quad \text{(ideal gas, any process)}, \quad (20\text{-}45)$$

in which the appropriate value of C_V must be substituted, according to the type of ideal gas.

Degrees of Freedom and C_V We find C_V itself by using the *equipartition of energy* theorem, which states that every *degree of freedom* of a molecule (that is, every independent way it can store energy) has associated with it—on average—an energy $\frac{1}{2}kT$ per molecule ($= \frac{1}{2}RT$ per mole). If f is the number of degrees of freedom, then $E_{\text{int}} = (f/2)nRT$ and

$$C_V = \left(\frac{f}{2}\right)R = 4.16f \text{ J/mol}\cdot\text{K}. \quad (20\text{-}51)$$

For monatomic gases $f = 3$ (three translational degrees); for diatomic gases $f = 5$ (three translational and two rotational degrees).

Adiabatic Process When an ideal gas undergoes a slow adiabatic volume change (a change for which $Q = 0$), its pressure and volume are related by

$$pV^\gamma = \text{a constant} \quad \text{(adiabatic process)}, \quad (20\text{-}53)$$

in which γ ($= C_p/C_V$) is the ratio of molar specific heats for the gas. For a free expansion, however, pV = a constant.

QUESTIONS

1. If the temperature of an ideal gas is changed from 20°C to 40°C while the volume is unchanged, is the pressure of the gas doubled, increased but less than doubled, or increased and more than doubled?

2. In Fig. 20-15*a*, three isothermal processes are shown for the same gas and for the same change in volume (V_i to V_f) but at different temperatures. Rank the processes according to (a) the work done by the gas, (b) the change in the internal energy of the gas, and (c) the energy transferred as heat to the gas, greatest first.

In Fig. 20-15*b*, three isothermal processes are shown along a

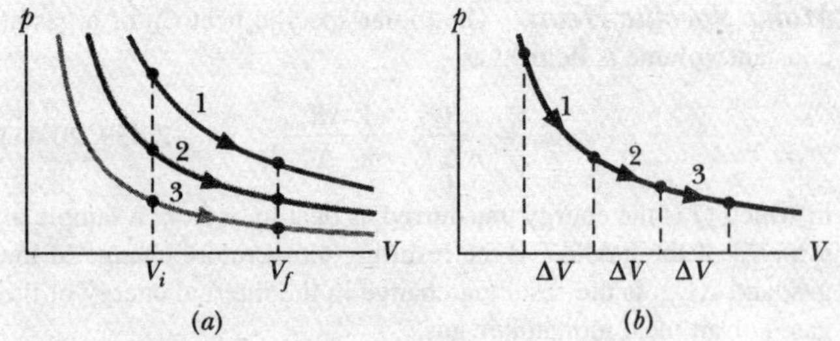

Fig. 20-15 Question 2.

single isotherm, for the same change ΔV in volume. Rank the processes according to (d) the work done by the gas, (e) the change in the internal energy of the gas, and (f) the energy transferred as heat to the gas, greatest first.

3. The volume of a gas and the number of gas molecules within that volume for four situations are (a) $2V_0$ and N_0, (b) $3V_0$ and $3N_0$, (c) $8V_0$ and $4N_0$, and (d) $3V_0$ and $9N_0$. Rank the situations according to the mean free path of the molecules, greatest first.

4. In Sample Problem 20-2, how much energy is transferred as heat during the expansion?

5. Figure 20-16 shows the initial state of an ideal gas and an isotherm through that state. Which of the paths shown result in a decrease in the temperature of the gas?

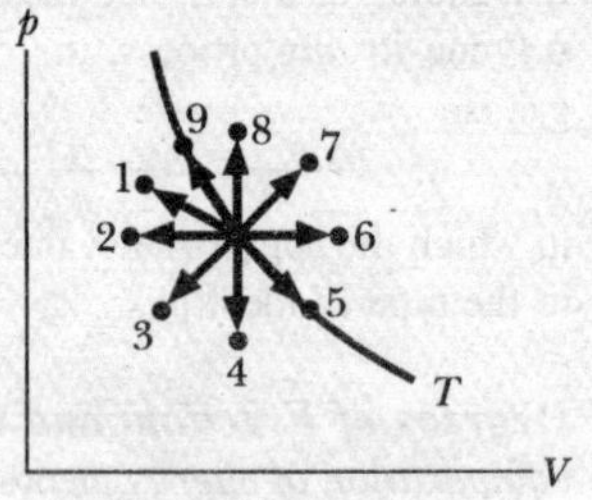

Fig. 20-16 Question 5.

6. For four situations for an ideal gas, the table gives the heat Q and either the work W done by the gas or the work W_{on} done on the gas, all in joules. Rank the four situations in terms of the temperature change of the gas, most positive first, most negative last.

	a	b	c	d
Q	−50	+35	−15	+20
W	−50	+35		
W_{on}			−40	+40

7. For a temperature increase of ΔT_1, a certain amount of an ideal gas requires 30 J when heated at constant volume and 50 J when heated at constant pressure. How much work is done by the gas in the second situation?

8. An ideal diatomic gas, with molecular rotation but not oscillation, loses energy as heat Q. Is the resulting decrease in the internal energy of the gas greater if the loss occurs in a constant-volume process or in a constant-pressure process?

9. A certain amount of energy is to be transferred as heat to 1 mol of a monatomic gas (a) at constant pressure and (b) at constant volume, and to 1 mol of a diatomic gas (c) at constant pressure and (d) at constant volume. Figure 20-17 shows four paths from an initial point to four final points on a p-V diagram. Which path goes with which process? (e) Are the molecules of the diatomic gas rotating?

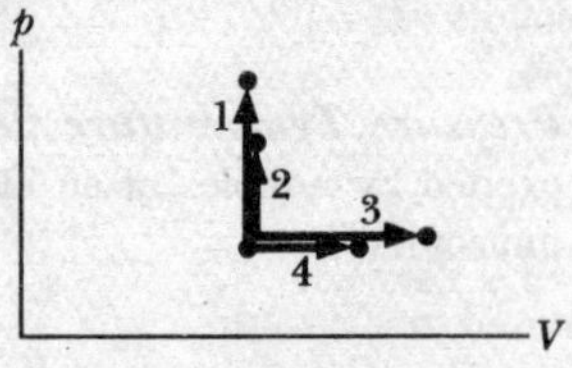

Fig. 20-17 Question 9.

10. Does the temperature of an ideal gas increase, decrease, or stay the same during (a) an isothermal expansion, (b) an expansion at constant pressure, (c) an adiabatic expansion, and (d) an increase in pressure at constant volume?

11. (a) Rank the four paths of Fig. 20-14 according to the work done by the gas, greatest first. (b) Rank paths 1, 2, and 3 according to the change in the internal energy of the gas, most positive first and most negative last.

12. In the p-V diagram of Fig. 20-18, the gas does 5 J of work along isotherm ab and 4 J along adiabat bc. What is the change in the internal energy of the gas if the gas traverses the straight path from a to c?

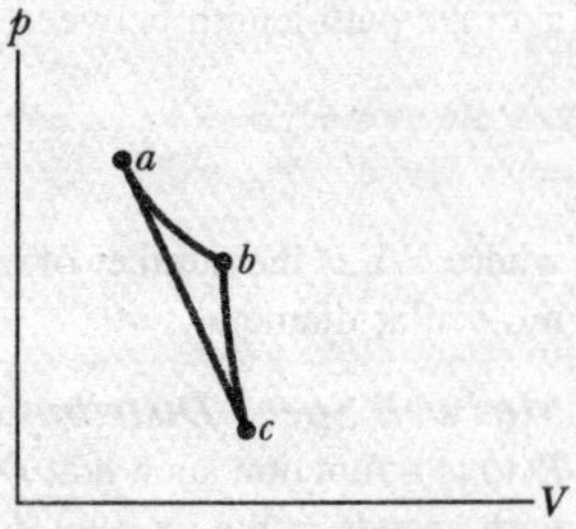

Fig. 20-18 Question 12.

EXERCISES & PROBLEMS

ssm Solution is in the Student Solutions Manual.
www Solution is available on the World Wide Web at: http://www.wiley.com/college/hrw
ilw Solution is available on the Interactive LearningWare.

SEC. 20-2 Avogadro's Number

1E. Find the mass in kilograms of 7.50×10^{24} atoms of arsenic, which has a molar mass of 74.9 g/mol. **ssm**

2E. Gold has a molar mass of 197 g/mol. (a) How many moles of gold are in a 2.50 g sample of pure gold? (b) How many atoms are in the sample?

3P. If the water molecules in 1.00 g of water were distributed uniformly over the surface of Earth, how many such molecules would there be on 1.00 cm^2 of the surface?

4P. A distinguished scientist has written: "There are enough molecules in the ink that makes one letter of this sentence to provide not only one for every inhabitant of Earth, but one for every creature if each star of our galaxy had a planet as populous as Earth." Check this statement. Assume the ink sample (molar mass = 18 g/mol) to have a mass of 1 μg, the population of Earth to be 5×10^9, and the number of stars in our galaxy to be 10^{11}.

SEC. 20-3 Ideal Gases

5E. Compute (a) the number of moles and (b) the number of molecules in 1.00 cm^3 of an ideal gas at a pressure of 100 Pa and a temperature of 220 K. **ssm**

6E. The best laboratory vacuum has a pressure of about 1.00×10^{-18} atm, or 1.01×10^{-13} Pa. How many gas molecules are there per cubic centimeter in such a vacuum at 293 K?

7E. Oxygen gas having a volume of 1000 cm^3 at 40.0°C and 1.01×10^5 Pa expands until its volume is 1500 cm^3 and its pressure is 1.06×10^5 Pa. Find (a) the number of moles of oxygen present and (b) the final temperature of the sample. ssm

8E. An automobile tire has a volume of 1.64×10^{-2} m^3 and contains air at a gauge pressure (pressure above atmospheric pressure) of 165 kPa when the temperature is 0.00°C. What is the gauge pressure of the air in the tires when its temperature rises to 27.0°C and its volume increases to 1.67×10^{-2} m^3? Assume atmospheric pressure is 1.01×10^5 Pa.

9E. A quantity of ideal gas at 10.0°C and 100 kPa occupies a volume of 2.50 m^3. (a) How many moles of the gas are present? (b) If the pressure is now raised to 300 kPa and the temperature is raised to 30.0°C, how much volume does the gas occupy? Assume no leaks.

10E. Calculate the work done by an external agent during an isothermal compression of 1.00 mol of oxygen from a volume of 22.4 L at 0°C and 1.00 atm pressure to 16.8 L.

11P. Pressure p, volume V, and temperature T for a certain material are related by

$$p = \frac{AT - BT^2}{V},$$

where A and B are constants. Find an expression for the work done by the material if the temperature changes from T_1 to T_2 while the pressure remains constant. ssm

12P. A container encloses two ideal gases. Two moles of the first gas are present, with molar mass M_1. The second gas has molar mass $M_2 = 3M_1$, and 0.5 mol of this gas is present. What fraction of the total pressure on the container wall is attributable to the second gas? (The kinetic theory explanation of pressure leads to the experimentally discovered law of partial pressures for a mixture of gases that do not react chemically: *The total pressure exerted by the mixture is equal to the sum of the pressures that the several gases would exert separately if each were to occupy the vessel alone.*)

13P. Air that initially occupies 0.14 m^3 at a gauge pressure of 103.0 kPa is expanded isothermally to a pressure of 101.3 kPa and then cooled at constant pressure until it reaches its initial volume. Compute the work done by the air. (Gauge pressure is the difference between the actual pressure and atmospheric pressure.) ssm ilw www

14P. A sample of an ideal gas is taken through the cyclic process *abca* shown in Fig. 20-19; at point a, T = 200 K. (a) How many moles of gas are in the sample? What are (b) the temperature of the gas at point b, (c) the temperature of the gas at point c, and (d) the net energy added to the gas as heat during the cycle?

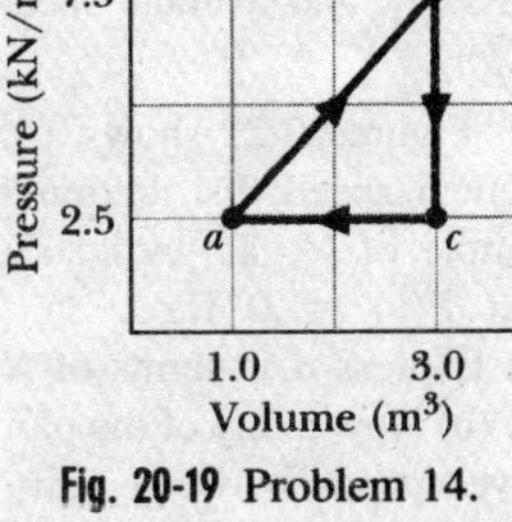

Fig. 20-19 Problem 14.

15P. An air bubble of 20 cm^3 volume is at the bottom of a lake 40 m deep where the temperature is 4.0°C. The bubble rises to the surface, which is at a temperature of 20°C. Take the temperature of the bubble's air to be the same as that of the surrounding water. Just as the bubble reaches the surface, what is its volume? ssm

16P. A pipe of length L = 25.0 m that is open at one end contains air at atmospheric pressure. It is thrust vertically into a freshwater lake until the water rises halfway up in the pipe, as shown in Fig. 20-20. What is the depth h of the lower end of the pipe? Assume that the temperature is the same everywhere and does not change.

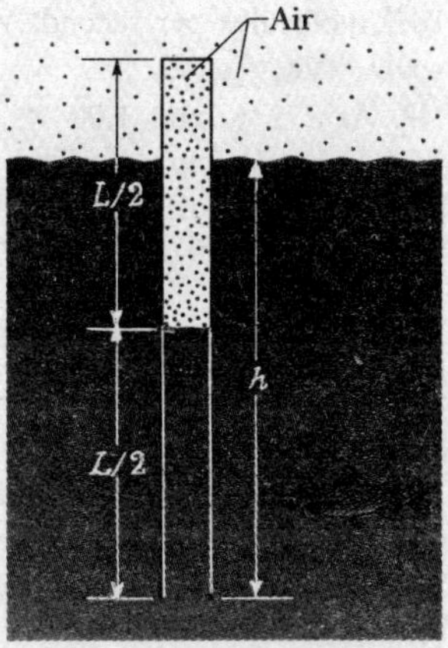

Fig. 20-20 Problem 16.

17P. Container A in Fig. 20-21 holds an ideal gas at a pressure of 5.0×10^5 Pa and a temperature of 300 K. It is connected by a thin tube (and a closed valve) to container B, with four times the volume of A. Container B holds the same ideal gas at a pressure of 1.0×10^5 Pa and a temperature of 400 K. The valve is opened to allow the pressures to equalize, but the temperature of each container is kept constant at its initial value. What then is the pressure in the two containers? ilw

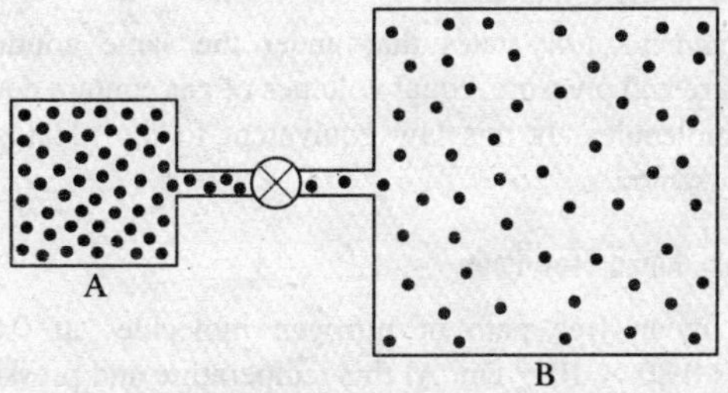

Fig. 20-21 Problem 17.

SEC. 20-4 Pressure, Temperature, and RMS Speed

18E. Calculate the rms speed of helium atoms at 1000 K. See Appendix F for the molar mass of helium atoms.

19E. The lowest possible temperature in outer space is 2.7 K. What is the root-mean-square speed of hydrogen molecules at this temperature? (The molar mass of hydrogen molecules (H_2) is given in Table 20-1.) ssm

20E. Find the rms speed of argon atoms at 313 K. See Appendix F for the molar mass of argon atoms.

21E. The temperature and pressure in the Sun's atmosphere are 2.00×10^6 K and 0.0300 Pa. Calculate the rms speed of free electrons (mass = 9.11×10^{-31} kg) there, assuming they are an ideal gas.

22E. (a) Compute the root-mean-square speed of a nitrogen molecule at 20.0°C. The molar mass of nitrogen molecules (N_2) is given in Table 20-1. At what temperatures will the root-mean-square speed be (b) half that value and (c) twice that value?

23P. A beam of hydrogen molecules (H_2) is directed toward a wall, at an angle of 55° with the normal to the wall. Each molecule in the beam has a speed of 1.0 km/s and a mass of 3.3×10^{-24} g. The beam strikes the wall over an area of 2.0 cm^2, at the rate of

10^{23} molecules per second. What is the beam's pressure on the wall? ssm

24P. At 273 K and 1.00×10^{-2} atm, the density of a gas is 1.24×10^{-5} g/cm^3. (a) Find v_{rms} for the gas molecules. (b) Find the molar mass of the gas and identify the gas. (*Hint:* The gas is listed in Table 20-1.)

SEC. 20-5 Translational Kinetic Energy

25E. What is the average translational kinetic energy of nitrogen molecules at 1600 K? ssm

26E. Determine the average value of the translational kinetic energy of the molecules of an ideal gas at (a) 0.00°C and (b) 100°C. What is the translational kinetic energy per mole of an ideal gas at (c) 0.00°C and (d) 100°C?

27P. Water standing in the open at 32.0°C evaporates because of the escape of some of the surface molecules. The heat of vaporization (539 cal/g) is approximately equal to εn, where ε is the average energy of the escaping molecules and n is the number of molecules per gram. (a) Find ε. (b) What is the ratio of ε to the average kinetic energy of H_2O molecules, assuming the latter is related to temperature in the same way as it is for gases? ssm www

28P. Show that the ideal gas equation, Eq. 20-5, can be written in the alternative form $p = \rho RT/M$, where ρ is the mass density of the gas and M is the molar mass.

29P. *Avogadro's law* states that under the same conditions of temperature and pressure, equal volumes of gas contain equal numbers of molecules. Is this law equivalent to the ideal gas law? Explain. ssm

SEC. 20-6 Mean Free Path

30E. The mean free path of nitrogen molecules at 0.0°C and 1.0 atm is 0.80×10^{-5} cm. At this temperature and pressure there are 2.7×10^{19} molecules/cm^3. What is the molecular diameter?

31E. At 2500 km above Earth's surface, the density of the atmosphere is about 1 molecule/cm^3. (a) What mean free path is predicted by Eq. 20-25 and (b) what is its significance under these conditions? Assume a molecular diameter of 2.0×10^{-8} cm. ssm

32E. At what frequency would the wavelength of sound in air be equal to the mean free path of oxygen molecules at 1.0 atm pressure and 0.00°C? Take the diameter of an oxygen molecule to be 3.0×10^{-8} cm.

33E. What is the mean free path for 15 spherical jelly beans in a bag that is vigorously shaken? The volume of the bag is 1.0 L, and the diameter of a jelly bean is 1.0 cm. (Consider bean–bean collisions, not bean–bag collisions.) ssm

34P. At 20°C and 750 torr pressure, the mean free paths for argon gas (Ar) and nitrogen gas (N_2) are $\lambda_{Ar} = 9.9 \times 10^{-6}$ cm and $\lambda_{N_2} = 27.5 \times 10^{-6}$ cm. (a) Find the ratio of the effective diameter of argon to that of nitrogen. What is the mean free path of argon at (b) 20°C and 150 torr, and (c) −40°C and 750 torr?

35P. In a certain particle accelerator, protons travel around a circular path of diameter 23.0 m in an evacuated chamber, whose residual gas is at 295 K and 1.00×10^{-6} torr pressure. (a) Calculate the number of gas molecules per cubic centimeter at this pressure. (b) What is the mean free path of the gas molecules if the molecular diameter is 2.00×10^{-8} cm? ssm

SEC. 20-7 The Distribution of Molecular Speeds

36E. Twenty-two particles have speeds as follows (N_i represents the number of particles that have speed v_i):

N_i	2	4	6	8	2
v_i (cm/s)	1.0	2.0	3.0	4.0	5.0

(a) Compute their average speed v_{avg}. (b) Compute their root-mean-square speed v_{rms}. (c) Of the five speeds shown, which is the most probable speed v_P?

37E. The speeds of 10 molecules are 2.0, 3.0, 4.0, . . . , 11 km/s. (a) What is their average speed? (b) What is their root-mean-square speed? ssm

38E. (a) Ten particles are moving with the following speeds: four at 200 m/s, two at 500 m/s, and four at 600 m/s. Calculate their average and root-mean-square speeds. Is $v_{rms} > v_{avg}$? (b) Make up your own speed distribution for the 10 particles and show that $v_{rms} \geq v_{avg}$ for your distribution. (c) Under what condition (if any) does $v_{rms} = v_{avg}$?

39P. (a) Compute the temperatures at which the rms speed for (a) molecular hydrogen and (b) molecular oxygen is equal to the speed of escape from Earth. (c) Do the same for the speed of escape from the Moon, assuming the gravitational acceleration on its surface to be $0.16g$. (d) The temperature high in Earth's upper atmosphere is about 1000 K. Would you expect to find much hydrogen there? Much oxygen? Explain. ssm www

40P. It is found that the most probable speed of molecules in a gas when it has (uniform) temperature T_2 is the same as the rms speed of the molecules in this gas when it has (uniform) temperature T_1. Calculate T_2/T_1.

41P. A molecule of hydrogen (diameter 1.0×10^{-8} cm), traveling with the rms speed, escapes from a furnace ($T = 4000$ K) into a chamber containing atoms of *cold* argon (diameter 3.0×10^{-8} cm) at a density of 4.0×10^{19} atoms/cm^3. (a) What is the speed of the hydrogen molecule? (b) If the H_2 molecule collides with an argon atom, what is the closest their centers can be, considering each as spherical? (c) What is the initial number of collisions per second experienced by the hydrogen molecule? (*Hint:* Assume that the cold argon atoms are stationary. Then the mean free path of the hydrogen molecule is given by Eq. 20-26 and not Eq. 20-25.) ssm

42P. Two containers are at the same temperature. The first contains gas with pressure p_1, molecular mass m_1, and root-mean-square speed v_{rms1}. The second contains gas with pressure $2p_1$, molecular mass m_2, and average speed $v_{avg2} = 2v_{rms1}$. Find the mass ratio m_1/m_2.

43P. Figure 20-22 shows a hypothetical speed distribution for a sample of N gas particles (note that $P(v) = 0$ for $v > 2v_0$). (a) Express a in terms of N and v_0. (b) How many of the particles have speeds between $1.5v_0$ and $2.0v_0$? (c) Express the average speed of the particles in terms of v_0. (d) Find v_{rms}. ssm

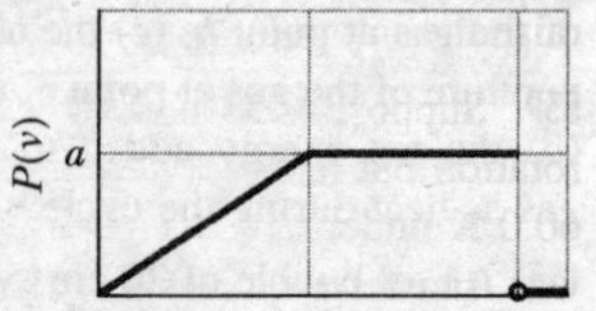

Fig. 20-22 Problem 43.

SEC. 20-8 The Molar Specific Heats of an Ideal Gas

44E. What is the internal energy of 1.0 mol of an ideal monatomic gas at 273 K?

45E. One mole of an ideal gas undergoes an isothermal expansion. Find the energy added to the gas as heat in terms of the initial and final volumes and the temperature. (*Hint:* Use the first law of thermodynamics.) ssm

46P. When 20.9 J was added as heat to a particular ideal gas, the volume of the gas changed from 50.0 cm^3 to 100 cm^3 while the pressure remained constant at 1.00 atm. (a) By how much did the internal energy of the gas change? If the quantity of gas present is 2.00×10^{-3} mol, find the molar specific heat of the gas at (b) constant pressure and (c) constant volume.

47P. A container holds a mixture of three nonreacting gases: n_1 moles of the first gas with molar specific heat at constant volume C_1, and so on. Find the molar specific heat at constant volume of the mixture, in terms of the molar specific heats and quantities of the separate gases. ssm

48P. One mole of an ideal diatomic gas goes from a to c along the diagonal path in Fig. 20-23. During the transition, (a) what is the change in internal energy of the gas, and (b) how much energy is added to the gas as heat? (c) How much heat is required if the gas goes from a to c along the indirect path abc?

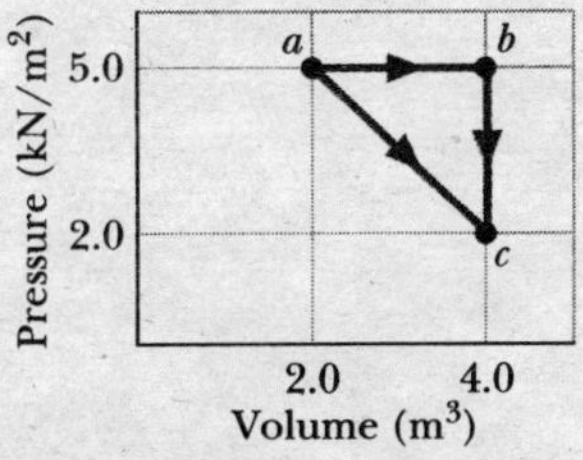

Fig. 20-23 Problem 48.

49P. The mass of a gas molecule can be computed from its specific heat at constant volume c_V. Take $c_V = 0.075$ cal/g · C° for argon and calculate (a) the mass of an argon atom and (b) the molar mass of argon. ilw

SEC. 20-9 Degrees of Freedom and Molar Specific Heats

50E. We give 70 J as heat to a diatomic gas, which then expands at constant pressure. The gas molecules rotate but do not oscillate. By how much does the internal energy of the gas increase?

51E. One mole of oxygen (O_2) is heated at constant pressure starting at 0°C. How much energy must be added to the gas as heat to double its volume? (The molecules rotate but do not oscillate.) ilw

52E. Suppose 12.0 g of oxygen (O_2) is heated at constant atmospheric pressure from 25.0°C to 125°C. (a) How many moles of oxygen are present? (See Table 20-1 for the molar mass.) (b) How much energy is transferred to the oxygen as heat? (The molecules rotate but do not oscillate.) (c) What fraction of the heat is used to raise the internal energy of the oxygen?

53P. Suppose 4.00 mol of an ideal diatomic gas, with molecular rotation but not oscillation, experienced a temperature increase of 60.0 K under constant-pressure conditions. (a) How much energy was transferred to the gas as heat? (b) How much did the internal energy of the gas increase? (c) How much work was done by the gas? (d) How much did the translational kinetic energy of the gas increase? ssm www

SEC. 20-11 The Adiabatic Expansion of an Ideal Gas

54E. (a) One liter of a gas with $\gamma = 1.3$ is at 273 K and 1.0 atm pressure. It is suddenly compressed adiabatically to half its original volume. Find its final pressure and temperature. (b) The gas is now cooled back to 273 K at constant pressure. What is its final volume?

55E. A certain gas occupies a volume of 4.3 L at a pressure of 1.2 atm and a temperature of 310 K. It is compressed adiabatically to a volume of 0.76 L. Determine (a) the final pressure and (b) the final temperature, assuming the gas to be an ideal gas for which $\gamma = 1.4$. ssm

56E. We know that for an adiabatic process pV^γ = a constant. Evaluate "a constant" for an adiabatic process involving exactly 2.0 mol of an ideal gas passing through the state having exactly $p = 1.0$ atm and $T = 300$ K. Assume a diatomic gas whose molecules have rotation but not oscillation.

57E. Let n moles of an ideal gas expand adiabatically from an initial temperature T_1 to a final temperature T_2. Prove that the work done by the gas is $nC_V(T_1 - T_2)$, where C_V is the molar specific heat at constant volume. (*Hint:* Use the first law of thermodynamics.) ssm

58E. For adiabatic processes in an ideal gas, show that (a) the bulk modulus is given by

$$B = -V\frac{dp}{dV} = \gamma p,$$

and therefore (b) the speed of sound in the gas is

$$v_s = \sqrt{\frac{\gamma p}{\rho}} = \sqrt{\frac{\gamma RT}{M}}.$$

See Eqs. 18-2 and 18-3.

59E. Air at 0.000°C and 1.00 atm pressure has a density of 1.29×10^{-3} g/cm^3, and the speed of sound in air is 331 m/s at that temperature. Use those data to compute the ratio γ of the molar specific heats of air. (*Hint:* See Exercise 58.)

60P. (a) An ideal gas initially at pressure p_0 undergoes a free expansion until its volume is 3.00 times its initial volume. What then is its pressure? (b) The gas is next slowly and adiabatically compressed back to its original volume. The pressure after compression is $(3.00)^{1/3}p_0$. Is the gas monatomic, diatomic, or polyatomic? (c) How does the average kinetic energy per molecule in this final state compare with that in the initial state?

61P. One mole of an ideal monatomic gas traverses the cycle of Fig. 20-24. Process 1 → 2 occurs at constant volume, process 2 → 3 is adiabatic, and process 3 → 1 occurs at constant pressure. (a) Compute the heat Q, the change in internal energy ΔE_{int}, and the work done W, for each of the three processes and for the cycle as a whole. (b) The initial pressure at point 1 is 1.00 atm. Find the pressure and the volume at points 2 and 3. Use 1.00 atm = 1.013×10^5 Pa and $R = 8.314$ J/mol · K. ssm

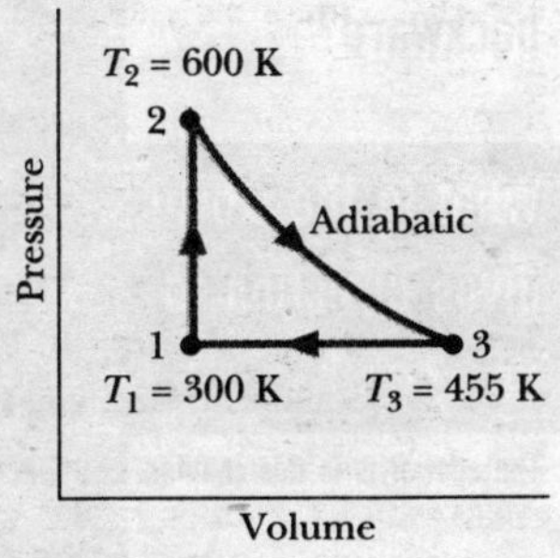

Fig. 20-24 Problem 61.

21 Entropy and the Second Law of Thermodynamics

An anonymous graffito on a wall of the Pecan Street Cafe in Austin, Texas, read: "Time is God's way of keeping things from happening all at once." Time also has direction—some things happen in a certain sequence and could never happen on their own in a reverse sequence. As an example, an accidentally dropped egg splatters in a cup. The reverse process, a splattered egg reforming into a whole egg and jumping up to an outstretched hand, will never happen on its own—but why not? Why can't that process be reversed, like a videotape run backward?

What in the world gives direction to time?

The answer is in this chapter.

21-1 Some One-Way Processes

Suppose you come indoors on a very cold day and wrap your cold hands around a warm mug of cocoa. Then your hands get warmer and the mug gets cooler. However, it never happens the other way around; that is, your cold hands never get still colder while the warm mug gets still warmer.

The system consisting of your hands and the mug is a *closed system,* one that is isolated from its environment. Here are some other one-way processes that occur in closed systems: (1) A crate sliding over an ordinary surface eventually stops—but you never see an initially stationary crate start to move all by itself. (2) If you drop a glob of putty, it falls to the floor—but an initially motionless glob of putty never leaps spontaneously into the air. (3) If you puncture a helium-filled balloon in a closed room, the helium gas spreads throughout the room—but the individual helium atoms will never clump up again into the shape of the balloon. We say that such one-way processes are **irreversible,** meaning that they cannot be reversed by means of only small changes in their environment.

The one-way character of such thermodynamic processes is so pervasive that we take it for granted. If these processes were to occur *spontaneously* (on their own) in the "wrong" direction, we would be astonished beyond belief. *Yet none of these "wrong-way" events would violate the law of conservation of energy.* In the cocoa mug example, that law would be obeyed even for a wrong-way transfer of energy as heat between hands and mug. It would be obeyed even if a stationary crate or a stationary glob of putty suddenly were to transfer some of its thermal energy to kinetic energy and begin to move. It would also be obeyed even if the helium atoms released from a balloon were, on their own, to clump together again.

In this chapter by closed, they mean isolated system, one that does not exchange even energy with its surroundings

Thus, changes in energy within a closed system do not set the direction of irreversible processes. Rather, that direction is set by another property that we shall discuss in this chapter—the *change in entropy* ΔS of the system. The change in entropy of a system is defined in the next section, but we can here state its central property, often called the *entropy postulate:*

> If an irreversible process occurs in a *closed* (isolated) system, the entropy S of the system always increases; it never decreases.

Entropy differs from energy in that it does *not* obey a conservation law. The *energy* of a closed system is conserved; it always remains constant. For irreversible processes, the *entropy* of a closed system always increases. Because of this property, the change in entropy is sometimes called "the arrow of time." For example, we associate the egg of our opening photograph, breaking irreversibly as it drops into a cup, with the forward direction of time and with an increase in entropy. The backward direction of time (a videotape run backward) would correspond to the broken egg re-forming into a whole egg and rising into the air. This backward process would result in an entropy decrease, so it never happens.

There are two equivalent ways to define the change in entropy of a system: (1) in terms of the system's temperature and the energy it gains or loses as heat, and (2) by counting the ways in which the atoms or molecules that make up the system can be arranged. We use the first approach in the next section, and the second in Section 21-7.

21-2 Change in Entropy

Let's approach this definition of *change in entropy* by looking again at a process that we described in Sections 19-10 and 20-11: the free expansion of an ideal gas.

Figure 21-1a shows the gas in its initial equilibrium state i, confined by a closed stopcock to the left half of a thermally insulated container. If we open the stopcock, the gas rushes to fill the entire container, eventually reaching the final equilibrium state f shown in Fig. 21-1b. This is an irreversible process; all the molecules of the gas will never return, by themselves, to the left half of the container.

The p-V plot of the process, in Fig. 21-2, shows the pressure and volume of the gas in its initial state i and final state f. Pressure and volume are *state properties,* properties that depend only on the state of the gas and not on how it reached that state. Other state properties are temperature and energy. We now assume that the gas has still another state property—its entropy. Furthermore, we define the **change in entropy** $S_f - S_i$ of a system during a process that takes the system from an initial state i to a final state f as

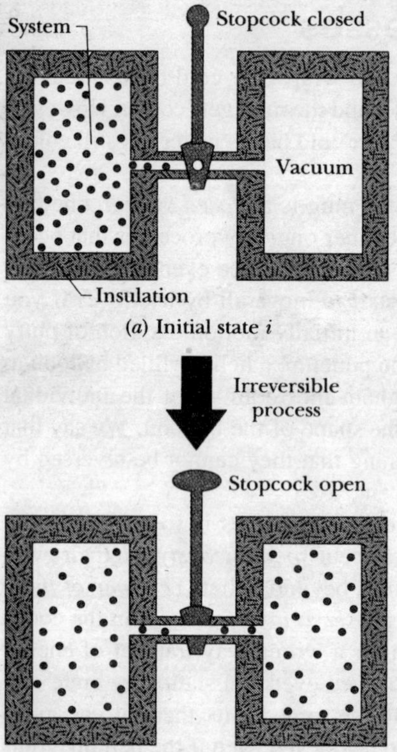

(a) Initial state i

(b) Final state f

Fig. 21-1 The free expansion of an ideal gas. (a) The gas is confined to the left half of an insulated container by a closed stopcock. (b) When the stopcock is opened, the gas rushes to fill the entire container. This process is irreversible; that is, it does not occur in reverse, with the gas spontaneously collecting itself in the left half of the container.

$$\Delta S = S_f - S_i = \int_i^f \frac{dQ}{T} \quad \text{(change in entropy defined).} \tag{21-1}$$

Here Q is the energy transferred as heat to or from the system during the process, and T is the temperature of the system in kelvins. Thus, an entropy change depends not only on the energy transferred as heat but also on the temperature at which the transfer takes place. Because T is always positive, the sign of ΔS is the same as that of Q. We see from Eq. 21-1 that the SI unit for entropy and entropy change is the joule per kelvin.

There is a problem, however, in applying Eq. 21-1 to the free expansion of Fig. 21-1. As the gas rushes to fill the entire container, the pressure, temperature, and volume of the gas fluctuate unpredictably. In other words, they do not have a sequence of well-defined equilibrium values during the intermediate stages of the change from initial equilibrium state i to final equilibrium state f. Thus, we cannot trace a pressure–volume path for the free expansion on the p-V plot of Fig. 21-2 and, more important, we cannot find a relation between Q and T that allows us to integrate as Eq. 21-1 requires.

However, if entropy is truly a state property, the difference in entropy between states i and f must depend *only on those states* and not at all on the way the system went from one state to the other. Suppose, then, that we replace the irreversible free expansion of Fig. 21-1 with a *reversible* process that connects states i and f. With a reversible process we can trace a pressure–volume path on a p-V plot, and we can find a relation between Q and T that allows us to use Eq. 21-1 to obtain the entropy change.

We saw in Section 20-11 that the temperature of an ideal gas does not change during a free expansion: $T_i = T_f = T$. Thus, points i and f in Fig. 21-2 must be on the same isotherm. A convenient replacement process is then a reversible isothermal expansion from state i to state f, which actually proceeds *along* that isotherm. Furthermore, because T is constant throughout a reversible isothermal expansion, the integral of Eq. 21-1 is greatly simplified.

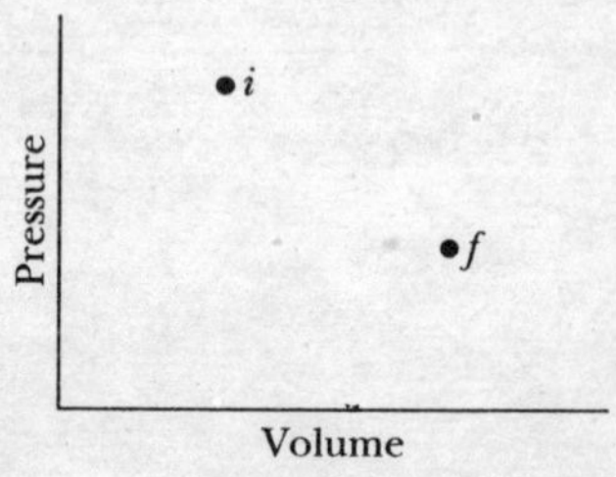

Fig. 21-2 A p-V diagram showing the initial state i and the final state f of the free expansion of Fig. 21-1. The intermediate states of the gas cannot be shown because they are not equilibrium states.

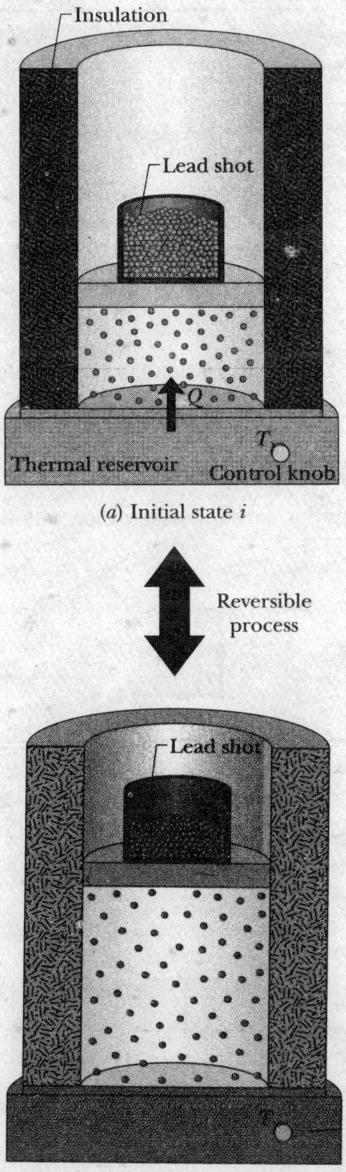

Fig. 21-3 The isothermal expansion of an ideal gas, done in a reversible way. The gas has the same initial state i and same final state f as in the irreversible process of Figs. 21-1 and 21-2.

Figure 21-3 shows how to produce such a reversible isothermal expansion. We confine the gas to an insulated cylinder that rests on a thermal reservoir maintained at the temperature T. We begin by placing just enough lead shot on the movable piston so that the pressure and volume of the gas are those of the initial state i of Fig. 21-1*a*. We then remove shot slowly (piece by piece) until the pressure and volume of the gas are those of the final state f of Fig. 21-1*b*. The temperature of the gas does not change because the gas remains in thermal contact with the reservoir throughout the process.

The reversible isothermal expansion of Fig. 21-3 is physically quite different from the irreversible free expansion of Fig. 21-1. However, *both processes have the same initial state and the same final state and thus must have the same change in entropy.* Because we removed the lead shot slowly, the intermediate states of the gas are equilibrium states, so we can plot them on a p-V diagram (Fig. 21-4).

To apply Eq. 21-1 to the isothermal expansion, we take the constant temperature T outside the integral, obtaining

$$\Delta S = S_f - S_i = \frac{1}{T}\int_i^f dQ.$$

Because $\int dQ = Q$, where Q is the total energy transferred as heat during the process, we have

$$\Delta S = S_f - S_i = \frac{Q}{T} \qquad \text{(change in entropy, isothermal process).} \qquad (21\text{-}2)$$

To keep the temperature T of the gas constant during the isothermal expansion of Fig. 21-3, heat Q must have been energy transferred *from* the reservoir *to* the gas. Thus, Q is positive and the entropy of the gas *increases* during the isothermal process and during the free expansion of Fig. 21-1.

To summarize:

> To find the entropy change for an irreversible process occurring in a *closed* system, replace that process with any reversible process that connects the same initial and final states. Calculate the entropy change for this reversible process with Eq. 21-1.

When the temperature change ΔT of a system is small relative to the temperature (in kelvins) before and after the process, the entropy change can be approximated as

$$\Delta S = S_f - S_i \approx \frac{Q}{T_{avg}}, \qquad (21\text{-}3)$$

where T_{avg} is the average temperature of the system in kelvins during the process.

CHECKPOINT 1: Water is heated on a stove. Rank the entropy changes of the water as its temperature rises (a) from 20°C to 30°C, (b) from 30°C to 35°C, and (c) from 80°C to 85°C, greatest first.

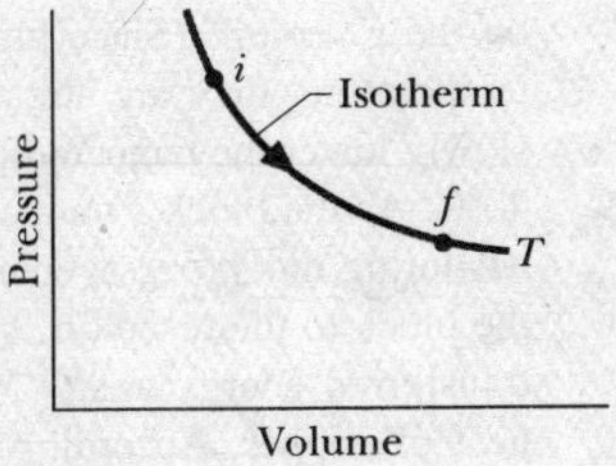

Fig. 21-4 A p-V diagram for the reversible isothermal expansion of Fig. 21-3. The intermediate states, which are now equilibrium states, are shown.

Sample Problem 21-1

One mole of nitrogen gas is confined to the left side of the container of Fig. 21-1*a*. You open the stopcock and the volume of the gas doubles. What is the entropy change of the gas for this irreversible process? Treat the gas as ideal.

SOLUTION: We need two **Key Ideas** here. One is that we can determine the entropy change for the irreversible process by calculating it for a reversible process that provides the same change in volume. The other is that the temperature of the gas does not change in the free expansion. Thus, the reversible process should be an isothermal expansion—namely, the one of Figs. 21-3 and 21-4.

From Table 20-4, the energy Q added as heat to the gas as it expands isothermally at temperature T from an initial volume V_i to a final volume V_f is

$$Q = nRT \ln \frac{V_f}{V_i},$$

in which n is the number of moles of gas present. From Eq. 21-2 the entropy change for this reversible process is

$$\Delta S_{rev} = \frac{Q}{T} = \frac{nRT \ln(V_f/V_i)}{T} = nR \ln \frac{V_f}{V_i}.$$

Substituting $n = 1.00$ mol and $V_f/V_i = 2$, we find

$$\Delta S_{rev} = nR \ln \frac{V_f}{V_i} = (1.00 \text{ mol})(8.31 \text{ J/mol} \cdot \text{K})(\ln 2)$$
$$= +5.76 \text{ J/K}.$$

Thus, the entropy change for the free expansion (and for all other processes that connect the initial and final states shown in Fig. 21-2) is

$$\Delta S_{irrev} = \Delta S_{rev} = +5.76 \text{ J/K}. \quad \text{(Answer)}$$

ΔS is positive, so the entropy increases, in accordance with the entropy postulate of Section 21-1.

CHECKPOINT 2: An ideal gas has temperature T_1 at the initial state i shown in the p-V diagram here. The gas has a higher temperature T_2 at final states a and b, which it can reach along the paths shown. Is the entropy change along the path to state a larger than, smaller than, or the same as that along the path to state b?

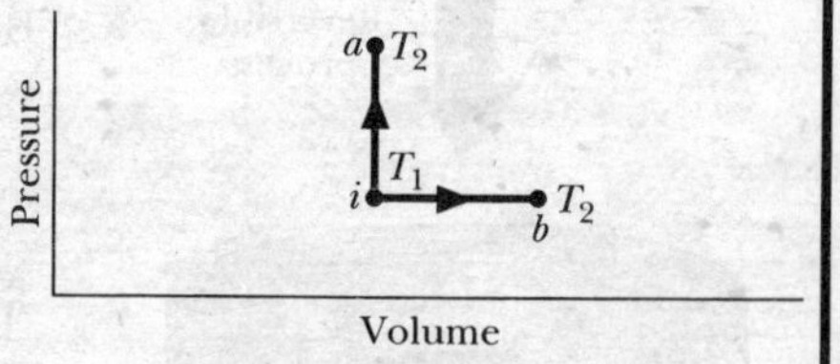

Sample Problem 21-2

Figure 21-5*a* shows two identical copper blocks of mass m = 1.5 kg: block L at temperature T_{iL} = 60°C and block R at temperature T_{iR} = 20°C. The blocks are in a thermally insulated box and are separated by an insulating shutter. When we lift the shutter, the blocks eventually come to the equilibrium temperature T_f = 40°C (Fig. 21-5*b*). What is the net entropy change of the two-block system during this irreversible process? The specific heat of copper is 386 J/kg · K.

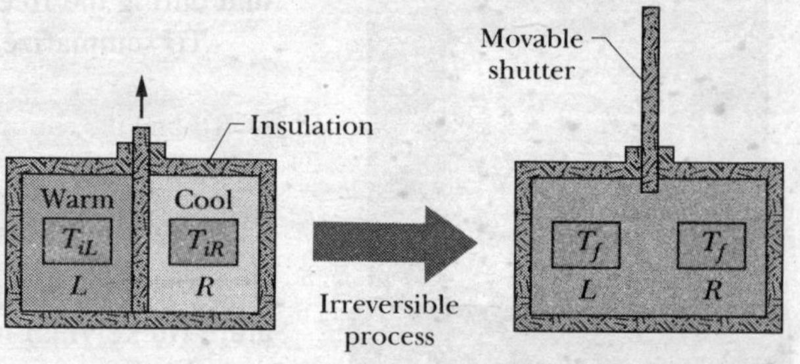

Fig. 21-5 Sample Problem 21-2. (*a*) In the initial state, two copper blocks L and R, identical except for their temperatures, are in an insulating box and are separated by an insulating shutter. (*b*) When the shutter is removed, the blocks exchange energy as heat and come to a final state, both with the same temperature T_f.

SOLUTION: The **Key Idea** here is that to calculate the entropy change, we must find a reversible process that takes the system from the initial state of Fig. 21-5*a* to the final state of Fig. 21-5*b*. We can calculate the net entropy change ΔS_{rev} of the reversible process using Eq. 21-1, and then the entropy change for the irreversible process is equal to ΔS_{rev}. For such a reversible process we need a thermal reservoir whose temperature can be changed slowly (say, by turning a knob). We then take the blocks through the following two steps, illustrated in Fig. 21-6.

Step 1. With the reservoir's temperature set at 60°C, put block L on the reservoir. (Since block and reservoir are at the same temperature, they are already in thermal equilibrium.) Then slowly lower the temperature of the reservoir and the block to 40°C. As the block's temperature changes by each increment dT during this process, energy dQ is transferred as heat *from* the block to the reservoir. Using Eq. 19-14, we can write this transferred energy as $dQ = mc\, dT$, where c is the specific heat of copper. According to Eq. 21-1, the entropy change

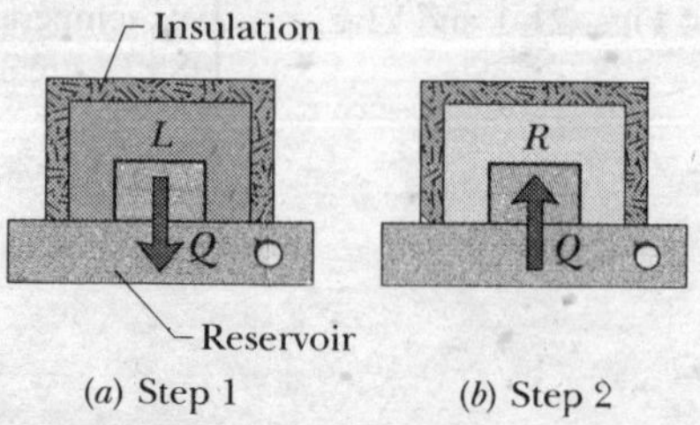

Fig. 21-6 The blocks of Fig. 21-5 can proceed from their initial state to their final state in a reversible way if we use a reservoir with a controllable temperature (*a*) to extract heat reversibly from block L and (*b*) to add heat reversibly to block R.

ΔS_L of block L during the full temperature change from initial temperature T_{iL} (= 60°C = 333 K) to final temperature T_f (= 40°C = 313 K) is

$$\Delta S_L = \int_i^f \frac{dQ}{T} = \int_{T_{iL}}^{T_f} \frac{mc\,dT}{T} = mc \int_{T_{iL}}^{T_f} \frac{dT}{T}$$
$$= mc \ln \frac{T_f}{T_{iL}}.$$

Inserting the given data yields

$$\Delta S_L = (1.5\text{ kg})(386\text{ J/kg}\cdot\text{K}) \ln \frac{313\text{ K}}{333\text{ K}}$$
$$= -35.86\text{ J/K}.$$

Step 2. With the reservoir's temperature now set at 20°C, put block R on the reservoir. Then slowly raise the temperature of the reservoir and the block to 40°C. With the same reasoning used to find ΔS_L, you can show that the entropy change ΔS_R of block R during this process is

$$\Delta S_R = (1.5\text{ kg})(386\text{ J/kg}\cdot\text{K}) \ln \frac{313\text{ K}}{293\text{ K}}$$
$$= +38.23\text{ J/K}.$$

The net entropy change ΔS_{rev} of the two-block system undergoing this two-step reversible process is then

$$\Delta S_{\text{rev}} = \Delta S_L + \Delta S_R$$
$$= -35.86\text{ J/K} + 38.23\text{ J/K} = 2.4\text{ J/K}.$$

Thus, the net entropy change ΔS_{irrev} for the two-block system undergoing the actual irreversible process is

$$\Delta S_{\text{irrev}} = \Delta S_{\text{rev}} = 2.4\text{ J/K}. \quad \text{(Answer)}$$

This result is positive, in accordance with the entropy postulate of Section 21-1.

Entropy as a State Function

We have assumed that entropy, like pressure, energy, and temperature, is a property of the state of a system and is independent of how that state is reached. That entropy is indeed a *state function* (as state properties are usually called) can only be deduced by experiment. However, we can prove it is a state function for the special and important case in which an ideal gas is taken through a reversible process.

To make the process reversible, it is done slowly in a series of small steps, with the gas in an equilibrium state at the end of each step. For each small step, the energy transferred as heat to or from the gas is dQ, the work done by the gas is dW, and the change in internal energy is dE_{int}. These are related by the first law of thermodynamics in differential form (Eq. 19-27):

$$dE_{\text{int}} = dQ - dW.$$

Because the steps are reversible, with the gas in equilibrium states, we can use Eq. 19-24 to replace dW with $p\,dV$ and Eq. 20-45 to replace dE_{int} with $nC_V\,dT$. Solving for dQ then leads to

$$dQ = p\,dV + nC_V\,dT.$$

Using the ideal gas law, we replace p in this equation with nRT/V. Then we divide each term in the resulting equation by T, obtaining

$$\frac{dQ}{T} = nR\frac{dV}{V} + nC_V\frac{dT}{T}.$$

Now let us integrate each term of this equation between an arbitrary initial state i and an arbitrary final state f to get

$$\int_i^f \frac{dQ}{T} = \int_i^f nR\frac{dV}{V} + \int_i^f nC_V\frac{dT}{T}.$$

The quantity on the left is the entropy change ΔS $(= S_f - S_i)$ defined by Eq. 21-1. Substituting this and integrating the quantities on the right yield

$$\Delta S = S_f - S_i = nR \ln \frac{V_f}{V_i} + nC_V \ln \frac{T_f}{T_i}. \qquad (21\text{-}4)$$

Note that we did not have to specify a particular reversible process when we integrated. Therefore, the integration must hold for all reversible processes that take the gas from state i to state f. Thus, the change in entropy ΔS between the initial and final states of an ideal gas depends only on properties of the initial state (V_i and T_i) and properties of the final state (V_f and T_f); ΔS does not depend on how the gas changes between the two states.

21-3 The Second Law of Thermodynamics

Here is a puzzle. We saw in Sample Problem 21-1 that if we cause the reversible process of Fig. 21-3 to proceed from (*a*) to (*b*) in that figure, the change in entropy of the gas—which we take as our system—is positive. However, because the process is reversible, we can just as easily make it proceed from (*b*) to (*a*), simply by slowly adding lead shot to the piston of Fig. 21-3*b* until the original volume of the gas is restored. In this reverse process, energy must be extracted as heat *from the gas* to keep its temperature from rising. Hence Q is negative and so, from Eq. 21-2, the entropy of the gas must decrease.

Doesn't this decrease in the entropy of the gas violate the entropy postulate of Section 21-1, which states that entropy always increases? No, because that postulate holds only for *irreversible* processes occurring in closed systems. The procedure suggested here does not meet these requirements. The process is *not* irreversible and (because energy is transferred as heat from the gas to the reservoir) the system—which is the gas alone—is *not* closed.

However, if we include the reservoir, along with the gas, as part of the system, then we do have a closed system. Let's check the change in entropy of the enlarged system *gas + reservoir* for the process that takes it from (*b*) to (*a*) in Fig. 21-3. During this reversible process, energy is transferred as heat from the gas to the reservoir—that is, from one part of the enlarged system to another. Let $|Q|$ represent the absolute value (or magnitude) of this heat. With Eq. 21-2, we can then calculate separately the entropy changes for the gas (which loses $|Q|$) and the reservoir (which gains $|Q|$). We get

$$\Delta S_{\text{gas}} = -\frac{|Q|}{T}$$

and

$$\Delta S_{\text{res}} = +\frac{|Q|}{T}.$$

The entropy change of the closed system is the sum of these two quantities, *which is zero.*

With this result, we can modify the entropy postulate of Section 21-1 to include both reversible and irreversible processes:

> If a process occurs in a closed system, the entropy of the system increases for irreversible processes and remains constant for reversible processes. It never decreases.

Although entropy may decrease in part of a closed system, there will always be an equal or larger entropy increase in another part of the system, so that the entropy of the system as a whole never decreases. This fact is one form of the **second law of thermodynamics** and can be written as

$$\Delta S \geq 0 \quad \text{(second law of thermodynamics)}, \tag{21-5}$$

where the greater-than sign applies to irreversible processes, and the equals sign to reversible processes. Equation 21-5 applies only to closed systems.

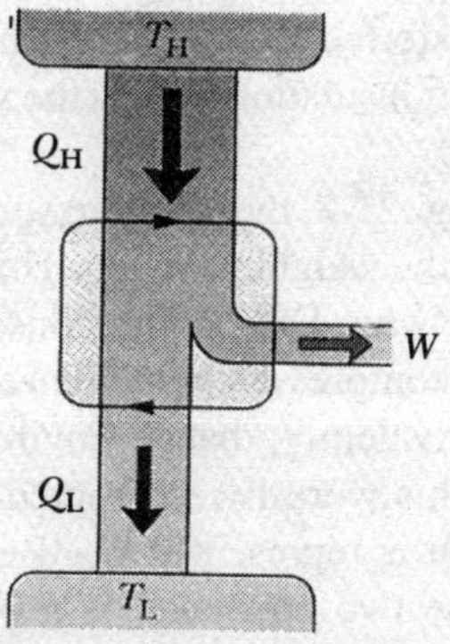

Fig. 21-7 The elements of an engine. The two black arrowheads on the central loop suggest the working substance operating in a cycle, as if on a *p-V* plot. Energy $|Q_H|$ is transferred as heat from the high-temperature reservoir at temperature T_H to the working substance. Energy $|Q_L|$ is transferred as heat from the working substance to the low-temperature reservoir at temperature T_L. Work *W* is done by the engine (actually by the working substance) on something in the environment.

In the real world almost all processes are irreversible to some extent because of friction, turbulence, and other factors, so the entropy of real closed systems undergoing real processes always increases. Processes in which the system's entropy remains constant are always idealizations.

21-4 Entropy in the Real World: Engines

A **heat engine,** or more simply, an **engine,** is a device that extracts energy from its environment in the form of heat and does useful work. At the heart of every engine is a *working substance*. In a steam engine, the working substance is water, in both its vapor and its liquid form. In an automobile engine the working substance is a gasoline–air mixture. If an engine is to do work on a sustained basis, the working substance must operate in a *cycle*; that is, the working substance must pass through a closed series of thermodynamic processes, called *strokes,* returning again and again to each state in its cycle. Let us see what the laws of thermodynamics can tell us about the operation of engines.

A Carnot Engine

We have seen that we can learn much about real gases by analyzing an ideal gas, which obeys the simple law $pV = nRT$. This is a useful plan because, although an ideal gas does not exist, any real gas approaches ideal behavior as closely as you wish if its density is low enough. In much the same spirit we choose to study real engines by analyzing the behavior of an **ideal engine.**

> In an ideal engine, all processes are reversible and no wasteful energy transfers occur due to, say, friction and turbulence.

We shall focus on a particular ideal engine called a **Carnot engine** after the French scientist and engineer N. L. Sadi Carnot (pronounced "car-no"), who first proposed the engine's concept in 1824. This ideal engine turns out to be the best (in principle) at using energy as heat to do useful work. Surprisingly, Carnot was able to analyze the performance of this engine before the first law of thermodynamics and the concept of entropy had been discovered.

Figure 21-7 shows schematically the operation of a Carnot engine. During each cycle of the engine, the working substance absorbs energy $|Q_H|$ as heat from a thermal reservoir at constant temperature T_H and discharges energy $|Q_L|$ as heat to a second thermal reservoir at a constant lower temperature T_L.

Figure 21-8 shows a *p-V* plot of the *Carnot cycle*—the cycle followed by the working substance. As indicated by the arrows, the cycle is traversed in the clockwise direction. Imagine the working substance to be a gas, confined to an insulating cylinder with a weighted, movable piston. The cylinder may be placed at will on either of the two thermal reservoirs, as in Fig. 21-3, or on an insulating slab. Figure 21-8 shows that, if we place the cylinder in contact with the high-temperature reservoir at temperature T_H, heat $|Q_H|$ is transferred *to* the working substance *from* this reservoir as the gas undergoes an isothermal *expansion* from volume V_a to volume V_b. Similarly, with the working substance in contact with the low-temperature reservoir at temperature T_L, heat $|Q_L|$ is transferred *from* the working substance *to* the low-temperature reservoir, as the gas undergoes an isothermal *compression* from volume V_c to volume V_d.

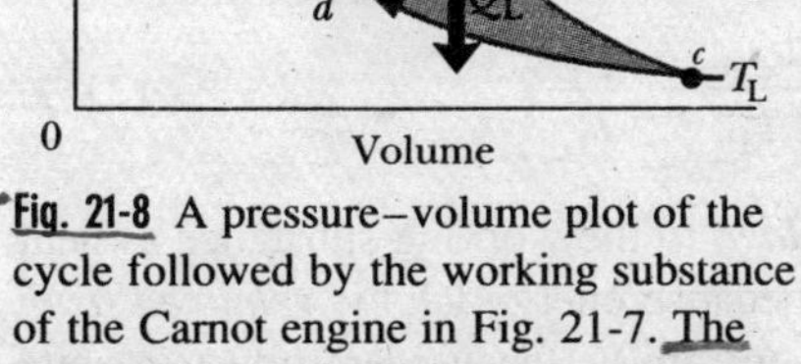

Fig. 21-8 A pressure–volume plot of the cycle followed by the working substance of the Carnot engine in Fig. 21-7. The cycle consists of two isotherm (*ab* and *cd*) and two adiabatic processes (*bc* and *da*). The shaded area enclosed by the cycle is equal to the work *W* per cycle done by the Carnot engine.

In the engine of Fig. 21-7, we assume that heat transfers to or from the working substance can take place *only* during the isothermal processes *ab* and *cd* of Fig. 21-8. Therefore, processes *bc* and *da* in that figure, which connect the two isotherms at temperatures T_H and T_L, must be (reversible) adiabatic processes; that is, they must be processes in which no energy is transferred as heat. To ensure this, during processes bc and da the cylinder is placed on an insulating slab as the volume of the working substance is changed.

During the consecutive processes ab and bc of Fig. 21-8, the working substance is expanding and thus doing positive work as it raises the weighted piston. This work is represented in Fig. 21-8 by the area under curve abc. During the consecutive processes cd and da, the working substance is being compressed, which means that it is doing negative work on its environment or, equivalently, that its environment is doing work on it as the loaded piston descends. This work is represented by the area under curve cda. The *net work per cycle*, which is represented by W in both Figs. 21-7 and 21-8, is the difference between these two areas and is a positive quantity equal to the area enclosed by cycle $abcda$ in Fig. 21-8. This work W is performed on some outside object, such as a load to be lifted.

Equation 21-1 ($\Delta S = \int dQ/T$) tells us that any energy transfer as heat must involve a change in entropy. To illustrate the entropy changes for a Carnot engine, we can plot the Carnot cycle on a temperature-entropy (T-S) diagram as shown in Fig. 21-9. The lettered points a, b, c, and d in Fig. 21-9 correspond to the lettered points in the p-V diagram in Fig. 21-8. The two horizontal lines in Fig. 21-9 correspond to the two isothermal processes of the Carnot cycle (because the temperature is constant). Process ab is the isothermal expansion of the cycle. As the working substance (reversibly) absorbs energy $|Q_H|$ as heat at constant temperature T_H during the expansion, its entropy increases. Similarly, during the isothermal compression cd, the working substance (reversibly) loses energy $|Q_L|$ as heat at constant temperature T_L, and its entropy decreases.

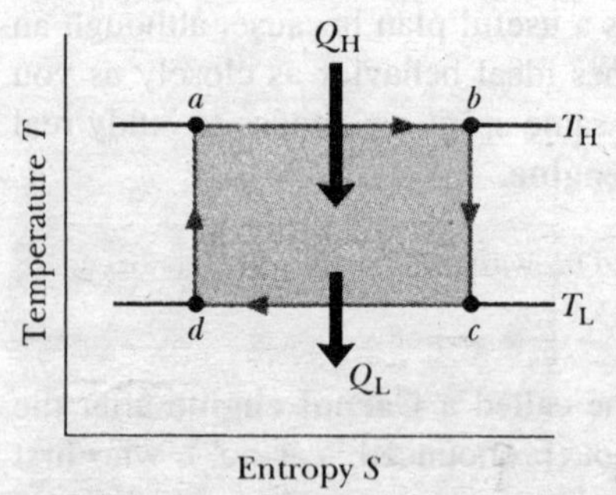

Fig. 21-9 The Carnot cycle of Fig. 21-8 plotted on a temperature–entropy diagram. During processes ab and cd the temperature remains constant. During processes bc and da the entropy remains constant.

The two vertical lines in Fig. 21-9 correspond to the two adiabatic processes of the Carnot cycle. Because no energy is transferred as heat during the two processes, the entropy of the working substance is constant during them.

The Work: To calculate the net work done by a Carnot engine during a cycle, let us apply Eq. 19-26, the first law of thermodynamics ($\Delta E_{int} = Q - W$), to the working substance. That substance must return again and again to any arbitrarily selected state in the cycle. Thus, if X represents any state property of the working substance, such as pressure, temperature, volume, internal energy, or entropy, we must have $\Delta X = 0$ for every cycle. It follows that $\Delta E_{int} = 0$ for a complete cycle of the working substance. Recalling that Q in Eq. 19-26 is the *net* heat transfer per cycle and W is the *net* work, we can write the first law of thermodynamics for the Carnot cycle as

$$W = |Q_H| - |Q_L|. \tag{21-6}$$

Entropy Changes: In a Carnot engine, there are *two* (and only two) reversible energy transfers as heat, and thus two changes in the entropy of the working substance—one at temperature T_H and one at T_L. The net entropy change per cycle is then

$$\Delta S = \Delta S_H + \Delta S_L = \frac{|Q_H|}{T_H} - \frac{|Q_L|}{T_L}. \tag{21-7}$$

Here ΔS_H is positive because energy $|Q_H|$ is *added to* the working substance as heat (an increase in entropy) and ΔS_L is negative because energy $|Q_L|$ is *removed from* the working substance as heat (a decrease in entropy). Because entropy is a state function, we must have $\Delta S = 0$ for a complete cycle. Putting $\Delta S = 0$ in Eq. 21-7 requires that

$$\frac{|Q_H|}{T_H} = \frac{|Q_L|}{T_L}. \tag{21-8}$$

Note that, because $T_H > T_L$, we must have $|Q_H| > |Q_L|$; that is, more energy is extracted as heat from the high-temperature reservoir than is delivered to the low-temperature reservoir.

We shall now use Eqs. 21-6 and 21-8 to derive an expression for the efficiency of a Carnot engine.

Efficiency of a Carnot Engine

The purpose of any engine is to transform as much of the extracted energy Q_H into work as possible. We measure its success in doing so by its **thermal efficiency** ε, defined as the work the engine does per cycle ("energy we get") divided by the energy it absorbs as heat per cycle ("energy we pay for"):

$$\varepsilon = \frac{\text{energy we get}}{\text{energy we pay for}} = \frac{|W|}{|Q_H|} \quad \text{(efficiency, any engine).} \tag{21-9}$$

For a Carnot engine we can substitute for W from Eq. 21-6 to write Eq. 21-9 as

$$\varepsilon_C = \frac{|Q_H| - |Q_L|}{|Q_H|} = 1 - \frac{|Q_L|}{|Q_H|}. \tag{21-10}$$

Using Eq. 21-8 we can write this as

$$\varepsilon_C = 1 - \frac{T_L}{T_H} \quad \text{(efficiency, Carnot engine),} \tag{21-11}$$

where the temperatures T_L and T_H are in kelvins. Because $T_L < T_H$, the Carnot engine necessarily has a thermal efficiency less than unity—that is, less than 100%. This is indicated in Fig. 21-7, which shows that only part of the energy extracted as heat from the high-temperature reservoir is available to do work, and the rest is delivered to the low-temperature reservoir. We will show in Section 21-6 that no real engine can have a thermal efficiency greater than that calculated from Eq. 21-11.

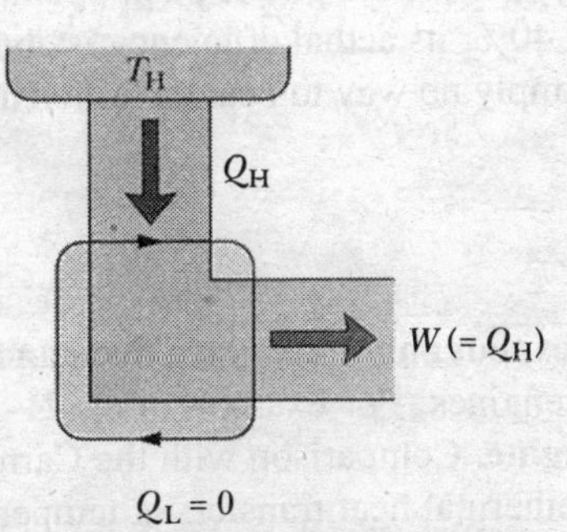

Fig. 21-10 The elements of a perfect engine—that is, one that converts heat Q_H from a high-temperature reservoir directly to work W with 100% efficiency.

Inventors continually try to improve engine efficiency by reducing the energy $|Q_L|$ that is "thrown away" during each cycle. The inventor's dream is to produce the *perfect engine*, diagrammed in Fig. 21-10, in which $|Q_L|$ is reduced to zero and $|Q_H|$ is converted completely into work. Such an engine on an ocean liner, for example, could extract energy as heat from the water and use it to drive the propellers, with no fuel cost. An automobile, fitted with such an engine, could extract energy as heat from the surrounding air and use it to drive the car, again with no fuel cost. Alas, a perfect engine is only a dream: Inspection of Eq. 21-11 shows that we can achieve 100% engine efficiency (that is, $\varepsilon = 1$) only if $T_L = 0$ or $T_H \to \infty$, requirements that are impossible to meet. Instead, decades of practical engineering experience have led to the following alternative version of the second law of thermodynamics:

> No series of processes is possible whose sole result is the transfer of energy as heat from a thermal reservoir and the complete conversion of this energy to work.

In short, *there are no perfect engines.*

To summarize: The thermal efficiency given by Eq. 21-11 applies only to Carnot engines. Real engines, in which the processes that form the engine cycle are not reversible, have lower efficiencies. If your car were powered by a Carnot engine, it

Fig. 21-11 The North Anna nuclear power plant near Charlottesville, Virginia, which generates electric energy at the rate of 900 MW. At the same time, by design, it discards energy into the nearby river at the rate of 2100 MW. This plant—and all others like it—throws away more energy than it delivers in useful form. It is a real counterpart to the ideal engine of Fig. 21-7.

would have an efficiency of about 55% according to Eq. 21-11; its actual efficiency is probably about 25%. A nuclear power plant (Fig. 21-11), taken in its entirety, is an engine. It extracts energy as heat from a reactor core, does work by means of a turbine, and discharges energy as heat to a nearby river. If the power plant operated as a Carnot engine, its efficiency would be about 40%; its actual efficiency is about 30%. In designing engines of any type, there is simply no way to beat the efficiency limitation imposed by Eq. 21-11.

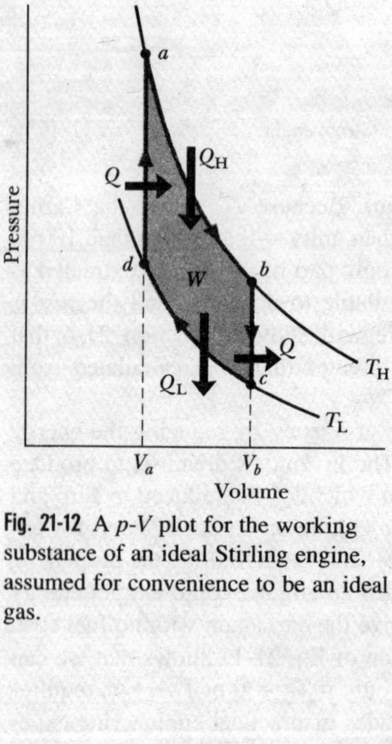

Fig. 21-12 A *p-V* plot for the working substance of an ideal Stirling engine, assumed for convenience to be an ideal gas.

Stirling Engine

Equation 21-11 does not apply to all ideal engines, but only to engines that can be represented as in Fig. 21-8—that is, to Carnot engines. For example, Fig. 21-12 shows the operating cycle of an ideal **Stirling engine.** Comparison with the Carnot cycle of Fig. 21-8 shows that each engine has isothermal heat transfers at temperatures T_H and T_L. However, the two isotherms of the Stirling engine cycle are connected, not by adiabatic processes as for the Carnot engine, but by constant-volume processes (Fig. 21-12). To increase the temperature of a gas at constant volume reversibly from T_L to T_H (process *da* of Fig. 21-12) requires a transfer of energy as heat to the working substance from a thermal reservoir whose temperature can be varied smoothly between those limits. Also, a reverse transfer is required in process *bc*. Thus, reversible heat transfers (and corresponding entropy changes) occur in all four of the processes that form the cycle of a Stirling engine, not just two processes as in a Carnot engine. Thus, the derivation that led to Eq. 21-11 does not apply to an ideal Stirling engine. More important, the efficiency of an ideal Stirling engine is lower than that of a Carnot engine operating between the same two temperatures. Real Stirling engines have even lower efficiencies.

The Stirling engine was developed in 1816 by Robert Stirling. This engine, long neglected, is now being developed for use in automobiles and spacecraft. A Stirling engine delivering 5000 hp (3.7 MW) has been built.

CHECKPOINT 3: Three Carnot engines operate between reservoir temperatures of (a) 400 and 500 K, (b) 600 and 800 K, and (c) 400 and 600 K. Rank the engines according to their thermal efficiencies, greatest first.

Sample Problem 21-3

Imagine a Carnot engine that operates between the temperatures $T_H = 850$ K and $T_L = 300$ K. The engine performs 1200 J of work each cycle, which takes 0.25 s.

(a) What is the efficiency of this engine?

SOLUTION: The **Key Idea** here is that the efficiency ε of a Carnot engine depends only on the ratio T_L/T_H of the temperatures (in kelvins) of the thermal reservoirs to which it is connected. Thus, from Eq. 21-11, we have

$$\varepsilon = 1 - \frac{T_L}{T_H} = 1 - \frac{300\text{ K}}{850\text{ K}} = 0.647 \approx 65\%. \quad \text{(Answer)}$$

(b) What is the average power of this engine?

SOLUTION: Here the **Key Idea** is that the average power P of an engine is the ratio of the work W it does per cycle to the time t that each cycle takes. For this Carnot engine, we find

$$P = \frac{W}{t} = \frac{1200\text{ J}}{0.25\text{ s}} = 4800\text{ W} = 4.8\text{ kW}. \quad \text{(Answer)}$$

(c) How much energy $|Q_H|$ is extracted as heat from the high-temperature reservoir every cycle?

SOLUTION: Now the **Key Idea** is that for any engine, including a Carnot engine, the efficiency ε is the ratio of the work W that is done per cycle to the energy $|Q_H|$ that is extracted as heat from the high-temperature reservoir per cycle ($\varepsilon = W/|Q_H|$). Thus,

$$|Q_H| = \frac{W}{\varepsilon} = \frac{1200\text{ J}}{0.647} = 1855\text{ J}. \quad \text{(Answer)}$$

(d) How much energy $|Q_L|$ is delivered as heat to the low-temperature reservoir every cycle?

SOLUTION: The **Key Idea** here is that for a Carnot engine, the work W done per cycle is equal to the difference in the energy transfers as heat: $|Q_H| - |Q_L|$, as in Eq. 21-6. Thus, we have

$$|Q_L| = |Q_H| - W = 1855\text{ J} - 1200\text{ J} = 655\text{ J}. \quad \text{(Answer)}$$

(e) What is the entropy change of the working substance for the energy transfer to it from the high-temperature reservoir? From it to the low-temperature reservoir?

SOLUTION: The **Key Idea** here is that the entropy change ΔS during a transfer of energy Q as heat at constant temperature T is given by Eq. 21-2 ($\Delta S = Q/T$). Thus, for the *positive* transfer of energy Q_H from the high-temperature reservoir at T_H, the change in the entropy of the working substance is

$$\Delta S_H = \frac{Q_H}{T_H} = \frac{1855\text{ J}}{850\text{ K}} = +2.18\text{ J/K}. \quad \text{(Answer)}$$

Similarly, for the *negative* transfer of energy Q_L to the low-temperature reservoir at T_L, we have

$$\Delta S_L = \frac{Q_L}{T_L} = \frac{-655\text{ J}}{300\text{ K}} = -2.18\text{ J/K}. \quad \text{(Answer)}$$

Note that the net entropy change of the working substance for one cycle is zero, as we discussed in deriving Eq. 21-8..

Sample Problem 21-4

OR L.H.S = 75/100 = 0.75
R.H.S = 100/373 ~ 0.2

An inventor claims to have constructed an engine that has an efficiency of 75% when operated between the boiling and freezing points of water. Is this possible?

SOLUTION: The **Key Idea** here is that the efficiency of a real engine (with its irreversible processes and wasteful energy transfers) must be less than the efficiency of a Carnot engine operating between the same two temperatures. From Eq. 21-11, we find that the efficiency of a Carnot engine operating between the boiling and freezing points of water is

$$\varepsilon = 1 - \frac{T_L}{T_H} = 1 - \frac{(0 + 273)\text{ K}}{(100 + 273)\text{ K}} = 0.268 \approx 27\%.$$

Thus, the claimed efficiency of 75% for a real engine operating between the given temperatures is impossible.

PROBLEM-SOLVING TACTICS

Tactic 1: *The Language of Thermodynamics*

A rich, but sometimes misleading, language is used in scientific and engineering studies of thermodynamics. You may see statements that say heat is added, absorbed, subtracted, extracted, rejected, discharged, discarded, withdrawn, delivered, gained, lost, transferred, or expelled, or that it flows from one body to another (as if it were a liquid). You may also see statements that describe a body as *having* heat (as if heat can be held or possessed), or that its heat is increased or decreased. You should always keep in mind what is meant by the term *heat:*

> Heat is energy that is transferred from one body to another body owing to a difference in the temperatures of the bodies.

When we identify one of the bodies as being our system of interest, any such transfer of energy into the system is positive heat Q, and any such transfer out of the system is negative heat Q.

The term *work* also requires close attention. You may see statements that say work is produced or generated, or combined with heat or changed from heat. Here is what is meant by the term *work:*

> Work is energy that is transferred from one body to another body owing to a force that acts between them.

When we identify one of the bodies as being our system of interest, any such transfer of energy out of the system is either positive work W done *by* the system or negative work W done *on* the system. Any such transfer of energy into the system is negative work done *by* the system or positive work done *on* the system. (The preposition that is used is important.) Obviously, this can be confusing—whenever you see the term *work,* you should read carefully to determine the intent.

21-5 Entropy in the Real World: Refrigerators

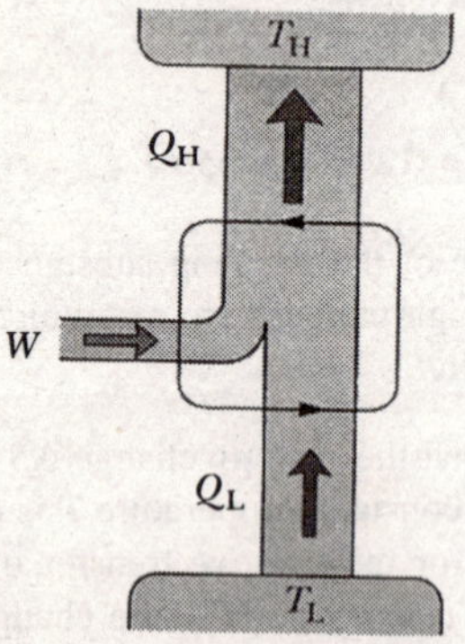

Fig. 21-13 The elements of a refrigerator. The two black arrowheads on the central loop suggest the working substance operating in a cycle, as if on a p-V plot. Energy Q_L is transferred as heat to the working substance from the low-temperature reservoir. Energy Q_H is transferred as heat to the high-temperature reservoir from the working substance. Work W is done on the refrigerator (on the working substance) by something in the environment.

A **refrigerator** is a device that uses work to transfer energy from a low-temperature reservoir to a high-temperature reservoir as it continuously repeats a set series of thermodynamic processes. In a household refrigerator, for example, work is done by an electrical compressor to transfer energy from the food storage compartment (a low-temperature reservoir) to the room (a high-temperature reservoir).

Air conditioners and heat pumps are also refrigerators. The differences are only in the nature of the high- and low-temperature reservoirs. For an air conditioner, the low-temperature reservoir is the room that is to be cooled, and the high-temperature reservoir is the (presumably warmer) outdoors. A heat pump is an air conditioner that can be operated in reverse to heat a room; the room is the high-temperature reservoir and heat is transferred to it from the (presumably cooler) outdoors.

Let us consider an *ideal refrigerator:*

> In an ideal refrigerator, all processes are reversible and no wasteful energy transfers occur due to, say, friction and turbulence.

Figure 21-13 shows the basic elements of an ideal refrigerator that operates in the reverse of the Carnot engine of Fig. 21-7. In other words, all the energy transfers, as either heat or work, are reversed from those of a Carnot engine. We can call such an ideal refrigerator a **Carnot refrigerator.**

The designer of a refrigerator would like to extract as much energy $|Q_L|$ as possible from the low-temperature reservoir (what we want) for the least amount of work $|W|$ (what we pay for). A measure of the efficiency of a refrigerator, then, is

$$K = \frac{\text{what we want}}{\text{what we pay for}} = \frac{|Q_L|}{|W|} \quad \text{(coefficient of performance, any refrigerator),} \tag{21-12}$$

where K is called the *coefficient of performance*. For a Carnot refrigerator, the first law of thermodynamics gives $|W| = |Q_H| - |Q_L|$, where $|Q_H|$ is the magnitude of the energy transferred as heat to the high-temperature reservoir. Equation 21-12 then becomes

$$K_C = \frac{|Q_L|}{|Q_H| - |Q_L|}. \tag{21-13}$$

Because a Carnot refrigerator is a Carnot engine operating in reverse, we can combine Eq. 21-8 with Eq. 21-13; after some algebra we find

$$K_C = \frac{T_L}{T_H - T_L} \quad \text{(coefficient of performance, Carnot refrigerator).} \tag{21-14}$$

For typical room air conditioners, $K \approx 2.5$. For household refrigerators, $K \approx 5$. Perversely, the value of K is higher the closer the temperatures of the two reservoirs are to each other. That is why heat pumps are more effective in temperate climates than in climates where the outside temperature varies widely.

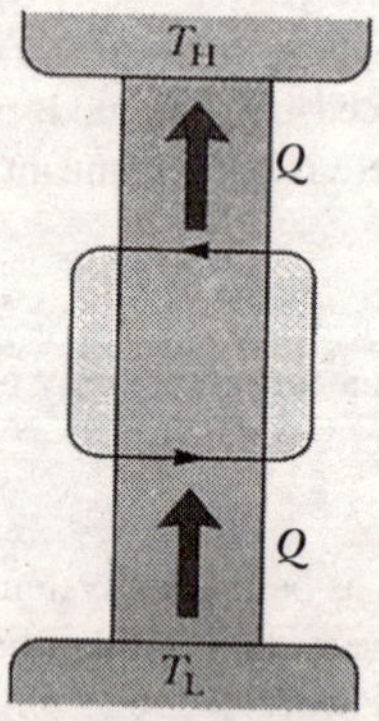

Fig. 21-14 The elements of a perfect refrigerator—that is, one that transfers energy from a low-temperature reservoir to a high-temperature reservoir without any input of work.

It would be nice to own a refrigerator that did not require some input of work—that is, one that would run without being plugged in. Figure 21-14 represents another "inventor's dream," a *perfect refrigerator* that transfers heat Q from a cold reservoir to a warm reservoir without the need for work. Because the unit operates in cycles, the entropy of the working substance does not change during a complete cycle. The entropies of the two reservoirs, however, do change: The entropy change for the cold reservoir is $-|Q|/T_L$, and that for the warm reservoir is $+|Q|/T_H$. Thus, the net entropy change for the entire system is

$$\Delta S = -\frac{|Q|}{T_L} + \frac{|Q|}{T_H}.$$

Because $T_H > T_L$, the right side of this equation is negative and thus the net change in entropy per cycle for the closed system *refrigerator + reservoirs* is also negative. Because such a decrease in entropy violates the second law of thermodynamics (Eq. 21-5), a perfect refrigerator does not exist. (If you want your refrigerator to operate, you must plug it in.)

This result leads us to another (equivalent) formulation of the second law of thermodynamics:

> No series of processes is possible whose sole result is the transfer of energy as heat from a reservoir at a given temperature to a reservoir at a higher temperature.

In short, *there are no perfect refrigerators.*

CHECKPOINT 4: You wish to increase the coefficient of performance of an ideal refrigerator. You can do so by (a) running the cold chamber at a slightly higher temperature, (b) running the cold chamber at a slightly lower temperature, (c) moving the unit to a slightly warmer room, or (d) moving it to a slightly cooler room. The magnitudes of the temperature changes are to be the same in all four cases. List the changes according to the resulting coefficients of performance, greatest first.

21-6 The Efficiencies of Real Engines

Let ε_C be the efficiency of a Carnot engine operating between two given temperatures. In this section we prove that no real engine operating between those temperatures can have an efficiency greater than ε_C. If it could, the engine would violate the second law of thermodynamics.

Let us assume that an inventor, working in her garage, has constructed an engine X, which she claims has an efficiency ε_X that is greater than ε_C:

$$\varepsilon_X > \varepsilon_C \quad \text{(a claim).} \tag{21-15}$$

Let us couple engine X to a Carnot refrigerator, as in Fig. 21-15*a*. We adjust the strokes of the Carnot refrigerator so that the work it requires per cycle is just equal

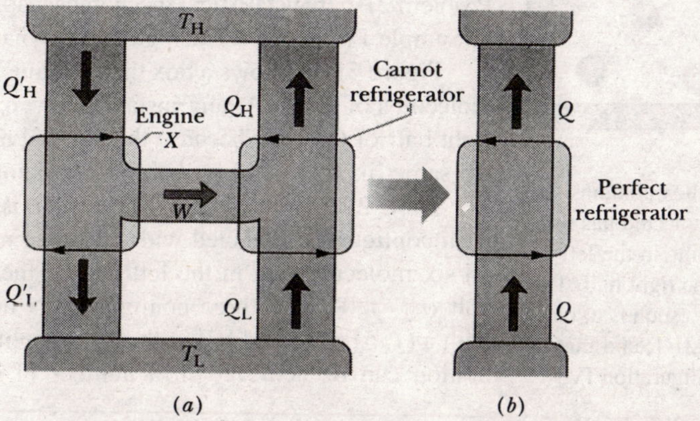

Fig. 21-15 (*a*) Engine X drives a Carnot refrigerator. (*b*) If, as claimed, engine X is more efficient than a Carnot engine, then the combination shown in (*a*) is equivalent to the perfect refrigerator shown here. This violates the second law of thermodynamics, so we conclude that engine X *cannot* be more efficient than a Carnot engine.

to that provided by engine X. Thus, no (external) work is performed on or by the combination *engine + refrigerator* of Fig. 21-15*a*, which we take as our system.

If Eq. 21-15 is true, from the definition of efficiency (Eq. 21-9), we must have

$$\frac{|W|}{|Q'_H|} > \frac{|W|}{|Q_H|},$$

where the prime refers to engine X and the right side of the inequality is the efficiency of the Carnot refrigerator when it operates as an engine. This inequality requires that

$$|Q_H| > |Q'_H|. \qquad (21\text{-}16)$$

Because the work done by engine X is equal to the work done on the Carnot refrigerator, we have, from the first law of thermodynamics (see Eq. 21-6),

$$|Q_H| - |Q_L| = |Q'_H| - |Q'_L|,$$

which we can write as

$$|Q_H| - |Q'_H| = |Q_L| - |Q'_L| = Q. \qquad (21\text{-}17)$$

Because of Eq. 21-16, the quantity Q in Eq. 21-17 must be positive.

Comparison of Eq. 21-17 with Fig. 21-15 shows that the net effect of engine X and the Carnot refrigerator, working in combination, is to transfer energy Q as heat from a low-temperature reservoir to a high-temperature reservoir without the requirement of work. Thus, the combination acts like the perfect refrigerator of Fig. 21-14, whose existence is a violation of the second law of thermodynamics.

Something must be wrong with one or more of our assumptions, and it can only be Eq. 21-15. We conclude that *no real engine can have an efficiency greater than that of a Carnot engine when both engines work between the same two temperatures.* At most, it can have an efficiency equal to that of a Carnot engine. In that case, engine X *is* a Carnot engine.

21-7 A Statistical View of Entropy

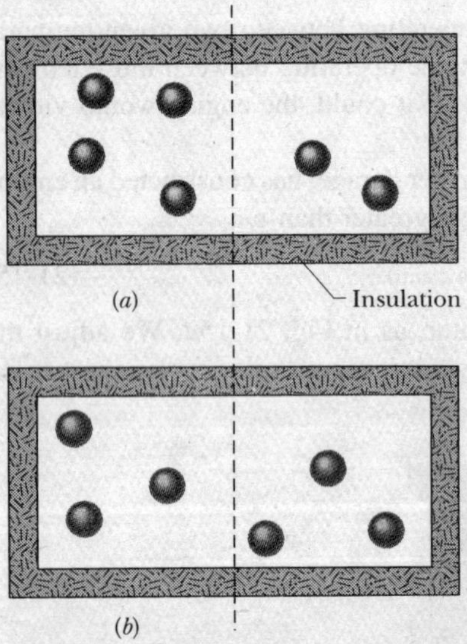

Fig. 21-16 An insulated box contains six gas molecules. Each molecule has the same probability of being in the left half of the box as in the right half. The arrangement in (*a*) corresponds to configuration III in Table 21-1, and that in (*b*) corresponds to configuration IV.

In Chapter 20 we saw that the macroscopic properties of gases can be explained in terms of their microscopic, or molecular, behavior. For one example, recall that we were able to account for the pressure exerted by a gas on the walls of its container in terms of the momentum transferred to those walls by rebounding gas molecules. Such explanations are part of a study called **statistical mechanics.**

Here we shall focus our attention on a single problem, involving the distribution of gas molecules between the two halves of an insulated box. This problem is reasonably simple to analyze, and it allows us to use statistical mechanics to calculate the entropy change for the free expansion of an ideal gas. You will see in Sample Problem 21-6 that statistical mechanics leads to the same entropy change we obtained in Sample Problem 21-1 using thermodynamics.

Figure 21-16 shows a box that contains six identical (and thus indistinguishable) molecules of a gas. At any instant, a given molecule will be in either the left or the right half of the box; because the two halves have equal volumes, the molecule has the same likelihood, or probability, of being in either half.

Table 21-1 shows four of the seven possible *configurations* of the six molecules, each configuration labeled with a Roman numeral. For example, in configuration I, all six molecules are in the left half of the box ($n_1 = 6$) and none are in the right half ($n_2 = 0$). The three configurations not shown are V with a (2, 4) split, VI with a (1, 5) split, and VII with a (0, 6) split. We see that, in general, a given configuration can be achieved in a number of different ways. We call these different

Table 21-1 Six Molecules in a Box

Configuration Label	n_1	n_2	Multiplicity W (number of microstates)	Calculation of W (Eq. 21-18)	Entropy 10^{-23} J/K (Eq. 21-19)
I	6	0	1	$6!/(6!\,0!) = 1$	0
II	5	1	6	$6!/(5!\,1!) = 6$	2.47
III	4	2	15	$6!/(4!\,2!) = 15$	3.74
IV	3	3	20	$6!/(3!\,3!) = 20$	4.13
Total number of microstates = 64					

arrangements of the molecules *microstates.* Let us see how to calculate the number of microstates that correspond to a given configuration.

Suppose we have N molecules, distributed with n_1 molecules in one half of the box and n_2 in the other. (Thus $n_1 + n_2 = N$.) Let us imagine that we distribute the molecules "by hand," one at a time. If $N = 6$, we can select the first molecule in six independent ways; that is, we can pick any one of the six molecules. We can pick the second molecule in five ways, by picking any one of the remaining five molecules; and so on. The total number of ways in which we can select all six molecules is the product of these independent ways, or $6 \times 5 \times 4 \times 3 \times 2 \times 1 = 720$. In mathematical shorthand we write this product as $6! = 720$, where $6!$ is pronounced "six factorial." Your hand calculator can probably calculate factorials. For later use you will need to know that $0! = 1$. (Check this on your calculator.)

However, because the molecules are indistinguishable, these 720 arrangements are not all different. In the case that $n_1 = 4$ and $n_2 = 2$ (which is configuration III in Table 21-1), for example, the order in which you put four molecules in one half of the box does not matter, because after you have put all four in, there is no way that you can tell the order in which you did so. The number of ways in which you can order the four molecules is $4!$ or 24. Similarly, the number of ways in which you can order two molecules for the other half of the box is simply $2!$ or 2. To get the number of *different* arrangements that lead to the (4, 2) split of configuration III, we must divide 720 by 24 and also by 2. We call the resulting quantity, which is the number of microstates that correspond to a given configuration, the *multiplicity* W of that configuration. Thus, for configuration III,

$$W_{\mathrm{III}} = \frac{6!}{4!\,2!} = \frac{720}{24 \times 2} = 15.$$

Thus, Table 21-1 tells us there are 15 independent microstates that correspond to configuration III. Note that, as the table also tells us, the total number of microstates for six molecules distributed over the seven configurations is 64.

Extrapolating from six molecules to the general case of N molecules, we have

$$W = \frac{N!}{n_1!\,n_2!} \quad \text{(multiplicity of configuration).} \qquad (21\text{-}18)$$

You should verify that Eq. 21-18 gives the multiplicities for all the configurations listed in Table 21-1.

The basic assumption of statistical mechanics is:

➤ All microstates are equally probable.

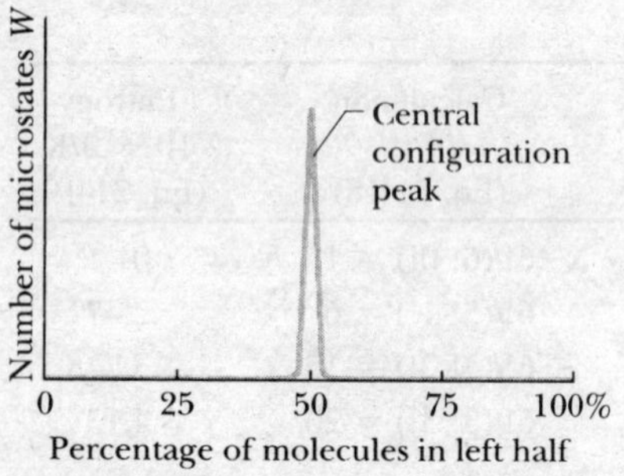

Fig. 21-17 For a *large* number of molecules in a box, a plot of the number of microstates that require various percentages of the molecules to be in the left half of the box. Nearly all the microstates correspond to an approximately equal sharing of the molecules between the two halves of the box; those microstates form the *central configuration peak* on the plot. For $N \approx 10^{22}$, the central configuration peak is much too narrow to be drawn on this plot.

In other words, if we were to take a great many snapshots of the six molecules as they jostle around in the box of Fig. 21-16 and then count the number of times each microstate occurred, we would find that all 64 microstates would occur equally often. In other words, the system will spend, on average, the same amount of time in each of the 64 microstates.

Because the microstates are equally probable, but different configurations have different numbers of microstates, the configurations are *not* equally probable. In Table 21-1 configuration IV, with 20 microstates, is the *most probable configuration,* with a probability of $20/64 = 0.313$. This result means that the system is in configuration IV 31.3% of the time. Configurations I and VII, in which all the molecules are in one half of the box, are the least probable, each with a probability of $1/64 = 0.016$ or 1.6%. It is not surprising that the most probable configuration is the one in which the molecules are evenly divided between the two halves of the box, because that is what we expect at thermal equilibrium. However, it *is* surprising that there is *any* probability, however small, of finding all six molecules clustered in half of the box, with the other half empty. In Sample Problem 21-5 we show that this state can occur because six molecules is an extremely small number.

For large values of N there are extremely large numbers of microstates, but nearly all the microstates belong to the configuration in which the molecules are divided equally between the two halves of the box, as Fig. 21-17 indicates. Even though the measured temperature and pressure of the gas remain constant, the gas is churning away endlessly as its molecules "visit" all probable microstates with equal probability. However, because so few microstates lie outside the very narrow central configuration peak of Fig. 21-17, we might as well assume that the gas molecules are always divided equally between the two halves of the box. As we shall see, this is the configuration with the greatest entropy.

Sample Problem 21-5

Suppose that there are 100 indistinguishable molecules in the box of Fig. 21-16. How many microstates are associated with the configuration $n_1 = 50$ and $n_2 = 50$? How many are associated with the configuration $n_1 = 100$ and $n_2 = 0$? Interpret the results in terms of the relative probabilities of the two configurations.

SOLUTION: The **Key Idea** here is that the multiplicity W of a configuration of indistinguishable molecules in a closed box is the number of independent microstates with that configuration, as given by Eq. 21-18. For the (n_1, n_2) configuration (50, 50), that equation yields

$$W = \frac{N!}{n_1!\,n_2!} = \frac{100!}{50!\,50!} = \frac{9.33 \times 10^{157}}{(3.04 \times 10^{64})(3.04 \times 10^{64})} = 1.01 \times 10^{29}. \quad \text{(Answer)}$$

Similarly, for the configuration of (100, 0), we have

$$W = \frac{N!}{n_1!\,n_2!} = \frac{100!}{100!\,0!} = \frac{1}{0!} = \frac{1}{1} = 1. \quad \text{(Answer)}$$

Thus, a 50–50 distribution is more likely than a 100–0 distribution by the enormous factor of about 1×10^{29}. If you could count, at one per nanosecond, the number of microstates that correspond to the 50–50 distribution, it would take you about 3×10^{12} years, which is about 750 times longer than the age of the universe. Even 100 molecules is *still* a very small number. Imagine what these calculated probabilities would be like for a mole of molecules, say about $N = 10^{24}$. Thus, you need never worry about suddenly finding all the air molecules clustering in one corner of your room!

Probability and Entropy

In 1877, Austrian physicist Ludwig Boltzmann (the Boltzmann of Boltzmann's constant k) derived a relationship between the entropy S of a configuration of a gas and the multiplicity W of that configuration. That relationship is

$$S = k \ln W \quad \text{(Boltzmann's entropy equation).} \qquad (21\text{-}19)$$

This famous formula is engraved on Boltzmann's tombstone.

It is natural that S and W should be related by a logarithmic function. The total entropy of two systems is the *sum* of their separate entropies. The probability of occurrence of two independent systems is the *product* of their separate probabilities. Because $\ln ab = \ln a + \ln b$, the logarithm seems the logical way to connect these quantities.

Table 21-1 displays the entropies of the configurations of the six-molecule system of Fig. 21-16, computed using Eq. 21-19. Configuration IV, which has the greatest multiplicity, also has the greatest entropy.

When you use Eq. 21-18 to calculate W, your calculator may signal "OVERFLOW" if you try to find the factorial of a number greater than a few hundred. Fortunately, there is a very good approximation, known as **Stirling's approximation,** not for $N!$ but for $\ln N!$, which is exactly what is needed in Eq. 21-19. Stirling's approximation is

$$\ln N! \approx N(\ln N) - N \quad \text{(Stirling's approximation).} \tag{21-20}$$

The Stirling of this approximation is not the Stirling of the Stirling engine.

CHECKPOINT 5: A box contains one mole of a gas. Consider two configurations: (a) each half of the box contains half the molecules, and (b) each third of the box contains one-third of the molecules. Which configuration has more microstates?

Sample Problem 21-6

In Sample Problem 21-1 we showed that when n moles of an ideal gas doubles its volume in a free expansion, the entropy increase from the initial state i to the final state f is $S_f - S_i = nR \ln 2$. Derive this result with statistical mechanics.

SOLUTION: One **Key Idea** here is that we can relate the entropy S of any given configuration of the molecules in the gas to the multiplicity W of microstates for that configuration, using Eq. 21-19 ($S = k \ln W$). We are interested in two configurations: the final configuration f (with the molecules occupying the full volume of their container in Fig. 21-1b) and the initial configuration i (with the molecules occupying the left half of the container).

A second **Key Idea** is that, because the molecules are in a closed container, we can calculate the multiplicity W of their microstates with Eq. 21-18. Here we have N molecules in the n moles of the gas. Initially, with the molecules all in the left half of the container, their (n_1, n_2) configuration is $(N, 0)$. Then, Eq. 21-18 gives their multiplicity as

$$W_i = \frac{N!}{N!\,0!} = 1.$$

Finally, with the molecules spread through the full volume, their (n_1, n_2) configuration is $(N/2, N/2)$. Then, Eq. 21-18 gives their multiplicity as

$$W_f = \frac{N!}{(N/2)!\,(N/2)!}.$$

From Eq. 21-19, the initial and final entropies are

$$S_i = k \ln W_i = k \ln 1 = 0$$

and

$$S_f = k \ln W_f = k \ln(N!) - 2k \ln[(N/2)!]. \tag{21-21}$$

In writing Eq. 21-21, we have used the relation

$$\ln \frac{a}{b^2} = \ln a - 2 \ln b.$$

Now, applying Eq. 21-20 to evaluate Eq. 21-21, we find that

$$\begin{aligned} S_f &= k \ln(N!) - 2k \ln[(N/2)!] \\ &= k[N(\ln N) - N] - 2k[(N/2)\ln(N/2) - (N/2)] \\ &= k[N(\ln N) - N - N\ln(N/2) + N] \\ &= k[N(\ln N) - N(\ln N - \ln 2)] = Nk \ln 2. \end{aligned} \tag{21-22}$$

From Section 20-3 we can substitute nR for Nk, where R is the universal gas constant. Equation 21-22 then becomes

$$S_f = nR \ln 2.$$

The change in entropy from the initial state to the final is thus

$$\begin{aligned} S_f - S_i &= nR \ln 2 - 0 \\ &= nR \ln 2, \end{aligned} \quad \text{(Answer)}$$

which is what we set out to show. In Sample Problem 21-1 we calculated this entropy increase for a free expansion with thermodynamics by finding an equivalent reversible process and calculating the entropy change for *that* process in terms of temperature and heat transfer. In this sample problem, we calculate the same increase in entropy with statistical mechanics using the fact that the system consists of molecules.

REVIEW & SUMMARY

One-Way Processes An **irreversible process** is one that cannot be reversed by means of small changes in the environment. The direction in which an irreversible process proceeds is set by the *change in entropy* ΔS of the system undergoing the process. Entropy S is a *state property* (or *state function*) of the system; that is, it depends only on the state of the system and not on how the system reached that state. The *entropy postulate* states (in part): *If an irreversible process occurs in a closed system, the entropy of the system always increases.*

Calculating Entropy Change The **entropy change** ΔS for an irreversible process that takes a system from an initial state i to a final state f is exactly equal to the entropy change ΔS for *any reversible process* that takes the system between those same two states. We can compute the latter (but not the former) with

$$\Delta S = S_f - S_i = \int_i^f \frac{dQ}{T}. \quad (21\text{-}1)$$

Here Q is the energy transferred as heat to or from the system during the process, and T is the temperature of the system in kelvins during the process.

For a reversible isothermal process, Eq. 21-1 reduces to

$$\Delta S = S_f - S_i = \frac{Q}{T}. \quad (21\text{-}2)$$

When the temperature change ΔT of a system is small relative to the temperature (in kelvins) before and after the process, the entropy change can be approximated as

$$\Delta S = S_f - S_i \approx \frac{Q}{T_{\text{avg}}}, \quad (21\text{-}3)$$

where T_{avg} is the system's average temperature during the process.

When an ideal gas changes reversibly from an initial state with temperature T_i and volume V_i to a final state with temperature T_f and volume V_f, the change ΔS in the entropy of the gas is

$$\Delta S = S_f - S_i = nR \ln\frac{V_f}{V_i} + nC_V \ln\frac{T_f}{T_i}. \quad (21\text{-}4)$$

The Second Law of Thermodynamics This law, which is an extension of the entropy postulate, states: *If a process occurs in a closed system, the entropy of the system increases for irreversible processes and remains constant for reversible processes. It never decreases.* In equation form,

$$\Delta S \geq 0. \quad (21\text{-}5)$$

Engines An **engine** is a device that, operating in a cycle, extracts energy $|Q_H|$ as heat from a high-temperature reservoir and does a certain amount of work $|W|$. The *efficiency* ε of any engine is defined as

$$\varepsilon = \frac{\text{energy we get}}{\text{energy we pay for}} = \frac{|W|}{|Q_H|}. \quad (21\text{-}9)$$

In an **ideal engine,** all processes are reversible and no wasteful energy transfers occur due to, say, friction and turbulence. A **Carnot engine** is an ideal engine that follows the cycle of Fig. 21-8. Its efficiency is

$$\varepsilon_C = 1 - \frac{|Q_L|}{|Q_H|} = 1 - \frac{T_L}{T_H}, \quad (21\text{-}10, 21\text{-}11)$$

in which T_H and T_L are the temperatures of the high- and low-temperature reservoirs, respectively. Real engines always have an efficiency lower than that given by Eq. 21-11. Ideal engines that are not Carnot engines also have efficiencies lower than that given by Eq. 21-11.

A *perfect engine* is an imaginary engine in which energy extracted as heat from the high-temperature reservoir is converted completely to work. Such would violate the second law of thermodynamics, which can be restated as follows: No series of processes is possible whose sole result is the absorption of energy as heat from a thermal reservoir and the complete conversion of this energy to work.

Refrigerators A refrigerator is a device that, operating in a cycle, has work W done on it as it extracts energy $|Q_L|$ as heat from a low-temperature reservoir. The coefficient of performance K of a refrigerator is defined as

$$K = \frac{\text{what we want}}{\text{what we pay for}} = \frac{|Q_L|}{|W|}. \quad (21\text{-}12)$$

A **Carnot refrigerator** is a Carnot engine operating in reverse. For a Carnot refrigerator, Eq. 21-12 becomes

$$K_C = \frac{|Q_L|}{|Q_H| - |Q_L|} = \frac{T_L}{T_H - T_L}. \quad (21\text{-}13, 21\text{-}14)$$

A *perfect refrigerator* is an imaginary refrigerator in which energy extracted as heat from the low-temperature reservoir is converted completely to heat discharged to the high-temperature reservoir, without any need for work. Such would violate the second law of thermodynamics, which can be restated as follows: No series of processes is possible whose sole result is the transfer of energy as heat from a reservoir at a given temperature to a reservoir at a higher temperature.

Entropy from a Statistical View The entropy of a system can be defined in terms of the possible distributions of its molecules. For identical molecules, each possible distribution of molecules is called a **microstate** of the system. All equivalent microstates are grouped into a **configuration** of the system. The number of microstates in a configuration is the **multiplicity** W of the configuration.

For a system of N molecules that may be distributed between the two halves of a box, the multiplicity is given by

$$W = \frac{N!}{n_1!\, n_2!}, \quad (21\text{-}18)$$

in which n_1 is the number of molecules in one half of the box and n_2 is the number in the other half. A basic assumption of *statistical mechanics* is that all the microstates are equally probable. Thus, configurations with a large multiplicity occur most often. When N is very large (say, $N = 10^{22}$ molecules or more), the molecules are nearly always in the configuration in which $n_1 = n_2$.

The multiplicity W of a configuration of a system and the entropy S of the system in that configuration are related by Boltzmann's entropy equation:

$$S = k \ln W, \qquad (21\text{-}19)$$

where $k = 1.38 \times 10^{-23}$ J/K is the Boltzmann constant.

When N is very large (the usual case), we can approximate $\ln N!$ with *Stirling's approximation:*

$$\ln N! \approx N(\ln N) - N. \qquad (21\text{-}20)$$

QUESTIONS

1. A gas, confined to an insulated cylinder, is compressed adiabatically to half its volume. Does the entropy of the gas increase, decrease, or remain unchanged during this process?

2. In four experiments, blocks A and B, starting at different initial temperatures, were brought together in an insulating box (as in Sample Problem 21-2) and allowed to reach a common final temperature. The entropy changes for the blocks in the four experiments had the following values (in Joules per kelvin), but not necessarily in the order given. Determine which values for A go with which values for B.

Block	Values			
A	8	5	3	9
B	−3	−8	−5	−2

3. Point i in Fig. 21-18 represents the initial state of an ideal gas at temperature T. Taking algebraic signs into account, rank the entropy changes that the gas undergoes as it moves, successively and reversibly, from point i to points a, b, c, and d, greatest first.

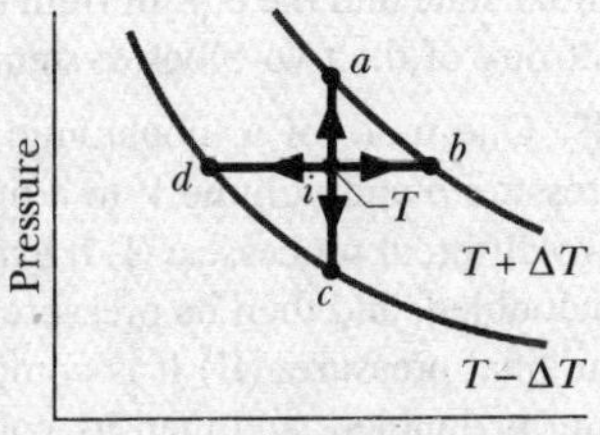

Fig. 21-18 Question 3.

4. An ideal gas, in contact with a controllable thermal reservoir, can be taken from initial state i to final state f along the four reversible paths in Fig. 21-19. Rank the paths according to the magnitudes of the resulting entropy changes of (a) the gas, (b) the reservoir, and (c) the gas–reservoir system, greatest first.

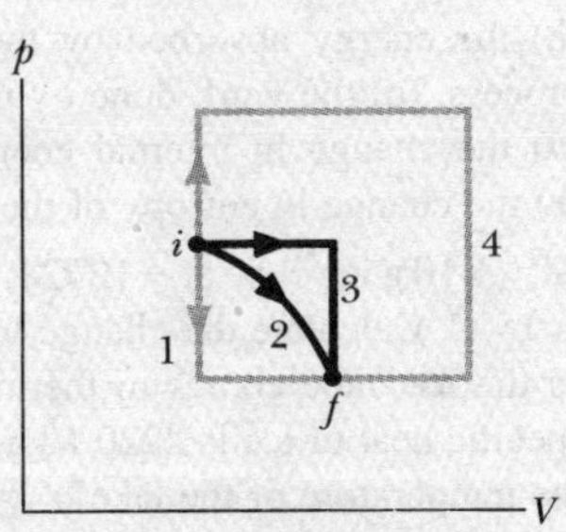

Fig. 21-19 Question 4.

5. You allow a gas to expand freely from volume V to volume $2V$. Later you allow that gas to expand freely from volume $2V$ to volume $3V$. Is the net entropy change for these two expansions greater than, less than, or equal to the entropy change that would occur if you allowed the gas to expand freely from volume V directly to volume $3V$?

6. Three Carnot engines operate between temperature limits of (a) 400 and 500 K, (b) 500 and 600 K, and (c) 400 and 600 K. Each engine extracts the same amount of energy per cycle from the high-temperature reservoir. Rank the magnitudes of the work done by the engines per cycle, greatest first.

7. Does the entropy per cycle increase, decrease, or remain the same for (a) a Carnot engine, (b) a real engine, and (c) a perfect engine (which is, of course, impossible to build)?

8. If you leave the door of your kitchen refrigerator open for several hours, does the temperature of the kitchen increase, decrease, or remain the same? Assume that the kitchen is closed and well insulated.

9. Does the entropy per cycle increase, decrease, or remain the same for (a) a Carnot refrigerator, (b) a real refrigerator, and (c) a perfect refrigerator (which is, of course, impossible to build)?

10. A box contains 100 atoms in a configuration, with 50 atoms in each half of the box. Suppose that you could count the different microstates associated with this configuration at the rate of 100 billion states per second, using a supercomputer. Without written calculation, guess how much computing time you would need: a day, a year, or much more than a year.

11. Figure 21-20 shows a snapshot at time $t = 0$ of molecules a and b in a box (similar to that of Fig. 21-16). The molecules have the same mass and speed v, and collisions between the molecules and the walls are elastic. What is the probability that snapshots taken at times (a) $t = 0.10L/v$ and (b) $t = 10L/v$ will show that a is in the left side of the box and b is in the right side of the box? (c) What is the probability that at some later time only half of the kinetic energy of the molecules will be in the right side of the box?

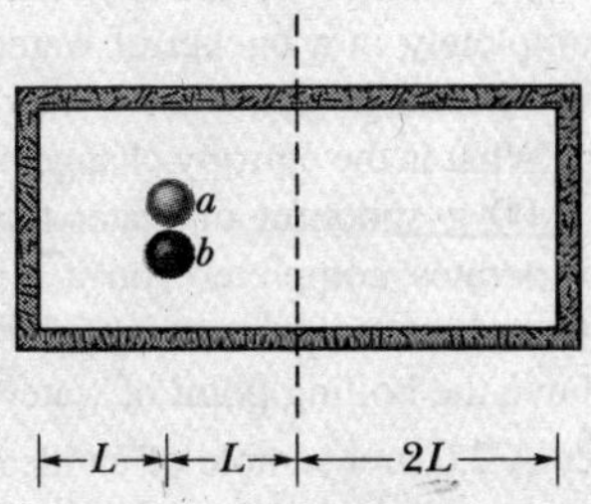

Fig. 21-20 Question 11.

EXERCISES & PROBLEMS

ssm Solution is in the Student Solutions Manual.
www Solution is available on the World Wide Web at: http://www.wiley.com/college/hrw
ilw Solution is available on the Interactive LearningWare.

SEC. 21-2 Change in Entropy

1E. A 2.50 mol sample of an ideal gas expands reversibly and isothermally at 360 K until its volume is doubled. What is the increase in entropy of the gas? **ilw**

2E. How much heat is required for a reversible isothermal expansion of an ideal gas at 132°C if the entropy of the gas increases by 46.0 J/K?

3E. Four moles of an ideal gas undergo a reversible isothermal expansion from volume V_1 to volume $V_2 = 2V_1$ at temperature $T = 400$ K. Find (a) the work done by the gas and (b) the entropy change of the gas. (c) If the expansion is reversible and adiabatic instead of isothermal, what is the entropy change of the gas? ssm

4E. An ideal gas undergoes a reversible isothermal expansion at 77.0°C, increasing its volume from 1.30 L to 3.40 L. The entropy change of the gas is 22.0 J/K. How many moles of gas are present?

5E. Find (a) the energy absorbed as heat and (b) the change in entropy of a 2.00 kg block of copper whose temperature is increased reversibly from 25°C to 100°C. The specific heat of copper is 386 J/kg · K. itw

6E. An ideal monatomic gas at initial temperature T_0 (in kelvins) expands from initial volume V_0 to volume $2V_0$ by each of the five

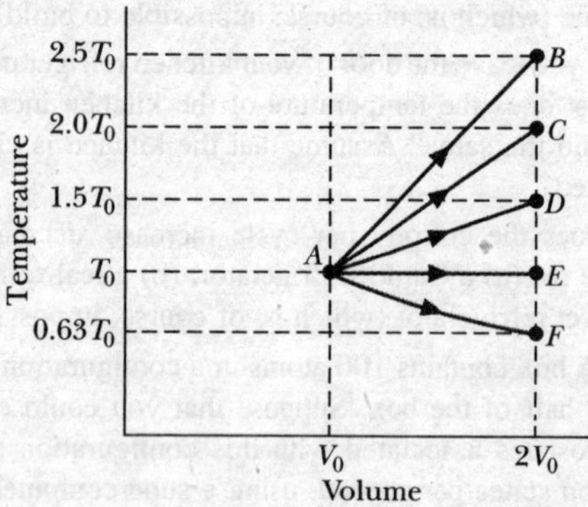

Fig. 21-21 Exercise 6.

processes indicated in the *T-V* diagram of Fig. 21-21. In which process is the expansion (a) isothermal, (b) isobaric (constant pressure), and (c) adiabatic? Explain your answers. (d) In which processes does the entropy of the gas decrease?

7E. (a) What is the entropy change of a 12.0 g ice cube that melts completely in a bucket of water whose temperature is just above the freezing point of water? (b) What is the entropy change of a 5.00 g spoonful of water that evaporates completely on a hot plate whose temperature is slightly above the boiling point of water?

8P. A 2.0 mol sample of an ideal monatomic gas undergoes the reversible process shown in Fig. 21-22. (a) How much energy is absorbed as heat by the gas? (b) What is the change in the internal energy of the gas? (c) How much work is done by the gas?

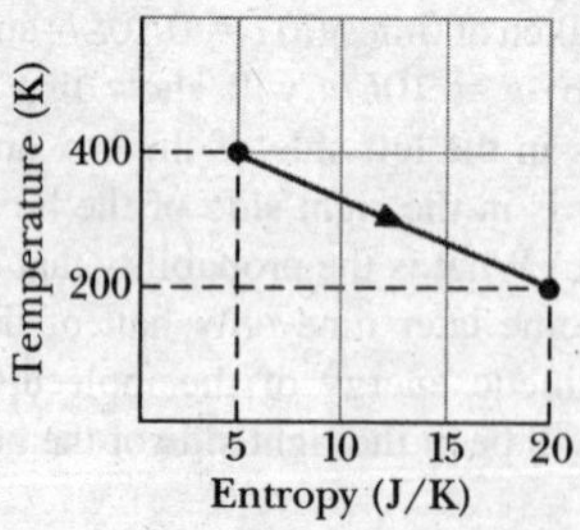

Fig. 21-22 Problem 8.

9P. In an experiment, 200 g of aluminum (with a specific heat of 900 J/kg · K) at 100°C is mixed with 50.0 g of water at 20.0°C, with the mixture thermally isolated. (a) What is the equilibrium temperature? What are the entropy changes of (b) the aluminum, (c) the water, and (d) the aluminum–water system? ssm www

10P. In the irreversible process of Fig. 21-5, let the initial temperatures of identical blocks *L* and *R* be 305.5 and 294.5 K, respectively, and let 215 J be the energy transfer between the blocks required to reach equilibrium. Then for the reversible processes of Fig. 21-6, what are the entropy changes of (a) block *L*, (b) its reservoir, (c) block *R*, (d) its reservoir, (e) the two-block system, and (f) the system of the two blocks and the two reservoirs?

11P. Use the reversible apparatus of Fig. 21-6 to show that, if the process of Fig. 21-5 happened in reverse, the entropy of the system would decrease, a violation of the second law of thermodynamics.

12P. An ideal diatomic gas, whose molecules are rotating but not oscillating, is taken through the cycle in Fig. 21-23. Determine for all three processes, in terms of p_1, V_1, T_1, and R: (a) p_2, p_3, and T_3 and (b) W, Q, ΔE_{int}, and ΔS per mole.

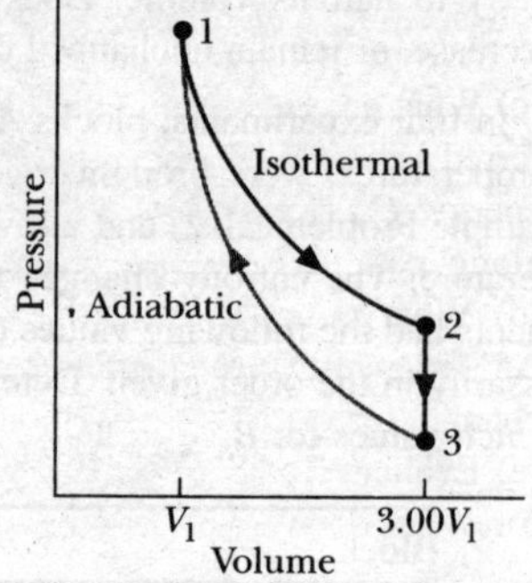

Fig. 21-23 Problem 12.

13P. A 50.0 g block of copper whose temperature is 400 K is placed in an insulating box with a 100 g block of lead whose temperature is 200 K. (a) What is the equilibrium temperature of the two-block system? (b) What is the change in the internal energy of the two-block system between the initial state and the equilibrium state? (c) What is the change in the entropy of the two-block system? (See Table 19-3.) itw

14P. One mole of a monatomic ideal gas is taken from an initial pressure p and volume V to a final pressure $2p$ and volume $2V$ by two different processes: (I) It expands isothermally until its volume is doubled, and then its pressure is increased at constant volume to the final pressure. (II) It is compressed isothermally until its pressure is doubled, and then its volume is increased at constant pressure to the final volume. (a) Show the path of each process on a *p-V* diagram. For each process calculate, in terms of p and V, (b) the energy absorbed by the gas as heat in each part of the process, (c) the work done by the gas in each part of the process, (d) the change in internal energy of the gas, $E_{\text{int},f} - E_{\text{int},i}$, and (e) the change in entropy of the gas, $S_f - S_i$.

15P. A 10 g ice cube at −10°C is placed in a lake whose temperature is 15°C. Calculate the change in entropy of the cube–lake system as the ice cube comes to thermal equilibrium with the lake. The specific heat of ice is 2220 J/kg · K. (*Hint:* Will the ice cube affect the temperature of the lake?) ssm

16P. An 8.0 g ice cube at −10°C is put into a Thermos flask containing 100 cm³ of water at 20°C. By how much has the entropy of the cube–water system changed when a final equilibrium state is reached? The specific heat of ice is 2220 J/kg · K.

17P. A mixture of 1773 g of water and 227 g of ice is in an initial equilibrium state at 0.00°C. The mixture is then, in a reversible process, brought to a second equilibrium state where the water–ice ratio, by mass, is 1:1 at 0.00°C. (a) Calculate the entropy change of the system during this process. (The heat of fusion for water is 333 kJ/kg.) (b) The system is then returned to the initial equilibrium state in an irreversible process (say, by using a Bunsen burner). Calculate the entropy change of the system during this process. (c) Are your answers consistent with the second law of thermodynamics? ssm

18P. A cylinder contains n moles of a monatomic ideal gas. If the gas undergoes a reversible isothermal expansion from initial volume V_i to final volume V_f along path I in Fig. 21-24, its change in entropy is $\Delta S = nR \ln (V_f/V_i)$. (See Sample Problem 21-1.) Now consider path II in Fig. 21-24, which takes the gas from the same initial state i to state x by a reversible adiabatic expansion, and then from that state x to the same final state f by a reversible constant-volume process. (a) Describe how you would carry out the two reversible processes for path II. (b) Show that the temperature of the gas in state x is

$$T_x = T_i(V_i/V_f)^{2/3}$$

(c) What are the energy Q_I transferred as heat along path I and the energy Q_{II} transferred as heat along path II? Are they equal? (d) What is the entropy change ΔS for path II? Is the entropy change for path I equal to it? (e) Evaluate T_x, Q_I, Q_{II}, and ΔS for $n = 1$, $T_i = 500$ K, and $V_f/V_i = 2$.

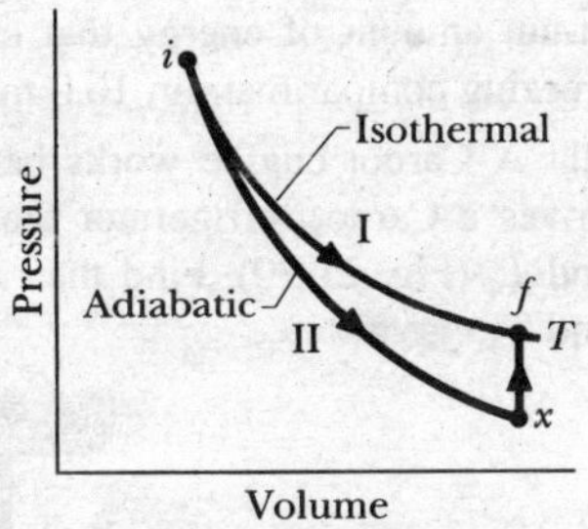

Fig. 21-24 Problem 18.

19P. One mole of an ideal monatomic gas is taken through the cycle in Fig. 21-25. (a) How much work is done by the gas in going from state a to state c along path abc? What are the changes in internal energy and entropy in going (b) from b to c and (c) through one complete cycle? Express all answers in terms of the pressure p_0, volume V_0, and temperature T_0 of state a. ssm

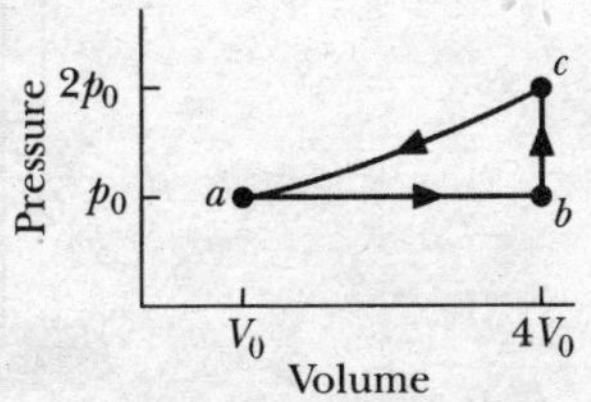

Fig. 21-25 Problem 19.

20P. One mole of an ideal monatomic gas, at an initial pressure of 5.00 kPa and initial temperature of 600 K, expands from initial volume $V_i = 1.00$ m^3 to final volume $V_f = 2.00$ m^3. During the expansion, the pressure p and volume V of the gas are related by $p = 5.00 \exp[(V_i - V)/a]$, where p is in kilopascals, V_i and V are in cubic meters, and $a = 1.00$ m^3. What are (a) the final pressure and (b) the final temperature of the gas? (c) How much work is done by the gas during the expansion? (d) What is the change in entropy of the gas during the expansion? (*Hint:* Use two simple reversible processes to find the entropy change.)

SEC. 21-4 Entropy in the Real World: Engines

21E. A Carnot engine absorbs 52 kJ as heat and exhausts 36 kJ as heat in each cycle. Calculate (a) the engine's efficiency and (b) the work done per cycle in kilojoules.

22E. A Carnot engine whose low-temperature reservoir is at 17°C has an efficiency of 40%. By how much should the temperature of the high-temperature reservoir be increased to increase the efficiency to 50%?

23E. A Carnot engine operates between 235°C and 115°C, absorbing 6.30×10^4 J per cycle at the higher temperature. (a) What is the efficiency of the engine? (b) How much work per cycle is this engine capable of performing? ssm

24E. In a hypothetical nuclear fusion reactor, the fuel is deuterium gas at a temperature of about 7×10^8 K. If this gas could be used to operate a Carnot engine with $T_L = 100$°C, what would be the engine's efficiency?

25E. A Carnot engine has an efficiency of 22.0%. It operates between constant-temperature reservoirs differing in temperature by 75.0 C°. What are the temperatures of the two reservoirs? ssm

26P. A Carnot engine has a power of 500 W. It operates between constant-temperature reservoirs at 100°C and 60.0°C. What are (a) the rate of heat input and (b) the rate of exhaust heat output, in kilojoules per second?

27P. One mole of a monatomic ideal gas is taken through the reversible cycle shown in Fig. 21-26. Process bc is an adiabatic expansion, with $p_b = 10.0$ atm and $V_b = 1.00 \times 10^{-3}$ m^3. Find (a) the energy added to the gas as heat, (b) the energy leaving the gas as heat, (c) the net work done by the gas, and (d) the efficiency of the cycle. ssm ilw

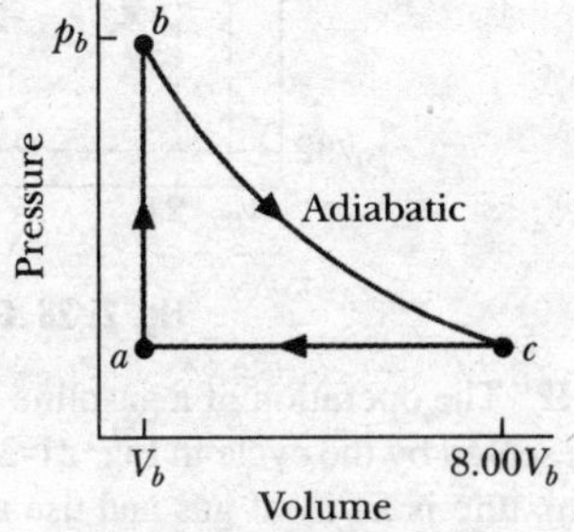

Fig. 21-26 Problem 27.

28P. Show that the area enclosed by the Carnot cycle on the temperature-entropy plot of Fig. 21-9 represents the net energy transfer per cycle as heat to the working substance.

29P. One mole of an ideal monatomic gas is taken through the cycle shown in Fig. 21-27. Assume that $p = 2p_0$, $V = 2V_0$, $p_0 = 1.01 \times 10^5$ Pa, and $V_0 = 0.0225$ m^3. Calculate (a) the work done during the cycle, (b) the energy added as heat during stroke abc, and (c) the efficiency of the cycle. (d) What is the efficiency of a Carnot engine operating between the highest and lowest temperatures that occur in the cycle? How does this compare to the efficiency calculated in (c)?

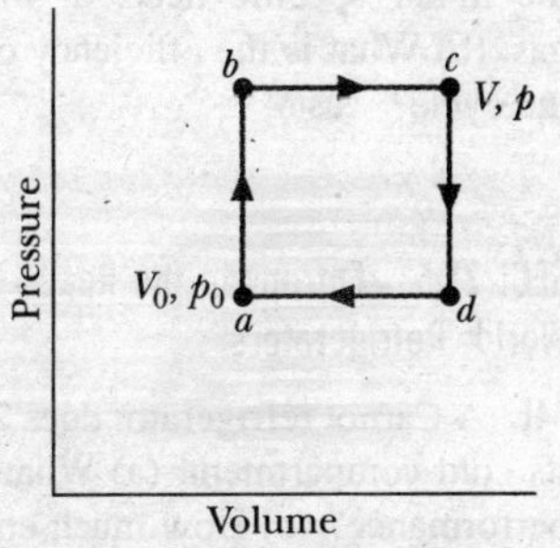

Fig. 21-27 Problem 29.

30P. In the first stage of a two-stage Carnot engine, energy Q_1 is absorbed as heat at temperature T_1, work W_1 is done, and energy Q_2 is expelled as heat at a lower temperature T_2. The second stage absorbs that energy Q_2, does work W_2, and expels energy Q_3 at a still lower temperature T_3. Prove that the efficiency of the two-stage engine is $(T_1 - T_3)/T_1$.

31P. Suppose that a deep shaft were drilled in Earth's crust near one of the poles, where the surface temperature is −40°C, to a depth where the temperature is 800°C. (a) What is the theoretical limit to the efficiency of an engine operating between these temperatures? (b) If all the energy released as heat into the low-temperature reservoir were used to melt ice that was initially at −40°C, at what rate could liquid water at 0°C be produced by a 100 MW power plant (treat it as an engine)? The specific heat of ice is 2220 J/kg · K; water's heat of fusion is 333 kJ/kg. (Note that the engine can operate only between 0°C and 800°C in this case. Energy exhausted at −40°C cannot be used to raise the temperature of anything above −40°C.) ssm www

32P. One mole of an ideal gas is used as the working substance of an engine that operates on the cycle shown in Fig. 21-28. *BC* and *DA* are reversible adiabatic processes. (a) Is the gas monatomic, diatomic, or polyatomic? (b) What is the efficiency of the engine?

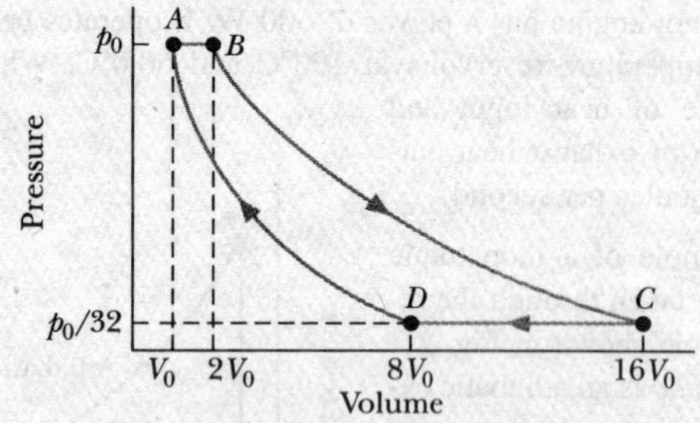

Fig. 21-28 Problem 32.

33P. The operation of a gasoline internal combustion engine is represented by the cycle in Fig. 21-29. Assume the gasoline–air intake mixture is an ideal gas and use a compression ratio of 4:1 ($V_4 = 4V_1$). Assume that $p_2 = 3p_1$. (a) Determine the pressure and temperature at each of the vertex points of the *p-V* diagram in terms of p_1, T_1, and the ratio γ of the molar specific heats of the gas. (b) What is the efficiency of the cycle? ssm

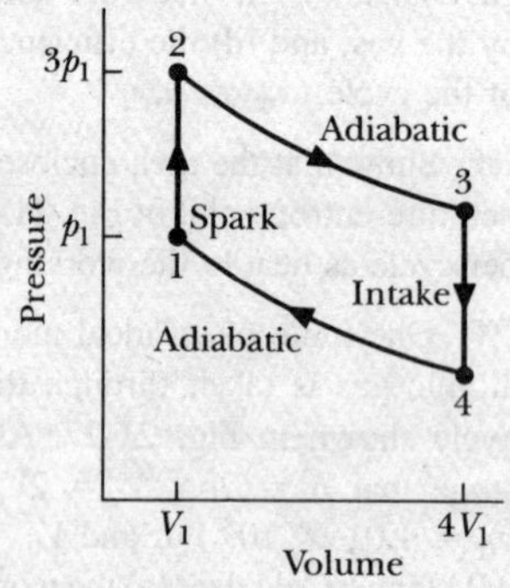

Fig. 21-29 Problem 33.

SEC. 21-5 Entropy in the Real World: Refrigerators

34E. A Carnot refrigerator does 200 J of work to remove 600 J from its cold compartment. (a) What is the refrigerator's coefficient of performance? (b) How much energy per cycle is exhausted to the kitchen as heat?

35E. A Carnot air conditioner takes energy from the thermal energy of a room at 70°F and transfers it to the outdoors, which is at 96°F. For each joule of electric energy required to operate the air conditioner, how many joules are removed from the room? ssm

36E. The electric motor of a heat pump transfers energy as heat from the outdoors, which is at −5.0°C, to a room, which is at 17°C. If the heat pump were a Carnot heat pump (a Carnot engine working in reverse), how many joules of heat would be transferred to the thermal energy of the room for each joule of electric energy consumed?

37E. A heat pump is used to heat a building. The outside temperature is −5.0°C, and the temperature inside the building is to be maintained at 22°C. The pump's coefficient of performance is 3.8, and the heat pump delivers 7.54 MJ as heat to the building each hour. If the heat pump is a Carnot engine working in reverse, at what rate must work be done to run the heat pump? ssm

38E. How much work must be done by a Carnot refrigerator to transfer 1.0 J as heat (a) from a reservoir at 7.0°C to one at 27°C, (b) from a reservoir at −73°C to one at 27°C, (c) from a reservoir at −173°C to one at 27°C, and (d) from a reservoir at −223°C to one at 27°C?

39P. An air conditioner operating between 93°F and 70°F is rated at 4000 Btu/h cooling capacity. Its coefficient of performance is 27% of that of a Carnot refrigerator operating between the same two temperatures. What horsepower is required of the air conditioner motor? ilw

40P. The motor in a refrigerator has a power of 200 W. If the freezing compartment is at 270 K and the outside air is at 300 K, and assuming the efficiency of a Carnot refrigerator, what is the maximum amount of energy that can be extracted as heat from the freezing compartment in 10.0 min?

41P. A Carnot engine works between temperatures T_1 and T_2. It drives a Carnot refrigerator that works between temperatures T_3 and T_4 (Fig. 21-30). Find the ratio Q_3/Q_1 in terms of T_1, T_2, T_3, and T_4. ssm

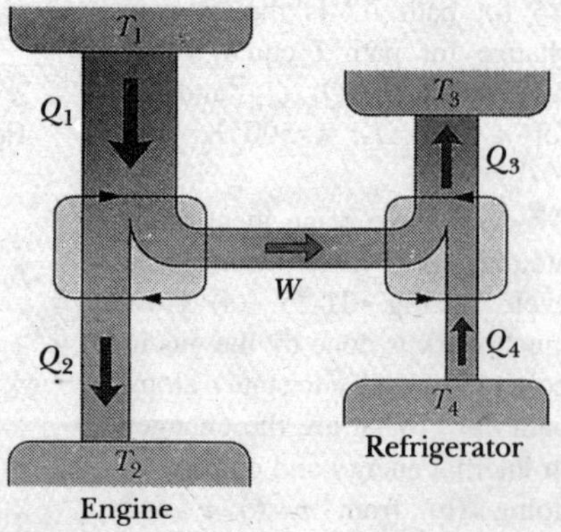

Fig. 21-30 Problem 41.

SEC. 21-7 A Statistical View of Entropy

42E. Construct a table like Table 21-1 for eight molecules.

43E. Show that for *N* molecules in a box, the number of possible microstates is 2^N when microstates are defined by whether a given molecule is in the left half of the box or the right half. Check this for the situation of Table 21-1.

44P. A box contains *N* gas molecules, equally divided between its two halves. For $N = 50$: (a) What is the multiplicity of this central configuration? (b) What is the total number of microstates for the system? (*Hint:* See Exercise 43.) (c) What percentage of the time does the system spend in its central configuration? (d) Repeat (a) through (c) for $N = 100$. (e) Repeat (a) through (c) for $N = 200$. (f) As *N* increases, you will find that the system spends *less* time (not more) in its central configuration. Explain why this is so.

45P. A box contains *N* gas molecules. Consider the box to be divided into three equal parts. (a) By extension of Eq. 21-18, write a formula for the multiplicity of any given configuration. (b) Consider two configurations: configuration *A* with equal numbers of molecules in all three thirds of the box, and configuration *B* with equal numbers of molecules in both halves of the box. What is the ratio W_A/W_B of the multiplicity of configuration *A* to that of configuration *B*? (c) Evaluate W_A/W_B for $N = 100$. (Because 100 is not evenly divisible by 3, put 34 molecules into one of the three box parts and 33 in each of the other parts for configuration *A*.) ssm www

22 Electric Charge

If you adapt your eyes to darkness for about 15 minutes and then have a friend chew a wintergreen LifeSaver, you will see a faint flash of blue light from your friend's mouth with each chomp. (To avoid wear on the teeth, you might crush the candy with pliers, as in the photograph.)

What causes this display of light, commonly called "sparking"?

The answer is in this chapter.

22-1 Electromagnetism

The early Greek philosophers knew that if you rubbed a piece of amber, it would attract bits of straw. This ancient observation is a direct ancestor of the electronic age in which we live. (The strength of the connection is indicated by our word *electron*, which is derived from the Greek word for amber.) The Greeks also recorded the observation that some naturally occurring "stones," known today as the mineral magnetite, would attract iron.

From these modest origins, the sciences of electricity and magnetism developed separately for centuries—until 1820, in fact, when Hans Christian Oersted found a connection between them: an electric current in a wire can deflect a magnetic compass needle. Interestingly enough, Oersted made this discovery while preparing a lecture demonstration for his physics students.

The new science of *electromagnetism* (the combination of electrical and magnetic phenomena) was developed further by workers in many countries. One of the best was Michael Faraday, a truly gifted experimenter with a talent for physical intuition and visualization. That talent is attested to by the fact that his collected laboratory notebooks do not contain a single equation. In the mid-19th century, James Clerk Maxwell put Faraday's ideas into mathematical form, introduced many new ideas of his own, and put electromagnetism on a sound theoretical basis.

Table 32-1 shows the basic laws of electromagnetism, now called Maxwell's equations. We plan to work our way through them in the chapters between here and there, but you might want to glance at them now, to see our goal.

22-2 Electric Charge

If you walk across a carpet in dry weather, you can produce a spark by bringing your finger close to a metal doorknob. Television advertising has alerted us to the problem of "static cling" in clothing (Fig. 22-1). On a grander scale, lightning is familiar to everyone. Each of these phenomena represents a tiny glimpse of the vast amount of *electric charge* that is stored in the familiar objects that surround us and—indeed—in our own bodies. **Electric charge** is an intrinsic characteristic of the fundamental particles making up those objects; that is, it is a characteristic that automatically accompanies those particles wherever they exist.

Fig. 22-1 Static cling, an electrical phenomenon that accompanies dry weather, causes these pieces of paper to stick to one another and to the plastic comb, and your clothing to stick to your body.

The vast amount of charge in an everyday object is usually hidden because the object contains equal amounts of the two kinds of charge: *positive charge* and *negative charge*. With such an equality—or *balance*—of charge, the object is said to be *electrically neutral*; that is, it contains no *net* charge. If the two types of charge are not in balance, then there *is* a net charge. We say that an object is *charged* to indicate that it has a charge imbalance, or net charge. The imbalance is always very small compared to the total amounts of positive charge and negative charge contained in the object.

Charged objects interact by exerting forces on one another. To show this, we first charge a glass rod by rubbing one end with silk. At points of contact between the rod and the silk, tiny amounts of charge are transferred from one to the other, slightly upsetting the electrical neutrality of each. (We *rub* the silk over the rod to increase the number of contact points and thus the amount, still tiny, of transferred charge.)

Suppose we now suspend the charged rod from a thread to *electrically isolate* it from its surroundings so that its charge cannot change. If we bring a second, similarly charged, glass rod nearby (Fig. 22-2*a*), the two rods *repel* each other; that is, each rod experiences a force directed away from the other rod. However, if we rub a plastic rod with fur and bring it near the suspended glass rod (Fig. 22-2*b*), the

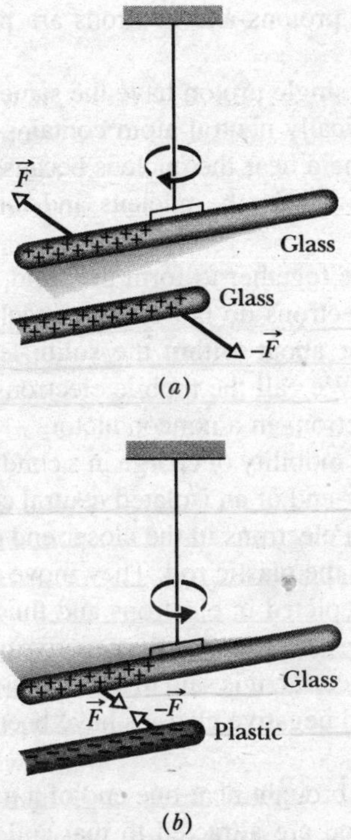

Fig. 22-2 (*a*) Two charged rods of the same signs repel each other. (*b*) Two charged rods of opposite signs attract each other. Plus signs indicate a positive net charge, and minus signs a negative net charge.

two rods *attract* each other; that is, each rod experiences a force directed toward the other rod.

We can understand these two demonstrations in terms of positive and negative charges. When a glass rod is rubbed with silk, the glass loses some of its negative charge and then has a small unbalanced positive charge (represented by the plus signs in Fig. 22-2*a*). When the plastic rod is rubbed with fur, the plastic gains a small unbalanced negative charge (represented by the minus signs in Fig. 22-2*b*). Our two demonstrations reveal the following:

> Charges with the same electrical sign repel each other, and charges with opposite electrical signs attract each other.

In Section 22-4, we shall put this rule into quantitative form as Coulomb's law of *electrostatic force* (or *electric force*) between charges. The term *electrostatic* is used to emphasize that, relative to each other, the charges are either stationary or moving only very slowly.

The "positive" and "negative" labels and signs for electric charge were chosen arbitrarily by Benjamin Franklin. He could easily have interchanged the labels or used some other pair of opposites to distinguish the two kinds of charge. (Franklin was a scientist of international reputation. It has even been said that Franklin's triumphs in diplomacy in France during the American War of Independence were facilitated, and perhaps even made possible, because he was so highly regarded as a scientist.)

The attraction and repulsion between charged bodies have many industrial applications, including electrostatic paint spraying and powder coating, fly-ash collection in chimneys, nonimpact ink-jet printing, and photocopying. Figure 22-3 shows a tiny carrier bead in a Xerox copying machine, covered with particles of black powder called *toner*, which stick to it by means of electrostatic forces. The negatively charged toner particles are eventually attracted from the carrier bead to a rotating drum, where a positively charged image of the document being copied has formed. A charged sheet of paper then attracts the toner particles from the drum to itself, after which they are heat-fused in place to produce the copy.

Fig. 22-3 A carrier bead from a Xerox copying machine; it is covered with toner particles that cling to it by electrostatic attraction. The diameter of the bead is about 0.3 mm.

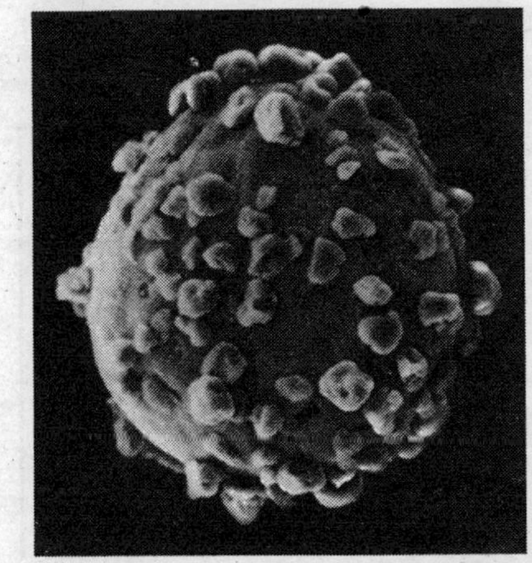

22-3 Conductors and Insulators

In some materials, such as metals, tap water, and the human body, some of the negative charge can move rather freely. We call such materials **conductors.** In other materials, such as glass, chemically pure water, and plastic, none of the charge can move freely. We call these materials **nonconductors** or **insulators.**

If you rub a copper rod with wool while holding the rod in your hand, you will not be able to charge the rod, because both you and the rod are conductors. The rubbing will cause a charge imbalance on the rod, but the excess charge will immediately move from the rod through you to the floor (which is connected to Earth's surface), and the rod will quickly be neutralized.

In thus setting up a pathway of conductors between an object and Earth's surface, we are said to *ground* the object, and in neutralizing the object (by eliminating an unbalanced positive or negative charge), we are said to *discharge* the object. If instead of holding the copper rod in your hand, you hold it by an insulating handle, you eliminate the conducting path to Earth, and the rod can then be charged by rubbing, as long as you do not touch it directly with your hand.

The properties of conductors and insulators are due to the structure and electrical nature of atoms. Atoms consist of positively charged *protons*, negatively charged

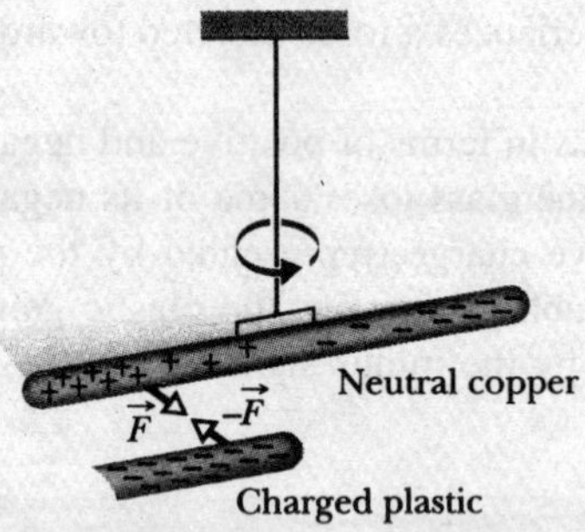

Fig. 22-4 A neutral copper rod is electrically isolated from its surroundings by being suspended on a nonconducting thread. Either end of the copper rod will be attracted by a charged rod. Here, conduction electrons in the copper rod are repelled to the far end of that rod by the negative charge on the plastic rod. Then that negative charge attracts the remaining positive charge on the near end of the copper rod, rotating the copper rod to bring that near end closer to the plastic rod.

electrons, and electrically neutral *neutrons*. The protons and neutrons are packed tightly together in a central *nucleus*.

The charge of a single electron and that of a single proton have the same magnitude but are opposite in sign. Hence, an electrically neutral atom contains equal numbers of electrons and protons. Electrons are held near the nucleus because they have the electrical sign opposite that of the protons in the nucleus and thus are attracted to the nucleus.

When atoms of a conductor like copper come together to form the solid, some of their outermost (and so most loosely held) electrons do not remain attached to the individual atoms but become free to wander about within the solid, leaving behind positively charged atoms (*positive ions*). We call the mobile electrons *conduction electrons*. There are few (if any) free electrons in a nonconductor.

The experiment of Fig. 22-4 demonstrates the mobility of charge in a conductor. A negatively charged plastic rod will attract either end of an isolated neutral copper rod. What happens is that many of the conduction electrons in the closer end of the copper rod are repelled by the negative charge on the plastic rod. They move to the far end of the copper rod, leaving the near end depleted in electrons and thus with an unbalanced positive charge. This positive charge is attracted to the negative charge in the plastic rod. Although the copper rod is still neutral, it is said to have an *induced charge*, which means that some of its positive and negative charges have been separated owing to the presence of a nearby charge.

Similarly, if a positively charged glass rod is brought near one end of a neutral copper rod, conduction electrons in the copper rod are attracted to that end. That end becomes negatively charged and the other end positively charged, so again an induced charge is set up in the copper rod. Although the copper rod is still neutral, it and the glass rod attract each other. (Figure 22-5 shows another demonstration of induced charge.) ∴ repulsion is the real test (of type of charge) not attraction

Note that only conduction electrons, with their negative charges, can move; positive ions are fixed in place. Thus, an object becomes positively charged only through the *removal of negative charges*.

Fig. 22-5 This is not a parlor stunt but a serious experiment carried out in 1774 to prove that the human body is a conductor of electricity. The etching shows a person suspended by nonconducting ropes while being charged by a charged rod (which probably touched flesh instead of the trousers). When the person brought his face, left hand, or the conducting ball and rod in his right hand near bits of paper on the plates, charge was induced on the paper, which then flew through the intermediate air to him.

Semiconductors, such as silicon and germanium, are materials that are intermediate between conductors and insulators. The microelectronic revolution that has transformed our lives in so many ways is due to devices constructed of semiconducting materials.

Finally, there are **superconductors,** so called because they present no resistance to the movement of electric charge through them. When charge moves through a material, we say that an **electric current** exists in the material. Ordinary materials, even good conductors, tend to resist the flow of charge through them. In a superconductor, however, the resistance is not just small; it is precisely zero. If you set up a current in a superconducting ring, it lasts "forever," with no battery or other source of energy needed to maintain it.

CHECKPOINT 1: The figure shows five pairs of plates: *A*, *B*, and *D* are charged plastic plates and *C* is an electrically neutral copper plate. The electrostatic forces between the pairs of plates are shown for three of the pairs. For the remaining two pairs, do the plates repel or attract each other?

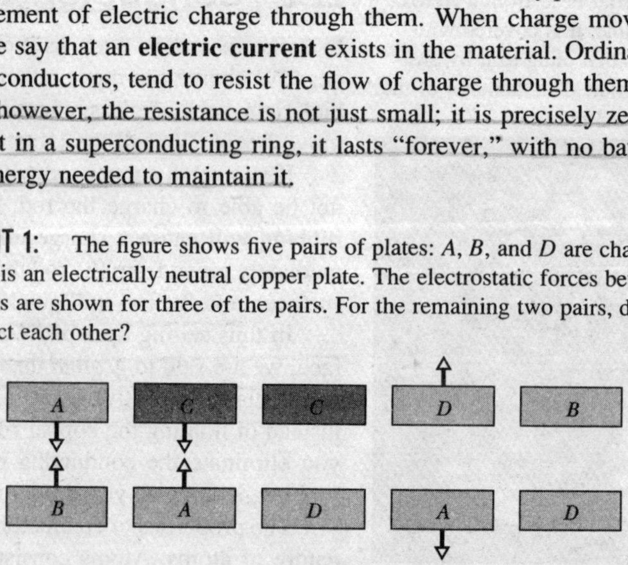

22-4 Coulomb's Law

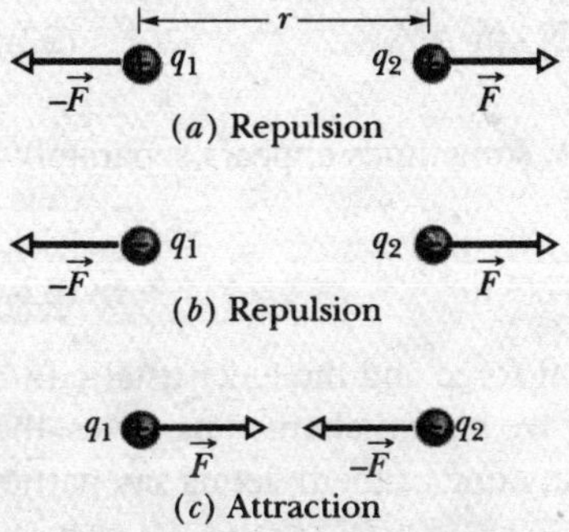

Fig. 22-6 Two charged particles, separated by distance r, repel each other if their charges are (a) both positive and (b) both negative. (c) They attract each other if their charges are of opposite signs. In each of the three situations, the force acting on one particle is equal in magnitude to the force acting on the other particle but has the opposite direction.

Let two charged particles (also called *point charges*) have charge magnitudes q_1 and q_2 and be separated by a distance r. The **electrostatic force** of attraction or repulsion between them has the magnitude

$$F = k\frac{|q_1||q_2|}{r^2} \quad \text{(Coulomb's law)}, \tag{22-1}$$

in which k is a constant. Each particle exerts a force of this magnitude on the other particle; the two forces form a third-law force pair. If the particles *repel* each other, the force on each particle is directed *away from* the other particle (as in Figs. 22-6a and b). If the particles *attract* each other, the force on each particle is directed *toward* the other particle (as in Fig. 22-6c).

Equation 22-1 is called **Coulomb's law** after Charles Augustin Coulomb, whose experiments in 1785 led him to it. Curiously, the form of Eq. 22-1 is the same as that of Newton's equation for the gravitational force between two particles with masses m_1 and m_2 that are separated by a distance r:

$$F = G\frac{m_1 m_2}{r^2}, \tag{22-2}$$

in which G is the gravitational constant.

The constant k in Eq. 22-1, by analogy with the gravitational constant G in Eq. 22-2, may be called the *electrostatic constant*. Both equations describe inverse square laws that involve a property of the interacting particles—the mass in one case and the charge in the other. The laws differ in that gravitational forces are always attractive but electrostatic forces may be either attractive or repulsive, depending on the signs of the two charges. This difference arises from the fact that, although there is only one kind of mass, there are two kinds of charge (and that is why absolute value signs are needed in Eq. 22-1 but not in Eq. 22-2).

Coulomb's law has survived every experimental test; no exceptions to it have ever been found. It holds even within the atom, correctly describing the force between the positively charged nucleus and each of the negatively charged electrons, even though classical Newtonian mechanics fails in that realm and is replaced there by quantum physics. This simple law also correctly accounts for the forces that bind atoms together to form molecules, and for the forces that bind atoms and molecules together to form solids and liquids.

For practical reasons having to do with the accuracy of measurements, the SI unit of charge is derived from the SI unit of electric current, the ampere (A). The SI unit of charge is the **coulomb** (C): *One coulomb is the amount of charge that is transferred through the cross section of a wire in 1 second when there is a current of 1 ampere in the wire.* In Section 30-2 we shall describe how the ampere is defined experimentally. In general, we can write

$$dq = i\,dt, \tag{22-3}$$

in which dq (in coulombs) is the charge transferred by a current i (in amperes) during the time interval dt (in seconds).

For historical reasons (and because doing so simplifies many other formulas), the electrostatic constant k of Eq. 22-1 is usually written $1/4\pi\varepsilon_0$. Then Coulomb's law becomes

$$F = \frac{1}{4\pi\varepsilon_0}\frac{|q_1||q_2|}{r^2} \quad \text{(Coulomb's law)}. \tag{22-4}$$

The constants in Eqs. 22-1 and 22-4 have the value

$$k = \frac{1}{4\pi\varepsilon_0} = 8.99 \times 10^9 \text{ N} \cdot \text{m}^2/\text{C}^2. \qquad (22\text{-}5)$$

The quantity ε_0, called the **permittivity constant,** sometimes appears separately in equations and is

$$\varepsilon_0 = 8.85 \times 10^{-12} \text{ C}^2/\text{N} \cdot \text{m}^2. \qquad (22\text{-}6)$$

Still another parallel between the gravitational force and the electrostatic force is that both obey the principle of superposition. If we have n charged particles, they interact independently in pairs, and the force on any one of them, let us say particle 1, is given by the vector sum

$$\vec{F}_{1,\text{net}} = \vec{F}_{12} + \vec{F}_{13} + \vec{F}_{14} + \vec{F}_{15} + \cdots + \vec{F}_{1n}, \qquad (22\text{-}7)$$

in which, for example, $\vec{F}_{14}$ is the force acting on particle 1 owing to the presence of particle 4. An identical formula holds for the gravitational force.

Finally, the two shell theorems that we found so useful in our study of gravitation have analogs in electrostatics:

spherical not necessary here

A shell of uniform charge attracts or repels a charged particle that is outside the shell as if all the shell's charge were concentrated at its center.

If a charged particle is located inside a shell of uniform charge, there is no net electrostatic force on the particle from the shell.

(In the first theorem, we assume that the charge on the shell is much greater than that of the particle. Then any redistribution of the charge on the shell due to the presence of the particle's charge can be neglected.)

Spherical Conductors

If excess charge is placed on a spherical shell that is made of conducting material, the excess charge spreads uniformly over the (external) surface. For example, if we place excess electrons on a spherical metal shell, those electrons repel one another and tend to move apart, spreading over the available surface until they are uniformly distributed. That arrangement maximizes the distances between all pairs of the excess electrons. According to the first shell theorem, the shell then will attract or repel an external charge as if all the excess charge on the shell were concentrated at its center.

If we remove negative charge from a spherical metal shell, the resulting positive charge of the shell is also spread uniformly over the surface of the shell. For example, if we remove n electrons, there are then n sites of positive charge (sites missing an electron) that are spread uniformly over the shell. According to the first shell theorem, the shell will again attract or repel an external charge as if all the shell's excess charge were concentrated at its center.

CHECKPOINT 2: The figure shows two protons (symbol p) and one electron (symbol e) on an axis. What are the directions of (a) the electrostatic force on the central proton due to the electron, (b) the electrostatic force on the central proton due to the other proton, and (c) the net electrostatic force on the central proton?

Sample Problem 22-1

(a) Figure 22-7*a* shows two positively charged particles fixed in place on an x axis. The charges are $q_1 = 1.60 \times 10^{-19}$ C and $q_2 = 3.20 \times 10^{-19}$ C, and the particle separation is $R = 0.0200$ m. What are the magnitude and direction of the electrostatic force $\vec{F}_{12}$ on particle 1 from particle 2?

SOLUTION: The **Key Idea** here is that, because both particles are positively charged, particle 1 is repelled by particle 2, with a force magnitude given by Eq. 22-4. Thus, the direction of force $\vec{F}_{12}$ on particle 1 is *away from* particle 2, in the negative direction of the x axis, as indicated in the free-body diagram of Fig. 22-7*b*. Using Eq. 22-4 with separation R substituted for r, we can write the magnitude F_{12} of this force as

$$\begin{aligned} F_{12} &= \frac{1}{4\pi\varepsilon_0}\frac{|q_1||q_2|}{R^2} \\ &= (8.99 \times 10^9\ \text{N}\cdot\text{m}^2/\text{C}^2) \\ &\quad \times \frac{(1.60 \times 10^{-19}\ \text{C})(3.20 \times 10^{-19}\ \text{C})}{(0.0200\ \text{m})^2} \\ &= 1.15 \times 10^{-24}\ \text{N}. \end{aligned}$$

Thus, force $\vec{F}_{12}$ has the following magnitude and direction (relative to the positive direction of the x axis):

$$1.15 \times 10^{-24}\ \text{N} \quad \text{and} \quad 180°. \qquad \text{(Answer)}$$

We can also write $\vec{F}_{12}$ in unit-vector notation as

$$\vec{F}_{12} = -(1.15 \times 10^{-24}\ \text{N})\hat{\text{i}}. \qquad \text{(Answer)}$$

(b) Figure 22-7*c* is identical to Fig. 22-7*a* except that particle 3 now lies on the x axis between particles 1 and 2. Particle 3 has charge $q_3 = -3.20 \times 10^{-19}$ C and is at a distance $\frac{3}{4}R$ from particle 1. What is the net electrostatic force $\vec{F}_{1,\text{net}}$ on particle 1 due to particles 2 and 3?

SOLUTION: One **Key Idea** here is that the presence of particle 3 does not alter the electrostatic force on particle 1 from particle 2. Thus, force $\vec{F}_{12}$ still acts on particle 1. Similarly, the force $\vec{F}_{13}$ that acts on particle 1 due to particle 3 is not affected by the presence of particle 2. Because particles 1 and 3 have charge of opposite sign, particle 1 is attracted to particle 3. Thus, force $\vec{F}_{13}$ is directed *toward* particle 3, as indicated in the free-body diagram of Fig. 22-7*d*.

To find the magnitude of $\vec{F}_{13}$, we can rewrite Eq. 22-4 as

$$\begin{aligned} F_{13} &= \frac{1}{4\pi\varepsilon_0}\frac{|q_1||q_3|}{(\frac{3}{4}R)^2} \\ &= (8.99 \times 10^9\ \text{N}\cdot\text{m}^2/\text{C}^2) \\ &\quad \times \frac{(1.60 \times 10^{-19}\ \text{C})(3.20 \times 10^{-19}\ \text{C})}{(\frac{3}{4})^2(0.0200\ \text{m})^2} \\ &= 2.05 \times 10^{-24}\ \text{N}. \end{aligned}$$

We can also write $\vec{F}_{13}$ in unit-vector notation:

$$\vec{F}_{13} = (2.05 \times 10^{-24}\ \text{N})\hat{\text{i}}.$$

A second **Key Idea** here is that the net force $\vec{F}_{1,\text{net}}$ on particle 1 is the vector sum of $\vec{F}_{12}$ and $\vec{F}_{13}$; that is, from Eq. 22-7, we can

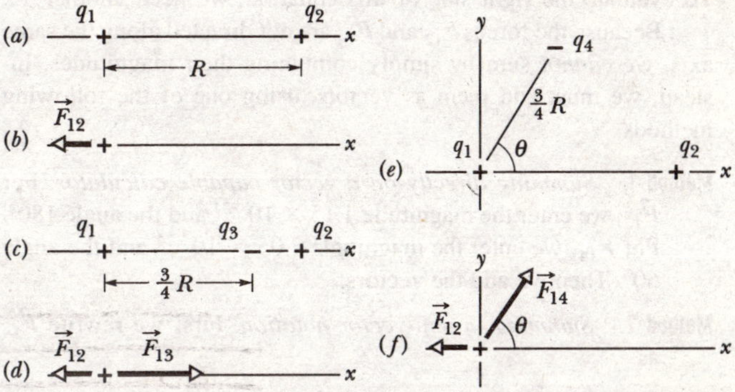

Fig. 22-7 Sample Problem 22-1. (*a*) Two charged particles of charges q_1 and q_2 are fixed in place on an x axis, with separation R. (*b*) The free-body diagram for particle 1, showing the electrostatic force on it from particle 2. (*c*) Particle 3 is now fixed in place on the x axis, along with particles 1 and 2. (*d*) The free-body diagram for particle 1. (*e*) Particle 4 is fixed in place on a line at angle θ to the x axis, again with particles 1 and 2. (*f*) The free-body diagram for particle 1.

write the net force $\vec{F}_{1,\text{net}}$ on particle 1 in unit-vector notation as

$$\begin{aligned} \vec{F}_{1,\text{net}} &= \vec{F}_{12} + \vec{F}_{13} \\ &= -(1.15 \times 10^{-24}\ \text{N})\hat{\text{i}} + (2.05 \times 10^{-24}\ \text{N})\hat{\text{i}} \\ &= (9.00 \times 10^{-25}\ \text{N})\hat{\text{i}}. \qquad \text{(Answer)} \end{aligned}$$

Thus, $\vec{F}_{1,\text{net}}$ has the following magnitude and direction (relative to the positive direction of the x axis):

$$9.00 \times 10^{-25}\ \text{N} \quad \text{and} \quad 0°. \qquad \text{(Answer)}$$

(c) Figure 22-7*e* is identical to Fig. 22-7*a* except that particle 4 is now positioned as shown. Particle 4 has charge $q_4 = -3.20 \times 10^{-19}$ C, is at a distance $\frac{3}{4}R$ from particle 1, and lies on a line that makes an angle $\theta = 60°$ with the x axis. What is the net electrostatic force $\vec{F}_{1,\text{net}}$ on particle 1 due to particles 2 and 4?

SOLUTION: The **Key Idea** here is that the net force $\vec{F}_{1,\text{net}}$ is the vector sum of $\vec{F}_{12}$ and a new force $\vec{F}_{14}$ acting on particle 1 due to particle 4. Because particles 1 and 4 have charges of opposite sign, particle 1 is attracted to particle 4. Thus, force $\vec{F}_{14}$ on particle 1 is directed *toward* particle 4, at angle $\theta = 60°$, as indicated in the free-body diagram of Fig. 22-7*f*.

To find the magnitude of $\vec{F}_{14}$, we can rewrite Eq. 22-4 as

$$\begin{aligned} F_{14} &= \frac{1}{4\pi\varepsilon_0}\frac{|q_1||q_4|}{(\frac{3}{4}R)^2} \\ &= (8.99 \times 10^9\ \text{N}\cdot\text{m}^2/\text{C}^2) \\ &\quad \times \frac{(1.60 \times 10^{-19}\ \text{C})(3.20 \times 10^{-19}\ \text{C})}{(\frac{3}{4})^2(0.0200\ \text{m})^2} \\ &= 2.05 \times 10^{-24}\ \text{N}. \end{aligned}$$

Then from Eq. 22-7, we can write the net force $\vec{F}_{1,\text{net}}$ on particle 1 as

$$\vec{F}_{1,\text{net}} = \vec{F}_{12} + \vec{F}_{14}.$$

To evaluate the right side of this equation, we need another : Because the forces $\vec{F}_{12}$ and $\vec{F}_{14}$ are not directed along the same axis, we *cannot* sum by simply combining their magnitudes. Instead, we must add them as vectors, using one of the following methods.

Method 1. *Summing directly on a vector-capable calculator.* For $\vec{F}_{12}$, we enter the magnitude 1.15×10^{-24} and the angle 180°. For $\vec{F}_{14}$, we enter the magnitude 2.05×10^{-24} and the angle 60°. Then we add the vectors.

Method 2. *Summing in unit-vector notation.* First we rewrite $\vec{F}_{14}$ as

$$\vec{F}_{14} = (F_{14} \cos \theta)\hat{\mathrm{i}} + (F_{14} \sin \theta)\hat{\mathrm{j}}.$$

Substituting 2.05×10^{-24} N for F_{14} and 60° for θ, this becomes

$$\vec{F}_{14} = (1.025 \times 10^{-24}\ \mathrm{N})\hat{\mathrm{i}} + (1.775 \times 10^{-24}\ \mathrm{N})\hat{\mathrm{j}}.$$

Then we sum:

$$\begin{aligned}\vec{F}_{1,\mathrm{net}} &= \vec{F}_{12} + \vec{F}_{14}\\ &= -(1.15 \times 10^{-24}\ \mathrm{N})\hat{\mathrm{i}}\\ &\quad + (1.025 \times 10^{-24}\ \mathrm{N})\hat{\mathrm{i}} + (1.775 \times 10^{-24}\ \mathrm{N})\hat{\mathrm{j}}\\ &\approx (-1.25 \times 10^{-25}\ \mathrm{N})\hat{\mathrm{i}} + (1.78 \times 10^{-24}\ \mathrm{N})\hat{\mathrm{j}}.\end{aligned}$$

(Answer)

Method 3. *Summing components axis by axis.* The sum of the x components gives us

$$\begin{aligned}F_{1,\mathrm{net},x} &= F_{12,x} + F_{14,x} = F_{12} + F_{14} \cos 60°\\ &= -1.15 \times 10^{-24}\ \mathrm{N} + (2.05 \times 10^{-24}\ \mathrm{N})(\cos 60°)\\ &= -1.25 \times 10^{-25}\ \mathrm{N}.\end{aligned}$$

The sum of the y components gives us

$$\begin{aligned}F_{1,\mathrm{net},y} &= F_{12,y} + F_{14,y} = 0 + F_{14} \sin 60°\\ &= (2.05 \times 10^{-24}\ \mathrm{N})(\sin 60°)\\ &= 1.78 \times 10^{-24}\ \mathrm{N}.\end{aligned}$$

The net force $\vec{F}_{1,\mathrm{net}}$ has the magnitude

$$F_{1,\mathrm{net}} = \sqrt{F^2_{1,\mathrm{net},x} + F^2_{1,\mathrm{net},y}} = 1.78 \times 10^{-24}\ \mathrm{N}. \quad \text{(Answer)}$$

To find the direction of $\vec{F}_{1,\mathrm{net}}$, we take

$$\theta = \tan^{-1} \frac{F_{1,\mathrm{net},y}}{F_{1,\mathrm{net},x}} = -86.0°.$$

However, this is an unreasonable result because $\vec{F}_{1,\mathrm{net}}$ must have a direction between the directions of $\vec{F}_{12}$ and $\vec{F}_{14}$. To correct θ, we add 180°, obtaining

$$-86.0° + 180° = 94.0°. \quad \text{(Answer)}$$

CHECKPOINT 3: The figure here shows three arrangements of an electron e and two protons p. (a) Rank the arrangements according to the magnitude of the net electrostatic force on the electron due to the protons, largest first. (b) In situation *c*, is the angle between the net force on the electron and the line labeled *d* less than or more than 45°?

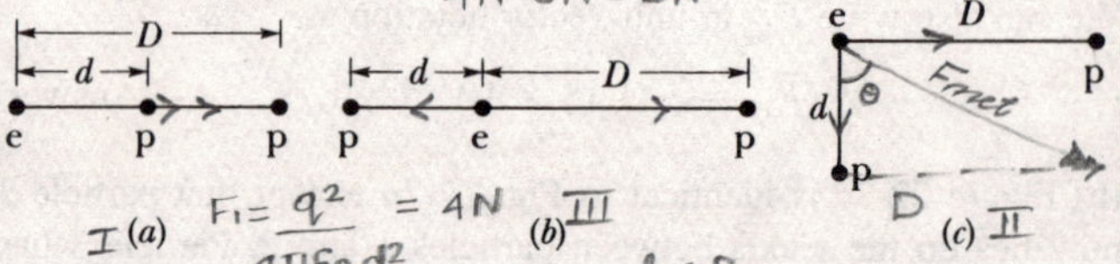

PROBLEM-SOLVING TACTICS

Tactic 1: *Symbols Representing Charge*
Here is a general guide to the symbols representing charge. If the symbol q, with or without a subscript, is used in a sentence when no electrical sign has been specified, the charge can be either positive or negative. Sometimes the sign is explicitly shown, as in the notation $+q$ or $-q$.

When more than one charged object is being considered, their charges might be given as multiples of a charge magnitude. As examples, the notation $+2q$ means a positive charge with magnitude twice that of some reference charge magnitude q, and $-3q$ means a negative charge with magnitude three times that of the reference charge magnitude q.

Sample Problem 22-2

Figure 22-8*a* shows two particles fixed in place: a particle of charge $q_1 = +8q$ at the origin and a particle of charge $q_2 = -2q$ at $x = L$. At what point (other than infinitely far away) can a proton be placed so that it is in *equilibrium* (meaning that the net force on it is zero)? Is that equilibrium *stable* or *unstable*?

SOLUTION: The **Key Idea** here is that, if $\vec{F}_1$ is the force on the proton due to charge q_1 and $\vec{F}_2$ is the force on the proton due to charge q_2, then the point we seek is where $\vec{F}_1 + \vec{F}_2 = 0$. This condition requires that

$$\vec{F}_1 = -\vec{F}_2. \tag{22-8}$$

This tells us that at the point we seek, the forces acting on the

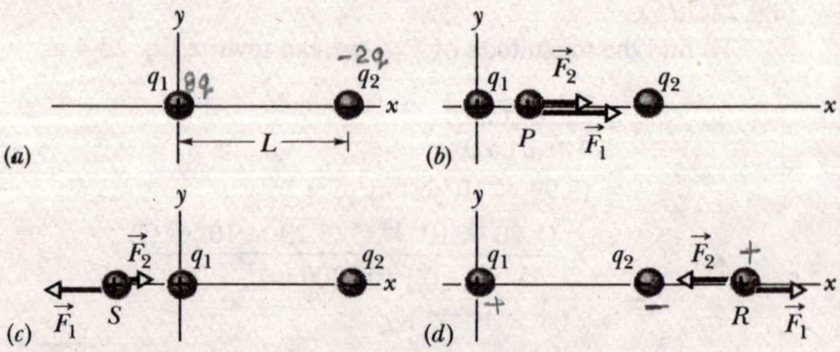

Fig. 22-8 Sample Problem 22-2. (*a*) Two particles of charges q_1 and q_2 are fixed in place on an x axis, with separation L. (*b*)–(*d*) Three possible locations P, S, and R for a proton. At each location, $\vec{F}_1$ is the force on the proton from particle 1 and $\vec{F}_2$ is the force on the proton from particle 2.

proton due to the other two particles must be of equal magnitudes,

$$F_1 = F_2, \qquad (22\text{-}9)$$

and that the forces must have opposite directions.

A proton has a positive charge. Thus, the proton and the particle of charge q_1 are of the same sign, and force $\vec{F}_1$ on the proton must point away from q_1. Also, the proton and the particle of charge q_2 are of opposite signs, so force $\vec{F}_2$ on the proton must point toward q_2. "Away from q_1" and "toward q_2" can be in opposite directions only if the proton is located on the x axis.

If the proton is on the x axis at any point between q_1 and q_2, such as P in Fig. 22-8b, then $\vec{F}_1$ and $\vec{F}_2$ are in the same direction and not in opposite directions as required. If the proton is at any point on the x axis to the left of q_1, such as point S in Fig. 22-8c, then $\vec{F}_1$ and $\vec{F}_2$ are in opposite directions. However, Eq. 22-4 tells us that $\vec{F}_1$ and $\vec{F}_2$ cannot have equal magnitudes there: F_1 must be greater than F_2, because F_1 is produced by a closer charge (with lesser r) of greater magnitude ($8q$ versus $2q$).

Finally, if the proton is at any point on the x axis to the right of q_2, such as point R in Fig. 22-8d, then $\vec{F}_1$ and $\vec{F}_2$ are again in opposite directions. However, because now the charge of greater magnitude (q_1) is *farther* away from the proton than the charge of lesser magnitude, there is a point at which F_1 is equal to F_2. Let x be the coordinate of this point, and let q_p be the charge of the proton. Then with the aid of Eq. 22-4, we can rewrite Eq. 22-9 as

$$\frac{1}{4\pi\varepsilon_0}\frac{8qq_p}{x^2} = \frac{1}{4\pi\varepsilon_0}\frac{2qq_p}{(x-L)^2}. \qquad (22\text{-}10)$$

(Note that only the charge magnitudes appear in Eq. 22-10.) Rearranging Eq. 22-10 gives us

$$\left(\frac{x-L}{x}\right)^2 = \frac{1}{4}.$$

After taking the square roots of both sides, we have

$$\frac{x-L}{x} = \frac{1}{2},$$

which gives us

$$x = 2L. \qquad \text{(Answer)}$$

The equilibrium at $x = 2L$ is unstable; that is, if the proton is displaced leftward from point R, then F_1 and F_2 both increase but F_2 increases more (because q_2 is closer than q_1), and a net force will drive the proton farther leftward. If the proton is displaced rightward, both F_1 and F_2 decrease but F_2 decreases more, and a net force will then drive the proton farther rightward. In a stable equilibrium, each time the proton is displaced slightly, it would return to the equilibrium position.

PROBLEM-SOLVING TACTICS

Tactic 2: *Drawing Electrostatic Force Vectors*

When you are given a diagram of charged particles, such as Fig. 22-7a, and are asked to find the net electrostatic force on one of them, you should usually draw a free-body diagram showing only the particle of concern and the forces *it* experiences, as in Fig. 22-7b. If, instead, you choose to superimpose those forces on the given diagram showing all the particles, be sure to draw the force vectors with either their tails (preferably) or their heads on the particle of concern. If you draw the vectors elsewhere in the diagram, you invite confusion—and confusion is guaranteed if you draw the vectors on the particles *causing* the forces on the particle of concern.

Sample Problem 22-3

In Fig. 22-9a, two identical, electrically isolated conducting spheres A and B are separated by a (center-to-center) distance a that is large compared to the spheres. Sphere A has a positive charge of $+Q$, and sphere B is electrically neutral. Initially, there is no electrostatic force between the spheres. (Assume that there is no induced charge on the spheres because of their large separation.)

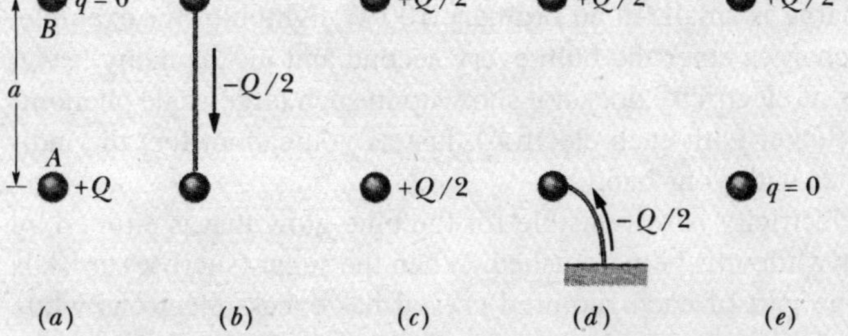

Fig. 22-9 Sample Problem 22-3. Two small conducting spheres A and B. (a) To start, sphere A is charged positively. (b) Negative charge is transferred between the spheres through a connecting wire. (c) Both spheres are then charged positively. (d) Negative charge is transferred through a grounding wire to sphere A. (e) Sphere A is then neutral.

(a) Suppose the spheres are connected for a moment by a conducting wire. The wire is thin enough so that any net charge on it is negligible. What is the electrostatic force between the spheres after the wire is removed?

SOLUTION: A **Key Idea** here is that when the spheres are wired together, the (negative) conduction electrons on sphere B, which always repel one another, have a way to move farther away from one another (along the wire to positively charged sphere A, which attracts them). (See Fig. 22-9b.) As sphere B loses negative charge, it becomes positively charged, and as A gains negative charge, it becomes *less* positively charged. A second **Key Idea** is that the spheres must end up with the same charge because they are identical. Thus, the transfer of charge stops when the excess charge on B has increased to $+Q/2$ and the excess charge on A has decreased to $+Q/2$. This condition occurs when a charge of $-Q/2$ has been transferred.

After the wire has been removed (Fig. 22-9c), we can assume that the charge on either sphere does not disturb the uniformity of the charge distribution on the other sphere, because the spheres are small relative to their separation. Thus, we can apply the first shell theorem to each sphere. By Eq. 22-4 with $q_1 = q_2 = Q/2$ and $r = a$, the electrostatic force between the spheres has a magnitude of

$$F = \frac{1}{4\pi\varepsilon_0}\frac{(Q/2)(Q/2)}{a^2} = \frac{1}{16\pi\varepsilon_0}\left(\frac{Q}{a}\right)^2. \quad \text{(Answer)}$$

The spheres, now positively charged, repel each other.

(b) Next, suppose sphere A is grounded momentarily, and then the ground connection is removed. What now is the electrostatic force between the spheres?

SOLUTION: The **Key Idea** here is that the ground connection allows electrons, with a total charge of $-Q/2$, to move from the ground to sphere A (Fig. 22-9d), neutralizing that sphere (Fig. 22-9e). With no charge on sphere A, there is no electrostatic force between the two spheres (just as initially, in Fig. 22-9a).

22-5 Charge Is Quantized

In Benjamin Franklin's day, electric charge was thought to be a continuous fluid—an idea that was useful for many purposes. However, we now know that fluids themselves, such as air and water, are not continuous but are made up of atoms and molecules; matter is discrete. Experiment shows that "electrical fluid" is also not continuous but is made up of multiples of a certain elementary charge. Any positive or negative charge q that can be detected can be written as

$$q = ne, \qquad n = \pm 1, \pm 2, \pm 3, \ldots, \tag{22-11}$$

in which e, the **elementary charge**, has the value

$$e = 1.60 \times 10^{-19}\ \text{C}. \tag{22-12}$$

The elementary charge e is one of the important constants of nature. The electron and proton both have a charge of magnitude e (Table 22-1). (Quarks, the constituent particles of protons and neutrons, have charges of $\pm e/3$ or $\pm 2e/3$, but they apparently cannot be detected individually. For this and for historical reasons, we do not take their charges to be the elementary charge.)

You often see phrases—such as "the charge on a sphere," "the amount of charge transferred," and "the charge carried by the electron"—that suggest that charge is a substance. (Indeed, such statements have already appeared in this chapter.) You should, however, keep in mind what is intended: *Particles* are the substance and charge happens to be one of their properties, just as mass is.

When a physical quantity such as charge can have only discrete values rather than any value, we say that the quantity is **quantized.** It is possible, for example, to find a particle that has no charge at all or a charge of $+10e$ or $-6e$, but not a particle with a charge of, say, $3.57e$.

The quantum of charge is small. In an ordinary 100 W lightbulb, for example, about 10^{19} elementary charges enter the bulb every second and just as many leave. However, the graininess of electricity does not show up in such large-scale phenomena (the bulb does not flicker with each electron), just as you cannot feel the individual molecules of water with your hand.

The graininess of electricity is responsible for the blue glow that is emitted by a wintergreen LifeSaver while it is being crushed. When the sugar (sucrose) crystals in the candy rupture, one part of each ruptured crystal has excess electrons while the other part has excess positive ions. Almost immediately, electrons and ions jump across the gap of the rupture to neutralize the two sides. During the jumps, the electrons and positive ions collide with nitrogen molecules in the air that is then flowing into the gap.

The collisions cause the nitrogen to emit ultraviolet light that you cannot see, as well as blue light (from the visible region of the spectrum) that is, however, too

TABLE 22-1 The Charges of Three Particles

Particle	Symbol	Charge
Electron	e or e^-	$-e$
Proton	p	$+e$
Neutron	n	0

dim to see: Oil of wintergreen in the crystals absorbs the ultraviolet light and immediately emits enough blue light to light up a mouth or a pair of pliers. However, if the candy is wet with saliva, the demonstration fails, because the conducting saliva neutralizes the two parts of a fractured crystal before sparking can occur.

CHECKPOINT 4: Initially, sphere A has a charge of $-50e$ and sphere B has a charge of $+20e$. The spheres are made of conducting material and are identical in size. If the spheres then touch, what is the resulting charge on sphere A? −15e

net charge of −30e is equally shared

Sample Problem 22-4

The nucleus in an iron atom has a radius of about 4.0×10^{-15} m and contains 26 protons.

(a) What is the magnitude of the repulsive electrostatic force between two of the protons that are separated by 4.0×10^{-15} m?

SOLUTION: The **Key Idea** here is that the protons can be treated as charged particles, so the magnitude of the electrostatic force on one from the other is given by Coulomb's law. Table 22-1 tells us that their charge is $+e$. Thus, Eq. 22-4 gives us

$$F = \frac{1}{4\pi\varepsilon_0}\frac{e^2}{r^2}$$

$$= \frac{(8.99 \times 10^9\ \mathrm{N \cdot m^2/C^2})(1.60 \times 10^{-19}\ \mathrm{C})^2}{(4.0 \times 10^{-15}\ \mathrm{m})^2}$$

$$= 14\ \mathrm{N}. \qquad \text{(Answer)}$$

This is a small force to be acting on a macroscopic object like a cantaloupe, but an enormous force to be acting on a proton. Such forces should blow apart the nucleus of any element but hydrogen (which has only one proton in its nucleus). However, they don't, not even in nuclei with a great many protons. Therefore, there must be some enormous attractive force to counter this enormous repulsive electrostatic force.

(b) What is the magnitude of the gravitational force between those same two protons?

SOLUTION: The **Key Idea** here is like that in part (a): Because the protons are particles, the magnitude of the gravitational force on one from the other is given by Newton's equation for the gravitational force (Eq. 22-2). With m_p ($= 1.67 \times 10^{-27}$ kg) representing the mass of a proton, Eq. 22-2 gives us

$$F = G\frac{m_p^2}{r^2}$$

$$= \frac{(6.67 \times 10^{-11}\ \mathrm{N \cdot m^2/kg^2})(1.67 \times 10^{-27}\ \mathrm{kg})^2}{(4.0 \times 10^{-15}\ \mathrm{m})^2}$$

$$= 1.2 \times 10^{-35}\ \mathrm{N}. \qquad \text{(Answer)}$$

This result tells us that the (attractive) gravitational force is far too weak to counter the repulsive electrostatic forces between protons in a nucleus. Instead, the protons are bound together by an enormous force called (aptly) the *strong nuclear force*—a force that acts between protons (and neutrons) when they are close together, as in a nucleus.

Although the gravitational force is many times weaker than the electrostatic force, it is more important in large-scale situations because it is always attractive. This means that it can collect many small bodies into huge bodies with huge masses, such as planets and stars, that then exert large gravitational forces. The electrostatic force, on the other hand, is repulsive for charges of the same sign, so it is unable to collect either positive charge or negative charge into large concentrations that would then exert large electrostatic forces.

22-6 Charge Is Conserved

If you rub a glass rod with silk, a positive charge appears on the rod. Measurement shows that a negative charge of equal magnitude appears on the silk. This suggests that rubbing does not create charge but only transfers it from one body to another, upsetting the electrical neutrality of each body during the process. This hypothesis of **conservation of charge,** first put forward by Benjamin Franklin, has stood up under close examination, both for large-scale charged bodies and for atoms, nuclei, and elementary particles. No exceptions have ever been found. Thus, we add electric charge to our list of quantities—including energy and both linear and angular momentum—that obey a conservation law.

Radioactive decay of nuclei, in which a nucleus spontaneously transforms into a different type of nucleus, gives us many instances of charge conservation at the nuclear level. For example, uranium-238, or ^{238}U, which is found in common uranium ore, can decay by emitting an alpha particle (which is a helium nucleus, ^{4}He)

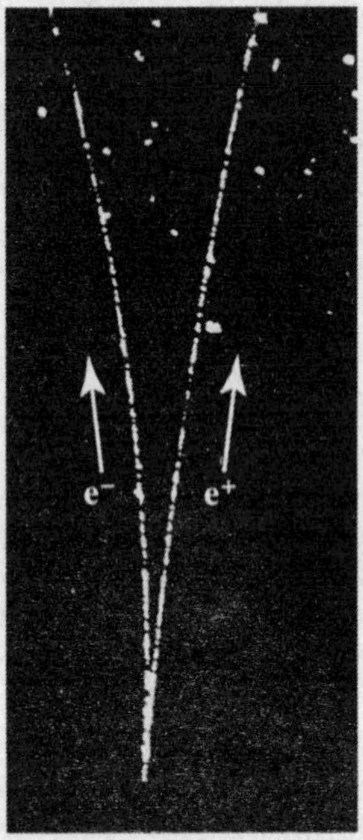

Fig. 22-10 A photograph of trails of bubbles left in a bubble chamber by an electron and a positron. The pair of particles was produced by a gamma ray that entered the chamber from the bottom. Being electrically neutral, the gamma ray did not generate a telltale trail of bubbles along its path, as the electron and positron did.

and transforming to thorium, ^{234}Th:

$$^{238}\text{U} \rightarrow {}^{234}\text{Th} + {}^{4}\text{He} \qquad \text{(radioactive decay).} \qquad (22\text{-}13)$$

The atomic number Z of the radioactive *parent* nucleus ^{238}U is 92, which tells us that this nucleus contains 92 protons and has a charge of $92e$. The emitted alpha particle has $Z = 2$, and the *daughter* nucleus ^{234}Th has $Z = 90$. Thus, the amount of charge present before the decay, $92e$, is equal to the total amount present after the decay, $90e + 2e$. Charge is conserved.

Another example of charge conservation occurs when an electron e^- (whose charge is $-e$) and its antiparticle, the *positron* e^+ (whose charge is $+e$), undergo an *annihilation process* in which they transform into two *gamma rays* (high-energy light):

$$e^- + e^+ \rightarrow \gamma + \gamma \qquad \text{(annihilation).} \qquad (22\text{-}14)$$

In applying the conservation-of-charge principle, we must add the charges algebraically, with due regard for their signs. In the annihilation process of Eq. 22-14 then, the net charge of the system is zero both before and after the event. Charge is conserved.

In *pair production*, the converse of annihilation, charge is also conserved. In this process a gamma ray transforms into an electron and a positron:

$$\gamma \rightarrow e^- + e^+ \qquad \text{(pair production).} \qquad (22\text{-}15)$$

Figure 22-10 shows such a pair-production event that occurred in a bubble chamber. A gamma ray entered the chamber from the bottom and at one point transformed into an electron and a positron. Because those new particles were charged and moving, each left a trail of tiny bubbles. (The trails were curved because a magnetic field had been set up in the chamber.) The gamma ray, being electrically neutral, left no trail. Still, you can tell exactly where it underwent pair production—at the tip of the curved **V**, where the trails of the electron and positron begin.

REVIEW & SUMMARY

Electric Charge The strength of a particle's electric interaction with objects around it depends on its **electric charge,** which can be either positive or negative. Charges with the same sign repel each other and charges with opposite signs attract each other. An object with equal amounts of the two kinds of charge is electrically neutral, whereas one with an imbalance is electrically charged.

Conductors are materials in which a significant number of charged particles (electrons in metals) are free to move. The charged particles in **nonconductors,** or **insulators,** are not free to move. When charge moves through a material, we say that an **electric current** exists in the material.

The Coulomb and Ampere The SI unit of charge is the **coulomb** (C). It is defined in terms of the unit of current, the ampere (A), as the charge passing a particular point in 1 second when there is a current of 1 ampere at that point.

Coulomb's Law *Coulomb's law* describes the **electrostatic force** between small (point) electric charges q_1 and q_2 at rest (or nearly at rest) and separated by a distance r:

$$F = \frac{1}{4\pi\varepsilon_0}\frac{|q_1||q_2|}{r^2} \qquad \text{(Coulomb's law).} \qquad (22\text{-}4)$$

Here $\varepsilon_0 = 8.85 \times 10^{-12}\ \text{C}^2/\text{N}\cdot\text{m}^2$ is the **permittivity constant,** and $1/4\pi\varepsilon_0 = k = 8.99 \times 10^9\ \text{N}\cdot\text{m}^2/\text{C}^2$.

The force of attraction or repulsion between point charges at rest acts along the line joining the two charges. If more than two charges are present, Eq. 22-4 holds for each pair of charges. The net force on each charge is then found, using the superposition principle, as the vector sum of the forces exerted on the charge by all the others.

The two shell theorems for electrostatics are

A shell of uniform charge attracts or repels a charged particle that is outside the shell as if all the shell's charge were concentrated at its center.

If a charged particle is located inside a shell of uniform charge, there is no net electrostatic force on the particle from the shell.

The Elementary Charge Electric charge is **quantized:** any charge can be written as ne, where n is a positive or negative integer, and e is a constant of nature called the **elementary charge** (approximately 1.60×10^{-19} C). Electric charge is **conserved:** the (algebraic) net charge of any isolated system cannot change.

QUESTIONS

1 Does Coulomb's law hold for all charged objects?

2 A particle of charge q is to be placed, in turn, outside four metal objects, each of uniform charge Q: (1) a large solid sphere, (2) a large spherical shell, (3) a small solid sphere, and (4) a small spherical shell. The distance between the particle and the center of the object is the same, and q is small enough not to alter significantly the uniform distribution of Q. Rank the objects according to the electrostatic force they exert on the particle, greatest first.

3 Figure 22-11 shows four situations in which charged particles are fixed in place on an axis. In which situations is there a point to the left of the particles where an electron will be in equilibrium?

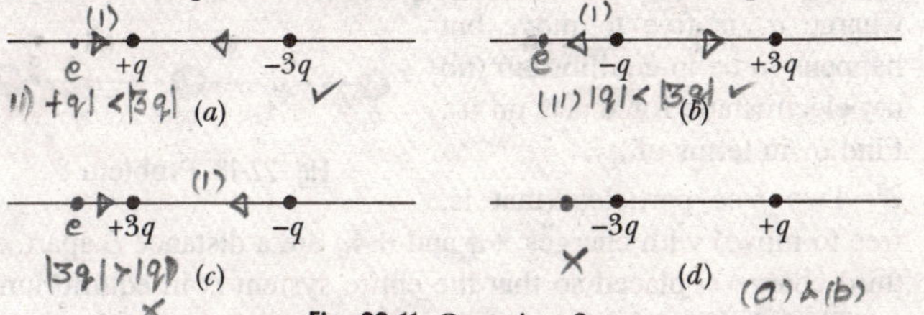

Fig. 22-11 Question 3.

4 Figure 22-12 shows two charged particles on an axis. The charges are free to move. At one point, however, a third charged particle can be placed such that all three particles are in equilibrium. (a) Is that point to the left of the first two particles, to their right, or between them? (b) Should the third particle be positively or negatively charged? (c) Is the equilibrium stable or unstable?

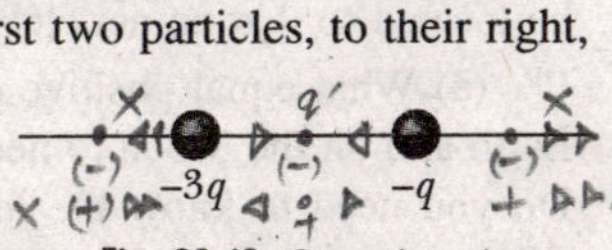

Fig. 22-12 Question 4.

5 In Fig. 22-13, a central particle of charge $-q$ is surrounded by two circular rings of charged particles, of radii r and R, with $R > r$. What are the magnitude and direction of the net electrostatic force on the central particle due to the other particles?

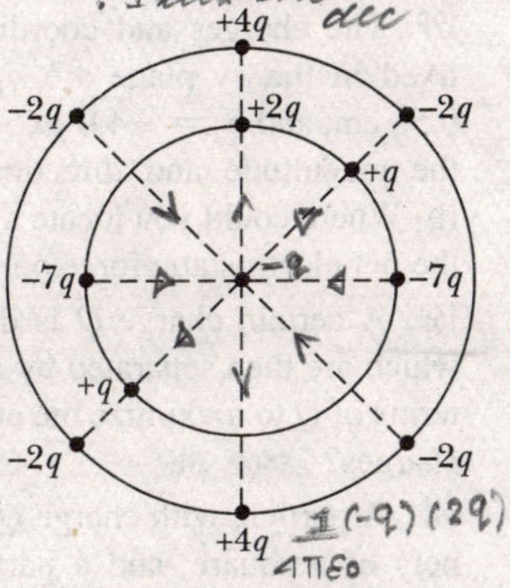

Fig. 22-13 Question 5.

6 Figure 22-14 shows four arrangements of charged particles. Rank the arrangements according to the magnitude of the net electrostatic force on the particle with charge $+Q$, greatest first.

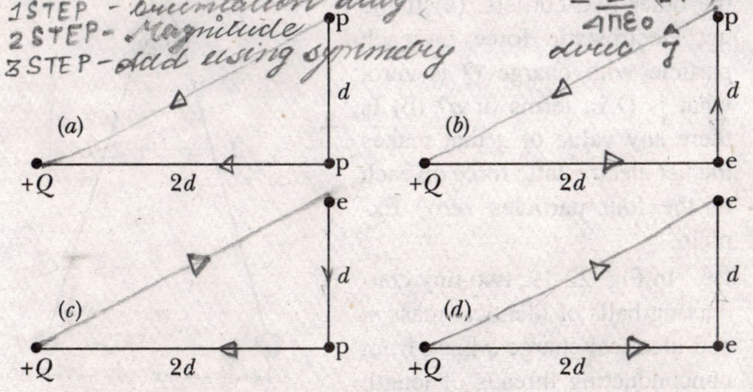

Fig. 22-14 Question 6.

7 Figure 22-15 shows four situations in which particles of charge $+q$ or $-q$ are fixed in place. In each, the particles on the x axis are equidistant from the y axis. First, consider the middle particle in situation 1; the middle particle experiences an electrostatic force from each of the other two particles. (a) Are the magnitudes F of those forces the same or different? (b) Is the magnitude of the net force on the middle particle equal to, greater than, or less than $2F$? (c) Do the x components of the two forces add or cancel? (d) Do their y components add or cancel? (e) Is the direction of the net force on the middle particle that of the canceling components or the adding components? (f) What is the direction of that net force? Now consider the remaining situations: What is the direction of the net force on the middle particle in (g) situation 2, (h) situation 3, and (i) situation 4? (In each, consider the symmetry of the charge distribution and determine the canceling components and the adding components.)

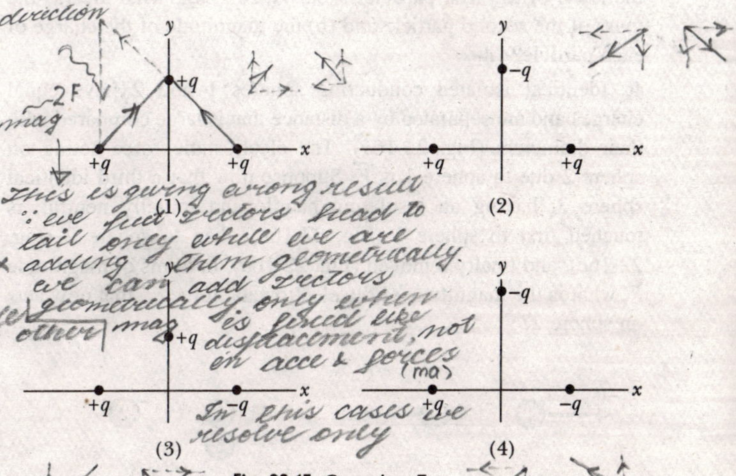

Fig. 22-15 Question 7.

8 A positively charged ball is brought close to a neutral isolated conductor. The conductor is then grounded while the ball is kept close. Is the conductor charged positively or negatively, or is it neutral, if (a) the ball is first taken away and then the ground connection is removed and (b) the ground connection is first removed and then the ball is taken away?

9 (a) A positively charged glass rod attracts an object suspended by a nonconducting thread. Is the object definitely negatively charged or only possibly negatively charged? (b) A positively charged glass rod repels a similarly suspended object. Is the object definitely positively charged or only possibly?

10 In Fig. 22-4, the nearby (negatively charged) plastic rod causes some of the conduction electrons in the copper rod to move to the far end of the copper rod. Why does the flow of the conduction electrons quickly cease? After all, a huge number of them are free to move to that far end.

11 A person standing on an electrically insulated platform touches a charged, electrically isolated conductor. Does this discharge the conductor completely?

EXERCISES & PROBLEMS

ssm Solution is in the Student Solutions Manual.
www Solution is available on the World Wide Web at: http://www.wiley.com/college/hrw
ilw Solution is available on the Interactive LearningWare.

SEC. 22-4 Coulomb's Law

1E. What must be the distance between point charge $q_1 = 26.0\ \mu C$ and point charge $q_2 = -47.0\ \mu C$ for the electrostatic force between them to have a magnitude of 5.70 N? ssm

2E. A point charge of $+3.00 \times 10^{-6}$ C is 12.0 cm distant from a second point charge of -1.50×10^{-6} C. Calculate the magnitude of the force on each charge.

3E. Two equally charged particles, held 3.2×10^{-3} m apart, are released from rest. The initial acceleration of the first particle is observed to be 7.0 m/s^2 and that of the second to be 9.0 m/s^2. If the mass of the first particle is 6.3×10^{-7} kg, what are (a) the mass of the second particle and (b) the magnitude of the charge of each particle? ilw

4E. Identical isolated conducting spheres 1 and 2 have equal charges and are separated by a distance that is large compared with their diameters (Fig. 22-16*a*). The electrostatic force acting on sphere 2 due to sphere 1 is $\vec{F}$. Suppose now that a third identical sphere 3, having an insulating handle and initially neutral, is touched first to sphere 1 (Fig. 22-16*b*), then to sphere 2 (Fig. 22-16*c*), and finally removed (Fig. 22-16*d*). In terms of magnitude F, what is the magnitude of the electrostatic force $\vec{F}'$ that now acts on sphere 2?

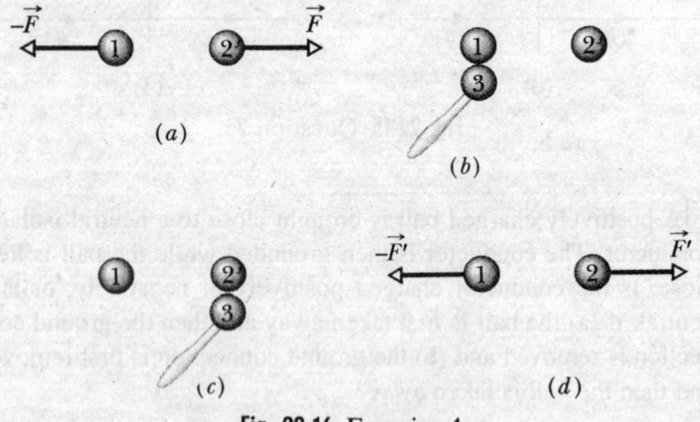

Fig. 22-16 Exercise 4.

5P. In Fig. 22-17, what are the (a) horizontal and (b) vertical components of the net electrostatic force on the charged particle in the lower left corner of the square if $q = 1.0 \times 10^{-7}$ C and $a = 5.0$ cm? ilw

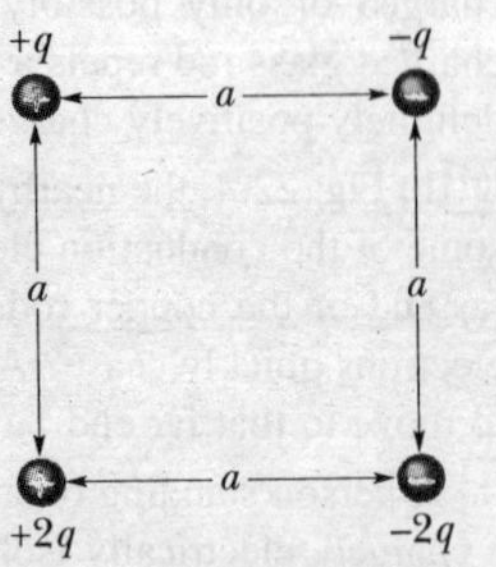

Fig. 22-17 Problem 5.

6P. Point charges q_1 and q_2 lie on the x axis at points $x = -a$ and $x = +a$, respectively. (a) How must q_1 and q_2 be related for the net electrostatic force on point charge $+Q$, placed at $x = +a/2$, to be zero? (b) Repeat (a) but with point charge $+Q$ now placed at $x = +3a/2$.

7P. Two identical conducting spheres, fixed in place, attract each other with an electrostatic force of 0.108 N when separated by 50.0 cm, center-to-center. The spheres are then connected by a thin conducting wire. When the wire is removed, the spheres repel each other with an electrostatic force of 0.0360 N. What were the initial charges on the spheres? ssm

8P. In Fig. 22-18, three charged particles lie on a straight line and are separated by distances d. Charges q_1 and q_2 are held fixed. Charge q_3 is free to move but happens to be in equilibrium (no net electrostatic force acts on it). Find q_1 in terms of q_2.

Fig. 22-18 Problem 8.

9P. Two *free* particles (that is, free to move) with charges $+q$ and $+4q$ are a distance L apart. A third charge is placed so that the entire system is in equilibrium. (a) Find the location, magnitude, and sign of the third charge. (b) Show that the equilibrium is unstable. ssm www

10P. Two fixed particles, of charges $q_1 = +1.0\ \mu C$ and $q_2 = -3.0\ \mu C$, are 10 cm apart. How far from each should a third charge be located so that no net electrostatic force acts on it?

11P. (a) What equal positive charges would have to be placed on Earth and on the Moon to neutralize their gravitational attraction? Do you need to know the lunar distance to solve this problem? Why or why not? (b) How many kilograms of hydrogen would be needed to provide the positive charge calculated in (a)? ssm

12P. The charges and coordinates of two charged particles held fixed in the xy plane are $q_1 = +3.0\ \mu C$, $x_1 = 3.5$ cm, $y_1 = 0.50$ cm, and $q_2 = -4.0\ \mu C$, $x_2 = -2.0$ cm, $y_2 = 1.5$ cm. (a) Find the magnitude and direction of the electrostatic force on q_2. (b) Where could you locate a third charge $q_3 = +4.0\ \mu C$ such that the net electrostatic force on q_2 is zero?

13P. A certain charge Q is divided into two parts q and $Q - q$, which are then separated by a certain distance. What must q be in terms of Q to maximize the electrostatic repulsion between the two charges? ssm ilw

14P. A particle with charge Q is fixed at each of two opposite corners of a square, and a particle with charge q is placed at each of the other two corners. (a) If the net electrostatic force on each particle with charge Q is zero, what is Q in terms of q? (b) Is there any value of q that makes the net electrostatic force on each of the four particles zero? Explain.

15P. In Fig. 22-19, two tiny conducting balls of identical mass m and identical charge q hang from nonconducting threads of length L. Assume that θ is so small that

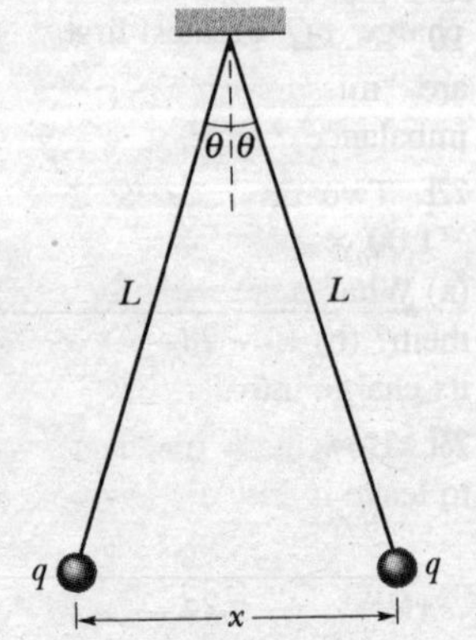

Fig. 22-19 Problem 15.

tan θ can be replaced by its approximate equal, sin θ. (a) Show that, for equilibrium,

$$x = \left(\frac{q^2L}{2\pi\varepsilon_0 mg}\right)^{1/3}$$

where x is the separation between the balls. (b) If L = 120 cm, m = 10 g, and x = 5.0 cm, what is q? ssm

16P. Explain what happens to the balls of Problem 15b if one of them is discharged (loses it charge q to, say, the ground), and find the new equilibrium separation x, using the given values of L and m and the computed value of q.

17P. Figure 22-20 shows a long, nonconducting, massless rod of length L, pivoted at its center and balanced with a block of weight W at a distance x from the left end. At the left and right ends of the rod are attached small conducting spheres with positive charges q and $2q$, respectively. A distance h directly beneath each of these spheres is a fixed sphere with positive charge Q. (a) Find the distance x when the rod is horizontal and balanced. (b) What value should h have so that the rod exerts no vertical force on the bearing when the rod is horizontal and balanced? ssm

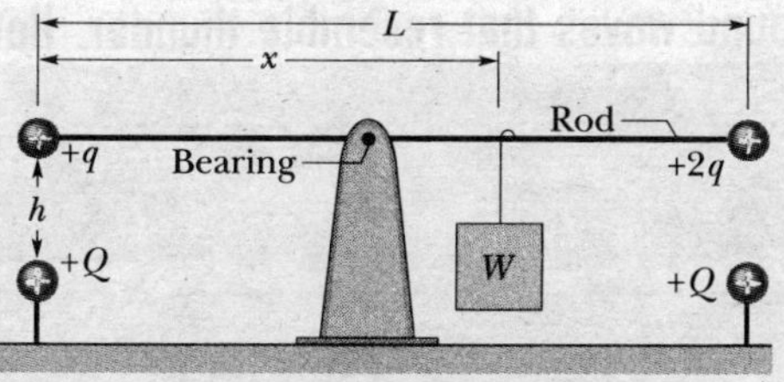

Fig. 22-20 Problem 17.

SEC. 22-5 Charge Is Quantized

18E. What is the magnitude of the electrostatic force between a singly charged sodium ion (Na^+, of charge $+e$) and an adjacent singly charged chlorine ion (Cl^-, of charge $-e$) in a salt crystal if their separation is 2.82×10^{-10} m?

19E. What is the total charge in coulombs of 75.0 kg of electrons? ssm

20E. How many megacoulombs of positive (or negative) charge are in 1.00 mol of neutral molecular-hydrogen gas (H_2)?

21E. The magnitude of the electrostatic force between two identical ions that are separated by a distance of 5.0×10^{-10} m is 3.7×10^{-9} N. (a) What is the charge of each ion? (b) How many electrons are "missing" from each ion (thus giving the ion its charge imbalance)? ssm

22E. Two tiny, spherical water drops, with identical charges of -1.00×10^{-16} C, have a center-to-center separation of 1.00 cm. (a) What is the magnitude of the electrostatic force acting between them? (b) How many excess electrons are on each drop, giving it its charge imbalance?

23E. How many electrons would have to be removed from a coin to leave it with a charge of $+1.0 \times 10^{-7}$ C? ilw

24E. An electron is in a vacuum near the surface of Earth. Where should a second electron be placed so that the electrostatic force it exerts on the first electron balances the gravitational force on the first electron due to Earth?

25P. Earth's atmosphere is constantly bombarded by *cosmic ray protons* that originate somewhere in space. If the protons all passed through the atmosphere, each square meter of Earth's surface would intercept protons at the average rate of 1500 protons per second. What would be the corresponding electric current intercepted by the total surface area of the planet? ilw

26P. Calculate the number of coulombs of positive charge in 250 cm^3 of (neutral) water (about a glassful).

27P. In the basic CsCl (cesium chloride) crystal structure, Cs^+ ions form the corners of a cube and a Cl^- ion is at the cube's center (Fig. 22-21). The edge length of the cube is 0.40 nm. The Cs^+ ions are each deficient by one electron (and thus each has a charge of $+e$), and the Cl^- ion has one excess electron (and thus has a charge of $-e$). (a) What is the magnitude of the net electrostatic force exerted on the Cl^- ion by the eight Cs^+ ions at the corners of the cube? (b) If one of the Cs^+ ions is missing, the crystal is said to have a *defect*; what is the magnitude of the net electrostatic force exerted on the Cl^- ion by the seven remaining Cs^+ ions? ssm www

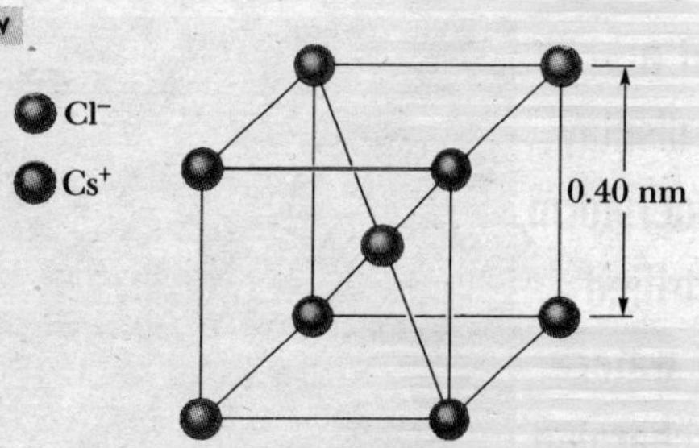

Fig. 22-21 Problem 27.

28P. We know that the negative charge on the electron and the positive charge on the proton are equal. Suppose, however, that these magnitudes differ from each other by 0.00010%. With what force would two copper coins, placed 1.0 m apart, repel each other? Assume that each coin contains 3×10^{22} copper atoms. (*Hint:* A neutral copper atom contains 29 protons and 29 electrons.) What do you conclude?

SEC. 22-6 Charge Is Conserved

29E. Identify X in the following nuclear reactions (in the first, n represents a neutron): (a) $^1H + {}^9Be \rightarrow X + n$; (b) $^{12}C + {}^1H \rightarrow X$; (c) $^{15}N + {}^1H \rightarrow {}^4He + X$. Appendix F will help. ssm

Additional Problem

30. In Problem 13, let $q = \alpha Q$. (a) Write an expression for the magnitude F of the force between the charges in terms of α, Q, and the charge separation d. (b) Graph F as a function of α. Graphically find the values of α that give (c) the maximum value of F and (d) half the maximum value of F.

23 Electric Fields

During the frequent eruptions of the Sakurajima volcano in Japan, multiple electrical discharges (sparks) flash over the volcano's crater, lighting up the sky and sending out sound waves that resemble thunder. However, this is not a lightning display in a thunderstorm, with electrified clouds of water drops discharging to the ground. This is something different.

How does the region above the volcano become electrified, and is there any way to tell whether the sparks travel up from the crater or down to it?

The answer is in this chapter.

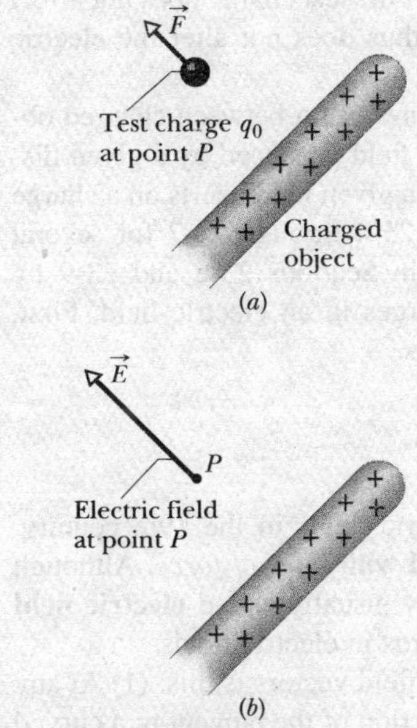

Fig. 23-1 (*a*) A positive test charge q_0 placed at point P near a charged object. An electrostatic force $\vec{F}$ acts on the test charge. (*b*) The electric field $\vec{E}$ at point P produced by the charged object.

23-1 Charges and Forces: A Closer Look

Suppose we fix a positive point charge q_1 in place and then put a second positive point charge q_2 near it. From Coulomb's law we know that q_1 exerts a repulsive electrostatic force on q_2 and, given enough data, we could determine the magnitude and direction of that force. Still, a nagging question remains: How does q_1 "know" of the presence of q_2? Since the charges do not touch, how can q_1 exert a force on q_2?

This question about *action at a distance* can be answered by saying that q_1 sets up an **electric field** in the space surrounding it. At any given point P in that space, the field has both magnitude and direction. The magnitude depends on the magnitude of q_1 and the distance between P and q_1. The direction depends on the direction from q_1 to P and the electrical sign of q_1. Thus when we place q_2 at P, q_1 interacts with q_2 through the electric field at P. The magnitude and direction of that electric field determine the magnitude and direction of the force acting on q_2.

Another action-at-a-distance problem arises if we move q_1, say, toward q_2. Coulomb's law tells us that when q_1 is closer to q_2, the repulsive electrostatic force acting on q_2 must be greater—and it is. However, here the nagging question is: Does the electric field at q_2, and thus the force acting on q_2, change immediately?

The answer is no. Instead, the information about the move by q_1 travels outward from q_1 (in all directions) as an electromagnetic wave at the speed of light c. The change in the electric field at q_2, and thus the change in the force acting on q_2, occurs when the wave finally reaches q_2.

23-2 The Electric Field

The temperature at every point in a room has a definite value. You can measure the temperature at any given point or combination of points by putting a thermometer there. We call the resulting distribution of temperatures a *temperature field*. In much the same way, you can imagine a *pressure field* in the atmosphere; it consists of the distribution of air pressure values, one for each point in the atmosphere. These two examples are of *scalar fields*, because temperature and air pressure are scalar quantities.

The electric field is a *vector field*; it consists of a distribution of *vectors*, one for each point in the region around a charged object, such as a charged rod. In principle, we can define the electric field at some point near the charged object, such as point P in Fig. 23-1*a*, as follows: We first place a *positive* charge q_0, called a *test charge*, at the point. We then measure the electrostatic force $\vec{F}$ that acts on the test charge. Finally, we define the electric field $\vec{E}$ at point P due to the charged object as

$$\vec{E} = \frac{\vec{F}}{q_0} \quad \text{(electric field).} \tag{23-1}$$

Thus, the magnitude of the electric field $\vec{E}$ at point P is $E = F/q_0$, and the direction of $\vec{E}$ is that of the force $\vec{F}$ that acts on the *positive* test charge. As shown in Fig. 23-1*b*, we represent the electric field at P with a vector whose tail is at P. To define the electric field within some region, we must similarly define it at all points in the region.

The SI unit for the electric field is the newton per coulomb (N/C). Table 23-1 shows the electric fields that occur in a few physical situations.

TABLE 23-1 Some Electric Fields

Field Location or Situation	Value (N/C)
At the surface of a uranium nucleus	3×10^{21}
Within a hydrogen atom, at a radius of 5.29×10^{-11} m	5×10^{11}
Electric breakdown occurs in air	3×10^{6}
Near the charged drum of a photocopier	10^{5}
Near a charged comb	10^{3}
In the lower atmosphere	10^{2}
Inside the copper wire of household circuits	10^{-2}

Although we use a positive test charge to define the electric field of a charged object, that field exists independently of the test charge. The field at point P in Figure 23-1*b* existed both before and after the test charge of Fig. 23-1*a* was put there. (We

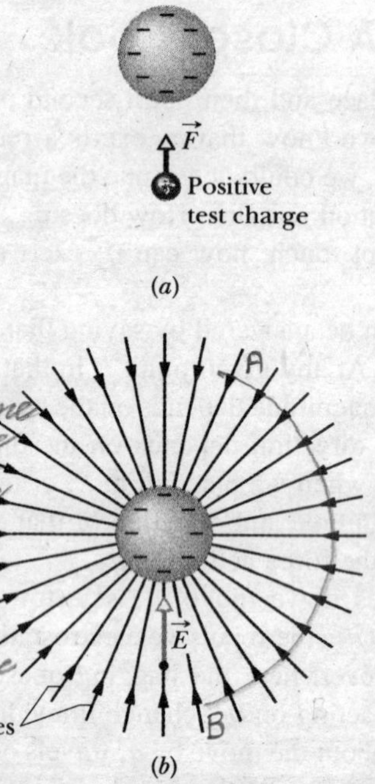

Fig. 23-2 *(a)* The electrostatic force $\vec{F}$ acting on a positive test charge near a sphere of uniform negative charge. *(b)* The electric field vector $\vec{E}$ at the location of the test charge, and the electric field lines in the space near the sphere. The field lines extend *toward* the negatively charged sphere. (They originate on distant positive charges.)

assume that in our defining procedure, the presence of the test charge does not affect the charge distribution on the charged object, and thus does not alter the electric field we are defining.)

To examine the role of an electric field in the interaction between charged objects, we have two tasks: (1) calculating the electric field produced by a given distribution of charge, and (2) calculating the force that a given field exerts on a charge placed in it. We perform the first task in Sections 23-4 through 23-7 for several charge distributions. We perform the second task in Sections 23-8 and 23-9 by considering a point charge and a pair of point charges in an electric field. First, however, we discuss a way to visualize electric fields.

23-3 Electric Field Lines

Michael Faraday, who introduced the idea of electric fields in the 19th century, thought of the space around a charged body as filled with *lines of force*. Although we no longer attach much reality to these lines, now usually called **electric field lines,** they still provide a nice way to visualize patterns in electric fields.

The relation between the field lines and electric field vectors is this: (1) At any point, the direction of a straight field line or the direction of the tangent to a curved field line gives the direction of $\vec{E}$ at that point, and (2) the field lines are drawn so that the number of lines per unit area, measured in a plane that is perpendicular to the lines, is proportional to the *magnitude* of $\vec{E}$. This second relation means that where the field lines are close together, E is large; and where they are far apart, E is small.

Figure 23-2*a* shows a sphere of uniform negative charge. If we place a *positive* test charge anywhere near the sphere, an electrostatic force pointing *toward* the center of the sphere will act on the test charge as shown. In other words, the electric field vectors at all points near the sphere are directed radially toward the sphere. This pattern of vectors is neatly displayed by the field lines in Fig. 23-2*b*, which point in the same directions as the force and field vectors. Moreover, the spreading of the field lines with distance from the sphere tells us that the magnitude of the electric field decreases with distance from the sphere.

If the sphere of Fig. 23-2 were of uniform *positive* charge, the electric field vectors at all points near the sphere would be directed radially *away from* the sphere. Thus, the electric field lines would also extend radially away from the sphere. We then have the following rule:

> Electric field lines extend away from positive charge (where they originate) and toward negative charge (where they terminate).

Figure 23-3*a* shows part of an infinitely large, nonconducting *sheet* (or plane) with a uniform distribution of positive charge on one side. If we were to place a positive test charge at any point near the sheet of Fig. 23-3*a*, the net electrostatic force acting on the test charge would be perpendicular to the sheet, because forces acting in all other directions would cancel one another as a result of the symmetry. Moreover, the net force on the test charge would point away from the sheet as shown. Thus, the electric field vector at any point in the space on either side of the sheet is also perpendicular to the sheet and directed away from it (Figs. 23-3*b* and *c*). Since the charge is uniformly distributed along the sheet, all the field vectors have the same magnitude. Such an electric field, with the same magnitude and direction at every point, is a *uniform electric field*.

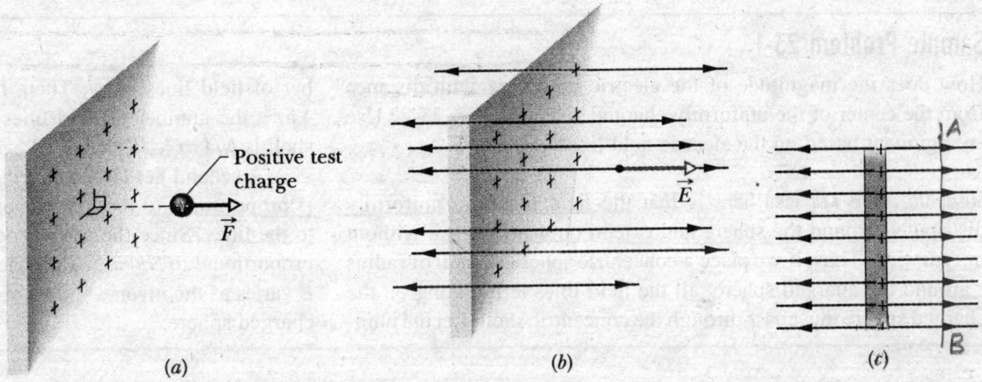

Fig. 23-3 (*a*) The electrostatic force $\vec{F}$ on a positive test charge near a very large, nonconducting sheet with uniformly distributed positive charge on one side. (*b*) The electric field vector $\vec{E}$ at the location of the test charge, and the electric field lines in the space near the sheet. The field lines extend *away from* the positively charged sheet. (*c*) Side view of (*b*).

Of course, no real nonconducting sheet (such as a flat expanse of plastic) is infinitely large, but if we consider a region that is near the middle of a real sheet and not near its edges, the field lines through that region are arranged as in Figs. 23-3*b* and *c*.

Figure 23-4 shows the field lines for two equal positive charges. Figure 23-5 shows the pattern for two charges that are equal in magnitude but of opposite sign, a configuration that we call an **electric dipole.** Although we do not often use field lines quantitatively, they are very useful to visualize what is going on. Can you not almost "see" the charges being pushed apart in Fig. 23-4 and pulled together in Fig. 23-5?

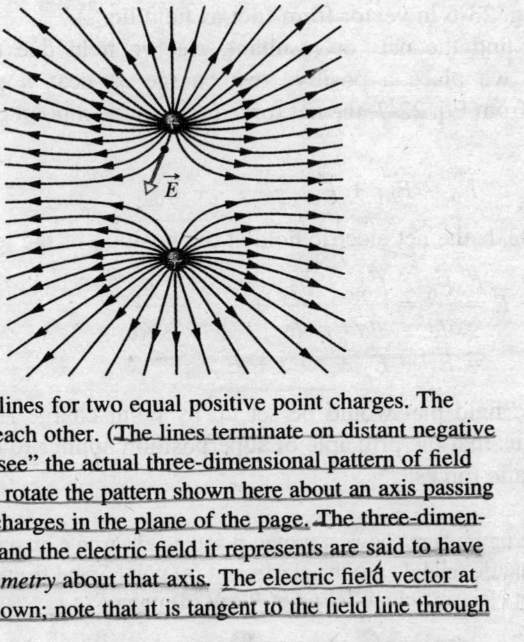

Fig. 23-4 Field lines for two equal positive point charges. The charges repel each other. (The lines terminate on distant negative charges.) To "see" the actual three-dimensional pattern of field lines, mentally rotate the pattern shown here about an axis passing through both charges in the plane of the page. The three-dimensional pattern and the electric field it represents are said to have *rotational symmetry* about that axis. The electric field vector at one point is shown; note that it is tangent to the field line through that point.

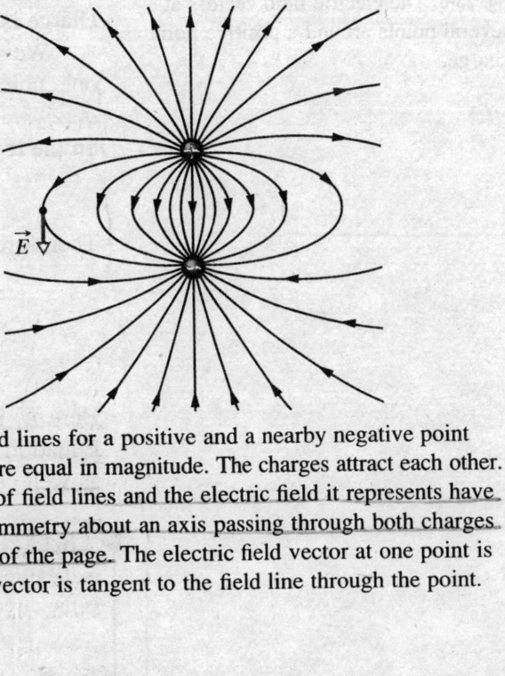

Fig. 23-5 Field lines for a positive and a nearby negative point charge that are equal in magnitude. The charges attract each other. The pattern of field lines and the electric field it represents have rotational symmetry about an axis passing through both charges in the plane of the page. The electric field vector at one point is shown; the vector is tangent to the field line through the point.

Sample Problem 23-1

How does the magnitude of the electric field vary with distance from the center of the uniformly charged sphere in Fig. 23-2? Use an argument based on the electric field lines.

SOLUTION: One **Key Idea** here is that the field lines are uniformly distributed around the sphere and extend outward from it without interruption. Thus, if we place a concentric spherical shell of radius r around the charged sphere, all the field lines terminating on the charged sphere must pass through the concentric shell. Let the number of field lines be N. Then, because the shell has surface area $4\pi r^2$, the number of field lines per unit area passing through the shell is $N/4\pi r^2$.

A second **Key Idea** is that the magnitude E of the electric field is proportional to the number of lines per unit area perpendicular to the lines. Since the shell is perpendicular to the field lines, E is proportional to $N/4\pi r^2$. Because r is the only variable in that term, E varies as the inverse square of the distance from the center of the charged sphere.

23-4 The Electric Field Due to a Point Charge

To find the electric field due to a point charge q (or charged particle) at any point a distance r from the point charge, we put a positive test charge q_0 at that point. From Coulomb's law (Eq. 22-4), the magnitude of the electrostatic force acting on q_0 is

$$F = \frac{1}{4\pi\varepsilon_0}\frac{|q||q_0|}{r^2}. \tag{23-2}$$

The direction of $\vec{F}$ is directly away from the point charge if q is positive, and directly toward the point charge if q is negative. The magnitude of the electric field vector is, from Eq. 23-1,

$$E = \frac{F}{q_0} = \frac{1}{4\pi\varepsilon_0}\frac{|q|}{r^2} \quad \text{(point charge)}. \tag{23-3}$$

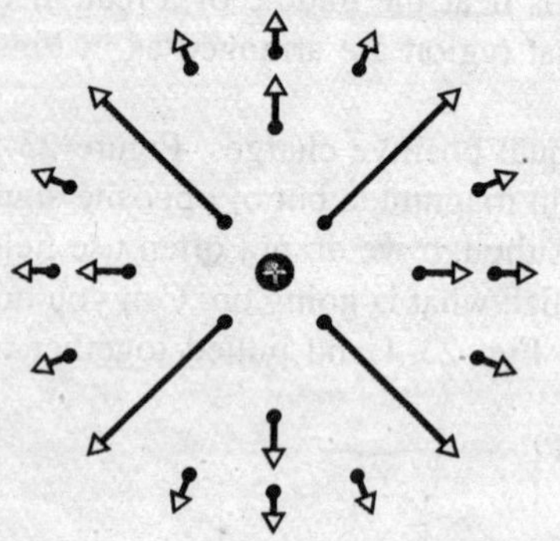

Fig. 23-6 The electric field vectors at several points around a positive point charge.

The direction of $\vec{E}$ is the same as that of the force on the positive test charge: directly away from the point charge if q is positive, and toward it if q is negative.

Because there is nothing special about the point we chose for q_0, Eq. 23-3 gives the field at every point around the point charge q. The field for a positive point charge is shown in Fig. 23-6 in vector form (not as field lines).

We can quickly find the net, or resultant, electric field due to more than one point charge. If we place a positive test charge q_0 near n point charges $q_1, q_2, \ldots, q_n$, then, from Eq. 22-7, the net force $\vec{F}_0$ from the n point charges acting on the test charge is

$$\vec{F}_0 = \vec{F}_{01} + \vec{F}_{02} + \cdots + \vec{F}_{0n}.$$

Therefore, from Eq. 23-1, the net electric field at the position of the test charge is

$$\begin{aligned}\vec{E} &= \frac{\vec{F}_0}{q_0} = \frac{\vec{F}_{01}}{q_0} + \frac{\vec{F}_{02}}{q_0} + \cdots + \frac{\vec{F}_{0n}}{q_0} \\ &= \vec{E}_1 + \vec{E}_2 + \cdots + \vec{E}_n.\end{aligned} \tag{23-4}$$

Here $\vec{E}_i$ is the electric field that would be set up by point charge i acting alone. Equation 23-4 shows us that the principle of superposition applies to electric fields as well as to electrostatic forces.

CHECKPOINT 1: The figure here shows a proton p and an electron e on an x axis. What is the direction of the electric field due to the electron at (a) point S and (b) point R? What is the direction of the net electric field at (c) point R and (d) point S?

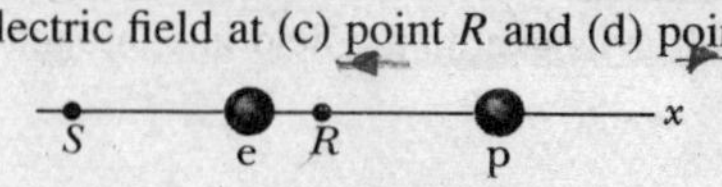

Sample Problem 23-2

Figure 23-7*a* shows three particles with charges $q_1 = +2Q$, $q_2 = -2Q$, and $q_3 = -4Q$, each a distance d from the origin. What net electric field $\vec{E}$ is produced at the origin?

SOLUTION: The **Key Idea** is that charges q_1, q_2, and q_3 produce electric field vectors $\vec{E}_1$, $\vec{E}_2$, and $\vec{E}_3$, respectively, at the origin, and the net electric field is the vector sum $\vec{E} = \vec{E}_1 + \vec{E}_2 + \vec{E}_3$. To find this sum, we first must find the magnitudes and orientations of the three field vectors. To find the magnitude of $\vec{E}_1$, which is due to q_1, we use Eq. 23-3, substituting d for r and $2Q$ for $|q|$ and obtaining

$$E_1 = \frac{1}{4\pi\varepsilon_0}\frac{2Q}{d^2}.$$

Similarly, we find the magnitudes of the fields $\vec{E}_2$ and $\vec{E}_3$ to be

$$E_2 = \frac{1}{4\pi\varepsilon_0}\frac{2Q}{d^2} \quad \text{and} \quad E_3 = \frac{1}{4\pi\varepsilon_0}\frac{4Q}{d^2}.$$

We next must find the orientations of the three electric field vectors at the origin. Because q_1 is a positive charge, the field vector it produces points directly *away* from it, and because q_2 and q_3 are both negative, the field vectors they produce point directly *toward* each of them. Thus, the three electric fields produced at the origin by the three charged particles are oriented as in Fig. 23-7*b*. (*Caution:* Note that we have placed the tails of the vectors at the point where the fields are to be evaluated; doing so decreases the chance of error.)

We can now add the fields vectorially as outlined for forces in Sample Problem 22-1c. However, here we can use symmetry to simplify the procedure. From Fig. 23-7*b*, we see that $\vec{E}_1$ and $\vec{E}_2$ have the same direction. Hence, their vector sum has that direction and has the magnitude

$$E_1 + E_2 = \frac{1}{4\pi\varepsilon_0}\frac{2Q}{d^2} + \frac{1}{4\pi\varepsilon_0}\frac{2Q}{d^2} = \frac{1}{4\pi\varepsilon_0}\frac{4Q}{d^2},$$

which happens to equal the magnitude of field $\vec{E}_3$.

We must now combine two vectors, $\vec{E}_3$ and the vector sum $\vec{E}_1 + \vec{E}_2$, that have the same magnitude and that are oriented symmetrically about the x axis, as shown in Fig. 23-7*c*. From the symmetry of Fig. 23-7*c*, we realize that the equal y components of our two vectors cancel and the equal x components add. Thus, the net electric field $\vec{E}$ at the origin is in the positive direction of the x axis and has the magnitude

$$E = 2E_{3x} = 2E_3 \cos 30^\circ = (2)\frac{1}{4\pi\varepsilon_0}\frac{4Q}{d^2}(0.866) = \frac{6.93Q}{4\pi\varepsilon_0 d^2}. \quad \text{(Answer)}$$

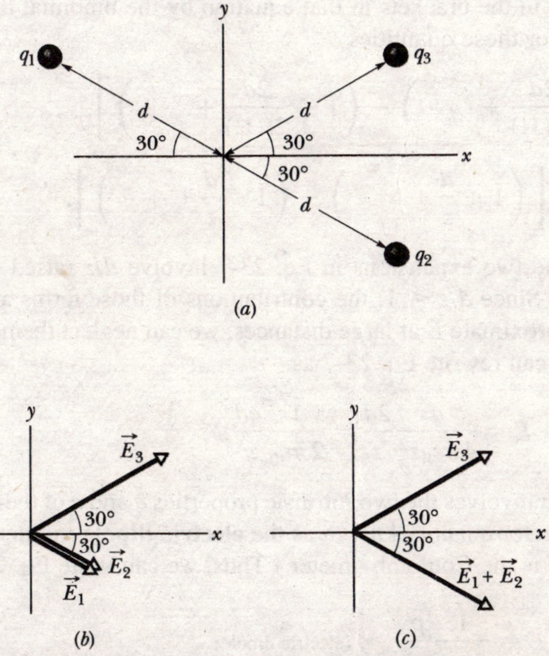

Fig. 23-7 Sample Problem 23-2. (*a*) Three particles with charges q_1, q_2, and q_3 are at the same distance d from the origin. (*b*) The electric field vectors $\vec{E}_1$, $\vec{E}_2$, and $\vec{E}_3$ at the origin due to the three particles. (*c*) The electric field vector $\vec{E}_3$ and the vector sum $\vec{E}_1 + \vec{E}_2$ at the origin.

CHECKPOINT 2: The figure here shows four situations in which charged particles are at equal distances from the origin. Rank the situations according to the magnitude of the net electric field at the origin, greatest first.

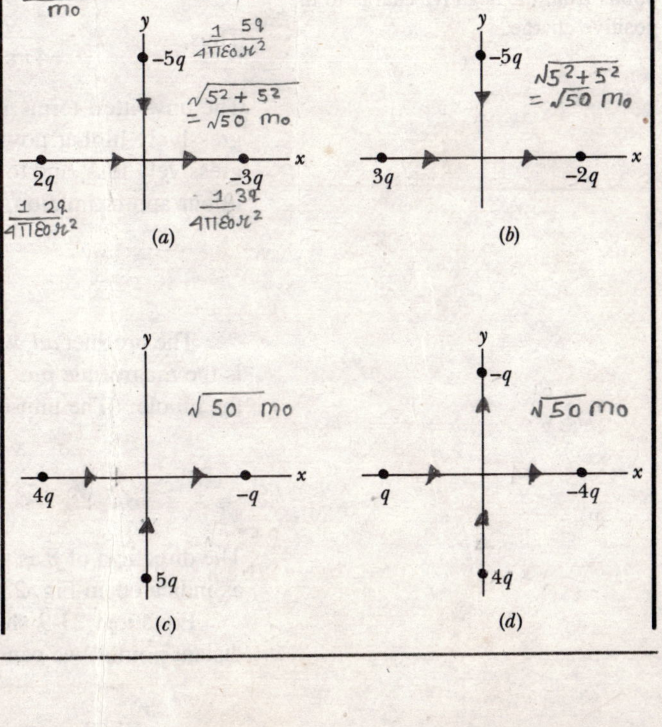

23-5 The Electric Field Due to an Electric Dipole

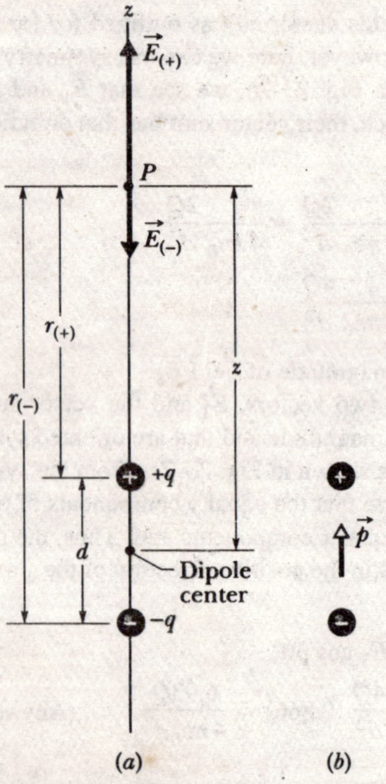

Fig. 23-8 (*a*) An electric dipole. The electric field vectors $\vec{E}_{(+)}$ and $\vec{E}_{(-)}$ at point *P* on the dipole axis result from the dipole's two charges. *P* is at distances $r_{(+)}$ and $r_{(-)}$ from the individual charges that make up the dipole. (*b*) The dipole moment $\vec{p}$ of the dipole points from the negative charge to the positive charge.

Figure 23-8*a* shows two charged particles of magnitude *q* but of opposite sign, separated by a distance *d*. As was noted in connection with Fig. 23-5, we call this configuration an *electric dipole*. Let us find the electric field due to the dipole of Fig. 23-8*a* at a point *P*, a distance *z* from the midpoint of the dipole and on the axis through the particles, which is called the *dipole axis*.

From symmetry, the electric field $\vec{E}$ at point *P*—and also the fields $\vec{E}_{(+)}$ and $\vec{E}_{(-)}$ due to the separate charges that make up the dipole—must lie along the dipole axis, which we have taken to be a *z* axis. Applying the superposition principle for electric fields, we find that the magnitude *E* of the electric field at *P* is

$$\begin{aligned} E &= E_{(+)} - E_{(-)} \\ &= \frac{1}{4\pi\varepsilon_0}\frac{q}{r_{(+)}^2} - \frac{1}{4\pi\varepsilon_0}\frac{q}{r_{(-)}^2} \\ &= \frac{q}{4\pi\varepsilon_0(z-\frac{1}{2}d)^2} - \frac{q}{4\pi\varepsilon_0(z+\frac{1}{2}d)^2}. \end{aligned} \tag{23-5}$$

After a little algebra, we can rewrite this equation as

$$E = \frac{q}{4\pi\varepsilon_0 z^2}\left[\left(1-\frac{d}{2z}\right)^{-2} - \left(1+\frac{d}{2z}\right)^{-2}\right]. \tag{23-6}$$

We are usually interested in the electrical effect of a dipole only at distances that are large compared with the dimensions of the dipole—that is, at distances such that $z \gg d$. At such large distances, we have $d/2z \ll 1$ in Eq. 23-6. We can then expand the two quantities in the brackets in that equation by the binomial theorem (Appendix E), obtaining for those quantities

$$\left[\left(1+\frac{2d}{2z(1!)}+\cdots\right) - \left(1-\frac{2d}{2z(1!)}+\cdots\right)\right].$$

Thus,

$$E = \frac{q}{4\pi\varepsilon_0 z^2}\left[\left(1+\frac{d}{z}+\cdots\right) - \left(1-\frac{d}{z}+\cdots\right)\right]. \tag{23-7}$$

The unwritten terms in the two expansions in Eq. 23-7 involve d/z raised to progressively higher powers. Since $d/z \ll 1$, the contributions of those terms are progressively less, and to approximate *E* at large distances, we can neglect them. Then, in our approximation, we can rewrite Eq. 23-7 as

$$E = \frac{q}{4\pi\varepsilon_0 z^2}\frac{2d}{z} = \frac{1}{2\pi\varepsilon_0}\frac{qd}{z^3}. \tag{23-8}$$

The product *qd*, which involves the two intrinsic properties *q* and *d* of the dipole, is the magnitude *p* of a vector quantity known as the **electric dipole moment** $\vec{p}$ of the dipole. (The unit of $\vec{p}$ is the Coulomb–meter.) Thus, we can write Eq. 23-8 as

$$E = \frac{1}{2\pi\varepsilon_0}\frac{p}{z^3} \quad \text{(electric dipole)}. \tag{23-9}$$

The direction of $\vec{p}$ is taken to be from the negative to the positive end of the dipole, as indicated in Fig. 23-8*b*. We can use $\vec{p}$ to specify the orientation of a dipole.

Equation 23-9 shows that, if we measure the electric field of a dipole only at distant points, we can never find *q* and *d* separately, only their product. The field at

TABLE 23-2 Some Measures of Electric Charge

Name	Symbol	SI Unit
Charge	q	C
Linear charge density	λ	C/m
Surface charge density	σ	C/m^2
Volume charge density	ρ	C/m^3

distant points would be unchanged if, for example, q were doubled and d simultaneously halved. Thus, the dipole moment is a basic property of a dipole.

Although Eq. 23-9 holds only for distant points along the dipole axis, it turns out that E for a dipole varies as $1/r^3$ for *all* distant points, regardless of whether they lie on the dipole axis; here r is the distance between the point in question and the dipole center.

Inspection of Fig. 23-8 and of the field lines in Fig. 23-5 shows that the direction of $\vec{E}$ for distant points on the dipole axis is always the direction of the dipole moment vector $\vec{p}$. This is true whether point P in Fig. 23-8a is on the upper or the lower part of the dipole axis.

Inspection of Eq. 23-9 shows that if you double the distance of a point from a dipole, the electric field at the point drops by a factor of 8. If you double the distance from a single point charge, however (see Eq. 23-3), the electric field drops only by a factor of 4. Thus the electric field of a dipole decreases more rapidly with distance than does the electric field of a single charge. The physical reason for this rapid decrease in electric field for a dipole is that from distant points a dipole looks like two equal but opposite charges that almost—but not quite—coincide. Thus, their electric fields at distant points almost—but not quite—cancel each other.

23-6 The Electric Field Due to a Line of Charge

So far we have considered the electric field that is produced by one or, at most, a few point charges. We now consider charge distributions that consist of a great many closely spaced point charges (perhaps billions) that are spread along a line, over a surface, or within a volume. Such distributions are said to be **continuous** rather than discrete. Since these distributions can include an enormous number of point charges, we find the electric fields that they produce by means of calculus rather than by considering the point charges one by one. In this section we discuss the electric field caused by a line of charge. We consider a charged surface in the next section. In the next chapter, we shall find the field inside a uniformly charged sphere.

When we deal with continuous charge distributions, it is most convenient to express the charge on an object as a *charge density* rather than as a total charge. For a line of charge, for example, we would report the linear charge density (or charge per unit length) λ, whose SI unit is the coulomb per meter. Table 23-2 shows the other charge densities we shall be using.

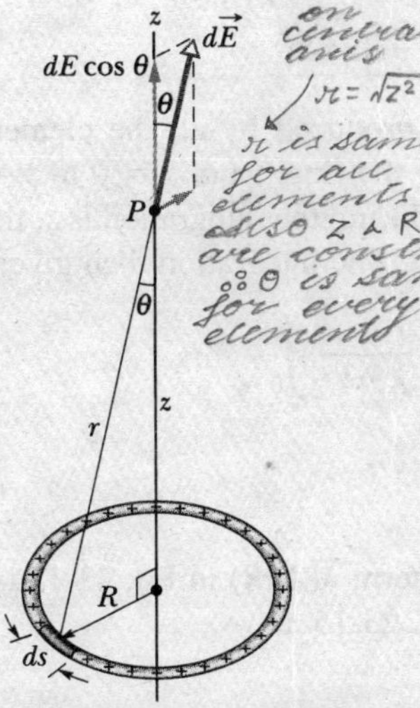

Fig. 23-9 A ring of uniform positive charge. A differential element of charge occupies a length ds (greatly exaggerated for clarity). This element sets up an electric field $d\vec{E}$ at point P. The component of $d\vec{E}$ along the central axis of the ring is $dE \cos\theta$.

Figure 23-9 shows a thin ring of radius R with a uniform positive linear charge density λ around its circumference. We may imagine the ring to be made of plastic or some other insulator, so that the charges can be regarded as fixed in place. What is the electric field $\vec{E}$ at point P, a distance z from the plane of the ring along its central axis?

To answer, we cannot just apply Eq. 23-3, which gives the electric field set up by a point charge, because the ring is obviously not a point charge. However, we can mentally divide the ring into differential elements of charge that are so small that they are like point charges, and then we can apply Eq. 23-3 to each of them. Next, we can add the electric fields set up at P by all the differential elements. The vector sum of all those fields gives us the field set up at P by the ring.

Let ds be the (arc) length of any differential element of the ring. Since λ is the charge per unit length, the element has a charge of magnitude

$$dq = \lambda\, ds. \tag{23-10}$$

This differential charge sets up a differential electric field $d\vec{E}$ at point P, which is a distance r from the element. Treating the element as a point charge, and using Eq. 23-10, we can rewrite Eq. 23-3 to express the magnitude of $d\vec{E}$ as

$$dE = \frac{1}{4\pi\varepsilon_0}\frac{dq}{r^2} = \frac{1}{4\pi\varepsilon_0}\frac{\lambda\, ds}{r^2}. \tag{23-11}$$

From Fig. 23-9, we can rewrite Eq. 23-11 as

$$dE = \frac{1}{4\pi\varepsilon_0}\frac{\lambda\, ds}{(z^2 + R^2)}. \tag{23-12}$$

Figure 23-9 shows us that $d\vec{E}$ is at an angle θ to the central axis (which we have taken to be a z axis) and has components perpendicular to and parallel to that axis.

Every charge element in the ring sets up a differential field $d\vec{E}$ at P, with magnitude given by Eq. 23-12. All the $d\vec{E}$ vectors have identical components parallel to the central axis, in both magnitude and direction. All these $d\vec{E}$ vectors have components perpendicular to the central axis as well; these perpendicular components are identical in magnitude but point in different directions. In fact, for any perpendicular component that points in a given direction, there is another one that points in the opposite direction. The sum of this pair of components, like the sum of all other pairs of oppositely directed components, is zero.

Thus, the perpendicular components cancel and we need not consider them further. This leaves the parallel components; they all have the same direction, so the net electric field at P is their sum.

The parallel component of $d\vec{E}$ shown in Fig. 23-9 has magnitude $dE \cos\theta$. The figure also shows us that

$$\cos\theta = \frac{z}{r} = \frac{z}{(z^2 + R^2)^{1/2}}. \tag{23-13}$$

Then Eqs. 23-13 and 23-12 give us, for the parallel component of $d\vec{E}$,

$$dE\cos\theta = \frac{z\lambda}{4\pi\varepsilon_0(z^2 + R^2)^{3/2}}\, ds. \tag{23-14}$$

To add the parallel components $dE \cos\theta$ produced by all the elements, we integrate Eq. 23-14 around the circumference of the ring, from $s = 0$ to $s = 2\pi R$. Since the only quantity in Eq. 23-14 that varies during the integration is s, the other quantities can be moved outside the integral sign. The integration then gives us

$$E = \int dE\cos\theta = \frac{z\lambda}{4\pi\varepsilon_0(z^2 + R^2)^{3/2}}\int_0^{2\pi R} ds$$

$$= \frac{z\lambda(2\pi R)}{4\pi\varepsilon_0(z^2 + R^2)^{3/2}}. \tag{23-15}$$

Since λ is the charge per length of the ring, the term $\lambda(2\pi R)$ in Eq. 23-15 is q, the total charge on the ring. We then can rewrite Eq. 23-15 as

or V due to ring on the central z axis

$V = \frac{1}{4\pi\varepsilon_0}\frac{q}{\sqrt{R^2+z^2}}$ ←

$$E = \frac{qz}{4\pi\varepsilon_0(z^2 + R^2)^{3/2}} \quad \text{(charged ring).} \tag{23-16}$$

find E: $E = -\frac{\partial V}{\partial z}$

If the charge on the ring is negative, instead of positive as we have assumed, the magnitude of the field at P is still given by Eq. 23-16. However, the electric field vector then points toward the ring instead of away from it.

Let us check Eq. 23-16 for a point on the central axis that is so far away that $z \gg R$. For such a point, the expression $z^2 + R^2$ in Eq. 23-16 can be approximated as z^2, and Eq. 23-16 becomes

$$E = \frac{1}{4\pi\varepsilon_0}\frac{q}{z^2} \quad \text{(charged ring at large distance).} \tag{23-17}$$

This is a reasonable result, because from a large distance, the ring "looks like" a point charge. If we replace z with r in Eq. 23-17, we indeed do have Eq. 23-3, the magnitude of the electric field due to a point charge.

Let us next check Eq. 23-16 for a point at the center of the ring—that is, for $z = 0$. At that point, Eq. 23-16 tells us that $E = 0$. This is a reasonable result, because if we were to place a test charge at the center of the ring, there would be no net electrostatic force acting on it; the force due to any element of the ring would be canceled by the force due to the element on the opposite side of the ring. By Eq. 23-1, if the force at the center of the ring were zero, the electric field there would also have to be zero.

Sample Problem 23-3

Field due to circular arc $E = \frac{1}{4\pi\varepsilon_0 r}\lambda[\sin\theta]_{\theta_1}^{\theta_2}$

Figure 23-10a shows a plastic rod having a uniformly distributed charge $-Q$. The rod has been bent in a 120° circular arc of radius r. We place coordinate axes such that the axis of symmetry of the rod lies along the x axis and the origin is at the center of curvature P of the rod. In terms of Q and r, what is the electric field $\vec{E}$ due to the rod at point P? Had rod been bent in a complete circle, electric field at P would have been zero

SOLUTION: The **Key Idea** here is that, because the rod has a continuous charge distribution, we must find an expression for the electric fields due to differential elements of the rod and then sum those fields via calculus. Consider a differential element having arc length ds and located at an angle θ above the x axis (Fig. 23-10b). If we let λ represent the linear charge density of the rod, our element ds has a differential charge of magnitude

$$dq = \lambda\, ds. \quad \text{— 1 step} \tag{23-18}$$

Our element produces a differential electric field $d\vec{E}$ at point P, which is a distance r from the element. Treating the element as a point charge, we can rewrite Eq. 23-3 to express the magnitude of $d\vec{E}$ as

$$dE = \frac{1}{4\pi\varepsilon_0}\frac{dq}{r^2} = \frac{1}{4\pi\varepsilon_0}\frac{\lambda\, ds}{r^2}. \quad \text{— 2 step} \tag{23-19}$$

The direction of $d\vec{E}$ is toward ds, because charge dq is negative. — 3 step

Our element has a symmetrically located (mirror image) element ds' in the bottom half of the rod. — 4 step The electric field $d\vec{E}'$ set up at P by ds' also has the magnitude given by Eq. 23-19, but the field vector points toward ds' as shown in Fig. 23-10b. If we resolve the electric field vectors of ds and ds' into x and y components as shown in Fig. 23-10b, we see that their y components cancel (because they have equal magnitudes and are in opposite directions). We also see that their x components have equal magnitudes and are in the same direction.

Thus, to find the electric field set up by the rod, we need sum (via integration) only the x components of the differential electric fields set up by all the differential elements of the rod. From

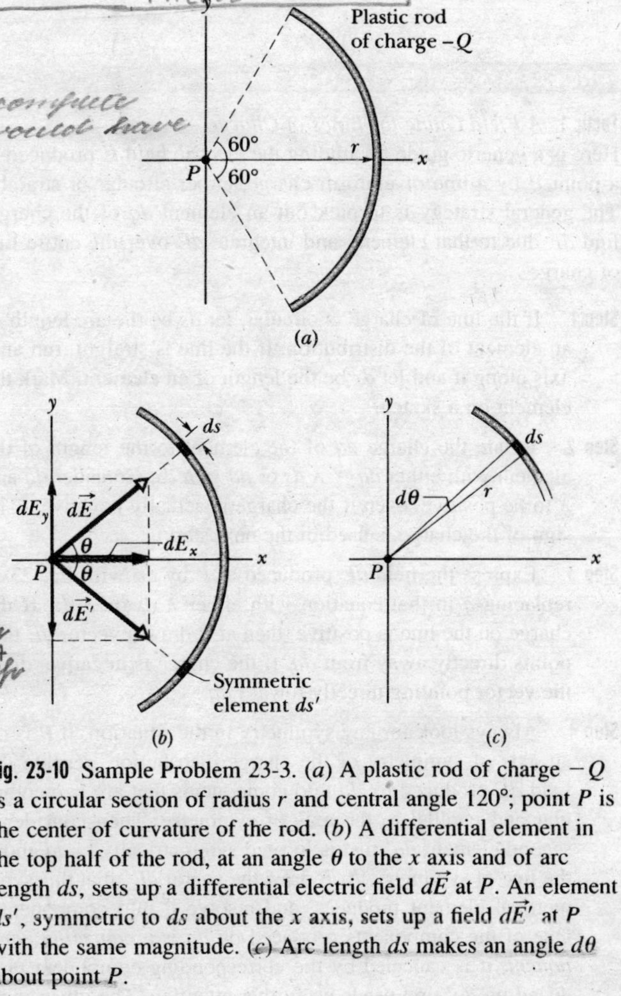

Fig. 23-10 Sample Problem 23-3. (a) A plastic rod of charge $-Q$ is a circular section of radius r and central angle 120°; point P is the center of curvature of the rod. (b) A differential element in the top half of the rod, at an angle θ to the x axis and of arc length ds, sets up a differential electric field $d\vec{E}$ at P. An element ds', symmetric to ds about the x axis, sets up a field $d\vec{E}'$ at P with the same magnitude. (c) Arc length ds makes an angle $d\theta$ about point P.

Fig. 23-10b and Eq. 23-19, we can write the component dE_x set up by ds as

$$dE_x = dE \cos\theta = \frac{1}{4\pi\varepsilon_0}\frac{\lambda}{r^2}\cos\theta\, ds. \tag{23-20}$$

Equation 23-20 has two variables, θ and s. Before we can integrate it, we must eliminate one variable. We do so by replacing ds, using the relation

$$ds = r\, d\theta,$$

in which $d\theta$ is the angle at P that includes arc length ds (Fig. 23-10c). With this replacement, we can integrate Eq. 23-20 over the angle made by the rod at P, from $\theta = -60°$ to $\theta = 60°$; that will give us the magnitude of the electric field at P due to the rod:

$$E = \int dE_x = \int_{-60°}^{60°} \frac{1}{4\pi\varepsilon_0}\frac{\lambda}{r^2}\cos\theta\, r\, d\theta$$

$$= \frac{\lambda}{4\pi\varepsilon_0 r}\int_{-60°}^{60°}\cos\theta\, d\theta = \frac{\lambda}{4\pi\varepsilon_0 r}\Big[\sin\theta\Big]_{-60°}^{60°}$$

$$= \frac{\lambda}{4\pi\varepsilon_0 r}[\sin 60° - \sin(-60°)]$$

$$= \frac{1.73\lambda}{4\pi\varepsilon_0 r}. \tag{23-21}$$

(If we had reversed the limits on the integration, we would have gotten the same result but with a minus sign. Since the integration gives only the magnitude of $\vec{E}$, we would then have discarded the minus sign.)

To evaluate λ, we note that the rod has an angle of 120° and so is one-third of a full circle. Its arc length is then $2\pi r/3$, and its linear charge density must be

$$\lambda = \frac{\text{charge}}{\text{length}} = \frac{Q}{2\pi r/3} = \frac{0.477Q}{r}.$$

Substituting this into Eq. 23-21 and simplifying give us

$$E = \frac{(1.73)(0.477Q)}{4\pi\varepsilon_0 r^2}$$

$$= \frac{0.83Q}{4\pi\varepsilon_0 r^2}. \qquad \text{(Answer)}$$

The direction of $\vec{E}$ is toward the rod, along the axis of symmetry of the charge distribution. We can write $\vec{E}$ in unit-vector notation as

$$\vec{E} = \frac{0.83Q}{4\pi\varepsilon_0 r^2}\,\hat{\mathrm{i}}.$$

PROBLEM-SOLVING TACTICS

Tactic 1: *A Field Guide for Lines of Charge*

Here is a generic guide for finding the electric field $\vec{E}$ produced at a point P by a line of uniform charge, either circular or straight. The general strategy is to pick out an element dq of the charge, find $d\vec{E}$ due to that element, and integrate $d\vec{E}$ over the entire line of charge.

Step 1. If the line of charge is circular, let ds be the arc length of an element of the distribution. If the line is straight, run an x axis along it and let dx be the length of an element. Mark the element on a sketch.

Step 2. Relate the charge dq of the element to the length of the element with either $dq = \lambda\, ds$ or $dq = \lambda\, dx$. Consider dq and λ to be positive, even if the charge is actually negative. (The sign of the charge is used in the next step.)

Step 3. Express the field $d\vec{E}$ produced at P by dq with Eq. 23-3, replacing q in that equation with either $\lambda\, ds$ or $\lambda\, dx$. If the charge on the line is positive, then at P draw a vector $d\vec{E}$ that points directly away from dq. If the charge is negative, draw the vector pointing directly toward dq.

Step 4. Always look for any symmetry in the situation. If P is on an axis of symmetry of the charge distribution, resolve the field $d\vec{E}$ produced by dq into components that are perpendicular and parallel to the axis of symmetry. Then consider a second element dq' that is located symmetrically to dq about the line of symmetry. At P draw the vector $d\vec{E}'$ that this symmetrical element produces, and resolve it into components. One of the components produced by dq is a *canceling component*; it is canceled by the corresponding component produced by dq' and needs no further attention. The other component produced by dq is an *adding component*; it adds to the corresponding component produced by dq'. Add the adding components of all the elements via integration.

Step 5. Here are four general types of uniform charge distributions, with strategies for simplifying the integral of step 4.

Ring, with point P on (central) axis of symmetry, as in Fig. 23-9. In the expression for dE, replace r^2 with $z^2 + R^2$, as in Eq. 23-12. Express the adding component of $d\vec{E}$ in terms of θ. That introduces $\cos\theta$, but θ is identical for all elements and thus is not a variable. Replace $\cos\theta$ as in Eq. 23-13. Integrate over s, around the circumference of the ring.

Circular arc, with point P at the center of curvature, as in Fig. 23-10. Express the adding component of $d\vec{E}$ in terms of θ. That introduces either $\sin\theta$ or $\cos\theta$. Reduce the resulting two variables s and θ to one, θ, by replacing ds with $r\, d\theta$. Integrate over θ, as in Sample Problem 23-3, from one end of the arc to the other end.

Straight line, with point P on an extension of the line, as in Fig. 23-11a. In the expression for dE, replace r with x. Integrate over x, from end to end of the line of charge.

Straight line, with point P at perpendicular distance y from the line of charge, as in Fig. 23-11b. In the expression for dE, replace r with an expression involving x and y. If P is on the perpendicular bisector of the line of charge, find an expression for the adding component of $d\vec{E}$. That will introduce either $\sin\theta$ or $\cos\theta$. Reduce the resulting two variables x and θ to one, x, by replacing the trigonometric function with an expression (its definition) involving x and y. Integrate over x from end to end of the line of charge. If P is not on a line

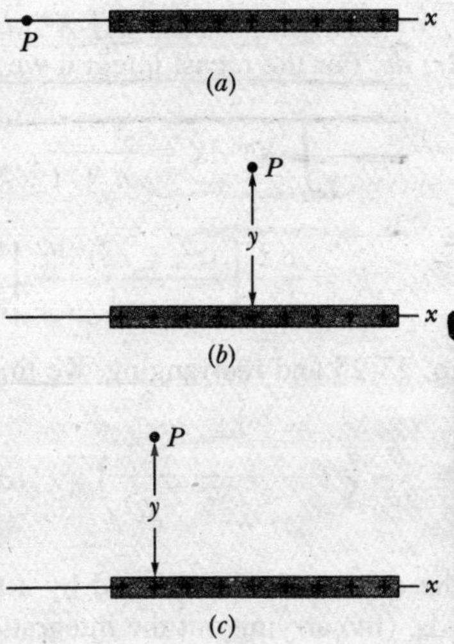

Fig. 23-11 (*a*) Point P is on an extension of the line of charge. (*b*) P is on a line of symmetry of the line of charge, at perpendicular distance y from that line. (*c*) Same as (*b*) except that P is not on a line of symmetry.

of symmetry, as in Fig. 23-11*c*, set up an integral to sum the components dE_x, and integrate over x to find E_x. Also set up an integral to sum the components dE_y, and integrate over x again to find E_y. Use the components E_x and E_y in the usual way to find the magnitude E and the orientation of $\vec{E}$.

Step 6. One arrangement of the integration limits gives a positive result. The reverse arrangement gives the same result with a minus sign; discard the minus sign. If the result is to be stated in terms of the total charge Q of the distribution, replace λ with Q/L, in which L is the length of the distribution. For a ring, L is the ring's circumference.

CHECKPOINT 3: The figure here shows three nonconducting rods, one circular and two straight. Each has a uniform charge of magnitude Q along its top half and another along its bottom half. For each rod, what is the direction of the net electric field at point P?

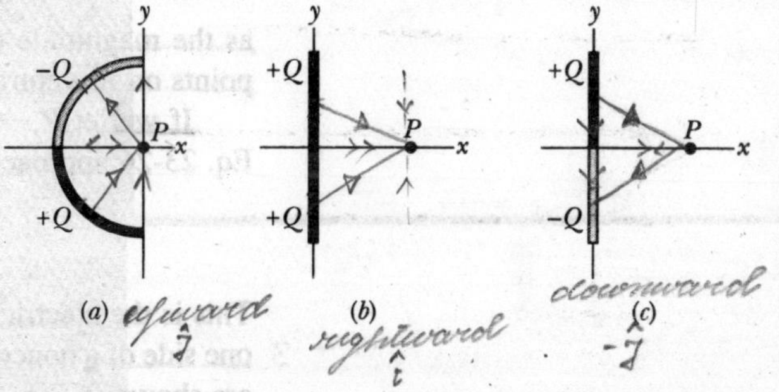

23-7 The Electric Field Due to a Charged Disk

Figure 23-12 shows a circular plastic disk of radius R that has a positive surface charge of uniform density σ on its upper surface (see Table 23-2). What is the electric field at point P, a distance z from the disk along its central axis?

Our plan is to divide the disk into concentric flat rings and then to calculate the electric field at point P by adding up (that is, by integrating) the contributions of all the rings. Figure 23-12 shows one such ring, with radius r and radial width dr. Since σ is the charge per unit area, the charge on the ring is

$$dq = \sigma\, dA = \sigma(2\pi r\, dr), \tag{23-22}$$

where dA is the differential area of the ring.

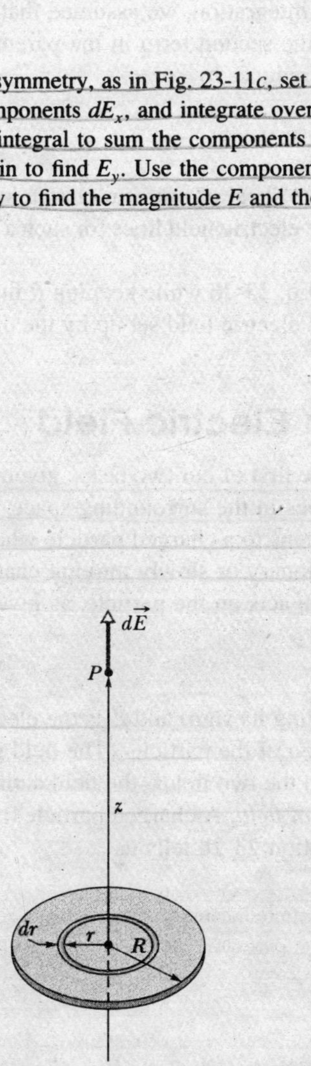

Fig. 23-12 A disk of radius R and uniform positive charge. The ring shown has radius r and radial width dr. It sets up a differential electric field $d\vec{E}$ at point P on its central axis.

We have already solved the problem of the electric field due to a ring of charge. Substituting dq from Eq. 23-22 for q in Eq. 23-16, and replacing R in Eq. 23-16 with r, we obtain an expression for the electric field dE at P due to our flat ring:

$$dE = \frac{z\sigma 2\pi r\, dr}{4\pi\varepsilon_0(z^2 + r^2)^{3/2}},$$

which we may write as

$$dE = \frac{\sigma z}{4\varepsilon_0}\frac{2r\, dr}{(z^2 + r^2)^{3/2}}. \tag{23-23}$$

We can now find E by integrating Eq. 23-23 over the surface of the disk—that is, by integrating with respect to the variable r from $r = 0$ to $r = R$. Note that z remains constant during this process. We get

$$E = \int dE = \frac{\sigma z}{4\varepsilon_0}\int_0^R (z^2 + r^2)^{-3/2}(2r)\, dr. \tag{23-24}$$

To solve this integral, we cast it in the form $\int X^m\,dX$ by setting $X = (z^2 + r^2)$, $m = -\frac{3}{2}$, and $dX = (2r)\,dr$. For the recast integral we have

$$\int X^m\,dX = \frac{X^{m+1}}{m+1},$$

so Eq. 23-24 becomes

$$E = \frac{\sigma z}{4\varepsilon_0}\left[\frac{(z^2+r^2)^{-1/2}}{-\frac{1}{2}}\right]_0^R. \tag{23-25}$$

Taking the limits in Eq. 23-25 and rearranging, we find

$$E = \frac{\sigma}{2\varepsilon_0}\left(1 - \frac{z}{\sqrt{z^2+R^2}}\right) \quad \text{(charged disk)} \tag{23-26}$$

...agnitude of the electric field produced by a flat, circular, charged disk at ... its central axis. (In carrying out the integration, we assumed that $z \ge 0$.) ...e let $R \to \infty$ while keeping z finite, the second term in the parentheses in ...6 approaches zero, and this equation reduces to

$$E = \frac{\sigma}{2\varepsilon_0} \quad \text{(infinite sheet).} \tag{23-27}$$

...e electric field produced by an infinite sheet of uniform charge located on ...f a nonconductor such as plastic. The electric field lines for such a situation ...n in Fig. 23-3.

...also get Eq. 23-27 if we let $z \to 0$ in Eq. 23-26 while keeping R finite. This ...at at points very close to the disk, the electric field set up by the disk is the ...f the disk were infinite in extent.

50. ∴ Infinite sheet approximation is quite good for a capacitor. The deviation from uniform electric field only becomes noticeable near the edges of the capacitor plates. Note however that $E_{conduc} \neq E_{non\,conduc}$ for capacitor plates

23-6 A Point Charge in an Electric Field

In the preceding four sections we worked at the first of our two tasks: given a charge distribution, to find the electric field it produces in the surrounding space. Here we begin the second task: to determine what happens to a charged particle when it is in an electric field that is produced by other stationary or slowly moving charges.

What happens is that an electrostatic force acts on the particle, as given by

$$\vec{F} = q\vec{E}, \tag{23-28}$$

in which q is the charge of the particle (including its sign) and $\vec{E}$ is the electric field that other charges have produced at the location of the particle. (The field is *not* the field set up by the particle itself; to distinguish the two fields, the field acting on the particle in Eq. 23-28 is often called the *external field*. A charged particle (or object) is not affected by its own electric field.) Equation 23-28 tells us

➤ The electrostatic force $\vec{F}$ acting on a charged particle located in an external electric field $\vec{E}$ has the direction of $\vec{E}$ if the charge q of the particle is positive and has the opposite direction if q is negative.

CHECKPOINT 4: (a) In the figure, what is the direction of the electrostatic force on the electron due to the electric field shown? (b) In which direction will the electron accelerate if it is moving parallel to the y axis before it encounters the electric field? (c) If, instead, the electron is initially moving rightward, will its speed increase, decrease, or remain constant?

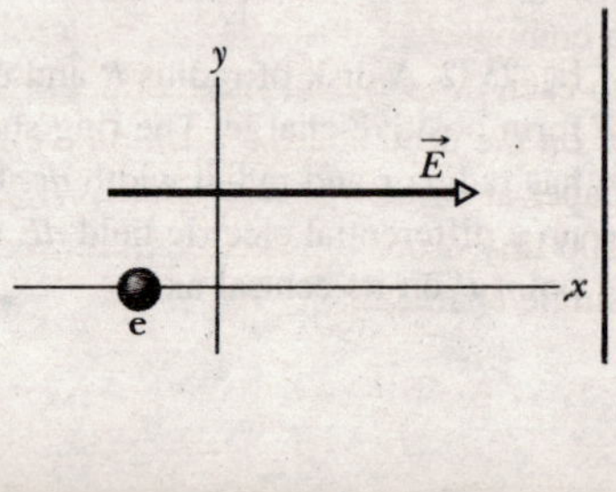

a) F = qE = −eE → F ⇒ (−î)
b) electron would accelerate in −î direction. In y no acc
c) decrease

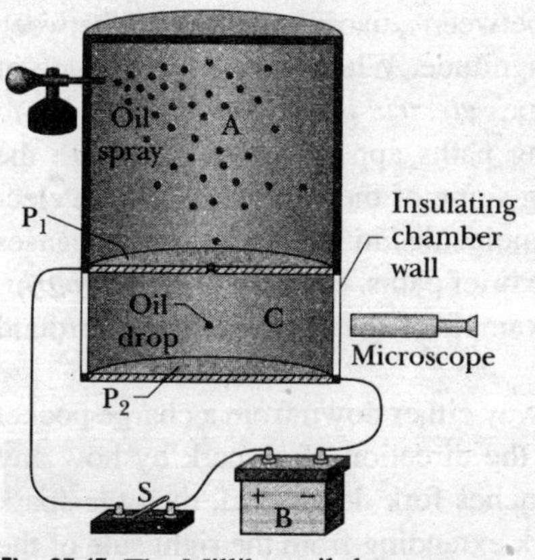

Fig. 23-13 The Millikan oil-drop apparatus for measuring the elementary charge *e*. When a charged oil drop drifted into chamber C through the hole in plate P_1, its motion could be controlled by closing and opening switch S and thereby setting up or eliminating an electric field in chamber C. The microscope was used to view the drop, to permit timing of its motion.

Discovered the value of fundamental charge e. Doesn't show that electron has charge e

Measuring the Elementary Charge

Equation 23-28 played a role in the measurement of the elementary charge *e* by American physicist Robert A. Millikan in 1910–1913. Figure 23-13 is a representation of his apparatus. When tiny oil drops are sprayed into chamber A, some of them become charged, either positively or negatively, in the process. Consider a drop that drifts downward through the small hole in plate P_1 and into chamber C. Let us assume that this drop has a negative charge *q*.

If switch S in Fig. 23-13 is open as shown, battery B has no electrical effect on chamber C. If the switch is closed (the connection between chamber C and the positive terminal of the battery is then complete), the battery causes an excess positive charge on conducting plate P_1 and an excess negative charge on conducting plate P_2. The charged plates set up a downward-directed electric field $\vec{E}$ in chamber C. According to Eq. 23-28, this field exerts an electrostatic force on any charged drop that happens to be in the chamber and affects its motion. In particular, our negatively charged drop will tend to drift upward.

By timing the motion of oil drops with the switch opened and with it closed and thus determining the effect of the charge *q*, Millikan discovered that the values of *q* were always given by

$$q = ne, \quad \text{for } n = 0, \pm 1, \pm 2, \pm 3, \ldots, \qquad (23\text{-}29)$$

in which *e* turned out to be the fundamental constant we call the *elementary charge*, 1.60×10^{-19} C. Millikan's experiment is convincing proof that charge is quantized, and he earned the 1923 Nobel prize in physics in part for this work. Modern measurements of the elementary charge rely on a variety of interlocking experiments, all more precise than the pioneering experiment of Millikan.

Ink-Jet Printing

The need for high-quality, high-speed printing has caused a search for an alternative to impact printing, such as occurs in a standard typewriter. Building up letters by squirting tiny drops of ink at the paper is one such alternative.

Figure 23-14 shows a negatively charged drop moving between two conducting deflecting plates, between which a uniform, downward-directed electric field $\vec{E}$ has been set up. The drop is deflected upward according to Eq. 23-28 and then strikes the paper at a position that is determined by the magnitudes of $\vec{E}$ and the charge *q* of the drop.

In practice, *E* is held constant and the position of the drop is determined by the charge *q* delivered to the drop in the charging unit, through which the drop must pass before entering the deflecting system. The charging unit, in turn, is activated by electronic signals that encode the material to be printed.

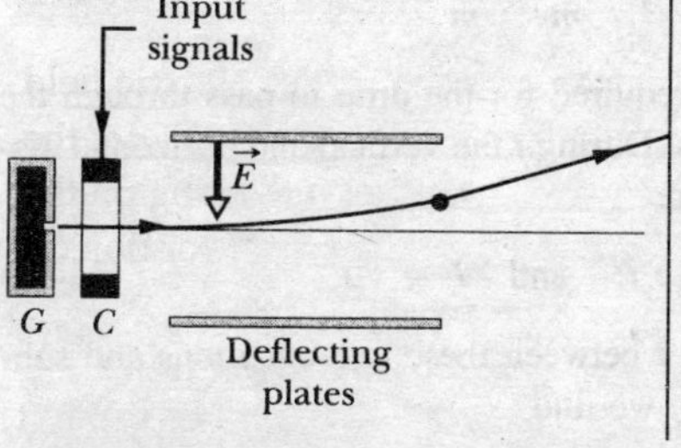

Fig. 23-14 The essential features of an ink-jet printer. Drops are shot out from generator *G* and receive a charge in charging unit *C*. An input signal from a computer controls the charge given to each drop and thus the effect of field $\vec{E}$ on the drop and the position on the paper at which the drop lands. About 100 tiny drops are needed to form a single character.

Volcanic Lightning

When the Sakurajima volcano erupts, as seen in this chapter's opening photograph, it spews ash into the air. That ash results when liquid water within the volcano, suddenly converted to steam by the flow of hot lava, shatters rock, which is then burnt. The liquid-to-steam conversion and the explosion of rock cause positive and negative charges to separate. Then, as the steam and ash are spewed into the air, they form a cloud that contains pockets of positive charge and pockets of negative charge.

As these pockets grow, the electric fields between adjacent pockets and between pockets and the volcano crater increase in magnitude. Whenever the magnitude of about 3×10^6 N/C is reached, the air undergoes *electric breakdown* and begins to conduct current. These momentary conducting paths appear in the air where the electric field has ionized air molecules, freeing some of their electrons. These electrons, propelled by the field, collide with air molecules in their way, which causes those molecules to emit light. We can see these brief paths, commonly called *sparks*, because of the light they emit. (A small-scale example of *sparking* can be seen around the charged metal cap in Fig. 23-15.)

The sparks above the volcano snake their way either down from a charge pocket to the crater wall or vice versa. You can tell the direction of a spark by how any dead-end branches on it are forked. If the branches fork downward, then the spark snaked its way downward. (See the bright spark extending from the right side of the photograph to the crater wall.) If the branches fork upward, then the spark snaked its way upward. (See the lower part of the central bright spark on the crater wall.) Sometimes a downward-snaking spark and an upward-snaking spark meet each other. Can you find an example in the photograph?

Fig. 23-15 The metal cap is so charged that the electric field it produces in the surrounding ... undergo electric breakdown. The visible sparks reveal where ...s are set up in the air, along which the electric field has re- ... molecules and then accelerated them into collisions with the

Is it reasonable to ignore the effect of gravity on the droplet in our calculations?

placement of the ink droplet is particularly important

Sample Problem 23-4

Figure 23-16 shows the deflecting plates of an ink-jet printer, with superimposed coordinate axes. An ink drop with a mass m of 1.3×10^{-10} kg and a negative charge of magnitude $Q = 1.5 \times 10^{-13}$ C enters the region between the plates, initially moving along the x axis with speed $v_x = 18$ m/s. The length L of the plates is 1.6 cm. The plates are charged and thus produce an electric field at all points between them. Assume that field $\vec{E}$ is downward directed, uniform, and has a magnitude of 1.4×10^6 N/C. What is the vertical deflection of the drop at the far edge of the plates? (The gravitational force on the drop is small relative to the electrostatic force acting on the drop and can be neglected.)

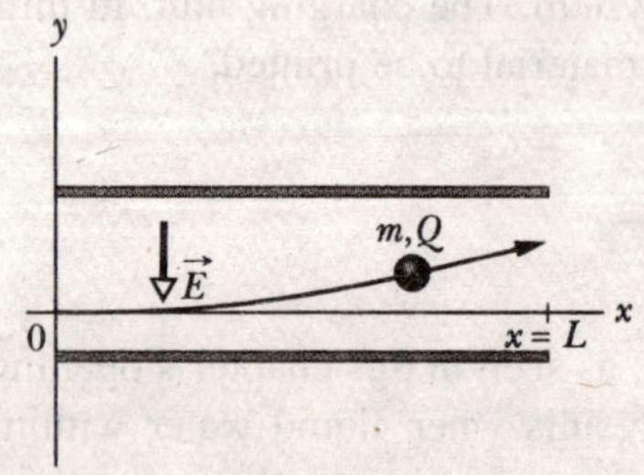

Fig. 23-16 Sample Problem 23-4. An ink drop of mass m and charge magnitude Q is deflected in the electric field of an ink-jet printer.

SOLUTION: The drop is negatively charged and the electric field is directed *downward*. The **Key Idea** here is that, from Eq. 23-28, a constant electrostatic force of magnitude QE acts *upward* on the charged drop. Thus, as the drop travels parallel to the x axis at constant speed v_x, it accelerates upward with some constant acceleration a_y. Applying Newton's second law ($F = ma$) for components along the y axis, we find that

$$a_y = \frac{F}{m} = \frac{QE}{m}. \qquad (23\text{-}30)$$

Let t represent the time required for the drop to pass through the region between the plates. During t the vertical and horizontal displacements of the drop are

$$y = \tfrac{1}{2}a_y t^2 \quad \text{and} \quad L = v_x t, \qquad (23\text{-}31)$$

respectively. Eliminating t between these two equations and substituting Eq. 23-30 for a_y, we find

$$y = \frac{QEL^2}{2mv_x^2}$$

$$= \frac{(1.5 \times 10^{-13}\ \text{C})(1.4 \times 10^6\ \text{N/C})(1.6 \times 10^{-2}\ \text{m})^2}{(2)(1.3 \times 10^{-10}\ \text{kg})(18\ \text{m/s})^2}$$

$$= 6.4 \times 10^{-4}\ \text{m}$$

$$= 0.64\ \text{mm}. \qquad \text{(Answer)}$$

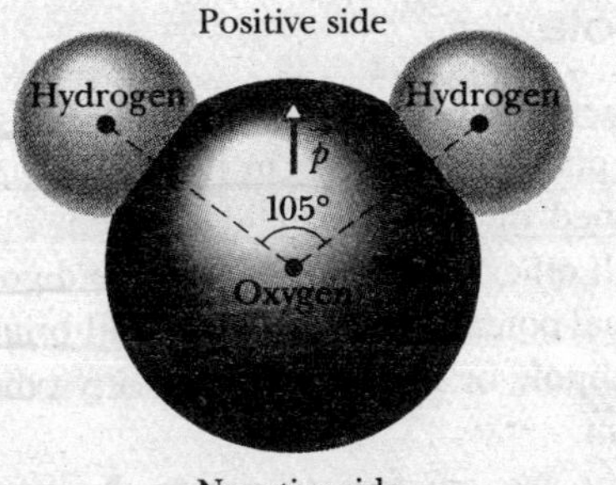

Fig. 23-17 A molecule of H_2O, showing the three nuclei (represented by dots) and the regions in which the electrons can be located. The electric dipole moment $\vec{p}$ points from the (negative) oxygen side to the (positive) hydrogen side of the molecule.

23-9 A Dipole in an Electric Field

We have defined the electric dipole moment $\vec{p}$ of an electric dipole to be a vector that points from the negative to the positive end of the dipole. As you will see, the behavior of a dipole in a uniform external electric field $\vec{E}$ can be described completely in terms of the two vectors $\vec{E}$ and $\vec{p}$, with no need of any details about the dipole's structure.

A molecule of water (H_2O) is an electric dipole; Fig. 23-17 shows why. There the black dots represent the oxygen nucleus (having eight protons) and the two hydrogen nuclei (having one proton each). The colored enclosed areas represent the regions in which electrons can be located around the nuclei.

In a water molecule, the two hydrogen atoms and the oxygen atom do not lie on a straight line but form an angle of about 105°, as shown in Fig. 23-17. As a result, the molecule has a definite "oxygen side" and "hydrogen side." Moreover, the 10 electrons of the molecule tend to remain closer to the oxygen nucleus than to the hydrogen nuclei. This makes the oxygen side of the molecule slightly more negative than the hydrogen side and creates an electric dipole moment $\vec{p}$ that points along the symmetry axis of the molecule as shown. If the water molecule is placed in an external electric field, it behaves as would be expected of the more abstract electric dipole of Fig. 23-8.

To examine this behavior, we now consider such an abstract dipole in a uniform external electric field $\vec{E}$, as shown in Fig. 23-18*a*. We assume that the dipole is a rigid structure that consists of two centers of opposite charge, each of magnitude q, separated by a distance d. The dipole moment $\vec{p}$ makes an angle θ with field $\vec{E}$.

Electrostatic forces act on the charged ends of the dipole. Because the electric field is uniform, those forces act in opposite directions (as shown in Fig. 23-18) and with the same magnitude $F = qE$. Thus, *because the field is uniform*, the net force on the dipole from the field is zero and the center of mass of the dipole does not move. However, the forces on the charged ends do produce a net torque $\vec{\tau}$ on the dipole about its center of mass. The center of mass lies on the line connecting the charged ends, at some distance x from one end and thus a distance $d - x$ from the other end. From Eq. 11-31 ($\tau = rF \sin \phi$), we can write the magnitude of the net torque $\vec{\tau}$ as

$$\tau = Fx \sin \theta + F(d - x) \sin \theta = Fd \sin \theta. \tag{23-32}$$

We can also write the magnitude of $\vec{\tau}$ in terms of the magnitudes of the electric field E and the dipole moment $p = qd$. To do so, we substitute qE for F and p/q for d in Eq. 23-32, finding that the magnitude of $\vec{\tau}$ is

$$\tau = pE \sin \theta. \tag{23-33}$$

We can generalize this equation to vector form as

$$\vec{\tau} = \vec{p} \times \vec{E} \quad \text{(torque on a dipole).} \tag{23-34}$$

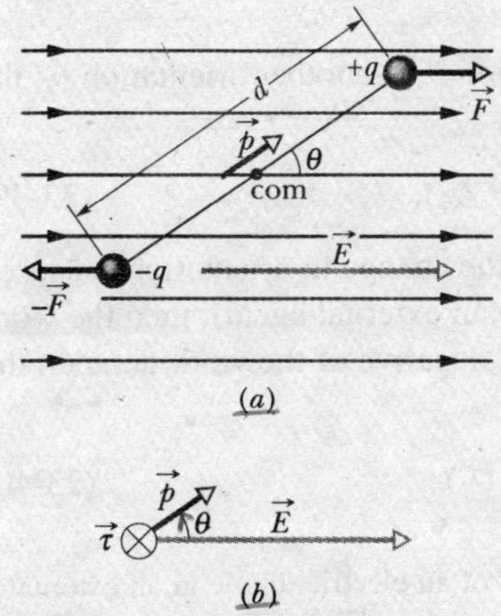

Fig. 23-18 (*a*) An electric dipole in a uniform electric field $\vec{E}$. Two centers of equal but opposite charge are separated by distance d. The line between them represents their rigid connection. (*b*) Field $\vec{E}$ causes a torque $\vec{\tau}$ on the dipole. The direction of $\vec{\tau}$ is into the page, as represented by the symbol ⊗.

Vectors $\vec{p}$ and $\vec{E}$ are shown in Fig. 23-18*b*. The torque acting on a dipole tends to rotate $\vec{p}$ (hence the dipole) into the direction of field $\vec{E}$, thereby reducing θ. In Fig. 23-18, such rotation is clockwise. As we discussed in Chapter 11, we can represent a torque that gives rise to a clockwise rotation by including a minus sign with the magnitude of the torque. With that notation, the torque of Fig. 23-18 is

$$\tau = -pE \sin \theta. \tag{23-35}$$

Potential Energy of an Electric Dipole

Potential energy can be associated with the orientation of an electric dipole in an electric field. The dipole has its least potential energy when it is in its equilibrium orientation, which is when its moment $\vec{p}$ is lined up with the field $\vec{E}$ (then $\vec{\tau} = \vec{p} \times \vec{E} = 0$). It has greater potential energy in all other orientations. Thus the dipole is like a pendulum, which has *its* least gravitational potential energy in *its* equilibrium orientation—at its lowest point. To rotate the dipole or the pendulum to any other orientation requires work by some external agent.

In any situation involving potential energy, we are free to define the zero-potential-energy configuration in a perfectly arbitrary way, because only differences in potential energy have physical meaning. It turns out that the expression for the potential energy of an electric dipole in an external electric field is simplest if we choose the potential energy to be zero when the angle θ in Fig. 23-18 is 90°. We then can find the potential energy U of the dipole at any other value of θ with Eq. 8-1 ($\Delta U = -W$) by calculating the work W done by the field on the dipole when the dipole is rotated to that value of θ from 90°. With the aid of Eq. 11-45 ($W = \int \tau\, d\theta$) and Eq. 23-35, we find that the potential energy U at any angle θ is

$$U = -W = -\int_{90^\circ}^{\theta} \tau\, d\theta = \int_{90^\circ}^{\theta} pE \sin\theta\, d\theta. \tag{23-36}$$

Evaluating the integral leads to

$$U = -pE\cos\theta. \tag{23-37}$$

We can generalize this equation to vector form as

$$U = -\vec{p} \cdot \vec{E} \quad \text{(potential energy of a dipole).} \tag{23-38}$$

Equations 23-37 and 23-38 show us that the potential energy of the dipole is least ($U = -pE$) when $\theta = 0$, which is when $\vec{p}$ and $\vec{E}$ are in the same direction; the potential energy is greatest ($U = pE$) when $\theta = 180°$, which is when $\vec{p}$ and $\vec{E}$ are in opposite directions.

When a dipole rotates from an initial orientation θ_i to another orientation θ_f, the work W done on the dipole by the electric field is

$$W = -\Delta U = -(U_f - U_i), \tag{23-39}$$

where U_f and U_i are calculated with Eq. 23-38. If the change in orientation is caused by an applied torque (commonly said to be due to an external agent), then the work W_a done on the dipole by the applied torque is the negative of the work done on the dipole by the field; that is,

$$W_a = -W = (U_f - U_i). \tag{23-40}$$

CHECKPOINT 5: The figure shows four orientations of an electric dipole in an external electric field. Rank the orientations according to (a) the magnitude of the torque on the dipole and (b) the potential energy of the dipole, greatest first.

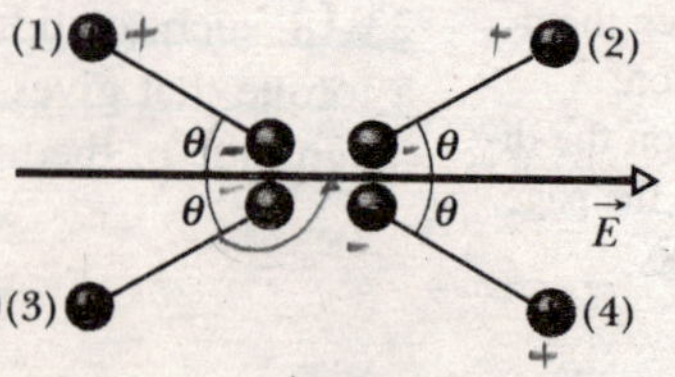

Sample Problem 23-5

A neutral water molecule (H_2O) in its vapor state has an electric dipole moment of magnitude 6.2×10^{-30} C · m.

(a) How far apart are the molecule's centers of positive and negative charge?

SOLUTION: The **Key Idea** here is that a molecule's dipole moment depends on the magnitude q of the molecule's positive or negative charge and the charge separation d. There are 10 electrons and 10 protons in a neutral water molecule, so the magnitude of its dipole moment is

$$p = qd = (10e)(d),$$

in which d is the separation we are seeking and e is the elementary charge. Thus,

$$d = \frac{p}{10e} = \frac{6.2 \times 10^{-30}\ \text{C}\cdot\text{m}}{(10)(1.60 \times 10^{-19}\ \text{C})}$$
$$= 3.9 \times 10^{-12}\ \text{m} = 3.9\ \text{pm}. \qquad \text{(Answer)}$$

This distance is not only small, but it is actually smaller than the radius of a hydrogen atom. 0·5 A° = 0·5×10⁻¹⁰ = 50pm

(b) If the molecule is placed in an electric field of 1.5×10^4 N/C, what maximum torque can the field exert on it? (Such a field can easily be set up in the laboratory.)

SOLUTION: The **Key Idea** here is that the torque on a dipole is maximum when the angle θ between $\vec{p}$ and $\vec{E}$ is 90°. Substituting this value in Eq. 23-33 yields

$$\tau = pE \sin\theta$$
$$= (6.2 \times 10^{-30}\ \text{C}\cdot\text{m})(1.5 \times 10^4\ \text{N/C})(\sin 90°)$$
$$= 9.3 \times 10^{-26}\ \text{N}\cdot\text{m}. \qquad \text{(Answer)}$$

(c) How much work must an *external agent* do to turn this molecule end for end in this field, starting from its fully aligned position, for which $\theta = 0$?

SOLUTION: The **Key Idea** here is that the work done by an external agent (by means of a torque applied to the molecule) is equal to the change in the molecule's potential energy due to the change in orientation. From Eq. 23-40, we find

$$W_a = U(180°) - U(0)$$
$$= (-pE \cos 180°) - (-pE \cos 0)$$
$$= 2pE = (2)(6.2 \times 10^{-30}\ \text{C}\cdot\text{m})(1.5 \times 10^4\ \text{N/C})$$
$$= 1.9 \times 10^{-25}\ \text{J}. \qquad \text{(Answer)}$$

REVIEW & SUMMARY

Electric Field One way to explain the electrostatic force between two charges is to assume that each charge sets up an electric field in the space around it. The electrostatic force acting on any one charge is then due to the electric field set up at its location by the other charge.

Definition of Electric Field The *electric field* $\vec{E}$ at any point is defined in terms of the electrostatic force $\vec{F}$ that would be exerted on a positive test charge q_0 placed there:

$$\vec{E} = \frac{\vec{F}}{q_0}. \qquad \text{(23-1)}$$

Electric Field Lines *Electric field lines* provide a means for visualizing the direction and magnitude of electric fields. The electric field vector at any point is tangent to a field line through that point. The density of field lines in any region is proportional to the magnitude of the electric field in that region. Field lines originate on positive charges and terminate on negative charges.

Field Due to a Point Charge The magnitude of the electric field $\vec{E}$ set up by a point charge q at a distance r from the charge is

$$E = \frac{1}{4\pi\varepsilon_0}\frac{|q|}{r^2}. \qquad \text{(23-3)}$$

The direction of $\vec{E}$ is away from the point charge if the charge is positive and toward the point charge if the charge is negative.

Field Due to an Electric Dipole An *electric dipole* consists of two particles with charges of equal magnitude q but opposite sign, separated by a small distance d. Their **dipole moment** $\vec{p}$ has magnitude qd and points from the negative charge to the positive charge. The magnitude of the electric field set up by the dipole at a distant point on the dipole axis (which runs through both charges) is

$$E = \frac{1}{2\pi\varepsilon_0}\frac{p}{z^3}, \qquad \text{(23-9)}$$

where z is the distance between the point and the dipole center.

Field Due to a Continuous Charge Distribution The electric field due to a *continuous charge distribution* is found by treating charge elements as point charges and then summing, via integration, the electric field vectors produced by all the charge elements.

Force on a Point Charge in an Electric Field When a point charge q is placed in an electric field $\vec{E}$ set up by other charges, the electrostatic force $\vec{F}$ that acts on the point charge is

$$\vec{F} = q\vec{E}. \qquad \text{(23-28)}$$

Force $\vec{F}$ has the same direction as $\vec{E}$ if q is positive and the opposite direction if q is negative.

Dipole in an Electric Field When an electric dipole of dipole moment $\vec{p}$ is placed in an electric field $\vec{E}$, the field exerts a torque $\vec{\tau}$ on the dipole:

$$\vec{\tau} = \vec{p} \times \vec{E}. \qquad (23\text{-}34)$$

The dipole has a potential energy U associated with its orientation in the field:

$$U = -\vec{p} \cdot \vec{E}. \qquad (23\text{-}38)$$

This potential energy is defined to be zero when $\vec{p}$ is perpendicular to $\vec{E}$; it is least ($U = -pE$) when $\vec{p}$ is aligned with $\vec{E}$, and most ($U = pE$) when $\vec{p}$ is directed opposite $\vec{E}$.

QUESTIONS

1. Figure 23-19 shows three electric field lines. What is the direction of the electrostatic force on a positive test charge placed at (a) point A and (b) point B? (c) At which point, A or B, will the acceleration of the test charge be greater if the charge is released?

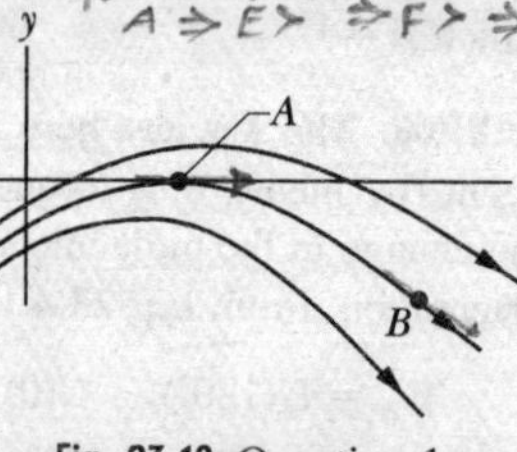

Fig. 23-19 Question 1.

2. Figure 23-20*a* shows two charged particles on an axis. (a) Where on the axis (other than at an infinite distance) is there a point at which their net electric field is zero: between the charges, to their left, or to their right? (b) Is there a point of zero electric field off the axis (other than at an infinite distance)?

Fig. 23-20 Questions 2 and 3.

3. Figure 23-20*b* shows two protons and an electron that are evenly spaced on an axis. Where on the axis (other than at an infinite distance) is there a point at which their net electric field is zero: to the left of the particles, to their right, between the two protons, or between the electron and the nearer proton?

4. Figure 23-21 shows two square arrays of charged particles. The squares, which are centered on point P, are misaligned. The particles are separated by either d or $d/2$ along the perimeters of the squares. What are the magnitude and direction of the net electric field at P?

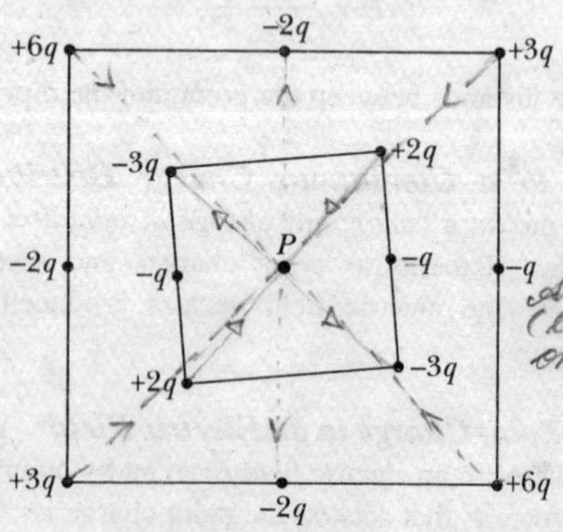

Fig. 23-21 Question 4.

5. In Fig. 23-22, two particles of charge $-q$ are arranged symmetrically about the y axis; each produces an electric field at point P on that axis. (a) Are the magnitudes of the fields at P equal? (b) Is each electric field directed toward or away from the charge producing it? (c) Is the magnitude of the net electric field at P equal to the sum of the magnitudes E of the two field vectors (is it equal to $2E$)? (d) Do the x components of those two field vectors add or cancel? (e) Do their y components add or cancel? (f) Is the direction of the net field at P that of the canceling components or the adding components? (g) What is the direction of the net field?

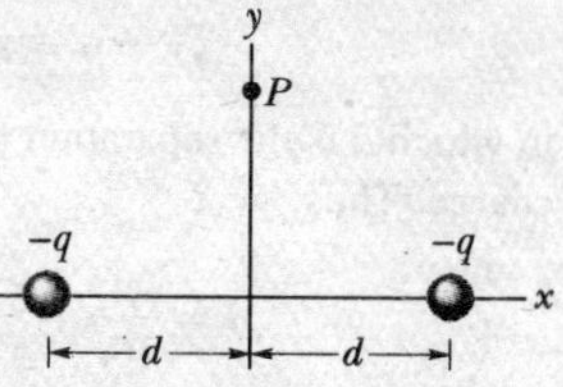

Fig. 23-22 Question 5.

6. Three circular nonconducting rods of the same radius of curvature have uniform charges. Rod A has charge $+2Q$ and subtends an arc of 30°, rod B has charge $+6Q$ and subtends 90°, and rod C has charge $+4Q$ and subtends 60°. Rank the rods according to their linear charge density, greatest first.

7. In Fig. 23-23*a*, a circular plastic rod with uniform charge $+Q$ produces an electric field of magnitude E at the center of curvature (at the origin). In Figs. 23-23*b*, *c*, and *d*, more circular rods with

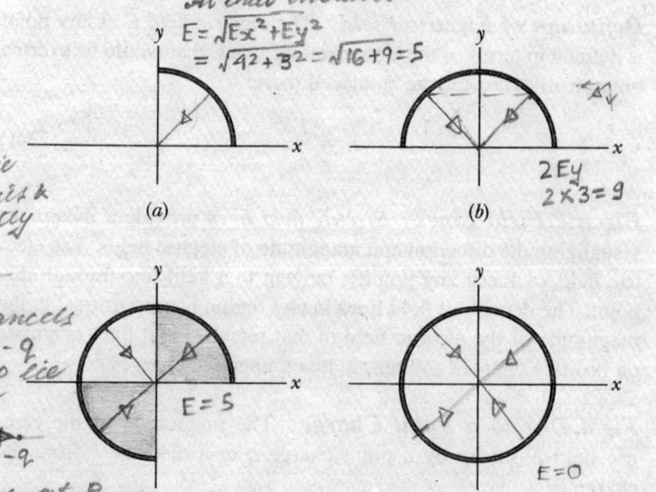

Fig. 23-23 Question 7.

identical uniform charges $+Q$ are added until the circle is complete. A fifth arrangement (which would be labeled *e*) is like that in *d* except that the rod in the fourth quadrant has charge $-Q$. Rank the five arrangements according to the magnitude of the electric field at the center of curvature, greatest first.

8. In Fig. 23-24, an electron e travels through a small hole in plate *A* and then toward plate *B*. A uniform electric field in the region between the plates then slows the electron without deflecting it. (a) What is the direction of the field? (b) Four other particles similarly travel through small holes in either plate *A* or plate *B* and then into the region between the plates. Three have charges $+q_1$, $+q_2$, and $-q_3$. The fourth (labeled n) is a neutron, which is electrically neutral. Does the speed of each of those four other particles increase, decrease, or remain the same in the region between the plates?

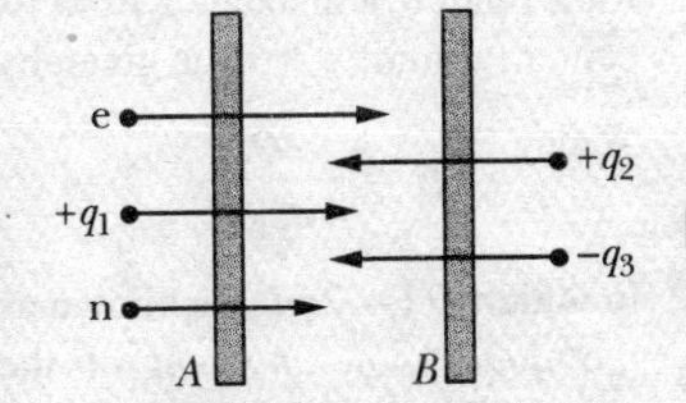

Fig. 23-24 Question 8.

9. Figure 23-25 shows the path of negatively charged particle 1 through a rectangular region of uniform electric field; the particle is deflected toward the top of the page. (a) Is the field directed leftward, rightward, toward the top of the page, or toward the bottom? (b) Three other charged particles are shown approaching the region of electric field. Which are deflected toward the top of the page and which toward the bottom?

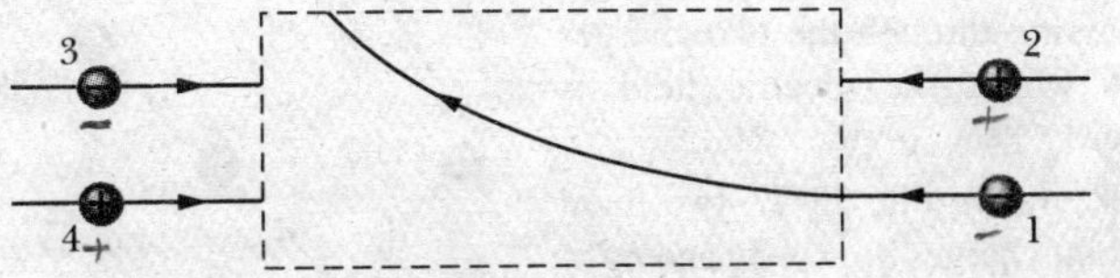

Fig. 23-25 Question 9.

10. (a) In Checkpoint 5, if the dipole rotates from orientation 1 to orientation 2, is the work done on the dipole by the field positive, negative, or zero? (b) If, instead, the dipole rotates from orientation 1 to orientation 4, is the work done by the field more than, less than, or the same as in (a)?

11. The potential energies associated with four orientations of an electric dipole in an electric field are (1) $-5U_0$, (2) $-7U_0$, (3) $3U_0$, and (4) $5U_0$, where U_0 is positive. Rank the orientations according to (a) the angle between the electric dipole moment $\vec{p}$ and the electric field $\vec{E}$, and (b) the magnitude of the torque on the electric dipole, greatest first.

12. If you walk across some types of carpet on a dry day and then reach for a metal doorknob or (for more fun) the back of someone's neck, you might produce a spark. Why does the spark occur? (You can increase the brightness and noise of the spark if you reach with a pointed finger or, even better, a metal key with the pointed end forward.)

EXERCISES & PROBLEMS

ssm Solution is in the Student Solutions Manual.
www Solution is available on the World Wide Web at: http://www.wiley.com/college/hrw
ilw Solution is available on the Interactive LearningWare.

SEC. 23-3 Electric Field Lines

1E. In Fig. 23-26 the electric field lines on the left have twice the separation of those on the right. (a) If the magnitude of the field at *A* is 40 N/C, what force acts on a proton at *A*? (b) What is the magnitude of the field at *B*?

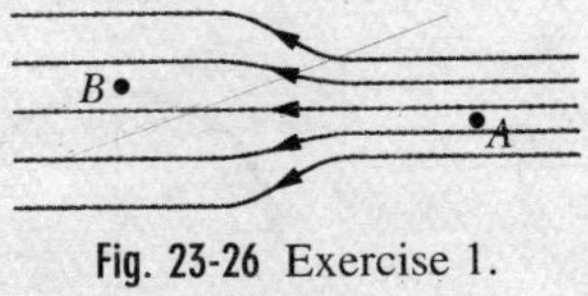

Fig. 23-26 Exercise 1.

2E. Sketch qualitatively the electric field lines both between and outside two concentric conducting spherical shells when a uniform positive charge q_1 is on the inner shell and a uniform negative charge $-q_2$ is on the outer. Consider the cases $q_1 > q_2$, $q_1 = q_2$, and $q_1 < q_2$.

3E. Sketch qualitatively the electric field lines for a thin, circular, uniformly charged disk of radius *R*. (*Hint:* Consider as limiting cases points very close to the disk, where the electric field is directed perpendicular to the surface, and points very far from it, where the electric field is like that of a point charge.) ssm

SEC. 23-4 The Electric Field Due to a Point Charge

4E. What is the magnitude of a point charge that would create an electric field of 1.00 N/C at points 1.00 m away?

5E. What is the magnitude of a point charge whose electric field 50 cm away has the magnitude 2.0 N/C? ssm

6E. Two particles with equal charge magnitudes 2.0×10^{-7} C but opposite signs are held 15 cm apart. What are the magnitude and direction of $\vec{E}$ at the point midway between the charges?

7E. An atom of plutonium-239 has a nuclear radius of 6.64 fm and the atomic number $Z = 94$. Assuming that the positive charge is distributed uniformly within the nucleus, what are the magnitude and direction of the electric field at the surface of the nucleus due to the positive charge? ssm

8P. In Fig. 23-27, two fixed point charges $q_1 = +1.0 \times 10^{-6}$ C and $q_2 = +3.0 \times 10^{-6}$ C are separated by a distance $d = 10$ cm. Plot their net electric field $E(x)$ as a function of x for both positive and negative values of x, taking E to be positive when the vector $\vec{E}$ points to the right and negative when $\vec{E}$ points to the left.

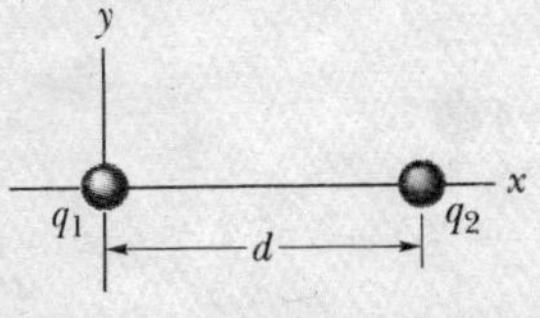

Fig. 23-27 Problems 8 and 10.

9P. Two point charges $q_1 = 2.1 \times 10^{-8}$ C and $q_2 = -4.0q_1$ are fixed in place 50 cm apart. Find the point along the straight line passing through the two charges at which the electric field is zero. ssm www

10P. (a) In Fig. 23-27, two fixed point charges $q_1 = -5q$ and $q_2 = +2q$ are separated by distance d. Locate the point (or points) at which the net electric field due to the two charges is zero. (b) Sketch the net electric field lines qualitatively.

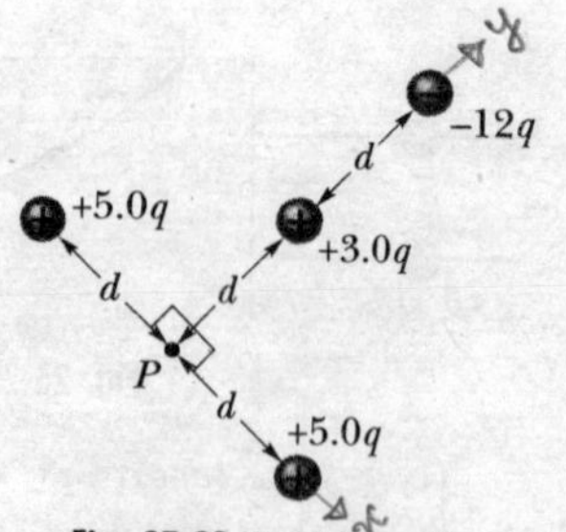

Fig. 23-28 Problem 11.

11P. In Fig. 23-28, what is the magnitude of the electric field at point P due to the four point charges shown?

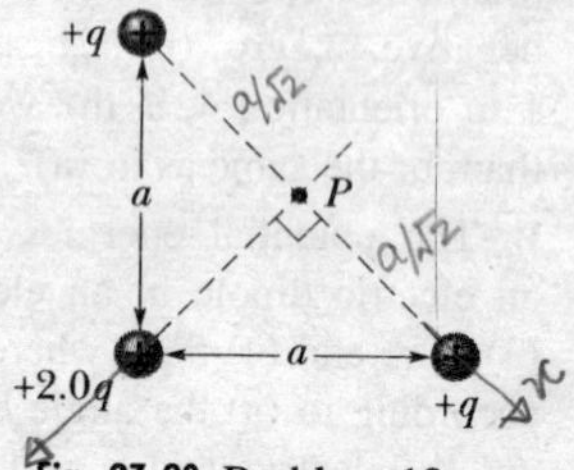

Fig. 23-29 Problem 12.

12P. Calculate the direction and magnitude of the electric field at point P in Fig. 23-29, due to the three point charges.

13P. What are the magnitude and direction of the electric field at the center of the square of Fig. 23-30 if $q = 1.0 \times 10^{-8}$ C and $a = 5.0$ cm? ssm ilw

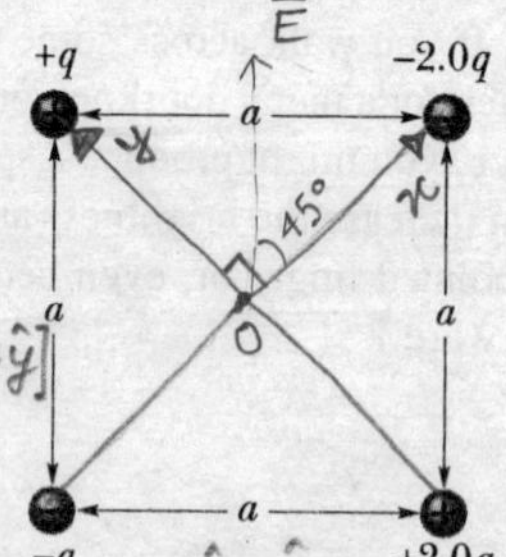

Fig. 23-30 Problem 13.

SEC. 23-5 The Electric Field Due to an Electric Dipole

14E. In Fig. 23-8, let both charges be positive. Assuming $z \gg d$, show that E at point P in that figure is then given by

$$E = \frac{1}{4\pi\varepsilon_0}\frac{2q}{z^2}.$$

15E. Calculate the electric dipole moment of an electron and a proton 4.30 nm apart. ssm

16P. Find the magnitude and direction of the electric field at point P due to the electric dipole in Fig. 23-31. P is located at a distance $r \gg d$ along the perpendicular bisector of the line joining the charges. Express your answer in terms of the magnitude and direction of the electric dipole moment $\vec{p}$.

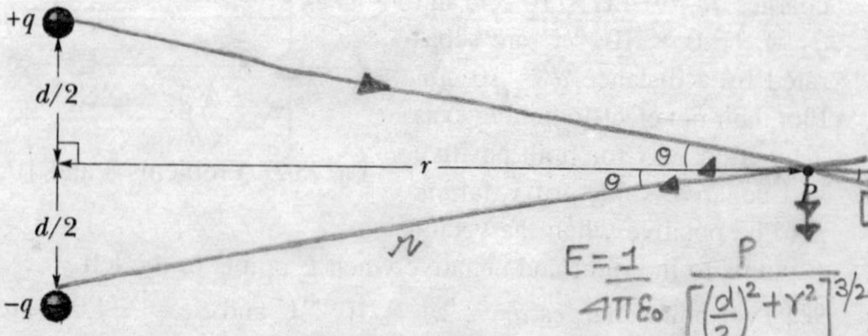

Fig. 23-31 Problem 16.

17P*. *Electric quadrupole.* Figure 23-32 shows an electric quadrupole. It consists of two dipoles with dipole moments that are equal in magnitude but opposite in direction. Show that the value of E on the axis of the quadrupole for a point P a distance z from its center (assume $z \gg d$) is given by

$$E = \frac{3Q}{4\pi\varepsilon_0 z^4},$$

in which $Q\ (= 2qd^2)$ is known as the *quadrupole moment* of the charge distribution. ssm

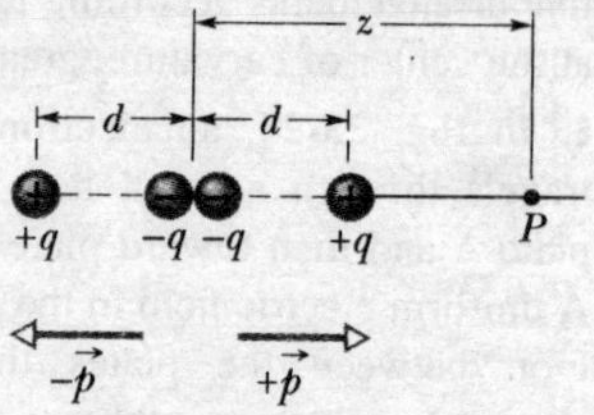

Fig. 23-32 Problem 17.

SEC. 23-6 The Electric Field Due to a Line of Charge

18E. Figure 23-33 shows two parallel nonconducting rings arranged with their central axes along a common line. Ring 1 has uniform charge q_1 and radius R; ring 2 has uniform charge q_2 and the same radius R. The rings are separated by a distance $3R$. The net electric field at point P on the common line, at distance R from ring 1, is zero. What is the ratio q_1/q_2?

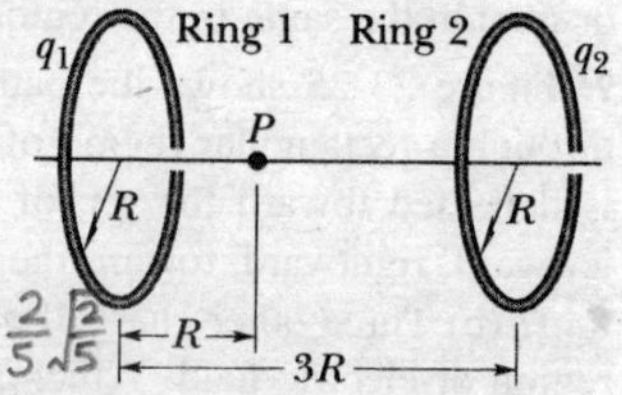

Fig. 23-33 Exercise 18.

19P. An electron is constrained to the central axis of the ring of charge of radius R discussed in Section 23-6. Show that the electrostatic force on the electron can cause it to oscillate through the center of the ring with an angular frequency

$$\omega = \sqrt{\frac{eq}{4\pi\varepsilon_0 m R^3}},$$

where q is the ring's charge and m is the electron's mass. ssm

20P. In Fig. 23-34*a*, two curved plastic rods, one of charge $+q$ and the other of charge $-q$, form a circle of radius R in an xy plane. The x axis passes through their connecting points, and the charge is distributed uniformly on both rods. What are the magnitude and direction of the electric field $\vec{E}$ produced at P, the center of the circle?

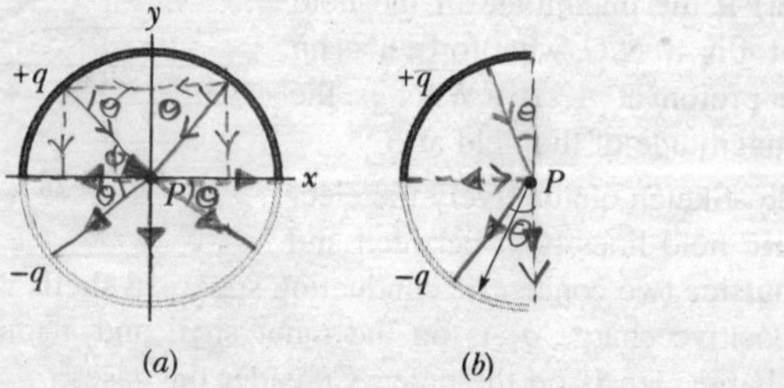

Fig. 23-34 Problems 20 and 21.

21P. A thin glass rod is bent into a semicircle of radius r. A charge $+q$ is uniformly distributed along the upper half, and a charge $-q$ is uniformly distributed along the lower half, as shown in Fig. 23-34*b*. Find the magnitude and direction of the electric field $\vec{E}$ at P, the center of the semicircle. ilw

22P. At what distance along the central axis of a ring of radius R and uniform charge is the magnitude of the electric field due to the ring's charge maximum?

23P. In Fig. 23-35, a nonconducting rod of length L has charge $-q$ uniformly distributed along its length. (a) What is the linear charge density of the rod? (b) What is the electric field at point P, a distance a from the end of the rod? (c) If P were very far from the rod compared to L, the rod would look like a point charge. Show that your answer to (b) reduces to the electric field of a point charge for $a \gg L$. ssm ilw www

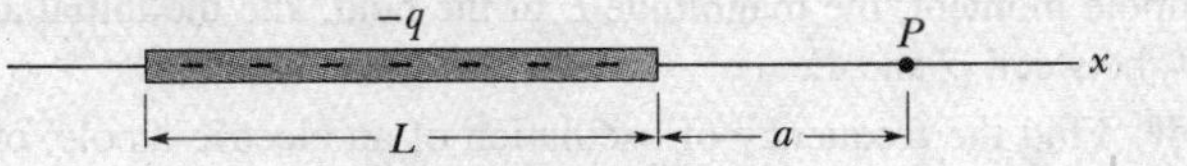

Fig. 23-35 Problem 23.

24P. A thin nonconducting rod of finite length L has a charge q spread uniformly along it. Show that

$$E = \frac{q}{2\pi\varepsilon_0 y}\frac{1}{(L^2 + 4y^2)^{1/2}}$$

gives the magnitude E of the electric field at point P on the perpendicular bisector of the rod (Fig. 23-36).

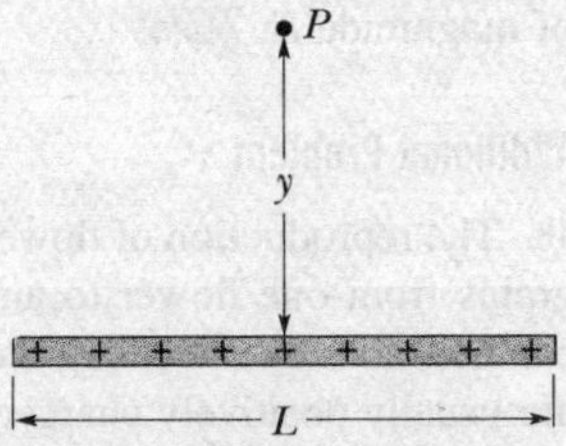

Fig. 23-36 Problem 24.

25P*. In Fig. 23-37, a "semi-infinite" nonconducting rod (that is, infinite in one direction only) has uniform linear charge density λ. Show that the electric field at point P makes an angle of 45° with the rod and that this result is independent of the distance R. (*Hint:* Separately find the parallel and perpendicular (to the rod) components of the electric field at P, and then compare those components.) ssm

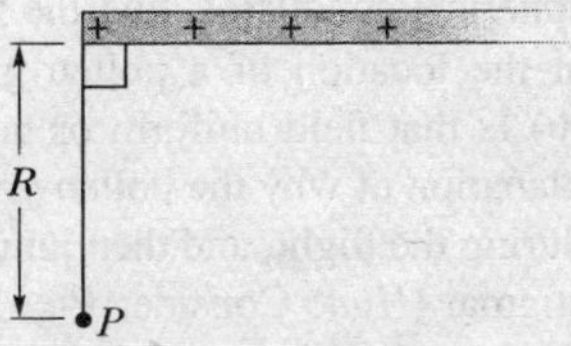

Fig. 23-37 Problem 25.

SEC. 23-7 The Electric Field Due to a Charged Disk

26E. A disk of radius 2.5 cm has a surface charge density of 5.3 μC/m^2 on its upper face. What is the magnitude of the electric field produced by the disk at a point on its central axis at distance z = 12 cm from the disk?

27P. At what distance along the central axis of a uniformly charged plastic disk of radius R is the magnitude of the electric field equal to one-half the magnitude of the field at the center of the surface of the disk? ssm

SEC. 23-8 A Point Charge in an Electric Field

28E. An electron is accelerated eastward at 1.80×10^9 m/s^2 by an electric field. Determine the magnitude and direction of the electric field.

29E. An electron is released from rest in a uniform electric field of magnitude 2.00×10^4 N/C. Calculate the acceleration of the electron. (Ignore gravitation.) ssm

30E. An alpha particle (the nucleus of a helium atom) has a mass of 6.64×10^{-27} kg and a charge of $+2e$. What are the magnitude and direction of the electric field that will balance the gravitational force on it?

31E. Calculate the magnitude of the force, due to an electric dipole of dipole moment 3.6×10^{-29} C · m, on an electron 25 nm from the center of the dipole, along the dipole axis. Assume that this distance is large relative to the dipole's charge separation. ilw

32E. Humid air breaks down (its molecules become ionized) in an electric field of 3.0×10^6 N/C. In that field, what is the magnitude of the electrostatic force on (a) an electron and (b) an ion with a single electron missing?

33E. A charged cloud system produces an electric field in the air near Earth's surface. A particle of charge -2.0×10^{-9} C is acted on by a downward electrostatic force of 3.0×10^{-6} N when placed in this field. (a) What is the magnitude of the electric field? (b) What are the magnitude and direction of the electrostatic force exerted on a proton placed in this field? (c) What is the gravitational force on the proton? (d) What is the ratio of the magnitude of the electrostatic force to the magnitude of the gravitational force in this case? ssm

34E. An electric field $\vec{E}$ with an average magnitude of about 150 N/C points downward in the atmosphere near Earth's surface. We wish to "float" a sulfur sphere weighing 4.4 N in this field by charging the sphere. (a) What charge (both sign and magnitude) must be used? (b) Why is the experiment impractical?

35E Beams of high-speed protons can be produced in "guns" using electric fields to accelerate the protons. (a) What acceleration would a proton experience if the gun's electric field were 2.00×10^4 N/C? (b) What speed would the proton attain if the field accelerated the proton through a distance of 1.00 cm? ssm

36E. An electron with a speed of 5.00×10^8 cm/s enters an electric field of magnitude 1.00×10^3 N/C, traveling along the field lines in the direction that retards its motion. (a) How far will the electron travel in the field before stopping momentarily and (b) how much time will have elapsed? (c) If the region with the electric field is only 8.00 mm long (too short for the electron to stop within it), what fraction of the electron's initial kinetic energy will be lost in that region?

37E. In Millikan's experiment, an oil drop of radius 1.64 μm and density 0.851 g/cm^3 is suspended in chamber C (Fig. 23-13) when a downward-directed electric field of 1.92×10^5 N/C is applied. Find the charge on the drop, in terms of e. ssm

38P In one of his experiments, Millikan observed that the following measured charges, among others, appeared at different times on a single drop:

6.563×10^{-19} C	13.13×10^{-19} C	19.71×10^{-19} C
8.204×10^{-19} C	16.48×10^{-19} C	22.89×10^{-19} C
11.50×10^{-19} C	18.08×10^{-19} C	26.13×10^{-19} C

What value for the elementary charge e can be deduced from these data?

39P A uniform electric field exists in a region between two oppositely charged plates. An electron is released from rest at the surface of the negatively charged plate and strikes the surface of the opposite plate, 2.0 cm away, in a time 1.5×10^{-8} s. (a) What is the speed of the electron as it strikes the second plate? (b) What is the magnitude of the electric field $\vec{E}$? **ilw**

40P. At some instant the velocity components of an electron moving between two charged parallel plates are $v_x = 1.5 \times 10^5$ m/s and $v_y = 3.0 \times 10^3$ m/s. Suppose that the electric field between the plates is given by $\vec{E} = (120 \text{ N/C})\hat{j}$. (a) What is the acceleration of the electron? (b) What will be the velocity of the electron after its x coordinate has changed by 2.0 cm?

41P. Two large parallel copper plates are 5.0 cm apart and have a uniform electric field between them as depicted in Fig. 23-38. An electron is released from the negative plate at the same time that a proton is released from the positive plate. Neglect the force of the particles on each other and find their distance from the positive plate when they pass each other. (Does it surprise you that you need not know the electric field to solve this problem?) **ssm** **www**

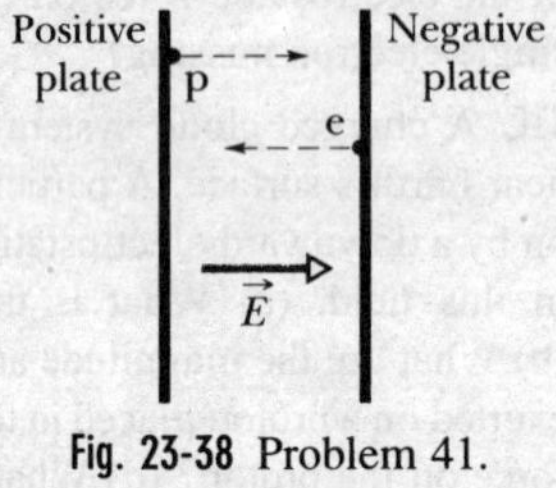

Fig. 23-38 Problem 41.

42P. A 10.0 g block with a charge of $+8.00 \times 10^{-5}$ C is placed in electric field $\vec{E} = (3.00 \times 10^3)\hat{i} - 600\hat{j}$, where $\vec{E}$ is in newtons per coulomb. (a) What are the magnitude and direction of the force on the block? (b) If the block is released from rest at the origin at $t = 0$, what will be its coordinates at $t = 3.00$ s?

43P. In Fig. 23-39, a uniform, upward-directed electric field $\vec{E}$ of magnitude 2.00×10^3 N/C has been set up between two horizontal plates by charging the lower plate positively and the upper plate negatively. The plates have length $L = 10.0$ cm and separation $d = 2.00$ cm. An electron is then shot between the plates from the left edge of the lower plate. The initial velocity $\vec{v}_0$ of the electron makes an angle $\theta = 45.0°$ with the lower plate and has a magnitude of 6.00×10^6 m/s. (a) Will the electron strike one of the plates? (b) If so, which plate and how far horizontally from the left edge will the electron strike? **ssm**

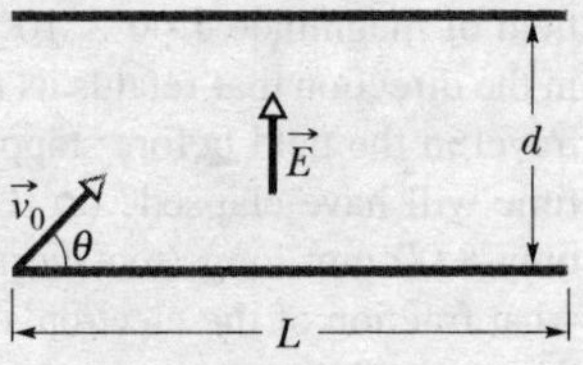

Fig. 23-39 Problem 43.

SEC. 23-9 A Dipole in an Electric Field

44E. An electric dipole, consisting of charges of magnitude 1.50 nC separated by 6.20 μm, is in an electric field of strength 1100 N/C. (a) What is the magnitude of the electric dipole moment? (b) What is the difference between the potential energies corresponding to dipole orientations parallel to and antiparallel to the field?

45E. An electric dipole consists of charges $+2e$ and $-2e$ separated by 0.78 nm. It is in an electric field of strength 3.4×10^6 N/C. Calculate the magnitude of the torque on the dipole when the dipole moment is (a) parallel to, (b) perpendicular to, and (c) antiparallel to the electric field.

46P. Find the work required to turn an electric dipole end for end in a uniform electric field $\vec{E}$, in terms of the magnitude p of the dipole moment, the magnitude E of the field, and the initial angle θ_0 between $\vec{p}$ and $\vec{E}$.

47P Find the frequency of oscillation of an electric dipole, of dipole moment $\vec{p}$ and rotational inertia I, for small amplitudes of oscillation about its equilibrium position in a uniform electric field of magnitude E. **ssm**

Additional Problem

48. The reproduction of flowers depends on insects carrying pollen grains from one flower to another. One way in which honeybees can do this is by collecting the grains electrically, because the bees are usually positively charged. When a bee hovers near a flower's anther (Fig. 23-40), which is electrically insulated, the pollen grains (which are moderately conducting) jump to the bee, where they cling during the flight to the next flower. As the bee nears that flower's stigma, which is electrically connected to ground through the flower's interior, the pollen grains jump from the bee to the stigma, fertilizing the flower.

(a) Assuming that a bee with a typical charge of 45 pC is a spherical conductor, find the magnitude of the bee's electric field at the location of a pollen grain 2.0 cm from the bee's center. (b) Is that field uniform or nonuniform? (c) Give a plausible explanation of why the pollen grains jump to the bee, cling to the bee during the flight, and then jump away from the bee to the grounded stigma. (*Hint:* Consider Fig. 22-5.) When a pollen grain reaches the bee, does it make electrical contact with it, so that the charge on the grain changes?

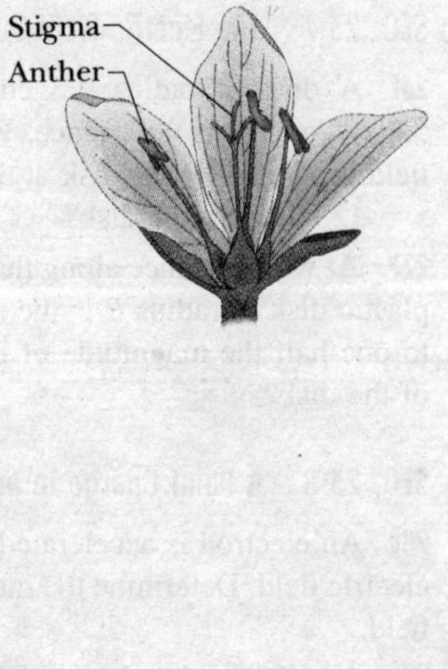

Fig. 23-40 Problem 48.

24 Gauss' Law

Lightning strikes Tucson in a brilliant display, each strike delivering about 10^{20} electrons from the cloud base to the ground.

How wide is a lightning strike? Since a strike can be seen from kilometers away, is it as wide as, say, a car?

The answer is in this chapter.

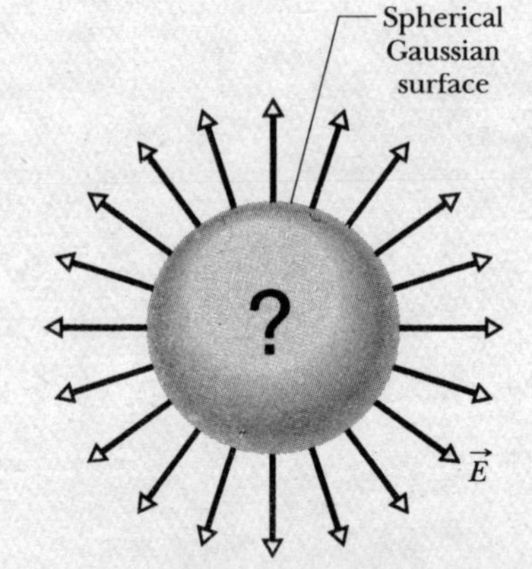

Fig. 24-1 A spherical Gaussian surface. If the electric field vectors are of uniform magnitude and point radially outward at all surface points, you can conclude that a net positive distribution of charge must lie within the surface and have spherical symmetry.

24-1 A New Look at Coulomb's Law

If you want to find the center of mass of a potato, you can do so by experiment or by laborious calculation, involving the numerical evaluation of a triple integral. However, if the potato happens to be a uniform ellipsoid, you know from its symmetry exactly where the center of mass is without calculation. Such are the advantages of symmetry. Symmetrical situations arise in all areas of physics; when possible, it makes sense to cast the laws of physics in forms that take full advantage of this fact.

Coulomb's law is the governing law in electrostatics, but it is not cast in a form that particularly simplifies the work in situations involving symmetry. In this chapter we introduce a new formulation of Coulomb's law, derived by German mathematician and physicist Carl Friedrich Gauss (1777–1855). This law, called **Gauss' law,** *can* be used to take advantage of special symmetry situations. For electrostatics problems, it is the full equivalent of Coulomb's law; which of them we choose to use depends only on the problem at hand.

Central to Gauss' law is a hypothetical closed surface called a **Gaussian surface.** The Gaussian surface can be of any shape you wish to make it, but the most useful surface is one that mimics the symmetry of the problem at hand. Thus, the Gaussian surface will often be a sphere, a cylinder, or some other symmetrical form. It must always be a *closed* surface, so that a clear distinction can be made between points that are inside the surface, on the surface, and outside the surface.

Imagine that you have established a Gaussian surface around a distribution of charges. Then Gauss' law comes into play:

> Gauss' law relates the electric fields at points on a (closed) Gaussian surface and the net charge enclosed by that surface.

Figure 24-1 shows a simple situation in which the Gaussian surface is a sphere. Suppose you know that there is an electric field at every point on the surface and that all the fields have the same magnitude and point radially outward. Without knowing anything about Gauss' law, you can guess that some net positive charge must be enclosed by the Gaussian surface. If you *do* know Gauss' law, you can calculate just how much net positive charge is enclosed. To make the calculation, you need know only "how much" electric field is intercepted by the surface—this "how much" involves the *flux* of the electric field through the surface.

24-2 Flux

Suppose that, as in Fig. 24-2*a*, you aim a wide airstream of uniform velocity $\vec{v}$ at a small square loop of area A. Let Φ represent the *volume flow rate* (volume per unit time) at which air flows through the loop. This rate depends on the angle between $\vec{v}$ and the plane of the loop. If $\vec{v}$ is perpendicular to the plane, the rate Φ is equal to vA.

If $\vec{v}$ is parallel to the plane of the loop, no air moves through the loop, so Φ is zero. For an intermediate angle θ, the rate Φ depends on the component of $\vec{v}$ that is normal to the plane (Fig. 24-2*b*). Since that component is $v\cos\theta$, the rate of volume flow through the loop is

$$\Phi = (v\cos\theta)A. \tag{24-1}$$

This rate of flow through an area is an example of a **flux**—a *volume flux* in this situation. Before we discuss a flux that is involved in electrostatics, we need to rewrite Eq. 24-1 in terms of vectors.

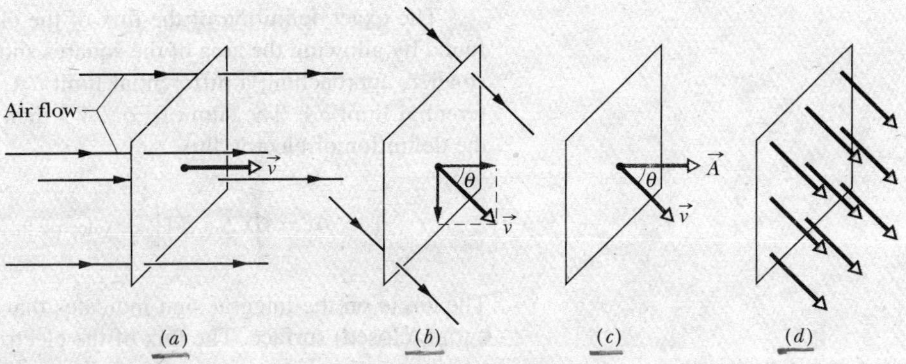

Fig. 24-2 (*a*) A uniform airstream of velocity $\vec{v}$ is perpendicular to the plane of a square loop of area A. (*b*) The component of $\vec{v}$ perpendicular to the plane of the loop is $v \cos\theta$, where θ is the angle between $\vec{v}$ and a normal to the plane. (*c*) The area vector $\vec{A}$ is perpendicular to the plane of the loop and makes an angle θ with $\vec{v}$. (*d*) The velocity field intercepted by the area of the loop.

To do this, we first define an *area vector* $\vec{A}$ as being a vector whose magnitude is equal to an area (here the area of the loop) and whose direction is normal to the plane of the area (Fig. 24-2*c*). We then rewrite Eq. 24-1 as the scalar (or dot) product of the velocity vector $\vec{v}$ of the airstream and the area vector $\vec{A}$ of the loop:

$$\Phi = vA\cos\theta = \vec{v}\cdot\vec{A}, \tag{24-2}$$

where θ is the angle between $\vec{v}$ and $\vec{A}$.

The word "flux" comes from the Latin word meaning "to flow." That meaning makes sense if we talk about the flow of air volume through the loop. However, Eq. 24-2 can be regarded in a more abstract way. To see it, note that we can assign a velocity vector to each point in the airstream passing through the loop (Fig. 24-2*d*). The composite of all those vectors is a *velocity field*, so we can interpret Eq. 24-2 as giving the *flux of the velocity field through the loop*. With this interpretation, flux no longer means the actual flow of something through an area—rather it means the product of an area and the field across that area.

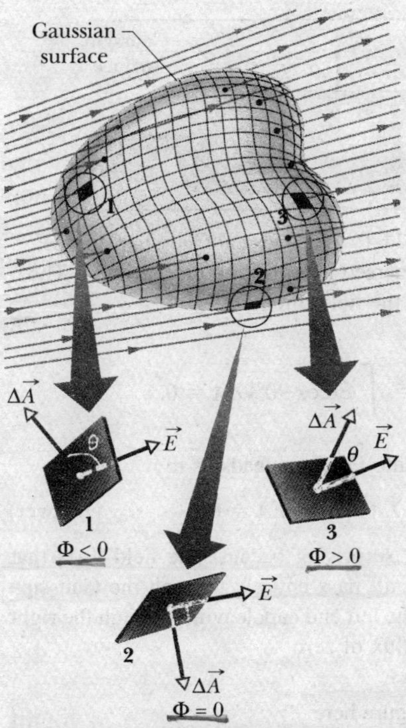

Fig. 24-3 A Gaussian surface of arbitrary shape immersed in an electric field. The surface is divided into small squares of area ΔA. The electric field vectors $\vec{E}$ and the area vectors $\Delta\vec{A}$ for three representative squares, marked 1, 2, and 3, are shown.

24-3 Flux of an Electric Field

To define the flux of an electric field, consider Fig. 24-3, which shows an arbitrary (asymmetric) Gaussian surface immersed in a nonuniform electric field. Let us divide the surface into small squares of area ΔA, each square being small enough to permit us to neglect any curvature and to consider the individual square to be flat. We represent each such element of area with an area vector $\Delta\vec{A}$, whose magnitude is the area ΔA. Each vector $\Delta\vec{A}$ is perpendicular to the Gaussian surface and directed away from the interior of the surface.

Because the squares have been taken to be arbitrarily small, the electric field $\vec{E}$ may be taken as constant over any given square. The vectors $\Delta\vec{A}$ and $\vec{E}$ for each square then make some angle θ with each other. Figure 24-3 shows an enlarged view of three squares (1, 2, and 3) on the Gaussian surface, and the angle θ for each.

A provisional definition for the flux of the electric field for the Gaussian surface of Fig. 24-3 is

$$\Phi = \sum \vec{E}\cdot\Delta\vec{A}. \tag{24-3}$$

This equation instructs us to visit each square on the Gaussian surface, to evaluate the scalar product $\vec{E}\cdot\Delta\vec{A}$ for the two vectors $\vec{E}$ and $\Delta\vec{A}$ that we find there, and to sum the results algebraically (that is, with signs included) for all the squares that make up the surface. The sign or a zero resulting from each scalar product determines whether the flux through its square is positive, negative, or zero. Squares like 1, in which $\vec{E}$ points inward, make a negative contribution to the sum of Eq. 24-3. Squares like 2, in which $\vec{E}$ lies in the surface, make zero contribution. Squares like 3, in which $\vec{E}$ points outward, make a positive contribution.

The exact definition of the flux of the electric field through a closed surface is found by allowing the area of the squares shown in Fig. 24-3 to become smaller and smaller, approaching a differential limit dA. The area vectors then approach a differential limit $d\vec{A}$. The sum of Eq. 24-3 then becomes an integral and we have, for the definition of electric flux,

$$\Phi = \oint \vec{E} \cdot d\vec{A} \quad \text{(electric flux through a Gaussian surface).} \tag{24-4}$$

The circle on the integral sign indicates that the integration is to be taken over the entire (closed) surface. The flux of the electric field is a scalar, and its SI unit is the newton–square-meter per coulomb ($N \cdot m^2/C$).

We can interpret Eq. 24-4 in the following way: First recall that we can use the density of electric field lines passing through an area as a proportional measure of an electric field $\vec{E}$ there. Specifically, the magnitude E is proportional to the number of electric field lines per unit area. Thus, the scalar product $\vec{E} \cdot d\vec{A}$ in Eq. 24-4 is proportional to the number of electric field lines passing through area $d\vec{A}$. Then, because the integration in Eq. 24-4 is carried out over a Gaussian surface, which is closed, we see that

> The electric flux Φ through a Gaussian surface is proportional to the net number of electric field lines passing through that surface.

Sample Problem 24-1

Figure 24-4 shows a Gaussian surface in the form of a cylinder of radius R immersed in a uniform electric field $\vec{E}$, with the cylinder axis parallel to the field. What is the flux Φ of the electric field through this closed surface?

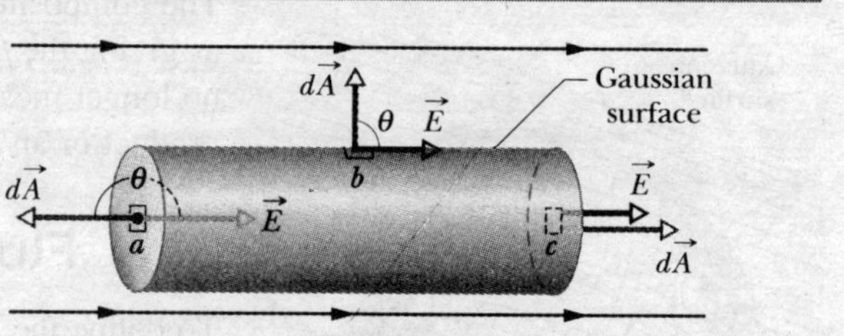

Fig. 24-4 Sample Problem 24-1. A cylindrical Gaussian surface, closed by end caps, is immersed in a uniform electric field. The cylinder axis is parallel to the field direction.

SOLUTION: The **Key Idea** here is that we can find the flux Φ through the surface by integrating the scalar product $\vec{E} \cdot d\vec{A}$ over the Gaussian surface. We can do this by writing the flux as the sum of three terms: integrals over the left cylinder cap a, the cylindrical surface b, and the right cap c. Thus, from Eq. 24-4,

Area vector always points outward from a surface

$$\Phi = \oint \vec{E} \cdot d\vec{A}$$

$$= \int_a \vec{E} \cdot d\vec{A} + \int_b \vec{E} \cdot d\vec{A} + \int_c \vec{E} \cdot d\vec{A}. \tag{24-5}$$

For all points on the left cap, the angle θ between $\vec{E}$ and $d\vec{A}$ is 180° and the magnitude E of the field is constant. Thus,

$$\int_a \vec{E} \cdot d\vec{A} = \int E(\cos 180°)\, dA = -E \int dA = -EA,$$

where $\int dA$ gives the cap's area, A ($= \pi R^2$). Similarly, for the right cap, where $\theta = 0$ for all points,

$$\int_c \vec{E} \cdot d\vec{A} = \int E(\cos 0)\, dA = EA.$$

Finally, for the cylindrical surface, where the angle θ is 90° at all points,

$$\int_b \vec{E} \cdot d\vec{A} = \int E(\cos 90°)\, dA = 0.$$

Substituting these results into Eq. 24-5 leads us to

$$\Phi = -EA + 0 + EA = 0. \quad \text{(Answer)}$$

This result is perhaps not surprising because the field lines that represent the electric field all pass entirely through the Gaussian surface, entering through the left end cap, leaving through the right end cap, and giving a net flux of zero.

CHECKPOINT 1: The figure here shows a Gaussian cube of face area A immersed in a uniform electric field $\vec{E}$ that has the positive direction of the z axis. In terms of E and A, what is the flux through (a) the front face (which is in the xy plane), (b) the rear face, (c) the top face, and (d) the whole cube?

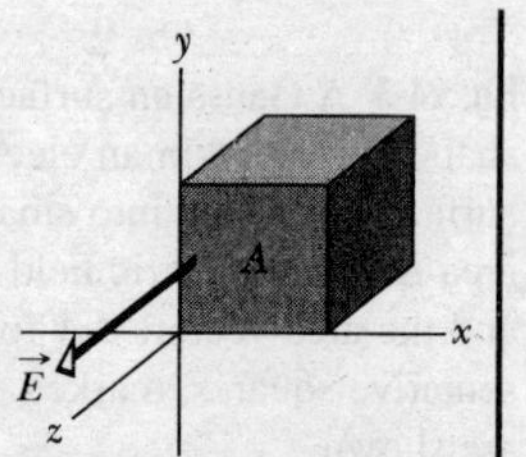

(a) $\int \vec{E} \cdot d\vec{A} = E \int \cos 0\, dA = EA$
(b) $\int \vec{E} \cdot d\vec{A} = E \int \cos 180°\, dA = -EA$
(c) $\int \vec{E} \cdot d\vec{A} = E \int \cos 90°\, dA = 0$
d) $EA - EA + 0 + 0 = 0$

Sample Problem 24-2

A *nonuniform* electric field given by $\vec{E} = 3.0x\hat{i} + 4.0\hat{j}$ pierces the Gaussian cube shown in Fig. 24-5. (E is in newtons per coulomb and x is in meters.) What is the electric flux through the right face, the left face, and the top face?

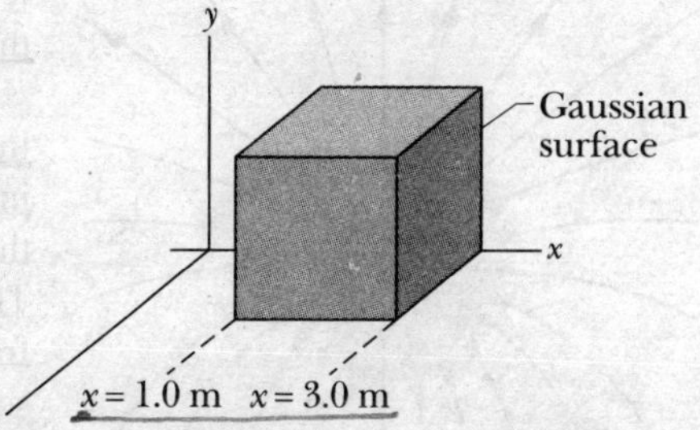

Fig. 24-5 Sample Problem 24-2. A Gaussian cube with one edge on the x axis lies within a nonuniform electric field.

SOLUTION: The **Key Idea** here is that we can find the flux Φ through the surface by integrating the scalar product $\vec{E} \cdot d\vec{A}$ over each face.

Right face: An area vector $\vec{A}$ is always perpendicular to its surface and always points away from the interior of a Gaussian surface. Thus, the vector $d\vec{A}$ for the right face of the cube must point in the positive x direction. In unit vector notation, then,

$$d\vec{A} = dA\hat{i}.$$

From Eq. 24-4, the flux Φ_r through the right face is then

$$\begin{aligned}\Phi_r &= \int \vec{E} \cdot d\vec{A} = \int (3.0x\hat{i} + 4.0\hat{j}) \cdot (dA\hat{i}) \\ &= \int [(3.0x)(dA)\hat{i} \cdot \hat{i} + (4.0)(dA)\hat{j} \cdot \hat{i}] \\ &= \int (3.0x\, dA + 0) = 3.0 \int x\, dA.\end{aligned}$$

We are about to integrate over the right face, but we note that x has the same value everywhere on that face—namely, $x = 3.0$ m. This means we can substitute that constant value for x. Then

$$\Phi_r = 3.0 \int (3.0)\, dA = 9.0 \int dA.$$

Now the integral merely gives us the area $A = 4.0\ \text{m}^2$ of the right face, so

$$\Phi_r = (9.0\ \text{N/C})(4.0\ \text{m}^2) = 36\ \text{N} \cdot \text{m}^2/\text{C}. \qquad \text{(Answer)}$$

Left face: The procedure for finding the flux through the left face is the same as that for the right face. However, two factors change. (1) The differential area vector $d\vec{A}$ points in the negative x direction and thus $d\vec{A} = -dA\hat{i}$. (2) The term x again appears in our integration, and it is again constant over the face being considered. However, on the left face, $x = 1.0$ m. With these two changes, we find that the flux Φ_l through the left face is

$$\Phi_l = -12\ \text{N} \cdot \text{m}^2/\text{C}. \qquad \text{(Answer)}$$

Top face: The differential area vector $d\vec{A}$ points in the positive y direction and thus $d\vec{A} = dA\hat{j}$. The flux Φ_t through the top face is then

$$\begin{aligned}\Phi_t &= \int (3.0x\hat{i} + 4.0\hat{j}) \cdot (dA\hat{j}) \\ &= \int [(3.0x)(dA)\hat{i} \cdot \hat{j} + (4.0)(dA)\hat{j} \cdot \hat{j}] \\ &= \int (0 + 4.0\, dA) = 4.0 \int dA \\ &= 16\ \text{N} \cdot \text{m}^2/\text{C}. \qquad \text{(Answer)}\end{aligned}$$

24-4 Gauss' Law

Gauss' law relates the net flux Φ of an electric field through a closed surface (a Gaussian surface) to the *net* charge q_{enc} that is *enclosed* by that surface. It tells us that

$$\varepsilon_0 \Phi = q_{enc} \qquad \text{(Gauss' law).} \qquad (24\text{-}6)$$

By substituting Eq. 24-4, the definition of flux, we can also write Gauss' law as

$$\varepsilon_0 \oint \vec{E} \cdot d\vec{A} = q_{enc} \qquad \text{(Gauss' law).} \qquad (24\text{-}7)$$

Equations 24-6 and 24-7 hold only when the net charge is located in a vacuum or (what is the same for most practical purposes) in air. In Section 26-8, we modify Gauss' law to include situations in which a material such as mica, oil, or glass is present.

In Eqs. 24-6 and 24-7, the net charge q_{enc} is the algebraic sum of all the *enclosed* positive and negative charges, and it can be positive, negative, or zero. We include the sign, rather than just use the magnitude of the enclosed charge, because the sign

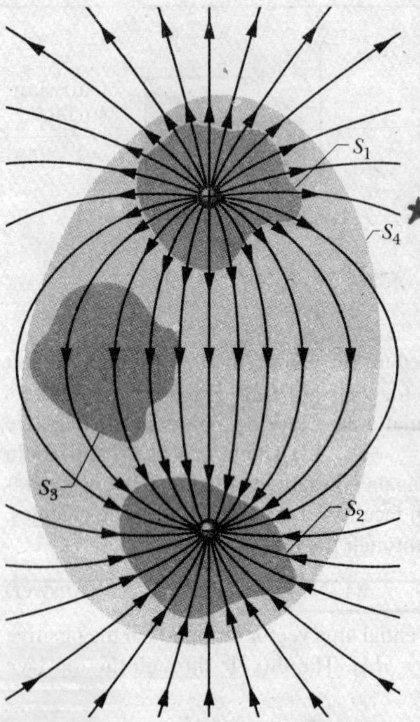

Fig. 24-6 Two point charges, equal in magnitude but opposite in sign, and the field lines that represent their net electric field. Four Gaussian surfaces are shown in cross section. Surface S_1 encloses the positive charge. Surface S_2 encloses the negative charge. Surface S_3 encloses no charge. Surface S_4 encloses both charges, and thus no net charge.

tells us something about the net flux through the Gaussian surface: If q_{enc} is positive, the net flux is *outward*; if q_{enc} is negative, the net flux is *inward*.

Charge outside the surface, no matter how large or how close it may be, is not included in the term q_{enc} in Gauss' law. The exact form or location of the charges inside the Gaussian surface is also of no concern; the only things that matter, on the right side of Eq. 24-7, are the magnitude and sign of the net enclosed charge. The quantity $\vec{E}$ on the left side of Eq. 24-7, however, is the electric field resulting from *all* charges, both those inside and those outside the Gaussian surface. This may seem to be inconsistent, but keep in mind what we saw in Sample Problem 24-1: The electric field due to a charge outside the Gaussian surface contributes zero net flux *through* the surface, because as many field lines due to that charge enter the surface as leave it.

Let us apply these ideas to Fig. 24-6, which shows two point charges, equal in magnitude but opposite in sign, and the field lines describing the electric fields that they set up in the surrounding space. Four Gaussian surfaces are also shown, in cross section. Let us consider each in turn.

Surface S_1. The electric field is outward for all points on this surface. Thus, the flux of the electric field through this surface is positive, and so is the net charge within the surface, as Gauss' law requires. (That is, in Eq. 24-6, if Φ is positive, q_{enc} must be also.)

Surface S_2. The electric field is inward for all points on this surface. Thus, the flux of the electric field is negative and so is the enclosed charge, as Gauss' law requires.

Surface S_3. This surface encloses no charge, and thus $q_{enc} = 0$. Gauss' law (Eq. 24-6) requires that the net flux of the electric field through this surface be zero. That is reasonable because all the field lines pass entirely through the surface, entering it at the top and leaving at the bottom.

Surface S_4. This surface encloses no *net* charge, because the enclosed positive and negative charges have equal magnitudes. Gauss' law requires that the net flux of the electric field through this surface be zero. That is reasonable because there are as many field lines leaving surface S_4 as entering it.

What would happen if we were to bring an enormous charge Q up close to surface S_4 in Fig. 24-6? The pattern of the field lines would certainly change, but the net flux for each of the four Gaussian surfaces would not change. We can understand this because the field lines associated with the added Q would pass entirely through each of the four Gaussian surfaces, making no contribution to the net flux through any of them. The value of Q would not enter Gauss' law in any way, because Q lies outside all four of the Gaussian surfaces that we are considering.

CHECKPOINT 2: The figure shows three situations in which a Gaussian cube sits in an electric field. The arrows and the values indicate the directions of the field lines and the magnitudes (in N · m²/C) of the flux through the six sides of each cube. (The lighter arrows are for the hidden faces.) In which situations does the cube enclose (a) a positive net charge, (b) a negative net charge, and (c) zero net charge?

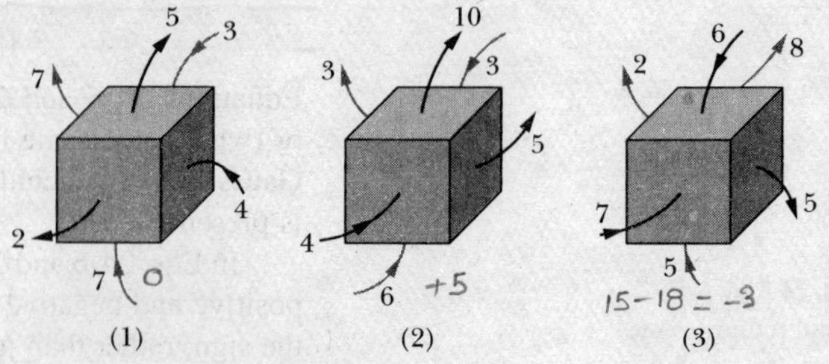

Sample Problem 24-3

Figure 24-7 shows five charged lumps of plastic and an electrically neutral coin. The cross section of a Gaussian surface S is indicated. What is the net electric flux through the surface if $q_1 = q_4 = +3.1$ nC, $q_2 = q_5 = -5.9$ nC, and $q_3 = -3.1$ nC?

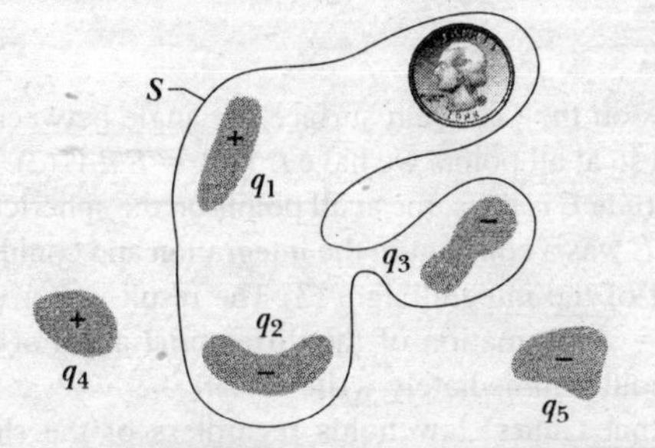

Fig. 24-7 Sample Problem 24-3. Five plastic objects, each with an electric charge, and a coin, which has no net charge. A Gaussian surface, shown in cross section, encloses three of the plastic objects and the coin.

SOLUTION: The **Key Idea** here is that the *net* flux Φ through the surface depends on the *net* charge q_{enc} enclosed by surface S. This means that the coin and charges q_4 and q_5 do not contribute to Φ. The coin does not contribute because it is neutral and thus contains equal amounts of positive and negative charge. Charges q_4 and q_5 do not contribute because they are outside surface S. Thus, q_{enc} is $q_1 + q_2 + q_3$ and Eq. 24-6 gives us

$$\Phi = \frac{q_{enc}}{\varepsilon_0} = \frac{q_1 + q_2 + q_3}{\varepsilon_0}$$
$$= \frac{+3.1 \times 10^{-9}\text{ C} - 5.9 \times 10^{-9}\text{ C} - 3.1 \times 10^{-9}\text{ C}}{8.85 \times 10^{-12}\text{ C}^2/\text{N}\cdot\text{m}^2}$$
$$= -670\text{ N}\cdot\text{m}^2/\text{C}. \quad \text{(Answer)}$$

The minus sign shows that the net flux through the surface is inward and thus that the net charge within the surface is negative.

24-5 Gauss' Law and Coulomb's Law

If Gauss' law and Coulomb's law are equivalent, we should be able to derive each from the other. Here we derive Coulomb's law from Gauss' law and some symmetry considerations.

Figure 24-8 shows a positive point charge q, around which we have drawn a concentric spherical Gaussian surface of radius r. Let us divide this surface into differential areas dA. By definition, the area vector $d\vec{A}$ at any point is perpendicular to the surface and directed outward from the interior. From the symmetry of the situation, we know that at any point the electric field $\vec{E}$ is also perpendicular to the surface and directed outward from the interior. Thus, since the angle θ between $\vec{E}$ and $d\vec{A}$ is zero, we can rewrite Eq. 24-7 for Gauss' law as

$$\varepsilon_0 \oint \vec{E}\cdot d\vec{A} = \varepsilon_0 \oint E\, dA = q_{enc}. \tag{24-8}$$

Here $q_{enc} = q$. Although E varies radially with the distance from q, it has the same value everywhere on the spherical surface. Since the integral in Eq. 24-8 is taken over that surface, E is a constant in the integration and can be brought out in front of the integral sign. That gives us

$$\varepsilon_0 E \oint dA = q. \tag{24-9}$$

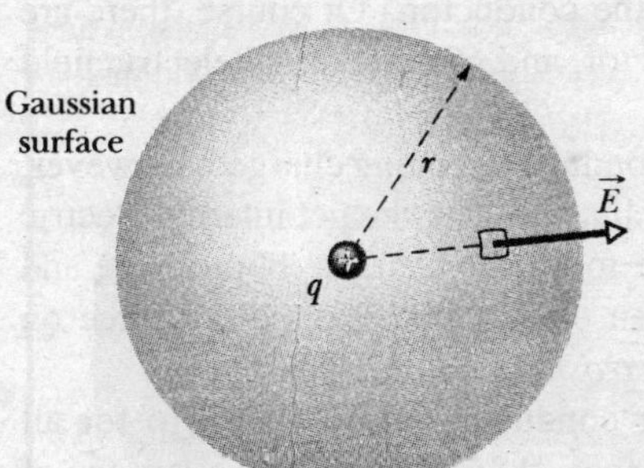

Fig. 24-8 A spherical Gaussian surface centered on a point charge q.

The integral is now merely the sum of all the differential areas dA on the sphere and thus is just the surface area, $4\pi r^2$. Substituting this, we have

$$\varepsilon_0 E(4\pi r^2) = q$$

or

$$E = \frac{1}{4\pi\varepsilon_0}\frac{q}{r^2}. \tag{24-10}$$

This is exactly the electric field due to a point charge (Eq. 23-3), which we found using Coulomb's law. Thus, Gauss' law is equivalent to Coulomb's law.

CHECKPOINT 3: There is a certain net flux Φ_i through a Gaussian sphere of radius r enclosing an isolated charged particle. Suppose the enclosing Gaussian surface is changed to (a) a larger Gaussian sphere, (b) a Gaussian cube with edge length equal to r, and (c) a Gaussian cube with edge length equal to $2r$. In each case, is the net flux through the new Gaussian surface greater than, less than, or equal to Φ_i?

PROBLEM-SOLVING TACTICS

Tactic 1: *Choosing a Gaussian Surface*
The derivation of Eq. 24-10 using Gauss' law is a warm-up for derivations of electric fields produced by other charge configurations, so let us go back over the steps involved. We started with a given positive point charge q; we know that electric field lines extend radially outward from q in a spherically symmetric pattern.

To find the magnitude E of the electric field at a distance r by Gauss' law (Eq. 24-7), we had to place a hypothetical closed Gaussian surface around q, through a point that is a distance r from q. Then we had to sum via integration the values of $\vec{E} \cdot d\vec{A}$ over the full Gaussian surface. To make this integration as simple as possible, we chose a spherical Gaussian surface (to mimic the spherical symmetry of the electric field). That choice produced three simplifying features. (1) The dot product $\vec{E} \cdot d\vec{A}$ became simple, because at all points on the Gaussian surface the angle between $\vec{E}$ and $d\vec{A}$ is zero, and so at all points we have $\vec{E} \cdot d\vec{A} = E\,dA$. (2) The electric field magnitude E is the same at all points on the spherical Gaussian surface, so E was a constant in the integration and could be brought out in front of the integral sign. (3) The result was a very simple integration—a summation of the differential areas of the sphere, which we could immediately write as $4\pi r^2$.

Note that Gauss' law holds regardless of the shape of the Gaussian surface we choose to place around charge q_{enc}. However, if we had chosen, say, a cubical Gaussian surface, our three simplifying features would have disappeared and the integration of $\vec{E} \cdot d\vec{A}$ over the cubical surface would have been very difficult. The moral here is to choose the Gaussian surface that most simplifies the integration in Gauss' law.

24-6 A Charged Isolated Conductor

Gauss' law permits us to prove an important theorem about isolated conductors:

> If an excess charge is placed on an isolated conductor, that amount of charge will move entirely to the surface of the conductor. None of the excess charge will be found within the body of the conductor.

This might seem reasonable, considering that charges with the same sign repel each other. You might imagine that, by moving to the surface, the added charges are getting as far away from each other as they can. We turn to Gauss' law for verification of this speculation.

Figure 24-9*a* shows, in cross section, an isolated lump of copper hanging from an insulating thread and having an excess charge q. We place a Gaussian surface just inside the actual surface of the conductor.

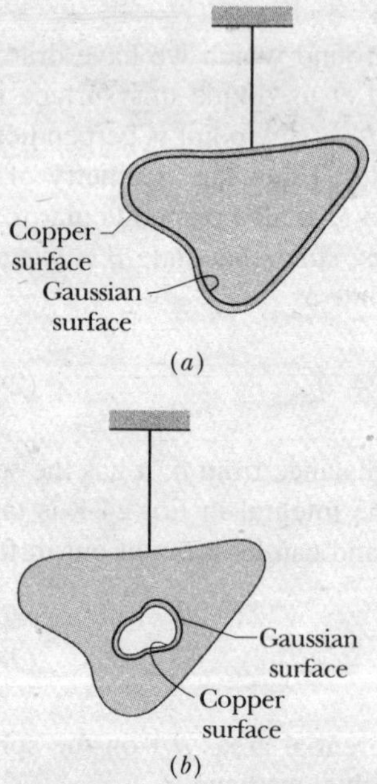

Fig. 24-9 (*a*) A lump of copper with a charge q hangs from an insulating thread. A Gaussian surface is placed within the metal, just inside the actual surface. (*b*) The lump of copper now has a cavity within it. A Gaussian surface lies within the metal, close to the cavity surface.

The electric field inside this conductor must be zero. If this were not so, the field would exert forces on the conduction (free) electrons, which are always present in a conductor, and thus current would always exist within a conductor. (That is, charge would flow from place to place within the conductor.) Of course, there are no such perpetual currents in an isolated conductor, and so the internal electric field is zero.

(An internal electric field *does* appear as a conductor is being charged. However, the added charge quickly distributes itself in such a way that the net internal electric field—the vector sum of the electric fields due to all the charges, both inside and outside—is zero. The movement of charge then ceases, because the net force on each charge is zero; the charges are then in *electrostatic equilibrium*.)

If $\vec{E}$ is zero everywhere inside our copper conductor, it must be zero for all points on the Gaussian surface because that surface, though close to the surface of the conductor, is definitely inside the conductor. This means that the flux through the Gaussian surface must be zero. Gauss' law then tells us that the net charge inside the Gaussian surface must also be zero. Then because the excess charge is not inside the Gaussian surface, it must be outside that surface, which means it must lie on the actual surface of the conductor.

An Isolated Conductor with a Cavity

Figure 24-9*b* shows the same hanging conductor, but now with a cavity that is totally within the conductor. It is perhaps reasonable to suppose that when we scoop out the electrically neutral material to form the cavity, we do not change the distribution of charge or the pattern of the electric field that exists in Fig. 24-9*a*. Again, we must turn to Gauss' law for a quantitative proof.

We draw a Gaussian surface surrounding the cavity, close to its surface but inside the conducting body. Because $\vec{E} = 0$ inside the conductor, there can be no flux through this new Gaussian surface. Therefore, from Gauss' law, that surface can enclose no net charge. We conclude that there is no net charge on the cavity walls; all the excess charge remains on the outer surface of the conductor, as in Fig. 24-9*a*.

The Conductor Removed

Suppose that, by some magic, the excess charges could be "frozen" into position on the conductor's surface, perhaps by embedding them in a thin plastic coating, and suppose that then the conductor could be removed completely. This is equivalent to enlarging the cavity of Fig. 24-9*b* until it consumes the entire conductor, leaving only the charges. The electric field would not change at all; it would remain zero inside the thin shell of charge and would remain unchanged for all external points. This shows us that the electric field is set up by the charges and not by the conductor. The conductor simply provides an initial pathway for the charges to take up their positions.

Spse conductor in consideration is thick, and we draw Gaussian surface coming from other side

Then $\varepsilon_0 \Phi = q_{enc}$
$\varepsilon_0 (EA + EA) = \sigma A + \sigma A$
$2\varepsilon_0 EA = 2\sigma A$
$E = \sigma/\varepsilon_0$

The External Electric Field

You have seen that the excess charge on an isolated conductor moves entirely to the conductor's surface. However, unless the conductor is spherical, the charge does not distribute itself uniformly. Put another way, the surface charge density σ (charge per unit area) varies over the surface of any nonspherical conductor. Generally, this variation makes the determination of the electric field set up by the surface charges very difficult.

However, the electric field just outside ... to determine using Gauss' law. To do this, we ... t is small enough to permit us to neglect any cu... to be flat. We then imagine a tiny cylindrical G... the section as in Fig. 24-10: One end cap is fully ... lly outside, and the cylinder is perpendicular to ...

Since the charge distribution doesn't change if we move in any direction of xy plane parallel to the sheet (sheet is in xy plane) ∴ E must be ⊥ to the sheet & must have the same magnitude at any given distance on either side of the sheet

The electric field $\vec{E}$ at and just outside ... be perpendicular to that surface. If it were not, ... ong the conductor's surface that would exert force ... em to move. However, such motion would viola... are dealing with electrostatic equilibrium. There... uc-tor's surface. at & just outside

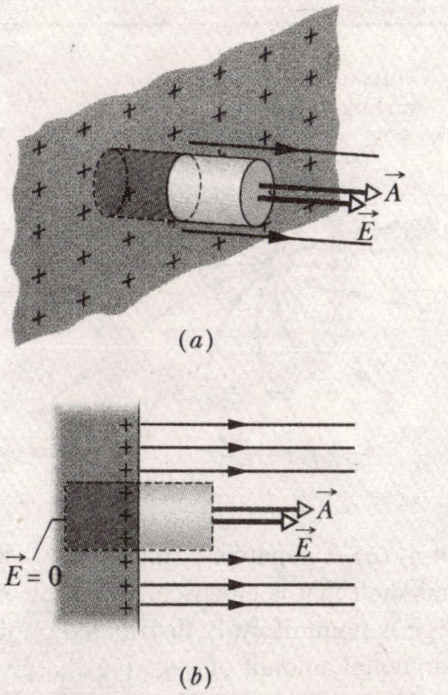

Fig. 24-10 Perspective view (*a*) and side view (*b*) of a tiny portion of a large, isolated conductor with excess positive charge on its surface. A (closed) cylindrical Gaussian surface, embedded perpendicularly in the conductor, encloses some of the charge. Electric field lines pierce the external end cap of the cylinder, but not the internal end cap. The external end cap has area *A* and area vector $\vec{A}$.

We now sum the flux through the Gaussian surface. There is no flux through the internal end cap, because the electric field within the conductor is zero. There is no flux through the curved surface of the cylinder, because internally (in the conductor) there is no electric field and externally the electric field is parallel to the curved portion of the Gaussian surface. The only flux through the Gaussian surface is that through the external end cap, where $\vec{E}$ is perpendicular to the plane of the cap. We assume that the cap area A is small enough that the field magnitude E is constant over the cap. Then the flux through the cap is EA, and that is the net flux Φ through the Gaussian surface.

The charge q_{enc} enclosed by the Gaussian surface lies on the conductor's surface in an area A. If σ is the charge per unit area, then q_{enc} is equal to σA. When we substitute σA for q_{enc} and EA for Φ, Gauss' law (Eq. 24-6) becomes

$$\varepsilon_0 EA = \sigma A,$$

from which we find

$$E = \frac{\sigma}{\varepsilon_0} \quad \text{(conducting surface).} \tag{24-11}$$

Thus, the magnitude of the electric field at a location just outside a conductor is proportional to the surface charge density at that location on the conductor. If the charge on the conductor is positive, the electric field is directed away from the conductor as in Fig. 24-10. It is directed toward the conductor if the charge is negative.

The field lines in Fig. 24-10 must terminate on negative charges somewhere in the environment. If we bring those charges near the conductor, the charge density at any given location on the conductor's surface changes, and so does the magnitude of the electric field. However, the relation between σ and E is still given by Eq. 24-11.

Sample Problem 24-4

Figure 24-11a shows a cross section of a spherical metal shell of inner radius R. A point charge of $-5.0\ \mu C$ is located at a distance $R/2$ from the center of the shell. If the shell is electrically neutral, what are the (induced) charges on its inner and outer surfaces? Are those charges uniformly distributed? What is the field pattern inside and outside the shell?

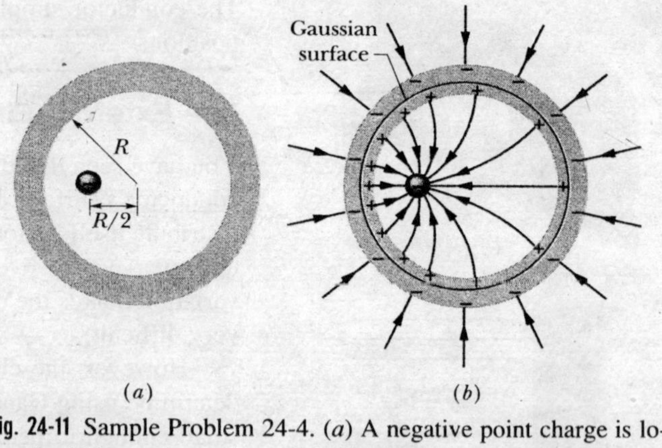

Fig. 24-11 Sample Problem 24-4. (a) A negative point charge is located within a spherical metal shell that is electrically neutral. (b) As a result, positive charge is nonuniformly distributed on the inner wall of the shell, and an equal amount of negative charge is uniformly distributed on the outer wall.

SOLUTION: Figure 24-11b shows a cross section of a spherical Gaussian surface within the metal, just outside the inner wall of the shell. One **Key Idea** here is that the electric field must be zero inside the metal (and thus on the Gaussian surface inside the metal). This means that the electric flux through the Gaussian surface must also be zero. Gauss' law then tells us that the *net* charge enclosed by the Gaussian surface must be zero. With a point charge of $-5.0\ \mu C$ within the shell, a charge of $+5.0\ \mu C$ must lie on the inner wall of the shell.

If the point charge were centered, this positive charge would be uniformly distributed along the inner wall. However, since the point charge is off-center, the distribution of positive charge is skewed, as suggested by Fig. 24-11b, because the positive charge tends to collect on the section of the inner wall nearest the (negative) point charge.

A second **Key Idea** is that because the shell is electrically neutral, its inner wall can have a charge of $+5.0\ \mu C$ only if electrons, with a total charge of $-5.0\ \mu C$, leave the inner wall and move to the outer wall. There they spread out uniformly, as is also suggested by Fig. 24-11b. This distribution of negative charge is uniform because the shell is spherical and because the skewed distribution of positive charge on the inner wall cannot produce an electric field in the shell to affect the distribution of charge on the outer wall.

The field lines inside and outside the shell are shown approximately in Fig. 24-11b. All the field lines intersect the shell and the point charge perpendicularly. Inside the shell the pattern of field lines is skewed owing to the skew of the positive charge distribution. Outside the shell the pattern is the same as if the point charge were centered and the shell were missing. In fact, this would be true no matter where inside the shell the point charge happened to be located.

CHECKPOINT 4: A ball of charge $-50e$ lies at the center of a hollow spherical metal shell that has a net charge of $-100e$. What is the charge on (a) the shell's inner surface and (b) its outer surface?

24-7 Applying Gauss' Law: Cylindrical Symmetry

Figure 24-12 shows a section of an infinitely long cylindrical plastic rod with a uniform positive linear charge density λ. Let us find an expression for the magnitude of the electric field $\vec{E}$ at a distance r from the axis of the rod.

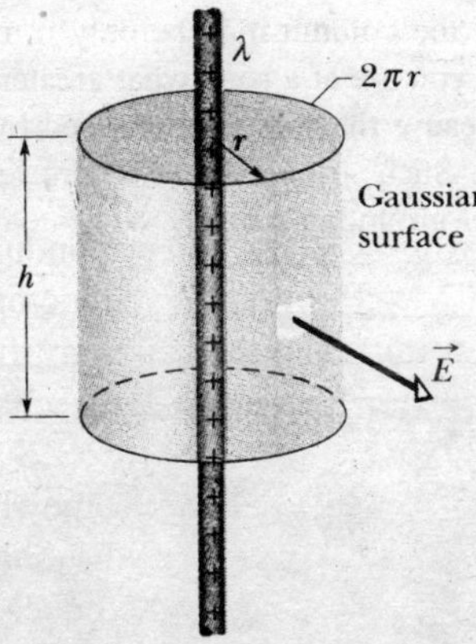

Fig. 24-12 A Gaussian surface in the form of a closed cylinder surrounds a section of a very long, uniformly charged, cylindrical plastic rod.

Our Gaussian surface should match the symmetry of the problem, which is cylindrical. We choose a circular cylinder of radius r and length h, coaxial with the rod. The Gaussian surface must be closed, so we include two end caps as part of the surface.

Imagine now that, while you are not watching, someone rotates the plastic rod around its longitudinal axis or turns it end for end. When you look again at the rod, you will not be able to detect any change. We conclude from this symmetry that the only uniquely specified direction in this problem is along a radial line. Thus, at every point on the cylindrical part of the Gaussian surface, $\vec{E}$ must have the same magnitude E and (for a positively charged rod) must be directed radially outward.

Since $2\pi r$ is the cylinder's circumference and h is its height, the area A of the cylindrical surface is $2\pi rh$. The flux of $\vec{E}$ through this cylindrical surface is then

$$\Phi = EA \cos\theta = E(2\pi rh)\cos 0 = E(2\pi rh).$$

There is no flux through the end caps because $\vec{E}$, being radially directed, is parallel to the end caps at every point.

The charge enclosed by the surface is λh, so Gauss' law,

$$\varepsilon_0 \Phi = q_{enc},$$

reduces to

$$\varepsilon_0 E(2\pi rh) = \lambda h,$$

yielding

$$E = \frac{\lambda}{2\pi\varepsilon_0 r} \quad \text{(line of charge).} \tag{24-12}$$

This is the electric field due to an infinitely long, straight line of charge, at a point that is a radial distance r from the line. The direction of $\vec{E}$ is radially outward from the line of charge if the charge is positive, and radially inward if it is negative. Equation 24-12 also approximates the field of a *finite* line of charge, at points that are not too near the ends (compared with the distance from the line).

Sample Problem 24-5

The visible portion of a lightning strike is preceded by an invisible stage in which a column of electrons extends downward from a cloud to the ground. These electrons come from the cloud and from air molecules that are ionized within the column. The linear charge density λ along the column is typically -1×10^{-3} C/m. Once the column reaches the ground, electrons within it are rapidly dumped to the ground. During the dumping, collisions between the moving electrons and the air within the column result in a brilliant flash of light. If air molecules break down (ionize) in an electric field exceeding 3×10^6 N/C, what is the radius of the column?

SOLUTION: One **Key Idea** here is that, although the column is not straight or infinitely long, we can approximate it as being a line of charge as in Fig. 24-12. (Since it contains a net negative charge,

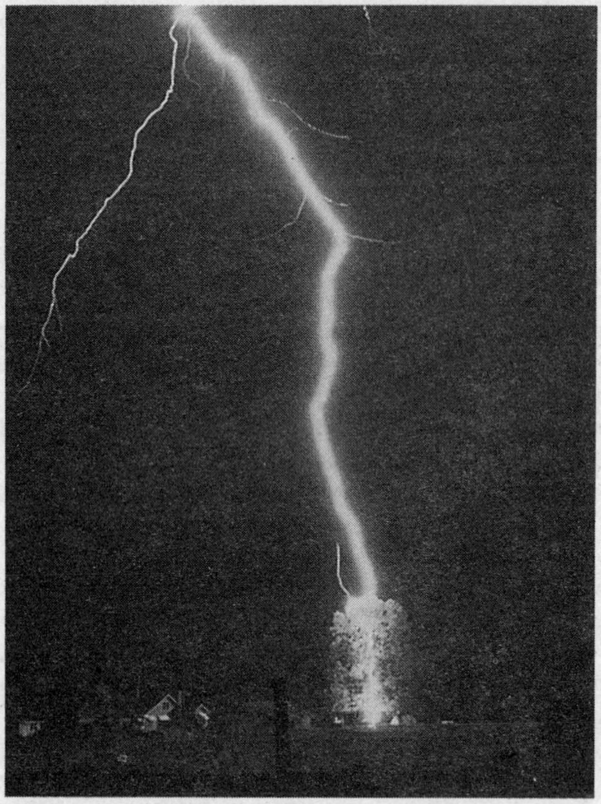

Fig. 24-13 Lightning strikes a 20-m-high sycamore. Because the tree was wet, most of the charge traveled through the water on it and the tree was unharmed.

its electric field $\vec{E}$ points radially inward.) Then, according to Eq. 24-12, the field's magnitude E decreases with distance from the axis of the column of charge.

A second **Key Idea** is that the surface of the column of charge must be at the radius r where the magnitude of $\vec{E}$ is 3×10^6 N/C, because air molecules within that radius ionize while those farther out do not. Solving Eq. 24-12 for r and inserting the known data, we find the radius of the column to be

$$r = \frac{\lambda}{2\pi\varepsilon_0 E}$$

$$= \frac{1 \times 10^{-3}\ \text{C/m}}{(2\pi)(8.85 \times 10^{-12}\ \text{C}^2/\text{N}\cdot\text{m}^2)(3 \times 10^6\ \text{N/C})}$$

$$= 6\ \text{m}. \qquad \text{(Answer)}$$

(The radius of the luminous portion of a lightning strike is smaller, perhaps only 0.5 m. You can get an idea of the width from Fig. 24-13.) Although the radius of the column may be only 6 m, do not assume that you are safe if you are at a somewhat greater distance from the strike point, because the electrons dumped by the strike travel along the ground. Such *ground currents* are lethal. Figure 24-14 shows evidence of ground currents.

Fig. 24-14 Ground currents from a lightning strike have burned grass off this golf green, exposing the soil.

24-8 Applying Gauss' Law: Planar Symmetry

Nonconducting Sheet

Sheet is so thin that Gaussian surface has to come out

Figure 24-15 shows a portion of a thin, infinite, nonconducting sheet with a uniform (positive) surface charge density σ. A sheet of thin plastic wrap, uniformly charged on one side, can serve as a simple model. Let us find the electric field $\vec{E}$ a distance r in front of the sheet.

A useful Gaussian surface is a closed cylinder with end caps of area A, arranged to pierce the sheet perpendicularly as shown. From symmetry, $\vec{E}$ must be perpendicular to the sheet and hence to the end caps. Furthermore, since the charge is positive, $\vec{E}$ is directed *away* from the sheet, and thus the electric field lines pierce the two Gaussian end caps in an outward direction. Because the field lines do not pierce the curved surface, there is no flux through this portion of the Gaussian

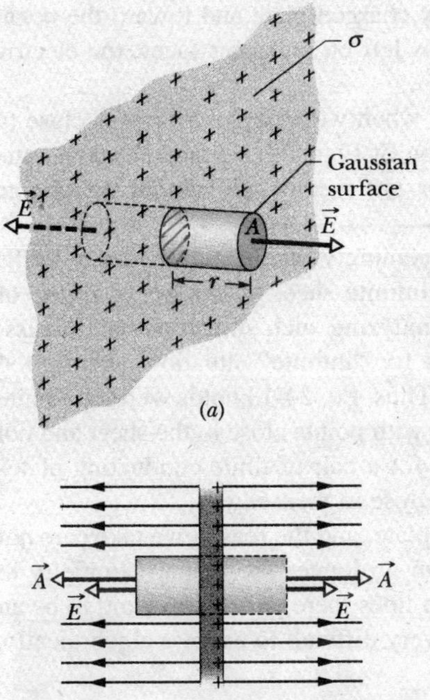

Fig. 24-15 Perspective view (*a*) and side view (*b*) of a portion of a very large, thin plastic sheet, uniformly charged on one side to surface charge density σ. A closed cylindrical Gaussian surface passes through the sheet and is perpendicular to it.

surface. Thus $\vec{E} \cdot d\vec{A}$ is simply $E\,dA$; then Gauss' law,

$$\varepsilon_0 \oint \vec{E} \cdot d\vec{A} = q_{enc},$$

becomes

$$\varepsilon_0(EA + EA) = \sigma A,$$

where σA is the charge enclosed by the Gaussian surface. This gives

$$E = \frac{\sigma}{2\varepsilon_0} \quad \text{(sheet of charge).} \tag{24-13}$$

Since we are considering an infinite sheet with uniform charge density, this result holds for any point at a finite distance from the sheet. Equation 24-13 agrees with Eq. 23-27, which we found by integration of the electric field components that are produced by individual charges. (Look back to that time-consuming and challenging integration, and note how much more easily we obtain the result with Gauss' law. That is one reason for devoting a whole chapter to that law: for certain symmetric arrangements of charge, it is very much easier to use than integration of field components.)

Two Conducting Plates

Figure 24-16*a* shows a cross section of a thin, infinite conducting plate with excess positive charge. From Section 24-6 we know that this excess charge lies on the surface of the plate. Since the plate is thin and very large, we can assume that essentially all the excess charge is on the two large faces of the plate.

If there is no external electric field to force the positive charge into some particular distribution, it will spread out on the two faces with a uniform surface charge density of magnitude σ_1. From Eq. 24-11 we know that just outside the plate this charge sets up an electric field of magnitude $E = \sigma_1/\varepsilon_0$. Because the excess charge is positive, the field is directed away from the plate.

Figure 24-16*b* shows an identical plate with excess negative charge having the same magnitude of surface charge density σ_1. The only difference is that now the electric field is directed toward the plate.

Suppose we arrange for the plates of Figs. 24-16*a* and *b* to be close to each other and parallel (Fig. 24-16*c*). Since the plates are conductors, when we bring them into this arrangement, the excess charge on one plate attracts the excess charge on the other plate, and all the excess charge moves onto the inner faces of the plates as in Fig. 24-16*c*. With twice as much charge now on each inner face, the new surface charge density (call it σ) on each inner face is twice σ_1. Thus, the electric field at any point between the plates has the magnitude

$$E = \frac{2\sigma_1}{\varepsilon_0} = \frac{\sigma}{\varepsilon_0}. \tag{24-14}$$

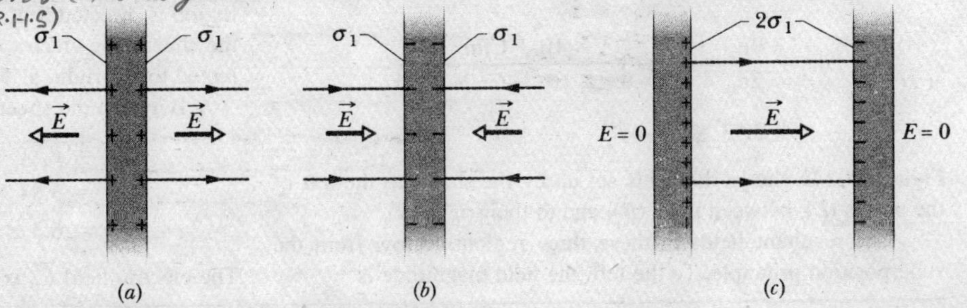

Fig. 24-16 (*a*) A thin, very large conducting plate with excess positive charge. (*b*) An identical plate with excess negative charge. (*c*) The two plates arranged so they are parallel and close.

This field is directed away from the positively charged plate and toward the negatively charged plate. Since no excess charge is left on the outer faces, the electric field to the left and right of the plates is zero.

Because the charges on the plates moved when we brought the plates close to each other, Fig. 24-16*c* is *not* the superposition of Figs. 24-16*a* and *b*; that is, the charge distribution of the two-plate system is not merely the sum of the charge distributions of the individual plates.

You may wonder why we discuss such seemingly unrealistic situations as the field set up by an infinite line of charge, an infinite sheet of charge, or a pair of infinite plates of charge. One reason is that analyzing such situations with Gauss' law is easy. More important is that analyses for "infinite" situations yield good approximations to many real-world problems. Thus, Eq. 24-13 holds well for a finite nonconducting sheet as long as we are dealing with points close to the sheet and not too near its edges. Equation 24-14 holds well for a pair of finite conducting plates as long as we consider points that are not too close to their edges.

The trouble with the edges of a sheet or a plate, and the reason we take care not to deal with them, is that near an edge we can no longer use planar symmetry to find expressions for the fields. In fact, the field lines there are curved (said to be an *edge effect* or *fringing*), and the fields can be very difficult to express algebraically.

Sample Problem 24-6

Had they been conducting sheets: a) σ_+/ε_0 b) $\frac{\sigma}{\varepsilon_0}$ c) 0

Figure 24-17*a* shows portions of two large, parallel, nonconducting sheets, each with a fixed uniform charge on one side. The magnitudes of the surface charge densities are $\sigma_{(+)} = 6.8\ \mu\text{C/m}^2$ for the positively charged sheet and $\sigma_{(-)} = 4.3\ \mu\text{C/m}^2$ for the negatively charged sheet.

Find the electric field $\vec{E}$ (a) to the left of the sheets, (b) between the sheets, and (c) to the right of the sheets.

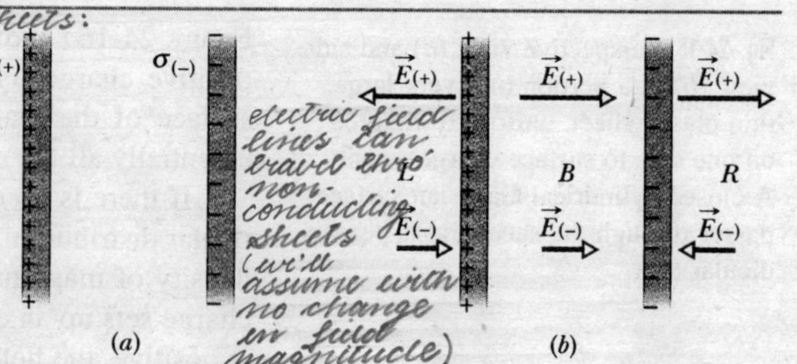

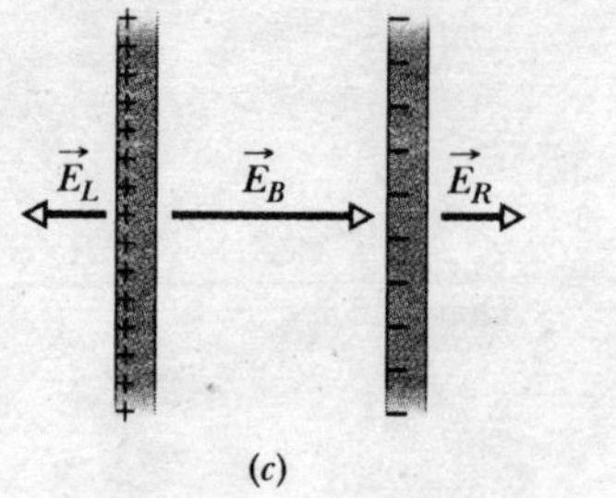

Fig. 24-17 Sample Problem 24-6. (*a*) Two large, parallel sheets, uniformly charged on one side. (*b*) The individual electric fields resulting from the two charged sheets. (*c*) The net field due to both charged sheets, found by superposition.

SOLUTION: The **Key Idea** here is that with the charges fixed in place, we can find the electric field of the sheets in Fig. 24-17*a* by (1) finding the field of each sheet as if that sheet were isolated and (2) algebraically adding the fields of the isolated sheets via the superposition principle. (We can add the fields algebraically because they are parallel to each other.) From Eq. 24-13, the magnitude $E_{(+)}$ of the electric field due to the positive sheet at any point is

$$E_{(+)} = \frac{\sigma_{(+)}}{2\varepsilon_0} = \frac{6.8 \times 10^{-6}\ \text{C/m}^2}{(2)(8.85 \times 10^{-12}\ \text{C}^2/\text{N}\cdot\text{m}^2)}$$
$$= 3.84 \times 10^5\ \text{N/C}.$$

Similarly, the magnitude $E_{(-)}$ of the electric field at any point due to the negative sheet is

$$E_{(-)} = \frac{\sigma_{(-)}}{2\varepsilon_0} = \frac{4.3 \times 10^{-6}\ \text{C/m}^2}{(2)(8.85 \times 10^{-12}\ \text{C}^2/\text{N}\cdot\text{m}^2)}$$
$$= 2.43 \times 10^5\ \text{N/C}.$$

Figure 24-17*b* shows the fields set up by the sheets to the left of the sheets (*L*), between them (*B*), and to their right (*R*).

The resultant fields in these three regions follow from the superposition principle. To the left, the field magnitude is

$$E_L = E_{(+)} - E_{(-)}$$
$$= 3.84 \times 10^5\ \text{N/C} - 2.43 \times 10^5\ \text{N/C}$$
$$= 1.4 \times 10^5\ \text{N/C}. \quad \text{(Answer)}$$

Because $E_{(+)}$ is larger than $E_{(-)}$, the net electric field $\vec{E}_L$ in this region is directed to the left, as Fig. 24-17*c* shows. To the right of the sheets, the electric field $\vec{E}_R$ has the same magnitude but is directed to the right, as Fig. 24-17*c* shows.

Between the sheets, the two fields add and we have

$$E_B = E_{(+)} + E_{(-)}$$
$$= 3.84 \times 10^5\ \text{N/C} + 2.43 \times 10^5\ \text{N/C}$$
$$= 6.3 \times 10^5\ \text{N/C}. \quad \text{(Answer)}$$

The electric field $\vec{E}_B$ is directed to the right.

24-9 Applying Gauss' Law: Spherical Symmetry

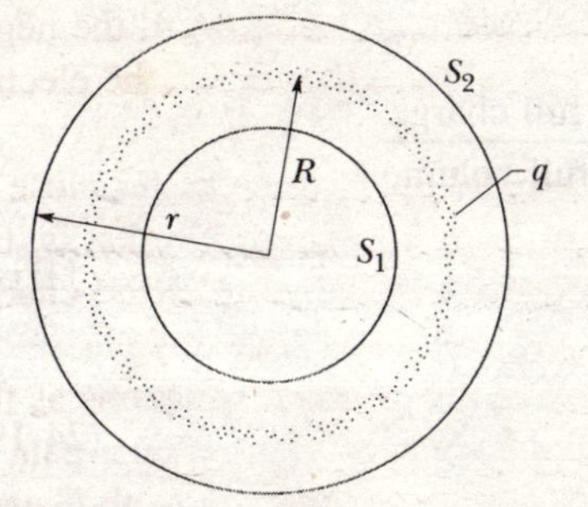

Fig. 24-18 A thin, uniformly charged, spherical shell with total charge q, in cross section. Two Gaussian surfaces S_1 and S_2 are also shown in cross section. Surface S_2 encloses the shell, and S_1 encloses only the empty interior of the shell.

Here we use Gauss' law to prove the two shell theorems presented without proof in Section 22-4:

> A shell of uniform charge attracts or repels a charged particle that is outside the shell as if all the shell's charge were concentrated at the center of the shell.

> A shell of uniform charge exerts no electrostatic force on a charged particle that is located inside the shell.

Figure 24-18 shows a charged spherical shell of total charge q and radius R and two concentric spherical Gaussian surfaces, S_1 and S_2. If we followed the procedure of Section 24-5 as we applied Gauss' law to surface S_2, for which $r \geq R$, we would find that

$$E = \frac{1}{4\pi\varepsilon_0}\frac{q}{r^2} \quad \text{(spherical shell, field at } r \geq R\text{).} \tag{24-15}$$

This is the same field that would be set up by a point charge q at the center of the shell of charge. Thus, a shell of charge q would produce the same force on a charged particle placed outside the shell as would a point charge q located at the center of the shell. This proves the first shell theorem.

Applying Gauss' law to surface S_1, for which $r < R$, leads directly to

$$E = 0 \quad \text{(spherical shell, field at } r < R\text{),} \tag{24-16}$$

because this Gaussian surface encloses no charge. Thus, if a charged particle were enclosed by the shell, the shell would exert no net electrostatic force on it. This proves the second shell theorem.

Any spherically symmetric charge distribution, such as that of Fig. 24-19, can be constructed with a nest of concentric spherical shells. For purposes of applying the two shell theorems, the volume charge density ρ should have a single value for each shell but need not be the same from shell to shell. Thus, for the charge distribution as a whole, ρ can vary, but only with r, the radial distance from the center. We can then examine the effect of the charge distribution "shell by shell."

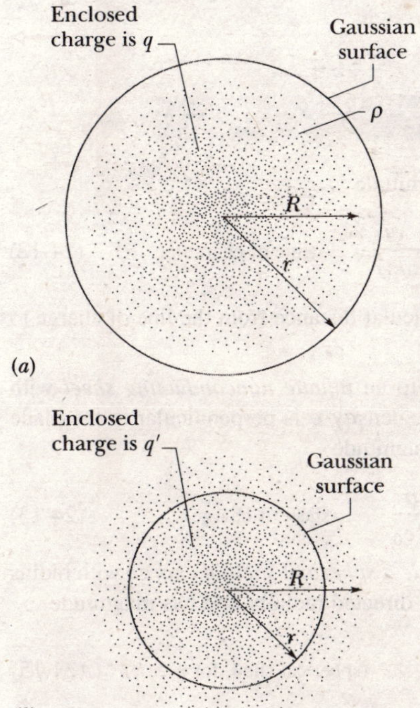

Fig. 24-19 The dots represent a spherically symmetric distribution of charge of radius R, whose volume charge density ρ is a function only of distance from the center. The charged object is not a conductor, and the charge is assumed to be fixed in position. A concentric spherical Gaussian surface with $r > R$ is shown in (*a*). A similar Gaussian surface with $r < R$ is shown in (*b*).

In Fig. 24-19*a* the entire charge lies within a Gaussian surface with $r > R$. The charge produces an electric field on the Gaussian surface as if the charge were a point charge located at the center, and Eq. 24-15 holds.

Figure 24-19*b* shows a Gaussian surface with $r < R$. To find the electric field at points on this Gaussian surface, we consider two sets of charged shells—one set inside the Gaussian surface and one set outside. Equation 24-16 says that the charge lying *outside* the Gaussian surface does not set up a net electric field on the Gaussian surface. Equation 24-15 says that the charge *enclosed* by the surface sets up an electric field as if that enclosed charge were concentrated at the center. Letting q' represent that enclosed charge, we can then rewrite Eq. 24-15 as

$$E = \frac{1}{4\pi\varepsilon_0}\frac{q'}{r^2} \quad \text{(spherical distribution, field at } r \leq R\text{).} \tag{24-17}$$

If the full charge q enclosed within radius R is uniform, then q' enclosed within radius r in Fig. 24-19b is proportional to q:

$$\frac{\text{charge enclosed by } r}{\text{volume enclosed by } r} = \frac{\text{full charge}}{\text{full volume}}$$

or

$$\frac{q'}{\frac{4}{3}\pi r^3} = \frac{q}{\frac{4}{3}\pi R^3}. \qquad (24\text{-}18)$$

This gives us

$$q' = q\frac{r^3}{R^3}. \qquad (24\text{-}19)$$

Substituting this into Eq. 24-17 yields

$$E = \left(\frac{q}{4\pi\varepsilon_0 R^3}\right)r \quad \text{(uniform charge, field at } r \leq R). \qquad (24\text{-}20)$$

CHECKPOINT 5: The figure shows two large, parallel, nonconducting sheets with identical (positive) uniform surface charge densities, and a sphere with a uniform (positive) volume charge density. Rank the four numbered points according to the magnitude of the net electric field there, greatest first.

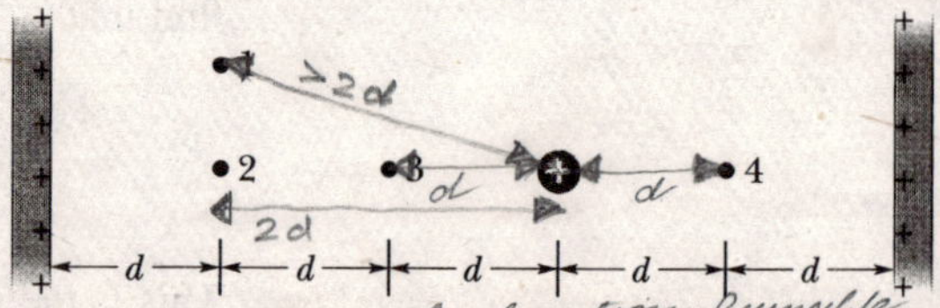

REVIEW & SUMMARY

Gauss' Law *Gauss' law* and Coulomb's law, although expressed in different forms, are equivalent ways of describing the relation between charge and electric field in static situations. Gauss' law is

$$\varepsilon_0\Phi = q_{enc} \quad \text{(Gauss' law)}, \qquad (24\text{-}6)$$

in which q_{enc} is the net charge inside an imaginary closed surface (a **Gaussian surface**) and Φ is the net **flux** of the electric field through the surface:

$$\Phi = \oint \vec{E}\cdot d\vec{A} \quad \text{(electric flux through a Gaussian surface)}. \qquad (24\text{-}4)$$

Coulomb's law can readily be derived from Gauss' law.

Applications of Gauss' Law Using Gauss' law and, in some cases, symmetry arguments, we can derive several important results in electrostatic situations. Among these are:

1. An excess charge on a *conductor* is located entirely on the outer surface of the conductor.
2. The external electric field near the *surface of a charged conductor* is perpendicular to the surface and has magnitude

$$E = \frac{\sigma}{\varepsilon_0} \quad \text{(conducting surface)}. \qquad (24\text{-}11)$$

Within the conductor, $E = 0$.

3. The electric field at any point due to an infinite *line of charge* with uniform linear charge density λ is perpendicular to the line of charge and has magnitude

$$E = \frac{\lambda}{2\pi\varepsilon_0 r} \quad \text{(line of charge)}, \qquad (24\text{-}12)$$

where r is the perpendicular distance from the line of charge to the point.

4. The electric field due to an *infinite nonconducting sheet* with uniform surface charge density σ is perpendicular to the plane of the sheet and has magnitude

$$E = \frac{\sigma}{2\varepsilon_0} \quad \text{(sheet of charge)}. \qquad (24\text{-}13)$$

5. The electric field *outside a spherical shell of charge* with radius R and total charge q is directed radially and has magnitude

$$E = \frac{1}{4\pi\varepsilon_0}\frac{q}{r^2} \quad \text{(spherical shell, for } r \geq R). \qquad (24\text{-}15)$$

Here r is the distance from the center of the shell to the point at which E is measured. (The charge behaves, for external points, as if it were all located at the center of the sphere.) The field *inside* a uniform spherical shell of charge is exactly zero:

$$E = 0 \quad \text{(spherical shell, for } r < R). \qquad (24\text{-}16)$$

6. The electric field *inside a uniform sphere of charge* is directed radially and has magnitude

$$E = \left(\frac{q}{4\pi\varepsilon_0 R^3}\right)r. \qquad (24\text{-}20)$$

QUESTIONS

1. A surface has the area vector $\vec{A} = (2\hat{i} + 3\hat{j})\ m^2$. What is the flux of an electric field through it if the field is (a) $\vec{E} = 4\hat{i}$ N/C and (b) $\vec{E} = 4\hat{k}$ N/C?

2. What is $\int dA$ for (a) a square of edge length a, (b) a circle of radius r, and (c) the curved surface of a cylinder of length h and radius r?

3. In Fig. 24-20, a full Gaussian surface encloses two of the four positively charged particles. (a) Which of the particles contribute to the electric field at point P on the surface? (b) Which net flux of electric field through the surface is greater (if either): that due to q_1 and q_2 or that due to all four charges?

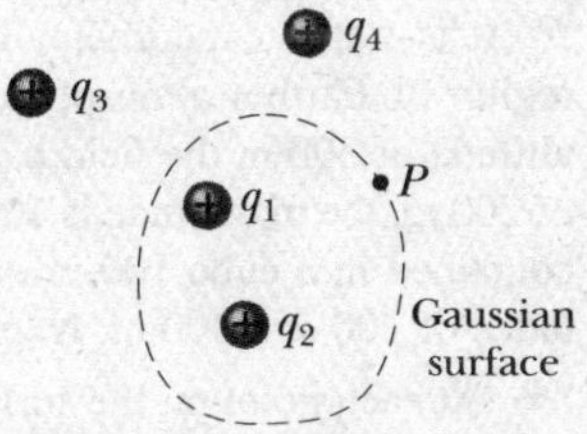

Fig. 24-20 Question 3.

4. Figure 24-21 shows, in cross section, a central metal ball, two spherical metal shells, and three spherical Gaussian surfaces of radii R, $2R$, and $3R$, all with the same center. The uniform charges on the three objects are: ball, Q; smaller shell, $3Q$; larger shell, $5Q$. Rank the Gaussian surfaces according to the magnitude of the electric field at any point on the surface, greatest first.

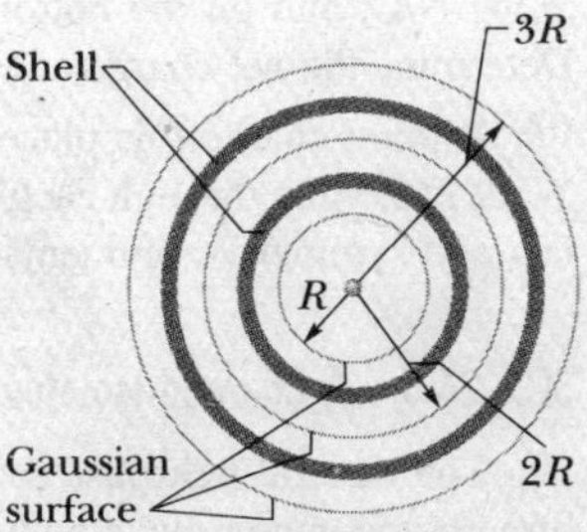

Fig. 24-21 Question 4.

5. Figure 24-22 shows three Gaussian surfaces, each half-submerged in a large, thick metal plate with a uniform surface charge density. Gaussian surface S_1 is the tallest and has the smallest square end caps; surface S_3 is shortest and has the largest square end caps; and S_2 has intermediate values. Rank the surfaces according to (a) the charge they enclose, (b) the magnitude of the electric field at points on their top end cap, (c) the net electric flux through that top end cap, and (d) the net electric flux through their bottom end cap, greatest first.

Fig. 24-22 Question 5.

6. Figure 24-23 shows, in cross section, three cylinders, each of uniform charge Q. Concentric with each cylinder is a cylindrical Gaussian surface, all three with the same radius. Rank the Gaussian surfaces according to the electric field at any point on the surface, greatest first.

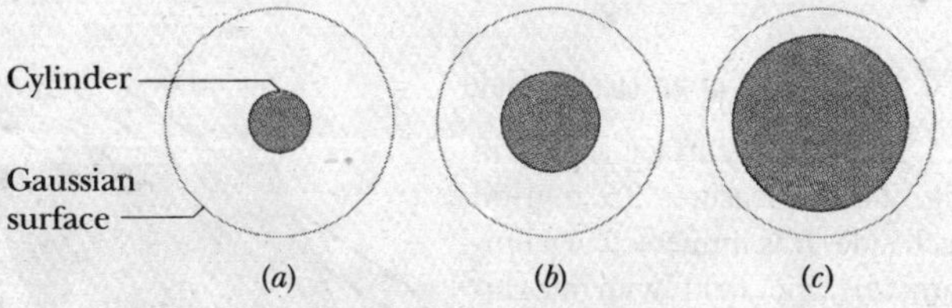

Fig. 24-23 Question 6.

7. Three infinite nonconducting sheets, with uniform surface charge densities σ, 2σ, and 3σ, are arranged to be parallel like the two sheets in Fig. 24-17a. What is their order, from left to right, if the electric field $\vec{E}$ produced by the arrangement has magnitude $E = 0$ in one region and $E = 2\sigma/\varepsilon_0$ in another region?

8. A small charged ball lies within the hollow of a metallic spherical shell of radius R. Here, for three situations, are the net charges on the ball and shell, respectively: (1) $+4q$, 0; (2) $-6q$, $+10q$; (3) $+16q$, $-12q$. Rank the situations according to the charge on (a) the inner surface of the shell and (b) the outer surface, most positive first.

9. Rank the situations of Question 8 according to the magnitude of the electric field (a) halfway through the shell and (b) at a point $2R$ from the center of the shell, greatest first.

10. Figure 24-24 shows four spheres, each with charge Q uniformly distributed through its volume. (a) Rank the spheres according to their volume charge density, greatest first. The figure also shows a point P for each sphere, all at the same distance from the center of the sphere. (b) Rank the spheres according to the magnitude of the electric field they produce at point P, greatest first.

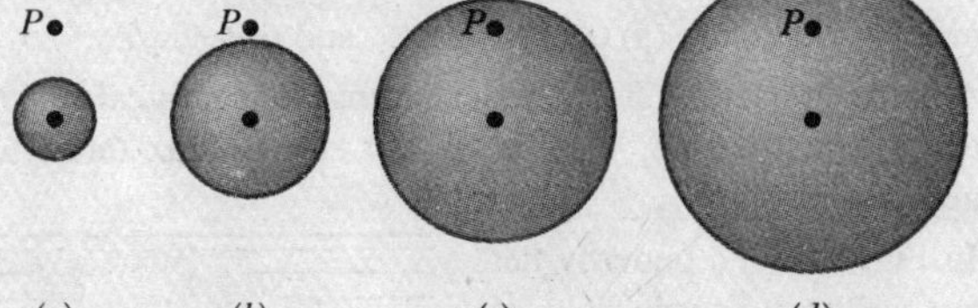

Fig. 24-24 Question 10.

EXERCISES & PROBLEMS

ssm Solution is in the Student Solutions Manual.
www Solution is available on the World Wide Web at: http://www.wiley.com/college/hrw
ilw Solution is available on the Interactive LearningWare.

SEC. 24-2 Flux

1E. Water in an irrigation ditch of width $w = 3.22$ m and depth $d = 1.04$ m flows with a speed of 0.207 m/s. The *mass flux* of the flowing water through an imaginary surface is the product of the water's density (1000 kg/m^3) and its volume flux through that surface. Find the mass flux through the following imaginary surfaces: (a) a surface of area wd, entirely in the water, perpendicular to the flow; (b) a surface with area $3wd/2$, of which wd is in the water, perpendicular to the flow; (c) a surface of area $wd/2$, entirely in the water, perpendicular to the flow; (d) a surface of area wd, half in the water and half out, perpendicular to the flow; (e) a surface of area wd, entirely in the water, with its normal 34° from the direction of flow.

SEC. 24-3 Flux of an Electric Field

2E. The square surface shown in Fig. 24-25 measures 3.2 mm on each side. It is immersed in a uniform electric field with magnitude $E = 1800$ N/C. The field lines make an angle of 35° with a normal to the surface, as shown. Take that normal to be directed "outward," as though the surface were one face of a box. Calculate the electric flux through the surface. ssm

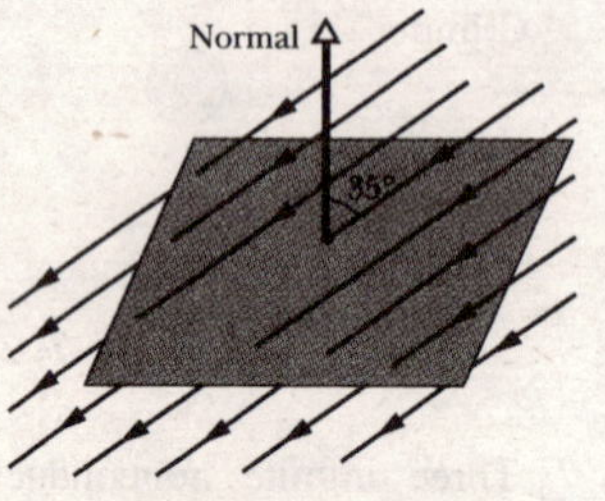

Fig. 24-25 Exercise 2.

3E. The cube in Fig. 24-26 has edge lengths of 1.40 m and is oriented as shown in a region of uniform electric field. Find the electric flux through the right face if the electric field, in newtons per coulomb, is given by (a) $6.00\hat{i}$, (b) $-2.00\hat{j}$, and (c) $-3.00\hat{i} + 4.00\hat{k}$. (d) What is the total flux through the cube for each of these fields?

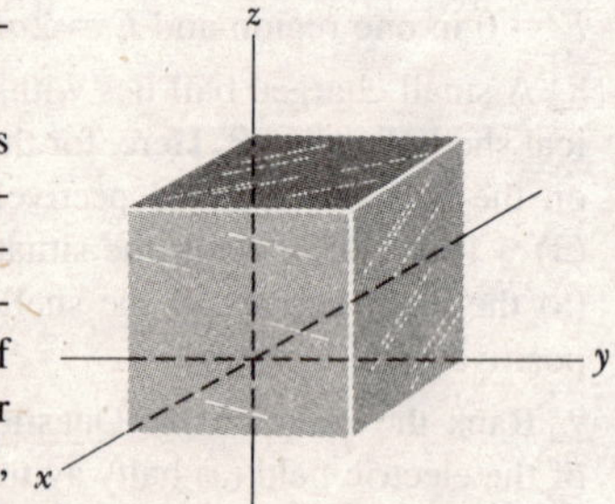

Fig. 24-26 Exercise 3 and Problems 7 and 10.

SEC. 24-4 Gauss' Law

4E. You have four point charges, $2q$, q, $-q$, and $-2q$. If possible, describe how you would place a closed surface that encloses at least the charge $2q$ (and perhaps other charges) and through which the net electric flux is (a) 0, (b) $+3q/\varepsilon_0$, and (c) $-2q/\varepsilon_0$.

5E. A point charge of 1.8 μC is at the center of a cubical Gaussian surface 55 cm on edge. What is the net electric flux through the surface? ssm

6E. In Fig. 24-27, a butterfly net is in a uniform electric field of magnitude E. The rim, a circle of radius a, is aligned perpendicular to the field. Find the electric flux through the netting.

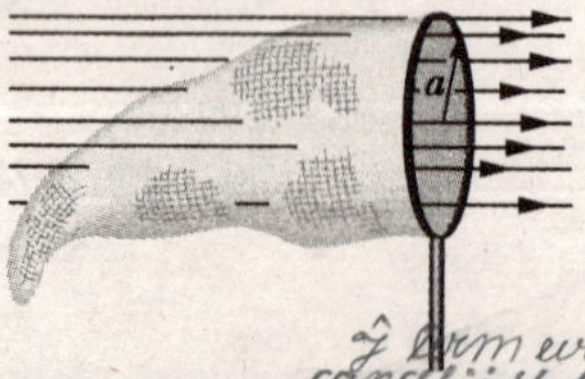

Fig. 24-27 Exercise 6.

7P. Find the net flux through the cube given in Exercise 3 and Fig. 24-26 if the electric field is given by (a) $\vec{E} = 3.00y\hat{j}$ and (b) $\vec{E} = -4.00\hat{i} + (6.00 + 3.00y)\hat{j}$. E is in newtons per coulomb, and y is in meters. (c) In each case, how much charge is enclosed by the cube? ilw

8P. When a shower is turned on in a closed bathroom, the splashing of the water on the bare tub can fill the room's air with negatively charged ions and produce an electric field in the air as great as 1000 N/C. Consider a bathroom with dimensions of 2.5 m × 3.0 m × 2.0 m. Along the ceiling, floor, and four walls, approximate the electric field in the air as being directed perpendicular to the surface and as having a uniform magnitude of 600 N/C. Also, treat those surfaces as forming a closed Gaussian surface around the room's air. What are (a) the volume charge density ρ and (b) the number of excess elementary charges e per cubic meter in the room's air?

9P. It is found experimentally that the electric field in a certain region of Earth's atmosphere is directed vertically down. At an altitude of 300 m the field has magnitude 60.0 N/C; at an altitude of 200 m, the magnitude is 100 N/C. Find the net amount of charge contained in a cube 100 m on edge, with horizontal faces at altitudes of 200 and 300 m. Neglect the curvature of Earth. ssm

10P. At each point on the surface of the cube shown in Fig. 24-26, the electric field is in the positive direction of y. The length of each edge of the cube is 3.0 m. On the top surface of the cube $\vec{E} = -34\hat{j}$ N/C, and on the bottom face of the cube $\vec{E} = +20\hat{j}$ N/C. Determine the net charge contained within the cube.

11P. A point charge q is placed at one corner of a cube of edge a. What is the flux through each of the cube faces? (*Hint:* Use Gauss' law and symmetry arguments.) ssm

SEC. 24-6 A Charged Isolated Conductor

12E. The electric field just above the surface of the charged drum of a photocopying machine has a magnitude E of 2.3×10^5 N/C. What is the surface charge density on the drum, assuming that the drum is a conductor?

13E. A uniformly charged conducting sphere of 1.2 m diameter has a surface charge density of 8.1 μC/m^2. (a) Find the net charge on the sphere. (b) What is the total electric flux leaving the surface of the sphere? ssm

14E. Space vehicles traveling through Earth's radiation belts can intercept a significant number of electrons. The resulting charge buildup can damage electronic components and disrupt operations. Suppose a spherical metallic satellite 1.3 m in diameter accumulates 2.4 μC of charge in one orbital revolution. (a) Find the resulting surface charge density. (b) Calculate the magnitude of the electric field just outside the surface of the satellite, due to the surface charge.

15P. An isolated conductor of arbitrary shape has a net charge of $+10 \times 10^{-6}$ C. Inside the conductor is a cavity within which is a point charge $q = +3.0 \times 10^{-6}$ C. What is the charge (a) on the cavity wall and (b) on the outer surface of the conductor? ssm www

SEC. 24-7 Applying Gauss' Law: Cylindrical Symmetry

16E. (a) The drum of the photocopying machine in Exercise 12 has a length of 42 cm and a diameter of 12 cm. What is the total charge on the drum? (b) The manufacturer wishes to produce a desktop version of the machine. This requires reducing the size of the drum to a length of 28 cm and a diameter of 8.0 cm. The electric field at the drum surface must remain unchanged. What must be the charge on this new drum?

17E. An infinite line of charge produces a field of 4.5×10^4 N/C at a distance of 2.0 m. Calculate the linear charge density. ssm

18P. Figure 24-28 shows a section of a long, thin-walled metal tube of radius R, with a charge per unit length λ on its surface. Derive expressions for E in terms of the distance r from the tube axis, considering both (a) $r > R$ and (b) $r < R$. Plot your results for the range $r = 0$ to $r = 5.0$ cm, assuming that $\lambda = 2.0 \times 10^{-8}$ C/m and $R = 3.0$ cm. (*Hint:* Use cylindrical Gaussian surfaces, coaxial with the metal tube.)

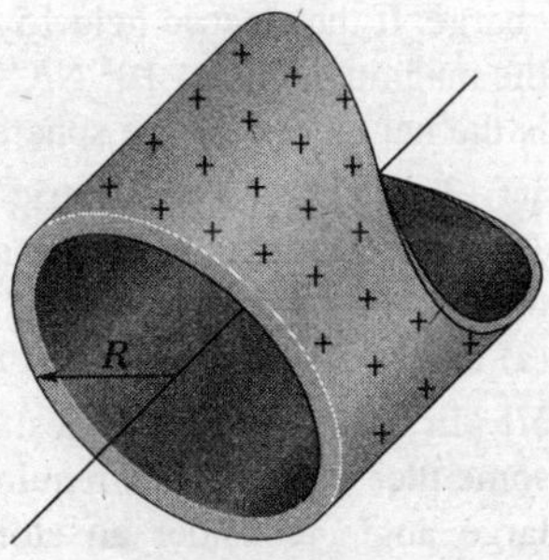

Fig. 24-28 Problem 18.

19P. A very long conducting cylindrical rod of length L with a total charge $+q$ is surrounded by a conducting cylindrical shell (also of length L) with total charge $-2q$, as shown in Fig. 24-29. Use Gauss' law to find (a) the electric field at points outside the conducting shell, (b) the distribution of charge on the shell, and (c) the electric field in the region between the shell and rod. ssm

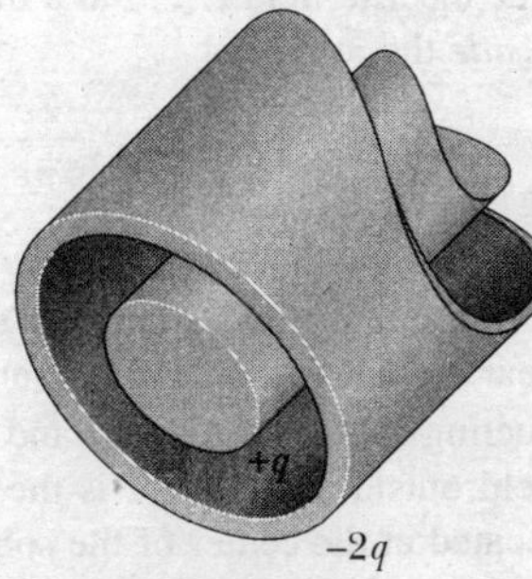

Fig. 24-29 Problem 19.

20P. A long, straight wire has fixed negative charge with a linear charge density of magnitude 3.6 nC/m. The wire is to be enclosed by a thin, nonconducting cylinder of outside radius 1.5 cm, coaxial with the wire. The cylinder is to have positive charge on its outside surface with a surface charge density σ such that the net external electric field is zero. Calculate the required σ.

21P. Two long, charged, concentric cylinders have radii of 3.0 and 6.0 cm. The charge per unit length is 5.0×10^{-6} C/m on the inner cylinder and -7.0×10^{-6} C/m on the outer cylinder. Find the electric field at (a) $r = 4.0$ cm and (b) $r = 8.0$ cm, where r is the radial distance from the common central axis. ilw

22P. A long, nonconducting, solid cylinder of radius 4.0 cm has a nonuniform volume charge density ρ that is a function of the radial distance r from the axis of the cylinder, as given by $\rho = Ar^2$, with $A = 2.5\ \mu\text{C/m}^5$. What is the magnitude of the electric field at a radial distance of (a) 3.0 cm and (b) 5.0 cm from the axis of the cylinder?

23P. Figure 24-30 shows a Geiger counter, a device used to detect ionizing radiation (radiation that causes ionization of atoms). The counter consists of a thin, positively charged central wire surrounded by a concentric, circular, conducting cylinder with an equal negative charge. Thus, a strong radial electric field is set up inside the cylinder. The cylinder contains a low-pressure inert gas. When a particle of radiation enters the device through the cylinder wall, it ionizes a few of the gas atoms. The resulting free electrons (label e) are drawn to the positive wire. However, the electric field is so intense that, between collisions with other gas atoms, the free electrons gain energy sufficient to ionize these atoms also. More free electrons are thereby created, and the process is repeated until the electrons reach the wire. The resulting "avalanche" of electrons is collected by the wire, generating a signal that is used to record the passage of the original particle of radiation. Suppose that the radius of the central wire is 25 μm, the radius of the cylinder 1.4 cm, and the length of the tube 16 cm. If the electric field at the cylinder's inner wall is 2.9×10^4 N/C, what is the total positive charge on the central wire? ssm

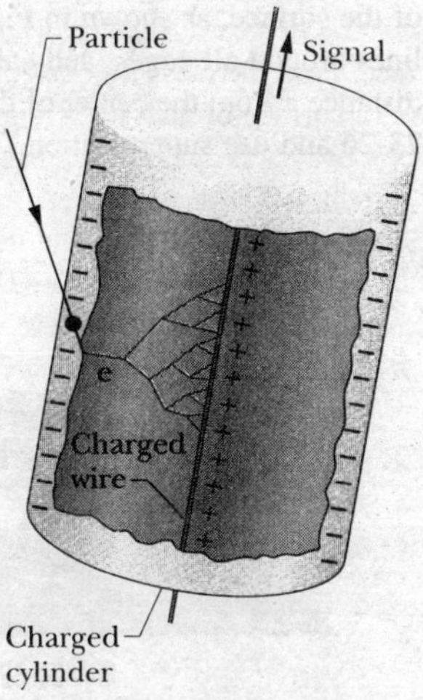

Fig. 24-30 Problem 23.

24P. A charge of uniform linear density 2.0 nC/m is distributed along a long, thin, nonconducting rod. The rod is coaxial with a long, hollow, conducting cylinder (inner radius = 5.0 cm, outer radius = 10 cm). The net charge on the conductor is zero. (a) What is the magnitude of the electric field 15 cm from the axis of the cylinder? What is the surface charge density on (b) the inner surface and (c) the outer surface of the conductor?

25P. Charge is distributed uniformly throughout the volume of an infinitely long cylinder of radius R. (a) Show that, at a distance r from the cylinder axis (for $r < R$),

$$E = \frac{\rho r}{2\varepsilon_0},$$

where ρ is the volume charge density. (b) Write an expression for E when $r > R$. ssm www

SEC. 24-8 Applying Gauss' Law: Planar Symmetry

26E. Figure 24-31 shows cross-sections through two large, parallel, nonconducting sheets with identical distributions of positive charge with surface charge density σ. What is $\vec{E}$ at points (a) above the sheets, (b) between them, and (c) below them?

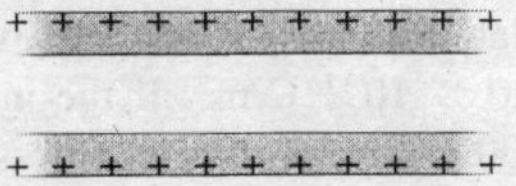

Fig. 24-31 Exercise 26.

27E. A square metal plate of edge length 8.0 cm and negligible thickness has a total charge of 6.0×10^{-6} C. (a) Estimate the magnitude E of the electric field just off the center of the plate (at, say, a distance of 0.50 mm) by assuming that the charge is spread uniformly over the two faces of the plate. (b) Estimate E at a distance of 30 m (large relative to the plate size) by assuming that the plate is a point charge. ssm

28E. A large, flat, nonconducting surface has a uniform charge density σ. A small circular hole of radius R has been cut in the middle of the surface, as shown in Fig. 24-32. Ignore fringing of the field lines around all edges, and calculate the electric field at point P, a distance z from the center of the hole along its axis. (*Hint:* See Eq. 23-26 and use superposition.)

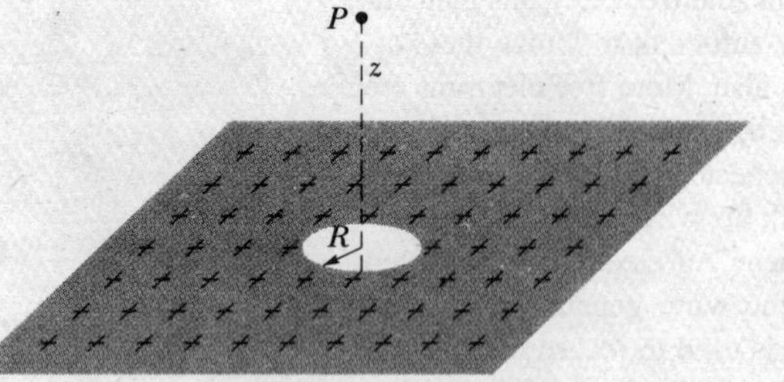

Fig. 24-32 Exercise 28.

29P. In Fig. 24-33, a small, nonconducting ball of mass m = 1.0 mg and charge q = 2.0 × 10^{-8} C (distributed uniformly through its volume) hangs from an insulating thread that makes an angle θ = 30° with a vertical, uniformly charged nonconducting sheet (shown in cross section). Considering the gravitational force on the ball and assuming that the sheet extends far vertically and into and out of the page, calculate the surface charge density σ of the sheet. ssm

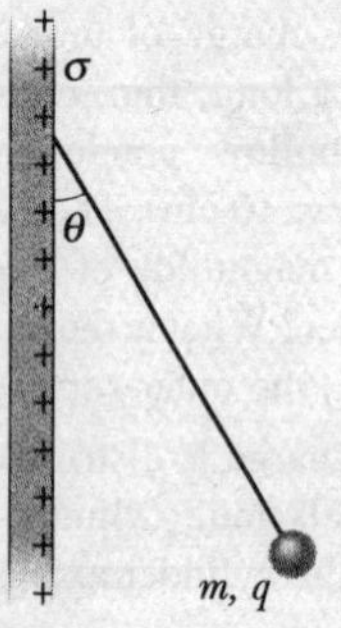

Fig. 24-33 Problem 29.

30P. Two large, thin metal plates are parallel and close to each other, as in Fig. 24-16*c*. On their inner faces, the plates have excess surface charge densities of opposite signs and with a magnitude 7.0 × 10^{-22} C/m², with the negatively charged plate on the left. What are the magnitude and direction of the electric field $\vec{E}$ (a) to the left of the plates, (b) to the right of the plates, and (c) between the plates?

31P. An electron is shot directly toward the center of a large metal plate that has excess negative charge with surface charge density 2.0 × 10^{-6} C/m². If the initial kinetic energy of the electron is 100 eV and if the electron is to stop (owing to electrostatic repulsion from the plate) just as it reaches the plate, how far from the plate must it be shot? ssm

32P. Two large metal plates of area 1.0 m² face each other. They are 5.0 cm apart and have equal but opposite charges on their inner surfaces. If the magnitude E of the electric field between the plates is 55 N/C, what is the magnitude of the charge on each plate? Neglect edge effects.

assume nonconductor unless stated otherwise

33P*. A planar slab of thickness d has a uniform volume charge density ρ. Find the magnitude of the electric field at all points in space both (a) inside and (b) outside the slab, in terms of x, the distance measured from the central plane of the slab. ssm

SEC. 24-9 Applying Gauss' Law: Spherical Symmetry

34E. A point charge causes an electric flux of −750 N · m²/C to pass through a spherical Gaussian surface of 10.0 cm radius centered on the charge. (a) If the radius of the Gaussian surface were doubled, how much flux would pass through the surface? (b) What is the value of the point charge?

35E. A conducting sphere of radius 10 cm has an unknown charge. If the electric field 15 cm from the center of the sphere has the magnitude 3.0 × 10^3 N/C and is directed radially inward, what is the net charge on the sphere? ssm

36E. Two charged concentric spheres have radii of 10.0 cm and 15.0 cm. The charge on the inner sphere is 4.00 × 10^{-8} C, and that on the outer sphere is 2.00 × 10^{-8} C. Find the electric field (a) at r = 12.0 cm and (b) at r = 20.0 cm.

37E. In a 1911 paper, Ernest Rutherford said: "In order to form some idea of the forces required to deflect an α particle through a large angle, consider an atom [as] containing a point positive charge Ze at its centre and surrounded by a distribution of negative electricity $-Ze$ uniformly distributed within a sphere of radius R. The electric field E . . . at a distance r from the center for a point *inside* the atom [is]

$$E = \frac{Ze}{4\pi\varepsilon_0}\left(\frac{1}{r^2} - \frac{r}{R^3}\right)."$$

Verify this equation. ssm

38E. Equation 24-11 ($E = \sigma/\varepsilon_0$) gives the electric field at points near a charged conducting surface. Apply this equation to a conducting sphere of radius r and charge q, and show that the electric field outside the sphere is the same as the field of a point charge located at the center of the sphere.

39P. A proton with speed v = 3.00 × 10^5 m/s orbits just outside a charged sphere of radius r = 1.00 cm. What is the charge on the sphere? ssm www

40P. A point charge $+q$ is placed at the center of an electrically neutral, spherical conducting shell with inner radius a and outer radius b. What charge appears on (a) the inner surface of the shell and (b) the outer surface? What is the net electric field at a distance r from the center of the shell if (c) $r < a$, (d) $b > r > a$, and (e) $r > b$? Sketch field lines for those three regions. For $r > b$, what is the net electric field due to (f) the central point charge plus the inner surface charge and (g) the outer surface charge? A point charge $-q$ is now placed outside the shell. Does this point charge change the charge distribution on (h) the outer surface and (i) the inner surface? Sketch the field lines now. (j) Is there an electrostatic force on the second point charge? (k) Is there a net electrostatic force on the first point charge? (l) Does this situation violate Newton's third law?

41P. A solid nonconducting sphere of radius R has a nonuniform charge distribution of volume charge density $\rho = \rho_s r/R$, where ρ_s is a constant and r is the distance from the center of the sphere. Show (a) that the total charge on the sphere is $Q = \pi\rho_s R^3$ and (b) that

$$E = \frac{1}{4\pi\varepsilon_0}\frac{Q}{R^4}r^2$$

gives the magnitude of the electric field inside the sphere. ilw

42P. A hydrogen atom can be considered as having a central point-like proton of positive charge $+e$ and an electron of negative charge $-e$ that is distributed about the proton according to the volume charge density $\rho = A\exp(-2r/a_0)$. Here A is a constant, $a_0 = 0.53 \times 10^{-10}$ m is the *Bohr radius,* and r is the distance from the center of the atom. (a) Using the fact that hydrogen is electrically neutral, find A. (b) Then find the electric field produced by the atom at the Bohr radius.

43P. In Fig. 24-34 a sphere, of radius a and charge $+q$ uniformly distributed throughout its volume, is concentric with a spherical conducting shell of inner radius b and outer radius c. This shell has a net charge of $-q$. Find expressions for the electric field, as a function of the radius r, (a) within the sphere ($r < a$), (b) between the sphere and the shell ($a < r < b$), (c) inside the shell ($b < r < c$), and (d) outside the shell ($r > c$). (e) What are the charges on the inner and outer surfaces of the shell? ssm

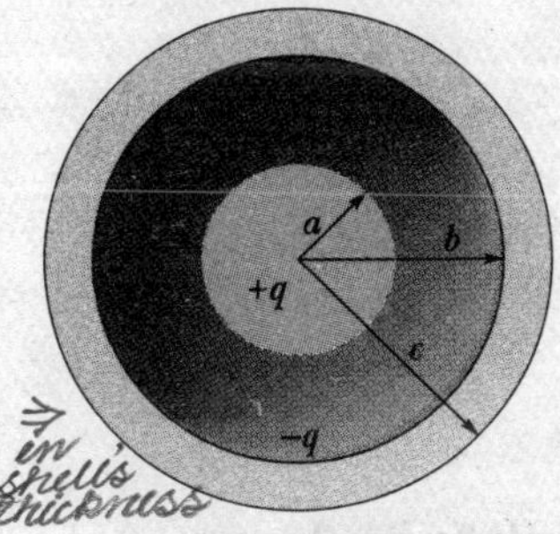

Fig. 24-34 Problem 43.

44P. Figure 24-35*a* shows a spherical shell of charge with uniform volume charge density ρ. Plot E due to the shell for distances r from the center of the shell ranging from zero to 30 cm. Assume that $\rho = 1.0 \times 10^{-6}$ C/m^3, $a = 10$ cm, and $b = 20$ cm.

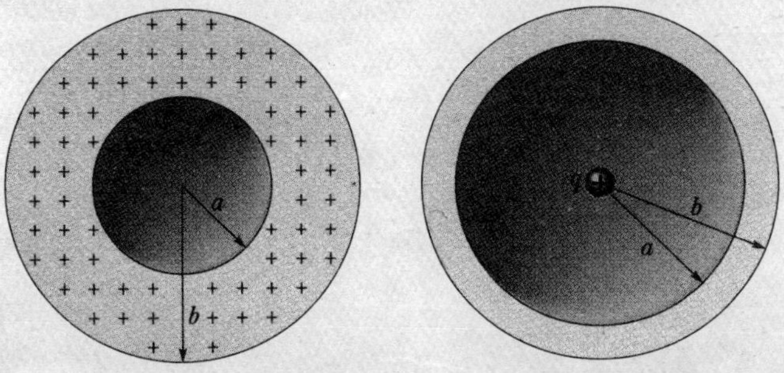

(*a*) (*b*)

Fig. 24-35 Problems 44 and 45.

45P. In Fig. 24-35*b*, a nonconducting spherical shell, of inner radius a and outer radius b, has a positive volume charge density $\rho = A/r$ (within its thickness), where A is a constant and r is the distance from the center of the shell. In addition, a positive point charge q is located at that center. What value should A have if the electric field in the shell ($a \le r \le b$) is to be uniform? (*Hint:* The constant A depends on a but not on b.) ssm

46P*. A nonconducting sphere has a uniform volume charge density ρ. Let $\vec{r}$ be the vector from the center of the sphere to a general point P within the sphere. (a) Show that the electric field at P is given by $\vec{E} = \rho\vec{r}/3\varepsilon_0$. (Note that the result is independent of the radius of the sphere.) (b) A spherical cavity is hollowed out of the sphere, as shown in Fig. 24-36. Using superposition concepts, show that the electric field at all points within the cavity is uniform and equal to $\vec{E} = \rho\vec{a}/3\varepsilon_0$, where $\vec{a}$ is the position vector from the center of the sphere to the center of the cavity. (Note that this result is independent of the radius of the sphere and the radius of the cavity.)

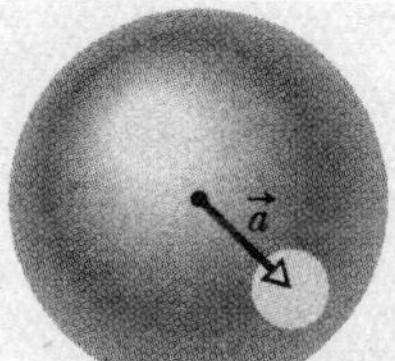

Fig. 24-36 Problem 46.

47P*. A spherically symmetrical but nonuniform volume distribution of charge produces an electric field of magnitude $E = Kr^4$, directed radially outward from the center of the sphere. Here r is the radial distance from that center, and K is a constant. What is the volume density ρ of the charge distribution?

Additional Problem

48. *The chocolate crumb mystery.* Explosions ignited by electrostatic discharges (sparks) constitute a serious danger in facilities handling grain or powder. Such an explosion occurred in chocolate crumb powder at a biscuit factory in the 1970s. At the factory, workers usually emptied newly delivered sacks of the powder into a loading bin, from which it was blown through grounded PVC pipes to a silo for storage. Somewhere along this route, two conditions for an explosion were met: (1) The magnitude of an electric field became 3.0×10^6 N/C or greater, so that electric breakdown and thus sparking could occur. (2) The energy of a spark was 150 mJ or greater so that it could ignite the powder explosively. Let us check for the first condition in the powder flow through the PVC pipes.

Suppose a stream of *negatively* charged chocolate crumb powder is blown through a cylindrical PVC pipe with radius $R = 5.0$ cm. Assume that the powder and its charge are spread uniformly through the pipe with a volume charge density ρ. (a) Using Gauss' law, find an expression for the magnitude of the electric field $\vec{E}$ in the pipe as a function of the radial distance r from the center of the pipe. (b) Does the magnitude increase or decrease with r? (c) Is the electric field $\vec{E}$ directed radially inward or outward? (d) Assuming a volume charge density ρ of magnitude 1.1×10^{-3} C/m^3 (which was typical at the biscuit factory), find the maximum magnitude of the electric field and determine where that maximum field occurs. (e) Could sparking occur, and if so, where? (The story continues with Problem 57 in Chapter 25.)

25 Electric Potential

While enjoying the Sequoia National Park from a lookout platform, this woman found her hair rising from her head. Amused, her brother took her photograph. Five minutes after they left, lightning struck the platform, killing one person and injuring seven.

What had caused the woman's hair to rise?

The answer is in this chapter.

25-1 Electric Potential Energy

Newton's law for the gravitational force and Coulomb's law for the electrostatic force are mathematically identical. Thus, the general features we have discussed for the gravitational force should apply to the electrostatic force.

The concept of PE holds only for conservative forces

In particular, we can infer, correctly, that the electrostatic force is a *conservative force*. Thus, when an electrostatic force acts between two or more charged particles within a system of particles, we can assign an **electric potential energy** U to the system. Moreover, if the system changes its configuration from an initial state i to a different final state f, the electrostatic force does work W on the particles. From Eq. 8-1, we then know that the resulting change ΔU in the potential energy of the system is

$$\boxed{\Delta U = U_f - U_i = -W.} \tag{25-1}$$

As with other conservative forces, the work done by the electrostatic force is *path independent*. Suppose a charged particle within the system moves from point i to point f while an electrostatic force between it and the rest of the system acts on it. Provided the rest of the system does not change, the work W done by the force is the same for *all* paths between points i and f.

For convenience, we usually take the *reference configuration* of a system of charged particles to be that in which the particles are all infinitely separated from each other. Also, we usually set the corresponding *reference potential energy* to be zero. Suppose that several charged particles come together from initially infinite separations (state i) to form a system of nearby particles (state f). Let the initial potential energy U_i be zero, and let W_∞ represent the work done by the electrostatic forces between the particles during the move in from infinity. Then from Eq. 25-1, the final potential energy U of the system is

$$\boxed{U = -W_\infty.} \tag{25-2}$$

As is true of other kinds of potential energy, electric potential energy is considered to be a type of mechanical energy. Recall from Chapter 8 that if only conservative forces act within a (closed) system, the mechanical energy of the system is conserved. We shall use this fact extensively in the rest of this chapter.

PROBLEM-SOLVING TACTICS

Tactic 1: *Electric Potential Energy; Work Done by a Field*
An electric potential energy is associated with a system of particles as a whole. However, you will see statements (starting with Sample Problem 25-1) that associate it with only one particle within a system. For example, you might read, "An electron in an electric field has a potential energy of 10^{-7} J." Such statements are often acceptable, but you should always keep in mind that the potential energy is actually associated with a system—here the electron plus the charged particles that set up the electric field. Also keep in mind that it makes sense to assign a particular potential energy value, such as 10^{-7} J here, to a particle or even a system *only* if the reference potential energy value is known.

When the potential energy is associated with only one particle within a system, you often will read that work is done on the particle *by the electric field*. What is meant is that work is done by the force on the particle due to the charges that set up the field.

Sample Problem 25-1

Electrons are continually being knocked out of air molecules in the atmosphere by cosmic-ray particles coming in from space. Once released, each electron experiences an electrostatic force $\vec{F}$ due to the electric field $\vec{E}$ that is produced in the atmosphere by charged particles already on Earth. Near Earth's surface the electric field has the magnitude $E = 150$ N/C and is directed downward. What is the change ΔU in the electric potential energy of a released electron when the electrostatic force causes it to move vertically upward through a distance $d = 520$ m (Fig. 25-1)?

SOLUTION: We need three **Key Ideas** here. One is that the change ΔU in the electric potential energy of the electron is related to the work

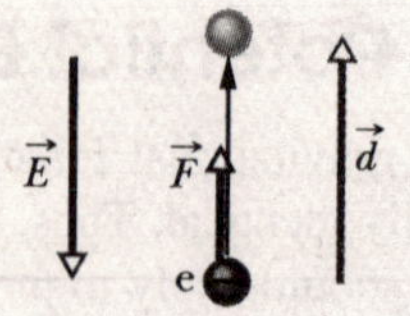

Fig. 25-1 Sample Problem 25-1. An electron in the atmosphere is moved upward through displacement $\vec{d}$ by an electrostatic force $\vec{F}$ due to an electric field $\vec{E}$.

W done on the electron by the electric field. Equation 25-1 ($\Delta U = -W$) gives the relation. A second **Key Idea** is that the work done by a constant force $\vec{F}$ on a particle undergoing a displacement $\vec{d}$ is

$$W = \vec{F} \cdot \vec{d}. \tag{25-3}$$

Finally, the third **Key Idea** is that the electrostatic force and the electric field are related by $\vec{F} = q\vec{E}$, where here q is the charge of an electron ($= -1.6 \times 10^{-19}$ C). Substituting for $\vec{F}$ in Eq. 25-3 and taking the dot product yield

$$W = q\vec{E} \cdot \vec{d} = qEd \cos \theta, \tag{25-4}$$

where θ is the angle between the directions of $\vec{E}$ and $\vec{d}$. The field $\vec{E}$ is directed downward and the displacement $\vec{d}$ is directed upward, so $\theta = 180°$. Substituting this and other data into Eq. 25-4, we find

$$W = (-1.6 \times 10^{-19}\text{ C})(150\text{ N/C})(520\text{ m}) \cos 180°$$
$$= 1.2 \times 10^{-14}\text{ J}.$$

Equation 25-1 then yields

$$\Delta U = -W = -1.2 \times 10^{-14}\text{ J}. \quad \text{(Answer)}$$

This result tells us that during the 520 m ascent, the electric potential energy of the electron *decreases* by 1.2×10^{-14} J.

CHECKPOINT 1: In the figure, a proton moves from point i to point f in a uniform electric field directed as shown. (a) Does the electric field do positive or negative work on the proton? (b) Does the electric potential energy of the proton increase or decrease?

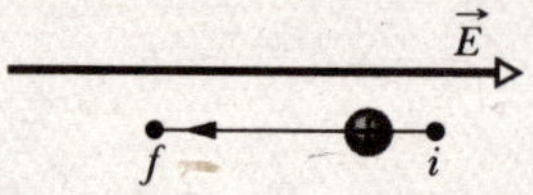

25-2 Electric Potential

As you can infer from Sample Problem 25-1, the potential energy of a charged particle in an electric field depends on the magnitude of the charge. However, the potential energy *per unit charge* has a unique value at any point in an electric field.

For an example of this, suppose we place a test particle of positive charge 1.60×10^{-19} C at a point in an electric field where the particle has an electric potential energy of 2.40×10^{-17} J. Then the potential energy per unit charge is

$$\frac{2.40 \times 10^{-17}\text{ J}}{1.60 \times 10^{-19}\text{ C}} = 150\text{ J/C}.$$

Next, suppose we replace that test particle with one having twice as much positive charge, 3.20×10^{-19} C. We would find that the second particle has an electric potential energy of 4.80×10^{-17} J, twice that of the first particle. However, the potential energy per unit charge would be the same, still 150 J/C.

Thus, the potential energy per unit charge, which can be symbolized as U/q, is independent of the charge q of the particle we happen to use and is *characteristic only of the electric field* we are investigating. The potential energy per unit charge at a point in an electric field is called the **electric potential** V (or simply the **potential**) at that point. Thus,

$$V = \frac{U}{q}. \tag{25-5}$$

Note that electric potential is a scalar, not a vector.

The *electric potential difference* ΔV between any two points i and f in an electric field is equal to the difference in potential energy per unit charge between the two points:

$$\Delta V = V_f - V_i = \frac{U_f}{q} - \frac{U_i}{q} = \frac{\Delta U}{q}. \tag{25-6}$$

Using Eq. 25-1 to substitute $-W$ for ΔU in Eq. 25-6, we can define the potential difference between points i and f as

$$\Delta V = V_f - V_i = -\frac{W}{q} \quad \text{(potential difference defined).} \tag{25-7}$$

The potential difference between two points is thus the negative of the work done by the electrostatic force to move a unit charge from one point to the other. A potential difference can be positive, negative, or zero, depending on the signs and magnitudes of q and W.

If we set $U_i = 0$ at infinity as our reference potential energy, then by Eq. 25-5, the electric potential must also be zero there. Then from Eq. 25-7, we can define the electric potential V at any point in an electric field to be

$$V = -\frac{W_\infty}{q} \quad \text{(potential defined),} \tag{25-8}$$

where W_∞ is the work done by the electric field on a charged particle as that particle moves in from infinity to point f. A potential V can be positive, negative, or zero, depending on the signs and magnitudes of q and W_∞.

The SI unit for potential that follows from Eq. 25-8 is the joule per coulomb. This combination occurs so often that a special unit, the *volt* (abbreviated V) is used to represent it. Thus,

$$1 \text{ volt} = 1 \text{ joule per coulomb.} \tag{25-9}$$

This new unit allows us to adopt a more conventional unit for the electric field $\vec{E}$, which we have measured up to now in newtons per coulomb. With two unit conversions, we obtain

$$1 \text{ N/C} = \left(1\frac{\text{N}}{\text{C}}\right)\left(\frac{1 \text{ V}\cdot\text{C}}{1 \text{ J}}\right)\left(\frac{1 \text{ J}}{1 \text{ N}\cdot\text{m}}\right)$$
$$= 1 \text{ V/m.} \tag{25-10}$$

The conversion factor in the second set of parentheses comes from Eq. 25-9; that in the third set of parentheses is derived from the definition of the joule. From now on, we shall express values of the electric field in volts per meter rather than in newtons per coulomb.

Finally, we can now define an energy unit that is a convenient one for energy measurements in the atomic and subatomic domain: One *electron-volt* (eV) is the energy equal to the work required to move a single elementary charge e, such as that of the electron or the proton, through a potential difference of exactly one volt. Equation 25-7 tells us that the magnitude of this work is $q\,\Delta V$, so

$$1 \text{ eV} = e(1 \text{ V})$$
$$= (1.60 \times 10^{-19} \text{ C})(1 \text{ J/C}) = 1.60 \times 10^{-19} \text{ J}.$$

PROBLEM-SOLVING TACTICS

Tactic 2: *Electric Potential and Electric Potential Energy*
Electric potential V and electric potential energy U are quite different quantities and should not be confused.

- *Electric potential* is a property of an electric field, regardless of whether a charged object has been placed in that field; it is measured in joules per coulomb, or volts.

- *Electric potential energy* is an energy of a charged object in an external electric field (or more precisely, an energy of the system consisting of the object and the external electric field); it is measured in joules.

Work Done by an Applied Force

Suppose we move a particle of charge q from point i to point f in an electric field by applying a force to it. During the move, our applied force does work W_{app} on the charge while the electric field does work W on it. By the work–kinetic energy theorem of Eq. 7-10, the change ΔK in the kinetic energy of the particle is

$$\Delta K = K_f - K_i = W_{app} + W. \tag{25-11}$$

Now suppose the particle is stationary before and after the move. Then K_f and K_i are both zero, and Eq. 25-11 reduces to

$$W_{app} = -W. \tag{25-12}$$

In words, the work W_{app} done by our applied force during the move is equal to the negative of the work W done by the electric field—provided there is no change in kinetic energy.

By using Eq. 25-12 to substitute W_{app} into Eq. 25-1, we can relate the work done by our applied force to the change in the potential energy of the particle during the move. We find

$$\Delta U = U_f - U_i = W_{app}. \tag{25-13}$$

By similarly using Eq. 25-12 to substitute W_{app} into Eq. 25-7, we can relate our work W_{app} to the electric potential difference ΔV between the initial and final locations of the particle. We find

$$W_{app} = q\,\Delta V. \tag{25-14}$$

W_{app} can be positive, negative, or zero depending on the signs and magnitudes of q and ΔV. It is the work we must do to move a particle of charge q through a potential difference ΔV with no change in the particle's kinetic energy.

CHECKPOINT 2: In the figure of Checkpoint 1, we move a proton from point i to point f in a uniform electric field directed as shown. (a) Does our force do positive or negative work? (b) Does the proton move to a point of higher or lower potential?

a) W_{app} is positive ∴ F_{app} & d is in the same direction
b) against (opp direc) to E ∴ higher potential

25-3 Equipotential Surfaces

Adjacent points that have the same electric potential form an **equipotential surface,** which can be either an imaginary surface or a real, physical surface. No net work W is done on a charged particle by an electric field when the particle moves between two points i and f on the same equipotential surface. This follows from Eq. 25-7, which tells us that W must be zero if $V_f = V_i$. Because of the path independence of work (and thus of potential energy and potential), $W = 0$ for *any* path connecting points i and f, regardless of whether that path lies entirely on the equipotential surface.

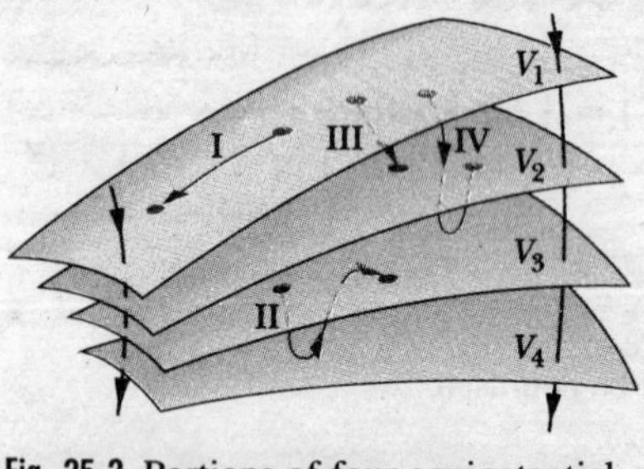

Fig. 25-2 Portions of four equipotential surfaces at electric potentials $V_1 = 100$ V, $V_2 = 80$ V, $V_3 = 60$ V, and $V_4 = 40$ V. Four paths along which a test charge may move are shown. Two electric field lines are also indicated.

Figure 25-2 shows a *family* of equipotential surfaces, associated with the electric field due to some distribution of charges. The work done by the electric field on a charged particle as the particle moves from one end to the other of paths I and II is zero because each of these paths begins and ends on the same equipotential surface. The work done as the charged particle moves from one end to the other of paths III and IV is not zero but has the same value for both these paths because the initial and final potentials are identical for the two paths; that is, paths III and IV connect the same pair of equipotential surfaces.

From symmetry, the equipotential surfaces produced by a point charge or a spherically symmetrical charge distribution are a family of concentric spheres. For a uniform electric field, the surfaces are a family of planes perpendicular to the field

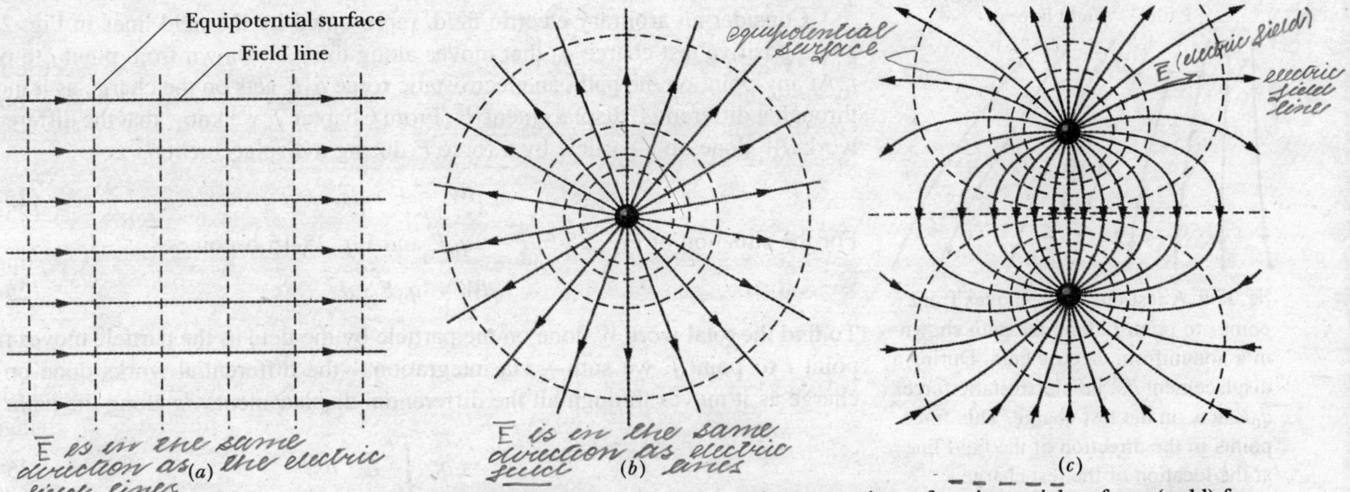

Fig. 25-3 Electric field lines (purple) and cross sections of equipotential surfaces (gold) for (*a*) a uniform field, (*b*) the field of a point charge, and (*c*) the field of an electric dipole.

Fig. 25-4 This enhancement of the chapter's opening photograph shows the result of overhead clouds creating a strong electric field $\vec{E}$ near a woman's head. Many of the hair strands extended along the field, which was perpendicular to the equipotential surfaces and greatest where those surfaces were closest, near the top of her head.

lines. In fact, equipotential surfaces are always perpendicular to electric field lines and thus to $\vec{E}$, which is always tangent to these lines. If $\vec{E}$ were *not* perpendicular to an equipotential surface, it would have a component lying along that surface. This component would then do work on a charged particle as it moved along the surface. However, by Eq. 25-7 work cannot be done if the surface is truly an equipotential surface; the only possible conclusion is that $\vec{E}$ must be everywhere perpendicular to the surface. Figure 25-3 shows electric field lines and cross sections of the equipotential surfaces for a uniform electric field and for the field associated with a point charge and with an electric dipole.

We now return to the woman in the opening photograph for this chapter. Because she was standing on a platform that was connected to the mountainside, she was at about the same potential as the mountainside. Overhead, a highly charged cloud system had moved in and created a strong electric field around her and the mountainside, with $\vec{E}$ pointing outward from her and the mountain. Electrostatic forces due to this field drove some of the conduction electrons in the woman downward through her body, leaving strands of her hair positively charged. The magnitude of $\vec{E}$ was apparently large, but less than the value of about 3×10^6 V/m that would have caused electrical breakdown of the air molecules. (That value was exceeded shortly later when lightning struck the platform.)

The equipotential surfaces surrounding the woman on the mountainside platform can be inferred from her hair; the strands are extended along the direction of $\vec{E}$ and thus are perpendicular to the equipotential surfaces, so the surfaces must be as drawn in Fig. 25-4. The field magnitude E was apparently greatest (the equipotential surfaces were apparently most closely spaced) just above her head, because the hair there extended farther than the hair around the sides.

The lesson here is simple. If an electric field causes the hairs on your head to stand up, you would do better to run for shelter than to pose for a snapshot.

25-4 Calculating the Potential from the Field

We can calculate the potential difference between any two points i and f in an electric field if we know the electric field vector $\vec{E}$ all along any path connecting those points. To make the calculation, we find the work done on a positive test charge by the field as the charge moves from i to f, and then use Eq. 25-7.

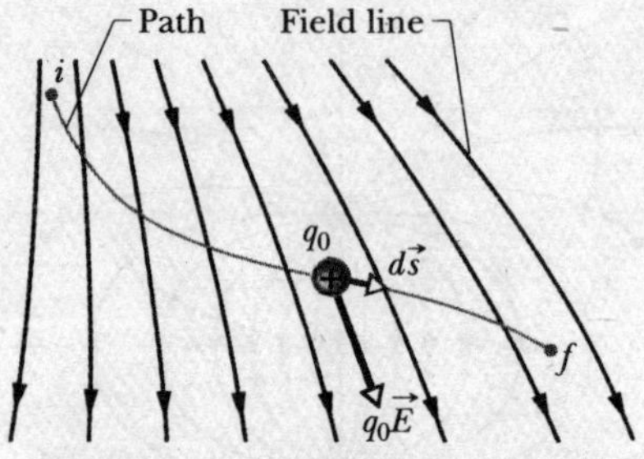

Fig. 25-5 A test charge q_0 moves from point i to point f along the path shown in a nonuniform electric field. During a displacement $d\vec{s}$, an electrostatic force $q_0\vec{E}$ acts on the test charge. This force points in the direction of the field line at the location of the test charge.

Consider an arbitrary electric field, represented by the field lines in Fig. 25-5, and a positive test charge q_0 that moves along the path shown from point i to point f. At any point on the path, an electrostatic force $q_0\vec{E}$ acts on the charge as it moves through a differential displacement $d\vec{s}$. From Chapter 7, we know that the differential work dW done on a particle by a force $\vec{F}$ during a displacement $d\vec{s}$ is

$$dW = \vec{F} \cdot d\vec{s}. \tag{25-15}$$

general case where F is not const (∵ E is not const)

For the situation of Fig. 25-5, $\vec{F} = q_0\vec{E}$ and Eq. 25-15 becomes

$$dW = q_0\vec{E} \cdot d\vec{s}. \tag{25-16}$$

To find the total work W done on the particle by the field as the particle moves from point i to point f, we sum—via integration—the differential works done on the charge as it moves through all the differential displacements $d\vec{s}$ along the path:

$$W = q_0 \int_i^f \vec{E} \cdot d\vec{s}. \tag{25-17}$$

If we substitute the total work W from Eq. 25-17 into Eq. 25-7, we find

$$V_f - V_i = -\int_i^f \vec{E} \cdot d\vec{s}. \tag{25-18}$$

Thus, the potential difference $V_f - V_i$ between any two points i and f in an electric field is equal to the negative of the *line integral* (meaning the integral along a particular path) of $\vec{E} \cdot d\vec{s}$ from i to f. However, because the electrostatic force is conservative, all paths (whether easy or difficult to use) yield the same result.

If the electric field is known throughout a certain region, Eq. 25-18 allows us to calculate the difference in potential between any two points in the field. If we choose the potential V_i at point i to be zero, then Eq. 25-18 becomes

$$V = -\int_i^f \vec{E} \cdot d\vec{s}, \tag{25-19}$$

in which we have dropped the subscript f on V_f. Equation 25-19 gives us the potential V at any point f in the electric field *relative to the zero potential* at point i. If we let point i be at infinity, then Eq. 25-19 gives us the potential V at any point f relative to the zero potential at infinity.

CHECKPOINT 3: The figure shows a family of parallel equipotential surfaces (in cross section) and five paths along which we shall move an electron from one surface to another. (a) What is the direction of the electric field associated with the surfaces? (b) For each path, is the work we do positive, negative, or zero? (c) Rank the paths according to the work we do, greatest first.

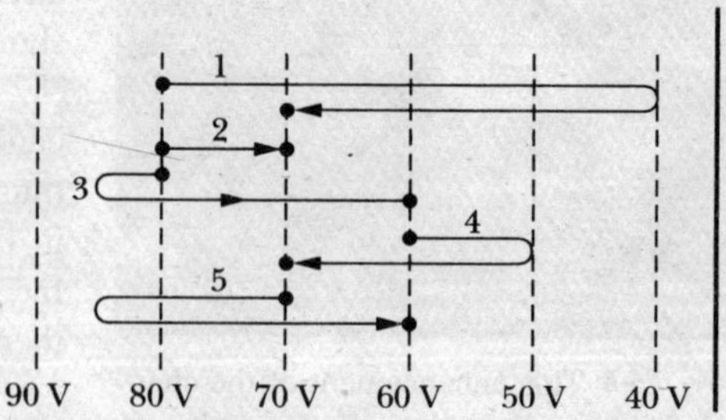

a) If the particle is left in the field all by itself, it will go from higher to lower potential. This is the direction of E which exerts force on it ∴ rightward

b) $W_{app} = q\Delta V = q(V_f - V_i) = -e(V_f - V_i)$

I $W_{app} = -e(70-80) = 10e$ (+)

∴ II, III, V +ve

IV −ve

c) 3, 1&2&5, 4

Sample Problem 25-2

(a) Figure 25-6*a* shows two points i and f in a uniform electric field $\vec{E}$. The points lie on the same electric field line (not shown) and are separated by a distance d. Find the potential difference $V_f - V_i$ by moving a positive test charge q_0 from i to f along the path shown, which is parallel to the field direction.

SOLUTION: The **Key Idea** here is that we can find the potential difference between any two points in an electric field by integrating $\vec{E} \cdot d\vec{s}$ along a path connecting those two points according to Eq. 25-18. We do this by mentally moving a test charge q_0 along that path, from initial point i to final point f. As we move such a test charge along the path in Fig. 25-6*a*, its differential displacement $d\vec{s}$ always has the same direction as $\vec{E}$. Thus, the angle θ between $\vec{E}$ and $d\vec{s}$ is zero and the dot product in Eq. 25-18 is

$$\vec{E} \cdot d\vec{s} = E\,ds \cos\theta = E\,ds. \qquad (25\text{-}20)$$

Equations 25-18 and 25-20 then give us

$$V_f - V_i = -\int_i^f \vec{E} \cdot d\vec{s} = -\int_i^f E\,ds. \qquad (25\text{-}21)$$

Since the field is uniform, E is constant over the path and can be moved outside the integral, giving us

$$V_f - V_i = -E \int_i^f ds = -Ed, \qquad \text{(Answer)}$$

in which the integral is simply the length d of the path. The minus sign in the result shows that the potential at point f in Fig. 25-6*a* is lower than the potential at point i. This is a general result: The potential always decreases along a path that extends in the direction of the electric field lines.

(b) Now find the potential difference $V_f - V_i$ by moving the positive test charge q_0 from i to f along the path icf shown in Fig. 25-6*b*.

SOLUTION: The **Key Idea** of (a) applies here too, except now we move the test charge along a path that consists of two lines: ic and cf. At all points along line ic, the displacement $d\vec{s}$ of the test charge is perpendicular to $\vec{E}$. Thus, the angle θ between $\vec{E}$ and $d\vec{s}$ is 90°, and the dot product $\vec{E} \cdot d\vec{s}$ is 0. Equation 25-18 then tells us that points i and c are at the same potential: $V_c - V_i = 0$.

For line cf we have $\theta = 45°$ and, from Eq. 25-18,

$$V_f - V_i = -\int_c^f \vec{E} \cdot d\vec{s} = -\int_c^f E(\cos 45°)\,ds$$
$$= -E(\cos 45°) \int_c^f ds.$$

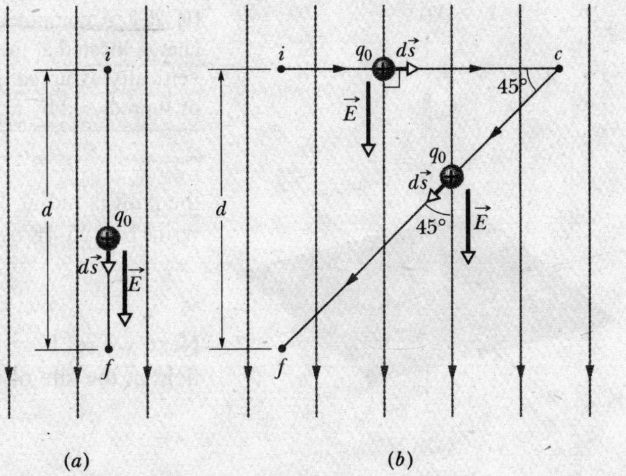

Fig. 25-6 Sample Problem 25-2. (*a*) A test charge q_0 moves in a straight line from point i to point f, along the direction of a uniform electric field. (*b*) Charge q_0 moves along path icf in the same electric field.

The integral in this equation is just the length of line cf; from Fig. 25-6*b*, that length is $d/\sin 45°$. Thus,

$$V_f - V_i = -E(\cos 45°)\frac{d}{\sin 45°} = -Ed. \qquad \text{(Answer)}$$

This is the same result we obtained in (a), as it must be; the potential difference between two points does not depend on the path connecting them. Moral: When you want to find the potential difference between two points by moving a test charge between them, you can save time and work by choosing a path that simplifies the use of Eq. 25-18.

25-5 Potential Due to a Point Charge

We will now use Eq. 25-18 to derive an expression for the electric potential V in the space around a charged particle, relative to the zero potential at infinity. Consider a point P at a distance R from a fixed particle of positive charge q (Fig. 25-7). To use Eq. 25-18, we imagine that we move a positive test charge q_0 from point P to infinity. Because the path we take does not matter, let us choose the simplest one—a line that extends radially from the fixed particle through P to infinity.

To use Eq. 25-18, we must evaluate the dot product

$$\vec{E} \cdot d\vec{s} = E \cos\theta\,ds. \qquad (25\text{-}22)$$

The electric field $\vec{E}$ in Fig. 25-7 is directed radially outward from the fixed particle. Thus, the differential displacement $d\vec{s}$ of the test particle along its path has the same direction as $\vec{E}$. That means that in Eq. 25-22, angle $\theta = 0$ and $\cos\theta = 1$. Because

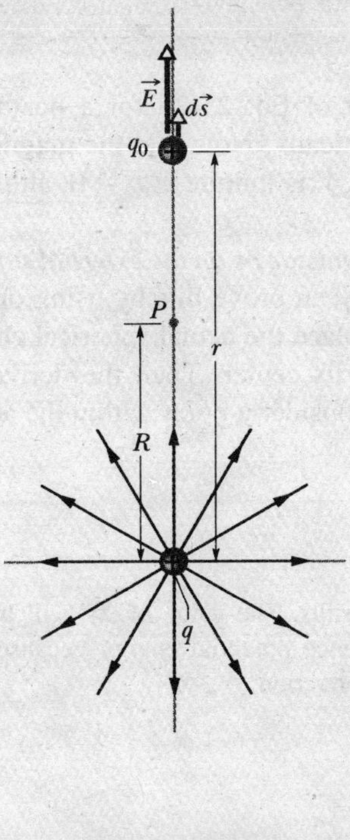

Fig. 25-7 The positive point charge q produces an electric field $\vec{E}$ and an electric potential V at point P. We find the potential by moving a test charge q_0 from P to infinity. The test charge is shown at distance r from the point charge, during differential displacement $d\vec{s}$.

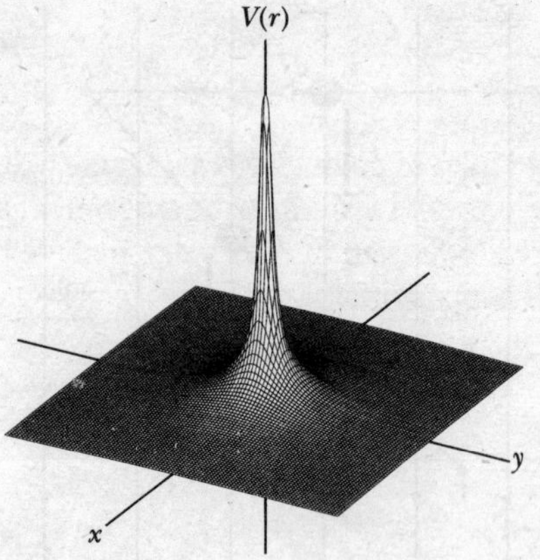

Fig. 25-8 A computer-generated plot of the electric potential $V(r)$ due to a positive point charge located at the origin of an xy plane. The potentials at points in that plane are plotted vertically. (Curved lines have been added to help you visualize the plot.) The infinite value of V predicted by Eq. 25-26 for $r = 0$ is not plotted.

the path is radial, let us write ds as dr. Then, substituting the limits R and ∞, we can write Eq. 25-18 as

$$V_f - V_i = -\int_R^\infty E\, dr. \tag{25-23}$$

Next, we set $V_f = 0$ (at ∞) and $V_i = V$ (at R). Then, for the magnitude of the electric field at the site of the test charge, we substitute from Eq. 23-3:

$$E = \frac{1}{4\pi\varepsilon_0}\frac{q}{r^2}. \tag{25-24}$$

With these changes, Eq. 25-23 then gives us

$$0 - V = -\frac{q}{4\pi\varepsilon_0}\int_R^\infty \frac{1}{r^2}\, dr = \frac{q}{4\pi\varepsilon_0}\left[\frac{1}{r}\right]_R^\infty$$

$$= -\frac{1}{4\pi\varepsilon_0}\frac{q}{R}. \tag{25-25}$$

Solving for V and switching R to r, we then have

$$V = \frac{1}{4\pi\varepsilon_0}\frac{q}{r} \tag{25-26}$$

ef $V_f=0$ at ∞

as the electric potential V due to a particle of charge q, at any radial distance r from the particle.

Although we have derived Eq. 25-26 for a positively charged particle, the derivation holds also for a negatively charged particle, in which case, q is a negative quantity. Note that the sign of V is the same as the sign of q:

> A positively charged particle produces a positive electric potential. A negatively charged particle produces a negative electric potential. When V is taken =0 at ∞

Figure 25-8 shows a computer-generated plot of Eq. 25-26 for a positively charged particle; the magnitude of V is plotted vertically. Note that the magnitude increases as $r \to 0$. In fact, according to Eq. 25-26, V is infinite at $r = 0$, although Fig. 25-8 shows a finite, smoothed-off value there.

Equation 25-26 also gives the electric potential *outside or on the external surface of* a spherically symmetric charge distribution. We can prove this by using one of the shell theorems of Sections 22-4 and 24-9 to replace the actual spherical charge distribution with an equal charge concentrated at its center. Then the derivation leading to Eq. 25-26 follows, provided we do not consider a point within the actual distribution.

PROBLEM-SOLVING TACTICS

Tactic 3: *Finding a Potential Difference*

To find the potential difference ΔV between any two points in the field of an isolated point charge, we can evaluate Eq. 25-26 at each point and then subtract the results. The value of ΔV will be the same for any choice of reference potential energy because that choice is eliminated by the subtraction.

25-6 Potential Due to a Group of Point Charges

We can find the net potential at a point due to a group of point charges with the help of the superposition principle. We calculate the potential resulting from each charge at the given point separately, using Eq. 25-26 with the sign of the charge included. Then we sum the potentials. For n charges, the net potential is

$$V = \sum_{i=1}^{n} V_i = \frac{1}{4\pi\varepsilon_0} \sum_{i=1}^{n} \frac{q_i}{r_i} \quad (n \text{ point charges}). \tag{25-27}$$

Here q_i is the value of the ith charge, and r_i is the radial distance of the given point from the ith charge. The sum in Eq. 25-27 is an *algebraic sum,* not a vector sum like the sum that would be used to calculate the electric field resulting from a group of point charges. Herein lies an important computational advantage of potential over electric field: It is a lot easier to sum several scalar quantities than to sum several vector quantities whose directions and components must be considered.

CHECKPOINT 4: The figure here shows three arrangements of two protons. Rank the arrangements according to the net electric potential produced at point P by the protons, greatest first.

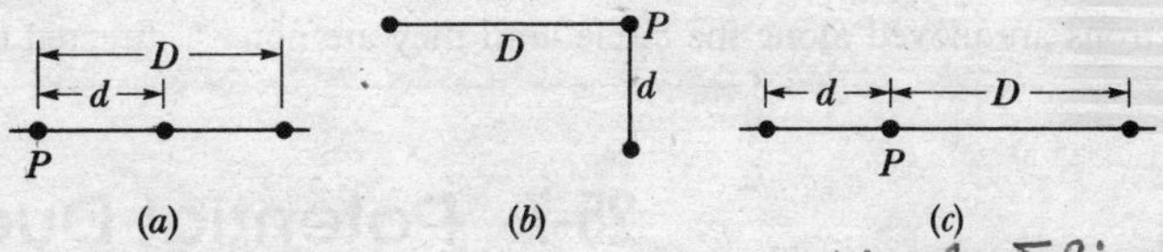

Sample Problem 25-3

What is the electric potential at point P, located at the center of the square of point charges shown in Fig. 25-9a? The distance d is 1.3 m, and the charges are

$$q_1 = +12 \text{ nC}, \qquad q_3 = +31 \text{ nC},$$
$$q_2 = -24 \text{ nC}, \qquad q_4 = +17 \text{ nC}.$$

SOLUTION: The **Key Idea** here is that the electric potential V at P is the algebraic sum of the electric potentials contributed by the four point charges. (Because electric potential is a scalar, the orientations of the point charges do not matter.) Thus, from Eq. 25-27, we have

$$V = \sum_{i=1}^{4} V_i = \frac{1}{4\pi\varepsilon_0}\left(\frac{q_1}{r} + \frac{q_2}{r} + \frac{q_3}{r} + \frac{q_4}{r}\right).$$

The distance r is $d/\sqrt{2}$, which is 0.919 m, and the sum of the charges is

$$q_1 + q_2 + q_3 + q_4 = (12 - 24 + 31 + 17) \times 10^{-9} \text{ C}$$
$$= 36 \times 10^{-9} \text{ C}.$$

Thus, $$V = \frac{(8.99 \times 10^9 \text{ N}\cdot\text{m}^2/\text{C}^2)(36 \times 10^{-9} \text{ C})}{0.919 \text{ m}}$$

$$\approx 350 \text{ V}. \qquad \text{(Answer)}$$

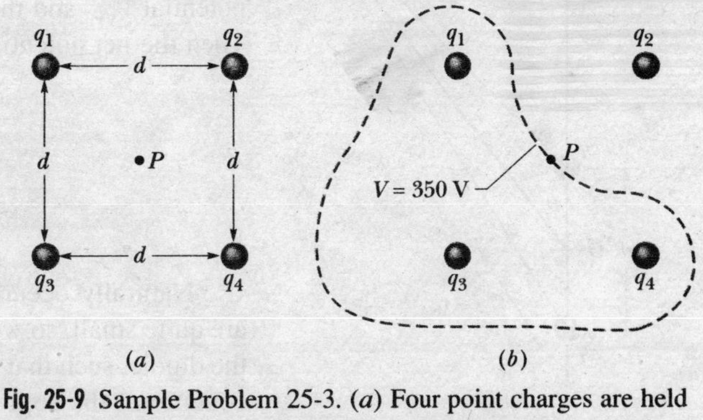

Fig. 25-9 Sample Problem 25-3. (a) Four point charges are held fixed at the corners of a square. (b) The closed curve is a cross section, in the plane of the figure, of the equipotential surface that contains point P. (The curve is only roughly drawn.)

Close to any of the three positive charges in Fig. 25-9a, the potential has very large positive values. Close to the single negative charge, the potential has very large negative values. Therefore, there must be points within the square that have the same intermediate potential as that at point P. The curve in Fig. 25-9b shows the intersection of the plane of the figure with the equipotential surface that contains point P. Any point along that curve has the same potential as point P.

Sample Problem 25-4

(a) In Fig. 25-10a, 12 electrons (of charge $-e$) are equally spaced and fixed around a circle of radius R. Relative to $V = 0$ at infinity, what are the electric potential and electric field at the center C of the circle due to these electrons?

SOLUTION: The **Key Idea** here is that the electric potential V at C is the algebraic sum of the electric potentials contributed by all the electrons. (Because electric potential is a scalar, the orientations of the electrons do not matter.) Because the electrons all have the same negative charge $-e$ and are all the same distance R from C, Eq. 25-27 gives us

$$V = -12\frac{1}{4\pi\varepsilon_0}\frac{e}{R}. \quad \text{(Answer)} \quad (25\text{-}28)$$

For the electric field at C, the **Key Idea** is that electric field is a vector quantity and thus the orientation of the electrons *is* important. Because of the symmetry of the arrangement in Fig. 25-10a, the electric field vector at C due to any given electron is canceled by the field vector due to the electron that is diametrically opposite it. Thus, at C,

$$\vec{E} = 0. \quad \text{(Answer)}$$

(b) If the electrons are moved along the circle until they are nonuniformly spaced over a 120° arc (Fig. 25-10b), what then is the potential at C? How does the electric field at C change (if at all)?

SOLUTION: The potential is still given by Eq. 25-28, because the distance between C and each electron is unchanged and orientation is irrelevant. The electric field is no longer zero, because the arrangement is no longer symmetric. There is now a net field that is directed toward the charge distribution.

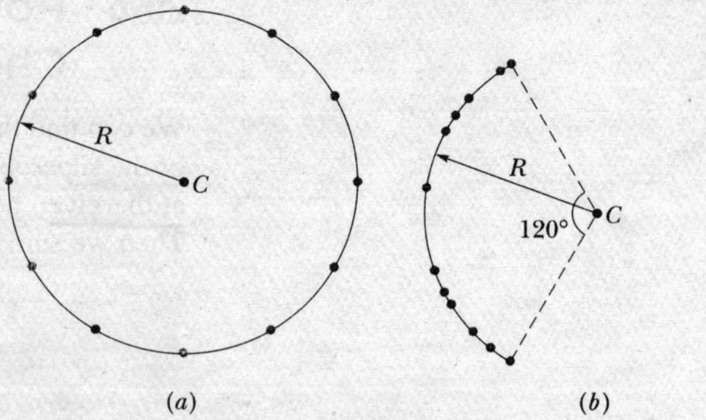

Fig. 25-10 Sample Problem 25-4. (a) Twelve electrons uniformly spaced around a circle. (b) Those electrons are now nonuniformly spaced along an arc of the original circle.

25-7 Potential Due to an Electric Dipole

Now let us apply Eq. 25-27 to an electric dipole to find the potential at an arbitrary point P in Fig. 25-11a. At P, the positive point charge (at distance $r_{(+)}$) sets up potential $V_{(+)}$ and the negative point charge (at distance $r_{(-)}$) sets up potential $V_{(-)}$. Then the net potential at P is given by Eq. 25-27 as

$$V = \sum_{i=1}^{2} V_i = V_{(+)} + V_{(-)} = \frac{1}{4\pi\varepsilon_0}\left(\frac{q}{r_{(+)}} + \frac{-q}{r_{(-)}}\right)$$

$$= \frac{q}{4\pi\varepsilon_0}\frac{r_{(-)} - r_{(+)}}{r_{(-)}r_{(+)}}. \quad (25\text{-}29)$$

Naturally occurring dipoles—such as those possessed by many molecules—are quite small, so we are usually interested only in points that are relatively far from the dipole, such that $r \gg d$, where d is the distance between the charges. Under those conditions, the approximations that follow from Fig. 25-11b are

$$r_{(-)} - r_{(+)} \approx d\cos\theta \quad \text{and} \quad r_{(-)}r_{(+)} \approx r^2.$$

If we substitute these quantities into Eq. 25-29, we can approximate V to be

$$V = \frac{q}{4\pi\varepsilon_0}\frac{d\cos\theta}{r^2},$$

where θ is measured from the dipole axis as shown in Fig. 25-11a. We can now

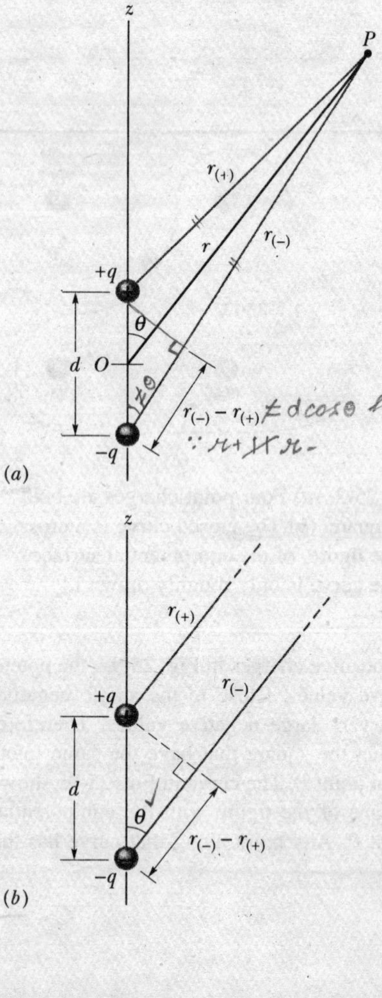

Fig. 25-11 (a) Point P is a distance r from the midpoint O of a dipole. The line OP makes an angle θ with the dipole axis. (b) If P is far from the dipole, the lines of lengths $r_{(+)}$ and $r_{(-)}$ are approximately parallel to the line of length r, and the dashed black line is approximately perpendicular to the line of length $r_{(-)}$.

write V as

$$V = \frac{1}{4\pi\varepsilon_0}\frac{p\cos\theta}{r^2} \quad \text{(electric dipole),} \tag{25-30}$$

in which p (= qd) is the magnitude of the electric dipole moment $\vec{p}$ defined in Section 23-5. The vector $\vec{p}$ is directed along the dipole axis, from the negative to the positive charge. (Thus, θ is measured from the direction of $\vec{p}$.)

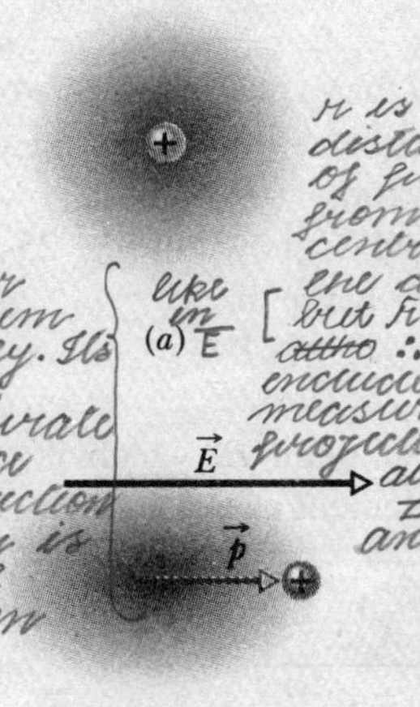

(b)

Fig. 25-12 (a) An atom, showing the positively charged nucleus (green) and the negatively charged electrons (gold shading). The centers of positive and negative charge coincide. (b) If the atom is placed in an external electric field $\vec{E}$, the electron orbits are distorted so that the centers of positive and negative charge no longer coincide. An induced dipole moment $\vec{p}$ appears. The distortion is greatly exaggerated here.

CHECKPOINT 5: Suppose that three points are set at equal (large) distances r from the center of the dipole in Fig. 25-11: Point a is on the dipole axis above the positive charge, point b is on the axis below the negative charge, and point c is on a perpendicular bisector through the line connecting the two charges. Rank the points according to the electric potential of the dipole there, greatest (most positive) first.

Induced Dipole Moment

Many molecules such as water have *permanent* electric dipole moments. In other molecules (called *nonpolar molecules*) and in every isolated atom, the centers of the positive and negative charges coincide (Fig. 25-12a) and thus no dipole moment is set up. However, if we place an atom or a nonpolar molecule in an external electric field, the field distorts the electron orbits and separates the centers of positive and negative charge (Fig. 25-12b). Because the electrons are negatively charged, they tend to be shifted in a direction opposite the field. This shift sets up a dipole moment $\vec{p}$ that points in the direction of the field. This dipole moment is said to be *induced* by the field, and the atom or molecule is then said to be *polarized* by the field (it has a positive side and a negative side). When the field is removed, the induced dipole moment and the polarization disappear.

25-8 Potential Due to a Continuous Charge Distribution

When a charge distribution q is continuous (as on a uniformly charged thin rod or disk), we cannot use the summation of Eq. 25-27 to find the potential V at a point P. Instead, we must choose a differential element of charge dq, determine the potential dV at P due to dq, and then integrate over the entire charge distribution.

Let us again take the zero of potential to be at infinity. If we treat the element of charge dq as a point charge, then we can use Eq. 25-26 to express the potential dV at point P due to dq:

$$dV = \frac{1}{4\pi\varepsilon_0}\frac{dq}{r} \quad \text{(positive or negative } dq\text{).} \tag{25-31}$$

Here r is the distance between P and dq. To find the total potential V at P, we integrate to sum the potentials due to all the charge elements:

$$V = \int dV = \frac{1}{4\pi\varepsilon_0}\int\frac{dq}{r}. \tag{25-32}$$

The integral must be taken over the entire charge distribution. Note that because the electric potential is a scalar, there are *no vector components* to consider in Eq. 25-32.

We now examine two continuous charge distributions, a line of charge and a charged disk.

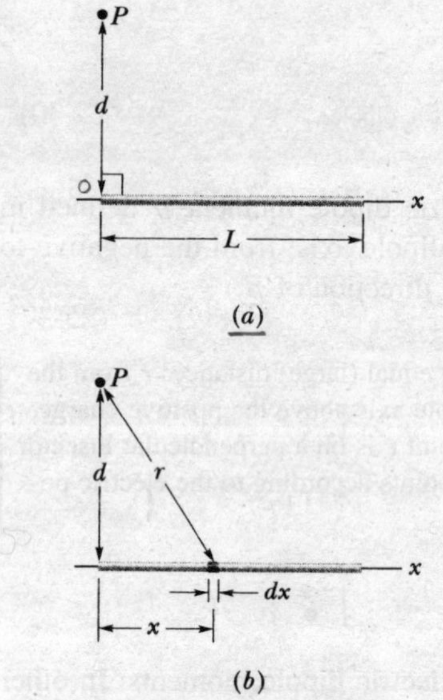

Fig. 25-13 (*a*) A thin, uniformly charged rod produces an electric potential V at point P. (*b*) An element of charge produces a differential potential dV at P.

Line of Charge

In Fig. 25-13*a*, a thin nonconducting rod of length L has a positive charge of uniform linear density λ. Let us determine the electric potential V due to the rod at point P, a perpendicular distance d from the left end of the rod.

We consider a differential element dx of the rod as shown in Fig. 25-13*b*. This (or any other) element of the rod has a differential charge of

$$dq = \lambda\, dx. \tag{25-33}$$

This element produces an electric potential dV at point P, which is a distance $r = (x^2 + d^2)^{1/2}$ from the element. Treating the element as a point charge, we can use Eq. 25-31 to write the potential dV as

$$dV = \frac{1}{4\pi\varepsilon_0}\frac{dq}{r} = \frac{1}{4\pi\varepsilon_0}\frac{\lambda\, dx}{(x^2 + d^2)^{1/2}}. \tag{25-34}$$

Since the charge on the rod is positive and we have taken $V = 0$ at infinity, we know from Section 25-5 that dV in Eq. 25-34 must be positive.

We now find the total potential V produced by the rod at point P by integrating Eq. 25-34 along the length of the rod, from $x = 0$ to $x = L$, using integral 17 in Appendix E. We find

$$\begin{aligned}
V &= \int dV = \int_0^L \frac{1}{4\pi\varepsilon_0}\frac{\lambda}{(x^2 + d^2)^{1/2}}\, dx \\
&= \frac{\lambda}{4\pi\varepsilon_0}\int_0^L \frac{dx}{(x^2 + d^2)^{1/2}} \\
&= \frac{\lambda}{4\pi\varepsilon_0}\left[\ln\left(x + (x^2 + d^2)^{1/2}\right)\right]_0^L \\
&= \frac{\lambda}{4\pi\varepsilon_0}\left[\ln\left(L + (L^2 + d^2)^{1/2}\right) - \ln d\right].
\end{aligned}$$

We can simplify this result by using the general relation $\ln A - \ln B = \ln(A/B)$. We then find

$$V = \frac{\lambda}{4\pi\varepsilon_0}\ln\left[\frac{L + (L^2 + d^2)^{1/2}}{d}\right]. \tag{25-35}$$

Because V is the sum of positive values of dV, it should be positive—but does Eq. 25-35 give a positive V? Since the argument of the logarithm is greater than one, the logarithm is a positive number and V is indeed positive.

Charged Disk

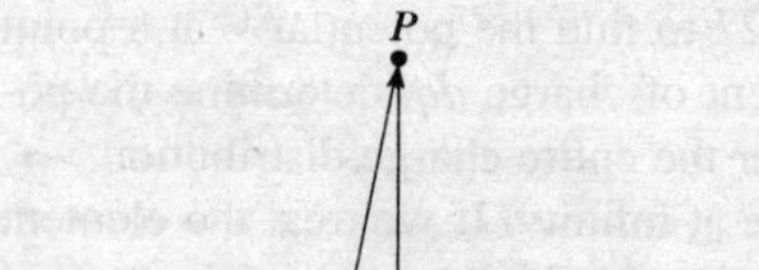

Fig. 25-14 A plastic disk of radius R, charged on its top surface to a uniform surface charge density σ. We wish to find the potential V at point P on the central axis of the disk.

In Section 23-7, we calculated the magnitude of the electric field at points on the central axis of a plastic disk of radius R that has a uniform charge density σ on one surface. Here we derive an expression for $V(z)$, the electric potential at any point on the central axis.

In Fig. 25-14, consider a differential element consisting of a flat ring of radius R' and radial width dR'. Its charge has magnitude

$$dq = \sigma(2\pi R')(dR'),$$

in which $(2\pi R')(dR')$ is the upper surface area of the ring. All parts of this charged element are the same distance r from point P on the disk's axis. With the aid of Fig. 25-14, we can use Eq. 25-31 to write the contribution of this ring to the electric potential at P as

$$dV = \frac{1}{4\pi\varepsilon_0}\frac{dq}{r} = \frac{1}{4\pi\varepsilon_0}\frac{\sigma(2\pi R')(dR')}{\sqrt{z^2 + R'^2}}. \tag{25-36}$$

We find the net potential at P by adding (via integration) the contributions of all the strips from $R' = 0$ to $R' = R$:

$$V = \int dV = \frac{\sigma}{2\varepsilon_0}\int_0^R \frac{R'\,dR'}{\sqrt{z^2 + R'^2}} = \frac{\sigma}{2\varepsilon_0}\left(\sqrt{z^2 + R^2} - z\right). \tag{25-37}$$

Note that the variable in the second integral of Eq. 25-37 is R' and not z, which remains constant while the integration over the surface of the disk is carried out. (Note also that, in evaluating the integral, we have assumed that $z \geq 0$.)

PROBLEM-SOLVING TACTICS

Tactic 4: *Signs of Trouble with Electric Potential*
When you calculate the potential V at some point P due to a line of charge or any other continuous charge configuration, the signs can cause you trouble. Here is a generic guide to sort out the signs.

If the charge is negative, should the symbols dq and λ represent negative quantities, or should you explicitly show the signs, using $-dq$ and $-\lambda$? You can do either as long as you remember what your notation means, so that when you get to the final step, you can correctly interpret the sign of V.

Another approach, which can be used when the entire charge distribution is of a single sign, is to let the symbols dq and λ represent magnitudes only. The result of the calculation will give you the magnitude of V at P. Then add a sign to V based on the sign of the charge. (If the zero potential is at infinity, positive charge gives a positive potential and negative charge gives a negative potential.)

If you happen to reverse the limits on the integral used to calculate a potential, you will obtain a negative value for V. The magnitude will be correct, but discard the minus sign. Then determine the proper sign for V from the sign of the charge. As an example, we would have obtained a minus sign in Eq. 25-35 if we had reversed the limits in the integral above that equation. We would then have discarded that minus sign and noted that the potential is positive because the charge producing it is positive.

25-9 Calculating the Field from the Potential

In Section 25-4, you saw how to find the potential at a point f if you know the electric field along a path from a reference point to point f. In this section, we propose to go the other way—that is, to find the electric field when we know the potential. As Fig. 25-3 shows, solving this problem graphically is easy: If we know the potential V at all points near an assembly of charges, we can draw in a family of equipotential surfaces. The electric field lines, sketched perpendicular to those surfaces, reveal the variation of $\vec{E}$. What we are seeking here is the mathematical equivalent of this graphical procedure.

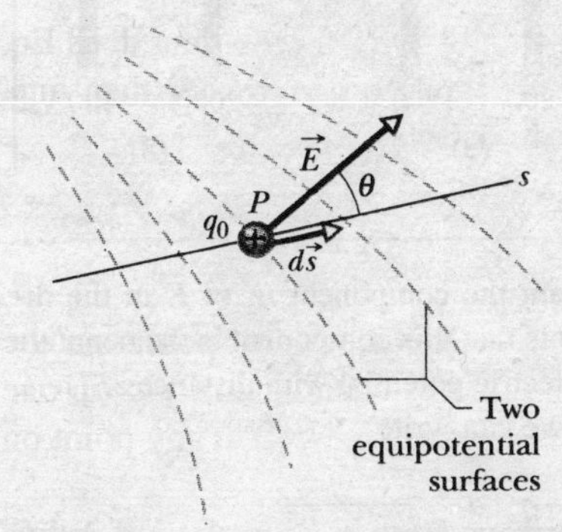

Fig. 25-15 A test charge q_0 moves a distance $d\vec{s}$ from one equipotential surface to another. (The separation between the surfaces has been exaggerated for clarity.) The displacement $d\vec{s}$ makes an angle θ with the direction of the electric field $\vec{E}$.

Figure 25-15 shows cross sections of a family of closely spaced equipotential surfaces, the potential difference between each pair of adjacent surfaces being dV. As the figure suggests, the field $\vec{E}$ at any point P is perpendicular to the equipotential surface through P.

Suppose that a positive test charge q_0 moves through a displacement $d\vec{s}$ from one equipotential surface to the adjacent surface. From Eq. 25-7, we see that the work the electric field does on the test charge during the move is $-q_0\,dV$. From Eq. 25-16 and Fig. 25-15, we see that the work done by the electric field may also be written as the scalar product $(q_0\vec{E})\cdot d\vec{s}$, or $q_0E(\cos\theta)\,ds$. Equating these two expressions for the work yields

$$-q_0\,dV = q_0E(\cos\theta)\,ds, \tag{25-38}$$

or

$$E\cos\theta = -\frac{dV}{ds}. \tag{25-39}$$

Since $E \cos \theta$ is the component of $\vec{E}$ in the direction of $d\vec{s}$, Eq. 25-39 becomes

$$E_s = -\frac{\partial V}{\partial s}. \tag{25-40}$$

We have added a subscript to E and switched to the partial derivative symbols to emphasize that Eq. 25-40 involves only the variation of V along a specified axis (here called the s axis) and only the component of $\vec{E}$ along that axis. In words, Eq. 25-40 (which is essentially the inverse of Eq. 25-18) states:

> The component of $\vec{E}$ in any direction is the negative of the rate of change of the electric potential with distance in that direction.

If we take the s axis to be, in turn, the x, y, and z axes, we find that the x, y, and z components of $\vec{E}$ at any point are

$$E_x = -\frac{\partial V}{\partial x}; \qquad E_y = -\frac{\partial V}{\partial y}; \qquad E_z = -\frac{\partial V}{\partial z}. \tag{25-41}$$

Thus, if we know V for all points in the region around a charge distribution—that is, if we know the function $V(x, y, z)$—we can find the components of $\vec{E}$, and thus $\vec{E}$ itself, at any point by taking partial derivatives.

For the simple situation in which the electric field $\vec{E}$ is uniform, Eq. 25-40 becomes

$$E = -\frac{\Delta V}{\Delta s}, \tag{25-42}$$

where s is perpendicular to the equipotential surfaces. The component of the electric field is zero in any direction parallel to the equipotential surfaces.

CHECKPOINT 6: The figure shows three pairs of parallel plates with the same separation, and the electric potential of each plate. The electric field between the plates is uniform and perpendicular to the plates. (a) Rank the pairs according to the magnitude of the electric field between the plates, greatest first. (b) For which pair is the electric field pointing rightward? (c) If an electron is released midway between the third pair of plates, does it remain there, move rightward at constant speed, move leftward at constant speed, accelerate rightward, or accelerate leftward?

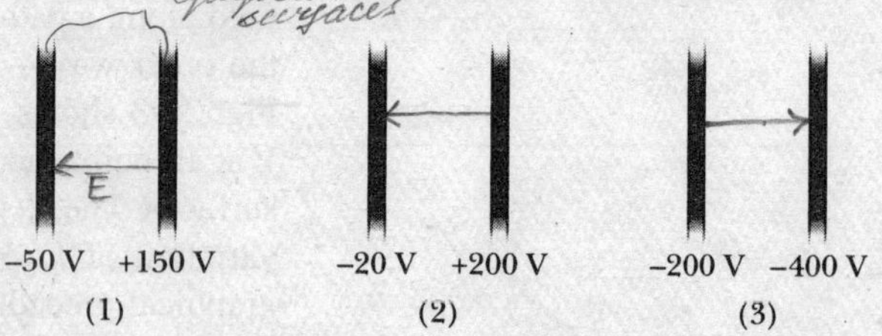

Sample Problem 25-5

The electric potential at any point on the central axis of a uniformly charged disk is given by Eq. 25-37,

$$V = \frac{\sigma}{2\varepsilon_0}\left(\sqrt{z^2 + R^2} - z\right).$$

Starting with this expression, derive an expression for the electric field at any point on the axis of the disk.

SOLUTION: We want the electric field $\vec{E}$ as a function of distance z along the axis of the disk. For any value of z, the direction of $\vec{E}$ must be along that axis because the disk has circular symmetry about that axis. Thus, we want the component E_z of $\vec{E}$ in the direction of z. Then the **Key Idea** is that this component is the negative of the rate of change of the electric potential with distance z. Thus, from the last of Eqs. 25-41, we can write

$$E_z = -\frac{\partial V}{\partial z} = -\frac{\sigma}{2\varepsilon_0}\frac{d}{dz}\left(\sqrt{z^2 + R^2} - z\right)$$

$$= \frac{\sigma}{2\varepsilon_0}\left(1 - \frac{z}{\sqrt{z^2 + R^2}}\right). \qquad \text{(Answer)}$$

This is the same expression that we derived in Section 23-7 by integration, using Coulomb's law.

25-10 Electric Potential Energy of a System of Point Charges

In Section 25-1, we discussed the electric potential energy of a charged particle as an electrostatic force does work on it. In that section, we assumed that the charges that produced the force were fixed in place, so that neither the force nor the corresponding electric field could be influenced by the presence of the test charge. In this section we can take a broader view, to find the electric potential energy of a *system* of charges due to the electric field produced *by* those same charges.

For a simple example, suppose you push together two bodies that have charges of the same electrical sign. The work that you must do is stored as electric potential energy in the two-body system (provided the kinetic energy of the bodies does not change). If you later release the charges, you can recover this stored energy, in whole or in part, as kinetic energy of the charged bodies as they rush away from each other.

We define the electric potential energy *of a system of point charges*, held in fixed positions by forces not specified, as follows:

> The electric potential energy of a system of fixed point charges is equal to the work that must be done by an external agent to assemble the system, bringing each charge in from an infinite distance.

We assume that the charges are stationary both in their initial infinitely distant positions and in their final assembled configuration.

Figure 25-16 shows two point charges q_1 and q_2, separated by a distance r. To find the electric potential energy of this two-charge system, we must mentally build the system, starting with both charges infinitely far away and at rest. When we bring q_1 in from infinity and put it in place we do no work, because no electrostatic force acts on q_1. However, when we next bring q_2 in from infinity and put it in place, we must do work, because q_1 exerts an electrostatic force on q_2 during the move.

We can calculate that work with Eq. 25-8 by dropping the minus sign (so that the equation gives the work *we* do rather than the field's work) and substituting q_2 for the general charge q. Our work is then equal to q_2V, where V is the potential that has been set up by q_1 at the point where we put q_2. From Eq. 25-26, that potential is

$$V = \frac{1}{4\pi\varepsilon_0}\frac{q_1}{r}.$$

Thus, from our definition, the electric potential energy of the pair of point charges of Fig. 25-16 is

$$U = W = q_2V = \frac{1}{4\pi\varepsilon_0}\frac{q_1q_2}{r}. \qquad (25\text{-}43)$$

If the charges have the same sign, we have to do positive work to push them together against their mutual repulsion. Hence, as Eq. 25-43 shows, the potential energy of the system is then positive. If the charges have opposite signs, we have to do negative work against their mutual attraction to bring them together if they are to be stationary. The potential energy of the system is then negative. Sample Problem 25-6 shows how to extend this process to more than two charges.

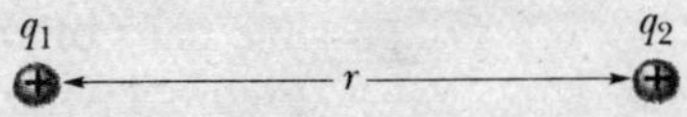

Fig. 25-16 Two charges held a fixed distance r apart.

Sample Problem 25-6

Figure 25-17 shows three point charges held in fixed positions by forces that are not shown. What is the electric potential energy U of this system of charges? Assume that $d = 12$ cm and that

$$q_1 = +q, \quad q_2 = -4q, \quad \text{and} \quad q_3 = +2q,$$

in which $q = 150$ nC.

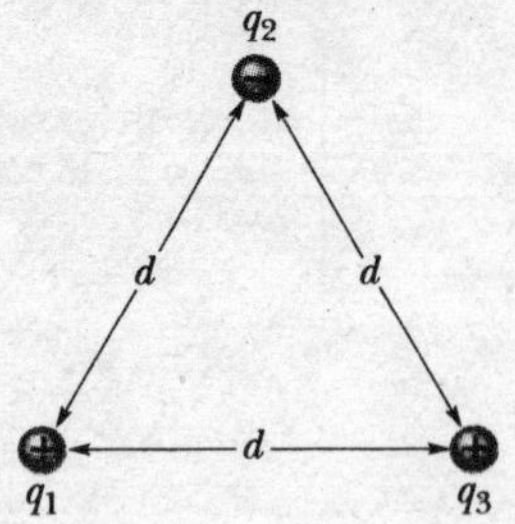

Fig. 25-17 Sample Problem 25-6. Three charges are fixed at the vertices of an equilateral triangle. What is the electric potential energy of the system?

SOLUTION: The **Key Idea** here is that the potential energy U of the system is equal to the work we must do to assemble the system, bringing in each charge from an infinite distance. Therefore, let's mentally build the system of Fig. 25-17, starting with one of the point charges, say q_1, in place and the others at infinity. Then we bring another one, say q_2, in from infinity and put it in place. From Eq. 25-43 with d substituted for r, the potential energy U_{12} associated with the pair of point charges q_1 and q_2 is

$$U_{12} = \frac{1}{4\pi\varepsilon_0}\frac{q_1 q_2}{d}.$$

negative sign with U would automatically come when we'll put sign of q_2

We then bring the last point charge q_3 in from infinity and put it in place. The work that we must do in this last step is equal to the sum of the work we must do to bring q_3 near q_1 and the work we must do to bring it near q_2. From Eq. 25-43, with d substituted for r, that sum is

$$W_{13} + W_{23} = U_{13} + U_{23} = \frac{1}{4\pi\varepsilon_0}\frac{q_1 q_3}{d} + \frac{1}{4\pi\varepsilon_0}\frac{q_2 q_3}{d}.$$

The total potential energy U of the three-charge system is the sum of the potential energies associated with the three pairs of charges. This sum (which is actually independent of the order in which the charges are brought together) is

$$U = U_{12} + U_{13} + U_{23}$$
$$= \frac{1}{4\pi\varepsilon_0}\left(\frac{(+q)(-4q)}{d} + \frac{(+q)(+2q)}{d} + \frac{(-4q)(+2q)}{d}\right)$$
$$= -\frac{10q^2}{4\pi\varepsilon_0 d}$$
$$= -\frac{(8.99 \times 10^9\ \text{N}\cdot\text{m}^2/\text{C}^2)(10)(150 \times 10^{-9}\ \text{C})^2}{0.12\ \text{m}}$$
$$= -1.7 \times 10^{-2}\ \text{J} = -17\ \text{mJ}. \quad \text{(Answer)}$$

The negative potential energy means that negative work would have to be done to assemble this structure, starting with the three charges infinitely separated and at rest. Put another way, an external agent would have to do 17 mJ of work to disassemble the structure completely, ending with the three charges infinitely far apart.

25-11 Potential of a Charged Isolated Conductor

In Section 24-6, we concluded that $\vec{E} = 0$ for all points inside an isolated conductor. We then used Gauss' law to prove that an excess charge placed on an isolated conductor lies entirely on its surface. (This is true even if the conductor has an empty internal cavity.) Here we use the first of these facts to prove an extension of the second:

> An excess charge placed on an isolated conductor will distribute itself on the surface of that conductor so that all points of the conductor—whether on the surface or inside—come to the same potential. This is true even if the conductor has an internal cavity and even if that cavity contains a net charge.

Our proof follows directly from Eq. 25-18, which is

$$V_f - V_i = -\int_i^f \vec{E}\cdot d\vec{s}.$$

Since $\vec{E} = 0$ for all points within a conductor, it follows directly that $V_f = V_i$ for all possible pairs of points i and f in the conductor.

Figure 25-18a is a plot of potential against radial distance r from the center for an isolated spherical conducting shell of 1.0 m radius, having a charge of 1.0 μC.

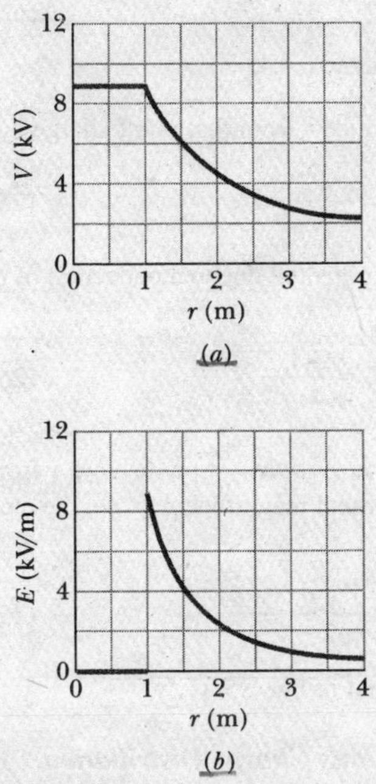

Fig. 25-18 (a) A plot of $V(r)$ both inside and outside a charged spherical shell of radius 1.0 m. (b) A plot of $E(r)$ for the same shell.

For points outside the shell, we can calculate $V(r)$ from Eq. 25-26 because the charge q behaves for such external points as if it were concentrated at the center of the shell. That equation holds right up to the surface of the shell. Now let us push a small test charge through the shell—assuming a small hole exists—to its center. No extra work is needed to do this because no net electric force acts on the test charge once it is inside the shell. Thus, the potential at all points inside the shell has the same value as that on the surface, as Fig. 25-18*a* shows.

Figure 25-18*b* shows the variation of electric field with radial distance for the same shell. Note that $E = 0$ everywhere inside the shell. The curves of Fig. 25-18*b* can be derived from the curve of Fig. 25-18*a* by differentiating with respect to r, using Eq. 25-40 (recall that the derivative of any constant is zero). The curve of Fig. 25-18*a* can be derived from the curves of Fig. 25-18*b* by integrating with respect to r, using Eq. 25-19.

On nonspherical conductors, a surface charge does not distribute itself uniformly over the surface of the conductor. At sharp points or edges, the surface charge density—and thus the external electric field, which is proportional to it—may reach very high values. The air around such sharp points may become ionized, producing the corona discharge that golfers and mountaineers see on the tips of bushes, golf clubs, and rock hammers when thunderstorms threaten. Such corona discharges, like hair that stands on end, are often the precursors of lightning strikes. In such circumstances, it is wise to enclose yourself in a cavity inside a conducting shell, where the electric field is guaranteed to be zero. A car (unless it is a convertible or made with a plastic body) is almost ideal (Fig. 25-19).

If an isolated conductor is placed in an *external electric field*, as in Fig. 25-20, all points of the conductor still come to a single potential regardless of whether the conductor has an excess charge. The free conduction electrons distribute themselves on the surface in such a way that the electric field they produce at interior points cancels the external electric field that would otherwise be there. Furthermore, the electron distribution causes the net electric field at all points on the surface to be perpendicular to the surface. If the conductor in Fig. 25-20 could be somehow removed, leaving the surface charges frozen in place, the pattern of the electric field would remain absolutely unchanged, for both exterior and interior points.

Fig. 25-19 A large spark jumps to a car's body and then exits by moving across the insulating left front tire (note the flash there), leaving the person inside unharmed.

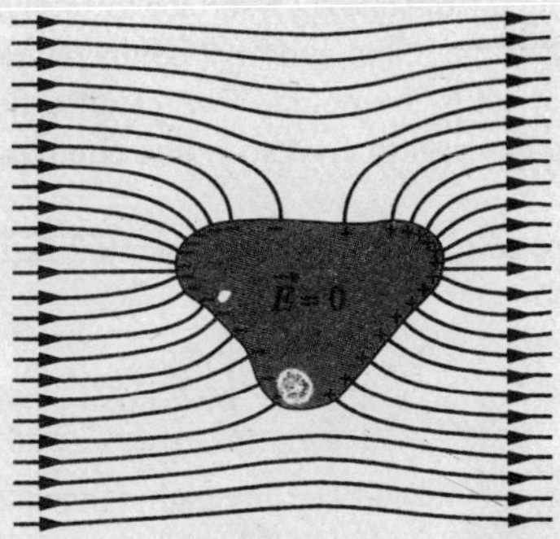

Fig. 25-20 An uncharged conductor is suspended in an external electric field. The free electrons in the conductor distribute themselves on the surface as shown, so as to reduce the net electric field inside the conductor to zero and make the net field at the surface perpendicular to the surface.

REVIEW & SUMMARY

Electric Potential Energy The change ΔU in the electric potential energy U of a point charge as the charge moves from an initial point i to a final point f in an electric field is

$$\Delta U = U_f - U_i = -W, \tag{25-1}$$

where W is the work done by the electrostatic force (due to the electric field) on the point charge during the move from i to f. If the potential energy is defined to be zero at infinity, the **electric potential energy** U of the point charge at a particular point is

$$U = -W_\infty. \tag{25-2}$$

Here W_∞ is the work done by the electrostatic force on the point charge as the charge moves from infinity to the particular point.

Electric Potential Difference and Electric Potential We define the **potential difference** ΔV between two points i and f in an electric field as

$$\Delta V = V_f - V_i = -\frac{W}{q}, \tag{25-7}$$

where q is the charge of a particle on which work is done by the field. The **potential** at a point is

$$V = -\frac{W_\infty}{q}. \tag{25-8}$$

The SI unit of potential is the *volt:* 1 volt = 1 joule per coulomb.

Potential and potential difference can also be written in terms of the electric potential energy U of a particle of charge q in an electric field:

$$V = \frac{U}{q} \tag{25-5}$$

$$\Delta V = V_f - V_i = \frac{U_f}{q} - \frac{U_i}{q} = \frac{\Delta U}{q}. \tag{25-6}$$

Equipotential Surfaces The points on an **equipotential surface** all have the same electric potential. The work done on a test charge in moving it from one such surface to another is independent of the locations of the initial and final points on these surfaces and of the path that joins the points. The electric field $\vec{E}$ is always directed perpendicularly to corresponding equipotential surfaces.

Finding V from $\vec{E}$ The electric potential difference between two points i and f is

$$V_f - V_i = -\int_i^f \vec{E}\cdot d\vec{s}, \tag{25-18}$$

where the integral is taken over any path connecting the points. If we choose $V_i = 0$ we have, for the potential at a particular point,

$$V = -\int_i^f \vec{E}\cdot d\vec{s}. \tag{25-19}$$

Potential Due to Point Charges The electric potential due to a single point charge at a distance r from that point charge is

$$V = \frac{1}{4\pi\varepsilon_0}\frac{q}{r}. \tag{25-26}$$

V has the same sign as q. The potential due to a collection of point charges is

$$V = \sum_{i=1}^{n} V_i = \frac{1}{4\pi\varepsilon_0}\sum_{i=1}^{n}\frac{q_i}{r_i}. \tag{25-27}$$

Potential Due to an Electric Dipole At a distance r from an electric dipole with dipole moment magnitude $p = qd$, the electric potential of the dipole is

$$V = \frac{1}{4\pi\varepsilon_0}\frac{p\cos\theta}{r^2} \tag{25-30}$$

for $r \gg d$; the angle θ is defined in Fig. 25-11.

Potential Due to a Continuous Charge Distribution For a continuous distribution of charge, Eq. 25-27 becomes

$$V = \frac{1}{4\pi\varepsilon_0}\int\frac{dq}{r}, \tag{25-32}$$

in which the integral is taken over the entire distribution.

Calculating $\vec{E}$ from V The component of $\vec{E}$ in any direction is the negative of the rate of change of the potential with distance in that direction:

$$E_s = -\frac{\partial V}{\partial s}. \tag{25-40}$$

The x, y, and z components of $\vec{E}$ may be found from

$$E_x = -\frac{\partial V}{\partial x};\quad E_y = -\frac{\partial V}{\partial y};\quad E_z = -\frac{\partial V}{\partial z}. \tag{25-41}$$

When $\vec{E}$ is uniform, Eq. 25-40 reduces to

$$E = -\frac{\Delta V}{\Delta s}, \tag{25-42}$$

where s is perpendicular to the equipotential surfaces. The electric field is zero in the direction parallel to an equipotential surface.

Electric Potential Energy of a System of Point Charges The electric potential energy of a system of point charges is equal to the work needed to assemble the system with the charges initially at rest and infinitely distant from each other. For two charges at separation r,

$$U = W = \frac{1}{4\pi\varepsilon_0}\frac{q_1q_2}{r}. \tag{25-43}$$

Potential of a Charged Conductor An excess charge placed on a conductor will, in the equilibrium state, be located entirely on the outer surface of the conductor. The charge will distribute itself so that the entire conductor, including interior points, is at a uniform potential.

QUESTIONS

1. Figure 25-21 shows three paths along which we can move positively charged sphere A closer to positively charged sphere B, which is fixed in place. (a) Would sphere A be moved to a higher or lower electric potential? Is the work done (b) by our force and (c) by the electric field (due to the second sphere) positive, negative, or zero? (d) Rank the paths according to the work our force does, greatest first.

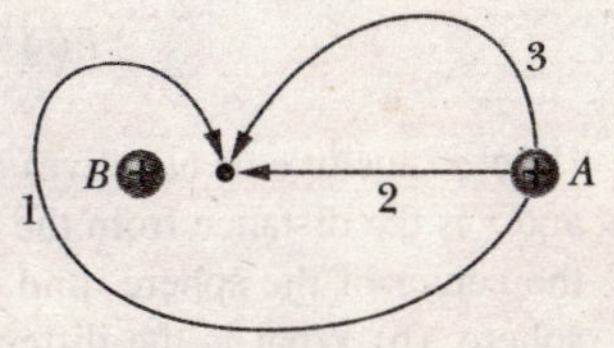

Fig. 25-21 Question 1.

2. Figure 25-22 shows four pairs of charged particles. Let $V = 0$ at infinity. For which pairs is there another point of zero net electric potential *on the axis shown,* (a) between the particles and (b) to their right? (c) Where such a zero potential point exists, is the net electric field $\vec{E}$ due to the particles equal to zero? (d) For each pair, are there points off the axis (other than at infinity, of course) where $V = 0$?

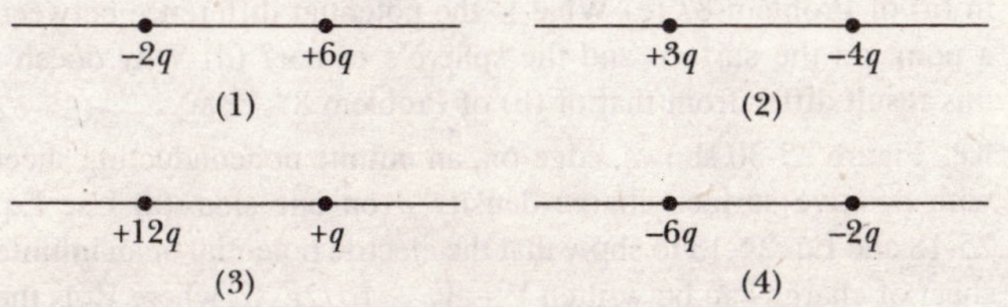

Fig. 25-22 Questions 2 and 8.

3. Figure 25-23 shows a square array of charged particles, with distance d between adjacent particles. What is the electric potential at point P at the center of the square if the electric potential is zero at infinity?

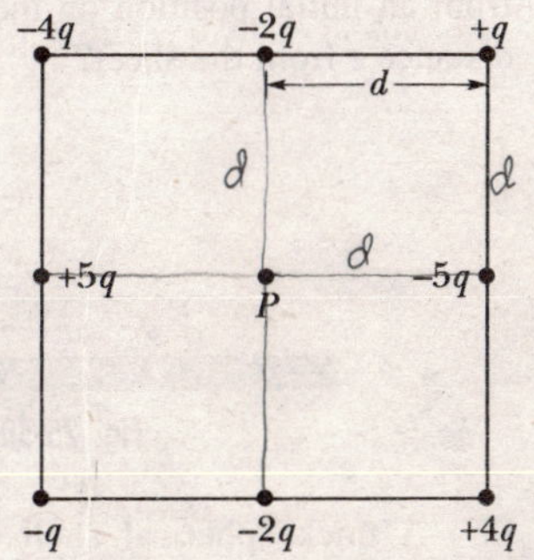

Fig. 25-23 Question 3.

4. Figure 25-24 shows four arrangements of charged particles, all the same distance from the origin. Rank the situations according to the net electric potential at the origin, most positive first. Take the potential to be zero at infinity.

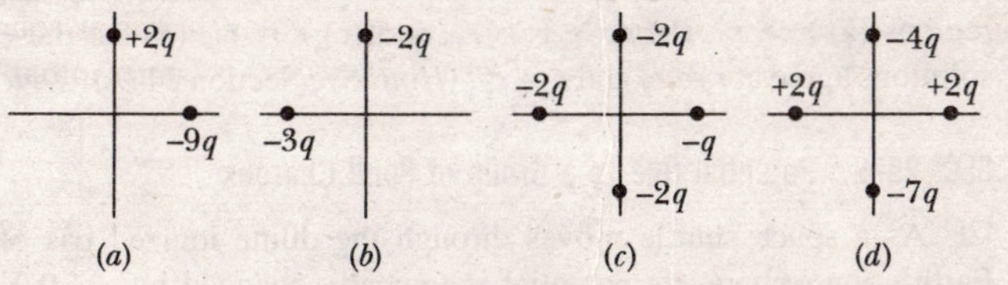

Fig. 25-24 Question 4.

5. (a) In Fig. 25-25*a*, what is the potential at point P due to charge Q at distance R from P? Set $V = 0$ at infinity. (b) In Fig. 25-25*b*, the same charge Q has been spread uniformly over a circular arc of radius R and central angle 40°. What is the potential at point P, the center of curvature of the arc? (c) In Fig. 25-25*c*, the same charge Q has been spread uniformly over a circle of radius R. What is the potential at point P, the center of the circle? (d) Rank the three situations according to the magnitude of the electric field that is set up at P, greatest first.

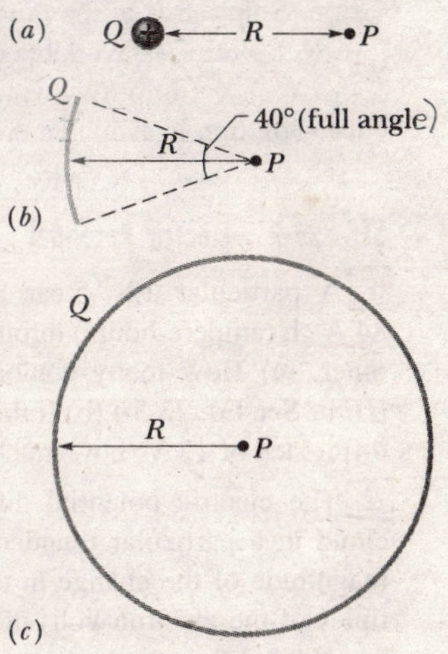

Fig. 25-25 Question 5.

6. Figure 25-26 shows three sets of cross sections of equipotential surfaces; all three cover the same size region of space. (a) Rank the arrangements according to the magnitude of the electric field present in the region, greatest first. (b) In which is the electric field directed down the page?

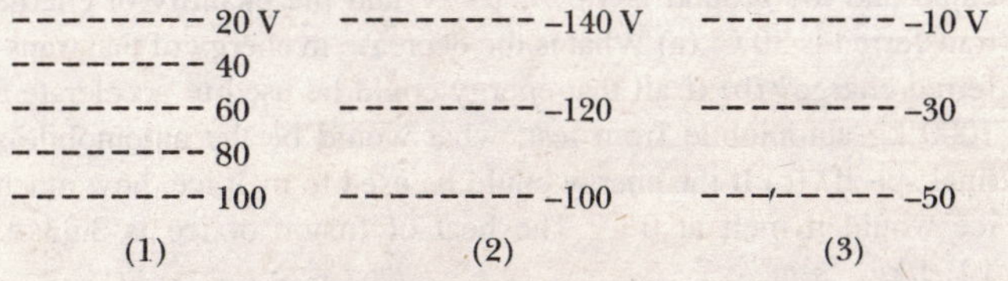

Fig. 25-26 Question 6.

7. Figure 25-27 gives the electric potential V as a function of x. (a) Rank the five regions according to the magnitude of the x component of the electric field within them, greatest first. What is the direction of the field along the x axis in (b) region 2 and (c) region 4?

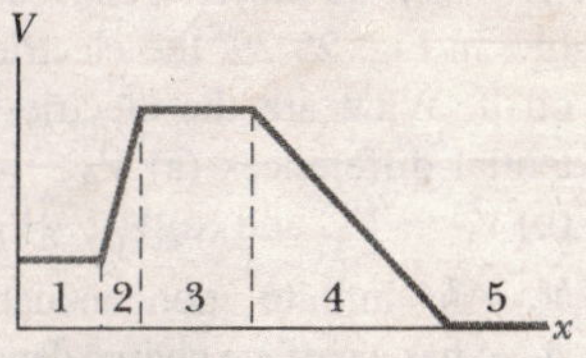

Fig. 25-27 Question 7.

8. Figure 25-22 shows four pairs of charged particles with identical separations. (a) Rank the pairs according to their electric potential energy, greatest (most positive) first. (b) For each pair, if the separation between the particles is increased, does the potential energy of the pair increase or decrease?

9. Figure 25-28 shows a system of three charged particles. If you move the particle of charge $+q$ from point A to point D, are the following positive, negative, or zero: (a) the change in the electric potential energy of the three-particle system, (b) the work done by the net electrostatic force on the particle you moved, and (c) the work done by your force? (d) What are the answers to (a) through (c) if, instead, the move is from point B to point C?

10. In the situation of Question 9, is the work done by your force positive, negative, or zero if the move is (a) from A to B, (b) from A to C, and (c) from B to D? (d) Rank those moves according to the magnitude of the work done by your force, greatest first.

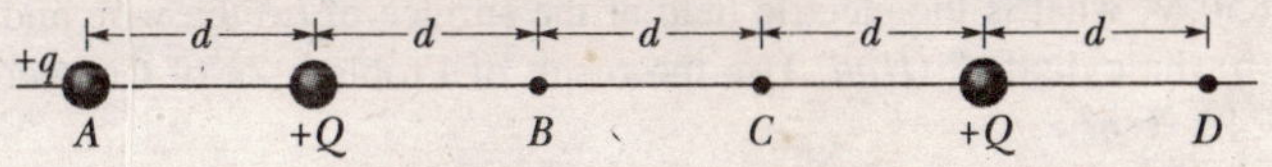

Fig. 25-28 Questions 9 and 10.

EXERCISES & PROBLEMS

ssm Solution is in the Student Solutions Manual.
www Solution is available on the World Wide Web at: http://www.wiley.com/college/hrw
ilw Solution is available on the Interactive LearningWare.

SEC. 25-2 Electric Potential

1E. A particular 12 V car battery can send a total charge of 84 A · h (ampere-hours) through a circuit, from one terminal to the other. (a) How many coulombs of charge does this represent? (*Hint:* See Eq. 22-3.) (b) If this entire charge undergoes a potential difference of 12 V, how much energy is involved? ssm

2E. The electric potential difference between the ground and a cloud in a particular thunderstorm is 1.2×10^9 V. What is the magnitude of the change in the electric potential energy (in multiples of the electron-volt) of an electron that moves between the ground and the cloud?

3P. In a given lightning flash, the potential difference between a cloud and the ground is 1.0×10^9 V and the quantity of charge transferred is 30 C. (a) What is the decrease in energy of that transferred charge? (b) If all that energy could be used to accelerate a 1000 kg automobile from rest, what would be the automobile's final speed? (c) If the energy could be used to melt ice, how much ice would it melt at 0°C? The heat of fusion of ice is 3.33×10^5 J/kg. ssm

SEC. 25-4 Calculating the Potential from the Field

4E. When an electron moves from A to B along an electric field line in Fig. 25-29, the electric field does 3.94×10^{-19} J of work on it. What are the electric potential differences (a) $V_B - V_A$, (b) $V_C - V_A$, and (c) $V_C - V_B$?

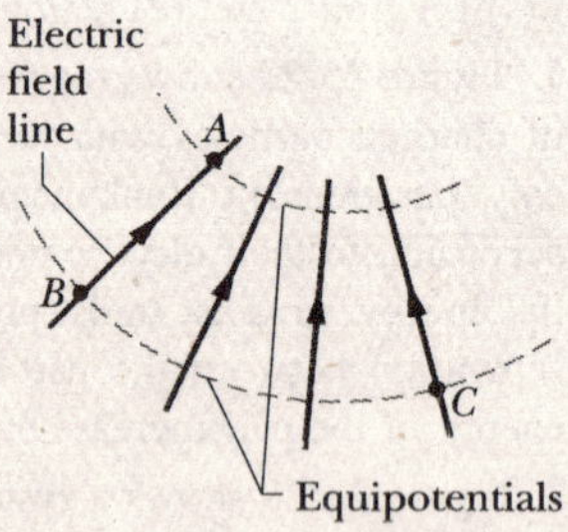

Fig. 25-29 Exercise 4.

5E. An infinite nonconducting sheet has a surface charge density $\sigma = 0.10\ \mu\text{C/m}^2$ on one side. How far apart are equipotential surfaces whose potentials differ by 50 V? ssm

6E. Two large, parallel, conducting plates are 12 cm apart and have charges of equal magnitude and opposite sign on their facing surfaces. An electrostatic force of 3.9×10^{-15} N acts on an electron placed anywhere between the two plates. (Neglect fringing.) (a) Find the electric field at the position of the electron. (b) What is the potential difference between the plates?

7P. A Geiger counter has a metal cylinder 2.00 cm in diameter along whose axis is stretched a wire 1.30×10^{-4} cm in diameter. If the potential difference between the wire and the cylinder is 850 V, what is the electric field at the surface of (a) the wire and (b) the cylinder? (*Hint:* Use the result of Problem 23 of Chapter 24.) ssm

8P. The electric field inside a nonconducting sphere of radius R, with charge spread uniformly throughout its volume, is radially directed and has magnitude

$$E(r) = \frac{qr}{4\pi\varepsilon_0 R^3}.$$

Here q (positive or negative) is the total charge within the sphere, and r is the distance from the sphere's center. (a) Taking $V = 0$ at the center of the sphere, find the electric potential $V(r)$ inside the sphere. (b) What is the difference in electric potential between a point on the surface and the sphere's center? (c) If q is positive, which of those two points is at the higher potential?

9P*. A charge q is distributed uniformly throughout a spherical volume of radius R. (a) Setting $V = 0$ at infinity, show that the potential at a distance r from the center, where $r < R$, is given by

$$V = \frac{q(3R^2 - r^2)}{8\pi\varepsilon_0 R^3}.$$

(*Hint:* See Section 24-9.) (b) Why does this result differ from that in (a) of Problem 8? (c) What is the potential difference between a point on the surface and the sphere's center? (d) Why doesn't this result differ from that of (b) of Problem 8? ssm

10P. Figure 25-30 shows, edge-on, an infinite nonconducting sheet with positive surface charge density σ on one side. (a) Use Eq. 25-18 and Eq. 24-13 to show that the electric potential of an infinite sheet of charge can be written $V = V_0 - (\sigma/2\varepsilon_0)z$, where V_0 is the electric potential at the surface of the sheet and z is the perpendicular distance from the sheet. (b) How much work is done by the electric field of the sheet as a small positive test charge q_0 is moved from an initial position on the sheet to a final position located a distance z from the sheet?

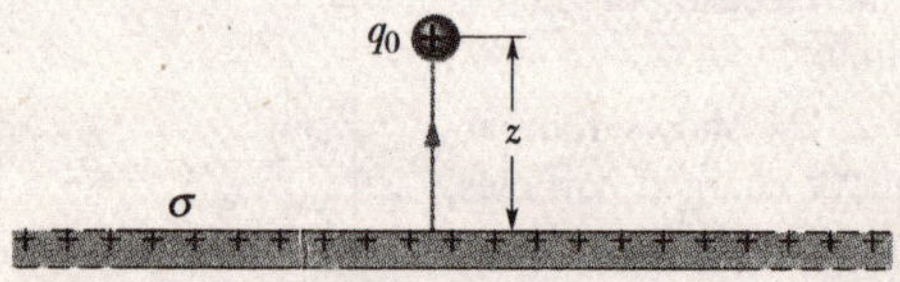

Fig. 25-30 Problem 10.

11P*. A thick spherical shell of charge Q and uniform volume charge density ρ is bounded by radii r_1 and r_2, where $r_2 > r_1$. With $V = 0$ at infinity, find the electric potential V as a function of the distance r from the center of the distribution, considering the regions (a) $r > r_2$, (b) $r_2 > r > r_1$, and (c) $r < r_1$. (d) Do these solutions agree at $r = r_2$ and $r = r_1$? (*Hint:* See Section 24-9.) ssm

SEC. 25-6 Potential Due to a Group of Point Charges

12E. As a space shuttle moves through the dilute ionized gas of Earth's ionosphere, its potential is typically changed by -1.0 V during one revolution. By assuming that the shuttle is a sphere of radius 10 m, estimate the amount of charge it collects.

13E. Consider a point charge $q = 1.0\ \mu$C, point A at distance $d_1 = 2.0$ m from q, and point B at distance $d_2 = 1.0$ m. (a) If these points are diametrically opposite each other, as in Fig. 25-31a, what is the electric potential difference $V_A - V_B$? (b) What is that electric potential difference if points A and B are located as in Fig. 25-31b?

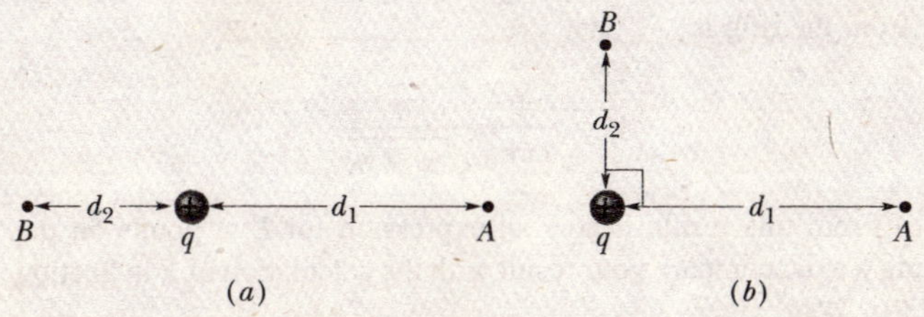

Fig. 25-31 Exercise 13.

14E. Figure 25-32 shows two charged particles on an axis. Sketch the electric field lines and the equipotential surfaces in the plane of the page for (a) $q_1 = +q$ and $q_2 = +2q$ and (b) $q_1 = +q$ and $q_2 = -3q$.

15E. In Fig. 25-32, set $V = 0$ at infinity and let the particles have charges $q_1 = +q$ and $q_2 = -3q$. Then locate (in terms of the separation distance d) any point on the x axis (other than at infinity) at which the net potential due to the two particles is zero.

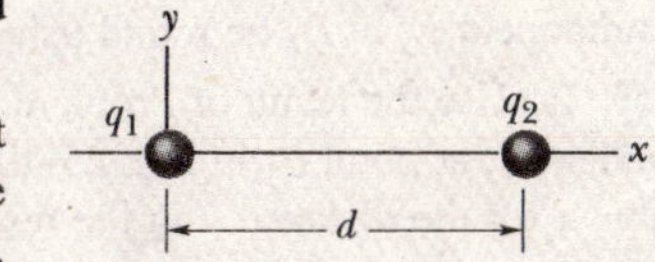

Fig. 25-32 Exercises 14, 15, and 16.

16E. Two particles, of charges q_1 and q_2, are separated by distance d in Fig. 25-32. The net electric field of the particles is zero at $x = d/4$. With $V = 0$ at infinity, locate (in terms of d) any point on the x axis (other than at infinity) at which the electric potential due to the two particles is zero.

17P. A spherical drop of water carrying a charge of 30 pC has a potential of 500 V at its surface (with $V = 0$ at infinity). (a) What is the radius of the drop? (b) If two such drops of the same charge and radius combine to form a single spherical drop, what is the potential at the surface of the new drop? ssm ilw www

18P. What are (a) the charge and (b) the charge density on the surface of a conducting sphere of radius 0.15 m whose potential is 200 V (with $V = 0$ at infinity)?

19P. An electric field of approximately 100 V/m is often observed near the surface of Earth. If this were the field over the entire surface, what would be the electric potential of a point on the surface? (Set $V = 0$ at infinity.) ssm

20P. In Fig. 25-33, point P is at the center of the rectangle. With $V = 0$ at infinity, what is the net electric potential at P due to the six charged particles?

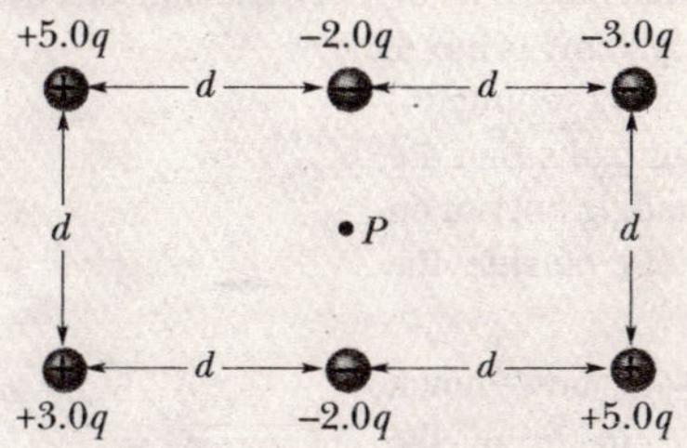

Fig. 25-33 Problem 20.

21P. In Fig. 25-34, what is the net potential at point P due to the four point charges, if $V = 0$ at infinity? ssm

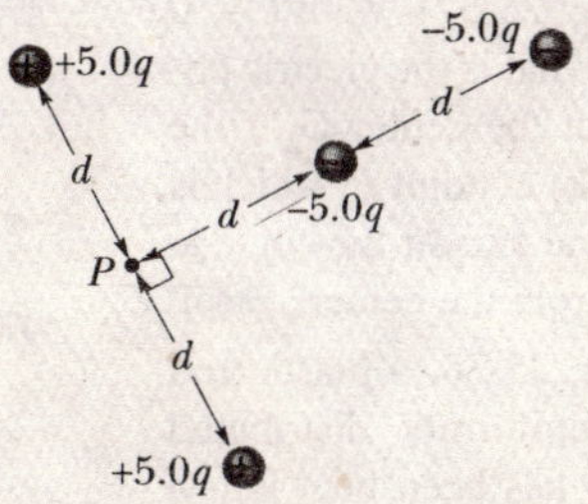

Fig. 25-34 Problem 21.

SEC. 25-7 Potential Due to an Electric Dipole

22E. The ammonia molecule NH_3 has a permanent electric dipole moment equal to 1.47 D, where 1 D = 1 debye unit = 3.34×10^{-30} C·m. Calculate the electric potential due to an ammonia molecule at a point 52.0 nm away along the axis of the dipole. (Set $V = 0$ at infinity.) ilw

23P. Figure 25-35 shows three charged particles located on a horizontal axis. For points (such as P) on the axis with $r \gg d$, show that the electric potential $V(r)$ is given by

$$V = \frac{1}{4\pi\varepsilon_0}\frac{q}{r}\left(1 + \frac{2d}{r}\right).$$

(*Hint:* The charge configuration can be viewed as the sum of an isolated charge and a dipole.) ssm www

or as sum of three isolated charges

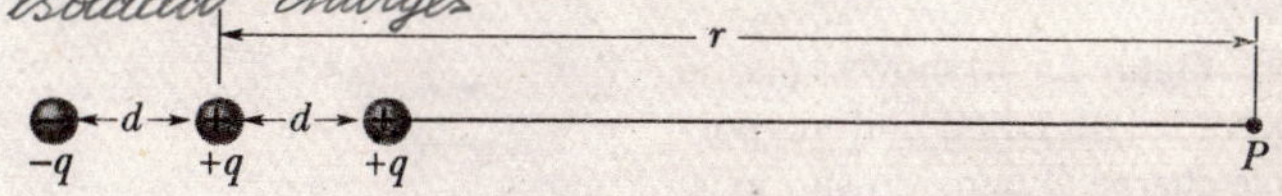

Fig. 25-35 Problem 23.

SEC. 25-8 Potential Due to a Continuous Charge Distribution

24E. (a) Figure 25-36a shows a positively charged plastic rod of length L and uniform linear charge density λ. Setting $V = 0$ at infinity and considering Fig. 25-13 and Eq. 25-35, find the electric potential at point P without written calculation. (b) Figure 25-36b shows an identical rod, except that it is split in half and the right half is negatively charged; the left and right halves have the same magnitude λ of uniform linear charge density. With V still zero at infinity, what is the electric potential at point P in Fig. 25-36b?

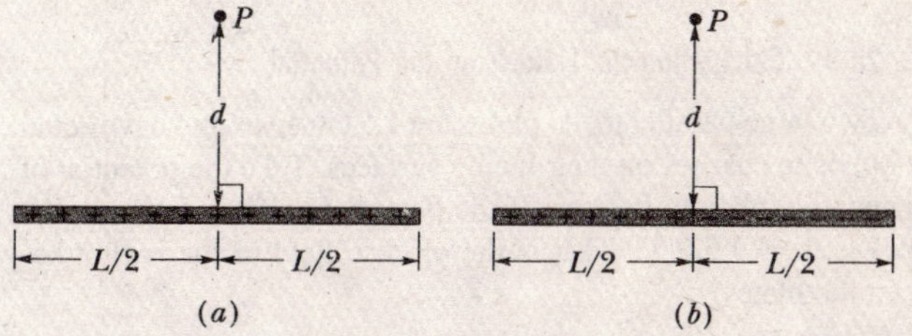

Fig. 25-36 Exercise 24.

25E. A plastic rod has been formed into a circle of radius R. It has a positive charge $+Q$ uniformly distributed along one-quarter of its circumference and a negative charge of $-6Q$ uniformly distributed along the rest of the circumference (Fig. 25-37). With $V = 0$ at infinity, what is the electric potential (a) at the center C of the circle and (b) at point P, which is on the central axis of the circle at distance z from the center? ssm

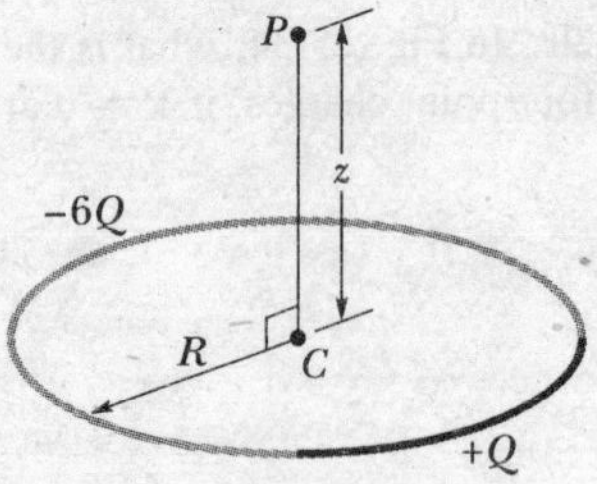

Fig. 25-37 Exercise 25.

26E. In Fig. 25-38, a plastic rod having a uniformly distributed charge $-Q$ has been bent into a circular arc of radius R and central angle 120°. With $V = 0$ at infinity, what is the electric potential at P, the center of curvature of the rod?

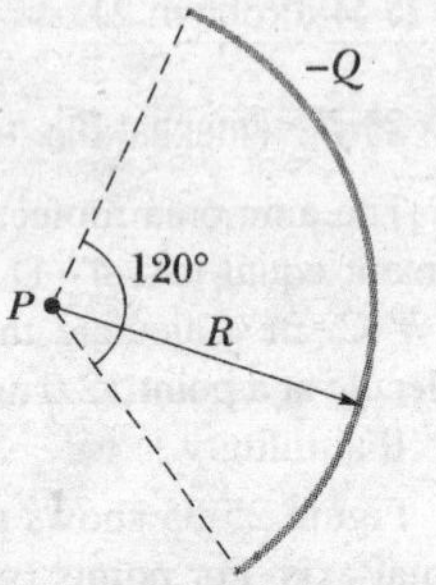

Fig. 25-38 Exercise 26.

27E. A plastic disk is charged on one side with a uniform surface charge density σ, and then three quadrants of the disk are removed. The remaining quadrant is shown in Fig. 25-39. With $V = 0$ at infinity, what is the potential due to the remaining quadrant at point P, which is on the central axis of the original disk at a distance z from the original center? ssm

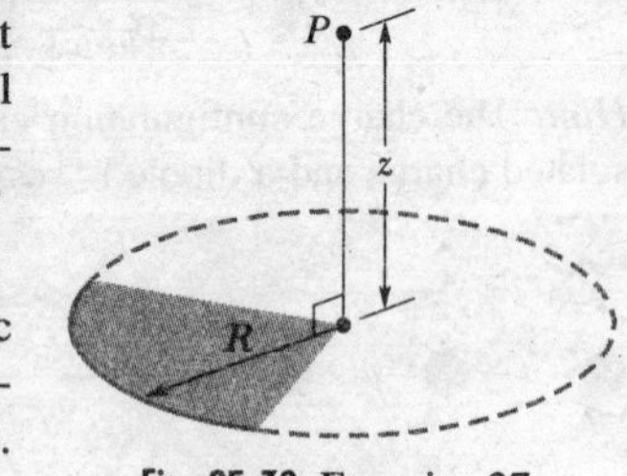

Fig. 25-39 Exercise 27.

28P. Figure 25-40 shows a plastic rod of length L and uniform positive charge Q lying on an x axis. With $V = 0$ at infinity, find the electric potential at point P_1 on the axis, at distance d from one end of the rod.

29P. The plastic rod shown in Fig. 25-40 has length L and a nonuniform linear charge density $\lambda = cx$, where c is a positive constant. With $V = 0$ at infinity, find the electric potential at point P_1 on the axis, at distance d from one end.

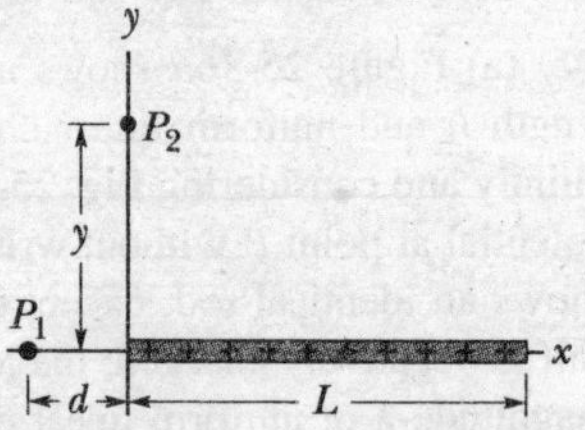

Fig. 25-40 Problems 28, 29, 34, and 35.

SEC. 25-9 Calculating the Field from the Potential

30E. Two large parallel metal plates are 1.5 cm apart and have equal but opposite charges on their facing surfaces. Take the potential of the negative plate to be zero. If the potential halfway between the plates is then +5.0 V, what is the electric field in the region between the plates?

31E. The electric potential at points in an xy plane is given by $V = (2.0\ \text{V/m}^2)x^2 - (3.0\ \text{V/m}^2)y^2$. What are the magnitude and direction of the electric field at the point (3.0 m, 2.0 m)?

32E. The electric potential V in the space between two flat parallel plates is given by $V = 1500x^2$, where V is in volts if x, the distance from one of the plates, is in meters. Calculate the magnitude and direction of the electric field at $x = 1.3$ cm.

33P. (a) Using Eq. 25-32, show that the electric potential at a point on the central axis of a thin ring of charge of radius R and a distance z from the ring is

$$V = \frac{1}{4\pi\varepsilon_0}\frac{q}{\sqrt{z^2 + R^2}}.$$

(b) From this result, derive an expression for E at points on the ring's axis; compare your result with the calculation of E in Section 23-6. ssm www

34P. The plastic rod of length L in Fig. 25-40 has the nonuniform linear charge density $\lambda = cx$, where c is a positive constant. (a) With $V = 0$ at infinity, find the electric potential at point P_2 on the y axis, a distance y from one end of the rod. (b) From that result, find the electric field component E_y at P_2. (c) Why cannot the field component E_x at P_2 be found using the result of (a)?

35P. (a) Use the result of Problem 28 to find the electric field component E_x at point P_1 in Fig. 25-40. (*Hint:* First substitute the variable x for the distance d in the result.) (b) Use symmetry to determine the electric field component E_y at P_1. ssm

SEC. 25-10 Electric Potential Energy of a System of Point Charges

36E. (a) What is the electric potential energy of two electrons separated by 2.00 nm? (b) If the separation increases, does the potential energy increase or decrease?

37E. Derive an expression for the work required to set up the four-charge configuration of Fig. 25-41, assuming the charges are initially infinitely far apart. ilw

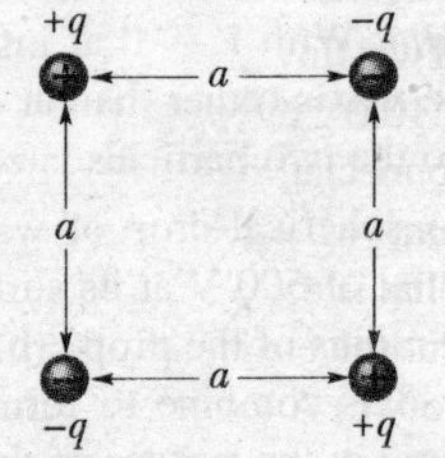

Fig. 25-41 Exercise 37.

38E. What is the electric potential energy of the charge configuration of Fig. 25-9*a*? Use the numerical values provided in Sample Problem 25-3.

39P. In the rectangle of Fig. 25-42, the sides have lengths 5.0 cm and 15 cm, $q_1 = -5.0\ \mu\text{C}$, and $q_2 = +2.0\ \mu\text{C}$. With $V = 0$ at infinity, what are the electric potentials (a) at corner A and (b) at corner B? (c) How much work is required to move a third charge $q_3 = +3.0\ \mu\text{C}$ from B to A along a diagonal of the rectangle? (d) Does this work increase or decrease the electric energy of the three-charge system? Is more, less, or the same work required if q_3 is moved along paths that are (e) inside the rectangle but not on a diagonal and (f) outside the rectangle? ssm

Fig. 25-42 Problem 39.

40P. In Fig. 25-43, how much work is required to bring the charge of $+5q$ in from infinity along the dashed line and place it

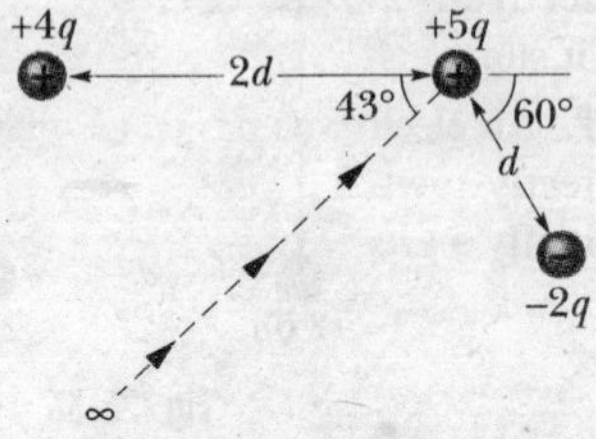

Fig. 25-43 Problem 40.

as shown near the two fixed charges $+4q$ and $-2q$? Take distance $d = 1.40$ cm and charge $q = 1.6 \times 10^{-19}$ C.

41P. A particle of positive charge Q is fixed at point P. A second particle of mass m and negative charge $-q$ moves at constant speed in a circle of radius r_1, centered at P. Derive an expression for the work W that must be done by an external agent on the second particle to increase the radius of the circle of motion to r_2. ssm www

42P. Calculate (a) the electric potential established by the nucleus of a hydrogen atom at the average distance ($r = 5.29 \times 10^{-11}$ m) of the atom's electron (take $V = 0$ at infinite distance), (b) the electric potential energy of the atom when the electron is at this radius, and (c) the kinetic energy of the electron, assuming it to be moving in a circular orbit of this radius centered on the nucleus. (d) How much energy is required to ionize the hydrogen atom (that is, to remove the electron from the nucleus so that the separation is effectively infinite)? Express all energies in electron-volts.

43P. A particle of charge q is fixed at point P, and a second particle of mass m and the same charge q is initially held a distance r_1 from P. The second particle is then released. Determine its speed when it is a distance r_2 from P. Let $q = 3.1\ \mu$C, $m = 20$ mg, $r_1 = 0.90$ mm, and $r_2 = 2.5$ mm. ilw

44P. A charge of -9.0 nC is uniformly distributed around a thin plastic ring of radius 1.5 m that lies in the yz plane with its center at the origin. A point charge of -6.0 pC is located on the x axis at $x = 3.0$ m. Calculate the work done on the point charge by an external force to move the point charge to the origin.

45P. Two tiny metal spheres A and B of mass $m_A = 5.00$ g and $m_B = 10.0$ g have equal positive charges $q = 5.00\ \mu$C. The spheres are connected by a massless nonconducting string of length $d = 1.00$ m, which is much greater than the radii of the spheres. (a) What is the electric potential energy of the system? (b) Suppose you cut the string. At that instant, what is the acceleration of each sphere? (c) A long time after you cut the string, what is the speed of each sphere? ssm

46P. A thin, spherical, conducting shell of radius R is mounted on an isolating support and charged to a potential of $-V$. An electron is then fired from point P at distance r from the center of the shell ($r \gg R$) with initial speed v_0 and directly toward the shell's center. What value of v_0 is needed for the electron to just reach the shell before reversing direction?

47P. Two electrons are fixed 2.0 cm apart. Another electron is shot from infinity and stops midway between the two. What is its initial speed? ssm

48P. Two charged, parallel, flat conducting surfaces are spaced $d = 1.00$ cm apart and produce a potential difference $\Delta V = 625$ V between them. An electron is projected from one surface directly toward the second. What is the initial speed of the electron if it stops just at the second surface?

49P. An electron is projected with an initial speed of 3.2×10^5 m/s directly toward a proton that is fixed in place. If the electron is initially a great distance from the proton, at what distance from the proton is the speed of the electron instantaneously equal to twice the initial value?

SEC. 25-11 Potential of a Charged Isolated Conductor

50E. An empty hollow metal sphere has a potential of $+400$ V with respect to ground (defined to be at $V = 0$) and has a charge of 5.0×10^{-9} C. Find the electric potential at the center of the sphere.

51E. What is the excess charge on a conducting sphere of radius $r = 0.15$ m if the potential of the sphere is 1500 V and $V = 0$ at infinity? ssm

52E. Consider two widely separated conducting spheres, 1 and 2, the second having twice the diameter of the first. The smaller sphere initially has a positive charge q, and the larger one is initially uncharged. You now connect the spheres with a long thin wire. (a) How are the final potentials V_1 and V_2 of the spheres related? $V_1 = V_2$ (b) What are the final charges q_1 and q_2 on the spheres, in terms of q? (c) What is the ratio of the final surface charge density of sphere 1 to that of sphere 2? 1:8 → $2q_1 = q_2$ & $q_1 + q_2 = q$ ⇒ $q_1 = q/3$ & $q_2 = 2q/3$

53P. Two metal spheres, each of radius 3.0 cm, have a center-to-center separation of 2.0 m. One has a charge of $+1.0 \times 10^{-8}$ C; the other has a charge of -3.0×10^{-8} C. Assume that the separation is large enough relative to the size of the spheres to permit us to consider the charge on each to be uniformly distributed (the spheres do not affect each other). With $V = 0$ at infinity, calculate (a) the potential at the point halfway between their centers and (b) the potential of each sphere. ssm

54P. A charged metal sphere of radius 15 cm has a net charge of 3.0×10^{-8} C. (a) What is the electric field at the sphere's surface? (b) If $V = 0$ at infinity, what is the electric potential at the sphere's surface? (c) At what distance from the sphere's surface has the electric potential decreased by 500 V?

55P. (a) If Earth had a net surface charge density of 1.0 electron per square meter (a very artificial assumption), what would its potential be? (Set $V = 0$ at infinity.) (b) What would be the electric field due to Earth just outside its surface?

56P. Two thin, isolated, concentric conducting spheres of radii R_1 and R_2 (with $R_1 < R_2$) have charges q_1 and q_2. With $V = 0$ at infinity, derive expressions for $E(r)$ and $V(r)$, where r is distance from the center of the spheres. Plot $E(r)$ and $V(r)$ from $r = 0$ to $r = 4.0$ m for $R_1 = 0.50$ m, $R_2 = 1.0$ m, $q_1 = +2.0\ \mu$C, and $q_2 = +1.0\ \mu$C.

Additional Problem

57. *The chocolate crumb mystery.* This story begins with Problem 48 in Chapter 24. (a) From the answer to part (a) of that problem, find an expression for the electric potential as a function of the radial distance r from the center of the pipe. (The electric potential is zero on the grounded pipe wall.) (b) For the typical volume charge density $\rho = -1.1 \times 10^{-3}$ C/m^3, what is the difference in the electric potential between the pipe's center and its inside wall? (The story continues with Problem 48 in Chapter 26.)

26 Capacitance

During ventricular fibrillation, a common type of heart attack, the chambers of the heart fail to pump blood because their muscle fibers randomly contract and relax. To save a victim of ventricular fibrillation, the heart muscle must be shocked to reestablish its normal rhythm. For that, 20 A of current must be sent through the chest cavity to transfer 200 J of electrical energy in about 2.0 ms. This requires about 100 kW of electric power. Such a requirement may easily be met in a hospital, but not by, say, the electrical system of an ambulance arriving to help the victim.

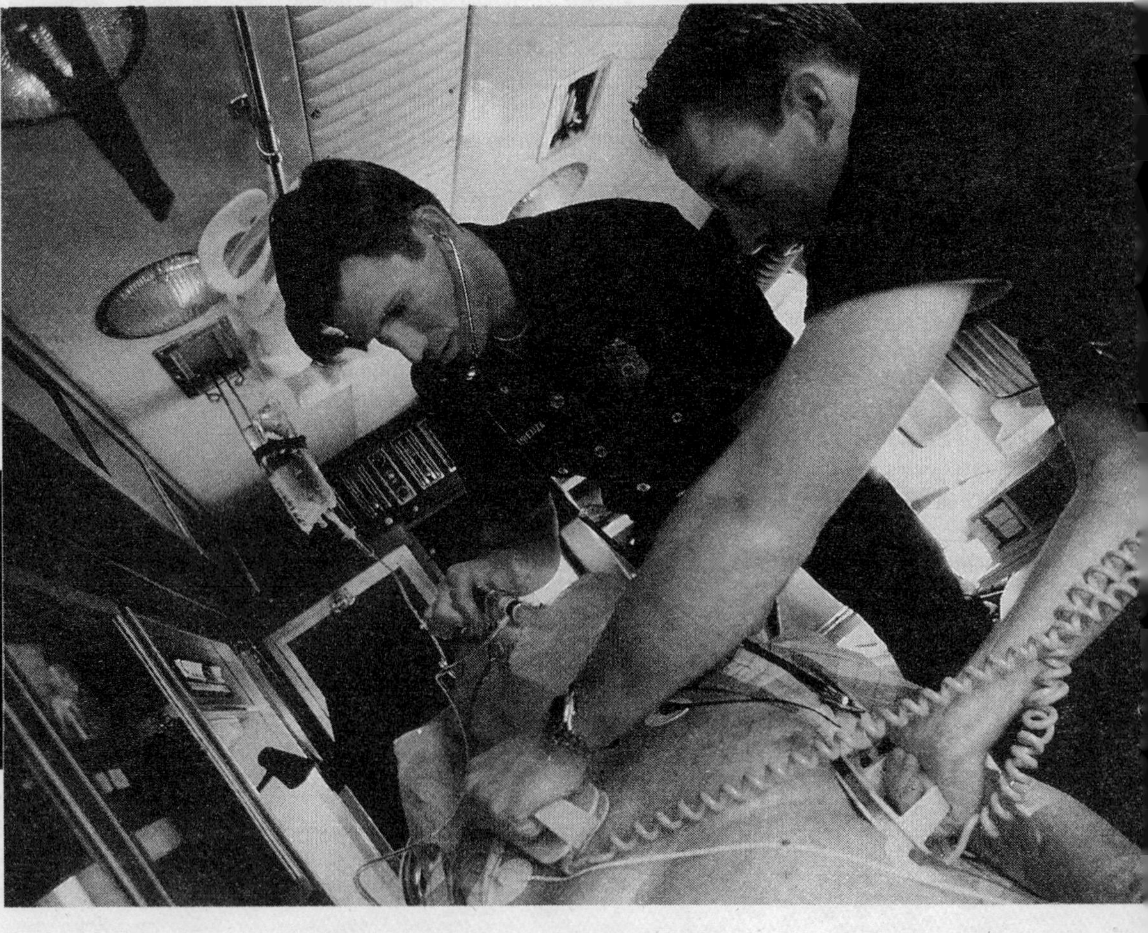

What, then, can provide the power needed for defibrillation at remote locations?

The answer is in this chapter.

Fig. 26-1 An assortment of capacitors.

26-1 The Uses of Capacitors

You can store energy as potential energy by pulling a bowstring, stretching a spring, compressing a gas, or lifting a book. You can also store energy as potential energy in an electric field, and a **capacitor** is a device you can use to do exactly that.

There is a capacitor in a portable battery-operated photoflash unit, for example. It accumulates charge relatively slowly during the readying process between flashes, building up an electric field as it does so. It holds this field and the associated energy until the energy is rapidly released to initiate the flash.

Capacitors have many uses in our electronic and microelectronic age beyond serving as storehouses for potential energy. As one example, they are vital elements in the circuits with which we tune radio and television transmitters and receivers. As another example, microscopic capacitors form the memory banks of computers. These tiny devices are not as important for their stored energy as for the ON–OFF information that the presence or absence of their electric fields provides.

26-2 Capacitance

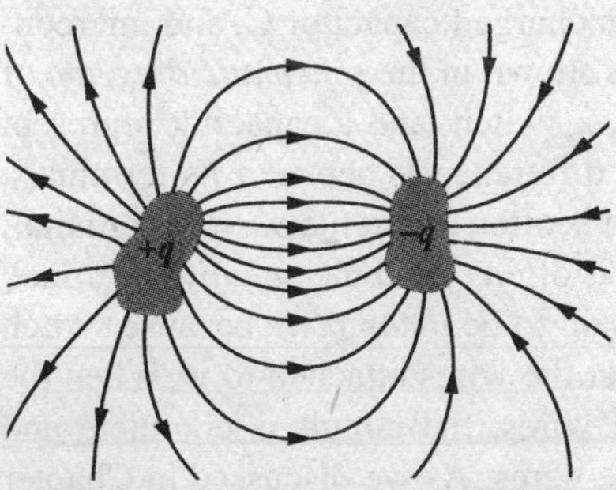

Fig. 26-2 Two conductors, isolated electrically from each other and from their surroundings, form a *capacitor*. When the capacitor is charged, the conductors, or *plates* as they are called, have equal but opposite charges of magnitude q.

Figure 26-1 shows some of the many sizes and shapes of capacitors. Figure 26-2 shows the basic elements of *any* capacitor—two isolated conductors of any shape. No matter what their geometry, flat or not, we call these conductors *plates*.

Figure 26-3a shows a less general but more conventional arrangement, called a *parallel-plate capacitor*, consisting of two parallel conducting plates of area A separated by a distance d. The symbol that we use to represent a capacitor (⊣⊢) is based on the structure of a parallel-plate capacitor but is used for capacitors of all geometries. We assume for the time being that no material medium (such as glass or plastic) is present in the region between the plates. In Section 26-6, we shall remove this restriction.

When a capacitor is *charged*, its plates have equal but opposite charges of $+q$ and $-q$. However, we refer to the *charge of a capacitor* as being q, the absolute value of these charges on the plates. (Note that q is not the net charge on the capacitor, which is zero.)

Because the plates are conductors, they are equipotential surfaces; all points on a plate are at the same electric potential. Moreover, there is a potential difference between the two plates. For historical reasons, we represent the absolute value of this potential difference with V rather than with ΔV as we would with previous notation.

The charge q and the potential difference V for a capacitor are proportional to each other; that is,

$$q = CV. \tag{26-1}$$

The proportionality constant C is called the **capacitance** of the capacitor. Its value depends only on the geometry of the plates and *not* on their charge or potential difference. The capacitance is a measure of how much charge must be put on the

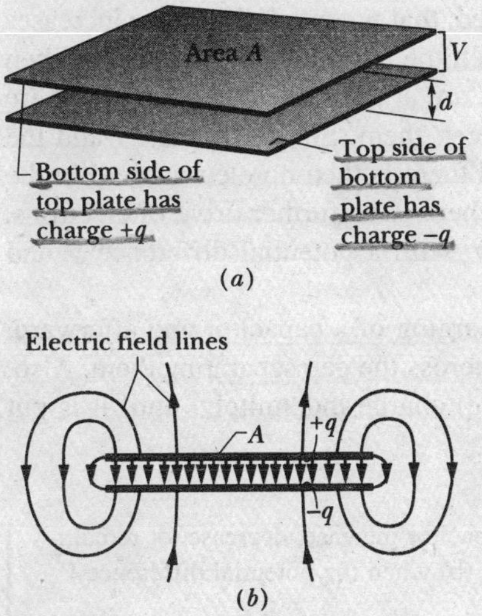

Fig. 26-3 (a) A parallel-plate capacitor, made up of two plates of area A separated by a distance d. The plates have equal and opposite charges of magnitude q on their facing surfaces. (b) As the field lines show, the electric field due to the charged plates is uniform in the central region between the plates. The field is not uniform at the edges of the plates, as indicated by the 'fringing' of the field lines there.

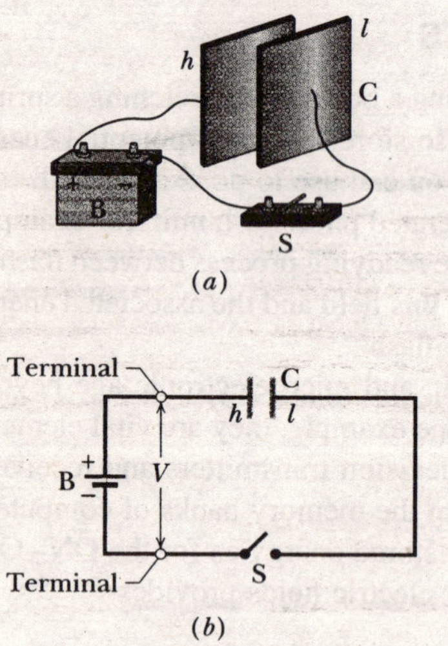

Fig. 26-4 (*a*) Battery B, switch S, and plates *h* and *l* of capacitor C, connected in a circuit. (*b*) A schematic diagram with the *circuit elements* represented by their symbols.

plates to produce a certain potential difference between them: The *greater the capacitance, the more charge is required*.

The SI unit of capacitance that follows from Eq. 26-1 is the coulomb per volt. This unit occurs so often that it is given a special name, the *farad* (F):

$$1 \text{ farad} = 1 \text{ F} = 1 \text{ coulomb per volt} = 1 \text{ C/V}. \tag{26-2}$$

As you will see, the farad is a very large unit. Submultiples of the farad, such as the microfarad ($1\ \mu F = 10^{-6}$ F) and the picofarad ($1\ pF = 10^{-12}$ F), are more convenient units in practice.

Charging a Capacitor

One way to charge a capacitor is to place it in an electric circuit with a battery. An *electric circuit* is a path through which charge can flow. A *battery* is a device that maintains a certain potential difference between its *terminals* (points at which charge can enter or leave the battery) by means of internal electrochemical reactions in which electric forces can move internal charge.

In Fig. 26-4*a*, a battery B, a switch S, an uncharged capacitor C, and interconnecting wires form a circuit. The same circuit is shown in the *schematic diagram* of Fig. 26-4*b*, in which the symbols for a battery, a switch, and a capacitor represent those devices. The battery maintains potential difference V between its terminals. The terminal of higher potential is labeled + and is often called the *positive* terminal; the terminal of lower potential is labeled − and is often called the *negative* terminal.

The circuit shown in Figs. 26-4*a* and *b* is said to be *incomplete* because switch S is *open*; that is, it does not electrically connect the wires attached to it. When the switch is *closed*, electrically connecting those wires, the circuit is complete and charge can then flow through the switch and the wires. As we discussed in Chapter 22, the charge that can flow through a conductor, such as a wire, is that of electrons. When the circuit of Fig. 26-4 is completed, electrons are driven through the wires by an electric field that the battery sets up in the wires. The field drives electrons from capacitor plate *h* to the positive terminal of the battery; thus, plate *h*, losing electrons, becomes positively charged. The field drives just as many electrons from the negative terminal of the battery to capacitor plate *l*; thus, plate *l*, gaining electrons, becomes negatively charged *just as much* as plate *h*, losing electrons, becomes positively charged.

Due to the PD b/w the two terminals, electric field is set up in the wire

negative terminal of battery has excess of electrons

Initially, when the plates are uncharged, the potential difference between them is zero. As the plates become oppositely charged, that potential difference increases until it equals the potential difference V between the terminals of the battery. Then plate *h* and the positive terminal of the battery are at the same potential, and there is no longer an electric field in the wire between them. Similarly, plate *l* and the negative terminal reach the same potential and there is then no electric field in the wire between them. Thus, with the field zero, there is no further drive of electrons. The capacitor is then said to be *fully charged*, with a potential difference V and charge q that are related by Eq. 26-1.

In this book we assume that during the charging of a capacitor and afterward, charge cannot pass from one plate to the other across the gap separating them. Also, we assume that a capacitor can retain (or *store*) charge indefinitely, until it is put into a circuit where it can be *discharged*.

CHECKPOINT 1 Does the capacitance C of a capacitor increase, decrease, or remain the same (a) when the charge q on it is doubled and (b) when the potential difference V across it is tripled?

C depends upon geometry of plates only

PROBLEM-SOLVING TACTICS

Tactic 1: *The Symbol V and Potential Difference*

In previous chapters, the symbol V represents an electric potential at a point or along an equipotential surface. However, in matters concerning electrical devices, V often represents a *potential difference* between two points or two equipotential surfaces. Equation 26-1 is an example of this second use of the symbol. In Section 26-3, you will see a mixture of the two meanings of V. There and in later chapters, you need to be alert as to the intent of this symbol.

You will also be seeing, in this book and elsewhere, a variety of phrases regarding potential difference. A potential difference or a "potential" or a "voltage" may be *applied* to a device, or it may be *across* a device. A capacitor can be charged to a potential difference, as in "a capacitor is charged to 12 V." Also, a battery can be characterized by the potential difference across it, as in "a 12 V battery." Always keep in mind what is meant by such phrases: There is a potential difference between two points, such as two points in a circuit or at the terminals of a device such as a battery.

26-3 Calculating the Capacitance

Our task here is to calculate the capacitance of a capacitor once we know its geometry. Because we will consider a number of different geometries, it seems wise to develop a general plan to simplify the work. In brief our plan is as follows: (1) Assume a charge q on the plates; (2) calculate the electric field $\vec{E}$ between the plates in terms of this charge, using Gauss' law; (3) knowing $\vec{E}$, calculate the potential difference V between the plates from Eq. 25-18; (4) calculate C from Eq. 26-1.

Before we start, we can simplify the calculation of both the electric field and the potential difference by making certain assumptions. We discuss each in turn.

Calculating the Electric Field

To relate the electric field $\vec{E}$ between the plates of a capacitor to the charge q on either plate, we shall use Gauss' law:

$$\varepsilon_0 \oint \vec{E} \cdot d\vec{A} = q. \tag{26-3}$$

Here q is the charge enclosed by a Gaussian surface, and $\oint \vec{E} \cdot d\vec{A}$ is the net electric flux through that surface. In all cases that we shall consider, the Gaussian surface will be such that whenever electric flux passes through it, $\vec{E}$ will have a uniform magnitude E and the vectors $\vec{E}$ and $d\vec{A}$ will be parallel. Equation 26-3 then will reduce to

$$q = \varepsilon_0 EA \quad \text{(special case of Eq. 26-3)}, \tag{26-4}$$

in which A is the area of that part of the Gaussian surface through which flux passes. For convenience, we shall always draw the Gaussian surface in such a way that it completely encloses the charge on the positive plate; see Fig. 26-5 for an example.

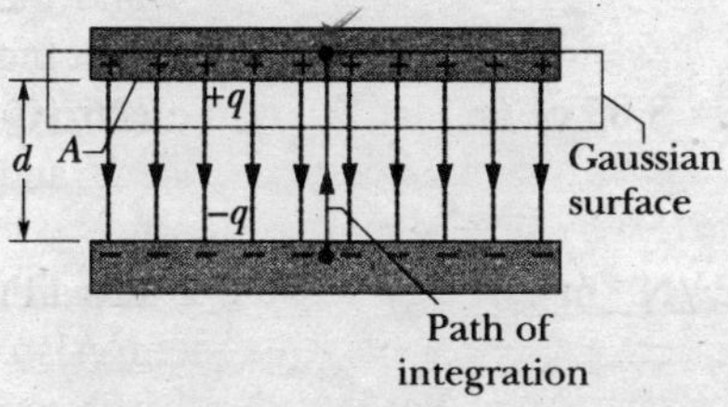

Fig. 26-5 A charged parallel-plate capacitor. A Gaussian surface encloses the charge on the positive plate. The integration of Eq. 26-6 is taken along a path extending directly from the negative plate to the positive plate.

Calculating the Potential Difference

In the notation of Chapter 25 (Eq. 25-18), the potential difference between the plates of a capacitor is related to the field $\vec{E}$ by

$$V_f - V_i = -\int_i^f \vec{E} \cdot d\vec{s}, \tag{26-5}$$

in which the integral is to be evaluated along any path that starts on one plate and ends on the other. We shall always choose a path that follows an electric field line, from the negative plate to the positive plate. For this path, the vectors $\vec{E}$ and $d\vec{s}$ will have opposite directions, so the dot product $\vec{E} \cdot d\vec{s}$ will be equal to $-E\,ds$. Thus, the right side of Eq. 26-5 will then be positive. Letting V represent the difference $V_f - V_i$, we can then recast Eq. 26-5 as

$$V = \int_{-}^{+} E\,ds \quad \text{(special case of Eq. 26-5)}, \tag{26-6}$$

in which the $-$ and $+$ remind us that our path of integration starts on the negative plate and ends on the positive plate.

We are now ready to apply Eqs. 26-4 and 26-6 to some particular cases.

A Parallel-Plate Capacitor

We assume, as Fig. 26-5 suggests, that the plates of our parallel-plate capacitor are so large and so close together that we can neglect the fringing of the electric field at the edges of the plates, taking $\vec{E}$ to be constant throughout the region between the plates.

We draw a Gaussian surface that encloses just the charge q on the positive plate, as in Fig. 26-5. From Eq. 26-4 we can then write

$$q = \varepsilon_0 EA, \tag{26-7}$$

where A is the area of the plate.

Equation 26-6 yields

$$V = \int_{-}^{+} E\,ds = E \int_{0}^{d} ds = Ed. \tag{26-8}$$

In Eq. 26-8, E can be placed outside the integral because it is a constant; the second integral then is simply the plate separation d.

If we now substitute q from Eq. 26-7 and V from Eq. 26-8 into the relation $q = CV$ (Eq. 26-1), we find

$$C = \frac{\varepsilon_0 A}{d} \quad \text{(parallel-plate capacitor)}. \tag{26-9}$$

Thus, the capacitance does indeed depend only on geometrical factors—namely, the plate area A and the plate separation d. Note that C increases as we increase the plate area A or decrease the separation d.

As an aside, we point out that Eq. 26-9 suggests one of our reasons for writing the electrostatic constant in Coulomb's law in the form $1/4\pi\varepsilon_0$. If we had not done so, Eq. 26-9—which is used more often in engineering practice than Coulomb's law—would have been less simple in form. We note further that Eq. 26-9 permits us to express the permittivity constant ε_0 in a unit more appropriate for use in problems involving capacitors; namely,

$$\varepsilon_0 = 8.85 \times 10^{-12}\ \text{F/m} = 8.85\ \text{pF/m}. \tag{26-10}$$

We have previously expressed this constant as

$$\varepsilon_0 = 8.85 \times 10^{-12}\ \text{C}^2/\text{N}\cdot\text{m}^2. \tag{26-11}$$

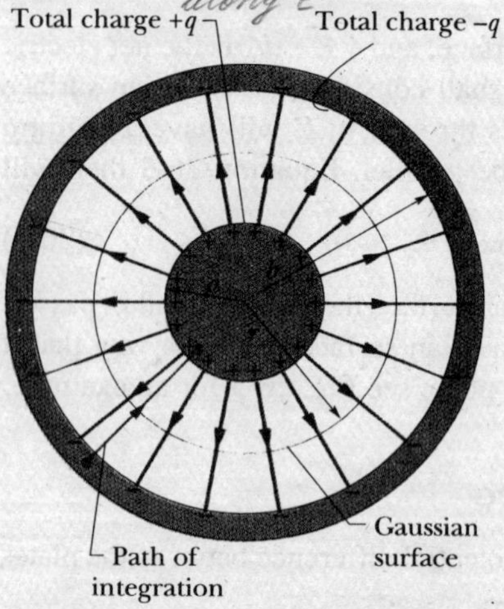

Fig. 26-6 A cross section of a long cylindrical capacitor, showing a cylindrical Gaussian surface of radius r (that encloses the positive plate) and the radial path of integration along which Eq. 26-6 is to be applied. This figure also serves to illustrate a spherical capacitor in a cross section through its center.

A Cylindrical Capacitor

Figure 26-6 shows, in cross section, a cylindrical capacitor of length L formed by two coaxial cylinders of radii a and b. We assume that $L \gg b$ so that we can neglect the fringing of the electric field that occurs at the ends of the cylinders. Each plate contains a charge of magnitude q.

As a Gaussian surface, we choose a cylinder of length L and radius r, closed by end caps and placed as is shown in Fig. 26-6. Equation 26-4 then yields

$$q = \varepsilon_0 EA = \varepsilon_0 E(2\pi rL),$$

in which $2\pi rL$ is the area of the curved part of the Gaussian surface. There is no flux through the end caps. Solving for E yields

$$E = \frac{q}{2\pi\varepsilon_0 Lr}. \tag{26-12}$$

Substitution of this result into Eq. 26-6 yields

$$V = \int_-^+ E\,ds = -\frac{q}{2\pi\varepsilon_0 L}\int_b^a \frac{dr}{r} = \frac{q}{2\pi\varepsilon_0 L}\ln\left(\frac{b}{a}\right), \tag{26-13}$$

where we have used the fact that here $ds = -dr$ (we integrated radially inward). From the relation $C = q/V$, we then have

$$C = 2\pi\varepsilon_0 \frac{L}{\ln(b/a)} \quad \text{(cylindrical capacitor)}. \tag{26-14}$$

We see that the capacitance of a cylindrical capacitor, like that of a parallel-plate capacitor, depends only on geometrical factors, in this case L, b, and a.

A Spherical Capacitor

Figure 26-6 can also serve as a central cross section of a capacitor that consists of two concentric spherical shells, of radii a and b. As a Gaussian surface we draw a sphere of radius r concentric with the two shells; then Eq. 26-4 yields

$$q = \varepsilon_0 EA = \varepsilon_0 E(4\pi r^2),$$

in which $4\pi r^2$ is the area of the spherical Gaussian surface. We solve this equation for E, obtaining

$$E = \frac{1}{4\pi\varepsilon_0}\frac{q}{r^2}, \tag{26-15}$$

which we recognize as the expression for the electric field due to a uniform spherical charge distribution (Eq. 24-15).

If we substitute this expression into Eq. 26-6, we find

$$V = \int_-^+ E\,ds = -\frac{q}{4\pi\varepsilon_0}\int_b^a \frac{dr}{r^2} = \frac{q}{4\pi\varepsilon_0}\left(\frac{1}{a} - \frac{1}{b}\right) = \frac{q}{4\pi\varepsilon_0}\frac{b-a}{ab}, \tag{26-16}$$

where again we have substituted $-dr$ for ds. If we now substitute Eq. 26-16 into Eq. 26-1 and solve for C, we find

$$C = 4\pi\varepsilon_0 \frac{ab}{b-a} \quad \text{(spherical capacitor)}. \tag{26-17}$$

An Isolated Sphere

We can assign a capacitance to a *single* isolated spherical conductor of radius R by assuming that the "missing plate" is a conducting sphere of infinite radius. After all, the field lines that leave the surface of a positively charged isolated conductor must end somewhere; the walls of the room in which the conductor is housed can serve effectively as our sphere of infinite radius.

To find the capacitance of the isolated conductor, we first rewrite Eq. 26-17 as

$$C = 4\pi\varepsilon_0 \frac{a}{1 - a/b}.$$

If we then let $b \rightarrow \infty$ and substitute R for a, we find

$$C = 4\pi\varepsilon_0 R \quad \text{(isolated sphere).} \tag{26-18}$$

Note that this formula and the others we have derived for capacitance (Eqs. 26-9, 26-14, and 26-17) involve the constant ε_0 multiplied by a quantity that has the dimensions of a length.

CHECKPOINT 2: For capacitors charged by the same battery, does the charge stored by the capacitor increase, decrease, or remain the same in each of the following situations? (a) The plate separation of a parallel-plate capacitor is increased. (b) The radius of the inner cylinder of a cylindrical capacitor is increased. (c) The radius of the outer spherical shell of a spherical capacitor is increased.

Sample Problem 26-1

A storage capacitor on a random access memory (RAM) chip has a capacitance of 55 fF. If the capacitor is charged to 5.3 V, how many excess electrons are on its negative plate?

SOLUTION: One **Key Idea** here is that we can find the number n of excess electrons on the negative plate if we know q, the total *amount* of excess charge on that plate. Then $n = q/e$, where e is the magnitude of the charge on each electron. A second **Key Idea** is that q is related to the potential difference V to which the capacitor is charged, according to Eq. 26-1 ($q = CV$). Combining these two ideas then gives us

$$n = \frac{q}{e} = \frac{CV}{e} = \frac{(55 \times 10^{-15}\ \text{F})(5.3\ \text{V})}{1.60 \times 10^{-19}\ \text{C}}$$
$$= 1.8 \times 10^{6} \text{ electrons.} \qquad \text{(Answer)}$$

For electrons, this is a very small number. A speck of household dust, so tiny that it essentially never settles, contains about 10^{17} electrons (and the same number of protons).

26-4 Capacitors in Parallel and in Series

When there is a combination of capacitors in a circuit, we can sometimes replace that combination with an **equivalent capacitor**—that is, a single capacitor that has the same capacitance as the actual combination of capacitors. With such a replacement, we can simplify the circuit, affording easier solutions for unknown quantities of the circuit. Here we discuss two basic combinations of capacitors that allow such a replacement.

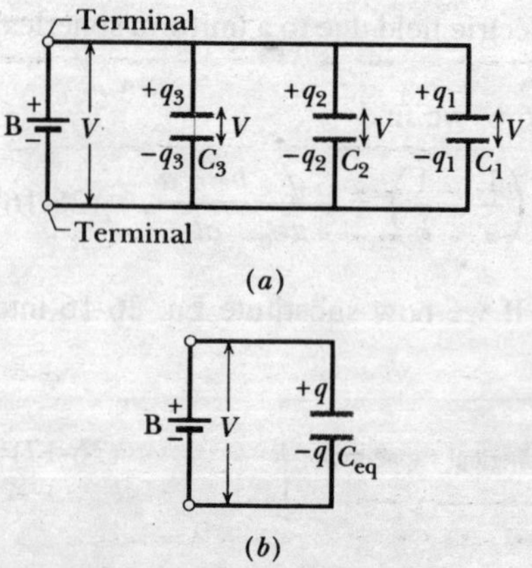

Fig. 26-7 (*a*) Three capacitors connected in parallel to battery B. The battery maintains potential difference V across its terminals and thus across *each* capacitor. (*b*) The equivalent capacitor, with capacitance C_{eq}, replaces the parallel combination.

Capacitors in Parallel

Figure 26-7*a* shows an electric circuit in which three capacitors are connected *in parallel* to battery B. This description has little to do with how the capacitor plates are drawn. Rather, "in parallel" means that the capacitors are directly wired together at one plate and directly wired together at the other plate, and that a potential difference V is applied across the two groups of wired-together plates. Thus, each capacitor has the same potential difference V, which produces charge on the capacitor. (In Fig. 26-7*a*, the applied potential V is maintained by the battery.) In general,

> When a potential difference V is applied across several capacitors connected in parallel, that potential difference V is applied across each capacitor. The total charge q stored on the capacitors is the sum of the charges stored on all the capacitors.

When we analyze a circuit of capacitors in parallel, we can simplify it with this mental replacement:

> Capacitors connected in parallel can be replaced with an equivalent capacitor that has the same *total* charge q and the same potential difference V as the actual capacitors.

(You might remember this result with the nonsense word "par-V," which is close to "party.") Figure 26-7*b* shows the equivalent capacitor (with equivalent capacitance C_{eq}) that has replaced the three capacitors (with actual capacitances C_1, C_2, and C_3) of Fig. 26-7*a*.

To derive an expression for C_{eq} in Fig. 26-7*b*, we first use Eq. 26-1 to find the charge on each actual capacitor:

$$q_1 = C_1V, \quad q_2 = C_2V, \quad \text{and} \quad q_3 = C_3V.$$

The total charge on the parallel combination of Fig. 26-7*a* is then

$$q = q_1 + q_2 + q_3 = (C_1 + C_2 + C_3)V.$$

The equivalent capacitance, with the same total charge q and applied potential difference V as the combination, is then

$$C_{eq} = \frac{q}{V} = C_1 + C_2 + C_3,$$

a result that we can easily extend to any number n of capacitors, as

$$C_{eq} = \sum_{j=1}^{n} C_j \quad (n \text{ capacitors in parallel}). \tag{26-19}$$

Thus, to find the equivalent capacitance of a parallel combination, we simply add the individual capacitances.

Capacitors in Series

Figure 26-8*a* shows three capacitors connected *in series* to battery B. This description has little to do with how the capacitors are drawn. Rather, "in series" means that the capacitors are wired serially, one after the other, and that a potential difference V is applied across the two ends of the series. (In Fig. 26-8*a*, this potential difference V is maintained by battery B.) The potential differences that then exist across the capacitors in the series produce identical charges q on them.

> When a potential difference V is applied across several capacitors connected in series, the capacitors have identical charges q. The sum of the potential differences across all the capacitors is equal to the applied potential difference V.

We can explain how the capacitors end up with identical charges by following a *chain reaction* of events, in which the charging of each capacitor causes the charging of the next capacitor. We start with capacitor 3 and work upward to capacitor 1. When the battery is first connected to the series of capacitors, it produces charge $-q$ on the bottom plate of capacitor 3. That charge then repels negative charge from the top plate of capacitor 3 (leaving it with charge $+q$). The repelled negative charge moves to the bottom plate of capacitor 2 (giving it charge $-q$). That charge on the bottom plate of capacitor 2 then repels negative charge from the top plate of capacitor 2 (leaving it with charge $+q$) to the bottom plate of capacitor 1 (giving it charge $-q$). Finally the charge on the bottom plate of capacitor 1 helps move negative

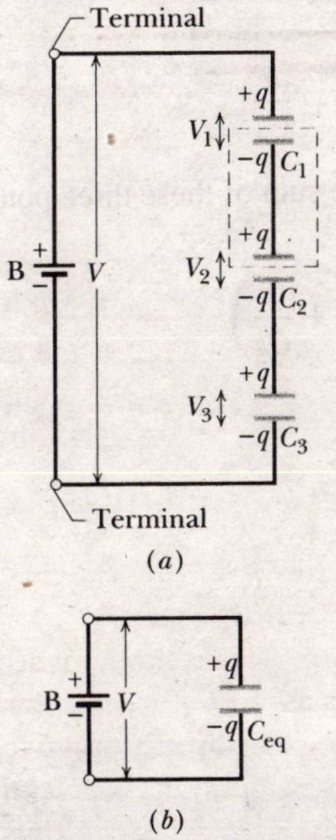

Fig. 26-8 (*a*) Three capacitors connected in series to battery B. The battery maintains potential difference V between the top and bottom plates of the series combination. (*b*) The equivalent capacitor, with capacitance C_{eq}, replaces the series combination.

charge from the top plate of capacitor 1 to the battery, leaving that top plate with charge $+q$.

Here are two important points about capacitors in series:

1. When charge is shifted from one capacitor to another in a series of capacitors, it can move along only one route, such as from capacitor 3 to capacitor 2 in Fig. 26-8*a*. If there are additional routes, the capacitors are not in series. An example is given in Sample Problem 26-2.
2. The battery directly produces charges on only the two plates to which it is connected (the bottom plate of capacitor 3 and the top plate of capacitor 1 in Fig. 26-8*a*). Charges that are produced on the other plates are due merely to the shifting of charge already there. For example, in Fig. 26-8, the part of the circuit enclosed by dashed lines is electrically isolated from the rest of the circuit. Thus, the net charge of that part cannot be changed by the battery—its charge can only be redistributed.

When we analyze a circuit of capacitors in series, we can simplify it with this mental replacement:

> Capacitors that are connected in series can be replaced with an equivalent capacitor that has the same charge q and the same *total* potential difference V as the actual series capacitors.

(You might remember this with the nonsense word "seri-q.") Figure 26-8*b* shows the equivalent capacitor (with equivalent capacitance C_{eq}) that has replaced the three actual capacitors (with actual capacitances C_1, C_2, and C_3) of Fig. 26-8*a*.

To derive an expression for C_{eq} in Fig. 26-8*b*, we first use Eq. 26-1 to find the potential difference of each actual capacitor:

$$V_1 = \frac{q}{C_1}, \quad V_2 = \frac{q}{C_2}, \quad \text{and} \quad V_3 = \frac{q}{C_3}.$$

The total potential difference V due to the battery is the sum of these three potential differences. Thus,

$$V = V_1 + V_2 + V_3 = q\left(\frac{1}{C_1} + \frac{1}{C_2} + \frac{1}{C_3}\right).$$

The equivalent capacitance is then

$$C_{eq} = \frac{q}{V} = \frac{1}{1/C_1 + 1/C_2 + 1/C_3},$$

or

$$\frac{1}{C_{eq}} = \frac{1}{C_1} + \frac{1}{C_2} + \frac{1}{C_3}.$$

We can easily extend this to any number n of capacitors as

$$\frac{1}{C_{eq}} = \sum_{j=1}^{n} \frac{1}{C_j} \quad (n \text{ capacitors in series}). \qquad (26\text{-}20)$$

Using Eq. 26-20 you can show that the equivalent of a series of capacitances is always *less* than the least capacitance in the series.

CHECKPOINT 3: A battery of potential V stores charge q on a combination of two identical capacitors. What are the potential difference across and the charge on either capacitor if the capacitors are (a) in parallel and (b) in series?

Sample Problem 26-2

(a) Find the equivalent capacitance for the combination of capacitances shown in Fig. 26-9*a*, across which potential difference V is applied. Assume

$$C_1 = 12.0\ \mu\text{F},\quad C_2 = 5.30\ \mu\text{F},\quad \text{and}\quad C_3 = 4.50\ \mu\text{F}.$$

SOLUTION: The **Key Idea** here is that any capacitors connected in series can be replaced with their equivalent capacitor, and any capacitors connected in parallel can be replaced with their equivalent capacitor. Therefore, we should first check whether any of the capacitors in Fig. 26-9*a* are in parallel or series.

Capacitors 1 and 3 are connected one after the other, but are they in series? No. The potential V that is applied to the capacitors produces charge on the bottom plate of capacitor 3. That charge causes charge to shift from the top plate of capacitor 3. However, note that the shifting charge can move to the bottom plates of both capacitor 1 and capacitor 2. Because there is more than one route for the shifting charge, capacitor 3 is *not* in series with capacitor 1 (or capacitor 2).

Are capacitor 1 and capacitor 2 in parallel? Yes. Their top plates are directly wired together and their bottom plates are directly wired together, and electric potential is applied between the top-plate pair and the bottom-plate pair. Thus, capacitor 1 and capacitor 2 are in parallel, and Eq. 26-19 tells us that their equivalent capacitance C_{12} is

$$C_{12} = C_1 + C_2 = 12.0\ \mu\text{F} + 5.30\ \mu\text{F} = 17.3\ \mu\text{F}.$$

In Fig. 26-9*b*, we have replaced capacitors 1 and 2 with their equivalent capacitor, call it capacitor 12 (say "one two"). (The connections at points A and B are exactly the same in Figs. 26-9*a* and *b*.)

Is capacitor 12 in series with capacitor 3? Again applying the test for series capacitances, we see that the charge that shifts from the top plate of capacitor 3 must entirely go to the bottom plate of capacitor 12. Thus, capacitor 12 and capacitor 3 are in series, and we can replace them with their equivalent C_{123}, as shown in Fig. 26-9*c*. From Eq. 26-20, we have

$$\frac{1}{C_{123}} = \frac{1}{C_{12}} + \frac{1}{C_3} = \frac{1}{17.3\ \mu\text{F}} + \frac{1}{4.50\ \mu\text{F}} = 0.280\ \mu\text{F}^{-1},$$

from which

$$C_{123} = \frac{1}{0.280\ \mu\text{F}^{-1}} = 3.57\ \mu\text{F}. \qquad \text{(Answer)}$$

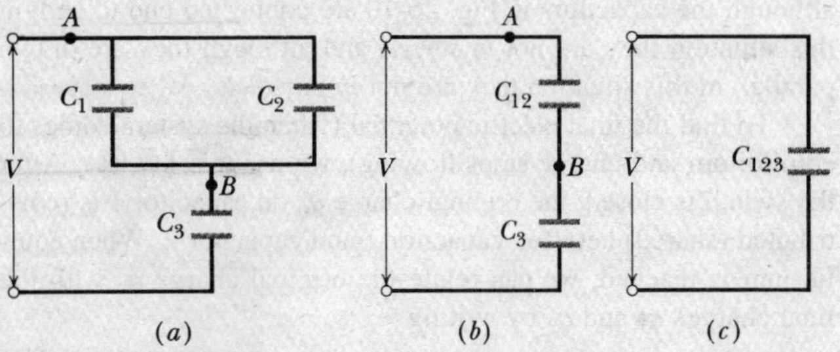

Fig. 26-9 Sample Problem 26-2. (*a*) Three capacitors. (*b*) C_1 and C_2, a parallel combination, are replaced by C_{12}. (*c*) C_{12} and C_3, a series combination, are replaced by the equivalent capacitance C_{123}.

(b) The potential difference that is applied to the input terminals in Fig. 26-9*a* is $V = 12.5$ V. What is the charge on C_1?

SOLUTION: One **Key Idea** here is that, to get the charge q_1 on capacitor 1, we now have to work backward to that capacitor, starting with the equivalent capacitor 123. Since the given potential difference V (= 12.5 V) is applied across the actual combination of three capacitors in Fig. 26-9*a*, it is also applied across capacitor 123 in Fig. 26-9*c*. Thus, Eq. 26-1 ($q = CV$) gives us

$$q_{123} = C_{123}V = (3.57\ \mu\text{F})(12.5\ \text{V}) = 44.6\ \mu\text{C}.$$

A second **Key Idea** is that the series capacitors 12 and 3 in Fig. 26-9*b* have the same charge as their equivalent capacitor 123 (recall "seri-q"). Thus, capacitor 12 has charge $q_{12} = q_{123} = 44.6\ \mu$C. From Eq. 26-1, the potential difference across capacitor 12 must be

$$V_{12} = \frac{q_{12}}{C_{12}} = \frac{44.6\ \mu\text{C}}{17.3\ \mu\text{F}} = 2.58\ \text{V}.$$

A third **Key Idea** is that the parallel capacitors 1 and 2 both have the same potential difference as their equivalent capacitor 12 (recall "par-V"). Thus, capacitor 1 has the potential difference $V_1 = V_{12} = 2.58$ V. Thus, from Eq. 26-1, the charge on capacitor 1 must be

$$q_1 = C_1V_1 = (12.0\ \mu\text{F})(2.58\ \text{V})$$
$$= 31.0\ \mu\text{C}. \qquad \text{(Answer)}$$

Sample Problem 26-3

Capacitor 1, with $C_1 = 3.55\ \mu$F, is charged to a potential difference $V_0 = 6.30$ V, using a 6.30 V battery. The battery is then removed and the capacitor is connected as in Fig. 26-10 to an uncharged capacitor 2, with $C_2 = 8.95\ \mu$F. When switch S is closed, charge flows between the capacitors until they have the same potential difference V. Find V.

SOLUTION: The situation here differs from the previous example because an applied electric potential is *not* maintained across a combination of capacitors by a battery or some other source. Here, just

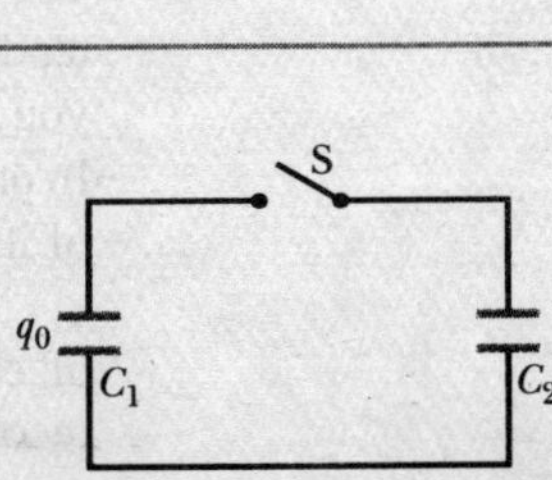

Fig. 26-10 Sample Problem 26-3. A potential difference V_0 is applied to capacitor 1 and the charging battery is removed. Switch S is then closed so that the charge on capacitor 1 is shared with capacitor 2.

after switch S is closed, the only applied electric potential is that of capacitor 1 on capacitor 2, and that potential is decreasing. Thus, although the capacitors in Fig. 26-10 are connected end to end, in this situation they are not *in series*; and although they are drawn parallel, in this situation they are not *in parallel*.

To find the final electric potential (when the system comes to equilibrium and charge stops flowing), we use this **Key Idea**: After the switch is closed, the original charge q_0 on capacitor 1 is redistributed (shared) between capacitor 1 and capacitor 2. When equilibrium is reached, we can relate the original charge q_0 with the final charges q_1 and q_2 by writing

$$q_0 = q_1 + q_2.$$

Applying the relation $q = CV$ to each term of this equation yields

$$C_1V_0 = C_1V + C_2V,$$

from which

$$V = V_0 \frac{C_1}{C_1 + C_2} = \frac{(6.30\ \text{V})(3.55\ \mu\text{F})}{3.55\ \mu\text{F} + 8.95\ \mu\text{F}}$$

$$= 1.79\ \text{V}. \qquad \text{(Answer)}$$

When the capacitors reach this value of electric potential difference, the charge flow stops.

CHECKPOINT 4: In this sample problem, suppose capacitor 2 is replaced by a series combination of capacitors 3 and 4. (a) After the switch is closed and charge has stopped flowing, what is the relation between the initial charge q_0, the charge q_1 then on capacitor 1, and the charge q_{34} then on the equivalent capacitor 34? (b) If $C_3 > C_4$, is the charge q_3 on capacitor 3 more than, less than, or equal to the charge q_4 on capacitor 4?

PROBLEM-SOLVING TACTICS

Tactic 2: *Multiple-Capacitor Circuits*

Let us review the procedure used in the solution of Sample Problem 26-2, in which several capacitors are connected to a battery. To find a single equivalent capacitance, we simplify the given arrangement of capacitances by replacing them, in steps, with equivalent capacitances, using Eq. 26-19 when we find capacitances in parallel and Eq. 26-20 when we find capacitances in series. Then, to find the charge stored by that single equivalent capacitance, we use Eq. 26-1 and the potential difference V applied by the battery.

That result tells us the net charge stored on the actual arrangement of capacitors. However, to find the charge on, or the potential difference across, any particular capacitor in the actual arrangement, we need to reverse our steps of simplification. With each reversed step, we use these two rules: When capacitances are in parallel, they have the same potential difference as their equivalent capacitance, and we use Eq. 26-1 to find the charge on each capacitance; when they are in series, they have the same charge as their equivalent capacitance, and we use Eq. 26-1 to find the potential difference across each capacitance.

Tactic 3: *Batteries and Capacitors*

A battery maintains a certain potential difference across its terminals. Thus, when capacitor 1 of Sample Problem 26-3 is connected to the 6.30 V battery, charge flows between the capacitor and the battery until the capacitor has the same potential difference across it as the battery.

A capacitor differs from a battery in that a capacitor lacks the internal electrochemical reactions needed to release charged particles (electrons) from internal atoms and molecules. Thus, when the charged capacitor 1 of Sample Problem 26-3 is disconnected from the battery and then connected to the uncharged capacitor 2 with switch S closed, the potential difference across capacitor 1 is not maintained. The quantity that *is* maintained is the charge q_0 of the two-capacitor system; that is, charge obeys a conservation law, *not* electric potential.

26-5 Energy Stored in an Electric Field

Work must be done by an external agent to charge a capacitor. Starting with an uncharged capacitor, for example, imagine that—using "magic tweezers"—you remove electrons from one plate and transfer them one at a time to the other plate. The electric field that builds up in the space between the plates has a direction that tends to oppose further transfer. Thus, as charge accumulates on the capacitor plates, you have to do increasingly larger amounts of work to transfer additional electrons. In practice, this work is done not by "magic tweezers" but by a battery, at the expense of its store of chemical energy.

We visualize the work required to charge a capacitor as being stored in the form of **electric potential energy** U in the electric field between the plates. You can recover this energy at will, by discharging the capacitor in a circuit, just as you can recover the potential energy stored in a stretched bow by releasing the bowstring to transfer the energy to the kinetic energy of an arrow.

Suppose that, at a given instant, a charge q' has been transferred from one plate of a capacitor to the other. The potential difference V' between the plates at that instant will be q'/C. If an extra increment of charge dq' is then transferred, the increment of work required will be, from Eq. 25-7,

$$dW = V'\,dq' = \frac{q'}{C}\,dq'.$$

The work required to bring the total capacitor charge up to a final value q is

$$W = \int dW = \frac{1}{C}\int_0^q q'\,dq' = \frac{q^2}{2C}.$$

This work is stored as potential energy U in the capacitor, so that

$$U = \frac{q^2}{2C} \quad \text{(potential energy).} \qquad (26\text{-}21)$$

From Eq. 26-1, we can also write this as

$$U = \tfrac{1}{2}CV^2 \quad \text{(potential energy).} \qquad (26\text{-}22)$$

Equations 26-21 and 26-22 hold no matter what the geometry of the capacitor is.

To gain some physical insight into energy storage, consider two parallel-plate capacitors that are identical except that capacitor 1 has twice the plate separation of capacitor 2. Then capacitor 1 has twice the volume between its plates and also, from Eq. 26-9, half the capacitance of capacitor 2. Equation 26-4 tells us that if both capacitors have the same charge q, the electric fields between their plates are identical. And Eq. 26-21 tells us that capacitor 1 has twice the stored potential energy of capacitor 2. Thus, of two otherwise identical capacitors with the same charge and same electric field, the one with twice the volume between its plates has twice the stored potential energy. Arguments like this tend to verify our earlier assumption:

The potential energy of a charged capacitor may be viewed as being stored in the electric field between its plates.

The Medical Defibrillator

The ability of a capacitor to store potential energy is the basis of *defibrillator* devices, which are used by emergency medical teams to stop the fibrillation of heart attack victims. In the portable version, a battery charges a capacitor to a high potential difference, storing a large amount of energy in less than a minute. The battery maintains only a modest potential difference; an electronic circuit repeatedly uses that potential difference to greatly increase the potential difference of the capacitor. The power, or rate of energy transfer, during this process is also modest.

Conducting leads ("paddles") are placed on the victim's chest. When a control switch is closed, the capacitor sends a portion of its stored energy from paddle to paddle through the victim. As an example, when a 70 μF capacitor in a defibrillator is charged to 5000 V, Eq. 26-22 gives the energy stored in the capacitor as

$$U = \tfrac{1}{2}CV^2 = \tfrac{1}{2}(70 \times 10^{-6}\ \text{F})(5000\ \text{V})^2 = 875\ \text{J}.$$

About 200 J of this energy is sent through the victim during a pulse of about 2.0 ms. The power of the pulse is

$$P = \frac{U}{t} = \frac{200\ \text{J}}{2.0 \times 10^{-3}\ \text{s}} = 100\ \text{kW},$$

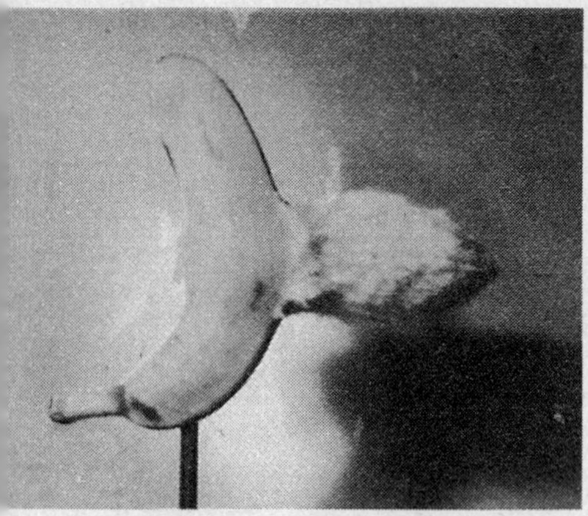

Fig. 26-11 To photograph a bullet blowing apart a banana, Harold Edgerton, the inventor of the stroboscope, used a capacitor to dump electrical energy into one of his stroboscopic lamps, which then brightly illuminated the banana for only 0.3 μs.

which is much greater than the power of the battery itself. This same technique of slowly charging a capacitor with a battery and then discharging the capacitor at a much higher power is commonly used in flash photography and stroboscopic photography (Fig. 26-11).

Energy Density

In a parallel-plate capacitor, neglecting fringing, the electric field has the same value at all points between the plates. Thus, the **energy density** u—that is, the potential energy per unit volume between the plates—should also be uniform. We can find u by dividing the total potential energy by the volume Ad of the space between the plates. Using Eq. 26-22, we obtain

$$u = \frac{U}{Ad} = \frac{CV^2}{2Ad}.$$

With Eq. 26-9 ($C = \varepsilon_0 A/d$), this result becomes

$$u = \tfrac{1}{2}\varepsilon_0\left(\frac{V}{d}\right)^2.$$

However, from Eq. 25-42, V/d equals the electric field magnitude E, so

$$u = \tfrac{1}{2}\varepsilon_0 E^2 \quad \text{(energy density).} \qquad (26\text{-}23)$$

Although we derived this result for the special case of a parallel-plate capacitor, it holds generally, whatever may be the source of the electric field. If an electric field $\vec{E}$ exists at any point in space, we can think of that point as a site of electric potential energy whose amount per unit volume is given by Eq. 26-23.

Sample Problem 26-4

An isolated conducting sphere whose radius R is 6.85 cm has a charge $q = 1.25$ nC.

(a) How much potential energy is stored in the electric field of this charged conductor?

SOLUTION: The **Key Idea** here is that the energy U stored in a capacitor depends on the charge q on the capacitor and the capacitance C of the capacitor, according to Eq. 26-21. Substituting from Eq. 26-18 for C, Eq. 26-21 gives us

$$U = \frac{q^2}{2C} = \frac{q^2}{8\pi\varepsilon_0 R}$$
$$= \frac{(1.25 \times 10^{-9}\ \text{C})^2}{(8\pi)(8.85 \times 10^{-12}\ \text{F/m})(0.0685\ \text{m})}$$
$$= 1.03 \times 10^{-7}\ \text{J} = 103\ \text{nJ}. \qquad \text{(Answer)}$$

(b) What is the energy density at the surface of the sphere?

SOLUTION: The **Key Idea** here is that the density u of the energy stored in an electric field depends on the magnitude E of the field, according to Eq. 26-23 ($u = \frac{1}{2}\varepsilon_0 E^2$), so we must first find E at the surface of the sphere. This is given by Eq. 24-15:

$$E = \frac{1}{4\pi\varepsilon_0}\frac{q}{R^2}.$$

The energy density is then

$$u = \tfrac{1}{2}\varepsilon_0 E^2 = \frac{q^2}{32\pi^2\varepsilon_0 R^4}$$
$$= \frac{(1.25 \times 10^{-9}\ \text{C})^2}{(32\pi^2)(8.85 \times 10^{-12}\ \text{C}^2/\text{N}\cdot\text{m}^2)(0.0685\ \text{m})^4}$$
$$= 2.54 \times 10^{-5}\ \text{J/m}^3 = 25.4\ \mu\text{J/m}^3. \qquad \text{(Answer)}$$

26-6 Capacitor with a Dielectric

If you fill the space between the plates of a capacitor with a *dielectric*, which is an insulating material such as mineral oil or plastic, what happens to the capacitance? Michael Faraday—to whom the whole concept of capacitance is largely due and for whom the SI unit of capacitance is named—first looked into this matter in 1837. Using simple equipment much like that shown in Fig. 26-12, he found that the

Fig. 26-12 The simple electrostatic apparatus used by Faraday. An assembled apparatus (second from left) forms a spherical capacitor consisting of a central brass ball and a concentric brass shell. Faraday placed dielectric materials in the space between the ball and the shell.

TABLE 26-1 Some Properties of Dielectrics[a]

Material	Dielectric Constant κ	Dielectric Strength (kV/mm)
Air (1 atm)	1.00054	3
Polystyrene	2.6	24
Paper	3.5	16
Transformer oil	4.5	
Pyrex	4.7	14
Ruby mica	5.4	
Porcelain	6.5	
Silicon	12	
Germanium	16	
Ethanol	25	
Water (20°C)	80.4	
Water (25°C)	78.5	
Titania ceramic	130	
Strontium titanate	310	8
For a vacuum, κ = unity.		

[a]Measured at room temperature, except for the water.

capacitance *increased* by a numerical factor κ, which he called the **dielectric constant** of the insulating material. Table 26-1 shows some dielectric materials and their dielectric constants. The dielectric constant of a vacuum is unity by definition. Because air is mostly empty space, its measured dielectric constant is only slightly greater than unity.

Another effect of the introduction of a dielectric is to limit the potential difference that can be applied between the plates to a certain value V_{max}, called the *breakdown potential.* If this value is substantially exceeded, the dielectric material will break down and form a conducting path between the plates. Every dielectric material has a characteristic *dielectric strength,* which is the maximum value of the electric field that it can tolerate without breakdown. A few such values are listed in Table 26-1.

As we discussed in connection with Eq. 26-18, the capacitance of any capacitor can be written in the form

$$C = \varepsilon_0 \mathscr{L}, \tag{26-24}$$

in which $\mathscr{L}$ has the dimensions of a length. For example, $\mathscr{L} = A/d$ for a parallel-plate capacitor. Faraday's discovery was that, with a dielectric *completely* filling the space between the plates, Eq. 26-24 becomes

$$C = \kappa\varepsilon_0 \mathscr{L} = \kappa C_{air}, \tag{26-25}$$

where C_{air} is the value of the capacitance with only air between the plates.

Figure 26-13 provides some insight into Faraday's experiments. In Fig. 26-13*a* the battery ensures that the potential difference V between the plates will remain

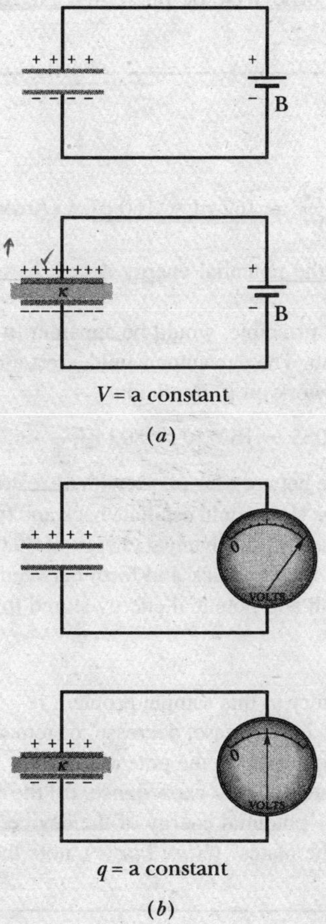

Fig. 26-13 (*a*) If the potential difference between the plates of a capacitor is maintained, as by battery B, the effect of a dielectric is to increase the charge on the plates. (*b*) If the charge on the capacitor plates is maintained, as in this case, the effect of a dielectric is to reduce the potential difference between the plates. The scale shown is that of a *potentiometer*, a device used to measure potential difference (here, between the plates). A capacitor cannot discharge through a potentiometer.

constant. When a dielectric slab is inserted between the plates, the charge q on the plates increases by a factor of κ; the additional charge is delivered to the capacitor plates by the battery. In Fig. 26-13b there is no battery and therefore the charge q must remain constant when the dielectric slab is inserted; then the potential difference V between the plates decreases by a factor of κ. Both these observations are consistent (through the relation $q = CV$) with the increase in capacitance caused by the dielectric.

Comparison of Eqs. 26-24 and 26-25 suggests that the effect of a dielectric can be summed up in more general terms:

> In a region completely filled by a dielectric material of dielectric constant κ, all electrostatic equations containing the permittivity constant ε_0 are to be modified by replacing ε_0 with $\kappa\varepsilon_0$.

Thus, a point charge inside a dielectric produces an electric field that, by Coulomb's law, has the magnitude

$$E = \frac{1}{4\pi\kappa\varepsilon_0}\frac{q}{r^2}. \tag{26-26}$$

Also, the expression for the electric field just outside an isolated conductor immersed in a dielectric (see Eq. 24-11) becomes

$$E = \frac{\sigma}{\kappa\varepsilon_0}. \tag{26-27}$$

Both these equations show that *for a fixed distribution of charges, the effect of a dielectric is to weaken the electric field* that would otherwise be present.

Sample Problem 26-5

A parallel-plate capacitor whose capacitance C is 13.5 pF is charged by a battery to a potential difference $V = 12.5$ V between its plates. The charging battery is now disconnected and a porcelain slab ($\kappa = 6.50$) is slipped between the plates. What is the potential energy of the capacitor–slab device, both before and after the slab is put into place?

SOLUTION: The **Key Idea** here is that we can relate the potential energy U of the capacitor to the capacitance C and either the potential V (with Eq. 26-22) or the charge q (with Eq. 26-21):

$$U_i = \tfrac{1}{2}CV^2 = \frac{q^2}{2C}.$$

Because we are given the initial potential V (= 12.5 V), we use Eq. 26-22 to find the initial stored energy:

$$U_i = \tfrac{1}{2}CV^2 = \tfrac{1}{2}(13.5 \times 10^{-12}\text{ F})(12.5\text{ V})^2$$
$$= 1.055 \times 10^{-9}\text{ J} = 1055\text{ pJ} \approx 1100\text{ pJ}. \quad \text{(Answer)}$$

To find the final potential energy U_f (after the slab is introduced), we need another **Key Idea**: Because the battery has been disconnected, the charge on the capacitor cannot change when the dielectric is inserted. However, the potential *does* change. Thus, we must now use Eq. 26-21 (based on q) to write the final potential energy U_f, but now that the slab is within the capacitor, the capacitance is κC. We then have

$$U_f = \frac{q^2}{2\kappa C} = \frac{U_i}{\kappa} = \frac{1055\text{ pJ}}{6.50} = 162\text{ pJ} \approx 160\text{ pJ}. \quad \text{(Answer)}$$

When the slab is introduced, the potential energy decreases by a factor of κ.

The "missing" energy, in principle, would be apparent to the person who introduced the slab. The capacitor would exert a tiny tug on the slab and would do work on it, in amount

$$W = U_i - U_f = (1055 - 162)\text{ pJ} = 893\text{ pJ}.$$

If the slab were allowed to slide between the plates with no restraint and if there were no friction, the slab would oscillate back and forth between the plates with a (constant) mechanical energy of 893 pJ, and this system energy would transfer back and forth between kinetic energy of the moving slab and potential energy stored in the electric field.

CHECKPOINT 5: If the battery in this sample problem remains connected, do the following increase, decrease, or remain the same when the slab is introduced: (a) the potential difference between the capacitor plates, (b) the capacitance, (c) the charge on the capacitor, (d) the potential energy of the device, (e) the electric field between the plates? (*Hint:* For (e), note that the charge is not fixed.)

26-7 Dielectrics: An Atomic View

What happens, in atomic and molecular terms, when we put a dielectric in an electric field? There are two possibilities, depending on the nature of the molecules:

1. *Polar dielectrics*. The molecules of some dielectrics, like water, have permanent electric dipole moments. In such materials (called *polar dielectrics*), the electric dipoles tend to line up with an external electric field as in Fig. 26-14. Because the molecules are continuously jostling each other as a result of their random thermal motion, this alignment is not complete, but it becomes more complete as the magnitude of the applied field is increased (or as the temperature, and thus the jostling, is decreased). The alignment of the electric dipoles produces an electric field that is directed opposite the applied field and smaller in magnitude.
2. *Nonpolar dielectrics*. Regardless of whether they have permanent electric dipole moments, molecules acquire dipole moments by induction when placed in an external electric field. In Section 25-7 (see Fig. 25-12), we saw that this occurs because the external field tends to "stretch" the molecules, slightly separating the centers of negative and positive charge.

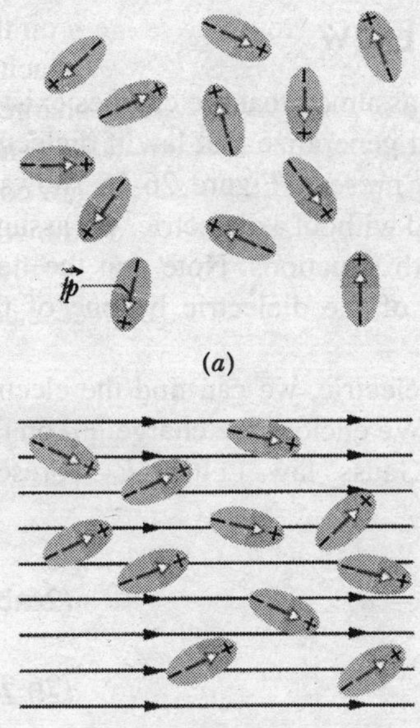

Fig. 26-14 (*a*) Molecules with a permanent electric dipole moment, showing their random orientation in the absence of an external electric field. (*b*) An electric field is applied, producing partial alignment of the dipoles. Thermal agitation prevents complete alignment.

Figure 26-15*a* shows a nonpolar dielectric slab with no external electric field applied. In Fig. 26-15*b*, an electric field $\vec{E}_0$ is applied via a capacitor, whose plates are charged as shown. The result is a slight separation of the centers of the positive and negative charge distributions within the slab, producing positive charge on one face of the slab (due to the positive ends of dipoles there) and negative charge on the opposite face (due to the negative ends of dipoles there). The slab as a whole remains electrically neutral and—within the slab—there is no excess charge in any volume element.

∴ volume charges sort of cancel one other

Figure 26-15*c* shows that the induced surface charges on the faces produce an electric field $\vec{E}'$ in the direction opposite that of the applied electric field $\vec{E}_0$. The resultant field $\vec{E}$ inside the dielectric (the vector sum of fields $\vec{E}_0$ and $\vec{E}'$) has the direction of $\vec{E}_0$ but is smaller in magnitude.

Both the field $\vec{E}'$ produced by the surface charges in Fig. 26-15*c* and the electric field produced by the permanent electric dipoles in Fig. 26-14 act in the same way—they oppose the applied field $\vec{E}$. Thus, the effect of both polar and nonpolar dielectrics is to weaken any applied field within them, as between the plates of a capacitor.

* We can now see why the dielectric porcelain slab in Sample Problem 26-5 is pulled into the capacitor: As it enters the space between the plates, the surface charge that appears on each slab face has the sign opposite that of the charge on the nearby capacitor plate. Thus, slab and plates attract each other.

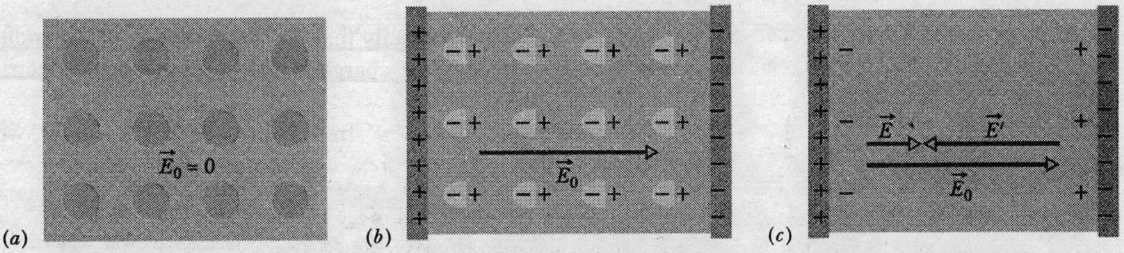

Fig. 26-15 (*a*) A nonpolar dielectric slab. The circles represent the electrically neutral atoms within the slab. (*b*) An electric field is applied via charged capacitor plates; the field slightly stretches the atoms, separating the centers of positive and negative charge. (*c*) The separation produces surface charges on the slab faces. These charges set up a field $\vec{E}'$, which opposes the applied field $\vec{E}_0$. The resultant field $\vec{E}$ inside the dielectric (the vector sum of $\vec{E}_0$ and $\vec{E}'$) has the same direction as $\vec{E}_0$ but less magnitude.

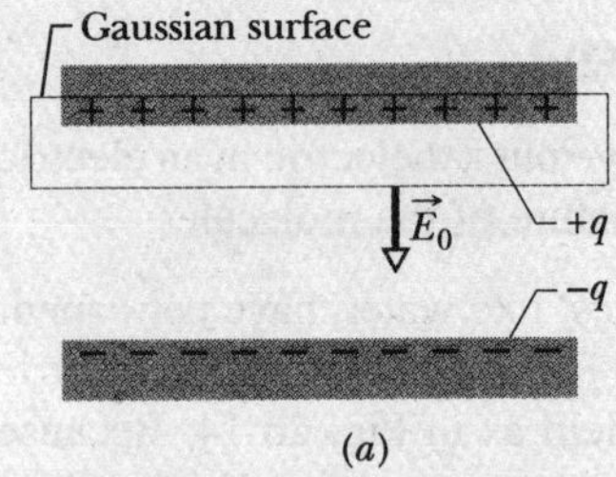

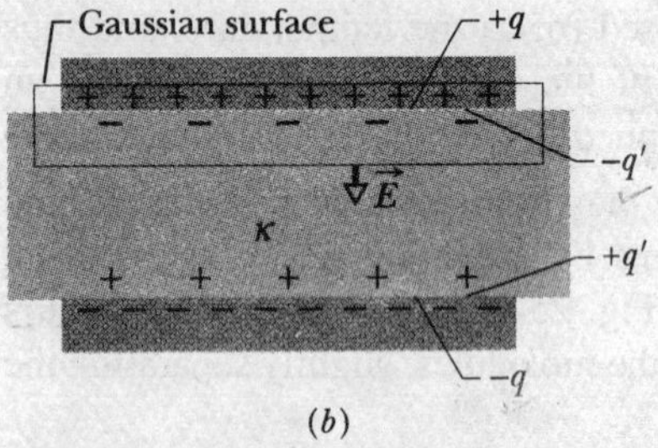

Fig. 26-16 A parallel-plate capacitor (*a*) without and (*b*) with a dielectric slab inserted. The charge q on the plates is assumed to be the same in both cases.

26-8 Dielectrics and Gauss' Law

In our discussion of Gauss' law in Chapter 24, we assumed that the charges existed in a vacuum. Here we shall see how to modify and generalize that law if dielectric materials, such as those listed in Table 26-1, are present. Figure 26-16 shows a parallel-plate capacitor of plate area A, both with and without a dielectric. We assume that the charge q on the plates is the same in both situations. Note that the field between the plates induces charges on the faces of the dielectric by one of the methods of Section 26-7.

For the situation of Fig. 26-16*a*, without a dielectric, we can find the electric field $\vec{E}_0$ between the plates as we did in Fig. 26-5: We enclose the charge $+q$ on the top plate with a Gaussian surface and then apply Gauss' law. Letting E_0 represent the magnitude of the field, we find

$$\varepsilon_0 \oint \vec{E} \cdot d\vec{A} = \varepsilon_0 E_0 A = q, \tag{26-28}$$

or

$$E_0 = \frac{q}{\varepsilon_0 A}. \tag{26-29}$$

In Fig. 26-16*b*, with the dielectric in place, we can find the electric field between the plates (and within the dielectric) by using the same Gaussian surface. However, now the surface encloses two types of charge: It still encloses charge $+q$ on the top plate, but it now also encloses the induced charge $-q'$ on the top face of the dielectric. The charge on the conducting plate is said to be *free charge* because it can move if we change the electric potential of the plate; the induced charge on the surface of the dielectric is not free charge because it cannot move from that surface.

The net charge enclosed by the Gaussian surface in Fig. 26-16*b* is $q - q'$, so Gauss' law now gives

$$\varepsilon_0 \oint \vec{E} \cdot d\vec{A} = \varepsilon_0 E A = q - q', \tag{26-30}$$

or

$$E = \frac{q - q'}{\varepsilon_0 A}. \tag{26-31}$$

The effect of the dielectric is to weaken the original field E_0 by a factor of κ, so we may write

$$E = \frac{E_0}{\kappa} = \frac{q}{\kappa \varepsilon_0 A}. \tag{26-32}$$

Comparison of Eqs. 26-31 and 26-32 shows that

$$q - q' = \frac{q}{\kappa}. \tag{26-33}$$

Equation 26-33 shows correctly that the magnitude q' of the induced surface charge is less than that of the free charge q and is zero if no dielectric is present (then, $\kappa = 1$ in Eq. 26-33).

By substituting for $q - q'$ from Eq. 26-33 in Eq. 26-30, we can write Gauss' law in the form

$$\varepsilon_0 \oint \kappa \vec{E} \cdot d\vec{A} = q \quad \text{(Gauss' law with dielectric).} \tag{26-34}$$

This important equation, although derived for a parallel-plate capacitor, is true generally and is the most general form in which Gauss' law can be written. Note the following:

1. The flux integral now involves $\kappa\vec{E}$, not just $\vec{E}$. (The vector $\varepsilon_0\kappa\vec{E}$ is sometimes called the *electric displacement* $\vec{D}$, so that Eq. 26-34 can be written in the form $\oint \vec{D}\cdot d\vec{A} = q$.)
2. The charge q enclosed by the Gaussian surface is now taken to be the *free charge only*. The induced surface charge is deliberately ignored on the right side of Eq. 26-34, having been taken fully into account by introducing the dielectric constant κ on the left side.
3. Equation 26-34 differs from Eq. 24-7, our original statement of Gauss' law, only in that ε_0 in the latter equation has been replaced by $\kappa\varepsilon_0$. We keep κ inside the integral of Eq. 26-34 to allow for cases in which κ is not constant over the entire Gaussian surface.

In a region that is completely filled by a dielectric, all electrostatic eqs must be modified by replacing ε_0 by $\kappa\varepsilon_0$

Sample Problem 26-6

Figure 26-17 shows a parallel-plate capacitor of plate area A and plate separation d. A potential difference V_0 is applied between the plates. The battery is then disconnected, and a dielectric slab of thickness b and dielectric constant κ is placed between the plates as shown. Assume

$A = 115\text{ cm}^2$, $d = 1.24\text{ cm}$, $V_0 = 85.5\text{ V}$,

$b = 0.780\text{ cm}$, $\kappa = 2.61$.

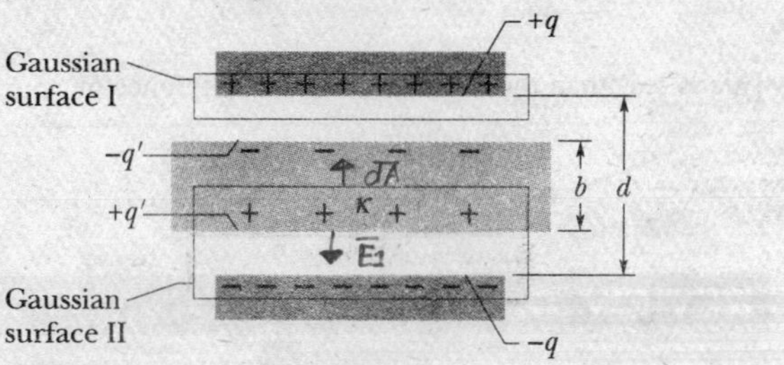

Fig. 26-17 Sample Problem 26-6. A parallel-plate capacitor containing a dielectric slab that only partially fills the space between the plates.

(a) What is the capacitance C_0 before the dielectric slab is inserted?

SOLUTION: From Eq. 26-9 we have

$$C_0 = \frac{\varepsilon_0 A}{d} = \frac{(8.85\times10^{-12}\text{ F/m})(115\times10^{-4}\text{ m}^2)}{1.24\times10^{-2}\text{ m}}$$
$$= 8.21\times10^{-12}\text{ F} = 8.21\text{ pF.} \qquad \text{(Answer)}$$

(b) What free charge appears on the plates?

SOLUTION: From Eq. 26-1,

$$q = C_0V_0 = (8.21\times10^{-12}\text{ F})(85.5\text{ V})$$
$$= 7.02\times10^{-10}\text{ C} = 702\text{ pC.} \qquad \text{(Answer)}$$

Because the charging battery was disconnected before the slab was introduced, the free charge remains unchanged as the slab is put into place.

(c) What is the electric field E_0 in the gaps between the plates and the dielectric slab?

SOLUTION: A **Key Idea** here is to apply Gauss' law, in the form of Eq. 26-34, to Gaussian surface I in Fig. 26-17—that surface passes through the gap, and so it encloses *only* the free charge on the upper capacitor plate. Because the area vector $d\vec{A}$ and the field vector $\vec{E}_0$ are both directed downward, the dot product in Eq. 26-34 becomes

$$\vec{E}_0\cdot d\vec{A} = E_0\,dA\cos 0° = E_0\,dA.$$

Equation 26-34 then becomes

$$\varepsilon_0\kappa E_0 \oint dA = q.$$

The integration now simply gives the surface area A of the plate. Thus, we obtain

$$\varepsilon_0\kappa E_0 A = q,$$

or

$$E_0 = \frac{q}{\varepsilon_0\kappa A}.$$

One more **Key Idea** is needed before we evaluate E_0; that is, we must put $\kappa = 1$ here because Gaussian surface I does not pass through the dielectric. Thus, we have

$$E_0 = \frac{q}{\varepsilon_0\kappa A} = \frac{7.02\times10^{-10}\text{ C}}{(8.85\times10^{-12}\text{ F/m})(1)(115\times10^{-4}\text{ m}^2)}$$
$$= 6900\text{ V/m} = 6.90\text{ kV/m.} \qquad \text{(Answer)}$$

Note that the value of E_0 does not change when the slab is introduced because the amount of charge enclosed by Gaussian surface I in Fig. 26-17 does not change.

(d) What is the electric field E_1 in the dielectric slab?

SOLUTION: The **Key Idea** here is to apply Eq. 26-34 to Gaussian surface II in Fig. 26-17. That surface encloses free charge $-q$ and induced charge $+q'$, but we ignore the latter when we use Eq. 26-34. We find

$$\varepsilon_0 \oint \kappa\vec{E}_1\cdot d\vec{A} = -\varepsilon_0\kappa E_1 A = -q. \qquad (26\text{-}35)$$

(The first minus sign in this equation comes from the dot product $\vec{E}_1\cdot d\vec{A}$, because now the field vector $\vec{E}_1$ is directed downward and the area vector $d\vec{A}$ is directed upward.) Equation 26-35 gives us

$$E_1 = \frac{q}{\varepsilon_0\kappa A} = \frac{E_0}{\kappa} = \frac{6.90\text{ kV/m}}{2.61} = 2.64\text{ kV/m.} \qquad \text{(Answer)}$$

$E_1 = \frac{E_0}{\kappa}$

(e) What is the potential difference V between the plates after the slab has been introduced?

SOLUTION: The **Key Idea** here is to find V by integrating along a straight-line path extending directly from the bottom plate to the top plate. Within the dielectric, the path length is b and the electric field is E_1. Within the two gaps above and below the dielectric, the total path length is $d - b$ and the electric field is E_0. Equation 26-6 then yields

$$V = \int_-^+ E\,ds = E_0(d - b) + E_1 b$$

$$= (6900\text{ V/m})(0.0124\text{ m} - 0.00780\text{ m})$$
$$+ (2640\text{ V/m})(0.00780\text{ m})$$
$$= 52.3\text{ V}. \quad \text{(Answer)}$$

This is less than the original potential difference of 85.5 V.

(f) What is the capacitance with the slab in place?

SOLUTION: The **Key Idea** now is that the capacitance C is related to the free charge q and the potential difference V via Eq. 26-1, just as when a dielectric is not in place. Taking q from (b) and V from (e), we have

$$C = \frac{q}{V} = \frac{7.02 \times 10^{-10}\text{ C}}{52.3\text{ V}}$$
$$= 1.34 \times 10^{-11}\text{ F} = 13.4\text{ pF}. \quad \text{(Answer)}$$

This is greater than the original capacitance of 8.21 pF.

CHECKPOINT 6: In this sample problem, if the thickness b of the slab increases, do the following increase, decrease, or remain the same: (a) the electric field E_1, (b) the potential difference between the plates, and (c) the capacitance of the capacitor?

REVIEW & SUMMARY

Capacitor; Capacitance A **capacitor** consists of two isolated conductors (the *plates*) with equal and opposite charges $+q$ and $-q$. Its **capacitance** C is defined from

$$q = CV, \tag{26-1}$$

where V is the potential difference between the plates. The SI unit of capacitance is the farad (1 farad = 1 F = 1 coulomb per volt).

Determining Capacitance We generally determine the capacitance of a particular capacitor configuration by (1) assuming a charge q to have been placed on the plates, (2) finding the electric field $\vec{E}$ due to this charge, (3) evaluating the potential difference V, and (4) calculating C from Eq. 26-1. Some specific results are the following:

A *parallel-plate capacitor* with flat parallel plates of area A and spacing d has capacitance

$$C = \frac{\varepsilon_0 A}{d}. \tag{26-9}$$

A *cylindrical capacitor* (two long coaxial cylinders) of length L and radii a and b has capacitance

$$C = 2\pi\varepsilon_0 \frac{L}{\ln(b/a)}. \tag{26-14}$$

A *spherical capacitor* with concentric spherical plates of radii a and b has capacitance

$$C = 4\pi\varepsilon_0 \frac{ab}{b - a}. \tag{26-17}$$

If we let $b \to \infty$ and $a = R$ in Eq. 26-17, we obtain the capacitance of an *isolated sphere* of radius R:

$$C = 4\pi\varepsilon_0 R. \tag{26-18}$$

Capacitors in Parallel and in Series The **equivalent capacitances** C_{eq} of combinations of individual capacitors connected in **parallel** and in **series** can be found from

$$C_{eq} = \sum_{j=1}^{n} C_j \quad (n \text{ capacitors in parallel}) \tag{26-19}$$

and

$$\frac{1}{C_{eq}} = \sum_{j=1}^{n} \frac{1}{C_j} \quad (n \text{ capacitors in series}). \tag{26-20}$$

Equivalent capacitances can be used to calculate the capacitances of more complicated series-parallel combinations.

Potential Energy and Energy Density The **electric potential energy** U of a charged capacitor,

$$U = \frac{q^2}{2C} = \tfrac{1}{2}CV^2, \tag{26-21, 26-22}$$

is equal to the work required to charge it. This energy can be associated with the capacitor's electric field $\vec{E}$. By extension we can associate stored energy with an electric field. In vacuum, the **energy density** u, or potential energy per unit volume, within an electric field of magnitude E is given by

$$u = \tfrac{1}{2}\varepsilon_0 E^2. \tag{26-23}$$

Capacitance with a Dielectric If the space between the plates of a capacitor is completely filled with a dielectric material, the capacitance C is increased by a factor κ, called the **dielectric constant,** which is characteristic of the material. In a region that is completely filled by a dielectric, all electrostatic equations containing ε_0 must be modified by replacing ε_0 with $\kappa\varepsilon_0$.

The effects of adding a dielectric can be understood physically in terms of the action of an electric field on the permanent or induced electric dipoles in the dielectric slab. The result is the formation of induced charges on the surfaces of the dielectric, which results in a weakening of the field within the dielectric for the same free charge on the plates.

Gauss' Law with a Dielectric When a dielectric is present, Gauss' law may be generalized to

$$\varepsilon_0 \oint \kappa \vec{E} \cdot d\vec{A} = q. \tag{26-34}$$

Here q is the free charge; any induced surface charge is accounted for by including the dielectric constant κ inside the integral.

QUESTIONS

1. Figure 26-18 shows plots of charge versus potential difference for three parallel-plate capacitors, which have the plate areas and separations given in the table. Which of the plots goes with which of the capacitors?

Capacitor	Area	Separation
1	A	d
2	$2A$	d
3	A	$2d$

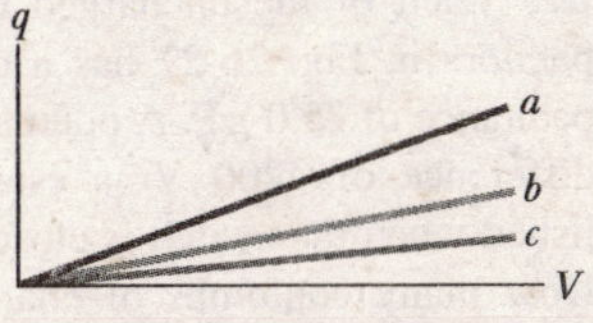

Fig. 26-18 Question 1.

2. Figure 26-19 shows an open switch, a battery of potential difference V, a current-measuring meter A, and three identical uncharged capacitors of capacitance C. When the switch is closed and the circuit reaches equilibrium, what are (a) the potential difference across each capacitor and (b) the charge on the left-hand plate of each capacitor? (c) During the charging process, what is the net charge that passes through the meter?

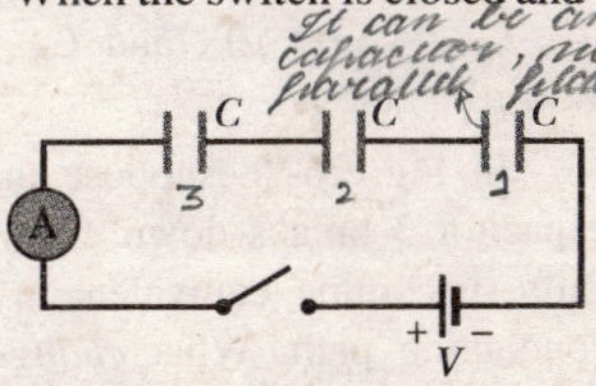

Fig. 26-19 Question 2.

3. For each circuit in Fig. 26-20, are the capacitors connected in series, in parallel, or in neither mode?

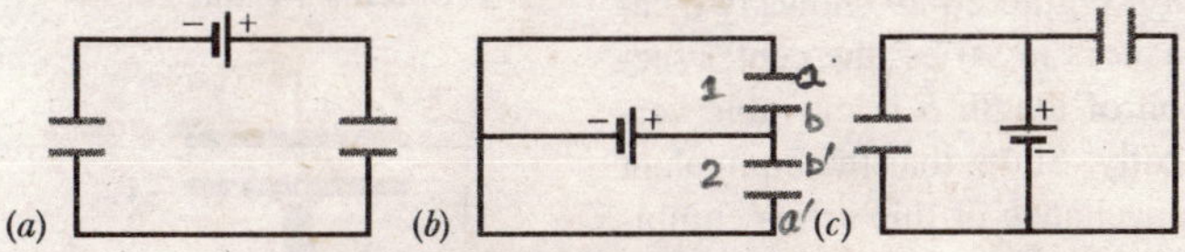

Fig. 26-20 Question 3.

4. (a) In Fig. 26-21*a*, are capacitors C_1 and C_3 in series? (b) In the same figure, are capacitors C_1 and C_2 in parallel? (c) Rank the equivalent capacitances of the four circuits shown in Fig. 26-21, greatest first.

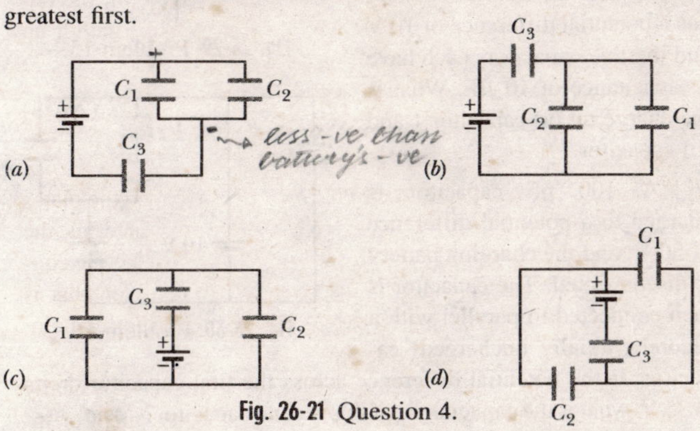

Fig. 26-21 Question 4.

5. What is the equivalent capacitance of three capacitors, each of capacitance C, if they are connected to a battery (a) in series with one another and (b) in parallel? (c) In which arrangement is there more charge on the equivalent capacitance?

6. You are to connect capacitances C_1 and C_2, with $C_1 > C_2$, to a battery, first individually, then in series, and then in parallel. Rank those arrangements according to the amount of charge stored, greatest first.

7. Initially, a single capacitance C_1 is wired to a battery. Then capacitance C_2 is added in parallel. Are (a) the potential difference across C_1 and (b) the charge q_1 on C_1 now more than, less than, or the same as previously? (c) Is the equivalent capacitance C_{12} of C_1 and C_2 more than, less than, or equal to C_1? (d) Is the total charge stored on C_1 and C_2 together more than, less than, or equal to the charge stored previously on C_1?

8. Repeat Question 7 for C_2 added in series, not in parallel.

9. Figure 26-22 shows three circuits, each consisting of a switch

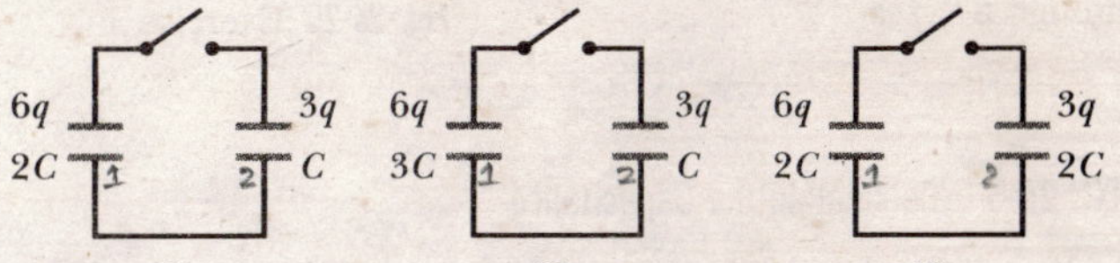

Fig. 26-22 Question 9.

and two capacitors, initially charged as indicated. After the switches have been closed, in which circuit (if any) will the charge on the left-hand capacitor (a) increase, (b) decrease, and (c) remain the same?

10. Two isolated metal spheres A and B have radii R and $2R$, respectively, and the same charge q. (a) Is the capacitance of A more than, less than, or equal to that of B? (b) Is the energy density just outside the surface of A more than, less than, or equal to that of B? (c) Is the energy density at distance $3R$ from the center of A more than, less than, or equal to that at the same distance from the center of B? (d) Is the total energy of the electric field due to A more than, less than, or equal to that of B?

11. When a dielectric slab is inserted between the plates of one of the two identical capacitors in Fig. 26-23, do the following properties of that capacitor increase, decrease, or remain the same: (a) capacitance, (b) charge, (c) potential difference, and (d) potential energy? (e) How about the same properties of the other capacitor?

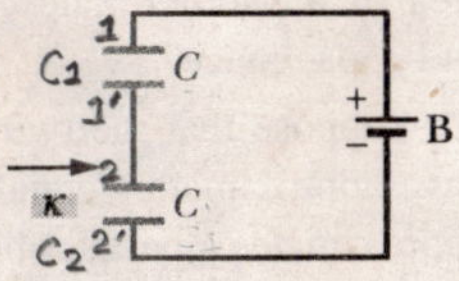

Fig. 26-23 Question 11.

EXERCISES & PROBLEMS

ssm Solution is in the Student Solutions Manual.
www Solution is available on the World Wide Web at: http://www.wiley.com/college/hrw
ilw Solution is available on the Interactive LearningWare.

SEC. 26-2 Capacitance

1E. An electrometer is a device used to measure static charge—an unknown charge is placed on the plates of the meter's capacitor, and the potential difference is measured. What minimum charge can be measured by an electrometer with a capacitance of 50 pF and a voltage sensitivity of 0.15 V?

2E. The two metal objects in Fig. 26-24 have net charges of +70 pC and −70 pC, which result in a 20 V potential difference between them. (a) What is the capacitance of the system? (b) If the charges are changed to +200 pC and −200 pC, what does the capacitance become? (c) What does the potential difference become?

Fig. 26-24 Exercise 2.

3E. The capacitor in Fig. 26-25 has a capacitance of 25 μF and is initially uncharged. The battery provides a potential difference of 120 V. After switch S is closed, how much charge will pass through it? ssm

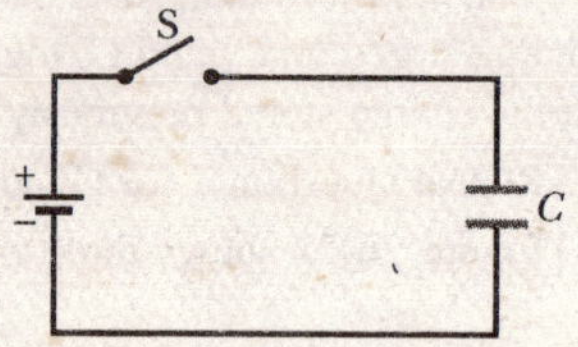

Fig. 26-25 Exercise 3.

SEC. 26-3 Calculating the Capacitance

4E. If we solve Eq. 26-9 for ε_0, we see that its SI unit is the farad per meter. Show that this unit is equivalent to that obtained earlier for ε_0—namely, the coulomb squared per newton-meter squared ($C^2/N \cdot m^2$).

5E. A parallel-plate capacitor has circular plates of 8.2 cm radius and 1.3 mm separation. (a) Calculate the capacitance. (b) What charge will appear on the plates if a potential difference of 120 V is applied? ssm

6E. You have two flat metal plates, each of area 1.00 m^2, with which to construct a parallel-plate capacitor. If the capacitance of the device is to be 1.00 F, what must be the separation between the plates? Could this capacitor actually be constructed?

7E. A spherical drop of mercury of radius R has a capacitance given by $C = 4\pi\varepsilon_0 R$. If two such drops combine to form a single larger drop, what is its capacitance? ssm

8E. The plates of a spherical capacitor have radii 38.0 mm and 40.0 mm. (a) Calculate the capacitance. (b) What must be the plate area of a parallel-plate capacitor with the same plate separation and capacitance?

9P. Suppose that the two spherical shells of a spherical capacitor have approximately equal radii. Under these conditions the device approximates a parallel-plate capacitor with $b - a = d$. Show that Eq. 26-17 does indeed reduce to Eq. 26-9 in this case. ssm

SEC. 26-4 Capacitors in Parallel and in Series

10E. In Fig. 26-26, find the equivalent capacitance of the combination. Assume that $C_1 = 10.0$ μF, $C_2 = 5.00$ μF, and $C_3 = 4.00$ μF.

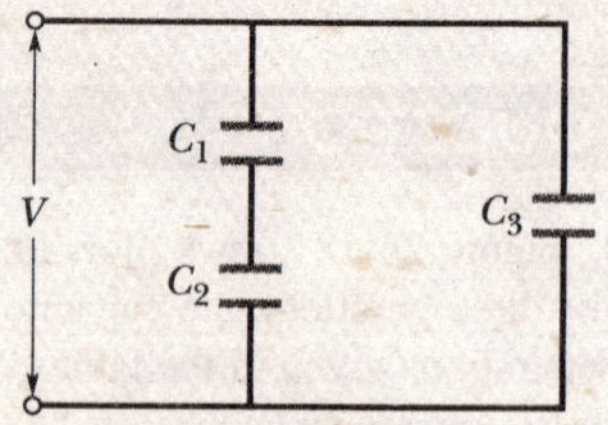

Fig. 26-26 Exercise 10 and Problem 30.

11E. How many 1.00 μF capacitors must be connected in parallel to store a charge of 1.00 C with a potential of 110 V across the capacitors? ssm

12E. Each of the uncharged capacitors in Fig. 26-27 has a capacitance of 25.0 μF. A potential difference of 4200 V is established when the switch is closed. How many coulombs of charge then pass through meter A?

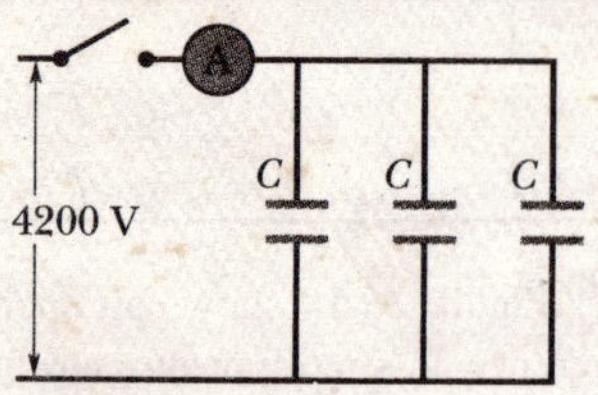

Fig. 26-27 Exercise 12.

13E. In Fig. 26-28 find the equivalent capacitance of the combination. Assume that $C_1 = 10.0$ μF, $C_2 = 5.00$ μF, and $C_3 = 4.00$ μF. ilw

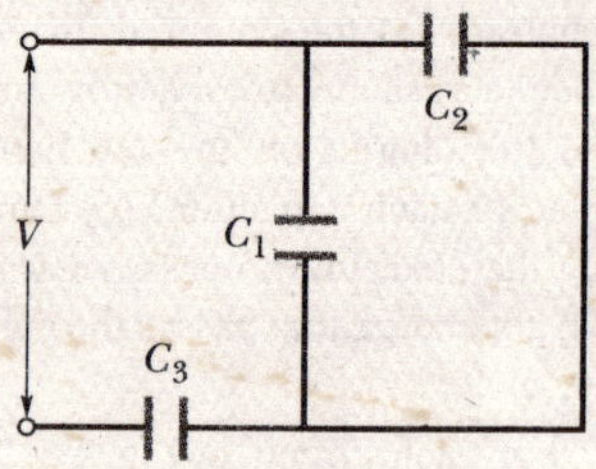

Fig. 26-28 Exercise 13, Problems 14 and 28.

14P. In Fig. 26-28 suppose that capacitor 3 breaks down electrically, becoming equivalent to a conducting path. What *changes* in (a) the charge and (b) the potential difference occur for capacitor 1? Assume that $V = 100$ V.

15P. Figure 26-29 shows two capacitors in series; the center section of length b is movable vertically. Show that the equivalent capacitance of this series combination is independent of the position of the center section and is given by $C = \varepsilon_0 A/(a - b)$, where A is the plate area. ssm www

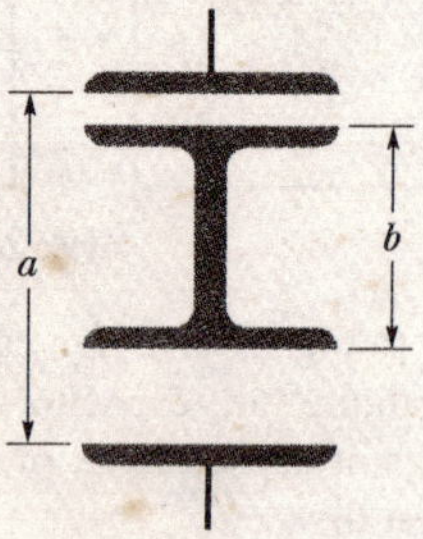

Fig. 26-29 Problem 15.

16P. In Fig. 26-30, the battery has a potential difference of 10 V and the five capacitors each have a capacitance of 10 μF. What is the charge on (a) capacitor 1 and (b) capacitor 2?

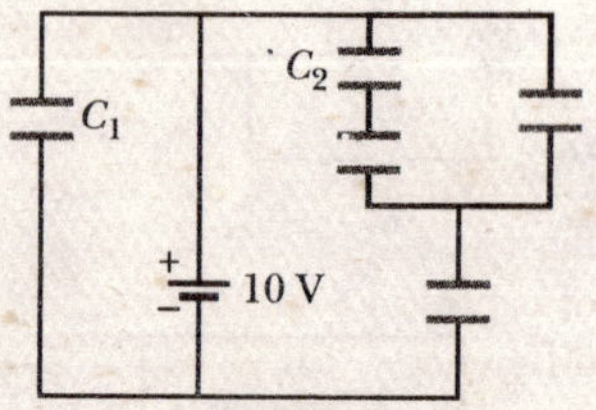

Fig. 26-30 Problem 16.

17P. A 100 pF capacitor is charged to a potential difference of 50 V, and the charging battery is disconnected. The capacitor is then connected in parallel with a second (initially uncharged) capacitor. If the potential difference across the first capacitor drops to 35 V, what is the capacitance of this second capacitor? ssm ilw

18P. In Fig. 26-31, the battery has a potential difference of 20 V. Find (a) the equivalent capacitance of all the capacitors and (b) the charge stored on that equivalent capacitance. Find the potential across and charge on (c) capacitor 1, (d) capacitor 2, and (e) capacitor 3.

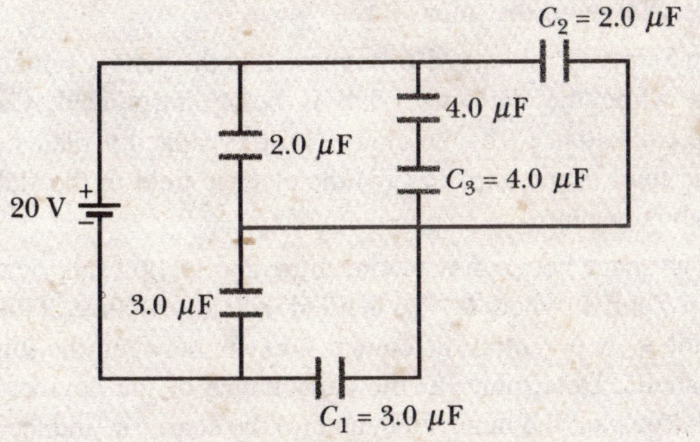

Fig. 26-31 Problem 18.

19P. In Fig. 26-32, the capacitances are $C_1 = 1.0\ \mu F$ and $C_2 = 3.0\ \mu F$ and both capacitors are charged to a potential difference of $V = 100$ V but with opposite polarity as shown. Switches S_1 and S_2 are now closed. (a) What is now the potential difference between points a and b? What are now the charges on capacitors (b) 1 and (c) 2? ssm www

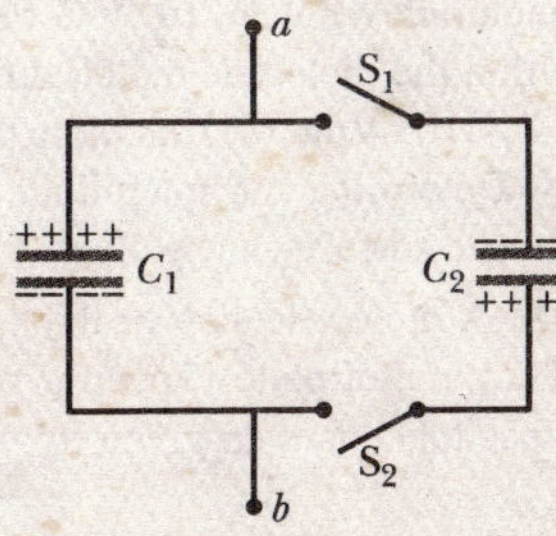

Fig. 26-32 Problem 19.

20P. In Fig. 26-33, battery B supplies 12 V. Find the charge on each capacitor (a) first when only switch S_1 is closed and (b) later when switch S_2 is also closed. Take $C_1 = 1.0\ \mu F$, $C_2 = 2.0\ \mu F$, $C_3 = 3.0\ \mu F$, and $C_4 = 4.0\ \mu F$.

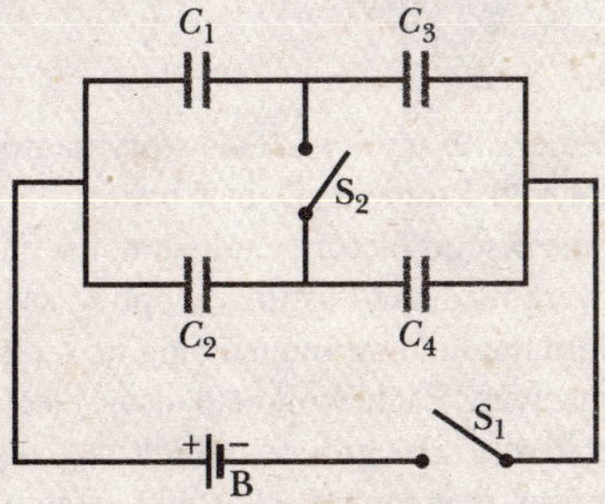

Fig. 26-33 Problem 20.

21P. When switch S is thrown to the left in Fig. 26-34, the plates of capacitor 1 acquire a potential difference V_0. Capacitors 2 and 3 are initially uncharged. The switch is now thrown to the right. What are the final charges q_1, q_2, and q_3 on the capacitors? ssm

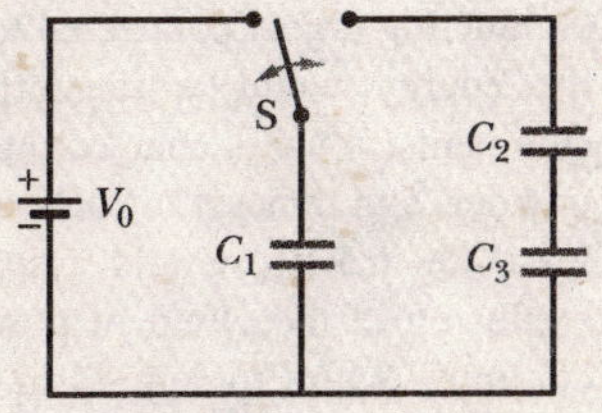

Fig. 26-34 Problem 21.

SEC. 26-5 Energy Stored in an Electric Field

22E. How much energy is stored in one cubic meter of air due to the "fair weather" electric field of magnitude 150 V/m?

23E. What capacitance is required to store an energy of 10 kW · h at a potential difference of 1000 V? ssm

24E. A parallel-plate air-filled capacitor having area 40 cm^2 and plate spacing 1.0 mm is charged to a potential difference of 600 V. Find (a) the capacitance, (b) the magnitude of the charge on each plate, (c) the stored energy, (d) the electric field between the plates, and (e) the energy density between the plates.

25E. Two capacitors, of 2.0 and 4.0 μF capacitance, are connected in parallel across a 300 V potential difference. Calculate the total energy stored in the capacitors. ssm

26P. A parallel-connected bank of 5.00 μF capacitors is used to store electric energy. What does it cost to charge the 2000 capacitors of the bank to 50 000 V, assuming 3.0¢/kW · h?

27P. One capacitor is charged until its stored energy is 4.0 J. A second uncharged capacitor is then connected to it in parallel. (a) If the charge distributes equally, what is now the total energy stored in the electric fields? (b) Where did the excess energy go? ilw

28P. In Fig. 26-28 find (a) the charge, (b) the potential difference, and (c) the stored energy for each capacitor. Assume the numerical values of Exercise 13, with $V = 100$ V.

29P. A parallel-plate capacitor has plates of area A and separation d and is charged to a potential difference V. The charging battery is then disconnected, and the plates are pulled apart until their separation is $2d$. Derive expressions in terms of A, d, and V for (a) the new potential difference; (b) the initial and final stored energies, U_i and U_f; and (c) the work required to separate the plates. ssm ilw

30P. In Fig. 26-26, find (a) the charge, (b) the potential difference, and (c) the stored energy for each capacitor. Assume the numerical values of Exercise 10, with $V = 100$ V.

31P. A cylindrical capacitor has radii a and b as in Fig. 26-6. Show that half the stored electric potential energy lies within a cylinder whose radius is $r = \sqrt{ab}$. ssm www

32P. A charged isolated metal sphere of diameter 10 cm has a potential of 8000 V relative to $V = 0$ at infinity. Calculate the energy density in the electric field near the surface of the sphere.

33P. (a) Show that the plates of a parallel-plate capacitor attract each other with a force given by $F = q^2/2\varepsilon_0 A$. Do so by calculating the work needed to increase the plate separation from x to $x + dx$, with the charge q remaining constant. (b) Next show that the force per unit area (the *electrostatic stress*) acting on either capacitor plate is given by $\frac{1}{2}\varepsilon_0 E^2$. (Actually, this is the force per unit area on *any* conductor of *any* shape with an electric field $\vec{E}$ at its surface.) ssm

SEC. 26-6 Capacitor with a Dielectric

34E. An air-filled parallel-plate capacitor has a capacitance of 1.3 pF. The separation of the plates is doubled and wax is inserted between them. The new capacitance is 2.6 pF. Find the dielectric constant of the wax.

35E. Given a 7.4 pF air-filled capacitor, you are asked to convert it to a capacitor that can store up to 7.4 μJ with a maximum potential difference of 652 V. What dielectric in Table 26-1 should you use to fill the gap in the air capacitor if you do not allow for a margin of error? ssm

36E. A parallel-plate air-filled capacitor has a capacitance of 50 pF. (a) If each of its plates has an area of 0.35 m^2, what is the separation? (b) If the region between the plates is now filled with material having $\kappa = 5.6$, what is the capacitance?

37E. A coaxial cable used in a transmission line has an inner radius of 0.10 mm and an outer radius of 0.60 mm. Calculate the capacitance per meter for the cable. Assume that the space between the conductors is filled with polystyrene. ssm

38P. You are asked to construct a capacitor having a capacitance near 1 nF and a breakdown potential in excess of 10 000 V. You think of using the sides of a tall Pyrex drinking glass as a dielectric, lining the inside and outside curved surfaces with aluminum foil to act as the plates. The glass is 15 cm tall with an inner radius of 3.6 cm and an outer radius of 3.8 cm. What are the (a) capacitance and (b) breakdown potential of this capacitor?

39P. A certain substance has a dielectric constant of 2.8 and a dielectric strength of 18 MV/m. If it is used as the dielectric material in a parallel-plate capacitor, what minimum area should the plates of the capacitor have to obtain a capacitance of 7.0×10^{-2} μF and to ensure that the capacitor will be able to withstand a potential difference of 4.0 kV? ssm ilw

40P. A parallel-plate capacitor of plate area A is filled with two dielectrics as in Fig. 26-35a. Show that the capacitance is

$$C = \frac{\varepsilon_0 A}{d}\,\frac{\kappa_1 + \kappa_2}{2}.$$

Check this formula for limiting cases. (*Hint:* Can you justify this arrangement as being two capacitors in parallel?)

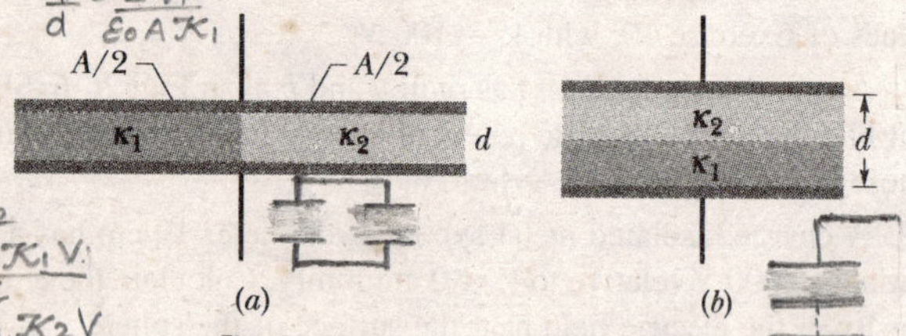

Fig. 26-35 Problems 40 and 41.

$E_1 = \frac{2q_1}{\varepsilon_0 A \kappa_1}$ $\frac{V}{d} = \frac{2q_1}{\varepsilon_0 A \kappa_1}$ $E_2 = \frac{2q_2}{\varepsilon_0 A \kappa_2}$ $\frac{V}{d} = \frac{2q_2}{\varepsilon_0 A \kappa_2}$ $q_1 = \frac{\varepsilon_0 A \kappa_1 V}{2d}$ $q_2 = \frac{\varepsilon_0 A \kappa_2 V}{2d}$ $q = q_1 + q_2$ $q = \frac{\varepsilon_0 A V(\kappa_1+\kappa_2)}{2d}$ $q = CV$ $C = \frac{\varepsilon_0 A(\kappa_1+\kappa_2)}{2d}$

41P. A parallel-plate capacitor of plate area A is filled with two dielectrics as in Fig. 26-35b. Show that the capacitance is

$$C = \frac{2\varepsilon_0 A}{d}\,\frac{\kappa_1 \kappa_2}{\kappa_1 + \kappa_2}.$$

Check this formula for limiting cases. (*Hint:* Can you justify this arrangement as being two capacitors in series?) ssm

42P. What is the capacitance of the capacitor, of plate area A, shown in Fig. 26-36? (*Hint:* See Problems 40 and 41.) 2d

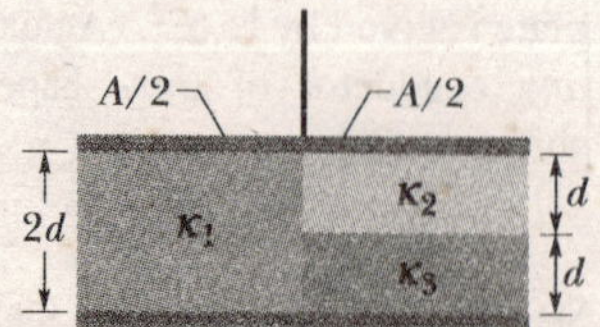

Fig. 26-36 Problem 42.

C_2 & C_3 in series & C_{series} & C_1 in parallel

$$C = \frac{A}{2}\frac{\varepsilon_0}{d}\left(\frac{\kappa_2\kappa_3}{\kappa_2+\kappa_3} + \frac{\kappa_1}{2}\right)$$

SEC. 26-8 Dielectrics and Gauss' Law

43E. A parallel-plate capacitor has a capacitance of 100 pF, a plate area of 100 cm^2, and a mica dielectric ($\kappa = 5.4$) completely filling the space between the plates. At 50 V potential difference, calculate (a) the electric field magnitude E in the mica, (b) the magnitude of the free charge on the plates, and (c) the magnitude of the induced surface charge on the mica. ssm

44E. In Sample Problem 26-6, suppose that the battery remains connected while the dielectric slab is being introduced. Calculate (a) the capacitance, (b) the charge on the capacitor plates, (c) the electric field in the gap, and (d) the electric field in the slab, after the slab is in place.

45P. The space between two concentric conducting spherical shells of radii b and a (where $b > a$) is filled with a substance of dielectric constant κ. A potential difference V exists between the inner and outer shells. Determine (a) the capacitance of the device, (b) the free charge q on the inner shell, and (c) the charge q' induced along the surface of the inner shell. ssm www

46P. Two parallel plates of area 100 cm^2 are given charges of equal magnitudes 8.9×10^{-7} C but opposite signs. The electric field within the dielectric material filling the space between the plates is 1.4×10^6 V/m. (a) Calculate the dielectric constant of the material. (b) Determine the magnitude of the charge induced on each dielectric surface.

47P. A dielectric slab of thickness b is inserted between the plates of a parallel-plate capacitor of plate separation d. Show that the capacitance is then given by

$$C = \frac{\kappa\varepsilon_0 A}{\kappa d - b(\kappa - 1)}.$$

(*Hint:* You can derive the formula following the procedure outlined in Sample Problem 26-6.) Does this formula predict the correct numerical result of Sample Problem 26-6? Verify that the formula gives reasonable results for the special cases of $b = 0$, $\kappa = 1$, and $b = d$. ssm

Additional Problem

48. *The chocolate crumb mystery.* This story begins with Problem 48 in Chapter 24 and Problem 57 in Chapter 25. As part of the investigation of the biscuit factory explosion, the electric potentials of the workers were measured as they emptied sacks of chocolate crumb powder into the loading bin, stirring up a cloud of the powder around themselves. Each worker had an electric potential of about 7.0 kV relative to the ground, which was taken as zero potential. (a) Assuming that each worker was effectively a capacitor with a typical capacitance of 200 pF, find the energy stored in that effective capacitor. If a single spark between the worker and any conducting object connected to the ground neutralized the worker, that energy would be transferred to the spark. According to measurements, a spark that could ignite a cloud of chocolate crumb powder, and thus set off an explosion, had to have an energy of at least 150 mJ. (b) Could a spark from a worker have set off an explosion in the cloud of powder in the loading bin? (The story continues with Problem 44 in Chapter 27.)

27 Current and Resistance

The pride of Germany and a wonder of its time, the zeppelin *Hindenburg*—almost the length of three football fields—was the largest flying machine that had ever been built. Although it was kept aloft by 16 cells of highly flammable hydrogen gas, it made many trans-Atlantic trips without incident. In fact, German zeppelins, which all depended on hydrogen, had never suffered an accident due to the hydrogen. However, shortly after 7:21 p.m. on May 6, 1937, as the *Hindenburg* was ready to land at the U.S. Naval Air Station at Lakehurst, New Jersey, the ship burst into flames. Its crew had been waiting for a rainstorm to diminish, and handling ropes had just been let down to a navy ground crew, when ripples were sighted on the outer fabric of the ship about one-third of the way forward from the stern. Seconds later a flame erupted from that region, and a red glow illuminated the interior of the ship. Within 32 seconds the burning ship fell to the ground.

After so many successful flights of hydrogen-floated zeppelins, why did this zeppelin burst into flames?

The answer is in this chapter.

27-1 Moving Charges and Electric Currents

Chapters 22 through 26 deal largely with *electrostatics*—that is, with charges at rest. With this chapter we begin to focus on **electric currents**—that is, charges in motion.

Examples of electric currents abound, ranging from the large currents that constitute lightning strokes to the tiny nerve currents that regulate our muscular activity. The currents in household wiring, in lightbulbs, and in electrical appliances are familiar to all. A beam of electrons—a current—moves through an evacuated space in the picture tube of a common television set. Charged particles of *both* signs flow in the ionized gases of fluorescent lamps, in the batteries of radios, and in car batteries. Electric currents can also be found in the semiconductors in calculators and in the chips that control microwave ovens and electric dishwashers.

On a global scale, charged particles trapped in the Van Allen radiation belts surge back and forth above the atmosphere between Earth's north and south magnetic poles. On the scale of the solar system, enormous currents of protons, electrons, and ions fly radially outward from the Sun as the *solar wind*. On the galactic scale, cosmic rays, which are largely energetic protons, stream through our Milky Way galaxy, some reaching Earth.

Although an electric current is a stream of moving charges, not all moving charges constitute an electric current. If there is to be an electric current through a given surface, there must be a net flow of charge through that surface. Two examples clarify our meaning.

1. The free electrons (conduction electrons) in an isolated length of copper wire are in random motion at speeds of the order of 10^6 m/s. If you pass a hypothetical plane through such a wire, conduction electrons pass through it *in both directions* at the rate of many billions per second—but there is *no net transport* of charge and thus *no current* through the wire. However, if you connect the ends of the wire to a battery, you slightly bias the flow in one direction, with the result that there now is a net transport of charge and thus an electric current through the wire.
2. The flow of water through a garden hose represents the directed flow of positive charge (the protons in the water molecules) at a rate of perhaps several million coulombs per second. There is no net transport of charge, however, because there is a parallel flow of negative charge (the electrons in the water molecules) of exactly the same amount moving in exactly the same direction.

In this chapter we restrict ourselves largely to the study—within the framework of classical physics—of *steady* currents of *conduction electrons* moving through *metallic conductors* such as copper wires.

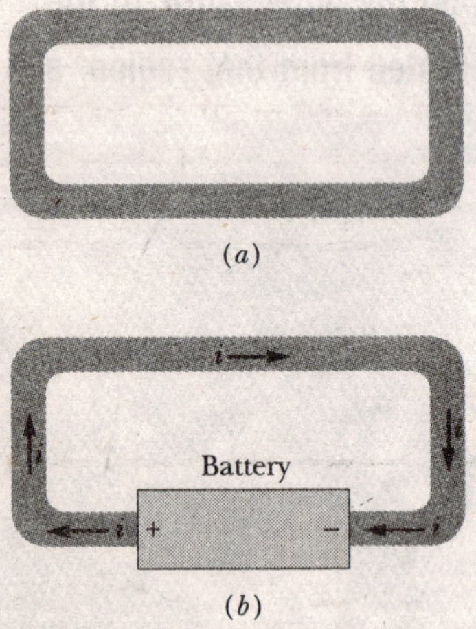

Fig. 27-1 (*a*) A loop of copper in electrostatic equilibrium. The entire loop is at a single potential, and the electric field is zero at all points inside the copper. (*b*) Adding a battery imposes an electric potential difference between the ends of the loop that are connected to the terminals of the battery. The battery thus produces an electric field within the loop, from terminal to terminal, and the field causes charges to move around the loop. This movement of charges is a current i.

27-2 Electric Current

As Fig. 27-1*a* reminds us, an isolated conducting loop—regardless of whether it has an excess charge—is all at the same potential. No electric field can exist within it or along its surface. Although conduction electrons are available, no net electric force acts on them and thus there is no current.

If, as in Fig. 27-1*b*, we insert a battery in the loop, the conducting loop is no longer at a single potential. Electric fields act inside the material making up the loop, exerting forces on the conduction electrons, causing them to move, and thus establishing a current. After a very short time, the electron flow reaches a constant value and the current is in its *steady state* (it does not vary with time).

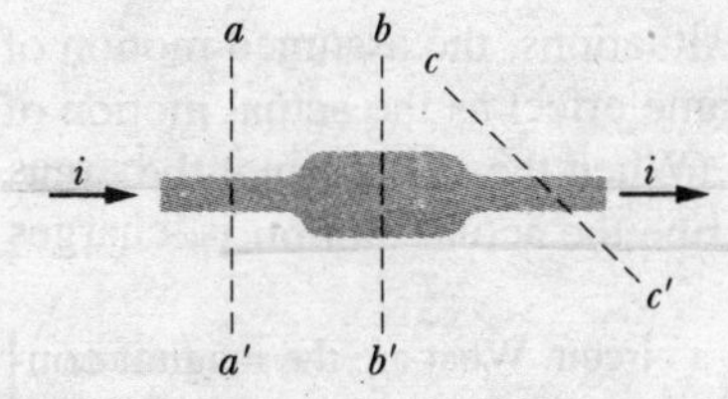

Fig. 27-2 The current i through the conductor has the same value at planes aa', bb', and cc'.

Figure 27-2 shows a section of a conductor, part of a conducting loop in which current has been established. If charge dq passes through a hypothetical plane (such as aa') in time dt, then the current i through that plane is defined as

$$i = \frac{dq}{dt} \quad \text{(definition of current).} \tag{27-1}$$

We can find the charge that passes through the plane in a time interval extending from 0 to t by integration:

$$q = \int dq = \int_0^t i\,dt, \tag{27-2}$$

in which the current i may vary with time.

Under steady-state conditions, the current is the same for planes aa', bb', and cc' and indeed for all planes that pass completely through the conductor, no matter what their location or orientation. This follows from the fact that charge is conserved. Under the steady-state conditions assumed here, an electron must pass through plane aa' for every electron that passes through plane cc'. In the same way, if we have a steady flow of water through a garden hose, a drop of water must leave the nozzle for every drop that enters the hose at the other end. The amount of water in the hose is a conserved quantity.

The SI unit for current is the coulomb per second, also called the *ampere* (A):

$$1 \text{ ampere} = 1 \text{ A} = 1 \text{ coulomb per second} = 1 \text{ C/s}.$$

The ampere is an SI base unit; the coulomb is defined in terms of the ampere, as we discussed in Chapter 22. The formal definition of the ampere is discussed in Chapter 30.

Current, as defined by Eq. 27-1, is a scalar because both charge and time in that equation are scalars. Yet, as in Fig. 27-1*b*, we often represent a current with an arrow to indicate that charge is moving. Such arrows are not vectors, however, and they do not require vector addition. Figure 27-3*a* shows a conductor with current i_0 splitting at a junction into two branches. Because charge is conserved, the magnitudes of the currents in the branches must add to yield the magnitude of the current in the original conductor, so that

$$i_0 = i_1 + i_2. \tag{27-3}$$

As Fig. 27-3*b* suggests, bending or reorienting the wires in space does not change the validity of Eq. 27-3. Current arrows show only a direction (or sense) of flow along a conductor, not a direction in space.

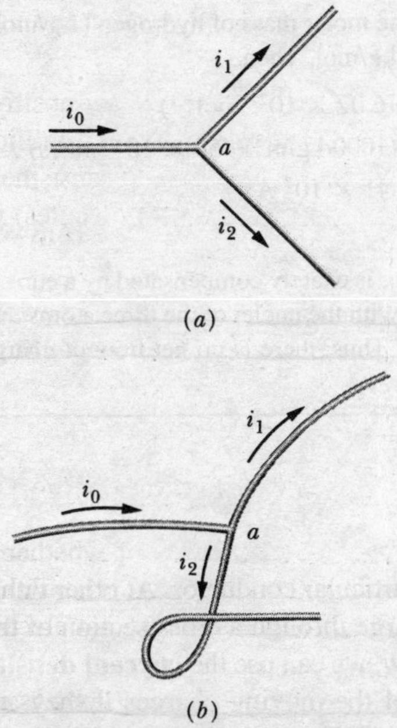

Fig. 27-3 The relation $i_0 = i_1 + i_2$ is true at junction a no matter what the orientation in space of the three wires. Currents are scalars, not vectors.

The Directions of Currents

In Fig. 27-1*b* we drew the current arrows in the direction in which positively charged particles would be forced to move through the loop by the electric field. Such positive *charge carriers*, as they are often called, would move away from the positive battery terminal and toward the negative terminal. Actually, the charge carriers in the copper loop of Fig. 27-1*b* are electrons and thus are negatively charged. The electric field forces them to move in the direction opposite the current arrows, from the negative terminal to the positive terminal. For historical reasons, however, we use the following convention:

A current arrow is drawn in the direction in which positive charge carriers would move, even if the actual charge carriers are negative and move in the opposite direction.

We can use this convention because in *most* situations, the assumed motion of positive charge carriers in one direction has the same effect as the actual motion of negative charge carriers in the opposite direction. (When the effect is not the same, we shall, of course, drop the convention and describe the actual motion.)

✓CHECKPOINT 1: The figure here shows a portion of a circuit. What are the magnitude and direction of the current i in the lower right-hand wire?

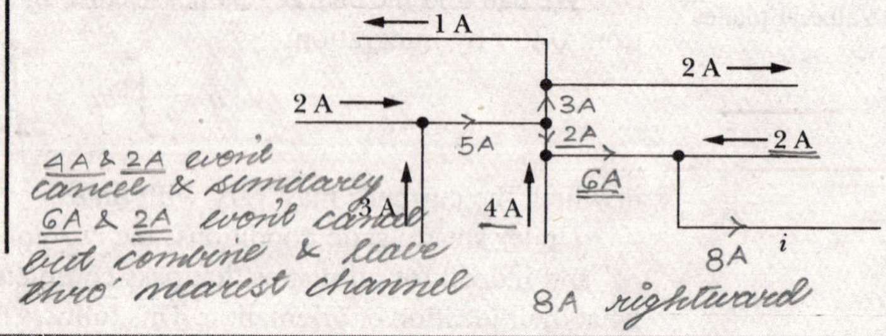

Sample Problem 27-1

Water flows through a garden hose at a volume flow rate dV/dt of 450 cm^3/s. What is the current of negative charge?

SOLUTION: The current i of negative charge is due to the electrons in the water molecules moving through the hose. The current is the rate at which that negative charge passes through any plane that cuts completely across the hose. Thus, the **Key Idea** here is that we can write the current in terms of the number of molecules that pass through such a plane per second as

$$i = \begin{pmatrix}\text{charge}\\ \text{per}\\ \text{electron}\end{pmatrix}\begin{pmatrix}\text{electrons}\\ \text{per}\\ \text{molecule}\end{pmatrix}\begin{pmatrix}\text{molecules}\\ \text{per}\\ \text{second}\end{pmatrix}$$

or

$$i = (e)(10)\frac{dN}{dt}.$$

We substitute 10 electrons per molecule because a water (H_2O) molecule contains 8 electrons in the single oxygen atom and 1 electron in each of the two hydrogen atoms.

We can express the rate dN/dt in terms of the given volume flow rate dV/dt by first writing

$$\begin{pmatrix}\text{molecules}\\ \text{per}\\ \text{second}\end{pmatrix} = \begin{pmatrix}\text{molecules}\\ \text{per}\\ \text{mole}\end{pmatrix}\begin{pmatrix}\text{moles}\\ \text{per unit}\\ \text{mass}\end{pmatrix}\begin{pmatrix}\text{mass}\\ \text{per unit}\\ \text{volume}\end{pmatrix}\begin{pmatrix}\text{volume}\\ \text{per}\\ \text{second}\end{pmatrix}.$$

"Molecules per mole" is Avogadro's number N_A. "Moles per unit mass" is the inverse of the mass per mole, which is the molar mass M of water. "Mass per unit volume" is the (mass) density ρ_{mass} of water. The volume per second is the volume flow rate dV/dt. Thus, we have

$$\frac{dN}{dt} = N_A\left(\frac{1}{M}\right)\rho_{mass}\left(\frac{dV}{dt}\right) = \frac{N_A\rho_{mass}}{M}\frac{dV}{dt}.$$

Substituting this into the equation for i, we find

$$i = 10eN_AM^{-1}\rho_{mass}\frac{dV}{dt}.$$

N_A is 6.02×10^{23} molecules/mol, or 6.02×10^{23} mol^{-1}, and ρ_{mass} is 1000 kg/m^3. We can get the molar mass of water from the molar masses listed in Appendix F: We add the molar mass of oxygen (16 g/mol) to twice the molar mass of hydrogen (1 g/mol), obtaining 18 g/mol = 0.018 kg/mol. Then

$$\begin{aligned} i &= (10)(1.6 \times 10^{-19}\ \text{C})(6.02 \times 10^{23}\ \text{mol}^{-1}) \\ &\quad \times (0.018\ \text{kg/mol})^{-1}(1000\ \text{kg/m}^3)(450 \times 10^{-6}\ \text{m}^3/\text{s}) \\ &= 2.41 \times 10^7\ \text{C/s} = 2.41 \times 10^7\ \text{A} \\ &= 24.1\ \text{MA}. \end{aligned}$$ (Answer)

This current of negative charge is exactly compensated by a current of positive charge associated with the nuclei of the three atoms that make up the water molecule. Thus, there is no net flow of charge through the hose.

27-3 Current Density

Sometimes we are interested in the current i in a particular conductor. At other times we take a localized view and study the flow of charge through a cross section of the conductor at a particular point. To describe this flow, we can use the **current density** $\vec{J}$, which has the same direction as the velocity of the moving charges if they are positive and the opposite direction if they are negative. For each element of the cross section, the magnitude J is equal to the current per unit area through that element. We can write the amount of current through the element as $\vec{J} \cdot d\vec{A}$, where $d\vec{A}$ is the area vector of the element, perpendicular to the element. The total current through the surface is then

$$i = \int \vec{J} \cdot d\vec{A}. \qquad (27\text{-}4)$$

If the current is uniform across the surface and parallel to $d\vec{A}$, then $\vec{J}$ is also uniform and parallel to $d\vec{A}$. Then Eq. 27-4 becomes

$$i = \int J\,dA = J \int dA = JA,$$

so

$$J = \frac{i}{A}, \qquad (27\text{-}5)$$

when J is uniform & parallel to dA

where A is the total area of the surface. From Eq. 27-4 or 27-5 we see that the SI unit for current density is the ampere per square meter (A/m^2).

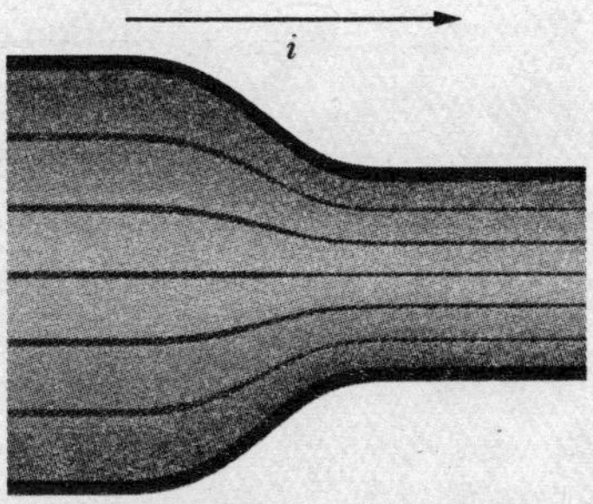

Fig. 27-4 Streamlines representing current density in the flow of charge through a constricted conductor.

In Chapter 23 we saw that we can represent an electric field with electric field lines. Figure 27-4 shows how current density can be represented with a similar set of lines, which we can call *streamlines*. The current, which is toward the right in Fig. 27-4, makes a transition from the wider conductor at the left to the narrower conductor at the right. Because charge is conserved during the transition, the amount of charge and thus the amount of current cannot change. However, the current density does change—it is greater in the narrower conductor. The spacing of the streamlines suggests this increase in current density; streamlines that are closer together imply greater current density.

Drift Speed

When a conductor does not have a current through it, its conduction electrons move randomly, with no net motion in any direction. When the conductor does have a current through it, these electrons actually still move randomly, but now they tend to *drift* with a **drift speed** v_d in the direction opposite that of the applied electric field that causes the current. The drift speed is tiny compared to the speeds in the random motion. For example, in the copper conductors of household wiring, electron drift speeds are perhaps 10^{-5} or 10^{-4} m/s, whereas the random-motion speeds are around 10^6 m/s.

We can use Fig. 27-5 to relate the drift speed v_d of the conduction electrons in a current through a wire to the magnitude J of the current density in the wire. For convenience, Fig. 27-5 shows the equivalent drift of *positive* charge carriers in the direction of the applied electric field $\vec{E}$. Let us assume that these charge carriers all move with the same drift speed v_d and that the current density J is uniform across the wire's cross-sectional area A. The number of charge carriers in a length L of the wire is nAL, where n is the number of carriers per unit volume. The total charge of the carriers in the length L, each with charge e, is then

$$q = (nAL)e.$$

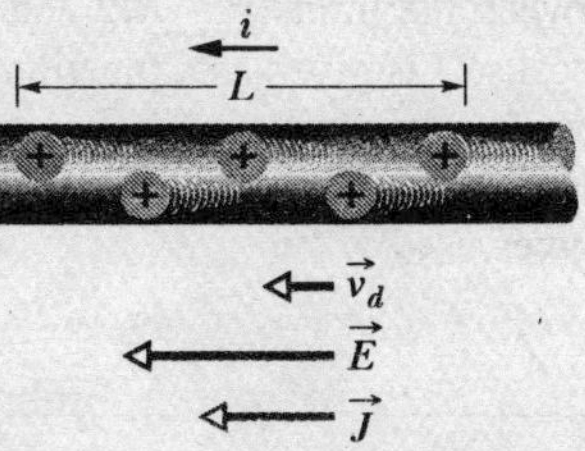

Fig. 27-5 Positive charge carriers drift at speed v_d in the direction of the applied electric field $\vec{E}$. By convention, the direction of the current density $\vec{J}$ and the sense of the current arrow are drawn in that same direction.

Because the carriers all move along the wire with speed v_d, this total charge moves through any cross section of the wire in the time interval

$$t = \frac{L}{v_d}.$$

Equation 27-1 tells us that the current i is the time rate of transfer of charge across a cross section, so here we have

$$i = \frac{q}{t} = \frac{nALe}{L/v_d} = nAev_d. \tag{27-6}$$

Solving for v_d and recalling Eq. 27-5 ($J = i/A$), we obtain

$$v_d = \frac{i}{nAe} = \frac{J}{ne}$$

or, extended to vector form,

$$\vec{J} = (ne)\vec{v}_d. \tag{27-7}$$

Here the product ne, whose SI unit is the coulomb per cubic meter (C/m^3), is the *carrier charge density*. For positive carriers, ne is positive and Eq. 27-7 predicts that $\vec{J}$ and $\vec{v}_d$ have the same direction. For negative carriers, ne is negative and $\vec{J}$ and $\vec{v}_d$ have opposite directions.

CHECKPOINT 2: The figure here shows conduction electrons moving leftward through a wire. Are the following leftward or rightward: (a) the current i, (b) the current density $\vec{J}$, (c) the electric field $\vec{E}$ in the wire?

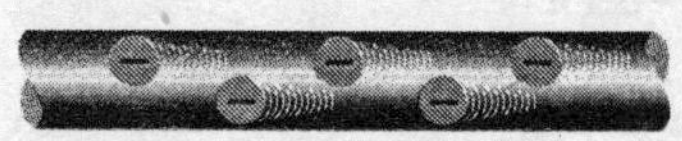

Sample Problem 27-2

(a) The current density in a cylindrical wire of radius $R = 2.0$ mm is uniform across a cross section of the wire and is $J = 2.0 \times 10^5$ A/m^2. What is the current through the outer portion of the wire between radial distances $R/2$ and R (Fig. 27-6*a*)?

SOLUTION: The **Key Idea** here is that, because the current density is uniform across the cross section, the current density J, the current i, and the cross-sectional area A are related by Eq. 27-5 ($J = i/A$). However, we want only the current through a reduced cross-sectional area A' of the wire (rather than the entire area), where

$$A' = \pi R^2 - \pi\left(\frac{R}{2}\right)^2 = \pi\left(\frac{3R^2}{4}\right)$$

$$= \frac{3\pi}{4}(0.002\text{ m})^2 = 9.424 \times 10^{-6}\text{ m}^2.$$

We now rewrite Eq. 27-5 as

$$i = JA'$$

and then substitute the data to find

$$i = (2.0 \times 10^5\text{ A/m}^2)(9.424 \times 10^{-6}\text{ m}^2)$$

$$= 1.9\text{ A.} \qquad \text{(Answer)}$$

(b) Suppose, instead, that the current density through a cross section varies with radial distance r as $J = ar^2$, in which $a = 3.0 \times 10^{11}$ A/m^4 and r is in meters. What now is the current through the same outer portion of the wire?

SOLUTION: The **Key Idea** here is that, because the current density is not uniform across a cross section of the wire, we must resort to Eq. 27-4 ($i = \int \vec{J} \cdot d\vec{A}$) and integrate the current density over the portion of the wire from $r = R/2$ to $r = R$. The current density vector $\vec{J}$ (along the wire's length) and the differential area vector $d\vec{A}$ (perpendicular to a cross section of the wire) have the same direction. Thus,

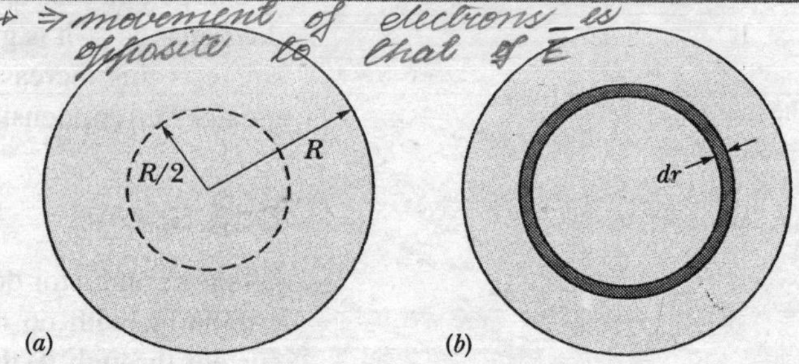

Fig. 27-6 Sample Problem 27-2. (*a*) Cross section of a wire of radius R. (*b*) A thin ring has width dr and circumference $2\pi r$, and thus a differential area $dA = 2\pi r\,dr$.

$$\vec{J} \cdot d\vec{A} = J\,dA\cos 0 = J\,dA.$$

We need to replace the differential area dA with something we can actually integrate between the limits $r = R/2$ and $r = R$. The simplest replacement (because J is given as a function of r) is the area $2\pi r\,dr$ of a thin ring of circumference $2\pi r$ and width dr (Fig. 27-6*b*). We can then integrate with r as the variable of integration. Equation 27-4 then gives us

$$i = \int \vec{J} \cdot d\vec{A} = \int J\,dA$$

$$= \int_{R/2}^{R} ar^2\, 2\pi r\,dr = 2\pi a \int_{R/2}^{R} r^3\,dr$$

$$= 2\pi a\left[\frac{r^4}{4}\right]_{R/2}^{R} = \frac{\pi a}{2}\left[R^4 - \frac{R^4}{16}\right] = \frac{15}{32}\pi a R^4$$

$$= \frac{15}{32}\pi(3.0 \times 10^{11}\text{ A/m}^4)(0.002\text{ m})^4 = 7.1\text{ A.} \qquad \text{(Answer)}$$

Sample Problem 27-3

What is the drift speed of the conduction electrons in a copper wire with radius $r = 900\ \mu m$ when it has a uniform current $i = 17$ mA? Assume that each copper atom contributes one conduction electron to the current and the current density is uniform across the wire's cross section.

SOLUTION: We need three **Key Ideas** here:

1. The drift speed v_d is related to the current density $\vec{J}$ and the number n of conduction electrons per unit volume according to Eq. 27-7, which we can write in magnitude form as $J = nev_d$.
2. Because the current density is uniform, its magnitude J is related to the given current i and wire size by Eq. 27-5 ($J = i/A$, where A is the cross-sectional area of the wire).
3. Because we assume one conduction electron per atom, the number n of conduction electrons per unit volume is the same as the number of atoms per unit volume.

Let us start with the third idea by writing

$$n = \begin{pmatrix}\text{atoms}\\ \text{per unit}\\ \text{volume}\end{pmatrix} = \begin{pmatrix}\text{atoms}\\ \text{per}\\ \text{mole}\end{pmatrix}\begin{pmatrix}\text{moles}\\ \text{per unit}\\ \text{mass}\end{pmatrix}\begin{pmatrix}\text{mass}\\ \text{per unit}\\ \text{volume}\end{pmatrix}.$$

The number of atoms per mole is just Avogadro's number N_A ($= 6.02 \times 10^{23}\ \text{mol}^{-1}$). Moles per unit mass is the inverse of the mass per mole, which here is the molar mass M of copper. The mass per unit volume is the (mass) density ρ_{mass} of copper. Thus,

$$n = N_A\left(\frac{1}{M}\right)\rho_{mass} = \frac{N_A \rho_{mass}}{M}.$$

Taking copper's molar mass M and density ρ_{mass} from Appendix F, we then ha

$n =$

$=$

or

Next let

There is only one pipe to raise water to height but two wires in the battery — one to raise electric PE of charge carrier & second to take them out as e⁻ cannot flow thro air

If ∃ wood instead of air, then water would be impermeable to it & we would need second pipe

$$\frac{i}{A} = nev_d.$$

Substituting for A with πr^2 ($= 2.54 \times 10^{-6}\ \text{m}^2$) and solving for v_d, we then find

$$v_d = \frac{i}{ne(\pi r^2)}$$

$$= \frac{17 \times 10^{-3}\ \text{A}}{(8.49 \times 10^{28}\ \text{m}^{-3})(1.6 \times 10^{-19}\ \text{C})(2.54 \times 10^{-6}\ \text{m}^2)}$$

$$= 4.9 \times 10^{-7}\ \text{m/s}, \qquad \text{(Answer)}$$

which is only 1.8 mm/h, slower than a sluggish snail.

You may well ask: "If the electrons drift so slowly, why do the room lights turn on so quickly when I throw the switch?" Confusion on this point results from not distinguishing between the drift speed of the electrons and the speed at which *changes* in the electric field configuration travel along wires. This latter speed is nearly that of light; electrons everywhere in the wire begin drifting almost at once, including into the lightbulbs. Similarly, when you open the valve on your garden hose, with the hose full of water, a pressure wave travels along the hose at the speed of sound in water. The speed at which the water itself moves through the hose—measured perhaps with a dye marker—is much slower.

If the current thro' the wire is ~ 1A, or 10A then v_d is of the order of 10^{-4} m/s or 10^{-3} m/s
It is ~ 0.1 mm/sec or 1 mm/sec

27-4 Resistance and Res

If we apply the same potential difference bet ... ilar rods of copper and of glass, very different c ... the conductor that enters here is its electrical re ... nce between any two points of a conductor by ap ... een those points and measuring the current i that

$$R = \frac{V}{i} \qquad \text{(defi} \qquad \text{8)}$$

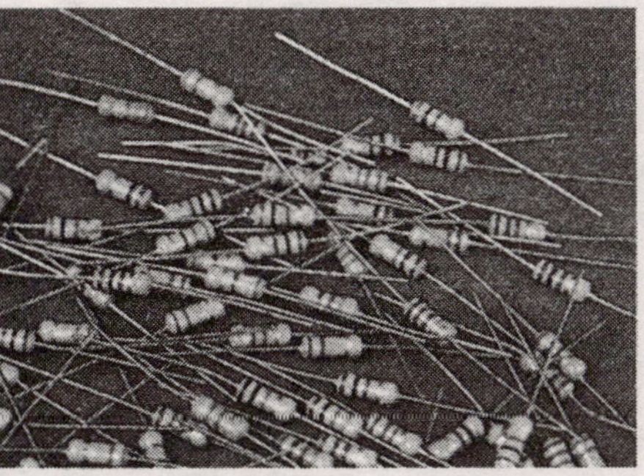

Fig. 27-7 An assortment of resistors. The circular bands are color-coding marks that identify the value of the resistance.

The SI unit for resistance that follows from Eq. 27-8 is the volt per ampere. This combination occurs so often that we give it a special name, the **ohm** (symbol Ω); that is,

$$1\ \text{ohm} = 1\ \Omega = 1\ \text{volt per ampere}$$
$$= 1\ \text{V/A}. \qquad (27\text{-}9)$$

A conductor whose function in a circuit is to provide a specified resistance is called a **resistor** (see Fig. 27-7). In a circuit diagram, we represent a resistor and a resistance

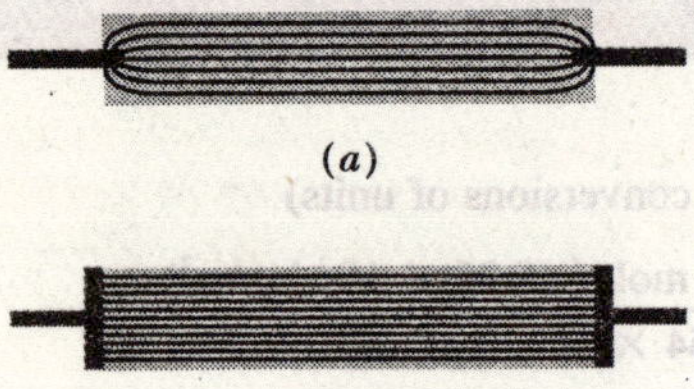

Fig. 27-8 Two ways of applying a potential difference to a conducting rod. The heavy gray connectors are assumed to have negligible resistance. When they are arranged as in (*a*), the measured resistance is larger than when they are arranged as in (*b*).

with the symbol -ᴡᴡ-. If we write Eq. 27-8 as

$$i = \frac{V}{R},$$

we see that "resistance" is aptly named. For a given potential difference, the greater the resistance (to current), the smaller the current.

★ The resistance of a conductor depends on the manner in which the potential difference is applied to it. Figure 27-8, for example, shows a given potential difference applied in two different ways to the same conductor. As the current density streamlines suggest, the currents in the two cases—hence the measured resistances—will be different. Unless otherwise stated, we shall assume that any given potential difference is applied as in Fig. 27-8*b*.

As we have done several times in other connections, we often wish to take a general view and deal not with particular objects but with materials. Here we do so by focusing not on the potential difference V across a particular resistor but on the electric field $\vec{E}$ at a point in a resistive material. Instead of dealing with the current i through the resistor, we deal with the current density $\vec{J}$ at the point in question. Instead of the resistance R of an object, we deal with the **resistivity** ρ of the *material*:

$$\rho = \frac{E}{J} \quad \text{(definition of } \rho\text{)}. \qquad (27\text{-}10)$$

(Compare this equation with Eq. 27-8.)

If we combine the SI units of E and J according to Eq. 27-10, we get, for the unit of ρ, the ohm-meter ($\Omega \cdot \text{m}$):

$$\frac{\text{unit } (E)}{\text{unit } (J)} = \frac{\text{V/m}}{\text{A/m}^2} = \frac{\text{V}}{\text{A}}\,\text{m} = \Omega \cdot \text{m}.$$

(Do not confuse the *ohm-meter*, the unit of resistivity, with the *ohmmeter*, which is an instrument that measures resistance.) Table 27-1 lists the resistivities of some materials.

We can write Eq. 27-10 in vector form as

$$\vec{E} = \rho \vec{J}. \qquad (27\text{-}11)$$

Equations 27-10 and 27-11 hold only for *isotropic* materials—materials whose electrical properties are the same in all directions.

We often speak of the **conductivity** σ of a material. This is simply the reciprocal of its resistivity, so

$$\sigma = \frac{1}{\rho} \quad \text{(definition of } \sigma\text{)}. \qquad (27\text{-}12)$$

The SI unit of conductivity is the reciprocal ohm-meter, $(\Omega \cdot \text{m})^{-1}$. The unit name mhos per meter is sometimes used (mho is ohm backwards). The definition of σ allows us to write Eq. 27-11 in the alternative form

$$\vec{J} = \sigma \vec{E}. \qquad (27\text{-}13)$$

Calculating Resistance from Resistivity

We have just made an important distinction:

Resistance is a property of an object. Resistivity is a property of a material.

TABLE 27-1 Resistivities of Some Materials at Room Temperature (20°C)

Material	Resistivity, ρ ($\Omega \cdot$ m)	Temperature Coefficient of Resistivity, α (K^{-1})
	Typical Metals	
Silver	1.62×10^{-8}	4.1×10^{-3}
Copper	1.69×10^{-8}	4.3×10^{-3}
Aluminum	2.75×10^{-8}	4.4×10^{-3}
Tungsten	5.25×10^{-8}	4.5×10^{-3}
Iron	9.68×10^{-8}	6.5×10^{-3}
Platinum	10.6×10^{-8}	3.9×10^{-3}
Manganin[a]	4.82×10^{-8}	0.002×10^{-3}
	Typical Semiconductors	
Silicon, pure	2.5×10^{3}	-70×10^{-3}
Silicon, *n*-type[b]	8.7×10^{-4}	
Silicon, *p*-type[c]	2.8×10^{-3}	
	Typical Insulators	
Glass	10^{10}–10^{14}	
Fused quartz	$\sim 10^{16}$	

[a]An alloy specifically designed to have a small value of α.

[b]Pure silicon doped with phosphorus impurities to a charge carrier density of 10^{23} m^{-3}.

[c]Pure silicon doped with aluminum impurities to a charge carrier density of 10^{23} m^{-3}.

If we know the resistivity of a substance such as copper, we can calculate the resistance of a length of wire made of that substance. Let A be the cross-sectional area of the wire, let L be its length, and let a potential difference V exist between its ends (Fig. 27-9). If the streamlines representing the current density are uniform throughout the wire, the electric field and the current density will be constant for all points within the wire and, from Eqs. 25-42 and 27-5, will have the values

$$E = V/L \quad \text{and} \quad J = i/A. \tag{27-14}$$

We can then combine Eqs. 27-10 and 27-14 to write

$$\rho = \frac{E}{J} = \frac{V/L}{i/A}. \tag{27-15}$$

However, V/i is the resistance R, which allows us to recast Eq. 27-15 as

$$R = \rho \frac{L}{A}. \tag{27-16}$$

Equation 27-16 can be applied only to a homogeneous isotropic conductor of uniform cross section, with the potential difference applied as in Fig. 27-8*b*.

The macroscopic quantities V, i, and R are of greatest interest when we are making electrical measurements on specific conductors. They are the quantities that we read directly on meters. We turn to the microscopic quantities E, J, and ρ when we are interested in the fundamental electrical properties of materials.

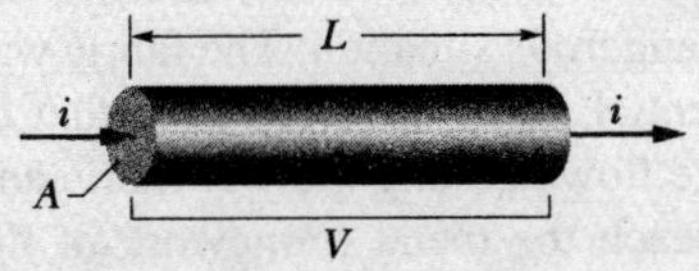

Fig. 27-9 A potential difference V is applied between the ends of a wire of length L and cross section A, establishing a current i.

CHECKPOINT 3: The figure here shows three cylindrical copper conductors along with their face areas and lengths. Rank them according to the current through them, greatest first, when the same potential difference V is placed across their lengths.

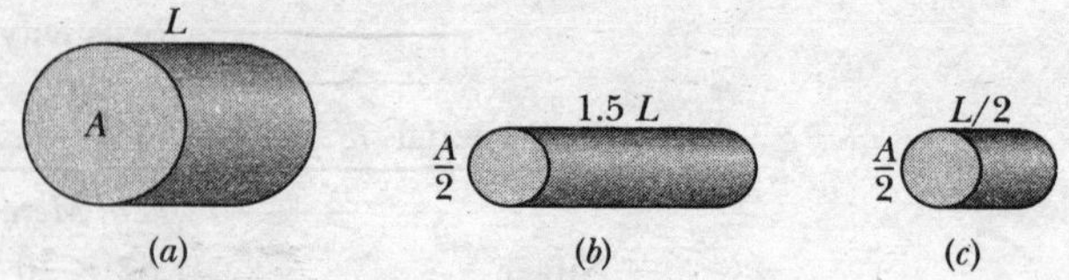

Variation with Temperature

The values of most physical properties vary with temperature, and resistivity is no exception. Figure 27-10, for example, shows the variation of this property for copper over a wide temperature range. The relation between temperature and resistivity for copper—and for metals in general—is fairly linear over a rather broad temperature range. For such linear relations we can write an empirical approximation that is good enough for most engineering purposes:

$$\rho - \rho_0 = \rho_0\alpha(T - T_0). \tag{27-17}$$

Here T_0 is a selected reference temperature and ρ_0 is the resistivity at that temperature. Usually $T_0 = 293$ K (room temperature), for which $\rho_0 = 1.69 \times 10^{-8}\ \Omega \cdot \text{m}$ for copper.

Because temperature enters Eq. 27-17 only as a difference, it does not matter whether you use the Celsius or Kelvin scale in that equation because the sizes of degrees on these scales are identical. The quantity α in Eq. 27-17, called the *temperature coefficient of resistivity,* is chosen so that the equation gives good agreement with experiment for temperatures in the chosen range. Some values of α for metals are listed in Table 27-1.

The *Hindenburg*

When the zeppelin *Hindenburg* was preparing to land, the handling ropes were let down to the ground crew. Exposed to the rain, the ropes became wet (and thus were able to conduct a current). In this condition, the ropes "grounded" the metal framework of the zeppelin to which they were attached; that is, the wet ropes formed a conducting path between the framework and the ground, making the electric potential of the framework the same as that of the ground. This should have also grounded the outer fabric of the zeppelin. The *Hindenburg*, however, had been the first zeppelin to have its outer fabric painted with a sealant of large electrical resistivity. Thus, the fabric remained at the electric potential of the atmosphere at the zeppelin's altitude of about 43 m. Due to the rainstorm, that potential was large relative to the potential at ground level.

The handling of the ropes apparently ruptured one of the hydrogen cells and released hydrogen between that cell and the zeppelin's outer fabric, causing the reported rippling of the fabric. There was then a dangerous situation: The fabric was wet with conducting rainwater and was at a potential much different from that of the framework of the zeppelin. Apparently, charge flowed along the wet fabric and then sparked through the released hydrogen to reach the metal framework of the zeppelin, igniting the hydrogen in the process. The burning rapidly ignited the cells of hydrogen in the zeppelin and brought the ship down. If the sealant on the outer fabric of the *Hindenburg* had been of less resistivity (like that of earlier and later zeppelins), the *Hindenburg* disaster probably would not have occurred.

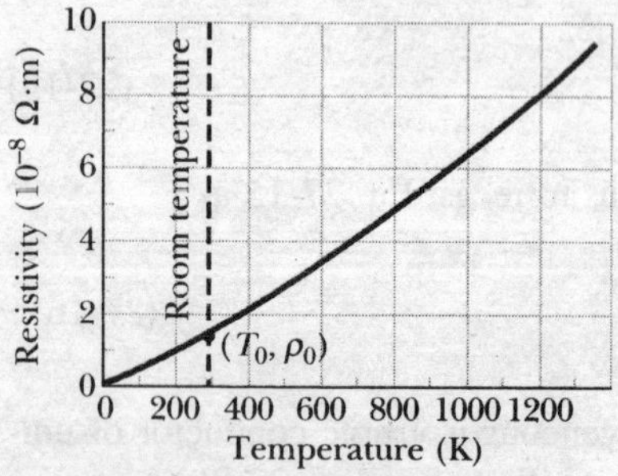

Fig. 27-10 The resistivity of copper as a function of temperature. The dot on the curve marks a convenient reference point at temperature $T_0 = 293$ K and resistivity $\rho_0 = 1.69 \times 10^{-8}\ \Omega \cdot \text{m}$.

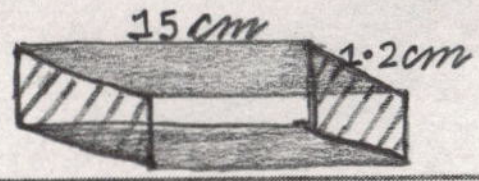

Sample Problem 27-4

A rectangular block of iron has dimensions 1.2 cm × 1.2 cm × 15 cm. A potential difference is to be applied to the block between parallel sides and in such a way that those sides are equipotential surfaces (as in Fig. 27-8*b*). What is the resistance of the block if the two parallel sides are (1) the square ends (with dimensions 1.2 cm × 1.2 cm) and (2) two rectangular sides (with dimensions 1.2 cm × 15 cm)?

SOLUTION: The **Key Idea** here is that the resistance R of an object depends on how the electric potential is applied to the object. In particular, it depends on the ratio L/A, according to Eq. 27-16 ($R = \rho L/A$), where A is the area of the surfaces to which the potential difference is applied and L is the distance between those surfaces. For arrangement 1, L = 15 cm = 0.15 m and

$$A = (1.2\text{ cm})^2 = 1.44 \times 10^{-4}\text{ m}^2.$$

Substituting into Eq. 27-16 with the resistivity ρ from Table 27-1, we then find that for arrangement 1,

$$R = \frac{\rho L}{A} = \frac{(9.68 \times 10^{-8}\ \Omega\cdot\text{m})(0.15\text{ m})}{1.44 \times 10^{-4}\text{ m}^2}$$
$$= 1.0 \times 10^{-4}\ \Omega = 100\ \mu\Omega. \qquad \text{(Answer)}$$

Similarly, for arrangement 2, with distance L = 1.2 cm and area A = (1.2 cm)(15 cm), we obtain

$$R = \frac{\rho L}{A} = \frac{(9.68 \times 10^{-8}\ \Omega\cdot\text{m})(1.2 \times 10^{-2}\text{ m})}{1.80 \times 10^{-3}\text{ m}^2}$$
$$= 6.5 \times 10^{-7}\ \Omega = 0.65\ \mu\Omega. \qquad \text{(Answer)}$$

27-5 Ohm's Law

As we just discussed in Section 27-4, a resistor is a conductor with a specified resistance. It has that same resistance no matter what the magnitude and direction (*polarity*) of the applied potential difference. Other conducting devices, however, might have resistances that change with the applied potential difference.

Figure 27-11*a* shows how to distinguish such devices. A potential difference V is applied across the device being tested, and the resulting current i through the device is measured as V is varied in both magnitude and polarity. The polarity of V is arbitrarily taken to be positive when the left terminal of the device is at a higher potential than the right terminal. The direction of the resulting current (from left to right) is arbitrarily assigned a plus sign. The reverse polarity of V (with the right terminal at a higher potential) is then negative; the current it causes is assigned a minus sign.

Figure 27-11*b* is a plot of i versus V for one device. This plot is a straight line passing through the origin, so the ratio i/V (which is the slope of the straight line) is the same for all values of V. This means that the resistance $R = V/i$ of the device is independent of the magnitude and polarity of the applied potential difference V.

Figure 27-11*c* is a plot for another conducting device. Current can exist in this device only when the polarity of V is positive and the applied potential difference is more than about 1.5 V. When current does exist, the relation between i and V is not linear; it depends on the value of the applied potential difference V.

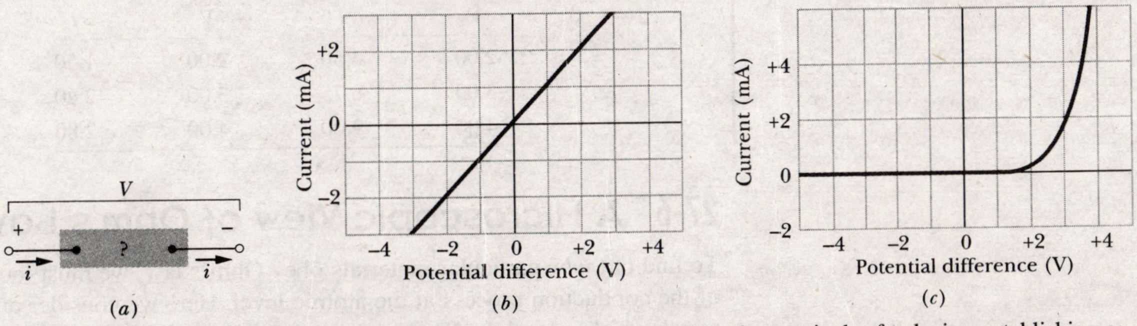

Fig. 27-11 (*a*) A potential difference V is applied to the terminals of a device, establishing a current i. (*b*) A plot of current i versus applied potential difference V when the device is a 1000 Ω resistor. (*c*) A plot when the device is a semiconducting *pn* junction diode.

We distinguish between the two types of device by saying that one obeys Ohm's law and the other does not.

> **Ohm's law** is an assertion that the current through a device is *always* directly proportional to the potential difference applied to the device.

(This assertion is correct only in certain situations; still, for historical reasons, the term "law" is used.) The device of Fig. 27-11*b*—which turns out to be a 1000 Ω resistor—obeys Ohm's law. The device of Fig. 27-11*c*—which turns out to be a so-called *pn* junction diode—does not.

> A conducting device obeys Ohm's law when the resistance of the device is independent of the magnitude and polarity of the applied potential difference.

Modern microelectronics—and therefore much of the character of our present technological civilization—depends almost totally on devices that do *not* obey Ohm's law. Your calculator, for example, is full of them.

It is often contended that $V = iR$ is a statement of Ohm's law. That is not true! This equation is the defining equation for resistance, and it applies to all conducting devices, whether they obey Ohm's law or not. If we measure the potential difference V across, and the current i through, any device, even a *pn* junction diode, we can find its resistance *at that value of* V as $R = V/i$. The essence of Ohm's law, however, is that a plot of i versus V is linear; that is, R is independent of V.

We can express Ohm's law in a more general way if we focus on conducting *materials* rather than on conducting *devices*. The relevant relation is then Eq. 27-11 ($\vec{E} = \rho\vec{J}$), which is the analog of $V = iR$.

> A conducting material obeys Ohm's law when the resistivity of the material is independent of the magnitude and direction of the applied electric field.

All homogeneous materials, whether they are conductors like copper or semiconductors like pure silicon or silicon containing special impurities, obey Ohm's law within some range of values of the electric field. If the field is too strong, however, there are departures from Ohm's law in all cases.

CHECKPOINT 4: The following table gives the current i (in amperes) through two devices for several values of potential difference V (in volts). From these data, determine which device does not obey Ohm's law.

Device 1		Device 2	
V	i	V	i
2.00	4.50	2.00	1.50
3.00	6.75	3.00	2.20
4.00	9.00	4.00	2.80

27-6 A Microscopic View of Ohm's Law

To find out *why* particular materials obey Ohm's law, we must look into the details of the conduction process at the atomic level. Here we consider only conduction in metals, such as copper. We base our analysis on the *free-electron model*, in which we assume that the conduction electrons in the metal are free to move throughout the volume of a sample, like the molecules of a gas in a closed container. We

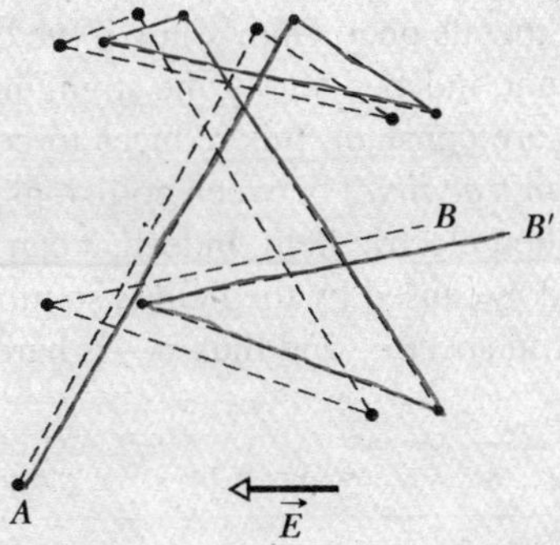

Fig. 27-12 The gray lines show an electron moving from A to B, making six collisions en route. The green lines show what its path might be in the presence of an applied electric field $\vec{E}$. Note the steady drift in the direction of $-\vec{E}$. (Actually, the green lines should be slightly curved, to represent the parabolic paths followed by the electrons between collisions, under the influence of an electric field.)

also assume that the electrons collide not with one another but only with atoms of the metal.

According to classical physics, the electrons should have a Maxwellian speed distribution somewhat like that of the molecules in a gas. In such a distribution (see Section 20-7), the average electron speed would be proportional to the square root of the absolute temperature. The motions of electrons are, however, governed not by the laws of classical physics but by those of quantum physics. As it turns out, an assumption that is much closer to the quantum reality is that conduction electrons in a metal move with a single effective speed v_{eff}, and this speed is essentially independent of the temperature. For copper, $v_{\text{eff}} \approx 1.6 \times 10^6$ m/s.

When we apply an electric field to a metal sample, the electrons modify their random motions slightly and drift very slowly—in a direction opposite that of the field—with an average drift speed v_d. As we saw in Sample Problem 27-3, the drift speed in a typical metallic conductor is about 5×10^{-7} m/s, less than the effective speed (1.6×10^6 m/s) by many orders of magnitude. Figure 27-12 suggests the relation between these two speeds. The gray lines show a possible random path for an electron in the absence of an applied field; the electron proceeds from A to B, making six collisions along the way. The green lines show how the same events *might* occur when an electric field $\vec{E}$ is applied. We see that the electron drifts steadily to the right, ending at B' rather than at B. Figure 27-12 was drawn with the assumption that $v_d \approx 0.02 v_{\text{eff}}$. However, because the actual value is more like $v_d \approx (10^{-13}) v_{\text{eff}}$, the drift displayed in the figure is greatly exaggerated.

The motion of the conduction electrons in an electric field $\vec{E}$ is thus a combination of the motion due to random collisions and that due to $\vec{E}$. When we consider all the free electrons, their random motions average to zero and make no contribution to the drift speed. Thus, the drift speed is due only to the effect of the electric field on the electrons.

If an electron of mass m is placed in an electric field of magnitude E, the electron will experience an acceleration given by Newton's second law:

$$a = \frac{F}{m} = \frac{eE}{m}. \tag{27-18}$$

The nature of the collisions experienced by conduction electrons is such that, after a typical collision, each electron will—so to speak—completely lose its memory of its previous drift velocity. Each electron will then start off fresh after every encounter, moving off in a random direction. In the average time τ between collisions, the average electron will acquire a drift speed of $v_d = a\tau$. Moreover, if we measure the drift speeds of all the electrons at any instant, we will find that their average drift speed is also $a\tau$. Thus, at any instant, on average, the electrons will have drift speed $v_d = a\tau$. Then Eq. 27-18 gives us

$$v_d = a\tau = \frac{eE\tau}{m}. \tag{27-19}$$

Combining this result with Eq. 27-7 ($\vec{J} = ne\vec{v}_d$), in magnitude form, yields

$$v_d = \frac{J}{ne} = \frac{eE\tau}{m},$$

which we can write as

$$E = \left(\frac{m}{e^2 n\tau}\right) J.$$

Comparing this with Eq. 27-11 ($\vec{E} = \rho\vec{J}$), in magnitude form, leads to

$$\rho = \frac{m}{e^2 n\tau}. \tag{27-20}$$

For any material

Equation 27-20 may be taken as a statement that metals obey Ohm's law if we can show that, for metals, their resistivity ρ is a constant, independent of the strength of the applied electric field $\vec{E}$. Because n, m, and e are constant, this reduces to convincing ourselves that τ, the average time (or *mean free time*) between collisions, is a constant, independent of the strength of the applied electric field. Indeed, τ can be considered to be a constant because the drift speed v_d caused by the field is so much smaller than the effective speed v_{eff} that the electron speed—and thus τ—is hardly affected by the field.

Sample Problem 27-5

(a) What is the mean free time τ between collisions for the conduction electrons in copper?

SOLUTION: **The Key Idea** here is that the mean free time τ of copper is approximately constant, and in particular does not depend on any electric field that might be applied to a sample of the copper. Thus, we need not consider any particular value of applied electric field. However, because the resistivity ρ displayed by copper under an electric field depends on τ, we can find the mean free time τ from Eq. 27-20 ($\rho = m/e^2 n\tau$). That equation gives us

$$\tau = \frac{m}{ne^2\rho}.$$

We take the value of n, the number of conduction electrons per unit volume in copper, from Sample Problem 27-3. We take the value of ρ from Table 27-1. The denominator then becomes

$$(8.49 \times 10^{28}\ \text{m}^{-3})(1.6 \times 10^{-19}\ \text{C})^2(1.69 \times 10^{-8}\ \Omega \cdot \text{m})$$
$$= 3.67 \times 10^{-17}\ \text{C}^2 \cdot \Omega/\text{m}^2 = 3.67 \times 10^{-17}\ \text{kg/s},$$

where we converted units as

$$\frac{\text{C}^2 \cdot \Omega}{\text{m}^2} = \frac{\text{C}^2 \cdot \text{V}}{\text{m}^2 \cdot \text{A}} = \frac{\text{C}^2 \cdot \text{J/C}}{\text{m}^2 \cdot \text{C/s}} = \frac{\text{kg} \cdot \text{m}^2/\text{s}^2}{\text{m}^2/\text{s}} = \frac{\text{kg}}{\text{s}}.$$

Using these results and substituting for the electron mass m, we then have

$$\tau = \frac{9.1 \times 10^{-31}\ \text{kg}}{3.67 \times 10^{-17}\ \text{kg/s}} = 2.5 \times 10^{-14}\ \text{s}. \quad \text{(Answer)}$$

(b) The mean free path λ of the conduction electrons in a conductor is the average distance traveled by an electron between collisions. (This definition parallels that in Section 20-6 for the mean free path of molecules in a gas.) What is λ for the conduction electrons in copper, assuming that their effective speed v_{eff} is 1.6×10^6 m/s?

SOLUTION: **The Key Idea** here is that the distance d any particle travels in a certain time t at a constant speed v is $d = vt$. For the electrons in copper, this gives us

$$\lambda = v_{\text{eff}}\tau = (1.6 \times 10^6\ \text{m/s})(2.5 \times 10^{-14}\ \text{s})$$
$$= 4.0 \times 10^{-8}\ \text{m} = 40\ \text{nm}. \quad \text{(Answer)}$$

This is about 150 times the distance between nearest-neighbor atoms in a copper lattice. Thus, on the average, each conduction electron passes many copper atoms before finally hitting one.

27-7 Power in Electric Circuits

Figure 27-13 shows a circuit consisting of a battery B that is connected by wires, which we assume have negligible resistance, to an unspecified conducting device. The device might be a resistor, a storage battery (a rechargeable battery), a motor, or some other electrical device. The battery maintains a potential difference of magnitude V across its own terminals, and thus (because of the wires) across the terminals of the unspecified device, with a greater potential at terminal a of the device than at terminal b.

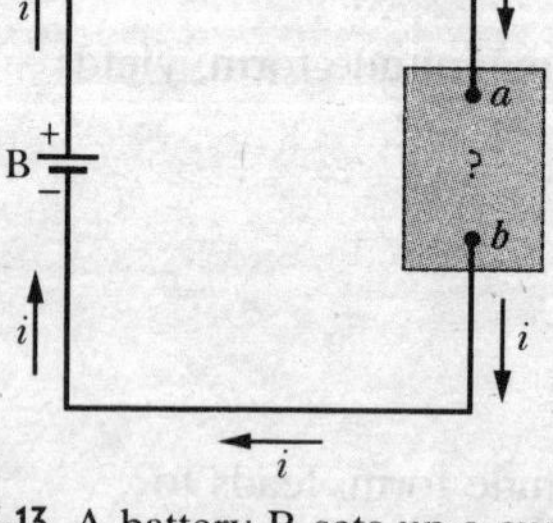

Fig. 27-13 A battery B sets up a current i in a circuit containing an unspecified conducting device.

Since there is an external conducting path between the two terminals of the battery, and since the potential differences set up by the battery are maintained, a steady current i is produced in the circuit, directed from terminal a to terminal b. The amount of charge dq that moves between those terminals in time interval dt is equal to $i\,dt$. This charge dq moves through a decrease in potential of magnitude V, and thus its electric potential energy decreases in magnitude by the amount

$$dU = dq\,V = i\,dt\,V.$$

The principle of conservation of energy tells us that the decrease in electric potential energy from a to b is accompanied by a transfer of energy to some other

The wire coils within a toaster have appreciable resistance. When there is a current through them, electric energy is transferred to thermal energy of the coils, increasing their temperature. The coils then emit infrared radiation and visible light that will toast (or burn) bread.

form. The power P associated with that transfer is the rate of transfer dU/dt, which is

$$P = iV \quad \text{(rate of electric energy transfer).} \tag{27-21}$$

Moreover, this power P is also the rate that energy is transferred from the battery to the unspecified device. If that device is a motor connected to a mechanical load, the energy is transferred as work done on the load. If the device is a storage battery that is being charged, the energy is transferred to stored chemical energy in the storage battery. If the device is a resistor, the energy is transferred to internal thermal energy, tending to increase the resistor's temperature.

The unit of power that follows from Eq. 27-21 is the volt-ampere (V · A). We can write it as

$$1 \text{ V} \cdot \text{A} = \left(1 \frac{\text{J}}{\text{C}}\right)\left(1 \frac{\text{C}}{\text{s}}\right) = 1 \frac{\text{J}}{\text{s}} = 1 \text{ W}.$$

The course of an electron moving through a resistor at constant drift speed is much like that of a stone falling through water at constant terminal speed. The average kinetic energy of the electron remains constant, and its lost electric potential energy appears as thermal energy in the resistor and the surroundings. On a microscopic scale this energy transfer is due to collisions between the electron and the molecules of the resistor, which leads to an increase in the temperature of the resistor lattice. The mechanical energy thus transferred to thermal energy is *dissipated* (lost), because the transfer cannot be reversed.

For a resistor or some other device with resistance R, we can combine Eqs. 27-8 ($R = V/i$) and 27-21 to obtain, for the rate of electric energy dissipation due to a resistance, either

$$P = i^2R \quad \text{(resistive dissipation)} \tag{27-22}$$

or

$$P = \frac{V^2}{R} \quad \text{(resistive dissipation).} \tag{27-23}$$

Caution: We must be careful to distinguish these two equations from Eq. 27-21: $P = iV$ applies to electric energy transfers of all kinds; $P = i^2R$ and $P = V^2/R$ apply only to the transfer of electric potential energy to thermal energy in a device with resistance.

CHECKPOINT 5: A potential difference V is connected across a device with resistance R, causing current i through the device. Rank the following variations according to the change in the rate at which electric energy is converted to thermal energy due to the resistance, greatest change first: (a) V is doubled with R unchanged, (b) i is doubled with R unchanged, (c) R is doubled with V unchanged, (d) R is doubled with i unchanged.

Sample Problem 27-6

You are given a length of uniform heating wire made of a nickel–chromium–iron alloy called Nichrome; it has a resistance R of 72 Ω. At what rate is energy dissipated in each of the following situations? (1) A potential difference of 120 V is applied across the full length of the wire. (2) The wire is cut in half, and a potential difference of 120 V is applied across the length of each half.

SOLUTION: The **Key Idea** is that a current in a resistive material produces a transfer of mechanical energy to thermal energy; the rate of transfer (dissipation) is given by Eqs. 27-21 to 27-23. Because we know the potential V and resistance R, we use Eq. 27-23, which yields, for situation 1,

$$P = \frac{V^2}{R} = \frac{(120 \text{ V})^2}{72\ \Omega} = 200 \text{ W}. \quad \text{(Answer)}$$

In situation 2, the resistance of each half of the wire is (72 Ω)/2, or 36 Ω. Thus, the dissipation rate for each half is

$$P' = \frac{(120\ \text{V})^2}{36\ \Omega} = 400\ \text{W},$$

and that for the two halves is

$$P = 2P' = 800\ \text{W}. \qquad \text{(Answer)}$$

This is four times the dissipation rate of the full length of wire. Thus, you might conclude that you could buy a heating coil, cut it in half, and reconnect it to obtain four times the heat output. Why is this unwise? (What would happen to the amount of current in the coil?)

27-8 Semiconductors

Semiconducting devices are at the heart of the microelectronic revolution that ushered in the information age. Table 27-2 compares the properties of silicon—a typical semiconductor—and copper—a typical metallic conductor. We see that silicon has many fewer charge carriers, a much higher resistivity, and a temperature coefficient of resistivity that is both large and negative. Thus, although the resistivity of copper increases with temperature, that of pure silicon decreases.

Pure silicon has such a high resistivity that it is effectively an insulator and thus not of much direct use in microelectronic circuits. However, its resistivity can be greatly reduced in a controlled way by adding minute amounts of specific "impurity" atoms in a process called *doping*. Table 27-1 gives typical values of resistivity for silicon before and after doping with two different impurities.

We can roughly explain the differences in resistivity (and thus in conductivity) between semiconductors, insulators, and metallic conductors in terms of the energies of their electrons. (We need quantum physics to explain in more detail.) In a metallic conductor such as copper wire, most of the electrons are firmly locked into place within the molecules; much energy would be required to free them so they could move and participate in an electric current. However, there are also some electrons that, roughly speaking, are only loosely held in place and that require only little energy to become free. Thermal energy can supply that energy, as can an electric field applied across the conductor. The field would not only free these loosely held electrons but would also propel them along the wire; thus, the field would drive a current through the conductor.

In an insulator, significantly greater energy is required to free electrons so they can move through the material. Thermal energy cannot supply enough energy, and neither can any reasonable electric field applied to the insulator. Thus, no electrons are available to move through the insulator, and hence no current occurs even with an applied electric field.

A semiconductor is like an insulator *except* that the energy required to free some electrons is not quite so great. More important, doping can supply electrons or positive charge carriers that are very loosely held within the material and thus are easy to get moving. Moreover, by controlling the doping of a semiconductor, we can

TABLE 27-2 Some Electrical Properties of Copper and Silicon[a]

Property	Copper	Silicon
Type of material	Metal	Semiconductor
Charge carrier density, m^{-3}	9×10^{28}	1×10^{16}
Resistivity, $\Omega \cdot m$	2×10^{-8}	3×10^{3}
Temperature coefficient of resistivity, K^{-1}	$+4 \times 10^{-3}$	-70×10^{-3}

[a] Rounded to one significant figure for easy comparison.

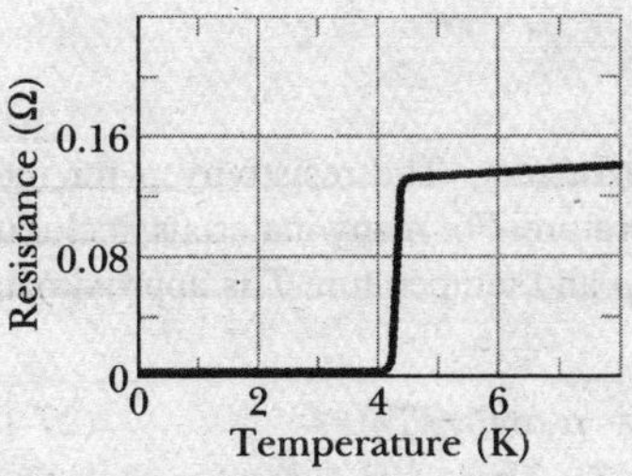

Fig. 27-14 The resistance of mercury drops to zero at a temperature of about 4 K.

control the density of charge carriers that can participate in a current, and thereby can control some of its electrical properties. Most semiconducting devices, such as transistors and junction diodes, are fabricated by the selective doping of different regions of the silicon with impurity atoms of different kinds.

Let us now look again at Eq. 27-20 for the resistivity of a conductor:

$$\rho = \frac{m}{e^2 n \tau}, \tag{27-24}$$

where n is the number of charge carriers per unit volume and τ is the mean time between collisions of the charge carriers. (We derived this equation for conductors, but it also applies to semiconductors.) Let us consider how the variables n and τ change as the temperature is increased.

In a conductor, n is large but very nearly constant with any change in temperature. The increase of resistivity with temperature for metals (Fig. 27-10) is due to an increase in the collision rate of the charge carriers, which shows up in Eq. 27-24 as a decrease in τ, the mean time between collisions.

In a semiconductor, n is small but increases very rapidly with temperature as the increased thermal agitation makes more charge carriers available. This causes a *decrease* of resistivity with increasing temperature, as indicated by the negative temperature coefficient of resistivity for silicon in Table 27-2. The same increase in collision rate that we noted for metals also occurs for semiconductors, but its effect is swamped by the rapid increase in the number of charge carriers.

27-9 Superconductors

In 1911, Dutch physicist Kamerlingh Onnes discovered that the resistivity of mercury absolutely disappears at temperatures below about 4 K (Fig. 27-14). This phenomenon of **superconductivity** is of vast potential importance in technology because it means that charge can flow through a superconducting conductor without producing thermal energy losses. Currents created in a superconducting ring, for example, have persisted for several years without diminution; the electrons making up the current require a force and a source of energy at start-up time, but not thereafter.

Prior to 1986, the technological development of superconductivity was throttled by the cost of producing the extremely low temperatures that were required to achieve the effect. In 1986, however, new ceramic materials were discovered that become superconducting at considerably higher (and thus cheaper to produce) temperatures. Practical application of superconducting devices at room temperature may eventually become feasible.

Superconductivity is a much different phenomenon from conductivity. In fact, the best of the normal conductors, such as silver and copper, cannot become superconducting at any temperature, and the new ceramic superconductors are actually good insulators when they are not at low enough temperatures to be in a superconducting state.

A disk-shaped magnet is levitated above a superconducting material that has been cooled by liquid nitrogen. The goldfish is along for the ride.

One explanation for superconductivity is that the electrons which make up the current move in coordinated pairs. One of the electrons in a pair may electrically distort the molecular structure of the superconducting material as it moves through, creating nearby a short-lived concentration of positive charge. The other electron in the pair may then be attracted toward this positive charge. According to the theory, such coordination between electrons would prevent them from colliding with the molecules of the material and thus would eliminate electrical resistance. The theory worked well to explain the pre-1986, lower temperature superconductors, but new theories appear to be needed for the newer, higher temperature superconductors.

REVIEW & SUMMARY

Current An **electric current** i in a conductor is defined by

$$i = \frac{dq}{dt}. \quad (27\text{-}1)$$

Here dq is the amount of (positive) charge that passes in time dt through a hypothetical surface that cuts across the conductor. By convention, the direction of electric current is taken as the direction in which positive charge carriers would move. The SI unit of electric current is the **ampere** (A): 1 A = 1 C/s.

Current Density Current (a scalar) is related to **current density** $\vec{J}$ (a vector) by

$$i = \int \vec{J} \cdot d\vec{A}, \quad (27\text{-}4)$$

where $d\vec{A}$ is a vector perpendicular to a surface element of area dA, and the integral is taken over any surface cutting across the conductor. $\vec{J}$ has the same direction as the velocity of the moving charges if they are positive and the opposite direction if they are negative.

Drift Speed of the Charge Carriers When an electric field $\vec{E}$ is established in a conductor, the charge carriers (assumed positive) acquire a **drift speed** v_d in the direction of $\vec{E}$; the velocity $\vec{v}_d$ is related to the current density by

$$\vec{J} = (ne)\vec{v}_d, \quad (27\text{-}7)$$

where ne is the carrier charge density.

Resistance of a Conductor The **resistance** R of a conductor is defined as

$$R = \frac{V}{i} \quad \text{(definition of } R\text{)}, \quad (27\text{-}8)$$

where V is the potential difference across the conductor and i is the current. The SI unit of resistance is the **ohm** (Ω): 1 Ω = 1 V/A. Similar equations define the **resistivity** ρ and **conductivity** σ of a material:

$$\rho = \frac{1}{\sigma} = \frac{E}{J} \quad \text{(definitions of } \rho \text{ and } \sigma\text{)}, \quad (27\text{-}12, 27\text{-}10)$$

where E is the magnitude of the applied electric field. The SI unit of resistivity is the ohm-meter ($\Omega \cdot$ m). Equation 27-10 corresponds to the vector equation

$$\vec{E} = \rho\vec{J}. \quad (27\text{-}11)$$

The resistance R of a conducting wire of length L and uniform cross section is

$$R = \rho\frac{L}{A}, \quad (27\text{-}16)$$

where A is the cross-sectional area.

Change of ρ with Temperature The resistivity ρ for most materials changes with temperature. For many materials, including metals, the relation between ρ and temperature T is approximated by the equation

$$\rho - \rho_0 = \rho_0\alpha(T - T_0). \quad (27\text{-}17)$$

Here T_0 is a reference temperature, ρ_0 is the resistivity at T_0, and α is the temperature coefficient of resistivity for the material.

Ohm's Law A given device (conductor, resistor, or any other electrical device) obeys *Ohm's law* if its resistance R, defined by Eq. 27-8 as V/i, is independent of the applied potential difference V. A given *material* obeys Ohm's law if its resistivity, defined by Eq. 27-10, is independent of the magnitude and direction of the applied electric field $\vec{E}$.

Resistivity of a Metal By assuming that the conduction electrons in a metal are free to move like the molecules of a gas, it is possible to derive an expression for the resistivity of a metal:

$$\rho = \frac{m}{e^2 n\tau}. \quad (27\text{-}20)$$

Here n is the number of free electrons per unit volume, and τ is the mean time between the collisions of an electron with the atoms of the metal. We can explain why metals obey Ohm's law by pointing out that τ is essentially independent of the magnitude E of any electric field applied to a metal.

Power The power P, or rate of energy transfer, in an electrical device across which a potential difference V is maintained is

$$P = iV \quad \text{(rate of electric energy transfer)}. \quad (27\text{-}21)$$

Resistive Dissipation If the device is a resistor, we can write Eq. 27-21 as

$$P = i^2R = \frac{V^2}{R} \quad \text{(resistive dissipation)}. \quad (27\text{-}22, 27\text{-}23)$$

In a resistor, electric potential energy is converted to internal thermal energy via collisions between charge carriers and atoms.

Semiconductors *Semiconductors* are materials with few conduction electrons but can become conductors when they are *doped* with other atoms that contribute free electrons.

Superconductors *Superconductors* are materials that lose all electrical resistance at low temperatures. Recent research has discovered materials that are superconducting at surprisingly high temperatures.

QUESTIONS

1. Figure 27-15 shows plots of the current i through a certain cross section of a wire over four different time periods. Rank the periods according to the net charge that passes through the cross section during each, greatest first.

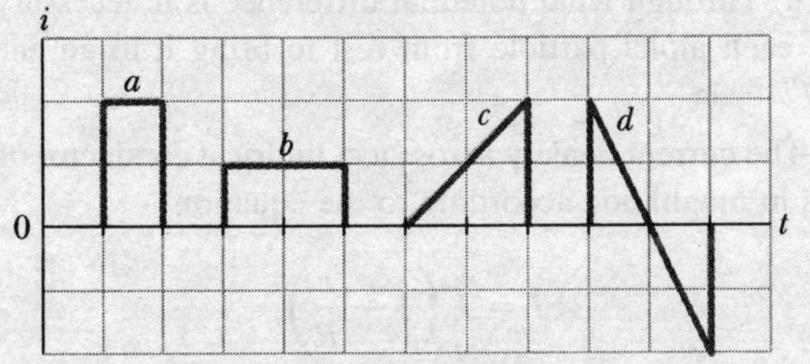

Fig. 27-15 Question 1.

2. Figure 27-16 shows four situations in which positive and negative charges move horizontally through a region and gives the rate at which each charge moves. Rank the situations according to the effective current through the regions, greatest first.

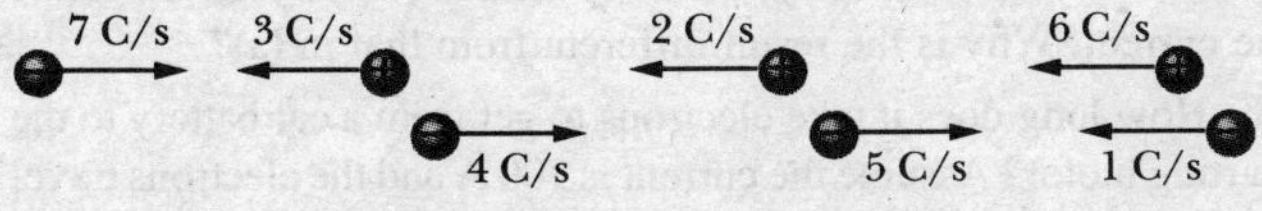

Fig. 27-16 Question 2.

3. Figure 27-17 shows cross sections through three wires of equal length and of the same material. The figure also gives the length of each side in millimeters. Rank the wires according to their resistances (measured end to end along each wire's length), greatest first.

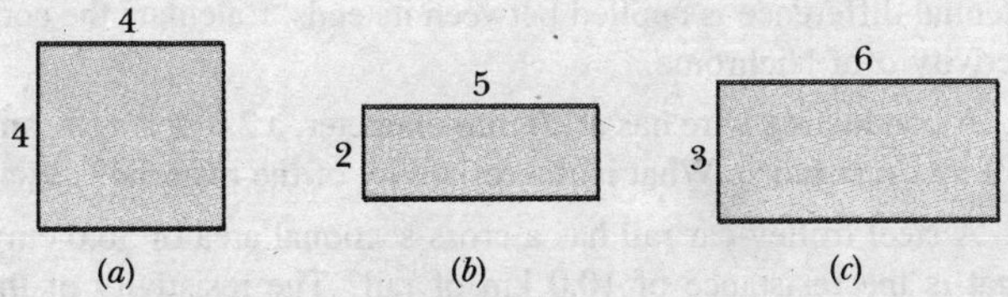

Fig. 27-17 Question 3.

4. If you stretch a cylindrical wire and it remains cylindrical, does the resistance of the wire (measured end to end along its length) increase, decrease, or remain the same?

5. Figure 27-18 shows cross sections through three long conductors of the same length and material, with square cross sections of edge lengths as shown. Conductor B will fit snugly within conductor A, and conductor C will fit snugly within conductor B. Rank the following according to their end-to-end resistances, greatest first: the individual conductors and the combinations of $A + B$, $B + C$, and $A + B + C$.

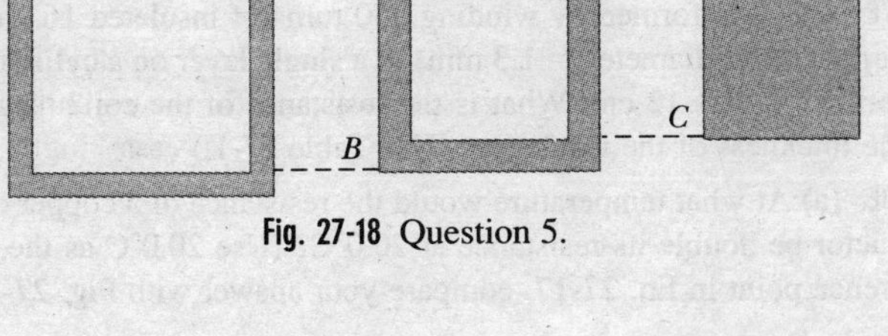

Fig. 27-18 Question 5.

6. Figure 27-19 shows a rectangular solid conductor of edge lengths L, $2L$, and $3L$. A certain potential difference V is to be applied between pairs of opposite faces of the conductor as in Fig. 27-8*b*: left–right, top–bottom, and front–back. Rank those pairs according to (a) the magnitude of the electric field within the conductor, (b) the current density within the conductor, (c) the current through the conductor, and (d) the drift speed of the electrons through the conductor, greatest first.

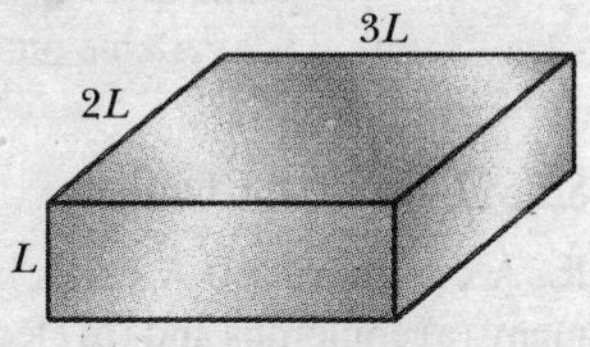

Fig. 27-19 Question 6.

7. The following table gives the lengths of three copper rods, their diameters, and the potential differences between their ends. Rank the rods according to (a) the magnitude of the electric field within them, (b) the current density within them, and (c) the drift speed of electrons through them, greatest first.

Rod	Length	Diameter	Potential Difference
1	L	$3d$	V
2	$2L$	d	$2V$
3	$3L$	$2d$	$2V$

8. The following table gives the conductivity and the density of conduction electrons for materials A, B, C, and D. Rank the materials according to the average time between collisions of the conduction electrons in the materials, greatest first.

	A	B	C	D
Conductivity	σ	2σ	2σ	σ
Electrons/m^3	n	$2n$	n	$2n$

9. Three wires, of the same diameter, are connected in turn between two points maintained at a constant potential difference. Their resistivities and lengths are ρ and L (wire A), 1.2ρ and $1.2L$ (wire B), and 0.9ρ and L (wire C). Rank the wires according to the rate at which energy is transferred to thermal energy within them, greatest first.

10. Figure 27-20 gives the resistivities of four materials as a function of temperature. (a) Which materials are conductors, and which are semiconductors? In which materials does an increase in temperature result in (b) an increase in the number of conduction electrons per unit volume and (c) an increase in the collision rate of conduction electrons?

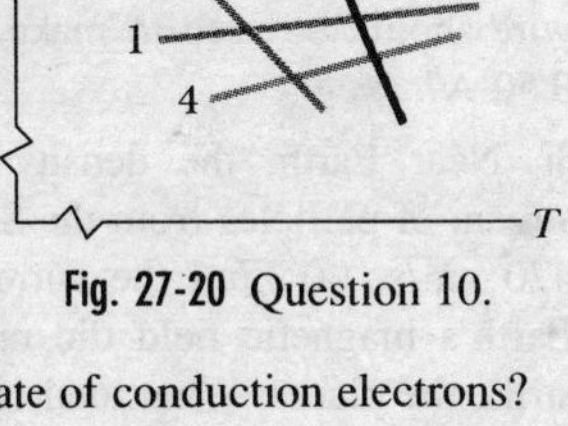

Fig. 27-20 Question 10.

EXERCISES & PROBLEMS

ssm Solution is in the Student Solutions Manual.
www Solution is available on the World Wide Web at: http://www.wiley.com/college/hrw
ilw Solution is available on the Interactive LearningWare.

SEC. 27-2 Electric Current

1E. A current of 5.0 A exists in a 10 Ω resistor for 4.0 min. How many (a) coulombs and (b) electrons pass through any cross section of the resistor in this time? ssm

2P. A charged belt, 50 cm wide, travels at 30 m/s between a source of charge and a sphere. The belt carries charge into the sphere at a rate corresponding to 100 μA. Compute the surface charge density on the belt.

3P. An isolated conducting sphere has a 10 cm radius. One wire carries a current of 1.000 002 0 A into it. Another wire carries a current of 1.000 000 0 A out of it. How long would it take for the sphere to increase in potential by 1000 V? ssm

SEC. 27-3 Current Density

4E. A small but measurable current of 1.2×10^{-10} A exists in a copper wire whose diameter is 2.5 mm. Assuming the current is uniform, calculate (a) the current density and (b) the electron drift speed. (See Sample Problem 27-3.)

5E. A beam contains 2.0×10^8 doubly charged positive ions per cubic centimeter, all of which are moving north with a speed of 1.0×10^5 m/s. (a) What are the magnitude and direction of the current density $\vec{J}$? (b) Can you calculate the total current i in this ion beam? If not, what additional information is needed? ssm

6E. The (United States) National Electric Code, which sets maximum safe currents for insulated copper wires of various diameters, is given (in part) in the table. Plot the safe current density as a function of diameter. Which wire gauge has the maximum safe current density? ("Gauge" is a way of identifying wire diameters, and 1 mil = 10^{-3} in.)

Gauge	4	6	8	10	12	14	16	18
Diameter, mils	204	162	129	102	81	64	51	40
Safe current, A	70	50	35	25	20	15	6	3

7E. A fuse in an electric circuit is a wire that is designed to melt, and thereby open the circuit, if the current exceeds a predetermined value. Suppose that the material to be used in a fuse melts when the current density rises to 440 A/cm^2. What diameter of cylindrical wire should be used to make a fuse that will limit the current to 0.50 A? ssm

8P. Near Earth, the density of protons in the solar wind (a stream of particles from the Sun) is 8.70 cm^{-3}, and their speed is 470 km/s. (a) Find the current density of these protons. (b) If Earth's magnetic field did not deflect them, the protons would strike the planet. What total current would Earth then receive?

9P. A steady beam of alpha particles ($q = +2e$) traveling with constant kinetic energy 20 MeV carries a current of 0.25 μA. (a) If the beam is directed perpendicular to a plane surface, how many alpha particles strike the surface in 3.0 s? (b) At any instant, how many alpha particles are there in a given 20 cm length of the beam? (c) Through what potential difference is it necessary to accelerate each alpha particle from rest to bring it to an energy of 20 MeV? ssm

10P. (a) The current density across a cylindrical conductor of radius R varies in magnitude according to the equation

$$J = J_0\left(1 - \frac{r}{R}\right),$$

where r is the distance from the central axis. Thus, the current density is a maximum J_0 at that axis ($r = 0$) and decreases linearly to zero at the surface ($r = R$). Calculate the current in terms of J_0 and the conductor's cross-sectional area $A = \pi R^2$. (b) Suppose that, instead, the current density is a maximum J_0 at the cylinder's surface and decreases linearly to zero at the axis: $J = J_0 r/R$. Calculate the current. Why is the result different from that in (a)?

11P. How long does it take electrons to get from a car battery to the starting motor? Assume the current is 300 A and the electrons travel through a copper wire with cross-sectional area 0.21 cm^2 and length 0.85 m. (See Sample Problem 27-3.) ilw

SEC. 27-4 Resistance and Resistivity

12E. A wire of Nichrome (a nickel–chromium–iron alloy commonly used in heating elements) is 1.0 m long and 1.0 mm^2 in cross-sectional area. It carries a current of 4.0 A when a 2.0 V potential difference is applied between its ends. Calculate the conductivity σ of Nichrome.

13E. A conducting wire has a 1.0 mm diameter, a 2.0 m length, and a 50 mΩ resistance. What is the resistivity of the material? ssm

14E. A steel trolley-car rail has a cross-sectional area of 56.0 cm^2. What is the resistance of 10.0 km of rail? The resistivity of the steel is 3.00×10^{-7} Ω · m.

15E. A human being can be electrocuted if a current as small as 50 mA passes near the heart. An electrician working with sweaty hands makes good contact with the two conductors he is holding, one in each hand. If his resistance is 2000 Ω, what might the fatal voltage be? ssm

16E. A wire 4.00 m long and 6.00 mm in diameter has a resistance of 15.0 mΩ. A potential difference of 23.0 V is applied between the ends. (a) What is the current in the wire? (b) What is the current density? (c) Calculate the resistivity of the wire material. Identify the material. (Use Table 27-1.)

17E. A coil is formed by winding 250 turns of insulated 16-gauge copper wire (diameter = 1.3 mm) in a single layer on a cylindrical form of radius 12 cm. What is the resistance of the coil? Neglect the thickness of the insulation. (Use Table 27-1.) ssm

18E. (a) At what temperature would the resistance of a copper conductor be double its resistance at 20.0°C? (Use 20.0°C as the reference point in Eq. 27-17; compare your answer with Fig. 27-10.)

(b) Does this same "doubling temperature" hold for all copper conductors, regardless of shape or size?

19E. A wire with a resistance of 6.0 Ω is drawn out through a die so that its new length is three times its original length. Find the resistance of the longer wire, assuming that the resistivity and density of the material are unchanged. ssm ilw

20E. A certain wire has a resistance R. What is the resistance of a second wire, made of the same material, that is half as long and has half the diameter?

21P. Two conductors are made of the same material and have the same length. Conductor A is a solid wire of diameter 1.0 mm. Conductor B is a hollow tube of outside diameter 2.0 mm and inside diameter 1.0 mm. What is the resistance ratio R_A/R_B, measured between their ends? ssm www

22P. An electrical cable consists of 125 strands of fine wire, each having 2.65 μΩ resistance. The same potential difference is applied between the ends of all the strands and results in a total current of 0.750 A. (a) What is the current in each strand? (b) What is the applied potential difference? (c) What is the resistance of the cable?

23P. When 115 V is applied across a wire that is 10 m long and has a 0.30 mm radius, the current density is 1.4×10^4 A/m^2. Find the resistivity of the wire. ssm

24P. A block in the shape of a rectangular solid has a cross-sectional area of 3.50 cm^2 across its width, a front-to-rear length of 15.8 cm, and a resistance of 935 Ω. The material of which the block is made has 5.33×10^{22} conduction electrons/m^3. A potential difference of 35.8 V is maintained between its front and rear faces. (a) What is the current in the block? (b) If the current density is uniform, what is its value? (c) What is the drift velocity of the conduction electrons? (d) What is the magnitude of the electric field in the block?

25P. A common flashlight bulb is rated at 0.30 A and 2.9 V (the values of the current and voltage under operating conditions). If the resistance of the bulb filament at room temperature (20°C) is 1.1 Ω, what is the temperature of the filament when the bulb is on? The filament is made of tungsten. ilw

26P. Earth's lower atmosphere contains negative and positive ions that are produced by radioactive elements in the soil and cosmic rays from space. In a certain region, the atmospheric electric field strength is 120 V/m, directed vertically down. This field causes singly charged positive ions, at a density of 620/cm^3, to drift downward and singly charged negative ions, at a density of 550/cm^3, to drift upward (Fig. 27-21). The measured conductivity of the air in that region is 2.70×10^{-14}/Ω · m. Calculate (a) the ion drift speed, assumed to be the same for positive and negative ions, and (b) the current density.

Fig. 27-21 Problem 26.

27P. When a metal rod is heated, not only its resistance but also its length and its cross-sectional area change. The relation $R = \rho L/A$ suggests that all three factors should be taken into account in measuring ρ at various temperatures. (a) If the temperature changes by 1.0 C°, what percentage changes in R, L, and A occur for a copper conductor? (b) The coefficient of linear expansion for copper is 1.7×10^{-5}/K. What conclusion do you draw? ssm www

28P. If the gauge number of a wire is increased by 6, the diameter is halved; if a gauge number is increased by 1, the diameter decreases by the factor $2^{1/6}$ (see the table in Exercise 6). Knowing this, and knowing that 1000 ft of 10-gauge copper wire has a resistance of approximately 1.00 Ω, estimate the resistance of 25 ft of 22-gauge copper wire.

29P. A resistor has the shape of a truncated right-circular cone (Fig. 27-22). The end radii are a and b, and the altitude is L. If the taper is small, we may assume that the current density is uniform across any cross section. (a) Calculate the resistance of this object. (b) Show that your answer reduces to $\rho(L/A)$ for the special case of zero taper (that is, for $a = b$). ssm

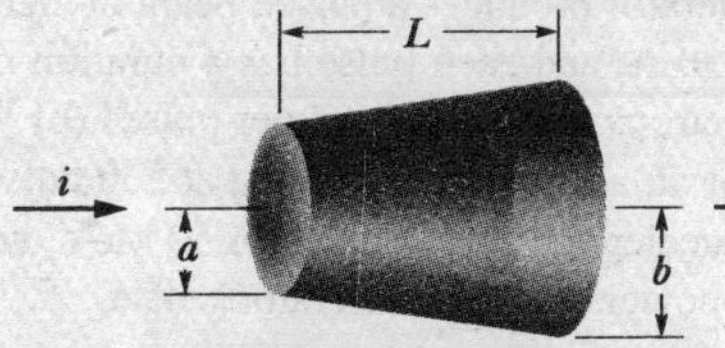

Fig. 27-22 Problem 29.

SEC. 27-6 A Microscopic View of Ohm's Law

30P. Show that, according to the free-electron model of electrical conduction in metals and classical physics, the resistivity of metals should be proportional to $\sqrt{T}$, where T is the temperature in kelvins. (See Eq. 20-31.)

SEC. 27-7 Power in Electric Circuits

31E. A certain x-ray tube operates at a current of 7.0 mA and a potential difference of 80 kV. What is its power in watts? ssm

32E. A student kept his 9.0 V, 7.0 W radio turned on at full volume from 9:00 p.m. until 2:00 a.m. How much charge went through it?

33E. A 120 V potential difference is applied to a space heater whose resistance is 14 Ω when hot. (a) At what rate is electric energy transferred to heat? (b) At 5.0¢/kW · h, what does it cost to operate the device for 5.0 h? ssm

34E. Thermal energy is produced in a resistor at a rate of 100 W when the current is 3.00 A. What is the resistance?

35E. An unknown resistor is connected between the terminals of a 3.00 V battery. Energy is dissipated in the resistor at the rate of 0.540 W. The same resistor is then connected between the terminals of a 1.50 V battery. At what rate is energy now dissipated? ilw

36E. A 120 V potential difference is applied to a space heater that dissipates 500 W during operation. (a) What is its resistance during operation? (b) At what rate do electrons flow through any cross section of the heater element?

37P. A 1250 W radiant heater is constructed to operate at 115 V. (a) What will be the current in the heater? (b) What is the resistance of the heating coil? (c) How much thermal energy is produced in 1.0 h by the heater? ssm ilw www

38P. A heating element is made by maintaining a potential difference of 75.0 V across the length of a Nichrome wire that has a $2.60 \times 10^{-6}\ \text{m}^2$ cross section. Nichrome has a resistivity of $5.00 \times 10^{-7}\ \Omega \cdot \text{m}$. (a) If the element dissipates 5000 W, what is its length? (b) If a potential difference of 100 V is used to obtain the same dissipation rate, what should the length be?

39P. A Nichrome heater dissipates 500 W when the applied potential difference is 110 V and the wire temperature is 800°C. What would be the dissipation rate if the wire temperature were held at 200°C by immersing the wire in a bath of cooling oil? The applied potential difference remains the same, and α for Nichrome at 800°C is 4.0×10^{-4}/K. ssm

40P. A 100 W lightbulb is plugged into a standard 120 V outlet. (a) How much does it cost per month to leave the light turned on continuously? Assume electric energy costs 6¢/kW · h. (b) What is the resistance of the bulb? (c) What is the current in the bulb? (d) Is the resistance different when the bulb is turned off?

41P. A linear accelerator produces a pulsed beam of electrons. The pulse current is 0.50 A, and each pulse has a duration of 0.10 μs. (a) How many electrons are accelerated per pulse? (b) What is the average current for an accelerator operating at 500 pulses/s? (c) If the electrons are accelerated to an energy of 50 MeV, what are the average and peak powers of the accelerator? ssm

42P. A cylindrical resistor of radius 5.0 mm and length 2.0 cm is made of material that has a resistivity of $3.5 \times 10^{-5}\ \Omega \cdot \text{m}$. What are (a) the current density and (b) the potential difference when the energy dissipation rate in the resistor is 1.0 W?

43P. A copper wire of cross-sectional area $2.0 \times 10^{-6}\ \text{m}^2$ and length 4.0 m has a current of 2.0 A uniformly distributed across that area. (a) What is the magnitude of the electric field along the wire? (b) How much electric energy is transferred to thermal energy in 30 min?

Additional Problems

44. *The chocolate crumb mystery.* This story begins with Problem 48 in Chapter 24 and continues through Chapters 25 and 26. The chocolate crumb powder moved to the silo through a pipe of radius R with uniform speed v and uniform charge density ρ. (a) Find an expression for the current i (the rate at which charge on the powder moved) through a perpendicular cross section of the pipe. (b) Evaluate i for the conditions at the factory: pipe radius $R = 5.0$ cm, speed $v = 2.0$ m/s, and charge density $\rho = 1.1 \times 10^{-3}\ \text{C/m}^3$.

If the powder were to flow through a change V in electric potential, its energy could be transferred to a spark at the rate $P = iV$. (c) Could there be such a transfer within the pipe due to the radial potential difference discussed in Problem 57 of Chapter 25?

As the powder flowed from the pipe into the silo, the electric potential of the powder changed. The magnitude of that change was at least equal to the radial potential difference within the pipe (as evaluated in Problem 57 of Chapter 25). (d) Assuming that value for the potential difference and using the current found in (b) above, find the rate at which energy could have been transferred from the powder to a spark as the powder exited the pipe. (e) If a spark did occur at the exit and lasted for 0.20 s (a reasonable expectation), how much energy would have been transferred to the spark?

Recall from Problem 48 in Chapter 24 that a minimum energy transfer of 150 mJ is needed to cause an explosion. (f) Where did the powder explosion most likely occur: in the powder cloud at the unloading bin (considered in Problem 48 of Chapter 26), within the pipe, or at the exit of the pipe into the silo?

45. *Heart attack or electrocution?* One morning a man walks barefooted away from a picnic onto moist ground near a tower that supports electric transmission lines. Suddenly he collapses. His relatives at the picnic table see him fall and, on reaching him a few seconds later, find that he is in ventricular fibrillation. The man dies before an emergency team can reach him with defibrillation equipment. Later the family files a lawsuit against the power company, claiming that the victim was electrocuted because of accidental current leakage from the tower. You are hired as part of the forensic team to investigate the death—was it due to heart attack or electrocution?

Investigation of the power company records reveals that there was indeed an *electrical fault* at the tower that morning—for about 1.0 s a current I leaked from a rod into the ground. Assume that the current spread uniformly (hemispherically) into the ground (Fig. 27-23). Let ρ be the resistivity of the ground and r be the distance from the rod. Find expressions for (a) the current density and (b) the electric field magnitude, both as functions of r. The lower end of the rod was spherical with a radius of b. (c) From your expression for the electric field magnitude, find an expression for the potential difference ΔV between the lower end of the rod and a point at distance r. Your investigation finds that $I = 100$ A, $\rho = 100\ \Omega \cdot \text{m}$, and $b = 1.0$ cm, and that the victim was located at $r = 10$ m. At the victim's location, what were (d) the current density, (e) the electric field magnitude, and (f) the potential difference ΔV? (This story continues with Problem 56 in Chapter 28.)

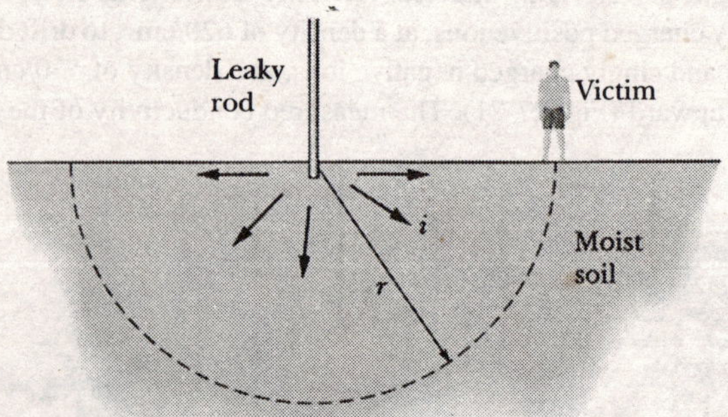

Fig. 27-23 Problem 45.

28 Circuits

The electric eel (*Electrophorus*) lurks in rivers of South America, killing the fish on which it preys with pulses of current. It does so by producing a potential difference of several hundred volts along its length; the resulting current in the surrounding water, from near the eel's head to the tail region, can be as much as one ampere. If you were to brush up against this eel while swimming, you might wonder (after recovering from the very painful stun):

How can the creature manage to produce a current that large without shocking itself?

The answer is in this chapter.

The world's largest battery, housed in Chino, California, has a power capability of 10 MW, which is put to use during peak power demands on the electric system served by Southern California Edison. Because the battery does work on charge carriers, it is an emf device.

28-1 "Pumping" Charges

If you want to make charge carriers flow through a resistor, you must establish a potential difference between the ends of the device. One way to do this is to connect each end of the resistor to one plate of a charged capacitor. The trouble with this scheme is that the flow of charge acts to discharge the capacitor, quickly bringing the plates to the same potential. When that happens, there is no longer an electric field in the resistor and thus the flow of charge stops.

To produce a steady flow of charge, you need a "charge pump," a device that—by doing work on the charge carriers—maintains a potential difference between a pair of terminals. We call such a device an **emf device,** and the device is said to provide an **emf** $\mathscr{E}$, which means that it does work on charge carriers. An emf device is sometimes called a *seat of emf.* The term *emf* comes from the outdated phrase *electromotive force,* which was adopted before scientists clearly understood the function of an emf device.

In Chapter 27, we discussed the motion of charge carriers through a circuit in terms of the electric field set up in the circuit—the field produces forces that move the charge carriers. In this chapter we take a different approach: We discuss the motion of the charge carriers in terms of the required energy—an emf device supplies the energy for the motion via the work it does.

A common emf device is the *battery,* used to power a wide variety of machines from wristwatches to submarines. The emf device that most influences our daily lives, however, is the *electric generator,* which, by means of electrical connections (wires) from a generating plant, creates a potential difference in our homes and workplaces. The emf devices known as *solar cells*, long familiar as the winglike panels on spacecraft, also dot the countryside for domestic applications. Less familiar emf devices are the *fuel cells* that power the space shuttles and the *thermopiles* that provide onboard electrical power for some spacecraft and for remote stations in Antarctica and elsewhere. An emf device does not have to be an instrument—living systems, ranging from electric eels and human beings to plants, have physiological emf devices.

Although the devices we have listed differ widely in their modes of operation, they all perform the same basic function—they do work on charge carriers and thus maintain a potential difference between their terminals.

28-2 Work, Energy, and Emf

Figure 28-1 shows an emf device (consider it to be a battery) that is part of a simple circuit containing a single resistance R (the symbol for resistance and a resistor is -WW-). The emf device keeps one of its terminals (called the positive terminal and often labeled +) at a higher electric potential than the other terminal (called the negative terminal and labeled −). We can represent the emf of the device with an arrow that points from the negative terminal toward the positive terminal as in Fig. 28-1. A small circle on the tail of the emf arrow distinguishes it from the arrows that indicate current direction.

When an emf device is not connected to a circuit, its internal chemistry does not cause any net flow of charge carriers within it. However, when it is connected to a circuit as in Fig. 28-1, its internal chemistry causes a net flow of positive charge carriers from the negative terminal to the positive terminal, in the direction of the emf arrow. This flow is part of the current that is set up around the circuit in that same direction (clockwise in Fig. 28-1).

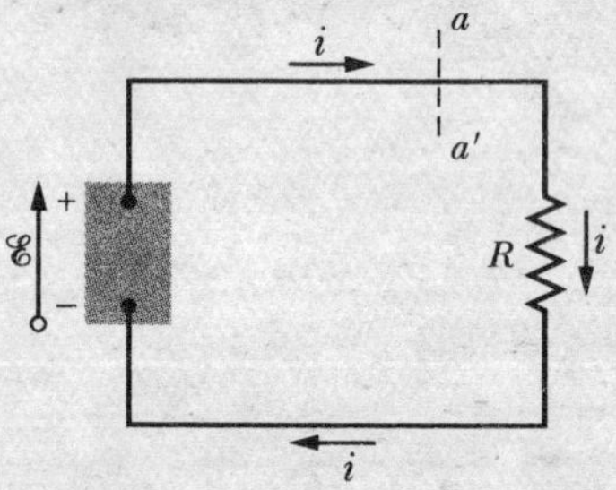

Fig. 28-1 A simple electric circuit, in which a device of emf $\mathscr{E}$ does work on the charge carriers and maintains a steady current i in a resistor of resistance R.

Within the emf device, positive charge carriers move from a region of low electric potential and thus low electric potential energy (at the negative terminal) to a region of higher electric potential and higher electric potential energy (at the positive terminal). This motion is just the opposite of what the electric field between the terminals (which is directed from the positive terminal toward the negative terminal) would cause the charge carriers to do.

Thus, there must be some source of energy within the device, enabling it to do work on the charges by forcing them to move as they do. The energy source may be chemical, as in a battery or a fuel cell. It may involve mechanical forces, as in an electric generator. Temperature differences may supply the energy, as in a thermopile; or the Sun may supply it, as in a solar cell.

Let us now analyze the circuit of Fig. 28-1 from the point of view of work and energy transfers. In any time interval dt, a charge dq passes through any cross section of this circuit, such as aa'. This same amount of charge must enter the emf device at its low-potential end and leave at its high-potential end. The device must do an amount of work dW on the charge dq to force it to move in this way. We define the emf of the emf device in terms of this work:

$$\mathscr{E} = \frac{dW}{dq} \quad \text{(definition of } \mathscr{E}\text{).} \tag{28-1}$$

In words, the emf of an emf device is the work per unit charge that the device does in moving charge from its low-potential terminal to its high-potential terminal. The SI unit for emf is the joule per coulomb; in Chapter 25 we defined that unit as the *volt*.

An **ideal emf device** is one that lacks any internal resistance to the internal movement of charge from terminal to terminal. The potential difference between the terminals of an ideal emf device is equal to the emf of the device. For example, an ideal battery with an emf of 12.0 V always has a potential difference of 12.0 V between its terminals.

A **real emf device,** such as any real battery, has internal resistance to the internal movement of charge. When a real emf device is not connected to a circuit, and thus does not have current through it, the potential difference between its terminals is equal to its emf. However, when that device has current through it, the potential difference between its terminals differs from its emf. We will discuss such real batteries in Section 28-4.

When an emf device is connected to a circuit, the device transfers energy to the charge carriers passing through it. This energy can then be transferred from the charge carriers to other devices in the circuit, for example, to light a bulb. Figure 28-2a shows a circuit containing two ideal rechargeable (*storage*) batteries A and B, a resistance R, and an electric motor M that can lift an object by using energy it obtains from charge carriers in the circuit. Note that the batteries are connected so that they tend to send charges around the circuit in opposite directions. The actual direction of the current in the circuit is determined by the battery with the larger emf, which happens to be battery B, so the chemical energy within battery B is decreasing as energy is transferred to the charge carriers passing through it. However, the chemical energy within battery A is increasing because the current in it is directed from the positive terminal to the negative terminal. Thus, battery B is charging battery A. Battery B is also providing energy to motor M and energy that is being dissipated by resistance R. Figure 28-2b shows all three energy transfers from battery B; each decreases that battery's chemical energy.

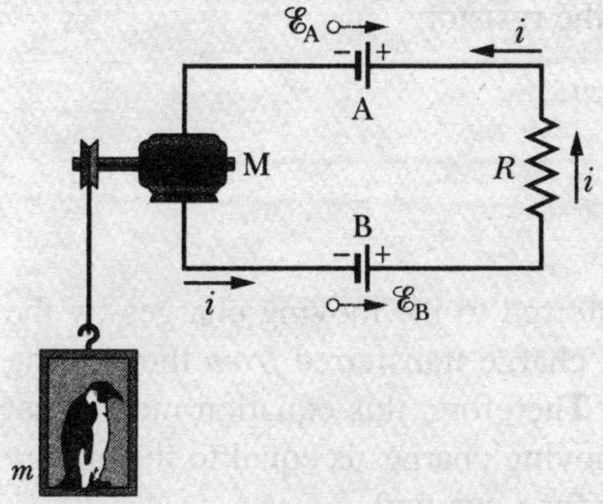

(a)

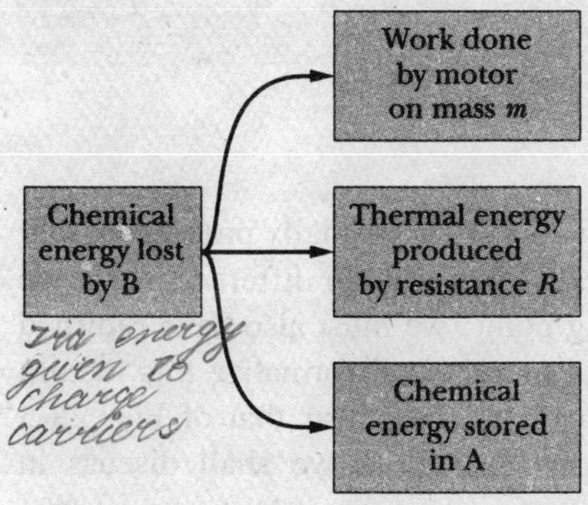

(b)

Fig. 28-2 (a) In the circuit, $\mathscr{E}_B > \mathscr{E}_A$; so battery B determines the direction of the current. (b) The energy transfers in the circuit, assuming that no dissipation occurs in the motor.

28-3 Calculating the Current in a Single-Loop Circuit

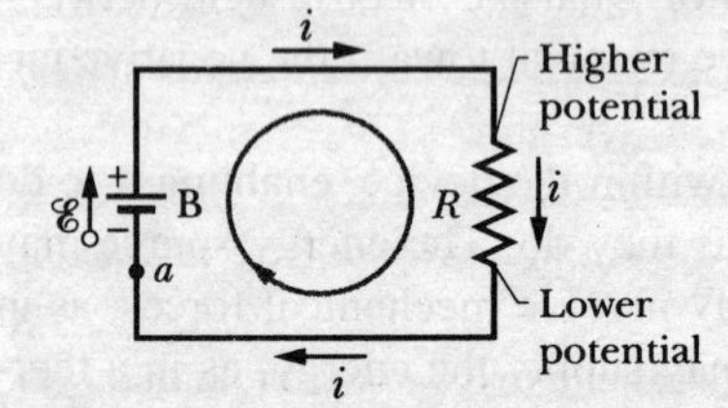

Fig. 28-3 A single-loop circuit in which a resistance R is connected across an ideal battery B with emf $\mathscr{E}$. The resulting current i is the same throughout the circuit.

We discuss here two equivalent ways to calculate the current in the simple *single-loop* circuit of Fig. 28-3; one method is based on energy conservation considerations, and the other on the concept of potential. The circuit consists of an ideal battery B with emf $\mathscr{E}$, a resistor of resistance R, and two connecting wires. (Unless otherwise indicated, we assume that wires in circuits have negligible resistance. Their function, then, is merely to provide pathways along which charge carriers can move.)

Energy Method

Equation 27-22 ($P = i^2R$) tells us that in a time interval dt an amount of energy given by $i^2R\,dt$ will appear in the resistor of Fig. 28-3 as thermal energy. (Since we assume the wires to have negligible resistance, no thermal energy will appear in them.) During the same interval, a charge $dq = i\,dt$ will have moved through battery B, and the work that the battery will have done on this charge, according to Eq. 28-1, is

$$dW = \mathscr{E}\,dq = \mathscr{E}i\,dt.$$

From the principle of conservation of energy, the work done by the (ideal) battery must equal the thermal energy that appears in the resistor:

$$\mathscr{E}i\,dt = i^2R\,dt.$$

This gives us

$$\mathscr{E} = iR.$$

The emf $\mathscr{E}$ is the energy per unit charge transferred to the moving charges by the battery. The quantity iR is the energy per unit charge transferred *from* the moving charges to thermal energy within the resistor. Therefore, this equation means that the energy per unit charge transferred to the moving charges is equal to the energy per unit charge transferred from them. Solving for i, we find

$$i = \frac{\mathscr{E}}{R}. \tag{28-2}$$

Potential Method

Suppose we start at any point in the circuit of Fig. 28-3 and mentally proceed around the circuit in either direction, adding algebraically the potential differences that we encounter. Then when we return to our starting point, we must also have returned to our starting potential. Before actually doing so, we shall formalize this idea in a statement that holds not only for single-loop circuits such as that of Fig. 28-3 but also for any complete loop in a *multiloop* circuit, as we shall discuss in Section 28-6:

> **LOOP RULE:** The algebraic sum of the changes in potential encountered in a complete traversal of any loop of a circuit must be zero.

This is often referred to as *Kirchhoff's loop rule* (or *Kirchhoff's voltage law*), after German physicist Gustav Robert Kirchhoff. This rule is equivalent to saying that each point on a mountain has only one elevation above sea level. If you start from any point and return to it after walking around the mountain, the algebraic sum of the changes in elevation that you encounter must be zero.

In Fig. 28-3, let us start at point a, whose potential is V_a, and mentally walk clockwise around the circuit until we are back at a, keeping track of potential changes as we move. Our starting point is at the low-potential terminal of the battery. Since the battery is ideal, the potential difference between its terminals is equal to $\mathscr{E}$. When we pass through the battery to the high-potential terminal, the change in potential is $+\mathscr{E}$.

As we walk along the top wire to the top end of the resistor, there is no potential change because the wire has negligible resistance; it is at the same potential as the high-potential terminal of the battery. So too is the top end of the resistor. When we pass through the resistor, however, the potential changes according to Eq. 27-8 (which we can rewrite as $V = iR$). Moreover, the potential must decrease because we are moving from the higher potential side of the resistor. Thus, the change in potential is $-iR$.

We return to point a by moving along the bottom wire. Since this wire also has negligible resistance, we again find no potential change. Back at point a, the potential is again V_a. Because we traversed a complete loop, our initial potential, as modified for potential changes along the way, must be equal to our final potential; that is,

$$V_a + \mathscr{E} - iR = V_a.$$

The value of V_a cancels from this equation, which becomes

$$\mathscr{E} - iR = 0.$$

Solving this equation for i gives us the same result, $i = \mathscr{E}/R$, as the energy method (Eq. 28-2).

If we apply the loop rule to a complete *counterclockwise* walk around the circuit, the rule gives us

$$-\mathscr{E} + iR = 0$$

and we again find that $i = \mathscr{E}/R$. Thus, you may mentally circle a loop in either direction to apply the loop rule.

To prepare for circuits more complex than that of Fig. 28-3, let us set down two rules for finding potential differences as we move around a loop:

RESISTANCE RULE: For a move through a resistance in the direction of the current, the change in potential is $-iR$; in the opposite direction it is $+iR$.

EMF RULE: For a move through an ideal emf device in the direction of the emf arrow, the change in potential is $+\mathscr{E}$; in the opposite direction it is $-\mathscr{E}$.

CHECKPOINT 1: The figure shows the current i in a single-loop circuit with a battery B and a resistance R (and wires of negligible resistance). (a) Should the emf arrow at B be drawn pointing leftward or rightward? At points a, b, and c, rank (b) the magnitude of the current, (c) the electric potential, and (d) the electric potential energy of the charge carriers, greatest first.

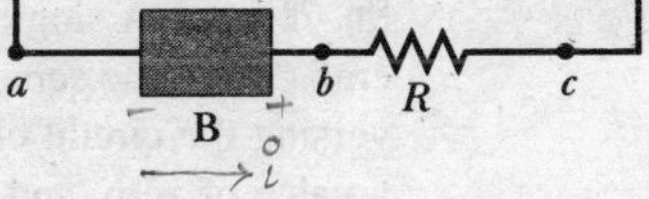

28-4 Other Single-Loop Circuits

In this section we extend the simple circuit of Fig. 28-3 in two ways.

Internal Resistance

Figure 28-4*a* shows a real battery, with internal resistance r, wired to an external resistor of resistance R. The internal resistance of the battery is the electrical resistance of the conducting materials of the battery and thus is an unremovable feature of the battery. In Fig. 28-4*a*, however, the battery is drawn as if it could be separated into an ideal battery with emf $\mathscr{E}$ and a resistor of resistance r. The order in which the symbols for these separated parts are drawn does not matter.

If we apply the loop rule clockwise beginning at point a, the *changes* in potential give us

$$\mathscr{E} - ir - iR = 0. \tag{28-3}$$

Solving for the current, we find

$$\boxed{i = \frac{\mathscr{E}}{R + r}}. \tag{28-4}$$

Note that this equation reduces to Eq. 28-2 if the battery is ideal—that is, if $r = 0$.

Figure 28-4*b* shows graphically the changes in electric potential around the circuit. (To better link Fig. 28-4*b* with the *closed circuit* in Fig. 28-4*a*, imagine curling the graph into a cylinder with point a at the left overlapping point a at the right.) Note how traversing the circuit is like walking around a (potential) mountain and returning to your starting point—you also return to the starting elevation.

In this book, when a battery is not described as real or if no internal resistance is indicated, you can generally assume that it is ideal—but, of course, in the real world batteries are always real and have internal resistance.

Resistances in Series

Figure 28-5*a* shows three resistances connected **in series** to an ideal battery with emf $\mathscr{E}$. This description has little to do with how the resistances are drawn. Rather, "in series" means that the resistances are wired one after another and that a potential difference V is applied across the two ends of the series. In Fig. 28-5*a*, the resistances are connected one after another between a and b, and a potential difference is main-

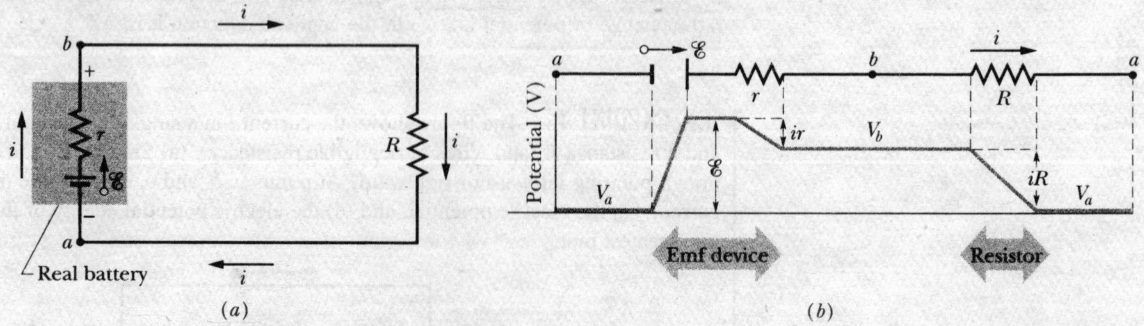

Fig. 28-4 (*a*) A single-loop circuit containing a real battery having internal resistance r and emf $\mathscr{E}$. (*b*) The same circuit, now spread out in a line. The potentials encountered in traversing the circuit clockwise from a are also shown. The potential V_a is arbitrarily assigned a value of zero, and other potentials in the circuit are graphed relative to V_a.

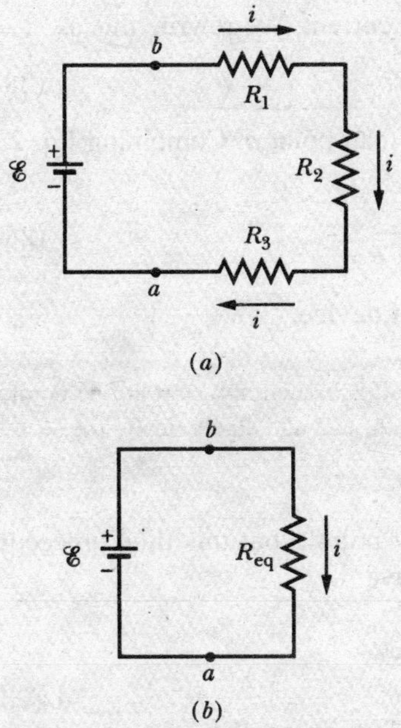

Fig. 28-5 (*a*) Three resistors are connected in series between points *a* and *b*. (*b*) An equivalent circuit, with the three resistors replaced with their equivalent resistance R_{eq}.

tained across *a* and *b* by the battery. The potential differences that then exist across the resistances in the series produce identical currents *i* in them. In general,

> When a potential difference *V* is applied across resistances connected in series, the resistances have identical currents *i*. The sum of the potential differences across the resistances is equal to the applied potential difference *V*.

Note that charge moving through the series resistances can move along only a single route. If there are additional routes, so that the currents in different resistances are different, the resistances are not connected in series.

> Resistances connected in series can be replaced with an equivalent resistance R_{eq} that has the same current *i* and the same *total* potential difference *V* as the actual resistances.

You might remember that R_{eq} and all the actual series resistances have the same current *i* with the nonsense word "ser-i." Figure 28-5*b* shows the equivalent resistance R_{eq} that can replace the three resistances of Fig. 28-5*a*.

To derive an expression for R_{eq} in Fig. 28-5*b*, we apply the loop rule to both circuits. For Fig. 28-5*a*, starting at terminal *a* and going clockwise around the circuit, we find

$$\mathscr{E} - iR_1 - iR_2 - iR_3 = 0,$$

or

$$i = \frac{\mathscr{E}}{R_1 + R_2 + R_3}. \tag{28-5}$$

For Fig. 28-5*b*, with the three resistances replaced with a single equivalent resistance R_{eq}, we find

$$\mathscr{E} - iR_{eq} = 0,$$

or

$$i = \frac{\mathscr{E}}{R_{eq}}. \tag{28-6}$$

Comparison of Eqs. 28-5 and 28-6 shows that

$$R_{eq} = R_1 + R_2 + R_3.$$

The extension to *n* resistances is straightforward and is

$$R_{eq} = \sum_{j=1}^{n} R_j \qquad (n \text{ resistances in series}). \tag{28-7}$$

Note that when resistances are in series, their equivalent resistance is greater than any of the individual resistances.

✔CHECKPOINT 2: In Fig. 28-5*a*, if $R_1 > R_2 > R_3$, rank the three resistances according to (a) the current through them and (b) the potential difference across them, greatest first.

all the

$V' = iR'$
$V_1 = iR_1$
$V_2 = iR_2$
$V_3 = iR_3$
$\because R_1 > R_2 > R_3$
$\Rightarrow V_1 > V_2 > V_3$

28-5 Potential Differences

We often want to find the potential difference between two points in a circuit. In Fig. 28-4*a*, for example, what is the potential difference between points *b* and *a*? To find out, let us start at point *b* and traverse the circuit clockwise to point *a*, passing through resistor *R*. If V_a and V_b are the potentials at *a* and *b*, respectively, we have

$$V_b - iR = V_a$$

because (according to our resistance rule) we experience a decrease in potential in going through a resistance in the direction of the current. We rewrite this as

$$V_b - V_a = +iR, \tag{28-8}$$

which tells us that point b is at greater potential than point a. Combining Eq. 28-8 with Eq. 28-4, we have

$$V_b - V_a = \mathscr{E}\frac{R}{R + r}, \tag{28-9}$$

where again r is the internal resistance of the emf device.

(6) ➤ **To find the potential difference between any two points in a circuit, start at one point and traverse the circuit to the other, following any path, and add algebraically the changes in potential that you encounter.**

Let us again calculate $V_b - V_a$, starting again at point b but this time proceeding counterclockwise to a through the battery. We have

$$V_b + ir - \mathscr{E} = V_a,$$

or

$$\Rightarrow V_b - V_a = \mathscr{E} - ir. \tag{28-10}$$

Combining this with Eq. 28-4 again leads to Eq. 28-9.

The quantity $V_b - V_a$ in Fig. 28-4 is the potential difference that the battery sets up across the battery terminals. As noted earlier, $V_b - V_a$ is equal to the emf $\mathscr{E}$ of the battery only if the battery has no internal resistance ($r = 0$ in Eq. 28-9) or if the circuit is open ($i = 0$ in Eq. 28-10).

Suppose that in Fig. 28-4, $\mathscr{E} = 12$ V, $R = 10\ \Omega$, and $r = 2.0\ \Omega$. Then Eq. 28-9 tells us that the potential across the battery's terminals is

$$V_b - V_a = (12\text{ V})\frac{10\ \Omega}{10\ \Omega + 2.0\ \Omega} = 10\text{ V}.$$

In "pumping" charge through itself, the battery does work per unit charge of $\mathscr{E}$ = 12 J/C, or 12 V. However, because of the internal resistance of the battery, it produces a potential difference of only 10 J/C, or 10 V, across its terminals.

Power, Potential, and Emf

When a battery or some other type of emf device does work on the charge carriers to establish a current i, it transfers energy from its source of energy (such as the chemical source in a battery) to the charge carriers. Because a real emf device has an internal resistance r, it also transfers energy to internal thermal energy via resistive dissipation, discussed in Section 27-7. Let us relate these transfers.

The net rate P of energy transfer from the emf device to the charge carriers is given by Eq. 27-21:

$$\boxed{P = iV,} \tag{28-11}$$

where V is the potential across the terminals of the emf device. From Eq. 28-10, we can substitute $V = \mathscr{E} - ir$ into Eq. 28-11 to find

$$P = i(\mathscr{E} - ir) = i\mathscr{E} - i^2r. \tag{28-12}$$

We see that the term i^2r in Eq. 28-12 is the rate P_r of energy transfer to thermal energy within the emf device:

$$\boxed{P_r = i^2r} \quad \text{(internal dissipation rate).} \tag{28-13}$$

Then the term $i\mathscr{E}$ in Eq. 28-12 must be the rate P_{emf} at which the emf device transfers energy to *both* the charge carriers and to internal thermal energy. Thus,

$$P_{emf} = i\mathscr{E} \quad \text{(power of emf device).} \tag{28-14}$$

If a battery is being *recharged*, with a "wrong way" current through it, the energy transfer is then *from* the charge carriers *to* the battery—both to the battery's chemical energy and to the energy dissipated in the internal resistance r. The rate of change of the chemical energy is given by Eq. 28-14, the rate of dissipation is given by Eq. 28-13, and the rate at which the carriers supply energy is given by Eq. 28-11.

Sample Problem 28-1

The emfs and resistances in the circuit of Fig. 28-6*a* have the following values:

$$\mathscr{E}_1 = 4.4\text{ V}, \quad \mathscr{E}_2 = 2.1\text{ V},$$

$$r_1 = 2.3\ \Omega, \quad r_2 = 1.8\ \Omega, \quad R = 5.5\ \Omega.$$

(a) What is the current i in the circuit?

SOLUTION: The **Key Idea** here is that we can get an expression involving the current i in this single-loop circuit by applying the loop rule. Although knowing the direction of i is not necessary, we can easily determine it from the emfs of the two batteries. Because $\mathscr{E}_1$ is greater than $\mathscr{E}_2$, battery 1 controls the direction of i, so the direction is clockwise. Let us then apply the loop rule by going counterclockwise—against the current—and starting at point a. We find

$$-\mathscr{E}_1 + ir_1 + iR + ir_2 + \mathscr{E}_2 = 0.$$

Check that this equation also results if we apply the loop rule clockwise or start at some point other than a. Also, take the time to compare this equation term by term with Fig. 28-6*b*, which shows the potential changes graphically (with the potential at point a arbitrarily taken to be zero).

Solving the above loop equation for the current i, we obtain

$$i = \frac{\mathscr{E}_1 - \mathscr{E}_2}{R + r_1 + r_2} = \frac{4.4\text{ V} - 2.1\text{ V}}{5.5\ \Omega + 2.3\ \Omega + 1.8\ \Omega}$$

$$= 0.2396\text{ A} \approx 240\text{ mA}. \qquad \text{(Answer)}$$

(b) What is the potential difference between the terminals of battery 1 in Fig. 28-6*a*?

SOLUTION: The **Key Idea** is to sum the potential differences between points a and b. Let us start at point b (effectively the negative terminal of battery 1) and travel clockwise through battery 1 to point a (effectively the positive terminal), keeping track of potential changes. We find that

$$V_b - ir_1 + \mathscr{E}_1 = V_a,$$

which gives us

$$V_a - V_b = -ir_1 + \mathscr{E}_1$$

$$= -(0.2396\text{ A})(2.3\ \Omega) + 4.4\text{ V}$$

$$= +3.84\text{ V} \approx 3.8\text{ V}, \qquad \text{(Answer)}$$

which is less than the emf of the battery. You can verify this result by starting at point b in Fig. 28-6*a* and traversing the circuit counterclockwise to point a.

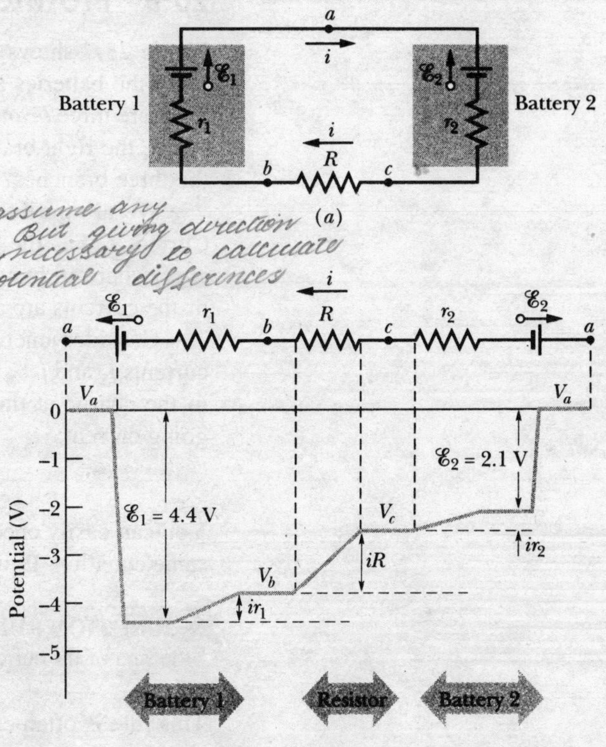

Fig. 28-6 Sample Problem 28-1. (*a*) A single-loop circuit containing two real batteries and a resistor. The batteries oppose each other; that is, they tend to send current in opposite directions through the resistor. (*b*) A graph of the potentials, counterclockwise from point a, with the potential at a arbitrarily taken to be zero. (To better link the circuit with the graph, mentally cut the circuit at a and then unfold the left side of the circuit toward the left and the right side of the circuit toward the right.)

CHECKPOINT 3: A battery has an emf of 12 V and an internal resistance of 2 Ω. Is the terminal-to-terminal potential difference greater than, less than, or equal to 12 V if the current in the battery is (a) from the negative to the positive terminal, (b) from the positive to the negative terminal, and (c) zero?

PROBLEM-SOLVING TACTICS

Tactic 1: *Assuming the Direction of a Current*

In solving circuit problems, you do not need to know the direction of a current in advance. Instead, you can just assume its direction, although that may take some physics courage. To show this, assume the current in Fig. 28-6*a* is counterclockwise; that is, reverse the direction of the current arrows shown. Applying the loop rule counterclockwise from point *a* now yields

$$-\mathscr{E}_1 - ir_1 - iR - ir_2 + \mathscr{E}_2 = 0,$$

or

$$i = -\frac{\mathscr{E}_1 - \mathscr{E}_2}{R + r_1 + r_2}.$$

Substituting the numerical values of Sample Problem 28-1 yields $i = -240$ mA for the current. The minus sign is a signal that the current is opposite the direction we initially assumed.

28-6 Multiloop Circuits

Figure 28-7 shows a circuit containing more than one loop. For simplicity, we assume the batteries are ideal. There are two *junctions* in this circuit, at *b* and *d*, and there are three *branches* connecting these junctions. The branches are the left branch (*bad*), the right branch (*bcd*), and the central branch (*bd*). What are the currents in the three branches?

We arbitrarily label the currents, using a different subscript for each branch. Current i_1 has the same value everywhere in branch *bad*, i_2 has the same value everywhere in branch *bcd*, and i_3 is the current through branch *bd*. The directions of the currents are assumed arbitrarily.

Consider junction *d* for a moment: Charge comes into that junction via incoming currents i_1 and i_3, and it leaves via outgoing current i_2. Because there is no variation in the charge at the junction, the total incoming current must equal the total outgoing current:

$$i_1 + i_3 = i_2. \qquad (28\text{-}15)$$

You can easily check that applying this condition to junction *b* leads to exactly the same equation. Equation 28-15 thus suggests a general principle:

> **JUNCTION RULE:** The sum of the currents entering any junction must be equal to the sum of the currents leaving that junction.

This rule is often called *Kirchhoff's junction rule* (or *Kirchhoff's current law*). It is simply a statement of the conservation of charge for a steady flow of charge—there is neither a build-up nor a depletion of charge at a junction. Thus, our basic tools for solving complex circuits are the *loop rule* (based on the conservation of energy) and the *junction rule* (based on the conservation of charge).

Equation 28-15 is a single equation involving three unknowns. To solve the circuit completely (that is, to find all three currents), we need two more equations involving those same unknowns. We obtain them by applying the loop rule twice. In the circuit of Fig. 28-7, we have three loops from which to choose: the left-hand loop (*badb*), the right-hand loop (*bcdb*), and the big loop (*badcb*). Which two loops we choose does not matter—let's choose the left-hand loop and the right-hand loop.

If we traverse the left-hand loop in a counterclockwise direction from point *b*, the loop rule gives us

$$\mathscr{E}_1 - i_1R_1 + i_3R_3 = 0. \qquad (28\text{-}16)$$

If we traverse the right-hand loop in a counterclockwise direction from point *b*, the loop rule gives us

$$-i_3R_3 - i_2R_2 - \mathscr{E}_2 = 0. \qquad (28\text{-}17)$$

We now have three equations (Eqs. 28-15, 28-16, and 28-17) in the three unknown currents, and they can be solved by a variety of techniques.

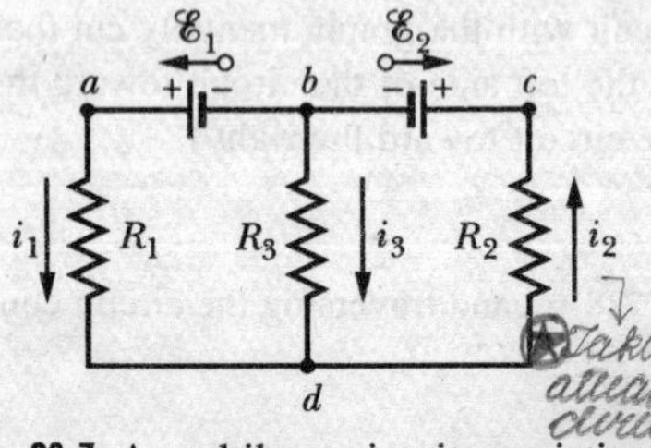

Fig. 28-7 A multiloop circuit consisting of three branches: left-hand branch *bad*, right-hand branch *bcd*, and central branch *bd*. The circuit also consists of three loops: left-hand loop *badb*, right-hand loop *bcdb*, and big loop *badcb*.

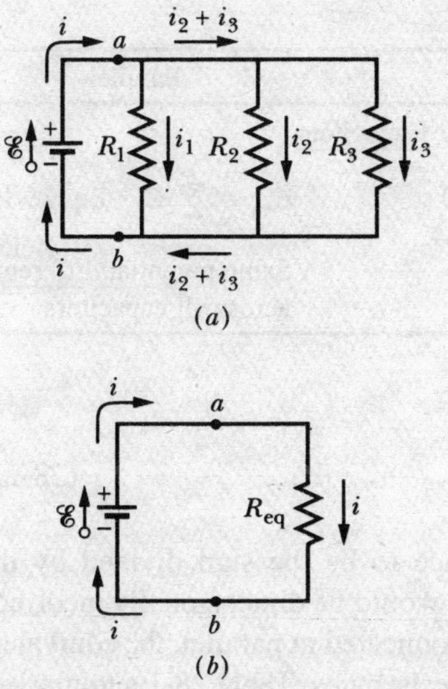

Fig. 28-8 (*a*) Three resistors connected in parallel across points *a* and *b*. (*b*) An equivalent circuit, with the three resistors replaced with their equivalent resistance R_{eq}.

Potential divides in series

PD across any resistor R′ in series

$V' = \left(\frac{R'}{Req}\right) V$

Req = Equivalent resistance of the series

V = Net PD across the series

Current divides in parallel

Current through any branch in parallel

$i' = i \left[\frac{\text{resistance of oppi branch}}{\text{total resistance}}\right]$

If we had applied the loop rule to the big loop, we would have obtained (moving counterclockwise from *b*) the equation

$$\mathscr{E}_1 - i_1R_1 - i_2R_2 - \mathscr{E}_2 = 0.$$

This equation may look like fresh information, but in fact it is only the sum of Eqs. 28-16 and 28-17. (It would, however, yield the proper results when used with Eq. 28-15 and either 28-16 or 28-17.)

Resistances in Parallel

Figure 28-8*a* shows three resistances connected *in parallel* to an ideal battery of emf $\mathscr{E}$. The term "in parallel" means that the resistances are directly wired together on one side and directly wired together on the other side, and that a potential difference V is applied across the pair of connected sides. Thus, all three resistances have the same potential difference V across them, producing a current through each. In general,

> When a potential difference V is applied across resistances connected in parallel, the resistances all have that same potential difference V.

In Fig. 28-8*a*, the applied potential difference V is maintained by the battery. In Fig. 28-8*b*, the three parallel resistances have been replaced with an equivalent resistance R_{eq}.

> Resistances connected in parallel can be replaced with an equivalent resistance R_{eq} that has the same potential difference V and the same *total* current i as the actual resistances.

You might remember that R_{eq} and all the actual parallel resistances have the same potential difference V with the nonsense word "par-V."

To derive an expression for R_{eq} in Fig. 28-8*b*, we first write the current in each actual resistance in Fig. 28-8*a* as

$$i_1 = \frac{V}{R_1}, \quad i_2 = \frac{V}{R_2}, \quad \text{and} \quad i_3 = \frac{V}{R_3},$$

where V is the potential difference between *a* and *b*. If we apply the junction rule at point *a* in Fig. 28-8*a* and then substitute these values, we find

$$i = i_1 + i_2 + i_3 = V\left(\frac{1}{R_1} + \frac{1}{R_2} + \frac{1}{R_3}\right). \tag{28-18}$$

If we replaced the parallel combination with the equivalent resistance R_{eq} (Fig. 28-8*b*), we would have

$$i = \frac{V}{R_{eq}}. \tag{28-19}$$

Comparing Eqs. 28-18 and 28-19 leads to

$$\frac{1}{R_{eq}} = \frac{1}{R_1} + \frac{1}{R_2} + \frac{1}{R_3}. \tag{28-20}$$

Extending this result to the case of n resistances, we have

$$\frac{1}{R_{eq}} = \sum_{j=1}^{n} \frac{1}{R_j} \quad (n \text{ resistances in parallel}). \tag{28-21}$$

For the case of two resistances, the equivalent resistance is their product divided by

TABLE 28-1 Series and Parallel Resistors and Capacitors

Series	Parallel	Series	Parallel
Resistors		Capacitors	
$R_{eq} = \sum_{j=1}^{n} R_j$ Eq. 28-7	$\frac{1}{R_{eq}} = \sum_{j=1}^{n} \frac{1}{R_j}$ Eq. 28-21	$\frac{1}{C_{eq}} = \sum_{j=1}^{n} \frac{1}{C_j}$ Eq. 26-20	$C_{eq} = \sum_{j=1}^{n} C_j$ Eq. 26-19
Same current through all resistors	Same potential difference across all resistors	Same charge on all capacitors	Same potential difference across all capacitors

their sum; that is,

$$R_{eq} = \frac{R_1 R_2}{R_1 + R_2}. \tag{28-22}$$

If you accidentally took the equivalent resistance to be the sum divided by the product, you would notice at once that this result would be dimensionally incorrect.

Note that when two or more resistances are connected in parallel, the equivalent resistance is smaller than any of the combining resistances. Table 28-1 summarizes the equivalence relations for resistors and capacitors in series and in parallel.

CHECKPOINT 4: A battery, with potential V across it, is connected to a combination of two identical resistors and then has current i through it. What are the potential difference across and the current through either resistor if the resistors are (a) in series and (b) in parallel?

Sample Problem 28-2

Figure 28-9*a* shows a multiloop circuit containing one ideal battery and four resistances with the following values:

$$R_1 = 20\ \Omega,\quad R_2 = 20\ \Omega,\quad \mathscr{E} = 12\ \text{V},$$
$$R_3 = 30\ \Omega,\quad R_4 = 8.0\ \Omega.$$

(a) What is the current through the battery?

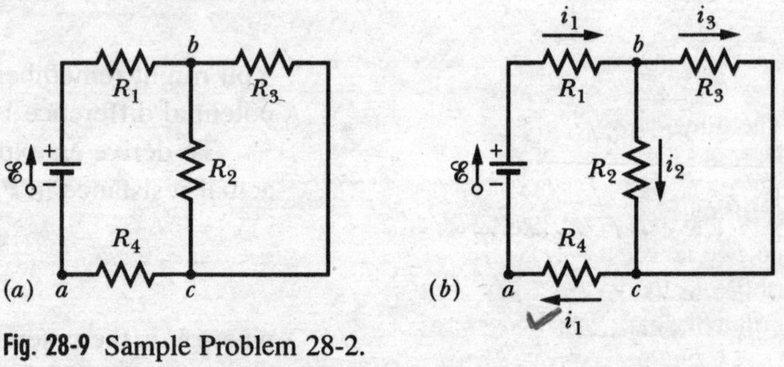

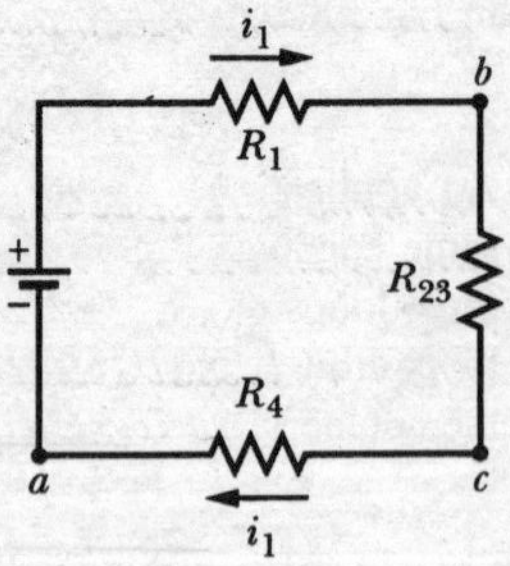

Fig. 28-9 Sample Problem 28-2. (*a*) A multiloop circuit with an ideal battery of emf $\mathscr{E}$ and four resistances. (*b*) Assumed currents through the resistances. (*c*) A simplification of the circuit, with resistances R_2 and R_3 replaced with their equivalent resistance R_{23}. The current through R_{23} is equal to that through R_1 and R_4.

SOLUTION: First note that the current through the battery must also be the current through R_1. Thus, one **Key Idea** here is that we might find that current by applying the loop rule to a loop that includes R_1 because the current would be included in the potential difference across R_1. Either the left-hand loop or the big loop will do. Noting that the emf arrow of the battery points upward, so the current the battery supplies is clockwise, we might apply the loop rule to the left-hand loop, clockwise from point a. With i being the current through the battery, we would get

$$+\mathscr{E} - iR_1 - iR_2 - iR_4 = 0 \qquad \text{(incorrect).}$$

However, this equation is incorrect because it assumes that R_1, R_2, and R_4 all have the same current i. Resistances R_1 and R_4 do have the same current, because the current passing through R_4 must pass through the battery and then through R_1 with no change in value. However, that current splits at junction point b—only part passes through R_2, the rest through R_3.

To distinguish the several currents in the circuit, we must label them individually as in Fig. 28-9*b*. Then, circling clockwise from a, we can write the loop rule for the left-hand loop as

$$+\mathscr{E} - i_1R_1 - i_2R_2 - i_1R_4 = 0.$$

Unfortunately, this equation contains two unknowns, i_1 and i_2; we would need at least one more equation to find them.

A second **Key Idea** is that a much easier option is to simplify the circuit of Fig. 28-9*b* by finding equivalent resistances. Note carefully that R_1 and R_2 are *not* in series and thus cannot be replaced with an equivalent resistance. However, R_2 and R_3 are in parallel, so we can use either Eq. 28-21 or Eq. 28-22 to find their equivalent resistance R_{23}. From the latter,

$$R_{23} = \frac{R_2R_3}{R_2 + R_3} = \frac{(20\ \Omega)(30\ \Omega)}{50\ \Omega} = 12\ \Omega.$$

We can now redraw the circuit as in Fig. 28-9*c*; note that the current through R_{23} must be i_1 because charge that moves through R_1 and R_4 must also move through R_{23}. For this simple one-loop circuit, the loop rule (applied clockwise from point *a*) yields

$$+\mathscr{E} - i_1R_1 - i_1R_{23} - i_1R_4 = 0.$$

Substituting the given data, we find

$$12\text{ V} - i_1(20\ \Omega) - i_1(12\ \Omega) - i_1(8.0\ \Omega) = 0,$$

which gives us

$$i_1 = \frac{12\text{ V}}{40\ \Omega} = 0.30\text{ A}. \qquad \text{(Answer)}$$

(b) What is the current i_2 through R_2?

SOLUTION: One **Key Idea** here is that we must work backward from the equivalent circuit of Fig. 28-9*c*, where R_{23} has replaced the parallel resistances R_2 and R_3. A second **Key Idea** is that, because R_2 and R_3 are in parallel, they both have the same potential difference across them as their equivalent R_{23}. We know the current through R_{23} is $i_1 = 0.30$ A. Thus, we can use Eq. 27-8 ($R = V/i$) to find the potential difference V_{23} across R_{23}:

$$V_{23} = i_1R_{23} = (0.30\text{ A})(12\ \Omega) = 3.6\text{ V}.$$

NOTE: PD across $R_{23} \neq \mathscr{E}$

The potential difference across R_2 is thus 3.6 V, so the current i_2 in R_2 must be, by Eq. 27-8,

$$i_2 = \frac{V_2}{R_2} = \frac{3.6\text{ V}}{20\ \Omega} = 0.18\text{ A}. \qquad \text{(Answer)}$$

(c) What is the current i_3 through R_3?

$i_3 = \frac{V_2}{R_3}$ ⇒

SOLUTION: We can answer by using the same technique as in (b), or we can use this **Key Idea**: The junction rule tells us that at point *b* in Fig. 28-9*b*, the incoming current i_1 and the outgoing currents i_2 and i_3 are related by

$$i_1 = i_2 + i_3.$$

This gives us

$$i_3 = i_1 - i_2 = 0.30\text{ A} - 0.18\text{ A}$$
$$= 0.12\text{ A}. \qquad \text{(Answer)}$$

Sample Problem 28-3

Figure 28-10 shows a circuit whose elements have the following values:

$$\mathscr{E}_1 = 3.0\text{ V}, \quad \mathscr{E}_2 = 6.0\text{ V},$$

$$R_1 = 2.0\ \Omega, \quad R_2 = 4.0\ \Omega.$$

The three batteries are ideal batteries. Find the magnitude and direction of the current in each of the three branches.

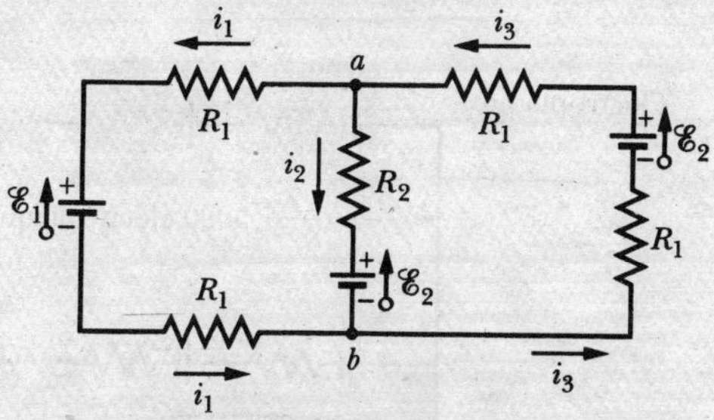

Fig. 28-10 Sample Problem 28-3. A multiloop circuit with three ideal batteries and five resistances.

SOLUTION: It is not worthwhile to try to simplify this circuit, because no two resistors are in parallel, and the resistors that are in series (those in the right branch or those in the left branch) present no problem. So, our **Key Idea** is to apply the junction and loop rules.

Using arbitrarily chosen directions for the currents as shown in Fig. 28-10, we apply the junction rule at point *a* by writing

$$i_3 = i_1 + i_2. \qquad (28\text{-}23)$$

An application of the junction rule at junction *b* gives only the same equation, so we next apply the loop rule to any two of the three loops of the circuit. We first arbitrarily choose the left-hand loop, arbitrarily start at point *a*, and arbitrarily traverse the loop in the counterclockwise direction, obtaining

$$-i_1R_1 - \mathscr{E}_1 - i_1R_1 + \mathscr{E}_2 + i_2R_2 = 0.$$

Substituting the given data and simplifying yield

$$i_1(4.0\ \Omega) - i_2(4.0\ \Omega) = 3.0\text{ V}. \qquad (28\text{-}24)$$

For our second application of the loop rule, we arbitrarily choose to traverse the right-hand loop clockwise from point *a*, finding

$$+i_3R_1 - \mathscr{E}_2 + i_3R_1 + \mathscr{E}_2 + i_2R_2 = 0.$$

Substituting the given data and simplifying yield

$$i_2(4.0\ \Omega) + i_3(4.0\ \Omega) = 0. \qquad (28\text{-}25)$$

$$y = \frac{\begin{vmatrix} a_1 & c_1 \\ a_2 & c_2 \end{vmatrix}}{\begin{vmatrix} a_1 & b_1 \\ a_2 & b_2 \end{vmatrix}} \qquad x = \frac{\begin{vmatrix} c_1 & b_1 \\ c_2 & b_2 \end{vmatrix}}{\begin{vmatrix} a_1 & b_1 \\ a_2 & b_2 \end{vmatrix}}$$

Using Eq. 28-23 to eliminate i_3 from Eq. 28-25 and simplifying give us

$$i_1(4.0\ \Omega) + i_2(8.0\ \Omega) = 0. \qquad (28\text{-}26)$$

We now have a system of two equations (Eqs. 28-24 and 28-26) in two unknowns (i_1 and i_2) to solve either "by hand" (which is easy enough here) or with a "math package." (One solution technique is Cramer's rule, given in Appendix E.) We find

$$i_2 = -0.25\text{ A}.$$

(The minus sign signals that our arbitrary choice of direction for i_2 in Fig. 28-10 is wrong; i_2 should point up through $\mathscr{E}_2$ and R_2.) Substituting $i_2 = -0.25$ A into Eq. 28-26 and solving for i_1 then give us

we include − sign

$$i_1 = 0.50\text{ A}. \qquad \text{(Answer)}$$

With Eq. 28-23 we then find that

$$i_3 = i_1 + i_2 = 0.25\text{ A}. \qquad \text{(Answer)}$$

The positive answers we obtained for i_1 and i_3 signal that our choices of directions for these currents are correct. We can now correct the direction for i_2 and write its magnitude as

$$i_2 = 0.25\text{ A}. \qquad \text{(Answer)}$$

Two simultaneous eqs in unknowns x & y can be solved by Cramer's rule

$a_1x + b_1y = c_1$
$a_2x + b_2y = c_2$

Sample Problem 28-4

Electric fish are able to generate current with biological cells called *electroplaques,* which are physiological emf devices. The electroplaques in the South American eel shown in the photograph that opens this chapter are arranged in 140 rows, each row stretching horizontally along the body and each containing 5000 electroplaques. The arrangement is suggested in Fig. 28-11*a*; each electroplaque has an emf $\mathscr{E}$ of 0.15 V and an internal resistance r of 0.25 Ω. The water surrounding the eel completes a circuit between the two ends of the electroplaque array, one end at the animal's head and the other near its tail.

(a) If the water surrounding the eel has resistance $R_w = 800\ \Omega$, how much current can the eel produce in the water?

SOLUTION: The **Key Idea** here is that we can simplify the circuit of Fig. 28-11*a* by replacing combinations of emfs and internal resistances with equivalent emfs and resistances. We first consider a single row. The total emf $\mathscr{E}_{row}$ along a row of 5000 electroplaques is the sum of the emfs:

$$\mathscr{E}_{row} = 5000\mathscr{E} = (5000)(0.15\text{ V}) = 750\text{ V}.$$

The total resistance R_{row} along a row is the sum of the internal resistances of the 5000 electroplaques:

$$R_{row} = 5000r = (5000)(0.25\ \Omega) = 1250\ \Omega.$$

We can now represent each of the 140 identical rows as having a single emf $\mathscr{E}_{row}$ and a single resistance R_{row}, as shown in Fig. 28-11*b*.

In Fig. 28-11*b*, the emf between point a and point b on any row is $\mathscr{E}_{row} = 750$ V. Because the rows are identical and because they are all connected together at the left in Fig. 28-11*b*, all points b in that figure are at the same electric potential. Thus, we can consider them to be connected so that there is only a single point b. The emf between point a and this single point b is $\mathscr{E}_{row} = 750$ V, so we can draw the circuit as shown in Fig. 28-11*c*.

Between points b and c in Fig. 28-11*c* are 140 resistances $R_{row} = 1250\ \Omega$, all in parallel. The equivalent resistance R_{eq} of this combination is given by Eq. 28-21 as

$$\frac{1}{R_{eq}} = \sum_{j=1}^{140} \frac{1}{R_j} = 140\frac{1}{R_{row}},$$

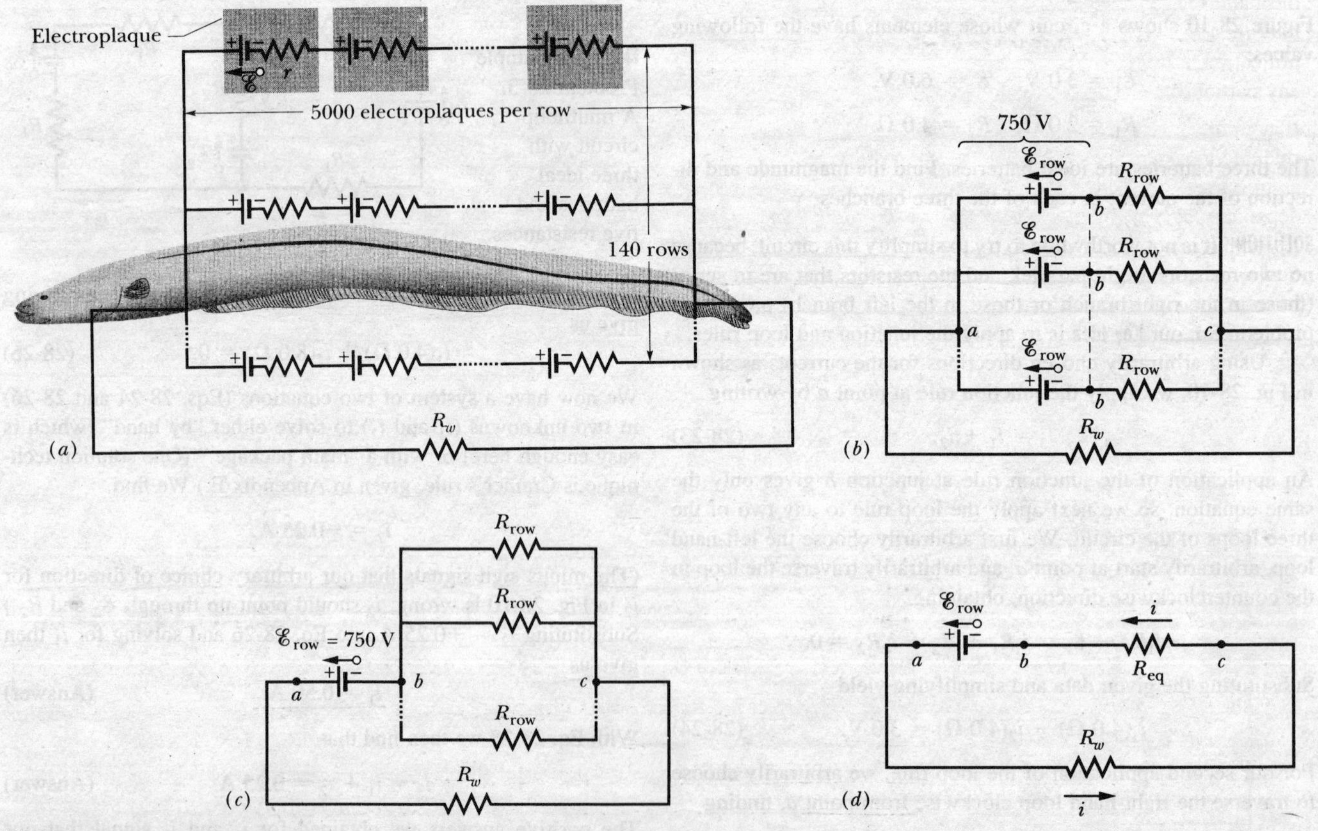

Fig. 28-11 Sample Problem 28-4. (*a*) A model of the electric circuit of an eel in water. Each electroplaque of the eel has an emf $\mathscr{E}$ and internal resistance r. Along each of 140 rows extending from the head to the tail of the eel, there are 5000 electroplaques. The surrounding water has resistance R_w. (*b*) The emf $\mathscr{E}_{row}$ and resistance R_{row} of each row. (*c*) The emf between points a and b is $\mathscr{E}_{row}$. Between points b and c are 140 parallel resistances R_{row}. (*d*) The simplified circuit, with R_{eq} replacing the parallel combination.

or
$$R_{eq} = \frac{R_{row}}{140} = \frac{1250\ \Omega}{140} = 8.93\ \Omega.$$

Replacing the parallel combination with R_{eq}, we obtain the simplified circuit of Fig. 28-11*d*. Applying the loop rule to this circuit counterclockwise from point *b*, we have

$$\mathscr{E}_{row} - iR_w - iR_{eq} = 0.$$

Solving for *i* and substituting the known data, we find

$$i = \frac{\mathscr{E}_{row}}{R_w + R_{eq}} = \frac{750\ \text{V}}{800\ \Omega + 8.93\ \Omega}$$
$$= 0.927\ \text{A} \approx 0.93\ \text{A}. \qquad \text{(Answer)}$$

If the head or tail of the eel is near a fish, some of this current could pass along a narrow path through the fish, stunning or killing it.

(b) How much current i_{row} travels through each row of Fig. 28-11*a*?

SOLUTION: The **Key Idea** here is that since the rows are identical, the current into and out of the eel is evenly divided among them:

$$i_{row} = \frac{i}{140} = \frac{0.927\ \text{A}}{140} = 6.6 \times 10^{-3}\ \text{A}. \qquad \text{(Answer)}$$

Thus, the current through each row is small, about two orders of magnitude smaller than the current through the water. This tends to spread the current through the eel's body, so that it need not stun or kill itself when it stuns or kills a fish.

PROBLEM-SOLVING TACTICS

Tactic 2: *Solving Circuits of Batteries and Resistors*

Here are two general techniques for solving circuits for unknown currents or potential differences.

1. If a circuit can be simplified by replacing resistors in series or in parallel with their equivalents, do so. If you can reduce the circuit to a single loop, then you can find the current through the battery with that loop, as in Sample Problem 28-2a. You may then have to "work backward," undoing the resistor simplification process, to find the current or potential difference for any particular resistor, as in Sample Problem 28-2b.
2. If a circuit cannot be simplified to a single loop, use the junction rule and the loop rule to write a set of simultaneous equations, as in Sample Problem 28-3. You need have only as many independent equations as there are unknowns in those equations. If you have to find the current or potential difference for a particular resistor, you can ensure that its current or potential difference appears in the equations by having at least one of the loops pass through it.

Tactic 3: *Arbitrary Choices in Solving Circuit Problems*

In Sample Problem 28-3, we made several arbitrary choices. (1) We assumed directions for the currents in Fig. 28-10 arbitrarily. (2) We chose which of the three possible loops to write equations for arbitrarily. (3) We chose the direction in which to traverse each loop arbitrarily. (4) We chose the starting and ending point for each traversal arbitrarily.

Such arbitrariness often worries a beginning circuit solver, but an experienced circuit solver knows that it does not matter. Just keep two rules firmly in mind. First, make sure you traverse each chosen loop completely. Second, once you have chosen a direction for a current, stick with it until you get numerical values for all the currents. If you were wrong about a direction, the algebra will signal you with a minus sign. Then you can make a correction by simply erasing the minus sign and reversing the arrow representing that current in the circuit diagram. However, *you should not make this correction until you have completed all the required calculations for the circuit*, as we did in Sample Problem 28-3.

28-7 The Ammeter and the Voltmeter

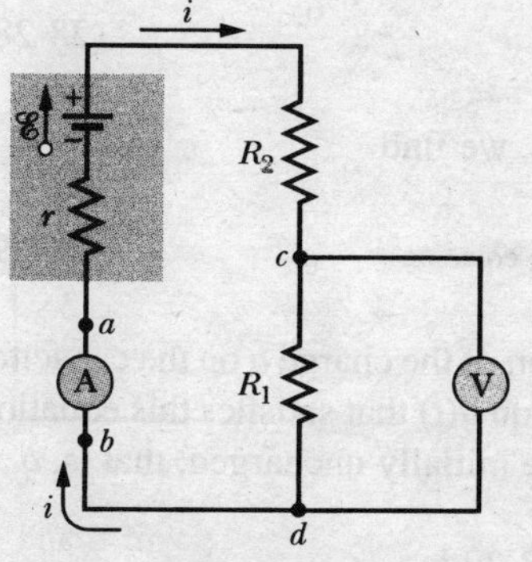

Fig. 28-12 A single-loop circuit, showing how to connect an ammeter (A) and a voltmeter (V).

An instrument used to measure currents is called an *ammeter*. To measure the current in a wire, you usually have to break or cut the wire and insert the ammeter so that the current to be measured passes through the meter. (In Fig. 28-12, ammeter A is set up to measure current *i*.)

It is essential that the resistance R_A of the ammeter be very small compared to other resistances in the circuit. Otherwise, the very presence of the meter will change the current to be measured.

A meter used to measure potential differences is called a *voltmeter*. To find the potential difference between any two points in the circuit, the voltmeter terminals are connected between those points, without breaking or cutting the wire. (In Fig. 28-12, voltmeter V is set up to measure the voltage across R_1.)

It is essential that the resistance R_V of a voltmeter be very large compared to the resistance of any circuit element across which the voltmeter is connected. Otherwise, the meter itself becomes an important circuit element and alters the potential difference that is to be measured.

Often a single meter is packaged so that, by means of a switch, it can be made to serve as either an ammeter or a voltmeter—and usually also as an *ohmmeter*, designed to measure the resistance of any element connected between its terminals. Such a versatile unit is called a *multimeter*.

28-8 *RC* Circuits

In preceding sections we dealt only with circuits in which the currents did not vary with time. Here we begin a discussion of time-varying currents.

Charging a Capacitor

The capacitor of capacitance C in Fig. 28-13 is initially uncharged. To charge it, we close switch S on point a. This completes an *RC series circuit* consisting of the capacitor, an ideal battery of emf $\mathscr{E}$, and a resistance R.

From Section 26-2, we already know that as soon as the circuit is complete, charge begins to flow (current exists) between a capacitor plate and a battery terminal on each side of the capacitor. This current increases the charge q on the plates and the potential difference V_C $(= q/C)$ across the capacitor. When that potential difference equals the potential difference across the battery (which here is equal to the emf $\mathscr{E}$), the current is zero. From Eq. 26-1 $(q = CV)$, the *equilibrium* (final) *charge* on the then fully charged capacitor is equal to $C\mathscr{E}$.

Here we want to examine the charging process. In particular we want to know how the charge $q(t)$ on the capacitor plates, the potential difference $V_C(t)$ across the capacitor, and the current $i(t)$ in the circuit vary with time during the charging process. We begin by applying the loop rule to the circuit, traversing it clockwise from the negative terminal of the battery. We find

$$\mathscr{E} - iR - \frac{q}{C} = 0. \tag{28-27}$$

The last term on the left side represents the potential difference across the capacitor. The term is negative because the capacitor's top plate, which is connected to the battery's positive terminal, is at a higher potential than the lower plate. Thus, there is a drop in potential as we move down through the capacitor.

We cannot immediately solve Eq. 28-27 because it contains two variables, i and q. However, those variables are not independent but are related by

$$i = \frac{dq}{dt}. \tag{28-28}$$

Substituting this for i in Eq. 28-27 and rearranging, we find

$$R\frac{dq}{dt} + \frac{q}{C} = \mathscr{E} \qquad \text{(charging equation).} \tag{28-29}$$

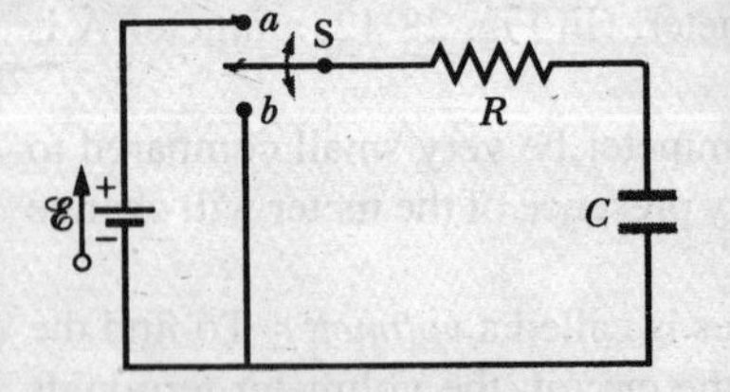

Fig. 28-13 When switch S is closed on a, the capacitor is *charged* through the resistor. When the switch is afterward closed on b, the capacitor *discharges* through the resistor.

This differential equation describes the time variation of the charge q on the capacitor in Fig. 28-13. To solve it, we need to find the function $q(t)$ that satisfies this equation and also satisfies the condition that the capacitor be initially uncharged; that is, $q = 0$ at $t = 0$.

We shall soon show that the solution to Eq. 28-29 is

$$q = C\mathscr{E}(1 - e^{-t/RC}) \qquad \text{(charging a capacitor).} \tag{28-30}$$

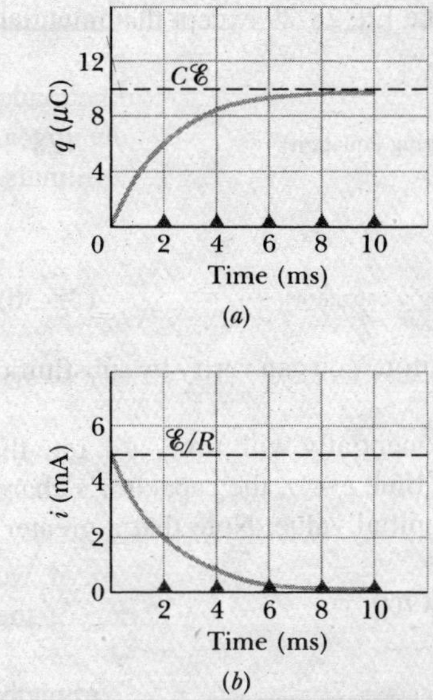

Fig. 28-14 (*a*) A plot of Eq. 28-30, which shows the buildup of charge on the capacitor of Fig. 28-13. (*b*) A plot of Eq. 28-31, which shows the decline of the charging current in the circuit of Fig. 28-13. The curves are plotted for $R = 2000\ \Omega$, $C = 1\ \mu\text{F}$, and $\mathscr{E} = 10$ V; the small triangles represent successive intervals of one time constant τ.

(Here e is the exponential base, 2.718 . . . , and not the elementary charge.) Note that Eq. 28-30 does indeed satisfy our required initial condition, because at $t = 0$ the term $e^{-t/RC}$ is unity; so the equation gives $q = 0$. Note also that as t goes to infinity (that is, a long time later), the term $e^{-t/RC}$ goes to zero; so the equation gives the proper value for the full (equilibrium) charge on the capacitor—namely, $q = C\mathscr{E}$. A plot of $q(t)$ for the charging process is given in Fig. 28-14*a*.

The derivative of $q(t)$ is the current $i(t)$ charging the capacitor:

$$i = \frac{dq}{dt} = \left(\frac{\mathscr{E}}{R}\right)e^{-t/RC} \quad \text{(charging a capacitor).} \qquad (28\text{-}31)$$

A plot of $i(t)$ for the charging process is given in Fig. 28-14*b*. Note that the current has the initial value $\mathscr{E}/R$ and that it decreases to zero as the capacitor becomes fully charged.

> A capacitor that is being charged initially acts like ordinary connecting wire relative to the charging current. A long time later, it acts like a broken wire.

By combining Eq. 26-1 ($q = CV$) and Eq. 28-30, we find that the potential difference $V_C(t)$ across the capacitor during the charging process is

$$V_C = \frac{q}{C} = \mathscr{E}(1 - e^{-t/RC}) \quad \text{(charging a capacitor).} \qquad (28\text{-}32)$$

This tells us that $V_C = 0$ at $t = 0$ and that $V_C = \mathscr{E}$ when the capacitor becomes fully charged as $t \to \infty$.

The Time Constant

The product RC that appears in Eqs. 28-30, 28-31, and 28-32 has the dimensions of time (both because the argument of an exponential must be dimensionless and because, in fact, $1.0\ \Omega \times 1.0\ \text{F} = 1.0$ s). RC is called the **capacitive time constant** of the circuit and is represented with the symbol τ:

$$\tau = RC \quad \text{(time constant).} \qquad (28\text{-}33)$$

From Eq. 28-30, we can now see that at time $t = \tau$ $(= RC)$, the charge on the initially uncharged capacitor of Fig. 28-13 has increased from zero to

$$q = C\mathscr{E}(1 - e^{-1}) = 0.63C\mathscr{E}. \qquad (28\text{-}34)$$

In words, during the first time constant τ the charge has increased from zero to 63% of its final value $C\mathscr{E}$. In Fig. 28-14, the small triangles along the time axes mark successive intervals of one time constant during the charging of the capacitor. The charging times for RC circuits are often stated in terms of τ; the greater τ is, the greater the charging time.

Discharging a Capacitor

Assume now that the capacitor of Fig. 28-13 is fully charged to a potential V_0 equal to the emf $\mathscr{E}$ of the battery. At a new time $t = 0$, switch S is thrown from a to b so that the capacitor can *discharge* through resistance R. How do the charge $q(t)$ on the capacitor and the current $i(t)$ through the discharge loop of capacitor and resistance now vary with time?

The differential equation describing $q(t)$ is like Eq. 28-29 except that now, with no battery in the discharge loop, $\mathscr{E} = 0$. Thus,

$$R\frac{dq}{dt} + \frac{q}{C} = 0 \quad \text{(discharging equation).} \tag{28-35}$$

The solution to this differential equation is

$$q = q_0 e^{-t/RC} \quad \text{(discharging a capacitor),} \tag{28-36}$$

where $q_0\,(= CV_0)$ is the initial charge on the capacitor. You can verify by substitution that Eq. 28-36 is indeed a solution of Eq. 28-35.

Equation 28-36 tells us that q decreases exponentially with time, at a rate that is set by the capacitive time constant $\tau = RC$. At time $t = \tau$, the capacitor's charge has been reduced to $q_0 e^{-1}$, or about 37% of the initial value. Note that a greater τ means a greater discharge time.

Differentiating Eq. 28-36 gives us the current $i(t)$:

$$i = \frac{dq}{dt} = -\left(\frac{q_0}{RC}\right)e^{-t/RC} \quad \text{(discharging a capacitor).} \tag{28-37}$$

This tells us that the current also decreases exponentially with time, at a rate set by τ. The initial current i_0 is equal to q_0/RC. Note that you can find i_0 by simply applying the loop rule to the circuit at $t = 0$; just then the capacitor's initial potential V_0 is connected across the resistance R, so the current must be $i_0 = V_0/R = (q_0/C)/R = q_0/RC$. The minus sign in Eq. 28-37 can be ignored; it merely means that the capacitor's charge q is decreasing.

Derivation of Eq. 28-30

To solve Eq. 28-29, we first rewrite it as

$$\frac{dq}{dt} + \frac{q}{RC} = \frac{\mathscr{E}}{R}. \tag{28-38}$$

The general solution to this differential equation is of the form

$$q = q_p + Ke^{-at}, \tag{28-39}$$

where q_p is a *particular solution* of the differential equation, K is a constant to be evaluated from the initial conditions, and $a = 1/RC$ is the coefficient of q in Eq. 28-38. To find q_p, we set $dq/dt = 0$ in Eq. 28-38 (corresponding to the final condition of no further charging), let $q = q_p$, and solve, obtaining

$$q_p = C\mathscr{E}. \tag{28-40}$$

To evaluate K, we first substitute this into Eq. 28-39 to get

$$q = C\mathscr{E} + Ke^{-at}.$$

Then substituting the initial conditions $q = 0$ and $t = 0$ yields

$$0 = C\mathscr{E} + K,$$

or $K = -C\mathscr{E}$. Finally, with the values of q_p, a, and K inserted, Eq. 28-39 becomes

$$q = C\mathscr{E} - C\mathscr{E}e^{-t/RC},$$

which, with a slight modification, is Eq. 28-30.

CHECKPOINT 5: The table gives four sets of values for the circuit elements in Fig. 28-13. Rank the sets according to (a) the initial current (as the switch is closed on *a*) and (b) the time required for the current to decrease to half its initial value, greatest first.

	1	2	3	4
$\mathscr{E}$ (V)	12	12	10	10
R (Ω)	2	3	10	5
C (μF)	3	2	0.5	2

Sample Problem 28-5

A capacitor of capacitance C is discharging through a resistor of resistance R.

(a) In terms of the time constant $\tau = RC$, when will the charge on the capacitor be half its initial value?

SOLUTION: The **Key Idea** here is that the charge on the capacitor varies according to Eq. 28-36,

$$q = q_0 e^{-t/RC},$$

in which q_0 is the initial charge. We are asked to find the time t at which $q = \frac{1}{2}q_0$, or at which

$$\tfrac{1}{2}q_0 = q_0 e^{-t/RC}. \tag{28-41}$$

After canceling q_0, we realize that the time t we seek is "buried" inside an exponential function. To expose the symbol t in Eq. 28-41, we take the natural logarithms of both sides of the equation. (The natural logarithm is the inverse function of the exponential function.) We find

$$\ln \tfrac{1}{2} = \ln(e^{-t/RC}) = -\frac{t}{RC},$$

or $t = (-\ln \tfrac{1}{2})RC = 0.69RC = 0.69\tau.$ (Answer)

(b) When will the energy stored in the capacitor be half its initial value?

SOLUTION: There are two **Key Ideas** here. First, the energy U stored in a capacitor is related to the charge q on the capacitor according to Eq. 26-21 ($U = Q^2/2C$). Second, that charge is decreasing according to Eq. 28-36. Combining these two ideas gives us

$$U = \frac{q^2}{2C} = \frac{q_0^2}{2C} e^{-2t/RC} = U_0 e^{-2t/RC},$$

in which U_0 is the initial stored energy. We are asked to find the time at which $U = \frac{1}{2}U_0$, or at which

$$\tfrac{1}{2}U_0 = U_0 e^{-2t/RC}.$$

Canceling U_0 and taking the natural logarithms of both sides, we obtain

$$\ln \tfrac{1}{2} = -\frac{2t}{RC},$$

or $t = -RC \dfrac{\ln \frac{1}{2}}{2} = 0.35RC = 0.35\tau.$ (Answer)

It takes longer (0.69τ versus 0.35τ) for the *charge* to fall to half its initial value than for the *stored energy* to fall to half its initial value. Doesn't this result surprise you?

REVIEW & SUMMARY

Emf An **emf device** does work on charges to maintain a potential difference between its output terminals. If dW is the work the device does to force positive charge dq from the negative to the positive terminal, then the **emf** (work per unit charge) of the device is

$$\mathscr{E} = \frac{dW}{dq} \quad \text{(definition of } \mathscr{E}). \tag{28-1}$$

The volt is the SI unit of emf as well as of potential difference. An **ideal emf device** is one that lacks any internal resistance. The potential difference between its terminals is equal to the emf. A **real emf device** has internal resistance. The potential difference between its terminals is equal to the emf only if there is no current through the device.

Analyzing Circuits The change in potential in traversing a resistance R in the direction of the current is $-iR$; in the opposite direction it is $+iR$. The change in potential in traversing an ideal emf device in the direction of the emf arrow is $+\mathscr{E}$; in the opposite direction it is $-\mathscr{E}$. Conservation of energy leads to the loop rule:

Loop Rule. *The algebraic sum of the changes in potential encountered in a complete traversal of any loop of a circuit must be zero.*

Conservation of charge gives us the junction rule:

Junction Rule. *The sum of the currents entering any junction must be equal to the sum of the currents leaving that junction.*

Single-Loop Circuits The current in a single-loop circuit containing a single resistance R and an emf device with emf $\mathscr{E}$ and internal resistance r is

$$i = \frac{\mathscr{E}}{R + r}, \qquad (28\text{-}4)$$

which reduces to $i = \mathscr{E}/R$ for an ideal emf device with $r = 0$.

Power When a real battery of emf $\mathscr{E}$ and internal resistance r does work on the charge carriers in a current i through it, the rate P of energy transfer to the charge carriers is

$$P = iV, \qquad (28\text{-}11)$$

where V is the potential across the terminals of the battery. The rate P_r of energy transfer to thermal energy within the battery is

$$P_r = i^2 r. \qquad (28\text{-}13)$$

The rate P_{emf} at which the chemical energy within the battery changes is

$$P_{emf} = i\mathscr{E}. \qquad (28\text{-}14)$$

Series Resistances When resistances are in **series,** they have the same current. The equivalent resistance that can replace a series combination of resistances is

$$R_{eq} = \sum_{j=1}^{n} R_j \quad (n \text{ resistances in series}). \qquad (28\text{-}7)$$

Parallel Resistances When resistances are in **parallel,** they have the same potential difference. The equivalent resistance that can replace a parallel combination of resistances is given by

$$\frac{1}{R_{eq}} = \sum_{j=1}^{n} \frac{1}{R_j} \quad (n \text{ resistances in parallel}). \qquad (28\text{-}21)$$

RC Circuits When an emf $\mathscr{E}$ is applied to a resistance R and capacitance C in series, as in Fig. 28-13 with the switch at a, the charge on the capacitor increases according to

$$q = C\mathscr{E}(1 - e^{-t/RC}) \quad \text{(charging a capacitor)}, \qquad (28\text{-}30)$$

in which $C\mathscr{E} = q_0$ is the equilibrium (final) charge and $RC = \tau$ is the **capacitive time constant** of the circuit. During the charging, the current is

$$i = \frac{dq}{dt} = \left(\frac{\mathscr{E}}{R}\right)e^{-t/RC} \quad \text{(charging a capacitor)}. \qquad (28\text{-}31)$$

When a capacitor discharges through a resistance R, the charge on the capacitor decays according to

$$q = q_0 e^{-t/RC} \quad \text{(discharging a capacitor)}. \qquad (28\text{-}36)$$

During the discharging, the current is

$$i = \frac{dq}{dt} = -\left(\frac{q_0}{RC}\right)e^{-t/RC} \quad \text{(discharging a capacitor)}. \qquad (28\text{-}37)$$

QUESTIONS

1. Figure 28-15 shows current i passing through a battery. The following table gives four sets of values for i and the battery's emf $\mathscr{E}$ and internal resistance r; it also gives the *polarity* (orientation of the terminals) of the battery. Rank the sets according to the rate at which energy is transferred between the battery and the charge carriers, greatest transfer *to* the carriers first and greatest transfer *from* the carriers last.

	$\mathscr{E}$	r	i	Polarity
(1)	$15\mathscr{E}_1$	0	i_1	+ at left
(2)	$10\mathscr{E}_1$	0	$2i_1$	+ at left
(3)	$10\mathscr{E}_1$	0	$2i_1$	− at left
(4)	$10\mathscr{E}_1$	r_1	$2i_1$	− at left

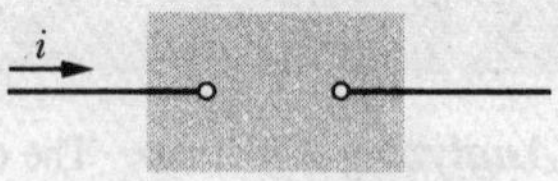

Fig. 28-15 Question 1.

2. For each circuit in Fig. 28-16, are the resistors connected in series, in parallel, or neither?

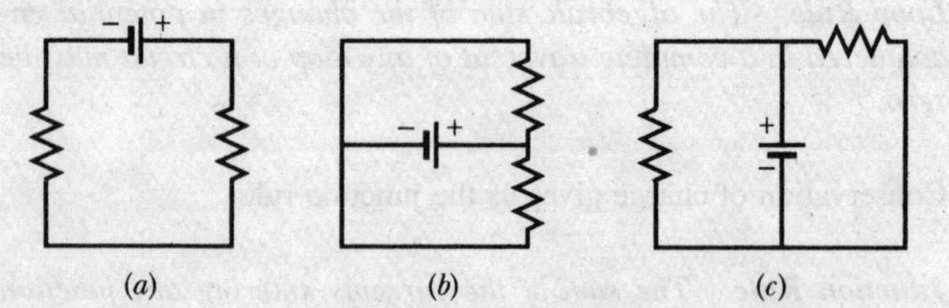

Fig. 28-16 Question 2.

Note: R_1 & R_2 are in || since they are connected end to end & same PD is applied across both ends

3. (a) In Fig. 28-17*a*, are resistors R_1 and R_3 in series? (b) Are resistors R_1 and R_2 in parallel? (c) Rank the equivalent resistances of the four circuits shown in Fig. 28-17, greatest first.

4. (a) In Fig. 28-17*a*, with $R_1 > R_2$, is the potential difference across R_2 more than, less than, or equal to that across R_1? (b) Is the current through resistor R_2 more than, less than, or equal to that through resistor R_1?

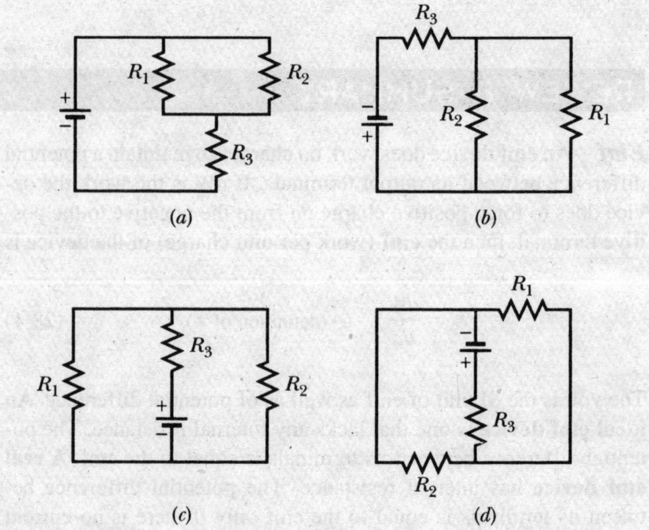

Fig. 28-17 Questions 3 and 4.

5. You are to connect resistors R_1 and R_2, with $R_1 > R_2$, to a battery, first individually, then in series, and then in parallel. Rank those arrangements according to the amount of current through the battery, greatest first.

6. *Res-monster maze.* In Fig. 28-18, all the resistors have a resistance of 4.0 Ω and all the (ideal) batteries have an emf of 4.0 V. What is the current through resistor R? (If you can find the proper loop through this maze, you can answer the question with a few seconds of mental calculation.)

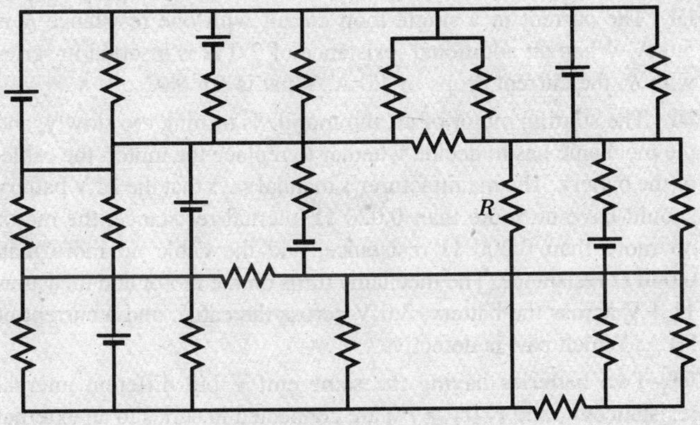

Fig. 28-18 Question 6.

7. Initially, a single resistor R_1 is wired to a battery. Then resistor R_2 is added in parallel. Are (a) the potential difference across R_1 and (b) the current i_1 through R_1 now more than, less than, or the same as previously? (c) Is the equivalent resistance R_{12} of R_1 and R_2 more than, less than, or equal to R_1? (d) Is the total current through R_1 and R_2 together more than, less than, or equal to the current through R_1 previously?

8. *Cap-monster maze.* In Fig. 28-19, all the capacitors have a capacitance of 6.0 μF, and all the batteries have an emf of 10 V. What is the charge on capacitor C? (If you can find the proper loop through this maze, you can answer the question with a few seconds of mental calculation.)

9. A resistor R_1 is wired to a battery, then resistor R_2 is added in series. Are (a) the potential difference across R_1 and (b) the current i_1 through R_1 now more than, less than, or the same as previously? (c) Is the equivalent resistance R_{12} of R_1 and R_2 more than, less than, or equal to R_1?

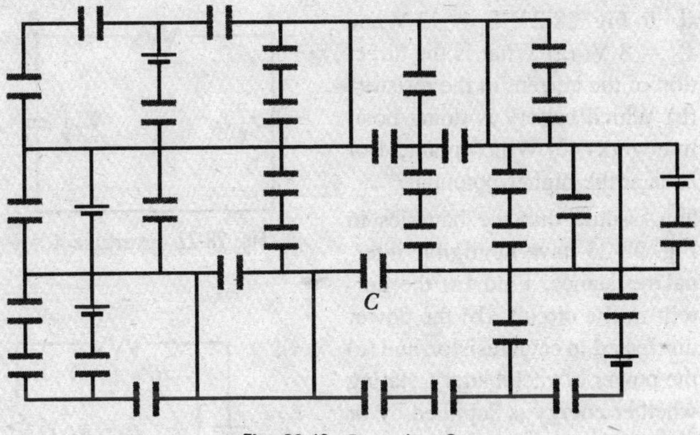

Fig. 28-19 Question 8.

10. Figure 28-20 shows three sections of circuit that are to be connected in turn to the same battery via a switch as in Fig. 28-13. The resistors are all identical, as are the capacitors. Rank the sections according to (a) the final (equilibrium) charge on the capacitor and (b) the time required for the capacitor to reach 50% of its final charge, greatest first.

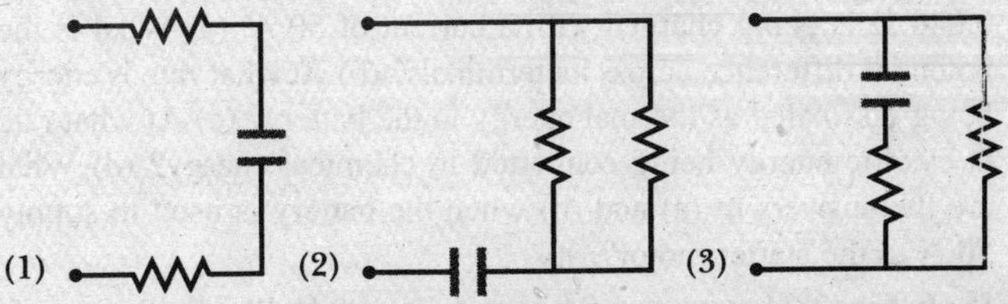

Fig. 28-20 Question 10.

11. Figure 28-21 shows plots of $V(t)$ for three capacitors that discharge (separately) through the same resistor. Rank the plots according to the capacitances of the capacitors, greatest first.

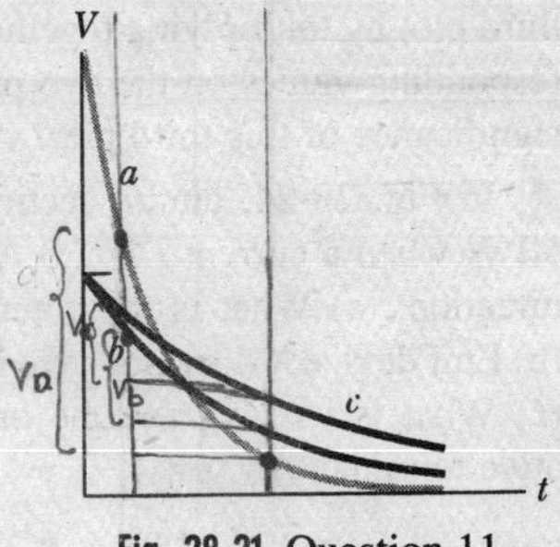

Fig. 28-21 Question 11.

EXERCISES & PROBLEMS

ssm Solution is in the Student Solutions Manual.
www Solution is available on the World Wide Web at: http://www.wiley.com/college/hrw
ilw Solution is available on the Interactive LearningWare.

SEC. 28-5 Potential Differences

1E. A standard flashlight battery can deliver about 2.0 W · h of energy before it runs down. (a) If a battery costs 80¢, what is the cost of operating a 100 W lamp for 8.0 h using batteries? (b) What is the cost if energy is provided at 6¢ per kilowatt-hour? ssm

2E. A 5.0 A current is set up in a circuit for 6.0 min by a rechargeable battery with a 6.0 V emf. By how much is the chemical energy of the battery reduced?

3E. A certain car battery with a 12 V emf has an initial charge of 120 A · h. Assuming that the potential across the terminals stays constant until the battery is completely discharged, for how long can it deliver energy at the rate of 100 W? ssm

4E. In Fig. 28-22, $\mathscr{E}_1 = 12$ V and $\mathscr{E}_2 = 8$ V. (a) What is the direction of the current in the resistor? (b) Which battery is doing positive work? (c) Which point, A or B, is at the higher potential?

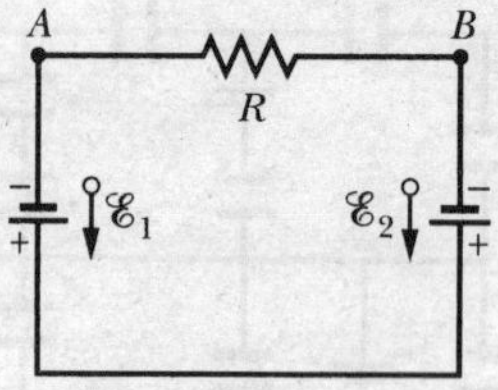

Fig. 28-22 Exercise 4.

5E. Assume that the batteries in Fig. 28-23 have negligible internal resistance. Find (a) the current in the circuit, (b) the power dissipated in each resistor, and (c) the power of each battery, stating whether energy is supplied by or absorbed by it. ssm

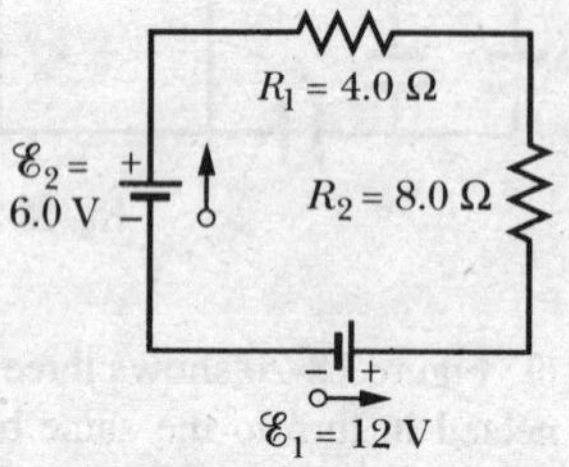

Fig. 28-23 Exercise 5.

6E. A wire of resistance 5.0 Ω is connected to a battery whose emf $\mathscr{E}$ is 2.0 V and whose internal resistance is 1.0 Ω. In 2.0 min, (a) how much energy is transferred from chemical to electrical form? (b) How much energy appears in the wire as thermal energy? (c) Account for the difference between (a) and (b).

7E. A car battery with a 12 V emf and an internal resistance of 0.040 Ω is being charged with a current of 50 A. (a) What is the potential difference across its terminals? (b) At what rate is energy being dissipated as thermal energy in the battery? (c) At what rate is electric energy being converted to chemical energy? (d) What are the answers to (a) and (b) when the battery is used to supply 50 A to the starter motor? Ilw

8E. In Fig. 28-4a, put $\mathscr{E} = 2.0$ V and $r = 100$ Ω. Plot (a) the current and (b) the potential difference across R, as functions of R over the range 0 to 500 Ω. Make both plots on the same graph. (c) Make a third plot by multiplying together, for various values of R, the corresponding values on the two plotted curves. What is the physical significance of this third plot?

9E. In Fig. 28-24, circuit section AB absorbs energy at a rate of 50 W when a current $i = 1.0$ A passes through it in the indicated direction. (a) What is the potential difference between A and B? (b) Emf device X does not have internal resistance. What is its emf? (c) What is its *polarity* (the orientation of its positive and negative terminals)? ssm

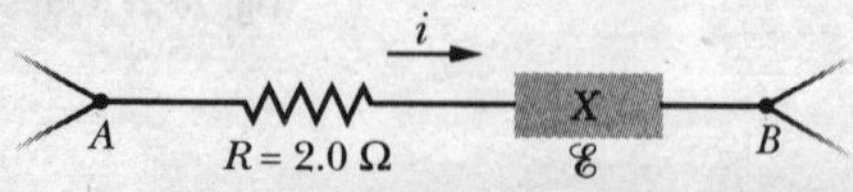

Fig. 28-24 Exercise 9.

10E. In Fig. 28-25, if the potential at point P is 100 V, what is the potential at point Q?

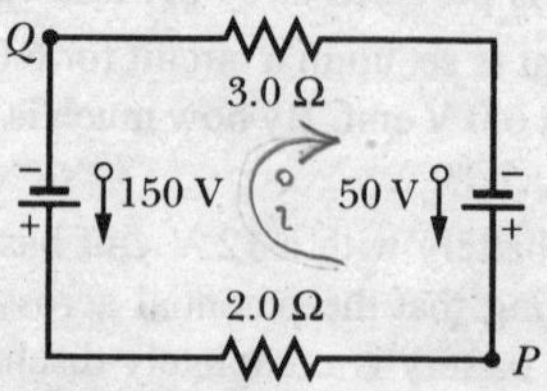

Fig. 28-25 Exercise 10.

11E. In Fig. 28-6a, calculate the potential difference between a and c by considering a path that contains R, r_1, and $\mathscr{E}_1$.

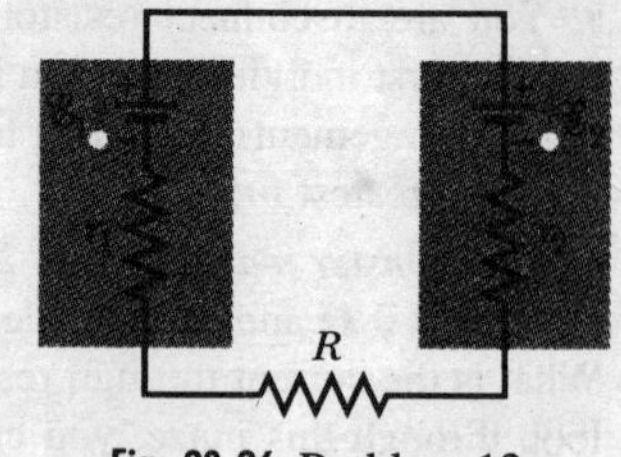

Fig. 28-26 Problem 12.

12P. (a) In Fig. 28-26, what value must R have if the current in the circuit is to be 1.0 mA? Take $\mathscr{E}_1 = 2.0$ V, $\mathscr{E}_2 = 3.0$ V, and $r_1 = r_2 = 3.0$ Ω. (b) What is the rate at which thermal energy appears in R?

13P. The current in a single-loop circuit with one resistance R is 5.0 A. When an additional resistance of 2.0 Ω is inserted in series with R, the current drops to 4.0 A. What is R? Ilw

14P. The starting motor of an automobile is turning too slowly, and the mechanic has to decide whether to replace the motor, the cable, or the battery. The manufacturer's manual says that the 12 V battery should have no more than 0.020 Ω internal resistance, the motor no more than 0.200 Ω resistance, and the cable no more than 0.040 Ω resistance. The mechanic turns on the motor and measures 11.4 V across the battery, 3.0 V across the cable, and a current of 50 A. Which part is defective?

15P. Two batteries having the same emf $\mathscr{E}$ but different internal resistances r_1 and r_2 ($r_1 > r_2$) are connected in series to an external resistance R. (a) Find the value of R that makes the potential difference zero between the terminals of one battery. (b) Which battery is it? ssm

16P. A solar cell generates a potential difference of 0.10 V when a 500 Ω resistor is connected across it, and a potential difference of 0.15 V when a 1000 Ω resistor is substituted. What are (a) the internal resistance and (b) the emf of the solar cell? (c) The area of the cell is 5.0 cm^2, and the rate per unit area at which it receives energy from light is 2.0 mW/cm^2. What is the efficiency of the cell for converting light energy to thermal energy in the 1000 Ω external resistor?

17P. (a) In Fig. 28-4a, show that the rate at which energy is dissipated in R as thermal energy is a maximum when $R = r$. (b) Show that this maximum power is $P = \mathscr{E}^2/4r$. ssm www

SEC. 28-6 Multiloop Circuits

18E. By using only two resistors—singly, in series, or in parallel—you are able to obtain resistances of 3.0, 4.0, 12, and 16 Ω. What are the two resistances?

19E. Four 18.0 Ω resistors are connected in parallel across a 25.0 V ideal battery. What is the current through the battery? ssm

20E. In Fig. 28-27, find the equivalent resistance between points D and E. (*Hint:* Imagine that a battery is connected between points D and E.)

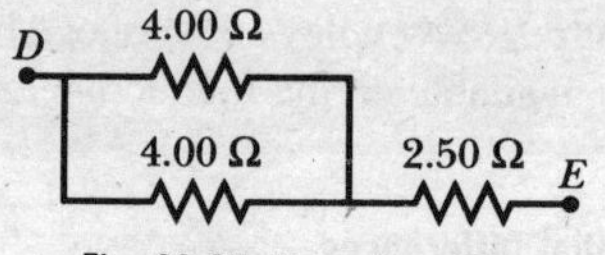

Fig. 28-27 Exercise 20.

21E. In Fig. 28-28 find the current in each resistor and the potential difference between points a and b. Put $\mathscr{E}_1 = 6.0$ V, $\mathscr{E}_2 = 5.0$ V, $\mathscr{E}_3 = 4.0$ V, $R_1 = 100\ \Omega$, and $R_2 = 50\ \Omega$. ssm

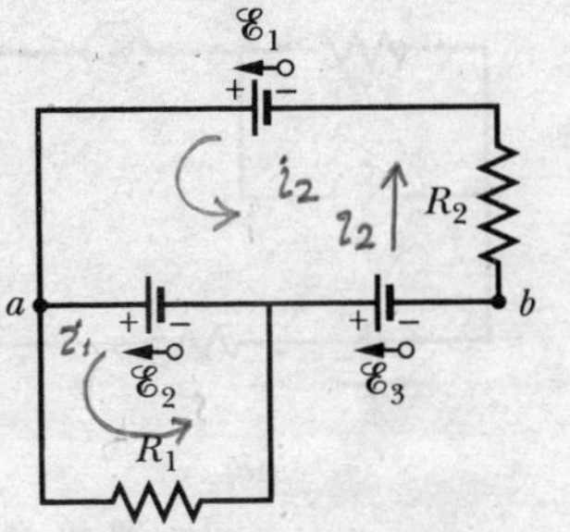

Fig. 28-28 Exercise 21.

22E. Figure 28-29 shows a circuit containing three switches, labeled S_1, S_2, and S_3. Find the current at a for all possible combinations of switch settings. Put $\mathscr{E} = 120$ V, $R_1 = 20.0\ \Omega$, and $R_2 = 10.0\ \Omega$. Assume that the battery has no resistance.

23E. Two lightbulbs, one of resistance R_1 and the other of resistance R_2, where $R_1 > R_2$, are connected to a battery (a) in parallel and (b) in series. Which bulb is brighter (dissipates more energy) in each case? ssm

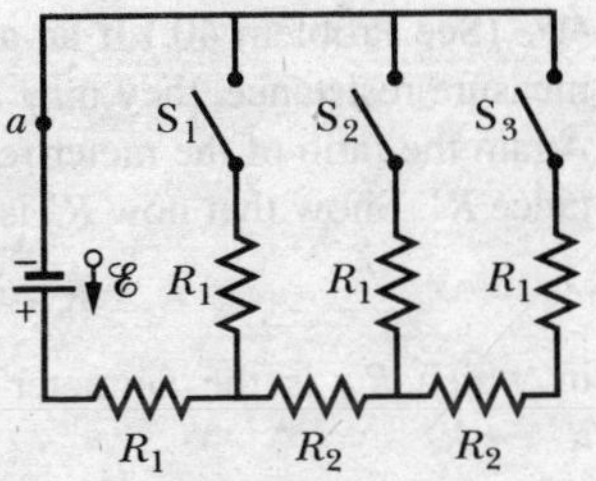

Fig. 28-29 Exercise 22.

24E. In Fig. 28-7, calculate the potential difference between points c and d by as many paths as possible. Assume that $\mathscr{E}_1 = 4.0$ V, $\mathscr{E}_2 = 1.0$ V, $R_1 = R_2 = 10\ \Omega$, and $R_3 = 5.0\ \Omega$.

25E. Nine copper wires of length l and diameter d are connected in parallel to form a single composite conductor of resistance R. What must be the diameter D of a single copper wire of length l if it is to have the same resistance? ssm

26P In Fig. 28-30, find the equivalent resistance between points (a) F and H and (b) F and G. (*Hint:* For each pair of points, imagine that a battery is connected across the pair.)

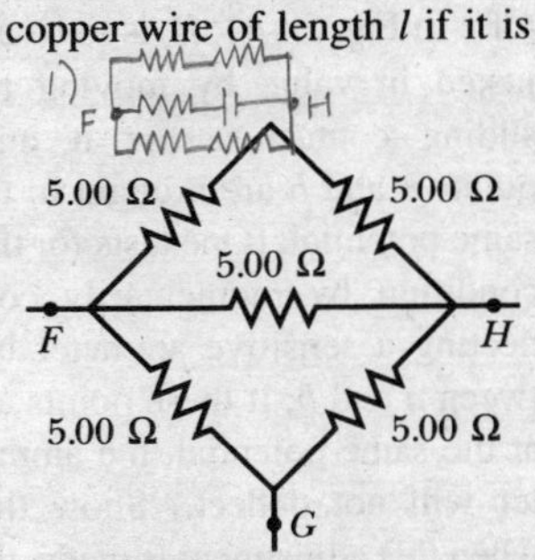

Fig. 28-30 Problem 26.

27P. You are given a number of $10\ \Omega$ resistors, each capable of dissipating only 1.0 W without being destroyed. What is the minimum number of such resistors that you need to combine in series or in parallel to make a $10\ \Omega$ resistance that is capable of dissipating at least 5.0 W? ssm

28P. (a) In Fig. 28-31, what is the equivalent resistance of the network shown? (b) What is the current in each resistor? Put $R_1 = 100\ \Omega$, $R_2 = R_3 = 50\ \Omega$, $R_4 = 75\ \Omega$, and $\mathscr{E} = 6.0$ V; assume the battery is ideal.

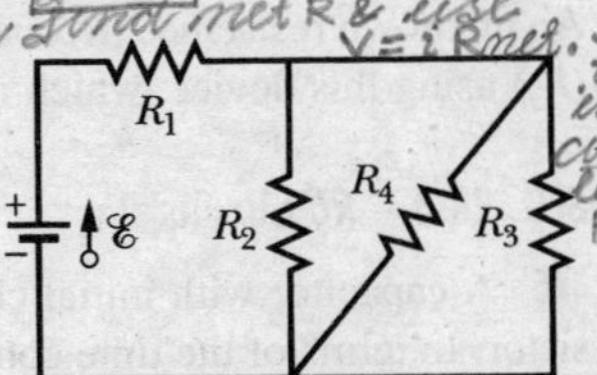

Fig. 28-31 Problem 28.

29P. Two batteries of emf $\mathscr{E}$ and internal resistance r are connected in parallel across a resistor R, as in Fig. 28-32a. (a) For what value of R is the rate of electrical energy dissipation by the resistor a maximum? (b) What is the maximum energy dissipation rate? ssm

30P. You are given two batteries of emf $\mathscr{E}$ and internal resistance r. They may be connected either in parallel (Fig. 28-32a) or in series (Fig. 28-32b) and are to be used to establish a current in a resistor R. (a) Derive expressions for the current in R for both arrangements. Which will yield the larger current (b) when $R > r$ and (c) when $R < r$?

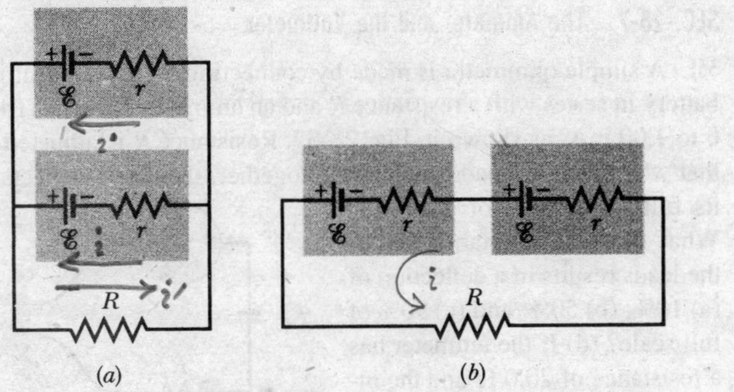

Fig. 28-32 Problems 29 and 30.

31P. In Fig. 28-33, $\mathscr{E}_1 = 3.00$ V, $\mathscr{E}_2 = 1.00$ V, $R_1 = 5.00\ \Omega$, $R_2 = 2.00\ \Omega$, $R_3 = 4.00\ \Omega$, and both batteries are ideal. What is the rate at which energy is dissipated in (a) R_1, (b) R_2, and (c) R_3? What is the power of (d) battery 1 and (e) battery 2? ssm www

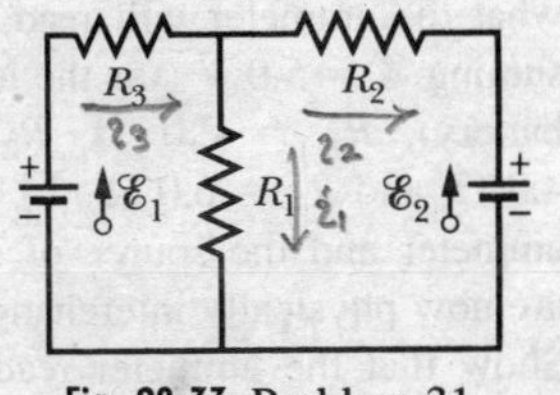

Fig. 28-33 Problem 31.

32P. In the circuit of Fig. 28-34, for what value of R will the ideal battery transfer energy to the resistors (a) at a rate of 60.0 W, (b) at the maximum possible rate, and (c) at the minimum possible rate? (d) What are those rates?

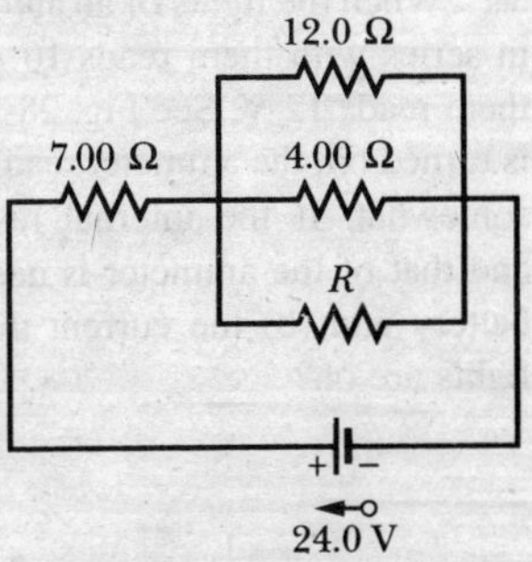

Fig. 28-34 Problem 32.

33P. (a) Calculate the current through each ideal battery in Fig. 28-35. Assume that $R_1 = 1.0\ \Omega$, $R_2 = 2.0\ \Omega$, $\mathscr{E}_1 = 2.0$ V, and $\mathscr{E}_2 = \mathscr{E}_3 = 4.0$ V. (b) Calculate $V_a - V_b$. ilw

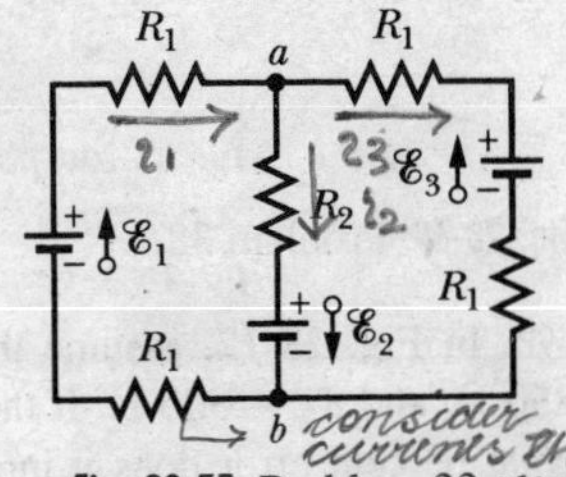

Fig. 28-35 Problem 33.

34P. In the circuit of Fig. 28-36, $\mathscr{E}$ has a constant value but R can be varied. Find the value of R that results in the maximum heating in that resistor. The battery is ideal.

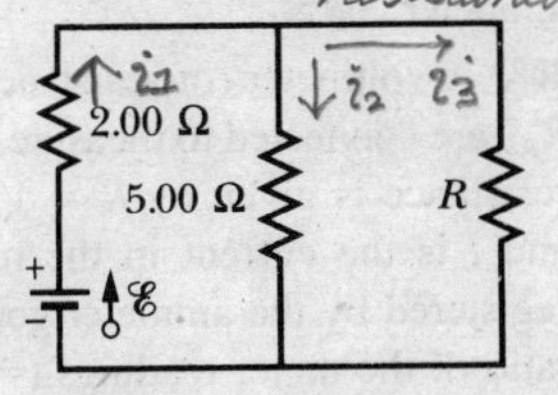

Fig. 28-36 Problem 34.

35P. A copper wire of radius $a = 0.250$ mm has an aluminum jacket of outer radius $b = 0.380$ mm. (a) There is a current $i = 2.00$ A in the composite wire. Using Table 27-1, calculate the current in each material. (b) If a potential difference $V = 12.0$ V between the ends maintains the current, what is the length of the composite wire? ssm

SEC. 28-7 The Ammeter and the Voltmeter

36E. A simple ohmmeter is made by connecting a 1.50 V flashlight battery in series with a resistance R and an ammeter that reads from 0 to 1.00 mA, as shown in Fig. 28-37. Resistance R is adjusted so that when the clip leads are shorted together, the meter deflects to its full-scale value of 1.00 mA. What external resistance across the leads results in a deflection of (a) 10%, (b) 50%, and (c) 90% of full scale? (d) If the ammeter has a resistance of 20.0 Ω and the internal resistance of the battery is negligible, what is the value of R?

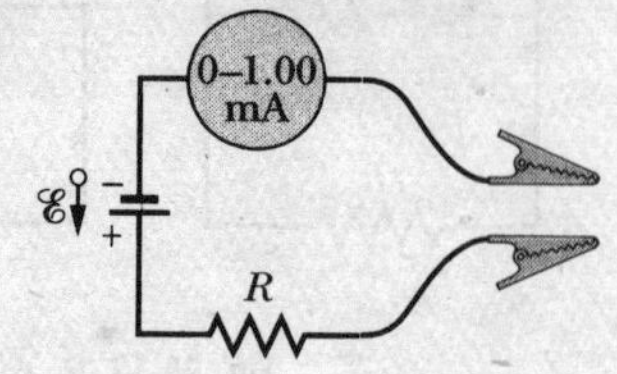

Fig. 28-37 Exercise 36.

37P. (a) In Fig. 28-38, determine what the ammeter will read, assuming $\mathscr{E} = 5.0$ V (for the ideal battery), $R_1 = 2.0\ \Omega$, $R_2 = 4.0\ \Omega$, and $R_3 = 6.0\ \Omega$. (b) The ammeter and the source of emf are now physically interchanged. Show that the ammeter reading remains unchanged. ilw

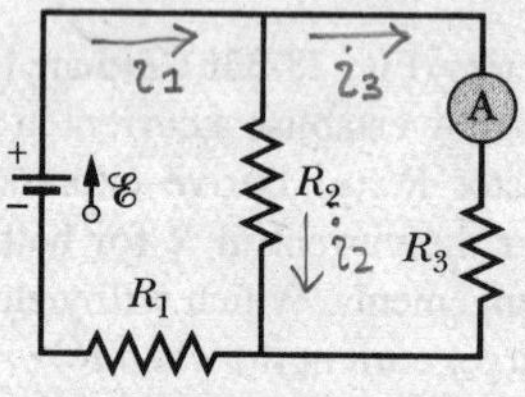

Fig. 28-38 Problem 37.

38P. When the lights of an automobile are switched on, an ammeter in series with them reads 10 A and a voltmeter connected across them reads 12 V. See Fig. 28-39. When the electric starting motor is turned on, the ammeter reading drops to 8.0 A and the lights dim somewhat. If the internal resistance of the battery is 0.050 Ω and that of the ammeter is negligible, what are (a) the emf of the battery and (b) the current through the starting motor when the lights are on?

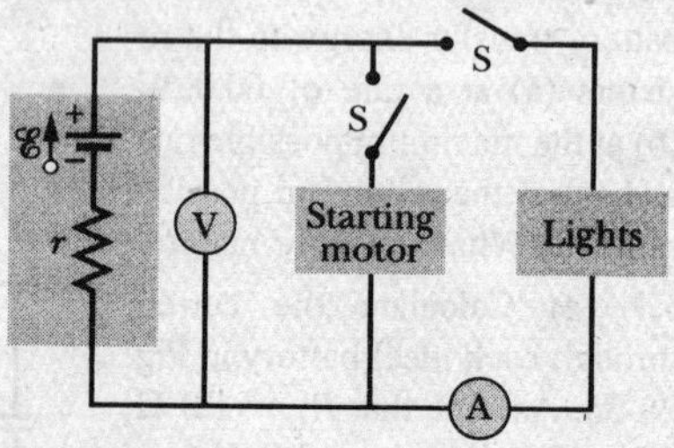

Fig. 28-39 Problem 38.

39P. In Fig. 28-12, assume that $\mathscr{E} = 3.0$ V, $r = 100\ \Omega$, $R_1 = 250\ \Omega$, and $R_2 = 300\ \Omega$. If the voltmeter resistance R_V is 5.0 kΩ, what percent error does it introduce into the measurement of the potential difference across R_1? Ignore the presence of the ammeter. ssm

40P. A voltmeter (of resistance R_V) and an ammeter (of resistance R_A) are connected to measure a resistance R, as in Fig. 28-40*a*. The resistance is given by $R = V/i$, where V is the voltmeter reading and i is the current in the resistance R. Some of the current i' registered by the ammeter goes through the voltmeter, so that the ratio of the meter readings ($= V/i'$) gives only an *apparent* resistance reading R'. Show that R and R' are related by

$$\frac{1}{R} = \frac{1}{R'} - \frac{1}{R_V}.$$

Note that as $R_V \to \infty$, $R' \to R$.

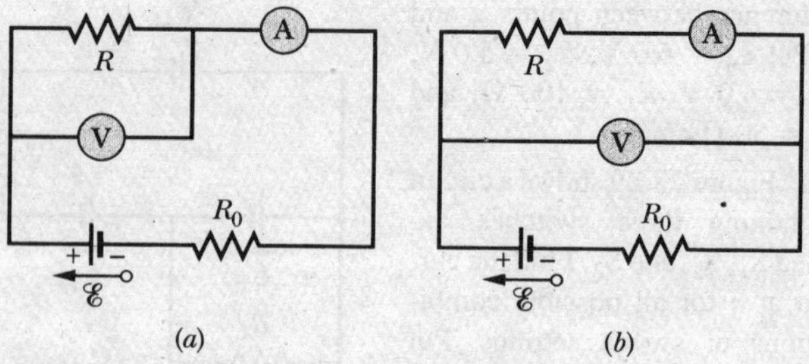

Fig. 28-40 Problems 40 to 42.

41P. (See Problem 40.) If an ammeter and a voltmeter are used to measure resistance, they may also be connected as in Fig. 28-40*b*. Again the ratio of the meter readings gives only an apparent resistance R'. Show that now R' is related to R by

$$R = R' - R_A,$$

in which R_A is the ammeter resistance. Note that as $R_A \to 0$, $R' \to R$.

42P. (See Problems 40 and 41.) In Fig. 28-40, the ammeter and voltmeter resistances are 3.00 and 300 Ω, respectively. Take $\mathscr{E} = 12.0$ V for the ideal battery and $R_0 = 100\ \Omega$. If $R = 85.0\ \Omega$, (a) what will the meters read for the two different connections (Figs. 28-40*a* and *b*)? (b) What apparent resistance R' will be computed in each case?

43P. In Fig. 28-41, R_s is to be adjusted in value by moving the sliding contact across it until points a and b are brought to the same potential. (One tests for this condition by momentarily connecting a sensitive ammeter between a and b; if these points are at the same potential, the ammeter will not deflect.) Show that when this adjustment is made, the following relation holds:

$$R_x = R_s\left(\frac{R_2}{R_1}\right).$$

An unknown resistance (R_x) can be measured in terms of a standard (R_s) using this device, which is called a Wheatstone bridge. ssm

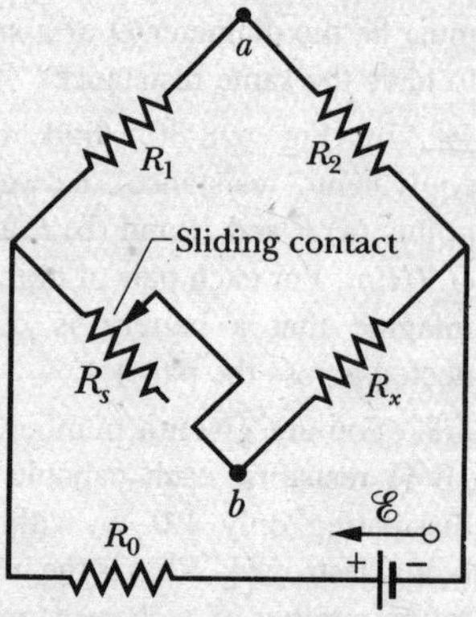

Fig. 28-41 Problem 43.

SEC. 28-8 *RC* Circuits

44E. A capacitor with initial charge q_0 is discharged through a resistor. In terms of the time constant τ, how long is required for the capacitor to lose (a) the first one-third of its charge and (b) two-thirds of its charge?

45E. How many time constants must elapse for an initially uncharged capacitor in an *RC* series circuit to be charged to 99.0% of its equilibrium charge? ssm

46E. In an *RC* series circuit, $\mathscr{E} = 12.0$ V, $R = 1.40$ MΩ, and $C = 1.80\ \mu$F. (a) Calculate the time constant. (b) Find the maximum charge that will appear on the capacitor during charging. (c) How long does it take for the charge to build up to 16.0 μC?

47E. A 15.0 kΩ resistor and a capacitor are connected in series and then a 12.0 V potential difference is suddenly applied across them. The potential difference across the capacitor rises to 5.00 V in 1.30 μs. (a) Calculate the time constant of the circuit. (b) Find the capacitance of the capacitor. ilw

48P. The potential difference between the plates of a leaky (meaning that charge leaks from one plate to the other) 2.0 μF capacitor drops to one-fourth its initial value in 2.0 s. What is the equivalent resistance between the capacitor plates?

49P. A 3.00 MΩ resistor and a 1.00 μF capacitor are connected in series with an ideal battery of emf $\mathscr{E}$ = 4.00 V. At 1.00 s after the connection is made, what are the rates at which (a) the charge of the capacitor is increasing, (b) energy is being stored in the capacitor, (c) thermal energy is appearing in the resistor, and (d) energy is being delivered by the battery? ssm

50P. An initially uncharged capacitor C is fully charged by a device of constant emf $\mathscr{E}$ connected in series with a resistor R. (a) Show that the final energy stored in the capacitor is half the energy supplied by the emf device. (b) By direct integration of i^2R over the charging time, show that the thermal energy dissipated by the resistor is also half the energy supplied by the emf device.

51P. A capacitor with an initial potential difference of 100 V is discharged through a resistor when a switch between them is closed at $t = 0$. At $t = 10.0$ s, the potential difference across the capacitor is 1.00 V. (a) What is the time constant of the circuit? (b) What is the potential difference across the capacitor at $t = 17.0$ s? ssm

52P. Figure 28-42 shows the circuit of a flashing lamp, like those attached to barrels at highway construction sites. The fluorescent lamp L (of negligible capacitance) is connected in parallel across the capacitor C of an RC circuit. There is a current through the lamp only when the potential difference across it reaches the breakdown voltage V_L; in this event, the capacitor discharges completely through the lamp and the lamp flashes briefly. Suppose that two flashes per second are needed. For a lamp with breakdown voltage $V_L = 72.0$ V, wired to a 95.0 V ideal battery and a 0.150 μF capacitor, what should be the resistance R?

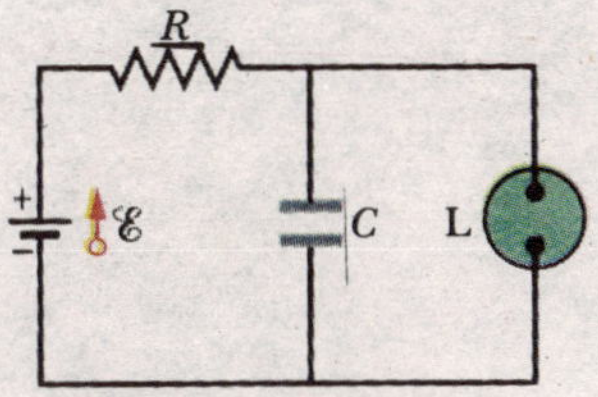

Fig. 28-42 Problem 52.

53P. A 1.0 μF capacitor with an initial stored energy of 0.50 J is discharged through a 1.0 MΩ resistor. (a) What is the initial charge on the capacitor? (b) What is the current through the resistor when the discharge starts? (c) Determine V_C, the potential difference across the capacitor, and V_R, the potential difference across the resistor, as functions of time. (d) Express the production rate of thermal energy in the resistor as a function of time. ssm www

54P. A controller on an electronic arcade game consists of a variable resistor connected across the plates of a 0.220 μF capacitor. The capacitor is charged to 5.00 V, then discharged through the resistor. The time for the potential difference across the plates to decrease to 0.800 V is measured by a clock inside the game. If the range of discharge times that can be handled effectively is from 10.0 μs to 6.00 ms, what should be the resistance range of the resistor?

55P*. In the circuit of Fig. 28-43, $\mathscr{E}$ = 1.2 kV, C = 6.5 μF, $R_1 = R_2 = R_3 = 0.73$ MΩ. With C completely uncharged, switch S is suddenly closed (at $t = 0$). (a) Determine the current through each resistor at $t = 0$ and as $t \to \infty$. (b) Draw qualitatively a graph of the potential difference V_2 across R_2 from $t = 0$ to $t = \infty$. (c) What are the numerical values of V_2 at $t = 0$ and as $t \to \infty$? (d) What is the physical meaning of "$t \to \infty$" in this case? ssm

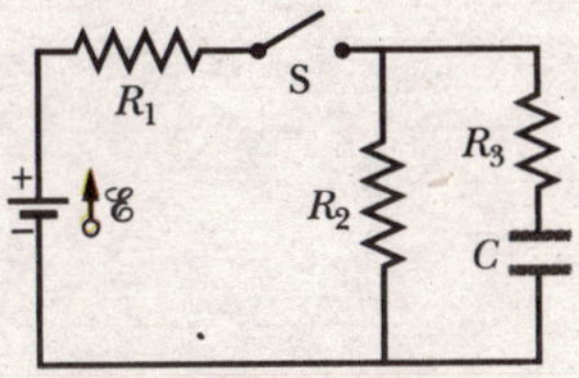

Fig. 28-43 Problem 55.

Additional Problem

56. *Heart attack or electrocution?* This story begins with Problem 45 in Chapter 27. Figure 28-44 shows the electrical pathway of the current up through one foot of the victim, across the torso (including the heart), and down through the other foot. (a) From the given data, find the potential difference between the man's feet, assuming that one foot was 0.50 m closer to the leaky rod than the other. (b) Assume that the resistance of a foot on wet soil is the typical value of 300 Ω and the resistance of the torso interior is the commonly accepted value of 1000 Ω. What then was the current across the victim's torso? (c) The human heart can be put into fibrillation by a current of 0.10 A to 1.0 A through the torso. Was the victim's fibrillation due to current leakage from the rod?

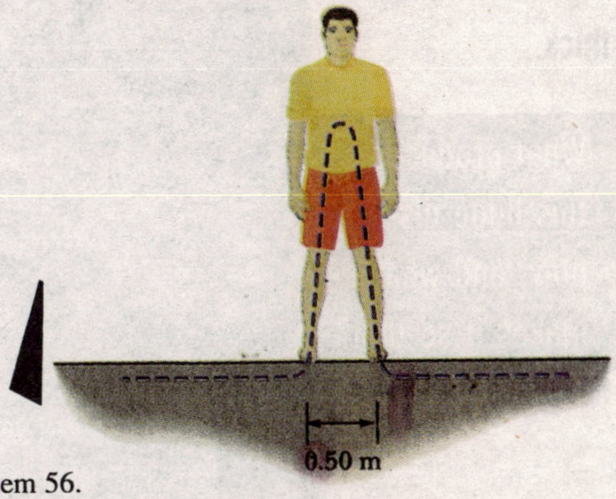

Fig. 28-44 Problem 56.

29 Magnetic Fields

If you are outside on a dark night in the middle to high latitudes, you might be able to see an aurora, a ghostly "curtain" of light that hangs down from the sky. This curtain is not only local; it may be several hundred kilometers high and several thousand kilometers long, stretching around Earth in an arc. However, it is less than 1 km thick.

What produces this huge display, and what makes it so thin?

The answer is in this chapter.

Fig. 29-1 Using an electromagnet to collect and transport scrap metal at a steel mill.

29-1 The Magnetic Field

We have discussed how a charged plastic rod produces a vector field—the electric field $\vec{E}$—at all points in the space around it. Similarly, a magnet produces a vector field—the **magnetic field** $\vec{B}$—at all points in the space around it. You get a hint of that magnetic field whenever you attach a note to a refrigerator door with a small magnet, or accidentally erase a computer disk by bringing it near a magnet. The magnet acts on the door or disk *by means of* its magnetic field.

In a familiar type of magnet, a wire coil is wound around an iron core and a current is sent through the coil; the strength of the magnetic field is determined by the size of the current. In industry, such **electromagnets** are used for sorting scrap metal (Fig. 29-1) among many other things. You are probably more familiar with **permanent magnets**—magnets, like the refrigerator-door type, that do not need current to have a magnetic field.

In Chapter 23 we saw that an *electric charge* sets up an electric field that can then affect other electric charges. Here, we might reasonably expect that a *magnetic charge* sets up a magnetic field that can then affect other magnetic charges. Although such magnetic charges, called *magnetic monopoles*, are predicted by certain theories, their existence has not been confirmed.

How then are magnetic fields set up? There are two ways. (1) Moving electrically charged particles, such as a current in a wire, create magnetic fields. (2) Elementary particles such as electrons have an *intrinsic* magnetic field around them; that is, this field is a basic characteristic of the particles, just as are their mass and electric charge (or lack of charge). As we shall discuss in Chapter 32, the magnetic fields of the electrons in certain materials add together to give a net magnetic field around the material. This is true for the material in permanent magnets (which is good, because they can then hold notes to a refrigerator door). In other materials, the magnetic fields of all the electrons cancel out, giving no net magnetic field surrounding the material. This is true for the material in your body (which is also good, because otherwise you might be slammed up against a refrigerator door every time you passed one).

Experimentally we find that when a charged particle (either alone or as part of a current) moves through a magnetic field, a force due to the field can act on the particle. In this chapter we focus on the relation between the magnetic field and this force.

29-2 The Definition of $\vec{B}$

We determined the electric field $\vec{E}$ at a point by putting a test particle of charge q at rest at that point and measuring the electric force $\vec{F}_E$ acting on the particle. We then defined $\vec{E}$ as

$$\vec{E} = \frac{\vec{F}_E}{q}. \tag{29-1}$$

If a magnetic monopole were available, we could define $\vec{B}$ in a similar way. Because such particles have not been found, we must define $\vec{B}$ in another way, in terms of the magnetic force $\vec{F}_B$ exerted on a moving electrically charged test particle.

In principle, we do this by firing a charged particle through the point at which $\vec{B}$ is to be defined, using various directions and speeds for the particle and determining the force $\vec{F}_B$ that acts on the particle at that point. After many such trials we would find that when the particle's velocity $\vec{v}$ is along a particular axis through the point, force $\vec{F}_B$ is zero. For all other directions of $\vec{v}$, the magnitude of $\vec{F}_B$ is always proportional to $v \sin \phi$, where ϕ is the angle between the zero-force axis and the direction of $\vec{v}$. Furthermore, the direction of $\vec{F}_B$ is always perpendicular to the direction of $\vec{v}$. (These results suggest that a cross product is involved.)

We can then define a magnetic field $\vec{B}$ to be a vector quantity that is directed along the zero-force axis. We can next measure the magnitude of $\vec{F}_B$ when $\vec{v}$ is directed perpendicular to that axis and then define the magnitude of $\vec{B}$ in terms of that force magnitude:

$$B = \frac{F_B}{|q|v},$$

where q is the charge of the particle.

We can summarize all these results with the following vector equation:

$$\vec{F}_B = q\vec{v} \times \vec{B}; \tag{29-2}$$

that is, the force $\vec{F}_B$ on the particle is equal to the charge q times the cross product of its velocity $\vec{v}$ and the field $\vec{B}$ (all measured in the same reference frame). Using Eq. 3-20 for the cross product, we can write the magnitude of $\vec{F}_B$ as

$$F_B = |q|vB \sin \phi, \tag{29-3}$$

where ϕ is the angle between the directions of velocity $\vec{v}$ and magnetic field $\vec{B}$.

Finding the Magnetic Force on a Particle

Equation 29-3 tells us that the magnitude of the force $\vec{F}_B$ acting on a particle in a magnetic field is proportional to the charge q and speed v of the particle. Thus, the force is equal to zero if the charge is zero or if the particle is stationary. Equation 29-3 also tells us that the magnitude of the force is zero if $\vec{v}$ and $\vec{B}$ are either parallel ($\phi = 0°$) or antiparallel ($\phi = 180°$), and the force is at its maximum when $\vec{v}$ and $\vec{B}$ are perpendicular to each other.

Equation 29-2 tells us all this plus the direction of $\vec{F}_B$. From Section 3-7, we know that the cross product $\vec{v} \times \vec{B}$ in Eq. 29-2 is a vector that is perpendicular to the two vectors $\vec{v}$ and $\vec{B}$. The right-hand rule (Fig. 29-2*a*) tells us that the thumb of the right hand points in the direction of $\vec{v} \times \vec{B}$ when the fingers sweep $\vec{v}$ into $\vec{B}$. If q is positive, then (by Eq. 29-2) the force $\vec{F}_B$ has the same sign as $\vec{v} \times \vec{B}$ and thus must be in the same direction; that is, for positive q, $\vec{F}_B$ is directed along the thumb (Fig. 29-2*b*). If q is negative, then the force $\vec{F}_B$ and cross product

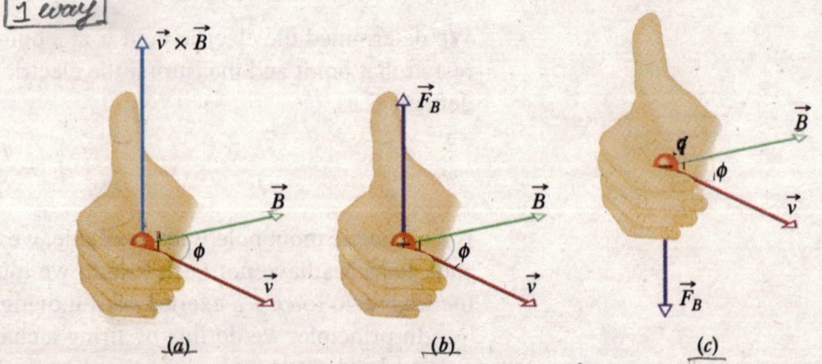

Fig. 29-2 (*a*) The right-hand rule (in which $\vec{v}$ is swept into $\vec{B}$ through the smaller angle ϕ between them) gives the direction of $\vec{v} \times \vec{B}$ as the direction of the thumb. (*b*) If q is positive, then the direction of $\vec{F}_B = q\vec{v} \times \vec{B}$ is in the direction of $\vec{v} \times \vec{B}$. (*c*) If q is negative, then the direction of $\vec{F}_B$ is opposite that of $\vec{v} \times \vec{B}$.

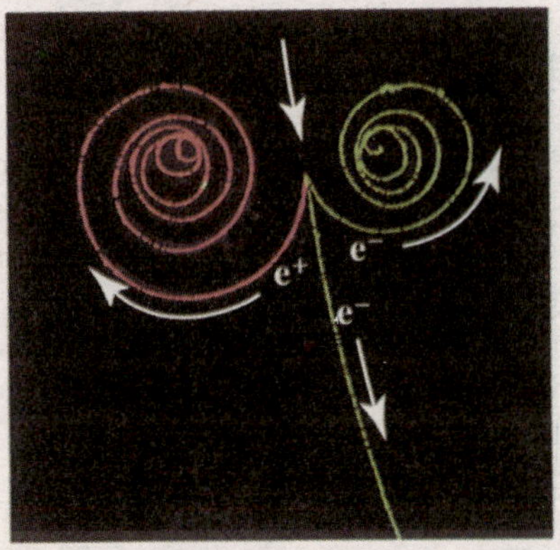

Fig. 29-3 The tracks of two electrons (e^-) and a positron (e^+) in a bubble chamber that is immersed in a uniform magnetic field that is directed out of the plane of the page.

$\vec{v} \times \vec{B}$ have opposite signs and thus must be in opposite directions. For negative q, $\vec{F}_B$ is directed opposite the thumb (Fig. 29-2*c*).

Regardless of the sign of the charge, however,

> The force $\vec{F}_B$ acting on a charged particle moving with velocity $\vec{v}$ through a magnetic field $\vec{B}$ is *always* perpendicular to $\vec{v}$ and $\vec{B}$.

Thus, $\vec{F}_B$ *never* has a component parallel to $\vec{v}$. This means that $\vec{F}_B$ cannot change the particle's speed v (and thus it cannot change the particle's kinetic energy). The force can change only the direction of $\vec{v}$ (and thus the direction of travel); only in this sense can $\vec{F}_B$ accelerate the particle.

To develop a feeling for Eq. 29-2, consider Fig. 29-3, which shows some tracks left by charged particles moving rapidly through a *bubble chamber* at the Lawrence Berkeley Laboratory. The chamber, which is filled with liquid hydrogen, is immersed in a strong uniform magnetic field that is directed out of the plane of the figure. An incoming gamma ray particle—which leaves no track because it is uncharged—transforms into an electron (spiral track marked e^-) and a positron (track marked e^+) while it knocks an electron out of a hydrogen atom (long track marked e^-). Check with Eq. 29-2 and Fig. 29-2 that the three tracks made by these two negative particles and one positive particle curve in the proper directions.

The SI unit for $\vec{B}$ that follows from Eqs. 29-2 and 29-3 is the newton per coulomb-meter per second. For convenience, this is called the **tesla** (T):

$$1 \text{ tesla} = 1 \text{ T} = 1 \frac{\text{newton}}{(\text{coulomb})(\text{meter/second})}.$$

Recalling that a coulomb per second is an ampere, we have

$$1 \text{ T} = 1 \frac{\text{newton}}{(\text{coulomb/second})(\text{meter})} = 1 \frac{\text{N}}{\text{A} \cdot \text{m}}. \tag{29-4}$$

An earlier (non-SI) unit for $\vec{B}$, still in common use, is the *gauss* (G), and

$$\boxed{1 \text{ tesla} = 10^4 \text{ gauss.}} \tag{29-5}$$

Table 29-1 lists the magnetic fields that occur in a few situations. Note that Earth's magnetic field near the planet's surface is about 10^{-4} T (= 100 μT or 1 gauss).

TABLE 29-1 Some Approximate Magnetic Fields

At the surface of a neutron star	10^8 T
Near a big electromagnet	1.5 T
Near a small bar magnet	10^{-2} T
At Earth's surface	10^{-4} T
In interstellar space	10^{-10} T
Smallest value in a magnetically shielded room	10^{-14} T

CHECKPOINT 1: The figure shows three situations in which a charged particle with velocity $\vec{v}$ travels through a uniform magnetic field $\vec{B}$. In each situation, what is the direction of the magnetic force $\vec{F}_B$ on the particle?

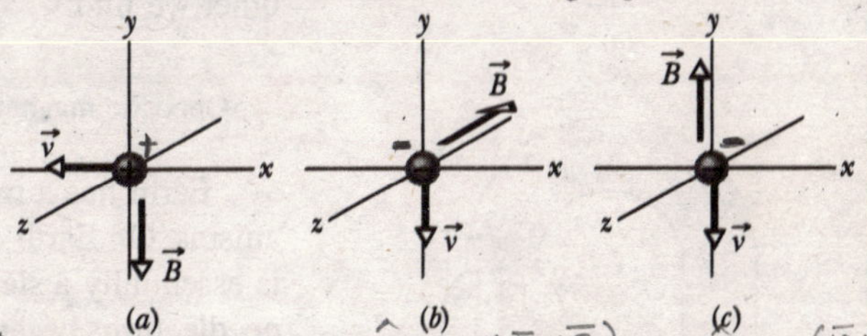

Magnetic Field Lines

We can represent magnetic fields with field lines, as we did for electric fields. Similar rules apply; that is, (1) the direction of the tangent to a magnetic field line at any point gives the direction of $\vec{B}$ at that point, and (2) the spacing of the lines represents the magnitude of $\vec{B}$—the magnetic field is stronger where the lines are closer together, and conversely.

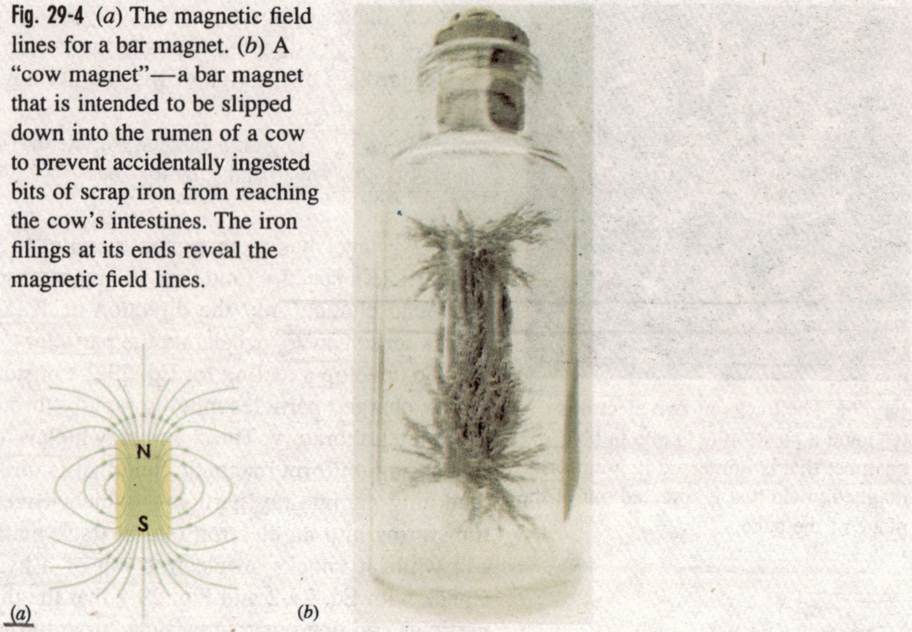

Fig. 29-4 (*a*) The magnetic field lines for a bar magnet. (*b*) A "cow magnet"—a bar magnet that is intended to be slipped down into the rumen of a cow to prevent accidentally ingested bits of scrap iron from reaching the cow's intestines. The iron filings at its ends reveal the magnetic field lines.

Figure 29-4*a* shows how the magnetic field near a *bar magnet* (a permanent magnet in the shape of a bar) can be represented by magnetic field lines. The lines all pass through the magnet, and they all form closed loops (even those that are not shown closed in the figure). The external magnetic effects of a bar magnet are strongest near its ends, where the field lines are most closely spaced. Thus, the magnetic field of the bar magnet in Fig. 29-4*b* collects the iron filings mainly near the two ends of the magnet.

The (closed) field lines enter one end of a magnet and exit the other end. The end of a magnet from which the field lines emerge is called the *north pole* of the magnet; the other end, where field lines enter the magnet, is called the *south pole*. The magnets we use to fix notes on refrigerators are short bar magnets. Figure 29-5 shows two other common shapes for magnets: a *horseshoe magnet* and a magnet that has been bent around into the shape of a **C** so that the *pole faces* are facing each other. (The magnetic field between the pole faces can then be approximately uniform.) Regardless of the shape of the magnets, if we place two of them near each other we find:

Opposite magnetic poles attract each other, and like magnetic poles repel each other.

Earth has a magnetic field that is produced in its core by still unknown mechanisms. On Earth's surface, we can detect this magnetic field with a compass, which is essentially a slender bar magnet on a low-friction pivot. This bar magnet, or this needle, turns because its north-pole end is attracted toward the Arctic region of Earth. Thus, the *south* pole of Earth's magnetic field must be located toward the Arctic. Logically, we then should call the pole there a south pole. However, because we call that direction north, we are trapped into the statement that Earth has a *geomagnetic north pole* in that direction.

With more careful measurement we would find that in the northern hemisphere, the magnetic field lines of Earth generally point down into Earth and toward the Arctic. In the southern hemisphere, they generally point up out of Earth and away from the Antarctic—that is, away from Earth's *geomagnetic south pole*.

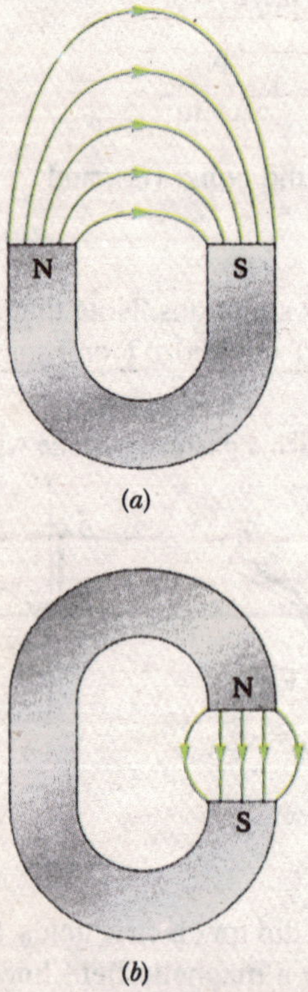

Fig. 29-5 (*a*) A horseshoe magnet and (*b*) a **C**-shaped magnet. (Only some of the external field lines are shown.)

Sample Problem 29-1

A uniform magnetic field $\vec{B}$, with magnitude 1.2 mT, is directed vertically upward throughout the volume of a laboratory chamber. A proton with kinetic energy 5.3 MeV enters the chamber, moving horizontally from south to north. What magnetic deflecting force acts on the proton as it enters the chamber? The proton mass is 1.67×10^{-27} kg. (Neglect Earth's magnetic field.)

SOLUTION: Because the proton is charged and moving through a magnetic field, a magnetic force $\vec{F}_B$ can act on it. The **Key Idea** here is that, because the initial direction of the proton's velocity is not along a magnetic field line, $\vec{F}_B$ is not simply zero. To find the magnitude of $\vec{F}_B$, we can use Eq. 29-3 provided we first find the proton's speed v. We can find v from the given kinetic energy, since $K = \frac{1}{2}mv^2$. Solving for v, we obtain

$$v = \sqrt{\frac{2K}{m}} = \sqrt{\frac{(2)(5.3\text{ MeV})(1.60 \times 10^{-13}\text{ J/MeV})}{1.67 \times 10^{-27}\text{ kg}}}$$

$$= 3.2 \times 10^7 \text{ m/s}.$$

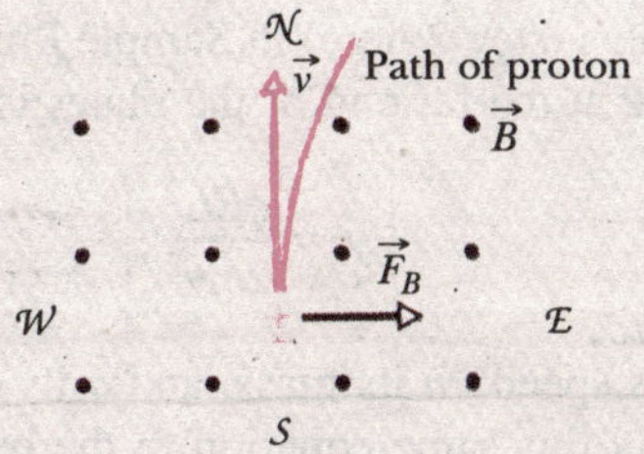

Fig. 29-6 Sample Problem 29-1. An overhead view of a proton moving from south to north with velocity $\vec{v}$ in a chamber. A magnetic field is directed vertically upward in the chamber, as represented by the array of dots (which resemble the tips of arrows). The proton is deflected toward the east.

Equation 29-3 then yields

$$F_B = |q|vB \sin \phi$$
$$= (1.60 \times 10^{-19}\text{ C})(3.2 \times 10^7\text{ m/s}) \times (1.2 \times 10^{-3}\text{ T})(\sin 90°)$$
$$= 6.1 \times 10^{-15}\text{ N.} \qquad \text{(Answer)}$$

This may seem like a small force, but it acts on a particle of small mass, producing a large acceleration; namely,

$$a = \frac{F_B}{m} = \frac{6.1 \times 10^{-15}\text{ N}}{1.67 \times 10^{-27}\text{ kg}} = 3.7 \times 10^{12}\text{ m/s}^2.$$

To find the direction of $\vec{F}_B$, we use the **Key Idea** that $\vec{F}_B$ has the direction of the cross product $q\vec{v} \times \vec{B}$. Because the charge q is positive, $\vec{F}_B$ must have the same direction as $\vec{v} \times \vec{B}$, which can be determined with the right-hand rule for cross products (as in Fig. 29-2*b*). We know that $\vec{v}$ is directed horizontally from south to north and $\vec{B}$ is directed vertically up. The right-hand rule shows us that the deflecting force $\vec{F}_B$ must be directed horizontally from west to east, as Fig. 29-6 shows. (The array of dots in the figure represents a magnetic field directed out of the plane of the figure. An array of Xs would have represented a magnetic field directed into that plane.)

If the charge of the particle were negative, the magnetic deflecting force would be directed in the opposite direction—that is, horizontally from east to west. This is predicted automatically by Eq. 29-2 if we substitute a negative value for q.

29-3 Crossed Fields: Discovery of the Electron

Both an electric field $\vec{E}$ and a magnetic field $\vec{B}$ can produce a force on a charged particle. When the two fields are perpendicular to each other, they are said to be *crossed fields*. Here we shall examine what happens to charged particles—namely, electrons—as they move through crossed fields. We use as our example the experiment that led to the discovery of the electron in 1897 by J. J. Thomson at Cambridge University.

Figure 29-7 shows a modern, simplified version of Thomson's experimental apparatus—a *cathode ray tube* (which is like the picture tube in a standard television

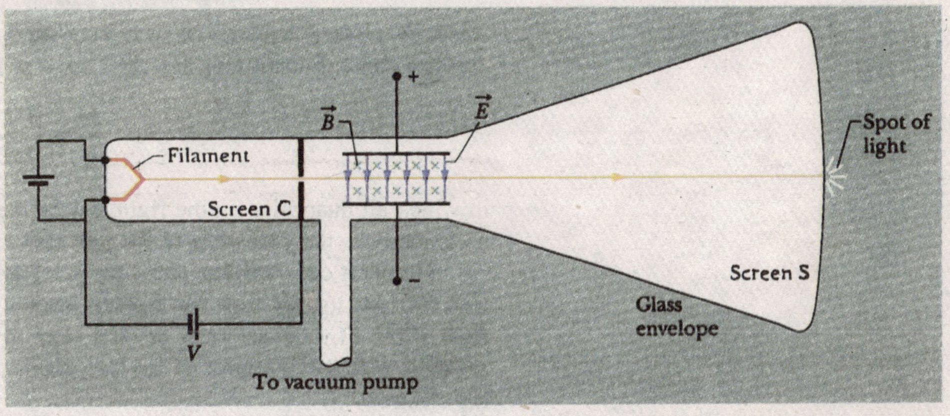

Fig. 29-7 A modern version of J. J. Thomson's apparatus for measuring the ratio of mass to charge for the electron. An electric field $\vec{E}$ is established by connecting a battery across the deflecting-plate terminals. A magnetic field $\vec{B}$ is set up by means of a current in a system of coils (not shown). The magnetic field shown is into the plane of the figure, as represented by the array of Xs (which resemble the feathered ends of arrows).

set). Charged particles (which we now know as electrons) are emitted by a hot filament at the rear of the evacuated tube and are accelerated by an applied potential difference V. After they pass through a slit in screen C, they form a narrow beam. They then pass through a region of crossed $\vec{E}$ and $\vec{B}$ fields, headed toward a fluorescent screen S, where they produce a spot of light (on a television screen the spot is part of the picture). The forces on the charged particles in the crossed-fields region can deflect them from the center of the screen. By controlling the magnitudes and directions of the fields, Thomson could thus control where the spot of light appeared on the screen. Recall that the force on a negatively charged particle due to an electric field is directed opposite the field. Thus, for the particular field arrangement of Fig. 29-7, electrons are forced up the page by the electric field $\vec{E}$ and down the page by the magnetic field $\vec{B}$; that is, the forces are *in opposition*. Thomson's procedure was equivalent to the following series of steps.

1. Set $E = 0$ and $B = 0$ and note the position of the spot on screen S due to the undeflected beam.
2. Turn on $\vec{E}$ and measure the resulting beam deflection.
3. Maintaining $\vec{E}$, now turn on $\vec{B}$ and adjust its value until the beam returns to the undeflected position. (With the forces in opposition, they can be made to cancel.)

We discussed the deflection of a charged particle moving through an electric field $\vec{E}$ between two plates (step 2 here) in Sample Problem 23-4. We found that the deflection of the particle at the far end of the plates is

$$y = \frac{qEL^2}{2mv^2}, \tag{29-6}$$

where v is the particle's speed, m its mass, and q its charge, and L is the length of the plates. We can apply this same equation to the beam of electrons in Fig. 29-7; if need be, we can calculate the deflection by measuring the deflection of the beam on screen S and then working back to calculate the deflection y at the end of the plates. (Because the direction of the deflection is set by the sign of the particle's charge, Thomson was able to show that the particles that were lighting up his screen were negatively charged.)

When the two fields in Fig. 29-7 are adjusted so that the two deflecting forces cancel (step 3), we have from Eqs. 29-1 and 29-3a

$$|q|E = |q|vB \sin(90°) = |q|vB$$

or

$$v = \frac{E}{B}. \tag{29-7}$$

Thus, the crossed fields allow us to measure the speed of the charged particles passing through them. Substituting Eq. 29-7 for v in Eq. 29-6 and rearranging yield

$$\frac{m}{q} = \frac{B^2L^2}{2yE}, \tag{29-8}$$

in which all quantities on the right can be measured. Thus, the crossed fields allow us to measure the ratio m/q of the particles moving through Thomson's apparatus.

Thomson claimed that these particles are found in all matter. He also claimed that they are lighter than the lightest known atom (hydrogen) by a factor of more than 1000. (The exact ratio proved later to be 1836.15.) His m/q measurement, coupled with the boldness of his two claims, is considered to be the "discovery of the electron."

CHECKPOINT 2: The figure shows four directions for the velocity vector $\vec{v}$ of a positively charged particle moving through a uniform electric field $\vec{E}$ (directed out of the page and represented with an encircled dot) and a uniform magnetic field $\vec{B}$. (a) Rank directions 1, 2, and 3 according to the magnitude of the net force on the particle, greatest first. (b) Of all four directions, which might result in a net force of zero?

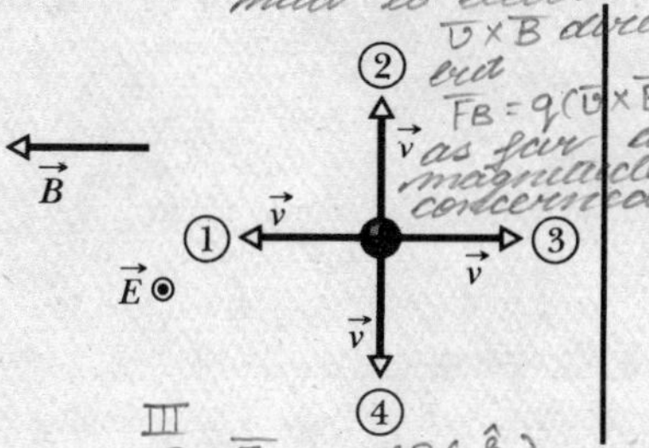

29-4 Crossed Fields: The Hall Effect

As we just discussed, a beam of electrons in a vacuum can be deflected by a magnetic field. Can the drifting conduction electrons in a copper wire also be deflected by a magnetic field? In 1879, Edwin H. Hall, then a 24-year-old graduate student at the Johns Hopkins University, showed that they can. This **Hall effect** allows us to find out whether the charge carriers in a conductor are positively or negatively charged. Beyond that, we can measure the number of such carriers per unit volume of the conductor.

Figure 29-8*a* shows a copper strip of width d, carrying a current i whose conventional direction is from the top of the figure to the bottom. The charge carriers are electrons and, as we know, they drift (with drift speed v_d) in the opposite direction, from bottom to top. At the instant shown in Fig. 29-8*a*, an external magnetic field $\vec{B}$, pointing into the plane of the figure, has just been turned on. From Eq. 29-2 we see that a magnetic deflecting force $\vec{F}_B$ will act on each drifting electron, pushing it toward the right edge of the strip.

As time goes on, electrons move to the right, mostly piling up on the right edge of the strip, leaving uncompensated positive charges in fixed positions at the left edge. The separation of positive and negative charges produces an electric field $\vec{E}$ within the strip, pointing from left to right in Fig. 29-8*b*. This field exerts an electric force $\vec{F}_E$ on each electron, tending to push it to the left.

An equilibrium quickly develops in which the electric force on each electron builds up until it just cancels the magnetic force. When this happens, as Fig. 29-8*b* shows, the force due to $\vec{B}$ and the force due to $\vec{E}$ are in balance. The drifting electrons then move along the strip toward the top of the page at velocity $\vec{v}_d$, with no further collection of electrons on the right edge of the strip and thus no further increase in the electric field $\vec{E}$.

A *Hall potential difference V* is associated with the electric field across strip width d. From Eq. 25-42, the magnitude of that potential difference is

$$V = Ed. \tag{29-9}$$

By connecting a voltmeter across the width, we can measure the potential difference between the two edges of the strip. Moreover, the voltmeter can tell us which edge is at higher potential. For the situation of Fig. 29-8*a*, we would find that the left

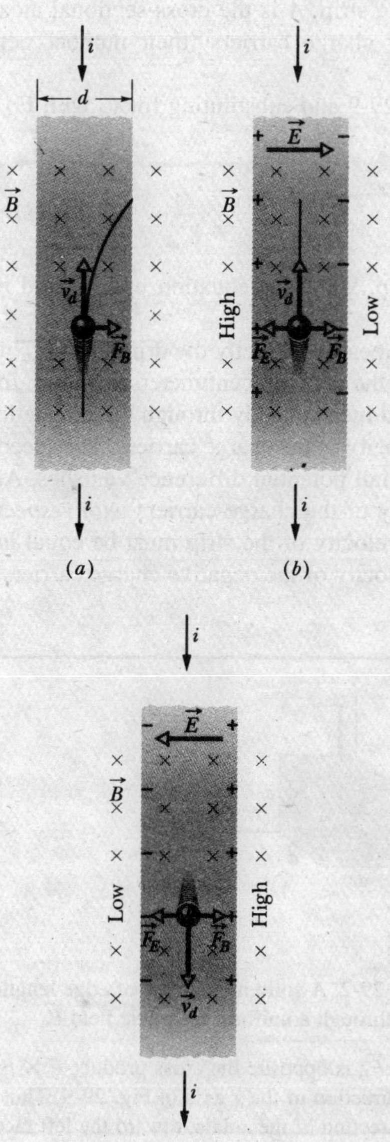

Fig. 29-8 A strip of copper carrying a current i is immersed in a magnetic field $\vec{B}$. (*a*) The situation immediately after the magnetic field is turned on. The curved path that will then be taken by an electron is shown. (*b*) The situation at equilibrium, which quickly follows. Note that negative charges pile up on the right side of the strip, leaving uncompensated positive charges on the left. Thus, the left side is at a higher potential than the right side. (*c*) For the same current direction, if the charge carriers were positively charged, *they* would pile up on the right side, and the right side would be at the higher potential.

edge is at higher potential, which is consistent with our assumption that the charge carriers are negatively charged.

For a moment, let us make the opposite assumption, that the charge carriers in current i are positively charged (Fig. 29-8*c*). Convince yourself that as these charge carriers move from top to bottom in the strip, they are pushed to the right edge by $\vec{F}_B$ and thus that the *right* edge is at higher potential. Because that last statement is contradicted by our voltmeter reading, the charge carriers must be negatively charged.

Now for the quantitative part. When the electric and magnetic forces are in balance (Fig. 29-8*b*), Eqs. 29-1 and 29-3 give us

$$eE = ev_dB. \qquad (29\text{-}10)$$

From Eq. 27-7, the drift speed v_d is

$$v_d = \frac{J}{ne} = \frac{i}{neA} \qquad (29\text{-}11)$$

in which J (= i/A) is the current density in the strip, A is the cross-sectional area of the strip, and n is the *number density* of charge carriers (their number per unit volume).

In Eq. 29-10, substituting for E with Eq. 29-9 and substituting for v_d with Eq. 29-11, we obtain

$$n = \frac{Bi}{Vle}, \qquad (29\text{-}12)$$

in which l (= A/d) is the thickness of the strip. With this equation we can find n from measurable quantities.

It is also possible to use the Hall effect to measure directly the drift speed v_d of the charge carriers, which you may recall is of the order of centimeters per hour. In this clever experiment, the metal strip is moved mechanically through the magnetic field in a direction opposite that of the drift velocity of the charge carriers. The speed of the moving strip is then adjusted until the Hall potential difference vanishes. At this condition, with no Hall effect, the velocity of the charge carriers *with respect to the laboratory frame* must be zero, so the velocity of the strip must be equal in magnitude but opposite the direction of the velocity of the negative charge carriers.

Sample Problem 29-2

Figure 29-9 shows a solid metal cube, of edge length d = 1.5 cm, moving in the positive y direction at a constant velocity $\vec{v}$ of magnitude 4.0 m/s. The cube moves through a uniform magnetic field $\vec{B}$ of magnitude 0.050 T directed toward positive z.

(a) Which cube face is at a lower electric potential and which is at a higher electric potential because of the motion through the field?

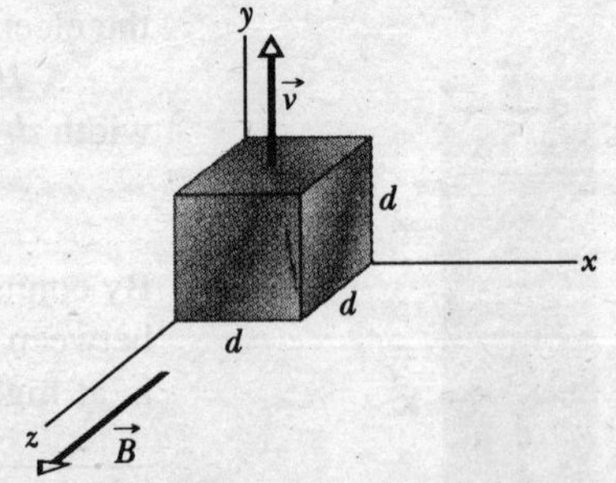

Fig. 29-9 Sample Problem 29-2. A solid metal cube of edge length d, at constant velocity $\vec{v}$ through a uniform magnetic field $\vec{B}$.

SOLUTION: One **Key Idea** here is that, because the cube is moving through a magnetic field $\vec{B}$, a magnetic force $\vec{F}_B$ acts on its charged particles, including its conduction electrons. A second **Key Idea** is how $\vec{F}_B$ causes an electric potential difference between certain faces of the cube. When the cube first begins to move through the magnetic field, its electrons do also. Because each electron has charge q and is moving through a magnetic field with velocity $\vec{v}$, the magnetic force $\vec{F}_B$ acting on it is given by Eq. 29-2. Because q is negative, the direction of $\vec{F}_B$ is opposite the cross product $\vec{v} \times \vec{B}$, which is in the positive direction of the x axis in Fig. 29-9. Thus, $\vec{F}_B$ acts in the negative direction of the x axis, toward the left face of the cube (which is hidden from view in Fig. 29-9).

Most of the electrons are fixed in place in the molecules of the cube. However, because the cube is a metal, it contains conduction electrons that are free to move. Some of those conduction electrons are deflected by $\vec{F}_B$ to the left cube face, making that face negatively charged and leaving the right face positively charged. This charge separation produces an electric field $\vec{E}$ directed from the positively charged right face to the negatively charged left face. Thus, the left face is at a lower electric potential, and the right face is at a higher electric potential.

(b) What is the potential difference between the faces of higher and lower electric potential?

SOLUTION: The **Key Ideas** here are these:

1. The electric field $\vec{E}$ created by the charge separation produces an electric force $\vec{F}_E = q\vec{E}$ on each electron. Because q is negative, this force is directed opposite the field $\vec{E}$—that is, toward the right. Thus on each electron, $\vec{F}_E$ acts toward the right and $\vec{F}_B$ acts toward the left.
2. When the cube had just begun to move through the magnetic field and the charge separation had just begun, the magnitude of $\vec{E}$ began to increase from zero. Thus, the magnitude of $\vec{F}_E$ also began to increase from zero and was initially smaller than the magnitude $\vec{F}_B$. During this early stage, the net force on any electron was dominated by $\vec{F}_B$, which continuously moved additional electrons to the left cube face, increasing the charge separation.
3. However, as the charge separation increased, eventually magnitude F_E became equal to magnitude F_B. The net force on any electron was then zero, and no additional electrons were moved to the left cube face. Thus, the magnitude of $\vec{F}_E$ could not increase further, and the electrons were then in equilibrium.

We seek the potential difference V between the left and right cube faces after equilibrium was reached (which occurred quickly). We can obtain V with Eq. 29-9 ($V = Ed$) provided we first find the magnitude E of the electric field at equilibrium. We can do so with the equation for the balance of forces ($F_E = F_B$).

For F_E, we substitute $|q|E$. For F_B, we substitute $|q|vB \sin \phi$ from Eq. 29-3. From Fig. 29-9, we see that the angle ϕ between vectors $\vec{v}$ and $\vec{B}$ is 90°; so $\sin \phi = 1$ and $F_E = F_B$ yields

$$|q|E = |q|vB \sin 90° = |q|vB.$$

This gives us $E = vB$, so Eq. 29-9 ($V = Ed$) becomes

$$V = vBd. \qquad (29\text{-}13)$$

Substituting known values gives us

$$V = (4.0\ \text{m/s})(0.050\ \text{T})(0.015\ \text{m})$$
$$= 0.0030\ \text{V} = 3.0\ \text{mV}. \qquad \text{(Answer)}$$

CHECKPOINT 3: The figure shows a metallic, rectangular solid that is to move at a certain speed v through the uniform magnetic field $\vec{B}$. Its dimensions are multiples of d, as shown. You have six choices for the direction of the velocity of the solid: it can be parallel to x, y, or z, in either the positive or negative direction. (a) Rank the six choices according to the potential difference that would be set up across the solid, greatest first. (b) For which choice is the front face at lower potential?

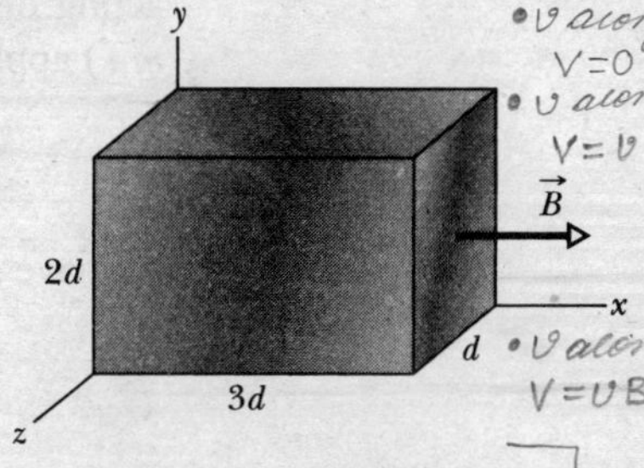

29-5 A Circulating Charged Particle

If a particle moves in a circle at constant speed, we can be sure that the net force acting on the particle is constant in magnitude and points toward the center of the circle, always perpendicular to the particle's velocity. Think of a stone tied to a string and whirled in a circle on a smooth horizontal surface, or of a satellite moving in a circular orbit around Earth. In the first case, the tension in the string provides the necessary force and centripetal acceleration. In the second case, Earth's gravitational attraction provides the force and acceleration.

Figure 29-10 shows another example: A beam of electrons is projected into a chamber by an *electron gun* G. The electrons enter in the plane of the page with speed v and then move in a region of uniform magnetic field $\vec{B}$ directed out of that plane. As a result, a magnetic force $\vec{F}_B = q\vec{v} \times \vec{B}$ continually deflects the electrons, and because $\vec{v}$ and $\vec{B}$ are always perpendicular to each other, this deflection causes the electrons to follow a circular path. The path is visible in the photo because atoms of gas in the chamber emit light when some of the circulating electrons collide with them.

Fig. 29-10 Electrons circulating in a chamber containing gas at low pressure (their path is the glowing circle). A uniform magnetic field $\vec{B}$, pointing directly out of the plane of the page, fills the chamber. Note the radially directed magnetic force $\vec{F}_B$; for circular motion to occur, $\vec{F}_B$ *must* point toward the center of the circle. Use the right-hand rule for cross products to confirm that $\vec{F}_B = q\vec{v} \times \vec{B}$ gives $\vec{F}_B$ the proper direction. (Don't forget the sign of q.)

We would like to determine the parameters that characterize the circular motion of these electrons, or of any particle of charge magnitude q and mass m moving perpendicular to a uniform magnetic field $\vec{B}$ at speed v. From Eq. 29-3, the force acting on the particle has a magnitude of qvB. From Newton's second law ($\vec{F} = m\vec{a}$) applied to uniform circular motion (Eq. 6-18),

$$F = m\frac{v^2}{r}, \tag{29-14}$$

we have

$$qvB = \frac{mv^2}{r}. \tag{29-15}$$

Solving for r, we find the radius of the circular path as

$$r = \frac{mv}{qB} \quad \text{(radius).} \tag{29-16}$$

★ The expression for T, f, ω would remain the same, regardless of ∠ b/w v & B provided ∠ ≠ 0 since then there would be no circulation

The period T (the time for one full revolution) is equal to the circumference divided by the speed:

$$T = \frac{2\pi r}{v} = \frac{2\pi}{v}\frac{mv}{qB} = \frac{2\pi m}{qB} \quad \text{(period).} \tag{29-17}$$

The frequency f (the number of revolutions per unit time) is

$$f = \frac{1}{T} = \frac{qB}{2\pi m} \quad \text{(frequency).} \tag{29-18}$$

The angular frequency ω of the motion is then

$$\omega = 2\pi f = \frac{qB}{m} \quad \text{(angular frequency).} \tag{29-19}$$

The quantities T, f, and ω do not depend on the speed of the particle (provided that speed is much less than the speed of light). Fast particles move in large circles and slow ones in small circles, but all particles with the same charge-to-mass ratio q/m

take the same time T (the period) to complete one round trip. Using Eq. 29-2, you can show that if you are looking in the direction of $\vec{B}$, the direction of rotation for a positive particle is always counterclockwise, and the direction for a negative particle is always clockwise.

Helical Paths

If the velocity of a charged particle has a component parallel to the (uniform) magnetic field, the particle will move in a helical path about the direction of the field vector. Figure 29-11*a*, for example, shows the velocity vector $\vec{v}$ of such a particle resolved into two components, one parallel to $\vec{B}$ and one perpendicular to it:

$$v_{\parallel} = v \cos \phi \quad \text{and} \quad v_{\perp} = v \sin \phi. \tag{29-20}$$

distance travelled by the particle parallel to the magnetic field during one period of circulation

The parallel component determines the *pitch p* of the helix—that is, the distance between adjacent turns (Fig. 29-11*b*). The perpendicular component determines the radius of the helix and is the quantity to be substituted for v in Eq. 29-16.

Figure 29-11*c* shows a charged particle spiraling in a nonuniform magnetic field. The more closely spaced field lines at the left and right sides indicate that the magnetic field is stronger there. When the field at an end is strong enough, the particle "reflects" from that end. If the particle reflects from both ends, it is said to be trapped in a *magnetic bottle*.

Electrons and protons are trapped in this way by the terrestrial magnetic field; the trapped particles form the *Van Allen radiation belts,* which loop well above Earth's atmosphere between Earth's north and south geomagnetic poles. These particles bounce back and forth, from one end of this magnetic bottle to the other, within a few seconds.

When a large solar flare shoots additional energetic electrons and protons into the radiation belts, an electric field is produced in the region where electrons normally reflect. This field eliminates the reflection and instead drives electrons down into the atmosphere, where they collide with atoms and molecules of air, causing

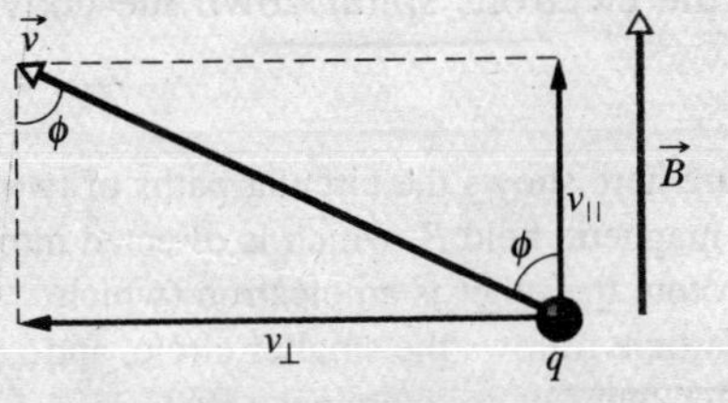

(*a*)

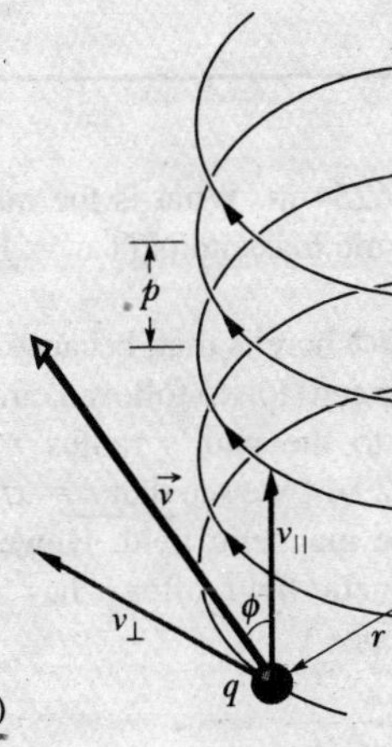

(*b*)

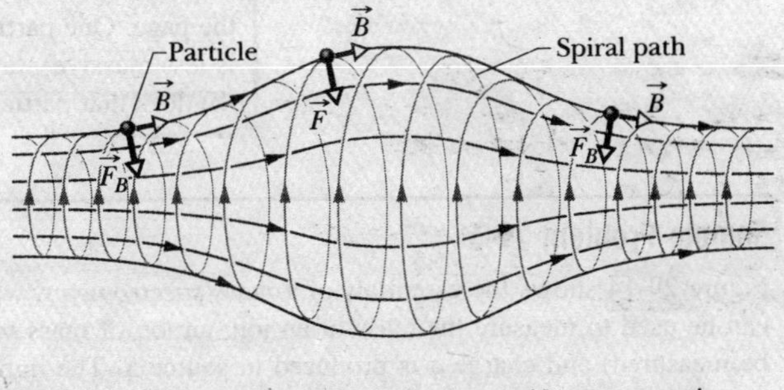

(*c*)

Fig. 29-11 (*a*) A charged particle moves in a uniform magnetic field $\vec{B}$, its velocity $\vec{v}$ making an angle ϕ with the field direction. (*b*) The particle follows a helical path, of radius r and pitch p. (*c*) A charged particle spiraling in a nonuniform magnetic field. (The particle can become trapped, spiraling back and forth between the strong field regions at either end.) Note that the magnetic force vectors at the left and right sides have a component pointing toward the center of the figure.

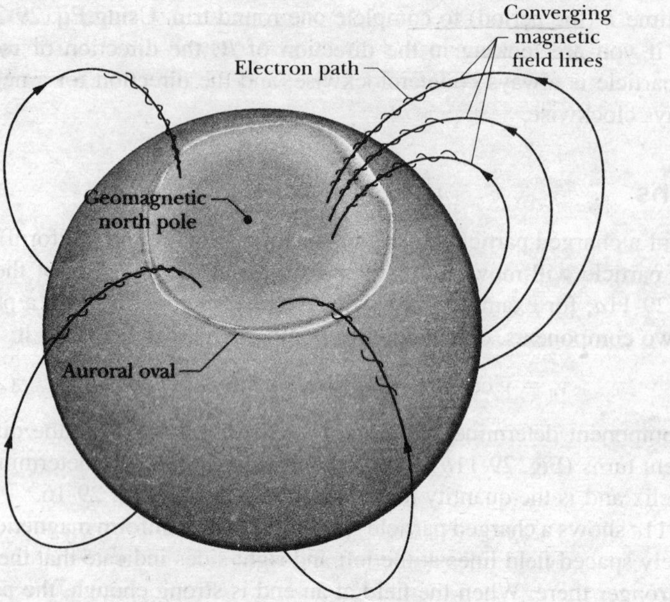

Fig. 29-12 The auroral oval surrounding Earth's geomagnetic north pole (in northwestern Greenland). Magnetic field lines converge toward that pole. Electrons moving toward Earth are "caught by" and spiral around these field lines, entering the terrestrial atmosphere at high latitudes and producing aurora within the oval.

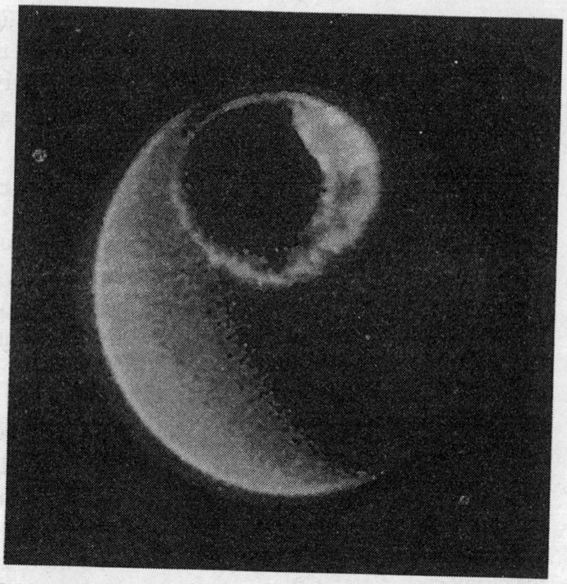

Fig. 29-13 A false-color image of aurora inside the north auroral oval, recorded by the satellite *Dynamic Explorer*, using ultraviolet light emitted by oxygen atoms excited in the aurora. The sun-lit portion of Earth is the crescent at the left.

that air to emit light. This light forms the aurora—a curtain of light that hangs down to an altitude of about 100 km. Green light is emitted by oxygen atoms, and pink light is emitted by nitrogen molecules, but often the light is so dim that we perceive only white light.

Aurora extend in arcs above Earth and can occur in a region called the *auroral oval* that is shown in Figs. 29-12 and 29-13 as seen from space. Although an aurora is long, it is less than 1 km thick (north to south) because the paths of the electrons producing it converge as the electrons spiral down the converging magnetic field lines (Fig. 29-12).

(b) x z y
• use either coordinates or sweep fingers from $\vec{v}$ to $\vec{B}$
consider point P, $\hat{B} = -\hat{k}$, $\hat{F} = \hat{j}$
proton $= +(\hat{v} \times -\hat{k}) = \hat{j}$
$\hat{v} = \hat{i}$ anticlockwise
electron $= -(\hat{v} \times -\hat{k}) = \hat{j}$
$= \hat{v} \times \hat{k} = \hat{j}$
$\hat{v} = -\hat{i}$ clockwise

CHECKPOINT 4: The figure here shows the circular paths of two particles that travel at the same speed in a uniform magnetic field $\vec{B}$, which is directed into the page. One particle is a proton; the other is an electron (which is less massive). (a) Which particle follows the smaller circle, and (b) does that particle travel clockwise or counterclockwise?

(a) Both electron & proton would experience the same force qvB
but $qvB = \frac{mv^2}{r}$ ∵ $m_e < m_p$ ⇒ $r_e < r_p$
ans = electron

Sample Problem 29-3

Figure 29-14 shows the essentials of a *mass spectrometer*, which can be used to measure the mass of an ion; an ion of mass m (to be measured) and charge q is produced in source S. The initially stationary ion is accelerated by the electric field due to a potential difference V. The ion leaves S and enters a separator chamber in which a uniform magnetic field $\vec{B}$ is perpendicular to the path of the ion. The magnetic field causes the ion to move in a semicircle, striking (and thus altering) a photographic plate at distance x from the entry slit. Suppose that in a certain trial $B = 80.000$ mT and $V = 1000.0$ V, and ions of charge $q = +1.6022 \times 10^{-19}$ C strike the plate at $x = 1.6254$ m. What is the mass m of the individual ions, in unified atomic mass units ($1\ \text{u} = 1.6605 \times 10^{-27}$ kg)?

SOLUTION: One **Key Idea** here is that, because the (uniform) magnetic field causes the (charged) ion to follow a circular path, we can relate the ion's mass m to the path's radius r with Eq. 29-16 ($r = mv/qB$). From Fig. 29-14 we see that $r = x/2$, and we are given the magnitude B of the magnetic field. However, we lack the ion's speed v in the magnetic field, after it has been accelerated due to the potential difference V.

ΔU=qΔV

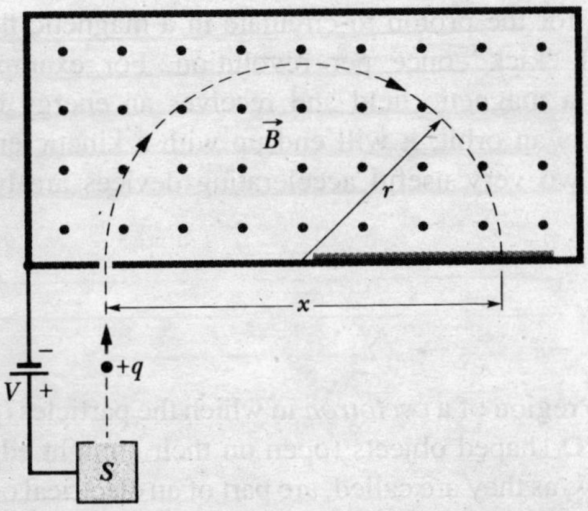

Fig. 29-14 Sample Problem 29-3. Essentials of a mass spectrometer. A positive ion, after being accelerated from its source S by potential difference V, enters a chamber of uniform magnetic field $\vec{B}$. There it travels through a semicircle of radius r and strikes a photographic plate at a distance x from where it entered the chamber.

To relate v and V, we use the **Key Idea** that mechanical energy ($E_{mec} = K + U$) is conserved during the acceleration. When the ion emerges from the source, its kinetic energy is approximately zero. At the end of the acceleration, its kinetic energy is $\frac{1}{2}mv^2$. Also, during the acceleration, the positive ion moves through a change in potential of $-V$. Thus, because the ion has positive charge q, its potential energy changes by $-qV$. If we now write the conservation of mechanical energy as

$$\Delta K + \Delta U = 0,$$

we get

$$\tfrac{1}{2}mv^2 - qV = 0$$

or

$$v = \sqrt{\frac{2qV}{m}}. \tag{29-21}$$

Substituting this into Eq. 29-16 gives us

$$r = \frac{mv}{qB} = \frac{m}{qB}\sqrt{\frac{2qV}{m}} = \frac{1}{B}\sqrt{\frac{2mV}{q}}.$$

Thus,

$$x = 2r = \frac{2}{B}\sqrt{\frac{2mV}{q}}.$$

Solving this for m and substituting the given data yield

$$m = \frac{B^2qx^2}{8V}$$
$$= \frac{(0.080000\ \text{T})^2(1.6022 \times 10^{-19}\ \text{C})(1.6254\ \text{m})^2}{8(1000.0\ \text{V})}$$
$$= 3.3863 \times 10^{-25}\ \text{kg} = 203.93\ \text{u}. \quad \text{(Answer)}$$

Sample Problem 29-4

An electron with a kinetic energy of 22.5 eV moves into a region of uniform magnetic field $\vec{B}$ of magnitude 4.55×10^{-4} T. The angle between the directions of $\vec{B}$ and the electron's velocity $\vec{v}$ is 65.5°. What is the pitch of the helical path taken by the electron?

SOLUTION: One **Key Idea** here is that the pitch p is the distance the electron travels parallel to the magnetic field $\vec{B}$ during one period T of circulation. A second **Key Idea** is that the period T is given by Eq. 29-17, regardless of the angle between the directions of $\vec{v}$ and $\vec{B}$ (provided the angle is not zero, for which there is no circulation of the electron). Thus, using Eqs. 29-20 and 29-17, we find

$$p = v_{\parallel}T = (v \cos \phi)\frac{2\pi m}{qB}. \tag{29-22}$$

We can calculate the electron's speed v from its kinetic energy as we did for the proton in Sample Problem 29-1. We find that $v = 2.81 \times 10^6$ m/s. Substituting this and known data in Eq. 29-22 gives us

$$p = (2.81 \times 10^6\ \text{m/s})(\cos 65.5°) \times \frac{2\pi(9.11 \times 10^{-31}\ \text{kg})}{(1.60 \times 10^{-19}\ \text{C})(4.55 \times 10^{-4}\ \text{T})}$$
$$= 9.16\ \text{cm}. \quad \text{(Answer)}$$

29-6 Cyclotrons and Synchrotrons

What is the structure of matter on the smallest scale? This question has always intrigued physicists. One way of getting at the answer is to allow an energetic charged particle (a proton, for example) to slam into a solid target. Better yet, allow two such energetic protons to collide head-on. Then analyze the debris from many such collisions to learn the nature of the subatomic particles of matter. The Nobel Prizes in physics for 1976 and 1984 were awarded for just such studies.

How can we give a proton enough kinetic energy for such an experiment? The direct approach is to allow the proton to "fall" through a potential difference V, thereby increasing its kinetic energy by eV. As we want higher and higher energies, however, it becomes more and more difficult to establish the necessary potential difference.

∵ proton/particle accelerates each time ⇒ KE inc

A better way is to arrange for the proton to circulate in a magnetic field, and to give it a modest electrical "kick" once per revolution. For example, if a proton circulates 100 times in a magnetic field and receives an energy boost of 100 keV every time it completes an orbit, it will end up with a kinetic energy of (100)(100 keV) or 10 MeV. Two very useful accelerating devices are based on this principle.

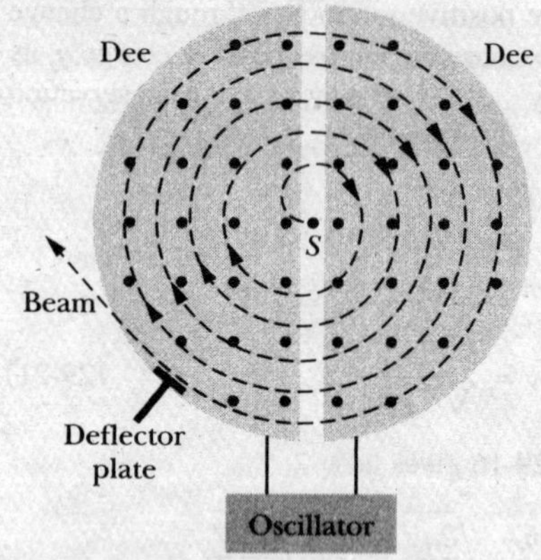

Fig. 29-15 The elements of a cyclotron, showing the particle source S and the dees. A uniform magnetic field is directed up from the plane of the page. Circulating protons spiral outward within the hollow dees, gaining energy every time they cross the gap between the dees.

The Cyclotron

Figure 29-15 is a top view of the region of a *cyclotron* in which the particles (protons, say) circulate. The two hollow **D**-shaped objects (open on their straight edges) are made of sheet copper. These *dees,* as they are called, are part of an electrical oscillator that alternates the electric potential difference across the gap between the dees. The electrical signs of the dees are alternated so that the electric field in the gap alternates in direction, first toward one dee and then the other dee, back and forth. The dees are immersed in a magnetic field ($B = 1.5$ T) whose direction is out of the plane of the page and that is set up by a large electromagnet.

Suppose that a proton, injected by source S at the center of the cyclotron in Fig. 29-15, initially moves toward a negatively charged dee. It will accelerate toward this dee and enter it. Once inside, it is shielded from electric fields by the copper walls of the dee; that is, the electric field does not enter the dee. The magnetic field, however, is not screened by the (nonmagnetic) copper dee, so the proton moves in a circular path whose radius, which depends on its speed, is given by Eq. 29-16 ($r = mv/qB$).

Let us assume that at the instant the proton emerges into the center gap from the first dee, the potential difference between the dees is reversed. Thus, the proton *again* faces a negatively charged dee and is *again* accelerated. This process continues, the circulating proton always being in step with the oscillations of the dee potential, until the proton has spiraled out to the edge of the dee system. There a deflector plate sends it out through a portal.

The key to the operation of the cyclotron is that the frequency f at which the proton circulates in the field (and that does *not* depend on its speed) must be equal to the fixed frequency f_{osc} of the electrical oscillator, or

$$f = f_{osc} \quad \text{(resonance condition).} \tag{29-23}$$

This *resonance condition* says that, if the energy of the circulating proton is to increase, energy must be fed to it at a frequency f_{osc} that is equal to the natural frequency f at which the proton circulates in the magnetic field.

Combining Eqs. 29-18 and 29-23 allows us to write the resonance condition as

$$qB = 2\pi m f_{osc}. \tag{29-24}$$

For the proton, q and m are fixed. The oscillator (we assume) is designed to work at a single fixed frequency f_{osc}. We then "tune" the cyclotron by varying B until Eq. 29-24 is satisfied and then many protons circulate through the magnetic field, to emerge as a beam.

The Proton Synchrotron

At proton energies above 50 MeV, the conventional cyclotron begins to fail because one of the assumptions of its design—that the frequency of revolution of a charged particle circulating in a magnetic field is independent of the particle's speed—is true only for speeds that are much less than the speed of light. At greater proton speeds (above about 10% of the speed of light), we must treat the problem relativistically. According to relativity theory, as the speed of a circulating proton approaches that of light, the proton's frequency of revolution decreases steadily. Thus, the protons get out of step with the cyclotron's oscillator—whose frequency remains fixed at f_{osc}—and eventually the energy of the circulating proton stops increasing.

There is another problem. For a 500 GeV proton in a magnetic field of 1.5 T, the path radius is 1.1 km. The corresponding magnet for a conventional cyclotron of the proper size would be impossibly expensive, the area of its pole faces being about 4×10^6 m^2.

The *proton synchrotron* is designed to meet these two difficulties. The magnetic field B and the oscillator frequency f_{osc}, instead of having fixed values as in the conventional cyclotron, are made to vary with time during the accelerating cycle. When this is done properly, (1) the frequency of the circulating protons remains in step with the oscillator at all times, and (2) the protons follow a circular—not a spiral—path. Thus, the magnet need extend only along that circular path, not over some 4×10^6 m^2. The circular path, however, still must be large if high energies are to be achieved. The proton synchrotron at the Fermi National Accelerator Laboratory (Fermilab) in Illinois has a circumference of 6.3 km and can produce protons with energies of about 1 TeV ($= 10^{12}$ eV).

Sample Problem 29-5

Suppose a cyclotron is operated at an oscillator frequency of 12 MHz and has a dee radius $R = 53$ cm.

(a) What is the magnitude of the magnetic field needed for deuterons to be accelerated in the cyclotron? A deuteron is the nucleus of deuterium, an isotope of hydrogen. It consists of a proton and a neutron and thus has the same charge as a proton. Its mass is $m = 3.34 \times 10^{-27}$ kg.

SOLUTION: The **Key Idea** here is that, for a given oscillator frequency f_{osc}, the magnetic field magnitude B required to accelerate any particle in a cyclotron depends on the ratio m/q of mass to charge for the particle, according to Eq. 29-24. For deuterons and the oscillator frequency $f_{osc} = 12$ MHz, we find

$$B = \frac{2\pi m f_{osc}}{q} = \frac{(2\pi)(3.34 \times 10^{-27}\ \text{kg})(12 \times 10^6\ \text{s}^{-1})}{1.60 \times 10^{-19}\ \text{C}}$$

$$= 1.57\ \text{T} \approx 1.6\ \text{T}. \qquad \text{(Answer)}$$

Note that, to accelerate protons, B would have to be reduced by a factor of 2, provided that the oscillator frequency remained fixed at 12 MHz.

(b) What is the resulting kinetic energy of the deuterons?

SOLUTION: One **Key Idea** here is that the kinetic energy ($\frac{1}{2}mv^2$) of a deuteron exiting the cyclotron is equal to the kinetic energy it had just before exiting, when it was traveling in a circular path with a radius approximately equal to the radius R of the cyclotron dees. A second **Key Idea** is that we can find the speed v of the deuteron in that circular path with Eq. 29-16 ($r = mv/qB$). Solving that equation for v, substituting R for r, and then substituting known data, we find

$$v = \frac{RqB}{m} = \frac{(0.53\ \text{m})(1.60 \times 10^{-19}\ \text{C})(1.57\ \text{T})}{3.34 \times 10^{-27}\ \text{kg}}$$

$$= 3.99 \times 10^7\ \text{m/s}.$$

This speed corresponds to a kinetic energy of

$$K = \tfrac{1}{2}mv^2$$

$$= \tfrac{1}{2}(3.34 \times 10^{-27}\ \text{kg})(3.99 \times 10^7\ \text{m/s})^2$$

$$= 2.7 \times 10^{-12}\ \text{J}, \qquad \text{(Answer)}$$

or about 17 MeV.

29-7 Magnetic Force on a Current-Carrying Wire

We have already seen (in connection with the Hall effect) that a magnetic field exerts a sideways force on electrons moving in a wire. This force must then be transmitted to the wire itself, because the conduction electrons cannot escape sideways out of the wire.

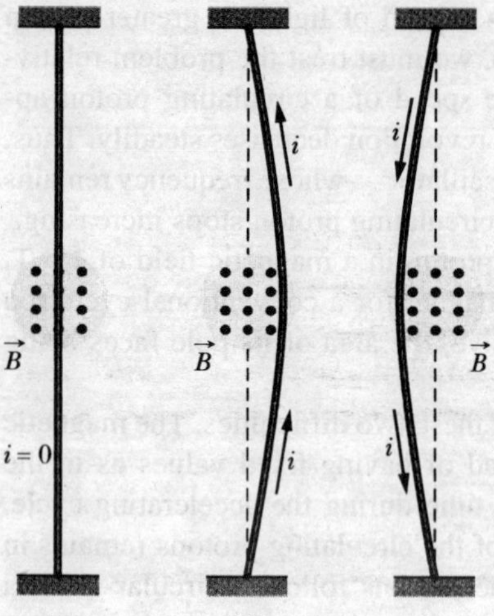

Fig. 29-16 A flexible wire passes between the pole faces of a magnet (only the farther pole face is shown). (*a*) Without current in the wire, the wire is straight. (*b*) With upward current, the wire is deflected rightward. (*c*) With downward current, the deflection is leftward. The connections for getting the current into the wire at one end and out of it at the other end are not shown.

In Fig. 29-16*a*, a vertical wire, carrying no current and fixed in place at both ends, extends through the gap between the vertical pole faces of a magnet. The magnetic field between the faces is directed outward from the page. In Fig. 29-16*b*, a current is sent upward through the wire; the wire deflects to the right. In Fig. 29-16*c*, we reverse the direction of the current and the wire deflects to the left.

Figure 29-17 shows what happens inside the wire of Fig. 29-16. We see one of the conduction electrons, drifting downward with an assumed drift speed v_d. Equation 29-3, in which we must put $\phi = 90°$, tells us that a force $\vec{F}_B$ of magnitude ev_dB must act on each such electron. From Eq. 29-2 we see that this force must be directed to the right. We expect then that the wire as a whole will experience a force to the right, in agreement with Fig. 29-16*b*.

If, in Fig. 29-17, we were to reverse *either* the direction of the magnetic field *or* the direction of the current, the force on the wire would reverse, being directed now to the left. Note too that it does not matter whether we consider negative charges drifting downward in the wire (the actual case) or positive charges drifting upward. The direction of the deflecting force on the wire is the same. We are safe then in dealing with a current of positive charge. ★ F is same for a particular current direction, no matter +ve or -ve carriers

NOTE Always remember:
1) Electron's v_d is opposite in direction w.r.t current
2) F_B has opposite direction if q is negative

Consider a length L of the wire in Fig. 29-17. All the conduction electrons in this section of wire will drift past plane xx in Fig. 29-17 in a time $t = L/v_d$. Thus, in that time a charge given by

$$q = it = i\frac{L}{v_d}$$

will pass through that plane. Substituting this into Eq. 29-3 yields

$$F_B = qv_dB\sin\phi = \frac{iL}{v_d}v_dB\sin 90°$$

or

$$F_B = iLB. \tag{29-25}$$

This equation gives the magnetic force that acts on a length L of straight wire carrying a current i and immersed in a magnetic field $\vec{B}$ that is perpendicular to the wire.

If the magnetic field is *not* perpendicular to the wire, as in Fig. 29-18, the magnetic force is given by a generalization of Eq. 29-25:

$$\vec{F}_B = i\vec{L} \times \vec{B} \quad \text{(force on a current).} \tag{29-26}$$

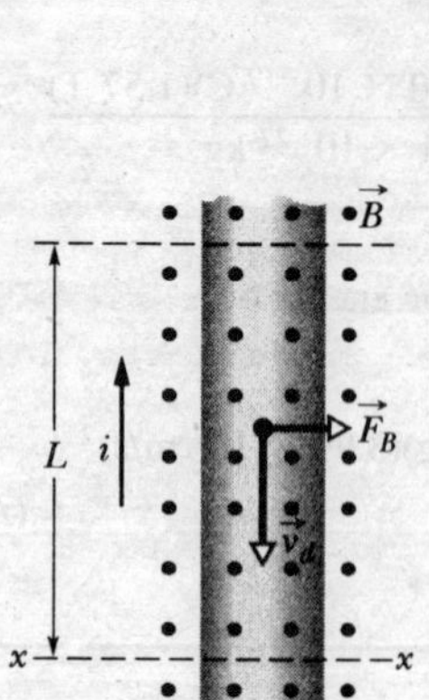

Fig. 29-17 A close-up view of a section of the wire of Fig. 29-16*b*. The current direction is upward, which means that electrons drift downward. A magnetic field that emerges from the plane of the page causes the electrons and the wire to be deflected to the right.

★Here $\vec{L}$ is a *length vector* that has magnitude L and is directed along the wire segment in the direction of the (conventional) current. The force magnitude F_B is

$$F_B = iLB\sin\phi, \tag{29-27}$$

where ϕ is the angle between the directions of $\vec{L}$ and $\vec{B}$. The direction of $\vec{F}_B$ is that of the cross product $\vec{L} \times \vec{B}$, because we take current i to be a positive quantity. Equation 29-26 tells us that $\vec{F}_B$ is always perpendicular to the plane defined by vectors $\vec{L}$ and $\vec{B}$, as indicated in Fig. 29-18.

Equation 29-26 is equivalent to Eq. 29-2 in that either can be taken as the

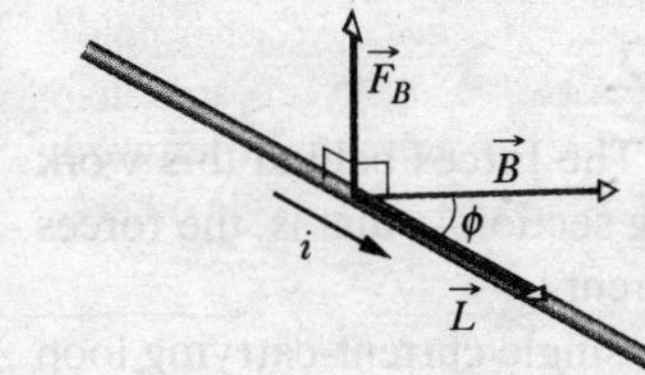

Fig. 29-18 A wire carrying current i makes an angle ϕ with magnetic field $\vec{B}$. The wire has length L in the field and length vector $\vec{L}$ (in the direction of the current). A magnetic force $\vec{F}_B = i\vec{L} \times \vec{B}$ acts on the wire.

defining equation for $\vec{B}$. In practice, we define $\vec{B}$ from Eq. 29-26. It is much easier to measure the magnetic force acting on a wire than that on a single moving charge.

If a wire is not straight or the field is not uniform, we can imagine it broken up into small straight segments and apply Eq. 29-26 to each segment. The force on the wire as a whole is then the vector sum of all the forces on the segments that make it up. In the differential limit, we can write

$$d\vec{F}_B = i\, d\vec{L} \times \vec{B}, \tag{29-28}$$

and we can find the resultant force on any given arrangement of currents by integrating Eq. 29-28 over that arrangement.

In using Eq. 29-28, bear in mind that there is no such thing as an isolated current-carrying wire segment of length dL. There must always be a way to introduce the current into the segment at one end and take it out at the other end.

CHECKPOINT 5: The figure shows a current i through a wire in a uniform magnetic field $\vec{B}$, as well as the magnetic force $\vec{F}_B$ acting on the wire. The field is oriented so that the force is maximum. In what direction is the field?

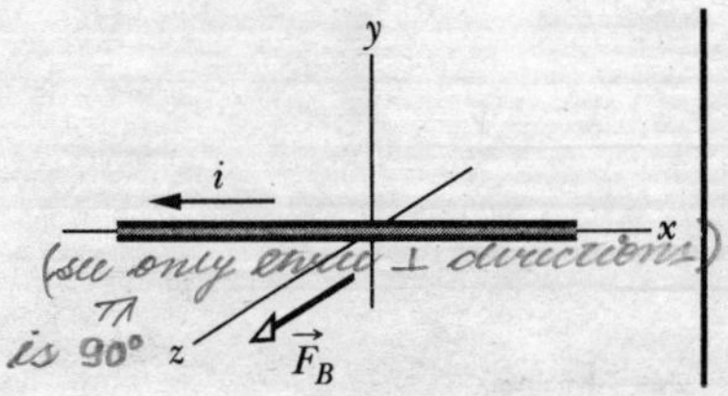

Sample Problem 29-6

A straight, horizontal length of copper wire has a current $i = 28$ A through it. What are the magnitude and direction of the minimum magnetic field $\vec{B}$ needed to suspend the wire—that is, to balance the gravitational force on it? The linear density (mass per unit length) of the wire is 46.6 g/m.

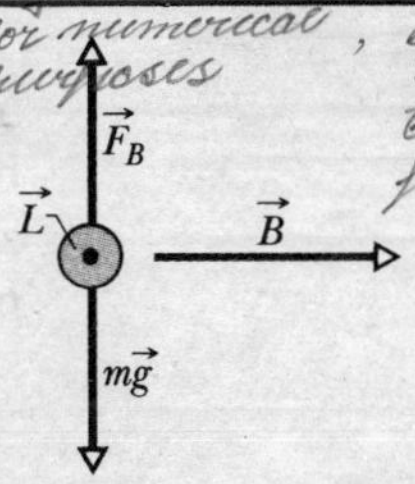

Fig. 29-19 Sample Problem 29-6. A current-carrying wire (shown in cross section) can be made to "float" in a magnetic field. The current in the wire emerges from the plane of the page, and the magnetic field is directed to the right.

SOLUTION: One **Key Idea** is that, because the wire carries a current, a magnetic force $\vec{F}_B$ can act on the wire if we place it in a magnetic field $\vec{B}$. To balance the downward gravitational force $\vec{F}_g$ on the wire, we want $\vec{F}_B$ to be directed upward (Fig. 29-19).

A second **Key Idea** is that the direction of $\vec{F}_B$ is related to the directions of $\vec{B}$ and the wire's length vector $\vec{L}$ by Eq. 29-26. Because $\vec{L}$ is directed horizontally (and the current is taken to be positive), Eq. 29-26 and the right-hand rule for cross products tell us that $\vec{B}$ must be horizontal and rightward (in Fig. 29-19) to give the required upward $\vec{F}_B$.

The magnitude of $\vec{F}_B$ is given by Eq. 29-27 ($F_B = iLB \sin \phi$). Because we want $\vec{F}_B$ to balance $\vec{F}_g$, we want

$$iLB \sin \phi = mg, \tag{29-29}$$

where mg is the magnitude of $\vec{F}_g$ and m is the mass of the wire. We also want the minimal field magnitude B for $\vec{F}_B$ to balance $\vec{F}_g$. Thus, we need to maximize $\sin \phi$ in Eq. 29-29. To do so, we set $\phi = 90°$, thereby arranging for $\vec{B}$ to be perpendicular to the wire. We then have $\sin \phi = 1$, so Eq. 29-29 yields

$$B = \frac{mg}{iL \sin \phi} = \frac{(m/L)g}{i}. \tag{29-30}$$

We write the result this way because we know m/L, the linear density of the wire. Substituting known data then gives us

$$B = \frac{(46.6 \times 10^{-3}\ \text{kg/m})(9.8\ \text{m/s}^2)}{28\ \text{A}}$$
$$= 1.6 \times 10^{-2}\ \text{T}. \qquad \text{(Answer)}$$

This is about 160 times the strength of Earth's magnetic field.

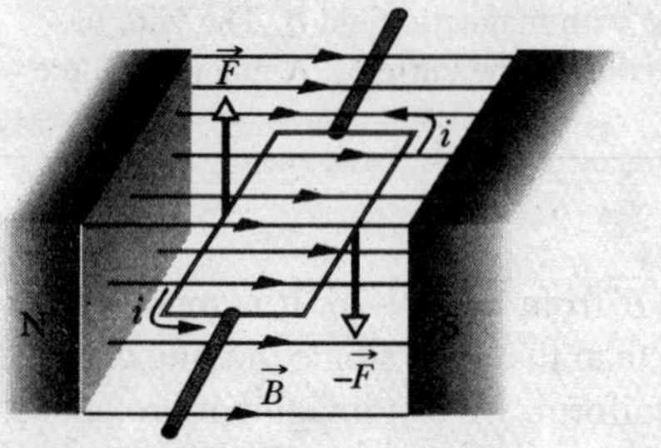

Fig. 29-20 The elements of an electric motor. A rectangular loop of wire, carrying a current and free to rotate about a fixed axis, is placed in a magnetic field. Magnetic forces on the wire produce a torque that rotates it. A commutator (not shown) reverses the direction of the current every half-revolution so that the torque always acts in the same direction.

29-8 Torque on a Current Loop

Much of the world's work is done by electric motors. The forces behind this work are the magnetic forces that we studied in the preceding section—that is, the forces that a magnetic field exerts on a wire that carries a current.

Figure 29-20 shows a simple motor, consisting of a single current-carrying loop immersed in a magnetic field $\vec{B}$. The two magnetic forces $\vec{F}$ and $-\vec{F}$ produce a torque on the loop, tending to rotate it about its central axis. Although many essential details have been omitted, the figure does suggest how the action of a magnetic field on a current loop produces rotary motion. Let us analyze that action.

Figure 29-21a shows a rectangular loop of sides a and b, carrying current i through uniform magnetic field $\vec{B}$. We place it in the field so that its long sides, labeled 1 and 3, are perpendicular to the field direction (which is into the page), but its short sides, labeled 2 and 4, are not. Wires to lead the current into and out of the loop are needed but, for simplicity, they are not shown.

To define the orientation of the loop in the magnetic field, we use a normal vector $\vec{n}$ that is perpendicular to the plane of the loop. Figure 29-21b shows a right-hand rule for finding the direction of $\vec{n}$. Point or curl the fingers of your right hand in the direction of the current at any point on the loop. Your extended thumb then points in the direction of the normal vector $\vec{n}$.

In Fig. 29-21c, the normal vector of the loop is shown at an arbitrary angle θ to the direction of the magnetic field $\vec{B}$. We wish to find the net force and net torque acting on the loop in this orientation.

The net force on the loop is the vector sum of the forces acting on its four sides. For side 2 the vector $\vec{L}$ in Eq. 29-26 points in the direction of the current and has magnitude b. The angle between $\vec{L}$ and $\vec{B}$ for side 2 (see Fig. 29-21c) is $90° - \theta$. Thus, the magnitude of the force acting on this side is

$$F_2 = ibB \sin(90° - \theta) = ibB \cos \theta. \tag{29-31}$$

You can show that the force $\vec{F}_4$ acting on side 4 has the same magnitude as $\vec{F}_2$ but the opposite direction. Thus, $\vec{F}_2$ and $\vec{F}_4$ cancel out exactly. Their net force is zero and, because their common line of action is through the center of the loop, their net torque is also zero.

The situation is different for sides 1 and 3. For them, $\vec{L}$ is perpendicular to $\vec{B}$,

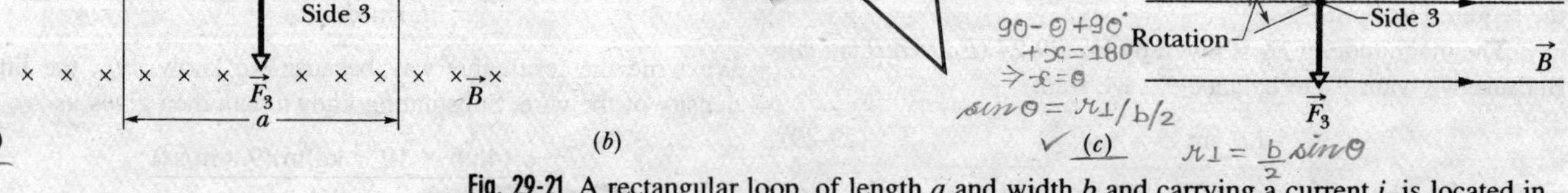

Fig. 29-21 A rectangular loop, of length a and width b and carrying a current i, is located in a uniform magnetic field. A torque $\vec{\tau}$ acts to align the normal vector $\vec{n}$ with the direction of the field. (a) The loop as seen by looking in the direction of the magnetic field. (b) A perspective of the loop showing how the right-hand rule gives the direction of $\vec{n}$, which is perpendicular to the plane of the loop. (c) A side view of the loop, from side 2. The loop rotates as indicated.

so the forces $\vec{F}_1$ and $\vec{F}_3$ have the common magnitude iaB. Because these two forces have opposite directions, they do not tend to move the loop up or down. However, as Fig. 29-21*c* shows, these two forces do *not* share the same line of action so they *do* produce a net torque. The torque tends to rotate the loop so as to align its normal vector $\vec{n}$ with the direction of the magnetic field $\vec{B}$. That torque has moment arm $(b/2)\sin\theta$ about the central axis of the loop. The magnitude τ' of the torque due to forces $\vec{F}_1$ and $\vec{F}_3$ is then (see Fig. 29-21*c*)

$$\tau' = \left(iaB\frac{b}{2}\sin\theta\right) + \left(iaB\frac{b}{2}\sin\theta\right) = iabB\sin\theta. \tag{29-32}$$

Suppose we replace the single loop of current with a *coil* of N loops, or *turns*. Further, suppose that the turns are wound tightly enough that they can be approximated as all having the same dimensions and lying in a plane. Then the turns form a *flat coil* and a torque τ' with the magnitude given in Eq. 29-32 acts on each of them. The total torque on the coil then has magnitude

$$\boxed{\tau = N\tau' = NiabB\sin\theta = (NiA)B\sin\theta,} \tag{29-33}$$

in which A $(= ab)$ is the area enclosed by the coil. The quantities in parentheses (NiA) are grouped together because they are all properties of the coil: its number of turns, its area, and the current it carries. Equation 29-33 holds for all flat coils, no matter what their shape, provided the magnetic field is uniform.

Instead of focusing on the motion of the coil, it is simpler to keep track of the vector $\vec{n}$, which is normal to the plane of the coil. Equation 29-33 tells us that a current-carrying flat coil placed in a magnetic field will tend to rotate so that $\vec{n}$ has the same direction as the field.

In a motor, the current in the coil is reversed as $\vec{n}$ begins to line up with the field direction, so that a torque continues to rotate the coil. This automatic reversal of the current is done via a commutator that electrically connects the rotating coil with the stationary contacts on the wires that supply the current from some source.

Sample Problem 29-7

Analog voltmeters and ammeters work by measuring the torque exerted by a magnetic field on a current-carrying coil. The reading is displayed by means of the deflection of a pointer over a scale. Figure 29-22 shows the essentials of a *galvanometer*, on which both analog ammeters and analog voltmeters are based. Assume the coil is 2.1 cm high and 1.2 cm wide, has 250 turns, and is mounted so that it can rotate about an axis (into the page) in a uniform *radial* magnetic field with $B = 0.23$ T. For any orientation of the coil, the net magnetic field through the coil is perpendicular to the normal vector of the coil (and thus parallel to the plane of the coil). A spring Sp provides a countertorque that balances the magnetic torque, so that a given steady current i in the coil results in a steady angular deflection ϕ. The greater the current is, the greater the deflection is, and thus the greater the torque required of the spring is. If a current of 100 μA produces an angular deflection of 28°, what must be the torsional constant κ of the spring, as used in Eq. 16-22 $(\tau = -\kappa\phi)$?

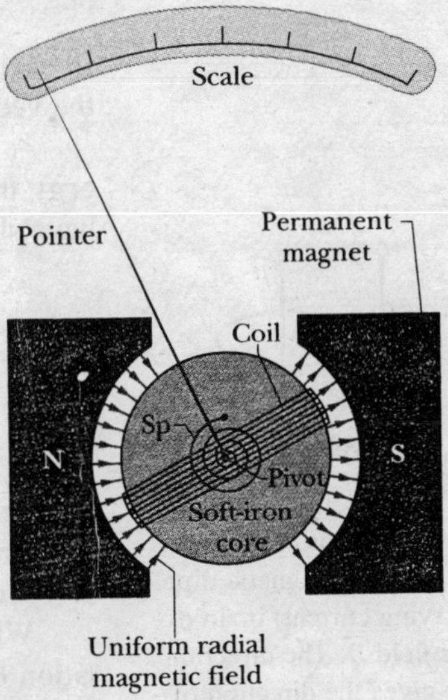

Fig. 29-22 Sample Problem 29-7. The elements of a galvanometer. Depending on the external circuit, this device can be wired up as either a voltmeter or an ammeter.

SOLUTION: The **Key Idea** here is that, with a constant current through the device, the resulting magnetic torque (Eq. 29-33) is balanced by the spring torque. Thus, the magnitudes of those torques are equal:

$$NiAB \sin \theta = \kappa\phi. \qquad (29\text{-}34)$$

Here ϕ is the angular deflection of the coil and pointer, and A ($= 2.52 \times 10^{-4}\ \text{m}^2$) is the area encircled by the coil. Since the net magnetic field through the coil is always perpendicular to the normal vector of the coil, $\theta = 90°$ for any orientation of the pointer.

Solving Eq. 29-34 for κ, we find

$$\kappa = \frac{NiAB \sin \theta}{\phi}$$
$$= (250)(100 \times 10^{-6}\ \text{A})(2.52 \times 10^{-4}\ \text{m}^2) \times \frac{(0.23\ \text{T})(\sin 90°)}{28°}$$
$$= 5.2 \times 10^{-8}\ \text{N} \cdot \text{m/degree}. \qquad \text{(Answer)}$$

Many modern ammeters and voltmeters are of the digital, direct-reading type and do not use a moving coil.

29-9 The Magnetic Dipole Moment

We can describe the current-carrying coil of the preceding section with a single vector $\vec{\mu}$, its **magnetic dipole moment.** We take the direction of $\vec{\mu}$ to be that of the normal vector $\vec{n}$ to the plane of the coil, as in Fig. 29-21c. We define the magnitude of $\vec{\mu}$ as

$$\mu = NiA \quad \text{(magnetic moment)}, \qquad (29\text{-}35)$$

in which N is the number of turns in the coil, i is the current through the coil, and A is the area enclosed by each turn of the coil. (Equation 29-35 tells us that the unit of $\vec{\mu}$ is the ampere-square meter.) Using $\vec{\mu}$, we can rewrite Eq. 29-33 for the torque on the coil due to a magnetic field as

$$\tau = \mu B \sin \theta, \qquad (29\text{-}36)$$

in which θ is the angle between the vectors $\vec{\mu}$ and $\vec{B}$.

We can generalize this to the vector relation

$$\vec{\tau} = \vec{\mu} \times \vec{B}, \qquad (29\text{-}37)$$

which reminds us very much of the corresponding equation for the torque exerted by an *electric* field on an *electric* dipole—namely, Eq. 23-34:

$$\vec{\tau} = \vec{p} \times \vec{E}.$$

In each case the torque due to the field—either magnetic or electric—is equal to the vector product of the corresponding dipole moment and the field vector.

A magnetic dipole in an external magnetic field has a **magnetic potential energy** that depends on the dipole's orientation in the field. For electric dipoles we have shown (Eq. 23-38) that

$$U(\theta) = -\vec{p} \cdot \vec{E}.$$

In strict analogy, we can write for the magnetic case

$$U(\theta) = -\vec{\mu} \cdot \vec{B}. \qquad (29\text{-}38)$$

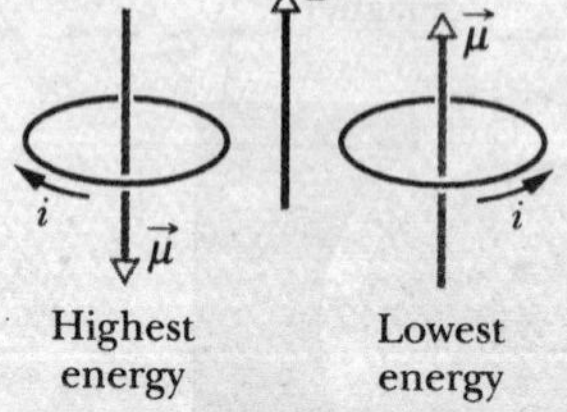

Fig. 29-23 The orientations of highest and lowest energy of a magnetic dipole (here a coil carrying current) in an external magnetic field $\vec{B}$. The direction of the current i gives the direction of the magnetic dipole moment $\vec{\mu}$ via the right-hand rule shown for $\vec{n}$ in Fig. 29-21b.

A magnetic dipole has its lowest energy ($= -\mu B \cos 0 = -\mu B$) when its dipole moment $\vec{\mu}$ is lined up with the magnetic field (Fig. 29-23). It has its highest energy ($= -\mu B \cos 180° = +\mu B$) when $\vec{\mu}$ is directed opposite the field.

When a magnetic dipole rotates from an initial orientation θ_i to another orientation θ_f, the work W done on the dipole by the magnetic field is

$$W = -\Delta U = -(U_f - U_i), \qquad (29\text{-}39)$$

where U_f and U_i are calculated with Eq. 29-38. If an applied torque (due to "an

TABLE 29-2 Some Magnetic Dipole Moments

A small bar magnet	5 J/T
Earth	8.0×10^{22} J/T
A proton	1.4×10^{-26} J/T
An electron	9.3×10^{-24} J/T

external agent") acts on the dipole during the change in its orientation, then work W_a is done on the dipole by the applied torque. *If the dipole is stationary* before and after the change in its orientation, then work W_a is the negative of the work done on the dipole by the field. Thus,

$$W_a = -W = U_f - U_i. \tag{29-40}$$

So far, we have identified only a current-carrying coil as a magnetic dipole. However, a simple bar magnet is also a magnetic dipole, as is a rotating sphere of charge. Earth itself is (approximately) a magnetic dipole. Finally, most subatomic particles, including the electron, the proton, and the neutron, have magnetic dipole moments. As you will see in Chapter 32, all these quantities can be viewed as current loops. For comparison, some approximate magnetic dipole moments are shown in Table 29-2.

CHECKPOINT 6: The figure shows four orientations, at angle θ, of a magnetic dipole moment $\vec{\mu}$ in a magnetic field. Rank the orientations according to (a) the magnitude of the torque on the dipole and (b) the potential energy of the dipole, greatest first.

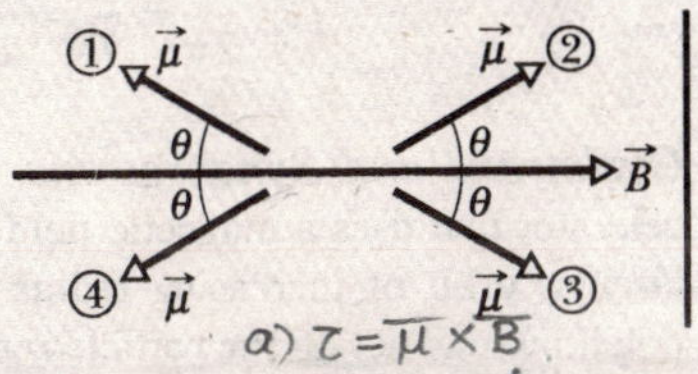

Sample Problem 29-8

Figure 29-24 shows a circular coil with 250 turns, an area A of 2.52×10^{-4} m^2, and a current of 100 μA. The coil is at rest in a uniform magnetic field of magnitude $B = 0.85$ T, with its magnetic dipole moment $\vec{\mu}$ initially aligned with $\vec{B}$.

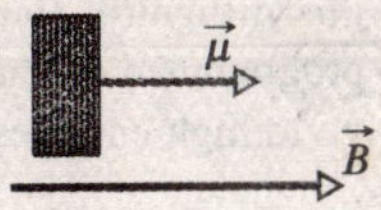

Fig. 29-24 Sample Problem 29-8. A side view of a circular coil carrying a current and oriented so that its magnetic dipole moment $\vec{\mu}$ is aligned with magnetic field $\vec{B}$.

(a) In Fig. 29-24, what is the direction of the current in the coil?

SOLUTION: The Key Idea here is to apply the following right-hand rule to the coil: Imagine cupping the coil with your right hand so that your right thumb is outstretched in the direction of $\vec{\mu}$. The direction in which your fingers curl around the coil is the direction of the current in the coil. Thus, in the wires on the near side of the coil—those we see in Fig. 29-24—the current is from top to bottom.

(b) How much work would the torque applied by an external agent have to do on the coil to rotate it 90° from its initial orientation, so that $\vec{\mu}$ is perpendicular to $\vec{B}$ and the coil is again at rest?

SOLUTION: The Key Idea here is that the work W_a done by the applied torque would be equal to the change in the coil's potential energy due to its change in orientation. From Eq. 29-40 ($W_a = U_f - U_i$), we find

$$\begin{aligned} W_a &= U(90°) - U(0°) \\ &= -\mu B \cos 90° - (-\mu B \cos 0°) = 0 + \mu B \\ &= \mu B. \end{aligned}$$

Substituting for μ from Eq. 29-35 ($\mu = NiA$), we find that

$$\begin{aligned} W_a &= (NiA)B \\ &= (250)(100 \times 10^{-6}\ \text{A})(2.52 \times 10^{-4}\ \text{m}^2)(0.85\ \text{T}) \\ &= 5.356 \times 10^{-6}\ \text{J} \approx 5.4\ \mu\text{J}. \end{aligned} \quad \text{(Answer)}$$

REVIEW & SUMMARY

Magnetic Field $\vec{B}$ A **magnetic field** $\vec{B}$ is defined in terms of the force $\vec{F}_B$ acting on a test particle with charge q moving through the field with velocity $\vec{v}$:

$$\vec{F}_B = q\vec{v} \times \vec{B}. \tag{29-2}$$

The SI unit for $\vec{B}$ is the **tesla** (T): 1 T = 1 N/(A · m) = 10^4 gauss.

The Hall Effect When a conducting strip of thickness l carrying a current i is placed in a uniform magnetic field $\vec{B}$, some charge carriers (with charge e) build up on the sides of the conductor, creating a potential difference V across the strip. The polarities of the sides indicate the sign of the charge carriers; the number density n of charge carriers can be calculated with

$$n = \frac{Bi}{Vle}. \tag{29-12}$$

A Charged Particle Circulating in a Magnetic Field A charged particle with mass m and charge magnitude q moving with velocity $\vec{v}$ perpendicular to a uniform magnetic field $\vec{B}$ will travel in a circle. Applying Newton's second law to the circular motion yields

$$qvB = \frac{mv^2}{r}, \qquad (29\text{-}15)$$

from which we find the radius r of the circle to be

$$r = \frac{mv}{qB}. \qquad (29\text{-}16)$$

The frequency of revolution f, the angular frequency ω, and the period of the motion T are given by

$$f = \frac{\omega}{2\pi} = \frac{1}{T} = \frac{qB}{2\pi m}. \qquad (29\text{-}19, 29\text{-}18, 29\text{-}17)$$

Cyclotrons and Synchrotrons A cyclotron is a particle accelerator that uses a magnetic field to hold a charged particle in a circular orbit of increasing radius so that a modest accelerating potential may act on the particle repeatedly, providing it with high energy. Because the moving particle gets out of step with the oscillator as its speed approaches that of light, there is an upper limit to the energy attainable with the cyclotron. A synchrotron avoids this difficulty. Here both B and the oscillator frequency f_{osc} are programmed to change cyclically so that the particle not only can go to high energies but can do so at a constant orbital radius.

Magnetic Force on a Current-Carrying Wire A straight wire carrying a current i in a uniform magnetic field experiences a sideways force

$$\vec{F}_B = i\vec{L} \times \vec{B}. \qquad (29\text{-}26)$$

The force acting on a current element $i\,d\vec{L}$ in a magnetic field is

$$d\vec{F}_B = i\,d\vec{L} \times \vec{B}. \qquad (29\text{-}28)$$

The direction of the length vector $\vec{L}$ or $d\vec{L}$ is that of the current i.

Torque on a Current-Carrying Coil A coil (of area A and carrying current i, with N turns) in a uniform magnetic field $\vec{B}$ will experience a torque $\vec{\tau}$ given by

$$\vec{\tau} = \vec{\mu} \times \vec{B}. \qquad (29\text{-}37)$$

Here $\vec{\mu}$ is the **magnetic dipole moment** of the coil, with magnitude $\mu = NiA$ and direction given by the right-hand rule.

Orientation Energy of a Magnetic Dipole The **magnetic potential energy** of a magnetic dipole in a magnetic field is

$$U(\theta) = -\vec{\mu} \cdot \vec{B}. \qquad (29\text{-}38)$$

If a magnetic dipole rotates from an initial orientation θ_i to another orientation θ_f, the work W done on the dipole by the magnetic field is

$$W = -\Delta U = -(U_f - U_i). \qquad (29\text{-}39)$$

QUESTIONS

1. Figure 29-25 shows three situations in which a positive particle of velocity $\vec{v}$ moves through a uniform magnetic field $\vec{B}$ and experiences a magnetic force $\vec{F}_B$. In each situation, determine whether the orientations of the vectors are physically reasonable.

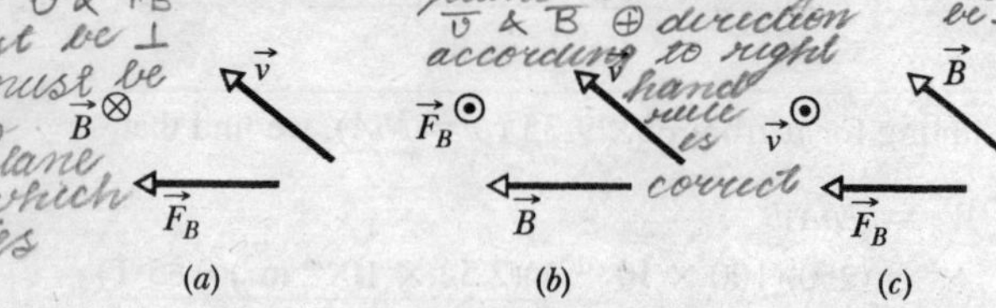

Fig. 29-25 Question 1.

2. For four situations, here is the velocity $\vec{v}$ of a proton at a certain instant as it moves through a uniform magnetic field $\vec{B}$:

(a) $\vec{v} = 2\hat{i} - 3\hat{j}$ and $\vec{B} = 4\hat{k}$
(b) $\vec{v} = 3\hat{i} + 2\hat{j}$ and $\vec{B} = -4\hat{k}$
(c) $\vec{v} = 3\hat{j} - 2\hat{k}$ and $\vec{B} = 4\hat{i}$
(d) $\vec{v} = 20\hat{i}$ and $\vec{B} = -4\hat{i}$.

Without written calculation, rank the situations according to the magnitude of the magnetic force on the proton, greatest first.

3. In Section 29-3, we discussed a charged particle moving through crossed fields with the forces $\vec{F}_E$ and $\vec{F}_B$ in opposition. We found that the particle moves in a straight line (that is, neither force dominates the motion) if its speed is given by Eq. 29-7 ($v = E/B$). Which of the two forces dominates if the speed of the particle is, instead, (a) $v < E/B$ and (b) $v > E/B$?

4. Figure 29-26 shows crossed and uniform electric and magnetic fields $\vec{E}$ and $\vec{B}$ and, at a certain instant, the velocity vectors of the 10 charged particles listed in Table 29-3. (The vectors are not drawn to scale.) The table gives the signs of the charges and the speeds of the particles; the speeds are given as either less than or greater than E/B (see Question 3). Which particles will move out of the page toward you after the instant of Fig. 29-26?

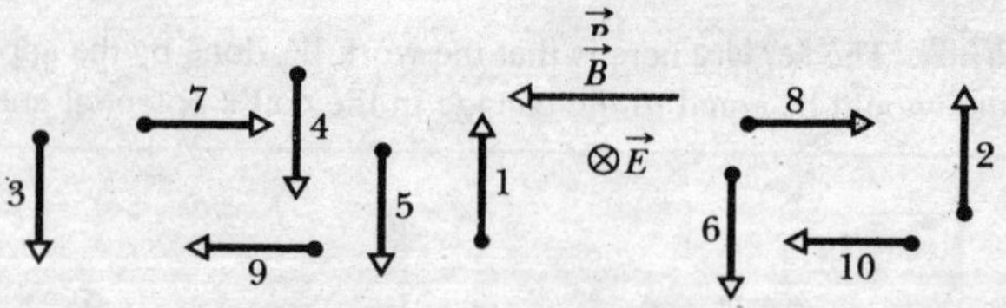

Fig. 29-26 Question 4.

TABLE 29-3 Question 4

Particle	Charge	Speed	Particle	Charge	Speed
1	+	Less	6	−	Greater
2	+	Greater	7	+	Less
3	+	Less	8	+	Greater
4	+	Greater	9	−	Less
5	−	Less	10	−	Greater

5. In Fig. 29-27, a charged particle enters a uniform magnetic field $\vec{B}$ with speed v_0, moves through a half-circle in time T_0, and then leaves the field. (a) Is the charge positive or negative? (b) Is the final speed of the particle greater than, less than, or equal to v_0? (c) If the initial speed had been $0.5v_0$, would the time spent in field $\vec{B}$ have been greater than, less than, or equal to T_0? (d) Would the path have been a half-circle, more than a half-circle, or less than a half-circle?

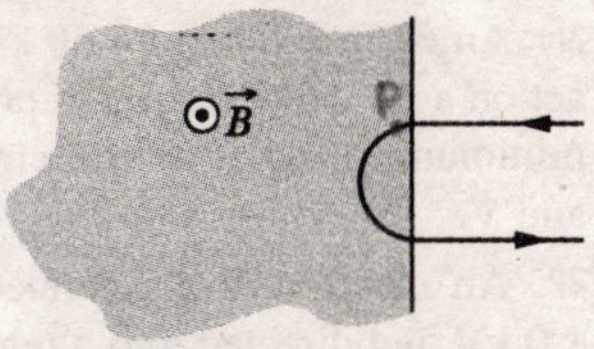

Fig. 29-27 Question 5.

6. Figure 29-28 shows the path of a particle through six regions of uniform magnetic field, where the path is either a half-circle or a quarter-circle. Upon leaving the last region, the particle travels between two charged, parallel plates and is deflected toward the plate of higher potential. What are the directions of the magnetic fields in the six regions? ⇒ particle is -vely charged

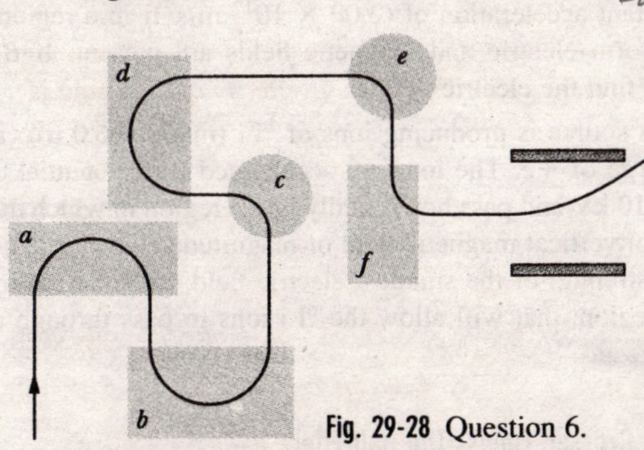

Fig. 29-28 Question 6.

7. Figure 29-29 shows the path of an electron that passes through two regions containing uniform magnetic fields of magnitudes B_1 and B_2. Its path in each region is a half-circle. (a) Which field is stronger? (b) What are the directions of the two fields? (c) Is the time spent by the electron in the $\vec{B}_1$ region greater than, less than, or the same as the time spent in the $\vec{B}_2$ region?

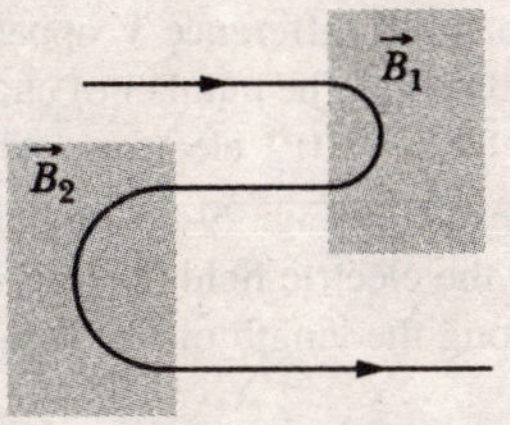

Fig. 29-29 Question 7.

8. *Particle Roundabout.* Figure 29-30 shows 11 paths through a region of uniform magnetic field. One path is a straight line; the rest are half-circles. Table 29-4 gives the masses, charges, and speeds of 11 particles that take these paths through the field in the directions shown. Which path in the figure corresponds to which particle in the table?

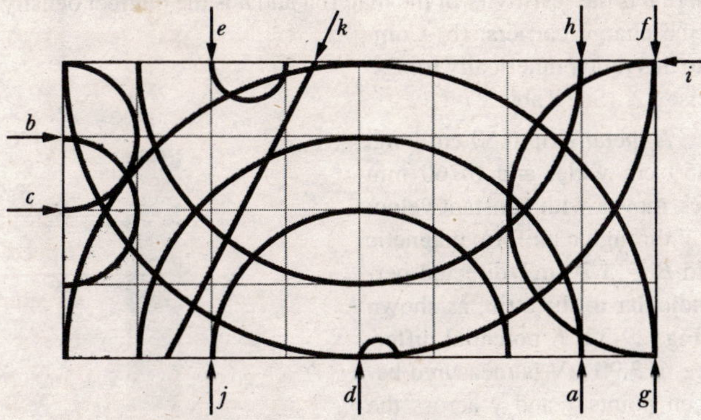

all 2ie **Fig. 29-30** Question 8.

TABLE 29-4 Question 8

Particle	Mass	Charge	Speed
1	$2m$	q	v
2	m	$2q$	v
3	$m/2$	q	$2v$
4	$3m$	$3q$	$3v$
5	$2m$	q	$2v$
6	m	$-q$	$2v$
7	m	$-4q$	v
8	m	$-q$	v
9	$2m$	$-2q$	$8v$
10	m	$-2q$	$8v$
11	$3m$	0	$3v$

9. Figure 29-31 shows eight wires that carry identical currents through the same uniform magnetic field (directed into the page) in eight separate experiments. Each wire consists of two straight sections (each of length L and either parallel or perpendicular to the x and y axes shown) and one curved section (with radius of curvature R). The directions of the currents through the wires are indicated by the arrows next to the wires. (a) Give the direction of the net magnetic force on each wire in terms of an angle measured counterclockwise from the positive direction of the x axis. (b) Rank wires 1 through 4 according to the magnitude of the net magnetic force on them, greatest first. (c) Do the same type of ranking for wires 5 through 8. ie from x

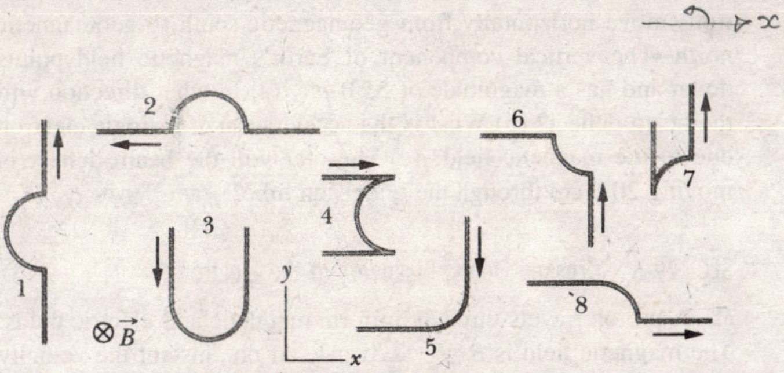

Fig. 29-31 Question 9.

10. (a) In Checkpoint 6, if the dipole moment $\vec{\mu}$ rotates from orientation 1 to orientation 2, is the work done on the dipole *by the magnetic field* positive, negative, or zero? (b) Rank the work done on the dipole by the magnetic field for rotations from orientation 1 to (1) orientation 2, (2) orientation 3, and (3) orientation 4, greatest first.

(a) $\Delta W_B = -(U_f - U_i) = -(U_f(\theta) - U_i(\theta))$

$U_f(\theta) = -\vec{\mu}\cdot\vec{B} = -\mu B\cos\theta$

$U_i(\theta) = -\vec{\mu}\cdot\vec{B} = -\mu B\cos(180-2\theta) = +\mu B\cos 2\theta$

$\Delta W = +\mu B(\cos\theta + \cos 2\theta)$ ΔW +ve

$\Delta W_{12} = \mu B(\cos\theta + \cos 2\theta)$

$\Delta W_{13} = \mu B(\cos\theta + \cos 2\theta)$

$\Delta W_{14} = \mu B(\cos\theta + \cos 2\theta)$

EXERCISES & PROBLEMS

ssm Solution is in the Student Solutions Manual.
www Solution is available on the World Wide Web at: http://www.wiley.com/college/hrw
ilw Solution is available on the Interactive LearningWare.

SEC. 29-2 The Definition of $\vec{B}$

1E. An alpha particle travels at a velocity $\vec{v}$ of magnitude 550 m/s through a uniform magnetic field $\vec{B}$ of magnitude 0.045 T. (An alpha particle has a charge of $+3.2 \times 10^{-19}$ C and a mass of 6.6×10^{-27} kg.) The angle between $\vec{v}$ and $\vec{B}$ is 52°. What are the magnitudes of (a) the force $\vec{F}_B$ acting on the particle due to the field and (b) the acceleration of the particle due to $\vec{F}_B$? (c) Does the speed of the particle increase, decrease, or remain equal to 550 m/s?

2E. An electron in a TV camera tube is moving at 7.20×10^6 m/s in a magnetic field of strength 83.0 mT. (a) Without knowing the direction of the field, what can you say about the greatest and least magnitudes of the force acting on the electron due to the field? (b) At one point the electron has an acceleration of magnitude 4.90×10^{14} m/s^2. What is the angle between the electron's velocity and the magnetic field?

3E. A proton traveling at 23.0° with respect to the direction of a magnetic field of strength 2.60 mT experiences a magnetic force of 6.50×10^{-17} N. Calculate (a) the proton's speed and (b) its kinetic energy in electron-volts. ssm ilw

4P. An electron that has velocity

$$\vec{v} = (2.0 \times 10^6 \text{ m/s})\hat{\text{i}} + (3.0 \times 10^6 \text{ m/s})\hat{\text{j}}$$

moves through the magnetic field $\vec{B} = (0.030 \text{ T})\hat{\text{i}} - (0.15 \text{ T})\hat{\text{j}}$. (a) Find the force on the electron. (b) Repeat your calculation for a proton having the same velocity.

5P. Each of the electrons in the beam of a television tube has a kinetic energy of 12.0 keV. The tube is oriented so that the electrons move horizontally from geomagnetic south to geomagnetic north. The vertical component of Earth's magnetic field points down and has a magnitude of 55.0 μT. (a) In what direction will the beam deflect? (b) What is the acceleration of a single electron due to the magnetic field? (c) How far will the beam deflect in moving 20.0 cm through the television tube? ssm www

SEC. 29-3 Crossed Fields: Discovery of the Electron

6E. A proton travels through uniform magnetic and electric fields. The magnetic field is $\vec{B} = -2.5\hat{\text{i}}$ mT. At one instant the velocity of the proton is $\vec{v} = 2000\hat{\text{j}}$ m/s. At that instant, what is the magnitude of the net force acting on the proton if the electric field is (a) $4.0\hat{\text{k}}$ V/m, (b) $-4.0\hat{\text{k}}$ V/m, and (c) $4.0\hat{\text{i}}$ V/m?

7E. An electron with kinetic energy 2.5 keV moves horizontally into a region of space in which there is a downward-directed uniform electric field of magnitude 10 kV/m. (a) What are the magnitude and direction of the (smallest) uniform magnetic field that will cause the electron to continue to move horizontally? Ignore the gravitational force, which is rather small. (b) Is it possible for a proton to pass through this combination of fields undeflected? If so, under what circumstances? ssm

8E. An electric field of 1.50 kV/m and a magnetic field of 0.400 T act on a moving electron to produce no net force. (a) Calculate the minimum speed v of the electron. (b) Draw the vectors $\vec{E}$, $\vec{B}$, and $\vec{v}$.

9P. An electron is accelerated through a potential difference of 1.0 kV and directed into a region between two parallel plates separated by 20 mm with a potential difference of 100 V between them. The electron is moving perpendicular to the electric field of the plates when it enters the region between the plates. What uniform magnetic field, applied perpendicular to both the electron path and the electric field, will allow the electron to travel in a straight line? ilw

(handwritten: only mag is asked; $-\frac{eV}{d} = -e(\vec{v}\times\vec{B}) \Rightarrow -evB = -\frac{eV}{d}$)

10P. An electron has an initial velocity of $(12.0\hat{\text{j}} + 15.0\hat{\text{k}})$ km/s and a constant acceleration of $(2.00 \times 10^{12} \text{ m/s}^2)\hat{\text{i}}$ in a region in which uniform electric and magnetic fields are present. If $\vec{B} = (400\ \mu\text{T})\hat{\text{i}}$, find the electric field $\vec{E}$.

11P. An ion source is producing ions of ^{6}Li (mass = 6.0 u), each with a charge of $+e$. The ions are accelerated by a potential difference of 10 kV and pass horizontally into a region in which there is a uniform vertical magnetic field of magnitude $B = 1.2$ T. Calculate the strength of the smallest electric field, to be set up over the same region, that will allow the ^{6}Li ions to pass through undeflected. ssm

(handwritten: $1u = 1.6\times10^{-27}$ kg = mass of proton)

SEC. 29-4 Crossed Fields: The Hall Effect

12E. A strip of copper 150 μm wide is placed in a uniform magnetic field $\vec{B}$ of magnitude 0.65 T, with $\vec{B}$ perpendicular to the strip. A current $i = 23$ A is then sent through the strip such that a Hall potential difference V appears across the width of the strip. Calculate V. (The number of charge carriers per unit volume for copper is 8.47×10^{28} electrons/m^3.)

13P. (a) In Fig. 29-8, show that the ratio of the Hall electric field E to the electric field E_C responsible for moving charge (the current) along the length of the strip is

$$\frac{E}{E_C} = \frac{B}{ne\rho},$$

where ρ is the resistivity of the material and n is the number density of the charge carriers. (b) Compute this ratio numerically for Exercise 12. (See Table 27-1.)

14P. A metal strip 6.50 cm long, 0.850 cm wide, and 0.760 mm thick moves with constant velocity $\vec{v}$ through a uniform magnetic field $B = 1.20$ mT directed perpendicular to the strip, as shown in Fig. 29-32. A potential difference of 3.90 μV is measured between points x and y across the strip. Calculate the speed v.

(handwritten: we always seek the PD after equilibrium had reached (which occurs quickly))

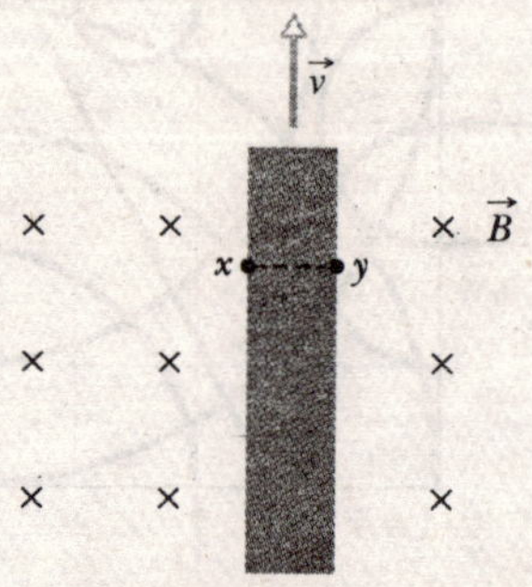

Fig. 29-32 Problem 14.

SEC. 29-5 A Circulating Charged Particle

15E. What uniform magnetic field, applied perpendicular to a beam of electrons moving at 1.3×10^6 m/s, is required to make the electrons travel in a circular arc of radius 0.35 m? ssm

16E. An electron is accelerated from rest by a potential difference of 350 V. It then enters a uniform magnetic field of magnitude 200 mT with its velocity perpendicular to the field. Calculate (a) the speed of the electron and (b) the radius of its path in the magnetic field.

17E. An electron with kinetic energy 1.20 keV circles in a plane perpendicular to a uniform magnetic field. The orbit radius is 25.0 cm. Find (a) the speed of the electron, (b) the magnetic field, (c) the frequency of circling, and (d) the period of the motion. ssm

18E. Physicist S. A. Goudsmit devised a method for measuring the masses of heavy ions by timing their periods of revolution in a known magnetic field. A singly charged ion of iodine makes 7.00 rev in a field of 45.0 mT in 1.29 ms. Calculate its mass, in unified atomic mass units. (Actually, the method allows mass measurements to be carried out to much greater accuracy than these approximate data suggest.)

19E. (a) Find the frequency of revolution of an electron with an energy of 100 eV in a uniform magnetic field of 35.0 μT. (b) Calculate the radius of the path of this electron if its velocity is perpendicular to the magnetic field. ilw

20E. An alpha particle ($q = +2e$, $m = 4.00$ u) travels in a circular path of radius 4.50 cm in a uniform magnetic field with $B =$ 1.20 T. Calculate (a) its speed, (b) its period of revolution, (c) its kinetic energy in electron-volts, and (d) the potential difference through which it would have to be accelerated to achieve this energy.

21E. A beam of electrons whose kinetic energy is K emerges from a thin-foil "window" at the end of an accelerator tube. There is a metal plate a distance d from this window and perpendicular to the direction of the emerging beam (Fig. 29-33). Show that we can prevent the beam from hitting the plate if we apply a uniform magnetic field $\vec{B}$ such that

$$B \geq \sqrt{\frac{2mK}{e^2d^2}},$$

in which m and e are the electron mass and charge. How should $\vec{B}$ be oriented? ssm

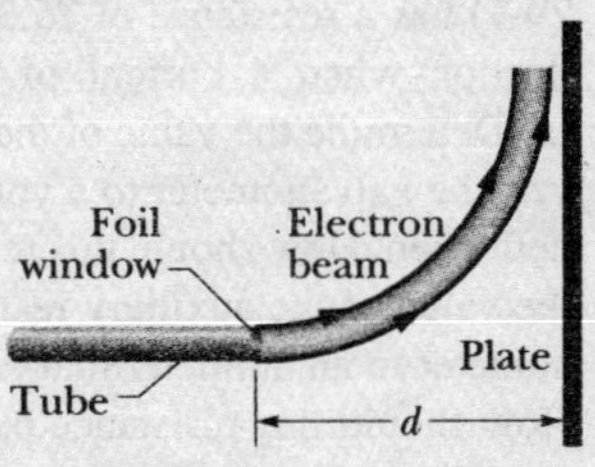

Fig. 29-33 Exercise 21.

22P. A source injects an electron of speed $v = 1.5 \times 10^7$ m/s into a uniform magnetic field of magnitude $B = 1.0 \times 10^{-3}$ T. The velocity of the electron makes an angle $\theta = 10°$ with the direction of the magnetic field. Find the distance d from the point of injection at which the electron next crosses the field line that passes through the injection point.

23P. In a nuclear experiment a proton with kinetic energy 1.0 MeV moves in a circular path in a uniform magnetic field. What energy must (a) an alpha particle ($q = +2e$, $m = 4.0$ u) and (b) a deuteron ($q = +e$, $m = 2.0$ u) have if they are to circulate in the same circular path?

24P. A proton, a deuteron ($q = +e$, $m = 2.0$ u), and an alpha particle ($q = +2e$, $m = 4.0$ u) with the same kinetic energies enter a region of uniform magnetic field $\vec{B}$, moving perpendicular to $\vec{B}$. Compare the radii of their circular paths.

25P. A certain commercial mass spectrometer (see Sample Problem 29-3) is used to separate uranium ions of mass 3.92×10^{-25} kg and charge 3.20×10^{-19} C from related species. The ions are accelerated through a potential difference of 100 kV and then pass into a uniform magnetic field, where they are bent in a path of radius 1.00 m. After traveling through 180° and passing through a slit of width 1.00 mm and height 1.00 cm, they are collected in a cup. (a) What is the magnitude of the (perpendicular) magnetic field in the separator? If the machine is used to separate out 100 mg of material per hour, calculate (b) the current of the desired ions in the machine and (c) the thermal energy produced in the cup in 1.00 h. ssm

26P. A proton of charge $+e$ and mass m enters a uniform magnetic field $\vec{B} = B\hat{i}$ with an initial velocity $\vec{v} = v_{0x}\hat{i} + v_{0y}\hat{j}$. Find an expression in unit-vector notation for its velocity $\vec{v}$ at any later time t.

27P. A positron with kinetic energy 2.0 keV is projected into a uniform magnetic field $\vec{B}$ of magnitude 0.10 T, with its velocity vector making an angle of 89° with $\vec{B}$. Find (a) the period, (b) the pitch p, and (c) the radius r of its helical path. ssm www

28P. In Fig. 29-34, a charged particle moves into a region of uniform magnetic field $\vec{B}$, goes through half a circle, and then exits that region. The particle is either a proton or an electron (you must decide which). It spends 130 ns within the region. (a) What is the magnitude of $\vec{B}$? (b) If the particle is sent back through the magnetic field (along the same initial path) but with 2.00 times its previous kinetic energy, how much time does it spend within the field?

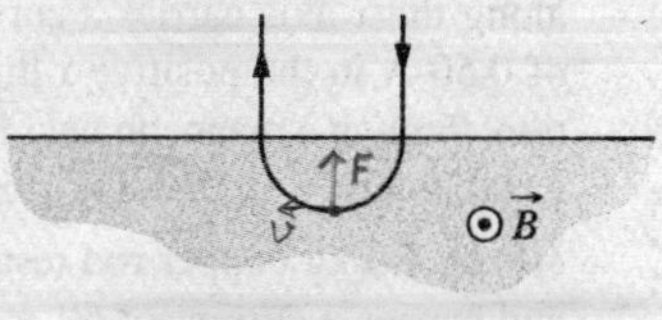

Fig. 29-34 Problem 28.

29P. A neutral particle is at rest in a uniform magnetic field $\vec{B}$. At time $t = 0$ it decays into two charged particles, each of mass m. (a) If the charge of one of the particles is $+q$, what is the charge of the other? (b) The two particles move off in separate paths, both of which lie in the plane perpendicular to $\vec{B}$. At a later time the particles collide. Express the time from decay until collision in terms of m, B, and q. ssm

SEC. 29-6 Cyclotrons and Synchrotrons

30E. In a certain cyclotron a proton moves in a circle of radius 0.50 m. The magnitude of the magnetic field is 1.2 T. (a) What is the oscillator frequency? (b) What is the kinetic energy of the proton, in electron-volts?

31P. Estimate the total path length traveled by a deuteron in the cyclotron of Sample Problem 29-5 during the (entire) acceleration process. Assume that the accelerating potential between the dees is 80 kV. ssm www

32P. The oscillator frequency of the cyclotron in Sample Problem 29-5 has been adjusted to accelerate deuterons ($q = +e$, $m =$ 2.0 u). (a) If protons are injected instead of deuterons, to what kinetic energy can the protons be accelerated, using the same oscillator frequency? (b) What magnetic field would be required? (c) What kinetic energy could be produced for protons if the magnetic field were left at the value used for deuterons? (d) What oscillator frequency would then be required? (e) Answer the same questions for alpha particles ($q = +2e$, $m = 4.0$ u).

SEC. 29-7 Magnetic Force on a Current-Carrying Wire

33E. A horizontal conductor that is part of a power line carries a current of 5000 A from south to north. Earth's magnetic field (60.0 μT) is directed toward the north and is inclined downward at 70° to the horizontal. Find the magnitude and direction of the magnetic force on 100 m of the conductor due to Earth's field. ssm

34E. A wire 1.80 m long carries a current of 13.0 A and makes an angle of 35.0° with a uniform magnetic field $B = 1.50$ T. Calculate the magnetic force on the wire.

35E. A wire of 62.0 cm length and 13.0 g mass is suspended by a pair of flexible leads in a uniform magnetic field of magnitude 0.440 T (Fig. 29-35). What are the magnitude and direction of the current required to remove the tension in the supporting leads? ssm ilw

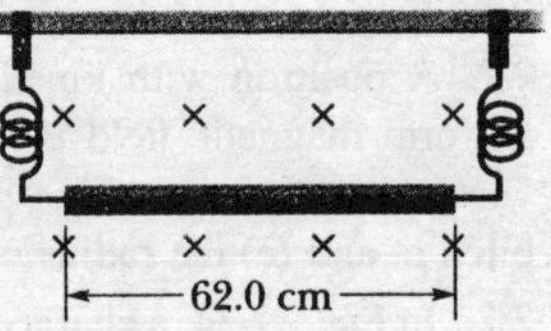

Fig. 29-35 Exercise 35.

36P. A wire 50 cm long lying along the x axis carries a current of 0.50 A in the positive x direction, through a magnetic field $\vec{B} = (0.0030\text{ T})\hat{j} + (0.010\text{ T})\hat{k}$. Find the magnetic force on the wire.

37P. A 1.0 kg copper rod rests on two horizontal rails 1.0 m apart and carries a current of 50 A from one rail to the other. The coefficient of static friction between rod and rails is 0.60. What is the smallest magnetic field (not necessarily vertical) that would cause the rod to slide? ssm

38P. Consider the possibility of a new design for an electric train. The engine is driven by the force on a conducting axle due to the vertical component of Earth's magnetic field. To produce the force, current is maintained down one rail, through a conducting wheel, through the axle, through another conducting wheel, and then back to the source via the other rail. (a) What current is needed to provide a modest 10 kN force? Take the vertical component of Earth's field to be 10 μT and the length of the axle to be 3.0 m. (b) At what rate would electric energy be lost for each ohm of resistance in the rails? (c) Is such a train totally or just marginally unrealistic?

SEC. 29-8 Torque on a Current Loop

39E. Figure 29-36 shows a rectangular 20-turn coil of wire, of dimensions 10 cm by 5.0 cm. It carries a current of 0.10 A and is hinged along one long side. It is mounted in the xy plane, at 30° to the direction of a uniform magnetic field of magnitude 0.50 T. Find the magnitude and direction of the torque acting on the coil about the hinge line. ssm

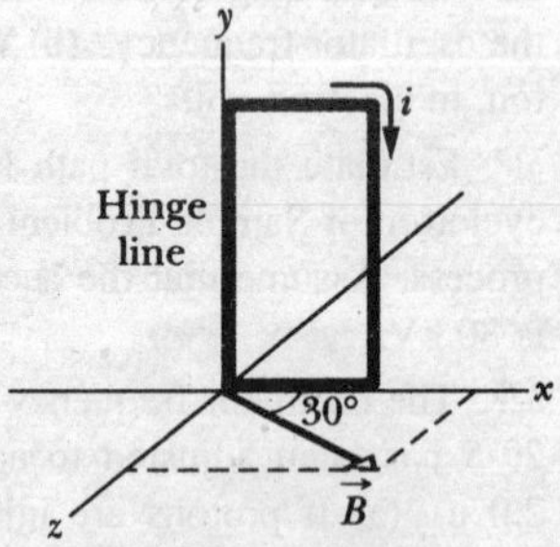

Fig. 29-36 Exercise 39.

40E. A single-turn current loop, carrying a current of 4.00 A, is in the shape of a right triangle with sides 50.0, 120, and 130 cm. The loop is in a uniform magnetic field of magnitude 75.0 mT whose direction is parallel to the current in the 130 cm side of the loop. (a) Find the magnitude of the magnetic force on each of the three sides of the loop. (b) Show that the total magnetic force on the loop is zero.

41E. A length L of wire carries a current i. Show that if the wire is formed into a circular coil, then the maximum torque in a given magnetic field is developed when the coil has one turn only, and that maximum torque has the magnitude $\tau = L^2iB/4\pi$. ssm ilw

42P. Prove that the relation $\tau = NiAB \sin\theta$ holds for closed loops of arbitrary shape and not only for rectangular loops as in Fig. 29-21. (*Hint:* Replace the loop of arbitrary shape with an assembly of adjacent long, thin, approximately rectangular loops that are nearly equivalent to the loop of arbitrary shape as far as the distribution of current is concerned.)

43P. Figure 29-37 shows a wire ring of radius a that is perpendicular to the general direction of a radially symmetric, diverging magnetic field. The magnetic field at the ring is everywhere of the same magnitude B, and its direction at the ring everywhere makes an angle θ with a normal to the plane of the ring. The twisted lead wires have no effect on the problem. Find the magnitude and direction of the force the field exerts on the ring if the ring carries a current i. ssm www

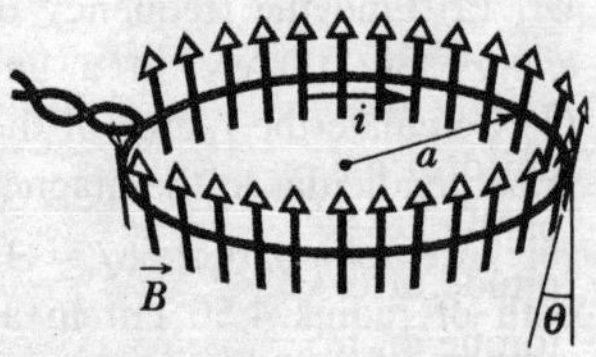

Fig. 29-37 Problem 43.

44P. A closed wire loop with current i is in a uniform magnetic field $\vec{B}$, with the plane of the loop at angle θ to the direction of $\vec{B}$. Show that the total magnetic force on the loop is zero. Does your proof also hold for a nonuniform magnetic field?

45P. The coil of a certain galvanometer (see Sample Problem 29-7) has a resistance of 75.3 Ω; its needle shows a full-scale deflection when a current of 1.62 mA passes through the coil. (a) Determine the value of the auxiliary resistance required to convert the galvanometer to a voltmeter that reads 1.00 V at full-scale deflection. How should this resistance be connected? (b) Determine the value of the auxiliary resistance required to convert the galvanometer to an ammeter that reads 50.0 mA at full-scale deflection. How should this resistance be connected? ssm

46P. A particle of charge q moves in a circle of radius a with speed v. Treating the circular path as a current loop with constant current equal to its average current, find the maximum torque exerted on the loop by a uniform magnetic field of magnitude B.

47P. Figure 29-38 shows a wood cylinder of mass $m = 0.250$ kg and length $L = 0.100$ m, with $N = 10.0$ turns of wire wrapped around it longitudinally, so that the plane of the wire coil contains the axis of the cylinder. What is the least current i through the coil

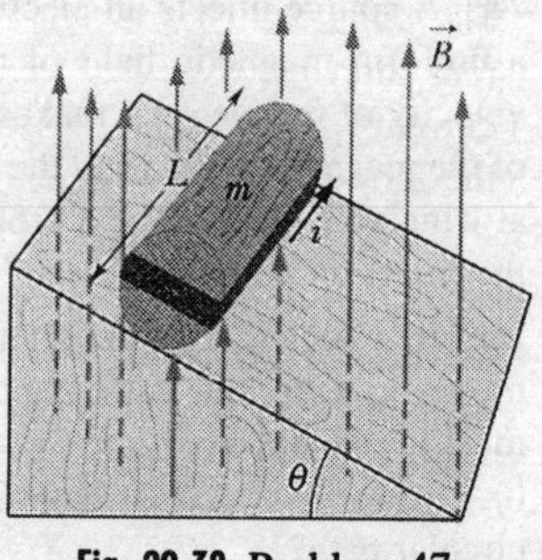

Fig. 29-38 Problem 47.

that will prevent the cylinder from rolling down a plane inclined at an angle θ to the horizontal, in the presence of a vertical, uniform magnetic field of magnitude 0.500 T, if the plane of the coil is parallel to the inclined plane? ssm

SEC. 29-9 The Magnetic Dipole Moment

48E. The magnetic dipole moment of Earth is 8.00×10^{22} J/T. Assume that this is produced by charges flowing in Earth's molten outer core. If the radius of their circular path is 3500 km, calculate the current they produce.

49E. A circular coil of 160 turns has a radius of 1.90 cm. (a) Calculate the current that results in a magnetic dipole moment of $2.30\ \mathrm{A \cdot m^2}$. (b) Find the maximum torque that the coil, carrying this current, can experience in a uniform 35.0 mT magnetic field. ssm

50E. A circular wire loop whose radius is 15.0 cm carries a current of 2.60 A. It is placed so that the normal to its plane makes an angle of 41.0° with a uniform magnetic field of 12.0 T. (a) Calculate the magnetic dipole moment of the loop. (b) What torque acts on the loop?

51E. A current loop, carrying a current of 5.0 A, is in the shape of a right triangle with sides 30, 40, and 50 cm. The loop is in a uniform magnetic field of magnitude 80 mT whose direction is parallel to the current in the 50 cm side of the loop. Find the magnitude of (a) the magnetic dipole moment of the loop and (b) the torque on the loop. ssm

52E. A stationary circular wall clock has a face with a radius of 15 cm. Six turns of wire are wound around its perimeter; the wire carries a current of 2.0 A in the clockwise direction. The clock is located where there is a constant, uniform external magnetic field of magnitude 70 mT (but the clock still keeps perfect time). At exactly 1:00 p.m., the hour hand of the clock points in the direction of the external magnetic field. (a) After how many minutes will the minute hand point in the direction of the torque on the winding due to the magnetic field? (b) Find the torque magnitude.

53E. Two concentric, circular wire loops, of radii 20.0 and 30.0 cm, are located in the xy plane; each carries a clockwise current of 7.00 A (Fig. 29-39). (a) Find the net magnetic dipole moment of this system. (b) Repeat for reversed current in the inner loop. ssm

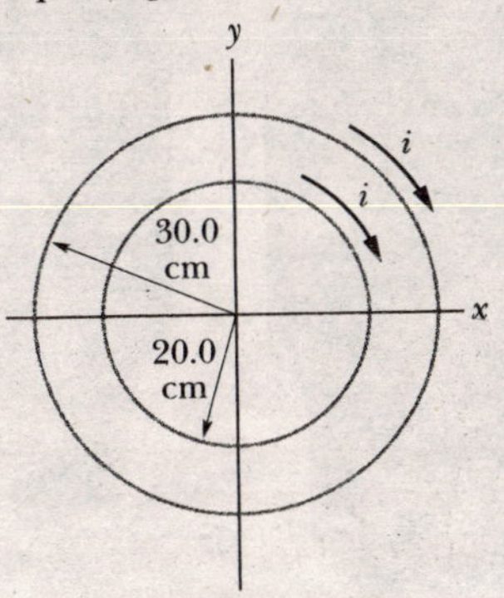

Fig. 29-39 Exercise 53.

54P. Figure 29-40 shows a current loop $ABCDEFA$ carrying a current $i = 5.00$ A. The sides of the loop are parallel to the coordinate axes, with $AB = 20.0$ cm, $BC = 30.0$ cm, and $FA = 10.0$ cm. Calculate the magnitude and direction of the magnetic dipole moment of this loop. (*Hint:* Imagine equal and opposite currents i in the line segment AD; then treat the two rectangular loops $ABCDA$ and $ADEFA$.)

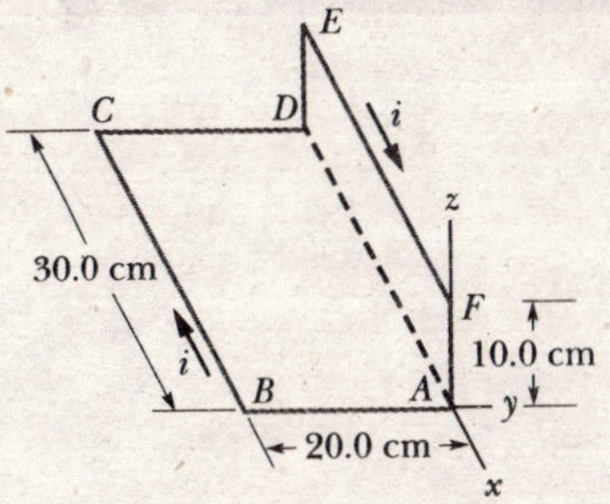

Fig. 29-40 Problem 54.

55P. A circular loop of wire having a radius of 8.0 cm carries a current of 0.20 A. A vector of unit length and parallel to the dipole moment $\vec{\mu}$ of the loop is given by $0.60\hat{\mathrm{i}} - 0.80\hat{\mathrm{j}}$. If the loop is located in a uniform magnetic field given by $\vec{B} = (0.25\ \mathrm{T})\hat{\mathrm{i}} + (0.30\ \mathrm{T})\hat{\mathrm{k}}$, find (a) the torque on the loop (in unit-vector notation) and (b) the magnetic potential energy of the loop. ssm

Additional Problems

56. A wire lying along a y axis from $y = 0$ to $y = 0.250$ m carries a current of 2.00 mA in the negative y direction. The wire lies in a nonuniform magnetic field given by

$$\vec{B} = (0.300\ \mathrm{T/m})y\hat{\mathrm{i}} + (0.400\ \mathrm{T/m})y\hat{\mathrm{j}}.$$

In unit-vector notation, what is the magnetic force on (a) an element dy of the wire at position y and (b) the entire wire?

57. A proton moves at a constant velocity of +50 m/s along an x axis, through crossed electric and magnetic fields. The magnetic field is $\vec{B} = (2.0\ \mathrm{mT})\hat{\mathrm{j}}$. What is the electric field?

58. A magnetic dipole with a dipole moment of magnitude 0.020 J/T is released from rest in a uniform magnetic field of magnitude 52 mT. The rotation of the dipole due to the magnetic force on it is unimpeded. When the dipole rotates through the orientation where its dipole moment is aligned with the magnetic field, its kinetic energy is 0.80 mJ. (a) What is the initial angle between the dipole moment and the magnetic field? (b) What is the angle when the dipole is next (momentarily) at rest?

59. An electron moves through a uniform magnetic field given by $\vec{B} = B_x\hat{\mathrm{i}} + (3B_x)\hat{\mathrm{j}}$. At a particular instant, the electron has the velocity $\vec{v} = (2.0\hat{\mathrm{i}} + 4.0\hat{\mathrm{j}})$ m/s and the magnetic force acting on it is $(6.4 \times 10^{-19}\ \mathrm{N})\hat{\mathrm{k}}$. Find B_x.

30 Magnetic Fields Due to Currents

This is the way we presently launch materials into space. However, when we begin mining the Moon and the asteroids, where we will not have a source of fuel for such conventional rockets, we shall need a more effective way. Electromagnetic launchers may be the answer. A small prototype, the *electromagnetic rail gun,* can presently accelerate a projectile from rest to a speed of 10 km/s (36 000 km/h) within 1 ms.

How can such rapid acceleration possibly be accomplished?

The answer is in this chapter.

30-1 Calculating the Magnetic Field Due to a Current

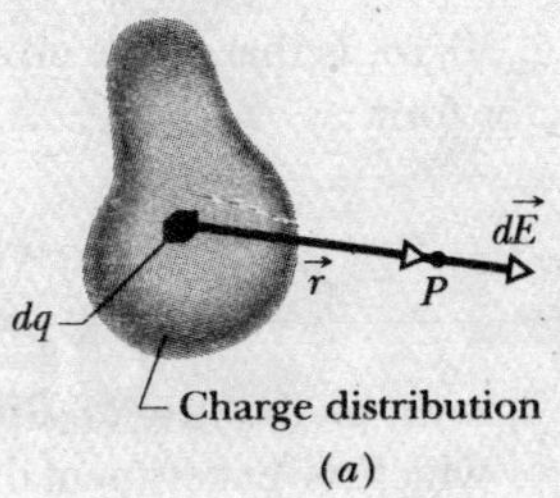

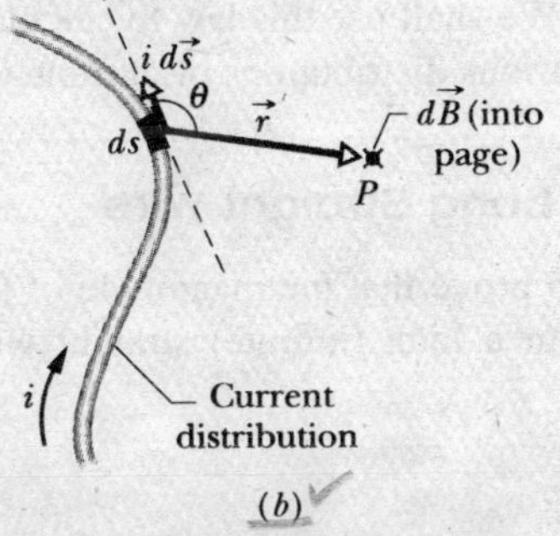

Fig. 30-1 (*a*) A charge element dq produces a differential electric field $d\vec{E}$ at point P. (*b*) A current-length element $i\,d\vec{s}$ produces a differential magnetic field $d\vec{B}$ at point P. The green × (the tail of an arrow) at the dot for point P indicates that $d\vec{B}$ is directed *into* the page there.

As we discussed in Section 29-1, one way to produce a magnetic field is with moving charges—that is, with a current. Our goal in this chapter is to calculate the magnetic field that is produced by a given distribution of currents. We shall use the same basic procedure we used in Chapter 23 to calculate the electric field produced by a given distribution of charged particles.

Let us quickly review that basic procedure. We first mentally divide the charge distribution into charge elements dq, as is done for a charge distribution of arbitrary shape in Fig. 30-1*a*. We then calculate the field $d\vec{E}$ produced at some point P by a typical charge element. Because the electric fields contributed by different elements can be superimposed, we calculate the net field $\vec{E}$ at P by summing, via integration, the contributions $d\vec{E}$ from all the elements.

Recall that we express the magnitude of $d\vec{E}$ as

$$dE = \frac{1}{4\pi\varepsilon_0}\frac{dq}{r^2}, \tag{30-1}$$

in which r is the distance from the charge element dq to point P. For a positively charged element, the direction of $d\vec{E}$ is that of $\vec{r}$, where $\vec{r}$ is the vector that extends from the charge element dq to the point P. Using $\vec{r}$, we can rewrite Eq. 30-1 in vector form as

$$d\vec{E} = \frac{1}{4\pi\varepsilon_0}\frac{dq}{r^3}\vec{r}, \tag{30-2}$$

which indicates that the direction of the vector $d\vec{E}$ produced by a positively charged element is the direction of the vector $\vec{r}$. Note that Eq. 30-2 is an inverse-square law ($d\vec{E}$ depends on inverse r^2) in spite of the exponent 3 in the denominator. That exponent is in the equation only because we added a factor of magnitude r in the numerator.

Now let us use the same basic procedure to calculate the magnetic field due to a current. Figure 30-1*b* shows a wire of arbitrary shape carrying a current i. We want to find the magnetic field $\vec{B}$ at a nearby point P. We first mentally divide the wire into differential elements ds and then define for each element a length vector $d\vec{s}$ that has length ds and whose direction is the direction of the current in ds. We can then define a differential *current-length element* to be $i\,d\vec{s}$; we wish to calculate the field $d\vec{B}$ produced at P by a typical current-length element. From experiment we find that magnetic fields, like electric fields, can be superimposed to find a net field. Thus, we can calculate the net field $\vec{B}$ at P by summing, via integration, the contributions $d\vec{B}$ from all the current-length elements. However, this summation is more challenging than the process associated with electric fields because of a complexity; whereas a charge element dq producing an electric field is a scalar, a current-length element $i\,d\vec{s}$ producing a magnetic field is the product of a scalar and a vector.

The magnitude of the field $d\vec{B}$ produced at point P by a current-length element $i\,d\vec{s}$ turns out to be

$$dB = \frac{\mu_0}{4\pi}\frac{i\,ds\sin\theta}{r^2}, \tag{30-3}$$

where θ is the angle between the directions of $d\vec{s}$ and $\vec{r}$, the vector that extends from ds to P. Symbol μ_0 is a constant, called the *permeability constant*, whose value is defined to be exactly

$$\mu_0 = 4\pi \times 10^{-7}\ \mathrm{T\cdot m/A} \approx 1.26 \times 10^{-6}\ \mathrm{T\cdot m/A}. \tag{30-4}$$

The direction of $d\vec{B}$, shown as being into the page in Fig. 30-1*b*, is that of the cross product $d\vec{s} \times \vec{r}$. We can therefore write Eq. 30-3 in vector form as

$$d\vec{B} = \frac{\mu_0}{4\pi} \frac{i\, d\vec{s} \times \vec{r}}{r^3} \quad \text{(Biot–Savart law).} \tag{30-5}$$

This vector equation and its scalar form, Eq. 30-3, are known as the **law of Biot and Savart** (rhymes with "Leo and bazaar"). The law, which is experimentally deduced, is an inverse-square law (the exponent in the denominator of Eq. 30-5 is 3 only because of the factor $\vec{r}$ in the numerator). We shall use this law to calculate the net magnetic field $\vec{B}$ produced at a point by various distributions of current.

Magnetic Field Due to a Current in a Long Straight Wire

Shortly we shall use the law of Biot and Savart to prove that the magnitude of the magnetic field at a perpendicular distance R from a long (infinite) straight wire carrying a current i is given by

$$B = \frac{\mu_0 i}{2\pi R} \quad \text{(long straight wire).} \tag{30-6}$$

The field magnitude B in Eq. 30-6 depends only on the current and the perpendicular distance R of the point from the wire. We shall show in our derivation that the field lines of $\vec{B}$ form concentric circles around the wire, as Fig. 30-2 shows and as the iron filings in Fig. 30-3 suggest. The increase in the spacing of the lines in Fig. 30-2 with increasing distance from the wire represents the $1/R$ decrease in the magnitude of $\vec{B}$ predicted by Eq. 30-6. The lengths of the two vectors $\vec{B}$ in the figure also show the $1/R$ decrease.

Here is a simple right-hand rule for finding the direction of the magnetic field set up by a current-length element, such as a section of a long wire:

> *Right-hand rule:* Grasp the element in your right hand with your extended thumb pointing in the direction of the current. Your fingers will then naturally curl around in the direction of the magnetic field lines due to that element.

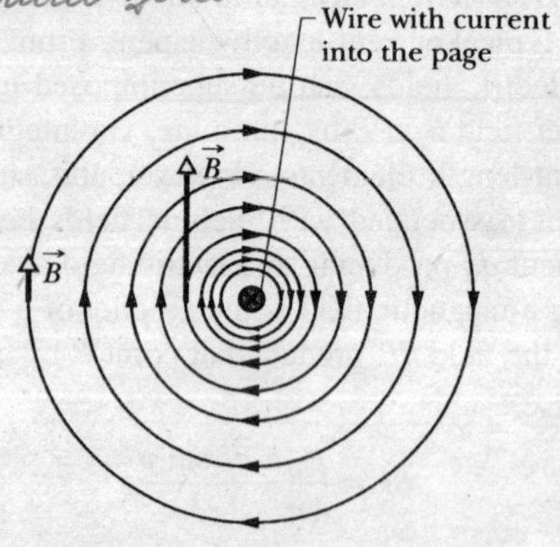

Fig. 30-2 The magnetic field lines produced by a current in a long straight wire form concentric circles around the wire. Here the current is into the page, as indicated by the ×.

Fig. 30-3 Iron filings that have been sprinkled onto cardboard collect in concentric circles when current is sent through the central wire. The alignment, which is along magnetic field lines, is caused by the magnetic field produced by the current.

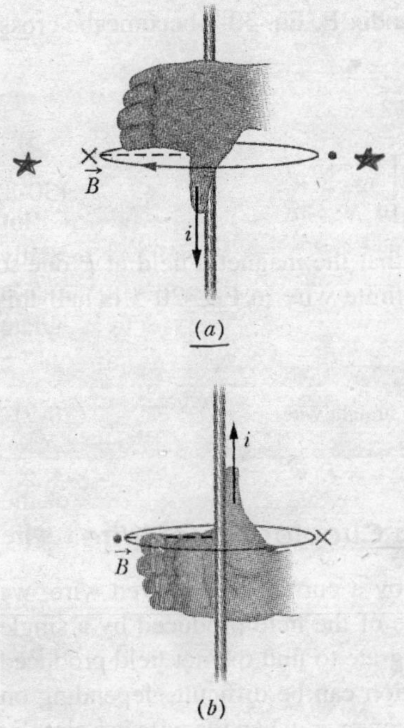

Fig. 30-4 A right-hand rule gives the direction of the magnetic field due to a current in a wire. (*a*) The situation of Fig. 30-2, seen from the side. The magnetic field $\vec{B}$ at any point to the left of the wire is perpendicular to the dashed radial line and directed into the page, in the direction of the fingertips, as indicated by the ×. (*b*) If the current is reversed, $\vec{B}$ at any point to the left is still perpendicular to the dashed radial line but now is directed out of the page, as indicated by the dot.

The result of applying this right-hand rule to the current in the straight wire of Fig. 30-2 is shown in a side view in Fig. 30-4*a*. To determine the direction of the magnetic field $\vec{B}$ set up at any particular point by this current, mentally wrap your right hand around the wire with your thumb in the direction of the current. Let your fingertips pass through the point; their direction is then the direction of the magnetic field at that point. In the view of Fig. 30-2, $\vec{B}$ at any point is *tangent to a magnetic field line;* in the view of Fig. 30-4, it is *perpendicular to a dashed radial line connecting the point and the current.*

Proof of Equation 30-6

Figure 30-5, which is just like Fig. 30-1*b* except that now the wire is straight and of infinite length, illustrates the task at hand; we seek the field $\vec{B}$ at point P, a perpendicular distance R from the wire. The magnitude of the differential magnetic field produced at P by the current-length element $i\,d\vec{s}$ located a distance r from P is given by Eq. 30-3:

$$dB = \frac{\mu_0}{4\pi}\frac{i\,ds\sin\theta}{r^2}.$$

The direction of $d\vec{B}$ in Fig. 30-5 is that of the vector $d\vec{s} \times \vec{r}$—namely, directly into the page.

Note that $d\vec{B}$ at point P has this same direction for all the current-length elements into which the wire can be divided. Thus, we can find the magnitude of the magnetic field produced at P by the current-length elements in the upper half of the infinitely long wire by integrating dB in Eq. 30-3 from 0 to ∞.

Now consider a current-length element in the lower half of the wire, one that is as far below P as $d\vec{s}$ is above P. By Eq. 30-5, the magnetic field produced at P by this current-length element has the same magnitude and direction as that from element $i\,d\vec{s}$ in Fig. 30-5. Further, the magnetic field produced by the lower half of the wire is exactly the same as that produced by the upper half. To find the magnitude of the *total* magnetic field $\vec{B}$ at P, we need only multiply the result of our integration by 2. We get

$$B = 2\int_0^\infty dB = \frac{\mu_0 i}{2\pi}\int_0^\infty \frac{\sin\theta\,ds}{r^2}. \qquad (30\text{-}7)$$

The variables θ, s, and r in this equation are not independent but (see Fig. 30-5) are related by

$$r = \sqrt{s^2 + R^2}$$

and

$$\sin\theta = \sin(\pi - \theta) = \frac{R}{\sqrt{s^2 + R^2}}.$$

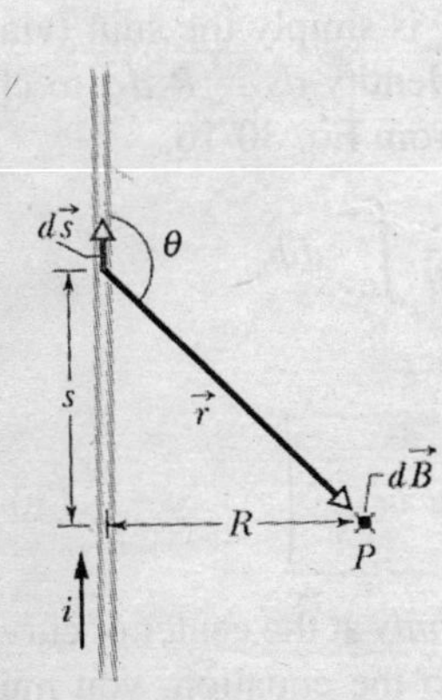

Fig. 30-5 Calculating the magnetic field produced by a current i in a long straight wire. The field $d\vec{B}$ at P associated with the current-length element $i\,d\vec{s}$ is into the page, as shown.

With these substitutions and integral 19 in Appendix E, Eq. 30-7 becomes

$$B = \frac{\mu_0 i}{2\pi} \int_0^{\infty} \frac{R\,ds}{(s^2 + R^2)^{3/2}}$$

$$= \frac{\mu_0 i}{2\pi R} \left[\frac{s}{(s^2 + R^2)^{1/2}} \right]_0^{\infty} = \frac{\mu_0 i}{2\pi R}, \tag{30-8}$$

which is the relation we set out to prove. Note that the magnetic field at P due to either the lower half or the upper half of the infinite wire in Fig. 30-5 is half this value; that is,

★R is the ⊥ distance of the point (where B is req) from the wire

$$B = \frac{\mu_0 i}{4\pi R} \quad \text{(semi-infinite straight wire)}. \tag{30-9}$$

Magnetic Field Due to a Current in a Circular Arc of Wire

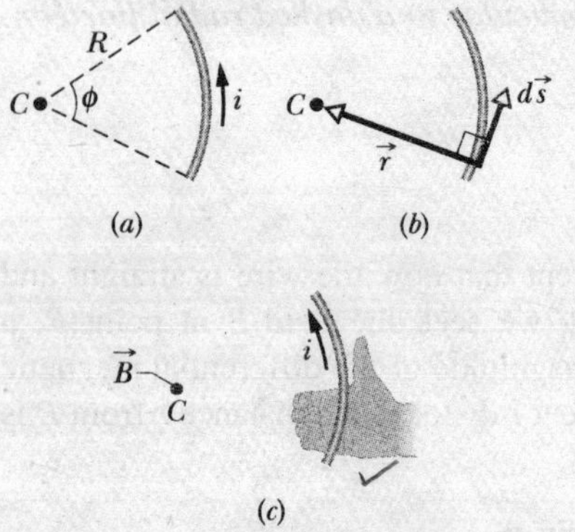

Fig. 30-6 (*a*) A wire in the shape of a circular arc with center C carries current i. (*b*) For any element of wire along the arc, the angle between the directions of $\vec{ds}$ and $\vec{r}$ is 90°. (*c*) Determining the direction of the magnetic field at the center C due to the current in the wire; the field is out of the page, in the direction of the fingertips, as indicated by the colored dot at C.

To find the magnetic field produced at a point by a current in a curved wire, we would again use Eq. 30-3 to write the magnitude of the field produced by a single current-length element, and we would again integrate to find the net field produced by all the current-length elements. That integration can be difficult, depending on the shape of the wire; it is fairly straightforward, however, when the wire is a circular arc and the point is the center of curvature.

Figure 30-6*a* shows such an arc-shaped wire with central angle ϕ, radius R, and center C, carrying current i. At C, each current-length element $i\,\vec{ds}$ of the wire produces a magnetic field of magnitude dB given by Eq. 30-3. Moreover, as Fig. 30-6*b* shows, no matter where the element is located on the wire, the angle θ between the vectors $\vec{ds}$ and $\vec{r}$ is 90°; also, $r = R$. Thus, by substituting R for r and 90° for θ, we obtain from Eq. 30-3,

$$dB = \frac{\mu_0}{4\pi} \frac{i\,ds \sin 90°}{R^2} = \frac{\mu_0}{4\pi} \frac{i\,ds}{R^2}. \tag{30-10}$$

The field at C due to each current-length element in the arc has this magnitude.

An application of the right-hand rule anywhere along the wire (as in Fig. 30-6*c*) will show that all the differential fields $\vec{dB}$ have the same direction at C—directly out of the page. Thus, the total field at C is simply the sum (via integration) of all the differential fields dB. We use the identity $ds = R\,d\phi$ to change the variable of integration from ds to $d\phi$ and obtain, from Eq. 30-10,

$$B = \int dB = \int_0^{\phi} \frac{\mu_0}{4\pi} \frac{iR\,d\phi}{R^2} = \frac{\mu_0 i}{4\pi R} \int_0^{\phi} d\phi.$$

Integrating, we find that

$$B = \frac{\mu_0 i \phi}{4\pi R} \quad \text{(at center of circular arc)}. \tag{30-11}$$

Note that this equation gives us the magnetic field *only* at the center of curvature of a circular arc of current. When you insert data into the equation, you must be careful to express ϕ in radians rather than degrees. For example, to find the magnitude of the magnetic field at the center of a full circle of current, you would substitute 2π rad for ϕ in Eq. 30-11, finding

$$B = \frac{\mu_0 i (2\pi)}{4\pi R} = \frac{\mu_0 i}{2R} \quad \text{(at center of full circle)}. \tag{30-12}$$

Sample Problem 30-1

The wire in Fig. 30-7*a* carries a current i and consists of a circular arc of radius R and central angle $\pi/2$ rad, and two straight sections whose extensions intersect the center C of the arc. What magnetic field $\vec{B}$ does the current produce at C?

SOLUTION: One **Key Idea** here is that we can find the magnetic field $\vec{B}$ at point C by applying the Biot–Savart law of Eq. 30-5 to the wire. A second **Key Idea** is that the application of Eq. 30-5 can be simplified by evaluating $\vec{B}$ separately for the three distinguishable sections of the wire—namely, (1) the straight section at the left, (2) the straight section at the right, and (3) the circular arc.

Straight sections. For any current-length element in section 1, the angle θ between $d\vec{s}$ and $\vec{r}$ is zero (Fig. 30-7*b*), so Eq. 30-3 gives us

$$dB_1 = \frac{\mu_0}{4\pi}\frac{i\,ds\sin\theta}{r^2} = \frac{\mu_0}{4\pi}\frac{i\,ds\sin 0}{r^2} = 0.$$

Thus, the current along the entire length of wire in straight section 1 contributes no magnetic field at C:

$$B_1 = 0.$$

The same situation prevails in straight section 2, where the angle θ between $d\vec{s}$ and $\vec{r}$ for any current-length element is 180°. Thus,

$$B_2 = 0.$$

Circular arc. The **Key Idea** here is that application of the Biot–Savart law to evaluate the magnetic field at the center of a circular arc leads to Eq. 30-11 ($B = \mu_0 i\phi/4\pi R$). Here the central angle ϕ of the arc is $\pi/2$ rad. Thus from Eq. 30-11, the magnitude of the magnetic field $\vec{B}_3$ at the arc's center C is

$$B_3 = \frac{\mu_0 i(\pi/2)}{4\pi R} = \frac{\mu_0 i}{8R}.$$

To find the direction of $\vec{B}_3$, we apply the right-hand rule displayed in Fig. 30-4. Mentally grasp the circular arc with your right hand as in Fig. 30-7*c*, with your thumb in the direction of the current. The direction in which your fingers curl around the wire indicates the direction of the magnetic field lines around the wire. In the region of point C (inside the circular arc), your fingertips point *into the plane* of the page. Thus, $\vec{B}_3$ is directed into that plane.

Net field. Generally, when we must combine two or more magnetic fields to find the net magnetic field, we must combine the fields as vectors and not simply add their magnitudes. Here, however, only the circular arc produces a magnetic field at point C.

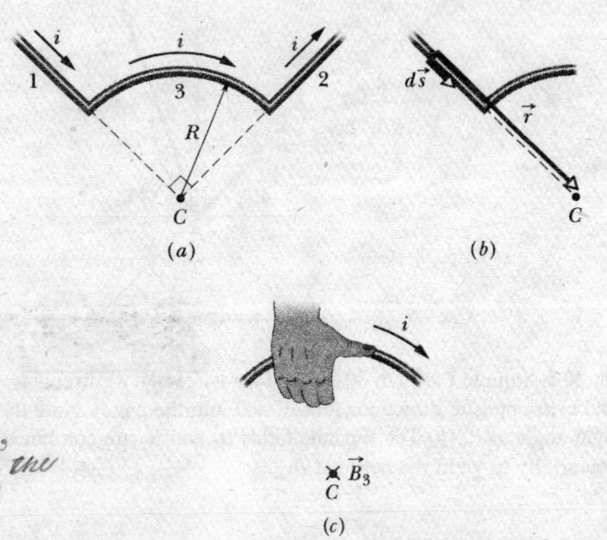

Fig. 30-7 Sample Problem 30-1. (*a*) A wire consists of two straight sections (1 and 2) and a circular arc (3), and carries current i. (*b*) For a current-length element in section 1, the angle between $d\vec{s}$ and $\vec{r}$ is zero. (*c*) Determining the direction of magnetic field $\vec{B}_3$ at C due to the current in the circular arc; the field is into the page there.

Thus, we can write the magnitude of the net field $\vec{B}$ as

$$B = B_1 + B_2 + B_3 = 0 + 0 + \frac{\mu_0 i}{8R} = \frac{\mu_0 i}{8R}. \quad \text{(Answer)}$$

The direction of $\vec{B}$ is the direction of $\vec{B}_3$—namely, into the plane of Fig. 30-7.

CHECKPOINT 1: The figure here shows three circuits consisting of concentric circular arcs (either half- or quarter-circles of radii r, $2r$, and $3r$) and radial lengths. The circuits carry the same current. Rank them according to the magnitude of the magnetic field produced at the center of curvature (the dot), greatest first.

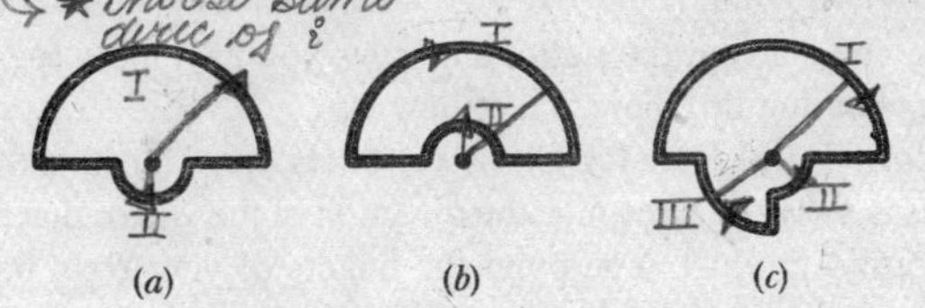

Sample Problem 30-2

Figure 30-8*a* shows two long parallel wires carrying currents i_1 and i_2 in opposite directions. What are the magnitude and direction of the net magnetic field at point P? Assume the following values: $i_1 = 15$ A, $i_2 = 32$ A, and $d = 5.3$ cm.

SOLUTION: One **Key Idea** here is that the net magnetic field $\vec{B}$ at point P is the vector sum of the magnetic fields due to the currents in the two wires. A second **Key Idea** is that we can find the magnetic field due to any current by applying the Biot–Savart law to the

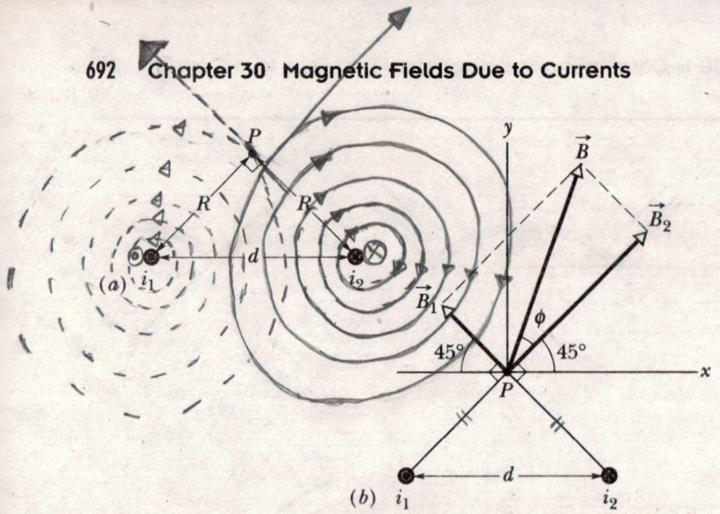

Fig. 30-8 Sample Problem 30-2. (*a*) Two wires carry currents i_1 and i_2 in opposite directions (out of and into the page). Note the right angle at P. (*b*) The separate fields $\vec{B}_1$ and $\vec{B}_2$ are combined vectorially to yield the net field $\vec{B}$.

current. For points near the current in a long straight wire, that law leads to Eq. 30-6.

In Fig. 30-8*a*, point P is distance R from both currents i_1 and i_2. Thus, Eq. 30-6 tells us that at point P those currents produce magnetic fields $\vec{B}_1$ and $\vec{B}_2$ with magnitudes

$$B_1 = \frac{\mu_0 i_1}{2\pi R} \quad \text{and} \quad B_2 = \frac{\mu_0 i_2}{2\pi R}.$$

In the right triangle of Fig. 30-8*a*, note that the base angles (between sides R and d) are both 45°. Thus, we may write $\cos 45° = R/d$ and replace R with $d \cos 45°$. Then the field magnitudes B_1 and B_2 become

$$B_1 = \frac{\mu_0 i_1}{2\pi d \cos 45°} \quad \text{and} \quad B_2 = \frac{\mu_0 i_2}{2\pi d \cos 45°}.$$

∵ R is not given

We want to combine $\vec{B}_1$ and $\vec{B}_2$ to find their vector sum, which is the net field $\vec{B}$ at P. To find the directions of $\vec{B}_1$ and $\vec{B}_2$, we apply the right-hand rule of Fig. 30-4 to each current in Fig. 30-8*a*. For wire 1, with current out of the page, we mentally grasp the wire with the right hand, with the thumb pointing out of the page. Then the curled fingers indicate that the field lines run counterclockwise. In particular, in the region of point P, they are directed upward to the left. Recall that the magnetic field at a point near a long, straight current-carrying wire must be directed perpendicular to a radial line between the point and the current. Thus, $\vec{B}_1$ must be directed upward to the left as drawn in Fig. 30-8*b*. (Note carefully the perpendicular symbol between vector $\vec{B}_1$ and the line connecting point P and wire 1.)

Repeating this analysis for the current in wire 2, we find that $\vec{B}_2$ is directed upward to the right as drawn in Fig. 30-8*b*. (Note the perpendicular symbol between vector $\vec{B}_2$ and the line connecting point P and wire 2.)

We can now vectorially add $\vec{B}_1$ and $\vec{B}_2$ to find the net magnetic field $\vec{B}$ at point P, either by using a vector-capable calculator or by resolving the vectors into components and then combining the components of $\vec{B}$. However, in Fig. 30-8*b*, there is a third method: Because $\vec{B}_1$ and $\vec{B}_2$ are perpendicular to each other, they form the legs of a right triangle, with $\vec{B}$ as the hypotenuse. The Pythagorean theorem then gives us

$$\begin{aligned} B &= \sqrt{B_1^2 + B_2^2} = \frac{\mu_0}{2\pi d(\cos 45°)}\sqrt{i_1^2 + i_2^2} \\ &= \frac{(4\pi \times 10^{-7}\ \text{T}\cdot\text{m/A})\sqrt{(15\ \text{A})^2 + (32\ \text{A})^2}}{(2\pi)(5.3 \times 10^{-2}\ \text{m})(\cos 45°)} \\ &= 1.89 \times 10^{-4}\ \text{T} \approx 190\ \mu\text{T}. \qquad \text{(Answer)} \end{aligned}$$

The angle ϕ between the directions of $\vec{B}$ and $\vec{B}_2$ in Fig. 30-8*b* follows from

$$\phi = \tan^{-1}\frac{B_1}{B_2},$$

which, with B_1 and B_2 as given above, yields

$$\phi = \tan^{-1}\frac{i_1}{i_2} = \tan^{-1}\frac{15\ \text{A}}{32\ \text{A}} = 25°.$$

The angle between the direction of $\vec{B}$ and the x axis shown in Fig. 30-8*b* is then

$$\phi + 45° = 25° + 45° = 70°. \qquad \text{(Answer)}$$

PROBLEM-SOLVING TACTICS

Tactic 1: *Right-Hand Rules*
To help you sort out the right-hand rules you have now seen (and the ones coming up), here is a review.

Right-Hand Rule for Cross Products. Introduced in Section 3-7, this is a way to determine the direction of the vector that results from a cross product. You point the fingers of your right hand so as to sweep the first vector expressed in the product into the second vector, through the smaller angle between the two vectors. Your outstretched thumb gives you the direction of the vector resulting from the cross product. In Chapter 12, we used this right-hand rule to find the directions of torque and angular momentum vectors; in Chapter 29, we used it to find the direction of the force on a current-carrying wire in a magnetic field.

Curled–Straight Right-Hand Rules for Magnetism. In many situations involving magnetism, you need to relate a "curled" element and a "straight" element. You can do so with the (curled) fingers and the (straight) thumb on your right hand. You have already seen an example in Section 29-8, in which we related the current around a loop (curled element) to the normal vector $\vec{n}$ (straight element) of the loop: You curl the fingers of your right hand around in the direction of the current along the loop; your outstretched thumb then gives the direction of $\vec{n}$. This is also the direction of the magnetic dipole moment $\vec{\mu}$ of the loop.

In this section you were introduced to a second curled–straight right-hand rule. To determine the direction of the magnetic field lines around a current-length element, you point the outstretched thumb of your right hand in the direction of the current. The fingers then curl around the current-length element in the direction of the field lines.

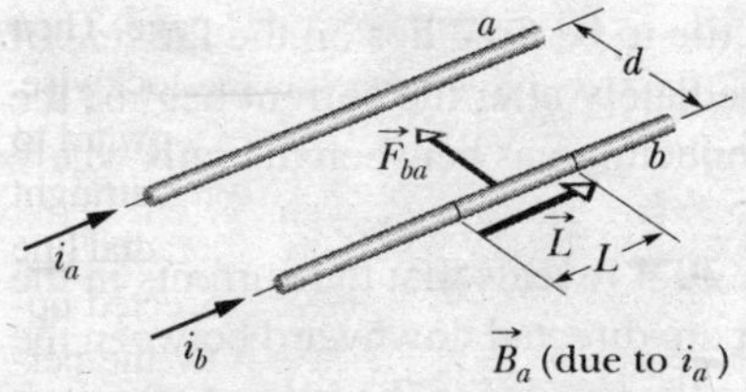

Fig. 30-9 Two parallel wires carrying currents in the same direction attract each other. $\vec{B}_a$ is the magnetic field at wire b produced by the current in wire a. $\vec{F}_{ba}$ is the resulting force acting on wire b because it carries current in $\vec{B}_a$.

30-2 Force Between Two Parallel Currents

Two long parallel wires carrying currents exert forces on each other. Figure 30-9 shows two such wires, separated by a distance d and carrying currents i_a and i_b. Let us analyze the forces on these wires due to each other.

We seek first the force on wire b in Fig. 30-9 due to the current in wire a. That current produces a magnetic field $\vec{B}_a$, and it is this magnetic field that actually causes the force we seek. To find the force, then, we need the magnitude and direction of the field $\vec{B}_a$ *at the site of wire b*. The magnitude of $\vec{B}_a$ at every point of wire b is, from Eq. 30-6,

$$B_a = \frac{\mu_0 i_a}{2\pi d}. \tag{30-13}$$

A (curled–straight) right-hand rule tells us that the direction of $\vec{B}_a$ at wire b is down, as Fig. 30-9 shows.

Now that we have the field, we can find the force it produces on wire b. Equation 29-27 tells us that the force $\vec{F}_{ba}$ on a length L of wire b due to the external magnetic field $\vec{B}_a$ is

$$\vec{F}_{ba} = i_b \vec{L} \times \vec{B}_a, \tag{30-14}$$

where $\vec{L}$ is the length vector of the wire. In Fig. 30-9, vectors $\vec{L}$ and $\vec{B}_a$ are perpendicular, so with Eq. 30-13, we can write

$$F_{ba} = i_b L B_a \sin 90° = \frac{\mu_0 L i_a i_b}{2\pi d}. \tag{30-15}$$

The direction of $\vec{F}_{ba}$ is the direction of the cross product $\vec{L} \times \vec{B}_a$. Applying the right-hand rule for cross products to $\vec{L}$ and $\vec{B}_a$ in Fig. 30-9, we see that $\vec{F}_{ba}$ is directly toward wire a, as shown.

The general procedure for finding the force on a current-carrying wire is this:

> **To find the force on a current-carrying wire due to a second current-carrying wire, first find the field due to the second wire at the site of the first wire. Then find the force on the first wire due to that field.**

We could now use this procedure to compute the force on wire a due to the current in wire b. We would find that the force is directly toward wire b; hence, the two wires with parallel currents attract each other. Similarly, if the two currents were antiparallel, we could show that the two wires repel each other. Thus,

> **Parallel currents attract, and antiparallel currents repel.**

The force acting between currents in parallel wires is the basis for the definition of the ampere, which is one of the seven SI base units. The definition, adopted in 1946, is this: The ampere is that constant current which, if maintained in two straight, parallel conductors of infinite length, of negligible circular cross section, and placed 1 m apart in vacuum, would produce on each of these conductors a force of magnitude 2×10^{-7} newton per meter of length.

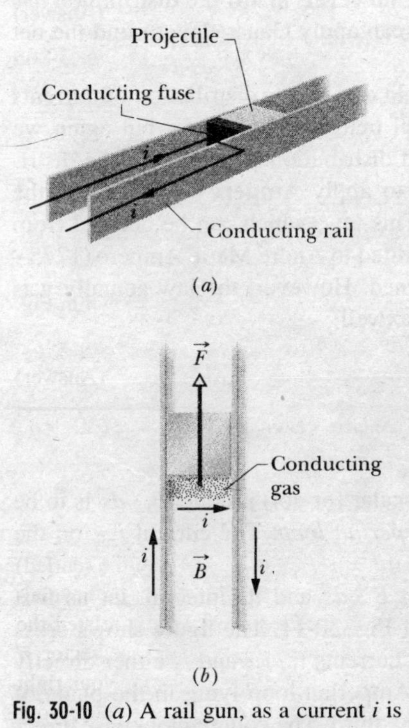

Fig. 30-10 (*a*) A rail gun, as a current i is set up in it. The current rapidly causes the conducting fuse to vaporize. (*b*) The current produces a magnetic field $\vec{B}$ between the rails, and the field causes a force $\vec{F}$ to act on the conducting gas, which is part of the current path. The gas propels the projectile along the rails, launching it.

Rail Gun

A rail gun is a device in which a magnetic force can accelerate a projectile to a high speed in a short time. The basics of a rail gun are shown in Fig. 30-10*a*. A large current is sent out along one of two parallel conducting rails, across a conducting "fuse" (such as a narrow piece of copper) between the rails, and then back to the current source along the second rail. The projectile to be fired lies on the far side of the fuse and fits loosely between the rails. Immediately after the current begins, the fuse element melts and vaporizes, creating a conducting gas between the rails where the fuse had been.

The curled–straight right-hand rule of Fig. 30-4 reveals that the currents in the rails of Fig. 30-10*a* produce magnetic fields that are directed downward between the rails. The net magnetic field $\vec{B}$ exerts a force $\vec{F}$ on the gas due to the current i through the gas (Fig. 30-10*b*). With Eq. 30-14 and the right-hand rule for cross products, we find that $\vec{F}$ points outward along the rails. As the gas is forced outward along the rails, it pushes the projectile, accelerating it by as much as $5 \times 10^6 g$, and then launches it with a speed of 10 km/s, all within 1 ms.

✔CHECKPOINT 2: The figure here shows three long, straight, parallel, equally spaced wires with identical currents either into or out of the page. Rank the wires according to the magnitude of the force on each due to the currents in the other two wires, greatest first.

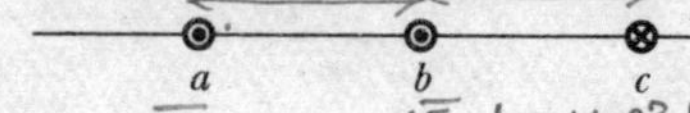

30-3 Ampere's Law

We can find the net electric field due to *any* distribution of charges with the inverse-square law for the differential field $d\vec{E}$ (Eq. 30-2), but if the distribution is complicated, we may have to use a computer. Recall, however, that if the distribution has planar, cylindrical, or spherical symmetry, we can apply Gauss' law to find the net electric field with considerably less effort.

Similarly, we can find the net magnetic field due to *any* distribution of currents with the inverse-square law for the differential field $d\vec{B}$ (Eq. 30-5), but again we may have to use a computer for a complicated distribution. However, if the distribution has some symmetry, we may be able to apply **Ampere's law** to find the magnetic field with considerably less effort. This law, which can be derived from the Biot–Savart law, has traditionally been credited to André Marie Ampere (1775–1836), for whom the SI unit of current is named. However, the law actually was advanced by English physicist James Clerk Maxwell.

Ampere's law is

$$\oint \vec{B} \cdot d\vec{s} = \mu_0 i_{enc} \quad \text{(Ampere's law).} \tag{30-16}$$

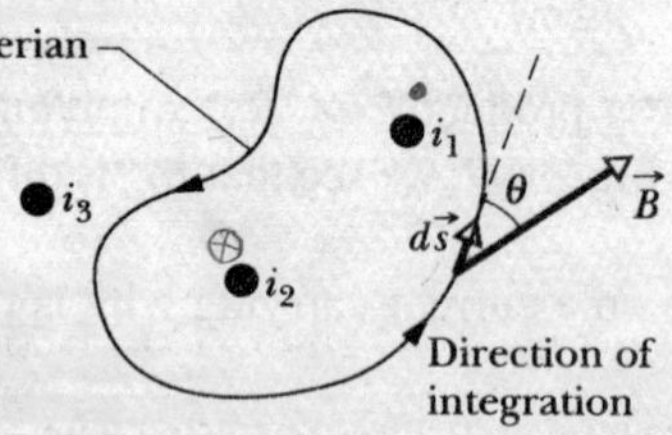

Fig. 30-11 Ampere's law applied to an arbitrary Amperian loop that encircles two long straight wires but excludes a third wire. Note the directions of the currents.

The circle on the integral sign means that the scalar (or dot) product $\vec{B} \cdot d\vec{s}$ is to be integrated around a *closed* loop, called an *Amperian loop*. The current i_{enc} on the right is the *net* current encircled by that loop.

To see the meaning of the scalar product $\vec{B} \cdot d\vec{s}$ and its integral, let us first apply Ampere's law to the general situation of Fig. 30-11. The figure shows cross sections of three long straight wires that carry currents i_1, i_2, and i_3 either directly into or directly out of the page. An arbitrary Amperian loop lying in the plane of the page encircles two of the currents but not the third. The counterclockwise direction marked on the loop indicates the arbitrarily chosen direction of integration for Eq. 30-16.

To apply Ampere's law, we mentally divide the loop into differential vector elements $d\vec{s}$ that are everywhere directed along the tangent to the loop in the direction of integration. Assume that at the location of the element $d\vec{s}$ shown in Fig. 30-11, the net magnetic field due to the three currents is $\vec{B}$. Because the wires are perpendicular to the page, we know that the magnetic field at $d\vec{s}$ due to each current is in the plane of Fig. 30-11; thus, their net magnetic field $\vec{B}$ at $d\vec{s}$ must also be in that plane. However, we do not know the orientation of $\vec{B}$ within the plane. In Fig. 30-11, $\vec{B}$ is arbitrarily drawn at an angle θ to the direction of $d\vec{s}$.

The scalar product $\vec{B} \cdot d\vec{s}$ on the left side of Eq. 30-16 is then equal to $B \cos \theta \, ds$. Thus, Ampere's law can be written as

$$\oint \vec{B} \cdot d\vec{s} = \oint B \cos \theta \, ds = \mu_0 i_{enc}. \tag{30-17}$$

We can now interpret the scalar product $\vec{B} \cdot d\vec{s}$ as being the product of a length ds of the Amperian loop and the field component $B \cos \theta$ that is tangent to the loop. Then we can interpret the integration as being the summation of all such products around the entire loop.

When we can actually perform this integration, we do not need to know the direction of $\vec{B}$ before integrating. Instead, we arbitrarily assume $\vec{B}$ to be generally in the direction of integration (as in Fig. 30-11). Then we use the following curled–straight right-hand rule to assign a plus sign or a minus sign to each of the currents that make up the net encircled current i_{enc}:

> Curl your right hand around the Amperian loop, with the fingers pointing in the direction of integration. A current through the loop in the general direction of your outstretched thumb is assigned a plus sign, and a current generally in the opposite direction is assigned a minus sign.

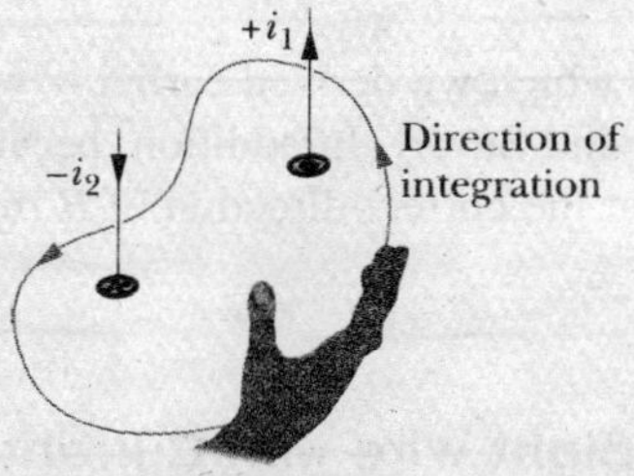

Fig. 30-12 A right-hand rule for Ampere's law, to determine the signs for currents encircled by an Amperian loop. The situation is that of Fig. 30-11.

Finally, we solve Eq. 30-17 for the magnitude of $\vec{B}$. If B turns out positive, then the direction we assumed for $\vec{B}$ is correct. If it turns out negative, we neglect the minus sign and redraw $\vec{B}$ in the opposite direction.

In Fig. 30-12 we apply the curled–straight rule for Ampere's law to the situation of Fig. 30-11. With the indicated counterclockwise direction of integration, the net current encircled by the loop is

$$i_{enc} = i_1 - i_2.$$

(Current i_3 is not encircled by the loop.) We can then rewrite Eq. 30-17 as

$$\oint B \cos \theta \, ds = \mu_0(i_1 - i_2). \tag{30-18}$$

You might wonder why, since current i_3 contributes to the magnetic-field magnitude B on the left side of Eq. 30-18, it is not needed on the right side. The answer is that the contributions of current i_3 to the magnetic field cancel out because the integration in Eq. 30-18 is made around the full loop. In contrast, the contributions of an encircled current to the magnetic field do not cancel out.

We cannot solve Eq. 30-18 for the magnitude B of the magnetic field, because for the situation of Fig. 30-11 we do not have enough information to simplify and solve the integral. However, we do know the outcome of the integration; it must be equal to the value of $\mu_0(i_1 - i_2)$, which is set by the net current passing through the loop.

We shall now apply Ampere's law to two situations in which symmetry does allow us to simplify and solve the integral, hence to find the magnetic field.

The Magnetic Field Outside a Long Straight Wire with Current

Figure 30-13 shows a long straight wire that carries current i directly out of the page. Equation 30-6 tells us that the magnetic field $\vec{B}$ produced by the current has the same

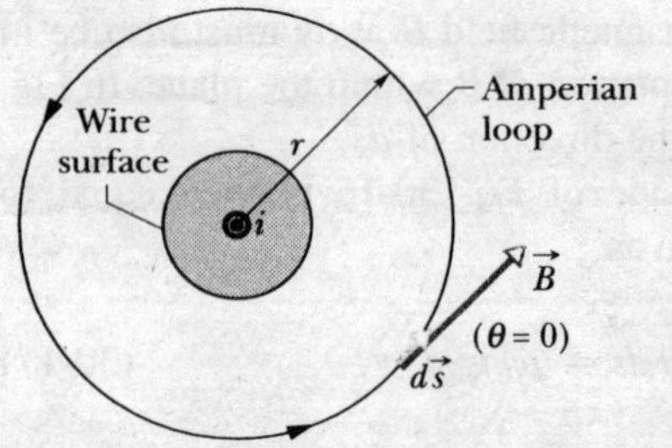

Fig. 30-13 Using Ampere's law to find the magnetic field produced by a current i in a long straight wire. The Amperian loop is a concentric circle that lies outside the wire.

magnitude at all points that are the same distance r from the wire; that is, the field $\vec{B}$ has cylindrical symmetry about the wire. We can take advantage of that symmetry to simplify the integral in Ampere's law (Eqs. 30-16 and 30-17) if we encircle the wire with a concentric circular Amperian loop of radius r, as in Fig. 30-13. The magnetic field $\vec{B}$ then has the same magnitude B at every point on the loop. We shall integrate counterclockwise, so that $d\vec{s}$ has the direction shown in Fig. 30-13.

We can further simplify the quantity $B \cos \theta$ in Eq. 30-17 by noting that $\vec{B}$ is tangent to the loop at every point along the loop, as is $d\vec{s}$. Thus, $\vec{B}$ and $d\vec{s}$ are either parallel or antiparallel at each point of the loop, and we shall arbitrarily assume the former. Then at every point the angle θ between $d\vec{s}$ and $\vec{B}$ is 0°, so $\cos \theta = \cos 0° = 1$. The integral in Eq. 30-17 then becomes

$$\oint \vec{B} \cdot d\vec{s} = \oint B \cos \theta \, ds = B \oint ds = B(2\pi r).$$

since mag field B is tangent to mag field lines

Note that $\oint ds$ above is the summation of all the line segment lengths ds around the circular loop; that is, it simply gives the circumference $2\pi r$ of the loop.

Our right-hand rule gives us a plus sign for the current of Fig. 30-13. The right side of Ampere's law becomes $+\mu_0 i$ and we then have

$$B(2\pi r) = \mu_0 i$$

or

$$B = \frac{\mu_0 i}{2\pi r}. \tag{30-19}$$

⇒ this law holds even if ∞, straight wire has some circular crossection

With a slight change in notation, this is Eq. 30-6, which we derived earlier—with considerably more effort—using the law of Biot and Savart. In addition, because the magnitude B turned out positive, we know that the correct direction of $\vec{B}$ must be the one shown in Fig. 30-13.

The Magnetic Field Inside a Long Straight Wire with Current

Figure 30-14 shows the cross section of a long straight wire of radius R that carries a uniformly distributed current i directly out of the page. Because the current is uniformly distributed over a cross section of the wire, the magnetic field $\vec{B}$ that it produces must be cylindrically symmetrical. Thus, to find the magnetic field at points inside the wire, we can again use an Amperian loop of radius r, as shown in Fig. 30-14, where now $r < R$. Symmetry again suggests that $\vec{B}$ is tangent to the loop, as shown, so the left side of Ampere's law again yields

$$\oint \vec{B} \cdot d\vec{s} = B \oint ds = B(2\pi r). \tag{30-20}$$

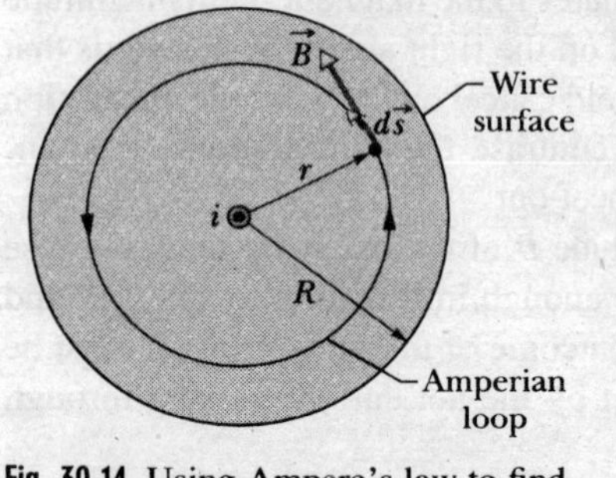

Fig. 30-14 Using Ampere's law to find the magnetic field that a current i produces inside a long straight wire of circular cross section. The current is uniformly distributed over the cross section of the wire and emerges from the page. An Amperian loop is drawn inside the wire.

To find the right side of Ampere's law, we note that because the current is uniformly distributed, the current i_{enc} encircled by the loop is proportional to the area encircled by the loop; that is,

uniformly distributed across the circular crossection ∴ current per unit area is same

$$i_{enc} = i \frac{\pi r^2}{\pi R^2}. \tag{30-21}$$

Our right-hand rule tells us that i_{enc} gets a plus sign. Then Ampere's law gives us

$$B(2\pi r) = \mu_0 i \frac{\pi r^2}{\pi R^2}$$

or

$$B = \left(\frac{\mu_0 i}{2\pi R^2}\right) r. \tag{30-22}$$

Thus, inside the wire, the magnitude B of the magnetic field is proportional to r; that magnitude is zero at the center and a maximum at the surface, where $r = R$. Note that Eqs. 30-19 and 30-22 give the same value for B at $r = R$; that is, the expressions for the magnetic field outside the wire and inside the wire yield the same result at the surface of the wire.

CHECKPOINT 3: The figure here shows three equal currents i (two parallel and one antiparallel) and four Amperian loops. Rank the loops according to the magnitude of $\oint \vec{B} \cdot d\vec{s}$ along each, greatest first.

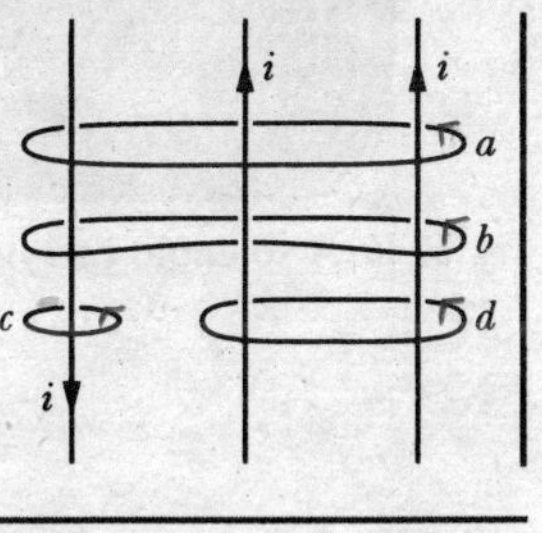

$\oint \vec{B} \cdot d\vec{s}$ is independent of the shape of the loop provided i_{enc} is the same

$\oint \vec{B} \cdot d\vec{s} = \mu_0 i_{enc}$

(a) $\mu_0(i+i-i) = \mu_0 i$ (d) $\mu_0(2i)$

(b) $\mu_0(i-i) = 0$

(c) $\mu_0(-i) = -\mu_0 i$ (d) > (a) & (c) > (b)

Sample Problem 30-3

Figure 30-15a shows the cross section of a long conducting cylinder with inner radius $a = 2.0$ cm and outer radius $b = 4.0$ cm. The cylinder carries a current out of the page, and the current density in the cross section is given by $J = cr^2$, with $c = 3.0 \times 10^6$ A/m^4 and r in meters. What is the magnetic field $\vec{B}$ at a point that is 3.0 cm from the central axis of the cylinder?

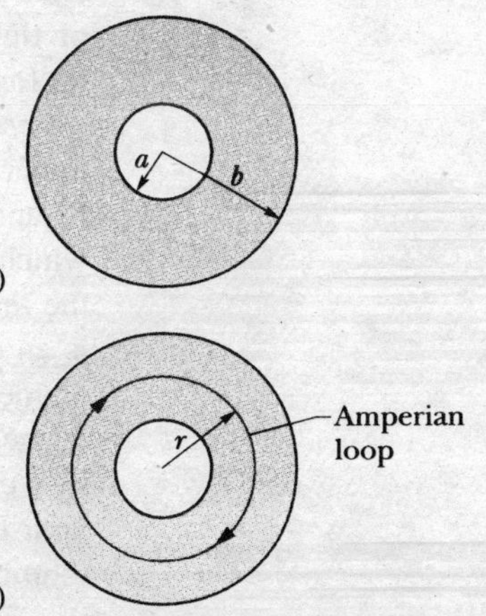

Fig. 30-15 Sample Problem 30-3. (a) Cross section of a conducting cylinder of inner radius a and outer radius b. (b) An Amperian loop of radius r is added to compute the magnetic field at points that are a distance r from the central axis.

SOLUTION: The point at which we want to evaluate $\vec{B}$ is inside the material of the conducting cylinder, between its inner and outer radii. We note that the current distribution has cylindrical symmetry (it is the same all around the cross section for any given radius). Thus, the **Key Idea** here is that the symmetry allows us to use Ampere's law to find $\vec{B}$ at the point. We first draw the Amperian loop shown in Fig. 30-15b. The loop is concentric with the cylinder and has radius $r = 3.0$ cm, because we want to evaluate $\vec{B}$ at that distance from the cylinder's central axis.

Next, we must compute the current i_{enc} that is encircled by the Amperian loop. However, a second **Key Idea** is that we *cannot* set up a proportionality as in Eq. 30-21, because here the current is not uniformly distributed. Instead, following the procedure of Sample Problem 27-2b, we must integrate the current density from the cylinder's inner radius a to the loop radius r:

$$i_{enc} = \int J\,dA = \int_a^r cr^2\,(2\pi r\,dr)$$

dA is area of crosssection ⇒ area ⊥ direction of flow of J

$$= 2\pi c \int_a^r r^3\,dr = 2\pi c\left[\frac{r^4}{4}\right]_a^r$$

$$= \frac{\pi c(r^4 - a^4)}{2}.$$

The direction of integration indicated in Fig. 30-15b is (arbitrarily) clockwise. Applying the right-hand rule for Ampere's law to that loop, we find that we should take i_{enc} as negative because the current is directed out of the page but our thumb is directed into the page.

We next evaluate the left side of Ampere's law exactly as we did in Fig. 30-14, and we again obtain Eq. 30-20. Then Ampere's law,

$$\oint \vec{B} \cdot d\vec{s} = \mu_0 i_{enc},$$

gives us

area enclosed by Amperian loop is say $2\pi r$

$$B(2\pi r) = -\frac{\mu_0 \pi c}{2}(r^4 - a^4).$$

Solving for B and substituting known data yield

$$B = -\frac{\mu_0 c}{4r}(r^4 - a^4)$$

$$= -\frac{(4\pi \times 10^{-7}\ \text{T}\cdot\text{m/A})(3.0 \times 10^6\ \text{A/m}^4)}{4(0.030\ \text{m})} \times [(0.030\ \text{m})^4 - (0.020\ \text{m})^4]$$

$$= -2.0 \times 10^{-5}\ \text{T}.$$

Thus, the magnetic field $\vec{B}$ at a point 3.0 cm from the central axis has the magnitude

$$B = 2.0 \times 10^{-5}\ \text{T} \qquad \text{(Answer)}$$

and forms magnetic field lines that are directed opposite our direction of integration, hence counterclockwise in Fig. 30-15b.

30-4 Solenoids and Toroids

Magnetic Field of a Solenoid

Fig. 30-16 A solenoid carrying current i.

We now turn our attention to another situation in which Ampere's law proves useful. It concerns the magnetic field produced by the current in a long, tightly wound helical coil of wire. Such a coil is called a **solenoid** (Fig. 30-16). We assume that the length of the solenoid is much greater than the diameter.

Figure 30-17 shows a section through a portion of a "stretched-out" solenoid. The solenoid's magnetic field is the vector sum of the fields produced by the individual turns (loops) that make up the solenoid. For points very close to each turn, the wire behaves magnetically almost like a long straight wire, and the lines of $\vec{B}$ there are almost concentric circles. Figure 30-17 suggests that the field tends to cancel between adjacent turns. It also suggests that, at points inside the solenoid and reasonably far from the wire, $\vec{B}$ is approximately parallel to the (central) solenoid axis. In the limiting case of an *ideal solenoid,* which is infinitely long and consists of tightly packed (*close-packed*) turns of square wire, the field inside the coil is uniform and parallel to the solenoid axis.

At points above the solenoid, such as P in Fig. 30-17, the field set up by the upper parts of the solenoid turns (marked ⊙) is directed to the left (as drawn near P) and tends to cancel the field set up by the lower parts of the turns (marked ⊗), which is directed to the right (not drawn). In the limiting case of an ideal solenoid, the magnetic field outside the solenoid is zero. Taking the external field to be zero is an excellent assumption for a real solenoid if its length is much greater than its diameter and if we consider external points such as point P that are not at either end of the solenoid. The direction of the magnetic field along the solenoid axis is given by a curled–straight right-hand rule: Grasp the solenoid with your right hand so that your fingers follow the direction of the current in the windings; your extended right thumb then points in the direction of the axial magnetic field.

Figure 30-18 shows the lines of $\vec{B}$ for a real solenoid. The spacing of the lines of $\vec{B}$ in the central region shows that the field inside the coil is fairly strong and uniform over the cross section of the coil. The external field, however, is relatively weak.

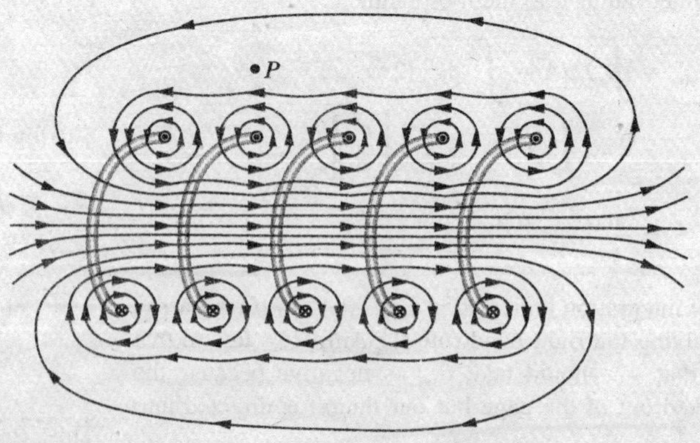

Fig. 30-17 A vertical cross section through the central axis of a "stretched-out" solenoid. The back portions of five turns are shown, as are the magnetic field lines due to a current through the solenoid. Each turn produces circular magnetic field lines near it. Near the solenoid's axis, the field lines combine into a net magnetic field that is directed along the axis. The closely spaced field lines there indicate a strong magnetic field. Outside the solenoid the field lines are widely spaced; the field there is very weak.

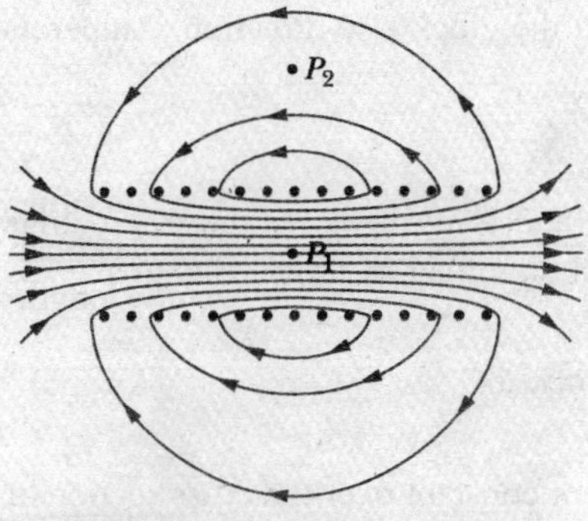

Fig. 30-18 Magnetic field lines for a real solenoid of finite length. The field is strong and uniform at interior points such as P_1 but relatively weak at external points such as P_2.

Let us now apply Ampere's law,

$$\oint \vec{B} \cdot d\vec{s} = \mu_0 i_{\text{enc}}, \tag{30-23}$$

to the ideal solenoid of Fig. 30-19, where $\vec{B}$ is uniform within the solenoid and zero outside it, using the rectangular Amperian loop *abcda*. We write $\oint \vec{B} \cdot d\vec{s}$ as the sum of four integrals, one for each loop segment:

$$\oint \vec{B} \cdot d\vec{s} = \int_a^b \vec{B} \cdot d\vec{s} + \int_b^c \vec{B} \cdot d\vec{s} + \int_c^d \vec{B} \cdot d\vec{s} + \int_d^a \vec{B} \cdot d\vec{s}. \tag{30-24}$$

The first integral on the right of Eq. 30-24 is Bh, where B is the magnitude of the uniform field $\vec{B}$ inside the solenoid and h is the (arbitrary) length of the segment from a to b. The second and fourth integrals are zero because for every element ds of these segments, $\vec{B}$ either is perpendicular to ds or is zero, and thus the product $\vec{B} \cdot d\vec{s}$ is zero. The third integral, which is taken along a segment that lies outside the solenoid, is zero because $B = 0$ at all external points. Thus, $\oint \vec{B} \cdot d\vec{s}$ for the entire rectangular loop has the value Bh.

The net current i_{enc} encircled by the rectangular Amperian loop in Fig. 30-19 is not the same as the current i in the solenoid windings because the windings pass more than once through this loop. Let n be the number of turns per unit length of the solenoid; then the loop encloses nh turns and

$$i_{\text{enc}} = i(nh).$$

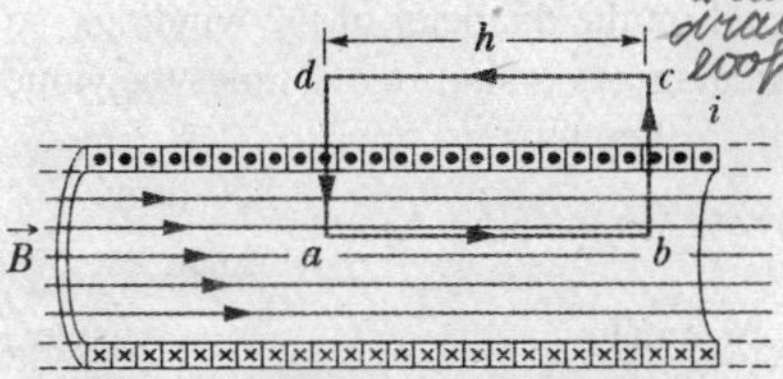

Fig. 30-19 Application of Ampere's law to a section of a long ideal solenoid carrying a current i. The Amperian loop is the rectangle *abcd*.

Ampere's law then gives us

$$Bh = \mu_0 inh$$

or

$$B = \mu_0 in \quad \text{(ideal solenoid).} \tag{30-25}$$

Although we derived Eq. 30-25 for an infinitely long ideal solenoid, it holds quite well for actual solenoids if we apply it only at interior points, well away from the solenoid ends. Equation 30-25 is consistent with the experimental fact that the magnetic field magnitude B within a solenoid does not depend on the diameter or the length of the solenoid and that B is uniform over the solenoidal cross section. A solenoid thus provides a practical way to set up a known uniform magnetic field for experimentation, just as a parallel-plate capacitor provides a practical way to set up a known uniform electric field.

Magnetic Field of a Toroid

Figure 30-20*a* shows a **toroid**, which we may describe as a solenoid bent into the shape of a hollow doughnut. What magnetic field $\vec{B}$ is set up at its interior points (within the hollow of the doughnut)? We can find out from Ampere's law and the symmetry of the doughnut.

From the symmetry, we see that the lines of $\vec{B}$ form concentric circles inside the toroid, directed as shown in Fig. 30-20*b*. Let us choose a concentric circle of

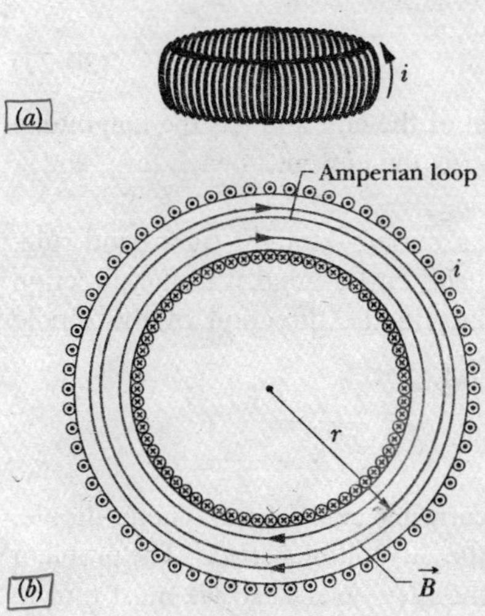

Fig. 30-20 (*a*) A toroid carrying a current i. (*b*) A horizontal cross section of the toroid. The interior magnetic field (inside the doughnut-shaped tube) can be found by applying Ampere's law with the Amperian loop shown.

radius r as an Amperian loop and traverse it in the clockwise direction. Ampere's law (Eq. 30-16) yields

$$(B)(2\pi r) = \mu_0 iN,$$

where i is the current in the toroid windings (and is positive for those windings enclosed by the Amperian loop) and N is the total number of turns. This gives

$$B = \frac{\mu_0 iN}{2\pi}\frac{1}{r}\ \hat{\phi} \quad \text{(toroid).} \tag{30-26}$$

In contrast to the situation for a solenoid, B is not constant over the cross section of a toroid. It is easy to show, with Ampere's law, that $B = 0$ for points outside an ideal toroid (as if the toroid were made from an ideal solenoid).

The direction of the magnetic field within a toroid follows from our curled–straight right-hand rule: Grasp the toroid with the fingers of your right hand curled in the direction of the current in the windings; your extended right thumb points in the direction of the magnetic field.

Sample Problem 30-4

A solenoid has length $L = 1.23$ m and inner diameter $d = 3.55$ cm, and it carries a current $i = 5.57$ A. It consists of five close-packed layers, each with 850 turns along length L. What is B at its center?

SOLUTION: One **Key Idea** here is that the magnitude B of the magnetic field along the solenoid's center is related to the solenoid's current i and number of turns per unit length n by Eq. 30-25. A second **Key Idea** is that B does not depend on the diameter of the windings, so the value of n for five identical layers is simply five times the value for each layer. Equation 30-25 then tells us

$$B = \mu_0 in = (4\pi \times 10^{-7}\ \text{T}\cdot\text{m/A})(5.57\ \text{A})\frac{5 \times 850\ \text{turns}}{1.23\ \text{m}}$$

$$= 2.42 \times 10^{-2}\ \text{T} = 24.2\ \text{mT}. \quad \text{(Answer)}$$

30-5 A Current-Carrying Coil as a Magnetic Dipole

So far we have examined the magnetic fields produced by current in a long straight wire, a solenoid, and a toroid. We turn our attention here to the field produced by a coil carrying a current. You saw in Section 29-9 that such a coil behaves as a magnetic dipole in that, if we place it in an external magnetic field $\vec{B}$, a torque $\vec{\tau}$ given by

$$\vec{\tau} = \vec{\mu} \times \vec{B} \tag{30-27}$$

acts on it. Here $\vec{\mu}$ is the magnetic dipole moment of the coil and has the magnitude NiA, where N is the number of turns (or loops), i is the current in each turn, and A is the area enclosed by each turn.

Recall that the direction of $\vec{\mu}$ is given by a curled–straight right-hand rule: Grasp the coil so that the fingers of your right hand curl around it in the direction of the current; your extended thumb then points in the direction of the dipole moment $\vec{\mu}$.

Magnetic Field of a Coil

We turn now to the other aspect of a current-carrying coil as a magnetic dipole. What magnetic field does *it* produce at a point in the surrounding space? The problem does not have enough symmetry to make Ampere's law useful, so we must turn to

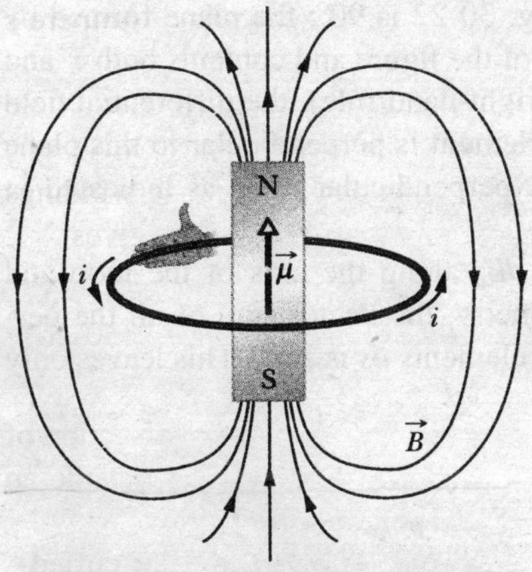

Fig. 30-21 A current loop produces a magnetic field like that of a bar magnet and thus has associated north and south poles. The magnetic dipole moment $\vec{\mu}$ of the loop, given by a curled–straight right-hand rule, points from the south pole to the north pole, in the direction of the field $\vec{B}$ within the loop.

the law of Biot and Savart. For simplicity, we first consider only a coil with a single circular loop and only points on its central axis, which we take to be a z axis. We shall show that the magnitude of the magnetic field at such points is

$$B(z) = \frac{\mu_0 i R^2}{2(R^2 + z^2)^{3/2}}, \tag{30-28}$$

in which R is the radius of the circular loop and z is the distance of the point in question from the center of the loop. Furthermore, the direction of the magnetic field $\vec{B}$ is the same as the direction of the magnetic dipole moment $\vec{\mu}$ of the loop.

For axial points far from the loop, we have $z \gg R$ in Eq. 30-28. With that approximation, the equation reduces to

$$B(z) \approx \frac{\mu_0 i R^2}{2z^3}.$$

Recalling that πR^2 is the area A of the loop and extending our result to include a coil of N turns, we can write this equation as

$$B(z) = \frac{\mu_0}{2\pi}\frac{NiA}{z^3}.$$

Further, since $\vec{B}$ and $\vec{\mu}$ have the same direction, we can write the equation in vector form, substituting from the identity $\mu = NiA$:

$$\vec{B}(z) = \frac{\mu_0}{2\pi}\frac{\vec{\mu}}{z^3} \quad \text{(current-carrying coil).} \tag{30-29}$$

Thus, we have two ways in which we can regard a current-carrying coil as a magnetic dipole: (1) it experiences a torque when we place it in an external magnetic field; (2) it generates its own intrinsic magnetic field, given, for distant points along its axis, by Eq. 30-29. Figure 30-21 shows the magnetic field of a current loop; one side of the loop acts as a north pole (in the direction of $\vec{\mu}$) and the other side as a south pole, as suggested by the lightly drawn magnet in the figure.

CHECKPOINT 4: The figure here shows four arrangements of circular loops of radius r or $2r$, centered on vertical axes (perpendicular to the loops) and carrying identical currents in the directions indicated. Rank the arrangements according to the magnitude of the net magnetic field at the dot, midway between the loops on the central axis, greatest first.

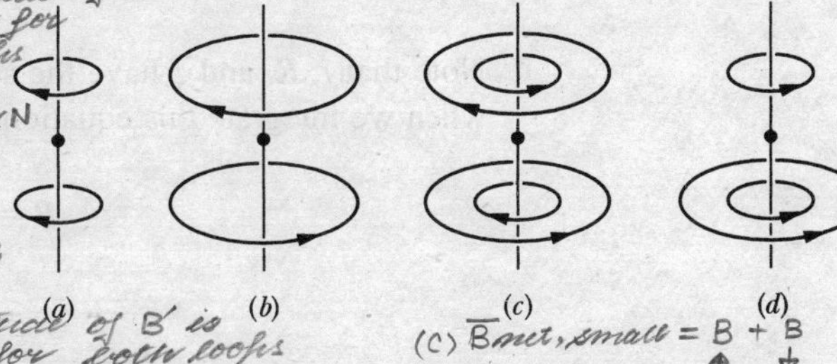

Proof of Equation 30-28

Figure 30-22 shows the back half of a circular loop of radius R carrying a current i. Consider a point P on the axis of the loop, a distance z from its plane. Let us apply the law of Biot and Savart to a differential element ds of the loop, located at the left side of the loop. The length vector $d\vec{s}$ for this element points perpendicularly out of

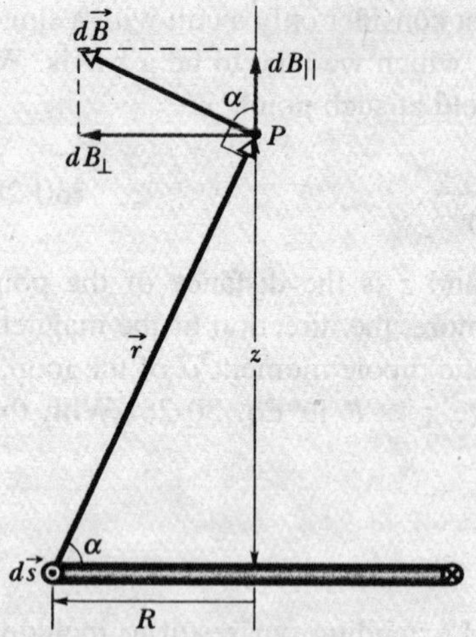

Fig. 30-22 A current loop of radius R. The plane of the loop is perpendicular to the page and only the back half of the loop is shown. We use the law of Biot and Savart to find the magnetic field at point P on the central axis of the loop.

the page. The angle θ between $d\vec{s}$ and $\vec{r}$ in Fig. 30-22 is 90°; the plane formed by these two vectors is perpendicular to the plane of the figure and contains both $\vec{r}$ and $d\vec{s}$. From the law of Biot and Savart (and the right-hand rule), the differential field $d\vec{B}$ produced at point P by the current in this element is perpendicular to this plane and thus is directed in the plane of the figure, perpendicular to $\vec{r}$, as indicated in Fig. 30-22.

Let us resolve $d\vec{B}$ into two components: $dB_{\parallel}$ along the axis of the loop and $dB_{\perp}$ perpendicular to this axis. From the symmetry, the vector sum of all the perpendicular components $dB_{\perp}$ due to all the loop elements ds is zero. This leaves only the axial components $dB_{\parallel}$ and we have

$$B = \int dB_{\parallel}.$$

For the element $d\vec{s}$ in Fig. 30-22, the law of Biot and Savart (Eq. 30-3) tells us that the magnetic field at distance r is

$$dB = \frac{\mu_0}{4\pi}\frac{i\,ds \sin 90°}{r^2}.$$

We also have

$$dB_{\parallel} = dB \cos \alpha.$$

Combining these two relations, we obtain

$$dB_{\parallel} = \frac{\mu_0 i \cos \alpha\, ds}{4\pi r^2}. \tag{30-30}$$

Figure 30-22 shows that r and α are not independent but are related to each other. Let us express each in terms of the variable z, the distance between point P and the center of the loop. The relations are

$$r = \sqrt{R^2 + z^2} \tag{30-31}$$

and

$$\cos \alpha = \frac{R}{r} = \frac{R}{\sqrt{R^2 + z^2}}. \tag{30-32}$$

Substituting Eqs. 30-31 and 30-32 into Eq. 30-30, we find

$$dB_{\parallel} = \frac{\mu_0 i R}{4\pi(R^2 + z^2)^{3/2}}\,ds.$$

Note that i, R, and z have the same values for all elements ds around the loop, so when we integrate this equation, we find that

$$B = \int dB_{\parallel}$$

$$= \frac{\mu_0 i R}{4\pi(R^2 + z^2)^{3/2}} \int ds$$

or, since $\int ds$ is simply the circumference $2\pi R$ of the loop,

$$B(z) = \frac{\mu_0 i R^2}{2(R^2 + z^2)^{3/2}}.$$

This is Eq. 30-28, the relation we sought to prove.

REVIEW & SUMMARY

The Biot–Savart Law The magnetic field set up by a current-carrying conductor can be found from the *Biot–Savart law*. This law asserts that the contribution $d\vec{B}$ to the field produced by a current-length element $i\,d\vec{s}$ at a point P, a distance r from the current element, is

$$d\vec{B} = \frac{\mu_0}{4\pi}\frac{i\,d\vec{s} \times \vec{r}}{r^3} \quad \text{(Biot–Savart law)}. \tag{30-5}$$

Here $\vec{r}$ is a vector that points from the element to P. The quantity μ_0, called the permeability constant, has the value $4\pi \times 10^{-7}\ \text{T}\cdot\text{m/A} \approx 1.26 \times 10^{-6}\ \text{T}\cdot\text{m/A}$.

Magnetic Field of a Long Straight Wire For a long straight wire carrying a current i, the Biot–Savart law gives, for the magnitude of the magnetic field at a perpendicular distance R from the wire,

$$B = \frac{\mu_0 i}{2\pi R} \quad \text{(long straight wire)}. \tag{30-6}$$

Magnetic Field of a Circular Arc The magnitude of the magnetic field at the center of a circular arc, of radius R and central angle ϕ (in radians), carrying current i, is

$$B = \frac{\mu_0 i \phi}{4\pi R} \quad \text{(at center of circular arc)}. \tag{30-11}$$

The Force Between Parallel Wires Carrying Currents Parallel wires carrying currents in the same direction attract each other, whereas parallel wires carrying currents in opposite directions repel each other. The magnitude of the force on a length L of either wire is

$$F_{ba} = i_b L B_a \sin 90^\circ = \frac{\mu_0 L i_a i_b}{2\pi d}, \tag{30-15}$$

where d is the wire separation, and i_a and i_b are the currents in the wires.

Ampere's Law **Ampere's law** states that

$$\oint \vec{B}\cdot d\vec{s} = \mu_0 i_{enc} \quad \text{(Ampere's law)}. \tag{30-16}$$

The line integral in this equation is evaluated around a closed loop called an *Amperian loop*. The current i is the *net* current encircled by the loop. For some current distributions, Eq. 30-16 is easier to use than Eq. 30-5 to calculate the magnetic field due to the currents.

Fields of a Solenoid and a Toroid Inside a *long solenoid* carrying current i, at points not near its ends, the magnitude B of the magnetic field is

$$B = \mu_0 i n \quad \text{(ideal solenoid)}, \tag{30-25}$$

where n is the number of turns per unit length. At a point inside a *toroid*, the magnitude B of the magnetic field is

$$B = \frac{\mu_0 i N}{2\pi}\frac{1}{r} \quad \text{(toroid)}, \tag{30-26}$$

where r is the distance from the center of the toroid to the point.

Field of a Magnetic Dipole The magnetic field produced by a current-carrying coil, which is a *magnetic dipole*, at a point P located a distance z along the coil's central axis is parallel to the axis and is given by

$$\vec{B}(z) = \frac{\mu_0}{2\pi}\frac{\vec{\mu}}{z^3}, \tag{30-29}$$

where $\vec{\mu}$ is the dipole moment of the coil. This equation applies only when z is much greater than the dimensions of the coil.

QUESTIONS

1 Figure 30-23 shows four arrangements in which long parallel wires carry equal currents directly into or out of the page at the corners of identical squares. Rank the arrangements according to the magnitude of the net magnetic field at the center of the square, greatest first. *see B direction at the centre of the sq*

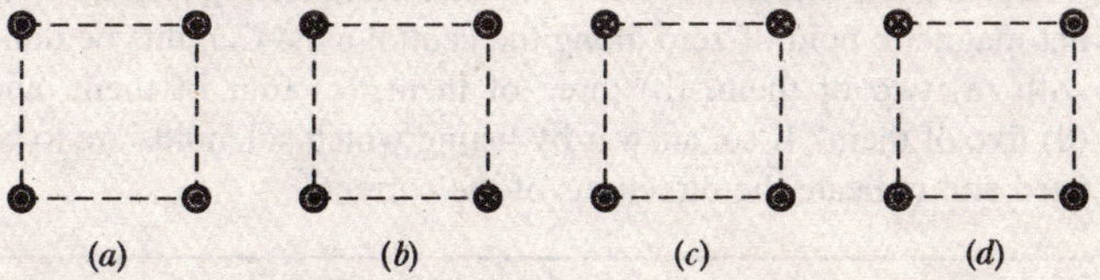

Fig. 30-23 Question 1.

2 Figure 30-24 shows cross sections of two long straight wires; the left-hand wire carries current i_1 directly out of the page. If the net magnetic field due to the two currents is to be zero at point P, (a) should the direction of current i_2 in the right-hand wire be directly into or out of the page and (b) should i_2 be greater than, less than, or equal to i_1?

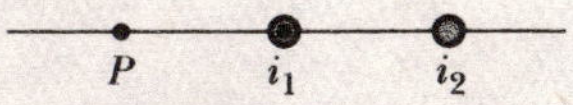

Fig. 30-24 Question 2.

3 Figure 30-25 shows three circuits, each consisting of two concentric circular arcs, one of radius r and the other of a larger radius

Fig. 30-25 Question 3. (a) (b) (c)

R, and two radial lengths. The circuits have the same current through them, and the radial lengths have the same angle between them. Rank the circuits according to the magnitude of the net magnetic field at the center, greatest first.

4. Figure 30-26 shows four arrangements in which long, parallel, equally spaced wires carry equal currents directly into or out of the page. Rank the arrangements according to the magnitude of the net force on the central wire due to the currents in the other wires, greatest first.

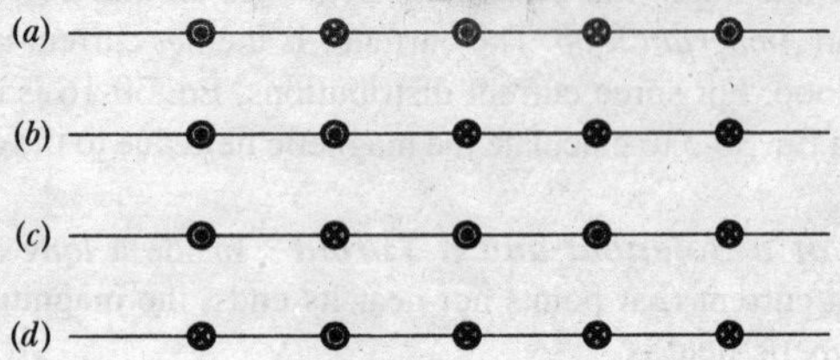

Fig. 30-26 Question 4.

5. Figure 30-27 shows three arrangements of three long straight wires, carrying equal currents directly into or out of the page. (a) Rank the arrangements according to the magnitude of the net force on the wire with the current directed out of the page due to the currents in the other wires, greatest first. (b) In arrangement 3, is the angle between the net force on that wire and the dashed line equal to, less than, or more than 45°?

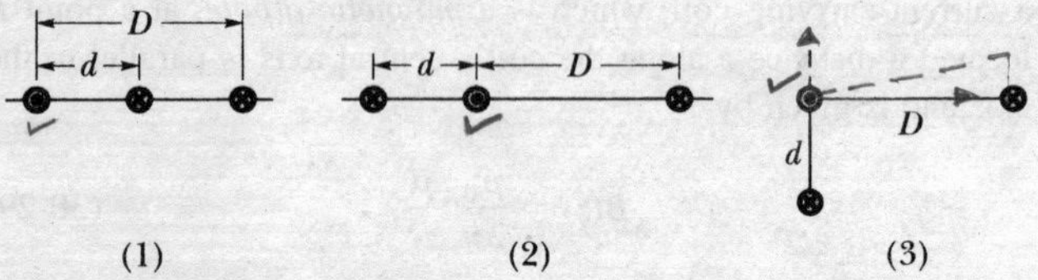

Fig. 30-27 Question 5.

6. Figure 30-28 shows a uniform magnetic field $\vec{B}$ and four straight-line paths of equal lengths. Rank the paths according to the magnitude of $\int \vec{B} \cdot d\vec{s}$ taken along the paths, greatest first.

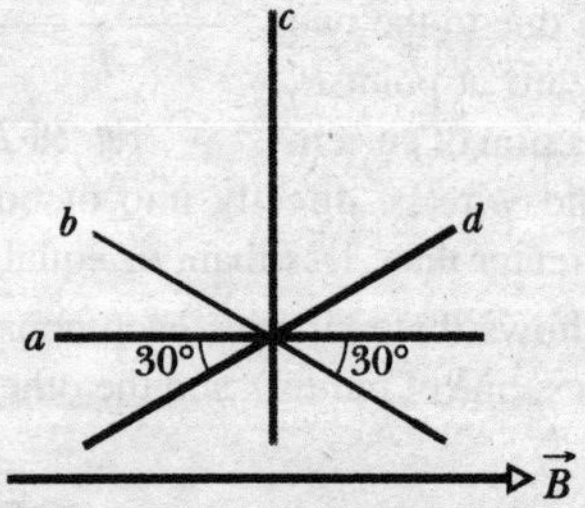

Fig. 30-28 Question 6.

7. Figure 30-29*a* shows four circular Amperian loops concentric with a wire whose current is directed out of the page. The current is uniform across the wire's circular cross section. Rank the loops according to the magnitude of $\oint \vec{B} \cdot d\vec{s}$ around each, greatest first.

8. Figure 30-29*b* shows four circular Amperian loops (red) and, in cross section, four long circular conductors (blue), all of which are concentric. Three of the conductors are hollow cylinders; the central conductor is a solid cylinder. The currents in the conductors are, from smallest radius to largest radius, 4 A out of the page, 9 A into the page, 5 A out of the page, and 3 A into the page. Rank the loops according to the magnitude of $\oint \vec{B} \cdot d\vec{s}$ around each, greatest first.

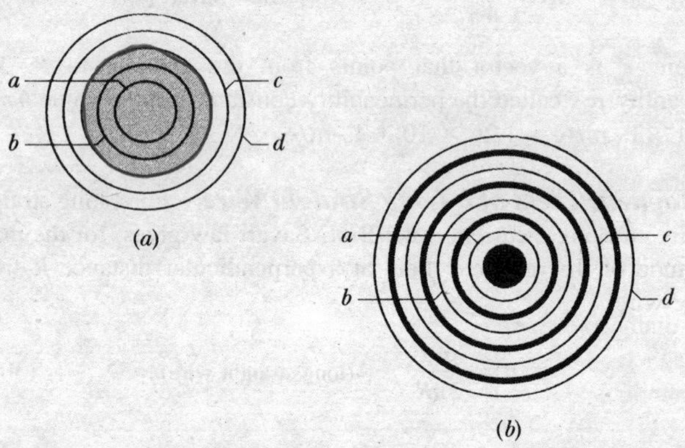

Fig. 30-29 Questions 7 and 8.

9. Figure 30-30 shows four identical currents *i* and five Amperian paths encircling them. Rank the paths according to the value of $\oint \vec{B} \cdot d\vec{s}$ taken in the directions shown, most positive first and most negative last.

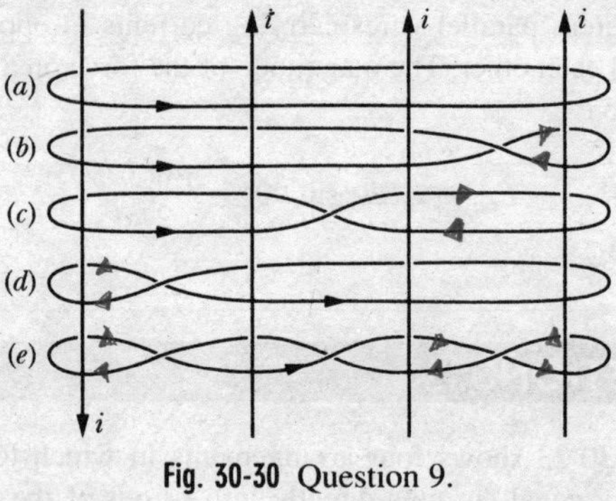

Fig. 30-30 Question 9.

10. The following table gives the number of turns per unit length *n* and the current *i* through six ideal solenoids of different radii. You want to combine several of them concentrically to produce a net magnetic field of zero along the central axis. Can this be done with (a) two of them, (b) three of them, (c) four of them, and (d) five of them? If so, answer by listing which solenoids are to be used and indicate the directions of the currents.

Solenoid:	1	2	3	4	5	6
n:	5	4	3	2	10	8
i:	5	3	7	6	2	3

EXERCISES & PROBLEMS

ssm Solution is in the Student Solutions Manual.
www Solution is available on the World Wide Web at: http://www.wiley.com/college/hrw
itw Solution is available on the Interactive LearningWare.

SEC. 30-1 Calculating the Magnetic Field Due to a Current

1E. A surveyor is using a magnetic compass 6.1 m below a power line in which there is a steady current of 100 A. (a) What is the magnetic field at the site of the compass due to the power line? (b) Will this interfere seriously with the compass reading? The horizontal component of Earth's magnetic field at the site is 20 μT. ssm

2E. The electron gun in a traditional television tube fires electrons of kinetic energy 25 keV at the screen in a circular beam 0.22 mm in diameter; 5.6×10^{14} electrons arrive each second. Calculate the magnetic field produced by the beam at a point 1.5 mm from the beam axis.

3E. At a certain position in the Philippines, Earth's magnetic field of 39 μT is horizontal and directed due north. Suppose the net field is zero exactly 8.0 cm above a long, straight, horizontal wire that carries a constant current. What are (a) the magnitude and (b) the direction of the current? ssm

4E. A long wire carrying a current of 100 A is placed in a uniform external magnetic field of 5.0 mT. The wire is perpendicular to this magnetic field. Locate the points at which the net magnetic field is zero.

5E. A particle with positive charge q is a distance d from a long straight wire that carries a current i; the particle is traveling with speed v perpendicular to the wire. What are the direction and magnitude of the force on the particle if it is moving (a) toward and (b) away from the wire? itw

6E. A straight conductor carrying a current i splits into identical semicircular arcs as shown in Fig. 30-31. What is the magnetic field at the center C of the resulting circular loop?

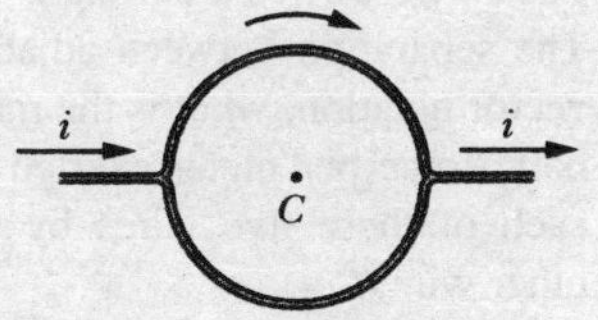

Fig. 30-31 Exercise 6.

7P. A wire carrying current i has the configuration shown in Fig. 30-32. Two semi-infinite straight sections, both tangent to the same circle, are connected by a circular arc, of central angle θ, along the circumference of the circle, with all sections lying in the same plane. What must θ be in order for B to be zero at the center of the circle? ssm

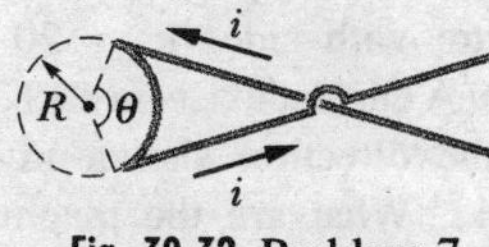

Fig. 30-32 Problem 7.

8P. Use the Biot–Savart law to calculate the magnetic field $\vec{B}$ at C, the common center of the semicircular arcs AD and HJ in Fig. 30-33a. The two arcs, of radii R_2 and R_1, respectively, form part of the circuit $ADJHA$ carrying current i.

9P. In the circuit of Fig. 30-33b, the curved segments are arcs of circles of radii a and b with common center P. The straight segments are along radii. Find the magnetic field $\vec{B}$ at point P, assuming a current i in the circuit. ssm

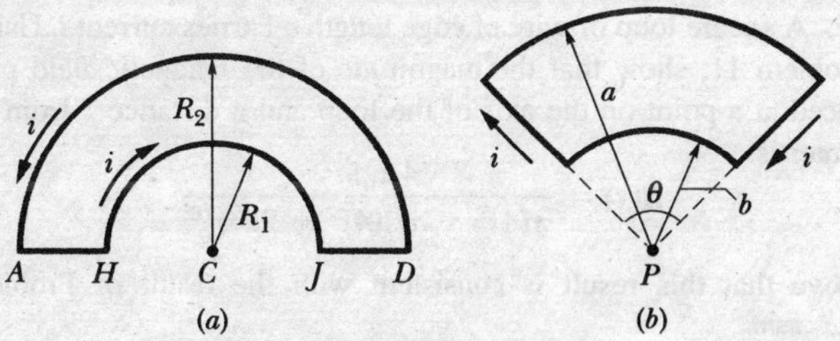

Fig. 30-33 Problems 8 and 9.

10P. The wire shown in Fig. 30-34 carries current i. What magnetic field $\vec{B}$ is produced at the center C of the semicircle by (a) each straight segment of length L, (b) the semicircular segment of radius R, and (c) the entire wire?

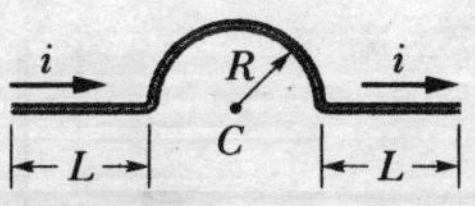

Fig. 30-34 Problem 10.

11P. In Fig. 30-35, a straight wire of length L carries current i. Show that the magnitude of the magnetic field $\vec{B}$ produced by this segment at P_1, a distance R from the segment along a perpendicular bisector, is

$$B = \frac{\mu_0 i}{2\pi R} \frac{L}{(L^2 + 4R^2)^{1/2}}.$$

Show that this expression for B reduces to an expected result as $L \to \infty$. ssm www

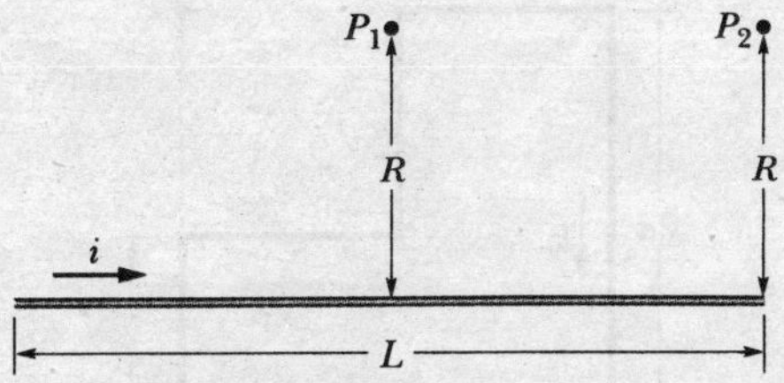

Fig. 30-35 Problems 11 and 13.

12P. A square loop of wire of edge length a carries current i. Using Problem 11, show that, at the center of the loop, the magnitude of the magnetic field produced by the current is

$$B = \frac{2\sqrt{2}\mu_0 i}{\pi a}.$$

13P. In Fig. 30-35, a straight wire of length L carries current i. Show that

$$B = \frac{\mu_0 i}{4\pi R} \frac{L}{(L^2 + R^2)^{1/2}}$$

gives the magnitude of the magnetic field $\vec{B}$ produced by the wire at P_2, a perpendicular distance R from one end of the wire. ssm

14P. Using Problem 11, show that the magnitude of the magnetic field produced at the center of a rectangular loop of wire of length L and width W, carrying a current i, is

$$B = \frac{2\mu_0 i}{\pi}\frac{(L^2 + W^2)^{1/2}}{LW}.$$

15P. A square loop of wire of edge length a carries current i. Using Problem 11, show that the magnitude of the magnetic field produced at a point on the axis of the loop and a distance x from its center is

$$B(x) = \frac{4\mu_0 i a^2}{\pi(4x^2 + a^2)(4x^2 + 2a^2)^{1/2}}.$$

Prove that this result is consistent with the result of Problem 12. ssm

16P. In Fig. 30-36, a straight wire of length a carries a current i. Show that the magnitude of the magnetic field produced by the current at point P is $B = \sqrt{2}\mu_0 i/8\pi a$.

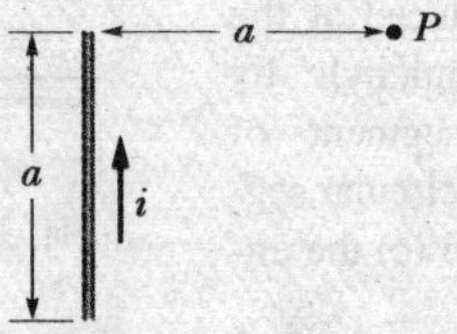

Fig. 30-36 Problem 16.

17P. Two wires, both of length L, are formed into a circle and a square, and each carries current i. Show that the square produces a greater magnetic field at its center than the circle produces at its center. (See Problem 12.) ssm

18P. Find the magnetic field $\vec{B}$ at point P in Fig. 30-37. (See Problem 16.)

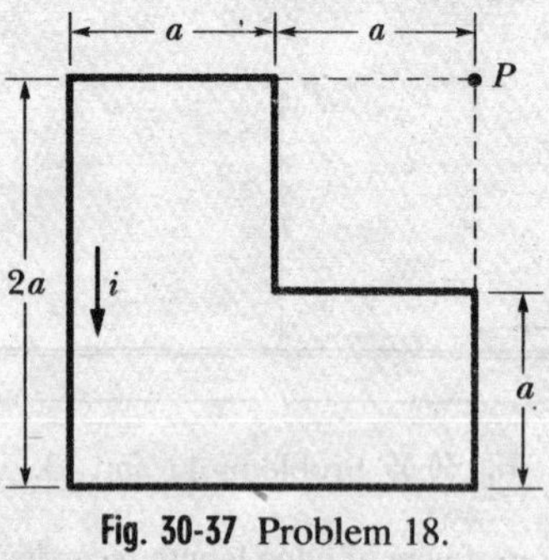

Fig. 30-37 Problem 18.

19P. Figure 30-38 shows a cross section of a long thin ribbon of width w that is carrying a uniformly distributed total current i into the page. Calculate the magnitude and direction of the magnetic field $\vec{B}$ at a point P in the plane of the ribbon at a distance d from its edge. (*Hint:* Imagine the ribbon to be constructed from many long, thin, parallel wires.) ilw

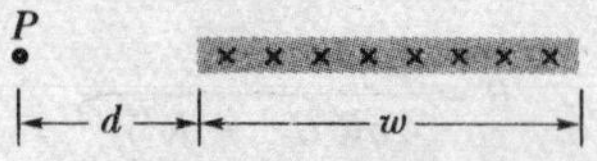

Fig. 30-38 Problem 19.

20P. Find the magnetic field $\vec{B}$ at point P in Fig. 30-39 for i = 10 A and a = 8.0 cm. (See Problems 13 and 16.)

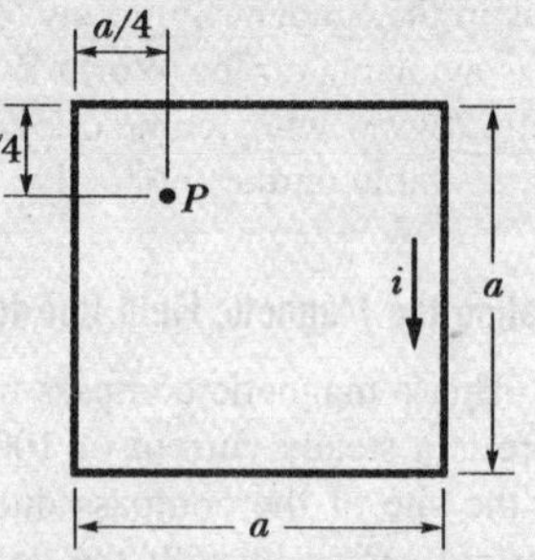

Fig. 30-39 Problem 20.

SEC. 30-2 Force Between Two Parallel Currents

21E. Two long parallel wires are 8.0 cm apart. What equal currents must be in the wires if the magnetic field halfway between them is to have a magnitude of 300 μT? Answer for both (a) parallel and (b) antiparallel currents. ssm

22E. Two long parallel wires a distance d apart carry currents of i and $3i$ in the same direction. Locate the point or points at which their magnetic fields cancel.

23E. Two long, straight, parallel wires, separated by 0.75 cm, are perpendicular to the plane of the page as shown in Fig. 30-40. Wire 1 carries a current of 6.5 A into the page. What must be the current (magnitude and direction) in wire 2 for the resultant magnetic field at point P to be zero?

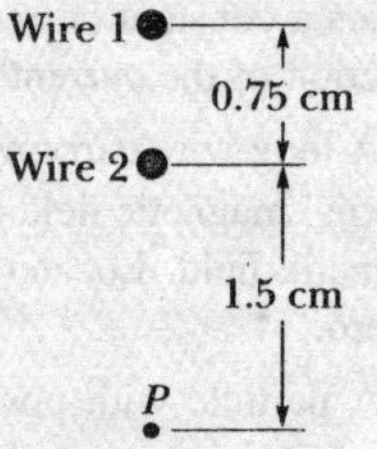

Fig. 30-40 Exercise 23.

24E. Figure 30-41 shows five long parallel wires in the xy plane. Each wire carries a current i = 3.00 A in the positive x direction. The separation between adjacent wires is d = 8.00 cm. In unit-vector notation, what is the magnetic force per meter exerted on each of these five wires by the other wires?

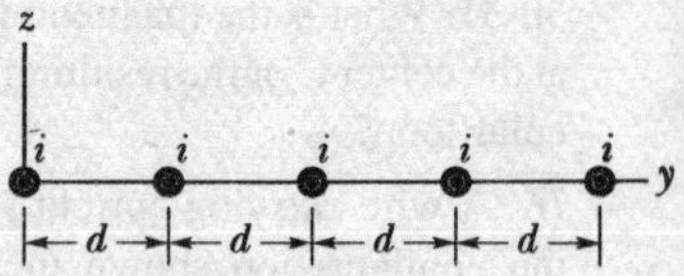

Fig. 30-41 Exercise 24.

25P. Four long copper wires are parallel to each other, their cross sections forming the corners of a square with sides a = 20 cm. A 20 A current exists in each wire in the direction shown in Fig. 30-42. What are the magnitude and direction of $\vec{B}$ at the center of the square? ssm www

26P. Four identical parallel currents i are arranged to form a square of edge length a as in Fig. 30-42, *except* that they are *all* out of the page. What is the force per unit length (magnitude and direction) on any one wire?

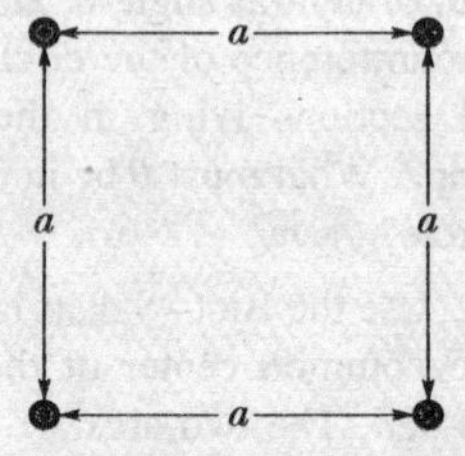

Fig. 30-42 Problems 25, 26, and 27.

27P. In Fig. 30-42, what is the force per unit length acting on the lower left wire, in magnitude and direction, with the current directions as shown? The currents are i.

28P. Figure 30-43 is an idealized schematic drawing of a rail gun. Projectile P sits between two wide rails of circular cross section; a source of current sends current through the rails and through the (conducting) projectile itself (a fuse is not used). (a) Let w be the distance between the rails, R the radius of the rails, and i the current. Show that the force on the projectile is directed to the right along the rails and is given approximately by

$$F = \frac{i^2 \mu_0}{2\pi} \ln \frac{w + R}{R}.$$

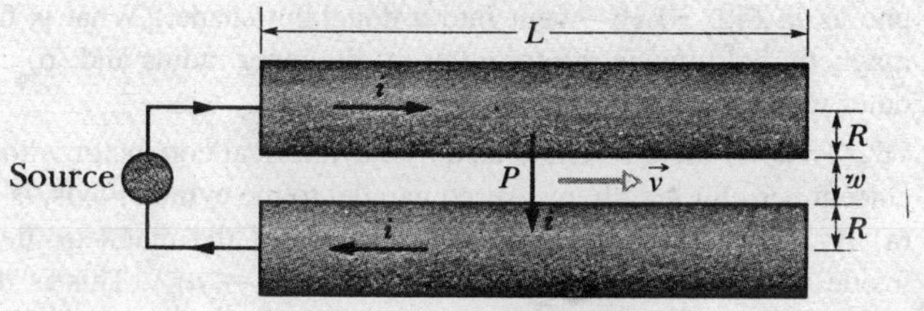

Fig. 30-43 Problem 28.

(b) If the projectile starts from the left end of the rails at rest, find the speed v at which it is expelled at the right. Assume that $i = 450$ kA, $w = 12$ mm, $R = 6.7$ cm, $L = 4.0$ m, and the mass of the projectile is $m = 10$ g.

29P. In Fig. 30-44, the long straight wire carries a current of 30 A and the rectangular loop carries a current of 20 A. Calculate the resultant force acting on the loop. Assume that $a = 1.0$ cm, $b = 8.0$ cm, and $L = 30$ cm. ilw

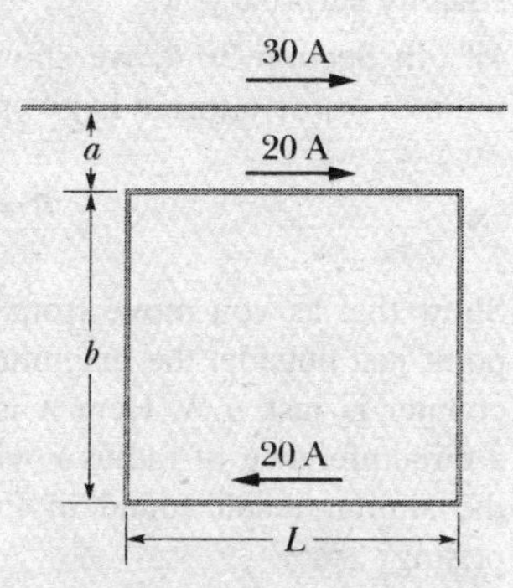

Fig. 30-44 Problem 29.

SEC. 30-3 Ampere's Law

30E. Eight wires cut the page perpendicularly at the points shown in Fig. 30-45. A wire labeled with the integer k ($k = 1, 2, \ldots, 8$) carries the current ki. For those with odd k, the current is out of the page; for those with even k, it is into the page. Evaluate $\oint \vec{B} \cdot d\vec{s}$ along the closed path in the direction shown.

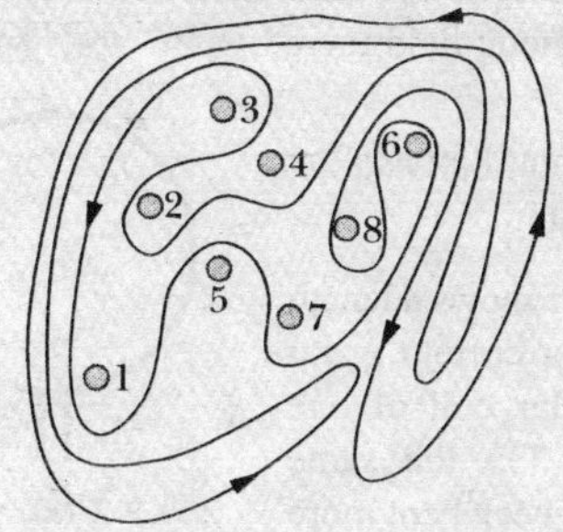

Fig. 30-45 Exercise 30.

31E. Each of the eight conductors in Fig. 30-46 carries 2.0 A of current into or out of the page. Two paths are indicated for the line integral $\oint \vec{B} \cdot d\vec{s}$. What is the value of the integral for the path (a) at the left and (b) at the right? ssm

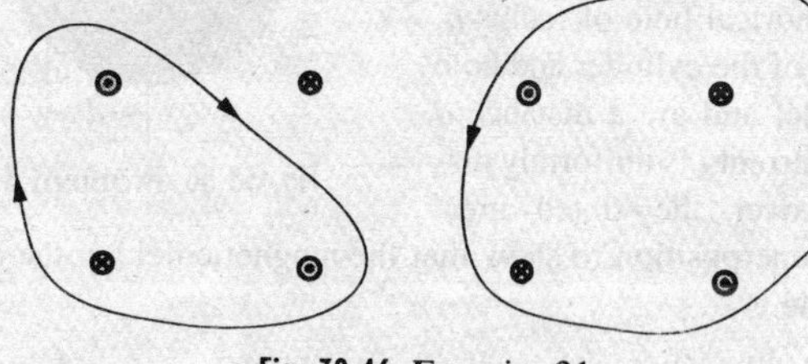
Fig. 30-46 Exercise 31.

32E. Figure 30-47 shows a cross section of a long cylindrical conductor of radius a, carrying a uniformly distributed current i. Assume that $a = 2.0$ cm and $i = 100$ A, and plot $B(r)$ over the range $0 < r < 6.0$ cm.

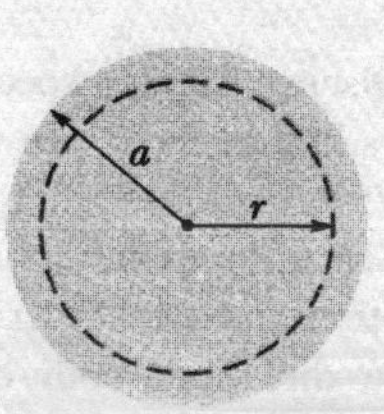

Fig. 30-47 Exercise 32.

33P. Show that a uniform magnetic field $\vec{B}$ cannot drop abruptly to zero (as is suggested by the lack of field lines to the right of point a in Fig. 30-48) as one moves perpendicular to $\vec{B}$, say along the horizontal arrow in the figure. (*Hint:* Apply Ampere's law to the rectangular path shown by the dashed lines.) In actual magnets "fringing" of the magnetic field lines always occurs, which means that $\vec{B}$ approaches zero in a gradual manner. Modify the field lines in the figure to indicate a more realistic situation. ssm

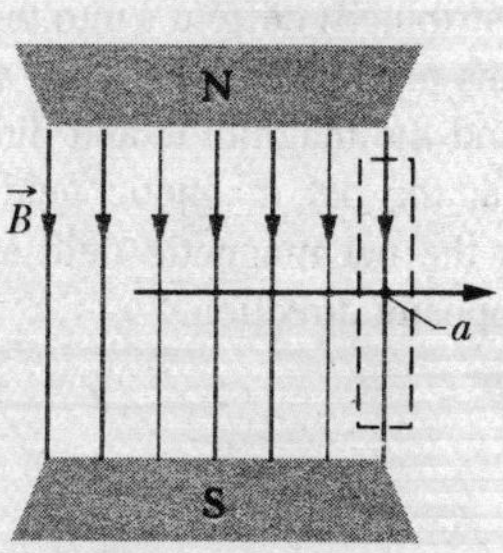

Fig. 30-48 Problem 33.

34P. Two square conducting loops carry currents of 5.0 and 3.0 A as shown in Fig. 30-49. What is the value of $\oint \vec{B} \cdot d\vec{s}$ for each of the two closed paths shown?

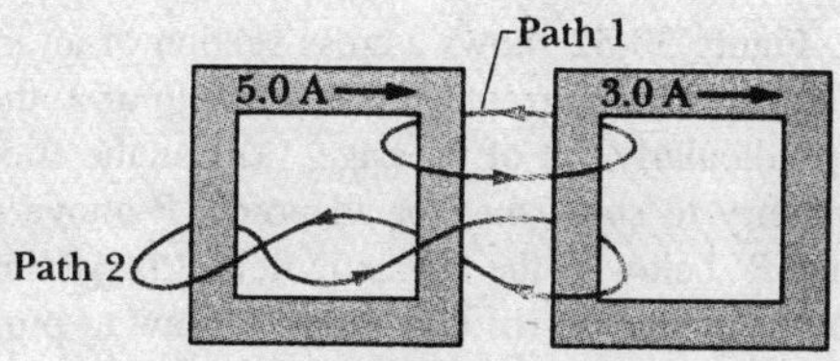

Fig. 30-49 Problem 34.

35P. The current density inside a long, solid, cylindrical wire of radius a is in the direction of the central axis and varies linearly with radial distance r from the axis according to $J = J_0 r/a$. Find the magnetic field inside the wire. ilw

36P. A long straight wire (radius = 3.0 mm) carries a constant current distributed uniformly over a cross section perpendicular to the axis of the wire. If the current density is 100 A/m^2, what are the magnitudes of the magnetic fields (a) 2.0 mm from the axis of the wire and (b) 4.0 mm from the axis of the wire?

37P. Figure 30-50 shows a cross section of a long cylindrical conductor of radius a containing a long cylindrical hole of radius b. The axes of the cylinder and hole are parallel and are a distance d apart; a current i is uniformly distributed over the tinted area.

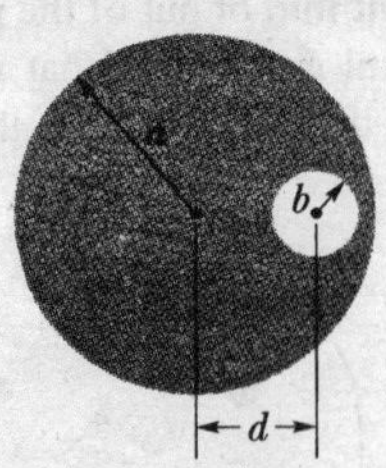

Fig. 30-50 Problem 37.

(a) Use superposition to show that the magnetic field at the center of the hole is

$$B = \frac{\mu_0 i d}{2\pi(a^2 - b^2)}.$$

(b) Discuss the two special cases $b = 0$ and $d = 0$. (c) Use Ampere's law to show that the magnetic field in the hole is uniform. (*Hint:* Regard the cylindrical hole as resulting from the superposition of a complete cylinder (no hole) carrying a current in one direction and a cylinder of radius b carrying a current in the opposite direction, both cylinders having the same current density.) ssm www

38P. A long circular pipe with outside radius R carries a (uniformly distributed) current i into the page as shown in Fig. 30-51. A wire runs parallel to the pipe at a distance of $3R$ from center to center. Find the magnitude and direction of the current in the wire such that the net magnetic field at point P has the same magnitude as the net magnetic field at the center of the pipe but is in the opposite direction.

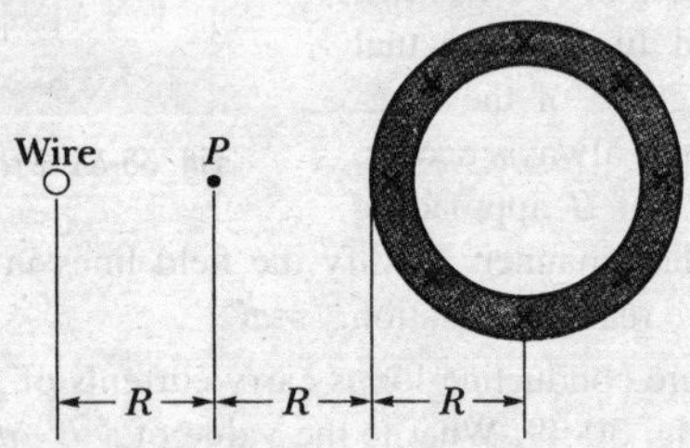

Fig. 30-51 Problem 38.

39P. Figure 30-52 shows a cross section of an infinite conducting sheet carrying a current per unit x-length of λ; the current emerges perpendicularly out of the page. (a) Use the Biot–Savart law and symmetry to show that for all points P above the sheet, and all points P' below it, the magnetic field $\vec{B}$ is parallel to the sheet and directed as shown. (b) Use Ampere's law to prove that $B = \frac{1}{2}\mu_0\lambda$ at all points P and P'. ssm

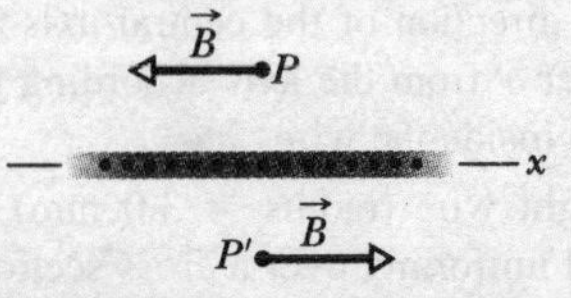

Fig. 30-52 Problems 39 and 44.

SEC. 30-4 Solenoids and Toroids

40E. A solenoid that is 95.0 cm long has a radius of 2.00 cm and a winding of 1200 turns; it carries a current of 3.60 A. Calculate the magnitude of the magnetic field inside the solenoid.

41E. A 200-turn solenoid having a length of 25 cm and a diameter of 10 cm carries a current of 0.30 A. Calculate the magnitude of the magnetic field $\vec{B}$ inside the solenoid. ssm

42E. A solenoid 1.30 m long and 2.60 cm in diameter carries a current of 18.0 A. The magnetic field inside the solenoid is 23.0 mT. Find the length of the wire forming the solenoid.

43E. A toroid having a square cross section, 5.00 cm on a side, and an inner radius of 15.0 cm has 500 turns and carries a current of 0.800 A. (It is made up of a square solenoid—instead of a round one as in Fig. 30-16—bent into a doughnut shape.) What is the magnetic field inside the toroid at (a) the inner radius and (b) the outer radius of the toroid? ssm

44P. Treat an ideal solenoid as a thin cylindrical conductor whose current per unit length, measured parallel to the cylinder axis, is λ. (a) By doing so, show that the magnitude of the magnetic field inside an ideal solenoid can be written as $B = \mu_0\lambda$. This is the value of the *change* in $\vec{B}$ that you encounter as you move from inside the solenoid to outside, through the solenoid wall. (b) Show that the same change occurs as you move through an infinite flat current sheet such as that of Fig. 30-52 (see Problem 39). Does this equality surprise you?

45P. In Section 30-4, we showed that the magnetic field at any radius r *inside* a toroid is given by

$$B = \frac{\mu_0 i N}{2\pi r}.$$

Show that as you move from any point just inside a toroid to a point just outside, the magnitude of the *change* in $\vec{B}$ that you encounter is just $\mu_0\lambda$. Here λ is the current per unit length along a circumference of radius r within the toroid. Compare this with the similar result found in Problem 44. Isn't the equality surprising? ssm

46P. A long solenoid has 100 turns/cm and carries current i. An electron moves within the solenoid in a circle of radius 2.30 cm perpendicular to the solenoid axis. The speed of the electron is $0.0460c$ (c = speed of light). Find the current i in the solenoid.

47P. A long solenoid with 10.0 turns/cm and a radius of 7.00 cm carries a current of 20.0 mA. A current of 6.00 A exists in a straight conductor located along the central axis of the solenoid. (a) At what radial distance from the axis will the direction of the resulting magnetic field be at 45.0° to the axial direction? (b) What is the magnitude of the magnetic field there? ssm ilw www

SEC. 30-5 A Current-Carrying Coil as a Magnetic Dipole

48E. Figure 30-53a shows a length of wire carrying a current i and bent into a circular coil of one turn. In Fig. 30-53b the same length of wire has been bent more sharply, to give a coil of two

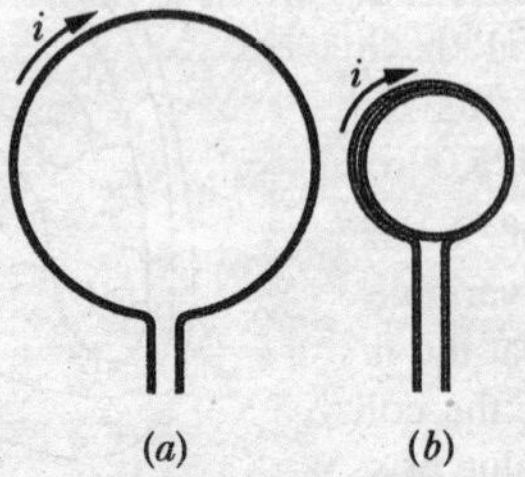

Fig. 30-53 Exercise 48.

turns, each of half the original radius. (a) If B_a and B_b are the magnitudes of the magnetic fields at the centers of the two coils, what is the ratio B_b/B_a? (b) What is the ratio of the dipole moments, μ_b/μ_a, of the coils?

49E. What is the magnetic dipole moment $\vec{\mu}$ of the solenoid described in Exercise 41? ssm

50E. Figure 30-54 shows an arrangement known as a Helmholtz coil. It consists of two circular coaxial coils, each of N turns and radius R, separated by a distance R. The two coils carry equal currents i in the same direction. Find the magnitude of the net magnetic field at P, midway between the coils.

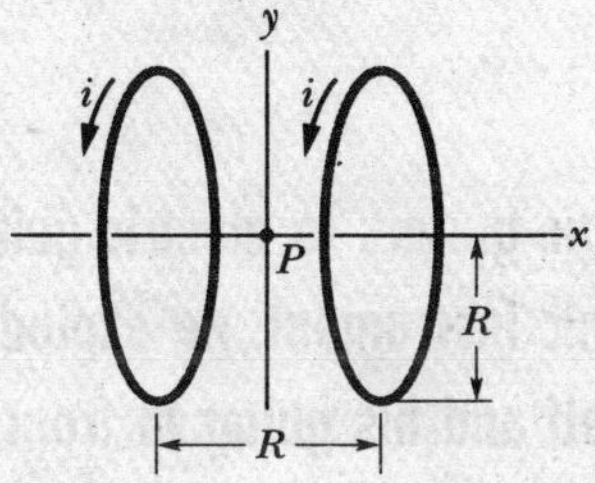

Fig. 30-54 Exercise 50; Problems 53 and 55.

51E. A student makes a short electromagnet by winding 300 turns of wire around a wooden cylinder of diameter $d = 5.0$ cm. The coil is connected to a battery producing a current of 4.0 A in the wire. (a) What is the magnetic moment of this device? (b) At what axial distance $z \gg d$ will the magnetic field of this dipole have the magnitude 5.0 μT (approximately one-tenth that of Earth's magnetic field)? ssm

52E. The magnitude $B(x)$ of the magnetic field at points on the axis of a square current loop of side a is given in Problem 15. (a) Show that the axial magnetic field of this loop, for $x \gg a$, is that of a magnetic dipole (see Eq. 30-29). (b) What is the magnetic dipole moment of this loop?

53P. Two 300-turn coils of radius R each carry a current i. They are arranged a distance R apart, as in Fig. 30-54. For $R = 5.0$ cm and $i = 50$ A, plot the magnitude B of the net magnetic field as a function of distance x along the common x axis over the range $x = -5$ cm to $x = +5$ cm, taking $x = 0$ at the midpoint P. (Such coils provide an especially uniform field $\vec{B}$ near point P.) (*Hint:* See Eq. 30-28.)

54P. A conductor carries a current of 6.0 A along the closed path *abcdefgha* involving 8 of the 12 edges of a cube of side 10 cm as shown in Fig. 30-55. (a) Why can one regard this as the superposition of three square loops: *bcfgb*, *abgha*, and *cdefc*? (*Hint:* Draw currents around those square loops.) (b) Use this superposition to find the magnetic dipole moment $\vec{\mu}$ (magnitude and direction) of the closed path. (c) Calculate $\vec{B}$ at the points $(x, y, z) =$ (0, 5.0 m, 0) and (5.0 m, 0, 0).

55P. In Exercise 50 (Fig. 30-54), let the separation of the coils be a variable s (not necessarily equal to the coil radius R). (a) Show that the first derivative of the magnitude of the net magnetic field of the coils (dB/dx) vanishes at the midpoint P regardless of the value of s. Why would you expect this to be true from symmetry? (b) Show that the second derivative (d^2B/dx^2) also vanishes at P, provided $s = R$. This accounts for the uniformity of B near P for this particular coil separation. ssm

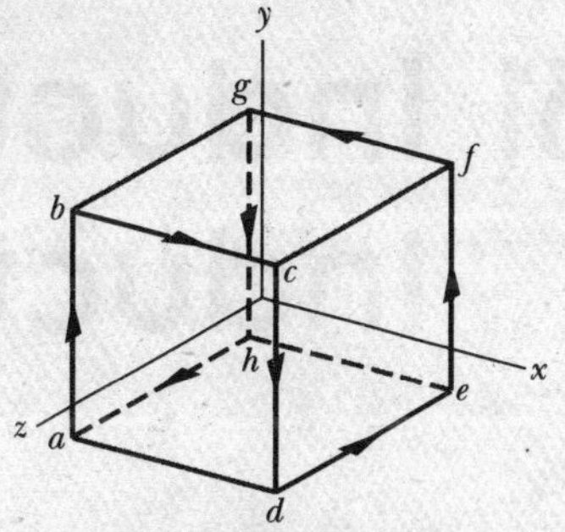

Fig. 30-55 Problem 54.

56P. A length of wire is formed into a closed circuit with radii a and b, as shown in Fig. 30-56, and carries a current i. (a) What are the magnitude and direction of $\vec{B}$ at point P? (b) Find the magnetic dipole moment of the circuit.

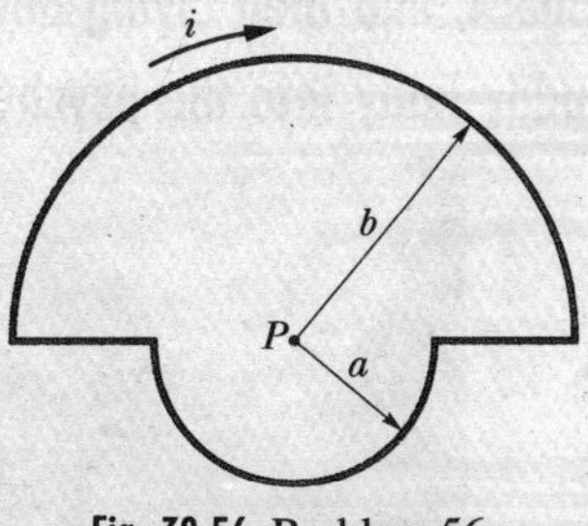

Fig. 30-56 Problem 56.

57P. A circular loop of radius 12 cm carries a current of 15 A. A flat coil of radius 0.82 cm, having 50 turns and a current of 1.3 A, is concentric with the loop. (a) What magnetic field $\vec{B}$ does the loop produce at its center? (b) What torque acts on the coil? Assume that the planes of the loop and coil are perpendicular and that the magnetic field due to the loop is essentially uniform throughout the volume occupied by the coil.

58P. (a) A long wire is bent into the shape shown in Fig. 30-57, without the wire actually touching itself at P. The radius of the circular section is R. Determine the magnitude and direction of $\vec{B}$ at the center C of the circular section when the current i is as indicated. (b) Suppose the circular section of the wire is rotated without distortion about the indicated diameter, until the plane of the circle is perpendicular to the straight sections of the wire. The magnetic dipole moment associated with the circular section is now in the direction of the current in the straight section of the wire. Determine $\vec{B}$ at C in this case.

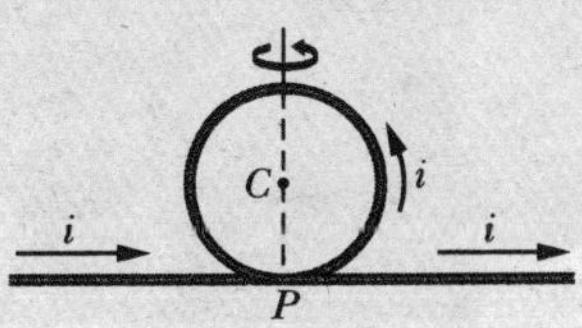

Fig. 30-57 Problem 58.

31 Induction and Inductance

Soon after rock began in the mid-1950s, guitarists switched from acoustic guitars to electric guitars—but it was Jimi Hendrix who first understood the electric guitar as an electronic instrument. He exploded on the scene in the 1960s, ripping his pick along the strings, positioning himself and his guitar in front of a speaker to sustain feedback, and then laying down chords on top of the feedback. He shoved rock forward from the melodies of Buddy Holly into the psychedelia of the late 1960s and into the early heavy metal of Led Zeppelin and the raw energy of Joy Division in the 1970s, and his ideas continue to influence rock today.

What is it about an electric guitar that distinguishes it from an acoustic guitar and enabled Hendrix to make so much broader use of this electronic instrument?

The answer is in this chapter.

31-1 Two Symmetric Situations

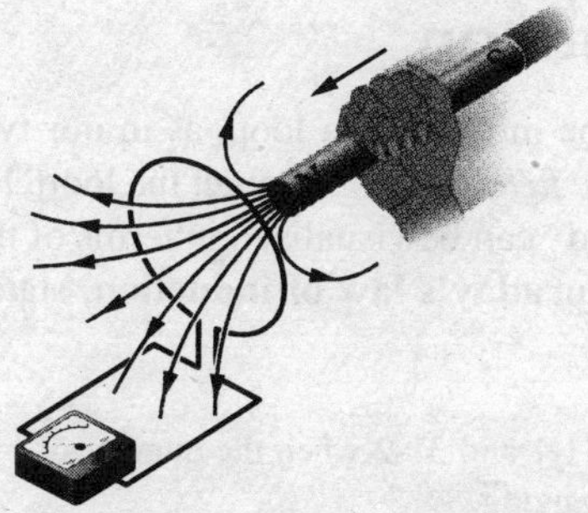

Fig. 31-1 A current meter registers a current in the wire loop when the magnet is moving with respect to the loop.

In Section 29-8, we saw that if we put a closed conducting loop in a magnetic field and then send current through the loop, forces due to the magnetic field create a torque to turn the loop:

$$\text{current loop} + \text{magnetic field} \Rightarrow \text{torque}. \tag{31-1}$$

Suppose that, instead, with the current off, we turn the loop by hand. Will the opposite of Eq. 31-1 occur? That is, will a current now appear in the loop:

$$\text{torque} + \text{magnetic field} \Rightarrow \text{current?} \tag{31-2}$$

The answer is yes—a current does appear. The situations of Eqs. 31-1 and 31-2 are symmetric. The physical law on which Eq. 31-2 depends is called *Faraday's law of induction*. Whereas Eq. 31-1 is the basis for the electric motor, Eq. 31-2 and Faraday's law are the basis for the electric generator. This chapter is concerned with that law and the process it describes.

31-2 Two Experiments

Let us examine two simple experiments to prepare for our discussion of Faraday's law of induction.

First Experiment. Figure 31-1 shows a conducting loop connected to a sensitive current meter. Since there is no battery or other source of emf included, there is no current in the circuit. However, if we move a bar magnet toward the loop, a current suddenly appears in the circuit. The current disappears when the magnet stops. If we then move the magnet away, a current again suddenly appears, but now in the opposite direction. If we experimented for a while, we would discover the following:

1. A current appears only if there is relative motion between the loop and the magnet (one must move relative to the other); the current disappears when the relative motion between them ceases.
2. Faster motion produces a greater current.
3. If moving the magnet's north pole toward the loop causes, say, clockwise current, then moving the north pole away causes counterclockwise current. Moving the south pole toward or away from the loop also causes currents, but in the reversed directions.

The current produced in the loop is called an **induced current;** the work done per unit charge to produce that current (to move the conduction electrons that constitute the current) is called an **induced emf;** and the process of producing the current and emf is called **induction.**

Fig. 31-2 The current meter registers a current in the left-hand wire loop just as switch S is closed (to turn on the current in the right-hand wire loop) or opened (to turn off the current in the right-hand loop). No motion of the coils is involved.

Second Experiment. For this experiment we use the apparatus of Fig. 31-2, with the two conducting loops close to each other but not touching. If we close switch S, to turn on a current in the right-hand loop, the meter suddenly and briefly registers a current—an induced current—in the left-hand loop. If we then open the switch, another sudden and brief induced current appears in the left-hand loop, but in the opposite direction. We get an induced current (and thus an induced emf) only when the current in the right-hand loop is changing (either turning on or turning off) and not when it is constant (even if it is large).

The induced emf and induced current in these experiments are apparently caused when something changes—but what is that "something"? Faraday knew.

31-3 Faraday's Law of Induction

Faraday realized that an emf and a current can be induced in a loop, as in our two experiments, by changing the *amount of magnetic field* passing through the loop. He further realized that the "amount of magnetic field" can be visualized in terms of the magnetic field lines passing through the loop. **Faraday's law of induction,** stated in terms of our experiments, is this:

> An emf is induced in the loop at the left in Figs. 31-1 and 31-2 when the number of magnetic field lines that pass through the loop is changing.

The actual number of field lines passing through the loop does not matter; the values of the induced emf and induced current are determined by the *rate* at which that number changes.

In our first experiment (Fig. 31-1), the magnetic field lines spread out from the north pole of the magnet. Thus, as we move the north pole closer to the loop, the number of field lines passing through the loop increases. That increase apparently causes conduction electrons in the loop to move (the induced current) and provides energy (the induced emf) for their motion. When the magnet stops moving, the number of field lines through the loop no longer changes and the induced current and induced emf disappear.

In our second experiment (Fig. 31-2), when the switch is open (no current), there are no field lines. However, when we turn on the current in the right-hand loop, the increasing current builds up a magnetic field around that loop and at the left-hand loop. While the field builds, the number of magnetic field lines through the left-hand loop increases. As in the first experiment, the increase in field lines through that loop apparently induces a current and an emf there. When the current in the right-hand loop reaches a final, steady value, the number of field lines through the left-hand loop no longer changes, and the induced current and induced emf disappear.

current carrying coil produces magnetic field

Faraday's law does not explain *why* a current and an emf are induced in either experiment; it is just a statement that helps us visualize the induction.

Does changing electric field also induces magnetic field?

A Quantitative Treatment

To put Faraday's law to work, we need a way to calculate the *amount of magnetic field* that passes through a loop. In Chapter 24, in a similar situation, we needed to calculate the amount of an electric field that passes through a surface. There we defined an electric flux $\Phi_E = \int \vec{E} \cdot d\vec{A}$. Here we define a *magnetic flux:* Suppose a loop enclosing an area A is placed in a magnetic field $\vec{B}$. Then the magnetic flux through the loop is

$$\Phi_B = \int \vec{B} \cdot d\vec{A} \quad \text{(magnetic flux through area } A\text{).} \tag{31-3}$$

As in Chapter 24, $d\vec{A}$ is a vector of magnitude dA that is perpendicular to a differential area dA.

As a special case of Eq. 31-3, suppose that the loop lies in a plane and that the magnetic field is perpendicular to the plane of the loop. Then we can write the dot product in Eq. 31-3 as $B\,dA \cos 0° = B\,dA$. If the magnetic field is also uniform, then B can be brought out in front of the integral sign. The remaining $\int dA$ then gives just the area A of the loop. Thus, Eq. 31-3 reduces to

$$\Phi_B = BA \quad (\vec{B} \perp \text{area } A, \vec{B} \text{ uniform}). \tag{31-4}$$

From Eqs. 31-3 and 31-4, we see that the SI unit for magnetic flux is the tesla–square meter, which is called the *weber* (abbreviated Wb):

$$1 \text{ weber} = 1 \text{ Wb} = 1 \text{ T} \cdot \text{m}^2. \tag{31-5}$$

With the notion of magnetic flux, we can state Faraday's law in a more quantitative and useful way:

> The magnitude of the emf $\mathscr{E}$ induced in a conducting loop is equal to the rate at which the magnetic flux Φ_B through that loop changes with time.

As you will see in the next section, the induced emf $\mathscr{E}$ tends to oppose the flux change, so Faraday's law is formally written as

$$\mathscr{E} = -\frac{d\Phi_B}{dt} \quad \text{(Faraday's law)}, \tag{31-6}$$

with the minus sign indicating that opposition. We often neglect the minus sign in Eq. 31-6, seeking only the magnitude of the induced emf.

If we change the magnetic flux through a coil of N turns, an induced emf appears in every turn and the total emf induced in the coil is the sum of these individual induced emfs. If the coil is tightly wound (*closely packed*), so that the same magnetic flux Φ_B passes through all the turns, the total emf induced in the coil is

$$\mathscr{E} = -N\frac{d\Phi_B}{dt} \quad \text{(coil of } N \text{ turns)}. \tag{31-7}$$

Here are the general means by which we can change the magnetic flux through a coil:

1. Change the magnitude B of the magnetic field within the coil.
2. Change the area of the coil, or the portion of that area that happens to lie within the magnetic field (for example, by expanding the coil or sliding it in or out of the field).
3. Change the angle between the direction of the magnetic field $\vec{B}$ and the area of the coil (for example, by rotating the coil so that field $\vec{B}$ is first perpendicular to the plane of the coil and then is along that plane).

CHECKPOINT 1: The graph gives the magnitude $B(t)$ of a uniform magnetic field that exists throughout a conducting loop, perpendicular to the plane of the loop. Rank the five regions of the graph according to the magnitude of the emf induced in the loop, greatest first. (we assume A = const)

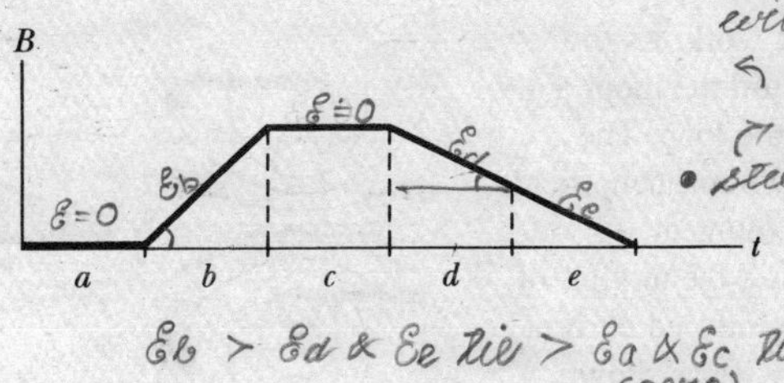

$\Phi_B = \int \vec{B} \cdot d\vec{A}$

$\frac{d\Phi_B}{dt} = \int \frac{d\vec{B}}{dt} \cdot d\vec{A}$

⇒ The rate at which B changes is ∝ rate at which Φ_B changes ∵ Actual change in Φ_B is $\frac{dB}{dt}$ multiplied by dA

• For slope, see the ∠ it makes with the x-axis; ↶ θ +ve; ↷ θ −ve
• steepness = |m| = |tan θ|

$\mathscr{E}_b > \mathscr{E}_d$ & $\mathscr{E}_e$ tie > $\mathscr{E}_a$ & $\mathscr{E}_c$ tie (zero)

Sample Problem 31-1

The long solenoid S shown (in cross section) in Fig. 31-3 has 220 turns/cm and carries a current $i = 1.5$ A; its diameter D is 3.2 cm. At its center we place a 130-turn closely packed coil C of diameter $d = 2.1$ cm. The current in the solenoid is reduced to zero at a steady rate in 25 ms. What is the magnitude of the emf that is induced in coil C while the current in the solenoid is changing?

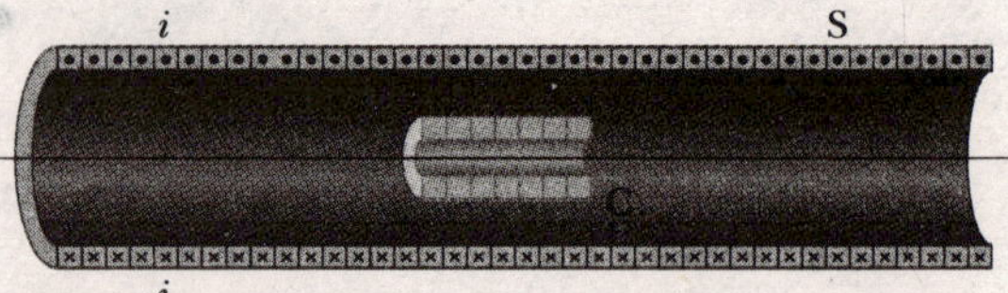

Fig. 31-3 Sample Problem 31-1. A coil C is located inside a solenoid S, which carries current i.

SOLUTION: The **Key Ideas** here are these:

1. Because coil C is located in the interior of the solenoid, it lies within the magnetic field produced by current i in the solenoid; thus, there is a magnetic flux Φ_B through coil C.
2. Because current i decreases, flux Φ_B also decreases.
3. As Φ_B decreases, emf $\mathscr{E}$ is induced in coil C, according to Faraday's law.

Because coil C consists of more than one turn, we apply Faraday's law in the form of Eq. 31-7 ($\mathscr{E} = -N\,d\Phi_B/dt$), where the number of turns N is 130 and $d\Phi_B/dt$ is the rate at which the flux in each turn changes.

Because the current in the solenoid decreases at a steady rate, flux Φ_B also decreases at a steady rate and we can write $d\Phi_B/dt$ as $\Delta\Phi_B/\Delta t$. Then, to evaluate $\Delta\Phi_B$, we need the final and initial flux. The final flux $\Phi_{B,f}$ is zero because the final current in the solenoid is zero. To find the initial flux $\Phi_{B,i}$, we need two more **Key Ideas**:

4. The flux through each turn of coil C depends on the area A and orientation of that turn in the solenoid's magnetic field $\vec{B}$. Because $\vec{B}$ is uniform and directed perpendicular to area A, the flux is given by Eq. 31-4 ($\Phi_B = BA$).
5. The magnitude B of the magnetic field in the interior of a solenoid depends on the solenoid's current i and its number n of turns per unit length, according to Eq. 30-25 ($B = \mu_0 in$).

For the situation of Fig. 31-3, A is $\frac{1}{4}\pi d^2$ ($= 3.46 \times 10^{-4}$ m^2) and n is 220 turns/cm, or 22 000 turns/m. Substituting Eq. 30-25 into Eq. 31-4 then leads to

$$\begin{aligned}\Phi_{B,i} &= BA = (\mu_0 in)A \\ &= (4\pi \times 10^{-7}\ \mathrm{T\cdot m/A})(1.5\ \mathrm{A})(22\,000\ \mathrm{turns/m}) \\ &\quad \times (3.46 \times 10^{-4}\ \mathrm{m^2}) \\ &= 1.44 \times 10^{-5}\ \mathrm{Wb}.\end{aligned}$$

Now we can write

$$\begin{aligned}\frac{d\Phi_B}{dt} &= \frac{\Delta\Phi_B}{\Delta t} = \frac{\Phi_{B,f} - \Phi_{B,i}}{\Delta t} \\ &= \frac{(0 - 1.44 \times 10^{-5}\ \mathrm{Wb})}{25 \times 10^{-3}\ \mathrm{s}} \\ &= -5.76 \times 10^{-4}\ \mathrm{Wb/s} = -5.76 \times 10^{-4}\ \mathrm{V}.\end{aligned}$$

We are interested only in magnitudes, so we ignore the minus signs here and in Eq. 31-7, writing

$$\begin{aligned}\mathscr{E} &= N\frac{d\Phi_B}{dt} = (130\ \mathrm{turns})(5.76 \times 10^{-4}\ \mathrm{V}) \\ &= 7.5 \times 10^{-2}\ \mathrm{V} = 75\ \mathrm{mV}. \qquad \text{(Answer)}\end{aligned}$$

31-4 Lenz's Law

Soon after Faraday proposed his law of induction, Heinrich Friedrich Lenz devised a rule—now known as **Lenz's law**—for determining the direction of an induced current in a loop:

> **An induced current has a direction such that the magnetic field due to *the current* opposes the change in the magnetic flux that induces the current.**

Furthermore, the direction of an induced emf is that of the induced current. To get a feel for Lenz's law, let us apply it in two different but equivalent ways to Fig. 31-4, where the north pole of a magnet is being moved toward a conducting loop.

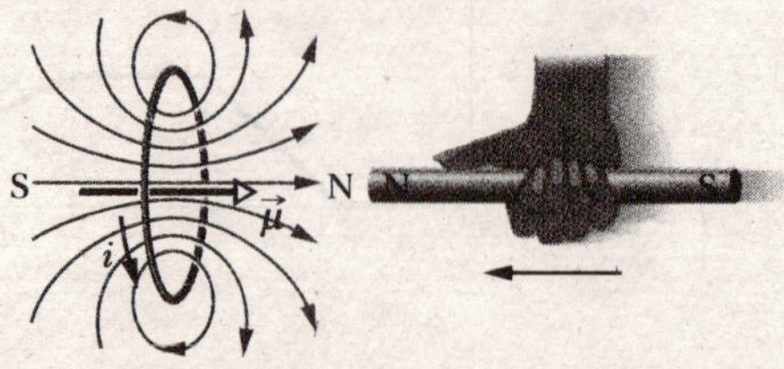

Fig. 31-4 Lenz's law at work. As the magnet is moved toward the loop, a current is induced in the loop. The current produces its own magnetic field, with magnetic dipole moment $\vec{\mu}$ oriented so as to oppose the motion of the magnet. Thus, the induced current must be counterclockwise as shown.

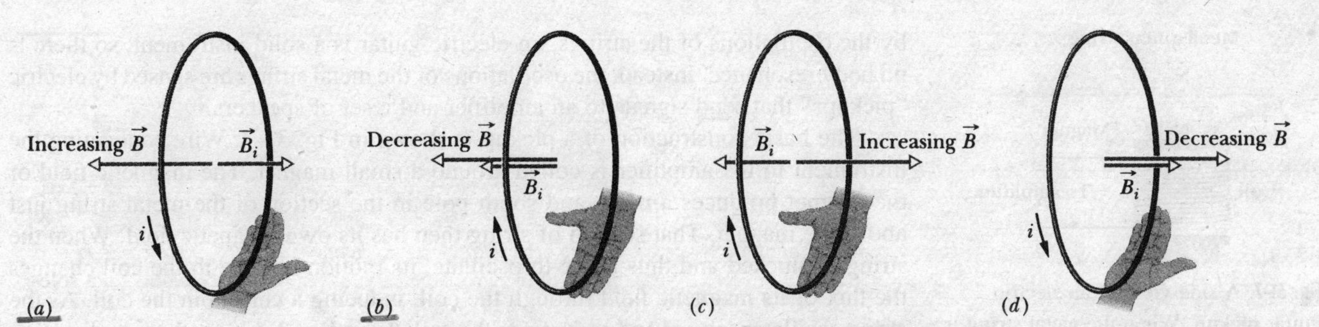

Fig. 31-5 The current i induced in a loop has the direction such that the current's magnetic field $\vec{B}_i$ opposes the *change* in the magnetic field $\vec{B}$ inducing i. The field $\vec{B}_i$ is always directed opposite an increasing field $\vec{B}$ (a, c) and in the same direction as a decreasing field $\vec{B}$ (b, d). The curled–straight right-hand rule gives the direction of the induced current based on the direction of the induced field.

Direction of induced current is such that

1. ***Opposition to Pole Movement.*** The approach of the magnet's north pole in Fig. 31-4 increases the magnetic flux in the loop and thereby induces a current in the loop. From Fig. 30-21, we know that the loop then acts as a magnetic dipole with a south pole and a north pole, and that its magnetic dipole moment $\vec{\mu}$ is directed from south to north. To *oppose* the magnetic flux increase being caused by the approaching magnet, the loop's north pole (and thus $\vec{\mu}$) must face *toward* the approaching north pole so as to repel it (Fig. 31-4). Then the curled–straight right-hand rule for $\vec{\mu}$ (Fig. 30-21) tells us that the current induced in the loop must be counterclockwise in Fig. 31-4.

   If we next pull the magnet away from the loop, a current will again be induced in the loop. Now, however, the loop will have a south pole facing the retreating north pole of the magnet, so as to oppose the retreat. Thus, the induced current will be clockwise.

2. ***Opposition to Flux Change.*** In Fig. 31-4, with the magnet initially distant, no magnetic flux passes through the loop. As the north pole of the magnet then nears the loop with its magnetic field $\vec{B}$ directed *toward the left,* the flux through the loop increases. To oppose this increase in flux, the induced current i must set up its own field $\vec{B}_i$ *directed toward the right* inside the loop, as shown in Fig. 31-5a; then the rightward flux of field $\vec{B}_i$ opposes the increasing leftward flux of field $\vec{B}$. The curled–straight right-hand rule of Fig. 30-21 then tells us that i must be counterclockwise in Fig. 31-5a.

   Note carefully that the flux of $\vec{B}_i$ always opposes the *change* in the flux of $\vec{B}$, but that does not always mean that $\vec{B}_i$ points opposite $\vec{B}$. For example, if we next pull the magnet away from the loop, the flux Φ_B from the magnet is still directed to the left through the loop, but it is now decreasing. The flux of $\vec{B}_i$ must now be to the left inside the loop, to oppose the *decrease* in Φ_B, as shown in Fig. 31-5b. Thus, $\vec{B}_i$ and $\vec{B}$ are now in the same direction.

   Figures 31-5c and d show the situations in which the south pole of the magnet approaches and retreats from the loop, respectively.

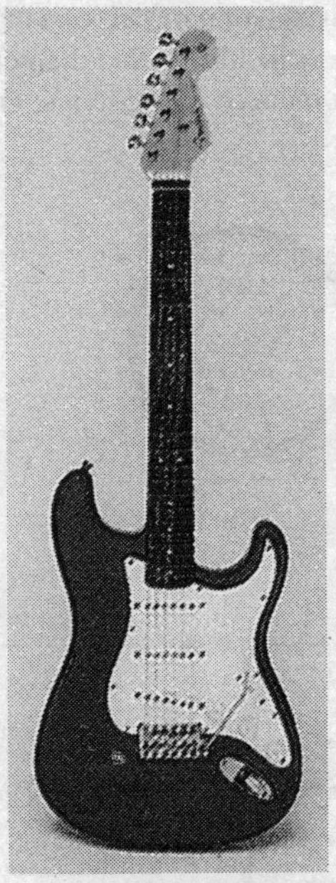

Fig. 31-6 A Fender Stratocaster has three groups of six electric pickups each (within the wide part of the body). A toggle switch (at the bottom of the guitar) allows the musician to determine which group of pickups sends signals to an amplifier and thus to a speaker system.

Electric Guitars

Figure 31-6 shows a Fender Stratocaster, the type of electric guitar that was used by Jimi Hendrix and by many other musicians. Whereas an acoustic guitar depends for its sound on the acoustic resonance produced in the hollow body of the instrument

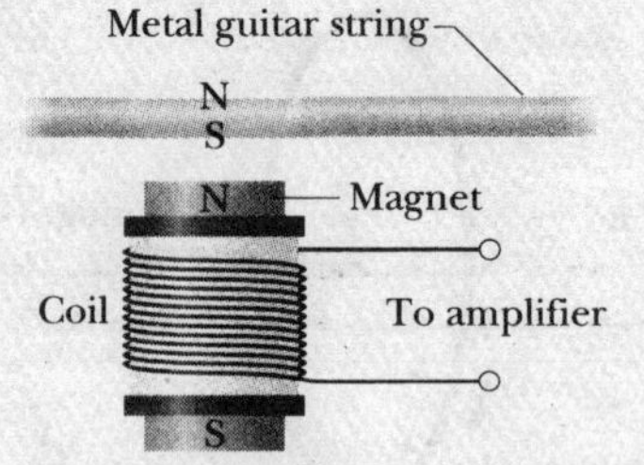

Fig. 31-7 A side view of an electric guitar pickup. When the metal string (which acts like a magnet) is made to oscillate, it causes a variation in magnetic flux that induces a current in the coil.

by the oscillations of the strings, an electric guitar is a solid instrument, so there is no body resonance. Instead, the oscillations of the metal strings are sensed by electric "pickups" that send signals to an amplifier and a set of speakers.

The basic construction of a pickup is shown in Fig. 31-7. Wire connecting the instrument to the amplifier is coiled around a small magnet. The magnetic field of the magnet produces a north and south pole in the section of the metal string just above the magnet. That section of string then has its own magnetic field. When the string is plucked and thus made to oscillate, its motion relative to the coil changes the flux of its magnetic field through the coil, inducing a current in the coil. As the string oscillates toward and away from the coil, the induced current changes direction at the same frequency as the string's oscillations, thus relaying the frequency of oscillation to the amplifier and speaker.

On a Stratocaster, there are three groups of pickups, placed at the near end of the strings (on the wide part of the body). The group closest to the near end better detects the high-frequency oscillations of the strings; the group farthest from the near end better detects the low-frequency oscillations. By throwing a toggle switch on the guitar, the musician can select which group or which pair of groups will send signals to the amplifier and speakers.

To gain further control over his music, Hendrix sometimes rewrapped the wire in the pickup coils of his guitar to change the number of turns. In this way, he altered the amount of emf induced in the coils and thus their relative sensitivity to string oscillations. Even without this additional measure, you can see that the electric guitar offers far more control over the sound that is produced than can be obtained with an acoustic guitar.

CHECKPOINT 2: The figure shows three situations in which identical circular conducting loops are in uniform magnetic fields that are either increasing (Inc) or decreasing (Dec) in magnitude at identical rates. In each, the dashed line coincides with a diameter. Rank the situations according to the magnitude of the current induced in the loops, greatest first.

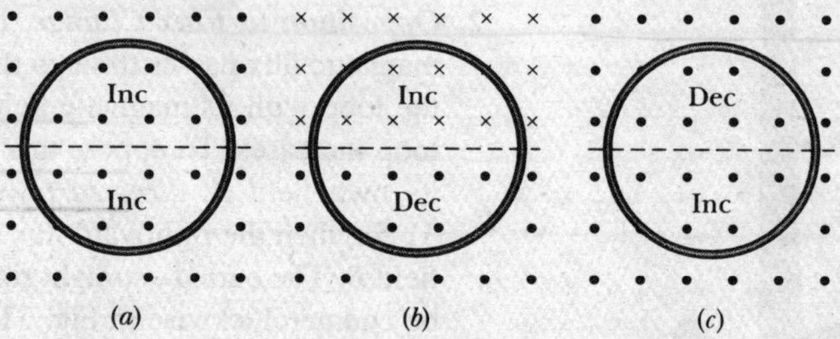

(a) & (b) tie then (c) zero

Sample Problem 31-2

Figure 31-8 shows a conducting loop consisting of a half-circle of radius $r = 0.20$ m and three straight sections. The half-circle lies in a uniform magnetic field $\vec{B}$ that is directed out of the page; the field magnitude is given by $B = 4.0t^2 + 2.0t + 3.0$, with B in teslas and t in seconds. An ideal battery with emf $\mathscr{E}_{bat} = 2.0$ V is connected to the loop. The resistance of the loop is 2.0 Ω.

(a) What are the magnitude and direction of the emf $\mathscr{E}_{ind}$ induced around the loop by field $\vec{B}$ at $t = 10$ s?

SOLUTION: One **Key Idea** here is that, according to Faraday's law, the magnitude of $\mathscr{E}_{ind}$ is equal to the rate $d\Phi_B/dt$ at which the magnetic

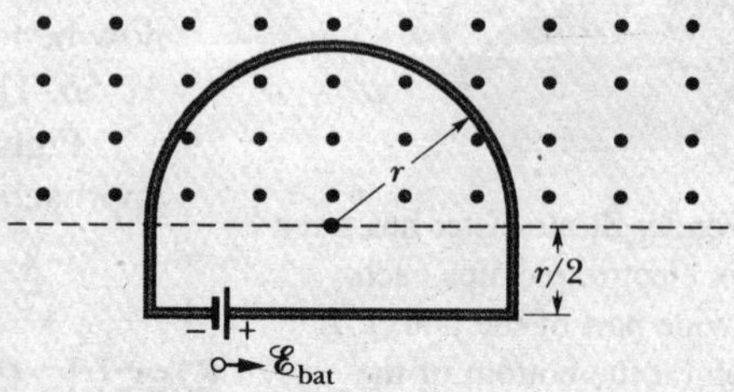

Fig. 31-8 Sample Problem 31-2. A battery is connected to a conducting loop consisting of a half-circle of radius r that lies in a uniform magnetic field. The field is directed out of the page; its magnitude is changing.

flux through the loop changes. A second **Key Idea** is that the flux through the loop depends on the loop's area A and its orientation in the magnetic field $\vec{B}$. Because $\vec{B}$ is uniform and is perpendicular to the plane of the loop, the flux is given by Eq. 31-4 ($\Phi_B = BA$). Using this equation and realizing that only the field magnitude B changes in time (not the area A), we rewrite Faraday's law, Eq. 31-6, as

$$\mathscr{E}_{ind} = \frac{d\Phi_B}{dt} = \frac{d(BA)}{dt} = A\frac{dB}{dt}.$$

A third **Key Idea** is that, because the flux penetrates the loop only within the half-circle, the area A in this equation is $\frac{1}{2}\pi r^2$. Substituting this and the given expression for B yields

$$\mathscr{E}_{ind} = A\frac{dB}{dt} = \frac{\pi r^2}{2}\frac{d}{dt}(4.0t^2 + 2.0t + 3.0)$$

$$= \frac{\pi r^2}{2}(8.0t + 2.0).$$

At $t = 10$ s, then,

$$\mathscr{E}_{ind} = \frac{\pi(0.20\text{ m})^2}{2}[8.0(10) + 2.0]$$

$$= 5.152\text{ V} \approx 5.2\text{ V}. \quad \text{(Answer)}$$

To find the direction of $\mathscr{E}_{ind}$, we first note that in Fig. 31-8 the flux through the loop is out of the page and increasing. Then the **Key Idea** here is that the induced field B_i (due to the induced current) must oppose that increase, and thus be *into* the page. Using the curled–straight right-hand rule (Fig. 30-7*c*), we find that the induced current must be clockwise around the loop. The induced emf $\mathscr{E}_{ind}$ must then also be clockwise.

(b) What is the current in the loop at $t = 10$ s?

SOLUTION: The **Key Idea** here is that two emfs tend to move charges around the loop. The induced emf $\mathscr{E}_{ind}$ tends to drive a current clockwise around the loop; the battery's emf $\mathscr{E}_{bat}$ tends to drive a current counterclockwise. Because $\mathscr{E}_{ind}$ is greater than $\mathscr{E}_{bat}$, the net emf $\mathscr{E}_{net}$ is clockwise, and thus so is the current. To find the current at $t = 10$ s, we use Eq. 28-2 ($i = \mathscr{E}/R$):

$$i = \frac{\mathscr{E}_{net}}{R} = \frac{\mathscr{E}_{ind} - \mathscr{E}_{bat}}{R}$$

$$= \frac{5.152\text{ V} - 2.0\text{ V}}{2.0\ \Omega} = 1.58\text{ A} \approx 1.6\text{ A}. \quad \text{(Answer)}$$

Sample Problem 31-3

Figure 31-9 shows a rectangular loop of wire immersed in a nonuniform and varying magnetic field $\vec{B}$ that is perpendicular to and directed into the page. The field's magnitude is given by $B = 4t^2x^2$, with B in teslas, t in seconds, and x in meters. The loop has width $W = 3.0$ m and height $H = 2.0$ m. What are the magnitude and direction of the induced emf $\mathscr{E}$ around the loop at $t = 0.10$ s?

SOLUTION: One **Key Idea** here is that because the magnitude of the magnetic field $\vec{B}$ is changing with time, the magnetic flux Φ_B through the loop is also changing. A second **Key Idea** is that the changing flux induces an emf $\mathscr{E}$ in the loop according to Faraday's law, which we can write as $\mathscr{E} = d\Phi_B/dt$.

To use that law, we need an expression for the flux Φ_B at any time t. However, a third **Key Idea** is that because B is *not* uniform over the area enclosed by the loop, we *cannot* use Eq. 31-4 ($\Phi_B = BA$) to find that expression; instead we must use Eq. 31-3 ($\Phi_B = \int \vec{B}\cdot d\vec{A}$).

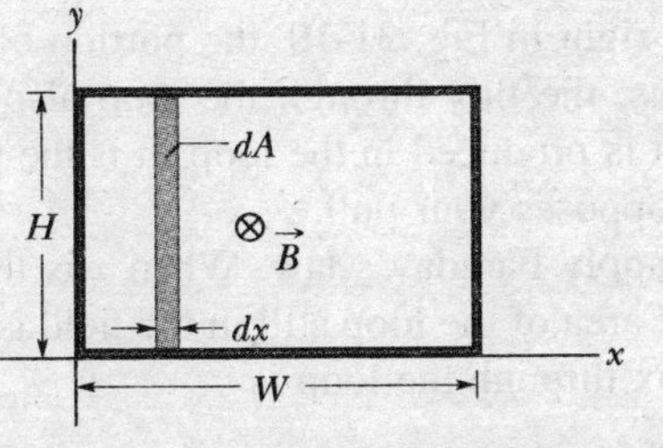

Fig. 31-9 Sample Problem 31-3. A closed conducting loop, of width W and height H, lies in a nonuniform, varying magnetic field that points directly into the page. To apply Faraday's law, we use the vertical strip of height H, width dx, and area dA.

In Fig. 31-9, $\vec{B}$ is perpendicular to the plane of the loop (and hence parallel to the differential area vector $d\vec{A}$), so the dot product in Eq. 31-3 gives $B\,dA$. Because the magnetic field varies with the coordinate x but not with the coordinate y, we can take the differential area dA to be the area of a vertical strip of height H and width dx (as shown in Fig. 31-9). Then $dA = H\,dx$, and the flux through the loop is

$$\Phi_B = \int \vec{B}\cdot d\vec{A} = \int B\,dA = \int BH\,dx = \int 4t^2x^2H\,dx.$$

Treating t as a constant for this integration and inserting the integration limits $x = 0$ and $x = 3.0$ m, we obtain

$$\Phi_B = 4t^2H\int_0^{3.0} x^2\,dx = 4t^2H\left[\frac{x^3}{3}\right]_0^{3.0} = 72t^2,$$

where we have substituted $H = 2.0$ m and Φ_B is in webers. Now we can use Faraday's law to find the magnitude of $\mathscr{E}$ at any time t:

$$\mathscr{E} = \frac{d\Phi_B}{dt} = \frac{d(72t^2)}{dt} = 144t,$$

in which $\mathscr{E}$ is in volts. At $t = 0.10$ s,

$$\mathscr{E} = (144\text{ V/s})(0.10\text{ s}) \approx 14\text{ V}. \quad \text{(Answer)}$$

The flux of $\vec{B}$ through the loop is into the page in Fig. 31-9 and is increasing in magnitude because B is increasing in magnitude with time. According to Lenz's law, the field B_i of the induced current must oppose this increase and so is directed out of the page. The curled–straight right-hand rule of Fig. 31-5 then tells us that the induced current is counterclockwise around the loop, and thus so is the induced emf $\mathscr{E}$.

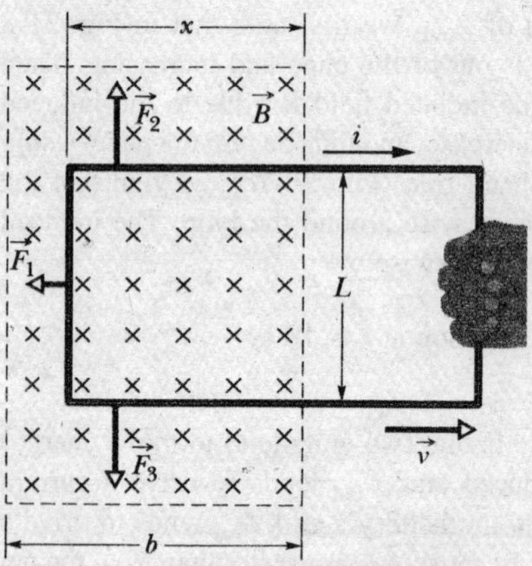

Fig. 31-10 You pull a closed conducting loop out of a magnetic field at constant velocity $\vec{v}$. While the loop is moving, a clockwise current i is induced in the loop, and the loop segments still within the magnetic field experience forces $\vec{F}_1$, $\vec{F}_2$, and $\vec{F}_3$.

31-5 Induction and Energy Transfers

By Lenz's law, whether you move the magnet toward or away from the loop in Fig. 31-1, a magnetic force resists the motion, requiring your applied force to do positive work. At the same time, thermal energy is produced in the material of the loop because of the material's electrical resistance to the current that is induced by the motion. The energy you transfer to the closed *loop + magnet* system via your applied force ends up in this thermal energy. (For now, we neglect energy that is radiated away from the loop as electromagnetic waves during the induction.) The faster you move the magnet, the more rapidly your applied force does work, and the greater the rate at which your energy is transferred to thermal energy in the loop; that is, the power of the transfer is greater.

Regardless of how current is induced in a loop, energy is always transferred to thermal energy during the process because of the electrical resistance of the loop (unless the loop is superconducting). For example, in Fig. 31-2, when switch S is closed and a current is briefly induced in the left-hand loop, energy is transferred from the battery to thermal energy in that loop.

Figure 31-10 shows another situation involving induced current. A rectangular loop of wire of width L has one end in a uniform external magnetic field that is directed perpendicularly into the plane of the loop. This field may be produced, for example, by a large electromagnet. The dashed lines in Fig. 31-10 show the assumed limits of the magnetic field; the fringing of the field at its edges is neglected. You are asked to pull this loop to the right at a constant velocity $\vec{v}$.

The situation of Fig. 31-10 does not differ in any essential way from that of Fig. 31-1. In each case a magnetic field and a conducting loop are in relative motion; in each case the flux of the field through the loop is changing with time. It is true that in Fig. 31-1 the flux is changing because $\vec{B}$ is changing and in Fig. 31-10 the flux is changing because the area of the loop still in the magnetic field is changing, but that difference is not important. The important difference between the two arrangements is that the arrangement of Fig. 31-10 makes calculations easier. Let us now calculate the rate at which you do mechanical work as you pull steadily on the loop in Fig. 31-10.

As you will see, to pull the loop at a constant velocity $\vec{v}$, you must apply a constant force $\vec{F}$ to the loop because a magnetic force of equal magnitude but opposite direction acts on the loop to oppose you. From Eq. 7-48, the rate at which you do work is then

$$P = Fv, \tag{31-8}$$

where F is the magnitude of your force. We wish to find an expression for P in terms of the magnitude B of the magnetic field and the characteristics of the loop—namely, its resistance R to current and its dimension L.

As you move the loop to the right in Fig. 31-10, the portion of its area within the magnetic field decreases. Thus, the flux through the loop also decreases and, according to Lenz's law, a current is produced in the loop. It is the presence of this current that causes the force that opposes your pull.

To find the current, we first apply Faraday's law. When x is the length of the loop still in the magnetic field, the area of the loop still in the field is Lx. Then from Eq. 31-4, the magnitude of the flux through the loop is

$$\Phi_B = BA = BLx. \tag{31-9}$$

As x decreases, the flux decreases. Faraday's law tells us that with this flux decrease, an emf is induced in the loop. Dropping the minus sign in Eq. 31-6 and using Eq.

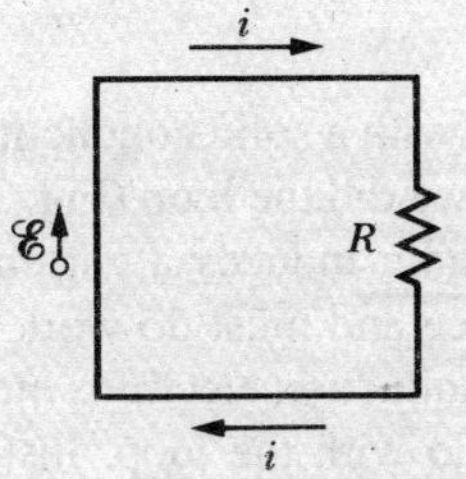

Fig. 31-11 A circuit diagram for the loop of Fig. 31-10 while the loop is moving.

31-9, we can write the magnitude of this emf as

$$\mathscr{E} = \frac{d\Phi_B}{dt} = \frac{d}{dt}BLx = BL\frac{dx}{dt} = BLv, \tag{31-10}$$

in which we have replaced dx/dt with v, the speed at which the loop moves.

Figure 31-11 shows the loop as a circuit: induced emf $\mathscr{E}$ is represented on the left, and the collective resistance R of the loop is represented on the right. The direction of the induced current i is obtained with a right-hand rule as in Fig. 31-5*b*; $\mathscr{E}$ must have the same direction.

To find the magnitude of the induced current, we cannot apply the loop rule for potential differences in a circuit because, as you will see in Section 31-6, we cannot define a potential difference for an induced emf. However, we can apply the equation $i = \mathscr{E}/R$, as we did in Sample Problem 31-2. With Eq. 31-10, this becomes

$$i = \frac{BLv}{R}. \tag{31-11}$$

Because three segments of the loop in Fig. 31-10 carry this current through the magnetic field, sideways deflecting forces act on those segments. From Eq. 29-26 we know that such a deflecting force is, in general notation,

$$\vec{F}_d = i\vec{L} \times \vec{B}. \tag{31-12}$$

In Fig. 31-10, the deflecting forces acting on the three segments of the loop are marked $\vec{F}_1$, $\vec{F}_2$, and $\vec{F}_3$. Note, however, that from the symmetry, forces $\vec{F}_2$ and $\vec{F}_3$ are equal in magnitude and cancel. This leaves only $\vec{F}_1$, which is directed opposite your force $\vec{F}$ on the loop and thus is the force opposing you. So, $\vec{F} = -\vec{F}_1$.

Using Eq. 31-12 to obtain the magnitude of $\vec{F}_1$ and noting that the angle between $\vec{B}$ and the length vector $\vec{L}$ for the left segment is 90°, we write

$$F = F_1 = iLB \sin 90° = iLB. \tag{31-13}$$

Substituting Eq. 31-11 for i in Eq. 31-13 then gives us

$$F = \frac{B^2L^2v}{R}. \tag{31-14}$$

Since B, L, and R are constants, the speed v at which you move the loop is constant if the magnitude F of the force you apply to the loop is also constant.

By substituting Eq. 31-14 into Eq. 31-8, we find the rate at which you do work on the loop as you pull it from the magnetic field:

$$P = Fv = \frac{B^2L^2v^2}{R} \quad \text{(rate of doing work)}. \tag{31-15}$$

To complete our analysis, let us find the rate at which thermal energy appears in the loop as you pull it along at constant speed. We calculate it from Eq. 27-22,

$$P = i^2R. \tag{31-16}$$

Substituting for i from Eq. 31-11, we find

$$P = \left(\frac{BLv}{R}\right)^2 R = \frac{B^2L^2v^2}{R} \quad \text{(thermal energy rate)}, \tag{31-17}$$

which is exactly equal to the rate at which you are doing work on the loop (Eq. 31-15). Thus, the work that you do in pulling the loop through the magnetic field appears as thermal energy in the loop.

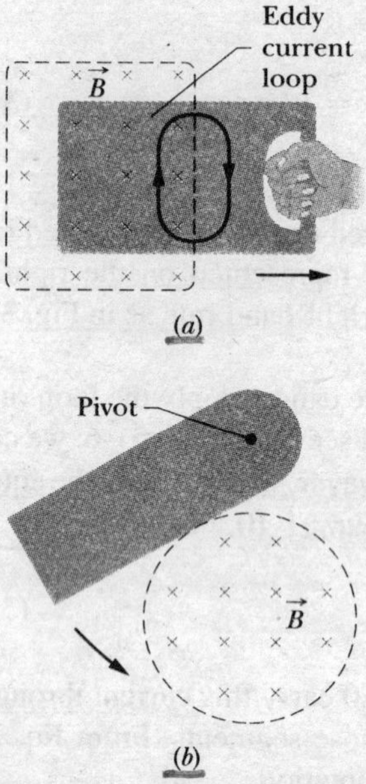

Fig. 31-12 (*a*) As you pull a solid conducting plate out of a magnetic field, *eddy currents* are induced in the plate. A typical loop of eddy current is shown. (*b*) A conducting plate is allowed to swing like a pendulum about a pivot and into a region of magnetic field. As it enters and leaves the field, eddy currents are induced in the plate.

Eddy Currents

Suppose we replace the conducting loop of Fig. 31-10 with a solid conducting plate. If we then move the plate out of the magnetic field as we did the loop (Fig. 31-12*a*), the relative motion of the field and the conductor again induces a current in the conductor. Thus, we again encounter an opposing force and must do work because of the induced current. With the plate, however, the conduction electrons making up the induced current do not follow one path as they do with the loop. Instead, the electrons swirl about within the plate as if they were caught in an eddy (or whirlpool) of water. Such a current is called an *eddy current* and can be represented as in Fig. 31-12*a as if* it followed a single path.

As with the conducting loop of Fig. 31-10, the current induced in the plate results in mechanical energy being dissipated as thermal energy. The dissipation is more apparent in the arrangement of Fig. 31-12*b*; a conducting plate, free to rotate about a pivot, is allowed to swing down through a magnetic field like a pendulum. Each time the plate enters and leaves the field, a portion of its mechanical energy is transferred to its thermal energy. After several swings, no mechanical energy remains and the warmed-up plate just hangs from its pivot.

CHECKPOINT 3: The figure shows four wire loops, with edge lengths of either L or $2L$. All four loops will move through a region of uniform magnetic field $\vec{B}$ (directed out of the page) at the same constant velocity. Rank the four loops according to the maximum magnitude of the emf induced as they move through the field, greatest first.

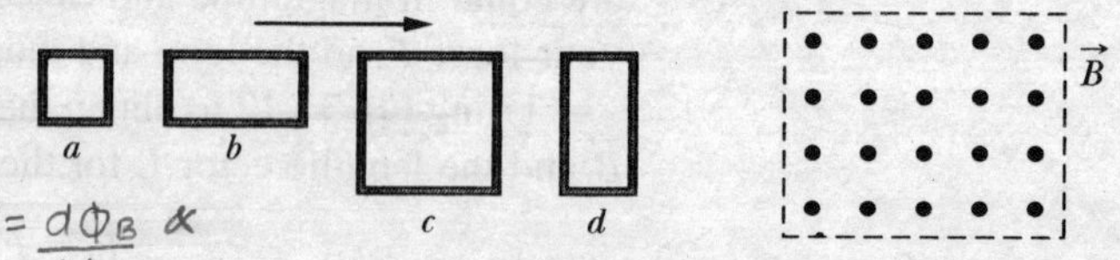

$\mathcal{E} = \frac{d\Phi_B}{dt} \propto$

$\Phi_B = \int \vec{B}\cdot d\vec{A} = BLx \Rightarrow \mathcal{E} = BL\frac{dx}{dt} = BLv$

B & v are same for all
c & d tie then
a & b tie

31-6 Induced Electric Fields

Let us place a copper ring of radius r in a uniform external magnetic field, as in Fig. 31-13*a*. The field—neglecting fringing—fills a cylindrical volume of radius R. Suppose that we increase the strength of this field at a steady rate, perhaps by increasing—in an appropriate way—the current in the windings of the electromagnet that produces the field. The magnetic flux through the ring will then change at a steady rate and—by Faraday's law—an induced emf and thus an induced current will appear in the ring. From Lenz's law we can deduce that the direction of the induced current is counterclockwise in Fig. 31-13*a*.

★ If there is a current in the copper ring, an electric field must be present along the ring; an electric field is needed to do the work of moving the conduction electrons. Moreover, the electric field must have been produced by the changing magnetic flux. This **induced electric field** $\vec{E}$ is just as real as an electric field produced by static charges; either field will exert a force $q_0\vec{E}$ on a particle of charge q_0.

By this line of reasoning, we are led to a useful and informative restatement of Faraday's law of induction:

> A changing magnetic field produces an electric field.

★ The striking feature of this statement is that the electric field is induced (⇒ produced) even if there is no copper ring.

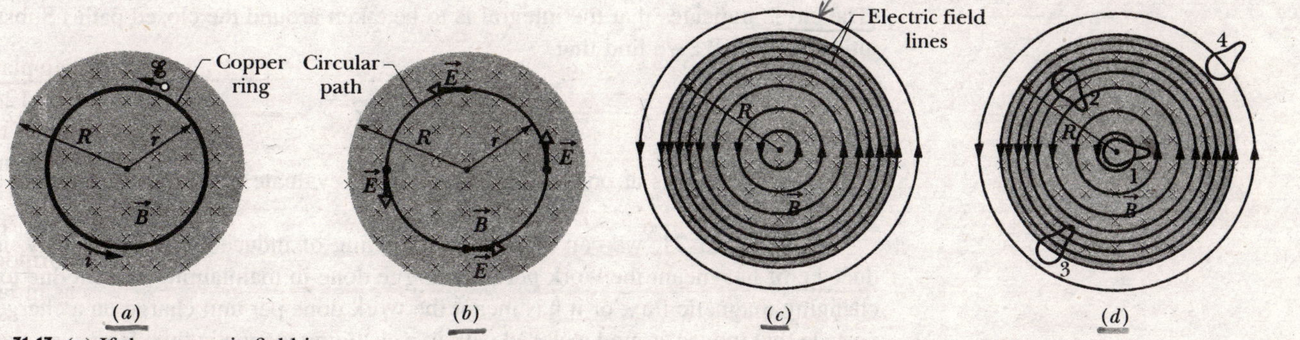

Fig. 31-13 (*a*) If the magnetic field increases at a steady rate, a constant induced current appears, as shown, in the copper ring of radius *r*. (*b*) An induced electric field exists even when the ring is removed; the electric field is shown at four points. (*c*) The complete picture of the induced electric field, displayed as field lines. (*d*) Four similar closed paths that enclose identical areas. Equal emfs are induced around paths 1 and 2, which lie entirely within the region of changing magnetic field. A smaller emf is induced around path 3, which only partially lies in that region. No emf is induced around path 4, which lies entirely outside the magnetic field.

To fix these ideas, consider Fig. 31-13*b*, which is just like Fig. 31-13*a* except the copper ring has been replaced by a hypothetical circular path of radius *r*. We assume, as previously, that the magnetic field $\vec{B}$ is increasing in magnitude at a constant rate dB/dt. The electric field induced at various points around the circular path must—from the symmetry—be tangent to the circle, as Fig. 31-13*b* shows.* Hence, the circular path is an electric field line. There is nothing special about the circle of radius *r*, so the electric field lines produced by the changing magnetic field must be a set of concentric circles, as in Fig. 31-13*c*.

As long as the magnetic field is *increasing* with time, the electric field represented by the circular field lines in Fig. 31-13*c* will be present. If the magnetic field remains *constant* with time, there will be no induced electric field and thus no electric field lines. If the magnetic field is *decreasing* with time (at a constant rate), the electric field lines will still be concentric circles as in Fig. 31-13*c*, but they will now have the opposite direction. All this is what we have in mind when we say: "A changing magnetic field produces an electric field."

A Reformulation of Faraday's Law

Consider a particle of charge q_0 moving around the circular path of Fig. 31-13*b*. The work *W* done on it in one revolution by the induced electric field is $\mathscr{E}q_0$, where $\mathscr{E}$ is the induced emf—that is, the work done per unit charge in moving the test charge around the path. From another point of view, the work is

$$\int \vec{F} \cdot d\vec{s} = (q_0 E)(2\pi r), \tag{31-18}$$

where q_0E is the magnitude of the force acting on the test charge and $2\pi r$ is the distance over which that force acts. Setting these two expressions for *W* equal to each other and canceling q_0, we find that

$$\mathscr{E} = 2\pi r E. \tag{31-19}$$

More generally, we can rewrite Eq. 31-18 to give the work done on a particle of charge q_0 moving along any closed path:

$$W = \oint \vec{F} \cdot d\vec{s} = q_0 \oint \vec{E} \cdot d\vec{s}. \tag{31-20}$$

*Arguments of symmetry would also permit the lines of $\vec{E}$ around the circular path to be *radial*, rather than tangential. However, such radial lines would imply that there are free charges, distributed symmetrically about the axis of symmetry, on which the electric field lines could begin or end; there are no such charges.

(The circle indicates that the integral is to be taken around the closed path.) Substituting $\mathscr{E}q_0$ for W, we find that

$$\mathscr{E} = \oint \vec{E} \cdot d\vec{s}. \tag{31-21}$$

This integral reduces at once to Eq. 31-19 if we evaluate it for the special case of Fig. 31-13*b*.

With Eq. 31-21, we can expand the meaning of induced emf. Previously, induced emf has meant the work per unit charge done in maintaining current due to a changing magnetic flux, or it has meant the work done per unit charge on a charged particle that moves around a closed path in a changing magnetic flux. However, with Fig. 31-13*b* and Eq. 31-21, an induced emf can exist without the need of a current or particle: An induced emf is the sum—via integration—of quantities $\vec{E} \cdot d\vec{s}$ around a closed path, where $\vec{E}$ is the electric field induced by a changing magnetic flux and $d\vec{s}$ is a differential length vector along the closed path.

If we combine Eq. 31-21 with Faraday's law in Eq. 31-6 ($\mathscr{E} = -d\Phi_B/dt$), we can rewrite Faraday's law as

$$\oint \vec{E} \cdot d\vec{s} = -\frac{d\Phi_B}{dt} \quad \text{(Faraday's law)}. \tag{31-22}$$

This equation says simply that a changing magnetic field induces an electric field. The changing magnetic field appears on the right side of this equation, the electric field on the left.

Faraday's law in the form of Eq. 31-22 can be applied to *any* closed path that can be drawn in a changing magnetic field. Figure 31-13*d*, for example, shows four such paths, all having the same shape and area but located in different positions in the changing field. For paths 1 and 2, the induced emfs $\mathscr{E}$ $(= \oint \vec{E} \cdot d\vec{s})$ are equal because these paths lie entirely in the magnetic field and thus have the same value of $d\Phi_B/dt$. This is true even though the electric field vectors at points along these paths are different, as indicated by the patterns of electric field lines in the figure. For path 3 the induced emf is smaller because the enclosed flux Φ_B (hence $d\Phi_B/dt$) is smaller, and for path 4 the induced emf is zero, even though the electric field is not zero at any point on the path. E would be there as long as B is changing even if path lies outside the region of B (or no emf)

A New Look at Electric Potential

Induced electric fields are produced not by static charges but by a changing magnetic flux. Although electric fields produced in either way exert forces on charged particles, there is an important difference between them. The simplest evidence of this difference is that the field lines of induced electric fields form closed loops, as in Fig. 31-13*c*. Field lines produced by static charges never do so but must start on positive charges and end on negative charges.

In a more formal sense, we can state the difference between electric fields produced by induction and those produced by static charges in these words:

> Electric potential has meaning only for electric fields that are produced by static charges; it has no meaning for electric fields that are produced by induction.

You can understand this statement qualitatively by considering what happens to a charged particle that makes a single journey around the circular path in Fig. 31-13*b*. It starts at a certain point and, on its return to that same point, has experienced an emf $\mathscr{E}$ of, let us say, 5 V; that is, work of 5 J/C has been done on the particle, and thus the particle should then be at a point that is 5 V greater in potential. However, that is impossible because the particle is back at the same point, which cannot have two different values of potential. We must conclude that potential has no meaning for electric fields that are set up by changing magnetic fields.

We can take a more formal look by recalling Eq. 25-18, which defines the potential difference between two points i and f in an electric field $\vec{E}$:

$$V_f - V_i = -\int_i^f \vec{E} \cdot d\vec{s}. \qquad (31\text{-}23)$$

In Chapter 25 we had not yet encountered Faraday's law of induction, so the electric fields involved in the derivation of Eq. 25-18 were those due to static charges. If i and f in Eq. 31-23 are the same point, the path connecting them is a closed loop, V_i and V_f are identical, and Eq. 31-23 reduces to

$$\oint \vec{E} \cdot d\vec{s} = 0. \qquad (31\text{-}24)$$

However, when a changing magnetic flux is present, this integral is *not* zero but is $-d\Phi_B/dt$, as Eq. 31-22 asserts. Thus, assigning electric potential to an induced electric field leads us to a contradiction. We must conclude that electric potential has no meaning for electric fields associated with induction.

CHECKPOINT 4: The figure shows five lettered regions in which a uniform magnetic field extends either directly out of the page (as in region a) or into the page. The field is increasing in magnitude at the same steady rate in all five regions; the regions are identical in area. Also shown are four numbered paths along which $\oint \vec{E} \cdot d\vec{s}$ has the magnitudes given below in terms of a quantity mag. Determine whether the magnetic fields in regions b through e are directed into or out of the page.

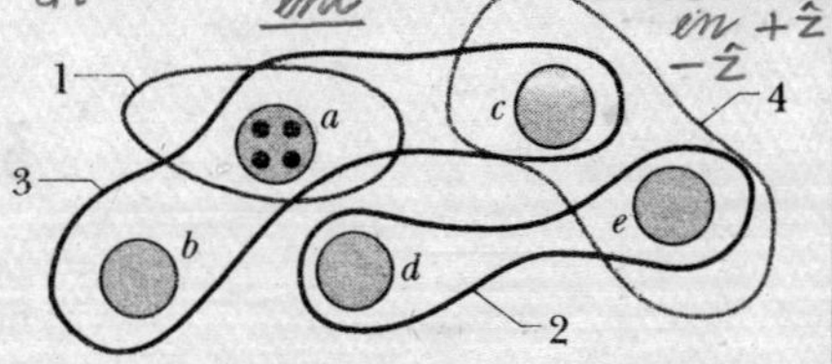

Path:	1	2	3	4
$\oint \vec{E} \cdot d\vec{s}$:	mag	2(mag)	3(mag)	0

Sample Problem 31-4

In Fig. 31-13b, take $R = 8.5$ cm and $dB/dt = 0.13$ T/s.

(a) Find an expression for the magnitude E of the induced electric field at points within the magnetic field, at radius r from the center of the magnetic field. Evaluate the expression for $r = 5.2$ cm.

SOLUTION: The **Key Idea** here is that an electric field is induced by the changing magnetic field, according to Faraday's law. To calculate the field magnitude E, we apply Faraday's law in the form of Eq. 31-22. We use a circular path of integration with radius $r \le R$ because we want E for points within the magnetic field. We assume from the symmetry that $\vec{E}$ in Fig. 31-13b is tangent to the circular path at all points. The path vector $d\vec{s}$ is also always tangent to the circular path, so the dot product $\vec{E} \cdot d\vec{s}$ in Eq. 31-22 must have the magnitude $E\,ds$ at all points on the path. We can also assume from the symmetry that E has the same value at all points along the circular path. Then the left side of Eq. 31-22 becomes

$$\oint \vec{E} \cdot d\vec{s} = \oint E\,ds = E \oint ds = E(2\pi r). \qquad (31\text{-}25)$$

(The integral $\oint ds$ is the circumference $2\pi r$ of the circular path.)

Next, we need to evaluate the right side of Eq. 31-22. Because $\vec{B}$ is uniform over the area A encircled by the path of integration and is directed perpendicular to that area, the magnetic flux is given by Eq. 31-4:

$$\Phi_B = BA = B(\pi r^2). \qquad (31\text{-}26)$$

Substituting this and Eq. 31-25 into Eq. 31-22 and dropping the minus sign, we find that

$$E(2\pi r) = (\pi r^2)\frac{dB}{dt}$$

or

$$E = \frac{r}{2}\frac{dB}{dt}. \qquad \text{(Answer)} \quad (31\text{-}27)$$

Equation 31-27 gives the magnitude of the electric field at any point for which $r \leq R$ (that is, within the magnetic field). Substituting given values yields, for the magnitude of $\vec{E}$ at $r = 5.2$ cm,

$$E = \frac{(5.2 \times 10^{-2}\text{ m})}{2}(0.13\text{ T/s})$$
$$= 0.0034\text{ V/m} = 3.4\text{ mV/m}. \qquad \text{(Answer)}$$

(b) Find an expression for the magnitude E of the induced electric field at points that are outside the magnetic field, at radius r. Evaluate the expression for $r = 12.5$ cm.

SOLUTION: The **Key Idea** of part (a) applies here also, except that we use a circular path of integration with radius $r \geq R$, because we want to evaluate E for points outside the magnetic field. Proceeding as in (a), we again obtain Eq. 31-25. However, we do not then obtain Eq. 31-26, because the new path of integration is now outside the magnetic field, and we need this **Key Idea**: The magnetic flux encircled by the new path is only that in the area πR^2 of the magnetic field region. Therefore,

$$\Phi_B = BA = B(\pi R^2). \qquad (31\text{-}28)$$

Substituting this and Eq. 31-25 into Eq. 31-22 (without the minus sign) and solving for E yield

$$E = \frac{R^2}{2r}\frac{dB}{dt}. \qquad \text{(Answer)} \quad (31\text{-}29)$$

Since E is not zero here, we know that an electric field is induced even at points that are outside the changing magnetic field, an important result that (as you shall see in Section 33-11) makes transformers possible. With the given data, Eq. 31-29 yields the magnitude of $\vec{E}$ at $r = 12.5$ cm:

$$E = \frac{(8.5 \times 10^{-2}\text{ m})^2}{(2)(12.5 \times 10^{-2}\text{ m})}(0.13\text{ T/s})$$
$$= 3.8 \times 10^{-3}\text{ V/m} = 3.8\text{ mV/m}. \qquad \text{(Answer)}$$

Equations 31-27 and 31-29 give the same result, as they must, for $r = R$. Figure 31-14 shows a plot of $E(r)$ based on these two equations.

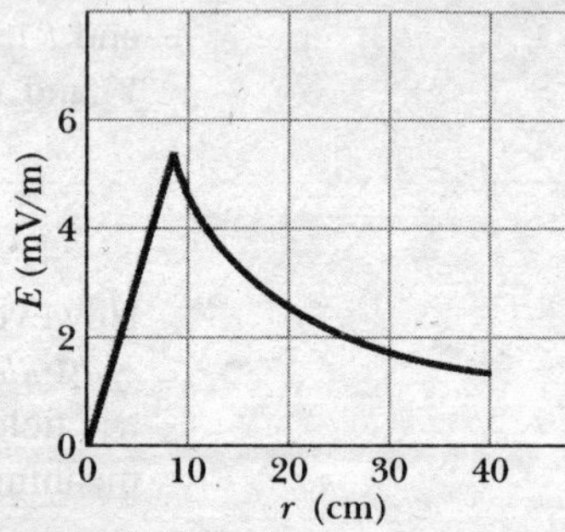

Fig. 31-14 A plot of the induced electric field $E(r)$ for the conditions of Sample Problem 31-4.

31-7 Inductors and Inductance

We found in Chapter 26 that a capacitor can be used to produce a desired electric field. We considered the parallel-plate arrangement as a basic type of capacitor. Similarly, an **inductor** (symbol ⅏) can be used to produce a desired magnetic field. We shall consider a long solenoid (more specifically, a short length near the middle of a long solenoid) as our basic type of inductor.

If we establish a current i in the windings (or turns) of an inductor (a solenoid), the current produces a magnetic flux Φ_B through the central region of the inductor. The **inductance** of the inductor is then

$$L = \frac{N\Phi_B}{i} \qquad \text{(inductance defined)}, \qquad (31\text{-}30)$$

in which N is the number of turns. The windings of the inductor are said to be *linked* by the shared flux, and the product $N\Phi_B$ is called the *magnetic flux linkage*. The inductance L is thus a measure of the flux linkage produced by the inductor per unit of current.

Because the SI unit of magnetic flux is the tesla–square meter, the SI unit of inductance is the tesla–square meter per ampere (T · m²/A). We call this the **henry** (H), after American physicist Joseph Henry, the codiscoverer of the law of induction and a contemporary of Faraday. Thus,

$$1\text{ henry} = 1\text{ H} = 1\text{ T}\cdot\text{m}^2/\text{A}. \qquad (31\text{-}31)$$

Through the rest of this chapter we assume that all inductors, no matter what their geometric arrangement, have no magnetic materials such as iron in their vicinity. Such materials would distort the magnetic field of an inductor.

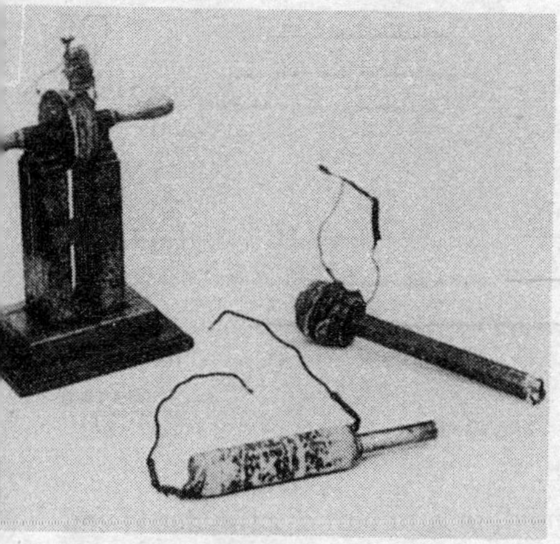

The crude inductors with which Michael Faraday discovered the law of induction. In those days amenities such as insulated wire were not commercially available. It is said that Faraday insulated his wires by wrapping them with strips cut from one of his wife's petticoats.

Inductance of a Solenoid

Consider a long solenoid of cross-sectional area A. What is the inductance per unit length near its middle?

To use the defining equation for inductance (Eq. 31-30), we must calculate the flux linkage set up by a given current in the solenoid windings. Consider a length l near the middle of this solenoid. The flux linkage for this section of the solenoid is

$$N\Phi_B = (nl)(BA)$$

in which n is the number of turns per unit length of the solenoid and B is the magnitude of the magnetic field within the solenoid.

The magnitude B is given by Eq. 30-25,

$$B = \mu_0 in,$$

so from Eq. 31-30,

$$L = \frac{N\Phi_B}{i} = \frac{(nl)(BA)}{i} = \frac{(nl)(\mu_0 in)(A)}{i}$$
$$= \mu_0 n^2 lA. \qquad (31\text{-}32)$$

Thus, the inductance per unit length for a long solenoid near its center is

$$\frac{L}{l} = \mu_0 n^2 A \quad \text{(solenoid)}. \qquad (31\text{-}33)$$

Inductance—like capacitance—depends only on the geometry of the device. The dependence on the square of the number of turns per unit length is to be expected. If you, say, triple n, you not only triple the number of turns (N) but you also triple the flux ($\Phi_B = BA = \mu_0 inA$) through each turn, multiplying the flux linkage $N\Phi_B$ and thus the inductance L by a factor of 9.

If the solenoid is very much longer than its radius, then Eq. 31-32 gives its inductance to a good approximation. This approximation neglects the spreading of the magnetic field lines near the ends of the solenoid, just as the parallel-plate capacitor formula ($C = \varepsilon_0 A/d$) neglects the fringing of the electric field lines near the edges of the capacitor plates.

From Eq. 31-32, and recalling that n is a number per unit length, we can see that an inductance can be written as a product of the permeability constant μ_0 and a quantity with the dimensions of a length. This means that μ_0 can be expressed in the unit henry per meter:

$$\mu_0 = 4\pi \times 10^{-7}\ \text{T}\cdot\text{m/A}$$
$$= 4\pi \times 10^{-7}\ \text{H/m}. \qquad (31\text{-}34)$$

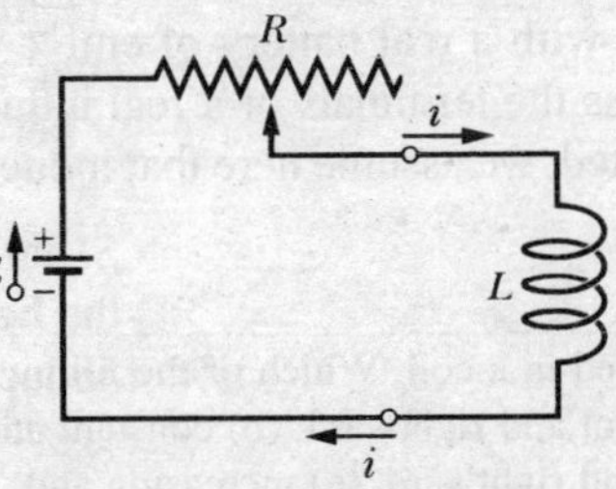

Fig. 31-15 If the current in a coil is changed by varying the contact position on a variable resistor, a self-induced emf $\mathscr{E}_L$ will appear in the coil *while the current is changing.*

31-8 Self-Induction

If two coils—which we can now call inductors—are near each other, a current i in one coil produces a magnetic flux Φ_B through the second coil. We have seen that if we change this flux by changing the current, an induced emf appears in the second coil according to Faraday's law. An induced emf appears in the first coil as well.

An induced emf $\mathscr{E}_L$ appears in any coil in which the current is changing.

This process (see Fig. 31-15) is called **self-induction,** and the emf that appears is called a **self-induced emf.** It obeys Faraday's law of induction just as other induced emfs do.

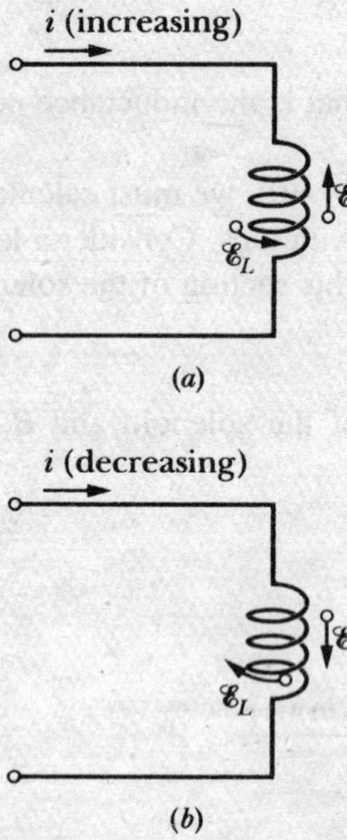

Fig. 31-16 (*a*) The current i is increasing and the self-induced emf $\mathscr{E}_L$ appears along the coil in a direction such that it opposes the increase. The arrow representing $\mathscr{E}_L$ can be drawn along a turn of the coil or alongside the coil. Both are shown. (*b*) The current i is decreasing and the self-induced emf appears in a direction such that it opposes the decrease.

For any inductor, Eq. 31-30 tells us that

$$N\Phi_B = Li. \tag{31-35}$$

Faraday's law tells us that

$$\mathscr{E}_L = -\frac{d(N\Phi_B)}{dt}. \tag{31-36}$$

By combining Eqs. 31-35 and 31-36 we can write:

$$\mathscr{E}_L = -L\frac{di}{dt} \quad \text{(self-induced emf)}. \tag{31-37}$$

Thus, in any inductor (such as a coil, a solenoid, or a toroid) a self-induced emf appears whenever the current changes with time. The magnitude of the current has no influence on the magnitude of the induced emf; only the rate of change of the current counts.

You can find the *direction* of a self-induced emf from Lenz's law. The minus sign in Eq. 31-37 indicates that—as the law states—the self-induced emf $\mathscr{E}_L$ has the orientation such that it opposes the change in current i. We can drop the minus sign when we want only the magnitude of $\mathscr{E}_L$.

Suppose that, as in Fig. 31-16*a*, you set up a current i in a coil and arrange to have it increase with time at a rate di/dt. In the language of Lenz's law, this increase in the current is the "change" that the self-induction must oppose. For such opposition to occur, a self-induced emf must appear in the coil, pointing—as the figure shows—so as to oppose the increase in the current. If you cause the current to decrease with time, as in Fig. 31-16*b*, the self-induced emf must point in a direction that tends to oppose the decrease in the current, as the figure shows.

In Section 31-6 we saw that we cannot define an electric potential for an electric field (and thus for an emf) that is induced by a changing magnetic flux. This means that when a self-induced emf is produced in the inductor of Fig. 31-15, we cannot define an electric potential within the inductor itself, where the flux is changing. However, potentials can still be defined at points of the circuit that are not within the inductor—points where the electric fields are due to charge distributions and their associated electric potentials.

Moreover, we can define a self-induced potential difference V_L *across an inductor* (between its terminals, which we assume to be outside the region of changing flux). If the inductor is an *ideal inductor* (its wire has negligible resistance), the magnitude of V_L is equal to the magnitude of the self-induced emf $\mathscr{E}_L$.

If, instead, the wire in the inductor has resistance r, we mentally separate the inductor into a resistance r (which we take to be outside the region of changing flux) and an ideal inductor of self-induced emf $\mathscr{E}_L$. As with a real battery of emf $\mathscr{E}$ and internal resistance r, the potential difference across the terminals of a real inductor then differs from the emf. Unless otherwise indicated, we assume here that inductors are ideal.

CHECKPOINT 5: The figure shows an emf $\mathscr{E}_L$ induced in a coil. Which of the following can describe the current through the coil: (a) constant and rightward, (b) constant and leftward, (c) increasing and rightward, (d) decreasing and rightward, (e) increasing and leftward, (f) decreasing and leftward?

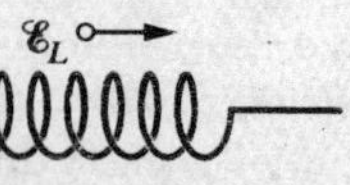

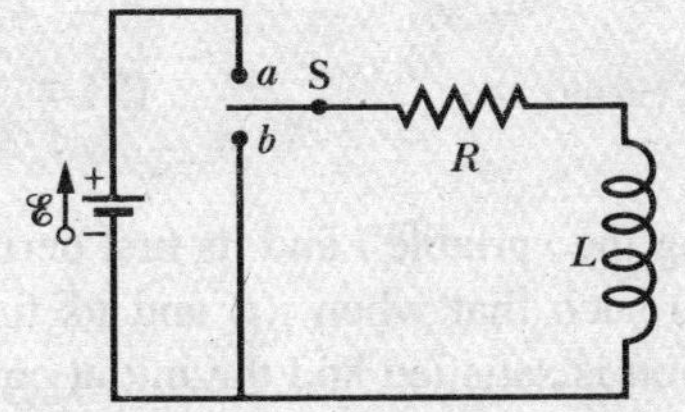

Fig. 31-17 An *RL* circuit. When switch S is closed on *a*, the current rises and approaches a limiting value of $\mathscr{E}/R$.

31-9 *RL* Circuits

In Section 28-8 we saw that if we suddenly introduce an emf $\mathscr{E}$ into a single-loop circuit containing a resistor R and a capacitor C, the charge on the capacitor does not build up immediately to its final equilibrium value $C\mathscr{E}$ but approaches it in an exponential fashion:

$$q = C\mathscr{E}(1 - e^{-t/\tau_C}). \tag{31-38}$$

The rate at which the charge builds up is determined by the capacitive time constant τ_C, defined in Eq. 28-33 as

$$\tau_C = RC. \tag{31-39}$$

If we suddenly remove the emf from this same circuit, the charge does not immediately fall to zero but approaches zero in an exponential fashion:

$$q = q_0 e^{-t/\tau_C}. \tag{31-40}$$

The time constant τ_C describes the fall of the charge as well as its rise.

An analogous slowing of the rise (or fall) of the current occurs if we introduce an emf $\mathscr{E}$ into (or remove it from) a single-loop circuit containing a resistor R and an inductor L. When the switch S in Fig. 31-17 is closed on *a*, for example, the current in the resistor starts to rise. If the inductor were not present, the current would rise rapidly to a steady value $\mathscr{E}/R$. Because of the inductor, however, a self-induced emf $\mathscr{E}_L$ appears in the circuit; from Lenz's law, this emf opposes the rise of the current, which means that it opposes the battery emf $\mathscr{E}$ in polarity. Thus, the current in the resistor responds to the difference between two emfs, a constant one $\mathscr{E}$ due to the battery and a variable one $\mathscr{E}_L$ $(= -L\, di/dt)$ due to self-induction. As long as $\mathscr{E}_L$ is present, the current in the resistor will be less than $\mathscr{E}/R$.

As time goes on, the rate at which the current increases becomes less rapid and the magnitude of the self-induced emf, which is proportional to di/dt, becomes smaller. Thus, the current in the circuit approaches $\mathscr{E}/R$ asymptotically.

We can generalize these results as follows:

> ► **Initially, an inductor acts to oppose changes in the current through it. A long time later, it acts like ordinary connecting wire.**

Now let us analyze the situation quantitatively. With the switch S in Fig. 31-17 thrown to *a*, the circuit is equivalent to that of Fig. 31-18. Let us apply the loop rule, starting at point x in this figure and moving clockwise around the loop along with current i.

1. *Resistor.* Because we move through the resistor in the direction of current i, the electric potential decreases by iR. Thus, as we move from point x to point y, we encounter a potential change of $-iR$.
2. *Inductor.* Because current i is changing, there is a self-induced emf $\mathscr{E}_L$ in the inductor. The magnitude of $\mathscr{E}_L$ is given by Eq. 31-37 as $L\, di/dt$. The direction of $\mathscr{E}_L$ is upward in Fig. 31-18 because current i is downward through the inductor *and* increasing. Thus, as we move from point y to point z, opposite the direction of $\mathscr{E}_L$, we encounter a potential change of $-L\, di/dt$.
3. *Battery.* As we move from point z back to starting point x, we encounter a potential change of $+\mathscr{E}$ due to the battery's emf.

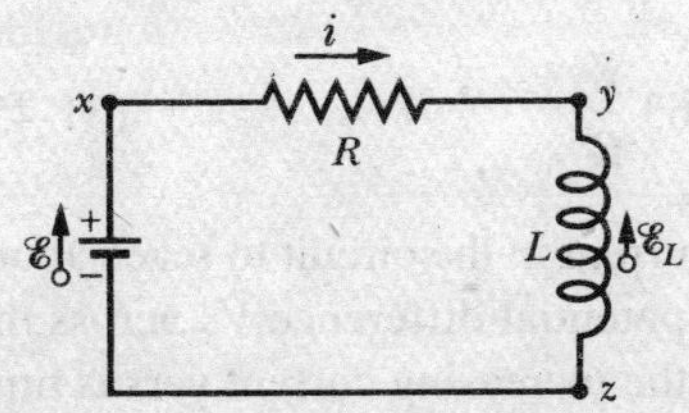

Fig. 31-18 The circuit of Fig. 31-17 with the switch closed on *a*. We apply the loop rule for circuits clockwise, starting at *x*.

Thus, the loop rule gives us

$$-iR - L\frac{di}{dt} + \mathscr{E} = 0$$

or

$$L\frac{di}{dt} + Ri = \mathscr{E} \quad \text{(RL circuit).} \tag{31-41}$$

Equation 31-41 is a differential equation involving the variable i and its first derivative di/dt. To solve it, we seek the function $i(t)$ such that when $i(t)$ and its first derivative are substituted in Eq. 31-41, the equation is satisfied and the initial condition $i(0) = 0$ is satisfied.

Equation 31-41 and its initial condition are of exactly the form of Eq. 28-29 for an RC circuit, with i replacing q, L replacing R, and R replacing $1/C$. The solution of Eq. 31-41 must then be of exactly the form of Eq. 28-30 with the same replacements. That solution is

$$i = \frac{\mathscr{E}}{R}(1 - e^{-Rt/L}), \tag{31-42}$$

which we can rewrite as

$$i = \frac{\mathscr{E}}{R}(1 - e^{-t/\tau_L}) \quad \text{(rise of current).} \tag{31-43}$$

Here τ_L, the **inductive time constant**, is given by

$$\tau_L = \frac{L}{R} \quad \text{(time constant).} \tag{31-44}$$

Let's examine Eq. 31-43 for when the switch is closed (at time $t = 0$) and for a time long after the switch is closed ($t \to \infty$). If we substitute $t = 0$ into Eq. 31-43, the exponential becomes $e^{-0} = 1$. Thus, Eq. 31-43 tells us that the current is initially $i = 0$, as we expected. Next, if we let t go to ∞, then the exponential goes to $e^{-\infty} = 0$. Thus, Eq. 31-43 tells us that the current goes to its equilibrium value of $\mathscr{E}/R$.

We can also examine the potential differences in the circuit. Figure 31-19 shows how the potential differences V_R $(= iR)$ across the resistor and V_L $(= L\,di/dt)$ across the inductor vary with time for particular values of $\mathscr{E}$, L, and R. Compare this figure carefully with the corresponding figure for an RC circuit (Fig. 28-14).

To show that the quantity τ_L $(= L/R)$ has the dimension of time, we convert from Henries per ohm as follows:

$$1\,\frac{\text{H}}{\Omega} = 1\,\frac{\text{H}}{\Omega}\left(\frac{1\,\text{V}\cdot\text{s}}{1\,\text{H}\cdot\text{A}}\right)\left(\frac{1\,\Omega\cdot\text{A}}{1\,\text{V}}\right) = 1\text{ s}.$$

The first quantity in parentheses is a conversion factor based on Eq. 31-37, and the second one is a conversion factor based on the relation $V = iR$.

The physical significance of the time constant follows from Eq. 31-43. If we put $t = \tau_L = L/R$ in this equation, it reduces to

$$i = \frac{\mathscr{E}}{R}(1 - e^{-1}) = 0.63\,\frac{\mathscr{E}}{R}. \tag{31-45}$$

Thus, the time constant τ_L is the time it takes the current in the circuit to reach about 63% of its final equilibrium value $\mathscr{E}/R$. Since the potential difference V_R across the resistor is proportional to the current i, a graph of the increasing current versus time has the same shape as that of V_R in Fig. 31-19a.

If the switch S in Fig. 31-17 is closed on a long enough for the equilibrium current $\mathscr{E}/R$ to be established, and then is thrown to b, the effect will be to remove the battery from the circuit. (The connection to b must actually be made an instant

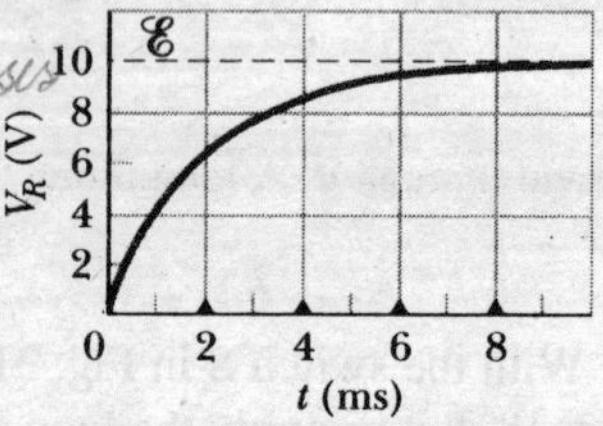

(a)

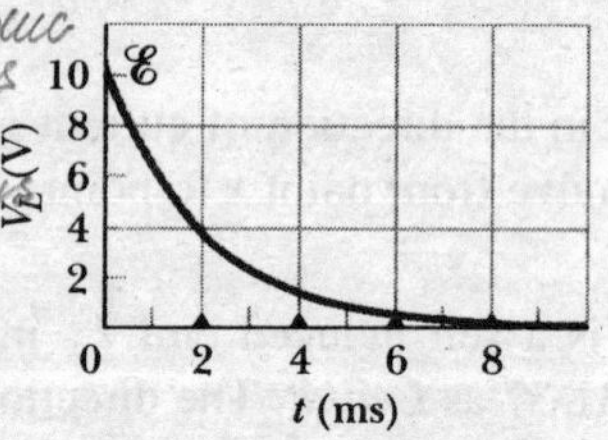

(b)

Fig. 31-19 The variation with time of (a) V_R, the potential difference across the resistor in the circuit of Fig. 31-18, and (b) V_L, the potential difference across the inductor in that circuit. The small triangles represent successive intervals of one inductive time constant $\tau_L = L/R$. The figure is plotted for $R = 2000\ \Omega$, $L = 4.0$ H, and $\mathscr{E} = 10$ V.

before the connection to *a* is broken. A switch that does this is called a *make-before-break* switch.)

With the battery gone, the current through the resistor will decrease. However, it cannot drop immediately to zero but must decay to zero over time. The differential equation that governs the decay can be found by putting $\mathscr{E} = 0$ in Eq. 31-41:

$$L\frac{di}{dt} + iR = 0. \tag{31-46}$$

By analogy with Eqs. 28-35 and 28-36, the solution of this differential equation that satisfies the initial condition $i(0) = i_0 = \mathscr{E}/R$ is

$$i = \frac{\mathscr{E}}{R}e^{-t/\tau_L} = i_0 e^{-t/\tau_L} \quad \text{(decay of current).} \tag{31-47}$$

We see that both current rise (Eq. 31-43) and current decay (Eq. 31-47) in an *RL* circuit are governed by the same inductive time constant, τ_L.

We have used i_0 in Eq. 31-47 to represent the current at time $t = 0$. In our case that happened to be $\mathscr{E}/R$, but it could be any other initial value.

Sample Problem 31-5

Figure 31-20*a* shows a circuit that contains three identical resistors with resistance $R = 9.0\ \Omega$, two identical inductors with inductance $L = 2.0$ mH, and an ideal battery with emf $\mathscr{E} = 18$ V.

(a) What is the current i through the battery just after the switch is closed?

SOLUTION: The **Key Idea** here is that just after the switch is closed, the inductor acts to oppose a change in the current through it. Because the current through each inductor is zero before the switch is closed, it will also be zero just afterward. Thus, immediately after the switch is closed, the inductors act as broken wires, as indicated in Fig. 31-20*b*. We then have a single-loop circuit for which the loop rule gives us

$$\mathscr{E} - iR = 0.$$

Substituting given data, we find that

$$i = \frac{\mathscr{E}}{R} = \frac{18\ \text{V}}{9.0\ \Omega} = 2.0\ \text{A}. \quad \text{(Answer)}$$

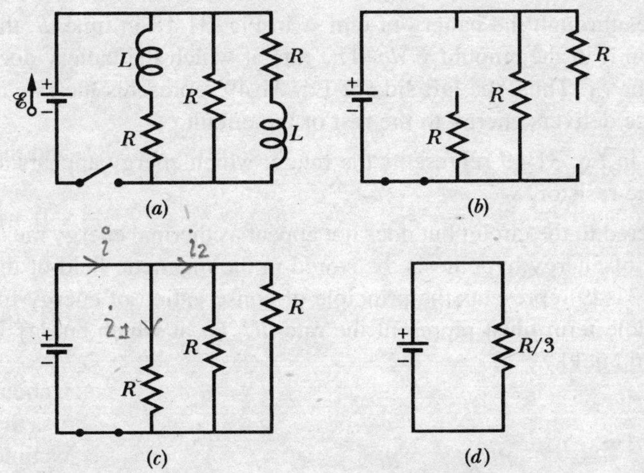

Fig. 31-20 Sample Problem 31-5. (*a*) A multiloop *RL* circuit with an open switch. (*b*) The equivalent circuit just after the switch has been closed. (*c*) The equivalent circuit a long time later. (*d*) The single-loop circuit that is equivalent to circuit (*c*).

(b) What is the current i through the battery long after the switch has been closed?

SOLUTION: The **Key Idea** here is that long after the switch has been closed, the currents in the circuit have reached their equilibrium values, and the inductors act as simple connecting wires, as indicated in Fig. 31-20*c*. We then have a circuit with three identical resistors in parallel; from Eq. 28-20, their equivalent resistance is $R_{eq} = R/3 = (9.0\ \Omega)/3 = 3.0\ \Omega$. The equivalent circuit shown in Fig. 31-20*d* then yields the loop equation $\mathscr{E} - iR_{eq} = 0$, or

$$i = \frac{\mathscr{E}}{R_{eq}} = \frac{18\ \text{V}}{3.0\ \Omega} = 6.0\ \text{A}. \quad \text{(Answer)}$$

CHECKPOINT 6: The figure shows three circuits with identical batteries, inductors, and resistors. Rank the circuits according to the current through the battery (a) just after the switch is closed and (b) a long time later, greatest first.

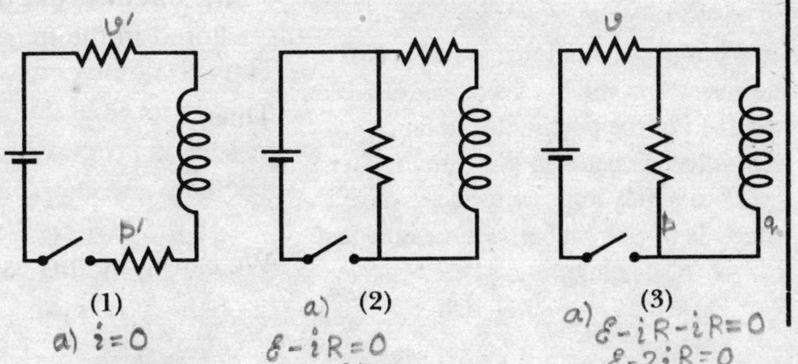

Sample Problem 31-6

A solenoid has an inductance of 53 mH and a resistance of 0.37 Ω. If it is connected to a battery, how long will the current take to reach half its final equilibrium value?

SOLUTION: One **Key Idea** here is that we can mentally separate the solenoid into a resistance and an inductance that are wired in series with a battery, as in Fig. 31-18. Then application of the loop rule leads to Eq. 31-41, which has the solution of Eq. 31-43 for the current i in the circuit.

The second **Key Idea** is that, according to that solution, current i increases exponentially from zero to its final equilibrium value of $\mathscr{E}/R$. Let t_0 be the time that current i takes to reach half its equilibrium value. Then Eq. 31-43 gives us

$$\frac{1}{2}\frac{\mathscr{E}}{R} = \frac{\mathscr{E}}{R}(1 - e^{-t_0/\tau_L}).$$

We solve for t_0 by canceling $\mathscr{E}/R$, isolating the exponential, and taking the natural logarithm of each side. We find

$$t_0 = \tau_L \ln 2 = \frac{L}{R}\ln 2 = \frac{53 \times 10^{-3}\text{ H}}{0.37\ \Omega}\ln 2$$
$$= 0.10\text{ s.} \qquad \text{(Answer)}$$

31-10 Energy Stored in a Magnetic Field

When we pull two particles with opposite signs of charge away from each other, we say that the resulting electric potential energy is stored in the electric field of the particles. We get it back from the field by letting the particles move closer together again. In the same way we can consider energy to be stored in a magnetic field.

To derive a quantitative expression for that stored energy, consider again Fig. 31-18, which shows a source of emf $\mathscr{E}$ connected to a resistor R and an inductor L. Equation 31-41, restated here for convenience,

$$\mathscr{E} = L\frac{di}{dt} + iR, \qquad (31\text{-}48)$$

is the differential equation that describes the growth of current in this circuit. Recall that this equation follows immediately from the loop rule and that the loop rule in turn is an expression of the principle of conservation of energy for single-loop circuits. If we multiply each side of Eq. 31-48 by i, we obtain

$$\mathscr{E}i = Li\frac{di}{dt} + i^2R, \qquad (31\text{-}49)$$

which has the following physical interpretation in terms of work and energy:

1. If a charge dq passes through the battery of emf $\mathscr{E}$ in Fig. 31-18 in time dt, the battery does work on it in the amount $\mathscr{E}\,dq$. The rate at which the battery does work is $(\mathscr{E}\,dq)/dt$, or $\mathscr{E}i$. Thus, the left side of Eq. 31-49 represents the rate at which the emf device delivers energy to the rest of the circuit.
2. The rightmost term in Eq. 31-49 represents the rate at which energy appears as thermal energy in the resistor.
3. Energy that is delivered to the circuit but does not appear as thermal energy must, by the conservation-of-energy hypothesis, be stored in the magnetic field of the inductor. Since Eq. 31-49 represents the principle of conservation of energy for RL circuits, the middle term must represent the rate dU_B/dt at which energy is stored in the magnetic field.

Thus

$$\frac{dU_B}{dt} = Li\frac{di}{dt}. \qquad (31\text{-}50)$$

We can write this as

$$dU_B = Li\,di.$$

Integrating yields

$$\int_0^{U_B} dU_B = \int_0^i Li\,di$$

or

$$U_B = \tfrac{1}{2}Li^2 \quad \text{(magnetic energy)}, \tag{31-51}$$

which represents the total energy stored by an inductor L carrying a current i. Note the similarity in form between this expression and the expression for the energy stored by a capacitor with capacitance C and charge q; namely,

$$U_E = \frac{q^2}{2C}. \tag{31-52}$$

(The variable i^2 corresponds to q^2, and the constant L corresponds to $1/C$.)

Sample Problem 31-7

A coil has an inductance of 53 mH and a resistance of 0.35 Ω.

(a) If a 12 V emf is applied across the coil, how much energy is stored in the magnetic field after the current has built up to its equilibrium value?

SOLUTION: The **Key Idea** here is that the energy stored in the magnetic field of a coil at any time depends on the current through the coil at that time, according to Eq. 31-51 ($U_B = \frac{1}{2}Li^2$). Thus, to find the energy $U_{B\infty}$ stored at equilibrium, we must first find the equilibrium current. From Eq. 31-43, the equilibrium current is

$$i_\infty = \frac{\mathscr{E}}{R} = \frac{12\text{ V}}{0.35\ \Omega} = 34.3\text{ A}. \tag{31-53}$$

Then substitution yields

$$U_{B\infty} = \tfrac{1}{2}Li_\infty^2 = (\tfrac{1}{2})(53 \times 10^{-3}\text{ H})(34.3\text{ A})^2$$
$$= 31\text{ J}. \quad \text{(Answer)}$$

(b) After how many time constants will half this equilibrium energy be stored in the magnetic field?

SOLUTION: The **Key Idea** of part (a) applies here also. Now we are being asked: At what time t will the relation

$$U_B = \tfrac{1}{2}U_{B\infty}$$

be satisfied? Using Eq. 31-51 twice allows us to rewrite this energy condition as

$$\tfrac{1}{2}Li^2 = (\tfrac{1}{2})\tfrac{1}{2}Li_\infty^2$$

or

$$i = \left(\frac{1}{\sqrt{2}}\right)i_\infty. \tag{31-54}$$

However, i is given by Eq. 31-43 and i_∞ (see Eq. 31-53) is $\mathscr{E}/R$, so Eq. 31-54 becomes

$$\frac{\mathscr{E}}{R}(1 - e^{-t/\tau_L}) = \frac{\mathscr{E}}{\sqrt{2}R}.$$

By canceling $\mathscr{E}/R$ and rearranging, this can be written as

$$e^{-t/\tau_L} = 1 - \frac{1}{\sqrt{2}} = 0.293,$$

which yields

$$\frac{t}{\tau_L} = -\ln 0.293 = 1.23$$

or

$$t \approx 1.2\tau_L. \quad \text{(Answer)}$$

Thus, the stored energy will reach half its equilibrium value 1.2 time constants after the emf is applied.

31-11 Energy Density of a Magnetic Field

Consider a length l near the middle of a long solenoid of cross-sectional area A carrying current i; the volume associated with this length is Al. The energy U_B stored by the length l of the solenoid must lie entirely within this volume because the magnetic field outside such a solenoid is approximately zero. Moreover, the stored energy must be uniformly distributed within the solenoid because the magnetic field is (approximately) uniform everywhere inside.

Thus, the energy stored per unit volume of the field is

$$u_B = \frac{U_B}{Al}$$

or, since

$$U_B = \tfrac{1}{2}Li^2,$$

we have

$$u_B = \frac{Li^2}{2Al} = \frac{L}{l}\frac{i^2}{2A}.$$

Here L is the inductance of length l of the solenoid.

Substituting for L/l from Eq. 31-33, we find

$$u_B = \tfrac{1}{2}\mu_0 n^2 i^2, \qquad (31\text{-}55)$$

where n is the number of turns per unit length. From Eq. 30-25 ($B = \mu_0 in$) we can write this *energy density* as

$$u_B = \frac{B^2}{2\mu_0} \quad \text{(magnetic energy density).} \qquad (31\text{-}56)$$

This equation gives the density of stored energy at any point where the magnetic field is B. Even though we derived it by considering the special case of a solenoid, Eq. 31-56 holds for all magnetic fields, no matter how they are generated. The equation is comparable to Eq. 26-23; namely,

$$u_E = \tfrac{1}{2}\varepsilon_0 E^2, \qquad (31\text{-}57)$$

which gives the energy density (in a vacuum) at any point in an electric field. Note that both u_B and u_E are proportional to the square of the appropriate field magnitude, B or E.

CHECKPOINT 7 The table lists the number of turns per unit length, current, and cross-sectional area for three solenoids. Rank the solenoids according to the magnetic energy density within them, greatest first.

Solenoid	Turns per Unit Length	Current	Area
a	$2n_1$	i_1	$2A_1$
b	n_1	$2i_1$	A_1
c	n_1	i_1	$6A_1$

B = ni μo
Ba = (2n1 i1) μo
Bb = (2n1 i1) μo
Bc = (n1 i1) μo
μa & μb tie then μc

Sample Problem 31-8

A long coaxial cable (Fig. 31-21) consists of two thin-walled concentric conducting cylinders with radii a and b. The inner cylinder carries a steady current i, the outer cylinder providing the return path for that current. The current sets up a magnetic field between the two cylinders.

(a) Calculate the energy stored in the magnetic field for a length ℓ of the cable.

SOLUTION: The **Key Ideas** for this challenging problem are these:

1. We can calculate the (total) energy U_B stored in the magnetic field from the energy density u_B of the field.
2. That energy density depends on the magnitude B of the magnetic field according to Eq. 31-56 ($u_B = B^2/2\mu_0$).

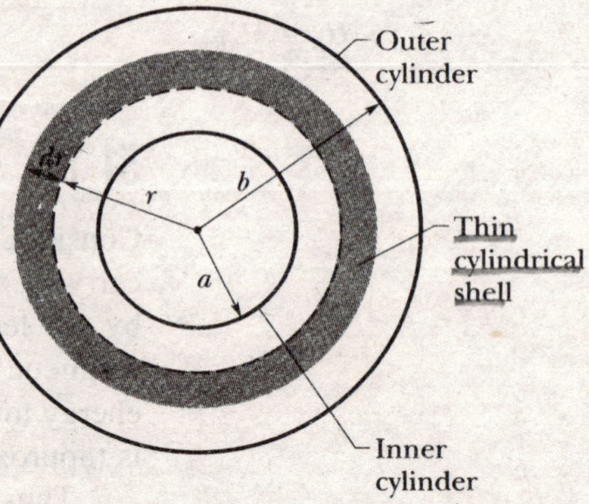

Fig. 31-21 Sample Problem 31-8. A cross section of a long coaxial cable consisting of two thin-walled conducting cylinders, the inner cylinder of radius a and the outer cylinder of radius b.

3. Because of the circular symmetry of the cable, we can find B by using Ampere's law with the given current i.

Finding B: To apply these ideas, we begin with Ampere's law, using a circular path of integration with radius r such that $a < r < b$ (between the two cylinders, as indicated by the dashed line in Fig. 31-21). The only current enclosed by this path is current i on the inner cylinder. Thus, we can write Ampere's law as

$$\oint \vec{B} \cdot d\vec{s} = \mu_0 i. \tag{31-58}$$

Next, we simplify the integral: Because of the circular symmetry, at all points along the circular path, $\vec{B}$ is tangent to the path and has the same magnitude B. Let us take the direction of integration along the path as the direction of the magnetic field around the path. Then we can replace $\vec{B} \cdot d\vec{s}$ with $B\,ds \cos 0 = B\,ds$, and then move magnitude B in front of the integration. The integral that remains is $\oint ds$, which just gives the circumference $2\pi r$ of the path. Thus, Eq. 31-58 simplifies to

$$B(2\pi r) = \mu_0 i$$

or

$$B = \frac{\mu_0 i}{2\pi r}. \tag{31-59}$$

Finding u_B: Next, to obtain the energy density, we substitute Eq. 31-59 into Eq. 31-56:

$$u_B = \frac{B^2}{2\mu_0} = \frac{\mu_0 i^2}{8\pi^2 r^2}. \tag{31-60}$$

Finding U_B: Note that u_B is not uniform in the volume between the two cylinders, but instead depends on the radial distance r. Thus, to find the total energy U_B stored between the cylinders, we must integrate u_B over that volume.

Because the volume between the cylinders has circular symmetry about the cable's central axis, we consider the volume dV of a cylindrical shell located between the cylinders; the shell has inner radius r, outer radius $r + dr$ (Fig. 31-21), and length ℓ. The shell's cross-sectional area (or face area) is the product of its circumference $2\pi r$ and thickness dr. Thus, the shell's volume dV is $(2\pi r)(dr)(\ell)$; that is, $dV = 2\pi r\ell\, dr$.

Because points within this shell are all at approximately the same radial distance r, they all have approximately the same energy density u_B. Thus, the total energy dU_B contained in the shell of volume dV is given by

$$\text{energy} = \begin{pmatrix}\text{energy per}\\ \text{unit volume}\end{pmatrix}(\text{volume})$$

or

$$dU_B = u_B\, dV.$$

Substituting Eq. 31-60 for u_B and $2\pi r\ell\, dr$ for dV, we obtain

$$dU_B = \frac{\mu_0 i^2}{8\pi^2 r^2}(2\pi r\ell)\, dr = \frac{\mu_0 i^2 \ell}{4\pi}\frac{dr}{r}.$$

To find the total energy contained between the two cylinders, we integrate this equation over the volume between the two cylinders:

$$U_B = \int dU_B = \frac{\mu_0 i^2 \ell}{4\pi}\int_a^b \frac{dr}{r}$$

$$= \frac{\mu_0 i^2 \ell}{4\pi} \ln\frac{b}{a}. \qquad \text{(Answer)} \quad (31\text{-}61)$$

No energy is stored outside the outer cylinder or inside the inner cylinder because the magnetic field is zero in both locations, as you can show with Ampere's law.

(b) What is the stored energy per unit length of the cable if $a = 1.2$ mm, $b = 3.5$ mm, and $i = 2.7$ A?

SOLUTION: From Eq. 31-61 we have

$$\frac{U_B}{\ell} = \frac{\mu_0 i^2}{4\pi}\ln\frac{b}{a}$$

$$= \frac{(4\pi \times 10^{-7}\ \text{H/m})(2.7\ \text{A})^2}{4\pi}\ln\frac{3.5\ \text{mm}}{1.2\ \text{mm}}$$

$$= 7.8 \times 10^{-7}\ \text{J/m} = 780\ \text{nJ/m}. \qquad \text{(Answer)}$$

31-12 Mutual Induction

In this section we return to the case of two interacting coils, which we first discussed in Section 31-2, and we treat it in a somewhat more formal manner. We saw earlier that if two coils are close together as in Fig. 31-2, a steady current i in one coil will set up a magnetic flux Φ through the other coil (*linking* the other coil). If we change i with time, an emf $\mathscr{E}$ given by Faraday's law appears in the second coil; we called this process *induction*. We could better have called it **mutual induction**, to suggest the mutual interaction of the two coils and to distinguish it from *self-induction*, in which only one coil is involved.

Let us look a little more quantitatively at mutual induction. Figure 31-22*a* shows two circular close-packed coils near each other and sharing a common central axis. There is a steady current i_1 in coil 1, produced by the battery in the external circuit. This current creates a magnetic field represented by the lines of $\vec{B}_1$ in the figure. Coil 2 is connected to a sensitive meter but contains no battery; a magnetic flux Φ_{21} (the flux through coil 2 associated with the current in coil 1) links the N_2 turns of coil 2.

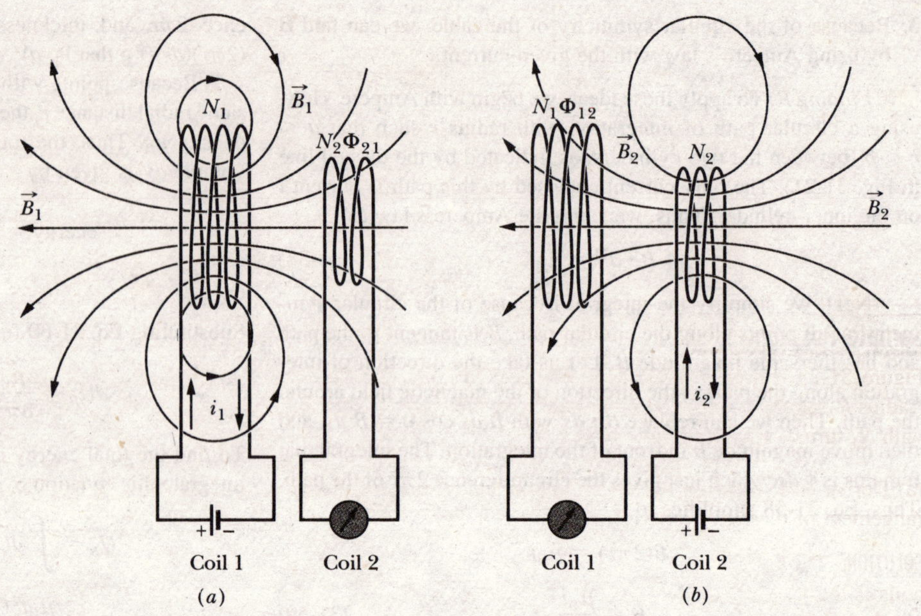

Fig. 31-22 Mutual induction. (*a*) If the current in coil 1 changes, an emf will be induced in coil 2. (*b*) If the current in coil 2 changes, an emf will be induced in coil 1.

We define the mutual inductance M_{21} of coil 2 with respect to coil 1 as

$$M_{21} = \frac{N_2\Phi_{21}}{i_1}, \tag{31-62}$$

which has the same form as Eq. 31-30 ($L = N\Phi/i$), the definition of (self) inductance. We can recast Eq. 31-62 as

$$M_{21}i_1 = N_2\Phi_{21}.$$

If, by external means, we cause i_1 to vary with time, we have

$$M_{21}\frac{di_1}{dt} = N_2\frac{d\Phi_{21}}{dt}.$$

The right side of this equation is, according to Faraday's law, just the magnitude of the emf $\mathscr{E}_2$ appearing in coil 2 due to the changing current in coil 1. Thus, with a minus sign to indicate direction,

$$\mathscr{E}_2 = -M_{21}\frac{di_1}{dt}, \tag{31-63}$$

which you should compare with Eq. 31-37 for self-induction ($\mathscr{E} = -L\, di/dt$).

Let us now interchange the roles of coils 1 and 2, as in Fig. 31-22*b*; that is, we set up a current i_2 in coil 2 by means of a battery, and this produces a magnetic flux Φ_{12} that links coil 1. If we change i_2 with time, we have, by the argument given above,

$$\mathscr{E}_1 = -M_{12}\frac{di_2}{dt}. \tag{31-64}$$

Thus, we see that the emf induced in either coil is proportional to the rate of change of current in the other coil. The proportionality constants M_{21} and M_{12} seem to be different. We assert, without proof, that they are in fact the same so that no subscripts are needed. (This conclusion is true but is in no way obvious.) Thus, we have

$$M_{21} = M_{12} = M, \tag{31-65}$$

and we can rewrite Eqs. 31-63 and 31-64 as

$$\mathscr{E}_2 = -M\frac{di_1}{dt} \tag{31-66}$$

and

$$\mathscr{E}_1 = -M\frac{di_2}{dt}. \tag{31-67}$$

The induction is indeed mutual. The SI unit for M (as for L) is the henry.

Sample Problem 31-9

Figure 31-23 shows two circular close-packed coils, the smaller (radius R_2, with N_2 turns) being coaxial with the larger (radius R_1, with N_1 turns) and in the same plane.

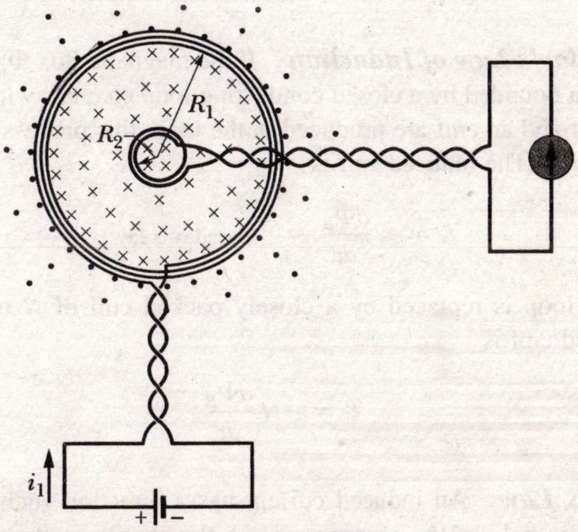

Fig. 31-23 Sample Problem 31-9. A small coil is located at the center of a large coil. The mutual inductance of the coils can be determined by sending current i_1 through the large coil.

(a) Derive an expression for the mutual inductance M for this arrangement of these two coils, assuming that $R_1 \gg R_2$.

SOLUTION: The **Key Idea** here is that the mutual inductance M for these coils is the ratio of the flux linkage ($N\Phi$) through one coil to the current i in the other coil, which produces that flux linkage. Thus, we need to assume that currents exist in the coils; then we need to calculate the flux linkage in one of the coils.

The magnetic field through the larger coil due to the smaller coil is nonuniform in both magnitude and direction, so the flux through the larger coil due to the smaller coil is nonuniform and difficult to calculate. However, the smaller coil is small enough for us to assume that the magnetic field through it due to the larger coil is approximately uniform. Thus, the flux through it due to the larger coil is also approximately uniform. Hence, to find M we shall assume a current i_1 in the larger coil and calculate the flux linkage $N_2\Phi_{21}$ in the smaller coil:

$$M = \frac{N_2\Phi_{21}}{i_1}. \tag{31-68}$$

A second **Key Idea** is that the flux Φ_{21} through each turn of the smaller coil is, from Eq. 31-4,

$$\Phi_{21} = B_1A_2,$$

where B_1 is the magnitude of the magnetic field at points within the small coil due to the larger coil, and A_2 ($= \pi R_2^2$) is the area enclosed by the turn. Thus, the flux linkage in the smaller coil (with its N_2 turns) is

$$N_2\Phi_{21} = N_2B_1A_2. \tag{31-69}$$

A third **Key Idea** is that to find B_1 at points within the smaller coil, we can use Eq. 30-28, with z set to 0 because the smaller coil is in the plane of the larger coil. That equation tells us that each turn of the larger coil produces a magnetic field of magnitude $\mu_0 i_1/2R_1$ at points within the smaller coil. Thus, the larger coil (with its N_1 turns) produces a total magnetic field of magnitude

$$B_1 = N_1\frac{\mu_0 i_1}{2R_1} \tag{31-70}$$

at points within the smaller coil.

Substituting Eq. 31-70 for B_1 and πR_2^2 for A_2 in Eq. 31-69 yields

$$N_2\Phi_{21} = \frac{\pi\mu_0 N_1 N_2 R_2^2 i_1}{2R_1}.$$

Substituting this result into Eq. 31-68, we find

$$M = \frac{N_2\Phi_{21}}{i_1} = \frac{\pi\mu_0 N_1 N_2 R_2^2}{2R_1}. \quad \text{(Answer)} \tag{31-71}$$

(b) What is the value of M for $N_1 = N_2 = 1200$ turns, $R_2 = 1.1$ cm, and $R_1 = 15$ cm?

SOLUTION: Equation 31-71 yields

$$M = \frac{(\pi)(4\pi \times 10^{-7}\ \text{H/m})(1200)(1200)(0.011\ \text{m})^2}{(2)(0.15\ \text{m})}$$

$$= 2.29 \times 10^{-3}\ \text{H} \approx 2.3\ \text{mH}. \quad \text{(Answer)}$$

Consider the situation if we reverse the roles of the two coils—that is, if we produce a current i_2 in the smaller coil and try to calculate M from Eq. 31-62 in the form

$$M = \frac{N_1\Phi_{12}}{i_2}.$$

The calculation of Φ_{12} (the nonuniform flux of the smaller coil's magnetic field encompassed by the larger coil) is not simple. If we were to do the calculation numerically using a computer, we would find M to be 2.3 mH, as above! This emphasizes that Eq. 31-65 ($M_{21} = M_{12} = M$) is not obvious.

REVIEW & SUMMARY

Magnetic Flux The *magnetic flux* Φ_B through an area A in a magnetic field $\vec{B}$ is defined as

$$\Phi_B = \int \vec{B} \cdot d\vec{A}, \tag{31-3}$$

where the integral is taken over the area. The SI unit of magnetic flux is the weber, where 1 Wb = 1 T · m². If $\vec{B}$ is perpendicular to the area and uniform over it, Eq. 31-3 becomes

$$\Phi_B = BA \qquad (\vec{B} \perp A, \vec{B} \text{ uniform}). \tag{31-4}$$

Faraday's Law of Induction If the magnetic flux Φ_B through an area bounded by a closed conducting loop changes with time, a current and an emf are produced in the loop; this process is called *induction*. The induced emf is

$$\mathscr{E} = -\frac{d\Phi_B}{dt} \qquad \text{(Faraday's law)}. \tag{31-6}$$

If the loop is replaced by a closely packed coil of N turns, the induced emf is

$$\mathscr{E} = -N\frac{d\Phi_B}{dt}. \tag{31-7}$$

Lenz's Law An induced current has a direction such that the magnetic field *of the current* opposes the change in the magnetic flux that produces the current. The induced emf has the same direction as the induced current.

Emf and the Induced Electric Field An emf is induced by a changing magnetic flux even if the loop through which the flux is changing is not a physical conductor but an imaginary line. The changing magnetic field induces an electric field $\vec{E}$ at every point of such a loop; the induced emf is related to $\vec{E}$ by

$$\mathscr{E} = \oint \vec{E} \cdot d\vec{s}, \tag{31-21}$$

where the integration is taken around the loop. From Eq. 31-21 we can write Faraday's law in its most general form,

$$\oint \vec{E} \cdot d\vec{s} = -\frac{d\Phi_B}{dt} \qquad \text{(Faraday's law)}. \tag{31-22}$$

The essence of this law is that *a changing magnetic field induces an electric field* $\vec{E}$.

Inductors An **inductor** is a device that can be used to produce a known magnetic field in a specified region. If a current i is established through each of the N windings of an inductor, a magnetic flux Φ_B links those windings. The **inductance** L of the inductor is

$$L = \frac{N\Phi_B}{i} \qquad \text{(inductance defined)}. \tag{31-30}$$

The SI unit of inductance is the **henry** (H), with

$$1 \text{ henry} = 1 \text{ H} = 1 \text{ T} \cdot \text{m}^2/\text{A}. \tag{31-31}$$

The inductance per unit length near the middle of a long solenoid of cross-sectional area A and n turns per unit length is

$$\frac{L}{l} = \mu_0 n^2 A \qquad \text{(solenoid)}. \tag{31-33}$$

Self-Induction If a current i in a coil changes with time, an emf is induced in the coil. This self-induced emf is

$$\mathscr{E}_L = -L\frac{di}{dt}. \tag{31-37}$$

The direction of $\mathscr{E}_L$ is found from Lenz's law: The self-induced emf acts to oppose the change that produces it.

Series RL Circuits If a constant emf $\mathscr{E}$ is introduced into a single-loop circuit containing a resistance R and an inductance L, the current rises to an equilibrium value of $\mathscr{E}/R$ according to

$$i = \frac{\mathscr{E}}{R}(1 - e^{-t/\tau_L}) \qquad \text{(rise of current)}. \tag{31-43}$$

Here τ_L ($= L/R$) governs the rate of rise of the current and is called the **inductive time constant** of the circuit. When the source of constant emf is removed, the current decays from a value i_0 according to

$$i = i_0 e^{-t/\tau_L} \qquad \text{(decay of current)}. \tag{31-47}$$

Magnetic Energy If an inductor L carries a current i, the inductor's magnetic field stores an energy given by

$$U_B = \tfrac{1}{2}Li^2 \qquad \text{(magnetic energy)}. \tag{31-51}$$

If B is the magnitude of a magnetic field at any point (in an inductor or anywhere else), the density of stored magnetic energy at that point is

$$u_B = \frac{B^2}{2\mu_0} \qquad \text{(magnetic energy density)}. \tag{31-56}$$

Mutual Induction If two coils (labeled 1 and 2) are near each other, a changing current in either coil can induce an emf in the other. This mutual induction is described by

$$\mathscr{E}_2 = -M\frac{di_1}{dt} \tag{31-66}$$

and

$$\mathscr{E}_1 = -M\frac{di_2}{dt}, \tag{31-67}$$

where M (measured in henries) is the mutual inductance for the coil arrangement.

QUESTIONS

1 In Fig. 31-24, a long straight wire with current i passes (without touching) three rectangular wire loops with edge lengths L, $1.5L$, and $2L$. The loops are widely spaced (so as to not affect one another). Loops 1 and 3 are symmetric about the long wire. Rank the loops according to the size of the current induced in them if current i is (a) constant and (b) increasing, greatest first.

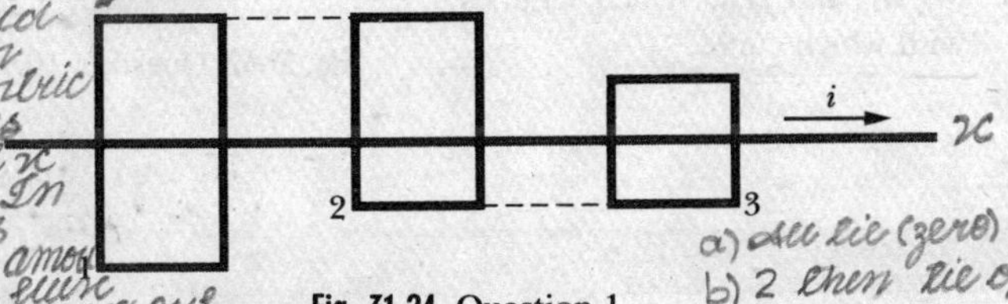

Fig. 31-24 Question 1.

2 If the circular conductor in Fig. 31-25 undergoes thermal expansion while it is in a uniform magnetic field, a current will be induced clockwise around it. Is the magnetic field directed into the page or out of it?

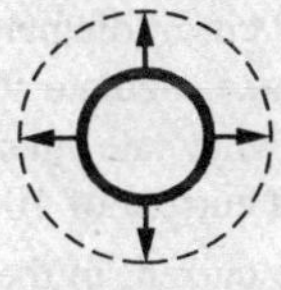

Fig. 31-25 Question 2.

3 Figure 31-26 shows two circuits in which a conducting bar is slid at the same speed v through the same uniform magnetic field and along a U-shaped wire. The parallel lengths of the wire are separated by $2L$ in circuit 1 and by L in circuit 2. The current induced in circuit 1 is counterclockwise. (a) Is the direction of the magnetic field into or out of the page? (b) Is the direction of the current induced in circuit 2 clockwise or counterclockwise? (c) Is the emf induced in circuit 1 larger than, smaller than, or the same as that in circuit 2?

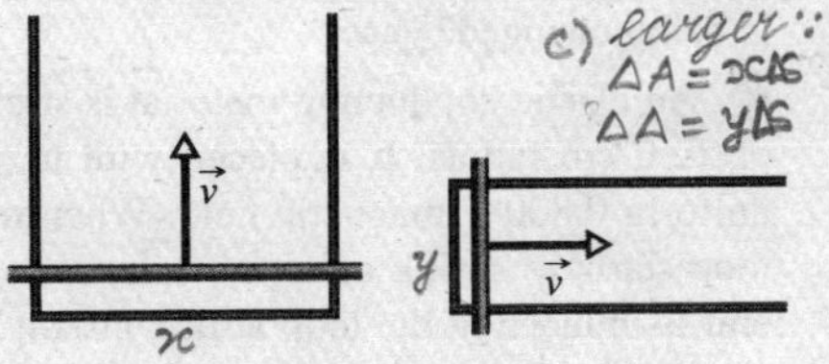

Fig. 31-26 Question 3.

4 Figure 31-27 shows two coils wrapped around nonconducting rods. Coil X is connected to a battery and a variable resistance. What is the direction of the induced current through the current meter connected to coil Y (a) when coil Y is moved toward coil X and (b) when the current in coil X is decreased without any change in the relative positions of the coils?

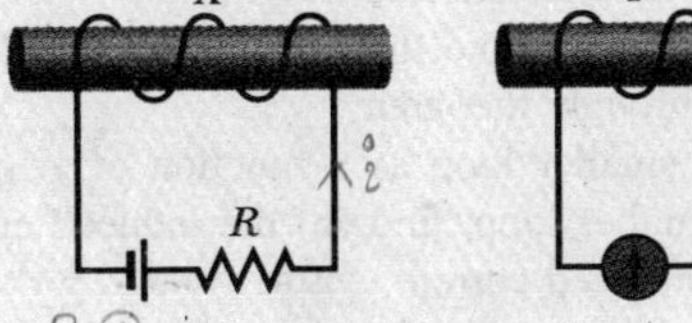

Fig. 31-27 Question 4.

5 Figure 31-28a shows a circular region in which an increasing uniform magnetic field is directed out of the page, as well as a concentric circular path along which $\oint \vec{E} \cdot d\vec{s}$ is to be evaluated. The table gives the initial magnitude of the magnetic field, the increase in that magnitude, and the time interval for the increase, in three situations. Rank the situations according to the magnitude of the electric field induced along the path, greatest first.

Situation	Initial Field	Increase	Time
a	B_1	ΔB_1	Δt_1
b	$2B_1$	$\Delta B_1/2$	Δt_1
c	$B_1/4$	ΔB_1	$\Delta t_1/2$

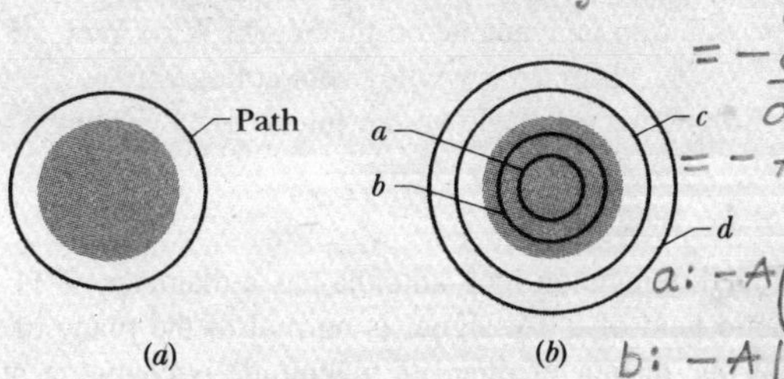

Fig. 31-28 Questions 5 and 6.

6 Figure 31-28b shows a circular region in which a decreasing uniform magnetic field is directed out of the page, as well as four concentric circular paths. Rank the paths according to the magnitude of $\oint \vec{E} \cdot d\vec{s}$ evaluated along them, greatest first.

7 Figure 31-29 gives the variation with time of the potential difference V_R across a resistor in three circuits wired as shown in Fig. 31-18. The circuits contain the same resistance R and emf $\mathscr{E}$ but differ in the inductance L. Rank the circuits according to the value of L, greatest first.

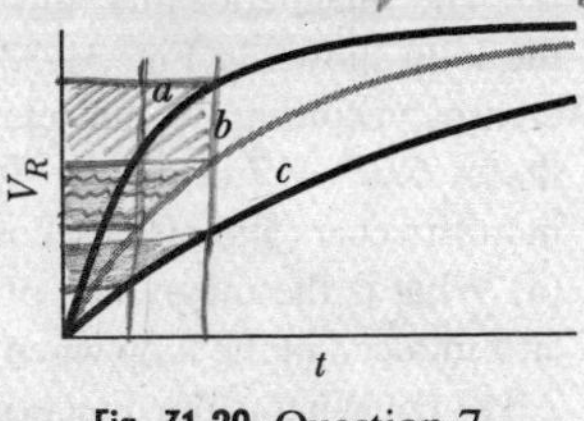

Fig. 31-29 Question 7.

8 Figure 31-30 shows three circuits with identical batteries, inductors, and resistors. Rank the circuits according to the time for the current to reach 50% of its equilibrium value after the switches are closed, greatest first.

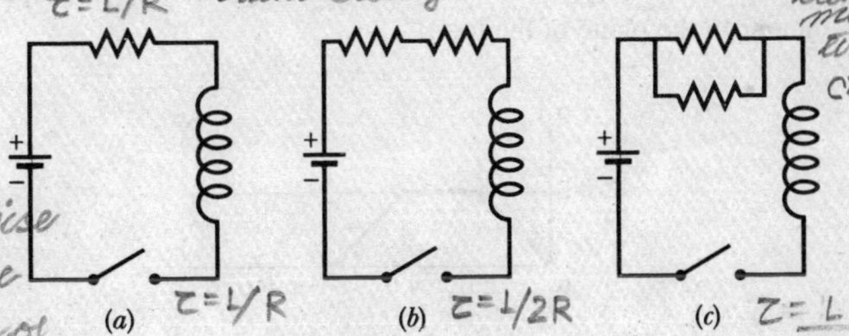

Fig. 31-30 Question 8.

9. Figure 31-31 shows a circuit with two identical resistors and an ideal inductor. Is the current through the central resistor more than, less than, or the same as that through the other resistor (a) just after the closing of switch S, (b) a long time after the closing of S, (c) just after S is reopened, a long time later, and (d) a long time after the reopening of S?

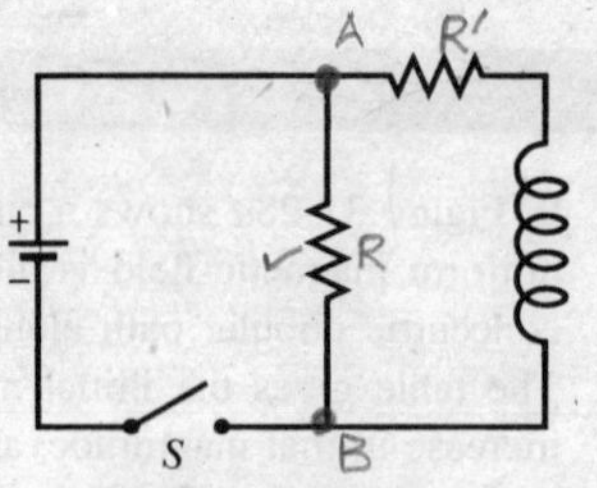

Fig. 31-31 Question 9.

10. The switch in the circuit of Fig. 31-17 has been closed on a for a very long time when it is then thrown to b. The resulting current through the inductor is indicated in Fig. 31-32 for four sets of values for the resistance R and inductance L: (1) R_0 and L_0, (2) $2R_0$ and L_0, (3) R_0 and $2L_0$, (4) $2R_0$ and $2L_0$. Which set goes with which curve?

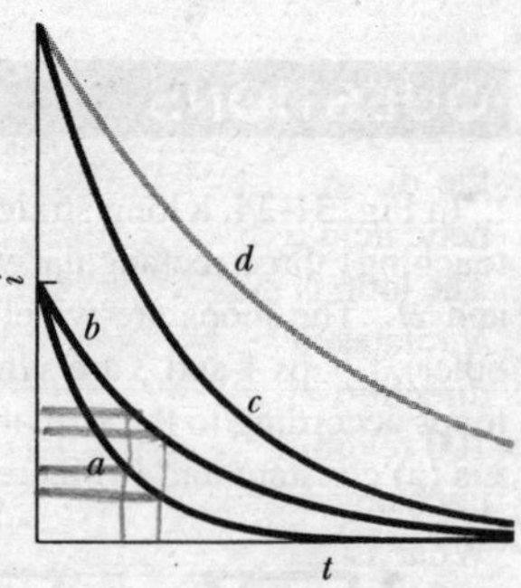

Fig. 31-32 Question 10.

EXERCISES & PROBLEMS

ssm Solution is in the Student Solutions Manual.
www Solution is available on the World Wide Web at: http://www.wiley.com/college/hrw
ilw Solution is available on the Interactive LearningWare.

SEC. 31-4 Lenz's Law

1E. A UHF television loop antenna has a diameter of 11 cm. The magnetic field of a TV signal is normal to the plane of the loop and, at one instant of time, its magnitude is changing at the rate 0.16 T/s. The magnetic field is uniform. What emf is induced in the antenna? **ssm**

2E. A small loop of area A is inside of, and has its axis in the same direction as, a long solenoid of n turns per unit length and current i. If $i = i_0 \sin \omega t$, find the emf induced in the loop.

3E. The magnetic flux through the loop shown in Fig. 31-33 increases according to the relation $\Phi_B = 6.0t^2 + 7.0t$, where Φ_B is in milliwebers and t is in seconds. (a) What is the magnitude of the emf induced in the loop when $t = 2.0$ s? (b) What is the direction of the current through R?

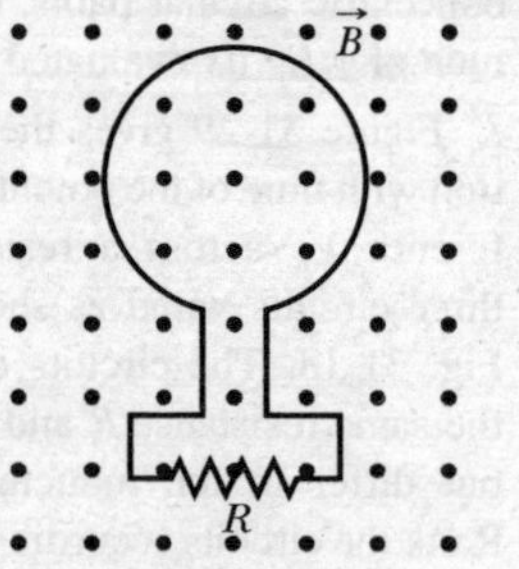

Fig. 31-33 Exercise 3 and Problem 11.

4E. The magnetic field through a single loop of wire, 12 cm in radius and of 8.5 Ω resistance, changes with time as shown in Fig. 31-34. Calculate the emf in the loop as a function of time. Consider the time intervals (a) $t = 0$ to $t = 2.0$ s, (b) $t = 2.0$ s to $t = 4.0$ s, (c) $t = 4.0$ s to $t = 6.0$ s. The (uniform) magnetic field is perpendicular to the plane of the loop.

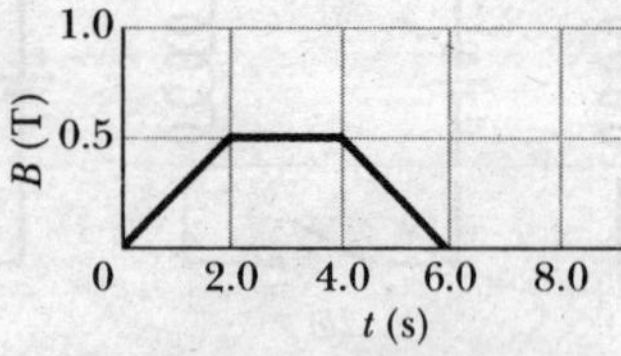

Fig. 31-34 Exercise 4.

5E. A uniform magnetic field is normal to the plane of a circular loop 10 cm in diameter and made of copper wire (of diameter 2.5 mm). (a) Calculate the resistance of the wire. (See Table 27-1.) (b) At what rate must the magnetic field change with time if an induced current of 10 A is to appear in the loop?

6P. The current in the solenoid of Sample Problem 31-1 changes, not as stated there, but according to $i = 3.0t + 1.0t^2$, where i is in amperes and t is in seconds. (a) Plot the induced emf in the coil from $t = 0$ to $t = 4.0$ s. (b) The resistance of the coil is 0.15 Ω. What is the current in the coil at $t = 2.0$ s?

7P. In Fig. 31-35 a 120-turn coil of radius 1.8 cm and resistance 5.3 Ω is placed *outside* a solenoid like that of Sample Problem 31-1. If the current in the solenoid is changed as in that sample problem, what current appears in the coil while the solenoid current is being changed? **ssm**

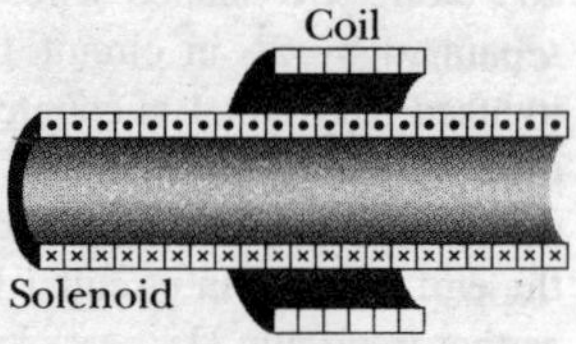

Fig. 31-35 Problem 7.

8P. An elastic conducting material is stretched into a circular loop of 12.0 cm radius. It is placed with its plane perpendicular to a uniform 0.800 T magnetic field. When released, the radius of the loop starts to shrink at an instantaneous rate of 75.0 cm/s. What emf is induced in the loop at that instant?

9P. Figure 31-36 shows two parallel loops of wire having a common axis. The smaller loop (radius r) is above the larger loop (radius R) by a distance $x \gg R$. Consequently, the magnetic field due to the current i in the larger loop is nearly constant throughout the smaller loop. Suppose that x is increasing at the constant rate of $dx/dt = v$. (a) Determine the magnetic flux through the area bounded by the smaller loop as a function of x. (*Hint:* See Eq. 30-29.) In the smaller loop, find (b) the induced emf and (c) the direction of the induced current. **ssm** **www**

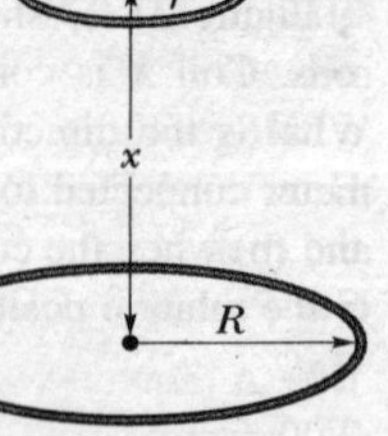

Fig. 31-36 Problem 9.

10P. In Fig. 31-37, a circular loop of wire 10 cm in diameter (seen edge-on) is placed with its normal $\vec{N}$ at an angle $\theta = 30°$ with the direction of a uniform magnetic field $\vec{B}$ of magnitude 0.50 T. The loop is then rotated such that $\vec{N}$ rotates in a cone about the field direction at the constant rate of 100 rev/min; the angle θ remains unchanged during the process. What is the emf induced in the loop?

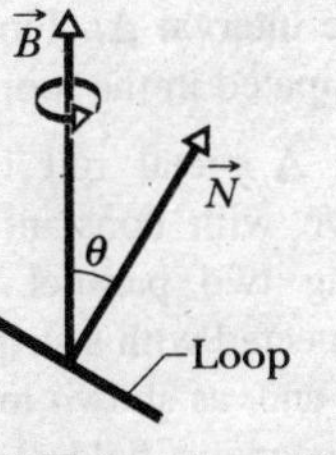

Fig. 31-37 Problem 10.

11P. In Fig. 31-33 let the flux through the loop be $\Phi_B(0)$ at time $t = 0$. Then let the magnetic field $\vec{B}$ vary in a continuous but unspecified way, in both magnitude and direction, so that at time t the flux is represented by $\Phi_B(t)$. (a) Show that the net charge $q(t)$ that has passed through resistor R in time t is

$$q(t) = \frac{1}{R}\,[\Phi_B(0) - \Phi_B(t)]$$

and is independent of the way $\vec{B}$ has changed. (b) If $\Phi_B(t) = \Phi_B(0)$ in a particular case, we have $q(t) = 0$. Is the induced current necessarily zero throughout the interval from 0 to t? ssm

12P. A small circular loop of area 2.00 cm^2 is placed in the plane of, and concentric with, a large circular loop of radius 1.00 m. The current in the large loop is changed uniformly from 200 A to -200 A (a change in direction) in a time of 1.00 s, beginning at $t = 0$. (a) What is the magnetic field at the center of the small circular loop due to the current in the large loop at $t = 0$, $t = 0.500$ s, and $t = 1.00$ s? (b) What emf is induced in the small loop at $t = 0.500$ s? (Since the inner loop is small, assume the field $\vec{B}$ due to the outer loop is uniform over the area of the smaller loop.)

13P. One hundred turns of insulated copper wire are wrapped around a wooden cylindrical core of cross-sectional area 1.20×10^{-3} m^2. The two terminals are connected to a resistor. The total resistance in the circuit is 13.0 Ω. If an externally applied uniform longitudinal magnetic field in the core changes from 1.60 T in one direction to 1.60 T in the opposite direction, how much charge flows through the circuit? (*Hint:* See Problem 11.) ilw

14P. At a certain place, Earth's magnetic field has magnitude $B = 0.590$ gauss and is inclined downward at an angle of 70.0° to the horizontal. A flat horizontal circular coil of wire with a radius of 10.0 cm has 1000 turns and a total resistance of 85.0 Ω. It is connected to a meter with 140 Ω resistance. The coil is flipped through a half-revolution about a diameter, so that it is again horizontal. How much charge flows through the meter during the flip? (*Hint:* See Problem 11.)

15P. A square wire loop with 2.00 m sides is perpendicular to a uniform magnetic field, with half the area of the loop in the field as shown in Fig. 31-38. The loop contains a 20.0 V battery with negligible internal resistance. If the magnitude of the field varies with time according to $B = 0.0420 - 0.870t$, with B in teslas and t in seconds, what are (a) the net emf in the circuit and (b) the direction of the current through the battery? ssm

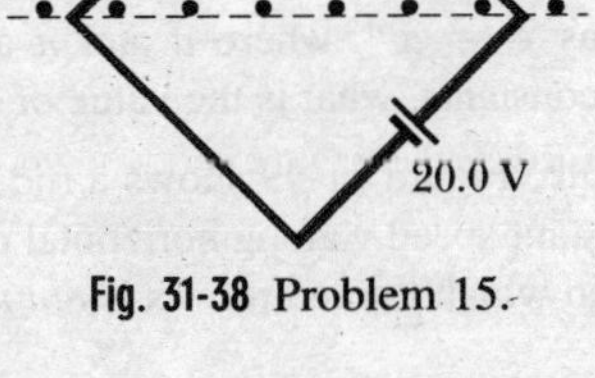

Fig. 31-38 Problem 15.

16P. A wire is bent into three circular segments, each of radius $r = 10$ cm, as shown in Fig. 31-39. Each segment is a quadrant of a circle, ab lying in the xy plane, bc lying in the yz plane, and ca lying in the zx plane. (a) If a uniform magnetic field $\vec{B}$ points in the positive x direction, what is the magnitude of the emf developed in the wire when B increases at the rate of 3.0 mT/s? (b) What is the direction of the current in segment bc?

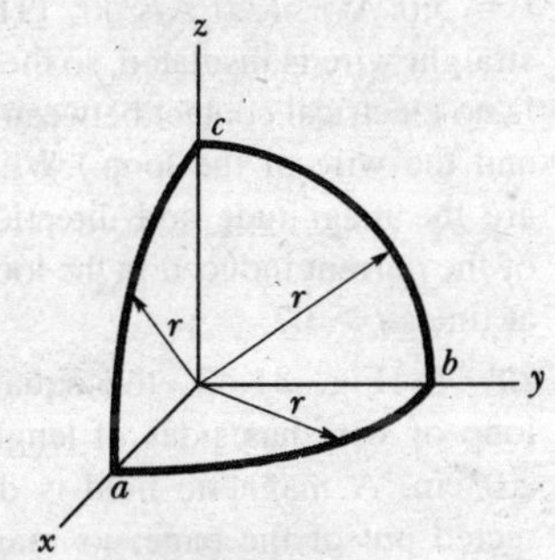

Fig. 31-39 Problem 16.

17P. A rectangular coil of N turns and of length a and width b is rotated at frequency f in a uniform magnetic field $\vec{B}$, as indicated in Fig. 31-40. The coil is connected to co-rotating cylinders, against which metal brushes slide to make contact. (a) Show that the emf induced in the coil is given (as a function of time t) by

$$\mathscr{E} = 2\pi f NabB \sin(2\pi f t) = \mathscr{E}_0 \sin(2\pi f t).$$

This is the principle of the commercial alternating-current generator. (b) Design a loop that will produce an emf with $\mathscr{E}_0 = 150$ V when rotated at 60.0 rev/s in a uniform magnetic field of 0.500 T. ssm www

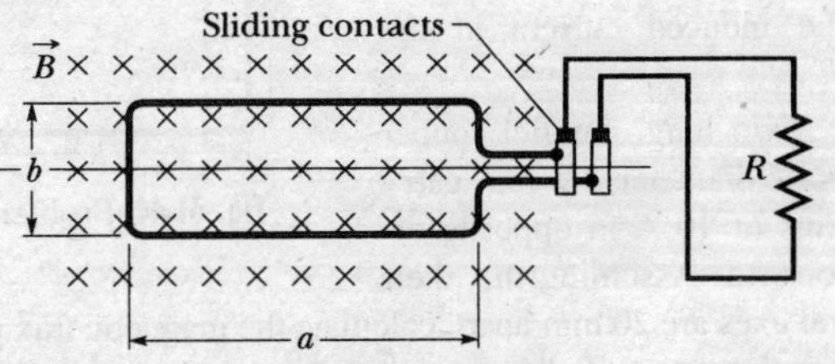

Fig. 31-40 Problem 17.

18P. A stiff wire bent into a semicircle of radius a is rotated with frequency f in a uniform magnetic field, as suggested in Fig. 31-41. What are (a) the frequency and (b) the amplitude of the varying emf induced in the loop?

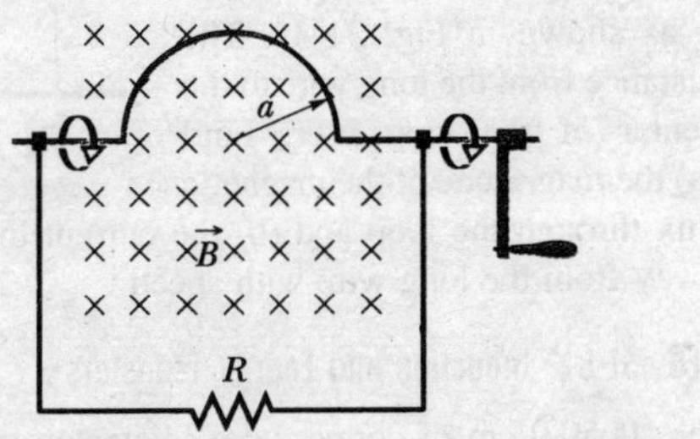

Fig. 31-41 Problem 18

19P. An electric generator consists of 100 turns of wire formed into a rectangular loop 50.0 cm by 30.0 cm, placed entirely in a uniform magnetic field with magnitude $B = 3.50$ T. What is the maximum value of the emf produced when the loop is spun at 1000 rev/min about an axis perpendicular to $\vec{B}$? ilw

20P. In Fig. 31-42, a wire forms a closed circular loop, with radius $R = 2.0$ m and resistance 4.0 Ω. The circle is centered on a long straight wire; at time $t = 0$, the current in the long straight wire is 5.0 A rightward. Thereafter, the current changes according to $i = 5.0\text{ A} - (2.0\text{ A/s}^2)t^2$. (The straight wire is insulated, so there is no electrical contact between it and the wire of the loop.) What are the magnitude and direction of the current induced in the loop at times $t > 0$?

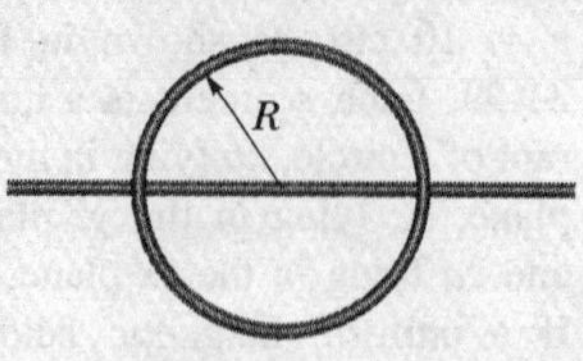

Fig. 31-42 Problem 20.

21P. In Fig. 31-43, the square loop of wire has sides of length 2.0 cm. A magnetic field is directed out of the page; its magnitude is given by $B = 4.0t^2y$, where B is in teslas, t is in seconds, and y is in meters. Determine the emf around the square at $t = 2.5$ s and give its direction. ssm ilw

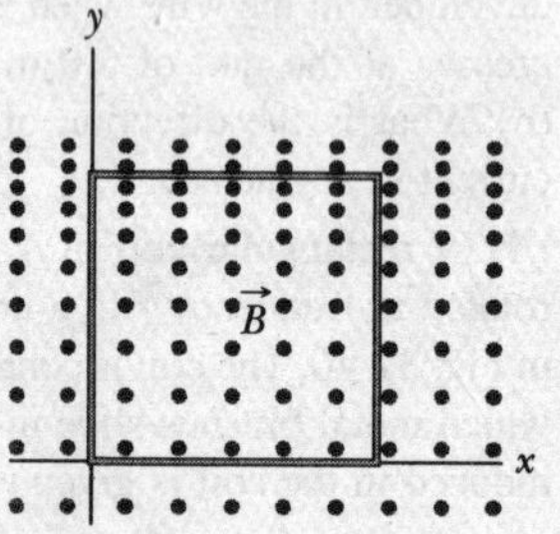

Fig. 31-43 Problem 21.

22P. For the situation shown in Fig. 31-44, $a = 12.0$ cm and $b = 16.0$ cm. The current in the long straight wire is given by $i = 4.50t^2 - 10.0t$, where i is in amperes and t is in seconds. (a) Find the emf in the square loop at $t = 3.00$ s. (b) What is the direction of the induced current in the loop?

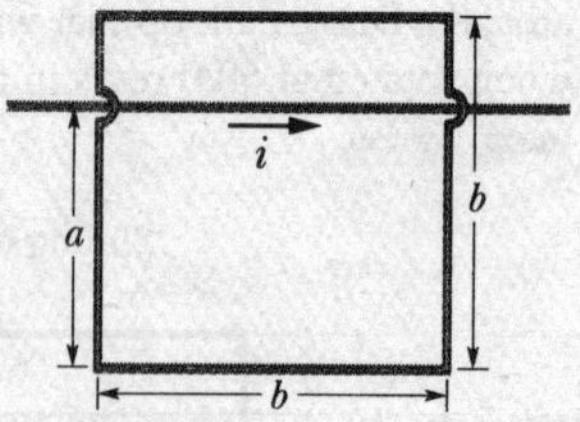

Fig. 31-44 Problem 22.

23P*. Two long, parallel copper wires of diameter 2.5 mm carry currents of 10 A in opposite directions. (a) Assuming that their central axes are 20 mm apart, calculate the magnetic flux per meter of wire that exists in the space between those axes. (b) What fraction of this flux lies inside the wires? (c) Repeat part (a) for parallel currents. ssm

24P. A rectangular loop of wire with length a, width b, and resistance R is placed near an infinitely long wire carrying current i, as shown in Fig. 31-45. The distance from the long wire to the center of the loop is r. Find (a) the magnitude of the magnetic flux through the loop and (b) the current in the loop as it moves away from the long wire with speed v.

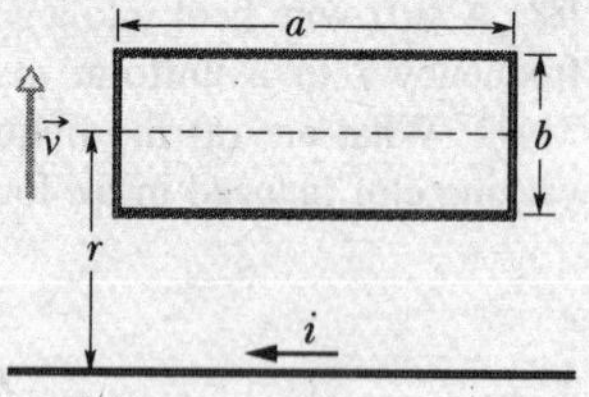

Fig. 31-45 Problem 24.

SEC. 31-5 Induction and Energy Transfers

25E. If 50.0 cm of copper wire (diameter = 1.00 mm) is formed into a circular loop and placed perpendicular to a uniform magnetic field that is increasing at the constant rate of 10.0 mT/s, at what rate is thermal energy generated in the loop? ssm ilw

26E. A loop antenna of area A and resistance R is perpendicular to a uniform magnetic field $\vec{B}$. The field drops linearly to zero in a time interval Δt. Find an expression for the total thermal energy dissipated in the loop.

27E. A metal rod is forced to move with constant velocity $\vec{v}$ along two parallel metal rails, connected with a strip of metal at one end, as shown in Fig. 31-46. A magnetic field $B = 0.350$ T points out of the page. (a) If the rails are separated by 25.0 cm and the speed of the rod is 55.0 cm/s, what emf is generated? (b) If the rod has a resistance of 18.0 Ω and the rails and connector have negligible resistance, what is the current in the rod? (c) At what rate is energy being transferred to thermal energy? ssm

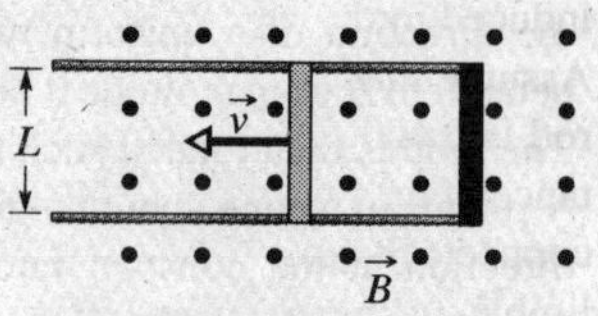

Fig. 31-46 Exercise 27 and Problem 29.

28P. In Fig. 31-47, a long rectangular conducting loop, of width L, resistance R, and mass m, is hung in a horizontal, uniform magnetic field $\vec{B}$ that is directed into the page and that exists only above line aa. The loop is then dropped; during its fall, it accelerates until it reaches a certain terminal speed v_t. Ignoring air drag, find that terminal speed.

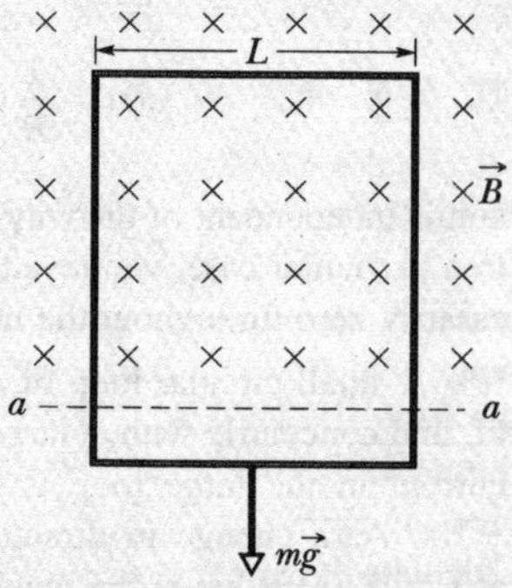

Fig. 31-47 Problem 28.

29P. The conducting rod shown in Fig. 31-46 has length L and is being pulled along horizontal, frictionless conducting rails at a constant velocity $\vec{v}$. The rails are connected at one end with a metal strip. A uniform magnetic field $\vec{B}$, directed out of the page, fills the region in which the rod moves. Assume that $L = 10$ cm, $v = 5.0$ m/s, and $B = 1.2$ T. (a) What are the magnitude and direction of the emf induced in the rod? (b) What is the current in the conducting loop? Assume that the resistance of the rod is 0.40 Ω and that the resistance of the rails and metal strip is negligibly small. (c) At what rate is thermal energy being generated in the rod? (d) What force must be applied to the rod by an external agent to maintain its motion? (e) At what rate does this external agent do work on the rod? Compare this answer with the answer to (c). ssm

30P. Two straight conducting rails form a right angle where their ends are joined. A conducting bar in contact with the rails starts at the vertex at time $t = 0$ and moves with a constant velocity of 5.20 m/s along them, as shown in Fig. 31-48. A magnetic field with $B = 0.350$ T is directed out of the page. Calculate (a) the flux through the triangle formed by the rails and bar at $t = 3.00$ s and (b) the emf around the triangle at that time. (c) If we write the emf as $\mathscr{E} = at^n$, where a and n are constants, what is the value of n?

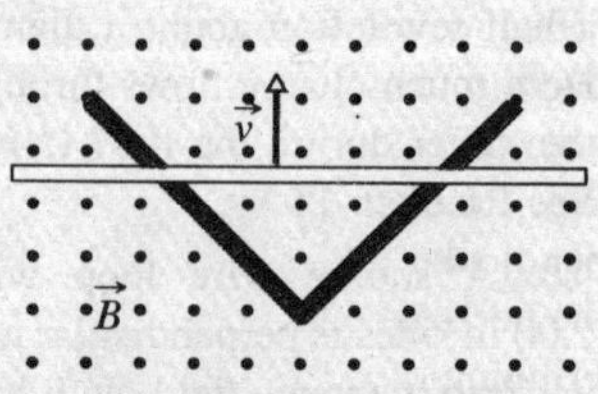

Fig. 31-48 Problem 30.

31P. Figure 31-49 shows a rod of length L caused to move at constant speed v along horizontal conducting rails. The magnetic field in which the rod moves is *not uniform* but is provided by a current

i in a long wire parallel to the rails. Assume that $v = 5.00$ m/s, $a = 10.0$ mm, $L = 10.0$ cm, and $i = 100$ A. (a) Calculate the emf induced in the rod. (b) What is the current in the conducting loop? Assume that the resistance of the rod is 0.400 Ω and that the resistance of the rails and the strip that connects them at the right is negligible. (c) At what rate is thermal energy being generated in the rod? (d) What force must be applied to the rod by an external agent to maintain its motion? (e) At what rate does this external agent do work on the rod? Compare this answer to that for (c). ssm

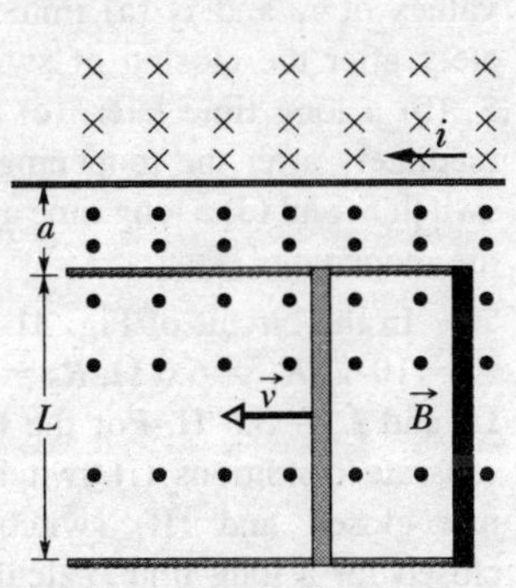

Fig. 31-49 Problem 31.

SEC. 31-6 Induced Electric Fields

32E. Figure 31-50 shows two circular regions R_1 and R_2 with radii $r_1 = 20.0$ cm and $r_2 = 30.0$ cm. In R_1 there is a uniform magnetic field $B_1 = 50.0$ mT into the page, and in R_2 there is a uniform magnetic field $B_2 = 75.0$ mT out of the page (ignore any fringing of these fields). Both fields are decreasing at the rate of 8.50 mT/s. Calculate the integral $\oint \vec{E} \cdot d\vec{s}$ for each of the three dashed paths.

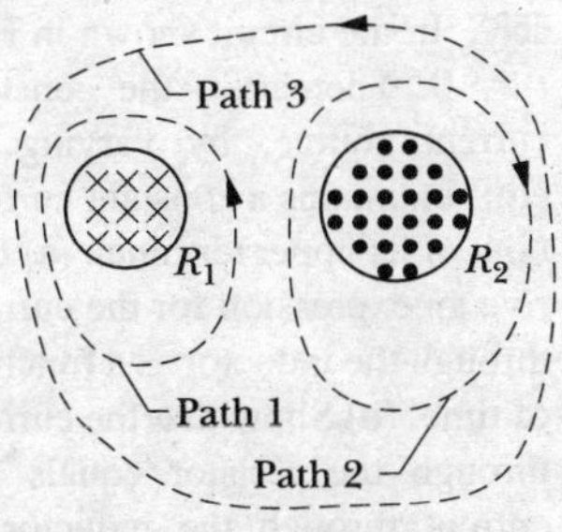

Fig. 31-50 Exercise 32.

33E. A long solenoid has a diameter of 12.0 cm. When a current i exists in its windings, a uniform magnetic field $B = 30.0$ mT is produced in its interior. By decreasing i, the field is caused to decrease at the rate of 6.50 mT/s. Calculate the magnitude of the induced electric field (a) 2.20 cm and (b) 8.20 cm from the axis of the solenoid. ssm ilw

34P. Early in 1981 the Francis Bitter National Magnet Laboratory at M.I.T. commenced operation of a 3.3-cm-diameter cylindrical magnet, which produces a 30 T field, then the world's largest steady-state field. The field can be varied sinusoidally between the limits of 29.6 and 30.0 T at a frequency of 15 Hz. When this is done, what is the maximum value of the induced electric field at a radial distance of 1.6 cm from the axis? (*Hint:* See Sample Problem 31-4.)

35P. Prove that the electric field $\vec{E}$ in a charged parallel-plate capacitor cannot drop abruptly to zero (as is suggested at point a in Fig. 31-51), as one moves perpendicular to the field, say, along the horizontal arrow in the figure. Fringing of the field lines always occurs in actual capacitors, which means that $\vec{E}$ approaches zero in a continuous and gradual way (see Problem 33 in Chapter 30). (*Hint:* Apply Faraday's law to the rectangular path shown by the dashed lines.) ssm

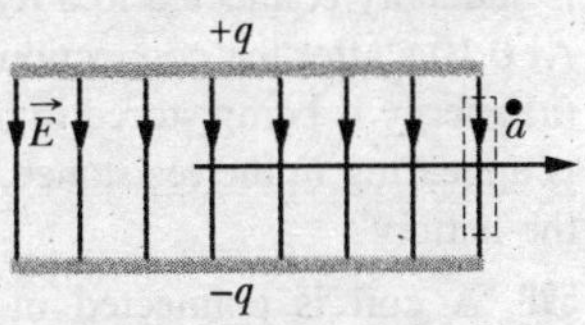

Fig. 31-51 Problem 35.

SEC. 31-7 Inductors and Inductance

36E. A circular coil has a 10.0 cm radius and consists of 30.0 closely wound turns of wire. An externally produced magnetic field of 2.60 mT is perpendicular to the coil. (a) If no current is in the coil, what magnetic flux links its turns? (b) When the current in the coil is 3.80 A in a certain direction, the net flux through the coil is found to vanish. What is the inductance of the coil?

37E. The inductance of a close-packed coil of 400 turns is 8.0 mH. Calculate the magnetic flux through the coil when the current is 5.0 mA. ssm

38P. A wide copper strip of width W is bent to form a tube of radius R with two parallel planar extensions, as shown in Fig. 31-52. There is a current i through the strip, distributed uniformly over its width. In this way a "one-turn solenoid" is formed. (a) Derive an expression for the magnitude of the magnetic field $\vec{B}$ in the tubular part (far away from the edges). (*Hint:* Assume that the magnetic field outside this one-turn solenoid is negligibly small.) (b) Find the inductance of this one-turn solenoid, neglecting the two planar extensions.

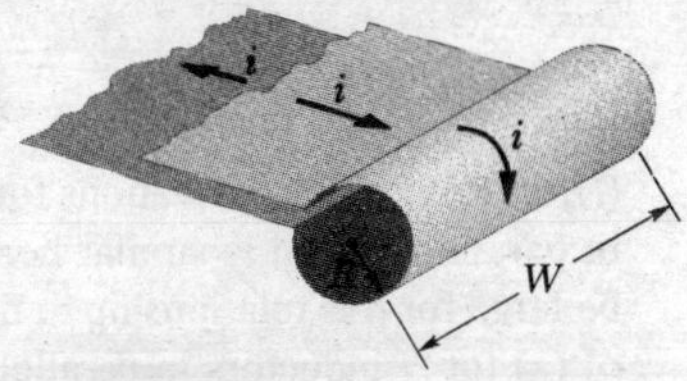

Fig. 31-52 Problem 38.

39P. Two long parallel wires, both of radius a and whose centers are a distance d apart, carry equal currents in opposite directions. Show that, neglecting the flux within the wires, the inductance of a length l of such a pair of wires is given by

$$L = \frac{\mu_0 l}{\pi} \ln \frac{d - a}{a}.$$

(*Hint:* Calculate the flux through a rectangle of which the wires form two opposite sides.) ssm www

SEC. 31-8 Self-Induction

40E. At a given instant the current and self-induced emf in an inductor are directed as indicated in Fig. 31-53. (a) Is the current increasing or decreasing? (b) The induced emf is 17 V and the rate of change of the current is 25 kA/s; find the inductance.

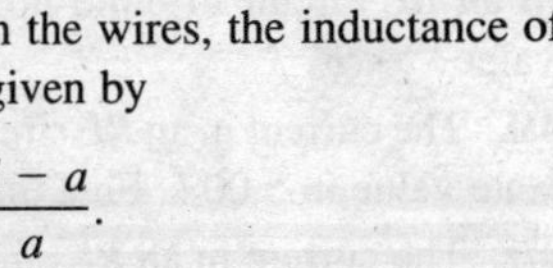

Fig. 31-53 Exercise 40.

41E. A 12 H inductor carries a steady current of 2.0 A. How can a 60 V self-induced emf be made to appear in the inductor? ssm

42P. The current i through a 4.6 H inductor varies with time t as shown by the graph of Fig. 31-54. The inductor has a resistance of

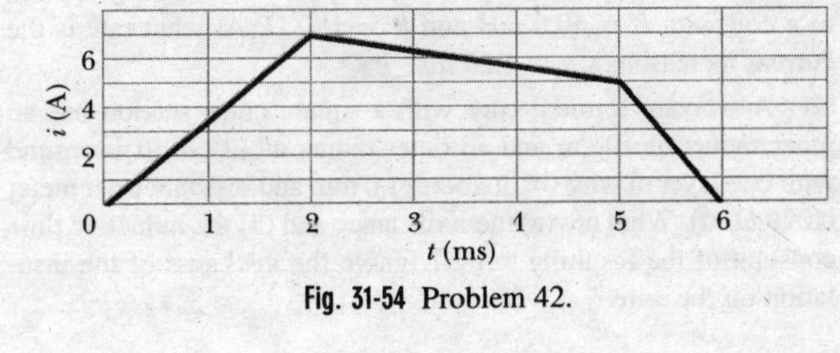

Fig. 31-54 Problem 42.

12 Ω. Find the magnitude of the induced emf $\mathscr{E}$ during the time intervals (a) $t = 0$ to $t = 2$ ms, (b) $t = 2$ ms to $t = 5$ ms, (c) $t = 5$ ms to $t = 6$ ms. (Ignore the behavior at the ends of the intervals.)

43P. *Inductors in series.* Two inductors L_1 and L_2 are connected in series and are separated by a large distance. (a) Show that the equivalent inductance is given by

$$L_{eq} = L_1 + L_2.$$

(*Hint:* Review the derivations for resistors in series and capacitors in series. Which is similar here?) (b) Why must their separation be large for this relationship to hold? (c) What is the generalization of (a) for N inductors in series?

44P. *Inductors in parallel.* Two inductors L_1 and L_2 are connected in parallel and separated by a large distance. (a) Show that the equivalent inductance is given by

$$\frac{1}{L_{eq}} = \frac{1}{L_1} + \frac{1}{L_2}.$$

(*Hint:* Review the derivations for resistors in parallel and capacitors in parallel. Which is similar here?) (b) Why must their separation be large for this relationship to hold? (c) What is the generalization of (a) for N inductors in parallel?

SEC. 31-9 *RL* Circuits

45E. In terms of τ_L, how long must we wait for the current in an *RL* circuit to build up to within 0.100% of its equilibrium value? ssm

46E. The current in an *RL* circuit builds up to one-third of its steady-state value in 5.00 s. Find the inductive time constant.

47E. The current in an *RL* circuit drops from 1.0 A to 10 mA in the first second following removal of the battery from the circuit. If L is 10 H, find the resistance R in the circuit. ilw

48E. Consider the *RL* circuit of Fig. 31-17. In terms of the battery emf $\mathscr{E}$, (a) what is the self-induced emf $\mathscr{E}_L$ when the switch has just been closed on a, and (b) what is $\mathscr{E}_L$ when $t = 2.0\tau_L$? (c) In terms of τ_L, when will $\mathscr{E}_L$ be just one-half the battery emf $\mathscr{E}$?

49E. A solenoid having an inductance of 6.30 μH is connected in series with a 1.20 kΩ resistor. (a) If a 14.0 V battery is switched across the pair, how long will it take for the current through the resistor to reach 80.0% of its final value? (b) What is the current through the resistor at time $t = 1.0\tau_L$? ssm

50P. Suppose the emf of the battery in the circuit of Fig. 31-18 varies with time t so that the current is given by $i(t) = 3.0 + 5.0t$, where i is in amperes and t is in seconds. Take $R = 4.0\ \Omega$ and $L = 6.0$ H, and find an expression for the battery emf as a function of time. (*Hint:* Apply the loop rule.)

51P. At time $t = 0$, a 45.0 V potential difference is suddenly applied to a coil with $L = 50.0$ mH and $R = 180\ \Omega$. At what rate is the current increasing at $t = 1.20$ ms? ilw

52P. A wooden toroidal core with a square cross section has an inner radius of 10 cm and an outer radius of 12 cm. It is wound with one layer of wire (of diameter 1.0 mm and resistance per meter 0.020 Ω/m). What are (a) the inductance and (b) the inductive time constant of the resulting toroid? Ignore the thickness of the insulation on the wire.

53P. In Fig. 31-55, $\mathscr{E} = 100$ V, $R_1 = 10.0\ \Omega$, $R_2 = 20.0\ \Omega$, $R_3 = 30.0\ \Omega$, and $L = 2.00$ H. Find the values of i_1 and i_2 (a) immediately after the closing of switch S, (b) a long time later, (c) immediately after the reopening of switch S, and (d) a long time after the reopening. ssm

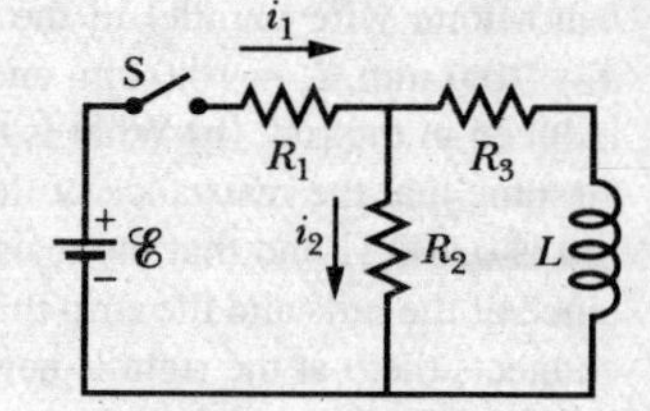

Fig. 31-55 Problem 53.

54P. In the circuit of Fig. 31-56, $\mathscr{E} = 10$ V, $R_1 = 5.0\ \Omega$, $R_2 = 10\ \Omega$, and $L = 5.0$ H. For the two separate conditions (I) switch S just closed and (II) switch S closed for a long time, calculate (a) the current i_1 through R_1, (b) the current i_2 through R_2, (c) the current i through the switch, (d) the potential difference across R_2, (e) the potential difference across L, and (f) the rate of change di_2/dt.

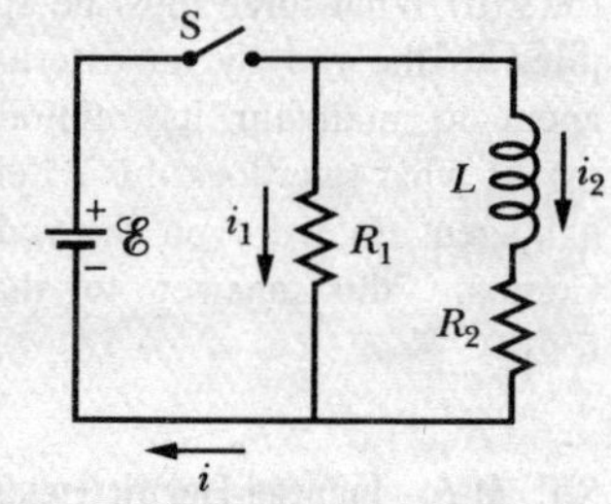

Fig. 31-56 Problem 54.

55P*. In the circuit shown in Fig. 31-57, switch S is closed at time $t = 0$. Thereafter, the constant current source, by varying its emf, maintains a constant current i out of its upper terminal. (a) Derive an expression for the current through the inductor as a function of time. (b) Show that the current through the resistor equals the current through the inductor at time $t = (L/R) \ln 2$. ssm www

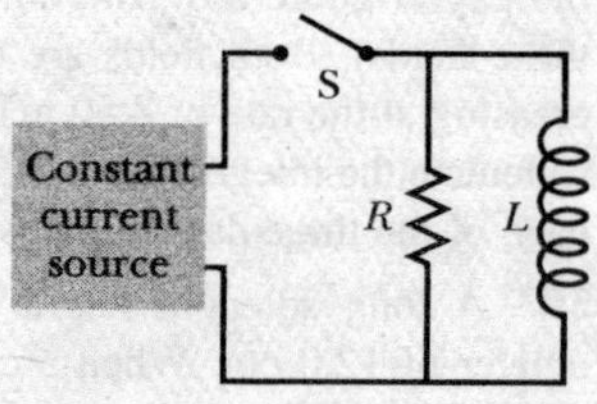

Fig. 31-57 Problem 55.

SEC. 31-10 Energy Stored in a Magnetic Field

56E. Consider the circuit of Fig. 31-18. In terms of the inductive time constant, at what instant after the battery is connected will the energy stored in the magnetic field of the inductor be half its steady-state value?

57E. Suppose that the inductive time constant for the circuit of Fig. 31-18 is 37.0 ms and the current in the circuit is zero at time $t = 0$. At what time does the rate at which energy is dissipated in the resistor equal the rate at which energy is being stored in the inductor? ilw

58E. A coil with an inductance of 2.0 H and a resistance of 10 Ω is suddenly connected to a resistanceless battery with $\mathscr{E} = 100$ V. At 0.10 s after the connection is made, what are the rates at which (a) energy is being stored in the magnetic field, (b) thermal energy is appearing in the resistance, and (c) energy is being delivered by the battery?

59P. A coil is connected in series with a 10.0 kΩ resistor. A 50.0 V battery is applied across the two devices, and the current reaches a value of 2.00 mA after 5.00 ms. (a) Find the inductance of the coil. (b) How much energy is stored in the coil at this same moment? ssm

60P. For the circuit of Fig. 31-18, assume that $\mathscr{E} = 10.0$ V, $R =$

6.70 Ω, and $L = 5.50$ H. The battery is connected at time $t = 0$. (a) How much energy is delivered by the battery during the first 2.00 s? (b) How much of this energy is stored in the magnetic field of the inductor? (c) How much of this energy is dissipated in the resistor?

61P. Prove that, after switch S in Fig. 31-17 has been thrown from a to b, all the energy stored in the inductor will ultimately appear as thermal energy in the resistor. ssm

SEC. 31-11 Energy Density of a Magnetic Field

62E. A toroidal inductor with an inductance of 90.0 mH encloses a volume of 0.0200 m^3. If the average energy density in the toroid is 70.0 J/m^3, what is the current through the inductor?

63E. A solenoid that is 85.0 cm long has a cross-sectional area of 17.0 cm^2. There are 950 turns of wire carrying a current of 6.60 A. (a) Calculate the energy density of the magnetic field inside the solenoid. (b) Find the total energy stored in the magnetic field there (neglect end effects). ssm

64E. The magnetic field in the interstellar space of our galaxy has a magnitude of about 10^{-10} T. How much energy is stored in this field in a cube 10 light-years on edge? (For scale, note that the nearest star is 4.3 light-years distant and the radius of our galaxy is about 8×10^4 light-years.)

65E. What must be the magnitude of a uniform electric field if it is to have the same energy density as that possessed by a 0.50 T magnetic field? Itw

66E. A circular loop of wire 50 mm in radius carries a current of 100 A. (a) Find the magnetic field strength at the center of the loop. (b) Calculate the energy density at the center of the loop.

67P. A length of copper wire carries a current of 10 A, uniformly distributed through its cross section. Calculate the energy density of (a) the magnetic field and (b) the electric field at the surface of the wire. The wire diameter is 2.5 mm, and its resistance per unit length is 3.3 Ω/km. ssm www

SEC. 31-12 Mutual Induction

68E. Coil 1 has $L_1 = 25$ mH and $N_1 = 100$ turns. Coil 2 has $L_2 = 40$ mH and $N_2 = 200$ turns. The coils are rigidly positioned with respect to each other; their mutual inductance M is 3.0 mH. A 6.0 mA current in coil 1 is changing at the rate of 4.0 A/s. (a) What magnetic flux Φ_{12} links coil 1, and what self-induced emf appears there? (b) What magnetic flux Φ_{21} links coil 2, and what mutually induced emf appears there?

69E. Two coils are at fixed locations. When coil 1 has no current and the current in coil 2 increases at the rate 15.0 A/s, the emf in coil 1 is 25.0 mV. (a) What is their mutual inductance? (b) When coil 2 has no current and coil 1 has a current of 3.60 A, what is the flux linkage in coil 2? ssm

70E. Two solenoids are part of the spark coil of an automobile. When the current in one solenoid falls from 6.0 A to zero in 2.5 ms, an emf of 30 kV is induced in the other solenoid. What is the mutual inductance M of the solenoids?

71P. Two coils, connected as shown in Fig. 31-58, separately have inductances L_1 and L_2. Their mutual inductance is M. (a) Show that this combination can be replaced by a single coil of equivalent inductance given by

$$L_{eq} = L_1 + L_2 + 2M.$$

(b) How could the coils in Fig. 31-58 be reconnected to yield an equivalent inductance of

$$L_{eq} = L_1 + L_2 - 2M?$$

(This problem is an extension of Problem 43, but the requirement that the coils be far apart has been removed.) ssm

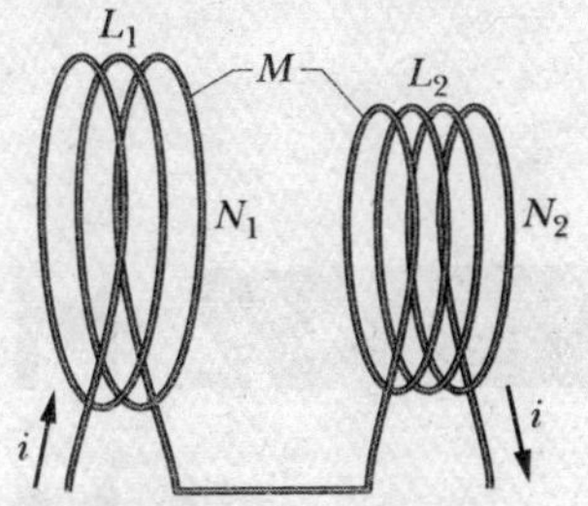

Fig. 31-58 Problem 71.

72P. A coil C of N turns is placed around a long solenoid S of radius R and n turns per unit length, as in Fig. 31-59. Show that the mutual inductance for the coil–solenoid combination is given by $M = \mu_0 \pi R^2 nN$. Explain why M does not depend on the shape, size, or possible lack of close-packing of the coil.

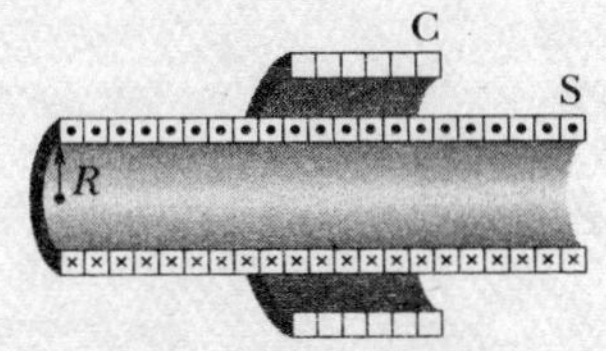

Fig. 31-59 Problem 72.

73P. Figure 31-60 shows, in cross section, two coaxial solenoids. Show that the mutual inductance M for a length l of this solenoid–solenoid combination is given by $M = \pi R_1^2 l \mu_0 n_1 n_2$, in which n_1 and n_2 are the respective numbers of turns per unit length and R_1 is the radius of the inner solenoid. Why does M depend on R_1 and not on R_2? ssm

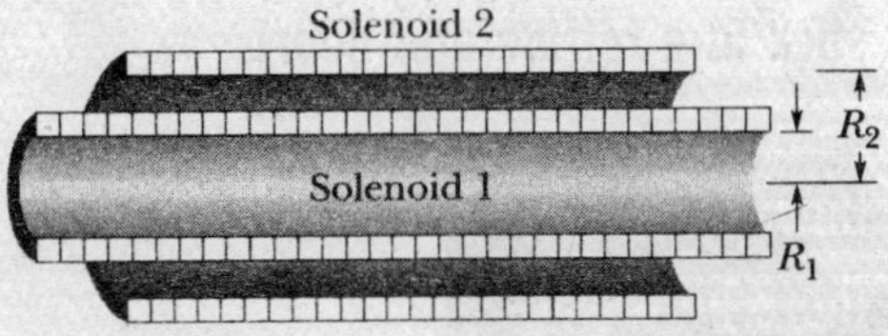

Fig. 31-60 Problem 73.

74P. Figure 31-61 shows a coil of N_2 turns wound as shown around part of a toroid of N_1 turns. The toroid's inner radius is a, its outer radius is b, and its height is h. Show that the mutual inductance M for the toroid–coil combination is

$$M = \frac{\mu_0 N_1 N_2 h}{2\pi} \ln \frac{b}{a}.$$

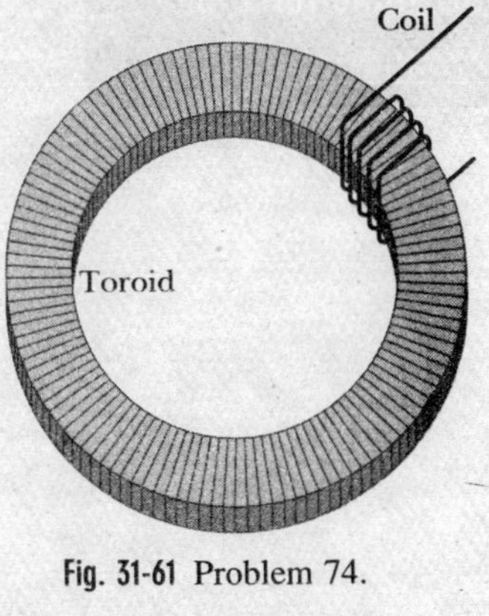

Fig. 31-61 Problem 74.

75P. A rectangular loop of N close-packed turns is positioned near a long straight wire as shown in Fig. 31-62. (a) What is the mutual inductance M for the loop–wire combination? (b) Evaluate M for $N = 100$, $a = 1.0$ cm, $b = 8.0$ cm, and $l = 30$ cm. Itw

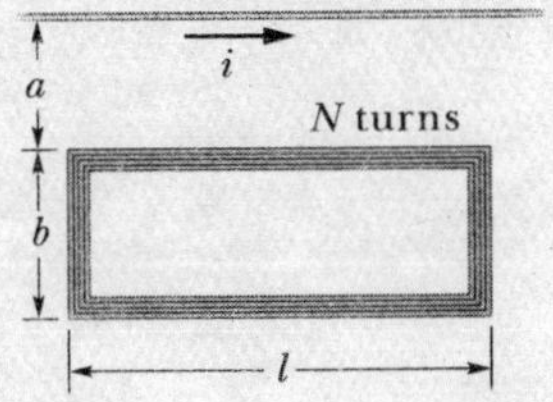

Fig. 31-62 Problem 75.

32 Magnetism of Matter: Maxwell's Equations

This is an overhead view of a frog that is being levitated in a magnetic field produced by current in a vertical solenoid below the frog. The solenoid's upward magnetic force on the frog balances the downward gravitational force on the frog. (The frog is not in discomfort; the sensation is like floating in water, which frogs like very much.) However, a frog is not magnetic (it would not, for example, stick to a refrigerator door).

How, then, can there be a magnetic force on the frog?

The answer is in this chapter.

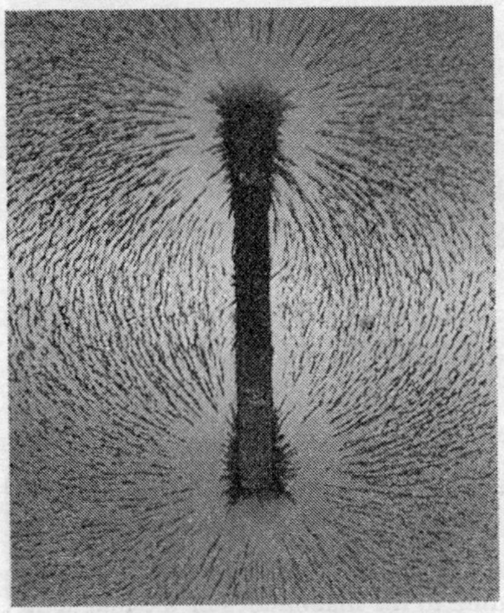

Fig. 32-1 A bar magnet is a magnetic dipole. The iron filings suggest the magnetic field lines. (The background is illuminated with colored light.)

32-1 Magnets

The first known magnets were *lodestones*, which are stones that have been *magnetized* (made magnetic) naturally. When the ancient Greeks and ancient Chinese discovered these rare stones, they were amused by the stones' ability to attract metal over a short distance, as if by magic. Only much later did they learn to use lodestones (and artificially magnetized pieces of iron) in compasses to determine direction.

Today, magnets and magnetic materials are ubiquitous. We find them in VCRs, audio cassettes, ATM and credit cards, audio headsets, and even in the inks for paper money. In fact, some breakfast cereals that are "iron fortified" contain small bits of magnetic materials (you can collect them from a slurry of cereal and water with a magnet). More important, the modern electronics industry as we know it (including the music and information sectors) would not exist without magnetic materials.

The magnetic properties of materials can be traced back to their atoms and electrons. We begin here, however, with the bar magnet in Fig. 32-1. As you have seen, iron filings sprinkled around such a magnet tend to align with the magnetic field of the magnet, and their pattern reveals the magnetic field lines. The clustering of the lines at the ends of the magnet suggests that one end is a *source* of the lines (the field diverges from it) and the other end is a *sink* of the lines (the field converges toward it). By convention, we call the source the *north pole* of the magnet and the opposite end the *south pole*, and we say that the magnet, with its two poles, is an example of a **magnetic dipole.**

Suppose we break apart a bar magnet the way we break a piece of chalk (Fig. 32-2). We should, it seems, be able to isolate a single pole, or *monopole*. However, we cannot—not even if we break the magnet down to its individual atoms and then to its electrons and nuclei. Each fragment has a north pole and a south pole. Thus:

> The simplest magnetic structure that can exist is a magnetic dipole. Magnetic monopoles do not exist (as far as we know).

32-2 Gauss' Law for Magnetic Fields

Gauss' law for magnetic fields is a formal way of saying that magnetic monopoles do not exist. The law asserts that the net magnetic flux Φ_B through any closed Gaussian surface is zero:

$$\Phi_B = \oint \vec{B} \cdot d\vec{A} = 0 \quad \text{(Gauss' law for magnetic fields).} \tag{32-1}$$

Contrast this with Gauss' law for electric fields,

$$\Phi_E = \oint \vec{E} \cdot d\vec{A} = \frac{q_{\text{enc}}}{\varepsilon_0} \quad \text{(Gauss' law for electric fields).}$$

In both equations, the integral is taken over a *closed* Gaussian surface. Gauss' law for electric fields says that this integral (the net electric flux through the surface) is proportional to the net electric charge q_{enc} enclosed by the surface. Gauss' law for magnetic fields says that there can be no net magnetic flux through the surface because there can be no net "magnetic charge" (individual magnetic poles) enclosed by the surface. The simplest magnetic structure that can exist and thus be enclosed by a Gaussian surface is a dipole, which consists of both a source and a sink for the field lines. Thus, there must always be as much magnetic flux into the surface as out of it, and the net magnetic flux must always be zero.

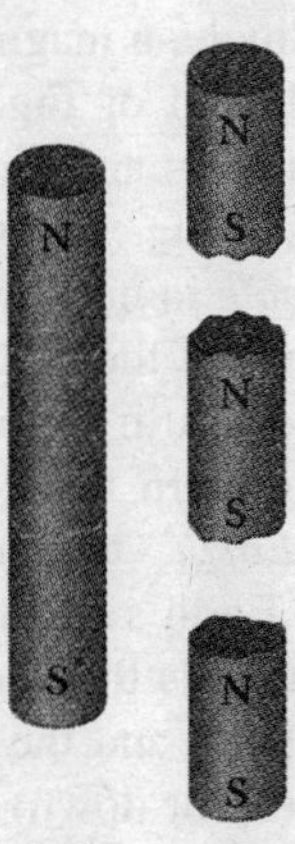

Fig. 32-2 If you break a magnet, each fragment becomes a separate magnet, with its own north and south poles.

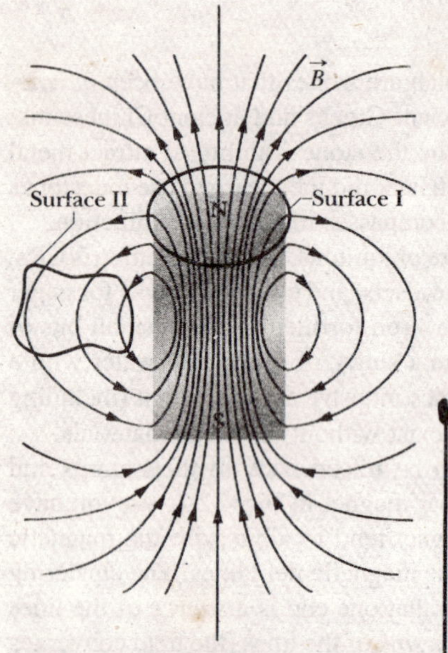

Fig. 32-3 The field lines for the magnetic field $\vec{B}$ of a short bar magnet. The red curves represent cross sections of closed, three-dimensional Gaussian surfaces.

Gauss' law for magnetic fields holds for more complicated structures than a magnetic dipole, and it holds even if the Gaussian surface does not enclose the entire structure. Gaussian surface II near the bar magnet of Fig. 32-3 encloses no poles, and we can easily conclude that the net magnetic flux through it is zero. Gaussian surface I is more difficult. It may seem to enclose only the north pole of the magnet because it encloses the label N and not the label S. However, a south pole must be associated with the lower boundary of the surface, because magnetic field lines enter the surface there. (The enclosed section is like one piece of the broken bar magnet in Fig. 32-2.) Thus, Gaussian surface I encloses a magnetic dipole and the net flux through the surface is zero.

CHECKPOINT 1: The figure here shows four closed surfaces with flat top and bottom faces and curved sides. The table gives the areas A of the faces and the magnitudes B of the uniform and perpendicular magnetic fields through those faces; the units of A and B are arbitrary but consistent. Rank the surfaces according to the magnitudes of the magnetic flux through their curved sides, greatest first.

Surface	A_{top}	B_{top}	A_{bot}	B_{bot}
a	2	6, outward	4	3, inward
b	2	1, inward	4	2, inward
c	2	6, inward	2	8, outward
d	2	3, outward	3	2, outward

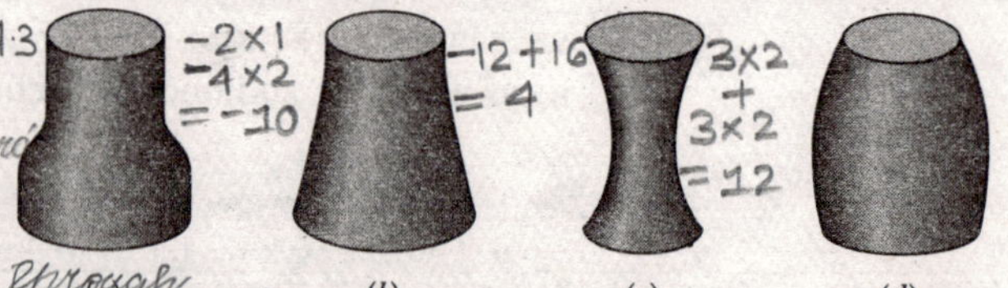

32-3 The Magnetism of Earth

Earth is a huge magnet; for points near Earth's surface, its magnetic field can be approximated as the field of a huge bar magnet—a magnetic dipole—that straddles the center of the planet. Figure 32-4 is an idealized symmetric depiction of the dipole field, without the distortion caused by passing charged particles from the Sun.

Because Earth's magnetic field is that of a magnetic dipole, a magnetic dipole moment $\vec{\mu}$ is associated with the field. For the idealized field of Fig. 32-4, the magnitude of $\vec{\mu}$ is 8.0×10^{22} J/T and the direction of $\vec{\mu}$ makes an angle of 11.5° with the rotation axis (RR) of Earth. The *dipole axis* (MM in Fig. 32-4) lies along $\vec{\mu}$ and intersects Earth's surface at the *geomagnetic north pole* in northwest Greenland and the *geomagnetic south pole* in Antarctica. The lines of the magnetic field $\vec{B}$ generally emerge in the southern hemisphere and reenter Earth in the northern hemisphere. Thus, the magnetic pole that is in Earth's northern hemisphere and known as a "north magnetic pole" *is really the south pole of Earth's magnetic dipole.*

The direction of the magnetic field at any location on Earth's surface is commonly specified in terms of two angles. The **field declination** is the angle (left or right) between geographic north (which is toward 90° latitude) and the horizontal component of the field. The **field inclination** is the angle (up or down) between a horizontal plane and the field's direction.

Magnetometers measure these angles and determine the field with much precision. However, you can do reasonably well with just a *compass* and a *dip meter*. A

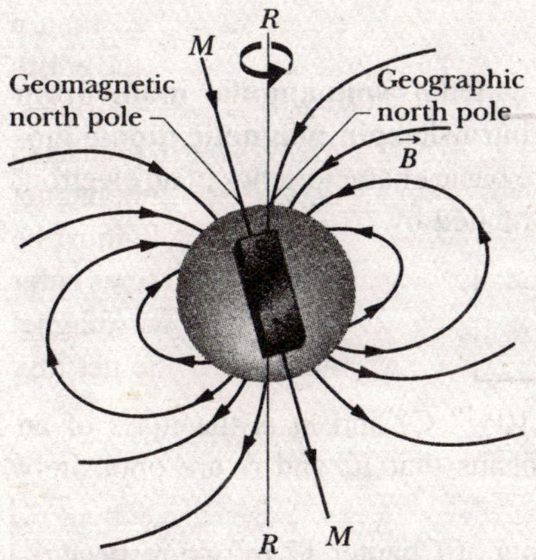

Fig. 32-4 Earth's magnetic field represented as a dipole field. The dipole axis *MM* makes an angle of 11.5° with Earth's rotational axis *RR*. The south pole of the dipole is in Earth's northern hemisphere.

compass is simply a needle-shaped magnet that is mounted so it can rotate freely about a vertical axis. When it is held in a horizontal plane, the north-pole end of the needle points, generally, toward the geomagnetic north pole (really a south magnetic pole, remember). The angle between the needle and geographic north is the field declination. A dip meter is a similar magnet that can rotate freely about a horizontal axis. When its vertical plane of rotation is aligned with the direction of the compass, the angle between the meter's needle and the horizontal is the field inclination.

At any point on Earth's surface, the measured magnetic field may differ appreciably, in both magnitude and direction, from the idealized dipole field of Fig. 32-4. In fact, the point where the field is actually perpendicular to Earth's surface and inward is not located at the geomagnetic north pole in Greenland as we would expect; instead, this so-called *dip north pole* is located in the Queen Elizabeth Islands in northern Canada, far from Greenland.

In addition, the field observed at any location on the surface of Earth varies with time, by measurable amounts over a period of a few years and by substantial amounts over, say, 100 years. For example, between 1580 and 1820 the direction indicated by compass needles in London changed by 35°.

In spite of these local variations, the average dipole field changes only slowly over such relatively short time periods. Variations over longer periods can be studied by measuring the weak magnetism of the ocean floor on either side of the Mid-Atlantic Ridge (Fig. 32-5). This floor has been formed by molten magma that oozed up through the ridge from Earth's interior, solidified, and was pulled away from the ridge (by the drift of tectonic plates) at the rate of a few centimeters per year. As the magma solidified, it became weakly magnetized with its magnetic field in the direction of Earth's magnetic field at the time of solidification. Study of this solidified magma across the ocean floor reveals that Earth's field has reversed its *polarity* (directions of the north pole and south pole) about every million years. The reason for the reversals is not known. In fact, the mechanism that produces Earth's magnetic field is only vaguely understood.

32-4 Magnetism and Electrons

Magnetic materials, from lodestones to videotapes, are magnetic because of the electrons within them. We have already seen one way in which electrons can generate a magnetic field: Send them through a wire as an electric current, and their motion produces a magnetic field around the wire. There are two more ways, each involving a magnetic dipole moment that produces a magnetic field in the surrounding space. However, their explanation requires quantum physics that is beyond the physics presented in this book, so here we shall only outline the results.

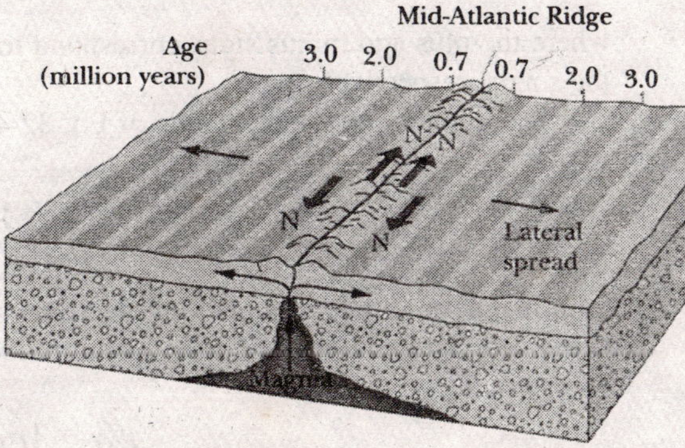

Fig. 32-5 A magnetic profile of the seafloor on either side of the Mid-Atlantic Ridge. The seafloor, extruded through the ridge and spreading out as part of the tectonic drift system, displays a record of the past magnetic history of Earth's core. The direction of the magnetic field produced by the core reverses about every million years.

Spin Magnetic Dipole Moment

An electron has an intrinsic angular momentum called its **spin angular momentum** (or just **spin**) $\vec{S}$; associated with this spin is an intrinsic **spin magnetic dipole moment** $\vec{\mu}_s$. (By *intrinsic*, we mean that $\vec{S}$ and $\vec{\mu}_s$ are basic characteristics of an electron, like its mass and electric charge.) $\vec{S}$ and $\vec{\mu}_s$ are related by

$$\vec{\mu}_s = -\frac{e}{m}\vec{S}, \tag{32-2}$$

in which e is the elementary charge (1.60×10^{-19} C) and m is the mass of an electron (9.11×10^{-31} kg). The minus sign means that $\vec{\mu}_s$ and $\vec{S}$ are oppositely directed.

Spin $\vec{S}$ is different from the angular momenta of Chapter 12 in two respects:

1. $\vec{S}$ itself cannot be measured. However, its component along any axis can be measured.
2. A measured component of $\vec{S}$ is *quantized*, which is a general term that means it is restricted to certain values. A measured component of $\vec{S}$ can have only two values, which differ only in sign.

Let us assume that the component of spin $\vec{S}$ is measured along the z axis of a coordinate system. Then the measured component S_z can have only the two values given by

$$S_z = m_s \frac{h}{2\pi}, \quad \text{for } m_s = \pm\tfrac{1}{2}, \tag{32-3}$$

where m_s is called the *spin magnetic quantum number* and h ($= 6.63 \times 10^{-34}$ J·s) is the Planck constant, the ubiquitous constant of quantum physics. The signs given in Eq. 32-3 have to do with the direction of S_z along the z axis. When S_z is parallel to the z axis, m_s is $+\frac{1}{2}$ and the electron is said to be *spin up*. When S_z is antiparallel to the z axis, m_s is $-\frac{1}{2}$ and the electron is said to be *spin down*.

The spin magnetic dipole moment $\vec{\mu}_s$ of an electron also cannot be measured; only its component along any axis can be measured, and that component too is quantized, with two possible values of the same magnitude but different signs. We can relate the component $\mu_{s,z}$ measured on the z axis to S_z by rewriting Eq. 32-2 in component form for the z axis as

$$\mu_{s,z} = -\frac{e}{m} S_z.$$

Substituting for S_z from Eq. 32-3 then gives us

$$\mu_{s,z} = \pm\frac{eh}{4\pi m}, \tag{32-4}$$

where the plus and minus signs correspond to $\mu_{s,z}$ being parallel and antiparallel to the z axis, respectively.

The quantity on the right side of Eq. 32-4 is called the *Bohr magneton* μ_B:

$$\mu_B = \frac{eh}{4\pi m} = 9.27 \times 10^{-24} \text{ J/T} \quad \text{(Bohr magneton).} \tag{32-5}$$

Spin magnetic dipole moments of electrons and other elementary particles can be expressed in terms of μ_B. For an electron, the magnitude of the measured z component of $\vec{\mu}_s$ is

$$\mu_{s,z} = 1\mu_B. \tag{32-6}$$

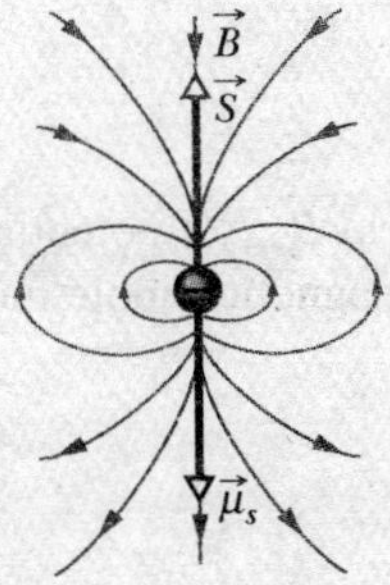

Fig. 32-6 The spin $\vec{S}$, spin magnetic dipole moment $\vec{\mu}_s$, and magnetic dipole field $\vec{B}$ of an electron represented as a microscopic sphere.

(The quantum physics of the electron, called *quantum electrodynamics*, or QED, reveals that $\mu_{s,z}$ is actually slightly greater than $1\,\mu_B$, but we shall neglect that fact.)

When an electron is placed in an external magnetic field $\vec{B}_{ext}$, a potential energy U can be associated with the orientation of the electron's spin magnetic dipole moment $\vec{\mu}_s$ just as a potential energy can be associated with the orientation of the magnetic dipole moment $\vec{\mu}$ of a current loop placed in $\vec{B}_{ext}$. From Eq. 29-38, the potential energy for the electron is

$$U = -\vec{\mu}_s \cdot \vec{B}_{ext} = -\mu_{s,z} B_{ext}, \tag{32-7}$$

where the z axis is taken to be in the direction of $\vec{B}_{ext}$.

If we imagine an electron to be a microscopic sphere (which it is not), we can represent the spin $\vec{S}$, the spin magnetic dipole moment $\vec{\mu}_s$, and the associated magnetic dipole field as in Fig. 32-6. Although we use the word "spin" here, electrons do not spin like tops. How, then, can something have angular momentum without actually rotating? Again, we would need quantum physics to provide the answer.

Protons and neutrons also have an intrinsic angular momentum called spin and an associated intrinsic spin magnetic dipole moment. For a proton those two vectors have the same direction, and for a neutron they have opposite directions. We shall not examine the contributions of these dipole moments to the magnetic fields of atoms because they are about a thousand times smaller than that due to an electron.

CHECKPOINT 2: The figure here shows the spin orientations of two particles in an external magnetic field $\vec{B}_{ext}$. (a) If the particles are electrons, which spin orientation is at lower potential energy? (b) If, instead, the particles are protons, which spin orientation is at lower potential energy?

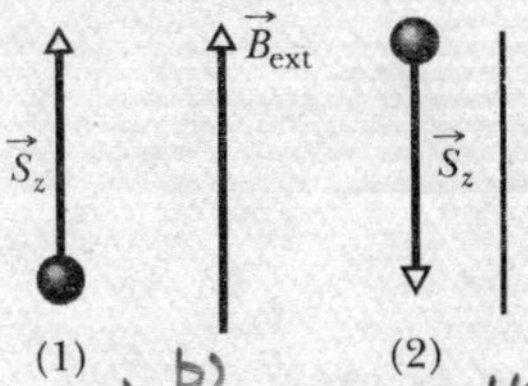

Orbital Magnetic Dipole Moment

When it is in an atom, an electron has an additional angular momentum called its **orbital angular momentum** $\vec{L}_{orb}$. Associated with $\vec{L}_{orb}$ is an **orbital magnetic dipole moment** $\vec{\mu}_{orb}$; the two are related by

$$\vec{\mu}_{orb} = -\frac{e}{2m}\vec{L}_{orb}. \tag{32-8}$$

The minus sign means that $\vec{\mu}_{orb}$ and $\vec{L}_{orb}$ have opposite directions.

Orbital angular momentum $\vec{L}_{orb}$ cannot be measured; only its component along any axis can be measured, and that component is quantized. The component along, say, a z axis can have only the values given by

$$L_{orb,z} = m_l \frac{h}{2\pi}, \quad \text{for } m_l = 0, \pm 1, \pm 2, \ldots, \pm(\text{limit}), \tag{32-9}$$

in which m_l is called the *orbital magnetic quantum number* and "limit" refers to some largest allowed integer value for m_l. The signs in Eq. 32-9 have to do with the direction of $L_{orb,z}$ along the z axis.

The orbital magnetic dipole moment $\vec{\mu}_{orb}$ of an electron also cannot itself be measured; only its component along an axis can be measured, and that component is again quantized. By writing Eq. 32-8 for a component along the same z axis as above and then substituting for $L_{orb,z}$ from Eq. 32-9, we can write the z component $\mu_{orb,z}$ of the orbital magnetic dipole moment as

$$\mu_{orb,z} = -m_l \frac{eh}{4\pi m} \tag{32-10}$$

and, in terms of the Bohr magneton, as

$$\mu_{orb,z} = -m_l \mu_B. \qquad (32\text{-}11)$$

When an atom is placed in an external magnetic field $\vec{B}_{ext}$, a potential energy U can be associated with the orientation of the orbital magnetic dipole moment of each electron in the atom. Its value is

$$U = -\vec{\mu}_{orb} \cdot \vec{B}_{ext} = -\mu_{orb,z} B_{ext}, \qquad (32\text{-}12)$$

where the z axis is taken in the direction of $\vec{B}_{ext}$.

Although we have used the words "orbit" and "orbital" here, electrons do not orbit the nucleus of an atom like planets orbiting the Sun. How can an electron have an orbital angular momentum without orbiting in the common meaning of the term? Once again, this can be explained only with quantum physics.

Loop Model for Electron Orbits

We can obtain Eq. 32-8 with the nonquantum derivation that follows, in which we assume that an electron moves along a circular path with a radius that is much larger than an atomic radius (hence the name "loop model"). However, the derivation does not apply to an electron within an atom (for which we need quantum physics).

We imagine an electron moving at constant speed v in a circular path of radius r, counterclockwise as shown in Fig. 32-7. The motion of the negative charge of the electron is equivalent to a conventional current i (of positive charge) that is clockwise, as also shown in Fig. 32-7. The magnitude of the orbital magnetic dipole moment of such a *current loop* is obtained from Eq. 29-35 with $N = 1$:

$$\mu_{orb} = iA, \qquad (32\text{-}13)$$

where A is the area enclosed by the loop. The direction of this magnetic dipole moment is, from the right-hand rule of Fig. 30-21, downward in Fig. 32-7.

To evaluate Eq. 32-13, we need the current i. Current is, generally, the rate at which charge passes some point in a circuit. Here, the charge of magnitude e takes a time $T = 2\pi r/v$ to circle from any point back through that point, so

$$i = \frac{\text{charge}}{\text{time}} = \frac{e}{2\pi r/v}. \qquad (32\text{-}14)$$

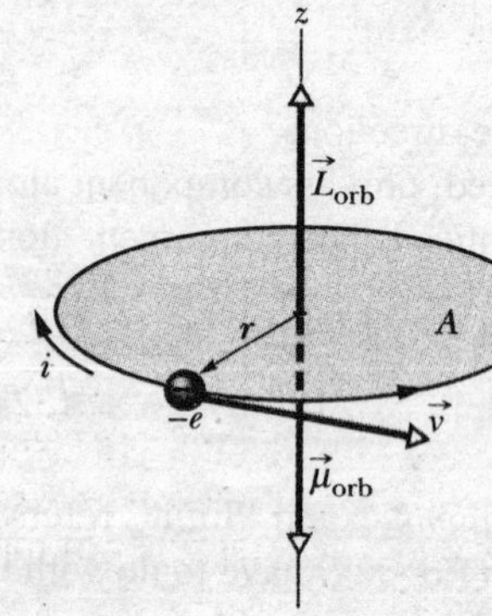

Fig. 32-7 An electron moving at constant speed v in a circular path of radius r that encloses an area A. The electron has an orbital angular momentum $\vec{L}_{orb}$ and an associated orbital magnetic dipole moment $\vec{\mu}_{orb}$. A clockwise current i (of positive charge) is equivalent to the counterclockwise circulation of the negatively charged electron.

Substituting this and the area $A = \pi r^2$ of the loop into Eq. 32-13 gives us

$$\mu_{orb} = \frac{e}{2\pi r/v}\,\pi r^2 = \frac{evr}{2}. \qquad (32\text{-}15)$$

To find the electron's orbital angular momentum $\vec{L}_{orb}$, we use Eq. 12-18, $\vec{\ell} = m(\vec{r} \times \vec{v})$. Because $\vec{r}$ and $\vec{v}$ are perpendicular, $\vec{L}_{orb}$ has the magnitude

$$L_{orb} = mrv \sin 90° = mrv. \qquad (32\text{-}16)$$

$\vec{L}_{orb}$ is directed upward in Fig. 32-7 (see Fig. 12-11). Combining Eqs. 32-15 and 32-16, generalizing to a vector formulation, and indicating the opposite directions of the vectors with a minus sign yield

$$\vec{\mu}_{orb} = -\frac{e}{2m}\vec{L}_{orb},$$

which is Eq. 32-8. Thus, by "classical" (nonquantum) analysis we have obtained the same result, in both magnitude and direction, given by quantum physics. You might

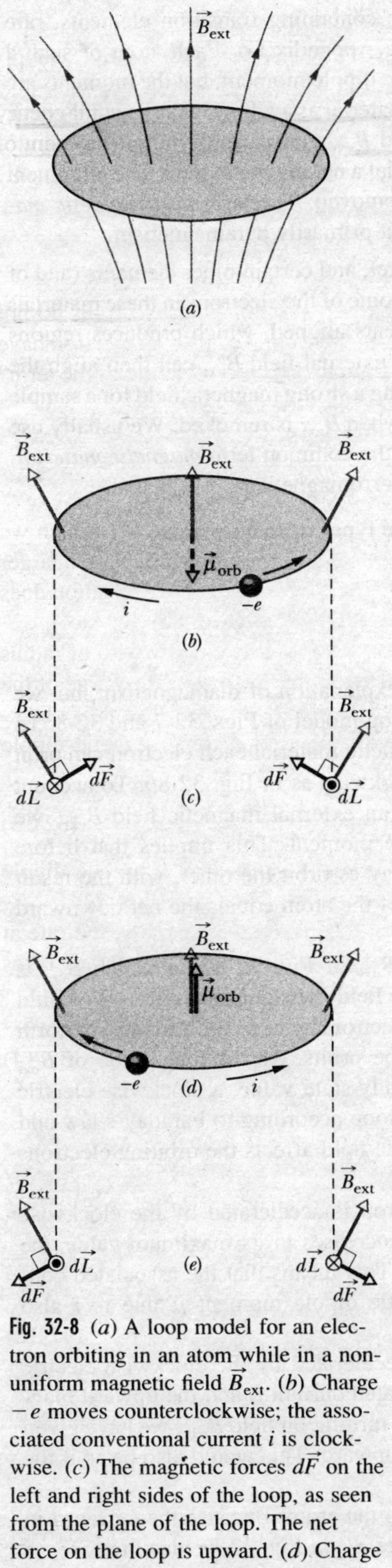

Fig. 32-8 (*a*) A loop model for an electron orbiting in an atom while in a nonuniform magnetic field $\vec{B}_{ext}$. (*b*) Charge $-e$ moves counterclockwise; the associated conventional current i is clockwise. (*c*) The magnetic forces $d\vec{F}$ on the left and right sides of the loop, as seen from the plane of the loop. The net force on the loop is upward. (*d*) Charge $-e$ now moves clockwise. (*e*) The net force on the loop is now downward.

wonder, since this derivation gives the correct result for an electron within an atom, why the derivation is invalid for that situation. The answer is that this line of reasoning yields other results that are contradicted by experiments.

Loop Model in a Nonuniform Field

We continue to consider an electron orbit as a current loop, as we did in Fig. 32-7. Now, however, we draw the loop in a nonuniform magnetic field $\vec{B}_{ext}$ as shown in Fig. 32-8*a*. (This field could be the diverging field near the north pole of the magnet in Fig. 32-3.) We make this change to prepare for the next several sections, in which we shall discuss the forces that act on magnetic materials when the materials are placed in a nonuniform magnetic field. We shall discuss these forces by assuming that the electron orbits in the materials are tiny current loops like that in Fig. 32-8*a*.

Here we assume that the magnetic field vectors all around the electron's circular path have the same magnitude and form the same angle with the vertical, as shown in Figs. 32-8*b* and *d*. We also assume that all the electrons in an atom move either counterclockwise (Fig. 32-8*b*) or clockwise (Fig. 32-8*d*). The associated conventional current i around the current loop and the orbital magnetic dipole moment $\vec{\mu}_{orb}$ produced by i are shown for each of these directions of motion.

Figures 32-8*c* and *e* show diametrically opposite views of a length element $d\vec{L}$ of the loop with the same direction as i, as seen from the plane of the orbit. Also shown are the field $\vec{B}_{ext}$ and the resulting magnetic force $d\vec{F}$ on $d\vec{L}$. Recall that a current along an element $d\vec{L}$ in a magnetic field $\vec{B}_{ext}$ experiences a magnetic force $d\vec{F}$ as given by Eq. 29-28:

$$d\vec{F} = i\,d\vec{L} \times \vec{B}_{ext}. \tag{32-17}$$

On the left side of Fig. 32-8*c*, Eq. 32-17 tells us that the force $d\vec{F}$ is directed upward and rightward. On the right side, the force $d\vec{F}$ is just as large and is directed upward and leftward. Because their angles are the same, the horizontal components of these two forces cancel and the vertical components add. The same is true at any other two symmetric points on the loop. Thus, the net force on the current loop of Fig. 32-8*b* must be upward. The same reasoning leads to a downward net force on the loop in Fig. 32-8*d*. We shall use these two results shortly when we examine the behavior of magnetic materials in nonuniform magnetic fields.

32-5 Magnetic Materials

Each electron in an atom has an orbital magnetic dipole moment and a spin magnetic dipole moment that combine vectorially. The resultant of these two vector quantities combines vectorially with similar resultants for all other electrons in the atom, and the resultant for each atom combines with those for all the other atoms in a sample of a material. If the combination of all these magnetic dipole moments produces a magnetic field, then the material is magnetic. There are three general types of magnetism: diamagnetism, paramagnetism, and ferromagnetism.

1. ***Diamagnetism*** is exhibited by all common materials but is so feeble that it is masked if the material also exhibits magnetism of either of the other two types. In diamagnetism, weak magnetic dipole moments are produced in the atoms of the material when the material is placed in an external magnetic field $\vec{B}_{ext}$; the combination of all those induced dipole moments gives the material as a whole only a feeble net magnetic field. The dipole moments and thus their net field disappear when $\vec{B}_{ext}$ is removed. The term *diamagnetic material* usually refers to materials that exhibit only diamagnetism.

2. ***Paramagnetism*** is exhibited by materials containing transition elements, rare earth elements, and actinide elements (see Appendix G). Each atom of such a material has a permanent resultant magnetic dipole moment, but the moments are randomly oriented in the material and the material as a whole lacks a net magnetic field. However, an external magnetic field $\vec{B}_{ext}$ can partially align the atomic magnetic dipole moments to give the material a net magnetic field. The alignment and thus its field disappear when $\vec{B}_{ext}$ is removed. The term *paramagnetic material* usually refers to materials that exhibit primarily paramagnetism.
3. ***Ferromagnetism*** is a property of iron, nickel, and certain other elements (and of compounds and alloys of these elements). Some of the electrons in these materials have their resultant magnetic dipole moments aligned, which produces regions with strong magnetic dipole moments. An external field $\vec{B}_{ext}$ can then align the magnetic moments of such regions, producing a strong magnetic field for a sample of the material; the field partially persists when $\vec{B}_{ext}$ is removed. We usually use the term *ferromagnetic material,* and even the common term *magnetic material,* to refer to materials that exhibit primarily ferromagnetism.

The next three sections explore these three types of magnetism.

32-6 Diamagnetism

We cannot yet discuss the quantum physical explanation of diamagnetism, but we can provide a classical explanation with the loop model of Figs. 32-7 and 32-8. To begin, we assume that in an atom of a diamagnetic material each electron can orbit only clockwise as in Fig. 32-8*d* or counterclockwise as in Fig. 32-8*b*. To account for the lack of magnetism in the absence of an external magnetic field $\vec{B}_{ext}$, we assume the atom lacks a net magnetic dipole moment. This implies that before $\vec{B}_{ext}$ is applied, as many electrons orbit one way as orbit the other, with the result that the net upward magnetic dipole moment of the atom equals the net downward magnetic dipole moment.

Now let's turn on the nonuniform field $\vec{B}_{ext}$ of Fig. 32-8*a*, in which $\vec{B}_{ext}$ is directed upward but is diverging (the magnetic field lines are diverging). We could do this by increasing the current through an electromagnet or by moving the north pole of a bar magnet closer to, and below, the orbits. As the magnitude of $\vec{B}_{ext}$ increases from zero to its final maximum, steady-state value, a clockwise electric field is induced around each electron's orbital loop according to Faraday's law and Lenz's law. Let us see how this induced electric field affects the orbiting electrons in Figs. 32-8*b* and *d*.

In Fig. 32-8*b*, the counterclockwise electron is accelerated by the clockwise electric field. Thus, as the magnetic field $\vec{B}_{ext}$ increases to its maximum value, the electron speed increases to a maximum value. This means that the associated conventional current i and the downward magnetic dipole moment $\vec{\mu}$ due to i also *increase.*

In Fig. 32-8*d*, the clockwise electron is decelerated by the clockwise electric field. Thus, here, the electron speed, the associated current i, and the upward magnetic dipole moment $\vec{\mu}$ due to i all *decrease.* By turning on field $\vec{B}_{ext}$, we have given the atom a *net* magnetic dipole moment that is upward. This would also be so if the magnetic field were uniform.

The nonuniformity of field $\vec{B}_{ext}$ also affects the atom. Because the current i in Fig. 32-8*b* increases, the upward magnetic forces $d\vec{F}$ in Fig. 32-8*c* also increase, as does the net upward force on the current loop. Because current i in Fig. 32-8*d* decreases, the downward magnetic forces $d\vec{F}$ in Fig. 32-8*e* also decrease, as does the net downward force on the current loop. Thus, by turning on the *nonuniform* field $\vec{B}_{ext}$, we have produced a net force on the atom; moreover, that force is directed *away* from the region of greater magnetic field.

We have argued with fictitious electron orbits (current loops), but we have ended up with exactly what happens to a diamagnetic material: If we apply the magnetic field of Fig. 32-8, the material develops a downward magnetic dipole moment and experiences an upward force. When the field is removed, both the dipole moment and the force disappear. The external field need not be positioned as shown; similar arguments can be made for other orientations of $\vec{B}_{ext}$. In general,

> A diamagnetic material placed in an external magnetic field $\vec{B}_{ext}$ develops a magnetic dipole moment directed opposite $\vec{B}_{ext}$. If the field is nonuniform, the diamagnetic material is repelled *from* a region of greater magnetic field *toward* a region of lesser field.

The frog in the photograph opening this chapter is diamagnetic (as is any other animal). When the frog was placed in the diverging magnetic field near the top end of a vertical current-carrying solenoid, every atom in the frog was repelled upward, away from the region of stronger magnetic field at that end of the solenoid. The frog moved upward into weaker and weaker magnetic field until the upward magnetic force balanced the gravitational force on it, and there it hung in midair. If we built a solenoid that was large enough, we could similarly levitate a person in midair owing to the person's diamagnetism.

CHECKPOINT 3: The figure shows two diamagnetic spheres located near the south pole of a bar magnet. Are (a) the magnetic forces on the spheres and (b) the magnetic dipole moments of the spheres directed toward or away from the bar magnet? (c) Is the magnetic force on sphere 1 greater than, less than, or equal to that on sphere 2?

a) away
b) away
c) less
(less lines of B)

32-7 Paramagnetism

In paramagnetic materials, the spin and orbital magnetic dipole moments of the electrons in each atom do not cancel but add vectorially to give the atom a net (and permanent) magnetic dipole moment $\vec{\mu}$. In the absence of an external magnetic field, these atomic dipole moments are randomly oriented, and the net magnetic dipole moment of the material is zero. However, if a sample of the material is placed in an external magnetic field $\vec{B}_{ext}$, the magnetic dipole moments tend to line up with the field, which gives the sample a net magnetic dipole moment. This alignment with the external field is the opposite of what we saw with diamagnetic materials.

Liquid oxygen is suspended between the two pole faces of a magnet because the liquid is paramagnetic and is magnetically attracted to the magnet.

> A paramagnetic material placed in an external magnetic field $\vec{B}_{ext}$ develops a magnetic dipole moment in the direction of $\vec{B}_{ext}$. If the field is nonuniform, the paramagnetic material is attracted *toward* a region of greater magnetic field *from* a region of lesser field.

A paramagnetic sample with N atoms would have a magnetic dipole moment of magnitude $N\mu$ if alignment of its atomic dipoles were complete. However, random collisions of atoms due to thermal agitation transfer energy among them, disrupting their alignment and thus reducing the sample's magnetic dipole moment.

The importance of thermal agitation may be measured by comparing two energies. One, from Eq. 20-24, is the mean translational kinetic energy K ($= \frac{3}{2}kT$) of an atom at temperature T, where k is the Boltzmann constant (1.38×10^{-23} J/K) and T is in kelvins (not Celsius degrees). The other, from Eq. 29-38, is the difference in energy ΔU_B ($= 2\mu B_{ext}$) between parallel alignment and antiparallel alignment of

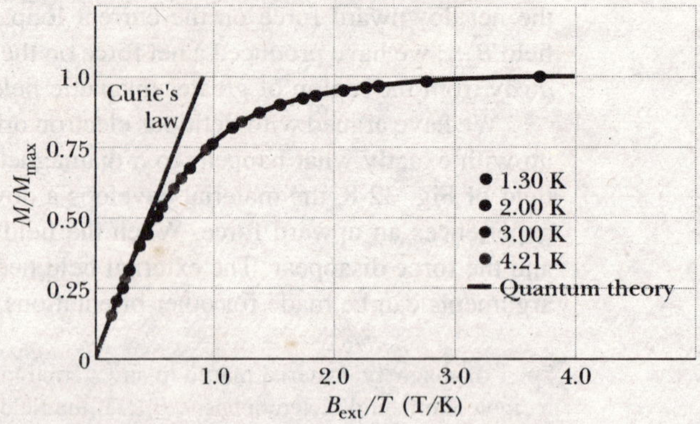

Fig. 32-9 A *magnetization curve* for potassium chromium sulfate, a paramagnetic salt. The ratio of magnetization M of the salt to the maximum possible magnetization M_{max} is plotted versus the ratio of the applied magnetic field B_{ext} to the temperature T. Curie's law fits the data at the left; quantum theory fits all the data. After W. E. Henry.

the magnetic dipole moment of an atom and the external field. As we shall show below, $K \gg \Delta U_B$, even for ordinary temperatures and field magnitudes. Thus, energy transfers during collisions among atoms can significantly disrupt the alignment of the atomic dipole moments, keeping the magnetic dipole moment of a sample much less than $N\mu$.

We can express the extent to which a given paramagnetic sample is magnetized by finding the ratio of its magnetic dipole moment to its volume V. This vector quantity, the magnetic dipole moment per unit volume, is the **magnetization** $\vec{M}$ of the sample, and its magnitude is

$$M = \frac{\text{measured magnetic moment}}{V}. \qquad (32\text{-}18)$$

The unit of $\vec{M}$ is the ampere–square meter per cubic meter, or ampere per meter (A/m). Complete alignment of the atomic dipole moments, called *saturation* of the sample, corresponds to the maximum value $M_{max} = N\mu/V$.

In 1895 Pierre Curie discovered experimentally that the magnetization of a paramagnetic sample is directly proportional to the external magnetic field $\vec{B}_{ext}$ and inversely proportional to the temperature T in kelvins; that is,

$$M = C\frac{B_{ext}}{T}. \qquad (32\text{-}19)$$

Equation 32-19 is known as *Curie's law*, and C is called the *Curie constant*. Curie's law is reasonable in that increasing B_{ext} tends to align the atomic dipole moments in a sample and thus to increase M, whereas increasing T tends to disrupt the alignment via thermal agitation and thus to decrease M. However, the law is actually an approximation that is valid only when the ratio B_{ext}/T is not too large.

Figure 32-9 shows the ratio M/M_{max} as a function of B_{ext}/T for a sample of the salt potassium chromium sulfate, in which chromium ions are the paramagnetic substance. The plot is called a *magnetization curve*. The straight line for Curie's law fits the experimental data at the left, for B_{ext}/T below about 0.5 T/K. The curve that fits all the data points is based on quantum physics. The data on the right side, near saturation, are very difficult to obtain because they require very strong magnetic fields (about 100 000 times Earth's field), even at very low temperatures.

CHECKPOINT 4: The figure here shows two paramagnetic spheres located near the south pole of a bar magnet. Are (a) the magnetic forces on the spheres and (b) the magnetic dipole moments of the spheres directed toward or away from the bar magnet? (c) Is the magnetic force on sphere 1 greater than, less than, or equal to that on sphere 2?

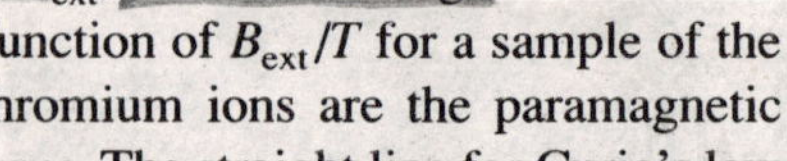

a) toward
b) toward
c) less

Sample Problem 32-1

A paramagnetic gas at room temperature ($T = 300$ K) is placed in an external uniform magnetic field of magnitude $B = 1.5$ T; the atoms of the gas have magnetic dipole moment $\mu = 1.0\mu_B$. Calculate the mean translational kinetic energy K of an atom of the gas and the energy difference ΔU_B between parallel alignment and antiparallel alignment of the atom's magnetic dipole moment with the external field.

SOLUTION: The first **Key Idea** here is that the mean translational kinetic energy K of an atom in a gas depends on the temperature of the gas. From Eq. 20-24, we have

$$K = \tfrac{3}{2}kT = \tfrac{3}{2}(1.38 \times 10^{-23}\ \text{J/K})(300\ \text{K})$$
$$= 6.2 \times 10^{-21}\ \text{J} = 0.039\ \text{eV}. \qquad \text{(Answer)}$$

The second **Key Idea** is that the potential energy U_B of a magnetic dipole $\vec{\mu}$ in an external magnetic field $\vec{B}$ depends on the angle θ between the directions of $\vec{\mu}$ and $\vec{B}$. From Eq. 29-38 ($U_B = -\vec{\mu} \cdot \vec{B}$), we can write the difference ΔU_B between parallel alignment ($\theta = 0°$) and antiparallel alignment ($\theta = 180°$) as

$$\Delta U_B = -\mu B \cos 180° - (-\mu B \cos 0°) = 2\mu B$$
$$= 2\mu_B B = 2(9.27 \times 10^{-24}\ \text{J/T})(1.5\ \text{T})$$
$$= 2.8 \times 10^{-23}\ \text{J} = 0.000\ 17\ \text{eV}. \qquad \text{(Answer)}$$

Here K is about 230 times ΔU_B, so energy exchanges among the atoms during their collisions with one another can easily reorient any magnetic dipole moments that might be aligned with the external magnetic field. The magnetic dipole moment exhibited by the paramagnetic gas must then be due to fleeting partial alignments of the atomic dipole moments.

32-8 Ferromagnetism

When we speak of magnetism in everyday conversation, we almost always have a mental picture of a bar magnet or a disk magnet (probably clinging to a refrigerator door). That is, we picture a ferromagnetic material having strong, permanent magnetism, and not a diamagnetic or paramagnetic material having weak, temporary magnetism.

Iron, cobalt, nickel, gadolinium, dysprosium, and alloys containing these elements exhibit ferromagnetism because of a quantum physical effect called *exchange coupling* in which the electron spins of one atom interact with those of neighboring atoms. The result is alignment of the magnetic dipole moments of the atoms, in spite of the randomizing tendency of atomic collisions. This persistent alignment is what gives ferromagnetic materials their permanent magnetism.

If the temperature of a ferromagnetic material is raised above a certain critical value, called the *Curie temperature,* the exchange coupling ceases to be effective. Most such materials then become simply paramagnetic; that is, the dipoles still tend to align with an external field but much more weakly, and thermal agitation can now more easily disrupt the alignment. The Curie temperature for iron is 1043 K (= 770°C).

The magnetization of a ferromagnetic material such as iron can be studied with an arrangement called a *Rowland ring* (Fig. 32-10). The material is formed into a thin toroidal core of circular cross section. A primary coil P having n turns per unit length is wrapped around the core and carries current i_P. (The coil is essentially a long solenoid bent into a circle.) If the iron core were not present, the magnitude of the magnetic field inside the coil would be, from Eq. 30-25,

$$B_0 = \mu_0 i_P n. \qquad (32\text{-}20)$$

However, with the iron core present, the magnetic field $\vec{B}$ inside the coil is greater than $\vec{B}_0$, usually by a large amount. We can write the magnitude of this field as

$$B = B_0 + B_M, \qquad (32\text{-}21)$$

where B_M is the magnitude of the magnetic field contributed by the iron core. This contribution results from the alignment of the atomic dipole moments within the iron, due to exchange coupling and to the applied magnetic field B_0, and is proportional to the magnetization M of the iron. That is, the contribution B_M is proportional

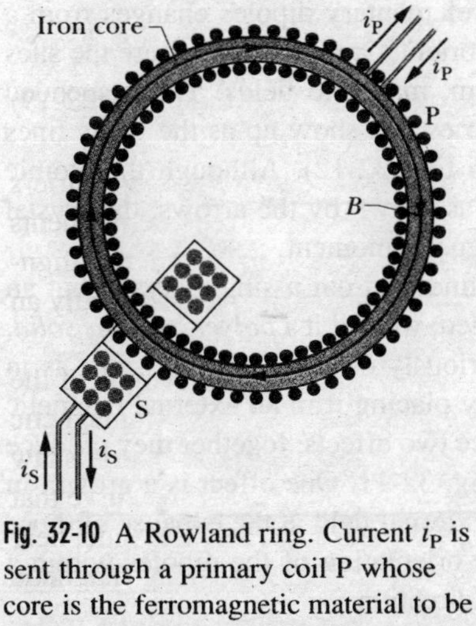

Fig. 32-10 A Rowland ring. Current i_P is sent through a primary coil P whose core is the ferromagnetic material to be studied (here iron) and that is magnetized by the current. (The turns of the coil are represented by dots.) The extent of magnetization of the core determines the total magnetic field $\vec{B}$ within coil P. Field $\vec{B}$ can be measured by means of a secondary coil S.

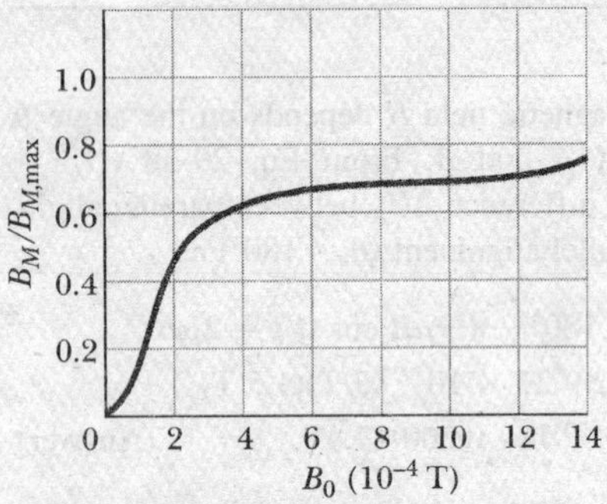

Fig. 32-11 A magnetization curve for a ferromagnetic core material in the Rowland ring of Fig. 32-10. On the vertical axis, 1.0 corresponds to complete alignment (saturation) of the atomic dipoles within the material.

to the magnetic dipole moment per unit volume of the iron. To determine B_M we use a secondary coil S to measure B, compute B_0 with Eq. 32-20, and subtract as suggested by Eq. 32-21.

Figure 32-11 shows a magnetization curve for a ferromagnetic material in a Rowland ring: the ratio $B_M/B_{M,\text{max}}$, where $B_{M,\text{max}}$ is the maximum possible value of B_M, corresponding to saturation, is plotted versus B_0. The curve is like Fig. 32-9, the magnetization curve for a paramagnetic substance: both curves show the extent to which an applied magnetic field can align the atomic dipole moments of a material.

For the ferromagnetic core yielding Fig. 32-11, the alignment of the dipole moments is about 70% complete for $B_0 \approx 1 \times 10^{-3}$ T. If B_0 were increased to 1 T, the alignment would be almost complete (but $B_0 = 1$ T, and thus almost complete saturation, is quite difficult to obtain).

Magnetic Domains

Exchange coupling produces strong alignment of adjacent atomic dipoles in a ferromagnetic material at a temperature below the Curie temperature. Why, then, isn't the material naturally at saturation even when there is no applied magnetic field B_0? That is, why isn't every piece of iron, such as an iron nail, a naturally strong magnet?

To understand this, consider a specimen of a ferromagnetic material such as iron that is in the form of a single crystal; that is, the arrangement of the atoms that make it up—its crystal lattice—extends with unbroken regularity throughout the volume of the specimen. Such a crystal will, in its normal state, be made up of a number of *magnetic domains*. These are regions of the crystal throughout which the alignment of the atomic dipoles is essentially perfect. The domains, however, are not all aligned. For the crystal as a whole, the domains are so oriented that they largely cancel each other as far as their external magnetic effects are concerned.

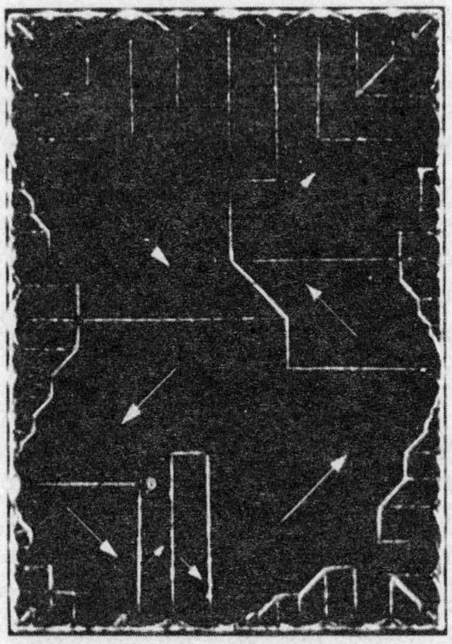

Fig. 32-12 A photograph of domain patterns within a single crystal of nickel; white lines reveal the boundaries of the domains. The white arrows superimposed on the photograph show the orientations of the magnetic dipoles within the domains and thus the orientations of the net magnetic dipoles of the domains. The crystal as a whole is unmagnetized if the net magnetic field (the vector sum over all the domains) is zero.

Figure 32-12 is a magnified photograph of such an assembly of domains in a single crystal of nickel. It was made by sprinkling a colloidal suspension of finely powdered iron oxide on the surface of the crystal. The domain boundaries, which are thin regions in which the alignment of the elementary dipoles changes from a certain orientation in one domain to a different orientation in the other, are the sites of intense, but highly localized and nonuniform, magnetic fields. The suspended colloidal particles are attracted to these boundaries and show up as the white lines (not all the domain boundaries are apparent in Fig. 32-12). Although the atomic dipoles in each domain are completely aligned as shown by the arrows, the crystal as a whole may have a very small resultant magnetic moment.

Actually, a piece of iron as we ordinarily find it is not a single crystal but an assembly of many tiny crystals, randomly arranged; we call it a *polycrystalline solid*. Each tiny crystal, however, has its array of variously oriented domains, just as in Fig. 32-12. If we magnetize such a specimen by placing it in an external magnetic field of gradually increasing strength, we produce two effects; together they produce a magnetization curve of the shape shown in Fig. 32-11. One effect is a growth in size of the domains that are oriented along the external field at the expense of those that are not. The second effect is a shift of the orientation of the dipoles within a domain, as a unit, to become closer to the field direction.

Exchange coupling and domain shifting give us the following result:

> A ferromagnetic material placed in an external magnetic field $\vec{B}_{\text{ext}}$ develops a strong magnetic dipole moment in the direction of $\vec{B}_{\text{ext}}$. If the field is nonuniform, the ferromagnetic material is attracted *toward* a region of greater magnetic field *from* a region of lesser field.

You can actually hear sound produced by shifting domains: Put an audio cassette player into its play mode without a cassette in place (or with a blank cassette) and turn the volume control to maximum. Then bring a strong magnet up to the play head (which is ferromagnetic). The magnetic field causes the domains in the play head to shift abruptly, which abruptly alters the magnetic field through a coil wrapped around the play head. The resulting suddenly induced currents in the coil are amplified and fed to the speaker, producing a fizzing sound.

Sample Problem 32-2

A compass needle made of pure iron (with density 7900 kg/m^3) has a length L of 3.0 cm, a width of 1.0 mm, and a thickness of 0.50 mm. The magnitude of the magnetic dipole moment of an iron atom is $\mu_{Fe} = 2.1 \times 10^{-23}$ J/T. If the magnetization of the needle is equivalent to the alignment of 10% of the atoms in the needle, what is the magnitude of the needle's magnetic dipole moment $\vec{\mu}$?

SOLUTION: One **Key Idea** here is that alignment of all N atoms in the needle would give a magnitude of $N\mu_{Fe}$ for the needle's magnetic dipole moment $\vec{\mu}$. However, the needle has only 10% alignment (the random orientation of the rest does not give any net contribution to $\vec{\mu}$). Thus,

$$\mu = 0.10 N \mu_{Fe}. \tag{32-22}$$

A second **Key Idea** is that we can find the number of atoms N in the needle from the needle's mass:

$$N = \frac{\text{needle's mass}}{\text{iron's atomic mass}}. \tag{32-23}$$

Iron's atomic mass is not listed in Appendix F, but its molar mass M is. Thus, we write

$$\text{iron's atomic mass} = \frac{\text{iron's molar mass } M}{\text{Avogadro's number } N_A}. \tag{32-24}$$

Equation 32-23 then becomes

$$N = \frac{mN_A}{M}. \tag{32-25}$$

The needle's mass m is the product of its density and its volume. The volume works out to be 1.5×10^{-8} m^3, so we can write

$$\begin{aligned} \text{needle's mass } m &= (\text{needle's density})(\text{needle's volume}) \\ &= (7900 \text{ kg/m}^3)(1.5 \times 10^{-8} \text{ m}^3) \\ &= 1.185 \times 10^{-4} \text{ kg}. \end{aligned}$$

Substituting into Eq. 32-25 with this value for m, and also 55.847 g/mole (= 0.055 847 kg/mole) for M and 6.02×10^{23} for N_A, we find

$$\begin{aligned} N &= \frac{(1.185 \times 10^{-4} \text{ kg})(6.02 \times 10^{23})}{0.055\,847 \text{ kg/mole}} \\ &= 1.2774 \times 10^{21}. \end{aligned}$$

Substituting this and the value of μ_{Fe} into Eq. 32-22 then yields

$$\begin{aligned} \mu &= (0.10)(1.2774 \times 10^{21})(2.1 \times 10^{-23} \text{ J/T}) \\ &= 2.682 \times 10^{-3} \text{ J/T} \approx 2.7 \times 10^{-3} \text{ J/T}. \quad \text{(Answer)} \end{aligned}$$

Hysteresis

Magnetization curves for ferromagnetic materials are not retraced as we increase and then decrease the external magnetic field B_0. Figure 32-13 is a plot of B_M versus B_0 during the following operations with a Rowland ring: (1) Starting with the iron unmagnetized (point a), increase the current in the toroid until B_0 ($= \mu_0 in$) has the value corresponding to point b; (2) reduce the current in the toroid winding (and thus B_0) back to zero (point c); (3) reverse the toroid current and increase it in magnitude until B_0 has the value corresponding to point d; (4) reduce the current to zero again (point e); (5) reverse the current once more until point b is reached again.

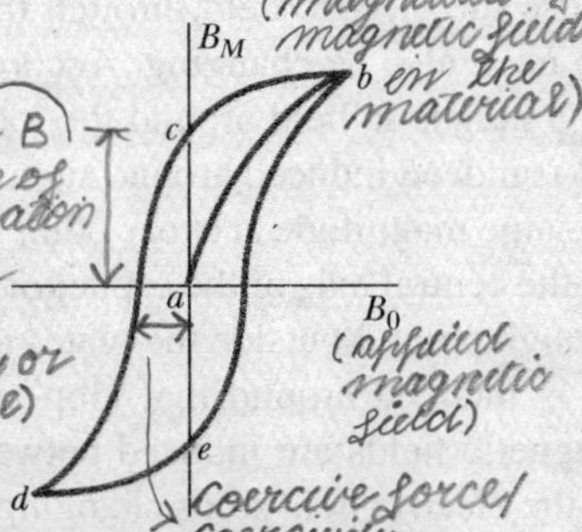

Fig. 32-13 A magnetization curve (ab) for a ferromagnetic specimen and an associated hysteresis loop ($bcdeb$).

The lack of retraceability shown in Fig. 32-13 is called **hysteresis,** and the curve $bcdeb$ is called a *hysteresis loop*. Note that at points c and e the iron core is magnetized, even though there is no current in the toroid windings; this is the familiar phenomenon of permanent magnetism.

Hysteresis can be understood through the concept of magnetic domains. Evidently the motions of the domain boundaries and the reorientations of the domain directions are not totally reversible. When the applied magnetic field B_0 is increased and then decreased back to its initial value, the domains do not return completely to their original configuration but retain some "memory" of their alignment after the initial increase. This memory of magnetic materials is essential for the magnetic storage of information, as on cassette tapes and computer disks.

This memory of the alignment of domains can also occur naturally. When lightning sends currents along multiple tortuous paths through the ground, the currents produce intense magnetic fields that can suddenly magnetize any ferromagnetic material in nearby rock. Because of hysteresis, such rock material retains some of that magnetization after the lightning strike (after the currents disappear). Pieces of the rock, later exposed, broken, and loosened by weathering, are then lodestones.

32-9 Induced Magnetic Fields

In Chapter 31 you saw that a changing magnetic flux induces an electric field, and we ended up with Faraday's law of induction in the form

$$\oint \vec{E} \cdot d\vec{s} = -\frac{d\Phi_B}{dt} \quad \text{(Faraday's law of induction).} \qquad (32\text{-}26)$$

Here $\vec{E}$ is the electric field induced along a closed loop by the changing magnetic flux Φ_B encircled by that loop. Because symmetry is often so powerful in physics, we should be tempted to ask whether induction can occur in the opposite sense; that is, can a changing electric flux induce a magnetic field?

The answer is that it can; furthermore, the equation governing the induction of a magnetic field is almost symmetric with Eq. 32-26. We often call it Maxwell's law of induction after James Clerk Maxwell, and we write it as

$$\oint \vec{B} \cdot d\vec{s} = \mu_0 \varepsilon_0 \frac{d\Phi_E}{dt} \quad \text{(Maxwell's law of induction).} \qquad (32\text{-}27)$$

Here $\vec{B}$ is the magnetic field induced along a closed loop by the changing electric flux Φ_E in the region encircled by that loop.

As an example of this sort of induction, we consider the charging of a parallel-plate capacitor with circular plates (Fig. 32-14*a*). (Although we shall focus on this particular arrangement, a changing electric flux will always induce a magnetic field whenever it occurs.) We assume that the charge on the capacitor is being increased at a steady rate by a constant current i in the connecting wires. Then the electric field magnitude between the plates must also be increasing at a steady rate.

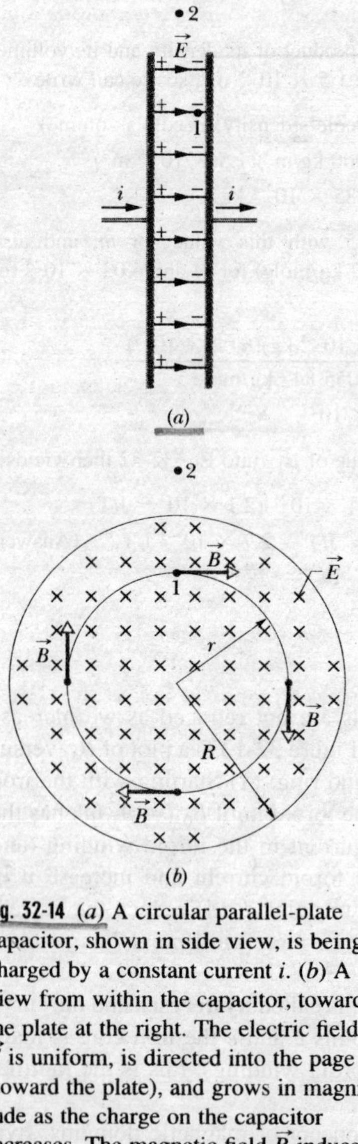

Fig. 32-14 (*a*) A circular parallel-plate capacitor, shown in side view, is being charged by a constant current i. (*b*) A view from within the capacitor, toward the plate at the right. The electric field $\vec{E}$ is uniform, is directed into the page (toward the plate), and grows in magnitude as the charge on the capacitor increases. The magnetic field $\vec{B}$ induced by this changing electric field is shown at four points on a circle with a radius r less than the plate radius R.

Figure 32-14*b* is a view of the right-hand plate of Fig. 32-14*a* from between the plates. The electric field is directed into the page. Let us consider a circular loop through point 1 in Figs. 32-14*a* and *b*, concentric with the capacitor plates and with a radius smaller than that of the plates. Because the electric field through the loop is changing, the electric flux through the loop must also be changing. According to Eq. 32-27, this changing electric flux induces a magnetic field around the loop.

Experiment proves that a magnetic field $\vec{B}$ *is* indeed induced around such a loop, directed as shown. This magnetic field has the same magnitude at every point around the loop and thus has circular symmetry about the central axis of the capacitor plates.

If we now consider a larger loop, say, through point 2 outside the plates in Figs. 32-14*a* and *b*, we find that a magnetic field is induced around that loop as well. Thus, while the electric field is changing, magnetic fields are induced between the plates, both inside and outside the gap. When the electric field stops changing, these induced magnetic fields disappear.

Although Eq. 32-27 is similar to Eq. 32-26, the equations differ in two ways. First, Eq. 32-27 has the two extra symbols, μ_0 and ε_0, but they appear only because

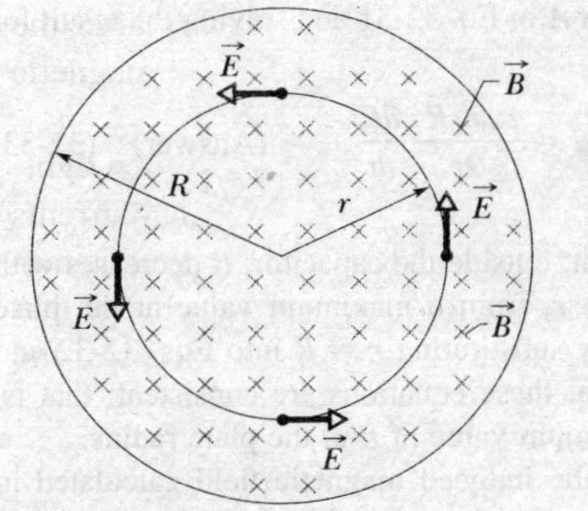

Fig. 32-15 A uniform magnetic field $\vec{B}$ in a circular region. The field, directed into the page, is increasing in magnitude. The electric field $\vec{E}$ induced by the changing magnetic field is shown at four points on a circle concentric with the circular region. Compare this situation with that of Fig. 32-14*b*.

we employ SI units. Second, Eq. 32-27 lacks the minus sign of Eq. 32-26, meaning that the induced electric field $\vec{E}$ and the induced magnetic field $\vec{B}$ have opposite directions when they are produced in otherwise similar situations. To see this opposition, examine Fig. 32-15, in which an increasing magnetic field $\vec{B}$, directed into the page, induces an electric field $\vec{E}$. The induced field $\vec{E}$ is counterclockwise, opposite the induced magnetic field $\vec{B}$ in Fig. 32-14*b*.

Ampere–Maxwell Law

Now recall that the left side of Eq. 32-27, the integral of the dot product $\vec{B}\cdot d\vec{s}$ around a closed loop, appears in another equation—namely, Ampere's law:

$$\oint \vec{B}\cdot d\vec{s} = \mu_0 i_{enc} \quad \text{(Ampere's law)}, \tag{32-28}$$

where i_{enc} is the current encircled by the closed loop. Thus, our two equations that specify the magnetic field $\vec{B}$ produced by means other than a magnetic material (that is, by a current and by a changing electric field) give the field in exactly the same form. We can combine the two equations into the single equation

$$\oint \vec{B}\cdot d\vec{s} = \mu_0\varepsilon_0 \frac{d\Phi_E}{dt} + \mu_0 i_{enc} \quad \text{(Ampere–Maxwell law)}. \tag{32-29}$$

When there is a current but no change in electric flux (such as with a wire carrying a constant current), the first term on the right side of Eq. 32-29 is zero, and so Eq. 32-29 reduces to Eq. 32-28, Ampere's law. When there is a change in electric flux but no current (such as inside or outside the gap of a charging capacitor), the second term on the right side of Eq. 32-29 is zero, and so Eq. 32-29 reduces to Eq. 32-27, Maxwell's law of induction.

Sample Problem 32-3

A parallel-plate capacitor with circular plates of radius R is being charged as in Fig. 32-14*a*.

(a) Derive an expression for the magnetic field at radii r for the case $r \le R$.

SOLUTION: The **Key Idea** here is that a magnetic field can be set up by a current and by induction due to a changing electric flux; both effects are included in Eq. 32-29. There is no current between the capacitor plates of Fig. 32-14, but the electric flux there is changing. Thus, Eq. 32-29 reduces to

$$\oint \vec{B}\cdot d\vec{s} = \mu_0\varepsilon_0 \frac{d\Phi_E}{dt}. \tag{32-30}$$

We shall separately evaluate the left and right sides of this equation.

Left side of Eq. 32-30: We choose a circular Amperian loop with a radius $r \le R$ as shown in Fig. 32-14, because we want to evaluate the magnetic field for $r \le R$—that is, inside the capacitor. The magnetic field $\vec{B}$ at all points along the loop is tangent to the loop, as is the path element $d\vec{s}$. Thus, $\vec{B}$ and $d\vec{s}$ are either parallel or antiparallel at each point of the loop. For simplicity, assume they are parallel (the choice does not alter our outcome here). Then

$$\oint \vec{B}\cdot d\vec{s} = \oint B\, ds \cos 0° = \oint B\, ds.$$

Due to the circular symmetry of the plates, we can also assume that $\vec{B}$ has the same magnitude at every point around the loop. Thus, B can be taken outside the integral on the right side of the above equation. The integral that remains is $\oint ds$, which simply gives the circumference $2\pi r$ of the loop. The left side of Eq. 32-30 is then $(B)(2\pi r)$.

Right side of Eq. 32-30: We assume that the electric field $\vec{E}$ is uniform between the capacitor plates and directed perpendicular to the plates. Then the electric flux Φ_E through the Amperian loop is EA, where A is the area encircled by the loop within the electric field. Thus, the right side of Eq. 32-30 is $\mu_0\varepsilon_0\, d(EA)/dt$.

Substituting our results for the left and right sides into Eq. 32-30, we get

$$(B)(2\pi r) = \mu_0\varepsilon_0 \frac{d(EA)}{dt}.$$

Because A is a constant, we write $d(EA)$ as $A\, dE$, so we have

$$(B)(2\pi r) = \mu_0\varepsilon_0 A \frac{dE}{dt}. \tag{32-31}$$

We next use this **Key Idea**: The area A that is encircled by the Amperian loop within the electric field is the full area πr^2 of the loop, because the loop's radius r is less than (or equal to) the plate radius R. Substituting πr^2 for A in Eq. 32-31 and solving the result for B give us, for $r \le R$,

$$B = \frac{\mu_0 \varepsilon_0 r}{2} \frac{dE}{dt}. \quad \text{(Answer)} \quad (32\text{-}32)$$

This equation tells us that, inside the capacitor, B increases linearly with increased radial distance r, from zero at the center of the plates to a maximum value at the plate edges (where $r = R$).

(b) Evaluate the field magnitude B for $r = R/5 = 11.0$ mm and $dE/dt = 1.50 \times 10^{12}$ V/m · s.

SOLUTION: From the answer to (a), we have

$$\begin{aligned} B &= \frac{1}{2} \mu_0 \varepsilon_0 r \frac{dE}{dt} \\ &= \tfrac{1}{2}(4\pi \times 10^{-7}\ \text{T}\cdot\text{m/A})(8.85 \times 10^{-12}\ \text{C}^2/\text{N}\cdot\text{m}^2) \\ &\quad \times (11.0 \times 10^{-3}\ \text{m})(1.50 \times 10^{12}\ \text{V/m}\cdot\text{s}) \\ &= 9.18 \times 10^{-8}\ \text{T}. \qquad \text{(Answer)} \end{aligned}$$

(c) Derive an expression for the induced magnetic field for the case $r \ge R$.

SOLUTION: Our procedure is the same as in (a) except we now use an Amperian loop with a radius r that is greater than the plate radius R, to evaluate B outside the capacitor. Evaluating the left and right sides of Eq. 32-30 again leads to Eq. 32-31. However, we then need this subtle **Key Idea**: The electric field exists only between the plates, not outside the plates. Thus, the area A that is encircled by the Amperian loop in the electric field is *not* the full area πr^2 of the loop. Rather, A is only the plate area πR^2.

Substituting πR^2 for A in Eq. 32-31 and solving the result for B give us, for $r \ge R$,

$$B = \frac{\mu_0 \varepsilon_0 R^2}{2r} \frac{dE}{dt}. \quad \text{(Answer)} \quad (32\text{-}33)$$

This equation tells us that, outside the capacitor, B decreases with increased radial distance r, from a maximum value at the plate edges (where $r = R$). By substituting $r = R$ into Eqs. 32-32 and 32-33, you can show that these equations are consistent; that is, they give the same maximum value of B at the plate radius.

The magnitude of the induced magnetic field calculated in (b) is so small that it can scarcely be measured with simple apparatus. This is in sharp contrast to the magnitudes of induced electric fields (Faraday's law), which can be measured easily. This experimental difference exists partly because induced emfs can easily be multiplied by using a coil of many turns. No technique of comparable simplicity exists for multiplying induced magnetic fields. In any case, the experiment suggested by this sample problem has been done, and the presence of the induced magnetic fields has been verified quantitatively.

CHECKPOINT 5: The figure shows graphs of the electric field magnitude E versus time t for four uniform electric fields, all contained within identical circular regions as in Fig. 32-14b. Rank the fields according to the magnitudes of the magnetic fields they induce at the edge of the region, greatest first.

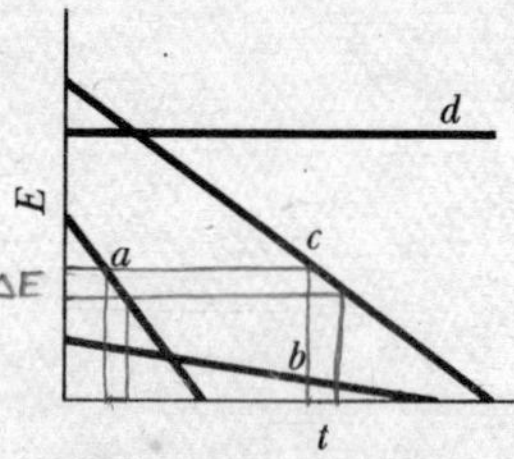

32-10 Displacement Current

If you compare the two terms on the right side of Eq. 32-29, you will see that the product $\varepsilon_0(d\Phi_E/dt)$ must have the dimension of a current. In fact, that product has been treated as being a fictitious current called the **displacement current** i_d:

$$i_d = \varepsilon_0 \frac{d\Phi_E}{dt} \quad \text{(displacement current).} \quad (32\text{-}34)$$

"Displacement" is poorly chosen in that nothing is being displaced, but we are stuck with the word. Nevertheless, we can now rewrite Eq. 32-29 as

$$\oint \vec{B} \cdot d\vec{s} = \mu_0 i_{d,\text{enc}} + \mu_0 i_{\text{enc}} \quad \text{(Ampere–Maxwell law),} \quad (32\text{-}35)$$

in which $i_{d,\text{enc}}$ is the displacement current that is encircled by the integration loop.

Let us again focus on a charging capacitor with circular plates, as in Fig. 32-16a. The real current i that is charging the plates changes the electric field $\vec{E}$ between the plates. The fictitious displacement current i_d between the plates is associated with that changing field $\vec{E}$. Let us relate these two currents.

The charge q on the plates at any time is related to the magnitude E of the field between the plates at that time by Eq. 26-4:

$$q = \varepsilon_0 A E, \quad (32\text{-}36)$$

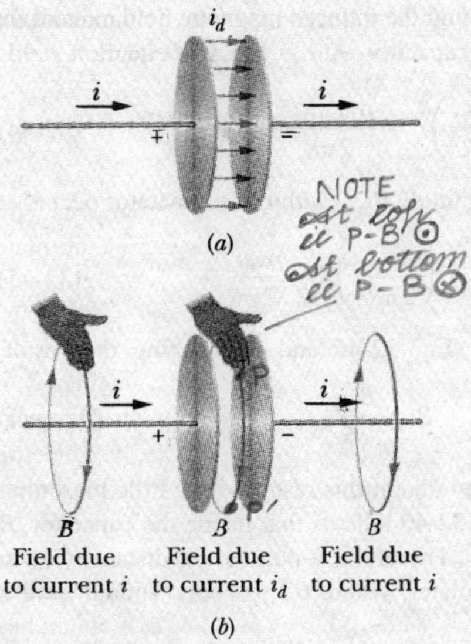

Fig. 32-16 (a) The displacement current i_d between the plates of a capacitor that is being charged by a current i. (b) The right-hand rule for finding the direction of the magnetic field around a wire with a real current (as at the left) also gives the direction of the magnetic field around a displacement current (as in the center).

in which A is the plate area. To get the real current i, we differentiate Eq. 32-36 with respect to time, finding

$$\frac{dq}{dt} = i = \varepsilon_0 A \frac{dE}{dt}. \tag{32-37}$$

To get the displacement current i_d, we can use Eq. 32-34. Assuming that the electric field $\vec{E}$ between the two plates is uniform (we neglect any fringing), we can replace the electric flux Φ_E in that equation with EA. Then Eq. 32-34 becomes

$$i_d = \varepsilon_0 \frac{d\Phi_E}{dt} = \varepsilon_0 \frac{d(EA)}{dt} = \varepsilon_0 A \frac{dE}{dt}. \tag{32-38}$$

Comparing Eqs. 32-37 and 32-38, we see that the real current i charging the capacitor and the fictitious displacement current i_d between the plates have the same magnitude:

$$i_d = i \quad \text{(displacement current in a capacitor).} \tag{32-39}$$

Thus, we can consider the fictitious displacement current i_d to be simply a continuation of the real current i from one plate, across the capacitor gap, to the other plate. Because the electric field is uniformly spread over the plates, the same is true of this fictitious displacement current i_d, as suggested by the spread of current arrows in Fig. 32-16a. Although no charge actually moves across the gap between the plates, the idea of the fictitious current i_d can help us to quickly find the direction and magnitude of an induced magnetic field, as follows.

Finding the Induced Magnetic Field

In Chapter 30 we found the direction of the magnetic field produced by a real current i by using the right-hand rule of Fig. 30-4. We can apply the same rule to find the direction of an induced magnetic field produced by a fictitious displacement current i_d, as is shown in the center of Fig. 32-16b for a capacitor.

We can also use i_d to find the magnitude of the magnetic field induced by a charging capacitor with parallel circular plates of radius R. We simply consider the space between the plates to be an imaginary circular wire of radius R carrying the imaginary current i_d. Then, from Eq. 30-22, the magnitude of the magnetic field at a point inside the capacitor at radius r from the center is

$$B = \left(\frac{\mu_0 i_d}{2\pi R^2}\right) r \quad \text{(inside a circular capacitor).} \tag{32-40}$$

Similarly, from Eq. 30-19, the magnitude of the magnetic field at a point outside the capacitor at radius r is

$$B = \frac{\mu_0 i_d}{2\pi r} \quad \text{(outside a circular capacitor).} \tag{32-41}$$

Sample Problem 32-4

The circular parallel-plate capacitor in Sample Problem 32-3 is being charged with a current i.

(a) Between the plates, what is the magnitude of $\oint \vec{B} \cdot d\vec{s}$, in terms of μ_0 and i, at a radius $r = R/5$ from their center?

SOLUTION: The first **Key Idea** of Sample Problem 32-3a holds here too. However, now we can replace the product $\varepsilon_0\, d\Phi_E/dt$ in Eq. 32-29 with a fictitious displacement current i_d. Then integral $\oint \vec{B} \cdot d\vec{s}$ is given by Eq. 32-35, but because there is no real current i between the capacitor plates, the equation reduces to

$$\oint \vec{B} \cdot d\vec{s} = \mu_0 i_{d,\text{enc}}. \tag{32-42}$$

Because we want to evaluate $\oint \vec{B} \cdot d\vec{s}$ at radius $r = R/5$ (within the capacitor), the integration loop encircles only a portion $i_{d,\text{enc}}$ of the total displacement current i_d. A second **Key Idea** is to assume that i_d is uniformly spread over the full plate area. Then the portion of the displacement current encircled by the loop is proportional to the area encircled by the loop:

$$\frac{\left(\begin{array}{c}\text{encircled displacement} \\ \text{current } i_{d,\text{enc}}\end{array}\right)}{\left(\begin{array}{c}\text{total displacement} \\ \text{current } i_d\end{array}\right)} = \frac{\text{encircled area } \pi r^2}{\text{full plate area } \pi R^2}.$$

This gives us

$$i_{d,\text{enc}} = i_d \frac{\pi r^2}{\pi R^2}.$$

Substituting this into Eq. 32-42, we obtain

$$\oint \vec{B} \cdot d\vec{s} = \mu_0 i_d \frac{\pi r^2}{\pi R^2}. \tag{32-43}$$

Now substituting $i_d = i$ (from Eq. 32-39) and $r = R/5$ into Eq. 32-43 leads to

$$\oint \vec{B} \cdot d\vec{s} = \mu_0 i \frac{(R/5)^2}{R^2} = \frac{\mu_0 i}{25}. \quad \text{(Answer)}$$

(b) In terms of the maximum induced magnetic field, what is the magnitude of the magnetic field induced at $r = R/5$, inside the capacitor?

SOLUTION: The **Key Idea** here is that, because the capacitor has parallel circular plates, we can treat the space between the plates as an imaginary wire of radius R carrying the imaginary current i_d. Then we can use Eq. 32-40 to find the induced magnetic field magnitude B at any point inside the capacitor. At $r = R/5$, that equation yields

$$B = \left(\frac{\mu_0 i_d}{2\pi R^2}\right) r = \frac{\mu_0 i_d (R/5)}{2\pi R^2} = \frac{\mu_0 i_d}{10\pi R}. \tag{32-44}$$

The maximum field magnitude $B_{\max}$ within the capacitor occurs at $r = R$. It is

$$B_{\max} = \left(\frac{\mu_0 i_d}{2\pi R^2}\right) R = \frac{\mu_0 i_d}{2\pi R}. \tag{32-45}$$

Dividing Eq. 32-44 by Eq. 32-45 and rearranging the result, we find

$$B = \frac{B_{\max}}{5}. \quad \text{(Answer)}$$

We should be able to obtain this result with a little reasoning and less work. Equation 32-40 tells us that inside the capacitor, B increases linearly with r. Therefore, a point $\frac{1}{5}$ the distance out to the full radius R of the plates, where $B_{\max}$ occurs, should have a field B that is $\frac{1}{5}B_{\max}$.

CHECKPOINT 6 The figure is a view of one plate of a parallel-plate capacitor from within the capacitor. The dashed lines show four integration paths (path b follows the edge of the plate). Rank the paths according to the magnitude of $\oint \vec{B} \cdot d\vec{s}$ along the paths during the discharging of the capacitor, greatest first.

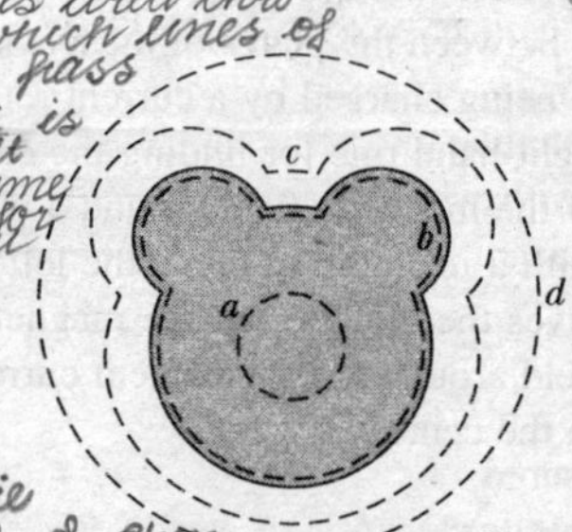

32-11 Maxwell's Equations

Equation 32-29 is the last of the four fundamental equations of electromagnetism, called *Maxwell's equations* and displayed in Table 32-1. These four equations explain a diverse range of phenomena, from why a compass needle points north to why a car starts when you turn the ignition key. They are the basis for the functioning of such electromagnetic devices as electric motors, cyclotrons, television transmitters and receivers, telephones, fax machines, radar, and microwave ovens.

Maxwell's equations are the basis from which many of the equations you have seen since Chapter 22 can be derived. They are also the basis of many of the equations you will see in Chapters 34 through 37, which introduce you to optics.

TABLE 32-1 Maxwell's Equations[a]

Name	Equation	
Gauss' law for electricity	$\oint \vec{E} \cdot d\vec{A} = q_{\text{enc}}/\varepsilon_0$	Relates net electric flux to net enclosed electric charge
Gauss' law for magnetism	$\oint \vec{B} \cdot d\vec{A} = 0$	Relates net magnetic flux to net enclosed magnetic charge
Faraday's law	$\oint \vec{E} \cdot d\vec{s} = -\dfrac{d\Phi_B}{dt}$	Relates induced electric field to changing magnetic flux
Ampere–Maxwell law	$\oint \vec{B} \cdot d\vec{s} = \mu_0\varepsilon_0 \dfrac{d\Phi_E}{dt} + \mu_0 i_{\text{enc}}$	Relates induced magnetic field to changing electric flux and to current

[a] Written on the assumption that no dielectric or magnetic materials are present.

REVIEW & SUMMARY

Gauss' Law for Magnetic Fields The simplest magnetic structures are magnetic dipoles. Magnetic monopoles do not exist (as far as we know). **Gauss' law** for magnetic fields,

$$\Phi_B = \oint \vec{B} \cdot d\vec{A} = 0, \qquad (32\text{-}1)$$

states that the net magnetic flux through any (closed) Gaussian surface is zero. It implies that magnetic monopoles don't exist.

Earth's Magnetic Field Earth's magnetic field can be approximated as being that of a magnetic dipole whose dipole moment makes an angle of 11.5° with Earth's rotation axis, and with the south pole of the dipole in the northern hemisphere. The direction of the local magnetic field at any point on Earth's surface is given by the *field declination* (the angle left or right from geographic north) and the *field inclination* (the angle up or down from the horizontal).

Spin Magnetic Dipole Moment An electron has an intrinsic angular momentum called *spin angular momentum* (or *spin*) $\vec{S}$, with which an intrinsic *spin magnetic dipole moment* $\vec{\mu}_s$ is associated:

$$\vec{\mu}_s = -\frac{e}{m}\vec{S}. \qquad (32\text{-}2)$$

Spin $\vec{S}$ cannot itself be measured, but any component can be measured. Assuming that the measurement is along the z axis of a coordinate system, the component S_z can have only the values given by

$$S_z = m_s \frac{h}{2\pi}, \quad \text{for } m_s = \pm\tfrac{1}{2}, \qquad (32\text{-}3)$$

where h (= 6.63×10^{-34} J · s) is the Planck constant. Similarly, the electron's spin magnetic dipole moment $\vec{\mu}_s$ cannot itself be measured but its component can be measured. Along a z axis, the component is

$$\mu_{s,z} = \pm\frac{eh}{4\pi m} = \pm\mu_B, \qquad (32\text{-}4, 32\text{-}6)$$

where μ_B is the *Bohr magneton:*

$$\mu_B = \frac{eh}{4\pi m} = 9.27 \times 10^{-24} \text{ J/T}. \qquad (32\text{-}5)$$

The potential energy U associated with the orientation of the spin magnetic dipole moment in an external magnetic field $\vec{B}_{ext}$ is

$$U = -\vec{\mu}_s \cdot \vec{B}_{ext} = -\mu_{s,z}B. \qquad (32\text{-}7)$$

Orbital Magnetic Dipole Moment An electron in an atom has an additional angular momentum called its *orbital angular momentum* $\vec{L}_{orb}$, with which an *orbital magnetic dipole moment* $\vec{\mu}_{orb}$ is associated:

$$\vec{\mu}_{orb} = -\frac{e}{2m}\vec{L}_{orb}. \qquad (32\text{-}8)$$

Orbital angular momentum is quantized and can have only values given by

$$L_{orb,z} = m_l \frac{h}{2\pi}, \quad \text{for } m_l = 0, \pm1, \pm2, \ldots, \pm(\text{limit}). \qquad (32\text{-}9)$$

Thus, the magnitude of the orbital angular momentum is

$$\mu_{orb,z} = -m_l \frac{eh}{4\pi m} = -m_l\mu_B. \qquad (32\text{-}10, 32\text{-}11)$$

The potential energy U associated with the orientation of the orbital magnetic dipole moment in an external magnetic field $\vec{B}_{ext}$ is

$$U = -\vec{\mu}_{orb} \cdot \vec{B}_{ext} = -\mu_{orb,z}B_{ext}. \qquad (32\text{-}12)$$

Diamagnetism *Diamagnetic materials* do not exhibit magnetism until they are placed in an external magnetic field $\vec{B}_{ext}$. They then develop a magnetic dipole moment directed opposite $\vec{B}_{ext}$. If the field is nonuniform, the diamagnetic material is repelled from regions of greater magnetic field. This property is called *diamagnetism.*

Paramagnetism In a *paramagnetic material,* each atom has a permanent magnetic dipole moment $\vec{\mu}$, but the dipole moments are randomly oriented and the material as a whole lacks a magnetic field. However, an external magnetic field $\vec{B}_{ext}$ can partially align the atomic dipole moments to give the material a net magnetic dipole moment in the direction of $\vec{B}_{ext}$. If $\vec{B}_{ext}$ is nonuniform, the material is attracted to regions of greater magnetic field. These properties are called *paramagnetism.*

The alignment of the atomic dipole moments increases with an increase in B_{ext} and decreases with an increase in temperature T. The extent to which a sample of volume V is magnetized is given by its *magnetization* $\vec{M}$, whose magnitude is

$$M = \frac{\text{measured magnetic moment}}{V}. \qquad (32\text{-}18)$$

Complete alignment of all N atomic magnetic dipoles in a sample, called *saturation* of the sample, corresponds to the maximum magnetization value $M_{max} = N\mu/V$. For low values of the ratio B_{ext}/T, we have the approximation

$$M = C\frac{B_{ext}}{T} \quad \text{(Curie's law)}, \qquad (32\text{-}19)$$

where C is called the *Curie constant.*

Ferromagnetism In the absence of an external magnetic field, some of the electrons in a ferromagnetic material have their magnetic dipole moments aligned by means of a quantum physical interaction called *exchange coupling,* producing regions (domains) within the material with strong magnetic dipole moments. An external field $\vec{B}_{ext}$ can align the magnetic dipole moments of those regions, producing a strong net magnetic dipole moment for the material as a whole, in the direction of $\vec{B}_{ext}$. This net magnetic dipole moment can partially persist when field $\vec{B}_{ext}$ is removed. If $\vec{B}_{ext}$ is nonuniform, the ferromagnetic material is attracted to regions of greater magnetic field. These properties are called *ferromagnetism.* Exchange coupling disappears when a sample's temperature exceeds its *Curie temperature,* and then the sample has only paramagnetism.

Maxwell's Extension of Ampere's Law A changing electric flux induces a magnetic field $\vec{B}$. Maxwell's law,

$$\oint \vec{B}\cdot d\vec{s} = \mu_0\varepsilon_0 \frac{d\Phi_E}{dt} \quad \text{(Maxwell's law of induction)}, \qquad (32\text{-}27)$$

relates the magnetic field induced along a closed loop to the changing electric flux Φ_E through the loop. Ampere's law, $\oint \vec{B}\cdot d\vec{s} = \mu_0 i_{enc}$ (Eq. 32-28), gives the magnetic field generated by a current i_{enc} encircled by a closed loop. Maxwell's law and Ampere's law can be written as the single equation

$$\oint \vec{B}\cdot d\vec{s} = \mu_0\varepsilon_0 \frac{d\Phi_E}{dt} + \mu_0 i_{enc} \quad \text{(Ampere–Maxwell law)}. \qquad (32\text{-}29)$$

Displacement Current We define the fictitious *displacement current* due to a changing electric field as

$$i_d = \varepsilon_0 \frac{d\Phi_E}{dt}. \qquad (32\text{-}34)$$

Equation 32-29 then becomes

$$\oint \vec{B}\cdot d\vec{s} = \mu_0 i_{d,enc} + \mu_0 i_{enc} \quad \text{(Ampere–Maxwell law)}, \qquad (32\text{-}35)$$

where $i_{d,enc}$ is the displacement current encircled by the integration loop. The idea of a displacement current allows us to retain the notion of continuity of current through a capacitor. However, displacement current is *not* a transfer of charge.

Maxwell's Equations Maxwell's equations, displayed in Table 32-1, summarize electromagnetism and form its foundation.

QUESTIONS

1. An electron in an external magnetic field $\vec{B}_{ext}$ has its spin angular momentum S_z antiparallel to $\vec{B}_{ext}$. If the electron undergoes a *spin-flip* so that S_z is then parallel with $\vec{B}_{ext}$, must energy be supplied to or lost by the electron?

2. Figure 32-17*a* shows a pair of opposite spin orientations for an electron in an external magnetic field $\vec{B}_{ext}$. Figure 32-17*b* gives three choices for the graph of the potential energies associated with those orientations as a function of the magnitude of $\vec{B}_{ext}$. Choices *b* and *c* consist of intersecting lines, choice *a* of parallel lines. Which is the correct choice?

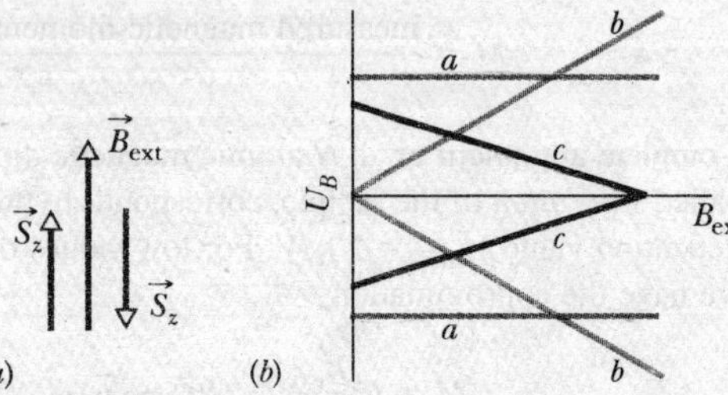

Fig. 32-17 Question 2. (*a*) (*b*)

3. Figure 32-18 shows three loop models of an electron orbiting counterclockwise within a magnetic field. The fields are nonuniform for models 1 and 2 and uniform for model 3. For each model, are (a) the magnetic dipole moment of the loop and (b) the magnetic force on the loop directed up, directed down, or zero?

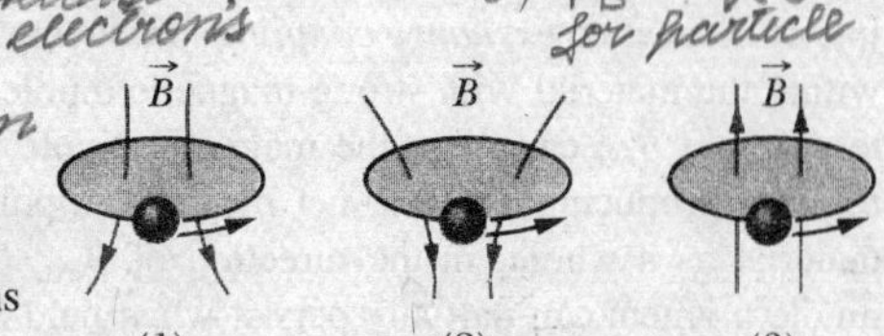

Fig. 32-18 Questions 3, 5, and 6. (1) (2) (3)

4. Does the magnitude of the net force on the loop of Figs. 32-8*a* and *b* increase, decrease, or remain the same if we increase (a) the magnitude of $\vec{B}_{ext}$ and (b) the divergence of $\vec{B}_{ext}$?

5. Replace the current loops of Question 3 and Fig. 32-18 with diamagnetic spheres. For each field, are (a) the magnetic dipole moment of the sphere and (b) the magnetic force on the sphere directed up, directed down, or zero?

6. Replace the current loops of Question 3 and Fig. 32-18 with paramagnetic spheres. For each field, are (a) the magnetic dipole moment of the sphere and (b) the magnetic force on the sphere up, down, or zero?

7. The magnetic dipoles in a diamagnetic material are represented, for three situations, in Fig. 32-19. (For simplicity, the dipole moments are assumed to be directed only up or down the page.) The three situations differ in the magnitude of a magnetic field applied to the material. (a) For each situation, is the applied field directed up or down the page? Rank the three situations according to (b) the magnitude of the applied field and (c) the magnetization of the material, greatest first.

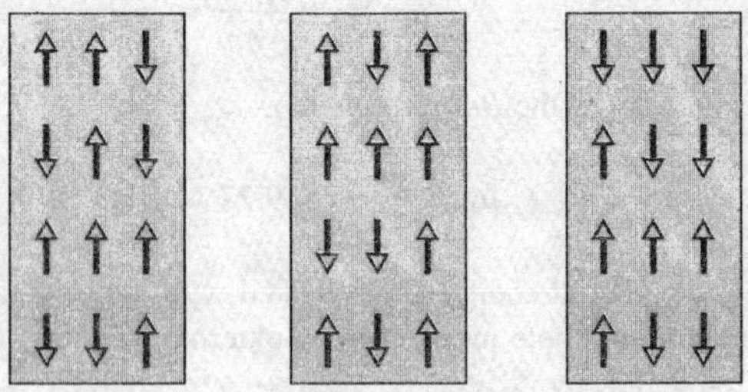

Fig. 32-19 Question 7. (1) (2) (3)

8. Figure 32-20 shows, in two situations, an electric field vector and an induced magnetic field line. In each, is the magnitude of $\vec{E}$ increasing or decreasing?

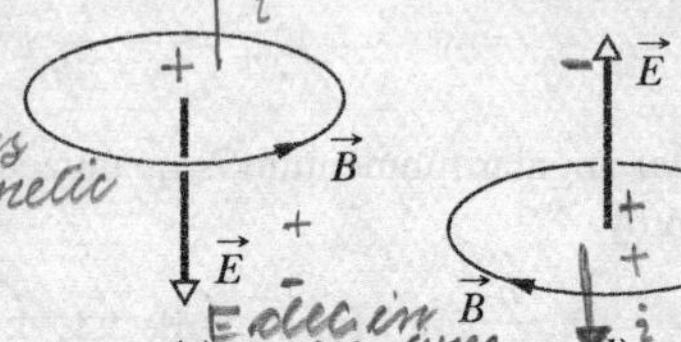

Fig. 32-20 Question 8. (*a*) (*b*)

9. Figure 32-21 shows a parallel-plate capacitor and the current in the connecting wires that is discharging the capacitor. Are the directions of (a) electric field $\vec{E}$ and (b) displacement current i_d leftward or rightward between the plates? (c) Is the magnetic field at point P into the page or out of the page?

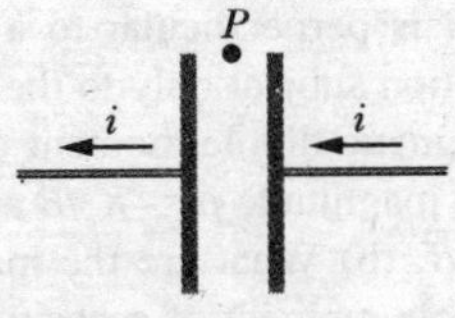

Fig. 32-21 Question 9.

10. A parallel-plate capacitor with rectangular plates is being discharged. A rectangular loop, centered on the plates and between them, measures L by $2L$; the plates measure $2L$ by $4L$. What fraction of the displacement current is encircled by the loop if that current is uniform?

11. Figure 32-22a shows a capacitor, with circular plates, that is being charged. Point a (near one of the connecting wires) and point b (inside the capacitor gap) are equidistant from the central axis, as are point c (not so near the wire) and point d (between the plates but outside the gap). In Fig. 32-22b, one curve gives the variation with distance r of the magnitude of the magnetic field inside and outside the wire. The other curve gives the variation with distance r of the magnitude of the magnetic field inside and outside the gap. The two curves partially overlap. Which of the three points on the curves correspond to which of the four points of Fig. 32-22a?

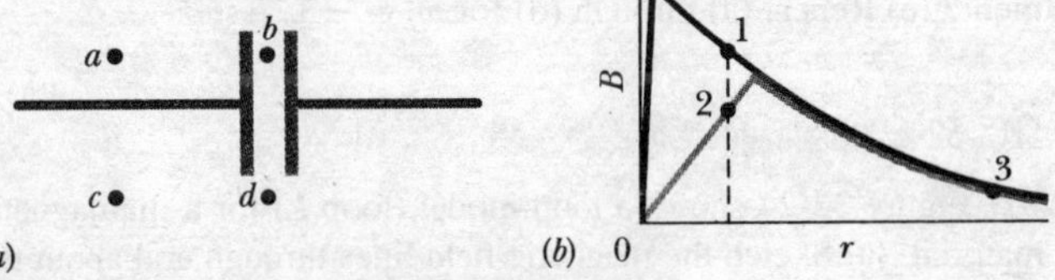

Fig. 32-22 Question 11.

EXERCISES & PROBLEMS

ssm Solution is in the Student Solutions Manual.
www Solution is available on the World Wide Web at: http://www.wiley.com/college/hrw
ilw Solution is available on the Interactive LearningWare.

SEC. 32-2 Gauss' Law for Magnetic Fields

1E. Imagine rolling a sheet of paper into a cylinder and placing a bar magnet near its end as shown in Fig. 32-23. (a) Sketch the magnetic field lines that pass through the surface of the cylinder. (b) What can you say about the sign of $\vec{B} \cdot d\vec{A}$ for every area $d\vec{A}$ on the surface? (c) Does this contradict Gauss' law for magnetism? Explain.

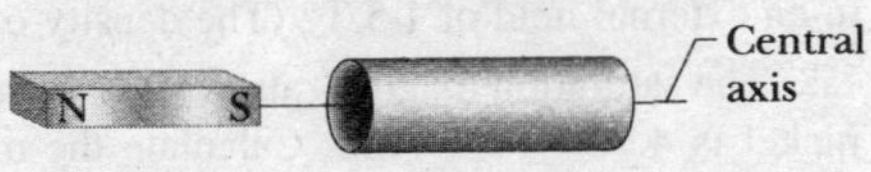

Fig. 32-23 Exercise 1.

2E. The magnetic flux through each of five faces of a die (singular of "dice") is given by $\Phi_B = \pm N$ Wb, where N (= 1 to 5) is the number of spots on the face. The flux is positive (outward) for N even and negative (inward) for N odd. What is the flux through the sixth face of the die?

3P. A Gaussian surface in the shape of a right circular cylinder with end caps has a radius of 12.0 cm and a length of 80.0 cm. Through one end there is an inward magnetic flux of 25.0 μWb. At the other end there is a uniform magnetic field of 1.60 mT, normal to the surface and directed outward. What is the net magnetic flux through the curved surface? ssm ilw www

SEC. 32-3 The Magnetism of Earth

4E. Assume the average value of the vertical component of Earth's magnetic field is 43 μT (downward) for all of Arizona, which has an area of 2.95×10^5 km^2, and calculate the net magnetic flux through the rest of Earth's surface (the entire surface excluding Arizona). Is that net magnetic flux outward or inward?

5E. In New Hampshire the average horizontal component of Earth's magnetic field in 1912 was 16 μT and the average inclination or "dip" was 73°. What was the corresponding magnitude of Earth's magnetic field? ssm

6P. The magnetic field of Earth can be approximated as the magnetic field of a dipole, with horizontal and vertical components, at a point a distance r from Earth's center, given by

$$B_h = \frac{\mu_0 \mu}{4\pi r^3} \cos \lambda_m, \qquad B_v = \frac{\mu_0 \mu}{2\pi r^3} \sin \lambda_m,$$

where λ_m is the *magnetic latitude* (this type of latitude is measured from the geomagnetic equator toward the north or south geomagnetic pole). Assume that Earth's magnetic dipole moment is $\mu = 8.00 \times 10^{22}$ A $\cdot$ m^2. (a) Show that the magnitude of Earth's field at latitude λ_m is given by

$$B = \frac{\mu_0 \mu}{4\pi r^3} \sqrt{1 + 3 \sin^2 \lambda_m}.$$

(b) Show that the inclination ϕ_i of the magnetic field is related to the magnetic latitude λ_m by

$$\tan \phi_i = 2 \tan \lambda_m.$$

7P. Use the results displayed in Problem 6 to predict Earth's magnetic field (both magnitude and inclination) at (a) the geomagnetic equator, (b) a point at geomagnetic latitude 60°, and (c) the north geomagnetic pole. ssm

8P. Using the approximations given in Problem 6, find (a) the altitude above Earth's surface where the magnitude of its magnetic field is 50% of the surface value at the same latitude; (b) the maximum magnitude of the magnetic field at the core–mantle boundary, 2900 km below Earth's surface; and (c) the magnitude and inclination of Earth's magnetic field at the north geographic pole. Suggest why the values you calculated for (c) differ from measured values.

SEC. 32-4 Magnetism and Electrons

9E. What is the measured component of the orbital magnetic dipole moment of an electron with (a) $m_l = 1$ and (b) $m_l = -2$? ssm

10E. What is the energy difference between parallel and antiparallel alignment of the z component of an electron's spin magnetic dipole moment with an external magnetic field of magnitude 0.25 T, directed parallel to the z axis?

11E. If an electron in an atom has an orbital angular momentum with $m_l = 0$, what are the components (a) $L_{orb,z}$ and (b) $\mu_{orb,z}$? If the atom is in an external magnetic field $\vec{B}$ of magnitude 35 mT and directed along the z axis, what are the potential energies associated with the orientations of (c) the electron's orbital magnetic dipole moment and (d) the electron's spin magnetic dipole moment? (e) Repeat (a) through (d) for $m_l = -3$. ssm

SEC. 32-6 Diamagnetism

12E. Figure 32-24 shows a loop model (loop L) for a diamagnetic material. (a) Sketch the magnetic field lines through and about the material due to the bar magnet. (b) What are the directions of the loop's net magnetic dipole moment $\vec{\mu}$ and the conventional current i in the loop? (c) What is the direction of the magnetic force on the loop?

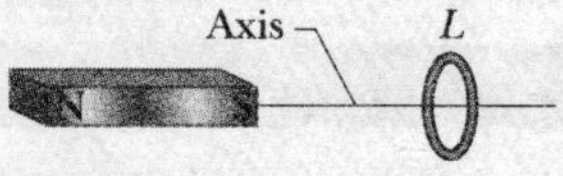

Fig. 32-24 Exercises 12 and 16.

13P*. Assume that an electron of mass m and charge magnitude e moves in a circular orbit of radius r about a nucleus. A uniform magnetic field $\vec{B}$ is then established perpendicular to the plane of the orbit. Assuming also that the radius of the orbit does not change and that the change in the speed of the electron due to field $\vec{B}$ is small, find an expression for the change in the orbital magnetic dipole moment of the electron due to the field. ssm

SEC. 32-7 Paramagnetism

14E. A 0.50 T magnetic field is applied to a paramagnetic gas whose atoms have an intrinsic magnetic dipole moment of 1.0×10^{-23} J/T. At what temperature will the mean kinetic energy of translation of the gas atoms be equal to the energy required to reverse such a dipole end for end in this magnetic field?

15E. A magnet in the form of a cylindrical rod has a length of 5.00 cm and a diameter of 1.00 cm. It has a uniform magnetization of 5.30×10^3 A/m. What is its magnetic dipole moment? ssm ilw

16E. Repeat Exercise 12 for the case in which loop L is the model for a paramagnetic material.

17E. A sample of the paramagnetic salt to which the magnetization curve of Fig. 32-9 applies is to be tested to see whether it obeys Curie's law. The sample is placed in a uniform 0.50 T magnetic field that remains constant throughout the experiment. The magnetization M is then measured at temperatures ranging from 10 to 300 K. Will it be found that Curie's law is valid under these conditions? ssm

18E. A sample of the paramagnetic salt to which the magnetization curve of Fig. 32-9 applies is held at room temperature (300 K). At what applied magnetic field will the degree of magnetic saturation of the sample be (a) 50% and (b) 90%? (c) Are these fields attainable in the laboratory?

19P. An electron with kinetic energy K_e travels in a circular path that is perpendicular to a uniform magnetic field, the electron's motion subject only to the force due to the field. (a) Show that the magnetic dipole moment of the electron due to its orbital motion has magnitude $\mu = K_e/B$ and that it is in the direction opposite that of $\vec{B}$. (b) What are the magnitude and direction of the magnetic dipole moment of a positive ion with kinetic energy K_i under the same circumstances? (c) An ionized gas consists of 5.3×10^{21} electrons/m^3 and the same number density of ions. Take the average electron kinetic energy to be 6.2×10^{-20} J and the average ion kinetic energy to be 7.6×10^{-21} J. Calculate the magnetization of the gas when it is in a magnetic field of 1.2 T. ssm www

SEC. 32-8 Ferromagnetism

20E. Measurements in mines and boreholes indicate that Earth's interior temperature increases with depth at the average rate of 30 C°/km. Assuming a surface temperature of 10°C, at what depth does iron cease to be ferromagnetic? (The Curie temperature of iron varies very little with pressure.)

21E. The exchange coupling mentioned in Section 32-8 as being responsible for ferromagnetism is *not* the mutual magnetic interaction between two elementary magnetic dipoles. To show this, calculate (a) the magnitude of the magnetic field a distance of 10 nm away, along the dipole axis, from an atom with magnetic dipole moment 1.5×10^{-23} J/T (cobalt), and (b) the minimum energy required to turn a second identical dipole end for end in this field. By comparing the latter with the results of Sample Problem 32-1, what can you conclude? ssm

22E. The dipole moment associated with an atom of iron in an iron bar is 2.1×10^{-23} J/T. Assume that all the atoms in the bar, which is 5.0 cm long and has a cross-sectional area of 1.0 cm^2, have their dipole moments aligned. (a) What is the dipole moment of the bar? (b) What torque must be exerted to hold this magnet perpendicular to an external field of 1.5 T? (The density of iron is 7.9 g/cm^3.)

23E. The saturation magnetization M_{max} of the ferromagnetic metal nickel is 4.70×10^5 A/m. Calculate the magnetic moment of a single nickel atom. (The density of nickel is 8.90 g/cm^3 and its molar mass is 58.71 g/mol.) ssm

24P. Figure 32-25 shows the apparatus used in a lecture demonstration of para- and diamagnetism. A sample of the magnetic material is suspended by a long string in the nonuniform field (d = 2 cm) between the poles of a powerful electromagnet. Pole P_1 is sharply pointed and pole P_2 is rounded as indicated. Any deflection of the string from the vertical is visible to the audience by means of an optical projection system (not shown). (a) First a bismuth (highly diamagnetic) sample is used. When the electromagnet is turned on, the sample is observed to deflect slightly (about 1 mm) toward one of the poles. What is the direction of this deflection? (b) Next an aluminum (paramagnetic, conducting) sample is used. When the electromagnet is turned on, the sample is observed to de-

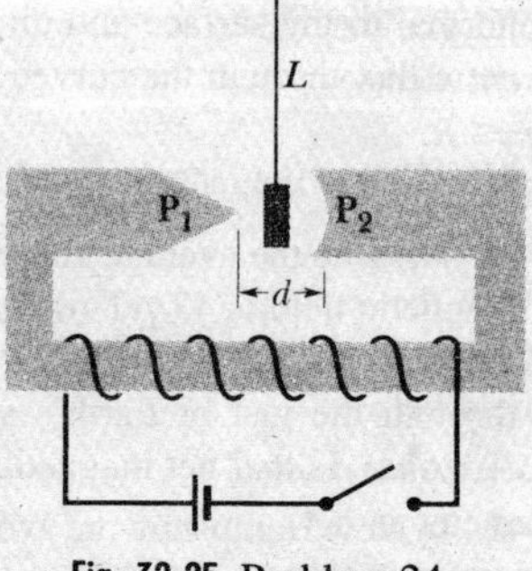

Fig. 32-25 Problem 24.

flect strongly (about 1 cm) toward one pole for about a second and then deflect moderately (a few millimeters) toward the other pole. Explain and indicate the direction of these deflections. (*Hint:* The aluminum sample is a conductor, for which Lenz's law applies.) (c) What would happen if a ferromagnetic sample were used?

25P. The magnetic dipole moment of Earth is 8.0×10^{22} J/T. (a) If the origin of this magnetism were a magnetized iron sphere at the center of Earth, what would be its radius? (b) What fraction of the volume of Earth would such a sphere occupy? Assume complete alignment of the dipoles. The density of Earth's inner core is 14 g/cm^3. The magnetic dipole moment of an iron atom is 2.1×10^{-23} J/T. (*Note:* Earth's inner core is in fact thought to be in both liquid and solid forms and partly iron, but a permanent magnet as the source of Earth's magnetism has been ruled out by several considerations. For one, the temperature is certainly above the Curie point.) ssm ilw

SEC. 32-9 Induced Magnetic Fields

26E. Sample Problem 32-3 describes the charging of a parallel-plate capacitor with circular plates of radius 55.0 mm. At what two radii r from the central axis of the capacitor is the magnitude of the induced magnetic field equal to 50% of its maximum value?

27E. The induced magnetic field 6.0 mm from the central axis of a circular parallel-plate capacitor and between the plates is 2.0×10^{-7} T. The plates have radius 3.0 mm. At what rate $d\vec{E}/dt$ is the electric field between the plates changing? ssm

28P. Suppose that a parallel-plate capacitor has circular plates with radius $R = 30$ mm and a plate separation of 5.0 mm. Suppose also that a sinusoidal potential difference with a maximum value of 150 V and a frequency of 60 Hz is applied across the plates; that is,

$$V = (150 \text{ V}) \sin[2\pi(60 \text{ Hz})t].$$

(a) Find $B_{max}(R)$, the maximum value of the induced magnetic field that occurs at $r = R$. (b) Plot $B_{max}(r)$ for $0 < r < 10$ cm.

SEC. 32-10 Displacement Current

29E. Prove that the displacement current in a parallel-plate capacitor of capacitance C can be written as $i_d = C(dV/dt)$, where V is the potential difference between the plates. ssm

30E. At what rate must the potential difference between the plates of a parallel-plate capacitor with a 2.0 μF capacitance be changed to produce a displacement current of 1.5 A?

31E. For the situation of Sample Problem 32-3, show that the current density of the displacement current is $J_d = \varepsilon_0(dE/dt)$ for $r \leq R$. ssm

32E. A parallel-plate capacitor with circular plates of radius 0.10 m is being discharged. A circular loop of radius 0.20 m is concentric with the capacitor and halfway between the plates. The displacement current through the loop is 2.0 A. At what rate is the electric field between the plates changing?

33P. As a parallel-plate capacitor with circular plates 20 cm in diameter is being charged, the current density of the displacement current in the region between the plates is uniform and has a magnitude of 20 A/m^2. (a) Calculate the magnitude B of the magnetic field at a distance $r = 50$ mm from the axis of symmetry of this region. (b) Calculate dE/dt in this region. ssm ilw

34P. The magnitude of the electric field between the two circular parallel plates in Fig. 32-26 is $E = (4.0 \times 10^5) - (6.0 \times 10^4 t)$, with E in volts per meter and t in seconds. At $t = 0$, the field is upward as shown. The plate area is 4.0×10^{-2} m^2. For $t \geq 0$, (a) what are the magnitude and direction of the displacement current between the plates and (b) is the direction of the induced magnetic field clockwise or counterclockwise around the plates?

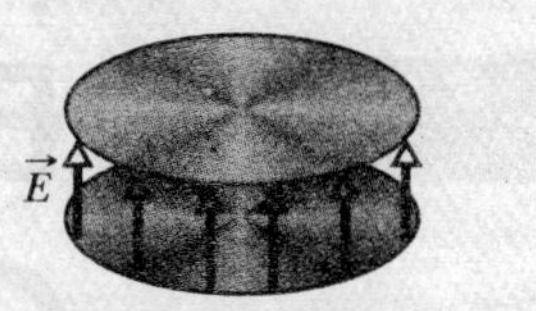

Fig. 32-26 Problem 34.

35P. A uniform electric field collapses to zero from an initial strength of 6.0×10^5 N/C in a time of 15 μs in the manner shown in Fig. 32-27. Calculate the magnitude of the displacement current, through a 1.6 m^2 area perpendicular to the field, during each of the time intervals a, b, and c shown on the graph. (Ignore the behavior at the ends of the intervals.) ssm ilw

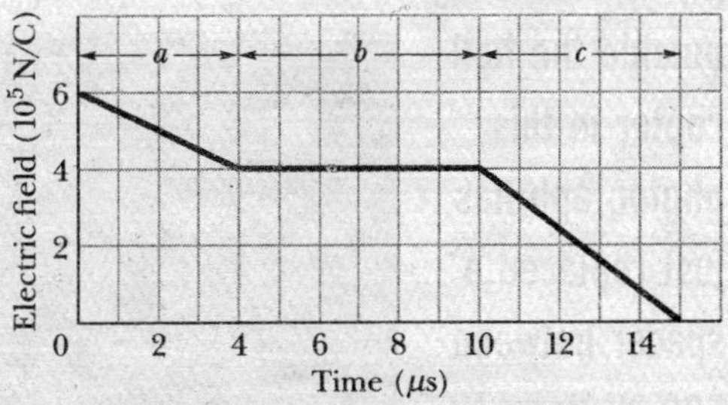

Fig. 32-27 Problem 35.

36P. A parallel-plate capacitor with circular plates is being charged. Consider a circular loop centered on the central axis between the plates. The loop radius is 0.20 m; the plate radius is 0.10 m; and the displacement current through the loop is 2.0 A. What is the rate at which the electric field between the plates is changing?

37P. A parallel-plate capacitor has square plates 1.0 m on a side as shown in Fig. 32-28. A current of 2.0 A charges the capacitor, producing a uniform electric field $\vec{E}$ between the plates, with $\vec{E}$ perpendicular to the plates. (a) What is the displacement current i_d through the region between the plates? (b) What is dE/dt in this region? (c) What is the displacement current through the square dashed path between the plates? (d) What is $\oint \vec{B} \cdot d\vec{s}$ around this square dashed path? ssm ilw www

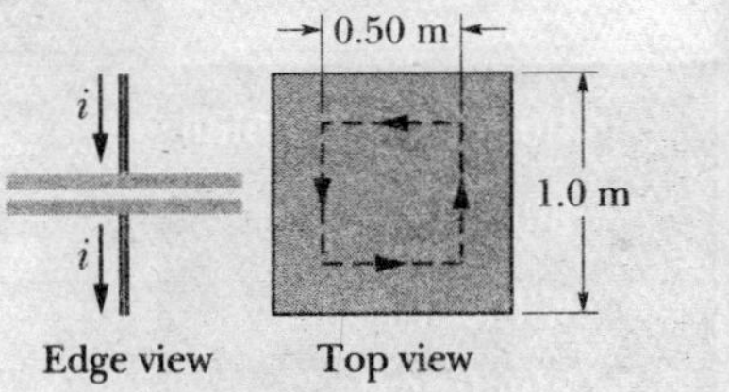

Fig. 32-28 Problem 37.

38P. A capacitor with parallel circular plates of radius R is discharging via a current of 12.0 A. Consider a loop of radius $R/3$ that is centered on the central axis between the plates. (a) How much displacement current is encircled by the loop? The maximum induced magnetic field has a magnitude of 12.0 mT. (b) At what radial distance from the central axis of the plate is the magnitude of the induced magnetic field 3.00 mT?

33 Electromagnetic Oscillations and Alternating Current

When a high-voltage power transmission line requires repair, a utility company cannot just shut it down, perhaps blacking out an entire city. Repairs must be made while the lines are electrically "hot." The man outside the helicopter in this photograph has just replaced a spacer between 500 kV lines *by hand,* a procedure that requires considerable expertise.

How does he manage this repair without being electrocuted?

The answer is in this chapter.

33-1 New Physics—Old Mathematics

In this chapter you will see how the electric charge q varies with time in a circuit made up of an inductor L, a capacitor C, and a resistor R. From another point of view, we shall discuss how energy shuttles back and forth between the magnetic field of the inductor and the electric field of the capacitor, while it is being gradually dissipated as thermal energy in the resistor.

We have discussed oscillations before, in another context. In Chapter 16 we saw how displacement x varies with time in a mechanical oscillating system made up of a block of mass m, a spring of spring constant k, and a viscous or frictional element such as oil; Fig. 16-15 shows such a system. We also saw how energy shuttles back and forth between the kinetic energy of the oscillating mass and the potential energy of the spring, being gradually dissipated as thermal energy.

The parallel between these two idealized systems is exact, and the controlling differential equations are identical. Thus, there is no new mathematics to be learned; we can simply change the symbols and give our full attention to the physics of the situation.

33-2 *LC* Oscillations, Qualitatively

Of the three circuit elements, resistance R, capacitance C, and inductance L, we have so far discussed the series combinations RC (in Section 28-8) and RL (in Section 31-9). In these two kinds of circuit we found that the charge, current, and potential difference grow and decay exponentially. The time scale of the growth or decay is given by a *time constant* τ, which is either capacitive or inductive.

We now examine the remaining two-element circuit combination LC. You will see that in this case the charge, current, and potential difference do not decay exponentially with time but vary sinusoidally (with period T and angular frequency ω). The resulting oscillations of the capacitor's electric field and the inductor's magnetic field are said to be **electromagnetic oscillations.** Such a circuit is said to oscillate.

Parts a through h of Fig. 33-1 show succeeding stages of the oscillations in a simple LC circuit. From Eq. 26-21, the energy stored in the electric field of the capacitor at any time is

$$U_E = \frac{q^2}{2C}, \qquad (33\text{-}1)$$

where q is the charge on the capacitor at that time. From Eq. 31-51, the energy stored in the magnetic field of the inductor at any time is

$$U_B = \frac{Li^2}{2}, \qquad (33\text{-}2)$$

where i is the current through the inductor at that time.

The method of repairing high-voltage lines shown in the opening photograph is patented by Scott H. Yenzer and is licensed exclusively to Haverfield Corporation of Gettysburg, PA. As the lineman approaches a hot line, the electric field surrounding the line brings his body to nearly the potential of the line. To match the two potentials, he then extends a conducting "wand" to the line. To avoid being electrocuted, he must be isolated from anything electrically connected to the ground. To ensure that his body is always at a single potential—that of the line he is working on—he wears a conducting suit, hood, and gloves, all of which are electrically connected to the line via the wand.

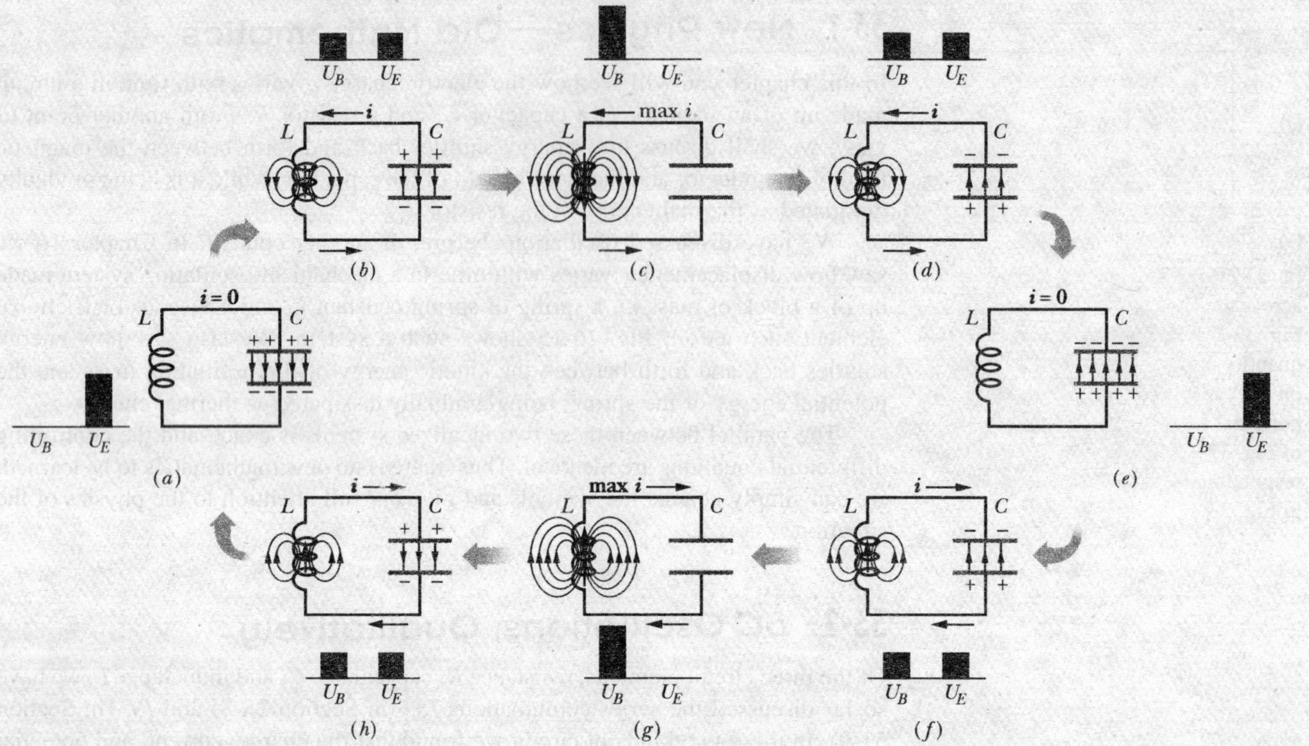

Fig. 33-1 Eight stages in a single cycle of oscillation of a resistanceless LC circuit. The bar graphs by each figure show the stored magnetic and electric energies. The magnetic field lines of the inductor and the electric field lines of the capacitor are shown. (*a*) Capacitor with maximum charge, no current. (*b*) Capacitor discharging, current increasing. (*c*) Capacitor fully discharged, current maximum. (*d*) Capacitor charging but with polarity opposite that in (*a*), current decreasing. (*e*) Capacitor with maximum charge having polarity opposite that in (*a*), no current. (*f*) Capacitor discharging, current increasing with direction opposite that in (*b*). (*g*) Capacitor fully discharged, current maximum. (*h*) Capacitor charging, current decreasing.

We now adopt the convention of representing *instantaneous values* of the electrical quantities of a sinusoidally oscillating circuit with small letters, such as q, and the *amplitudes* of those quantities with capital letters, such as Q. With this convention in mind, let us assume that initially the charge q on the capacitor in Fig. 33-1 is at its maximum value Q and that the current i through the inductor is zero. This initial state of the circuit is shown in Fig. 33-1*a*. The bar graphs for energy included there indicate that at this instant, with zero current through the inductor and maximum charge on the capacitor, the energy U_B of the magnetic field is zero and the energy U_E of the electric field is a maximum.

The capacitor now starts to discharge through the inductor, positive charge carriers moving counterclockwise, as shown in Fig. 33-1*b*. This means that a current i, given by dq/dt and pointing down in the inductor, is established. As the capacitor's charge decreases, the energy stored in the electric field within the capacitor also decreases. This energy is transferred to the magnetic field that appears around the inductor because of the current i that is building up there. Thus, the electric field decreases and the magnetic field builds up as energy is transferred from the electric field to the magnetic field.

The capacitor eventually loses all its charge (Fig. 33-1*c*) and thus also loses its electric field and the energy stored in that field. The energy has then been fully transferred to the magnetic field of the inductor. The magnetic field is then at its maximum magnitude, and the current through the inductor is then at its maximum value I.

Although the charge on the capacitor is now zero, the counterclockwise current must continue because the inductor does not allow it to change suddenly to zero. The current continues to transfer positive charge from the top plate to the bottom plate through the circuit (Fig. 33-1*d*). Energy now flows from the inductor back to the capacitor as the electric field within the capacitor builds up again. The current

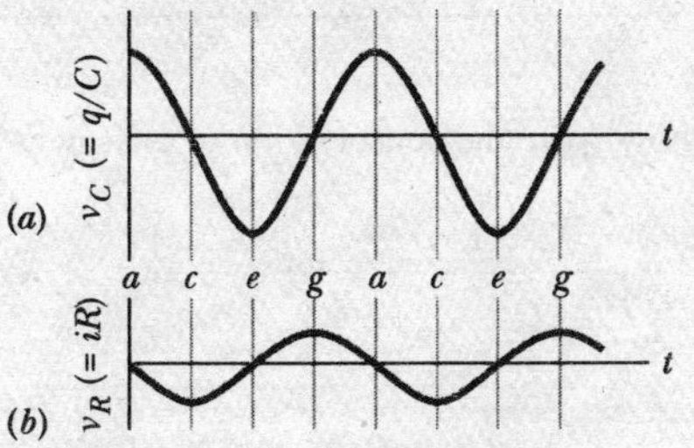

Fig. 33-2 (*a*) The potential difference across the capacitor of the circuit of Fig. 33-1 as a function of time. This quantity is proportional to the charge on the capacitor. (*b*) A potential proportional to the current in the circuit of Fig. 33-1. The letters refer to the correspondingly labeled oscillation stages in Fig. 33-1.

gradually decreases during this energy transfer. When, eventually, the energy has been transferred completely back to the capacitor (Fig. 33-1*e*), the current has decreased to zero (momentarily). The situation of Fig. 33-1*e* is like the initial situation, except that the capacitor is now charged oppositely.

The capacitor then starts to discharge again but now with a clockwise current (Fig. 33-1*f*). Reasoning as before, we see that the clockwise current builds to a maximum (Fig. 33-1*g*) and then decreases (Fig. 33-1*h*), until the circuit eventually returns to its initial situation (Fig. 33-1*a*). The process then repeats at some frequency f and thus at an angular frequency $\omega = 2\pi f$. In the ideal *LC* circuit with no resistance, there are no energy transfers other than that between the electric field of the capacitor and the magnetic field of the inductor. Owing to the conservation of energy, the oscillations continue indefinitely. The oscillations need not begin with the energy all in the electric field; the initial situation could be any other stage of the oscillation.

To determine the charge q on the capacitor as a function of time, we can put in a voltmeter to measure the time-varying potential difference (or *voltage*) v_C that exists across the capacitor C. From Eq. 26-1 we can write

$$v_C = \left(\frac{1}{C}\right)q,$$

which allows us to find q. To measure the current, we can connect a small resistance R in series with the capacitor and inductor and measure the time-varying.potential difference v_R across it; v_R is proportional to i through the relation

$$v_R = iR.$$

We assume here that R is so small that its effect on the behavior of the circuit is negligible. The variations in time of v_C and v_R, and thus of q and i, are shown in Fig. 33-2. All four quantities vary sinusoidally.

In an actual *LC* circuit, the oscillations will not continue indefinitely because there is always some resistance present that will drain energy from the electric and magnetic fields and dissipate it as thermal energy (the circuit may become warmer). The oscillations, once started, will die away as Fig. 33-3 suggests. Compare this figure with Fig. 16-16, which shows the decay of mechanical oscillations caused by frictional damping in a block–spring system.

✔CHECKPOINT 1: A charged capacitor and an inductor are connected in series at time $t = 0$. In terms of the period T of the resulting oscillations, determine how much later the following reach their maximums: (a) the charge on the capacitor; (b) the voltage across the capacitor, with its original polarity; (c) the energy stored in the electric field; and (d) the current.

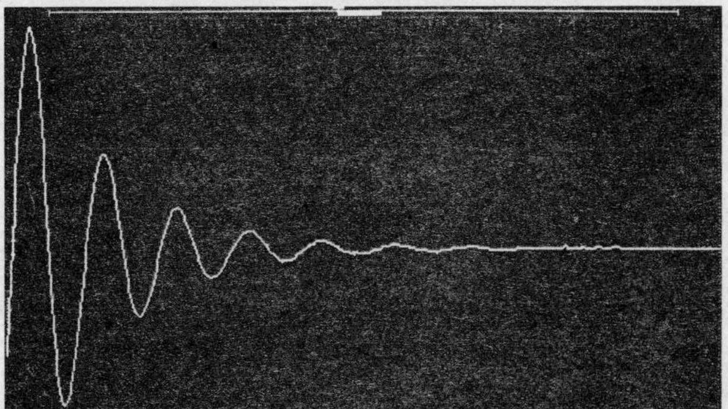

Fig. 33-3 An oscilloscope trace showing how the oscillations in an *RLC* circuit actually die away because energy is dissipated in the resistor as thermal energy.

Sample Problem 33-1

A 1.5 μF capacitor is charged to 57 V. The charging battery is then disconnected, and a 12 mH coil is connected in series with the capacitor so that LC oscillations occur. What is the maximum current in the coil? Assume that the circuit contains no resistance.

SOLUTION: The **Key Ideas** here are these:

1. Because the circuit contains no resistance, the electromagnetic energy of the circuit is conserved as the energy is transferred back and forth between the electric field of the capacitor and the magnetic field of the coil (inductor).
2. At any time t, the energy $U_B(t)$ of the magnetic field is related to the current $i(t)$ through the coil by Eq. 33-2 ($U_B = Li^2/2$). When all the energy is stored as magnetic energy, the current is at its maximum value I and that energy is $U_{B,\max} = LI^2/2$.
3. At any time t, the energy $U_E(t)$ of the electric field is related to the charge $q(t)$ on the capacitor by Eq. 33-1 ($U_E = q^2/2C$). When all the energy is stored as electric energy, the charge is at its maximum value Q and that energy is $U_{E,\max} = Q^2/2C$.

With these ideas, we can now write the conservation of energy as

$$U_{B,\max} = U_{E,\max}$$

or

$$\frac{LI^2}{2} = \frac{Q^2}{2C}.$$

Solving for I gives us

$$I = \sqrt{\frac{Q^2}{LC}}.$$

We know L and C, but not Q. However, with Eq. 26-1 ($q = CV$) we can relate Q to the maximum potential difference V across the capacitor, which is the initial potential difference of 57 V. Thus, substituting $Q = CV$ leads to

$$I = V\sqrt{\frac{C}{L}} = (57\ \text{V})\sqrt{\frac{1.5 \times 10^{-6}\ \text{F}}{12 \times 10^{-3}\ \text{H}}}$$

$$= 0.637\ \text{A} \approx 640\ \text{mA}. \qquad \text{(Answer)}$$

33-3 The Electrical–Mechanical Analogy

Let us look a little closer at the analogy between the oscillating LC system of Fig. 33-1 and an oscillating block–spring system. Two kinds of energy are involved in the block–spring system. One is potential energy of the compressed or extended spring; the other is kinetic energy of the moving block. These two energies are given by the familiar formulas in the left energy column in Table 33-1.

The table also shows, in the right energy column, the two kinds of energy involved in LC oscillations. By looking across the table, we can see an analogy between the forms of the two pairs of energies—the mechanical energies of the block–spring system and the electromagnetic energies of the LC oscillator. The equations for v and i at the bottom of the table help us see the details of the analogy. They tell us that q corresponds to x, and i corresponds to v (in both equations, the former is differentiated to obtain the latter). These correspondences then suggest that, in the energy expressions, $1/C$ corresponds to k and L corresponds to m. Thus,

q corresponds to x, $1/C$ corresponds to k,
i corresponds to v, and L corresponds to m.

These correspondences suggest that in an LC oscillator, the capacitor is mathematically like the spring in a block–spring system, and the inductor is like the block.

In Section 16-3 we saw that the angular frequency of oscillation of a (friction-

TABLE 33-1 The Energy in Two Oscillating Systems Compared

Block–Spring System		LC Oscillator	
Element	Energy	Element	Energy
Spring	Potential, $\frac{1}{2}kx^2$	Capacitor	Electric, $\frac{1}{2}(1/C)q^2$
Block	Kinetic, $\frac{1}{2}mv^2$	Inductor	Magnetic, $\frac{1}{2}Li^2$
$v = dx/dt$		$i = dq/dt$	

less) block–spring system is

$$\omega = \sqrt{\frac{k}{m}} \qquad \text{(block–spring system).} \tag{33-3}$$

The correspondences listed above suggest that to find the angular frequency of oscillation for a (resistanceless) *LC* circuit, *k* should be replaced by $1/C$ and *m* by *L*, yielding

$$\omega = \frac{1}{\sqrt{LC}} \qquad (LC\ \text{circuit}). \tag{33-4}$$

We derive this result in the next section.

33-4 *LC* Oscillations, Quantitatively

Here we want to show explicitly that Eq. 33-4 for the angular frequency of *LC* oscillations is correct. At the same time, we want to examine even more closely the analogy between *LC* oscillations and block–spring oscillations. We start by extending somewhat our earlier treatment of the mechanical block–spring oscillator.

The Block–Spring Oscillator

We analyzed block–spring oscillations in Chapter 16 in terms of energy transfers and did not—at that early stage—derive the fundamental differential equation that governs those oscillations. We do so now.

We can write, for the total energy *U* of a block–spring oscillator at any instant,

$$U = U_b + U_s = \tfrac{1}{2}mv^2 + \tfrac{1}{2}kx^2, \tag{33-5}$$

where U_b and U_s are, respectively, the kinetic energy of the moving block and the potential energy of the stretched or compressed spring. If there is no friction—which we assume—the total energy *U* remains constant with time, even though *v* and *x* vary. In more formal language, $dU/dt = 0$. This leads to

$$\frac{dU}{dt} = \frac{d}{dt}\left(\tfrac{1}{2}mv^2 + \tfrac{1}{2}kx^2\right) = mv\frac{dv}{dt} + kx\frac{dx}{dt} = 0. \tag{33-6}$$

However, $v = dx/dt$ and $dv/dt = d^2x/dt^2$. With these substitutions, Eq. 33-6 becomes

$$m\frac{d^2x}{dt^2} + kx = 0 \qquad \text{(block–spring oscillations).} \tag{33-7}$$

Equation 33-7 is the fundamental *differential equation* that governs the frictionless block–spring oscillations.

The general solution to Eq. 33-7—that is, the function $x(t)$ that describes the block–spring oscillations—is (as we saw in Eq. 16-3)

$$x = X\cos(\omega t + \phi) \qquad \text{(displacement),} \tag{33-8}$$

in which *X* is the amplitude of the mechanical oscillations (represented by x_m in Chapter 16), ω is the angular frequency of the oscillations, and ϕ is a phase constant.

The *LC* Oscillator

Now let us analyze the oscillations of a resistanceless *LC* circuit, proceeding exactly as we just did for the block–spring oscillator. The total energy *U* present at any instant in an oscillating LC circuit is given by

$$U = U_B + U_E = \frac{Li^2}{2} + \frac{q^2}{2C}, \tag{33-9}$$

in which U_B is the energy stored in the magnetic field of the inductor and U_E is the energy stored in the electric field of the capacitor. Since we have assumed the circuit resistance to be zero, no energy is transferred to thermal energy and U remains constant with time. In more formal language, dU/dt must be zero. This leads to

$$\frac{dU}{dt} = \frac{d}{dt}\left(\frac{Li^2}{2} + \frac{q^2}{2C}\right) = Li\frac{di}{dt} + \frac{q}{C}\frac{dq}{dt} = 0. \tag{33-10}$$

However, $i = dq/dt$ and $di/dt = d^2q/dt^2$. With these substitutions, Eq. 33-10 becomes

$$L\frac{d^2q}{dt^2} + \frac{1}{C}q = 0 \quad \text{(LC oscillations).} \tag{33-11}$$

This is the *differential equation* that describes the oscillations of a resistanceless LC circuit. Equations 33-11 and 33-7 are exactly of the same mathematical form.

Charge and Current Oscillations

Since the differential equations are mathematically identical, their solutions must also be mathematically identical. Because q corresponds to x, we can write the general solution of Eq. 33-11, by analogy to Eq. 33-8, as

$$q = Q\cos(\omega t + \phi) \quad \text{(charge),} \tag{33-12}$$

where Q is the amplitude of the charge variations, ω is the angular frequency of the electromagnetic oscillations, and ϕ is the phase constant.

Taking the first derivative of Eq. 33-12 with respect to time gives us the current of the LC oscillator:

$$i = \frac{dq}{dt} = -\omega Q\sin(\omega t + \phi) \quad \text{(current).} \tag{33-13}$$

The amplitude I of this sinusoidally varying current is

$$I = \omega Q, \tag{33-14}$$

so we can rewrite Eq. 33-13 as

$$i = -I\sin(\omega t + \phi). \tag{33-15}$$

Angular Frequencies

We can test whether Eq. 33-12 is a solution of Eq. 33-11 by substituting it and its second derivative with respect to time into Eq. 33-11. The first derivative of Eq. 33-12 is Eq. 33-13. The second derivative is then

$$\frac{d^2q}{dt^2} = -\omega^2 Q\cos(\omega t + \phi).$$

Substituting for q and d^2q/dt^2 into Eq. 33-11, we obtain

$$-L\omega^2 Q\cos(\omega t + \phi) + \frac{1}{C}Q\cos(\omega t + \phi) = 0.$$

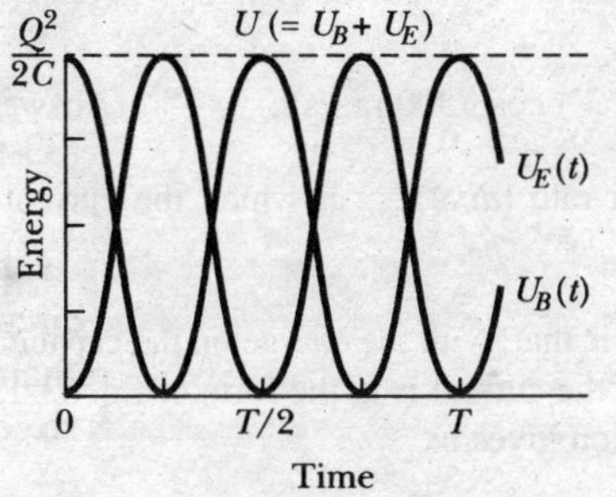

Fig. 33-4 The stored magnetic energy and electric energy in the circuit of Fig. 33-1 as a function of time. Note that their sum remains constant. T is the period of oscillation.

Canceling $Q\cos(\omega t + \phi)$ and rearranging lead to

$$\omega = \frac{1}{\sqrt{LC}}.$$

Thus, Eq. 33-12 is indeed a solution of Eq. 33-11 if ω has the constant value $1/\sqrt{LC}$. Note that this expression for ω is exactly that given by Eq. 33-4, which we arrived at by examining correspondences.

The phase constant ϕ in Eq. 33-12 is determined by the conditions that exist at any certain time, say, $t = 0$. If the conditions yield $\phi = 0$ at $t = 0$, Eq. 33-12 requires that $q = Q$ and Eq. 33-13 requires that $i = 0$; these are the initial conditions represented by Fig. 33-1*a*.

Electric and Magnetic Energy Oscillations

The electric energy stored in the *LC* circuit at any time t is, from Eqs. 33-1 and 33-12,

$$U_E = \frac{q^2}{2C} = \frac{Q^2}{2C}\cos^2(\omega t + \phi). \qquad (33\text{-}16)$$

The magnetic energy is, from Eqs. 33-2 and 33-13,

$$U_B = \tfrac{1}{2}Li^2 = \tfrac{1}{2}L\omega^2Q^2\sin^2(\omega t + \phi).$$

Substituting for ω from Eq. 33-4 then gives us

$$U_B = \frac{Q^2}{2C}\sin^2(\omega t + \phi). \qquad (33\text{-}17)$$

Figure 33-4 shows plots of $U_E(t)$ and $U_B(t)$ for the case of $\phi = 0$. Note that

1. The maximum values of U_E and U_B are both $Q^2/2C$.
2. At any instant the sum of U_E and U_B is equal to $Q^2/2C$, a constant.
3. When U_E is maximum, U_B is zero, and conversely.

✓CHECKPOINT 2: A capacitor in an *LC* oscillator has a maximum potential difference of 17 V and a maximum energy of 160 μJ. When the capacitor has a potential difference of 5 V and an energy of 10 μJ, what are (a) the emf across the inductor and (b) the energy stored in the magnetic field?

Sample Problem 33-2

For the situation described in Sample Problem 33-1, let the coil (inductor) be connected to the charged capacitor at time $t = 0$. The result is an *LC* circuit like that in Fig. 33-1.

(a) What is the potential difference $v_L(t)$ across the inductor as a function of time?

SOLUTION: One **Key Idea** here is that the current and potential differences of the circuit undergo sinusoidal oscillations. Another **Key Idea** is that we can still apply the loop rule to this oscillating circuit—just as we did for the nonoscillating circuits of Chapter 28. At any time t during the oscillations, the loop rule and Fig. 33-1 give us

$$v_L(t) = v_C(t); \qquad (33\text{-}18)$$

that is, the potential difference v_L across the inductor must always be equal to the potential difference v_C across the capacitor, so that the net potential difference around the circuit is zero. Thus, we will find $v_L(t)$ if we can find $v_C(t)$, and we can find $v_C(t)$ from $q(t)$ with Eq. 26-1 ($q = CV$).

Because the potential difference $v_C(t)$ is maximum when the oscillations begin at time $t = 0$, the charge q on the capacitor must also be maximum then. Thus, phase constant ϕ must be zero, so that Eq. 33-12 gives us

$$q = Q \cos \omega t. \qquad (33\text{-}19)$$

(Note that this cosine function does indeed yield maximum q (= Q) when $t = 0$.) To get the potential difference $v_C(t)$, we divide both sides of Eq. 33-19 by C to write

$$\frac{q}{C} = \frac{Q}{C} \cos \omega t,$$

and then use Eq. 26-1 to write

$$v_C = V_C \cos \omega t. \qquad (33\text{-}20)$$

Here, V_C is the amplitude of the oscillations in the potential difference v_C across the capacitor.

Next, substituting $v_C = v_L$ from Eq. 33-18, we find

$$v_L = V_C \cos \omega t. \qquad (33\text{-}21)$$

We can evaluate the right side of this equation by first noting that the amplitude V_C is equal to the initial (maximum) potential difference of 57 V across the capacitor. Then, using the values of L and C from Sample Problem 33-1, we find ω with Eq. 33-4:

$$\omega = \frac{1}{\sqrt{LC}} = \frac{1}{[(0.012 \text{ H})(1.5 \times 10^{-6} \text{ F})]^{0.5}}$$
$$= 7454 \text{ rad/s} \approx 7500 \text{ rad/s}.$$

Thus, Eq. 33-21 becomes

$$v_L = (57 \text{ V}) \cos(7500 \text{ rad/s})t. \qquad \text{(Answer)}$$

(b) What is the maximum rate $(di/dt)_{max}$ at which the current i changes in the circuit?

SOLUTION: The **Key Idea** here is that, with the charge on the capacitor oscillating as in Eq. 33-12, the current is in the form of Eq. 33-13. Because $\phi = 0$, that equation gives us

$$i = -\omega Q \sin \omega t.$$

Then $$\frac{di}{dt} = \frac{d}{dt}(-\omega Q \sin \omega t) = -\omega^2 Q \cos \omega t.$$

We can simplify this equation by substituting CV_C for Q (because we know C and V_C but not Q) and $1/\sqrt{LC}$ for ω according to Eq. 33-4. We get

$$\frac{di}{dt} = -\frac{1}{LC} CV_C \cos \omega t = -\frac{V_C}{L} \cos \omega t.$$

This tells us that the current changes at a varying (sinusoidal) rate, with its maximum rate of change being

$$\frac{V_C}{L} = \frac{57 \text{ V}}{0.012 \text{ H}} = 4750 \text{ A/s} \approx 4800 \text{ A/s}. \qquad \text{(Answer)}$$

33-5 Damped Oscillations in an *RLC* Circuit

A circuit containing resistance, inductance, and capacitance is called an *RLC circuit.* We shall here discuss only *series RLC circuits* like that shown in Fig. 33-5. With a resistance R present, the total *electromagnetic energy U* of the circuit (the sum of the electric energy and magnetic energy) is no longer constant; instead, it decreases with time as energy is transferred to thermal energy in the resistance. Because of this loss of energy, the oscillations of charge, current, and potential difference continuously decrease in amplitude, and the oscillations are said to be *damped.* As you will see, they are damped in exactly the same way as those of the damped block–spring oscillator of Section 16-8.

To analyze the oscillations of this circuit, we write an equation for the total electromagnetic energy U in the circuit at any instant. Because the resistance does not store electromagnetic energy, we can use Eq. 33-9:

$$U = U_B + U_E = \frac{Li^2}{2} + \frac{q^2}{2C}. \qquad (33\text{-}22)$$

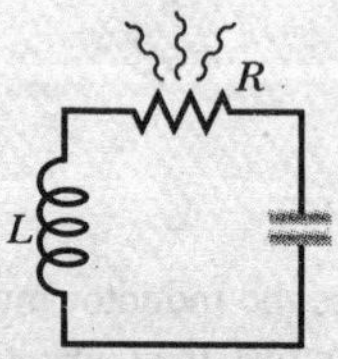

Fig. 33-5 A series *RLC* circuit. As the charge contained in the circuit oscillates back and forth through the resistance, electromagnetic energy is dissipated as thermal energy, damping (decreasing the amplitude of) the oscillations.

Now, however, this total energy decreases as energy is transferred to thermal energy. The rate of that transfer is, from Eq. 27-22,

$$\frac{dU}{dt} = -i^2 R, \qquad (33\text{-}23)$$

where the minus sign indicates that U decreases. By differentiating Eq. 33-22 with respect to time and then substituting the result in Eq. 33-23, we obtain

$$\frac{dU}{dt} = Li\frac{di}{dt} + \frac{q}{C}\frac{dq}{dt} = -i^2 R.$$

Substituting dq/dt for i and d^2q/dt^2 for di/dt, we obtain

$$L\frac{d^2q}{dt^2} + R\frac{dq}{dt} + \frac{1}{C}q = 0 \qquad (RLC \text{ circuit}), \qquad (33\text{-}24)$$

which is the differential equation for damped oscillations in an *RLC* circuit.

The solution to Eq. 33-24 is

$$q = Qe^{-Rt/2L}\cos(\omega' t + \phi), \qquad (33\text{-}25)$$

in which

$$\omega' = \sqrt{\omega^2 - (R/2L)^2}, \qquad (33\text{-}26)$$

where $\omega = 1/\sqrt{LC}$, as with an undamped oscillator. Equation 33-25 tells us how the charge on the capacitor oscillates in a damped *RLC* circuit; that equation is the electromagnetic counterpart of Eq. 16-40, which gives the displacement of a damped block–spring oscillator.

Equation 33-25 describes a sinusoidal oscillation (the cosine function) with an *exponentially decaying amplitude* $Qe^{-Rt/2L}$ (the factor that multiplies the cosine). The angular frequency ω' of the damped oscillations is always less than the angular frequency ω of the undamped oscillations; however, we shall here consider only situations in which R is small enough for us to replace ω' with ω.

Let us next find an expression for the total electromagnetic energy U of the circuit as a function of time. One way to do so is to monitor the energy of the electric field in the capacitor, which is given by Eq. 33-1 ($U_E = q^2/2C$). By substituting Eq. 33-25 into Eq. 33-1, we obtain

$$U_E = \frac{q^2}{2C} = \frac{[Qe^{-Rt/2L}\cos(\omega' t + \phi)]^2}{2C} = \frac{Q^2}{2C}e^{-Rt/L}\cos^2(\omega' t + \phi). \qquad (33\text{-}27)$$

Thus, the energy of the electric field oscillates according to a cosine-squared term and the amplitude of that oscillation decreases exponentially with time.

Sample Problem 33-3

A series *RLC* circuit has inductance $L = 12$ mH, capacitance $C = 1.6\ \mu$F, and resistance $R = 1.5\ \Omega$.

(a) At what time t will the amplitude of the charge oscillations in the circuit be 50% of its initial value?

SOLUTION: The **Key Idea** here is that the amplitude of the charge oscillations decreases exponentially with time t: According to Eq. 33-25, the charge amplitude at any time t is $Qe^{-Rt/2L}$, in which Q is the amplitude at time $t = 0$. We want the time when the charge amplitude has decreased to $0.500Q$, that is, when

$$Qe^{-Rt/2L} = 0.50Q.$$

Canceling Q and taking the natural logarithms of both sides, we have

$$-\frac{Rt}{2L} = \ln 0.50.$$

Solving for t and then substituting given data yield

$$t = -\frac{2L}{R}\ln 0.50 = -\frac{(2)(12\times10^{-3}\ \text{H})(\ln 0.50)}{1.5\ \Omega}$$

$$= 0.0111\ \text{s} \approx 11\ \text{ms}. \qquad \text{(Answer)}$$

(b) How many oscillations are completed within this time?

SOLUTION: The **Key Idea** here is that the time for one complete oscillation is the period $T = 2\pi/\omega$, where the angular frequency for *LC* oscillations is given by Eq. 33-4 ($\omega = 1/\sqrt{LC}$). Thus, in the time interval $\Delta t = 0.0111$ s, the number of complete oscillations is

$$\frac{\Delta t}{T} = \frac{\Delta t}{2\pi\sqrt{LC}}$$

$$= \frac{0.0111\ \text{s}}{2\pi[(12\times10^{-3}\ \text{H})(1.6\times10^{-6}\ \text{F})]^{1/2}} \approx 13. \qquad \text{(Answer)}$$

Thus, the amplitude decays by 50% in about 13 complete oscillations. This damping is less severe than that shown in Fig. 33-3, where the amplitude decays by a little more than 50% in one oscillation.

33-6 Alternating Current

The oscillations in an *RLC* circuit will not damp out if an external emf device supplies enough energy to make up for the energy dissipated as thermal energy in the resistance *R*. Circuits in homes, offices, and factories, including countless *RLC* circuits, receive such energy from local power companies. In most countries the energy is supplied via oscillating emfs and currents—the current is said to be an **alternating current,** or **ac** for short. (The nonoscillating current from a battery is said to be a **direct current,** or **dc**.) These oscillating emfs and currents vary sinusoidally with time, reversing direction (in North America) 120 times per second and thus having frequency $f = 60$ Hz.

At first sight this may seem to be a strange arrangement. We have seen that the drift speed of the conduction electrons in household wiring may typically be 4×10^{-5} m/s. If we now reverse their direction every $\frac{1}{120}$ s, such electrons can move only about 3×10^{-7} m in a half-cycle. At this rate, a typical electron can drift past no more than about 10 atoms in the wiring before it is required to reverse its direction. How, you may wonder, can the electron ever get anywhere?

Although this question may be worrisome, it is a needless concern. The conduction electrons do not have to "get anywhere." When we say that the current in a wire is one ampere, we mean that charge passes through any plane cutting across that wire at the rate of one coulomb per second. The speed at which the charge carriers cross that plane does not matter directly; one ampere may correspond to many charge carriers moving very slowly or to a few moving very rapidly. Furthermore, the signal to the electrons to reverse directions—which originates in the alternating emf provided by the power company's generator—is propagated along the conductor at a speed close to that of light. All electrons, no matter where they are located, get their reversal instructions at about the same instant. Finally, we note that for many devices, such as lightbulbs and toasters, the direction of motion is unimportant as long as the electrons do move so as to transfer energy to the device via collisions with atoms in the device.

The basic advantage of alternating current is this: *As the current alternates, so does the magnetic field that surrounds the conductor*. This makes possible the use of Faraday's law of induction, which, among other things, means that we can step up (increase) or step down (decrease) the magnitude of an alternating potential difference at will, using a device called a transformer, as you will see later in this chapter. Moreover, alternating current is more readily adaptable to rotating machinery such as generators and motors than is (nonalternating) direct current.

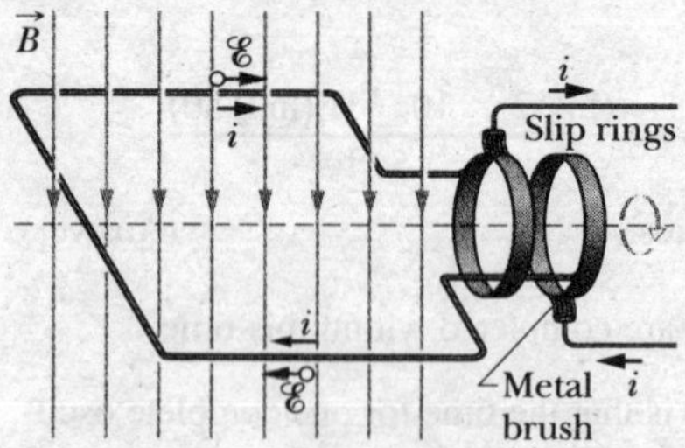

Fig. 33-6 The basic mechanism of an alternating-current generator is a conducting loop rotated in an external magnetic field. In practice, the alternating emf induced in a coil of many turns of wire is made accessible by means of slip rings attached to the rotating loop. Each ring is connected to one end of the loop wire and is electrically connected to the rest of the generator circuit by a conducting brush against which it slips as the loop (and it) rotates.

Figure 33-6 shows a simple model of an ac generator. As the conducting loop is forced to rotate through the external magnetic field $\vec{B}$, a sinusoidally oscillating emf $\mathscr{E}$ is induced in the loop:

$$\mathscr{E} = \mathscr{E}_m \sin \omega_d t. \tag{33-28}$$

The *angular frequency* ω_d of the emf is equal to the angular speed with which the loop rotates in the magnetic field; the *phase* of the emf is $\omega_d t$; and the *amplitude* of the emf is $\mathscr{E}_m$ (where the subscript stands for maximum). When the rotating loop is part of a closed conducting path, this emf produces (*drives*) a sinusoidal (alternating) current along the path with the same angular frequency ω_d, which then is called the **driving angular frequency.** We can write the current as

$$i = I \sin(\omega_d t - \phi), \tag{33-29}$$

in which I is the amplitude of the driven current. (The phase $\omega_d t - \phi$ of the current is traditionally written with a minus sign instead of as $\omega_d t + \phi$.) We include a phase

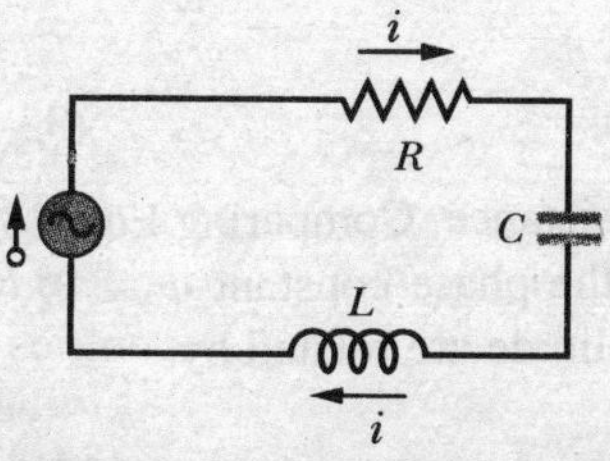

Fig. 33-7 A single-loop circuit containing a resistor, a capacitor, and an inductor. A generator, represented by a sine wave in a circle, produces an alternating emf that establishes an alternating current; the directions of the emf and current are indicated here at only one instant.

constant ϕ in Eq. 33-29 because the current i may not be in phase with the emf $\mathscr{E}$. (As you will see, the phase constant depends on the circuit to which the generator is connected.) We can also write the current i in terms of the **driving frequency** f_d of the emf, by substituting $2\pi f_d$ for ω_d in Eq. 33-29.

33-7 Forced Oscillations

We have seen that once started, the charge, potential difference, and current in both undamped LC circuits and damped RLC circuits (with small enough R) oscillate at angular frequency $\omega = 1/\sqrt{LC}$. Such oscillations are said to be *free oscillations* (free of any external emf), and the angular frequency ω is said to be the circuit's **natural angular frequency.**

When the external alternating emf of Eq. 33-28 is connected to an RLC circuit, the oscillations of charge, potential difference, and current are said to be *driven oscillations* or *forced oscillations*. These oscillations always occur at the driving angular frequency ω_d:

> Whatever the natural angular frequency ω of a circuit may be, forced oscillations of charge, current, and potential difference in the circuit always occur at the driving angular frequency ω_d.

However, as you will see in Section 33-9, the amplitudes of the oscillations very much depend on how close ω_d is to ω. When the two angular frequencies match—a condition known as **resonance**—the amplitude I of the current in the circuit is maximum.

33-8 Three Simple Circuits

Later in this chapter, we shall connect an external alternating emf device to a series RLC circuit as in Fig. 33-7. We shall then find expressions for the amplitude I and phase constant ϕ of the sinusoidally oscillating current in terms of the amplitude $\mathscr{E}_m$ and angular frequency ω_d of the external emf. First, however, let us consider three simpler circuits, each having an external emf and only one other circuit element: R, C, or L. We start with a resistive element (a purely *resistive load*).

A Resistive Load

Figure 33-8*a* shows a circuit containing a resistance element of value R and an ac generator with the alternating emf of Eq. 33-28. By the loop rule, we have

$$\mathscr{E} - v_R = 0.$$

With Eq. 33-28, this gives us

$$v_R = \mathscr{E}_m \sin \omega_d t.$$

Because the amplitude V_R of the alternating potential difference (or voltage) across the resistance is equal to the amplitude $\mathscr{E}_m$ of the alternating emf, we can write this as

$$v_R = V_R \sin \omega_d t. \tag{33-30}$$

From the definition of resistance ($R = V/i$), we can now write the current i_R in the resistance as

$$i_R = \frac{v_R}{R} = \frac{V_R}{R} \sin \omega_d t. \tag{33-31}$$

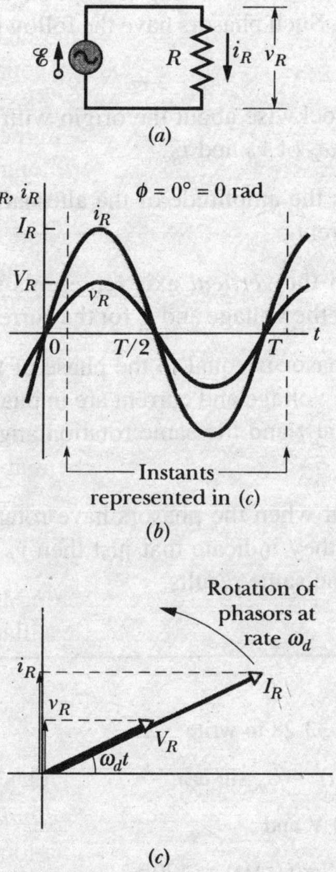

Fig. 33-8 (*a*) A resistor is connected across an alternating-current generator. (*b*) The current i_R and the potential difference v_R across the resistor are plotted on the same graph, both versus time t. They are in phase and complete one cycle in one period T. (*c*) A phasor diagram shows the same thing as (*b*).

From Eq. 33-29, we can also write this current as

$$i_R = I_R \sin(\omega_d t - \phi), \tag{33-32}$$

where I_R is the amplitude of the current i_R in the resistance. Comparing Eqs. 33-31 and 33-32, we see that for a purely resistive load the phase constant $\phi = 0°$. We also see that the voltage amplitude and current amplitude are related by

$$V_R = I_R R \quad \text{(resistor).} \tag{33-33}$$

Although we found this relation for the circuit of Fig. 33-8*a*, it applies to any resistance in any ac circuit.

By comparing Eqs. 33-30 and 33-31, we see that the time-varying quantities v_R and i_R are both functions of $\sin \omega_d t$ with $\phi = 0°$. Thus, these two quantities are *in phase,* which means that their corresponding maxima (and minima) occur at the same times. Figure 33-8*b*, which is a plot of $v_R(t)$ and $i_R(t)$, illustrates this fact. Note that v_R and i_R do not decay here, because the generator supplies energy to the circuit to make up for the energy dissipated in R.

The time-varying quantities v_R and i_R can also be represented geometrically by *phasors.* Recall from Section 17-10 that phasors are vectors that rotate around an origin. Those that represent the voltage across and current in the resistor of Fig. 33-8*a* are shown in Fig. 33-8*c* at an arbitrary time t. Such phasors have the following properties:

Angular speed: Both phasors rotate counterclockwise about the origin with an angular speed equal to the angular frequency ω_d of v_R and i_R.

Length: The length of each phasor represents the amplitude of the alternating quantity: V_R for the voltage and I_R for the current.

Projection: The projection of each phasor on the *vertical* axis represents the value of the alternating quantity at time t: v_R for the voltage and i_R for the current.

Rotation angle: The rotation angle of each phasor is equal to the phase of the alternating quantity at time t. In Fig. 33-8*c*, the voltage and current are in phase, so their phasors always have the same phase $\omega_d t$ and the same rotation angle, and thus they rotate together.

Mentally follow the rotation. Can you see that when the phasors have rotated so that $\omega_d t = 90°$ (they point vertically upward), they indicate that just then $v_R = V_R$ and $i_R = I_R$? Equations 33-30 and 33-32 give the same results.

Sample Problem 33-4

Purely resistive load. In Fig. 33-8*a*, resistance R is 200 Ω and the sinusoidal alternating emf device operates at amplitude $\mathscr{E}_m = 36.0$ V and frequency $f_d = 60.0$ Hz.

(a) What is the potential difference $v_R(t)$ across the resistance as a function of time t, and what is the amplitude V_R of $v_R(t)$?

SOLUTION: When we applied the loop rule to the circuit of Fig. 33-8*a*, we found this **Key Idea**: In a circuit with a purely resistive load, the potential difference $v_R(t)$ across the resistance is always equal to the potential difference $\mathscr{E}(t)$ across the emf device. Thus, $v_R(t) = \mathscr{E}(t)$ and $V_R = \mathscr{E}_m$. Since $\mathscr{E}_m$ is given, we can write

$$V_R = \mathscr{E}_m = 36.0 \text{ V}. \quad \text{(Answer)}$$

To find $v_R(t)$, we use Eq. 33-28 to write

$$v_R(t) = \mathscr{E}(t) = \mathscr{E}_m \sin \omega_d t, \tag{33-34}$$

and then substitute $\mathscr{E}_m = 36.0$ V and

$$\omega_d = 2\pi f_d = 2\pi(60 \text{ Hz}) = 120\pi$$

to obtain

$$v_R = (36.0 \text{ V}) \sin(120\pi t). \quad \text{(Answer)}$$

We can leave the argument of the sine in this form for convenience, or we can write it as (377 rad/s)t or as (377 s^{-1})t.

(b) What are the current $i_R(t)$ in the resistance and the amplitude I_R of $i_R(t)$?

SOLUTION: The **Key Idea** here is that in an ac circuit with a purely resistive load, the alternating current $i_R(t)$ in the resistance is *in phase* with the alternating potential difference $v_R(t)$ across the resistance; that is, the phase constant ϕ for the current is zero. Thus, we can write Eq. 33-29 as

$$i_R = I_R \sin(\omega_d t - \phi) = I_R \sin \omega_d t. \qquad (33\text{-}35)$$

From Eq. 33-33, the amplitude I_R is

$$I_R = \frac{V_R}{R} = \frac{36.0 \text{ V}}{200\ \Omega} = 0.180 \text{ A}. \qquad \text{(Answer)}$$

Substituting this and $\omega_d = 2\pi f_d = 120\pi$ into Eq. 33-35, we have

$$i_R = (0.180 \text{ A}) \sin(120\pi t). \qquad \text{(Answer)}$$

✔CHECKPOINT 3: If we increase the driving frequency in a circuit with a purely resistive load, do (a) amplitude V_R and (b) amplitude I_R increase, decrease, or remain the same?

A Capacitive Load

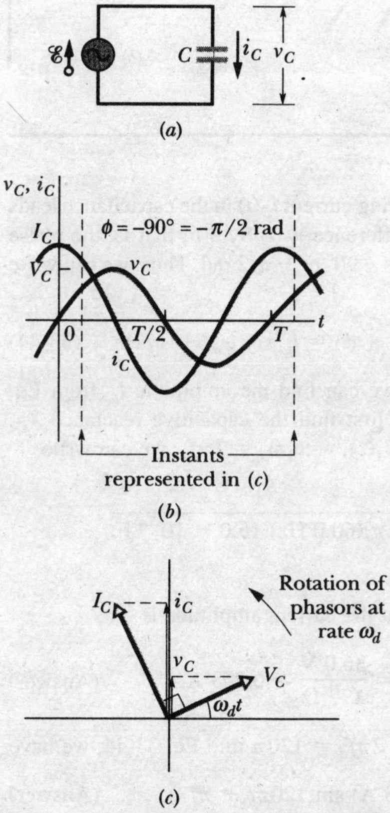

Fig. 33-9 (*a*) A capacitor is connected across an alternating-current generator. (*b*) The current in the capacitor leads the voltage by 90° (= $\pi/2$ rad). (*c*) A phasor diagram shows the same thing.

Figure 33-9*a* shows a circuit containing a capacitance and a generator with the alternating emf of Eq. 33-28. Using the loop rule and proceeding as we did when we obtained Eq. 33-30, we find that the potential difference across the capacitor is

$$v_C = V_C \sin \omega_d t, \qquad (33\text{-}36)$$

where V_C is the amplitude of the alternating voltage across the capacitor. From the definition of capacitance we can also write

$$q_C = C v_C = C V_C \sin \omega_d t. \qquad (33\text{-}37)$$

Our concern, however, is with the current rather than the charge. Thus, we differentiate Eq. 33-37 to find

$$i_C = \frac{dq_C}{dt} = \omega_d C V_C \cos \omega_d t. \qquad (33\text{-}38)$$

We now modify Eq. 33-38 in two ways. First, for reasons of symmetry of notation, we introduce the quantity X_C, called the **capacitive reactance** of a capacitor, defined as

$$X_C = \frac{1}{\omega_d C} \quad \text{(capacitive reactance).} \qquad (33\text{-}39)$$

Its value depends not only on the capacitance but also on the driving angular frequency ω_d. We know from the definition of the capacitive time constant ($\tau = RC$) that the SI unit for C can be expressed as seconds per ohm. Applying this to Eq. 33-39 shows that the SI unit of X_C is the *ohm,* just as for resistance R.

Second, we replace $\cos \omega_d t$ in Eq. 33-38 with a phase-shifted sine:

$$\cos \omega_d t = \sin(\omega_d t + 90°).$$

You can verify this identity by shifting a sine curve in the negative direction by 90°.

With these two modifications, Eq. 33-38 becomes

$$i_C = \left(\frac{V_C}{X_C}\right) \sin(\omega_d t + 90°). \qquad (33\text{-}40)$$

From Eq. 33-29, we can also write the current i_C in C as

$$i_C = I_C \sin(\omega_d t - \phi), \qquad (33\text{-}41)$$

where I_C is the amplitude of i_C. Comparing Eqs. 33-40 and 33-41, we see that for a purely capacitive load the phase constant ϕ for the current is $-90°$. We also see that the voltage amplitude and current amplitude are related by

$$V_C = I_C X_C \quad \text{(capacitor).} \qquad (33\text{-}42)$$

Although we found this relation for the circuit of Fig. 33-9*a*, it applies to any capacitance in any ac circuit.

Comparison of Eqs. 33-36 and 33-40, or inspection of Fig. 33-9*b*, shows that the quantities v_C and i_C are 90°, or one-quarter cycle, out of phase. Furthermore, we see that i_C *leads* v_C, which means that, if you monitored the current i_C and the potential difference v_C in the circuit of Fig. 33-9*a*, you would find that i_C reaches its maximum *before* v_C does, by one-quarter cycle.

This relation between i_C and v_C is illustrated by the phasor diagram of Fig. 33-9*c*. As the phasors representing these two quantities rotate counterclockwise together, the phasor labeled I_C does indeed lead that labeled V_C, and by an angle of 90°; that is, the phasor I_C coincides with the vertical axis one-quarter cycle before the phasor V_C does. Be sure to convince yourself that the phasor diagram of Fig. 33-9*c* is consistent with Eqs. 33-36 and 33-40.

CHECKPOINT 4: The figure shows, in (*a*), a sine curve $S(t) = \sin(\omega_d t)$ and three other sinusoidal curves $A(t)$, $B(t)$, and $C(t)$, each of the form $\sin(\omega_d t - \phi)$. (a) Rank the three other curves according to the value of ϕ, most positive first and most negative last. (b) Which curve corresponds to which phasor in (*b*) of the figure? (c) Which curve leads the others?

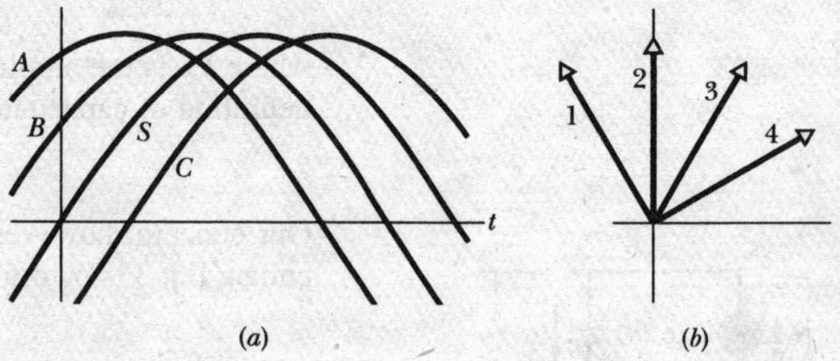

Sample Problem 33-5

Purely capacitive load. In Fig. 33-9*a*, capacitance C is 15.0 μF and the sinusoidal alternating emf device operates at amplitude $\mathscr{E}_m = 36.0$ V and frequency $f_d = 60.0$ Hz.

(a) What are the potential difference $v_C(t)$ across the capacitance and the amplitude V_C of $v_C(t)$?

SOLUTION: If we apply the loop rule to the circuit of Fig. 33-9*a*, we find this **Key Idea**: In a circuit with a purely capacitive load, the potential difference $v_C(t)$ across the capacitance is always equal to the potential difference $\mathscr{E}(t)$ across the emf device. Thus, $v_C(t) = \mathscr{E}(t)$ and $V_C = \mathscr{E}_m$. Since $\mathscr{E}_m$ is given, we have

$$V_C = \mathscr{E}_m = 36.0 \text{ V.} \qquad \text{(Answer)}$$

To find $v_C(t)$, we use Eq. 33-28 to write

$$v_C(t) = \mathscr{E}(t) = \mathscr{E}_m \sin \omega_d t. \qquad (33\text{-}43)$$

Then, substituting $\mathscr{E}_m = 36.0$ V and $\omega_d = 2\pi f_d = 120\pi$ into Eq. 33-43, we have

$$v_C = (36.0 \text{ V}) \sin(120\pi t). \qquad \text{(Answer)}$$

(b) What are the current $i_C(t)$ in the circuit as a function of time and the amplitude I_C of $i_C(t)$?

SOLUTION: The **Key Idea** here is that in an ac circuit with a purely capacitive load, the alternating current $i_C(t)$ in the capacitance leads the alternating potential difference $v_C(t)$ by 90°; that is, the phase constant ϕ for the current is $-90°$ or $-\pi/2$ rad. Thus, we can write Eq. 33-29 as

$$i_C = I_C \sin(\omega_d t - \phi) = I_C \sin(\omega_d t + \pi/2). \qquad (33\text{-}44)$$

A second **Key Idea** is that we can find the amplitude I_C from Eq. 33-42 ($V_C = I_C X_C$) if we first find the capacitive reactance X_C. From Eq. 33-39 ($X_C = 1/\omega_d C$), with $\omega_d = 2\pi f_d$, we can write

$$X_C = \frac{1}{2\pi f_d C} = \frac{1}{(2\pi)(60.0 \text{ Hz})(15.0 \times 10^{-6} \text{ F})} = 177\ \Omega.$$

Then Eq. 33-42 tells us that the current amplitude is

$$I_C = \frac{V_C}{X_C} = \frac{36.0 \text{ V}}{177\ \Omega} = 0.203 \text{ A.} \qquad \text{(Answer)}$$

Substituting this and $\omega_d = 2\pi f_d = 120\pi$ into Eq. 33-44, we have

$$i_C = (0.203 \text{ A}) \sin(120\pi t + \pi/2). \qquad \text{(Answer)}$$

CHECKPOINT 5: If we increase the driving frequency in a circuit with a purely capacitive load, do (a) amplitude V_C and (b) amplitude I_C increase, decrease, or remain the same?

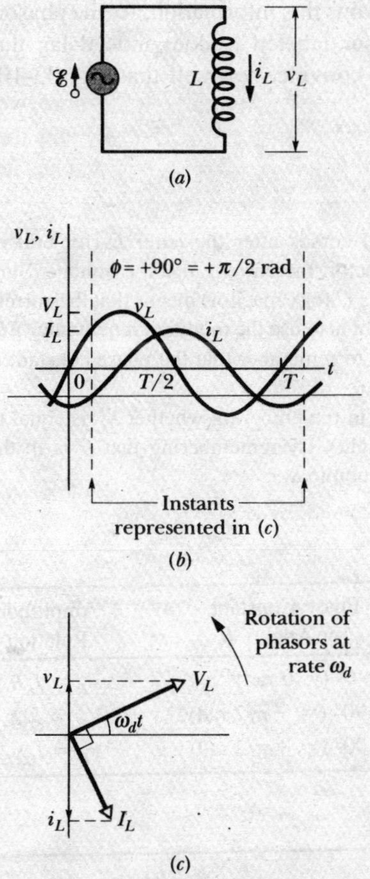

Fig. 33-10 (*a*) An inductor is connected across an alternating-current generator. (*b*) The current in the inductor lags the voltage by 90° ($= \pi/2$ rad). (*c*) A phasor diagram shows the same thing.

An Inductive Load

Figure 33-10*a* shows a circuit containing an inductance and a generator with the alternating emf of Eq. 33-28. Using the loop rule and proceeding as we did to obtain Eq. 33-30, we find that the potential difference across the inductance is

$$v_L = V_L \sin \omega_d t, \tag{33-45}$$

where V_L is the amplitude of v_L. From Eq. 31-37, we can write the potential difference across an inductance L, in which the current is changing at the rate di_L/dt, as

$$v_L = L\frac{di_L}{dt}. \tag{33-46}$$

If we combine Eqs. 33-45 and 33-46, we have

$$\frac{di_L}{dt} = \frac{V_L}{L} \sin \omega_d t. \tag{33-47}$$

Our concern, however, is with the current rather than with its time derivative. We find the former by integrating Eq. 33-47, obtaining

$$i_L = \int di_L = \frac{V_L}{L}\int \sin \omega_d t\, dt = -\left(\frac{V_L}{\omega_d L}\right) \cos \omega_d t. \tag{33-48}$$

We now modify this equation in two ways. First, for reasons of symmetry of notation, we introduce the quantity X_L, called the **inductive reactance** of an inductor, which is defined as

$$X_L = \omega_d L \qquad \text{(inductive reactance).} \tag{33-49}$$

The value of X_L depends on the driving angular frequency ω_d. The unit of the inductive time constant τ_L indicates that the SI unit of X_L is the *ohm,* just as it is for X_C and for R.

Second, we replace $-\cos \omega_d t$ in Eq. 33-48 with a phase-shifted sine:

$$-\cos \omega_d t = \sin(\omega_d t - 90°).$$

You can verify this identity by shifting a sine curve in the positive direction by 90°.

With these two changes, Eq. 33-48 becomes

$$i_L = \left(\frac{V_L}{X_L}\right) \sin(\omega_d t - 90°). \tag{33-50}$$

From Eq. 33-29, we can also write this current in the inductance as

$$i_L = I_L \sin(\omega_d t - \phi), \tag{33-51}$$

where I_L is the amplitude of the current i_L. Comparing Eqs. 33-50 and 33-51, we see that for a purely inductive load the phase constant ϕ for the current is $+90°$. We also see that the voltage amplitude and current amplitude are related by

$$V_L = I_L X_L \qquad \text{(inductor).} \tag{33-52}$$

Although we found this relation for the circuit of Fig. 33-10*a*, it applies to any inductance in any ac circuit.

Comparison of Eqs. 33-45 and 33-50, or inspection of Fig. 33-10*b*, shows that the quantities i_L and v_L are 90° out of phase. In this case, however, i_L *lags* v_L; that is, if you monitored the current i_L and the potential difference v_L in the circuit of Fig. 33-10*a*, you would find that i_L reaches its maximum value *after* v_L does, by one-quarter cycle.

The phasor diagram of Fig. 33-10*c* also contains this information. As the phasors rotate counterclockwise in the figure, the phasor labeled I_L does indeed lag that labeled V_L, and by an angle of 90°. Be sure to convince yourself that Fig. 33-10*c* represents Eqs. 33-45 and 33-50.

PROBLEM-SOLVING TACTICS

Tactic 1: *Leading and Lagging in AC Circuits*
Table 33-2 summarizes the relations between the current i and the voltage v for each of the three kinds of circuit elements we have considered. When an applied alternating voltage produces an alternating current in them, the current is in phase with the voltage across a resistor, leads the voltage across a capacitor, and lags the voltage across an inductor.

Many students remember these results with the mnemonic "*ELI* the *ICE* man." *ELI* contains the letter *L* (for inductor), and in it the letter *I* (for current) comes after the letter *E* (for emf or voltage). Thus, for an inductor, the current *lags* the voltage. Similarly *ICE* (which contains a *C* for capacitor) means that the current *leads* the voltage. You might also use the modified mnemonic "*ELI positively* is the *ICE* man" to remember that the phase constant ϕ is positive for an inductor.

If you have difficulty in remembering whether X_C is equal to $\omega_d C$ (wrong) or $1/\omega_d C$ (right), try remembering that *C* is in the "cellar"—that is, in the denominator.

TABLE 33-2 Phase and Amplitude Relations for Alternating Currents and Voltages

Circuit Element	Symbol	Resistance or Reactance	Phase of the Current	Phase Constant (or Angle) ϕ	Amplitude Relation
Resistor	R	R	In phase with v_R	0° (= 0 rad)	$V_R = I_R R$
Capacitor	C	$X_C = 1/\omega_d C$	Leads v_C by 90° (= $\pi/2$ rad)	−90° (= $-\pi/2$ rad)	$V_C = I_C X_C$
Inductor	L	$X_L = \omega_d L$	Lags v_L by 90° (= $\pi/2$ rad)	+90° (= $+\pi/2$ rad)	$V_L = I_L X_L$

Sample Problem 33-6

Purely inductive load. In Fig. 33-10*a*, inductance L is 230 mH and the sinusoidal alternating emf device operates at amplitude $\mathscr{E}_m$ = 36.0 V and frequency f_d = 60.0 Hz.

(a) What are the potential difference $v_L(t)$ across the inductance and the amplitude V_L of $v_L(t)$?

SOLUTION: If we apply the loop rule to the circuit in Fig. 33-10*a*, we find this **Key Idea:** In a circuit with a purely inductive load, the potential difference $v_L(t)$ across the inductance is always equal to the potential difference $\mathscr{E}(t)$ across the emf device. Thus, $v_L(t) = \mathscr{E}(t)$ and $V_L = \mathscr{E}_m$. Since $\mathscr{E}_m$ is given, we have

$$V_L = \mathscr{E}_m = 36.0 \text{ V}. \quad \text{(Answer)}$$

To find $v_L(t)$, we use Eq. 33-28 to write

$$v_L(t) = \mathscr{E}(t) = \mathscr{E}_m \sin \omega_d t. \quad (33\text{-}53)$$

Then, substituting $\mathscr{E}_m$ = 36.0 V and $\omega_d = 2\pi f_d = 120\pi$ into Eq. 33-53, we have

$$v_L = (36.0 \text{ V}) \sin(120\pi t). \quad \text{(Answer)}$$

(b) What are the current $i_L(t)$ in the circuit as a function of time and the amplitude I_L of $i_L(t)$?

SOLUTION: The **Key Idea** here is that in an ac circuit with a purely inductive load, the alternating current $i_L(t)$ in the inductance lags the alternating potential difference $v_L(t)$ by 90°. (In the mnemonic of Problem-Solving Tactic 1, this circuit is "positively an *ELI* circuit," which tells us that the emf *E* leads the current *I* and that ϕ is *positive*.) Thus, the phase constant ϕ for the current is +90° or $+\pi/2$ rad, and we can write Eq. 33-29 as

$$i_L = I_L \sin(\omega_d t - \phi) = I_L \sin(\omega_d t - \pi/2). \quad (33\text{-}54)$$

A second **Key Idea** is that we can find the amplitude I_L from Eq. 33-52 ($V_L = I_L X_L$) if we first find the inductive reactance X_L. From Eq. 33-49 ($X_L = \omega_d L$), with $\omega_d = 2\pi f_d$, we can write

$$X_L = 2\pi f_d L = (2\pi)(60.0 \text{ Hz})(230 \times 10^{-3} \text{ H})$$
$$= 86.7 \ \Omega.$$

Then Eq. 33-52 tells us that the current amplitude is

$$I_L = \frac{V_L}{X_L} = \frac{36.0 \text{ V}}{86.7 \ \Omega} = 0.415 \text{ A}. \quad \text{(Answer)}$$

Substituting this and $\omega_d = 2\pi f_d = 120\pi$ into Eq. 33-54, we have

$$i_L = (0.415 \text{ A}) \sin(120\pi t - \pi/2). \quad \text{(Answer)}$$

✔CHECKPOINT 6: If we increase the driving frequency in a circuit with a purely inductive load, do (a) amplitude V_L and (b) amplitude I_L increase, decrease, or remain the same?

33-9 The Series *RLC* Circuit

We are now ready to apply the alternating emf of Eq. 33-28,

$$\mathscr{E} = \mathscr{E}_m \sin \omega_d t \quad \text{(applied emf)}, \tag{33-55}$$

to the full *RLC* circuit of Fig. 33-7. Because *R*, *L*, and *C* are in series, the same current

$$i = I \sin(\omega_d t - \phi) \tag{33-56}$$

is driven in all three of them. We wish to find the current amplitude *I* and the phase constant ϕ. The solution is simplified by the use of phasor diagrams.

The Current Amplitude

We start with Fig. 33-11*a*, which shows the phasor representing the current of Eq. 33-56 at an arbitrary time *t*. The length of the phasor is the current amplitude *I*, the projection of the phasor on the vertical axis is the current *i* at time *t*, and the angle of rotation of the phasor is the phase $\omega_d t - \phi$ of the current at time *t*.

Figure 33-11*b* shows the phasors representing the voltages across *R*, *L*, and *C* at the same time *t*. Each phasor is oriented relative to the angle of rotation of current phasor *I* in Fig. 33-11*a*, based on the information in Table 33-2:

Resistor: Here current and voltage are in phase, so the angle of rotation of voltage phasor V_R is the same as that of phasor *I*.

Capacitor: Here current leads voltage by 90°, so the angle of rotation of voltage phasor V_C is 90° less than that of phasor *I*.

Inductor: Here current lags voltage by 90°, so the angle of rotation of voltage phasor v_L is 90° greater than that of phasor *I*.

Figure 33-11*b* also shows the instantaneous voltages v_R, v_C, and v_L across *R*, *C*, and *L* at time *t*; those voltages are the projections of the corresponding phasors on the vertical axis of the figure.

Figure 33-11*c* shows the phasor representing the applied emf of Eq. 33-55. The length of the phasor is the emf amplitude $\mathscr{E}_m$, the projection of the phasor on the vertical axis is the emf $\mathscr{E}$ at time *t*, and the angle of rotation of the phasor is the phase $\omega_d t$ of the emf at time *t*.

From the loop rule we know that at any instant the sum of the voltages v_R, v_C, and v_L is equal to the applied emf $\mathscr{E}$:

$$\mathscr{E} = v_R + v_C + v_L. \tag{33-57}$$

Thus, at time *t* the projection $\mathscr{E}$ in Fig. 33-11*c* is equal to the algebraic sum of the projections v_R, v_C, and v_L in Fig. 33-11*b*. In fact, as the phasors rotate together, this equality always holds. This means that phasor $\mathscr{E}_m$ in Fig. 33-11*c* must be equal to the vector sum of the three voltage phasors V_R, V_C, and V_L in Fig. 33-11*b*.

Fig. 33-11 (*a*) A phasor representing the alternating current in the driven *RLC* circuit of Fig. 33-7 at time *t*. The amplitude *I*, the instantaneous value *i*, and the phase $(\omega_d t - \phi)$ are shown. (*b*) Phasors representing the voltages across the inductor, resistor, and capacitor, oriented with respect to the current phasor in (*a*). (*c*) A phasor representing the alternating emf that drives the current of (*a*). (*d*) The emf phasor is equal to the vector sum of the three voltage phasors of (*b*). Here, voltage phasors V_L and V_C have been added to yield their net phasor $(V_L - V_C)$.

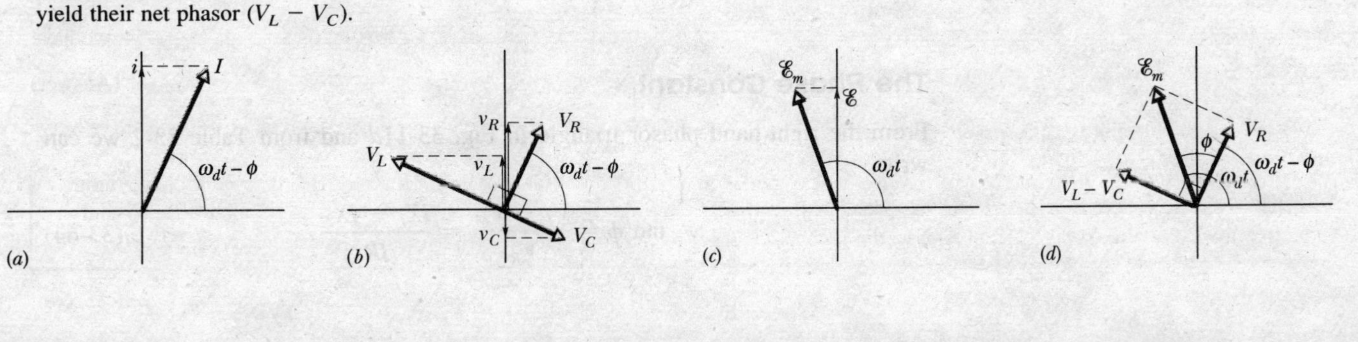

That requirement is indicated in Fig. 33-11d, where phasor $\mathscr{E}_m$ is drawn as the sum of phasors V_R, V_L, and V_C. Because phasors V_L and V_C have opposite directions in the figure, we simplify the vector sum by first combining V_L and V_C to form the single phasor $V_L - V_C$. Then we combine that single phasor with V_R to find the net phasor. Again, the net phasor must coincide with phasor $\mathscr{E}_m$, as shown.

Both triangles in Fig. 33-11d are right triangles. Applying the Pythagorean theorem to either one yields

$$\mathscr{E}_m^2 = V_R^2 + (V_L - V_C)^2. \tag{33-58}$$

From the amplitude information displayed in Table 33-2 we can rewrite this as

$$\mathscr{E}_m^2 = (IR)^2 + (IX_L - IX_C)^2, \tag{33-59}$$

and then rearrange it to the form

$$I = \frac{\mathscr{E}_m}{\sqrt{R^2 + (X_L - X_C)^2}}. \tag{33-60}$$

The denominator in Eq. 33-60 is called the **impedance** Z of the circuit for the driving angular frequency ω_d:

$$Z = \sqrt{R^2 + (X_L - X_C)^2} \quad \text{(impedance defined).} \tag{33-61}$$

We can then write Eq. 33-60 as

$$I = \frac{\mathscr{E}_m}{Z}. \tag{33-62}$$

If we substitute for X_C and X_L from Eqs. 33-39 and 33-49, we can write Eq. 33-60 more explicitly as

$$I = \frac{\mathscr{E}_m}{\sqrt{R^2 + (\omega_d L - 1/\omega_d C)^2}} \quad \text{(current amplitude).} \tag{33-63}$$

We have now accomplished half our goal: We have obtained an expression for the current amplitude I in terms of the sinusoidal driving emf and the circuit elements in a series RLC circuit.

The value of I depends on the difference between $\omega_d L$ and $1/\omega_d C$ in Eq. 33-63 or, equivalently, the difference between X_L and X_C in Eq. 33-60. In either equation, it does not matter which of the two quantities is greater because the difference is always squared.

The current that we have been describing in this section is the *steady-state current* that occurs after the alternating emf has been applied for some time. When the emf is first applied to a circuit, a brief *transient current* occurs. Its duration (before settling down into the steady-state current) is determined by the time constants $\tau_L = L/R$ and $\tau_C = RC$ as the inductive and capacitive elements "turn on." This transient current can be large and can, for example, destroy a motor on start-up if it is not properly taken into account in the motor's circuit design.

The Phase Constant

From the right-hand phasor triangle in Fig. 33-11d and from Table 33-2 we can write

$$\tan \phi = \frac{V_L - V_C}{V_R} = \frac{IX_L - IX_C}{IR}, \tag{33-64}$$

which gives us

$$\tan\phi = \frac{X_L - X_C}{R} \qquad \text{(phase constant).} \tag{33-65}$$

This is the other half of our goal: an equation for the phase constant ϕ in a sinusoidally driven series *RLC* circuit. In essence, it gives us three different results for the phase constant, depending on the relative values of X_L and X_C:

$X_L > X_C$: The circuit is said to be *more inductive than capacitive*. Equation 33-65 tells us that ϕ is positive for such a circuit, which means that phasor I rotates behind phasor $\mathscr{E}_m$ (Fig. 33-12*a*). A plot of $\mathscr{E}$ and i versus time is like that in Fig. 33-12*b*. (Figures 33-11*c* and *d* were drawn assuming $X_L > X_C$.)

$X_C > X_L$: The circuit is said to be *more capacitive than inductive*. Equation 33-65 tells us that ϕ is negative for such a circuit, which means that phasor I rotates ahead of phasor $\mathscr{E}_m$ (Fig. 33-12*c*). A plot of $\mathscr{E}$ and i versus time is like that in Fig. 33-12*d*.

$X_C = X_L$: The circuit is said to be in *resonance*, a state that is discussed next. Equation 33-65 tells us that $\phi = 0°$ for such a circuit, which means that phasors $\mathscr{E}_m$ and I rotate together (Fig. 33-12*e*). A plot of $\mathscr{E}$ and i versus time is like that in Fig. 33-12*f*.

As illustration, let us reconsider two extreme circuits: In the *purely inductive circuit* of Fig. 33-10*a*, where X_L is nonzero and $X_C = R = 0$, Eq. 33-65 tells us that $\phi = +90°$ (the greatest value of ϕ), consistent with Fig. 33-10*c*. In the *purely capacitive circuit* of Fig. 33-9*a*, where X_C is nonzero and $X_L = R = 0$, Eq. 33-65 tells us that $\phi = -90°$ (the least value of ϕ), consistent with Fig. 33-9*c*.

Resonance

Equation 33-63 gives the current amplitude I in an *RLC* circuit as a function of the driving angular frequency ω_d of the external alternating emf. For a given resistance R, that amplitude is a maximum when the quantity $\omega_d L - 1/\omega_d C$ in the denominator

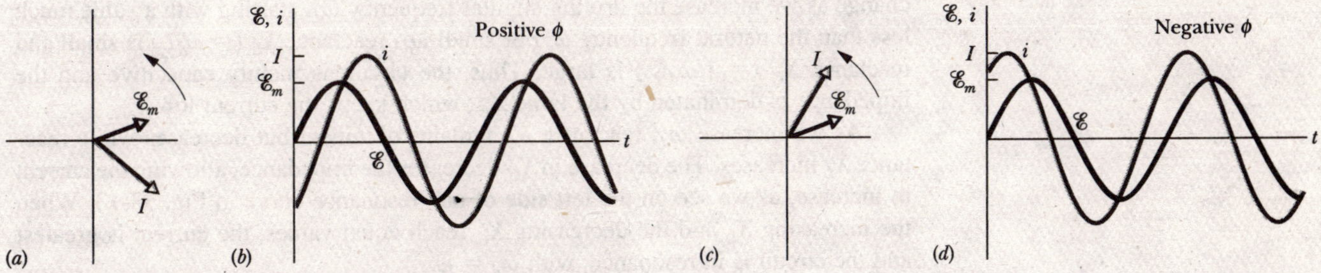

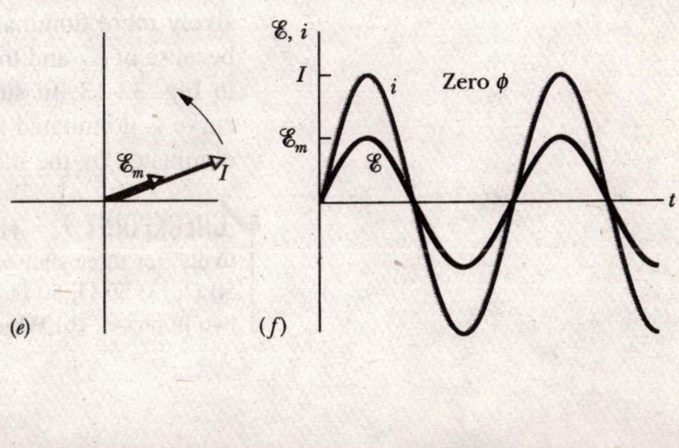

Fig. 33-12 Phasor diagrams and graphs of the alternating emf $\mathscr{E}$ and current i for the driven *RLC* circuit of Fig. 33-7. In the phasor diagram of (*a*) and the graph of (*b*), the current i lags the driving emf $\mathscr{E}$ and the current's phase constant ϕ is positive. In (*c*) and (*d*), the current i leads the driving emf $\mathscr{E}$ and its phase constant ϕ is negative. In (*e*) and (*f*), the current i is in phase with the driving emf $\mathscr{E}$ and its phase constant ϕ is zero.

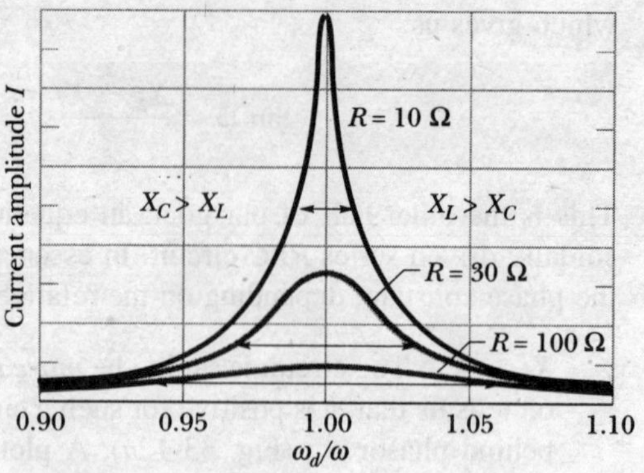

Fig. 33-13 *Resonance curves* for the driven *RLC* circuit of Fig. 33-7 with $L = 100\ \mu\text{H}$, $C = 100$ pF, and three values of R. The current amplitude I of the alternating current depends on how close the driving angular frequency ω_d is to the natural angular frequency ω. The horizontal arrow on each curve measures the curve's width at the half-maximum level, a measure of the sharpness of the resonance. To the left of $\omega_d/\omega = 1.00$, the circuit is mainly capacitive, with $X_C > X_L$; to the right, it is mainly inductive, with $X_L > X_C$.

is zero—that is, when

$$\omega_d L = \frac{1}{\omega_d C}$$

or

$$\omega_d = \frac{1}{\sqrt{LC}} \quad \text{(maximum } I\text{)}. \tag{33-66}$$

Because the natural angular frequency ω of the *RLC* circuit is also equal to $1/\sqrt{LC}$, the maximum value of I occurs when the driving angular frequency matches the natural angular frequency—that is, at resonance. Thus, in an *RLC* circuit, resonance and maximum current amplitude I occur when

$$\omega_d = \omega = \frac{1}{\sqrt{LC}} \quad \text{(resonance)}. \tag{33-67}$$

Figure 33-13 shows three *resonance curves* for sinusoidally driven oscillations in three series *RLC* circuits differing only in R. Each curve peaks at its maximum current amplitude I when the ratio ω_d/ω is 1.00, but the maximum value of I decreases with increasing R. (The maximum I is always $\mathscr{E}_m/R$; to see why, combine Eqs. 33-61 and 33-62.) In addition, the curves increase in width (measured in Fig. 33-13 at half the maximum value of I) with increasing R.

To make physical sense of Fig. 33-13, consider how the reactances X_L and X_C change as we increase the driving angular frequency ω_d, starting with a value much less than the natural frequency ω. For small ω_d, reactance X_L $(= \omega_d L)$ is small and reactance X_C $(= 1/\omega_d C)$ is large. Thus, the circuit is mainly capacitive and the impedance is dominated by the large X_C, which keeps the current low.

As we increase ω_d, reactance X_C remains dominant but decreases while reactance X_L increases. The decrease in X_C decreases the impedance, allowing the current to increase, as we see on the left side of any resonance curve in Fig. 33-13. When the increasing X_L and the decreasing X_C reach equal values, the current is greatest and the circuit is in resonance, with $\omega_d = \omega$.

As we continue to increase ω_d, the increasing reactance X_L becomes progressively more dominant over the decreasing reactance X_C. The impedance increases because of X_L and the current decreases, as on the right side of any resonance curve in Fig. 33-13. In summary, then: The low-angular-frequency side of a resonance curve is dominated by the capacitor's reactance, the high-angular-frequency side is dominated by the inductor's reactance, and resonance occurs in the middle.

CHECKPOINT 7: Here are the capacitive reactance and inductive reactance, respectively, for three sinusoidally driven series *RLC* circuits: (1) 50 Ω, 100 Ω; (2) 100 Ω, 50 Ω; (3) 50 Ω, 50 Ω. (a) For each, does the current lead or lag the applied emf, or are the two in phase? (b) Which circuit is in resonance?

Sample Problem 33-7

In Fig. 33-7 let $R = 200\ \Omega$, $C = 15.0\ \mu F$, $L = 230$ mH, $f_d = 60.0$ Hz, and $\mathscr{E}_m = 36.0$ V. (These parameters are those used in Sample Problems 33-4, 33-5, and 33-6.)

(a) What is the current amplitude I?

SOLUTION: The **Key Idea** here is that current amplitude I depends on the amplitude $\mathscr{E}_m$ of the driving emf and on the impedance Z of the circuit, according to Eq. 33-62 ($I = \mathscr{E}_m/Z$). Thus, we need to find Z, which depends on the circuit's resistance R, capacitive reactance X_C, and inductive reactance X_L.

The circuit's only resistance is the given resistance R. Its only capacitive reactance is due to the given capacitance and, from Sample Problem 33-5, $X_C = 177\ \Omega$. Its only inductive reactance is due to the given inductance and, from Sample Problem 33-6, $X_L = 86.7\ \Omega$. Thus, the circuit's impedance is

$$Z = \sqrt{R^2 + (X_L - X_C)^2}$$
$$= \sqrt{(200\ \Omega)^2 + (86.7\ \Omega - 177\ \Omega)^2}$$
$$= 219\ \Omega.$$

We then find

$$I = \frac{\mathscr{E}_m}{Z} = \frac{36.0\ \text{V}}{219\ \Omega} = 0.164\ \text{A}. \qquad \text{(Answer)}$$

(b) What is the phase constant ϕ of the current in the circuit relative to the driving emf?

SOLUTION: The **Key Idea** here is that the phase constant depends on the inductive reactance, the capacitive reactance, and the resistance of the circuit, according to Eq. 33-65. Solving that equation for ϕ leads to

$$\phi = \tan^{-1}\frac{X_L - X_C}{R} = \tan^{-1}\frac{86.7\ \Omega - 177\ \Omega}{200\ \Omega}$$
$$= -24.3° = -0.424\ \text{rad}. \qquad \text{(Answer)}$$

The negative phase constant is consistent with the fact that the load is mainly capacitive; that is, $X_C > X_L$. In the mnemonic of Problem-Solving Tactic 1, this circuit is an *ICE* circuit—the current *leads* the driving emf.

33-10 Power in Alternating-Current Circuits

In the *RLC* circuit of Fig. 33-7, the source of energy is the alternating-current generator. Some of the energy that it provides is stored in the electric field in the capacitor, some is stored in the magnetic field in the inductor, and some is dissipated as thermal energy in the resistor. In steady-state operation—which we assume—the average energy stored in the capacitor and inductor together remains constant. The net transfer of energy is thus from the generator to the resistor, where electromagnetic energy is dissipated as thermal energy.

The instantaneous rate at which energy is dissipated in the resistor can be written, with the help of Eqs. 27-22 and 33-29, as

$$P = i^2R = [I\sin(\omega_d t - \phi)]^2 R = I^2R\sin^2(\omega_d t - \phi). \qquad (33\text{-}68)$$

The *average* rate at which energy is dissipated in the resistor, however, is the average of Eq. 33-68 over time. Over one complete cycle, the average value of $\sin\theta$, where θ is any variable, is zero (Fig. 33-14*a*) but the average value of $\sin^2\theta$ is $\frac{1}{2}$ (Fig. 33-14*b*). (Note in Fig. 33-14*b* how the shaded areas under the curve but above the horizontal line marked $+\frac{1}{2}$ exactly fill in the unshaded spaces below that line.) Thus, we can write, from Eq. 33-68,

$$P_{\text{avg}} = \frac{I^2R}{2} = \left(\frac{I}{\sqrt{2}}\right)^2 R. \qquad (33\text{-}69)$$

The quantity $I/\sqrt{2}$ is called the **root-mean-square,** or **rms,** value of the current i:

$$I_{\text{rms}} = \frac{I}{\sqrt{2}} \quad \text{(rms current)}. \qquad (33\text{-}70)$$

We can now rewrite Eq. 33-69 as

$$P_{\text{avg}} = I_{\text{rms}}^2 R \quad \text{(average power)}. \qquad (33\text{-}71)$$

Equation 33-71 looks much like Eq. 27-22 ($P = i^2R$); the message is that if we

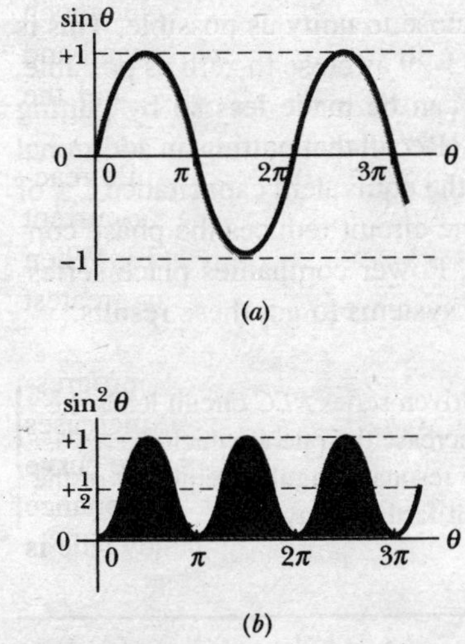

Fig. 33-14 (*a*) A plot of $\sin\theta$ versus θ. The average value over one cycle is zero. (*b*) A plot of $\sin^2\theta$ versus θ. The average value over one cycle is $\frac{1}{2}$.

switch to the rms current, we can compute the average rate of energy dissipation for alternating-current circuits just as for direct-current circuits.

We can also define rms values of voltages and emfs for alternating-current circuits:

$$V_{\text{rms}} = \frac{V}{\sqrt{2}} \quad \text{and} \quad \mathscr{E}_{\text{rms}} = \frac{\mathscr{E}_m}{\sqrt{2}} \quad \text{(rms voltage; rms emf).} \tag{33-72}$$

Alternating-current instruments, such as ammeters and voltmeters, are usually calibrated to read I_{rms}, V_{rms}, and $\mathscr{E}_{\text{rms}}$. Thus, if you plug an alternating-current voltmeter into a household electric outlet and it reads 120 V, that is an rms voltage. The *maximum* value of the potential difference at the outlet is $\sqrt{2} \times (120\ \text{V})$, or 170 V.

Because the proportionality factor $1/\sqrt{2}$ in Eqs. 33-70 and 33-72 is the same for all three variables, we can write Eqs. 33-62 and 33-60 as

$$I_{\text{rms}} = \frac{\mathscr{E}_{\text{rms}}}{Z} = \frac{\mathscr{E}_{\text{rms}}}{\sqrt{R^2 + (X_L - X_C)^2}}, \tag{33-73}$$

and, indeed, this is the form that we almost always use.

We can use the relationship $I_{\text{rms}} = \mathscr{E}_{\text{rms}}/Z$ to recast Eq. 33-71 in a useful equivalent way. We write

$$P_{\text{avg}} = \frac{\mathscr{E}_{\text{rms}}}{Z} I_{\text{rms}} R = \mathscr{E}_{\text{rms}} I_{\text{rms}} \frac{R}{Z}. \tag{33-74}$$

From Fig. 33-11*d*, Table 33-2, and Eq. 33-62, however, we see that R/Z is just the cosine of the phase constant ϕ:

$$\cos\phi = \frac{V_R}{\mathscr{E}_m} = \frac{IR}{IZ} = \frac{R}{Z}. \tag{33-75}$$

Equation 33-74 then becomes

$$P_{\text{avg}} = \mathscr{E}_{\text{rms}} I_{\text{rms}} \cos\phi \quad \text{(average power),} \tag{33-76}$$

in which the term $\cos\phi$ is called the **power factor.** Because $\cos\phi = \cos(-\phi)$, Eq. 33-76 is independent of the sign of the phase constant ϕ.

To maximize the rate at which energy is supplied to a resistive load in an *RLC* circuit, we should keep the power factor $\cos\phi$ as close to unity as possible. This is equivalent to keeping the phase constant ϕ in Eq. 33-29 as close to zero as possible. If, for example, the circuit is highly inductive, it can be made less so by putting more capacitance in the circuit, connected in series. (Recall that putting an additional capacitance into a series of capacitances decreases the equivalent capacitance C_{eq} of the series.) Thus, the resulting decrease in C_{eq} in the circuit reduces the phase constant and increases the power factor in Eq. 33-76. Power companies place series-connected capacitors throughout their transmission systems to get these results.

✔CHECKPOINT 8: (a) If the current in a sinusoidally driven series *RLC* circuit leads the emf, would we increase or decrease the capacitance to increase the rate at which energy is supplied to the resistance? (b) Will this change bring the resonant angular frequency of the circuit closer to the angular frequency of the emf or put it farther away?

Sample Problem 33-8

A series *RLC* circuit, driven with $\mathscr{E}_{\text{rms}} = 120$ V at frequency $f_d = 60.0$ Hz, contains a resistance $R = 200\ \Omega$, an inductance with $X_L = 80.0\ \Omega$, and a capacitance with $X_C = 150\ \Omega$.

(a) What are the power factor $\cos\phi$ and phase constant ϕ of the circuit?

SOLUTION: The **Key Idea** here is that the power factor $\cos\phi$ can be found from the resistance R and impedance Z via Eq. 33-75

($\cos \phi = R/Z$). To calculate Z, we use Eq. 33-61:

$$Z = \sqrt{R^2 + (X_L - X_C)^2}$$
$$= \sqrt{(200\ \Omega)^2 + (80.0\ \Omega - 150\ \Omega)^2} = 211.90\ \Omega.$$

Equation 33-75 then gives us

$$\cos \phi = \frac{R}{Z} = \frac{200\ \Omega}{211.90\ \Omega} = 0.9438 \approx 0.944. \quad \text{(Answer)}$$

Taking the inverse cosine then yields

$$\phi = \cos^{-1} 0.944 = \pm 19.3°.$$

Both $+19.3°$ and $-19.3°$ have a cosine of 0.944. To determine which sign is correct, we must consider whether the current leads or lags the driving emf. Because $X_C > X_L$, this circuit is mainly capacitive, with the current leading the emf. Thus, ϕ must be negative:

$$\phi = -19.3°. \quad \text{(Answer)}$$

We could, instead, have found ϕ with Eq. 33-65. A calculator would then have given us the complete answer, with the minus sign.

(b) What is the average rate P_{avg} at which energy is dissipated in the resistance?

SOLUTION: One way to answer this question is to use this **Key Idea**: Because the circuit is assumed to be in steady-state operation, the rate at which energy is dissipated in the resistance is equal to the rate at which energy is supplied to the circuit, as given by Eq. 33-76 ($P_{avg} = \mathscr{E}_{rms} I_{rms} \cos \phi$).

We are given the rms driving emf $\mathscr{E}_{rms}$ and we know $\cos \phi$ from part (a). To find I_{rms} we use the **Key Idea** that the rms current is determined by the rms value of the driving emf and the circuit's impedance Z (which we know), according to Eq. 33-73:

$$I_{rms} = \frac{\mathscr{E}_{rms}}{Z}.$$

Substituting this into Eq. 33-76 then leads to

$$P_{avg} = \mathscr{E}_{rms} I_{rms} \cos \phi = \frac{\mathscr{E}^2_{rms}}{Z} \cos \phi$$
$$= \frac{(120\ \text{V})^2}{211.90\ \Omega}(0.9438) = 64.1\ \text{W}. \quad \text{(Answer)}$$

A second way to answer the question is to use the **Key Idea** that the rate at which energy is dissipated in a resistance R depends on the square of the rms current I_{rms} through it, according to Eq. 33-71. We then find

$$P_{avg} = I^2_{rms} R = \frac{\mathscr{E}^2_{rms}}{Z^2} R$$
$$= \frac{(120\ \text{V})^2}{(211.90\ \Omega)^2}(200\ \Omega) = 64.1\ \text{W}. \quad \text{(Answer)}$$

(c) What new capacitance C_{new} is needed to maximize P_{avg} if the other parameters of the circuit are not changed?

SOLUTION: One **Key Idea** here is that the average rate P_{avg} at which energy is supplied and dissipated is maximized if the circuit is brought into resonance with the driving emf. A second **Key Idea** is that resonance occurs when $X_C = X_L$. From the given data, we have $X_C > X_L$. Thus, we must decrease X_C to reach resonance. From Eq. 33-39 ($X_C = 1/\omega_d C$), we see that this means we must increase C to the new value C_{new}.

Using Eq. 33-39, we can write the condition $X_C = X_L$ as

$$\frac{1}{\omega_d C_{new}} = X_L.$$

Substituting $2\pi f_d$ for ω_d (because we are given f_d and not ω_d) and then solving for C_{new}, we find

$$C_{new} = \frac{1}{2\pi f_d X_L} = \frac{1}{(2\pi)(60\ \text{Hz})(80.0\ \Omega)}$$
$$= 3.32 \times 10^{-5}\ \text{F} = 33.2\ \mu\text{F}. \quad \text{(Answer)}$$

Following the procedure of part (b), you can show that with C_{new}, P_{avg} would then be at its maximum value of 72.0 W.

33-11 Transformers

Energy Transmission Requirements

When an ac circuit has only a resistive load, the power factor in Eq. 33-76 is $\cos 0° = 1$ and the applied rms emf $\mathscr{E}_{rms}$ is equal to the rms voltage V_{rms} across the load. Thus, with an rms current I_{rms} in the load, energy is supplied and dissipated at the average rate of

$$P_{avg} = \mathscr{E}I = IV. \quad (33\text{-}77)$$

(In Eq. 33-77 and the rest of this section, we follow conventional practice and drop the subscripts identifying rms quantities. Engineers and scientists assume that all time-varying currents and voltages are reported as rms values; that is what the meters read.) Equation 33-77 tells us that, to satisfy a given power requirement, we have a range of choices, from a relatively large current I and a relatively small voltage V to just the reverse, provided only that the product IV is as required.

At 5:17 p.m. on November 9, 1965, a faulty relay in the power system near Niagara Falls opened a circuit breaker on a transmission line, automatically causing the current to switch to other lines, which overloaded those lines and made other circuit breakers open. Within minutes a runaway shutdown had blacked out much of New York, New England, and Ontario.

In electric power distribution systems it is desirable for reasons of safety and for efficient equipment design to deal with relatively low voltages at both the generating end (the electric power plant) and the receiving end (the home or factory). Nobody wants an electric toaster or a child's electric train to operate at, say, 10 kV. On the other hand, in the transmission of electric energy from the generating plant to the consumer, we want the lowest practical current (hence the largest practical voltage) to minimize I^2R losses (often called *ohmic losses*) in the transmission line.

As an example, consider the 735 kV line used to transmit electric energy from the La Grande 2 hydroelectric plant in Quebec to Montreal, 1000 km away. Suppose that the current is 500 A and the power factor is close to unity. Then from Eq. 33-77, energy is supplied at the average rate

$$P_{avg} = \mathscr{E}I = (7.35 \times 10^5\ \text{V})(500\ \text{A}) = 368\ \text{MW}.$$

The resistance of the transmission line is about 0.220 Ω/km; thus, there is a total resistance of about 220 Ω for the 1000 km stretch. Energy is dissipated owing to that resistance at a rate of about

$$P_{avg} = I^2R = (500\ \text{A})^2(220\ \Omega) = 55.0\ \text{MW},$$

which is nearly 15% of the supply rate.

Imagine what would happen if we doubled the current and halved the voltage. Energy would be supplied by the plant at the same average rate of 368 MW as previously, but now energy would be dissipated at the rate of about

$$P_{avg} = I^2R = (1000\ \text{A})^2(220\ \Omega) = 220\ \text{MW},$$

which is *almost 60% of the supply rate*. Hence the general energy transmission rule: Transmit at the highest possible voltage and the lowest possible current.

The Ideal Transformer

The transmission rule leads to a fundamental mismatch between the requirement for efficient high-voltage transmission and the need for safe low-voltage generation and consumption. We need a device with which we can raise (for transmission) and lower (for use) the ac voltage in a circuit, keeping the product current × voltage essentially constant. The **transformer** is such a device. It has no moving parts, operates by Faraday's law of induction, and has no simple direct-current counterpart.

The *ideal transformer* in Fig. 33-15 consists of two coils, with different numbers of turns, wound around an iron core. (The coils are insulated from the core.) In use, the primary winding, of N_p turns, is connected to an alternating-current generator whose emf $\mathscr{E}$ at any time t is given by

$$\mathscr{E} = \mathscr{E}_m \sin \omega t. \tag{33-78}$$

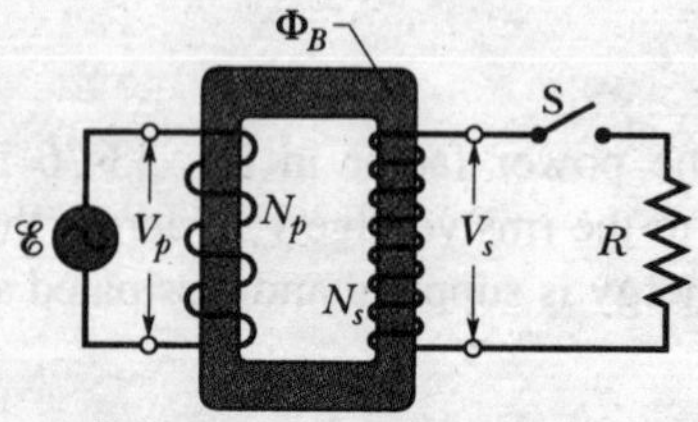

Fig. 33-15 An ideal transformer (two coils wound on an iron core) in a basic transformer circuit. An ac generator produces current in the coil at the left (the *primary*). The coil at the right (the *secondary*) is connected to the resistive load R when switch S is closed.

The secondary winding, of N_s turns, is connected to load resistance R, but its circuit is an open circuit as long as switch S is open (which we assume for the present). Thus, there can be no current through the secondary coil. We assume further for this ideal transformer that the resistances of the primary and secondary windings are negligible, as are energy losses due to magnetic hysteresis in the iron core. Well-designed, high-capacity transformers can have energy losses as low as 1%, so our assumptions are reasonable.

For the assumed conditions, the primary winding (or *primary*) is a pure inductance, and the primary circuit is like that in Fig. 33-10*a*. Thus, the (very small) primary current, also called the *magnetizing current* I_{mag}, lags the primary voltage V_p by 90°; the primary's power factor (= cos ϕ in Eq. 33-76) is zero, so no power is delivered from the generator to the transformer.

However, the small alternating primary current I_{mag} induces an alternating magnetic flux Φ_B in the iron core. Because the core extends through the secondary winding (or *secondary*), this induced flux also extends through the turns of the secondary. From Faraday's law of induction (Eq. 31-6), the induced emf per turn $\mathscr{E}_{turn}$ is the same for both the primary and the secondary. Also, the voltage V_p across the primary is equal to the emf induced in the primary, and the voltage V_s across the secondary is equal to the emf induced in the secondary. Thus, we can write

$$\mathscr{E}_{turn} = \frac{d\Phi_B}{dt} = \frac{V_p}{N_p} = \frac{V_s}{N_s}$$

and thus,

$$V_s = V_p \frac{N_s}{N_p} \quad \text{(transformation of voltage).} \qquad (33\text{-}79)$$

If $N_s > N_p$, the transformer is called a *step-up transformer* because it steps the primary's voltage V_p *up* to a higher voltage V_s. Similarly, if $N_s < N_p$, the device is a *step-down transformer*.

So far, with switch S open, no energy is transferred from the generator to the rest of the circuit. Now let us close S to connect the secondary to the resistive load R. (In general, the load would also contain inductive and capacitive elements, but here we consider just resistance R.) We find that now energy *is* transferred from the generator. Let us see why.

Several things happen when we close switch S.

1. An alternating current I_s appears in the secondary circuit, with corresponding energy dissipation rate I_s^2R $(= V_s^2/R)$ in the resistive load.
2. This current produces its own alternating magnetic flux in the iron core, and this flux induces (from Faraday's law and Lenz's law) an opposing emf in the primary windings.
3. The voltage V_p of the primary, however, cannot change in response to this opposing emf because it must always be equal to the emf $\mathscr{E}$ that is provided by the generator; closing switch S cannot change this fact.
4. To maintain V_p, the generator now produces (in addition to I_{mag}) an alternating current I_p in the primary circuit; the magnitude and phase constant of I_p are just those required for the emf induced by I_p in the primary to exactly cancel the emf induced there by I_s. Because the phase constant of I_p is not 90° like that of I_{mag}, this current I_p can transfer energy to the primary.

We want to relate I_s to I_p. However, rather than analyze the foregoing complex process in detail, let us just apply the principle of conservation of energy. The rate at which the generator transfers energy to the primary is equal to I_pV_p. The rate at which the primary then transfers energy to the secondary (via the alternating magnetic field linking the two coils) is I_sV_s. Because we assume that no energy is lost along the way, conservation of energy requires that

$$I_pV_p = I_sV_s.$$

Substituting for V_s from Eq. 33-79, we find that

$$I_s = I_p \frac{N_p}{N_s} \quad \text{(transformation of currents).} \qquad (33\text{-}80)$$

This equation tells us that the current I_s in the secondary can differ from the current I_p in the primary, depending on the *turns ratio* N_p/N_s.

Current I_p appears in the primary circuit because of the resistive load R in the secondary circuit. To find I_p, we substitute $I_s = V_s/R$ into Eq. 33-80 and then we substitute for V_s from Eq. 33-79. We find

$$I_p = \frac{1}{R}\left(\frac{N_s}{N_p}\right)^2 V_p. \tag{33-81}$$

This equation has the form $I_p = V_p/R_{eq}$, where equivalent resistance R_{eq} is

$$R_{eq} = \left(\frac{N_p}{N_s}\right)^2 R. \tag{33-82}$$

This R_{eq} is the value of the load resistance as "seen" by the generator; the generator produces the current I_p and voltage V_p as if it were connected to a resistance R_{eq}.

Impedance Matching

Equation 33-82 suggests still another function for the transformer. For maximum transfer of energy from an emf device to a resistive load, the resistance of the emf device and the resistance of the load must be equal. The same relation holds for ac circuits except that the *impedance* (rather than just the resistance) of the generator must be matched to that of the load. Often this condition is not met. For example, in a music-playing system, the amplifier has high impedance and the speaker set has low impedance. We can match the impedances of the two devices by coupling them through a transformer with a suitable turns ratio N_p/N_s.

✔CHECKPOINT 9: An alternating-current emf device has a smaller resistance than that of the resistive load; to increase the transfer of energy from the device to the load, a transformer will be connected between the two. (a) Should N_s be greater than or less than N_p? (b) Will that make it a step-up or step-down transformer?

Sample Problem 33-9

A transformer on a utility pole operates at $V_p = 8.5$ kV on the primary side and supplies electric energy to a number of nearby houses at $V_s = 120$ V, both quantities being rms values. Assume an ideal step-down transformer, a purely resistive load, and a power factor of unity.

(a) What is the turns ratio N_p/N_s of the transformer?

SOLUTION: The **Key Idea** here is that the turns ratio N_p/N_s is related to the (given) rms primary and secondary voltages via Eq. 33-79, which we can write as

$$\frac{V_s}{V_p} = \frac{N_s}{N_p}.$$

(Note that the right side of this equation is the *inverse* of the turns ratio.) Inverting both sides then gives us

$$\frac{N_p}{N_s} = \frac{V_p}{V_s} = \frac{8.5 \times 10^3 \text{ V}}{120 \text{ V}} = 70.83 \approx 71. \quad \text{(Answer)}$$

(b) The average rate of energy consumption (or dissipation) in the houses served by the transformer is 78 kW. What are the rms currents in the primary and secondary of the transformer?

SOLUTION: The **Key Idea** here is that for a purely resistive load, the power factor cos ϕ is unity; thus, the average rate at which energy is supplied and dissipated is given by Eq. 33-77. In the primary circuit, with $V_p = 8.5$ kV, Eq. 33-77 yields

$$I_p = \frac{P_{avg}}{V_p} = \frac{78 \times 10^3 \text{ W}}{8.5 \times 10^3 \text{ V}} = 9.176 \text{ A} \approx 9.2 \text{ A}. \quad \text{(Answer)}$$

Similarly, in the secondary circuit,

$$I_s = \frac{P_{avg}}{V_s} = \frac{78 \times 10^3 \text{ W}}{120 \text{ V}} = 650 \text{ A}. \quad \text{(Answer)}$$

You can check that $I_s = I_p(N_p/N_s)$ as required by Eq. 33-80.

(c) What is the resistive load R_s in the secondary circuit? What is the corresponding resistive load R_p in the primary circuit?

SOLUTION: For both circuits, the **Key Idea** here is that we can relate the resistive load to the rms voltage and current with $V = IR$. For the secondary circuit, we find

$$R_s = \frac{V_s}{I_s} = \frac{120 \text{ V}}{650 \text{ A}} = 0.1846\ \Omega \approx 0.18\ \Omega. \quad \text{(Answer)}$$

Similarly, for the primary circuit we find

$$R_p = \frac{V_p}{I_p} = \frac{8.5 \times 10^3 \text{ V}}{9.176 \text{ A}} = 926\ \Omega \approx 930\ \Omega. \quad \text{(Answer)}$$

Another **Key Idea** that we can use to find R_p is that R_p is the equivalent resistive load "seen" from the primary side of the transformer, as given by Eq. 33-82. If we substitute R_p for R_{eq} and R_s for R, that equation yields

$$R_p = \left(\frac{N_p}{N_s}\right)^2 R_s = (70.83)^2(0.1846\ \Omega)$$

$$= 926\ \Omega \approx 930\ \Omega. \quad \text{(Answer)}$$

REVIEW & SUMMARY

***LC* Energy Transfers** In an oscillating *LC* circuit, energy is shuttled periodically between the electric field of the capacitor and the magnetic field of the inductor; instantaneous values of the two forms of energy are

$$U_E = \frac{q^2}{2C} \quad \text{and} \quad U_B = \frac{Li^2}{2}, \qquad (33\text{-}1, 33\text{-}2)$$

where q is the instantaneous charge on the capacitor and i is the instantaneous current through the inductor. The total energy $U\ (= U_E + U_B)$ remains constant.

***LC* Charge and Current Oscillations** The principle of conservation of energy leads to

$$L\frac{d^2q}{dt^2} + \frac{1}{C}q = 0 \quad (LC \text{ oscillations}) \qquad (33\text{-}11)$$

as the differential equation of *LC* oscillations (with no resistance). The solution of Eq. 33-11 is

$$q = Q\cos(\omega t + \phi) \quad \text{(charge)}, \qquad (33\text{-}12)$$

in which Q is the *charge amplitude* (maximum charge on the capacitor) and the angular frequency ω of the oscillations is

$$\omega = \frac{1}{\sqrt{LC}}. \qquad (33\text{-}4)$$

The phase constant ϕ in Eq. 33-12 is determined by the initial conditions (at $t = 0$) of the system.

The current i in the system at any time t is

$$i = -\omega Q \sin(\omega t + \phi) \quad \text{(current)}, \qquad (33\text{-}13)$$

in which ωQ is the *current amplitude* I.

Damped Oscillations Oscillations in an *LC* circuit are damped when a dissipative element R is also present in the circuit. Then

$$L\frac{d^2q}{dt^2} + R\frac{dq}{dt} + \frac{1}{C}q = 0 \quad (RLC \text{ circuit}). \qquad (33\text{-}24)$$

The solution of this differential equation is

$$q = Qe^{-Rt/2L}\cos(\omega' t + \phi), \qquad (33\text{-}25)$$

where

$$\omega' = \sqrt{\omega^2 - (R/2L)^2}. \qquad (33\text{-}26)$$

We consider only situations with small R and thus small damping; then $\omega' \approx \omega$.

Alternating Currents; Forced Oscillations A series *RLC* circuit may be set into *forced oscillation* at a *driving angular frequency* ω_d by an external alternating emf

$$\mathscr{E} = \mathscr{E}_m \sin \omega_d t. \qquad (33\text{-}28)$$

The current driven in the circuit by the emf is

$$i = I\sin(\omega_d t - \phi), \qquad (33\text{-}29)$$

where ϕ is the phase constant of the current.

Resonance The current amplitude I in a series *RLC* circuit driven by a sinusoidal external emf is a maximum ($I = \mathscr{E}_m/R$) when the driving angular frequency ω_d equals the natural angular frequency ω of the circuit (that is, at *resonance*). Then $X_C = X_L$, $\phi = 0$, and the current is in phase with the emf.

Single Circuit Elements The alternating potential difference across a resistor has amplitude $V_R = IR$; the current is in phase with the potential difference.

For a *capacitor*, $V_C = IX_C$, in which $X_C = 1/\omega_d C$ is the **capacitive reactance;** the current here leads the potential difference by 90° ($\phi = -90° = -\pi/2$ rad).

For an *inductor*, $V_L = IX_L$, in which $X_L = \omega_d L$ is the **inductive reactance;** the current here lags the potential difference by 90° ($\phi = +90° = +\pi/2$ rad).

Series *RLC* Circuits For a series *RLC* circuit with external emf given by Eq. 33-28 and current given by Eq. 33-29,

$$I = \frac{\mathscr{E}_m}{\sqrt{R^2 + (X_L - X_C)^2}}$$

$$= \frac{\mathscr{E}_m}{\sqrt{R^2 + (\omega_d L - 1/\omega_d C)^2}} \quad \text{(current amplitude)} \qquad (33\text{-}60, 33\text{-}63)$$

and

$$\tan\phi = \frac{X_L - X_C}{R} \quad \text{(phase constant)}. \qquad (33\text{-}65)$$

Defining the impedance Z of the circuit as

$$Z = \sqrt{R^2 + (X_L - X_C)^2} \quad \text{(impedance)} \qquad (33\text{-}61)$$

allows us to write Eq. 33-60 as $I = \mathscr{E}_m/Z$.

Power In a series *RLC* circuit, the **average power** P_{avg} of the generator is equal to the production rate of thermal energy in the resistor:

$$P_{avg} = I_{rms}^2 R = \mathscr{E}_{rms} I_{rms}\cos\phi. \qquad (33\text{-}71, 33\text{-}76)$$

Here rms stands for **root-mean-square;** the rms quantities are related to the maximum quantities by $I_{rms} = I/\sqrt{2}$, $V_{rms} = V_m/\sqrt{2}$, and $\mathscr{E}_{rms} = \mathscr{E}_m/\sqrt{2}$. The term $\cos\phi$ is called the **power factor** of the circuit.

Transformers A *transformer* (assumed to be ideal) is an iron core on which are wound a primary coil of N_p turns and a secondary coil of N_s turns. If the primary coil is connected across an alternating-current generator, the primary and secondary voltages are related by

$$V_s = V_p \frac{N_s}{N_p} \quad \text{(transformation of voltage).} \qquad (33\text{-}79)$$

The currents through the coils are related by

$$I_s = I_p \frac{N_p}{N_s} \quad \text{(transformation of currents),} \qquad (33\text{-}80)$$

and the equivalent resistance of the secondary circuit, as seen by the generator, is

$$R_{eq} = \left(\frac{N_p}{N_s}\right)^2 R, \qquad (33\text{-}82)$$

where R is the resistive load in the secondary circuit. The ratio N_p/N_s is called the transformer's *turns ratio*.

QUESTIONS

1. A charged capacitor and an inductor are connected at time $t = 0$. In terms of the period T of the resulting oscillations, what is the first later time at which the following reach a maximum: (a) U_B, (b) the magnetic flux through the inductor, (c) di/dt, and (d) the emf of the inductor?

2. What values of phase constant ϕ in Eq. 33-12 allow situations (*a*), (*c*), (*e*), and (*g*) of Fig. 33-1 to occur at $t = 0$?

3. Figure 33-16 shows three oscillating LC circuits with identical inductors and capacitors. Rank the circuits according to the time taken to fully discharge the capacitors during the oscillations, greatest first.

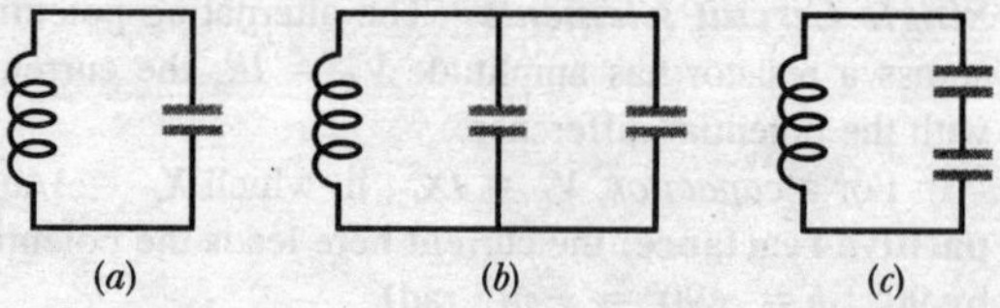

Fig. 33-16 Question 3.

4. Figure 33-17 shows graphs of capacitor voltage v_C for LC circuits 1 and 2, which contain identical capacitances and have the same maximum charge Q. Are (a) the inductance L and (b) the maximum current I in circuit 1 greater than, less than, or the same as those in circuit 2?

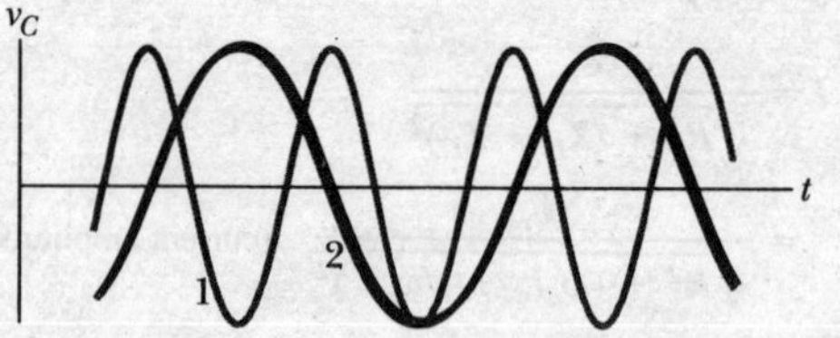

Fig. 33-17 Question 4.

5. Charges on the capacitors in three oscillating LC circuits vary as follows: (1) $q = 2\cos 4t$, (2) $q = 4\cos t$, (3) $q = 3\cos 4t$ (with q in coulombs and t in seconds). Rank the circuits according to (a) the current amplitude and (b) the period, greatest first.

6. If you increase the inductance L in an oscillating LC circuit having a given maximum charge Q, do (a) the current magnitude I and (b) the maximum magnetic energy U_B increase, decrease, or stay the same?

7. An alternating emf source with a certain emf amplitude is connected, in turn, to a resistor, a capacitor, and then an inductor. Once connected to one of the devices, the driving frequency f_d is varied and the amplitude I of the resulting current through the device is measured and plotted. Which of the three plots in Fig. 33-18 corresponds to which of the three devices?

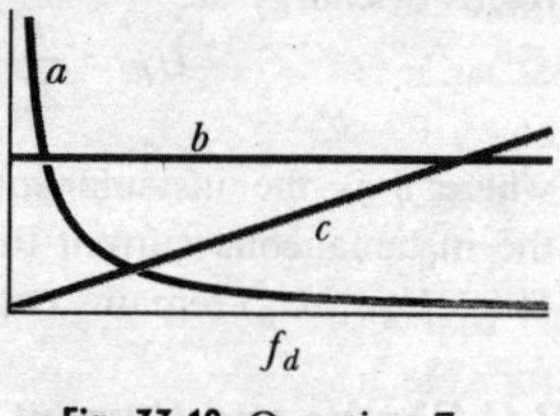

Fig. 33-18 Question 7.

8. The values of the phase constant ϕ for four sinusoidally driven series RLC circuits are (1) $-15°$, (2) $+35°$, (3) $\pi/3$ rad, and (4) $-\pi/6$ rad. (a) In which is the load primarily capacitive? (b) In which does the current lag the alternating emf?

9. Figure 33-19 shows the current i and driving emf $\mathscr{E}$ for a series RLC circuit. (a) Does the current lead or lag the emf? (b) Is the circuit's load mainly capacitive or mainly inductive? (c) Is the angular frequency ω_d of the emf greater than or less than the natural angular frequency ω?

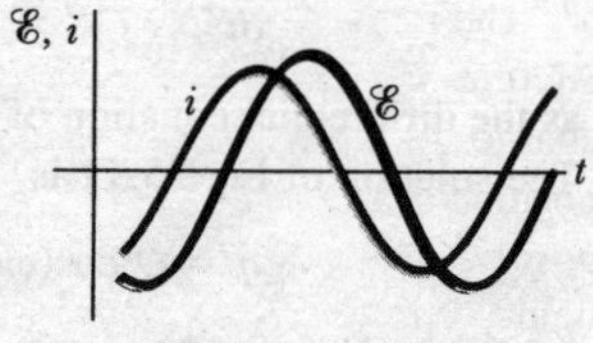

Fig. 33-19 Questions 9 and 11.

10. Figure 33-20 shows three situations like those of Fig. 33-12. For each situation, is the driving angular frequency greater than, less than, or equal to the resonant angular frequency of the circuit?

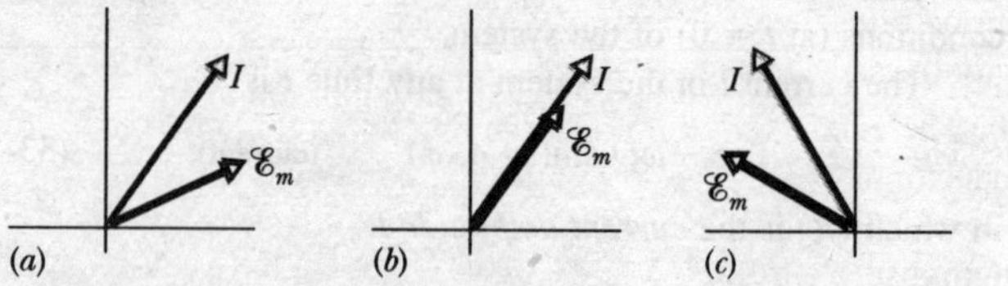

Fig. 33-20 Question 10.

11. Figure 33-19 shows the current i and driving emf $\mathscr{E}$ for a series RLC circuit. Relative to the emf curve, does the current curve shift leftward or rightward and does the amplitude of that curve increase or decrease if we slightly increase (a) L, (b) C, and (c) ω_d?

12. Figure 33-21 shows the current i and driving emf $\mathscr{E}$ for a series RLC circuit. (a) Is the phase constant positive or negative? (b) To increase the rate at which energy is transferred to the resistive load, should L be increased or decreased? (c) Should, instead, C be increased or decreased?

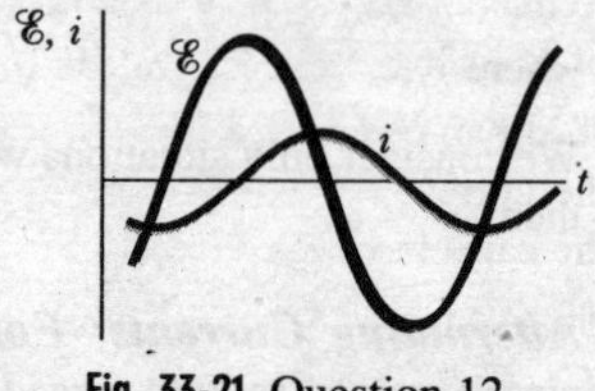

Fig. 33-21 Question 12.

EXERCISES & PROBLEMS

ssm Solution is in the Student Solutions Manual.
www Solution is available on the World Wide Web at: http://www.wiley.com/college/hrw
ilw Solution is available on the Interactive LearningWare.

SEC. 33-2 *LC* Oscillations, Qualitatively

1E. What is the capacitance of an oscillating *LC* circuit if the maximum charge on the capacitor is 1.60 μC and the total energy is 140 μJ?

2E. In an oscillating *LC* circuit, $L = 1.10$ mH and $C = 4.00$ μF. The maximum charge on the capacitor is 3.00 μC. Find the maximum current.

3E. An oscillating *LC* circuit consists of a 75.0 mH inductor and a 3.60 μF capacitor. If the maximum charge on the capacitor is 2.90 μC, (a) what is the total energy in the circuit and (b) what is the maximum current? ssm

4E. In a certain oscillating *LC* circuit the total energy is converted from electric energy in the capacitor to magnetic energy in the inductor in 1.50 μs. (a) What is the period of oscillation? (b) What is the frequency of oscillation? (c) How long after the magnetic energy is a maximum will it be a maximum again?

5P. The frequency of oscillation of a certain *LC* circuit is 200 kHz. At time $t = 0$, plate *A* of the capacitor has maximum positive charge. At what times $t > 0$ will (a) plate *A* again have maximum positive charge, (b) the other plate of the capacitor have maximum positive charge, and (c) the inductor have maximum magnetic field?

SEC. 33-3 The Electrical–Mechanical Analogy

6E. A 0.50 kg body oscillates in SHM on a spring that, when extended 2.0 mm from its equilibrium, has an 8.0 N restoring force. (a) What is the angular frequency of oscillation? (b) What is the period of oscillation? (c) What is the capacitance of an *LC* circuit with the same period if *L* is chosen to be 5.0 H?

7P. The energy in an oscillating *LC* circuit containing a 1.25 H inductor is 5.70 μJ. The maximum charge on the capacitor is 175 μC. Find (a) the mass, (b) the spring constant, (c) the maximum displacement, and (d) the maximum speed for a mechanical system with the same period. ssm

SEC. 33-4 *LC* Oscillations, Quantitatively

8E. *LC* oscillators have been used in circuits connected to loudspeakers to create some of the sounds of electronic music. What inductance must be used with a 6.7 μF capacitor to produce a frequency of 10 kHz, which is near the middle of the audible range of frequencies?

9E. In an oscillating *LC* circuit with $L = 50$ mH and $C = 4.0$ μF, the current is initially a maximum. How long will it take before the capacitor is fully charged for the first time? ssm ilw

10E. A single loop consists of inductors ($L_1, L_2, \ldots$), capacitors ($C_1, C_2, \ldots$), and resistors ($R_1, R_2, \ldots$) connected in series as shown, for example, in Fig. 33-22*a*. Show that regardless of the sequence of these circuit elements in the loop, the behavior of this circuit is identical to that of the simple *LC* circuit shown in Fig. 33-22*b*. (*Hint:* Consider the loop rule and see Problem 43 in Chapter 31.)

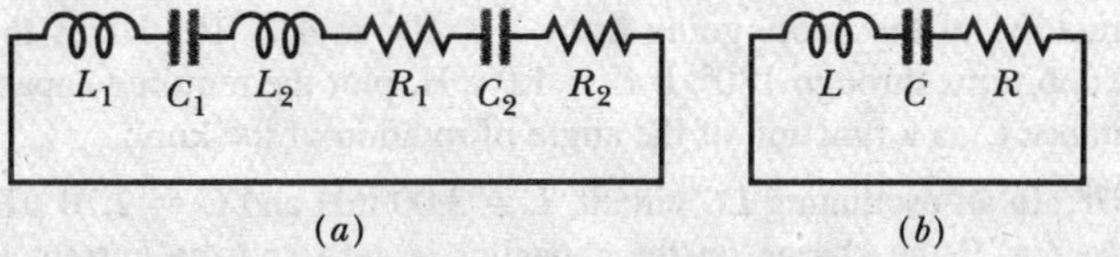

Fig. 33-22 Exercise 10.

11P. An oscillating *LC* circuit consisting of a 1.0 nF capacitor and a 3.0 mH coil has a maximum voltage of 3.0 V. (a) What is the maximum charge on the capacitor? (b) What is the maximum current through the circuit? (c) What is the maximum energy stored in the magnetic field of the coil? ilw

12P. In an oscillating *LC* circuit in which $C = 4.00$ μF, the maximum potential difference across the capacitor during the oscillations is 1.50 V and the maximum current through the inductor is 50.0 mA. (a) What is the inductance *L*? (b) What is the frequency of the oscillations? (c) How much time is required for the charge on the capacitor to rise from zero to its maximum value?

13P. In the circuit shown in Fig. 33-23 the switch is kept in position *a* for a long time. It is then thrown to position *b*. (a) Calculate the frequency of the resulting oscillating current. (b) What is the amplitude of the current oscillations? ssm ilw

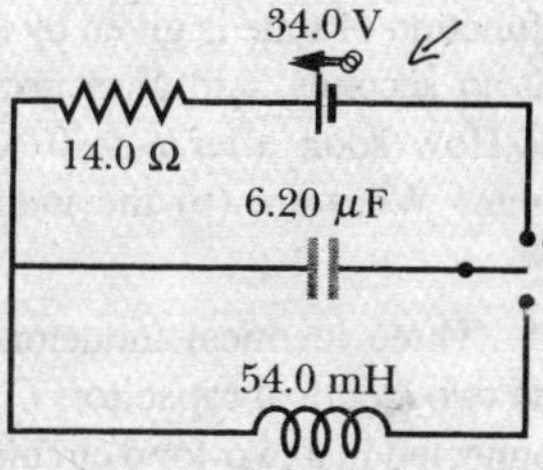

Fig. 33-23 Problem 13.

14P. You are given a 10 mH inductor and two capacitors, of 5.0 μF and 2.0 μF capacitance. List the oscillation frequencies that can be generated by connecting these elements in various combinations.

15P. A variable capacitor with a range from 10 to 365 pF is used with a coil to form a variable-frequency *LC* circuit to tune the input to a radio. (a) What ratio of maximum to minimum frequencies may be obtained with such a capacitor? (b) If this circuit is to obtain frequencies from 0.54 MHz to 1.60 MHz, the ratio computed in (a) is too large. By adding a capacitor in parallel to the variable capacitor, this range may be adjusted. What should be the capacitance of this added capacitor, and what inductance should be used to obtain the desired range of frequencies? ssm www

16P. In an oscillating *LC* circuit, 75.0% of the total energy is stored in the magnetic field of the inductor at a certain instant. (a) In terms of the maximum charge on the capacitor, what is the charge there at that instant? (b) In terms of the maximum current in the inductor, what is the current there at that instant?

17P. In an oscillating *LC* circuit, $L = 25.0$ mH and $C = 7.80$ μF. At time $t = 0$ the current is 9.20 mA, the charge on the capacitor is 3.80 μC, and the capacitor is charging. (a) What is the total energy in the circuit? (b) What is the maximum charge on the capacitor? (c) What is the maximum current? (d) If the charge on the capacitor is given by $q = Q \cos(\omega t + \phi)$, what is the phase angle ϕ? (e) Suppose the data are the same, except that the capacitor is discharging at $t = 0$. What then is ϕ? ssm

18P. An inductor is connected across a capacitor whose capacitance can be varied by turning a knob. We wish to make the frequency of oscillation of this *LC* circuit vary linearly with the angle of rotation of the knob, going from 2×10^5 to 4×10^5 Hz as the knob turns through 180°. If $L = 1.0$ mH, plot the required capacitance *C* as a function of the angle of rotation of the knob.

19P. In an oscillating *LC* circuit, $L = 3.00$ mH and $C = 2.70\ \mu$F. At $t = 0$ the charge on the capacitor is zero and the current is 2.00 A. (a) What is the maximum charge that will appear on the capacitor? (b) In terms of the period *T* of oscillation, how much time will elapse after $t = 0$ until the energy stored in the capacitor will be increasing at its greatest rate? (c) What is this greatest rate at which energy is transferred to the capacitor? ssm

20P. A series circuit containing inductance L_1 and capacitance C_1 oscillates at angular frequency ω. A second series circuit, containing inductance L_2 and capacitance C_2, oscillates at the same angular frequency. In terms of ω, what is the angular frequency of oscillation of a series circuit containing all four of these elements? Neglect resistance. (*Hint:* Use the formulas for equivalent capacitance and equivalent inductance; see Section 26-4 and Problem 43 in Chapter 31.)

21P. In an oscillating *LC* circuit with $C = 64.0\ \mu$F, the current as a function of time is given by $i = (1.60) \sin(2500t + 0.680)$, where *t* is in seconds, *i* in amperes, and the phase constant in radians. (a) How soon after $t = 0$ will the current reach its maximum value? What are (b) the inductance *L* and (c) the total energy? ilw

22P. Three identical inductors *L* and two identical capacitors *C* are connected in a two-loop circuit as shown in Fig. 33-24. (a) Suppose the currents are as shown in Fig. 33-24*a*. What is the current in the middle inductor? Write the loop equations and show that they are satisfied if the current oscillates with angular frequency $\omega = 1/\sqrt{LC}$. (b) Now suppose the currents are as shown in Fig. 33-24*b*. What is the current in the middle inductor? Write the loop equations and show that they are satisfied if the current oscillates with angular frequency $\omega = 1/\sqrt{3LC}$. Because the circuit can oscillate at two different frequencies, we cannot find an equivalent single-loop *LC* circuit to replace it.

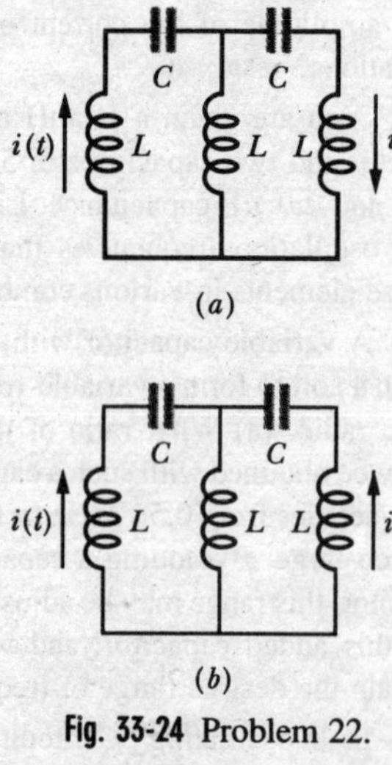

Fig. 33-24 Problem 22.

23P*. In Fig. 33-25, capacitor 1 with $C_1 = 900\ \mu$F is initially charged to 100 V and capacitor 2 with $C_2 = 100\ \mu$F is uncharged. The inductor has an inductance of 10.0 H. Describe in detail how one might charge capacitor 2 to 300 V by manipulating switches S_1 and S_2. ssm www

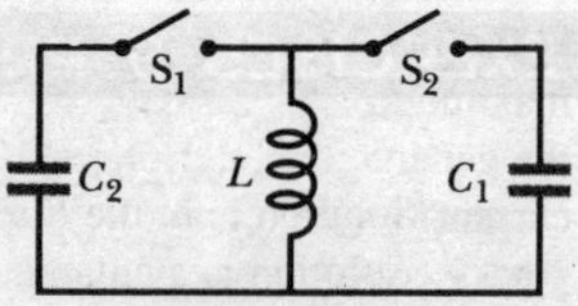

Fig. 33-25 Problem 23.

SEC. 33-5 Damped Oscillations in an *RLC* Circuit

24E. Consider a damped *LC* circuit. (a) Show that the damping term $e^{-Rt/2L}$ (which involves *L* but not *C*) can be rewritten in a more symmetric manner (involving *L* and *C*) as $e^{-\pi R(\sqrt{C/L})t/T}$. Here *T* is the period of oscillation (neglecting resistance). (b) Using (a), show that the SI unit of $\sqrt{L/C}$ is the ohm. (c) Using (a), show that the condition that the fractional energy loss per cycle be small is $R \ll \sqrt{L/C}$.

25E. What resistance *R* should be connected in series with an inductance $L = 220$ mH and capacitance $C = 12.0\ \mu$F for the maximum charge on the capacitor to decay to 99.0% of its initial value in 50.0 cycles? (Assume $\omega' \approx \omega$.) ssm ilw

26P. A single-loop circuit consists of a 7.20 Ω resistor, a 12.0 H inductor, and a 3.20 μF capacitor. Initially the capacitor has a charge of 6.20 μC and the current is zero. Calculate the charge on the capacitor *N* complete cycles later for $N = 5$, 10, and 100.

27P. In an oscillating series *RLC* circuit, find the time required for the maximum energy present in the capacitor during an oscillation to fall to half its initial value. Assume $q = Q$ at $t = 0$. ssm

28P. At time $t = 0$ there is no charge on the capacitor of a series *RLC* circuit but there is current *I* through the inductor. (a) Find the phase constant ϕ in Eq. 33-25 for the circuit. (b) Write an expression for the charge *q* on the capacitor as a function of time *t* and in terms of the current amplitude and angular frequency ω' of the oscillations.

29P*. In an oscillating series *RLC* circuit, show that the fraction of the energy lost per cycle of oscillation, $\Delta U/U$, is given to a close approximation by $2\pi R/\omega L$. The quantity $\omega L/R$ is often called the *Q* of the circuit (for *quality*). A high-*Q* circuit has low resistance and a low fractional energy loss ($= 2\pi/Q$) per cycle. ssm www

SEC. 33-8 Three Simple Circuits

30E. A 1.50 μF capacitor is connected as in Fig. 33-9*a* to an ac generator with $\mathscr{E}_m = 30.0$ V. What is the amplitude of the resulting alternating current if the frequency of the emf is (a) 1.00 kHz and (b) 8.00 kHz?

31E. A 50.0 mH inductor is connected as in Fig. 33-10*a* to an ac generator with $\mathscr{E}_m = 30.0$ V. What is the amplitude of the resulting alternating current if the frequency of the emf is (a) 1.00 kHz and (b) 8.00 kHz? ssm ilw

32E. A 50 Ω resistor is connected as in Fig. 33-8*a* to an ac generator with $\mathscr{E}_m = 30.0$ V. What is the amplitude of the resulting alternating current if the frequency of the emf is (a) 1.00 kHz and (b) 8.00 kHz?

33E. (a) At what frequency would a 6.0 mH inductor and a 10 μF capacitor have the same reactance? (b) What would the reactance be? (c) Show that this frequency would be the natural frequency of an oscillating circuit with the same *L* and *C*. ssm

34P. An ac generator has emf $\mathscr{E} = \mathscr{E}_m \sin \omega_d t$, with $\mathscr{E}_m = 25.0$ V and $\omega_d = 377$ rad/s. It is connected to a 12.7 H inductor. (a) What is the maximum value of the current? (b) When the current is a maximum, what is the emf of the generator? (c) When the emf of the generator is -12.5 V and increasing in magnitude, what is the current?

35P. An ac generator has emf $\mathscr{E} = \mathscr{E}_m \sin(\omega_d t - \pi/4)$, where $\mathscr{E}_m = 30.0$ V and $\omega_d = 350$ rad/s. The current produced in a connected circuit is $i(t) = I \sin(\omega_d t - 3\pi/4)$, where $I = 620$ mA. (a) At what time after $t = 0$ does the generator emf first reach a maximum? (b) At what time after $t = 0$ does the current first reach a maximum? (c) The circuit contains a single element other than the generator. Is it a capacitor, an inductor, or a resistor? Justify your answer. (d) What is the value of the capacitance, inductance, or resistance, as the case may be? ssm

36P. The ac generator of Problem 34 is connected to a 4.15 μF capacitor. (a) What is the maximum value of the current? (b) When the current is a maximum, what is the emf of the generator? (c) When the emf of the generator is -12.5 V and increasing in magnitude, what is the current?

SEC. 33-9 The Series *RLC* Circuit

37E. (a) Find Z, ϕ, and I for the situation of Sample Problem 33-7 with the capacitor removed from the circuit, all other parameters remaining unchanged. (b) Draw to scale a phasor diagram like that of Fig. 33-11*d* for this new situation.

38E. (a) Find Z, ϕ, and I for the situation of Sample Problem 33-7 with the inductor removed from the circuit, all other parameters remaining unchanged. (b) Draw to scale a phasor diagram like that of Fig. 33-11*d* for this new situation.

39E. (a) Find Z, ϕ, and I for the situation of Sample Problem 33-7 with $C = 70.0$ μF, the other parameters remaining unchanged. (b) Draw a phasor diagram like that of Fig. 33-11*d* for this new situation and compare the two diagrams closely. ssm www

40P. In Fig. 33-26, a generator with an adjustable frequency of oscillation is connected to a variable resistance R, a capacitor of $C = 5.50$ μF, and an inductor of inductance L. The amplitude of the current produced in the circuit by the generator is at half-maximum level when the generator's frequency is 1.30 or 1.50 kHz. (a) What is L? (b) If R is increased, what happens to the frequencies at which the current amplitude is at half-maximum level?

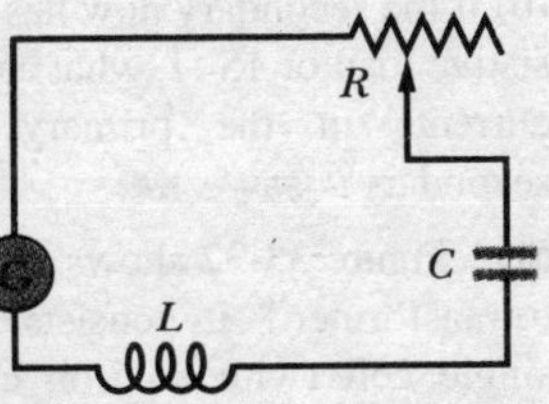

Fig. 33-26 Problem 40.

41P. In an *RLC* circuit, can the amplitude of the voltage across an inductor be greater than the amplitude of the generator emf? Consider an *RLC* circuit with $\mathscr{E}_m = 10$ V, $R = 10$ Ω, $L = 1.0$ H, and $C = 1.0$ μF. Find the amplitude of the voltage across the inductor at resonance. ssm ilw

42P. When the generator emf in Sample Problem 33-7 is a maximum, what is the voltage across (a) the generator, (b) the resistance, (c) the capacitance, and (d) the inductance? (e) By summing these with appropriate signs, verify that the loop rule is satisfied.

43P. A coil of inductance 88 mH and unknown resistance and a 0.94 μF capacitor are connected in series with an alternating emf of frequency 930 Hz. If the phase constant between the applied voltage and the current is 75°, what is the resistance of the coil? ssm ilw

44P. An ac generator with $\mathscr{E}_m = 220$ V and operating at 400 Hz causes oscillations in a series *RLC* circuit having $R = 220$ Ω, $L = 150$ mH, and $C = 24.0$ μF. Find (a) the capacitive reactance X_C, (b) the impedance Z, and (c) the current amplitude I. A second capacitor of the same capacitance is then connected in series with the other components. Determine whether the values of (d) X_C, (e) Z, and (f) I increase, decrease, or remain the same.

45P. An *RLC* circuit such as that of Fig. 33-7 has $R = 5.00$ Ω, $C = 20.0$ μF, $L = 1.00$ H, and $\mathscr{E}_m = 30.0$ V. (a) At what angular frequency ω_d will the current amplitude have its maximum value, as in the resonance curves of Fig. 33-13? (b) What is this maximum value? (c) At what two angular frequencies ω_{d1} and ω_{d2} will the current amplitude be half this maximum value? (d) What is the fractional half-width $[= (\omega_{d1} - \omega_{d2})/\omega]$ of the resonance curve for this circuit? ssm

46P. An ac generator is to be connected in series with an inductor of $L = 2.00$ mH and a capacitance C. You are to produce C by using capacitors of capacitances $C_1 = 4.00$ μF and $C_2 = 6.00$ μF, either singly or together. What resonant frequencies can the circuit have, depending on how you use C_1 and C_2?

47P. Show that the fractional half-width (see Problem 45) of a resonance curve is given by

$$\frac{\Delta\omega_d}{\omega} = \sqrt{\frac{3C}{L}}\,R,$$

in which ω is the angular frequency at resonance and $\Delta\omega_d$ is the width of the resonance curve at half-amplitude. Note that $\Delta\omega_d/\omega$ increases with R, as Fig. 33-13 shows. Use this formula to check the answer to Problem 45d. ssm

48P. In Fig. 33-27, a generator with an adjustable frequency of oscillation is connected to resistance $R = 100$ Ω, inductances $L_1 = 1.70$ mH and $L_2 = 2.30$ mH, and capacitances $C_1 = 4.00$ μF, $C_2 = 2.50$ μF, and $C_3 = 3.50$ μF. (a) What is the resonant frequency of the circuit? (*Hint:* See Problem 43 in Chapter 31.) What happens to the resonant frequency if (b) the value of R is increased, (c) the value of L_1 is increased, and (d) capacitance C_3 is removed from the circuit?

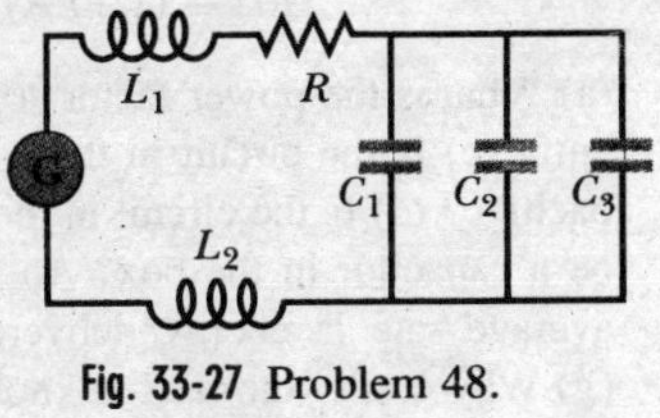

Fig. 33-27 Problem 48.

SEC. 33-10 Power in Alternating-Current Circuits

49E. What direct current will produce the same amount of thermal energy, in a particular resistor, as an alternating current that has a maximum value of 2.60 A? ssm

50E. An ac voltmeter with large impedance is connected in turn across the inductor, the capacitor, and the resistor in a series circuit having an alternating emf of 100 V (rms); it gives the same reading in volts in each case. What is this reading?

51E. What is the maximum value of an ac voltage whose rms value is 100 V?

52E. (a) For the conditions in Problem 34c, is the generator supplying energy to or taking energy from the rest of the circuit? (b) Repeat for the conditions of Problem 36c.

53E. Calculate the average rate of energy dissipation in the circuits of Exercises 31, 32, 37, and 38.

54E. Show that the average rate at which energy is supplied to the circuit of Fig. 33-7 can also be written as $P_{avg} = \mathscr{E}^2_{rms}R/Z^2$. Show that this expression for average power gives reasonable results for a purely resistive circuit, for an *RLC* circuit at resonance, for a purely capacitive circuit, and for a purely inductive circuit.

55E. An air conditioner connected to a 120 V rms ac line is equivalent to a 12.0 Ω resistance and a 1.30 Ω inductive reactance in series. (a) Calculate the impedance of the air conditioner. (b) Find the average rate at which energy is supplied to the appliance. ssm

56P. In a series oscillating *RLC* circuit, $R = 16.0\ \Omega$, $C = 31.2\ \mu\text{F}$, $L = 9.20$ mH, and $\mathscr{E} = \mathscr{E}_m \sin \omega_d t$ with $\mathscr{E}_m = 45.0$ V and $\omega_d = 3000$ rad/s. For time $t = 0.442$ ms find (a) the rate at which energy is being supplied by the generator, (b) the rate at which the energy in the capacitor is changing, (c) the rate at which the energy in the inductor is changing, and (d) the rate at which energy is being dissipated in the resistor. (e) What is the meaning of a negative result for any of (a), (b), and (c)? (f) Show that the results of (b), (c), and (d) sum to the result of (a).

57P. Figure 33-28 shows an ac generator connected to a "black box" through a pair of terminals. The box contains an *RLC* circuit, possibly even a multiloop circuit, whose elements and connections we do not know. Measurements outside the box reveal that

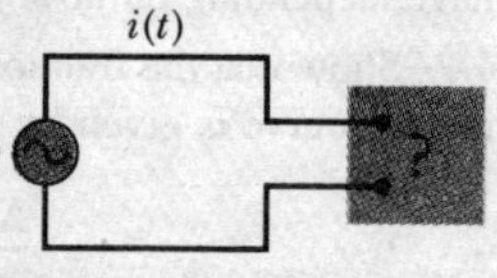

Fig. 33-28 Problem 57.

$$\mathscr{E}(t) = (75.0\ \text{V}) \sin \omega_d t$$

and

$$i(t) = (1.20\ \text{A}) \sin(\omega_d t + 42.0°).$$

(a) What is the power factor? (b) Does the current lead or lag the emf? (c) Is the circuit in the box largely inductive or largely capacitive? (d) Is the circuit in the box in resonance? (e) Must there be a capacitor in the box? An inductor? A resistor? (f) At what average rate is energy delivered to the box by the generator? (g) Why don't you need to know the angular frequency ω_d to answer all these questions? ssm www

58P. In Fig. 33-29 show that the average rate at which energy is dissipated in resistance *R* is a maximum when *R* is equal to the internal resistance *r* of the ac generator. (In the text discussion we have tacitly assumed that $r = 0$.)

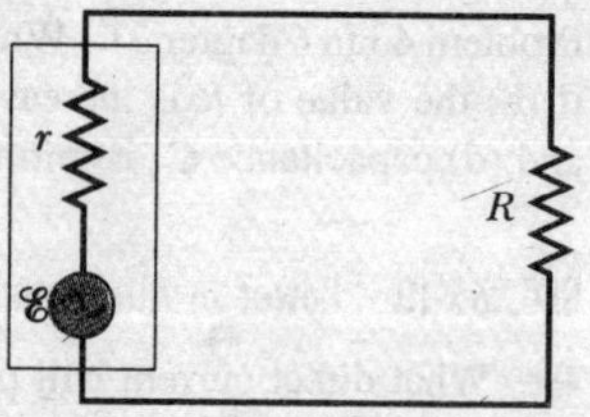

Fig. 33-29 Problems 58 and 65.

59P. In an *RLC* circuit such as that of Fig. 33-7 assume that $R = 5.00\ \Omega$, $L = 60.0$ mH, $f_d = 60.0$ Hz, and $\mathscr{E}_m = 30.0$ V. For what values of the capacitance would the average rate at which energy is dissipated in the resistance be (a) a maximum and (b) a minimum? (c) What are these maximum and minimum energy dissipation rates? What are (d) the corresponding phase angles and (e) the corresponding power factors? ssm

60P. A typical "light dimmer" used to dim the stage lights in a theater consists of a variable inductor *L* (whose inductance is adjustable between zero and L_{max}) connected in series with the lightbulb B as shown in Fig. 33-30. The electrical supply is 120 V (rms) at 60.0 Hz; the lightbulb is rated as "120 V, 1000 W." (a) What L_{max} is required if the rate of energy dissipation in the lightbulb is to be varied by a factor of 5 from its upper limit of 1000 W? Assume that the resistance of the lightbulb is independent of its temperature. (b) Could one use a variable resistor (adjustable between zero and R_{max}) instead of an inductor? If so, what R_{max} is required? Why isn't this done?

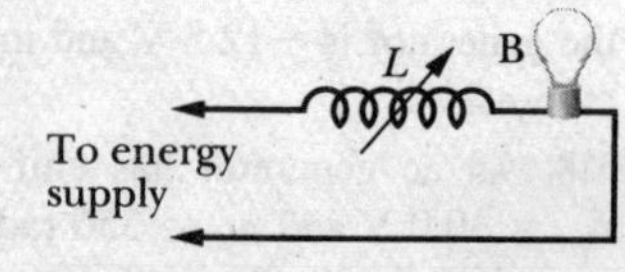

Fig. 33-30 Problem 60.

61P. In Fig. 33-31, $R = 15.0\ \Omega$, $C = 4.70\ \mu\text{F}$, and $L = 25.0$ mH. The generator provides a sinusoidal voltage of 75.0 V (rms) and frequency $f = 550$ Hz. (a) Calculate the rms current. (b) Find the rms voltages V_{ab}, V_{bc}, V_{cd}, V_{bd}, V_{ad}. (c) At what average rate is energy dissipated by each of the three circuit elements? ilw

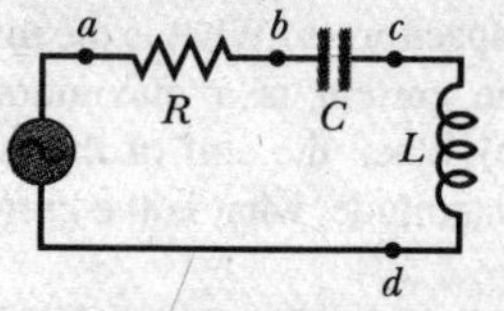

Fig. 33-31 Problem 61.

SEC. 33-11 Transformers

62E. A generator supplies 100 V to the primary coil of a transformer of 50 turns. If the secondary coil has 500 turns, what is the secondary voltage?

63E. A transformer has 500 primary turns and 10 secondary turns. (a) If V_p is 120 V (rms), what is V_s with an open circuit? (b) If the secondary now has a resistive load of 15 Ω, what are the currents in the primary and secondary? ssm ilw

64E. Figure 33-32 shows an "autotransformer." It consists of a single coil (with an iron core). Three taps T_i are provided. Between taps T_1 and T_2 there are 200 turns, and between taps T_2 and T_3 there are 800 turns. Any two taps can be considered the "primary terminals" and any two taps can be considered the "secondary terminals." List all the ratios by which the primary voltage may be changed to a secondary voltage.

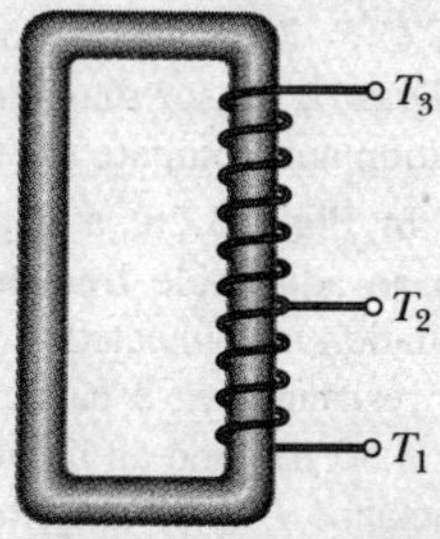

Fig. 33-32 Exercise 64.

65P. In Fig. 33-29 let the rectangular box on the left represent the (high-impedance) output of an audio amplifier, with $r = 1000\ \Omega$. Let $R = 10\ \Omega$ represent the (low-impedance) coil of a loudspeaker. For maximum transfer of energy to the load *R* we must have $R = r$, and that is not true in this case. However, a transformer can be used to "transform" resistances, making them behave electrically as if they were larger or smaller than they actually are. Sketch the primary and secondary coils of a transformer that can be introduced between the amplifier and the speaker in Fig. 33-29 to match the impedances. What must be the turns ratio? ssm

34 Electromagnetic Waves

As a comet swings around the Sun, ice on its surface vaporizes, releasing trapped dust and charged particles. The electrically charged "solar wind" forces the charged particles into a straight "tail" that points radially away from the Sun. However, the dust is unaffected by the solar wind and seemingly should continue to travel along the comet's orbit.

Why, instead, does much of the dust fashion the curved lower tail seen in the photograph?

The answer is in this chapter.

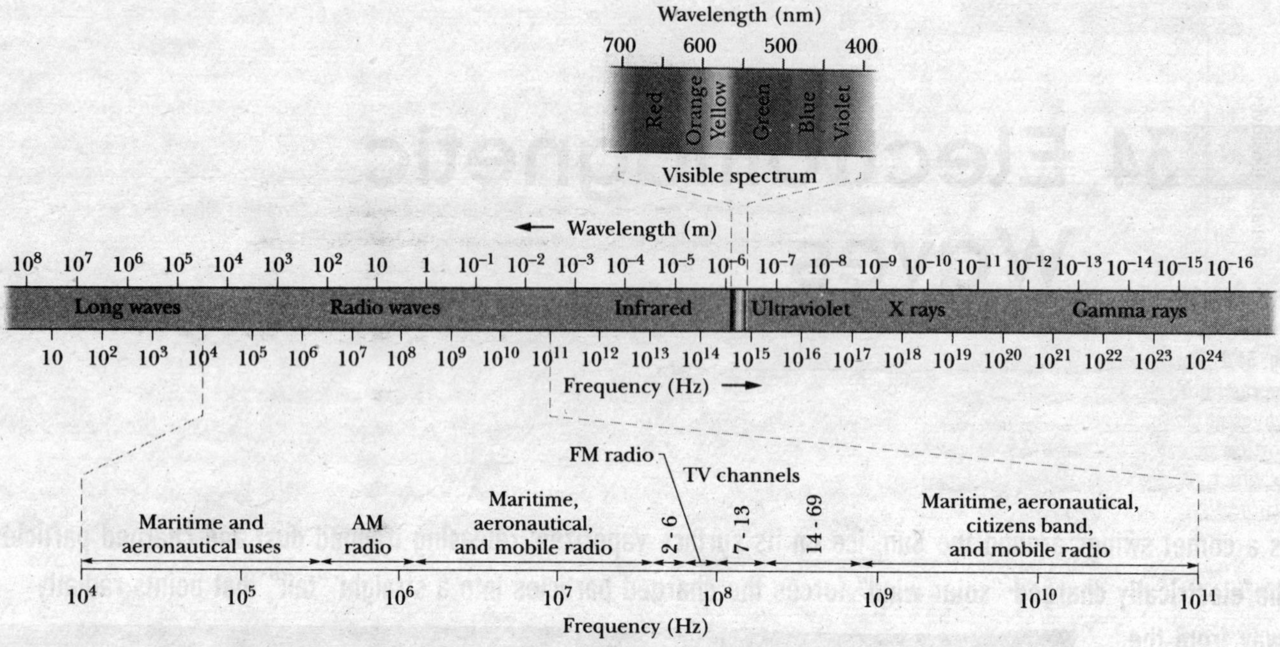

Fig. 34-1 The electromagnetic spectrum.

34-1 Maxwell's Rainbow

James Clerk Maxwell's crowning achievement was to show that a beam of light is a traveling wave of electric and magnetic fields—an **electromagnetic wave**—and thus that optics, the study of visible light, is a branch of electromagnetism. In this chapter we move from one to the other: we conclude our discussion of strictly electric and magnetic phenomena, and we build a foundation for optics.

In Maxwell's time (the mid 1800s), the visible, infrared, and ultraviolet forms of light were the only electromagnetic waves known. Spurred on by Maxwell's work, however, Heinrich Hertz discovered what we now call radio waves and verified that they move through the laboratory at the same speed as visible light.

As Fig. 34-1 shows, we now know a wide *spectrum* (or range) of electromagnetic waves, referred to by one imaginative writer as "Maxwell's rainbow." Consider the extent to which we are bathed in electromagnetic waves throughout this spectrum. The Sun, whose radiations define the environment in which we as a species have evolved and adapted, is the dominant source. We are also crisscrossed by radio and television signals. Microwaves from radar systems and from telephone relay systems may reach us. There are electromagnetic waves from lightbulbs, from the heated engine blocks of automobiles, from x-ray machines, from lightning flashes, and from buried radioactive materials. Beyond this, radiation reaches us from stars and other objects in our galaxy and from other galaxies. Electromagnetic waves also travel in the other direction. Television signals, transmitted from Earth since about 1950, have now taken news about us (along with episodes of *I Love Lucy,* albeit *very* faintly) to whatever technically sophisticated inhabitants there may be on whatever planets may encircle the nearest 400 or so stars.

In the wavelength scale in Fig. 34-1 (and similarly the corresponding frequency scale), each scale marker represents a change in wavelength (and correspondingly in frequency) by a factor of 10. The scale is open-ended; the wavelengths of electromagnetic waves have no inherent upper or lower bounds.

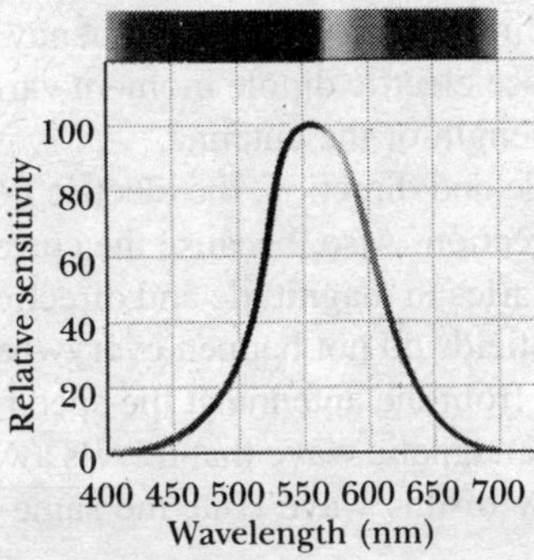

Fig. 34-2 The relative sensitivity of the average human eye to electromagnetic waves at different wavelengths. This portion of the electromagnetic spectrum to which the eye is sensitive is called *visible light.*

Certain regions of the electromagnetic spectrum in Fig. 34-1 are identified by familiar labels, such as *x rays* and *radio waves*. These labels denote roughly defined wavelength ranges within which certain kinds of sources and detectors of electromagnetic waves are in common use. Other regions of Fig. 34-1, such as those labeled television and AM radio, represent specific wavelength bands assigned by law for certain commercial or other purposes. There are no gaps in the electromagnetic spectrum—and all electromagnetic waves, no matter where they lie in the spectrum, travel through *free space* (vacuum) with the same speed *c*.

The visible region of the spectrum is of course of particular interest to us. Figure 34-2 shows the relative sensitivity of the human eye to light of various wavelengths. The center of the visible region is about 555 nm, which produces the sensation that we call yellow-green.

The limits of this visible spectrum are not well defined because the eye sensitivity curve approaches the zero-sensitivity line asymptotically at both long and short wavelengths. If we take the limits, arbitrarily, as the wavelengths at which eye sensitivity has dropped to 1% of its maximum value, these limits are about 430 and 690 nm; however, the eye can detect electromagnetic waves somewhat beyond these limits if they are intense enough.

34-2 The Traveling Electromagnetic Wave, Qualitatively

Some electromagnetic waves, including x rays, gamma rays, and visible light, are *radiated* (emitted) from sources that are of atomic or nuclear size, where quantum physics rules. Here we discuss how other electromagnetic waves are generated. To simplify matters, we restrict ourselves to that region of the spectrum (wavelength $\lambda \approx 1$ m) in which the source of the *radiation* (the emitted waves) is both macroscopic and of manageable dimensions.

Figure 34-3 shows, in broad outline, the generation of such waves. At its heart is an *LC oscillator,* which establishes an angular frequency ω $(= 1/\sqrt{LC})$. Charges and currents in this circuit vary sinusoidally at this frequency, as depicted in Fig. 33-1. An external source—possibly an ac generator—must be included to supply energy to compensate both for thermal losses in the circuit and for energy carried away by the radiated electromagnetic wave.

The *LC* oscillator of Fig. 34-3 is coupled by a transformer and a transmission line to an *antenna,* which consists essentially of two thin, solid, conducting rods. Through this coupling, the sinusoidally varying current in the oscillator causes charge to oscillate sinusoidally along the rods of the antenna at the angular frequency ω of the *LC* oscillator. The current in the rods associated with this movement of

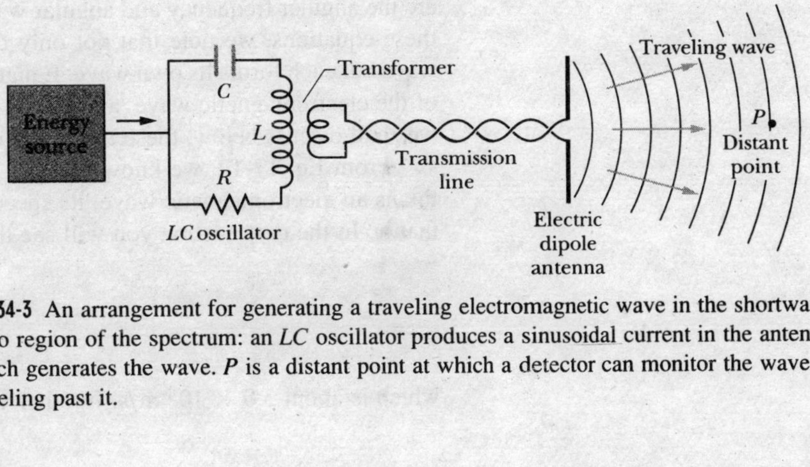

Fig. 34-3 An arrangement for generating a traveling electromagnetic wave in the shortwave radio region of the spectrum: an *LC* oscillator produces a sinusoidal current in the antenna, which generates the wave. *P* is a distant point at which a detector can monitor the wave traveling past it.

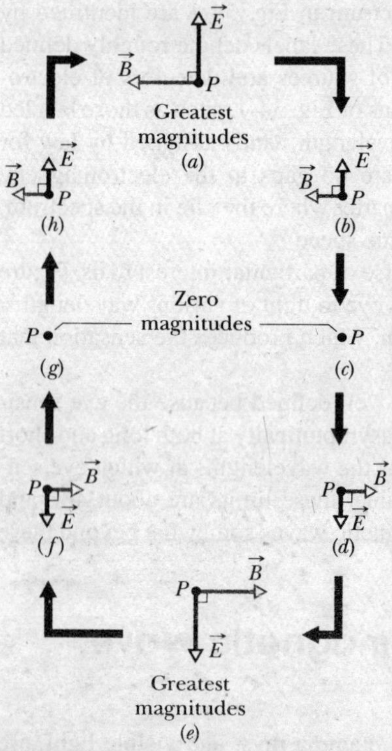

Fig. 34-4 (*a*)–(*h*) The variation in the electric field $\vec{E}$ and the magnetic field $\vec{B}$ at the distant point *P* of Fig. 34-3 as one wavelength of the electromagnetic wave travels past it. In this perspective, the wave is traveling directly out of the page. The two fields vary sinusoidally in magnitude and direction. Note that they are always perpendicular to each other and to the direction of travel of the wave.

charge also varies sinusoidally, in magnitude and direction, at angular frequency ω. The antenna has the effect of an electric dipole whose electric dipole moment varies sinusoidally in magnitude and direction along the length of the antenna.

Because the dipole moment varies in magnitude and direction, the electric field produced by the dipole varies in magnitude and direction. Also, because the current varies, the magnetic field produced by the current varies in magnitude and direction. However, the changes in the electric and magnetic fields do not happen everywhere instantaneously; rather, the changes travel outward from the antenna at the speed of light c. Together the changing fields form an electromagnetic wave that travels away from the antenna at speed c. The angular frequency of this wave is ω, the same as that of the *LC* oscillator.

Figure 34-4 shows how the electric field $\vec{E}$ and the magnetic field $\vec{B}$ change with time as one wavelength of the wave sweeps past the distant point *P* of Fig. 34-3; in each part of Fig. 34-4, the wave is traveling directly out of the page. (We choose a distant point so that the curvature of the waves suggested in Fig. 34-3 is small enough to neglect. At such points, the wave is said to be a *plane wave,* and discussion of the wave is much simplified.) Note several key features in Fig. 34-4; they are present regardless of how the wave is created:

For an e-m wave:

1. The electric and magnetic fields $\vec{E}$ and $\vec{B}$ are always perpendicular to the direction of travel of the wave. Thus, the wave is a *transverse wave,* as discussed in Chapter 17.
2. The electric field is always perpendicular to the magnetic field.
3. The cross product $\vec{E} \times \vec{B}$ always gives the direction of travel of the wave.
4. The fields (E & B) always vary sinusoidally, just like the transverse waves discussed in Chapter 17. Moreover, the fields vary with the same frequency and *in phase* (in step) with each other.

In keeping with these features, we can assume that the electromagnetic wave is traveling toward *P* in the positive direction of an *x* axis, that the electric field in Fig. 34-4 is oscillating parallel to the *y* axis, and that the magnetic field is then oscillating parallel to the *z* axis (using a right-handed coordinate system, of course). Then we can write the electric and magnetic fields as sinusoidal functions of position *x* (along the path of the wave) and time *t*:

$$E = E_m \sin(kx - \omega t), \tag{34-1}$$

$$B = B_m \sin(kx - \omega t), \tag{34-2}$$

in which E_m and B_m are the amplitudes of the fields and, as in Chapter 17, ω and k are the angular frequency and angular wave number of the wave, respectively. From these equations, we note that not only do the two fields form the electromagnetic wave but each forms its own wave. Equation 34-1 gives the *electric wave component* of the electromagnetic wave, and Eq. 34-2 gives the *magnetic wave component.* As we shall discuss below, these two wave components cannot exist independently.

From Eq. 17-12, we know that the speed of the wave is ω/k. However, since this is an electromagnetic wave, its speed (in vacuum) is given the symbol c rather than v. In the next section you will see that c has the value

$$c = \frac{1}{\sqrt{\mu_0 \varepsilon_0}} \quad \text{(wave speed)}, \tag{34-3}$$

which is about 3.0×10^8 m/s. In other words,

All electromagnetic waves, including visible light, have the same speed c in vacuum.

You will also see that the wave speed c and the amplitudes of the electric and magnetic fields are related by

$$\frac{E_m}{B_m} = c \qquad \text{(amplitude ratio).} \tag{34-4}$$

If we divide Eq. 34-1 by Eq. 34-2 and then substitute with Eq. 34-4, we find that the magnitudes of the fields at every instant and at any point are related by

$$\frac{E}{B} = c \qquad \text{(magnitude ratio).} \tag{34-5}$$

We can represent the electromagnetic wave as in Fig. 34-5*a*, with a *ray* (a directed line showing the wave's direction of travel) or with *wavefronts* (imaginary surfaces over which the wave has the same magnitude of electric field), or both. The two wavefronts shown in Fig. 34-5*a* are separated by one wavelength λ $(= 2\pi/k)$ of the wave. (Waves traveling in approximately the same direction form a *beam*, such as a laser beam, which can also be represented with a ray.)

We can also represent the wave as in Fig. 34-5*b*, which shows the electric and magnetic field vectors in a "snapshot" of the wave at a certain instant. The curves through the tips of the vectors represent the sinusoidal oscillations given by Eqs. 34-1 and 34-2; the wave components $\vec{E}$ and $\vec{B}$ are in phase, perpendicular to each other, and perpendicular to the wave's direction of travel.

Interpretation of Fig. 34-5*b* requires some care. The similar drawings for a transverse wave on a taut string that we discussed in Chapter 17 represented the up and down displacement of sections of the string as the wave passed (*something actually moved*). Figure 34-5*b* is more abstract. At the instant shown, the electric and magnetic fields each have a certain magnitude and direction (but always perpendicular to the x axis) at each point along the x axis. We choose to represent these vector quantities with a pair of arrows for each point, so we must draw arrows of different lengths for different points, all directed away from the x axis, like thorns on a rose stem. However, the arrows represent only field values at points that are on the x axis. Neither the arrows nor the sinusoidal curves represent a sideways motion of anything, nor do the arrows connect points on the x axis with points off the axis.

Drawings like Fig. 34-5 help us visualize what is actually a very complicated situation. First consider the magnetic field. Because it varies sinusoidally, it induces (via Faraday's law of induction) a perpendicular electric field that also varies sinusoidally. However, because that electric field is varying sinusoidally, it induces (via

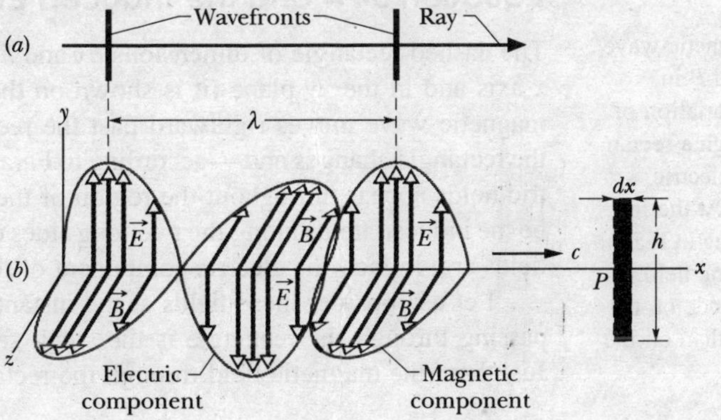

Fig. 34-5 (*a*) An electromagnetic wave represented with a ray and two wavefronts; the wavefronts are separated by one wavelength λ. (*b*) The same wave represented in a "snapshot" of its electric field $\vec{E}$ and magnetic field $\vec{B}$ at points on the x axis, along which the wave travels at speed c. As it travels past point P, the fields vary as shown in Fig. 34-4. The electric component of the wave consists of only the electric fields; the magnetic component consists of only the magnetic fields. The dashed rectangle at P is used in Fig. 34-6.

Maxwell's law of induction) a perpendicular magnetic field that also varies sinusoidally. And so on. The two fields continuously create each other via induction, and the resulting sinusoidal variations in the fields travel as a wave—the electromagnetic wave. Without this amazing result, we could not see; indeed, because we need electromagnetic waves from the Sun to maintain Earth's temperature, without this result we could not even exist.

A Most Curious Wave

The waves we discussed in Chapters 17 and 18 require a *medium* (some material) through which or along which to travel. We had waves traveling along a string, through Earth, and through the air. However, an electromagnetic wave (let's use the term *light wave* or *light*) is curiously different in that it requires no medium for its travel. It can, indeed, travel through a medium such as air or glass, but it can also travel through the vacuum of space between a star and us.

Once the special theory of relativity became accepted, long after Einstein published it in 1905, the speed of light waves was realized to be special. One reason is that light has the same speed regardless of the frame of reference from which it is measured. If you send a beam of light along an axis and ask several observers to measure its speed while they move at different speeds along that axis, either in the direction of the light or opposite it, they will all measure the *same speed* for the light. This result is an amazing one and quite different from what would have been found if those observers had measured the speed of any other type of wave; for other waves, the speed of the observers relative to the wave would have affected their measurements.

The meter has now been defined so that the speed of light (any electromagnetic wave) in vacuum has the exact value

$$c = 299\,792\,458 \text{ m/s},$$

which can be used as a standard. In fact, if you now measure the travel time of a pulse of light from one point to another, you are not really measuring the speed of the light but rather the distance between those two points.

34-3 The Traveling Electromagnetic Wave, Quantitatively

We shall now derive Eqs. 34-3 and 34-4 and, even more important, explore the dual induction of electric and magnetic fields that gives us light.

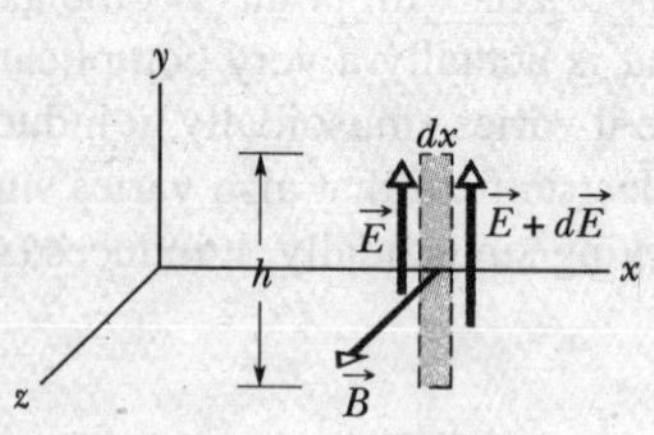

Fig. 34-6 As the electromagnetic wave travels rightward past point P in Fig. 34-5, the sinusoidal variation of the magnetic field $\vec{B}$ through a rectangle centered at P induces electric fields along the rectangle. At the instant shown, $\vec{B}$ is decreasing in magnitude and the induced electric field is therefore greater in magnitude on the right side of the rectangle than on the left.

Equation 34-4 and the Induced Electric Field

The dashed rectangle of dimensions dx and h in Fig. 34-6 is fixed at point P on the x axis and in the xy plane (it is shown on the right in Fig. 34-5b). As the electromagnetic wave moves rightward past the rectangle, the magnetic flux Φ_B through the rectangle changes and—according to Faraday's law of induction—induced electric fields appear throughout the region of the rectangle. We take $\vec{E}$ and $\vec{E} + d\vec{E}$ to be the induced fields along the two long sides of the rectangle. These induced electric fields are, in fact, the electric component of the electromagnetic wave.

Let us consider these fields at the instant when the magnetic wave component passing through the rectangle is the small section marked with red in Fig. 34-5b. Just then, the magnetic field through the rectangle points in the positive z direction and is decreasing in magnitude (the magnitude was greater just before the red section arrived). Because the magnetic field is decreasing, the magnetic flux Φ_B through the rectangle is also decreasing. According to Faraday's law, this change in flux is opposed by induced electric fields, which produce a magnetic field $\vec{B}$ in the positive z direction.

According to Lenz's law, this in turn means that if we imagine the boundary of the rectangle to be a conducting loop, a counterclockwise induced current would have to appear in it. There is, of course, no conducting loop; but this analysis shows that the induced electric field vectors $\vec{E}$ and $\vec{E} + d\vec{E}$ are indeed oriented as shown in Fig. 34-6, with the magnitude of $\vec{E} + d\vec{E}$ greater than that of $\vec{E}$. Otherwise, the net induced electric field would not act counterclockwise around the rectangle.

Let us now apply Faraday's law of induction,

$$\oint \vec{E} \cdot d\vec{s} = -\frac{d\Phi_B}{dt}, \qquad (34\text{-}6)$$

counterclockwise around the rectangle of Fig. 34-6. There is no contribution to the integral from the top or bottom of the rectangle because $\vec{E}$ and $d\vec{s}$ are perpendicular there. The integral then has the value

$$\oint \vec{E} \cdot d\vec{s} = (E + dE)h - Eh = h\, dE. \qquad (34\text{-}7)$$

The flux Φ_B through this rectangle is

$$\Phi_B = (B)(h\, dx), \qquad (34\text{-}8)$$

where B is the magnitude of $\vec{B}$ within the rectangle and $h\, dx$ is the area of the rectangle. Differentiating Eq. 34-8 with respect to t gives

$$\frac{d\Phi_B}{dt} = h\, dx \frac{dB}{dt}. \qquad (34\text{-}9)$$

If we substitute Eqs. 34-7 and 34-9 into Eq. 34-6, we find

$$h\, dE = -h\, dx \frac{dB}{dt}$$

or

$$\frac{dE}{dx} = -\frac{dB}{dt}. \qquad (34\text{-}10)$$

Actually, both B and E are functions of *two* variables, x and t, as Eqs. 34-1 and 34-2 imply. However, in evaluating dE/dx, we must assume that t is constant because Fig. 34-6 is an "instantaneous snapshot." Also, in evaluating dB/dt we must assume that x is constant because we are dealing with the time rate of change of B at a particular place, the point P in Fig. 34-5*b*. The derivatives under these circumstances are *partial derivatives,* and Eq. 34-10 must be written

$$\frac{\partial E}{\partial x} = -\frac{\partial B}{\partial t}. \qquad (34\text{-}11)$$

The minus sign in this equation is appropriate and necessary because, although E is increasing with x at the site of the rectangle in Fig. 34-6, B is decreasing with t.

From Eq. 34-1 we have

$$\frac{\partial E}{\partial x} = kE_m \cos(kx - \omega t)$$

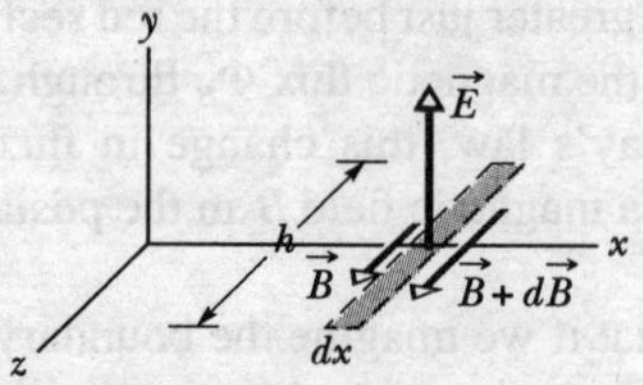

Fig. 34-7 The sinusoidal variation of the electric field through this rectangle, located (but not shown) at point P in Fig. 34-5, induces magnetic fields along the rectangle. The instant shown is that of Fig. 34-6: $\vec{E}$ is decreasing in magnitude and the induced magnetic field is greater in magnitude on the right side of the rectangle than on the left.

and from Eq. 34-2

$$\frac{\partial B}{\partial t} = -\omega B_m \cos(kx - \omega t).$$

Then Eq. 34-11 reduces to

$$kE_m \cos(kx - \omega t) = \omega B_m \cos(kx - \omega t). \tag{34-12}$$

The ratio ω/k for a traveling wave is its speed, which we are calling c. Equation 34-12 then becomes

$$\frac{E_m}{B_m} = c \quad \text{(amplitude ratio)}, \tag{34-13}$$

which is just Eq. 34-4.

Equation 34-3 and the Induced Magnetic Field

Figure 34-7 shows another dashed rectangle at point P of Fig. 34-5; this one is in the xz plane. As the electromagnetic wave moves rightward past this new rectangle, the electric flux Φ_E through the rectangle changes and—according to Maxwell's law of induction—induced magnetic fields appear throughout the region of the rectangle. These induced magnetic fields are, in fact, the magnetic component of the electromagnetic wave.

We see from Fig. 34-5 that at the instant chosen for the magnetic field in Fig. 34-6, the electric field through the rectangle of Fig. 34-7 is directed as shown. Recall that at the chosen instant, the magnetic field in Fig. 34-6 is decreasing. Because the two fields are in phase, the electric field in Fig. 34-7 must also be decreasing, and so must the electric flux Φ_E through the rectangle. By applying the same reasoning we applied to Fig. 34-6, we see that the changing flux Φ_E will induce a magnetic field with vectors $\vec{B}$ and $\vec{B} + d\vec{B}$ oriented as shown in Fig. 34-7, where $\vec{B} + d\vec{B}$ is greater than $\vec{B}$.

Let us apply Maxwell's law of induction,

$$\oint \vec{B} \cdot d\vec{s} = \mu_0 \varepsilon_0 \frac{d\Phi_E}{dt}, \tag{34-14}$$

by proceeding counterclockwise around the dashed rectangle of Fig. 34-7. Only the long sides of the rectangle contribute to the integral, whose value is

$$\oint \vec{B} \cdot d\vec{s} = -(B + dB)h + Bh = -h\, dB. \tag{34-15}$$

The flux Φ_E through the rectangle is

$$\Phi_E = (E)(h\, dx), \tag{34-16}$$

where E is the average magnitude of $\vec{E}$ within the rectangle. Differentiating Eq. 34-16 with respect to t gives

$$\frac{d\Phi_E}{dt} = h\, dx \frac{dE}{dt}.$$

If we substitute this and Eq. 34-15 into Eq. 34-14, we find

$$-h\, dB = \mu_0 \varepsilon_0 \left(h\, dx \frac{dE}{dt} \right)$$

or, changing to partial-derivative notation as we did before (Eq. 34-11),

$$-\frac{\partial B}{\partial x} = \mu_0 \varepsilon_0 \frac{\partial E}{\partial t}. \tag{34-17}$$

Again, the minus sign in this equation is necessary because, although B is increasing with x at point P in the rectangle in Fig. 34-7, E is decreasing with t.

Evaluating Eq. 34-17 by using Eqs. 34-1 and 34-2 leads to

$$-kB_m \cos(kx - \omega t) = -\mu_0\varepsilon_0\omega E_m \cos(kx - \omega t),$$

which we can write as

$$\frac{E_m}{B_m} = \frac{1}{\mu_0\varepsilon_0(\omega/k)} = \frac{1}{\mu_0\varepsilon_0 c}.$$

Combining this with Eq. 34-13 leads at once to

$$c = \frac{1}{\sqrt{\mu_0\varepsilon_0}} \quad \text{(wave speed)}, \tag{34-18}$$

which is exactly Eq. 34-3.

CHECKPOINT 1: The magnetic field $\vec{B}$ through the rectangle of Fig. 34-6 is shown at a different instant in part 1 of the figure here; $\vec{B}$ is directed in the xz plane, parallel to the z axis, and its magnitude is increasing. (a) Complete part 1 by drawing the induced electric fields, indicating both directions and relative magnitudes (as in Fig. 34-6). (b) For the same instant, complete part 2 of the figure by drawing the electric field of the electromagnetic wave. Also draw the induced magnetic fields, indicating both directions and relative magnitudes (as in Fig. 34-7).

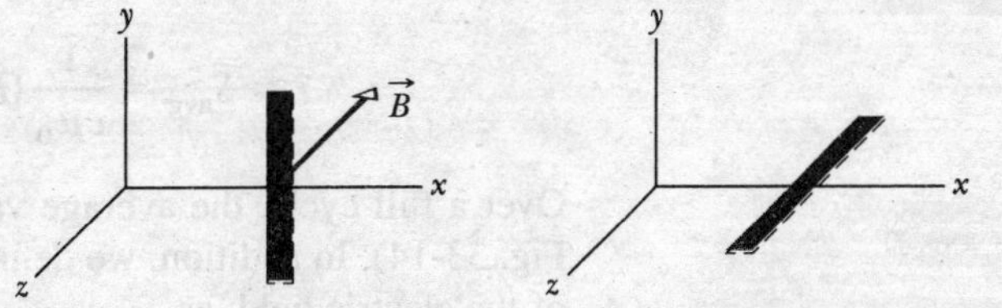

(1) (2)

34-4 Energy Transport and the Poynting Vector

All sunbathers know that an electromagnetic wave can transport energy and deliver it to a body on which it falls. The rate of energy transport per unit area in such a wave is described by a vector $\vec{S}$, called the **Poynting vector** after physicist John Henry Poynting (1852–1914), who first discussed its properties. $\vec{S}$ is defined as

$$\vec{S} = \frac{1}{\mu_0}\vec{E} \times \vec{B} \quad \text{(Poynting vector)}. \tag{34-19}$$

Its magnitude S is related to the rate at which energy is transported by a wave across a unit area at any instant (inst):

$$S = \left(\frac{\text{energy/time}}{\text{area}}\right)_{\text{inst}} = \left(\frac{\text{power}}{\text{area}}\right)_{\text{inst}}. \tag{34-20}$$

From this we can see that the SI unit for $\vec{S}$ is the watt per square meter (W/m^2).

The direction of the Poynting vector $\vec{S}$ of an electromagnetic wave at any point gives the wave's direction of travel and the direction of energy transport at that point.

Because $\vec{E}$ and $\vec{B}$ are perpendicular to each other in an electromagnetic wave, the magnitude of $\vec{E} \times \vec{B}$ is EB. Then the magnitude of $\vec{S}$ is

$$S = \frac{1}{\mu_0} EB, \tag{34-21}$$

in which S, E, and B are instantaneous values. E and B are so closely coupled to each other that we need to deal with only one of them; we choose E, largely because most instruments for detecting electromagnetic waves deal with the electric component of the wave rather than the magnetic component. Using $B = E/c$ from Eq. 34-5, we can rewrite Eq. 34-21 as

$$S = \frac{1}{c\mu_0} E^2 \quad \text{(instantaneous energy flow rate).} \tag{34-22}$$

By substituting $E = E_m \sin(kx - \omega t)$ into Eq. 34-22, we could obtain an equation for the energy transport rate as a function of time. More useful in practice, however, is the average energy transported over time; for that, we need to find the time-averaged value of S, written S_{avg} and also called the **intensity** I of the wave. Thus from Eq. 34-20, the intensity I is

$$I = S_{avg} = \left(\frac{\text{energy/time}}{\text{area}}\right)_{avg} = \left(\frac{\text{power}}{\text{area}}\right)_{avg}. \tag{34-23}$$

From Eq. 34-22, we find

$$I = S_{avg} = \frac{1}{c\mu_0} [E^2]_{avg} = \frac{1}{c\mu_0} [E_m^2 \sin^2(kx - \omega t)]_{avg}. \tag{34-24}$$

Over a full cycle, the average value of $\sin^2 \theta$, for any angular variable θ, is $\frac{1}{2}$ (see Fig. 33-14). In addition, we define a new quantity E_{rms}, the *root-mean-square* value of the electric field, as

$$E_{rms} = \frac{E_m}{\sqrt{2}}. \tag{34-25}$$

We can then rewrite Eq. 34-24 as

$$I = \frac{1}{c\mu_0} E_{rms}^2. \tag{34-26}$$

Because $E = cB$ and c is such a very large number, you might conclude that the energy associated with the electric field is much greater than that associated with the magnetic field. That conclusion is incorrect; the two energies are exactly equal. To show this, we start with Eq. 26-23, which gives the energy density $u\ (= \frac{1}{2}\varepsilon_0 E^2)$ within an electric field, and substitute cB for E; then we can write

$$u_E = \tfrac{1}{2}\varepsilon_0 E^2 = \tfrac{1}{2}\varepsilon_0 (cB)^2.$$

If we now substitute for c with Eq. 34-3, we get

$$u_E = \tfrac{1}{2}\varepsilon_0 \frac{1}{\mu_0 \varepsilon_0} B^2 = \frac{B^2}{2\mu_0}.$$

However, Eq. 31-56 tells us that $B^2/2\mu_0$ is the energy density u_B of a magnetic field $\vec{B}$, so we see that $u_E = u_B$ everywhere along an electromagnetic wave.

Variation of Intensity with Distance

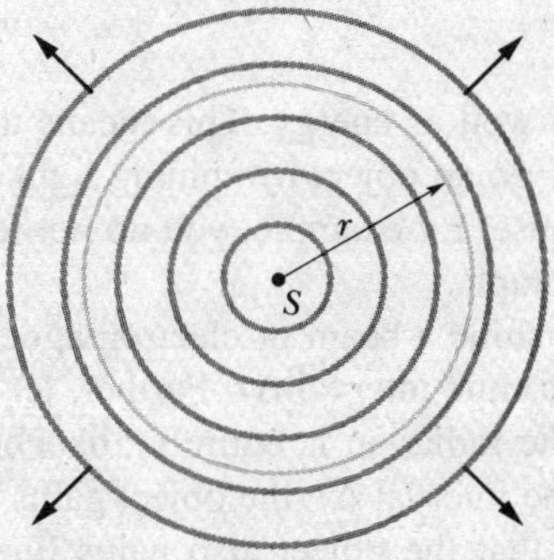

Fig. 34-8 A point source S emits electromagnetic waves uniformly in all directions. The spherical wavefronts pass through an imaginary sphere of radius r that is centered on S.

How intensity varies with distance from a real source of electromagnetic radiation is often complex—especially when the source (like a searchlight at a movie premier) beams the radiation in a particular direction. However, in some situations we can assume that the source is a *point source* that emits the light *isotropically*—that is, with equal intensity in all directions. The spherical wavefronts spreading from such an isotropic point source S at a particular instant are shown in cross section in Fig. 34-8.

Let us assume that the energy of the waves is conserved as they spread from this source. Let us also center an imaginary sphere of radius r on the source, as shown in Fig. 34-8. All the energy emitted by the source must pass through the sphere. Thus, the rate at which energy is transferred through the sphere by the radiation must equal the rate at which energy is emitted by the source—that is, the power P_s of the source. The intensity I at the sphere must then be

$$I = \frac{P_s}{4\pi r^2}, \tag{34-27}$$

where $4\pi r^2$ is the area of the sphere. Equation 34-27 tells us that the intensity of the electromagnetic radiation from an isotropic point source decreases with the square of the distance r from the source.

CHECKPOINT 2: The figure here gives the electric field of an electromagnetic wave at a certain point and a certain instant. The wave is transporting energy in the negative z direction. What is the direction of the magnetic field of the wave at that point and instant?

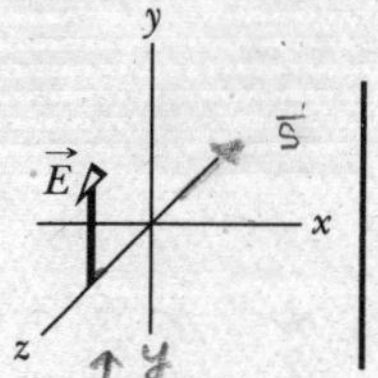

$\vec{S} = \frac{1}{\mu_0} \vec{E} \times \vec{B}$

$-\hat{z} = \frac{1}{\mu_0} (\hat{y} \times \vec{B})$

$\hat{B} = \hat{x}$

Sample Problem 34-1

An observer is 1.8 m from an isotropic point light source whose power P_s is 250 W. Calculate the rms values of the electric and magnetic fields due to the source at the position of the observer.

SOLUTION: The first two **Key Ideas** here are these:

1. The rms value E_{rms} of the electric field in light is related to the intensity I of the light via Eq. 34-26 ($I = E_{rms}^2/c\mu_0$).
2. Because the source is a point source emitting light with equal intensity in all directions, the intensity I at any distance r from the source is related to the source's power P_s via Eq. 34-27 ($I = P_s/4\pi r^2$).

Putting these two ideas together gives us

$$I = \frac{P_s}{4\pi r^2} = \frac{E_{rms}^2}{c\mu_0},$$

which leads to

$$E_{rms} = \sqrt{\frac{P_s c\mu_0}{4\pi r^2}}$$

$$= \sqrt{\frac{(250\text{ W})(3.00 \times 10^8\text{ m/s})(4\pi \times 10^{-7}\text{ H/m})}{(4\pi)(1.8\text{ m})^2}}$$

$$= 48.1\text{ V/m} \approx 48\text{ V/m}. \qquad \text{(Answer)}$$

The third **Key Idea** here is that magnitudes of the electric field and magnetic field of an electromagnetic wave at any instant and at any point in the wave are related by the speed of light c according to Eq. 34-5 ($E/B = c$). Thus, the rms values of those fields are also related by Eq. 34-5 and we can write

$$B_{rms} = \frac{E_{rms}}{c}$$

$$= \frac{48.1\text{ V/m}}{3.00 \times 10^8\text{ m/s}}$$

$$= 1.6 \times 10^{-7}\text{ T}. \qquad \text{(Answer)}$$

Note that E_{rms} ($= 48$ V/m) is appreciable as judged by ordinary laboratory standards, but B_{rms} ($= 1.6 \times 10^{-7}$ T) is quite small. This difference helps to explain why most instruments used for the detection and measurement of electromagnetic waves are designed to respond to the electric component of the wave. It is wrong, however, to say that the electric component of an electromagnetic wave is "stronger" than the magnetic component. You cannot compare quantities that are measured in different units. As we have seen, the electric and magnetic components are on an equal basis as far as the propagation of the wave is concerned, because their average energies, which *can* be compared, are exactly equal.

34-5 Radiation Pressure

Electromagnetic waves have linear momentum as well as energy. This means that we can exert a pressure—a **radiation pressure**—on an object by shining light on it. However, the pressure must be very small because, for example, you do not feel a camera flash when it is used to take your photograph.

To find an expression for the pressure, let us shine a beam of electromagnetic radiation—light, for example—on an object for a time interval Δt. Further, let us assume that the object is free to move and that the radiation is entirely **absorbed** (taken up) by the object. This means that during the interval Δt, the object gains an energy ΔU from the radiation. Maxwell showed that the object also gains linear momentum. The magnitude Δp of the momentum change of the object is related to the energy change ΔU by

$$\Delta p = \frac{\Delta U}{c} \quad \text{(total absorption)}, \tag{34-28}$$

where c is the speed of light. The direction of the momentum change of the object is the direction of the *incident* (incoming) beam that the object absorbs.

Instead of being absorbed, the radiation can be **reflected** by the object; that is, the radiation can be sent off in a new direction as if it bounced off the object. If the radiation is entirely reflected back along its original path, the magnitude of the momentum change of the object is twice that given above, or

$$\Delta p = \frac{2\,\Delta U}{c} \quad \text{(total reflection back along path)}. \tag{34-29}$$

In the same way, an object undergoes twice as much momentum change when a perfectly elastic tennis ball is bounced from it as when it is struck by a perfectly inelastic ball (a lump of wet putty, say) of the same mass and velocity. If the incident radiation is partly absorbed and partly reflected, the momentum change of the object is between $\Delta U/c$ and $2\,\Delta U/c$.

From Newton's second law, we know that a change in momentum is related to a force by

$$F = \frac{\Delta p}{\Delta t}. \tag{34-30}$$

To find expressions for the force exerted by radiation in terms of the intensity I of the radiation, suppose that a flat surface of area A, perpendicular to the path of the radiation, intercepts the radiation. In time interval Δt, the energy intercepted by area A is

$$\Delta U = IA\,\Delta t. \tag{34-31}$$

If the energy is completely absorbed, then Eq. 34-28 tells us that $\Delta p = IA\,\Delta t/c$ and, from Eq. 34-30, the magnitude of the force on the area A is

$$F = \frac{IA}{c} \quad \text{(total absorption)}. \tag{34-32}$$

Similarly, if the radiation is totally reflected back along its original path, Eq. 34-29 tells us that $\Delta p = 2IA\,\Delta t/c$ and, from Eq. 34-30,

$$F = \frac{2IA}{c} \quad \text{(total reflection back along path)}. \tag{34-33}$$

If the radiation is partly absorbed and partly reflected, the magnitude of the force on area A is between the values of IA/c and $2IA/c$.

The force per unit area on an object due to radiation is the radiation pressure p_r. We can find it for the situations of Eqs. 34-32 and 34-33 by dividing both sides of each equation by A. We obtain

$$p_r = \frac{I}{c} \quad \text{(total absorption)} \tag{34-34}$$

and

$$p_r = \frac{2I}{c} \quad \text{(total reflection back along path).} \tag{34-35}$$

Be careful not to confuse the symbol p_r for radiation pressure with the symbol p for momentum. Just as with fluid pressure in Chapter 15, the SI unit of radiation pressure is the newton per square meter (N/m^2), which is called the pascal (Pa).

The development of laser technology has permitted researchers to achieve radiation pressures much greater than, say, that due to a camera flashlamp. This comes about because a beam of laser light—unlike a beam of light from a small lamp filament—can be focused to a tiny spot only a few wavelengths in diameter. This permits the delivery of great amounts of energy to small objects placed at that spot.

CHECKPOINT 3: Light of uniform intensity shines perpendicularly on a totally absorbing surface, fully illuminating the surface. If the area of the surface is decreased, do (a) the radiation pressure and (b) the radiation force on the surface increase, decrease, or stay the same?

$P_r = I/c$ I same ∴ same

$P_r = \frac{F}{A}$ A ↓ ⇒ F = P_r A ⇒ F dec

Sample Problem 34-2

When dust is released by a comet, it does not continue along the comet's orbit because radiation pressure from sunlight pushes it radially outward from the Sun. Assume that a dust particle is spherical with radius R, has density $\rho = 3.5 \times 10^3$ kg/m^3, and totally absorbs the sunlight it intercepts. For what value of R does the gravitational force $\vec{F}_g$ on the dust particle due to the Sun just balance the radiation force $\vec{F}_r$ on it from the sunlight?

SOLUTION: We can assume that the Sun is far enough from the particle to act as an isotropic point source of light. Then because we are told that the radiation pressure pushes the particle radially outward from the Sun, we know that the radiation force $\vec{F}_r$ on the particle is directed radially outward from the center of the Sun. At the same time, the gravitational force $\vec{F}_g$ on the particle is directed radially inward *toward* the center of the Sun. Thus, we can write the balance of these two forces as

$$F_r = F_g. \tag{34-36}$$

Let us consider these forces separately.

Radiation force: To evaluate the left side of Eq. 34-36, we use these three **Key Ideas.**

1. Because the particle is totally absorbing, the force magnitude F_r can be found from the intensity I of sunlight at the particle's location and the particle's cross-sectional area A, via Eq. 34-32 ($F = IA/c$).
2. Because we assume that the Sun is an isotropic point source of light, we can use Eq. 34-27 ($I = P_s/4\pi r^2$) to relate the Sun's power P_S to the intensity I of the sunlight at the particle's distance r from the Sun.
3. Because the particle is spherical, its cross-sectional area A is πR^2 (*not* half its surface area).

Putting these three ideas together gives us

$$F_r = \frac{IA}{c} = \frac{P_S \pi R^2}{4\pi r^2 c} = \frac{P_S R^2}{4r^2 c}. \tag{34-37}$$

Gravitational force: The **Key Idea** here is Newton's law of gravitation (Eq. 14-1), which gives us the magnitude of the gravitational force on the particle as

$$F_g = \frac{GM_S m}{r^2}, \tag{34-38}$$

where M_S is the Sun's mass and m is the particle's mass. Next, the particle's mass is related to its density ρ and volume V ($= \frac{4}{3}\pi R^3$, for a sphere) by

$$\rho = \frac{m}{V} = \frac{m}{\frac{4}{3}\pi R^3}.$$

Solving this for m and substituting the result into Eq. 34-38 give us

$$F_g = \frac{GM_S \rho(\frac{4}{3}\pi R^3)}{r^2}. \tag{34-39}$$

Then substituting Eqs. 34-37 and 34-39 into Eq. 34-36 and solving for R yield

$$R = \frac{3P_S}{16\pi c \rho G M_S}.$$

Using the given value of ρ and the known values of G (Appendix B) and M_S (Appendix C), we can evaluate the denominator:

$$(16\pi)(3 \times 10^8 \text{ m/s})(3.5 \times 10^3 \text{ kg/m}^3) \times (6.67 \times 10^{-11} \text{ N} \cdot \text{m}^2/\text{kg}^2)(1.99 \times 10^{30} \text{ kg}) = 7.0 \times 10^{33} \text{ N/s}.$$

Using P_S from Appendix C, we then have

$$R = \frac{(3)(3.9 \times 10^{26} \text{ W})}{7.0 \times 10^{33} \text{ N/s}} = 1.7 \times 10^{-7} \text{ m}. \quad \text{(Answer)}$$

Note that this result is independent of the particle's distance r from the Sun.

Dust particles with radius $R \approx 1.7 \times 10^{-7}$ m follow an approximately straight path like path b in Fig. 34-9. For larger values of R, comparison of Eqs. 34-37 and 34-39 shows that, because F_g varies with R^3 and F_r varies with R^2, the gravitational force F_g dominates the radiation force F_r. Thus, such particles follow a path that is curved toward the Sun like path c in Fig. 34-9. Similarly, for smaller values of R, the radiation force dominates, and the dust follows a path that is curved away from the Sun like path a. The composite of these dust particles is the dust tail of the comet.

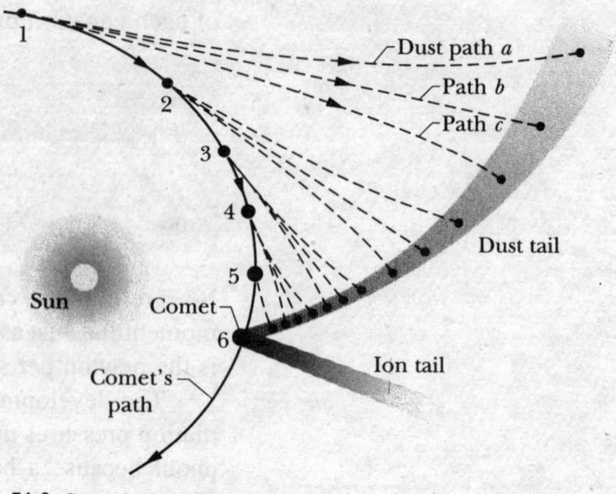

Fig. 34-9 Sample Problem 34-2. A comet is now at position 6. Dust it has released at five previous positions has been pushed outward by radiation pressure from sunlight, has taken the dashed paths, and now forms the comet's curved dust tail.

34-6 Polarization

VHF (very high frequency) television antennas in England are oriented vertically, but those in North America are horizontal. The difference is due to the direction of oscillation of the electromagnetic waves carrying the TV signal. In England, the transmitting equipment is designed to produce waves that are **polarized** vertically; that is, their electric field oscillates vertically. Thus, for the electric field of the incident television waves to drive a current along an antenna (and provide a signal to a television set), the antenna must be vertical. In North America, the waves are polarized horizontally .

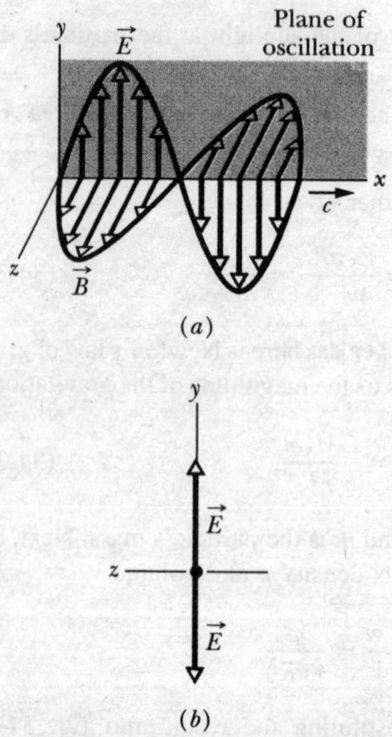

Fig. 34-10 (a) The plane of oscillation of a polarized electromagnetic wave. (b) To represent the polarization, we view the plane of oscillation "head-on" and indicate the directions of the oscillating electric field with a double arrow.

Figure 34-10a shows an electromagnetic wave with its electric field oscillating parallel to the vertical y axis. The plane containing the $\vec{E}$ vectors is called the **plane of oscillation** of the wave (hence, the wave is said to be *plane-polarized* in the y direction). We can represent the wave's *polarization* (state of being polarized) by showing the directions of the electric field oscillations in a "head-on" view of the plane of oscillation, as in Fig. 34-10b. The vertical "double arrow" in that figure indicates that as the wave travels past us, its electric field oscillates vertically—it continuously changes between being directed up and down along the y axis.

Polarized Light

The electromagnetic waves emitted by a television station all have the same polarization, but the electromagnetic waves emitted by any common source of light (such as the Sun or a bulb) are **polarized randomly** or **unpolarized;** that is, the electric field at any given point is always perpendicular to the direction of travel of the waves but changes directions randomly. Thus, if we try to represent a head-on view of the oscillations over some time period, we do not have a simple drawing with a single double arrow like that of Fig. 34-10b; instead we have a mess of double arrows like that in Fig. 34-11a.

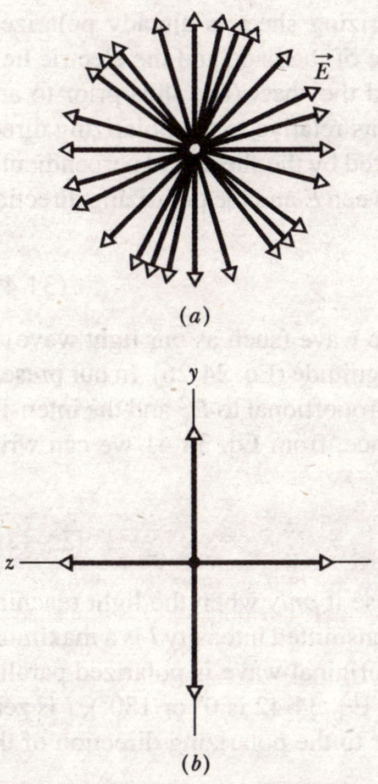

Fig. 34-11 (*a*) Unpolarized light consists of waves with randomly directed electric fields. Here the waves are all traveling along the same axis, directly out of the page, and all have the same amplitude *E*. (*b*) A second way of representing unpolarized light—the light is the superposition of two polarized waves whose planes of oscillation are perpendicular to each other.

In principle, we can simplify the mess by resolving each electric field of Fig. 34-11*a* into *y* and *z* components. Then as the wave travels past us, the net *y* component oscillates parallel to the *y* axis and the net *z* component oscillates parallel to the *z* axis. We can then represent the unpolarized light with a pair of double arrows as shown in Fig. 34-11*b*. The double arrow along the *y* axis represents the oscillations of the net *y* component of the electric field. The double arrow along the *z* axis represents the oscillations of the net *z* component of the electric field. In doing all this, we effectively change unpolarized light into the superposition of two polarized waves whose planes of oscillation are perpendicular to each other—one plane contains the *y* axis and the other contains the *z* axis. One reason to make this change is that drawing Fig. 34-11*b* is a lot easier than drawing Fig. 34-11*a*.

We can draw similar figures to represent light that is **partially polarized** (its field oscillations are not completely random as in Fig. 34-11*a* nor are they parallel to a single axis as in Fig. 34-10*b*). For this situation, we can draw one of the double arrows in a perpendicular pair of double arrows longer than the other one.

We can transform unpolarized visible light into polarized light by sending it through a *polarizing sheet,* as is shown in Fig. 34-12. Such sheets, commercially known as Polaroids or Polaroid filters, were invented in 1932 by Edwin Land while he was an undergraduate student. A polarizing sheet consists of certain long molecules embedded in plastic. When the sheet is manufactured, it is stretched to align the molecules in parallel rows, like rows in a plowed field. When light is then sent through the sheet, electric field components along one direction pass through the sheet, while components perpendicular to that direction are absorbed by the molecules and disappear.

We shall not dwell on the molecules but, instead, shall assign to the sheet a *polarizing direction,* along which electric field components are passed:

An electric field component parallel to the polarizing direction is passed (*transmitted*) by a polarizing sheet; a component perpendicular to it is absorbed.

Thus, the electric field of the light emerging from the sheet consists of only the components that are parallel to the polarizing direction of the sheet; hence the light is polarized in that direction. In Fig. 34-12, the vertical electric field components are transmitted by the sheet; the horizontal components are absorbed. The transmitted waves are then vertically polarized.

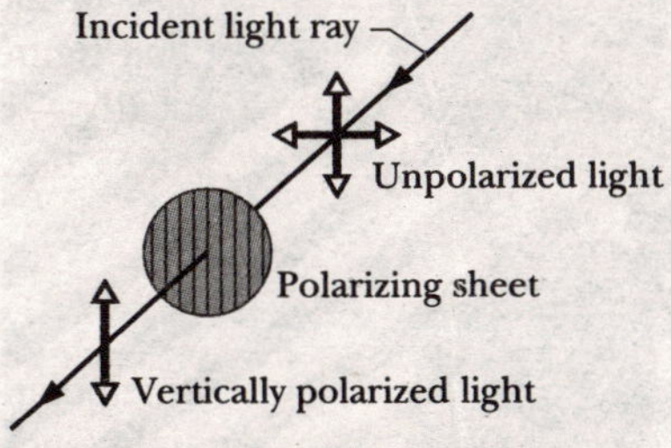

Fig. 34-12 Unpolarized light becomes polarized when it is sent through a polarizing sheet. Its direction of polarization is then parallel to the polarizing direction of the sheet, which is represented here by the vertical lines drawn in the sheet.

Intensity of Transmitted Polarized Light

We now consider the intensity of light transmitted by a polarizing sheet. We start with unpolarized light, whose electric field oscillations we can resolve into *y* and *z* components as represented in Fig. 34-11*b*. Further, we can arrange for the *y* axis to be parallel to the polarizing direction of the sheet. Then only the *y* components of the light's electric field are passed by the sheet; the *z* components are absorbed. As suggested by Fig. 34-11*b*, if the original waves are randomly oriented, the sum of the *y* components and the sum of the *z* components are equal. When the *z* components are absorbed, half the intensity I_0 of the original light is lost. The intensity *I* of the emerging polarized light is then

$$I = \tfrac{1}{2}I_0. \qquad (34\text{-}40)$$

Let us call this the *one-half rule;* we can use it *only* when the light reaching a polarizing sheet is unpolarized.

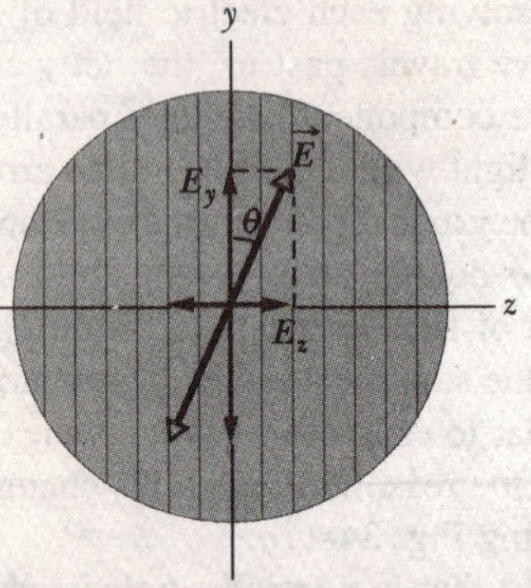

Fig. 34-13 Polarized light approaching a polarizing sheet. The electric field $\vec{E}$ of the light can be resolved into components E_y (parallel to the polarizing direction of the sheet) and E_z (perpendicular to that direction). Component E_y will be transmitted by the sheet; component E_z will be absorbed.

Suppose now that the light reaching a polarizing sheet is already polarized. Figure 34-13 shows a polarizing sheet in the plane of the page and the electric field $\vec{E}$ of such a polarized light wave traveling toward the sheet (and thus prior to any absorption). We can resolve $\vec{E}$ into two components relative to the polarizing direction of the sheet: parallel component E_y is transmitted by the sheet, and perpendicular component E_z is absorbed. Since θ is the angle between $\vec{E}$ and the polarizing direction of the sheet, the transmitted parallel component is

$$E_y = E\cos\theta. \tag{34-41}$$

Recall that the intensity of an electromagnetic wave (such as our light wave) is proportional to the square of the electric field's magnitude (Eq. 34-26). In our present case then, the intensity I of the emerging wave is proportional to E_y^2 and the intensity I_0 of the original wave is proportional to E^2. Hence, from Eq. 34-41 we can write $I/I_0 = \cos^2\theta$, or

$$I = I_0\cos^2\theta. \tag{34-42}$$

Let us call this the *cosine-squared rule;* we can use it *only* when the light reaching a polarizing sheet is already polarized. Then the transmitted intensity I is a maximum and is equal to the original intensity I_0 when the original wave is polarized parallel to the polarizing direction of the sheet (when θ in Eq. 34-42 is 0° or 180°). I is zero when the original wave is polarized perpendicular to the polarizing direction of the sheet (when θ is 90°).

Figure 34-14 shows an arrangement in which initially unpolarized light is sent through two polarizing sheets P_1 and P_2. (Often, the first sheet is called the *polarizer*, and the second the *analyzer*.) Because the polarizing direction of P_1 is vertical, the light transmitted by P_1 to P_2 is polarized vertically. If the polarizing direction of P_2 is also vertical, then all the light transmitted by P_1 is transmitted by P_2. If the polarizing direction of P_2 is horizontal, none of the light transmitted by P_1 is transmitted by P_2. We reach the same conclusions by considering only the *relative* orientations of the two sheets: If their polarizing directions are parallel, all the light passed by the first sheet is passed by the second sheet. If those directions are perpendicular (the sheets are said to be *crossed*), no light is passed by the second sheet. These two extremes are displayed with polarized sunglasses in Fig. 34-15.

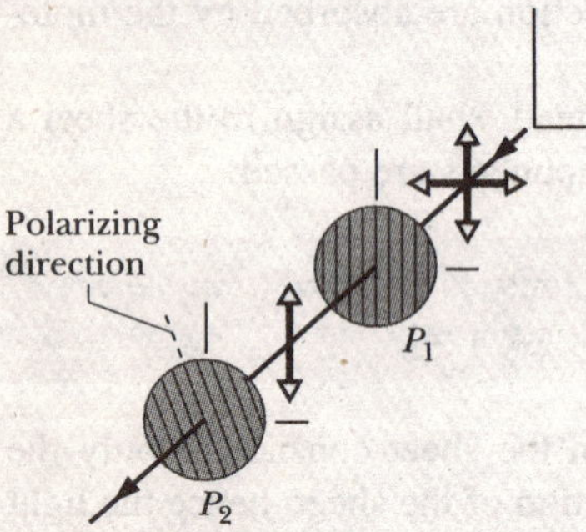

Fig. 34-14 The light transmitted by polarizing sheet P_1 is vertically polarized, as represented by the vertical double arrow. The amount of that light that is then transmitted by polarizing sheet P_2 depends on the angle between the polarization direction of that light and the polarizing direction of P_2 (indicated by the lines drawn in the sheet and by the dashed line).

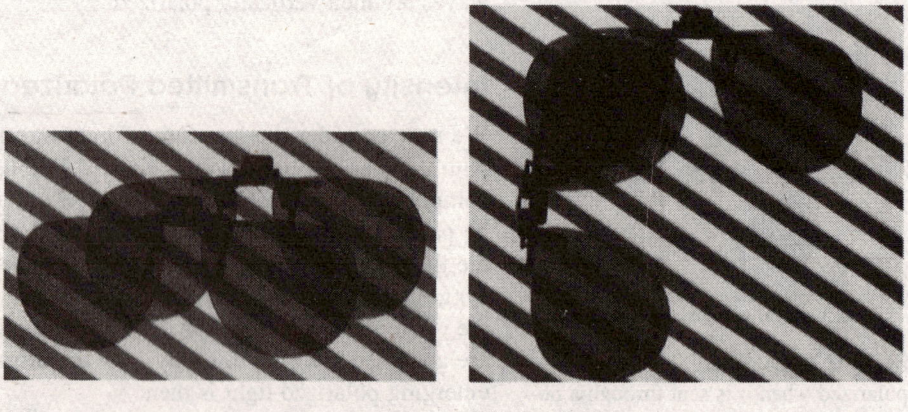

(*a*) (*b*)

Fig. 34-15 Polarizing sunglasses consist of sheets whose polarizing directions are vertical when the sunglasses are worn. (*a*) Overlapping sunglasses transmit light fairly well when their polarizing directions have the same orientation, but (*b*) they block most of the light when they are crossed.

Finally, if the two polarizing directions of Fig. 34-14 make an angle between 0° and 90°, some of the light transmitted by P_1 will be transmitted by P_2. The intensity of that light is determined by Eq. 34-42.

Light can be polarized by means other than polarizing sheets, such as by reflection (discussed in Section 34-9) and by scattering from atoms or molecules. In *scattering,* light that is intercepted by an object, such as a molecule, is sent off in many, perhaps random, directions. An example is the scattering of sunlight by molecules in the atmosphere, which gives the sky its general glow.

Although direct sunlight is unpolarized, light from much of the sky is at least partially polarized by such scattering. Bees use the polarization of sky light in navigating to and from their hives. Similarly, the Vikings used it to navigate across the North Sea when the daytime Sun was below the horizon (because of the high latitude of the North Sea). These early seafarers had discovered certain crystals (now called cordierite) that changed color when rotated in polarized light. By looking at the sky through such a crystal while rotating it about their line of sight, they could locate the hidden Sun and thus determine which way was south.

Sample Problem 34-3

Figure 34-16*a* shows a system of three polarizing sheets in the path of initially unpolarized light. The polarizing direction of the first sheet is parallel to the *y* axis, that of the second sheet is 60° counterclockwise from the *y* axis, and that of the third sheet is parallel to the *x* axis. What fraction of the initial intensity I_0 of the light emerges from the system, and how is that light polarized?

SOLUTION: The **Key Ideas** here are these:

1. We work through the system sheet by sheet, from the first one encountered by the light to the last one.
2. To find the intensity transmitted by any sheet, we apply either the one-half rule or the cosine-squared rule, depending on whether the light reaching the sheet is unpolarized or already polarized.
3. The light that is transmitted by a polarizing sheet is always polarized parallel to the polarizing direction of the sheet.

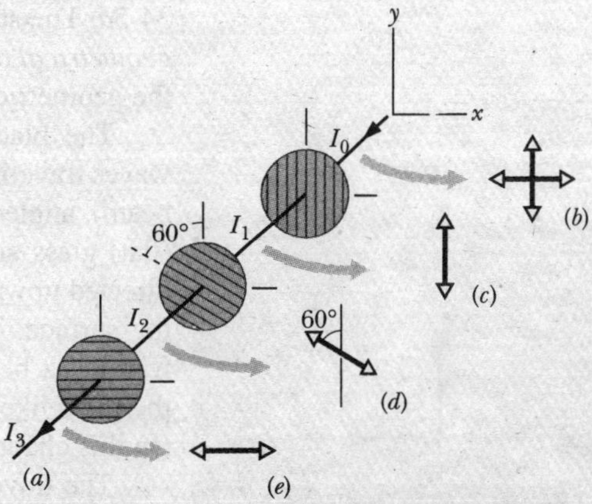

Fig. 34-16 Sample Problem 34-3. (*a*) Initially unpolarized light of intensity I_0 is sent into a system of three polarizing sheets. The intensities I_1, I_2, and I_3 of the light transmitted by the sheets are labeled. Shown also are the polarizations, from head-on views, of (*b*) the initial light and the light transmitted by (*c*) the first sheet, (*d*) the second sheet, and (*e*) the third sheet.

First sheet: The original light wave is represented in Fig. 34-16*b*, using the head-on, double-arrow representation of Fig. 34-11*b*. Because the light is initially unpolarized, the intensity I_1 of the light transmitted by the first sheet is given by the one-half rule (Eq. 34-40):

$$I_1 = \tfrac{1}{2}I_0.$$

Because the polarizing direction of the first sheet is parallel to the *y* axis, the polarization of the light transmitted by it is also, as shown in the head-on view of Fig. 34-16*c*.

Second sheet: Since the light reaching the second sheet is polarized, the intensity I_2 of the light transmitted by that sheet is given by the cosine-squared rule (Eq. 34-42). The angle θ in the rule is the angle between the polarization direction of the entering light (parallel to the *y* axis) and the polarizing direction of the second sheet (60° counterclockwise from the *y* axis), and so θ is 60°. Then

$$I_2 = I_1 \cos^2 60°.$$

The polarization of this transmitted light is parallel to the polarizing direction of the sheet transmitting it—that is, 60° counterclockwise from the *y* axis, as shown in the head-on view of Fig. 34-16*d*.

Third sheet: Because the light reaching the third sheet is polarized, the intensity I_3 of the light transmitted by that sheet is given by the cosine-squared rule. The angle θ is now the angle between the polarization direction of the entering light (Fig. 34-16*d*) and the polarizing direction of the third sheet (parallel to the x axis), and so $\theta = 30°$. Thus,

$$I_3 = I_2 \cos^2 30°.$$

This final transmitted light is polarized parallel to the x axis (Fig. 34-16*e*). We find its intensity by substituting first for I_2 and then for I_1 in the equation above:

$$I_3 = I_2 \cos^2 30° = (I_1 \cos^2 60°) \cos^2 30°$$
$$= (\tfrac{1}{2} I_0) \cos^2 60° \cos^2 30° = 0.094 I_0.$$

Thus,
$$\frac{I_3}{I_0} = 0.094. \qquad \text{(Answer)}$$

That is to say, 9.4% of the initial intensity emerges from the three-sheet system. (If we now remove the second sheet, what fraction of the initial intensity emerges from the system?)

✔CHECKPOINT 4: The figure shows four pairs of polarizing sheets, seen face-on. Each pair is mounted in the path of initially unpolarized light (like the three sheets in Fig. 34-16*a*). The polarizing direction of each sheet (indicated by the dashed line) is referenced to either a horizontal x axis or a vertical y axis. Rank the pairs according to the fraction of the initial intensity that they pass, greatest first.

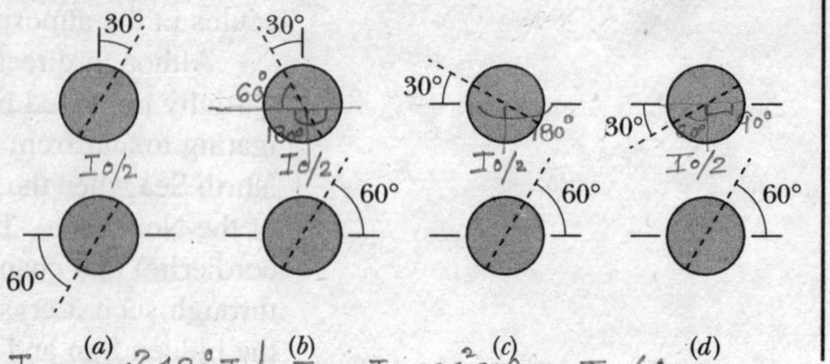

34-7 Reflection and Refraction

Although a light wave spreads as it moves away from its source, we can often approximate its travel as being in a straight line; we did so for the light wave in Fig. 34-5*a*. The study of the properties of light waves under that approximation is called *geometrical optics*. For the rest of this chapter and all of Chapter 35, we shall discuss the geometrical optics of visible light.

The black-and-white photograph in Fig. 34-17*a* shows an example of light waves traveling in approximately straight lines. A narrow beam of light (the *incident* beam), angled downward from the left and traveling through air, encounters a *plane* (flat) glass surface. Part of the light is **reflected** by the surface, forming a beam directed upward toward the right, traveling as if the original beam had bounced from the surface. The rest of the light travels through the surface and into the glass, forming a beam directed downward to the right. Because light can travel through the glass like this, the glass is said to be *transparent;* that is, we can see through it. (In this chapter we shall consider only transparent materials.)

The travel of light through a surface (or *interface*) that separates two media is called **refraction,** and the light is said to be *refracted*. Unless an incident beam of light is perpendicular to a surface, refraction by the surface changes the light's direction of travel. For this reason, the beam is said to be "bent" by the refraction. Note in Fig. 34-17*a* that the bending occurs only at the surface; within the glass, the light travels in a straight line.

In Figure 34-17*b*, the beams of light in the photograph are represented with an *incident ray,* a *reflected ray,* and a *refracted ray* (and wavefronts). Each ray is oriented with respect to a line, called the *normal,* that is perpendicular to the surface at the point of reflection and refraction. In Fig. 34-17*b*, the **angle of incidence** is θ_1, the **angle of reflection** is θ_1', and the **angle of refraction** is θ_2, all measured *relative to the normal* as shown. The plane containing the incident ray and the normal is the *plane of incidence,* which is in the plane of the page in Fig. 34-17*b*.

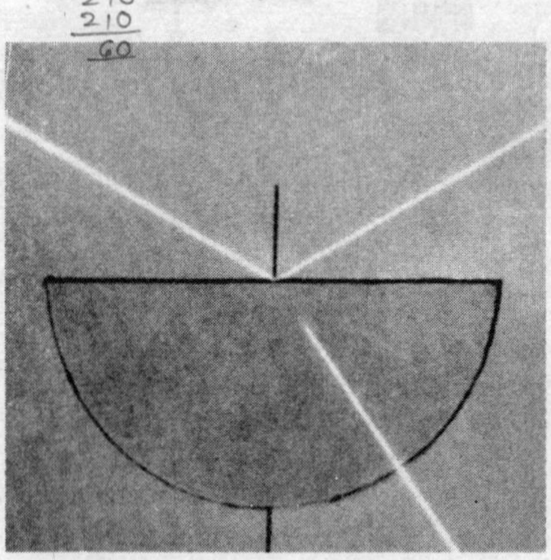

(*a*)

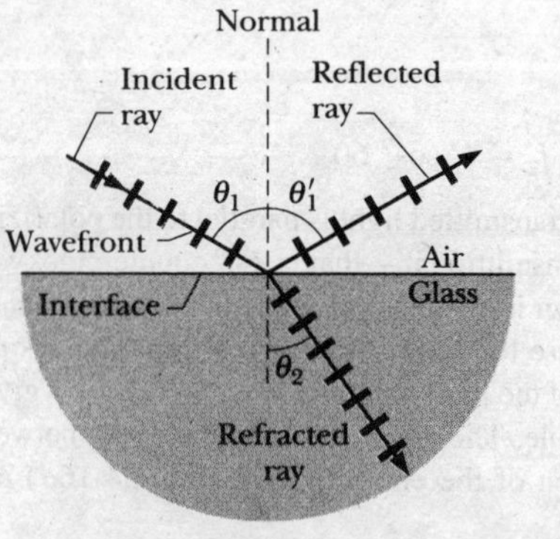

(*b*)

Fig. 34-17 (*a*) A black-and-white photograph showing the reflection and refraction of an incident beam of light by a horizontal plane glass surface. (A portion of the refracted beam within the glass was not well photographed.) At the bottom surface, which is curved, the beam is perpendicular to the surface; so the refraction there does not bend the beam. (*b*) A representation of (*a*) using rays. The angles of incidence (θ_1), of reflection (θ_1'), and of refraction (θ_2) are marked.

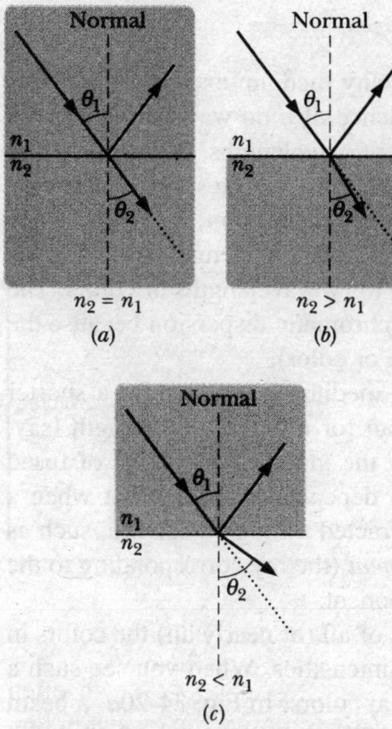

Fig. 34-18 Light refracting from a medium with an index of refraction n_1 and into a medium with an index of refraction n_2. (*a*) The beam does not bend when $n_2 = n_1$; the refracted light then travels in the *undeflected direction* (the dotted line), which is the same as the direction of the incident beam. The beam bends (*b*) toward the normal when $n_2 > n_1$ and (*c*) away from the normal when $n_2 < n_1$.

Experiment shows that reflection and refraction are governed by two laws:

Law of reflection: A reflected ray lies in the plane of incidence and has an angle of reflection equal to the angle of incidence. In Fig. 34-17*b*, this means that

$$\theta_1' = \theta_1 \quad \text{(reflection)}. \tag{34-43}$$

(We shall now usually drop the prime on the angle of reflection.)

Law of refraction: A refracted ray lies in the plane of incidence and has an angle of refraction θ_2 that is related to the angle of incidence θ_1 by

$$n_2 \sin \theta_2 = n_1 \sin \theta_1 \quad \text{(refraction)}. \tag{34-44}$$

Here each of the symbols n_1 and n_2 is a dimensionless constant, called the **index of refraction,** that is associated with a medium involved in the refraction. We derive this equation, called Snell's law, in Chapter 36. As we shall discuss there, the index of refraction of a medium is equal to c/v, where v is the speed of light in that medium and c is its speed in vacuum.

Table 34-1 gives the indexes of refraction of vacuum and some common substances. For vacuum, n is defined to be exactly 1; for air, n is very close to 1.0 (an approximation we shall often make). Nothing has an index of refraction below 1.

We can rearrange Eq. 34-44 as

$$\sin \theta_2 = \frac{n_1}{n_2} \sin \theta_1 \tag{34-45}$$

to compare the angle of refraction θ_2 with the angle of incidence θ_1. We can then see that the relative value of θ_2 depends on the relative values of n_2 and n_1. In fact, we can have three basic results:

1. If n_2 is equal to n_1, then θ_2 is equal to θ_1. In this case, refraction does not bend the light beam, which continues in the *undeflected direction,* as in Fig. 34-18*a*.
2. If n_2 is greater than n_1, then θ_2 is less than θ_1. In this case, refraction bends the light beam away from the undeflected direction and toward the normal, as in Fig. 34-18*b*.
3. If n_2 is less than n_1, then θ_2 is greater than θ_1. In this case, refraction bends the light beam away from the undeflected direction and away from the normal, as in Fig. 34-18*c*.

Refraction *cannot* bend a beam so much that the refracted ray is on the same side of the normal as the incident ray.

Table 34-1 Some Indexes of Refraction[a]

Medium	Index	Medium	Index
Vacuum	Exactly 1	Typical crown glass	1.52
Air (STP)[b]	1.00029	Sodium chloride	1.54
Water (20°C)	1.33	Polystyrene	1.55
Acetone	1.36	Carbon disulfide	1.63
Ethyl alcohol	1.36	Heavy flint glass	1.65
Sugar solution (30%)	1.38	Sapphire	1.77
Fused quartz	1.46	Heaviest flint glass	1.89
Sugar solution (80%)	1.49	Diamond	2.42

[a]For a wavelength of 589 nm (yellow sodium light).
[b]STP means "standard temperature (0°C) and pressure (1 atm)."

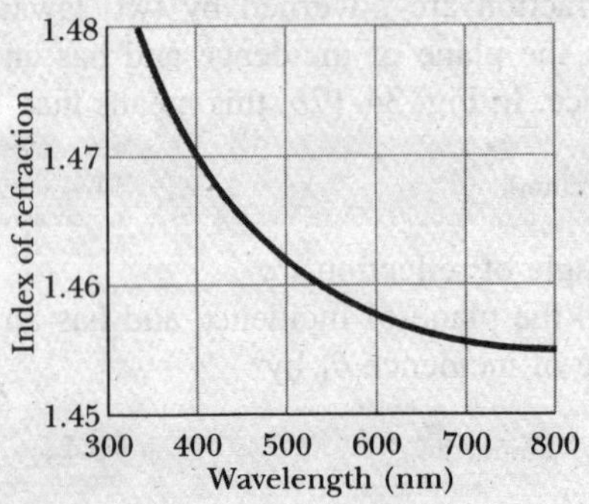

Fig. 34-19 The index of refraction as a function of wavelength for fused quartz. The graph indicates that a beam of short-wavelength light, for which the index of refraction is higher, is bent more upon entering or leaving quartz than a beam of long-wavelength light.

Chromatic Dispersion

The index of refraction n encountered by light in any medium except vacuum depends on the wavelength of the light. The dependence of n on wavelength implies that when a light beam consists of rays of different wavelengths, the rays will be refracted at different angles by a surface; that is, the light will be spread out by the refraction. This spreading of light is called **chromatic dispersion,** in which "chromatic" refers to the colors associated with the individual wavelengths and "dispersion" refers to the spreading of the light according to its wavelengths or colors. The refractions of Figs. 34-17 and 34-18 do not show chromatic dispersion because the beams are *monochromatic* (of a single wavelength or color).

Generally, the index of refraction of a given medium is *greater* for a shorter wavelength (corresponding to, say, blue light) than for a longer wavelength (say, red light). As an example, Fig. 34-19 shows how the index of refraction of fused quartz depends on the wavelength of light. Such dependence means that when a beam with waves of both blue and red light is refracted through a surface, such as from air into quartz or vice versa, the blue *component* (the ray corresponding to the wave of blue light) bends more than the red component.

A beam of *white light* consists of components of all (or nearly all) the colors in the visible spectrum with approximately uniform intensities. When you see such a beam, you perceive white rather than the individual colors. In Fig. 34-20*a*, a beam of white light in air is incident on a glass surface. (Because the pages of this book are white, a beam of white light is represented with a gray ray here. Also, a beam of monochromatic light is generally represented with a red ray.) Of the refracted light in Fig. 34-20*a*, only the red and blue components are shown. Because the blue component is bent more than the red component, the angle of refraction θ_{2b} for the blue component is *smaller* than the angle of refraction θ_{2r} for the red component. (Remember, angles are measured relative to the normal.) In Fig. 34-20*b*, a ray of white light in glass is incident on a glass–air interface. Again, the blue component is bent more than the red component, but now θ_{2b} is greater than θ_{2r}.

To increase the color separation, we can use a solid glass prism with a triangular cross section, as in Fig. 34-21*a*. The dispersion at the first surface (on the left in Fig. 34-21*a*, *b*) is then enhanced by that at the second surface.

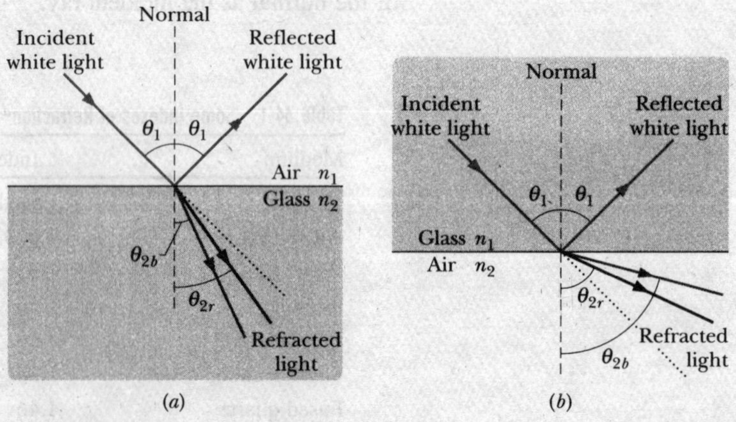

Fig. 34-20 Chromatic dispersion of white light. The blue component is bent more than the red component. (*a*) Passing from air to glass, the blue component ends up with the smaller angle of refraction. (*b*) Passing from glass to air, the blue component ends up with the greater angle of refraction.

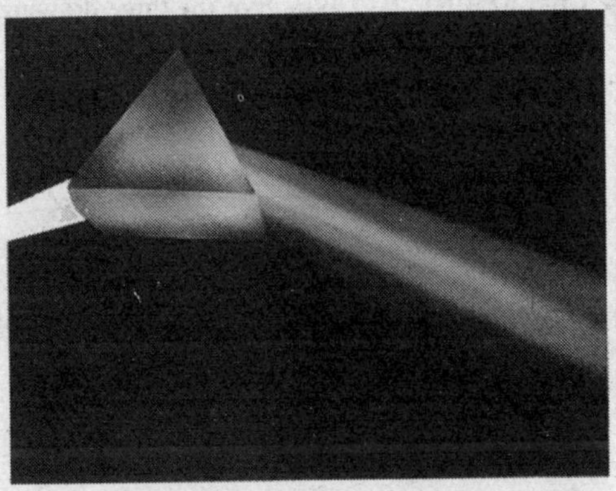

(a)

(b)

Fig. 34-21 (a) A triangular prism separating white light into its component colors. (b) Chromatic dispersion occurs at the first surface and is increased at the second surface.

The most charming example of chromatic dispersion is a rainbow. When white sunlight is intercepted by a falling raindrop, some of the light refracts into the drop, reflects from the drop's inner surface, and then refracts out of the drop (Fig. 34-22). As with a prism, the first refraction separates the sunlight into its component colors, and the second refraction increases the separation.

The rainbow you see is formed by light refracted by many such drops; the red comes from drops angled slightly higher in the sky, the blue from drops angled slightly lower, and the intermediate colors from drops at intermediate angles. All the drops sending separated colors to you are angled at about 42° from a point that is directly opposite the Sun in your view. If the rainfall is extensive and brightly lit, you see a circular arc of color, with red on top and blue on bottom. Your rainbow is a personal one, because another observer intercepts light from other drops.

(a)

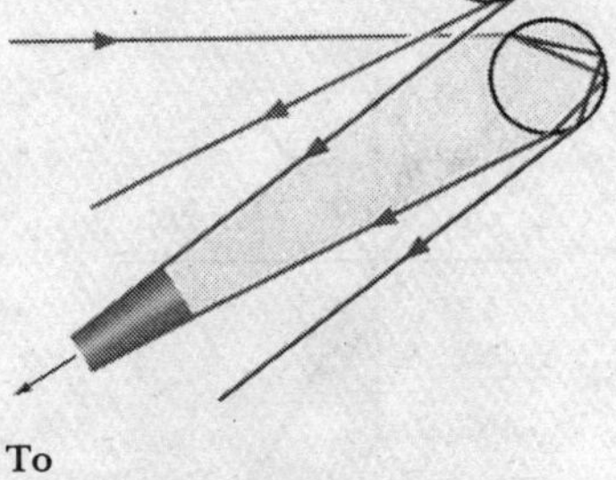

(b)

Fig. 34-22 (a) A rainbow is always a circular arc that is centered on the direction you would look if you looked directly away from the Sun. Under normal conditions, you are lucky if you see a long arc, but if you are looking downward from an elevated position, you might actually see a full circle. (b) The separation of colors when sunlight refracts into and out of falling raindrops leads to a rainbow. The figure shows the situation for the Sun on the horizon (the rays of sunlight are then horizontal). The paths of red and blue rays from two drops are indicated. Many other drops also contribute red and blue rays, as well as the intermediate colors of the visible spectrum.

CHECKPOINT 5: Which of the three drawings here (if any) show physically possible refraction?

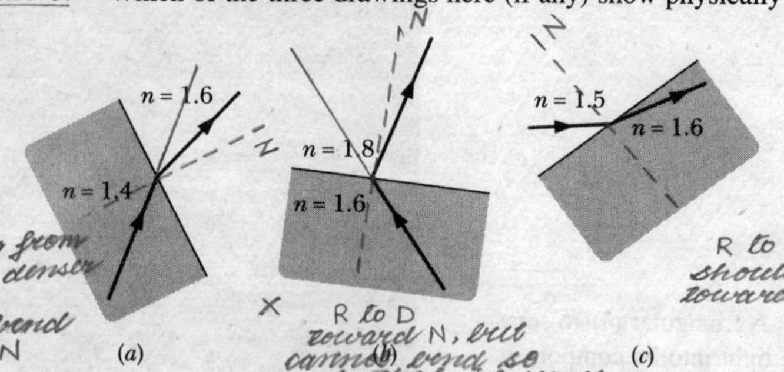

Ray is travelling from rarer to denser medium should bend toward N ✓ — X R to D toward N, but cannot bend so much that it comes on the same side as incident ray — X R to D should bend toward N

Sample Problem 34-4

(a) In Fig. 34-23*a*, a beam of monochromatic light reflects and refracts at point *A* on the interface between material 1 with index of refraction $n_1 = 1.33$ and material 2 with index of refraction $n_2 = 1.77$. The incident beam makes an angle of 50° with the interface. What is the angle of reflection at point *A*? What is the angle of refraction there?

SOLUTION: The **Key Idea** of any reflection is that the angle of reflection is equal to the angle of incidence. Further, both angles are measured between the corresponding light ray and a normal to the interface at the point of reflection. In Fig. 34-23*a*, the normal at point *A* is drawn as a dashed line through the point. Note that the angle of incidence θ_1 is not the given 50° but rather is 90° − 50° = 40°. Thus, the angle of reflection is

$$\theta_1' = \theta_1 = 40°. \quad \text{(Answer)}$$

The light that passes from material 1 into material 2 undergoes refraction at point *A* on the interface between the two materials. The **Key Idea** of any refraction is that we can relate the angle of incidence, the angle of refraction, and the indexes of refraction of the two materials via Eq. 34-44:

$$n_2 \sin \theta_2 = n_1 \sin \theta_1. \quad (34\text{-}46)$$

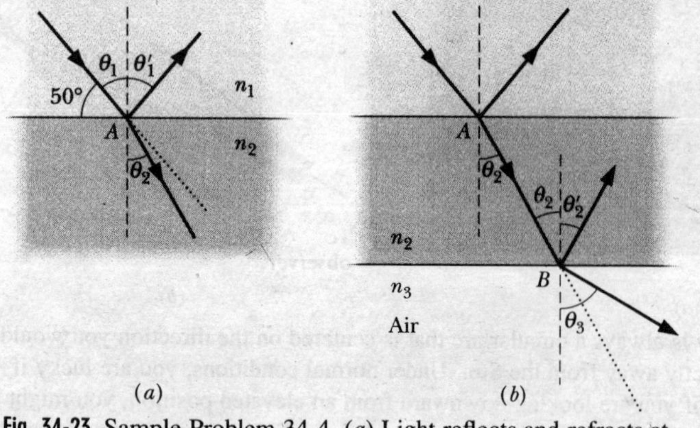

Fig. 34-23 Sample Problem 34-4. (*a*) Light reflects and refracts at point *A* on the interface between materials 1 and 2. (*b*) The light that passes through material 2 reflects and refracts at point *B* on the interface between materials 2 and 3 (air).

Again we measure angles between light rays and a normal, here at the point of refraction. Thus, in Fig. 34-23*a*, the angle of refraction is the angle marked θ_2. Solving Eq. 34-46 for θ_2 gives us

$$\theta_2 = \sin^{-1}\left(\frac{n_1}{n_2}\sin\theta_1\right) = \sin^{-1}\left(\frac{1.33}{1.77}\sin 40°\right) \quad \text{—}*$$

$$= 28.88° \approx 29°. \quad \text{(Answer)}$$

This result means that the beam swings toward the normal (it was at 40° to the normal and is now at 29°). The reason is that when the light travels across the interface, it moves into a material with a greater index of refraction.

(b) The light that enters material 2 at point *A* then reaches point *B* on the interface between material 2 and material 3, which is air, as shown in Fig. 34-23*b*. The interface through *B* is parallel to that through *A*. At *B*, some of the light reflects and the rest enters the air. What is the angle of reflection? What is the angle of refraction into the air?

SOLUTION: We first need to relate one of the angles at point *B* with a known angle at point *A*. Because the interface through point *B* is parallel to that through point *A*, the incident angle at *B* must be equal to the angle of refraction θ_2, as shown in Fig. 34-23*b*. Then for reflection, we use the same **Key Idea** as in (a): the law of reflection. Thus, the angle of reflection at *B* is

$$\theta_2' = \theta_2 = 28.88° \approx 29°. \quad \text{(Answer)}$$

Next, the light that passes from material 2 into the air undergoes refraction at point *B*, with refraction angle θ_3. Thus, the **Key Idea** here is again to apply the law of refraction, but this time by writing Eq. 34-46 as

$$n_3 \sin \theta_3 = n_2 \sin \theta_2.$$

Solving for θ_3 then leads to

$$\theta_3 = \sin^{-1}\left(\frac{n_2}{n_3}\sin\theta_2\right) = \sin^{-1}\left(\frac{1.77}{1.00}\sin 28.88°\right)$$

$$= 58.75° \approx 59°. \quad \text{(Answer)}$$

This result means that the beam swings away from the normal (it was at 29° to the normal and is now at 59°). The reason is that when the light travels across the interface, it moves into a material (air) with a lower index of refraction.

Had $n_3 = n_1$, $\theta_3 = \theta_1 = 40°$ (by eq *)

34-8 Total Internal Reflection

Figure 34-24 shows rays of monochromatic light from a point source S in glass incident on the interface between the glass and air. For ray a, which is perpendicular to the interface, part of the light reflects at the interface and the rest travels through it with no change in direction.

For rays b through e, which have progressively larger angles of incidence at the interface, there are also both reflection and refraction at the interface. As the angle of incidence increases, the angle of refraction increases; for ray e it is 90°, which means that the refracted ray points directly along the interface. The angle of incidence giving this situation is called the **critical angle** θ_c. For angles of incidence larger than θ_c, such as for rays f and g, there is no refracted ray and *all* the light is reflected; this effect is called **total internal reflection.**

To find θ_c, we use Eq. 34-44; we arbitrarily associate subscript 1 with the glass and subscript 2 with the air, and then we substitute θ_c for θ_1 and 90° for θ_2, finding

$$n_1 \sin \theta_c = n_2 \sin 90°,$$

which gives us

$$\theta_c = \sin^{-1} \frac{n_2}{n_1} \quad \text{(critical angle).} \tag{34-47}$$

Because the sine of an angle cannot exceed unity, n_2 cannot exceed n_1 in this equation. This restriction tells us that total internal reflection cannot occur when the incident light is in the medium of lower index of refraction. If source S were in the air in Fig. 34-24, all its rays that are incident on the air–glass interface (including f and g) would be both reflected *and* refracted at the interface.

Total internal reflection has found many applications in medical technology. For example, a physician can search for an ulcer in the stomach of a patient by running two thin bundles of *optical fibers* (Fig. 34-25) down the patient's throat. Light introduced at the outer end of one bundle undergoes repeated total internal reflection within the fibers so that, even though the bundle provides a curved path, most of the light ends up exiting the other end and illuminating the interior of the stomach. Some of the light reflected from the interior then comes back up the second bundle in a similar way, to be detected and converted to an image on a monitor's screen for the physician to view.

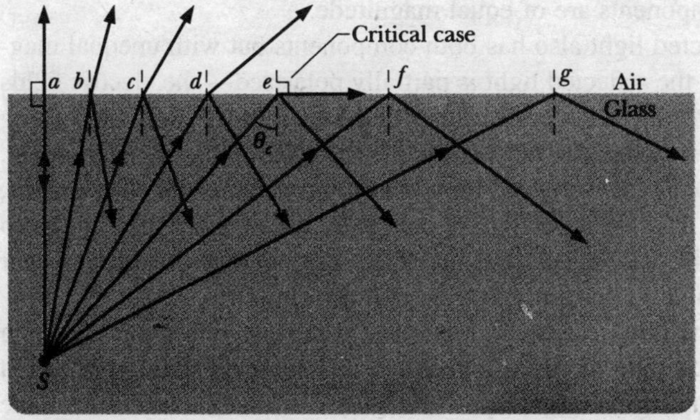

Fig. 34-24 Total internal reflection of light from a point source S in glass occurs for all angles of incidence greater than the critical angle θ_c. At the critical angle, the refracted ray points along the air–glass interface.

Fig. 34-25 Light sent into one end of an optical fiber like those shown here is transmitted to the opposite end with little loss of light through the sides of the fiber.

Sample Problem 34-5

Figure 34-26 shows a triangular prism of glass in air; an incident ray enters the glass perpendicular to one face and is totally reflected at the far glass–air interface as indicated. If θ_1 is 45°, what can you say about the index of refraction n of the glass?

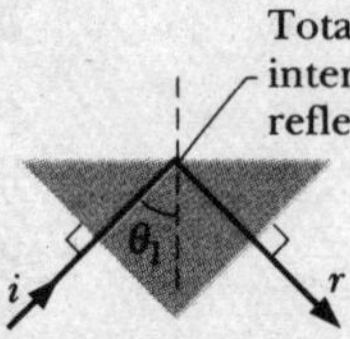

Fig. 34-26 Sample Problem 34-5. The incident ray i is totally internally reflected at the glass–air interface, becoming the reflected ray r.

SOLUTION: One **Key Idea** here is that because the light ray is totally reflected at the interface, the critical angle θ_c for that interface must be less than the incident angle of 45°. A second **Key Idea** is that we can relate the index of refraction n of the glass to θ_c with the law of refraction, which leads to Eq. 34-47. Substituting $n_2 = 1$ (for the air) and $n_1 = n$ (for the glass) into that equation yields

$$\theta_c = \sin^{-1}\frac{n_2}{n_1} = \sin^{-1}\frac{1}{n}.$$

Because θ_c must be less than the incident angle of 45°, we have

$$\sin^{-1}\frac{1}{n} < 45°,$$

which gives us

$$\frac{1}{n} < \sin 45°$$

or

$$n > \frac{1}{\sin 45°} = 1.4. \quad \text{(Answer)}$$

The index of refraction of the glass must be greater than 1.4; otherwise, total internal reflection would not occur for the incident ray shown.

CHECKPOINT 6: Suppose the prism in this sample problem has the index of refraction $n = 1.4$. Does the light still totally internally reflect if we keep the incident ray horizontal but rotate the prism (a) 10° clockwise and (b) 10° counterclockwise in Fig. 34-26?

$n = 1.4 \Rightarrow \theta_c = \sin^{-1}\frac{1}{1.4} = 45°$

a) $\theta_i \downarrow$ $\theta_i < \theta_1 (=45°)$ NO TIR $\because \theta_i < \theta_c$

b) $\theta_i \uparrow$ $\theta_i > \theta_1 (=45°)$ TIR $\because \theta_i > \theta_c$

34-9 Polarization by Reflection

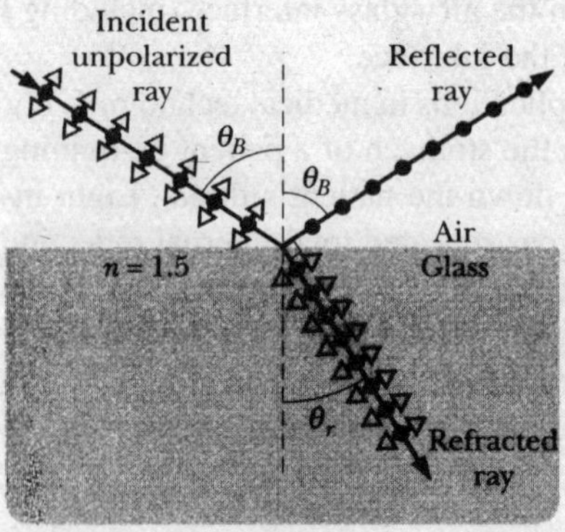

• Component perpendicular to page
<-> Component parallel to page

Fig. 34-27 A ray of unpolarized light in air is incident on a glass surface at the Brewster angle θ_B. The electric fields along that ray have been resolved into components perpendicular to the page (the plane of incidence, reflection, and refraction) and components parallel to the page. The reflected light consists only of components perpendicular to the page and is thus polarized in that direction. The refracted light consists of the original components parallel to the page and weaker components perpendicular to the page; this light is partially polarized.

You can vary the glare you see in sunlight that has been reflected from, say, water by looking through a polarizing sheet (such as a polarizing sunglass lens) and then rotating the sheet's polarizing axis around your line of sight. You can do so because reflected light is fully or partially polarized by the reflection from a surface.

Figure 34-27 shows a ray of unpolarized light incident on a glass surface. Let us resolve the electric field vectors of the light into two components. The *perpendicular components* are perpendicular to the plane of incidence and thus also to the page in Fig. 34-27; these components are represented with dots (as if we see the tips of the vectors). The *parallel components* are parallel to the plane of incidence and the page; they are represented with double-headed arrows. Because the light is unpolarized, these two components are of equal magnitude.

In general, the reflected light also has both components but with unequal magnitudes. This means that the reflected light is partially polarized—the electric fields oscillating along one direction have greater amplitudes than those oscillating along other directions. However, when the light is incident at a particular incident angle, called the *Brewster angle* θ_B, the reflected light has only perpendicular components, as shown in Fig. 34-27. The reflected light is then fully polarized perpendicular to the plane of incidence. The parallel components of the incident light do not disappear but (with the perpendicular components) refract into the glass.

Glass, water, and the other dielectric materials discussed in Section 26-7 can partially and fully polarize light by reflection. When you intercept sunlight reflected from such a surface, you see a bright spot (the glare) on the surface where the reflection takes place. If the surface is horizontal as in Fig. 34-27, the reflected light is partially or fully polarized horizontally. To eliminate such glare from horizontal surfaces, the lenses in polarizing sunglasses are mounted with their polarizing direction vertical.

Brewster's Law

For light incident at the Brewster angle θ_B, we find experimentally that the reflected and refracted rays are perpendicular to each other. Because the reflected ray is reflected at the angle θ_B in Fig. 34-27 and the refracted ray is at an angle θ_r, we have

$$\theta_B + \theta_r = 90° \qquad (34\text{-}48)$$

These two angles can also be related with Eq. 34-44. Arbitrarily assigning subscript 1 in Eq. 34-44 to the material through which the incident and reflected rays travel, we have, from that equation,

$$n_1 \sin \theta_B = n_2 \sin \theta_r.$$

Combining these equations leads to

$$n_1 \sin \theta_B = n_2 \sin(90° - \theta_B) = n_2 \cos \theta_B,$$

which gives us

$$\theta_B = \tan^{-1} \frac{n_2}{n_1} \quad \text{(Brewster angle).} \qquad (34\text{-}49)$$

(Note carefully that the subscripts in Eq. 34-49 are *not* arbitrary because of our decision as to their meanings.) If the incident and reflected rays travel *in air*, we can approximate n_1 as unity and let n represent n_2 in order to write Eq. 34-49 as

$$\theta_B = \tan^{-1} n \quad \text{(Brewster's law).} \qquad (34\text{-}50)$$

This simplified version of Eq. 34-49 is known as **Brewster's law.** Like θ_B, it is named after Sir David Brewster, who found both experimentally in 1812.

REVIEW & SUMMARY

Electromagnetic Waves An electromagnetic wave consists of oscillating electric and magnetic fields. The various possible frequencies of electromagnetic waves form a *spectrum,* a small part of which is visible light. An electromagnetic wave traveling along an x axis has an electric field $\vec{E}$ and a magnetic field $\vec{B}$ with magnitudes that depend on x and t:

$$E = E_m \sin(kx - \omega t)$$

and

$$B = B_m \sin(kx - \omega t), \qquad (34\text{-}1, 34\text{-}2)$$

where E_m and B_m are the amplitudes of $\vec{E}$ and $\vec{B}$. The electric field induces the magnetic field and vice versa. The speed of any electromagnetic wave in vacuum is c, which can be written as

$$c = \frac{E}{B} = \frac{1}{\sqrt{\mu_0 \varepsilon_0}}, \qquad (34\text{-}5, 34\text{-}3)$$

where E and B are the simultaneous magnitudes of the fields.

Energy Flow The rate per unit area at which energy is transported via an electromagnetic wave is given by the Poynting vector $\vec{S}$:

$$\vec{S} = \frac{1}{\mu_0} \vec{E} \times \vec{B}. \qquad (34\text{-}19)$$

The direction of $\vec{S}$ (and thus of the wave's travel and the energy transport) is perpendicular to the directions of both $\vec{E}$ and $\vec{B}$. The time-averaged rate per unit area at which energy is transported is S_{avg}, which is called the *intensity* I of the wave:

$$I = \frac{1}{c\mu_0} E_{rms}^2, \qquad (34\text{-}26)$$

in which $E_{rms} = E_m/\sqrt{2}$. A *point source* of electromagnetic waves emits the waves *isotropically*—that is, with equal intensity in all directions. The intensity of the waves at distance r from a point source of power P_s is

$$I = \frac{P_s}{4\pi r^2}. \qquad (34\text{-}27)$$

Radiation Pressure When a surface intercepts electromagnetic radiation, a force and a pressure are exerted on the surface. If the radiation is totally absorbed by the surface, the force is

$$F = \frac{IA}{c} \quad \text{(total absorption),} \qquad (34\text{-}32)$$

in which I is the intensity of the radiation and A is the area of the surface perpendicular to the path of the radiation. If the radiation is totally reflected back along its original path, the force is

$$F = \frac{2IA}{c} \quad \text{(total reflection back along path).} \qquad (34\text{-}33)$$

The radiation pressure p_r is the force per unit area:

$$p_r = \frac{I}{c} \quad \text{(total absorption)} \qquad (34\text{-}34)$$

and

$$p_r = \frac{2I}{c} \quad \text{(total reflection back along path).} \qquad (34\text{-}35)$$

Polarization Electromagnetic waves are **polarized** if their electric field vectors are all in a single plane, called the *plane of oscillation.* Light waves from common sources are not polarized; that is, they are **unpolarized** or **randomly polarized.**

Polarizing Sheets When a polarizing sheet is placed in the path of light, only electric field components of the light parallel to the sheet's **polarizing direction** are *transmitted* by the sheet; components perpendicular to the polarizing direction are absorbed. The light that emerges from a polarizing sheet is polarized parallel to the polarizing direction of the sheet.

If the original light is initially unpolarized, the transmitted intensity I is half the original intensity I_0:

$$I = \tfrac{1}{2}I_0. \qquad (34\text{-}40)$$

If the original light is initially polarized, the transmitted intensity depends on the angle θ between the polarization direction of the original light and the polarizing direction of the sheet:

$$I = I_0 \cos^2 \theta. \qquad (34\text{-}42)$$

Geometrical Optics *Geometrical optics* is an approximate treatment of light in which light waves are represented as straight-line rays.

Reflection and Refraction When a light ray encounters a boundary between two transparent media, a **reflected** ray and a **refracted** ray generally appear. Both rays remain in the plane of incidence. The **angle of reflection** is equal to the angle of incidence, and the **angle of refraction** is related to the angle of incidence by

$$n_2 \sin \theta_2 = n_1 \sin \theta_1 \quad \text{(refraction),} \qquad (34\text{-}44)$$

where n_1 and n_2 are the indexes of refraction of the media in which the incident and refracted rays travel.

Total Internal Reflection A wave encountering a boundary across which the index of refraction decreases will experience **total internal reflection** if the angle of incidence exceeds a **critical angle** θ_c, where

$$\theta_c = \sin^{-1}\frac{n_2}{n_1} \quad \text{(critical angle).} \qquad (34\text{-}47)$$

Polarization by Reflection A reflected wave will be fully **polarized,** with its $\vec{E}$ vectors perpendicular to the plane of incidence, if it strikes a boundary at the **Brewster angle** θ_B, where

$$\theta_B = \tan^{-1}\frac{n_2}{n_1} \quad \text{(Brewster angle).} \qquad (34\text{-}49)$$

QUESTIONS

1. If the magnetic field of a light wave oscillates parallel to a y axis and is given by $B_y = B_m \sin(kz - \omega t)$, (a) in what direction does the wave travel and (b) parallel to which axis does the associated electric field oscillate?

2. Figure 34-28 shows the electric and magnetic fields of an electromagnetic wave at a certain instant. Is the wave traveling into the page or out of it?

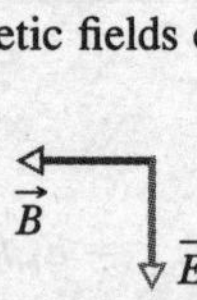

Fig. 34-28 Question 2.

3. (a) Figure 34-29 shows light reaching a polarizing sheet whose polarizing direction is parallel to a y axis. We shall rotate the sheet 40° clockwise about the light's indicated line of travel. During this rotation, does the fraction of the initial light intensity passed by the sheet increase, decrease, or remain the same if the light is (a) initially unpolarized, (b) initially polarized parallel to the x axis, and (c) initially polarized parallel to the y axis?

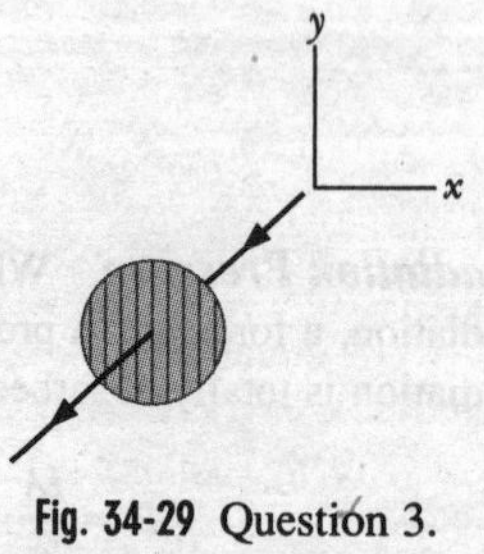

Fig. 34-29 Question 3.

4. In Fig. 34-16a, start with light that is initially polarized parallel to the x axis, and write the ratio of its final intensity I_3 to its initial intensity I_0 as $I_3/I_0 = A \cos^n \theta$. What are A, n, and θ if we rotate the polarizing direction of the first sheet (a) 60° counterclockwise and (b) 90° clockwise from what is shown?

5. Suppose we rotate the second sheet in Fig. 34-16a, starting with its polarization direction aligned with the y axis ($\theta = 0$) and ending with its polarization direction aligned with the x axis ($\theta = 90°$). Which of the three curves in Fig. 34-30 best shows the intensity of the light through the three-sheet system during this 90° rotation?

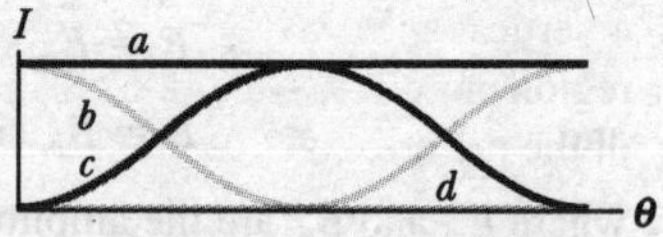

Fig. 34-30 Question 5.

6. Figure 34-31 shows the multiple reflections of a light ray along a glass corridor where the walls are either parallel or perpendicular to one another. If the angle of incidence at point a is 30°, what are the angles of reflection of the light ray at points b, c, d, e, and f?

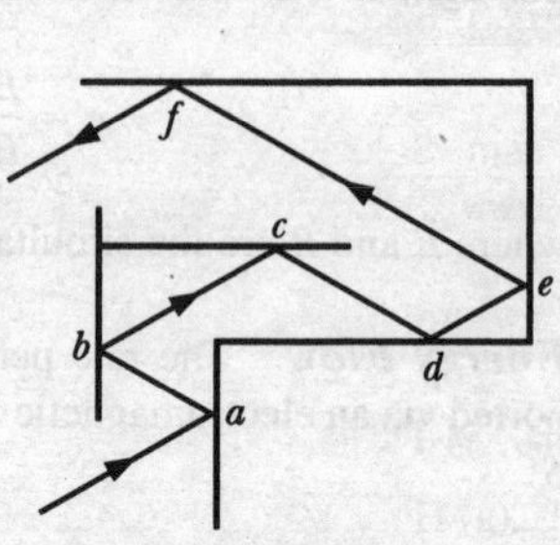

Fig. 34-31 Question 6.

7. Figure 34-32 shows rays of monochromatic light passing through three materials a, b, and c. Rank the materials according to their indexes of refraction, greatest first.

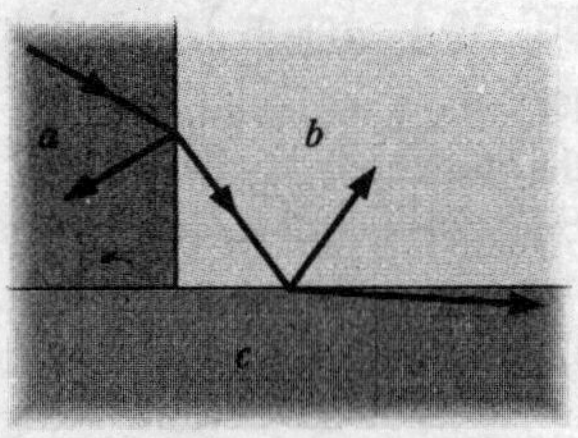

Fig. 34-32 Question 7.

8. In Fig. 34-33, light travels from material a, through three layers of other materials with surfaces parallel to one another, and then back into another layer of material a. The refractions (but not the associated reflections) at the surfaces are shown. Rank the materials according to their indexes of refraction, greatest first.

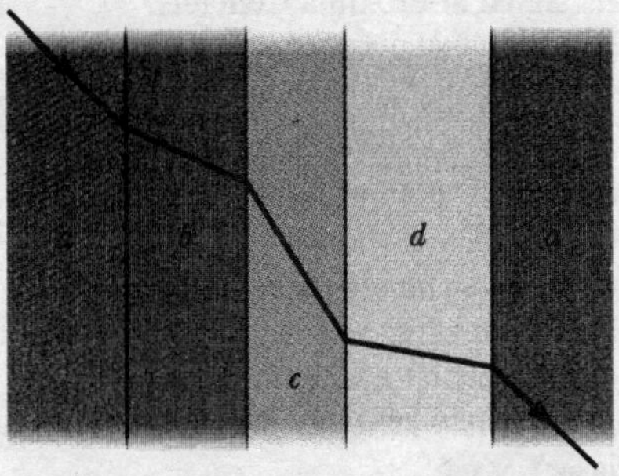

Fig. 34-33 Question 8.

9. Each part of Fig. 34-34 shows light that refracts through an interface between two materials. The incident ray (shown gray in the figure) consists of red and blue light. The approximate index of refraction for visible light is indicated for each material. Which of the three parts show physically possible refraction?

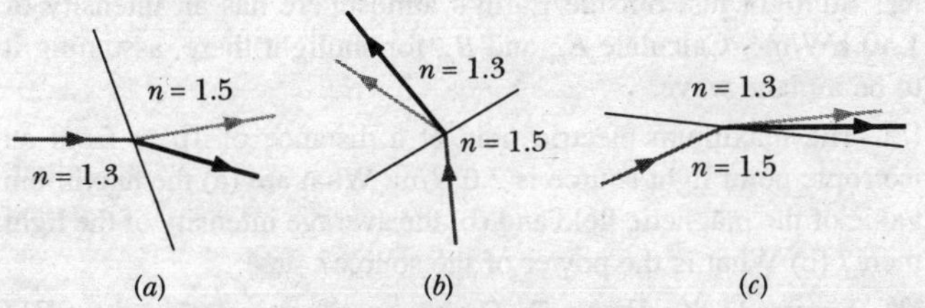

Fig. 34-34 Question 9.

10. (a) Figure 34-35a shows a ray of sunlight that just barely passes a vertical stick in a pool of water. Does that ray end in the general region of point a or point b? (b) Does the red or the blue component of the ray end up closer to the stick? (c) Figure 34-35b shows a flat object (such as a double-edged razor blade) floating in shallow water and illuminated vertically. The gravitational pull on the object and the cohesion of the water cause the water surface to curve as shown. In which general region (a, b, or c) is the edge of the object's shadow? (To the right of the edge of the shadow, many rays of sunlight are concentrated and produce an especially bright region, said to be a *caustic*.)

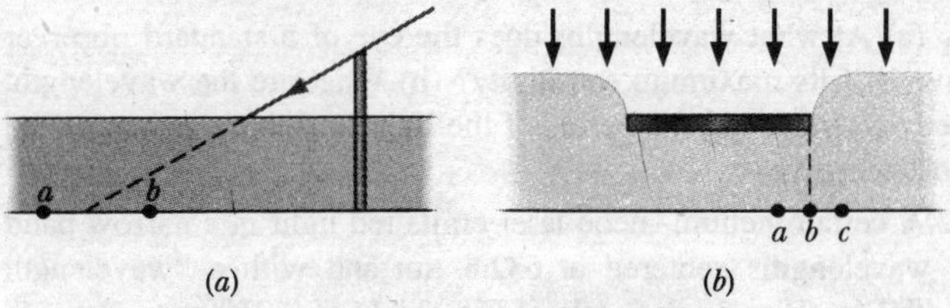

Fig. 34-35 Question 10.

11. Figure 34-22 shows some of the rays of sunlight responsible for the *primary rainbow* (which involves one reflection inside each water drop). A fainter, less frequent *secondary rainbow* (involving two reflections inside each water drop) can appear above a primary rainbow, formed by rays entering and exiting water drops as shown in Fig. 34-36 (without color indicated). Which ray, a or b, corresponds to red light?

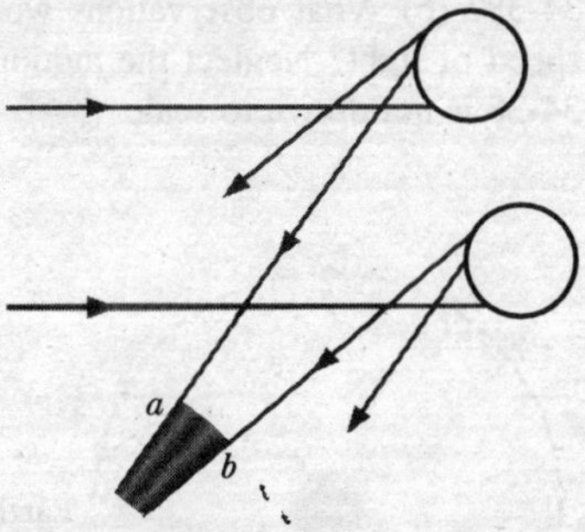

Fig. 34-36 Question 11.

12. Figure 34-37 shows four long horizontal layers of different materials, with air above and below them. The index of refraction of each material is given. Rays of light are sent into the left ends of each layer as shown. In which layer (give the index of refraction) is there the possibility of totally trapping the light in that layer so that, after many reflections, all the light reaches the right end of the layer?

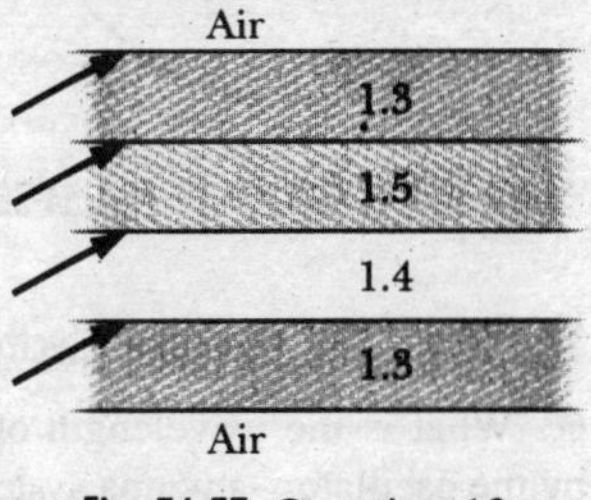

Fig. 34-37 Question 12.

EXERCISES & PROBLEMS

ssm Solution is in the Student Solutions Manual.
www Solution is available on the World Wide Web at: http://www.wiley.com/college/hrw
ilw Solution is available on the Interactive LearningWare.

SEC. 34-1 Maxwell's Rainbow

1E. (a) How long does it take a radio signal to travel 150 km from a transmitter to a receiving antenna? (b) We see a full Moon by reflected sunlight. How much earlier did the light that enters our eye leave the Sun? The Earth–Moon and Earth–Sun distances are 3.8×10^5 km and 1.5×10^8 km. (c) What is the round-trip travel time for light between Earth and a spaceship orbiting Saturn, 1.3×10^9 km distant? (d) The Crab nebula, which is about 6500 light-years (ly) distant, is thought to be the result of a supernova explosion recorded by Chinese astronomers in A.D. 1054. In approximately what year did the explosion actually occur? ssm

2E. Project Seafarer was an ambitious program to construct an enormous antenna, buried underground on a site about 10 000 km^2 in area. Its purpose was to transmit signals to submarines while they were deeply submerged. If the effective wavelength were 1.0×10^4 Earth radii, what would be (a) the frequency and (b) the period of the radiations emitted? Ordinarily, electromagnetic radiations do not penetrate very far into conductors such as seawater.

3E. (a) At what wavelengths does the eye of a standard observer have half its maximum sensitivity? (b) What are the wavelength, the frequency, and the period of the light for which the eye is the most sensitive?

4E. A certain helium–neon laser emits red light in a narrow band of wavelengths centered at 632.8 nm and with a "wavelength width" (such as on the scale of Fig. 34-1) of 0.0100 nm. What is the corresponding "frequency width" for the emission?

5P. One method for measuring the speed of light, based on observations by Roemer in 1676, consisted of observing the apparent times of revolution of one of the moons of Jupiter. The true period of revolution is 42.5 h. (a) Taking into account the finite speed of light, how would you expect the apparent time for one revolution to change as Earth moves in its orbit from point x to point y in Fig. 34-38? (b) What observations would be needed to compute the speed of light? Neglect the motion of Jupiter in its orbit. Figure 34-38 is not drawn to scale. ssm

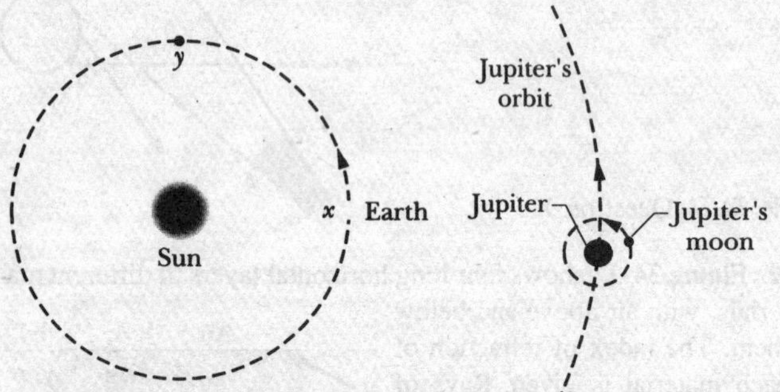

Fig. 34-38 Problem 5.

SEC. 34-2 The Traveling Electromagnetic Wave, Qualitatively

6E. What is the wavelength of the electromagnetic wave emitted by the oscillator–antenna system of Fig. 34-3 if $L = 0.253\ \mu H$ and $C = 25.0$ pF?

7E. What inductance must be connected to a 17 pF capacitor in an oscillator capable of generating 550 nm (i.e., visible) electromagnetic waves? Comment on your answer. ssm

SEC. 34-3 The Traveling Electromagnetic Wave, Quantitatively

8E. A plane electromagnetic wave has a maximum electric field of 3.20×10^{-4} V/m. Find the maximum magnetic field.

9E. The electric field of a certain plane electromagnetic wave is given by $E_x = 0$; $E_y = 0$; $E_z = 2.0 \cos[\pi \times 10^{15}(t - x/c)]$, with $c = 3.0 \times 10^8$ m/s and all quantities in SI units. The wave is propagating in the positive x direction. Write expressions for the components of the magnetic field of the wave. ilw

SEC. 34-4 Energy Transport and the Poynting Vector

10E. Show, by finding the direction of the Poynting vector $\vec{S}$, that the directions of the electric and magnetic fields at all points in Figs. 34-4 to 34-7 are consistent at all times with the assumed directions of propagation.

11E. Some neodymium–glass lasers can provide 100 TW of power in 1.0 ns pulses at a wavelength of 0.26 μm. How much energy is contained in a single pulse? ssm ilw

12E. Our closest stellar neighbor, Proxima Centauri, is 4.3 ly away. It has been suggested that TV programs from our planet have reached this star and may have been viewed by the hypothetical inhabitants of a hypothetical planet orbiting it. Suppose a television station on Earth has a power of 1.0 MW. What is the intensity of its signal at Proxima Centauri?

13E. The radiation emitted by a laser spreads out in the form of a narrow cone with circular cross section. The angle θ of the cone (see Fig. 34-39) is called the *full-angle beam divergence*. An argon laser, radiating at 514.5 nm, is aimed at the Moon in a ranging experiment. If the beam has a full-angle beam divergence of 0.880 μrad, what area on the Moon's surface is illuminated by the laser? ssm

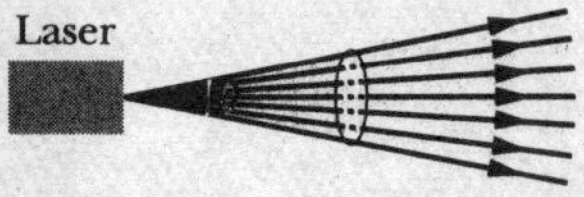

Fig. 34-39 Exercise 13.

14E. What is the intensity of a plane traveling electromagnetic wave if B_m is 1.0×10^{-4} T?

15E. In a plane radio wave the maximum value of the electric field component is 5.00 V/m. Calculate (a) the maximum value of the magnetic field component and (b) the wave intensity.

16P. Sunlight just outside Earth's atmosphere has an intensity of 1.40 kW/m^2. Calculate E_m and B_m for sunlight there, assuming it to be a plane wave.

17P. The maximum electric field at a distance of 10 m from an isotropic point light source is 2.0 V/m. What are (a) the maximum value of the magnetic field and (b) the average intensity of the light there? (c) What is the power of the source? ilw

18P. Figure 34-40: Frank D. Drake, an investigator in the SETI (Search for Extra-Terrestrial Intelligence) program, once said that

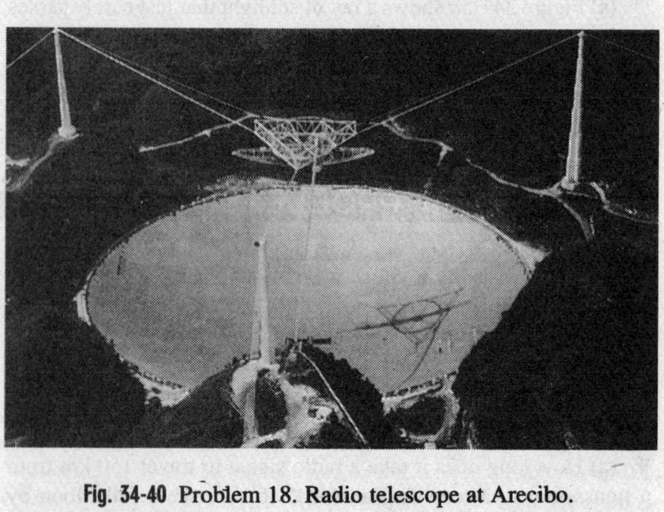

Fig. 34-40 Problem 18. Radio telescope at Arecibo.

the large radio telescope in Arecibo, Puerto Rico, "can detect a signal which lays down on the entire surface of the earth a power of only one picowatt." (a) What is the power that would be received by the Arecibo antenna for such a signal? The antenna diameter is 300 m. (b) What would be the power of a source at the center of our galaxy that could provide such a signal? The galactic center is 2.2×10^4 ly away. Take the source as radiating uniformly in all directions.

19P. An airplane flying at a distance of 10 km from a radio transmitter receives a signal of intensity 10 μW/m^2. Calculate (a) the amplitude of the electric field at the airplane due to this signal, (b) the amplitude of the magnetic field at the airplane, and (c) the total power of the transmitter, assuming the transmitter to radiate uniformly in all directions. ssm www

SEC. 34-5 Radiation Pressure

20E. A black, totally absorbing piece of cardboard of area A = 2.0 cm^2 intercepts light with an intensity of 10 W/m^2 from a camera strobe light. What radiation pressure is produced on the cardboard by the light?

21E. High-power lasers are used to compress a plasma (a gas of charged particles) by radiation pressure. A laser generating pulses of radiation of peak power 1.5×10^3 MW is focused onto 1.0 mm^2 of high-electron-density plasma. Find the pressure exerted on the plasma if the plasma reflects all the light beams directly back along their paths. ssm

22E. Radiation from the Sun reaching Earth (just outside the atmosphere) has an intensity of 1.4 kW/m^2. (a) Assuming that Earth (and its atmosphere) behaves like a flat disk perpendicular to the Sun's rays and that all the incident energy is absorbed, calculate the force on Earth due to radiation pressure. (b) Compare it with the force due to the Sun's gravitational attraction.

23E. What is the radiation pressure 1.5 m away from a 500 W lightbulb? Assume that the surface on which the pressure is exerted faces the bulb and is perfectly absorbing and that the bulb radiates uniformly in all directions. ssm ilw

24P. A helium–neon laser of the type often found in physics laboratories has a beam power of 5.00 mW at a wavelength of 633 nm. The beam is focused by a lens to a circular spot whose effective diameter may be taken to be equal to 2.00 wavelengths. Calculate (a) the intensity of the focused beam, (b) the radiation pressure exerted on a tiny perfectly absorbing sphere whose diameter is that of the focal spot, (c) the force exerted on this sphere, and (d) the magnitude of the acceleration imparted to it. Assume a sphere density of 5.00×10^3 kg/m^3.

25P. A plane electromagnetic wave, with wavelength 3.0 m, travels in vacuum in the positive x direction with its electric field $\vec{E}$, of amplitude 300 V/m, directed along the y axis. (a) What is the frequency f of the wave? (b) What are the direction and amplitude of the magnetic field associated with the wave? (c) What are the values of k and ω if $E = E_m \sin(kx - \omega t)$? (d) What is the time-averaged rate of energy flow in watts per square meter associated with this wave? (e) If the wave falls on a perfectly absorbing sheet of area 2.0 m^2, at what rate is momentum delivered to the sheet and what is the radiation pressure exerted on the sheet? ssm www

26P. In Fig. 34-41, a laser beam of power 4.60 W and diameter 2.60 mm is directed upward at one circular face (of diameter $d <$ 2.60 mm) of a perfectly reflecting cylinder, which is made to "hover" by the beam's radiation pressure. The cylinder's density is 1.20 g/cm^3. What is the cylinder's height H?

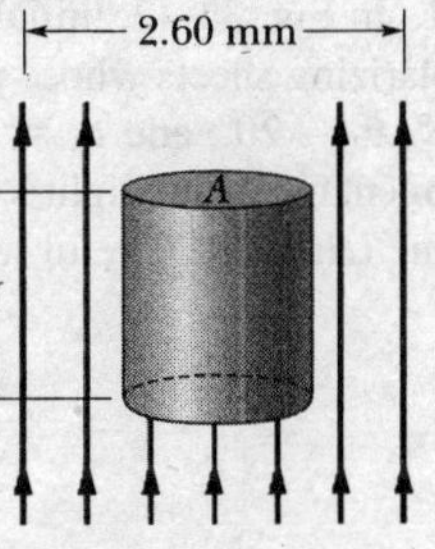

Fig. 34-41 Problem 26.

27P. Prove, for a plane electromagnetic wave that is normally incident on a plane surface, that the radiation pressure on the surface is equal to the energy density in the incident beam. (This relation between pressure and energy density holds no matter what fraction of the incident energy is reflected.) ssm

28P. Prove that the average pressure of a stream of bullets striking a plane surface perpendicularly is twice the kinetic energy density in the stream outside the surface. Assume that the bullets are completely absorbed by the surface. Contrast this with Problem 27.

29P. A small spaceship whose mass is 1.5×10^3 kg (including an astronaut) is drifting in outer space with negligible gravitational forces acting on it. If the astronaut turns on a 10 kW laser beam, what speed will the ship attain in 1.0 day because of the momentum carried away by the beam? ssm

30P. It has been proposed that a spaceship might be propelled in the solar system by radiation pressure, using a large sail made of foil. How large must the sail be if the radiation force is to be equal in magnitude in the Sun's gravitational attraction? Assume that the mass of the ship + sail is 1500 kg, that the sail is perfectly reflecting, and that the sail is oriented perpendicular to the Sun's rays. See Appendix C for needed data. (With a larger sail, the ship is continually driven away from the Sun.)

31P. A particle in the solar system is under the combined influence of the Sun's gravitational attraction and the radiation force due to the Sun's rays. Assume that the particle is a sphere of density 1.0×10^3 kg/m^3 and that all the incident light is absorbed. (a) Show that, if its radius is less than some critical radius R, the particle will be blown out of the solar system. (b) Calculate the critical radius. ssm

SEC. 34-6 Polarization

32E. The magnetic field equations for an electromagnetic wave in vacuum are $B_x = B \sin(ky + \omega t)$, $B_y = B_z = 0$. (a) What is the direction of propagation? (b) Write the electric field equations. (c) Is the wave polarized? If so, in what direction?

33E. A beam of unpolarized light of intensity 10 mW/m^2 is sent through a polarizing sheet as in Fig. 34-12. (a) Find the maximum value of the electric field of the transmitted beam. (b) What radiation pressure is exerted on the polarizing sheet? ssm

34E. In Fig. 34-42, initially unpolarized light is sent through three polarizing sheets whose polarizing directions make angles of $\theta_1 = \theta_2 = \theta_3 = 50°$ with the direction of the y axis. What percentage of the initial intensity is transmitted by the system of the three sheets? (*Hint:* Be careful with the angles.)

35E. In Fig. 34-42, initially unpolarized light is sent through three polarizing sheets whose polarizing directions make angles of $\theta_1 = 40°$, $\theta_2 = 20°$, and $\theta_3 = 40°$ with the direction of the y axis. What percentage of the light's initial intensity is transmitted by the system? (*Hint:* Be careful with the angles.) ssm

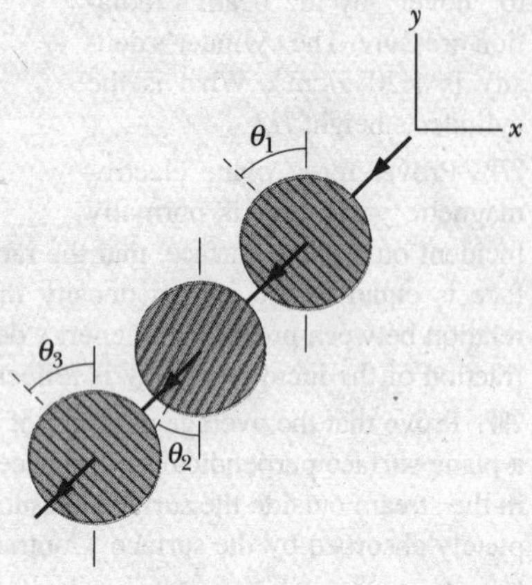

Fig. 34-42 Exercises 34 and 35.

36P. A beam of polarized light is sent through a system of two polarizing sheets. Relative to the polarization direction of that incident light, the polarizing directions of the sheets are at angles θ for the first sheet and 90° for the second sheet. If 0.10 of the incident intensity is transmitted by the two sheets, what is θ?

37P. A horizontal beam of vertically polarized light of intensity 43 W/m^2 is sent through two polarizing sheets. The polarizing direction of the first is at 70° to the vertical, and that of the second is horizontal. What is the intensity of the light transmitted by the pair of sheets? ssm ilw

38P. Suppose that in Problem 37 the initial beam is unpolarized. What then is the intensity of the transmitted light?

39P. A beam of partially polarized light can be considered to be a mixture of polarized and unpolarized light. Suppose we send such a beam through a polarizing filter and then rotate the filter through 360° while keeping it perpendicular to the beam. If the transmitted intensity varies by a factor of 5.0 during the rotation, what fraction of the intensity of the original beam is associated with the beam's polarized light? ssm www

40P. At a beach the light is generally partially polarized owing to reflections off sand and water. At a particular beach on a particular day near sundown, the horizontal component of the electric field vector is 2.3 times the vertical component. A standing sunbather puts on polarizing sunglasses; the glasses eliminate the horizontal field component. (a) What fraction of the light intensity received before the glasses were put on now reaches the sunbather's eyes? (b) The sunbather, still wearing the glasses, lies on his side. What fraction of the light intensity received before the glasses were put on now reaches his eyes?

41P. We want to rotate the direction of polarization of a beam of polarized light through 90° by sending the beam through one or more polarizing sheets. (a) What is the minimum number of sheets required? (b) What is the minimum number of sheets required if the transmitted intensity is to be more than 60% of the original intensity? ssm

SEC. 34-7 Reflection and Refraction

42E. Figure 34-43 shows light reflecting from two perpendicular reflecting surfaces A and B. Find the angle between the incoming ray i and the outgoing ray r'.

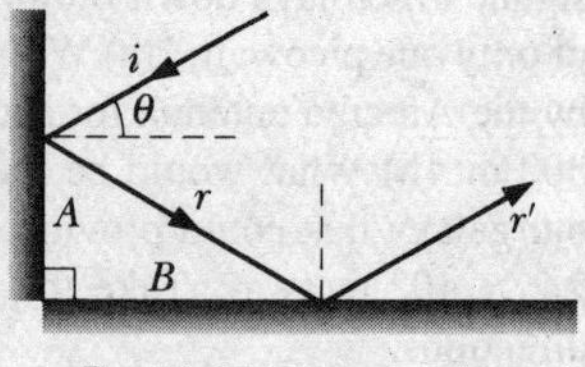

Fig. 34-43 Exercise 42.

43E. Light in vacuum is incident on the surface of a glass slab. In the vacuum the beam makes an angle of 32.0° with the normal to the surface, while in the glass it makes an angle of 21.0° with the normal. What is the index of refraction of the glass? ssm

44E. In about A.D. 150, Claudius Ptolemy gave the following measured values for the angle of incidence θ_1 and the angle of refraction θ_2 for a light beam passing from air to water:

θ_1	θ_2	θ_1	θ_2
10°	8°	50°	35°
20°	15°30′	60°	40°30′
30°	22°30′	70°	45°30′
40°	29°	80°	50°

(a) Are these data consistent with the law of refraction? (b) If so, what index of refraction results? These data are interesting as perhaps the oldest recorded physical measurements.

45E. When the rectangular metal tank in Fig. 34-44 is filled to the top with an unknown liquid, an observer with eyes level with the top of the tank can just see the corner E; a ray that refracts toward the observer at the top surface of the liquid is shown. Find the index of refraction of the liquid. ssm

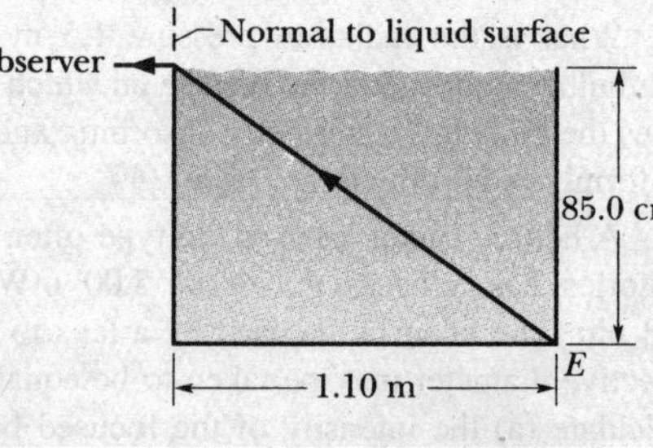

Fig. 34-44 Exercise 45.

46P. In Fig. 34-45, light is incident at angle $\theta_1 = 40.1°$ on a boundary between two transparent materials. Some of the light then travels down through the next three layers of transparent materials, while some of it reflects upward and then escapes into the air. What are the values of (a) θ_5 and (b) θ_4?

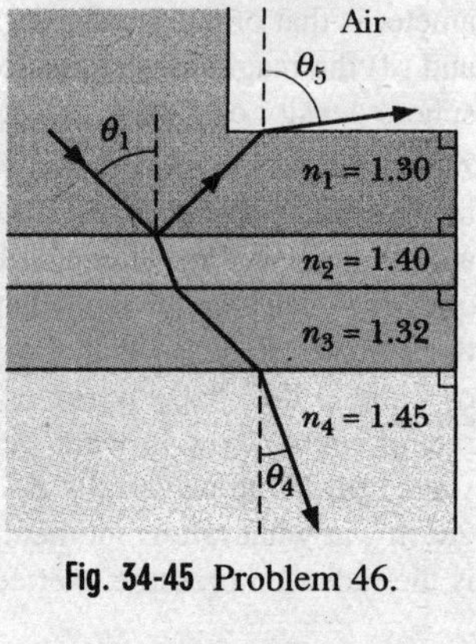

Fig. 34-45 Problem 46.

47P. In Fig. 34-46, a 2.00-m-long vertical pole extends from the bottom of a swimming pool to a point 50.0 cm above the water. Sunlight is incident at 55.0° above the horizon. What is the length of the shadow of the pole on the level bottom of the pool? ssm

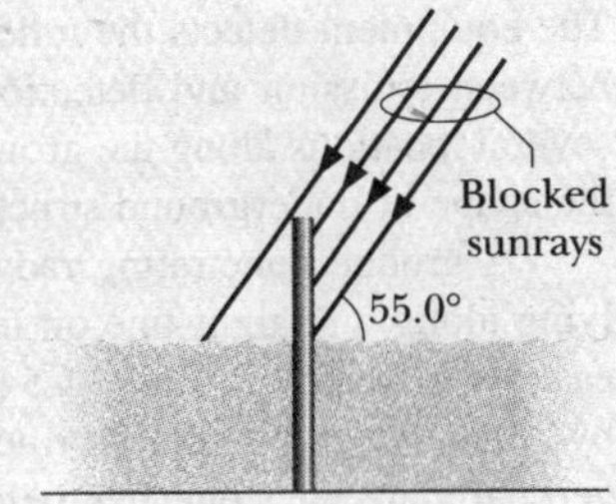

Fig. 34-46 Problem 47.

48P. A ray of white light makes an angle of incidence of 35° on one face of a prism of fused quartz; the prism's cross section is an equilateral triangle. Sketch the light as it passes through the prism, showing the paths traveled by rays representing (a) blue light, (b) yellow-green light, and (c) red light.

49P. Prove that a ray of light incident on the surface of a sheet of plate glass of thickness t emerges from the opposite face parallel to its initial direction but displaced sideways, as in Fig. 34-47. Show that, for small angles of incidence θ, this displacement is given by

$$x = t\theta\,\frac{n-1}{n},$$

where n is the index of refraction of the glass and θ is measured in radians. ssm www

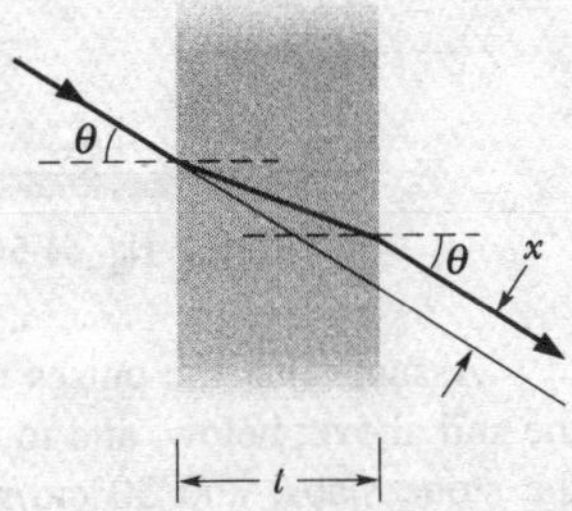

Fig. 34-47 Problem 49.

50P. In Fig. 34-48, two perpendicular mirrors form the sides of a vessel filled with water. (a) A light ray is incident from above, normal to the water surface. Show that the emerging ray is parallel to the incident ray. Assume that there are two reflections at the mirror surfaces. (b) Repeat the analysis for the case of oblique incidence, with the incident ray in the plane of the figure.

Fig. 34-48 Problem 50.

51P. In Fig. 34-49, a ray is incident on one face of a triangular glass prism in air. The angle of incidence θ is chosen so that the emerging ray also makes the same angle θ with the normal to the other face. Show that the index of refraction n of the glass prism is given by

$$n = \frac{\sin\frac{1}{2}(\psi + \phi)}{\sin\frac{1}{2}\phi},$$

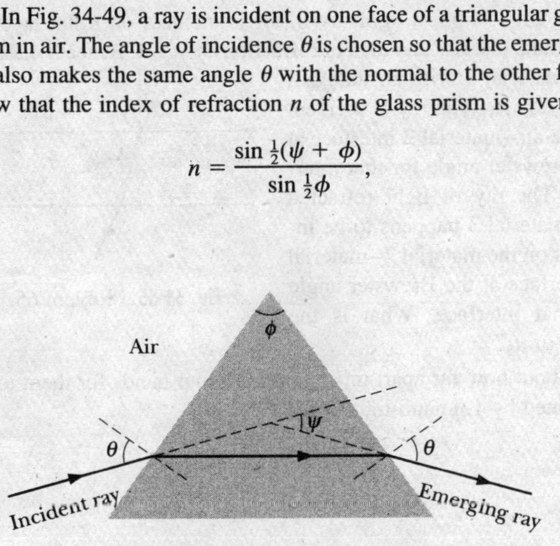

Fig. 34-49 Problems 51 and 58.

where ϕ is the vertex angle of the prism and ψ is the *deviation angle,* the total angle through which the beam is turned in passing through the prism. (Under these conditions the deviation angle ψ has the smallest possible value, which is called the *angle of minimum deviation.*) ilw

SEC. 34-8 Total Internal Reflection

52E. The index of refraction of benzene is 1.8. What is the critical angle for a light ray traveling in benzene toward a plane layer of air above the benzene?

53E. In Fig. 34-50, a light ray enters a glass slab at point A and then undergoes total internal reflection at point B. What minimum value for the index of refraction of the glass can be inferred from this information? ssm

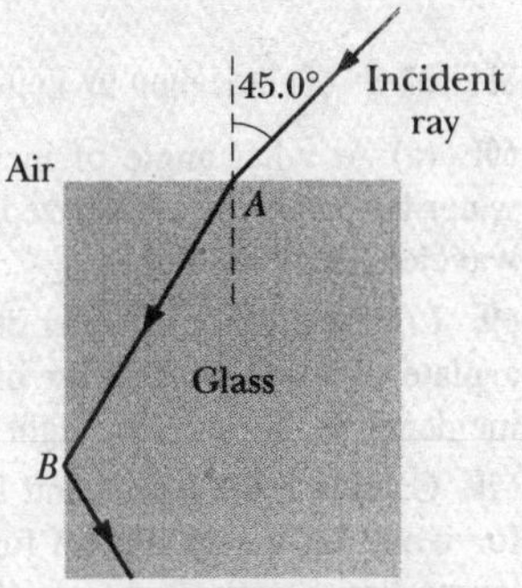

Fig. 34-50 Exercise 53.

54E. A point source of light is 80.0 cm below the surface of a body of water. Find the diameter of the circle at the surface through which light emerges from the water.

55E. In Fig. 34-51, a ray of light is perpendicular to the face ab of a glass prism ($n = 1.52$). Find the largest value for the angle ϕ so that the ray is totally reflected at face ac if the prism is immersed (a) in air and (b) in water. ssm ilw

Fig. 34-51 Exercise 55.

56P. A ray of white light travels through fused quartz that is surrounded by air. If all the color components of the light undergo total internal reflection at the surface, then the reflected light forms a reflected ray of white light. However, if the color component at one end of the visible range (either blue or red) partially refracts through the surface into the air, there is less of that component in the reflected light. Then the reflected light is not white but has the tint of the opposite end of the visible range. (If blue were partially lost to refraction, then the reflected beam would be reddish, and vice versa.) Is it possible for the reflected light to be (a) bluish or (b) reddish? (c) If so, what must be the angle of incidence of the original white light on the quartz surface? (See Fig. 34-19.)

57P. A solid glass cube, of edge length 10 mm and index of refraction 1.5, has a small spot at its center. (a) What parts of each cube face must be covered to prevent the spot from being seen, no matter what the direction of viewing? (Neglect light that reflects inside the cube and then refracts out into the air.) (b) What fraction of the cube surface must be so covered? ssm

58P. Suppose the prism of Fig. 34-49 has apex angle $\phi = 60.0°$ and index of refraction $n = 1.60$. (a) What is the smallest angle of incidence θ for which a ray can enter the left face of the prism and exit the right face? (b) What angle of incidence θ is required for the ray to exit the prism with an identical angle θ for its refraction, as it does in Fig. 34-49? (See Problem 51.)

59P. In Fig. 34-52, light enters a 90° triangular prism at point P with incident angle θ and then some of it refracts at point Q with an angle of refraction of 90°. (a) What is the index of refraction of the prism in terms of θ? (b) What, numerically, is the maximum value that the index of refraction can have? Explain what happens to the light at Q if the incident angle at Q is (c) increased slightly and (d) decreased slightly. ssm

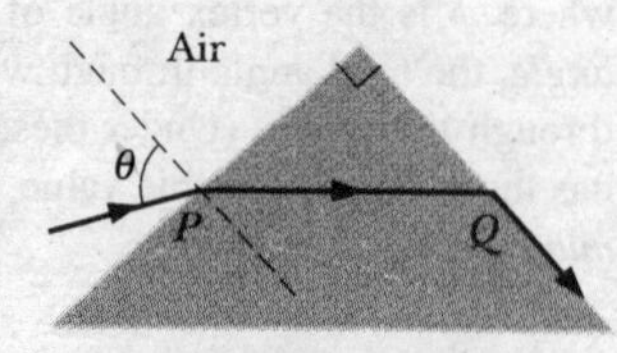

Fig. 34-52 Problem 59.

SEC. 34-9 Polarization by Reflection

60E. (a) At what angle of incidence will the light reflected from water be completely polarized? (b) Does this angle depend on the wavelength of the light?

61E. Light traveling in water of refractive index 1.33 is incident on a plate of glass with index of refraction 1.53. At what angle of incidence is the reflected light fully polarized? ssm

62E. Calculate the upper and lower limits of the Brewster angles for white light incident on fused quartz. Assume that the wavelength limits of the light are 400 and 700 nm.

Additional Problems

63. In Fig. 34-53, an albatross glides at a constant 15 m/s horizontally above level ground, moving in a vertical plane that contains the Sun. It glides toward a solid wall of height h = 2.0 m, which it will just barely clear. At that time of day, the angle of the Sun relative to the ground is θ = 30°. At what speed does the shadow of the albatross move (a) across the level ground and then (b) up the wall? Suppose that later a hawk happens to glide along the same path, also at 15 m/s. You see that when its shadow reaches the wall, the speed of the shadow noticeably increases. (c) Is the Sun now higher or lower in the sky than when the albatross flew by earlier? (d) If the speed of the hawk's shadow on the wall is 45 m/s, what is the angle θ of the Sun just then?

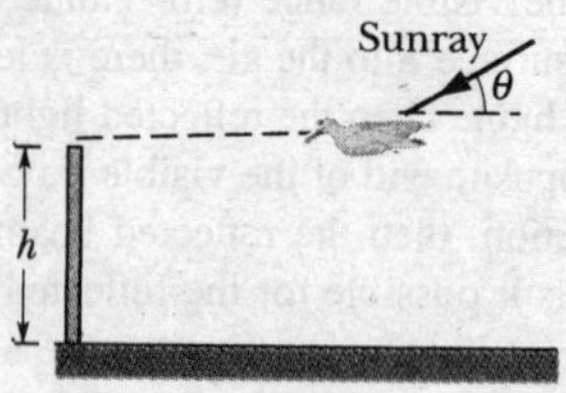

Fig. 34-53 Problem 63.

64. *Searching for graves.* In an archaeological investigation, unmarked graves and underground tombs can be located and mapped by *ground-penetrating radar* without disruption of the grave site. The radar unit emits a wave pulse directly downward into the ground; the pulse is then partially reflected upward by any underground interface. That is to say, a pulse is reflected upward by any horizontal boundary across which the speed of the pulse changes. The equipment detects the reflection and records the time interval between emission and detection. If this procedure is repeated at several locations along the ground, an archaeologist can determine the shape of underground structures.

A ground-penetrating radar unit was used at eight locations lying along a straight line on level ground and numbered west to east, as indicated in Fig. 34-54. Adjacent locations are separated by 2.0 m. An empty tomb formed with equally thick horizontal and vertical slabs of stone lies below these locations; the horizontal slabs form the ceiling and floor of the tomb, the vertical slabs form walls. The following table gives the time intervals Δt (in nanoseconds) recorded for the pulses at the eight locations. For example, at location 4, the original pulse emitted into the ground resulted in four reflected pulses, the first one returning 63.00 ns after the emission and the last one returning 86.54 ns after the emission.

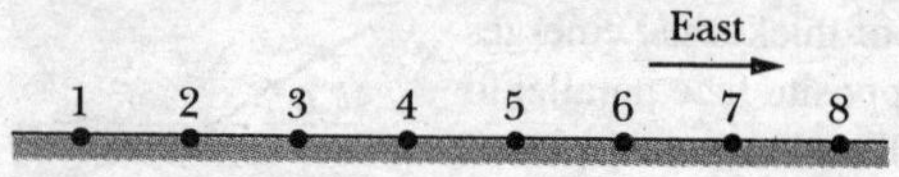

Fig. 34-54 Problem 64.

Assume that the pulses have a wave speed of 10.0 cm/ns in the soil above, below, and to the sides of the tomb; 10.6 cm/ns in the stone slabs; and 30 cm/ns in the air within the tomb. Find (a) the depth of the top surface of the tomb's ceiling, (b) the horizontal length of the tomb along the west–east line containing the eight locations, and (c) the vertical dimensions of the tomb's interior.

Location	1	2	3	4	5	6	7	8
Δt	None	63.00	63.00	63.00	63.00	63.00	63.00	None
		115.8	66.77	66.77	66.77	66.77	93.19	
			82.77	82.77	74.77	74.77		
			86.54	86.54	101.2	78.54		

65. In Fig. 34-55, a light ray in air is incident on a flat layer of material 2 that has an index of refraction n_2 = 1.5. Beneath material 2 is material 3 with an index of refraction n_3. The ray is incident on the air–material 2 interface at the Brewster angle for that interface. The ray of light refracted into material 3 happens to be incident on the material 2–material 3 interface at the Brewster angle for that interface. What is the value of n_3?

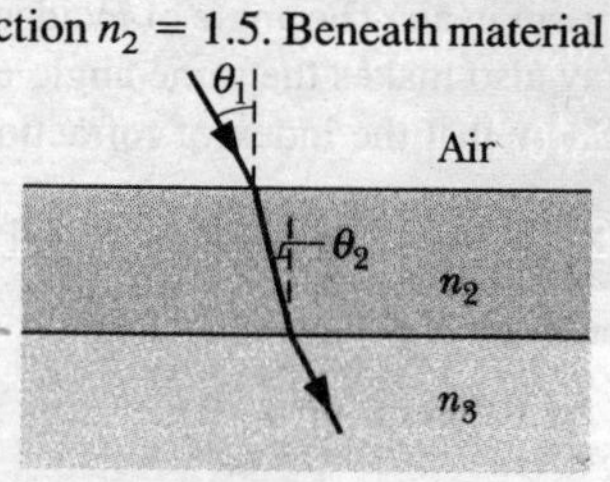

Fig. 34-55 Problem 65.

66. About how far apart must you hold your hands for them to be separated by 1.0 nano-light-second?

35 Images

Edouard Manet's *A Bar at the Folies-Bergère* has enchanted viewers ever since it was painted in 1882. Part of its appeal lies in the contrast between an audience ready for entertainment and a bartender whose eyes betray her fatigue. Its appeal also depends on subtle distortions of reality that Manet hid in the painting— distortions that give an eerie feel to the scene even before you recognize what is "wrong."

Can you find those subtle distortions of reality?

The answer is in this chapter.

35-1 Two Types of Image

For you to see, say, a penguin, your eye must intercept some of the light rays spreading from the penguin and then redirect them onto the retina at the rear of the eye. Your visual system, starting with the retina and ending with the visual cortex at the rear of your brain, automatically and subconsciously processes the information provided by the light. That system identifies edges, orientations, textures, shapes, and colors and then rapidly brings to your consciousness an **image** (a reproduction derived from light) of the penguin; you perceive and recognize the penguin as being in the direction from which the light rays came and at the proper distance.

Your visual system goes through this processing and recognition even if the light rays do not come directly from the penguin, but instead reflect toward you from a mirror or refract through the lenses in a pair of binoculars. However, you now see the penguin in the direction from which the light rays came after they reflected or refracted, and the distance you perceive may be quite different from the penguin's true distance.

For example, if the light rays have been reflected toward you from a standard flat mirror, the penguin appears to be behind the mirror because the rays you intercept come from that direction. Of course, the penguin is not back there. This type of image, which is called a **virtual image,** truly exists only within the brain but nevertheless is *said* to exist at the perceived location.

A **real image** differs in that it can be formed on a surface, such as a card or a movie screen. You can see a real image (otherwise movie theaters would be empty), but the existence of the image does not depend on your seeing it and it is present even if you are not.

In this chapter we explore several ways in which virtual and real images are formed by reflection (as with mirrors) and refraction (as with lenses). We also distinguish between the two types of image more clearly, but here first is an example of a natural virtual image.

A Common Mirage

A common example of a virtual image is a pool of water that appears to lie on the road some distance ahead of you on a sunny day, but that you can never reach. The pool is a *mirage* (a type of illusion), formed by light rays coming from the low section of the sky in front of you (Fig. 35-1*a*). As the rays approach the road, they travel through progressively warmer air that has been heated by the road, which is usually relatively warm. With an increase in air temperature, the speed of light in air increases slightly and, correspondingly, the index of refraction of the air decreases

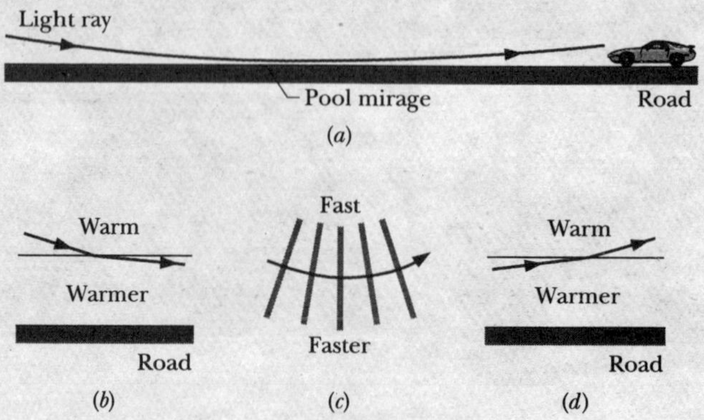

Fig. 35-1 (*a*) A ray from a low section of the sky refracts through air that is heated by a road (without reaching the road). An observer who intercepts the light perceives it to be from a pool of water on the road. (*b*) Bending (exaggerated) of a light ray descending across an imaginary boundary from warm air to warmer air. (*c*) Shifting of wavefronts and associated bending of a ray, which occurs because the lower ends of wavefronts move faster in warmer air. (*d*) Bending of a ray ascending across an imaginary boundary to warm air from warmer air.

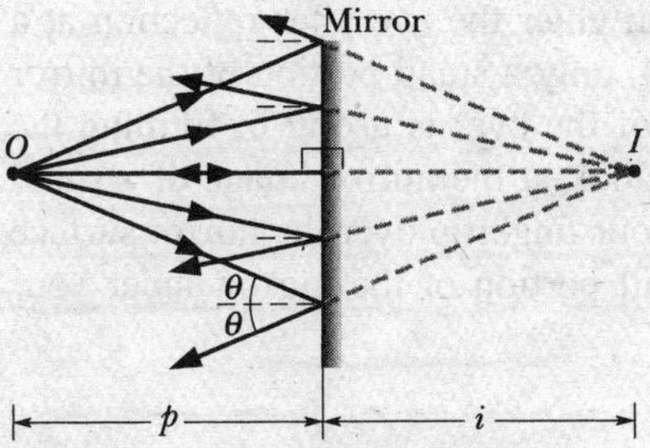

Fig. 35-2 A point source of light O, called the *object*, is a perpendicular distance p in front of a plane mirror. Light rays reaching the mirror from O reflect from the mirror. If your eye intercepts some of the reflected rays, you perceive a point source of light I to be behind the mirror, at a perpendicular distance i. The perceived source I is a virtual image of object O.

slightly. Thus, as the rays descend, encountering progressively smaller indexes of refraction, they continuously bend toward the horizontal (Fig. 35-1*b*).

Once a ray is horizontal, somewhat above the road's surface, it still bends because the lower portion of each associated wavefront is in slightly warmer air and is moving slightly faster than the upper portion of the wavefront (Fig. 35-1*c*). This nonuniform motion of the wavefronts bends the ray upward. As the ray then ascends, it continues to bend upward through progressively greater indexes of refraction (Fig. 35-1*d*).

If you intercept some of this light, your visual system automatically infers that it originated along a backward extension of the rays you have intercepted and, to make sense of the light, assumes that it came from the road surface. If the light happens to be bluish from blue sky, the mirage appears bluish, like water. Because the air is probably turbulent due to the heating, the mirage shimmies, as if water waves were present. The bluish coloring and the shimmy enhance the illusion of a pool of water, but you are actually seeing a virtual image of a low section of the sky.

35-2 Plane Mirrors

A **mirror** is a surface that can reflect a beam of light in one direction instead of either scattering it widely in many directions or absorbing it. A shiny metal surface acts as a mirror; a concrete wall does not. In this section we examine the images that a **plane mirror** (a flat reflecting surface) can produce.

POINT OBJECT

Figure 35-2 shows a point source of light O, which we shall call the *object*, at a perpendicular distance p in front of a plane mirror. The light that is incident on the mirror is represented with rays spreading from O. The reflection of that light is represented with reflected rays spreading from the mirror. If we extend the reflected rays backward (behind the mirror), we find that the extensions intersect at a point that is a perpendicular distance i behind the mirror.

If you look into the mirror of Fig. 35-2, your eyes intercept some of the reflected light. To make sense of what you see, you perceive a point source of light located at the point of intersection of the extensions. This point source is the image I of object O. It is called a *point image* because it is a point, and it is a virtual image because the rays do not actually pass through it. (As you will see, rays *do* pass through a point of intersection for a real image.)

Figure 35-3 shows two rays selected from the many rays in Fig. 35-2. One reaches the mirror at point b, perpendicularly. The other reaches it at an arbitrary point a, with an angle of incidence θ. The extensions of the two reflected rays are also shown. The right triangles $aOba$ and $aIba$ have a common side and three equal angles and are thus congruent (equal in size), so their horizontal sides have the same length. That is,

$$Ib = Ob, \tag{35-1}$$

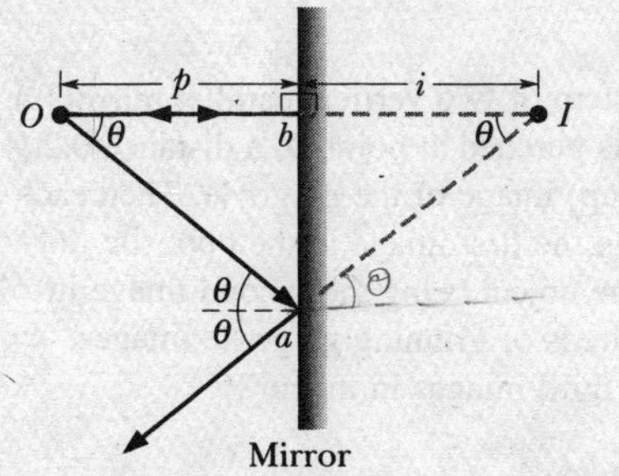

Fig. 35-3 Two rays from Fig. 35-2. Ray Oa makes an arbitrary angle θ with the normal to the mirror surface. Ray Ob is perpendicular to the mirror.

where Ib and Ob are the distances from the mirror to the image and the object, respectively. Equation 35-1 tells us that the image is as far behind the mirror as the object is in front of it. By convention (that is, to get our equations to work out), *object distances p* are taken to be positive quantities, and *image distances i* for virtual images (as here) are taken to be negative quantities. Thus, Eq. 35-1 can be written as $|i| = p$, or as

$$i = -p \quad \text{(plane mirror).} \tag{35-2}$$

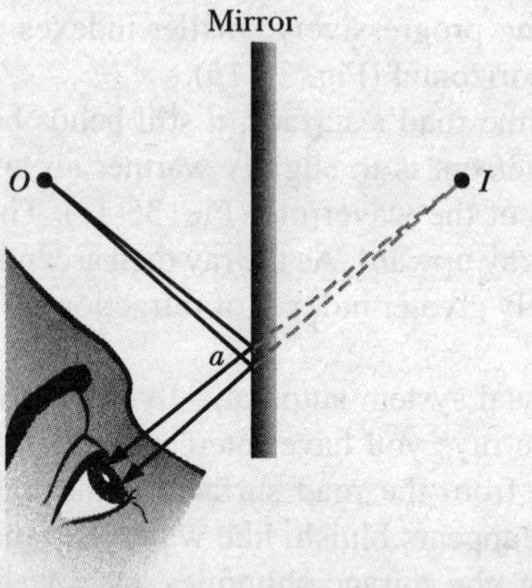

Fig. 35-4 A "pencil" of rays from O enters the eye after reflection at the mirror. Only a small portion of the mirror near a is involved in this reflection. The light appears to originate at point I behind the mirror.

Only rays that are fairly close together can enter the eye after reflection at a mirror. For the eye position shown in Fig. 35-4, only a small portion of the mirror near point a (a portion smaller than the pupil of the eye) is useful in forming the image. To find this portion, close one eye and look at the mirror image of a small object such as the tip of a pencil. Then move your fingertip over the mirror surface until you cannot see the image. Only that small portion of the mirror under your fingertip produced the image.

Extended Objects

In Fig. 35-5, an extended object O, represented by an upright arrow, is at perpendicular distance p in front of a plane mirror. Each small portion of the object that faces the mirror acts like the point source O of Figs. 35-2 and 35-3. If you intercept the light reflected by the mirror, you perceive a virtual image I that is a composite of the virtual point images of all those portions of the object and seems to be at distance i behind the mirror. Distances i and p are related by Eq. 35-2.

We can also locate the image of an extended object as we did for a point object in Fig. 35-2: we draw some of the rays that reach the mirror from the top of the object, draw the corresponding reflected rays, and then extend those reflected rays behind the mirror until they intersect to form an image of the top of the object. We then do the same for rays from the bottom of the object. As shown in Fig. 35-5, we find that virtual image I has the same orientation and *height* (measured parallel to the mirror) as object O.

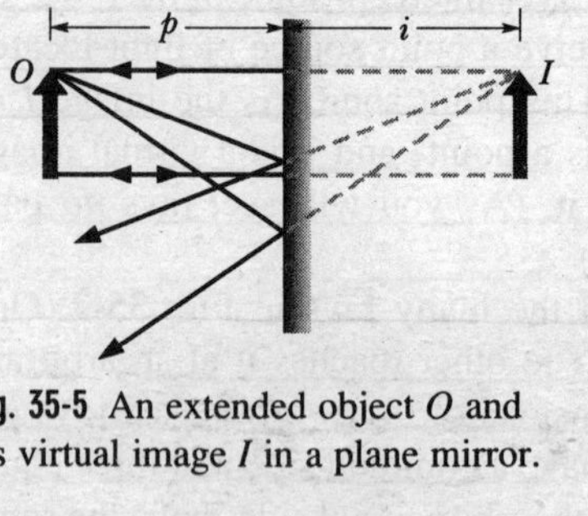

Fig. 35-5 An extended object O and its virtual image I in a plane mirror.

Manet's "Folies-Bergère"

In *A Bar at the Folies-Bergère* you see the barroom via reflection by a large mirror on the wall behind the woman tending bar, but the reflection is subtly wrong in three ways. First note the bottles at the left. Manet painted their reflections in the mirror but misplaced them, painting them farther toward the front of the bar than they should be.

Now note the reflection of the woman. Since your view is from directly in front of the woman, her reflection should be behind her, with only a little of it (if any) visible to you; yet Manet painted her reflection well off to the right. Finally, note the reflection of the man facing her. He must be you, because the reflection shows that he is directly in front of the woman, and thus he must be the viewer of the painting. You are looking into Manet's work and seeing your reflection well off to your right. The effect is eerie because it is not what we expect from a painting or from a mirror.

CHECKPOINT 1: In the figure you look into a system of two vertical parallel mirrors A and B separated by distance d. A grinning gargoyle is perched at point O, a distance $0.2d$ from mirror A. Each mirror produces a *first* (least deep) image of the gargoyle. Then each mirror produces a *second* image with the object being the first image in the opposite mirror. Then each mirror produces a *third* image with the object being the second image in the opposite mirror, and so on—you might see hundreds of grinning gargoyle images. How deep behind mirror A are the first, second, and third images in mirror A?

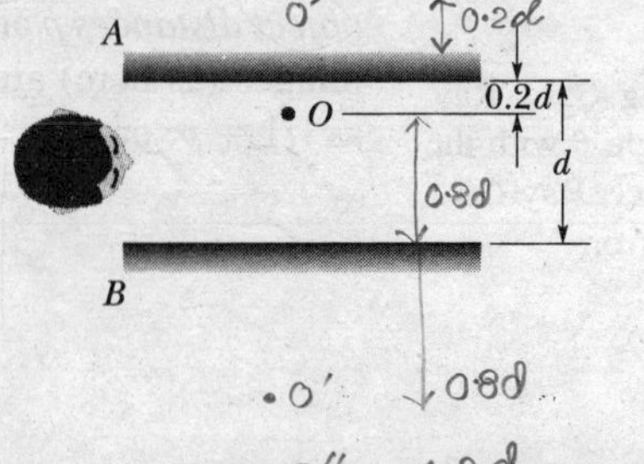

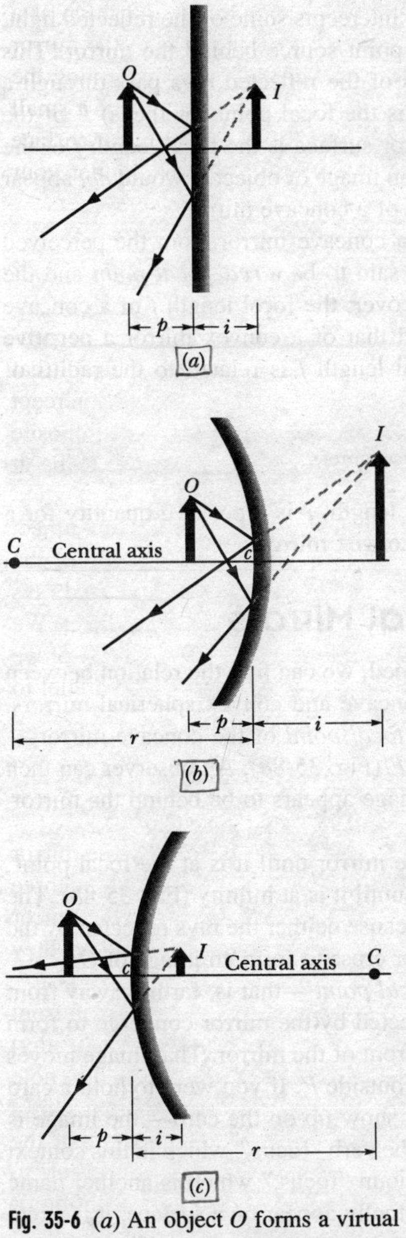

Fig. 35-6 (*a*) An object *O* forms a virtual image *I* in a plane mirror. (*b*) If the mirror is bent so that it becomes *concave*, the image moves farther away and becomes larger. (*c*) If the plane mirror is bent so that it becomes *convex*, the image moves closer and becomes smaller.

35-3 Spherical Mirrors

We turn now from images produced by plane mirrors to images produced by mirrors with curved surfaces. In particular, we shall consider spherical mirrors, which are simply mirrors in the shape of a small section of the surface of a sphere. A plane mirror is in fact a spherical mirror with an infinitely large *radius of curvature*.

Making a Spherical Mirror

We start with the plane mirror of Fig. 35-6*a*, which faces leftward toward an object *O* that is shown and an observer that is not shown. We make a **concave mirror** by curving the mirror's surface so it is *concave* ("caved in") as in Fig. 35-6*b*. Curving the surface in this way changes several characteristics of the mirror and the image it produces of the object:

1. The *center of curvature C* (the center of the sphere of which the mirror's surface is part) was infinitely far from the plane mirror; it is now closer but still in front of the concave mirror.
2. The *field of view*—the extent of the scene that is reflected to the observer—was wide; it is now smaller.
3. The image of the object was as far behind the plane mirror as the object was in front; the image is farther behind the concave mirror; that is, $|i|$ is greater.
4. The height of the image was equal to the height of the object; the height of the image is now greater. This feature is why many makeup mirrors and shaving mirrors are concave—they produce a larger image of a face.

We can make a **convex mirror** by curving a plane mirror so its surface is *convex* ("flexed out") as in Fig. 35-6*c*. Curving the surface in this way (1) moves the center of curvature *C* to *behind* the mirror and (2) *increases* the field of view. It also (3) moves the image of the object *closer* to the mirror and (4) *shrinks* it. Store surveillance mirrors are usually convex to take advantage of the increase in the field of view—more of the store can then be monitored with a single mirror.

Focal Points of Spherical Mirrors

For a plane mirror, the magnitude of the image distance *i* is always equal to the object distance *p*. Before we can determine how these two distances are related for a spherical mirror, we must consider the reflection of light from an object *O* located an effectively infinite distance in front of a spherical mirror, on the mirror's *central axis*. That axis extends through the center of curvature *C* and the center *c* of the mirror. Because of the great distance between the object and the mirror, the light waves spreading from the object are plane waves when they reach the mirror along the central axis. This means that the rays representing the light waves are all parallel to the central axis when they reach the mirror.

When these parallel rays reach a concave mirror like that of Fig. 35-7*a*, those near the central axis are reflected through a common point *F*; two of these reflected rays are shown in the figure. If we placed a (small) card at *F*, a point image of the infinitely distant object *O* would appear on the card. (This would occur for any infinitely distant object.) Point *F* is called the **focal point** (or **focus**) of the mirror, and its distance from the center of the mirror is the **focal length** *f* of the mirror.

If we now substitute a convex mirror for the concave mirror, we find that the parallel rays are no longer reflected through a common point. Instead, they diverge

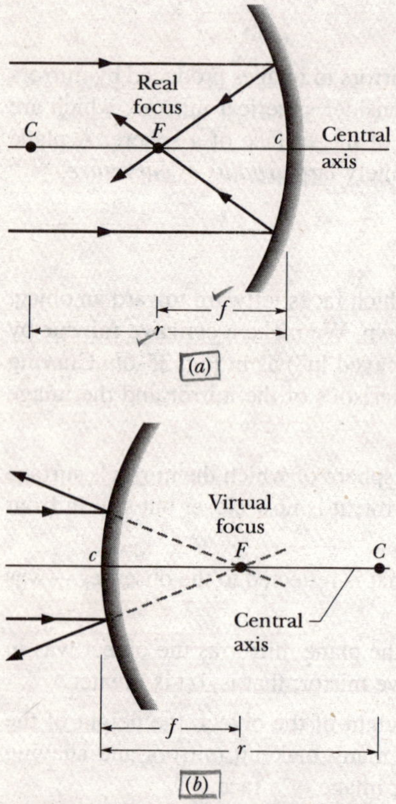

Fig. 35-7 (*a*) In a concave mirror, incident parallel light rays are brought to a real focus at *F*, on the same side of the mirror as the light rays. (*b*) In a convex mirror, incident parallel light rays seem to diverge from a virtual focus at *F*, on the side of the mirror opposite the light rays.

as shown in Fig. 35-7*b*. However, if your eye intercepts some of the reflected light, you perceive the light as originating from a point source behind the mirror. This perceived source is located where extensions of the reflected rays pass through a common point (*F* in Fig. 35-7*b*). That point is the focal point (or focus) *F* of the convex mirror, and its distance from the mirror surface is the focal length *f* of the mirror. If we placed a card at this focal point, an image of object *O* would *not* appear on the card, so this focal point is not like that of a concave mirror.

To distinguish the actual focal point of a concave mirror from the perceived focal point of a convex mirror, the former is said to be a *real focal point* and the latter is said to be a *virtual focal point.* Moreover, the focal length *f* of a concave mirror is taken to be a positive quantity, and that of a convex mirror a negative quantity. For mirrors of both types, the focal length *f* is related to the radius of curvature *r* of the mirror by

$$f = \tfrac{1}{2}r \qquad \text{(spherical mirror)}, \qquad (35\text{-}3)$$

where, consistent with the signs for the focal length, *r* is a positive quantity for a concave mirror and a negative quantity for a convex mirror.

35-4 Images from Spherical Mirrors

With the focal point of a spherical mirror defined, we can find the relation between image distance *i* and object distance *p* for concave and convex spherical mirrors. We begin by placing the object *O inside the focal point* of the concave mirror—that is, between the mirror and its focal point *F* (Fig. 35-8*a*). An observer can then see a virtual image of *O* in the mirror: The image appears to be behind the mirror, and it has the same orientation as the object.

If we now move the object away from the mirror until it is at the focal point, the image moves farther back from the mirror until it is at infinity (Fig. 35-8*b*). The image is then ambiguous and imperceptible because neither the rays reflected by the mirror nor the ray extensions behind the mirror cross to form an image of *O*.

If we next move the object *outside the focal point*—that is, farther away from the mirror than the focal point—the rays reflected by the mirror converge to form an *inverted* image of object *O* (Fig. 35-8*c*) in front of the mirror. That image moves in from infinity as we move the object farther outside *F*. If you were to hold a card at the position of the image, the image would show up on the card—the image is said to be *focused* on the card by the mirror. (The verb "focus," which in this context means to produce an image, differs from the noun "focus," which is another name for the focal point.) Because this image can actually appear on a surface, it is a real image—the rays actually intersect to create the image, regardless of whether an observer is present. The image distance *i* of a real image is a positive quantity, in contrast to that for a virtual image. We also see that

Fig. 35-8 (*a*) An object *O* inside the focal point of a concave mirror, and its virtual image *I*. (*b*) The object at the focal point *F*. (*c*) The object outside the focal point, and its real image *I*.

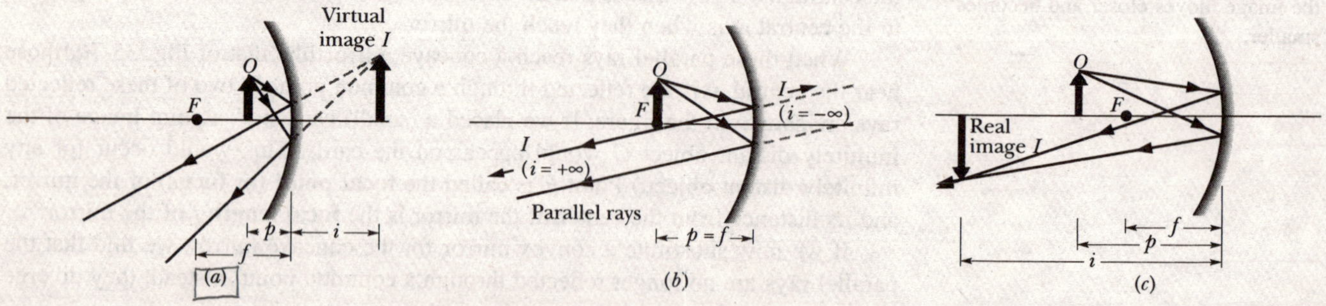

➤ Real images form on the side of a mirror where the object is, and virtual images form on the opposite side.

As we shall prove in Section 35-8, when light rays from an object make only small angles with the central axis of a spherical mirror, a simple equation relates the object distance p, the image distance i, and the focal length f:

$$\frac{1}{p}+\frac{1}{i}=\frac{1}{f} \quad \text{(spherical mirror).} \tag{35-4}$$

We assume such small angles in figures such as Fig. 35-8, but for clarity the rays are drawn with exaggerated angles. With that assumption, Eq. 35-4 applies to any concave, convex, or plane mirror. For a convex or plane mirror, only a virtual image can be formed, regardless of the object's location on the central axis. As shown in the example of a convex mirror in Fig. 35-6*c*, the image is always on the opposite side of the mirror from the object and has the same orientation as the object.

The size of an object or image, as measured *perpendicular* to the mirror's central axis, is called the object or image *height*. Let h represent the height of the object, and h' the height of the image. Then the ratio h'/h is called the **lateral magnification** m produced by the mirror. However, by convention, the lateral magnification always includes a plus sign when the image orientation is that of the object and a minus sign when the image orientation is opposite that of the object. For this reason, we write the formula for m as

$$|m|=\frac{h'}{h} \quad \text{(lateral magnification).} \tag{35-5}$$

We shall soon prove that the lateral magnification can also be written as

$$m=-\frac{i}{p} \quad \text{(lateral magnification).} \tag{35-6}$$

For a plane mirror, for which $i=-p$, we have $m=+1$. The magnification of 1 means that the image is the same size as the object. The plus sign means that the image and the object have the same orientation. For the concave mirror of Fig. 35-8*c*, $m \approx -1.5$.

Equations 35-3 through 35-6 hold for all plane mirrors, concave spherical mirrors, and convex spherical mirrors. In addition to those equations, you have been asked to absorb a lot of information about these mirrors, and you should organize it for yourself by filling in Table 35-1. Under Image Location, note whether the image is on the *same* side of the mirror as the object or on the *opposite* side. Under Image Type, note whether the image is *real* or *virtual*. Under Image Orientation, note whether the image has the *same* orientation as the object or is *inverted*. Under Sign,

Table 35-1 Your Organizing Table for Mirrors

Mirror Type	Object Location	Image Location	Image Type	Image Orientation	Sign of f	Sign of r	Sign of m
Plane	Anywhere						
Concave	Inside F						
	Outside F						
Convex	Anywhere						

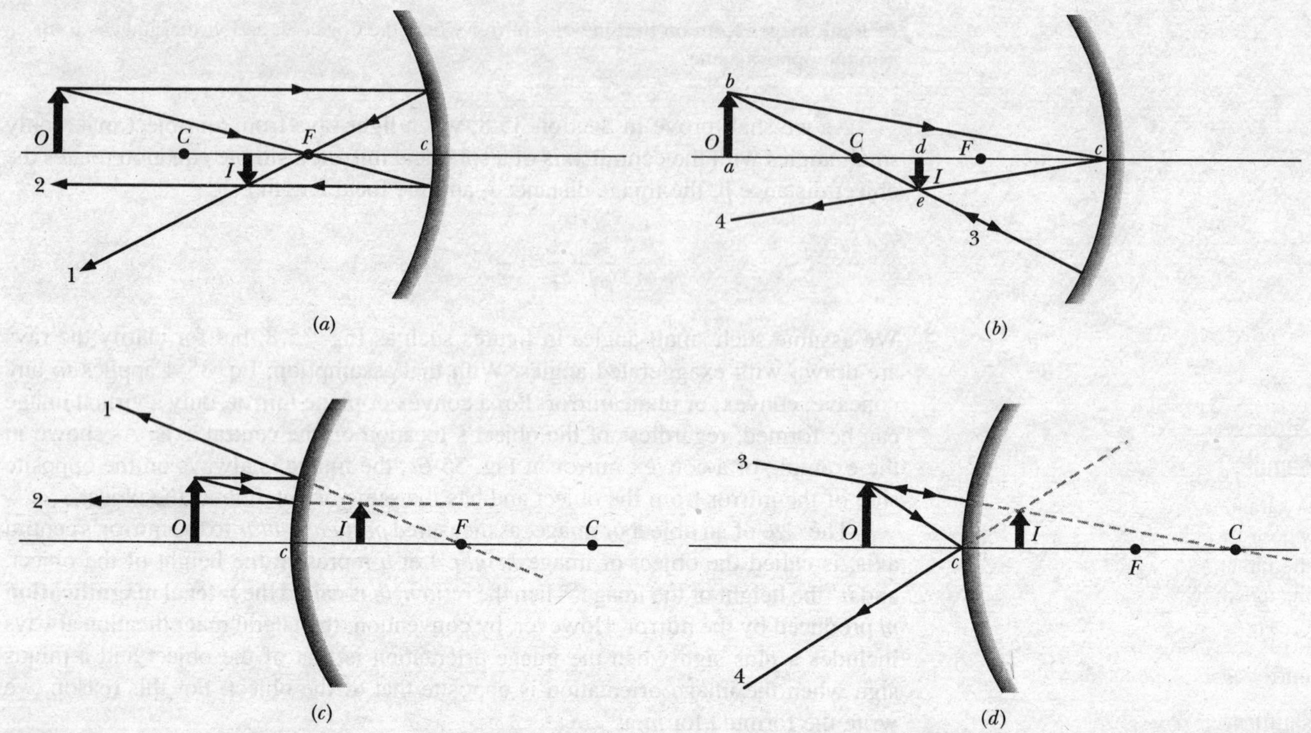

Fig. 35-9 (*a*, *b*) Four rays that may be drawn to find the image of an object in a concave mirror. For the object position shown, the image is real, inverted, and smaller than the object. (*c*, *d*) Four similar rays for the case of a convex mirror. For a convex mirror, the image is always virtual, oriented like the object, and smaller than the object. [In (*c*), ray 2 is initially directed toward focal point *F*. In (*d*), ray 3 is initially directed toward center of curvature *C*.]

give the sign of the quantity or fill in ± if the sign is ambiguous. You will need this organization to tackle homework or a test.

Locating Images by Drawing Rays

Figures 35-9*a* and *b* show an object *O* in front of a concave mirror. We can graphically locate the image of any off-axis point of the object by drawing a *ray diagram* with any two of four special rays through the point:

1. A ray that is initially parallel to the central axis reflects through the focal point *F* (ray 1 in Fig. 35-9*a*).
2. A ray that reflects from the mirror after passing through the focal point emerges parallel to the central axis (ray 2 in Fig. 35-9*a*).
3. A ray that reflects from the mirror after passing through the center of curvature *C* returns along itself (ray 3 in Fig. 35-9*b*).
4. A ray that reflects from the mirror at its intersection *c* with the central axis is reflected symmetrically about that axis (ray 4 in Fig. 35-9*b*).

The image of the point is at the intersection of the two special rays you choose. The image of the object can then be found by locating the images of two or more of its off-axis points. You need to modify the descriptions of the rays slightly to apply them to convex mirrors, as in Figs. 35-9*c* and *d*.

Proof of Equation 35-6

We are now in a position to derive Eq. 35-6 ($m = -i/p$), the equation for the lateral magnification of an object reflected in a mirror. Consider ray 4 in Fig. 35-9*b*. It is reflected at point *c* so that the incident and reflected rays make equal angles with the axis of the mirror at that point.

The two right triangles *abc* and *dec* in the figure are similar (have the same set of angles), so we can write

$$\frac{de}{ab} = \frac{cd}{ca}.$$

The quantity on the left (apart from the question of sign) is the lateral magnification m produced by the mirror. Since we indicate an inverted image as a *negative* magnification, we symbolize this as $-m$. However, $cd = i$ and $ca = p$, so we have

$$m = -\frac{i}{p} \quad \text{(magnification)}, \qquad (35\text{-}7)$$

which is the relation we set out to prove.

Sample Problem 35-1

A tarantula of height h sits cautiously before a spherical mirror whose focal length has absolute value $|f| = 40$ cm. The image of the tarantula produced by the mirror has the same orientation as the tarantula and has height $h' = 0.20h$.

(a) Is the image real or virtual, and is it on the same side of the mirror as the tarantula or the opposite side?

SOLUTION: The **Key Idea** here is that because the image has the same orientation as the tarantula (the object), it must be virtual and on the opposite side of the mirror. (You can easily see this result if you have filled out Table 35-1.)

(b) Is the mirror concave or convex, and what is its focal length f, sign included?

SOLUTION: We *cannot* tell the type of mirror from the type of image, because both types of mirror can produce virtual images. Similarly, we cannot tell the type of mirror from the sign of the focal length f, as obtained from Eq 35-3 or Eq. 35-4, because we lack enough information to use either equation. However—and this is the **Key Idea** here—we can make use of the magnification information. We know that the ratio of image height h' to object height h is 0.20. Thus, from Eq. 35-5 we have

$$|m| = \frac{h'}{h} = 0.20.$$

Because the object and image have the same orientation, we know that m must be positive: $m = +0.20$. Substituting this into Eq. 35-6 and solving for, say, i gives us

$$i = -0.20p,$$

which does not appear to be of help in finding f. However, it is helpful if we substitute it into Eq. 35-4. That equation gives us

$$\frac{1}{f} = \frac{1}{i} + \frac{1}{p} = \frac{1}{-0.20p} + \frac{1}{p} = \frac{1}{p}(-5 + 1),$$

from which we find

$$f = -p/4.$$

Now we have it: Because p is positive, f must be negative, which means that the mirror is convex with

$$f = -40 \text{ cm}. \qquad \text{(Answer)}$$

✔CHECKPOINT 2: A Central American vampire bat, dozing on the central axis of a spherical mirror, is magnified by $m = -4$. Is its image (a) real or virtual, (b) inverted or of the same orientation as the bat, and (c) on the same side of the mirror as the bat or on the opposite side?

35-5 Spherical Refracting Surfaces

We now turn from images formed by reflections to images formed by refraction through surfaces of transparent materials, such as glass. We shall consider only spherical surfaces, with radius of curvature r and center of curvature C. The light will be emitted by a point object O in a medium with index of refraction n_1; it will refract through a spherical surface into a medium of index of refraction n_2.

Our concern is whether the light rays, after refracting through the surface, form a real image (no observer necessary) or a virtual image (assuming that an observer intercepts the rays). The answer depends on the relative values of n_1 and n_2 and on the geometry of the situation.

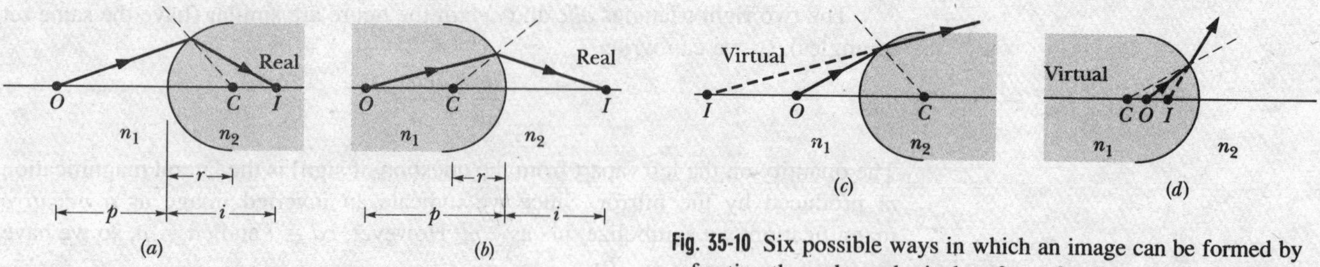

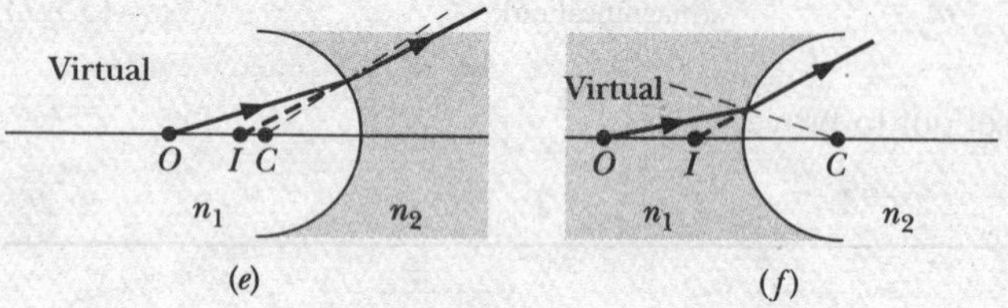

Fig. 35-10 Six possible ways in which an image can be formed by refraction through a spherical surface of radius r and center of curvature C. The surface separates a medium with index of refraction n_1 from a medium with index of refraction n_2. The point object O is always in the medium with n_1, to the left of the surface. The material with the lesser index of refraction is unshaded (think of it as being air, and the other material as being glass). Real images are formed in (*a*) and (*b*); virtual images are formed in the other four situations.

Six possible results are shown in Fig. 35-10. In each part of the figure, the medium with the greater index of refraction is shaded, and object O is always in the medium with index of refraction n_1, to the left of the refracting surface. In each part, a representative ray is shown refracting through the surface. (That ray and a ray along the central axis suffice to determine the position of the image in each case.)

At the point of refraction of each ray, the normal to the refracting surface is a radial line through the center of curvature C. Because of the refraction, the ray bends toward the normal if it is entering a medium of greater index of refraction, and away from the normal if it is entering a medium of lesser index of refraction. If the refracted ray is then directed toward the central axis, it and other (undrawn) rays will form a real image on that axis. If it is directed away from the central axis, it cannot form a real image; however, backward extensions of it and other refracted rays can form a virtual image, provided (as with mirrors) some of those rays are intercepted by an observer.

Real images I are formed (at image distance i) in parts a and b of Fig. 35-10, where the refraction directs the ray *toward* the central axis. Virtual images are formed in parts c and d, where the refraction directs the ray *away* from the central axis. Note, in these four parts, that real images are formed when the object is relatively far from the refracting surface, and virtual images are formed when the object is nearer the refracting surface. In the final situations (Figs. 35-10e and f), refraction always directs the ray away from the central axis and virtual images are always formed, regardless of the object distance.

Note the following major difference from reflected images:

Real images form on the side of a refracting surface that is opposite the object, and virtual images form on the same side as the object.

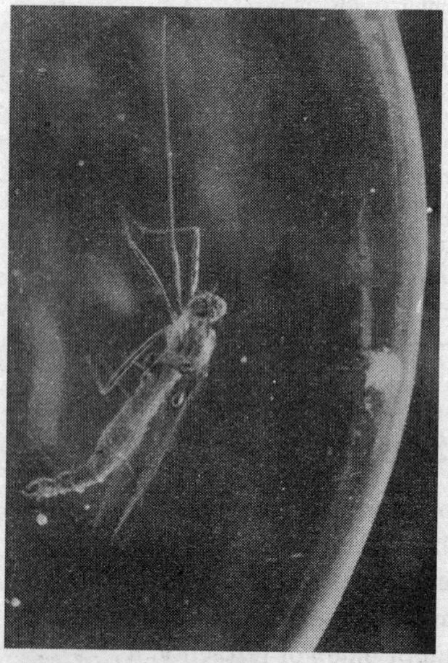

This insect has been entombed in amber for about 25 million years. Because we view the insect through a curved refracting surface, the image we see does not coincide with the insect.

In Section 35-8, we shall show that (for light rays making only small angles with the central axis)

$$\frac{n_1}{p} + \frac{n_2}{i} = \frac{n_2 - n_1}{r}. \tag{35-8}$$

Just as with mirrors, the object distance p is positive, and the image distance i is positive for a real image and negative for a virtual image. However, to keep all the signs correct in Eq. 35-8, we must use the following rule for the sign of the radius of curvature r:

> When the object faces a convex refracting surface, the radius of curvature r is positive. When it faces a concave surface, r is negative.

Be careful: This is just the reverse of the sign convention we have for mirrors.

CHECKPOINT 3: A bee is hovering in front of the concave spherical refracting surface of a glass sculpture. (a) Which of the general situations of Fig. 35-10 is like this situation? (b) Is the image produced by the surface real or virtual, and is it on the same side as the bee or the opposite side?

Sample Problem 35-2

A Jurassic mosquito is discovered embedded in a chunk of amber, which has index of refraction 1.6. One surface of the amber is spherically convex with radius of curvature 3.0 mm (Fig. 35-11). The mosquito head happens to be on the central axis of that surface and, when viewed along the axis, appears to be buried 5.0 mm into the amber. How deep is it really?

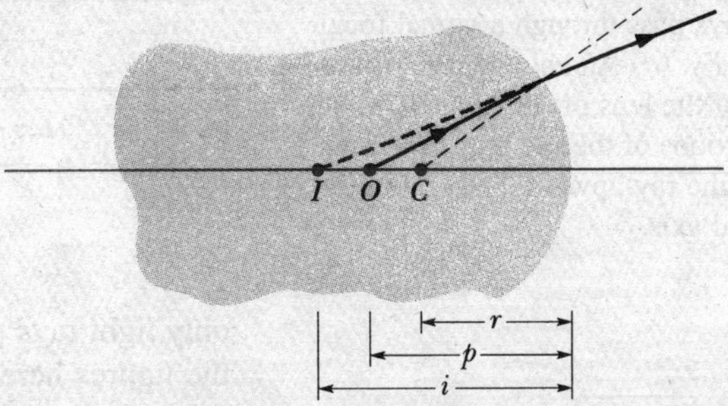

Fig. 35-11 Sample Problem 35-2. A piece of amber with a mosquito from the Jurassic period, with the head buried at point O. The spherical refracting surface at the right end, with center of curvature C, provides an image I to an observer intercepting rays from the object at O.

SOLUTION: The **Key Idea** here is that the head only appears to be 5.0 mm into the amber because the light rays that the observer intercepts are bent by refraction at the convex amber surface. The image distance i differs from the actual object distance p according to Eq. 35-8. To use that equation to find the actual object distance, we first note:

1. Because the object (the head) and its image are on the same side of the refracting surface, the image must be virtual and so $i = -5.0$ mm.
2. Because the object is always taken to be in the medium of index of refraction n_1, we must have $n_1 = 1.6$ and $n_2 = 1.0$.
3. Because the object faces a concave refracting surface, the radius of curvature r is negative and so $r = -3.0$ mm.

Making these substitutions in Eq. 35-8,

$$\frac{n_1}{p} + \frac{n_2}{i} = \frac{n_2 - n_1}{r},$$

yields

$$\frac{1.6}{p} + \frac{1.0}{-5.0\text{ mm}} = \frac{1.0 - 1.6}{-3.0\text{ mm}}$$

and

$$p = 4.0\text{ mm}. \qquad \text{(Answer)}$$

35-6 Thin Lenses

A **lens** is a transparent object with two refracting surfaces whose central axes coincide. The common central axis is the central axis of the lens. When a lens is surrounded by air, light refracts from the air into the lens, crosses through the lens, and then refracts back into the air. Each refraction can change the direction of travel of the light.

A lens that causes light rays initially parallel to the central axis to converge is (reasonably) called a **converging lens.** If, instead, it causes such rays to diverge, the lens is a **diverging lens.** When an object is placed in front of a lens of either type, refraction by the lens's surface of light rays from the object can produce an image of the object.

Now the rays spreading from the object, enter the lens.

We shall consider only the special case of a **thin lens**—that is, a lens in which the thickest part is thin compared to the object distance p, the image distance i, and the radii of curvature r_1 and r_2 of the two surfaces of the lens. We shall also consider

The rays may not be parallel to the central axis

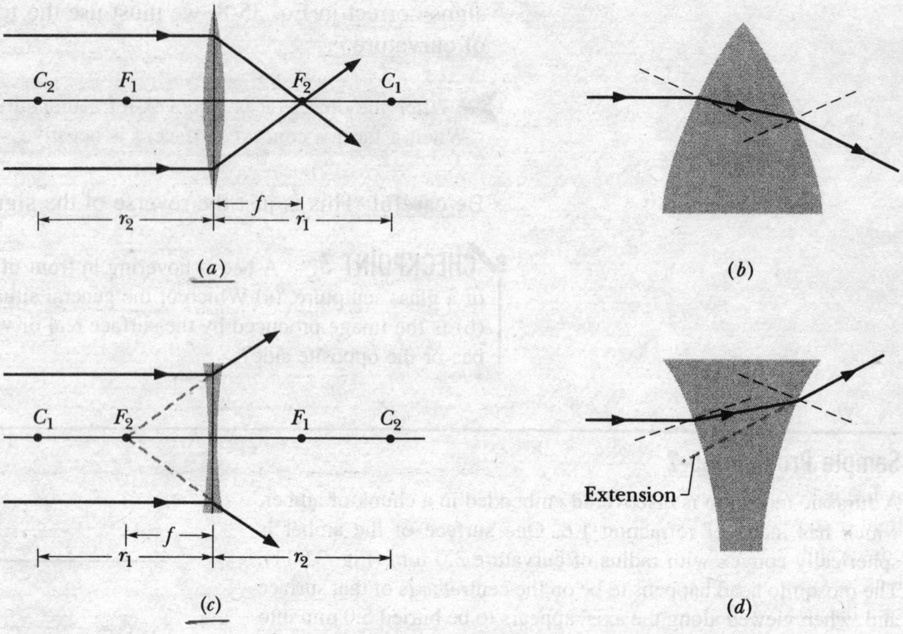

Fig. 35-12 (*a*) Rays initially parallel to the central axis of a converging lens are made to converge to a real focal point F_2 by the lens. The lens is thinner than drawn, with a width like that of the vertical line through it, where we shall consider all the bending of rays to occur. (*b*) An enlargement of the top part of the lens of (*a*); normals to the surfaces are shown dashed. Note that both refractions of the ray at the surfaces bend the ray downward, toward the central axis. (*c*) The same initially parallel rays are made to diverge by a diverging lens. Extensions of the diverging rays pass through a virtual focal point F_2. (*d*) An enlargement of the top part of the lens of (*c*). Note that both refractions of the ray at the surfaces bend the ray upward, away from the central axis.

only light rays that make small angles with the central axis (they are exaggerated in the figures here). In Section 35-8 we shall prove that for such rays, a thin lens has a focal length f. Moreover, i and p are related to each other by

$$\frac{1}{f} = \frac{1}{p} + \frac{1}{i} \quad \text{(thin lens)}, \qquad (35\text{-}9)$$

which is the same as we had for mirrors. We shall also prove that when a thin lens with index of refraction n is surrounded by air, this focal length f is given by

$$\frac{1}{f} = (n-1)\left(\frac{1}{r_1} - \frac{1}{r_2}\right) \quad \text{(thin lens in air)}, \qquad (35\text{-}10)$$

which is often called the *lens maker's equation*. Here r_1 is the radius of curvature of the lens surface nearer the object, and r_2 is that of the other surface. The signs of these radii are found with the rules in Section 35-5 for the radii of spherical refracting surfaces. If the lens is surrounded by some medium other than air (say, corn oil) with index of refraction n_{medium}, we replace n in Eq. 35-10 with n/n_{medium}. Keep in mind the basis of Eqs. 35-9 and 35-10:

A lens can produce an image of an object only because it can bend light rays; but it can bend light rays only if its index of refraction differs from that of the surrounding medium.

Figure 35-12*a* shows a thin lens with convex refracting surfaces, or *sides*. When rays that are parallel to the central axis of the lens are sent through the lens, they refract twice, as is shown enlarged in Fig. 35-12*b*. This double refraction causes the rays to converge and pass through a common point F_2 at a distance f from the center of the lens. Hence, this lens is a converging lens; further, a *real* focal point (or focus) exists at F_2 (because the rays really do pass through it), and the associated focal length is f. When rays parallel to the central axis are sent in the opposite direction through the lens, we find another real focal point at F_1 on the other side of the lens. For a thin lens, these two focal points are equidistant from the lens.

A fire is being started by focusing sunlight onto newspaper by means of a converging lens made of clear ice. The lens was made by melting both sides of a flat piece of ice into a convex shape in the shallow vessel (which has a curved bottom).

If object is kept at F in either case, images are formed at ∞

Because the focal points of a converging lens are real, we take the associated focal lengths f to be positive, just as we do with a real focus of a concave mirror. However, signs in optics can be tricky, so we had better check this in Eq. 35-10. The left side of that equation is positive if f is positive; how about the right side? We examine it term by term. Because the index of refraction n of glass or any other material is greater than 1, the term $(n - 1)$ must be positive. Because the source of the light (which is the object) is at the left and faces the convex left side of the lens, the radius of curvature r_1 of that side must be positive according to the sign rule for refracting surfaces. Similarly, because the object faces a concave right side of the lens, the radius of curvature r_2 of that side must be negative according to that rule. Thus, the term $(1/r_1 - 1/r_2)$ is positive, the whole right side of Eq. 35-10 is positive, and all the signs are consistent.

Figure 35-12c shows a thin lens with concave sides. When rays that are parallel to the central axis of the lens are sent through this lens, they refract twice, as is shown enlarged in Fig. 35-12d; these rays *diverge*, never passing through any common point, and so this lens is a diverging lens. However, extensions of the rays do pass through a common point F_2 at a distance f from the center of the lens. Hence, the lens has a *virtual* focal point at F_2. (If your eye intercepts some of the diverging rays, you perceive a bright spot to be at F_2, as if it is the source of the light.) Another virtual focus exists on the opposite side of the lens at F_1, symmetrically placed if the lens is thin. Because the focal points of a diverging lens are virtual, we take the focal length f to be negative.

Images from Thin Lenses

We now consider the types of image formed by converging and diverging lenses. Figure 35-13a shows an object O outside the focal point F_1 of a converging lens. The two rays drawn in the figure show that the lens forms a real, inverted image I of the object on the side of the lens opposite the object.

When the object is placed inside the focal point F_1, as in Fig. 35-13b, the lens forms a virtual image I on the same side of the lens as the object and with the same orientation. Hence, a converging lens can form either a real image or a virtual image, depending on whether the object is outside or inside the focal point, respectively.

Figure 35-13c shows an object O in front of a diverging lens. Regardless of the object distance (regardless of whether O is inside or outside the virtual focal point), this lens produces a virtual image that is on the same side of the lens as the object and has the same orientation.

As with mirrors, we take the image distance i to be positive when the image is real and negative when the image is virtual. However, the locations of real and virtual images from lenses are the reverse of those from mirrors:

> Real images form on the side of a lens that is opposite the object, and virtual images form on the side where the object is.

Fig. 35-13 (a) A real, inverted image I is formed by a converging lens when the object O is outside the focal point F_1. (b) The image I is virtual and has the same orientation as O when O is inside the focal point. (c) A diverging lens forms a virtual image I, with the same orientation as the object O, whether O is inside or outside the focal point of the lens.

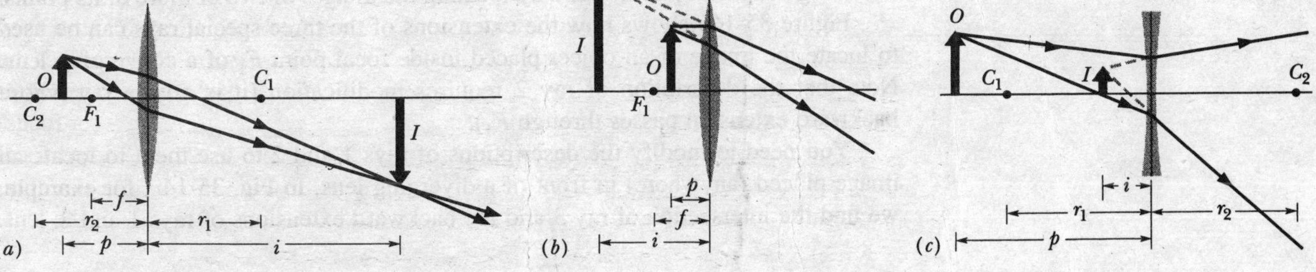

TABLE 35-2 Your Organizing Table for Thin Lenses

Lens Type	Object Location	Image Location	Image Type	Image Orientation	Sign of f	Sign of i	Sign of m
Converging	Inside F						
	Outside F						
Diverging	Anywhere						

The lateral magnification m produced by converging and diverging lenses is given by Eqs. 35-5 and 35-6, the same as for mirrors.

You have been asked to absorb a lot of information in this section, and you should organize it for yourself by filling in Table 35-2 for thin *symmetric lenses* (both sides are convex or both sides are concave). Under Image Location note whether the image is on the *same* side of the lens as the object or on the *opposite* side. Under Image Type note whether the image is *real* or *virtual*. Under Image Orientation note whether the image has the *same* orientation as the object or is *inverted*.

PROBLEM-SOLVING TACTICS

Tactic 1: *Signs of Trouble with Mirrors and Lenses*
Be careful: A mirror with a convex surface has a negative focal length f, just the opposite of a lens with convex surfaces. A mirror with a concave surface has a positive focal length f, just the opposite of a lens with concave surfaces. Confusing lens properties with mirror properties is a common mistake.

Locating Images of Extended Objects by Drawing Rays

Figure 35-14*a* shows an object O outside focal point F_1 of a converging lens. We can graphically locate the image of any off-axis point on such an object (such as the tip of the arrow in Fig. 35-14*a*) by drawing a ray diagram with any two of three special rays through the point. These special rays, chosen from all those that pass through the lens to form the image, are the following:

1. A ray that is initially parallel to the central axis of the lens will pass through focal point F_2 (ray 1 in Fig. 35-14*a*).
2. A ray that initially passes through focal point F_1 will emerge from the lens parallel to the central axis (ray 2 in Fig. 35-14*a*).
3. A ray that is initially directed toward the center of the lens will emerge from the lens with no change in its direction (ray 3 in Fig. 35-14*a*) because the ray encounters the two sides of the lens where they are almost parallel.

The image of the point is located where the rays intersect on the far side of the lens. The image of the object is found by locating the images of two or more of its points.

Figure 35-14*b* shows how the extensions of the three special rays can be used to locate the image of an object placed inside focal point F_1 of a converging lens. Note that the description of ray 2 requires modification (it is now a ray whose backward extension passes through F_1).

You need to modify the descriptions of rays 1 and 2 to use them to locate an image placed (anywhere) in front of a diverging lens. In Fig. 35-14*c*, for example, we find the intersection of ray 3 and the backward extensions of rays 1 and 2.

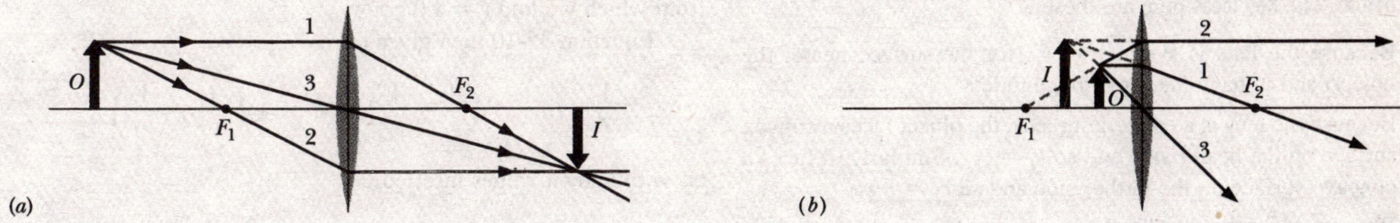

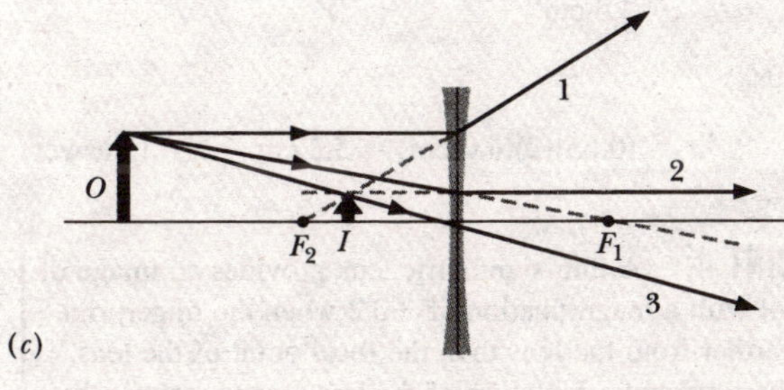

Fig. 35-14 Three special rays allow us to locate an image formed by a thin lens whether the object O is (*a*) outside or (*b*) inside the focal point of a converging lens, or (*c*) anywhere in front of a diverging lens.

Two-Lens Systems

When an object O is placed in front of a system of two lenses whose central axes coincide, we can locate the final image of the system (that is, the image produced by the lens farther from the object) by working in steps. Let lens 1 be the nearer lens and lens 2 the farther lens.

Step 1. We let p_1 represent the distance of object O from lens 1. We then find the distance i_1 of the image produced by lens 1, either by use of Eq. 35-9 or by drawing rays.

Step 2. Now, ignoring the presence of lens 1, we treat the image found in step 1 *as the object* for lens 2. If this new object is located beyond lens 2, the object distance p_2 for lens 2 is taken to be negative. (Note this exception to the rule that says the object distance is positive; the exception occurs because the object here is on the side opposite the source of light.) Otherwise, p_2 is taken to be positive as usual. We then find the distance i_2 of the (final) image produced by lens 2 by use of Eq. 35-9 or by drawing rays.

A similar step-by-step solution can be used for any number of lenses or if a mirror is substituted for lens 2.

The overall lateral magnification M produced by a system of two lenses is the product of the lateral magnifications m_1 and m_2 produced by the two lenses:

$$M = m_1 m_2. \tag{35-11}$$

Sample Problem 35-3

A praying mantis preys along the central axis of a thin symmetric lens, 20 cm from the lens. The lateral magnification of the mantis provided by the lens is $m = -0.25$, and the index of refraction of the lens material is 1.65.

(a) Determine the type of image produced by the lens; the type of lens; whether the object (mantis) is inside or outside the focal point; on which side of the lens the image appears; and whether the image is inverted.

SOLUTION: The **Key Idea** here is that we can tell a lot about the lens and the image from the given value of m. From it and Eq. 35-6 ($m = -i/p$), we see that

$$i = -mp = 0.25p.$$

Even without finishing the calculation, we can answer the questions. Because p is positive, i here must be positive. That means we have a real image, which means we have a converging lens (the only lens that can produce a real image). The object must be outside the focal point (the only way a real image can be produced). Also, the image is inverted and on the side of the lens opposite the object. (That is how a converging lens makes a real image.)

(b) What are the two radii of curvature of the lens?

SOLUTION: The **Key Ideas** here are these:

1. Because the lens is symmetric, r_1 (for the surface nearer the object) and r_2 have the same magnitude r.
2. Because the lens is a converging lens, the object faces a convex surface on the nearer side and so $r_1 = +r$. Similarly, it faces a concave surface on the farther side and so $r_2 = -r$.
3. We can relate these radii of curvature to the focal length f via the lens maker's equation, Eq. 35-10 (our only equation involving the radii of curvature of a lens).
4. We can relate f to the object distance p and image distance i via Eq. 35-9.

We know p but we do not know i. Thus, our starting point is to finish the calculation for i in part (a); we obtain

$$i = (0.25)(20 \text{ cm}) = 5.0 \text{ cm}.$$

i is +ve ∴ real image

Now Eq. 35-9 gives us

$$\frac{1}{f} = \frac{1}{p} + \frac{1}{i} = \frac{1}{20 \text{ cm}} + \frac{1}{5.0 \text{ cm}},$$

from which we find $f = 4.0$ cm.

Equation 35-10 then gives us

$$\frac{1}{f} = (n-1)\left(\frac{1}{r_1} - \frac{1}{r_2}\right) = (n-1)\left(\frac{1}{+r} - \frac{1}{-r}\right)$$

or, with known values inserted,

$$\frac{1}{4.0 \text{ cm}} = (1.65 - 1)\frac{2}{r},$$

which yields

$$r = (0.65)(2)(4.0 \text{ cm}) = 5.2 \text{ cm}. \quad \text{(Answer)}$$

✔CHECKPOINT 4: A thin symmetric lens provides an image of a fingerprint with a magnification of +0.2 when the fingerprint is 1.0 cm farther from the lens than the focal point of the lens. What are the type and orientation of the image, and what is the type of lens?

Sample Problem 35-4

Figure 35-15*a* shows a jalapeño seed O_1 that is placed in front of two thin symmetrical coaxial lenses 1 and 2, with focal lengths $f_1 = +24$ cm and $f_2 = +9.0$ cm, respectively, and with lens separation $L = 10$ cm. The seed is 6.0 cm from lens 1. Where does the system of two lenses produce an image of the seed?

SOLUTION: We could locate the image produced by the system of lenses by tracing light rays from the seed through the two lenses. However, the **Key Idea** here is that we can, instead, calculate the location of that image by working through the system in steps, lens by lens. We begin with the lens closer to the seed. The image we seek is the final one—that is, image I_2 produced by lens 2.

Lens 1. Ignoring lens 2, we locate the image I_1 produced by lens 1 by applying Eq. 35-9 to lens 1 alone:

$$\frac{1}{p_1} + \frac{1}{i_1} = \frac{1}{f_1}.$$

Take f of concave lens as + & f of convex lens as −, even if in Q, it's not given

The object O_1 for lens 1 is the seed, which is 6.0 cm from the lens; thus, we substitute $p_1 = +6.0$ cm. Also substituting the given value of f_1, we then have

$$\frac{1}{+6.0 \text{ cm}} + \frac{1}{i_1} = \frac{1}{+24 \text{ cm}},$$

which yields $i_1 = -8.0$ cm.

This tells us that image I_1 is 8.0 cm from lens 1 and virtual. (We could have guessed that it is virtual by noting that the seed is inside the focal point of lens 1.) Since I_1 is virtual, it is on the same

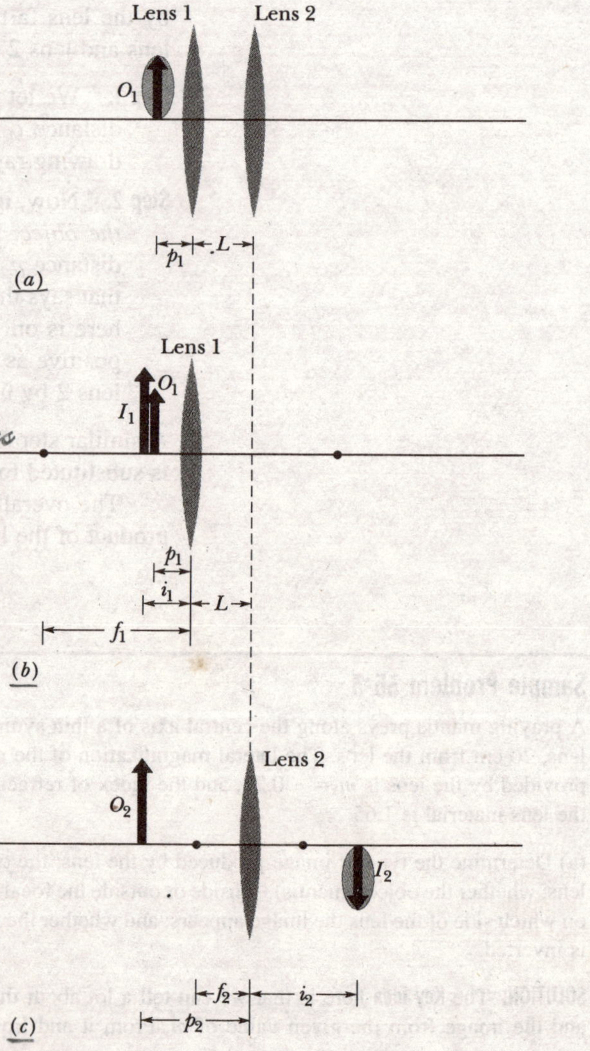

Fig. 35-15 Sample Problem 35-4. (*a*) Seed O_1 is distance p_1 from a two-lens system with lens separation L. We use the arrow to orient the seed. (*b*) The image I_1 produced by lens 1 alone. (*c*) Image I_1 acts as object O_2 for lens 2 alone, which produces the final image I_2.

side of the lens as object O_1 and has the same orientation as the seed, as shown in Fig. 35-15*b*.

Lens 2. In the second step of our solution, the **Key Idea** is that we can treat image I_1 as an object O_2 for the second lens and now ignore lens 1. We first note that this object O_2 is outside the focal point of lens 2. So the image I_2 produced by lens 2 must be real, inverted, and on the side of the lens opposite O_2. Let us see.

The distance p_2 between this object O_2 and lens 2 is, from Fig. 35-15*c*,

$$p_2 = L + |i_1| = 10\text{ cm} + 8.0\text{ cm} = 18\text{ cm}.$$

Then Eq. 35-9, now written for lens 2, yields

$$\frac{1}{+18\text{ cm}} + \frac{1}{i_2} = \frac{1}{+9.0\text{ cm}}.$$

Hence,

$$i_2 = +18\text{ cm}. \qquad \text{(Answer)}$$

The plus sign confirms our guess: Image I_2 produced by lens 2 is real, inverted, and on the side of lens 2 opposite O_2, as shown in Fig. 35-15*c*.

35-7 Optical Instruments

The human eye is a remarkably effective organ, but its range can be extended in many ways by optical instruments such as eyeglasses, simple magnifying lenses, motion picture projectors, cameras (including TV cameras), microscopes, and telescopes. Many such devices extend the scope of our vision beyond the visible range; satellite-borne infrared cameras and x-ray microscopes are just two examples.

The mirror and thin-lens formulas can be applied only as approximations to most sophisticated optical instruments. The lenses in typical laboratory microscopes are by no means "thin." In most optical instruments the lenses are compound lenses; that is, they are made of several components, the interfaces rarely being exactly spherical. Now we discuss three optical instruments, assuming, for simplicity, that the thin-lens formulas apply.

Simple Magnifying Lens

The normal human eye can focus a sharp image of an object on the retina (at the rear of the eye) if the object is located anywhere from infinity to a certain point called the *near point* P_n. If you move the object closer to the eye than the near point, the perceived retinal image becomes fuzzy. The location of the near point normally varies with age. We have all heard about people who claim not to need glasses but

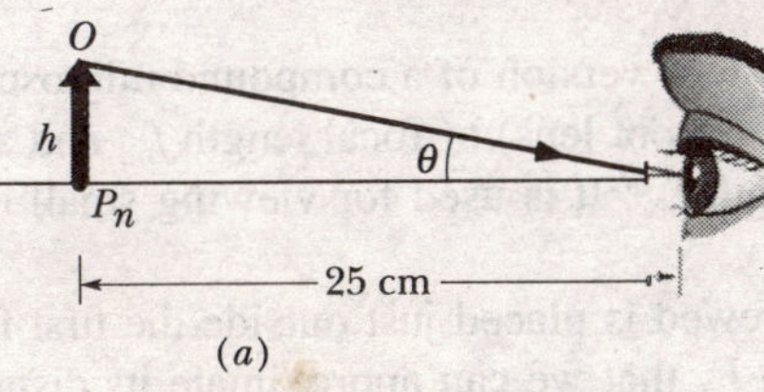

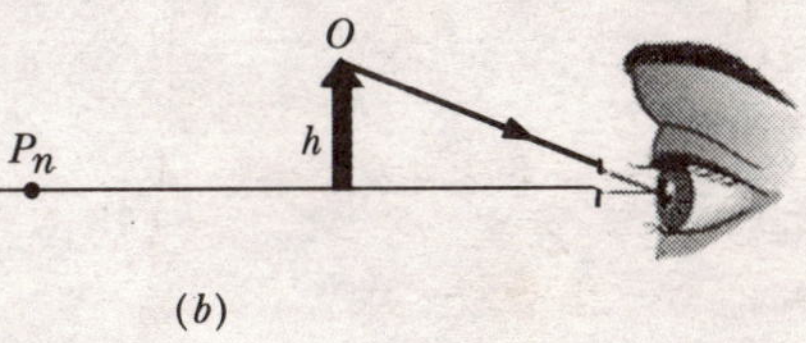

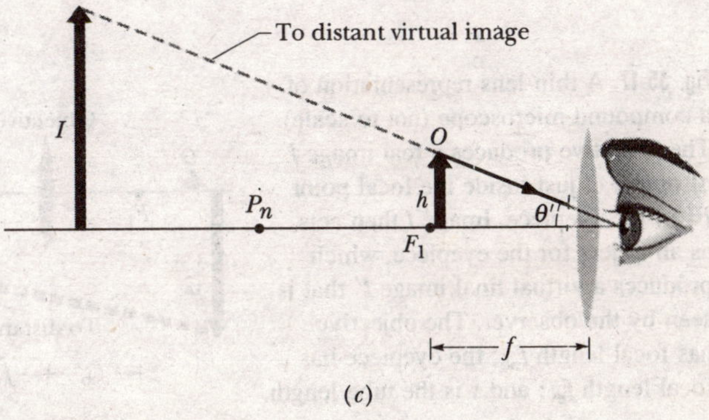

Fig. 35-16 (*a*) An object O of height h, placed at the near point of a human eye, occupies angle θ in the eye's view. (*b*) The object is moved closer to increase the angle, but now the observer cannot bring the object into focus. (*c*) A converging lens is placed between the object and the eye, with the object just inside the focal point F_1 of the lens. The image produced by the lens is then far enough away to be focused by the eye, and the image occupies a larger angle θ' than object O does in (*a*).

read their newspapers at arm's length; their near points are receding. To find your own near point, remove your glasses or contacts if you wear any, close one eye, and then bring this page closer to your open eye until it becomes indistinct. In what follows, we take the near point to be 25 cm from the eye, a bit more than the typical value for 20-year-olds.

Figure 35-16*a* shows an object O placed at the near point P_n of an eye. The size of the image of the object produced on the retina depends on the angle θ that the object occupies in the field of view from that eye. By moving the object closer to the eye, as in Fig. 35-16*b*, you can increase the angle and, hence, the possibility of distinguishing details of the object. However, because the object is then closer than the near point, it is no longer *in focus*; that is, the image is no longer clear.

You can restore the clarity by looking at O through a converging lens, placed so that O is just inside the focal point F_1 of the lens, which is at focal length f (Fig. 35-16*c*). What you then see is the virtual image of O produced by the lens. That image is farther away than the near point; thus, the eye can see it clearly.

Moreover, the angle θ' occupied by the virtual image is larger than the largest angle θ that the object alone can occupy and still be seen clearly. The *angular magnification* m_θ (not to be confused with lateral magnification m) of what is seen is

$$m_\theta = \theta'/\theta.$$

In words, the angular magnification of a simple magnifying lens is a comparison of the angle occupied by the image the lens produces with the angle occupied by the object when the object is moved to the near point of the viewer.

From Fig. 35-16, assuming that O is at the focal point of the lens, and approximating $\tan\theta$ as θ and $\tan\theta'$ as θ' for small angles, we have

$$\theta \approx h/25 \text{ cm} \quad \text{and} \quad \theta' \approx h/f.$$

We then find that

$$m_\theta \approx \frac{25 \text{ cm}}{f} \quad \text{(simple magnifier).} \qquad (35\text{-}12)$$

Compound Microscope

Figure 35-17 shows a thin-lens version of a compound microscope. The instrument consists of an *objective* (the front lens) of focal length f_{ob} and an *eyepiece* (the lens near the eye) of focal length f_{ey}. It is used for viewing small objects that are very close to the objective.

The object O to be viewed is placed just outside the first focal point F_1 of the objective, close enough to F_1 that we can approximate its distance p from the lens

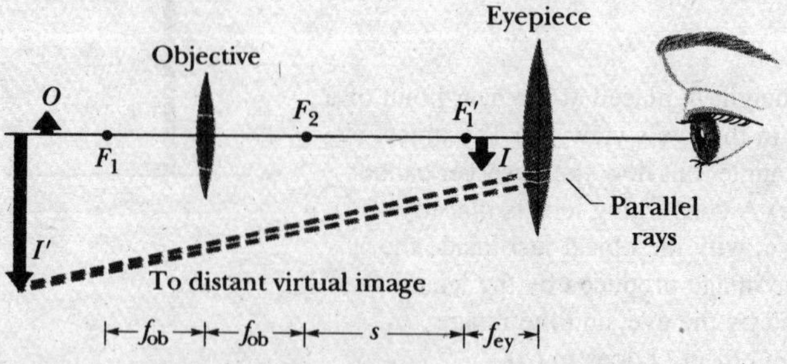

Fig. 35-17 A thin-lens representation of a compound microscope (not to scale). The objective produces a real image I of object O just inside the focal point F_1' of the eyepiece. Image I then acts as an object for the eyepiece, which produces a virtual final image I' that is seen by the observer. The objective has focal length f_{ob}; the eyepiece has focal length f_{ey}; and s is the tube length.

as being f_{ob}. The separation between the lenses is then adjusted so that the enlarged, inverted, real image I produced by the objective is located just inside the first focal point F_1' of the eyepiece. The *tube length s* shown in Fig. 35-17 is actually large relative to f_{ob}, and we can approximate the distance i between the objective and the image I as being length s.

From Eq. 35-6, and using our approximations for p and i, we can write the lateral magnification produced by the objective as

$$m = -\frac{i}{p} = -\frac{s}{f_{ob}}. \quad (35\text{-}13)$$

Since the image I is located just inside the focal point F_1' of the eyepiece, the eyepiece acts as a simple magnifying lens, and an observer sees a final (virtual, inverted) image I' through it. The overall magnification of the instrument is the product of the lateral magnification m produced by the objective, given by Eq. 35-13, and the angular magnification m_θ produced by the eyepiece, given by Eq. 35-12; that is,

$$M = mm_\theta = -\frac{s}{f_{ob}}\frac{25\text{ cm}}{f_{ey}} \quad \text{(microscope).} \quad (35\text{-}14)$$

Refracting Telescope

Telescopes come in a variety of forms. The form we describe here is the simple refracting telescope that consists of an objective and an eyepiece; both are represented in Fig. 35-18 with simple lenses, although in practice, as is also true for most microscopes, each lens is actually a compound lens system.

The lens arrangements for telescopes and for microscopes are similar, but telescopes are designed to view large objects, such as galaxies, stars, and planets, at large distances, whereas microscopes are designed for just the opposite purpose. This difference requires that in the telescope of Fig. 35-18 the second focal point of the objective F_2 coincide with the first focal point of the eyepiece F_1', whereas in the microscope of Fig. 35-17 these points are separated by the tube length s.

In Fig. 35-18*a*, parallel rays from a distant object strike the objective, making an angle θ_{ob} with the telescope axis and forming a real, inverted image at the common focal point F_2, F_1'. This image I acts as an object for the eyepiece, through which

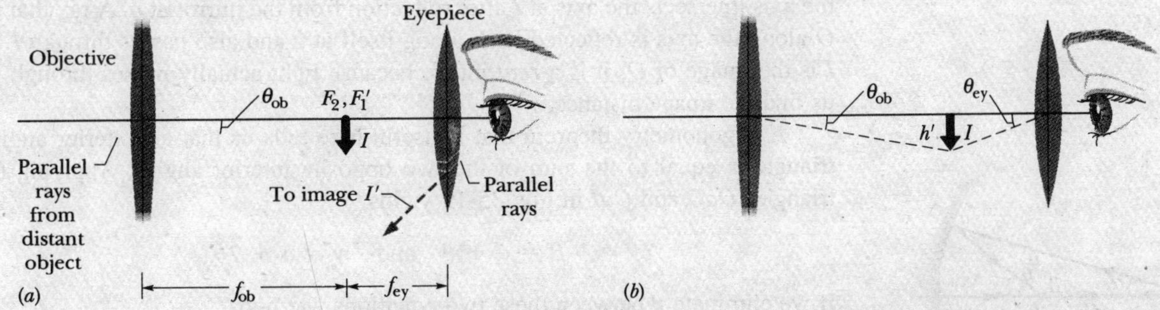

Fig. 35-18 (*a*) A thin-lens representation of a refracting telescope. The objective produces a real image I of a distant source of light (the object), with approximately parallel light rays at the objective. (One end of the object is assumed to lie on the central axis.) Image I, formed at the common focal points F_2 and F_1', acts as an object for the eyepiece, which produces a virtual final image I' at a great distance from the observer. The objective has focal length f_{ob}; the eyepiece has focal length f_{ey}. (*b*) Image I has height h' and takes up angle θ_{ob} measured from the objective and angle θ_{ey} measured from the eyepiece.

an observer sees a distant (still inverted) virtual image I'. The rays defining the image make an angle θ_{ey} with the telescope axis.

The angular magnification m_θ of the telescope is θ_{ey}/θ_{ob}. From Fig. 35-18*b*, for rays close to the central axis, we can write $\theta_{ob} = h'/f_{ob}$ and $\theta_{ey} \approx h'/f_{ey}$, which gives us

$$m_\theta = -\frac{f_{ob}}{f_{ey}} \quad \text{(telescope)}, \qquad (35\text{-}15)$$

where the minus sign indicates that I' is inverted. In words, the angular magnification of a telescope is a comparison of the angle occupied by the image the telescope produces with the angle occupied by the distant object as seen without the telescope.

Magnification is only one of the design factors for an astronomical telescope and is indeed easily achieved. A good telescope needs *light-gathering power*, which determines how bright the image is. This is important for viewing faint objects such as distant galaxies and is accomplished by making the objective diameter as large as possible. A telescope also needs *resolving power*, which is the ability to distinguish between two distant objects (stars, say) whose angular separation is small. *Field of view* is another important design parameter. A telescope designed to look at galaxies (which occupy a tiny field of view) is much different from one designed to track meteors (which move over a wide field of view).

The telescope designer must also take into account the difference between real lenses and the ideal thin lenses we have discussed. A real lens with spherical surfaces does not form sharp images, a flaw called *spherical aberration*. Also, because refraction by the two surfaces of a real lens depends on wavelength, a real lens does not focus light of different wavelengths to the same point, a flaw called *chromatic aberration*.

This brief discussion by no means exhausts the design parameters of astronomical telescopes—many others are involved. We could make a similar listing for any other high-performance optical instrument.

35-8 Three Proofs

The Spherical Mirror Formula (Eq. 35-4)

Figure 35-19 shows a point object O placed on the central axis of a concave spherical mirror, outside its center of curvature C. A ray from O that makes an angle α with the axis intersects the axis at I after reflection from the mirror at a. A ray that leaves O along the axis is reflected back along itself at c and also passes through I. Thus, I is the image of O; it is a *real* image because light actually passes through it. Let us find the image distance i.

A trigonometry theorem that is useful here tells us that an exterior angle of a triangle is equal to the sum of the two opposite interior angles. Applying this to triangles OaC and OaI in Fig. 35-19 yields

$$\beta = \alpha + \theta \quad \text{and} \quad \gamma = \alpha + 2\theta.$$

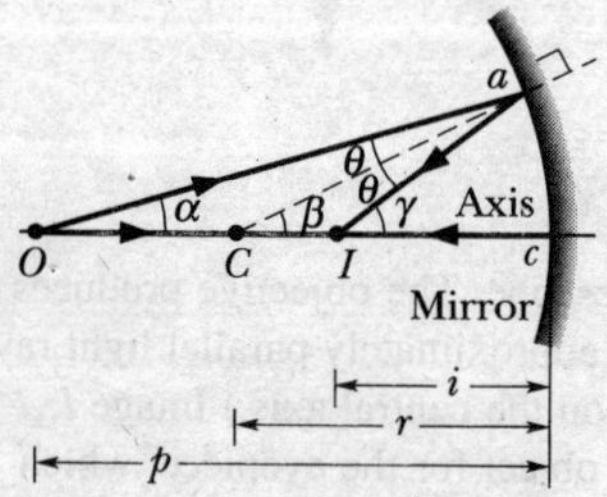

Fig. 35-19 A concave spherical mirror forms a real point image I by reflecting light rays from a point object O.

If we eliminate θ between these two equations, we find

$$\alpha + \gamma = 2\beta. \qquad (35\text{-}16)$$

We can write angles α, β, and γ, in radian measure, as

$$\alpha \approx \frac{\widehat{ac}}{cO} = \frac{\widehat{ac}}{p}, \qquad \beta = \frac{\widehat{ac}}{cC} = \frac{\widehat{ac}}{r},$$

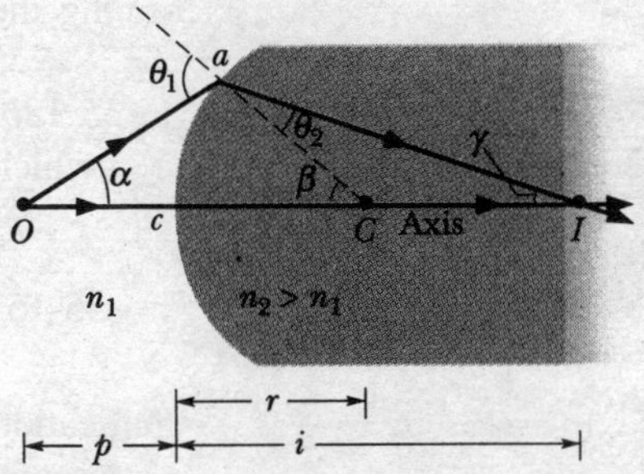

Fig. 35-20 A real point image I of a point object O is formed by refraction at a spherical convex surface between two media.

and
$$\gamma \approx \frac{\widehat{ac}}{cI} = \frac{\widehat{ac}}{i}. \tag{35-17}$$

Only the equation for β is exact, because the center of curvature of arc ac is at C. However, the equations for α and γ are approximately correct if these angles are small enough (that is, for rays close to the central axis). Substituting Eqs. 35-17 into Eq. 35-16, using Eq. 35-3 to replace r with $2f$, and canceling $\widehat{ac}$ lead exactly to Eq. 35-4, the relation that we set out to prove.

The Refracting Surface Formula (Eq. 35-8)

The incident ray from point object O in Fig. 35-20 that falls on point a of a spherical refracting surface is refracted there according to Eq. 34-44,

$$n_1 \sin \theta_1 = n_2 \sin \theta_2.$$

If α is small, θ_1 and θ_2 will also be small and we can replace the sines of these angles with the angles themselves. Thus, the equation above becomes

$$n_1\theta_1 \approx n_2\theta_2. \tag{35-18}$$

We again use the fact that an exterior angle of a triangle is equal to the sum of the two opposite interior angles. Applying this to triangles COa and ICa yields

$$\theta_1 = \alpha + \beta \quad \text{and} \quad \beta = \theta_2 + \gamma. \tag{35-19}$$

If we use Eqs. 35-19 to eliminate θ_1 and θ_2 from Eq. 35-18, we find

$$n_1\alpha + n_2\gamma = (n_2 - n_1)\beta. \tag{35-20}$$

In radian measure the angles α, β, and γ are

$$\alpha \approx \frac{\widehat{ac}}{p}; \qquad \beta = \frac{\widehat{ac}}{r}; \qquad \gamma \approx \frac{\widehat{ac}}{i}. \tag{35-21}$$

Only the second of these equations is exact. The other two are approximate because I and O are not the centers of circles of which $\widehat{ac}$ is a part. However, for α small enough (for rays close to the axis), the inaccuracies in Eqs. 35-21 are small. Substituting Eqs. 35-21 into Eq. 35-20 leads directly to Eq. 35-8, as we wanted.

The Thin-Lens Formulas (Eqs. 35-9 and 35-10)

Our plan is to consider each lens surface as a separate refracting surface, and to use the image formed by the first surface as the object for the second.

We start with the thick glass "lens" of length L in Fig. 35-21a whose left and right refracting surfaces are ground to radii r' and r''. A point object O' is placed near the left surface as shown. A ray leaving O' along the central axis is not deflected on entering or leaving the lens.

A second ray leaving O' at an angle α with the central axis intersects the left surface at point a', is refracted, and intersects the second (right) surface at point a''. The ray is again refracted and crosses the axis at I'', which, being the intersection of two rays from O', is the image of point O', formed after refraction at two surfaces.

Figure 35-21b shows that the first (left) surface also forms a virtual image of O' at I'. To locate I', we use Eq. 35-8,

$$\frac{n_1}{p} + \frac{n_2}{i} = \frac{n_2 - n_1}{r}.$$

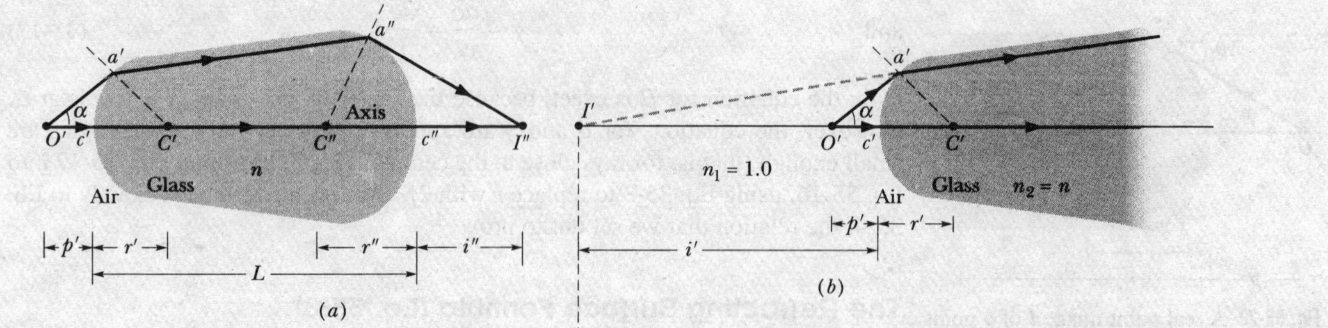

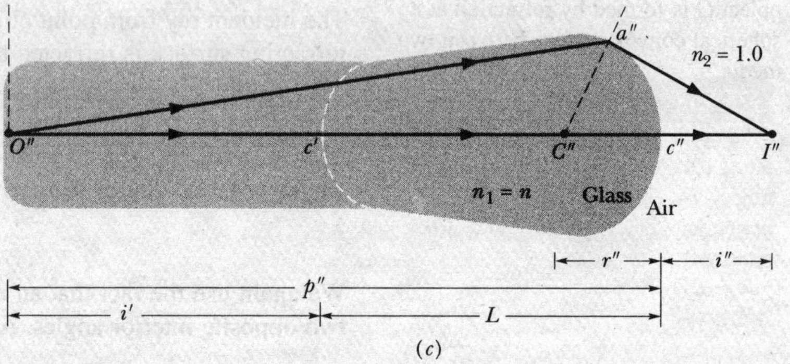

Fig. 35-21 (*a*) Two rays from point object O' form a real image I'' after refracting through two spherical surfaces of a "lens." The object faces a convex surface at the left side of the lens and a concave surface at the right side. The ray traveling through points a' and a'' is actually close to the central axis through the lens. (*b*) The left side and (*c*) the right side of the "lens" in (*a*), shown separately.

Putting $n_1 = 1$ for air and $n_2 = n$ for lens glass and bearing in mind that the image distance is negative (that is, $i = -i'$ in Fig. 35-21*b*), we obtain

$$\frac{1}{p'} - \frac{n}{i'} = \frac{n-1}{r'}. \tag{35-22}$$

In this equation i' will be a positive number because we have already introduced the minus sign appropriate to a virtual image.

Figure 35-21*c* shows the second surface again. Unless an observer at point a'' were aware of the existence of the first surface, the observer would think that the light striking that point originated at point I' in Fig. 35-21*b* and that the region to the left of the surface was filled with glass as indicated. Thus, the (virtual) image I' formed by the first surface serves as a real object O'' for the second surface. The distance of this object from the second surface is

$$p'' = i' + L. \tag{35-23}$$

To apply Eq. 35-8 to the second surface, we must insert $n_1 = n$ and $n_2 = 1$ because the object now is effectively imbedded in glass. If we substitute with Eq. 35-23, then Eq. 35-8 becomes

$$\frac{n}{i' + L} + \frac{1}{i''} = \frac{1-n}{r''}. \tag{35-24}$$

Let us now assume that the thickness L of the "lens" in Fig. 35-21*a* is so small that we can neglect it in comparison with our other linear quantities (such as p', i', p'', i'', r', and r''). In all that follows we make this *thin-lens approximation*. Putting $L = 0$ in Eq. 35-24 and rearranging the right side lead to

$$\frac{n}{i'} + \frac{1}{i''} = -\frac{n-1}{r''}. \tag{35-25}$$

Adding Eqs. 35-22 and 35-25 leads to

$$\frac{1}{p'} + \frac{1}{i''} = (n-1)\left(\frac{1}{r'} - \frac{1}{r''}\right).$$

Finally, calling the original object distance simply p and the final image distance simply i leads to

$$\frac{1}{p} + \frac{1}{i} = (n-1)\left(\frac{1}{r'} - \frac{1}{r''}\right), \tag{35-26}$$

which, with a small change in notation, is Eqs. 35-9 and 35-10, the relations we set out to prove.

REVIEW & SUMMARY

Real and Virtual Images An *image* is a reproduction of an object via light. If the image can form on a surface, it is a *real image* and can exist even if no observer is present. If the image requires the visual system of an observer, it is a *virtual image.*

Image Formation *Spherical mirrors, spherical refracting surfaces,* and *thin lenses* can form images of a source of light—the object—by redirecting rays emerging from the source. The image occurs where the redirected rays cross (forming a real image) or where backward extensions of those rays cross (forming a virtual image). If the rays are sufficiently close to the *central axis* through the spherical mirror, refracting surface, or thin lens, we have the following relations between the *object distance* p (which is positive) and the *image distance* i (which is positive for real images and negative for virtual images):

1. Spherical Mirror:

$$\frac{1}{p} + \frac{1}{i} = \frac{1}{f} = \frac{2}{r}, \tag{35-4, 35-3}$$

where f is the mirror's focal length and r is the mirror's radius of curvature. A *plane mirror* is a special case for which $r \rightarrow \infty$, so that $p = -i$. Real images form on the side of a mirror where the object is located, and virtual images form on the opposite side.

2. Spherical Refracting Surface:

$$\frac{n_1}{p} + \frac{n_2}{i} = \frac{n_2 - n_1}{r} \quad \text{(single surface)}, \tag{35-8}$$

where n_1 is the index of refraction of the material where the object is located, n_2 is the index of refraction of the material on the other side of the refracting surface, and r is the radius of curvature of the surface. When the object faces a convex refracting surface, the radius r is positive. When it faces a concave surface, r is negative. Real images form on the side of a refracting surface that is opposite the object, and virtual images form on the same side as the object.

3. Thin Lens:

$$\frac{1}{p} + \frac{1}{i} = \frac{1}{f} = (n-1)\left(\frac{1}{r_1} - \frac{1}{r_2}\right), \tag{35-9, 35-10}$$

where f is the lens's focal length, n is the index of refraction of the lens material, and r_1 and r_2 are the radii of curvature of the two sides of the lens, which are spherical surfaces. A convex lens surface that faces the object has a positive radius of curvature; a concave lens surface that faces the object has a negative radius of curvature. Real images form on the side of a lens that is opposite the object, and virtual images form on the same side as the object.

Lateral Magnification The *lateral magnification m* produced by a spherical mirror or a thin lens is

$$m = -\frac{i}{p}. \tag{35-6}$$

The magnitude of m is given by

$$|m| = \frac{h'}{h}, \tag{35-5}$$

where h and h' are the heights (measured perpendicular to the central axis) of the object and image, respectively.

Optical Instruments Three optical instruments that extend human vision are:

1. The *simple magnifying lens,* which produces an *angular magnification* m_θ given by

$$m_\theta = \frac{25\text{ cm}}{f}, \tag{35-12}$$

where f is the focal length of the magnifying lens.

2. The *compound microscope,* which produces an *overall magnification M* given by

$$M = mm_\theta = -\frac{s}{f_{ob}}\frac{25\text{ cm}}{f_{ey}}, \tag{35-14}$$

where m is the lateral magnification produced by the objective, m_θ is the angular magnification produced by the eyepiece, s is the tube length, and f_{ob} and f_{ey} are the focal lengths of the objective and eyepiece, respectively.

3. The *refracting telescope,* which produces an *angular magnification* m_θ given by

$$m_\theta = -\frac{f_{ob}}{f_{ey}}. \tag{35-15}$$

QUESTIONS

1. Lake monsters, mermen, and mermaids have long been "sighted" by observers located either on a shore or on a low deck of a ship. From such a low point, an observer can intercept rays of light that leave a floating object (say, a log or a porpoise) and bend slightly back downward toward the observer (one such refracted ray, exaggerated, is shown in Fig. 35-22*a*). The observer then perceives the object as being elongated upward from the water (and probably oscillating because of air turbulence) in a mirage that might easily resemble one of the fabled creatures. Figure 35-22*b* gives several plots of height from the water surface versus air temperature. Which one of the plots best illustrates the air-temperature conditions that can bend the light rays so as to create this mirage?

(*a*) (*b*)

Fig. 35-22 Question 1.

2. Figure 35-23 shows a fish and a fish stalker in water. (a) Does the stalker see the fish in the general region of point *a* or point *b*? (b) Does the fish see the (wild) eyes of the stalker in the general region of point *c* or point *d*?

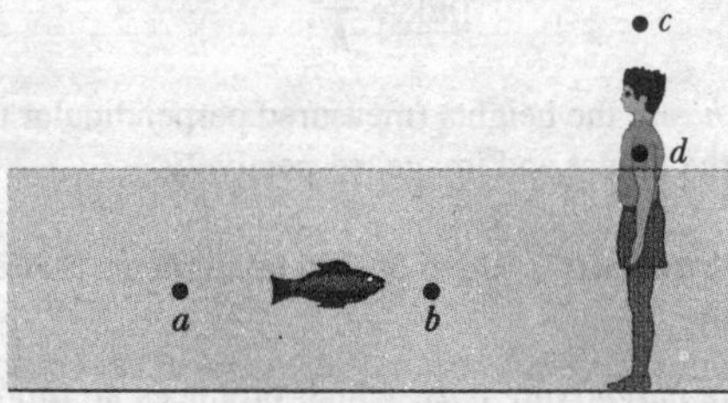

Fig. 35-23 Question 2.

3. In the mirror maze of Fig. 35-24*a*, many "virtual hallways" seem to extend away from you because you see multiple reflections from the mirrors that form the walls of the maze. Those mirrors are placed along some sides of repeated equilateral triangles on the floor. The floor plan for a similar but different maze is shown in Fig. 35-24*b*; every wall section within this maze is mirrored. If you stand at entrance *x*, (a) which of the maze monsters *a*, *b*, and *c* hiding in the maze can you see along the virtual hallways extending from entrance *x*; (b) how many times does each visible monster appear in a hallway; and (c) what is at the far end of a hallway? (*Hint:* The two rays shown are coming down virtual hallways; follow them back into the maze, using the law of reflection at each mirror along the path of each ray. Do they pass through a triangle with a monster? If so, how many times? For additional analysis, see J. Walker, "The Amateur Scientist," *Scientific American*, Vol. 254, pages 120–126, June 1986.)

(*a*)

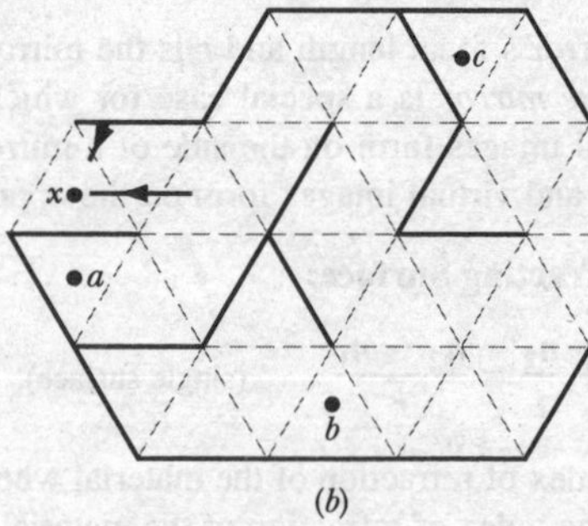

(*b*)

Fig. 35-24 Question 3.

4. A penguin waddles along the central axis of a concave mirror, from the focal point to an effectively infinite distance. (a) How does its image move? (b) Does the height of its image increase continually, decrease continually, or change in some more complicated manner?

5. When a *T. rex* pursues a jeep in the movie *Jurassic Park*, we see a reflected image of the *T. rex* via a side-view mirror, on which is printed the (then darkly humorous) warning: "Objects in mirror are closer than they appear." Is the mirror flat, convex, or concave?

6. Figure 35-25 shows four thin lenses, all of the same material, with sides that either are flat or have a radius of curvature of magnitude 10 cm. Without written calculation, rank the lenses according to the magnitude of the focal length, greatest first.

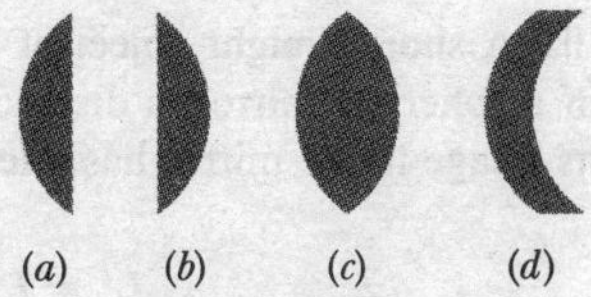

Fig. 35-25 Question 6.

7. An object lies before a thin, symmetric, converging lens. Does the image distance increase, decrease, or remain the same if we increase (a) the index of refraction n of the lens, (b) the magnitude of the radius of curvature of the two sides, and (c) the index of refraction n_{med} of the surrounding medium, keeping n_{med} less than n?

8. A concave mirror and a converging lens (glass with $n = 1.5$) both have a focal length of 3 cm when in air. When they are in water ($n = 1.33$), are their focal lengths greater than, less than, or equal to 3 cm?

9. The table details six variations of the basic arrangement of two thin lenses represented in Fig. 35-26. (The points labeled F_1 and F_2 are the focal points of lenses 1 and 2.) An object is distance p_1 to the left of lens 1, as in Fig. 35-15. (a) For which variations can we tell, *without calculation*, whether the final image (that due to lens 2) is to the left or right of lens 2 and whether it has the same orientation as the object? (b) For those "easy" variations, give the image location as "left" or "right" and the orientation as "same" or "inverted."

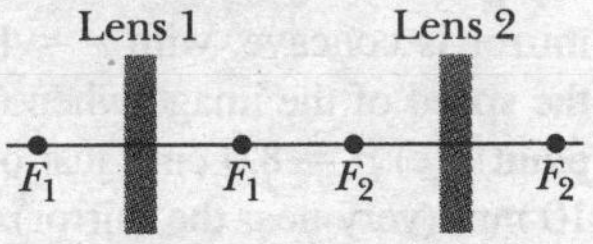

Fig. 35-26 Question 9.

Variation	Lens 1	Lens 2	
1	Converging	Converging	$p_1 < f_1$
2	Converging	Converging	$p_1 > f_1$
3	Diverging	Converging	$p_1 < f_1$
4	Diverging	Converging	$p_1 > f_1$
5	Diverging	Diverging	$p_1 < f_1$
6	Diverging	Diverging	$p_1 > f_1$

10. Much of the bending of light rays necessary for human vision occurs at the cornea (at the air–eye interface). The cornea has an index of refraction somewhat greater than that of water. (a) When your eye is submerged in a swimming pool, is the bending of light rays at the cornea greater than, less than, or the same as in air? (b) The Central American fish *Anableps anableps* can see simultaneously above and below water because it swims with its eyes partially extending above the water surface. To provide clear sight in both media, is the radius of curvature of the submerged portion of the cornea greater than, less than, or equal to that of the exposed portion?

EXERCISES & PROBLEMS

ssm Solution is in the Student Solutions Manual.
www Solution is available on the World Wide Web at: http://www.wiley.com/college/hrw
ilw Solution is available on the Interactive LearningWare.

SEC. 35-2 Plane Mirrors

1E. A moth at about eye level is 10 cm in front of a plane mirror; you are behind the moth, 30 cm from the mirror. What is the distance between your eyes and the apparent position of the moth's image in the mirror? ssm ilw

2E. You look through a camera toward an image of a hummingbird in a plane mirror. The camera is 4.30 m in front of the mirror. The bird is at camera level, 5.00 m to your right and 3.30 m from the mirror. What is the distance between the camera and the apparent position of the bird's image in the mirror?

3E. Figure 35-27*a* is an overhead view of two vertical plane mirrors with an object O placed between them. If you look into the mirrors, you see multiple images of O. You can find them by drawing the reflection in each mirror of the angular region between the mirrors, as is done for the left-hand mirror in Fig. 35-27*b*. Then draw the reflection of the reflection. Continue this on the left and on the right until the reflections meet or overlap at the rear of the mirrors. Then you can count the number of images of O. (a) If $\theta = 90°$, how many images of O would you see? (b) Draw their locations and orientations (as in Fig. 35-27*b*). ssm

4P. Repeat Exercise 3 for the mirror angle θ equal to (a) 45°, (b) 60°, and (c) 120°. (d) Explain why there are several possible answers for (c).

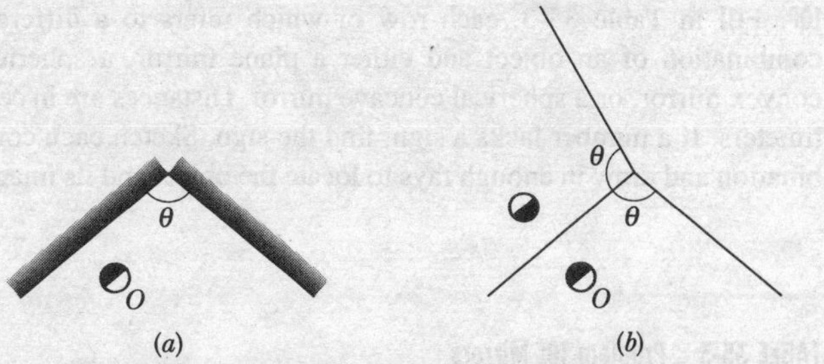

Fig. 35-27 Exercise 3 and Problem 4.

5P. Prove that if a plane mirror is rotated through an angle α, the reflected beam is rotated through an angle 2α. Show that this result is reasonable for $\alpha = 45°$. ssm

6P. Figure 35-28 shows an overhead view of a corridor with a

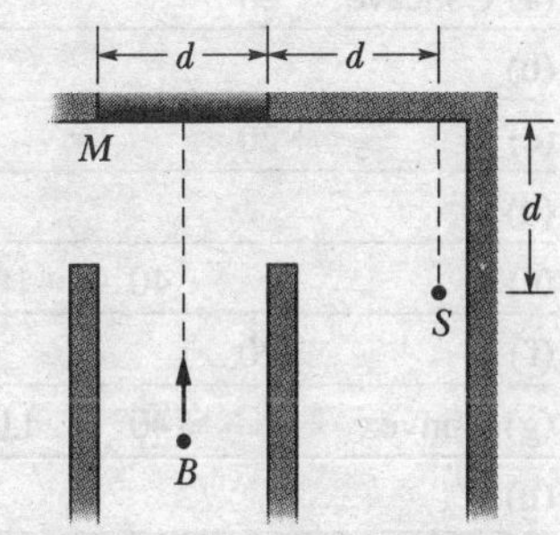

Fig. 35-28 Problem 6.

plane mirror M mounted at one end. A burglar B sneaks along the corridor directly toward the center of the mirror. If $d = 3.0$ m, how far from the mirror will she be when the security guard S can first see her in the mirror?

7P. You put a point source of light S a distance d in front of a screen A. How is the light intensity at the center of the screen changed if you put a completely reflecting mirror M a distance d behind the source, as in Fig. 35-29? (*Hint:* Use Eq. 34-27.) ssm

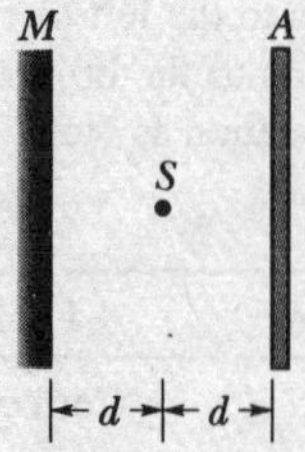

Fig. 35-29 Problem 7.

8P. Figure 35-30 shows a small lightbulb suspended above the surface of the water in a swimming pool. The bottom of the pool is a large mirror. How far below the mirror's surface is the image of the bulb? (*Hint:* Construct a diagram of two rays like that of Fig. 35-3, but take into account the bending of light rays by refraction. Assume that the rays are close to a vertical axis through the bulb, and use the small-angle approximation that $\sin\theta \approx \tan\theta \approx \theta$.)

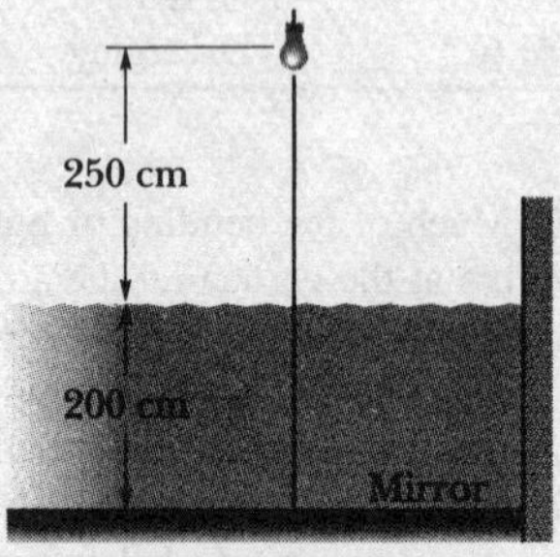

Fig. 35-30 Problem 8.

SEC. 35-4 Images from Spherical Mirrors

9E. A concave shaving mirror has a radius of curvature of 35.0 cm. It is positioned so that the (upright) image of a man's face is 2.50 times the size of the face. How far is the mirror from the face?

10P. Fill in Table 35-3, each row of which refers to a different combination of an object and either a plane mirror, a spherical convex mirror, or a spherical concave mirror. Distances are in centimeters. If a number lacks a sign, find the sign. Sketch each combination and draw in enough rays to locate the object and its image.

11P. A short straight object of length L lies along the central axis of a spherical mirror, a distance p from the mirror. (a) Show that its image in the mirror has a length L' where

$$L' = L\left(\frac{f}{p-f}\right)^2.$$

(*Hint:* Locate the two ends of the object.) (b) Show that the *longitudinal magnification* m' $(= L'/L)$ is equal to m^2, where m is the lateral magnification. ssm www

12P. (a) A luminous point is moving at speed v_O toward a spherical mirror with radius of curvature r, along the central axis of the mirror. Show that the image of this point is moving at speed

$$v_I = -\left(\frac{r}{2p-r}\right)^2 v_O,$$

where p is the distance of the luminous point from the mirror at any given time. (*Hint:* Start with Eq. 35-4.) Now assume that the mirror is concave, with $r = 15$ cm, and let $v_O = 5.0$ cm/s. Find the speed of the image when (b) $p = 30$ cm (far outside the focal point), (c) $p = 8.0$ cm (just outside the focal point), and (d) $p = 10$ mm (very near the mirror).

SEC. 35-5 Spherical Refracting Surfaces

13P. A beam of parallel light rays from a laser is incident on a solid transparent sphere of index of refraction n (Fig. 35-31). (a) If a point image is produced at the back of the sphere, what is the index of refraction of the sphere? (b) What index of refraction, if any, will produce a point image at the center of the sphere?

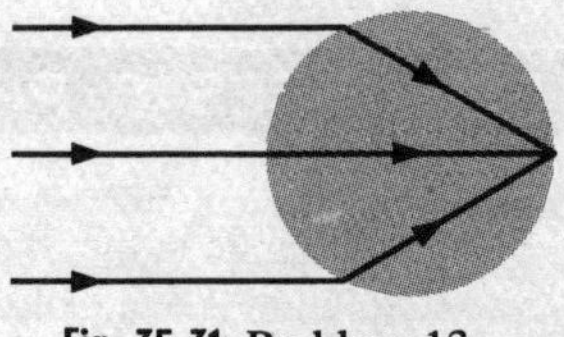

Fig. 35-31 Problem 13.

14P. Fill in Table 35-4, each row of which refers to a different combination of a point object and a spherical refracting surface separating two media with different indexes of refraction. Distances are in centimeters. If a number lacks a sign, find the sign. Sketch

TABLE 35-3 Problem 10: Mirrors

Type	f	r	i	p	m	Real Image?	Inverted Image?
(a) Concave	20			+10			
(b)				+10	+1.0	No	
(c)	+20			+30			
(d)				+60	−0.50		
(e)		−40	−10				
(f)	20				+0.10		
(g) Convex		40	4.0				
(h)				+24	0.50		Yes

TABLE 35-4 Problem 14: Spherical Refracting Surfaces

	n_1	n_2	p	i	r	Inverted Image?
(a)	1.0	1.5	+10		+30	
(b)	1.0	1.5	+10	−13		
(c)	1.0	1.5		+600	+30	
(d)	1.0		+20	−20	−20	
(e)	1.5	1.0	+10	−6.0		
(f)	1.5	1.0		−7.5	−30	
(g)	1.5	1.0	+70		+30	
(h)	1.5		+100	+600	−30	

TABLE 35-5 Problem 24: Thin Lenses

	Type	f	r_1	r_2	i	p	n	m	Real Image?	Inverted Image?
(a)	C	10				+20				
(b)		+10				+5.0				
(c)		10				+5.0		>1.0		
(d)		10				+5.0		<1.0		
(e)			+30	−30		+10	1.5			
(f)			−30	+30		+10	1.5			
(g)			−30	−60		+10	1.5			
(h)						+10		0.50		No
(i)						+10		−0.50		

each combination and draw in enough rays to locate the object and image.

15P. You look downward at a coin that lies at the bottom of a pool of liquid with depth d and index of refraction n (Fig. 35-32). Because you view with two eyes, which intercept different rays of light from the coin, you perceive the coin to be where extensions of the intercepted rays cross, at depth d_a instead of d. Assuming that the intercepted rays in Fig. 35-32 are close to a vertical axis through the coin, show that $d_a = d/n$. (*Hint:* Use the small-angle approximation that $\sin\theta \approx \tan\theta \approx \theta$.)

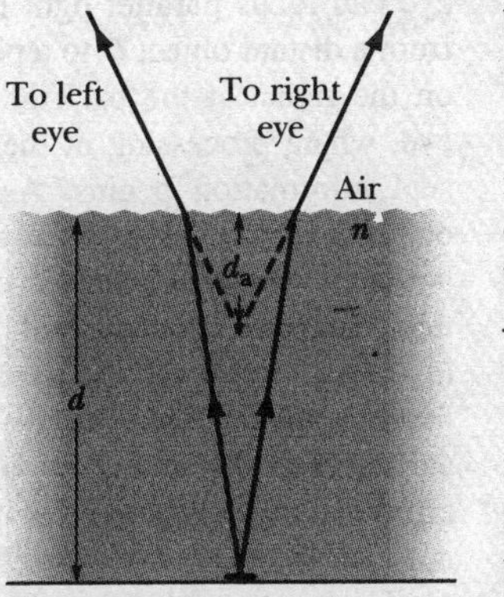

Fig. 35-32 Problem 15.

16P. A 20-mm-thick layer of water (n = 1.33) floats on a 40-mm-thick layer of carbon tetrachloride (n = 1.46) in a tank. A coin lies at the bottom of the tank. At what depth below the top water surface do you perceive the coin? (*Hint:* Use the result and assumptions of Problem 15 and work with a ray diagram of the situation.)

SEC. 35-6 Thin Lenses

17E. An object is 20 cm to the left of a thin diverging lens having a 30 cm focal length. What is the image distance i? Find the image position with a ray diagram. ssm ilw

18E. You produce an image of the Sun on a screen, using a thin lens whose focal length is 20.0 cm. What is the diameter of the image? (See Appendix C for needed data on the Sun.)

19E. A double-convex lens is to be made of glass with an index of refraction of 1.5. One surface is to have twice the radius of curvature of the other and the focal length is to be 60 mm. What are the radii? ssm

20E. A lens is made of glass having an index of refraction of 1.5. One side of the lens is flat, and the other is convex with a radius of curvature of 20 cm. (a) Find the focal length of the lens. (b) If an object is placed 40 cm in front of the lens, where will the image be located?

21E. The formula

$$\frac{1}{p} + \frac{1}{i} = \frac{1}{f}$$

is called the *Gaussian* form of the thin-lens formula. Another form of this formula, the *Newtonian* form, is obtained by considering the distance x from the object to the first focal point and the distance x' from the second focal point to the image. Show that

$$xx' = f^2$$

is the Newtonian form of the thin-lens formula. ssm

22E. A movie camera with a (single) lens of focal length 75 mm takes a picture of a 180-cm-high person standing 27 m away. What is the height of the image of the person on the film?

23P. An illuminated slide is held 44 cm from a screen. How far from the slide must a lens of focal length 11 cm be placed to form an image of the slide's picture on the screen? ilw

24P. To the extent possible, fill in Table 35-5, each row of which refers to a different combination of an object and a thin lens. Distances are in centimeters. For the type of lens, use C for converging and D for diverging. If a number (except for the index of refraction) lacks a sign, find the sign. Sketch each combination and draw in enough rays to locate the object and image.

25P. Show that the distance between an object and its real image formed by a thin converging lens is always greater than or equal to four times the focal length of the lens. ssm www

26P. A diverging lens with a focal length of −15 cm and a converging lens with a focal length of 12 cm have a common central axis. Their separation is 12 cm. An object of height 1.0 cm is 10 cm in front of the diverging lens, on the common central axis. (a) Where does the lens combination produce the final image of the object (the one produced by the second, converging lens)? (b) What is the height of that image? (c) Is the image real or virtual? (d) Does the image have the same orientation as the object or is it inverted?

27P. A converging lens with a focal length of +20 cm is located 10 cm to the left of a diverging lens having a focal length of

−15 cm. If an object is located 40 cm to the left of the converging lens, locate and describe completely the final image formed by the diverging lens.

28P. An object is 20 cm to the left of a lens with a focal length of +10 cm. A second lens of focal length +12.5 cm is 30 cm to the right of the first lens. (a) Find the location and relative size of the final image. (b) Verify your conclusions by drawing the lens system to scale and constructing a ray diagram. (c) Is the final image real or virtual? (d) Is it inverted?

29P. Two thin lenses of focal lengths f_1 and f_2 are in contact. Show that they are equivalent to a single thin lens with

$$f = \frac{f_1 f_2}{f_1 + f_2}$$

as its focal length. ssm

30P. In Fig. 35-33, a real inverted image I of an object O is formed by a certain lens (not shown); the object–image separation is d = 40.0 cm, measured along the central axis of the lens. The image is just half the size of the object. (a) What kind of lens must be used to produce this image? (b) How far from the object must the lens be placed? (c) What is the focal length of the lens?

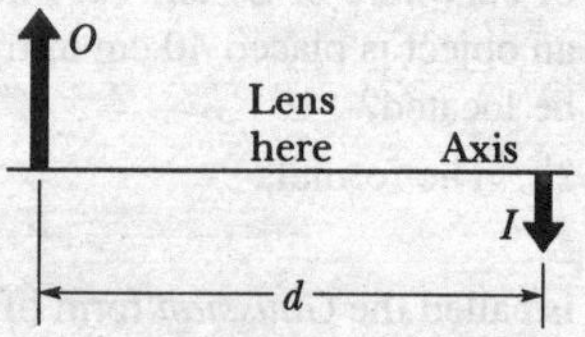

Fig. 35-33 Problem 30.

31P. A luminous object and a screen are a fixed distance D apart. (a) Show that a converging lens of focal length f, placed between object and screen, will form a real image on the screen for two lens positions that are separated by a distance

$$d = \sqrt{D(D - 4f)}.$$

(b) Show that

$$\left(\frac{D - d}{D + d}\right)^2$$

gives the ratio of the two image sizes for these two positions of the lens. ssm www

SEC. 35-7 Optical Instruments

32E. If the angular magnification of an astronomical telescope is 36 and the diameter of the objective is 75 mm, what is the minimum diameter of the eyepiece required to collect all the light entering the objective from a distant point source on the telescope axis?

33E. In a microscope of the type shown in Fig. 35-17, the focal length of the objective is 4.00 cm, and that of the eyepiece is 8.00 cm. The distance between the lenses is 25.0 cm. (a) What is the tube length s? (b) If image I in Fig. 35-17 is to be just inside focal point F_1', how far from the objective should the object be? What then are (c) the lateral magnification m of the objective, (d) the angular magnification m_θ of the eyepiece, and (e) the overall magnification M of the microscope? ssm

34P. A simple magnifying lens of focal length f is placed near the eye of someone whose near point P_n is 25 cm from the eye. An object is positioned so that its image in the magnifying lens appears at P_n. (a) What is the lens's angular magnification? (b) What is the angular magnification if the object is moved so that its image appears at infinity? (c) Evaluate the angular magnifications of (a) and (b) for f = 10 cm. (Viewing an image at P_n requires effort by muscles in the eye, whereas for many people viewing an image at infinity requires no effort.)

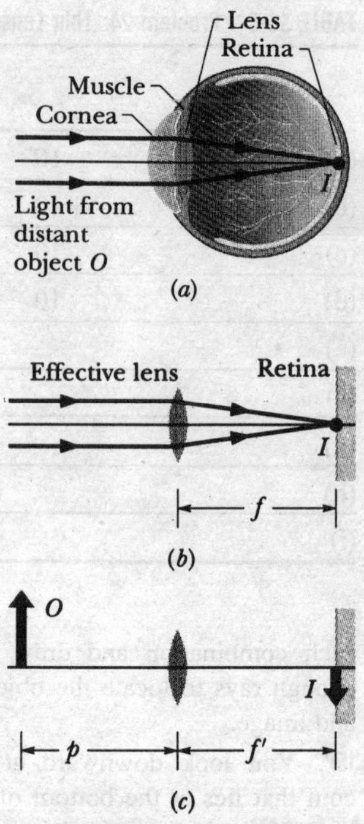

Fig. 35-34 Problem 35.

35P. Figure 35-34a shows the basic structure of a human eye. Light refracts into the eye through the cornea and is then further redirected by a lens whose shape (and thus ability to focus the light) is controlled by muscles. We can treat the cornea and eye lens as a single effective thin lens (Fig. 35-34b). A "normal" eye can focus parallel light rays from a distant object O to a point on the retina at the back of the eye, where processing of the visual information begins. As an object is brought close to the eye, however, the muscles must change the shape of the lens so that rays form an inverted real image on the retina (Fig. 35-34c). (a) Suppose that for the parallel rays of Figs. 35-34a and b, the focal length f of the effective thin lens of the eye is 2.50 cm. For an object at distance p = 40.0 cm, what focal length f' of the effective lens is required for it to be seen clearly? (b) Must the eye muscles increase or decrease the radii of curvature of the eye lens to give focal length f'? ssm

36P. An object is 10.0 mm from the objective of a certain compound microscope. The lenses are 300 mm apart and the intermediate image is 50.0 mm from the eyepiece. What overall magnification is produced by the instrument?

37P. Figure 35-35a shows the basic structure of a camera. A lens can be moved forward or back to produce an image on film at the back of the camera. For a certain camera, with the distance i between the lens and the film set at f = 5.0 cm, parallel light rays from a very distant object O converge to a point image on the film, as shown. The object is now brought closer, to a distance of p = 100 cm, and the lens–film distance is adjusted so that an inverted real image forms on the film (Fig. 35-35b). (a) What is the lens–film distance i now? (b) By how much was i changed?

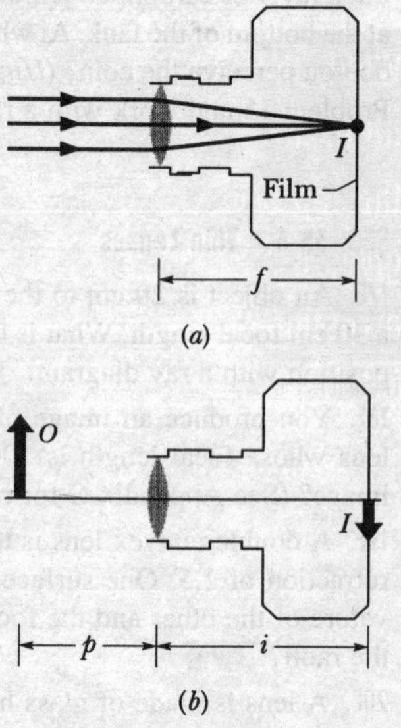

Fig. 35-35 Problem 37.

36 Interference

At first glance, the top surface of the *Morpho* butterfly's wing is simply a beautiful blue-green. There is something strange about the color, however, for it almost glimmers, unlike the colors of most objects—and if you change your perspective, or if the wing moves, the tint of the color changes. The wing is said to be iridescent, and the blue-green we see hides the wing's "true" dull brown color that appears on the bottom surface.

What, then, is so different about the top surface that gives us this arresting display?

The answer is in this chapter.

36-1 Interference

Sunlight, as the rainbow shows us, is a composite of all the colors of the visible spectrum. The colors reveal themselves in the rainbow because the incident wavelengths are bent through different angles as they pass through raindrops that produce the bow. However, soap bubbles and oil slicks can also show striking colors, produced not by refraction but by constructive and destructive **interference** of light. The interfering waves combine either to enhance or to suppress certain colors in the spectrum of the incident sunlight. Interference of light waves is thus a superposition phenomenon like those we discussed in Chapter 17.

This selective enhancement or suppression of wavelengths has many applications. When light encounters an ordinary glass surface, for example, about 4% of the incident energy is reflected, thus weakening the transmitted beam by that amount. This unwanted loss of light can be a real problem in optical systems with many components. A thin, transparent "interference film," deposited on the glass surface, can reduce the amount of reflected light (and thus enhance the transmitted light) by destructive interference. The bluish cast of a camera lens reveals the presence of such a coating. Interference coatings can also be used to enhance—rather than reduce—the ability of a surface to reflect light.

To understand interference, we must go beyond the restrictions of geometrical optics and employ the full power of wave optics. In fact, as you will see, the existence of interference phenomena is perhaps our most convincing evidence that light is a wave—because interference cannot be explained other than with waves.

36-2 Light as a Wave

Is there some other case in which the surface/wavefront is not a plane?

The first person to advance a convincing wave theory for light was Dutch physicist Christian Huygens, in 1678. Although much less comprehensive than the later electromagnetic theory of Maxwell, Huygens' theory was simpler mathematically and remains useful today. Its great advantages are that it accounts for the laws of reflection and refraction in terms of waves and gives physical meaning to the index of refraction.

Huygens' wave theory is based on a geometrical construction that allows us to tell where a given wavefront will be at any time in the future if we know its present position. This construction is based on **Huygens' principle,** which is:

> All points on a wavefront serve as point sources of spherical secondary wavelets. After a time t, the new position of the wavefront will be that of a surface tangent to these secondary wavelets.

For a plane wave, wavefront at any time is a plane

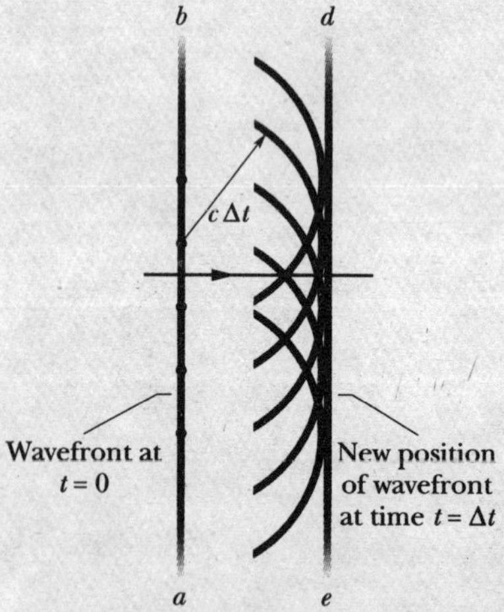

Fig. 36-1 The propagation of a plane wave in vacuum, as portrayed by Huygens' principle.

Here is a simple example. At the left in Fig. 36-1, the present location of a wavefront of a plane wave traveling to the right in vacuum is represented by plane *ab*, perpendicular to the page. Where will the wavefront be at time Δt later? We let several points on plane *ab* (the dots) serve as sources of spherical secondary wavelets that are emitted at $t = 0$. At time Δt, the radius of all these spherical wavelets will have grown to $c\,\Delta t$, where c is the speed of light in vacuum. We draw plane *de* tangent to these wavelets at time Δt. This plane represents the wavefront of the plane wave at time Δt; it is parallel to plane *ab* and a perpendicular distance $c\,\Delta t$ from it.

The Law of Refraction

We now use Huygens' principle to derive the law of refraction, Eq. 34-44 (Snell's law). Figure 36-2 shows three stages in the refraction of several wavefronts at a plane interface between air (medium 1) and glass (medium 2). We arbitrarily choose the wavefronts in the incident light beam to be separated by λ_1, the wavelength in medium 1. Let the speed of light in air be v_1 and that in glass be v_2. We assume that $v_2 < v_1$, which happens to be true.

Angle θ_1 in Fig. 36-2*a* is the angle between the wavefront and the interface; it has the same value as the angle between the *normal* to the wavefront (that is, the incident ray) and the *normal* to the interface. Thus, θ_1 is the angle of incidence.

As the wave moves into the glass, a Huygens wavelet at point *e* will expand to pass through point *c*, at a distance of λ_1 from point *e*. The time interval required for this expansion is that distance divided by the speed of the wavelet, or λ_1/v_1. Now note that in this same time interval, a Huygens wavelet at point *h* will expand to pass through point *g*, at the reduced speed v_2 and with wavelength λ_2. Thus, this time interval must also be equal to λ_2/v_2. By equating these times of travel, we obtain the relation

$$\frac{\lambda_1}{\lambda_2} = \frac{v_1}{v_2}, \tag{36-1}$$

which shows that the wavelengths of light in two media are proportional to the speeds of light in those media.

By Huygens' principle, the refracted wavefront must be tangent to an arc of radius λ_2 centered on *h*, say at point *g*. The refracted wavefront must also be tangent to an arc of radius λ_1 centered on *e*, say at *c*. Then the refracted wavefront must be oriented as shown. Note that θ_2, the angle between the refracted wavefront and the interface, is actually the angle of refraction.

For the right triangles *hce* and *hcg* in Fig. 36-2*b* we may write

$$\sin\theta_1 = \frac{\lambda_1}{hc} \quad \text{(for triangle } hce\text{)}$$

and

$$\sin\theta_2 = \frac{\lambda_2}{hc} \quad \text{(for triangle } hcg\text{)}.$$

Dividing the first of these two equations by the second and using Eq. 36-1, we find

$$\frac{\sin\theta_1}{\sin\theta_2} = \frac{\lambda_1}{\lambda_2} = \frac{v_1}{v_2}. \tag{36-2}$$

We can define an **index of refraction** *n* for each medium as the ratio of the speed of light in vacuum to the speed of light *v* in the medium. Thus,

$$n = \frac{c}{v} \quad \text{(index of refraction).} \tag{36-3}$$

In particular, for our two media, we have

$$n_1 = \frac{c}{v_1} \quad \text{and} \quad n_2 = \frac{c}{v_2}. \tag{36-4}$$

If we combine Eqs. 36-2 and 36-4, we find

$$\frac{\sin\theta_1}{\sin\theta_2} = \frac{c/n_1}{c/n_2} = \frac{n_2}{n_1} \tag{36-5}$$

or

$$n_1 \sin\theta_1 = n_2 \sin\theta_2 \quad \text{(law of refraction),} \tag{36-6}$$

as introduced in Chapter 34.

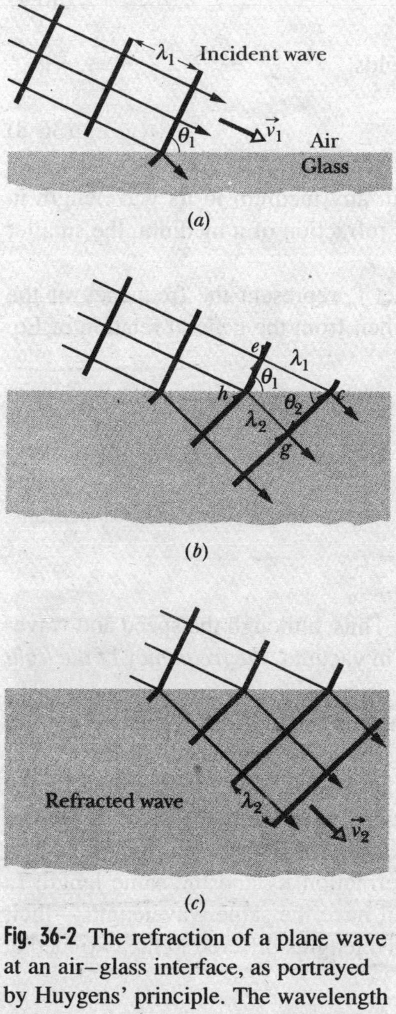

Fig. 36-2 The refraction of a plane wave at an air–glass interface, as portrayed by Huygens' principle. The wavelength in glass is smaller than that in air. For simplicity, the reflected wave is not shown. Parts (*a*) through (*c*) represent three successive stages of the refraction.

$v = \frac{c}{n}$
velocity of light in any medium is represented by v
$v_b > v_c > v_a$

CHECKPOINT 1: The figure shows a monochromatic ray of light traveling across parallel interfaces, from an original material *a*, through layers of materials *b* and *c*, and then back into material *a*. Rank the materials according to the speed of light in them, greatest first.

$n_b < n_c < n_a$ —Given

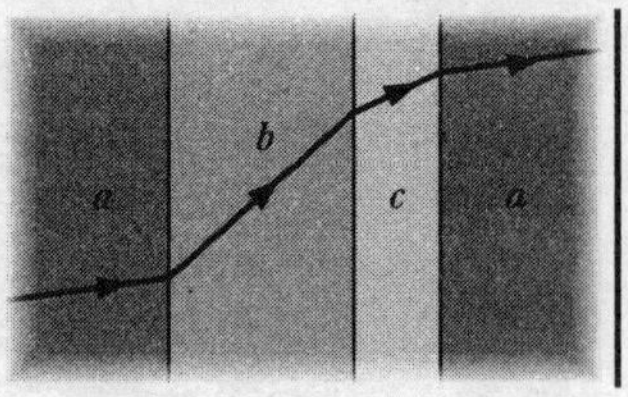

Wavelength and Index of Refraction

We have now seen that the wavelength of light changes when the speed of the light changes, as happens when light crosses an interface from one medium into another. Further, the speed of light in any medium depends on the index of refraction of the medium, according to Eq. 36-3. Thus, the wavelength of light in any medium depends on the index of refraction of the medium. Let a certain monochromatic light have wavelength λ and speed c in vacuum and wavelength λ_n and speed v in a medium with an index of refraction n. Now we can rewrite Eq. 36-1 as

$$\lambda_n = \lambda \frac{v}{c}. \tag{36-7}$$

Using Eq. 36-3 to substitute $1/n$ for v/c then yields

$$\lambda_n = \frac{\lambda}{n}. \tag{36-8}$$

This equation relates the wavelength of light in any medium to its wavelength in vacuum. It tells us that the greater the index of refraction of a medium, the smaller the wavelength of light in that medium.

What about the frequency of the light? Let f_n represent the frequency of the light in a medium with index of refraction n. Then from the general relation of Eq. 17-12 ($v = \lambda f$), we can write

$$f_n = \frac{v}{\lambda_n}.$$

Substituting Eqs. 36-3 and 36-8 then gives us

$$f_n = \frac{c/n}{\lambda/n} = \frac{c}{\lambda} = f,$$

where f is the frequency of the light in vacuum. Thus, although the speed and wavelength of light are different in the medium than in vacuum, *the frequency of the light in the medium is the same as it is in vacuum.*

The fact that the wavelength of light depends on the index of refraction via Eq. 36-8 is important in certain situations involving the interference of light waves. For example, in Fig. 36-3, the *waves of the rays* (that is, the waves represented by the rays) have identical wavelengths λ and are initially in phase in air ($n \approx 1$). One of the waves travels through medium 1 of index of refraction n_1 and length L. The other travels through medium 2 of index of refraction n_2 and the same length L. When the waves leave the two media, they will have the same wavelength—their wavelength λ in air. However, because their wavelengths differed in the two media, the two waves may no longer be in phase.

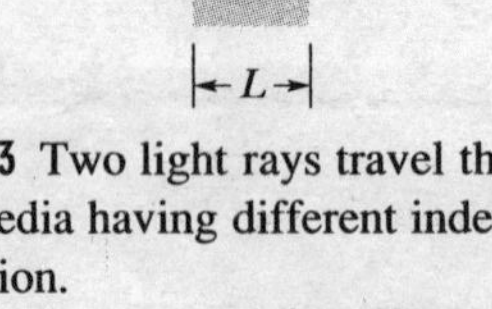

Fig. 36-3 Two light rays travel through two media having different indexes of refraction.

> The phase difference between two light waves can change if the waves travel through different materials having different indexes of refraction.

As we shall discuss soon, this change in the phase difference can determine how the light waves will interfere if they reach some common point.

To find their new phase difference in terms of wavelengths, we first count the number N_1 of wavelengths there are in the length L of medium 1. From Eq. 36-8, the wavelength in medium 1 is $\lambda_{n1} = \lambda/n_1$, so

$$N_1 = \frac{L}{\lambda_{n1}} = \frac{Ln_1}{\lambda}. \tag{36-9}$$

Similarly, we count the number N_2 of wavelengths there are in the length L of medium 2, where the wavelength is $\lambda_{n2} = \lambda/n_2$:

$$N_2 = \frac{L}{\lambda_{n2}} = \frac{Ln_2}{\lambda}. \tag{36-10}$$

To find the new phase difference between the waves, we subtract the smaller of N_1 and N_2 from the larger. Assuming $n_2 > n_1$, we obtain

$$N_2 - N_1 = \frac{Ln_2}{\lambda} - \frac{Ln_1}{\lambda} = \frac{L}{\lambda}(n_2 - n_1). \tag{36-11}$$

Path lengths of waves in two media are identical

Suppose Eq. 36-11 tells us that the waves now have a phase difference of 45.6 wavelengths. That is equivalent to taking the initially in-phase waves and shifting one of them by 45.6 wavelengths. However, a shift of an integer number of wavelengths (such as 45) would put the waves back in phase, so it is only the decimal fraction (here, 0.6) that is important. A phase difference of 45.6 wavelengths is equivalent to an *effective phase difference* of 0.6 wavelength.

A phase difference of 0.5 wavelength puts two waves exactly out of phase. If the waves had equal amplitudes and were to reach some common point, they would then undergo fully destructive interference, producing darkness at that point. With a phase difference of 0.0 or 1.0 wavelength, they would, instead, undergo fully constructive interference, resulting in brightness at the common point. Our phase difference of 0.6 wavelength is an intermediate situation, but closer to destructive interference, and the waves would produce a dimly illuminated common point.

We can also express phase difference in terms of radians and degrees, as we have done already. A phase difference of one wavelength is equivalent to phase differences of 2π rad and 360°.

Sample Problem 36-1

In Fig. 36-3, the two light waves that are represented by the rays have wavelength 550.0 nm before entering media 1 and 2. They also have equal amplitudes and are in phase. Medium 1 is now just air, and medium 2 is a transparent plastic layer of index of refraction 1.600 and thickness 2.600 μm.

(a) What is the phase difference of the emerging waves in wavelengths, radians, and degrees? What is their effective phase difference (in wavelengths)?

SOLUTION: One **Key Idea** here is that the phase difference of two light waves can change if they travel through different media, with different indexes of refraction. The reason is that their wavelengths are different in the different media. We can calculate the change in phase difference by counting the number of wavelengths that fits into each medium and then subtracting those numbers. When the path lengths of the waves in the two media are identical, Eq. 36-11 gives the result. Here we have $n_1 = 1.000$ (for the air), $n_2 = 1.600$, $L = 2.600\ \mu$m, and $\lambda = 550.0$ nm. Thus, Eq. 36-11 yields

$$N_2 - N_1 = \frac{L}{\lambda}(n_2 - n_1)$$
$$= \frac{2.600 \times 10^{-6}\ \text{m}}{5.500 \times 10^{-7}\ \text{m}}(1.600 - 1.000)$$
$$= 2.84. \qquad \text{(Answer)}$$

Thus, the phase difference of the emerging waves is 2.84 wavelengths. Because 1.0 wavelength is equivalent to 2π rad and 360°, you can show that this phase difference is equivalent to

$$\text{phase difference} = 17.8\ \text{rad} \approx 1020°. \qquad \text{(Answer)}$$

A second **Key Idea** is that the effective phase difference is the decimal part of the actual phase difference *expressed in wavelengths*. Thus, we have

effective phase difference = 0.84 wavelength. (Answer)

You can show that this is equivalent to 5.3 rad and about 300°. *Caution:* We do *not* find the effective phase difference by taking the decimal part of the actual phase difference as expressed in radians or degrees. For example, we do *not* take 0.8 rad from the actual phase difference of 17.8 rad.

(b) If the rays of the waves were angled slightly so that the waves reached the same point on a distant viewing screen, what type of interference would the waves produce at that point?

SOLUTION: The **Key Idea** here is to compare the effective phase difference of the waves with the phase differences that give the extreme types of interference. Here the effective phase difference of 0.84 wavelength is between 0.5 wavelength (for fully destructive interference, or the darkest possible result) and 1.0 wavelength (for fully constructive interference, or the brightest possible result), but closer to 1.0 wavelength. Thus, the waves would produce intermediate interference that is closer to fully constructive interference—they would produce a relatively bright spot.

✔CHECKPOINT 2: The light waves of the rays in Fig. 36-3 have the same wavelength and amplitude and are initially in phase. (a) If 7.60 wavelengths fit within the length of the top material and 5.50 wavelengths fit within that of the bottom material, which material has the greater index of refraction? (b) If the rays are angled slightly so that they meet at the same point on a distant screen, will the interference there result in the brightest possible illumination, bright intermediate illumination, dark intermediate illumination, or darkness?

36-3 Diffraction

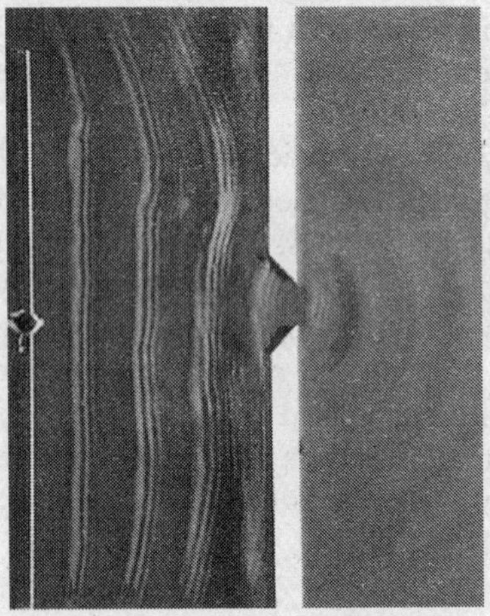

Fig. 36-4 The diffraction of water waves in a ripple tank. The waves are produced by an oscillating paddle at the left. As they move from left to right, they flare out through an opening in a barrier along the water surface.

In the next section we shall discuss the experiment that first proved that light is a wave. To prepare for that discussion, we must introduce the idea of **diffraction** of waves, a phenomenon that we explore much more fully in Chapter 37. Its essence is this: If a wave encounters a barrier that has an opening of dimensions similar to the wavelength, the part of the wave that passes through the opening will flare (spread) out —will *diffract*—into the region beyond the barrier. The flaring is consistent with the spreading of wavelets in the Huygens construction of Fig. 36-1. Diffraction occurs for waves of all types, not just light waves; Fig. 36-4 shows the diffraction of water waves traveling across the surface of water in a shallow tank.

Figure 36-5*a* shows the situation schematically for an incident plane wave of wavelength λ encountering a slit that has width $a = 6.0\lambda$ and extends into and out of the page. The wave flares out on the far side of the slit. Figures 36-5*b* (with $a = 3.0\lambda$) and 36-5*c* ($a = 1.5\lambda$) illustrate the main feature of diffraction: the narrower the slit, the greater the diffraction.

Diffraction limits geometrical optics, in which we represent an electromagnetic wave with a ray. If we actually try to form a ray by sending light through a narrow slit, or through a series of narrow slits, diffraction will always defeat our effort because it always causes the light to spread. Indeed, the narrower we make the slits (in the hope of producing a narrower beam), the greater the spreading is. Thus, geometrical optics holds only when slits or other apertures that might be located in the path of light do not have dimensions comparable to or smaller than the wavelength of the light.

36-4 Young's Interference Experiment

In 1801, Thomas Young experimentally proved that light is a wave, contrary to what most other scientists then thought. He did so by demonstrating that light undergoes interference, as do water waves, sound waves, and waves of all other types. In addition, he was able to measure the average wavelength of sunlight; his value, 570 nm, is impressively close to the modern accepted value of 555 nm. We shall here examine Young's experiment as an example of the interference of light waves.

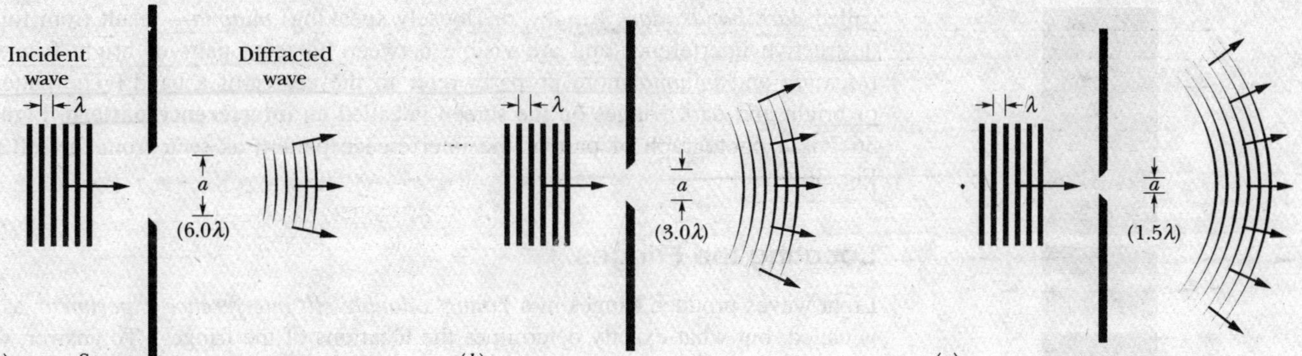

Fig. 36-5 Diffraction represented schematically. For a given wavelength λ, the diffraction is more pronounced the smaller the slit width a. The figures show the cases for (*a*) slit width $a = 6.0\lambda$, (*b*) slit width $a = 3.0\lambda$, and (*c*) slit width $a = 1.5\lambda$. In all three cases, the screen and the length of the slit extend well into and out of the page, perpendicular to it.

Figure 36-6 gives the basic arrangement of Young's experiment. Light from a distant monochromatic source illuminates slit S_0 in screen A. The emerging light then spreads via diffraction to illuminate two slits S_1 and S_2 in screen B. Diffraction of the light by these two slits sends overlapping circular waves into the region beyond screen B, where the waves from one slit interfere with the waves from the other slit.

The "snapshot" of Fig. 36-6 depicts the interference of the ovelapping waves. However, we cannot see evidence for the interference except where a viewing screen C intercepts the light. Where it does so, points of interference maxima form visible bright rows—called *bright bands, bright fringes*, or (loosely speaking) *maxima*—that extend across the screen (into and out of the page in Fig. 36-6). Dark regions—

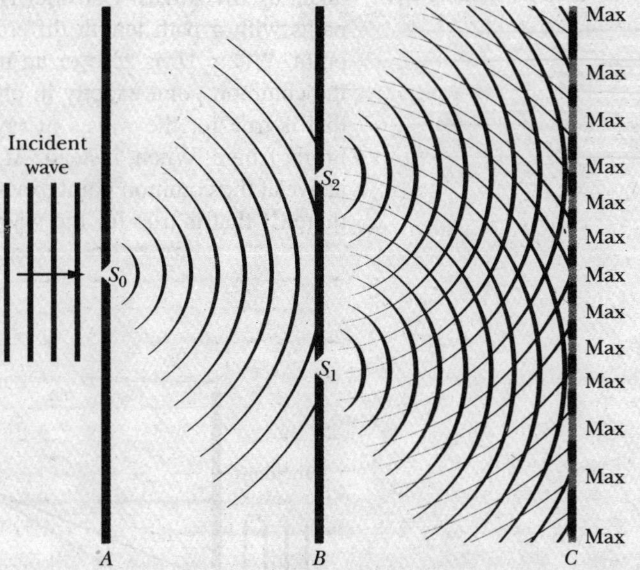

Fig. 36-6 In Young's interference experiment, incident monochromatic light is diffracted by slit S_0, which then acts as a point source of light that emits semicircular wavefronts. As that light reaches screen B, it is diffracted by slits S_1 and S_2, which then act as two point sources of light. The light waves traveling from slits S_1 and S_2 overlap and undergo interference, forming an interference pattern of maxima and minima on viewing screen C. This figure is a cross section; the screens, slits, and interference pattern extend into and out of the page. Between screens B and C, the semicircular wavefronts centered on S_2 depict the waves that would be there if only S_2 were open. Similarly, those centered on S_1 depict waves that would be there if only S_1 were open.

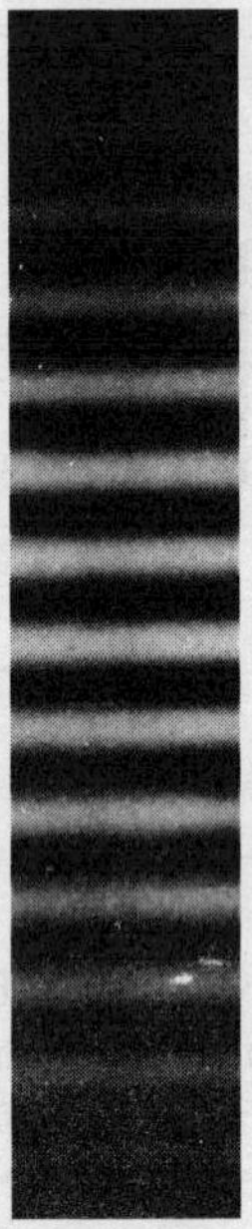

Fig. 36-7 A photograph of the interference pattern produced by the arrangement shown in Fig. 36-6. (The photograph is a front view of part of screen *C*.) The alternating maxima and minima are called *interference fringes* (because they resemble the decorative fringe sometimes used on clothing and rugs).

called *dark bands, dark fringes,* or (loosely speaking) *minima*—result from fully destructive interference and are visible between adjacent pairs of bright fringes. (*Maxima* and *minima* more properly refer to the center of a band.) The pattern of bright and dark fringes on the screen is called an **interference pattern.** Figure 36-7 is a photograph of part of the interference pattern as seen from the left in Fig. 36-6.

Locating the Fringes

Light waves produce fringes in a *Young's double-slit interference experiment,* as it is called, but what exactly determines the locations of the fringes? To answer, we shall use the arrangement in Fig. 36-8*a*. There, a plane wave of monochromatic light is incident on two slits S_1 and S_2 in screen *B*; the light diffracts through the slits and produces an interference pattern on screen *C*. We draw a central axis from the point halfway between the slits to screen *C* as a reference. We then pick, for discussion, an arbitrary point *P* on the screen, at angle θ to the central axis. This point intercepts the wave of ray r_1 from the bottom slit and the wave of ray r_2 from the top slit.

These waves are in phase when they pass through the two slits because there they are just portions of the same incident wave. However, once they have passed the slits, the two waves must travel different distances to reach *P*. We saw a similar situation in Section 18-4 with sound waves and concluded that

> The phase difference between two waves can change if the waves travel paths of different lengths.

The change in phase difference is due to the *path length difference* ΔL in the paths taken by the waves. Consider two waves initially exactly in phase, traveling along paths with a path length difference ΔL, and then passing through some common point. When ΔL is zero or an integer number of wavelengths, the waves arrive at the common point exactly in phase and they interfere fully constructively there. If that is true for the waves of rays r_1 and r_2 in Fig. 36-8, then point *P* is part of a bright fringe. When, instead, ΔL is an odd multiple of half a wavelength, the waves arrive at the common point exactly out of phase and they interfere fully destructively there. If that is true for the waves of rays r_1 and r_2, then point *P* is part of a dark

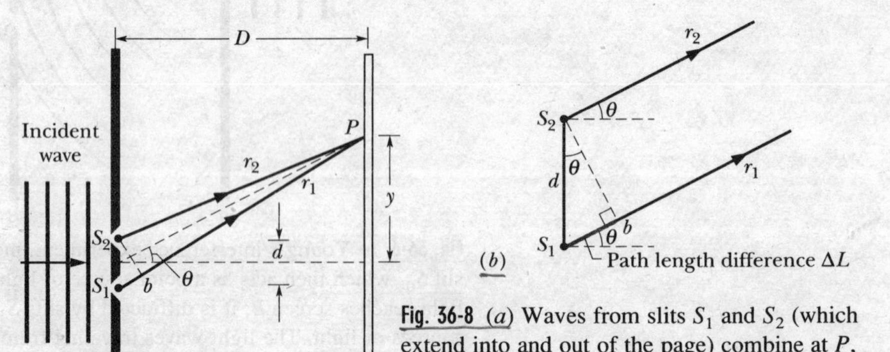

Fig. 36-8 (*a*) Waves from slits S_1 and S_2 (which extend into and out of the page) combine at *P*, an arbitrary point on screen *C* at distance *y* from the central axis. The angle θ serves as a convenient locator for *P*. (*b*) For $D \gg d$, we can approximate rays r_1 and r_2 as being parallel, at angle θ to the central axis.

fringe. (And, of course, we can have intermediate situations of interference and thus intermediate illumination at P.) Thus,

> What appears at each point on the viewing screen in a Young's double-slit interference experiment is determined by the path length difference ΔL of the rays reaching that point.

We can specify where each bright or dark fringe is located on the screen by giving the angle θ from the central axis to that fringe. To find θ, we must relate it to ΔL. We start with Fig. 36-8*a* by finding a point b along ray r_1 such that the path length from b to P equals the path length from S_2 to P. Then the path length difference ΔL between the two rays is the distance from S_1 to b.

The relation between this S_1-to-b distance and θ is complicated, but we can simplify it considerably if we arrange for the distance D from the slits to the screen to be much greater than the slit separation d. Then we can approximate rays r_1 and r_2 as being parallel to each other and at angle θ to the central axis (Fig. 36-8*b*). We can also approximate the triangle formed by S_1, S_2, and b as being a right triangle, and approximate the angle inside that triangle at S_2 as being θ. Then, for that triangle, $\sin\theta = \Delta L/d$ and thus

$$\Delta L = d \sin\theta \qquad \text{(path length difference).} \tag{36-12}$$

For a bright fringe, we saw that ΔL must be zero or an integer number of wavelengths. Using Eq. 36-12, we can write this requirement as

$$\Delta L = d \sin\theta = (\text{integer})(\lambda), \tag{36-13}$$

or as

$$d \sin\theta = m\lambda, \quad \text{for } m = 0, 1, 2, \ldots \qquad \text{(maxima—bright fringes).} \tag{36-14}$$

For a dark fringe, ΔL must be an odd multiple of half a wavelength. Again using Eq. 36-12, we can write this requirement as

$$\Delta L = d \sin\theta = (\text{odd number})(\tfrac{1}{2}\lambda), \tag{36-15}$$

or as

$$d \sin\theta = (m + \tfrac{1}{2})\lambda, \quad \text{for } m = 0, 1, 2, \ldots \qquad \text{(minima—dark fringes).} \tag{36-16}$$

With Eqs. 36-14 and 36-16, we can find the angle θ to any fringe and thus locate that fringe; further, we can use the values of m to label the fringes. For the value and label $m = 0$, Eq. 36-14 tells us that a bright fringe is at $\theta = 0$—that is, on the central axis. This *central maximum* is the point at which waves arriving from the two slits have a path length difference $\Delta L = 0$, hence zero phase difference.

For, say, $m = 2$, Eq. 36-14 tells us that *bright* fringes are at the angle

$$\theta = \sin^{-1}\left(\frac{2\lambda}{d}\right)$$

above and below the central axis. Waves from the two slits arrive at these two fringes with $\Delta L = 2\lambda$ and with a phase difference of two wavelengths. These fringes are said to be the *second-order fringes* (meaning $m = 2$) or the *second side maxima* (the second maxima to the side of the central maximum), or they are described as being the second fringes from the central maximum.

For $m = 1$, Eq. 36-16 tells us that *dark* fringes are at the angle

$$\theta = \sin^{-1}\left(\frac{1.5\lambda}{d}\right)$$

above and below the central axis. Waves from the two slits arrive at these two fringes with $\Delta L = 1.5\lambda$ and with a phase difference, in wavelengths, of 1.5. These fringes are called the *second dark fringes* or *second minima* because they are the second dark fringes from the central axis. (The first dark fringes, or first minima, are at locations for which $m = 0$ in Eq. 36-16.)

We derived Eqs. 36-14 and 36-16 for the situation $D \gg d$. However, they also apply if we place a converging lens between the slits and the viewing screen and then move the viewing screen closer to the slits, to the focal point of the lens. (The screen is then said to be in the *focal plane* of the lens; that is, it is in the plane perpendicular to the central axis at the focal point.) One property of a converging lens is that it focuses all rays that are parallel to one another to the same point on its focal plane. Thus, the rays that now arrive at any point on the screen (in the focal plane) were exactly parallel (rather than approximately) when they left the slits. They are like the initially parallel rays in Fig. 35-12*a* that are directed to a point (the focal point) by a lens.

CHECKPOINT 3: In Fig. 36-8*a*, what are ΔL (as a multiple of the wavelength) and the phase difference (in wavelengths) for the two rays if point P is (a) a third side maximum and (b) a third minimum?

Sample Problem 36-2

What is the distance on screen C in Fig. 36-8*a* between adjacent maxima near the center of the interference pattern? The wavelength λ of the light is 546 nm, the slit separation d is 0.12 mm, and the slit–screen separation D is 55 cm. Assume that θ in Fig. 36-8 is small enough to permit use of the approximations $\sin\theta \approx \tan\theta \approx \theta$, in which θ is expressed in radian measure.

SOLUTION: First, let us pick a maximum with a low value of m to ensure that it is near the center of the pattern. Then one **Key Idea** is that, from the geometry of Fig. 36-8*a*, the maximum's vertical distance y_m from the center of the pattern is related to its angle θ from the central axis by

$$\tan\theta \approx \theta = \frac{y_m}{D}.$$

A second **Key Idea** is that, from Eq. 36-14, this angle θ for the mth maximum is given by

$$\sin\theta \approx \theta = \frac{m\lambda}{d}.$$

If we equate these two expressions for θ and solve for y_m, we find

$$y_m = \frac{m\lambda D}{d}. \qquad (36\text{-}17)$$

For the next farther out maximum, we have

$$y_{m+1} = \frac{(m+1)\lambda D}{d}. \qquad (36\text{-}18)$$

We find the distance between these adjacent maxima by subtracting Eq. 36-17 from Eq. 36-18:

$$\Delta y = y_{m+1} - y_m = \frac{\lambda D}{d}$$

$$= \frac{(546 \times 10^{-9}\ \text{m})(55 \times 10^{-2}\ \text{m})}{0.12 \times 10^{-3}\ \text{m}}$$

$$= 2.50 \times 10^{-3}\ \text{m} \approx 2.5\ \text{mm}. \qquad \text{(Answer)}$$

As long as d and θ in Fig. 36-8*a* are small, the separation of the interference fringes is independent of m; that is, the fringes are evenly spaced.

36-5 Coherence

For the interference pattern to appear on viewing screen C in Fig. 36-6, the light waves reaching any point P on the screen must have a phase difference that does not vary in time. That is the case in Fig. 36-6, because the waves passing through slits S_1 and S_2 are portions of the single light wave that illuminates the slits. Because the phase difference remains constant, the light from slits S_1 and S_2 is said to be completely **coherent**.

Direct sunlight is partially coherent; that is, sunlight waves intercepted at two points have a constant phase difference only if the points are very close. If you look closely at your fingernail in bright sunlight, you can see a faint interference pattern called *speckle* that causes the nail to appear to be covered with specks. You see this effect because light waves scattering from very close points on the nail are sufficiently coherent to interfere with one another at your eye. The slits in a double-slit experiment, however, are not close enough, and in direct sunlight, the light at the slits would be **incoherent.** To get coherent light, we would have to send the sunlight through a single slit as in Fig. 36-6; because that single slit is small, light that passes through it is coherent. In addition, the smallness of the slit causes the coherent light to spread via diffraction to illuminate both slits in the double-slit experiment.

If we replace the double slits with two similar but independent monochromatic light sources, such as two fine incandescent wires, the phase difference between the waves emitted by the sources varies rapidly and randomly. (This occurs because the light is emitted by vast numbers of atoms in the wires, acting randomly and independently for extremely short times—of the order of nanoseconds.) As a result, at any given point on the viewing screen, the interference between the waves from the two sources varies rapidly and randomly between fully constructive and fully destructive. The eye (and most common optical detectors) cannot follow such changes, and no interference pattern can be seen. The fringes disappear, and the screen is seen as being uniformly illuminated.

A *laser* differs from common light sources in that its atoms emit light in a cooperative manner, thereby making the light coherent. Moreover, the light is almost monochromatic, is emitted in a thin beam with little spreading, and can be focused to a width that almost matches the wavelength of the light.

36-6 Intensity in Double-Slit Interference

Equations 36-14 and 36-16 tell us how to locate the maxima and minima of the double-slit interference pattern on screen C of Fig. 36-8 as a function of the angle θ in that figure. Here we wish to derive an expression for the intensity I of the fringes as a function of θ.

The light leaving the slits is in phase. However, let us assume that the light waves from the two slits are not in phase when they arrive at point P. Instead, the electric field components of those waves at point P are not in phase and vary with time as

$$E_1 = E_0 \sin \omega t \tag{36-19}$$

and

$$E_2 = E_0 \sin(\omega t + \phi), \tag{36-20}$$

where ω is the angular frequency of the waves and ϕ is the phase constant of wave E_2. Note that the two waves have the same amplitude E_0 and a phase difference of ϕ. Because that phase difference does not vary, the waves are coherent. We shall show that these two waves will combine at P to produce an intensity I given by

$$I = 4I_0 \cos^2 \tfrac{1}{2}\phi, \tag{36-21}$$

and that

$$\phi = \frac{2\pi d}{\lambda} \sin \theta. \tag{36-22}$$

In Eq. 36-21, I_0 is the intensity of the light that arrives on the screen from one slit when the other slit is temporarily covered. We assume that the slits are so narrow in comparison to the wavelength that this single-slit intensity is essentially uniform over the region of the screen in which we wish to examine the fringes.

Equations 36-21 and 36-22, which together tell us how the intensity I of the fringe pattern varies with the angle θ in Fig. 36-8, necessarily contain information about the location of the maxima and minima. Let us see if we can extract that information, to find equations about those locations.

Study of Eq. 36-21 shows that intensity maxima will occur when

$$\tfrac{1}{2}\phi = m\pi, \qquad \text{for } m = 0, 1, 2, \ldots . \tag{36-23}$$

If we put this result into Eq. 36-22, we find

$$2m\pi = \frac{2\pi d}{\lambda}\sin\theta, \qquad \text{for } m = 0, 1, 2, \ldots$$

or

$$d\sin\theta = m\lambda, \qquad \text{for } m = 0, 1, 2, \ldots \quad \text{(maxima)}, \tag{36-24}$$

which is exactly Eq. 36-14, the expression that we derived earlier for the locations of the maxima.

The minima in the fringe pattern occur when

$$\tfrac{1}{2}\phi = (m + \tfrac{1}{2})\pi, \qquad \text{for } m = 0, 1, 2, \ldots .$$

If we combine this relation with Eq. 36-22, we are led at once to

$$d\sin\theta = (m + \tfrac{1}{2})\lambda \qquad \text{for } m = 0, 1, 2, \ldots \quad \text{(minima)}, \tag{36-25}$$

which is just Eq. 36-16, the expression we derived earlier for the locations of the fringe minima.

Figure 36-9, which is a plot of Eq. 36-21, shows the intensity of double-slit interference patterns as a function of the phase difference ϕ between the waves at the screen. The horizontal solid line is I_0, the (uniform) intensity on the screen when one of the slits is covered up. Note in Eq. 36-21 and the graph that the intensity I varies from zero at the fringe minima to $4I_0$ at the fringe maxima.

If the waves from the two sources (slits) were *incoherent*, so that no enduring phase relation existed between them, there would be no fringe pattern and the intensity would have the uniform value $2I_0$ for all points on the screen; the horizontal dashed line in Fig. 36-9 shows this uniform value.

Interference cannot create or destroy energy but merely redistributes it over the screen. Thus, the *average* intensity on the screen must be the same $2I_0$ regardless of whether the sources are coherent. This follows at once from Eq. 36-21; if we substitute $\frac{1}{2}$, the average value of the cosine-squared function, this equation reduces to $I_{avg} = 2I_0$.

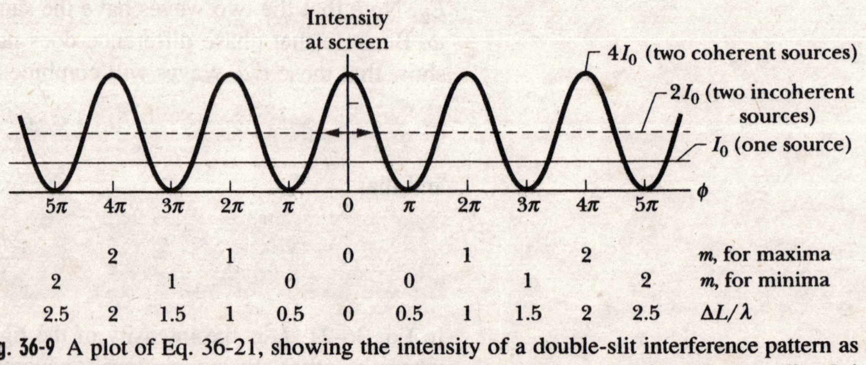

Fig. 36-9 A plot of Eq. 36-21, showing the intensity of a double-slit interference pattern as a function of the phase difference between the waves when they arrive from the two slits. I_0 is the (uniform) intensity that would appear on the screen if one slit were covered. The average intensity of the fringe pattern is $2I_0$, and the *maximum* intensity (for coherent light) is $4I_0$.

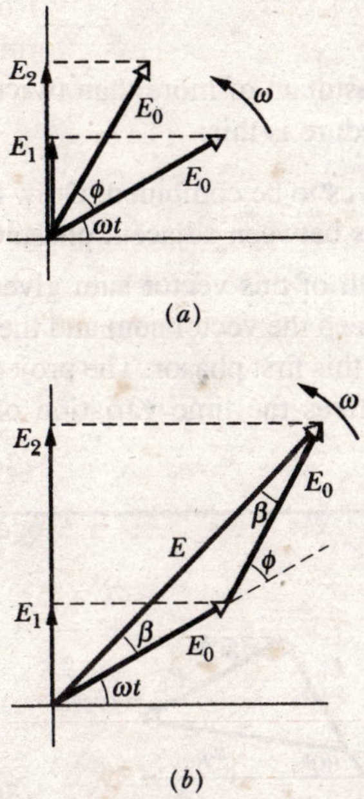

Fig. 36-10 (*a*) Phasors representing, at time t, the electric field components given by Eqs. 36-19 and 36-20. Both phasors have magnitude E_0 and rotate with angular speed ω. Their phase difference is ϕ. (*b*) Vector addition of the two phasors gives the phasor representing the resultant wave, with amplitude E and phase constant β.

Proof of Eqs. 36-21 and 36-22

We shall combine the electric field components E_1 and E_2, given by Eqs. 36-19 and 36-20, respectively, by the method of phasors as is discussed in Section 17-10. In Fig. 36-10*a*, the waves with components E_1 and E_2 are represented by phasors of magnitude E_0 that rotate around the origin at angular speed ω. The values of E_1 and E_2 at any time are the projections of the corresponding phasors on the vertical axis. Figure 36-10*a* shows the phasors and their projections at an arbitrary time t. Consistent with Eqs. 36-19 and 36-20, the phasor for E_1 has a rotation angle ωt and the phasor for E_2 has a rotation angle $\omega t + \phi$.

To combine the field components E_1 and E_2 at any point P in Fig. 36-8, we add their phasors vectorially, as shown in Fig. 36-10*b*. The magnitude of the vector sum is the amplitude E of the resultant wave at point P, and that wave has a certain phase constant β. To find the amplitude E in Fig. 36-10*b*, we first note that the two angles marked β are equal because they are opposite equal-length sides of a triangle. From the theorem (for triangles) that an exterior angle (here ϕ, as shown in Fig. 36-10*b*) is equal to the sum of the two opposite interior angles (here that sum is $\beta + \beta$), we see that $\beta = \frac{1}{2}\phi$. Thus, we have

$$E = 2(E_0 \cos \beta)$$
$$= 2E_0 \cos \tfrac{1}{2}\phi. \tag{36-26}$$

If we square each side of this relation, we obtain

$$E^2 = 4E_0^2 \cos^2 \tfrac{1}{2}\phi. \tag{36-27}$$

Now, from Eq. 34-24, we know that the intensity of an electromagnetic wave is proportional to the square of its amplitude. Therefore, the waves we are combining in Fig. 36-10*b*, whose amplitudes are E_0, each has an intensity I_0 that is proportional to E_0^2, and the resultant wave, with amplitude E, has an intensity I that is proportional to E^2. Thus,

$$\frac{I}{I_0} = \frac{E^2}{E_0^2}.$$

Substituting Eq. 36-27 into this equation and rearranging then yield

$$I = 4I_0 \cos^2 \tfrac{1}{2}\phi,$$

which is Eq. 36-21, which we set out to prove.

It remains to prove Eq. 36-22, which relates the phase difference ϕ between the waves arriving at any point P on the screen of Fig. 36-8 to the angle θ that serves as a locator of that point.

The phase difference ϕ in Eq. 36-20 is associated with the path length difference S_1b in Fig. 36-8*b*. If S_1b is $\frac{1}{2}\lambda$, then ϕ is π; if S_1b is λ, then ϕ is 2π, and so on. This suggests

$$\begin{pmatrix}\text{phase}\\ \text{difference}\end{pmatrix} = \frac{2\pi}{\lambda}\begin{pmatrix}\text{path length}\\ \text{difference}\end{pmatrix}. \tag{36-28}$$

The path length difference S_1b in Fig. 36-8*b* is $d \sin \theta$, so Eq. 36-28 becomes

$$\phi = \frac{2\pi d}{\lambda} \sin \theta,$$

which is Eq. 36-22, the other equation that we set out to prove.

Combining More Than Two Waves

In a more general case, we might want to find the resultant of more than two sinusoidally varying waves at a point. The general procedure is this:

1. Construct a series of phasors representing the waves to be combined. Draw them end to end, maintaining the proper phase relations between adjacent phasors.
2. Construct the vector sum of this array. The length of this vector sum gives the amplitude of the resultant phasor. The angle between the vector sum and the first phasor is the phase of the resultant with respect to this first phasor. The projection of this vector-sum phasor on the vertical axis gives the time variation of the resultant wave.

Sample Problem 36-3

Three light waves combine at a certain point where their electric field components are

$$E_1 = E_0 \sin \omega t,$$
$$E_2 = E_0 \sin(\omega t + 60°),$$
$$E_3 = E_0 \sin(\omega t - 30°).$$

Find their resultant component $E(t)$ at that point.

SOLUTION: The resultant wave is

$$E(t) = E_1(t) + E_2(t) + E_3(t).$$

The **Key Idea** here is a two-fold idea: We can use the method of phasors to find this sum and we are free to evaluate the phasors at any time t. To simplify the solution we choose $t = 0$, for which the phasors representing the three waves are shown in Fig. 36-11. We can add these three phasors either directly on a vector-capable calculator or by components. For the component approach, we first write the sum of their horizontal components as

$$\sum E_h = E_0 \cos 0 + E_0 \cos 60° + E_0 \cos(-30°) = 2.37E_0.$$

The sum of their vertical components, which is the value of E at $t = 0$, is

$$\sum E_v = E_0 \sin 0 + E_0 \sin 60° + E_0 \sin(-30°) = 0.366E_0.$$

The resultant wave $E(t)$ thus has an amplitude E_R of

$$E_R = \sqrt{(2.37E_0)^2 + (0.366E_0)^2} = 2.4E_0,$$

and a phase angle β relative to the phasor representing E_1 of

$$\beta = \tan^{-1}\left(\frac{0.366E_0}{2.37E_0}\right) = 8.8°.$$

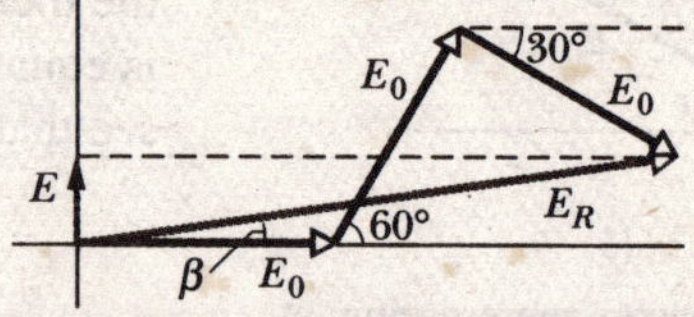

Fig. 36-11 Sample Problem 36-3. Three phasors, representing waves with equal amplitudes E_0 and with phase constants 0°, 60°, and −30°, shown at time $t = 0$. The phasors combine to give a resultant phasor with magnitude E_R, at angle β.

We can now write, for the resultant wave $E(t)$,

$$E = E_R \sin(\omega t + \beta)$$
$$= 2.4E_0 \sin(\omega t + 8.8°). \quad \text{(Answer)}$$

Be careful to interpret the angle β correctly in Fig. 36-11: It is the constant angle between E_R and the phasor representing E_1 as the four phasors rotate as a single unit around the origin. The angle between E_R and the horizontal axis in Fig. 36-11 does not remain equal to β.

✔CHECKPOINT 4: Each of four pairs of light waves arrives at a certain point on a screen. The waves have the same wavelength. At the arrival point, their amplitudes and phase differences are (a) $2E_0$, $6E_0$, and π rad; (b) $3E_0$, $5E_0$, and π rad; (c) $9E_0$, $7E_0$, and 3π rad; (d) $2E_0$, $2E_0$, and 0 rad. Rank the four pairs according to the intensity of the light at those points, greatest first. (*Hint:* Draw phasors.)

36-7 Interference from Thin Films

The colors we see when sunlight illuminates a soap bubble or an oil slick are caused by the interference of light waves reflected from the front and back surfaces of a thin transparent film. The thickness of the soap or oil film is typically of the order of magnitude of the wavelength of the (visible) light involved. (Greater thicknesses spoil the coherence of the light needed to produce the colors.)

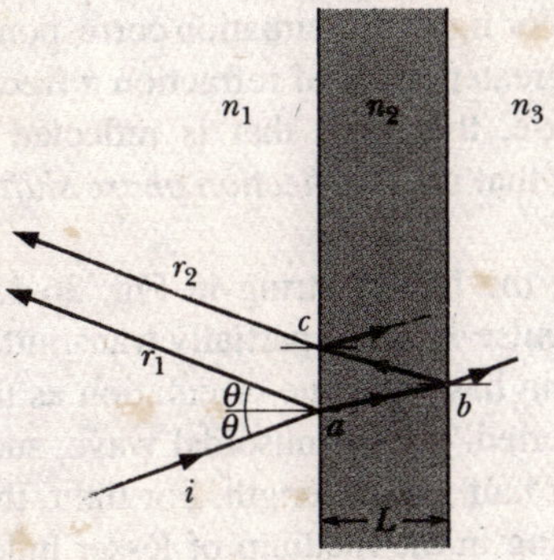

Fig. 36-12 Light waves, represented with ray i, are incident on a thin film of thickness L and index of refraction n_2. Rays r_1 and r_2 represent light waves that have been reflected by the front and back surfaces of the film, respectively. (All three rays are actually nearly perpendicular to the film.) The interference of the waves of r_1 and r_2 with each other depends on their phase difference. The index of refraction n_1 of the medium at the left can differ from the index of refraction n_3 of the medium at the right, but for now we assume that both media are air, with $n_1 = n_3 = 1.0$, which is less than n_2.

Figure 36-12 shows a thin transparent film of uniform thickness L and index of refraction n_2, illuminated by bright light of wavelength λ from a distant point source. For now, we assume that air lies on both sides of the film and thus that $n_1 = n_3$ in Fig. 36-12. For simplicity, we also assume that the light rays are almost perpendicular to the film ($\theta \approx 0$). We are interested in whether the film is bright or dark to an observer viewing it almost perpendicularly. (Since the film is brightly illuminated, how could it possibly be dark? You will see.)

The incident light, represented by ray i, intercepts the front (left) surface of the film at point a and undergoes both reflection and refraction there. The reflected ray r_1 is intercepted by the observer's eye. The refracted light crosses the film to point b on the back surface, where it undergoes both reflection and refraction. The light reflected at b crosses back through the film to point c, where it undergoes both reflection and refraction. The light refracted at c, represented by ray r_2, is intercepted by the observer's eye.

If the light waves of rays r_1 and r_2 are exactly in phase at the eye, they produce an interference maximum, and region ac on the film is bright to the observer. If they are exactly out of phase, they produce an interference minimum, and region ac is dark to the observer, *even though it is illuminated*. If there is some intermediate phase difference, there are intermediate interference and intermediate brightness.

Thus, the key to what the observer sees is the phase difference between the waves of rays r_1 and r_2. Both rays are derived from the same ray i, but the path involved in producing r_2 involves light traveling twice across the film (a to b, and then b to c), whereas the path involved in producing r_1 involves no travel through the film. Because θ is about zero, we approximate the path length difference between the waves of r_1 and r_2 as $2L$. However, to find the phase difference between the waves, we cannot just find the number of wavelengths λ that is equivalent to a path length difference of $2L$. This simple approach is impossible for two reasons: (1) the path length difference occurs in a medium other than air, and (2) reflections are involved, which can change the phase.

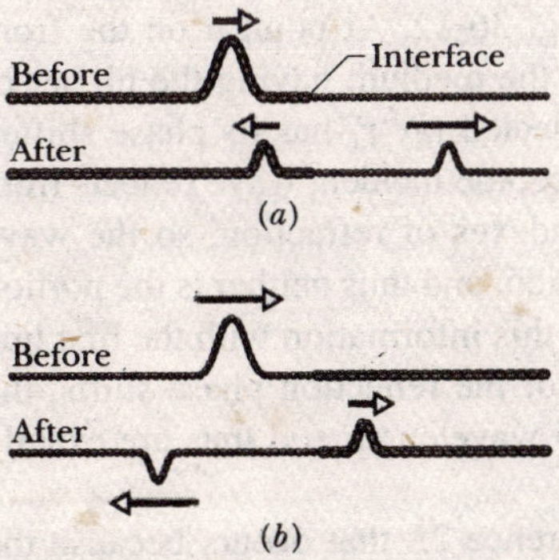

Fig. 36-13 Phase changes when a pulse is reflected at the interface between two stretched strings of different linear densities. The wave speed is greater in the lighter string. (*a*) The incident pulse is in the denser string. (*b*) The incident pulse is in the lighter string. Only here is there a phase change, and only in the reflected wave.

> The phase difference between two waves can change if one or both are reflected.

Before we continue our discussion of interference from thin films, we must discuss changes in phase that are caused by reflections.

Reflection Phase Shifts

Refraction at an interface never causes a phase change—but reflection can, depending on the indexes of refraction on the two sides of the interface. Figure 36-13 shows what happens when reflection causes a phase change, using as an example pulses on a denser string (along which pulse travel is relatively slow) and a lighter string (along which pulse travel is relatively fast).

When a pulse traveling relatively slowly along the denser string in Fig. 36-13a reaches the interface with the lighter string, the pulse is partially transmitted and partially reflected, with no change in orientation. For light, this situation corresponds to the incident wave traveling in the medium of greater index of refraction n (recall that greater n means slower speed). In that case, the wave that is reflected at the interface does not undergo a change in phase; that is, its *reflection phase shift* is zero.

When a pulse traveling more quickly along the lighter string in Fig. 36-13*b* reaches the interface with the denser string, the pulse is again partially transmitted and partially reflected. The transmitted pulse again has the same orientation as the incident pulse, but now the reflected pulse is inverted. For a sinusoidal wave, such an inversion involves a phase change of π rad, or half a wavelength. For light, this situation corresponds to the incident wave traveling in the medium of lesser index of refraction (with greater speed). In that case, the wave that is reflected at the interface undergoes a phase shift of π rad, or half a wavelength.

We can summarize these results for light in terms of the index of refraction of the medium off which (or from which) the light reflects:

Reflection	Reflection phase shift
Off lower index	0
Off higher index	0.5 wavelength

This might be remembered as "higher means half."

Equations for Thin-Film Interference

In this chapter we have now seen three ways in which the phase difference between two waves can change:

1. by reflection
2. by the waves traveling along paths of different lengths
3. by the waves traveling through media of different indexes of refraction

When light reflects from a thin film, producing the waves of rays r_1 and r_2 shown in Fig. 36-12, all three ways are involved. Let us consider them one by one.

We first reexamine the two reflections in Fig. 36-12. At point a on the front interface, the incident wave (in air) reflects from the medium having the higher of the two indexes of refraction, so the wave of reflected ray r_1 has its phase shifted by 0.5 wavelength. At point b on the back interface, the incident wave reflects from the medium (air) having the lower of the two indexes of refraction, so the wave reflected there is not shifted in phase by the reflection, and thus neither is the portion of it that exits the film as ray r_2. We can organize this information with the first line in Table 36-1. It tells us that, so far, as a result of the reflection phase shifts, the waves of r_1 and r_2 have a phase difference of 0.5 wavelength and thus are exactly out of phase.

Now we must consider the path length difference $2L$ that occurs because the wave of ray r_2 crosses the film twice. (This difference $2L$ is shown on the second line in Table 36-1.) If the waves of r_1 and r_2 are to be exactly in phase so that they produce fully constructive interference, the path length $2L$ must cause an additional phase difference of 0.5, 1.5, 2.5, . . . wavelengths. Only then will the net phase difference be an integer number of wavelengths. Thus, for a bright film, we must have

$$2L = \frac{\text{odd number}}{2} \times \text{wavelength} \quad \text{(in-phase waves).} \qquad (36\text{-}29)$$

TABLE 36-1 An Organizing Table for Thin-Film Interference in Air[a]

Reflection	r_1	r_2
phase shifts	0.5 wavelength	0
Path length difference	$2L$	
Index in which path length difference occurs	n_2	
In phase[a]:	$2L = \frac{\text{odd number}}{2} \times \frac{\lambda}{n_2}$	
Out of phase[a]:	$2L = \text{integer} \times \frac{\lambda}{n_2}$	

[a]Valid for $n_2 > n_1$ and $n_2 > n_3$.

The wavelength we need here is the wavelength λ_{n2} of the light in the medium containing path length $2L$—that is, in the medium with index of refraction n_2. Thus, we can rewrite Eq. 36-29 as

$$2L = \frac{\text{odd number}}{2} \times \lambda_{n2} \quad \text{(in-phase waves).} \qquad (36\text{-}30)$$

If, instead, the waves are to be exactly out of phase so that there is fully destructive interference, the path length $2L$ must cause either no additional phase difference or a phase difference of 1, 2, 3, . . . wavelengths. Only then will the net phase difference be an odd number of half-wavelengths. For a dark film, we must have

$$2L = \text{integer} \times \text{wavelength}, \qquad (36\text{-}31)$$

where, again, the wavelength is the wavelength λ_{n2} in the medium containing $2L$. Thus, this time we have

$$2L = \text{integer} \times \lambda_{n2} \quad \text{(out-of-phase waves).} \qquad (36\text{-}32)$$

Now we can use Eq. 36-8 ($\lambda_n = \lambda/n$) to write the wavelength of the wave of ray r_2 inside the film as

$$\lambda_{n2} = \frac{\lambda}{n_2}, \qquad (36\text{-}33)$$

where λ is the wavelength of the incident light in vacuum (and approximately also in air). Substituting Eq. 36-33 into Eq. 36-30 and replacing "odd number/2" with $(m + \frac{1}{2})$ give us

$$2L = (m + \tfrac{1}{2})\frac{\lambda}{n_2}, \quad \text{for } m = 0, 1, 2, \ldots \quad \text{(maxima—bright film in air).} \qquad (36\text{-}34)$$

Similarly, with m replacing "integer," Eq. 36-32 yields

$$2L = m\frac{\lambda}{n_2}, \quad \text{for } m = 0, 1, 2, \ldots \quad \text{(minima—dark film in air).} \qquad (36\text{-}35)$$

For a given film thickness L, Eqs. 36-34 and 36-35 tell us the wavelengths of light for which the film appears bright and dark, respectively, one wavelength for each value of m. Intermediate wavelengths give intermediate brightnesses. For a given wavelength λ, Eqs. 36-34 and 36-35 tell us the thicknesses of the films that appear bright and dark in that light, respectively, one thickness for each value of m. Intermediate thicknesses give intermediate brightnesses.

A special situation arises when a film is so thin that L is much less than λ, say, $L < 0.1\lambda$. Then the path length difference $2L$ can be neglected, and the phase difference between r_1 and r_2 is due *only* to reflection phase shifts. If the film of Fig. 36-12, where the reflections cause a phase difference of 0.5 wavelength, has thickness $L < 0.1\lambda$, then r_1 and r_2 are exactly out of phase, and thus the film is dark, regardless of the wavelength and even the intensity of the light that illuminates it. This special situation corresponds to $m = 0$ in Eq. 36-35. We shall count any thickness $L < 0.1\lambda$ as being the least thickness specified by Eq. 36-35 to make the film of Fig. 36-12 dark. (Every such thickness will correspond to $m = 0$.) The next greater thickness that will make the film dark is that corresponding to $m = 1$.

Figure 36-14 shows a vertical soap film whose thickness increases from top to bottom because gravitation has caused the film to slump. Bright white light illumi-

Fig. 36-14 The reflection of light from a soapy water film spanning a vertical loop. The top portion is so thin that the light reflected there undergoes destructive interference, making that portion dark. Colored interference fringes, or bands, decorate the rest of the film but are marred by circulation of liquid within the film as the liquid is gradually pulled downward by gravitation.

nates the film. However, the top portion is so thin that it is dark. In the (somewhat thicker) middle we see fringes, or bands, whose color depends primarily on the wavelength at which reflected light undergoes fully constructive interference for a particular thickness. Toward the (thickest) bottom of the film the fringes become progressively narrower and the colors begin to overlap and fade.

Iridescence of a *Morpho* Butterfly Wing

A surface that displays colors due to thin-film interference is said to be *iridescent* because the tints of the colors change as you change your view of the surface. The iridescence of the top surface of a *Morpho* butterfly wing is due to thin-film interference of light reflected by thin terraces of transparent cuticle-like material on the wing. These terraces are arranged like wide, flat branches on a tree-like structure that extends perpendicular to the wing.

Suppose you look directly down on these terraces as white light shines directly down on the wing. Then the light reflected back up to you from the terraces undergoes fully constructive interference in the blue-green region of the visible spectrum. Light in the yellow and red regions, at the opposite end of the spectrum, is weaker because it undergoes only intermediate interference. Thus, the top surface of the wing looks blue-green to you.

If you intercept light that reflects from the wing in some other direction, the light has traveled along a slanted path through the terraces. Then the wavelength at which there is fully constructive interference is somewhat different from that for light reflected directly upward. Thus, if the wing moves in your view so that the angle at which you view it changes, the color at which the wing is brightest changes somewhat, producing the iridescence of the wing.

PROBLEM-SOLVING TACTICS

Tactic 1: *Thin-Film Equations*
Some students believe that Eq. 36-34 gives the maxima and Eq. 36-35 gives the minima for *all* thin-film situations. This is not true. These relations were derived only for the situation in which $n_2 > n_1$ and $n_2 > n_3$ in Fig. 36-12.

The appropriate equations for other relative values of the indexes of refraction can be derived by following the reasoning of this section and constructing new versions of Table 36-1. In each case you will end up with Eqs. 36-34 and 36-35, but sometimes Eq. 36-34 will give the minima and Eq. 36-35 will give the maxima—the opposite of what we found here. Which equation gives which depends on whether the reflections at the two interfaces give the same reflection phase shift.

CHECKPOINT 5: The figure shows four situations in which light reflects perpendicularly from a thin film of thickness L (as in Fig. 36-12), with indexes of refraction as given. (a) For which situations does reflection at the film interfaces cause a zero phase difference for the two reflected rays? (b) For which situations will the film be dark if the path length difference $2L$ causes a phase difference of 0.5 wavelength?

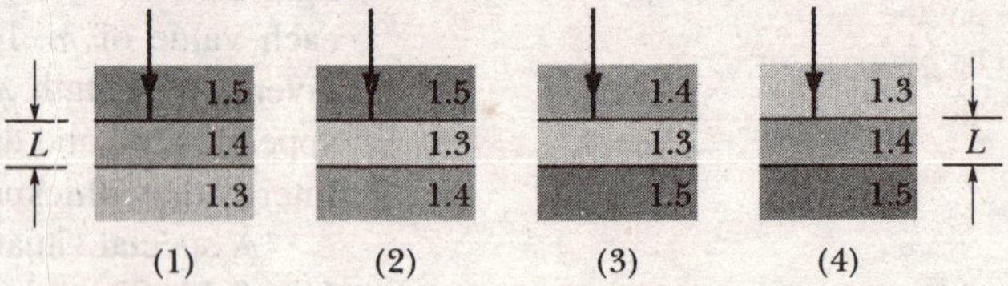

★ Sample Problem 36-4

White light, with a uniform intensity across the visible wavelength range of 400 to 690 nm, is perpendicularly incident on a water film, of index of refraction $n_2 = 1.33$ and thickness $L = 320$ nm, that is suspended in air. At what wavelength λ is the light reflected by the film brightest to an observer?

SOLUTION: The **Key Idea** here is that the reflected light from the film is brightest at the wavelengths λ for which the reflected rays are in phase with one another. The equation relating these wavelengths λ to the given film thickness L and film index of refraction n_2 is either Eq. 36-34 or Eq. 36-35, depending on the reflection phase shifts for this particular film.

To determine which equation is needed, we should fill out an organizing table like Table 36-1. However, because there is air on both sides of the water film, the situation here is exactly like that in Fig. 36-12, and thus the table would be exactly like Table 36-1. Then from Table 36-1, we see that the reflected rays are in phase (and thus the film is brightest) when

$$2L = \frac{\text{odd number}}{2} \times \frac{\lambda}{n_2},$$

which leads to Eq. 36-34:

$$2L = (m + \tfrac{1}{2})\frac{\lambda}{n_2}.$$

Solving for λ and substituting for L and n_2, we find

$$\lambda = \frac{2n_2L}{m+\frac{1}{2}} = \frac{(2)(1.33)(320\text{ nm})}{m+\frac{1}{2}} = \frac{851\text{ nm}}{m+\frac{1}{2}}.$$

For $m = 0$, this gives us $\lambda = 1700$ nm, which is in the infrared region. For $m = 1$, we find $\lambda = 567$ nm, which is yellow-green light, near the middle of the visible spectrum. For $m = 2$, $\lambda = 340$ nm, which is in the ultraviolet region. Thus, the wavelength at which the light seen by the observer is brightest is

$$\lambda = 567\text{ nm}. \quad \text{(Answer)}$$

★Sample Problem 36-5

In Fig. 36-15, a glass lens is coated on one side with a thin film of magnesium fluoride (MgF_2) to reduce reflection from the lens surface. The index of refraction of MgF_2 is 1.38; that of the glass is 1.50. What is the least coating thickness that eliminates (via interference) the reflections at the middle of the visible spectrum ($\lambda = 550$ nm)? Assume that the light is approximately perpendicular to the lens surface.

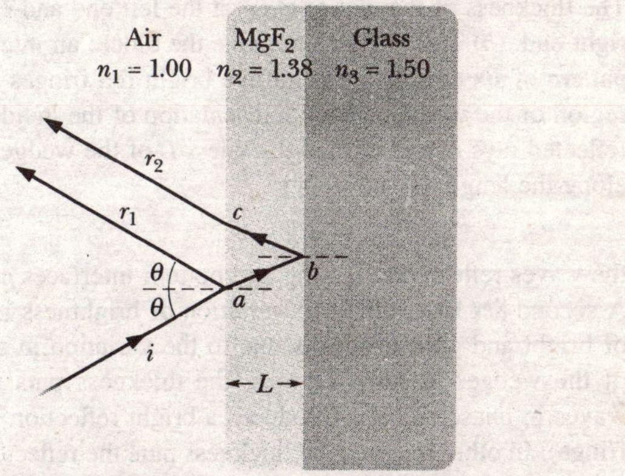

Fig. 36-15 Sample Problem 36-5. Unwanted reflections from glass can be suppressed (at a chosen wavelength) by coating the glass with a thin transparent film of magnesium fluoride of the properly chosen thickness.

SOLUTION: The **Key Idea** here is that reflection is eliminated if the film thickness L is such that light waves reflected from the two film interfaces are exactly out of phase. The equation relating L to the given wavelength λ and the index of refraction n_2 of the thin film is either Eq. 36-34 or Eq. 36-35, depending on the reflection phase shifts at the interfaces.

To determine which equation is needed, we fill out an organizing table like Table 36-1. At the first interface, the incident light is in air, which has a lesser index of refraction than the MgF_2 (the thin film). Thus, we fill in 0.5 wavelength under r_1 in our organizing table (meaning that the waves of ray r_1 are shifted by 0.5λ at the first interface). At the second interface, the incident light is in the MgF_2, which has a lesser index of refraction than the glass on the other side of the interface. Thus, we fill in 0.5 wavelength under r_2 in our table.

Because both reflections cause the same phase shift, they tend to put the waves of r_1 and r_2 in phase. Since we want those waves to be *out of phase,* their path length difference $2L$ must be an odd number of half-wavelengths:

$$2L = \frac{\text{odd number}}{2} \times \frac{\lambda}{n_2}.$$

This leads to Eq. 36-34. Solving that equation for L then gives us the film thicknesses that will eliminate reflection from the lens and coating:

$$L = (m + \tfrac{1}{2})\frac{\lambda}{2n_2}, \quad \text{for } m = 0, 1, 2, \ldots. \quad (36\text{-}36)$$

We want the least thickness for the coating—that is, the least L. Thus, we choose $m = 0$, the least possible value of m. Substituting it and the given data in Eq. 36-36, we obtain

$$L = \frac{\lambda}{4n_2} = \frac{550\text{ nm}}{(4)(1.38)} = 99.6\text{ nm}. \quad \text{(Answer)}$$

★Sample Problem 36-6

Figure 36-16*a* shows a transparent plastic block with a thin wedge of air at the right. (The wedge thickness is exaggerated in the figure.) A broad beam of red light, with wavelength $\lambda = 632.8$ nm, is directed downward through the top of the block (at an incidence angle of 0°). Some of the light is reflected back up from the top and bottom surfaces of the wedge, which acts as a thin film (of air) with a thickness that varies uniformly and gradually from L_L at the left-hand end to L_R at the right-hand end. (The plastic layers above and below the wedge of air are too thick to act as thin films.) An observer looking down on the block sees an interference pattern consisting of six dark fringes and five bright red fringes along the wedge. What is the change in thickness ΔL ($= L_R - L_L$) along the wedge?

SOLUTION: One **Key Idea** here is that the brightness at any point along the left–right length of the air wedge is due to the interference of

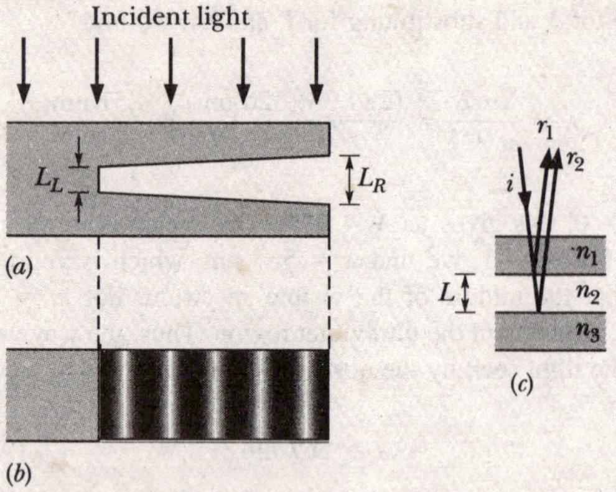

Fig. 36-16 Sample Problem 36-6. (*a*) Red light is incident on a thin, air-filled wedge in the side of a transparent plastic block. The thickness of the wedge is L_L at the left end and L_R at the right end. (*b*) The view from above the block: an interference pattern of six dark fringes and five bright red fringes lies over the region of the wedge. (*c*) A representation of the incident ray i, reflected rays r_1 and r_2, and thickness L of the wedge anywhere along the length of the wedge.

the waves reflected at the top and bottom interfaces of the wedge. A second **Key Idea** is that the variation of brightness in the pattern of bright and dark fringes is due to the variation in the thickness of the wedge. In some regions, the thickness puts the reflected waves in phase and thus produces a bright reflection (a bright red fringe). In other regions, the thickness puts the reflected waves out of phase and thus produces no reflection (a dark fringe).

Because the observer sees more dark fringes than bright fringes, we can assume that a dark fringe is produced at both the left and right ends of the wedge. Thus, the interference pattern is that shown in Fig. 36-16*b*, which we can use to determine the change in thickness ΔL of the wedge.

Another **Key Idea** is that we can represent the reflection of light at the top and bottom interfaces of the wedge, at any point along its length, with Fig. 36-16*c*, in which L is the wedge thickness at that point. Let us apply this figure to the left end of the wedge, where the reflections give a dark fringe.

We know that, for a dark fringe, the waves of rays r_1 and r_2 in Fig. 36-16*c* must be out of phase. We also know that the equation relating the film thickness L to the light's wavelength λ and the film's index of refraction n_2 is either Eq. 36-34 or Eq. 36-35, depending on the reflection phase shifts. To determine which equation gives a dark fringe at the left end of the wedge, we should fill out an organizing table like Table 36-1.

At the top interface of the wedge, the incident light is in the plastic, which has a greater index of refraction than the air beneath that interface. Thus, we fill in 0 under r_1 in our organizing table. At the bottom interface of the wedge, the incident light is in air, which has a lesser index of refraction than the plastic beneath that interface. Thus, we fill in 0.5 wavelength under r_2 in our organizing table. Therefore, the reflections alone tend to put the waves of r_1 and r_2 out of phase.

Since the waves are, in fact, out of phase at the left end of the air wedge, the path length difference $2L$ at that end of the wedge must be given by

$$2L = \text{integer} \times \frac{\lambda}{n_2},$$

which leads to Eq. 36-35:

$$2L = m\frac{\lambda}{n_2}, \qquad \text{for } m = 0, 1, 2, \ldots. \tag{36-37}$$

Here is another **Key Idea**: Eq. 36-37 holds not only for the left end of the wedge but also at any point along the wedge where a dark fringe is observed, including the right end—with a different integer value of m for each fringe. The least value of m is associated with the least thickness of the wedge where a dark fringe is observed. Progressively greater values of m are associated with progressively greater thicknesses of the wedge where a dark fringe is observed. Let m_L be the value at the left end. Then the value at the right end must be $m_L + 5$ because, from Fig. 36-16*b*, the right end is located at the fifth dark fringe from the left end.

We want the change ΔL in thickness, from the left end to the right end of the wedge. To find it we first solve Eq. 36-37 twice—once for the thickness L_L at the left end and once for the thickness L_R at the right end:

$$L_L = (m_L)\frac{\lambda}{2n_2}, \qquad L_R = (m_L + 5)\frac{\lambda}{2n_2}. \tag{36-38}$$

To find the change in thickness ΔL, we can now subtract L_L from L_R and substitute known data, including $n_2 = 1.00$ for the air within the wedge:

$$\Delta L = L_R - L_L = \frac{(m_L + 5)\lambda}{2n_2} - \frac{m_L\lambda}{2n_2} = \frac{5}{2}\frac{\lambda}{n_2}$$

$$= \frac{5}{2}\frac{632.8 \times 10^{-9}\text{ m}}{1.00}$$

$$= 1.58 \times 10^{-6}\text{ m}. \qquad \text{(Answer)}$$

36-8 Michelson's Interferometer

An **interferometer** is a device that can be used to measure lengths or changes in length with great accuracy by means of interference fringes. We describe the form originally devised and built by A. A. Michelson in 1881.

Consider light that leaves point P on extended source S in Fig. 36-17 and encounters *beam splitter M*. A beam splitter is a mirror that transmits half the incident light and reflects the other half. In the figure we have assumed, for convenience, that

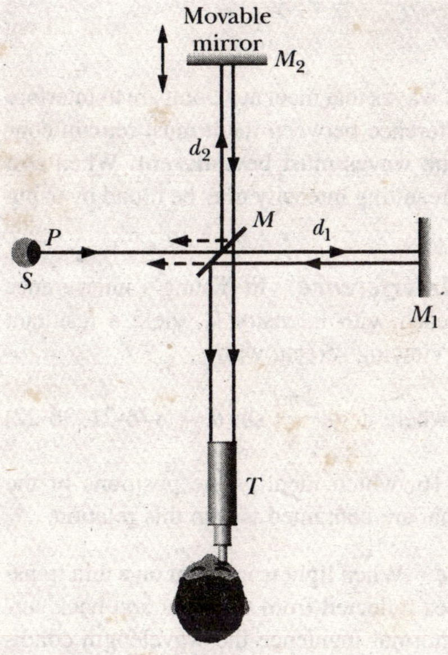

Fig. 36-17 Michelson's interferometer, showing the path of light originating at point P of an extended source S. Mirror M splits the light into two beams, which reflect from mirrors M_1 and M_2 back to M and then to telescope T. In the telescope an observer sees a pattern of interference fringes.

this mirror possesses negligible thickness. At M the light thus divides into two waves. One proceeds by transmission toward mirror M_1; the other proceeds by reflection toward mirror M_2. The waves are entirely reflected at these mirrors and are sent back along their directions of incidence, each wave eventually entering telescope T. What the observer sees is a pattern of curved or approximately straight interference fringes; in the latter case the fringes resemble the stripes on a zebra.

The path length difference for the two waves when they recombine at the telescope is $2d_2 - 2d_1$, and anything that changes this path length difference will cause a change in the phase difference between these two waves at the eye. As an example, if mirror M_2 is moved by a distance $\frac{1}{2}\lambda$, the path length difference is changed by λ and the fringe pattern is shifted by one fringe (as if each dark stripe on a zebra had moved to where the adjacent dark stripe had been). Similarly, moving mirror M_2 by $\frac{1}{4}\lambda$ causes a shift by half a fringe (each dark zebra stripe shifts to where the adjacent white stripe was).

A shift in the fringe pattern can also be caused by the insertion of a thin transparent material into the optical path of one of the mirrors, say, M_1. If the material has thickness L and index of refraction n, then the number of wavelengths along the light's to-and-fro path through the material is, from Eq. 36-9,

$$N_m = \frac{2L}{\lambda_n} = \frac{2Ln}{\lambda}. \tag{36-39}$$

The number of wavelengths in the same thickness $2L$ of air before the insertion of the material is

$$N_a = \frac{2L}{\lambda}. \tag{36-40}$$

When the material is inserted, the light returned by mirror M_1 undergoes a phase change (in terms of wavelengths) of

$$N_m - N_a = \frac{2Ln}{\lambda} - \frac{2L}{\lambda} = \frac{2L}{\lambda}(n-1). \tag{36-41}$$

For each phase change of one wavelength, the fringe pattern is shifted by one fringe. Thus, by counting the number of fringes through which the material causes the pattern to shift, and substituting that number for $N_m - N_a$ in Eq. 36-41, you can determine the thickness L of the material in terms of λ.

By such techniques the lengths of objects can be expressed in terms of the wavelengths of light. In Michelson's day, the standard of length—the meter—was chosen by international agreement to be the distance between two fine scratches on a certain metal bar preserved at Sèvres, near Paris. Michelson was able to show, using his interferometer, that the standard meter was equivalent to 1 553 163.5 wavelengths of a certain monochromatic red light emitted from a light source containing cadmium. For this careful measurement, Michelson received the 1907 Nobel prize in physics. His work laid the foundation for the eventual abandonment (in 1961) of the meter bar as a standard of length and for the redefinition of the meter in terms of the wavelength of light. By 1983, even this wavelength standard was not precise enough to meet the growing requirements of science and technology, and it was replaced with a new standard based on a defined value for the speed of light.

REVIEW & SUMMARY

Huygens' Principle The three-dimensional transmission of waves, including light, may often be predicted by *Huygens' principle,* which states that all points on a wavefront serve as point sources of spherical secondary wavelets. After a time t, the new position of the wavefront will be that of a surface tangent to these secondary wavelets.

The law of refraction can be derived from Huygens' principle by assuming that the index of refraction of any medium is $n = c/v$, in which v is the speed of light in the medium and c is the speed of light in vacuum.

Wavelength and Index of Refraction The wavelength λ_n of light in a medium depends on the index of refraction n of the medium:

$$\lambda_n = \frac{\lambda}{n}, \qquad (36\text{-}8)$$

in which λ is the wavelength of the light in vacuum. Because of this dependency, the phase difference between two waves can change if they pass through different materials with different indexes of refraction.

Young's Experiment In **Young's interference experiment,** light passing through a single slit falls on two slits in a screen. The light leaving these slits flares out (by diffraction), and interference occurs in the region beyond the screen. A fringe pattern, due to the interference, forms on a viewing screen.

The light intensity at any point on the viewing screen depends in part on the difference in the path lengths from the slits to that point. If this difference is an integer number of wavelengths, the waves interfere constructively and an intensity maximum results. If it is an odd number of half-wavelengths, there is destructive interference and an intensity minimum occurs. The conditions for maximum and minimum intensity are

$$d \sin \theta = m\lambda, \quad \text{for } m = 0, 1, 2, \ldots \quad \text{(maxima—bright fringes)}, \qquad (36\text{-}14)$$

$$d \sin \theta = (m + \tfrac{1}{2})\lambda, \quad \text{for } m = 0, 1, 2, \ldots \quad \text{(minima—dark fringes)}, \qquad (36\text{-}16)$$

where θ is the angle the light path makes with a central axis and d is the slit separation.

Coherence If two light waves that meet at a point are to interfere perceptibly, the phase difference between them must remain constant with time; that is, the waves must be **coherent.** When two coherent waves meet, the resulting intensity may be found by using phasors.

Intensity in Two-Slit Interference In Young's interference experiment, two waves, each with intensity I_0, yield a resultant wave of intensity I at the viewing screen, with

$$I = 4I_0 \cos^2 \tfrac{1}{2}\phi, \quad \text{where } \phi = \frac{2\pi d}{\lambda} \sin \theta. \qquad (36\text{-}21, 36\text{-}22)$$

Equations 36-14 and 36-16, which identify the positions of the fringe maxima and minima, are contained within this relation.

Thin-Film Interference When light is incident on a thin transparent film, the light waves reflected from the front and back surfaces interfere. For near-normal incidence the wavelength conditions for maximum and minimum intensity of the light reflected from a *film in air* are

$$2L = (m + \tfrac{1}{2})\frac{\lambda}{n_2}, \quad \text{for } m = 0, 1, 2, \ldots \quad \text{(maxima—bright film in air)}, \qquad (36\text{-}34)$$

$$2L = m\frac{\lambda}{n_2}, \quad \text{for } m = 0, 1, 2, \ldots \quad \text{(minima—dark film in air)}, \qquad (36\text{-}35)$$

where n_2 is the index of refraction of the film, L is its thickness, and λ is the wavelength of the light in air.

If the light incident at an interface between media with different indexes of refraction is in the medium with the smaller index of refraction, the reflection causes a phase change of π rad, or half a wavelength, in the reflected wave. Otherwise, there is no phase change due to the reflection. Refraction at an interface does not cause a phase shift.

The Michelson Interferometer In *Michelson's interferometer* a light wave is split into two beams that, after traversing paths of different lengths, are recombined so they interfere and form a fringe pattern. Varying the path length of one of the beams allows distances to be accurately expressed in terms of wavelengths of light, by counting the number of fringes through which the pattern shifts because of the change.

QUESTIONS

1. In Fig. 36-18, three pulses of light—a, b, and c—of the same wavelength are sent through layers of plastic whose indexes of refraction are given. Rank the pulses according to their travel time through the plastic, greatest first.

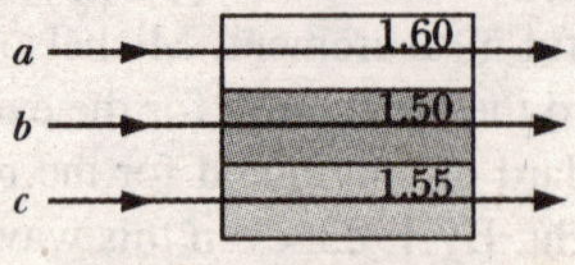

Fig. 36-18 Question 1.

2. Light travels along the length of a 1500-nm-long nanostructure. When a peak of the wave is at one end of the nanostructure, is there a peak or a valley at the other end if the wavelength is (a) 500 nm and (b) 1000 nm?

3. Figure 36-19 shows two rays of light, of wavelength 600 nm, that reflect from glass surfaces separated by 150 nm. The rays are initially in phase. (a) What is the path length difference of the rays?

(b) When they have cleared the reflection region, are the rays exactly in phase, exactly out of phase, or in some intermediate state?

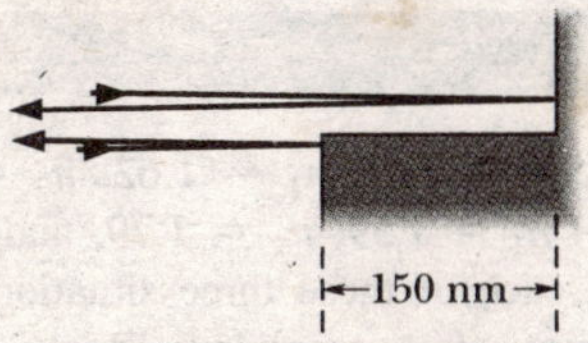

Fig. 36-19 Question 3.

4. Figure 36-20 shows two light rays that are initially exactly in phase and that reflect from several glass surfaces. Neglect the slight slant in the path of the light in the second arrangement. (a) What is the path length difference of the rays? In wavelengths λ, (b) what should that path length difference equal if the rays are to be exactly out of phase when they emerge, and (c) what is the smallest value of d that will allow that final phase difference?

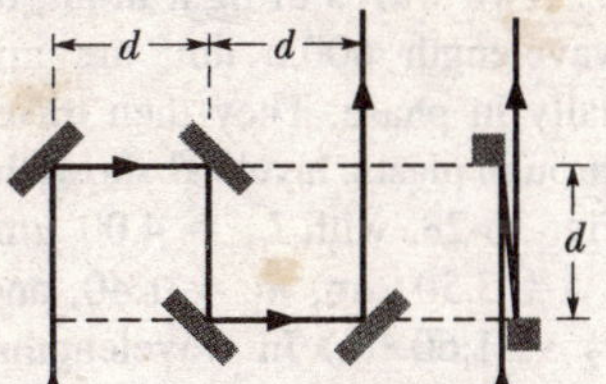

Fig. 36-20 Question 4.

5. Is there an interference maximum, a minimum, an intermediate state closer to a maximum, or an intermediate state closer to a minimum at point P in Fig. 36-8 if the path length difference of the two rays is (a) 2.2λ, (b) 3.5λ, (c) 1.8λ, and (d) 1.0λ? For each situation, give the value of m associated with the maximum or minimum involved.

6. (a) If you move from one bright fringe in a two-slit interference pattern to the next one farther out, (a) does the path length difference ΔL increase or decrease and (b) by how much does it change, in wavelengths λ?

7. Does the spacing between fringes in a two-slit interference pattern increase, decrease, or stay the same if (a) the slit separation is increased, (b) the color of the light is switched from red to blue, and (c) the whole apparatus is submerged in cooking sherry? (d) If the slits are illuminated with white light, then at any side maximum, does the blue component or the red component peak closer to the central maximum?

8. Each part of Fig. 36-21 shows phasors representing the two light waves in a double-slit interference experiment. Further, each part represents a different point on the viewing screen, at a different time. Assuming all eight phasors have the same length, rank the points according to the intensity of the light there, greatest first.

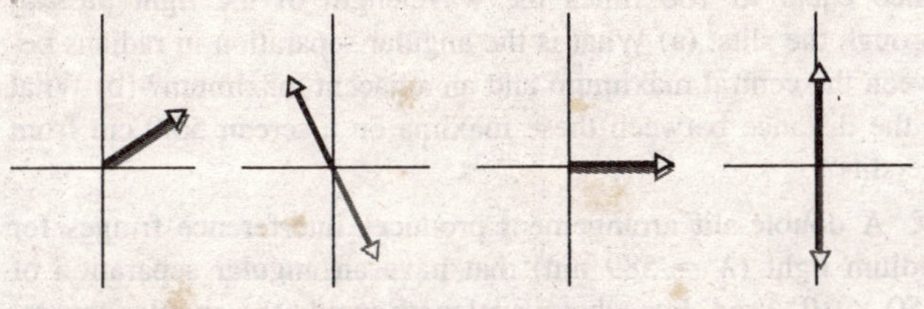

Fig. 36-21 Question 8.

9. Figure 36-22 shows two sources S_1 and S_2 that emit radio waves of wavelength λ in all directions. The sources are exactly in phase and are separated by a distance equal to 1.5λ. The vertical broken line is the perpendicular bisector of the distance between the sources. (a) If we start at the indicated start point and travel along path 1, does the interference produce a maximum all along the path, a minimum all along the path, or alternating maxima and minima? Repeat for (b) path 2 and (c) path 3.

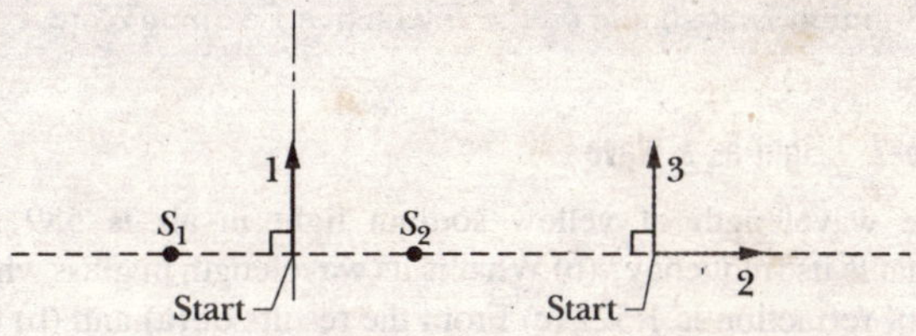

Fig. 36-22 Question 9.

10. Figure 36-23 shows two rays of light encountering interfaces, where they reflect and refract. Which of the resulting waves are shifted in phase at the interface?

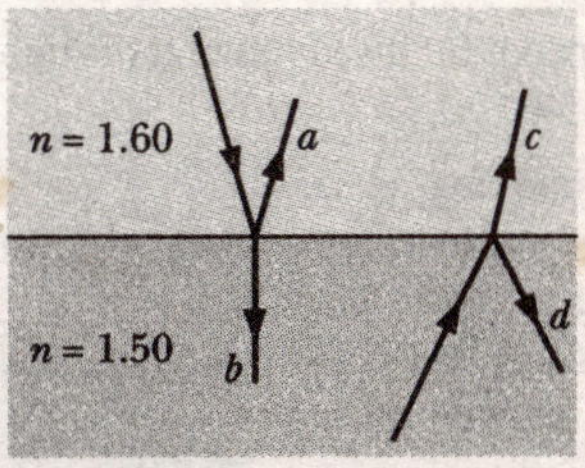

Fig. 36-23 Question 10.

11. Figure 36-24a shows the cross section of a vertical thin film whose width increases downward because gravitation causes slumping. Figure 36-24b is a face-on view of the film, showing four bright interference fringes that result when the film is illuminated with a perpendicular beam of red light. Points in the cross section corresponding to the bright fringes are labeled. In terms of the wavelength of the light inside the film, what is the difference in film thickness between (a) points a and b and (b) points b and d?

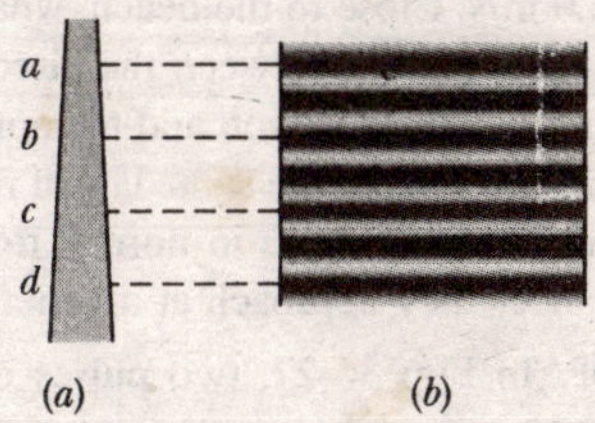

Fig. 36-24 Question 11.

12. Figure 36-25 shows the transmission of light through a thin film in air by a perpendicular beam (tilted in the figure for clarity). (a) Did ray r_3 undergo a phase shift due to reflection? (b) In wavelengths, what is the reflection phase shift for ray r_4? (c) If the film thickness is L, what is the path length difference between rays r_3 and r_4?

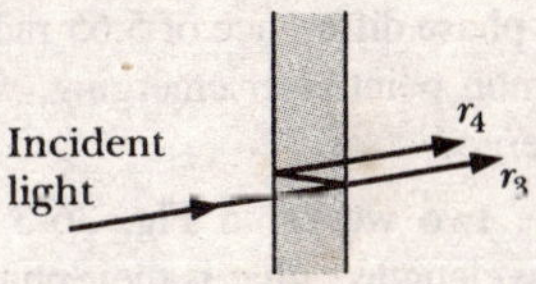

Fig. 36-25 Question 12.

EXERCISES & PROBLEMS

ssm Solution is in the Student Solutions Manual.
www Solution is available on the World Wide Web at: http://www.wiley.com/college/hrw
ilw Solution is available on the Interactive LearningWare.

SEC. 36-2 Light as a Wave

1E. The wavelength of yellow sodium light in air is 589 nm. (a) What is its frequency? (b) What is its wavelength in glass whose index of refraction is 1.52? (c) From the results of (a) and (b) find its speed in this glass.

2E. How much faster, in meters per second, does light travel in sapphire than in diamond? See Table 34-1.

3E. The speed of yellow light (from a sodium lamp) in a certain liquid is measured to be 1.92×10^8 m/s. What is the index of refraction of this liquid for the light?

4E. What is the speed in fused quartz of light of wavelength 550 nm? (See Fig. 34-19.)

5P. Ocean waves moving at a speed of 4.0 m/s are approaching a beach at an angle of 30° to the normal, as shown from above in Fig. 36-26. Suppose the water depth changes abruptly at a certain distance from the beach and the wave speed there drops to 3.0 m/s. Close to the beach, what is the angle θ between the direction of wave motion and the normal? (Assume the same law of refraction as for light.) Explain why most waves come in normal to a shore even though at large distances they approach at a variety of angles.

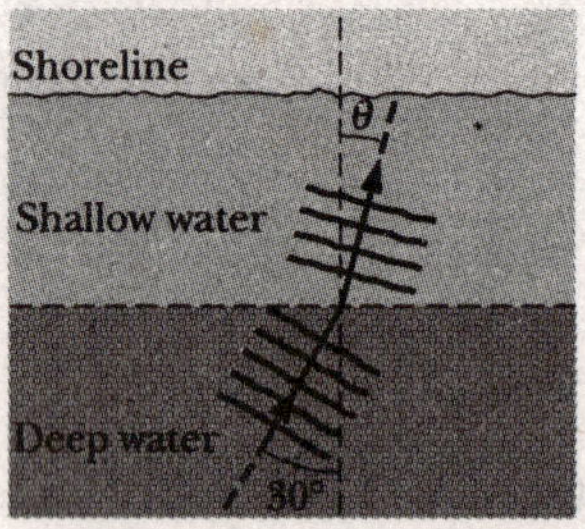

Fig. 36-26 Problem 5.

6P. In Fig. 36-27, two pulses of light are sent through layers of plastic with the indexes of refraction indicated and with thicknesses of either L or $2L$ as shown. (a) Which pulse travels through the plastic in less time? (b) In terms of L/c, what is the difference in the traversal times of the pulses?

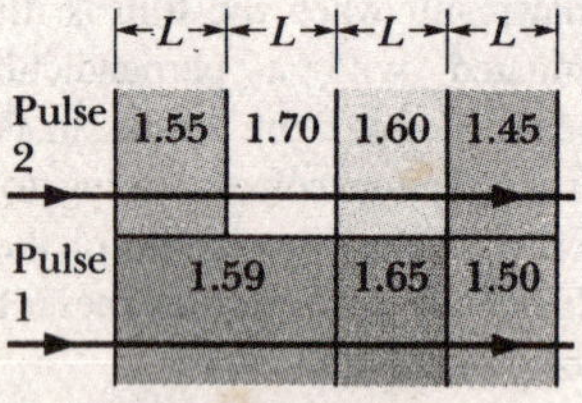

Fig. 36-27 Problem 6.

7P. In Fig. 36-3, assume that two waves of light in air, of wavelength 400 nm, are initially in phase. One travels through a glass layer of index of refraction $n_1 = 1.60$ and thickness L. The other travels through an equally thick plastic layer of index of refraction $n_2 = 1.50$. (a) What is the least value L should have if the waves are to end up with a phase difference of 5.65 rad? (b) If the waves arrive at some common point after emerging, what type of interference do they undergo? ssm

8P. Suppose that the two waves in Fig. 36-3 have wavelength 500 nm in air. In wavelengths, what is their phase difference after traversing media 1 and 2 if (a) $n_1 = 1.50$, $n_2 = 1.60$, and $L = 8.50$ μm; (b) $n_1 = 1.62$, $n_2 = 1.72$, and $L = 8.50$ μm; and (c) $n_1 = 1.59$, $n_2 = 1.79$, and $L = 3.25$ μm? (d) Suppose that in each of these three situations the waves arrive at a common point after emerging. Rank the situations according to the brightness the waves produce at the common point.

9P. Two waves of light in air, of wavelength 600.0 nm, are initially in phase. They then travel through plastic layers as shown in Fig. 36-28, with $L_1 = 4.00$ μm, $L_2 = 3.50$ μm, $n_1 = 1.40$, and $n_2 = 1.60$. (a) In wavelengths, what is their phase difference after they both have emerged from the layers? (b) If the waves later arrive at some common point, what type of interference do they undergo? ilw

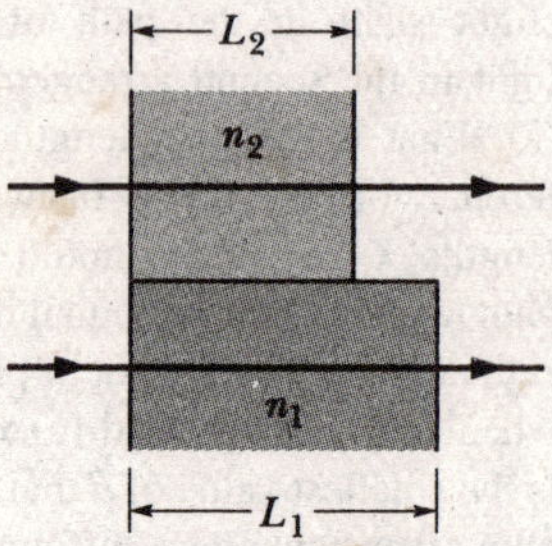

Fig. 36-28 Problem 9.

10P. In Fig. 36-3, assume that the two light waves, of wavelength 620 nm in air, are initially out of phase by π rad. The indexes of refraction of the media are $n_1 = 1.45$ and $n_2 = 1.65$. (a) What is the least thickness L that will put the waves exactly in phase once they pass through the two media? (b) What is the next greater L that will do this?

SEC. 36-4 Young's Interference Experiment

11E. Monochromatic green light, of wavelength 550 nm, illuminates two parallel narrow slits 7.70 μm apart. Calculate the angular deviation (θ in Fig. 36-8) of the third-order (for $m = 3$) bright fringe (a) in radians and (b) in degrees.

12E. What is the phase difference of the waves from the two slits when they arrive at the mth dark fringe in a Young's double-slit experiment?

13E. Suppose that Young's experiment is performed with blue-green light of wavelength 500 nm. The slits are 1.20 mm apart, and the viewing screen is 5.40 m from the slits. How far apart are the bright fringes? ssm ilw

14E. In a double-slit arrangement the slits are separated by a distance equal to 100 times the wavelength of the light passing through the slits. (a) What is the angular separation in radians between the central maximum and an adjacent maximum? (b) What is the distance between these maxima on a screen 50.0 cm from the slits?

15E. A double-slit arrangement produces interference fringes for sodium light ($\lambda = 589$ nm) that have an angular separation of 3.50×10^{-3} rad. For what wavelength would the angular separation be 10.0% greater? ssm

16E. A double-slit arrangement produces interference fringes for sodium light ($\lambda = 589$ nm) that are 0.20° apart. What is the angular fringe separation if the entire arrangement is immersed in water ($n = 1.33$)?

17E. Two radio-frequency point sources separated by 2.0 m are radiating in phase with $\lambda = 0.50$ m. A detector moves in a circular path around the two sources in a plane containing them. Without written calculation, find how many maxima it detects. ssm

18E. Sources A and B emit long-range radio waves of wavelength 400 m, with the phase of the emission from A ahead of that from source B by 90°. The distance r_A from A to a detector is greater than the corresponding distance r_B by 100 m. What is the phase difference at the detector?

19P. In a double-slit experiment the distance between slits is 5.0 mm and the slits are 1.0 m from the screen. Two interference patterns can be seen on the screen: one due to light with wavelength 480 nm, and the other due to light with wavelength 600 nm. What is the separation on the screen between the third-order ($m = 3$) bright fringes of the two interference patterns? ssm

20P. In Fig. 36-29, S_1 and S_2 are identical radiators of waves that are in phase and of the same wavelength λ. The radiators are separated by distance $d = 3.00\lambda$. Find the greatest distance from S_1, along the x axis, for which fully destructive interference occurs. Express this distance in wavelengths.

Fig. 36-29 Problems 20, 27, and 59.

21P. A thin flake of mica ($n = 1.58$) is used to cover one slit of a double-slit interference arrangement. The central point on the viewing screen is now occupied by what had been the seventh bright side fringe ($m = 7$) before the mica was used. If $\lambda = 550$ nm, what is the thickness of the mica? (*Hint:* Consider the wavelength of the light within the mica.) ssm www

22P. Laser light of wavelength 632.8 nm passes through a double-slit arrangement at the front of a lecture room, reflects off a mirror 20.0 m away at the back of the room, and then produces an interference pattern on a screen at the front of the room. The distance between adjacent bright fringes is 10.0 cm. (a) What is the slit separation? (b) What happens to the pattern when the lecturer places a thin cellophane sheet over one slit, thereby increasing by 2.50 the number of wavelengths along the path that includes the cellophane?

SEC. 36-6 Intensity in Double-Slit Interference

23E. Two waves of the same frequency have amplitudes 1.00 and 2.00. They interfere at a point where their phase difference is 60.0°. What is the resultant amplitude? ssm

24E. Find the sum y of the following quantities:

$$y_1 = 10 \sin \omega t \quad \text{and} \quad y_2 = 8.0 \sin(\omega t + 30°).$$

25E. Add the quantities

$$y_1 = 10 \sin \omega t$$
$$y_2 = 15 \sin(\omega t + 30°)$$
$$y_3 = 5.0 \sin(\omega t - 45°)$$

using the phasor method. ilw

26E. Light of wavelength 600 nm is incident normally on two parallel narrow slits separated by 0.60 mm. Sketch the intensity pattern observed on a distant screen as a function of angle θ from the pattern's center for the range of values $0 \le \theta \le 0.0040$ rad.

27P. S_1 and S_2 in Fig. 36-29 are point sources of electromagnetic waves of wavelength 1.00 m. They are in phase and separated by $d = 4.00$ m, and they emit at the same power. (a) If a detector is moved to the right along the x axis from source S_1, at what distances from S_1 are the first three interference maxima detected? (b) Is the intensity of the nearest minimum exactly zero? (*Hint:* Does the intensity of a wave from a point source remain constant with an increase in distance from the source?) ssm

28P. The double horizontal arrow in Fig. 36-9 marks the points on the intensity curve where the intensity of the central fringe is half the maximum intensity. Show that the angular separation $\Delta\theta$ between the corresponding points on the viewing screen is

$$\Delta\theta = \frac{\lambda}{2d}$$

if θ in Fig. 36-8 is small enough so that $\sin \theta \approx \theta$.

29P*. Suppose that one of the slits of a double-slit interference experiment is wider than the other, so the amplitude of the light reaching the central part of the screen from one slit, acting alone, is twice that from the other slit, acting alone. Derive an expression for the light intensity I at the screen as a function of θ, corresponding to Eqs. 36-21 and 36-22. ssm www

SEC. 36-7 Interference from Thin Films

30E. In Fig. 36-30, light wave W_1 reflects once from a reflecting surface while light wave W_2 reflects twice from that surface and once from a reflecting sliver at distance L from the mirror. The waves are initially in phase and have a wavelength of 620 nm. Neglect the slight tilt of the rays. (a) For what least value of L are the reflected waves exactly out of phase? (b) How far must the sliver be moved to put the waves exactly out of phase again?

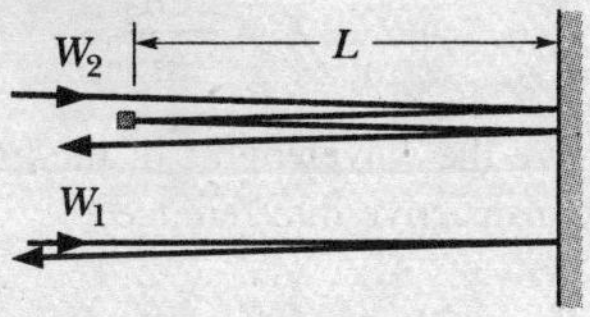

Fig. 36-30 Exercises 30 and 32.

31E. Bright light of wavelength 585 nm is incident perpendicularly on a soap film ($n = 1.33$) of thickness 1.21 μm, suspended in air. Is the light reflected by the two surfaces of the film closer to interfering fully destructively or fully constructively? ssm

32E. Suppose the light waves of Exercise 30 are initially exactly out of phase. Find an expression for the values of L (in terms of the wavelength λ) that put the reflected waves exactly in phase.

33E. Light of wavelength 624 nm is incident perpendicularly on a soap film (with $n = 1.33$) suspended in air. What are the least two thicknesses of the film for which the reflections from the film undergo fully constructive interference? ilw

34E. A camera lens with index of refraction greater than 1.30 is coated with a thin transparent film of index of refraction 1.25 to eliminate by interference the reflection of light at wavelength λ that is incident perpendicularly on the lens. In terms of λ, what minimum film thickness is needed?

35E. The rhinestones in costume jewelry are glass with index of refraction 1.50. To make them more reflective, they are often coated with a layer of silicon monoxide of index of refraction 2.00. What is the minimum coating thickness needed to ensure that light of wavelength 560 nm and of perpendicular incidence will be reflected from the two surfaces of the coating with fully constructive interference? ssm

36E. In Fig. 36-31, light of wavelength 600 nm is incident perpendicularly on five sections of a transparent structure suspended in air. The structure has index of refraction 1.50. The thickness of each section is given in terms of $L = 4.00\ \mu m$. For which sections will the light that is reflected from the top and bottom surfaces of that section undergo fully constructive interference?

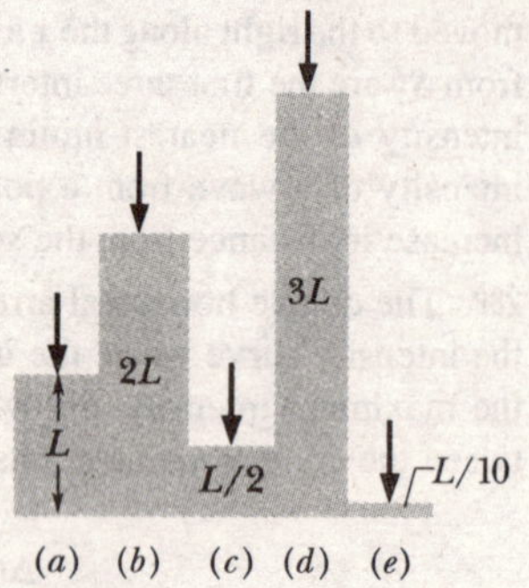

Fig. 36-31 Exercise 36.

37E. We wish to coat flat glass ($n = 1.50$) with a transparent material ($n = 1.25$) so that reflection of light at wavelength 600 nm is eliminated by interference. What minimum thickness can the coating have to do this? ssm www

38P. In Fig. 36-32, light is incident perpendicularly on four thin layers of thickness L. The indexes of refraction of the thin layers and of the media above and below these layers are given. Let λ represent the wavelength of the light in air, and n_2 represent the index of refraction of the thin layer in each situation. Consider only the transmission of light that undergoes no reflection or two reflections, as in Fig. 36-32*a*. For which of the situations does the expression

$$\lambda = \frac{2Ln_2}{m}, \quad \text{for } m = 1, 2, 3, \ldots,$$

give the wavelengths of the transmitted light that undergoes fully constructive interference?

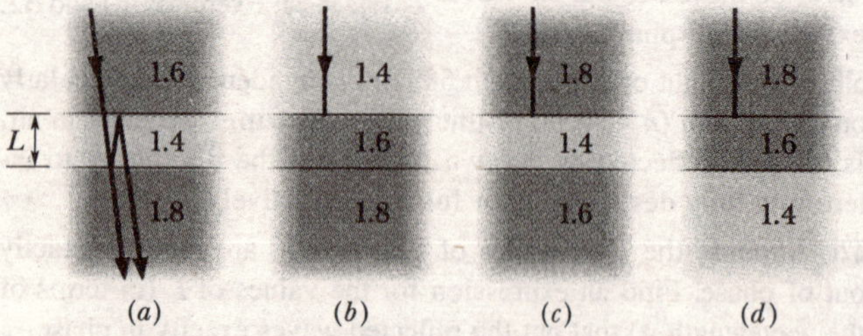

Fig. 36-32 Problems 38 and 39.

39P. A disabled tanker leaks kerosene ($n = 1.20$) into the Persian Gulf, creating a large slick on top of the water ($n = 1.30$). (a) If you are looking straight down from an airplane, while the Sun is overhead, at a region of the slick where its thickness is 460 nm, for which wavelength(s) of visible light is the reflection brightest because of constructive interference? (b) If you are scuba diving directly under this same region of the slick, for which wavelength(s) of visible light is the transmitted intensity strongest? (*Hint:* Use Fig. 36-32*a* with appropriate indexes of refraction.)

40P. A plane wave of monochromatic light is incident normally on a uniform thin film of oil that covers a glass plate. The wavelength of the source can be varied continuously. Fully destructive interference of the reflected light is observed for wavelengths of 500 and 700 nm and for no wavelengths in between. If the index of refraction of the oil is 1.30 and that of the glass is 1.50, find the thickness of the oil film.

41P. A plane monochromatic light wave in air is perpendicularly incident on a thin film of oil that covers a glass plate. The wavelength of the source may be varied continuously. Fully destructive interference of the reflected light is observed for wavelengths of 500 and 700 nm and for no wavelength in between. The index of refraction of the glass is 1.50. Show that the index of refraction of the oil must be less than 1.50. ssm

42P. The reflection of perpendicularly incident white light by a soap film in air has an interference maximum at 600 nm and a minimum at 450 nm, with no minimum in between. If $n = 1.33$ for the film, what is the film thickness, assumed uniform?

43P. In Fig. 36-33, a broad beam of light of wavelength 683 nm is sent directly downward through the top plate of a pair of glass plates. The plates are 120 mm long, touch at the left end, and are separated by a wire of diameter 0.048 mm at the right end. The air between the plates acts as a thin film. How many bright fringes will be seen by an observer looking down through the top plate? ssm

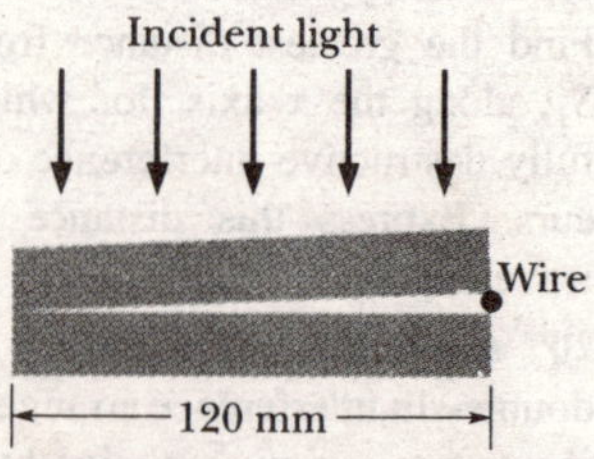

Fig. 36-33 Problems 43 and 44.

44P. In Fig. 36-33, white light is sent directly downward through the top plate of a pair of glass plates. The plates touch at the left end and are separated by a wire of diameter 0.048 mm at the right end; the air between the plates acts as a thin film. An observer looking down through the top plate sees bright and dark fringes due to that film. (a) Is a dark fringe or a bright fringe seen at the left end? (b) To the right of that end, fully destructive interference occurs at different locations for different wavelengths of the light. Does it occur first for the red end or the blue end of the visible spectrum?

45P. A broad beam of light of wavelength 630 nm is incident at 90° on a thin, wedge-shaped film with index of refraction 1.50. An observer intercepting the light transmitted by the film sees 10 bright and 9 dark fringes along the length of the film. By how much does the film thickness change over this length? ssm www

46P. A thin film of acetone ($n = 1.25$) coats a thick glass plate ($n = 1.50$). White light is incident normal to the film. In the reflections, fully destructive interference occurs at 600 nm and fully constructive interference at 700 nm. Calculate the thickness of the acetone film.

47P. Two glass plates are held together at one end to form a wedge of air that acts as a thin film. A broad beam of light of wavelength 480 nm is directed through the plates, perpendicular to the first plate. An observer intercepting light reflected from the plates sees on the plates an interference pattern that is due to the wedge of air. How much thicker is the wedge at the sixteenth bright fringe than it is at the sixth bright fringe, counting from where the plates touch?

48P. A broad beam of monochromatic light is directed perpendicularly through two glass plates that are held together at one end to create a wedge of air between them. An observer intercepting light reflected from the wedge of air, which acts as a thin film, sees 4001 dark fringes along the length of the wedge. When the air between the plates is evacuated, only 4000 dark fringes are seen. Calculate the index of refraction of air from these data.

49P. Figure 36-34*a* shows a lens with radius of curvature R lying on a plane glass plate and illuminated from above by light with wavelength λ. Figure 36-34*b* (a photograph taken from above the lens) shows that circular interference fringes (called *Newton's rings*) appear, associated with the variable thickness d of the air film between the lens and the plate. Find the radii r of the interference maxima assuming $r/R \ll 1$. ssm ilw

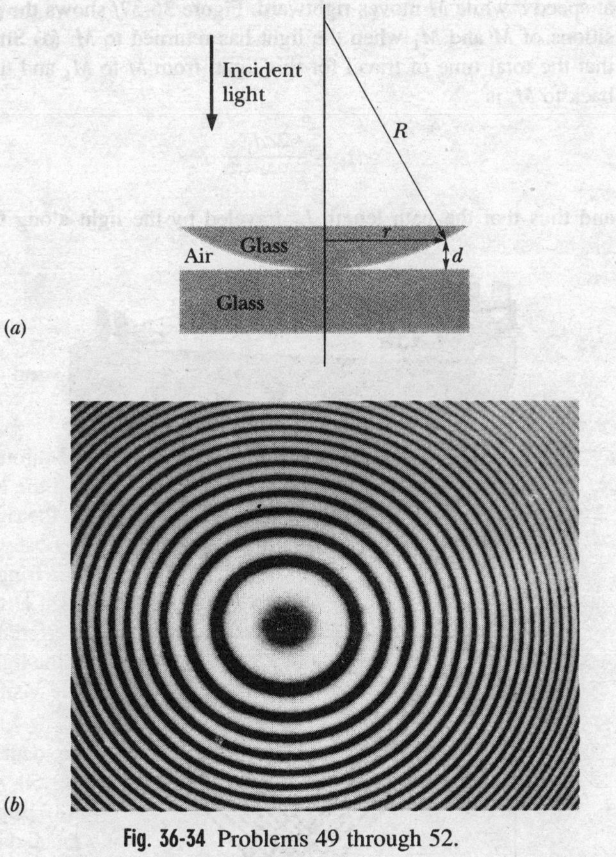

Fig. 36-34 Problems 49 through 52.

50P. In a Newton's rings experiment (see Problem 49), the radius of curvature R of the lens is 5.0 m and the lens diameter is 20 mm. (a) How many bright rings are produced? Assume that λ = 589 nm. (b) How many bright rings would be produced if the arrangement were immersed in water (n = 1.33)?

51P. A Newton's rings apparatus is to be used to determine the radius of curvature of a lens (see Fig. 36-34 and Problem 49). The radii of the nth and $(n + 20)$th bright rings are measured and found to be 0.162 and 0.368 cm, respectively, in light of wavelength 546 nm. Calculate the radius of curvature of the lower surface of the lens.

52P. (a) Use the result of Problem 49 to show that, in a Newton's rings experiment, the difference in radius between adjacent bright rings (maxima) is given by

$$\Delta r = r_{m+1} - r_m \approx \tfrac{1}{2}\sqrt{\lambda R/m},$$

assuming $m \gg 1$. (b) Now show that the *area* between adjacent bright rings is given by

$$A = \pi \lambda R,$$

assuming $m \gg 1$. Note that this area is independent of m.

53P. In Fig. 36-35, a microwave transmitter at height a above the water level of a wide lake transmits microwaves of wavelength λ toward a receiver on the opposite shore, a distance x above the water level. The microwaves reflecting from the water interfere with the microwaves arriving directly from the transmitter. Assuming that the lake width D is much greater than a and x, and that $\lambda \geq a$, at what values of x is the signal at the receiver maximum? (*Hint:* Does the reflection cause a phase change?) ssm

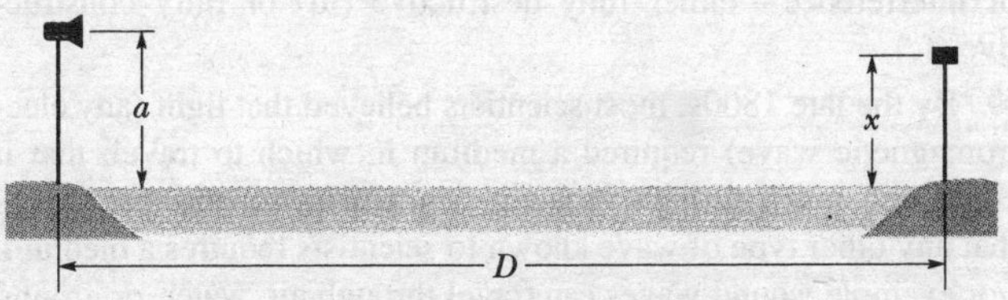

Fig. 36-35 Problem 53.

SEC. 36-8 Michelson's Interferometer

54E. A thin film with index of refraction n = 1.40 is placed in one arm of a Michelson interferometer, perpendicular to the optical path. If this causes a shift of 7.0 fringes of the pattern produced by light of wavelength 589 nm, what is the film thickness?

55E. If mirror M_2 in a Michelson interferometer (Fig. 36-17) is moved through 0.233 mm, a shift of 792 fringes occurs. What is the wavelength of the light producing the fringe pattern? ssm

56P. The element sodium can emit light at two wavelengths, λ_1 = 589.10 nm and λ_2 = 589.59 nm. Light from sodium is being used in a Michelson interferometer (Fig. 36-17). Through what distance must mirror M_2 be moved to shift the fringe pattern for one wavelength by 1.00 fringe more than the fringe pattern for the other wavelength?

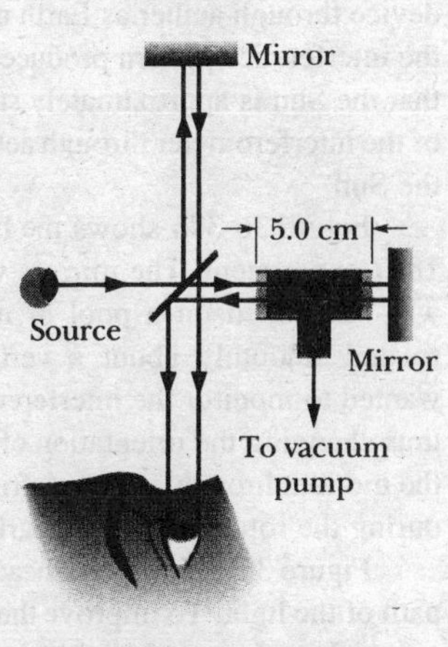

Fig. 36-36 Problem 57.

57P. In Fig. 36-36, an airtight chamber 5.0 cm long with glass windows is placed in one arm of a Michelson interferometer. Light of wavelength λ = 500 nm is used. Evacuating the air from the chamber causes a shift of 60 fringes. From these data, find the index of refraction of air at atmospheric pressure. ssm

58P. Write an expression for the intensity observed in a Michelson interferometer (Fig. 36-17) as a function of the position of the movable mirror. Measure the position of the mirror from the point at which $d_2 = d_1$.

Additional Problems

59. Figure 36-29 shows two point sources S_1 and S_2 that emit light of wavelength $\lambda = 500$ nm. The emissions are isotropic and in phase, and the separation between the sources is $d = 2.00\ \mu$m. At any point P on the x axis, the wave from S_1 and the wave from S_2 interfere. When P is very far away ($x \approx \infty$), what are (a) the phase difference between the waves arriving from S_1 and S_2 and (b) the type of interference they produce (approximately fully constructive or fully destructive)? (c) As we then move P along the x axis toward S_1, does the phase difference between the waves from S_1 and S_2 increase or decrease? (d) Produce a table that gives the positions x at which the phase differences are 0, 0.50λ, 1.00λ, . . . , 2.50λ, and for each indicate the corresponding type of interference—either fully destructive (fd) or fully constructive (fc).

60. By the late 1800s, most scientists believed that light (any electromagnetic wave) required a medium in which to travel, that it could not travel through vacuum. One reason for this belief was that any other type of wave known to scientists requires a medium. For example, sound waves can travel through air, water, or ground but not through vacuum. Thus, reasoned the scientists, when light travels from the Sun or any other star to Earth, it cannot be traveling through vacuum; instead, it must be traveling through a medium that fills all of space and through which Earth slips. Presumably, light has a certain speed c through this medium, which was called *aether* (or *ether*).

In 1887, Michelson and Edward Morely used a version of Michelson's interferometer to test for the effects of aether on the travel of light within the device. Specifically, the motion of the device through aether as Earth moves around the Sun should affect the interference pattern produced by the device. Scientists assumed that the Sun is approximately stationary in aether; hence the speed of the interferometer through aether should be Earth's speed v about the Sun.

Figure 36-37*a* shows the basic arrangement of mirrors in the 1887 experiment. The mirrors were mounted on a heavy slab that was suspended on a pool of mercury so that the slab could be rotated smoothly about a vertical axis. Michelson and Morely wanted to monitor the interference pattern as they rotated the slab, thus changing the orientation of the interferometer arms relative to the motion through aether. A fringe shift in the interference pattern during the rotation would clearly signal the presence of aether.

Figure 36-37*b*, an overhead view of the equipment, shows the path of the light. To improve the possibility of fringe shift, the light was reflected several times along the arms of the interferometer, instead of only once along each arm as indicated in the basic interferometer of Fig. 36-17. This repeated reflection increased the effective length of each arm to about 10 m. In spite of the added complexity, the interferometer of Figs. 36-37*a* and *b* functions just like the simpler interferometer of Fig. 36-17; so we can use Fig. 36-17 in our discussion here by merely taking the arm lengths d_1 and d_2 to be 10 m each.

Let us assume that there is aether through which light has speed c. Figure 36-37*c* shows a side view of the arm of length d_1 from the aether reference frame as the interferometer moves rightward through it with velocity $\vec{v}$. (For simplicity, the beam splitter M of Fig. 36-17 is drawn parallel to the mirror M_1 at the far end of the arm.) Figure 36-37*d* shows the arm just as a particular portion of the light (represented by a dot) begins its travel along the arm. We shall follow this light to find the path length along the arm, from the beam splitter to M_1 and then back to the beam splitter.

As the light moves at speed c rightward through aether and toward mirror M_1, that mirror moves rightward at speed v. Figure 36-37*e* shows the positions of M and M_1 when the light reaches M_1, reflecting there. The light now moves leftward through aether at speed c while M moves rightward. Figure 36-37*f* shows the positions of M and M_1 when the light has returned to M. (a) Show that the total time of travel for this light, from M to M_1 and then back to M, is

$$t_1 = \frac{2cd_1}{c^2 - v^2}$$

and thus that the path length L_1 traveled by the light along this

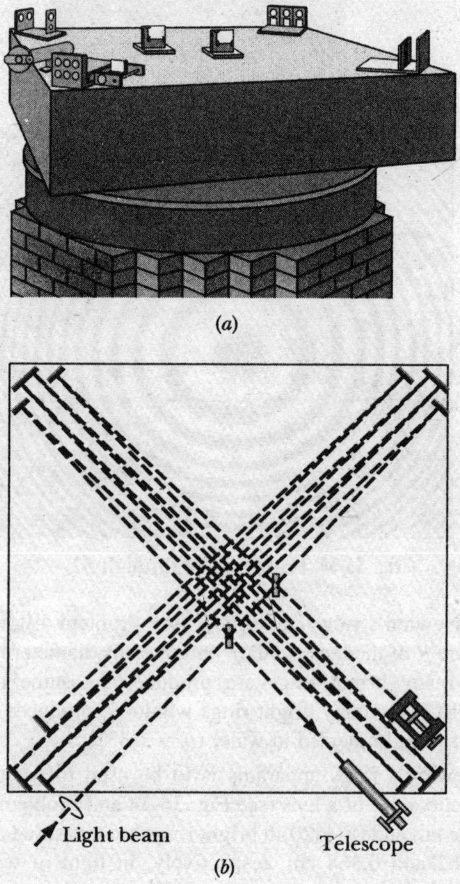

Fig. 36-37 Problem 60.

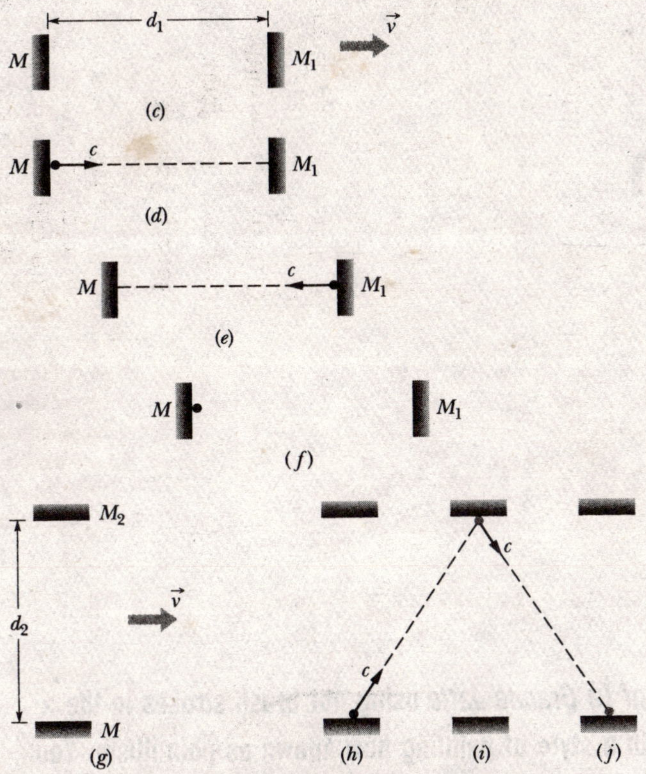

Fig. 36-37 Problem 60 (*continued*).

arm is

$$L_1 = ct_1 = \frac{2c^2 d_1}{c^2 - v^2}.$$

Figure 36-37*g* shows a view of the arm of length d_2; that arm also moves rightward with velocity $\vec{v}$ through the aether. For simplicity, the beam splitter M of Fig. 36-17 is now drawn parallel to the mirror M_2 at the far end of this arm. Figure 36-37*h* shows the arm just as a particular portion of the light (the dot) begins its travel along the arm. Because the arm moves rightward during the flight of the light, the path of the light is angled rightward toward the position that M_2 will have when the light reaches that mirror (Fig. 36-37*i*). The reflection of the light from M_2 sends the light angled rightward toward the position that M will have when the light returns to it (Fig. 36-37*j*). (b) Show that the total time of travel for the light, from M to M_2 and then back to M, is

$$t_2 = \frac{2d_2}{\sqrt{c^2 - v^2}}$$

and thus that the path length L_2 traveled by the light along this arm is

$$L_2 = ct_2 = \frac{2cd_2}{\sqrt{c^2 - v^2}}.$$

Substitute d for d_1 and d_2 in the expressions for L_1 and L_2. Then expand the two expressions by using the binomial expansion (given in Appendix E); retain the first two terms in each expansion. (c) Show that path length L_1 is greater than path length L_2 and that their difference ΔL is

$$\Delta L = \frac{dv^2}{c^2}.$$

(d) Next show that, at the telescope, the phase difference (in terms of wavelengths) between the light traveling along L_1 and that along L_2 is

$$\frac{\Delta L}{\lambda} = \frac{dv^2}{\lambda c^2},$$

where λ is the wavelength of the light. This phase difference determines the fringe pattern produced by the light arriving at the telescope in the interferometer.

Now rotate the interferometer by 90° so that the arm of length d_2 is along the direction of motion through the aether and the arm of length d_1 is perpendicular to that direction. (e) Show that the shift in the fringe pattern due to this rotation is

$$\text{shift} = \frac{2dv^2}{\lambda c^2}.$$

(f) Evaluate the shift, setting $c = 3.0 \times 10^8$ m/s, $d = 10$ m, and $\lambda = 500$ nm and using data about Earth given in Appendix C.

This expected fringe shift would have been easily observable. However, Michelson and Morely observed no fringe shift, which cast grave doubt on the existence of aether. In fact, the idea of aether soon disappeared. Moreover, the null result of Michelson and Morely led, at least indirectly, to Einstein's special theory of relativity.

37 Diffraction

Georges Seurat painted *Sunday Afternoon on the Island of La Grande Jatte* using not brush strokes in the usual sense, but rather a myriad of small colored dots, in a style of painting now known as pointillism. You can see the dots if you stand close enough to the painting, but as you move away from it, they eventually blend and cannot be distinguished. Moreover, the color that you see at any given place on the painting changes as you move away—which is why Seurat painted with the dots.

What causes this change in color?

The answer is in this chapter.

37-1 Diffraction and the Wave Theory of Light

Fig. 37-1 This diffraction pattern appeared on a viewing screen when light that had passed through a narrow vertical slit reached the screen. Diffraction causes light to flare out perpendicular to the long sides of the slit. That produces an interference pattern consisting of a broad central maximum and less intense and narrower secondary (or side) maxima, with minima between them.

In Chapter 36 we defined diffraction rather loosely as the flaring of light as it emerges from a narrow slit. More than just flaring occurs, however, because the light produces an interference pattern called a **diffraction pattern.** For example, when monochromatic light from a distant source (or a laser) passes through a narrow slit and is then intercepted by a viewing screen, the light produces on the screen a diffraction pattern like that in Fig. 37-1. This pattern consists of a broad and intense (very bright) central maximum and a number of narrower and less intense maxima (called **secondary** or **side** maxima) to both sides. In between the maxima are minima.

Such a pattern would be totally unexpected in geometrical optics: If light traveled in straight lines as rays, then the slit would allow some of those rays through and they would form a sharp, bright rendition of the slit on the viewing screen. As in Chapter 36, we must conclude that geometrical optics is only an approximation.

Diffraction of light is not limited to situations of light passing through a narrow opening (such as a slit or pinhole). It also occurs when light passes an edge, such as the edges of the razor blade whose diffraction pattern is shown in Fig. 37-2. Note the lines of maxima and minima that run approximately parallel to the edges, at both the inside edges of the blade and the outside edges. As the light passes, say, the vertical edge at the left, it flares left and right and undergoes interference, producing the pattern along the left edge. The rightmost portion of that pattern actually lies within what would have been the shadow of the blade if geometrical optics prevailed.

You encounter a common example of diffraction when you look at a clear blue sky and see tiny specks and hairlike structures floating in your view. These *floaters,* as they are called, are produced when light passes the edges of tiny deposits in the vitreous humor, the transparent material filling most of the eyeball. What you are seeing when a floater is in your field of vision is the diffraction pattern produced on the retina by one of these deposits. If you sight through a pinhole in an otherwise opaque sheet so as to make the light entering your eye approximately a plane wave, you can distinguish individual maxima and minima in the patterns.

The Fresnel Bright Spot

Diffraction finds a ready explanation in the wave theory of light. However, this theory, originally advanced in the late 1600s by Huygens and used 123 years later by Young to explain double-slit interference, was very slow in being adopted, largely because it ran counter to Newton's theory that light was a stream of particles.

Newton's view was the prevailing view in French scientific circles of the early nineteenth century, when Augustin Fresnel was a young military engineer. Fresnel, who believed in the wave theory of light, submitted a paper to the French Academy of Sciences describing his experiments with light and his wave-theory explanations of them.

In 1819, the Academy, dominated by supporters of Newton and thinking to challenge the wave point of view, organized a prize competition for an essay on the subject of diffraction. Fresnel won. The Newtonians, however, were neither converted nor silenced. One of them, S. D. Poisson, pointed out the "strange result" that if Fresnel's theories were correct, then light waves should flare into the shadow region of a sphere as they pass the edge of the sphere, producing a bright spot at the center of the shadow. The prize committee arranged a test of the famous mathe-

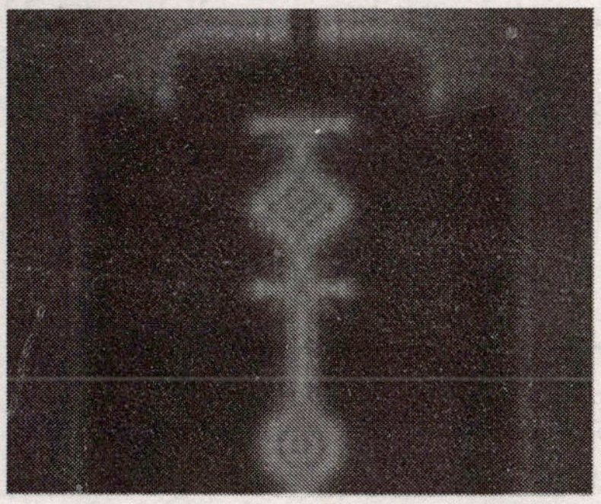

Fig. 37-2 The diffraction pattern produced by a razor blade in monochromatic light. Note the lines of alternating maximum and minimum intensity.

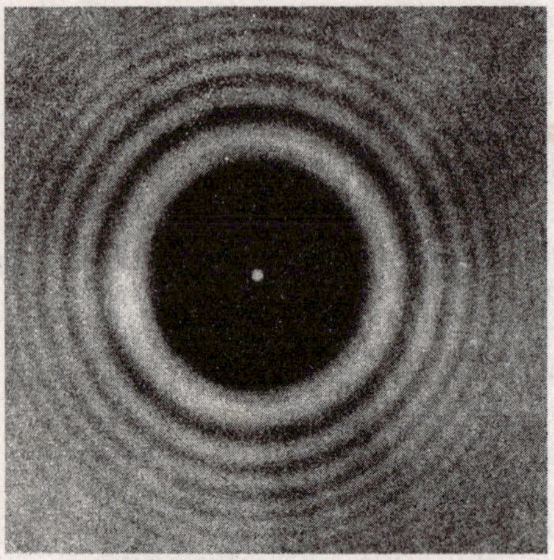

Fig. 37-3 A photograph of the diffraction pattern of a disk. Note the concentric diffraction rings and the Fresnel bright spot at the center of the pattern. This experiment is essentially identical to that arranged by the committee testing Fresnel's theories, because both the sphere they used and the disk used here have a cross section with a circular edge.

matician's prediction and discovered (see Fig. 37-3) that the predicted *Fresnel bright spot,* as we call it today, was indeed there! Nothing builds confidence in a theory so much as having one of its unexpected and counterintuitive predictions verified by experiment.

37-2 Diffraction by a Single Slit: Locating the Minima

Let us now examine the diffraction pattern of plane waves of light of wavelength λ that are diffracted by a single, long, narrow slit of width a in an otherwise opaque screen B, as shown in cross section in Fig. 37-4*a*. (In that figure, the slit's length extends into and out of the page, and the incoming wavefronts are parallel to screen B.) When the diffracted light reaches viewing screen C, waves from different points within the slit undergo interference and produce a diffraction pattern of bright and dark fringes (interference maxima and minima) on the screen. To locate the fringes, we shall use a procedure somewhat similar to the one we used to locate the fringes in a two-slit interference pattern. However, diffraction is more mathematically challenging, and here we shall be able to find equations for only the dark fringes.

Before we do that, however, we can justify the central bright fringe seen in Fig. 37-1 by noting that the Huygens wavelets from all points in the slit travel about the same distance to reach the center of the pattern and thus are in phase there. As for the other bright fringes, we can say only that they are approximately halfway between adjacent dark fringes.

To find the dark fringes, we shall use a clever (and simplifying) strategy that involves pairing up all the rays coming through the slit and then finding what conditions cause the wavelets of the rays in each pair to cancel each other. We apply this strategy in Fig. 37-4*a* to locate the first dark fringe, at point P_1. First, we mentally divide the slit into two *zones* of equal widths $a/2$. Then we extend to P_1 a light ray r_1 from the top point of the top zone and a light ray r_2 from the top point of the bottom zone. A central axis is drawn from the center of the slit to screen C, and P_1 is located at an angle θ to that axis.

The wavelets of the pair of rays r_1 and r_2 are in phase within the slit because they originate from the same wavefront passing through the slit, along the width of the slit. However, to produce the first dark fringe they must be out of phase by $\lambda/2$ when they reach P_1; this phase difference is due to their path length difference, with the wavelet of r_2 traveling a longer path to reach P_1 than the wavelet of r_1. To display this path length difference, we find a point b on ray r_2 such that the path length from b to P_1 matches the path length of ray r_1. Then the path length difference between the two rays is the distance from the center of the slit to b.

When viewing screen C is near screen B, as in Fig. 37-4*a*, the diffraction pattern on C is difficult to describe mathematically. However, we can simplify the mathematics considerably if we arrange for the screen separation D to be much larger than the slit width a. Then we can approximate rays r_1 and r_2 as being parallel, at angle θ to the central axis (Fig. 37-4*b*). We can also approximate the triangle formed by point b, the top point of the slit, and the center point of the slit as being a right triangle, and one of the angles inside that triangle as being θ. The path length difference between rays r_1 and r_2 (which is still the distance from the center of the slit to point b) is then equal to $(a/2)\sin\theta$.

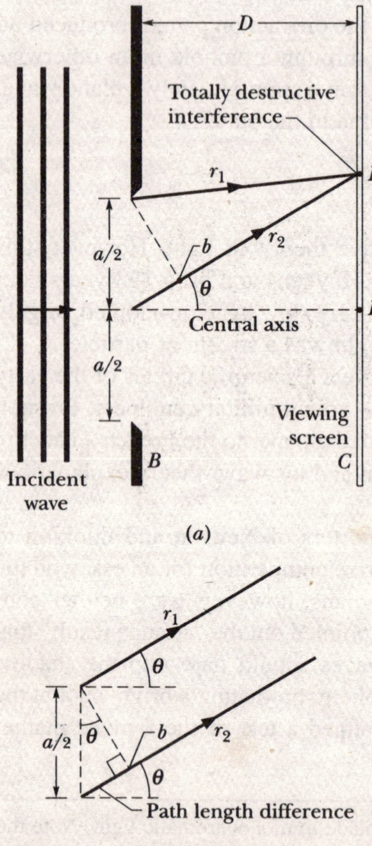

Fig. 37-4 (*a*) Waves from the top points of two zones of width $a/2$ undergo totally destructive interference at point P_1 on viewing screen C. (*b*) For $D \gg a$, we can approximate rays r_1 and r_2 as being parallel, at angle θ to the central axis.

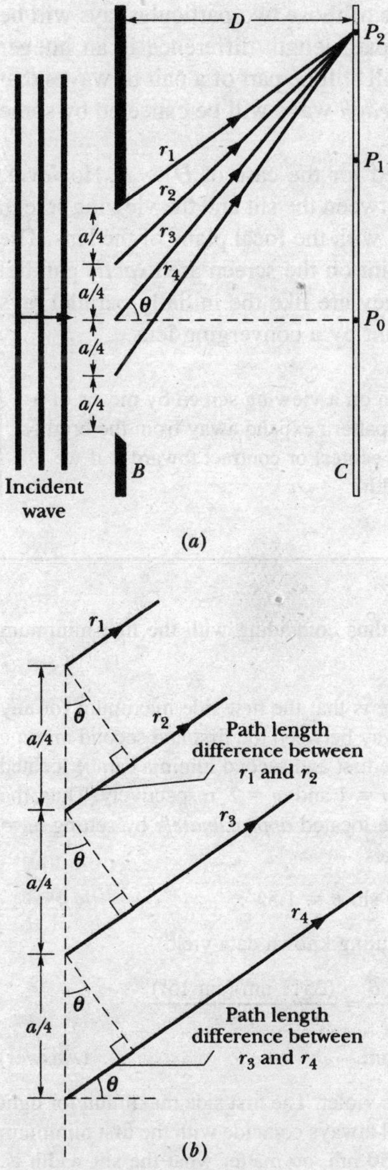

Fig. 37-5 (*a*) Waves from the top points of four zones of width $a/4$ undergo totally destructive interference at point P_2. (*b*) For $D \gg a$, we can approximate rays r_1, r_2, r_3, and r_4 as being parallel, at angle θ to the central axis.

We can repeat this analysis for any other pair of rays originating at corresponding points in the two zones (say, at the midpoints of the zones) and extending to point P_1. Each such pair of rays has the same path length difference $(a/2) \sin \theta$. Setting this common path length difference equal to $\lambda/2$ (our condition for the first dark fringe), we have

$$\frac{a}{2} \sin \theta = \frac{\lambda}{2},$$

which gives us

$$a \sin \theta = \lambda \quad \text{(first minimum).} \qquad (37\text{-}1)$$

Given slit width a and wavelength λ, Eq. 37-1 tells us the angle θ of the first dark fringe above and (by symmetry) below the central axis.

Note that if we begin with $a > \lambda$ and then narrow the slit while holding the wavelength constant, we increase the angle at which the first dark fringes appear; that is, the extent of the diffraction (the extent of the flaring and the width of the pattern) is *greater* for a *narrower* slit. When we have reduced the slit width to the wavelength (that is, $a = \lambda$), the angle of the first dark fringes is 90°. Since the first dark fringes mark the two edges of the central bright fringe, that bright fringe must then cover the entire viewing screen.

We find the second dark fringes above and below the central axis as we found the first dark fringes, except that we now divide the slit into *four* zones of equal widths $a/4$, as shown in Fig. 37-5*a*. We then extend rays r_1, r_2, r_3, and r_4 from the top points of the zones to point P_2, the location of the second dark fringe above the central axis. To produce that fringe, the path length difference between r_1 and r_2, that between r_2 and r_3, and that between r_3 and r_4 must all be equal to $\lambda/2$.

For $D \gg a$, we can approximate these four rays as being parallel, at angle θ to the central axis. To display their path length differences, we extend a perpendicular line through each adjacent pair of rays, as shown in Fig. 37-5*b*, to form a series of right triangles, each of which has a path length difference as one side. We see from the top triangle that the path length difference between r_1 and r_2 is $(a/4) \sin \theta$. Similarly, from the bottom triangle, the path length difference between r_3 and r_4 is also $(a/4) \sin \theta$. In fact, the path length difference for any two rays that originate at corresponding points in two adjacent zones is $(a/4) \sin \theta$. Since in each such case the path length difference is equal to $\lambda/2$, we have

$$\frac{a}{4} \sin \theta = \frac{\lambda}{2},$$

which gives us

$$a \sin \theta = 2\lambda \quad \text{(second minimum).} \qquad (37\text{-}2)$$

We could now continue to locate dark fringes in the diffraction pattern by splitting up the slit into more zones of equal width. We would always choose an even number of zones so that the zones (and their waves) could be paired as we have been doing. We would find that the dark fringes above and below the central axis can be located with the following general equation:

$$a \sin \theta = m\lambda, \quad \text{for } m = 1, 2, 3, \ldots \quad \text{(minima—dark fringes).} \qquad (37\text{-}3)$$

You can remember this result in the following way. Draw a triangle like the one in Fig. 37-4*b*, but for the full slit width a, and note that the path length difference between the top and bottom rays from the slit equals $a \sin \theta$. Thus, Eq. 37-3 says:

In a single-slit diffraction experiment, dark fringes are produced where the path length differences ($a \sin \theta$) between the top and bottom rays are equal to $\lambda, 2\lambda, 3\lambda, \ldots$.

This may seem to be wrong, because the waves of those two particular rays will be exactly in phase with each other when their path length difference is an integer number of wavelengths. However, they each will still be part of a pair of waves that are exactly out of phase with each other; thus, *each* wave will be canceled by some other wave, resulting in darkness.

Equations 37-1, 37-2, and 37-3 are derived for the case of $D \gg a$. However, they also apply if we place a converging lens between the slit and the viewing screen and then move the screen in so that it coincides with the focal plane of the lens. The lens ensures that rays which now reach any point on the screen are *exactly* parallel (rather than approximately) back at the slit. They are like the initially parallel rays of Fig. 35-12*a* that are directed to the focal point by a converging lens.

CHECKPOINT 1: We produce a diffraction pattern on a viewing screen by means of a long narrow slit illuminated by blue light. Does the pattern expand away from the bright center (the maxima and minima shift away from the center) or contract toward it if we (a) switch to yellow light or (b) decrease the slit width?

expand expand

Sample Problem 37-1

A slit of width a is illuminated by white light (which consists of all the wavelengths in the visible range).

(a) For what value of a will the first minimum for red light of wavelength $\lambda = 650$ nm appear at $\theta = 15°$?

SOLUTION: The **Key Idea** here is that diffraction occurs separately for each wavelength in the range of wavelengths passing through the slit, with the locations of the minima for each wavelength given by Eq. 37-3 ($a \sin\theta = m\lambda$). When we set $m = 1$ (for the first minimum) and substitute the given values of θ and λ, Eq. 37-3 yields

$$a = \frac{m\lambda}{\sin\theta} = \frac{(1)(650 \text{ nm})}{\sin 15°}$$
$$= 2511 \text{ nm} \approx 2.5\ \mu\text{m}. \qquad \text{(Answer)}$$

For the incident light to flare out that much ($\pm 15°$ to the first minima) the slit has to be very fine indeed—about four times the wavelength. For comparison, note that a fine human hair may be about 100 μm in diameter.

(b) What is the wavelength λ' of the light whose first side diffraction maximum is at 15°, thus coinciding with the first minimum for the red light?

SOLUTION: The **Key Idea** here is that the first side maximum for any wavelength is about halfway between the first and second minima for that wavelength. Those first and second minima can be located with Eq. 37-3 by setting $m = 1$ and $m = 2$, respectively. Thus, the first side maximum can be located *approximately* by setting $m = 1.5$. Then Eq. 37-3 becomes

$$a \sin\theta = 1.5\lambda'.$$

Solving for λ' and substituting known data yield

$$\lambda' = \frac{a \sin\theta}{1.5} = \frac{(2511 \text{ nm})(\sin 15°)}{1.5}$$
$$= 430 \text{ nm}. \qquad \text{(Answer)}$$

Light of this wavelength is violet. The first side maximum for light of wavelength 430 nm will always coincide with the first minimum for light of wavelength 650 nm, no matter what the slit width is. If the slit is relatively narrow, the angle θ at which this overlap occurs will be relatively large, and conversely.

37-3 Intensity in Single-Slit Diffraction, Qualitatively

In Section 37-2 we saw how to find the positions of the minima and the maxima in a single-slit diffraction pattern. Now we turn to a more general problem: find an expression for the intensity I of the pattern as a function of θ, the angular position of a point on a viewing screen.

To do this, we divide the slit of Fig. 37-4*a* into N zones of equal widths Δx small enough that we can assume each zone acts as a source of Huygens wavelets. We wish to superimpose the wavelets arriving at an arbitrary point P on the viewing screen, at angle θ to the central axis, so that we can determine the amplitude E_θ of

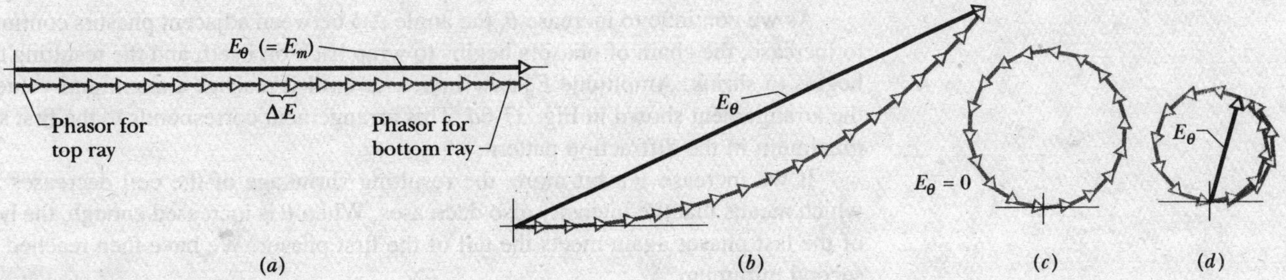

Fig. 37-6 Phasor diagrams for $N = 18$ phasors, corresponding to the division of a single slit into 18 zones. Resultant amplitudes E_θ are shown for (*a*) the central maximum at $\theta = 0$, (*b*) a point on the screen lying at a small angle θ to the central axis, (*c*) the first minimum, and (*d*) the first side maximum.

the electric component of the resultant wave at P. The intensity of the light at P is then proportional to the square of that amplitude.

To find E_θ, we need the phase relationships among the arriving wavelets. The phase difference between wavelets from adjacent zones is given by

$$\begin{pmatrix}\text{phase}\\ \text{difference}\end{pmatrix} = \left(\frac{2\pi}{\lambda}\right)\begin{pmatrix}\text{path length}\\ \text{difference}\end{pmatrix}.$$

For point P at angle θ, the path length difference between wavelets from adjacent zones is $\Delta x \sin\theta$, so the phase difference $\Delta\phi$ between wavelets from adjacent zones is

$$\Delta\phi = \left(\frac{2\pi}{\lambda}\right)(\Delta x \sin\theta). \tag{37-4}$$

We assume that the wavelets arriving at P all have the same amplitude ΔE. To find the amplitude E_θ of the resultant wave at P, we add the amplitudes ΔE via phasors. To do this, we construct a diagram of N phasors, one corresponding to the wavelet from each zone in the slit.

For point P_0 at $\theta = 0$ on the central axis of Fig. 37-4*a*, Eq. 37-4 tells us that the phase difference $\Delta\phi$ between the wavelets is zero; that is, the wavelets all arrive in phase. Figure 37-6*a* is the corresponding phasor diagram; adjacent phasors represent wavelets from adjacent zones and are arranged head to tail. Because there is zero phase difference between the wavelets, there is zero angle between each pair of adjacent phasors. The amplitude E_θ of the net wave at P_0 is the vector sum of these phasors. This arrangement of the phasors turns out to be the one that gives the greatest value for the amplitude E_θ. We call this value E_m; that is, E_m is the value of E_θ for $\theta = 0$.

We next consider a point P that is at a small angle θ to the central axis. Equation 37-4 now tells us that the phase difference $\Delta\phi$ between wavelets from adjacent zones is no longer zero. Figure 37-6*b* shows the corresponding phasor diagram; as before, the phasors are arranged head to tail, but now there is an angle $\Delta\phi$ between adjacent phasors. The amplitude E_θ at this new point is still the vector sum of the phasors, but it is smaller than that in Fig. 37-6*a*, which means that the intensity of the light is less at this new point P than at P_0.

If we continue to increase θ, the angle $\Delta\phi$ between adjacent phasors increases, and eventually the chain of phasors curls completely around so that the head of the last phasor just reaches the tail of the first phasor (Fig. 37-6*c*). The amplitude E_θ is now zero, which means that the intensity of the light is also zero. We have reached the first minimum, or dark fringe, in the diffraction pattern. The first and last phasors now have a phase difference of 2π rad, which means that the path length difference between the top and bottom rays through the slit equals one wavelength. Recall that this is the condition we determined for the first diffraction minimum.

As we continue to increase θ, the angle $\Delta\phi$ between adjacent phasors continues to increase, the chain of phasors begins to wrap back on itself, and the resulting coil begins to shrink. Amplitude E_θ now increases until it reaches a maximum value in the arrangement shown in Fig. 37-6*d*. This arrangement corresponds to the first side maximum in the diffraction pattern.

If we increase θ a bit more, the resulting shrinkage of the coil decreases E_θ, which means that the intensity also decreases. When θ is increased enough, the head of the last phasor again meets the tail of the first phasor. We have then reached the second minimum.

We could continue this qualitative method of determining the maxima and minima of the diffraction pattern but, instead, we shall now turn to a quantitative method.

✔CHECKPOINT 2: The figures represent, in smoother form (with more phasors) than Fig. 37-6, the phasor diagrams for two points of a diffraction pattern that are on opposite sides of a certain diffraction maximum. (a) Which maximum is it? (b) What is the approximate value of m (in Eq. 37-3) that corresponds to this maximum?

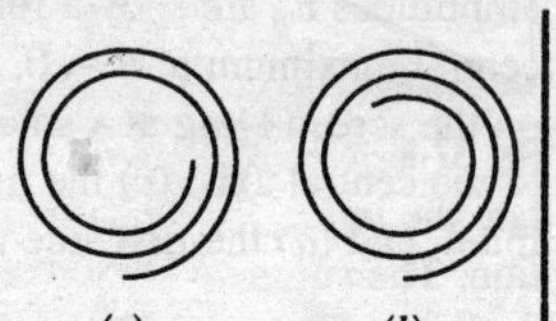

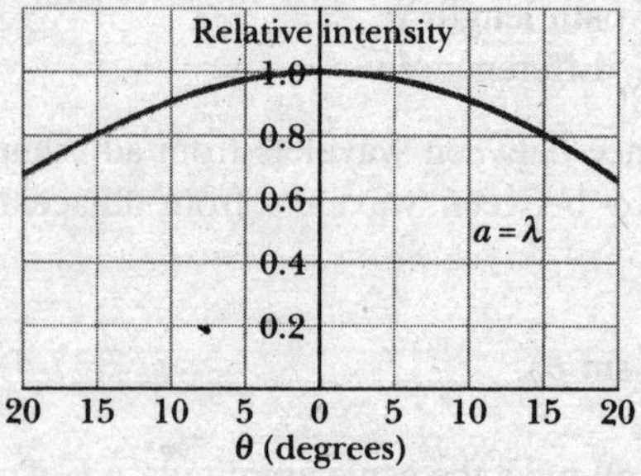

(*a*)

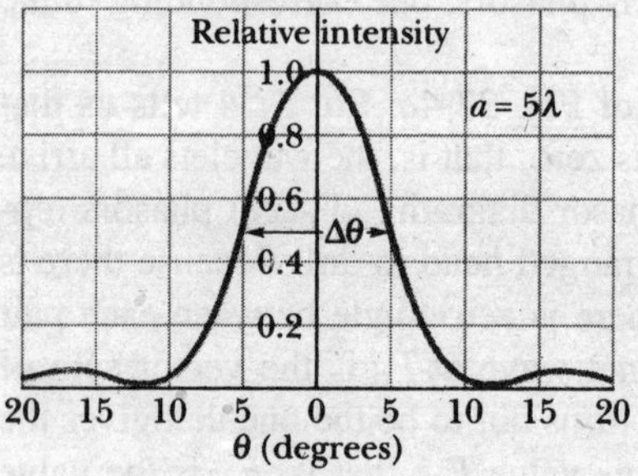

(*b*)

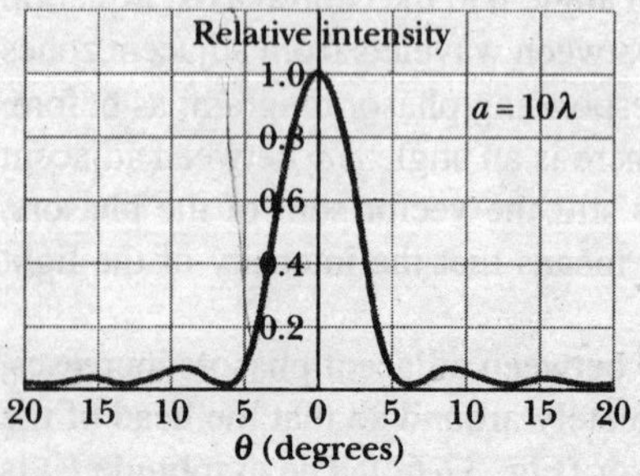

(*c*)

Fig. 37-7 The relative intensity in single-slit diffraction for three values of the ratio a/λ. The wider the slit is, the narrower is the central diffraction maximum.

37-4 Intensity in Single-Slit Diffraction, Quantitatively

Equation 37-3 tells us how to locate the minima of the single-slit diffraction pattern on screen C of Fig. 37-4*a* as a function of the angle θ in that figure. Here we wish to derive an expression for the intensity $I(\theta)$ of the pattern as a function of θ. We state, and shall prove below, that the intensity is given by

$$I(\theta) = I_m\left(\frac{\sin\alpha}{\alpha}\right)^2, \tag{37-5}$$

where

$$\alpha = \tfrac{1}{2}\phi = \frac{\pi a}{\lambda}\sin\theta. \tag{37-6}$$

The symbol α is just a convenient connection between the angle θ that locates a point on the viewing screen and the light intensity $I(\theta)$ at that point. I_m is the greatest value of the intensities $I(\theta)$ in the pattern and occurs at the central maximum (where $\theta = 0$), and ϕ is the phase difference (in radians) between the top and bottom rays from the slit width a.

Study of Eq. 37-5 shows that intensity minima will occur where

$$\alpha = m\pi, \qquad \text{for } m = 1, 2, 3, \ldots. \tag{37-7}$$

If we put this result into Eq. 37-6, we find

$$m\pi = \frac{\pi a}{\lambda}\sin\theta, \qquad \text{for } m = 1, 2, 3, \ldots$$

or

$$a\sin\theta = m\lambda, \qquad \text{for } m = 1, 2, 3, \ldots \quad \text{(minima—dark fringes)}, \tag{37-8}$$

which is exactly Eq. 37-3, the expression that we derived earlier for the location of the minima.

Figure 37-7 shows plots of the intensity of a single-slit diffraction pattern, calculated with Eqs. 37-5 and 37-6 for three slit widths: $a = \lambda$, $a = 5\lambda$, and $a = 10\lambda$. Note that as the slit width increases (relative to the wavelength), the width of the *central diffraction maximum* (the central hill-like region of the graphs) decreases;

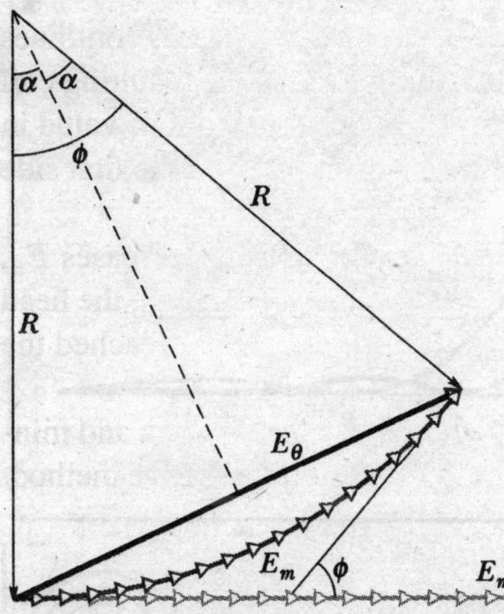

Fig. 37-8 A construction used to calculate the intensity in single-slit diffraction. The situation shown corresponds to that of Fig. 37-6*b*.

that is, the light undergoes less flaring by the slit. The secondary maxima also decrease in width (and become weaker). In the limit of slit width a being much greater than wavelength λ, the secondary maxima due to the slit disappear; we then no longer have single-slit diffraction (but we still have diffraction due to the edges of the wide slit, like that produced by the edges of the razor blade in Fig. 37-2).

Proof of Eqs. 37-5 and 37-6

The arc of phasors in Fig. 37-8 represents the wavelets that reach an arbitrary point P on the viewing screen of Fig. 37-4, corresponding to a particular small angle θ. The amplitude E_θ of the resultant wave at P is the vector sum of these phasors. If we divide the slit of Fig. 37-4 into infinitesimal zones of width Δx, the arc of phasors in Fig. 37-8 approaches the arc of a circle; we call its radius R as indicated in that figure. The length of the arc must be E_m, the amplitude at the center of the diffraction pattern, because if we straightened out the arc we would have the phasor arrangement of Fig. 37-6*a* (shown lightly in Fig. 37-8).

The angle ϕ in the lower part of Fig. 37-8 is the difference in phase between the infinitesimal vectors at the left and right ends of arc E_m. From the geometry, ϕ is also the angle between the two radii marked R in Fig. 37-8. The dashed line in that figure, which bisects ϕ, then forms two congruent right triangles. From either triangle we can write

$$\sin \tfrac{1}{2}\phi = \frac{E_\theta}{2R}. \tag{37-9}$$

In radian measure, ϕ is (with E_m considered to be a circular arc)

$$\phi = \frac{E_m}{R}.$$

Solving this equation for R and substituting in Eq. 37-9 lead to

$$E_\theta = \frac{E_m}{\frac{1}{2}\phi} \sin \tfrac{1}{2}\phi. \tag{37-10}$$

In Section 34-4 we saw that the intensity of an electromagnetic wave is proportional to the square of the amplitude of its electric field. Here, this means that the maximum intensity I_m (which occurs at the center of the diffraction pattern) is proportional to E_m^2 and the intensity $I(\theta)$ at angle θ is proportional to E_θ^2. Thus, we may write

$$\frac{I(\theta)}{I_m} = \frac{E_\theta^2}{E_m^2}. \tag{37-11}$$

Substituting for E_θ with Eq. 37-10 and then substituting $\alpha = \frac{1}{2}\phi$, we are led to the following expression for the intensity as a function of θ:

$$I(\theta) = I_m\left(\frac{\sin \alpha}{\alpha}\right)^2.$$

This is exactly Eq. 37-5, one of the two equations we set out to prove.

The second equation we wish to prove relates α to θ. The phase difference ϕ between the rays from the top and bottom of the entire slit may be related to a path length difference with Eq. 37-4; it tells us that

$$\phi = \left(\frac{2\pi}{\lambda}\right)(a \sin \theta),$$

where a is the sum of the widths Δx of the infinitesimal zones. However, $\phi = 2\alpha$, so this equation reduces to Eq. 37-6.

CHECKPOINT 3: Two wavelengths, 650 and 430 nm, are used separately in a single-slit diffraction experiment. The figure shows the results as graphs of intensity I versus angle θ for the two diffraction patterns. If both wavelengths are then used simultaneously, what color will be seen in the combined diffraction pattern at (a) angle A and (b) angle B?

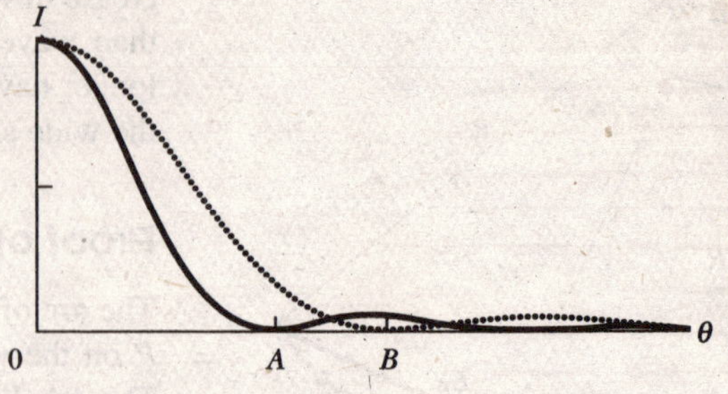

Sample Problem 37-2

Find the intensities of the first three secondary maxima (side maxima) in the single-slit diffraction pattern of Fig. 37-1, measured relative to the intensity of the central maximum.

SOLUTION: One **Key Idea** here is that the secondary maxima lie approximately halfway between the minima, whose angular locations are given by Eq. 37-7 ($\alpha = m\pi$). The locations of the secondary maxima are then given (approximately) by

$$\alpha = (m + \tfrac{1}{2})\pi, \qquad \text{for } m = 1, 2, 3, \ldots,$$

with α in radian measure.

A second **Key Idea** is that we can relate the intensity I at any point in the diffraction pattern to the intensity I_m of the central maximum via Eq. 37-5. Thus, we can substitute the approximate values of α for the secondary maxima into Eq. 37-5 to obtain the relative intensities at those maxima. We get

$$\frac{I}{I_m} = \left(\frac{\sin\alpha}{\alpha}\right)^2 = \left(\frac{\sin(m + \frac{1}{2})\pi}{(m + \frac{1}{2})\pi}\right)^2, \qquad \text{for } m = 1, 2, 3, \ldots.$$

The first of the secondary maxima occurs for $m = 1$, and its relative intensity is

$$\frac{I_1}{I_m} = \left(\frac{\sin(1 + \frac{1}{2})\pi}{(1 + \frac{1}{2})\pi}\right)^2 = \left(\frac{\sin 1.5\pi}{1.5\pi}\right)^2$$

$$= 4.50 \times 10^{-2} \approx 4.5\%. \qquad \text{(Answer)}$$

For $m = 2$ and $m = 3$ we find that

$$\frac{I_2}{I_m} = 1.6\% \quad \text{and} \quad \frac{I_3}{I_m} = 0.83\%. \qquad \text{(Answer)}$$

Successive secondary maxima decrease rapidly in intensity. Figure 37-1 was deliberately overexposed to reveal them.

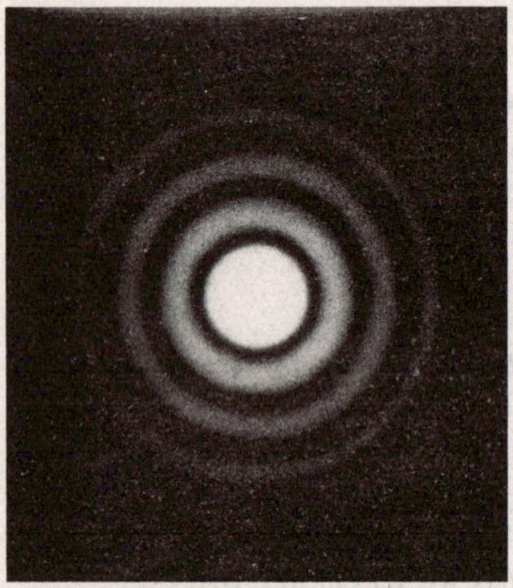

Fig. 37-9 The diffraction pattern of a circular aperture. Note the central maximum and the circular secondary maxima. The figure has been overexposed to bring out these secondary maxima, which are much less intense than the central maximum.

37-5 Diffraction by a Circular Aperture

Here we consider diffraction by a circular aperture—that is, a circular opening such as a circular lens, through which light can pass. Figure 37-9 shows the image of a distant point source of light (a star, for instance) formed on photographic film placed in the focal plane of a converging lens. This image is not a point, as geometrical optics would suggest, but a circular disk surrounded by several progressively fainter secondary rings. Comparison with Fig. 37-1 leaves little doubt that we are dealing with a diffraction phenomenon. Here, however, the aperture is a circle of diameter d rather than a rectangular slit.

The analysis of such patterns is complex. It shows, however, that the first minimum for the diffraction pattern of a circular aperture of diameter d is located by

$$\sin\theta = 1.22\frac{\lambda}{d} \qquad \text{(first minimum; circular aperture).} \qquad (37\text{-}12)$$

The angle θ here is the angle from the central axis to any point on that (circular) minimum. Compare this with Eq. 37-1,

$$\sin\theta = \frac{\lambda}{a} \qquad \text{(first minimum; single slit),} \qquad (37\text{-}13)$$

which locates the first minimum for a long narrow slit of width a. The main difference is the factor 1.22, which enters because of the circular shape of the aperture.

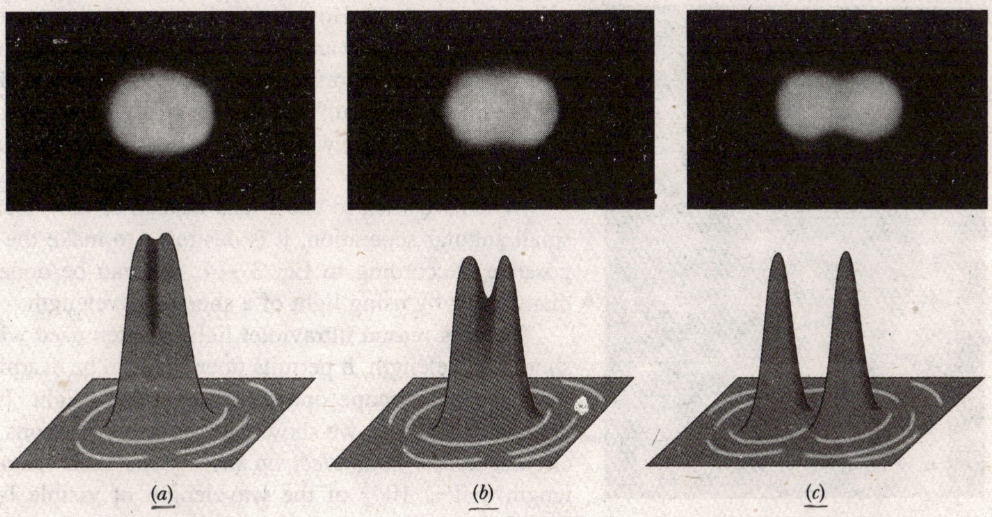

Fig. 37-10 At the top, the images of two point sources (stars), formed by a converging lens. At the bottom, representations of the image intensities. In (*a*) the angular separation of the sources is too small for them to be distinguished, in (*b*) they can be marginally distinguished, and in (*c*) they are clearly distinguished. Rayleigh's criterion is satisfied in (*b*), with the central maximum of one diffraction pattern coinciding with the first minimum of the other.

Resolvability

The fact that lens images are diffraction patterns is important when we wish to *resolve* (distinguish) two distant point objects whose angular separation is small. Figure 37-10 shows, in three different cases, the visual appearance and corresponding intensity pattern for two distant point objects (stars, say) with small angular separation. In Figure 37-10*a*, the objects are not resolved because of diffraction; that is, their diffraction patterns (mainly their central maxima) overlap so much that the two objects cannot be distinguished from a single point object. In Fig. 37-10*b* the objects are barely resolved, and in Fig. 37-10*c* they are fully resolved.

In Fig. 37-10*b* the angular separation of the two point sources is such that the central maximum of the diffraction pattern of one source is centered on the first minimum of the diffraction pattern of the other, a condition called **Rayleigh's criterion** for resolvability. From Eq. 37-12, two objects that are barely resolvable by this criterion must have an angular separation θ_R of

$$\theta_R = \sin^{-1} \frac{1.22\lambda}{d}.$$

Since the angles are small, we can replace $\sin \theta_R$ with θ_R expressed in radians:

$$\theta_R = 1.22 \frac{\lambda}{d} \quad \text{(Rayleigh's criterion).} \qquad (37\text{-}14)$$

Rayleigh's criterion for resolvability is only an approximation, because resolvability depends on many factors, such as the relative brightness of the sources and their surroundings, turbulence in the air between the sources and the observer, and the functioning of the observer's visual system. Experimental results show that the least angular separation that can actually be resolved by a person is generally somewhat greater than the value given by Eq. 37-14. However, for the sake of calculations here, we shall take Eq. 37-14 as being a precise criterion: If the angular separation θ between the sources is greater than θ_R, we can resolve the sources; if it is less, we cannot.

Rayleigh's criterion can explain the colors in Seurat's *Sunday Afternoon on the Island of La Grande Jatte* (or any other pointillistic painting). When you stand close enough to the painting, the angular separations θ of adjacent dots are greater than θ_R and thus the dots can be seen individually. Their colors are the colors of the

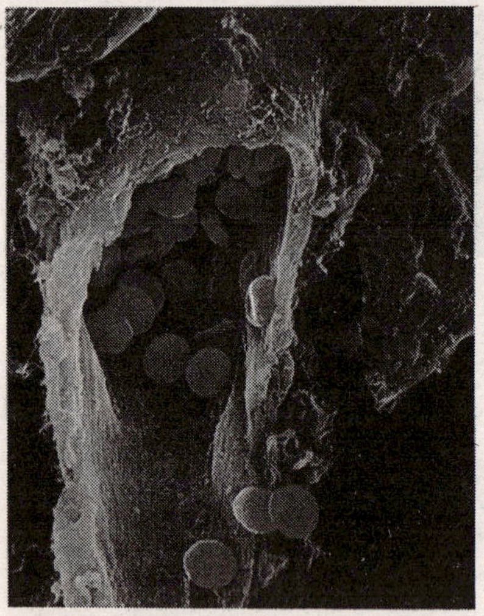

Fig. 37-11 A false-color scanning electron micrograph of a vein containing red blood cells.

paints Seurat used. However, when you stand far enough from the painting, the angular separations θ are less than θ_R and the dots cannot be seen individually. The resulting blend of colors coming into your eye from any group of dots can then cause your brain to "make up" a color for that group—a color that may not actually exist in the group. In this way, Seurat uses your visual system to create the colors of his art.

When we wish to use a lens instead of our visual system to resolve objects of small angular separation, it is desirable to make the diffraction pattern as small as possible. According to Eq. 37-14, this can be done either by increasing the lens diameter or by using light of a shorter wavelength.

For this reason ultraviolet light is often used with microscopes; because of its shorter wavelength, it permits finer detail to be examined than would be possible for the same microscope operated with visible light. In Chapter 39 of the extended version of this text, we show that beams of electrons behave like waves under some circumstances. In an *electron microscope* such beams may have an effective wavelength that is 10^{-5} of the wavelength of visible light. They permit the detailed examination of tiny structures, like that in Fig. 37-11, that would be blurred by diffraction if viewed with an optical microscope.

CHECKPOINT 4: Suppose that you can barely resolve two red dots, owing to diffraction by the pupil of your eye. If we increase the general illumination around you so that the pupil decreases in diameter, does the resolvability of the dots improve or diminish? Consider only diffraction. (You might experiment to check your answer.)

Sample Problem 37-3

A circular converging lens, with diameter d = 32 mm and focal length f = 24 cm, forms images of distant point objects in the focal plane of the lens. Light of wavelength λ = 550 nm is used.

(a) Considering diffraction by the lens, what angular separation must two distant point objects have to satisfy Rayleigh's criterion?

SOLUTION: Figure 37-12 shows two distant point objects P_1 and P_2, the lens, and a viewing screen in the focal plane of the lens. It also shows, on the right, plots of light intensity I versus position on the screen for the central maxima of the images formed by the lens. Note that the angular separation θ_o of the objects equals the angular separation θ_i of the images. Thus, the **Key Idea** here is that if the images are to satisfy Rayleigh's criterion for resolvability, the angular separations on both sides of the lens must be given by Eq. 37-14 (assuming small angles). Substituting the given data, we obtain from Eq. 37-14

$$\theta_o = \theta_i = \theta_R = 1.22\frac{\lambda}{d}$$

$$= \frac{(1.22)(550 \times 10^{-9}\ \text{m})}{32 \times 10^{-3}\ \text{m}} = 2.1 \times 10^{-5}\ \text{rad}. \quad \text{(Answer)}$$

At this angular separation, each central maximum in the two intensity curves of Fig. 37-12 is centered on the first minimum of the other curve.

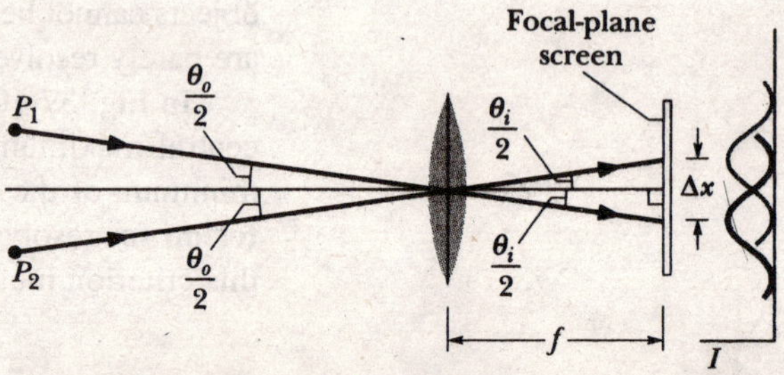

Fig. 37-12 Sample Problem 37-3. Light from two distant point objects P_1 and P_2 passes through a converging lens and forms images on a viewing screen in the focal plane of the lens. Only one representative ray from each object is shown. The images are not points but diffraction patterns, with intensities approximately as plotted at the right. The angular separation of the objects is θ_o and that of the images is θ_i; the central maxima of the images have a separation Δx.

(b) What is the separation Δx of the centers of the *images* in the focal plane? (That is, what is the separation of the *central* peaks in the two curves?)

SOLUTION: The **Key Idea** here is to relate the separation Δx to the angle θ_i, which we now know. From either triangle between the lens and the screen in Fig. 37-12, we see that $\tan \theta_i/2 = \Delta x/2f$. Rearranging this and making the approximation $\tan\theta \approx \theta$, we find

$$\Delta x = f\theta_i, \quad (37\text{-}15)$$

where θ_i is in radian measure. Substituting known data then yields

$$\Delta x = (0.24\ \text{m})(2.1 \times 10^{-5}\ \text{rad}) = 5.0\ \mu\text{m}. \quad \text{(Answer)}$$

37-6 Diffraction by a Double Slit

In the double-slit experiments of Chapter 36, we implicitly assumed that the slits were narrow compared to the wavelength of the light illuminating them; that is, $a \ll \lambda$. For such narrow slits, the central maximum of the diffraction pattern of either slit covers the entire viewing screen. Moreover, the interference of light from the two slits produces bright fringes with approximately the same intensity (Fig. 36-9).

In practice with visible light, however, the condition $a \ll \lambda$ is often not met. For relatively wide slits, the interference of light from two slits produces bright fringes that do not all have the same intensity. That is, the intensities of the fringes produced by double-slit interference (as discussed in Chapter 36) are modified by diffraction of the light passing through each slit (as discussed in this chapter).

As an example, the intensity plot of Fig. 37-13*a* suggests the double-slit interference pattern that would occur if the slits were infinitely narrow (and thus $a \ll \lambda$); all the bright interference fringes would have the same intensity. The intensity plot of Fig. 37-13*b* is that for diffraction by a single actual slit; the diffraction pattern has a broad central maximum and weaker secondary maxima at $\pm 17°$. The plot of Fig. 37-13*c* suggests the interference pattern for two actual slits. That plot was constructed by using the curve of Fig. 37-13*b* as an *envelope* on the intensity plot in Fig. 37-13*a*. The positions of the fringes are not changed; only the intensities are affected.

Figure 37-14*a* shows an actual pattern in which both double-slit interference and diffraction are evident. If one slit is covered, the single-slit diffraction pattern of Fig. 37-14*b* results. Note the correspondence between Figs. 37-14*a* and 37-13*c*, and between Figs. 37-14*b* and 37-13*b*. In comparing these figures, bear in mind that Fig. 37-14 has been deliberately overexposed to bring out the faint secondary maxima and that two secondary maxima (rather than one) are shown.

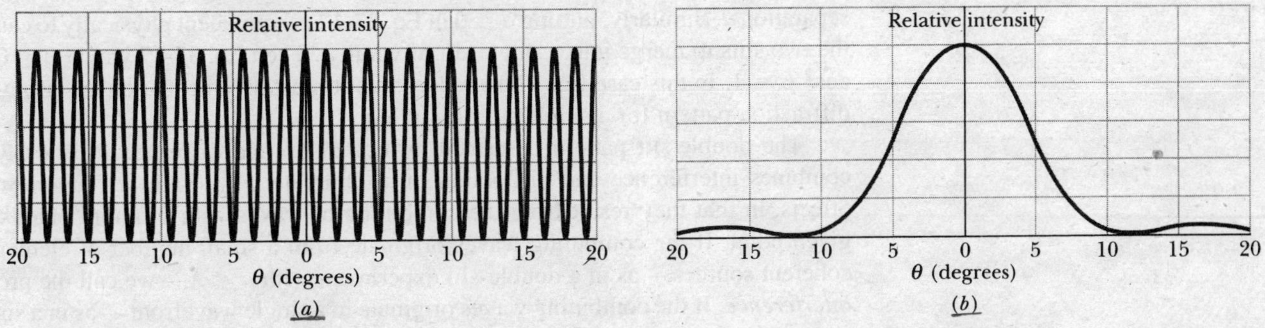

Fig. 37-13 (*a*) The intensity plot to be expected in a double-slit interference experiment with vanishingly narrow slits. (*b*) The intensity plot for diffraction by a typical slit of width a (not vanishingly narrow). (*c*) The intensity plot to be expected for two slits of width a. The curve of (*b*) acts as an envelope, limiting the intensity of the double-slit fringes in (*a*). Note that the first minima of the diffraction pattern of (*b*) eliminate the double-slit fringes that would occur near 12° in (*c*).

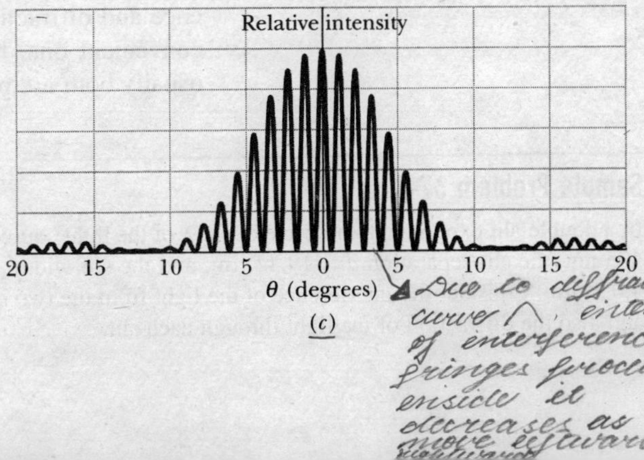

Fig. 37-14 (*a*) Interference fringes for an actual double-slit system; compare with Fig. 37-13*c*. (*b*) The diffraction pattern of a single slit; compare with Fig. 37-13*b*.

secondary maxima decrease in width as well as in intensity

With diffraction effects taken into account, the intensity of a double-slit interference pattern is given by

$$I(\theta) = I_m(\cos^2 \beta)\left(\frac{\sin \alpha}{\alpha}\right)^2 \quad \text{(double slit)}, \tag{37-16}$$

in which

$$\beta = \frac{\pi d}{\lambda} \sin \theta \tag{37-17}$$

and

$$\alpha = \frac{\pi a}{\lambda} \sin \theta. \tag{37-18}$$

Here d is the distance between the centers of the slits, and a is the slit width. Note carefully that the right side of Eq. 37-16 is the product of I_m and two factors. (1) The *interference factor* $\cos^2 \beta$ is due to the interference between two slits with slit separation d (as given by Eqs. 36-17 and 36-18). (2) The *diffraction factor* $[(\sin \alpha)/\alpha]^2$ is due to diffraction by a single slit of width a (as given by Eqs. 37-5 and 37-6).

Let us check these factors. If we let $a \to 0$ in Eq. 37-18, for example, then $\alpha \to 0$ and $(\sin \alpha)/\alpha \to 1$. Equation 37-16 then reduces, as it must, to an equation describing the interference pattern for a pair of vanishingly narrow slits with slit separation d. Similarly, putting $d = 0$ in Eq. 37-17 is equivalent physically to causing the two slits to merge into a single slit of width a. Then Eq. 37-17 yields $\beta = 0$ and $\cos^2 \beta = 1$. In this case Eq. 37-16 reduces, as it must, to an equation describing the diffraction pattern for a single slit of width a.

The double-slit pattern described by Eq. 37-16 and displayed in Fig. 37-14*a* combines interference and diffraction in an intimate way. Both are superposition effects, in that they result from the combining of waves with different phases at a given point. If the combining waves originate from a small number of elementary coherent sources—as in a double-slit experiment with $a \ll \lambda$—we call the process *interference*. If the combining waves originate in a single wavefront—as in a single-slit experiment—we call the process *diffraction*. This distinction between interference and diffraction (which is somewhat arbitrary and not always adhered to) is a convenient one, but we should not forget that both are superposition effects and usually both are present simultaneously (as in Fig. 37-14*a*).

Sample Problem 37-4

In a double-slit experiment, the wavelength λ of the light source is 405 nm, the slit separation d is 19.44 μm, and the slit width a is 4.050 μm. Consider the interference of the light from the two slits and also the diffraction of the light through each slit.

(a) How many bright interference fringes are within the central peak of the diffraction envelope?

SOLUTION: Let us first analyze the two basic mechanisms responsible for the optical pattern produced in the experiment:

Single-slit diffraction: The Key Idea here is that the limits of the central peak are the first minima in the diffraction pattern due to either slit, individually. (See Fig. 37-13.) The angular locations of those minima are given by Eq. 37-3 ($a \sin \theta = m\lambda$). Let us write this equation as $a \sin \theta = m_1 \lambda$, with the subscript 1 referring to the one-slit diffraction. For the first minima in the diffraction pattern, we substitute $m_1 = 1$, obtaining

$$a \sin \theta = \lambda. \qquad (37\text{-}19)$$

Double-slit interference: The Key Idea here is that the angular locations of the bright fringes of the double-slit interference pattern are given by Eq. 36-14, which we can write as

$$d \sin \theta = m_2 \lambda, \qquad \text{for } m_2 = 0, 1, 2, \ldots. \qquad (37\text{-}20)$$

Here the subscript 2 refers to the double-slit interference.

We can locate the first diffraction minimum within the double-slit fringe pattern by dividing Eq. 37-20 by Eq. 37-19 and solving for m_2. By doing so and then substituting the given data, we obtain

$$m_2 = \frac{d}{a} = \frac{19.44\ \mu\text{m}}{4.050\ \mu\text{m}} = 4.8.$$

This tells us that the bright interference fringe for $m_2 = 4$ fits into the central peak of the one-slit diffraction pattern, but the fringe for $m_2 = 5$ does not fit. Within the central diffraction peak we have the central bright fringe ($m_2 = 0$), and four bright fringes (up to $m_2 = 4$) on each side of it. Thus, a total of nine bright fringes of the double-slit interference pattern are within the central peak of the diffraction envelope. The bright fringes to one side of the central bright fringe are shown in Fig. 37-15.

(b) How many bright fringes are within either of the first side peaks of the diffraction envelope?

SOLUTION: The Key Idea here is that the outer limits of the first side diffraction peaks are the second diffraction minima, each of which is at the angle θ given by $a \sin \theta = m_1 \lambda$ with $m_1 = 2$:

$$a \sin \theta = 2\lambda. \qquad (37\text{-}21)$$

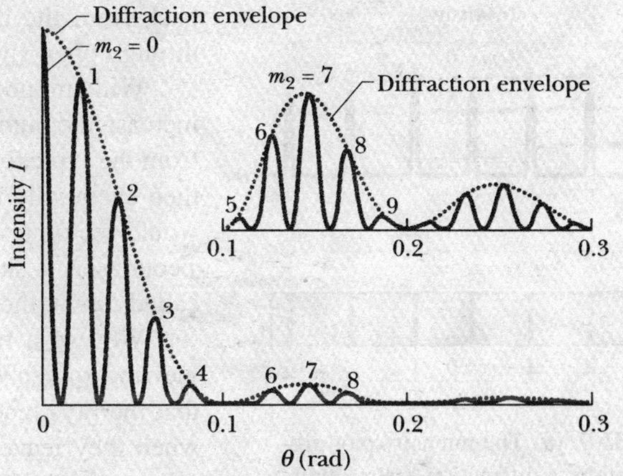

Fig. 37-15 Sample Problem 37-4. One side of the intensity plot for a two-slit interference experiment; the diffraction envelope is indicated by the dotted curve. The smaller inset shows (vertically expanded) the intensity plot within the first and second side peaks of the diffraction envelope.

Dividing Eq. 37-20 by Eq. 37-21, we find

$$m_2 = \frac{2d}{a} = \frac{(2)(19.44\ \mu\text{m})}{4.050\ \mu\text{m}} = 9.6.$$

This tells us that the second diffraction minimum occurs just before the bright interference fringe for $m_2 = 10$ in Eq. 37-20. Within either first side diffraction peak we have the fringes from $m_2 = 5$ to $m_2 = 9$ for a total of five bright fringes of the double-slit interference pattern (shown in the inset of Fig. 37-15). However, if the $m_2 = 5$ bright fringe, which is almost eliminated by the first diffraction minimum, is considered too dim to count, then only four bright fringes are in the first side diffraction peak.

CHECKPOINT 5: If we increase the wavelength of the light source in this sample problem to 550 nm, do (a) the width of the central diffraction peak and (b) the number of bright interference fringes within that peak increase, decrease, or remain the same?

37-7 Diffraction Gratings

One of the most useful tools in the study of light and of objects that emit and absorb light is the **diffraction grating.** This device is somewhat like the double-slit arrangement of Fig. 36-8 but has a much greater number N of slits, often called *rulings*, perhaps as many as several thousand per millimeter. An idealized grating consisting of only five slits is represented in Fig. 37-16. When monochromatic light is sent through the slits, it forms narrow interference fringes that can be analyzed to determine the wavelength of the light. (Diffraction gratings can also be opaque surfaces with narrow parallel grooves arranged like the slits in Fig. 37-16. Light then scatters

Here interference fringes do not have the same intensity

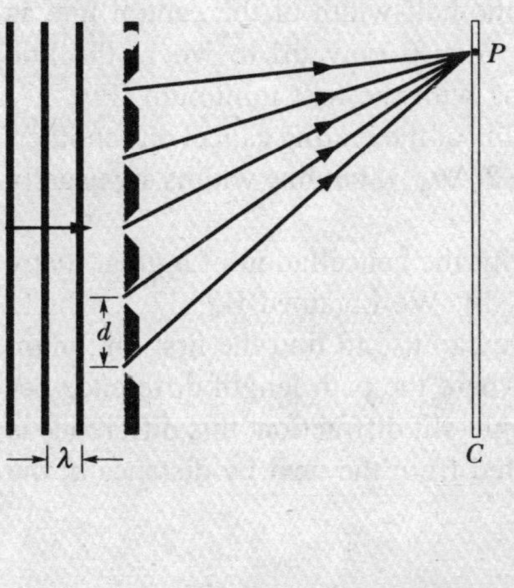

Fig. 37-16 An idealized diffraction grating, consisting of only five rulings, that produces an interference pattern on a distant viewing screen C.

back from the grooves to form interference fringes rather than being transmitted through open slits.)

With monochromatic light incident on a diffraction grating, if we gradually increase the number of slits from two to a large number N, the intensity plot changes from the typical double-slit plot of Fig. 37-13c to a much more complicated one and then eventually to a simple graph like that shown in Fig. 37-17a. The pattern you would see on a viewing screen using monochromatic red light from, say, a helium–neon laser, is shown in Fig. 37-17b. The maxima are now very narrow (and so are called *lines*); they are separated by relatively wide dark regions.

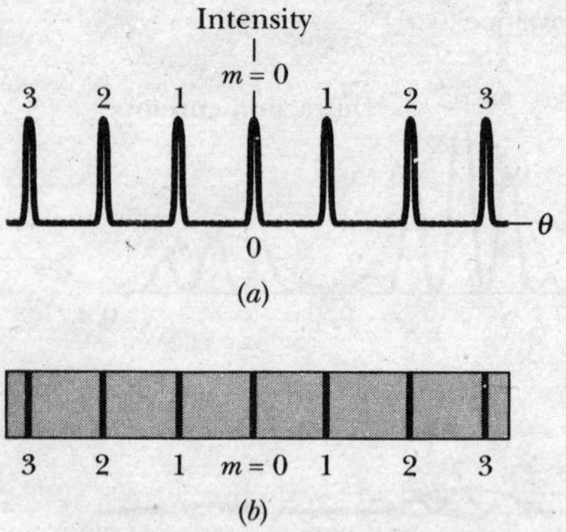

Fig. 37-17 (*a*) The intensity plot produced by a diffraction grating with a great many rulings consists of narrow peaks, here labeled with their order numbers m. (*b*) The corresponding bright fringes seen on the screen are called lines and are here also labeled with order numbers m. Lines of the zeroth, first, second, and third orders are shown.

We use a familiar procedure to find the locations of the bright lines on the viewing screen. We first assume that the screen is far enough from the grating so that the rays reaching a particular point P on the screen are approximately parallel when they leave the grating (Fig. 37-18). Then we apply to each pair of adjacent rulings the same reasoning we used for double-slit interference. The separation d between rulings is called the *grating spacing*. (If N rulings occupy a total width w, then $d = w/N$.) The path length difference between adjacent rays is again $d \sin \theta$ (Fig. 37-18), where θ is the angle from the central axis of the grating (and of the diffraction pattern) to point P. A line will be located at P if the path length difference between adjacent rays is an integer number of wavelengths—that is, if

$$d \sin \theta = m\lambda, \qquad \text{for } m = 0, 1, 2, \ldots \qquad \text{(maxima—lines)}, \qquad (37\text{-}22)$$

where λ is the wavelength of the light. Each integer m represents a different line; hence these integers can be used to label the lines, as in Fig. 37-17. The integers are then called the *order numbers*, and the lines are called the zeroth-order line (the central line, with $m = 0$), the first-order line, the second-order line, and so on.

If we rewrite Eq. 37-22 as $\theta = \sin^{-1}(m\lambda/d)$, we see that, for a given diffraction grating, the angle from the central axis to any line (say, the third-order line) depends on the wavelength of the light being used. Thus, when light of an unknown wavelength is sent through a diffraction grating, measurements of the angles to the higher-order lines can be used in Eq. 37-22 to determine the wavelength. Even light of several unknown wavelengths can be distinguished and identified in this way. We cannot do that with the double-slit arrangement of Section 36-4, even though the same equation and wavelength dependence apply there. In double-slit interference, the bright fringes due to different wavelengths overlap too much to be distinguished.

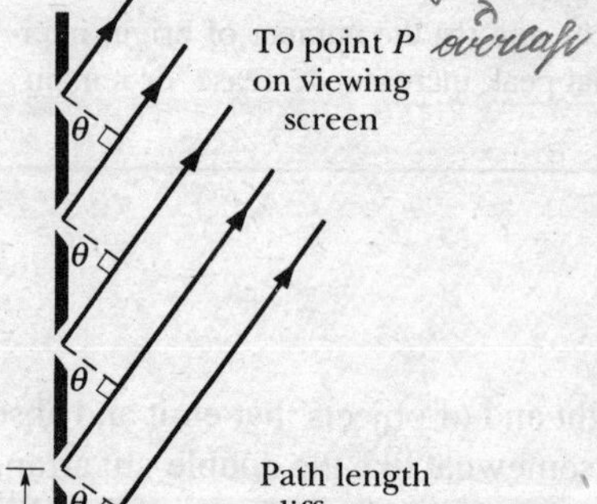

Fig. 37-18 The rays from the rulings in a diffraction grating to a distant point P are approximately parallel. The path length difference between each two adjacent rays is $d \sin \theta$, where θ is measured as shown. (The rulings extend into and out of the page.)

Width of the Lines

A grating's ability to resolve (separate) lines of different wavelengths depends on the width of the lines. We shall here derive an expression for the *half-width* of the central line (the line for which $m = 0$) and then state an expression for the half-widths of the higher-order lines. We measure the half-width of the central line as the angle $\Delta\theta_{hw}$ from the center of the line at $\theta = 0$ outward to where the line effectively ends and darkness effectively begins with the first minimum (Fig. 37-19). At such a minimum, the N rays from the N slits of the grating cancel one another. (The actual width of the central line is, of course $2(\Delta\theta_{hw})$, but line widths are usually compared via half-widths.)

In Section 37-2 we were also concerned with the cancellation of a great many rays, there due to diffraction through a single slit. We obtained Eq. 37-3, which, owing to the similarity of the two situations, we can use to find the first minimum here. It tells us that the first minimum occurs where the path length difference between the top and bottom rays equals λ. For single-slit diffraction, this difference is $a \sin \theta$. For a grating of N rulings, each separated from the next by distance d, the

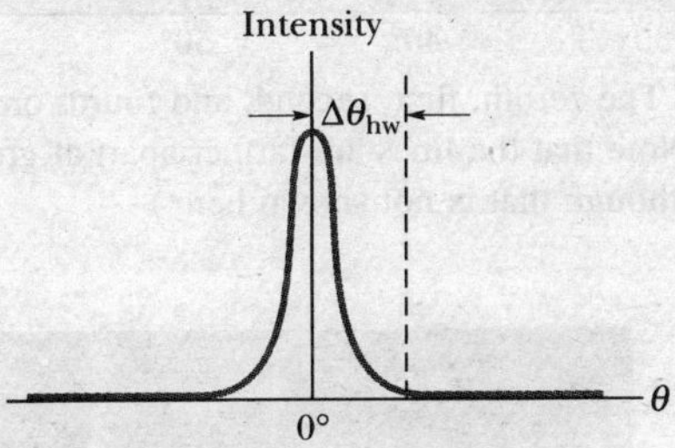

Fig. 37-19 The half-width $\Delta\theta_{hw}$ of the central line is measured from the center of that line to the adjacent minimum on a plot of I versus θ like Fig. 37-17*a*.

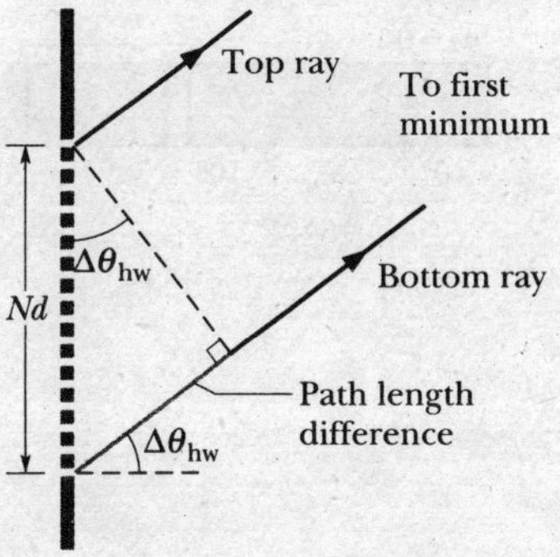

Fig. 37-20 The top and bottom rulings of a diffraction grating of N rulings are separated by distance Nd. The top and bottom rays passing through these rulings have a path length difference of $Nd \sin \Delta\theta_{hw}$, where $\Delta\theta_{hw}$ is the angle to the first minimum. (The angle is here greatly exaggerated for clarity.)

distance between the top and bottom rulings is Nd (Fig. 37-20), so the path length difference between the top and bottom rays here is $Nd \sin \Delta\theta_{hw}$. Thus, the first minimum occurs where

$$Nd \sin \Delta\theta_{hw} = \lambda. \tag{37-23}$$

Because $\Delta\theta_{hw}$ is small, $\sin \Delta\theta_{hw} = \Delta\theta_{hw}$ (in radian measure). Substituting this in Eq. 37-23 gives the half-width of the central line as

$$\Delta\theta_{hw} = \frac{\lambda}{Nd} \quad \text{(half-width of central line).} \tag{37-24}$$

We state without proof that the half-width of any other line depends on its location relative to the central axis and is

$$\Delta\theta_{hw} = \frac{\lambda}{Nd \cos \theta} \quad \text{(half-width of line at } \theta\text{).} \tag{37-25}$$

Note that for light of a given wavelength λ and a given ruling separation d, the widths of the lines decrease with an increase in the number N of rulings. Thus, of two diffraction gratings, the grating with the larger value of N is better able to distinguish between wavelengths because its diffraction lines are narrower and so produce less overlap.

An Application of Diffraction Gratings

Diffraction gratings are widely used to determine the wavelengths that are emitted by sources of light ranging from lamps to stars. Figure 37-21 shows a simple *grating spectroscope* in which a grating is used for this purpose. Light from source S is focused by lens L_1 on a vertical slit S_1 placed in the focal plane of lens L_2. The light emerging from tube C (called a *collimator*) is a plane wave and is incident perpendicularly on grating G, where it is diffracted into a diffraction pattern, with the $m = 0$ order diffracted at angle $\theta = 0$ along the central axis of the grating.

We can view the diffraction pattern that would appear on a viewing screen at any angle θ simply by orienting telescope T in Fig. 37-21 to that angle. Lens L_3 of the telescope then focuses the light diffracted at angle θ (and at slightly smaller and

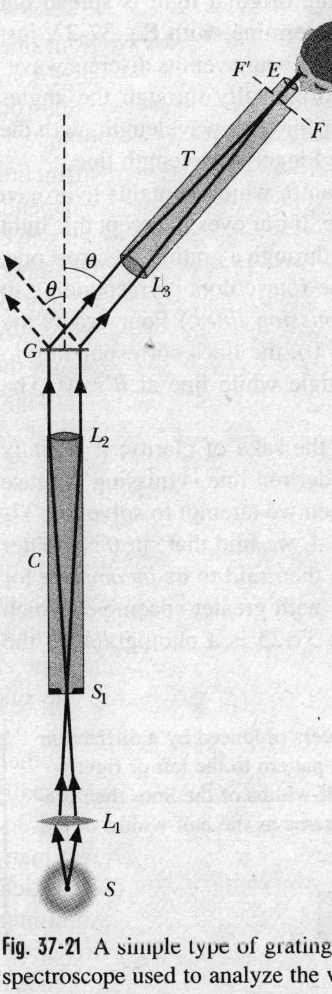

Fig. 37-21 A simple type of grating spectroscope used to analyze the wavelengths of light emitted by source S.

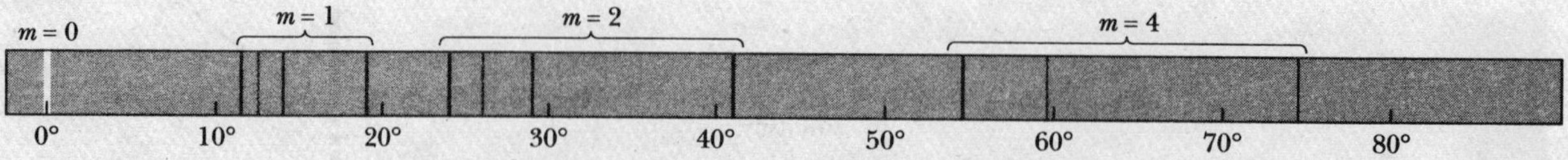

Fig. 37-22 The zeroth, first, second, and fourth orders of the visible emission lines from hydrogen. Note that the lines are farther apart at greater angles. (They are also dimmer and wider, although that is not shown here.)

Fig. 37-23 The visible emission lines of cadmium, as seen through a grating spectroscope.

larger angles) onto a focal plane FF' within the telescope. When we look through eyepiece E, we see a magnified view of this focused image.

By changing the angle θ of the telescope, we can examine the entire diffraction pattern. For any order number other than $m = 0$, the original light is spread out according to wavelength (or color) so that we can determine, with Eq. 37-22, just what wavelengths are being emitted by the source. If the source emits discrete wavelengths, what we see as we rotate the telescope horizontally through the angles corresponding to an order m is a vertical line of color for each wavelength, with the shorter-wavelength line at a smaller angle θ than the longer-wavelength line.

For example, the light emitted by a hydrogen lamp, which contains hydrogen gas, has four discrete wavelengths in the visible range. If our eyes intercept this light directly, it appears to be white. If, instead, we view it through a grating spectroscope, we can distinguish, in several orders, the lines of the four colors corresponding to these visible wavelengths. (Such lines are called *emission lines*.) Four orders are represented in Fig. 37-22. In the central order ($m = 0$), the lines corresponding to all four wavelengths are superimposed, giving a single white line at $\theta = 0$. The colors are separated in the higher orders.

The third order is not shown in Fig. 37-22 for the sake of clarity; it actually overlaps the second and fourth orders. The fourth-order red line is missing because it is not formed by the grating used here. That is, when we attempt to solve Eq. 37-22 for the angle θ for the red wavelength when $m = 4$, we find that $\sin\theta$ is greater than unity, which is not possible. The fourth order is then said to be *incomplete* for this grating; it might not be incomplete for a grating with greater spacing d, which will spread the lines less than in Fig. 37-22. Figure 37-23 is a photograph of the visible emission lines produced by cadmium.

CHECKPOINT 6: The figure shows lines of different orders produced by a diffraction grating in monochromatic red light. (a) Is the center of the pattern to the left or right? (b) If we switch to monochromatic green light, will the half-widths of the lines then produced in the same orders be greater than, less than, or the same as the half-widths of the lines shown?

37-8 Gratings: Dispersion and Resolving Power

The fine rulings, each 0.5 μm wide, on a compact disc function as a diffraction grating. When a small source of white light illuminates a disc, the diffracted light forms colored "lanes" that are the composite of the diffraction patterns from the rulings.

Dispersion

To be useful in distinguishing wavelengths that are close to each other (as in a grating spectroscope), a grating must spread apart the diffraction lines associated with the various wavelengths. This spreading, called **dispersion,** is defined as

$$D = \frac{\Delta\theta}{\Delta\lambda} \quad \text{(dispersion defined).} \tag{37-26}$$

Here $\Delta\theta$ is the angular separation of two lines whose wavelengths differ by $\Delta\lambda$. The greater D is, the greater is the distance between two emission lines whose wavelengths differ by $\Delta\lambda$. We show below that the dispersion of a grating at angle θ is given by

$$D = \frac{m}{d \cos \theta} \quad \text{(dispersion of a grating).} \tag{37-27}$$

Thus, to achieve higher dispersion we must use a grating of smaller grating spacing d and work in a higher order m. Note that the dispersion does not depend on the number of rulings N in the grating. The SI unit for D is the degree per meter or the radian per meter.

Resolving Power

To *resolve* lines whose wavelengths are close together (that is, to make the lines distinguishable), the line should also be as narrow as possible. Expressed otherwise, the grating should have a high **resolving power** R, defined as

$$R = \frac{\lambda_{avg}}{\Delta\lambda} \quad \text{(resolving power defined).} \tag{37-28}$$

Here λ_{avg} is the mean wavelength of two emission lines that can barely be recognized as separate, and $\Delta\lambda$ is the wavelength difference between them. The greater R is, the closer two emission lines can be and still be resolved. We shall show below that the resolving power of a grating is given by the simple expression

$$R = Nm \quad \text{(resolving power of a grating).} \tag{37-29}$$

To achieve high resolving power, we must use many rulings (large N in Eq. 37-29).

Proof of Eq. 37-27

Let us start with Eq. 37-22, the expression for the locations of the lines in the diffraction pattern of a grating:

$$d \sin \theta = m\lambda.$$

Let us regard θ and λ as variables and take differentials of this equation. We find

$$d \cos \theta \, d\theta = m \, d\lambda.$$

For small enough angles, we can write these differentials as small differences, obtaining

$$d \cos \theta \, \Delta\theta = m \, \Delta\lambda \tag{37-30}$$

or

$$\frac{\Delta\theta}{\Delta\lambda} = \frac{m}{d \cos \theta}.$$

The ratio on the left is simply D (see Eq. 37-26), so we have indeed derived Eq. 37-27.

Proof of Eq. 37-29

We start with Eq. 37-30, which was derived from Eq. 37-22, the expression for the locations of the lines in the diffraction pattern formed by a grating. Here $\Delta\lambda$ is the small wavelength difference between two waves that are diffracted by the grating, and $\Delta\theta$ is the angular separation between them in the diffraction pattern. If $\Delta\theta$ is to be the smallest angle that will permit the two lines to be resolved, it must (by Rayleigh's criterion) be equal to the half-width of each line, which is given by Eq. 37-25:

$$\Delta\theta_{\text{hw}} = \frac{\lambda}{Nd \cos \theta}.$$

If we substitute $\Delta\theta_{\text{hw}}$ as given here for $\Delta\theta$ in Eq. 37-30, we find that

$$\frac{\lambda}{N} = m \, \Delta\lambda,$$

from which it readily follows that

$$R = \frac{\lambda}{\Delta\lambda} = Nm.$$

This is Eq. 37-29, which we set out to derive.

Dispersion and Resolving Power Compared

The resolving power of a grating must not be confused with its dispersion. Table 37-1 shows the characteristics of three gratings, all illuminated with light of wavelength $\lambda = 589$ nm, whose diffracted light is viewed in the first order ($m = 1$ in Eq. 37-22). You should verify that the values of D and R as given in the table can be calculated with Eqs. 37-27 and 37-29, respectively. (In the calculations for D, you will need to convert radians per meter to degrees per micrometer.)

For the conditions noted in Table 37-1, gratings A and B have the same *dispersion* and A and C have the same *resolving power*.

Figure 37-24 shows the intensity patterns (also called *line shapes*) that would be produced by these gratings for two lines of wavelengths λ_1 and λ_2, in the vicinity of $\lambda = 589$ nm. Grating B, with the higher resolving power, produces narrower lines and thus is capable of distinguishing lines that are much closer together in wavelength than those in the figure. Grating C, with the higher dispersion, produces the greater angular separation between the lines.

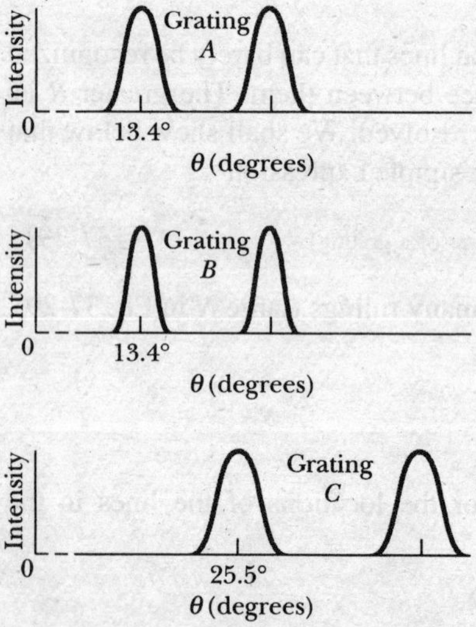

Fig. 37-24 The intensity patterns for light of two wavelengths sent through the gratings of Table 37-1. Grating B has the highest resolving power, and grating C the highest dispersion.

TABLE 37-1 Three Gratings[a]

Grating	N	d (nm)	θ	D (°/μm)	R
A	10 000	2540	13.4°	23.2	10 000
B	20 000	2540	13.4°	23.2	20 000
C	10 000	1370	25.5°	46.3	10 000

[a]Data are for $\lambda = 589$ nm and $m = 1$.

Sample Problem 37-5 ★

A diffraction grating has 1.26×10^4 rulings uniformly spaced over width $w = 25.4$ mm. It is illuminated at normal incidence by yellow light from a sodium vapor lamp. This light contains two closely spaced emission lines (known as the sodium doublet) of wavelengths 589.00 nm and 589.59 nm.

(a) At what angle does the first-order maximum occur (on either side of the center of the diffraction pattern) for the wavelength of 589.00 nm?

SOLUTION: The **Key Idea** here is that the maxima produced by the diffraction grating can be located with Eq. 37-22 ($d \sin \theta = m\lambda$). The grating spacing d for this diffraction grating is

$$d = \frac{w}{N} = \frac{25.4 \times 10^{-3}\text{ m}}{1.26 \times 10^4} = 2.016 \times 10^{-6}\text{ m} = 2016\text{ nm}.$$

The first-order maximum corresponds to $m = 1$. Substituting these values for d and m into Eq. 37-22 leads to

$$\theta = \sin^{-1}\frac{m\lambda}{d} = \sin^{-1}\frac{(1)(589.00\text{ nm})}{2016\text{ nm}} = 16.99° \approx 17.0°. \quad \text{(Answer)}$$

(b) Using the dispersion of the grating, calculate the angular separation between the two lines in the first order.

SOLUTION: One **Key Idea** here is that the angular separation $\Delta\theta$ between the two lines in the first order depends on their wavelength difference $\Delta\lambda$ and the dispersion D of the grating, according to Eq. 37-26 ($D = \Delta\theta/\Delta\lambda$). A second **Key Idea** is that the dispersion D depends on the angle θ at which it is to be evaluated. We can assume that, in the first order, the two sodium lines occur close enough to each other for us to evaluate D at the angle $\theta = 16.99°$ we found in part (a) for one of those lines. Then Eq. 37-27 gives the dispersion as

$$D = \frac{m}{d \cos \theta} = \frac{1}{(2016\text{ nm})(\cos 16.99°)} = 5.187 \times 10^{-4}\text{ rad/nm}.$$

From Eq. 37-26, we then have

$$\Delta\theta = D\,\Delta\lambda = (5.187 \times 10^{-4}\text{ rad/nm})(589.59\text{ nm} - 589.00\text{ nm}) = 3.06 \times 10^{-4}\text{ rad} = 0.0175°. \quad \text{(Answer)}$$

You can show that this result depends on the grating spacing d but not on the number of rulings there are in the grating.

(c) What is the least number of rulings a grating can have and still be able to resolve the sodium doublet in the first order?

SOLUTION: One **Key Idea** here is that the resolving power of a grating in any order m is physically set by the number of rulings N in the grating according to Eq. 37-29 ($R = Nm$). A second **Key Idea** is that the least wavelength difference $\Delta\lambda$ that can be resolved depends on the average wavelength involved and the resolving power R of the grating, according to Eq. 37-28 ($R = \lambda_{avg}/\Delta\lambda$). For the sodium doublet to be barely resolved, $\Delta\lambda$ must be their wavelength separation of 0.59 nm, and λ_{avg} must be their average wavelength of 589.30 nm.

Putting these ideas together, we find that the least number of rulings for a grating to resolve the sodium doublet is

$$N = \frac{R}{m} = \frac{\lambda_{avg}}{m\,\Delta\lambda} = \frac{589.30\text{ nm}}{(1)(0.59\text{ nm})} = 999\text{ rulings}. \quad \text{(Answer)}$$

37-9 X-Ray Diffraction

X rays are electromagnetic radiation whose wavelengths are of the order of 1 Å ($= 10^{-10}$ m). Compare this with a wavelength of 550 nm ($= 5.5 \times 10^{-7}$ m) at the center of the visible spectrum. Figure 37-25 shows that x rays are produced when electrons escaping from a heated filament F are accelerated by a potential difference V and strike a metal target T.

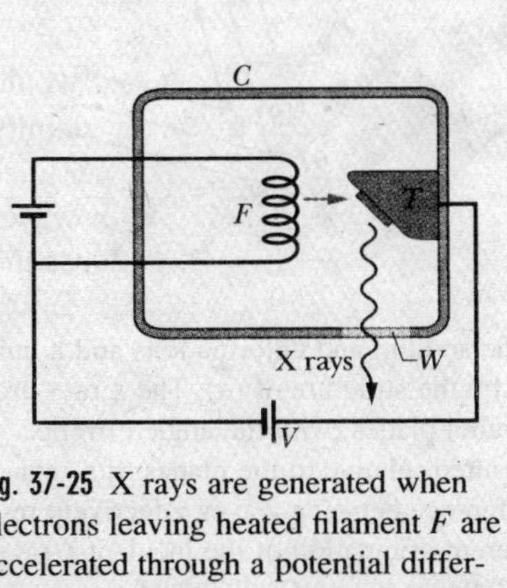

Fig. 37-25 X rays are generated when electrons leaving heated filament F are accelerated through a potential difference V and strike a metal target T. The "window" W in the evacuated chamber C is transparent to x rays.

A standard optical diffraction grating cannot be used to discriminate between different wavelengths in the x-ray wavelength range. For $\lambda = 1$ Å ($= 0.1$ nm) and $d = 3000$ nm, for example, Eq. 37-22 shows that the first-order maximum occurs at

$$\theta = \sin^{-1}\frac{m\lambda}{d} = \sin^{-1}\frac{(1)(0.1\text{ nm})}{3000\text{ nm}} = 0.0019°.$$

This is too close to the central maximum to be practical. A grating with $d \approx \lambda$ is desirable, but, since x-ray wavelengths are about equal to atomic diameters, such gratings cannot be constructed mechanically.

In 1912, it occurred to German physicist Max von Laue that a crystalline solid, which consists of a regular array of atoms, might form a natural three-dimensional "diffraction grating" for x rays. The idea is that, in a crystal such as sodium chloride (NaCl), a basic unit of atoms (called the *unit cell*) repeats itself throughout the array. In NaCl four sodium ions and four chlorine ions are associated with each unit cell. Figure 37-26*a* represents a section through a crystal of NaCl and identifies this basic unit. The unit cell is a cube measuring a_0 on each side.

When an x-ray beam enters a crystal such as NaCl, x rays are *scattered*—that is, redirected—in all directions by the crystal structure. In some directions the scattered waves undergo destructive interference, resulting in intensity minima; in other directions the interference is constructive, resulting in intensity maxima. This process of scattering and interference is a form of diffraction, although it is unlike the diffraction of light traveling through a slit or past an edge as we discussed earlier.

Although the process of diffraction of x rays by a crystal is complicated, the maxima turn out to be in directions *as if* the x rays were reflected by a family of parallel *reflecting planes* (or *crystal planes*) that extend through the atoms within the crystal and that contain regular arrays of the atoms. (The x rays are not actually reflected; we use these fictional planes only to simplify the analysis of the actual diffraction process.)

Figure 37-26*b* shows three of the family of planes, with *interplanar spacing d*, from which the incident rays shown are said to reflect. Rays 1, 2, and 3 reflect from the first, second, and third planes, respectively. At each reflection the angle of incidence and the angle of reflection are represented with θ. Contrary to the custom

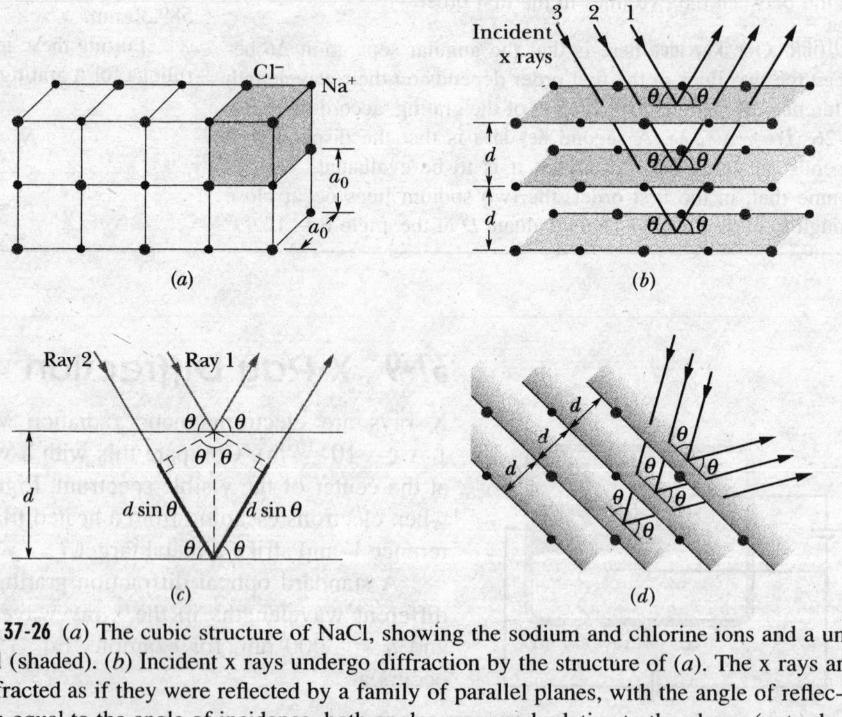

Fig. 37-26 (*a*) The cubic structure of NaCl, showing the sodium and chlorine ions and a unit cell (shaded). (*b*) Incident x rays undergo diffraction by the structure of (*a*). The x rays are diffracted as if they were reflected by a family of parallel planes, with the angle of reflection equal to the angle of incidence, both angles measured relative to the planes (not relative to a normal as in optics). (*c*) The path length difference between waves effectively reflected by two adjacent planes is $2d \sin \theta$. (*d*) A different orientation of the incident x rays relative to the structure. A different family of parallel planes now effectively reflects the x rays.

in optics, these angles are defined relative to the *surface* of the reflecting plane rather than a normal to that surface. For the situation of Fig. 37-26*b*, the interplanar spacing happens to be equal to the unit cell dimension a_0.

Figure 37-26*c* shows an edge-on view of reflection from an adjacent pair of planes. The waves of rays 1 and 2 arrive at the crystal in phase. After they are reflected, they must again be in phase, because the reflections and the reflecting planes have been defined solely to explain the intensity maxima in the diffraction of x rays by a crystal. Unlike light rays, the x rays do not refract upon entering the crystal; moreover, we do not define an index of refraction for this situation. Thus, the relative phase between the waves of rays 1 and 2 as they leave the crystal is set solely by their path length difference. For these rays to be in phase, the path length difference must be equal to an integer multiple of the wavelength λ of the x rays.

By drawing the dashed perpendiculars in Fig. 37-26*c*, we find that the path length difference is $2d \sin \theta$. In fact, this is true for any pair of adjacent planes in the family of planes represented in Fig. 37-26*b*. Thus, we have, as the criterion for intensity maxima for x-ray diffraction,

$$2d \sin \theta = m\lambda, \quad \text{for } m = 1, 2, 3, \ldots \quad \text{(Bragg's law)}, \tag{37-31}$$

where m is the order number of an intensity maximum. Equation 37-31 is called **Bragg's law** after British physicist W. L. Bragg, who first derived it. (He and his father shared the 1915 Nobel prize for their use of x rays to study the structures of crystals.) The angle of incidence and reflection in Eq. 37-31 is called a *Bragg angle.*

Regardless of the angle at which x rays enter a crystal, there is always a family of planes from which they can be said to reflect so that we can apply Bragg's law. In Fig. 37-26*d*, notice that the crystal structure has the same orientation as it does in Fig. 37-26*a*, but the angle at which the beam enters the structure differs from that shown in Fig. 37-26*b*. This new angle requires a new family of reflecting planes, with a different interplanar spacing d and different Bragg angle θ, in order to explain the x-ray diffraction via Bragg's law.

Figure 37-27 shows how the interplanar spacing d can be related to the unit cell dimension a_0. For the particular family of planes shown there, the Pythagorean theorem gives

$$5d = \sqrt{5}a_0,$$

or

$$d = \frac{a_0}{\sqrt{5}}. \tag{37-32}$$

Figure 37-27 suggests how the dimensions of the unit cell can be found once the interplanar spacing has been measured by means of x-ray diffraction.

X-ray diffraction is a powerful tool for studying both x-ray spectra and the arrangement of atoms in crystals. To study spectra, a particular set of crystal planes, having a known spacing d, is chosen. These planes effectively reflect different wavelengths at different angles. A detector that can discriminate one angle from another can then be used to determine the wavelength of radiation reaching it. The crystal itself can be studied with a monochromatic x-ray beam, to determine not only the spacing of various crystal planes but also the structure of the unit cell.

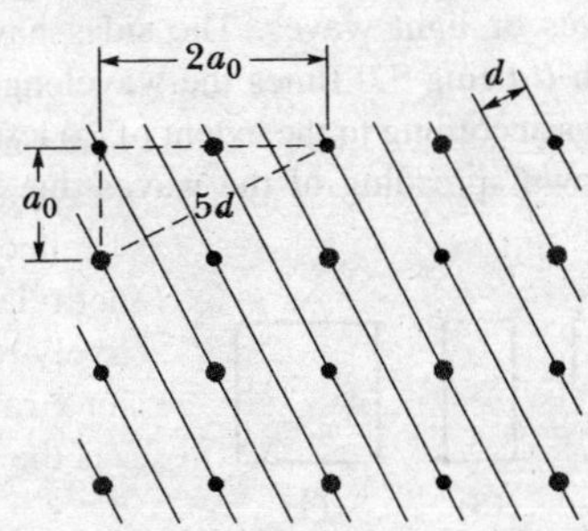

Fig. 37-27 A family of planes through the structure of Fig. 37-26*a*, and a way to relate the edge length a_0 of a unit cell to the interplanar spacing d.

REVIEW & SUMMARY

Diffraction When waves encounter an edge or an obstacle or an aperture with a size comparable to the wavelength of the waves, those waves spread out as they travel and, as a result, undergo interference. This is called **diffraction.**

Single-Slit Diffraction Waves passing through a long narrow slit of width a produce, on a viewing screen, a **single-slit diffraction pattern** that includes a central maximum and other maxima, separated by minima located at angles θ to the central axis that satisfy

$$a \sin \theta = m\lambda, \quad \text{for } m = 1, 2, 3, \ldots \quad \text{(minima)}. \quad (37\text{-}3)$$

The intensity of the diffraction pattern at any given angle θ is

$$I(\theta) = I_m\left(\frac{\sin \alpha}{\alpha}\right)^2, \quad \text{where} \quad \alpha = \frac{\pi a}{\lambda} \sin \theta \quad (37\text{-}5, 37\text{-}6)$$

and I_m is the intensity at the center of the pattern.

Circular-Aperture Diffraction Diffraction by a circular aperture or a lens with diameter d produces a central maximum and concentric maxima and minima, with the first minimum at an angle θ given by

$$\sin \theta = 1.22 \frac{\lambda}{d} \quad \text{(first minimum; circular aperture)}. \quad (37\text{-}12)$$

Rayleigh's Criterion *Rayleigh's criterion* suggests that two objects are on the verge of resolvability if the central diffraction maximum of one is at the first minimum of the other. Their angular separation must then be at least

$$\theta_R = 1.22 \frac{\lambda}{d} \quad \text{(Rayleigh's criterion)}, \quad (37\text{-}14)$$

in which d is the diameter of the aperture through which the light passes.

Double-Slit Diffraction Waves passing through two slits, each of width a, whose centers are a distance d apart, display diffraction patterns whose intensity I at angle θ is

$$I(\theta) = I_m(\cos^2 \beta)\left(\frac{\sin \alpha}{\alpha}\right)^2 \quad \text{(double slit)}, \quad (37\text{-}16)$$

with $\beta = (\pi d/\lambda) \sin \theta$ and α the same as for the case of single-slit diffraction.

Multiple-Slit Diffraction Diffraction by N (multiple) slits results in maxima (lines) at angles θ such that

$$d \sin \theta = m\lambda, \quad \text{for } m = 0, 1, 2, \ldots \quad \text{(maxima)}, \quad (37\text{-}22)$$

with the half-widths of the lines given by

$$\Delta\theta_{hw} = \frac{\lambda}{Nd \cos \theta} \quad \text{(half-widths)}. \quad (37\text{-}25)$$

Diffraction Gratings A *diffraction grating* is a series of "slits" used to separate an incident wave into its component wavelengths by separating and displaying their diffraction maxima. A grating is characterized by its dispersion D and resolving power R:

$$D = \frac{\Delta\theta}{\Delta\lambda} = \frac{m}{d \cos \theta} \quad (37\text{-}26, 37\text{-}27)$$

$$R = \frac{\lambda_{avg}}{\Delta\lambda} = Nm. \quad (37\text{-}28, 37\text{-}29)$$

X-Ray Diffraction The regular array of atoms in a crystal is a three-dimensional diffraction grating for short-wavelength waves such as x rays. For analysis purposes, the atoms can be visualized as being arranged in planes with characteristic interplanar spacing d. Diffraction maxima (due to constructive interference) occur if the incident direction of the wave, measured from the surfaces of these planes, and the wavelength λ of the radiation satisfy **Bragg's law:**

$$2d \sin \theta = m\lambda, \quad \text{for } m = 1, 2, 3, \ldots \quad \text{(Bragg's law)}. \quad (37\text{-}31)$$

QUESTIONS

1. Light of frequency f illuminating a long narrow slit produces a diffraction pattern. (a) If we switch to light of frequency $1.3f$, does the pattern expand away from the center or contract toward the center? (b) Does the pattern expand or contract if, instead, we submerge the equipment in clear corn syrup?

2. You are conducting a single-slit diffraction experiment with light of wavelength λ. What appears, on a distant viewing screen, at a point at which the top and bottom rays through the slit have a path length difference equal to (a) 5λ and (b) 4.5λ?

3. If you speak with the same intensity with and without a megaphone in front of your mouth, in which situation do you sound louder to someone directly in front of you?

4. Figure 37-28 shows four choices for the rectangular opening of a source of either sound waves or light waves. The sides have lengths of either L or $2L$, with L being 3.0 times the wavelength of the waves. Rank the openings according to the extent of (a) left–right spreading and (b) up–down spreading of the waves due to diffraction, greatest first.

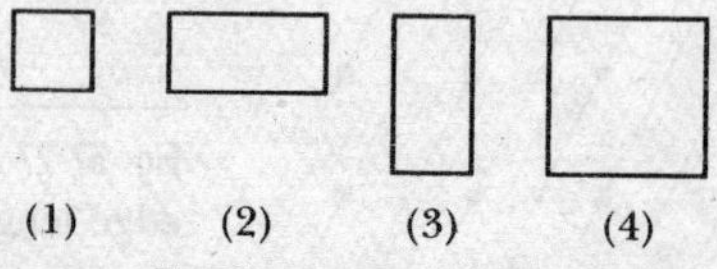

Fig. 37-28 Question 4.

5. In a single-slit diffraction experiment, the top and bottom rays through the slit arrive at a certain point on the viewing screen with a path length difference of 4.0 wavelengths. In a phasor representation like those in Fig 37-6, how many overlapping circles does the chain of phasors make?

6. At night many people see rings (called *entoptic halos*) surrounding bright outdoor lamps in otherwise dark surroundings. The rings are the first of the side maxima in diffraction patterns produced by structures that are thought to be within the cornea (or possibly the lens) of the observer's eye. (The central maxima of such patterns overlap the lamp.) (a) Would a particular ring become smaller or larger if the lamp were switched from blue to red light? (b) If a lamp emits white light, is blue or red on the outside edge of the ring?

7. Figure 37-29 shows the bright fringes that lie within the central diffraction envelope in two double-slit diffraction experiments using the same wavelength of light. Are (a) the slit width a, (b) the slit separation d, and (c) the ratio d/a in experiment B greater than, less than, or the same as those in experiment A?

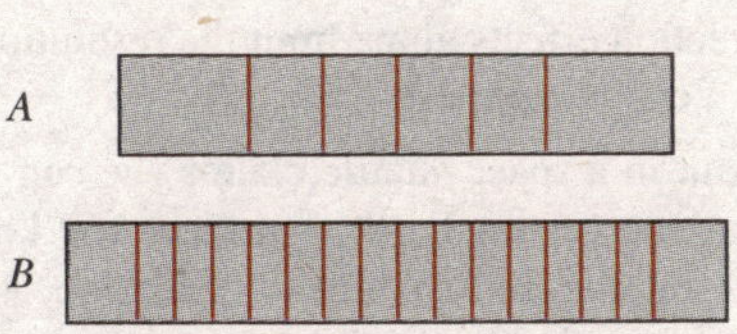

Fig. 37-29 Question 7.

8. Figure 37-30 shows a red line and a green line of the same order in the pattern produced by a diffraction grating. If we increased the number of rulings in the grating, say, by removing tape that had covered half the rulings, would (a) the half-widths of the lines and (b) the separation of the lines increase, decrease, or remain the same? (c) Would the lines shift to the right, shift to the left, or remain in place?

Fig. 37-30 Questions 8 and 9.

9. For the situation of Question 8 and Fig. 37-30, if instead we increased the grating spacing, would (a) the half-widths of the lines and (b) the separation of the lines increase, decrease, or remain the same? (c) Would the lines shift to the right, shift to the left, or remain in place?

10. (a) Figure 37-31a shows the lines produced by diffraction gratings A and B using light of the same wavelength; the lines are of the same order and appear at the same angles θ. Which grating has the greater number of rulings? (b) Figure 37-31b shows lines of two orders produced by a single diffraction grating using light of two wavelengths, both in the red region of the spectrum. Which lines, the left pair or right pair, are in the order with greater m? Is the center of the diffraction pattern to the left or to the right in (c) Fig. 37-31a and (d) Fig. 37-31b?

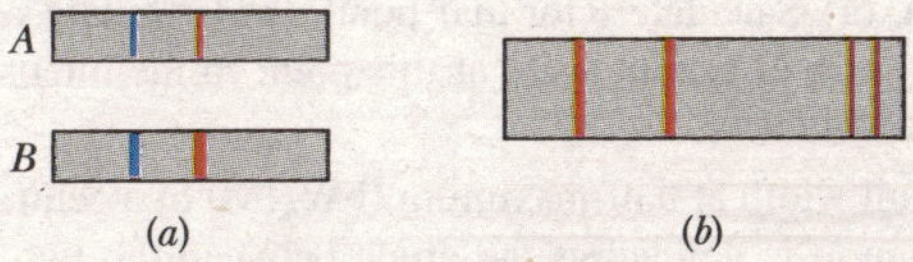

Fig. 37-31 Question 10.

11. (a) For a given diffraction grating, does the least difference $\Delta\lambda$ in two wavelengths that can be resolved increase, decrease, or remain the same as the wavelength increases? (b) For a given wavelength region (say, around 500 nm), is $\Delta\lambda$ greater in the first order or in the third order?

EXERCISES & PROBLEMS

ssm Solution is in the Student Solutions Manual.
www Solution is available on the World Wide Web at: http://www.wiley.com/college/hrw
ilw Solution is available on the Interactive LearningWare.

SEC. 37-2 Diffraction by a Single Slit: Locating the Minima

1E. Light of wavelength 633 nm is incident on a narrow slit. The angle between the first diffraction minimum on one side of the central maximum and the first minimum on the other side is 1.20°. What is the width of the slit? ssm

2E. Monochromatic light of wavelength 441 nm is incident on a narrow slit. On a screen 2.00 m away, the distance between the second diffraction minimum and the central maximum is 1.50 cm. (a) Calculate the angle of diffraction θ of the second minimum. (b) Find the width of the slit.

3E. A single slit is illuminated by light of wavelengths λ_a and λ_b, chosen so the first diffraction minimum of the λ_a component coincides with the second minimum of the λ_b component. (a) What relationship exists between the two wavelengths? (b) Do any other minima in the two diffraction patterns coincide? ssm

4E. The distance between the first and fifth minima of a single-slit diffraction pattern is 0.35 mm with the screen 40 cm away from the slit, when light of wavelength 550 nm is used. (a) Find the slit width. (b) Calculate the angle θ of the first diffraction minimum.

5E. A plane wave of wavelength 590 nm is incident on a slit with a width of $a = 0.40$ mm. A thin converging lens of focal length +70 cm is placed between the slit and a viewing screen and focuses the light on the screen. (a) How far is the screen from the lens? (b) What is the distance on the screen from the center of the diffraction pattern to the first minimum? ssm

6P. Sound waves with frequency 3000 Hz and speed 343 m/s diffract through the rectangular opening of a speaker cabinet and into a large auditorium. The opening, which has a horizontal width of 30.0 cm, faces a wall 100 m away (Fig. 37-32). Where along that wall will a listener be at the first diffraction minimum and thus have difficulty hearing the sound? (Neglect reflections.)

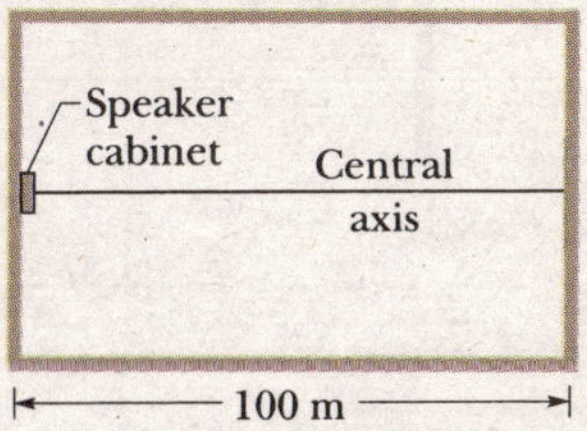

Fig. 37-32 Problem 6.

7P. A slit 1.00 mm wide is illuminated by light of wavelength 589 nm. We see a diffraction pattern on a screen 3.00 m away. What is the distance between the first two diffraction minima on the same side of the central diffraction maximum? ssm ilw

SEC. 37-4 Intensity in Single-Slit Diffraction, Quantitatively

8E. A 0.10-mm-wide slit is illuminated by light of wavelength 589 nm. Consider a point P on a viewing screen on which the diffraction pattern of the slit is viewed; the point is at 30° from the central axis of the slit. What is the phase difference between the Huygens wavelets arriving at point P from the top and midpoint of the slit? (*Hint:* See Eq. 37-4.)

9E. If you double the width of a single slit, the intensity of the central maximum of the diffraction pattern increases by a factor of 4, even though the energy passing through the slit only doubles. Explain this quantitatively. ssm

10E. Monochromatic light with wavelength 538 nm is incident on a slit with width 0.025 mm. The distance from the slit to a screen is 3.5 m. Consider a point on the screen 1.1 cm from the central maximum. (a) Calculate θ for that point. (b) Calculate α. (c) Calculate the ratio of the intensity at this point to the intensity at the central maximum.

11P. The full width at half-maximum (FWHM) of a central diffraction maximum is defined as the angle between the two points in the pattern where the intensity is one-half that at the center of the pattern. (See Fig. 37-7*b*.) (a) Show that the intensity drops to one-half the maximum value when $\sin^2 \alpha = \alpha^2/2$. (b) Verify that $\alpha = 1.39$ rad (about 80°) is a solution to the transcendental equation of (a). (c) Show that the FWHM is $\Delta\theta = 2\sin^{-1}(0.443\lambda/a)$, where a is the slit width. (d) Calculate the FWHM of the central maximum for slits whose widths are 1.0, 5.0, and 10 wavelengths. ssm www

12P. *Babinet's Principle.* A monochromatic beam of parallel light is incident on a "collimating" hole of diameter $x \gg \lambda$. Point P lies in the geometrical shadow region on a *distant* screen (Fig. 37-33*a*). Two diffracting objects, shown in Fig. 37-33*b*, are placed in turn over the collimating hole. A is an opaque circle with a hole in it and B is the "photographic negative" of A. Using superposition concepts, show that the intensity at P is identical for the two diffracting objects A and B.

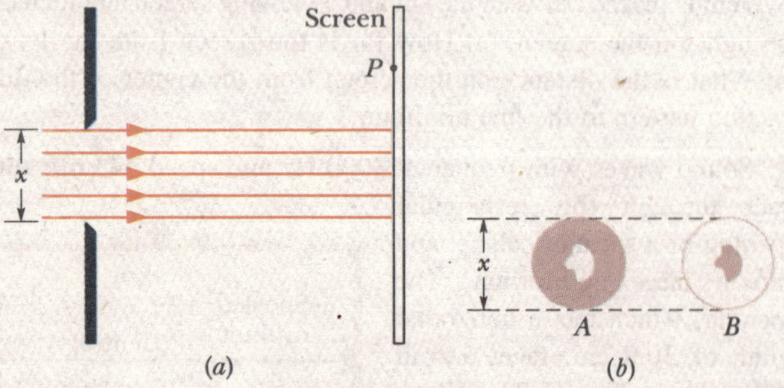

Fig. 37-33 Problem 12.

13P. (a) Show that the values of α at which intensity maxima for single-slit diffraction occur can be found exactly by differentiating Eq. 37-5 with respect to α and equating the result to zero, obtaining the condition $\tan \alpha = \alpha$. (b) Find the values of α satisfying this relation by plotting the curve $y = \tan \alpha$ and the straight line $y = \alpha$ and finding their intersections or by using a calculator to find an appropriate value of α by trial and error. (c) Find the (noninteger) values of m corresponding to successive maxima in the single-slit pattern. Note that the secondary maxima do not lie exactly halfway between minima. ssm

SEC. 37-5 Diffraction by a Circular Aperture

14E. Assume that the lamp in Question 6 emits light at wavelength 550 nm. If a ring has an angular diameter of 2.5°, approximately what is the (linear) diameter of the structure in the eye that causes the ring?

15E. The two headlights of an approaching automobile are 1.4 m apart. At what (a) angular separation and (b) maximum distance will the eye resolve them? Assume that the pupil diameter is 5.0 mm, and use a wavelength of 550 nm for the light. Also assume that diffraction effects alone limit the resolution so that Rayleigh's criterion can be applied. ssm

16E. An astronaut in a space shuttle claims she can just barely resolve two point sources on Earth's surface, 160 km below. Calculate their (a) angular and (b) linear separation, assuming ideal conditions. Take $\lambda = 540$ nm and the pupil diameter of the astronaut's eye to be 5.0 mm.

17E. Find the separation of two points on the Moon's surface that can just be resolved by the 200 in. (= 5.1 m) telescope at Mount Palomar, assuming that this separation is determined by diffraction effects. The distance from Earth to the Moon is 3.8×10^5 km. Assume a wavelength of 550 nm for the light. ilw

18E. The wall of a large room is covered with acoustic tile in which small holes are drilled 5.0 mm from center to center. How far can a person be from such a tile and still distinguish the individual holes, assuming ideal conditions, the pupil diameter of the observer's eye to be 4.0 mm, and the wavelength of the room light to be 550 nm?

19E. Estimate the linear separation of two objects on the planet Mars that can just be resolved under ideal conditions by an observer on Earth (a) using the naked eye and (b) using the 200 in. (= 5.1 m) Mount Palomar telescope. Use the following data: distance to Mars = 8.0×10^7 km, diameter of pupil = 5.0 mm, wavelength of light = 550 nm. ssm

20E. The radar system of a navy cruiser transmits at a wavelength of 1.6 cm, from a circular antenna with a diameter of 2.3 m. At a range of 6.2 km, what is the smallest distance that two speedboats can be from each other and still be resolved as two separate objects by the radar system?

21P. The wings of tiger beetles (Fig. 37-34) are colored by interference due to thin cuticle-like layers. In addition, these layers are arranged in patches that are 60 μm across and produce different colors. The color you see is a pointillistic mixture of thin-film interference colors that varies with perspective. Approximately what viewing distance from a wing puts you at the limit of resolving the different colored patches according to Rayleigh's criterion? Use

Fig. 37-34 Problem 21. Tiger beetles are colored by pointillistic mixtures of thin-film interference colors.

Fig. 37-35 Problem 24. The corona around the Moon is a composite of the diffraction patterns of airborne water drops.

550 nm as the wavelength of light and 3.00 mm as the diameter of your pupil.

22P. In June 1985, a laser beam was sent out from the Air Force Optical Station on Maui, Hawaii, and reflected back from the shuttle *Discovery* as it sped by, 354 km overhead. The diameter of the central maximum of the beam at the shuttle position was said to be 9.1 m, and the beam wavelength was 500 nm. What is the effective diameter of the laser aperture at the Maui ground station? (*Hint:* A laser beam spreads only because of diffraction; assume a circular exit aperture.)

23P. Millimeter-wave radar generates a narrower beam than conventional microwave radar, making it less vulnerable to antiradar missiles. (a) Calculate the angular width of the central maximum, from first minimum to first minimum, produced by a 220 GHz radar beam emitted by a 55.0-cm-diameter circular antenna. (The frequency is chosen to coincide with a low-absorption atmospheric "window.") (b) Calculate the same quantity for the ship's radar described in Exercise 20. ssm www

24P. A circular obstacle produces the same diffraction pattern as a circular hole of the same diameter (except very near $\theta = 0$). Airborne water drops are examples of such obstacles. When you see the Moon through suspended water drops, such as in a fog, you intercept the diffraction pattern from many drops. The composite of the central diffraction maxima of those drops forms a white region that surrounds the Moon and may obscure it. Figure 37-35 is a photograph in which the Moon is obscured. There are two, faint, colored rings around the Moon (the larger one may be too faint to be seen in your copy of the photograph). The smaller ring is on the outer edge of the central maxima from the drops; the somewhat larger ring is on the outer edge of the smallest of the secondary maxima from the drops (see Fig. 37-9). The color is visible because the rings are adjacent to the diffraction minima (dark rings) in the patterns. (Colors in other parts of the pattern overlap too much to be visible.)

(a) What is the color of these rings on the outer edges of the diffraction maxima? (b) The colored ring around the central maxima in Fig. 37-35 has an angular diameter that is 1.35 times the angular diameter of the Moon, which is 0.50°. Assume that the drops all have about the same diameter. Approximately what is that diameter?

25P. (a) What is the angular separation of two stars if their images are barely resolved by the Thaw refracting telescope at the Allegheny Observatory in Pittsburgh? The lens diameter is 76 cm and its focal length is 14 m. Assume $\lambda = 550$ nm. (b) Find the distance between these barely resolved stars if each of them is 10 light-years distant from Earth. (c) For the image of a single star in this telescope, find the diameter of the first dark ring in the diffraction pattern, as measured on a photographic plate placed at the focal plane of the telescope lens. Assume that the structure of the image is associated entirely with diffraction at the lens aperture and not with lens "errors."

26P. In a joint Soviet–French experiment to monitor the Moon's surface with a light beam, pulsed radiation from a ruby laser ($\lambda = 0.69\ \mu$m) was directed to the Moon through a reflecting telescope with a mirror radius of 1.3 m. A reflector on the Moon behaved like a circular plane mirror with radius 10 cm, reflecting the light directly back toward the telescope on Earth. The reflected light was then detected after being brought to a focus by this telescope. What fraction of the original light energy was picked up by the detector? Assume that for each direction of travel all the energy is in the central diffraction peak.

SEC. 37-6 Diffraction by a Double Slit

27E. Suppose that the central diffraction envelope of a double-slit diffraction pattern contains 11 bright fringes and the first diffraction minima eliminate (are coincident with) bright fringes. How many bright fringes lie between the first and second minima of the diffraction envelope? ssm

28E. In a double-slit experiment, the slit separation d is 2.00 times the slit width w. How many bright interference fringes are in the central diffraction envelope?

29P. (a) In a double-slit experiment, what ratio of d to a causes diffraction to eliminate the fourth bright side fringe? (b) What other bright fringes are also eliminated?

30P. Two slits of width a and separation d are illuminated by a coherent beam of light of wavelength λ. What is the linear separation of the bright interference fringes observed on a screen that is at a distance D away?

31P. (a) How many bright fringes appear between the first diffraction-envelope minima to either side of the central maximum in a double-slit pattern if $\lambda = 550$ nm, $d = 0.150$ mm, and $a = 30.0\ \mu$m? (b) What is the ratio of the intensity of the third bright fringe to the intensity of the central fringe? ssm

32P. Light of wavelength 440 nm passes through a double slit, yielding a diffraction pattern whose graph of intensity I versus angular position θ is shown in Fig. 37-36. Calculate (a) the slit width and (b) the slit separation. (c) Verify the displayed intensities of the $m = 1$ and $m = 2$ interference fringes.

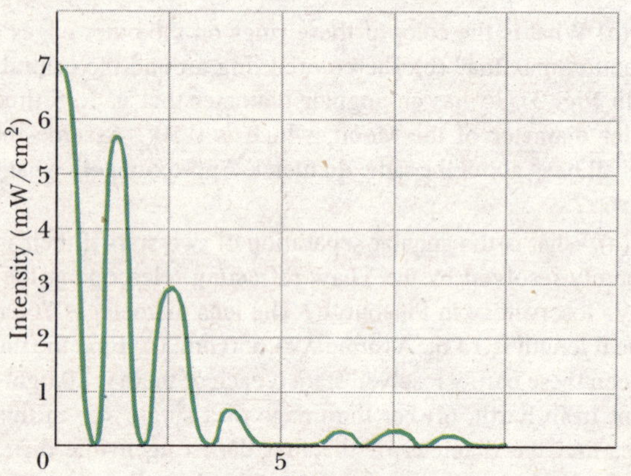

Fig. 37-36 Problem 32.

SEC. 37-7 Diffraction Gratings

33E. A diffraction grating 20.0 mm wide has 6000 rulings. (a) Calculate the distance d between adjacent rulings. (b) At what angles θ will intensity maxima occur on a viewing screen if the radiation incident on the grating has a wavelength of 589 nm?

34E. A grating has 315 rulings/mm. For what wavelengths in the visible spectrum can fifth-order diffraction be observed when this grating is used in a diffraction experiment?

35E. A grating has 400 lines/mm. How many orders of the entire visible spectrum (400–700 nm) can it produce in a diffraction experiment, in addition to the $m = 0$ order? ssm ilw

36E. Perhaps to confuse a predator, some tropical gyrinid beetles (whirligig beetles) are colored by optical interference that is due to scales whose alignment forms a diffraction grating (which scatters light instead of transmiting it). When the incident light rays are perpendicular to the grating, the angle between the first-order maxima (on opposite sides of the zeroth-order maximum) is about 26° in light with a wavelength of 550 nm. What is the grating spacing of the beetle?

37P. Light of wavelength 600 nm is incident normally on a diffraction grating. Two adjacent maxima occur at angles given by $\sin\theta = 0.2$ and $\sin\theta = 0.3$. The fourth-order maxima are missing. (a) What is the separation between adjacent slits? (b) What is the smallest slit width this grating can have? (c) Which orders of intensity maxima are produced by the grating, assuming the values derived in (a) and (b)? ssm

38P. A diffraction grating is made up of slits of width 300 nm with separation 900 nm. The grating is illuminated by monochromatic plane waves of wavelength $\lambda = 600$ nm at normal incidence. (a) How many maxima are there in the full diffraction pattern? (b) What is the width of a spectral line observed in the first order if the grating has 1000 slits?

39P. Assume that the limits of the visible spectrum are arbitrarily chosen as 430 and 680 nm. Calculate the number of rulings per millimeter of a grating that will spread the first-order spectrum through an angle of 20°. ssm www

40P. With light from a gaseous discharge tube incident normally on a grating with slit separation 1.73 μm, sharp maxima of green light are produced at angles $\theta = \pm 17.6°$, 37.3°, −37.1°, 65.2°, and −65.0°. Compute the wavelength of the green light that best fits these data.

41P. Light is incident on a grating at an angle ψ as shown in Fig. 37-37. Show that bright fringes occur at angles θ that satisfy the equation

$$d(\sin\psi + \sin\theta) = m\lambda, \quad \text{for } m = 0, 1, 2, \ldots.$$

(Compare this equation with Eq. 37-22.) Only the special case $\psi = 0$ has been treated in this chapter.

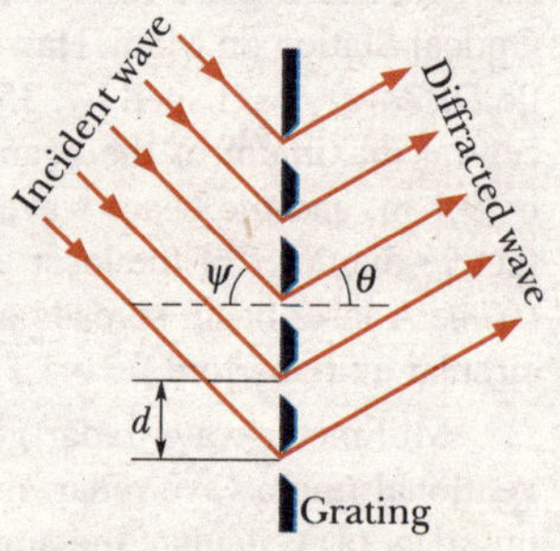

Fig. 37-37 Problem 41.

42P. A grating with $d = 1.50\ \mu$m is illuminated at various angles of incidence by light of wavelength 600 nm. Plot, as a function of the angle of incidence (0 to 90°), the angular deviation of the first-order maximum from the incident direction. (See Problem 41.)

43P. Derive Eq. 37-25; the expression for the half-widths of lines in a grating's diffraction pattern. ssm

44P. A grating has 350 rulings per millimeter and is illuminated at normal incidence by white light. A spectrum is formed on a screen 30 cm from the grating. If a hole 10 mm square is cut in the screen, its inner edge being 50 mm from the central maximum and parallel to it, what is the range in the wavelengths of the light that passes through the hole?

45P*. Derive this expression for the intensity pattern for a three-slit "grating":

$$I = \tfrac{1}{9} I_m (1 + 4\cos\phi + 4\cos^2\phi),$$

where $\phi = (2\pi d \sin\theta)/\lambda$. Assume that $a \ll \lambda$; be guided by the derivation of the corresponding double-slit formula (Eq. 36-21). ssm

SEC. 37-8 Gratings: Dispersion and Resolving Power

46E. The D line in the spectrum of sodium is a doublet with wavelengths 589.0 and 589.6 nm. Calculate the minimum number of lines needed in a grating that will resolve this doublet in the second-order spectrum. See Sample Problem 37-5.

47E. A source containing a mixture of hydrogen and deuterium atoms emits red light at two wavelengths whose mean is 656.3 nm and whose separation is 0.180 nm. Find the minimum number of lines needed in a diffraction grating that can resolve these lines in the first order. ssm ilw

48E. A grating has 600 rulings/mm and is 5.0 mm wide. (a) What is the smallest wavelength interval it can resolve in the third order at $\lambda = 500$ nm? (b) How many higher orders of maxima can be seen?

49E. Show that the dispersion of a grating is $D = (\tan\theta)/\lambda$. ssm

50E. With a particular grating the sodium doublet (see Sample Problem 37-5) is viewed in the third order at 10° to the normal and is barely resolved. Find (a) the grating spacing and (b) the total width of the rulings.

51P. A diffraction grating has resolving power $R = \lambda_{avg}/\Delta\lambda = Nm$. (a) Show that the corresponding frequency range Δf that can just be resolved is given by $\Delta f = c/Nm\lambda$. (b) From Fig. 37-18, show that the times required for light to travel along the ray at the bottom of the figure and the ray at the top differ by an amount $\Delta t = (Nd/c)\sin\theta$. (c) Show that $(\Delta f)(\Delta t) = 1$, this relation being independent of the various grating parameters. Assume $N \gg 1$. ssm

52P. (a) In terms of the angle θ locating a line produced by a grating, find the product of that line's half-width and the resolving power of the grating. (b) Evaluate that product for the grating of Problem 38, for the first order.

SEC. 37-9 X-Ray Diffraction

53E. X rays of wavelength 0.12 nm are found to undergo second-order reflection at a Bragg angle of 28° from a lithium fluoride crystal. What is the interplanar spacing of the reflecting planes in the crystal? ssm

54E. Figure 37-38 is a graph of intensity versus angular position θ for the diffraction of an x-ray beam by a crystal. The beam consists of two wavelengths, and the spacing between the reflecting planes is 0.94 nm. What are the two wavelengths?

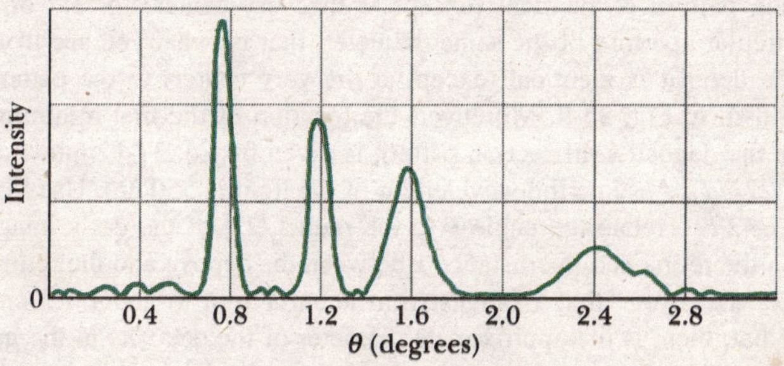

Fig. 37-38 Exercise 54.

55E. An x-ray beam of a certain wavelength is incident on a NaCl crystal, at 30.0° to a certain family of reflecting planes of spacing 39.8 pm. If the reflection from those planes is of the first order, what is the wavelength of the x rays?

56E. An x-ray beam of wavelength A undergoes first-order reflection from a crystal when its angle of incidence to a crystal face is 23°, and an x-ray beam of wavelength 97 pm undergoes third-order reflection when its angle of incidence to that face is 60°. Assuming that the two beams reflect from the same family of reflecting planes, find (a) the interplanar spacing and (b) the wavelength A.

57P. Prove that it is not possible to determine both wavelength of incident radiation and spacing of reflecting planes in a crystal by measuring the Bragg angles for several orders. ssm

58P. In Fig. 37-39, first-order reflection from the reflection planes shown occurs when an x-ray beam of wavelength 0.260 nm makes an angle of 63.8° with the top face of the crystal. What is the unit cell size a_0?

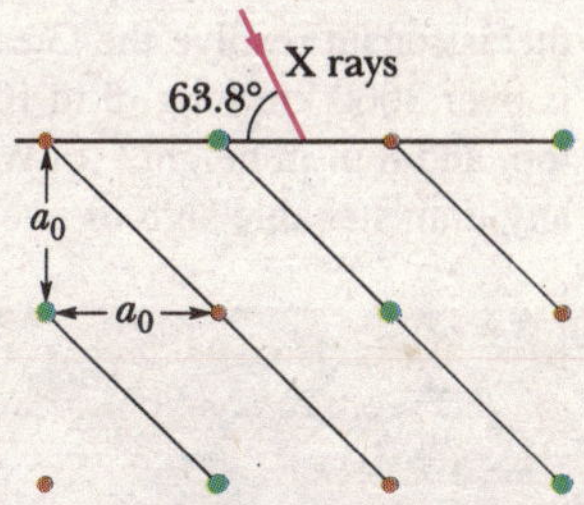

Fig. 37-39 Problem 58.

59P. Consider a two-dimensional square crystal structure, such as one side of the structure shown in Fig. 37-26*a*. One interplanar spacing of reflecting planes is the unit cell size a_0. (a) Calculate and sketch the next five smaller interplanar spacings. (b) Show that your results in (a) are consistent with the general formula

$$d = \frac{a_0}{\sqrt{h^2 + k^2}},$$

where h and k are relatively prime integers (they have no common factor other than unity). ssm www

60P. In Fig. 37-40, an x-ray beam of wavelengths from 95.0 pm to 140 pm is incident at 45° to a family of reflecting planes with spacing $d = 275$ pm. At which wavelengths will these planes produce intensity maxima in their reflections?

61P. In Fig. 37-40, let a beam of x rays of wavelength 0.125 nm be incident on an NaCl crystal at an angle of 45.0° to the top face of the crystal and a family of reflecting planes. Let the reflecting planes have separation $d = 0.252$ nm. Through what angles must the crystal be turned about an axis that is perpendicular to the plane of the page for these reflecting planes to give intensity maxima in their reflections? ssm

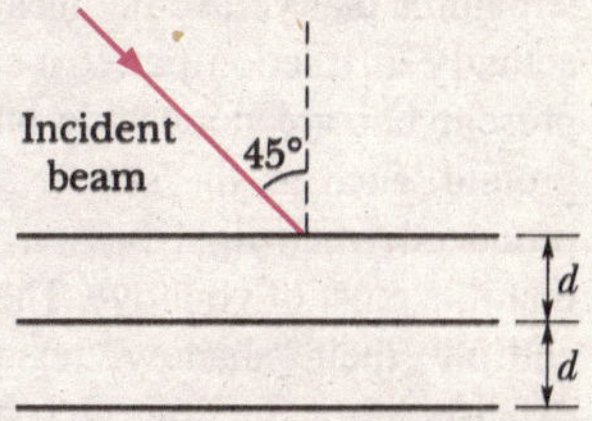

Fig. 37-40 Problems 60 and 61.

Additional Problems

62. In conventional television, signals are broadcast from towers to home receivers. Even when a receiver is not in direct view of a tower because of a hill or building, it can still intercept a signal if the signal diffracts enough around the obstacle, into the obstacle's "shadow region." Current television signals have a wavelength of about 50 cm, but future digital television signals that are to be transmitted from towers will have a wavelength of about 10 mm. (a) Will this change in wavelength increase or decrease the diffraction of the signals into the shadow regions of obstacles? Assume that a signal passes through an opening of 5.0 m width between two adjacent buildings. What is the angular spread of the central diffraction maximum (out to the first minima) for wavelengths of (b) 50 cm and (c) 10 mm?

63. Assume that Rayleigh's criterion gives the limit of resolution of an astronaut's eye looking down on Earth's surface from a typical space shuttle altitude of 400 km. (a) Under that idealized assumption, estimate the least linear width on Earth's surface that the astronaut can resolve. Take the astronaut's pupil diameter to be 5 mm and the wavelength of visible light to be 550 nm. (b) Can the astronaut resolve the Great Wall of China (Fig. 37-41), which is over 3000 km long, 5 to 10 m thick at its base, 4 m thick at its top, and 8 m in height? (c) Would the astronaut be able to resolve any unmistakable sign of intelligent life on Earth's surface?

Fig. 37-41 Problem 63. The Great Wall of China.

64. *Floaters.* As described in Section 37-1, the specks and hairlike structures that you sometimes see floating in your field of view are actually diffraction patterns cast on your retina. They are always present but are noticeable only if you view a featureless background, such as the sky or a brightly lit wall. The patterns are produced when light passes deposits in the gel (*vitreous humor*) that fills most of your eye. The light diffracts around these deposits and into their "shadow" region, much as light did in the Fresnel experiment of Section 37-1. You perceive not the deposits themselves, but their diffraction patterns on your retina. The patterns are called "floaters" because when you move your eye, the gel shimmies (somewhat like a gelatin dessert when shook), causing the diffraction patterns to move around on your retina. As you age, the gel can shimmy more because its attachment to the interior wall of the eye weakens; thus, with age, your floaters will be more noticeable (and a frequent reminder of diffraction physics).

To study the patterns, you can make them more distinct by looking through a pinhole, because the pinhole acts as a single point source of light (as in Fig. 36-5c). Then you can tell that floaters can be circular with a bright center and one or more dark rings (Fig. 37-42a); they can also be in the shape of a hair, with a bright interior and one or more dark bands running along the sides (Fig. 37-42b).

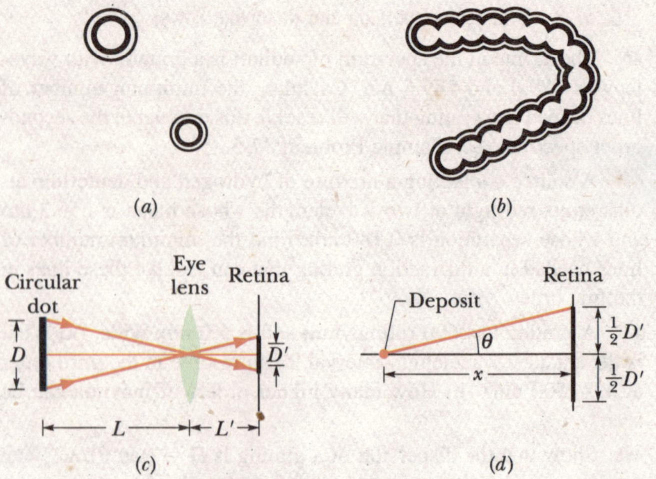

Fig. 37-42 Problem 64.

Estimate the size of the deposits in your eye's gel with the following procedure. Punch a pinhole through an opaque sheet of cardboard, about as distant from one edge as your nose is distant from the center of your eye. Draw a circular dot of diameter $D = 2$ mm on another sheet of cardboard. Position the pinhole immediately in front of your right eye and hold the circular dot in front of your left eye. Simultaneously look at the sky through the pinhole with your right eye and at the dot with your left eye. With a little practice, you can mentally merge the two views so that the dot mentally appears among the floaters.

Adjust the distance of the dot from your left eye until the dot's size approximates the size of one of the circular floaters. Have someone measure the distance L between the dot and your left eye (an estimate will do). Figure 37-42c shows a simplified schematic of your view of the dot: Rays pass straight through an eye lens to form the image of the dot on the retina, at a distance $L' = 2.0$ cm behind the lens. From this view of the dot and the value of L, find the diameter D' of the dot's image (and the floater's pattern) on the retina.

Let us approximate the deposit as spherical. Then its diffraction pattern is identical (except at the very center) to that of a circular aperture of the same diameter; that is, what you see from the deposit is identical (except at the very center) to the pattern shown in Fig. 37-9. Moreover, the location of the first minimum in the deposit's diffraction pattern is given by Eq. 37-12 ($\sin\theta = 1.22\lambda/d$). Assume the wavelength of the light is 550 nm. Use Fig. 37-42d to relate the angle θ to the radius $\frac{1}{2}D'$ of the dot's image on the retina and the distance x between the deposit and the retina. Let us assume that x ranges from about 1 mm to about 1.5 cm. What, then, is the approximate diameter of the deposits in the gel of your eye?

38 Relativity

In modern long-range navigation, the precise location and speed of moving craft are continuously monitored and updated. A system of navigation satellites called NAVSTAR permits locations and speeds anywhere on Earth to be determined to within about 16 m and 2 cm/s. However, if relativity effects were not taken into account, speeds could not be determined any closer than about 20 cm/s, which is unacceptable for modern navigation systems.

How can something as abstract as Einstein's special theory of relativity be involved in something as practical as navigation?

The answer is in this chapter.

38-1 What Is Relativity All About?

One principal focus of **relativity** has to do with measurements of events (things that happen): where and when they happen, and by how much any two events are separated in space and in time. In addition, relativity has to do with transforming such measurements and others between reference frames that move relative to each other. (Hence the name *relativity*.) We discussed such matters in Sections 4-8 and 4-9.

Transformations and moving reference frames were well understood and quite routine to physicists in 1905. Then Albert Einstein (Fig. 38-1) published his **special theory of relativity.** The adjective *special* means that the theory deals only with **inertial reference frames,** which are frames in which Newton's laws are valid. This means that the frames do not accelerate; instead they can move only at constant velocities relative to one another. (Einstein's *general theory of relativity* treats the more challenging situation in which reference frames accelerate; in this chapter the term *relativity* implies only inertial reference frames.)

Starting with two deceivingly simple postulates, Einstein stunned the scientific world by showing that the old ideas about relativity were wrong, even though everyone was so accustomed to them that they seemed to be unquestionable common sense. This supposed common sense, however, was derived from experience only with things that move rather slowly. Einstein's relativity, which turns out to be correct for all possible speeds, predicted many effects that were, at first study, bizarre because no one had experienced them.

In particular, Einstein demonstrated that space and time are entangled; that is, the time between two events depends on how far apart they occur, and vice versa. Also, the entanglement is different for observers who move relative to each other. One result is that time does not pass at a fixed rate, as if it were ticked off with mechanical regularity on some master grandfather clock that controls the universe. Rather, that rate is adjustable: Relative motion can change the rate at which time passes. Prior to 1905, no one but a few daydreamers would have thought that. Now,

Fig. 38-1 Einstein in the early 1900s, at his desk at the patent office in Bern, Switzerland, where he was employed when he published his special theory of relativity.

engineers and scientists take it for granted because their experience with special relativity has reshaped their common sense.

Special relativity has the reputation of being difficult. It is not difficult mathematically, at least not here. However, it is difficult in that we must be very careful about *who* measures *what* about an event and just *how* that measurement is made—and it can be difficult because it can contradict experience.

38-2 The Postulates

We now examine the two postulates of relativity, on which Einstein's theory is based:

1. The Relativity Postulate: The laws of physics are the same for observers in all inertial reference frames. No frame is preferred.

Galileo assumed that the laws of *mechanics* were the same in all inertial reference frames. (Newton's first law of motion is one important consequence.) Einstein extended that idea to include *all* the laws of physics, especially electromagnetism and optics. This postulate does *not* say that the measured values of all physical quantities are the same for all inertial observers; most are not the same. It is the *laws of physics*, which relate these measurements to each other, that are the same.

2. The Speed of Light Postulate: The speed of light in vacuum has the same value c in all directions and in all inertial reference frames.

We can also phrase this postulate to say that there is in nature an *ultimate speed c*, the same in all directions and in all inertial reference frames. Light happens to travel at this ultimate speed, as do any massless particles (neutrinos might be an example). However, no entity that carries energy or information can exceed this limit. Moreover, no particle that does have mass can actually reach speed c, no matter how much or how long it is accelerated.

Both postulates have been exhaustively tested, and no exceptions have ever been found.

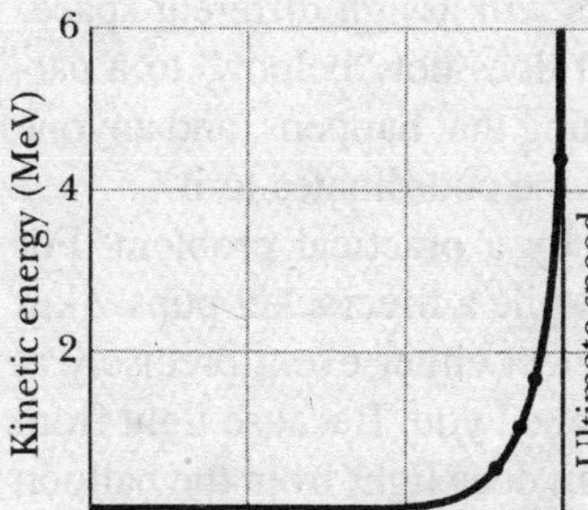

Fig. 38-2 The dots show measured values of the kinetic energy of an electron plotted against its measured speed. No matter how much energy is given to an electron (or to any other particle having mass), its speed can never equal or exceed the ultimate limiting speed c. (The plotted curve through the dots shows the predictions of Einstein's special theory of relativity.)

The Ultimate Speed

The existence of a limit to the speed of accelerated electrons was shown in a 1964 experiment of W. Bertozzi. He accelerated electrons to various measured speeds (see Fig. 38-2) and—by an independent method—also measured their kinetic energies. He found that as the force on a very fast electron is increased, the electron's measured kinetic energy increases toward very large values but its speed does not increase appreciably. Electrons have been accelerated to at least 0.999 999 999 95 times the speed of light but—close though it may be—that speed is still less than the ultimate speed c.

This ultimate speed has been defined to be exactly

$$c = 299\,792\,458 \text{ m/s}. \tag{38-1}$$

So far in this book we have (appropriately) approximated c as 3.0×10^8 m/s, but in this chapter we shall approximate it as 2.998×10^8 m/s. You might want to store the exact value in your calculator's memory (if it is not there already), to be called up when needed.

Testing the Speed of Light Postulate

If the speed of light is the same in all inertial reference frames, then the speed of light that is emitted by a moving source should be the same as the speed of light that is emitted by a source at rest in the laboratory. This claim has been tested directly, in an experiment of high precision. The "light source" was the *neutral pion* (symbol π^0), an unstable, short-lived particle that can be produced by collisions in a particle accelerator. It decays into two gamma rays by the process

$$\pi^0 \rightarrow \gamma + \gamma. \tag{38-2}$$

Gamma rays are part of the electromagnetic spectrum (at very high frequencies) and so obey the speed of light postulate, just as visible light does.

In a 1964 experiment, physicists at CERN, the European particle-physics laboratory near Geneva, generated a beam of pions moving at a speed of 0.999 75c with respect to the laboratory. The experimenters then measured the speed of the gamma rays emitted from these very rapidly moving sources. They found that the speed of the light emitted by the pions was the same as would be measured if the pions were at rest in the laboratory.

38-3 Measuring an Event

An **event** is something that happens, to which an observer can assign three space coordinates and one time coordinate. Among many possible events are (1) the turning on or off of a tiny lightbulb, (2) the collision of two particles, (3) the passage of a pulse of light through a specified point, (4) an explosion, and (5) the coincidence of the hand of a clock with a marker on the rim of the clock. A certain observer, fixed in a certain inertial reference frame, might, for example, assign to an event A the coordinates given in Table 38-1. Because space and time are entangled with each other in relativity, we can describe these coordinates collectively as *spacetime* coordinates. The coordinate system itself is part of the reference frame of the observer.

A given event may be recorded by any number of observers, each in a different inertial reference frame. In general, different observers will assign different spacetime coordinates to the same event. Note that an event does not "belong" to a particular inertial reference frame. An event is just something that happens, and anyone in any reference frame may detect it and assign spacetime coordinates to it.

Making such an assignment can be complicated by a practical problem. For example, suppose a balloon bursts 1 km to your right while a firecracker pops 2 km to your left, both at 9:00 a.m. However, you do not detect either event precisely at 9:00 a.m. because light from the events has not yet reached you. Because light from the pop has farther to go, it arrives at your eyes later than does light from the balloon burst, and thus the pop will seem to have occurred later than the burst. To sort out the actual times and to assign 9:00 a.m. to both events, you must calculate the travel times of the light and then subtract them from the arrival times.

This procedure can be very messy in more challenging situations, and we need an easier procedure that automatically eliminates any concern about the travel time from an event to an observer. To set up such a procedure, we shall construct an imaginary array of measuring rods and clocks throughout the observer's inertial frame (the array moves rigidly with the observer). This construction may seem contrived, but it spares us much confusion and calculation and allows us to find the space coordinates, the time coordinate, and the spacetime coordinates, as follows.

TABLE 38-1 Record of Event *A*

Coordinate	Value
x	3.58 m
y	1.29 m
z	0 m
t	34.5 s

1. **The Space Coordinates.** We imagine the observer's coordinate system fitted with a close-packed, three-dimensional array of measuring rods, one set of rods parallel

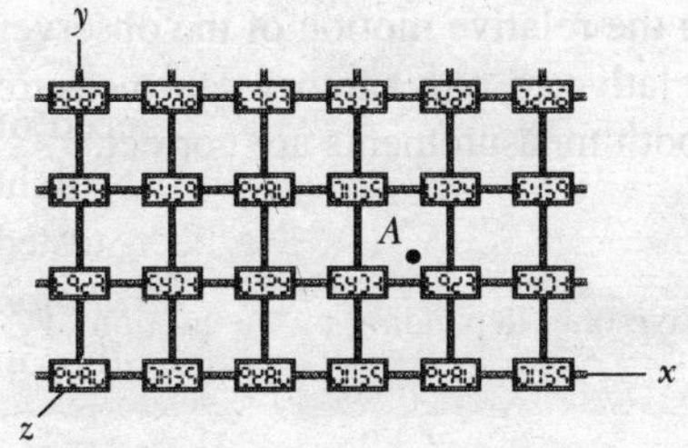

Fig. 38-3 One section of a three-dimensional array of clocks and measuring rods by which an observer can assign spacetime coordinates to an event, such as a flash of light at point A. The event's space coordinates are approximately $x = 3.7$ rod lengths, $y = 1.2$ rod lengths, and $z = 0$. The time coordinate is whatever time appears on the clock closest to A at the instant of the flash.

to each of the three coordinate axes. These rods provide a way to determine coordinates along the axes. Thus, if the event is, say, the turning on of a small lightbulb, the observer, in order to locate the position of the event, need only read the three space coordinates at the bulb's location.

2. **The Time Coordinate.** For the time coordinate, we imagine that every point of intersection in the array of measuring rods includes a tiny clock, which the observer can read by the light generated by the event. Figure 38-3 suggests one plane in the "jungle gym" of clocks and measuring rods that we have described.

   The array of clocks must be synchronized properly. It is not enough to assemble a set of identical clocks, set them all to the same time, and then move them to their assigned positions. We do not know, for example, whether moving the clocks will change their rates. (Actually, it will.) We must put the clocks in place and *then* synchronize them.

   If we had a method of transmitting signals at infinite speed, synchronization would be a simple matter. However, no known signal has this property. We therefore choose light (interpreted broadly to include the entire electromagnetic spectrum) to send out our synchronizing signals because, in vacuum, light travels at the greatest possible speed, the limiting speed c.

   Here is one of many ways in which an observer might synchronize an array of clocks using light signals: The observer enlists the help of a great number of temporary helpers, one for each clock. The observer then stands at a point selected as the origin and sends out a pulse of light when the origin clock reads $t = 0$. When the light pulse reaches the location of a helper, that helper sets the clock there to read $t = r/c$, where r is the distance between the helper and the origin. The clocks are then synchronized.

3. **The Spacetime Coordinates.** The observer can now assign spacetime coordinates to an event by simply recording the time on the clock nearest the event and the position as measured on the nearest measuring rods. If there are two events, the observer computes their separation in time as the difference of the times on clocks near each, and their separation in space from the differences of coordinates on rods near each. We thus avoid the practical problem of calculating the travel times of signals that reach the observer from the events.

38-4 The Relativity of Simultaneity

Suppose that one observer (Sam) notes that two independent events (event Red and event Blue) occur at the same time. Suppose also that another observer (Sally), who is moving at a constant velocity $\vec{v}$ with respect to Sam, also records these same two events. Will Sally also find that they occur at the same time?

The answer is that in general she will not:

If two observers are in relative motion, they will not, in general, agree as to whether two events are simultaneous. If one observer finds them to be simultaneous, the other generally will not.

We cannot say that one observer is right and the other wrong. Their observations are equally valid, and there is no reason to favor one over the other.

The realization that two contradictory statements about the same natural event can be correct is a seemingly strange outcome of Einstein's theory. However, in Chapter 18 we discussed another way in which motion can affect measurement, without balking at the contradictory results: In the Doppler effect, the frequency an observer measures for a sound wave depends on the relative motion of the observer and the source. Thus, two observers moving relative to one another can measure different frequencies for the same wave—and both measurements are correct.

We conclude the following:

> Simultaneity is not an absolute concept but a relative one, depending on the motion of the observer.

If the relative speed of the observers is very much less than the speed of light, then measured departures from simultaneity are so small that they are not noticeable. Such is the case for all our experiences of daily living; that is why the relativity of simultaneity is unfamiliar.

A Closer Look at Simultaneity

Let us clarify the relativity of simultaneity with an example based on the postulates of relativity, no clocks or measuring rods being directly involved. Figure 38-4 shows two long spaceships (the SS *Sally* and the SS *Sam*), which can serve as inertial reference frames for observers Sally and Sam. The two observers are stationed at the midpoints of their ships. The ships are separating along a common x axis, the relative velocity of *Sally* with respect to *Sam* being $\vec{v}$. Figure 38-4*a* shows the ships with the two observer stations momentarily aligned opposite each other.

Two large meteorites strike the ships, one setting off a red flare (event Red) and the other a blue flare (event Blue), not necessarily simultaneously. Each event leaves a permanent mark on each ship, at positions R,R' and B,B'.

Let us suppose that the expanding wavefronts from the two events happen to reach Sam at the same time, as Fig. 38-4*c* shows. Let us further suppose that, after the episode, Sam finds, by measurements of the marks on his spaceship, that he was indeed stationed exactly halfway between the markers B and R on his ship when the two events occurred. He will say:

Sam: Light from event Red and light from event Blue reached me at the same time. From the marks on my spaceship, I find that I was standing halfway between the two sources when the light from them reached me. Therefore, event Red and event Blue were simultaneous events.

As study of Fig. 38-4 shows, Sally and the expanding wavefront from event Red are moving *toward* each other, while she and the expanding wavefront from event Blue

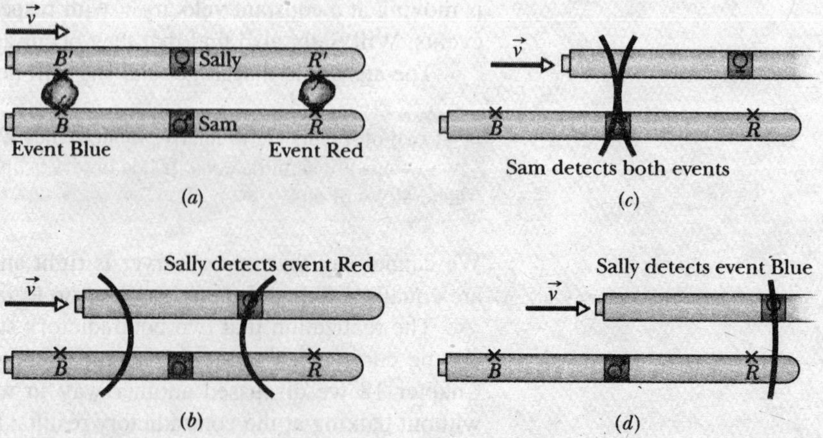

Fig. 38-4 The spaceships of Sally and Sam and the occurrences of events from Sam's view. Sally's ship moves rightward with velocity $\vec{v}$. (*a*) Event Red occurs at positions R,R' and event Blue occurs at positions B,B'; each event sends out a wave of light. (*b*) Sally detects the wave from event Red. (*c*) Sam simultaneously detects the waves from event Red and event Blue. (*d*) Sally detects the wave from event Blue.

are moving in the *same direction*. Thus, the wavefront from event Red will reach Sally *before* the wavefront from event Blue does. She will say:

Sally: Light from event Red reached me before light from event Blue did. From the marks on my spaceship, I found that I too was standing halfway between the two sources. Therefore, the events were *not* simultaneous; event Red occurred first, followed by event Blue.

These reports do not agree. Nevertheless, *both* observers are correct.

Note carefully that there is only one wavefront expanding from the site of each event and that *this wavefront travels with the same speed c in both reference frames*, exactly as the speed of light postulate requires.

It *might* have happened that the meteorites struck the ships in such a way that the two hits appeared to Sally to be simultaneous. If that had been the case, then Sam would have declared them not to be simultaneous.

38-5 The Relativity of Time

If observers who move relative to each other measure the time interval (or *temporal separation*) between two events, they generally will find different results. Why? Because the spatial separation of the events can affect the time intervals measured by the observers.

> The time interval between two events depends on how far apart they occur, in both space and time; that is, their spatial and temporal separations are entangled.

In this section we discuss this entanglement by means of an example; however, the example is restricted in a crucial way: *To one of two observers, the two events occur at the same location.* We shall not get to more general examples until Section 38-7.

Figure 38-5*a* shows the basics of an experiment Sally conducts while she and her equipment ride in a train moving with constant velocity $\vec{v}$ relative to a station.

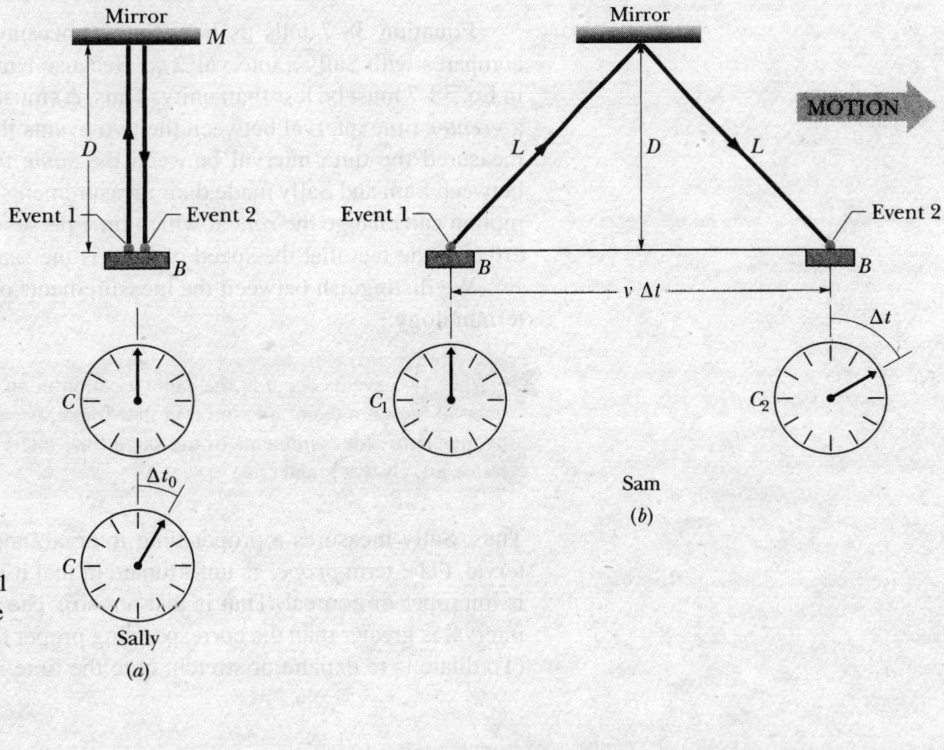

Fig. 38-5 (*a*) Sally, on the train, measures the time interval Δt_0 between events 1 and 2 using a single clock C on the train. That clock is shown twice: first for event 1 and then for event 2. (*b*) Sam, watching from the station as the events occur, requires two synchronized clocks, C_1 at event 1 and C_2 at event 2, to measure the time interval between the two events; his measured time interval is Δt.

A pulse of light leaves a light source B (event 1), travels vertically upward, is reflected vertically downward by a mirror, and then is detected back at the source (event 2). Sally measures a certain time interval Δt_0 between the two events, related to the distance D from source to mirror by

$$\Delta t_0 = \frac{2D}{c} \quad \text{(Sally).} \tag{38-3}$$

The two events occur at the same location in Sally's reference frame, and she needs only one clock C at that location to measure the time interval. Clock C is shown twice in Fig. 38-5*a*, at the beginning and end of the interval.

Consider now how these same two events are measured by Sam, who is standing on the station platform as the train passes. Because the equipment moves with the train during the travel time of the light, Sam sees the path of the light as shown in Fig. 38-5*b*. For him, the two events occur at different places in his reference frame, so to measure the time interval between events, Sam must use *two* synchronized clocks, C_1 and C_2, one at each event. According to Einstein's speed of light postulate, the light travels at the same speed c for Sam as for Sally. Now, however, the light travels distance $2L$ between events 1 and 2. The time interval measured by Sam between the two events is

$$\Delta t = \frac{2L}{c} \quad \text{(Sam),} \tag{38-4}$$

in which

$$L = \sqrt{(\tfrac{1}{2}v\,\Delta t)^2 + D^2}\,. \tag{38-5}$$

From Eq. 38-3, we can write this as

$$L = \sqrt{(\tfrac{1}{2}v\,\Delta t)^2 + (\tfrac{1}{2}c\,\Delta t_0)^2}\,. \tag{38-6}$$

If we eliminate L between Eqs. 38-4 and 38-6 and solve for Δt, we find

$$\Delta t = \frac{\Delta t_0}{\sqrt{1 - (v/c)^2}}\,. \tag{38-7}$$

Equation 38-7 tells us how Sam's measured interval Δt between the events compares with Sally's interval Δt_0. Because v must be less than c, the denominator in Eq. 38-7 must be less than unity. Thus, Δt must be greater than Δt_0: Sam measures a *greater* time interval between the two events than does Sally. Sam and Sally have measured the time interval between the *same* two events, but the relative motion between Sam and Sally made their measurements *different*. We conclude that relative motion can change the *rate* at which time passes between two events; the key to this effect is the fact that the speed of light is the same for both observers.

We distinguish between the measurements of Sam and Sally with the following terminology:

➤ When two events occur at the same location in an inertial reference frame, the time interval between them, measured in that frame, is called the **proper time interval** or the **proper time.** Measurements of the same time interval from any other inertial reference frame are always greater.

Thus, Sally measures a proper time interval, and Sam measures a greater time interval. (The term *proper* is unfortunate in that it implies that any other measurement is improper or nonreal. That is just not so.) The amount by which a measured time interval is greater than the corresponding proper time interval is called **time dilation.** (To dilate is to expand or stretch; here the time interval is expanded or stretched.)

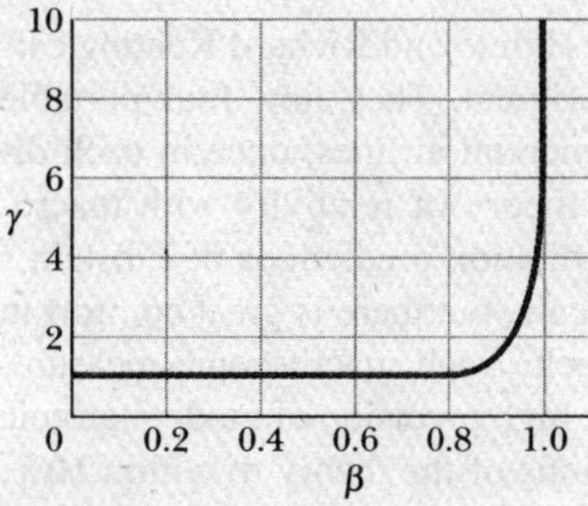

Fig. 38-6 A plot of the Lorentz factor γ as a function of the speed parameter $\beta\ (= v/c)$.

Often the dimensionless ratio v/c in Eq. 38-7 is replaced with β, called the **speed parameter,** and the dimensionless inverse square root in Eq. 38-7 is often replaced with γ, called the **Lorentz factor:**

$$\gamma = \frac{1}{\sqrt{1-\beta^2}} = \frac{1}{\sqrt{1-(v/c)^2}}. \tag{38-8}$$

With these replacements, we can rewrite Eq. 38-7 as

$$\Delta t = \gamma\,\Delta t_0 \quad \text{(time dilation).} \tag{38-9}$$

The speed parameter β is always less than unity and, provided v is not zero, γ is always greater than unity. However, the difference between γ and 1 is not significant unless $v > 0.1c$. Thus, in general, "old relativity" works well enough for $v < 0.1c$, but we must use special relativity for greater values of v. As shown in Fig. 38-6, γ increases rapidly in magnitude as β approaches 1 (as v approaches c). Therefore, the greater the relative speed between Sally and Sam is, the greater will be the time interval measured by Sam, until at a great enough speed, the interval takes "forever."

You might wonder what Sally says about Sam's having measured a greater time interval than she did. His measurement comes as no surprise to her, because to her, he failed to synchronize his clocks C_1 and C_2 in spite of his insistence that he did. Recall that observers in relative motion generally do not agree about simultaneity. Here, Sam insists that his two clocks simultaneously read the same time when event 1 occurred. To Sally, however, Sam's clock C_2 was erroneously set ahead during the synchronization process. Thus, when Sam read the time of event 2 on it, to Sally he was reading off a time that was too large, and that is why the time interval he measured between the two events was greater than the interval she measured.

Two Tests of Time Dilation

1. **Microscopic Clocks.** Subatomic particles called *muons* are unstable; that is, when a muon is produced, it lasts for only a short time before it *decays* (transforms into particles of other types). The *lifetime* of a muon is the time interval between its production (event 1) and its decay (event 2). When muons are stationary and their lifetimes are measured with stationary clocks (say, in a laboratory), their average lifetime is 2.200 μs. This is a proper time interval because, for each muon, events 1 and 2 occur at the same location in the reference frame of the muon, namely at the muon itself. We can represent this proper time interval with Δt_0; moreover, we can call the reference frame in which it is measured the *rest frame* of the muon.

   If, instead, the muons are moving, say, through a laboratory, then measurements of their lifetimes made with the laboratory clocks should yield a greater average lifetime (a dilated average lifetime). To check this conclusion, measurements were made of the average lifetime of muons moving with a speed of $0.9994c$ relative to laboratory clocks. From Eq. 38-8, with $\beta = 0.9994$, the Lorentz factor for this speed is

   $$\gamma = \frac{1}{\sqrt{1-\beta^2}} = \frac{1}{\sqrt{1-(0.9994)^2}} = 28.87.$$

   Equation 38-9 then yields, for the average dilated lifetime,

   $$\Delta t = \gamma\,\Delta t_0 = (28.87)(2.200\ \mu\text{s}) = 63.51\ \mu\text{s}.$$

   The actual measured value matched this result within experimental error.

2. Macroscopic Clocks. In October 1977, Joseph Hafele and Richard Keating carried out what must have been a grueling experiment. They flew four portable atomic clocks twice around the world on commercial airlines, once in each direction. Their purpose was "to test Einstein's theory of relativity with macroscopic clocks." As we have just seen, the time dilation predictions of Einstein's theory have been confirmed on a microscopic scale, but there is great comfort in seeing a confirmation made with an actual clock. Such macroscopic measurements became possible only because of the very high precision of modern atomic clocks. Hafele and Keating verified the predictions of the theory to within 10%. (Einstein's *general* theory of relativity, which predicts that the rate at which time passes on a clock is influenced by the gravitational force on the clock, also plays a role in this experiment.)

A few years later, physicists at the University of Maryland carried out a similar experiment with improved precision. They flew an atomic clock round and round over Chesapeake Bay for flights lasting 15 h and succeeded in checking the time dilation prediction to better than 1%. Today, when atomic clocks are transported from one place to another for calibration or other purposes, the time dilation caused by their motion is always taken into account.

✔CHECKPOINT 1: Standing beside railroad tracks, we are suddenly startled by a relativistic boxcar traveling past us as shown in the figure. Inside, a well-equipped hobo fires a laser pulse from the front of the boxcar to its rear. (a) Is our measurement of the speed of the pulse greater than, less than, or the same as that measured by the hobo? (b) Is his measurement of the flight time of the pulse a proper time? (c) Are his measurement and our measurement of the flight time related by Eq. 38-9?

Sample Problem 38-1

Your starship passes Earth with a relative speed of $0.9990c$. After traveling 10.0 y (your time), you stop at lookout post LP13, turn, and then travel back to Earth with the same relative speed. The trip back takes another 10.0 y (your time). How long does the round trip take according to measurements made on Earth? (Neglect any effects due to the accelerations involved with stopping, turning, and getting back up to speed.)

SOLUTION: We begin by analyzing the outward trip only, with these **Key Ideas:**

1. This problem involves measurements made from two (inertial) reference frames, one attached to Earth and the other (your reference frame) attached to your ship.
2. The outward trip involves two events: the start of the trip at Earth and the end of the trip at LP13.
3. Your measurement of 10 y for the outward trip is the proper time Δt_0 between those two events, because the events occur at the same location in your reference frame, namely on your ship.
4. The Earth-frame measurement of the time interval Δt for the outward trip must be greater than Δt_0, according to Eq. 38-9 ($\Delta t = \gamma\,\Delta t_0$) for time dilation.

Using Eq. 38-8 to substitute for γ in Eq. 38-9, we find

$$\Delta t = \frac{\Delta t_0}{\sqrt{1-(v/c)^2}}$$

$$= \frac{10.0 \text{ y}}{\sqrt{1-(0.9990c/c)^2}} = (22.37)(10.0 \text{ y}) = 224 \text{ y}.$$

On the return trip, we have the same situation and the same data. Thus, the round trip requires 20 y of your time but

$$\Delta t_{\text{total}} = (2)(224 \text{ y}) = 448 \text{ y} \qquad \text{(Answer)}$$

of Earth time. In other words, you have aged 20 y while the Earth has aged 448 y. Although you cannot travel into the past (as far as we know), you can travel into the future of, say, Earth, by using high-speed relative motion to adjust the rate at which time passes.

Sample Problem 38-2

The elementary particle known as the *positive kaon* (K^+) has, on average, a lifetime of 0.1237 μs when stationary—that is, when the lifetime is measured in the rest frame of the kaon. If a positive kaon has a speed of 0.990c relative to a laboratory reference frame when it is produced, how far can it travel in that frame during its lifetime according to *classical physics* (which is a reasonable approximation for speeds much less than c) and according to special relativity (which is correct for all physically possible speeds)?

SOLUTION: We begin with these **Key Ideas:**

1. This problem involves measurements made from two (inertial) reference frames, one attached to the kaon and the other attached to the laboratory.
2. This problem also involves two events: the start of the kaon's travel (when the kaon is produced) and the end of that travel (at the end of the kaon's lifetime).
3. The distance traveled by the kaon between those two events is related to its speed v and the time interval for the travel by

$$v = \frac{\text{distance}}{\text{time interval}}. \tag{38-10}$$

With these ideas in mind, let us solve for the distance first with classical physics and then with special relativity.

Classical physics: In classical physics, we have this **Key Idea:** We would find the same distance and time interval (in Eq. 38-10) whether we measured them from the kaon frame or from the laboratory frame. Thus, we need not be careful about the frame in which the measurements are made. To find the kaon's travel distance d_{cp} according to classical physics, we first rewrite Eq. 38-10 as

$$d_{cp} = v\,\Delta t, \tag{38-11}$$

where Δt is the time interval between the two events in either frame. Then, substituting 0.990c for v and 0.1237 μs for Δt in Eq. 38-11, we find

$$\begin{aligned} d_{cp} &= (0.990c)\,\Delta t \\ &= (0.990)(2.998 \times 10^8 \text{ m/s})(0.1237 \times 10^{-6} \text{ s}) \\ &= 36.7 \text{ m.} \qquad \text{(Answer)} \end{aligned}$$

This is how far the kaon would travel if classical physics were correct at speeds close to c.

Special relativity: In special relativity we have this **Key Idea:** We must be very careful that both the distance and the time interval in Eq. 38-10 are measured in the *same* reference frame—especially when the speed is close to c, as here. Thus, to find the actual travel distance d_{sr} of the kaon *as measured from the laboratory frame* and according to special relativity, we rewrite Eq. 38-10 as

$$d_{sr} = v\,\Delta t, \tag{38-12}$$

where Δt is the time interval between the two events *as measured from the laboratory frame.*

Before we can evaluate d_{sr} in Eq. 38-12, we must find Δt by using this **Key Idea**: the 0.1237 μs time interval is a proper time, because the two events occur at the same location in the kaon frame—namely, at the kaon itself. Therefore, let Δt_0 represent this proper time interval. Then we can use Eq. 38-9 ($\Delta t = \gamma\,\Delta t_0$) for time dilation to find the time interval Δt as measured from the laboratory frame. Using Eq. 38-8 to substitute for γ in Eq. 38-9 leads to

$$\Delta t = \frac{\Delta t_0}{\sqrt{1 - (v/c)^2}} = \frac{0.1237 \times 10^{-6} \text{ s}}{\sqrt{1 - (0.990c/c)^2}} = 8.769 \times 10^{-7} \text{ s.}$$

This is about seven times longer than the kaon's proper lifetime. That is, the kaon's lifetime is about seven times longer in the laboratory frame than in its own frame—the kaon's lifetime is dilated. We can now evaluate Eq. 38-12 for the travel distance d_{sr} in the laboratory frame as

$$\begin{aligned} d_{sr} &= v\,\Delta t = (0.990c)\,\Delta t \\ &= (0.990)(2.998 \times 10^8 \text{ m/s})(8.769 \times 10^{-7} \text{ s}) \\ &= 260 \text{ m.} \qquad \text{(Answer)} \end{aligned}$$

This is about seven times d_{cp}. Experiments like the one outlined here, which verify special relativity, became routine in physics laboratories decades ago. The engineering design and the construction of any scientific or medical facility that employs high-speed particles must take relativity into account.

38-6 The Relativity of Length

If you want to measure the length of a rod that is at rest with respect to you, you can—at your leisure—note the positions of its end points on a long stationary scale and subtract one reading from the other. If the rod is moving, however, you must note the positions of the end points *simultaneously* (in your reference frame) or your measurement cannot be called a length. Figure 38-7 suggests the difficulty of trying to measure the length of a moving penguin by locating its front and back at different times. Because simultaneity is relative and it enters into length measurements, length should also be a relative quantity. It is.

Let L_0 be the length of a rod that you measure when the rod is stationary (meaning you and it are in the same reference frame, the rod's rest frame). If, instead,

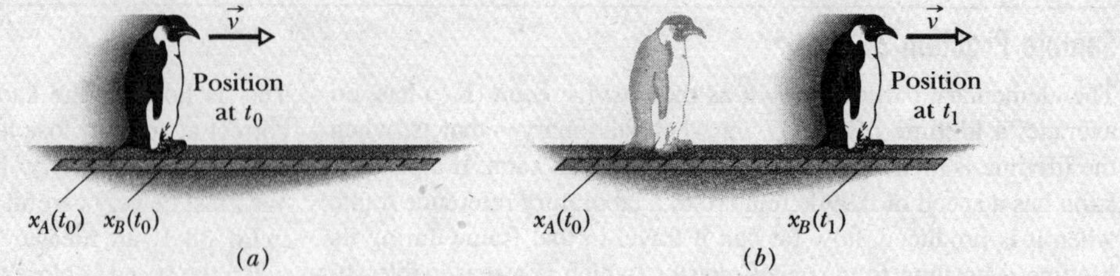

Fig. 38-7 If you want to measure the front-to-back length of a penguin while it is moving, you must mark the positions of its front and back simultaneously (in your reference frame), as in (*a*), rather than at different times, as in (*b*).

there is relative motion at speed v between you and the rod *along the length of the rod,* then with simultaneous measurement you obtain a length L given by

$$L = L_0\sqrt{1-\beta^2} = \frac{L_0}{\gamma} \quad \text{(length contraction).} \tag{38-13}$$

Because the Lorentz factor γ is always greater than unity if there is relative motion, L is less than L_0. The relative motion causes a *length contraction,* and L is called a *contracted length.* Because γ increases with speed v, the length contraction also increases with v.

➤ The length L_0 of an object measured in the rest frame of the object is its **proper length** or **rest length.** Measurements of the length from any reference frame that is in relative motion parallel to that length are always less than the proper length.

Be careful: Length contraction occurs only along the direction of relative motion. Also, the length that is measured does not have to be that of an object like a rod or a circle. Instead, it can be the length (or distance) between two objects in the same rest frame—for example, the Sun and a nearby star (which are, at least approximately, at rest relative to each other).

Does a moving object *really* shrink? Reality is based on observations and measurements; if the results are always consistent and if no error can be determined, then what is observed and measured is real. In that sense, the object really does shrink. However, a more precise statement is that the object *is really measured* to shrink—motion affects that measurement and thus reality.

When you measure a contracted length for, say, a rod, what does an observer moving with the rod say of your measurement? To that observer, you did not locate the two ends of the rod simultaneously. (Recall that observers in motion relative to each other do not agree about simultaneity.) To the observer, you first located the rod's front end and then, slightly later, its rear end, and that is why you measured a length that is less than the proper length.

Proof of Eq. 38-13

Length contraction is a direct consequence of time dilation. Consider once more our two observers. This time, both Sally, seated on a train moving through a station, and Sam, again on the station platform, want to measure the length of the platform. Sam, using a tape measure, finds the length to be L_0, a proper length because the platform is at rest with respect to him. Sam also notes that Sally, on the train, moves through this length in a time $\Delta t = L_0/v$, where v is the speed of the train; that is,

$$L_0 = v\,\Delta t \quad \text{(Sam).} \tag{38-14}$$

This time interval Δt is not a proper time interval because the two events that define it (Sally passes the back of the platform and Sally passes the front of the platform) occur at two different places and Sam must use two synchronized clocks to measure the time interval Δt.

For Sally, however, the platform is moving past her. She finds that the two events measured by Sam occur *at the same place* in her reference frame. She can time them with a single stationary clock, so the interval Δt_0 that she measures is a proper time interval. To her, the length L of the platform is given by

$$L = v\,\Delta t_0 \quad \text{(Sally)}. \tag{38-15}$$

If we divide Eq. 38-15 by Eq. 38-14 and apply Eq. 38-9, the time dilation equation, we have

$$\frac{L}{L_0} = \frac{v\,\Delta t_0}{v\,\Delta t} = \frac{1}{\gamma},$$

or

$$L = \frac{L_0}{\gamma}, \tag{38-16}$$

which is Eq. 38-13, the length contraction equation.

Sample Problem 38-3

In Fig. 38-8, Sally (at point A) and Sam's spaceship (of proper length $L_0 = 230$ m) pass each other with constant relative speed v. Sally measures a time interval of 3.57 μs for the ship to pass her (from the passage of point B to the passage of point C). In terms of c, what is the relative speed v between Sally and the ship?

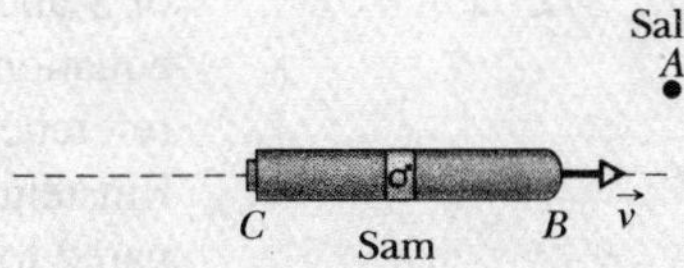

Fig. 38-8 Sample Problem 38-3. Sally, at point A, measures the time a spaceship takes to pass her.

SOLUTION: Let us assume that speed v is near c. Then we begin with these **Key Ideas**:

1. This problem involves measurements made from two (inertial) reference frames, one attached to Sally and the other attached to Sam and his spaceship.
2. This problem also involves two events: the first is the passage of point B and the second is the passage of point C.
3. From either reference frame, the other reference frame passes at speed v and moves a certain distance in the time interval between the two events:

$$v = \frac{\text{distance}}{\text{time interval}}. \tag{38-17}$$

Because speed v is assumed to be near the speed of light, we must be careful that the distance and the time interval in Eq. 38-17 are measured in the *same* reference frame.

We are free to use either frame for the measurements. Because we know that the time interval Δt between the two events measured from Sally's frame is 3.57 μs, let us also use the distance L between the two events measured from her frame. Equation 38-17 then becomes

$$v = \frac{L}{\Delta t}. \tag{38-18}$$

We do not know L but we can relate it to the given L_0 with this additional **Key Idea**: The distance between the two events as measured from Sam's frame is the ship's proper length L_0. Thus, the distance L measured from Sally's frame must be less than L_0, as given by Eq. 38-13 ($L = L_0/\gamma$) for length contraction. Substituting L_0/γ for L in Eq. 38-18 and then substituting Eq. 38-8 for γ, we find

$$v = \frac{L_0/\gamma}{\Delta t} = \frac{L_0\sqrt{(1-(v/c)^2}}{\Delta t}.$$

Solving this equation for v leads us to

$$v = \frac{L_0 c}{\sqrt{(c\,\Delta t)^2 + L_0^2}}$$

$$= \frac{(230\text{ m})c}{\sqrt{(2.998\times 10^8\text{ m/s})^2(3.57\times 10^{-6}\text{ s})^2 + (230\text{ m})^2}}$$

$$= 0.210c. \qquad \text{(Answer)}$$

Thus, the relative speed between Sally and the ship is 21% of the speed of light. Note that only the relative motion of Sally and Sam matters here; whether either is stationary relative to, say, a space station is irrelevant. In Fig. 38-8 we took Sally to be stationary, but we could instead have taken the ship to be stationary, with Sally moving past it. Nothing would have changed in our result.

✔CHECKPOINT 2: In this sample problem, Sally measures the passage time of the ship. If Sam does also, (a) which measurement, if either, is a proper time and (b) which measurement is smaller?

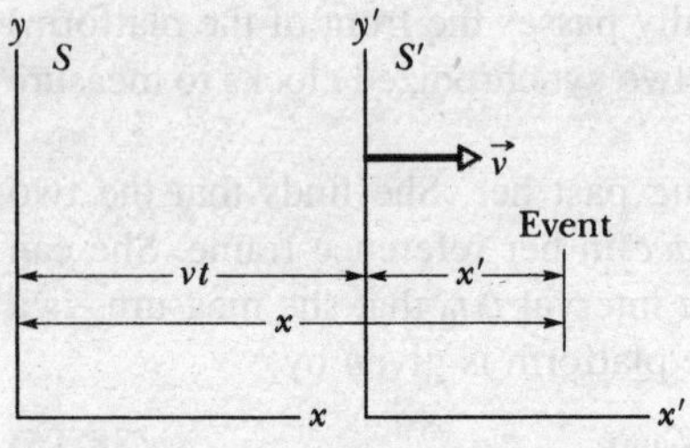

Fig. 38-9 Two inertial reference frames: frame S' has velocity $\vec{v}$ relative to frame S.

38-7 The Lorentz Transformation

Figure 38-9 shows inertial reference frame S' moving with speed v relative to frame S, in the common positive direction of their horizontal axes (marked x and x'). An observer in S reports spacetime coordinates x, y, z, t for an event, and an observer in S' reports x', y', z', t' for the same event. How are these sets of numbers related?

We claim at once (although it requires proof) that the y and z coordinates, which are perpendicular to the motion, are not affected by the motion; that is, $y = y'$ and $z = z'$. Our interest then reduces to the relation between x and x' and between t and t'.

The Galilean Transformation Equations

Prior to Einstein's publication of his special theory of relativity, the four coordinates of interest were assumed to be related by the *Galilean transformation equations:*

$$\begin{aligned} x' &= x - vt \\ t' &= t \end{aligned} \qquad \text{(Galilean transformation equations; approximately valid at low speeds).} \qquad (38\text{-}19)$$

(These equations are written with the assumption that $t = t' = 0$ when the origins of S and S' coincide.) You can verify the first equation with Fig. 38-9. The second equation effectively claims that time passes at the same rate for observers in both reference frames. That would have been so obviously true to a scientist prior to Einstein that it would not even have been mentioned. When speed v is small compared to c, Eqs. 38-19 generally work well.

The Lorentz Transformation Equations

We state without proof that the correct transformation equations, which remain valid for all speeds up to the speed of light, can be derived from the postulates of relativity. The results, called the **Lorentz transformation equations*** or sometimes (more loosely) just the Lorentz transformations, are

$$\begin{aligned} x' &= \gamma(x - vt), \\ y' &= y, \\ z' &= z, \\ t' &= \gamma(t - vx/c^2) \end{aligned} \qquad \text{(Lorentz transformation equations; valid at all physically possible speeds).} \qquad (38\text{-}20)$$

(The equations are written with the assumption that $t = t' = 0$ when the origins of S and S' coincide.) Note that the spatial values x and the temporal values t are bound together in the first and last equations. This entanglement of space and time was a prime message of Einstein's theory, a message that was long rejected by many of his contemporaries.

It is a formal requirement of relativistic equations that they should reduce to familiar classical equations if we let c approach infinity. That is, if the speed of light were infinitely great, *all* finite speeds would be "low" and classical equations would

*You may wonder why we do not call these the *Einstein transformation equations* (and why not the *Einstein factor* for γ). H. A. Lorentz actually derived these equations before Einstein did, but as the great Dutch physicist graciously conceded, he did not take the further bold step of interpreting these equations as describing the true nature of space and time. It is this interpretation, first made by Einstein, that is at the heart of relativity.

TABLE 38-2 The Lorentz Transformation Equations for Pairs of Events

1. $\Delta x = \gamma(\Delta x' + v\,\Delta t')$	1′. $\Delta x' = \gamma(\Delta x - v\,\Delta t)$
2. $\Delta t = \gamma(\Delta t' + v\,\Delta x'/c^2)$	2′. $\Delta t' = \gamma(\Delta t - v\,\Delta x/c^2)$

$$\gamma = \frac{1}{\sqrt{1-(v/c)^2}} = \frac{1}{\sqrt{1-\beta^2}}$$

Frame S' moves at velocity v relative to frame S.

never fail. If we let $c \rightarrow \infty$ in Eqs. 38-20, $\gamma \rightarrow 1$ and these equations reduce—as we expect—to the Galilean equations (Eqs. 38-19). You should check this.

Equations 38-20 are written in a form that is useful if we are given x and t and wish to find x' and t'. We may wish to go the other way, however. In that case we simply solve Eqs. 38-20 for x and t, obtaining

$$x = \gamma(x' + vt') \quad \text{and} \quad t = \gamma(t' + vx'/c^2). \tag{38-21}$$

Comparison shows that, starting from either Eqs. 38-20 or Eqs. 38-21, you can find the other set by interchanging primed and unprimed quantities and reversing the sign of the relative velocity v.

Equations 38-20 and 38-21 relate the coordinates of a single event as seen by two observers. Sometimes we want to know not the coordinates of a single event but the differences between coordinates for a pair of events. That is, if we label our events 1 and 2, we may want to relate

$$\Delta x = x_2 - x_1 \quad \text{and} \quad \Delta t = t_2 - t_1,$$

as measured by an observer in S, and

$$\Delta x' = x_2' - x_1' \quad \text{and} \quad \Delta t' = t_2' - t_1',$$

as measured by an observer in S'.

Table 38-2 displays the Lorentz equations in difference form, suitable for analyzing pairs of events. The equations in the table were derived by simply substituting differences (such as Δx and $\Delta x'$) for the four variables in Eqs. 38-20 and 38-21.

Be careful: When substituting values for these differences, you must be consistent and not mix the values for the first event with those for the second event. Also, if, say, Δx is a negative quantity, you must be certain to include the minus sign in a substitution.

✓CHECKPOINT 3: The figure here shows three situations in which a blue reference frame and a green reference frame are in relative motion along the common direction of their x and x' axes, as indicated by the velocity vector attached to one of the frames. For each situation, if we choose the blue frame to be stationary, then is v in the equations of Table 38-2 a positive or negative quantity?

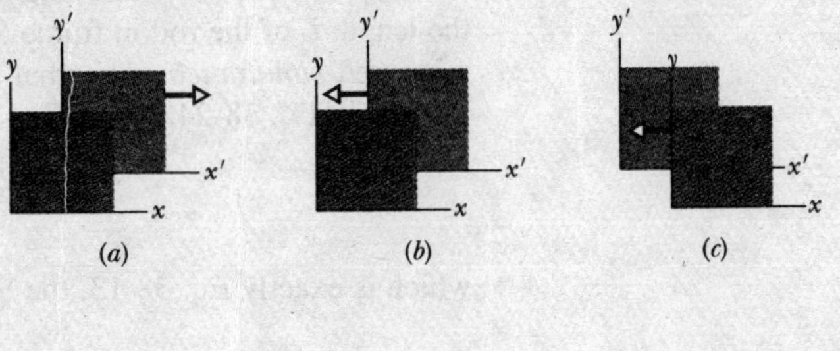

38-8 Some Consequences of the Lorentz Equations

Here we use the transformation equations of Table 38-2 to affirm some of the conclusions that we reached earlier by arguments based directly on the postulates.

Simultaneity

Consider Eq. 2 of Table 38-2,

$$\Delta t = \gamma\left(\Delta t' + \frac{v\,\Delta x'}{c^2}\right). \tag{38-22}$$

If two events occur at different places in reference frame S' of Fig. 38-9, then $\Delta x'$ in this equation is not zero. It follows that even if the events are simultaneous in S' (so $\Delta t' = 0$), they will not be simultaneous in frame S. (This is in accord with our conclusion in Section 38-4.) The time interval between the events in S will be

$$\Delta t = \gamma \frac{v\,\Delta x'}{c^2} \qquad \text{(simultaneous events in } S'\text{)}.$$

Time Dilation

Suppose now that two events occur at the same place in S' (so $\Delta x' = 0$) but at different times (so $\Delta t' \neq 0$). Equation 38-22 then reduces to

$$\Delta t = \gamma\,\Delta t' \qquad \text{(events in same place in } S'\text{)}. \tag{38-23}$$

This confirms time dilation. Because the two events occur at the same place in S', the time interval $\Delta t'$ between them can be measured with a single clock, located at that place. Under these conditions, the measured interval is a proper time interval, and we can label it Δt_0. Thus, Eq. 38-23 becomes

$$\Delta t = \gamma\,\Delta t_0 \qquad \text{(time dilation)},$$

which is exactly Eq. 38-9, the time dilation equation.

Length Contraction

Consider Eq. 1′ of Table 38-2,

$$\Delta x' = \gamma(\Delta x - v\,\Delta t). \tag{38-24}$$

If a rod lies parallel to the x and x' axes of Fig. 38-9 and is at rest in reference frame S', an observer in S' can measure its length at leisure. One way to do so is by subtracting the coordinates of the end points of the rod. The value of $\Delta x'$ that is obtained will be the proper length L_0 of the rod.

Suppose the rod is moving in frame S. This means that Δx can be identified as the length L of the rod in frame S only if the coordinates of the rod's end points are measured *simultaneously*—that is, if $\Delta t = 0$. If we put $\Delta x' = L_0$, $\Delta x = L$, and $\Delta t = 0$ in Eq. 38-24, we find

$$L = \frac{L_0}{\gamma} \qquad \text{(length contraction)}, \tag{38-25}$$

which is exactly Eq. 38-13, the length contraction equation.

Sample Problem 38-4

An Earth starship has been sent to check an Earth outpost on the planet P1407, whose moon houses a battle group of the often hostile Reptulians. As the ship follows a straight-line course first past the planet and then past the moon, it detects a high-energy microwave burst at the Reptulian moon base and then, 1.10 s later, an explosion at the Earth outpost, which is 4.00×10^8 m from the Reptulian base as measured from the ship's reference frame. The Reptulians have obviously attacked the Earth outpost, so the starship begins to prepare for a confrontation with them.

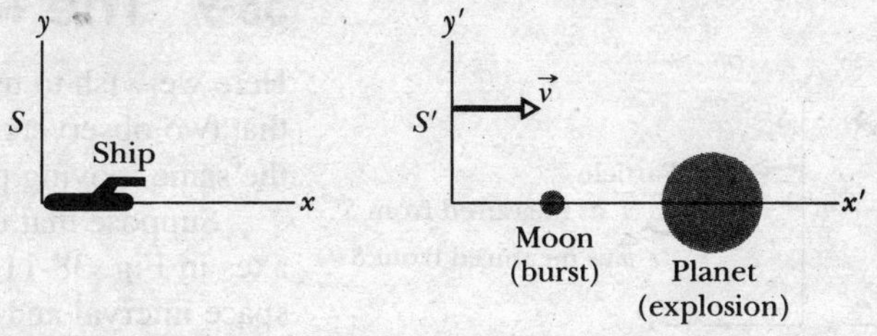

Fig. 38-10 Sample Problem 38-4. A planet and its moon in reference frame S' move rightward with speed v relative to a starship in reference frame S.

(a) The speed of the ship relative to the planet and its moon is $0.980c$. What are the distance and time interval between the burst and the explosion as measured in the planet–moon inertial frame (and thus according to the occupants of the stations)?

SOLUTION: We begin with these **Key Ideas:**

1. This problem involves measurements made from two reference frames, the planet–moon frame and the starship frame.
2. This problem involves two events: the burst and the explosion.
3. We need to transform the given data about the time and distance between the two events as measured in the starship frame to the corresponding data as measured in the planet–moon frame.

Before we get to the transformation, we need to carefully choose our notation. We begin with a sketch of the situation as shown in Fig. 38-10. There, we have chosen the ship's frame S to be stationary and the planet–moon frame S' to be moving with positive velocity (rightward). (This is an arbitrary choice; we could, instead, have chosen the planet–moon frame to be stationary. Then we would redraw $\vec{v}$ in Fig. 38-10 as being attached to the S frame and indicating leftward motion; v would then be a negative quantity. The results would be the same.) Let subscripts e and b represent the explosion and burst, respectively. Then the given data, all in the unprimed (starship) reference frame, are

$$\Delta x = x_e - x_b = +4.00 \times 10^8 \text{ m}$$

and

$$\Delta t = t_e - t_b = +1.10 \text{ s}.$$

Here, Δx is a positive quantity because in Fig. 38-10, the coordinate x_e for the explosion is greater than the coordinate x_b for the burst; Δt is also a positive quantity because the time t_e of the explosion is greater (later) than the time t_b of the burst.

We seek $\Delta x'$ and $\Delta t'$, which we shall get by transforming the given S-frame data to the planet–moon frame S'. Because we are considering a pair of events, we choose transformation equations from Table 38-2, namely Eqs. 1′ and 2′:

$$\Delta x' = \gamma(\Delta x - v\,\Delta t) \qquad (38\text{-}26)$$

and

$$\Delta t' = \gamma\left(\Delta t - \frac{v\,\Delta x}{c^2}\right). \qquad (38\text{-}27)$$

Here, $v = +0.980c$, and the Lorentz factor is

$$\gamma = \frac{1}{\sqrt{1-(v/c)^2}} = \frac{1}{\sqrt{1-(+0.980c/c)^2}} = 5.0252.$$

Equation 38-26 then becomes

$$\Delta x' = (5.0252) \times [4.00 \times 10^8 \text{ m} - (+0.980)(2.998 \times 10^8 \text{ m/s})(1.10 \text{ s})]$$
$$= 3.86 \times 10^8 \text{ m}, \qquad \text{(Answer)}$$

and Eq. 38-27 becomes

$$\Delta t' = (5.0252) \times \left[(1.10 \text{ s}) - \frac{(+0.980)(2.998 \times 10^8 \text{ m/s})(4.00 \times 10^8 \text{ m})}{(2.998 \times 10^8 \text{ m/s})^2}\right]$$
$$= -1.04 \text{ s}. \qquad \text{(Answer)}$$

(b) What is the meaning of the minus sign in the value for $\Delta t'$?

SOLUTION: The **Key Idea** here is to be consistent with the notation we set up in part (a). Recall how we originally defined the time interval between burst and explosion: $\Delta t = t_e - t_b = +1.10$ s. To be consistent with that choice of notation, our definition of $\Delta t'$ must be $t'_e - t'_b$; thus, we have found that

$$\Delta t' = t'_e - t'_b = -1.04 \text{ s}.$$

The minus sign here tells us that $t'_b > t'_e$; that is, in the planet–moon reference frame, the burst occurred 1.04 s *after* the explosion, not 1.10 s *before* the explosion as detected in the ship frame.

(c) Did the burst cause the explosion, or vice versa?

SOLUTION: The sequence of events measured in the planet–moon reference frame is the reverse of that measured in the ship frame. The **Key Idea** here is that in either situation, if there is a causal relationship between the two events, information must travel from the location of one event to the location of the other to cause it. Let us check the required speed of the information. In the ship frame, this speed is

$$v_{info} = \frac{\Delta x}{\Delta t} = \frac{4.00 \times 10^8 \text{ m}}{1.10 \text{ s}} = 3.64 \times 10^8 \text{ m/s},$$

but that speed is impossible because it exceeds c. In the planet–moon frame, the speed comes out to be 3.70×10^8 m/s, also impossible. Therefore, neither event could possibly have caused the other event; that is, they are *unrelated* events. Thus, the starship should not confront the Reptulians.

38-9 The Relativity of Velocities

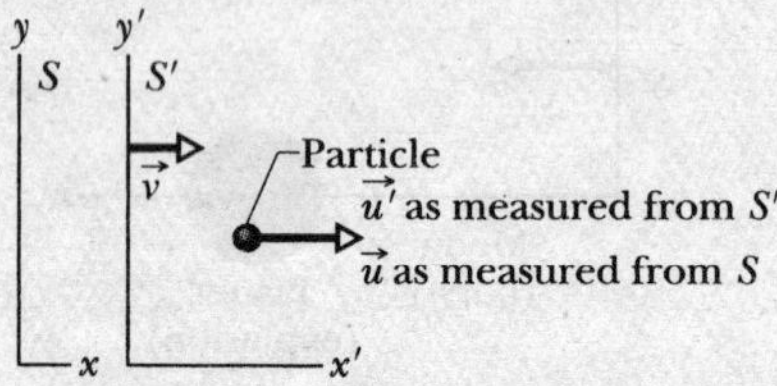

Fig. 38-11 Reference frame S' moves with velocity $\vec{v}$ relative to frame S. A particle has velocity $\vec{u}'$ relative to reference frame S' and velocity $\vec{u}$ relative to reference frame S.

Here we wish to use the Lorentz transformation equations to compare the velocities that two observers in different inertial reference frames S and S' would measure for the same moving particle. Let S' move with velocity v relative to S.

Suppose that the particle, moving with constant velocity parallel to the x and x' axes in Fig. 38-11, sends out two signals as it moves. Each observer measures the space interval and the time interval between these two events. These four measurements are related by Eqs. 1 and 2 of Table 38-2,

$$\Delta x = \gamma(\Delta x' + v\,\Delta t')$$

and

$$\Delta t = \gamma\left(\Delta t' + \frac{v\,\Delta x'}{c^2}\right).$$

If we divide the first of these equations by the second, we find

$$\frac{\Delta x}{\Delta t} = \frac{\Delta x' + v\,\Delta t'}{\Delta t' + v\,\Delta x'/c^2}.$$

Dividing the numerator and denominator of the right side by $\Delta t'$, we find

$$\frac{\Delta x}{\Delta t} = \frac{\Delta x'/\Delta t' + v}{1 + v(\Delta x'/\Delta t')/c^2}.$$

However, in the differential limit, $\Delta x/\Delta t$ is u, the velocity of the particle as measured in S, and $\Delta x'/\Delta t'$ is u', the velocity of the particle as measured in S'. Then we have, finally,

$$u = \frac{u' + v}{1 + u'v/c^2} \qquad \text{(relativistic velocity transformation)} \qquad (38\text{-}28)$$

as the relativistic velocity transformation equation. This equation reduces to the classical, or Galilean, velocity transformation equation,

$$u = u' + v \qquad \text{(classical velocity transformation)}, \qquad (38\text{-}29)$$

when we apply the formal test of letting $c \to \infty$. In other words, Eq. 38-28 is correct for all physically possible speeds while Eq. 38-29 is approximately correct for speeds much less than c.

38-10 Doppler Effect for Light

In Section 18-8 we discussed the Doppler effect (a shift in detected frequency) for sound waves traveling in air. For such waves, the Doppler effect depends on two velocities—namely, the velocities of the source and detector with respect to the air. (Air is the medium that transmits the waves.)

That is not the situation with light waves, for they (and other electromagnetic waves) require no medium, being able to travel even through vacuum. The Doppler effect for light waves depends on only one velocity, the relative velocity $\vec{v}$ between source and detector, as measured from the reference frame of either. Let f_0 represent the **proper frequency** of the source—that is, the frequency that is measured by an observer in the rest frame of the source. Let f represent the frequency detected by an observer moving with velocity $\vec{v}$ relative to that rest frame. Then, when the direction of $\vec{v}$ is directly away from the source,

$$f = f_0 \sqrt{\frac{1-\beta}{1+\beta}} \quad \text{(source and detector separating)}, \tag{38-30}$$

where $\beta = v/c$. When the direction of $\vec{v}$ is directly toward the source, we must change the signs in front of both β symbols in Eq. 38-30.

Low-Speed Doppler Effect

For low speeds ($\beta \ll 1$), Eq. 38-30 can be expanded in a power series in β and approximated as

$$f = f_0(1 - \beta + \tfrac{1}{2}\beta^2) \quad \text{(source and detector separating, } \beta \ll 1). \tag{38-31}$$

The corresponding low-speed equation for the Doppler effect with sound waves (or any waves except light waves) has the same first two terms but a different coefficient in the third term. Thus, the relativistic effect for low-speed light sources and detectors shows up only with the β^2 term.

A police radar unit employs the Doppler effect with microwaves to measure the speed v of a car. A source in the radar unit emits a microwave beam at a certain (proper) frequency f_0 along the road. A car that is moving toward the unit intercepts that beam but at a frequency that is shifted upward by the Doppler effect, due to the car's motion toward the radar unit. The car reflects the beam back toward the radar unit. Because the car is moving toward the radar unit, the detector in the unit intercepts a reflected beam that is further shifted up in frequency. The unit compares that detected frequency with f_0 and computes the speed v of the car.

Astronomical Doppler Effect

In astronomical observations of stars, galaxies, and other sources of light, we can determine how fast the sources are moving, either directly away from us or directly toward us, by measuring the *Doppler shift* of the light that reaches us. If a certain star were at rest relative to us, we would detect light from it with a certain proper frequency f_0. However, if the star is moving either directly away from us or directly toward us, the light we detect has a frequency f that is shifted from f_0 by the Doppler effect. This Doppler shift is due only to the *radial* motion of the star (its motion directly toward us or away from us), and the speed we can determine by measuring this Doppler shift is only the *radial speed* v of the star—that is, only the radial component of the star's velocity relative to us.

Let us assume that the radial speed v of a certain light source is low enough (β is small enough) for us to neglect the β^2 term in Eq. 38-31. Also, let us explicitly show a $\pm$ option in front of the β term—the minus sign corresponding to radial motion away from us and the plus sign corresponding to radial motion toward us. Then Eq. 38-31 becomes

$$f = f_0(1 \pm \beta). \tag{38-32}$$

Astronomical measurements involving light are usually done in wavelengths rather than frequencies, so let us replace f with c/λ and f_0 with c/λ_0, where λ is the measured wavelength and λ_0 is the **proper wavelength.** Also replacing β with v/c in Eq. 38-32, we have

$$\frac{c}{\lambda} = \frac{c}{\lambda_0}\left(1 \pm \frac{v}{c}\right),$$

which leads to

$$v = \pm \frac{\lambda - \lambda_0}{\lambda} c.$$

This is conventionally written as

$$v = \frac{\Delta\lambda}{\lambda} c \quad \text{(radial speed of light source, } v \ll c\text{)}, \tag{38-33}$$

where $\Delta\lambda$ $(= |\lambda - \lambda_0|)$ is the *wavelength* Doppler shift of the light source. If the source is moving away from us, λ is greater than λ_0 and the Doppler shift is called a *red shift*. (The term does not mean that the detected light is red or even visible; it merely means the wavelength increased.) Similarly, if the source is moving toward us, λ is less than λ_0 and the Doppler shift is called a *blue shift*.

CHECKPOINT 4: The figure shows a source that emits light of proper frequency f_0 while moving directly toward the right with speed $c/4$ as measured from reference frame S. The figure also shows a light detector, which measures a frequency $f > f_0$ for the emitted light. (a) Is the detector moving toward the left or the right? (b) Is the speed of the detector as measured from reference frame S more than $c/4$, less than $c/4$, or equal to $c/4$?

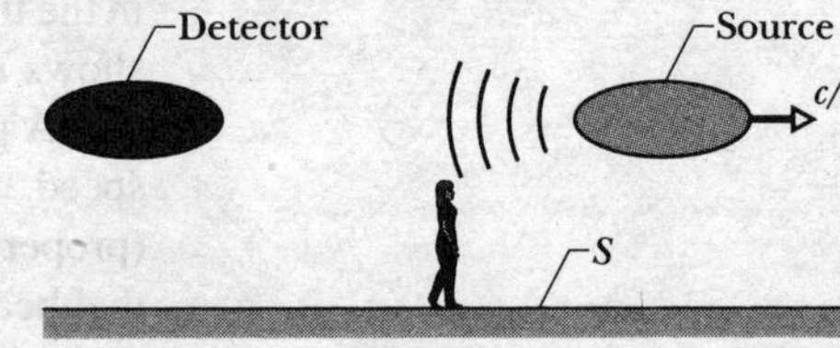

Transverse Doppler Effect

So far, we have discussed the Doppler effect, here and in Chapter 18, only for situations in which the source and the detector move either directly toward or directly away from each other. Figure 38-12 shows a different arrangement, in which a source S moves past a detector D. When S reaches point P, its velocity is perpendicular to the line joining S and D and, at that instant, it is moving neither toward nor away from D. If the source is emitting sound waves of frequency f_0, D detects that frequency (with no Doppler effect) when it intercepts the waves that were emitted at point P. However, if the source is emitting light waves, there is still a Doppler effect, called the **transverse Doppler effect.** In this situation, the detected frequency of the light emitted when the source is at point P is

$$f = f_0\sqrt{1 - \beta^2} \quad \text{(transverse Doppler effect)}. \tag{38-34}$$

For low speeds ($\beta \ll 1$), Eq. 38-34 can be expanded in a power series in β and approximated as

$$f = f_0(1 - \tfrac{1}{2}\beta^2) \quad \text{(low speeds)}. \tag{38-35}$$

Here the first term is what we would expect for sound waves and, again, the relativistic effect for low-speed light sources and detectors appears with the β^2 term.

In principle, a police radar unit can determine the speed of a car even when the path of the radar beam is perpendicular (transverse) to the path of the car. However, Eq. 38-35 tells us that because β is small even for a fast car, the relativistic term $\beta^2/2$ in the transverse Doppler effect is extremely small. Thus, $f \approx f_0$ and the radar unit computes a speed of zero. For this reason, police officers always try to direct the radar beam along the car's path to get a Doppler shift that gives the car's actual speed. Any deviation from that alignment works in favor of the motorist, because it reduces the measured speed.

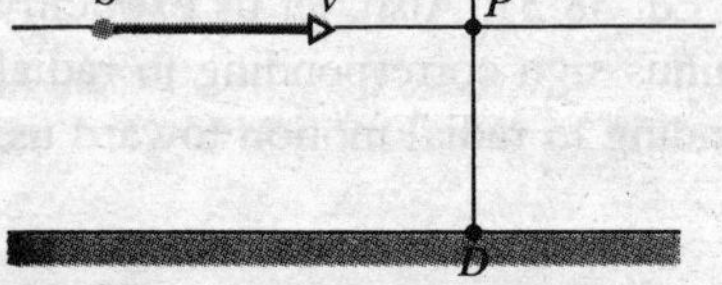

Fig. 38-12 A light source S travels with velocity $\vec{v}$ past a detector at D. The special theory of relativity predicts a transverse Doppler effect as the source passes through point P, where the direction of travel is perpendicular to the line extending through D. Classical theory predicts no such effect.

The transverse Doppler effect is really another test of time dilation. If we rewrite Eq. 38-34 in terms of the period T of oscillation of the emitted light wave instead of the frequency, we have, since $T = 1/f$,

$$T = \frac{T_0}{\sqrt{1-\beta^2}} = \gamma T_0, \qquad (38\text{-}36)$$

in which T_0 ($= 1/f_0$) is the **proper period** of the source. As comparison with Eq. 38-9 shows, Eq. 38-36 is simply the time dilation formula, since a period is a time interval.

The NAVSTAR Navigation System

Each NAVSTAR satellite continually broadcasts radio signals giving its location, at a set frequency that is controlled by precise atomic clocks. When the signal is sensed by the detector on, say, a commercial aircraft, the frequency has been Doppler-shifted. By detecting the signals from several NAVSTAR satellites simultaneously, the detector can determine the direction to any one of them and the direction of the velocity of that satellite. From the Doppler shift of the signal, the detector then determines the speed of the aircraft.

Let us use some rough numbers to see how well this can be done. The speed of a NAVSTAR satellite relative to the center of Earth is about 1.0×10^4 m/s. The associated β is about 3.0×10^{-5}. Thus, the term $\beta^2/2$ in Eqs. 38-31 and 38-35 (that is, the relativity term) is about 4.5×10^{-10}. In other words, relativity changes the Doppler shift of the detected signal by about 4.5 parts in 10^{10}, which seems hardly worth considering.

However, it is indeed important. The atomic clocks in the satellites are so precise that the variation in the frequency of the satellite signal is only 2 parts in 10^{12}. From Eq. 38-35, we see that β (hence v) depends on the square root of f/f_0. Thus, the clock's frequency variation of 2×10^{-12} causes a variation of

$$\sqrt{2 \times 10^{-12}} = 1.4 \times 10^{-6}$$

in the measured value of the relative speed v between satellite and aircraft.

Since v is due primarily to the satellite's great speed, 1.0×10^4 m/s, this means that v (hence the aircraft's speed) can be determined to an accuracy of about

$$(1.4 \times 10^{-6})(1.0 \times 10^4 \text{ m/s}) = 1.4 \text{ cm/s}.$$

Suppose the aircraft flies for 1 h (3600 s). Knowing the speed to about 1.4 cm/s allows the location at the end of that hour to be predicted to about

$$(0.014 \text{ m/s})(3600 \text{ s}) = 50 \text{ m},$$

which is acceptable in modern navigation.

If relativity effects were not taken into account, the speed of the aircraft could not be known any closer than 21 cm/s, and its location after an hour's flight could not be predicted any better than within 760 m.

Sample Problem 38-5

Figure 38-13a shows curves of intensity versus wavelength for light reaching us from interstellar gas on two opposite sides of galaxy M87 (Fig. 38-13b). One curve peaks at 499.8 nm; the other at 501.6 nm. The gas orbits the core of the galaxy at a radius r = 100 light-years, apparently moving toward us on one side of the core and moving away from us on the opposite side.

(a) Which curve corresponds to the gas moving toward us? What is the speed of the gas relative to us (and relative to the galaxy's core)?

SOLUTION: The **Key Ideas** here are these:

1. If the gas were not moving around the galaxy's core, the light from it would be detected at a certain wavelength.
2. The motion of the gas changes the detected wavelength via the Doppler effect, increasing the wavelength for the gas moving away from us and decreasing it for the gas moving toward us.

Thus, the curve peaking at 501.6 nm corresponds to motion away from us, and that peaking at 499.8 nm corresponds to motion toward us.

Let us assume that the increase and the decrease in wavelength due to the motion of the gas are equal in magnitude. Then the unshifted wavelength, which we shall take as the proper wavelength λ_0, must be the average of the two shifted wavelengths:

$$\lambda_0 = \frac{501.6 \text{ nm} + 499.8 \text{ nm}}{2} = 500.7 \text{ nm}.$$

The Doppler shift $\Delta\lambda$ of the light from the gas moving away from us is then

$$\Delta\lambda = |\lambda - \lambda_0| = 501.6 \text{ nm} - 500.7 \text{ nm}$$
$$= 0.90 \text{ nm}.$$

Substituting this and $\lambda = 501.6$ nm into Eq. 38-33, we find that the speed of the gas is

$$v = \frac{\Delta\lambda}{\lambda} c = \frac{0.90 \text{ nm}}{501.6 \text{ nm}} 2.998 \times 10^8 \text{ m/s}$$
$$= 5.38 \times 10^5 \text{ m/s}. \quad \text{(Answer)}$$

(b) The gas orbits the core of the galaxy because it experiences a gravitational force due to the mass M of the core. What is that mass in multiples of the Sun's mass M_S ($= 1.99 \times 10^{30}$ kg)?

SOLUTION: There are two **Key Ideas** here:

1. From Eq. 14-1, the magnitude F of the gravitational force on an orbiting gas element of mass m at orbital radius r is

$$F = \frac{GMm}{r^2}.$$

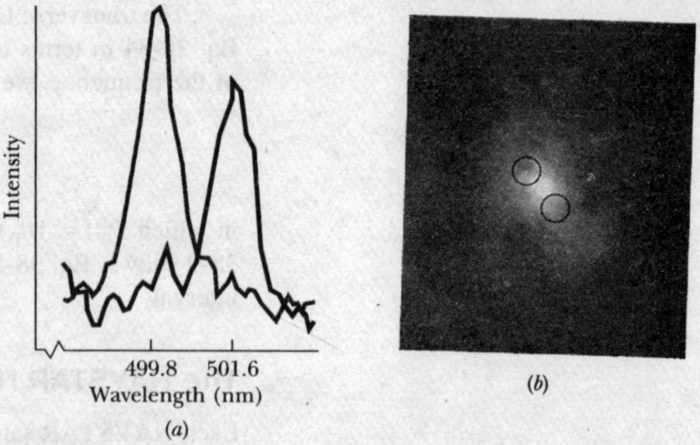

Fig. 38-13 Sample Problem 38-5. (*a*) Plots of intensity versus wavelength for light emitted by gas on opposite sides of galaxy M87 and detected on Earth. (*b*) The central region of M87. The circles indicate the locations of the gas whose intensity is given in (*a*). The core of M87 is halfway between the circles.

2. If the gas element orbits the galaxy core in a circle, then it must have a centripetal acceleration of magnitude $a = v^2/r$, directed toward the core.
3. Newton's second law, written for a radial axis extending from the core to the gas element, tells us that $F = ma$.

Putting these three ideas together, we have

$$\frac{GMm}{r^2} = m\frac{v^2}{r}.$$

Solving this for M and substituting known data, we find

$$M = \frac{v^2 r}{G}$$
$$= \frac{(5.38 \times 10^5 \text{ m/s})^2(100 \text{ ly})(9.46 \times 10^{15} \text{ m/ly})}{6.67 \times 10^{-11} \text{ N} \cdot \text{m}^2/\text{kg}^2}$$
$$= 4.11 \times 10^{39} \text{ kg} = (2.1 \times 10^9)M_S. \quad \text{(Answer)}$$

This result tells us that a mass equivalent to two billion suns has been compacted into the core of the galaxy, strongly suggesting that a "supermassive" black hole occupies the core.

38-11 A New Look at Momentum

Suppose that a number of observers, each in a different inertial reference frame, watch an isolated collision between two particles. In classical mechanics, we have seen that—even though the observers measure different velocities for the colliding particles—they all find that the law of conservation of momentum holds. That is, they find that the total momentum of the system of particles after the collision is the same as it was before the collision.

How is this situation affected by relativity? We find that if we continue to define the momentum $\vec{p}$ of a particle as $m\vec{v}$, the product of its mass and its velocity, total momentum is *not* conserved for the observers in different inertial frames. We have two choices: (1) Give up the law of conservation of momentum or (2) see if we can redefine the momentum of a particle in some new way so that the law of conservation of momentum still holds. The correct choice is the second one.

Consider a particle moving with constant speed v in the positive x direction. Classically, its momentum has magnitude

$$p = mv = m\frac{\Delta x}{\Delta t} \quad \text{(classical momentum)}, \tag{38-37}$$

in which Δx is the distance it travels in time Δt. To find a relativistic expression for momentum, we start with the new definition

$$p = m\frac{\Delta x}{\Delta t_0}.$$

Here, as before, Δx is the distance traveled by a moving particle as viewed by an observer watching that particle. However, Δt_0 is the time required to travel that distance, measured not by the observer watching the moving particle but by an observer moving with the particle. The particle is at rest with respect to this second observer, with the result that the time this observer measures is a proper time.

Using the time dilation formula (Eq. 38-9), we can then write

$$p = m\frac{\Delta x}{\Delta t_0} = m\frac{\Delta x}{\Delta t}\frac{\Delta t}{\Delta t_0} = m\frac{\Delta x}{\Delta t}\gamma.$$

However, since $\Delta x/\Delta t$ is just the particle velocity v,

$$p = \gamma mv \quad \text{(momentum)}. \tag{38-38}$$

Note that this differs from the classical definition of Eq. 38-37 only by the Lorentz factor γ. However, that difference is important: Unlike classical momentum, relativistic momentum approaches an infinite value as v approaches c.

We can generalize the definition of Eq. 38-38 to vector form as

$$\vec{p} = \gamma m\vec{v} \quad \text{(momentum)}. \tag{38-39}$$

This equation gives the correct definition of momentum for all physically possible speeds. For a speed much less than c, it reduces to the classical definition of momentum ($\vec{p} = m\vec{v}$).

38-12 A New Look at Energy

Mass Energy

The science of chemistry was initially developed with the assumption that in chemical reactions, energy and mass are conserved separately. In 1905, Einstein showed that as a consequence of his theory of special relativity, mass can be considered to be another form of energy. Thus, the law of conservation of energy is really the law of conservation of mass–energy.

In a *chemical reaction* (a process in which atoms or molecules interact), the amount of mass that is transferred into other forms of energy (or vice versa) is such a tiny fraction of the total mass involved that there is no hope of measuring the mass change with even the best laboratory balances. Mass and energy truly *seem* to be separately conserved. However, in a *nuclear reaction* (in which nuclei or fundamental particles interact), the energy released is often about a million times greater

TABLE 38-3 The Energy Equivalents of a Few Objects

Object	Mass (kg)	Energy Equivalent	
Electron	9.11×10^{-31}	8.19×10^{-14} J	(= 511 keV)
Proton	1.67×10^{-27}	1.50×10^{-10} J	(= 938 MeV)
Uranium atom	3.95×10^{-25}	3.55×10^{-8} J	(= 225 GeV)
Dust particle	1×10^{-13}	1×10^{4} J	(= 2 kcal)
U.S. penny	3.1×10^{-3}	2.8×10^{14} J	(= 78 GW · h)

than in a chemical reaction, and the change in mass can easily be measured. Taking mass–energy transfers into account in nuclear reactions became routine long ago.

An object's mass m and the energy equivalent E_0 of that mass are related by

$$E_0 = mc^2, \tag{38-40}$$

which, without the subscript 0, is the best-known science equation of all time. This energy that is associated with the mass of an object is called **mass energy** or **rest energy.** The second phrase suggests that E_0 is an energy that the object has even when it is at rest, simply because it has mass. (If you continue your study of physics beyond this book, you will see more refined discussions of the relation between mass and energy. You might even encounter disagreements about just what that relation is and means.)

Table 38-3 shows the mass energy or rest energy of a few objects. The mass energy of, say, a U.S. penny is enormous; the equivalent amount of electric energy would cost well over a million dollars. On the other hand, the entire annual U.S. electric energy production corresponds to a mass of only a few hundred kilograms of matter (stones, burritos, or anything else).

In practice, SI units are rarely used with Eq. 38-40 because they are too large to be convenient. Masses are usually measured in atomic mass units, where

$$1 \text{ u} = 1.66 \times 10^{-27} \text{ kg}, \tag{38-41}$$

and energies are usually measured in electron-volts or multiples of it, where

$$1 \text{ eV} = 1.60 \times 10^{-19} \text{ J}. \tag{38-42}$$

In the units of Eqs. 38-41 and 38-42, the multiplying constant c^2 has the values

$$\begin{aligned} c^2 &= 9.315 \times 10^8 \text{ eV/u} = 9.315 \times 10^5 \text{ keV/u} \\ &= 931.5 \text{ MeV/u}. \end{aligned} \tag{38-43}$$

Total Energy

Equation 38-40 gives an object's mass energy (or rest energy) E_0 that is associated with the object's mass m, regardless of whether the object is at rest or moving. If the object is moving, it has additional energy in the form of kinetic energy K. If we assume that its potential energy is zero, then its total energy E is the sum of its mass energy and its kinetic energy:

$$E = E_0 + K = mc^2 + K. \tag{38-44}$$

Although we shall not prove it, the total energy E can also be written as

$$E = \gamma mc^2, \tag{38-45}$$

where γ is the Lorentz factor for the object's motion.

Since Chapter 7, we have discussed many examples involving changes in the total energy of a particle or a system of particles. However, we did not include mass energy in the discussions because the changes in mass energy were either zero or small enough to be neglected. The law of conservation of total energy still applies even if changes in mass energy are significant. Thus, regardless of what happens to the mass energy, the following statement from Section 8-7 is still true:

> The total energy E of an *isolated system* cannot change.

For example, if the total mass energy of two interacting particles in an isolated system decreases, some other type of energy in the system must increase because the total energy cannot change.

In a system undergoing a chemical or nuclear reaction, a change in the total mass energy of the system due to the reaction is often given as a Q value. The Q value for a reaction is obtained from the relation

$$\begin{pmatrix}\text{system's initial}\\ \text{total mass energy}\end{pmatrix} = \begin{pmatrix}\text{system's final}\\ \text{total mass energy}\end{pmatrix} + Q$$

or

$$E_{0i} = E_{0f} + Q. \tag{38-46}$$

Using Eq. 38-40 ($E_0 = mc^2$), we can rewrite this in terms of the initial *total* mass M_i and the final *total* mass M_f as

$$2M_i c^2 = M_f c^2 + Q$$

or

$$Q = M_i c^2 - M_f c^2 = -\Delta M\, c^2, \tag{38-47}$$

where the change in mass due to the reaction is $\Delta M = M_f - M_i$.

If a reaction results in the transfer of energy from mass energy to, say, kinetic energy of the reaction products, the system's total mass energy E_0 (and total mass M) decreases and Q is positive. If, instead, a reaction requires that energy be transferred to mass energy, the system's total mass energy E_0 (and its total mass M) increases and Q is negative.

For example, suppose two hydrogen nuclei undergo a *fusion reaction* in which they join together to form a single nucleus and release two particles. The total mass energy (and total mass) of the resultant single nucleus and two released particles is less than the total mass energy (and total mass) of the initial hydrogen nuclei. Thus, the Q of the fusion reaction is positive, and energy is said to be *released* (transferred from mass energy) by the reaction. This release is important to you, because the fusion of hydrogen nuclei in the Sun is one part of the process that results in sunshine on Earth and makes life here possible.

Kinetic Energy

In Chapter 7 we defined the kinetic energy K of an object of mass m and with a speed v well below c to be

$$K = \tfrac{1}{2}mv^2. \tag{38-48}$$

However, this classical equation is only an approximation that is good enough when the speed is well below the speed of light.

Let us now find an expression for kinetic energy that is correct for *all* physically possible speeds, including speeds close to c. Solving Eq. 38-44 for K and then

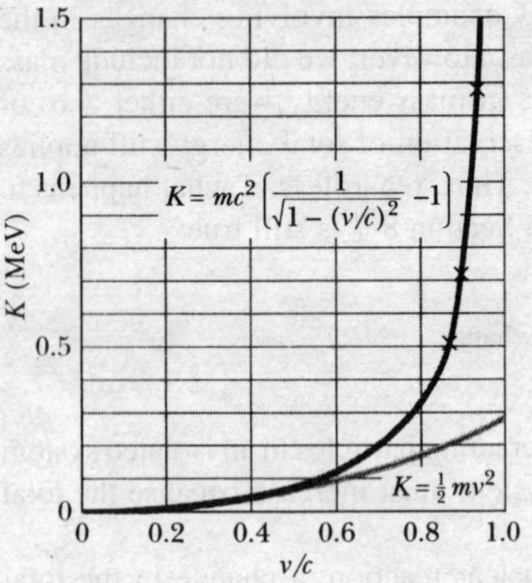

Fig. 38-14 The relativistic (Eq. 38-49) and classical (Eq. 38-48) equations for the kinetic energy of an electron, plotted as a function of v/c, where v is the speed of the electron and c is the speed of light. Note that the two curves blend together at low speeds and diverge widely at high speeds. Experimental data (at the × marks) show that at high speeds the relativistic curve agrees with experiment but the classical curve does not.

substituting for E_0 from Eq. 38-45 lead to

$$K = E - mc^2 = \gamma mc^2 - mc^2$$
$$= mc^2(\gamma - 1) \quad \text{(kinetic energy)}, \qquad (38\text{-}49)$$

where $\gamma\ (= 1/\sqrt{1-(v/c)^2})$ is the Lorentz factor for the object's motion.

Figure 38-14 shows plots of the kinetic energy of an electron as calculated with the correct definition (Eq. 38-49) and the classical approximation (Eq. 38-48), both as functions of v/c. Note that on the left side of the graph the two plots coincide; this is the part of the graph—at lower speeds—where we have calculated kinetic energies so far in this book. That part of the graph tells us that we have been justified in calculating kinetic energy with the classical expression of Eq. 38-48. However, on the right side of the graph—at speeds near c—the two plots differ significantly. As v/c approaches 1.0, the plot for the classical definition of kinetic energy increases only moderately while the plot for the correct definition of kinetic energy increases dramatically, approaching an infinite value as v/c approaches 1.0. Thus, when an object's speed v is near c, we *must* use Eq. 38-49 to calculate its kinetic energy.

Figure 38-14 also tells us something about the work we must do on an object to increase the object's speed by, say, 1%. The required work W is equal to the resulting change ΔK in the object's kinetic energy. If the change is to occur on the low-speed left side of Fig. 38-14, the required work might be modest. However, if the change is to occur on the high-speed right side of Fig. 38-14, the required work could be enormous, because the kinetic energy K increases so rapidly there with an increase in speed v. To increase an object's speed to c would require, in principle, an infinite amount of energy; thus, doing so is impossible.

The kinetic energies of electrons, protons, and other particles are often stated with the unit electron-volt or one of its multiples used as an adjective. For example, an electron with a kinetic energy of 20 MeV may be described as a 20 MeV electron.

Momentum and Kinetic Energy

In classical mechanics, the momentum p of a particle is mv and its kinetic energy K is $\frac{1}{2}mv^2$. If we eliminate v between these two expressions, we find a direct relation between momentum and kinetic energy:

$$p^2 = 2Km \quad \text{(classical)}. \qquad (38\text{-}50)$$

We can find a similar connection in relativity by eliminating v between the relativistic definition of momentum (Eq. 38-38) and the relativistic definition of kinetic energy (Eq. 38-49). Doing so leads, after some algebra, to

$$(pc)^2 = K^2 + 2Kmc^2. \qquad (38\text{-}51)$$

With the aid of Eq. 38-44, we can transform Eq. 38-51 into a relation between the momentum p and the total energy E of a particle:

$$E^2 = (pc)^2 + (mc^2)^2. \qquad (38\text{-}52)$$

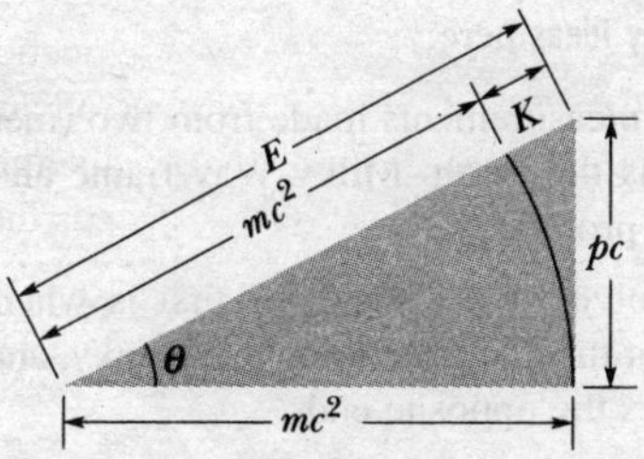

Fig. 38-15 A useful mnemonic device for remembering the relativistic relations among the total energy E, the rest energy or mass energy mc^2, the kinetic energy K, and the momentum p.

The right triangle of Fig. 38-15 can help you keep these useful relations in mind. You can also show that, in that triangle,

$$\sin\theta = \beta \quad \text{and} \quad \cos\theta = 1/\gamma. \qquad (38\text{-}53)$$

With Eq. 38-52 we can see that the product pc must have the same unit as energy E; thus, we can express the unit of momentum p as an energy unit divided by c. In fact, momentum in fundamental particle physics is often reported in the units MeV/c or GeV/c.

✔CHECKPOINT 5: Are (a) the kinetic energy and (b) the total energy of a 1 GeV electron more than, less than, or equal to those of a 1 GeV proton?

Sample Problem 38-6

(a) What is the total energy E of a 2.53 MeV electron?

SOLUTION: The **Key Idea** here is that, from Eq. 38-44, the total energy E is the sum of the electron's mass energy (or rest energy) mc^2 and its kinetic energy:

$$E = mc^2 + K. \qquad (38\text{-}54)$$

The adjective "2.53 MeV" in the problem statement means that the electron's kinetic energy is 2.53 MeV. To evaluate the electron's mass energy mc^2, we substitute the electron's mass m from Appendix B, obtaining

$$mc^2 = (9.109 \times 10^{-31}\ \text{kg})(2.998 \times 10^8\ \text{m/s})^2 = 8.187 \times 10^{-14}\ \text{J}.$$

Then dividing this result by 1.602×10^{-13} J/MeV gives us 0.511 MeV as the electron's mass energy (confirming the value in Table 38-3). Equation 38-54 then yields

$$E = 0.511\ \text{MeV} + 2.53\ \text{MeV} = 3.04\ \text{MeV}. \qquad \text{(Answer)}$$

(b) What is the magnitude p of the electron's momentum, in the unit MeV/c?

SOLUTION: The **Key Idea** here is that we can find p from the total energy E and the mass energy mc^2 via Eq. 38-52,

$$E^2 = (pc)^2 + (mc^2)^2.$$

Solving for pc gives us

$$pc = \sqrt{E^2 - (mc^2)^2} = \sqrt{(3.04\ \text{MeV})^2 - (0.511\ \text{MeV})^2} = 3.00\ \text{MeV}.$$

Finally, dividing both sides by c we find

$$p = 3.00\ \text{MeV}/c. \qquad \text{(Answer)}$$

Sample Problem 38-7

The most energetic proton ever detected in the cosmic rays coming to Earth from space had an astounding kinetic energy of 3.0×10^{20} eV (enough energy to warm a teaspoon of water by a few degrees).

(a) What were the proton's Lorentz factor γ and speed v (both relative to the ground-based detector)?

SOLUTION: One **Key Idea** here is that the proton's Lorentz factor γ relates its total energy E to its mass energy mc^2 via Eq. 38-45 ($E = \gamma mc^2$). A second **Key Idea** is that the proton's total energy is the sum of its mass energy mc^2 and its (given) kinetic energy K. Putting these ideas together we have

$$\gamma = \frac{E}{mc^2} = \frac{mc^2 + K}{mc^2} = 1 + \frac{K}{mc^2}. \qquad (38\text{-}55)$$

We can calculate the proton's mass energy mc^2 from its mass given in Appendix B, as we did for the electron in Sample Problem 38-6a. We find that mc^2 is 938 MeV (as listed in Table 38-3). Substituting this and the given kinetic energy into Eq. 38-55, we obtain

$$\gamma = 1 + \frac{3.0 \times 10^{20}\ \text{eV}}{938 \times 10^6\ \text{eV}} = 3.198 \times 10^{11} \approx 3.2 \times 10^{11}. \qquad \text{(Answer)}$$

This computed value for γ is so large that we cannot use the definition of γ (Eq. 38-8) to find v. Try it; your calculator will tell you that β is effectively equal to 1 and thus that v is effectively equal to c. Actually, v is almost c, but we want a more accurate answer, which we can obtain by first solving Eq. 38-8 for $1 - \beta$. To begin we write

$$\gamma = \frac{1}{\sqrt{1-\beta^2}} = \frac{1}{\sqrt{(1-\beta)(1+\beta)}} \approx \frac{1}{\sqrt{2(1-\beta)}},$$

where we have used the fact that β is so close to unity that $1 + \beta$ is very close to 2. The velocity we seek is contained in the $1 - \beta$

term. Solving for $1 - \beta$ then yields

$$1 - \beta = \frac{1}{2\gamma^2} = \frac{1}{(2)(3.198 \times 10^{11})^2}$$
$$= 4.9 \times 10^{-24} \approx 5 \times 10^{-24}.$$

Thus, $$\beta = 1 - 5 \times 10^{-24}$$

and, since $v = \beta c$,

$$v \approx 0.999\,999\,999\,999\,999\,999\,999\,995c. \quad \text{(Answer)}$$

(b) Suppose that the proton travels along a diameter (9.8×10^4 ly) of the Milky Way galaxy. Approximately how long does the proton take to travel that diameter as measured from the common reference frame of Earth and the galaxy?

SOLUTION: We just saw that this *ultrarelativistic* proton is traveling at a speed barely less than c. Then the **Key Idea** here is that by the definition of light-year, light takes 1 y to travel 1 ly, so light should take 9.8×10^4 y to travel 9.8×10^4 ly, and this proton should take almost the same time. Thus, from our Earth–Milky Way reference frame, the proton's trip takes

$$\Delta t = 9.8 \times 10^4 \text{ y}. \quad \text{(Answer)}$$

(c) How long does the trip take as measured in the reference frame of the proton?

SOLUTION: We need four **Key Ideas** here:

1. This problem involves measurements made from two (inertial) reference frames, one is the Earth–Milky Way frame and the other is attached to the proton.
2. This problem also involves two events: the first is when the proton passes one end of the diameter along the Galaxy, and the second is when it passes the opposite end.
3. The time interval between those two events as measured in the proton's reference frame is the proper time interval Δt_0 because the events occur at the same location in that frame—namely, at the proton itself.
4. We can find the proper time interval Δt_0 from the time interval Δt measured in the Earth–Milky Way frame by using Eq. 38-9 ($\Delta t = \gamma\,\Delta t_0$) for time dilation.

Solving Eq. 38-9 for Δt_0 and substituting γ from (a) and Δt from (b), we find

$$\Delta t_0 = \frac{\Delta t}{\gamma} = \frac{9.8 \times 10^4 \text{ y}}{3.198 \times 10^{11}}$$
$$= 3.06 \times 10^{-7} \text{ y} = 9.7 \text{ s}. \quad \text{(Answer)}$$

In our frame, the trip takes 98 000 y. In the proton's frame, it takes 9.7 s! As promised at the start of this chapter, relative motion can alter the rate at which time passes, and we have here an extreme example.

REVIEW & SUMMARY

The Postulates Einstein's **special theory of relativity** is based on two postulates:

1. The laws of physics are the same for observers in all inertial reference frames. No frame is preferred.
2. The speed of light in vacuum has the same value c in all directions and in all inertial reference frames.

The speed of light c in vacuum is an ultimate speed that cannot be exceeded by any entity carrying either energy or information.

Coordinates of an Event Three space coordinates and one time coordinate specify an **event.** One task of special relativity is to relate these coordinates as assigned by two observers who are in uniform motion with respect to each other.

Simultaneous Events If two observers are in relative motion, they will not, in general, agree as to whether two events are simultaneous. If one of the observers finds two events at different locations to be simultaneous, the other will not, and conversely. Simultaneity is *not* an absolute concept but a relative one, depending on the motion of the observer. The relativity of simultaneity is a direct consequence of the finite ultimate speed c.

Time Dilation If two successive events occur at the same place in an inertial reference frame, the time interval Δt_0 between them, measured on a single clock where they occur, is the **proper time** between the events. *Observers in frames moving relative to that frame will measure a larger value for this interval.* For an observer moving with relative speed v, the measured time interval is

$$\Delta t = \frac{\Delta t_0}{\sqrt{1 - (v/c)^2}} = \frac{\Delta t_0}{\sqrt{1 - \beta^2}}$$
$$= \gamma\,\Delta t_0 \quad \text{(time dilation)}. \quad \text{(38-7 to 38-9)}$$

Here $\beta = v/c$ is the **speed parameter** and $\gamma = 1/\sqrt{1 - \beta^2}$ is the **Lorentz factor.** An important consequence of time dilation is that moving clocks run slow as measured by an observer at rest.

Length Contraction The length L_0 of an object measured by an observer in an inertial reference frame in which the object is at rest is called its **proper length.** *Observers in frames moving relative to that frame and parallel to that length will measure a shorter length.* For an observer moving with relative speed v, the measured length is

$$L = L_0\sqrt{1 - \beta^2} = \frac{L_0}{\gamma} \quad \text{(length contraction)}. \quad \text{(38-13)}$$

The Lorentz Transformation The *Lorentz transformation* equations relate the spacetime coordinates of a single event as seen by observers in two inertial frames, S and S', where S' is moving relative to S with velocity v in the positive x, x' direction. The four coordinates are related by

$$\begin{aligned} x' &= \gamma(x - vt), \\ y' &= y, \\ z' &= z, \\ t' &= \gamma(t - vx/c^2) \end{aligned} \qquad \begin{array}{l}\text{(Lorentz transformation equations;} \\ \text{valid at all physically possible} \\ \text{speeds)}\end{array} \qquad (38\text{-}20)$$

Relativity of Velocities When a particle is moving with speed u' in the positive x' direction in an inertial reference frame S' that itself is moving with speed v parallel to the x direction of a second inertial frame S, the speed u of the particle as measured in S is

$$u = \frac{u' + v}{1 + u'v/c^2} \quad \text{(relativistic velocity).} \qquad (38\text{-}28)$$

Relativistic Doppler Effect If a source emitting light waves of frequency f_0 moves directly away from a detector with relative radial speed v (and speed parameter $\beta = v/c$), the frequency f measured by the detector is

$$f = f_0\sqrt{\frac{1 - \beta}{1 + \beta}}. \qquad (38\text{-}30)$$

If the source moves directly toward the detector, the signs in Eq. 38-30 are reversed.

For astronomical observations, the Doppler effect is measured in wavelengths. For speeds much less than c, Eq. 38-30 leads to

$$v = \frac{\Delta\lambda}{\lambda}c, \qquad (38\text{-}33)$$

where $\Delta\lambda$ is the *Doppler shift* in wavelength (the magnitude of the change in wavelength) due to the motion.

Transverse Doppler Effect If the relative motion of the light source is perpendicular to a line joining the source and detector, the Doppler frequency formula is

$$f = f_0\sqrt{1 - \beta^2}. \qquad (38\text{-}34)$$

This **transverse Doppler effect** is due to time dilation.

Momentum and Energy The following definitions of linear momentum $\vec{p}$, kinetic energy K, and total energy E for a particle of mass m are valid at any physically possible speed:

$$\vec{p} = \gamma m\vec{v} \quad \text{(momentum),} \qquad (38\text{-}39)$$

$$E = mc^2 + K = \gamma mc^2 \quad \text{(total energy),} \qquad (38\text{-}44, 38\text{-}45)$$

$$K = mc^2(\gamma - 1) \quad \text{(kinetic energy).} \qquad (38\text{-}49)$$

Here γ is the Lorentz factor for the particle's motion, and mc^2 is the *mass energy* or *rest energy* associated with the mass of the particle. These equations lead to the relationships

$$(pc)^2 = K^2 + 2Kmc^2, \qquad (38\text{-}51)$$

and

$$E^2 = (pc)^2 + (mc^2)^2. \qquad (38\text{-}52)$$

When a system of particles undergoes a chemical or nuclear reaction, the Q of the reaction is the negative of the change in the system's total mass energy:

$$Q = M_ic^2 - M_fc^2 = -\Delta M\,c^2, \qquad (38\text{-}47)$$

where M_i is the system's total mass before the reaction and M_f is its total mass after the reaction.

QUESTIONS

1. In Fig. 38-16, ship A sends a laser pulse to an oncoming ship B, while scout ship C races away. The indicated speeds of the ships are all measured from the same reference frame. Rank the ships according to the speed of the pulse as measured from each ship, greatest first.

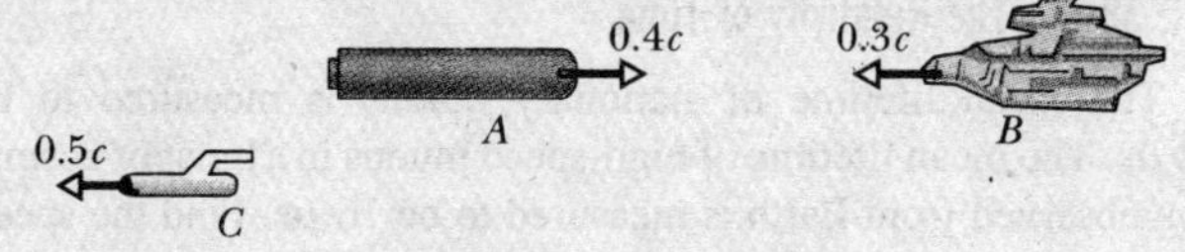

Fig. 38-16 Questions 1 and 7.

2. Figure 38-17a shows two clocks in stationary frame S (they are synchronized in that frame) and one clock in moving frame S'. Clocks C_1 and C_1' read zero when they pass each other. When clocks C_1' and C_2 pass each other, (a) which clock has the smaller reading and (b) which clock measures a proper time?

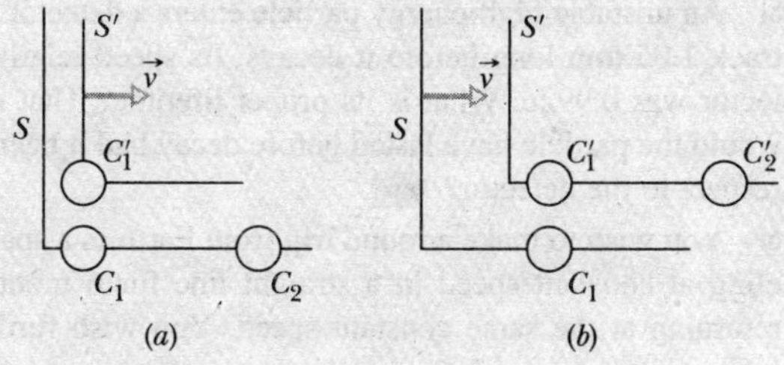

Fig. 38-17 Questions 2 and 3.

3. Figure 38-17b shows two clocks in stationary frame S' (they are synchronized in that frame) and one clock in moving frame S. Clocks C_1 and C_1' read zero when they pass each other. When clocks C_1 and C_2' pass each other, (a) which clock has the smaller reading and (b) which clock measures a proper time?

4. Sam leaves Venus in a spaceship to Mars and passes Sally, who is on Earth, with a relative speed of $0.5c$. (a) Each measures the Venus–Mars voyage time. Who measures a proper time: Sam, Sally, or neither? (b) On the way, Sam sends a pulse of light to Mars. Each measures the travel time of the pulse. Who measures a proper time?

5. Figure 38-18 shows a ship (with on-board reference frame S') passing us (with reference frame S). A proton is fired at nearly the speed of light along the length of the ship, from the front to the rear. (a) Is the spatial separation $\Delta x'$ between the firing of

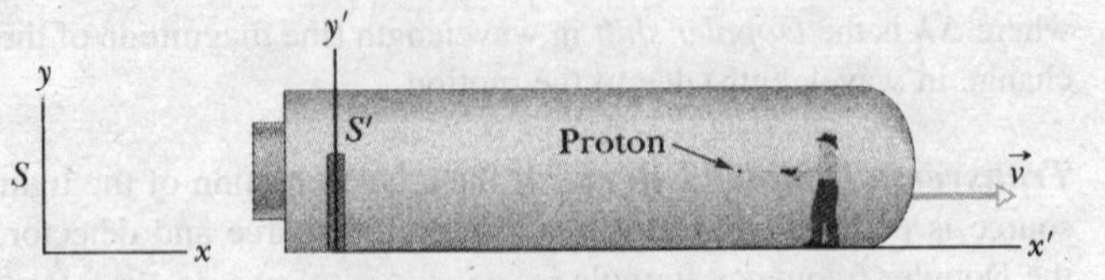

Fig. 38-18 Question 5.

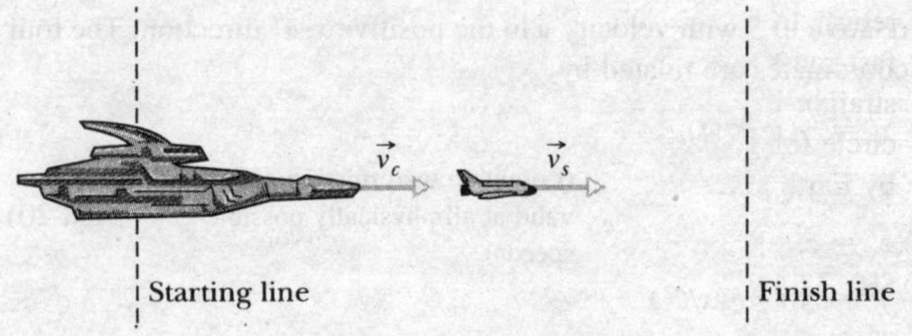

Fig. 38-19 Question 8.

the proton and its impact a positive or negative quantity? (b) Is the temporal separation $\Delta t'$ between those events a positive or negative quantity?

6. (a) In Fig. 38-9, suppose an observer in frame S' measures two events to be at the same location (say, at x') but not at the same time. Can an observer in frame S possibly measure them to be at the same location? (b) If two events occur simultaneously at the same place for one observer, will they be simultaneous for all other observers? (c) Will they occur at the same place for all other observers?

7. Ships A and B in Fig. 38-16 are moving directly toward each other; the velocities indicated are all measured from the same reference frame. Is the speed of ship A relative to ship B more than $0.7c$, less than $0.7c$, or equal to $0.7c$?

8. Figure 38-19 shows one of four star cruisers that are in a race. As each cruiser passes the starting line, a shuttle craft leaves the cruiser and races toward the finish line. You, judging the race, are stationary relative to the starting and finish lines. The speeds v_c of the cruisers relative to you and the speeds v_s of the shuttle craft relative to their starships are, in that order, (1) $0.70c$, $0.40c$; (2) $0.40c$, $0.70c$; (3) $0.20c$, $0.90c$; (4) $0.50c$, $0.60c$. (a) Without written calculation, rank the shuttle craft according to their speeds relative to you, greatest first. (b) Still without written calculation, rank the shuttle craft according to the distances their pilots measure from the starting line to the finish line, greatest first. (c) Each starship sends a signal to its shuttle craft at a certain frequency f_0 as measured on board the starship. Again without written calculation, rank the shuttle craft according to the frequencies they detect, greatest first.

9. While on board a starship, you intercept signals from four shuttle craft that are moving either directly toward or directly away from you. The signals have the same proper frequency f_0. The speed and direction (both relative to you) of the shuttle craft are (a) $0.3c$ toward, (b) $0.6c$ toward, (c) $0.3c$ away, and (d) $0.6c$ away. Rank the shuttle craft according to the frequency you receive, greatest first.

10. The rest energy and total energy, respectively, of three particles, expressed in terms of a basic amount A are (1) A, $2A$; (2) A, $3A$; (3) $3A$, $4A$. Without written calculation, rank the particles according to (a) their mass, (b) their kinetic energy, (c) their Lorentz factor, and (d) their speed, greatest first.

EXERCISES & PROBLEMS

ssm Solution is in the Student Solutions Manual.
www Solution is available on the World Wide Web at: http://www.wiley.com/college/hrw
ilw Solution is available on the Interactive LearningWare.

SEC. 38-2 The Postulates

1E. Quite apart from effects due to Earth's rotational and orbital motions, a laboratory reference frame is not strictly an inertial frame because a particle placed at rest there will not, in general, remain at rest; it will fall. Often, however, events happen so quickly that we can ignore the gravitational acceleration and treat the frame as inertial. Consider, for example, an electron of speed $v = 0.992c$, projected horizontally into a laboratory test chamber and moving through a distance of 20 cm. (a) How long would that journey take, and (b) how far would the electron fall during this interval? What can you conclude about the suitability of the laboratory as an inertial frame in this case? ssm

2E. What fraction of the speed of light does each of the following speeds represent; that is, what is the associated speed parameter β? (a) A typical rate of continental drift (3 cm/y). (b) A highway speed limit of 90 km/h. (c) A supersonic plane flying at Mach 2.5 (1200 km/h). (d) The escape speed of a projectile from the surface of Earth. (e) A typical recession speed of a distant quasar (3.0×10^4 km/s).

SEC. 38-5 The Relativity of Time

3E. The mean lifetime of stationary muons is measured to be 2.2 μs. The mean lifetime of high-speed muons in a burst of cosmic rays observed from Earth is measured to be 16 μs. Find the speed of these cosmic-ray muons relative to Earth. ssm

4E. What must be the speed parameter β if the Lorentz factor γ is (a) 1.01, (b) 10.0, (c) 100, and (d) 1000?

5P. An unstable high-energy particle enters a detector and leaves a track 1.05 mm long before it decays. Its speed relative to the detector was $0.992c$. What is its proper lifetime? That is, how long would the particle have lasted before decay had it been at rest with respect to the detector? ilw

6P. You wish to make a round trip from Earth in a spaceship, traveling at constant speed in a straight line for 6 months and then returning at the same constant speed. You wish further, on your return, to find Earth as it will be 1000 years in the future. (a) How fast must you travel? (b) Does it matter whether you travel in a straight line on your journey? If, for example, you traveled in a circle for 1 year, would you still find that 1000 years had elapsed by Earth clocks when you returned?

SEC. 38-6 The Relativity of Length

7E. A rod lies parallel to the x axis of reference frame S, moving along this axis at a speed of $0.630c$. Its rest length is 1.70 m. What will be its measured length in frame S? ssm

8E. An electron of $\beta = 0.999\ 987$ moves along the axis of an evacuated tube that has a length of 3.00 m as measured by a laboratory observer S at rest relative to the tube. An observer S' at rest relative to the electron, however, would see this tube moving with speed v ($= \beta c$). What length would observer S' measure for the tube?

9E. A meter stick in frame S' makes an angle of 30° with the x' axis. If that frame moves parallel to the x axis of frame S with speed $0.90c$ relative to frame S, what is the length of the stick as measured from S?

10E. The length of a spaceship is measured to be exactly half its rest length. (a) In terms of c, what is the speed of the spaceship relative to the observer's frame? (b) By what factor do the spaceship's clocks run slow, compared to clocks in the observer's frame?

11E. A spaceship of rest length 130 m races past a timing station at a speed of $0.740c$. (a) What is the length of the spaceship as measured by the timing station? (b) What time interval will the station clock record between the passage of the front and back ends of the ship? ssm

12P. (a) Can a person, in principle, travel from Earth to the galactic center (which is about 23 000 ly distant) in a normal lifetime? Explain, using either time-dilation or length-contraction arguments. (b) What constant speed is needed to make the trip in 30 y (proper time)?

13P. A space traveler takes off from Earth and moves at speed $0.99c$ toward the star Vega, which is 26 ly distant. How much time will have elapsed by Earth clocks (a) when the traveler reaches Vega and (b) when Earth observers receive word from the traveler that she has arrived? (c) How much older will Earth observers calculate the traveler to be (measured from her frame) when she reaches Vega than she was when she started the trip? ssm www

SEC. 38-8 Some Consequences of the Lorentz Equations

14E. Observer S reports that an event occurred on the x axis of his reference frame at $x = 3.00 \times 10^8$ m at time $t = 2.50$ s. (a) Observer S' and her frame are moving in the positive direction of x at a speed of $0.400c$. Further, $x = x' = 0$ at $t = t' = 0$. What coordinates does observer S' report for the event? (b) What coordinates would observer S' report if she were moving in the *negative* direction of x at this same speed?

15E. Observer S assigns the spacetime coordinates

$$x = 100 \text{ km} \quad \text{and} \quad t = 200\ \mu\text{s}$$

to an event. What are the coordinates of this event in frame S', which moves in the positive direction of x with speed $0.950c$ relative to S? Assume $x = x' = 0$ at $t = t' = 0$. ssm

16E. Inertial frame S' moves at a speed of $0.60c$ with respect to frame S (Fig. 38-9). Further, $x = x' = 0$ at $t = t' = 0$. Two events are recorded. In frame S, event 1 occurs at the origin at $t = 0$ and event 2 occurs on the x axis at $x = 3.0$ km at $t = 4.0\ \mu$s. What times of occurrence does observer S' record for these same events? Explain the difference in the time order.

17E. An experimenter arranges to trigger two flashbulbs simultaneously, producing a big flash located at the origin of his reference frame and a small flash at $x = 30.0$ km. An observer, moving at a speed of $0.250c$ in the positive direction of x, also views the flashes. (a) What is the time interval between them according to her? (b) Which flash does she say occurs first? ssm www

18P. An observer S sees a big flash of light 1200 m from his position and a small flash of light 720 m closer to him directly in line with the big flash. He determines that the time interval between the flashes is 5.00 μs and the big flash occurs first. (a) What is the relative velocity $\vec{v}$ (give both magnitude and direction) of a second observer S' for whom these flashes occur at the same place in the S' reference frame? (b) From the point of view of S', which flash occurs first? (c) What time interval between them does S' measure?

19P. A clock moves along the x axis at a speed of $0.600c$ and reads zero as it passes the origin. (a) Calculate the clock's Lorentz factor. (b) What time does the clock read as it passes $x = 180$ m? ssm www

20P. In Problem 18, observer S sees the two flashes in the same positions as before, but they now occur closer together in time. How close together in time can they be in the frame of S and still allow the possibility of finding a frame S' in which they occur at the same place?

SEC. 38-9 The Relativity of Velocities

21E. A particle moves along the x' axis of frame S' with a speed of $0.40c$. Frame S' moves with a speed of $0.60c$ with respect to frame S. What is the measured speed of the particle in frame S? ssm

22E. Frame S' moves relative to frame S at $0.62c$ in the positive direction of x. In frame S' a particle is measured to have a velocity of $0.47c$ in the positive direction of x'. (a) What is the velocity of the particle with respect to frame S? (b) What would be the velocity of the particle with respect to S if the particle moved (at $0.47c$) in the *negative* direction of x' in the S' frame? In each case, compare your answers with the predictions of the classical velocity transformation equation.

23E. Galaxy A is reported to be receding from us with a speed of $0.35c$. Galaxy B, located in precisely the opposite direction, is also found to be receding from us at this same speed. What recessional speed would an observer on Galaxy A find (a) for our galaxy and (b) for Galaxy B? ssm

24E. It is concluded from measurements of the red shift of the emitted light that quasar Q_1 is moving away from us at a speed of $0.800c$. Quasar Q_2, which lies in the same direction in space but is closer to us, is moving away from us at a speed $0.400c$. What velocity for Q_2 would be measured by an observer on Q_1?

25P. A spaceship whose rest length is 350 m has a speed of $0.82c$ with respect to a certain reference frame. A micrometeorite, also with a speed of $0.82c$ in this frame, passes the spaceship on an antiparallel track. How long does it take this object to pass the spaceship as measured on the ship? ssm ilw www

26P. An armada of spaceships that is 1.00 ly long (in its rest frame) moves with speed $0.800c$ relative to ground station S. A messenger travels from the rear of the armada to the front with a speed of $0.950c$ relative to S. How long does the trip take as measured (a) in the messenger's rest frame, (b) in the armada's rest frame, and (c) by an observer in frame S?

SEC. 38-10 Doppler Effect for Light

27E. A spaceship, moving away from Earth at a speed of $0.900c$, reports back by transmitting at a frequency (measured in the spaceship frame) of 100 MHz. To what frequency must Earth receivers be tuned to receive the report? ssm

28E. Figure 38-20 is a graph of intensity versus wavelength for light reaching Earth from galaxy NGC 7319, which is about 3×10^8 light-years away. The most intense light is emitted by the oxygen in NGC 7319. In a laboratory that emission is at wavelength $\lambda = 513$ nm, but in the light from NGC 7319 it has been shifted to 525 nm due to the Doppler effect (all the emissions from NGC 7319 have been shifted). (a) What is the radial speed of NGC 7319 relative to Earth? (b) Is the relative motion toward or away from our planet?

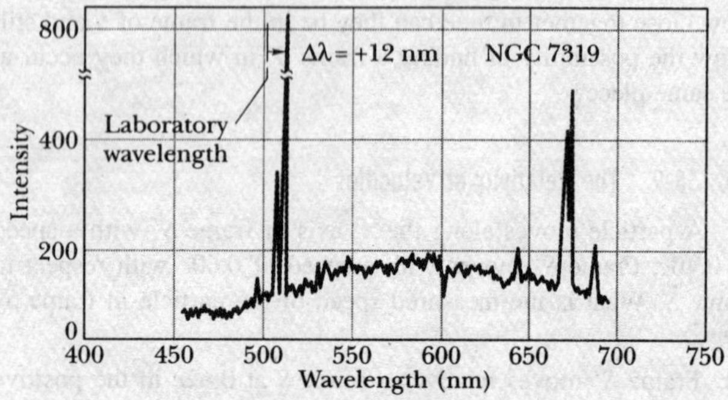

Fig. 38-20 Exercise 28.

29E. Certain wavelengths in the light from a galaxy in the constellation Virgo are observed to be 0.4% longer than the corresponding light from Earth sources. What is the radial speed of this galaxy with respect to Earth? Is it approaching or receding?

30E. Assuming that Eq. 38-33 holds, find how fast you would have to go through a red light to have it appear green. Take 620 nm as the wavelength of red light and 540 nm as the wavelength of green light.

31P. A spaceship is moving away from Earth at a speed of $0.20c$. A light source on the rear of the ship appears blue ($\lambda = 450$ nm) to passengers on the ship. What color would that source appear to an observer on Earth monitoring the receding spaceship? ssm ilw www

SEC. 38-12 A New Look at Energy

32E. How much work must be done to increase the speed of an electron from rest to (a) $0.50c$, (b) $0.990c$, and (c) $0.9990c$?

33E. Find the speed parameter β and Lorentz factor γ for an electron that has a kinetic energy of (a) 1.00 keV, (b) 1.00 MeV, and (c) 1.00 GeV.

34E. Find the speed parameter β and Lorentz factor γ for a particle whose kinetic energy is 10.0 MeV if the particle is (a) an electron, (b) a proton, and (c) an alpha particle.

35E. In terms of c, what is the speed of an electron whose kinetic energy is 100 MeV? ssm

36E. The precise masses in the reaction

$$p + {}^{19}F \rightarrow \alpha + {}^{16}O$$

have been determined to be

$$m(p) = 1.007825 \text{ u}, \qquad m(\alpha) = 4.002603 \text{ u},$$
$$m(F) = 18.998405 \text{ u}, \qquad m(O) = 15.994915 \text{ u}.$$

Calculate the Q of the reaction from these data.

37P. Quasars are thought to be the nuclei of active galaxies in the early stages of their formation. A typical quasar radiates energy at the rate of 10^{41} W. At what rate is the mass of this quasar being reduced to supply this energy? Express your answer in solar mass units per year, where one solar mass unit (1 smu $= 2.0 \times 10^{30}$ kg) is the mass of our Sun. ssm

38P. How much work must be done to increase the speed of an electron from (a) $0.18c$ to $0.19c$ and (b) $0.98c$ to $0.99c$? Note that the speed increase is $0.01c$ in both cases.

39P. A certain particle of mass m has momentum of magnitude mc. What are (a) its speed, (b) its Lorentz factor, and (c) its kinetic energy? ssm

40P. What is the speed of a particle (a) whose kinetic energy is equal to twice its rest energy and (b) whose total energy is equal to twice its rest energy?

41P. What must be the momentum of a particle with mass m so that the total energy of the particle is 3 times its rest energy? ilw

42P. (a) If the kinetic energy K and the momentum p of a particle can be measured, it should be possible to find its mass m and thus identify the particle. Show that

$$m = \frac{(pc)^2 - K^2}{2Kc^2}.$$

(b) Show that this expression reduces to an expected result as $u/c \rightarrow 0$, in which u is the speed of the particle. (c) Find the mass of a particle whose kinetic energy is 55.0 MeV and whose momentum is 121 MeV/c. Express your answer in terms of the mass m_e of the electron.

43P. A 5.00 grain aspirin tablet has a mass of 320 mg. For how many kilometers would the energy equivalent of this mass power an automobile? Assume 12.75 km/L and a heat of combustion of 3.65×10^7 J/L for the gasoline used in the automobile. ssm

44P. The average lifetime of muons at rest is 2.20 μs. A laboratory measurement on muons traveling in a beam emerging from a particle accelerator yields an average muon lifetime of 6.90 μs. What are (a) the speed of these muons in the laboratory, (b) their kinetic energy, and (c) their momentum? The mass of a muon is 207 times that of an electron.

45P. In a high-energy collision between a cosmic-ray particle and a particle near the top of Earth's atmosphere, 120 km above sea level, a pion is created. The pion has a total energy E of 1.35×10^5 MeV and is traveling vertically downward. In the pion's rest frame, the pion decays 35.0 ns after its creation. At what altitude above sea level, as measured from Earth's reference frame, does the decay occur? The rest energy of a pion is 139.6 MeV. ssm www

46P. In Section 29-5 we showed that a particle of charge q and mass m moving with speed v perpendicular to a uniform magnetic field B moves in a circle of radius r given by Eq. 29-16:

$$r = \frac{mv}{qB}.$$

Also, it was demonstrated that the period T of the circular motion is independent of the speed of the particle. These results hold only if $v \ll c$. For particles moving faster, the radius of the circular path must be obtained with

$$r = \frac{p}{qB} = \frac{\gamma mv}{qB} = \frac{mv}{qB\sqrt{1-\beta^2}}.$$

This equation is valid at all speeds. Compute the radius of the path of a 10.0 MeV electron moving perpendicular to a uniform 2.20 T magnetic field, using the (a) classical and (b) relativistic formulas. (c) Calculate the period $T = 2\pi r/v$ of the circular motion using the relativistic formula for r. Is the result independent of the speed of the electron?

47P. Ionization measurements show that a certain low-mass nuclear particle has a charge of $2e$ and is moving with a speed of $0.710c$. The radius of curvature of its path in a magnetic field of 1.00 T is 6.28 m. (The path is a circle whose plane is perpendicular to the magnetic field.) Find the mass of the particle and identify it. [*Hint:* Low-mass nuclear particles are made up of neutrons (which have no charge) and protons (charge $= +e$), in roughly equal numbers. Take the mass of each of these particles to be 1.00 u. Also, see Problem 46.] ssm

48P. A 10 GeV proton in cosmic radiation moves in Earth's magnetic field $\vec{B}$, with its velocity $\vec{v}$ perpendicular to $\vec{B}$, in a region over which the field's average magnitude is 55 μT. What is the radius of the proton's curved path in that region? (See Problem 46.)

49P. A 2.50 MeV electron moves perpendicular to a magnetic field in a path whose radius of curvature is 3.0 cm. What is the magnitude B of the magnetic field? (See Problem 46.)

50P. The proton synchrotron at Fermilab accelerates protons to a kinetic energy of 500 GeV. At such a great energy, relativistic effects are important; in particular, as the speed of the proton increases, the time the proton takes to make a trip around its circular orbit in the synchrotron also increases. In a cyclotron, where the magnetic field magnitude and the oscillator frequency are fixed, this effect of time dilation will put the proton's circling out of synchronization with the oscillator. That eliminates repeated acceleration; hence, the proton will not reach an energy as great as 500 GeV. However, in a synchrotron, both the magnitude of the magnetic field and the oscillation frequency are varied to allow for the change due to time dilation.

At the energy of 500 GeV, calculate (a) the Lorentz factor, (b) the speed parameter, and (c) the magnetic field magnitude at the proton orbit, which has a radius of curvature of 750 m. (See Problem 46; use 938.3 MeV as the proton's rest energy.)

51P*. An alpha particle with kinetic energy 7.70 MeV collides with an ^{14}N nucleus at rest, and the two transform into an ^{17}O nucleus and a proton. The proton is emitted at 90° to the direction of the incident alpha particle and has a kinetic energy of 4.44 MeV. The masses of the various particles are: alpha particle, 4.00260 u; ^{14}N, 14.00307 u; proton, 1.007825 u; and ^{17}O, 16.99914 u. In mega-electron volts, what are (a) the kinetic energy of the oxygen nucleus and (b) the Q of the reaction? (*Hint:* The speeds of the particles are much less than c.)

Additional Problems

52. *The car-in-the-garage problem.* Carman has just purchased the world's longest stretch limo, which has a proper length of $L_c = 30.5$ m. In Fig. 38-21*a*, it is shown parked in front of a garage with a proper length of $L_g = 6.00$ m. The garage has a front door (shown open) and a back door (shown closed). The limo is obviously longer than the garage. Still, Garageman, who owns the garage and knows something about relativistic length contraction, makes a bet with Carman that the limo can fit in the garage with both doors closed. Carman, who dropped the physics course before reaching special relativity, says such a thing, even in principle, is impossible.

To analyze Garageman's scheme, an x_c axis is attached to the limo, with $x_c = 0$ at the rear bumper, and an x_g axis is attached to the garage, with $x_g = 0$ at the (now open) front door. Then Carman is to drive the limo directly toward the front door at a velocity of $0.9980c$ (which is, of course, both technically and financially impossible). Carman is stationary in the x_c reference frame; Garageman is stationary in the x_g reference frame.

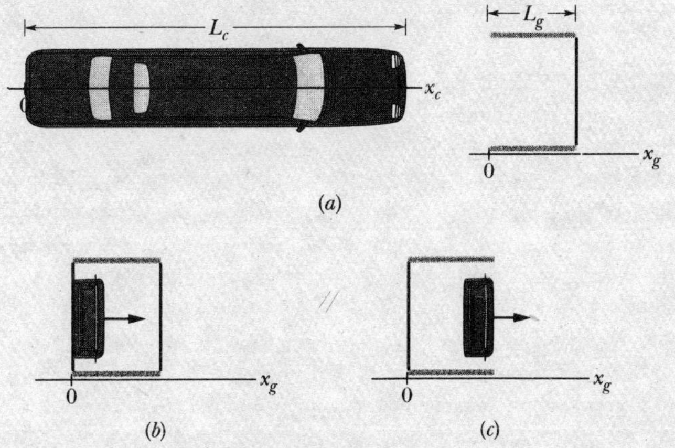

Fig. 38-21 Problem 52.

There are two events to consider. *Event 1:* When the rear bumper clears the front door, the front door is closed. Let the time of this event be zero to both Carman and Garageman: $t_{g1} = t_{c1} = 0$. The event occurs at $x_c = x_g = 0$. Figure 38-21*b* shows event 1 according to the x_g reference frame. *Event 2:* When the front bumper reaches the back door, that door opens. Figure 38-21*c* shows event 2 according to the x_g reference frame.

According to Garageman, (a) what is the length of the limo and (b) what are the spacetime coordinates x_{g2} and t_{g2} of event 2? (c) For how long is the limo temporarily "trapped" inside the garage, with both doors shut?

Now consider the situation from the x_c reference frame, in which the garage comes racing past the limo at a velocity of $-0.9980c$. According to Carman, (d) what is the length of the passing garage, (e) what are the spacetime coordinates x_{c2} and t_{c2} of event 2, (f) is the limo ever in the garage with both doors shut, and (g) which event occurs first? (h) Sketch events 1 and 2 as seen by Carman. (Are the events causally related; that is, does one of them cause the other?) (i) Finally, who wins the bet?

53. *Superluminal jets.* Figure 38-22*a* shows the path taken by a knot in a jet of ionized gas that has been expelled from a galaxy. The knot travels at constant velocity $\vec{v}$ at angle θ from the direction of Earth. The knot occasionally emits a burst of light, which is eventually detected on Earth. Two bursts are indicated in Fig. 38-22*a*, separated by time t as measured in a stationary frame near the bursts. The bursts are shown in Fig. 38-22*b* as if they were photographed on the same piece of film, first when light from burst 1 arrived on Earth and then later when light from burst 2 arrived. The apparent distance D_{app} traveled by the knot between the two bursts is the distance across an Earth-observer's view of the knot's path. The apparent time T_{app} between the bursts is the difference in the arrival times of the light from them. The apparent speed of the knot is then $V_{app} = D_{app}/T_{app}$. In terms of v, t, and θ, what are (a) D_{app} and (b) T_{app}? (c) Evaluate V_{app} for $v = 0.980c$ and $\theta = 30.0°$. When superluminal (faster than light) jets were first observed, they seemed to defy special relativity—at least until the geometry of viewing, such as in Fig. 38-22*a*, was understood.

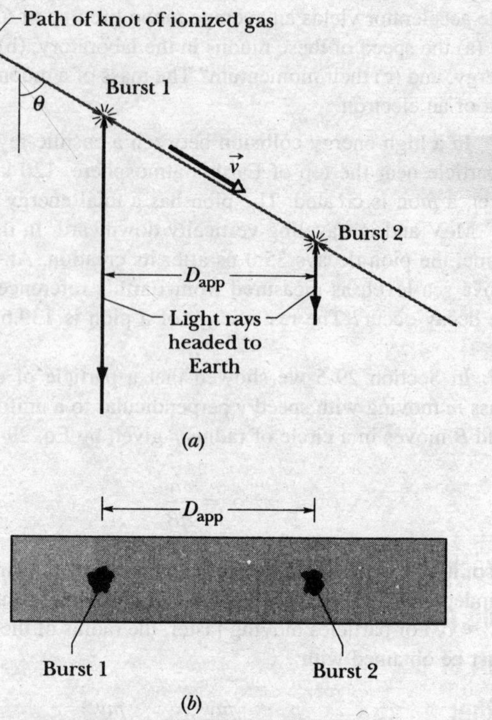

Fig. 38-22 Problem 53.

39 Photons and Matter Waves

Tracks of tiny vapor bubbles in this bubble-chamber image reveal where electrons (tracks color-coded green) and positrons (red) moved. A gamma ray (which left no track when it entered at the top) kicked an electron out of one of the hydrogen atoms filling the chamber and then converted to an electron–positron pair. Another gamma ray underwent another pair production farther down. These tracks (curved because of a magnetic field) clearly show that electrons and positrons are particles that move along narrow paths. Yet, those particles can also be interpreted in terms of waves.

Can a particle be a wave?

The answer is in this chapter.

39-1 A New Direction

Our discussion of Einstein's theory of relativity took us into a world far beyond that of ordinary experience—the world of objects moving at speeds close to the speed of light. Among other surprises, Einstein's theory predicts that the rate at which a clock runs depends on how fast the clock is moving relative to the observer: the faster the motion, the slower the clock rate. This and other predictions of the theory have passed every experimental test devised thus far, and relativity theory has led us to a deeper and more satisfying view of the nature of space and time.

Now you are about to explore a second world that is outside ordinary experience—the subatomic world. You will encounter a new set of surprises that, though they may sometimes seem bizarre, have led physicists step by step to a deeper view of reality.

Quantum physics, as our new subject is called, answers such questions as: Why do the stars shine? Why do the elements exhibit the order that is so apparent in the periodic table? How do transistors and other microelectronic devices work? Why does copper conduct electricity but glass does not? Because quantum physics accounts for all of chemistry, including biochemistry, we need to understand it if we are to understand life itself.

Some of the predictions of quantum physics seem strange even to the physicists and philosophers who study its foundations. Still, experiment after experiment has proved the theory correct, and many have exposed even stranger aspects of the theory. The quantum world is an amusement park full of wonderful rides that are guaranteed to shake up the commonsense world view you have developed since childhood. We begin our exploration of that quantum park with the photon.

39-2 The Photon, the Quantum of Light

Quantum physics (which is also known as *quantum mechanics* and *quantum theory*) is largely the study of the microscopic world. There many quantities are found only in certain minimum (*elementary*) amounts, or integer multiples of those elementary amounts; they are then said to be *quantized*. The elementary amount that is associated with such a quantity is called the **quantum** of that quantity (*quanta* is the plural).

In a loose sense, U.S. currency is quantized because the coin of least value is the penny, or \$0.01 coin, and the values of all other coins and bills are restricted to integer multiples of that least amount. In other words, the currency quantum is \$0.01, and all greater amounts of currency are of the form n(\$0.01), where n is a positive integer. For example, you cannot hand someone \$0.755 = 75.5(\$0.01).

In 1905, Einstein proposed that electromagnetic radiation (or simply *light*) is quantized and exists in elementary amounts (quanta) that we now call **photons.** This proposal should seem strange to you because we have just spent several chapters discussing the classical idea that light is a sinusoidal wave, with a wavelength λ, a frequency f, and a speed c such that

$$f = \frac{c}{\lambda}. \tag{39-1}$$

Furthermore, in Chapter 34 we discussed the classical light wave as being an interdependent combination of electric and magnetic fields, each oscillating at frequency f. How can this wave of oscillating fields consist of an elementary amount of something—the light quantum? What *is* a photon?

The concept of a light quantum, or a photon, turns out to be far more subtle and mysterious than Einstein imagined. Indeed, it is still very poorly understood. In this book, we shall discuss only some of the basic aspects of the photon concept, somewhat along the lines of Einstein's proposal.

According to that proposal, the quantum of a light wave of frequency f has the energy

$$E = hf \quad \text{(photon energy).} \tag{39-2}$$

Here h is the **Planck constant,** which has the value

$$h = 6.63 \times 10^{-34}\ \text{J}\cdot\text{s} = 4.14 \times 10^{-15}\ \text{eV}\cdot\text{s}. \tag{39-3}$$

The least energy a light wave of frequency f can have is hf, the energy of a single photon. If the wave has more energy, its total energy must be an integer multiple of hf, just as the currency in our previous example must be an integer multiple of \$0.01. The light cannot have an energy of $0.6hf$ or $75.5hf$.

Einstein further proposed that when light is absorbed or emitted by an object (matter), the absorption or emission event occurs at the atoms of the object. When light of frequency f is absorbed by an atom, the energy hf of one photon is transferred from the light to the atom. In this *absorption event,* the photon vanishes and the atom is said to absorb it. When light of frequency f is emitted by an atom, an energy hf is transferred from the atom to the light. In this *emission event,* a photon suddenly appears and the atom is said to emit it. Thus, we can have *photon absorption* and *photon emission* by atoms in an object.

For an object consisting of many atoms, there can be many photon absorptions (such as with sunglasses) or photon emissions (such as with lamps). However, each absorption or emission event still involves the transfer of energy equal to that of a single photon of the light.

When we discussed the absorption or emission of light in previous chapters, our examples involved so much light that we had no need of quantum physics, and we got by with classical physics. However, in the late twentieth century, technology became advanced enough that single-photon experiments could be conducted and put to practical use. Since then quantum physics has become part of standard engineering practice, especially in optical engineering.

✔CHECKPOINT 1: Rank the following radiations according to their associated photon energies, greatest first: (a) yellow light from a sodium vapor lamp, (b) a gamma ray emitted by a radioactive nucleus, (c) a radio wave emitted by the antenna of a commercial radio station, (d) a microwave beam emitted by airport traffic control radar.

Sample Problem 39-1

A sodium vapor lamp is placed at the center of a large sphere that absorbs all the light reaching it. The rate at which the lamp emits energy is 100 W; assume that the emission is entirely at a wavelength of 590 nm. At what rate are photons absorbed by the sphere?

SOLUTION: We assume that all the light emitted by the lamp reaches (and thus is absorbed by) the sphere. Then the **Key Idea** is that the light is emitted and absorbed as photons. The rate R at which photons are absorbed by the sphere is equal to the rate R_{emit} at which photons are emitted by the lamp. That rate is

$$R_{\text{emit}} = \frac{\text{rate of energy emission}}{\text{energy per emitted photon}} = \frac{P_{\text{emit}}}{E}.$$

We then have, from Eq. 39-2 ($E = hf$),

$$R = R_{\text{emit}} = \frac{P_{\text{emit}}}{hf}.$$

Using Eq. 39-1 ($f = c/\lambda$) to substitute for f and then entering known data, we obtain

$$R = \frac{P_{\text{emit}}\lambda}{hc}$$

$$= \frac{(100\ \text{W})(590 \times 10^{-9}\ \text{m})}{(6.63 \times 10^{-34}\ \text{J}\cdot\text{s})(3.0 \times 10^{8}\ \text{m/s})}$$

$$= 2.97 \times 10^{20}\ \text{photons/s}. \qquad \text{(Answer)}$$

39-3 The Photoelectric Effect

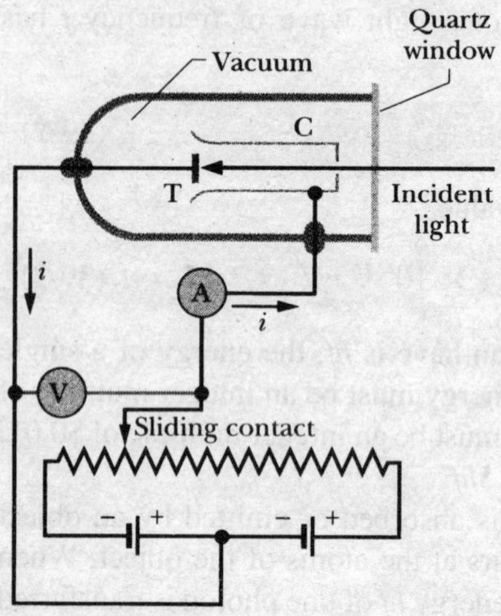

Fig. 39-1 An apparatus used to study the photoelectric effect. The incident light shines on target T, ejecting electrons, which are collected by collector cup C. The electrons move in the circuit in a direction opposite the conventional current arrows. The batteries and the variable resistor are used to produce and adjust the electric potential difference between T and C.

If you direct a beam of light of short enough wavelength onto a clean metal surface, the light will cause electrons to leave that surface (the light will *eject* the electrons from the surface). This **photoelectric effect** is used in many devices, including TV cameras, camcorders, and night vision viewers. Einstein supported his photon concept by using it to explain this effect, which simply cannot be understood without quantum physics.

Let us analyze two basic photoelectric experiments, each using the apparatus of Fig. 39-1 in which light of frequency f is directed onto target T and ejects electrons from it. A potential difference V is maintained between target T and collector cup C to sweep up these electrons, said to be **photoelectrons.** This collection produces a **photoelectric current** i that is measured with meter A.

First Photoelectric Experiment

We adjust the potential difference V by moving the sliding contact in Fig. 39-1 so that collector C is slightly negative with respect to target T. This potential difference acts to slow down the ejected electrons. We then vary V until it reaches a certain value, called the **stopping potential** V_{stop}, at which the reading of meter A has just dropped to zero. When $V = V_{stop}$, the most energetic ejected electrons are turned back just before reaching the collector. Then K_{max}, the kinetic energy of these most energetic electrons, is

$$K_{max} = eV_{stop}, \tag{39-4}$$

where e is the elementary charge.

Measurements show that for light of a given frequency, K_{max} *does not depend on the intensity of the light source.* Whether the source is dazzling bright or so feeble that you can scarcely detect it (or has some intermediate brightness), the maximum kinetic energy of the ejected electrons always has the same value.

This experimental result is a puzzle for classical physics. Classically, the incident light is a sinusoidally oscillating electromagnetic wave. An electron in the target should oscillate sinusoidally due to the oscillating electric force on it from the wave's electric field. If the amplitude of the electron's oscillation is great enough, the electron should break free of the target's surface—that is, be ejected from the target. Thus, if we increase the amplitude of the wave and its oscillating electric field, the electron should get a more energetic "kick" as it is being ejected. *However, that is not what happens.* For a given frequency, intense light beams and feeble light beams give exactly the same maximum kick to ejected electrons.

The actual result follows naturally if we think in terms of photons. Now the energy that can be transferred from the incident light to an electron in the target is that of a single photon. Increasing the light intensity increases the *number* of photons in the light, but the photon energy, given by Eq. 39-2, is unchanged because the frequency is unchanged. Thus, the energy transferred to the kinetic energy of an electron is also unchanged.

Second Photoelectric Experiment

Now we vary the frequency f of the incident light and measure the associated stopping potential V_{stop}. Figure 39-2 is a plot of V_{stop} versus f. Note that the photoelectric effect does not occur if the frequency is below a certain **cutoff frequency** f_0 or, equivalently, if the wavelength is greater than the corresponding **cutoff wavelength** $\lambda_0 = c/f_0$. This is so *no matter how intense the incident light is.*

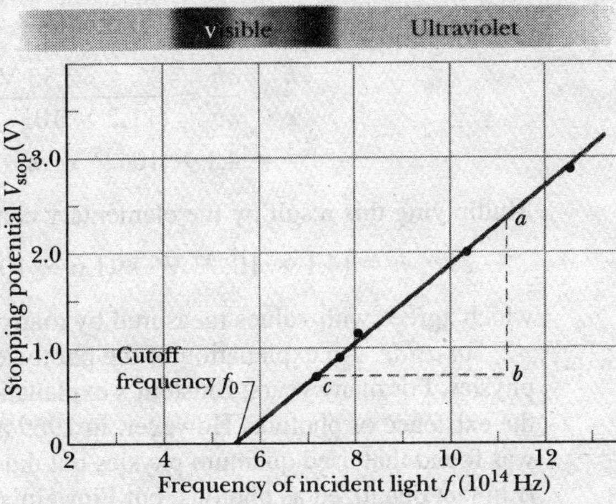

Fig. 39-2 The stopping potential V_{stop} as a function of the frequency f of the incident light for a sodium target T in the apparatus of Fig. 39-1. (Data reported by R. A. Millikan in 1916.)

This is another puzzle for classical physics. If you view light as an electromagnetic wave, you must expect that no matter how low the frequency, electrons can always be ejected by light if you supply them with enough energy—that is, if you use a light source that is bright enough. *That is not what happens.* For light below the cutoff frequency f_0, the photoelectric effect does not occur, no matter how bright the light source.

The existence of a cutoff frequency is, however, just what we should expect if the energy is transferred via photons. The electrons within the target are held there by electric forces. (If they weren't, they would drip out of the target due to the gravitational force on them.) To just escape from the target, an electron must pick up a certain minimum energy Φ, where Φ is a property of the target material called its **work function.** If the energy hf transferred to an electron by a photon exceeds the work function of the material (if $hf > \Phi$), the electron can escape the target. If the energy transferred does not exceed the work function (that is, if $hf \leq \Phi$), the electron cannot escape. This is what Fig. 39-2 shows.

The Photoelectric Equation

Einstein summed up the results of such photoelectric experiments in the equation

$$hf = K_{max} + \Phi \quad \text{(photoelectric equation).} \tag{39-5}$$

This is a statement of the conservation of energy for a single photon absorption by a target with work function Φ. Energy equal to the photon's energy hf is transferred to a single electron in the material of the target. If the electron is to escape from the target, it must pick up energy at least equal to Φ. Any additional energy $(hf - \Phi)$ that the electron acquires from the photon appears as kinetic energy K of the electron. In the most favorable circumstance, the electron can escape through the surface without losing any of this kinetic energy in the process; it then appears outside the target with the maximum possible kinetic energy K_{max}.

Let us rewrite Eq. 39-5 by substituting for K_{max} from Eq. 39-4. After a little rearranging we get

$$V_{stop} = \left(\frac{h}{e}\right) f - \frac{\Phi}{e}. \tag{39-6}$$

The ratios h/e and Φ/e are constants, so we would expect a plot of the measured stopping potential V_{stop} versus the frequency f of the light to be a straight line, as it is in Fig. 39-2. Further, the slope of that straight line should be h/e. As a check, we measure ab and bc in Fig. 39-2 and write

$$\frac{h}{e} = \frac{ab}{bc} = \frac{2.35\text{ V} - 0.72\text{ V}}{(11.2 \times 10^{14} - 7.2 \times 10^{14})\text{ Hz}}$$
$$= 4.1 \times 10^{-15}\text{ V}\cdot\text{s}.$$

Multiplying this result by the elementary charge e, we find

$$h = (4.1 \times 10^{-15}\text{ V}\cdot\text{s})(1.6 \times 10^{-19}\text{ C}) = 6.6 \times 10^{-34}\text{ J}\cdot\text{s},$$

which agrees with values measured by many other methods.

An aside: An explanation of the photoelectric effect certainly requires quantum physics. For many years, Einstein's explanation was also a compelling argument for the existence of photons. However, in 1969 an alternative explanation for the effect was found that used quantum physics but did not need the concept of photons. Light *is* in fact quantized as photons, but Einstein's explanation of the photoelectric effect is not the best argument for that fact.

✔CHECKPOINT 2: The figure shows data like those of Fig. 39-2 for targets of cesium, potassium, sodium, and lithium. The plots are parallel. (a) Rank the targets according to their work functions, greatest first. (b) Rank the plots according to the value of h they yield, greatest first.

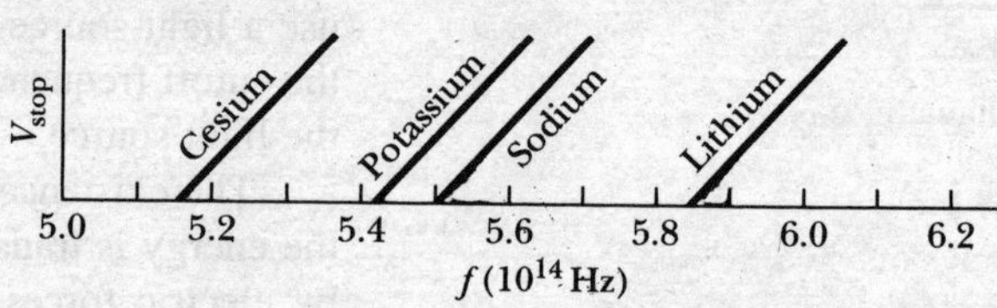

Sample Problem 39-2

A potassium foil is a distance $r = 3.5$ m from an isotropic light source that emits energy at the rate $P = 1.5$ W. The work function Φ of potassium is 2.2 eV. Suppose that the energy transported by the incident light were transferred to the target foil continuously and smoothly (that is, if classical physics prevailed instead of quantum physics). How long would it take for the foil to absorb enough energy to eject an electron? Assume that the foil totally absorbs all the energy reaching it and that the to-be-ejected electron collects energy from a circular patch of the foil whose radius is 5.0×10^{-11} m, about that of a typical atom.

SOLUTION: The **Key Ideas** here are these:

1. The time interval Δt required for the patch to absorb energy ΔE depends on the rate P_{abs} at which the energy is absorbed:

$$\Delta t = \frac{\Delta E}{P_{abs}}.$$

2. If the electron is to be ejected from the foil, the least energy ΔE it must gain from the light is equal to the work function Φ of potassium. Thus,

$$\Delta t = \frac{\Phi}{P_{abs}}.$$

3. Because the patch is totally absorbing, the rate of absorption P_{abs} is equal to the rate P_{arr} at which energy arrives at the patch; that is,

$$\Delta t = \frac{\Phi}{P_{arr}}.$$

4. With Eq. 34-23, we can relate the energy arrival rate P_{arr} to the intensity I of the light at the patch and the area A of the patch:

$$P_{arr} = IA.$$

Then

$$\Delta t = \frac{\Phi}{IA}.$$

5. Because the light source is isotropic, the light intensity I at distance r from the source depends on the rate P_{emit} at which energy is emitted by the source, according to Eq. 34-27:

$$I = \frac{P_{emit}}{4\pi r^2}.$$

Thus, finally, we have

$$\Delta t = \frac{4\pi r^2 \Phi}{P_{emit} A}.$$

The detection area A is $\pi(5.0 \times 10^{-11}\text{ m})^2 = 7.85 \times 10^{-21}\text{ m}^2$, and the work function Φ is 2.2 eV $= 3.5 \times 10^{-19}$ J. Substituting these and other data, we find that

$$\Delta t = \frac{4\pi(3.5\text{ m})^2(3.5 \times 10^{-19}\text{ J})}{(1.5\text{ W})(7.85 \times 10^{-21}\text{ m}^2)}$$
$$= 4580\text{ s} \approx 1.3\text{ h}. \qquad \text{(Answer)}$$

Thus, classical physics tells us that we would have to wait more than an hour after turning on the light source for a photoelectron to be ejected. The actual waiting time is less than 10^{-9} s. Apparently, then, an electron does *not* gradually absorb energy from the light arriving at the patch containing the electron. Rather, either the electron does not absorb any energy at all or it absorbs a quantum of energy instantaneously, by absorbing a photon from the light.

Sample Problem 39-3

Find the work function Φ of sodium from Fig. 39-2.

SOLUTION: The **Key Idea** here is that we can find the work function Φ from the cutoff frequency f_0 (which we can measure on the plot). The reasoning is this: At the cutoff frequency, the kinetic energy K_{max} in Eq. 39-5 is zero. Thus, all the energy hf that is transferred from a photon to an electron goes into the electron's escape, which requires an energy of Φ. Equation 39-5 then gives us, with $f = f_0$,

$$hf_0 = 0 + \Phi = \Phi.$$

In Fig. 39-2, the cutoff frequency f_0 is the frequency at which the plotted line intercepts the horizontal frequency axis, about 5.5×10^{14} Hz. We then have

$$\Phi = hf_0 = (6.63 \times 10^{-34}\ \text{J}\cdot\text{s})(5.5 \times 10^{14}\ \text{Hz})$$
$$= 3.6 \times 10^{-19}\ \text{J} = 2.3\ \text{eV}. \qquad \text{(Answer)}$$

39-4 Photons Have Momentum

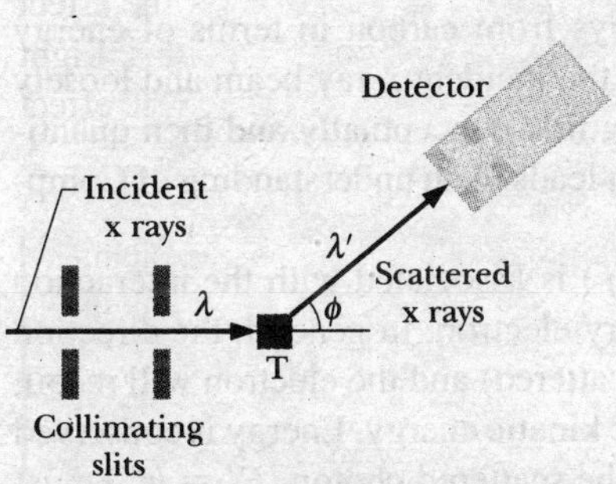

Fig. 39-3 Compton's apparatus. A beam of x rays of wavelength $\lambda = 71.1$ pm is directed onto a carbon target T. The x rays scattered from the target are observed at various angles ϕ to the direction of the incident beam. The detector measures both the intensity of the scattered x rays and their wavelength.

In 1916, Einstein extended his concept of light quanta (photons) by proposing that a quantum of light has linear momentum. For a photon with energy hf, the magnitude of that momentum is

$$p = \frac{hf}{c} = \frac{h}{\lambda} \quad \text{(photon momentum)}, \qquad (39\text{-}7)$$

where we have substituted for f from Eq. 39-1 ($f = c/\lambda$). Thus, when a photon interacts with matter, energy *and* momentum are transferred, *as if* there were a collision between the photon and matter in the classical sense (as in Chapter 10).

In 1923, Arthur Compton at Washington University in St. Louis carried out an experiment that supported the view that both momentum and energy are transferred via photons. He arranged for a beam of x rays of wavelength λ to be directed onto a target made of carbon, as shown in Fig. 39-3. An x ray is a form of electromagnetic radiation, at high frequency and thus small wavelength. Compton measured the wavelengths and intensities of the x rays that were scattered in various directions from his carbon target.

Figure 39-4 shows his results. Although there is only a single wavelength ($\lambda = 71.1$ pm) in the incident x-ray beam, we see that the scattered x rays contain a range

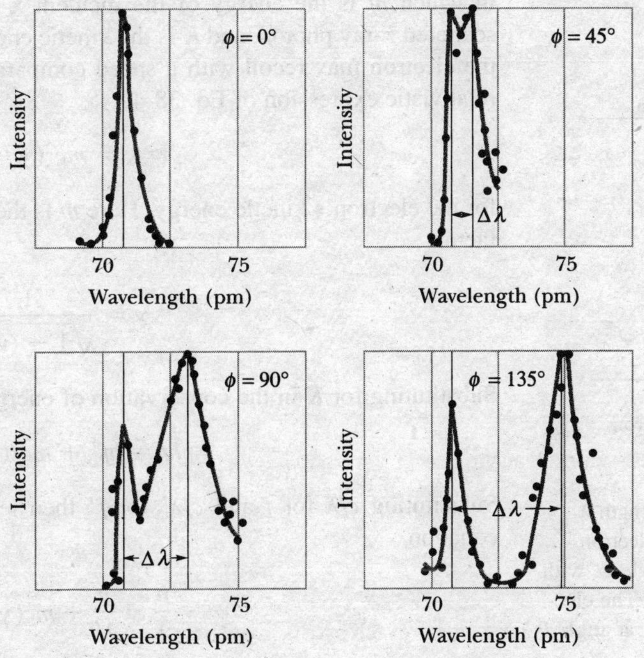

Fig. 39-4 Compton's results for four values of the scattering angle ϕ. Note that the Compton shift $\Delta\lambda$ increases as the scattering angle increases.

of wavelengths with two prominent intensity peaks. One peak is centered about the incident wavelength λ, the other about a wavelength λ' that is longer than λ by an amount $\Delta\lambda$, which is called the **Compton shift.** The value of the Compton shift varies with the angle at which the scattered x rays are detected.

Figure 39-4 is still another puzzle for classical physics. Classically, the incident x-ray beam is a sinusoidally oscillating electromagnetic wave. An electron in the carbon target should oscillate sinusoidally due to the oscillating electric force on it from the wave's electric field. Further, the electron should oscillate at the same frequency as the wave and should send out waves *at this same frequency,* as if it were a tiny transmitting antenna. Thus, the x rays scattered by the electron should have the same frequency, and the same wavelength, as the x rays in the incident beam—but they don't.

Compton interpreted the scattering of x rays from carbon in terms of energy and momentum transfers, via photons, between the incident x-ray beam and loosely bound electrons in the carbon target. Let us see, first conceptually and then quantitatively, how this quantum physics interpretation leads to an understanding of Compton's results.

Suppose a single photon (of energy $E = hf$) is associated with the interaction between the incident x-ray beam and a stationary electron. In general, the direction of travel of the x ray will change (the x ray is scattered) and the electron will recoil, which means that the electron has obtained some kinetic energy. Energy is conserved in this isolated interaction. Thus, the energy of the scattered photon ($E' = hf'$) must be less than that of the incident photon. The scattered x rays must then have a lower frequency f' and thus a longer wavelength λ' than the incident x rays, just as Compton's experimental results in Fig. 39-4 show.

For the quantitative part, we first apply the law of conservation of energy. Figure 39-5 suggests a "collision" between an x ray and an initially stationary free electron in the target. As a result of the collision, an x ray of wavelength λ' moves off at an angle ϕ and the electron moves off at an angle θ, as shown. The conservation of energy then gives us

$$hf = hf' + K,$$

in which hf is the energy of the incident x-ray photon, hf' is the energy of the scattered x-ray photon, and K is the kinetic energy of the recoiling electron. Because the electron may recoil with a speed comparable to that of light, we must use the relativistic expression of Eq. 38-49,

$$K = mc^2(\gamma - 1),$$

for the electron's kinetic energy. Here m is the electron's mass and γ is the Lorentz factor

$$\gamma = \frac{1}{\sqrt{1 - (v/c)^2}}.$$

Substituting for K in the conservation of energy equation yields

$$hf = hf' + mc^2(\gamma - 1).$$

Substituting c/λ for f and c/λ' for f' then leads to the new energy conservation equation

$$\frac{h}{\lambda} = \frac{h}{\lambda'} + mc(\gamma - 1). \tag{39-8}$$

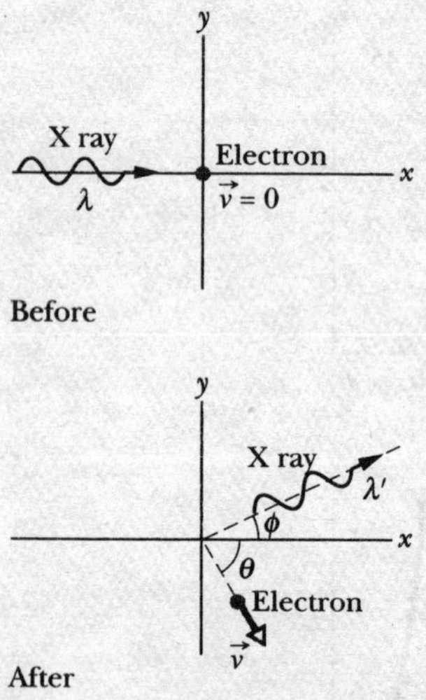

Fig. 39-5 An x ray of wavelength λ interacts with a stationary electron. The x ray is scattered at angle ϕ, with an increased wavelength λ'. The electron moves off with speed v at angle θ.

Next we apply the law of conservation of momentum to the x-ray–electron collision of Fig. 39-5. From Eq. 39-7, the magnitude of the momentum of the incident photon is h/λ, and that of the scattered photon is h/λ'. From Eq. 38-38, the magnitude for the recoiling electron's momentum is γmv. Because we have a two-dimensional situation, we write separate equations for the conservation of momentum along the x and y axes, obtaining

$$\frac{h}{\lambda} = \frac{h}{\lambda'}\cos\phi + \gamma mv\cos\theta \qquad (x \text{ axis}) \tag{39-9}$$

and

$$0 = \frac{h}{\lambda'}\sin\phi - \gamma mv\sin\theta \qquad (y \text{ axis}). \tag{39-10}$$

We want to find $\Delta\lambda$ $(= \lambda' - \lambda)$, the Compton shift of the scattered x rays. Of the five collision variables (λ, λ', v, ϕ, and θ) that appear in Eqs. 39-8, 39-9, and 39-10, we choose to eliminate v and θ, which deal only with the recoiling electron. Carrying out the algebra (it is somewhat complicated) leads to an equation for the Compton shift as a function of the scattering angle ϕ:

$$\Delta\lambda = \frac{h}{mc}(1 - \cos\phi) \qquad \text{(Compton shift).} \tag{39-11}$$

Equation 39-11 agrees exactly with Compton's experimental results.

The quantity h/mc in Eq. 39-11 is a constant called the **Compton wavelength.** Its value depends on the mass m of the particle from which the x rays scatter. Here that particle is a loosely bound electron, and thus we would substitute the mass of an electron for m to evaluate the *Compton wavelength for Compton scattering from an electron.*

A Loose End

The peak at the incident wavelength λ (= 71.1 pm) in Fig. 39-4 still needs to be explained. This peak arises not from interactions between x rays and the very loosely bound electrons in the target but from interactions between x rays and the electrons that are *tightly* bound to the carbon atoms making up the target. Effectively, each of these latter collisions occurs between an incident x ray and an entire carbon atom. If we substitute for m in Eq. 39-11 the mass of a carbon atom (which is about 22 000 times that of an electron), we see that $\Delta\lambda$ becomes about 22 000 times smaller than the Compton shift for an electron—too small to detect. Thus, the x rays scattered in these collisions have the same wavelength as the incident x rays.

Sample Problem 39-4

X rays of wavelength $\lambda = 22$ pm (photon energy = 56 keV) are scattered from a carbon target, and the scattered rays are detected at 85° to the incident beam.

(a) What is the Compton shift of the scattered rays?

SOLUTION: The Key Idea here is that the Compton shift is the wavelength change of the x rays due to scattering from loosely bound electrons in a target. Further, that shift depends on the angle at which the scattered x rays are detected, according to Eq. 39-11. Substituting 85° for that angle and 9.11×10^{-31} kg for the electron mass (because the scattering is from electrons) in Eq. 39-11 gives us

$$\Delta\lambda = \frac{h}{mc}(1 - \cos\phi)$$

$$= \frac{(6.63 \times 10^{-34}\ \text{J}\cdot\text{s})(1 - \cos 85°)}{(9.11 \times 10^{-31}\ \text{kg})(3.00 \times 10^{8}\ \text{m/s})}$$

$$= 2.21 \times 10^{-12}\ \text{m} \approx 2.2\ \text{pm}. \qquad \text{(Answer)}$$

(b) What percentage of the initial x-ray photon energy is transferred to an electron in such scattering?

SOLUTION: The **Key Idea** here is to find the *fractional energy loss* (let us call it *frac*) for photons that scatter from the electrons:

$$frac = \frac{\text{energy loss}}{\text{initial energy}} = \frac{E - E'}{E}.$$

From Eq. 39-2 ($E = hf$), we can substitute for the initial energy E and the detected energy E' of the x rays in terms of frequencies. Then, from Eq. 39-1 ($f = c/\lambda$), we can substitute for those frequencies in terms of the wavelengths. We find

$$frac = \frac{hf - hf'}{hf} = \frac{c/\lambda - c/\lambda'}{c/\lambda} = \frac{\lambda' - \lambda}{\lambda'}$$
$$= \frac{\Delta\lambda}{\lambda + \Delta\lambda}. \qquad (39\text{-}12)$$

Substitution of data yields

$$frac = \frac{2.21 \text{ pm}}{22 \text{ pm} + 2.21 \text{ pm}} = 0.091 \quad \text{or} \quad 9.1\%. \quad \text{(Answer)}$$

Although the Compton shift $\Delta\lambda$ is independent of the wavelength λ of the incident x rays (see Eq. 39-11), the *fractional* photon energy loss of the x rays does depend on λ, increasing as the wavelength of the incident radiation decreases, as indicated by Eq. 39-12.

✔CHECKPOINT 3: Compare Compton scattering for x rays ($\lambda \approx 20$ pm) and visible light ($\lambda \approx 500$ nm) at a particular angle of scattering. Which has the greater (a) Compton shift, (b) fractional wavelength shift, (c) fractional photon energy change, and (d) energy imparted to the electron?

39-5 Light as a Probability Wave

A fundamental mystery in physics is how light can be a wave (which spreads out over a region) in classical physics, whereas it is emitted and absorbed as photons (which originate and vanish at points) in quantum physics. The double-slit experiment of Section 36-4 lies at the heart of this mystery. Let us discuss three versions of that experiment.

The Standard Version

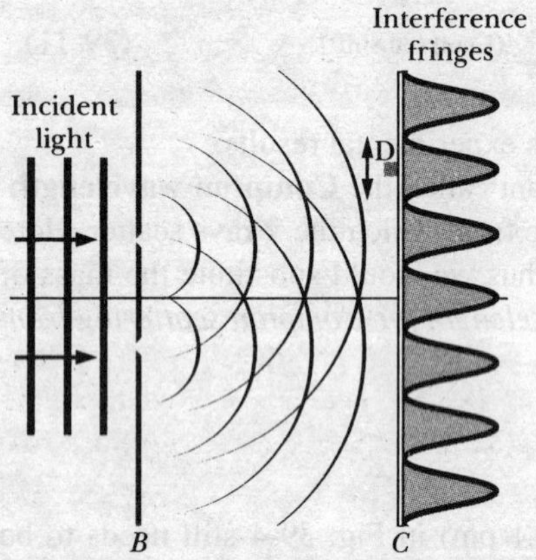

Fig. 39-6 Light is directed onto screen B, which contains two parallel slits. Light emerging from these slits spreads out by diffraction. The two diffracted waves overlap at screen C and form a pattern of interference fringes. A small photon detector D in the plane of screen C generates a sharp click for each photon that it absorbs.

Figure 39-6 is a sketch of the original experiment carried out by Thomas Young in 1801 (see also Fig. 36-6). Light shines on screen B, which contains two narrow parallel slits. The light waves emerging from the two slits spread out by diffraction and overlap on screen C where, by interference, they form a pattern of alternating intensity maxima and minima. In Section 36-4 we took the existence of these interference fringes as compelling evidence for the wave nature of light.

Let us place a tiny photon detector D at one point in the plane of screen C. Let the detector be a photoelectric device that clicks when it absorbs a photon. We would find that the detector produces a series of clicks, randomly spaced in time, each click signaling the transfer of energy from the light wave to the screen via a photon absorption.

If we moved the detector very slowly up or down as indicated by the black arrow in Fig. 39-6, we would find that the click rate increases and decreases, passing through alternate maxima and minima that correspond exactly to the maxima and minima of the interference fringes.

The point of this thought experiment is as follows. We cannot predict when a photon will be detected at any particular point on screen C; photons are detected at individual points at random times. We can, however, predict that the relative *probability* that a single photon will be detected at a particular point in a specified time interval is proportional to the intensity of the incident light at that point.

We saw in Section 34-4 that the intensity I of a light wave at any point is proportional to the square of E_m, the amplitude of the oscillating electric field vector of the wave at that point. Thus,

➤ The probability (per unit time interval) that a photon will be detected in any small volume centered on a given point in a light wave is proportional to the square of the amplitude of the wave's electric field vector at that point.

We now have a probabilistic description of a light wave, hence another way to view light. It is not only an electromagnetic wave but it is also a **probability wave.** That is, to every point in a light wave we can attach a numerical probability (per unit time interval) that a photon can be detected in any small volume centered on that point.

The Single-Photon Version

A single-photon version of the double-slit experiment was first carried out by G. I. Taylor in 1909 and has been repeated many times since. It differs from the standard version in that the light source is so extremely feeble that it emits only one photon at a time, at random intervals. Astonishingly, interference fringes still build up on screen C if the experiment runs long enough (several months for Taylor's early experiment).

What explanation can we offer for the result of this single-photon double-slit experiment? Before we can even consider the result, we are compelled to ask questions like these: If the photons move through the apparatus one at a time, through which of the two slits in screen B does a given photon pass? How does a given photon even "know" that there is another slit present so that interference is a possibility? Can a single photon somehow pass through both slits and interfere with itself?

Bear in mind that we can only know when photons interact with matter—we have no way of detecting them without an interaction with matter, such as with a detector or a screen. Thus, in the experiment of Fig. 39-6, we can only know that photons originate at the light source and vanish at the screen. Between source and screen, we cannot know what the photon is or does. However, because an interference pattern eventually builds up on the screen, we can speculate that each photon travels from source to screen *as a wave* that fills up the space between those two objects and then vanishes in a photon absorption, with a transfer of energy and momentum, at some point on the screen.

We *cannot* predict where this transfer will occur (where a photon will be detected) for any given photon originating at the source. However, we *can* predict the probability that a transfer will occur at any given point on the screen. Transfers will tend to occur (and thus photons will tend to be absorbed) in the regions of the bright fringes in the interference pattern that builds up on the screen. Transfers will tend *not* to occur (and thus photons will tend *not* to be absorbed) in the regions of the dark fringes in the built-up pattern. Thus, we can say that the wave traveling from the source to the screen is a *probability wave,* which produces a pattern of "probability fringes" on the screen.

The Single-Photon, Wide-Angle Version

In the past, physicists tried to explain the single-photon double-slit experiment in terms of small packets of classical light waves that are individually sent toward the slits. They would define these small packets as photons. However, modern experiments invalidate this explanation and definition. Figure 39-7 shows the arrangement of one of these experiments, reported in 1992 by Ming Lai and Jean-Claude Diels of the University of New Mexico. Source S contains molecules that emit photons at well separated times. Mirrors M_1 and M_2 are positioned to reflect light that the source emits along two distinct paths, 1 and 2, that are separated by an angle θ, which is close to 180°. This arrangement differs from the standard two-slit experiment, in which the angle between the paths of the light reaching two slits is very small.

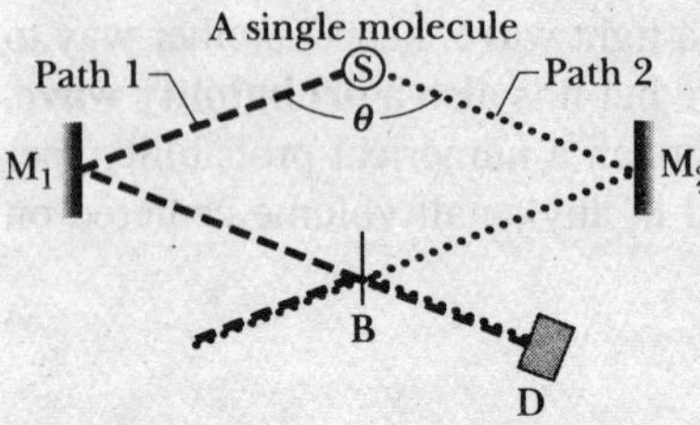

Fig. 39-7 The light from a single photon emission in source S travels over two widely separated paths and interferes with itself at detector D after being recombined by beam splitter B. (After Ming Lai and Jean-Claude Diels, *Journal of the Optical Society of America B*, **9**, 2290–2294, December 1992.)

After reflection from mirrors M_1 and M_2, the light waves traveling along paths 1 and 2 meet at beam splitter B. (A beam splitter is an optical device that transmits half the light incident upon it and reflects the other half.) On the right side of the beam splitter in Fig. 39-7, the light wave traveling along path 2 and reflected by B combines with the light wave traveling along path 1 and transmitted by B. These two waves then interfere with each other as they arrive at detector D (a *photomultiplier tube* that can detect individual photons).

The output of the detector is a randomly spaced series of electronic pulses, one for each detected photon. In the experiment, the beam splitter is moved slowly in a horizontal direction (in the reported experiment, a distance of only about 50 μm maximum), and the detector output is recorded on a chart recorder. Moving the beam splitter changes the lengths of paths 1 and 2, producing a phase shift between the light waves arriving at detector D. Interference maxima and minima appear in the detector's output signal.

This experiment is difficult to understand in traditional terms. For example, when a molecule in the source emits a single photon, does that photon travel along path 1 or path 2 in Fig. 39-7 (or along any other path)? How can it move in both directions at once? To answer, we assume that when a molecule emits a photon, a probability wave radiates in all directions from it. The experiment samples this wave in two of those directions, chosen to be nearly opposite each other.

We see that we can interpret all three versions of the double-slit experiment if we assume that (1) light is generated in the source as photons, (2) light is absorbed in the detector as photons, and (3) light travels between source and detector as a probability wave.

39-6 Electrons and Matter Waves

In 1924 French physicist Louis de Broglie made the following appeal to symmetry: A beam of light is a wave, but it transfers energy and momentum to matter only at points, via photons. Why can't a beam of particles have the same properties? That is, why can't we think of a moving electron—or any other particle, for that matter—as a **matter wave** that transfers energy and momentum to other matter at points?

In particular, de Broglie suggested that Eq. 39-7 ($p = h/\lambda$) might apply not only to photons but also to electrons. We used that equation in Section 39-4 to assign a momentum p to a photon of light with wavelength λ. We now use it, in the form

$$\lambda = \frac{h}{p} \quad \text{(de Broglie wavelength)}, \qquad (39\text{-}13)$$

to assign a wavelength λ to a particle with momentum of magnitude p. The wavelength calculated from Eq. 39-13 is called the **de Broglie wavelength** of the moving particle. De Broglie's prediction of the existence of matter waves was first verified experimentally in 1927, by C. J. Davisson and L. H. Germer of the Bell Telephone Laboratories and by George P. Thomson of the University of Aberdeen in Scotland.

Figure 39-8 shows photographic proof of matter waves in a more recent experiment. In the experiment, an interference pattern was built up when electrons were sent, *one by one,* through a double-slit apparatus. The apparatus was like the ones we have previously used to demonstrate optical interference, except that the viewing screen was similar to a conventional television screen. When an electron hit the screen, it caused a flash of light whose position was recorded.

The first several electrons (top two photos) revealed nothing interesting and seemingly hit the screen at random points. However, after many thousands of elec-

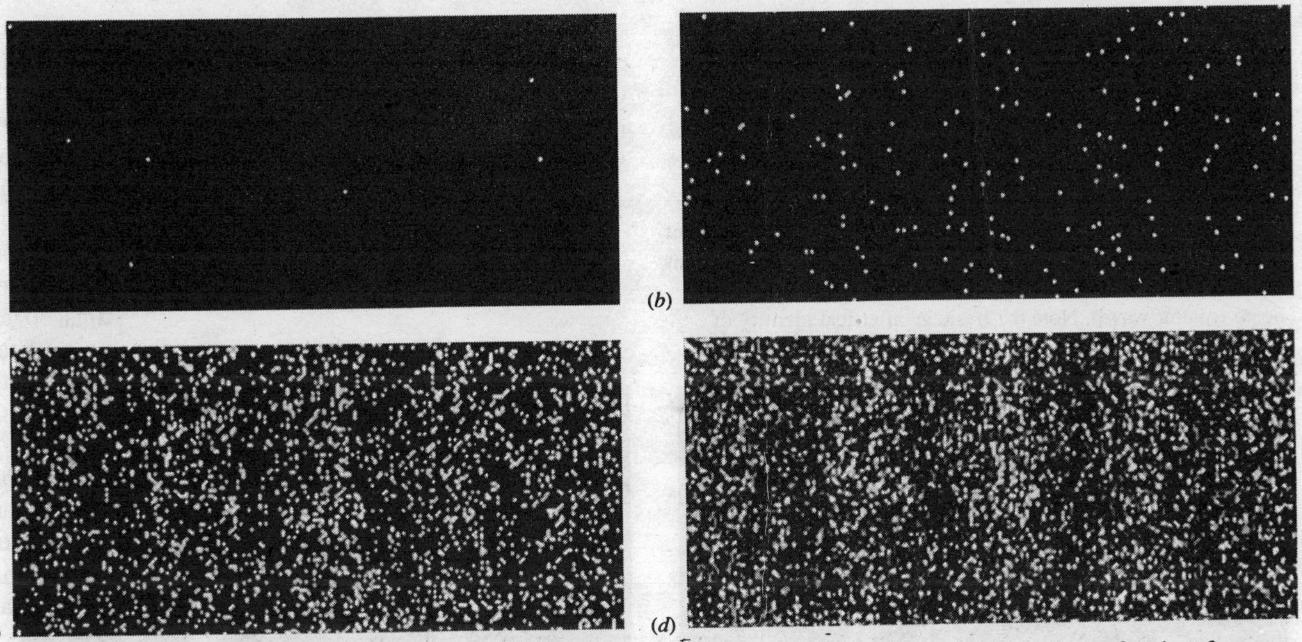

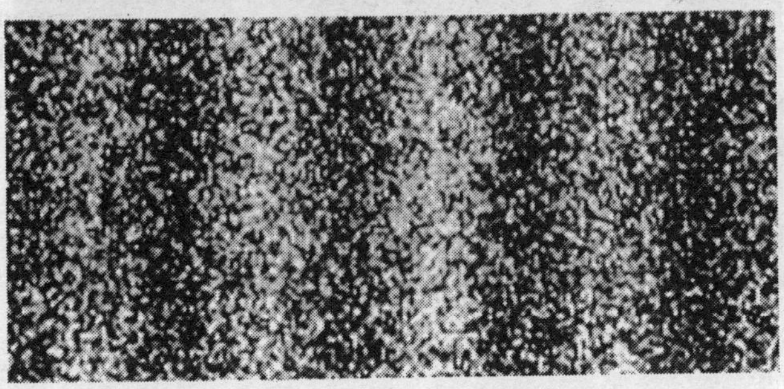

Fig. 39-8 Photographs showing the buildup of an interference pattern by a beam of electrons in a two-slit interference experiment like that of Fig. 39-6. Matter waves, like light waves, are *probability waves*. From top to bottom the approximate numbers of electrons involved are 7, 100, 3000, 20 000, and 70 000.

trons were sent through the apparatus, a pattern appeared on the screen, revealing fringes where many electrons had hit the screen and fringes where few had hit the screen. The pattern is exactly what we would expect for wave interference. Thus, *each* electron passed through the apparatus as a matter wave—the portion that traveled through one slit interfered with the portion that traveled through the other slit. That interference then determined the probability that the electron would materialize at a given point on the screen, hitting the screen there. Many electrons materialized in regions corresponding to bright fringes in optical interference, and few electrons materialized in regions corresponding to dark fringes.

Similar interference has been demonstrated with protons, neutrons, and various atoms. In 1994, it was demonstrated with iodine molecules I_2, which are not only 500 000 times more massive than electrons but far more complex. In 1999, it was demonstrated with the even more complex *fullerenes* (or *buckyballs*) C_{60} and C_{70}. (Fullerenes are soccer-ball-like molecules of carbon atoms, 60 in C_{60} and 70 in C_{70}.) Apparently, such small objects as electrons, protons, atoms, and molecules travel as matter waves. However, as we consider larger and more complex objects, there must come a point at which we are no longer justified in considering the wave nature of an object. At that point, we are back in our familiar nonquantum world, with the physics of earlier chapters of this book. In short, an electron is a matter wave and can undergo interference with itself, but a cat is not a matter wave and cannot undergo interference with itself (which must be a relief to cats).

Fig. 39-9 (*a*) An experimental arrangement used to demonstrate, by diffraction techniques, the wavelike character of the incident beam. Photographs of the diffraction patterns when the incident beam is (*b*) an x-ray beam (light wave) and (*c*) an electron beam (matter wave). Note the basic geometrical identity of the patterns.

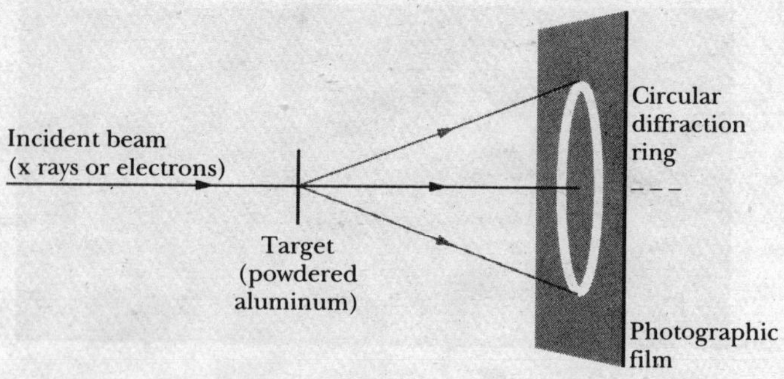

(*a*)

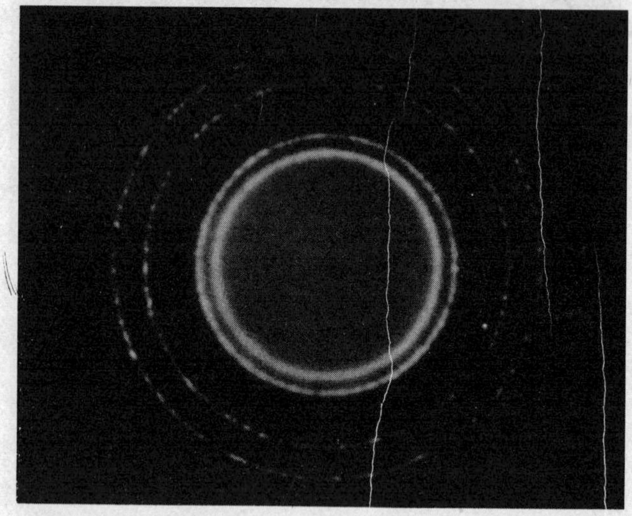

(*b*)

(*c*)

The wave nature of particles and atoms is now taken for granted in many scientific and engineering fields. For example, electron and neutron diffraction are used to study the atomic structures of solids and liquids, and electron diffraction is used to study the atomic features of surfaces on solids.

Figure 39-9*a* shows an arrangement that can be used to demonstrate the scattering of either x rays or electrons by crystals. A beam of one or the other is directed onto a target consisting of a powder of tiny aluminum crystals. The x rays have a certain wavelength λ. The electrons are given enough energy so that their de Broglie wavelength is the same wavelength λ. The scatter of x rays or electrons by the crystals produces a circular interference pattern on a photographic film. Figure 39-9*b* shows the pattern for the scatter of x rays, whereas Fig. 39-9*c* shows the pattern for the scatter of electrons. The patterns are the same—both x rays and electrons are waves.

Waves and Particles

Figures 39-8 and 39-9 are convincing evidence of the *wave* nature of matter, but we have at least as many experiments that suggest the *particle* nature of matter. Consider the tracks generated by electrons and displayed in the opening photo of this chapter. Surely these tracks—which are strings of bubbles left in the liquid hydrogen that

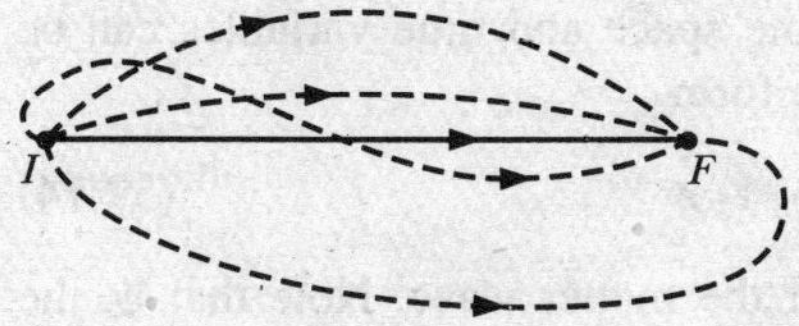

Fig. 39-10 A few of the many paths that connect two particle detection points I and F. Only matter waves that follow paths close to the straight line between these points interfere constructively. For all other paths, the waves following neighboring paths interfere destructively. Thus, a matter wave leaves a straight track.

fills the bubble chamber—strongly suggest the passage of a particle. Where is the wave?

To simplify the situation, let us turn off the magnetic field so that the strings of bubbles will be straight. We can view each bubble as a detection point for the electron. Matter waves traveling between detection points such as I and F in Fig. 39-10 will explore all possible paths, a few of which are shown.

In general, for every path connecting I and F (except the straight-line path), there will be a neighboring path such that matter waves following the two paths cancel each other by interference. This is not true, however, for the straight-line path joining I and F; in this case, matter waves traversing all neighboring paths reinforce the wave following the direct path. You can think of the bubbles that form the track as a series of detection points at which the matter wave undergoes constructive interference.

Sample Problem 39-5

What is the de Broglie wavelength of an electron with a kinetic energy of 120 eV?

SOLUTION: One **Key Idea** here is that we can find the electron's de Broglie wavelength λ from Eq. 39-13 ($\lambda = h/p$) if we first find the magnitude of its momentum p. A second **Key Idea** is that we find p from the given kinetic energy K of the electron. That kinetic energy is much less than the rest energy of an electron (0.511 MeV, from Table 38-3). Thus, we can get by with the classical approximations for momentum p $(= mv)$ and kinetic energy K $(= \frac{1}{2}mv^2)$.

Eliminating the speed v between these two equations yields

$$p = \sqrt{2mK}$$
$$= \sqrt{(2)(9.11 \times 10^{-31}\ \text{kg})(120\ \text{eV})(1.60 \times 10^{-19}\ \text{J/eV})}$$
$$= 5.91 \times 10^{-24}\ \text{kg} \cdot \text{m/s}.$$

From Eq. 39-13 then

$$\lambda = \frac{h}{p}$$
$$= \frac{6.63 \times 10^{-34}\ \text{J} \cdot \text{s}}{5.91 \times 10^{-24}\ \text{kg} \cdot \text{m/s}}$$
$$= 1.12 \times 10^{-10}\ \text{m} = 112\ \text{pm}. \qquad \text{(Answer)}$$

This is about the size of a typical atom. If we increase the kinetic energy, the wavelength becomes even smaller.

CHECKPOINT 4: An electron and a proton can have the same (a) kinetic energy, (b) momentum, or (c) speed. In each case, which particle has the shorter de Broglie wavelength?

39-7 Schrödinger's Equation

A simple traveling wave of any kind, be it a wave on a string, a sound wave, or a light wave, is described in terms of some quantity that varies in a wavelike fashion. For light waves, for example, this quantity is $\vec{E}(x, y, z, t)$, the electric field component of the wave. Its observed value at any point depends on the location of that point and on the time at which the observation is made.

What varying quantity should we use to describe a matter wave? We should expect this quantity, which we call the **wave function** $\Psi(x, y, z, t)$, to be more complicated than the corresponding quantity for a light wave because a matter wave, in addition to energy and momentum, transports mass and (often) electric charge. It turns out that Ψ, the uppercase Greek letter psi, usually represents a function that is complex in the mathematical sense; that is, we can always write its values in the form $a + ib$, in which a and b are real numbers and $i^2 = -1$.

In all the situations you will meet here, the space and time variables can be grouped separately and Ψ can be written in the form

$$\Psi(x, y, z, t) = \psi(x, y, z)\, e^{-i\omega t}, \tag{39-14}$$

where ω ($= 2\pi f$) is the angular frequency of the matter wave. Note that ψ, the lowercase Greek letter psi, represents only the space-dependent part of the complete, time-dependent wave function Ψ. We shall deal almost exclusively with ψ. Two questions arise: What is meant by the wave function, and how do we find it?

What does the wave function mean? It has to do with the fact that a matter wave, like a light wave, is a probability wave. Suppose that a matter wave reaches a particle detector that is small; then the probability that a particle will be detected in a specified time interval is proportional to $|\psi|^2$, where $|\psi|$ is the absolute value of the wave function at the location of the detector. Although ψ is usually a complex quantity, $|\psi|^2$ is always both real and positive. It is, then, $|\psi|^2$, which we call the **probability density,** and not ψ, that has *physical* meaning. Speaking loosely, the meaning is this:

The probability (per unit time) of detecting a particle in a small volume centered on a given point in a matter wave is proportional to the value of $|\psi|^2$ at that point.

Because ψ is usually a complex quantity, we find the square of its absolute value by multiplying ψ by ψ^*, the *complex conjugate* of ψ. (To find ψ^* we replace the imaginary number i in ψ with $-i$, wherever it occurs.)

How do we find the wave function? Sound waves and waves on strings are described by the equations of Newtonian mechanics. Light waves are described by Maxwell's equations. Matter waves are described by **Schrödinger's equation,** advanced in 1926 by Austrian physicist Erwin Schrödinger.

Many of the situations that we shall discuss involve a particle traveling in the x direction through a region in which forces acting on the particle cause it to have a potential energy $U(x)$. In this special case, Schrödinger's equation reduces to

$$\frac{d^2\psi}{dx^2} + \frac{8\pi^2 m}{h^2}[E - U(x)]\psi = 0 \quad \text{(Schrödinger's equation, one-dimensional motion),} \tag{39-15}$$

in which E is the total mechanical energy (potential energy plus kinetic energy) of the moving particle. (We do not consider mass energy in this nonrelativistic equation.) We cannot derive Schrödinger's equation from more basic principles; it *is* the basic principle.

If $U(x)$ in Eq. 39-15 is zero, that equation describes a **free particle**—that is, a moving particle on which no net force acts. The particle's total energy in this case is all kinetic, and thus E in Eq. 39-15 is $\frac{1}{2}mv^2$. That equation then becomes

$$\frac{d^2\psi}{dx^2} + \frac{8\pi^2 m}{h^2}\left(\frac{mv^2}{2}\right)\psi = 0,$$

which we can recast as

$$\frac{d^2\psi}{dx^2} + \left(2\pi\frac{p}{h}\right)^2\psi = 0.$$

To obtain this equation, we replaced mv with the momentum p and regrouped terms.

From Eq. 39-13 we recognize p/h in the equation above as $1/\lambda$, where λ is the de Broglie wavelength of the moving particle. We further recognize $2\pi/\lambda$ as the *angular wave number* k, which we defined in Eq. 17-5. With this substitution, the equation above becomes

$$\frac{d^2\psi}{dx^2} + k^2\psi = 0 \quad \text{(Schrödinger's equation, free particle).} \tag{39-16}$$

The most general solution of Eq. 39-16 is

$$\psi(x) = Ae^{ikx} + Be^{-ikx}, \tag{39-17}$$

in which A and B are arbitrary constants. You can show that this equation is indeed a solution of Eq. 39-16 by substituting $\psi(x)$ and its second derivative into that equation and noting that an identity results.

If we combine Eqs. 39-14 and 39-17, we find, for the time-dependent wave function Ψ of a free particle traveling in the x direction,

$$\Psi(x, t) = \psi(x)e^{-i\omega t} = (Ae^{ikx} + Be^{-ikx})e^{-i\omega t}$$
$$= Ae^{i(kx-\omega t)} + Be^{-i(kx+\omega t)}. \tag{39-18}$$

Finding the Probability Density $|\psi|^2$

In Section 17-5 we saw that *any function F* of the form $F(kx \pm \omega t)$ represents a traveling wave. This applies to exponential functions like those in Eq. 39-18 as well as to the sinusoidal functions we have used to describe waves on strings. In fact, these two representations of functions are related by

$$e^{i\theta} = \cos\theta + i\sin\theta \quad \text{and} \quad e^{-i\theta} = \cos\theta - i\sin\theta,$$

where θ is any angle.

The first term on the right in Eq. 39-18 thus represents a wave traveling in the direction of increasing x and the second a wave traveling in the negative direction of x. However, we have assumed that the free particle we are considering travels only in the positive direction of x. To reduce the general solution (Eq. 39-18) to our case of interest, we choose the arbitrary constant B in Eqs. 39-18 and 39-17 to be zero. At the same time, we relabel the constant A as ψ_0. Equation 39-17 then becomes

$$\psi(x) = \psi_0\, e^{ikx}. \tag{39-19}$$

To calculate the probability density, we take the square of the absolute value of $\psi(x)$. We get

$$|\psi|^2 = |\psi_0\, e^{ikx}|^2 = (\psi_0^2)\,|e^{ikx}|^2.$$

Now, because

$$|e^{ikx}|^2 = (e^{ikx})(e^{ikx})^* = e^{ikx}\, e^{-ikx} = e^{ikx-ikx} = e^0 = 1,$$

we get

$$|\psi|^2 = (\psi_0^2)(1)^2 = \psi_0^2 \quad \text{(a constant).}$$

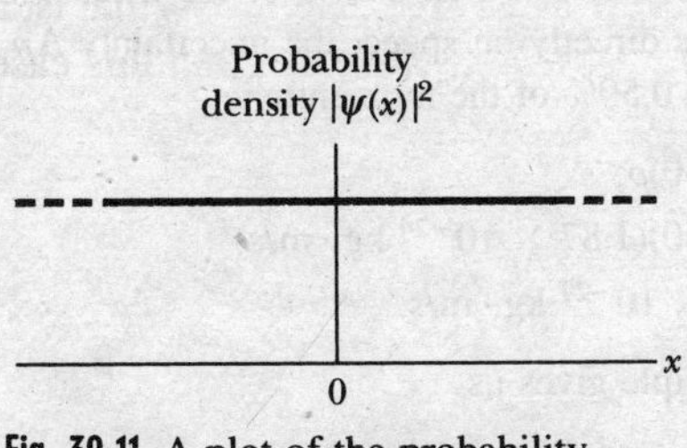

Fig. 39-11 A plot of the probability density $|\psi|^2$ for a free particle moving in the positive x direction. Since $|\psi|^2$ has the same constant value for all values of x, the particle has the same probability of detection at all points along its path.

Figure 39-11 is a plot of the probability density $|\psi|^2$ versus x for a free particle—a straight line parallel to the x axis from $-\infty$ to $+\infty$. We see that the probability density $|\psi|^2$ is the same for all values of x, which means that the particle has equal probabilities of being *anywhere* along the x axis. There is no distinguishing feature by which we can predict a most likely position for the particle. That is, all positions are equally likely.

We'll see what this means in the next section.

39-8 Heisenberg's Uncertainty Principle

Our inability to predict the position of a free particle, as indicated by Fig. 39-11, is our first example of **Heisenberg's uncertainty principle,** proposed in 1927 by German physicist Werner Heisenberg. It states that measured values cannot be assigned to the position $\vec{r}$ and the momentum $\vec{p}$ of a particle simultaneously with unlimited precision.

For the components of $\vec{r}$ and $\vec{p}$, Heisenberg's principle gives the following limits in terms of $\hbar = h/2\pi$ (called "h-bar"):

$$\Delta x \cdot \Delta p_x \geq \hbar$$
$$\Delta y \cdot \Delta p_y \geq \hbar \quad \text{(Heisenberg's uncertainty principle).} \qquad (39\text{-}20)$$
$$\Delta z \cdot \Delta p_z \geq \hbar$$

Here Δx and Δp_x, as examples, represent the intrinsic uncertainties in the measurements of the x components of $\vec{r}$ and $\vec{p}$. Even with the best measuring instruments that technology could ever provide, each product of a position uncertainty and a momentum uncertainty in Eq. 39-20 will be greater than $\hbar$; it can *never* be less.

The particle whose probability density is plotted in Fig. 39-11 is a free particle; that is, no force acts on it, so its momentum $\vec{p}$ must be constant. We implied—without making a point of it—that we can determine $\vec{p}$ with absolute precision; we assumed that $\Delta p_x = \Delta p_y = \Delta p_z = 0$ in Eq. 39-20. That assumption then requires $\Delta x \rightarrow \infty$, $\Delta y \rightarrow \infty$, and $\Delta z \rightarrow \infty$. With such infinitely great uncertainties, the position of the particle is completely unspecified, as Fig. 39-11 shows.

Do not think that the particle *really has* a sharply defined position that is, for some reason, hidden from us. If its momentum can be specified with absolute precision, the words "position of the particle" simply lose all meaning. The particle in Fig. 39-11 can be found *with equal probability* anywhere along the x axis.

Sample Problem 39-6

Assume that an electron is moving along an x axis and that you measure its speed to be 2.05×10^6 m/s, which can be known with a precision of 0.50%. What is the minimum uncertainty (as allowed by the uncertainty principle in quantum theory) with which you can simultaneously measure the position of the electron along the x axis?

SOLUTION: The **Key Idea** here is that the minimum uncertainty allowed by quantum theory is given by Heisenberg's uncertainty principle in Eq. 39-20. We need only consider components along the x axis because we have motion only along that axis and want the uncertainty Δx in location along that axis. Since we want the minimum allowed uncertainty, we use the equality instead of the inequality in the x-axis part of Eq. 39-20, writing

$$\Delta x \cdot \Delta p_x = \hbar.$$

To evaluate the uncertainty Δp_x in the momentum, we must first evaluate the momentum component p_x. Because the electron's speed v_x is much less than the speed of light c, we can evaluate p_x with the classical expression for momentum instead of using a relativistic expression. We find

$$p_x = mv_x = (9.11 \times 10^{-31}\ \text{kg})(2.05 \times 10^6\ \text{m/s})$$
$$= 1.87 \times 10^{-24}\ \text{kg} \cdot \text{m/s}.$$

The uncertainty in the speed is given as 0.50% of the measured speed. Because p_x depends directly on speed, the uncertainty Δp_x in the momentum must be 0.50% of the momentum:

$$\Delta p_x = (0.0050)p_x$$
$$= (0.0050)(1.87 \times 10^{-24}\ \text{kg} \cdot \text{m/s})$$
$$= 9.35 \times 10^{-27}\ \text{kg} \cdot \text{m/s}.$$

Then the uncertainty principle gives us

$$\Delta x = \frac{\hbar}{\Delta p_x} = \frac{(6.63 \times 10^{-34}\ \text{J} \cdot \text{s})/2\pi}{9.35 \times 10^{-27}\ \text{kg} \cdot \text{m/s}}$$
$$= 1.13 \times 10^{-8}\ \text{m} \approx 11\ \text{nm}, \qquad \text{(Answer)}$$

which is about 100 atomic diameters. Given your measurement of the electron's speed, it makes no sense to try to pin down the electron's position to any greater precision.

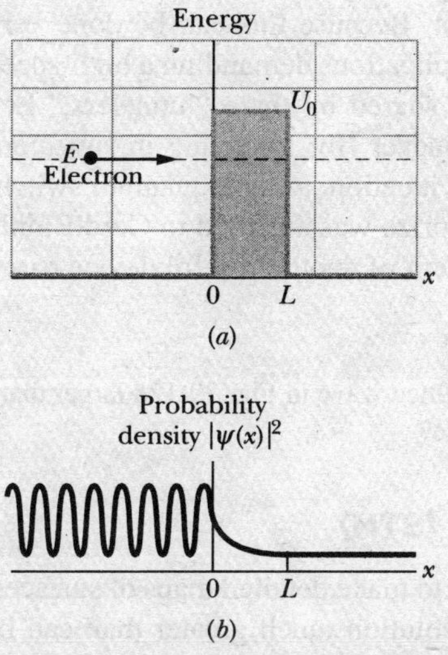

Fig. 39-12 (*a*) An energy diagram showing a potential energy barrier of height U_0 and thickness L. An electron with total energy E approaches the barrier from the left. (*b*) The probability density $|\psi|^2$ of the matter wave representing the electron, showing the tunneling of the electron through the barrier. The curve to the left of the barrier represents a standing matter wave that results from the superposition of the incident and reflected matter waves.

39-9 Barrier Tunneling

Suppose you repeatedly flip a jelly bean along a tabletop on which a book is positioned somewhere along the jelly bean's path. You would be very surprised to see the jelly bean appear on the other side of the book instead of bouncing back from it. Don't expect this surprising result from jelly beans. However, something very much like it, called **barrier tunneling,** *does* happen for electrons and other particles with small masses.

Figure 39-12*a* shows an electron of total energy E moving parallel to the x axis. Forces act on the electron such that its potential energy is zero except when it is in the region $0 < x < L$, where its potential energy has the constant value U_0. We define this region as a **potential energy barrier** (often loosely called a **potential barrier**) of height U_0 and thickness L.

Classically, because $E < U_0$, an electron approaching the barrier from the left would be reflected from the barrier and would move back in the direction from which it came. In quantum physics, however, the electron is a matter wave and there is a finite chance that it will "leak through" the barrier and appear on the other side. This means that there is a finite probability that the electron will end up on the far side of the barrier, moving to the right.

The wave function $\psi(x)$ describing the electron can be found by solving Schrödinger's equation (Eq. 39-15) separately for the three regions in Fig. 39-12*a*: (1) to the left of the barrier, (2) within the barrier, and (3) to the right of the barrier. The arbitrary constants that appear in the solutions can then be chosen so that the values of $\psi(x)$ and its derivative with respect to x join smoothly (no jumps, no kinks) at $x = 0$ and at $x = L$. Squaring the absolute value of $\psi(x)$ then yields the probability density.

Figure 39-12*b* shows a plot of the result. The oscillating curve to the left of the barrier (for $x < 0$) is a combination of the incident matter wave and the reflected matter wave (which has a smaller amplitude than the incident wave). The oscillations occur because these two waves, traveling in opposite directions, interfere with each other, setting up a standing wave pattern.

Within the barrier (for $0 < x < L$) the probability density decreases exponentially with x. However, provided L is small, the probability density is not quite zero at $x = L$.

To the right of the barrier of Fig. 39-12 (for $x > L$), the probability density plot describes a transmitted (through the barrier) wave with low but constant amplitude. Thus, the electron can be detected in this region but with a relatively small probability. (Compare this part of the figure with Fig. 39-11 for a free particle.)

We can assign a *transmission coefficient* T to the incident matter wave and the barrier in Fig. 39-12*a*. This coefficient gives the probability with which an approaching electron will be transmitted through the barrier—that is, that tunneling will occur. As an example, if $T = 0.020$, then of every 1000 electrons fired at the barrier, 20 (on average) will tunnel through it and 980 will be reflected.

The **transmission coefficient** T is approximately

$$T \approx e^{-2kL}, \tag{39-21}$$

in which

$$k = \sqrt{\frac{8\pi^2 m(U_0 - E)}{h^2}}. \tag{39-22}$$

Because of the exponential form of Eq. 39-21, the value of T is very sensitive to the three variables on which it depends: particle mass m, barrier thickness L, and energy difference $U_0 - E$.

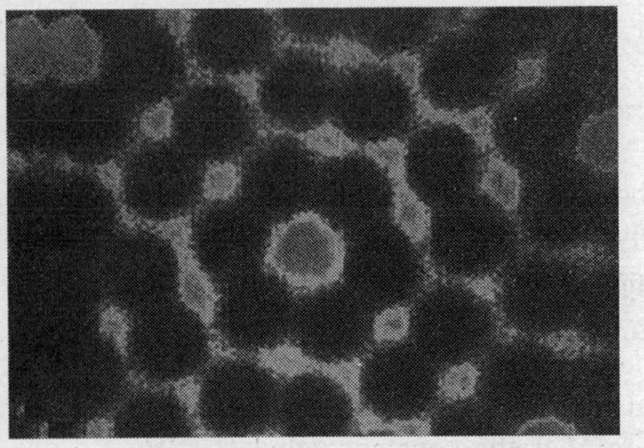

Fig. 39-13 An array of silicon atoms as revealed by a scanning tunneling microscope.

Barrier tunneling finds many applications in technology, among them the tunnel diode, in which a flow of electrons (by tunneling through a device) can be rapidly turned on or off by controlling the barrier height. Because this can be done very quickly (within 5 ps), the device is suitable for applications demanding a high-speed response. The 1973 Nobel prize in physics was shared by three "tunnelers," Leo Esaki (for tunneling in semiconductors), Ivar Giaever (for tunneling in superconductors), and Brian Josephson (for the Josephson junction, a rapid quantum switching device based on tunneling). The 1986 Nobel prize was awarded to Gerd Binnig and Heinrich Rohrer to recognize their development of another useful device based on tunneling, the scanning tunneling microscope.

CHECKPOINT 5: Is the wavelength of the transmitted wave in Fig. 39-12*b* larger than, smaller than, or the same as that of the incident wave?

The Scanning Tunneling Microscope (STM)

A device based on tunneling, the STM allows one to make detailed maps of surfaces, revealing features on the atomic scale with a resolution much greater than can be obtained with an optical or electron microscope. Figure 39-13 shows an example, the individual atoms of the surface being readily apparent.

Figure 39-14 shows the heart of the scanning tunneling microscope. A fine metallic tip, mounted at the intersection of three mutually perpendicular quartz rods, is placed close to the surface to be examined. A small potential difference, perhaps only 10 mV, is applied between tip and surface.

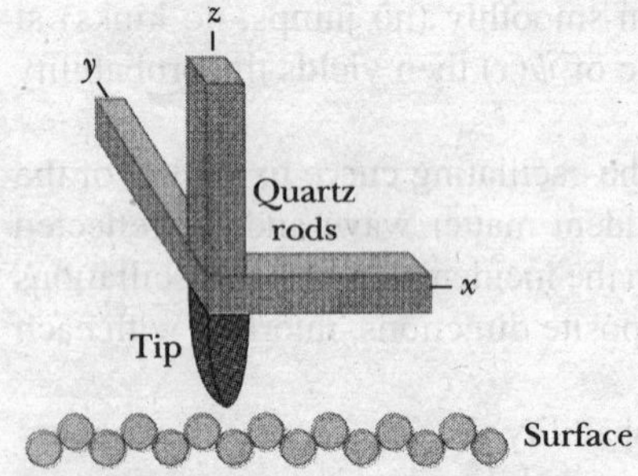

Fig. 39-14 The essence of a scanning tunneling microscope (STM). Three quartz rods are used to scan a sharply pointed conducting tip across the surface of interest and to maintain a constant separation between tip and surface. The tip thus moves up and down to match the contours of the surface, and a record of its movement is a map like that of Fig. 39-13.

Crystalline quartz has an interesting property called *piezoelectricity:* When an electric potential difference is applied across a sample of crystalline quartz, the dimensions of the sample change slightly. This property is used to change the length of each of the three rods in Fig. 39-14, smoothly and by tiny amounts, so that the tip can be scanned back and forth over the surface (in the x and y directions) and also lowered or raised with respect to the surface (in the z direction).

The space between the surface and the tip forms a potential energy barrier, much like that of Fig. 39-12*a*. If the tip is close enough to the surface, electrons from the sample can tunnel through this barrier from the surface to the tip, forming a tunneling current.

In operation, an electronic feedback arrangement adjusts the vertical position of the tip to keep the tunneling current constant as the tip is scanned over the surface. This means that the tip–surface separation also remains constant during the scan. The output of the device—for example, Fig. 39-13—is a video display of the varying vertical position of the tip, hence of the surface contour, as a function of the tip position in the xy plane.

Scanning tunneling microscopes are available commercially and are used in laboratories all over the world.

Sample Problem 39-7

Suppose that the electron in Fig. 39-12*a*, having a total energy E of 5.1 eV, approaches a barrier of height $U_0 = 6.8$ eV and thickness $L = 750$ pm.

(a) What is the approximate probability that the electron will be transmitted through the barrier, to appear (and be detectable) on the other side of the barrier?

SOLUTION: The **Key Idea** here is that the probability we seek is the transmission coefficient T as given by Eq. 39-21 ($T \approx e^{-2kL}$), where k is

$$k = \sqrt{\frac{8\pi^2 m(U_0 - E)}{h^2}}.$$

The numerator of the fraction under the square-root sign is

$$(8\pi^2)(9.11 \times 10^{-31}\text{ kg})(6.8\text{ eV} - 5.1\text{ eV}) \times (1.60 \times 10^{-19}\text{ J/eV}) = 1.956 \times 10^{-47}\text{ J}\cdot\text{kg}.$$

Thus, $k = \sqrt{\frac{1.956 \times 10^{-47}\text{ J}\cdot\text{kg}}{(6.63 \times 10^{-34}\text{ J}\cdot\text{s})^2}} = 6.67 \times 10^9\text{ m}^{-1}.$

The (dimensionless) quantity $2kL$ is then

$$2kL = (2)(6.67 \times 10^9\text{ m}^{-1})(750 \times 10^{-12}\text{ m}) = 10.0$$

and, from Eq. 39-21, the transmission coefficient is

$$T \approx e^{-2kL} = e^{-10.0} = 45 \times 10^{-6}. \quad \text{(Answer)}$$

Thus, of every million electrons that strike the barrier, about 45 will tunnel through it.

(b) What is the approximate probability that a proton with the same total energy of 5.1 eV will be transmitted through the barrier, to appear (and be detectable) on the other side of the barrier?

SOLUTION: The **Key Idea** here is that the transmission coefficient T (and thus the probability of transmission) depends on the mass of the particle. Indeed, because mass m is one of the factors in the exponent of e in the equation for T, the probability of transmission is very sensitive to the mass of the particle. This time, the mass is that of a proton (1.67×10^{-27} kg), which is significantly greater than that of the electron in (a). By substituting the proton's mass for the mass in (a) and then continuing as we did there, we find that $T \approx 10^{-186}$. Thus, the probability that the proton will be transmitted is not zero, but barely more than zero. For even more massive particles with the same total energy of 5.1 eV, the probability of transmission is exponentially lower.

REVIEW & SUMMARY

Light Quanta—Photons An electromagnetic wave (light) is quantized, and its quanta are called *photons*. For a light wave of frequency f and wavelength λ, the energy E and momentum magnitude p of a photon are

$$E = hf \quad \text{(photon energy)} \quad (39\text{-}2)$$

and

$$p = \frac{hf}{c} = \frac{h}{\lambda} \quad \text{(photon momentum).} \quad (39\text{-}7)$$

Photoelectric Effect When light of high enough frequency falls on a clean metal surface, electrons are emitted from the surface by photon–electron interactions within the metal. The governing relation is

$$hf = K_{max} + \Phi, \quad (39\text{-}5)$$

in which hf is the photon energy, K_{max} is the kinetic energy of the most energetic emitted electrons, and Φ is the **work function** of the target material—that is, the minimum energy an electron must have if it is to emerge from the surface of the target. If hf is less than Φ, the photoelectric effect does not occur.

Compton Shift When x rays are scattered by loosely bound electrons in a target, some of the scattered x rays have a longer wavelength than do the incident x rays. This **Compton shift** (in wavelength) is given by

$$\Delta\lambda = \frac{h}{mc}(1 - \cos\phi), \quad (39\text{-}11)$$

in which ϕ is the angle at which the x rays are scattered.

Light Waves and Photons When light interacts with matter, energy and momentum are transferred via photons. When light is in transit, however, we interpret the light wave as a **probability wave,** in which the probability (per unit time) that a photon can be detected is proportional to E_m^2, where E_m is the amplitude of the oscillating electric field of the light wave at the detector.

Matter Waves A moving particle such as an electron or a proton can be described as a **matter wave;** its wavelength (called the **de Broglie wavelength**) is given by $\lambda = h/p$, where p is the momentum of the particle.

The Wave Function A matter wave is described by its **wave function** $\Psi(x, y, z, t)$, which can be separated into a space-dependent part $\psi(x, y, z)$ and a time-dependent part $e^{-i\omega t}$. For a particle of mass m moving in the x direction with constant total energy E through a region in which its potential energy is $U(x)$, $\psi(x)$ can be found by solving the simplified **Schrödinger equation:**

$$\frac{d^2\psi}{dx^2} + \frac{8\pi^2 m}{h^2}[E - U(x)]\psi = 0. \quad (39\text{-}15)$$

A matter wave, like a light wave, is a probability wave in the sense that if a particle detector is inserted into the wave, the probability that the detector will register a particle during any specified time interval is proportional to $|\psi|^2$, a quantity called the **probability density.**

For a free particle—that is, a particle for which $U(x) = 0$—moving in the x direction, $|\psi|^2$ has a constant value for all positions along the x axis.

Heisenberg's Uncertainty Principle The probabilistic nature of quantum physics places an important limitation on detecting a particle's position and momentum. That is, it is not possible to measure the position $\vec{r}$ and the momentum $\vec{p}$ of a particle simultaneously with unlimited precision. The uncertainties in the components of these quantities are given by

$$\Delta x \cdot \Delta p_x \geq \hbar$$
$$\Delta y \cdot \Delta p_y \geq \hbar \quad (39\text{-}20)$$
$$\Delta z \cdot \Delta p_z \geq \hbar.$$

Barrier Tunneling According to classical physics, an incident particle will be reflected from a potential energy barrier whose height is greater than the particle's kinetic energy. According to quantum physics, however, the particle has a finite probability of tunneling through such a barrier. The probability that a given particle of mass m and energy E will tunnel through a barrier of height U_0 and thickness L is given by the transmission coefficient T:

$$T \approx e^{-2kL}, \tag{39-21}$$

where

$$k = \sqrt{\frac{8\pi^2 m(U_0 - E)}{h^2}}. \tag{39-22}$$

QUESTIONS

1. Of the electromagnetic waves generated in a microwave oven and in your dentist's x-ray machine, which has (a) the greater wavelength, (b) the greater frequency, and (c) the greater photon energy?

2. Of the following statements about the photoelectric effect, which are true and which are false? (a) The greater the frequency of the incident light is, the greater is the stopping potential. (b) The greater the intensity of the incident light is, the greater is the cutoff frequency. (c) The greater the work function of the target material is, the greater is the stopping potential. (d) The greater the work function of the target material is, the greater is the cutoff frequency. (e) The greater the frequency of the incident light is, the greater is the maximum kinetic energy of the ejected electrons. (f) The greater the energy of the photons is, the smaller is the stopping potential.

3. According to the figure for Checkpoint 2, is the maximum kinetic energy of the ejected electrons greater for a target made of sodium or of potassium for a given frequency of incident light?

4. In the photoelectric effect (for a given target and a given frequency of the incident light), which of these quantities, if any, depend on the intensity of the incident light beam: (a) the maximum kinetic energy of the electrons, (b) the maximum photoelectric current, (c) the stopping potential, (d) the cutoff frequency?

5. If you shine ultraviolet light on an isolated metal plate, the plate emits electrons for a while. Why does it eventually stop?

6. A metal plate is illuminated with light of a certain frequency. Which of the following determine whether or not electrons are ejected: (a) the intensity of the light, (b) the length of time of exposure to the light, (c) the thermal conductivity of the plate, (d) the area of the plate, (e) the material of the plate?

7. In a Compton-shift experiment, an x-ray photon is scattered in the forward direction, at $\phi = 0$ in Fig. 39-3. How much energy does the electron acquire during this interaction?

8. According to Eq. 39-11 the Compton shift is the same for x rays and for visible light. Why is it that the Compton shift for x rays can be measured readily but that for visible light cannot?

9. Photon A has twice the energy of photon B. (a) Is the momentum of A less than, equal to, or greater than that of B? (b) Is the wavelength of A less than, equal to, or greater than that of B?

10. Photon A is from an ultraviolet tanning lamp and photon B is from a television transmitter. Which has the greater (a) wavelength, (b) energy, (c) frequency, and (d) momentum?

11. The data shown in Fig. 39-4 were taken by directing x rays onto a carbon target. In what essential way, if any, would these data differ if the target were sulfur instead of carbon?

12. An electron and a proton have the same kinetic energy. Which has the greater de Broglie wavelength?

13. (a) If you double the kinetic energy of a nonrelativistic particle, how does its de Broglie wavelength change? (b) What if you double the speed of the particle?

14. The following nonrelativistic particles all have the same kinetic energy. Rank them in order of their de Broglie wavelengths, greatest first: electron, alpha particle, neutron.

15. Figure 39-15 shows four situations in which an electron is moving through a field. It is moving (a) opposite an electric field, (b) in the same direction as an electric field, (c) in the same direction as a magnetic field, (d) perpendicular to a magnetic field. For each situation, is the de Broglie wavelength of the electron increasing, decreasing, or remaining the same?

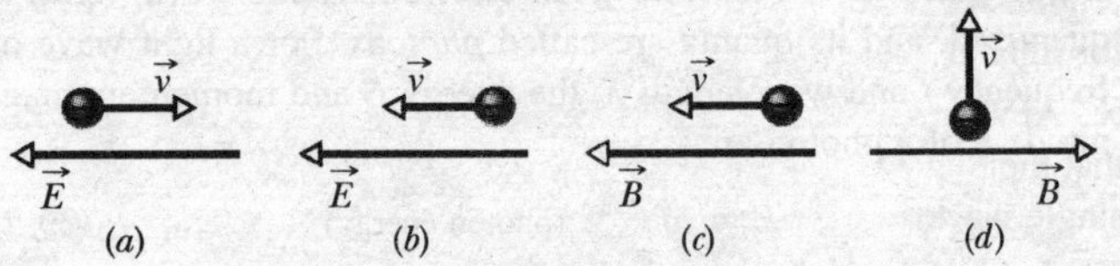

Fig. 39-15 Question 15.

16. A proton and a deuteron, each having a kinetic energy of 3 MeV, approach a potential energy barrier whose height U_0 is 10 MeV. Which particle has the greater chance of tunneling through the barrier? (A deuteron is twice as massive as a proton.)

17. Which has the greater effect on the transmission coefficient T for electron tunneling through a potential energy barrier: (a) raising the barrier height U_0 by 1% or (b) lowering the kinetic energy E of the incident electron by 1%?

18. At the left in Fig. 39-12*b*, why are the minima in the values of $|\psi|^2$ greater than zero?

19. Suppose that the height of the potential energy barrier in Fig. 39-12*a* is infinite. (a) What value would you expect for the transmission coefficient of electrons approaching the barrier? (b) Does Eq. 39-21 predict your expected result?

20. The table gives relative values for three situations for the barrier tunneling experiment of Fig. 39-12. Rank the situations according to the probability of the electron tunneling through the barrier, greatest first.

	Electron Energy	Barrier Height	Barrier Thickness
(a)	E	$5E$	L
(b)	E	$17E$	$L/2$
(c)	E	$2E$	$2L$

EXERCISES & PROBLEMS

ssm Solution is in the Student Solutions Manual.
www Solution is available on the World Wide Web at: http://www.wiley.com/college/hrw
ilw Solution is available on the Interactive LearningWare.

SEC. 39-2 The Photon, the Quantum of Light

1E. Express the Planck constant h in terms of the unit electron-volt-femtoseconds.

2E. Monochromatic light (that is, light of a single wavelength) is to be absorbed by a sheet of photographic film and thus recorded on the film. Photon absorption will occur if the photon energy equals or exceeds the least energy of 0.6 eV needed to dissociate an AgBr molecule in the film. What is the greatest wavelength of light that can be recorded by the film? In what region of the electromagnetic spectrum is this wavelength located?

3E. Show that, for light of wavelength λ in nanometers, the photon energy hf in electron-volts is $1240/\lambda$. ssm

4E. The yellow-colored light from a highway sodium lamp is brightest at a wavelength of 589 nm. What is the photon energy for light at that wavelength?

5E. At what rate does the Sun emit photons? For simplicity, assume that the Sun's entire emission at the rate of 3.9×10^{26} W is at the single wavelength of 550 nm.

6E. A helium–neon laser emits red light at wavelength λ = 633 nm, in a beam of diameter 3.5 mm, and at an energy-emission rate of 5.0 mW. A detector in the beam's path totally absorbs the beam. At what rate per unit area does the detector absorb photons?

7E. A spectral emission line is electromagnetic radiation that is emitted in a wavelength range narrow enough to be taken as a single wavelength. One such emission line that is important in astronomy has a wavelength of 21 cm. What is the photon energy in the electromagnetic wave at that wavelength?

8E. How fast must an electron move to have a kinetic energy equal to the photon energy of sodium light at wavelength 590 nm?

9E. The meter was once defined as 1 650 763.73 wavelengths of the orange light emitted by a source containing krypton-86 atoms. What is the photon energy of that light?

10P. Under ideal conditions, a visual sensation can occur in the human visual system if light of wavelength 550 nm is absorbed by the eye's retina at a rate as low as 100 photons per second. What is the corresponding rate at which energy is absorbed by the retina?

11P. A special kind of lightbulb emits monochromatic light of wavelength 630 nm. Electric energy is supplied to it at the rate of 60 W, and the bulb is 93% efficient at converting that energy to light energy. How many photons are emitted by the bulb during its lifetime of 730 h?

12P. The beam emerging from a 1.5 W argon laser (λ = 515 nm) has a diameter d of 3.0 mm. The beam is focused by a lens system with an effective focal length f_L of 2.5 mm. The focused beam strikes a totally absorbing screen, where it forms a circular diffraction pattern whose central disk has a radius R given by $1.22 f_L\lambda/d$. It can be shown that 84% of the incident energy ends up within this central disk. At what rate are photons absorbed by the screen in the central disk of the diffraction pattern?

13P. An ultraviolet lamp emits light of wavelength 400 nm, at the rate (power) of 400 W. An infrared lamp emits light of wavelength 700 nm, also at the rate of 400 W. (a) Which lamp emits photons at the greater rate and (b) what is that greater rate? ssm

14P. A satellite in Earth orbit maintains a panel of solar cells of area 2.60 m^2 perpendicular to the direction of the Sun's light rays. The intensity of the light at the panel is 1.39 kW/m^2. (a) At what rate does solar energy arrive at the panel? (b) At what rate are solar photons absorbed by the panel? Assume that the solar radiation is monochromatic, with a wavelength of 550 nm, and that all the solar radiation striking the panel is absorbed. (c) How long would it take for a "mole of photons" to be absorbed by the panel?

15P. A 100 W sodium lamp (λ = 589 nm) radiates energy uniformly in all directions. (a) At what rate are photons emitted by the lamp? (b) At what distance from the lamp will a totally absorbing screen absorb photons at the rate of 1.00 photon/$cm^2 \cdot s$? (c) What is the photon flux (photons per unit area per unit time) on a small screen 2.00 m from the lamp? ssm www

SEC. 39-3 The Photoelectric Effect

16E. (a) The least energy needed to eject an electron from metallic sodium is 2.28 eV. Does sodium show a photoelectric effect for red light, with λ = 680 nm? (b) What is the cutoff wavelength for photoelectric emission from sodium? To what color does that correspond?

17E. You wish to pick an element for a photocell that will operate via the photoelectric effect with visible light. Which of the following are suitable (work functions are in parentheses): tantalum (4.2 eV), tungsten (4.5 eV), aluminum (4.2 eV), barium (2.5 eV), lithium (2.3 eV)?

18E. The work functions for potassium and cesium are 2.25 and 2.14 eV, respectively. (a) Will the photoelectric effect occur for either of these elements with incident light of wavelength 565 nm? (b) With light of wavelength 518 nm?

19E. Light strikes a sodium surface, causing photoelectric emission. The stopping potential for the ejected electrons is 5.0 V, and the work function of sodium is 2.2 eV. What is the wavelength of the incident light? ssm

20E. Find the maximum kinetic energy of electrons ejected from a certain material if the material's work function is 2.3 eV and the frequency of the incident radiation is 3.0×10^{15} Hz.

21E. The work function of tungsten is 4.50 eV. Calculate the speed of the fastest electrons ejected from a tungsten surface when light whose photon energy is 5.80 eV shines on the surface.

22P. (a) If the work function for a certain metal is 1.8 eV, what is the stopping potential for electrons ejected from the metal when light of wavelength 400 nm shines on the metal? (b) What is the maximum speed of the ejected electrons?

23P. Light of wavelength 200 nm shines on an aluminum surface. In aluminum, 4.20 eV is required to eject an electron. What is the kinetic energy of (a) the fastest and (b) the slowest ejected electrons? (c) What is the stopping potential for this situation? (d) What is the cutoff wavelength for aluminum? ssm

24P. The wavelength associated with the cutoff frequency for silver is 325 nm. Find the maximum kinetic energy of electrons ejected from a silver surface by ultraviolet light of wavelength 254 nm.

25P. An orbiting satellite can become charged by the photoelectric effect when sunlight ejects electrons from its outer surface. Satellites must be designed to minimize such charging. Suppose a satellite is coated with platinum, a metal with a very large work function ($\Phi = 5.32$ eV). Find the longest wavelength of incident sunlight that can eject an electron from the platinum.

26P. In a photoelectric experiment using a sodium surface, you find a stopping potential of 1.85 V for a wavelength of 300 nm and a stopping potential of 0.820 V for a wavelength of 400 nm. From these data find (a) a value for the Planck constant, (b) the work function Φ for sodium, and (c) the cutoff wavelength λ_0 for sodium.

27P. The stopping potential for electrons emitted from a surface illuminated by light of wavelength 491 nm is 0.710 V. When the incident wavelength is changed to a new value, the stopping potential is 1.43 V. (a) What is this new wavelength? (b) What is the work function for the surface? ssm www

28P. In about 1916, R. A. Millikan found the following stopping-potential data for lithium in his photoelectric experiments:

Wavelength (nm)	433.9	404.7	365.0	312.5	253.5
Stopping potential (V)	0.55	0.73	1.09	1.67	2.57

Use these data to make a plot like Fig. 39-2 (which is for sodium) and then use the plot to find (a) the Planck constant and (b) the work function for lithium.

29P. Suppose the *fractional efficiency* of a cesium surface (with work function 1.80 eV) is 1.0×10^{-16}; that is, on average one electron is ejected for every 10^{16} photons that reach the surface. What would be the current of electrons ejected from such a surface if it were illuminated with 600 nm light from a 2.00 mW laser and all the ejected electrons took part in the charge flow?

30P. X rays with a wavelength of 71 pm are directed onto a gold foil and eject tightly bound electrons from the gold atoms. The ejected electrons then move in circular paths of radius r in a region of uniform magnetic field $\vec{B}$, with $Br = 1.88 \times 10^{-4}$ T · m. Find (a) the maximum kinetic energy of those electrons and (b) the work done in removing them from the gold atoms.

SEC. 39-4 Photons Have Momentum

31E. Light of wavelength 2.4 pm is directed onto a target containing free electrons. (a) Find the wavelength of light scattered at 30° from the incident direction. (b) Do the same for a scattering angle of 120°. ssm

32E. (a) What is the momentum of a photon whose energy equals the rest energy of an electron? What are (b) the wavelength and (c) the frequency of the corresponding radiation?

33E. A certain x-ray beam has a wavelength of 35.0 pm. (a) What is the corresponding frequency? Calculate the corresponding (b) photon energy and (c) photon momentum.

34P. X rays of wavelength 0.010 nm are directed onto a target containing loosely bound electrons. For Compton scattering from one of those electrons, at an angle of 180°, what are (a) the Compton shift, (b) the corresponding change in photon energy, (c) the kinetic energy of the recoiling electron, and (d) the electron's direction of motion?

35P. Show, by analyzing a collision between a photon and a free electron (using relativistic mechanics), that it is impossible for a photon to transfer all its energy to a free electron (and thus for the photon to vanish).

36P. Gamma rays of photon energy 0.511 MeV are directed onto an aluminum target and are scattered in various directions by loosely bound electrons there. (a) What is the wavelength of the incident gamma rays? (b) What is the wavelength of gamma rays scattered at 90.0° to the incident beam? (c) What is the photon energy of the rays scattered in this direction?

37P. Calculate the Compton wavelength for (a) an electron and (b) a proton. What is the photon energy for an electromagnetic wave with a wavelength equal to the Compton wavelength of (c) the electron and (d) the proton? ssm

38P. What is the maximum wavelength shift for a Compton collision between a photon and a free *proton*?

39P. What percentage increase in wavelength leads to a 75% loss of photon energy in a photon–free electron collision? ssm www

40P. Calculate the percentage change in photon energy during a collision like that in Fig. 39-5 for $\phi = 90°$ and for radiation in (a) the microwave range, with $\lambda = 3.0$ cm; (b) the visible range, with $\lambda = 500$ nm; (c) the x-ray range, with $\lambda = 25$ pm; and (d) the gamma-ray range, with a gamma photon energy of 1.0 MeV. (e) What are your conclusions about the feasibility of detecting the Compton shift in these various regions of the electromagnetic spectrum, judging solely by the criterion of energy loss in a single photon–electron encounter?

41P. An electron of mass m and speed v "collides" with a gamma-ray photon of initial energy hf_0, as measured in the laboratory frame. The photon is scattered in the electron's direction of travel. Verify that the energy of the scattered photon, as measured in the laboratory frame, is

$$E = hf_0\left(1 + \frac{2hf_0}{mc^2}\sqrt{\frac{1 + v/c}{1 - v/c}}\right)^{-1}.$$

42P. Show that $\Delta E/E$, the fractional loss of energy of a photon during a collision with a particle of mass m, is given by

$$\frac{\Delta E}{E} = \frac{hf'}{mc^2}(1 - \cos\phi),$$

where E is the energy of the incident photon, f' is the frequency of the scattered photon, and ϕ is defined as in Fig. 39-5.

43P. Consider a collision between an x-ray photon of initial energy 50.0 keV and an electron at rest, in which the photon is scattered backward and the electron is knocked forward. (a) What is the energy of the back-scattered photon? (b) What is the kinetic energy of the electron?

44P. What would be (a) the Compton shift, (b) the fractional Compton shift, and (c) the change in photon energy for light of wavelength 590 nm scattering from a free, initially stationary electron if the scattering is at 90° to the direction of the incident beam? (d) Calculate the same quantities for x rays whose photon energy is 50.0 keV.

45P. What is the maximum kinetic energy of electrons knocked out of a thin copper foil by Compton scattering of an incident beam of 17.5 keV x rays?

46P. Derive Eq. 39-11, the equation for the Compton shift, from Eqs. 39-8, 39-9, and 39-10 by eliminating v and θ.

47P. Through what angle must a 200 keV photon be scattered by a free electron so that the photon loses 10% of its energy?

48P. Show that when a photon of energy E is scattered from a free electron at rest, the maximum kinetic energy of the recoiling electron is given by

$$K_{max} = \frac{E^2}{E + mc^2/2}.$$

SEC. 39-6 Electrons and Matter Waves

49E. Using the classical equations for momentum and kinetic energy, show that an electron's de Broglie wavelength in nanometers can be written as $\lambda = 1.226/\sqrt{K}$, in which K is the electron's kinetic energy in electron-volts. ssm

50E. A bullet of mass 40 g travels at 1000 m/s. Although the bullet is clearly too large to be treated as a matter wave, determine what Eq. 39-13 predicts for its de Broglie wavelength.

51E. In an ordinary television set, electrons are accelerated through a potential difference of 25.0 kV. What is the de Broglie wavelength of such electrons? (Relativity is not needed.) ssm

52E. Calculate the de Broglie wavelength of (a) a 1.00 keV electron, (b) a 1.00 keV photon, and (c) a 1.00 keV neutron.

53P. The wavelength of the yellow spectral emission line of sodium is 590 nm. At what kinetic energy would an electron have that wavelength as its de Broglie wavelength? ssm

54P. If the de Broglie wavelength of a proton is 100 fm, (a) what is the speed of the proton and (b) through what electric potential would the proton have to be accelerated to acquire this speed?

55P. Neutrons in thermal equilibrium with matter have an average kinetic energy of $(3/2)kT$, where k is the Boltzmann constant and T, which may be taken to be 300 K, is the temperature of the environment of the neutrons. (a) What is the average kinetic energy of such a neutron? (b) What is the corresponding de Broglie wavelength?

56P. An electron and a photon each have a wavelength of 0.20 nm. Calculate (a) their momenta and (b) their energies.

57P. (a) A photon has an energy of 1.00 eV, and an electron has a kinetic energy of that same amount. What are their wavelengths? (b) Repeat for an energy of 1.00 GeV. ssm www

58P. Consider a balloon filled with helium gas at room temperature and pressure. Calculate (a) the average de Broglie wavelength of the helium atoms and (b) the average distance between atoms under these conditions. The average kinetic energy of an atom is equal to $(3/2)kT$, where k is the Boltzmann constant. (c) Can the atoms be treated as particles under these conditions?

59P. Singly charged sodium ions are accelerated through a potential difference of 300 V. (a) What is the momentum acquired by such an ion? (b) What is its de Broglie wavelength? ssm

60P. (a) A photon and an electron both have a wavelength of 1.00 nm. What are the energy of the photon and the kinetic energy of the electron? (b) Repeat for a wavelength of 1.00 fm.

61P. The large electron accelerator at Stanford University provides a beam of electrons with kinetic energies of 50 GeV. Electrons with this energy have small wavelengths, suitable for probing the fine details of nuclear structure via scattering. What de Broglie wavelength does a 50 GeV electron have? How does this wavelength compare with the radius of an average nucleus, taken to be about 5.0 fm? (At this energy you can use the extreme relativistic relationship between momentum and energy, namely, $p = E/c$. This relationship, used for light, is justified when the kinetic energy of a particle is much greater than its rest energy, as in this case.)

62P. The existence of the atomic nucleus was discovered in 1911 by Ernest Rutherford, who properly interpreted some experiments in which a beam of alpha particles was scattered from a metal foil of atoms such as gold. (a) If the alpha particles had a kinetic energy of 7.5 MeV, what was their de Broglie wavelength? (b) Should the wave nature of the incident alpha particles have been taken into account in interpreting these experiments? The mass of an alpha particle is 4.00 u (atomic mass units), and its distance of closest approach to the nuclear center in these experiments was about 30 fm. (The wave nature of matter was not postulated until more than a decade after these crucial experiments were first performed.)

63P. A nonrelativistic particle is moving three times as fast as an electron. The ratio of the de Broglie wavelength of the particle to that of the electron is 1.813×10^{-4}. By calculating its mass, identify the particle. ssm

64P. The highest achievable resolving power of a microscope is limited only by the wavelength used; that is, the smallest item that can be distinguished has dimensions about equal to the wavelength. Suppose one wishes to "see" inside an atom. Assuming the atom to have a diameter of 100 pm, this means that one must be able to resolve a width of, say, 10 pm. (a) If an electron microscope is used, what minimum electron energy is required? (b) If a light microscope is used, what minimum photon energy is required? (c) Which microscope seems more practical? Why?

65P. What accelerating voltage would be required for the electrons of an electron microscope if the microscope is to have the same resolving power as could be obtained using 100 keV gamma rays? (See Problem 64.) ssm

SEC. 39-7 Schrödinger's Equation

66E. (a) Let $n = a + ib$ be a complex number, where a and b are real (positive or negative) numbers. Show that the product nn^* is always a positive real number. (b) Let $m = c + id$ be another complex number. Show that $|nm| = |n|\,|m|$.

67P. Show that Eq. 39-17 is indeed a solution of Eq. 39-16 by substituting $\psi(x)$ and its second derivative into Eq. 39-16 and noting that an identity results.

68P. (a) Write the wave function $\psi(x)$ displayed in Eq. 39-19 in the form $\psi(x) = a + ib$, where a and b are real quantities. (Assume that ψ_0 is real.) (b) Write the time-dependent wave function $\Psi(x, t)$ that corresponds to $\psi(x)$.

69P. Show that the angular wave number k for a nonrelativistic free particle of mass m can be written as

$$k = \frac{2\pi\sqrt{2mK}}{h},$$

in which K is the particle's kinetic energy. ssm

70P. Show that $|\psi|^2 = |\Psi|^2$, with ψ and Ψ related as in Eq. 39-14. That is, show that the probability density does not depend on the time variable.

71P. The function $\psi(x)$ displayed in Eq. 39-19 describes a free particle, for which we assumed that $U(x) = 0$ in Schrödinger's equation (Eq. 39-15). Assume now that $U(x) = U_0$ = a constant in that equation. Show that Eq. 39-19 is still a solution of Schrödinger's equation, with

$$k = \frac{2\pi}{h}\sqrt{2m(E - U_0)}$$

now giving the angular wave number k of the particle. ssm www

72P. Suppose that we had put $A = 0$ in Eq. 39-17 and relabeled B as ψ_0. What would the resulting wave function then describe? How, if at all, would Fig. 39-11 be altered?

73P. In Eq. 39-18 keep both terms, putting $A = B = \psi_0$. The equation then describes the superposition of two matter waves of equal amplitude, traveling in opposite directions. (Recall that this is the condition for a standing wave.) (a) Show that $|\Psi(x, t)|^2$ is then given by

$$|\Psi(x, t)|^2 = 2\psi_0^2[1 + \cos 2kx].$$

(b) Plot this function, and demonstrate that it describes the square of the amplitude of a standing matter wave. (c) Show that the nodes of this standing wave are located at

$$x = (2n + 1)(\tfrac{1}{4}\lambda), \qquad \text{where } n = 0, 1, 2, 3, \ldots$$

and λ is the de Broglie wavelength of the particle. (d) Write an expression for the most probable locations of the particle.

SEC. 39-8 Heisenberg's Uncertainty Principle

74E. Figure 39-11 shows that because of Heisenberg's uncertainty principle, it is not possible to assign an x coordinate to the position of the electron. (a) Can you assign a y or a z coordinate? (*Hint:* The momentum of the electron has no y or z component.) (b) Describe the extent of the matter wave in three dimensions.

75E. Imagine playing baseball in a universe (not ours!) where the Planck constant is 0.60 J · s. What would be the uncertainty in the position of a 0.50 kg baseball that is moving at 20 m/s along an axis if the uncertainty in the speed is 1.0 m/s? ssm

76E. The uncertainty in the position of an electron is given as 50 pm, which is about equal to the radius of a hydrogen atom. What is the least uncertainty in any simultaneous measurement of the momentum of this electron?

77P. Figure 39-11 shows a case in which the momentum p_x of a particle is fixed so that $\Delta p_x = 0$; then, from Heisenberg's uncertainty principle (Eq. 39-20), the position x of the particle is completely unknown. From the same principle it follows that the opposite is also true; that is, if the position of a particle is exactly known ($\Delta x = 0$), the uncertainty in its momentum is infinite.

Consider an intermediate case, in which the position of a particle is measured, not to infinite precision, but to within a distance of $\lambda/2\pi$, where λ is the particle's de Broglie wavelength. Show that the uncertainty in the (simultaneously measured) momentum is then equal to the momentum itself; that is, $\Delta p_x = p$. Under these circumstances, would a measured momentum of zero surprise you? What about a measured momentum of $0.5p$? Of $2p$? Of $12p$? ssm

78P. You will find in Chapter 40 that electrons cannot move in definite orbits within atoms, like the planets in our solar system. To see why, let us try to "observe" such an orbiting electron by using a light microscope to measure the electron's presumed orbital position with a precision of, say, 10 pm (a typical atom has a radius of about 100 pm). The wavelength of the light used in the microscope must then be about 10 pm. (a) What would be the photon energy of this light? (b) How much energy would such a photon impart to an electron in a head-on collision? (c) What do these results tell you about the possibility of "viewing" an atomic electron at two or more points along its presumed orbital path? (*Hint:* The outer electrons of atoms are bound to the atom by energies of only a few electron-volts.)

SEC. 39-9 Barrier Tunneling

79P. A proton and a deuteron (the latter has the same charge as a proton but twice the mass) strike a potential energy barrier that is 10 fm thick and 10 MeV high. Each particle has a kinetic energy of 3.0 MeV before it strikes the barrier. (a) What is the transmission coefficient for each? (b) What are their respective kinetic energies after they pass through the barrier (assuming that they do so)? (c) What are their respective kinetic energies if they are reflected from the barrier? ssm

80P. Consider a potential energy barrier like that of Fig. 39-12*a* but whose height U_0 is 6.0 eV and whose thickness L is 0.70 nm. What is the energy of an incident electron whose transmission coefficient is 0.0010?

81P. Consider the barrier-tunneling situation in Sample Problem 39-7. What percentage change in the transmission coefficient T occurs for a 1.0% change in (a) the barrier height, (b) the barrier thickness, and (c) the kinetic energy of the incident electron? ssm

82P. (a) Suppose a beam of 5.0 eV protons strikes a potential energy barrier of height 6.0 eV and thickness 0.70 nm, at a rate equivalent to a current of 1000 A. How long would you have to wait—on average—for one proton to be transmitted? (b) How long would you have to wait if the beam consisted of electrons rather than protons?

83P. A 1500 kg car moving at 20 m/s approaches a hill that is 24 m high and 30 m long. Although the car and hill are clearly too large to be treated as matter waves, determine what Eq. 39-21 predicts for the transmission coefficient of the car, as if it could tunnel through the hill as a matter wave. Treat the hill as a potential energy barrier where the potential energy is gravitational.

40 More About Matter Waves

This spectacular computer image was produced in 1993 at IBM's Almaden Research Center in California. The 48 peaks forming the circle mark the positions of individual atoms of iron on a specially prepared copper surface. The circle, which is about 14 nm in diameter, is called a *quantum corral*.

How do these atoms come to be arranged in a circle, and what are the ripples that are trapped within the corral?

The answer is in this chapter.

40-1 Atom Building

Early in the twentieth century nobody knew how the electrons in an atom are arranged, what their motions are, how atoms emit or absorb light, or even why atoms are stable. Without this knowledge it is not possible to understand how atoms combine to form molecules or stack up to form solids. As a consequence, the foundations of chemistry—including biochemistry, which underlies the nature of life itself—were more or less a mystery.

In 1926 all these questions and many others were answered with the development of quantum physics. Its basic premise is that moving electrons, protons, and particles of any kind are best viewed as matter waves, whose motions are governed by Schrödinger's equation. Although quantum theory also applies to massive particles, there is no point in treating baseballs, automobiles, planets, and such objects with quantum theory. For such massive, slow-moving objects, Newtonian physics and quantum physics yield the same answers.

Before we can apply quantum physics to the problem of atomic structure, we need to develop some insights by applying quantum ideas in a few simpler situations. These "practice problems" may seem artificial but, as you will see, they provide a firm foundation for understanding a very real problem that we shall analyze in Section 40-8—the structure of the hydrogen atom.

40-2 Waves on Strings and Matter Waves

In Chapter 17 we saw that waves of two kinds can be set up on a stretched string. If the string is so long that we can take it to be infinitely long, we can set up a *traveling wave* of essentially any frequency. However, if the stretched string has only a finite length, perhaps because it is rigidly clamped at both ends, we can set up only *standing waves* on it; further, these standing waves can have only discrete frequencies. In other words, confining the wave to a finite region of space leads to *quantization* of the motion—to the existence of discrete *states* for the wave, each state with a sharply defined frequency.

This observation applies to waves of all kinds, including matter waves. For matter waves, however, it is more convenient to deal with the energy E of the associated particle than with the frequency f of the wave. In all that follows we shall focus on the matter wave associated with an electron, but the results apply to any confined matter wave.

Consider the matter wave associated with an electron moving in the positive x direction and subject to no net force—a so-called *free particle*. The energy of such an electron can have any reasonable value, just as a wave traveling along a stretched string of infinite length can have any reasonable frequency.

Consider next the matter wave associated with an atomic electron, perhaps the *valence* (least tightly bound) electron in a sodium atom. Such an electron—held within the atom by the attractive Coulomb force between it and the positively charged nucleus—is *not* a free particle. It can exist only in a set of discrete states, each having a discrete energy E. This sounds much like the discrete states and quantized frequencies that are available to a stretched string of finite length. For matter waves, then, as for waves of all kinds, we may state a **confinement principle:**

> Confinement of a wave leads to quantization—that is, to the existence of discrete states with discrete energies. The wave can have only those energies.

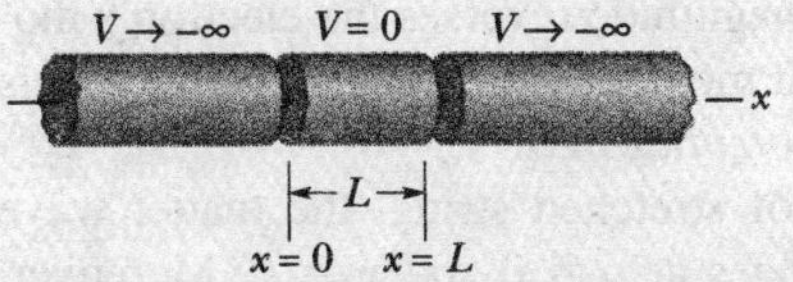

Fig. 40-1 The elements of an idealized "trap" designed to confine an electron to the central cylinder. We take the semi-infinitely long end cylinders to be at an infinitely great negative potential and the central cylinder to be at zero potential.

40-3 Energies of a Trapped Electron

One-Dimensional Traps

Here we examine the matter wave associated with an electron confined to a limited region of space. We do so by analogy with standing waves on a string of finite length, stretched along an x axis and confined between rigid supports. Because the supports are rigid, the two ends of the string are nodes, or points at which the string is always at rest. There may be other nodes along the string, but these two must always be present, as Fig. 17-21 shows.

The states, or discrete standing wave patterns in which the string can oscillate, are those for which the length L of the string is equal to an integer number of half-wavelengths. That is, the string can occupy only states for which

$$L = \frac{n\lambda}{2}, \qquad \text{for } n = 1, 2, 3, \ldots. \qquad (40\text{-}1)$$

Each value of n identifies a state of the oscillating string; using the language of quantum physics, we can call the integer n a **quantum number.**

For each state of the string permitted by Eq. 40-1, the transverse displacement of the string at any position x along the string is given by

$$y_n(x) = A \sin\left(\frac{n\pi}{L}x\right), \qquad \text{for } n = 1, 2, 3, \ldots, \qquad (40\text{-}2)$$

in which the quantum number n identifies the oscillation pattern, and A depends on the time at which you inspect the string. (Equation 40-2 is a short version of Eq. 17-47.) We see that for all values of n and for all times, there is a point of zero displacement (a node) at $x = 0$ and at $x = L$, as there must be. Figure 17-20 shows time exposures of such a stretched string for $n = 2$, 3, and 4.

Now let us turn our attention to matter waves. Our first problem is to physically confine an electron that is moving along the x axis so that it remains within a finite segment of that axis. Figure 40-1 shows a conceivable one-dimensional *electron trap*. It consists of two semi-infinitely long cylinders, each of which has an electric potential approaching $-\infty$; between them is a hollow cylinder of length L, which has an electric potential of zero. We put a single electron into this central cylinder to trap it.

The trap of Fig. 40-1 is easy to analyze but is not very practical. Single electrons *can*, however, be trapped in the laboratory with traps that are more complex in design but similar in concept. At the University of Washington, for example, a single electron has been held in a trap for months on end, permitting scientists to make extremely precise measurements of its properties.

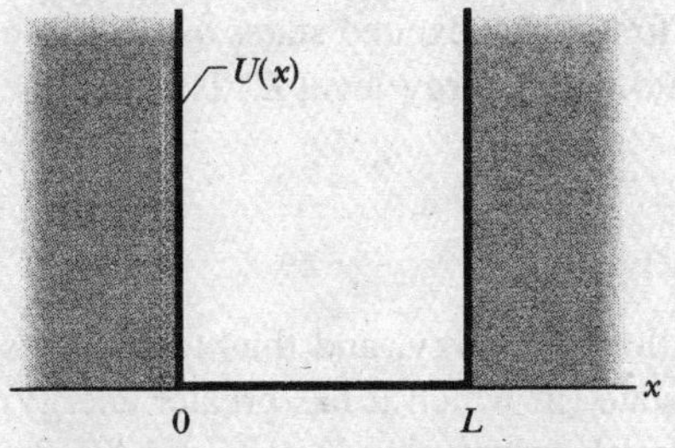

Fig. 40-2 The electric potential energy $U(x)$ of an electron confined to the central cylinder of the idealized trap of Fig. 40-1. We see that $U = 0$ for $0 < x < L$, and $U \to \infty$ for $x < 0$ and $x > L$.

Finding the Quantized Energies

Figure 40-2 shows the potential energy of the electron as a function of its position along the x axis of the idealized trap of Fig. 40-1. When the electron is in the central cylinder, its potential energy $U\ (= -eV)$ is zero because there the potential V is zero. If the electron could get outside this region, its potential energy would be positive and of infinite magnitude, because there $V \to -\infty$. We call the potential energy pattern of Fig. 40-2 an **infinitely deep potential energy well** or, for short, an *infinite potential well*. It is a "well" because an electron placed in the central cylinder of Fig. 40-1 cannot escape from it. As the electron approaches either end of the cylinder, a force of essentially infinite magnitude reverses the electron's motion, thus trapping it. Because the electron can move along only a single axis, this trap can be called a *one-dimensional infinite potential well.*

Just like the standing wave in a length of stretched string, the matter wave describing the confined electron must have nodes at $x = 0$ and $x = L$. Moreover, Eq. 40-1 applies to such a matter wave if we interpret λ in that equation as the de Broglie wavelength associated with the moving electron.

The de Broglie wavelength λ is defined in Eq. 39-13 as $\lambda = h/p$, where p is the magnitude of the electron's momentum. This magnitude p is related to the electron's kinetic energy K by $p = \sqrt{2mK}$, where m is the mass of the electron. For an electron moving within the central cylinder of Fig. 40-1, where $U = 0$, the total (mechanical) energy E is equal to the kinetic energy. Hence, we can write the de Broglie wavelength of this electron as

$$\lambda = \frac{h}{p} = \frac{h}{\sqrt{2mE}}. \tag{40-3}$$

If we substitute Eq. 40-3 into Eq. 40-1 and solve for the energy E, we find that E depends on n according to

$$E_n = \left(\frac{h^2}{8mL^2}\right)n^2, \quad \text{for } n = 1, 2, 3, \ldots. \tag{40-4}$$

The integer n here is the quantum number of the electron's quantum state in the trap.

Equation 40-4 tells us something important: Because the electron is confined to the trap, it can have only the energies given by the equation. It *cannot* have an energy that is, say, halfway between the values for $n = 1$ and $n = 2$. Why this restriction? Because an electron is a matter wave. Were it, instead, a particle as assumed in classical physics, it could have *any* value of energy while it is confined to the trap.

Figure 40-3 is a graph showing the lowest five allowed energy values for an electron in an infinite well with $L = 100$ pm (about the size of a typical atom). The values are called *energy levels,* and they are drawn in Fig. 40-3 as levels, or steps, on a ladder, in an *energy-level diagram.* Energy is plotted vertically; nothing is plotted horizontally.

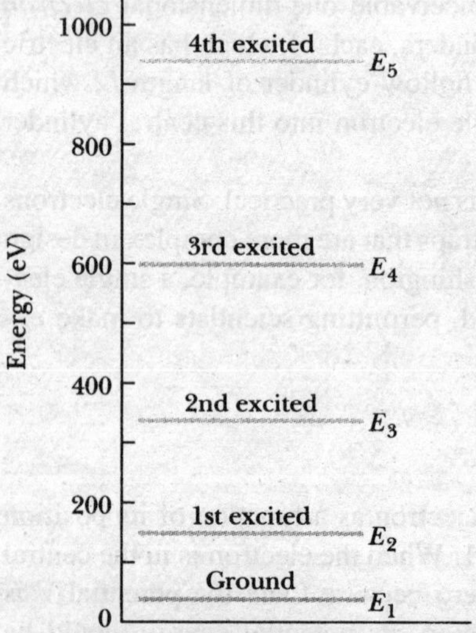

Fig. 40-3 Several of the allowed energies given by Eq. 40-4 for an electron confined to the infinite well of Fig. 40-2. Here width $L = 100$ pm. Such a plot is called an *energy-level diagram.*

The quantum state with the lowest possible energy level E_1 allowed by Eq. 40-4, with quantum number $n = 1$, is called the *ground state* of the electron. The electron tends to be in this lowest energy state. All the quantum states with greater energies (corresponding to quantum numbers $n = 2$ or greater), are called *excited states* of the electron. The state with energy level E_2, for quantum number $n = 2$, is called the *first excited state* because it is the first of the excited states as we move up the energy-level diagram. Similarly, the state with energy level E_3 is called the *second excited state.*

Energy Changes

A trapped electron tends to have the lowest allowed energy, and thus to be in its ground state. It can be changed to an excited state (in which it has greater energy) only if an external source provides the additional energy that is required for the change. Let E_{low} be the initial energy of the electron, and E_{high} be the greater energy in a state that is higher on its energy-level diagram. Then the amount of energy that is required for the electron's change of state is

$$\Delta E = E_{\text{high}} - E_{\text{low}}. \tag{40-5}$$

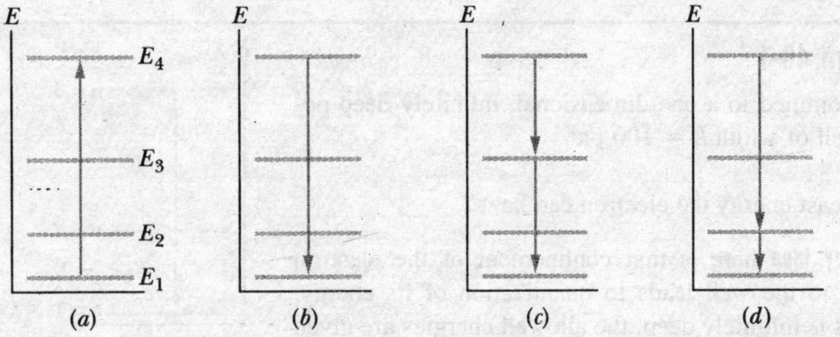

Fig. 40-4 *(a)* Excitation of a trapped electron from the energy level of its ground state to the level of its third excited state. *(b)*–*(d)* Three of four possible ways the electron can de-excite to return to the energy level of its ground state. (Which way is not shown?)

An electron that receives such energy is said to make a *quantum jump* (or *transition*), or to be *excited* from the lower-energy state to the higher-energy state. Figure 40-4*a* represents a quantum jump from the ground state (with energy level E_1) to the third excited state (with energy level E_4). As shown, the jump *must* be from one energy level to another but it can bypass one or more intermediate energy levels.

One way an electron can gain energy to make a quantum jump up to a greater energy level is to absorb a photon. However, this absorption and quantum jump can occur only if the following condition is met:

If a confined electron is to absorb a photon, the energy hf of the photon must equal the energy difference ΔE between the initial energy level of the electron and a higher level.

Thus, excitation by the absorption of light requires that

$$hf = \Delta E = E_{\text{high}} - E_{\text{low}}. \tag{40-6}$$

When an electron reaches an excited state, it does not stay there but quickly *de-excites* by decreasing its energy. Figures 40-4*b* to *d* represent some of the possible quantum jumps down from the energy level of the third excited state. The electron can reach its ground-state level either with one direct quantum jump (Fig. 40-4*b*) or with shorter jumps via intermediate energy levels (Figs. 40-4*c* and *d*).

One way in which an electron can decrease its energy is by emitting a photon, but only if the following condition is met:

If a confined electron emits a photon, the energy hf of that photon must equal the energy difference ΔE between the initial energy level of the electron and a lower level.

Thus, Eq. 40-6 applies to both the absorption and the emission of light by a confined electron. That is, the absorbed or emitted light can have only certain values of hf, and thus only certain values of frequency f and wavelength λ.

Aside: Although Eq. 40-6 and what we have discussed about photon absorption and emission can be applied to physical (real) electron traps, they actually cannot be applied to one-dimensional (unreal) electron traps. The reason involves the need to conserve angular momentum in a photon absorption or emission process. In this book, we shall neglect that need and use Eq. 40-6 even for one-dimensional traps.

CHECKPOINT 1: Rank the following pairs of quantum states for an electron confined to an infinite well according to the energy differences between the states, greatest first: (a) $n = 3$ to $n = 1$, (b) $n = 5$ to $n = 4$, (c) $n = 4$ to $n = 3$.

Sample Problem 40-1

An electron is confined to a one-dimensional, infinitely deep potential energy well of width $L = 100$ pm.

(a) What is the least energy the electron can have?

SOLUTION: The **Key Idea** here is that confinement of the electron (a matter wave) to the well leads to quantization of its energy. Because the well is infinitely deep, the allowed energies are given by Eq. 40-4 ($E_n = (h^2/8mL^2)n^2$), with the quantum number n a positive integer. Here, the collection of constants in front of n^2 in Eq. 40-4 is evaluated as

$$\frac{h^2}{8mL^2} = \frac{(6.63 \times 10^{-34}\ \text{J}\cdot\text{s})^2}{(8)(9.11 \times 10^{-31}\ \text{kg})(100 \times 10^{-12}\ \text{m})^2}$$
$$= 6.031 \times 10^{-18}\ \text{J}. \qquad (40\text{-}7)$$

The least energy of the electron corresponds to the least quantum number, which is $n = 1$ for the ground state of the electron. Thus, Eqs. 40-4 and 40-7 give us

$$E_1 = \left(\frac{h^2}{8mL^2}\right)n^2 = (6.031 \times 10^{-18}\ \text{J})(1^2)$$
$$\approx 6.03 \times 10^{-18}\ \text{J} = 37.7\ \text{eV}. \qquad \text{(Answer)}$$

(b) How much energy must be transferred to the electron if it is to make a quantum jump from its ground state to its second excited state?

SOLUTION: *First a caution:* Note that, from Fig. 40-3, the *second* excited state corresponds to the *third* energy level, with quantum number $n = 3$. Then one **Key Idea** is that if the electron is to jump from the $n = 1$ level to the $n = 3$ level, the required change in its energy is, from Eq. 40-5,

$$\Delta E_{31} = E_3 - E_1. \qquad (40\text{-}8)$$

A second **Key Idea** is that the energies E_3 and E_1 depend on the quantum number n, according to Eq. 40-4. Therefore, substituting that equation into Eq. 40-8 for energies E_3 and E_1 and using Eq. 40-7 lead to

$$\Delta E_{31} = \left(\frac{h^2}{8mL^2}\right)(3)^2 - \left(\frac{h^2}{8mL^2}\right)(1)^2 = \frac{h^2}{8mL^2}(3^2 - 1^2)$$
$$= (6.031 \times 10^{-18}\ \text{J})(8)$$
$$= 4.83 \times 10^{-17}\ \text{J} = 302\ \text{eV}. \qquad \text{(Answer)}$$

(c) If the electron gains the energy for the jump from energy level E_1 to energy level E_3 by absorbing light, what light wavelength is required?

SOLUTION: One **Key Idea** here is that if light is to transfer energy to the electron, the transfer must be by photon absorption. A second **Key Idea** is that the photon's energy must equal the energy difference ΔE between the initial energy level of the electron and a higher level, according to Eq. 40-6 ($hf = \Delta E$). Otherwise, a photon *cannot* be absorbed. Substituting c/λ for f, we can rewrite Eq. 40-6 as

$$\lambda = \frac{hc}{\Delta E}. \qquad (40\text{-}9)$$

For the energy difference ΔE_{31} we found in (b), this equation gives us

$$\lambda = \frac{hc}{\Delta E_{31}}$$
$$= \frac{(6.63 \times 10^{-34}\ \text{J}\cdot\text{s})(3.0 \times 10^8\ \text{m/s})}{4.83 \times 10^{-17}\ \text{J}}$$
$$= 4.12 \times 10^{-9}\ \text{m}. \qquad \text{(Answer)}$$

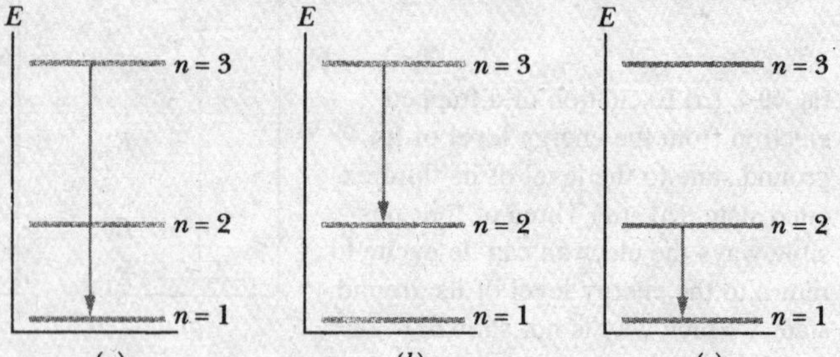

Fig. 40-5 Sample Problem 40-1. De-excitation from the second excited state to the ground state either directly (*a*) or via the first excited state (*b, c*).

(d) Once the electron has been excited to the second excited state, what wavelengths of light can it emit by de-excitation?

SOLUTION: We have three **Key Ideas** here:

1. The electron tends to de-excite, rather than remain in an excited state, until it reaches the ground state ($n = 1$).
2. If the electron is to de-excite, it must lose just enough energy to jump to a lower energy level.
3. If it is to lose energy by emitting light, then the loss of energy must be by emission of a photon.

Starting in the second excited state (at the $n = 3$ level), the electron can reach the ground state ($n = 1$) by *either* making a quantum jump directly to the ground-state energy level (Fig. 40-5*a*) or by making two *separate* jumps by way of the $n = 2$ level (Figs. 40-5*b* and *c*).

The direct jump involves the same energy difference ΔE_{31} we found in (c). Then the wavelength is the same as we calculated in (c)—except now the wavelength is for light that is emitted, not absorbed. Thus, the electron can jump directly to the ground state by emitting light of wavelength

$$\lambda = 4.12 \times 10^{-9}\ \text{m}. \qquad \text{(Answer)}$$

Following the procedure of part (b), you can show that the energy differences for the jumps of Figs. 40-5*b* and *c* are

$$\Delta E_{32} = 3.016 \times 10^{-17}\ \text{J} \quad \text{and} \quad \Delta E_{21} = 1.809 \times 10^{-17}\ \text{J}.$$

From Eq. 40-9, we then find that the wavelength of the light emitted in the first of these jumps (from $n = 3$ to $n = 2$) is

$$\lambda = 6.60 \times 10^{-9}\ \text{m}, \qquad \text{(Answer)}$$

and the wavelength of the light emitted in the second of these jumps (from $n = 2$ to $n = 1$) is

$$\lambda = 1.10 \times 10^{-8}\ \text{m}. \qquad \text{(Answer)}$$

40-4 Wave Functions of a Trapped Electron

If we solve Schrödinger's equation for an electron trapped in a one-dimensional infinite potential well of width L, we find that the wave functions for the electron are given by

$$\psi_n(x) = A \sin\left(\frac{n\pi}{L}x\right), \qquad \text{for } n = 1, 2, 3, \ldots, \tag{40-10}$$

for $0 \leq x \leq L$ (the wave function is zero outside that range). We shall soon evaluate the amplitude constant A in this equation.

Note that the wave functions $\psi_n(x)$ have the same form as the displacement functions $y_n(x)$ for a standing wave on a string stretched between rigid supports (see Eq. 40-2). We can picture an electron trapped in a one-dimensional well between infinite-potential walls as being a standing matter wave.

Probability of Detection

The wave function $\psi_n(x)$ cannot be detected or directly measured in any way—we cannot simply look inside the well to see the wave, like we can see a wave in a bathtub of water. All we can do is insert a probe of some kind, to try to detect the electron. At the instant of detection, the electron would materialize at the point of detection, at some position along the x axis within the well.

If we repeated this detection procedure at many positions throughout the well, we would find that the probability of detecting the electron is related to the probe's position x in the well. In fact, they are related by the *probability density* $\psi_n^2(x)$. Recall from Section 39-7 that in general the probability that a particle can be detected in a specified infinitesimal volume centered on a specified point is proportional to $|\psi_n^2|$. Here, with the electron trapped in a one-dimensional well, we are concerned only with detection of the electron along the x axis. Thus, the probability density $\psi_n^2(x)$ here is a probability per unit length along the x axis. (We can omit the absolute value sign here, because $\psi_n(x)$ in Eq. 40-10 is a real quantity, not a complex one.) The probability $p(x)$ that an electron can be detected at position x within the well is

$$\begin{pmatrix}\text{probability } p(x) \\ \text{of detection in width } dx \\ \text{centered on position } x\end{pmatrix} = \begin{pmatrix}\text{probability density } \psi_n^2(x) \\ \text{at position } x\end{pmatrix}(\text{width } dx),$$

or

$$p(x) = \psi_n^2(x)\, dx. \tag{40-11}$$

From Eq. 40-10, we see that the probability density $\psi_n^2(x)$ for the trapped electron is

$$\psi_n^2(x) = A^2 \sin^2\left(\frac{n\pi}{L}x\right), \qquad \text{for } n = 1, 2, 3, \ldots, \tag{40-12}$$

for the range $0 \leq x \leq L$ (the probability density is zero outside that range). Figure 40-6 shows $\psi_n^2(x)$ for $n = 1$, 2, 3, and 15 for an electron in an infinite well whose width L is 100 pm.

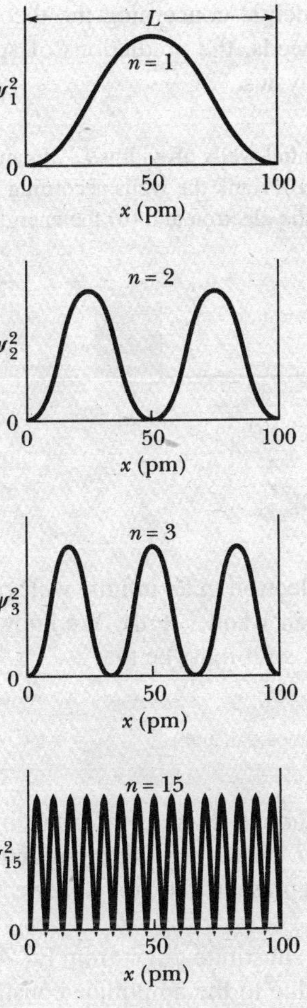

Fig. 40-6 The probability density $\psi_n^2(x)$ for four states of an electron trapped in a one-dimensional infinite well; their quantum numbers are $n = 1$, 2, 3, and 15. The electron is most likely to be found where $\psi_n^2(x)$ is greatest, and least likely to be found where $\psi_n^2(x)$ is least.

To find the probability that the electron can be detected in any finite section of the well—say, between point x_1 and point x_2—we must integrate $p(x)$ between those points. Thus, from Eqs. 40-11 and 40-12,

$$\left(\begin{matrix}\text{probability of detection}\\ \text{between } x_1 \text{ and } x_2\end{matrix}\right) = \int_{x_1}^{x_2} p(x)$$

$$= \int_{x_1}^{x_2} A^2 \sin^2\left(\frac{n\pi}{L}x\right) dx. \qquad (40\text{-}13)$$

If classical physics prevailed, we would expect the trapped electron to be detectable with equal probabilities in all parts of the well. From Fig. 40-6 we see that it is not. For example, inspection of that figure or of Eq. 40-12 shows that for the state with $n = 2$, the electron is most likely to be detected near $x = 25$ pm and $x = 75$ pm. It can be detected with near-zero probability near $x = 0$, $x = 50$ pm, and $x = 100$ pm.

The case of $n = 15$ in Fig. 40-6 suggests that as n increases, the probability of detection becomes more and more uniform across the well. This result is an instance of a general principle called the **correspondence principle:**

> At large enough quantum numbers, the predictions of quantum physics merge smoothly with those of classical physics.

This principle, first advanced by Danish physicist Niels Bohr, holds for all quantum predictions. It should remind you of a similar principle concerning the theory of relativity—namely, that at low enough particle speeds, the predictions of special relativity merge smoothly with those of classical physics.

CHECKPOINT 2: The figure shows three infinite potential wells of widths L, $2L$, and $3L$; each contains an electron in the state for which $n = 10$. Rank the wells according to (a) the number of maxima for the probability density of the electron and (b) the energy of the electron, greatest first.

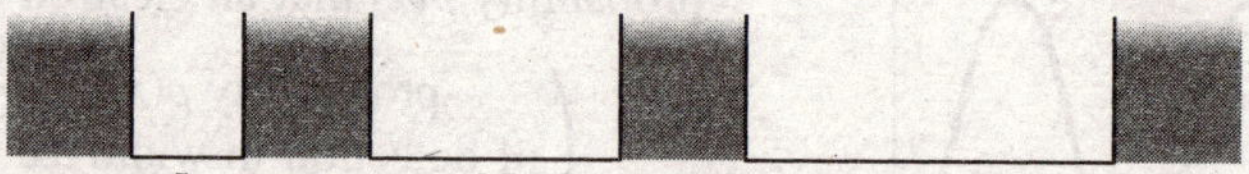

Normalization

The product $\psi_n^2(x)\, dx$ gives the probability that an electron in an infinite well can be detected in the interval of the x axis that lies between x and $x + dx$. We know that the electron must be *somewhere* in the infinite well, so it must be that

$$\int_{-\infty}^{+\infty} \psi_n^2(x)\, dx = 1 \quad \text{(normalization equation)}, \qquad (40\text{-}14)$$

since the probability 1 corresponds to certainty. Although the integral is taken over the entire x axis, only the region from $x = 0$ to $x = L$ makes any contribution to the probability. Graphically, the integral in Eq. 40-14 represents the area under each of the plots of Fig. 40-6.

In Sample Problem 40-2 we shall see that if we substitute $\psi_n^2(x)$ from Eq. 40-12 into Eq. 40-14, it is possible to assign a specific value to the amplitude constant A

that appears in Eq. 40-12; namely, $A = \sqrt{2/L}$. This process of using Eq. 40-14 to evaluate the amplitude of a wave function is called **normalizing** the wave function. The process applies to *all* one-dimensional wave functions.

Zero-Point Energy

Substituting $n = 1$ in Eq. 40-4 defines the state of lowest energy for an electron in an infinite potential well, the ground state. That is the state the confined electron will occupy unless energy is supplied to it to raise it to an excited state.

The question arises: Why can't we include $n = 0$ among the possibilities listed for n in Eq. 40-4? Putting $n = 0$ in this equation would indeed yield a ground-state energy of zero. However, putting $n = 0$ in Eq. 40-12 would also yield $\psi_n^2(x) = 0$ for all x, which we can interpret only to mean that there is no electron in the well. We know that there is, so $n = 0$ is not a possible quantum number.

It is an important conclusion of quantum physics that confined systems cannot exist in states with zero energy. They must always have a certain minimum energy called the **zero-point energy.**

We can make the zero-point energy as small as we like by making the infinite well wider—that is, by increasing L in Eq. 40-4 for $n = 1$. In the limit as $L \to \infty$, the zero-point energy E_1 approaches zero. In this limit, however, with an infinitely wide well, the electron is a free particle, no longer confined in the x direction. Also, because the energy of a free particle is not quantized, that energy can have any value, including zero. Only a confined particle must have a finite zero-point energy and can never be at rest.

CHECKPOINT 3: Each of the following particles is confined to an infinite well, and all four wells have the same width: (a) an electron, (b) a proton, (c) a deuteron, and (d) an alpha particle. Rank their zero-point energies, greatest first. The particles are listed in order of increasing mass.

Sample Problem 40-2

Evaluate the amplitude constant A in Eq. 40-10 for an infinite potential well extending from $x = 0$ to $x = L$.

SOLUTION: The **Key Idea** here is that the wave functions of Eq. 40-10 must satisfy the normalization requirement of Eq. 40-14, which states that the probability that the electron can be detected somewhere along the x axis is 1. Substituting Eq. 40-10 into Eq. 40-14 and taking the constant A outside the integral yield

$$A^2 \int_0^L \sin^2\left(\frac{n\pi}{L}x\right) dx = 1. \quad (40\text{-}15)$$

We have changed the limits of the integral from $-\infty$ and $+\infty$ to 0 and L because the wave function is zero outside these new limits (so there's no need to integrate out there).

We can simplify the indicated integration by changing the variable from x to the dimensionless variable y, where

$$y = \frac{n\pi}{L}x, \quad (40\text{-}16)$$

hence

$$dx = \frac{L}{n\pi}dy.$$

When we change the variable, we must also change the integration limits (again). Equation 40-16 tells us that $y = 0$ when $x = 0$ and that $y = n\pi$ when $x = L$, so 0 and $n\pi$ are our new limits. With all these substitutions, Eq. 40-15 becomes

$$A^2 \frac{L}{n\pi} \int_0^{n\pi} (\sin^2 y)\, dy = 1.$$

We can use integral 11 in Appendix E to evaluate the integral, obtaining the equation

$$\frac{A^2 L}{n\pi}\left[\frac{y}{2} - \frac{\sin 2y}{4}\right]_0^{n\pi} = 1.$$

Evaluating at the limits yields

$$\frac{A^2 L}{n\pi}\frac{n\pi}{2} = 1,$$

so

$$A = \sqrt{\frac{2}{L}}. \quad \text{(Answer)} \quad (40\text{-}17)$$

This result tells us that the dimension for A^2, and thus for $\psi_n^2(x)$, is an inverse length. This is appropriate because the probability density of Eq. 40-12 is a probability *per unit length.*

Sample Problem 40-3

A ground-state electron is trapped in the one-dimensional infinite potential well of Fig. 40-2, with width $L = 100$ pm.

(a) What is the probability that the electron can be detected in the left one-third of the well (between $x_1 = 0$ and $x_2 = L/3$)?

SOLUTION: One **Key Idea** here is that if we probe the left one-third of the well, there is no guarantee that we will detect the electron. However, we can calculate the probability of detecting it with the integral of Eq. 40-13. A second **Key Idea** is that the probability very much depends on which state the electron is in—that is, the value of the electron's quantum number n. Because here the electron is in the ground state, we set $n = 1$ in Eq. 40-13.

We also set the limits of integration as the positions $x_1 = 0$ and $x_2 = L/3$ and, from Sample Problem 40-2, set the amplitude constant A as $\sqrt{2/L}$. We then see that

$$\begin{pmatrix}\text{probability of detection}\\ \text{in left one-third}\end{pmatrix} = \int_0^{L/3} \frac{2}{L}\sin^2\left(\frac{1\pi}{L}x\right) dx.$$

We could find this probability by substituting 100×10^{-12} m for L and then using a graphing calculator or a computer math package to evaluate the integral. Instead, we shall follow the steps of Sample Problem 40-2. From Eq. 40-16, we obtain for the new integration variable y,

$$y = \frac{\pi}{L}x \quad \text{and} \quad dx = \frac{L}{\pi}dy.$$

From the first of these equations, we find the new limits of integration to be $y_1 = 0$ for $x_1 = 0$ and $y_2 = \pi/3$ for $x_2 = L/3$. We then must evaluate

$$\text{probability} = \left(\frac{2}{L}\right)\left(\frac{L}{\pi}\right)\int_0^{\pi/3} (\sin^2 y)\, dy.$$

Using integral 11 in Appendix E, we then find

$$\text{probability} = \frac{2}{\pi}\left(\frac{y}{2} - \frac{\sin 2y}{4}\right)_0^{\pi/3} = 0.20.$$

Thus, we have

$$\begin{pmatrix}\text{probability of detection}\\ \text{in left one-third}\end{pmatrix} = 0.20. \qquad \text{(Answer)}$$

That is, if we repeatedly probe the left one-third of the well, then on average we can detect the electron with 20% of the probes.

(b) What is the probability that the electron can be detected in the middle one-third of the well (between $x_1 = L/3$ and $x_2 = 2L/3$)?

SOLUTION: We now know that the probability of detection in the left one-third of the well is 0.20. A **Key Idea** here is that by symmetry, the probability of detection in the right one-third of the well is also 0.20. A second **Key Idea** is that because the electron is certainly in the well, the probability of detection in the entire well is 1.0. Thus, the probability of detection in the middle one-third of the well is

$$\begin{pmatrix}\text{probability of detection}\\ \text{in middle one-third}\end{pmatrix} = 1 - 0.20 - 0.20$$

$$= 0.60. \qquad \text{(Answer)}$$

40-5 An Electron in a Finite Well

A potential energy well of infinite depth is an idealization. Figure 40-7 shows a realizable potential energy well—one in which the potential energy of an electron outside the well is not infinitely great but has a finite positive value U_0, called the **well depth.** The analogy between waves on a stretched string and matter waves fails us for wells of finite depth because we can no longer be sure that matter wave nodes exist at $x = 0$ and at $x = L$. (As we shall see, they don't.)

To find the wave functions describing the quantum states of an electron in the finite well of Fig. 40-7, we *must* resort to Schrödinger's equation, the basic equation of quantum physics. From Section 39-7 recall that, for motion in one dimension, we use Schrödinger's equation in the form of Eq. 39-15:

$$\frac{d^2\psi}{dx^2} + \frac{8\pi^2 m}{h^2}[E - U(x)]\psi = 0. \qquad (40\text{-}18)$$

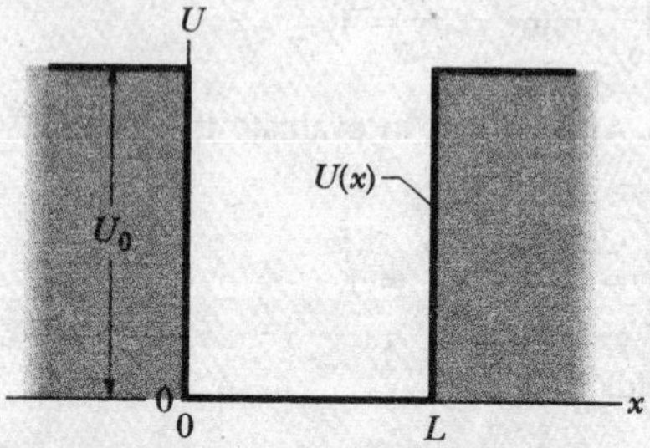

Fig. 40-7 A *finite* potential energy well. The depth of the well is U_0 and its width is L. As in the infinite potential well of Fig. 40-2, the motion of the trapped electron is restricted to the x direction.

Rather than attempting to solve this equation for the finite well, we simply state the results for particular numerical values of U_0 and L. Figure 40-8 shows these results as graphs of $\psi_n^2(x)$, the probability density, for a well with $U_0 = 450$ eV and $L = 100$ pm.

The probability density $\psi_n^2(x)$ for each graph in Fig. 40-8 satisfies Eq. 40-14, the normalization equation, so we know that the areas under all three probability density plots are numerically equal to 1.

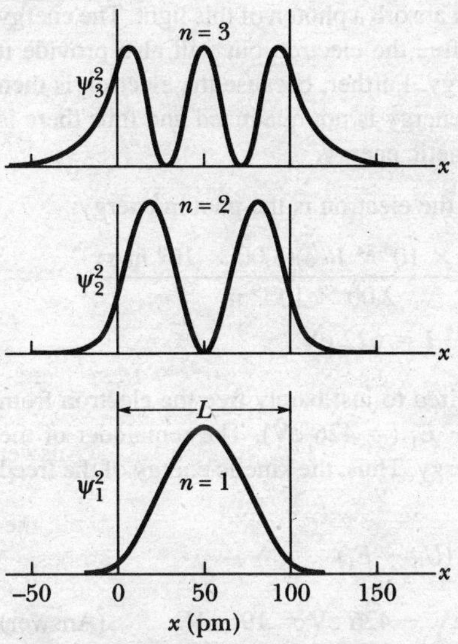

Fig. 40-8 Probability densities $\psi_n^2(x)$ for an electron confined to a finite potential well of depth $U_0 = 450$ eV and width $L = 100$ pm. The only quantum states the electron can have in this well are those that have quantum numbers $n = 1, 2,$ and 3.

If you compare Fig. 40-8, for a finite well, with Fig. 40-6, for an infinite well, you will see one striking difference: For a finite well, the electron matter wave penetrates the walls of the well—into a region in which Newtonian mechanics says the electron cannot exist. This penetration should not be surprising, because we saw in Section 39-9 that an electron can tunnel through a potential energy barrier. "Leaking" into the walls of a finite potential energy well is a similar phenomenon. From the plots of ψ^2 in Fig. 40-8, we see that the leakage is greater for greater values of quantum number n.

Because a matter wave *does* leak into the walls of a finite well, the wavelength λ for any given quantum state is greater when the electron is trapped in a finite well than when it is trapped in an infinite well. Equation 40-3 then tells us that the energy E for an electron in any given state is less in the finite well than in the infinite well.

That fact allows us to approximate the energy-level diagram for an electron trapped in a finite well. As an example, we can approximate the diagram for the finite well of Fig. 40-8, which has width $L = 100$ pm and depth $U_0 = 450$ J. The energy-level diagram for an *infinite* well of that width is shown in Fig. 40-3. First we remove the portion of Fig. 40-3 above 450 J. Then we shift the remaining three energy levels down, shifting the level for $n = 3$ the most because the wave leakage into the walls is greatest for $n = 3$. The result is approximately the energy-level diagram for the finite well. The actual diagram is shown in Fig. 40-9.

In that figure, an electron with an energy greater than U_0 (= 450 J) has too much energy to be trapped in the finite well. Thus, it is not confined and its energy is not quantized; that is, its energy is not restricted to certain values. To reach this *nonquantized* portion of the energy-level diagram and thus to be free, a trapped electron must somehow obtain enough energy to have a mechanical energy of 450 J or greater.

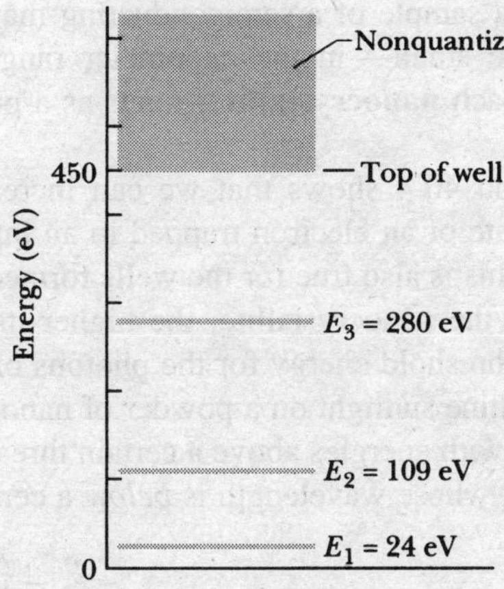

Fig. 40-9 The energy-level diagram corresponding to the probability densities of Fig. 40-8. If an electron is trapped in the finite potential well, it can have only the energies corresponding to $n = 1, 2,$ and 3. If it has an energy of 450 eV or greater, it is not trapped and its energy is not quantized.

Sample Problem 40-4

Suppose a finite well with $U_0 = 450$ eV and $L = 100$ pm confines a single electron in its ground state.

(a) What wavelength of light is needed to barely free the electron from the potential well by a single photon absorption?

SOLUTION: One **Key Idea** here is that for the electron to escape from the potential well, it must receive enough energy to put it into the nonquantized energy region of Fig. 40-9. Thus, it must end up with an energy of at least U_0 (= 450 eV). A second **Key Idea** is that the electron is initially in its ground state, with an energy of $E_1 = 24$ eV. Thus, to barely become free, it must receive an energy of

$$U_0 - E_1 = 450 \text{ eV} - 24 \text{ eV} = 426 \text{ eV}.$$

If it receives this energy from light, then it must absorb a photon with that much energy. From Eq. 40-6, with c/λ substituted for f, we can then write

$$\frac{hc}{\lambda} = U_0 - E_1,$$

from which we find

$$\lambda = \frac{hc}{U_0 - E_1}$$
$$= \frac{(6.63 \times 10^{-34}\ \text{J}\cdot\text{s})(3.00 \times 10^8\ \text{m/s})}{(426\ \text{eV})(1.60 \times 10^{-19}\ \text{J/eV})}$$
$$= 2.92 \times 10^{-9}\ \text{m} = 2.92\ \text{nm}. \qquad \text{(Answer)}$$

Thus, if the electron absorbs a photon from light of wavelength 2.92 nm, it just barely escapes the potential well.

(b) Can the electron, initially in the ground state, absorb light with a wavelength of 2.00 nm? If so, what then is the electron's energy?

SOLUTION: The **Key Ideas** here are these:

1. In (a) we found that light of 2.92 nm will just barely free the electron from the potential well.
2. We are now considering light with a shorter wavelength of 2.00 nm and thus a greater energy per photon ($hf = hc/\lambda$).
3. Hence, the electron *can* absorb a photon of this light. The energy transfer will not only free the electron but will also provide it with more kinetic energy. Further, because the electron is then no longer trapped, its energy is not quantized and thus there is no restriction on its kinetic energy.

The energy transferred to the electron is the photon energy:

$$hf = h\frac{c}{\lambda} = \frac{(6.63 \times 10^{-34}\ \text{J}\cdot\text{s})(3.00 \times 10^8\ \text{m/s})}{2.00 \times 10^{-9}\ \text{m}}$$
$$= 9.95 \times 10^{-17}\ \text{J} = 622\ \text{eV}.$$

From (a), the energy required to just barely free the electron from the potential well is $U_0 - E_1$ (= 426 eV). The remainder of the 622 eV goes to kinetic energy. Thus, the kinetic energy of the freed electron is

$$K = hf - (U_0 - E_1)$$
$$= 622\ \text{eV} - 426\ \text{eV} = 196\ \text{eV}. \qquad \text{(Answer)}$$

40-6 More Electron Traps

Here we discuss three types of artificial electron traps.

Nanocrystallites

Fig. 40-10 Two samples of powdered cadmium selenide, a semiconductor, differing only in the size of their granules. Each granule serves as an electron trap. The upper sample has the larger granules and consequently the smaller spacing between energy levels and the lower photon energy threshold for the absorption of light. Light not absorbed is scattered, causing the sample to scatter light of greater wavelength and appear red. The lower sample, because of its smaller granules, and consequently its larger level spacing and its larger energy threshold for absorption, appears yellow.

Perhaps the most direct way to construct a potential energy well in the laboratory is to prepare a sample of a semiconducting material in the form of a powder whose granules are small—in the nanometer range—and of uniform size. Each such granule—each **nanocrystallite**—acts as a potential well for the electrons trapped within it.

Equation 40-4 shows that we can increase the energy of the least energetic quantum state of an electron trapped in an infinite well by reducing the width L of that well. This is also true for the wells formed by individual nanocrystallites. Thus, the smaller the nanocrystallite, the higher its lowest available level—that is, the higher the threshold energy for the photons of light that it can absorb.

If we shine sunlight on a powder of nanocrystallites, the crystallites can absorb all photons with energies above a certain threshold energy E_t (= hf_t). Thus, they can absorb light whose wavelength is *below* a certain threshold λ_t, where

$$\lambda_t = \frac{c}{f_t} = \frac{ch}{E_t}. \qquad (40\text{-}19)$$

Since light not absorbed is scattered, our powder of nanocrystallites will scatter all wavelengths above λ_t.

We see the powder sample by the light it scatters back to our eyes. Thus, by controlling the size of the nanocrystallites in a sample, we can control the wavelengths of the light scattered by the sample, and hence the sample's color.

Figure 40-10 shows two samples of the semiconductor cadmium selenide, each consisting of a powder of nanocrystallites of uniform size. The upper sample scatters light at the red end of the spectrum. The lower sample differs from the upper sample *only* in that the lower sample is composed of smaller nanocrystallites. For this reason its threshold energy E_t is greater and, from Eq. 40-19, its threshold wavelength λ_t is shorter. The sample takes on a color of shorter wavelength—in this case yellow.

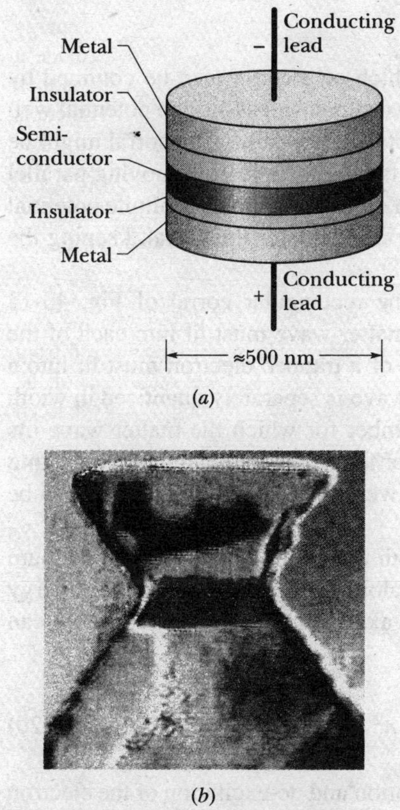

Fig. 40-11 A quantum dot, or "artificial atom." (*a*) A central semiconducting layer forms a potential energy well in which electrons are trapped. The lower insulating layer is thin enough to allow electrons to be added to or removed from the central layer by barrier tunneling if an appropriate voltage is applied between the leads. (*b*) A photograph of an actual quantum dot. The central purple band is the electron confinement region.

The striking contrast in color between the two samples is compelling evidence of the quantization of the energies of trapped electrons and the dependence of these energies on the size of the electron trap.

Quantum Dots

The highly developed techniques used to fabricate computer chips can be used to construct, atom by atom, individual potential energy wells that behave, in many respects, like artificial atoms. These **quantum dots,** as they are usually called, have promising applications in electron optics and computer technology.

In one such arrangement, a "sandwich" is fabricated in which a thin layer of a semiconducting material, shown in purple in Fig. 40-11*a*, is deposited between two insulating layers, one of which is much thinner than the other. Metal end caps with conducting leads are added at both ends. The materials are chosen to ensure that the potential energy of an electron in the central layer is less than it is in the two insulating layers, causing the central layer to act as a potential energy well. Figure 40-11*b* is a photograph of an actual quantum dot; the well in which individual electrons can be trapped is the purple region.

The lower (but not the upper) insulating layer in Fig. 40-11*a* is thin enough to permit electrons to tunnel through it if an appropriate potential difference is applied between the leads. In this way the number of electrons confined to the well can be controlled. The arrangement does indeed behave like an artificial atom with the property that the number of electrons it contains can be controlled. Quantum dots can be constructed in two-dimensional arrays that could well form the basis for computing systems of great speed and storage capacity.

Quantum Corrals

When a scanning tunneling microscope (described in Section 39-9 and Fig. 39-14) is in operation, its tip exerts a small force on isolated atoms that may be located on an otherwise smooth surface. By careful manipulation of the position of the tip, such isolated atoms can be "dragged" across the surface and deposited at another location. Using this technique, scientists at IBM's Almaden Research Center moved iron atoms across a carefully prepared copper surface, forming the atoms into a circle, which they named a **quantum corral.** The result is shown in the photograph that opens this chapter. Each iron atom in the circle is nestled in a hollow in the copper surface, equidistant from three nearest-neighbor copper atoms. The corral was fabricated at a low temperature (about 4 K) to minimize the tendency of the iron atoms to move randomly about on the surface because of their thermal energies.

The ripples within the corral are due to matter waves associated with electrons that can move over the copper surface but are largely trapped in the potential well of the corral. The dimensions of the ripples are in excellent agreement with the predictions of quantum theory.

40-7 Two- and Three-Dimensional Electron Traps

In the next section, we shall discuss the hydrogen atom as being a three-dimensional finite potential well. As a warm-up for the hydrogen atom, let us extend our discussion of infinite potential wells to two and three dimensions.

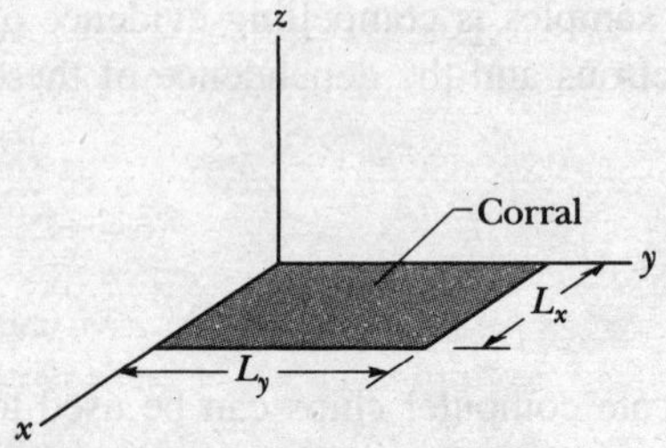

Fig. 40-12 A rectangular corral—a two-dimensional version of the infinite potential well of Fig. 40-2—with widths L_x and L_y.

Rectangular Corral

Figure 40-12 shows the rectangular area to which an electron can be confined by the two-dimensional version of Fig. 40-2—a two-dimensional infinite potential well of widths L_x and L_y. Such a well is called a rectangular *corral*. The corral might be on the surface of a body that somehow prevents the electron from moving parallel to the z axis and thus from leaving the surface. You have to imagine infinite potential energy functions (like $U(x)$ in Fig. 40-2) along each side of the corral, keeping the electron within the corral.

Solution of Schrödinger's equation for the rectangular corral of Fig. 40-12 shows that, for the electron to be trapped, its matter wave must fit into each of the two widths separately, just as the matter wave of a trapped electron must fit into a one-dimensional infinite well. This means the wave is separately quantized in width L_x and in width L_y. Let n_x be the quantum number for which the matter wave fits into width L_x, and let n_y be the quantum number for which the matter wave fits into width L_y. As with a one-dimensional potential well, these quantum numbers can be only positive integers.

The energy of the electron depends on both quantum numbers and is the sum of the energy it would have if it were confined along the x axis alone and the energy it would have if it were confined along the y axis alone. From Eq. 40-4, we can write this sum as

$$E_{nx,ny} = \left(\frac{h^2}{8mL_x^2}\right)n_x^2 + \left(\frac{h^2}{8mL_y^2}\right)n_y^2 = \frac{h^2}{8m}\left(\frac{n_x^2}{L_x^2} + \frac{n_y^2}{L_y^2}\right). \tag{40-20}$$

Excitation of the electron by photon absorption and de-excitation of the electron by photon emission have the same requirements as for one-dimensional traps. The only major difference for the two-dimensional corral is that the energy of any given state depends on two quantum numbers (n_x and n_y) instead of just one (n). In general, different states (with different pairs of values for n_x and n_y) have different energies. However, in some situations, different states can have the same energy. Such states (and their energy levels) are said to be *degenerate*. Degenerate states cannot occur in a one-dimensional well.

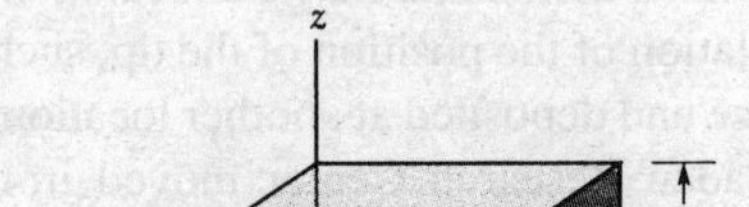

Fig. 40-13 A rectangular box—a three-dimensional version of the infinite potential well of Fig. 40-2—with widths L_x, L_y, and L_z.

Rectangular Box

An electron can also be trapped in a three-dimensional infinite potential well—a *box*. If the box is rectangular as in Fig. 40-13, then Schrödinger's equation shows us that we can write the energy of the electron as

$$E_{nx,ny,nz} = \frac{h^2}{8m}\left(\frac{n_x^2}{L_x^2} + \frac{n_y^2}{L_y^2} + \frac{n_z^2}{L_z^2}\right). \tag{40-21}$$

Here n_z is a third quantum number, for fitting the matter wave into width L_z.

CHECKPOINT 4: In the notation of Eq. 40-20, is $E_{0,0}$, $E_{1,0}$, $E_{0,1}$, or $E_{1,1}$ the ground-state energy of an electron in a rectangular corral?

Sample Problem 40-5

An electron is trapped in a square corral that is a two-dimensional infinite potential well (Fig. 40-12) with widths $L_x = L_y$. (a) Find the energies of the lowest five energy levels for the electron, and construct an energy-level diagram.

SOLUTION: The **Key Idea** here is that because the electron is trapped in a two-dimensional well that is rectangular, its energy depends on two quantum numbers, n_x and n_y, according to Eq. 40-20. Because the well is square, we can let the widths be $L_x = L_y = L$.

TABLE 40-1 Energy Levels

n_x	n_y	Energy*	n_x	n_y	Energy*
1	3	10	2	4	20
3	1	10	4	2	20
2	2	8	3	3	18
1	2	5	1	4	17
2	1	5	4	1	17
1	1	2	2	3	13
			3	2	13

*In multiples of $h^2/8mL^2$

Then Eq. 40-20 simplifies to

$$E_{nx,ny} = \frac{h^2}{8mL^2}(n_x^2 + n_y^2). \tag{40-22}$$

The lowest energy states correspond to low values of the quantum numbers n_x and n_y, which are the positive integers 1, 2, . . . , ∞. Substituting those integers for n_x and n_y in Eq. 40-22, starting with the lowest value 1, we can obtain the energy values as listed in Table 40-1. There we can see that several of the pairs of quantum numbers (n_x, n_y) give the same energy. For example, the (1, 2) and (2, 1) states both have an energy of $5(h^2/8mL^2)$. Each such pair is associated with degenerate energy levels. Note also that, perhaps surprisingly, the (4, 1) and (1, 4) states have less energy than the (3, 3) state.

From Table 40-1 (carefully keeping track of degenerate levels), we can construct the energy-level diagram of Fig. 40-14.

(b) As a multiple of $h^2/8mL^2$, what is the energy difference between the ground state and the third excited state of the electron?

SOLUTION: From Fig. 40-14, we see that the ground state is the (1, 1) state, with an energy of $2(h^2/8mL^2)$. We also see that the third excited state (the third state up from the ground state in the energy-level diagram) is the degenerate (1, 3) and (3, 1) states, with an energy of $10(h^2/8mL^2)$. Thus, the difference ΔE between these two states is

$$\Delta E = 10\left(\frac{h^2}{8mL^2}\right) - 2\left(\frac{h^2}{8mL^2}\right) = 8\left(\frac{h^2}{8mL^2}\right). \quad \text{(Answer)}$$

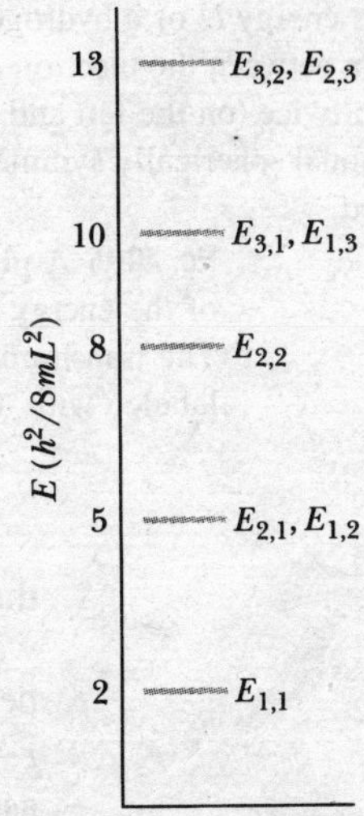

Fig. 40-14 Sample Problem 40-5. Energy-level diagram.

40-8 The Hydrogen Atom

We now move from artificial and fictitious electron traps to natural ones, using the simplest atom—hydrogen—as our example. This atom consists of a single electron (charge $-e$) bound to its central nucleus, a single proton (charge $+e$), by the attractive Coulomb force that acts between them. The hydrogen atom, like all atoms, is an electron trap; it confines its single electron to a region of space. From the confinement principle, we then expect that the electron can exist only in a discrete set of quantum states, each with a certain energy. We wish to identify the energies and the wave functions of these states.

The Energies of the Hydrogen Atom States

In Chapter 25 we wrote Eq. 25-43 for the (electric) potential energy of a two-particle system with charges q_1 and q_2:

$$U = \frac{1}{4\pi\varepsilon_0}\frac{q_1 q_2}{r},$$

where r is the distance between the particles. For the two-particle system of a hydrogen atom, we can write the potential energy as

$$U = \frac{1}{4\pi\varepsilon_0}\frac{(e)(-e)}{r} = -\frac{1}{4\pi\varepsilon_0}\frac{e^2}{r}. \tag{40-23}$$

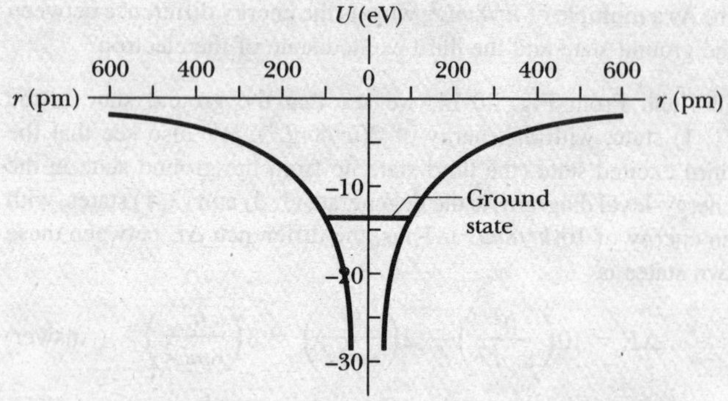

Fig. 40-15 The potential energy U of a hydrogen atom as a function of the separation r between the electron and the central proton. The plot is shown twice (on the left and on the right) to suggest the three-dimensional spherically symmetric trap to which the electron is confined.

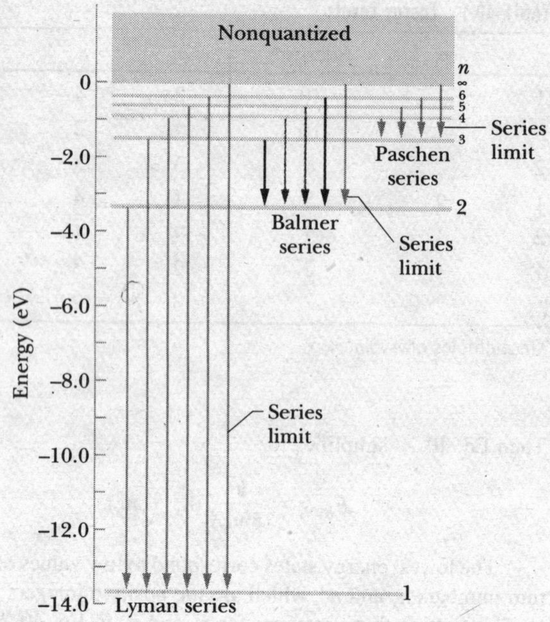

Fig. 40-16 A plot of Eq. 40-24, showing a few of the energy levels of the hydrogen atom. The transitions are grouped into series, each labeled with the name of a person.

The plot of Fig. 40-15 suggests the three-dimensional potential well in which the hydrogen atom's electron is trapped. This well differs from the finite potential well of Fig. 40-7 in that, for the hydrogen atom, U is negative for all values of r because we have (arbitrarily) chosen our zero of potential energy to correspond to $r = \infty$. For the finite well of Fig. 40-7, however, we (equally arbitrarily) chose to assign the zero of potential energy to the region inside the well.

To find the energies of the quantum states of the hydrogen atom, we must solve Schrödinger's equation, with Eq. 40-23 substituted for U in that equation. However, because the electron in the hydrogen atom is trapped in a three-dimensional well, we must use a three-dimensional form of Schrödinger's equation.

Solving that equation reveals that the energies of the electron's quantum states are given by

$$E_n = -\frac{me^4}{8\varepsilon_0^2 h^2}\frac{1}{n^2} = -\frac{13.6\ \text{eV}}{n^2}, \qquad \text{for } n = 1, 2, 3, \ldots, \qquad (40\text{-}24)$$

where n is a quantum number and m is the mass of an electron. The lowest energy, which is for the ground state with $n = 1$, is indicated in Fig. 40-15. Figure 40-16 shows the energy levels of the ground state and five excited states, each labeled with its quantum number n. It also shows the energy level for the greatest value of n—namely, $n = \infty$—for which $E_n = 0$. For any greater energy, the electron and proton are not bound together (there is no hydrogen atom), and the corresponding region in Fig. 40-16 is like the nonquantized region for the finite well of Fig. 40-9.

The quantized potential energy values given by Eq. 40-24 are actually those of the hydrogen atom—that is, of the *electron + proton* system. However, we can usually attribute the energy to the electron alone because its mass is much less than that of the proton. (Similarly, we can attribute the energy of a *ball + Earth* system to the ball alone.) Thus, we can say that when an electron is trapped in a hydrogen atom, the *electron* can have only energy values given by Eq. 40-24.

As we have seen with electrons in other potential wells, the electron in a hydrogen atom tends to be in the lowest energy level—that is, in its ground state. It can make a quantum jump up to a higher level, at greater energy, only if it is given the required energy to reach the higher level. One way it can receive that energy is by photon absorption. As we have discussed, this absorption can occur only if the photon's energy hf is equal to the energy difference ΔE between the electron's initial energy level and another level, as given by Eq. 40-6. To decrease its energy, the electron can make a quantum jump down to a lower energy level. If it does so by emitting a photon, the photon's energy hf must again equal the difference ΔE.

Because hf must equal a difference ΔE between the energies of two quantum levels, which can have only certain energy values, a hydrogen atom can emit and absorb light at only certain frequencies f—and thus also at only certain wavelengths λ. Any such wavelength is often called a *line* because of the way it is detected with a spectroscope; thus, a hydrogen atom has *absorption lines* and *emission lines*. A collection of such lines, such as in those in the visible range, is called a **spectrum** of the hydrogen atom.

The lines for hydrogen are said to be grouped into *series,* according to the level that upward jumps start on and downward jumps end on. For example, the emission and absorption lines for all possible jumps up from the $n = 1$ level and down to the $n = 1$ level are said to be in the *Lyman series,* named after the person who first studied those lines. Further, we can say that the Lyman series in the hydrogen spectrum has a *home-base level* of $n = 1$. Similarly, the *Balmer series* has a home-base level of $n = 2$, and the *Paschen series* has a home-base level of $n = 3$.

Some of the downward quantum jumps for these three series are shown in Fig. 40-16. Four lines in the Balmer series are in the visible range, and they are represented in Fig. 40-16 with arrows corresponding to their colors. The shortest of those arrows represents the shortest jump in the series, from the $n = 3$ level to the $n = 2$ level. Thus, that jump involves the least change in the electron's energy and the least emitted photon energy for the series. The emitted light is red. The next jump in the series, from $n = 4$ to $n = 2$, is longer, the photon energy is greater, the wavelength of the emitted light is shorter, and the light is green. The third, fourth, and fifth arrows represent longer jumps and shorter wavelengths. For the fifth jump, the emitted light is in the ultraviolet range and thus is not visible.

The *series limit* of a series is the line produced by the jump between the home-base level and the highest energy level, which is the level with quantum number $n = \infty$. Thus, the series limit is the shortest wavelength in the series. Figure 40-17 is a photograph of the Balmer emission lines taken with a spectroscope (as in Figs. 37-22 and 37-23). The series limit for the series is marked with a small triangle.

Bohr's Theory of the Hydrogen Atom

In 1913, some 13 years before the formulation of Schrödinger's equation, Bohr proposed a model of the hydrogen atom based on a clever combination of classical

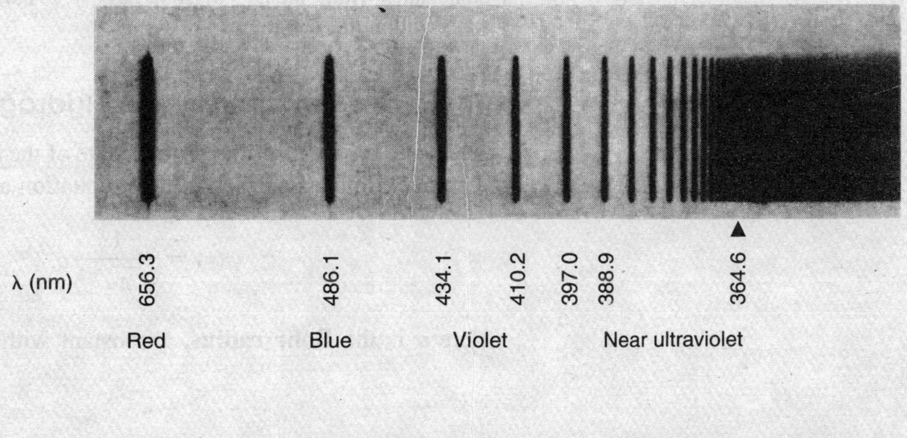

Fig. 40-17 The spectrum of emission lines of the Balmer series of the hydrogen atom. Whereas Fig. 40-16 shows four transitions of this series, along with the series limit, this figure shows about a dozen lines of this series; note how they are progressively closer toward the series limit, which is marked with a triangle.

TABLE 40-2 Quantum Numbers for the Hydrogen Atom

Symbol	Name	Allowed Values
n	Principal quantum number	$1, 2, 3, \ldots$
l	Orbital quantum number	$0, 1, 2, \ldots, n-1$
m_l	Orbital magnetic quantum number	$-l, -(l-1), \ldots, +(l-1), +l$

and early quantum concepts. His basic assumption—that atoms exist in discrete quantum states of well-defined energy—was a bold break with classical ideas; it carries over today as an indispensable concept in modern quantum physics. With this assumption, Bohr made skillful use of the correspondence principle (see Section 40-4), not only to derive Eq. 40-24 for the energies of the quantum states of the hydrogen atom but also to derive a numerical value (the *Bohr radius*) for the effective radius of that atom. In spite of its successes, Bohr's specific model of the hydrogen atom, based on the assumption that the electron is a particle that moves in planet-like orbits around the nucleus, was inconsistent with the uncertainty principle and was replaced by the probability density model derived from Schrödinger's work. For Bohr's brilliant achievements, which greatly stimulated progress toward the modern quantum theory, he was awarded the Nobel prize in physics in 1922.

Quantum Numbers for the Hydrogen Atom

Although the energies of the hydrogen atom states can be described by the single quantum number n, the wave functions describing these states require three quantum numbers, corresponding to the three dimensions in which the electron can move. The three quantum numbers, along with their names and the values that they may have, are shown in Table 40-2.

Each set of quantum numbers (n, l, m_l) identifies the wave function of a particular quantum state. The quantum number n, called the **principal quantum number,** appears in Eq. 40-24 for the energy of the state. The **orbital quantum number** l is a measure of the magnitude of the angular momentum associated with the quantum state. The **orbital magnetic quantum number** m_l is related to the orientation in space of this angular momentum vector. The restrictions on the values of the quantum numbers for the hydrogen atom, as listed in Table 40-2, are not arbitrary but come out of the solution to Schrödinger's equation. Note that for the ground state ($n = 1$), the restrictions require that $l = 0$ and $m_l = 0$. That is, the hydrogen atom in its ground state has zero angular momentum.

CHECKPOINT 5: (a) A group of quantum states of the hydrogen atom has $n = 5$. How many values of l are possible for states within this group? (b) A subgroup of hydrogen atom states within the $n = 5$ group has $l = 3$. How many values of m_l are possible for states within this subgroup?

The Wave Function of the Hydrogen Atom's Ground State

The wave function for the ground state of the hydrogen atom, as obtained by solving the three-dimensional Schrödinger equation and normalizing the result, is

$$\psi(r) = \frac{1}{\sqrt{\pi}a^{3/2}} e^{-r/a} \quad \text{(ground state).} \tag{40-25}$$

Here a is the **Bohr radius,** a constant with the dimension *length*. This radius is loosely taken to be the effective radius of a hydrogen atom and turns out to be a convenient unit of length for other situations involving atomic dimensions. Its value is

$$a = \frac{h^2\varepsilon_0}{\pi m e^2} = 5.29 \times 10^{-11}\text{ m} = 52.9\text{ pm}. \qquad (40\text{-}26)$$

As with other wave functions, $\psi(r)$ in Eq. 40-25 does not have physical meaning but $\psi^2(r)$ does. It is the probability density—the probability per unit volume—that the electron can be detected. Specifically, $\psi^2(r)\,dV$ is the probability that the electron can be detected in any given (infinitesimal) volume element dV located at radius r from the center of the atom:

$$\begin{pmatrix}\text{probability of detection}\\ \text{in volume } dV\\ \text{at radius } r\end{pmatrix} = \begin{pmatrix}\text{volume probability}\\ \text{density } \psi^2(r)\\ \text{at radius } r\end{pmatrix}(\text{volume } dV). \qquad (40\text{-}27)$$

Because $\psi^2(r)$ here depends only on r, it makes sense to choose, as a volume element dV, the volume between two concentric spherical shells whose radii are r and $r + dr$. That is, we take the volume element dV to be

$$dV = (4\pi r^2)\,dr, \qquad (40\text{-}28)$$

in which $4\pi r^2$ is the area of the inner shell and dr is the radial distance between the two shells. Then, combining Eqs. 40-25, 40-27, and 40-28 gives us

$$\begin{pmatrix}\text{probability of detection}\\ \text{in volume } dV\\ \text{at radius } r\end{pmatrix} = \psi^2(r)\,dV = \frac{4}{a^3}\,e^{-2r/a}r^2\,dr. \qquad (40\text{-}29)$$

Describing the probability of detection is easier if we work with a **radial probability density** $P(r)$ instead of a volume probability density $\psi^2(r)$. This $P(r)$ is a linear probability density such that

$$\begin{pmatrix}\text{radial probability}\\ \text{density } P(r)\\ \text{at radius } r\end{pmatrix}\begin{pmatrix}\text{radial}\\ \text{width } dr\end{pmatrix} = \begin{pmatrix}\text{volume probability}\\ \text{density } \psi^2(r)\\ \text{at radius } r\end{pmatrix}(\text{volume } dV)$$

or

$$P(r)\,dr = \psi^2(r)\,dV. \qquad (40\text{-}30)$$

Substituting for $\psi^2(r)\,dV$ from Eq. 40-29, we obtain

$$P(r) = \frac{4}{a^3}\,r^2e^{-2r/a} \quad \text{(radial probability density, hydrogen atom ground state).} \qquad (40\text{-}31)$$

Figure 40-18 is a plot of Eq. 40-31. The area under the plot is unity; that is,

$$\int_0^\infty P(r)\,dr = 1. \qquad (40\text{-}32)$$

This equation simply states that in a normal hydrogen atom, the electron must be *somewhere* in the space surrounding the nucleus.

The triangular marker on the horizontal axis of Fig. 40-18 is located one Bohr radius from the origin. The graph tells us that in the ground state of the hydrogen atom, the electron is most likely to be found at about this distance from the center of the atom.

Figure 40-18 conflicts sharply with the popular view that electrons in atoms follow well-defined orbits like planets moving around the Sun. *This popular view, however familiar, is incorrect.* Figure 40-18 shows us all that we can ever know

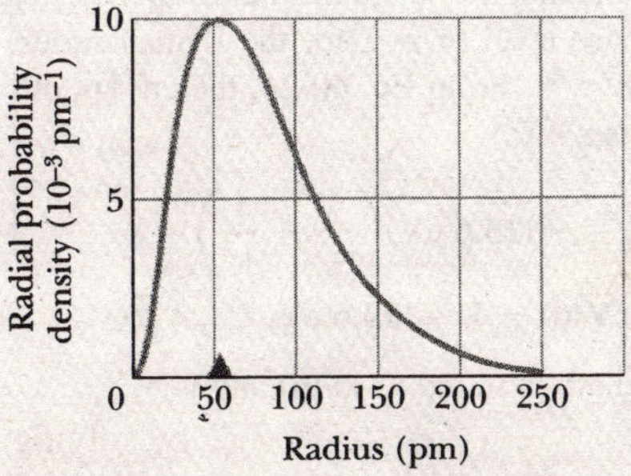

Fig. 40-18 A plot of the radial probability density $P(r)$ for the ground state of the hydrogen atom. The triangular marker is located at one Bohr radius from the origin, and the origin represents the center of the atom.

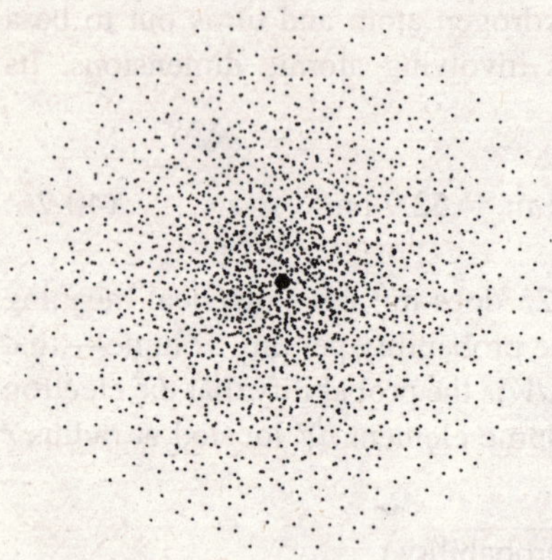

Fig. 40-19 A "dot plot" showing the probability density $\psi^2(r)$—not the *radial* probability density $P(r)$—for the ground state of the hydrogen atom. The density of dots drops exponentially with increasing distance from the nucleus, which is represented here by a red spot. Such dot plots provide a mental image of the "electron cloud" of an atom.

about the location of the electron in the ground state of the hydrogen atom. The appropriate question is not "When will the electron arrive at such-and-such a point?" but "What are the odds that the electron will be detected in a small volume centered on such-and-such a point?" Figure 40-19, which we call a dot plot, suggests the probabilistic nature of the wave function and provides a useful mental model of the hydrogen atom in its ground state. Think of the atom in this state as a fuzzy ball with no sharply defined boundary and no hint of orbits.

It is not easy for a beginner to envision subatomic particles in this probabilistic way. The difficulty is our natural impulse to regard an electron as something like a tiny jelly bean, located at certain places at certain times and following a well-defined path. Electrons and other subatomic particles simply do not behave in this way.

The energy of the ground state, found by putting $n = 1$ in Eq. 40-24, is $E_1 = -13.6$ eV. The wave function of Eq. 40-25 results if you solve Schrödinger's equation with this value of the energy. Actually, you can find a solution of Schrödinger's equation for *any* value of the energy, say $E = -11.6$ eV or -14.3 eV. This may suggest that the energies of the hydrogen atom states are not quantized—but we know that they are.

The puzzle was solved when physicists realized that such solutions of Schrödinger's equation are not physically acceptable because they yield increasingly large values as $r \rightarrow \infty$. These "wave functions" tell us that the electron is more likely to be found very far from the nucleus than closer to it, which makes no sense. We get rid of these unwanted solutions by imposing a so-called **boundary condition,** in which we agree to accept only solutions of Schrödinger's equation for which $\psi(r) \rightarrow 0$ as $r \rightarrow \infty$; that is, we agree to deal only with *confined* electrons. With this restriction, the solutions of Schrödinger's equation form a discrete set, with quantized energies given by Eq. 40-24.

Sample Problem 40-6

(a) What is the wavelength of light for the least energetic photon emitted in the Lyman series of the hydrogen atom spectrum lines?

SOLUTION: One **Key Idea** here is that for any series, the transition that produces the least energetic photon is the transition between the home-base level that defines the series and the level immediately above it. A second **Key Idea** is that for the Lyman series, the home-base level is at $n = 1$ (Fig. 40-16). Thus, the transition that produces the least energetic photon is the transition from the $n = 2$ level to the $n = 1$ level. From Eq. 40-24 the energy difference is

$$\Delta E = E_2 - E_1 = -(13.6 \text{ eV})\left(\frac{1}{2^2} - \frac{1}{1^2}\right) = 10.2 \text{ eV}.$$

Then from Eq. 40-6 ($\Delta E = hf$), with c/λ replacing f, we have

$$\lambda = \frac{hc}{\Delta E} = \frac{(6.63 \times 10^{-34} \text{ J}\cdot\text{s})(3.00 \times 10^8 \text{ m/s})}{(10.2 \text{ eV})(1.60 \times 10^{-19} \text{ J/eV})}$$
$$= 1.22 \times 10^{-7} \text{ m} = 122 \text{ nm}. \quad \text{(Answer)}$$

Light with this wavelength is in the ultraviolet range.

(b) What is the wavelength of the series limit for the Lyman series?

SOLUTION: The **Key Idea** here is that the series limit corresponds to a jump between the home-base level ($n = 1$ for the Lyman series) and the level at the limit $n = \infty$. From Eq. 40-24, the energy difference for this transition is

$$\Delta E = E_\infty - E_1 = -(13.6 \text{ eV})\left(\frac{1}{\infty^2} - \frac{1}{1^2}\right)$$
$$= -(13.6 \text{ eV})(0 - 1) = 13.6 \text{ eV}.$$

The corresponding wavelength is found as in (a) and is

$$\lambda = \frac{hc}{\Delta E}$$
$$= \frac{(6.63 \times 10^{-34} \text{ J}\cdot\text{s})(3.00 \times 10^8 \text{ m/s})}{(13.6 \text{ eV})(1.60 \times 10^{-19} \text{ J/eV})}$$
$$= 9.14 \times 10^{-8} \text{ m} = 91.4 \text{ nm}. \quad \text{(Answer)}$$

Light with this wavelength is also in the ultraviolet range.

Sample Problem 40-7

Show that the radial probability density for the ground state of the hydrogen atom has a maximum at $r = a$.

SOLUTION: One **Key Idea** here is that the radial probability density for a ground-state hydrogen atom is given by Eq. 40-31,

$$P(r) = \frac{4}{a^3} r^2 e^{-2r/a}.$$

A second **Key Idea** is that to find the maximum (or minimum) of any function, we must differentiate it and set the result equal to zero. If we differentiate $P(r)$ with respect to r, using derivative 7 of Appendix E and the chain rule for differentiating products, we get

$$\frac{dP}{dr} = \frac{4}{a^3} r^2 \left(\frac{-2}{a}\right) e^{-2r/a} + \frac{4}{a^3} 2r\, e^{-2r/a}$$

$$= \frac{8r}{a^3} e^{-2r/a} - \frac{8r^2}{a^4} e^{-2r/a}$$

$$= \frac{8}{a^4} r(a - r)e^{-2r/a}.$$

If we set the right side equal to zero, we obtain an equation that is true if $r = a$. In other words, dP/dr is equal to zero when $r = a$. (Note that we also have $dP/dr = 0$ at $r = 0$ and at $r = \infty$. However, these conditions correspond to a *minimum* in $P(r)$, as you can see in Fig. 40-18.)

Sample Problem 40-8

It can be shown that the probability $p(r)$ that the electron in the ground state of the hydrogen atom will be detected inside a sphere of radius r is given by

$$p(r) = 1 - e^{-2x}(1 + 2x + 2x^2),$$

in which x, a dimensionless quantity, is equal to r/a. Find r for $p(r) = 0.90$.

SOLUTION: The **Key Idea** here is that there is no guarantee of detecting the electron at any particular radial distance r from the center of the hydrogen atom. However, with the given function, we can calculate the probability the electron will be detected *somewhere* within a sphere of radius r. We seek the radius of a sphere for which $p(r) = 0.90$. Substituting that value in the expression for $p(r)$, we have

$$0.90 = 1 - e^{-2x}(1 + 2x + 2x^2)$$

or

$$10e^{-2x}(1 + 2x + 2x^2) = 1.$$

We must find the value of x that satisfies this equality. It is not possible to solve explicitly for x, but an equation solver on a calculator yields $x = 2.66$. This means that the radius of a sphere such that the electron will be detected inside it 90% of the time is $2.66a$. Mark this position on the horizontal axis of Fig. 40-18—is it a reasonable answer?

TABLE 40-3 Quantum Numbers for Hydrogen Atom States with $n = 2$

n	l	m_l
2	0	0
2	1	+1
2	1	0
2	1	−1

Hydrogen Atom States with $n = 2$

According to the requirements of Table 40-2, there are four states of the hydrogen atom with $n = 2$; their quantum numbers are listed in Table 40-3. Consider first the state with $n = 2$ and $l = m_l = 0$; its probability density is represented by the dot plot of Fig. 40-20. Note that this plot, like the plot for the ground state shown in Fig. 40-19, is spherically symmetric. That is, in a spherical coordinate system like that defined in Fig. 40-21, the probability density is a function of the radial coordinate r only and is independent of the angular coordinates θ and ϕ.

It turns out that all quantum states with $l = 0$ have spherically symmetric wave functions. This is reasonable because the quantum number l is a measure of the angular momentum associated with a given state. If $l = 0$, the angular momentum is also zero, which requires that the probability density representing the state have no preferred axis of symmetry.

Dot plots of ψ^2 for the three states with $n = 2$ and $l = 1$ are shown in Fig. 40-22. The probability densities for the states with $m_l = +1$ and $m_l = -1$ are identical. Although these plots are symmetric about the z axis, they are *not* spherically symmetric. That is, the probability densities for these three states are functions of both r and the angular coordinate θ.

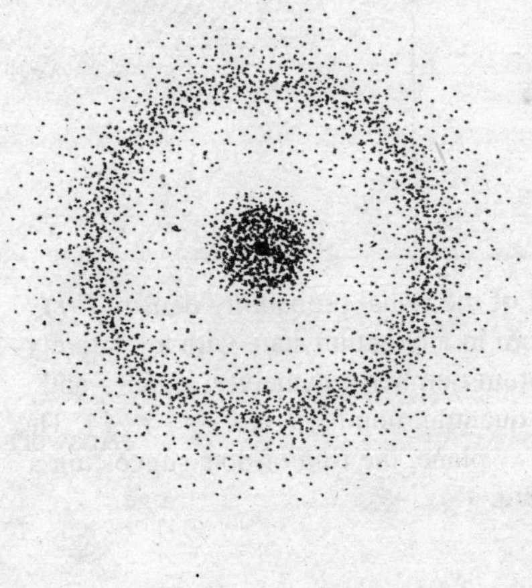

Fig. 40-20 A dot plot showing the probability density $\psi^2(r)$ for the hydrogen atom in the quantum state with $n = 2$, $l = 0$, and $m_l = 0$. The plot has spherical symmetry about the central nucleus. The gap in the dot density pattern marks a spherical surface over which $\psi^2(r) = 0$.

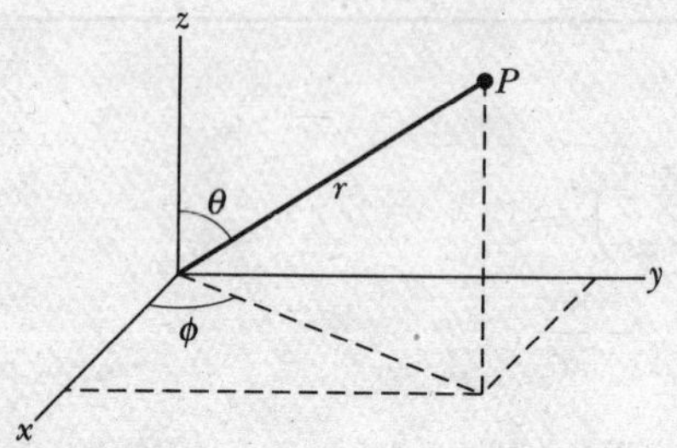

Fig. 40-21 The relationship between the coordinates x, y, and z of the rectangular coordinate system and the coordinates r, θ, and ϕ of the spherical coordinate system. The latter are more appropriate for analyzing situations involving spherical symmetry, such as the hydrogen atom.

Here is a puzzle: What is there about the hydrogen atom that establishes the axis of symmetry that is so obvious in Fig. 40-22? The answer: *absolutely nothing*.

The solution to this puzzle comes about when we realize that all three states shown in Fig. 40-22 have the same energy. Recall that the energy of a state, given by Eq. 40-24, depends only on the principal quantum number n and is independent of l and m_l. In fact, for an *isolated* hydrogen atom there is no way to differentiate experimentally among the three states of Fig. 40-22.

If we add the probability densities for the three states, $n = 2$ and $l = 1$, the combined probability density turns out to be spherically symmetrical, with no unique axis. One can, then, think of the electron as spending one-third of its time in each of the three states of Fig. 40-22, and one can think of the weighted sum of the three independent wave functions as defining a spherically symmetric **subshell,** specified by the quantum numbers $n = 2, l = 1$. The individual states will display their separate existence only if we place the hydrogen atom in an external electric or magnetic field. The three states of the $n = 2, l = 1$ subshell will then have different energies, and the field direction will establish the necessary symmetry axis.

The $n = 2, l = 0$ state, whose probability density is shown in Fig. 40-20, *also* has the same energy as each of the three states of Fig. 40-22. We can view all four states whose quantum numbers are listed in Table 40-3 as forming a spherically symmetric **shell,** specified by the single quantum number n. The importance of shells and subshells will become evident in Chapter 41, where we discuss atoms having more than one electron.

To round out our picture of the hydrogen atom, we display in Fig. 40-23 a dot plot of the radial probability density for a hydrogen atom state with a relatively high quantum number ($n = 45$) and the highest orbital quantum number that the restrictions of Table 40-2 permit ($l = n - 1 = 44$). The probability density forms a ring that is symmetrical about the z axis and lies very close to the xy plane. The mean radius of the ring is n^2a, where a is the Bohr radius. This mean radius is more than 2000 times the effective radius of the hydrogen atom in its ground state.

Figure 40-23 suggests the electron orbit of classical physics. Thus, we have another illustration of Bohr's correspondence principle—namely, that at large quan-

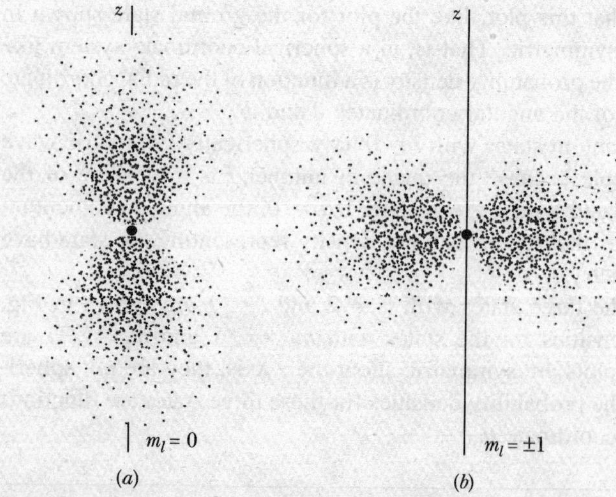

Fig. 40-22 Dot plots of the probability density $\psi^2(r, \theta)$ for the hydrogen atom in states with $n = 2$ and $l = 1$. (*a*) Plot for $m_l = 0$. (*b*) Plot for $m_l = +1$ and $m_l = -1$. Both plots show that the probability density is symmetric about the z axis.

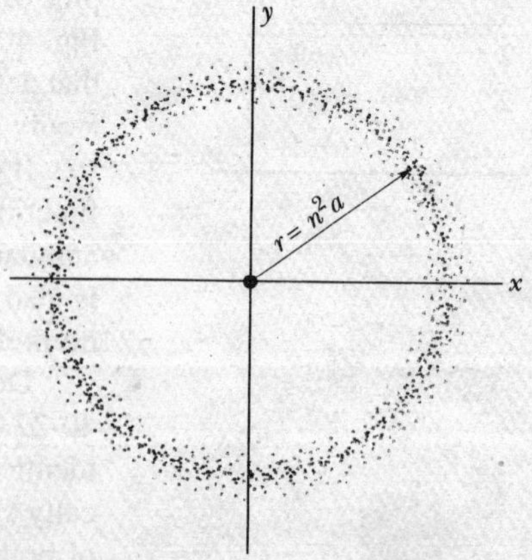

Fig. 40-23 A dot plot of the radial probability density $P(r)$ for the hydrogen atom in a quantum state with a relatively large principal quantum number—namely, $n = 45$—and angular momentum quantum number $l = n - 1 = 44$. The dots lie close to the xy plane, the ring of dots suggesting a classical electron orbit.

tum numbers the predictions of quantum mechanics merge smoothly with those of classical physics. Imagine what a dot plot like that of Figure 40-23 would look like for *really* large values of n and l, say, $n = 1000$ and $l = 999$.

REVIEW & SUMMARY

The Confinement Principle The **confinement principle** applies to waves of all kinds, including waves on a string and the matter waves of quantum physics. It states that confinement leads to quantization—that is, to the existence of discrete states with discrete energies.

An Electron in an Infinite Potential Well An infinite potential well is a device for confining an electron. From the confinement principle we expect that the matter wave representing a trapped electron can exist only in a set of discrete states. For a one-dimensional infinite potential well, the energies associated with these *quantum states* are

$$E_n = \left(\frac{h^2}{8mL^2}\right)n^2, \quad \text{for } n = 1, 2, 3, \ldots, \tag{40-4}$$

in which L is the width of the well and n is a **quantum number.** If the electron is to change from one state to another, its energy must change by the amount

$$\Delta E = E_{\text{high}} - E_{\text{low}}, \tag{40-5}$$

where E_{high} is the higher energy and E_{low} is the lower energy. If the change is done by photon absorption or emission, the energy of the photon must be

$$hf = \Delta E = E_{\text{high}} - E_{\text{low}}. \tag{40-6}$$

The **wave functions** associated with the quantum states are

$$\psi_n(x) = A \sin\left(\frac{n\pi}{L}x\right), \quad \text{for } n = 1, 2, 3, \ldots. \tag{40-10}$$

The **probability density** $\psi_n^2(x)$ for an allowed state has the physical meaning that $\psi_n^2(x)\,dx$ is the probability that the electron will be detected in the interval between x and $x + dx$. For an electron in an infinite well, the probability densities are

$$\psi_n^2(x) = A^2 \sin^2\left(\frac{n\pi}{L}x\right), \quad \text{for } n = 1, 2, 3, \ldots. \tag{40-12}$$

At high quantum numbers n, the electron tends toward classical behavior in that it tends to occupy all parts of the well with equal probability. This fact leads to the **correspondence principle:** At large enough quantum numbers, the predictions of quantum physics merge smoothly with those of classical physics.

Normalization and Zero-Point Energy The amplitude A^2 in Eq. 40-12 can be found from the **normalizing equation,**

$$\int_{-\infty}^{+\infty} \psi_n^2(x)\,dx = 1, \tag{40-14}$$

which asserts that the electron must be *somewhere* within the well, because the probability 1 implies certainty.

From Eq. 40-4 we see that the lowest permitted energy for the electron is not zero but the energy that corresponds to $n = 1$. This lowest energy is called the **zero-point energy** of the electron–well system.

An Electron in a Finite Potential Well A finite potential well is one for which the potential energy of an electron inside the well is less than that for one outside the well by a finite amount U_0. The wave function for an electron trapped in such a well extends into the walls of the well.

Two- and Three-Dimensional Electron Traps The quantized energies for an electron trapped in a two-dimensional infinite potential well that forms a rectangular corral are

$$E_{nx,ny} = \frac{h^2}{8m}\left(\frac{n_x^2}{L_x^2} + \frac{n_y^2}{L_y^2}\right), \tag{40-20}$$

where n_x is a quantum number for which the electron's matter wave fits in well width L_x and n_y is a quantum number for which the electron's matter wave fits in well width L_y. Similarly, the energies for an electron trapped in a three-dimensional infinite potential well that forms a rectangular box are

$$E_{nx,ny,nz} = \frac{h^2}{8m}\left(\frac{n_x^2}{L_x^2} + \frac{n_y^2}{L_y^2} + \frac{n_z^2}{L_z^2}\right). \tag{40-21}$$

Here n_z is a third quantum number, for which the matter wave fits in well width L_z.

The Hydrogen Atom The potential energy function for the hydrogen atom is

$$U = -\frac{1}{4\pi\varepsilon_0}\frac{e^2}{r}. \tag{40-23}$$

The energies of the quantum states of the hydrogen atom are found from the three-dimensional form of Schrödinger's equation to be

$$E_n = -\frac{me^4}{8\varepsilon_0^2h^2}\frac{1}{n^2} = -\frac{13.6\text{ eV}}{n^2}, \quad n = 1, 2, 3, \ldots, \tag{40-24}$$

in which n is the **principal quantum number.** The hydrogen atom requires three quantum numbers for its complete description; their names and allowed values are shown in Table 40-2.

The **radial probability density** $P(r)$ for a state of the hydrogen atom is defined so that $P(r)\,dr$ is the probability that the electron will be detected between two concentric shells, centered on the atom's nucleus, whose radii are r and $r + dr$. For the hydrogen atom's ground state,

$$P(r) = \frac{4}{a^3}r^2e^{-2r/a}, \tag{40-31}$$

in which a, the **Bohr radius,** is a length unit whose value is 52.9 pm. Figure 40-18 is a plot of $P(r)$ for the ground state.

Figures 40-20 and 40-22 represent the probability densities (not the *radial* probability densities) for the four hydrogen atom states with $n = 2$. The plot of Fig. 40-20 ($n = 2$, $l = 0$, $m_l = 0$) is spherically symmetric. The plots of Fig. 40-22 ($n = 2$, $l = 1$, $m_l = 0, +1, -1$) are symmetric about the z axis but, when added together, are also spherically symmetric.

All four states with $n = 2$ have the same energy and may be usefully regarded as a **shell,** identified as the $n = 2$ shell. The three states of Fig. 40-22, taken together, may be regarded as the $n = 2$, $l = 1$ **subshell.** It is not possible to separate the four $n = 2$ states experimentally unless the hydrogen atom is placed in an electric or magnetic field, to permit the establishment of a definite symmetry axis.

QUESTIONS

1. If you double the width of a one-dimensional infinite potential well, (a) is the energy of the ground state of the trapped electron multiplied by 4, 2, $\frac{1}{2}$, $\frac{1}{4}$, or some other number? (b) Are the energies of the higher energy states multiplied by this factor or by some other factor, depending on their quantum number?

2. Three electrons are trapped in three different one-dimensional infinite potential wells of widths (a) 50 pm, (b) 200 pm, and (c) 100 pm. Rank the electrons according to their ground-state energies, greatest first.

3. If you wanted to use the idealized trap of Fig. 40-1 to trap a positron, would you need to change (a) the geometry of the trap, (b) the electric potential of the central cylinder, or (c) the electric potentials of the two semi-infinite end cylinders? (A positron has the same mass as an electron but is positively charged.)

4. An electron is trapped in a one-dimensional infinite potential well in a state with $n = 17$. How many points of (a) zero probability and (b) maximum probability does its matter wave have?

5. Figure 40-24 shows three infinite potential wells, each on an x axis. Without written calculation, determine the wave function ψ for a ground-state electron trapped in each well.

Fig. 40-24 Question 5.

6. Figure 40-25 indicates the lowest energy levels (in electron-volts) for five situations in which an electron is trapped in a one-dimensional infinite potential well. In wells B, C, D, and E, the electron is in the ground state. We shall excite the electron in well A to the fourth excited state (at 25 eV). The electron can then de-excite to the ground state by emitting one or more photons, corresponding to one long jump or several short jumps. What photon *emission* energies of this de-excitation match a photon *absorption* energy (from the ground state) of the other four electrons? Give the corresponding quantum numbers.

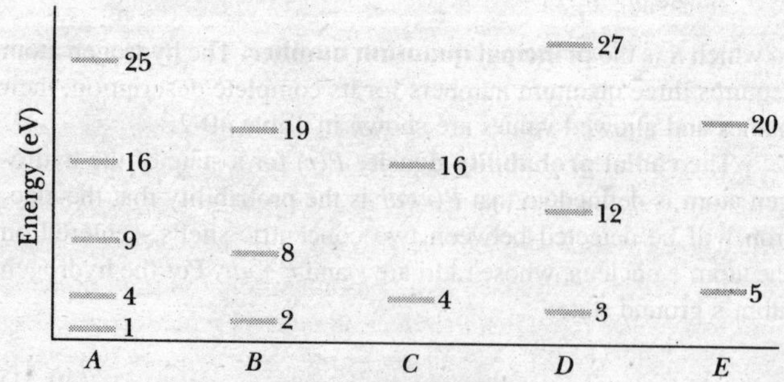

Fig. 40-25 Question 6.

7. Is the ground-state energy of a proton trapped in a one-dimensional infinite potential well greater than, less than, or equal to that of an electron trapped in the same potential well?

8. A proton and an electron are trapped in identical one-dimensional infinite potential wells; each particle is in its ground state. At the center of the wells, is the probability density for the proton greater than, less than, or equal to that of the electron?

9. You want to modify the finite potential well of Fig. 40-7 to allow its trapped electron to exist in more than three quantum states. Could you do so by making the well (a) wider or narrower, (b) deeper or shallower?

10. An electron is trapped in a finite potential well that is deep enough to allow the electron to exist in a state with $n = 4$. How many points of (a) zero probability and (b) probability maximum does its matter wave have within the well?

11. From a visual inspection of Fig. 40-8, rank the quantum numbers of the three quantum states according to the de Broglie wavelength of the electron, greatest first.

12. An electron, trapped in a finite potential energy well such as that of Fig. 40-7, is in its state of lowest energy. Are (a) its de Broglie wavelength, (b) the magnitude of its momentum, and (c) its energy greater than, the same as, or less than they would be if the potential well were infinite, as in Fig. 40-2?

13. The table lists the quantum numbers for five proposed hydrogen atom states. Which of them are not possible?

	n	l	m_l
(a)	3	2	0
(b)	2	3	1
(c)	4	3	−4
(d)	5	5	0
(e)	5	3	−2

14. In 1996 physicists working at an accelerator laboratory succeeded in producing atoms of antihydrogen. Such atoms consist of a positron moving in the electric field of an antiproton. A positron has the same mass as an electron but the opposite charge. An antiproton has the same mass as a proton but the opposite charge. Would you expect the spectrum of antihydrogen to be the same as that of normal hydrogen or different?

15. (a) From Fig. 40-16, the energy level diagram for the hydrogen atom, you can show that the photon energy of the second spectral line of the Lyman series is equal to the sum of the photon energies of two other lines. What are those lines? (b) The photon energy of the second spectral line of the Lyman series is also equal to the *difference* between the photon energies of two other lines. What are *those* lines?

16. A hydrogen atom is in the third excited state. To what state (give the quantum number n) should it jump to (a) emit light with the longest possible wavelength, (b) emit light with the shortest possible wavelength, and (c) absorb light with the longest possible wavelength?

EXERCISES & PROBLEMS

ssm Solution is in the Student Solutions Manual.
www Solution is available on the World Wide Web at: http://www.wiley.com/college/hrw
ilw Solution is available on the Interactive LearningWare.

SEC. 40-3 Energies of a Trapped Electron

1E. What is the ground-state energy of (a) an electron and (b) a proton if each is trapped in a one-dimensional infinite potential well that is 100 pm wide?

2E. You wish to reduce by one-half the ground-state energy of an electron trapped in a one-dimensional infinite potential well. By what factor must you change the width of the potential well?

3E. Consider an atomic nucleus to be equivalent to a one-dimensional infinite potential well with $L = 1.4 \times 10^{-14}$ m, a typical nuclear diameter. What would be the ground-state energy of an electron if it were trapped in such a potential well? (*Note:* Nuclei do not contain electrons.) ssm

4E. What must be the width of a one-dimensional infinite potential well if an electron trapped in it in the $n = 3$ state is to have an energy of 4.7 eV?

5E. A proton is confined to a one-dimensional infinite potential well 100 pm wide. What is its ground-state energy?

6E. The ground-state energy of an electron trapped in a one-dimensional infinite potential well is 2.6 eV. What will this quantity be if the width of the potential well is doubled?

7E. An electron, trapped in a one-dimensional infinite potential well 250 pm wide, is in its ground state. How much energy must it absorb if it is to jump up to the state with $n = 4$?

8P. An electron is trapped in a one-dimensional infinite potential well. (a) What pair of adjacent energy levels (if any) has an energy difference equal to the energy of the electron in the state with $n = 5$? (b) With $n = 6$?

9P. An electron is trapped in a one-dimensional infinite potential well. Show that the energy difference ΔE between its quantum levels n and $n + 2$ is $(h^2/2mL^2)(n + 1)$.

10P. An electron is trapped in a one-dimensional infinite potential well. (a) What pair of adjacent energy levels (if any) will have three times the energy difference that exists between levels $n = 3$ and $n = 4$? (b) What pair (if any) will have twice that energy difference?

11P. An electron is trapped in a one-dimensional infinite well of width 250 pm and is in its ground state. What are the four longest wavelengths of light that can excite the electron from the ground state via a single photon absorption? ssm www

12P. Suppose that an electron trapped in a one-dimensional infinite well of width 250 pm is excited from its first excited state to its third excited state. (a) In electron-volts, what energy must be transferred to the electron for this quantum jump? If the electron then de-excites by emitting light, (b) what wavelengths can it emit and (c) in which groupings (and orders) can they be emitted? (d) Show the several possible ways the electron can de-excite on an energy-level diagram.

13P. An electron is confined to a narrow evacuated tube of length 3.0 m; the tube functions as a one-dimensional infinite potential well. (a) In electron-volts, what is the energy difference between the electron's ground state and its first excited state? (b) At what quantum number n would the energy difference between adjacent energy levels be 1.0 eV—which is measurable, unlike the result of (a)? At that quantum number, (c) what would be the energy of the electron and (d) would the electron be relativistic?

SEC. 40-4 Wave Functions of a Trapped Electron

14E. An electron that is trapped in a one-dimensional infinite potential well of width L is excited from the ground state to the first excited state. (a) Does that increase, decrease, or have no effect on the probability of detecting the electron in a small length of the x axis (a) at the center of the well and (b) near one of the well walls?

15E. Let ΔE_{adj} be the energy difference between two adjacent energy levels for an electron trapped in a one-dimensional infinite potential well. Let E be the energy of either of the two levels. (a) Show that the ratio $\Delta E_{adj}/E$ approaches the value $2/n$ at large values of the quantum number n. As $n \rightarrow \infty$, does (b) ΔE_{adj}, (c) E, or (d) $\Delta E_{adj}/E$ approach zero? (e) What do these results mean in terms of the correspondence principle? ssm

16P. A particle is confined to the one-dimensional infinite potential well of Fig. 40-2. If the particle is in its ground state, what is its probability of detection between (a) $x = 0$ and $x = 0.25L$, (b) $x = 0.75L$ and $x = L$, and (c) $x = 0.25L$ and $x = 0.75L$?

17P. An electron is trapped in a one-dimensional infinite potential well that is 100 pm wide; the electron is in its ground state. What is the probability that you can detect the electron in an interval of width $\Delta x = 5.0$ pm centered at $x =$ (a) 25 pm, (b) 50 pm, and

(c) 90 pm? (*Hint:* The interval Δx is so narrow that you can take the probability density to be constant within it.) ssm

SEC. 40-5 An Electron in a Finite Well

18E. (a) Show that the terms in Schrödinger's equation (Eq. 40-18) have the same dimensions. (b) What is the common SI unit for each of these terms?

19E. An electron in the $n = 2$ state in the finite potential well of Fig. 40-7 absorbs 400 eV of energy from an external source. What is its kinetic energy after this absorption, assuming that the electron moves to a position for which $x > L$? ssm

20E. Figure 40-9 gives the energy levels for an electron trapped in a finite potential energy well 450 eV deep. If the electron is in the $n = 3$ state, what is its kinetic energy?

21P. As Fig. 40-8 suggests, the probability density for the region $x > L$ in the finite potential well of Fig. 40-7 drops off exponentially according to

$$\psi^2(x) = Ce^{-2kx},$$

where C is a constant. (a) Show that the wave function $\psi(x)$ that may be found from this equation is a solution of Schrödinger's equation in its one-dimensional form. (b) What must be the value of k for this to be true?

22P. As Fig. 40-8 suggests, the probability density for an electron in the region $0 < x < L$ for the finite potential well of Fig. 40-7 is sinusoidal, being given by

$$\psi^2(x) = B\sin^2 kx,$$

in which B is a constant. (a) Show that the wave function $\psi(x)$ that may be found from this equation is a solution of Schrödinger's equation in its one-dimensional form. (b) What must be the value of k for this to be true?

23P. Show that for the region $x > L$ in the finite potential well of Fig. 40-7, $\psi(x) = De^{2kx}$ is a solution of Schrödinger's equation in its one-dimensional form, where D is a constant and k is positive. On what basis do we find this mathematically acceptable solution to be physically unacceptable? ssm www

SEC. 40-7 Two- and Three-Dimensional Electron Traps

24E. An electron is contained in the rectangular corral of Fig. 40-12, with widths $L_x = 800$ pm and $L_y = 1600$ pm. What is the electron's ground-state energy in electron-volts?

25E. An electron is contained in the rectangular box of Fig. 40-13, with widths $L_x = 800$ pm, $L_y = 1600$ pm, and $L_z = 400$ pm. What is the electron's ground-state energy in electron-volts?

26P. A rectangular corral of widths $L_x = L$ and $L_y = 2L$ contains an electron. What multiple of $h^2/8mL^2$, where m is the electron's mass, are (a) the energy of the electron's ground state, (b) the energy of its first excited state, (c) the energy of its lowest degenerate states, and (d) the difference between the energies of its second and third excited states?

27P. For Problem 26, at what frequencies can light be absorbed or emitted by the electron for transitions between the lowest five energy levels? Answer in multiples of $h/8mL^2$. ssm www

28P. A cubical box of widths $L_x = L_y = L_z = L$ contains an electron. What multiple of $h^2/8mL^2$, where m is the electron's mass, are (a) the energy of the electron's ground state, (b) the energy of its second excited state, and (c) the difference between the energies of its second and third excited states? How many degenerate states have the energy of (d) the first excited state and (e) the fifth excited state?

29P. For the situation of Problem 28, at what frequencies can light be absorbed or emitted by the electron for transitions between the lowest five energy levels? Answer in multiples of $h/8mL^2$.

SEC. 40-8 The Hydrogen Atom

30E. Verify that the constant appearing in Eq. 40-24 is 13.6 eV.

31E. An atom (not a hydrogen atom) absorbs a photon whose associated frequency is 6.2×10^{14} Hz. By what amount does the energy of the atom increase?

32E. An atom (not a hydrogen atom) absorbs a photon whose associated wavelength is 375 nm and then immediately emits a photon whose associated wavelength is 580 nm. How much net energy is absorbed by the atom in this process?

33E. What is the ratio of the shortest wavelength of the Balmer series to the shortest wavelength of the Lyman series? ssm

34E. (a) What is the energy E of the hydrogen atom electron whose probability density is represented by the dot plot of Fig. 40-20? (b) What minimum energy is needed to remove this electron from the atom?

35E. What are (a) the energy, (b) the magnitude of the momentum, and (c) the wavelength of the photon emitted when a hydrogen atom undergoes a transition from a state with $n = 3$ to a state with $n = 1$? ssm

36E. Repeat Sample Problem 40-6 for the Balmer series of the hydrogen atom.

37E. A neutron, with a kinetic energy of 6.0 eV, collides with a stationary hydrogen atom in its ground state. Explain why the collision must be elastic—that is, why kinetic energy must be conserved. (*Hint:* Show that the hydrogen atom cannot be excited as a result of the collision.) ssm

38E. For the hydrogen atom in its ground state, calculate (a) the probability density $\psi^2(r)$ and (b) the radial probability density $P(r)$ for $r = a$, where a is the Bohr radius.

39E. Calculate the radial probability density $P(r)$ for the hydrogen atom in its ground state at (a) $r = 0$, (b) $r = a$, and (c) $r = 2a$, where a is the Bohr radius.

40E. A hydrogen atom is excited from its ground state to the state with $n = 4$. (a) How much energy must be absorbed by the atom? (b) Calculate and display on an energy-level diagram the different photon energies that may be emitted as the atom returns to its ground state.

41P. How much work must be done to pull apart the electron and the proton that make up the hydrogen atom if the atom is initially in (a) its ground state and (b) the state with $n = 2$? ssm

42P. A hydrogen atom, initially at rest in the $n = 4$ quantum state, undergoes a transition to the ground state, emitting a photon in the process. What is the speed of the recoiling hydrogen atom?

43P. Light of wavelength 486.1 nm is emitted by a hydrogen atom. (a) What transition of the atom is responsible for this radiation? (b) To what series does this transition belong?

44P. What are the widths of the wavelength intervals over which (a) the Lyman series and (b) the Balmer series extend? (Each width begins at the longest wavelength and ends at the series limit.) (c) What are the widths of the corresponding frequency intervals? Express the frequency intervals in terahertz (1 THz = 10^{12} Hz).

45P. In the ground state of the hydrogen atom, the electron has a total energy of -13.6 eV. What are (a) its kinetic energy and (b) its potential energy if the electron is one Bohr radius from the central nucleus? ssm

46P. (a) Find, using the energy-level diagram of Fig. 40-16, the quantum numbers corresponding to a transition in which the wavelength of the emitted radiation is 121.6 nm. (b) To what series does this transmission belong?

47P. A hydrogen atom in a state having a *binding energy* (the energy required to remove an electron) of 0.85 eV makes a transition to a state with an *excitation energy* (the difference between the energy of the state and that of the ground state) of 10.2 eV. (a) What is the energy of the photon emitted as a result of the transition? (b) Identify this transition, using the energy-level diagram of Fig. 40-16.

48P. Verify the wavelengths given in Fig. 40-17 for the visible spectral lines of the Balmer series.

49P. What is the probability that in the ground state of the hydrogen atom, the electron will be found at a radius greater than the Bohr radius? (*Hint:* See Sample Problem 40-8.) ssm www

50P. A hydrogen atom emits light of wavelength 102.6 nm. What are the initial and final quantum numbers for this transition?

51P. Schrödinger's equation for states of the hydrogen atom for which the orbital quantum number l is zero is

$$\frac{1}{r^2}\frac{d}{dr}\left(r^2\frac{d\psi}{dr}\right)+\frac{8\pi^2 m}{h^2}[E-U(r)]\psi=0.$$

Verify that Eq. 40-25, which describes the ground state of the hydrogen atom, is a solution of this equation. ssm

52P. Calculate the probability that the electron in the hydrogen atom, in its ground state, will be found between spherical shells whose radii are a and $2a$, where a is the Bohr radius. (*Hint:* See Sample Problem 40-8.)

53P. Verify that Eq. 40-31, the radial probability density for the ground state of the hydrogen atom, is normalized. That is, verify that

$$\int_0^\infty P(r)\,dr=1$$

is true. ssm

54P. (a) For a given value of the principal quantum number n, how many values of the orbital quantum number l are possible? (b) For a given value of l, how many values of the orbital magnetic quantum number m_l are possible? (c) For a given value of n, how many values of m_l are possible?

55P. What is the probability that an electron in the ground state of the hydrogen atom will be found between two spherical shells whose radii are r and $r+\Delta r$, (a) if $r=0.500a$ and $\Delta r=0.010a$ and (b) if $r=1.00a$ and $\Delta r=0.01a$, where a is the Bohr radius? (*Hint:* Δr is small enough to permit the radial probability density to be taken to be constant between r and $r+\Delta r$.) ssm

56P. For what value of the principal quantum number n would the effective radius, as shown in a probability density dot plot for the hydrogen atom, be 1.0 mm? Assume that l has its maximum value of $n-1$. (*Hint:* Be guided by Fig. 40-23.)

57P*. In Sample Problem 40-7 we showed that the radial probability density for the ground state of the hydrogen atom is a maximum when $r=a$, where a is the Bohr radius. Show that the *average* value of r, defined as

$$r_{avg}=\int P(r)\,r\,dr,$$

has the value $1.5a$. In this expression for r_{avg}, each value of $P(r)$ is weighted with the value of r at which it occurs. Note that the average value of r is greater than the value of r for which $P(r)$ is a maximum. ssm

58P*. The wave function for the hydrogen-atom quantum state with the dot plot shown in Fig. 40-20, which has $n=2$ and $l=m_l=0$, is

$$\psi_{200}(r)=\frac{1}{4\sqrt{2\pi}}a^{-3/2}\left(2-\frac{r}{a}\right)e^{-r/2a},$$

in which a is the Bohr radius and the subscript on $\psi(r)$ gives the values of the quantum numbers n, l, m_l. (a) Plot $\psi^2_{200}(r)$ and show that your plot is consistent with the dot plot of Fig. 40-20. (b) Show analytically that $\psi^2_{200}(r)$ has a maximum at $r=4a$. (c) Find the radial probability density $P_{200}(r)$ for this state. (d) Show that

$$\int_0^\infty P_{200}(r)\,dr=1,$$

and thus that the expression above for the wave function $\psi_{200}(r)$ has been properly normalized.

59P. The wave functions for the three states with the dot plots shown in Fig. 40-22, which have $n=2$, $l=1$, and $m_l=0$, $+1$, and -1, are

$$\psi_{210}(r,\theta)=(1/4\sqrt{2\pi})(a^{-3/2})(r/a)e^{-r/2a}\cos\theta,$$

$$\psi_{21+1}(r,\theta)=(1/8\sqrt{\pi})(a^{-3/2})(r/a)e^{-r/2a}(\sin\theta)e^{+i\phi},$$

$$\psi_{21-1}(r,\theta)=(1/8\sqrt{\pi})(a^{-3/2})(r/a)e^{-r/2a}(\sin\theta)e^{-i\phi},$$

in which the subscripts on $\psi(r,\theta)$ give the values of the quantum numbers n, l, m_l and the angles θ and ϕ are defined in Fig. 40-21. Note that the first wave function is real but the others, which involve the imaginary number i, are complex. (a) Find the probability density for each wave function and show that each is consistent with its dot plot in Fig. 40-22. (b) Add the three probability densities derived in (a) and show that their sum is spherically symmetric, depending only on the radial coordinate r. ssm

41 All About Atoms

Soon after lasers were invented in the 1960s, they became novel sources of light in research laboratories. Today, lasers are ubiquitous and are found in such diverse applications as voice and data transmission, surveying, welding, and grocery-store price scanning. The photograph shows surgery being performed with laser light transmitted via optical fibers. Light from a laser and light from any other source are both due to emissions by atoms.

What, then, is so different about the light from a laser?

The answer is in this chapter.

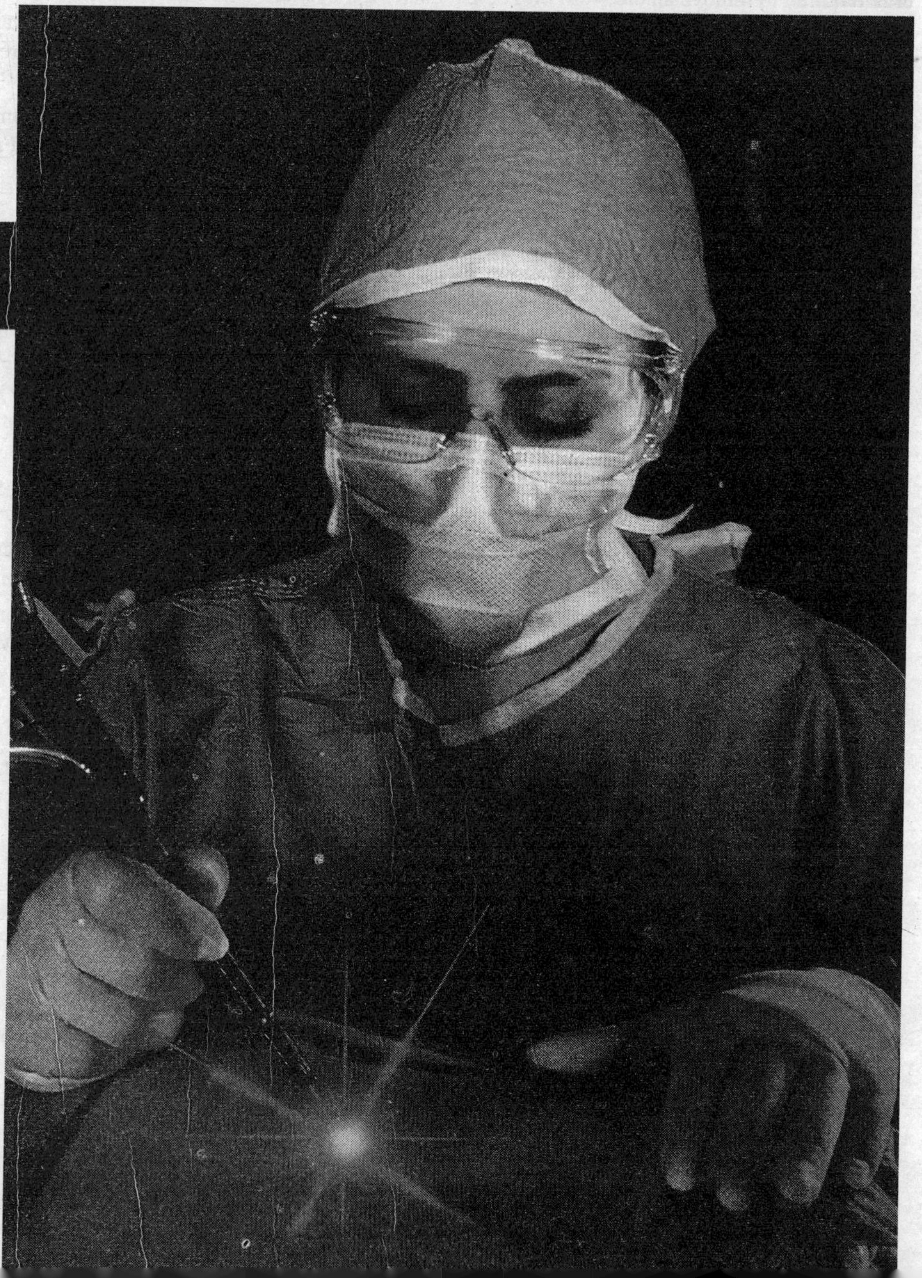

Fig. 41-1 The blue dot is a photograph of the light emitted from a single barium ion held for a long time in a trap at the University of Washington. Special techniques caused the ion to emit light over and over again as it underwent transitions between the same pair of energy levels. The dot represents the cumulative emission of many photons.

41-1 Atoms and the World Around Us

In the early years of the twentieth century many prominent scientists doubted the very existence of atoms. Today, however, every well-informed person believes that atoms exist and are the building blocks of the material world. Today, we can even pick up individual atoms and move them around. That's how the quantum corral on the opening page of Chapter 40 was formed. You can easily count the 48 iron atoms that make up the circle in that image. We can even photograph single atoms by the light they emit. For example, the faint blue dot in Figure 41-1 is due to light emitted by a single barium ion held in a trap at the University of Washington.

41-2 Some Properties of Atoms

You may think the details of atomic physics are remote from your daily life. However, consider how the following properties of atoms—so basic that we rarely think about them—affect the way we live in our world.

Atoms are stable. Essentially all the atoms that form our tangible world have existed without change for billions of years. What would the world be like if atoms continually changed into other forms, perhaps every few weeks or every few years?

Atoms combine with each other. They stick together to form stable molecules and stack up to form rigid solids. An atom is mostly empty space, but you can stand on a floor—made up of atoms—without falling through it.

These basic properties of atoms can be explained by quantum physics, as can the three less apparent properties that follow.

Atoms Are Put Together Systematically

Figure 41-2 shows an example of a repetitive property of the elements as a function of their position in the periodic table (Appendix G). The figure is a plot of the **ionization energy** of the elements; the energy required to remove the most loosely

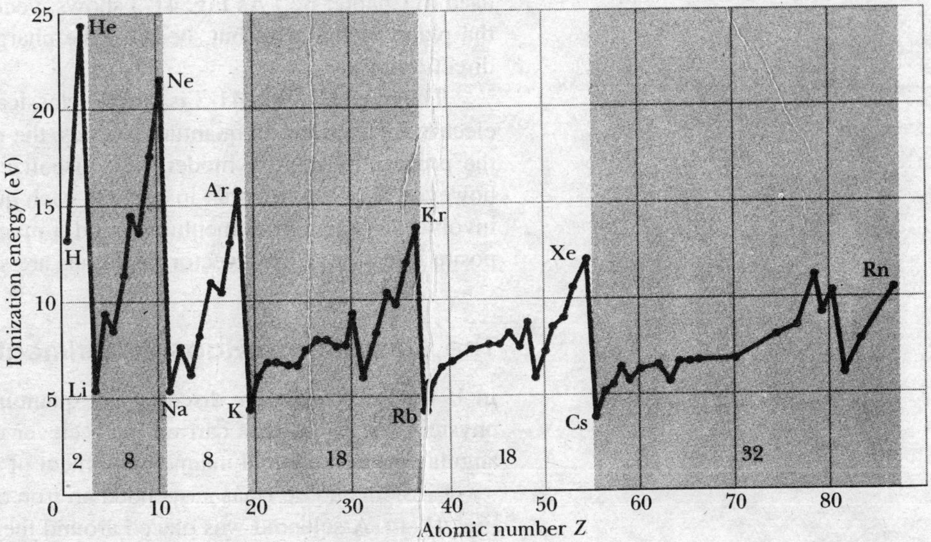

Fig. 41-2 A plot of the ionization energies of the elements as a function of atomic number, showing the periodic repetition of properties through the six complete horizontal periods of the periodic table. The number of elements in each of these periods is indicated.

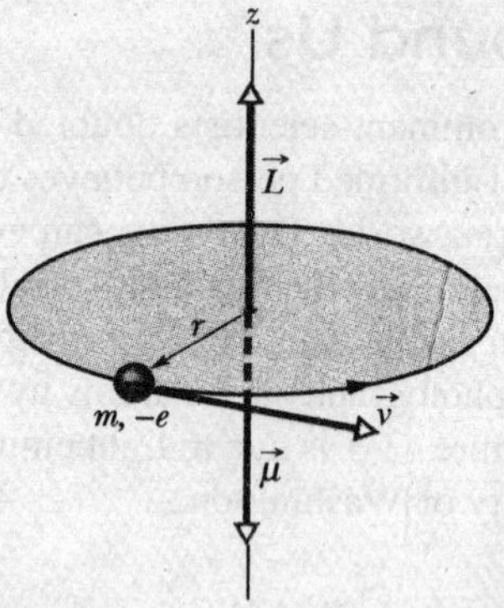

Fig. 41-3 A classical model showing a particle of mass m and charge $-e$ moving with speed v in a circle of radius r. The moving particle has an angular momentum $\vec{L}$ given by $\vec{r} \times \vec{p}$, where $\vec{p}$ is its linear momentum $m\vec{v}$. The particle's motion is equivalent to a current loop that has an associated magnetic momentum $\vec{\mu}$ which is directed opposite $\vec{L}$.

bound electron from a neutral atom is plotted as a function of the position in the periodic table of the element to which the atom belongs. The remarkable similarities in the chemical and physical properties of the elements in each vertical column of the periodic table are evidence enough that the atoms are constructed according to systematic rules.

The elements are arranged in the periodic table in six horizontal **periods;** except for the first, each period starts at the left with a highly reactive alkali metal (lithium, sodium, potassium, and so on) and ends at the right with a chemically inert noble gas (neon, argon, krypton, and so on). Quantum physics accounts for the chemical properties of these elements. The numbers of elements in the six periods are

2, 8, 8, 18, 18, and 32.

Quantum physics predicts these numbers.

Atoms Emit and Absorb Light

We have already seen that atoms can exist only in discrete quantum states, each state having a certain energy. An atom can make a transition from one state to another by emitting light (to jump to a lower energy level E_{low}) or by absorbing light (to jump to a higher energy level E_{high}). As we first discussed in Section 40-3, the light is emitted or absorbed as a photon with energy

$$hf = E_{high} - E_{low}. \tag{41-1}$$

Thus, the problem of finding the frequencies of light emitted or absorbed by an atom reduces to the problem of finding the energies of the quantum states of that atom. Quantum physics allows us—in principle at least—to calculate these energies.

Atoms Have Angular Momentum and Magnetism

Figure 41-3 shows a negatively charged particle moving in a circular orbit around a fixed center. As we discussed in Section 32-4, the orbiting particle has both an angular momentum $\vec{L}$ and (since its path is equivalent to a tiny current loop) a magnetic dipole moment $\vec{\mu}$. (Here, for brevity, we drop the subscript orb that we used in Chapter 32.) As Fig. 41-3 shows, vectors $\vec{L}$ and $\vec{\mu}$ are both perpendicular to the plane of the orbit but, because the charge is negative, they point in opposite directions.

The model of Fig. 41-3 is strictly classical and does not accurately represent an electron in an atom. In quantum physics, the rigid orbit model has been replaced by the probability density model, best visualized as a dot plot. In quantum physics, however, it is still true that in general, each quantum state of an electron in an atom involves an angular momentum $\vec{L}$ and a magnetic dipole moment $\vec{\mu}$ that have opposite directions (those vector quantities are said to be *coupled*).

The Einstein–de Haas Experiment

In 1915, well before the discovery of quantum physics, Albert Einstein and Dutch physicist W. J. de Haas carried out a clever experiment designed to show that the angular momentum and magnetic moment of individual atoms are coupled.

Einstein and de Haas suspended an iron cylinder from a thin fiber, as shown in Fig. 41-4*a*. A solenoid was placed around the cylinder but not touching it. Initially,

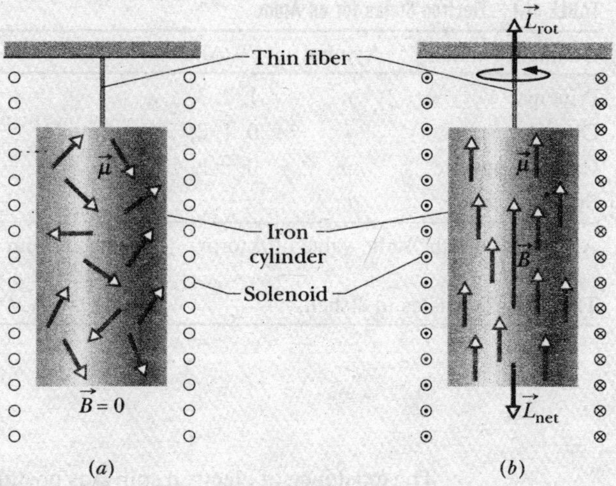

Fig. 41-4 The Einstein–de Haas experimental setup. (*a*) Initially, the magnetic field in the iron cylinder is zero and the magnetic dipole moment vectors $\vec{\mu}$ of its atoms are randomly oriented. The atomic angular momentum vectors (not shown) are directed opposite the magnetic dipole moment vectors and thus are also randomly oriented. (*b*) When a magnetic field $\vec{B}$ is set up along the cylinder's axis, the magnetic dipole moment vectors line up parallel to $\vec{B}$, which means that the angular momentum vectors line up opposite $\vec{B}$. Because the cylinder is initially isolated from external torques, its angular momentum is conserved and the cylinder as a whole must begin to rotate as shown.

the magnetic dipole moments $\vec{\mu}$ of the atoms of the cylinder point in random directions, so their external magnetic effects cancel (Fig. 41-4*a*). However, when a current is switched on in the solenoid (Fig. 41-4*b*) so that a magnetic field $\vec{B}$ is set up parallel to the axis of the cylinder, the magnetic dipole moments of the atoms of the cylinder reorient themselves, lining up with that field. If the angular momentum $\vec{L}$ of each atom is coupled to its magnetic moment $\vec{\mu}$, then this alignment of the atomic magnetic moments must cause an alignment of the atomic angular momenta opposite the magnetic field.

No external torques initially act on the cylinder; thus, its angular momentum must remain at its initial zero value. However, when $\vec{B}$ is turned on and the atomic angular momenta line up antiparallel to $\vec{B}$, they tend to give a net angular momentum $\vec{L}_{net}$ to the cylinder as a whole (directed downward in Fig. 41-4*b*). To maintain zero angular momentum, the cylinder begins to rotate around its central axis to produce an angular momentum $\vec{L}_{rot}$ in the opposite direction (upward in Fig. 41-4*b*).

Were it not for the fiber, the cylinder would continue to rotate for as long as the magnetic field is present. However, the twisting of the fiber quickly produces a torque that momentarily stops the cylinder's rotation and then rotates the cylinder in the opposite direction as the twisting is undone. Thereafter, the fiber will twist and untwist as the cylinder oscillates about its initial orientation in angular simple harmonic motion.

Observation of the cylinder's rotation verified that the angular momentum and the magnetic dipole moment of an atom are coupled in opposite directions. Moreover, it dramatically demonstrated that the angular momenta associated with quantum states of atoms can result in *visible* rotation of an object of everyday size.

41-3 Electron Spin

As we discussed in Section 32-4, whether an electron is *trapped* in an atom or is *free*, it has an intrinsic **spin angular momentum** $\vec{S}$, often called simply **spin.** (Recall that *intrinsic* means that $\vec{S}$ is a basic characteristic of an electron, like its mass and electric charge.) As we shall discuss in the next section, the magnitude of $\vec{S}$ is quantized and depends on a **spin quantum number** s, which is always $\frac{1}{2}$ for electrons (and for protons and neutrons). In addition, the component of $\vec{S}$ measured along any axis is quantized and depends on a **spin magnetic quantum number** m_s, which can have only the value $+\frac{1}{2}$ or $-\frac{1}{2}$.

TABLE 41-1 Electron States for an Atom

Quantum Number	Symbol	Allowed Values	Related to
Principal	n	$1, 2, 3, \ldots$	Distance from the nucleus
Orbital	l	$0, 1, 2, \ldots, (n-1)$	Orbital angular momentum
Orbital magnetic	m_l	$0, \pm 1, \pm 2, \ldots, \pm l$	Orbital angular momentum (z component)
Spin magnetic	m_s	$\pm\frac{1}{2}$	Spin angular momentum (z component)

All states with the same value of n form a **shell.**
There are $2n^2$ states in a shell.

All states with the same values of n and l form a **subshell.**
All states in a subshell have the same energy.
There are $2(2l + 1)$ states in a subshell.

The existence of electron spin was postulated on an empirical basis by two Dutch graduate students, George Uhlenbeck and Samuel Goudsmit, from their studies of atomic spectra. The quantum physics basis for electron spin was provided a few years later, by British physicist P. A. M. Dirac, who developed (in 1929) a relativistic quantum theory of the electron.

It is tempting to account for electron spin by thinking of the electron as a tiny sphere spinning about an axis. However, that classical model, like the classical model of orbits, does not hold up. In quantum physics, spin angular momentum is best thought of as a measurable intrinsic property of the electron; you simply can't visualize it with a classical model.

Table 41-1, an extension of Table 40-2, shows the four quantum numbers n, l, m_l, and m_s that completely specify the quantum states of the electron in a hydrogen atom. (Quantum number s is not included because all electrons have the value $s = \frac{1}{2}$.) The same quantum numbers also specify the allowed states of any single electron in a multielectron atom.

41-4 Angular Momenta and Magnetic Dipole Moments

Every quantum state of an electron in an atom has an associated orbital angular momentum and a corresponding orbital magnetic dipole moment. Every electron, whether trapped in an atom or free, has a spin angular momentum and a corresponding spin magnetic dipole moment. We discuss these quantities separately first, and then in combination.

Orbital Angular Momentum and Magnetism

The magnitude L of the **orbital angular momentum** $\vec{L}$ of an electron *in an atom* is quantized; that is, it can have only certain values. These values are

$$L = \sqrt{l(l+1)}\hbar, \tag{41-2}$$

in which l is the orbital quantum number and $\hbar$ is $h/2\pi$. According to Table 41-1, l must be either zero or a positive integer no greater than $n - 1$. For a state with $n = 3$, for example, only $l = 2$, $l = 1$, and $l = 0$ are permitted.

As we discussed in Section 32-4, a magnetic dipole is associated with the orbital angular momentum $\vec{L}$ of an electron in an atom. This magnetic dipole has an **orbital**

magnetic dipole moment $\vec{\mu}_{orb}$, which is related to the angular momentum by Eq. 32-8:

$$\vec{\mu}_{orb} = -\frac{e}{2m}\vec{L}. \tag{41-3}$$

The minus sign in this relation means that $\vec{\mu}_{orb}$ is directed opposite $\vec{L}$. Because the magnitude of $\vec{L}$ is quantized (Eq. 41-2), the magnitude of $\vec{\mu}_{orb}$ must also be quantized and given by

$$\mu_{orb} = \frac{e}{2m}\sqrt{l(l+1)}\hbar. \tag{41-4}$$

Neither $\vec{\mu}_{orb}$ nor $\vec{L}$ can be measured in any way. However, we *can* measure the components of those two vectors along a given axis. Let us imagine that the atom is located in a magnetic field $\vec{B}$; assume that a z axis extends in the direction of the field lines. Then we can measure the z components of $\vec{\mu}_{orb}$ and $\vec{L}$ along that axis.

The components $\mu_{orb,z}$ of the orbital magnetic dipole moment are quantized and given by

$$\mu_{orb,z} = -m_l\mu_B. \tag{41-5}$$

Here m_l is the orbital magnetic quantum number of Table 41-1 and μ_B is the **Bohr magneton:**

$$\mu_B = \frac{eh}{4\pi m} = \frac{e\hbar}{2m} = 9.274 \times 10^{-24}\ \text{J/T} \qquad \text{(Bohr magneton)}, \tag{41-6}$$

where m is the electron mass.

The components L_z of the angular momentum are also quantized, and they are given by

$$L_z = m_l\hbar. \tag{41-7}$$

Figure 41-5 shows the five quantized components L_z of the orbital angular momentum for an electron with $l = 2$, as well as the associated orientations of the angular momentum $\vec{L}$. However, *do not take the figure literally* because we cannot detect $\vec{L}$ in any way. Thus, drawing it in a figure like Fig. 41-5 is merely a visual aide. We can extend that visual aide by saying that $\vec{L}$ makes a certain angle θ with the z axis, such that

$$\cos\theta = \frac{L_z}{L}. \tag{41-8}$$

We can call θ the *semi-classical angle* between vector $\vec{L}$ and the z axis, because it is a classical measurement of something that quantum theory tells us cannot be measured.

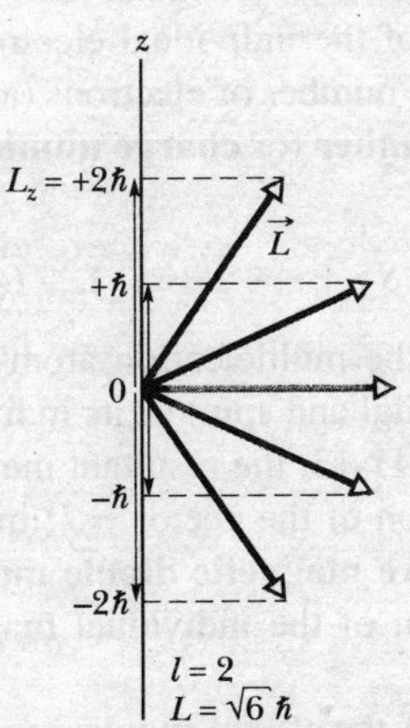

Fig. 41-5 The allowed values of L_z for an electron in a quantum state with $l = 2$. For every orbital angular momentum vector $\vec{L}$ in the figure, there is a vector pointing in the opposite direction, representing the magnitude and direction of the orbital magnetic dipole moment $\vec{\mu}_{orb}$.

Spin Angular Momentum and Spin Magnetic Dipole Moment

The magnitude S of the spin angular momentum $\vec{S}$ of any electron, whether *free or trapped,* has the single value given by

$$S = \sqrt{s(s+1)}\hbar$$
$$= \sqrt{(\tfrac{1}{2})(\tfrac{1}{2}+1)}\hbar = 0.866\hbar, \tag{41-9}$$

where s $(= \tfrac{1}{2})$ is the spin quantum number of the electron.

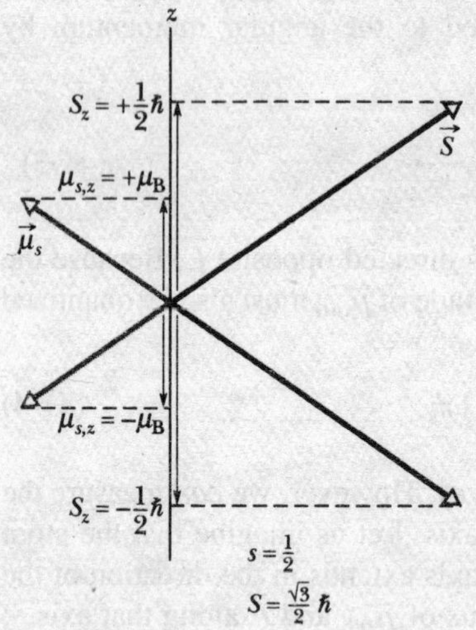

Fig. 41-6 The allowed values of S_z and μ_z for an electron.

As we discussed in Section 32-4, an electron has an intrinsic magnetic dipole that is associated with its spin angular momentum $\vec{S}$, whether the electron is confined to an atom or free. This magnetic dipole has a **spin magnetic dipole moment** $\vec{\mu}_s$, which is related to the spin angular momentum by Eq. 32-2:

$$\vec{\mu}_s = -\frac{e}{m}\vec{S}. \tag{41-10}$$

The minus sign in this relation means that $\vec{\mu}_s$ is directed opposite $\vec{S}$. Because the magnitude of $\vec{S}$ is quantized (Eq. 41-9), the magnitude of $\vec{\mu}_s$ must also be quantized and given by

$$\mu_s = \frac{e}{m}\sqrt{s(s+1)}\hbar. \tag{41-11}$$

Neither $\vec{S}$ nor $\vec{\mu}_s$ can be measured in any way. However, we *can* measure their components along any given axis—call it the z axis. The components S_z of the spin angular momentum are quantized and given by

$$S_z = m_s\hbar, \tag{41-12}$$

where m_s is the spin magnetic quantum number of Table 41-1. That quantum number can have only two values: $m_s = +\frac{1}{2}$ (the electron is said to be *spin up*) and $m_s = -\frac{1}{2}$ (the electron is said to be *spin down*).

The components $\mu_{s,z}$ of the spin magnetic dipole moment are also quantized, and they are given by

$$\mu_{s,z} = -2m_s\mu_B. \tag{41-13}$$

Figure 41-6 shows the two quantized components S_z of the spin angular momentum for an electron and the associated orientations of vector $\vec{S}$. It also shows the quantized components $\mu_{s,z}$ of the spin magnetic dipole moment and the associated orientations of $\vec{\mu}_s$.

Orbital and Spin Angular Momenta Combined

For an atom containing more than one electron, we define a total angular momentum $\vec{J}$, which is the vector sum of the angular momenta of the individual electrons—both their orbital and their spin angular momenta. The number of electrons (and the number of protons) in a neutral atom is the **atomic number** (or **charge number**) Z. Thus, for a neutral atom,

$$\vec{J} = (\vec{L}_1 + \vec{L}_2 + \vec{L}_3 + \cdots + \vec{L}_Z) + (\vec{S}_1 + \vec{S}_2 + \vec{S}_3 + \cdots + \vec{S}_Z). \tag{41-14}$$

Similarly, the total magnetic dipole moment of the multielectron atom is the vector sum of the magnetic dipole moments (both orbital and spin) of its individual electrons. However, because of the factor of 2 in Eq. 41-13, the resultant magnetic dipole moment for the atom does not have the direction of the vector $-\vec{J}$; instead, it makes a certain angle with that vector. The **effective magnetic dipole moment** $\vec{\mu}_{\text{eff}}$ for the atom is the component of the vector sum of the individual magnetic dipole moments in the direction of $-\vec{J}$ (Fig. 41-7).

As you will see in the next section, in typical atoms the orbital angular momenta and the spin angular momenta of most of the electrons sum vectorially to zero. Then $\vec{J}$ and $\vec{\mu}_{\text{eff}}$ of those atoms are due to a relatively small number of electrons, often only a single valence electron.

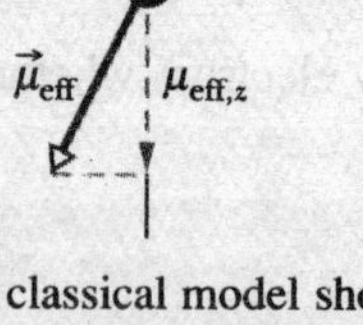

Fig. 41-7 A classical model showing the total angular momentum vector $\vec{J}$ and the effective magnetic moment vector $\vec{\mu}_{\text{eff}}$.

CHECKPOINT 1: An electron is in a quantum state for which the magnitude of the electron's orbital angular momentum $\vec{L}$ is $2\sqrt{3}\hbar$. How many projections of the electron's orbital magnetic dipole moment on a z axis are allowed?

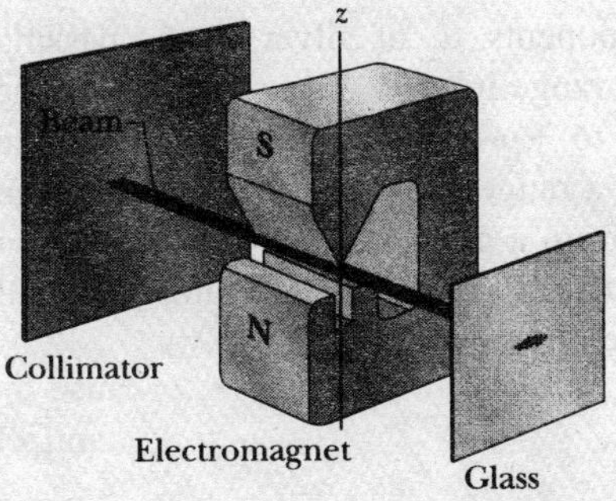

Fig. 41-8 Apparatus used by Stern and Gerlach.

41-5 The Stern–Gerlach Experiment

In 1922, Otto Stern and Walther Gerlach at the University of Hamburg in Germany showed experimentally that the magnetic moment of cesium atoms is quantized. In the Stern–Gerlach experiment, as it is now known, silver is vaporized in an oven, and some of the atoms in that vapor escape through a narrow slit in the oven wall, into an evacuated tube. Some of those escaping atoms then pass through a second narrow slit, to form a narrow beam of atoms (Fig. 41-8). (The atoms are said to be *collimated*—made into a beam—and the second slit is called a *collimator*.) The beam passes between the poles of an electromagnet and then lands on a glass detector plate where it forms a silver deposit.

When the electromagnet is off, the silver deposit is a narrow spot. However, when the electromagnet is turned on, the silver deposit is spread vertically. The spreading occurs because silver atoms are magnetic dipoles, so vertical magnetic forces act on them as they pass through the vertical magnetic field of the electromagnet; these forces deflect them slightly up or down. Thus, by analyzing the silver deposit on the plate, we can determine what deflections the atoms underwent in the magnetic field. When Stern and Gerlach analyzed the pattern of silver on their detector plate, they found a surprise. However, before we discuss that surprise and its quantum implications, let us discuss the magnetic deflecting force acting on the silver atoms.

The Magnetic Deflecting Force on a Silver Atom

We have not previously discussed the type of magnetic force that deflects the silver atoms in a Stern–Gerlach experiment. It is *not* the magnetic deflecting force that acts on a moving charged particle, as given by Eq. 29-2 ($\vec{F} = q\vec{v} \times \vec{B}$). The reason is simple: A silver atom is electrically neutral (its net charge q is zero) and thus this type of magnetic force is also zero.

The type of magnetic force we seek is due to an interaction between the magnetic field $\vec{B}$ of the electromagnet and the magnetic dipole of the individual silver atom. We can derive an expression for the force in this interaction by starting with the potential energy U of the dipole in the magnetic field. Equation 29-38 tells us that

$$U = -\vec{\mu} \cdot \vec{B}, \qquad (41\text{-}15)$$

where $\vec{\mu}$ is the magnetic dipole moment of a silver atom. In Fig. 41-8, the positive direction of the z axis and the direction of $\vec{B}$ are vertically upward. Thus, we can write Eq. 41-15 in terms of the component μ_z of the atom's magnetic dipole moment along the direction of $\vec{B}$:

$$U = -\mu_z B. \qquad (41\text{-}16)$$

Then, using Eq. 8-20 ($F = -dU/dx$) for the z axis shown in Fig. 41-8, we obtain

$$F_z = -\frac{dU}{dz} = \mu_z \frac{dB}{dz}. \qquad (41\text{-}17)$$

This is what we sought—an equation for the magnetic force that deflects a silver atom as the atom passes through a magnetic field.

The term dB/dz in Eq. 41-17 is the *gradient* of the magnetic field along the z axis. If the magnetic field does not change along the z axis (as in a uniform magnetic field or no magnetic field), then $dB/dz = 0$ and a silver atom is not deflected as it moves between the magnet's poles. In the Stern–Gerlach experiment, the poles are designed to maximize the gradient dB/dz, to vertically deflect the silver atoms passing between the poles as much as possible, so that their deflections show up in the deposit on the glass plate.

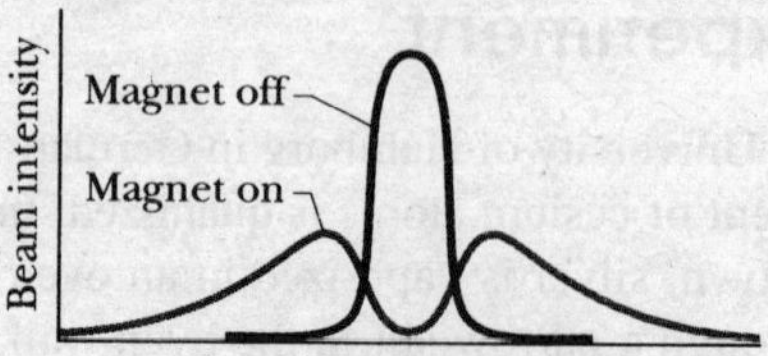

Fig. 41-9 Results of a modern repetition of the Stern–Gerlach experiment. With the electromagnet turned off, there is only a single beam; with the electromagnet turned on, the original beam splits into two subbeams. The two subbeams correspond to parallel and antiparallel alignment of the magnetic moments of cesium atoms with the external magnetic field.

According to classical physics, the components μ_z of silver atoms passing through the magnetic field in Fig. 41-8 should range in value from $-\mu$ (the dipole moment $\vec{\mu}$ is directed straight down the z axis) to $+\mu$ ($\vec{\mu}$ is directed straight up the z axis). Thus, from Eq. 41-17, there should be a range of forces on the atoms, and therefore a range of deflections of the atoms, from a greatest downward deflection to a greatest upward deflection. This means that we should expect the atoms to land along a vertical line on the glass plate. However, this does not happen.

The Experimental Surprise

What Stern and Gerlach found was that the atoms formed two distinct spots on the glass plate, one spot above the point where they would have landed with no deflection and the other spot just as far below that point. This two-spot result can be seen in the plots of Fig. 41-9, which shows the outcome of a more recent version of the Stern–Gerlach experiment. In that version, a beam of cesium atoms (magnetic dipoles like the silver atoms in the original Stern–Gerlach experiment) was sent through a magnetic field with a large vertical gradient dB/dz. The field could be turned on and off, and a detector could be moved up and down through the beam.

When the field was turned off, the beam was, of course, undeflected and the detector recorded the central-peak pattern shown in Fig. 41-9. When the field was turned on, the original beam was split vertically by the magnetic field into two smaller beams, one beam higher than the previously undeflected beam and the other beam lower. As the detector moved vertically up through these two smaller beams, it recorded the two-peak pattern shown in Fig. 41-9.

The Meaning of the Results

In the original Stern–Gerlach experiment, two spots of silver were formed on the glass plate, not a vertical line of silver. This means that the component μ_z along $\vec{B}$ (and z) could not have any value between $-\mu$ and $+\mu$ as classical physics predicts. Instead, μ_z is restricted to only two values, one for each spot on the glass. Thus, the original Stern–Gerlach experiment showed that μ_z is quantized, implying (correctly) that $\vec{\mu}$ is also. Moreover, because the angular momentum $\vec{L}$ of an atom is associated with $\vec{\mu}$, that angular momentum and its component L_z are also quantized.

With modern quantum theory, we can add to the explanation of the two-spot result in the Stern–Gerlach experiment. We now know that a silver atom consists of many electrons, each with a spin magnetic moment and an orbital magnetic moment. We also know that all those moments vectorially cancel out *except* for a single electron, and the orbital dipole moment of that electron is zero. Thus, the combined dipole moment $\vec{\mu}$ of a silver atom is the *spin* magnetic dipole moment of that single electron. According to Eq. 41-13, that means that μ_z can have only two components along the z axis in Fig. 41-8. One component is for quantum number $m_s = +\frac{1}{2}$ (the single electron is spin up), and the other component is for quantum number $m_s = -\frac{1}{2}$ (the single electron is spin down). Substituting into Eq. 41-13 gives us

$$\mu_{s,z} = -2(+\tfrac{1}{2})\mu_B = -\mu_B \quad \text{and} \quad \mu_{s,z} = -2(-\tfrac{1}{2})\mu_B = +\mu_B. \qquad (41\text{-}18)$$

Then substituting these expressions for μ_z in Eq. 41-17, we find that the force component F_z deflecting the silver atoms as they pass through the magnetic field can have only the two values

$$F_z = -\mu_B\left(\frac{dB}{dz}\right) \quad \text{and} \quad F_z = +\mu_B\left(\frac{dB}{dz}\right), \qquad (41\text{-}19)$$

which result in the two spots of silver on the glass.

Sample Problem 41-1

In the Stern–Gerlach experiment of Fig. 41-8, a beam of silver atoms passes through a magnetic field gradient dB/dz of magnitude 1.4 T/mm that is set up along the z axis. This region has a length w of 3.5 cm in the direction of the original beam. The speed of the atoms is 750 m/s. By what distance d have the atoms been deflected when they leave the region of the magnetic field gradient? The mass M of a silver atom is 1.8×10^{-25} kg.

SOLUTION: One **Key Idea** here is that the deflection of a silver atom in the beam is due to an interaction between the magnetic dipole of the atom and the magnetic field, because of the gradient dB/dz. The deflecting force is directed along the field gradient (along the z axis) and is given by Eqs. 41-17. Let us consider only deflection in the positive direction of z; thus, we shall use $F_z = \mu_B(dB/dz)$ from Eqs. 41-19.

A second **Key Idea** is that we assume the field gradient dB/dz has the same value throughout the region through which the silver atoms travel. Thus, force component F_z is constant in that region, and from Newton's second law, the acceleration a_z of an atom along the z axis due to F_z is also constant and is given by

$$a_z = \frac{F_z}{M} = \frac{\mu_B(dB/dz)}{M}.$$

Because this acceleration is constant, we can use Eq. 2-15 (from Table 2-1) to write the deflection d parallel to the z axis as

$$d = v_{0z}t + \tfrac{1}{2}a_z t^2 = 0t + \tfrac{1}{2}\left(\frac{\mu_B(dB/dz)}{M}\right)t^2. \tag{41-20}$$

Because the deflecting force on the atom acts perpendicular to the atom's original direction of travel, the component v of the atom's velocity along the original direction of travel is not changed by the force. Thus, the atom requires time $t = w/v$ to travel through length w in that direction. Substituting w/v for t into Eq. 41-20, we find

$$d = \tfrac{1}{2}\left(\frac{\mu_B(dB/dz)}{M}\right)\left(\frac{w}{v}\right)^2 = \frac{\mu_B(dB/dz)w^2}{2Mv^2}$$

$$= (9.27 \times 10^{-24}\ \text{J/T})(1.4 \times 10^3\ \text{T/m}) \times \frac{(3.5 \times 10^{-2}\ \text{m})^2}{(2)(1.8 \times 10^{-25}\ \text{kg})(750\ \text{m/s})^2}$$

$$= 7.85 \times 10^{-5}\ \text{m} \approx 0.08\ \text{mm}. \qquad \text{(Answer)}$$

The separation between the two subbeams is twice this, or 0.16 mm. This separation is not large but is easily measured.

41-6 Magnetic Resonance

As we discussed briefly in Section 32-4, a proton has an intrinsic spin angular momentum $\vec{S}$ and an associated spin magnetic dipole moment $\vec{\mu}$ that are in the same direction (because the proton is positively charged). If a proton is located in a uniform magnetic field $\vec{B}$ directed along a z axis, the z component μ_z of the spin magnetic dipole moment can have only two quantized orientations: either parallel to $\vec{B}$ or antiparallel to $\vec{B}$, as shown in Fig. 41-10a. From Eq. 29-38, we know that these two orientations differ in energy by $2\mu_z B$, which is the energy involved in reversing a magnetic dipole in a uniform magnetic field. The lower energy state is the one with μ_z parallel to $\vec{B}$, and the higher energy state has μ_z antiparallel to $\vec{B}$.

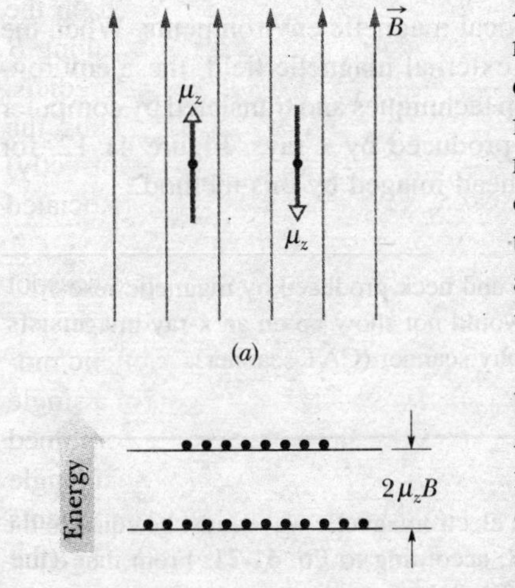

Fig. 41-10 (a) A proton (red dot), whose spin component in the direction of an applied magnetic field is $\frac{1}{2}\hbar$, can occupy either of two quantized orientations in an external magnetic field. If Eq. 41-21 is satisfied, the protons in the sample can be induced to flip from one orientation to the other. (b) Normally, there are more protons in the lower energy state than in the higher energy state.

Let us place a drop of water in a uniform magnetic field $\vec{B}$; then the protons in the hydrogen of the water molecules each have μ_z either parallel or antiparallel to $\vec{B}$. If we next apply to the drop an alternating electromagnetic field of a certain frequency f, the protons in the lower energy state can undergo reversal of their μ_z orientation. This process of reversal is called *spin-flipping* (because the reversal of a proton's magnetic dipole moment requires a reversal of the proton's spin). The frequency f required for the spin-flipping is given by

$$hf = 2\mu_z B, \tag{41-21}$$

a condition called **magnetic resonance** (or, as originally, **nuclear magnetic resonance**). In words, if an alternating electromagnetic field is to cause protons to spin-flip in the magnetic field, the photons associated with that field must have an energy hf equal to the energy difference $2\mu_z B$ between the two possible orientations of μ_z (and thus proton spin) in that field.

Once a proton is spin-flipped to the higher energy state, it can drop back to the lower energy state by emitting a photon of the same energy hf given by Eq. 41-21.

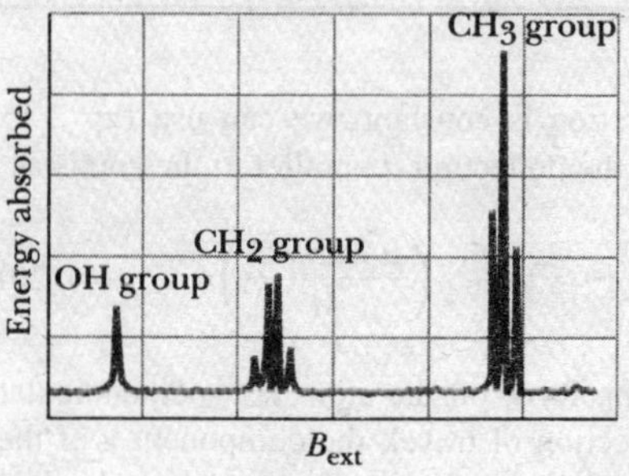

Fig. 41-11 A nuclear magnetic resonance spectrum for ethanol. The spectral lines represent the absorption of energy associated with spin flips of protons. The three groups of lines correspond, as indicated, to protons in the OH group, the CH_2 group, and the CH_3 group of the ethanol molecule. Note that the two protons in the CH_2 group occupy four different local environments. The entire horizontal axis covers less than 10^{-4} T.

Normally more protons are in the lower energy state than in the higher energy state, as Fig. 41-10*b* suggests. This means that there will be a detectable net *absorption* of energy from the alternating electromagnetic field.

The constant field whose magnitude $\vec{B}$ appears in Eq. 41-21 is actually *not* the imposed external field $\vec{B}_{ext}$ in which the water drop is placed; rather, it is that field as modified by the small, local, internal magnetic field $\vec{B}_{local}$ due to the magnetic moments of the atoms and nuclei near a given proton. Thus, we can rewrite Eq. 41-21 as

$$hf = 2\mu_z(B_{ext} + B_{local}). \tag{41-22}$$

To achieve magnetic resonance, it is customary to leave the frequency f of the electromagnetic oscillations fixed and to vary B_{ext} until Eq. 41-22 is satisfied and an absorption peak is recorded.

Nuclear magnetic resonance is a property that is the basis for a valuable analytical tool, particularly for the identification of unknown compounds. Figure 41-11 shows a **nuclear magnetic resonance spectrum,** as it is called, for ethanol, whose formula we may write as CH_3-CH_2-OH. The various resonance peaks all represent spin flips of protons. They occur at different values of B_{ext}, however, because the local environments of the six protons within the ethanol molecule differ from one another. The spectrum of Fig. 41-11 is a unique signature for ethanol.

Spin technology, called **magnetic resonance imaging** (MRI), has been applied to medical diagnostics with great success. The protons of the various tissues of the human body are situated in many different local magnetic environments. When the body, or part of it, is immersed in a strong external magnetic field, these environmental differences can be detected by spin-flip techniques and translated by computer processing into an image resembling those produced by x rays. Figure 41-12, for example, shows a cross section of a human head imaged by this method.

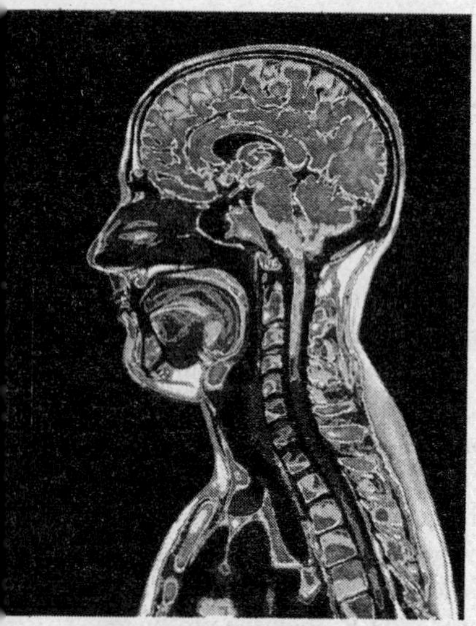

Fig. 41-12 A cross-sectional view of a human head and neck produced by magnetic resonance imaging. Some of the details visible here would not show up on an x-ray image, even with a modern computerized axial tomography scanner (CAT scanner).

Sample Problem 41-2

A drop of water is suspended in a magnetic field $\vec{B}$ of magnitude 1.80 T and an alternating electromagnetic field is applied, its frequency adjusted to produce spin flips of the protons in the water. The component μ_z of the magnetic dipole moment of a proton, measured along the direction of $\vec{B}$, is 1.41×10^{-26} J/T. Assume that the local magnetic fields are negligible compared to $\vec{B}$. What are the frequency f and wavelength λ of the alternating field?

SOLUTION: One **Key Idea** here is that when a proton is located in a magnetic field $\vec{B}$, it has a potential energy because it is a magnetic dipole. A second **Key Idea** is that this potential energy is restricted to two values, with a difference of $2\mu_z B$. The third **Key Idea** is that if the proton is to jump between these two energies (spin-flip), the photon energy hf of the electromagnetic wave must be equal to the energy difference $2\mu_z B$, according to Eq. 41-21. From that equation, we then find

$$f = \frac{2\mu_z B}{h} = \frac{(2)(1.41 \times 10^{-26}\ \text{J/T})(1.80\ \text{T})}{6.63 \times 10^{-34}\ \text{J}\cdot\text{s}}$$

$$= 7.66 \times 10^{7}\ \text{Hz} = 76.6\ \text{MHz}. \qquad \text{(Answer)}$$

The corresponding wavelength is

$$\lambda = \frac{c}{f} = \frac{3.00 \times 10^{8}\ \text{m/s}}{7.66 \times 10^{7}\ \text{Hz}} = 3.92\ \text{m}. \qquad \text{(Answer)}$$

This frequency and wavelength are in the short radio wave region of the electromagnetic spectrum.

41-7 The Pauli Exclusion Principle

In Chapter 40 we considered a variety of electron traps, from fictional one-dimensional traps to the real three-dimensional trap of a hydrogen atom. In all those examples, we trapped only one electron. However, when we discuss traps containing two or more electrons (as we shall in the next two sections), we must consider a principle that governs any particle whose spin quantum number s is not zero or an integer. This principle applies not only to electrons but also to protons and neutrons, all of which have $s = \frac{1}{2}$. The principle is known as the **Pauli exclusion principle** after Wolfgang Pauli, who formulated it in 1925. For electrons, it states that

> No two electrons confined to the same trap can have the same set of values for its quantum numbers.

As we shall discuss in Section 41-9, this principle means that no two electrons in an atom can have the same four values for the quantum numbers n, l, m_l, and m_s. In other words, the quantum numbers of any two electrons in an atom must differ in at least one quantum number. Were this not true, atoms would collapse, and thus you and the world as you know it could not exist.

41-8 Multiple Electrons in Rectangular Traps

To prepare for our discussion of multiple electrons in atoms, let us discuss two electrons confined to the rectangular traps of Chapter 40. We shall again use the quantum numbers we found for those traps when only one electron was confined. However, here we shall also include the spin angular momenta of the two electrons. To do this, we assume that the traps are located in a uniform magnetic field. Then according to Eq. 41-12, an electron can be either spin up with $m_s = \frac{1}{2}$ or spin down with $m_s = -\frac{1}{2}$. (We shall assume that the magnetic field is very weak so that we can neglect the potential energies of the electrons due to the field.)

As we confine the two electrons to one of the traps, we must keep the Pauli exclusion principle in mind; that is, the electrons cannot have the same set of values for their quantum numbers.

1. *One-dimensional trap.* In the one-dimensional trap of Fig. 40-2, fitting an electron wave to the trap's width L requires the single quantum number n. Therefore, any electron confined to the trap must have a certain value of n, and its quantum number m_s can be either $+\frac{1}{2}$ or $-\frac{1}{2}$. The two electrons could have different values of n, or they could have the same value of n if one of them is spin up and the other is spin down.
2. *Rectangular corral.* In the rectangular corral of Fig. 40-12, fitting an electron wave to the corral's widths L_x and L_y requires the two quantum numbers n_x and n_y. Thus, any electron confined to the trap must have certain values for those two quantum numbers, and its quantum number m_s can be either $+\frac{1}{2}$ or $-\frac{1}{2}$—so now there are three quantum numbers. According to the Pauli exclusion principle, two electrons confined to the trap must have different values for at least one of those three quantum numbers.
3. *Rectangular box.* In the rectangular box of Fig. 40-13, fitting an electron wave to the box's widths L_x, L_y, and L_z requires the three quantum numbers n_x, n_y, and n_z. Thus, any electron confined to the trap must have certain values for these three quantum numbers, and its quantum number m_s can be either $+\frac{1}{2}$ or $-\frac{1}{2}$—so now there are four quantum numbers. According to the Pauli exclusion

principle, two electrons confined to the trap must have different values for at least one of those four quantum numbers.

Suppose we add more than two electrons, one by one, to a rectangular trap in the preceding list. The first electrons naturally go into the lowest possible energy level—they are said to *occupy* that level. However, eventually the Pauli exclusion principle disallows any more electrons from occupying that lowest energy level, and the next electron must occupy the next higher level. When an energy level cannot be occupied by more electrons because of the Pauli exclusion principle, we say that level is **full** or **fully occupied.** In contrast, a level that is not occupied by any electrons is **empty** or **unoccupied.** For intermediate situations, the level is **partially occupied.** The *electron configuration* of a system of trapped electrons is a listing or drawing of the energy levels the electrons occupy, or the set of the quantum numbers of the electrons.

Finding the Total Energy

We shall later want to find the energy of a *system* of two or more electrons confined to a rectangular trap. That is, we shall want to find the total energy for any configuration of the trapped electrons.

For simplicity, we shall assume that the electrons do not electrically interact with one another; that is, we shall neglect the electric potential energies of pairs of electrons. In that case, we can calculate the total energy for any electron configuration by calculating the energy of each electron as we did in Chapter 40, and then summing those energies. (In Sample Problem 41-3 we do so for seven electrons confined to a rectangular corral.)

A good way to organize the energy values of a given system of electrons is with an energy-level diagram *for the system,* just as we did for a single electron in the traps of Chapter 40. The lowest level, with energy E_{gr}, corresponds to the ground state of the system. The next higher level, with energy E_{fe}, corresponds to the first excited state of the system. The next level, with energy E_{se}, corresponds to the second excited state of the system. And so on.

Sample Problem 41-3

Seven electrons are confined to the square corral of Sample Problem 40-5, where the corral is a two-dimensional infinite potential well with widths $L_x = L_y = L$ (Fig. 40-12). Assume that the electrons do not electrically interact with one another.

(a) What is the electron configuration for the ground state of the system of seven electrons?

SOLUTION: We can determine the electron configuration of the system by placing the seven electrons in the corral one by one, to build up the system. One **Key Idea** here is that because we assume the electrons do not electrically interact with one another, we can use the energy-level diagram for a single trapped electron in order to keep track of how we place the seven electrons in the corral. That *one-electron energy-level diagram* is given in Fig. 40-14 and partially reproduced here as Fig. 41-13*a*. Recall that the levels are labeled as $E_{nx,ny}$ for their associated energy. For example, the lowest level is for energy $E_{1,1}$, where quantum number n_x is 1 and quantum number n_y is 1.

A second **Key Idea** here is that the trapped electrons must obey the Pauli exclusion principle; that is, no two electrons can have the same set of values for their quantum numbers n_x, n_y, and m_s.

The first electron goes into energy level $E_{1,1}$ and can have $m_s = \frac{1}{2}$ or $m_s = -\frac{1}{2}$. We arbitrarily choose the latter and draw a down arrow (to represent spin down) on the $E_{1,1}$ level in Fig. 41-13*a*. The second electron also goes into the $E_{1,1}$ level but must have $m_s = +\frac{1}{2}$ so that one of its quantum numbers differs from those of the first electron. We represent this second electron with an up arrow (for spin up) on the $E_{1,1}$ level in Fig. 41-13*b*.

Another **Key Idea** now comes into play: The level for energy $E_{1,1}$ is fully occupied, and thus the third electron cannot have that energy. Therefore, the third electron goes into the next higher level, which is for the equal energies $E_{2,1}$ and $E_{1,2}$ (the level is degenerate). This third electron can have quantum numbers n_x and n_y of either 1 and 2 or 2 and 1, respectively. It can also have a quantum number m_s of either $+\frac{1}{2}$ or $-\frac{1}{2}$. Let us arbitrarily assign it the quantum numbers $n_x = 2$, $n_y = 1$, and $m_s = -\frac{1}{2}$. We then represent it with a down arrow on the level for $E_{1,2}$ and $E_{2,1}$ in Fig. 41-13*c*.

You can show that the next three electrons can also go into

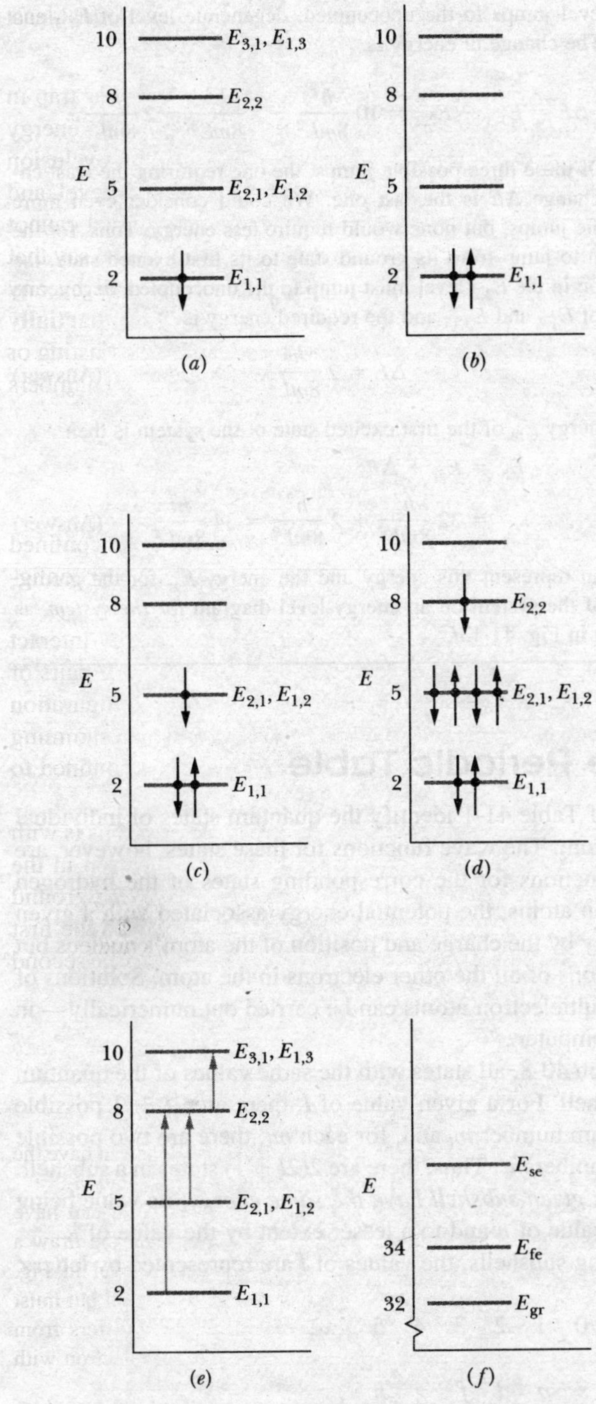

Fig. 41-13 (*a*) Energy-level diagram for one electron in a square corral of widths L. (Energy E is in multiples of $h^2/8mL^2$.) A spin-down electron occupies the lowest level. (*b*) Two electrons (one spin down, the other spin up) occupy the lowest level of the one-electron energy-level diagram. (*c*) A third electron occupies the next energy level. (*d*) The system's ground-state configuration, for all 7 electrons. (*e*) Three transitions to consider as possibly taking the 7-electron system to its first excited state. (*f*) The system's energy-level diagram, for the lowest three total energies of the system (in multiples of $h^2/8mL^2$).

TABLE 41-2 Ground-State Configuration and Energies

n_x	n_y	m_s	Energy*
2	2	$-\frac{1}{2}$	8
2	1	$+\frac{1}{2}$	5
2	1	$-\frac{1}{2}$	5
1	2	$+\frac{1}{2}$	5
1	2	$-\frac{1}{2}$	5
1	1	$+\frac{1}{2}$	2
1	1	$-\frac{1}{2}$	2
		Total	32

*In multiples of $h^2/8mL^2$

the level for energies $E_{2,1}$ and $E_{1,2}$, provided that no set of three quantum numbers is completely duplicated. That level then contains four electrons, with quantum numbers (n_x, n_y, m_s) of

$$(2, 1, -\tfrac{1}{2}),\ (2, 1, +\tfrac{1}{2}),\ (1, 2, -\tfrac{1}{2}),\ (1, 2, +\tfrac{1}{2}),$$

and the level is fully occupied. Thus, the seventh electron goes into the next higher level, which is the $E_{2,2}$ level. Let us arbitrarily assume it is spin down, with $m_s = -\frac{1}{2}$.

Figure 41-13*d* shows all seven electrons on a one-electron energy-level diagram. We now have seven electrons in the corral, and they are in the configuration with the lowest energy that satisfies the Pauli exclusion principle. Thus, the ground-state configuration of the system is that shown in Fig. 41-13*d* and listed in Table 41-2.

(b) What is the total energy of the seven-electron system in its ground state, as a multiple of $h^2/8mL^2$?

SOLUTION: The **Key Idea** here is that the total energy E_{gr} of the system in its ground state is the sum of the energies of the individual electrons in the system's ground-state configuration. The energy of each electron can be read from Table 40-1, which is partially reproduced in Table 41-2, or from Fig. 41-13*d*. Because there are two electrons in the first (lowest) level, four in the second level, and one in the third level, we have

$$E_{gr} = 2\left(2\frac{h^2}{8mL^2}\right) + 4\left(5\frac{h^2}{8mL^2}\right) + 1\left(8\frac{h^2}{8mL^2}\right)$$

$$= 32\frac{h^2}{8mL^2}. \qquad \text{(Answer)}$$

(c) How much energy must be transferred to the system for it to jump to its first excited state, and what is the energy of that state?

SOLUTION: The **Key Ideas** here are these:

1. If the system is to be excited, one of the seven electrons must make a quantum jump up the one-electron energy-level diagram of Fig. 41-13*d*.
2. If that jump is to occur, the energy change ΔE of the electron (and thus the system) must be $\Delta E = E_{high} - E_{low}$ (Eq. 40-5),

where E_{low} is the energy of the level where the jump begins and E_{high} is the energy of the level where the jump ends.

3. The Pauli exclusion principle must still apply; in particular, an electron *cannot* jump to a level that is fully occupied.

Let us consider the three jumps shown in Fig. 41-13*e*; all are allowed by the Pauli exclusion principle because they are jumps to empty or partially occupied states. In one of those possible jumps, an electron jumps from the $E_{1,1}$ level to the partially occupied $E_{2,2}$ level. The change in the energy is

$$\Delta E = E_{2,2} - E_{1,1} = 8\frac{h^2}{8mL^2} - 2\frac{h^2}{8mL^2} = 6\frac{h^2}{8mL^2}.$$

(We shall assume that the spin orientation of the electron making the jump can change as needed.)

In another of the possible jumps in Fig. 41-13*e*, an electron jumps from the degenerate level of $E_{2,1}$ and $E_{1,2}$ to the partially occupied $E_{2,2}$ level. The change in the energy is

$$\Delta E = E_{2,2} - E_{2,1} = 8\frac{h^2}{8mL^2} - 5\frac{h^2}{8mL^2} = 3\frac{h^2}{8mL^2}.$$

In the third possible jump in Fig. 41-13*e*, the electron in the $E_{2,2}$ level jumps to the unoccupied, degenerate level of $E_{1,3}$ and $E_{3,1}$. The change in energy is

$$\Delta E = E_{1,3} - E_{2,2} = 10\frac{h^2}{8mL^2} - 8\frac{h^2}{8mL^2} = 2\frac{h^2}{8mL^2}.$$

Of these three possible jumps, the one requiring the least energy change ΔE is the last one. We could consider even more possible jumps, but none would require less energy. Thus, for the system to jump from its ground state to its first excited state, the electron in the $E_{2,2}$ level must jump to the unoccupied, degenerate level of $E_{1,3}$ and $E_{3,1}$, and the required energy is

$$\Delta E = 2\frac{h^2}{8mL^2}. \qquad \text{(Answer)}$$

The energy E_{fe} of the first excited state of the system is then

$$E_{\text{fe}} = E_{\text{gr}} + \Delta E$$
$$= 32\frac{h^2}{8mL^2} + 2\frac{h^2}{8mL^2} = 34\frac{h^2}{8mL^2}. \qquad \text{(Answer)}$$

We can represent this energy and the energy E_{gr} for the ground state of the system on an energy-level diagram *for the system,* as shown in Fig. 41-13*f*.

41-9 Building the Periodic Table

The four quantum numbers of Table 41-1 identify the quantum states of individual electrons in a multielectron atom. The wave functions for these states, however, are not the same as the wave functions for the corresponding states of the hydrogen atom because, in multielectron atoms, the potential energy associated with a given electron is determined not only by the charge and position of the atom's nucleus but also by the charges and positions of all the other electrons in the atom. Solutions of Schrödinger's equation for multielectron atoms can be carried out numerically—in principle at least—using a computer.

As we discussed in Section 40-8, all states with the same values of the quantum numbers n and l form a subshell. For a given value of l, there are $2l + 1$ possible values of the magnetic quantum number m_l and, for each m_l, there are two possible values for the spin quantum number m_s. Thus, there are $2(2l + 1)$ states in a subshell. It turns out that *all states in a given subshell have the same energy,* its value being determined primarily by the value of n and to a lesser extent by the value of l.

For the purpose of labeling subshells, the values of l are represented by letters:

$$l = 0 \quad 1 \quad 2 \quad 3 \quad 4 \quad 5 \quad \ldots$$

$$s \quad p \quad d \quad f \quad g \quad h \quad \ldots$$

For example, the $n = 3$, $l = 2$ subshell would be labeled the $3d$ subshell.

When we assign electrons to states in a multielectron atom, we must be guided by the Pauli exclusion principle of Section 41-7; that is, no two electrons in an atom can have the same set of the quantum numbers n, l, m_l, and m_s. If this important principle did not hold, *all* the electrons in any atom could jump to the atom's lowest energy level, which would eliminate the chemistry of atoms and molecules, and thus also biochemistry. Let us examine the atoms of a few elements to see how the Pauli exclusion principle operates in the building up of the periodic table.

Neon

The neon atom has 10 electrons. Only two of them fit into the lowest energy subshell, the 1s subshell. These two electrons both have $n = 1$, $l = 0$, and $m_l = 0$, but one has $m_s = +\frac{1}{2}$ and the other has $m_s = -\frac{1}{2}$. The 1s subshell, according to Table 41-1, contains $2(2l + 1) = 2$ states. Because this subshell then contains all the electrons permitted by the Pauli principle, it is said to be **closed.**

Two of the remaining eight electrons fill the next lowest energy subshell, the 2s subshell. The last six electrons just fill the 2p subshell which, with $l = 1$, holds $2(2l + 1) = 6$ states.

In a closed subshell, all allowed z projections of the orbital angular momentum vector $\vec{L}$ are present and, as you can verify from Fig. 41-5, these projections cancel for the subshell as a whole; for every positive projection there is a corresponding negative projection of the same magnitude. Similarly, the z projections of the spin angular momenta also cancel. Thus, a closed subshell has no angular momentum and no magnetic moment of any kind. Furthermore, its probability density is spherically symmetric. Then neon with its three closed subshells (1s, 2s, and 2p) has no "loosely dangling electrons" to encourage chemical interaction with other atoms. Neon, like the other **noble gases** that form the right-hand column of the periodic table, is chemically inert.

Sodium

Next after neon in the periodic table comes sodium, with 11 electrons. Ten of them form a closed neonlike core, which, as we have seen, has zero angular momentum. The remaining electron is largely outside this inert core, in the 3s subshell—the next lowest energy subshell. Because this **valence electron** of sodium is in a state with $l = 0$ (that is, an s state), the sodium atom's angular momentum and magnetic dipole moment must be due entirely to the spin of this single electron.

Sodium readily combines with other atoms that have a "vacancy" into which sodium's loosely bound valence electron can fit. Sodium, like the other **alkali metals** that form the left-hand column of the periodic table, is chemically active.

Chlorine

The chlorine atom, which has 17 electrons, has a closed 10-electron, neonlike core, with 7 electrons left over. Two of them fill the 3s subshell, leaving five to be assigned to the 3p subshell, which is the subshell next lowest in energy. This subshell, which has $l = 1$, can hold $2(2l + 1) = 6$ electrons, so there is a vacancy, or a "hole," in this subshell.

Chlorine is receptive to interacting with other atoms that have a valence electron that might fill this hole. Sodium chloride (NaCl), for example, is a very stable compound. Chlorine, like the other **halogens** that form column VIIA of the periodic table, is chemically active.

Iron

The arrangement of the 26 electrons of the iron atom can be represented as follows:

$$\underbrace{1s^2\ \ 2s^2\ 2p^6\ \ 3s^2\ 3p^6}\ 3d^6\ \ 4s^2.$$

The subshells are listed in numerical order and, following convention, a superscript gives the number of electrons in each subshell. From Table 41-1 we can see that an

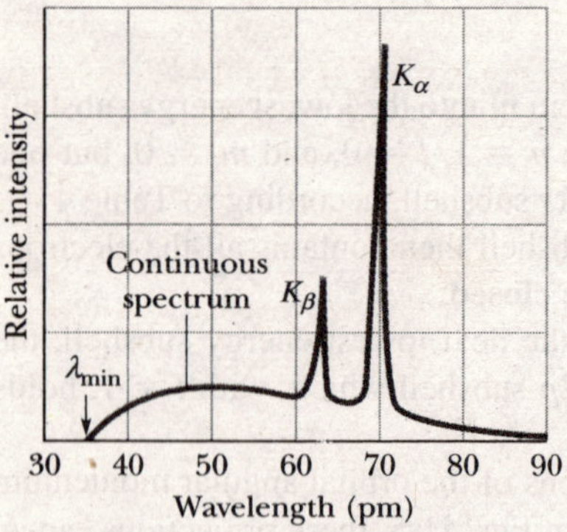

Fig. 41-14 The distribution by wavelength of the x rays produced when 35 keV electrons strike a molybdenum target. The sharp peaks and the continuous spectrum from which they rise are produced by different mechanisms.

s-subshell can hold 2 electrons, a *p*-subshell 6, and a *d*-subshell 10. Thus, iron's first 18 electrons form the five filled subshells that are marked off by the bracket, leaving 8 electrons to be accounted for. Six of the eight go into the 3*d* subshell and the remaining two go into the 4*s* subshell.

The last two electrons do not also go into the 3*d* subshell (which can hold 10 electrons) because the $3d^6\ 4s^2$ configuration results in a lower energy state for the atom as a whole than would the $3d^8$ configuration. An iron atom with 8 electrons (rather than 6) in the 3*d* subshell would quickly make a transition to the $3d^6\ 4s^2$ configuration, emitting electromagnetic radiation in the process. The lesson here is that except for the simplest elements, the states may not be filled in what we might think of as their "logical" sequence.

41-10 X Rays and the Numbering of the Elements

When a solid target, such as solid copper or tungsten, is bombarded with electrons whose kinetic energies are in the kiloelectron-volt range, electromagnetic radiation called **x rays** is emitted. Our concern here is what these rays—whose medical, dental, and industrial usefulness is so well known and widespread—can teach us about the atoms that absorb or emit them. Figure 41-14 shows the wavelength spectrum of the x rays produced when a beam of 35 keV electrons falls on a molybdenum target. We see a broad, continuous spectrum of radiation on which are superimposed two peaks of sharply defined wavelengths. The continuous spectrum and the peaks arise in different ways, which we next discuss separately.

The Continuous X-Ray Spectrum

Here we examine the continuous x-ray spectrum of Fig. 41-14, ignoring for the time being the two prominent peaks that rise from it. Consider an electron of initial kinetic energy K_0 that collides (interacts) with one of the target atoms, as in Fig. 41-15. The electron may lose an amount of energy ΔK, which will appear as the energy of an x-ray photon that is radiated away from the site of the collision. (Very little energy is transferred to the recoiling atom because of the relatively large mass of the atom; here we neglect that transfer.)

The scattered electron in Fig. 41-15, whose energy is now less than K_0, may have a second collision with a target atom, generating a second photon, whose energy will in general be different from the energy of the photon produced in the first collision. This electron-scattering process can continue until the electron is approximately stationary. All the photons generated by these collisions form part of the continuous x-ray spectrum.

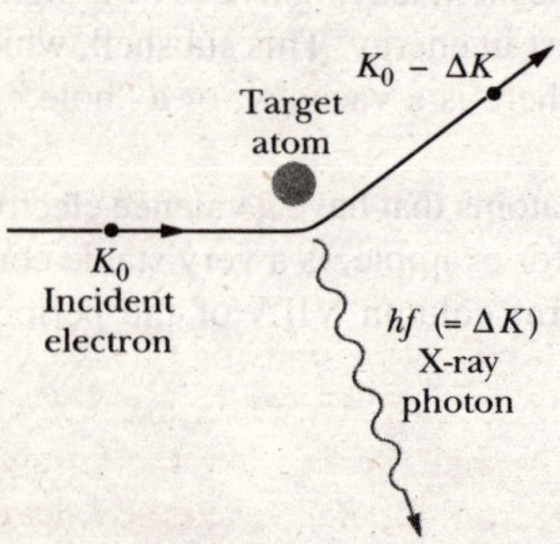

Fig. 41-15 An electron of kinetic energy K_0 passing near an atom in the target may generate an x-ray photon, the electron losing part of its energy in the process. The continuous x-ray spectrum arises in this way.

A prominent feature of that spectrum in Fig. 41-14 is the sharply defined **cutoff wavelength** λ_{min}, below which the continuous spectrum does not exist. This minimum wavelength corresponds to a collision in which an incident electron loses *all* its initial kinetic energy K_0 in a single head-on collision with a target atom. Essentially all this energy appears as the energy of a single photon, whose associated wavelength—the minimum possible x-ray wavelength—is found from

$$K_0 = hf = \frac{hc}{\lambda_{min}},$$

or

$$\lambda_{min} = \frac{hc}{K_0} \quad \text{(cutoff wavelength).} \qquad (41\text{-}23)$$

The cutoff wavelength is totally independent of the target material. If we were to switch from a molybdenum target to a copper target, for example, all features of the x-ray spectrum of Fig. 41-14 would change *except* the cutoff wavelength.

CHECKPOINT 2: Does the cutoff wavelength λ_{min} of the continuous x-ray spectrum increase, decrease, or remain the same if you (a) increase the kinetic energy of the electrons that strike the x-ray target, (b) allow the electrons to strike a thin foil rather than a thick block of the target material, (c) change the target to an element of higher atomic number?

Sample Problem 41-4

A beam of 35.0 keV electrons strikes a molybdenum target, generating the x rays whose spectrum is shown in Fig. 41-14. What is the cutoff wavelength?

SOLUTION: The **Key Idea** here is that the cutoff wavelength λ_{min} corresponds to an electron transferring (approximately) all of its energy to an x-ray photon, thus producing a photon with the greatest possible frequency and least possible wavelength. From Eq. 41-23, we have

$$\lambda_{min} = \frac{hc}{K_0} = \frac{(4.14 \times 10^{-15}\ \text{eV} \cdot \text{s})(3.00 \times 10^8\ \text{m/s})}{35.0 \times 10^3\ \text{eV}}$$

$$= 3.55 \times 10^{-11}\ \text{m} = 35.5\ \text{pm}. \quad \text{(Answer)}$$

The Characteristic X-Ray Spectrum

We now turn our attention to the two peaks of Fig. 41-14, labeled K_α and K_β. These (and other peaks that appear at wavelengths beyond the wavelength range displayed in Fig. 41-14) form the **characteristic x-ray spectrum** of the target material.

The peaks arise in a two-part process. (1) An energetic electron strikes an atom in the target and, while it is being scattered, the incident electron knocks out one of the atom's deep-lying (low n value) electrons. If the deep-lying electron is in the shell defined by $n = 1$ (called, for historical reasons, the K shell), there remains a vacancy, or *hole,* in this shell. (2) An electron in one of the shells with a higher energy jumps to the K shell, filling the hole in this shell. During this jump, the atom emits a characteristic x-ray photon. If the electron that fills the K-shell vacancy jumps from the shell with $n = 2$ (called the L shell), the emitted radiation is the K_α line of Fig. 41-14; if it jumps from the shell with $n = 3$ (called the M shell), it produces the K_β line, and so on. The hole left in either the L or M shell will be filled by an electron from still farther out in the atom.

In studying x rays, it is more convenient to keep track of the hole created deep in the atom's "electron cloud" than to record the changes in the quantum state of the electrons that jump to fill that hole. Figure 41-16 does exactly that; it is an energy-level diagram for molybdenum, the element to which Fig. 41-14 refers. The baseline ($E = 0$) represents the neutral atom in its ground state. The level marked K (at $E =$ 20 keV) represents the energy of the molybdenum atom with a hole in its K shell. Similarly, the level marked L (at $E = 2.7$ keV) represents the atom with a hole in its L shell, and so on.

The transitions marked K_α and K_β in Fig. 41-16 are the ones that produce the two x-ray peaks in Fig. 41-14. The K_α spectral line, for example, originates when an electron from the L shell fills a hole in the K shell. In Fig. 41-16, this jump corresponds to a *downward* transition of the hole, from the K level to the L level.

Fig. 41-16 A simplified atomic energy-level diagram for molybdenum, showing the transitions (of holes rather than electrons) that give rise to some of the characteristic x rays of that element. Each horizontal line represents the energy of the atom with a hole (a missing electron) in the shell indicated.

Numbering the Elements

In 1913 British physicist H. G. J. Moseley generated characteristic x rays for as many elements as he could find—he found 38—by using them as targets for electron bombardment in an evacuated tube of his own design. By means of a trolley manipulated by strings, Moseley was able to move the individual targets into the path of an electron beam. He measured the wavelengths of the emitted x rays by the crystal diffraction method described in Section 37-9.

Moseley then sought (and found) regularities in these spectra as he moved from element to element in the periodic table. In particular, he noted that if, for a given spectral line such as K_α, he plotted for each element the square root of the frequency f against the position of the element in the periodic table, a straight line resulted. Figure 41-17 shows a portion of his extensive data. Moseley's conclusion was this:

> We have here a proof that there is in the atom a fundamental quantity, which increases by regular steps as we pass from one element to the next. This quantity can only be the charge on the central nucleus.

Owing to Moseley's work, the characteristic x-ray spectrum became the universally accepted signature of an element, permitting the solution of a number of periodic table puzzles. Prior to that time (1913), the positions of elements in the table were assigned in order of atomic *weight,* although it was necessary to invert this order for several pairs of elements because of compelling chemical evidence; Moseley showed that it is the nuclear charge (that is, the atomic *number* Z) that is the real basis for numbering the elements.

In 1913 the periodic table had several empty squares, and a surprising number of claims for new elements had been advanced. The x-ray spectrum provided a conclusive test of such claims. The lanthanide elements, often called the rare earth elements, had been sorted out only imperfectly because their similar chemical properties made sorting difficult. Once Moseley's work was reported, these elements were properly organized. In more recent times, the identities of some elements beyond uranium were pinned down beyond dispute when the elements became available in quantities large enough to permit a study of their individual x-ray spectra.

It is not hard to see why the characteristic x-ray spectrum shows such impressive regularities from element to element whereas the optical spectrum in the visible and near-visible region does not: The key to the identity of an element is the charge on its nucleus. Gold, for example, is what it is because its atoms have a nuclear charge of $+79e$ (that is, $Z = 79$). An atom with one more elementary charge on its nucleus is mercury; with one fewer, it is platinum. The K electrons, which play such a large role in the production of the x-ray spectrum, lie very close to the nucleus and are thus sensitive probes of its charge. The optical spectrum, on the other hand, involves transitions of the outermost electrons, which are heavily screened from the nucleus by the remaining electrons of the atom and thus are *not* sensitive probes of nuclear charge.

Accounting for the Moseley Plot

Moseley's experimental data, of which the Moseley plot of Fig. 41-17 is but a part, can be used directly to assign the elements to their proper places in the periodic table. This can be done even if no theoretical basis for Moseley's results can be established. However, there is such a basis.

According to Eq. 40-24 the energy of the hydrogen atom is

$$E_n = -\frac{me^4}{8\varepsilon_0^2 h^2}\frac{1}{n^2} = -\frac{13.6\text{ eV}}{n^2}, \qquad \text{for } n = 1, 2, 3, \ldots . \qquad (41\text{-}24)$$

Consider now one of the two innermost electrons in the K shell of a multielectron atom. Because of the presence of the other K-shell electron, our electron "sees" an effective nuclear charge of approximately $(Z - 1)e$, where e is the elementary charge and Z is the atomic number of the element. The factor e^4 in Eq. 41-24 is the product

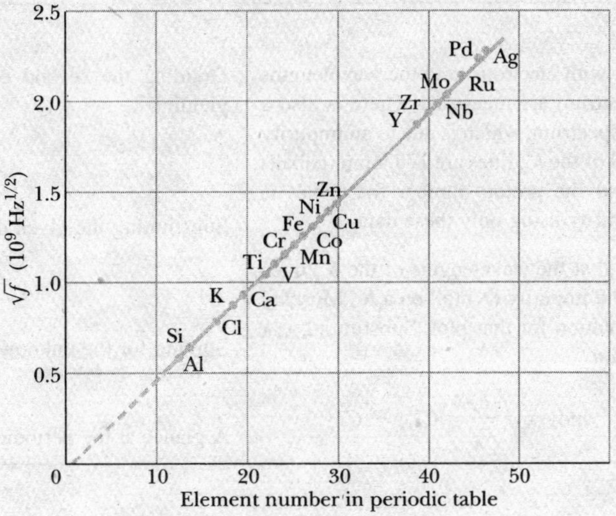

Fig. 41-17 A Moseley plot of the K_α line of the characteristic x-ray spectra of 21 elements. The frequency is calculated from the measured wavelength.

of e^2—the square of hydrogen's nuclear charge—and $(-e)^2$—the square of an electron's charge. For a multielectron atom, we can approximate the effective energy of the atom by replacing the factor e^4 in Eq. 41-24 with $(Z-1)^2e^2 \times (-e)^2$, or $e^4(Z-1)^2$. That gives us

$$E_n = -\frac{(13.6\text{ eV})(Z-1)^2}{n^2}. \tag{41-25}$$

We saw that the K_α x-ray photon (of energy hf) arises when an electron makes a transition from the L shell (with $n = 2$ and energy E_2) to the K shell (with $n = 1$ and energy E_1). Thus, using Eq. 41-25, we may write the energy change as

$$\begin{aligned}\Delta E &= E_2 - E_1 \\ &= \frac{-(13.6\text{ eV})(Z-1)^2}{2^2} - \frac{-(13.6\text{ eV})(Z-1)^2}{1^2} \\ &= (10.2\text{ eV})(Z-1)^2.\end{aligned}$$

Then the frequency f of the K_α line is

$$\begin{aligned}f &= \frac{\Delta E}{h} = \frac{(10.2\text{ eV})(Z-1)^2}{(4.14 \times 10^{-15}\text{ eV}\cdot\text{s})} \\ &= (2.46 \times 10^{15}\text{ Hz})(Z-1)^2.\end{aligned} \tag{41-26}$$

Taking the square root of both sides yields

$$\sqrt{f} = CZ - C, \tag{41-27}$$

in which C is a constant ($= 4.96 \times 10^7\text{ Hz}^{1/2}$). Equation 41-27 is the equation of a straight line. It shows that if we plot the square root of the frequency of the K_α x-ray spectral line against the atomic number Z, we should obtain a straight line. As Fig. 41-17 shows, that is exactly what Moseley found.

CHECKPOINT 3: The K_α x rays arising from a cobalt ($Z = 27$) target have a wavelength of about 179 pm. Is the wavelength of the K_α x rays arising from a nickel ($Z = 28$) target greater than or less than 179 pm?

Sample Problem 41-5

A cobalt target is bombarded with electrons, and the wavelengths of its characteristic x-ray spectrum are measured. There is also a second, fainter characteristic spectrum, which is due to an impurity in the cobalt. The wavelengths of the K_α lines are 178.9 pm (cobalt) and 143.5 pm (impurity), and the proton number for cobalt is $Z_{Co} = 27$. Determine the impurity using only these data.

SOLUTION: The **Key Idea** here is that the wavelengths of the K_α lines for both the cobalt (Co) and the impurity (X) fall on a K_α Moseley plot, and Eq. 41-27 is the equation for that plot. Substituting c/λ for f in that equation, we obtain

$$\sqrt{\frac{c}{\lambda_{Co}}} = CZ_{Co} - C \quad \text{and} \quad \sqrt{\frac{c}{\lambda_X}} = CZ_X - C.$$

Dividing the second equation by the first neatly eliminates C, yielding

$$\sqrt{\frac{\lambda_{Co}}{\lambda_X}} = \frac{Z_X - 1}{Z_{Co} - 1}.$$

Substituting the given data yields

$$\sqrt{\frac{178.9 \text{ pm}}{143.5 \text{ pm}}} = \frac{Z_X - 1}{27 - 1}.$$

Solving for the unknown, we find that

$$Z_X = 30.0. \quad \text{(Answer)}$$

A glance at the periodic table identifies the impurity as zinc.

41-11 Lasers and Laser Light

In the late 1940s and again in the early 1960s, quantum physics made two enormous contributions to technology: the **transistor,** which ushered in the computer revolution, and the **laser.** Laser light, like the light from an ordinary lightbulb, is emitted when atoms make a transition from one quantum state to a quantum state of lower energy. In a laser, however—but not in other light sources—the atoms act together to produce light with several special characteristics:

1. ***Laser light is highly monochromatic.*** Light from an ordinary incandescent lightbulb is spread over a continuous range of wavelengths and is certainly not monochromatic. The radiation from a fluorescent neon sign *is* monochromatic, to about 1 part in about 10^6. However, the sharpness of definition of laser light can be many times greater, as much as 1 part in 10^{15}.
2. ***Laser light is highly coherent.*** Individual long waves (*wave trains*) for laser light can be several hundred kilometers long. When two separated beams that have traveled such distances over separate paths are recombined, they "remember" their common origin and are able to form a pattern of interference fringes. The corresponding *coherence length* for wave trains emitted by a lightbulb is typically less than a meter.
3. ***Laser light is highly directional.*** A laser beam spreads very little; it departs from strict parallelism only because of diffraction at the exit aperture of the laser. For example, a laser pulse used to measure the distance to the Moon generates a spot on the Moon's surface with a diameter of only a few meters. Light from an ordinary bulb can be made into an approximately parallel beam by a lens, but the beam divergence is much greater than for laser light. Each point on a lightbulb's filament forms its own separate beam, and the angular divergence of the overall composite beam is set by the size of the filament.
4. ***Laser light can be sharply focused.*** If two light beams transport the same amount of energy, the beam that can be focused to the smaller spot will have the greater intensity at that spot. For laser light, the focused spot can be so small that an intensity of 10^{17} W/cm^2 is readily obtained. An oxyacetylene flame, by contrast, has an intensity of only about 10^3 W/cm^2.

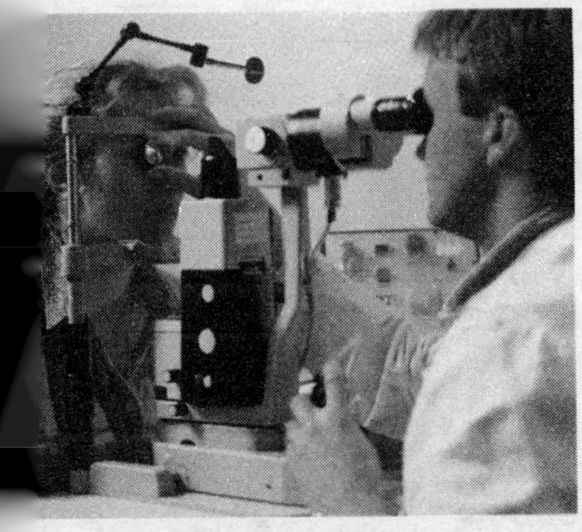

Fig. 41-18 A patient's loose retina is being welded into place by a laser directed into her eye.

Lasers Have Many Uses

The smallest lasers, used for voice and data transmission over optical fibers, have as their active medium a semiconducting crystal about the size of a pinhead. Small as they are, such lasers can generate about 200 mW of power. The largest lasers, used for nuclear fusion research and for astronomical and military applications, fill a large building. The largest such laser can generate brief pulses of laser light with a power level, during the pulse, of about 10^{14} W. This is a few hundred times greater than the total electric power generating capacity of the United States. To avoid a brief national power blackout during a pulse, the energy required for each pulse is stored up at a steady rate during the relatively long interpulse interval.

Among the many uses of lasers are reading bar codes, manufacturing and reading compact discs, performing surgery of many kinds (see the opening photo of this chapter and Fig. 41-18), surveying, cutting cloth in the garment industry (several hundred layers at a time), welding auto bodies, and generating holograms.

41-12 How Lasers Work

The word "laser" is an acronym for "light amplification by the stimulated emission of radiation," so you should not be surprised that **stimulated emission** is the key to laser operation. Einstein introduced this concept in 1917. Although the world had to wait until 1960 to see an operating laser, the groundwork for its development was put in place decades earlier.

Consider an isolated atom that can exist either in its state of lowest energy (its ground state), whose energy is E_0, or in a state of higher energy (an excited state), whose energy is E_x. Here are three processes by which the atom can move from one of these states to the other:

1. ***Absorption.*** Figure 41-19*a* shows the atom initially in its ground state. If the atom is placed in an electromagnetic field that is alternating at frequency *f*, the atom can absorb an amount of energy *hf* from that field and move to the higher energy state. From the principle of conservation of energy we have

$$hf = E_x - E_0. \tag{41-28}$$

   We call this process **absorption.**

2. ***Spontaneous emission.*** In Fig. 41-19*b* the atom is in its excited state and no external radiation is present. After a time, the atom will move *of its own accord* to its ground state, emitting a photon of energy *hf* in the process. We call this process **spontaneous emission**—*spontaneous* because the event was not triggered by any outside influence. The light from the filament of an ordinary lightbulb is generated in this way.

   Normally, the mean life of excited atoms before spontaneous emission occurs is about 10^{-8} s. However, for some excited states, this mean life is perhaps as much as 10^5 times longer. We call such long-lived states **metastable;** they play an important role in laser operation.

3. ***Stimulated emission.*** In Fig. 41-19*c* the atom is again in its excited state, but this time radiation with a frequency given by Eq. 41-28 is present. A photon of energy *hf* can stimulate the atom to move to its ground state, during which process the atom emits an additional photon, whose energy is also *hf*. We call this process **stimulated emission**—*stimulated* because the event is triggered by the external photon. The emitted photon is in every way identical to the stimulating photon.

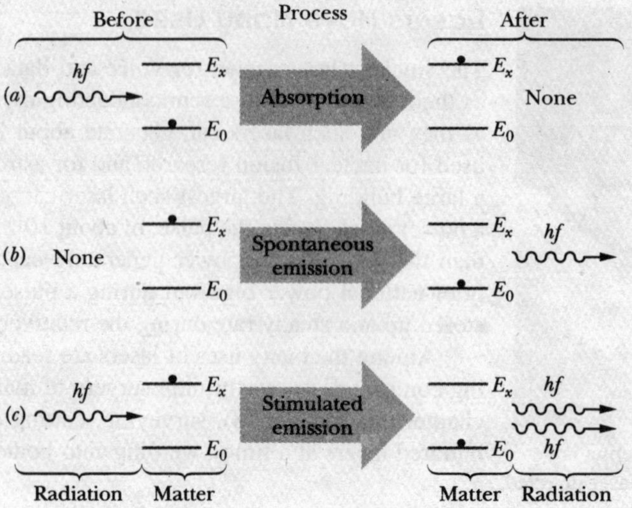

Fig. 41-19 The interaction of radiation and matter in the processes of (*a*) absorption, (*b*) spontaneous emission, and (*c*) stimulated emission. An atom (matter) is represented by the red dot; the atom is in either a lower quantum state with energy E_0 or a higher quantum state with energy E_x. In (*a*) the atom absorbs a photon of energy *hf* from a passing light wave. In (*b*) it emits a light wave by emitting a photon of energy *hf*. In (*c*) a passing light wave with photon energy *hf* causes the atom to emit a photon of the same energy, increasing the energy of the light wave.

Thus, the waves associated with the photons have the same energy, phase, polarization, and direction of travel.

Figure 41-19*c* describes stimulated emission for a single atom. Suppose now that a sample contains a large number of atoms in thermal equilibrium at temperature T. Before any radiation is directed at the sample, a number N_0 of these atoms are in their ground state with energy E_0, and a number N_x are in a state of higher energy E_x. Ludwig Boltzmann showed that N_x is given in terms of N_0 by

$$N_x = N_0 e^{-(E_x - E_0)/kT}, \tag{41-29}$$

in which k is Boltzmann's constant. This equation seems reasonable. The quantity kT is the mean kinetic energy of an atom at temperature T. The higher the temperature, the more atoms—on average—will have been "bumped up" by thermal agitation (that is, by atom–atom collisions) to the higher energy state E_x. Also, because $E_x > E_0$, Eq. 41-29 requires that $N_x < N_0$; that is, there will always be fewer atoms in the excited state than in the ground state. This is what we expect if the level populations N_0 and N_x are determined only by the action of thermal agitation. Figure 41-20*a* illustrates this situation.

If we now flood the atoms of Fig. 41-20*a* with photons of energy $E_x - E_0$, photons will disappear via absorption by ground-state atoms, and photons will be generated largely via stimulated emission of excited-state atoms. Einstein showed that the probabilities per atom for these two processes are identical. Thus, because there are more atoms in the ground state, the *net* effect will be the absorption of photons.

To produce laser light, we must have more photons emitted than absorbed; that is, we must have a situation in which stimulated emission dominates. The direct way to bring this about is to start with more atoms in the excited state than in the ground state, as in Fig. 41-20*b*. However, since such a **population inversion** is not consistent with thermal equilibrium, we must think up clever ways to set up and maintain one.

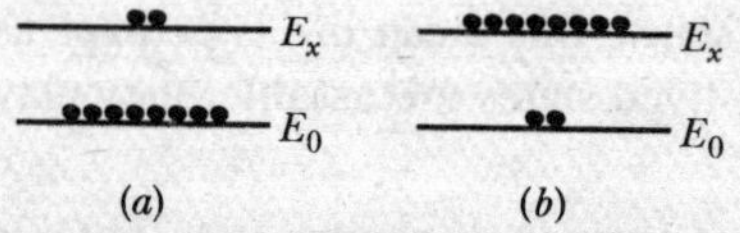

Fig. 41-20 (*a*) The equilibrium distribution of atoms between the ground state E_0 and excited state E_x, accounted for by thermal agitation. (*b*) An inverted population, obtained by special methods. Such an inverted population is essential for laser action.

The Helium–Neon Gas Laser

Figure 41-21 shows a type of laser commonly found in student laboratories. It was developed in 1961 by Ali Javan and his coworkers. The glass discharge tube is filled

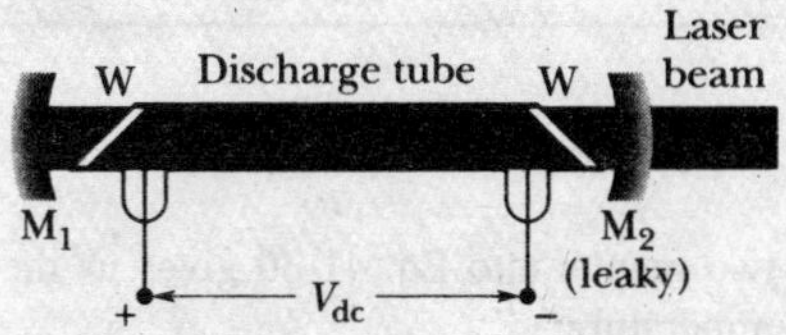

Fig. 41-21 The elements of a helium–neon gas laser. An applied potential V_{dc} sends electrons through a discharge tube containing a mixture of helium gas and neon gas. Electrons collide with helium atoms, which then collide with neon atoms, which emit light along the length of the tube. The light passes through transparent windows W and reflects back and forth through the tube from mirrors M_1 and M_2 to cause more neon atom emissions. Some of the light leaks through mirror M_2 to form the laser beam.

with a 20 : 80 mixture of helium and neon gases, neon being the medium in which laser action occurs.

Figure 41-22 shows simplified energy-level diagrams for the two atoms. An electric current passed through the helium–neon gas mixture serves—through collisions between helium atoms and electrons of the current—to raise many helium atoms to state E_3, which is metastable.

The energy of helium state E_3 (20.61 eV) is very close to the energy of neon state E_2 (20.66 eV). Thus, when a metastable (E_3) helium atom and a ground-state (E_0) neon atom collide, the excitation energy of the helium atom is often transferred to the neon atom, which then moves to state E_2. In this manner, neon level E_2 in Fig. 41-22 can become more heavily populated than neon level E_1.

This population inversion is relatively easy to set up because (1) initially there are essentially no neon atoms in state E_1, (2) the metastability of helium level E_3 ensures a ready supply of neon atoms in level E_2, and (3) atoms in level E_1 decay rapidly (through intermediate levels not shown) to the neon ground state E_0.

Suppose now that a single photon is spontaneously emitted as a neon atom transfers from state E_2 to state E_1. Such a photon can trigger a stimulated emission event, which, in turn, can trigger other stimulated emission events. Through such a chain reaction, a coherent beam of red laser light, moving parallel to the tube axis, can build up rapidly. This light, of wavelength 632.8 nm, moves through the discharge tube many times by successive reflections from mirrors M_1 and M_2 (Fig. 41-21), accumulating additional stimulated emission photons with each passage. M_1 is totally reflecting but M_2 is slightly "leaky" so that a small fraction of the laser light escapes to form a useful external beam.

CHECKPOINT 4: The wavelength of light from laser *A* (a helium–neon gas laser) is 632.8 nm; that from laser *B* (a carbon dioxide gas laser) is 10.6 μm. That from laser *C* (a gallium arsenide semiconductor laser) is 840 nm. Rank these lasers according to the energy interval between the two quantum states responsible for laser action, greatest first.

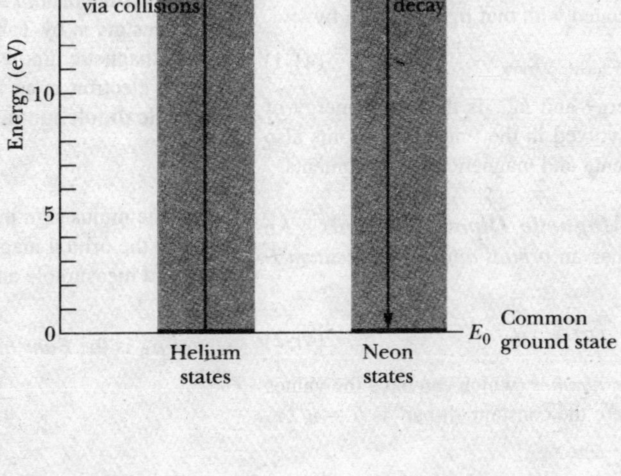

Fig. 41-22 Four essential energy levels for helium and neon atoms in a helium–neon gas laser. Laser action occurs between levels E_2 and E_1 of neon when more atoms are at the E_2 level than at the E_1 level.

Sample Problem 41-6

In the helium–neon laser of Fig. 41-21, laser action occurs between two excited states of the neon atom. However, in many lasers, laser action (*lasing*) occurs between the ground state and an excited state, as suggested in Fig. 41-20.

(a) Consider such a laser that emits at wavelength $\lambda = 550$ nm. If a population inversion is not generated, what is the ratio of the population of atoms in state E_x to the population in the ground state E_0, with the atoms at room temperature?

SOLUTION: One **Key Idea** here is that the naturally occurring population ratio N_x/N_0 of the two states is due to thermal agitation among the gas atoms, according to Eq. 41-29, which we can write as

$$N_x/N_0 = e^{-(E_x - E_0)/kT}. \qquad (41\text{-}30)$$

To find N_x/N_0 with Eq. 41-30, we need to find the energy separation $E_x - E_0$ between the two states. Here we use another **Key Idea**: We can obtain $E_x - E_0$ from the given wavelength of 550 nm for the lasing between those two states. We find

$$E_x - E_0 = hf = \frac{hc}{\lambda}$$
$$= \frac{(6.63 \times 10^{-34}\ \text{J}\cdot\text{s})(3.00 \times 10^8\ \text{m/s})}{(550 \times 10^{-9}\ \text{m})(1.60 \times 10^{-19}\ \text{J/eV})}$$
$$= 2.26\ \text{eV}.$$

To solve Eq. 41-30, we also need the mean energy of thermal agitation kT for an atom at room temperature (assumed to be 300 K), which is

$$kT = (8.62 \times 10^{-5}\ \text{eV/K})(300\ \text{K}) = 0.0259\ \text{eV}.$$

Substituting the last two results into Eq. 41-30 gives us the population ratio at room temperature:

$$N_x/N_0 = e^{-(2.26\ \text{eV})/(0.0259\ \text{eV})}$$
$$\approx 1.3 \times 10^{-38}. \qquad \text{(Answer)}$$

This is an extremely small number. It is not unreasonable, however. An atom whose mean thermal agitation energy is only 0.0259 eV will not often impart an energy of 2.26 eV to another atom in a collision.

(b) For the conditions of (a), at what temperature would the ratio N_x/N_0 be 1/2?

SOLUTION: The two **Key Ideas** of (a) apply here but this time we want the temperature T such that thermal agitation has bumped enough neon atoms up to the higher energy state to give $N_x/N_0 = 1/2$. Substituting that ratio into Eq. 41-30, taking the natural logarithm of both sides, and solving for T yield

$$T = \frac{E_x - E_0}{k(\ln 2)} = \frac{2.26\ \text{eV}}{(8.62 \times 10^{-5}\ \text{eV/K})(\ln 2)}$$
$$= 38\ 000\ \text{K}. \qquad \text{(Answer)}$$

This is much hotter than the surface of the Sun. It is clear that if we are to invert the populations of these two levels, some specific mechanism for bringing this about is needed—that is, we must "pump" the atoms. No temperature, however high, will naturally generate a population inversion by thermal agitation.

REVIEW & SUMMARY

Some Properties of Atoms The energies of atoms are quantized; that is, the atoms have only certain specific values of energy associated with different quantum states. Atoms can make transitions between different quantum states by emitting or absorbing a photon; the frequency f associated with that light is given by

$$hf = E_{\text{high}} - E_{\text{low}}, \qquad (41\text{-}1)$$

where E_{high} is the higher energy and E_{low} is the lower energy of the pair of quantum states involved in the transition. Atoms also have quantized angular momenta and magnetic dipole moments.

Angular Momenta and Magnetic Dipole Moments An electron trapped in an atom has an *orbital angular momentum* $\vec{L}$ with a magnitude given by

$$L = \sqrt{l(l+1)}\,\hbar, \qquad (41\text{-}2)$$

where l is the *orbital quantum number* (which can have the values given by Table 41-1) and where the constant "h-bar" is $\hbar = h/2\pi$.

The projection L_z of $\vec{L}$ on an arbitrary z axis is quantized and measurable and can have the values

$$L_z = m_l\hbar, \qquad (41\text{-}7)$$

where m_l is the *orbital magnetic quantum number* (which can have the values given by Table 41-1).

A magnetic dipole is associated with the angular momentum $\vec{L}$ of an electron in an atom. This magnetic dipole has an **orbital magnetic dipole moment** $\vec{\mu}_{\text{orb}}$ that is directed opposite $\vec{L}$:

$$\vec{\mu}_{\text{orb}} = -\frac{e}{2m}\vec{L}, \qquad (41\text{-}3)$$

where the minus sign indicates opposite directions. The projection $\mu_{\text{orb},z}$ of the orbital magnetic dipole moment on the z axis is quantized and measurable and can have the values

$$\mu_{\text{orb},z} = -m_l\mu_B, \qquad (41\text{-}5)$$

where μ_B is the *Bohr magneton:*

$$\mu_B = \frac{eh}{4\pi m} = 9.274 \times 10^{-24}\ \text{J/T}. \qquad (41\text{-}6)$$

An electron, whether trapped or free, has an intrinsic *spin angular momentum* (or just *spin*) $\vec{S}$ with a magnitude given by

$$S = \sqrt{s(s+1)}\hbar, \qquad (41\text{-}9)$$

where s is the *spin quantum number* of the electron, which is always $\frac{1}{2}$. The projection S_z of $\vec{S}$ on an arbitrary z axis is quantized and measurable and can have the values

$$S_z = m_s\hbar, \qquad (41\text{-}12)$$

where m_s is the *spin magnetic quantum number* of the electron, which can be $+\frac{1}{2}$ or $-\frac{1}{2}$.

An electron has an intrinsic magnetic dipole that is associated with its spin angular momentum $\vec{S}$, whether the electron is confined to an atom or free. This magnetic dipole has a **spin magnetic dipole moment** $\vec{\mu}_s$ that is directed opposite $\vec{S}$:

$$\vec{\mu}_s = -\frac{e}{m}\vec{S}. \qquad (41\text{-}10)$$

The projection $\mu_{s,z}$ of the spin magnetic dipole moment $\vec{\mu}_s$ on an arbitrary z axis is quantized and measurable and can have the values

$$\mu_{s,z} = -2m_s\mu_B. \qquad (41\text{-}13)$$

Spin and Magnetic Resonance A proton has an intrinsic spin angular momentum $\vec{S}$ and an associated spin magnetic dipole moment $\vec{\mu}$ that is always in the *same* direction as $\vec{S}$. If a proton is located in an external magnetic field $\vec{B}$, the projection μ_z of $\vec{\mu}$ on an axis z (defined to be along the direction of $\vec{B}$) can have only two quantized orientations: parallel to $\vec{B}$ or antiparallel to $\vec{B}$. The energy difference between these orientations is $2\mu_z B$. The energy required of a photon to *spin-flip* the proton between the two orientations is

$$hf = 2\mu_z(B_{ext} + B_{local}), \qquad (41\text{-}22)$$

where B_{ext} now represents the external field and B_{local} is the local magnetic field set up by the atoms and nuclei surrounding the proton. Detection of such spin flips can lead to *nuclear magnetic resonance spectra* by which specific substances can be identified.

Pauli Exclusion Principle Electrons in atoms and other traps obey the **Pauli exclusion principle,** which requires that *no two electrons in the same atom or any other type of trap can have the same set of quantum numbers.*

Building the Periodic Table The elements are listed in the periodic table in order of increasing atomic number Z; the nuclear charge is Ze, and Z is both the number of protons in the nucleus and the number of electrons in the neutral atom.

States with the same value of n form a **shell,** and those with the same values of both n and l form a **subshell.** In *closed* shells and subshells, which are those that contain the maximum number of electrons, the angular momenta and the magnetic moments of the individual electrons sum to zero.

X Rays and the Numbering of the Elements A **continuous spectrum** of x rays is emitted when high-energy electrons lose some of their energy in a collision with atomic nuclei. The **cutoff wavelength** λ_{min} is the wavelength emitted when such electrons lose *all* their initial energy in a single such encounter and is

$$\lambda_{min} = \frac{hc}{K_0}, \qquad (41\text{-}23)$$

in which K_0 is the initial kinetic energy of the electrons that strike the target.

The **characteristic x ray spectrum** arises when high-energy electrons eject electrons from deep within the atom; when a resulting "hole" is filled by an electron from farther out in the atom, a photon of the characteristic x-ray spectrum is generated.

In 1913 British physicist H. G. J. Moseley measured the frequencies of the characteristic x rays from a number of elements. He noted that when the square root of the frequency is plotted against the position of the element in the periodic table, a straight line results, as in the **Moseley plot** of Fig. 41-17. This allowed Moseley to conclude that the property that determines the position of an element in the periodic table is not its atomic mass but its **atomic number** Z—that is, the number of protons in its nucleus.

Lasers and Laser Light Laser light arises by **stimulated emission.** That is, radiation of a frequency given by

$$hf = E_x - E_0 \qquad (41\text{-}28)$$

can cause an atom to undergo a transition from an upper energy level (of energy E_x) to a lower energy level with a photon of frequency f being emitted. The stimulating photon and the emitted photon are identical in every respect and combine to form laser light.

For the emission process to predominate, there must normally be a **population inversion;** that is, there must be more atoms in the upper energy level than in the lower one.

QUESTIONS

1. An electron in an atom of gold is in a state with $n = 4$. Which of these values of l are possible for it: $-3, 0, 2, 3, 4, 5$?

2. An atom of silver has closed $3d$ and $4d$ subshells. Which subshell has the greater number of electrons, or do they have the same number?

3. An atom of uranium has closed $6p$ and $7s$ subshells. Which subshell has the greater number of electrons?

4. An electron in a mercury atom is in the $3d$ subshell. Which values of m_l are possible for it: $-3, -1, 0, 1, 2$?

5. (a) How many subshells are there in the $n = 2$ shell? How many electron states? (b) Repeat (a) for the $n = 5$ shell.

6. From which atom of each of the following pairs is it easier to remove an electron? (a) Krypton or bromine? (b) Rubidium or cerium? (c) Helium or hydrogen?

7. On what quantum numbers does the energy of an electron depend in (a) a hydrogen atom and (b) a vanadium atom?

8. Label these statements as true or false: (a) One (and only one) of these subshells cannot exist: $2p$, $4f$, $3d$, $1p$. (b) The number of values of m_l that are allowed depends only on l and not on n. (c) There are four subshells with $n = 4$. (d) The least value of n for a given value of l is $l + 1$. (e) All states with $l = 0$ also have $m_l = 0$. (f) There are n subshells for each value of n.

9. Which (if any) of these statements about the Einstein–de Haas experiment or its results are true? (a) Atoms have angular momentum. (b) The angular momentum of atoms is quantized. (c) Atoms have magnetic moments. (d) The magnetic moments of atoms are quantized. (e) The angular momentum of an atom is strongly coupled to its magnetic moment. (f) The experiment relies on the conservation of angular momentum.

10. Consider the elements krypton and rubidium. (a) Which is more suitable for use in a Stern–Gerlach experiment of the kind described in connection with Fig. 41-8? (b) Which, if either, would not work at all?

11. The x-ray spectrum of Fig. 41-14 is for 35.0 keV electrons striking a molybdenum ($Z = 42$) target. If you substitute a silver ($Z = 47$) target for the molybdenum target, will (a) λ_{min}, (b) the wavelength for the K_α line, and (c) the wavelength for the K_β line increase, decrease, or remain unchanged?

12. The K_α x-ray line for any element arises because of a transition between the K shell ($n = 1$) and the L shell ($n = 2$). Figure 41-14 shows this line (for a molybdenum target) occurring at a single wavelength. With higher resolution, however, the line splits into several wavelength components because the L shell does not have a unique energy. (a) How many components does the K_α line have? (b) Similarly, how many components does the K_β line have?

13. Which (if any) of the following is essential for laser action to occur between two energy levels of an atom? (a) There are more atoms in the upper level than in the lower. (b) The upper level is metastable. (c) The lower level is metastable. (d) The lower level is the ground state of the atom. (e) The lasing medium is a gas.

14. Figure 41-22 shows partial energy-level diagrams for the helium and neon atoms that are involved in the operation of a helium–neon laser. It is said that a helium atom in state E_3 can collide with a neon atom in its ground state and raise the neon atom to state E_2. The energy of helium state E_3 (20.61 eV) is close to, but not exactly equal to, the energy of neon state E_2 (20.66 eV). How can the energy transfer take place if these energies are not *exactly* equal?

EXERCISES & PROBLEMS

ssm Solution is in the Student Solutions Manual.
www Solution is available on the World Wide Web at: http://www.wiley.com/college/hrw
ilw Solution is available on the Interactive LearningWare.

SEC. 41-4 Angular Momenta and Magnetic Dipole Moments

1E. Show that $\hbar = 1.06 \times 10^{-34}$ J · s $= 6.59 \times 10^{-16}$ eV · s.

2E. How many electron states are in these subshells: (a) $n = 4$, $l = 3$; (b) $n = 3$, $l = 1$; (c) $n = 4$, $l = 1$; (d) $n = 2$, $l = 0$?

3E. (a) How many l values are associated with $n = 3$? (b) How many m_l values are associated with $l = 1$? ssm

4E. (a) What is the magnitude of the orbital angular momentum in a state with $l = 3$? (b) What is the magnitude of its largest projection on an imposed z axis?

5E. How many electron states are there in the following shells: (a) $n = 4$, (b) $n = 1$, (c) $n = 3$, (d) $n = 2$?

6E. Write down all the quantum numbers for states that form the subshell with $n = 4$ and $l = 3$.

7E. An electron in a hydrogen atom is in a state with $l = 5$. What is the minimum possible angle between $\vec{L}$ and L_z? ssm

8E. An electron in a multielectron atom has a maximum m_l value of +4. What can you say about the rest of its quantum numbers?

9E. An electron in a multielectron atom is known to have the quantum number $l = 3$. What are its possible n, m_l, and m_s quantum numbers?

10E. How many electron states are there in a shell defined by the quantum number $n = 5$?

11P. An electron is in a state with quantum number $l = 3$. What are the magnitudes of (a) $\vec{L}$ and (b) $\vec{\mu}$? (c) Construct a table of the allowed values of m_l, L_z (in terms of $\hbar$), $\mu_{orb,z}$ (in terms of μ_B), and the semi-classical angle θ between $\vec{L}$ and the positive direction of the z axis. ssm www

12P. An electron is in a state with $n = 3$. What are (a) the number of possible values of l, (b) the number of possible values of m_l, (c) the number of possible values of m_s, (d) the number of states in the $n = 3$ shell, and (e) the number of subshells in the $n = 3$ shell?

13P. If orbital angular momentum $\vec{L}$ is measured along, say, the z axis to obtain a value for L_z, show that

$$(L_x^2 + L_y^2)^{1/2} = [l(l+1) - m_l^2]^{1/2}\hbar$$

is the most that can be said about the other two components of the orbital angular momentum. ssm

14P. (A correspondence principle problem.) Estimate (a) the quantum number l for the orbital motion of Earth around the Sun and (b) the number of allowed orientations of the plane of Earth's orbit, according to the rules of space quantization. (c) Find θ_{min}, the half-angle of the smallest cone that can be swept out by a perpendicular to Earth's orbit as Earth revolves around the Sun.

SEC. 41-5 The Stern–Gerlach Experiment

15E. Calculate the two possible angles between the electron spin angular momentum vector and the magnetic field in Sample Problem 41-1. Bear in mind that the orbital angular momentum of the valence electron in the silver atom is zero. ssm

16E. Assume that in the Stern–Gerlach experiment as described for neutral silver atoms, the magnetic field $\vec{B}$ has a magnitude of 0.50 T. (a) What is the energy difference between the magnetic moment orientations of the silver atoms in the two subbeams? (b) What is the frequency of the radiation that would induce a transition between these two states? (c) What is its wavelength, and to what part of the electromagnetic spectrum does it belong? The magnetic moment of a neutral silver atom is 1 Bohr magneton.

17E. What is the acceleration of a silver atom as it passes through the deflecting magnet in the Stern–Gerlach experiment of Sample Problem 41-1? ssm

18P. Suppose that a hydrogen atom in its ground state moves 80 cm through and perpendicular to a vertical magnetic field that has a magnetic field gradient $dB/dz = 1.6 \times 10^2$ T/m. (a) What magnitude of force does the field gradient exert on the atom due to the magnetic moment of its electron, which we take to be 1 Bohr magneton? (b) What is the vertical displacement of the atom in the 80 cm of travel if its speed is 1.2×10^5 m/s?

SEC. 41-6 Magnetic Resonance

19E. What is the wavelength associated with a photon that will induce a transition of an electron spin from parallel to antiparallel orientation in a magnetic field of magnitude 0.200 T? Assume that $l = 0$. ssm

20E. The proton, like the electron, has a spin quantum number s of $\frac{1}{2}$. In the hydrogen atom in its ground state ($n = 1$ and $l = 0$), there are two energy levels, depending on whether the electron and proton spins are parallel or antiparallel. If an atom has a spin flip from the state of higher energy to that of lower energy, a photon of wavelength 21 cm is emitted. Radio astronomers observe this 21 cm radiation coming from deep space. What is the effective magnetic field (due to the magnetic dipole moment of the proton) experienced by the electron emitting this radiation?

21E. Excited sodium atoms emit two closely spaced spectrum lines (the *sodium doublet;* see Fig. 41-23) with wavelengths 588.995 nm and 589.592 nm. (a) What is the difference in energy between the two upper energy levels? (b) This energy difference occurs because the electron's spin magnetic moment (= 1 Bohr magneton) can be oriented either parallel or antiparallel to the internal magnetic field associated with the electron's orbital motion. Use your result in (a) to find the strength of this internal magnetic field.

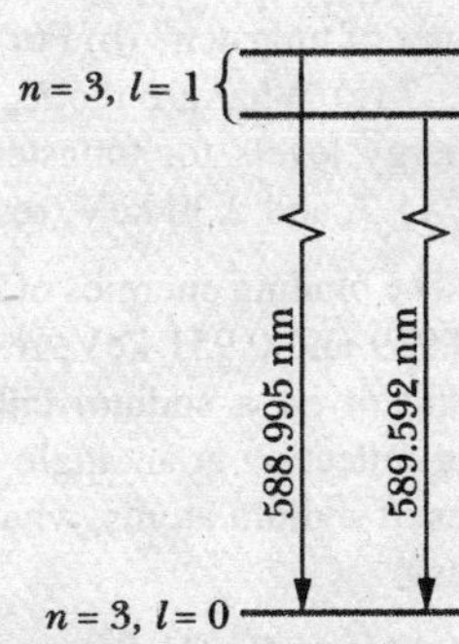

Fig. 41-23 Exercise 21.

22E. An external oscillating magnetic field of frequency 34 MHz is applied to a sample that contains hydrogen atoms. Resonance is observed when the strength of the constant external magnetic field equals 0.78 T. Calculate the strength of the local magnetic field at the site of the protons that are undergoing spin flips, assuming the external and local fields are parallel there. The protons have $\mu_z = 1.41 \times 10^{-26}$ J/T.

SEC. 41-8 Multiple Electrons in Rectangular Traps

23E. Seven electrons are trapped in a one-dimensional infinite potential well of width L. As a multiple of $h^2/8mL^2$, what is the energy of the ground state of the system of seven electrons? Assume that the electrons do not interact with one another, and do not neglect spin.

24E. A rectangular corral of widths $L_x = L$ and $L_y = 2L$ contains seven electrons. As a multiple of $h^2/8mL^2$, what is the energy of the ground state of the system of seven electrons? Assume that the electrons do not interact with one another, and do not neglect spin.

25P. For the situation of Exercise 23, and as multiples of $h^2/8mL^2$, what are the energies of (a) the first excited state, (b) the second excited state, and (c) the third excited state of the system of seven electrons? (d) Construct an energy-level diagram for the lowest four energy levels of the system.

26P. For Exercise 24, and as multiples of $h^2/8mL^2$, what are the energies of (a) the first excited state, (b) the second excited state, and (c) the third excited state of the system of seven electrons? (d) Construct an energy-level diagram for the lowest four energy levels of the system.

27P. A cubical box of widths $L_x = L_y = L_z = L$ contains eight electrons. As a multiple of $h^2/8mL^2$, what is the energy of the ground state of the system of eight electrons? Assume that the electrons do not interact with one another, and do not neglect spin. ssm www

28P. For the situation of Problem 27, and as multiples of $h^2/8mL^2$, what are the energies of (a) the first excited state, (b) the second excited state, and (c) the third excited state of the system of eight electrons? (d) Construct an energy-level diagram for the lowest four energy levels of the system.

SEC. 41-9 Building the Periodic Table

29E. Show that if the 63 electrons in an atom of europium were assigned to shells according to the "logical" sequence of quantum numbers, this element would be chemically similar to sodium.

30E. Consider the elements selenium (Z = 34), bromine (Z = 35), and krypton (Z = 36). In their part of the periodic table, the subshells of the electronic states are filled in the sequence

$$1s \quad 2s \;\; 2p \quad 3s \;\; 3p \;\; 3d \quad 4s \;\; 4p \;\cdots$$

For each of these elements, identify the highest occupied subshell and state how many electrons are in it.

31E. Suppose that the electron had no spin and that the Pauli exclusion principle still held. Which, if any, of the present noble gases would remain in that category?

32E. What are the four quantum numbers for the two electrons of the helium atom in its ground state?

33P. Two electrons in lithium (Z = 3) have the quantum numbers $n = 1$, $l = 0$, $m_l = 0$, and $m_s = \pm\frac{1}{2}$. What quantum numbers can the third electron have if the atom is to be in (a) its ground state and (b) its first excited state? ssm

34P. Suppose there are two electrons in the same atom, both of which have $n = 2$ and $l = 1$. (a) If the Pauli exclusion principle did not apply, how many combinations of states would conceivably be possible? (b) How many states does the exclusion principle forbid? Which are they?

35P. Show that the number of states with the same quantum number n is $2n^2$. ssm

SEC. 41-10 X Rays and the Numbering of the Elements

36E. Through what minimum potential difference must an electron in an x-ray tube be accelerated so that it can produce x rays with a wavelength of 0.100 nm?

37E. Knowing that the minimum x-ray wavelength produced by 40.0 keV electrons striking a target is 31.1 pm, determine the Planck constant h.

38E. Show that the cutoff wavelength (in picometers) in the continuous x-ray spectrum from any target is given by $\lambda_{min} = 1240/V$, where V is the potential difference (in kilovolts) through which the electrons are accelerated before they strike the target.

39P. X rays are produced in an x-ray tube by electrons accelerated through an electric potential difference of 50.0 kV. An electron makes three collisions in the target before coming to rest and loses half its remaining kinetic energy in each of the first two collisions. Determine the wavelengths of the resulting photons. (Neglect the recoil of the heavy target atoms.) ssm www

40P. A 20 keV electron is brought to rest by undergoing two successive nuclear encounters such as that of Fig. 41-15, thus transferring its kinetic energy to the energy of two photons. The wavelength associated with the second photon is 130 pm greater than the wavelength associated with the first photon. (a) Find the kinetic energy of the electron after its first encounter. (b) What are the associated wavelengths and energies of the two photons?

41P. Show that a moving electron cannot spontaneously change into an x-ray photon in free space. A third body (atom or nucleus) must be present. Why is it needed? (*Hint:* Examine the conservation of energy and momentum.) ssm

42P. When electrons bombard a molybdenum target, they produce both continuous and characteristic x rays as shown in Fig. 41-14. In that figure the kinetic energy of the incident electrons is 35.0 keV. If the accelerating potential is increased to 50.0 keV, what mean values of (a) λ_{min}, (b) the wavelength of the K_α line, and (c) the wavelength of the K_β line result?

43P. In Fig. 41-14, the x rays shown are produced when 35.0 keV electrons strike a molybdenum ($Z = 42$) target. If the accelerating potential is maintained at this value but a silver ($Z = 47$) target is used instead, what values of (a) λ_{min}, (b) the wavelength of the K_α line, and (c) the wavelength of the K_β line result? The K, L, and M atomic x-ray levels for silver (compare Fig. 41-16) are 25.51, 3.56, and 0.53 keV. ssm

44P. The wavelength of the K_α line from iron is 193 pm. What is the energy difference between the two states of the iron atom that give rise to this transition?

45P. Calculate the ratio of the wavelength of the K_α line for niobium (Nb) to that for gallium (Ga). Take needed data from the periodic table of Appendix G. ssm

46P. From Fig. 41-14, calculate approximately the energy difference $E_L - E_M$ for molybdenum. Compare it with the value that may be obtained from Fig. 41-16.

47P. Here are the K_α wavelengths of a few elements:

Element	λ (pm)	Element	λ (pm)
Ti	275	Co	179
V	250	Ni	166
Cr	229	Cu	154
Mn	210	Zn	143
Fe	193	Ga	134

Make a Moseley plot (like that in Fig. 41-17) from these data and verify that its slope agrees with the value given for C in Section 41-10.

48P. A molybdenum ($Z = 42$) target is bombarded with 35.0 keV electrons and the x-ray spectrum of Fig. 41-14 results. The K_β and K_α wavelengths are 63.0 and 71.0 pm, respectively. (a) What are the corresponding photon energies? (b) It is desired to filter these radiations through a material that will absorb the K_β line much more strongly than it will absorb the K_α line. What substance would you use? The ionization energies of the K electrons in molybdenum and in four neighboring elements are as follows:

	Zr	Nb	Mo	Tc	Ru
Z	40	41	42	43	44
E_K (keV)	18.00	18.99	20.00	21.04	22.12

(*Hint:* A substance will absorb one x radiation more strongly than another if the photons of the first have enough energy to eject a K electron from the atom of the substance but the photons of the second do not.)

49P. A tungsten ($Z = 74$) target is bombarded by electrons in an x-ray tube. (a) What is the minimum value of the accelerating potential that will permit the production of the characteristic K_α and K_β lines of tungsten? (b) For this same accelerating potential, what is λ_{min}? (c) What are the K_α and K_β wavelengths? The K, L, and M energy levels for tungsten (see Fig. 41-16) have the energies 69.5, 11.3, and 2.30 keV, respectively. ssm

50P. The binding energies of K-shell and L-shell electrons in copper are 8.979 and 0.951 keV, respectively. If a K_α x ray from copper is incident on a sodium chloride crystal and gives a first-order Bragg reflection at an angle of 74.1° measured relative to parallel planes of sodium atoms, what is the spacing between these parallel planes?

51P. (a) Using Eq. 41-26, estimate the ratios of photon energies due to K_α transitions in two atoms whose atomic numbers are Z and Z'. (b) What is this ratio for uranium and aluminum? (c) For uranium and lithium?

52P. Determine how close the theoretical K_α x-ray photon energies, as obtained from Eq. 41-27, are to the measured energies of the low-mass elements from lithium to magnesium. To do this, (a) first determine the constant C in Eq. 41-27 to five significant figures by finding C in terms of the fundamental constants in Eq. 41-24 and then using data from Appendix B to evaluate those constants. (b) Next, calculate the percentage deviations of the theoretical from the measured energies. (c) Finally, plot the deviations and comment on the trend. The measured energies (eV) of the K_α photons for these elements are as follows:

Li	54.3	O	524.9
Be	108.5	F	676.8
B	183.3	Ne	848.6
C	277	Na	1041
N	392.4	Mg	1254

(There is actually more than one K_α ray because of the splitting of the L energy level, but that effect is negligible for the elements listed here.)

SEC. 41-12 How Lasers Work

53E. Lasers can be used to generate pulses of light whose durations are as short as 10 fs. (a) How many wavelengths of light (λ = 500 nm) are contained in such a pulse? (b) Supply the missing quantity X (in years):

$$\frac{10\text{ fs}}{1\text{ s}} = \frac{1\text{ s}}{X}.$$

54E. For the conditions of Sample Problem 41-6a, how many moles of neon are needed to put 10 atoms in the excited state E_x?

55E. A hypothetical atom has energy levels uniformly separated by 1.2 eV. At a temperature of 2000 K, what is the ratio of the number of atoms in the 13th excited state to the number in the 11th excited state? ssm

56E. By measuring the go-and-return time for a laser pulse to travel from an Earth-bound observatory to a reflector on the Moon, it is possible to measure the separation between these bodies. (a) What is the predicted value of this time? (b) The separation can be measured to a precision of about 15 cm. To what uncertainty in travel time does this correspond? (c) If the laser beam forms a spot on the Moon 3 km in diameter, what is the angular divergence of the beam?

57E. A hypothetical atom has only two atomic energy levels, separated by 3.2 eV. Suppose that at a certain altitude in the atmosphere of a star there are $6.1 \times 10^{13}/\text{cm}^3$ of these atoms in the higher energy state and $2.5 \times 10^{15}/\text{cm}^3$ in the lower energy state. What is the temperature of the star's atmosphere at that altitude?

58E. A population inversion for two energy levels is often described by assigning a negative Kelvin temperature to the system. What negative temperature would describe a system in which the population of the upper energy level exceeds that of the lower level by 10% and the energy difference between the two levels is 2.1 eV?

59E. A pulsed laser emits light at a wavelength of 694.4 nm. The pulse duration is 12 ps and the energy per pulse is 0.150 J. (a) What is the length of the pulse? (b) How many photons are emitted in each pulse? ssm

60E. A helium–neon laser emits laser light at a wavelength of 632.8 nm and a power of 2.3 mW. At what rate are photons emitted by this device?

61E. A high-powered laser beam (λ = 600 nm) with a beam diameter of 12 cm is aimed at the Moon, 3.8×10^5 km distant. The beam spreads only because of diffraction. The angular location of the edge of the central diffraction disk (see Eq. 37-12) is given by

$$\sin\theta = \frac{1.22\lambda}{d},$$

where d is the diameter of the beam aperture. What is the diameter of the central diffraction disk on the Moon's surface?

62E. Assume that lasers are available whose wavelengths can be precisely "tuned" to anywhere in the visible range—that is, in the range 450 nm < λ < 650 nm. If every television channel occupies a bandwidth of 10 MHz, how many channels could be accommodated within this wavelength range?

63E. The active volume of a laser constructed of the semiconductor GaAlAs is only 200 μm^3 (smaller than a grain of sand) and yet the laser can continuously deliver 5.0 mW of power at a wavelength of 0.80 μm. At what rate does it generate photons?

64P. The mirrors in the laser of Fig. 41-21, which are separated by 8.0 cm, form an optical cavity in which standing waves of laser light can be set up. Each standing wave has an integral number n of half wavelengths in the 8.0 cm length, where n is large and the waves differ slightly in wavelength. Near λ = 533 nm, how far apart in wavelength are the standing waves?

65P. The active medium in a particular laser that generates laser light at a wavelength of 694 nm is 6.00 cm long and 1.00 cm in diameter. (a) Treat the medium as an optical resonance cavity analogous to a closed organ pipe. How many standing wave nodes are there along the laser axis? (b) By what amount Δf would the beam frequency have to shift to increase this number by one? (c) Show that Δf is just the inverse of the travel time of laser light for one round trip back and forth along the laser axis. (d) What is the corresponding fractional frequency shift $\Delta f/f$? The appropriate index of refraction of the lasing medium (a ruby crystal) is 1.75. ssm www

66P. A hypothetical atom has two energy levels, with a transition wavelength between them of 580 nm. In a particular sample at 300 K, 4.0×10^{20} such atoms are in the state of lower energy. (a) How many atoms are in the upper state, assuming conditions of thermal equilibrium? (b) Suppose, instead, that 3.0×10^{20} of these atoms are "pumped" into the upper state by an external process, with 1.0×10^{20} atoms remaining in the lower state. What is the maximum energy that could be released by the atoms in a single laser pulse if each atom jumps once between those two states (either via absorption or stimulated emission).

67P. Can an incoming intercontinental ballistic missile be destroyed by an intense laser beam? A beam of intensity 10^8 W/m^2 would probably burn into and destroy a hardened (nonspinning) missile in 1 s. (a) If the laser had 5.0 MW power, 3.0 μm wavelength, and a 4.0 m beam diameter (a very powerful laser indeed), would it destroy a missile at a distance of 3000 km? (b) If the wavelength could be changed, what maximum value would work? Use the equation for the central disk given in Exercise 61. ssm

68P. The beam from an argon laser (of wavelength 515 nm) has a diameter d of 3.00 mm and a continuous energy output rate of 5.00 W. The beam is focused onto a diffuse surface by a lens whose focal length f is 3.50 cm. A diffraction pattern such as that of Fig. 37-9 is formed, the radius of the central disk being given by

$$R = \frac{1.22 f \lambda}{d}$$

(see Eq. 37-12 and Sample Problem 37-3). The central disk can be shown to contain 84% of the incident power. (a) What is the radius of the central disk? (b) What is the average intensity (power per unit area) in the incident beam? (c) What is the average intensity in the central disk?

Additional Problems

69. *Martian CO_2 laser.* Where sunlight shines on the atmosphere of Mars, carbon dioxide molecules at an altitude of about 75 km undergo natural laser action. The energy levels involved in the action are shown in Fig. 41-24; population inversion occurs between energy levels E_2 and E_1. (a) What wavelength of sunlight excites the molecules in the lasing action? (b) At what wavelength does lasing occur? (c) In what region of the electromagnetic spectrum do the excitation and lasing wavelengths lie?

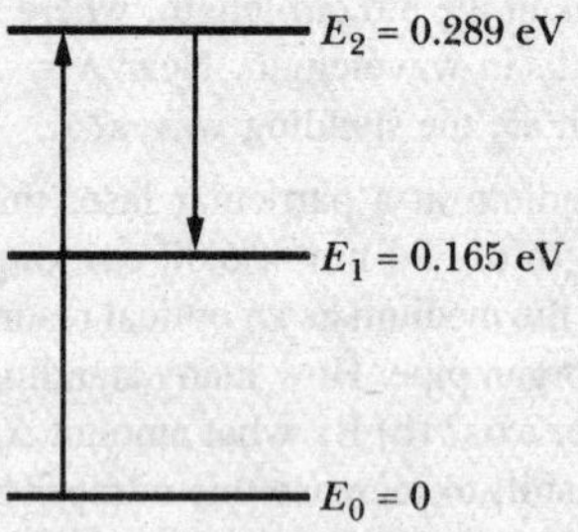

Fig. 41-24 Problem 69.

70. *Comet stimulated emission.* When a comet approaches the Sun, the increased warmth evaporates water from the frozen ice on the surface of the comet nucleus, producing a thin atmosphere of water vapor around the nucleus. Sunlight can then dissociate the water vapor into H and OH. The sunlight can also excite the OH molecules into higher energy levels, two of which are represented in Fig. 41-25.

When the comet is still relatively far from the Sun, the sunlight causes equal excitation to the E_2 and E_1 levels (Fig. 41-25*a*). Hence, there is no population inversion between the two levels. However, as the comet approaches the Sun, the excitation to the E_1 level decreases and population inversion occurs. The reason has to do with one of the many wavelengths—said to be *Fraunhofer lines*—that are missing in sunlight because, as the light travels outward through the Sun's atmosphere, those particular wavelengths are absorbed by the atmosphere.

As a comet approaches the Sun, the Doppler effect due to the comet's speed relative to the Sun shifts the Fraunhofer lines in wavelength, apparently overlapping one of them with the wavelength required for excitation to the E_1 level in OH molecules. Population inversion then occurs in those molecules, and they radiate stimulated emission (Fig. 41-25*b*). For example, as comet Kouhoutek approached the Sun in December 1973 and January 1974, it radiated stimulated emission at about 1666 MHz during mid-January. (a) What was the energy difference $E_2 - E_1$ for that emission? (b) In what region of the electromagnetic spectrum was the emission?

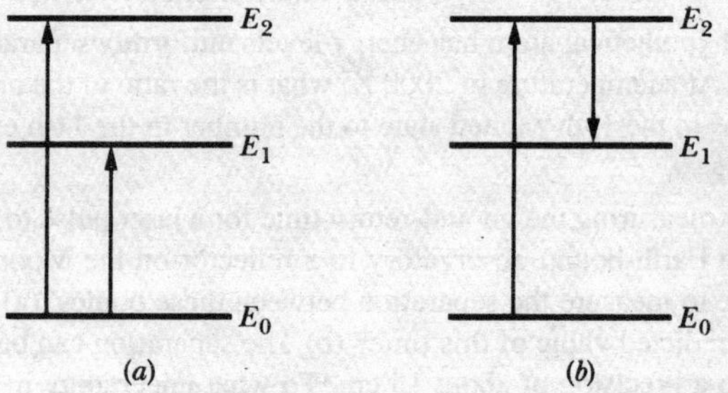

Fig. 41-25 Problem 70.

42 Conduction of Electricity in Solids

A few of the workers at the Fab 11 factory at Rio Rancho, New Mexico. The plant, which represents an investment of $2.5 billion, has a floor area equivalent to that of about two dozen football fields. According to the *New York Times,* the plant "on a high desert mesa in New Mexico is probably the most productive factory in the world, in terms of the value of the goods that it makes."

What do these workers manufacture that requires them to be suited up like astronauts?

The answer is in this chapter.

42-1 Solids

You have seen how well quantum physics works when we apply it to questions involving individual atoms. In this chapter we hope to show, with a single broad example, that this theory works just as well when we apply it to questions involving assemblies of atoms in the form of solids.

Every solid has an enormous range of properties that we can choose to examine. Is it transparent? Can it be hammered out into a thin sheet? At what speeds do sound waves travel through it? Is it magnetic? Is it a good heat conductor? . . . The list goes on and on. However, we choose to focus this entire chapter on a single question: *What are the mechanisms by which a solid conducts, or does not conduct, electricity?* As you will see, quantum physics provides the answer.

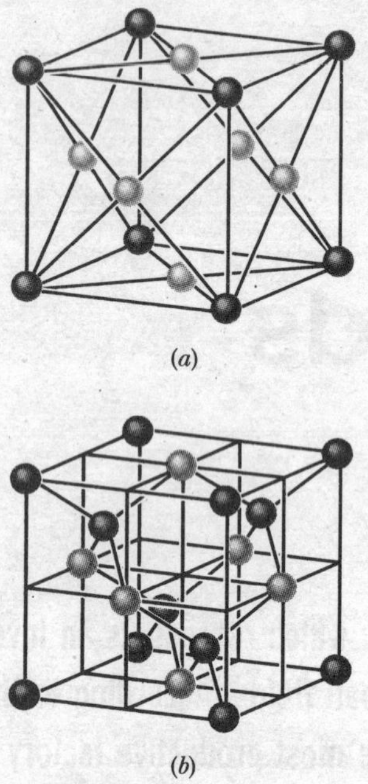

Fig. 42-1 (*a*) The unit cell for copper is a cube. There is one copper atom (darker) at each corner of the cube and one copper atom (lighter) at the center of each face of the cube. The arrangement is called *face-centered cubic.* (*b*) The unit cell for silicon and diamond is also a cube, the atoms being arranged in a so-called *diamond lattice*. There is one atom (darkest) at each corner of the cube and one atom (lightest) at the center of each cube face; in addition, four atoms (medium color) lie within the cube. Every atom is bonded to its four nearest neighbors by a two-electron covalent bond (only the four atoms within the cube show all four *nearest* neighbors).

42-2 The Electrical Properties of Solids

We shall examine only **crystalline solids**—that is, solids whose atoms are arranged in a repetitive three-dimensional structure called a **lattice.** We shall not consider such solids as wood, plastic, glass, and rubber, whose atoms are not arranged in such repetitive patterns. Figure 42-1 shows the basic repetitive units (the **unit cells**) of the lattice structures of copper, our prototype of a metal, and silicon and diamond, our prototypes of a semiconductor and an insulator, respectively.

We can classify solids electrically according to three basic properties:

1. Their **resistivity** ρ at room temperature, with the SI unit ohm-meter ($\Omega \cdot m$); resistivity is defined in Section 27-4.
2. Their **temperature coefficient of resistivity** α, defined as $\alpha = (1/\rho)(d\rho/dT)$ in Eq. 27-17 and having the SI unit inverse kelvin (K^{-1}). We can evaluate α for any solid by measuring ρ over a range of temperatures.
3. Their **number density of charge carriers** n. This quantity, the number of charge carriers per unit volume, can be found from measurements of the Hall effect, as discussed in Section 29-4, and from other measurements. It has the SI unit inverse cubic meter (m^{-3}).

From measurements of room-temperature resistivity alone, we discover that there are some materials—we call them **insulators**—that for all practical purposes do not conduct electricity at all. These are materials with very high resistivity. Diamond, an excellent example, has a resistivity greater than that of copper by the enormous factor of about 10^{24}.

We can then use measurements of ρ, α, and n to divide most noninsulators, at least at low temperatures, into two major categories: **metals** and **semiconductors.**

Semiconductors have a considerably greater resistivity ρ than metals.

Semiconductors have a temperature coefficient of resistivity α that is both high and negative. That is, the resistivity of a semiconductor *decreases* with temperature, whereas that of a metal *increases*.

Semiconductors have a considerably lower number density of charge carriers n than metals.

Table 42-1 shows values of these quantities for copper, our prototype metal, and silicon, our prototype semiconductor.

Now, with measurements of ρ, α, and n in hand, we have an experimental basis for refining our central question about the conduction of electricity in solids: *What*

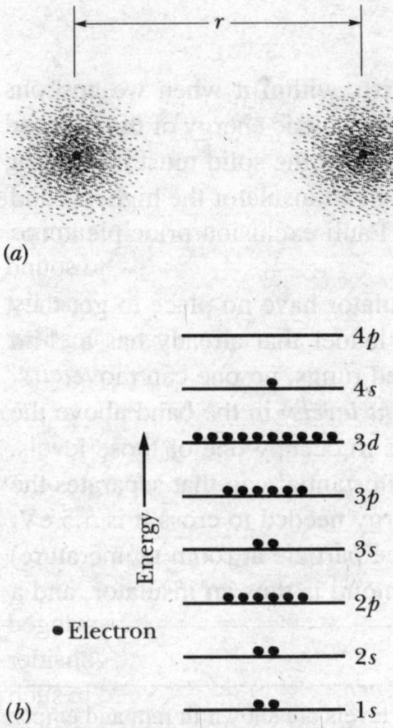

Fig. 42-2 (*a*) Two copper atoms separated by a large distance; their electron distributions are represented by dot plots. (*b*) Each copper atom has 29 electrons distributed among a set of subshells. In the neutral atom in its ground state, all subshells up through the 3*d* level are filled, the 4*s* subshell contains one electron (it can hold two), and higher subshells are empty. For simplicity, the subshells are shown as being evenly spaced in energy.

TABLE 42-1 Some Electrical Properties of Two Materials[a]

Property	Unit	Material: Copper	Material: Silicon
Type of conductor		Metal	Semiconductor
Resistivity, ρ	$\Omega \cdot m$	2×10^{-8}	3×10^{3}
Temperature coefficient of resistivity, α	K^{-1}	$+4 \times 10^{-3}$	-70×10^{-3}
Number density of charge carriers, n	m^{-3}	9×10^{28}	1×10^{16}

[a] All values are for room temperature.

features make diamond an insulator, copper a metal, and silicon a semiconductor? Again, quantum physics provides the answers.

42-3 Energy Levels in a Crystalline Solid

The distance between adjacent copper atoms in solid copper is 260 pm. Figure 42-2*a* shows two isolated copper atoms separated by a distance r that is much greater than that. As Fig. 42-2*b* shows, each of these isolated neutral atoms stacks up its 29 electrons in an array of discrete subshells, as follows:

$$1s^2\ 2s^2\ 2p^6\ 3s^2\ 3p^6\ 3d^{10}\ 4s^1.$$

Here we use the shorthand notation of Section 41-9 to identify the subshells. Recall, for example, that the subshell with principal quantum number $n = 3$ and orbital quantum number $l = 1$ is called the 3*p* subshell; it can hold up to $2(2l + 1) = 6$ electrons; the number it actually contains is indicated by a numerical superscript. We see above that the first six subshells in copper are filled, but the (outermost) 4*s* subshell, which can hold 2 electrons, holds only one.

If we bring the atoms of Fig. 42-2*a* closer together, they will—speaking loosely—begin to sense each other's presence. In the language of quantum physics, their wave functions will start to overlap, beginning with those of the outermost electrons.

When the wave functions of the two atoms overlap, we speak not of two independent atoms but of a single two-atom system; here the system contains $2 \times 29 = 58$ electrons. The Pauli exclusion principle also applies to this larger system and requires that each of these 58 electrons occupy a different quantum state. In fact, 58 quantum states are available because each energy level of the isolated atom splits into *two* levels for the two-atom system.

If we bring up more atoms, we gradually assemble a lattice of solid copper. If, say, our lattice contains N atoms, then each level of an isolated copper atom must split into N levels in the solid. Thus, the individual energy levels of the solid form energy **bands,** adjacent bands being separated by an energy **gap,** which represents a range of energies that no electron can possess. A typical band is only a few electron-volts wide. Since N may be of the order of 10^{24}, we see that the individual levels within a band are very close together indeed, and there are a vast number of levels.

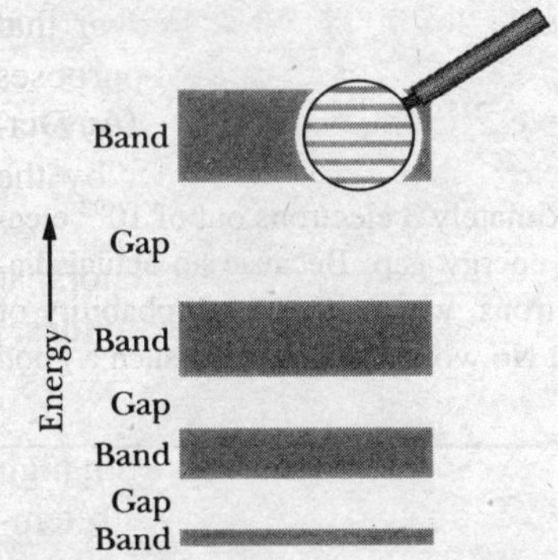

Fig. 42-3 The band–gap pattern of energy levels for an idealized crystalline solid. As the magnified view suggests, each band consists of a very large number of very closely spaced energy levels. (In many solids, adjacent bands may overlap; for clarity, we have not shown this condition.)

Figure 42-3 suggests the band–gap structure of the energy levels in a generalized crystalline solid. Note that bands of lower energy are narrower than those of higher energy. This occurs because electrons that occupy the lower energy bands spend most of their time deep within the atom's electron cloud. The wave functions of these core electrons do not overlap as much as the wave functions of the outer electrons. Hence the splitting of these levels is not as great as it is for the higher energy levels normally occupied by the outer electrons.

42-4 Insulators

A solid is said to be an insulator if no current exists within it when we apply a potential difference across it. For a current to exist, the kinetic energy of the average electron must increase. In other words, some electrons in the solid must move to a higher energy level. However, as Fig. 42-4*a* shows, in an insulator the highest band containing any electrons is fully occupied, and the Pauli exclusion principle keeps electrons from moving to occupied levels.

Thus, the electrons in the filled band of an insulator have no place to go; they are in gridlock. It is as if a child tries to climb a ladder that already has a child standing on each rung; since there are no unoccupied rungs, no one can move.

There are plenty of unoccupied levels (or *vacant levels*) in the band above the filled band in Fig. 42-4*a*. However, if an electron is to occupy one of those levels, it must acquire enough energy to jump across the substantial gap that separates the two bands. In diamond, this gap is so wide (the energy needed to cross it is 5.5 eV, about 140 times the average thermal energy of a free particle at room temperature) that essentially no electron can jump across it. Diamond is thus an insulator, and a very good one.

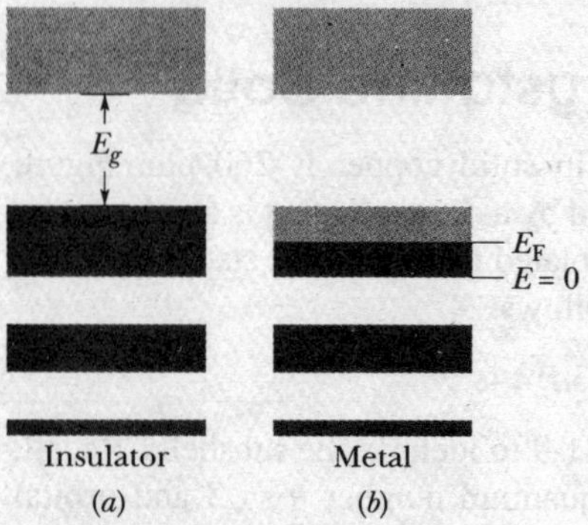

Fig. 42-4 (*a*) The band–gap pattern for an insulator; filled levels are shown in red, and empty levels in blue. Note that the highest filled level lies at the top of a band and the next higher vacant level is separated from it by a relatively large energy gap E_g. (*b*) The band–gap pattern for a metal. The highest filled level, called the Fermi level, lies near the middle of a band. Since vacant levels are available within that band, electrons in the band can easily change levels, and conduction can take place.

Sample Problem 42-1

Approximately what is the probability that, at room temperature (300 K), an electron at the top of the highest filled band in diamond (an insulator) will jump the energy gap E_g in Fig. 42-4*a*. For diamond, E_g is 5.5 eV.

SOLUTION: In Chapter 41 we used Eq. 41-29,

$$\frac{N_x}{N_0} = e^{-(E_x - E_0)/kT}, \tag{42-1}$$

to relate the population N_x of atoms at energy level E_x to the population N_0 at energy level E_0, where the atoms are part of a system at temperature T (measured in kelvins); k is the Boltzmann constant (8.62×10^{-5} eV/K).

A **Key Idea** here is that we can use Eq. 42-1 to *approximate* the probability P that an electron in an insulator will jump the energy gap E_g in Fig. 42-4*a*. To do so, we first set the energy difference $E_x - E_0$ to E_g. Then the probability P of the jump is approximately equal to the ratio N_x/N_0 of the number of electrons just above the energy gap to the number of electrons just below the gap.

For diamond, the exponent in Eq. 42-1 is

$$-\frac{E_g}{kT} = -\frac{5.5 \text{ eV}}{(8.62 \times 10^{-5} \text{ eV/K})(300 \text{ K})} = -213.$$

The required probability is then

$$P = \frac{N_x}{N_0} = e^{-(E_g/kT)} = e^{-213} \approx 3 \times 10^{-93}. \qquad \text{(Answer)}$$

This result tells us that approximately 3 electrons out of 10^{93} electrons would jump across the energy gap. Because an actual diamond has less than 10^{23} electrons, we see that the probability of the jump is vanishingly small. No wonder diamond is such a good insulator.

42-5 Metals

The feature that defines a metal is that, as Fig. 42-4*b* shows, the highest occupied energy level falls somewhere near the middle of an energy band. If we apply a potential difference across a metal, a current can exist because there are plenty of vacant levels at nearby higher energies into which electrons (the charge carriers in a metal) can jump. Thus, a metal can conduct electricity because electrons in its highest occupied band can easily move into higher energy levels within that band.

In Section 27-6 we introduced the **free-electron model** of a metal, in which the **conduction electrons** are free to move throughout the volume of the sample like the molecules of a gas in a closed container. We used this model to derive an expression for the resistivity of a metal, assuming that the electrons follow the laws of Newtonian mechanics. Here we use that same model to explain the behavior of the electrons—called the conduction electrons—in the partially filled band of Fig. 42-4*b*. However, we follow the laws of quantum physics by assuming the energies of these electrons to be quantized and the Pauli exclusion principle to hold.

We assume too that the electric potential energy of a conduction electron has the same constant value at all points within the lattice. If we choose this value of the potential energy to be zero, as we are free to do, then the mechanical energy E of the conduction electrons is entirely kinetic.

The level at the bottom of the partially filled band of Fig. 42-4*b* corresponds to $E = 0$. The highest occupied level in this band at absolute zero ($T = 0$ K) is called the **Fermi level,** and the energy corresponding to it is called the **Fermi energy** E_F; for copper, $E_F = 7.0$ eV.

The electron speed corresponding to the Fermi energy is called the **Fermi speed** v_F. For copper the Fermi speed is 1.6×10^6 m/s. This fact should be enough to shatter the popular misconception that all motion ceases at absolute zero; at that temperature—and solely because of the Pauli exclusion principle—the conduction electrons are stacked up in the partially filled band of Fig. 42-4*b* with energies that range from zero to the Fermi energy.

How Many Conduction Electrons Are There?

If we could bring individual atoms together to form a sample of a metal, we would find that the conduction electrons in the metal are the *valence electrons* of the atoms (the electrons in the outer shells of the individual atoms). A *monovalent* atom contributes one such electron to the conduction electrons in a metal; a *bivalent* atom contributes two such electrons. Thus, the total number of conduction electrons

$$\begin{pmatrix}\text{number of conduction}\\ \text{electrons in sample}\end{pmatrix} = \begin{pmatrix}\text{number of atoms}\\ \text{in sample}\end{pmatrix}\begin{pmatrix}\text{number of valence}\\ \text{electrons per atom}\end{pmatrix}. \tag{42-2}$$

(In this chapter, we shall write several equations largely in words because the symbols we have previously used for the quantities in them now represent other quantities.) The *number density* n of conduction electrons in a sample is the number of conduction electrons per unit volume:

$$n = \frac{\text{number of conduction electrons in sample}}{\text{sample volume } V}. \tag{42-3}$$

We can relate the number of atoms in a sample to various other properties of the sample and the material making up the sample with the following equation:

$$\begin{aligned}\begin{pmatrix}\text{number of atoms}\\ \text{in sample}\end{pmatrix} &= \frac{\text{sample mass } M_{sam}}{\text{atomic mass}} = \frac{\text{sample mass } M_{sam}}{(\text{molar mass } M)/N_A}\\ &= \frac{(\text{material's density})(\text{sample volume } V)}{(\text{molar mass } M)/N_A},\end{aligned} \tag{42-4}$$

where the molar mass M is the mass of one mole of the material in the sample and N_A is Avogadro's number (6.02×10^{23} mol^{-1}).

Sample Problem 42-2

How many conduction electrons are in a cube of magnesium with a volume of $2.00 \times 10^{-6}\ m^3$? Magnesium atoms are bivalent.

SOLUTION: The **Key Ideas** here are these:

1. Because magnesium atoms are bivalent, each magnesium atom contributes two conduction electrons.
2. The number of conduction electrons in the cube is related to the number of magnesium atoms in the cube by Eq. 42-2.
3. We can find the number of atoms with Eq. 42-4 and known data about the cube's volume and magnesium's properties.

We can write Eq. 42-4 as

$$\begin{pmatrix}\text{number}\\ \text{of atoms}\\ \text{in sample}\end{pmatrix} = \frac{(\text{material's density})(\text{sample volume } V)N_A}{\text{molar mass } M}.$$

Magnesium has a density of $1.738\ g/cm^3$ ($= 1.738 \times 10^3\ kg/m^3$) and a molar mass of 24.312 g/mol ($= 24.312 \times 10^{-3}$ kg/mol) (see Appendix F). The numerator gives us

$$(1.738 \times 10^3\ kg/m^3)(2.00 \times 10^{-6}\ m^3)(6.02 \times 10^{23}\ mol^{-1}) = 2.0926 \times 10^{21}\ kg/mol.$$

Thus, we have

$$\begin{pmatrix}\text{number of atoms}\\ \text{in sample}\end{pmatrix} = \frac{2.0926 \times 10^{21}\ kg/mol}{24.312 \times 10^{-3}\ kg/mol} = 8.61 \times 10^{22}.$$

Using this result and the fact that magnesium atoms are bivalent, we find that Eq. 42-2 yields

$$\begin{pmatrix}\text{number of}\\ \text{conduction electrons}\\ \text{in sample}\end{pmatrix} = (8.61 \times 10^{22}\ \text{atoms})\left(2\ \frac{\text{electrons}}{\text{atom}}\right) = 1.72 \times 10^{23}\ \text{electrons.} \quad \text{(Answer)}$$

Conductivity at $T > 0$

Our practical interest in the conduction of electricity in metals is at temperatures above absolute zero. What happens to the electron distribution of Fig. 42-4*b* at such higher temperatures? As we shall see, surprisingly little.

Of the electrons in the partially filled band of Fig. 42-4*b*, only those that are close to the Fermi energy find unoccupied levels above them, and only those electrons are free to be boosted to these higher levels by thermal agitation. Even at $T = 1000$ K, a temperature at which copper would glow brightly in a dark room, the distribution of electrons among the available levels does not differ much from the distribution at $T = 0$ K.

Let us see why. The quantity kT, where k is the Boltzmann constant, is a convenient measure of the energy that may be given to a conduction electron by the random thermal motions of the lattice. At $T = 1000$ K, we have $kT = 0.086$ eV. No electron can hope to have its energy changed by more than a few times this relatively small amount by thermal agitation alone, so at best, only those few conduction electrons whose energies are close to the Fermi energy are likely to jump to higher energy levels due to thermal agitation. Poetically stated, thermal agitation normally causes only ripples on the surface of the Fermi sea of electrons; the vast depths of that sea lie undisturbed.

How Many Quantum States Are There?

The ability of a metal to conduct electricity depends on how many quantum states are available to its electrons and what the energies of those states are. Thus, a question arises: What are the energies of the individual states in the partially filled band of Fig. 42-4*b*? This question is too difficult to answer because we cannot possibly list the energies of so many states individually. We ask instead: How many states in a unit volume of a sample have energies in the energy range E to $E + dE$? We write this number as $N(E)\,dE$, where $N(E)$ is called the **density of states** at energy E. The conventional unit for $N(E)\,dE$ is states per cubic meter (states/m^3, or simply m^{-3}); the unit for $N(E)$ is states per cubic meter per electron-volt ($m^{-3}\ eV^{-1}$).

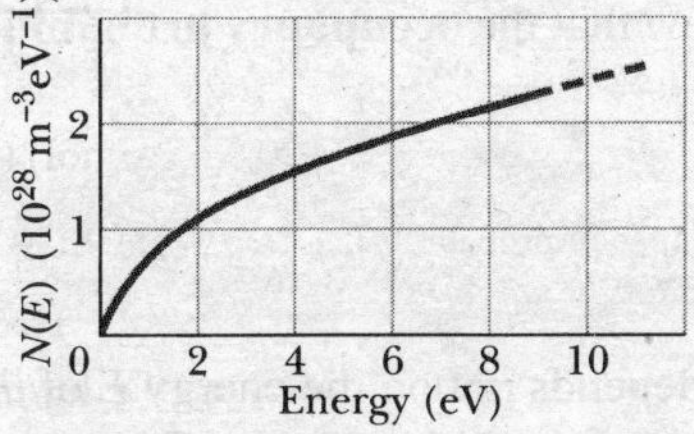

Fig. 42-5 The density of states $N(E)$—that is, the number of electron energy levels per unit energy interval and per unit volume—plotted as a function of electron energy. The density of states function simply counts the available states; it says nothing about whether these states are occupied by electrons.

We can find an expression for the density of states by counting the number of standing electron matter waves that can fit into a box the size of the metal sample we are considering. This is analogous to counting the number of standing waves of sound that can exist in a closed organ pipe. The differences are that our problem is three-dimensional (the organ pipe problem is one-dimensional) and the waves are matter waves (the organ-pipe waves are sound waves). The result of such counting can be shown to be

$$N(E) = \frac{8\sqrt{2}\pi m^{3/2}}{h^3} E^{1/2} \quad \text{(density of states)}, \tag{42-5}$$

where m is the mass of the electron and E is the energy at which $N(E)$ is to be evaluated. Note that nothing in this equation involves the shape of the sample, its temperature, or the material of which it is made. Equation 42-5 is plotted in Fig. 42-5.

✔CHECKPOINT 1: (a) Is the spacing between adjacent energy levels at $E = 4$ eV in copper larger than, the same as, or smaller than the spacing at $E = 6$ eV? (b) Is the spacing between adjacent energy levels at $E = 4$ eV in copper larger than, the same as, or smaller than the spacing for an identical volume of aluminum at that same energy?

Sample Problem 42-3

(a) Using Fig. 42-5, determine the number of states per electron-volt at 7 eV in a metal sample with a volume V of 2×10^{-9} m^3.

SOLUTION: The **Key Idea** is that we can obtain the number of states per electron-volt at a given energy by using the density of states $N(E)$ at that energy and the sample's volume V. At an energy of 7 eV, this means that

$$\begin{pmatrix}\text{number of states}\\ \text{per eV at 7 eV}\end{pmatrix} = \begin{pmatrix}\text{density of states}\\ N(E)\text{ at 7 eV}\end{pmatrix}\begin{pmatrix}\text{volume } V\\ \text{of sample}\end{pmatrix}.$$

From Fig. 42-5, we see that at an energy E of 7 eV, the density of states is about 2×10^{28} m^{-3} eV^{-1}. Thus,

$$\begin{pmatrix}\text{number of states}\\ \text{per eV at 7 eV}\end{pmatrix} = (2 \times 10^{28}\text{ m}^{-3}\text{ eV}^{-1})(2 \times 10^{-9}\text{ m}^3)$$
$$= 4 \times 10^{19}\text{ eV}^{-1}. \quad \text{(Answer)}$$

(b) Next, determine the number of states N in the sample within a *small* energy range ΔE of 0.003 eV, centered at 7 eV.

SOLUTION: From Eq. 42-5 and Fig. 42-5, we know that the density of states is a function of energy E. However, for an energy range ΔE that is small relative to E, we can approximate the density of states (and thus the number of states per electron-volt) to be constant. Thus, at an energy of 7 eV, we find the number of states N in the energy range ΔE of 0.003 eV as

$$\begin{pmatrix}\text{number of states } N\\ \text{in range } \Delta E \text{ at 7 eV}\end{pmatrix} = \begin{pmatrix}\text{number of states}\\ \text{per eV at 7 eV}\end{pmatrix} \times (\text{energy range } \Delta E)$$

or

$$N = (4 \times 10^{19}\text{ eV}^{-1})(0.003\text{ eV})$$
$$= 1.2 \times 10^{17} \approx 1 \times 10^{17}. \quad \text{(Answer)}$$

The Occupancy Probability $P(E)$

The ability of a metal to conduct electricity depends on the probability that available vacant levels will actually be occupied. Thus, another question arises: If an energy level is available at energy E, what is the probability $P(E)$ that it is actually occupied by an electron? At $T = 0$ K, we know that for all levels with energies below the Fermi energy, $P(E) = 1$, corresponding to a certainty that the level is occupied. We also know that, at $T = 0$ K, for all levels with energies above the Fermi energy, $P(E) = 0$, corresponding to a certainty that the level is *not* occupied. Figure 42-6*a* illustrates this situation.

To find $P(E)$ at temperatures above absolute zero, we must use a set of quantum counting rules called **Fermi–Dirac statistics,** named for the physicists who intro-

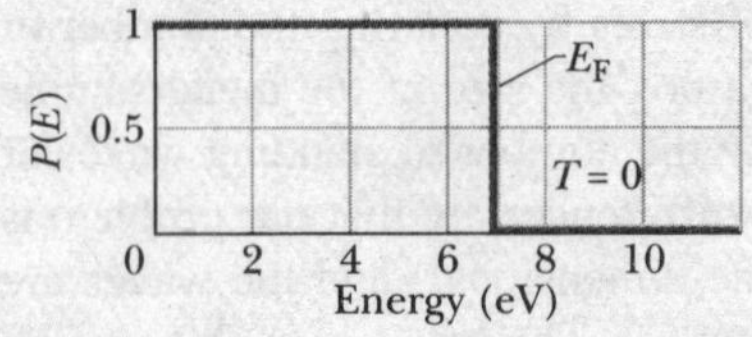

(*a*)

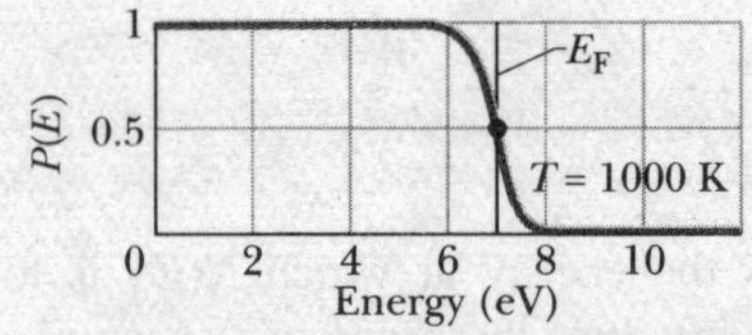

(*b*)

Fig. 42-6 The occupancy probability $P(E)$ is the probability that an energy level will be occupied by an electron. (*a*) At $T = 0$ K, $P(E)$ is unity for levels with energies E up to the Fermi energy E_F and zero for levels with higher energies. (*b*) At $T = 1000$ K, a few electrons whose energies were slightly less than the Fermi energy at $T = 0$ K move up to states with energies slightly greater than the Fermi energy. The dot on the curve shows that, for $E = E_F$, $P(E) = 0.5$.

duced them. Using these rules, it is possible to show that the **occupancy probability** $P(E)$ is

$$P(E) = \frac{1}{e^{(E - E_F)/kT} + 1} \quad \text{(occupancy probability)}, \tag{42-6}$$

in which E_F is the Fermi energy. Note that $P(E)$ depends not on the energy E of the level but only on the difference $E - E_F$, which may be positive or negative.

To see whether Eq. 42-6 describes Fig. 42-6*a*, we substitute $T = 0$ K in it. Then,

For $E < E_F$, the exponential term in Eq. 42-6 is $e^{-\infty}$, or zero, so $P(E) = 1$, in agreement with Fig. 42-6*a*.

For $E > E_F$, the exponential term is $e^{+\infty}$, so $P(E) = 0$, again in agreement with Fig. 42-6*a*.

Figure 42-6*b* is a plot of $P(E)$ for $T = 1000$ K. It shows that, as stated above, changes in the distribution of electrons among the available states involve only states whose energies are near the Fermi energy E_F. Note that if $E = E_F$ (no matter what the temperature T), the exponential term in Eq. 42-6 is $e^0 = 1$ and $P(E) = 0.5$. This leads us to a more useful definition of the Fermi energy:

> The Fermi energy of a given material is the energy of a quantum state that has the probability 0.5 of being occupied by an electron.

Figures 42-6*a* and *b* are plotted for copper, which has a Fermi energy of 7.0 eV. Thus, for copper both at $T = 0$ K and at $T = 1000$ K, a state at energy $E = 7.0$ eV has a probability of 0.5 of being occupied.

Sample Problem 42-4

(a) What is the probability that a quantum state whose energy is 0.10 eV above the Fermi energy will be occupied? Assume a sample temperature of 800 K.

SOLUTION: The **Key Idea** here is that the occupancy probability of any state in a metal can be found from Fermi–Dirac statistics according to Eq. 42-6. To apply that equation, let us first calculate its dimensionless exponent:

$$\frac{E - E_F}{kT} = \frac{0.10 \text{ eV}}{(8.62 \times 10^{-5} \text{ eV/K})(800 \text{ K})} = 1.45.$$

Inserting this exponent into Eq. 42-6 yields

$$P(E) = \frac{1}{e^{1.45} + 1} = 0.19 \text{ or } 19\%. \quad \text{(Answer)}$$

(b) What is the probability of occupancy for a state that is 0.10 eV *below* the Fermi energy?

SOLUTION: The **Key Idea** of part (a) applies here also except that now the state has an energy *below* the Fermi energy. Thus, the exponent in Eq. 42-6 has the same magnitude we found in part (a) but is negative, so Eq. 42-6 now yields

$$P(E) = \frac{1}{e^{-1.45} + 1} = 0.81 \text{ or } 81\%. \quad \text{(Answer)}$$

For states below the Fermi energy, we are often more interested in the probability that the state is *not* occupied. This probability is just $1 - P(E)$, or 19%. Note that it is the same as the probability of occupancy in (a).

How Many *Occupied* States Are There?

Equation 42-5 and Fig. 42-5 tell us how the available states are distributed in energy. The occupancy probability of Eq. 42-6 gives us the probability that any given state will actually be occupied by an electron. To find $N_o(E)$, the density of *occupied* states, we must weight each available state by the appropriate value of the occupancy

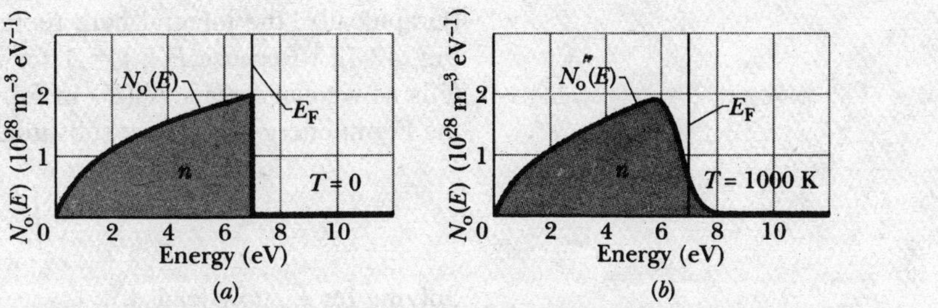

Fig. 42-7 (*a*) The density of occupied states $N_o(E)$ for copper at absolute zero. The area under the curve is the number density of electrons *n*. Note that all states with energies up to the Fermi energy $E_F = 7$ eV are occupied, and all those with energies above the Fermi energy are vacant. (*b*) The same for copper at $T = 1000$ K. Note that only electrons with energies near the Fermi energy have been affected and redistributed.

probability; that is,

$$\begin{pmatrix}\text{density of occupied states}\\ N_o(E)\text{ at energy }E\end{pmatrix} = \begin{pmatrix}\text{density of states}\\ N(E)\text{ at energy }E\end{pmatrix}\begin{pmatrix}\text{occupancy probability}\\ P(E)\text{ at energy }E\end{pmatrix}$$

or

$$N_o(E) = N(E)\,P(E) \quad \text{(density of occupied states).} \tag{42-7}$$

Figure 42-7*a* is a plot of Eq. 42-7 for copper at $T = 0$ K. It is found by multiplying, at each energy, the value of the density of states function (Fig. 42-5) by the value of the occupancy probability for absolute zero (Fig. 42-6*a*). Figure 42-7*b*, calculated similarly, shows the density of occupied states for copper at $T = 1000$ K.

Sample Problem 42-5

If the sample in Sample Problem 42-3 is copper, which has a Fermi energy of 7.0 eV, how many occupied states per electron-volt lie in a narrow energy range around 7.0 eV?

SOLUTION: The **Key Idea** of Sample Problem 42-3a applies here also, except that now we use the density of *occupied* states $N_o(E)$ as given by Eq. 42-7 ($N_o(E) = N(E)\,P(E)$). A second **Key Idea** is that because we want to evaluate quantities for a narrow energy range around 7.0 eV (the Fermi energy for copper), the occupancy probability $P(E)$ is 0.50. From Fig. 42-5, we see that the density of states at 7 eV is 2×10^{18} m^{-3} eV^{-1}. Thus, Eq. 42-7 tells us that the density of occupied states is

$$N_o(E) = N(E)\,P(E) = (2 \times 10^{28}\ \text{m}^{-3}\ \text{eV}^{-1})(0.50)$$
$$= 1 \times 10^{28}\ \text{m}^{-3}\ \text{eV}^{-1}.$$

Next, we rewrite the equation in Sample Problem 42-3a in terms of occupied states:

$$\begin{pmatrix}\text{number of } occupied\\ \text{states per eV at 7 eV}\end{pmatrix} = \begin{pmatrix}\text{density of } occupied\\ \text{states } N_o(E)\text{ at 7 eV}\end{pmatrix} \times \begin{pmatrix}\text{volume } V\\ \text{of sample}\end{pmatrix}.$$

Substituting our result for $N_o(E)$ and the previously given volume 2×10^{-9} m^3 for V then gives us

$$\begin{pmatrix}\text{number of occupied}\\ \text{states per eV}\\ \text{at 7 eV}\end{pmatrix} = (1 \times 10^{28}\ \text{m}^{-3}\ \text{eV}^{-1})(2 \times 10^{-9}\ \text{m}^3)$$
$$= 2 \times 10^{19}\ \text{eV}^{-1}. \quad \text{(Answer)}$$

Calculating the Fermi Energy

Suppose we add up (via integration) the number of occupied states per unit volume in Fig. 42-7*a* at all energies between $E = 0$ and $E = E_F$. The result must equal *n*, the number of conduction electrons per unit volume for the metal. In equation form, we have

$$n = \int_0^{E_F} N_o(E)\,dE. \tag{42-8}$$

(Graphically, the integral here represents the area under the distribution curve of Fig. 42-7*a*.) Because $P(E) = 1$ for all energies below the Fermi energy, Eq. 42-7 tells us we can replace $N_o(E)$ in Eq. 42-8 with $N(E)$ and then use Eq. 42-8 to find the Fermi energy E_F. If we substitute Eq. 42-5 into Eq. 42-8, we find that

$$n = \frac{8\sqrt{2}\pi m^{3/2}}{h^3} \int_0^{E_F} E^{1/2}\, dE = \frac{8\sqrt{2}\pi m^{3/2}}{h^3} \frac{2E_F^{3/2}}{3}.$$

Solving for E_F now leads to

$$E_F = \left(\frac{3}{16\sqrt{2}\pi}\right)^{2/3} \frac{h^2}{m} n^{2/3} = \frac{0.121h^2}{m} n^{2/3}. \tag{42-9}$$

Thus, when we know n, the number of conduction electrons per unit volume for a metal, we can find the Fermi energy for that metal.

42-6 Semiconductors

If you compare Fig. 42-8*a* with Fig. 42-4*a*, you can see that the band structure of a semiconductor is like that of an insulator. The main difference is that the semiconductor has a much smaller energy gap E_g between the top of the highest filled band (called the **valence band**) and the bottom of the vacant band just above it (called the **conduction band**). Thus, there is no doubt that silicon ($E_g = 1.1$ eV) is a semiconductor and diamond ($E_g = 5.5$ eV) is an insulator. In silicon—but not in diamond—there is a real possibility that thermal agitation at room temperature will cause electrons to jump the gap from the valence band to the conduction band.

In Table 42-1 we compared three basic electrical properties of copper, our prototype metallic conductor, and silicon, our prototype semiconductor. Let us look again at that table, one row at a time, to see how a semiconductor differs from a metal.

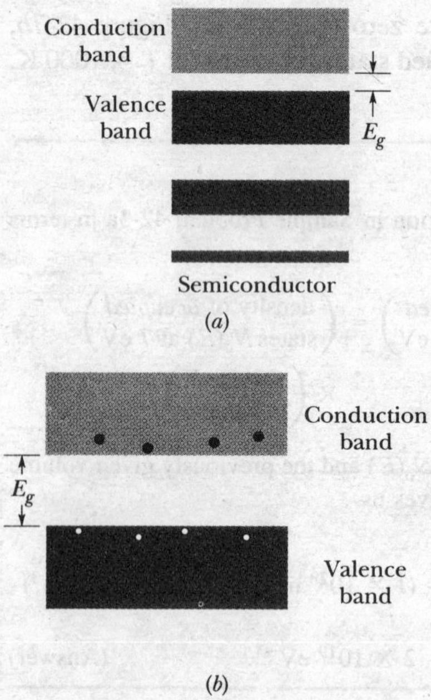

Fig. 42-8 (*a*) The band–gap pattern for a semiconductor. It resembles that of an insulator (see Fig. 42-4*a*) except that here the energy gap E_g is much smaller; thus electrons, because of their thermal agitation, have some reasonable probability of being able to jump the gap. (*b*) Thermal agitation has caused a few electrons to jump the gap from the valence band to the conduction band, leaving an equal number of holes in the valence band.

Number Density of Charge Carriers *n*

The bottom row of Table 42-1 shows that copper has far more charge carriers per unit volume than silicon, by a factor of about 10^{13}. For copper, each atom contributes one electron, its single valence electron, to the conduction process. Charge carriers in silicon arise only because, at thermal equilibrium, thermal agitation causes a certain (very small) number of valence-band electrons to jump the energy gap into the conduction band, leaving an equal number of unoccupied energy states, called **holes,** in the valence band. Figure 42-8*b* shows the situation.

Both the electrons in the conduction band and the holes in the valence band serve as charge carriers. The holes do so by permitting a certain freedom of movement to electrons in the valence band that, in the absence of holes, would be gridlocked. If an electric field $\vec{E}$ is set up in a semiconductor, the electrons in the valence band, being negatively charged, tend to drift in the direction opposite $\vec{E}$. This causes the positions of the holes to drift in the direction of $\vec{E}$. In effect, the holes behave like moving particles of charge $+e$.

It may help to think of a row of cars parked bumper to bumper, with the leading car at one car's length from a barrier. If the leading car moves forward to the barrier, it opens up a car's length space behind it. The second car can then move up to fill that space, allowing the third car to move up, and so on. The motions of the many cars toward the barrier are most simply analyzed by focusing attention on the drift of the single "hole" (parking space) away from the barrier.

In semiconductors, conduction by holes is just as important as conduction by electrons. In thinking about hole conduction, it is well to imagine that all unoccupied states in the valence band are occupied by particles of charge $+e$, and that all electrons in the valence band have been removed, so that these positive charge carriers can move freely throughout the band.

Resistivity ρ

From Chapter 27 recall that the resistivity ρ of a material is $m/e^2n\tau$, where m is the electron mass, e is the fundamental charge, n is the number of charge carriers per unit volume, and τ is the mean time between collisions of the charge carriers. Table 42-1 shows that, at room temperature, the resistivity of silicon is higher than that of copper, by a factor of about 10^{11}. This vast difference can be accounted for by the vast difference in n. Other factors enter, but their effect on the resistivity is swamped by the enormous difference in n.

Temperature Coefficient of Resistivity α

Recall that α (see Eq. 27-17) is the fractional change in resistivity per unit change in temperature:

$$\alpha = \frac{1}{\rho}\frac{d\rho}{dT}. \tag{42-10}$$

The resistivity of copper *increases* with temperature (that is, $d\rho/dT > 0$) because collisions of copper's charge carriers occur more frequently at higher temperatures. Thus, α is *positive* for copper.

The collision frequency also increases with temperature for silicon. However, the resistivity of silicon actually *decreases* with temperature ($d\rho/dT < 0$) because the number of charge carriers n (electrons in the conduction band and holes in the valence band) increases so rapidly with temperature. (More electrons jump the gap from the valence band to the conduction band.) Thus, the fractional change α is *negative* for silicon.

✔CHECKPOINT 2: The research laboratory of a large corporation developed three new solid materials whose electrical properties are shown here. Anticipating patent applications, the laboratory identified these materials with code names. Classify each material as a metal, an insulator, a semiconductor, or none of the above:

Material (Code Name)	n (m^{-3})	ρ ($\Omega \cdot$ m)	α (K^{-1})
Cleveland	10^{29}	10^{-8}	$+10^{-3}$
Boca Raton	10^{28}	10^{-9}	-10^{-3}
Seattle	10^{15}	10^{3}	-10^{-2}

42-7 Doped Semiconductors

The usefulness of semiconductors in technology can be greatly improved by introducing a small number of suitable replacement atoms (called impurities) into the semiconductor lattice—a process called **doping.** Typically, only about 1 silicon atom in 10^7 is replaced by a dopant atom in the doped semiconductor. Essentially all modern semiconducting devices are based on doped material. Such materials are of two types, called ***n*-type** and ***p*-type**; we discuss each in turn.

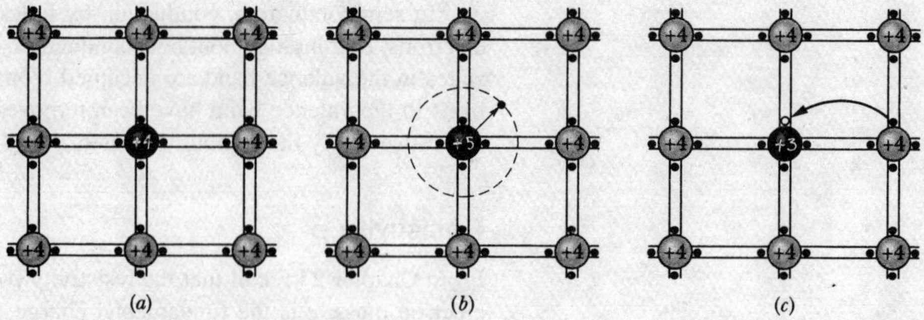

Fig. 42-9 (*a*) A flattened-out representation of the lattice structure of pure silicon. Each silicon ion is coupled to its four nearest neighbors by a two-electron covalent bond (represented by a pair of red dots between two parallel black lines). The electrons belong to the bond—not to the individual atoms—and form the valence band of the sample. (*b*) One silicon atom is replaced by a phosphorus atom (valence = 5). The "extra" electron is only loosely bound to its ion core and may easily be elevated to the conduction band, where it is free to wander through the volume of the lattice. (*c*) One silicon atom is replaced by an aluminum atom (valence = 3). There is now a hole in one of the covalent bonds and thus in the valence band of the sample. The hole can easily migrate through the lattice as electrons from neighboring bonds move in to fill it. Here the hole migrates rightward.

n-Type Semiconductors

The electrons in an isolated silicon atom are arranged in subshells according to the scheme

$$1s^2\ 2s^2\ 2p^6\ 3s^2\ 3p^2,$$

in which, as usual, the superscripts (which add to 14, the atomic number of silicon) represent the numbers of electrons in the specified subshells.

Figure 42-9*a* is a flattened-out representation of a portion of the lattice of pure silicon in which the portion has been projected onto a plane; compare the figure with Fig. 42-1*b*, which represents the unit cell of the lattice in three dimensions. Each silicon atom contributes its pair of 3*s* electrons and its pair of 3*p* electrons to form a rigid two-electron covalent bond with each of its four nearest neighbors. (A covalent bond is a link between two atoms in which the atoms share a pair of electrons.) The four atoms that lie within the unit cell in Fig. 42-1*b* show these four bonds.

The electrons that form the silicon–silicon bonds constitute the valence band of the silicon sample. If an electron is torn from one of these bonds so that it becomes free to wander throughout the lattice, we say that the electron has been raised from the valence band to the conduction band. The minimum energy required to do this is the gap energy E_g.

Because four of its electrons are involved in bonds, each silicon "atom" is actually an ion consisting of an inert neonlike electron cloud (containing 10 electrons) surrounding a nucleus whose charge is $+14e$, where 14 is the atomic number of silicon. The net charge of each of these ions is thus $+4e$, and the ions are said to have a *valence number* of 4.

In Fig. 42-9*b* the central silicon ion has been replaced by an atom of phosphorus (valence = 5). Four of the valence electrons of the phosphorus form bonds with the four surrounding silicon ions. The fifth ("extra") electron is only loosely bound to the phosphorus ion core. On an energy-band diagram, we usually say that such an electron occupies a localized energy state that lies within the energy gap, at an average energy interval E_d below the bottom of the conduction band; this is indicated in Fig. 42-10*a*. Because $E_d \ll E_g$, the energy required to excite electrons from *these* levels into the conduction band is much less than that required to excite silicon valence electrons into the conduction band.

The phosphorus atom is called a **donor** atom because it readily *donates* an electron to the conduction band. In fact, at room temperature virtually *all* the electrons contributed by the donor atoms are in the conduction band. By adding donor atoms, it is possible to increase greatly the number of electrons in the conduction band, by a factor very much larger than Fig. 42-10*a* suggests.

Semiconductors doped with donor atoms are called ***n*-type semiconductors;** the *n* stands for *negative,* to imply that the negative charge carriers introduced into the conduction band greatly outnumber the positive charge carriers, which are the holes

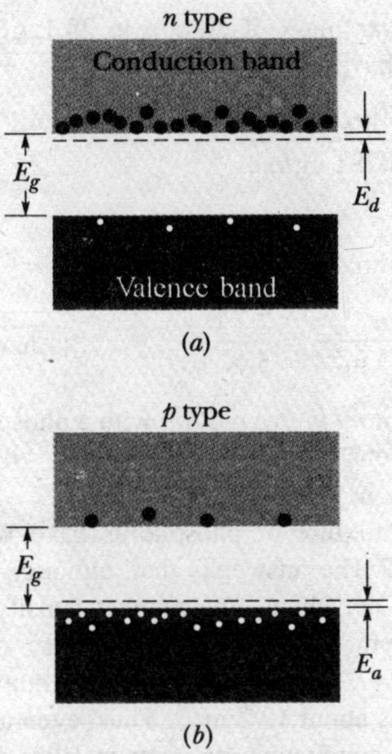

Fig. 42-10 (*a*) In a doped *n*-type semiconductor, the energy levels of donor electrons lie a small interval E_d below the bottom of the conduction band. Because donor electrons can be easily excited to the conduction band, there are now many more electrons in that band. The valence band contains the same small number of holes as before. (*b*) In a doped *p*-type semiconductor, the acceptor levels lie a small energy interval E_a above the top of the valence band. There are now relatively many more holes in the valence band. The conduction band contains the same small number of electrons as before. The ratio of majority carriers to minority carriers in both (*a*) and (*b*) is very much greater than is suggested by these diagrams.

TABLE 42-2 Properties of Two Doped Semiconductors

	Type of Semiconductor	
Property	*n*	*p*
Matrix material	Silicon	Silicon
Matrix nuclear charge	$+14e$	$+14e$
Matrix energy gap	1.2 eV	1.2 eV
Dopant	Phosphorus	Aluminum
Type of dopant	Donor	Acceptor
Majority carriers	Electrons	Holes
Minority carriers	Holes	Electrons
Dopant energy gap	0.045 eV	0.067 eV
Dopant valence	5	3
Dopant nuclear charge	$+15e$	$+13e$
Dopant net ion charge	$+e$	$-e$

in the valence band. In *n*-type semiconductors, the electrons are called the **majority carriers,** and the holes the **minority carriers.**

p-Type Semiconductors

Now consider Fig. 42-9*c*, in which one of the silicon atoms (valence = 4) has been replaced by an atom of aluminum (valence = 3). The aluminum atom can bond covalently with only three silicon atoms, so there is now a "missing" electron (a hole) in one aluminum–silicon bond. With a small expenditure of energy, an electron can be torn from a neighboring silicon–silicon bond to fill this hole, thereby creating a hole in *that* bond. Similarly, an electron from some other bond can be moved to fill the second hole. In this way, the hole can migrate through the lattice.

The aluminum atom is called an **acceptor** atom because it readily *accepts* an electron from a neighboring bond—that is, from the valence band of silicon. As Fig. 42-10*b* suggests, this electron occupies a localized acceptor state that lies within the energy gap, at an average energy interval E_a above the top of the valence band. By adding acceptor atoms, it is possible to increase very greatly the number of holes in the valence band, by a factor much larger than Fig. 42-10*b* suggests. In silicon at room temperature, virtually *all* the acceptor levels are occupied by electrons.

Semiconductors doped with acceptor atoms are called ***p*-type semiconductors;** the *p* stands for *positive* to imply that the holes introduced into the valence band, which behave like positive charge carriers, greatly outnumber the electrons in the conduction band. In *p*-type semiconductors, holes are the majority carriers and electrons are the minority carriers.

Table 42-2 summarizes the properties of a typical *n*-type and a typical *p*-type semiconductor. Note particularly that the donor and acceptor ion cores, although they are charged, are not charge *carriers* because at normal temperatures they remain fixed in their lattice sites.

Sample Problem 42-6

The number density n_0 of conduction electrons in pure silicon at room temperature is about 10^{16} m^{-3}. Assume that, by doping the silicon lattice with phosphorus, we want to increase this number by a factor of a million (10^6). What fraction of silicon atoms must we replace with phosphorus atoms? (Recall that at room temperature, thermal agitation is so effective that essentially every phosphorus atom donates its "extra" electron to the conduction band.)

SOLUTION: One **Key Idea** here is that, because each phosphorus atom contributes one conduction electron and because we want the total number density of conduction electrons to be $10^6 n_0$, then the number density of phosphorus atoms n_P must be given by

$$10^6 n_0 = n_0 + n_P.$$

Then
$$n_P = 10^6 n_0 - n_0 \approx 10^6 n_0$$
$$= (10^6)(10^{16}\ \text{m}^{-3}) = 10^{22}\ \text{m}^{-3}.$$

This tells us that we must add 10^{22} atoms of phosphorus per cubic meter of silicon.

A second **Key Idea** is that we can find the number density n_{Si} of silicon atoms in pure silicon (before the doping) from Eq. 42-4, which we can write as

$$\begin{pmatrix}\text{number of atoms}\\ \text{in sample}\end{pmatrix} = \frac{(\text{silicon density})(\text{sample volume } V)}{(\text{silicon molar mass } M_{Si})/N_A}.$$

Dividing both sides by the sample volume V to get the number density of silicon atoms n_{Si} on the left, we then have

$$n_{Si} = \frac{(\text{silicon density})N_A}{M_{Si}}.$$

Appendix F tells us that the density of silicon is 2.33 g/cm³ (= 2330 kg/m³) and the molar mass of silicon is 28.1 g/mol (= 0.0281 kg/mol). Thus, we have

$$n_{Si} = \frac{(2330\ \text{kg/m}^3)(6.02 \times 10^{23}\ \text{mol}^{-1})}{0.0281\ \text{kg/mol}}$$
$$= 5 \times 10^{28}\ \text{m}^{-3}.$$

The fraction we seek is approximately

$$\frac{n_P}{n_{Si}} = \frac{10^{22}\ \text{m}^{-3}}{5 \times 10^{28}\ \text{m}^{-3}} = \frac{1}{5 \times 10^6}. \quad \text{(Answer)}$$

If we replace only *one silicon atom in five million* with a phosphorus atom, the number of electrons in the conduction band will be increased by a factor of a million.

How can such a tiny admixture of phosphorus have what seems to be such a big effect? The answer is that, although the effect is very significant, it is not "big." The number density of conduction electrons was 10^{16} m^{-3} before doping and 10^{22} m^{-3} after doping. For copper, however, the conduction-electron number density (given in Table 42-1) is about 10^{29} m^{-3}. Thus, even after doping, the number density of conduction electrons in silicon remains much less than that of a typical metal, such as copper, by a factor of about 10^7.

42-8 The *p-n* Junction

A ***p-n* junction** (Fig. 42-11*a*) is a single semiconductor crystal that has been selectively doped so that one region is *n*-type material and the adjacent region is *p*-type material. Such junctions are at the heart of essentially all semiconductor devices.

We assume, for simplicity, that the junction has been formed mechanically, by jamming together a bar of *n*-type semiconductor and a bar of *p*-type semiconductor. Thus, the transition from one region to the other is perfectly sharp, occurring at a single **junction plane.**

Let us discuss the motions of electrons and holes just after the *n*-type bar and the *p*-type bar, both electrically neutral, have been jammed together to form the junction. We first examine the majority carriers, which are electrons in the *n*-type material and holes in the *p*-type material.

Motions of the Majority Carriers

If you burst a helium-filled balloon, helium atoms will diffuse (spread) outward into the surrounding air. This happens because there are very few helium atoms in normal air. In more formal language, there is a helium *density gradient* at the balloon–air interface (the number density of helium atoms varies across the interface); the helium atoms move so as to reduce the gradient.

In the same way, electrons on the *n* side of Fig. 42-11*a* that are close to the junction plane tend to diffuse across it (from right to left in the figure) and into the *p* side, where there are very few free electrons. Similarly, holes on the *p* side that are close to the junction plane tend to diffuse across that plane (from left to right) and into the *n* side, where there are very few holes. The motions of both the electrons and the holes contribute to a **diffusion current** I_{diff}, conventionally directed from left to right as indicated in Fig. 42-11*d*.

Recall that the *n*-side is studded throughout with positively charged donor ions,

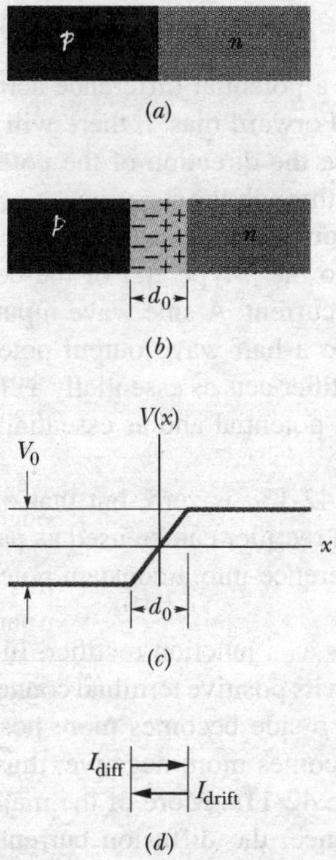

Fig. 42-11 (*a*) A *p-n* junction. (*b*) Motions of the majority charge carriers across the junction plane uncover a space charge associated with uncompensated donor ions (to the right of the plane) and acceptor ions (to the left). (*c*) Associated with the space charge is a contact potential difference V_0 across d_0. (*d*) The diffusion of majority carriers (both electrons and holes) across the junction plane produces a diffusion current I_{diff}. (In a real *p-n* junction, the boundaries of the depletion zone would not be sharp, as shown here, and the contact potential curve (*c*) would be smooth, with no sharp corners.)

fixed firmly in their lattice sites. Normally, the excess positive charge of each of these ions is compensated electrically by one of the conduction-band electrons. When an *n*-side electron diffuses across the junction plane, however, the diffusion "uncovers" one of these donor ions, thus introducing a fixed positive charge near the junction plane on the *n* side. When the diffusing electron arrives on the *p* side, it quickly combines with an acceptor ion (which lacks one electron), thus introducing a fixed negative charge near the junction plane on the *p* side.

In this way electrons diffusing through the junction plane from right to left in Fig. 42-11*a* result in a buildup of **space charge** on each side of the junction plane, as indicated in Fig. 42-11*b*. Holes diffusing through the junction plane from left to right have exactly the same effect. (Take the time now to convince yourself of that.) The motions of both majority carriers—electrons and holes—contribute to the buildup of these two space charge regions, one positive and one negative. These two regions form a **depletion zone,** so named because it is relatively free of *mobile* charge carriers; its width is shown as d_0 in Fig. 42-11*b*.

The buildup of space charge generates an associated **contact potential difference** V_0 across the depletion zone, as Fig. 42-11*c* shows. This potential difference limits further diffusion of electrons and holes across the junction plane. Negative charges tend to avoid regions of low potential. Thus, an electron approaching the junction plane from the right in Fig. 42-11*b* is moving toward a region of low potential and would tend to turn back into the *n* side. Similarly, a positive charge (a hole) approaching the junction plane from the left is moving toward a region of high potential and would tend to turn back into the *p* side.

Motions of the Minority Carriers

As Fig. 42-10*a* shows, although the majority carriers in *n*-type material are electrons, there are nevertheless a few holes. Likewise in *p*-type material (Fig. 42-10*b*), although the majority carriers are holes, there are also a few electrons. These few holes and electrons are the minority carriers in the corresponding materials.

Although the potential difference V_0 in Fig. 42-11*c* acts as a barrier for the majority carriers, it is a downhill trip for the minority carriers, be they electrons on the *p* side or holes on the *n* side. Positive charges (holes) tend to seek regions of low potential; negative charges (electrons) tend to seek regions of high potential. Thus, both types of carriers are *swept across* the junction plane by the contact potential difference and, together, constitute a **drift current** I_{drift} across the junction plane from right to left, as Fig. 42-11*d* indicates.

Thus, an isolated *p-n* junction is in an equilibrium state in which a contact potential difference V_0 exists between its ends. At equilibrium, the average diffusion current I_{diff} that moves through the junction plane from the *p* side to the *n* side is just balanced by an average drift current I_{drift} that moves in the opposite direction. These two currents cancel because the net current through the junction plane must be zero; otherwise charge would be transferred without limit from one end of the junction to the other.

CHECKPOINT 3: Which of the following five currents across the junction plane of Fig. 42-11*a* must be zero?
(a) the net current due to holes, both majority and minority carriers included
(b) the net current due to electrons, both majority and minority carriers included
(c) the net current due to both holes and electrons, both majority and minority carriers included
(d) the net current due to majority carriers, both holes and electrons included
(e) the net current due to minority carriers, both holes and electrons included

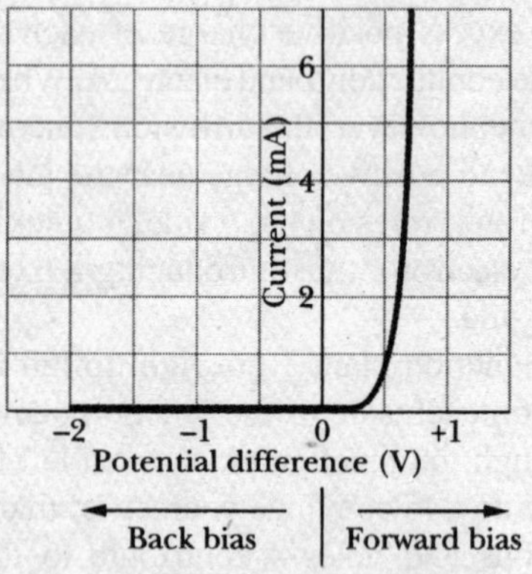

Fig. 42-12 A current–voltage plot for a *p-n* junction, showing that the junction is highly conducting when forward-biased and essentially nonconducting when back-biased.

42-9 The Junction Rectifier

Look now at Fig. 42-12. It shows that, if we place a potential difference across a *p-n* junction in one direction (here labeled + and "Forward bias"), there will be a current through the junction. However, if we reverse the direction of the potential difference, there will be approximately zero current through the junction.

One application of this property is the **junction rectifier,** whose symbol is shown in Fig. 42-13*b*; the arrowhead corresponds to the *p*-type end of the device and points in the allowed direction of conventional current. A sine wave input potential to the device (Fig. 42-13*a*) is transformed to a half-wave output potential (Fig. 42-13*c*) by the junction rectifier; that is, the rectifier acts as essentially a closed switch (zero resistance) for one polarity of the input potential and as essentially an open switch (infinite resistance) for the other.

The average value of the input voltage in Fig. 42-13*a* is zero, but that of the output voltage in Fig. 42-13*c* is not. Thus, a junction rectifier can be used as part of an apparatus to convert an alternating potential difference into a constant potential difference, as for an electronic power supply.

Figure 42-14 shows why a *p-n* junction operates as a junction rectifier. In Fig. 42-14*a*, a battery is connected across the junction with its positive terminal connected at the *p* side. In this **forward-bias connection,** the *p* side becomes more positive than it was before the connection and the *n* side becomes more negative, thus *decreasing* the height of the potential barrier V_0 of Fig. 42-11*c*. More of the majority carriers can now surmount this smaller barrier; hence, the diffusion current I_{diff} increases markedly.

The minority carriers that form the drift current, however, sense no barrier, so the drift current I_{drift} is not affected by the external battery. The nice current balance that existed at zero bias (see Fig. 42-11*d*) is thus upset and, as shown in Fig. 42-14*a*, a large net forward current I_F appears in the circuit.

Another effect of forward bias is to narrow the depletion zone, as a comparison of Figs. 42-11*b* and Fig. 42-14*a* shows. The depletion zone narrows because the reduced potential barrier associated with forward bias must be associated with a smaller space charge. Because the ions producing the space charge are fixed in their lattice sites, a reduction in their number can come about only through a reduction in the width of the depletion zone.

Because the depletion zone normally contains very few charge carriers, it is normally a region of high resistivity. However, when its width is substantially re-

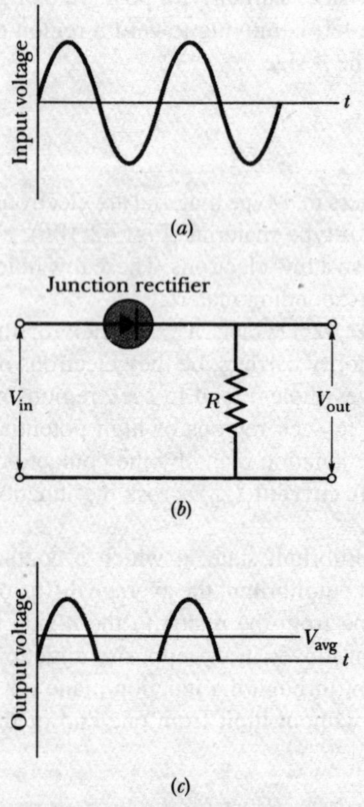

Fig. 42-13 A *p-n* junction connected as a junction rectifier. The action of the circuit in (*b*) is to pass the positive half of the input wave form (*a*) but to suppress the negative half. The average potential of the input wave form is zero; that of the output wave form (*c*) has a positive value V_{avg}.

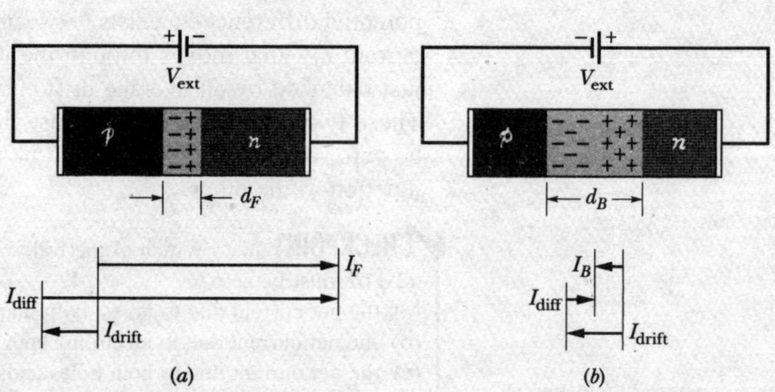

Fig. 42-14 (*a*) The forward-bias connection of a *p-n* junction, showing the narrowed depletion zone and the large forward current I_F. (*b*) The back-bias connection, showing the widened depletion zone and the small back current I_B.

duced by a forward bias, its resistance is also reduced substantially, as is consistent with the large forward current.

Figure 42-14*b* shows the **back-bias** connection, in which the negative terminal of the battery is connected at the *p*-type end of the *p-n* junction. Now the applied emf *increases* the contact potential difference, the diffusion current *decreases* substantially while the drift current remains unchanged, and a relatively *small* back current I_B results. The depletion zone *widens*, its *high* resistance being consistent with the *small* back current I_B.

42-10 The Light-Emitting Diode (LED)

Nowadays, we can hardly avoid the brightly colored "electronic" numbers that glow at us from cash registers and gasoline pumps, microwave ovens and alarm clocks, and we cannot seem to do without the invisible infrared beams that control elevator doors and operate television sets via remote control. In nearly all cases this light is emitted from a *p-n* junction operating as a **light-emitting diode** (LED). How can a *p-n* junction generate light?

Consider first a simple semiconductor. When an electron from the bottom of the conduction band falls into a hole at the top of the valence band, an energy E_g equal to the gap width is released. In silicon, germanium, and many other semiconductors, this energy is largely transformed into thermal energy of the vibrating lattice, and as a result, no light is emitted.

In some semiconductors, however, including gallium arsenide, the energy can be emitted as a photon of energy hf at wavelength

$$\lambda = \frac{c}{f} = \frac{c}{E_g/h} = \frac{hc}{E_g}. \qquad (42\text{-}11)$$

To emit enough light to be useful as an LED, the material must have a suitably large number of electron–hole transitions. This condition is *not* satisfied by a pure semiconductor because, at room temperature, there are simply not enough electron–hole pairs. As Fig. 42-10 suggests, doping will not help. In doped *n*-type material the number of conduction electrons is greatly increased, but there are not enough holes for them to combine with; in doped *p*-type material there are plenty of holes but not enough electrons to combine with them. Thus, neither a pure semiconductor nor a doped semiconductor can provide enough electron–hole transitions to serve as a practical LED.

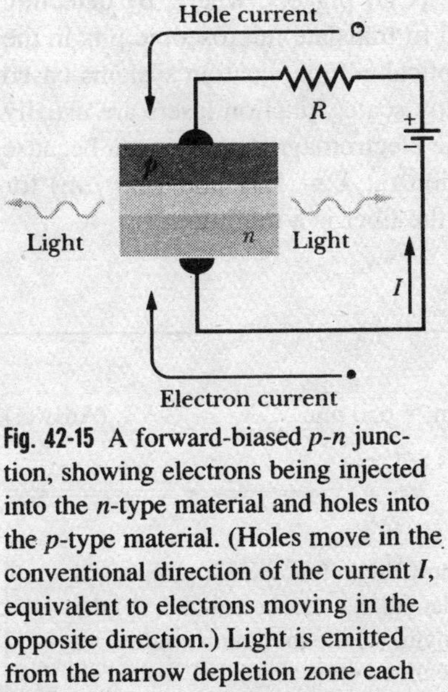

Fig. 42-15 A forward-biased *p-n* junction, showing electrons being injected into the *n*-type material and holes into the *p*-type material. (Holes move in the conventional direction of the current *I*, equivalent to electrons moving in the opposite direction.) Light is emitted from the narrow depletion zone each time an electron and a hole combine across that zone.

What we need is a semiconductor material with a very large number of electrons in the conduction band *and* a correspondingly large number of holes in the valence band. A device with this property can be fabricated by placing a strong forward bias on a heavily doped *p-n* junction, as in Fig. 42-15. In such an arrangement the current *I* through the device serves to inject electrons into the *n*-type material and to inject holes into the *p*-type material. If the doping is heavy enough and the current is great enough, the depletion zone can become very narrow, perhaps only a few micrometers wide. The result is a great number density of electrons in the *n*-type material facing a correspondingly great number density of holes in the *p*-type material, across the narrow depletion zone. With such great number densities so near, many electron–hole combinations occur, causing light to be emitted from that zone. Figure 42-16 shows the construction of an actual LED.

Commercial LEDs designed for the visible region are commonly based on gallium, suitably doped with arsenic and phosphorus atoms. An arrangement in which 60% of the nongallium sites are occupied by arsenic ions and 40% by phosphorus ions results in a gap width E_g of about 1.8 eV, corresponding to red light. Other

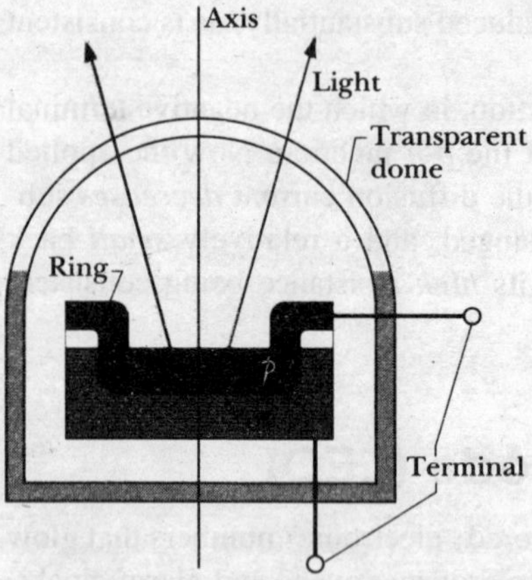

Fig. 42-16 Cross section of an LED (the device has rotational symmetry about the central axis). The *p*-type material, which is thin enough to transmit light, is in the form of a circular disk. A connection is made to the *p*-type material through a circular metal ring that touches the disk at its periphery. The depletion zone between the *n*-type material and the *p*-type material is not shown.

doping and transition level arrangements make it possible to construct LEDs that emit light in essentially any desired region of the visible and near-visible spectra.

The Photo-Diode

Passing a current through a suitably arranged *p-n* junction can generate light. The reverse is also true; that is, shining light on a suitably arranged *p-n* junction can produce a current in a circuit that includes the junction. This is the basis for the **photo-diode.**

When you click your television remote control, an LED in the device sends out a coded sequence of pulses of infrared light. The receiving device in your television set is an elaboration of the simple (two-terminal) photo-diode that not only detects the infrared signals but also amplifies them and transforms them into electrical signals that change the channel or adjust the volume, among other tasks.

The Junction Laser

In the arrangement of Fig. 42-15 there are many electrons in the conduction band of the *n*-type material and many holes in the valence band of the *p*-type material. Thus, there is a **population inversion** for the electrons; that is, there are more electrons in higher energy levels than in lower energy levels. As we discussed in Section 41-12, this is normally a necessary—but not a sufficient—condition for laser action.

When a single electron moves from the conduction band to the valence band, it can release its energy as a photon. This photon can stimulate a second electron to fall into the valence band, producing a second photon by stimulated emission. In this way, if the current through the junction is great enough, a chain reaction of stimulated emission events can occur and laser light can be generated. To bring this about, opposite faces of the *p-n* junction crystal must be flat and parallel, so that light can be reflected back and forth within the crystal. (Recall that in the helium–neon laser of Fig. 41-21, a pair of mirrors served this purpose.) Thus, a *p-n* junction can act as a **junction laser,** its light output being highly coherent and much more sharply defined in wavelength than light from an LED.

Fig. 42-17 A junction laser developed at the AT&T Bell Laboratories. The cube at the right is a grain of salt.

Junction lasers are built into compact disc (CD) players, where, by detecting reflections from the rotating disc, they are used to translate microscopic pits in the disc into sound. They are also much used in optical communication systems based on optical fibers. Figure 42-17 suggests their tiny scale. Junction lasers are usually designed to operate in the infrared region of the electromagnetic spectrum because optical fibers have two "windows" in that region (at $\lambda = 1.31$ and $1.55\ \mu$m) for which the energy absorption per unit length of the fiber is a minimum.

Sample Problem 42-7

An LED is constructed from a *p-n* junction based on a certain Ga-As-P semiconducting material whose energy gap is 1.9 eV. What is the wavelength of the emitted light?

SOLUTION: The **Key Idea** here is to assume that the transitions are from the bottom of the conduction band to the top of the valence band; then Eq. 42-11 holds. From this equation

$$\lambda = \frac{hc}{E_g} = \frac{(6.63 \times 10^{-34}\ \text{J}\cdot\text{s})(3.00 \times 10^8\ \text{m/s})}{(1.9\ \text{eV})(1.60 \times 10^{-19}\ \text{J/eV})}$$

$$= 6.5 \times 10^{-7}\ \text{m} = 650\ \text{nm}. \quad \text{(Answer)}$$

Light of this wavelength is red.

CHECKPOINT 4: For the LED in this sample problem, is 650 nm (a) the only wavelength that can be emitted, (b) the maximum emitted wavelength, (c) the minimum emitted wavelength, or (d) the average emitted wavelength? (Consider the quantum jump assumed in the solution.)

42-11 The Transistor

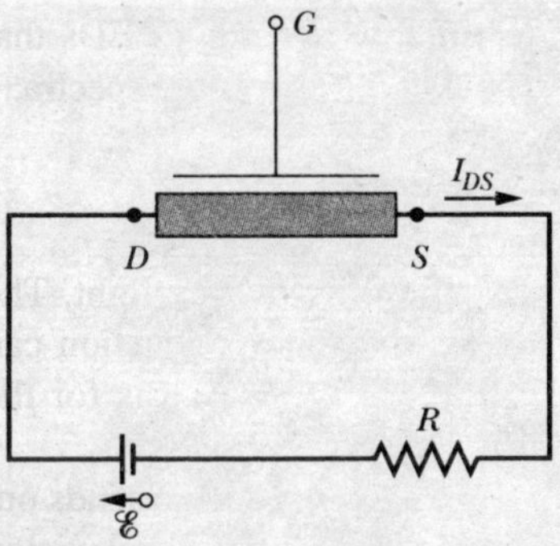

Fig. 42-18 A circuit containing a generalized field effect transistor, in which electrons flow through the device from the source terminal S to the drain terminal D. (The conventional current I_{DS} is in the opposite direction.) The magnitude of I_{DS} is controlled by the electric field set up within the body of the device by a potential applied to G, the gate terminal.

A **transistor** is a three-terminal semiconducting device that can be used to amplify input signals. Figure 42-18 shows a generalized **f**ield-**e**ffect **t**ransistor (FET); in it, the flow of electrons from terminal S (the *source*) to terminal D (the *drain*) can be controlled by an electric field (hence field effect) set up within the device by a suitable electric potential applied to terminal G (the *gate*). Transistors are available in many types; we shall discuss only a particular FET called a MOSFET, or **m**etal-**o**xide-**s**emiconductor-**f**ield-**e**ffect **t**ransistor. The MOSFET has been described as the workhorse of the modern electronics industry.

For many applications the MOSFET is operated in only two states: with the drain-to-source current I_{DS} ON (gate open) or with it OFF (gate closed). The first of these can represent a 1 and the other a 0 in the binary arithmetic on which digital logic is based, and therefore MOSFETs can be used in digital logic circuits. Switching between the ON and OFF states can occur at high speed, so that binary logic data can be moved through MOSFET-based circuits very rapidly. MOSFETs about 500 nm in length—about the same as the wavelength of yellow light—are routinely fabricated for use in electronic devices of all kinds.

Figure 42-19 shows the basic structure of a MOSFET. A single crystal of silicon or other semiconductor is lightly doped to form p-type material. Embedded in this substrate, by heavily "overdoping" with n-type dopants, are two "islands" of n-type material, forming the drain D and the source S. The drain and source are connected by a thin channel of n-type material, called the ***n* channel.** A thin insulating layer of silicon dioxide (hence the O in MOSFET) is deposited on the crystal and penetrated by two metallic terminals (hence the M) at D and S, so that electrical contact can be made with the drain and the source. A thin metallic layer—the gate G—is deposited facing the n channel. Note that the gate makes no electrical contact with the transistor proper, being separated from it by the insulating oxide layer.

Consider first that the source and p-type substrate are grounded (at zero potential) and the gate is "floating"; that is, the gate is not connected to an external source of emf. Let a potential V_{DS} be applied between the drain and the source, such that the drain is positive. Electrons will then flow through the n channel from source to drain, and the conventional current I_{DS}, as shown in Fig. 42-19, will be from drain to source through the n-type material.

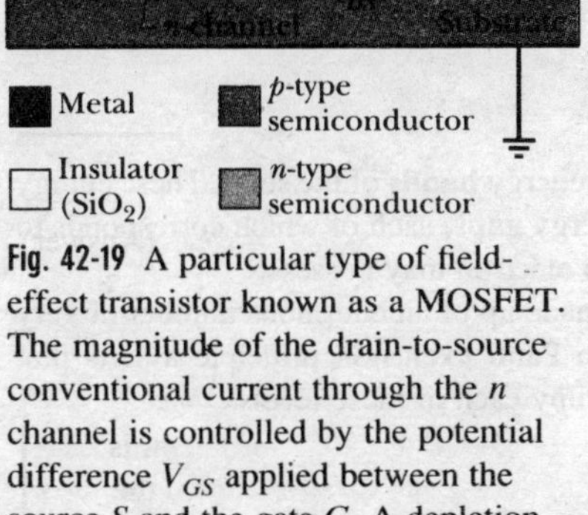

Fig. 42-19 A particular type of field-effect transistor known as a MOSFET. The magnitude of the drain-to-source conventional current through the n channel is controlled by the potential difference V_{GS} applied between the source S and the gate G. A depletion zone that exists between the n-type material and the p-type material is not shown.

Now let a potential V_{GS} be applied to the gate, making it negative with respect to the source. The negative gate sets up within the device an electric field (hence the "field effect") that tends to repel electrons from the n channel into the substrate. This electron movement widens the (naturally occurring) depletion zone between the n channel and the substrate, at the expense of the n channel. The reduced width of the n channel, coupled with a reduction in the number of charge carriers in that channel, increases the resistance of that channel and thus decreases the current I_{DS}. With the proper value of V_{GS}, this current can be shut off completely; hence, by controlling V_{GS}, the MOSFET can be switched between its ON and OFF modes.

Charge carriers do not flow through the *substrate* because the substrate (1) is lightly doped, (2) is not a good conductor, and (3) is separated from the n-channel and the two n-type islands by an insulating depletion zone, not specifically shown in Fig. 42-19. Such a depletion zone always exists at a boundary between n-type material and p-type material, as Fig. 42-11b shows.

Integrated Circuits

Computers and other electronic devices employ thousands (if not millions) of transistors and other electronic components such as capacitors and resistors. These are

Fig. 42-20 An integrated circuit for the Intel Pentium chip, used mainly in computers. It will be encapsulated in a ceramic coating for installation and use.

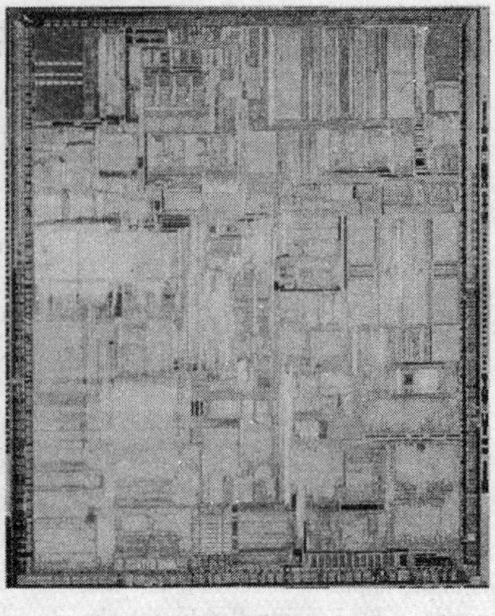

Fig. 42-21 Enlarged photograph of the layout of an Intel chip.

not assembled as separate units but are crafted into a single semiconducting **chip,** forming an **integrated circuit.**

Figure 42-20 shows a Pentium microprocessor chip, manufactured by Intel Corporation. It contains almost 7 million transistors, along with many other electronic components. Figure 42-21 shows a greatly enlarged view of part of the layout of another chip, the different colors identifying different layers of the chip.

At Intel's Rio Rancho plant, chips are fabricated in a 140-step process on 20 cm silicon wafers, each wafer holding about 300 chips. The individual electronic chip components are so small that the tiniest speck of dust can ruin a chip. Precautions are taken to maintain a dust-free atmosphere in the plant's clean rooms, where manufacturing takes place, and which are thousands of times more pristine than a hospital operating room. That is the reason for the workers' protective clothing, shown in the photograph that opens this chapter. As part of the cleanliness program, highly filtered air circulates through the perforated floor at about 30 m/min. There are also air showers and wipedown stations for removing cosmetics from employees.

REVIEW & SUMMARY

Conductors, Semiconductors, and Insulators Three electrical properties that can be used to distinguish among crystalline solids are the **resistivity** ρ, the **temperature coefficient of resistivity** α, and the **number density of charge carriers** n. Solids can be broadly divided into **conductors** (with small ρ) and **insulators** (with large ρ). Conductors can be further divided into **metals** (with small ρ, positive α, large n) and **semiconductors** (with larger ρ, negative α, and smaller n).

Energy Levels and Gaps in a Crystalline Solid An isolated atom can exist in only a discrete set of energy levels. As atoms come together to form a solid, the levels of the individual atoms merge to form the discrete energy **bands** of the solid. These energy bands are separated by energy **gaps,** each of which corresponds to a range of energies that no electron may possess.

Any energy band is made up of an enormous number of very closely spaced levels. The Pauli exclusion principle asserts that only one electron may occupy each of these levels.

Insulators In an insulator, the highest band containing electrons is completely filled and is separated from the vacant band above it by an energy gap so large that electrons can essentially never become thermally agitated enough to jump across the gap.

Metals In a **metal,** the highest band that contains any electrons is only partially filled. The energy of the highest filled level at a temperature of 0 K is called the **Fermi energy** E_F for the metal; for copper, $E_F = 7.0$ eV.

The electrons in the partially filled band are the *conduction electrons* and their number is

$$\begin{pmatrix}\text{number of conduction}\\ \text{electrons in sample}\end{pmatrix} = \begin{pmatrix}\text{number of atoms}\\ \text{in sample}\end{pmatrix} \times \begin{pmatrix}\text{number of valence}\\ \text{electrons per atom}\end{pmatrix}. \qquad (42\text{-}2)$$

The number of atoms in a sample is given by

$$\begin{pmatrix}\text{number of atoms}\\ \text{in sample}\end{pmatrix} = \frac{\text{sample mass } M_{\text{sam}}}{\text{atomic mass}} = \frac{\text{sample mass } M_{\text{sam}}}{(\text{molar mass } M)/N_A} = \frac{\begin{pmatrix}\text{material's}\\ \text{density}\end{pmatrix}\begin{pmatrix}\text{sample}\\ \text{volume } V\end{pmatrix}}{(\text{molar mass } M)/N_A}. \qquad (42\text{-}4)$$

The number density n of the conduction electrons is

$$n = \frac{\text{number of conduction electrons in sample}}{\text{sample volume } V}. \qquad (42\text{-}3)$$

The **density of states** function $N(E)$ is the number of available energy levels per unit volume of the sample and per unit energy interval and is given by

$$N(E) = \frac{8\sqrt{2}\pi m^{3/2}}{h^3} E^{1/2} \quad \text{(density of states)}, \qquad (42\text{-}5)$$

where E is the energy at which $N(E)$ is evaluated.

The **occupancy probability** $P(E)$ (the probability that a given available state will be occupied by an electron) is

$$P(E) = \frac{1}{e^{(E - E_F)/kT} + 1} \quad \text{(occupancy probability)}. \qquad (42\text{-}6)$$

The **density of occupied states** $N_o(E)$ is given by the product of the two quantities in Eqs. (42-5) and (42-6):

$$N_o(E) = N(E)\, P(E) \quad \text{(density of occupied states)}. \qquad (42\text{-}7)$$

The Fermi energy for a metal can be found by integrating $N_o(E)$ for $T = 0$ from $E = 0$ to $E = E_F$. The result is

$$E_F = \left(\frac{3}{16\sqrt{2}\pi}\right)^{2/3} \frac{h^2}{m} n^{2/3} = \frac{0.121 h^2}{m} n^{2/3}. \qquad (42\text{-}9)$$

Semiconductors The band structure of a **semiconductor** is like that of an insulator except that the gap width E_g is much smaller in the semiconductor. For silicon (a semiconductor) at room temperature, thermal agitation raises a few electrons to the **conduction band,** leaving an equal number of **holes** in the **valence band.** Both electrons and holes serve as charge carriers.

The number of electrons in the conduction band of silicon can be increased greatly by doping with small amounts of phosphorus, thus forming ***n*-type material.** The number of holes in the valence band can be greatly increased by doping with aluminum, thus forming ***p*-type material.**

The p-n Junction A ***p-n* junction** is a single semiconducting crystal with one end doped to form *p*-type material and the other end doped to form *n*-type material, the two types meeting at a **junction plane.** At thermal equilibrium, the following occurs at that plane:

The **majority carriers** (electrons on the *n* side and holes on the *p* side) diffuse across the junction plane, producing a **diffusion current** I_{diff}.

The **minority carriers** (holes on the *n* side and electrons on the *p* side) are swept across the junction plane, forming a **drift current** I_{drift}. These two currents are equal in magnitude, so the net current is zero.

A **depletion zone,** consisting largely of charged donor and acceptor ions, forms across the junction plane.

A **contact potential difference** V_0 develops across the depletion zone.

Applications of the p-n Junction When a potential difference is applied across a *p-n* junction, the device conducts electricity more readily for one polarity of the applied potential difference than for the other. Thus, a *p-n* junction can serve as a **junction rectifier.**

When a *p-n* junction is forward biased, it can emit light, hence can serve as a **light-emitting diode** (LED). The wavelength of the emitted light is

$$\lambda = \frac{c}{f} = \frac{hc}{E_g}. \qquad (42\text{-}11)$$

A strongly forward-biased *p-n* junction with parallel end faces can operate as a **junction laser,** emitting light of a sharply defined wavelength.

MOSFETS In a MOSFET, a type of three-terminal transistor, a potential applied to the **gate** terminal *G* controls the internal flow of electrons from the **source** terminal *S* to the **drain** terminal *D*. Commonly, a MOSFET is operated only in its ON (conducting) or OFF (not conducting) condition. Installed by the thousands and millions on silicon wafers (**chips**) to form **integrated circuits,** MOSFETs form the basis for computer hardware.

QUESTIONS

1. Figure 42-1*a* shows 14 atoms that represent the unit cell of copper. However, since each of these atoms is shared with one or more adjoining unit cells, only a fraction of each atom belongs to the unit cell shown. What is the number of atoms per unit cell for copper? (To answer, count up the fractional atoms belonging to a single unit cell.)

2. Figure 42-1*b* shows 18 atoms that represent the unit cell of silicon. Fourteen of these atoms, however, are shared with one or more adjoining unit cells. What is the number of atoms per unit cell for silicon? (See Question 1.)

3. Does the interval between adjacent energy levels in the highest occupied band of a metal depend on (a) the material of which the sample is made, (b) the size of the sample, (c) the position of the level in the band, (d) the temperature of the sample, or (e) the Fermi energy of the metal?

4. Compare the drift speed v_d of the conduction electrons in a current-carrying copper wire with the Fermi speed v_F for copper. Is v_d (a) about equal to v_F, (b) much greater than v_F, or (c) much less than v_F?

5. In a silicon lattice, where should you look if you want to find (a) a conduction electron, (b) a valence electron, and (c) an electron associated with the 2*p* subshell of the isolated silicon atom?

6. Which of the following statements, if any, are true? (a) At low enough temperatures, silicon behaves like an insulator. (b) At high enough temperatures, silicon becomes a good conductor. (c) At high enough temperatures, silicon behaves like a metal.

7. The energy gaps E_g for the semiconductors silicon and germanium are, respectively, 1.12 and 0.67 eV. Which of the following statements, if any, are true? (a) Both substances have the same number density of charge carriers at room temperature. (b) At room temperature, germanium has a greater number density of charge carriers than silicon. (c) Both substances have a greater number density of conduction electrons than holes. (d) For each substance, the number density of electrons equals that of holes.

8. An isolated atom of germanium has 32 electrons, arranged in subshells according to this scheme:

$$1s^2\ 2s^2\ 2p^6\ 3s^2\ 3p^6\ 3d^{10}\ 4s^2\ 4p^2.$$

This element has the same crystal structure as silicon and, like silicon, is a semiconductor. Which of these electrons form the valence band of crystalline germanium?

9. Germanium ($Z = 32$) has the same crystal structure and the same bonding pattern as silicon. Is the net charge on a germanium ion within its lattice $+e$, $+2e$, $+4e$, $+28e$, or $+32e$?

10. (a) Of the elements arsenic, indium, tin, gallium, antimony, and boron, which would produce *n*-type material if used as a dopant in silicon? (b) Which would produce *p*-type material? (c) Which would be unsuitable as a dopant? (*Hint*: Consult the periodic table in Appendix G.)

11. A sample of silicon is doped with phosphorus. Which of the following statements, if any, are true? (a) The number of holes in the sample is slightly increased. (b) The sample's resistivity is increased. (c) The sample becomes positively charged. (d) The sample becomes negatively charged. (e) The gap between the valence band and the conduction band decreases slightly.

12. To fabricate an *n*-type semiconductor, would you use (a) silicon doped with arsenic or (b) germanium doped with indium? (*Hint:* Consult the periodic table.)

13. In the biased *p-n* junctions shown in Fig. 42-14, there is an electric field $\vec{E}$ in each of the two depletion zones, associated with the potential difference that exists across that zone. (a) Is $\vec{E}$ directed from left to right or from right to left? (b) Is its magnitude greater for forward bias or for back bias?

14. A certain isolated *p-n* junction develops a contact potential difference V_0 of 0.78 V across its depletion zone. A voltmeter is connected across the terminals of the junction, the positive terminal of the meter being connected to the *p* side of the junction. Will the meter read (a) +0.78 V, (b) −0.78 V, (c) zero, or (d) something else? (*Hint:* Contact potentials appear at the connections between the *p-n* junction and the voltmeter leads.)

15. Which of the following obey Ohm's law: (a) a bar of pure silicon, (b) a bar of *n*-type silicon, (c) a bar of *p*-type silicon, (d) a *p-n* junction?

16. An LED based on a gallium–arsenic–phosphorus semiconducting crystal emits red light. If you look at a white surface through such a crystal, will you see (a) red, (b) blue, (c) nothing, because the crystal is opaque, or (d) white?

EXERCISES & PROBLEMS

ssm Solution is in the Student Solutions Manual.
www Solution is available on the World Wide Web at: http://www.wiley.com/college/hrw
ilw Solution is available on the Interactive LearningWare.

SEC. 42-5 Metals

1E. Copper, a monovalent metal, has molar mass 63.54 g/mol and density 8.96 g/cm³. What is the number density *n* of conduction electrons in copper? **ssm**

2E. Verify the numerical factor 0.121 in Eq. 42-9.

3E. At what pressure, in atmospheres, would the number of molecules per unit volume in an ideal gas be equal to the number density of the conduction electrons in copper, with both gas and copper at temperature $T = 300$ K?

4E. Use Eq. 42-9 to verify 7.0 eV as copper's Fermi energy.

5E. Calculate $d\rho/dT$ at room temperature for (a) copper and (b) silicon, using data from Table 42-1.

6E. What is the number density of conduction electrons in gold, which is a monovalent metal? Use the molar mass and density provided in Appendix F.

7E. (a) Show that Eq. 42-5 can be written as $N(E) = CE^{1/2}$. (b) Evaluate C in terms of meters and electron-volts. (c) Calculate $N(E)$ for $E = 5.00$ eV. **ssm**

8E. The Fermi energy of copper is 7.0 eV. Verify that the corresponding Fermi speed is 1600 km/s.

9E. What is the probability that a state 0.062 eV above the Fermi energy will be occupied at (a) $T = 0$ K and (b) $T = 320$ K? **ssm**

10E. Calculate the density of states $N(E)$ for a metal at energy $E =$

8.0 eV and show that your result is consistent with the curve of Fig. 42-5.

11E. Show that Eq. 42-9 can be written as $E_F = An^{2/3}$, where the constant A has the value $3.65 \times 10^{-19}\ m^2 \cdot eV$. ssm

12E. Use the result of Exercise 6 to calculate the Fermi energy of gold.

13E. A state 63 meV above the Fermi level has a probability of occupancy of 0.090. What is the probability of occupancy for a state 63 meV *below* the Fermi level?

14P. The Fermi energy for copper is 7.0 eV. For copper at 1000 K, (a) find the energy of the energy level whose probability of being occupied by an electron is 0.90. For this energy, evaluate (b) the density of states $N(E)$ and (c) the density of occupied states $N_o(E)$.

15P. In Eq. 42-6 let $E - E_F = \Delta E = 1.00$ eV. (a) At what temperature does the result of using this equation differ by 1.0% from the result of using the classical Boltzmann equation $P(E) = e^{-\Delta E/kT}$ (which is Eq. 42-1 with two changes in notation)? (b) At what temperature do the results from these two equations differ by 10%? ssm www

16P. Show that $P(E)$, the occupancy probability in Eq. 42-6, is symmetrical about the value of the Fermi energy; that is, show that

$$P(E_F + \Delta E) + P(E_F - \Delta E) = 1.$$

17P. Assume that the total volume of a metal sample is the sum of the volume occupied by the metal ions making up the lattice and the (separate) volume occupied by the conduction electrons. The density and molar mass of sodium (a metal) are 971 kg/m^3 and 23.0 g/mol, respectively; the radius of the Na^+ ion is 98 pm. (a) What percent of the volume of a sample of metallic sodium is occupied by its conduction electrons? (b) Carry out the same calculation for copper, which has density, molar mass, and ionic radius of 8960 kg/m^3, 63.5 g/mol, and 135 pm, respectively. (c) For which of these metals do you think the conduction electrons behave more like a free-electron gas?

18P. Calculate $N_o(E)$, the density of occupied states, for copper at $T = 1000$ K for the energies $E = 4.00$, 6.75, 7.00, 7.25, and 9.00 eV. Compare your results with the graph of Fig. 42-7*b*. The Fermi energy for copper is 7.00 eV.

19P. Calculate the number density (number per unit volume) for (a) molecules of oxygen gas at 0°C and 1.0 atm pressure and (b) conduction electrons in copper. (c) What is the ratio of the latter to the former? (d) What is the average distance between particles in each case? Assume this distance is the edge length of a cube whose volume is equal to the available volume per particle.

20P. What is the probability that an electron will jump across the energy gap E_g (see Fig. 42-4*a*) in a diamond whose mass is equal to the mass of Earth? Use the result of Sample Problem 42-1 and the molar mass of carbon in Appendix F; assume that in diamond there is one valence electron per carbon atom.

21P. The Fermi energy for silver is 5.5 eV. (a) At $T = 0$°C, what are the probabilities that states with the following energies are occupied: 4.4, 5.4, 5.5, 5.6, and 6.4 eV? (b) At what temperature is the probability 0.16 that a state with energy $E = 5.6$ eV is occupied? ssm www

22P. Show that the probability $P(E)$ that an energy level at energy E is not occupied is

$$P(E) = \frac{1}{e^{-\Delta E/kT} + 1},$$

where $\Delta E = E - E_F$.

23P. The Fermi energy of aluminum is 11.6 eV; its density and molar mass are 2.70 g/cm^3 and 27.0 g/mol, respectively. From these data, determine the number of conduction electrons per atom. ssm

24P. At $T = 300$ K, how close to the Fermi energy will we find a state whose probability of occupation by a conduction electron is 0.10?

25P. Silver is a monovalent metal. Calculate (a) the number density of conduction electrons, (b) the Fermi energy, (c) the Fermi speed, and (d) the de Broglie wavelength corresponding to this electron speed. See Appendix F for the needed data on silver. ssm

26P. Zinc is a bivalent metal. Calculate (a) the number density of conduction electrons, (b) the Fermi energy, (c) the Fermi speed, and (d) the de Broglie wavelength corresponding to this electron speed. See Appendix F for the needed data on zinc.

27P. (a) Show that the density of states at the Fermi energy is given by

$$N(E_F) = \frac{(4)(3^{1/3})(\pi^{2/3})mn^{1/3}}{h^2}$$
$$= (4.11 \times 10^{18}\ m^{-2}\ eV^{-1})n^{1/3},$$

in which n is the number density of conduction electrons. (b) Calculate $N(E_F)$ for copper using the result of Exercise 1, and verify your calculation with the curve of Fig. 42-5, recalling that $E_F = 7.0$ eV for copper.

28P. (a) Show that the slope dP/dE of Eq. 42-6 at $E = E_F$ is $-1/4kT$. (b) Show that the tangent line to the curve of Fig. 42-6*b* at $E = E_F$ intercepts the horizontal axis at $E = E_F + 2kT$.

29P. Show that, at $T = 0$ K, the average energy E_{avg} of the conduction electrons in a metal is equal to $\frac{3}{5}E_F$. (*Hint:* By definition of average, $E_{avg} = (1/n) \int E\, N_o(E)\, dE$, where n is the number density of charge carriers.) ssm

30P. Use the result of Problem 29 to calculate the total translational kinetic energy of the conduction electrons in 1.0 cm^3 of copper at $T = 0$ K.

31P. (a) Using the result of Problem 29, estimate how much energy would be released by the conduction electrons in a copper coin with mass 3.1 g if we could suddenly turn off the Pauli exclusion principle. (b) For how long would this amount of energy light a 100 W lamp? (*Note:* There is no way to turn off the Pauli principle!)

32P. At 1000 K, the fraction of the conduction electrons in a metal that have energies greater than the Fermi energy is equal to the area under the curve of Fig. 42-7*b* beyond E_F divided by the area under the entire curve. It is difficult to find these areas by direct integration. However, an approximation to this fraction at any temperature T is

$$frac = \frac{3kT}{2E_F}.$$

Note that $frac = 0$ for $T = 0$ K, just as we would expect. What is this fraction for copper at (a) 300 K and (b) 1000 K? For copper,

$E_F = 7.0$ eV. (c) Check your answers by numerical integration using Eq. 42-7.

33P. At what temperature do 1.3% of the conduction electrons in lithium (a metal) have energies greater than the Fermi energy E_F, which is 4.7 eV? (See Problem 32.) ssm

34P. Silver melts at 961°C. At the melting point, what fraction of the conduction electrons are in states with energies greater than the Fermi energy of 5.5 eV? (See Problem 32.)

SEC. 42-6 Semiconductors

35E. (a) What is the maximum wavelength of the light that will excite an electron in the valence band of diamond to the conduction band? The energy gap is 5.5 eV. (b) In what part of the electromagnetic spectrum does this wavelength lie? ssm

36P. The compound gallium arsenide is a commonly used semiconductor, having an energy gap E_g of 1.43 eV. Its crystal structure is like that of silicon, except that half the silicon atoms are replaced by gallium atoms and half by arsenic atoms. Draw a flattened-out sketch of the gallium arsenide lattice, following the pattern of Fig. 42-9*a*. (a) What are the net charges of the gallium and arsenic ion cores? (b) How many electrons per bond are there? (*Hint:* Consult the periodic table in Appendix G.)

37P. (a) Find the angle θ between adjacent nearest-neighbor bonds in the silicon lattice. Recall that each silicon atom is bonded to four of its nearest neighbors. The four neighbors form a regular tetrahedron—a three-sided pyramid whose sides and base are equilateral triangles. (b) Find the bond length, given that the atoms at the corners of the tetrahedron are 388 pm apart.

38P. The occupancy probability function (Eq. 42-6) can be applied to semiconductors as well as to metals. In semiconductors the Fermi energy is close to the midpoint of the gap between the valence band and the conduction band. For germanium, the gap width is 0.67 eV. What is the probability that (a) a state at the bottom of the conduction band is occupied and (b) a state at the top of the valence band is not occupied. Assume that $T = 290$ K. (*Note:* Figure 42-4*b* shows that, in a metal, the Fermi energy lies symmetrically between the population of conduction electrons and the population of holes. To match this scheme in a semiconductor, the Fermi energy must lie near the center of the gap. There need not be an available state at the location of the Fermi energy.)

39P. In a simplified model of an undoped semiconductor, the actual distribution of energy states may be replaced by one in which there are N_v states in the valence band, all these states having the same energy E_v, and N_c states in the conduction band, all these states having the same energy E_c. The number of electrons in the conduction band equals the number of holes in the valence band. (a) Show that this last condition implies that

$$\frac{N_c}{\exp(\Delta E_c/kT) + 1} = \frac{N_v}{\exp(\Delta E_v/kT) + 1},$$

in which

$$\Delta E_c = E_c - E_F \quad \text{and} \quad \Delta E_v = -(E_v - E_F).$$

(*Hint:* See Problem 22.) (b) If the Fermi level is in the gap between the two bands and is far from both bands compared with kT, then the exponentials dominate in the denominators. Under these conditions show that

$$E_F = \frac{(E_c + E_v)}{2} + \frac{kT \ln(N_v/N_c)}{2}$$

and that, if $N_v \approx N_c$, the Fermi level for the undoped semiconductor is close to the gap's center, as stated in Problem 38.

SEC. 42-7 Doped Semiconductors

40P. Pure silicon at room temperature has an electron number density in the conduction band of about 5×10^{15} m^{-3} and an equal density of holes in the valence band. Suppose that one of every 10^7 silicon atoms is replaced by a phosphorus atom. (a) Which type will the doped semiconductor be, *n* or *p*? (b) What charge carrier number density will the phosphorus add? (c) What is the ratio of the charge carrier number density (electrons in the conduction band and holes in the valence band) in the doped silicon to that in pure silicon?

41P. What mass of phosphorus is needed to dope 1.0 g of silicon to the extent described in Sample Problem 42-6? ssm

42P. A silicon sample is doped with atoms having donor states 0.110 eV below the bottom of the conduction band. (The energy gap in silicon is 1.11 eV.) (a) If each of these donor states is occupied with a probability of 5.00×10^{-5} at $T = 300$ K, where is the Fermi level with respect to the top of the silicon valence band? (b) What then is the probability that a state at the bottom of the silicon conduction band is occupied?

43P. Doping changes the Fermi energy of a semiconductor. Consider silicon, with a gap of 1.11 eV between the top of the valence band and the bottom of the conduction band. At 300 K the Fermi level of the pure material is nearly at the midpoint of the gap. Suppose that silicon is doped with donor atoms, each of which has a state 0.15 eV below the bottom of the silicon conduction band, and suppose further that doping raises the Fermi level to 0.11 eV below the bottom of that band (Fig. 42-22). (a) For both pure and doped silicon, calculate the probability that a state at the bottom of the silicon conduction band is occupied. (b) Calculate the probability that a donor state in the doped material is occupied. ssm www

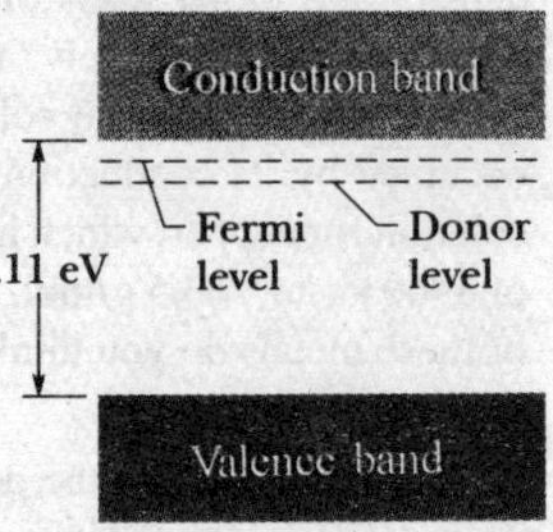

Fig. 42-22 Problem 43.

SEC. 42-9 The Junction Rectifier

44E. For an ideal *p-n* junction rectifier with a sharp boundary between its two semiconducting sides, the current I is related to the potential difference V across the rectifier by

$$I = I_0(e^{eV/kT} - 1),$$

where I_0, which depends on the materials but not on the current or the potential difference, is called the *reverse saturation current*. V is positive if the rectifier is forward-biased and negative if it is back-biased. (a) Verify that this expression predicts the behavior of a junction rectifier by graphing I versus V over the range -0.12 V to $+0.12$ V. Take $T = 300$ K and $I_0 = 5.0$ nA. (b) For the same temperature, calculate the ratio of the current for a 0.50 V forward-bias to the current for a 0.50 V back-bias.

45E. When a photon enters the depletion zone of a *p-n* junction, it can scatter from the valence electrons there, transferring part of its energy to each electron, which then jumps to the conduction band. Thus, the photon creates electron–hole pairs. For this reason, the junctions are often used as light detectors, especially in the x-ray and gamma-ray regions of the electromagnetic spectrum. Suppose a single 662 keV gamma-ray photon transfers its energy to electrons in multiple scattering events inside a semiconductor with an energy gap of 1.1 eV, until all the energy is transferred. Assuming that each of those electrons jumps the gap from the top of the valence band to the bottom of the conduction band, find the number of electron–hole pairs created by the process. ssm

SEC. 42-10 The Light-Emitting Diode (LED)

46P. A potassium chloride crystal has an energy band gap of 7.6 eV above the topmost occupied band, which is full. Is this crystal opaque or transparent to light of wavelength 140 nm?

47P. In a particular crystal, the highest occupied band is full. The crystal is transparent to light of wavelengths longer than 295 nm but opaque at shorter wavelengths. Calculate, in electron-volts, the gap between the highest occupied band and the next higher (empty) band for this material. ssm

SEC. 42-11 The Transistor

48P. A Pentium computer chip, which is about the size of a postage stamp (2.54 cm × 2.22 cm), contains about 3.5 million transistors. If the transistors are square, what must be their *maximum* dimension? (*Note:* Devices other than transistors are also on the chip, and there must be room for the interconnections among the circuit elements. Transistors smaller than 0.7 μm are now commonly and inexpensively fabricated.)

49P. A silicon-based MOSFET has a square gate 0.50 μm on edge. The insulating silicon oxide layer that separates it from the *p*-type substrate is 0.20 μm thick and has a dielectric constant of 4.5. (a) What is the equivalent gate–substrate capacitance (treating the gate as one plate and the substrate as the other plate)? (b) How many elementary charges *e* appear in the gate when there is a gate–source potential difference of 1.0 V?

43 Nuclear Physics

Radioactive nuclei that are injected into a patient collect at certain sites within the patient's body, undergo radioactive decay, and emit gamma rays. These gamma rays can be recorded by a detector, and a color-coded image of the patient's body produced on a video monitor. In the images reproduced here (the left one is a front view of a patient and the right one is a back view), you can tell just where the radioactive nuclei have collected (spine, pelvis, and ribs) by the color-coding of brown and orange.

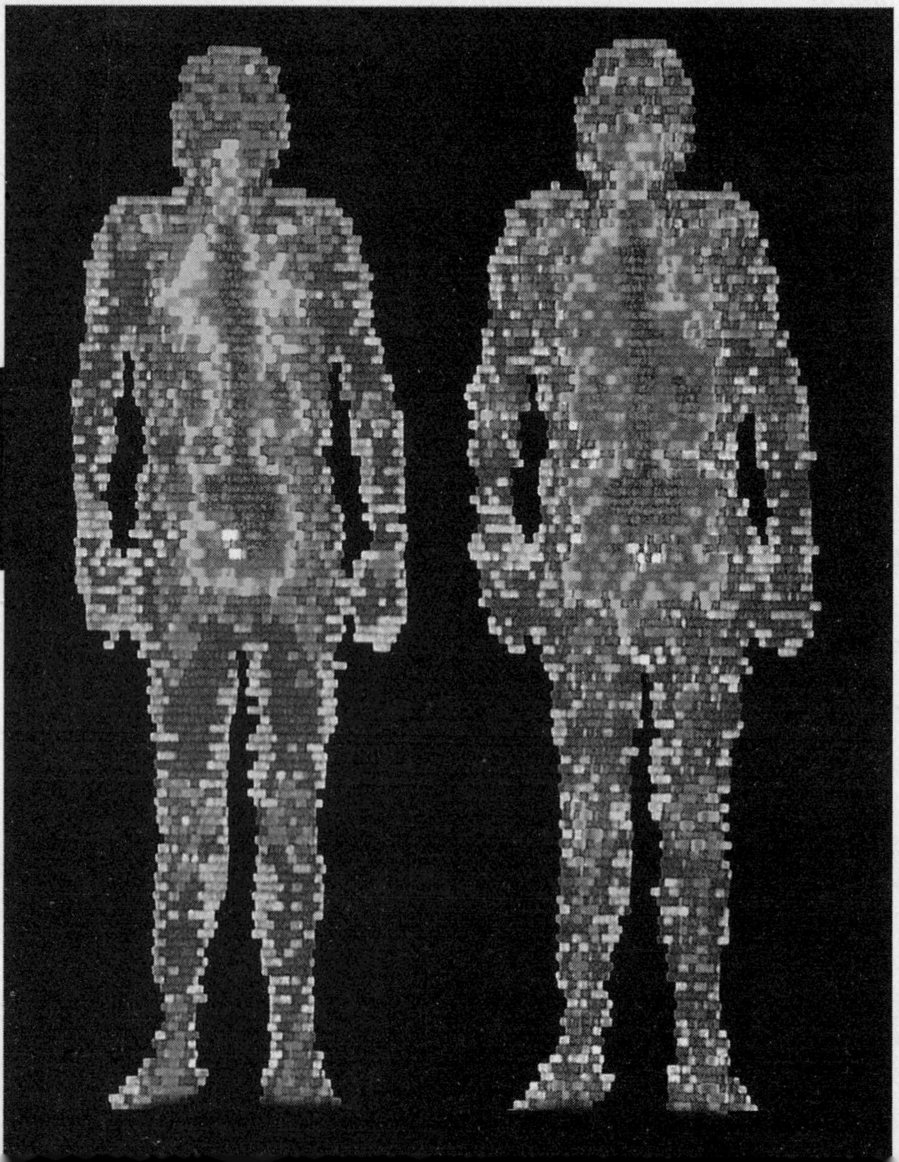

But what actually happens to radioactive nuclei when they undergo decay, and what exactly does "decay" mean?

The answer is in this chapter.

43-1 Discovering the Nucleus

In the first years of the twentieth century not much was known about the structure of atoms beyond the fact that they contain electrons. The electron had been discovered (by J. J. Thomson) in 1897, and its mass was unknown in those early days. Thus, it was not possible even to say how many negatively charged electrons a given atom contained. Atoms were electrically neutral so they must also contain some positive charge, but nobody knew what form this compensating positive charge took.

In 1911 Ernest Rutherford proposed that the positive charge of the atom is densely concentrated at the center of the atom, forming its **nucleus,** and that, furthermore, the nucleus is responsible for most of the mass of the atom. Rutherford's proposal was no mere conjecture but was based firmly on the results of an experiment suggested by him and carried out by his collaborators, Hans Geiger (of Geiger counter fame) and Ernest Marsden, a 20-year-old student who had not yet earned his bachelor's degree.

In Rutherford's day it was known that certain elements, called **radioactive,** transform into other elements spontaneously, emitting particles in the process. One such element is radon, which emits alpha (α) particles with energies of about 5.5 MeV. We now know that these particles are the nuclei of atoms of helium.

Rutherford's idea was to direct energetic alpha particles at a thin target foil and measure the extent to which they were deflected as they passed through the foil. Alpha particles, which are about 7300 times more massive than electrons, have a charge of $+2e$.

Figure 43-1 shows the experimental arrangement of Geiger and Marsden. Their alpha source was a thin-walled glass tube of radon gas. The experiment involves counting the number of alpha particles that are deflected through various scattering angles ϕ.

Figure 43-2 shows their results. Note especially that the vertical scale is logarithmic. We see that most of the particles are scattered through rather small angles

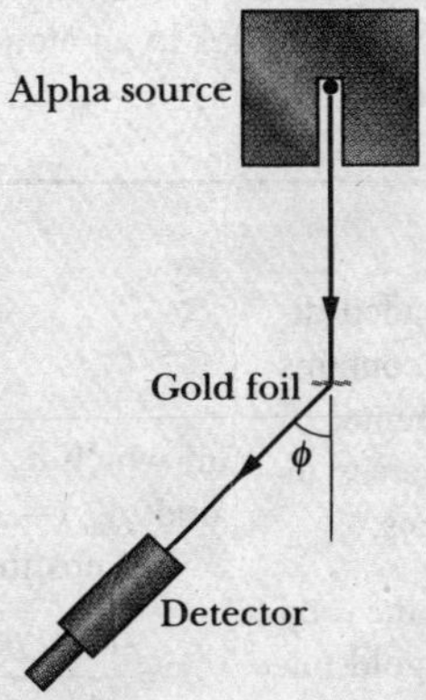

Fig. 43-1 An arrangement (top view) used in Rutherford's laboratory in 1911–1913 to study the scattering of α particles by thin metal foils. The detector can be rotated to various values of the scattering angle ϕ. The alpha source was radon gas, a decay product of radium. With this simple "tabletop" apparatus, the atomic nucleus was discovered.

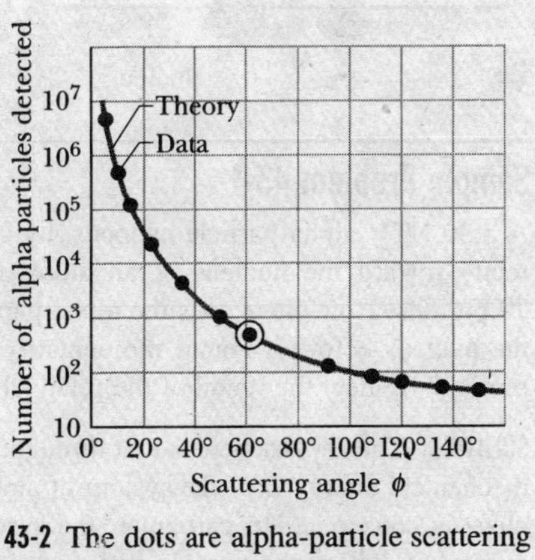

Fig. 43-2 The dots are alpha-particle scattering data for a gold foil, obtained by Geiger and Marsden using the apparatus of Fig. 43-1. The solid curve is the theoretical prediction, based on the assumption that the atom has a small, massive, positively charged nucleus. Note that the vertical scale is logarithmic, covering six orders of magnitude. The data have been adjusted to fit the theoretical curve at the experimental point that is enclosed in a circle.

but—and this was the big surprise—a very small fraction of them are scattered through very large angles, approaching 180°. In Rutherford's words: "It was quite the most incredible event that ever happened to me in my life. It was almost as incredible as if you had fired a 15-inch shell at a piece of tissue paper and it came back and hit you."

Why was Rutherford so surprised? At the time of these experiments, most physicists believed in the so-called plum pudding model of the atom, which had been advanced by J. J. Thomson. In this view the positive charge of the atom was thought to be spread out through the entire volume of the atom. The electrons (the "plums") were thought to vibrate about fixed points within this sphere of positive charge (the "pudding").

The maximum deflecting force that could act on an alpha particle as it passed through such a large positive sphere of charge would be far too small to deflect the alpha particle by even as much as 1°. (The expected deflection has been compared to what you would observe if you fired a bullet through a sack of snowballs.) The electrons in the atom would also have very little effect on the massive, energetic alpha particle. They would, in fact, be themselves strongly deflected, much as a swarm of gnats would be brushed aside by a stone thrown through them.

Rutherford saw that, to deflect the alpha particle backward, there must be a large force; this force could be provided if the positive charge, instead of being spread throughout the atom, were concentrated tightly at its center. Then the incoming alpha particle could get very close to the positive charge without penetrating it; such a close encounter would result in a large deflecting force.

Figure 43-3 shows possible paths taken by typical alpha particles as they pass through the atoms of the target foil. As we see, most are either undeflected or only slightly deflected, but a few (those whose incoming paths pass, by chance, very close to a nucleus) are deflected through large angles. From an analysis of the data, Rutherford concluded that the radius of the nucleus must be smaller than the radius of an atom by a factor of about 10^4. In other words, the atom is mostly empty space.

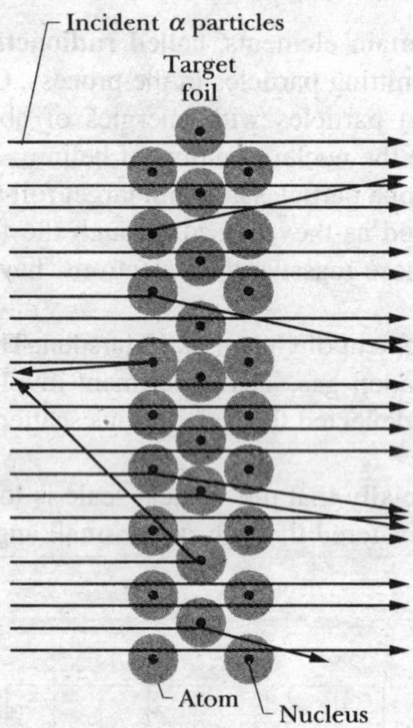

Fig. 43-3 The angle through which an incident alpha particle is scattered depends on how close the particle's path lies to an atomic nucleus. Large deflections result only from very close encounters.

Sample Problem 43-1

A 5.30 MeV alpha particle happens, by chance, to be headed directly toward the nucleus of an atom of gold, which contains 79 protons. How close does the alpha particle get to the center of the nucleus before it comes momentarily to rest and reverses its motion? Neglect the recoil of the relatively massive nucleus.

SOLUTION: The **Key Idea** here is that throughout this process, the total mechanical energy E of the system of alpha particle and gold nucleus is conserved. In particular, the system's initial mechanical energy E_i, before the particle and nucleus interact, is equal to its mechanical energy E_f when the alpha particle momentarily stops. The initial energy E_i is just the kinetic energy K_α of the incoming alpha particle. The final energy E_f is just the electric potential energy U of the system (the kinetic energy is then zero). We can find U with Eq. 25-43 ($U = q_1q_2/4\pi\varepsilon_0 r$).

Let d be the center-to-center distance between the alpha particle and the gold nucleus when the alpha particle is at its stopping point. Then we can write the conservation of energy $E_i = E_f$ as

$$K_\alpha = \frac{1}{4\pi\varepsilon_0}\frac{q_\alpha q_{\text{Au}}}{d},$$

in which q_α (= $2e$) is the charge of the alpha particle (2 protons) and q_{Au} (= $79e$) is the charge of the gold nucleus (79 protons).

Substituting for the charges and solving for d yield

$$d = \frac{(2e)(79e)}{4\pi\varepsilon_0 K_\alpha}$$

$$= \frac{(2 \times 79)(1.60 \times 10^{-19}\ \text{C})^2}{(4\pi)(8.85 \times 10^{-12}\ \text{F/m})(5.30\ \text{MeV})(1.60 \times 10^{-13}\ \text{J/MeV})}$$

$$= 4.29 \times 10^{-14}\ \text{m}. \qquad \text{(Answer)}$$

This is a small distance by atomic standards but not by nuclear standards. It is, in fact, considerably larger than the sum of the radii of the gold nucleus and the alpha particle. Thus, this alpha particle reverses its motion without ever actually "touching" the gold nucleus.

43-2 Some Nuclear Properties

Table 43-1 shows some properties of a few atomic nuclei. When we are interested primarily in their properties as specific nuclear species (rather than as parts of atoms), we call these particles **nuclides.**

Some Nuclear Terminology

Nuclei are made up of protons and neutrons. The number of protons in a nucleus (called the **atomic number** or **proton number** of the nucleus) is represented by the symbol Z; the number of neutrons (the **neutron number**) is represented by the symbol N. The total number of neutrons and protons in a nucleus is called its **mass number** A, so

$$A = Z + N. \tag{43-1}$$

Neutrons and protons, when considered collectively, are called **nucleons.**

We represent nuclides with symbols such as those displayed in the first column of Table 43-1. Consider ^{197}Au, for example. The superscript 197 is the mass number A. The chemical symbol Au tells us that this element is gold, whose atomic number is 79. From Eq. 43-1, the neutron number of this nuclide is 197 − 79, or 118.

Nuclides with the same atomic number Z but different neutron numbers N are called **isotopes** of each other. The element gold has 32 isotopes, ranging from ^{173}Au to ^{204}Au. Only one of them (^{197}Au) is stable; the remaining 31 are radioactive. Such **radionuclides** undergo **decay** (or **disintegration**) by emitting a particle and thereby transforming to a different nuclide.

Organizing the Nuclides

The neutral atoms of all isotopes of an element (all with the same Z) have the same number of electrons and the same chemical properties, and they fit into the same box in the periodic table of the elements. The *nuclear* properties of the isotopes of a given element, however, are very different. Thus, the periodic table is of limited use to the nuclear physicist, the nuclear chemist, or the nuclear engineer.

TABLE 43-1 Some Properties of Selected Nuclides

Nuclide	Z	N	A	Stability[a]	Mass[b] (u)	Spin[c]	Binding Energy (MeV/nucleon)
^{1}H	1	0	1	99.985%	1.007 825	$\frac{1}{2}$	—
^{7}Li	3	4	7	92.5%	7.016 003	$\frac{3}{2}$	5.60
^{31}P	15	16	31	100%	30.973 762	$\frac{1}{2}$	8.48
^{84}Kr	36	48	84	57.0%	83.911 507	0	8.72
^{120}Sn	50	70	120	32.4%	119.902 199	0	8.51
^{157}Gd	64	93	157	15.7%	156.923 956	$\frac{3}{2}$	8.21
^{197}Au	79	118	197	100%	196.966 543	$\frac{3}{2}$	7.91
^{227}Ac	89	138	227	21.8 y	227.027 750	$\frac{3}{2}$	7.65
^{239}Pu	94	145	239	24 100 y	239.052 158	$\frac{1}{2}$	7.56

[a]For stable nuclides, the **isotopic abundance** is given; this is the fraction of atoms of this type found in a typical sample of the element. For radioactive nuclides, the half-life is given.

[b]Following standard practice, the reported mass is that of the neutral atom, not that of the bare nucleus.

[c]Spin angular momentum in units of $\hbar$.

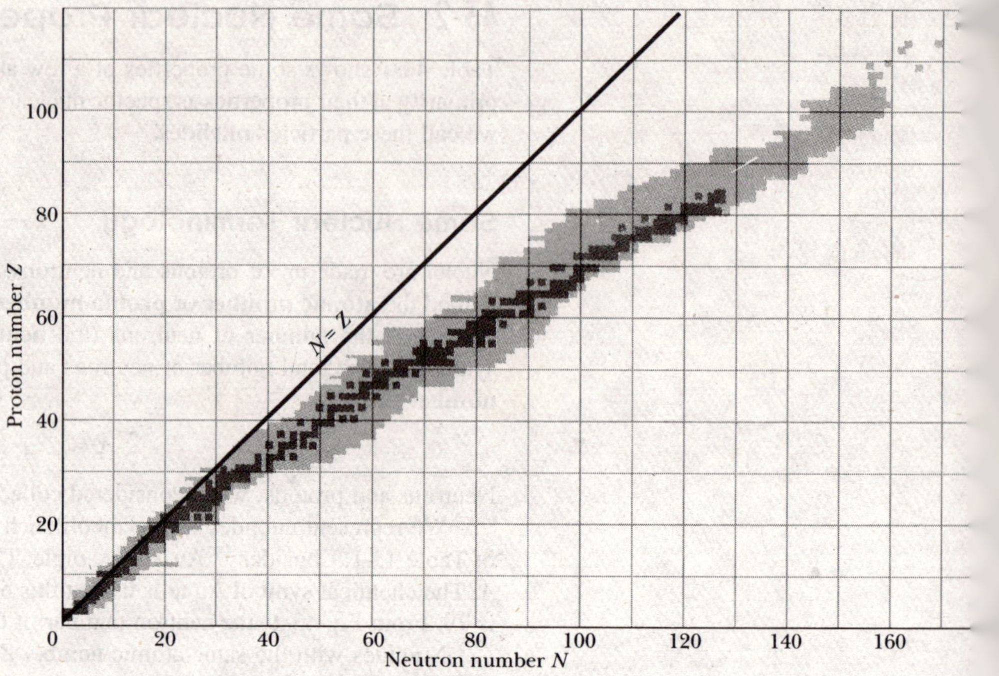

Fig. 43-4 A plot of the known nuclides. The green shading identifies the band of stable nuclides, the beige shading the radionuclides. Low-mass, stable nuclides have essentially equal numbers of neutrons and protons, but more massive nuclides have an increasing excess of neutrons. The figure shows that there are no stable nuclides with $Z > 83$ (bismuth).

We organize the nuclides on a **nuclidic chart** like that in Fig. 43-4, in which a nuclide is represented by plotting its proton number against its neutron number. The stable nuclides in this figure are represented by the green, the radionuclides by the yellow. As you can see, the radionuclides tend to lie on either side of—and at the upper end of—a well-defined band of stable nuclides. Note too that light stable nuclides tend to lie close to the line $N = Z$, which means that they have about the same numbers of neutrons and protons. Heavier nuclides, however, tend to have many more neutrons than protons. As an example, we saw that ^{197}Au has 118 neutrons and only 79 protons, a *neutron excess* of 39.

Nuclidic charts are available as wall charts, in which each small box on the chart is filled with data about the nuclide it represents. Figure 43-5 shows a section of such a chart, centered on ^{197}Au. Relative abundances (usually, as found on Earth) are shown for stable nuclides, and half-lives (a measure of decay rate) are shown for radionuclides. The sloping line points out a line of **isobars**—nuclides of the same mass number, $A = 198$ in this case.

As of early 2000, nuclides with atomic numbers as high as $Z = 118$ ($A = 293$) had been found in laboratory experiments (no elements with Z greater than 92 occur naturally). Although large nuclides generally should be highly unstable and last only a very brief time, certain supermassive nuclides are relatively stable, with fairly long lifetimes. These stable supermassive nuclides form an *island of stability* at high Z and N on a nuclidic chart like Fig. 43-4.

✔CHECKPOINT 1: Based on Fig. 43-4, which of the following nuclides do you conclude are not likely to be detected: ^{52}Fe ($Z = 26$), ^{90}As ($Z = 33$), ^{158}Nd ($Z = 60$), ^{175}Lu ($Z = 71$), ^{208}Pb ($Z = 82$)?

Nuclear Radii

A convenient unit for measuring distances on the scale of nuclei is the *femtometer* ($= 10^{-15}$ m). This unit is often called the *fermi;* the two names share the same

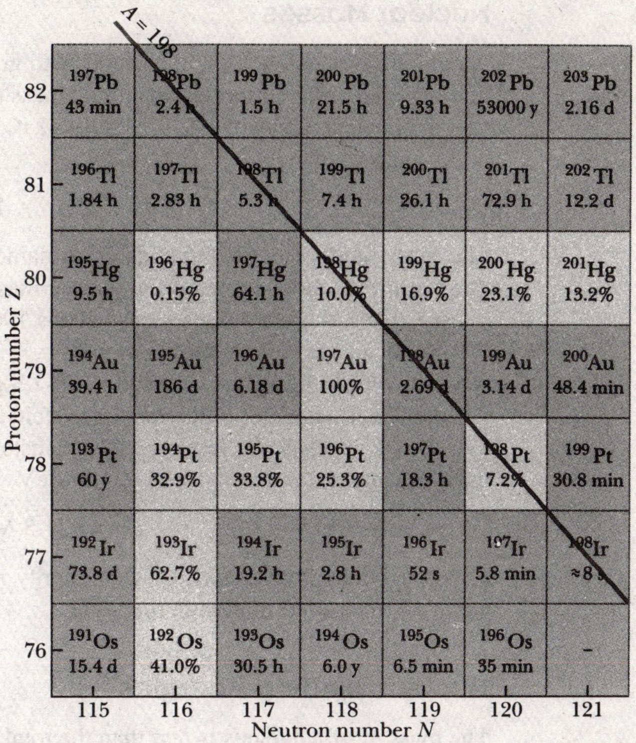

Fig. 43-5 An enlarged and detailed section of the nuclidic chart of Fig. 43-4, centered on ^{197}Au. Green squares represent stable nuclides, for which relative isotopic abundances are given. Beige squares represent radionuclides, for which half-lives are given. Isobaric lines of constant mass number A slope as shown by the example line for $A = 198$.

abbreviation. Thus,

$$1 \text{ femtometer} = 1 \text{ fermi} = 1 \text{ fm} = 10^{-15} \text{ m}. \qquad (43\text{-}2)$$

We can learn about the size and structure of nuclei by bombarding them with a beam of high-energy electrons and observing how the nuclei deflect the incident electrons. The electrons must be energetic enough (at least 200 MeV) to have de Broglie wavelengths that are smaller than the nuclear structures they are to probe.

The nucleus, like the atom, is not a solid object with a well-defined surface. Furthermore, although most nuclides are spherical, some are notably ellipsoidal. Nevertheless, electron-scattering experiments (as well as experiments of other kinds) allow us to assign to each nuclide an effective radius given by

$$r = r_0 A^{1/3}, \qquad (43\text{-}3)$$

in which A is the mass number and $r_0 \approx 1.2$ fm. We see that the volume of a nucleus, which is proportional to r^3, is directly proportional to the mass number A and is independent of the separate values of Z and N.

Equation 43-3 does not apply to *halo nuclides,* which are neutron-rich nuclides that were first produced in laboratories in the 1980s. These nuclides are larger than predicted by Eq. 43-3, because some of the neutrons form a *halo* around a spherical core of the protons and the rest of the neutrons. Lithium isotopes give an example. When a neutron is added to ^{8}Li to form ^{9}Li, neither of which are halo nuclides, the effective radius increases by about 4%. However, when two neutrons are added to ^{9}Li to form the neutron-rich isotope ^{11}Li (the largest of the lithium isotopes), they do not join that existing nucleus but instead form a halo around it, increasing the effective radius by about 30%. Apparently this halo configuration involves less energy than a core containing all 11 nucleons. (In this chapter we shall generally assume that Eq. 43-3 applies.)

Nuclear Masses

Atomic masses can be now measured with great precision. Recall from Section 1-6 that such masses are reported in atomic mass units u, chosen so that the atomic mass (not the nuclear mass) of ^{12}C is exactly 12 u. The relation of this unit to the SI mass unit is, approximately,

$$1 \text{ u} = 1.661 \times 10^{-27} \text{ kg}. \tag{43-4}$$

The mass number A of a nuclide is so named because the number represents the mass of the nuclide, expressed in atomic mass units and rounded off to the nearest integer. Thus, the atomic mass of ^{197}Au is 196.966573 u, which we round to 197 u.

In nuclear reactions, the relation $Q = -\Delta m\, c^2$ (Eq. 38-47) is an indispensable workaday tool. As we saw in Section 38-12, Q is the energy released (or absorbed) when the mass of a closed interacting system of particles changes by an amount Δm.

As we also saw in Section 38-12, Einstein's relation $E = mc^2$ tells us that the mass energy of a mass of 1 u is 931.5 MeV. Thus, from Eq. 38-43, we can use

$$c^2 = 931.5 \text{ MeV/u} \tag{43-5}$$

as a convenient conversion between energy measured in millions of electron-volts and mass measured in atomic mass units.

Nuclear Binding Energies

The mass M of a nucleus is *less* than the total mass Σm of its individual protons and neutrons. That means that the mass energy Mc^2 of a nucleus is *less* than the total mass energy $\Sigma(mc^2)$ of its individual protons and neutrons. The difference between these two energies is called the **binding energy** of the nucleus

$$\Delta E_{be} = \Sigma(mc^2) - Mc^2 \qquad \text{(binding energy).} \tag{43-6}$$

Caution: Binding energy is not an energy that resides in the nucleus. Rather, it is a *difference* in mass energy between a nucleus and its individual nucleons. If we were able to separate a nucleus into its nucleons, we would have to transfer a total energy equal to ΔE_{be} to those particles during the separating process. Although we cannot actually tear apart a nucleus in this way, the nuclear binding energy is still a convenient measure of how well a nucleus is held together.

A better measure is the **binding energy per nucleon** ΔE_{ben}, which is the ratio of the binding energy ΔE_{be} of a nucleus to the number A of nucleons in that nucleus:

$$\Delta E_{ben} = \frac{\Delta E_{be}}{A} \qquad \text{(binding energy per nucleon).} \tag{43-7}$$

We can think of the binding energy per nucleon as the average energy needed to separate a nucleus into its individual nucleons.

Figure 43-6 is a plot of the binding energy per nucleon ΔE_{ben} versus mass number A for a large number of nuclei. Those high on the plot are very tightly bound; that is, we would have to supply a great amount of energy per nucleon to break apart one of those nuclei. The nuclei that are lower on the plot, at the left and right sides, are less tightly bound, and less energy per nucleon would be required to break them apart.

These simple statements about Fig. 43-6 have profound consequences. The nucleons in a nucleus on the right side of the plot would be more tightly bound if that nucleus were to split into two nuclei that lie near the top of the plot. Such a process, called **fission,** occurs naturally with large (high mass number A) nuclei such as uranium, which can undergo fission spontaneously (that is, without an external cause

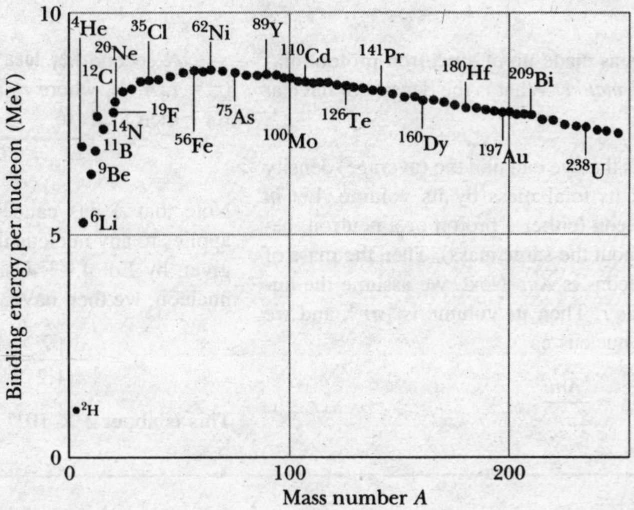

Fig. 43-6 The binding energy per nucleon for some representative nuclides. The nickel nuclide ^{62}Ni has the highest binding energy per nucleon (about 8.794 60 MeV/nucleon) of any known stable nuclide. Note that the alpha particle (^{4}He) has a higher binding energy per nucleon than its neighbors in the periodic table and thus is also particularly stable.

or source of energy). The process can also occur in nuclear weapons in which many uranium or plutonium nuclei are made to fission all at once, to create an explosion.

The nucleons in any pair of nuclei on the left side of the plot would be more tightly bound if the pair were to combine to form a single nucleus that lies near the top of the plot. Such a process, called **fusion,** occurs naturally in stars. Were this not true, the Sun would not shine and thus life could not exist on Earth.

Nuclear Energy Levels

The energy of nuclei, like that of atoms, is quantized. That is, nuclei can exist only in discrete quantum states, each with a well-defined energy. Figure 43-7 shows some of these energy levels for ^{28}Al, a typical low-mass nuclide. Note that the energy scale is in millions of electron-volts, rather than the electron-volts used for atoms. When a nucleus makes a transition from one level to a level of lower energy, the emitted photon is typically in the gamma-ray region of the electromagnetic spectrum.

Nuclear Spin and Magnetism

Many nuclides have an intrinsic *nuclear angular momentum,* or spin, and an associated intrinsic *nuclear magnetic moment*. Although nuclear angular momenta are roughly of the same magnitude as the angular momenta of atomic electrons, nuclear magnetic moments are much smaller than typical atomic magnetic moments.

The Nuclear Force

The force that controls the motions of atomic electrons is the familiar electromagnetic force. To bind the nucleus together, however, there must be a strong attractive nuclear force of a totally different kind, strong enough to overcome the repulsive force between the (positively charged) nuclear protons and to bind both protons and neutrons into the tiny nuclear volume. The nuclear force must also be of short range because its influence does not extend very far beyond the nuclear "surface."

The present view is that the nuclear force that binds neutrons and protons in the nucleus is not a fundamental force of nature but is a secondary, or "spillover," effect of the **strong force** that binds quarks together to form neutrons and protons. In much the same way, the attractive force between certain neutral molecules is a spillover effect of the Coulomb electric force that acts within each molecule to bind it together.

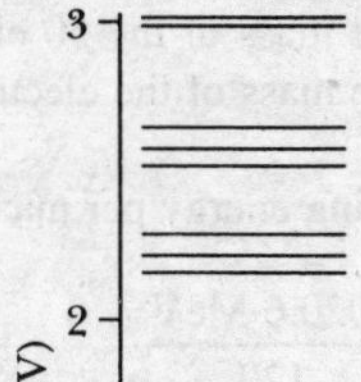

Fig. 43-7 Energy levels for the nuclide ^{28}Al, deduced from nuclear reaction experiments.

Sample Problem 43-2

We can think of all nuclides as made up of a neutron-proton mixture that we can call *nuclear matter*. What is the density of nuclear matter?

SOLUTION: One **Key Idea** here is that we can find the (average) density ρ of a nucleus by dividing its total mass by its volume. Let m represent the mass of a nucleon (either a proton or a neutron, because those particles have about the same mass). Then the mass of a nucleus containing A nucleons is Am. Next, we assume the nucleus is spherical with radius r. Then its volume is $\frac{4}{3}\pi r^3$, and we can write the density of the nucleus as

$$\rho = \frac{Am}{\frac{4}{3}\pi r^3}.$$

A second **Key Idea** is that the radius r is given by Eq. 43-3 ($r = r_0 A^{1/3}$), where r_0 is 1.2 fm ($= 1.2 \times 10^{-15}$ m). Substituting for r then leads to

$$\rho = \frac{Am}{\frac{4}{3}\pi r_0^3 A} = \frac{m}{\frac{4}{3}\pi r_0^3}.$$

Note that A has canceled out; thus, this equation for density ρ applies to any nucleus that can be treated as spherical with a radius given by Eq. 43-3. Using 1.67×10^{-27} kg for the mass m of a nucleon, we then have

$$\rho = \frac{1.67 \times 10^{-27}\ \text{kg}}{\frac{4}{3}\pi(1.2 \times 10^{-15}\ \text{m})^3} \approx 2 \times 10^{17}\ \text{kg/m}^3. \quad \text{(Answer)}$$

This is about 2×10^{14} times the density of water.

Sample Problem 43-3

What is the binding energy per nucleon for ^{120}Sn?

SOLUTION: We need two **Key Ideas** here:

1. We can find the binding energy per nucleon ΔE_{ben} if we first find the binding energy ΔE_{be} and then divide by the number of nucleons A in the nucleus, according to Eq. 43-7 ($\Delta E_{\text{ben}} = \Delta E_{\text{be}}/A$).
2. We can find ΔE_{be} by finding the difference between the mass energy Mc^2 of the nucleus and the total mass energy $\Sigma(mc^2)$ of the individual nucleons that make up the nucleus, according to Eq. 43-6 ($\Delta E_{\text{be}} = \Sigma(mc^2) - Mc^2$).

From Table 43-1, we see that a ^{120}Sn nucleus consists of 50 protons ($Z = 50$) and 70 neutrons ($N = A - Z = 120 - 50 = 70$). Thus, we need to imagine a ^{120}Sn nucleus being separated into its 50 protons and 70 neutrons,

$$(^{120}\text{Sn nucleus}) \rightarrow 50\begin{pmatrix}\text{separate}\\ \text{protons}\end{pmatrix} + 70\begin{pmatrix}\text{separate}\\ \text{neutrons}\end{pmatrix}, \quad (43\text{-}8)$$

and then compute the resulting change in mass energy.

For that computation, we need the masses of a ^{120}Sn nucleus, a proton, and a neutron. However, because the mass of a neutral atom (nucleus *plus* electrons) is much easier to measure than the mass of a bare nucleus, calculations of binding energies are traditionally done with atomic masses. Thus, let's modify Eq. 43-8 so that it has a neutral ^{120}Sn atom on the left side. To do that, we include 50 electrons on the left side (to match the 50 protons in the ^{120}Sn nucleus). We must also add 50 electrons on the right side to balance Eq. 43-8. Those 50 electrons can be combined with the 50 protons, to form 50 neutral hydrogen atoms. We then have

$$(^{120}\text{Sn atom}) \rightarrow 50\begin{pmatrix}\text{separate}\\ \text{H atoms}\end{pmatrix} + 70\begin{pmatrix}\text{separate}\\ \text{neutrons}\end{pmatrix}. \quad (43\text{-}9)$$

In Table 43-1, the mass M_{Sn} of a ^{120}Sn atom is 119.902 199 u and the mass m_{H} of a hydrogen atom is 1.007 825 u; the mass m_{n} of a neutron is 1.008 665 u. Thus, Eq. 43-6 yields

$$\begin{aligned}\Delta E_{\text{be}} &= \Sigma(mc^2) - Mc^2\\ &= 50(m_{\text{H}}c^2) + 70(m_{\text{n}}c^2) - M_{\text{Sn}}c^2\\ &= 50(1.007\,825\ \text{u})c^2 + 70(1.008\,665\ \text{u})c^2\\ &\quad - (119.902\,199\ \text{u})c^2\\ &= (1.095\,601\ \text{u})c^2 = (1.095\,601\ \text{u})(931.5\ \text{MeV/u})\\ &= 1020.6\ \text{MeV},\end{aligned}$$

where Eq. 43-5 ($c^2 = 931.5$ MeV/u) provides an easy unit conversion. Note that using atomic masses instead of nuclear masses does not affect the result because the mass of the 50 electrons in the ^{120}Sn atom subtracts out from the mass of the electrons in the 50 hydrogen atoms.

Now Eq. 43-7 gives us the binding energy per nucleon as

$$\begin{aligned}\Delta E_{\text{ben}} &= \frac{\Delta E_{\text{be}}}{A} = \frac{1020.6\ \text{MeV}}{120}\\ &= 8.51\ \text{MeV/nucleon}. \quad \text{(Answer)}\end{aligned}$$

43-3 Radioactive Decay

As Fig. 43-4 shows, most of the nuclides that have been identified are radioactive. A radioactive nuclide spontaneously emits a particle, transforming itself in the process into a different nuclide, occupying a different square on the nuclidic chart.

Radioactive decay provided the first evidence that the laws that govern the subatomic world are statistical. Consider, for example, a 1 mg sample of uranium metal. It contains 2.5×10^{18} atoms of the very long-lived radionuclide ^{238}U.

The nuclei of these particular atoms have existed without decaying since they were created—well before the formation of our solar system. During any given second only about 12 of the nuclei in our sample will happen to decay by emitting an alpha particle, transforming themselves into nuclei of ^{234}Th. However,

There is absolutely no way to predict whether any given nucleus in a radioactive sample will be among the small number of nuclei that decay during the next second. All have the same chance.

Although we cannot predict which nuclei in a sample will decay, we can say that if a sample contains N radioactive nuclei, then the rate $(= -dN/dt)$ at which nuclei will decay is proportional to N:

$$-\frac{dN}{dt} = \lambda N, \tag{43-10}$$

in which λ, the **disintegration constant** (or **decay constant**) has a characteristic value for every radionuclide. Its SI unit is the inverse second (s^{-1}).

To find N as a function of time t, we first rearrange Eq. 43-10 as

$$\frac{dN}{N} = -\lambda\, dt, \tag{43-11}$$

and then integrate both sides, obtaining

$$\int_{N_0}^{N} \frac{dN}{N} = -\lambda \int_{t_0}^{t} dt,$$

or

$$\ln N - \ln N_0 = -\lambda(t - t_0). \tag{43-12}$$

Here N_0 is the number of radioactive nuclei in the sample at some arbitrary initial time t_0. Setting $t_0 = 0$ and rearranging Eq. 43-12 give us

$$\ln \frac{N}{N_0} = -\lambda t. \tag{43-13}$$

Taking the exponential of both sides (the exponential function is the antifunction of the natural logarithm) leads to

$$\frac{N}{N_0} = e^{-\lambda t}$$

or

$$N = N_0 e^{-\lambda t} \quad \text{(radioactive decay)}, \tag{43-14}$$

in which N_0 is the number of radioactive nuclei in the sample at $t = 0$ and N is the number remaining at any subsequent time t. Note that lightbulbs (for one example) follow no such exponential decay law. If we life-test 1000 bulbs, we expect that they will all "decay" (that is, burn out) at more or less the same time. The decay of radionuclides follows quite a different law.

We are often more interested in the decay rate R $(= -dN/dt)$ than in N itself. Differentiating Eq. 43-14, we find

$$R = -\frac{dN}{dt} = \lambda N_0 e^{-\lambda t}$$

or

$$R = R_0 e^{-\lambda t} \quad \text{(radioactive decay)}, \tag{43-15}$$

an alternative form of the law of radioactive decay (Eq. 43-14). Here R_0 is the decay rate at time $t = 0$, and R is the rate at any subsequent time t. We can now rewrite Eq. 43-10 in terms of the decay rate R of the sample as

$$R = \lambda N, \tag{43-16}$$

where R and the number of radioactive nuclei N that have not yet undergone decay must be evaluated at the same instant.

The total decay rate R of a sample of one or more radionuclides is called the **activity** of that sample. The SI unit for activity is the **becquerel,** named for Henri Becquerel, the discoverer of radioactivity:

$$1 \text{ becquerel} = 1 \text{ Bq} = 1 \text{ decay per second.}$$

An older unit, the **curie,** is still in common use:

$$1 \text{ curie} = 1 \text{ Ci} = 3.7 \times 10^{10} \text{ Bq.}$$

Here is an example using these units: "The activity of spent reactor fuel rod #5658 on January 15, 2000, was 3.5×10^{15} Bq ($= 9.5 \times 10^4$ Ci)." Thus, on that day 3.5×10^{15} radioactive nuclei in the rod decayed each second. The identities of the radionuclides in the fuel rod, their disintegration constants λ, and the types of radiation they emit have no bearing on this measure of activity.

Often a radioactive sample will be placed near a detector that, for reasons of geometry or detector inefficiency, does not record all the disintegrations that occur in the sample. The reading of the detector under these circumstances is proportional to (and smaller than) the true activity of the sample. Such proportional activity measurements are reported not in becquerel units but simply in counts per unit time.

There are two common time measures of how long any given type of radionuclides lasts. One measure is the **half-life** $T_{1/2}$ of a radionuclide, which is the time at which both N and R have been reduced to one-half their initial values. The other measure is the **mean life** τ, which is the time at which both N and R have been reduced to e^{-1} of their initial values.

To relate $T_{1/2}$ to the disintegration constant λ, we put $R = \frac{1}{2}R_0$ in Eq. 43-15 and substitute $T_{1/2}$ for t. We obtain

$$\tfrac{1}{2}R_0 = R_0 e^{-\lambda T_{1/2}}.$$

Taking the natural logarithm of both sides and solving for $T_{1/2}$, we find

$$T_{1/2} = \frac{\ln 2}{\lambda}.$$

Similarly, to relate τ to λ, we put $R = e^{-1}R_0$ in Eq. 43-15, substitute τ for t, and solve for τ, finding

$$\tau = \frac{1}{\lambda}.$$

We summarize these results with the following:

$$T_{1/2} = \frac{\ln 2}{\lambda} = \tau \ln 2. \tag{43-17}$$

✔CHECKPOINT 2: The nuclide ^{131}I is radioactive, with a half-life of 8.04 days. At noon on January 1, the activity of a certain sample is 600 Bq. Using the concept of half-life, without written calculation, determine whether the activity at noon on January 24 will be a little less than 200 Bq, a little more than 200 Bq, a little less than 75 Bq, or a little more than 75 Bq.

Sample Problem 43-4

The table that follows shows some measurements of the decay rate of a sample of ^{128}I, a radionuclide often used medically as a tracer to measure the rate at which iodine is absorbed by the thyroid gland.

Time (min)	R (counts/s)	Time (min)	R (counts/s)
4	392.2	132	10.9
36	161.4	164	4.56
68	65.5	196	1.86
100	26.8	218	1.00

Find the disintegration constant λ and the half-life $T_{1/2}$ for this radionuclide.

SOLUTION: One **Key Idea** here is that the disintegration constant λ determines the exponential rate at which the decay rate R decreases with time t (as indicated by Eq. 43-15). Therefore, we should be able to determine λ by plotting the measurements of R against the measurement times t.

However, obtaining λ from a plot of R versus t is difficult because R decreases exponentially with t, according to Eq. 43-15. Thus, a second **Key Idea** is to transform Eq. 43-15 into a linear function of t, so that we can easily find λ. To do so, we take the natural logarithms of both sides of Eq. 43-15. We obtain

$$\ln R = \ln(R_0 e^{-\lambda t}) = \ln R_0 + \ln(e^{-\lambda t}) = \ln R_0 - \lambda t. \qquad (43\text{-}18)$$

Because Eq. 43-18 is of the form $y = b + mx$, with b and m constants, it is a linear equation giving the quantity $\ln R$ as a function of t. Thus, if we plot $\ln R$ (instead of R) versus t, we should get a straight line. Further, the slope of the line should be equal to $-\lambda$.

Figure 43-8 shows a plot of $\ln R$ versus time t for the given measurements. The slope of the straight line that fits through the plotted points is

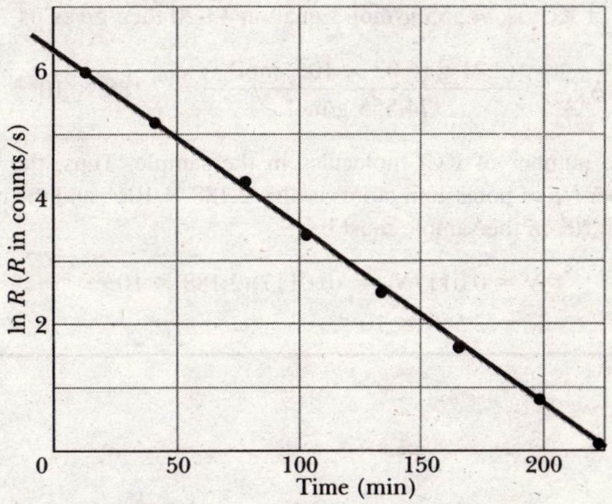

Fig. 43-8 Sample Problem 43-4. A semilogarithmic plot of the decay of a sample of ^{128}I, based on the data in the table.

$$\text{slope} = \frac{0 - 6.2}{225 \text{ min} - 0} = -0.0275 \text{ min}^{-1}.$$

Thus,

$$-\lambda = -0.0275 \text{ min}^{-1}$$

or

$$\lambda = 0.0275 \text{ min}^{-1} \approx 1.7 \text{ h}^{-1}. \qquad \text{(Answer)}$$

To find the half-life $T_{1/2}$ of the radionuclide, we use the **Key Idea** that the time for the decay rate R to decrease by 1/2 is related to the disintegration constant λ via Eq. 43-17 ($T_{1/2} = (\ln 2)/\lambda$). From that equation, we find

$$T_{1/2} = \frac{\ln 2}{\lambda} = \frac{\ln 2}{0.0275 \text{ min}^{-1}} \approx 25 \text{ min}. \qquad \text{(Answer)}$$

Sample Problem 43-5

A 2.71 g sample of KCl from the chemistry stockroom is found to be radioactive, and it is decaying at a constant rate of 4490 Bq. The decays are traced to the element potassium and in particular to the isotope ^{40}K, which constitutes 1.17% of normal potassium. Calculate the half-life of this nuclide.

SOLUTION: One **Key Idea** here is that because the activity R of the sample is apparently constant, we cannot find the half-life $T_{1/2}$ by plotting $\ln R$ versus time t as we did in Sample Problem 43-4. (We would just get a horizontal plot.) However, we can use the following two **Key Ideas**:

1. We can relate the half-life $T_{1/2}$ to the disintegration constant λ via Eq. 43-17 ($T_{1/2} = (\ln 2)/\lambda$).
2. We can then relate λ to the given activity R of 4490 Bq by means of Eq. 43-16 ($R = \lambda N$), where N is the number of ^{40}K nuclei (and thus atoms) in the sample.

Combining Eqs. 43-17 and 43-16 yields

$$T_{1/2} = \frac{N \ln 2}{R}. \qquad (43\text{-}19)$$

We know that N in this equation is 1.17% of the total number N_K of potassium atoms in the sample. We also know that N_K must equal the number N_{KCl} of molecules in the sample. We can obtain N_{KCl} from the molar mass M_{KCl} of KCl (the mass of one mole of KCl) and the given mass M_{sam} of the sample by combining Eqs. 20-2 and 20-3 to write

$$N_{KCl} = \left(\begin{matrix}\text{number of moles}\\ \text{in sample}\end{matrix}\right) N_A = \frac{M_{sam}}{M_{KCl}} N_A, \qquad (43\text{-}20)$$

where N_A is Avogadro's number (6.02×10^{23} mol^{-1}). From Appendix F, we see that the molar mass of potassium is 39.102 g/mol and the molar mass of chlorine is 35.453 g/mol; thus, the molar mass of KCl is 74.555 g/mol. Equation 43-20 then gives us

$$N_{KCl} = \frac{(2.71 \text{ g})(6.02 \times 10^{23} \text{ mol}^{-1})}{74.555 \text{ g/mol}} = 2.188 \times 10^{22}$$

as the number of KCl molecules in the sample. Thus, the total number N_K of potassium atoms is also 2.188×10^{22}, and the number of ^{40}K in the sample must be

$$N = 0.0117 N_K = (0.0117)(2.188 \times 10^{22}) = 2.560 \times 10^{20}.$$

Substituting this value for N and the given activity of 4490 Bq ($= 4490 \text{ s}^{-1}$) for R into Eq. 43-19 leads to

$$T_{1/2} = \frac{(2.560 \times 10^{20}) \ln 2}{4490 \text{ s}^{-1}} = 3.95 \times 10^{16} \text{ s} = 1.25 \times 10^{9} \text{ y}. \quad \text{(Answer)}$$

This half-life of ^{40}K turns out to have the same order of magnitude as the age of the universe. Thus, the activity of ^{40}K in the stockroom sample decreases *very* slowly, too slowly for us to detect during a few days of observation or even an entire lifetime. A portion of the potassium in our bodies consists of this radioisotope, which means that we are all slightly radioactive.

43-4 Alpha Decay

When a nucleus undergoes **alpha decay,** it transforms to a different nuclide by emitting an alpha particle (a helium nucleus, ^{4}He). For example, when uranium ^{238}U undergoes alpha decay, it transforms to thorium ^{234}Th:

$$^{238}\text{U} \rightarrow {}^{234}\text{Th} + {}^{4}\text{He}. \quad (43\text{-}21)$$

This alpha decay of ^{238}U can occur spontaneously (without an external source of energy) because the total mass of the decay products ^{234}Th and ^{4}He is less than the mass of the original ^{238}U. Thus, the total mass energy of the decay products is less than the mass energy of the original nuclide. As defined by Eq. 38-47, in such a process the difference between the initial mass energy and the total final mass energy is called the Q of the process.

For a nuclear decay, we say that the difference in mass energy is the decay's *disintegration energy* Q. The Q for the decay in Eq. 43-21 is 4.25 MeV—that amount of energy is said to be released by the alpha decay of ^{238}U, with the energy transferred from mass energy to the kinetic energy of the two products.

The half-life of ^{238}U for this decay process is 4.5×10^9 y. Why so long? If ^{238}U can decay in this way, why doesn't every ^{238}U nuclide in a sample of ^{238}U atoms simply decay at once? To answer the questions, we must examine the process of alpha decay.

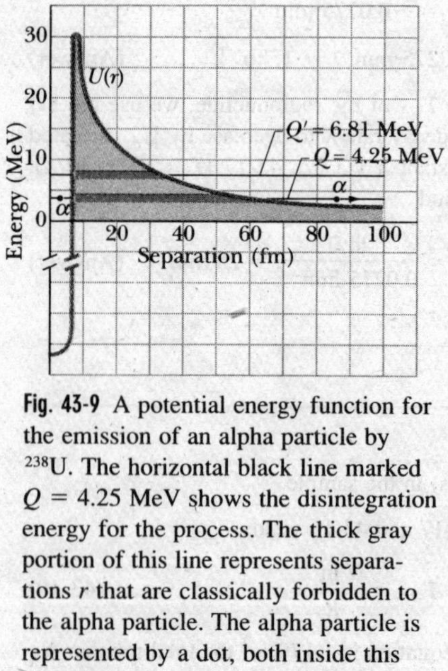

Fig. 43-9 A potential energy function for the emission of an alpha particle by ^{238}U. The horizontal black line marked $Q = 4.25$ MeV shows the disintegration energy for the process. The thick gray portion of this line represents separations r that are classically forbidden to the alpha particle. The alpha particle is represented by a dot, both inside this potential energy barrier (at the left) and outside it (at the right), after the particle has tunneled through. The horizontal black line marked $Q' = 6.81$ MeV shows the disintegration energy for the alpha decay of ^{228}U. (Both isotopes have the same potential energy function because they have the same nuclear charge.)

We choose a model in which the alpha particle is imagined to exist (already formed) inside the nucleus before it escapes from the nucleus. Figure 43-9 shows the approximate potential energy $U(r)$ of the system consisting of the alpha particle and the residual ^{234}Th nucleus, as a function of their separation r. This energy is a combination of (1) the potential energy associated with the (attractive) strong nuclear force that acts in the nuclear interior and (2) a Coulomb potential associated with the (repulsive) electric force that acts between the two particles before and after the decay has occurred.

The horizontal black line marked $Q = 4.25$ MeV shows the disintegration energy for the process. If we assume that this represents the total energy of the alpha particle during the decay process, then the part of the $U(r)$ curve above this line constitutes a potential energy barrier like that in Fig. 39-12. This barrier cannot be surmounted. If the alpha particle were able to be at some separation r within the barrier, its potential energy U would exceed its total energy E. This would mean, classically, that its kinetic energy K (which equals $E - U$) would be negative, an impossible situation.

We can see now why the alpha particle is not immediately emitted from the ^{238}U nucleus. That nucleus is surrounded by an impressive potential barrier, occupying—if you think of it in three dimensions—the volume lying between two spherical shells (of radii about 8 and 60 fm). This argument is so convincing that we now change our last question and ask: How, since the particle seems permanently trapped inside the nucleus by the barrier, can the ^{238}U nucleus *ever* emit an alpha particle? The answer is that, as you learned in Section 39-9, there is a finite probability that a particle can tunnel through an energy barrier that is classically insurmountable. In fact, alpha decay occurs as a result of barrier tunneling.

Since the half-life of ^{238}U is very long, the barrier is apparently not very "leaky." The alpha particle, presumed to be rattling back and forth within the nucleus, must arrive at the inner surface of the barrier about 10^{38} times before it succeeds in tunneling through the barrier. This is about 10^{21} times per second for about 4×10^9 years (the age of Earth)! We, of course, are waiting on the outside, able to count only the alpha particles that *do* manage to escape.

We can test this explanation of alpha decay by examining other alpha emitters. For an extreme contrast, consider the alpha decay of another uranium isotope, ^{228}U, which has a disintegration energy Q' of 6.81 MeV, about 60% higher than that of ^{238}U. (The value of Q' is also shown as a horizontal black line in Fig. 43-9.) Recall from Section 39-9 that the transmission coefficient of a barrier is very sensitive to small changes in the total energy of the particle seeking to penetrate it. Thus, we expect alpha decay to occur more readily for this nuclide than for ^{238}U. Indeed it does. As Table 43-2 shows, its half-life is only 9.1 min! An increase in Q by a factor of only 1.6 produces a decrease in half-life (that is, in the effectiveness of the barrier) by a factor of 3×10^{14}. This is sensitivity indeed.

TABLE 43-2 Two Alpha Emitters Compared

Radionuclide	Q	Half-Life
^{238}U	4.25 MeV	4.5×10^9 y
^{228}U	6.81 MeV	9.1 min

Sample Problem 43-6

We are given the following atomic masses:

^{238}U	238.050 79 u	^{4}He	4.002 60 u
^{234}Th	234.043 63 u	^{1}H	1.007 83 u
^{237}Pa	237.051 21 u		

Here Pa is the symbol for the element protactinium ($Z = 91$).

(a) Calculate the energy released during the alpha decay of ^{238}U. The decay process is

$$^{238}U \rightarrow {}^{234}Th + {}^{4}He.$$

Note, incidentally, how nuclear charge is conserved in this equation: The atomic numbers of thorium (90) and helium (2) add up to the atomic number of uranium (92). The number of nucleons is also conserved: $238 = 234 + 4$.

SOLUTION: The **Key Idea** here is that the energy released in the decay is the disintegration energy Q, which we can calculate from the change in mass ΔM due to the ^{238}U decay. We use Eq. 38-47,

$$Q = M_i c^2 - M_f c^2, \tag{43-22}$$

where the initial mass M_i is that of ^{238}U and the final mass M_f is the sum of the ^{234}Th and ^{4}He masses. As in Sample Problem 43-3, we must do this calculation for neutral atoms—that is, with atomic masses. Using the atomic masses given in the problem statement, Eq. 43-22 becomes

$$Q = (238.050\ 79\ u)c^2 - (234.043\ 63\ u + 4.002\ 60\ u)c^2$$
$$= (0.004\ 56\ u)c^2 = (0.004\ 56\ u)(931.5\ MeV/u)$$
$$= 4.25\ MeV. \quad \text{(Answer)}$$

Note that using atomic masses instead of nuclear masses does not affect the result because the total mass of the electrons in the products subtracts out from the mass of the nucleons + electrons in the original ^{238}U.

(b) Show that ^{238}U cannot spontaneously emit a proton.

SOLUTION: If this happened, the decay process would be

$$^{238}U \rightarrow {}^{237}Pa + {}^{1}H.$$

(You should verify that both nuclear charge and the number of nucleons are conserved in this process.) Using the same Key Idea as in part (a) and proceeding as we did there, we would find that the mass of the two decay products (= 237.051 21 u + 1.007 83 u) would *exceed* the mass of ^{238}U by $\Delta m = 0.008\ 25$ u, with disintegration energy $Q = -7.68$ MeV. The minus sign indicates that we must *add* 7.68 MeV to a ^{238}U nucleus before it will emit a proton; it will certainly not do so spontaneously.

43-5 Beta Decay

A nucleus that decays spontaneously by emitting an electron or a positron (a positively charged particle with the mass of an electron) is said to undergo **beta decay.** Like alpha decay, this is a spontaneous process, with a definite disintegration energy and half-life. Again like alpha decay, beta decay is a statistical process, governed by Eqs. 43-14 and 43-15. In *beta-minus* (β^-) decay, an electron is emitted by a nucleus, as in the decay

$$^{32}P \rightarrow {}^{32}S + e^- + \nu \quad (T_{1/2} = 14.3 \text{ d}). \tag{43-23}$$

In *beta-plus* (β^+) decay, a positron is emitted by a nucleus, as in the decay

$$^{64}Cu \rightarrow {}^{64}Ni + e^+ + \nu \quad (T_{1/2} = 12.7 \text{ h}). \tag{43-24}$$

The symbol ν represents a **neutrino,** a neutral particle, with very little or no mass, that is emitted from the nucleus along with the electron or positron during the decay process. Neutrinos interact only very weakly with matter and—for that reason—are so extremely difficult to detect that their presence long went unnoticed.*

Both charge and nucleon number are conserved in the above two processes. In the decay of Eq. 43-23, for example, we can write for charge conservation

$$(+15e) = (+16e) + (-e) + (0),$$

because ^{32}P has 15 protons, ^{32}S has 16 protons, and the neutrino ν has zero charge. Similarly, for nucleon conservation, we can write

$$(32) = (32) + (0) + (0),$$

because ^{32}P and ^{32}S each have 32 nucleons and neither the electron nor the neutrino is a nucleon.

It may seem surprising that nuclei can emit electrons, positrons, and neutrinos, since we have said that nuclei are made up of neutrons and protons only. However, we saw earlier that atoms emit photons, and we certainly do not say that atoms "contain" photons. We say that the photons are created during the emission process.

It is the same with the electrons, positrons, and neutrinos emitted from nuclei during beta decay. They are created during the emission process. For beta-minus decay, a neutron transforms into a proton within the nucleus according to

$$n \rightarrow p + e^- + \nu. \tag{43-25}$$

For beta-plus decay, a proton transforms into a neutron via

$$p \rightarrow n + e^+ + \nu. \tag{43-26}$$

Both of these beta-decay processes provide evidence that—as was pointed out—neutrons and protons are not truly fundamental particles. These processes show why the mass number A of a nuclide undergoing beta decay does not change; one of its constituent nucleons simply changes its character according to Eq. 43-25 or 43-26.

*Beta decay also includes *electron capture,* in which a nucleus decays by absorbing one of its atomic electrons, emitting a neutrino in the process. We do not consider that process here. Also, the neutral particle emitted in the decay process of Eq. 43-23 is actually an *antineutrino,* a distinction we shall not make in this introductory treatment.

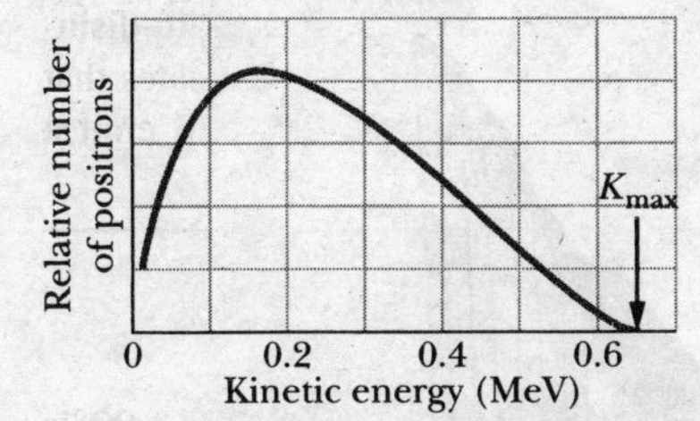

Fig. 43-10 The distribution of the kinetic energies of positrons emitted in the beta decay of ^{64}Cu. The maximum kinetic energy of the distribution (K_{max}) is 0.653 MeV. In all ^{64}Cu decay events, this energy is shared between the positron and the neutrino, in varying proportions. The *most probable* energy for an emitted positron is about 0.15 MeV.

In both alpha decay and beta decay, the same amount of energy is released in every individual decay of a particular radionuclide. In the alpha decay of a particular radionuclide, every emitted alpha particle has the same sharply defined kinetic energy. However, in the beta-minus decay of Eq. 43-25 with electron emission, the disintegration energy Q is shared—in varying proportions—between the emitted electron and neutrino. Sometimes the electron gets nearly all the energy, sometimes the neutrino does. In every case, however, the sum of the electron's energy and the neutrino's energy gives the same value Q. A similar sharing of energy, with a sum equal to Q, occurs in beta-plus decay (Eq. 43-26).

Thus, in beta decay the energy of the emitted electrons or positrons may range from zero up to a certain maximum K_{max}. Figure 43-10 shows the distribution of positron energies for the beta decay of ^{64}Cu (see Eq. 43-24). The maximum positron energy K_{max} must equal the disintegration energy Q because the neutrino carries away approximately zero energy when the positron carries away K_{max}; that is,

$$Q = K_{max}. \tag{43-27}$$

The Neutrino

Wolfgang Pauli first suggested the existence of neutrinos in 1930. His neutrino hypothesis not only permitted an understanding of the energy distribution of electrons or positrons in beta decay but also solved another early beta-decay puzzle involving "missing" angular momentum.

The neutrino is a truly elusive particle; the mean free path of an energetic neutrino in water has been calculated as no less than several thousand light-years. At the same time, neutrinos left over from the big bang that presumably marked the creation of the universe are the most abundant particles of physics. Billions of them pass through our bodies every second, leaving no trace.

In spite of their elusive character, neutrinos have been detected in the laboratory. This was first done in 1953 by F. Reines and C. L. Cowan, using neutrinos generated in a high-power nuclear reactor. (In 1995 Reines, the surviving member of the pair, received a Nobel prize for this work.) In spite of the difficulties of detection, experimental neutrino physics is now a well-developed branch of experimental physics, with avid practitioners at laboratories throughout the world.

The Sun emits neutrinos copiously from the nuclear furnace at its core and, at night, these messengers from the center of the Sun come up at us from below, Earth being almost totally transparent to them. In February 1987, light from an exploding star in the Large Magellanic Cloud (a nearby galaxy) reached Earth after having traveled for 170 000 years. Enormous numbers of neutrinos were generated in this explosion, and about 10 of them were picked up by a sensitive neutrino detector in Japan; Fig. 43-11 shows a record of their passage.

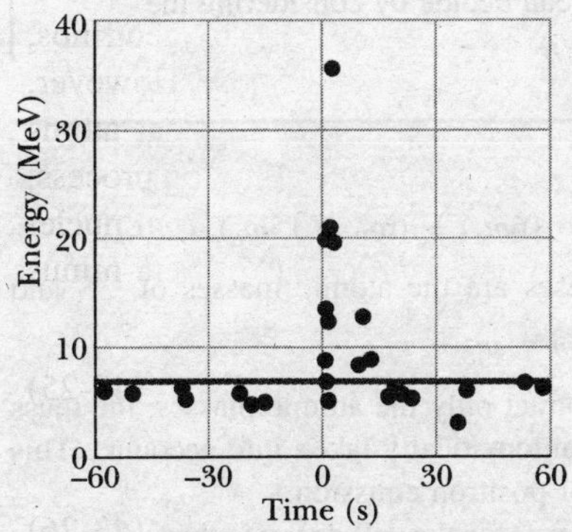

Fig. 43-11 A burst of neutrinos from the supernova SN 1987A, which occurred at (relative) time 0, stands out from the usual *background* of neutrinos. (For neutrinos, 10 is a "burst.") The particles were detected by an elaborate detector housed deep in a mine in Japan. The supernova was visible only in the Southern Hemisphere, so the neutrinos had to penetrate Earth (a trifling barrier for them) to reach the detector.

Radioactivity and the Nuclidic Chart

We can increase the information of the nuclidic chart of Fig. 43-4 by plotting the **mass excess** of each nuclide in a direction perpendicular to the *N-Z* plane. The mass excess of a nuclide is (in spite of its name) an energy that approximates the nuclide's *total* binding energy. It is defined as $(m - A)c^2$, where m is the atomic mass of the nuclide and A is its mass number, both expressed in atomic mass units, and c^2 is 931.5 MeV/u.

The surface so formed gives a graphic representation of nuclear stability. As Fig. 43-12 shows (for the low mass nuclides), this surface describes a "valley of the

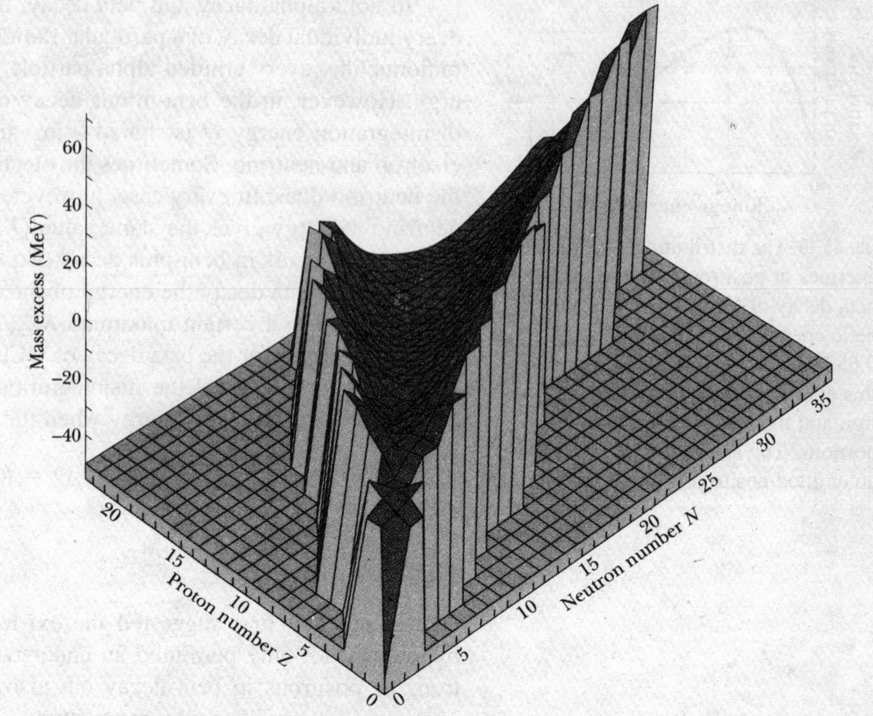

Fig. 43-12 A portion of the valley of the nuclides, showing only the nuclides of low mass. Deuterium, tritium, and helium lie at the nearest end of the plot, with helium at the high point. The valley stretches away from us, with the plot stopping at about $Z = 22$ and $N = 35$. Nuclides with large values of A, which would be plotted much beyond the valley, can decay into the valley by repeated alpha emissions and by fission (splitting of a nuclide).

nuclides," with the stability band of Fig. 43-4 running along its bottom. Nuclides on the proton-rich side of the valley decay into it by emitting positrons, and those on the neutron-rich side do so by emitting electrons.

✔CHECKPOINT 3: ^{238}U decays into ^{234}Th by the emission of an alpha particle. There follows a chain of further radioactive decays, either by alpha decay or by beta decay. Eventually a stable nuclide is reached and after that, no further radioactive decay is possible. Which of the following stable nuclides is the end product of the ^{238}U radioactive decay chain: ^{206}Pb, ^{207}Pb, ^{208}Pb, or ^{209}Pb? (*Hint:* You can decide by considering the changes in mass number A for the two types of decay.)

Sample Problem 43-7

Calculate the disintegration energy Q for the beta decay of ^{32}P, as described by Eq. 43-23. The needed atomic masses are 31.973 91 u for ^{32}P and 31.972 07 u for ^{32}S.

SOLUTION: The **Key Idea** here is that the disintegration energy Q for the beta decay is the amount by which the mass energy is changed by the decay. Q is given by Eq. 38-47 ($Q = -\Delta M\, c^2$). However, because an individual electron is emitted (and not an electron bound up in an atom), we must be careful to distinguish between nuclear masses (which we do not know) and atomic masses (which we do know). Let the boldface symbols $\mathbf{m_P}$ and $\mathbf{m_S}$ represent the nuclear masses of ^{32}P and ^{32}S, and let the italic symbols m_P and m_S represent their atomic masses. Then we can write the change in mass for the decay of Eq. 43-23 as

$$\Delta m = (\mathbf{m_S} + m_e) - \mathbf{m_P},$$

in which m_e is the mass of the electron. If we add and subtract $15m_e$ on the right side of this equation, we obtain

$$\Delta m = (\mathbf{m_S} + 16m_e) - (\mathbf{m_P} + 15m_e).$$

The quantities in parentheses are the atomic masses of ^{32}S and ^{32}P, so

$$\Delta m = m_S - m_P.$$

We thus see that if we subtract only the atomic masses, the mass of the emitted electron is automatically taken into account. (This procedure will not work for positron emission.)

The disintegration energy for the ^{32}P decay is then

$$Q = -\Delta m\, c^2$$
$$= -(31.972\,07\text{ u} - 31.973\,91\text{ u})(931.5\text{ MeV/u})$$
$$= 1.71\text{ MeV}. \qquad \text{(Answer)}$$

Experimentally, this calculated quantity proves to be equal to K_{max}, the maximum energy the emitted electrons can have. Although 1.71 MeV is released every time a ^{32}P nucleus decays, in essentially every case the electron carries away less energy than this. The neutrino gets all the rest, carrying it stealthily out of the laboratory.

A fragment of the Dead Sea scrolls and the caves from which the scrolls were recovered.

43-6 Radioactive Dating

If you know the half-life of a given radionuclide, you can in principle use the decay of that radionuclide as a clock to measure time intervals. The decay of very long-lived nuclides, for example, can be used to measure the age of rocks—that is, the time that has elapsed since they were formed. Such measurements for rocks from Earth and the Moon, and for meteorites, yield a consistent maximum age of about 4.5×10^9 y for these bodies.

The radionuclide ^{40}K, for example, decays to ^{40}Ar, a stable isotope of the noble gas argon. The half-life for this decay is 1.25×10^9 y. A measurement of the ratio of ^{40}K to ^{40}Ar, as found in the rock in question, can be used to calculate the age of that rock. Other long-lived decays, such as that of ^{235}U to ^{207}Pb (involving a number of intermediate stages), can be used to verify this calculation.

For measuring shorter time intervals, in the range of historical interest, radiocarbon dating has proved invaluable. The radionuclide ^{14}C (with $T_{1/2} = 5730$ y) is produced at a constant rate in the upper atmosphere as atmospheric nitrogen is bombarded by cosmic rays. This radiocarbon mixes with the carbon that is normally present in the atmosphere (as CO_2) so that there is about one atom of ^{14}C for every 10^{13} atoms of ordinary stable ^{12}C. Through biological activity such as photosynthesis and breathing, the atoms of atmospheric carbon trade places randomly, one atom at a time, with the atoms of carbon in every living thing, including broccoli, mushrooms, penguins, and humans. Eventually an exchange equilibrium is reached at which the carbon atoms of every living thing contain a fixed small fraction of the radioactive nuclide ^{14}C.

This equilibrium persists as long as the organism is alive. When the organism dies, the exchange with the atmosphere stops and the amount of radiocarbon trapped in the organism, since it is no longer being replenished, dwindles away with a half-life of 5730 y. By measuring the amount of radiocarbon per gram of organic matter, it is possible to measure the time that has elapsed since the organism died. Charcoal from ancient campfires, the Dead Sea scrolls, and many prehistoric artifacts have been dated in this way. The age of the scrolls was determined by radiocarbon dating a sample of the cloth used to plug the jars in which the scrolls were sealed.

Sample Problem 43-8

Mass spectrometric analysis of potassium and argon atoms in a Moon rock sample shows that the ratio of the number of (stable) ^{40}Ar atoms present to the number of (radioactive) ^{40}K atoms is 10.3. Assume that all the argon atoms were produced by the decay of potassium atoms, with a half-life of 1.25×10^9 y. How old is the rock?

SOLUTION: The **Key Idea** here is that if N_0 potassium atoms were present at the time the rock was formed by solidification from a molten form, the number of potassium atoms remaining at the time of analysis is, from Eq. 43-14,

$$N_K = N_0 e^{-\lambda t}, \tag{43-28}$$

in which t is the age of the rock. For every potassium atom that decays, an argon atom is produced. Thus, the number of argon atoms present at the time of the analysis is

$$N_{Ar} = N_0 - N_K. \tag{43-29}$$

We cannot measure N_0, so let's eliminate it from Eqs. 43-28 and 43-29. We find, after some algebra, that

$$\lambda t = \ln\left(1 + \frac{N_{Ar}}{N_K}\right), \tag{43-30}$$

in which N_{Ar}/N_K *can* be measured. Solving for t and using Eq. 43-17 to replace λ with $(\ln 2)/T_{1/2}$ yield

$$t = \frac{T_{1/2}\ln(1 + N_{Ar}/N_K)}{\ln 2} = \frac{(1.25 \times 10^9\ \text{y})[\ln(1 + 10.3)]}{\ln 2} = 4.37 \times 10^9\ \text{y}. \quad \text{(Answer)}$$

Lesser ages may be found for other lunar or terrestrial rock samples, but no substantially greater ones. Thus, the solar system must be about 4 billion years old.

43-7 Measuring Radiation Dosage

The effect of radiation such as gamma rays, electrons, and alpha particles on living tissue (particularly our own) is a matter of public interest. Such radiation is found in nature in cosmic rays and arises from radioactive elements in Earth's crust. Radiation associated with some human activities, such as using x rays and radionuclides in medicine and in industry, also contributes.

It is not our task here to explore the various sources of radiation but simply to describe the units in which the properties and effects of such radiations are expressed. We have already discussed the *activity* of a radioactive source. There are two remaining quantities of interest.

1. *Absorbed Dose*. This is a measure of the radiation dose (as energy per unit mass) actually absorbed by a specific object, such as a patient's hand or chest. Its SI unit is the **gray** (Gy). An older unit, the **rad** (from **r**adiation **a**bsorbed **d**ose) is still in common use. The terms are defined and related as follows:

$$1 \text{ Gy} = 1 \text{ J/kg} = 100 \text{ rad}. \tag{43-31}$$

A typical dose-related statement is: "A whole-body, short-term gamma-ray dose of 3 Gy (= 300 rad) will cause death in 50% of the population exposed to it." Thankfully, our present average absorbed dose per year, from sources of both natural and human origin, is only about 2 mGy (= 0.2 rad).

2. *Dose Equivalent*. Although different types of radiation (gamma rays and neutrons, say) may deliver the same amount of energy to the body, they do not have the same biological effect. The dose equivalent allows us to express the biological effect by multiplying the absorbed dose (in grays or rads) by a numerical **RBE** factor (from **r**elative **b**iological **e**ffectiveness). For x rays and electrons, for example, RBE = 1; for slow neutrons RBE = 5; for alpha particles RBE = 10; and so on. Personnel-monitoring devices such as film badges register the dose equivalent.

The SI unit of dose equivalent is the **sievert** (Sv). An earlier unit, the **rem,** is still in common use. Their relationship is

$$1 \text{ Sv} = 100 \text{ rem}. \tag{43-32}$$

An example of the correct use of these terms is: "The recommendation of the National Council on Radiation Protection is that no individual who is (nonoccupationally) exposed to radiation should receive a dose equivalent greater than 5 mSv (= 0.5 rem) in any one year." This includes radiation of all kinds; of course the appropriate RBE factor must be used for each kind.

Sample Problem 43-9

We have seen that a gamma-ray dose of 3 Gy is lethal to half the people exposed to it. If the equivalent energy were absorbed as heat, what rise in body temperature would result?

SOLUTION: One **Key Idea** here is that we can relate an absorbed energy Q and the resulting temperature increase ΔT with Eq. 19-14 ($Q = cm\,\Delta T$). In that equation, m is the mass of the material absorbing the energy and c is the specific heat of that material. Another **Key Idea** is that an absorbed dose of 3 Gy corresponds to an absorbed energy per unit mass of 3 J/kg. Let us assume that c, the specific heat of the human body, is the same as that of water, 4180 J/kg · K. Then we find that

$$\Delta T = \frac{Q/m}{c} = \frac{3 \text{ J/kg}}{4180 \text{ J/kg} \cdot \text{K}} = 7.2 \times 10^{-4} \text{ K} \approx 700\ \mu\text{K}.$$

(Answer)

Obviously the damage done by ionizing radiation has nothing to do with thermal heating. The harmful effects arise because the radiation damages DNA and thus interferes with the normal functioning of tissues in which it is absorbed.

43-8 Nuclear Models

Nuclei are more complicated than atoms. For atoms, the basic force law (Coulomb's law) is simple in form and there is a natural force center, the nucleus. For nuclei the force law is complicated and cannot, in fact, be written down explicitly in full detail. Furthermore, the nucleus—a jumble of protons and neutrons—has no natural force center to simplify the calculations.

In the absence of a comprehensive nuclear *theory,* we turn to the construction of nuclear *models*. A nuclear model is simply a way of looking at the nucleus that gives a physical insight into as wide a range of its properties as possible. The usefulness of a model is tested by its ability to provide predictions that can be verified experimentally in the laboratory.

Two models of the nucleus have proved useful. Although based on assumptions that seem flatly to exclude each other, each accounts very well for a selected group of nuclear properties. After describing them separately, we shall see how these two models may be combined to form a single coherent picture of the atomic nucleus.

The Collective Model

In the *collective model,* formulated by Niels Bohr, the nucleons, moving around within the nucleus at random, are imagined to interact strongly with each other, like the molecules in a drop of liquid. A given nucleon collides frequently with other nucleons in the nuclear interior, its mean free path as it moves about being substantially less than the nuclear radius.

The collective model permits us to correlate many facts about nuclear masses and binding energies; it is useful (as you will see later) in explaining nuclear fission. It is also useful for understanding a large class of nuclear reactions.

Consider, for example, a generalized nuclear reaction of the form

$$X + a \rightarrow C \rightarrow Y + b. \tag{43-33}$$

We imagine that projectile a enters target nucleus X, forming a **compound nucleus** C and conveying to it a certain amount of excitation energy. The projectile, perhaps a neutron, is at once caught up by the random motions that characterize the nuclear interior. It quickly loses its identity—so to speak—and the excitation energy it carried into the nucleus is quickly shared with all the other nucleons in C.

The quasi-stable state represented by C in Eq. 43-33 may have a mean life of 10^{-16} s before it decays to Y and b. By nuclear standards, this is a very long time, being about one million times longer than the time required for a nucleon with a few million electron-volts of energy to travel across a nucleus.

The central feature of this compound-nucleus concept is that the formation of the compound nucleus and its eventual decay are totally independent events. At the time of its decay, the compound nucleus has "forgotten" how it was formed. Hence, its mode of decay is not influenced by its mode of formation. As an example, Fig. 43-13 shows three possible ways in which the compound nucleus ^{20}Ne might be formed and three in which it might decay. Any of the three formation modes can lead to any of the three decay modes.

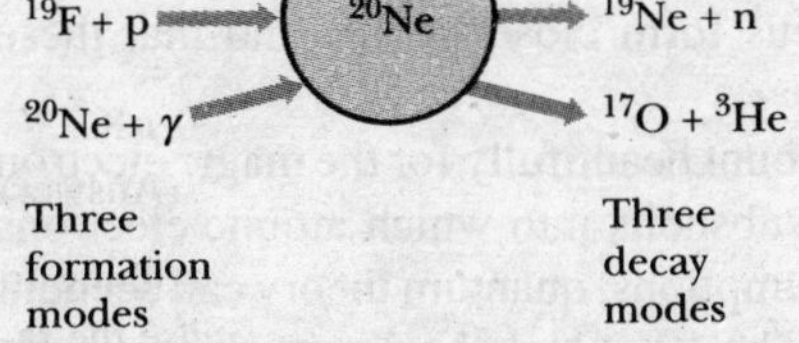

Fig. 43-13 The formation modes and the decay modes of the compound nucleus ^{20}Ne.

The Independent Particle Model

In the collective model, we assume that the nucleons move around at random and bump into each other frequently. The *independent particle model,* however, is based on just the opposite assumption—namely, that each nucleon remains in a well-defined quantum state within the nucleus and hardly makes any collisions at all! The nucleus, unlike the atom, has no fixed center of charge; we assume in this model that each nucleon moves in a potential well that is determined by the smeared-out (time-averaged) motions of all the other nucleons.

A nucleon in a nucleus, like an electron in an atom, has a set of quantum numbers that defines its state of motion. Also, nucleons obey the Pauli exclusion principle, just as electrons do; that is, no two nucleons in a nucleus may occupy the same quantum state at the same time. In this regard, the neutrons and the protons are treated separately, each particle type with its own set of quantum states.

The fact that nucleons obey the Pauli exclusion principle helps us to understand the relative stability of nucleon states. If two nucleons within the nucleus are to collide, the energy of each of them after the collision must correspond to the energy of an *unoccupied* state. If no such state is available, the collision simply cannot occur. Thus, any given nucleon experiencing repeated "frustrated collision opportunities" will maintain its state of motion long enough to give meaning to the statement that it exists in a quantum state with a well-defined energy.

In the atomic realm, the repetitions of physical and chemical properties that we find in the periodic table are associated with a property of atomic electrons—namely, they arrange themselves in shells that have a special stability when fully occupied. We can take the atomic numbers of the noble gases,

$$2, 10, 18, 36, 54, 86, \ldots,$$

as *magic electron numbers* that mark the completion (or closure) of such shells.

Nuclei also show such closed-shell effects, associated with certain **magic nucleon numbers:**

$$2, 8, 20, 28, 50, 82, 126, \ldots.$$

Any nuclide whose proton number Z or neutron number N has one of these values turns out to have a special stability that may be made apparent in a variety of ways.

Examples of "magic" nuclides are ^{18}O ($Z = 8$), ^{40}Ca ($Z = 20, N = 20$), ^{92}Mo ($N = 50$), and ^{208}Pb ($Z = 82, N = 126$). Both ^{40}Ca and ^{208}Pb are said to be "doubly magic" because they contain both filled shells of protons *and* filled shells of neutrons.

The magic number 2 shows up in the exceptional stability of the alpha particle (^{4}He), which, with $Z = N = 2$, is doubly magic. For example, on the binding energy curve of Fig. 43-6, the binding energy per nucleon for this nuclide stands well above those of its periodic-table neighbors hydrogen, lithium, and beryllium. The alpha particle is so tightly bound, in fact, that it is impossible to add another particle to it; there is no stable nuclide with $A = 5$.

The central idea of a closed shell is that a single particle outside a closed shell can be relatively easily removed, but considerably more energy must be expended to remove a particle from the shell itself. The sodium atom, for example, has one (valence) electron outside a closed electron shell. Only about 5 eV is required to strip the valence electron away from a sodium atom; however, to remove a *second* electron (which must be plucked out of a closed shell) requires 22 eV. As a nuclear case, consider ^{121}Sb ($Z = 51$), which contains a single proton outside a closed shell of 50 protons. To remove this lone proton requires 5.8 MeV; to remove a *second* proton, however, requires an energy of 11 MeV. There is much additional experimental evidence that the nucleons in a nucleus form closed shells and that these shells exhibit stable properties.

We have seen that quantum theory can account beautifully for the magic electron numbers—that is, for the populations of the subshells into which atomic electrons are grouped. It turns out that, under certain assumptions, quantum theory can account equally well for the magic nucleon numbers! The 1963 Nobel prize in physics was, in fact, awarded to Maria Mayer and Hans Jensen "for their discoveries concerning nuclear shell structure."

A Combined Model

Consider a nucleus in which a small number of neutrons (or protons) exist outside a core of closed shells that contains magic numbers of neutrons or protons. The outside nucleons occupy quantized states in a potential well established by the central core, thus preserving the central feature of the independent-particle model. These outside nucleons also interact with the core, deforming it and setting up "tidal wave" motions of rotation or vibration within it. These collective motions of the core preserve the central feature of that model. Such a model of nuclear structure thus succeeds in combining the seemingly irreconcilable points of view of the collective and independent-particle models. It has been remarkably successful in explaining observed nuclear properties.

Sample Problem 43-10

Consider the neutron capture reaction

$$^{109}\text{Ag} + \text{n} \rightarrow {}^{110}\text{Ag} \rightarrow {}^{110}\text{Ag} + \gamma, \tag{43-34}$$

in which a compound nucleus (^{110}Ag) is formed. Figure 43-14 shows the relative rate at which such events take place, plotted against the energy of the incoming neutron. Find the mean lifetime of this compound nucleus by using the uncertainty principle in the form

$$\Delta E \cdot \Delta t \approx \hbar. \tag{43-35}$$

Here ΔE is a measure of the uncertainty with which the energy of a state can be defined. The quantity Δt is a measure of the time available to measure this energy. In fact, here Δt is just t_{avg}, the average life of the compound nucleus before it decays to its ground state.

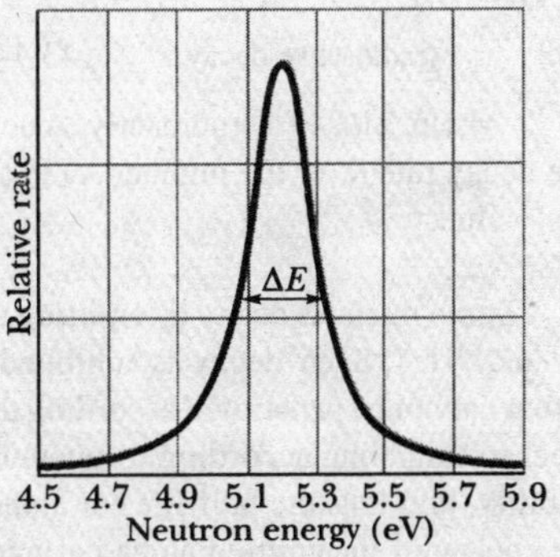

Fig. 43-14 Sample Problem 43-10. A plot of the relative number of reaction events of the type described by Eq. 43-34, as a function of the energy of the incident neutron. The half-width ΔE of the resonance peak is about 0.20 eV.

SOLUTION: We see that the relative reaction rate peaks sharply at a neutron energy of about 5.2 eV. This suggests that we are dealing with a single excited energy level of the compound nucleus ^{110}Ag. When the available energy (of the incoming neutron) just matches the energy of this level above the ^{110}Ag ground state, we have "resonance" and the reaction of Eq. 43-34 really "goes."

However, the resonance peak is not infinitely sharp but has an approximate half-width (ΔE in the figure) of about 0.20 eV. The **Key Idea** here is that we can account for this resonance-peak width by saying that the excited level is not sharply defined in energy, having an energy uncertainty ΔE of about 0.20 eV. Thus, Eq. 43-35 gives us

$$\Delta t = t_{avg} \approx \frac{\hbar}{\Delta E} = \frac{(4.14 \times 10^{-15}\ \text{eV} \cdot \text{s})/2\pi}{0.20\ \text{eV}}$$

$$\approx 3 \times 10^{-15}\ \text{s}. \qquad \text{(Answer)}$$

This is several hundred times greater than the time a 0.20 eV neutron takes to cross the diameter of a ^{109}Ag nucleus. Therefore, the neutron is spending this time of 3×10^{-15} s *as part of* the nucleus.

REVIEW & SUMMARY

The Nuclides Approximately 2000 **nuclides** are known to exist. Each is characterized by an **atomic number** Z (the number of protons), a **neutron number** N, and a **mass number** A (the total number of **nucleons**—protons and neutrons). Thus, $A = Z + N$. Nuclides with the same atomic number but different neutron numbers are **isotopes** of each other. Nuclei have a mean radius r given by

$$r = r_0 A^{1/3}, \tag{43-3}$$

where $r_0 \approx 1.2$ fm.

Mass–Energy Exchanges The energy equivalent of one mass unit (u) is 931.5 MeV. The binding energy curve shows that middle-mass nuclides are the most stable and that energy can be released both by fission of high-mass nuclei and by fusion of low-mass nuclei.

The Nuclear Force Nuclei are held together by an attractive force acting among the nucleons. It is thought to be a secondary effect of the **strong force** acting between the quarks that make up the nucleons. Nuclei can exist in a number of discrete energy states, each with a characteristic intrinsic angular momentum and magnetic moment.

Radioactive Decay Most known nuclides are radioactive; they spontaneously decay at a rate $R\ (= -dN/dt)$ that is proportional to the number N of radioactive atoms present, the proportionality constant being the **disintegration constant** λ. This leads to the law of exponential decay:

$$N = N_0 e^{-\lambda t}, \qquad R = \lambda N = R_0 e^{-\lambda t} \quad \text{(radioactive decay).} \qquad (43\text{-}14, 43\text{-}15, 43\text{-}16)$$

The half-life $T_{1/2} = (\ln 2)/\lambda$ of a radioactive nuclide is the time required for the decay rate R (or the number N) in a sample to drop to half its initial value.

Alpha Decay Some nuclides decay by emitting an alpha particle (a helium nucleus, ^{4}He). Such decay is inhibited by a potential energy barrier that cannot be penetrated according to classical physics but is subject to tunneling according to quantum physics. The barrier penetrability, and thus the half-life for alpha decay, is very sensitive to the energy of the emitted alpha particle.

Beta Decay In **beta decay** either an electron or a positron is emitted by a nucleus, along with a neutrino. The emitted particles share the available disintegration energy. The electrons and positrons emitted in beta decay have a continuous spectrum of energies from near zero up to a limit K_{max} $(= Q = -\Delta m\, c^2)$.

Radioactive Dating Naturally occurring radioactive nuclides provide a means for estimating the dates of historic and prehistoric events. For example, the ages of organic materials can often be found by measuring their ^{14}C content; rock samples can be dated using the radioactive isotope ^{40}K.

Radiation Dosage Three units are used to describe exposure to ionizing radiation. The **becquerel** (1 Bq = 1 decay per second) measures the **activity** of a source. The amount of energy actually absorbed is measured in **grays,** with 1 Gy corresponding to 1 J/kg. The estimated biological effect of the absorbed energy is measured in **sieverts;** a dose equivalent of 1 Sv causes the same biological effect regardless of the radiation type by which it was acquired.

Nuclear Models The **collective** model of nuclear structure assumes that nucleons collide constantly and that relatively long-lived **compound nuclei** are formed when a projectile is captured. The formation of a compound nucleus and the eventual decay of that nucleus are totally independent events.

The **independent particle** model of nuclear structure assumes that each nucleon moves, essentially without collisions, in a quantized state within the nucleus. The model predicts nucleon levels and **magic numbers** of nucleons (2, 8, 20, 28, 50, 82, and 126) associated with closed shells of nucleons; nuclides with any of these numbers of neutrons or protons are particularly stable.

The **combined** model, in which extra nucleons occupy quantized states outside a central core of closed shells, is highly successful in predicting many nuclear properties.

QUESTIONS

1. Suppose the alpha particle of Sample Problem 43-1 is replaced with a proton of the same initial kinetic energy and also headed directly toward the nucleus of the gold atom. Will the distance from the center of the nucleus at which the proton stops be greater than, less than, or the same as that of the alpha particle?

2. In your body are there more protons than neutrons, more neutrons than protons, or about the same number of each?

3. The nuclide ^{244}Pu ($Z = 94$) is an alpha-particle emitter. Into which of the following nuclides does it decay: ^{240}Np ($Z = 93$), ^{240}U ($Z = 92$), ^{248}Cm ($Z = 96$), or ^{244}Am ($Z = 95$)?

4. A certain nuclide is said to be particularly stable. Does its binding energy per nucleon lie slightly above or slightly below the binding energy curve of Fig. 43-6?

5. Is the mass excess of an alpha particle (use a straightedge on Fig. 43-12) greater than or less than the particle's total binding energy (use the binding energy per nucleon from Fig. 43-6)?

6. The radionuclide ^{196}Ir decays by emitting an electron. (a) Into which square in Fig. 43-5 is it transformed? (b) Do further decays then occur?

7. A lead nuclide contains 82 protons. (a) If it also contained 82 neutrons, where would it be located on the plot of Fig. 43-4? (b) If such a nucleus could be formed, would it emit positrons, emit electrons, or be stable? (c) From Fig. 43-4, about how many neutrons do you expect to find in a stable lead nuclide?

8. The nuclide ^{238}U ($Z = 92$) can fission into two parts that have identical atomic numbers and mass numbers. (a) Is the nuclide ^{238}U above or below the $N = Z$ line of Fig. 43-4? (b) Are the two fragments above or below this line? (c) Are these fragments stable or radioactive?

9. Radionuclides decay exponentially, as in Eq. 43-15. Batteries, stars, and even students also decay, where "decay" stands for "burn out." Do these items decay exponentially?

10. At $t = 0$, a sample of radionuclide A has the same decay rate as a sample of radionuclide B has at $t = 30$ min. The disintegration

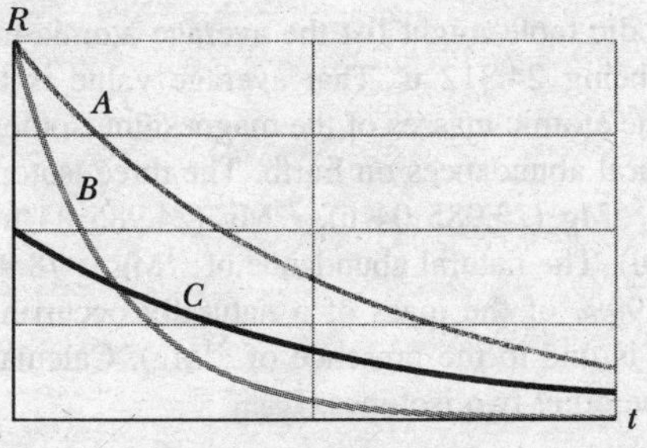

Fig. 43-15 Question 12.

constants are λ_A and λ_B, with $\lambda_A < \lambda_B$. Will the two samples ever have (simultaneously) the same decay rate? (*Hint:* Sketch a graph of their activities.)

11. At $t = 0$, a sample of radionuclide A has twice the decay rate as a sample of radionuclide B. The disintegration constants are λ_A and λ_B, with $\lambda_A > \lambda_B$. Will the two samples ever have (simultaneously) the same decay rate?

12. Figure 43-15 gives the activities of three radioactive samples versus time. Rank the samples according to their (a) half-life and (b) disintegration constant, greatest first. (*Hint:* For (a), use a straightedge on the graph.)

13. If the mass of a radioactive sample is doubled, do (a) the activity of the sample and (b) the disintegration constant of the sample increase, decrease, or remain the same?

14. At $t = 0$ we begin to observe two identical radioactive nuclei with a half-life of 5 min. At $t = 1$ min, one of the nuclei decays. Does that event increase or decrease the chance of the second nucleus decaying in the next 4 min, or is there no effect on the second nucleus?

15. The radionuclide ^{49}Sc has a half-life of 57.0 min. The counting rate of a sample of this nuclide at $t = 0$ is 6000 counts/min above the general background activity, which is 30 counts/min. Without actual computation, determine whether the counting rate of the sample will be about equal to the background rate in about 3 h, 7 h, 10 h, or a time much longer than 10 h.

16. The radionuclides ^{209}At and ^{209}Po emit alpha particles with energies of 5.65 and 4.88 MeV, respectively. Which nuclide has the longer half-life?

17. The magic numbers for nuclei are given in Section 43-8 as 2, 8, 20, 28, 50, 82, and 126. Are nuclides magic (that is, especially stable) when (a) only the mass number A, (b) only the atomic number Z, (c) only the neutron number N, or (d) either Z or N (or both) is equal to one of these numbers? Pick all correct phrases.

18. (a) Which of the following nuclides are magic: ^{122}Sn, ^{132}Sn, ^{98}Cd, ^{198}Au, ^{208}Pb? (b) Which, if any, are doubly magic?

EXERCISES & PROBLEMS

ssm Solution is in the Student Solutions Manual.
www Solution is available on the World Wide Web at: http://www.wiley.com/college/hrw
ilw Solution is available on the Interactive LearningWare.

SEC. 43-1 Discovering the Nucleus

1E. Assume that a gold nucleus has a radius of 6.23 fm and an alpha particle has a radius of 1.80 fm. What energy must an incident alpha particle have in order to "touch" the gold nucleus according to the type of calculation in Sample Problem 43-1?

2E. Calculate the distance of closest approach for a head-on collision between a 5.30 MeV alpha particle and the nucleus of a copper atom.

3P. When an alpha particle collides elastically with a nucleus, the nucleus recoils. Suppose a 5.00 MeV alpha particle has a head-on elastic collision with a gold nucleus that is initially at rest. What is the kinetic energy of (a) the recoiling nucleus and (b) the rebounding alpha particle?

SEC. 43-2 Some Nuclear Properties

4E. The radius of a spherical nucleus is measured, by electron-scattering methods, to be 3.6 fm. What is the likely mass number of the nucleus?

5E. Locate the nuclides displayed in Table 43-1 on the nuclidic chart of Fig. 43-4. Verify that they lie in the stability zone.

6E. A neutron star is a stellar object whose density is about that of nuclear matter, as calculated in Sample Problem 43-2. Suppose that the Sun were to collapse and become such a star without losing any of its present mass. What would be its radius?

7E. The nuclide ^{14}C contains (a) how many protons and (b) how many neutrons?

8E. Using a nuclidic chart, write the symbols for (a) all stable isotopes with $Z = 60$, (b) all radioactive nuclides with $N = 60$, and (c) all nuclides with $A = 60$.

9E. Make a nuclidic chart similar to Fig. 43-5 for the 25 nuclides $^{118-122}$Te, $^{117-121}$Sb, $^{116-120}$Sn, $^{115-119}$In, and $^{114-118}$Cd. Draw in and label (a) all isobaric (constant A) lines and (b) all lines of constant neutron excess, defined as $N - Z$.

10E. The strong neutron excess (defined as $N - Z$) of high-mass nuclei is illustrated by noting that most high-mass nuclides could never fission into two stable nuclei without neutrons being left over. For example, consider the spontaneous fission of a ^{235}U nucleus into two stable *daughter nuclei* with atomic numbers 39 and 53. (a) From Appendix F, what are the daughter elements? From Fig. 43-4, approximately how many neutrons are (b) in the daughter isotopes and (c) left over?

11E. The electric potential energy of a uniform sphere of charge q and radius r is

$$U = \frac{3q^2}{20\pi\varepsilon_0 r}.$$

(a) Find the electric potential energy of the nuclide ^{239}Pu, assumed to be spherical with radius 6.64 fm. (b) For this nuclide, compare the electric potential energy per nucleon, and also per proton, with the binding energy per nucleon of 7.56 MeV. (c) What do you conclude?

12E. Calculate and compare (a) the nuclear mass density ρ_m and (b) the nuclear charge density ρ_q for the fairly low-mass nuclide ^{55}Mn and for the fairly high-mass nuclide ^{209}Bi. (c) Are the differences what you would expect? Explain.

13E. Verify the binding energy per nucleon given in Table 43-1 for ^{239}Pu. The atomic masses you will need are 239.052 16 u (^{239}Pu), 1.007 83 u (^{1}H), and 1.008 67 u (neutron). ssm

14E. (a) Show that an approximate formula for the mass M of an atom is $M = Am_p$, where A is the mass number and m_p is the proton mass. (b) What percent error is committed in using this formula to calculate the masses of the atoms in Table 43-1? The mass of the bare proton is 1.007 276 u. (c) Is this formula accurate enough to be used for calculations of nuclear binding energy?

15E. Nuclear radii may be measured by scattering high-energy electrons from nuclei. (a) What is the de Broglie wavelength for 200 MeV electrons? (b) Are these electrons suitable probes for this purpose? ssm

16E. The characteristic nuclear time is a useful but loosely defined quantity, taken to be the time required for a nucleon with a few million electron-volts of kinetic energy to travel a distance equal to the diameter of a middle-mass nuclide. What is the order of magnitude of this quantity? Consider 5 MeV neutrons traversing a nuclear diameter of ^{197}Au; use Eq. 43-3.

17E. Because a nucleon is confined to a nucleus, we can take the uncertainty in its position to be approximately the nuclear radius r. What does the uncertainty principle say about the kinetic energy of a nucleon in a nucleus with, say, $A = 100$? (*Hint:* Take the uncertainty in momentum Δp to be the actual momentum p.)

18E. The atomic masses of ^{1}H, ^{12}C, and ^{238}U are 1.007 825 u, 12.000 000 u (this one is exact by definition), and 238.050 785 u, respectively. (a) What would these masses be if the mass unit were defined to give the mass of ^{1}H as (exactly) 1.000 000 u? (b) Use your result to suggest why this definition was not made.

19P. (a) Show that the energy associated with the strong force between nucleons in a nucleus is proportional to A, the mass number of the nucleus in question. (b) Show that the energy associated with the Coulomb force between protons in a nucleus is proportional to $Z(Z-1)$. (c) Show that, as we move to larger and larger nuclei (see Fig. 43-4), the importance of the Coulomb force increases more rapidly than does that of the strong force. ssm

20P. You are asked to pick apart an alpha particle (^{4}He) by removing, in sequence, a proton, a neutron, and a proton. Calculate (a) the work required for each step, (b) the total binding energy of the alpha particle, and (c) the binding energy per nucleon. Some needed atomic masses are

^{4}He	4.002 60 u	^{2}H	2.014 10 u
^{3}H	3.016 05 u	^{1}H	1.007 83 u
n	1.008 67 u		

21P. A periodic table might list the average atomic mass of magnesium as being 24.312 u. That average value is the result of *weighting* the atomic masses of the magnesium isotopes according to their natural abundances on Earth. The three isotopes and their masses are ^{24}Mg (23.985 04 u), ^{25}Mg (24.985 84 u), and ^{26}Mg (25.982 59 u). The natural abundance of ^{24}Mg is 78.99% by mass (that is, 78.99% of the mass of a naturally occurring sample of magnesium is due to the presence of ^{24}Mg). Calculate the abundances of the other two isotopes. ssm

22P. To simplify calculations, atomic masses are sometimes tabulated not as the actual atomic mass m but as $(m - A)c^2$, where A is the mass number expressed in atomic mass units. This quantity, usually reported in millions of electron-volts, is called the *mass excess,* represented with symbol Δ. Using data from Sample Problem 43-3, find the mass excesses for (a) ^{1}H, (b) the neutron, and (c) ^{120}Sn.

23P. A penny has a mass of 3.0 g. Calculate the nuclear energy that would be required to separate all the neutrons and protons in this coin from one another. For simplicity assume that the penny is made entirely of ^{63}Cu atoms (of mass 62.929 60 u). The masses of the proton and the neutron are 1.007 83 u and 1.008 67 u, respectively.

24P. Because the neutron has no charge, its mass must be found in some way other than by using a mass spectrometer. When a neutron and a proton meet (assume both to be almost stationary), they combine and form a deuteron, emitting a gamma ray whose energy is 2.2233 MeV. The masses of the proton and the deuteron are 1.007 825 035 u and 2.014 101 9 u, respectively. Find the mass of the neutron from these data, to as many significant figures as the data warrant. (A value of the mass–energy conversion factor c^2 that is more precise than the one presented in the text is 931.502 MeV/u.)

25P. Show that the total binding energy E_{be} of a nuclide is

$$E_{be} = Z\Delta_H + N\Delta_n - \Delta,$$

where Δ_H, Δ_n, and Δ are the appropriate mass excesses (see Problem 22). Using this method, calculate the binding energy per nucleon for ^{197}Au. Compare your result with the value listed in Table 43-1. The needed mass excesses, rounded to three significant figures, are $\Delta_H = +7.29$ MeV, $\Delta_n = +8.07$ MeV, and $\Delta_{197} = -31.2$ MeV. Note the economy of calculation that results when mass excesses are used in place of the actual masses. ssm www

SEC. 43-3 Radioactive Decay

26E. A radioactive nuclide has a half-life of 30 y. What fraction of an initially pure sample of this nuclide will remain undecayed at the end of (a) 60 y and (b) 90 y?

27E. The half-life of a radioactive isotope is 140 d. How many days would it take for the decay rate of a sample of this isotope to fall to one-fourth of its initial value?

28E. The half-life of a particular radioactive isotope is 6.5 h. If there are initially 48×10^{19} atoms of this isotope, how many remain at the end of 26 h?

29E. Consider an initially pure 3.4 g sample of ^{67}Ga, an isotope that has a half-life of 78 h. (a) What is its initial decay rate? (b) What is its decay rate 48 h later? ssm

30E. From data presented in the first few paragraphs of Section 43-3, find (a) the disintegration constant λ and (b) the half-life of ^{238}U.

31E. A radioactive isotope of mercury, ^{197}Hg, decays into gold, ^{197}Au, with a disintegration constant of 0.0108 h^{-1}. (a) Calculate its half-life. What fraction of a sample will remain at the end of (b) three half-lives and (c) 10.0 days? ssm

32E. The plutonium isotope ^{239}Pu is produced as a by-product in nuclear reactors and hence is accumulating in our environment. It is radioactive, decaying with a half-life of 2.41×10^4 y. (a) How many nuclei of Pu constitute a chemically lethal dose of 2 mg? (b) What is the decay rate of this amount?

33E. Cancer cells are more vulnerable to x and gamma radiation than are healthy cells. In the past, the standard source for radiation therapy was radioactive ^{60}Co, which decays, with a half-life of 5.27 y, into an excited nuclear state of ^{60}Ni. That nickel isotope then immediately emits two gamma-ray photons, each with an approximate energy of 1.2 MeV. How many radioactive ^{60}Co nuclei are present in a 6000 Ci source of the type used in hospitals? (Energetic particles from linear accelerators are now used in radiation therapy.) ssm

34P. The radionuclide ^{64}Cu has a half-life of 12.7 h. If a sample contains 5.50 g of initially pure ^{64}Cu at $t = 0$, how much of it will decay between $t = 14.0$ h and $t = 16.0$ h?

35P. After long effort, in 1902 Marie and Pierre Curie succeeded in separating from uranium ore the first substantial quantity of radium, one decigram of pure $RaCl_2$. The radium was the radioactive isotope ^{226}Ra, which has a half-life of 1600 y. (a) How many radium nuclei had the Curies isolated? (b) What was the decay rate of their sample, in disintegrations per second? ssm www

36P. The radionuclide ^{32}P ($T_{1/2} = 14.28$ d) is often used as a tracer to follow the course of biochemical reactions involving phosphorus. (a) If the counting rate in a particular experimental setup is initially 3050 counts/s, how much time will the rate take to fall to 170 counts/s? (b) A solution containing ^{32}P is fed to the root system of an experimental tomato plant and the ^{32}P activity in a leaf is measured 3.48 days later. By what factor must this reading be multiplied to correct for the decay that has occurred since the experiment began?

37P. A source contains two phosphorus radionuclides, ^{32}P ($T_{1/2} =$ 14.3 d) and ^{33}P ($T_{1/2} = 25.3$ d). Initially, 10.0% of the decays come from ^{33}P. How long must one wait until 90.0% do so?

38P. Plutonium isotope ^{239}Pu decays by alpha decay with a half-life of 24 100 y. How many milligrams of helium are produced by an initially pure 12.0 g sample of ^{239}Pu at the end of 20 000 y? (Consider only the helium produced directly by the plutonium and not by any by-products of the decay process.)

39P. A 1.00 g sample of samarium emits alpha particles at a rate of 120 particles/s. The responsible isotope is ^{147}Sm, whose natural abundance in bulk samarium is 15.0%. Calculate the half-life for the decay process. ssm

40P. After a brief neutron irradiation of silver, two isotopes are present: ^{108}Ag ($T_{1/2} = 2.42$ min) with an initial decay rate of 3.1×10^5/s, and ^{110}Ag ($T_{1/2} = 24.6$ s) with an initial decay rate of 4.1×10^6/s. Make a semilog plot similar to Fig. 43-8 showing the total combined decay rate of the two isotopes as a function of time from $t = 0$ until $t = 10$ min. We used Fig. 43-8 to illustrate the extraction of the half-life for simple (one isotope) decays. Given only your plot of total decay rate for the two-isotope system here, suggest a way to analyze it in order to find the half-lives of both isotopes.

41P. A certain radionuclide is being manufactured, say, in a cyclotron, at a constant rate R. It is also decaying, with disintegration constant λ. Assume that the production process has been going on for a time that is long compared to the half-life of the radionuclide. Show that the number of radioactive nuclei present after such time remains constant and is given by $N = R/\lambda$. Now show that this result holds no matter how many radioactive nuclei were present initially. The nuclide is said to be in *secular equilibrium* with its source; in this state its decay rate is just equal to its production rate. ssm www

42P. Calculate the mass of a sample of (initially pure) ^{40}K with an initial decay rate of 1.70×10^5 disintegrations/s. The isotope has a half-life of 1.28×10^9 y.

43P. (See Problem 41.) The radionuclide ^{56}Mn has a half-life of 2.58 h and is produced in a cyclotron by bombarding a manganese target with deuterons. The target contains only the stable manganese isotope ^{55}Mn, and the manganese–deuteron reaction that produces ^{56}Mn is

$$^{55}\text{Mn} + \text{d} \rightarrow {}^{56}\text{Mn} + \text{p}.$$

After bombardment for a time much longer than 2.58 h, the activity of the target, due to ^{56}Mn, is 8.88×10^{10} s^{-1}. (a) At what constant rate R are ^{56}Mn nuclei being produced in the cyclotron during the bombardment? (b) At what rate are they decaying (also during the bombardment)? (c) How many ^{56}Mn nuclei are present at the end of the bombardment? (d) What is their total mass? ssm

44P. (See Problems 41 and 43.) A radium source contains 1.00 mg of ^{226}Ra, which decays with a half-life of 1600 y to produce ^{222}Rn, a noble gas. This radon isotope in turn decays by alpha emission with a half-life of 3.82 d. (a) What is the rate of disintegration of ^{226}Ra in the source? (b) How long does it take for the radon to come to secular equilibrium with its radium parent? (c) At what rate is the radon then decaying? (d) How much radon is in equilibrium with its radium parent?

45P. One of the dangers of radioactive fallout from a nuclear bomb is its ^{90}Sr, which decays with a 29 year half-life. Because it has chemical properties much like those of calcium, the strontium, if ingested by a cow, becomes concentrated in the cow's milk. Some of the ^{90}Sr ends up in the bones of whoever drinks the milk. The energetic electrons emitted in the beta decay of ^{90}Sr damage the bone marrow and thus impair the production of red blood cells. A 1 megaton bomb produces approximately 400 g of ^{90}Sr. If the fallout spreads uniformly over a 2000 km^2 area, what ground area would hold an amount of radioactivity equal to the "allowed" limit for one person, which is 74 000 counts/s?

SEC. 43-4 Alpha Decay

46E. Consider a ^{238}U nucleus to be made up of an alpha particle (^{4}He) and a residual nucleus (^{234}Th). Plot the electrostatic potential energy $U(r)$, where r is the distance between these particles. Cover the approximate range $10 \text{ fm} < r < 100 \text{ fm}$ and compare your plot with that of Fig. 43-9.

47E. Generally, more massive nuclides tend to be more unstable to alpha decay. For example, the most stable isotope of uranium, ^{238}U, has an alpha decay half-life of 4.5×10^9 y. The most stable isotope of plutonium is ^{244}Pu with an 8.0×10^7 y half-life, and for curium we have ^{248}Cm and 3.4×10^5 y. When half of an original sample of ^{238}U has decayed, what fractions of the original samples of these isotopes of plutonium and curium are left? ssm

48P. Consider that a ^{238}U nucleus emits (a) an alpha particle or (b) a sequence of neutron, proton, neutron, proton. Calculate the energy released in each case. (c) Convince yourself both by reasoned argument and by direct calculation that the difference between these two numbers is just the total binding energy of the alpha particle. Find that binding energy. Some needed atomic and particle masses are

^{238}U	238.050 79 u	^{234}Th	234.043 63 u
^{237}U	237.048 73 u	^{4}He	4.002 60 u
^{236}Pa	236.048 91 u	^{1}H	1.007 83 u
^{235}Pa	235.045 44 u	n	1.008 67 u

49P. A ^{238}U nucleus emits a 4.196 MeV alpha particle. Calculate the disintegration energy Q for this process, taking the recoil energy of the residual ^{234}Th nucleus into account. ssm

50P. Large radionuclides emit an alpha particle rather than other combinations of nucleons because the alpha particle has such a stable, tightly bound structure. To confirm this statement, calculate the disintegration energies for these hypothetical decay processes and discuss the meaning of your findings:

(a) $^{235}U \rightarrow {}^{232}Th + {}^{3}He$,

(b) $^{235}U \rightarrow {}^{231}Th + {}^{4}He$,

(c) $^{235}U \rightarrow {}^{230}Th + {}^{5}He$.

The needed atomic masses are

^{232}Th	232.0381 u	^{3}He	3.0160 u
^{231}Th	231.0363 u	^{4}He	4.0026 u
^{230}Th	230.0331 u	^{5}He	5.0122 u
^{235}U	235.0439 u		

51P. Under certain rare circumstances, a nucleus can decay by emitting a particle more massive than an alpha particle. Consider the decays

$$^{223}Ra \rightarrow {}^{209}Pb + {}^{14}C$$

and

$$^{223}Ra \rightarrow {}^{219}Rn + {}^{4}He.$$

(a) Calculate the Q values for these decays and determine that both are energetically possible. (b) The Coulomb barrier height for alpha-particle emission is 30.0 MeV. What is the barrier height for ^{14}C emission? The needed atomic masses are

^{223}Ra	223.018 50 u	^{14}C	14.003 24 u
^{209}Pb	208.981 07 u	^{4}He	4.002 60 u
^{219}Rn	219.009 48 u		

SEC. 43-5 Beta Decay

52E. Large-mass radionuclides, which may be either alpha or beta emitters, belong to one of four decay chains, depending on whether their mass numbers A are of the form $4n$, $4n + 1$, $4n + 2$, or $4n + 3$, where n is a positive integer. (a) Justify this statement and show that if a nuclide belongs to one of these families, all its decay products belong to the same family. (b) Classify these nuclides as to family: ^{235}U, ^{236}U, ^{238}U, ^{239}Pu, ^{240}Pu, ^{245}Cm, ^{246}Cm, ^{249}Cf, and ^{253}Fm.

53E. A certain stable nuclide, after absorbing a neutron, emits an electron, and the new nuclide splits spontaneously into two alpha particles. Identify the nuclide. ssm

54E. An electron is emitted from a middle-mass nuclide (A = 150, say) with a kinetic energy of 1.0 MeV. (a) What is its de Broglie wavelength? (b) Calculate the radius of the emitting nucleus. (c) Can such an electron be confined as a standing wave in a "box" of such dimensions? (d) Can you use these numbers to disprove the (abandoned) argument that electrons actually exist in nuclei?

55E. The cesium isotope ^{137}Cs is present in the fallout from aboveground detonations of nuclear bombs. Because it decays with a slow (30.2 y) half-life into ^{137}Ba, releasing considerable energy in the process, it is of environmental concern. The atomic masses of the Cs and Ba are 136.9071 and 136.9058 u, respectively; calculate the total energy released in such a decay. ssm

56P. Some radionuclides decay by capturing one of their own atomic electrons, a K-shell electron, say. An example is

$$^{49}V + e^- \rightarrow {}^{49}Ti + \nu, \qquad T_{1/2} = 331 \text{ d}.$$

Show that the disintegration energy Q for this process is given by

$$Q = (m_V - m_{Ti})c^2 - E_K,$$

where m_V and m_{Ti} are the atomic masses of ^{49}V and ^{49}Ti, respectively, and E_K is the binding energy of the vanadium K-shell electron. (*Hint:* Put $\mathbf{m}_V$ and $\mathbf{m}_{Ti}$ as the corresponding nuclear masses and proceed as in Sample Problem 43-7.)

57P. A free neutron decays according to Eq. 43-26. If the neutron–hydrogen atom mass difference is 840 μu, what is the maximum kinetic energy K_{max} of the electron energy spectrum? ssm

58P. Find the disintegration energy Q for the decay of ^{49}V by K-electron capture, as described in Problem 56. The needed data are $m_V = 48.948\,52$ u, $m_{Ti} = 48.947\,87$ u, and $E_K = 5.47$ keV.

59P. The radionuclide ^{11}C decays according to

$$^{11}C \rightarrow {}^{11}B + e^+ + \nu, \qquad T_{1/2} = 20.3 \text{ min}.$$

The maximum energy of the emitted positrons is 0.960 MeV.

(a) Show that the disintegration energy Q for this process is given by

$$Q = (m_C - m_B - 2m_e)c^2,$$

where m_C and m_B are the atomic masses of ^{11}C and ^{11}B, respectively, and m_e is the mass of a positron. (b) Given the mass values $m_C = 11.011\ 434$ u, $m_B = 11.009\ 305$ u, and $m_e = 0.000\ 548\ 6$ u, calculate Q and compare it with the maximum energy of the emitted positron given above. (*Hint:* Let $\mathbf{m}_C$ and $\mathbf{m}_B$ be the nuclear masses and proceed as in Sample Problem 43-7 for beta decay. Note that beta-plus decay is an exception to the general rule that if atomic masses are used in nuclear decay calculations, the mass of the emitted electron is automatically taken care of.)

60P. Two radioactive materials that are unstable with regard to alpha decay, ^{238}U and ^{232}Th, and one that is unstable with regard to beta decay, ^{40}K, are sufficiently abundant in granite to contribute significantly to the heating of Earth through the decay energy produced. The alpha-unstable isotopes give rise to decay chains that stop when stable lead isotopes are formed. The isotope ^{40}K has a single beta decay. Here is the information:

Parent	Decay Mode	Half-Life (y)	Stable End Point	Q (MeV)	f (ppm)
^{238}U	α	4.47×10^9	^{206}Pb	51.7	4
^{232}Th	α	1.41×10^{10}	^{208}Pb	42.7	13
^{40}K	β	1.28×10^9	^{40}Ca	1.31	4

In the table Q is the *total* energy released in the decay of one parent nucleus to the *final* stable end point and f is the abundance of the isotope in kilograms per kilogram of granite; ppm means parts per million. (a) Show that these materials produce energy as heat at the rate of 1.0×10^{-9} W for each kilogram of granite. (b) Assuming that there is 2.7×10^{22} kg of granite in a 20-km-thick spherical shell at the surface of Earth, estimate the power of this decay process over all of Earth. Compare this power with the total solar power intercepted by Earth, 1.7×10^{17} W.

61P*. The radionuclide ^{32}P decays to ^{32}S as described by Eq. 43-23. In a particular decay event, a 1.71 MeV electron is emitted, the maximum possible value. What is the kinetic energy of the recoiling ^{32}S atom in this event? (*Hint:* For the electron it is necessary to use the relativistic expressions for kinetic energy and linear momentum. Newtonian mechanics may safely be used for the relatively slow-moving ^{32}S atom.) ssm www

SEC. 43-6 Radioactive Dating

62E. A 5.00 g charcoal sample from an ancient fire pit has a ^{14}C activity of 63.0 disintegrations/min. A living tree has a ^{14}C activity of 15.3 disintegrations/min per 1.00 g. The half-life of ^{14}C is 5730 y. How old is the charcoal sample?

63E. ^{238}U decays to ^{206}Pb with a half-life of 4.47×10^9 y. Although the decay occurs in many individual steps, the first step has by far the longest half-life; therefore, one can often consider the decay to go directly to lead. That is,

$$^{238}U \rightarrow {}^{206}Pb + \text{various decay products.}$$

A rock is found to contain 4.20 mg of ^{238}U and 2.135 mg of ^{206}Pb. Assume that the rock contained no lead at formation, so all the lead now present arose from the decay of uranium. (a) How many atoms of ^{238}U and ^{206}Pb does the rock now contain? (b) How many atoms of ^{238}U did the rock contain at formation? (c) What is the age of the rock? ssm

64P. A particular rock is thought to be 260 million years old. If it contains 3.70 mg of ^{238}U, how much ^{206}Pb should it contain? See Exercise 63.

65P. A rock, recovered from far underground, is found to contain 0.86 mg of ^{238}U, 0.15 mg of ^{206}Pb, and 1.6 mg of ^{40}Ar. How much ^{40}K will it likely contain? Needed half-lives are listed in Problem 60.

SEC. 43-7 Measuring Radiation Dosage

66E. A Geiger counter records 8700 counts in 1 min. Calculate the activity of the source in becquerels and in curies, assuming that the counter records all decays.

67E. The nuclide ^{198}Au, with a half-life of 2.70 d, is used in cancer therapy. What mass of this nuclide is required to produce an activity of 250 Ci? ssm

68E. An airline pilot spends an average of 20 h per week flying at an altitude of 10 km, at which the dose equivalent due to cosmic radiation is 7.0 μSv/h. What is the annual (52 week) dose equivalent from this source alone? Note that the maximum permitted yearly dose equivalent (from all sources) for the general population is 5 mSv, and for radiation workers it is 50 mSv.

69P. A typical chest x-ray radiation dose is 250 μSv, delivered by x rays with an RBE factor of 0.85. Assuming that the mass of the exposed tissue is one-half the patient's mass of 88 kg, calculate the energy absorbed in joules. ssm

70P. A 75 kg person receives a whole-body radiation dose of 2.4×10^{-4} Gy, delivered by alpha particles for which the RBE factor is 12. Calculate (a) the absorbed energy in joules and (b) the dose equivalent in sieverts and in rem.

71P. An 85 kg worker at a breeder reactor plant accidentally ingests 2.5 mg of ^{239}Pu dust. ^{239}Pu has a half-life of 24 100 y, decaying by alpha decay. The energy of the emitted alpha particles is 5.2 MeV, with an RBE factor of 13. Assume that the plutonium resides in the worker's body for 12 h and that 95% of the emitted alpha particles are stopped within the body. Calculate (a) the number of plutonium atoms ingested, (b) the number that decay during the 12 h, (c) the energy absorbed by the body, (d) the resulting physical dose in grays, and (e) the dose equivalent in sieverts.

SEC. 43-8 Nuclear Models

72E. A typical kinetic energy for a nucleon in a middle-mass nucleus may be taken as 5.00 MeV. To what effective nuclear temperature does this correspond, based on the assumptions of the collective model of nuclear structure?

73E. An intermediate nucleus in a particular nuclear reaction decays within 10^{-22} s of its formation. (a) What is the uncertainty ΔE in our knowledge of this intermediate state? (b) Can this state be called a compound nucleus? (See Sample Problem 43-10.)

74E. In the following list of nuclides, identify (a) those with filled nucleon shells, (b) those with one nucleon outside a filled shell, and (c) those with one vacancy in an otherwise filled shell: ^{13}C, ^{18}O, ^{40}K, ^{49}Ti, ^{60}Ni, ^{91}Zr, ^{92}Mo, ^{121}Sb, ^{143}Nd, ^{144}Sm, ^{205}Tl, and ^{207}Pb.

75P. Consider the three formation processes shown for the compound nucleus ^{20}Ne in Fig. 43-13. Here are some of the masses:

^{20}Ne	19.992 44 u	α	4.002 60 u
^{19}F	18.998 40 u	p	1.007 83 u
^{16}O	15.994 91 u		

What energy must (a) the alpha particle, (b) the proton, and (c) the γ-ray photon have to provide 25.0 MeV of excitation energy to the compound nucleus? ssm www

Additional Problems

76. At the end of World War II, Dutch authorities arrested Dutch artist Hans van Meegeren for treason because, during the war, he sold a masterpiece painting to the infamous Nazi Hermann Goering. The painting, *Christ and His Disciples at Emmaus* by Dutch master Johannes Vermeer (1632–1675), had been discovered in 1937 by van Meegeren, after it had been lost for almost 300 years. Soon after the discovery, art experts proclaimed that *Emmaus* was possibly the best Vermeer ever seen. Selling such a Dutch national treasure to the enemy was unthinkable treason.

However, shortly after being imprisoned, van Meegeren suddenly announced that he, not Vermeer, had painted *Emmaus*. He explained that he had carefully mimicked Vermeer's style, using a 300-year-old canvas and Vermeer's choice of pigments; he had then signed Vermeer's name to the work and baked the painting to give it an authentically old look.

Was van Meegeren lying to avoid a conviction of treason, hoping to be convicted of only the lesser crime of fraud? To art experts, *Emmaus* certainly looked like a Vermeer but, at the time of van Meegeren's trial in 1947, there was no scientific way to answer the question. However, in 1968 Bernard Keisch of Carnegie-Mellon University was able to answer the question with newly developed techniques of radioactive analysis.

Specifically, he analyzed a small sample of white lead-bearing pigment removed from *Emmaus*. This pigment is refined from lead ore, in which the lead is produced by a long radioactive decay series that starts with unstable ^{238}U and ends with stable ^{206}Pb. To follow the spirit of Keisch's analysis, focus on the following abbreviated portion of that decay series, in which intermediate, relatively short-lived radionuclides have been omitted:

$$^{230}\text{Th} \xrightarrow[75.4\text{ ky}]{} {}^{226}\text{Ra} \xrightarrow[1.60\text{ ky}]{} {}^{210}\text{Pb} \xrightarrow[22.6\text{ y}]{} {}^{206}\text{Pb}.$$

The longer and more important half-lives in this portion of the decay series are indicated.

(a) Show that in a unit sample of lead ore, the rate at which the number of ^{210}Pb nuclei changes is given by

$$\frac{dN_{210}}{dt} = \lambda_{226}N_{226} - \lambda_{210}N_{210},$$

where N_{210} and N_{226} are the numbers of ^{210}Pb nuclei and ^{226}Ra nuclei in the unit sample and λ_{210} and λ_{226} are the corresponding disintegration constants.

Because the decay series has been active for billions of years and because the half-life of ^{210}Pb is much less than that of ^{226}Ra, the nuclides ^{226}Ra and ^{210}Pb are in *equilibrium;* that is, their numbers, or concentrations, in the sample do not change. (b) What is the ratio R_{226}/R_{210} of the activities of these nuclides in the unit sample of lead ore? (c) What is the ratio N_{226}/N_{210} of their numbers?

When lead pigment is refined from the ore, most of the ^{226}Ra is eliminated. Assume that only 1.00% remains. Just after the pigment is produced, what are the ratios (d) R_{226}/R_{210} and (e) N_{226}/N_{210}?

Keisch realized that with time the ratio R_{226}/R_{210} of the pigment would gradually change from the value of freshly refined pigment back to the value of the ore as equilibrium between the ^{210}Pb and the remaining ^{226}Ra is established in the pigment. If *Emmaus* were painted by Vermeer and the sample of pigment taken from it were 300 years old when examined in 1968, the ratio would be close to the answer of (b). If *Emmaus* were painted by van Meegeren in the 1930s and the sample were only about 30 years old, the ratio would be close to the answer of (d). Keisch found a ratio of 0.09. (f) Is *Emmaus* a Vermeer?

77. When aboveground nuclear tests were conducted, the explosions shot radioactive dust into the upper atmosphere. Global air circulations then spread the dust worldwide before it settled out on ground and water. One such test was conducted in October 1976. What fraction of the ^{90}Sr produced by that explosion will still exist in October 2006? The half-life of ^{90}Sr is 29 y.

78. A radioactive sample intended for irradiation of a hospital patient is prepared at a nearby laboratory. The sample has a half-life of 83.61 h. What should its initial activity be if its activity is to be 7.4×10^8 Bq when it is used to irradiate the patient 24 h later?

79. The radioactive nuclide ^{99}Tc can be injected into a patient's blood system in order to monitor the blood flow, to measure the blood volume, or to find a tumor, among other goals. The nuclide is produced in a hospital by a "cow" containing ^{99}Mo, a radioactive nuclide that decays to ^{99}Tc with a half-life of 67 h. Once a day, the cow is "milked" for its ^{99}Tc, which is produced in an excited state by the ^{99}Mo; the ^{99}Tc de-excites to its lowest energy state by emitting a gamma-ray photon, which is recorded by detectors placed around the patient. The de-excitation has a half-life of 6.0 h. (a) By what process does ^{99}Mo decay to ^{99}Tc? (b) If a patient is injected with a 8.2×10^7 Bq sample of ^{99}Tc, how many gamma-ray photons are initially produced within the patient each second? (c) If the emission rate of gamma-ray photons from a small tumor that has collected ^{99}Tc is 38 per second at a certain time, how many excited state ^{99}Tc are located in the tumor at that time?

80. Because of the 1986 explosion and fire in a reactor at the Chernobyl nuclear power plant in northern Ukraine, part of the Ukraine

is contaminated with ^{137}Ce, which undergoes beta-minus decay with a half-life of 30.2 y. In 1996, the total activity of this contamination over an area of 2.6×10^5 km^2 was estimated to be 1×10^{16} Bq. Assume that the ^{137}Ce is uniformly spread over that area and that the beta-decay electrons travel either directly upward or directly downward. How many beta-decay electrons would you intercept were you to lie on the ground in that area for 1 h (a) in 1996 and (b) today? (You need to estimate your cross-sectional area that intercepts those electrons.)

81. In October 1992, Swiss police arrested two men who were attempting to smuggle osmium out of Eastern Europe for a clandestine sale. However, by error, the smugglers had picked up ^{137}Ce. Reportedly, each smuggler was carrying a 1.0 g sample of ^{137}Ce in a pocket. In bequerels and curies, what was the activity of each sample? ^{137}Ce has a half-life of 30.2 y. (The activities of radioisotopes commonly used in hospitals range up to a few millicuries.)

82. Figure 43-16 shows part of the decay scheme of ^{237}Np on a plot of mass number A versus proton number Z; five lines that represent either alpha decay or beta-minus decay connect dots that represent isotopes. What is the isotope at the end of the five decays (as marked with a question mark in Fig. 43-16)?

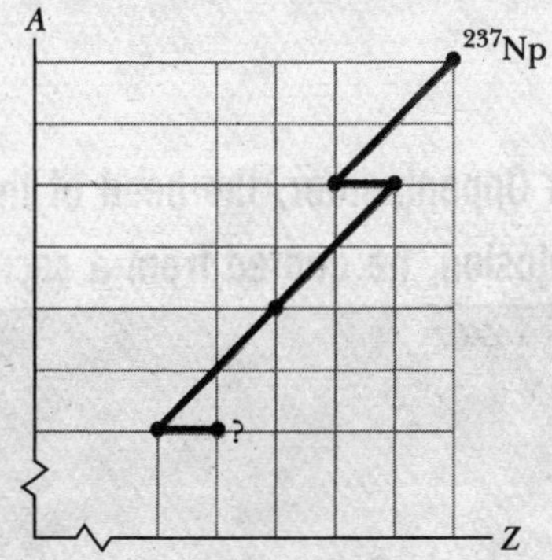

Fig. 43-16 Problem 82.

83. The isotope ^{40}K, with a half-life of 1.26×10^9 y, can decay to either ^{40}Ca or ^{40}Ar. The ratio of the Ca produced to the Ar produced is $8.54/1 = 8.54$. A sample originally had only ^{40}K. It now has equal amounts of ^{40}K and ^{40}Ar; that is, the ratio of K to Ar is $1/1 = 1$. How old is the sample? (*Hint:* Work this like other radioactive-dating problems, except that this decay has two products instead of just one.)

84. The air in some caves has a significant amount of radon gas, which can lead to lung cancer if breathed over a prolonged time. In British caves, the air in the cave with the greatest amount of the gas has an activity per volume of 1.55×10^5 Bq/m^3. Suppose that you spend two full days exploring (and sleeping in) that cave. Approximately how many ^{222}Rn atoms would you take in and out of your lungs during your two-day stay? The radionuclide ^{222}Rn in radon gas has a half-life of 3.82 days. You need to estimate your lung capacity and average breathing rate.

85. A ^{7}Li nucleus with a kinetic energy of 3.00 MeV is sent toward a ^{232}Th nucleus. What is the least center-to-center separation between the two nuclei, assuming that the (more massive) ^{232}Th nucleus does not move?

86. What is the binding energy per nucleon of ^{262}Bh? The mass of the atom is 262.1231 u.

87. How many years are needed to reduce the activity of ^{14}C to 0.020 of its original activity? The half-life of ^{14}C is 5730 y.

88. What is the activity of a 20 ng sample of ^{92}Kr, which has a half-life of 1.84 s?

44 Energy from the Nucleus

This image has transfixed the world since World War II. When Robert Oppenheimer, the head of the scientific team that developed the atomic bomb, witnessed the first atomic explosion, he quoted from a sacred Hindu text: "Now I am become Death, the destroyer of worlds."

What is the physics behind this image that has so horrified the world?

The answer is in this chapter.

44-1 The Atom and Its Nucleus

When we get energy from coal by burning the fuel in a furnace, we are tinkering with atoms of carbon and oxygen, rearranging their outer *electrons* into more stable combinations. When we get energy from uranium in a nuclear reactor, we are again burning a fuel, but then we are tinkering with its nucleus, rearranging its *nucleons* into more stable combinations.

Electrons are held in atoms by the electromagnetic Coulomb force, and it takes only a few electron-volts to pull one of them out. On the other hand, nucleons are held in nuclei by the strong force, and it takes a few *million* electron-volts to pull one of *them* out. This factor of a few million is reflected in the fact that we can extract about that much more energy from a kilogram of uranium than we can from a kilogram of coal.

In both atomic and nuclear burning, the release of energy is accompanied by a decrease in mass, according to the equation $Q = -\Delta m c^2$. The central difference between burning uranium and burning coal is that, in the former case, a much larger fraction of the available mass (again, by a factor of a few million) is consumed.

The different processes that can be used for atomic or nuclear burning provide different levels of power, or rates at which the energy is delivered. In the nuclear case, we can burn a kilogram of uranium explosively in a bomb or slowly in a power reactor. In the atomic case, we might consider exploding a stick of dynamite or digesting a jelly doughnut.

Table 44-1 shows how much energy can be extracted from 1 kg of matter by doing various things to it. Instead of reporting the energy directly, the table shows how long the extracted energy could operate a 100 W lightbulb. Only processes in the first three rows of the table have actually been carried out; the remaining three represent theoretical limits that may not be attainable in practice. The bottom row, the total mutual annihilation of matter and antimatter, is an ultimate energy production goal. In that process, *all* the mass energy is transferred to other forms of energy.

Keep in mind that the comparisons of Table 44-1 are computed on a per-unit-mass basis. Kilogram for kilogram, you get several million times more energy from uranium than you do from coal or from falling water. On the other hand, there is a lot of coal in Earth's crust, and water is easily backed up behind a dam.

44-2 Nuclear Fission: The Basic Process

In 1932 English physicist James Chadwick discovered the neutron. A few years later Enrico Fermi in Rome found that when various elements are bombarded by neutrons, new radioactive elements are produced. Fermi had predicted that the neutron, being uncharged, would be a useful nuclear projectile; unlike the proton or the alpha par-

TABLE 44-1 Energy Released by 1 kg of Matter

Form of Matter	Process	Time[a]
Water	A 50 m waterfall	5 s
Coal	Burning	8 h
Enriched UO_2	Fission in a reactor	690 y
^{235}U	Complete fission	3×10^4 y
Hot deuterium gas	Complete fusion	3×10^4 y
Matter and antimatter	Complete annihilation	3×10^7 y

[a]This column shows the time for which the generated energy could power a 100 W lightbulb.

ticle, it experiences no repulsive Coulomb force when it nears a nuclear surface. Even *thermal neutrons,* which are slowly moving neutrons in thermal equilibrium with the surrounding matter at room temperature, with a mean kinetic energy of only about 0.04 eV, are useful projectiles in nuclear studies.

In the late 1930s physicist Lise Meitner and chemists Otto Hahn and Fritz Strassmann, working in Berlin and following up the work of Fermi and co-workers, bombarded solutions of uranium salts with such thermal neutrons. They found that after the bombardment a number of new radionuclides were present. In 1939 one of the radionuclides produced in this way was positively identified, by repeated tests, as barium. But how, Hahn and Strassmann wondered, could this middle-mass element ($Z = 56$) be produced by bombarding uranium ($Z = 92$) with neutrons?

The puzzle was solved within a few weeks by Meitner and her nephew Otto Frisch. They suggested the mechanism by which a uranium nucleus, having absorbed a thermal neutron, could split, with the release of energy, into two roughly equal parts, one of which might well be barium. Frisch named the process **fission.**

Meitner's central role in the discovery of fission was not fully recognized until recent historical research brought it to light. She did not share in the Nobel prize in chemistry that was awarded to Otto Hahn in 1944. However, both Hahn and Meitner have been honored by having elements named after them: hahnium (symbol Ha, $Z = 105$) and meitnerium (symbol Mt, $Z = 109$).

A Closer Look at Fission

Figure 44-1 shows the distribution by mass number of the fragments produced when ^{235}U is bombarded with thermal neutrons. The most probable mass numbers, occurring in about 7% of the events, are centered around $A \approx 95$ and $A \approx 140$. Curiously, the "double-peaked" character of Fig. 44-1 is still not understood.

In a typical ^{235}U fission event, a ^{235}U nucleus absorbs a thermal neutron, producing a compound nucleus ^{236}U in a highly excited state. It is *this* nucleus that actually undergoes fission, splitting into two fragments. These fragments—between them—rapidly emit two neutrons, leaving (in a typical case) ^{140}Xe ($Z = 54$) and ^{94}Sr ($Z = 38$) as fission fragments. Thus, the overall fission equation for this event is

$$^{235}U + n \rightarrow {}^{236}U \rightarrow {}^{140}Xe + {}^{94}Sr + 2n. \qquad (44\text{-}1)$$

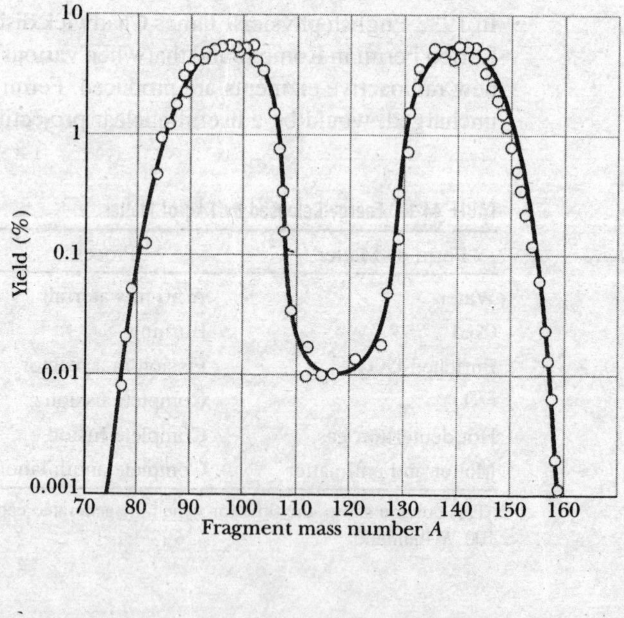

Fig. 44-1 The distribution by mass number of the fragments that are found when many fission events of ^{235}U are examined. Note that the vertical scale is logarithmic.

Note that during the formation and fission of the compound nucleus, there is conservation of the number of protons and of the number of neutrons involved in the process (and thus their total number and the net charge).

In Eq. 44-1, the fragments ^{140}Xe and ^{94}Sr are both highly unstable, undergoing beta decay (with the conversion of a neutron to a proton and the emission of an electron and a neutrino) until each reaches a stable end product. For xenon, the decay chain is

	$^{140}Xe \rightarrow$	$^{140}Cs \rightarrow$	$^{140}Ba \rightarrow$	$^{140}La \rightarrow$	^{140}Ce
$T_{1/2}$	14 s	64 s	13 d	40 h	Stable
Z	54	55	56	57	58

(44-2)

For strontium, it is

	$^{94}Sr \rightarrow$	$^{94}Y \rightarrow$	^{94}Zr
$T_{1/2}$	75 s	19 min	Stable
Z	38	39	40

(44-3)

As we should expect from Section 43-5, the mass numbers (140 and 94) of the fragments remain unchanged during these beta-decay processes, and the atomic numbers (initially 54 and 38) increase by unity at each step.

Inspection of the stability band on the nuclidic chart of Fig. 43-4 shows why the fission fragments are unstable. The nuclide ^{236}U, which is the fissioning nucleus in the reaction of Eq. 44-1, has 92 protons and 236 − 92, or 144, neutrons, for a neutron/proton ratio of about 1.6. The primary fragments formed immediately after the fission reaction have about this same neutron/proton ratio. However, stable nuclides in the middle-mass region have smaller neutron/proton ratios, in the range of 1.3 to 1.4. The primary fragments are thus *neutron rich* (they have too many neutrons) and will eject a few neutrons, two in the case of the reaction of Eq. 44-1. The fragments that remain are still too neutron rich to be stable. Beta decay offers a mechanism for getting rid of the excess neutrons—namely, by changing them into protons within the nucleus.

We can estimate the energy released by the fission of a high-mass nuclide by examining the total binding energy per nucleon ΔE_{ben} before and after the fission. The idea is that fission can occur because the total mass energy will decrease; that is, ΔE_{ben} will *increase* so that the products of the fission are *more* tightly bound. Thus, the energy Q released by the fission is

$$Q = \begin{pmatrix}\text{total final}\\ \text{binding energy}\end{pmatrix} - \begin{pmatrix}\text{initial}\\ \text{binding energy}\end{pmatrix}. \qquad (44\text{-}4)$$

For our estimate, let us assume that fission transforms an initial high-mass nucleus to two middle-mass nuclei with the same number of nucleons. Then we have

$$Q = \begin{pmatrix}\text{final}\\ \Delta E_{\text{ben}}\end{pmatrix}\begin{pmatrix}\text{final number}\\ \text{of nucleons}\end{pmatrix} - \begin{pmatrix}\text{initial}\\ \Delta E_{\text{ben}}\end{pmatrix}\begin{pmatrix}\text{initial number}\\ \text{of nucleons}\end{pmatrix}. \qquad (44\text{-}5)$$

From Fig. 43-6, we see that for a high-mass nuclide ($A \approx 240$), the binding energy per nucleon is about 7.6 MeV/nucleon. For middle-mass nuclides ($A \approx 120$), it is about 8.5 MeV/nucleon. Thus, the energy released by fission of a high-mass nuclide to two middle-mass nuclides is

$$Q = \left(8.5\,\frac{\text{MeV}}{\text{nucleon}}\right)(2\text{ nuclei})\left(120\,\frac{\text{nucleons}}{\text{nucleus}}\right) - \left(7.6\,\frac{\text{MeV}}{\text{nucleon}}\right)(240\text{ nucleons}) \approx 200\text{ MeV}. \qquad (44\text{-}6)$$

CHECKPOINT 1: A generic fission event is

$$^{235}\text{U} + \text{n} \rightarrow X + Y + 2\text{n}.$$

Which of the following pairs *cannot* represent X and Y: (a) ^{141}Xe and ^{93}Sr; (b) ^{139}Cs and ^{95}Rb; (c) ^{156}Nd and ^{79}Ge; (d) ^{121}In and ^{113}Ru?

Sample Problem 44-1

Find the disintegration energy Q for the fission event of Eq. 44-1, taking into account the decay of the fission fragments as displayed in Eqs. 44-2 and 44-3. Some needed atomic and particle masses are

^{235}U	235.0439 u	^{140}Ce	139.9054 u
n	1.008 67 u	^{94}Zr	93.9063 u

SOLUTION: The **Key Ideas** here are (1) that the disintegration energy Q is the energy transferred from mass energy to kinetic energy of the decay products, and (2) that $Q = -\Delta m\, c^2$, where Δm is the change in mass. Because we are to include the decay of the fission fragments, we combine Eqs. 44-1, 44-2, and 44-3 to write the overall transformation as

$$^{235}\text{U} \rightarrow {}^{140}\text{Ce} + {}^{94}\text{Zr} + \text{n}. \qquad (44\text{-}7)$$

Only the single neutron appears here because the initiating neutron on the left side of Eq. 44-1 cancels one of the two neutrons on the right of that equation. The mass difference for the reaction of Eq. 44-7 is

$$\Delta m = (139.9054\text{ u} + 93.9063\text{ u} + 1.008\,67\text{ u}) - (235.0439\text{ u}) = -0.223\,53\text{ u},$$

and the corresponding disintegration energy is

$$Q = -\Delta m\, c^2 = (-0.223\,53\text{ u})(931.5\text{ MeV/u}) = 208\text{ MeV}, \qquad \text{(Answer)}$$

which is in good agreement with our estimate of Eq. 44-6.

If the fission event takes place in a bulk solid, most of this disintegration energy, which first goes into kinetic energy of the decay products, appears eventually as an increase in the internal energy of that body, revealing itself as a rise in temperature. Five or six percent or so of the disintegration energy, however, is associated with neutrinos that are emitted during the beta decay of the primary fission fragments. This energy is carried out of the system and is lost.

44-3 A Model for Nuclear Fission

Soon after the discovery of fission, Niels Bohr and John Wheeler used the collective model of the nucleus (Section 43-8), based on the analogy between a nucleus and a charged liquid drop, to explain its main features. Figure 44-2 suggests how the fission

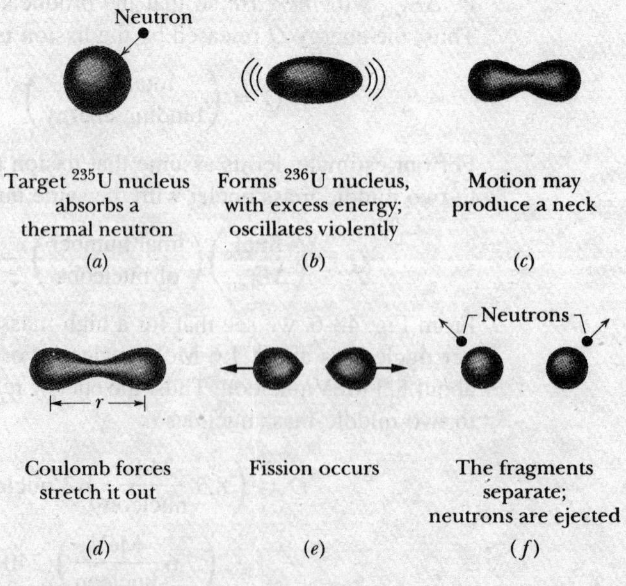

Fig. 44-2 The stages of a typical fission process, according to the collective model of Bohr and Wheeler.

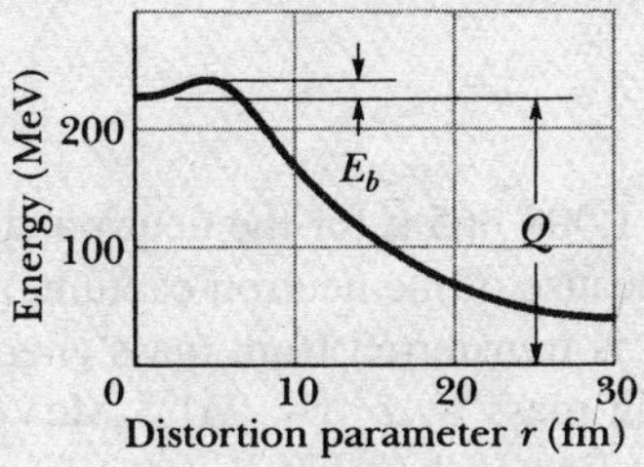

Fig. 44-3 The potential energy at various stages in the fission process, as predicted from the collective model of Bohr and Wheeler. The Q of the reaction (about 200 MeV) and the fission barrier height E_b are both indicated.

process proceeds from this point of view. When a high-mass nucleus—let us say ^{235}U—absorbs a slow (thermal) neutron, as in Fig. 44-2*a*, that neutron falls into the potential well associated with the strong forces that act in the nuclear interior. The neutron's potential energy is then transformed into internal excitation energy of the nucleus, as Fig. 44-2*b* suggests. The amount of excitation energy that a slow neutron carries into a nucleus is equal to the binding energy E_n of the neutron in that nucleus, which is the change in mass energy of the neutron–nucleus system due to the neutron's capture.

Figures 44-2*c* and *d* show that the nucleus, behaving like an energetically oscillating charged liquid drop, will sooner or later develop a short "neck" and will begin to separate into two charged "globs." If the electric repulsion between these two globs forces them far enough apart to break the neck, the two fragments, each still carrying some residual excitation energy, will fly apart (Figs. 44-2*e* and *f*). Fission has occurred.

This model gave a good qualitative picture of the fission process. What remained to be seen, however, was whether it could answer a hard question: Why are some high-mass nuclides (^{235}U and ^{239}Pu, say) readily fissionable by thermal neutrons when other, equally massive nuclides (^{238}U and ^{243}Am, say) are not?

Bohr and Wheeler were able to answer this question. Figure 44-3 shows a graph of the potential energy of the fissioning nucleus at various stages, derived from their model for the fission process. This energy is plotted against the *distortion parameter* r, which is a rough measure of the extent to which the oscillating nucleus departs from a spherical shape. Figure 44-2*d* suggests how this parameter is defined just before fission occurs. When the fragments are far apart, this parameter is simply the distance between their centers.

The energy difference between the initial state and the final state of the fissioning nucleus—that is, the disintegration energy Q—is labeled in Fig. 44-3. The central feature of that figure, however, is that the potential energy curve passes through a maximum at a certain value of r. Thus, there is a *potential barrier* of height E_b that must be surmounted (or tunneled through) before fission can occur. This reminds us of alpha decay (Fig. 43-9), which is also a process that is inhibited by a potential barrier.

We see then that fission will occur only if the absorbed neutron provides an excitation energy E_n great enough to overcome the barrier. This energy E_n need not be *quite* as great as the barrier height E_b because of the possibility of quantum-physics tunneling.

Table 44-2 shows this test of whether capture of a thermal neutron can cause fissioning, for four high-mass nuclides. For each nuclide, the table shows both the barrier height E_b of the nucleus that is formed by the neutron capture and the excitation energy E_n due to the capture. The values of E_b are calculated from the theory of Bohr and Wheeler. The values of E_n are calculated from the change in mass energy due to the neutron capture.

For an example of the calculation of E_n, we can go to the first line in the table,

TABLE 44-2 Test of the Fissionability of Four Nuclides

Target Nuclide	Nuclide Being Fissioned	E_n (MeV)	E_b (MeV)	Fission by Thermal Neutrons?
^{235}U	^{236}U	6.5	5.2	Yes
^{238}U	^{239}U	4.8	5.7	No
^{239}P	^{240}Pu	6.4	4.8	Yes
^{243}Am	^{244}Am	5.5	5.8	No

which represents the neutron capture process

$$^{235}\text{U} + \text{n} \rightarrow {}^{236}\text{U}.$$

The masses involved are 235.043 923 u for ^{235}U, 1.008 665 u for the neutron, and 236.045 562 u for ^{236}U. It is easy to show that, because of the neutron capture, the mass decreases by 7.026×10^{-3} u. Thus, energy is transferred from mass energy to excitation energy E_n. Multiplying the change in mass by c^2 (= 931.5 MeV/u) gives us $E_n = 6.5$ MeV, which is listed on the first line of the table.

The first and third results in Table 44-2 are historically profound, because they are the reasons why the two atomic bombs used in World War II contained ^{235}U (first bomb) and ^{239}Pu (second bomb). That is, for ^{235}U and ^{239}Pu, $E_n > E_b$. This means that fission by absorption of a thermal neutron is predicted to occur for these nuclides. For the other two nuclides (^{238}U and ^{243}Am), we have $E_n < E_b$; thus, there is not enough energy from a thermal neutron for the excited nucleus to surmount the barrier or to tunnel through it effectively. Instead of fissioning, the nucleus gets rid of its excitation energy by emitting a gamma-ray photon.

^{238}U and ^{243}Am *can* be made to fission, however, if they absorb a substantially energetic (rather than a thermal) neutron. For ^{238}U, for example, the absorbed neutron must have at least 1.3 MeV of energy for this *fast fission* process to have a good chance of occurring.

The two atomic bombs used in World War II depended on the ability of thermal neutrons to cause many high-mass nuclides in the cores of the bombs to fission nearly all at once, so that the fissioning would result in an explosive and devastating output of energy. The first bomb used ^{235}U because enough of it had been refined from uranium ore to make that bomb and a test bomb. (The ore consists mainly of ^{238}U, which, as we have seen, is not caused to fission by thermal neutrons.) The second bomb used ^{239}Pu, based only on theoretical calculations as summarized in Table 44-2, because not enough additional ^{235}U was available when the second bomb was ordered.

44-4 The Nuclear Reactor

For large-scale energy release due to fission, one fission event must trigger others, so that the process spreads throughout the nuclear fuel like flame through a log. The fact that more neutrons are produced in fission than are consumed raises the possibility of just such a **chain reaction,** with each neutron that is produced potentially triggering another fission. The reaction can be either rapid (as in a nuclear bomb) or controlled (as in a nuclear reactor).

Suppose that we wish to design a reactor based on the fission of ^{235}U by thermal neutrons. Natural uranium contains 0.7% of this isotope, the remaining 99.3% being ^{238}U, which is not fissionable by thermal neutrons. Let us give ourselves an edge by artificially *enriching* the uranium fuel so that it contains perhaps 3% ^{235}U. Three difficulties still stand in the way of a working reactor.

1. *The Neutron Leakage Problem.* Some of the neutrons produced by fission will leak out of the reactor and so not be part of the chain reaction. Leakage is a surface effect; its magnitude is proportional to the square of a typical reactor dimension (the surface area of a cube of edge length a is $6a^2$). Neutron production, however, occurs throughout the volume of the fuel and is thus proportional to the cube of a typical dimension (the volume of the same cube is a^3). We can make the fraction of neutrons lost by leakage as small as we wish by making the reactor core large enough, thereby reducing the surface-to-volume ratio (= $6/a$ for a cube).

The scene 20 m from the Chernobyl reactor unit 4 (near Kiev), after it exploded in April 1986. Nearly all the volatile radionuclides inside the reactor were released into the air.

2. *The Neutron Energy Problem.* The neutrons produced by fission are fast, with kinetic energies of about 2 MeV. However, fission is induced most effectively by thermal neutrons. The fast neutrons can be slowed down by mixing the uranium fuel with a substance—called a **moderator**—that has two properties: It is effective in slowing down neutrons via elastic collisions, and it does not remove neutrons from the core by absorbing them so that they do not result in fission. Most power reactors in North America use water as a moderator; the hydrogen nuclei (protons) in the water are the effective component. We saw in Chapter 10 that, if a moving particle has a head-on elastic collision with a stationary particle, the moving particle loses *all* its kinetic energy if the two particles have the same mass. Thus, protons form an effective moderator because they have approximately the same mass as the fast neutrons whose speed we wish to reduce.

3. *The Neutron Capture Problem.* As the fast (2 MeV) neutrons generated by fission are slowed down in the moderator to thermal energies (about 0.04 eV), they must pass through a critical energy interval (from 1 to 100 eV) in which they are particularly susceptible to nonfission capture by ^{238}U nuclei. Such *resonance capture,* which results in the emission of a gamma ray, removes the neutron from the fission chain. To minimize such nonfission capture, the uranium fuel and the moderator are not intimately mixed but are "clumped together," occupying different regions of the reactor volume.

In a typical reactor, the uranium fuel is in the form of uranium oxide pellets, which are inserted end to end into long hollow metal tubes. The liquid moderator surrounds bundles of these **fuel rods,** forming the reactor **core.** This geometric arrangement increases the probability that a fast neutron, produced in a fuel rod, will find itself in the moderator when it passes through the critical energy interval. Once the neutron has reached thermal energies, it may *still* be captured in ways that do not result in fission (called *thermal capture*). However, it is much more likely that the thermal neutron will wander back into a fuel rod and produce a fission event.

Figure 44-4 shows the neutron balance in a typical power reactor operating at constant power. Let us trace a sample of 1000 thermal neutrons through one complete

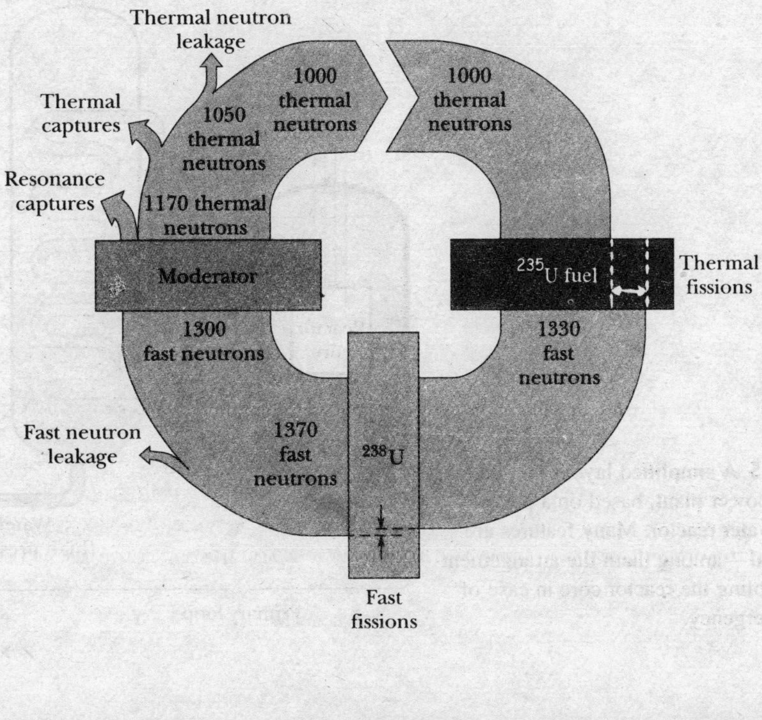

Fig. 44-4 Neutron bookkeeping in a reactor. A generation of 1000 thermal neutrons interacts with the ^{235}U fuel, the ^{238}U matrix, and the moderator. They produce 1370 neutrons by fission; 370 of these are lost by nonfission capture or by leakage; so 1000 thermal neutrons are left to form the next generation. The figure is drawn for a reactor running at a steady power level.

cycle, or *generation,* in the reactor core. They produce 1330 neutrons by fission in the ^{235}U fuel and 40 neutrons by fast fission in ^{238}U, which gives 370 neutrons more than the original 1000, all of them fast. When the reactor is operating at a steady power level, exactly the same number of neutrons (370) is then lost by leakage from the core and by nonfission capture, leaving 1000 thermal neutrons to start the next generation. In this cycle, of course, each of the 370 neutrons produced by fission events represents a deposit of energy in the reactor core, heating up the core.

The *multiplication factor k*—an important reactor parameter—is the ratio of the number of neutrons present at the beginning of a particular generation to the number present at the beginning of the next generation. In Fig. 44-4, the multiplication factor is 1000/1000, or exactly unity. For $k = 1$, the operation of the reactor is said to be exactly *critical,* which is what we wish it to be for steady-power operation. Reactors are actually designed so that they are inherently *supercritical* ($k > 1$); the multiplication factor is then adjusted to critical operation ($k = 1$) by inserting **control rods** into the reactor core. These rods, containing a material such as cadmium that absorbs neutrons readily, can be inserted farther to reduce the operating power level and withdrawn to increase the power level or to compensate for the tendency of reactors to go *subcritical* as (neutron-absorbing) fission products build up in the core during continued operation.

If you pulled out one of the control rods rapidly, how fast would the reactor power level increase? This *response time* is controlled by the fascinating circumstance that a small fraction of the neutrons generated by fission do not escape promptly from the newly formed fission fragments but are emitted from these fragments later, as the fragments decay by beta emission. Of the 370 "new" neutrons produced in Fig. 44-4, for example, perhaps 16 are delayed, being emitted from fragments following beta decays whose half-lives range from 0.2 to 55 s. These delayed neutrons are few in number, but they serve the essential purpose of slowing the reactor response time to match practical mechanical reaction times.

Figure 44-5 shows the broad outlines of an electric power plant based on a *pressurized-water reactor* (PWR), a type in common use in North America. In such

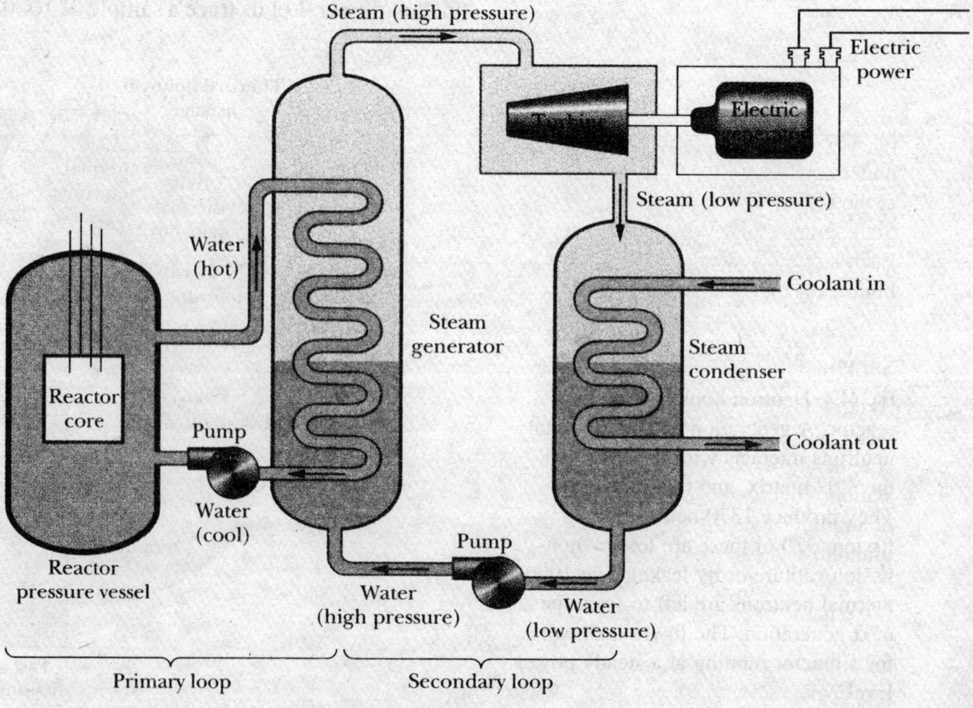

Fig. 44-5 A simplified layout of a nuclear power plant, based on a pressurized-water reactor. Many features are omitted—among them the arrangement for cooling the reactor core in case of an emergency.

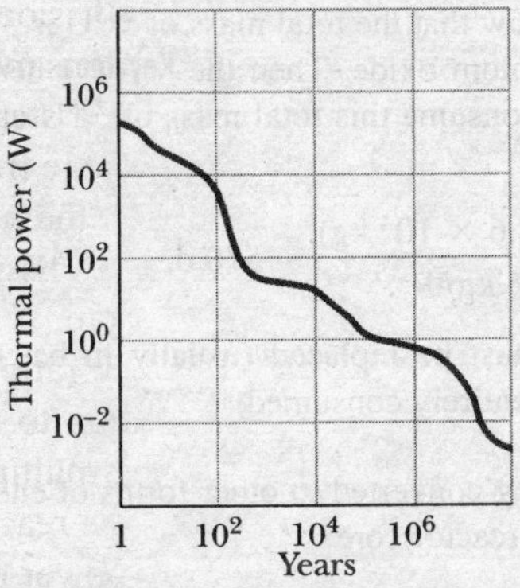

Fig. 44-6 The thermal power released by the radioactive wastes from one year's operation of a typical large nuclear power plant, shown as a function of time. The curve is the superposition of the effects of many radionuclides, with a wide variety of half-lives. Note that both scales are logarithmic.

a reactor, water is used both as the moderator and as the heat transfer medium. In the *primary loop,* water is circulated through the reactor vessel and transfers energy at high temperature and pressure (possibly 600 K and 150 atm) from the hot reactor core to the steam generator, which is part of the *secondary loop*. In the steam generator, evaporation provides high-pressure steam to operate the turbine that drives the electric generator. To complete the *secondary loop,* low-pressure steam from the turbine is cooled and condensed to water and forced back into the steam generator by a pump. To give some idea of scale, a typical reactor vessel for a 1000 MW (electric) plant may be 12 m high and weigh 4 MN. Water flows through the primary loop at a rate of about 1 ML/min.

An unavoidable feature of reactor operation is the accumulation of radioactive wastes, including both fission products and heavy *transuranic* nuclides such as plutonium and americium. One measure of their radioactivity is the rate at which they release energy in thermal form. Figure 44-6 shows the thermal power generated by such wastes from one year's operation of a typical large nuclear plant. Note that both scales are logarithmic. Most "spent" fuel rods from power reactor operation are stored on site, immersed in water; permanent secure storage facilities for reactor waste have yet to be completed.

Much weapons-derived radioactive waste accumulated during World War II and in subsequent years is also still in on-site storage. For example, Fig. 44-7 shows an underground storage tank farm under construction at the Hanford Site in Washington State; each large tank holds 1 ML of highly radioactive liquid waste. There are now 152 such tanks at the site. In addition, much solid waste, both low-level radioactive waste (contaminated clothing, for example) and high-level waste (reactor cores from decommissioned nuclear submarines, for example) is buried in trenches.

Fig. 44-7 An underground tank farm under construction during World War II at the Hanford Site in Washington State. Note the trucks and the workers. Each large tank now holds 1 ML of high-level radioactive waste.

Sample Problem 44-2

A large electric generating station is powered by a pressurized-water nuclear reactor. The thermal power produced in the reactor core is 3400 MW, and 1100 MW of electricity is generated by the station. The fuel charge is 8.60×10^4 kg of uranium, in the form of uranium oxide, distributed among 5.70×10^4 fuel rods. The uranium is enriched to 3.0% ^{235}U.

(a) What is the station's efficiency?

SOLUTION: The **Key Idea** here is the definition of efficiency for this power plant or any other energy device: Efficiency is the ratio of the output power (rate at which useful energy is provided) to the input power (rate at which energy must be supplied). Here the efficiency (eff) is

$$\text{eff} = \frac{\text{useful output}}{\text{energy input}} = \frac{1100 \text{ MW (electric)}}{3400 \text{ MW (thermal)}}$$
$$= 0.32, \text{ or } 32\%. \qquad \text{(Answer)}$$

The efficiency—as for all power plants—is controlled by the second law of thermodynamics. To run this plant, energy at the rate of 3400 MW − 1100 MW, or 2300 MW, must be discharged as thermal energy to the environment.

(b) At what rate R do fission events occur in the reactor core?

SOLUTION: The **Key Ideas** here are (1) that the fission events provide the input power P of 3400 MW ($= 3.4 \times 10^9$ J/s) and (2) from Eq. 44-6, the energy Q released by each event is about 200 MeV. Thus, for steady-state operation (P is constant), we find

$$R = \frac{P}{Q} = \left(\frac{3.4 \times 10^9 \text{ J/s}}{200 \text{ MeV/fission}}\right)\left(\frac{1 \text{ MeV}}{1.60 \times 10^{-13} \text{ J}}\right)$$
$$= 1.06 \times 10^{20} \text{ fissions/s}$$
$$\approx 1.1 \times 10^{20} \text{ fissions/s.} \qquad \text{(Answer)}$$

(c) At what rate (in kilograms per day) is the ^{235}U fuel disappearing? Assume conditions at start-up.

SOLUTION: Here the **Key Idea** is that ^{235}U disappears due to two processes: (1) the fission process with the rate calculated in part (b) and (2) the nonfission capture of neutrons at about one-fourth that rate. Thus, the total rate at which ^{235}U disappears is

$$(1 + 0.25)(1.06 \times 10^{20} \text{ atoms/s}) = 1.33 \times 10^{20} \text{ atoms/s.}$$

We next need the mass of each ^{235}U atom. We cannot use the molar mass for uranium listed in Appendix F, because that molar mass is for ^{238}U, the most common uranium isotope. Instead, we shall assume that the mass of each ^{235}U atom in atomic mass units is equal to the mass number A. Thus, the mass of each ^{235}U atom is 235 u ($= 3.90 \times 10^{-25}$ kg). Then the rate at which the ^{235}U fuel disappears is

$$\frac{dM}{dt} = (1.33 \times 10^{20} \text{ atoms/s})(3.90 \times 10^{-25} \text{ kg})$$
$$= 5.19 \times 10^{-5} \text{ kg/s} \approx 4.5 \text{ kg/d.} \qquad \text{(Answer)}$$

(d) At this rate of fuel consumption, how long would the fuel supply of ^{235}U last?

SOLUTION: At start-up, we know that the total mass of ^{235}U is 3.0% of the 8.6×10^4 kg of uranium oxide. Then the **Key Idea** here is that the time T required to consume this total mass of ^{235}U at the steady rate of 4.5 kg/d is

$$T = \frac{(0.030)(8.6 \times 10^4 \text{ kg})}{4.5 \text{ kg/d}} \approx 570 \text{ d.} \qquad \text{(Answer)}$$

In practice, the fuel rods must be replaced (usually in batches) before their ^{235}U content is entirely consumed.

(e) At what rate is mass being converted to other forms of energy by the fission of ^{235}U in the reactor core?

SOLUTION: The **Key Idea** here is that the conversion of mass energy to other forms of energy is linked only to the fissioning that produces the input power (3400 MW), and not to the nonfission capture of neutrons (although both these processes affect the rate at which ^{235}U is consumed). Thus, from Einstein's relation $E = mc^2$, we can write

$$\frac{dm}{dt} = \frac{dE/dt}{c^2} = \frac{3.4 \times 10^9 \text{ W}}{(3.00 \times 10^8 \text{ m/s})^2}$$
$$= 3.8 \times 10^{-8} \text{ kg/s} = 3.3 \text{ g/d.} \qquad \text{(Answer)}$$

We see that the mass conversion rate is about the mass of one common coin per day, considerably less than the fuel consumption rate calculated in (c).

✔CHECKPOINT 2: In this sample problem, we saw that the generated power of the nuclear power plant ($P_{gen} = 1100$ MW) was less than the power discharged to the environment ($P_{dis} =$ 2300 MW). Does the second law of thermodynamics: (a) require that P_{gen} always be less than P_{dis}; (b) permit P_{gen} to be greater than P_{dis}; and (c) permit P_{dis} to be zero, assuming optimum reactor design?

44-5 A Natural Nuclear Reactor

On December 2, 1942, when their reactor first became operational (Fig. 44-8), Enrico Fermi and his associates had every right to assume that they had put into operation the first fission reactor that had ever existed on this planet. About 30 years later it was discovered that, if they did in fact think that, they were wrong.

Fig. 44-8 A painting of the first nuclear reactor, assembled during World War II on a squash court at the University of Chicago by a team headed by Enrico Fermi. This reactor, which went critical on December 2, 1942, was built of lumps of uranium embedded in blocks of graphite. It served as a prototype for later reactors whose purpose was to manufacture plutonium for the construction of nuclear weapons.

Some two billion years ago, in a uranium deposit now being mined in Gabon, West Africa, a natural fission reactor apparently went into operation and ran for perhaps several hundred thousand years before shutting down. We can test whether this could actually have happened by considering two questions:

1. *Was There Enough Fuel?* The fuel for a uranium-based fission reactor must be the easily fissionable isotope ^{235}U, which constitutes only 0.72% of natural uranium. This isotopic ratio has been measured for terrestrial samples, in Moon rocks, and in meteorites; in all cases the abundance values are the same. The clue to the discovery in West Africa was that the uranium in that deposit was deficient in ^{235}U, some samples having abundances as low as 0.44%. Investigation led to the speculation that this deficit in ^{235}U could be accounted for if, at some earlier time, the ^{235}U was partially consumed by the operation of a natural fission reactor.

   The serious problem remains that, with an isotopic abundance of only 0.72%, a reactor can be assembled (as Fermi and his team learned) only after thoughtful design and with scrupulous attention to detail. There seems no chance that a nuclear reactor could go critical "naturally."

   However, things were different in the distant past. Both ^{235}U and ^{238}U are radioactive, with half-lives of 7.04×10^8 y and 44.7×10^8 y, respectively. Thus, the half-life of the readily fissionable ^{235}U is about 6.4 times shorter than that of ^{238}U. Because ^{235}U decays faster, there was more of it, relative to ^{238}U, in the past. Two billion years ago, in fact, this abundance was not 0.72%, as it is now, but 3.8%. This abundance happens to be just about the abundance to which natural uranium is artificially enriched to serve as fuel in modern power reactors.

   With this readily fissionable fuel available, the presence of a natural reactor (provided certain other conditions are met) is less surprising. The fuel was there. Two billion years ago, incidentally, the highest order of life-form to have evolved was the blue-green alga.

2. *What Is the Evidence?* The mere depletion of ^{235}U in an ore deposit does not prove the existence of a natural fission reactor. One looks for more convincing evidence.

   If there was a reactor, there must now be fission products. Of the 30 or so elements whose stable isotopes are produced in a reactor, some must still remain. Study of their isotopic abundances could provide the evidence we need.

   Of the several elements investigated, the case of neodymium is spectacularly convincing. Figure 44-9*a* shows the isotopic abundances of the seven stable neodymium isotopes as they are normally found in nature. Figure 44-9*b* shows these abundances as they appear among the ultimate stable fission products of the

Fig. 44-9 The distribution by mass number of the isotopes of neodymium as they occur in (*a*) natural terrestrial deposits of the ores of this element and (*b*) the spent fuel of a power reactor. (*c*) The distribution (after several corrections) found for neodymium from the uranium mine in Gabon, West Africa. Note that (*b*) and (*c*) are virtually identical and are quite different from (*a*).

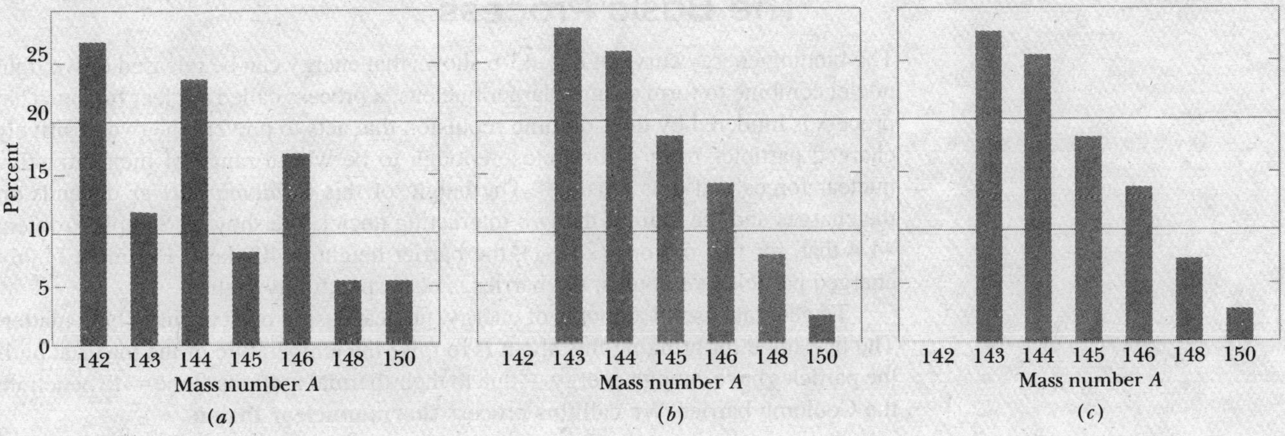

fission of ^{235}U. The clear differences are not surprising, considering the totally different origins of the two sets of isotopes. Note particularly that ^{142}Nd, the dominant isotope in the natural element, is absent from the fission products.

The big question is: What do the neodymium isotopes found in the uranium ore body in West Africa look like? If a natural reactor operated there, we would expect to find isotopes from *both* sources (that is, natural isotopes as well as fission-produced isotopes). Figure 44-9*c* shows the abundances after dual-source and other corrections have been made to the data. Comparison of Figs. 44-9*b* and *c* indicates that there was indeed a natural fission reactor at work.

The failure of the fission products of the West African natural reactor to migrate far from their region of production over about 2 billion years may support the idea of long-term storage of radioactive waste in suitably chosen geological environments.

Sample Problem 44-3

The ratio of ^{235}U to ^{238}U in natural uranium deposits today is 0.0072. What was this ratio 2.0×10^9 y ago? The half-lives of the two isotopes are 7.04×10^8 y and 44.7×10^8 y, respectively.

SOLUTION: The **Key Idea** here is that the ratio N_5/N_8 of ^{235}U to ^{238}U at time $t = 0$ was not equal to 0.0072 (the ratio today, at time $t = 2.0 \times 10^9$ y), because those two isotopes have been decaying at different rates. Let $N_5(0)$ and $N_8(0)$ be the numbers of the isotopes in a sample of uranium at $t = 0$, and let $N_5(t)$ and $N_8(t)$ be the numbers of the isotopes at some later time t. Then, for each isotope, we can use Eq. 43-14 to write the number at time t in terms of the number at time $t = 0$:

$$N_5(t) = N_5(0)e^{-\lambda_5 t} \quad \text{and} \quad N_8(t) = N_8(0)e^{-\lambda_8 t},$$

in which λ_5 and λ_8 are the corresponding disintegration constants. Dividing gives

$$\frac{N_5(t)}{N_8(t)} = \frac{N_5(0)}{N_8(0)} e^{-(\lambda_5-\lambda_8)t}.$$

Because we want the ratio $N_5(0)/N_8(0)$, we rearrange this to be

$$\frac{N_5(0)}{N_8(0)} = \frac{N_5(t)}{N_8(t)} e^{(\lambda_5-\lambda_8)t}. \qquad (44\text{-}8)$$

The disintegration constants are related to the half-lives by Eq. 43-17, which yields

$$\lambda_5 = \frac{\ln 2}{T_{1/2,5}} = \frac{\ln 2}{7.04 \times 10^8 \text{ y}} = 9.85 \times 10^{-10} \text{ y}^{-1}$$

and

$$\lambda_8 = \frac{\ln 2}{T_{1/2,8}} = \frac{\ln 2}{44.7 \times 10^8 \text{ y}} = 1.55 \times 10^{-10} \text{ y}^{-1}.$$

The exponent in Eq. 44-8 is then

$$(\lambda_5 - \lambda_8)t = [(9.85 - 1.55) \times 10^{-10} \text{ y}^{-1}](2 \times 10^9 \text{ y}) = 1.66.$$

Equation 44-8 then yields

$$\frac{N_5(0)}{N_8(0)} = \frac{N_5(t)}{N_8(t)} e^{(\lambda_5-\lambda_8)t} = (0.0072)(e^{1.66}) = 0.0379 \approx 3.8\%. \qquad \text{(Answer)}$$

The ratio of ^{235}U to ^{238}U was much greater than this (about 30%) when Earth was formed 4.5 billion years ago.

44-6 Thermonuclear Fusion: The Basic Process

The binding energy curve of Fig. 43-6 shows that energy can be released if two light nuclei combine to form a single larger nucleus, a process called nuclear **fusion.** That process is hindered by the Coulomb repulsion that acts to prevent the two positively charged particles from getting close enough to be within range of their attractive nuclear forces and thus "fusing." The height of this *Coulomb barrier* depends on the charges and the radii of the two interacting nuclei. We show in Sample Problem 44-4 that, for two protons ($Z = 1$), the barrier height is 400 keV. For more highly charged particles, of course, the barrier is correspondingly higher.

To generate useful amounts of energy, nuclear fusion must occur in bulk matter. The best hope for bringing this about is to raise the temperature of the material until the particles have enough energy—due to their thermal motions alone—to penetrate the Coulomb barrier. We call this process **thermonuclear fusion.**

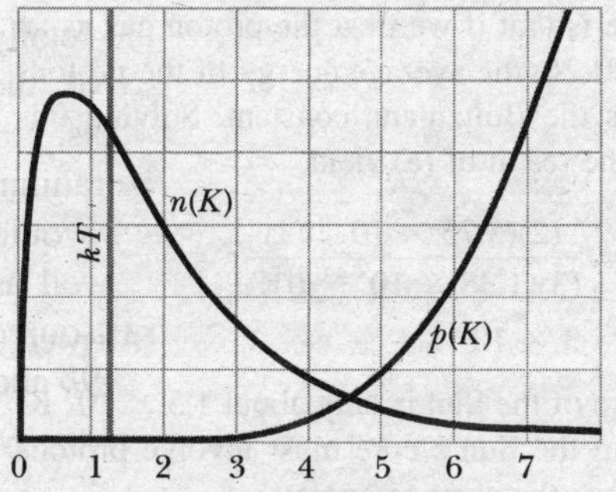

Fig. 44-10 The curve marked $n(K)$ gives the number density per unit energy for protons at the center of the Sun. The curve marked $p(K)$ gives the probability of barrier penetration (and hence fusion) for proton-proton collisions at the Sun's core temperature. The vertical line marks the value of kT at this temperature. Note that the two curves are drawn to (separate) arbitrary vertical scales.

In thermonuclear studies, temperatures are reported in terms of the kinetic energy K of interacting particles via the relation

$$K = kT, \tag{44-9}$$

in which K is the kinetic energy corresponding to the *most probable speed* of the interacting particles, k is the Boltzmann constant, and the temperature T is in kelvins. Thus, rather than saying, "The temperature at the center of the Sun is 1.5×10^7 K," it is more common to say, "The temperature at the center of the Sun is 1.3 keV."

Room temperature corresponds to $K \approx 0.03$ eV; a particle with only this amount of energy could not hope to overcome a barrier as high as, say, 400 keV Even at the center of the Sun, where $kT = 1.3$ keV, the outlook for thermonuclear fusion does not seem promising at first glance. Yet we know that thermonuclear fusion not only occurs in the core of the Sun but is the dominant feature of that body and of all other stars.

The puzzle is solved when we realize two facts: (1) The energy calculated with Eq. 44-9 is that of the particles with the *most probable* speed, as defined in Section 20-7; there is a long tail of particles with much higher speeds and, correspondingly, much higher energies. (2) The barrier heights that we have calculated represent the *peaks* of the barriers. Barrier tunneling can occur at energies considerably below those peaks, as we saw in the case of alpha decay in Section 43-4.

Figure 44-10 sums things up. The curve marked $n(K)$ in this figure is a Maxwell distribution curve for the protons in the Sun's core, drawn to correspond to the Sun's central temperature. This curve differs from the Maxwell distribution curve given in Fig. 20-7 in that here the curve is drawn in terms of energy and not of speed. Specifically, for any kinetic energy K, the expression $n(K)\,dK$ gives the probability that a proton will have a kinetic energy lying between K and $K + dK$. The value of kT in the core of the Sun is indicated by the vertical line in the figure; note that many of the Sun's core protons have energies greater than this value.

The curve marked $p(K)$ in Fig. 44-10 is the probability of barrier penetration by two colliding protons. The two curves in Fig. 44-10 suggest that there is a particular proton energy at which proton-proton fusion events occur at a maximum rate. At energies much above this value, the barrier is transparent enough but too few protons have these energies, so the fusion reaction cannot be sustained. At energies much below this value, plenty of protons have these energies but the Coulomb barrier is too formidable.

CHECKPOINT 3: Which of these potential fusion reactions will *not* result in the net release of energy: (a) ^{6}Li + ^{6}Li, (b) ^{4}He + ^{4}He, (c) ^{12}C + ^{12}C, (d) ^{20}Ne + ^{20}Ne, (e) ^{35}Cl + ^{35}Cl, and (f) ^{14}N + ^{35}Cl? (*Hint:* Consult the binding energy curve of Fig. 43-6.)

Sample Problem 44-4

Assume a proton is a sphere of radius $R \approx 1$ fm. Two protons are fired at each other with the same kinetic energy K.

(a) What must K be if the particles are brought to rest by their mutual Coulomb repulsion when they are just "touching" each other? We can take this value of K as a representative measure of the height of the Coulomb barrier.

SOLUTION: The **Key Idea** here is that the mechanical energy E of the two-proton system is conserved as the protons move toward each other and momentarily stop. In particular, the initial mechanical energy E_i is equal to the mechanical energy E_f when they stop. The initial energy E_i consists only of the total kinetic energy $2K$ of the two protons. When the protons stop, energy E_f consists only of the electric potential energy U of the system, as given by Eq. 25-43 ($U = q_1q_2/4\pi\varepsilon_0 r$). Here the distance r between the protons when they stop is their center-to-center distance $2R$, and their charges q_1 and q_2 are both e. Then we can write the conservation of energy $E_i = E_f$ as

$$2K = \frac{1}{4\pi\varepsilon_0}\frac{e^2}{2R}.$$

This yields, with known values,

$$K = \frac{e^2}{16\pi\varepsilon_0 R}$$

$$= \frac{(1.60 \times 10^{-19}\ \text{C})^2}{(16\pi)(8.85 \times 10^{-12}\ \text{F/m})(1 \times 10^{-15}\ \text{m})}$$

$$= 5.75 \times 10^{-14}\ \text{J} = 360\ \text{keV} \approx 400\ \text{keV}. \quad \text{(Answer)}$$

(b) At what temperature would a proton in a gas of protons have the average kinetic energy calculated in (a), and thus have energy equal to the height of the Coulomb barrier?

SOLUTION: The **Key Idea** here is that if we treat the proton gas as an ideal gas, then from Eq. 20-24, the average energy of the protons is $K_{avg} = \frac{3}{2}kT$, where k is the Boltzmann constant. Solving this equation for T and using the result of (a) yield

$$T = \frac{2K_{avg}}{3k} = \frac{(2)(5.75 \times 10^{-14}\ \text{J})}{(3)(1.38 \times 10^{-23}\ \text{J/K})}$$

$$\approx 3 \times 10^9\ \text{K}. \quad \text{(Answer)}$$

The temperature of the core of the Sun is only about 1.5×10^7 K, so it is clear that fusion in the Sun's core must involve protons whose energies are *far* above the average energy.

44-7 Thermonuclear Fusion in the Sun and Other Stars

The Sun radiates energy at the rate of 3.9×10^{26} W and has been doing so for several billion years. Where does all this energy come from? Chemical burning is ruled out; if the Sun had been made of coal and oxygen—in the right proportions for combustion—it would have lasted for only about 1000 y. Another possibility is that the Sun is slowly shrinking, under the action of its own gravitational forces. By transferring gravitational potential energy to thermal energy, the Sun might maintain its temperature and continue to radiate. Calculation shows, however, that this mechanism also fails; it produces a solar lifetime that is too short by a factor of at least 500. That leaves only thermonuclear fusion. The Sun, as you will see, burns not coal but hydrogen, and in a nuclear furnace, not an atomic or chemical one.

The fusion reaction in the Sun is a multistep process in which hydrogen is burned into helium, hydrogen being the "fuel" and helium the "ashes." Figure 44-11 shows the **proton-proton** (p-p) **cycle** by which this occurs.

The p-p cycle starts with the collision of two protons (^{1}H + ^{1}H) to form a deuteron (^{2}H), with the simultaneous creation of a positron (e^+) and a neutrino (ν). The positron very quickly encounters a free electron (e^-) in the Sun and both particles annihilate (see Section 22-6), their mass energy appearing as two gamma-ray photons (γ).

A pair of such events is shown in the top row of Fig. 44-11. These events are actually extremely rare. In fact, only once in about 10^{26} proton-proton collisions is a deuteron formed; in the vast majority of cases, the two protons simply rebound elastically from each other. It is the slowness of this "bottleneck" process that regulates the rate of energy production and keeps the Sun from exploding. In spite of this slowness, there are so very many protons in the huge and dense volume of the Sun's core that deuterium is produced in just this way at the rate of 10^{12} kg/s.

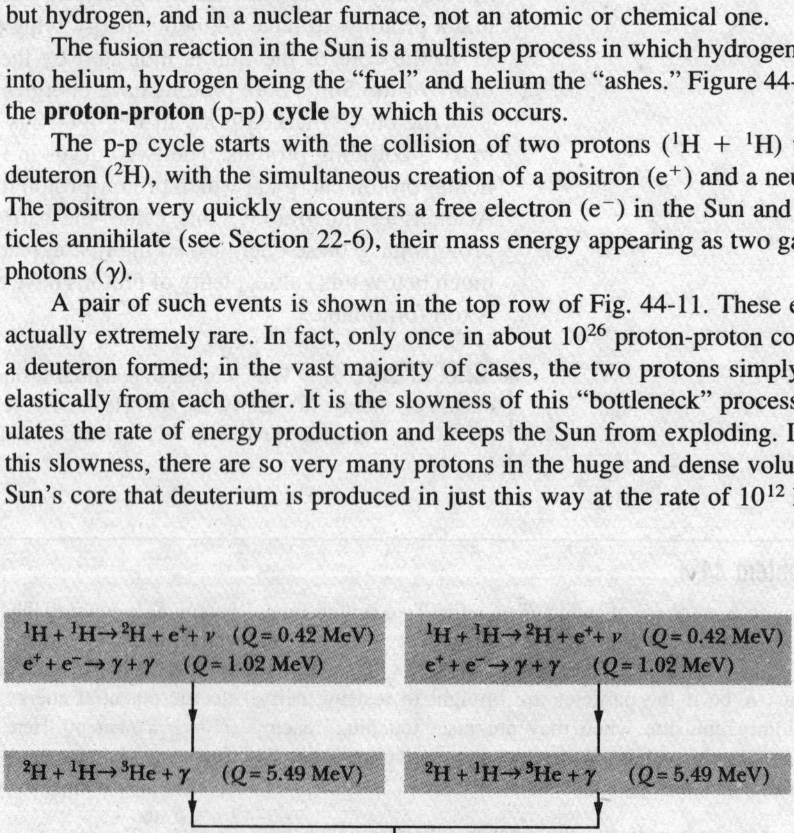

Fig. 44-11 The proton-proton mechanism that accounts for energy production in the Sun. In this process, protons fuse to form an alpha particle (^{4}He), with a net energy release of 26.7 MeV for each event.

Once a deuteron has been produced, it quickly collides with another proton and forms a ^{3}He nucleus, as the middle row of Fig. 44-11 shows. Two such ^{3}He nuclei may eventually (within 10^5 y; there is plenty of time) find each other, forming an alpha particle (^{4}He) and two protons, as the bottom row in the figure shows.

Overall, we see from Fig. 44-11 that the p-p cycle amounts to the combination of four protons and two electrons to form an alpha particle, two neutrinos, and six gamma-ray photons. That is,

$$4\,{}^1\text{H} + 2e^- \rightarrow {}^4\text{He} + 2\nu + 6\gamma. \tag{44-10}$$

Let us now add two electrons to each side of Eq. 44-10, obtaining

$$(4\,{}^1\text{H} + 4e^-) \rightarrow ({}^4\text{He} + 2e^-) + 2\nu + 6\gamma. \tag{44-11}$$

The quantities in the two sets of parentheses then represent *atoms* (not bare nuclei) of hydrogen and of helium. That allows us to compute the energy release in the overall reaction of Eq. 44-10 (and Eq. 44-11) as

$$\begin{aligned} Q &= -\Delta m\, c^2 \\ &= -[4.002\,603\text{ u} - (4)(1.007\,825\text{ u})][931.5\text{ MeV/u}] \\ &= 26.7\text{ MeV}, \end{aligned}$$

in which 4.002 603 u is the mass of a helium atom and 1.007 825 u is the mass of a hydrogen atom. Neutrinos have at most a negligibly small mass, and gamma-ray photons have no mass; thus, they do not enter into the calculation of the disintegration energy.

This same value of Q follows (as it must) from adding up the Q values for the separate steps of the proton-proton cycle in Fig. 44-11. Thus,

$$\begin{aligned} Q &= (2)(0.42\text{ MeV}) + (2)(1.02\text{ MeV}) + (2)(5.49\text{ MeV}) + 12.86\text{ MeV} \\ &= 26.7\text{ MeV}. \end{aligned}$$

About 0.5 MeV of this energy is carried out of the Sun by the two neutrinos indicated in Eqs. 44-10 and 44-11; the rest (= 26.2 MeV) is deposited in the core of the Sun as thermal energy. That thermal energy is then gradually transported to the Sun's surface where it is radiated away from the Sun as electromagnetic waves, including visible light.

The burning of hydrogen in the Sun's core is alchemy on a grand scale in the sense that one element is turned into another. The medieval alchemists, however, were more interested in changing lead into gold than in changing hydrogen into helium. In a sense, they were on the right track, except that their furnaces were not hot enough. Instead of being at a temperature of, say, 600 K, the ovens should have been at least as hot as 10^8 K.

Hydrogen burning has been going on in the Sun for about 5×10^9 y, and calculations show that there is enough hydrogen left to keep the Sun going for about the same length of time into the future. In 5 billion years, however, the Sun's core, which by that time will be largely helium, will begin to cool and the Sun will start to collapse under its own gravity. This will raise the core temperature and cause the outer envelope to expand, turning the Sun into what is called a *red giant*.

If the core temperature increases to about 10^8 K again, energy can be produced through fusion once more—this time by burning helium to make carbon. As a star evolves further and becomes still hotter, other elements can be formed by other fusion reactions. However, elements more massive than those near the peak of the binding energy curve of Fig. 43-6 cannot be produced by further fusion processes.

Elements with mass numbers beyond the peak of that curve are thought to be formed by neutron capture during cataclysmic stellar explosions that we call *super-*

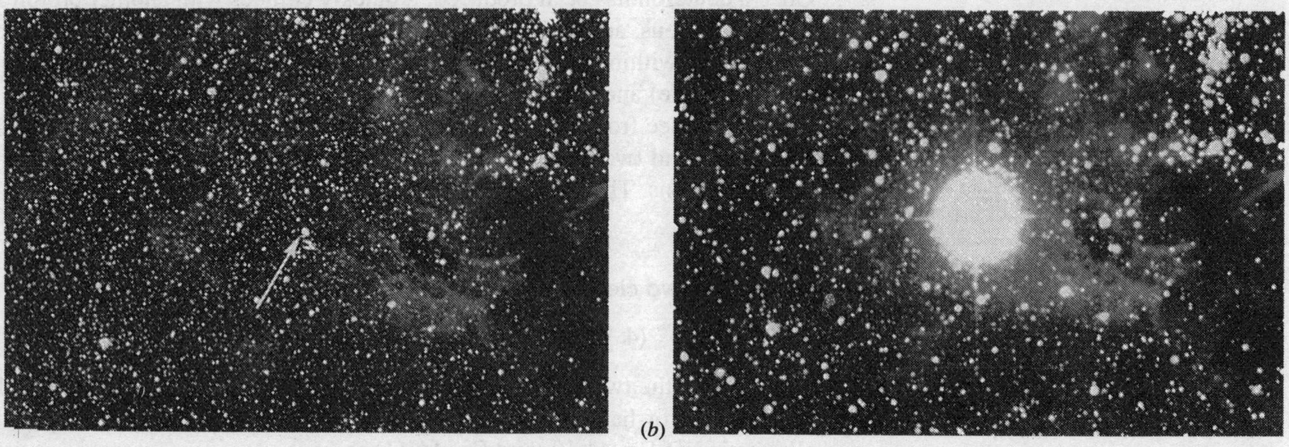

Fig. 44-12 (*a*) The star known as Sanduleak, as it appeared until 1987. (*b*) We then began to intercept light from the supernova of that star; the explosion was 100 million times brighter than our Sun and could be seen with the unaided eye. The explosion took place 155 000 light-years away and thus actually occurred 155 000 years ago.

novas (Fig. 44-12). In such an event the outer shell of the star is blown outward into space, where it mixes with—and becomes part of—the tenuous medium that fills the space between the stars. It is from this medium, continually enriched by debris from stellar explosions, that new stars form, by condensation under the influence of the gravitational force.

The abundance on Earth of elements heavier than hydrogen and helium suggests that our solar system has condensed out of interstellar material that contained the remnants of such explosions. Thus, all the elements around us—including those in our own bodies—were manufactured in the interiors of stars that no longer exist. As one scientist put it: "In truth, we are the children of the stars."

Sample Problem 44-5

At what rate dm/dt is hydrogen being consumed in the core of the Sun by the p-p cycle of Fig. 44-11?

SOLUTION: The **Key Idea** is that the rate dE/dt at which energy is produced by hydrogen (proton) consumption within the Sun is equal to the rate P at which energy is radiated by the Sun:

$$P = \frac{dE}{dt}.$$

To bring the mass consumption rate dm/dt into this equation, we can rewrite it as

$$P = \frac{dE}{dt} = \frac{dE}{dm}\frac{dm}{dt} \approx \frac{\Delta E}{\Delta m}\frac{dm}{dt}, \qquad (44\text{-}12)$$

where ΔE is the energy produced when protons of mass Δm are consumed. From our discussion in this section, we know that 26.2 MeV ($= 4.20 \times 10^{-12}$ J) of thermal energy is produced when four protons are consumed. That is, $\Delta E = 4.20 \times 10^{-12}$ J for a mass consumption of $\Delta m = 4(1.67 \times 10^{-27}$ kg). Substituting these data into Eq. 44-12 and using the power P of the Sun given in Appendix C, we find that

$$\frac{dm}{dt} = \frac{\Delta m}{\Delta E}P = \frac{4(1.67 \times 10^{-27}\text{ kg})}{4.20 \times 10^{-12}\text{ J}}(3.90 \times 10^{26}\text{ W})$$

$$= 6.2 \times 10^{11}\text{ kg/s}. \qquad \text{(Answer)}$$

Thus, a huge amount of hydrogen is consumed by the Sun every second. However, you need not worry too much about the Sun running out of hydrogen, because its mass of 2×10^{30} kg will keep it burning for a long, long time.

44-8 Controlled Thermonuclear Fusion

The first thermonuclear reaction on Earth occurred at Eniwetok Atoll on November 1, 1952, when the United States exploded a fusion device, generating an energy release equivalent to 10 million tons of TNT. The high temperatures and densities needed to initiate the reaction were provided by using a fission bomb as a trigger.

A sustained and controllable source of fusion power—a fusion reactor as part of, say, an electric generating plant—is considerably more difficult to achieve. That goal is nonetheless being pursued vigorously in many countries around the world, because many people look to the fusion reactor as the power source of the future, at least for the generation of electricity.

The p-p scheme displayed in Fig. 44-11 is not suitable for an Earth-bound fusion reactor because it is hopelessly slow. The process succeeds in the Sun only because of the enormous density of protons in the center of the Sun. The most attractive reactions for terrestrial use appear to be two deuteron-deuteron (d-d) reactions,

$$^2\text{H} + {}^2\text{H} \rightarrow {}^3\text{He} + \text{n} \qquad (Q = +3.27\ \text{MeV}), \tag{44-13}$$

$$^2\text{H} + {}^2\text{H} \rightarrow {}^3\text{H} + {}^1\text{H} \qquad (Q = +4.03\ \text{MeV}), \tag{44-14}$$

and the deuteron-triton (d-t) reaction*

$$^2\text{H} + {}^3\text{H} \rightarrow {}^4\text{He} + \text{n} \qquad (Q = +17.59\ \text{MeV}). \tag{44-15}$$

Deuterium, the source of deuterons for these reactions, has an isotopic abundance of only 1 part in 6700, but is available in unlimited quantities as a component of seawater. Proponents of power from the nucleus have described our ultimate power choice—when we have burned up all our fossil fuels—as either "burning rocks" (fission of uranium extracted from ores) or "burning water" (fusion of deuterium extracted from water).

There are three requirements for a successful thermonuclear reactor:

1. *A High Particle Density n.* The (number) density of interacting particles (the number of, say, deuterons per unit volume) must be great enough to ensure that the d-d collision rate is high enough. At the high temperatures required, the deuterium would be completely ionized, forming a neutral **plasma** (ionized gas) of deuterons and electrons.
2. *A High Plasma Temperature T.* The plasma must be hot. Otherwise the colliding deuterons will not be energetic enough to penetrate the Coulomb barrier that tends to keep them apart. A plasma ion temperature of 35 keV, corresponding to 4×10^8 K, has been achieved in the laboratory. This is about 30 times higher than the Sun's central temperature.
3. *A Long Confinement Time τ.* A major problem is containing the hot plasma long enough to maintain it at a density and a temperature sufficiently high to ensure the fusion of enough of the fuel. It is clear that no solid container can withstand the high temperatures that are necessary, so clever confining techniques are called for; we shall shortly discuss two of them.

It can be shown that, for the successful operation of a thermonuclear reactor using the d-t reaction, it is necessary to have

$$n\tau > 10^{20}\ \text{s/m}^3. \tag{44-16}$$

This condition, known as **Lawson's criterion,** tells us that we have a choice between confining a lot of particles for a short time or fewer particles for a longer time. Beyond meeting this criterion, it is still necessary that the plasma temperature be high enough.

Two approaches to controlled nuclear power generation are currently under study. Although neither approach has yet been successful, both are being pursued because of their promise and because of the potential importance of controlled fusion to the world's energy problems.

*The nucleus of the hydrogen isotope ^{3}H (tritium) is called the *triton.* It is a radionuclide, with a half-life of 12.3 y.

Fig. 44-13 The Tokamak Fusion Test Reactor at Princeton University.

Magnetic Confinement

In one version of this approach, a suitably shaped magnetic field is used to confine a hot plasma in an evacuated doughnut-shaped chamber called a **tokamak** (the name is an abbreviation consisting of parts of three Russian words). The magnetic forces acting on the charged particles that make up the hot plasma keep the plasma from touching the walls of the chamber. Figure 44-13 shows such a device at the Plasma Physics Laboratory of Princeton University.

The plasma is heated by inducing a current to flow in it and by bombarding the plasma with an externally accelerated beam of particles. The first goal of this approach is to achieve **breakeven,** which occurs when the Lawson criterion is met or exceeded. The ultimate goal is **ignition,** which corresponds to a self-sustaining thermonuclear reaction, with a net generation of energy. As of 2000, ignition has not been achieved, either in tokamaks or in other magnetic confinement devices.

Inertial Confinement

Fig. 44-14 The small spheres on the quarter are deuterium–tritium fuel pellets, designed to be used in a laser fusion chamber.

This approach to confining and heating fusion fuel so that a thermonuclear reaction can occur involves "zapping" a solid fuel pellet from all sides with intense laser beams, evaporating some material from the surface of the pellet. This boiled-off material causes an inward-moving shock wave that compresses the core of the pellet, increasing both its particle density and its temperature. The process is called *inertial confinement* because (a) the fuel is *confined* to the pellet and (b) the particles do not escape from the heated pellet during the very short zapping interval because of their *inertia* (their mass).

Laser fusion, using the inertial confinement approach, is being investigated in many laboratories in the United States and elsewhere. At the Lawrence Livermore Laboratory, for example, deuterium–tritium fuel pellets, each smaller than a grain of sand (Fig. 44-14), are to be zapped by 10 synchronized high-power laser pulses symmetrically arranged around the pellet. The laser pulses are designed to deliver, in total, some 200 kJ of energy to each fuel pellet in less than a nanosecond. This is a delivered power of about 2×10^{14} W during the pulse, which is roughly 100 times the total installed (sustained) electric power generating capacity of the world!

In an operating thermonuclear reactor of the laser-fusion type, fuel pellets are to be exploded—like miniature hydrogen bombs—at a rate of perhaps 10 to 100 per second. The feasibility of laser fusion as the basis of a thermonuclear power reactor has not been demonstrated as of 2000, but research is continuing at a vigorous pace.

Sample Problem 44-6

Suppose a fuel pellet in a laser fusion device contains equal numbers of deuterium and tritium atoms (and no other material). The density $d = 200\ \text{kg/m}^3$ of the pellet is increased by a factor of 10^3 by the action of the laser pulses.

(a) How many particles per unit volume (both deuterons and tritons) does the pellet contain in its compressed state? The molar mass M_d of deuterium atoms is 2.0×10^{-3} kg/mol, and the molar mass M_t of tritium atoms is 3.0×10^{-3} kg/mol.

SOLUTION: The **Key Idea** here is that, for a system consisting of only one type of particle, we can write the (mass) density of the system in terms of the particles' masses and their number density (the number of particles per unit volume):

$$\begin{pmatrix}\text{density,}\\ \text{kg/m}^3\end{pmatrix} = \begin{pmatrix}\text{number density,}\\ \text{m}^{-3}\end{pmatrix}\begin{pmatrix}\text{particle mass,}\\ \text{kg}\end{pmatrix}. \quad (44\text{-}17)$$

Let n be the total number of particles per unit volume in the compressed pellet. Then the number of deuterium atoms per unit volume is $n/2$, and the number of tritium atoms per unit volume is also $n/2$.

Next, we can extend Eq. 44-17 to the system consisting of the two types of particles by writing the density d^* of the compressed pellet as the sum of the individual densities:

$$d^* = \frac{n}{2} m_d + \frac{n}{2} m_t, \tag{44-18}$$

where m_d and m_t are the masses of a deuterium atom and a tritium atom, respectively. We can replace those masses with the given molar masses by substituting

$$m_d = \frac{M_d}{N_A} \quad \text{and} \quad m_t = \frac{M_t}{N_A},$$

where N_A is Avogadro's number. After making those replacements and substituting $1000d$ for the compressed density d^*, we solve Eq. 44-18 for n to obtain

$$n = \frac{2000 d N_A}{M_d + M_t},$$

which gives us

$$n = \frac{(2000)(200\ \text{kg/m}^3)(6.02 \times 10^{23}\ \text{mol}^{-1})}{2.0 \times 10^{-3}\ \text{kg/mol} + 3.0 \times 10^{-3}\ \text{kg/mol}}$$
$$= 4.8 \times 10^{31}\ \text{m}^{-3}. \quad \text{(Answer)}$$

(b) According to Lawson's criterion, how long must the pellet maintain this particle density if breakeven operation is to take place?

SOLUTION: The **Key Idea** here is that, if breakeven operation is to occur, the compressed density must be maintained for a time period τ given by Eq. 44-16 ($n\tau > 10^{20}$ s/m^3). Thus, we have

$$\tau > \frac{10^{20}\ \text{s/m}^3}{4.8 \times 10^{31}\ \text{m}^{-3}} \approx 10^{-12}\ \text{s}. \quad \text{(Answer)}$$

(The plasma temperature must also be suitably high.)

REVIEW & SUMMARY

Energy from the Nucleus Nuclear processes are about a million times more effective, per unit mass, than chemical processes in transforming mass into other forms of energy.

Nuclear Fission Equation 44-1 shows a **fission** of ^{236}U induced by thermal neutrons bombarding ^{235}U. Equations 44-2 and 44-3 show the beta-decay chains of the primary fragments. The energy released in such a fission event is $Q \approx 200$ MeV.

Fission can be understood in terms of the collective model, in which a nucleus is likened to a charged liquid drop carrying a certain excitation energy. A potential barrier must be tunneled if fission is to occur. Fissionability depends on the relationship between the barrier height E_b and the excitation energy E_n.

The neutrons released during fission make possible a fission **chain reaction.** Figure 44-4 shows the neutron balance for one cycle of a typical reactor. Figure 44-5 suggests the layout of a complete nuclear power plant.

Nuclear Fusion The release of energy by the **fusion** of two light nuclei is inhibited by their mutual Coulomb barrier. Fusion can occur in bulk matter only if the temperature is high enough (that is, if the particle energy is high enough) for appreciable barrier tunneling to occur.

The Sun's energy arises mainly from the thermonuclear burning of hydrogen to form helium by the **proton-proton cycle** outlined in Fig. 44-11. Elements up to $A \approx 56$ (the peak of the binding energy curve) can be built up by other fusion processes once the hydrogen fuel supply of a star has been exhausted.

Controlled Fusion Controlled **thermonuclear fusion** for energy generation has not yet been achieved. The d-d and d-t reactions are the most promising mechanisms. A successful fusion reactor must satisfy **Lawson's criterion,**

$$n\tau > 10^{20}\ \text{s/m}^3, \tag{44-16}$$

and must have a suitably high plasma temperature T.

In a **tokamak** the plasma is confined by a magnetic field. In **laser fusion** inertial confinement is used.

QUESTIONS

1. In Table 44-1, does the relation $Q = -\Delta m\, c^2$ apply to (a) all the processes, (b) all the processes except the waterfall, (c) the fission processes only, (d) the fission and fusion processes only?

2. According to Fig. 44-1, fission of ^{235}U by thermal neutrons into two equally massive fragments occurs in about one event in (a) 10 000, (b) 1000, (c) 100, (d) 10.

3. Do the initial fragments formed by fission have (a) more protons than neutrons, (b) more neutrons than protons, or (c) about the same number of each?

4. Consider the fission reaction

$$^{235}\text{U} + \text{n} \rightarrow \text{X} + \text{Y} + 2\text{n}.$$

Rank the following possible nuclides for representation by X (or Y), most likely first: (a) ^{152}Nd, (b) ^{140}I, (c) ^{128}In, (d) ^{115}Pd, (e) ^{105}Mo. (*Hint:* See Fig. 44-1.)

5. Pick the most likely member of each of these pairs to be one of the initial fragments formed by a fission event: (a) ^{93}Sr or ^{93}Ru, (b) ^{140}Gd or ^{140}I, (c) ^{155}Nd or ^{155}Lu. (*Hint:* See Fig. 43-4 and the periodic table.)

6. Suppose that a ^{238}U nucleus "swallows" a neutron and then decays not by fission but by beta decay, emitting an electron and a neutrino. Which nuclide remains after this decay: (a) ^{239}Pu, (b) ^{238}Np, (c) ^{239}Np, or (d) ^{238}Pa?

7. A nuclear reactor is operating at a certain power level, with its multiplication factor k adjusted to unity. If the control rods are used to reduce the power output of the reactor to 25% of its former value, is the multiplication factor now (a) a little less than unity, (b) substantially less than unity, or (c) still equal to unity?

8. A nuclear reactor core should have the smallest possible surface-to-volume ratio. Order the following solids according to their surface-to-volume ratios, greatest first: (a) a cube of edge a, (b) a sphere of radius a, (c) a cone of height a and base radius a, and (d) a cylinder of radius a and height a. (The area of the *curved* surface of the cone is $\sqrt{2}\pi a^2$ and its volume is $\pi a^3/3$.)

9. Figure 44-6 shows how the heat generated by the nuclear waste from one year's operation of a large nuclear power plant decays over time. By what approximate factor has this thermal energy output decreased at the end of 100 years: (a) 20, (b) 200, (c) 2000, (d) more than 2000?

10. Which of these elements is *not* "cooked up" by thermonuclear fusion processes in stellar interiors: (a) carbon, (b) silicon, (c) chromium, (d) bromine?

11. About 2% of the energy generated in the Sun's core by the p-p reaction is carried out of the Sun by neutrinos. Is the energy associated with this neutrino flux (a) equal to, (b) greater than, or (c) less than the energy radiated from the Sun's surface as electromagnetic radiation?

12. Lawson's criterion for the d-t reaction (Eq. 44-16) is $n\tau > 10^{20}$ s/m^3. For the d-d reaction, do you expect the number on the right-hand side to be (a) the same, (b) smaller, or (c) larger?

EXERCISES & PROBLEMS

ssm Solution is in the Student Solutions Manual.
www Solution is available on the World Wide Web at: http://www.wiley.com/college/hrw
ilw Solution is available on the Interactive LearningWare.

SEC. 44-2 Nuclear Fission: The Basic Process

1E. (a) How many atoms are contained in 1.0 kg of pure ^{235}U? (b) How much energy, in joules, is released by the complete fissioning of 1.0 kg of ^{235}U? Assume Q = 200 MeV. (c) For how long would this energy light a 100 W lamp? **ssm**

2E. Complete the following table, which refers to the generalized fission reaction $^{235}\text{U} + \text{n} \rightarrow \text{X} + \text{Y} + b\text{n}$.

X	Y	b
^{140}Xe	—	1
^{139}I	—	2
—	^{100}Zr	2
^{141}Cs	^{92}Rb	—

3E. At what rate must ^{235}U nuclei undergo fission by neutron bombardment to generate energy at the rate of 1.0 W? Assume that Q = 200 MeV. **ssm**

4E. The fission properties of the plutonium isotope ^{239}Pu are very similar to those of ^{235}U. The average energy released per fission is 180 MeV. How much energy, in MeV, is released if all the atoms in 1.00 kg of pure ^{239}Pu undergo fission?

5E. Verify that, as stated in Section 44-2, neutrons in equilibrium with matter at room temperature, 300 K, have an average kinetic energy of about 0.04 eV.

6E. Calculate the disintegration energy Q for the fission of ^{98}Mo into two equal parts. The masses you will need are 97.905 41 u for ^{98}Mo and 48.950 02 u for ^{49}Sc. If Q turns out to be positive, discuss why this process does not occur spontaneously.

7E. Calculate the disintegration energy Q for the fission of ^{52}Cr into two equal fragments. The masses you will need are 51.940 51 u for ^{52}Cr and 25.982 59 u for ^{26}Mg. **ssm**

8E. ^{235}U decays by alpha emission with a half-life of 7.0×10^8 y. It also decays (rarely) by spontaneous fission, and if the alpha decay did not occur, its half-life due to this process alone would be 3.0×10^{17} y. (a) At what rate do spontaneous fission decays occur in 1.0 g of ^{235}U? (b) How many ^{235}U alpha-decay events are there for every spontaneous fission event?

9E. Calculate the energy released in the fission reaction

$$^{235}\text{U} + \text{n} \rightarrow {}^{141}\text{Cs} + {}^{93}\text{Rb} + 2\text{n}.$$

Needed atomic and particle masses are

^{235}U	235.043 92 u	^{93}Rb	92.921 57 u
^{141}Cs	140.919 63 u	n	1.008 67 u

10P. Verify that, as reported in Table 44-1, fissioning of the ^{235}U in 1.0 kg of UO_2 (enriched so that ^{235}U is 3.0% of the total uranium) could keep a 100 W lamp burning for 690 y.

11P. In a particular fission event in which ^{235}U is fissioned by slow neutrons, no neutron is emitted and one of the primary fission fragments is ^{83}Ge. (a) What is the other fragment? (b) How is the disintegration energy Q = 170 MeV split between the two fragments? (c) Calculate the initial speed (just after the fission) of each fragment. **ssm** **www**

12P Consider the fission of ^{238}U by fast neutrons. In one fission event no neutrons are emitted and the final stable end products, after the beta decay of the primary fission fragments, are ^{140}Ce and ^{99}Ru. (a) How many beta-decay events are there in the two beta-decay chains, considered together? (b) Calculate Q for this fission process. The relevant atomic and particle masses are

^{238}U	238.050 79 u	^{140}Ce	139.905 43 u
n	1.008 67 u	^{99}Ru	98.905 94 u

13P. Assume that immediately after the fission of ^{236}U according to Eq. 44-1, the resulting ^{140}Xe and ^{94}Sr nuclei are just touching at their surfaces. (a) Assuming the nuclei to be spherical, calculate the electric potential energy (in MeV) associated with the repulsion between the two fragments. (*Hint:* Use Eq. 43-3 to calculate the radii of the fragments.) (b) Compare this energy with the energy released in a typical fission event. ssm

14P. A ^{236}U nucleus undergoes fission and breaks into two middle-mass fragments, ^{140}Xe and ^{96}Sr. (a) By what percentage does the surface area of the fission products differ from that of the original ^{236}U nucleus? (b) By what percentage does the volume change? (c) By what percentage does the electric potential energy change? The electric potential energy of a uniformly charged sphere of radius r and charge Q is given by

$$U = \frac{3}{5}\left(\frac{Q^2}{4\pi\varepsilon_0 r}\right).$$

SEC. 44-4 The Nuclear Reactor

15E. A 200 MW fission reactor consumes half its fuel in 3.00 y. How much ^{235}U did it contain initially? Assume that all the energy generated arises from the fission of ^{235}U and that this nuclide is consumed only by the fission process. ssm

16E. Repeat Exercise 15 taking into account nonfission neutron capture by the ^{235}U.

17E. ^{238}Np requires 4.2 MeV for fission. To remove a neutron from this nuclide requires an energy expenditure of 5.0 MeV. Is ^{237}Np fissionable by thermal neutrons?

18P. In an atomic bomb, energy release is due to the uncontrolled fission of plutonium ^{239}Pu (or ^{235}U). The bomb's rating is the magnitude of the released energy, specified in terms of the mass of TNT required to produce the same energy release. One megaton (10^6 tons) of TNT releases 2.6×10^{28} MeV of energy. (a) Calculate the rating, in tons of TNT, of an atomic bomb containing 95 kg of ^{239}Pu, of which 2.5 kg actually undergoes fission. (See Exercise 4.) (b) Why is the other 92.5 kg of ^{239}Pu needed if it does not fission?

19P. The thermal energy generated when radiations from radionuclides are absorbed in matter can serve as the basis for a small power source for use in satellites, remote weather stations, and other isolated locations. Such radionuclides are manufactured in abundance in nuclear reactors and may be separated chemically from the spent fuel. One suitable radionuclide is ^{238}Pu ($T_{1/2}$ = 87.7 y), which is an alpha emitter with Q = 5.50 MeV. At what rate is thermal energy generated in 1.00 kg of this material? ssm

20P. (See Problem 19.) Among the many fission products that may be extracted chemically from the spent fuel of a nuclear reactor is ^{90}Sr ($T_{1/2}$ = 29 y). This isotope is produced in typical large reactors at the rate of about 18 kg/y. By its radioactivity it generates thermal energy at the rate of 0.93 W/g. (a) Calculate the effective disintegration energy Q_{eff} associated with the decay of a ^{90}Sr nucleus. (Q_{eff} includes contributions from the decay of the ^{90}Sr daughter products in its decay chain but not from neutrinos, which escape totally from the sample.) (b) It is desired to construct a power source generating 150 W (electric) to use in operating electronic equipment in an underwater acoustic beacon. If the power source is based on the thermal energy generated by ^{90}Sr and if the efficiency of the thermal–electric conversion process is 5.0%, how much ^{90}Sr is needed?

21P. Many fear that helping additional nations develop nuclear power reactor technology will increase the likelihood of nuclear war because reactors can be used not only to produce electrical energy but, as a by-product through neutron capture with inexpensive ^{238}U, to make ^{239}Pu, which is a "fuel" for nuclear bombs. What simple series of reactions involving neutron capture and beta decay would yield this plutonium isotope?

22P. The neutron generation time t_{gen} in a reactor is the average time needed for a fast neutron emitted in one fission to be slowed to thermal energies by the moderator and to initiate another fission. Suppose that the power output of a reactor at time $t = 0$ is P_0. Show that the power output a time t later is $P(t)$, where

$$P(t) = P_0 k^{t/t_{\text{gen}}},$$

and k is the multiplication factor. For constant power output, $k = 1$.

23P. A 66 kiloton atomic bomb (see Problem 18) is fueled with pure ^{235}U (Fig. 44-15), 4.0% of which actually undergoes fission. (a) How much uranium is in the bomb? (b) How many primary fission fragments are produced? (c) How many neutrons generated in the fissions are released to the environment? (On average, each fission produces 2.5 neutrons.) ssm

Fig. 44-15 Problem 23. A "button" of ^{235}U, ready to be recast and machined for a warhead.

24P. The neutron generation time t_{gen} (see Problem 22) in a particular reactor is 1.0 ms. If the reactor is operating at a power level of 500 MW, about how many free neutrons are present in the reactor at any moment?

25P. The neutron generation time (see Problem 22) of a particular reactor is 1.3 ms. The reactor is generating energy at the rate of 1200 MW. To perform certain maintenance checks, the power level must temporarily be reduced to 350 MW. It is desired that the transition to the reduced power level take 2.6 s. To what (constant) value should the multiplication factor be set to effect the transition in the desired time? ssm www

26P. A reactor operates at 400 MW with a neutron generation time (see Problem 22) of 30.0 ms. If its power increases for 5.00 min with a multiplication factor of 1.0003, what is the power output at the end of the 5.00 min?

27P. (a) A neutron of mass m_n and kinetic energy K makes a head-on elastic collision with a stationary atom of mass m. Show that the fractional kinetic energy loss of the neutron is given by

$$\frac{\Delta K}{K} = \frac{4m_n m}{(m + m_n)^2}.$$

(b) Find $\Delta K/K$ for each of the following, acting as the stationary atom: hydrogen, deuterium, carbon, and lead. (c) If $K = 1.00$ MeV initially, how many such head-on collisions would it take to reduce the neutron's kinetic energy to a thermal value (0.025 eV) if the stationary atoms it collides with are deuterium, a commonly used moderator? (In actual moderators, most collisions are not head-on.) ssm

SEC. 44-5 A Natural Nuclear Reactor

28E. How long ago was the ratio $^{235}U/^{238}U$ in natural uranium deposits equal to 0.15?

29E. The natural fission reactor discussed in Section 44-5 is estimated to have generated 15 gigawatt-years of energy during its lifetime. (a) If the reactor lasted for 200 000 y, at what average power level did it operate? (b) How many kilograms of ^{235}U did it consume during its lifetime?

30P. Some uranium samples from the natural reactor site described in Section 44-5 were found to be slightly *enriched* in ^{235}U, rather than depleted. Account for this in terms of neutron absorption by the abundant isotope ^{238}U and the subsequent beta and alpha decay of its products.

31P. Mixed in with ^{238}U, uranium mined today contains 0.72% of fissionable ^{235}U, too little to make reactor fuel for thermal-neutron fission. For this reason, the natural uranium must be enriched with ^{235}U. Both ^{235}U ($T_{1/2} = 7.0 \times 10^8$ y) and ^{238}U ($T_{1/2} = 4.5 \times 10^9$ y) are radioactive. How far back in time would natural uranium have been a practical reactor fuel, with a $^{235}U/^{238}U$ ratio of 3.0%? ssm www

SEC. 44-6 Thermonuclear Fusion: The Basic Process

32E. From information given in the text, collect and write down the approximate heights of the Coulomb barriers for (a) the alpha decay of ^{238}U and (b) the fission of ^{235}U by thermal neutrons.

33E. Calculate the height of the Coulomb barrier for the head-on collision of two deuterons. Take the effective radius of a deuteron to be 2.1 fm. ssm

34E. Verify that the fusion of 1.0 kg of deuterium by the reaction

$$^2H + {}^2H \rightarrow {}^3He + n \qquad (Q = +3.27 \text{ MeV})$$

could keep a 100 W lamp burning for 3×10^4 y.

35E. Methods other than heating the material have been suggested for overcoming the Coulomb barrier for fusion. For example, one might consider particle accelerators. If you were to use two of them to accelerate two beams of deuterons directly toward each other so as to collide head-on, (a) what voltage would each accelerator require for the colliding deuterons to overcome the Coulomb barrier? (b) Why do you suppose this method is not presently used?

36P. Calculate the Coulomb barrier height for two ^7Li nuclei that are fired at each other with the same initial kinetic energy K. (*Hint:* Use Eq. 43-3 to calculate the radii of the nuclei.)

37P. In Fig. 44-10, the equation for $n(K)$, the number density per unit energy for particles, is

$$n(K) = 1.13n \frac{K^{1/2}}{(kT)^{3/2}} e^{-K/kT},$$

where n is the total number density of particles. At the center of the Sun the temperature is 1.50×10^7 K and the average proton energy K_{avg} is 1.94 keV. Find the ratio of the number density of protons at 5.00 keV to that at the mean proton energy.

38P. Expressions for the Maxwell speed distribution for molecules in a gas are given in Chapter 20. (a) Show that the *most probable energy* is given by

$$K_p = \tfrac{1}{2}kT.$$

Verify this result with the energy distribution curve of Fig. 44-10, for which $T = 1.5 \times 10^7$ K. (b) Show that the *most probable speed* is given by

$$v_p = \sqrt{\frac{2kT}{m}}.$$

Find its value for protons at $T = 1.5 \times 10^7$ K. (c) Show that the *energy corresponding to the most probable speed* (which is not the same as the most probable energy) is

$$K_{v,p} = kT.$$

Locate this quantity on the curve of Fig. 44-10.

SEC. 44-7 Thermonuclear Fusion in the Sun and Other Stars

39E. Show that the energy released when three alpha particles fuse to form ^{12}C is 7.27 MeV. The atomic mass of ^4He is 4.0026 u, and that of ^{12}C is 12.0000 u. ssm

40E. We have seen that Q for the overall proton-proton fusion cycle is 26.7 MeV. How can you relate this number to the Q values for the reactions that make up this cycle, as displayed in Fig. 44-11?

41E. At the center of the Sun the density is 1.5×10^5 kg/m^3 and the composition is essentially 35% hydrogen by mass and 65% helium. (a) What is the density of protons at the center of the Sun? (b) How much greater is this than the density of particles in an ideal gas at standard conditions of temperature (0°C) and pressure (1.01×10^5 Pa)?

42P. Verify the three Q values reported for the reactions given in Fig. 44-11. The needed atomic and particle masses are

^1H	1.007 825 u	^4He	4.002 603 u
^2H	2.014 102 u	$e^{\pm}$	0.000 548 6 u
^3He	3.016 029 u		

(*Hint:* Distinguish carefully between atomic and nuclear masses, and take the positrons properly into account.)

43P. The Sun has a mass of 2.0×10^{30} kg and radiates energy at the rate of 3.9×10^{26} W. (a) At what rate does the Sun transfer its mass to other forms of energy? (b) What fraction of its original mass has the Sun lost in this way since it began to burn hydrogen, about 4.5×10^9 y ago? ssm

44P. Calculate and compare the energy released by (a) the fusion of 1.0 kg of hydrogen deep within the Sun and (b) the fission of 1.0 kg of ^{235}U in a fission reactor.

45P. (a) Calculate the rate at which the Sun generates neutrinos. Assume that energy production is entirely by the proton-proton fusion cycle. (b) At what rate do solar neutrinos reach Earth?

46P. In certain stars the *carbon cycle* is more likely than the proton-proton cycle to be effective in generating energy. This cycle is

$$\begin{aligned}
{}^{12}\mathrm{C} + {}^{1}\mathrm{H} &\rightarrow {}^{13}\mathrm{N} + \gamma, & Q_1 &= 1.95\ \mathrm{MeV},\\
{}^{13}\mathrm{N} &\rightarrow {}^{13}\mathrm{C} + e^+ + \nu, & Q_2 &= 1.19,\\
{}^{13}\mathrm{C} + {}^{1}\mathrm{H} &\rightarrow {}^{14}\mathrm{N} + \gamma, & Q_3 &= 7.55,\\
{}^{14}\mathrm{N} + {}^{1}\mathrm{H} &\rightarrow {}^{15}\mathrm{O} + \gamma, & Q_4 &= 7.30,\\
{}^{15}\mathrm{O} &\rightarrow {}^{15}\mathrm{N} + e^+ + \nu, & Q_5 &= 1.73,\\
{}^{15}\mathrm{N} + {}^{1}\mathrm{H} &\rightarrow {}^{12}\mathrm{C} + {}^{4}\mathrm{He}, & Q_6 &= 4.97.
\end{aligned}$$

(a) Show that this cycle of reactions is exactly equivalent in its overall effects to the proton-proton cycle of Fig. 44-11. (b) Verify that the two cycles, as expected, have the same Q value.

47P. Coal burns according to the reaction $C + O_2 \rightarrow CO_2$. The heat of combustion is 3.3×10^7 J/kg of atomic carbon consumed. (a) Express this in terms of energy per carbon atom. (b) Express it in terms of energy per kilogram of the initial reactants, carbon and oxygen. (c) Suppose that the Sun (mass $= 2.0 \times 10^{30}$ kg) were made of carbon and oxygen in combustible proportions and that it continued to radiate energy at its present rate of 3.9×10^{26} W. How long would the Sun last? ssm

48P. Assume that the core of the Sun has one-eighth of the Sun's mass and is compressed within a sphere whose radius is one-fourth of the solar radius. Assume further that the composition of the core is 35% hydrogen by mass and that essentially all the Sun's energy is generated there. If the Sun continues to burn hydrogen at the rate calculated in Sample Problem 44-5, how long will it be before the hydrogen is entirely consumed? The Sun's mass is 2.0×10^{30} kg.

49P. A star converts all its hydrogen to helium, achieving a 100% helium composition. Next it converts the helium to carbon via the triple-alpha process,

$$^{4}\mathrm{He} + {}^{4}\mathrm{He} + {}^{4}\mathrm{He} \rightarrow {}^{12}\mathrm{C} + 7.27\ \mathrm{MeV}.$$

The mass of the star is 4.6×10^{32} kg, and it generates energy at the rate of 5.3×10^{30} W. How long will it take to convert all the helium to carbon at this rate? ssm

50P. The effective Q for the proton-proton cycle of Fig. 44-11 is 26.2 MeV. (a) Express this as energy per kilogram of hydrogen consumed. (b) The power of the Sun is 3.9×10^{26} W. If its energy derives from the proton-proton cycle, at what rate is it losing hydrogen? (c) At what rate is it losing mass? Account for the difference in the results for (b) and (c). (d) The mass of the Sun is 2.0×10^{30} kg. If it loses mass at the constant rate calculated in (c), how long will it take to lose 0.10% of its mass?

51P. Figure 44-16 shows an early proposal for a hydrogen bomb. The fusion fuel is deuterium, ^{2}H. The high temperature and particle density needed for fusion are provided by an atomic bomb "trigger," which involves a ^{235}U or ^{239}Pu fission fuel that is arranged to impress an imploding, compressive shock wave on the deuterium. The operative fusion reaction is

$$5\ {}^{2}\mathrm{H} \rightarrow {}^{3}\mathrm{He} + {}^{4}\mathrm{He} + {}^{1}\mathrm{H} + 2\mathrm{n}.$$

(a) Calculate Q for the fusion reaction. For needed atomic masses, see Problem 42. (b) Calculate the rating (see Problem 18) of the fusion part of the bomb if it contains 500 kg of deuterium, 30.0% of which undergoes fusion.

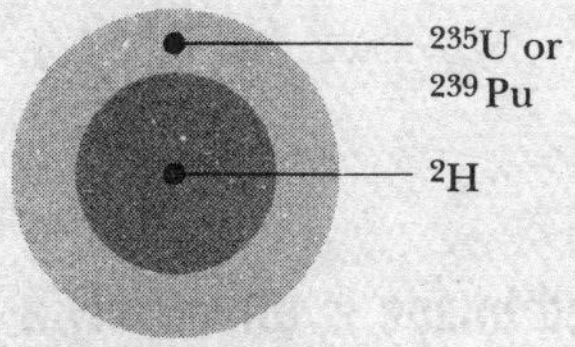

Fig. 44-16 Problem 51.

SEC. 44-8 Controlled Thermonuclear Fusion

52E. Verify the Q values reported in Eqs. 44-13, 44-14, and 44-15. The needed masses are

^{1}H	1.007 825 u	^{4}He	4.002 603 u
^{2}H	2.014 102 u	n	1.008 665 u
^{3}H	3.016 049 u		

53P. Ordinary water consists of roughly 0.0150% by mass of "heavy water," in which one of the two hydrogens is replaced with deuterium, ^{2}H. How much average fusion power could be obtained if we "burned" all the ^{2}H in 1 liter of water in 1 day by somehow causing the deuterium to fuse via the reaction $^2\mathrm{H} + {}^2\mathrm{H} \rightarrow {}^3\mathrm{He} + \mathrm{n}$? ssm

54P. In the deuteron-triton fusion reaction of Eq. 44-15, how is the reaction energy Q shared between the alpha particle and the neutron? Neglect the relatively small kinetic energies of the two combining particles.

45 Quarks, Leptons, and the Big Bang

This color-coded image is effectively a photograph of the universe when it was only 300 000 y old, which was about 15 $\times$ 10^9 y ago. This is what you would have seen then as you looked away in all directions (the view has been condensed to this oval). Patches of light from atoms stretch across the "sky," but galaxies, stars, and planets have not yet formed.

How can such a photograph of the early universe be taken?

The answer is in this chapter.

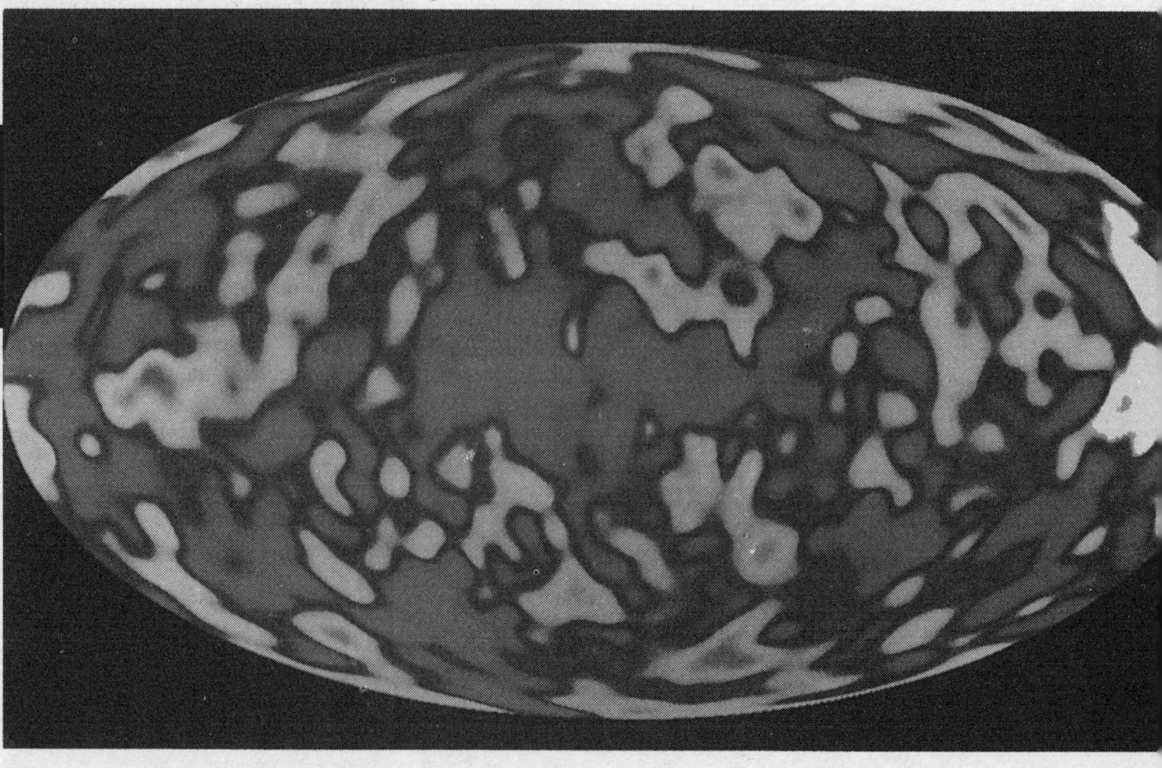

45-1 Life at the Cutting Edge

Physicists often refer to the theories of relativity and quantum physics as "modern physics," to distinguish them from the theories of Newtonian mechanics and Maxwellian electromagnetism, which are lumped together as "classical physics." As the years go by, the word "modern" seems less and less appropriate for theories whose foundations were laid down in the opening years of the twentieth century. Nevertheless, the label hangs on.

In this closing chapter we consider two lines of investigation that are truly "modern" but at the same time have the most ancient of roots. They center around two deceptively simple questions:

What is the universe made of?

How did the universe come to be the way it is?

Progress in answering these questions has been rapid in the last few decades.

Many new insights are based on experiments carried out with large particle accelerators. However, as physicists bang particles together at higher and higher energies using larger and larger accelerators, they come to realize that no conceivable Earth-bound accelerator can generate particles with energies great enough to test their ultimate theories. There has been only one source of particles with these energies, and that was the universe itself within the first millisecond of its existence.

In this chapter you will encounter a host of new terms and a veritable flood of particles with names that you should not try to remember. If you are temporarily bewildered, you are sharing the bewilderment of the physicists who lived through these developments and who at times saw nothing but increasing complexity with little hope of understanding. If you stick with it, however, you will come to share the excitement physicists felt as marvelous new accelerators poured out new results, as the theorists put forth ideas each more daring than the last, and as clarity finally sprang from obscurity.

45-2 Particles, Particles, Particles

In the 1930s, there were many scientists who thought that the problem of the ultimate structure of matter was well on the way to being solved. The atom could be understood in terms of only three particles—the electron, the proton, and the neutron. Quantum physics accounted well for the structure of the atom and for radioactive alpha decay. The neutrino had been postulated and, although not yet observed, had been incorporated by Enrico Fermi into a successful theory of beta decay. There was hope that quantum theory, applied to protons and neutrons, would soon account for the structure of the nucleus. What else was there?

The euphoria did not last. The end of that same decade saw the beginning of a period of discovery of new particles that continues to this day. The new particles have names and symbols such as *muon* (μ), *pion* (π), *kaon* (K), and *sigma* (Σ). All the new particles are unstable; that is, they spontaneously transform into other types of particles according to the same functions of time that apply to unstable nuclei. Thus, if N_0 particles of any one type are present in a sample at time $t = 0$, then the number N of those particles present at some later time t is given by Eq. 43-14,

$$N = N_0 e^{-\lambda t}; \qquad (45\text{-}1)$$

the rate of decay R, from an initial value of R_0, is given by Eq. 43-15,

$$R = R_0 e^{-\lambda t}; \qquad (45\text{-}2)$$

Fig. 45-1 The OPAL (omni-purpose apparatus) particle detector at CERN, the European high-energy particle physics laboratory near Geneva, Switzerland. OPAL is designed to measure the energies of particles produced in electron–positron collisions at energies of 50 GeV. Although the detector is huge, it is small compared with the accelerator itself, which is a ring with a circumference of 27 km.

and the half-life $T_{1/2}$, decay constant λ, and mean life τ are related by Eq. 43-17,

$$T_{1/2} = \frac{\ln 2}{\lambda} = \tau \ln 2. \tag{45-3}$$

The half-lives of the new particles range from about 10^{-6} s to 10^{-23} s. Indeed, some of the particles last so briefly that they cannot be detected directly but can only be inferred from indirect evidence.

These new particles are commonly produced in head-on collisions between protons or electrons accelerated to high energies in accelerators at places like Fermilab (near Chicago), CERN (near Geneva), SLAC (at Stanford), and DESY (near Hamburg, Germany). They are discovered with particle detectors that have grown in sophistication until (see Fig. 45-1) they rival the size and complexity of entire accelerators of only a few decades ago.

Today there are several hundred known particles. Naming them has strained the resources of the Greek alphabet, and most are known only by an assigned number in a periodically issued compilation. To make sense of this array of particles, we look for simple physical criteria. We can make a first rough cut among the particles in at least the following three ways.

Fermion or Boson?

All particles have an intrinsic angular momentum called **spin,** as we discussed for electrons, protons, and neutrons in Section 32-4. Generalizing the notation of that section, we can write the component of spin $\vec{S}$ in any direction (assume it to be along a z axis) as

$$S_z = m_s\hbar \qquad \text{for } m_s = s, s-1, \ldots, -s, \tag{45-4}$$

in which $\hbar$ is $h/2\pi$, m_s is the *spin magnetic quantum number,* and s is the *spin quantum number*. The latter can have either positive half-integer values ($\frac{1}{2}, \frac{3}{2}, \ldots$) or nonnegative integer values (0, 1, 2, . . .). For example, an electron has the value $s = \frac{1}{2}$. Hence the spin of an electron (measured along any direction) can have the values

$$S_z = \tfrac{1}{2}\hbar \qquad \text{(spin up)}$$

or

$$S_z = -\tfrac{1}{2}\hbar \qquad \text{(spin down)}.$$

Confusingly, the term *spin* is actually used in two ways: It properly means a particle's intrinsic angular momentum $\vec{S}$, but it is often used loosely to mean the particle's spin quantum number s. In the latter case, for example, an electron is said to be a spin-$\frac{1}{2}$ particle.

Particles with half-integer spin quantum numbers (like electrons) are called **fermions,** after Fermi, who (simultaneously with Paul Dirac) discovered the statistical laws that govern their behavior. Like electrons, protons and neutrons also have $s = \frac{1}{2}$ and are fermions.

Particles with zero or integer spin quantum numbers are called **bosons,** after Indian physicist Satyendra Nath Bose, who (simultaneously with Albert Einstein) discovered the governing statistical laws for *those* particles. Photons, which have $s = 1$, are bosons; you will soon meet other particles in this class.

This may seem a trivial way to classify particles, but it is very important for this reason:

> Fermions obey the Pauli exclusion principle, which asserts that only a single particle can be assigned to a given quantum state. Bosons *do not* obey this principle. Any number of bosons can occupy a given quantum state.

We saw how important the Pauli exclusion principle is when we "built up" the atoms by assigning (spin-$\frac{1}{2}$) electrons to individual quantum states. Using that principle led to a full accounting of the structure and properties of atoms of different types and of solids such as metals and semiconductors.

Because bosons do *not* obey the Pauli principle, those particles tend to pile up in the quantum state of lowest energy. In 1995 a group in Boulder, Colorado, succeeded in producing a condensate of about 2000 rubidium-87 atoms—they are bosons—in a single quantum state of approximately zero energy.

For this to happen, the rubidium has to be a vapor with a temperature so low and a density so great that the de Broglie wavelengths of the individual atoms are greater than the average separation between the atoms. When this condition is met, the wave functions of the individual atoms overlap and the entire assembly becomes a single quantum system, called a *Bose–Einstein condensate*. Figure 45-2 shows that, as the temperature of the rubidium vapor is lowered to about 1.70×10^{-7} K, this system does indeed "collapse" into a single sharply defined state corresponding to approximately zero speed for its atoms.

Hadron or Lepton?

We can also classify particles in terms of the four fundamental forces that act on them. The *gravitational force* acts on *all* particles, but its effects at the level of subatomic particles are so weak that we need not consider that force (at least not in today's research). The *electromagnetic force* acts on all *electrically charged* particles; its effects are well known and we can take them into account when we need to; we largely ignore this force in this chapter.

We are left with the *strong force,* which is the force that binds nucleons together, and the *weak force,* which is involved in beta decay and similar processes. The weak force acts on all particles, the strong force only on some.

We can, then, roughly classify particles on the basis of whether the strong force acts on them. Particles on which the *strong force* acts are called **hadrons.** Particles

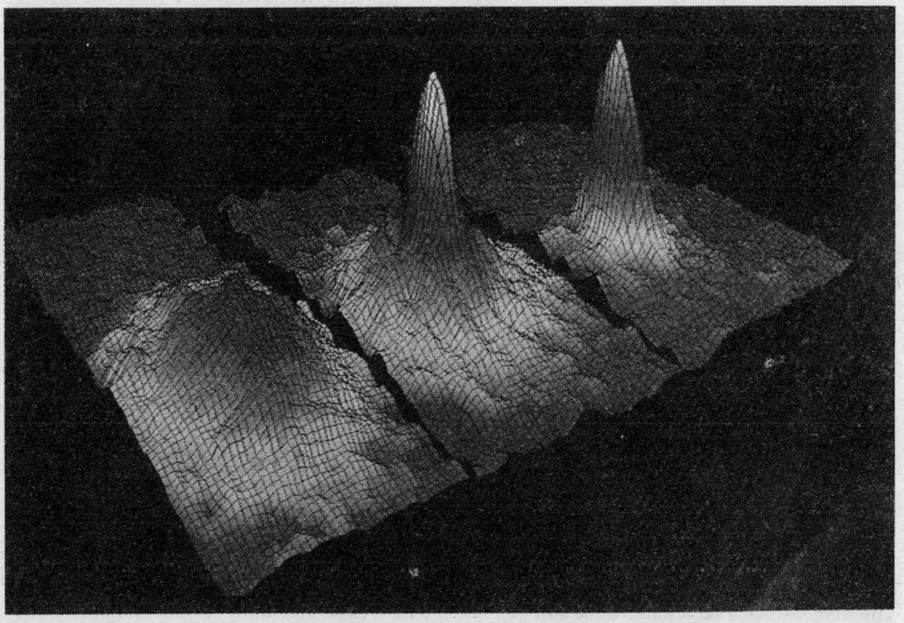

Fig. 45-2 Three plots of the particle speed distribution in a vapor of rubidium-87 atoms. The temperature of the vapor is successively reduced from plot (*a*) to plot (*c*). Plot (*c*) shows a sharp peak centered around zero speed; that is, all the atoms are in the same quantum state. The achievement of such a Bose–Einstein condensate, often called the Holy Grail of atomic physics, was finally recorded in 1995.

on which the strong force does *not* act, leaving the weak force as the dominant force, are called **leptons.** Protons, neutrons, and pions are hadrons; electrons and neutrinos are leptons. You will soon meet other members of these classes.

We can make a further distinction among the hadrons because some of them are bosons (we call them **mesons**); the pion is an example. The other hadrons are fermions (we call them **baryons**); the proton is an example.

Particle or Antiparticle?

In 1928 Dirac predicted that the electron e^- should have a positively charged counterpart of the same mass and spin. The counterpart, the *positron* e^+, was discovered in cosmic radiation in 1932 by Carl Anderson. Physicists then gradually realized that *every* particle has a corresponding **antiparticle.** The members of such pairs have the same mass and spin but opposite signs of electric charge (if they are charged) and opposite signs of quantum numbers that we have not yet discussed.

At first, *particle* was used to refer to the common particles such as electrons, protons, and neutrons, and *antiparticle* referred to their rarely detected counterparts. Later, for the less common particles, the assignment of *particle* and *antiparticle* was made so as to be consistent with certain conservation laws that we shall discuss later in this chapter. (Confusingly, both particles and antiparticles are sometimes called particles when no distinction is needed.) We often, but not always, represent an antiparticle by putting a bar over the symbol for the particle. Thus, p is the symbol for the proton, and $\bar{p}$ (pronounced "p bar") is the symbol for the antiproton.

When a particle meets its antiparticle, the two can *annihilate* each other. That is, the particle and antiparticle disappear and their combined energies reappear in other forms. For an electron annihilating with a positron, this energy reappears as two gamma-ray photons:

$$e^- + e^+ \rightarrow \gamma + \gamma. \qquad (45\text{-}5)$$

If the electron and positron are stationary when they annihilate, their total energy is their total mass energy, and that energy is then shared equally by the two photons. To conserve momentum and because photons cannot be stationary, the photons fly off in opposite directions.

In 1996 physicists at CERN succeeded in producing, for a few fleeting nanoseconds, a handful of antihydrogen atoms, each consisting of a positron and an antiproton bound together (presumably just as the electron and proton of a hydrogen atom are bound together). Such an assembly of antiparticles is called *antimatter* to distinguish it from an assembly of common particles (*matter*).

One can speculate that there are galaxies of antimatter, complete with atoms, molecules, and even physicists. One can even contemplate the disaster that would occur if, say, an asteroid freed from such a galaxy collided with (hence annihilating) a section of Earth. Thankfully, however, the present view is that not only our galaxy but the universe as a whole consists largely of matter rather than antimatter. (This lack of symmetry is disturbing to a physicist, who normally expects to find symmetry in nature.)

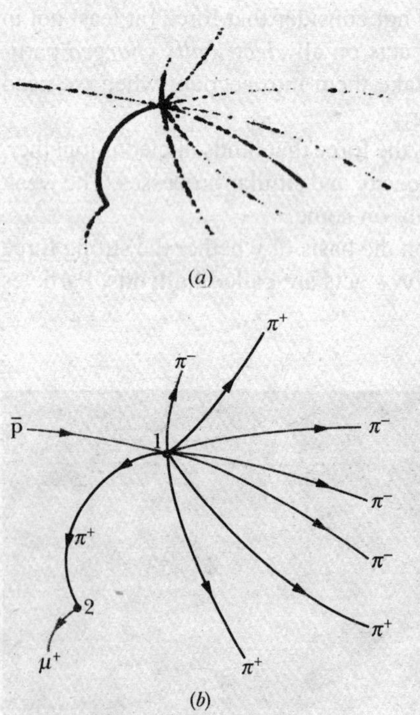

Fig. 45-3 (*a*) A bubble-chamber photograph of a series of events initiated by an antiproton that enters the chamber from the left. (*b*) The tracks redrawn and labeled for clarity. The dots at points 1 and 2 indicate the sites of specific secondary events that are described in the text. The tracks are curved because a magnetic field present in the chamber exerts a deflecting force on each moving charged particle.

45-3 An Interlude

Before pressing on with the task of classifying the particles, let us step aside for a moment and capture some of the spirit of particle research by analyzing a typical particle event—namely, that shown in the bubble-chamber photograph of Fig. 45-3*a*.

The tracks in this figure consist of bubbles formed along the paths of electrically charged particles as they move through a chamber filled with liquid hydrogen. We

TABLE 45-1 The Particles or Antiparticles Involved in the Event of Fig. 45-3

Particle	Symbol	Charge q	Mass (MeV/c^2)	Spin s	Identity	Mean Life (s)	Antiparticle
Neutrino	ν	0	0	$\frac{1}{2}$	Lepton	Stable	$\bar{\nu}$
Electron	e^-	-1	0.511	$\frac{1}{2}$	Lepton	Stable	e^+
Muon	μ^-	-1	105.7	$\frac{1}{2}$	Lepton	2.2×10^{-6}	μ^+
Pion	μ^+	$+1$	139.6	0	Meson	2.6×10^{-8}	π^-
Proton	p	$+1$	938.3	$\frac{1}{2}$	Baryon	Stable	$\bar{p}$

can identify the particle that makes a particular track by—among other means—measuring the relative spacing between the bubbles. The chamber lies in a uniform magnetic field that deflects the tracks of positively charged particles counterclockwise and the tracks of negatively charged particles clockwise. By measuring the radius of curvature of a track, we can calculate the momentum of the particle that made it. Table 45-1 shows some properties of the particles and antiparticles that participated in the event of Fig. 45-3*a*, including those that did not make tracks. Following common practice, we express the masses of the particles listed in Table 45-1—and in all other tables in this chapter—in the unit MeV/c^2. The reason for this notation is that the rest energy of a particle is needed more often than its mass. Thus, the mass of a proton is shown in Table 45-1 to be 938.3 MeV/c^2. To find the proton's rest energy, multiply this mass by c^2 to obtain 938.3 MeV.

The general tools used for the analysis of photographs like Fig. 45-3*a* are the laws of conservation of energy, linear momentum, angular momentum, and electric charge, along with other conservation laws that we have not yet discussed. Figure 45-3*a* is actually one of a stereo pair of photographs so that, in practice, these analyses are carried out in three dimensions.

The event of Fig. 45-3*a* is triggered by an energetic antiproton ($\bar{p}$) that, generated in an accelerator at the Lawrence Berkeley Laboratory, enters the chamber from the left. There are three separate subevents; two occur at points 1 and 2 in Fig. 45-3*b*, and the third occurs out of the frame of the figure. Let's examine each:

1. *Proton-Antiproton Annihilation.* At point 1 in Fig. 45-3*b*, the initiating antiproton (blue track) slams into a proton of the liquid hydrogen in the chamber, and the result is mutual annihilation. We can tell that annihilation occurred while the incoming antiproton was in flight because most of the particles generated in the encounter move in the forward direction—that is, toward the right in Fig. 45-3. From the principle of conservation of linear momentum, the incoming antiproton must have had a forward momentum when it underwent annihilation.

The total energy involved in the collision of the antiproton and the proton is the sum of the antiproton's kinetic energy and the two (identical) rest energies of those two particles (2 × 938.3 MeV, or 1876.6 MeV). This is enough energy to create a number of lighter particles and give them kinetic energy. In this case, the annihilation produces four positive pions (red tracks in Fig. 45-3*b*) and four negative pions (green tracks). (For simplicity, we assume that no gamma-ray photons, which would leave no tracks because they lack electric charge, are produced.) Then the annihilation process is

$$p + \bar{p} \rightarrow 4\pi^+ + 4\pi^-. \tag{45-6}$$

We see from Table 45-1 that the positive pions (π^+) are *particles* and the negative pions (π^-) are *antiparticles*. The reaction of Eq. 45-6 is a *strong interaction* (it involves the strong force), because all the particles involved are hadrons.

Let us check whether electric charge is conserved in the reaction. To do so, we can write the electric charge of a particle as qe, in which q is a **charge quantum number.** Then determining whether electric charge is conserved in a process amounts to determining whether the initial net charge quantum number is equal to the final net charge quantum number. In the process of Eq. 45-6, the initial net charge number is $1 + (-1)$, or 0, and the final net charge number is $4(1) + 4(-1)$, or 0. Thus, charge *is* conserved.

For the energy balance, note from above that the energy available from the p-p̄ annihilation process is at least the sum of the proton and antiproton rest energies, 1876.6 MeV. The rest energy of a pion is 139.6 MeV, so the rest energies of the eight pions amount to 8×139.6 MeV, or 1116.8 MeV. This leaves at least about 760 MeV to distribute among the eight pions as kinetic energy. Thus, the requirement of energy conservation is easily met.

2. *Pion Decay*. Pions are unstable particles; charged pions decay with a mean life of 2.6×10^{-8} s. At point 2 in Fig. 45-3*b*, one of the positive pions comes to rest in the chamber and decays spontaneously into an antimuon μ^+ (purple track) and a neutrino ν:

$$\pi^+ \rightarrow \mu^+ + \nu. \qquad (45\text{-}7)$$

The neutrino, being uncharged, leaves no track. Both the antimuon and the neutrino are leptons; that is, they are particles on which the strong force does not act. Thus, the decay process of Eq. 45-7, which is governed by the weak force, is described as a *weak interaction*. The rest energy of an antimuon is 105.7 MeV, so an energy of 139.6 MeV − 105.7 MeV, or 33.9 MeV, is available to share between the antimuon and the neutrino as kinetic energy.

Let us check whether spin angular momentum is conserved by the process of Eq. 45-7. This amounts to determining whether the net component S_z of spin angular momentum along some arbitrary z axis can be conserved by the process. The spin quantum numbers s of the particles in the process are 0 for the pion π^+ and $\frac{1}{2}$ for both the antimuon μ^+ and the neutrino ν. Thus, for π^+, the component S_z must be $0\hbar$, and for μ^+ and ν, it can be either $+\frac{1}{2}\hbar$ or $-\frac{1}{2}\hbar$.

S_z is conserved by the process of Eq. 45-7 if there is *any* way in which the initial S_z ($= 0\hbar$) can be equal to the final net S_z. We see that if one of the products, either μ^+ or ν, has $S_z = +\frac{1}{2}\hbar$ and the other has $S_z = -\frac{1}{2}\hbar$, then their final net value is $0\hbar$. Thus, because S_z can be conserved, the decay process of Eq. 45-7 *can* occur.

From Eq. 45-7, we also see that the net charge is conserved by the process, because before the process the net charge quantum number is +1, and after the process it is $+1 + 0 = +1$.

3. *Muon Decay*. Muons (whether μ^- or μ^+) are also unstable, decaying with a mean life of 2.2×10^{-6} s. Although the decay products are not shown in Fig. 45-3*b*, the antimuon produced in the reaction of Eq. 45-7 comes to rest and decays spontaneously according to

$$\mu^+ \rightarrow e^+ + \nu + \bar{\nu}. \qquad (45\text{-}8)$$

The rest energy of the antimuon is 105.7 MeV and that of the positron is only 0.511 MeV, leaving 105.2 MeV to be shared as kinetic energy among the three particles produced in the decay process of Eq. 45-8.

You may wonder: Why *two* neutrinos in Eq. 45-8? Why not just one, as in the pion decay in Eq. 45-7? One answer is that the spin quantum numbers of the antimuon, the positron, and the neutrino are each $\frac{1}{2}$; with only one neutrino, the net component S_z of spin angular momentum could not be conserved in the antimuon decay of Eq. 45-8. In Section 45-4 we shall discuss another reason.

Sample Problem 45-1

A stationary positive pion can decay according to

$$\pi^+ \rightarrow \mu^+ + \nu.$$

What is the kinetic energy of the antimuon μ^+? What is the kinetic energy of the neutrino?

SOLUTION: The **Key Idea** here is that the pion decay process must conserve both total energy and total linear momentum. Let us first write the conservation of total energy (rest energy mc^2 plus kinetic energy K) for the decay process as

$$m_\pi c^2 + K_\pi = m_\mu c^2 + K_\mu + m_\nu c^2 + K_\nu.$$

Because the pion was stationary, its kinetic energy K_π is zero. Then, using the masses listed for m_π, m_μ, and m_ν in Table 45-1, we find

$$\begin{aligned} K_\mu + K_\nu &= m_\pi c^2 - m_\mu c^2 - m_\nu c^2 \\ &= 139.6 \text{ MeV} - 105.7 \text{ MeV} - 0 \\ &= 33.9 \text{ MeV}. \end{aligned} \quad (45\text{-}9)$$

We cannot solve Eq. 45-9 for either K_μ or K_ν separately, so let us next apply the principle of conservation of linear momentum to the decay process. Because the pion is stationary when it decays, that principle requires that the muon and neutrino move in opposite directions after the decay. Assume that their motion is along an axis. Then, for components along that axis, we can write the conservation of linear momentum for the decay as

$$p_\pi = p_\mu + p_\nu,$$

which, with $p_\pi = 0$, gives us

$$p_\mu = -p_\nu. \quad (45\text{-}10)$$

We want to relate these momenta p_μ and $-p_\nu$ to the kinetic energies K_μ and K_ν so that we can solve for the kinetic energies. Because we have no reason to believe that classical physics can be applied to the motion of the muon and neutrino, we use Eq. 38-51, the momentum–kinetic energy relation from special relativity:

$$(pc)^2 = K^2 + 2Kmc^2. \quad (45\text{-}11)$$

From Eq. 45-10, we know that

$$(p_\mu c)^2 = (p_\nu c)^2. \quad (45\text{-}12)$$

Substituting from Eq. 45-11 for the left side and then for the right side of Eq. 45-12 yields

$$K_\mu^2 + 2K_\mu m_\mu c^2 = K_\nu^2 + 2K_\nu m_\nu c^2.$$

Setting the neutrino mass $m_\nu = 0$ (as assumed in Table 45-1), substituting $K_\nu = 33.9$ MeV $- K_\mu$ from Eq. 45-9, and then solving for K_μ, we find

$$\begin{aligned} K_\mu &= \frac{(33.9 \text{ MeV})^2}{(2)(33.9 \text{ MeV} + m_\mu c^2)} \\ &= \frac{(33.9 \text{ MeV})^2}{(2)(33.9 \text{ MeV} + 105.7 \text{ MeV})} \\ &= 4.12 \text{ MeV}. \end{aligned} \quad \text{(Answer)}$$

The kinetic energy of the neutrino is then, from Eq. 45-9,

$$\begin{aligned} K_\nu = 33.9 \text{ MeV} - K_\mu &= 33.9 \text{ MeV} - 4.12 \text{ MeV} \\ &= 29.8 \text{ MeV}. \end{aligned} \quad \text{(Answer)}$$

We see that, although the magnitudes of the momenta of the two recoiling particles are the same, the neutrino gets the larger share (88%) of the kinetic energy.

Sample Problem 45-2

Stationary protons in a bubble chamber are bombarded by energetic negative pions, and the following reaction occurs:

$$\pi^- + \text{p} \rightarrow \text{K}^- + \Sigma^+.$$

The rest energies of these particles are

π^-	139.6 MeV	K^-	493.7 MeV
p	938.3 MeV	Σ^+	1189.4 MeV

What is the Q of the reaction?

SOLUTION: The **Key Idea** here is that the Q of a reaction is

$$Q = \begin{pmatrix}\text{initial total} \\ \text{mass energy}\end{pmatrix} - \begin{pmatrix}\text{final total} \\ \text{mass energy}\end{pmatrix}.$$

For the given reaction, we find

$$\begin{aligned} Q &= (m_\pi c^2 + m_\text{p} c^2) - (m_\text{K} c^2 + m_\Sigma c^2) \\ &= (139.6 \text{ MeV} + 938.3 \text{ MeV}) \\ &\quad -(493.7 \text{ MeV} + 1189.4 \text{ MeV}) \\ &= -605 \text{ MeV}. \end{aligned} \quad \text{(Answer)}$$

The minus sign means that the reaction is *endothermic;* that is, the incoming pion (π^-) must have a kinetic energy greater than a certain threshold value to cause the reaction to occur. The threshold energy is greater than 605 MeV because linear momentum must be conserved, which means that the kaon (K^-) and the sigma (Σ^+) must not only be created but also must be given some kinetic energy. A relativistic calculation whose details are beyond our scope shows that the threshold energy for the reaction is 907 MeV.

45-4 The Leptons

In this and the next section, we discuss some of the particles of one of our classification schemes: lepton or hadron. We begin with the leptons, those particles on which the strong force does *not* act. So far, we have encountered, as leptons, the familiar electron and the neutrino that accompanies it in beta decay. The muon, whose decay is described in Eq. 45-8, is another member of this family. Physicists gradually learned that the neutrino that appears in Eq. 45-7, associated with the production of a muon, is *not the same particle* as the neutrino produced in beta decay, associated with the appearance of an electron. We call the former the **muon neutrino** (symbol ν_μ) and the latter the **electron neutrino** (symbol ν_e) when it is necessary to distinguish between them.

These two types of neutrino are known to be different particles because, if a beam of muon neutrinos (produced from pion decay as in Eq. 45-7) strikes a solid target, *only muons*—and never electrons—are produced. On the other hand, if electron neutrinos (produced by the beta decay of fission products in a nuclear reactor) strike a solid target, *only electrons*—and never muons—are produced.

Another lepton, the **tau,** was discovered at SLAC in 1975; its discoverer, Martin Perl, shared the 1995 Nobel prize in physics. The tau has its own associated neutrino, different still from the other two. Table 45-2 lists all the leptons (both particles and antiparticles); all have a spin quantum number s of $\frac{1}{2}$.

There are reasons for dividing the leptons into three families, each consisting of a particle (electron, muon, or tau), its associated neutrino, and the corresponding antiparticles. Furthermore, there are reasons to believe that there are *only* the three families of leptons shown in Table 45-2. Leptons have no internal structure and no measurable dimensions; they are believed to be truly pointlike fundamental particles when they interact with other particles or with electromagnetic waves.

The Conservation of Lepton Number

According to experiment, particle interactions involving leptons obey a certain conservation law for a quantum number called the **lepton number** L. Each (normal) particle in Table 45-2 is assigned $L = +1$ and each antiparticle is assigned $L = -1$. All other particles, which are not leptons, are assigned $L = 0$. Also according to experiment,

> **In all particle interactions, the net lepton number *for each family* is separately conserved.**

Thus, there are actually three lepton numbers L_e, L_μ, L_ν, and the net of *each* must

TABLE 45-2 The Leptons[a]

Family	Particle	Symbol	Mass (MeV/c^2)	Charge q	Antiparticle
Electron	Electron	e^-	0.511	-1	e^+
	Electron neutrino[b]	ν_e	0	0	$\bar{\nu}_e$
Muon	Muon	μ^-	105.7	-1	μ^+
	Muon neutrino[b]	ν_μ	0	0	$\bar{\nu}_\mu$
Tau	Tau	τ^-	1777	-1	τ^+
	Tau neutrino[b]	ν_τ	0	0	$\bar{\nu}_\tau$

[a] All leptons have a spin quantum number of $\frac{1}{2}$ and are thus fermions.

[b] If the neutrino masses are not zero, they are very small. As of 2000, their values have not been well determined.

remain unchanged during any particle interaction. This experimental fact is called the law of **conservation of lepton number.**

We can illustrate this law by reconsidering the antimuon decay process shown in Eq. 45-8, which we now write more fully as

$$\mu^+ \rightarrow e^+ + \nu_e + \bar{\nu}_\mu. \tag{45-13}$$

Consider this first in terms of the muon family of leptons. The μ^+ is an antiparticle (see Table 45-2) and thus has the muon lepton number $L_\mu = -1$. The two particles e^+ and ν_e do not belong to the muon family and thus have $L_\mu = 0$. This leaves $\bar{\nu}_\mu$ on the right which, being an antiparticle, also has the muon lepton number $L_\mu = -1$. Thus, both sides of Eq. 45-13 have the same net muon lepton number—namely, $L_\mu = -1$; if they did not, the μ^+ would not decay by this process.

No members of the electron family appear on the left in Eq. 45-13, so there the net electron lepton number must be $L_e = 0$. On the right side of Eq. 45-13, the positron, being an antiparticle (again see Table 45-2), has the electron lepton number $L_e = -1$. The electron neutrino ν_e, being a particle, has the electron number $L_e = +1$. Thus, the net electron lepton number for these two particles on the right in Eq. 45-13 is also zero; the electron lepton number is also conserved in the process.

No members of the tau family appear on either side of Eq. 45-13, so we must have $L_\tau = 0$ on each side. Thus, each of the lepton quantum numbers L_μ, L_e, and L_τ remains unchanged during the decay process of Eq. 45-13, their constant values being -1, 0, and 0, respectively. This example is but one illustration of the conservation of lepton number; it holds for all particle interactions.

CHECKPOINT 1: (a) The π^+ meson decays by the process $\pi^+ \rightarrow \mu^+ + \nu$. To what lepton family does the neutrino ν belong? (b) Is this neutrino a particle or an antiparticle? (c) What is its lepton number?

45-5 The Hadrons

We are now ready to consider hadrons (baryons and mesons), those particles whose interactions are governed by the strong force. We start by adding another conservation law to our list: conservation of baryon number.

To develop this conservation law, let us consider the proton decay process

$$p \rightarrow e^+ + \nu_e. \tag{45-14}$$

This process *never* happens. We should be glad that it does not because otherwise all protons in the universe would gradually change into positrons, with disastrous consequences. Yet this decay process violates none of the conservation laws we have so far discussed, including the conservation of lepton number.

We account for the apparent stability of the proton—and for the absence of many other processes that might otherwise occur—by introducing a new quantum number, the **baryon number** B, and a new conservation law, the **conservation of baryon number:**

> To every baryon we assign $B = +1$. To every antibaryon we assign $B = -1$. To all particles of other types we assign $B = 0$. A particle process cannot occur if it changes the net baryon number.

In the process of Eq. 45-14, the proton has a baryon number of $B = +1$ and the positron and neutrino both have a baryon number of $B = 0$. Thus, the process does not conserve baryon number and cannot occur.

CHECKPOINT 2: This mode of decay for a neutron is *not* observed:

$$n \rightarrow p + e^-.$$

Which of the following conservation laws does this process violate: (a) energy, (b) angular momentum, (c) linear momentum, (d) charge, (e) lepton number, (f) baryon number? The masses are $m_n = 939.6\ \text{MeV}/c^2$, $m_p = 938.3\ \text{MeV}/c^2$, and $m_e = 0.511\ \text{MeV}/c^2$.

Sample Problem 45-3

Determine whether a stationary proton can decay according to the proposed scheme

$$p \rightarrow \pi^0 + \pi^+.$$

Properties of the proton and the π^+ pion are listed in Table 45-1. The π^0 pion has zero charge, zero spin, and a mass energy of 135.0 MeV.

SOLUTION: The **Key Idea** here is to see whether the proposed decay violates any of the conservation laws we have discussed. For electric charge, we see that the net charge quantum number is initially +1 and finally is 0 + 1, or +1. Thus, charge is conserved by the decay. Lepton number is also conserved, because none of the three particles is a lepton and thus their lepton numbers are all zero.

Linear momentum can also be conserved: Because the proton is stationary, with zero linear momentum, the two pions must merely move in opposite directions with equal magnitudes of linear momentum (so that their total linear momentum is also zero) to conserve linear momentum. The fact that linear momentum *can* be conserved means that the process does not violate the conservation of linear momentum.

Is there energy for the decay? Because the proton is stationary, that question amounts to asking whether the proton's mass energy is sufficient to produce the mass energies and kinetic energies of the pions. To answer, we evaluate the Q of the decay:

$$\begin{aligned} Q &= \begin{pmatrix}\text{initial total}\\ \text{mass energy}\end{pmatrix} - \begin{pmatrix}\text{final total}\\ \text{mass energy}\end{pmatrix} \\ &= m_p c^2 - (m_0 c^2 + m_+ c^2) \\ &= 938.3\ \text{MeV} - (135.0\ \text{MeV} + 139.6\ \text{MeV}) \\ &= 663.7\ \text{MeV}. \end{aligned}$$

The fact that Q is positive indicates that the initial mass energy exceeds the final mass energy. Thus, the proton *does* have enough mass energy to create the pair of pions.

Is spin angular momentum conserved by the decay? This amounts to determining whether the net component S_z of spin angular momentum along some arbitrary z axis can be conserved by the decay. The spin quantum numbers s of the particles in the process are $\frac{1}{2}$ for the proton and 0 for both pions. Thus, for the proton the component S_z can be either $+\frac{1}{2}\hbar$ or $-\frac{1}{2}\hbar$ and for each pion it is $0\hbar$. We see that there is no way that S_z can be conserved. Hence, spin angular momentum is not conserved and the proposed decay of the proton cannot occur.

The decay also violates the conservation of baryon number: The proton has a baryon number of $B = +1$ and both pions have a baryon number of $B = 0$. Thus, nonconservation of baryon number is another reason why the proposed decay cannot occur.

Sample Problem 45-4

A particle called xi-minus and having the symbol Ξ^- decays as follows:

$$\Xi^- \rightarrow \Lambda^0 + \pi^-.$$

The Λ^0 particle (called lambda-zero) and the π^- particle are both unstable. The following decay processes occur in *cascade* until only relatively stable products remain:

$$\Lambda^0 \rightarrow p + \pi^- \qquad \pi^- \rightarrow \mu^- + \bar{\nu}_\mu \qquad \mu^- \rightarrow e^- + \nu_\mu + \bar{\nu}_e.$$

(a) Is the Ξ^- particle a lepton or a hadron? If the latter, is it a baryon or a meson?

SOLUTION: The **Key Idea** for answering the first question is that only three families of leptons exist (Table 45-2) and none include the Ξ^- particle. Thus, the Ξ^- must be a hadron.

The **Key Idea** for answering the second question is to determine the baryon number of the Ξ^- particle. If it is +1 or −1, then the Ξ^- is a baryon. If, instead, it is 0, then the Ξ^- is a meson. To see, let us the write the overall decay scheme, from the initial Ξ^- to the final relatively stable products, as

$$\Xi^- \rightarrow p + 2(e^- + \bar{\nu}_e) + 2(\nu_\mu + \bar{\nu}_\mu). \qquad (45\text{-}15)$$

On the right side, the proton has a baryon number of +1 and the electrons and neutrinos have baryon numbers of 0. Thus, the net baryon number of the right side is +1. That must then be the baryon number of the lone Ξ^- particle on the left side. We conclude that the Ξ^- particle is a baryon.

(b) Does the decay process conserve the three lepton numbers?

SOLUTION: The **Key Idea** here is that any process must separately conserve the net lepton number for each lepton family of Table 45-2. Let us first consider the electron lepton number L_e, which is +1 for the electron e^-, −1 for the anti-electron neutrino $\bar{\nu}_e$, and 0 for the other particles in the overall decay of Eq. 45-15. We see that the net L_e is 0 before the decay and $2[+1 + (-1)] + 2(0 + 0) = 0$ after the decay. Thus, the net electron lepton number *is* conserved. You can similarly show that the net muon lepton number and the net tau lepton number are also conserved.

(c) What can you say about the spin of the Ξ^- particle?

SOLUTION: The **Key Idea** here is that the overall decay scheme of Eq. 45-15 must conserve the net spin component S_z. Thus, we can determine the spin component S_z of the Ξ^- particle on the left side of Eq. 45-15 by considering the S_z components of the nine particles on the right side. All nine of those particles are spin-$\frac{1}{2}$ particles and thus can have S_z of either $+\frac{1}{2}\hbar$ of $-\frac{1}{2}\hbar$. No matter how we choose between those two possible values of S_z, the net S_z for those nine particles must be a *half-integer* times $\hbar$. Thus, the Ξ^- particle must have S_z of a *half-integer* times $\hbar$, and that means that its spin quantum number s must be a half-integer. (Actually, the quantum number is $\frac{1}{2}$; the Ξ^- particle is listed with other spin-$\frac{1}{2}$ baryons in Table 45-3.)

45-6 Still Another Conservation Law

Particles have intrinsic properties in addition to the ones we have listed so far: mass, charge, spin, lepton number, and baryon number. The first of these additional properties was discovered when researchers observed that certain new particles, such as the kaon (K) and the sigma (Σ), always seemed to be produced in pairs. It seemed impossible to produce only one of them at a time. Thus, if a beam of energetic pions interacts with the protons in a bubble chamber, the reaction

$$\pi^+ + p \rightarrow K^+ + \Sigma^+ \tag{45-16}$$

often occurs. The reaction

$$\pi^+ + p \rightarrow \pi^+ + \Sigma^+, \tag{45-17}$$

which violates no conservation law known in the early days of particle physics, never occurs.

It was eventually proposed (by Murray Gell-Mann in the United States and independently by K. Nishijima in Japan) that certain particles possess a new property, called **strangeness,** with its own quantum number S and its own conservation law. (Be careful not to confuse the symbol S here with spin.) The name *strangeness* arises from the fact that, before the identities of these particles were pinned down, they were known as "strange particles," and the label stuck.

The proton, neutron, and pion have $S = 0$; that is, they are not "strange." It was proposed, however, that the K^+ particle has strangeness $S = +1$ and that Σ^+ has $S = -1$. In the reaction of Eq. 45-16, the net strangeness is initially zero and finally zero; thus, the reaction conserves strangeness. However, in the reaction shown in Eq. 45-17, the final net strangeness is -1; thus, that reaction does not conserve strangeness and cannot occur. Apparently, then, we must add one more conservation law to our list—the **conservation of strangeness:**

➤ Strangeness is conserved in interactions involving the strong force.

It may seem heavy-handed to invent a new property of particles just to account for a little puzzle like that posed by Eqs. 45-16 and 45-17. However, strangeness and its quantum number soon revealed themselves in many other areas of particle physics, and strangeness is now fully accepted as a legitimate particle attribute, on a par with charge and spin.

Do not be misled by the whimsical character of the name. Strangeness is no more mysterious a property of particles than is charge. Both are properties that particles may (or may not) have; each is described by an appropriate quantum number. Each obeys a conservation law. Still other properties of particles have been discovered and given even more whimsical names, such as *charm* and *bottomness,* but all are perfectly legitimate properties. Let us see, as an example, how the new property of strangeness "earns its keep" by leading us to uncover important regularities in the properties of the particles.

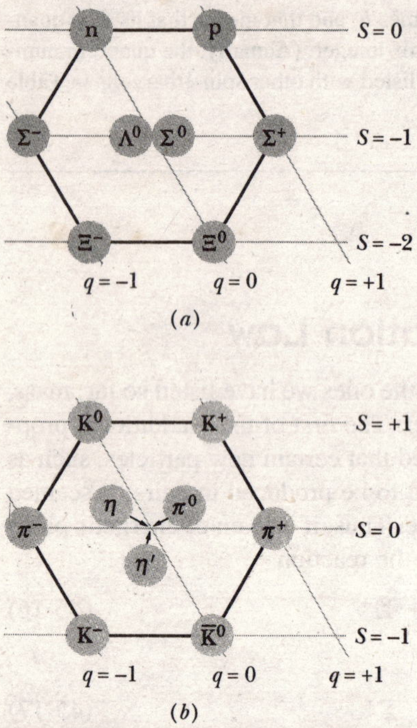

Fig. 45-4 (*a*) The Eightfold Way pattern for the eight spin-$\frac{1}{2}$ baryons listed in Table 45-3. The particles are represented as disks on a strangeness–charge plot, using a sloping axis for the charge quantum number. (*b*) A similar pattern for the nine spin-zero mesons listed in Table 45-4.

45-7 The Eightfold Way

There are eight baryons—the neutron and the proton among them—that have a spin quantum number of $\frac{1}{2}$. Table 45-3 shows some of their other properties. Figure 45-4*a* shows the fascinating pattern that emerges if we plot the strangeness of these baryons against their charge quantum number, using a sloping axis for the charge quantum numbers. Six of the eight form a hexagon with the two remaining baryons at its center.

Let us turn now from the hadrons called baryons to the hadrons called mesons. Nine with a spin of zero are listed in Table 45-4. If we plot them on a sloping strangeness–charge diagram, as in Fig. 45-4*b*, the same fascinating pattern emerges! These and related plots, called the *Eightfold Way* patterns,* were proposed independently in 1961 by Murray Gell-Mann at the California Institute of Technology and by Yuval Ne'eman at Imperial College, London. The two patterns of Fig. 45-4 are representative of a larger number of symmetrical patterns in which groups of baryons and mesons can be displayed.

The symmetry of the Eightfold Way pattern for the spin-$\frac{3}{2}$ baryons (not shown here) calls for *ten* particles arranged in a pattern like that of the tenpins in a bowling alley. However, when the pattern was first proposed, only *nine* such particles were known; the "headpin" was missing. In 1962, guided by theory and the symmetry of the pattern, Gell-Mann made a prediction in which he essentially said:

> There exists a spin-$\frac{3}{2}$ baryon with a charge of -1, a strangeness of -3, and a rest energy of about 1680 MeV. If you look for this *omega minus* particle (as I propose to call it), I think you will find it.

A team of physicists headed by Nicholas Samios of the Brookhaven National Laboratory took up the challenge and found the "missing" particle, confirming all its predicted properties. Nothing beats prompt experimental confirmation for building confidence in a theory!

The Eightfold Way patterns bear the same relationship to particle physics that the periodic table does to chemistry. In each case, there is a pattern of organization

*A borrowing from Eastern mysticism. The "Eight" refers to the eight quantum numbers (only a few of which we have defined here) that are involved in the symmetry-based theory that predicts the existence of the patterns.

TABLE 45-3 Eight Spin-$\frac{1}{2}$ Baryons

Particle	Symbol	Mass (MeV/c^2)	Quantum Numbers: Charge q	Quantum Numbers: Strangeness S
Proton	p	938.3	+1	0
Neutron	n	939.6	0	0
Lambda	Λ^0	1115.6	0	−1
Sigma	Σ^+	1189.4	+1	−1
Sigma	Σ^0	1192.5	0	−1
Sigma	Σ^-	1197.3	−1	−1
Xi	Ξ^0	1314.9	0	−2
Xi	Ξ^-	1321.3	−1	−2

TABLE 45-4 Nine Spin-Zero Mesons[a]

Particle	Symbol	Mass (MeV/c^2)	Quantum Numbers: Charge q	Quantum Numbers: Strangeness S
Pion	π^0	135.0	0	0
Pion	π^+	139.6	+1	0
Pion	π^-	139.6	−1	0
Kaon	K^+	493.7	+1	+1
Kaon	K^-	493.7	−1	−1
Kaon	K^0	497.7	0	+1
Kaon	$\bar{K}^0$	497.7	0	−1
Eta	η	547.5	0	0
Eta prime	η'	957.8	0	0

[a]All mesons are bosons, having spins of 0, 1, 2, ... The ones listed here all have a spin of 0.

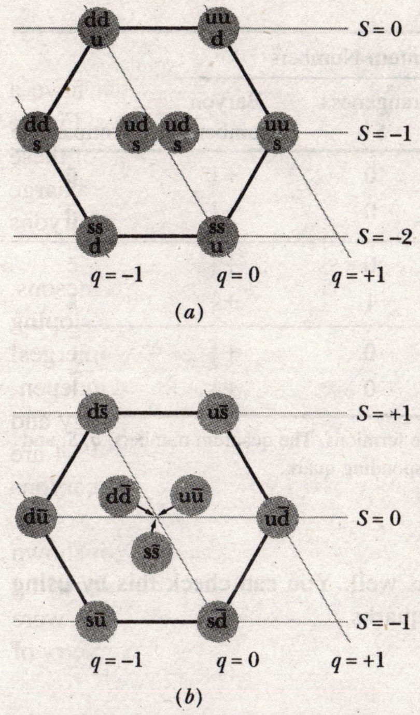

Fig. 45-5 (*a*) The quark compositions of the eight spin-$\frac{1}{2}$ baryons plotted in Fig. 45-4*a*. (Although the two central baryons share the same quark structure, the sigma is an excited state of the lambda, decaying into the lambda by emission of a gamma-ray photon.) (*b*) The quark compositions of the nine spin-zero mesons plotted in Fig. 45-4*b*.

in which vacancies (missing particles or missing elements) stick out like sore thumbs, guiding experimenters in their searches. In the case of the periodic table, its very existence strongly suggests that the atoms of the elements are not fundamental particles but have an underlying structure. Similarly, the Eightfold Way patterns strongly suggest that the mesons and the baryons must have an underlying structure, in terms of which their properties can be understood. That structure can be explained in terms of the *quark model,* which we now discuss.

45-8 The Quark Model

In 1964 Gell-Mann and George Zweig independently pointed out that the Eightfold Way patterns can be understood in a simple way if the mesons and the baryons are built up out of subunits that Gell-Mann called **quarks.** We deal first with three of them, called the *up quark* (symbol u), the *down quark* (symbol d), and the *strange quark* (symbol s), and we assign to them the properties displayed in Table 45-5. (The names of the quarks, along with those assigned to three other quarks that we shall meet later, have no meanings other than as convenient labels. Collectively, these names are called the *quark flavors*. We could just as well call them vanilla, chocolate, and strawberry instead of up, down, and strange.)

The fractional charge quantum numbers of the quarks may jar you a little. However, withhold judgment until you see how neatly these fractional charges account for the observed integral charges of the mesons and the baryons. In all normal situations, whether here on Earth or in any astronomical process, quarks are always bound up together in twos or threes for reasons that are still not well understood. However, in 2000, certain accelerators were finally able to fire atoms into targets with enough energy that the collision freed quarks from one another.

We have seen how we can put atoms together by combining electrons and nuclei. Now let us see how we can put mesons and baryons together by combining quarks. As it turns out,

> There is no known meson or baryon whose properties cannot be understood in terms of an appropriate combination of quarks. Conversely, there is no possible quark combination that does not correspond to an observed meson or baryon.

Let us look first at the baryons of Table 45-3.

Quarks and Baryons

Each baryon is a combination of three quarks; some of the combinations are given in Fig. 45-5*a*. With regard to baryon number, we see that any three quarks (each with $B = +\frac{1}{3}$) yield a proper baryon (with $B = +1$).

Charges also work out, as we can see from three examples. The proton has a quark composition of uud, so its charge quantum number is

$$q(\text{uud}) = \tfrac{2}{3} + \tfrac{2}{3} + (-\tfrac{1}{3}) = +1.$$

The neutron has a quark composition of udd and its charge quantum number is

$$q(\text{udd}) = \tfrac{2}{3} + (-\tfrac{1}{3}) + (-\tfrac{1}{3}) = 0.$$

The Σ^- (sigma-minus) particle has a quark composition of dds and its charge quantum number is

$$q(\text{dds}) = -\tfrac{1}{3} + (-\tfrac{1}{3}) + (-\tfrac{1}{3}) = -1.$$

TABLE 45-5 The Quarks[a]

Particle	Symbol	Mass (MeV/c^2)	Quantum Numbers: Charge q	Strangeness S	Baryon Number B	Antiparticle
Up	u	5	$+\frac{2}{3}$	0	$+\frac{1}{3}$	$\bar{u}$
Down	d	10	$-\frac{1}{3}$	0	$+\frac{1}{3}$	$\bar{d}$
Charm	c	1500	$+\frac{2}{3}$	0	$+\frac{1}{3}$	$\bar{c}$
Strange	s	200	$-\frac{1}{3}$	−1	$+\frac{1}{3}$	$\bar{s}$
Top	t	175 000	$+\frac{2}{3}$	0	$+\frac{1}{3}$	$\bar{t}$
Bottom	b	4300	$-\frac{1}{3}$	0	$+\frac{1}{3}$	$\bar{b}$

[a]All quarks (including antiquarks) have spin $\frac{1}{2}$ and thus are fermions. The quantum numbers, q, S, and B for an antiquark are the negatives of those for the corresponding quark.

The strangeness quantum numbers work out as well. You can check this by using Table 45-3 for the Σ^- and Table 45-5 for the quarks.

Quarks and Mesons

Mesons are quark–antiquark pairs; some of their compositions are given in Fig. 45-5*b*. The quark–antiquark model is consistent with the fact that mesons are not baryons; that is, mesons have a baryon number $B = 0$. The baryon number for a quark is $+\frac{1}{3}$ and for an antiquark is $-\frac{1}{3}$; thus, the combination of baryon numbers in a meson is zero.

Consider the meson π^+, which consists of an up quark u and an antidown quark $\bar{d}$. We see from Table 45-5 that the charge quantum number of the up quark is $+\frac{2}{3}$ and that of the antidown quark is $+\frac{1}{3}$ (the sign is opposite that of the down quark). This adds nicely to a charge quantum number of +1 for the π^+ meson; that is,

$$q(u\bar{d}) = \tfrac{2}{3} + \tfrac{1}{3} = +1.$$

All the charge and strangeness quantum numbers of Fig. 45-5*b* agree with those of Table 45-4 and Fig. 45-4*b*. Convince yourself that all possible up, down, and strange quark–antiquark combinations are used and that all known spin-zero mesons are accounted for. Everything fits.

CHECKPOINT 3: Is a combination of a down quark (d) and an antiup quark ($\bar{u}$) called (a) a π^0 meson, (b) a proton, (c) a π^- meson, (d) a π^+ meson, or (e) a neutron?

A New Look at Beta Decay

Let us see how beta decay appears from the quark point of view. In Eq. 43-23, we presented a typical example of this process:

$$^{32}P \rightarrow {}^{32}S + e^- + \nu.$$

After the neutron was discovered and Fermi had worked out his theory of beta decay, physicists came to view the fundamental beta-decay process as the changing of a neutron into a proton inside the nucleus, according to

$$n \rightarrow p + e^- + \bar{\nu}_e,$$

in which the neutrino is identified more completely. Today we look deeper and see that a neutron (udd) can change into a proton (uud) by changing a down quark into an up quark. We now view the fundamental beta-decay process as

$$d \rightarrow u + e^- + \bar{\nu}_e.$$

Thus, as we come to know more and more about the fundamental nature of matter, we can examine familiar processes at deeper and deeper levels. We see too that the quark model not only helps us to understand the structure of particles but also clarifies their interactions.

Still More Quarks

There are other particles and other Eightfold Way patterns that we have not discussed. To account for them, it turns out that we need to postulate three more quarks, the *charm quark* c, the *top quark* t, and the *bottom quark* b. Thus, a total of six quarks exist, as listed in Table 45-5.

Note that three quarks are exceptionally massive, the most massive of them (top) being almost 170 times more massive than a proton. To generate particles that contain such quarks, with such large mass energies, we must go to higher and higher energies, which is the reason that these three quarks were not discovered earlier.

The first particle containing a charm quark to be observed was the J/Ψ meson, whose quark structure is $c\bar{c}$. It was discovered simultaneously and independently in 1974 by groups headed by Samuel Ting at the Brookhaven National Laboratory and Burton Richter at Stanford University.

The top quark defied all efforts to generate it in the laboratory until 1995, when its existence was finally demonstrated in the Tevatron, a large particle accelerator at Fermilab. In this accelerator, protons and antiprotons, each with an energy of 0.9 TeV ($= 9 \times 10^{11}$ eV), are made to collide at the centers of two large particle detectors. In a very few cases, the colliding protons generate a top–antitop ($t\bar{t}$) quark pair, which *very* quickly decays into particles that can be detected and thus can be used to infer the existence of the top–antitop pair.

Look back for a moment at Table 45-5 (the quark family) and Table 45-2 (the lepton family) and notice the neat symmetry of these two "six-packs" of particles, each dividing naturally into three corresponding two-particle families. In terms of what we know today, the quarks and the leptons seem to be truly fundamental particles with no internal structure.

Sample Problem 45-5

The Ξ^- particle has a spin quantum number s of $\frac{1}{2}$, a charge quantum number q of -1, and a strangeness quantum number S of -2. Also, it is known not to contain a bottom quark. What combination of quarks makes up the Ξ^-?

SOLUTION: From Sample Problem 45-4, we know that the Ξ^- is a baryon. One **Key Idea** then is that it must consist of three quarks (not two as for a meson).

Let us next consider the strangeness $S = -2$ of the Ξ^-. A **Key Idea** here is that only the strange quark s and the antistrange quark $\bar{s}$ have nonzero values of strangeness (see Table 45-5). Further, because only the strange quark s has a *negative* value of strangeness, Ξ^- must contain that quark. In fact, for Ξ^- to have a strangeness of -2, it must contain two strange quarks.

To determine the third quark, call it x, we can consider the other known properties of Ξ^-. Its charge quantum number q is -1, and the charge quantum number q of each strange quark is $-\frac{1}{3}$. Thus, the third quark x must have a charge quantum number of $-\frac{1}{3}$, so that we can have

$$q(\Xi^-) = q(\text{ssx}) = -\tfrac{1}{3} + (-\tfrac{1}{3}) + (-\tfrac{1}{3}) = -1.$$

Besides quark s, the only quarks with $q = -\frac{1}{3}$ are the down quark d and bottom quark b. Because the problem statement ruled out a bottom quark, the third quark must be a down quark. This conclusion is also consistent with the baryon quantum numbers:

$$B(\Xi^-) = B(\text{ssd}) = \tfrac{1}{3} + \tfrac{1}{3} + \tfrac{1}{3} = +1.$$

Thus, the quark composition of the Ξ^- particle is ssd.

45-9 The Basic Forces and Messenger Particles

We turn now from cataloging the particles to considering the forces between them.

The Electromagnetic Force

At the atomic level, we say that two electrons exert electromagnetic forces on each other according to Coulomb's law. At a deeper level, this interaction is described by a highly successful theory called **quantum electrodynamics** (QED). From this point of view we say that each electron senses the presence of the other by exchanging photons with it.

We cannot detect these photons because they are emitted by one electron and absorbed by the other a very short time later. Because of their undetectable existence, we call them **virtual photons.** Because they communicate between the two interacting charged particles, we sometimes call these photons *messenger particles*.

If a stationary electron emits a photon and remains itself unchanged, energy is not conserved. The principle of conservation of energy is saved, however, by the uncertainty principle, written in the form

$$\Delta E \cdot \Delta t \approx \hbar, \tag{45-18}$$

as discussed in Sample Problem 43-10. Here we interpret this relation to mean that you can "overdraw" an amount of energy ΔE, violating conservation of energy, *provided* you "return" it within an interval Δt given by $\hbar/\Delta E$ so that the violation cannot be detected. The virtual photons do just that. When, say, electron A emits a virtual photon, the overdraw in energy is quickly set right when that electron receives a virtual photon from electron B, and the violation of the principle of conservation of energy for the electron pair is hidden by the inherent uncertainty.

The Weak Force

A theory of the weak force, which acts on all particles, was developed by analogy with the theory of the electromagnetic force. The messenger particles that transmit the weak force between particles, however, are not (massless) photons but massive particles, identified by the symbols W and Z. The theory was so successful that it revealed the electromagnetic force and the weak force as being different aspects of a single **electroweak force.** This accomplishment is a logical extension of the work of Maxwell, who revealed the electric and magnetic forces as being different aspects of a single *electromagnetic* force.

The electroweak theory was specific in predicting the properties of the messenger particles. Their charges and masses, for example, were predicted to be

Particle	Charge	Mass
W	$\pm e$	80.6 GeV/c^2
Z	0	91.2 GeV/c^2

Recall that the proton mass is only 0.938 GeV/c^2; these are massive particles! The 1979 Nobel prize in physics was awarded to Sheldon Glashow, Steven Weinberg, and Abdus Salam for their development of the electroweak theory.

The theory was confirmed in 1983 by Carlo Rubbia and his group at CERN, who experimentally verified both messenger particles and found that their masses agreed with the predicted values. The 1984 Nobel prize in physics went to Rubbia and Simon van der Meer for this brilliant experimental work.

Some notion of the complexity of particle physics in this day and age can be found by looking at an earlier Nobel prize particle physics experiment—the discovery of the neutron. This vitally important discovery was a "tabletop" experiment, employing particles emitted by naturally occurring radioactive materials as projectiles; it was reported in 1932 under the title "Possible Existence of a Neutron," the single author being James Chadwick.

The discovery of the W and Z messenger particles in 1983, by contrast, was carried out at a large particle accelerator, about 7 km in circumference and operating in the range of several hundred billion electron-volts. The principal particle detector alone weighed 20 MN. The experiment employed more than 130 physicists from 12 institutions in 8 countries, along with a large support staff.

The Strong Force

A theory of the strong force—that is, the force that acts between quarks to bind hadrons together—has also been developed. The messenger particles in this case are called **gluons** and, like the photon, they are predicted to be massless. The theory assumes that each "flavor" of quark comes in three varieties that, for convenience, have been labeled *red, yellow,* and *blue*. Thus, there are three up quarks, one of each color, and so on. The antiquarks also come in three colors, which we call *antired, antiyellow,* and *antiblue*. You must not think that quarks are actually colored, like tiny jelly beans. The names are labels of convenience but (for once) they do have a certain formal justification, as you shall see.

The force acting between quarks is called a **color force** and the underlying theory, by analogy with quantum electrodynamics (QED), is called **quantum chromodynamics** (QCD). Apparently, quarks can be assembled only in combinations that are *color-neutral*.

There are two ways to bring about color neutrality. In the theory of actual colors, red + yellow + blue yields white, which is color-neutral; thus, we can assemble three quarks to form a baryon, provided one is a yellow quark, one is a red quark, and one is a blue quark. Antired + antiyellow + antiblue is also white, so that we can assemble three antiquarks (of the proper anticolors) to form an antibaryon. Finally, red + antired, or yellow + antiyellow, or blue + antiblue also yields white. Thus, we can assemble a quark–antiquark combination to form a meson. The color-neutral rule does not permit any other combination of quarks, and none are observed.

The color force not only acts to bind together quarks as baryons and mesons, but it also acts between such particles, in which case it has traditionally been called the strong force. Hence, not only does the color force bind together quarks to form protons and neutrons, but it also binds together the protons and neutrons to form nuclei.

Einstein's Dream

The unification of the fundamental forces of nature into a single force—which occupied Einstein's attention for much of his later life—is very much a current focus of research. We have seen that the weak force has been successfully combined with electromagnetism so that they may be jointly viewed as aspects of a single *electroweak force*. Theories that attempt to add the strong force to this combination—called *grand unification theories* (GUTs)—are being pursued actively. Theories that seek to complete the job by adding gravity—sometimes called *theories of everything* (TOE)—are at an encouraging but speculative stage at this time.

45-10 A Pause for Reflection

Let us put what you have just learned in perspective. If all we are interested in is the structure of the world around us, we can get along nicely with the electron, the neutrino, the neutron, and the proton. As someone has said, we can operate "Spaceship Earth" quite well with just these particles. We can see a few of the more exotic particles by looking for them in the cosmic rays; however, to see most of them, we must build massive accelerators and look for them at great effort and expense.

The reason we must go to such effort is that—measured in energy terms—we live in a world of very low temperatures. Even at the center of the Sun, the value of kT is only about 1 keV. To produce the exotic particles, we must be able to accelerate protons or electrons to energies in the GeV and TeV range and higher.

Once upon a time the temperature everywhere *was* high enough to provide just such energies, and far greater. That time of extremely high temperatures occurred in the **big bang** beginning of the universe, when the universe (and both space and time) first existed. Thus, one reason why scientists study particles at high energies is to understand what the universe was like just after it began.

As we shall discuss shortly, *all* of space within the universe was initially tiny in extent, and the temperature of the particles within that space was incredibly high. With time, however, the universe expanded and cooled to lower temperatures, eventually to the size and temperature we see today.

Actually, the phrase "we see today" is complicated: When we look out into space, we are actually looking back in time, because the light from the stars and galaxies has taken a long time to reach us. The most distant objects that we can detect are **quasars** (*qua*si*s*tell*ar* objects), which are the extremely bright cores of galaxies that are as much as 14×10^9 ly from us. Each such core contains a gigantic black hole; as material (gas and even stars) is pulled into one of those black holes, the material heats up and radiates a tremendous amount of light, enough for us to detect in spite of the huge distance. We "see" a quasar as it once was, when that light began its journey to us billions of years ago.

45-11 The Universe Is Expanding

As we saw in Section 38-10, it is possible to measure the relative speeds at which galaxies are approaching us or receding from us by measuring the shifts in the wavelength of the light they emit. If we look only at distant galaxies, beyond our immediate galactic neighbors, we find an astonishing fact. They are *all* moving away (receding) from us!

In 1929 Edwin P. Hubble established a connection between the apparent speed of recession v of a galaxy and its distance r from us—namely, that they are directly proportional. That is,

$$v = Hr \quad \text{(Hubble's law)}, \tag{45-19}$$

in which H, the proportionality constant, is called the **Hubble constant.** The value of H is usually measured in the unit kilometers per second-megaparsec (km/s · Mpc), where megaparsec is a length unit commonly used in astrophysics and astronomy:

$$1 \text{ Mpc} = 3.084 \times 10^{19} \text{ km} = 3.260 \times 10^{6} \text{ ly}. \tag{45-20}$$

The Hubble constant H has not had the same value since the universe began. Determining its current value is extremely difficult because it involves measurements for very distant galaxies. Recently, one study put the current value of H at 70 ± 7 km/s · Mpc, whereas another study strongly argued for a value of 58 km/s · Mpc.

In this chapter, we shall use a "compromise value":

$$H = 63.0 \text{ km/s} \cdot \text{Mpc} = 19.3 \text{ mm/s} \cdot \text{ly}. \qquad (45\text{-}21)$$

We interpret the recession of the galaxies to mean that the universe is expanding, much as the raisins in what is to be a loaf of raisin bread grow farther apart as the dough expands. Observers on all other galaxies would find that distant galaxies were rushing away from them also, in accordance with Hubble's law. In keeping with our analogy, we can say that no raisin (galaxy) has a unique or preferred view.

Hubble's law is consistent with the hypothesis that the universe began with the big bang and has been expanding ever since. If we assume that the rate of expansion has been constant (that is, the value of H has been constant), then we can estimate the age T of the universe by using Eq. 45-19. Let us also assume that since the big bang, any given part of the universe (say, a galaxy) has been receding from our location at a speed v given by Eq. 45-19. Then the time required for the given part to recede a distance r is

$$T = \frac{r}{v} = \frac{r}{Hr} = \frac{1}{H} \quad \text{(estimated age of universe).} \qquad (45\text{-}22)$$

For our compromise value of H in Eq. 45-21, T works out to be 15×10^9 y. Much more sophisticated studies of the expansion of the universe put T between 12×10^9 y and 15×10^9 y.

Sample Problem 45-6

The wavelength shift in the light from a particular quasar indicates that the quasar has a recessional speed of 2.8×10^8 m/s (which is 93% of the speed of light). Approximately how far from us is the quasar?

SOLUTION: The **Key Idea** here is to apply Hubble's law to the given speed v. From Eqs. 45-19 and 45-21, we find

$$r = \frac{v}{H} = \frac{2.8 \times 10^8 \text{ m/s}}{19.3 \text{ mm/s} \cdot \text{ly}} (1000 \text{ mm/m})$$

$$= 14.5 \times 10^9 \text{ ly}. \qquad \text{(Answer)}$$

This is only an approximation because the quasar has not always been receding from our location at the same speed v; that is, H has not had its current value throughout the expansion of the universe.

Sample Problem 45-7

A particular emission line, detected in the light from a galaxy, has a wavelength $\lambda_{det} = 1.1\lambda$, where λ is the proper wavelength of the line. What is the galaxy's distance from us?

SOLUTION: One **Key Idea** here is to assume that Hubble's law ($v = Hr$) applies to the recession of the galaxy. A second **Key Idea** is to assume that the astronomical Doppler shift of Eq. 38-33 ($v = c\,\Delta\lambda/\lambda$) applies to the shift in wavelength due to the recession. We can then set the right side of these two equations equal to each other to write

$$Hr = \frac{c\,\Delta\lambda}{\lambda}, \qquad (45\text{-}23)$$

which leads us to

$$r = \frac{c\,\Delta\lambda}{H\lambda}. \qquad (45\text{-}24)$$

In this equation,

$$\Delta\lambda = \lambda_{det} - \lambda = 1.1\lambda - \lambda = 0.1\lambda.$$

Substituting this into Eq. 45-24 then gives us

$$r = \frac{c(0.1\lambda)}{H\lambda} = \frac{0.1c}{H}$$

$$= \frac{(0.1)(3.0 \times 10^8 \text{ m/s})}{19.3 \text{ mm/s} \cdot \text{ly}} (1000 \text{ mm/m})$$

$$= 1.6 \times 10^9 \text{ ly}. \qquad \text{(Answer)}$$

45-12 The Cosmic Background Radiation

In 1965 Arno Penzias and Robert Wilson, of what was then the Bell Telephone Laboratories, were testing a sensitive microwave receiver used for communications research. They discovered a faint background "hiss" that remained unchanged in intensity no matter where their antenna was pointed. It soon became clear that Penzias and Wilson were observing a **cosmic background radiation,** generated in the early universe and filling all space almost uniformly. This radiation, whose maximum intensity occurs at a wavelength of 1.1 mm, has the same distribution in wavelength as does radiation in an enclosure whose walls are held at a temperature of 2.7 K. In this situation the enclosure is the entire universe. Penzias and Wilson were awarded the 1978 Nobel prize in physics for their discovery.

This radiation originated about 300 000 years after the big bang, when the universe suddenly became transparent to electromagnetic waves (the waves were not immediately absorbed by particles). The radiation at that time corresponded to radiation in an enclosure at a temperature of perhaps 10^5 K. As the universe expanded, however, the temperature dropped to its present value of 2.7 K.

45-13 Dark Matter

At the Kitt Peak National Observatory in Arizona, Vera Rubin and her co-worker Kent Ford measured the rotational rates of a number of distant galaxies. They did so by measuring the Doppler shifts of bright clusters of stars located within each galaxy at various distances from the galactic center. As Fig. 45-6 shows, their results were surprising: The orbital speed of stars at the outer visible edge of the galaxy is about the same as that of stars close to the galactic center.

As the solid curve in Fig. 45-6 attests, that is not what we would expect to find if all the mass of the galaxy were represented by visible light. Nor is the pattern found by Rubin and Ford what we find in the solar system. For example, the orbital speed of Pluto (the planet most distant from the Sun) is only about one-tenth that of Mercury (the planet closest to the Sun).

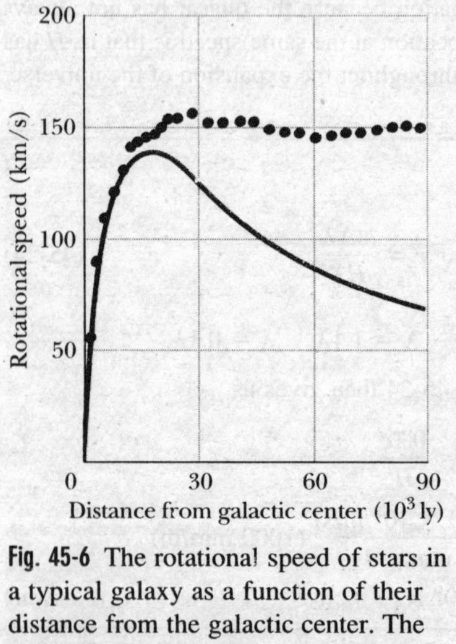

Fig. 45-6 The rotational speed of stars in a typical galaxy as a function of their distance from the galactic center. The theoretical solid curve shows that if a galaxy contained only the mass that is visible, the observed rotational speed would drop off with distance at large distances. The dots are the experimental data, which show that the rotational speed is approximately constant at large distances.

The only explanation for the findings of Rubin and Ford that is consistent with Newtonian mechanics is that a typical galaxy contains much more matter than what we can actually see. In fact, the visible portion of a galaxy represents only about 5 to 10% of the total mass of the galaxy. In addition to these studies of galactic rotation, many other observations lead to the conclusion that the universe abounds in matter that we cannot see.

What then is this **dark matter** that permeates and surrounds a typical galaxy like a gigantic halo whose diameter is perhaps 30 times the diameter of the visible galaxy? Dark matter candidates fall into two classes, whimsically called WIMPs (*w*eakly *i*nteracting *m*assive *p*articles) and MACHOs (*ma*ssive *c*ompact *h*alo *o*bjects). If neutrinos have mass, they are possible WIMP candidates. MACHOs can include objects such as black holes, white dwarf stars, and brown dwarf stars; the latter are Jupiter-size objects that are not massive enough to become actual stars, which shine (and hence are visible) because of fusion.

As of 2000, there is convincing evidence that MACHOs do indeed exist in our own galaxy. Assume that an (invisible) MACHO in our galaxy passes, by chance, between Earth and a star of a nearby galaxy. Einstein, in his general theory of relativity, predicted that light rays passing near any massive object will be deflected by the mass of that object (see Section 14-9). Thus, if star, MACHO, and Earth are aligned, the MACHO will act as a *gravitational lens,* focusing the light rays from the star that pass near it and causing the image of the star to brighten while the MACHO is eclipsing it.

Enough such events have been observed to convince astronomers that MACHOs can account for a substantial fraction (some say 50%) of the dark matter in our own galaxy. Observations are ongoing.

45-14 The Big Bang

In 1985, a physicist remarked at a scientific meeting:

> It is as certain that the universe started with a Big Bang about 15 billion years ago as it is that the Earth goes around the Sun.

This strong statement suggests the level of confidence in which the big bang theory, first advanced by Belgian physicist Georges Lemaître, is held by those who study these matters.

You must not imagine that the big bang was like the explosion of some gigantic firecracker and that, in principle at least, you could have stood to one side and watched. There was no "one side" because the big bang represents the beginning of spacetime itself. From the point of view of our present universe, there is no position in space to which you can point and say, "The big bang happened there." It happened everywhere.

Moreover, there was no "before the big bang," because time *began* with that creation event. In this context, the word "before" loses its meaning. We can, however, conjecture about what went on during succeeding intervals of time after the big bang.

$t \approx 10^{-43}$ **s.** This is the earliest time at which we can say anything meaningful about the development of the universe. It is at this moment that the concepts of space and time come to have their present meanings and the laws of physics as we know them become applicable. At this instant, the entire universe is much smaller than a proton and its temperature is about 10^{32} K.

$t \approx 10^{-34}$ **s.** By this moment the universe has undergone a tremendously rapid inflation, increasing in size by a factor of about 10^{30}. It has become a hot soup of photons, quarks, and leptons, at a temperature of about 10^{27} K.

$t \approx 10^{-4}$ **s.** Quarks can now combine to form protons and neutrons and their antiparticles. The universe has now cooled to such an extent by continued (but much slower) expansion that photons lack the energy needed to break up these new particles. Particles of matter and antimatter collide and annihilate each other. There is a slight excess of matter, which, failing to find annihilation partners, survives to form the world of matter that we know today.

$t \approx 1$ **min.** The universe has now cooled enough so that protons and neutrons, in colliding, can stick together to form the low-mass nuclei ^2H, ^3He, ^4He, and ^7Li. The predicted relative abundances of these nuclides are just what we observe in the universe today. There is plenty of radiation present, but light cannot travel far before it interacts with a nucleus. Thus the universe is opaque.

$t \approx 300\,000$ **y.** The temperature has now fallen to about 10^4 K, and electrons can stick to bare nuclei when they collide, forming atoms. Because light does not interact appreciably with (uncharged) particles such as neutral atoms, the light is now free to travel great distances. This radiation forms the cosmic *background radiation* discussed in Section 45-12. Atoms of hydrogen and helium, under the influence of gravity, begin to clump together, starting the formation of galaxies and stars.

Early measurements suggested that the cosmic background radiation is uniform in all directions, implying that 300 000 y after the big bang all matter in the universe was uniformly distributed. This finding was most puzzling because matter in the present universe is not uniformly distributed, but instead is collected in galaxies, clusters of galaxies, and superclusters of galactic clusters. There are also vast *voids* in which there is relatively little matter, and there are regions so crowded with matter that they are called *walls*. If the big bang theory of the beginning of the universe is even approximately correct, the seeds for this nonuniform distribution of matter must have been in place before the universe was 300 000 y old and now should show up as a nonuniform distribution of the microwave background radiation.

In 1992, measurements made by NASA's Cosmic Background Explorer (COBE) satellite revealed that the background radiation is, in fact, not perfectly uniform. The image shown on this chapter's opening page was made from those measurements and shows us the universe when it was only 300 000 y old. As you can see, large-scale collecting of matter had already begun; thus, the big bang theory is, in principle, on the right track.

45-15 A Summing Up

Let us, in these closing paragraphs, consider where our rapidly accumulating store of knowledge about the universe is leading us. That it provides satisfaction to a host of curiosity-motivated physicists and astronomers is beyond dispute. However, some view it as a humbling experience in that each increase in knowledge seems to reveal more clearly our own relative insignificance in the grand scheme of things. Thus, in roughly chronological order, we humans have come to realize that

Our Earth is not the center of the solar system.

Our Sun is but one star among many in our galaxy.

Our galaxy is but one of many, and our Sun is an insignificant star in it.

Our Earth has existed for perhaps only a third of the age of the universe and will surely disappear when our Sun burns up its fuel and becomes a red giant.

Our species has inhabited Earth for less than a million years—a blink in cosmological time.

Although our position in the universe may be insignificant, the laws of physics that we have discovered (uncovered?) seem to hold throughout the universe and—as far as we know—since the universe began and for all future time. At least, there is no evidence that other laws hold in other parts of the universe. Thus, until someone complains, we are entitled to stamp the laws of physics "Discovered on Earth." Much remains to be discovered: *"The universe is full of magical things, patiently waiting for our wits to grow sharper."*

REVIEW & SUMMARY

Leptons and Quarks Current research supports the view that all matter is made of six kinds of **leptons** (Table 45-2), six kinds of **quarks** (Table 45-5), and 12 **antiparticles,** one corresponding to each of the leptons and quarks. All these particles have spin quantum numbers equal to $\frac{1}{2}$ and are thus **fermions** (particles with half-integer spin quantum numbers).

The Interactions Particles with electric charge interact through the electromagnetic force by exchanging **virtual photons.** Leptons can interact with each other and with quarks through the **weak force,** via massive W and Z particles as messengers. In addition, quarks interact with each other through the **color force.** The electromagnetic and weak forces are different manifestations of the same force, called the **electroweak force.**

Leptons Three of the leptons (the **electron, muon,** and **tau**) have electric charge equal to $-1e$. There are also three uncharged **neutrinos** (also leptons), one corresponding to each of the charged leptons. The neutrinos have very small, possibly zero, mass. The antiparticles for the charged leptons have positive charge.

Quarks The six quarks (up, down, strange, charm, bottom, and top, in order of increasing mass) each have baryon number $+\frac{1}{3}$ and charge equal to either $+\frac{2}{3}e$ or $-\frac{1}{3}e$. The strange quark has strangeness −1, whereas the others all have strangeness 0. These four algebraic signs are reversed for the antiquarks.

Hadrons: Baryons and Mesons Quarks combine into strongly interacting particles called **hadrons. Baryons** are hadrons with half-integer spin quantum numbers ($\frac{1}{2}$ or $\frac{3}{2}$). **Mesons** are hadrons with integer spin quantum numbers (0 or 1) and thus are **bosons.** Baryons are fermions. Mesons have baryon number equal to zero; baryons have baryon number equal to +1 or −1. **Quantum chromodynamics** predicts that the possible combinations of quarks are either a quark with an antiquark, three quarks, or three antiquarks (this prediction is consistent with experiment).

Expansion of the Universe Current evidence strongly suggests that the universe is expanding, with the distant galaxies moving away from us at a rate v given by **Hubble's law:**

$$v = Hr \qquad \text{(Hubble's law)}. \qquad (45\text{-}19)$$

Here we take H, the **Hubble constant,** to have the value

$$H = 63.0 \text{ km/s} \cdot \text{Mpc} = 19.3 \text{ mm/s} \cdot \text{ly}. \qquad (45\text{-}21)$$

The expansion described by Hubble's law and the presence of ubiquitous background microwave radiation suggest that the universe began in a "big bang" between 12 and 15 billion years ago.

QUESTIONS

1. Not only particles such as electrons and protons but also entire atoms can be classified as fermions or bosons, depending on whether their overall spin quantum numbers are, respectively, half-integral or integral. Consider the helium isotopes ^{3}He and ^{4}He. Which of the following statements is correct? (a) Both are fermions. (b) Both are bosons. (c) ^{4}He is a fermion, and ^{3}He is a boson. (d) ^{3}He is a fermion, and ^{4}He is a boson. (The two helium electrons form a closed shell and play no role in this determination.)

2. Is the direction of the magnetic field in Fig. 45-3*b* out of the page or into the page?

3. Which of the eight pions in Fig. 45-3*b* has the least kinetic energy?

4. An electron cannot decay into two neutrinos. Which of the following conservation laws would be violated if it did: (a) energy, (b) angular momentum, (c) charge, (d) lepton number, (e) linear momentum, (f) baryon number?

5. A proton cannot decay into a neutron and a neutrino. Which of the following conservation laws would be violated if it did: (a) energy (assume the proton is stationary), (b) angular momentum, (c) charge, (d) lepton number, (e) linear momentum, (f) baryon number?

6. A proton has enough mass energy to decay into a shower made up of electrons, neutrinos, and their antiparticles. Which of the following conservation laws would be violated if it did: (a) electron lepton number, (b) angular momentum, (c) charge, (d) muon lepton number, (e) linear momentum, (f) baryon number?

7. As we have seen, the π^- meson has the quark structure d$\bar{\text{u}}$. Which of the following conservation laws would be violated if a π^- was formed, instead, from a d quark and a u quark: (a) energy, (b) angular momentum, (c) charge, (d) lepton number, (e) linear momentum, (f) baryon number?

8. A Σ^+ particle has these quantum numbers: strangeness $S = -1$, charge $q = +1$, and spin $s = \frac{1}{2}$. Which of the following quark combinations produce it: (a) dds, (b) s$\bar{\text{s}}$, (c) uus, (d) ssu, or (e) uu$\bar{\text{s}}$?

9. The left column that follows lists ideas from atomic physics, and the right column from particle physics. Match the items in the two columns.

1. chemistry	a. Eightfold Way patterns
2. electrons	b. missing hadrons
3. periodic table	c. quantum chromodynamics
4. missing elements	d. particle physics
5. quantum mechanics	e. quarks

10. Consider the neutrino whose symbol is $\bar{\nu}_\tau$. (a) Is it a quark, a lepton, a meson, or a baryon? (b) Is it a particle or an antiparticle? (c) Is it a boson or a fermion? (d) Is it stable against spontaneous decay?

11. Match the items in these two columns:

1. tau	a. quark
2. pion	b. lepton
3. proton	c. meson
4. positron	d. baryon
5. charm	e. antiparticle

12. List these particles in order of mass, least first: (a) proton, (b) neutrino, (c) π^+ meson, (d) strange quark, (e) tau, (f) electron, and (g) Σ^-.

13. What are the lepton numbers of these particles: (a) π^-, (b) e^-, (c) μ^+, (d) τ^-, (e) $\bar{\nu}_\mu$?

EXERCISES & PROBLEMS

ssm Solution is in the Student Solutions Manual.
www Solution is available on the World Wide Web at: http://www.wiley.com/college/hrw
ilw Solution is available on the Interactive LearningWare.

SEC. 45-3 An Interlude

1E. Calculate the difference in mass, in kilograms, between the muon and pion of Sample Problem 45-1.

2E. An electron and a positron are separated by a distance r. Find the ratio of the gravitational force to the electric force between them. From the result, what can you conclude concerning the forces acting between particles detected in a bubble chamber?

3E. A neutral pion decays into two gamma rays: $\pi^0 \rightarrow \gamma + \gamma$. Calculate the wavelength of the gamma rays produced by the decay of a neutral pion at rest. ssm

4E. A positively charged pion decays by Eq. 45-7: $\pi^+ \rightarrow \mu^+ + \nu$. What then must be the decay scheme of the negatively charged pion? (*Hint:* The π^- is the antiparticle of the π^+.)

5E. How much energy would be released if Earth were annihilated by collision with an anti-Earth? ssm

6P. Certain theories predict that the proton is unstable, with a half-life of about 10^{32} years. Assuming that this is true, calculate the number of proton decays you would expect to occur in one year in the water of an Olympic-sized swimming pool holding 4.32×10^5 L of water.

7P. Observations of neutrinos emitted by the supernova SN1987a (Fig. 44-12) in the Large Magellanic Cloud place an upper limit of 20 eV on the rest energy of the electron neutrino. Suppose that the rest energy of this neutrino, rather than being zero, is in fact equal to 20 eV. How much slower than the speed of light would be the speed of a 1.5 MeV neutrino that was emitted in a beta decay?

8P. A neutral pion has a rest energy of 135 MeV and a mean life of 8.3×10^{-17} s. If it is produced with an initial kinetic energy of 80 MeV and decays after one mean lifetime, what is the longest possible track this particle could leave in a bubble chamber? Use relativistic time dilation.

9P. The rest energies of many short-lived particles cannot be measured directly but must be inferred from the measured momenta and known rest energies of the decay products. Consider the ρ^0 meson, which decays by the reaction $\rho^0 \rightarrow \pi^+ + \pi^-$. Calculate the rest energy of the ρ^0 meson given that the oppositely directed momenta of the created pions each have magnitude 358.3 MeV/c. See Table 45-4 for the rest energies of the pions. ssm

10P. A positive tau (τ^+, rest energy = 1777 MeV) is moving with 2200 MeV of kinetic energy in a circular path perpendicular to a uniform 1.20 T magnetic field. (a) Calculate the momentum of the tau in kilogram-meters per second. Relativistic effects must be considered. (b) Find the radius of the circular path.

11P. (a) A stationary particle 1 decays into particles 2 and 3, which move off with equal but oppositely directed momenta. Show that the kinetic energy K_2 of particle 2 is given by

$$K_2 = \frac{1}{2E_1}[(E_1 - E_2)^2 - E_3^2],$$

where m_1, m_2, and m_3 are masses and E_1, E_2, and E_3 are the corresponding rest energies. (*Hint:* Follow the arguments of Sample Problem 45-1 except that, in this case, neither of the created particles has zero mass.) (b) Show that the result in (a) yields the kinetic energy of the muon as calculated in Sample Problem 45-1.

SEC. 45-6 Still Another Conservation Law

12E. Verify that the hypothetical proton decay scheme in Eq. 45-14 does not violate the conservation laws of (a) charge, (b) energy, (c) linear momentum, and (d) angular momentum.

13E. Which conservation law is violated in each of these proposed decays? Assume that the initial particle is stationary and the decay products have zero orbital angular momentum. (a) $\mu^- \rightarrow e^- + \nu_\mu$; (b) $\mu^- \rightarrow e^+ + \nu_e + \bar{\nu}_\mu$; (c) $\mu^+ \rightarrow \pi^+ + \nu_\mu$. ssm

14P. The A_2^+ particle and its products decay according to the following scheme:

$$A_2^+ \rightarrow \rho^0 + \pi^+, \qquad \mu^+ \rightarrow e^+ + \nu + \bar{\nu},$$
$$\rho^0 \rightarrow \pi^+ + \pi^-, \qquad \pi^- \rightarrow \mu^- + \bar{\nu},$$
$$\pi^+ \rightarrow \mu^+ + \nu, \qquad \mu^- \rightarrow e^- + \nu + \bar{\nu}.$$

(a) What are the final stable decay products? (b) From the evidence, is the A_2^+ particle a fermion or a boson? Is that particle a meson or a baryon? What is its baryon number? (*Hint:* See Sample Problem 45-4.)

SEC. 45-7 The Eightfold Way

15E. The reaction $\pi^+ + p \rightarrow p + p + \bar{n}$ proceeds via the strong interaction. By applying the conservation laws, deduce the charge quantum number, baryon number, and strangeness of the antineutron. ssm

16E. By examining strangeness, determine which of the following decays or reactions proceed via the strong interaction: (a) $K^0 \rightarrow \pi^+ + \pi^-$; (b) $\Lambda^0 + p \rightarrow \Sigma^+ + n$; (c) $\Lambda^0 \rightarrow p + \pi^-$; (d) $K^- + p \rightarrow \Lambda^0 + \pi^0$.

17E. Which conservation law is violated in each of these proposed reactions and decays? (Assume that the products have zero orbital angular momentum.) (a) $\Lambda^0 \rightarrow p + K^-$; (b) $\Omega^- \rightarrow \Sigma^- + \pi^0$ ($S = -3$, $q = -1$, and $m = 1672$ MeV/c^2 for Ω^-); (c) $K^- + p \rightarrow \Lambda^0 + \pi^+$. ssm

18E. Calculate the disintegration energy of the reactions (a) $\pi^+ + p \rightarrow \Sigma^+ + K^+$ and (b) $K^- + p \rightarrow \Lambda^0 + \pi^0$.

19E. A Σ^- particle moving with 220 MeV of kinetic energy decays according to $\Sigma^- \rightarrow \pi^- + n$. Calculate the total kinetic energy of the decay products.

20P. Show that if, instead of plotting strangeness S versus charge q for the spin-$\frac{1}{2}$ baryons in Fig. 45-4*a* and for the spin-zero mesons in Fig. 45-4*b*, the quantity $Y = B + S$ is plotted against the quantity $T_z = q - \frac{1}{2}B$, then the hexagonal patterns emerge with the use of nonsloping (perpendicular) axes. (The quantity Y is called *hypercharge* and T_z is related to a quantity called *isospin*.)

21P. Use the conservation laws to identify the particle labeled x in each of the following reactions, which proceed by means of the strong interaction: (a) $p + p \rightarrow p + \Lambda^0 + x$; (b) $p + \bar{p} \rightarrow n + x$; (c) $\pi^- + p \rightarrow \Xi^0 + K^0 + x$. ssm www

22P. Consider the decay $\Lambda^0 \rightarrow p + \pi^-$ with the Λ^0 at rest. (a) Calculate the disintegration energy. (b) Find the kinetic energy of the proton. (c) What is the kinetic energy of the pion? (*Hint:* See Problem 11.)

SEC. 45-8 The Quark Model

23E. The quark makeups of the proton and neutron are uud and udd, respectively. What are the quark makeups of (a) the antiproton and (b) the antineutron?

24E. From Tables 45-3 and 45-5, determine the identities of the baryons formed from the following combinations of quarks. Check your answers with the baryon octet shown in Fig. 45-4*a*: (a) ddu; (b) uus; (c) ssd.

25E. Using the up, down, and strange quarks only, construct, if possible, a baryon (a) with $q = +1$ and strangeness $S = -2$ and (b) with $q = +2$ and strangeness $S = 0$. ssm

26E. What quark combinations are needed to form (a) a Λ^0 and (b) a Ξ^0?

27E. There are 10 baryons with spin $\frac{3}{2}$. Their symbols and quantum numbers for charge q and strangeness S are as follows:

	q	S		q	S
Δ^-	−1	0	Σ^{*0}	0	−1
Δ^0	0	0	Σ^{*+}	+1	−1
Δ^+	+1	0	Ξ^{*-}	−1	−2
Δ^{++}	+2	0	Ξ^{*0}	0	−2
Σ^{*-}	−1	−1	Ω^-	−1	−3

Make a charge–strangeness plot for these baryons, using the sloping coordinate system of Fig. 45-4. Compare your plot with this figure.

28P. There is no known meson with charge quantum number $q = +1$ and strangeness $S = -1$ or with $q = -1$ and $S = +1$. Explain why, in terms of the quark model.

29P. The spin-$\frac{3}{2}$ Σ^{*0} baryon (see Exercise 27) has a rest energy of 1385 MeV (with an intrinsic uncertainty ignored here); the spin-$\frac{1}{2}$ Σ^0 baryon has a rest energy of 1192.5 MeV. If each of these particles has a kinetic energy of 1000 MeV, which, if either, is moving faster and by how much?

SEC. 45-11 The Universe Is Expanding

30E. If Hubble's law can be extrapolated to very large distances, at what distance would the apparent recessional speed become equal to the speed of light?

31E. What is the observed wavelength of the 656.3 nm (first Balmer) line of hydrogen emitted by a galaxy at a distance of 2.40×10^8 ly? Assume that the Doppler shift of Eq. 38-33 and Hubble's law apply. ssm www

32E. In the laboratory, one of the lines of sodium is emitted at a wavelength of 590.0 nm. In the light from a particular galaxy, however, this line is seen at a wavelength of 602.0 nm. Calculate the distance to the galaxy, assuming that Hubble's law holds and that the Doppler shift of Eq. 38-33 applies.

33P. Will the universe continue to expand forever? To attack this question, make the (reasonable?) assumption that the recessional speed v of a galaxy a distance r from us is determined only by the matter that lies inside a sphere of radius r centered on us. If the total mass inside this sphere is M, the escape speed v_e from the sphere is $v_e = \sqrt{2GM/r}$ (Eq. 14-27). (a) Show that to prevent unlimited expansion, the average density ρ inside the sphere must be at least equal to

$$\rho = \frac{3H^2}{8\pi G}.$$

(b) Evaluate this "critical density" numerically; express your answer in terms of hydrogen atoms per cubic meter. Measurements of the actual density are difficult and are complicated by the presence of dark matter.

34P. The apparent recessional speeds of galaxies and quasars at great distances are close to the speed of light, so the relativistic Doppler shift formula (Eq. 38-30) must be used. The red shift is reported as fractional red shift $z = \Delta\lambda/\lambda_0$. (a) Show that, in terms of z, the recessional speed parameter $\beta = v/c$ is given by

$$\beta = \frac{z^2 + 2z}{z^2 + 2z + 2}.$$

(b) A quasar detected in 1987 has $z = 4.43$. Calculate its speed parameter. (c) Find the distance to the quasar, assuming that Hubble's law is valid to these distances.

SEC. 45-12 The Cosmic Background Radiation

35P. Due to the presence everywhere of the microwave background radiation, the minimum possible temperature of a gas in interstellar or intergalactic space is not 0 K but 2.7 K. This implies that a significant fraction of the molecules in space that can occupy excited states of low excitation energy may, in fact, be in those excited states. Subsequent de-excitation would lead to the emission of radiation that could be detected. Consider a (hypothetical) molecule with just one excited state. (a) What would the excitation energy have to be for 25% of the molecules to be in the excited state? (*Hint:* See Eq. 41-29.) (b) What would be the wavelength of the photon emitted in a transition back to the ground state?

SEC. 45-13 Dark Matter

36E. What would the mass of the Sun have to be if Pluto (the outermost planet most of the time) were to have the same orbital speed that Mercury (the innermost planet) has now? Use data from Appendix C, and express your answer in terms of the Sun's current mass M. (Assume circular orbits.)

37P. Suppose that the radius of the Sun were increased to 5.90×10^{12} m (the average radius of the orbit of the planet Pluto, the outermost planet), that the density of this expanded Sun were uniform, and that the planets revolved within this tenuous object. (a) Calculate Earth's orbital speed in this new configuration and compare this with its present orbital speed of 29.8 km/s. Assume that the radius of Earth's orbit remains unchanged. (b) What would be the new period of revolution of Earth? (The Sun's mass remains unchanged.) ssm www

38P. Suppose that the matter (stars, gas, dust) of a particular galaxy, of total mass M, is distributed uniformly throughout a sphere of radius R. A star of mass m is revolving about the center of the galaxy in a circular orbit of radius $r < R$. (a) Show that the orbital speed v of the star is given by

$$v = r\sqrt{GM/R^3},$$

and therefore that the period T of revolution is

$$T = 2\pi\sqrt{R^3/GM},$$

independent of r. Ignore any resistive forces. (b) What is the corresponding formula for the orbital period, assuming that the mass of the galaxy is strongly concentrated toward the center of the galaxy, so that essentially all the mass is at distances from the center less than r?

SEC. 45-14 The Big Bang

39E. The wavelength at which a thermal radiator at temperature T radiates electromagnetic waves most intensely is given by Wien's law: $\lambda_{max} = (2898\ \mu m \cdot K)/T$. (a) Show that the energy E of a photon corresponding to that wavelength can be computed from

$$E = (4.28 \times 10^{-10}\ \text{MeV/K})T.$$

(b) At what minimum temperature can this photon create an electron–positron pair (as discussed in Section 22-6)? ssm

40E. Use Wien's law (see Exercise 39) to answer the following questions: (a) The microwave background radiation peaks in intensity at a wavelength of 1.1 mm. To what temperature does this correspond? (b) About 300 000 y after the big bang, the universe became transparent to electromagnetic radiation. Its temperature then was about 10^5 K. What was the wavelength at which the background radiation was then most intense?

Additional Problems

41. Figure 45-7 shows part of the experimental arrangement in which antiprotons were discovered in the 1950s. A beam of 6.2 GeV protons emerged from a particle accelerator and collided with nuclei in a copper target. According to theoretical predictions at the time, collisions with the protons and neutrons in those nuclei should produce antiprotons via the reactions

$$p + p \rightarrow p + p + p + \bar{p}$$

and $$p + n \rightarrow p + n + p + \bar{p}.$$

However, even if these reactions did occur, they would be rare compared to the reactions

$$p + p \rightarrow p + p + \pi^+ + \pi^-$$

and $$p + n \rightarrow p + n + \pi^+ + \pi^-.$$

Thus, most of the particles produced by the collisions between the 6.2 GeV protons and the copper target were pions.

To prove that antiprotons existed and were also produced by the collisions, particles leaving the target were sent into a series of magnetic fields and detectors as shown in Fig. 45-7. The first magnetic field M1 curved the paths of any charged particles passing through it; moreover, the field was arranged so that the only particles that emerged from it to reach the second magnetic field (Q1) had to be negatively charged (either a $\bar{p}$ or a π^-) and have a momentum of 1.19 GeV/c. Q1 was a special type of magnetic field (a *quadrapole field*) that focused the particles reaching it into a beam, allowing them to pass through a hole in thick shielding to a *scintillation counter* S1. The travel of a charged particle through such a counter triggers a signal (much like a conventional television screen emits a pulse of light when struck by an electron). Thus, each signal indicated the passage of either a 1.19 GeV/c π^- or (presumably) a 1.19 GeV/c $\bar{p}$.

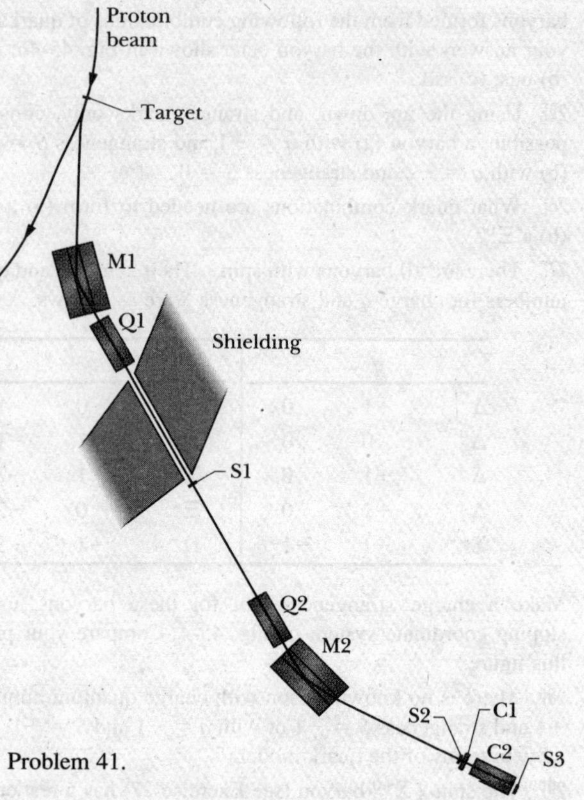

Fig. 45-7 Problem 41.

After being refocused by magnetic field Q2, the particles were directed by magnetic field M2 through a second scintillation counter S2 and then through two *Cerenkov counters* C1 and C2. These latter detectors can be manufactured so that they send a signal only when a particle passes through them with a speed in a certain range. In the experiment, a particle with a speed greater than $0.79c$ would trigger C1 and a particle with a speed between $0.75c$ and $0.78c$ would trigger C2.

There were then two ways to distinguish the predicted rare antiprotons from the abundant negative pions. Both ways involved the fact that the speed of a 1.19 GeV/c $\bar{p}$ would differ from that of a 1.19 GeV/c π^-: (1) According to calculations, a $\bar{p}$ would trigger one of the Cerenkov counters and a π^- would trigger the other. (2) Also, the time interval Δt between signals from S1 and S2, which were separated by 12 m, would be one value for a $\bar{p}$ and another value for a π^-. Thus, if the correct Cerenkov counter were triggered and the time interval Δt had the correct value, the experiment would prove the existence of antiprotons.

What is the speed of (a) an antiproton with a momentum of 1.19 GeV/c and (b) a negative pion with that same momentum? The speed of an antiproton through the Cerenkov detectors would actually be slightly less than calculated here because an antiproton would lose a little energy within the detectors. Which Cerenkov detector was triggered by (c) an antiproton and (d) a negative pion? What time interval Δt indicated the passage of (e) an antiproton and (f) a negative pion? [Problem adapted from O. Chamberlain, E. Segrè, C. Wiegand, and T. Ypsilantis, "Observation of Antiprotons," *Physical Review*, Vol. 100, pp. 947–950 (1955).]

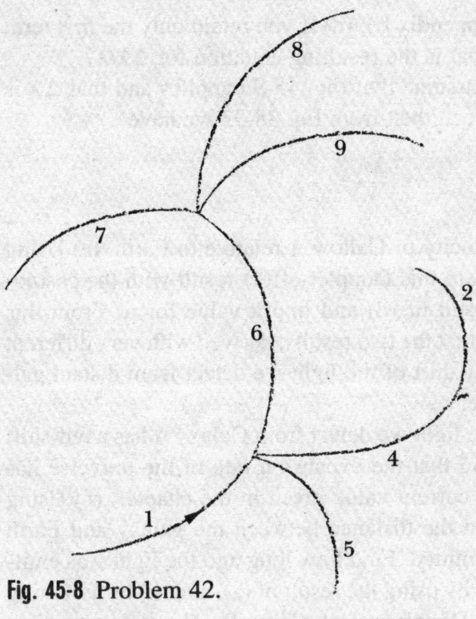

Fig. 45-8 Problem 42.

42. *A particle game*. Figure 45-8 is a sketch of the tracks made by particles in a *fictional* cloud chamber experiment (with a uniform magnetic field directed perpendicular to the page), and the following table gives *fictional* quantum numbers associated with the particles making the tracks. Particle *A* entered the chamber, leaving track 1 and decaying into three particles. Then the particle creating track 6 decayed into three other particles, and the particle creating track 4 decayed into two other particles, one of which was electrically uncharged—the path of that uncharged particle is represented by the dashed straight line. The particle that created track 8 is known to have a seriousness quantum number of zero.

By conserving the fictional quantum numbers at each decay point and by noting the directions of curvature of the tracks, identify which particle goes with which track. One of the listed particles is not formed; the others appear only once each.

Particle	Charge	Whimsy	Seriousness	Cuteness
A	1	1	−2	−2
B	0	4	3	0
C	1	2	−3	−1
D	−1	−1	0	1
E	−1	0	−4	−2
F	1	0	0	0
G	−1	−1	1	−1
H	3	3	1	0
I	0	6	4	6
J	1	−6	−4	−6

43. *Cosmological red-shift*. The expansion of the universe is often represented with a drawing like Fig. 45-9*a*. In that figure, we are located at the symbol labeled MW (for the Milky Way galaxy), at the origin of an r axis that extends radially away from us in any direction. Other, very distant galaxies are also represented. Superimposed on their symbols are their velocity vectors as inferred from the red-shift of the light reaching us from the galaxies. In accord with Hubble's law, the velocity of each galaxy is proportional to its distance from us. Such drawings can be misleading because they imply (1) that the red-shifts are due to the motions of galaxies relative to us, as they rush away from us through static (stationary) space, and (2) that we are at the center of all this motion.

Actually, the expansion of the universe and the increased separation of the galaxies are due not to an outward rush of the galaxies into pre-existing space but to an expansion of space itself throughout the universe. *Space is dynamic, not static.*

Figures 45-9*b*, *c*, and *d* show a different way of representing the universe and its expansion. Each part of the figure gives part of a one-dimensional section of the universe (along an r axis); the other two spatial dimensions of the universe are not shown. Each of the three parts of the figure show the Milky Way and six other galaxies (represented by dots); the parts are positioned along a time

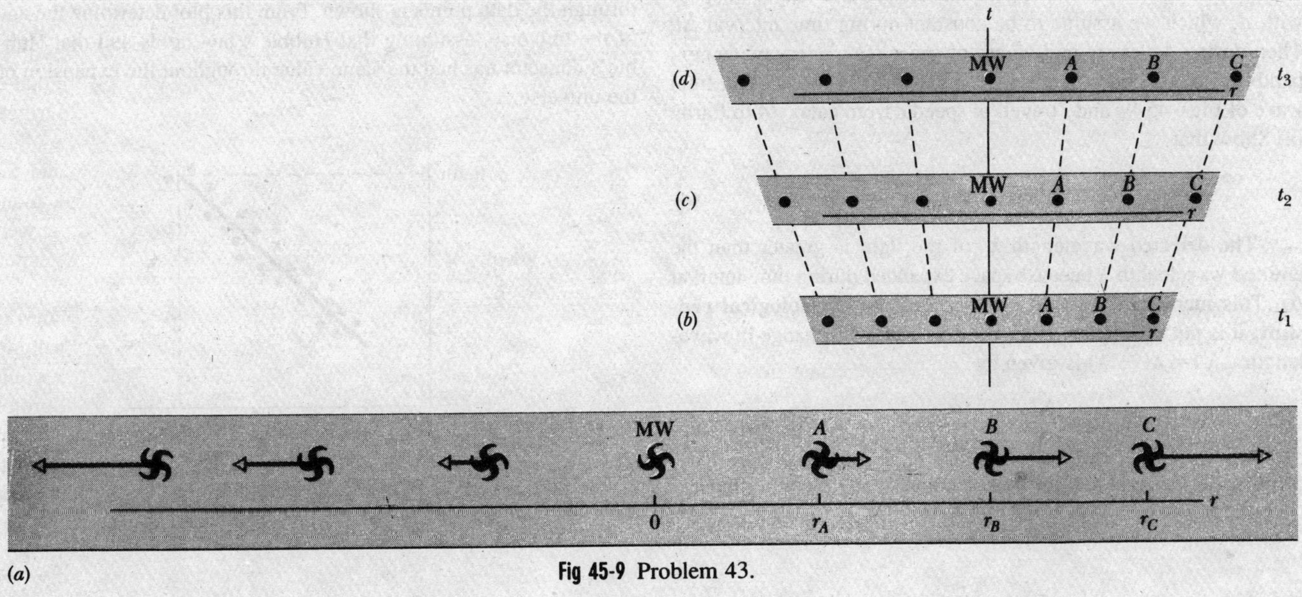

Fig 45-9 Problem 43.

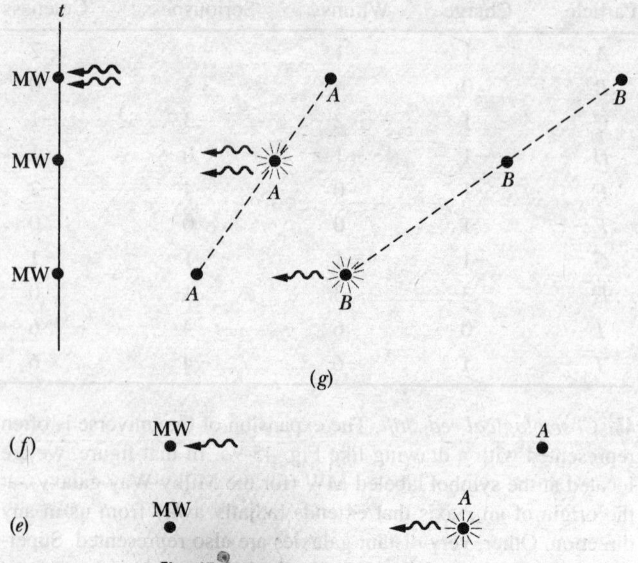

Fig. 45-9 Problem 43 (continued).

axis, with time increasing upward. In part *b*, at the earliest time of the three parts, the Milky Way and the six other galaxies are represented as being relatively close. As time progresses upward in the figures, space expands, causing the galaxies to move apart. Note that the figure parts are drawn relative to the Milky Way, and from that observation point all the other galaxies move away from it because of the expansion. However, there is nothing special about the Milky Way—the galaxies also move away from any other choice of observation point.

Figures 45-9*e* and *f* focus on just the Milky Way galaxy and one of the other galaxies, galaxy *A*, at two particular times during the expansion. In part *e*, galaxy *A* is a distance r from the Milky Way and is emitting a light wave of wavelength λ. In part *f*, after a time interval Δt, that light wave is being detected at Earth. Let us represent the universe's expansion rate per unit length of space with α, which we assume to be constant during time interval Δt. Then during Δt, every unit length of space (say, every meter) expands by $\alpha\,\Delta t$; hence, a distance r expands by $r\alpha\,\Delta t$. The light wave of Figs. 45-9*e* and *f* travels at speed c from galaxy *A* to Earth. (a) Show that

$$\Delta t = \frac{r}{c - r\alpha}.$$

The detected wavelength λ' of the light is greater than the emitted wavelength λ because space expanded during time interval Δt. This increase in wavelength is called the **cosmological red-shift;** it is not a Doppler effect. (b) Show that the change in wavelength $\Delta\lambda$ ($= \lambda' - \lambda$) is given by

$$\frac{\Delta\lambda}{\lambda} = \frac{r\alpha}{c - r\alpha}.$$

(c) Expand the right side of this equation, using the binomial expansion (given in Appendix E). (d) If you retain only the first term of the expansion, what is the resulting equation for $\Delta\lambda/\lambda$?

If, instead, we assume that Fig. 45-9*a* applies and that $\Delta\lambda$ is due to a Doppler effect, then from Eq. 38-33 we have

$$\frac{\Delta\lambda}{\lambda} = \frac{v}{c},$$

where v is radial velocity of Galaxy *A* relative to Earth. (e) Using Hubble's law, compare this Doppler-effect result with the cosmological-expansion result of (d) and find a value for α. From this analysis you can see that the two results, derived with very different models about the red-shift of the light we detect from distant galaxies, are compatible.

Suppose that the light we detect from Galaxy *A* has a red-shift of $\Delta\lambda/\lambda = 0.050$ and that the expansion rate of the universe has been constant at the current value given in the chapter. (f) Using the result of (b), find the distance between the galaxy and Earth when the light was emitted. Find how long ago the light was emitted by the galaxy (g) by using the result of (a), and (h) by assuming that the red-shift is a Doppler effect. (*Hint:* For (h), the time is just the distance at the time of emission divided by the speed of light, because if the red-shift is just a Doppler effect, the distance does not change during the light's travel to us. Here the two models about the red-shift of the light differ in their results.) (i) At the time of detection, what is the distance between Earth and Galaxy *A*? (We make the assumption that Galaxy *A* still exists; if it ceased to exist, humans would not know about its death until the last light emitted by the galaxy reached Earth.)

Now suppose that the light we detect from Galaxy *B* (Fig. 45-9*g*) has a red-shift of $\Delta\lambda/\lambda = 0.080$. (j) Using the result of (b), find the distance between Galaxy *B* and Earth when the light was emitted. (k) Using the result of (a), find how long ago the light was emitted by Galaxy *B*. (l) When the light that we detect from Galaxy *A* was emitted, what was the distance between Galaxy *A* and Galaxy *B*?

44. Figure 45-10 is a hypothetical plot of the recessional speeds v of galaxies against their distance r from us; the best-fit straight line through the data points is shown. From this plot determine the age of the universe, assuming that Hubble's law holds and that Hubble's constant has had the same value throughout the expansion of the universe.

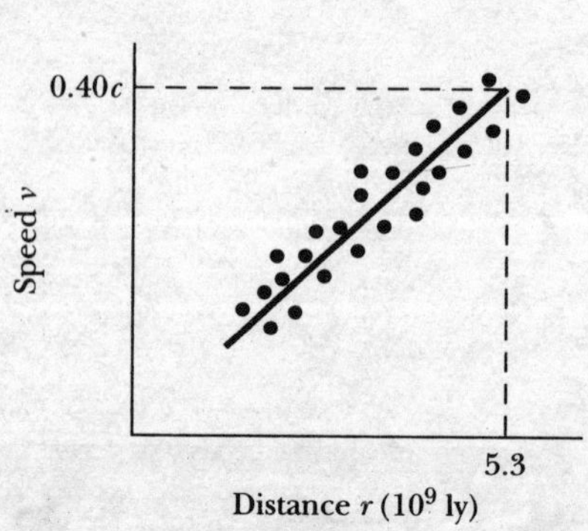

Fig. 45-10 Problem 44.

APPENDIX A

The International System of Units (SI)*

1. The SI Base Units

Quantity	Name	Symbol	Definition
length	meter	m	". . . the length of the path traveled by light in vacuum in 1/299,792,458 of a second." (1983)
mass	kilogram	kg	". . . this prototype [a certain platinum–iridium cylinder] shall henceforth be considered to be the unit of mass." (1889)
time	second	s	". . . the duration of 9,192,631,770 periods of the radiation corresponding to the transition between the two hyperfine levels of the ground state of the cesium-133 atom." (1967)
electric current	ampere	A	". . . that constant current which, if maintained in two straight parallel conductors of infinite length, of negligible circular cross section, and placed 1 meter apart in vacuum, would produce between these conductors a force equal to 2×10^{-7} newton per meter of length." (1946)
thermodynamic temperature	kelvin	K	". . . the fraction 1/273.16 of the thermodynamic temperature of the triple point of water." (1967)
amount of substance	mole	mol	". . . the amount of substance of a system which contains as many elementary entities as there are atoms in 0.012 kilogram of carbon-12." (1971)
luminous intensity	candela	cd	". . . the luminous intensity, in the perpendicular direction, of a surface of 1/600,000 square meter of a blackbody at the temperature of freezing platinum under a pressure of 101.325 newtons per square meter." (1967)

*Adapted from "The International System of Units (SI)," National Bureau of Standards Special Publication 330, 1972 edition. The definitions above were adopted by the General Conference of Weights and Measures, an international body, on the dates shown. In this book we do not use the candela.

2. Some SI Derived Units

Quantity	Name of Unit	Symbol	
area	square meter	m^2	
volume	cubic meter	m^3	
frequency	hertz	Hz	s^{-1}
mass density (density)	kilogram per cubic meter	kg/m^3	
speed, velocity	meter per second	m/s	
angular velocity	radian per second	rad/s	
acceleration	meter per second per second	m/s^2	
angular acceleration	radian per second per second	rad/s^2	
force	newton	N	$kg \cdot m/s^2$
pressure	pascal	Pa	N/m^2
work, energy, quantity of heat	joule	J	$N \cdot m$
power	watt	W	J/s
quantity of electric charge	coulomb	C	$A \cdot s$
potential difference, electromotive force	volt	V	W/A
electric field strength	volt per meter (or newton per coulomb)	V/m	N/C
electric resistance	ohm	Ω	V/A
capacitance	farad	F	$A \cdot s/V$
magnetic flux	weber	Wb	$V \cdot s$
inductance	henry	H	$V \cdot s/A$
magnetic flux density	tesla	T	Wb/m^2
magnetic field strength	ampere per meter	A/m	
entropy	joule per kelvin	J/K	
specific heat	joule per kilogram kelvin	$J/(kg \cdot K)$	
thermal conductivity	watt per meter kelvin	$W/(m \cdot K)$	
radiant intensity	watt per steradian	W/sr	

3. The SI Supplementary Units

Quantity	Name of Unit	Symbol
plane angle	radian	rad
solid angle	steradian	sr

APPENDIX B
Some Fundamental Constants of Physics*

Constant	Symbol	Computational Value	Best (1998) Value	
			Value[a]	Uncertainty[b]
Speed of light in a vacuum	c	3.00×10^{8} m/s	2.997 924 58	exact
Elementary charge	e	1.60×10^{-19} C	1.602 176 462	0.039
Gravitational constant	G	6.67×10^{-11} m³/s² · kg	6.673	1500
Universal gas constant	R	8.31 J/mol · K	8.314 472	1.7
Avogadro constant	N_A	6.02×10^{23} mol⁻¹	6.022 141 99	0.079
Boltzmann constant	k	1.38×10^{-23} J/K	1.380 650 3	1.7
Stefan–Boltzmann constant	σ	5.67×10^{-8} W/m² · K⁴	5.670 400	7.0
Molar volume of ideal gas at STP[d]	V_m	2.27×10^{-2} m³/mol	2.271 098 1	1.7
Permittivity constant	ϵ_0	8.85×10^{-12} F/m	8.854 187 817 62	exact
Permeability constant	μ_0	1.26×10^{-6} H/m	1.256 637 061 43	exact
Planck constant	h	6.63×10^{-34} J · s	6.626 068 76	0.078
Electron mass[c]	m_e	9.11×10^{-31} kg	9.109 381 88	0.079
		5.49×10^{-4} u	5.485 799 110	0.0021
Proton mass[c]	m_p	1.67×10^{-27} kg	1.672 621 58	0.079
		1.0073 u	1.007 276 466 88	1.3×10^{-4}
Ratio of proton mass to electron mass	m_p/m_e	1840	1836.152 667 5	0.0021
Electron charge-to-mass ratio	e/m_e	1.76×10^{11} C/kg	1.758 820 174	0.040
Neutron mass[c]	m_n	1.68×10^{-27} kg	1.674 927 16	0.079
		1.0087 u	1.008 664 915 78	5.4×10^{-4}
Hydrogen atom mass[c]	$m_{^1H}$	1.0078 u	1.007 825 031 6	0.0005
Deuterium atom mass[c]	$m_{^2H}$	2.0141 u	2.014 101 777 9	0.0005
Helium atom mass[c]	$m_{^4He}$	4.0026 u	4.002 603 2	0.067
Muon mass	m_μ	1.88×10^{-28} kg	1.883 531 09	0.084
Electron magnetic moment	μ_e	9.28×10^{-24} J/T	9.284 763 62	0.040
Proton magnetic moment	μ_p	1.41×10^{-26} J/T	1.410 606 663	0.041
Bohr magneton	μ_B	9.27×10^{-24} J/T	9.274 008 99	0.040
Nuclear magneton	μ_N	5.05×10^{-27} J/T	5.050 783 17	0.040
Bohr radius	r_B	5.29×10^{-11} m	5.291 772 083	0.0037
Rydberg constant	R	1.10×10^{7} m⁻¹	1.097 373 156 854 8	7.6×10^{-6}
Electron Compton wavelength	λ_C	2.43×10^{-12} m	2.426 310 215	0.0073

[a]Values given in this column should be given the same unit and power of 10 as the computational value.

[b]Parts per million.

[c]Masses given in u are in unified atomic mass units, where 1 u = $1.660\ 538\ 73 \times 10^{-27}$ kg.

[d]STP means standard temperature and pressure: 0°C and 1.0 atm (0.1 MPa).

*The values in this table were selected from the 1998 CODATA recommended values (www.physics.nist.gov).

APPENDIX C

Some Astronomical Data

Some Distances from Earth

To the Moon*	3.82×10^{8} m	To the center of our galaxy	2.2×10^{20} m
To the Sun*	1.50×10^{11} m	To the Andromeda Galaxy	2.1×10^{22} m
To the nearest star (Proxima Centauri)	4.04×10^{16} m	To the edge of the observable universe	$\sim 10^{26}$ m

*Mean distance.

The Sun, Earth, and the Moon

Property	Unit	Sun	Earth	Moon
Mass	kg	1.99×10^{30}	5.98×10^{24}	7.36×10^{22}
Mean radius	m	6.96×10^{8}	6.37×10^{6}	1.74×10^{6}
Mean density	kg/m^3	1410	5520	3340
Free-fall acceleration at the surface	m/s^2	274	9.81	1.67
Escape velocity	km/s	618	11.2	2.38
Period of rotation[a]	—	37 d at poles[b] 26 d at equator[b]	23 h 56 min	27.3 d
Radiation power[c]	W	3.90×10^{26}		

[a]Measured with respect to the distant stars.
[b]The Sun, a ball of gas, does not rotate as a rigid body.
[c]Just outside Earth's atmosphere solar energy is received, assuming normal incidence, at the rate of 1340 W/m^2.

Some Properties of the Planets

	Mercury	Venus	Earth	Mars	Jupiter	Saturn	Uranus	Neptune	Pluto
Mean distance from Sun, 10^6 km	57.9	108	150	228	778	1430	2870	4500	5900
Period of revolution, y	0.241	0.615	1.00	1.88	11.9	29.5	84.0	165	248
Period of rotation,[a] d	58.7	-243[b]	0.997	1.03	0.409	0.426	-0.451[b]	0.658	6.39
Orbital speed, km/s	47.9	35.0	29.8	24.1	13.1	9.64	6.81	5.43	4.74
Inclination of axis to orbit	<28°	≈3°	23.4°	25.0°	3.08°	26.7°	97.9°	29.6°	57.5°
Inclination of orbit to Earth's orbit	7.00°	3.39°		1.85°	1.30°	2.49°	0.77°	1.77°	17.2°
Eccentricity of orbit	0.206	0.0068	0.0167	0.0934	0.0485	0.0556	0.0472	0.0086	0.250
Equatorial diameter, km	4880	12 100	12 800	6790	143 000	120 000	51 800	49 500	2300
Mass (Earth = 1)	0.0558	0.815	1.000	0.107	318	95.1	14.5	17.2	0.002
Density (water = 1)	5.60	5.20	5.52	3.95	1.31	0.704	1.21	1.67	2.03
Surface value of g,[c] m/s^2	3.78	8.60	9.78	3.72	22.9	9.05	7.77	11.0	0.5
Escape velocity,[c] km/s	4.3	10.3	11.2	5.0	59.5	35.6	21.2	23.6	1.1
Known satellites	0	0	1	2	16 + ring	18 + rings	17 + rings	8 + rings	1

[a]Measured with respect to the distant stars.
[b]Venus and Uranus rotate opposite their orbital motion.
[c]Gravitational acceleration measured at the planet's equator.

APPENDIX D
Conversion Factors

Conversion factors may be read directly from these tables. For example, 1 degree = 2.778×10^{-3} revolutions, so $16.7° = 16.7 \times 2.778 \times 10^{-3}$ rev. The SI units are fully capitalized. Adapted in part from G. Shortley and D. Williams, *Elements of Physics*, 1971, Prentice-Hall, Englewood Cliffs, NJ.

Plane Angle

	°	′	″	RADIAN	rev
1 degree =	1	60	3600	1.745×10^{-2}	2.778×10^{-3}
1 minute =	1.667×10^{-2}	1	60	2.909×10^{-4}	4.630×10^{-5}
1 second =	2.778×10^{-4}	1.667×10^{-2}	1	4.848×10^{-6}	7.716×10^{-7}
1 RADIAN =	57.30	3438	2.063×10^{5}	1	0.1592
1 revolution =	360	2.16×10^{4}	1.296×10^{6}	6.283	1

Solid Angle

1 sphere = 4π steradians = 12.57 steradians

Length

	cm	METER	km	in.	ft	mi
1 centimeter =	1	10^{-2}	10^{-5}	0.3937	3.281×10^{-2}	6.214×10^{-6}
1 METER =	100	1	10^{-3}	39.37	3.281	6.214×10^{-4}
1 kilometer =	10^{5}	1000	1	3.937×10^{4}	3281	0.6214
1 inch =	2.540	2.540×10^{-2}	2.540×10^{-5}	1	8.333×10^{-2}	1.578×10^{-5}
1 foot =	30.48	0.3048	3.048×10^{-4}	12	1	1.894×10^{-4}
1 mile =	1.609×10^{5}	1609	1.609	6.336×10^{4}	5280	1

1 angström = 10^{-10} m
1 nautical mile = 1852 m = 1.151 miles = 6076 ft
1 fermi = 10^{-15} m
1 light-year = 9.460×10^{12} km
1 parsec = 3.084×10^{13} km
1 fathom = 6 ft
1 Bohr radius = 5.292×10^{-11} m
1 yard = 3 ft
1 rod = 16.5 ft
1 mil = 10^{-3} in.
1 nm = 10^{-9} m

Area

	METER^2	cm^2	ft^2	in.^{2}
1 SQUARE METER =	1	10^{4}	10.76	1550
1 square centimeter =	10^{-4}	1	1.076×10^{-3}	0.1550
1 square foot =	9.290×10^{-2}	929.0	1	144
1 square inch =	6.452×10^{-4}	6.452	6.944×10^{-3}	1

1 square mile = 2.788×10^{7} ft^2 = 640 acres
1 barn = 10^{-28} m^2
1 acre = 43 560 ft^2
1 hectare = 10^{4} m^2 = 2.471 acres

Volume

METER^3	cm^3	L	ft^3	$in.^3$
1 CUBIC METER = 1	10^6	1000	35.31	6.102×10^4
1 cubic centimeter = 10^{-6}	1	1.000×10^{-3}	3.531×10^{-5}	6.102×10^{-2}
1 liter = 1.000×10^{-3}	1000	1	3.531×10^{-2}	61.02
1 cubic foot = 2.832×10^{-2}	2.832×10^4	28.32	1	1728
1 cubic inch = 1.639×10^{-5}	16.39	1.639×10^{-2}	5.787×10^{-4}	1

1 U.S. fluid gallon = 4 U.S. fluid quarts = 8 U.S. pints = 128 U.S. fluid ounces = 231 $in.^3$

1 British imperial gallon = 277.4 $in.^3$ = 1.201 U.S. fluid gallons

Mass

Quantities in the colored areas are not mass units but are often used as such. When we write, for example, 1 kg "=" 2.205 lb, this means that a kilogram is a *mass* that *weighs* 2.205 pounds at a location where g has the standard value of 9.80665 m/s^2.

g	KILOGRAM	slug	u	oz	lb	ton
1 gram = 1	0.001	6.852×10^{-5}	6.022×10^{23}	3.527×10^{-2}	2.205×10^{-3}	1.102×10^{-6}
1 KILOGRAM = 1000	1	6.852×10^{-2}	6.022×10^{26}	35.27	2.205	1.102×10^{-3}
1 slug = 1.459×10^4	14.59	1	8.786×10^{27}	514.8	32.17	1.609×10^{-2}
1 atomic mass unit = 1.661×10^{-24}	1.661×10^{-27}	1.138×10^{-28}	1	5.857×10^{-26}	3.662×10^{-27}	1.830×10^{-30}
1 ounce = 28.35	2.835×10^{-2}	1.943×10^{-3}	1.718×10^{25}	1	6.250×10^{-2}	3.125×10^{-5}
1 pound = 453.6	0.4536	3.108×10^{-2}	2.732×10^{26}	16	1	0.0005
1 ton = 9.072×10^5	907.2	62.16	5.463×10^{29}	3.2×10^4	2000	1

1 metric ton = 1000 kg

Density

Quantities in the colored areas are weight densities and, as such, are dimensionally different from mass densities. See note for mass table.

slug/ft^3	KILOGRAM/ METER^3	g/cm^3	lb/ft^3	lb/$in.^3$
1 slug per foot^3 = 1	515.4	0.5154	32.17	1.862×10^{-2}
1 KILOGRAM per METER^3 = 1.940×10^{-3}	1	0.001	6.243×10^{-2}	3.613×10^{-5}
1 gram per centimeter^3 = 1.940	1000	1	62.43	3.613×10^{-2}
1 pound per foot^3 = 3.108×10^{-2}	16.02	16.02×10^{-2}	1	5.787×10^{-4}
1 pound per inch^3 = 53.71	2.768×10^4	27.68	1728	1

Time

y	d	h	min	SECOND
1 year = 1	365.25	8.766×10^3	5.259×10^5	3.156×10^7
1 day = 2.738×10^{-3}	1	24	1440	8.640×10^4
1 hour = 1.141×10^{-4}	4.167×10^{-2}	1	60	3600
1 minute = 1.901×10^{-6}	6.944×10^{-4}	1.667×10^{-2}	1	60
1 SECOND = 3.169×10^{-8}	1.157×10^{-5}	2.778×10^{-4}	1.667×10^{-2}	1

Speed

	ft/s	km/h	METER/SECOND	mi/h	cm/s
1 foot per second =	1	1.097	0.3048	0.6818	30.48
1 kilometer per hour =	0.9113	1	0.2778	0.6214	27.78
1 METER per SECOND =	3.281	3.6	1	2.237	100
1 mile per hour =	1.467	1.609	0.4470	1	44.70
1 centimeter per second =	3.281×10^{-2}	3.6×10^{-2}	0.01	2.237×10^{-2}	1

1 knot = 1 nautical mi/h = 1.688 ft/s 1 mi/min = 88.00 ft/s = 60.00 mi/h

Force

Force units in the colored areas are now little used. To clarify: 1 gram-force (= 1 gf) is the force of gravity that would act on an object whose mass is 1 gram at a location where g has the standard value of 9.80665 m/s^2.

	dyne	NEWTON	lb	pdl	gf	kgf
1 dyne =	1	10^{-5}	2.248×10^{-6}	7.233×10^{-5}	1.020×10^{-3}	1.020×10^{-6}
1 NEWTON =	10^5	1	0.2248	7.233	102.0	0.1020
1 pound =	4.448×10^5	4.448	1	32.17	453.6	0.4536
1 poundal =	1.383×10^4	0.1383	3.108×10^{-2}	1	14.10	1.410×10^2
1 gram-force =	980.7	9.807×10^{-3}	2.205×10^{-3}	7.093×10^{-2}	1	0.001
1 kilogram-force =	9.807×10^5	9.807	2.205	70.93	1000	1

1 ton = 2000 lb

Pressure

	atm	dyne/cm^2	inch of water	cm Hg	PASCAL	lb/in.^{2}	lb/ft^2
1 atmosphere =	1	1.013×10^6	406.8	76	1.013×10^5	14.70	2116
1 dyne per centimeter^2 =	9.869×10^{-7}	1	4.015×10^{-4}	7.501×10^{-5}	0.1	1.405×10^{-5}	2.089×10^{-3}
1 inch of water^a at 4°C =	2.458×10^{-3}	2491	1	0.1868	249.1	3.613×10^{-2}	5.202
1 centimeter of mercury^a at 0°C =	1.316×10^{-2}	1.333×10^4	5.353	1	1333	0.1934	27.85
1 PASCAL =	9.869×10^{-6}	10	4.015×10^{-3}	7.501×10^{-4}	1	1.450×10^{-4}	2.089×10^{-2}
1 pound per inch^2 =	6.805×10^{-2}	6.895×10^4	27.68	5.171	6.895×10^3	1	144
1 pound per foot^2 =	4.725×10^{-4}	478.8	0.1922	3.591×10^{-2}	47.88	6.944×10^{-3}	1

^{a}Where the acceleration of gravity has the standard value of 9.80665 m/s^2.

1 bar = 10^6 dyne/cm^2 = 0.1 MPa 1 millibar = 10^3 dyne/cm^2 = 10^2 Pa 1 torr = 1 mm Hg

Energy, Work, Heat

Quantities in the colored areas are not energy units but are included for convenience. They arise from the relativistic mass–energy equivalence formula $E = mc^2$ and represent the energy released if a kilogram or unified atomic mass unit (u) is completely converted to energy (bottom two rows) or the mass that would be completely converted to one unit of energy (rightmost two columns).

	Btu	erg	ft · lb	hp · h	JOULE	cal	kW · h	eV	MeV	kg	u
1 British thermal unit =	1	1.055×10^{10}	777.9	3.929×10^{-4}	1055	252.0	2.930×10^{-4}	6.585×10^{21}	6.585×10^{15}	1.174×10^{-14}	7.070×10^{12}
1 erg =	9.481×10^{-11}	1	7.376×10^{-8}	3.725×10^{-14}	10^{-7}	2.389×10^{-8}	2.778×10^{-14}	6.242×10^{11}	6.242×10^{5}	1.113×10^{-24}	670.2
1 foot-pound =	1.285×10^{-3}	1.356×10^{7}	1	5.051×10^{-7}	1.356	0.3238	3.766×10^{-7}	8.464×10^{18}	8.464×10^{12}	1.509×10^{-17}	9.037×10^{9}
1 horsepower-hour =	2545	2.685×10^{13}	1.980×10^{6}	1	2.685×10^{6}	6.413×10^{5}	0.7457	1.676×10^{25}	1.676×10^{19}	2.988×10^{-11}	1.799×10^{16}
1 JOULE =	9.481×10^{-4}	10^{7}	0.7376	3.725×10^{-7}	1	0.2389	2.778×10^{-7}	6.242×10^{18}	6.242×10^{12}	1.113×10^{-17}	6.702×10^{9}
1 calorie =	3.969×10^{-3}	4.186×10^{7}	3.088	1.560×10^{-6}	4.186	1	1.163×10^{-6}	2.613×10^{19}	2.613×10^{13}	4.660×10^{-17}	2.806×10^{10}
1 kilowatt-hour =	3413	3.600×10^{13}	2.655×10^{6}	1.341	3.600×10^{6}	8.600×10^{5}	1	2.247×10^{25}	2.247×10^{19}	4.007×10^{-11}	2.413×10^{16}
1 electron-volt =	1.519×10^{-22}	1.602×10^{-12}	1.182×10^{-19}	5.967×10^{-26}	1.602×10^{-19}	3.827×10^{-20}	4.450×10^{-26}	1	10^{-6}	1.783×10^{-36}	1.074×10^{-9}
1 million electron-volts =	1.519×10^{-16}	1.602×10^{-6}	1.182×10^{-13}	5.967×10^{-20}	1.602×10^{-13}	3.827×10^{-14}	4.450×10^{-20}	10^{-6}	1	1.783×10^{-30}	1.074×10^{-3}
1 kilogram =	8.521×10^{13}	8.987×10^{23}	6.629×10^{16}	3.348×10^{10}	8.987×10^{16}	2.146×10^{16}	2.497×10^{10}	5.610×10^{35}	5.610×10^{29}	1	6.022×10^{26}
1 unified atomic mass unit =	1.415×10^{-13}	1.492×10^{-3}	1.101×10^{-10}	5.559×10^{-17}	1.492×10^{-10}	3.564×10^{-11}	4.146×10^{-17}	9.320×10^{8}	932.0	1.661×10^{-27}	1

Power

	Btu/h	ft · lb/s	hp	cal/s	kW	WATT
1 British thermal unit per hour =	1	0.2161	3.929×10^{-4}	6.998×10^{-2}	2.930×10^{-4}	0.2930
1 foot-pound per second =	4.628	1	1.818×10^{-3}	0.3239	1.356×10^{-3}	1.356
1 horsepower =	2545	550	1	178.1	0.7457	745.7
1 calorie per second =	14.29	3.088	5.615×10^{-3}	1	4.186×10^{-3}	4.186
1 kilowatt =	3413	737.6	1.341	238.9	1	1000
1 WATT =	3.413	0.7376	1.341×10^{-3}	0.2389	0.001	1

Magnetic Field

	gauss	TESLA	milligauss
1 gauss =	1	10^{-4}	1000
1 TESLA =	10^{4}	1	10^{7}
1 milligauss =	0.001	10^{-7}	1

1 tesla = 1 weber/meter^2

Magnetic Flux

	maxwell	WEBER
1 maxwell =	1	10^{-8}
1 WEBER =	10^{8}	1

APPENDIX E

Mathematical Formulas

Geometry

Circle of radius r: circumference $= 2\pi r$; area $= \pi r^2$.

Sphere of radius r: area $= 4\pi r^2$; volume $= \frac{4}{3}\pi r^3$.

Right circular cylinder of radius r and height h:
area $= 2\pi r^2 + 2\pi rh$; volume $= \pi r^2 h$.

Triangle of base a and altitude h: area $= \frac{1}{2}ah$.

Quadratic Formula

If $ax^2 + bx + c = 0$, then $x = \dfrac{-b \pm \sqrt{b^2 - 4ac}}{2a}$.

Trigonometric Functions of Angle θ

$\sin\theta = \dfrac{y}{r}$ $\cos\theta = \dfrac{x}{r}$

$\tan\theta = \dfrac{y}{x}$ $\cot\theta = \dfrac{x}{y}$

$\sec\theta = \dfrac{r}{x}$ $\csc\theta = \dfrac{r}{y}$

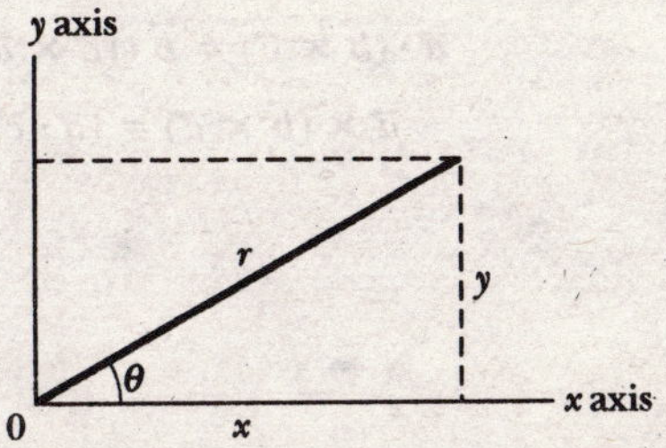

Pythagorean Theorem

In this right triangle,
$a^2 + b^2 = c^2$

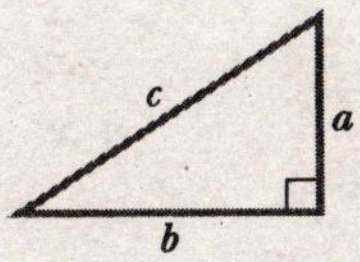

Triangles

Angles are A, B, C

Opposite sides are a, b, c

Angles $A + B + C = 180°$

$\dfrac{\sin A}{a} = \dfrac{\sin B}{b} = \dfrac{\sin C}{c}$

$c^2 = a^2 + b^2 - 2ab\cos C$

Exterior angle $D = A + C$

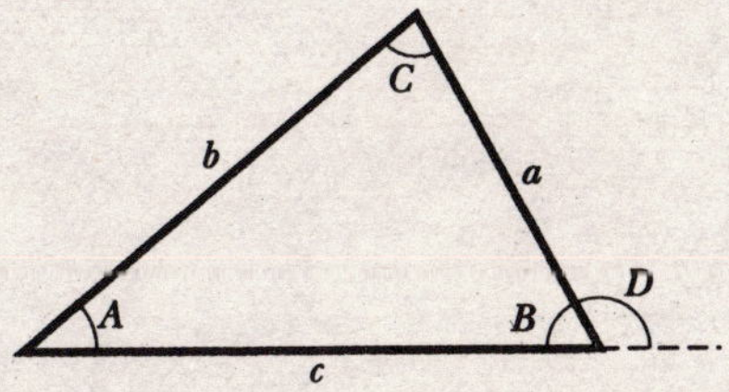

Mathematical Signs and Symbols

$=$ equals

$\approx$ equals approximately

$\sim$ is the order of magnitude of

$\neq$ is not equal to

$\equiv$ is identical to, is defined as

$>$ is greater than ($\gg$ is much greater than)

$<$ is less than ($\ll$ is much less than)

$\geq$ is greater than or equal to (or, is no less than)

$\leq$ is less than or equal to (or, is no more than)

$\pm$ plus or minus

$\propto$ is proportional to

Σ the sum of

x_{avg} the average value of x

Trigonometric Identities

$\sin(90° - \theta) = \cos\theta$

$\cos(90° - \theta) = \sin\theta$

$\sin\theta/\cos\theta = \tan\theta$

$\sin^2\theta + \cos^2\theta = 1$

$\sec^2\theta - \tan^2\theta = 1$

$\csc^2\theta - \cot^2\theta = 1$

$\sin 2\theta = 2\sin\theta\cos\theta$

$\cos 2\theta = \cos^2\theta - \sin^2\theta = 2\cos^2\theta - 1 = 1 - 2\sin^2\theta$

$\sin(\alpha \pm \beta) = \sin\alpha\cos\beta \pm \cos\alpha\sin\beta$

$\cos(\alpha \pm \beta) = \cos\alpha\cos\beta \mp \sin\alpha\sin\beta$

$\tan(\alpha \pm \beta) = \dfrac{\tan\alpha \pm \tan\beta}{1 \mp \tan\alpha\tan\beta}$

$\sin\alpha \pm \sin\beta = 2\sin\frac{1}{2}(\alpha \pm \beta)\cos\frac{1}{2}(\alpha \mp \beta)$

$\cos\alpha + \cos\beta = 2\cos\frac{1}{2}(\alpha + \beta)\cos\frac{1}{2}(\alpha - \beta)$

$\cos\alpha - \cos\beta = -2\sin\frac{1}{2}(\alpha + \beta)\sin\frac{1}{2}(\alpha - \beta)$

Binomial Theorem

$$(1+x)^n = 1 + \frac{nx}{1!} + \frac{n(n-1)x^2}{2!} + \cdots \qquad (x^2 < 1)$$

Exponential Expansion

$$e^x = 1 + x + \frac{x^2}{2!} + \frac{x^3}{3!} + \cdots$$

Logarithmic Expansion

$$\ln(1+x) = x - \tfrac{1}{2}x^2 + \tfrac{1}{3}x^3 - \cdots \qquad (|x| < 1)$$

Trigonometric Expansions (θ in radians)

$$\sin\theta = \theta - \frac{\theta^3}{3!} + \frac{\theta^5}{5!} - \cdots$$

$$\cos\theta = 1 - \frac{\theta^2}{2!} + \frac{\theta^4}{4!} - \cdots$$

$$\tan\theta = \theta + \frac{\theta^3}{3} + \frac{2\theta^5}{15} + \cdots$$

Cramer's Rule

Two simultaneous equations in unknowns x and y,

$$a_1x + b_1y = c_1 \quad \text{and} \quad a_2x + b_2y = c_2,$$

have the solutions

$$x = \frac{\begin{vmatrix} c_1 & b_1 \\ c_2 & b_2 \end{vmatrix}}{\begin{vmatrix} a_1 & b_1 \\ a_2 & b_2 \end{vmatrix}} = \frac{c_1b_2 - c_2b_1}{a_1b_2 - a_2b_1}$$

and

$$y = \frac{\begin{vmatrix} a_1 & c_1 \\ a_2 & c_2 \end{vmatrix}}{\begin{vmatrix} a_1 & b_1 \\ a_2 & b_2 \end{vmatrix}} = \frac{a_1c_2 - a_2c_1}{a_1b_2 - a_2b_1}.$$

Products of Vectors

Let $\hat{\mathrm{i}}$, $\hat{\mathrm{j}}$, and $\hat{\mathrm{k}}$ be unit vectors in the x, y, and z directions. Then

$$\hat{\mathrm{i}}\cdot\hat{\mathrm{i}} = \hat{\mathrm{j}}\cdot\hat{\mathrm{j}} = \hat{\mathrm{k}}\cdot\hat{\mathrm{k}} = 1, \qquad \hat{\mathrm{i}}\cdot\hat{\mathrm{j}} = \hat{\mathrm{j}}\cdot\hat{\mathrm{k}} = \hat{\mathrm{k}}\cdot\hat{\mathrm{i}} = 0,$$

$$\hat{\mathrm{i}}\times\hat{\mathrm{i}} = \hat{\mathrm{j}}\times\hat{\mathrm{j}} = \hat{\mathrm{k}}\times\hat{\mathrm{k}} = 0,$$

$$\hat{\mathrm{i}}\times\hat{\mathrm{j}} = \hat{\mathrm{k}}, \qquad \hat{\mathrm{j}}\times\hat{\mathrm{k}} = \hat{\mathrm{i}}, \qquad \hat{\mathrm{k}}\times\hat{\mathrm{i}} = \hat{\mathrm{j}}.$$

Any vector $\vec{a}$ with components a_x, a_y, and a_z along the x, y, and z axes can be written as

$$\vec{a} = a_x\hat{\mathrm{i}} + a_y\hat{\mathrm{j}} + a_z\hat{\mathrm{k}}.$$

Let $\vec{a}$, $\vec{b}$, and $\vec{c}$ be arbitrary vectors with magnitudes a, b, and c. Then

$$\vec{a}\times(\vec{b}+\vec{c}) = (\vec{a}\times\vec{b}) + (\vec{a}\times\vec{c})$$

$$(s\vec{a})\times\vec{b} = \vec{a}\times(s\vec{b}) = s(\vec{a}\times\vec{b}) \qquad (s = \text{a scalar}).$$

Let θ be the smaller of the two angles between $\vec{a}$ and $\vec{b}$. Then

$$\vec{a}\cdot\vec{b} = \vec{b}\cdot\vec{a} = a_xb_x + a_yb_y + a_zb_z = ab\cos\theta$$

$$\vec{a}\times\vec{b} = -\vec{b}\times\vec{a} = \begin{vmatrix} \hat{\mathrm{i}} & \hat{\mathrm{j}} & \hat{\mathrm{k}} \\ a_x & a_y & a_z \\ b_x & b_y & b_z \end{vmatrix}$$

$$= \hat{\mathrm{i}}\begin{vmatrix} a_y & a_z \\ b_y & b_z \end{vmatrix} - \hat{\mathrm{j}}\begin{vmatrix} a_x & a_z \\ b_x & b_z \end{vmatrix} + \hat{\mathrm{k}}\begin{vmatrix} a_x & a_y \\ b_x & b_y \end{vmatrix}$$

$$= (a_yb_z - b_ya_z)\hat{\mathrm{i}} + (a_zb_x - b_za_x)\hat{\mathrm{j}} + (a_xb_y - b_xa_y)\hat{\mathrm{k}}$$

$$|\vec{a}\times\vec{b}| = ab\sin\theta$$

$$\vec{a}\cdot(\vec{b}\times\vec{c}) = \vec{b}\cdot(\vec{c}\times\vec{a}) = \vec{c}\cdot(\vec{a}\times\vec{b})$$

$$\vec{a}\times(\vec{b}\times\vec{c}) = (\vec{a}\cdot\vec{c})\vec{b} - (\vec{a}\cdot\vec{b})\vec{c}$$

Derivatives and Integrals

In what follows, the letters u and v stand for any functions of x, and a and m are constants. To each of the indefinite integrals should be added an arbitrary constant of integration. The *Handbook of Chemistry and Physics* (CRC Press Inc.) gives a more extensive tabulation.

1. $\dfrac{dx}{dx} = 1$
2. $\dfrac{d}{dx}(au) = a\dfrac{du}{dx}$
3. $\dfrac{d}{dx}(u + v) = \dfrac{du}{dx} + \dfrac{dv}{dx}$
4. $\dfrac{d}{dx}x^m = mx^{m-1}$
5. $\dfrac{d}{dx}\ln x = \dfrac{1}{x}$
6. $\dfrac{d}{dx}(uv) = u\dfrac{dv}{dx} + v\dfrac{du}{dx}$
7. $\dfrac{d}{dx}e^x = e^x$
8. $\dfrac{d}{dx}\sin x = \cos x$
9. $\dfrac{d}{dx}\cos x = -\sin x$
10. $\dfrac{d}{dx}\tan x = \sec^2 x$
11. $\dfrac{d}{dx}\cot x = -\csc^2 x$.
12. $\dfrac{d}{dx}\sec x = \tan x \sec x$
13. $\dfrac{d}{dx}\csc x = -\cot x \csc x$
14. $\dfrac{d}{dx}e^u = e^u\dfrac{du}{dx}$
15. $\dfrac{d}{dx}\sin u = \cos u\dfrac{du}{dx}$
16. $\dfrac{d}{dx}\cos u = -\sin u\dfrac{du}{dx}$

1. $\int dx = x$
2. $\int au\,dx = a\int u\,dx$
3. $\int (u + v)\,dx = \int u\,dx + \int v\,dx$
4. $\int x^m\,dx = \dfrac{x^{m+1}}{m+1} \quad (m \neq -1)$
5. $\int \dfrac{dx}{x} = \ln|x|$
6. $\int u\dfrac{dv}{dx}dx = uv - \int v\dfrac{du}{dx}dx$
7. $\int e^x\,dx = e^x$
8. $\int \sin x\,dx = -\cos x$
9. $\int \cos x\,dx = \sin x$
10. $\int \tan x\,dx = \ln|\sec x|$
11. $\int \sin^2 x\,dx = \frac{1}{2}x - \frac{1}{4}\sin 2x$
12. $\int e^{-ax}\,dx = -\dfrac{1}{a}e^{-ax}$
13. $\int xe^{-ax}\,dx = -\dfrac{1}{a^2}(ax + 1)e^{-ax}$
14. $\int x^2e^{-ax}\,dx = -\dfrac{1}{a^3}(a^2x^2 + 2ax + 2)e^{-ax}$
15. $\int_0^\infty x^n e^{-ax}\,dx = \dfrac{n!}{a^{n+1}}$
16. $\int_0^\infty x^{2n}e^{-ax^2}\,dx = \dfrac{1\cdot 3\cdot 5\cdots(2n-1)}{2^{n+1}a^n}\sqrt{\dfrac{\pi}{a}}$
17. $\int \dfrac{dx}{\sqrt{x^2 + a^2}} = \ln(x + \sqrt{x^2 + a^2})$
18. $\int \dfrac{x\,dx}{(x^2 + a^2)^{3/2}} = -\dfrac{1}{(x^2 + a^2)^{1/2}}$
19. $\int \dfrac{dx}{(x^2 + a^2)^{3/2}} = \dfrac{x}{a^2(x^2 + a^2)^{1/2}}$
20. $\int_0^\infty x^{2n+1}e^{-ax^2}\,dx = \dfrac{n!}{2a^{n+1}} \quad (a > 0)$
21. $\int \dfrac{x\,dx}{x + d} = x - d\ln(x + d)$

APPENDIX F

Properties of the Elements

All physical properties are for a pressure of 1 atm unless otherwise specified.

Element	Symbol	Atomic Number Z	Molar Mass, g/mol	Density, g/cm^3 at 20°C	Melting Point, °C	Boiling Point, °C	Specific Heat, J/(g · °C) at 25°C
Actinium	Ac	89	(227)	10.06	1323	(3473)	0.092
Aluminum	Al	13	26.9815	2.699	660	2450	0.900
Americium	Am	95	(243)	13.67	1541	—	—
Antimony	Sb	51	121.75	6.691	630.5	1380	0.205
Argon	Ar	18	39.948	1.6626×10^{-3}	−189.4	−185.8	0.523
Arsenic	As	33	74.9216	5.78	817 (28 atm)	613	0.331
Astatine	At	85	(210)	—	(302)	—	—
Barium	Ba	56	137.34	3.594	729	1640	0.205
Berkelium	Bk	97	(247)	14.79	—	—	—
Beryllium	Be	4	9.0122	1.848	1287	2770	1.83
Bismuth	Bi	83	208.980	9.747	271.37	1560	0.122
Bohrium	Bh	107	262.12	—	—	—	—
Boron	B	5	10.811	2.34	2030	—	1.11
Bromine	Br	35	79.909	3.12 (liquid)	−7.2	58	0.293
Cadmium	Cd	48	112.40	8.65	321.03	765	0.226
Calcium	Ca	20	40.08	1.55	838	1440	0.624
Californium	Cf	98	(251)	—	—	—	—
Carbon	C	6	12.01115	2.26	3727	4830	0.691
Cerium	Ce	58	140.12	6.768	804	3470	0.188
Cesium	Cs	55	132.905	1.873	28.40	690	0.243
Chlorine	Cl	17	35.453	3.214×10^{-3} (0°C)	−101	−34.7	0.486
Chromium	Cr	24	51.996	7.19	1857	2665	0.448
Cobalt	Co	27	58.9332	8.85	1495	2900	0.423
Copper	Cu	29	63.54	8.96	1083.40	2595	0.385
Curium	Cm	96	(247)	13.3	—	—	—
Dubnium	Db	105	262.114	—	—	—	—
Dysprosium	Dy	66	162.50	8.55	1409	2330	0.172
Einsteinium	Es	99	(254)	—	—	—	—
Erbium	Er	68	167.26	9.15	1522	2630	0.167
Europium	Eu	63	151.96	5.243	817	1490	0.163
Fermium	Fm	100	(237)	—	—	—	—
Fluorine	F	9	18.9984	1.696×10^{-3} (0°C)	−219.6	−188.2	0.753
Francium	Fr	87	(223)	—	(27)	—	—
Gadolinium	Gd	64	157.25	7.90	1312	2730	0.234
Gallium	Ga	31	69.72	5.907	29.75	2237	0.377

Element	Symbol	Atomic Number Z	Molar Mass, g/mol	Density, g/cm³ at 20°C	Melting Point, °C	Boiling Point, °C	Specific Heat, J/(g · °C) at 25°C
Germanium	Ge	32	72.59	5.323	937.25	2830	0.322
Gold	Au	79	196.967	19.32	1064.43	2970	0.131
Hafnium	Hf	72	178.49	13.31	2227	5400	0.144
Hassium	Hs	108	(265)	—	—	—	—
Helium	He	2	4.0026	0.1664×10^{-3}	−269.7	−268.9	5.23
Holmium	Ho	67	164.930	8.79	1470	2330	0.165
Hydrogen	H	1	1.00797	0.08375×10^{-3}	−259.19	−252.7	14.4
Indium	In	49	114.82	7.31	156.634	2000	0.233
Iodine	I	53	126.9044	4.93	113.7	183	0.218
Iridium	Ir	77	192.2	22.5	2447	(5300)	0.130
Iron	Fe	26	55.847	7.874	1536.5	3000	0.447
Krypton	Kr	36	83.80	3.488×10^{-3}	−157.37	−152	0.247
Lanthanum	La	57	138.91	6.189	920	3470	0.195
Lawrencium	Lr	103	(257)	—	—	—	—
Lead	Pb	82	207.19	11.35	327.45	1725	0.129
Lithium	Li	3	6.939	0.534	180.55	1300	3.58
Lutetium	Lu	71	174.97	9.849	1663	1930	0.155
Magnesium	Mg	12	24.312	1.738	650	1107	1.03
Manganese	Mn	25	54.9380	7.44	1244	2150	0.481
Meitnerium	Mt	109	(266)	—	—	—	—
Mendelevium	Md	101	(256)	—	—	—	—
Mercury	Hg	80	200.59	13.55	−38.87	357	0.138
Molybdenum	Mo	42	95.94	10.22	2617	5560	0.251
Neodymium	Nd	60	144.24	7.007	1016	3180	0.188
Neon	Ne	10	20.183	0.8387×10^{-3}	−248.597	−246.0	1.03
Neptunium	Np	93	(237)	20.25	637	—	1.26
Nickel	Ni	28	58.71	8.902	1453	2730	0.444
Niobium	Nb	41	92.906	8.57	2468	4927	0.264
Nitrogen	N	7	14.0067	1.1649×10^{-3}	−210	−195.8	1.03
Nobelium	No	102	(255)	—	—	—	—
Osmium	Os	76	190.2	22.59	3027	5500	0.130
Oxygen	O	8	15.9994	1.3318×10^{-3}	−218.80	−183.0	0.913
Palladium	Pd	46	106.4	12.02	1552	3980	0.243
Phosphorus	P	15	30.9738	1.83	44.25	280	0.741
Platinum	Pt	78	195.09	21.45	1769	4530	0.134
Plutonium	Pu	94	(244)	19.8	640	3235	0.130
Polonium	Po	84	(210)	9.32	254	—	—
Potassium	K	19	39.102	0.862	63.20	760	0.758
Praseodymium	Pr	59	140.907	6.773	931	3020	0.197
Promethium	Pm	61	(145)	7.22	(1027)	—	—
Protactinium	Pa	91	(231)	15.37 (estimated)	(1230)	—	—
Radium	Ra	88	(226)	5.0	700	—	—
Radon	Rn	86	(222)	9.96×10^{-3} (0°C)	(−71)	−61.8	0.092
Rhenium	Re	75	186.2	21.02	3180	5900	0.134

Element	Symbol	Atomic Number Z	Molar Mass, g/mol	Density, g/cm^3 at 20°C	Melting Point, °C	Boiling Point, °C	Specific Heat, J/(g · °C) at 25°C
Rhodium	Rh	45	102.905	12.41	1963	4500	0.243
Rubidium	Rb	37	85.47	1.532	39.49	688	0.364
Ruthenium	Ru	44	101.107	12.37	2250	4900	0.239
Rutherfordium	Rf	104	261.11	—	—	—	—
Samarium	Sm	62	150.35	7.52	1072	1630	0.197
Scandium	Sc	21	44.956	2.99	1539	2730	0.569
Seaborgium	Sg	106	263.118	—	—	—	—
Selenium	Se	34	78.96	4.79	221	685	0.318
Silicon	Si	14	28.086	2.33	1412	2680	0.712
Silver	Ag	47	107.870	10.49	960.8	2210	0.234
Sodium	Na	11	22.9898	0.9712	97.85	892	1.23
Strontium	Sr	38	87.62	2.54	768	1380	0.737
Sulfur	S	16	32.064	2.07	119.0	444.6	0.707
Tantalum	Ta	73	180.948	16.6	3014	5425	0.138
Technetium	Tc	43	(99)	11.46	2200	—	0.209
Tellurium	Te	52	127.60	6.24	449.5	990	0.201
Terbium	Tb	65	158.924	8.229	1357	2530	0.180
Thallium	Tl	81	204.37	11.85	304	1457	0.130
Thorium	Th	90	(232)	11.72	1755	(3850)	0.117
Thulium	Tm	69	168.934	9.32	1545	1720	0.159
Tin	Sn	50	118.69	7.2984	231.868	2270	0.226
Titanium	Ti	22	47.90	4.54	1670	3260	0.523
Tungsten	W	74	183.85	19.3	3380	5930	0.134
Un-named	Uun	110	(269)	—	—	—	—
Un-named	Uuu	111	(272)	—	—	—	—
Un-named	Uub	112	(264)	—	—	—	—
Un-named	Uut	113	—	—	—	—	—
Un-named	Unq	114	(285)	—	—	—	—
Un-named	Uup	115	—	—	—	—	—
Un-named	Uuh	116	(289)	—	—	—	—
Un-named	Uus	117	—	—	—	—	—
Un-named	Uuo	118	(293)	—	—	—	—
Uranium	U	92	(238)	18.95	1132	3818	0.117
Vanadium	V	23	50.942	6.11	1902	3400	0.490
Xenon	Xe	54	131.30	5.495×10^{-3}	−111.79	−108	0.159
Ytterbium	Yb	70	173.04	6.965	824	1530	0.155
Yttrium	Y	39	88.905	4.469	1526	3030	0.297
Zinc	Zn	30	65.37	7.133	419.58	906	0.389
Zirconium	Zr	40	91.22	6.506	1852	3580	0.276

The values in parentheses in the column of molar masses are the mass numbers of the longest-lived isotopes of those elements that are radioactive. Melting points and boiling points in parentheses are uncertain.

The data for gases are valid only when these are in their usual molecular state, such as H_2, He, O_2, Ne, etc. The specific heats of the gases are the values at constant pressure.

Source: Adapted from J. Emsley, *The Elements,* 3rd ed., 1998, Clarendon Press, Oxford. See also www.webelements.com for the latest values and newest elements.

APPENDIX G

Periodic Table of the Elements

Metals
Metalloids
Nonmetals

Transition metals

THE HORIZONTAL PERIODS	Alkali metals IA	IIA	IIIB	IVB	VB	VIB	VIIB	VIIIB	VIIIB	VIIIB	IB	IIB	IIIA	IVA	VA	VIA	VIIA	Noble gases 0
1	1 H																	2 He
2	3 Li	4 Be											5 B	6 C	7 N	8 O	9 F	10 Ne
3	11 Na	12 Mg											13 Al	14 Si	15 P	16 S	17 Cl	18 Ar
4	19 K	20 Ca	21 Sc	22 Ti	23 V	24 Cr	25 Mn	26 Fe	27 Co	28 Ni	29 Cu	30 Zn	31 Ga	32 Ge	33 As	34 Se	35 Br	36 Kr
5	37 Rb	38 Sr	39 Y	40 Zr	41 Nb	42 Mo	43 Tc	44 Ru	45 Rh	46 Pd	47 Ag	48 Cd	49 In	50 Sn	51 Sb	52 Te	53 I	54 Xe
6	55 Cs	56 Ba	57-71 *	72 Hf	73 Ta	74 W	75 Re	76 Os	77 Ir	78 Pt	79 Au	80 Hg	81 Tl	82 Pb	83 Bi	84 Po	85 At	86 Rn
7	87 Fr	88 Ra	89-103 †	104 Rf	105 Db	106 Sg	107 Bh	108 Hs	109 Mt	110	111	112	113	114	115	116	117	118

Inner transition metals

Lanthanide series *	57 La	58 Ce	59 Pr	60 Nd	61 Pm	62 Sm	63 Eu	64 Gd	65 Tb	66 Dy	67 Ho	68 Er	69 Tm	70 Yb	71 Lu
Actinide series †	89 Ac	90 Th	91 Pa	92 U	93 Np	94 Pu	95 Am	96 Cm	97 Bk	98 Cf	99 Es	100 Fm	101 Md	102 No	103 Lr

The names for elements 104 through 109 (Rutherfordium, Dubnium, Seaborgium, Bohrium, Hassium, and Meitnerium, respectively) were adopted by the International Union of Pure and Applied Chemistry (IUPAC) in 1997. Elements 110, 111, 112, 114, 116, and 118 have been discovered but, as of 2000, have not yet been named. See www.webelements.com for the latest information and newest elements.

ANSWERS

to Checkpoints and Odd-Numbered Questions, Exercises, and Problems

CHAPTER 1

EP **1.** (a) 10^9; (b) 10^{-4}; (c) 9.1×10^5 **3.** (a) 160 rods; (b) 40 chains **5.** (a) 4.00×10^4 km; (b) 5.10×10^8 km^2; (c) 1.08×10^{12} km^3 **7.** 1.9×10^{22} cm^3 **9.** 1.1×10^3 acre-feet **11.** (a) 0.98 ft/ns; (b) 0.30 mm/ps **13.** C, D, A, B, E; the important criterion is the constancy of the daily variation, not its magnitude **15.** 0.12 AU/min **17.** 2.1 h **19.** 9.0×10^{49} **21.** (a) 10^3 kg; (b) 158 kg/s **23.** (a) 1.18×10^{-29} m^3; (b) 0.282 nm **25.** (a) 60.8 W; (b) 43.3 Z **27.** 89 km **29.** $\approx 1 \times 10^{36}$

CHAPTER 2

CP **1.** b and c **2.** zero (zero displacement for entire trip) **3.** (check the derivative dx/dt) (a) 1 and 4; (b) 2 and 3 **4.** (see Tactic 5) (a) plus; (b) minus; (c) minus; (d) plus **5.** 1 and 4 ($a = d^2x/dt^2$ must be a constant) **6.** (a) plus (upward displacement on y axis); (b) minus (downward displacement on y axis); (c) $a = -g = -9.8$ m/s^2 **Q** **1.** (a) all tie; (b) 4, tie of 1 and 2, then 3 **3.** E **5.** a and c **7.** $x = t^2$ and $x = 8(t-2) + (1.5)(t-2)^2$ **9.** same **EP** **1.** 414 ms **3.** (a) +40 km/h; (b) 40 km/h **5.** (a) 73 km/h; (b) 68 km/h; (c) 70 km/h; (d) 0 **7.** (a) 0, −2, 0, 12 m; (b) +12 m; (c) +7 m/s **9.** 1.4 m **11.** (a) −6 m/s; (b) negative x direction; (c) 6 m/s; (d) first smaller, then zero, and then larger; (e) yes ($t = 2$ s); (f) no **13.** 100 m **15.** (a) velocity squared; (b) acceleration; (c) m^2/s^2, m/s^2 **17.** 20 m/s^2, in the direction opposite to its initial velocity **19.** (a) 80 m/s; (b) 110 m/s; (c) 20 m/s^2 **21.** (a) m/s^2, m/s^3; (b) 1.0 s; (c) 82 m; (d) −80 m; (e) 0, −12, −36, −72 m/s; (f) −6, −18, −30, −42 m/s^2 **23.** 0.10 m **25.** (a) 1.6 m/s; (b) 18 m/s **27.** (a) 3.1×10^6 s = 1.2 months; (b) 4.6×10^{13} m **29.** 1.62×10^{15} m/s^2 **31.** 2.5 s **33.** (a) 3.56 m/s^2; (b) 8.43 m/s **35.** (a) 5.00 m/s; (b) 1.67 m/s^2; (c) 7.50 m **37.** (a) 0.74 s; (b) −6.2 m/s^2 **39.** (a) 10.6 m; (b) 41.5 s **41.** (a) 29.4 m; (b) 2.45 s **43.** (a) 31 m/s; (b) 6.4 s **45.** (a) 3.2 s; (b) 1.3 s **47.** (a) 3.70 m/s; (b) 1.74 m/s; (c) 0.154 m **49.** 4.0 m/s **51.** 857 m/s^2, upward **53.** 1.26×10^3 m/s^2, upward **55.** 22 cm and 89 cm below the nozzle **57.** 1.5 s **59.** (a) 5.4 s; (b) 41 m/s **61.** (a) 76 m; (b) 4.2 s **63.** (a) 1.23 cm; (b) 4 times, 9 times, 16 times, 25 times **65.** 2.34 m

CHAPTER 3

CP **1.** (a) 7 m ($\vec{a}$ and $\vec{b}$ are in same direction); (b) 1 m ($\vec{a}$ and $\vec{b}$ are in opposite directions) **2.** c, d, f (components must be head-to-tail; $\vec{a}$ must extend from tail of one component to head of the other) **3.** (a) +, +; (b) +, −; (c) +, + (draw vector from tail of $\vec{d}_1$ to head of $\vec{d}_2$) **4.** (a) 90°; (b) 0° (vectors are parallel—same direction); (c) 180° (vectors are antiparallel—opposite directions) **5.** (a) 0° or 180°; (b) 90° **Q** **1.** $\vec{A}$ and $\vec{B}$ **3.** no, but $\vec{a}$ and $-\vec{b}$ are commutative: $\vec{a} + (-\vec{b}) = (-\vec{b}) + \vec{a}$ **5.** (a) $\vec{a}$ and $\vec{b}$ are parallel; (b) $\vec{b} = 0$; (c) $\vec{a}$ and $\vec{b}$ are perpendicular **7.** all but (e) **9.** (a) 0 (vectors are parallel); (b) 0 (vectors are antiparallel) **EP** **1.** The displacements should be (a) parallel, (b) antiparallel, (c) perpendicular **3.** (a) −2.5 m; (b) −6.9 m **5.** (a) 47.2 m; (b) 122° **7.** (a) 168 cm; (b) 32.5° above the floor **9.** (a) 6.42 m; (b) no; (c) yes; (d) yes; (e) a possible answer: $(4.30\text{ m})\hat{i} + (3.70\text{ m})\hat{j} + (3.00\text{ m})\hat{k}$; (f) 7.96 m **11.** (a) 370 m; (b) 36° north of east; (c) 425 m; (d) the distance **13.** (a) $(-9\text{ m})\hat{i} + (10\text{ m})\hat{j}$; (b) 13 m; (c) +132° **15.** (a) 4.2 m; (b) 40° east of north; (c) 8.0 m; (d) 24° north of west **17.** (a) $(3.0\text{ m})\hat{i} - (2.0\text{ m})\hat{j} + (5.0\text{ m})\hat{k}$; (b) $(5.0\text{ m})\hat{i} - (4.0\text{ m})\hat{j} - (3.0\text{ m})\hat{k}$; (c) $(-5.0\text{ m})\hat{i} + (4.0\text{ m})\hat{j} + (3.0\text{ m})\hat{k}$ **19.** (a) 38 m; (b) 320°; (c) 130 m; (d) 1.2°; (e) 62 m; (f) 130° **21.** (a) 1.59 m; (b) 12.1 m; (c) 12.2 m; (d) 82.5° **27.** (a) Put axes along cube edges, with the origin at one corner. Diagonals are $a\hat{i} + a\hat{j} + a\hat{k}$, $a\hat{i} + a\hat{j} - a\hat{k}$, $a\hat{i} - a\hat{j} - a\hat{k}$, $a\hat{i} - a\hat{j} + a\hat{k}$; (b) 54.7°; (c) $\sqrt{3}a$ **29.** (a) 30; (b) 52 **31.** 22° **35.** (b) $a^2b \sin \phi$ **37.** (a) 3.0 m; (b) 0; (c) 3.46 m; (d) 2.00 m; (e) −5.00 m; (f) 8.66 m; (g) −6.67; (h) 4.33

CHAPTER 4

CP **1.** (a) $(8\hat{i} - 6\hat{j})$ m; (b) yes, the xy plane (no z component) **2.** (draw $\vec{v}$ tangent to path, tail on path) (a) first; (b) third **3.** (take second derivative with respect to time) (1) and (3) a_x and a_y are both constant and thus $\vec{a}$ is constant; (2) and (4) a_y is constant but a_x is not, thus $\vec{a}$ is not **4.** 4 m/s^3, −2 m/s, 3 m **5.** (a) v_x constant; (b) v_y initially positive, decreases to zero, and then becomes progressively more negative; (c) $a_x = 0$ throughout; (d) $a_y = -g$ throughout **6.** (a) $-(4\text{ m/s})\hat{i}$; (b) $-(8\text{ m/s}^2)\hat{j}$ **7.** (a) 0, distance not changing; (b) +70 km/h, distance increasing; (c) +80 km/h, distance decreasing **8.** (a)–(c) increase **Q** **1.** (a) $(7\text{ m})\hat{i} + (1\text{ m})\hat{j} + (-2\text{ m})\hat{k}$; (b) $(5\text{ m})\hat{i} + (-3\text{ m})\hat{j} + (1\text{ m})\hat{k}$; (c) $(-2\text{ m})\hat{i}$ **3.** a, b, c **5.** (a) all tie; (b) 1 and 2 tie (the rocket is shot upward), then 3 and 4 tie (it is shot into the ground!) **7.** (a) 3, 2, 1; (b) 1, 2, 3; (c) all tie; (d) 6, 5, 4 **9.** (a) less; (b) unanswerable; (c) equal; (d) unanswerable **11.** (a) 2; (b) 3; (c) 1; (d) 2; (e) 3; (f) 1 **13.** (a) yes; (b) no; (c) yes **EP** **1.** (a) $(-5.0\hat{i} + 8.0\hat{j})$ m; (b) 9.4 m; (c) 122°; (e) $(8\hat{i} - 8\hat{j})$ m; (f) 11 m; (g) −45° **3.** (a) $(-7.0\hat{i} + 12\hat{j})$ m; (b) xy plane **5.** 7.59 km/h, 22.5° east of north **7.** (a) $(3.00\hat{i} - 8.00t\hat{j})$ m/s; (b) $(3.00\hat{i} - 16.0\hat{j})$ m/s; (c) 16.3 m/s; (d) −79.4° **9.** (a) $(8t\hat{j} + \hat{k})$ m/s; (b) $8\hat{j}$ m/s^2 **11.** (a) $(6.00\hat{i} - 106\hat{j})$ m;

(b) $(19.0\hat{i} - 224\hat{j})$ m/s; (c) $(24.0\hat{i} - 336\hat{j})$ m/s^2; (d) $-85.2°$ to $+x$ **13.** (a) $(-1.5\hat{j})$ m/s; (b) $(4.5\hat{i} - 2.25\hat{j})$ m **15.** (a) 45 m; (b) 22 m/s **17.** (a) 62 ms; (b) 480 m/s **19.** (a) 0.205 s; (b) 0.205 s; (c) 20.5 cm; (d) 61.5 cm **21.** (a) 2.00 ns; (b) 2.00 mm; (c) 1.00×10^7 m/s; (d) 2.00×10^6 m/s **23.** (a) 16.9 m; (b) 8.21 m; (c) 27.6 m; (d) 7.26 m; (e) 40.2 m; (f) 0 **25.** 4.8 cm **29.** (a) 11 m; (b) 23 m; (c) 17 m/s; (d) 63° below horizontal **31.** (a) 24 m/s; (b) 65° above the horizontal **33.** (a) 10 s; (b) 897 m **35.** the third **37.** (a) 202 m/s; (b) 806 m; (c) 161 m/s; (d) -171 m/s **39.** (a) yes; (b) 2.56 m **41.** between the angles 31° and 63° above the horizontal **43.** (a) 7.49 km/s; (b) 8.00 m/s^2 **45.** (a) 19 m/s; (b) 35 rev/min; (c) 1.7 s **47.** (a) 0.034 m/s^2; (b) 84 min **49.** (a) 12 s; (b) 4.1 m/s^2, down; (c) 4.1 m/s^2, up **51.** 160 m/s^2 **53.** (a) 13 m/s^2, eastward; (b) 13 m/s^2, eastward **55.** 36 s, no **57.** 60° **59.** 32 m/s **61.** (a) 38 knots, 1.5° east of north; (b) 4.2 h; (c) 1.5° west of south **63.** (a) 37° west of north; (b) 62.6 s

CHAPTER 5

CP **1.** *c*, *d*, and *e* ($\vec{F}_1$ and $\vec{F}_2$ must be head-to-tail, $\vec{F}_{net}$ must be from tail of one of them to head of the other) **2.** (a) and (b) 2 N, leftward (acceleration is zero in each situation) **3.** (a) and (b) 1, 4, 3, 2 **4.** (a) equal; (b) greater (acceleration is upward, thus net force on body must be upward) **5.** (a) equal; (b) greater; (c) less **6.** (a) increase; (b) yes; (c) same; (d) yes **7.** (a) $F \sin \theta$; (b) increase **8.** 0 (because now $a = -g$)
Q **1.** (a) 5; (b) 7; (c) $(2\text{ N})\hat{i}$; (d) $-(6\text{ N})\hat{j}$; (e) fourth; (f) fourth **3.** (a) 2 and 4; (b) 2 and 4 **5.** (a) 2, 3, 4; (b) 1, 3, 4; (c) 1, $+y$; 2, $+x$; 3, fourth quadrant; 4, third quadrant **7.** (a) less; (b) greater **9.** (a) 20 kg; (b) 18 kg; (c) 10 kg; (d) all tie; (e) 3, 2, 1 **11.** (a) 4 or 5, choose 4; (b) 2; (c) 1; (d) 4 or 5, choose 5; (e) 3; (f) 6; (g) 3 and 6; 1, 2, and 5; (h) 3 and 6; (i) 1, 2, and 5
EP **1.** (a) $F_x = 1.88$ N; (b) $F_y = 0.684$ N; (c) $(1.88\hat{i} + 0.684\hat{j})$ N **3.** 2.9 m/s^2 **5.** $(3\hat{i} - 11\hat{j} + 4\hat{k})$ N **7.** (a) $(-32\hat{i} - 21\hat{j})$ N; (b) 38 N; (c) 213° from $+x$ **9.** (a) 108 N; (b) 108 N; (c) 108 N **11.** (a) 11 N; (b) 2.2 kg; (c) 0; (d) 2.2 kg **13.** 16 N **15.** (a) 42 N; (b) 72 N; (c) 4.9 m/s^2 **17.** (a) 0.02 m/s^2; (b) 8×10^4 km; (c) 2×10^3 m/s **19.** 1.2×10^5 N **21.** 1.5 mm **23.** (a) $(285\hat{i} + 705\hat{j})$ N; (b) $285\hat{i} - 115\hat{j})$ N; (c) 307 N; (d) $-22°$ from $+x$; (e) 3.67 m/s^2; (f) $-22°$ from $+x$ **25.** (a) 0.62 m/s^2; (b) 0.13 m/s^2; (c) 2.6 m **27.** (a) 494 N, up; (b) 494 N, down **29.** (a) 2.2×10^{-3} N; (b) 3.7×10^{-3} N **31.** (a) 1.1 N **33.** 1.8×10^4 N **35.** (a) 620 N; (b) 580 N **37.** (a) 3260 N; (b) 2.7×10^3 kg; (c) 1.2 m/s^2 **39.** (a) 180 N; (b) 640 N **41.** (a) 1.23 N; (b) 2.46 N; (c) 3.69 N; (d) 4.92 N; (e) 6.15 N; (f) 0.25 N **43.** (a) 0.735 m/s^2; (b) downward; (c) 20.8 N **45.** (a) 1.18 m; (b) 0.674 s; (c) 3.50 m/s **47.** (a) 4.9 m/s^2; (b) 2.0 m/s^2; (c) upward; (d) 120 N **49.** (a) 2.18 m/s^2; (b) 116 N; (c) 21.0 m/s^2 **51.** (b) $F/(m + M)$; (c) $MF/(m + M)$; (d) $F(m + 2M)/2(m + M)$ **53.** $2Ma/(a + g)$ **55.** (a) 31.3 kN; (b) 24.3 kN

CHAPTER 6

CP **1.** (a) zero (because there is no attempt at sliding); (b) 5 N; (c) no; (d) yes; (e) 8 N **2.** (a) same (10 N); (b) decreases; (c) decreases (because N decreases) **3.** greater (from Sample Problem 6-5, v_t depends on $\sqrt{R}$) **4.** ($\vec{a}$ is directed toward center of circular path) (a) $\vec{a}$ downward, $\vec{N}$ upward; (b) $\vec{a}$ and $\vec{N}$ upward **5.** (a) same (must still match the gravitational force on the rider); (b) increases ($N = mv^2/R$); (c) increases ($f_{s,\max} = \mu_s N$) **6.** (a) $4R_1$; (b) $4R_1$ Q **1.** (a) F_1, F_2, F_3; (b) all tie **3.** (a) same; (b) increases; (c) increases; (d) no **5.** (a) decrease; (b) decrease; (c) increase; (d) increase; (e) increase **7.** (a) the block's mass m; (b) equal (they are a third-law pair); (c) that on the slab is in the direction of the applied force; that on the block is in the opposite direction; (d) the slab's mass M **9.** 4, 3, then 1, 2, and 5 tie EP **1.** (a) 200 N; (b) 120 N **3.** 0.61 **5.** (a) 190 N; (b) 0.56 m/s^2 **7.** (a) 0.13 N; (b) 0.12 **9.** (a) no; (b) $(-12\hat{i} + 5\hat{j})$ N **13.** (a) 300 N; (b) 1.3 m/s^2 **15.** (a) 66 N; (b) 2.3 m/s^2 **17.** (b) 3.0×10^7 N **19.** 100 N **21.** (a) 0; (b) 3.9 m/s^2 down the incline; (c) 1.0 m/s^2 down the incline **23.** (a) 3.5 m/s^2; (b) 0.21 N; (c) blocks move independently **25.** 490 N **27.** (a) 6.1 m/s^2, leftward; (b) 0.98 m/s^2, leftward **29.** $g(\sin \theta - \sqrt{2}\mu_k \cos \theta)$ **31.** 9.9 s **33.** 6200 N **35.** 2.3 **37.** about 48 km/h **39.** 21 m **41.** $\sqrt{Mgr/m}$ **43.** (a) light; (b) 778 N; (c) 223 N **45.** 2.2 km **47.** (b) 8.74 N; (c) 37.9 N, radially inward; (d) 6.45 m/s

CHAPTER 7

CP **1.** (a) decrease; (b) same; (c) negative, zero **2.** *d*, *c*, *b*, *a* **3.** (a) same; (b) smaller **4.** (a) positive; (b) negative; (c) zero **5.** zero Q **1.** all tie **3.** (a) positive; (b) zero; (c) negative; (d) negative; (e) zero; (f) positive **5.** (a) A, B, C; (b) C, B, A; (c) C, B, A; (d) A, 2; B, 3; C, 1 **7.** all tie **9.** *c*, *d*, then *a* and *b* tie, then *f*, *e* **11.** (a) $2F_1$; (b) $2W_1$ **13.** B, C, A
EP **1.** 1.2×10^6 m/s **3.** (a) 3610 J; (b) 1900 J; (c) 1.1×10^{10} J **5.** (a) 2.9×10^7 m/s; (b) 2.1×10^{-13} J **7.** (a) 590 J; (b) 0; (c) 0; (d) 590 J **9.** (a) 170 N; (b) 340 m; (c) -5.8×10^4 J; (d) 340 N; (e) 170 m; (f) -5.8×10^4 J **11.** (a) 1.50 J; (b) increases **13.** 15.3 J **15.** (a) 98 N; (b) 4.0 cm; (c) 3.9 J; (d) -3.9 J **17.** (a) 1.2×10^4 J; (b) -1.1×10^4 J; (c) 1100 J; (d) 5.4 m/s **19.** (a) $-3Mgd/4$; (b) Mgd; (c) $Mgd/4$; (d) $\sqrt{gd/2}$ **21.** (a) -0.043 J; (b) -0.13 J **23.** (a) 6.6 m/s; (b) 4.7 m **25.** 800 J **27.** 0, by both methods **29.** -6 J **31.** 490 W **33.** (a) 0.83 J; (b) 2.5 J; (c) 4.2 J; (d) 5.0 W **35.** 740 W **37.** 68 kW **39.** (a) 1.8×10^5 ft · lb; (b) 0.55 hp

CHAPTER 8

CP **1.** no (consider round trip on the small loop) **2.** 3, 1, 2 (see Eq. 8-6) **3.** (a) all tie; (b) all tie **4.** (a) CD, AB, BC (zero) (check slope magnitudes); (b) positive direction of x **5.** all tie
Q **1.** (a) 12 J; (b) -2 J **3.** (a) all tie; (b) all tie **5.** (a) 4; (b) returns to its starting point and repeats the trip; (c) 1; (d) 1 **7.** (a) fL; (b) 0.50; (c) 1.25; (d) 2.25; (e) b, center; c, right; d, left **9.** (a) increasing; (b) decreasing; (c) decreasing; (d) constant in AB and BC, decreasing in CD EP **1.** 89 N/cm **3.** (a) 4.31 mJ; (b) -4.31 mJ; (c) 4.31 mJ; (d) -4.31 mJ; (e) all increase **5.** (a) mgL; (b) $-mgL$; (c) 0; (d) $-mgL$; (e) mgL; (f) 0; (g) same **7.** (a) 184 J; (b) -184 J; (c) -184 J

9. (a) 2.08 m/s; (b) 2.08 m/s; (c) increase **11.** (a) $\sqrt{2gL}$; (b) $2\sqrt{gL}$; (c) $\sqrt{2gL}$; (d) all the same **13.** (a) 260 m; (b) same; (c) decrease **15.** (a) 21.0 m/s; (b) 21.0 m/s; (c) 21.0 m/s **17.** (a) 0.98 J; (b) −0.98 J; (c) 3.1 N/cm **19.** (a) 39.2 J; (b) 39.2 J; (c) 4.00 m **21.** (a) 35 cm; (b) 1.7 m/s **23.** (a) 4.8 m/s; (b) 2.4 m/s **25.** 10 cm **27.** 1.25 cm **31.** (a) $2\sqrt{gL}$; (b) $5mg$; (c) 71° **33.** $mgL/32$ **37.** (a) $1.12(A/B)^{1/6}$; (b) repulsive; (c) attractive **39.** (a) 5.6 J; (b) 3.5 J **41.** (a) 30.1 J; (b) 30.1 J; (c) 0.22 **43.** (a) −2900 J; (b) 390 J; (c) 210 N **45.** 11 kJ **47.** 20 ft · lb **49.** (a) 1.5 MJ; (b) 0.51 MJ; (c) 1.0 MJ; (d) 63 m/s **51.** (a) 67 J; (b) 67 J; (c) 46 cm **53.** (a) 31.0 J; (b) 5.35 m/s; (c) conservative **55.** (a) 44 m/s; (b) 0.036 **57.** (a) −0.90 J; (b) 0.46 J; (c) 1.0 m/s **59.** 1.2 m **63.** in the center of the flat part **65.** (a) 216 J; (b) 1180 N; (c) 432 J; (d) motor also supplies thermal energy to crate and belt **67.** (b) $\rho(L - x)/2$; (c) $v = v_0[2(\rho L + m_f)/(\rho L + 2m_f - \rho x)]^{0.5}$; (e) 35 m/s

CHAPTER 9

CP **1.** (a) origin; (b) fourth quadrant; (c) on y axis below origin; (d) origin; (e) third quadrant; (f) origin **2.** (a) to (c) at the center of mass, still at the origin (their forces are internal to the system and cannot move the center of mass) **3.** (Consider slopes and Eq. 9-23.) (a) 1, 3, and then 2 and 4 tie (zero force); (b) 3 **4.** (No net external force; $\vec{P}$ conserved.) (a) 0; (b) no; (c) $-x$ **5.** (a) 500 km/h; (b) 2600 km/h; (c) 1600 km/h **6.** (a) yes; (b) no (because of net force along y) Q **1.** (a) to (d) at the origin **3.** (a) at the center of the sled; (b) $L/4$, to the right; (c) not at all (no net external force); (d) $L/4$, to the left; (e) L; (f) $L/2$; (g) $L/2$ **5.** (a) ac, cd, and bc; (b) bc; (c) bd and ad **7.** c, d, and then a and b tie **9.** b, c, a EP **1.** (a) 4600 km; (b) $0.73R_e$ **3.** (a) 1.1 m; (b) 1.3 m; (c) shifts toward topmost particle **5.** (a) −0.25 m; (b) 0 **7.** 6.8×10^{-12} m from the nitrogen atom, along axis of symmetry **9.** (a) $H/2$; (b) $H/2$; (c) descends to lowest point and then ascends to $H/2$; (d) $\frac{HM}{m}\left(\sqrt{1 + \frac{m}{M}} - 1\right)$ **11.** 72 km/h **13.** (a) 28 cm; (b) 2.3 m/s **15.** 53 m **17.** (a) halfway between the containers; (b) 26 mm toward the heavier container; (c) down; (d) -1.6×10^{-2} m/s² **19.** 4.2 m **21.** 24 km/h **23.** (a) 7.5×10^4 J; (b) 3.8×10^4 kg · m/s; (c) 38° south of east **25.** (a) $(-4.0 \times 10^4\hat{i})$ kg · m/s; (b) west; (c) 0 **27.** 3.0 mm/s, away from the stone **29.** increases by 4.4 m/s **31.** 4400 km/h **33.** (a) 7290 m/s; (b) 8200 m/s; (c) 1.271×10^{10} J; (d) 1.275×10^{10} J **35.** (a) 1.4×10^{-22} kg · m/s; (b) 150°; (c) 120°; (d) 1.6×10^{-19} J **37.** (a) 1010 m/s, 9.48° clockwise from the $+x$ direction; (b) 3.23 MJ **39.** 14 m/s, 135° from the other pieces **41.** 108 m/s **43.** (a) 1.57×10^6 N; (b) 1.35×10^5 kg; (c) 2.08 km/s **45.** 2.2×10^{-3} **47.** (a) 46 N; (b) none **49.** (a) 0.2 to 0.3 MJ; (b) same amount **51.** (a) 8.8 m/s; (b) 2600 J; (c) 1.6 kW **53.** 24 W **55.** (a) 860 N; (b) 2.4 m/s **57.** (a) 2.1×10^6 kg; (b) $\sqrt{100 + 1.5t}$ m/s; (c) $(1.5 \times 10^6)/\sqrt{100 + 1.5t}$ N; (d) 6.7 km **59.** 1.5 cm/s downward (the bubbles rise but the layers descend)

CHAPTER 10

CP **1.** (a) unchanged; (b) unchanged (see Eq. 10-4); (c) decrease (see Eq. 10-8) **2.** (a) zero; (b) positive (initial p_y down y; final p_y up y); (c) positive direction of y **3.** (a) 10 kg · m/s; (b) 14 kg · m/s; (c) 6 kg · m/s **4.** (a) 4 kg · m/s; (b) 8 kg · m/s; (c) 3 J **5.** (a) 2 kg · m/s (conserve momentum along x); (b) 3 kg · m/s (conserve momentum along y) Q **1.** all tie **3.** b and c **5.** (a) rightward; (b) rightward; (c) smaller **7.** (a) one was stationary; (b) 2; (c) 5; (d) equal (pool player's result) **9.** (a) 2; (b) 1; (c) 3; (d) yes; (e) no EP **1.** 2.5 m/s **3.** 3000 N **5.** 67 m/s, in opposite direction **7.** (a) 42 N · s; (b) 2100 N **9.** (a) $(7.4 \times 10^3\hat{i} - 7.4 \times 10^3\hat{j})$ N · s; (b) $(-7.4 \times 10^3\hat{i})$ N · s; (c) 2.3×10^3 N; (d) 2.1×10^4 N; (e) −45° **11.** 10 m/s **13.** (a) 1.0 kg · m/s; (b) 250 J; (c) 10 N; (d) 1700 N; (e) answer for (c) includes time between pellet collisions **15.** 41.7 cm/s **17.** (a) 1.8 N · s, upward in figure; (b) 180 N, downward in figure **19.** (a) 9.0 kg · m/s; (b) 3000 N; (c) 4500 N; (d) 20 m/s **21.** 3.0 m/s **23.** ≈2 mm/y **25.** (a) 4.6 m/s; (b) 3.9 m/s; (c) 7.5 m/s **27.** (a) $mR(\sqrt{2gh} + gt)$; (b) 5.06 kg **29.** 1.18×10^4 kg **31.** (a) $mv_i/(m + M)$; (b) $M/(m + M)$ **33.** 25 cm **35.** (a) 1.9 m/s, to the right; (b) yes; (c) no, total kinetic energy would have increased **37.** (a) 99 g; (b) 1.9 m/s; (c) 0.93 m/s **39.** 7.8% **41.** (a) 1.2 kg; (b) 2.5 m/s **43.** (a) 100 g; (b) 1.0 m/s **45.** (a) 1/3; (b) $4h$ **47.** (a) 4.15×10^5 m/s; (b) 4.84×10^5 m/s **49.** (a) 41°; (b) 4.76 m/s; (c) no **51.** 120° **53.** (a) 6.9 m/s, 30° to $+x$ direction; (b) 6.9 m/s, −30° to $+x$ direction; (c) 2.0 m/s, $-x$ direction **57.** (a) $5mg$; (b) $7mg$; (c) 5 m

CHAPTER 11

CP **1.** (b) and (c) **2.** (a) and (d) ($\alpha = d^2\theta/dt^2$ must be a constant) **3.** (a) yes; (b) no; (c) yes; (d) yes **4.** all tie **5.** 1, 2, 4, 3 (see Eq. 11-29) **6.** (see Eq. 11-32) 1 and 3 tie, 4, then 2 and 5 tie (zero) **7.** (a) downward in the figure ($\tau_{net} = 0$); (b) less (consider moment arms) Q **1.** (a) positive; (b) zero; (c) negative; (d) negative **3.** finite angular displacements are not commutative **5.** (a) c, a, then b and d tie; (b) b, then a and c tie, then d **7.** 3, 1, 2 **9.** 90°, then 70° and 110° tie **11.** (a) decrease; (b) clockwise; (c) counterclockwise EP **1.** (a) $a + 3bt^2 - 4ct^3$; (b) $6bt - 12ct^2$ **3.** (a) 5.5×10^{15} s; (b) 26 **5.** (a) 2 rad; (b) 0; (c) 130 rad/s; (d) 32 rad/s²; (e) no **7.** 11 rad/s **9.** (a) −67 rev/min²; (b) 8.3 rev **11.** 200 rev/min **13.** 8.0 s **15.** (a) 44 rad; (b) 5.5 s, 32 s; (c) −2.1 s, 40 s **17.** (a) 340 s; (b) -4.5×10^{-3} rad/s²; (c) 98 s **19.** 1.8 m/s², toward the center **21.** 0.13 rad/s **23.** (a) 3.0 rad/s; (b) 30 m/s; (c) 6.0 m/s²; (d) 90 m/s² **25.** (a) 3.8×10^3 rad/s; (b) 190 m/s **27.** (a) 7.3×10^{-5} rad/s; (b) 350 m/s; (c) 7.3×10^{-5} rad/s; (d) 460 m/s **29.** 16 s **31.** (a) -2.3×10^{-9} rad/s²; (b) 2600 y; (c) 24 ms **33.** 12.3 kg · m² **35.** (a) 1100 J; (b) 9700 J **37.** (a) $5md^2 + \frac{8}{3}Md^2$; (b) $(\frac{5}{2}m + \frac{4}{3}M)d^2\omega^2$ **39.** 0.097 kg · m² **41.** $\frac{1}{3}M(a^2 + b^2)$ **45.** 4.6 N · m **47.** (a) $r_1F_1 \sin\theta_1 - r_2F_2 \sin\theta_2$; (b) −3.8 N · m **49.** (a) 28.2 rad/s²; (b) 338 N · m **51.** (a) 155 kg · m²; (b) 64.4 kg **53.** 130 N **55.** (a) 6.00 cm/s²; (b) 4.87 N; (c) 4.54 N; (d) 1.20 rad/s²; (e) 0.0138 kg · m² **57.** (a) 1.73 m/s²; (b) 6.92 m/s² **59.** 396 N · m **61.** (a) $mL^2\omega^2/6$; (b) $L^2\omega^2/6g$ **63.** 5.42 m/s **65.** $\sqrt{9g/4L}$

67. (a) $[(3g/H)(1 - \cos\theta)]^{0.5}$; (b) $3g(1 - \cos\theta)$; (c) $\frac{3}{2}g \sin\theta$; (d) 41.8° **69.** 17

CHAPTER 12

CP **1.** (a) same; (b) less **2.** less (consider the transfer of energy from rotational kinetic energy to gravitational potential energy) **3.** (draw the vectors, use right-hand rule) (a) $\pm z$; (b) $+y$; (c) $-x$ **4.** (see Eq. 12-21) (a) 1 and 3 tie, then 2 and 4 tie, then 5 (zero); (b) 2 and 3 **5.** (see Eqs. 12-23 and 12-16) (a) 3, 1; then 2 and 4 tie (zero); (b) 3 **6.** (a) all tie (same τ, same t, thus same ΔL); (b) sphere, disk, hoop (reverse order of I) **7.** (a) decreases; (b) same ($\tau_{net} = 0$, so L conserved) (c) increases **Q** **1.** (a) same; (b) block; (c) block **3.** (a) L; (b) $1.5L$ **5.** b, then c and d tie, then a and e tie (zero) **7.** a, then b and c tie, then e, d (zero) **9.** (a) same; (b) increase; (c) decrease; (d) same, decrease, increase **11.** (a) 1, 2, 3 (zero); (b) 1 and 2 tie, then 3; (c) 1 and 3 tie, then 2 **13.** (a) 3, 1, 2; (b) 3, 1, 2 **EP** **1.** (a) 59.3 rad/s; (b) 9.31 rad/s^2; (c) 70.7 m **3.** -3.15 J **5.** 1/50 **7.** (a) 8.0°; (b) more **9.** 4.8 m **11.** (a) 63 rad/s; (b) 4.0 m **13.** (a) 8.0 J; (b) 3.0 m/s; (c) 6.9 J; (d) 1.8 m/s **15.** (a) 13 cm/s^2; (b) 4.4 s; (c) 55 cm/s; (d) 1.8×10^{-2} J; (e) 1.4 J; (f) 27 rev/s **19.** (a) 10 N · m, parallel to yz plane, at 53° to $+y$; (b) 22 N · m, $-x$ **21.** (a) $50\hat{k}$ N · m; (b) 90° **23.** 9.8 kg · m^2/s **25.** (a) 0; (b) $(8.0\hat{i} + 8.0\hat{k})$ N · m **27.** (a) mvd; (b) no; (c) 0, yes **29.** (a) $-170\hat{k}$ kg · m^2/s; (b) $+56\hat{k}$ N · m; (c) $+56\hat{k}$ kg · m^2/s^2 **31.** (a) 0; (b) $8t$ N · m, in $-z$ direction; (c) $2/\sqrt{t}$ N · m, $-z$; (d) $8/t^3$ N · m, $+z$ **33.** (a) -1.47 N · m; (b) 20.4 rad; (c) -29.9 J; (d) 19.9 W **35.** (a) $14md^2$; (b) $4md^2\omega$; (c) $14md^2\omega$ **37.** $\omega_0 R_1 R_2 I_1/(I_1 R_2^2 + I_2 R_1^2)$ **39.** (a) 3.6 rev/s; (b) 3.0; (c) in moving the bricks in, the forces on them from the man transferred energy from internal energy of the man to kinetic energy **41.** (a) 267 rev/min; (b) 2/3 **43.** (a) 149 kg · m^2; (b) 158 kg · m^2/s; (c) 0.746 rad/s **45.** $\dfrac{m}{M + m}\left(\dfrac{v}{R}\right)$ **47.** (a) $(mRv - I\omega_0)/(I + mR^2)$; (b) no, energy transferred to internal energy of cockroach **49.** 3.4 rad/s **51.** (a) 0.148 rad/s; (b) 0.0123; (c) 181° **53.** the day would be longer by about 0.8 s **55.** (a) 18 rad/s; (b) 0.92 **57.** (a) 0.24 kg · m^2; (b) 1800 m/s **59.** $\theta = \cos^{-1}\left[1 - \dfrac{6m^2h}{d(2m + M)(3m + M)}\right]$

CHAPTER 13

CP **1.** c, e, f **2.** directly below the rod (torque on the apple due to $\vec{F}_g$, about the suspension, is zero) **3.** (a) no; (b) at site of $\vec{F}_1$, perpendicular to plane of figure; (c) 45 N **4.** (a) at C (to eliminate forces there from a torque equation); (b) plus; (c) minus; (d) equal **5.** d **6.** (a) equal; (b) B; (c) B **Q** **1.** (a) yes; (b) yes; (c) yes; (d) no **3.** a and c (forces and torques balance) **5.** $m_2 = 12$ kg, $m_3 = 3$ kg, $m_4 = 1$ kg **7.** (a) 15 N (the key is the pulley with the 10 N piñata); (b) 10 N **9.** A, then tie of B and C **EP** **1.** (a) 2; (b) 7 **3.** (a) $(-27\hat{i} + 2\hat{j})$ N; (b) 176° counterclockwise from $+x$ direction **5.** 7920 N **7.** (a) $(mg/L)\sqrt{L^2 + r^2}$; (b) mgr/L **9.** (a) 1160 N, down; (b) 1740 N, up; (c) left; (d) right **11.** 74 g **13.** (a) 280 N; (b) 880 N, 71° above the horizontal **15.** (a) 8010 N; (b) 3.65 kN; (c) 5.66 kN **17.** 71.7 N **19.** (a) 5.0 N; (b) 30 N; (c) 1.3 m **21.** $mg\dfrac{\sqrt{2rh - h^2}}{r - h}$ **23.** (a) 192 N; (b) 96.1 N; (c) 55.5 N **25.** (a) 6630 N; (b) 5740 N; (c) 5960 N **27.** 2.20 m **29.** 0.34 **31.** (a) 211 N; (b) 534 N; (c) 320 N **33.** (a) 445 N; (b) 0.50; (c) 315 N **35.** (a) slides at 31°; (b) tips at 34° **37.** (a) 6.5×10^6 N/m^2; (b) 1.1×10^{-5} m **39.** (a) 867 N; (b) 143 N; (c) 0.165 **41.** (a) 51°; (b) $0.64Mg$

CHAPTER 14

CP **1.** all tie **2.** (a) 1, tie of 2 and 4, then 3; (b) line d **3.** negative y direction **4.** (a) increase; (b) negative **5.** (a) 2; (b) 1 **6.** (a) path 1 (decreased E (more negative) gives decreased a); (b) less (decreased a gives decreased T) **Q** **1.** (a) between, closer to less massive particle; (b) no; (c) no (other than infinity) **3.** $3GM^2/d^2$, leftward **5.** (a) 1 and 2 tie, then 3 and 4 tie; (b) 1, 2, 3, 4 **7.** $U_i/4$ **9.** (a) all tie; (b) all tie **11.** (a)–(d) zero **EP** **1.** 19 m **3.** 29 pN **5.** 1/2 **7.** 2.60×10^5 km **9.** 0.017 N, toward the 300 kg sphere **11.** 3.2×10^{-7} N **13.** $\dfrac{GmM}{d^2}\left[1 - \dfrac{1}{8(1 - R/2d)^2}\right]$ **15.** 2.6×10^6 m **17.** (b) 1.9 h **21.** 4.7×10^{24} kg **23.** (a) $(3.0 \times 10^{-7}$ N/kg$)m$; (b) $(3.3 \times 10^{-7}$ N/kg$)m$; (c) $(6.7 \times 10^{-7}$ N/kg · m$)mr$ **25.** (a) 9.83 m/s^2; (b) 9.84 m/s^2; (c) 9.79 m/s^2 **27.** (a) -1.3×10^{-4} J; (b) less; (c) positive; (d) negative **29.** (a) 0.74; (b) 3.7 m/s^2; (c) 5.0 km/s **31.** (a) 5.0×10^{-11} J; (b) -5.0×10^{-11} J **35.** (a) 1700 m/s; (b) 250 km; (c) 1400 m/s **37.** (a) 82 km/s; (b) 1.8×10^4 km/s **39.** 2.5×10^4 km **41.** 6.5×10^{23} kg **43.** 5×10^{10} **45.** (a) 7.82 km/s; (b) 87.5 min **47.** (a) 6640 km; (b) 0.0136 **49.** (a) 1.9×10^{13} m; (b) $3.5R_P$ **53.** 0.71 y **55.** $\sqrt{GM/L}$ **57.** (a) 2.8 y; (b) 1.0×10^{-4} **61.** (a) no; (b) same; (c) yes **63.** (a) 7.5 km/s; (b) 97 min; (c) 410 km; (d) 7.7 km/s; (e) 92 min; (f) 3.2×10^{-3} N; (g) no; (h) yes, if the satellite–Earth system is considered isolated

CHAPTER 15

CP **1.** all tie **2.** (a) all tie (the gravitational force on the penguin is the same); (b) $0.95\rho_0$, ρ_0, $1.1\rho_0$ **3.** 13 cm^3/s, outward **4.** (a) all tie; (b) 1, then 2 and 3 tie, 4 (wider means slower); (c) 4, 3, 2, 1 (wider and lower mean more pressure) **Q** **1.** e, then b and d tie, then a and c tie **3.** (a) 2; (b) 1, less; 3, equal; 4, greater **5.** (a) moves downward; (b) moves downward **7.** all tie **9.** (a) downward; (b) downward; (c) same **EP** **1.** 1.1×10^5 Pa or 1.1 atm **3.** 2.9×10^4 N **5.** 0.074 **7.** (b) 26 kN **9.** 5.4×10^4 Pa **11.** (a) 5.3×10^6 N; (b) 2.8×10^5 N; (c) 7.4×10^5 N; (d) no **13.** 7.2×10^5 N **15.** $\frac{1}{4}\rho g A(h_2 - h_1)^2$ **17.** 1.7 km **19.** (a) $\rho g W D^2/2$; (b) $\rho g W D^3/6$; (c) $D/3$ **21.** (a) 7.9 km; (b) 16 km **23.** 4.4 mm **25.** (a) 2.04×10^{-2} m^3; (b) 1570 N **27.** (a) 670 kg/m^3; (b) 740 kg/m^3 **29.** (a) 1.2 kg; (b) 1300 kg/m^3 **31.** 57.3 cm **33.** 0.126 m^3 **35.** (a) 45 m^2; (b) car should be over center of slab if slab is to be level **37.** (a) 9.4 N; (b) 1.6 N **39.** 8.1 m/s **41.** 66 W **43.** (a) 2.5 m/s; (b) 2.6×10^5 Pa **45.** (a) 3.9 m/s; (b) 88 kPa **47.** (a) 1.6×10^{-3} m^3/s; (b) 0.90 m **49.** 116 m/s **51.** (a) 6.4 m^3; (b) 5.4 m/s; (c) 9.8×10^4 Pa **53.** (a) 74 N; (b) 150 m^3 **55.** (b) 2.0×10^{-2} m^3/s **57.** (b) 63.3 m/s

59. (a) 180 kN; (b) 81 kN; (c) 20 kN; (d) 0; (e) 78 kPa; (f) no **61.** (a) 0.050; (b) 0.41; (c) no; (d) Lay back on the surface, slowly pull your legs free, and then roll over to the shore.

CHAPTER 16

CP 1. (sketch x versus t) (a) $-x_m$; (b) $+x_m$; (c) 0 **2.** a (F must have form of Eq. 6-10) **3.** (a) 5 J; (b) 2 J; (c) 5 J **4.** all tie (in Eq. 16-29, m is included in I) **5.** 1, 2, 3 (the ratio m/b matters; k does not) **Q 1.** c **3.** (a) 2; (b) positive; (c) between 0 and $+x_m$ **5.** (a) toward $-x_m$; (b) toward $+x_m$; (c) between $-x_m$ and 0; (d) between $-x_m$ and 0; (e) decreasing; (f) increasing **7.** (a) π rad; (b) π rad; (c) $\pi/2$ rad **9.** (a) varies; (b) varies; (c) $x = \pm x_m$; (d) more likely **11.** b (infinite period; does not oscillate), c, a **13.** one system: k = 1500 N/m, m = 500 kg; other system: k = 1200 N/m, m = 400 kg; the same ratio k/m = 3 gives resonance for both systems **EP 1.** (a) 0.50 s; (b) 2.0 Hz; (c) 18 cm **3.** (a) 0.500 s; (b) 2.00 Hz; (c) 12.6 rad/s; (d) 79.0 N/m; (e) 4.40 m/s; (f) 27.6 N **5.** $f > 500$ Hz **7.** (a) 6.28×10^5 rad/s; (b) 1.59 mm **9.** (a) 1.0 mm; (b) 0.75 m/s; (c) 570 m/s^2 **11.** (a) 1.29×10^5 N/m; (b) 2.68 Hz **13.** 7.2 m/s **15.** 2.08 h **17.** 3.1 cm **19.** (a) 5.58 Hz; (b) 0.325 kg; (c) 0.400 m **21.** (a) 2.2 Hz; (b) 56 cm/s; (c) 0.10 kg; (d) 20.0 cm below y_i **23.** (a) $0.183A$; (b) same direction **29.** (a) $(n+1)k/n$; (b) $(n+1)k$; (c) $\sqrt{(n+1)/n}\,f$; (d) $\sqrt{n+1}\,f$ **31.** 37 mJ **33.** (a) 2.25 Hz; (b) 125 J; (c) 250 J; (d) 86.6 cm **35.** (a) 130 N/m; (b) 0.62 s; (c) 1.6 Hz; (d) 5.0 cm; (e) 0.51 m/s **37.** (a) $\frac{3}{4}$; (b) $\frac{1}{4}$; (c) $x_m/\sqrt{2}$ **39.** (a) 16.7 cm; (b) 1.23% **41.** (a) 39.5 rad/s; (b) 34.2 rad/s; (c) 124 rad/s^2 **43.** 99 cm **45.** 5.6 cm **47.** (a) $2\pi\sqrt{\dfrac{L^2+12d^2}{12gd}}$; (b) increases for $d < L/\sqrt{12}$, decreases for $d > L/\sqrt{12}$; (c) increases; (d) no change **49.** (a) 0.205 kg · m^2; (b) 47.7 cm; (c) 1.50 s **53.** $2\pi\sqrt{m/3k}$ **55.** (a) 0.35 Hz; (b) 0.39 Hz; (c) 0 **57.** (b) smaller **59.** 0.39 **61.** (a) 14.3 s; (b) 5.27 **63.** (a) $F_m/b\omega$; (b) F_m/b

CHAPTER 17

CP 1. a, 2; b, 3; c, 1 (compare with phase in Eq. 17-2, then see Eq. 17-5) **2.** (a) 2, 3, 1 (see Eq. 17-12); (b) 3, then 1 and 2 tie (find amplitude of dy/dt) **3.** (a) same (independent of f); (b) decrease ($\lambda = v/f$); (c) increase; (d) increase **4.** (a) increase; (b) increase; (c) increase **5.** 0.20 and 0.80 tie, then 0.60, 0.45 **6.** (a) 1; (b) 3; (c) 2 **7.** (a) 75 Hz; (b) 525 Hz **Q 1.** 7d **3.** (a) $\pi/2$ rad and 0.25 wavelength; (b) π rad and 0.5 wavelength; (c) $3\pi/2$ rad and 0.75 wavelength; (d) 2π rad and 1.0 wavelength; (e) $3T/4$; (f) $T/2$ **5.** (a) 4; (b) 4; (c) 3 **7.** a and d tie, then b and c tie **9.** d **11.** (a) decrease; (b) disappears **EP 1.** (a) 3.49 m^{-1}; (b) 31.5 m/s **3.** (a) 0.68 s; (b) 1.47 Hz; (c) 2.06 m/s **7.** (a) $y(x, t) = 2.0 \sin 2\pi(0.10x - 400t)$, with x and y in cm and t in s; (b) 50 m/s; (c) 40 m/s **9.** (a) 11.7 cm; (b) π rad **11.** 129 m/s **13.** (a) 15 m/s; (b) 0.036 N **15.** $y(x, t) = 0.12 \sin(141x + 628t)$, with y in mm, x in m, and t in s **17.** (a) $2\pi y_m/\lambda$; (b) no **19.** (a) 5.0 cm; (b) 40 cm; (c) 12 m/s; (d) 0.033 s; (e) 9.4 m/s; (f) $5.0 \sin(16x + 190t + 0.93)$, with x in m, y in cm, and t in s **21.** 2.63 m from the end of the wire from which the later pulse originates **25.** (a) 3.77 m/s; (b) 12.3 N; (c) zero; (d) 46.3 W; (e) zero; (f) zero; (g) ± 0.50 cm **27.** $1.4y_m$ **29.** 5.0 cm **31.** (a) $0.83y_1$; (b) 37° **33.** (a) 140 m/s; (b) 60 cm; (c) 240 Hz **35.** (a) 82.0 m/s; (b) 16.8 m; (c) 4.88 Hz **37.** 7.91 Hz, 15.8 Hz, 23.7 Hz **39.** (a) 105 Hz; (b) 158 m/s **41.** (a) 0.25 cm; (b) 120 cm/s; (c) 3.0 cm; (d) zero **43.** (a) 50 Hz; (b) $y = 0.50 \sin[\pi(x \pm 100t)]$, with x in m, y in cm, and t in s **45.** (a) 1.3 m; (b) $y = 0.002 \sin(9.4x) \cos(3800t)$, with x and y in m and t in s **47.** (a) 2.0 Hz; (b) 200 cm; (c) 400 cm/s; (d) 50 cm, 150 cm, 250 cm, etc.; (e) 0, 100 cm, 200 cm, etc. **51.** (a) 323 Hz; (b) eight

CHAPTER 18

CP 1. beginning to decrease (example: mentally move the curves of Fig. 18-7 rightward past the point at x = 42 m) **2.** (a) 0, fully constructive; (b) 4λ, fully constructive **3.** (a) 1 and 2 tie, then 3 (see Eq. 18-28); (b) 3, then 1 and 2 tie (see Eq. 18-26) **4.** second (see Eqs. 18-39 and 18-41) **5.** loosen **6.** a, greater; b, less; c, can't tell; d, can't tell; e, greater; f, less **7.** (measure speeds relative to the air) (a) 222 m/s; (b) 222 m/s **Q 1.** pulse along path 2 **3.** (a) 2.0 wavelengths; (b) 1.5 wavelengths; (c) fully constructive, fully destructive **5.** (a) exactly out of phase; (b) exactly out of phase **7.** (a) one; (b) nine **9.** (a) increase; (b) decrease **11.** all odd harmonics **13.** d, e, b, c, a **EP 1.** divide the time by 3 **3.** (a) 79 m, 41 m; (b) 89 m **5.** 1900 km **7.** 40.7 m **9.** (a) 0.0762 mm; (b) 0.333 mm **11.** (a) 1.50 Pa; (b) 158 Hz; (c) 2.22 m; (d) 350 m/s **13.** (a) $343(1 + 2m)$ Hz, with m being an integer from 0 to 28; (b) $686m$ Hz, with m being an integer from 1 to 29 **15.** (a) 143 Hz, 429 Hz, 715 Hz; (b) 286 Hz, 572 Hz, 858 Hz **17.** 15.0 mW **19.** 36.8 nm **21.** (a) 1000; (b) 32 **23.** (a) 59.7; (b) 2.81×10^{-4} **25.** (b) 5.76×10^{-17} J/m^3 **27.** (b) length^2 **29.** (a) 5200 Hz; (b) amplitude$_{SAD}$/amplitude$_{SBD}$ = 2 **31.** (a) 57.2 cm; (b) 42.9 cm **33.** (a) 405 m/s; (b) 596 N; (c) 44.0 cm; (d) 37.3 cm **35.** (a) 1129, 1506, and 1882 Hz **37.** 12.4 m **39.** (a) node; (c) 22 s **41.** 45.3 N **43.** 387 Hz **45.** 0.02 **47.** 17.5 kHz **49.** (a) 526 Hz; (b) 555 Hz **51.** (a) 1.02 kHz; (b) 1.04 kHz **53.** 155 Hz **55.** (a) 485.8 Hz; (b) 500.0 Hz; (c) 486.2 Hz; (d) 500.0 Hz **57.** (a) 598 Hz; (b) 608 Hz; (c) 589 Hz **59.** (a) 42°; (b) 11 s

CHAPTER 19

CP 1. (a) all tie; (b) 50°X, 50°Y, 50°W **2.** (a) 2 and 3 tie, then 1, then 4; (b) 3, 2, then 1 and 4 tie (from Eqs. 19-9 and 19-10, assume that change in area is proportional to initial area) **3.** A (see Eq. 19-14) **4.** c and e (maximize area enclosed by a clockwise cycle) **5.** (a) all tie (ΔE_{int} depends on i and f, not on path); (b) 4, 3, 2, 1 (compare areas under curves); (c) 4, 3, 2, 1 (see Eq. 19-26) **6.** (a) zero (closed cycle); (b) negative (W_{net} is negative; see Eq. 19-26) **7.** b and d tie, then a, c (P_{cond} identical; see Eq. 19-32) **Q 1.** 25 S°, 25 U°, 25 R° **3.** A and B tie, then C, D **5.** (a) both clockwise; (b) both clockwise **7.** c, a, b **9.** sphere, hemisphere, cube **11.** (a) at freezing point; (b) no liquid freezes; (c) ice partly melts **EP 1.** 0.05 kPa, nitrogen **3.** 348 K **5.** (a) −40°; (b) 575°; (c) Celsius and Kelvin cannot give the same reading **7.** (a) Dimensions are inverse time. **9.** −92.1°X **11.** 960 μm **13.** 2.731 cm **15.** 29 cm^3 **17.** 0.26 cm^3 **19.** 360°C **23.** 0.68 s/h, fast **25.** 7.5 cm **27.** (a) 523 J/kg · K; (b) 26.2 J/mol · K; (c) 0.600 mole **29.** 42.7 kJ **31.** 1.9 times as great **33.** (a) 33.9 Btu; (b) 172 F°

35. 160 s **37.** 2.8 days **39.** 742 kJ **41.** 82 cal **43.** 33 g **45.** (a) 0°C; (b) 2.5°C **47.** 8.72 g **49.** *A*: 120 J, *B*: 75 J, *C*: 30 J **51.** −30 J **53.** (a) 6.0 cal; (b) −43 cal; (c) 40 cal; (d) 18 cal, 18 cal **55.** (a) 0.13 m; (b) 2.3 km **57.** 1660 J/s **59.** (a) 16 J/s; (b) 0.048 g/s **61.** 0.50 min **63.** (a) 17 kW/m^2; (b) 18 W/m^2 **65.** 0.40 cm/h **67.** (a) 90 W; (b) 230 W; (c) 330 W

CHAPTER 20

CP **1.** all but *c* **2.** (a) all tie; (b) 3, 2, 1 **3.** gas *A* **4.** 5 (greatest change in *T*), then tie of 1, 2, 3, and 4 **5.** 1, 2, 3 ($Q_3 = 0$, Q_2 goes into work W_2, but Q_1 goes into greater work W_1 and increases gas temperature) Q **1.** increased but less than doubled **3.** tie of a and c, then b, then d **5.** 1–4 **7.** 20 J **9.** (a) 3; (b) 1; (c) 4; (d) 2; (e) yes **11.** (a) 1, 2, 3, 4; (b) 1, 2, 3 EP **1.** 0.933 kg **3.** 6560 **5.** (a) 5.47×10^{-8} mol; (b) 3.29×10^{16} **7.** (a) 0.0388 mol; (b) 220°C **9.** (a) 106; (b) 0.892 m^3 **11.** $A(T_2 - T_1) - B(T_2^2 - T_1^2)$ **13.** 5600 J **15.** 100 cm^3 **17.** 2.0×10^5 Pa **19.** 180 m/s **21.** 9.53×10^6 m/s **23.** 1.9 kPa **25.** 3.3×10^{-20} J **27.** (a) 6.75×10^{-20} J; (b) 10.7 **31.** (a) 6×10^9 km **33.** 15 cm **35.** (a) 3.27×10^{10}; (b) 172 m **37.** (a) 6.5 km/s; (b) 7.1 km/s **39.** (a) 1.0×10^4 K; (b) 1.6×10^5 K; (c) 440 K, 7000 K; (d) hydrogen, no; oxygen, yes **41.** (a) 7.0 km/s; (b) 2.0×10^{-8} cm; (c) 3.5×10^{10} collisions/s **43.** (a) $\frac{3}{4}v_0$; (b) $N/3$; (c) $122v_0$; (d) $1.31v_0$ **45.** $RT \ln(V_f/V_i)$ **47.** $(n_1C_1 + n_2C_2 + n_3C_3)/(n_1 + n_2 + n_3)$ **49.** (a) 6.6×10^{-26} kg; (b) 40 g/mol **51.** 8000 J **53.** (a) 6980 J; (b) 4990 J; (c) 1990 J; (d) 2990 J **55.** (a) 14 atm; (b) 620 K **59.** 1.40 **61.** (a) In joules, in the order Q, ΔE_{int}, W: 1 → 2: 3740, 3740, 0; 2 → 3: 0, −1810, 1810; 3 → 1: −3220, −1930, −1290; Cycle: 520, 0, 520; (b) $V_2 = 0.0246$ m^3, $p_2 = 2.00$ atm, $V_3 = 0.0373$ m^3, $p_3 = 1.00$ atm

CHAPTER 21

CP **1.** a, b, c **2.** smaller (*Q* is smaller) **3.** c, b, a **4.** a, d, c, b **5.** b Q **1.** unchanged **3.** *b*, *a*, *c*, *d* **5.** equal **7.** (a) same; (b) increase; (c) decrease **9.** (a) same; (b) increase; (c) decrease **11.** (a) 0; (b) 0.25; (c) 0.50 EP **1.** 14.4 J/K **3.** (a) 9220 J; (b) 23.0 J/K; (c) 0 **5.** (a) 5.79×10^4 J; (b) 173 J/K **7.** (a) 14.6 J/K; (b) 30.2 J/K **9.** (a) 57.0°C; (b) −22.1 J/K; (c) +24.9 J/K; (d) +2.8 J/K **13.** (a) 320 K; (b) 0; (c) +1.72 J/K **15.** +0.75 J/K **17.** (a) −943 J/K; (b) +943 J/K; (c) yes **19.** (a) $3p_0V_0$; (b) $\Delta E_{\text{int}} = 6RT_0$, $\Delta S = \frac{3}{2}R \ln 2$; (c) both are zero **21.** (a) 31%; (b) 16 kJ **23.** (a) 23.6%; (b) 1.49×10^4 J **25.** 266 K and 341 K **27.** (a) 1470 J; (b) 554 J; (c) 918 J; (d) 62.4% **29.** (a) 2270 J; (b) 14 800 J; (c) 15.4%; (d) 75.0%, greater **31.** (a) 78%; (b) 81 kg/s **33.** (a) $T_2 = 3T_1$, $T_3 = 3T_1/4^{\gamma-1}$, $T_4 = T_1/4^{\gamma-1}$, $p_2 = 3p_1$, $p_3 = 3p_1/4^{\gamma}$, $p_4 = p_1/4^{\gamma}$; (b) $1 - 4^{1-\gamma}$ **35.** 21 J **37.** 440 W **39.** 0.25 hp **41.** $[1 - (T_2/T_1)]/[1 - (T_4/T_3)]$ **45.** (a) $W = N!/(n_1!\, n_2!\, n_3!)$; (b) $[(N/2)!\,(N/2)!]/[(N/3)!\,(N/3)!\,(N/3)!]$; (c) 4.2×10^{16}

CHAPTER 22

CP **1.** *C* and *D* attract; *B* and *D* attract **2.** (a) leftward; (b) leftward; (c) leftward **3.** (a) *a*, *c*, *b*; (b) less than **4.** $-15e$ (net charge of $-30e$ is equally shared) Q **1.** no, only for charged particles, charged particle-like objects, and spherical shells (including solid spheres) of uniform charge **3.** a and b **5.** $2q^2/4\pi\varepsilon_0 r^2$, up the page **7.** (a) same; (b) less than; (c) cancel; (d) add; (e) the adding components; (f) positive direction of *y*; (g) negative direction of *y*; (h) positive direction of *x*; (i) negative direction of *x* **9.** (a) possibly; (b) definitely **11.** no (the person and the conductor share the charge) EP **1.** 1.38 m **3.** (a) 4.9×10^{-7} kg; (b) 7.1×10^{-11} C **5.** (a) 0.17 N; (b) −0.046 N **7.** either −1.00 μC and +3.00 μC or +1.00 μC and −3.00 μC **9.** (a) charge $-4q/9$ must be located on the line joining the two positive charges, a distance $L/3$ from charge $+q$. **11.** (a) 5.7×10^{13} C, no; (b) 6.0×10^5 kg **13.** $q = Q/2$ **15.** (b) $\pm 2.4 \times 10^{-8}$ C **17.** (a) $\frac{L}{2}\left(1 + \frac{1}{4\pi\varepsilon_0}\frac{qQ}{Wh^2}\right)$; (b) $\sqrt{3qQ/4\pi\varepsilon_0 W}$ **19.** -1.32×10^{13} C **21.** (a) 3.2×10^{-19} C; (b) two **23.** 6.3×10^{11} **25.** 122 mA **27.** (a) 0; (b) 1.9×10^{-9} N **29.** (a) ^{9}B; (b) ^{13}N; (c) ^{12}C

CHAPTER 23

CP **1.** (a) rightward; (b) leftward; (c) leftward; (d) rightward (p and e have same charge magnitude, and p is farther) **2.** all tie **3.** (a) toward positive *y*; (b) toward positive *x*; (c) toward negative *y* **4.** (a) leftward; (b) leftward; (c) decrease **5.** (a) all tie; (b) 1 and 3 tie, then 2 and 4 tie Q **1.** (a) toward positive *x*; (b) downward and to the right; (c) *A* **3.** two points: one to the left of the particles, the other between the protons **5.** (a) yes; (b) toward; (c) no (the field vectors are not along the same line); (d) cancel; (e) add; (f) adding components; (g) toward negative *y* **7.** *e*, *b*, then *a* and *c* tie, then *d* (zero) **9.** (a) toward the bottom; (b) 2 and 4 toward the bottom, 3 toward the top **11.** (a) 4, 3, 1, 2; (b) 3, then 1 and 4 tie, then 2 EP **1.** (a) 6.4×10^{-18} N; (b) 20 N/C **5.** 56 pC **7.** 3.07×10^{21} N/C, radially outward **9.** 50 cm from q_1 and 100 cm from q_2 **11.** 0 **13.** 1.02×10^5 N/C, upward **15.** 6.88×10^{-28} C · m **21.** $q/\pi^2\varepsilon_0 r^2$, vertically downward **23.** (a) $-q/L$; (b) $q/4\pi\varepsilon_0 a(L + a)$ **27.** $R/\sqrt{3}$ **29.** 3.51×10^{15} m/s^2 **31.** 6.6×10^{-15} N **33.** (a) 1.5×10^3 N/C; (b) 2.4×10^{-16} N, up; (c) 1.6×10^{-26} N; (d) 1.5×10^{10} **35.** (a) 1.92×10^{12} m/s^2; (b) 1.96×10^5 m/s **37.** $-5e$ **39.** (a) 2.7×10^6 m/s; (b) 1000 N/C **41.** 27 μm **43.** (a) yes; (b) upper plate, 2.73 cm **45.** (a) 0; (b) 8.5×10^{-22} N · m; (c) 0 **47.** $(1/2\pi)\sqrt{pE/I}$

CHAPTER 24

CP **1.** (a) $+EA$; (b) $-EA$; (c) 0; (d) 0 **2.** (a) 2; (b) 3; (c) 1 **3.** (a) equal; (b) equal; (c) equal **4.** (a) $+50e$; (b) $-150e$ **5.** 3 and 4 tie, then 2, 1 Q **1.** (a) 8 N · m^2/C; (b) 0 **3.** (a) all four; (b) neither (they are equal) **5.** (a) S_3, S_2, S_1; (b) all tie; (c) S_3, S_2, S_1; (d) all tie (zero) **7.** 2σ, σ, 3σ; or 3σ, σ, 2σ **9.** (a) all tie ($E = 0$); (b) all tie EP **1.** (a) 693 kg/s; (b) 693 kg/s; (c) 347 kg/s; (d) 347 kg/s; (e) 575 kg/s **3.** (a) 0; (b) −3.92 N · m^2/C; (c) 0; (d) 0 for each field **5.** 2.0×10^5 N · m^2/C **7.** (a) 8.23 N · m^2/C; (b) 8.23 N · m^2/C; (c) 72.8 pC in each case **9.** 3.54 μC **11.** 0 through each of the three faces meeting at *q*, $q/24\varepsilon_0$ through each of the other faces **13.** (a) 37 μC; (b) 4.1×10^6 N · m^2/C **15.** (a) -3.0×10^{-6} C; (b) $+1.3 \times 10^{-5}$ C **17.** 5.0 μC/m **19.** (a) $E = q/2\pi\varepsilon_0 LR$, radially inward; (b) $-q$ on both inner and outer surfaces; (c) $E = q/2\pi\varepsilon_0 Lr$, radially outward **21.** (a) 2.3×10^6 N/C, radially out; (b) 4.5×10^5 N/C, radially in **23.** 3.6 nC **25.** (b) $\rho R^2/2\varepsilon_0 r$ **27.** (a) 5.3×10^7 N/C; (b) 60 N/C **29.** 5.0 nC/m² **31.** 0.44 mm **33.** (a) $\rho x/\varepsilon_0$; (b) $\rho d/2\varepsilon_0$ **35.** -7.5 nC **39.** -1.04 nC **43.** (a) $E = (q/4\pi\varepsilon_0 a^3)r$; (b) $E = q/4\pi\varepsilon_0 r^2$; (c) 0; (d) 0; (e) inner, $-q$; outer, 0 **45.** $q/2\pi a^2$ **47.** $6K\varepsilon_0 r^3$

CHAPTER 25

CP **1.** (a) negative; (b) increase **2.** (a) positive; (b) higher **3.** (a) rightward; (b) 1, 2, 3, 5: positive; 4, negative; (c) 3, then 1, 2, and 5 tie, then 4 **4.** all tie **5.** a, c (zero), b **6.** (a) 2, then 1 and 3 tie; (b) 3; (c) accelerate leftward Q **1.** (a) higher; (b) positive; (c) negative; (d) all tie **3.** $-4q/4\pi\varepsilon_0 d$ **5.** (a)–(c) $Q/4\pi\varepsilon_0 R$; (d) a, b, c **7.** (a) 2, 4, and then a tie of 1, 3, and 5 (where $E = 0$); (b) negative x direction; (c) positive x direction **9.** (a)–(d) zero EP **1.** (a) 3.0×10^5 C; (b) 3.6×10^6 J **3.** (a) 3.0×10^{10} J; (b) 7.7 km/s; (c) 9.0×10^4 kg **5.** 8.8 mm **7.** (a) 136 MV/m; (b) 8.82 kV/m **9.** (b) because $V = 0$ point is chosen differently; (c) $q/(8\pi\varepsilon_0 R)$; (d) potential differences are independent of the choice of $V = 0$ point **11.** (a) $Q/4\pi\varepsilon_0 r$; (b) $\frac{\rho}{3\varepsilon_0}\left(\frac{3}{2}r_2^2 - \frac{1}{2}r^2 - \frac{r_1^3}{r}\right)$, $\rho = \frac{Q}{\frac{4\pi}{3}(r_2^3 - r_1^3)}$; (c) $\frac{\rho}{2\varepsilon_0}(r_2^2 - r_1^2)$, with ρ as in (b); (d) yes **13.** (a) -4.5 kV; (b) -4.5 kV **15.** $x = d/4$ and $x = -d/2$ **17.** (a) 0.54 mm; (b) 790 V **19.** 6.4×10^8 V **21.** $2.5q/4\pi\varepsilon_0 d$ **25.** (a) $-5Q/4\pi\varepsilon_0 R$; (b) $-5Q/4\pi\varepsilon_0(z^2 + R^2)^{1/2}$ **27.** $(\sigma/8\varepsilon_0)[(z^2 + R^2)^{1/2} - z]$ **29.** $(c/4\pi\varepsilon_0)[L - d\ln(1 + L/d)]$ **31.** 17 V/m at 135° counterclockwise from $+x$ **35.** (a) $\frac{Q}{4\pi\varepsilon_0 d(d + L)}$, leftward; (b) 0 **37.** $-0.21q^2/\varepsilon_0 a$ **39.** (a) $+6.0 \times 10^4$ V; (b) -7.8×10^5 V; (c) 2.5 J; (d) increase; (e) same; (f) same **41.** $W = \frac{qQ}{8\pi\varepsilon_0}\left(\frac{1}{r_1} - \frac{1}{r_2}\right)$ **43.** 2.5 km/s **45.** (a) 0.225 J; (b) A, 45.0 m/s²; B, 22.5 m/s²; (c) A, 7.75 m/s; B, 3.87 m/s **47.** 0.32 km/s **49.** 1.6×10^{-9} m **51.** 2.5×10^{-8} C **53.** (a) -180 V; (b) 2700 V, -8900 V **55.** (a) -0.12 V; (b) 1.8×10^{-8} N/C, radially inward

CHAPTER 26

CP **1.** (a) same; (b) same **2.** (a) decreases; (b) increases; (c) decreases **3.** (a) V, $q/2$; (b) $V/2$, q **4.** (a) $q_0 = q_1 + q_{34}$; (b) equal (C_3 and C_4 are in series) **5.** (a) same; (b)–(d) increase; (e) same (same potential difference across same plate separation) **6.** (a) same; (b) decrease; (c) increase Q **1.** a, 2; b, 1; c, 3 **3.** a, series; b, parallel; c, parallel **5.** (a) $C/3$; (b) $3C$; (c) parallel **7.** (a) same; (b) same; (c) more; (d) more **9.** (a) 2; (b) 3; (c) 1 **11.** (a) increases; (b) increases; (c) decreases; (d) decreases; (e) same, increases, increases, increases EP **1.** 7.5 pC **3.** 3.0 mC **5.** (a) 140 pF; (b) 17 nC **7.** $5.04\pi\varepsilon_0 R$ **11.** 9090 **13.** 3.16 μF **17.** 43 pF **19.** (a) 50 V; (b) 5.0×10^{-5} C; (c) 1.5×10^{-4} C **21.** $q_1 = \frac{C_1C_2 + C_1C_3}{C_1C_2 + C_1C_3 + C_2C_3}C_1V_0$, $q_2 = q_3 = \frac{C_2C_3}{C_1C_2 + C_1C_3 + C_2C_3}C_1V_0$ **23.** 72 F **25.** 0.27 J **27.** (a) 2.0 J **29.** (a) $2V$; (b) $U_i = \varepsilon_0 AV^2/2d$, $U_f = 2U_i$; (c) $\varepsilon_0 AV^2/2d$ **35.** Pyrex **37.** 81 pF/m **39.** 0.63 m² **43.** (a) 10 kV/m; (b) 5.0 nC; (c) 4.1 nC **45.** (a) $C = 4\pi\varepsilon_0\kappa\left(\frac{ab}{b - a}\right)$; (b) $q = 4\pi\varepsilon_0\kappa V\left(\frac{ab}{b - a}\right)$; (c) $q' = q(1 - 1/\kappa)$

CHAPTER 27

CP **1.** 8 A, rightward **2.** (a)–(c) rightward **3.** a and c tie, then b **4.** Device 2 **5.** (a) and (b) tie, then (d), then (c) Q **1.** a, b, and c tie, then d (zero) **3.** b, a, c **5.** tie of A, B, and C, then tie of $A + B$ and $B + C$, then $A + B + C$ **7.** (a)–(c) 1 and 2 tie, then 3 **9.** C, A, B EP **1.** (a) 1200 C; (b) 7.5×10^{21} **3.** 5.6 ms **5.** (a) 6.4 A/m², north; (b) no, cross-sectional area **7.** 0.38 mm **9.** (a) 2×10^{12}; (b) 5000; (c) 10 MV **11.** 13 min **13.** 2.0×10^{-8} Ω · m **15.** 100 V **17.** 2.4 Ω **19.** 54 Ω **21.** 3.0 **23.** 8.2×10^{-4} Ω · m **25.** 2000 K **27.** (a) 0.43%, 0.0017%, 0.0034% **29.** (a) $R = \rho L/\pi ab$ **31.** 560 W **33.** (a) 1.0 kW; (b) 25¢ **35.** 0.135 W **37.** (a) 10.9 A; (b) 10.6 Ω; (c) 4.5 MJ **39.** 660 W **41.** (a) 3.1×10^{11}; (b) 25 μA; (c) 1300 W, 25 MW **43.** (a) 17 mV/m; (b) 243 J **45.** (a) $J = I/2\pi r^2$; (b) $E = \rho I/2\pi r^2$; (c) $\Delta V = \rho I(1/r - 1/b)/2\pi$; (d) 0.16 A/m²; (e) 16 V/m; (f) 0.16 MV

CHAPTER 28

CP **1.** (a) rightward; (b) all tie; (c) b, then a and c tie; (d) b, then a and c tie **2.** (a) all tie; (b) R_1, R_2, R_3 **3.** (a) less; (b) greater; (c) equal **4.** (a) $V/2$, i; (b) V, $i/2$ **5.** (a) 1, 2, 4, 3; (b) 4, tie of 1 and 2; then 3 Q **1.** 3, 4, 1, 2 **3.** (a) no; (b) yes; (c) all tie **5.** parallel, R_2, R_1, series **7.** (a) same; (b) same; (c) less; (d) more **9.** (a) less; (b) less; (c) more **11.** c, b, a EP **1.** (a) $320; (b) 4.8 cents **3.** 14 h 24 min **5.** (a) 0.50 A; (b) $P_1 = 1.0$ W, $P_2 = 2.0$ W; (c) $P_1 = 6.0$ W supplied, $P_2 = 3.0$ W absorbed **7.** (a) 14 V; (b) 100 W; (c) 600 W; (d) 10 V, 100 W **9.** (a) 50 V; (b) 48 V; (c) B is connected to the negative terminal **11.** 2.5 V **13.** 8.0 Ω **15.** (a) $r_1 - r_2$; (b) battery with r_1 **19.** 5.56 A **21.** $i_1 =$ 50 mA, $i_2 = 60$ mA, $V_{ab} = 9.0$ V **23.** (a) bulb 2; (b) bulb 1 **25.** $3d$ **27.** nine **29.** (a) $R = r/2$; (b) $P_{max} = \mathscr{E}^2/2r$ **31.** (a) 0.346 W; (b) 0.050 W; (c) 0.709 W; (d) 1.26 W; (e) -0.158 W **33.** (a) battery 1, 0.67 A down; battery 2, 0.33 A up; battery 3, 0.33 A up; (b) 3.3 V **35.** (a) Cu: 1.11 A, Al: 0.893 A; (b) 126 m **37.** 0.45 A **39.** -3.0% **45.** 4.6 **47.** (a) 2.41 μs; (b) 161 pF **49.** (a) 0.955 μC/s; (b) 1.08 μW; (c) 2.74 μW; (d) 3.82 μW **51.** (a) 2.17 s; (b) 39.6 mV **53.** (a) 1.0×10^{-3} C; (b) 10^{-3} A; (c) $V_C = 10^3 e^{-t}$ V, $V_R = 10^3 e^{-t}$ V; (d) $P = e^{-2t}$ W **55.** (a) at $t = 0$, $i_1 = 1.1$ mA, $i_2 = i_3 = 0.55$ mA; at $t = \infty$, $i_1 = i_2 = 0.82$ mA, $i_3 = 0$; (c) at $t = 0$, $V_2 = 400$ V; at $t = \infty$, $V_2 = 600$ V; (d) after several time constants ($\tau = 7.1$ s) have elapsed

CHAPTER 29

CP **1.** *a*, $+z$; *b*, $-x$; *c*, $\vec{F}_B = 0$ **2.** (a) 2, then tie of 1 and 3 (zero); (b) 4 **3.** (a) $+z$ and $-z$ tie, then $+y$ and $-y$ tie, then $+x$ and $-x$ tie (zero); (b) $+y$ **4.** (a) electron; (b) clockwise **5.** $-y$ **6.** (a) all tie; (b) 1 and 4 tie, then 2 and 3 tie **Q** **1.** (a) no, $\vec{v}$ and $\vec{F}_B$ must be perpendicular; (b) yes; (c) no, $\vec{B}$ and $\vec{F}_B$ must be perpendicular **3.** (a) $\vec{F}_E$; (b) $\vec{F}_B$ **5.** (a) negative; (b) equal; (c) equal; (d) half-circle **7.** (a) $\vec{B}_1$; (b) B_1 into page, B_2 out of page; (c) less **9.** (a) 1, 180°; 2, 270°; 3, 90°; 4, 0°; 5, 315°; 6, 225°; 7, 135°; 8, 45°; (b) 1 and 2 tie, then 3 and 4 tie; (c) 8, then 5 and 6 tie, then 7 **EP** **1.** (a) 6.2×10^{-18} N; (b) 9.5×10^{8} m/s²; (c) remains equal to 550 m/s **3.** (a) 400 km/s; (b) 835 eV **5.** (a) east; (b) 6.28×10^{14} m/s²; (c) 2.98 mm **7.** (a) 3.4×10^{-4} T, horizontal and to the left as viewed along $\vec{v}_0$; (b) yes, if its velocity is the same as the electron's velocity **9.** 0.27 mT **11.** 680 kV/m **13.** (b) 2.84×10^{-3} **15.** 21 μT **17.** (a) 2.05×10^{7} m/s; (b) 467 μT; (c) 13.1 MHz; (d) 76.3 ns **19.** (a) 0.978 MHz; (b) 96.4 cm **23.** (a) 1.0 MeV; (b) 0.5 MeV **25.** (a) 495 mT; (b) 22.7 mA; (c) 8.17 MJ **27.** (a) 0.36 ns; (b) 0.17 mm; (c) 1.5 mm **29.** (a) $-q$; (b) $\pi m/qB$ **31.** 240 m **33.** 28.2 N, horizontally west **35.** 467 mA, from left to right **37.** 0.10 T, at 31° from the vertical **39.** 4.3×10^{-3} N·m, negative y **43.** $2\pi aiB \sin\theta$, normal to the plane of the loop (up) **45.** (a) 540 Ω, connected in series with the galvanometer; (b) 2.52 Ω, connected in parallel **47.** 2.45 A **49.** (a) 12.7 A; (b) 0.0805 N·m **51.** (a) 0.30 J/T; (b) 0.024 N·m **53.** (a) 2.86 A·m²; (b) 1.10 A·m² **55.** (a) $(8.0 \times 10^{-4}$ N·m$)(-1.2\hat{i} - 0.90\hat{j} + 1.0\hat{k})$; (b) -6.0×10^{-4} J **57.** $-(0.10$ V/m$)\hat{k}$ **59.** -2.0 T

CHAPTER 30

CP **1.** *a*, *c*, *b* **2.** *b*, *c*, *a* **3.** *d*, tie of *a* and *c*, then *b* **4.** *d*, *a*, tie of *b* and *c* (zero) **Q** **1.** *c*, *d*, then *a* and *b* tie **3.** *c*, *a*, *b* **5.** (a) 1, 3, 2; (b) less **7.** *c* and *d* tie, then *b*, *a* **9.** *d*, then tie of *a* and *e*, then *b*, *c* **EP** **1.** (a) 3.3 μT; (b) yes **3.** (a) 16 A; (b) west to east **5.** (a) $\mu_0 qvi/2\pi d$, antiparallel to i; (b) same magnitude, parallel to i **7.** 2 rad **9.** $\dfrac{\mu_0 i\theta}{4\pi}\left(\dfrac{1}{b} - \dfrac{1}{a}\right)$, out of page **19.** $(\mu_0 i/2\pi w)\ln(1 + w/d)$, up **21.** (a) it is impossible to have other than $B = 0$ midway between them; (b) 30 A **23.** 4.3 A, out of page **25.** 80 μT, up the page **27.** $0.791\mu_0 i^2/\pi a$, 162° counterclockwise from the horizontal **29.** 3.2 mN, toward the wire **31.** (a) $(-2.0\text{ A})\mu_0$; (b) 0 **35.** $\mu_0 J_0 r^2/3a$ **41.** 0.30 mT **43.** (a) 533 μT; (b) 400 μT **47.** (a) 4.77 cm; (b) 35.5 μT **49.** 0.47 A·m² **51.** (a) 2.4 A·m²; (b) 46 cm **57.** (a) 79 μT; (b) 1.1×10^{-6} N·m

CHAPTER 31

CP **1.** *b*, then *d* and *e* tie, and then *a* and *c* tie (zero) **2.** *a* and *b* tie, then *c* (zero) **3.** *c* and *d* tie, then *a* and *b* tie **4.** *b*, out; *c*, out; *d*, into; *e*, into **5.** d and e **6.** (a) 2, 3, 1 (zero); (b) 2, 3, 1 **7.** *a* and *b* tie, then *c* **Q** **1.** (a) all tie (zero); (b) 2, then tie of 1 and 3 (zero) **3.** (a) into; (b) counterclockwise; (c) larger **5.** *c*, *a*, *b* **7.** *c*, *b*, *a* **9.** (a) more; (b) same; (c) same; (d) same (zero) **EP** **1.** 1.5 mV **3.** (a) 31 mV; (b) right to left **5.** (a) 1.1×10^{-3} Ω; (b) 1.4 T/s **7.** 30 mA **9.** (a) $\mu_0 iR^2\pi r^2/2x^3$; (b) $3\mu_0 i\pi R^2 r^2 v/2x^4$; (c) in the same direction as the current in the large loop **11.** (b) no **13.** 29.5 mC **15.** (a) 21.7 V; (b) counterclockwise **17.** (b) design it so that $Nab = (5/2\pi)$ m² **19.** 5.50 kV **21.** 80 μV, clockwise **23.** (a) 13 μWb/m; (b) 17%; (c) 0 **25.** 3.66 μW **27.** (a) 48.1 mV; (b) 2.67 mA; (c) 0.128 mW **29.** (a) 600 mV, up the page; (b) 1.5 A, clockwise; (c) 0.90 W; (d) 0.18 N; (e) same as (c) **31.** (a) 240 μV; (b) 0.600 mA; (c) 0.144 μW; (d) 2.88×10^{-8} N; (e) same as (c) **33.** (a) 71.5 μV/m; (b) 143 μV/m **37.** 0.10 μWb **41.** let the current change at 5.0 A/s **43.** (b) so that the changing magnetic field of one does not induce current in the other; (c) $L_{eq} = \sum_{j=1}^{N} L_j$ **45.** $6.91\tau_L$ **47.** 46 Ω **49.** (a) 8.45 ns; (b) 7.37 mA **51.** 12.0 A/s **53.** (a) $i_1 = i_2 = 3.33$ A; (b) $i_1 = 4.55$ A, $i_2 = 2.73$ A; (c) $i_1 = 0$, $i_2 = 1.82$ A (reversed); (d) $i_1 = i_2 = 0$ **55.** (a) $i(1 - e^{-Rt/L})$ **57.** 25.6 ms **59.** (a) 97.9 H; (b) 0.196 mJ **63.** (a) 34.2 J/m³; (b) 49.4 mJ **65.** 1.5×10^{8} V/m **67.** (a) 1.0 J/m³; (b) 4.8×10^{-15} J/m³ **69.** (a) 1.67 mH; (b) 6.00 mWb **71.** (b) have the turns of the two solenoids wrapped in opposite directions **73.** magnetic field exists only within the cross section of solenoid 1 **75.** (a) $\dfrac{\mu_0 Nl}{2\pi}\ln\left(1 + \dfrac{b}{a}\right)$; (b) 13 μH

CHAPTER 32

CP **1.** *d*, *b*, *c*, *a* (zero) **2.** (a) 2; (b) 1 **3.** (a) away; (b) away; (c) less **4.** (a) toward; (b) toward; (c) less **5.** *a*, *c*, *b*, *d* (zero) **6.** tie of *b*, *c*, and *d*, then *a* **Q** **1.** supplied **3.** (a) all down; (b) 1 up, 2 down, 3 zero **5.** (a) 1 up, 2 up, 3 down; (b) 1 down, 2 up, 3 zero **7.** (a) 1, up; 2, up; 3, down; (b) and (c) 2, then 1 and 3 tie **9.** (a) rightward; (b) leftward; (c) into **11.** 1, *a*; 2, *b*; 3, *c* and *d* **EP** **1.** (b) sign is minus; (c) no, there is compensating positive flux through open end near magnet **3.** 47.4 μWb, inward **5.** 55 μT **7.** (a) 31.0 μT, 0°; (b) 55.9 μT, 73.9°; (c) 62.0 μT, 90° **9.** (a) -9.3×10^{-24} J/T; (b) 1.9×10^{-23} J/T **11.** (a) 0; (b) 0; (c) 0; (d) $\pm 3.2 \times 10^{-25}$ J; (e) -3.2×10^{-34} J·s, 2.8×10^{-23} J/T, $+9.7 \times 10^{-25}$ J, $\pm 3.2 \times 10^{-25}$ J **13.** $\Delta\mu = e^2 r^2 B/4m$ **15.** 20.8 mJ/T **17.** yes **19.** (b) K_i/B, opposite to the field; (c) 310 A/m **21.** (a) 3.0 μT; (b) 5.6×10^{-10} eV **23.** 5.15×10^{-24} A·m² **25.** (a) 180 km; (b) 2.3×10^{-5} **27.** 2.4×10^{13} V/m·s **33.** (a) 0.63 μT; (b) 2.3×10^{12} V/m·s **35.** (a) 710 mA; (b) 0; (c) 1.1 A **37.** (a) 2.0 A; (b) 2.3×10^{11} V/m·s; (c) 0.50 A; (d) 0.63 μT·m

CHAPTER 33

CP **1.** (a) $T/2$, (b) T, (c) $T/2$, (d) $T/4$ **2.** (a) 5 V; (b) 150 μJ **3.** (a) remains the same; (b) remains the same **4.** (a) *C*, *B*, *A*; (b) 1, *A*; 2, *B*; 3, *S*; 4, *C*; (c) *A* **5.** (a) remains the same; (b) increases **6.** (a) remains the same; (b) decreases **7.** (a) 1, lags; 2, leads; 3, in phase; (b) 3 ($\omega_d = \omega$ when $X_L = X_C$) **8.** (a) increase (circuit is mainly capacitive; increase C to decrease X_C to be closer to resonance for maximum P_{avg}); (b) closer **9.** (a) greater; (b) step-up **Q** **1.** (a) $T/4$; (b) $T/4$; (c) $T/2$ (see Fig. 33-2); (d) $T/2$ (see Eq. 31-37) **3.** *b*, *a*, *c* **5.** (a) 3, 1, 2;

(b) 2, tie of 1 and 3 **7.** *a*, inductor; *b*, resistor; *c*, capacitor **9.** (a) leads; (b) capacitive; (c) less **11.** (a) rightward, increase (X_L increases, closer to resonance); (b) rightward, increase (X_C decreases, closer to resonance); (c) rightward, increase (ω_d/ω increases, closer to resonance) **EP** **1.** 9.14 nF **3.** (a) 1.17 μJ; (b) 5.58 mA **5.** with *n* a positive integer: (a) $t = n(5.00\ \mu s)$; (b) $t = (2n - 1)(2.50\ \mu s)$; (c) $t = (2n - 1)(1.25\ \mu s)$ **7.** (a) 1.25 kg; (b) 372 N/m; (c) 1.75×10^{-4} m; (d) 3.02 mm/s **9.** 7.0×10^{-4} s **11.** (a) 3.0 nC; (b) 1.7 mA; (c) 4.5 nJ **13.** (a) 275 Hz; (b) 364 mA **15.** (a) 6.0:1; (b) 36 pF, 0.22 mH **17.** (a) 1.98 μJ; (b) 5.56 μC; (c) 12.6 mA; (d) $-46.9°$; (e) $+46.9°$ **19.** (a) 0.180 mC; (b) $T/8$; (c) 66.7 W **21.** (a) 356 μs; (b) 2.50 mH; (c) 3.20 mJ **23.** Let T_2 (= 0.596 s) be the period of the inductor plus the 900 μF capacitor and let T_1 (= 0.199 s) be the period of the inductor plus the 100 μF capacitor. Close S_2, wait $T_2/4$; quickly close S_1, then open S_2; wait $T_1/4$ and then open S_1. **25.** 8.66 mΩ **27.** (L/R) ln 2 **31.** (a) 0.0955 A; (b) 0.0119 A **33.** (a) 0.65 kHz; (b) 24 Ω **35.** (a) 6.73 ms; (b) 11.2 ms; (c) inductor; (d) 138 mH **37.** (a) $X_C = 0$, $X_L = 86.7\ \Omega$, $Z = 182\ \Omega$, $I = 198$ mA, $\phi = 28.5°$ **39.** (a) $X_C = 37.9\ \Omega$, $X_L = 86.7\ \Omega$, $Z = 167\ \Omega$, $I = 216$ mA, $\phi = 17.1°$ **41.** 1000 V **43.** 89 Ω **45.** (a) 224 rad/s; (b) 6.00 A; (c) 228 rad/s, 219 rad/s; (d) 0.040 **49.** 1.84 A **51.** 141 V **53.** 0, 9.00 W, 3.14 W, 1.82 W **55.** (a) 12.1 Ω; (b) 1.19 kW **57.** (a) 0.743; (b) leads; (c) capacitive; (d) no; (e) yes, no, yes; (f) 33.4 W **59.** (a) 117 μF; (b) 0; (c) 90.0 W, 0; (d) 0°, 90°; (e) 1, 0 **61.** (a) 2.59 A; (b) 38.8 V, 159 V, 224 V, 64.2 V, 75.0 V; (c) 100 W for *R*, 0 for *L* and *C* **63.** (a) 2.4 V; (b) 3.2 mA, 0.16 A **65.** 10

CHAPTER 34

CP **1.** (a) (Use Fig. 34-5.) On right side of rectangle, $\vec{E}$ is in negative *y* direction; on left side, $\vec{E} + d\vec{E}$ is greater and in same direction; (b) $\vec{E}$ is downward. On right side, $\vec{B}$ is in negative *z* direction; on left side, $\vec{B} + d\vec{B}$ is greater and in same direction. **2.** positive direction of *x* **3.** (a) same; (b) decrease **4.** *a*, *d*, *b*, *c* (zero) **5.** *a* **6.** (a) no; (b) yes **Q** **1.** (a) positive direction of *z*; (b) *x* **3.** (a) same; (b) increase; (c) decrease **5.** *c* **7.** *a*, *b*, *c* **9.** none **11.** *b* **EP** **1.** (a) 0.50 ms; (b) 8.4 min; (c) 2.4 h; (d) 5500 B.C. **3.** (a) 515 nm, 610 nm; (b) 555 nm, 5.41×10^{14} Hz, 1.85×10^{-15} s **5.** it would steadily increase; (b) the summed discrepancies between the apparent time of eclipse and those observed from *x*; the radius of Earth's orbit **7.** 5.0×10^{-21} H **9.** $B_x = 0$, $B_y = -6.7 \times 10^{-9} \cos[\pi \times 10^{15}(t - x/c)]$, $B_z = 0$ in SI units **11.** 0.10 MJ **13.** 8.88×10^4 m^2 **15.** (a) 16.7 nT; (b) 33.1 mW/m^2 **17.** (a) 6.7 nT; (b) 5.3 mW/m^2; (c) 6.7 W **19.** (a) 87 mV/m; (b) 0.30 nT; (c) 13 kW **21.** 1.0×10^7 Pa **23.** 5.9×10^{-8} Pa **25.** (a) 100 MHz; (b) 1.0 μT along the *z* axis; (c) 2.1 m^{-1}, 6.3×10^8 rad/s; (d) 120 W/m^2; (e) 8.0×10^{-7} N, 4.0×10^{-7} Pa **29.** 1.9 mm/s **31.** (b) 580 nm **33.** (a) 1.9 V/m; (b) 1.7×10^{-11} Pa **35.** 3.1% **37.** 4.4 W/m^2 **39.** 2/3 **41.** (a) 2 sheets; (b) 5 sheets **43.** 1.48 **45.** 1.26 **47.** 1.07 m **53.** 1.22 **55.** (a) 49°; (b) 29° **57.** (a) cover the center of each face with an opaque disk of radius 4.5 mm; (b) about 0.63 **59.** (a) $\sqrt{1 + \sin^2 \theta}$; (b) $\sqrt{2}$; (c) light emerges at the right; (d) no light emerges at the right **61.** 49.0° **63.** (a) 15 m/s; (b) 8.7 m/s; (c) higher; (d) 72° **65.** 1.0

CHAPTER 35

CP **1.** 0.2*d*, 1.8*d*, 2.2*d* **2.** (a) real; (b) inverted; (c) same **3.** (a) *e*; (b) virtual, same **4.** virtual, same as object, diverging **Q** **1.** *c* **3.** (a) *a* and *c*; (b) three times; (c) you **5.** convex **7.** (a) decrease; (b) increase; (c) increase **9.** (a) all but variation 2; (b) for 1, 3, and 4: right, inverted; for 5 and 6: left, same **EP** **1.** 40 cm **3.** (a) 3 **7.** new illumination is 10/9 of the old **9.** 10.5 cm **13.** (a) 2.00; (b) none **17.** $i = -12$ cm **19.** 45 mm, 90 mm **23.** 22 cm **27.** same orientation, virtual, 30 cm to the left of the second lens; $m = 1$ **33.** (a) 13.0 cm; (b) 5.23 cm; (c) -3.25; (d) 3.13; (e) -10.2 **35.** (a) 2.35 cm; (b) decrease **37.** (a) 5.3 cm; (b) 3.0 mm

CHAPTER 36

CP **1.** *b* (least *n*), *c*, *a* **2.** (a) top; (b) bright intermediate illumination (phase difference is 2.1 wavelengths) **3.** (a) 3λ, 3; (b) 2.5λ, 2.5 **4.** a and d tie (amplitude of resultant wave is $4E_0$), then b and c tie (amplitude of resultant wave is $2E_0$) **5.** (a) 1 and 4; (b) 1 and 4 **Q** **1.** *a*, *c*, *b* **3.** (a) 300 nm; (b) exactly out of phase **5.** (a) intermediate closer to maximum, $m = 2$; (b) minimum, $m = 3$; (c) intermediate closer to maximum, $m = 2$; (d) maximum, $m = 1$ **7.** (a)–(c) decrease; (d) blue **9.** (a) maximum; (b) minimum; (c) alternates **11.** (a) 0.5 wavelength; (b) 1 wavelength **EP** **1.** (a) 5.09×10^{14} Hz; (b) 388 nm; (c) 1.97×10^8 m/s **3.** 1.56 **5.** 22°, refraction reduces θ **7.** (a) 3.60 μm; (b) intermediate, closer to fully constructive interference **9.** (a) 0.833; (b) intermediate, closer to fully constructive interference **11.** (a) 0.216 rad; (b) 12.4° **13.** 2.25 mm **15.** 648 nm **17.** 16 **19.** 0.072 mm **21.** 6.64 μm **23.** 2.65 **25.** $y = 27 \sin(\omega t + 8.5°)$ **27.** (a) 1.17 m, 3.00 m, 7.50 m; (b) no **29.** $I = \frac{1}{9}I_m[1 + 8\cos^2(\pi d \sin\theta/\lambda)]$, I_m = intensity of central maximum **31.** fully constructively **33.** 0.117 μm, 0.352 μm **35.** 70.0 nm **37.** 120 nm **39.** (a) 552 nm; (b) 442 nm **43.** 140 **45.** 1.89 μm **47.** 2.4 μm **49.** $\sqrt{(m + \frac{1}{2})\lambda R}$, for $m = 0, 1, 2, \ldots$ **51.** 1.00 m **53.** $x = (D/2a)(m + \frac{1}{2})\lambda$, for $m = 0, 1, 2, \ldots$ **55.** 588 nm **57.** 1.00030 **59.** (a) 0; (b) fully constructive; (c) increase;

(d)

Phase Difference	Position x (μm)	Type
0	$\approx\infty$	fc
0.50λ	7.88	fd
1.00λ	3.75	fc
1.50λ	2.29	fd
2.00λ	1.50	fc
2.50λ	0.975	fd

CHAPTER 37

CP **1.** (a) expand; (b) expand **2.** (a) second side maximum; (b) 2.5 **3.** (a) red; (b) violet **4.** diminish **5.** (a) increase;

(b) same **6.** (a) left; (b) less **Q** **1.** (a) contract; (b) contract **3.** with megaphone (larger opening, less diffraction) **5.** four **7.** (a) less; (b) greater; (c) greater **9.** (a) decrease; (b) decrease; (c) to the right **11.** (a) increase; (b) first order **EP** **1.** 60.4 μm **3.** (a) $\lambda_a = 2\lambda_b$; (b) coincidences occur when $m_b = 2m_a$ **5.** (a) 70 cm; (b) 1.0 mm **7.** 1.77 mm **11.** (d) 53°, 10°, 5.1° **13.** (b) 0, 4.493 rad, etc.; (c) −0.50, 0.93, etc. **15.** (a) 1.3×10^{-4} rad; (b) 10 km **17.** 50 m **19.** (a) 1.1×10^4 km; (b) 11 km **21.** 27 cm **23.** (a) 0.347°; (b) 0.97° **25.** (a) 8.7×10^{-7} rad; (b) 8.4×10^7 km; (c) 0.025 mm **27.** five **29.** (a) 4; (b) every fourth bright fringe **31.** (a) nine; (b) 0.255 **33.** (a) 3.33 μm; (b) 0, ±10.2°, ±20.7°, ±32.0°, ±45.0°, ±62.2° **35.** three **37.** (a) 6.0 μm; (b) 1.5 μm; (c) m = 0, 1, 2, 3, 5, 6, 7, 9 **39.** 1100 **47.** 3650 **53.** 0.26 nm **55.** 39.8 pm **59.** (a) $a_0/\sqrt{2}$, $a_0/\sqrt{5}$, $a_0/\sqrt{10}$, $a_0/\sqrt{13}$, $a_0/\sqrt{17}$ **61.** 30.6°, 15.3° (clockwise); 3.08°, 37.8° (counterclockwise) **63.** (a) 50 m; (b) no, the width of 10 m is too narrow to resolve; (c) not during daylight, but the light pollution during the night would be a sure sign

CHAPTER 38

CP **1.** (a) same (speed of light postulate); (b) no (the start and end of the flight are spatially separated); (c) no (because his measurement is not a proper time) **2.** (a) Sally's; (b) Sally's **3.** a, positive; b, negative; c, positive **4.** (a) right; (b) more **5.** (a) equal; (b) less **Q** **1.** all tie (pulse speed is c) **3.** (a) C_1; (b) C_1 **5.** (a) negative; (b) positive **7.** less **9.** b, a, c, d **EP** **1.** (a) 6.7×10^{-10} s; (b) 2.2×10^{-18} m **3.** 0.99c **5.** 0.445 ps **7.** 1.32 m **9.** 0.63 m **11.** (a) 87.4 m; (b) 394 ns **13.** (a) 26 y; (b) 52 y; (c) 3.7 y **15.** x' = 138 km, t' = −374 μs **17.** (a) 25.8 μs; (b) small flash **19.** (a) 1.25; (b) 0.800 μs **21.** 0.81c **23.** (a) 0.35c; (b) 0.62c **25.** 1.2 μs **27.** 22.9 MHz **29.** 1×10^6 m/s, receding **31.** yellow (550 nm) **33.** (a) 0.0625, 1.00196; (b) 0.941, 2.96; (c) 0.999 999 87, 1960 **35.** 0.999 987c **37.** 18 smu/y **39.** (a) 0.707c; (b) 1.41; (c) 0.414mc^2 **41.** $\sqrt{8}mc$ **43.** 1.01×10^7 km, or 250 Earth circumferences **45.** 110 km **47.** 4.00 u, probably a helium nucleus **49.** 330 mT **51.** (a) 2.08 MeV; (b) −1.18 MeV **53.** (a) $vt \sin\theta$; (b) $t[1 - (v/c)\cos\theta]$; (c) 3.24c

CHAPTER 39

CP **1.** b, a, d, c **2.** (a) lithium, sodium, potassium, cesium; (b) all tie **3.** (a) same; (b)–(d) x rays **4.** (a) proton; (b) same; (c) proton **5.** same **Q** **1.** (a) microwave; (b) x ray; (c) x ray **3.** potassium **5.** positive charge builds up on the plate, inhibiting further electron emission **7.** none **9.** (a) greater; (b) less **11.** no essential change **13.** (a) decreases by a factor of $1/\sqrt{2}$; (b) decreases by a factor of 1/2 **15.** (a) decreasing; (b) increasing; (c) same; (d) same **17.** a **19.** (a) zero; (b) yes **EP** **1.** 4.14 eV · fs **5.** 1.0×10^{45} photons/s **7.** 5.9 μeV **9.** 2.047 eV **11.** 4.7×10^{26} photons **13.** (a) infrared lamp; (b) 1.4×10^{21} photons/s **15.** (a) 2.96×10^{20} photons/s; (b) 48 600 km; (c) 5.89×10^{18} photons/m² · s **17.** barium and lithium **19.** 170 nm **21.** 676 km/s **23.** (a) 2.00 eV; (b) 0; (c) 2.00 eV; (d) 295 nm **25.** 233 nm **27.** (a) 382 nm; (b) 1.82 eV **29.** 9.68×10^{-20} A **31.** (a) 2.7 pm; (b) 6.05 pm **33.** (a) 8.57×10^{18} Hz; (b) 35.4 keV; (c) 1.89×10^{-23} kg · m/s = 35.4 keV/c **37.** (a) 2.43 pm; (b) 1.32 fm; (c) 0.511 MeV; (d) 938 MeV **39.** 300% **43.** (a) 41.8 keV; (b) 8.2 keV **45.** 1.12 keV **47.** 44° **51.** 7.75 pm **53.** 4.3 μeV **55.** (a) 38.8 meV; (b) 146 pm **57.** (a) photon: 1.24 μm; electron: 1.22 nm; (b) 1.24 fm for each **59.** (a) 1.9×10^{-21} kg · m/s; (b) 346 fm **61.** 0.025 fm, about 200 times smaller than a nuclear radius **63.** neutron **65.** 9.70 kV (relativistic calculation), 9.76 kV (classical calculation) **73.** (d) $x = n(\lambda/2)$, where n = 0, 1, 2, 3, . . . **75.** 0.19 m **79.** (a) proton: 9.02×10^{-6}, deuteron: 7.33×10^{-8}; (b) 3.0 MeV for each; (c) 3.0 MeV for each **81.** (a) −20%; (b) −10%; (c) +15% **83.** $T = 10^{-x}$, where $x = 7.2 \times 10^{39}$ (T is very small)

CHAPTER 40

CP **1.** b, a, c **2.** (a) all tie; (b) a, b, c **3.** a, b, c, d **4.** $E_{1,1}$ (neither n_x nor n_y can be zero) **5.** (a) 5; (b) 7 **Q** **1.** (a) 1/4; (b) same factor **3.** c **5.** (a) $(\sqrt{1/L})\sin(\pi/2L)x$; (b) $(\sqrt{4/L})\sin(2\pi/L)x$; (c) $(\sqrt{2/L})\cos(\pi/L)x$ **7.** less **9.** (a) wider; (b) deeper **11.** n = 1, n = 2, n = 3 **13.** b, c, and d **15.** (a) first Lyman plus first Balmer; (b) Lyman series limit minus Paschen series limit **EP** **1.** (a) 37.7 eV; (b) 0.0206 eV **3.** 1900 MeV **5.** 0.020 eV **7.** 90.3 eV **11.** 68.7 nm, 25.8 nm, 13.7 nm, and 8.59 nm **13.** (a) 1.3×10^{-19} eV; (b) about $n = 1.2 \times 10^{19}$; (c) 0.95 J = 5.9×10^{18} eV; (d) yes **15.** (b) no; (c) no; (d) yes **17.** (a) 0.050; (b) 0.10; (c) 0.0095 **19.** 59 eV **21.** (b) $k = (2\pi/h)[2m(U_0 - E)]^{1/2}$ **25.** 3.08 eV **27.** 0.75, 1.00, 1.25, 1.75, 2.00, 2.25, 3.00, 3.75 **29.** 1.00, 2.00, 3.00, 5.00, 6.00, 8.00, 9.00 **31.** 2.6 eV **33.** 4.0 **35.** (a) 12 eV; (b) 6.5×10^{-27} kg · m/s; (c) 103 nm **39.** (a) 0; (b) 10.2 nm⁻¹; (c) 5.54 nm⁻¹ **41.** (a) 13.6 eV; (b) 3.40 eV **43.** (a) n = 4 to n = 2; (b) Balmer series **45.** (a) 13.6 eV; (b) −27.2 eV **47.** (a) 2.6 eV; (b) n = 4 to n = 2 **49.** 0.68 **55.** (a) 0.0037; (b) 0.0054 **59.** (a) $P_{210} = (r^4/8a^5)e^{-r/a}\cos^2\theta$; $P_{21+1} = P_{21-1} = (r^4/16a^5)e^{-r/a}\sin^2\theta$

CHAPTER 41

CP **1.** 7 **2.** (a) decrease; (b)–(c) remain the same **3.** less **4.** A, C, B **Q** **1.** 0, 2, and 3 **3.** 6p **5.** (a) 2, 8; (b) 5, 50 **7.** (a) n; (b) n and l **9.** a, c, e, f **11.** (a) unchanged; (b) decrease; (c) decrease **13.** a and b **EP** **3.** (a) 3; (b) 3 **5.** (a) 32; (b) 2; (c) 18; (d) 8 **7.** 24.1° **9.** $n > 3$; m_l = +3, +2, +1, 0, −1, −2, −3; $m_s = \pm\frac{1}{2}$ **11.** (a) $\sqrt{12}\hbar$; (b) $\sqrt{12}\mu_B$;

(c)

m_l	L_z	$\mu_{orb,z}$	θ
−3	$-3\hbar$	$+3\mu_B$	150°
−2	$-2\hbar$	$+2\mu_B$	125°
−1	$-\hbar$	$+\mu_B$	107°
0	0	0	90°
+1	$+\hbar$	$-\mu_B$	73.2°
+2	$+2\hbar$	$-2\mu_B$	54.7°
+3	$+3\hbar$	$-3\mu_B$	30.0°

15. 54.7° and 125° **17.** 73 km/s² **19.** 5.35 cm **21.** (a) 2.13 meV; (b) 18 T **23.** $44(h^2/8mL^2)$

25. (a) $51(h^2/8mL^2)$; (b) $56(h^2/8mL^2)$; (c) $59(h^2/8mL^2)$ **27.** $42(h^2/8mL^2)$ **31.** argon **33.** (a) $(n, l, m_l, m_s) = (2, 0, 0, \pm\frac{1}{2})$; (b) $n = 2, l = 1, m_l = 1, 0,$ or $-1, m_s = \pm\frac{1}{2}$ **39.** 49.6 pm, 99.2 pm **43.** (a) 35.4 pm, as for molybdenum; (b) 57 pm; (c) 50 pm **45.** 9/16 **49.** (a) 69.5 kV; (b) 17.9 pm; (c) K_α: 21.4 pm, K_β: 18.5 pm **51.** (a) $(Z-1)^2/(Z'-1)^2$; (b) 57.5; (c) 2070 **53.** (a) 6; (b) 3.2×10^6 years **55.** 9.1×10^{-7} **57.** 10 000 K **59.** (a) 3.60 mm; (b) 5.25×10^{17} **61.** 4.7 km **63.** $2.0 \times 10^{16}\ s^{-1}$ **65.** (a) 3.03×10^5; (b) 1430 MHz; (d) 3.30×10^{-6} **67.** (a) no; (b) 140 nm **69.** (a) 4.3 μm; (b) 10 μm; (c) infrared

CHAPTER 42

CP **1.** (a) larger; (b) same **2.** Cleveland, metal; Boca Raton, none; Seattle, semiconductor **3.** a, b, and c **4.** b Q **1.** 4 **3.** b and c, yes **5.** (a) anywhere in the lattice; (b) in any silicon–silicon bond; (c) in a silicon ion core, at a lattice site **7.** b and d **9.** $+4e$ **11.** none **13.** (a) right to left; (b) back bias **15.** a, b, and c EP **1.** $8.49 \times 10^{28}\ m^{-3}$ **3.** 3490 atm **5.** (a) $+8.0 \times 10^{-11}\ \Omega \cdot m/K$; (b) $-210\ \Omega \cdot m/K$ **7.** (b) $6.81 \times 10^{27}\ m^{-3}\ eV^{-3/2}$; (c) $1.52 \times 10^{28}\ m^{-3}\ eV^{-1}$ **9.** (a) 0; (b) 0.0955 **13.** 0.91 **15.** (a) 2500 K; (b) 5300 K **17.** (a) 90.0%; (b) 12.5%; (c) sodium **19.** (a) $2.7 \times 10^{25}\ m^{-3}$; (b) $8.43 \times 10^{28}\ m^{-3}$; (c) 3100; (d) molecules: 3.3 nm; electrons: 0.228 nm **21.** (a) 1.0, 0.99, 0.50, 0.014, 2.5×10^{-17}; (b) 700 K **23.** 3 **25.** (a) $5.86 \times 10^{28}\ m^{-3}$; (b) 5.52 eV; (c) 1390 km/s; (d) 0.522 nm **27.** (b) $1.80 \times 10^{28}\ m^{-3}\ eV^{-1}$ **31.** (a) 19.8 kJ; (b) 197 s **33.** 200° C **35.** (a) 225 nm; (b) ultraviolet **37.** (a) 109.5°; (b) 235 pm **41.** 0.22 μg **43.** (a) pure: 4.78×10^{-10}; doped: 0.0141; (b) 0.824 **45.** 6.02×10^5 **47.** 4.20 eV **49.** (a) 5.0×10^{-17} F; (b) about $300e$

CHAPTER 43

CP **1.** ^{90}As and ^{158}Nd **2.** a little more than 75 Bq (elapsed time is a little less than three half-lives) **3.** ^{206}Pb Q **1.** less **3.** ^{240}U **5.** less **7.** (a) on the $N = Z$ line; (b) positrons; (c) about 120 **9.** no **11.** yes **13.** (a) increases; (b) remains the same **15.** 7 h **17.** d EP **1.** 28.3 MeV **3.** (a) 0.390 MeV; (b) 4.61 MeV **7.** (a) six; (b) eight **11.** (a) 1150 MeV; (b) 4.81 MeV/nucleon, 12.2 MeV/proton **15.** (a) 6.2 fm; (b) yes **17.** $K \approx 30$ MeV **21.** ^{25}Mg: 9.303%; ^{26}Mg: 11.71% **23.** 1.6×10^{25} MeV **25.** 7.92 MeV **27.** 280 d **29.** (a) $7.6 \times 10^{16}\ s^{-1}$; (b) $4.9 \times 10^{16}\ s^{-1}$ **31.** (a) 64.2 h; (b) 0.125; (c) 0.0749 **33.** 5.3×10^{22} **35.** (a) 2.0×10^{20}; (b) $2.8 \times 10^9\ s^{-1}$ **37.** 209 d **39.** 1.13×10^{11} y **43.** (a) $8.88 \times 10^{10}\ s^{-1}$; (b) $8.88 \times 10^{10}\ s^{-1}$; (c) 1.19×10^{15}; (d) 0.111 μg **45.** 730 cm² **47.** Pu: 1.2×10^{-17}, Cm: $e^{-9173} \approx 0$ **49.** 4.269 MeV **51.** (a) 31.8 MeV, 5.98 MeV; (b) 86 MeV **53.** ^{7}Li **55.** 1.21 MeV **57.** 0.782 MeV **59.** (b) 0.961 MeV **61.** 78.4 eV **63.** (a) U: 1.06×10^{19}, Pb: 0.624×10^{19}; (b) 1.69×10^{19}; (c) 2.98×10^9 y **65.** 1.8 mg **67.** 1.02 mg **69.** 13 mJ **71.** (a) 6.3×10^{18}; (b) 2.5×10^{11}; (c) 0.20 J; (d) 2.3 mGy; (e) 30 mSv **73.** (a) 6.6 MeV; (b) no **75.** (a) 25.4 MeV; (b) 12.8 MeV; (c) 25.0 MeV **77.** (a) 3.85 MeV, 7.95 MeV; (b) 3.98 MeV, 7.33 MeV **79.** (a) 7.21 MeV; (b) 12.0 MeV; (c) 8.69 MeV **81.** 0.49 **83.** (a) beta-minus decay; (b) 8.2×10^7; (c) 1.2×10^6 **85.** 3.2×10^{12} Bq = 86 Ci **87.** 4.28×10^9 y **89.** 1.3×10^{-13} m **91.** 3.2×10^4 y

CHAPTER 44

CP **1.** c and d **2.** (a) no; (b) yes; (c) no **3.** e Q **1.** a **3.** b **5.** (a) ^{93}Sr; (b) ^{140}I; (c) ^{155}Nd **7.** c **9.** a **11.** c EP **1.** (a) 2.6×10^{24}; (b) 8.2×10^{13} J; (c) 2.6×10^4 y **3.** $3.1 \times 10^{10}\ s^{-1}$ **7.** -23.0 MeV **9.** 181 MeV **11.** (a) ^{153}Nd; (b) 110 MeV to ^{83}Ge, 60 MeV to ^{153}Nd; (c) 1.6×10^7 m/s for ^{83}Ge, 8.7×10^6 m/s for ^{153}Nd **13.** (a) 252 MeV; (b) typical fission energy is 200 MeV **15.** 461 kg **17.** yes **19.** 557 W **21.** $^{238}U + n \rightarrow {}^{239}U \rightarrow {}^{239}Np + e, {}^{239}Np \rightarrow {}^{239}Pu + e$ **23.** (a) 84 kg; (b) 1.7×10^{25}; (c) 1.3×10^{25} **25.** 0.99938 **27.** (b) 1.0, 0.89, 0.28, 0.019; (c) 8 **29.** (a) 75 kW; (b) 5800 kg **31.** 1.7×10^9 y **33.** 170 keV **35.** (a) 170 kV **37.** 0.151 **41.** (a) 3.1×10^{31} protons/m³; (b) 1.2×10^6 times **43.** (a) 4.3×10^9 kg/s; (b) 3.1×10^{-4} **45.** (a) $1.83 \times 10^{38}\ s^{-1}$; (b) $8.25 \times 10^{28}\ s^{-1}$ **47.** (a) 4.1 eV/atom; (b) 9.0 MJ/kg; (c) 1500 y **49.** 1.6×10^8 y **51.** (a) 24.9 MeV; (b) 8.65 megatons **53.** 14.4 kW

CHAPTER 45

CP **1.** (a) the muon family; (b) a particle; (c) $L_\mu = +1$ **2.** b and e **3.** c Q **1.** d **3.** the leftmost π^+ pion whose track curves downward **5.** a, b, c, d **7.** c, f **9.** 1d, 2e, 3a, 4b, 5c **11.** 1b, 2c, 3d, 4e, 5a **13.** (a) 0; (b) +1; (c) −1; (d) +1; (e) −1 EP **1.** 6.03×10^{-29} kg **3.** 18.4 fm **5.** 1.08×10^{42} J **7.** 2.7 cm/s **9.** 769 MeV **13.** (a) L_e, spin angular momentum; (b) L_μ, charge; (c) energy, L_μ **15.** $q = 0, B = -1, S = 0$ **17.** (a) energy; (b) strangeness; (c) charge **19.** 338 MeV **21.** (a) K^+; (b) $\bar{n}$; (c) K^0 **23.** (a) $\bar{u}\bar{u}\bar{d}$; (b) $\bar{u}\bar{d}\bar{d}$ **25.** (a) not possible; (b) uuu **29.** Σ^0, 7530 km/s **31.** 666 nm **33.** (b) 4.5 H-atoms/m³ **35.** (a) 256 μeV; (b) 4.84 mm **37.** (a) 122 m/s; (b) 246 y **39.** (b) 2.38×10^9 K **41.** (a) $0.785c$; (b) $0.993c$; (c) C2; (d) C1; (e) 51 ns; (f) 40 ns **43.** (c) $r\alpha/c + (r\alpha/c)^2 + (r\alpha/c)^3 + \cdots$; (d) $\Delta\lambda/\lambda = r\alpha/c$; (e) $\alpha = H$; (f) 7.4×10^8 ly; (g) 7.8×10^8 y; (h) 7.4×10^8 y; (i) 7.8×10^8 ly; (j) 1.2×10^9 ly; (k) 1.2×10^9 y; (l) 4.4×10^8 ly

PHOTO CREDITS

CHAPTER 1

Page 1: George D. Lepp/Science Photo Source/Photo Researchers. Page 6: Steven Pitkin. Page 7: Courtesy Bureau International des Poids et Mesures, France.

CHAPTER 2

Page 11: James Neiss/Niagara Gazette/Corbis-Sygma. Pages 18–19: Courtesy U.S. Air Force. Page 23: James Sugar/Black Star.

CHAPTER 3

Page 31: Courtesy David des Marais, © 1976 The Cave Research Foundation.

CHAPTER 4

Page 48: J. Yulsman/The Image Bank. Page 55: Courtesy C. E. Miller. Page 56: Richard Megna/Fundamental Photographs. Page 57: Jamie Budge/Gamma Liaison. Page 68: Mike Powell/Allsport.

CHAPTER 5

Page 72: AP/Wide World Photos. Page 96: Hartwell/Corbis-Sygma.

CHAPTER 6

Page 98: Gerard Lucz/© Natural History Photographic Agency. Page 104: Tony Duffy/Allsport. Page 105: Didier Klein/AgenceVandystadt/Allsport. Page 108: Photograph reproduced with permission of Circus World Museum. Page 112: Jerry Schad/Photo Researchers. Page 113: Susan Copen Oken/Dot Pictures.

CHAPTER 7

Page 116: Photo by Terry Schmitt/UPI/Presslink.com. Page 118: Courtesy Library of Congress. Page 124: AP/Wide World Photos. Page 132: Photri, Inc. Page 136: Photo by James Baker, courtesy Dennis Milon.

CHAPTER 8

Page 139: Malcolm S. Kirk/Peter Arnold, Inc. Page 146: Jeff Schultz/Alaska Stock Images. Page 156: David Stoecklein/The Stock Market.

CHAPTER 9

Page 168: Lois Greenfield. Page 169: Courtesy C. E. Miller. Page 181: Courtesy NASA.

CHAPTER 10

Page 193: Courtesy C. E. Miller. Page 194 (center left): Bruce Kent/Earth Scenes/Animals Animals. Page 194 (center right): Science Photo Library/Photo Researchers. Page 194 (bottom): Russ Kinne/Comstock, Inc. Page 198: Courtesy Mercedes-Benz of North America. Page 209: George Lang/Sports Illustrated/© Time, Inc. Page 210: Superman #48, October 1990. © D. C. Comics. All rights reserved. Reprinted with permission. Page 214: Stephen Dalton/Photo Researchers.

CHAPTER 11

Page 215: Amwell/Tony Stone Images/New York, Inc. Page 216 (top): Chris Trotman/Duomo Photography, Inc. Page 216 (bottom): Steven Sutton/Duomo Photography, Inc. Page 225: Roger Ressmeyer/Corbis-Bettmann. Page 229: Courtesy Test Devices, Inc. Page 241: Courtesy Lick Observatory. Page 243: Courtesy Lawrence Livermore Laboratory, University of California.

CHAPTER 12

Page 245: Courtesy Ringling Brothers and Barnum & Bailey Circus. Page 246: Richard Megna/Fundamental Photographs. Page 247: Courtesy Alice Halliday. Page 259: From *Shepp's World's Fair*, photographed by James W. Shepp and Daniel P. Shepp. Globe Publishing Co., Chicago and Philadelphia, 1893. Photo scan provided courtesy Jeffery Howe.

CHAPTER 13

Page 273: Pascal Tournaire/Vandystadt/Allsport. Page 274: Fred Hirschmann/Allstock/Tony Stone Images/New York, Inc. Page 275: Andy Levin/Photo Researchers. Page 285: Courtesy Measurements Group, Inc., Raleigh, NC, USA. Page 289: Hideo Kurihara/Tony Stone Images/New York, Inc.

CHAPTER 14

Page 294: Courtesy Lund Observatory. Pages 295, 310 and 319: Courtesy NASA. Page 313: Courtesy National Radio Astronomy Observatory.

CHAPTER 15

Page 321: Colin Prior/Tony Stone Images/New York, Inc. Page 330: T. Orban/Sygma. Page 332: Will McIntyre/Photo Researchers. Page 333 (left): Courtesy D.H. Peregrine, University of Bristol. Page 333 (right): Courtesy Volvo North America Corporation.

CHAPTER 16

Page 346: T Campion/Sygma. Page 366: Courtesy NASA.

CHAPTER 17

Page 370: John Visser/Bruce Coleman, Inc. Page 390: Richard Megna/Fundamental Photographs. Page 391: Courtesy T.D. Rossing, Northern Illinois University.

CHAPTER 18

Page 398: Stephen Dalton/Animals Animals. Page 399: Howard Sochurak/The Stock Market. Page 407: Ben Rose/The Image Bank. Page 409: Bob Gruen/Star File. Page 410: John Eastcott/ Yva Momativk/DRK Photo. Page 419: U.S. Navy photo by Ensign John Gay.

CHAPTER 19

Page 425: ©Dr. Mosato Ono, Tamagawa University. Reproduced with permission. Page 431: AP/Wide World Photos. Page 445: Courtesy Daedalus Enterprises, Inc. Page 453: Kim Westerskov/ Tony Stone Images/New York, Inc.

CHAPTER 20

Page 454: Tom Branch.

CHAPTER 21

Page 482: Steven Dalton/Photo Researchers. Page 492: Richard Ustinich/The Image Bank.

CHAPTER 22

Page 505: Michael Watson. Page 506: ©Fundamental Photographs. Page 507: Courtesy Xerox Corporation. Page 508: Johann Gabriel Doppelmayr, *Neuentdeckte Phaenomena von Bewünderswurdigen Würckungen der Natur*, Nuremberg, 1744. Page 516: Courtesy Lawrence Berkeley Laboratory.

CHAPTER 23

Page 520: Tsuyoshi Nishiinoue/Orion Press. Page 534: Russ Kinne/Comstock, Inc.

CHAPTER 24

Page 543: Ralph H. Wetmore II/Tony Stone Images/New York, Inc. Page 554 (left): ©C. Johnny Autery. Page 554 (right): Courtesy E. Philip Krider, Institute for Atmospheric Physics, University of Arizona, Tucson.

CHAPTER 25

Pages 564 and 569: Courtesy NOAA. Page 581: Courtesy Westinghouse Corporation.

CHAPTER 26

Page 588: Bruce Ayres/Tony Stone Images/New York, Inc. Page 589: Paul Silvermann/Fundamental Photographs. Page 600: ©Harold & Ester Edgerton Foundation, 1999, courtesy of Palm Press, Inc. Page 601: Courtesy The Royal Institute, England.

CHAPTER 27

Page 611: ©UPI/Corbis Images. Page 617: The Image Works. Page 625: ©Laurie Rubin. Page 627: Courtesy Shoji Tonaka, International Superconductivity Technology Center, Tokyo, Japan.

CHAPTER 28

Page 633: Hans Reinhard/Bruce Coleman, Inc. Page 634: Courtesy Southern California Edison Company.

CHAPTER 29

Page 658: Johnny Johnson/Tony Stone Images/New York, Inc. Page 659: Ray Pfortner/Peter Arnold, Inc. Page 661: Lawrence Berkeley Laboratory/Photo Researchers. Page 662: Courtesy Dr. Richard Cannon, Southeast Missouri State University, Cape Girardeau. Page 668: Courtesy John Le P. Webb, Sussex University, England. Page 670: Courtesy Dr. L. A. Frank, University of Iowa.

CHAPTER 30

Page 686: Michael Brown/Florida Today/Gamma Liaison. Page 688: Courtesy Education Development Center.

CHAPTER 31

Page 710: Dan McCoy/Black Star. Page 715: Courtesy Fender Musical Instruments Corporation. Page 725: Courtesy The Royal Institute, England.

CHAPTER 32

Page 744: Courtesy A. K. Geim, High Field Magnet Laboratory, University of Nijmegen, The Netherlands. Page 745: Runk/ Schoenberger/Grant Heilman Photography. Page 753: Peter Lerman. Page 756: Courtesy Ralph W. DeBlois.

CHAPTER 33

Page 768: Photo by Rick Diaz, provided courtesy Haverfield Helicopter Co. Page 771: Courtesy Agilent Technologies. Page 792: Ted Cowell/Black Star.

CHAPTER 34

Page 801: John Chumack/Photo Researchers. Page 816: Diane Schiumo/Fundamental Photographs. Page 818: *PSSC Physics*, 2nd edition; ©1975 D. C. Heath and Co. with Education Development Center, Newton, MA. Reproduced with permission of Education Development Center. Page 821 (top): Courtesy Bausch & Lomb. Page 821 (bottom): Barbara Filet/Tony Stone Images/New York, Inc. Page 823: Greg Pease/Tony Stone Images/New York, Inc. Page 828: Courtesy Cornell University.

CHAPTER 35

Page 833: Courtesy Courtauld Institute Galleries, London. Page 842: Dr. Paul A. Zahl/Photo Researchers. Page 845: Courtesy Matthew J. Wheeler. Page 856: Piergiorgio Scharandis/Black Star.

CHAPTER 36

Page 861: David Julian/Phototake. Page 866: Runk Schoenberger/Grant Heilman Photography. Page 868: From *Atlas of Optical Phenomena* by M. Cagnet et al., Springer-Verlag, Prentice Hall, 1962. Page 878: Richard Megna/Fundamental Photographs. Page 887: Courtesy Bausch & Lomb.

CHAPTER 37

Page 890: Georges Seurat, *A Sunday on La Grande Jatte*, 1884; oil on canvas (207.5 × 308 cm). Helen Birch Bartlett Memorial Collection, 1926; photograph ©1996, The Art Institute of Chicago. All rights reserved. Page 891: Ken Kay/Fundamental

Photographs. Pages 892, 898, and 902: From *Atlas of Optical Phenomena* by Cagnet, Francon, Thierr, Springer-Verlag, Berlin, 1962. Reproduced with permission. Page 900: Warren Rosenberg/BPS/Tony Stone Images/New York, Inc. Page 906: Department of Physics, Imperial College/Science Photo Library/Photo Researchers. Page 907: Kristen Brochmann/Fundamental Photographs. Page 915 (left): Kjell B. Sandved/Bruce Coleman, Inc. Page 915 (right): Pekka Parvianen/Photo Researchers. Page 918: AP/Wide World Photos.

CHAPTER 38

Page 919: Jeffrey Zaruba/Tony Stone Images/New York, Inc. Page 920: Courtesy of the Albert Einstein Archives, The Jewish National & University Library, The Hebrew University of Jerusalem, Israel. Page 940: Courtesy NASA.

CHAPTER 39

Page 953: Lawrence Berkeley Laboratory/Science Photo Library/Photo Researchers. Page 965: Courtesy A. Tonomura, J. Endo, T. Matsuda, and T. Kawasaki/Advanced Research Laboratory, Hitachi, Ltd., Kokubinju, Tokyo; H. Ezawa, Department of Physics, Gakushuin University, Mejiro, Tokyo. Page 966 (left): Courtesy Riber Division of Instruments, Inc. Page 966 (right): From PSSC film "Matter Waves," courtesy Education Development Center, Newton, Massachusetts. Page 972: ©IBMRL/Visuals Unlimited.

CHAPTER 40

Page 979: Courtesy International Business Machines Corporation, Almaden Research Center, CA. Page 990: From *Scientific American,* January 1993, page 122. Reproduced with permission of Michael Steigerwald, Bell Labs–Lucent Technologies. Page 991: From *Scientific American,* September 1995, page 67. Image reproduced with permission of H. Temkin, Texas Tech University. Page 995: From W. Finkelnburg, *Structure of Matter*, Springer-Verlag, 1964. Reproduced with permission.

CHAPTER 41

Page 1006: Larry Mulvehill/Photo Researchers. Page 1007: Courtesy Warren Nagourney. Page 1016: David Job/Tony Stone Images/New York, Inc. Page 1027: Michael Rosenfeld/Tony Stone Images/New York, Inc.

CHAPTER 42

Page 1037: Steve Northrup/New York Times Pictures. Page 1054: Courtesy AT&T. Page 1056: Courtesy Intel.

CHAPTER 43

Page 1062: Elscint/Science Photo Library/Photo Researchers. Page 1079 (top inset): R. Perry/Corbis Sygma. Page 1079 (top): George Rockwin/Bruce Coleman, Inc.

CHAPTER 44

Pages 1092 and 1101: Courtesy U.S. Department of Energy. Page 1099: V. Ivelva/Magnum Photos, Inc. Page 1102: Gary Sheehan, *Birth of the Atomic Age,* 1957. Reproduced courtesy Chicago Historical Society. Page 1108: Courtesy Anglo Australian Telescope Board. Page 1110 (top): Courtesy Princeton University Physics Laboratory. Page 1110 (center): Courtesy Los Alamos National Laboratory, New Mexico. Page 1113: Courtesy Martin Marietta Energy Systems/U.S. Department of Energy.

CHAPTER 45

Page 1116: Courtesy NASA. Page 1118: David Parker/Photo Researchers. Page 1119: Courtesy Michael Mathews. Page 1120: Courtesy Lawrence Berkeley Laboratory.

INDEX

Figures are noted by page numbers in *italics;* tables are indicated by t following the page number.

D

F

N

X

Y

Z

Problem Solving Tactics

Some Physical Constants*

Speed of light	c	3.00×10^8 m/s
Gravitational constant	G	6.67×10^{-11} N·m²/kg²
Avogadro constant	N_A	6.02×10^{23} mol^{-1}
Universal gas constant	R	8.31 J/mol·K
Mass-energy relation	c^2	8.99×10^{16} J/kg
		931.5 MeV/u
Permittivity constant	ε_0	8.85×10^{-12} F/m
Permeability constant	μ_0	1.26×10^{-6} H/m
Planck constant	h	6.63×10^{-34} J·s
		4.14×10^{-15} eV·s
Boltzmann constant	κ	1.38×10^{-23} J/K
		8.62×10^{-5} eV/K
Elementary charge	e	1.60×10^{-19} C
Electron mass	m_e	9.11×10^{-31} kg
Proton mass	m_p	1.67×10^{-27} kg
Neutron mass	m_n	1.68×10^{-27} kg
Deuteron mass	m_d	3.34×10^{-27} kg
Bohr radius	a	5.29×10^{-11} m
Bohr magneton	μ_B	9.27×10^{-24} J/T
		5.79×10^{-5} eV/T
Rydberg constant	R	0.01097 nm^{-1}

* For a more complete list, showing also the best experimental values, see Appendix B.

Some Conversion Factors*

Mass and Density

1 kg = 1000 g = 6.02×10^{26} u

1 slug = 14.6 kg

1 u = 1.66×10^{-27} kg

1 kg/m³ = 10^{-3} g/cm³

Length and Volume

1 m = 100 cm = 39.4 in. = 3.28 ft

1 mi = 1.61 km = 5280 ft

1 in. = 2.54 cm

1 nm = 10^{-9} m = 10 Å

1 pm = 10^{-12} m = 1000 fm

1 light-year = 9.46×10^{15} m

1 m³ = 1000 L = 35.3 ft³ = 264 gal

Time

1 d = 86 400 s

1 y = $365\frac{1}{4}$ d = 3.16×10^7 s

Angular Measure

1 rad = 57.3° = 0.159 rev

π rad = 180° = $\frac{1}{2}$ rev

Speed

1 m/s = 3.28 ft/s = 2.24 mi/h

1 km/h = 0.621 mi/h = 0.278 m/s

Force and Pressure

1 N = 10^5 dyne = 0.225 lb

1 lb = 4.45 N

1 ton = 2000 lb

1 Pa = 1 N/m² = 10 dyne/cm²
= 1.45×10^{-4} lb/in.²

1 atm = 1.01×10^5 Pa = 14.7 lb/in.²
= 76 cm-Hg

Energy and Power

1 J = 10^7 erg = 0.239 cal = 0.738 ft·lb

1 kW·h = 3.6×10^6 J

1 cal = 4.19 J

1 eV = 1.60×10^{-19} J

1 horsepower = 746 W = 550 ft·lb/s

Magnetism

1 T = 1 Wb/m² = 10^4 gauss

* See Appendix D for a more complete list.

ALSO AVAILABLE:

US $4.50 (Rs. 210 apx.)
ISBN: 9971513323

US $5.00 (Rs. 235 apx.)
ISBN: 9971513331